Organic Geochemistry

Principles and Applications

TOPICS IN GEOBIOLOGY
Series Editors: F. G. Stehli, DOSECC, Inc., Gainesville, Florida
D. S. Jones, University of Florida, Gainesville, Florida

Organic Geochemistry

Principles and Applications

Edited by

Michael H. Engel

School of Geology and Geophysics
The University of Oklahoma
Norman, Oklahoma

and

Stephen A. Macko

Department of Environmental Sciences
University of Virginia
Charlottesville, Virginia

Springer Science+Business Media, LLC

Library of Congress Cataloging in Publication Data

Organic geochemistry: principles and applications / edited by Michael H. Engel and Stephen A. Macko.
p. cm.—(Topics in geobiology; v. 11)
Includes bibliographical references and index.
ISBN 978-1-4613-6252-4 ISBN 978-1-4615-2890-6 (eBook)
DOI 10.1007/978-1-4615-2890-6
1. Organic geochemistry. I. Engel, Michael (Michael H.). II. Macko, Stephen A. III. Series.
QE516.5.07 1993 93-28298
551.9—dc20 CIP

ISBN 978-1-4613-6252-4

Originally published by Plenum Press in 1993
Softcover reprint of the hardcover 1st edition 1993

*To Christopher, Daniel, Nikolas, Rebekah,
and Sarah—may they always
seek answers to their questions*

Contributors

John A. Baross School of Oceanography, University of Washington, Seattle, Washington 98195

Stuart A. Bradshaw Organic Geochemistry Unit, University of Bristol, Bristol B58 1TS, England; *present address*: Mobil Oil Company, Limited, Information Systems Department, Mobil House, London SW1E 6QB, England

Simon C. Brassell Department of Geology, Stanford University, Stanford, California 94305-2115; *present address*: Department of Geological Science, Indiana University, Bloomington, Indiana 47405-5101

K. Brew Department of Biochemistry and Molecular Biology, University of Miami School of Medicine, Miami, Florida 33101-9990

James M. Brooks Geochemical and Environmental Research Group (GERG), Texas A&M University, College Station, Texas 77845

David J. Carlson TOGA/COARE International Project Office, UCAR, Boulder, Colorado 80307

Luis A. Cifuentes Department of Oceanography, Texas A&M University, College Station, Texas 77843

H. Benjamin Cox Radian Corporation, Austin, Texas 78759

Joseph A. Curiale Unocal Energy Resources, Brea, California 92621

Jody W. Deming School of Oceanography, University of Washington, Seattle, Washington 98195

Geoffrey Eglinton Organic Geochemistry Unit, School of Chemistry, University of Bristol, Bristol BS8 1TS, England

Michael H. Engel School of Geology and Geophysics, The University of Oklahoma, Norman, Oklahoma 73019

Robert B. Finkelman U.S. Geological Survey, Reston, Virginia 22092

Marilyn L. Fogel Geophysical Laboratory, Carnegie Institution of Washington, Washington, D.C. 20015

Brian Fry The Ecosystems Center, Marine Biological Laboratory, Woods Hole Oceanographic Institution, Woods Hole, Massachusetts 02543

Andrew P. Gize Department of Geology, University of Manchester, Manchester M13 9PL, England

John I. Hedges School of Oceanography, University of Washington, Seattle, Washington 98195

Susan M. Henrichs Institute of Marine Science, University of Alaska, Fairbanks, Alaska 99775-1080

B. Horsfield Institute of Petroleum and Organic Geochemistry (ICG-4), KFA-Jülich GmbH, D-5170 Jülich, Germany

Scott W. Imbus School of Geology and Geophysics, The University of Oklahoma, Norman, Oklahoma 73019; *present address*: Energy and Petroleum Technology Department, Texaco Inc., Houston, Texas 77042

Ryoshi Ishiwatari Department of Chemistry, Tokyo Metropolitan University, Tokyo 192-03, Japan

James F. Kasting Department of Geosciences, The Pennsylvania State University, University Park, Pennsylvania 16802

Mahlon C. Kennicutt II Geochemical and Environmental Research Group (GERG), Texas A&M University, College Station, Texas 77845

Charles R. Landis Arco Oil Exploration Research and Technical Services, Plano, Texas 75075

C. Largeau Laboratoire de Chimie Bioorganique et Organique Physique, UA CNRS 456, École Nationale

S. R. Larter Newcastle Research Group in Fossil Fuels & Environmental Geochemistry, University of Newcastle, Newcastle upon Tyne NE1 7RU, England

Cindy Lee Marine Sciences Research Center, State University of New York, Stony Brook, New York 11794

J. W. de Leeuw Organic Geochemistry Unit, Faculty of Chemical Engineering and Materials Science, Delft University of Technology, 2628 RZ Delft, The Netherlands; *present address*: Division of Marine Biogeochemistry, Netherlands Institute of Sea Research (NIOZ), Texel, The Netherlands

Joel Leventhal U.S. Geological Survey, Federal Center MS 973, Denver, Colorado 80225

M. D. Lewan Amoco Production Company, Research Center, Tulsa, Oklahoma 74102; *present address*: U.S. Geological Survey, Federal Center, Denver, Colorado 80225

C. Anthony Lewis Centre for Petroleum and Environmental Organic Geochemistry, School of Applied Chemistry, Curtin University of Technology, Perth 6001, Western Australia, Australia; *present address*: Department of Environmental Sciences, University of Plymouth, Plymouth, Devon PL4 8AA, England

Jerold M. Lowenstein Department of Medicine, University of California–San Francisco, San Francisco, California 94143-0568

Stephen A. Macko Department of Environmental Sciences, The University of Virginia, Charlottesville, Virginia 22903

David A. C. Manning Department of Geology, University of Manchester, Manchester M13 9PL, England

Lawrence M. Mayer Department of Oceanography, Darling Marine Center, University of Maine, Walpole, Maine 04573

David McKirdy Department of Geology and Geophysics, The University of Adelaide, Adelaide, South Australia 5001, Australia

Renee McLaughlin Core Laboratories, Carrollton, Texas 75006

Philip A. Meyers Department of Geological Sciences and Center for Great Lakes and Aquatic Science, The University of Michigan, Ann Arbor, Michigan 48109

Stanley L. Miller Department of Chemistry, University of California–San Diego, La Jolla, California 92093-0317

Richard M. Mitterer Department of Geosciences, University of Texas at Dallas, Richardson, Texas 75083-0688

G. Muyzer Department of Biochemistry, Leiden University, 2333 CC Leiden, The Netherlands; *present address*: Molecular Ecology Unit, Max-Planck-Institute for Marine Microbiology, D-2800 Bremen 33, Germany

Bartholomew Nagy Department of Geosciences, The University of Arizona, Tucson, Arizona 85721

Peggy H. Ostrom Department of Geological Sciences, Michigan State University, East Lansing, Michigan 48824-1115

Susan E. Palmer Amoco Production Company, Research Center, Tulsa, Oklahoma 74102

Patrick L. Parker Department of Marine Sciences, The University of Texas at Austin, Austin, Texas 78712

R. P. Philp School of Geology and Geophysics, The University of Oklahoma, Norman, Oklahoma 73019

Fredrick G. Prahl College of Oceanography, Oregon State University, Corvallis, Oregon 97331

L. L. Robbins Department of Geology, University of South Florida, Tampa, Florida 33620

Jürgen Rullkötter Institut für Chemie und Biologie des Meeres (ICBM), Carl von Ossietzky Universität Oldenburg, D-2611 Oldenburg, Germany

Manfred Schidlowski Max-Planck-Institut für Chemie (Otto-Hahn-Institut), D-55020 Mainz, Germany

Joseph T. Senftle ARCO Oil Exploration Research and Technical Services, Plano, Texas 75075

Bernd R. T. Simoneit Petroleum Research Group, College of Oceanic and Atmospheric Sciences, Oregon State University, Corvallis, Oregon 97331

Roger E. Summons Australian Geological Survey Organisation, Canberra, ACT 2601, Australia

Carolyn L. Thompson-Rizer Conoco, Inc., Ponca City, Oklahoma 74602-1267

Stuart G. Wakeham Skidaway Institute of Oceanography, Savannah, Georgia 31416

John F. Wehmiller Department of Geology, University of Delaware, Newark, Delaware 19716

Jean K. Whelan Department of Marine Chemistry and Geochemistry, Woods Hole Oceanographic Institution, Woods Hole, Massachusetts 02543

John E. Zumberge GeoMark Research, Inc., Houston, Texas 77095

Foreword

The publication of *Organic Geochemistry* by Professors Engel and Macko is welcomed by geochemists and geologists working in the disciplinary field of geochemistry. The forty chapters and fifty-eight authors represent the many specialties in geochemistry as well as the interdisciplinary and international nature of the field. Geophysicists, mathematicians, exobiologists, paleontologists, microbiologists, and stable isotope chemists also contribute to the diverse investigations of organic matter as it appears in the biomass of the earth and solar system.

This volume appears almost twenty-five years after the publication of *Organic Geochemistry* by Professors Eglinton and Murphy. Since that time, biomarkers in ancient and Recent sediments have been a significant development. New analytical methods and instrumentation to identify chemicals that come from life or could support life have extended the capability to detect smaller amounts of more complex organic molecules. Most instrument detector signals are now sent directly to a computer that is programmed to print out results. Modeling has also contributed to the interpretation of analytical data and increased our knowledge and theory of older sediments.

Since 1969, new sample materials and remote sensing data have become available, including lunar rocks, Antarctic meteorites, Viking data from Mars, the Voyager results from the planetary atmospheres such as Titan, the mapping of Venus, the Hubble telescope pictures, and deeper and older source rocks from off-shore ocean cores.

The world has also changed dramatically. As a result, cooperation among nations is greater now, and the cooperative spirit common among scientists will enhance future geochemical investigations around the world.

As was perhaps true in 1969, needs still exist for basic research as well as synthesis of model compounds and standards. Environmental problems now considered on a worldwide basis require fundamental research needing the expertise of geochemistry. The nature, origin, and occurrence of biomass and fossil fuels in the environment still influence quality of life, economics, and human values.

As geochemists continue to search for new solutions, we acknowledge the interdependence of academe, government, and industry in the development of organic geochemistry. Each has a unique contribution to make to basic and applied research. Special thanks are due to educators, both faculty and mentors, who have initiated further basic research and patiently guided the next generation of geochemists to take their rightful place in the profession.

Government, with its service agencies and financial support of basic research, has provided support for academic institutions. Industry has been committed to the development of new and improved technologies, under constraints of time, money, and personnel, to meet the challenges of unsolved problems.

In a practical way, this book will help to provide background for the various interdisciplinary fields in organic geochemistry. We are very grateful to professors Michael Engel and Stephen Macko for their determined efforts to prepare this volume in time for the twenty-fifth anniversary of the publication of *Organic Geochemistry* by Professors Eglinton and Murphy. May it meet the expectations of the researcher in geochemistry and serve as a significant text for students for the next twenty-five years.

S. Mary Ellen Murphy, RSM, Ph.D.

Saint Joseph's College, Windham, Maine

Preface

With the passage of nearly twenty-five years since the publication of the primary text and reference in the field, it has become clear that the organic geochemistry community needs an updated work on this subject. Much of the methodology, interpretations, and findings reported in that primary work have drastically changed, and whole new fields of research have been introduced. This book represents a community effort to present the current understanding in this diverse field. We have encouraged the authors to introduce as many details as possible within the constraints of their chapters, as well as to provide direction for future study to students using this text. With this in mind, chemical terms in the Index are cross-listed with *Chemical Abstracts* to assist the reader in future literature searches.

The first three chapters of the book provide a general introduction to organic geochemistry, and focus on the composition of the biosphere, both chemically and isotopically, thus affording a perspective on the nature and fate of organic compounds that may be preserved in the geosphere. The second section of the book presents an overview of processes of organic matter and approaches to interpreting and understanding the geochemical record of these compounds. In Chapters 4 to 12, the early diagenesis and preservation of organic components are discussed. Chapters 13 to 18 address the macromolecular organic matter (e.g., kerogen) commonly found in rocks and sediments, the methods used to characterize it, and its conversion to fossil fuels. Chapters 19 to 24 focus on application of organic geochemical methods for hydrocarbon exploration. Chapters 25 to 28 address the associations of organic matter with metalliferous deposits.

The origin and composition of organic matter on the early Earth are discussed in Chapters 29 to 33. Applications of organic geochemistry in Quaternary research, ranging from stratigraphic correlation to chemotaxonomy, are presented in Chapters 34 to 39. Finally, in Chapter 40, Professor G. Eglinton presents his assessment of present-day problems and future perspectives in organic geochemistry.

Since the first reports in 1934 by the chemist Treibs of organic compounds in oils and shales that could be associated with living organisms, research in organic geochemistry has grown at an exponential rate. The field has expanded from a small community interested primarily in petroleum to one encompassing such diverse interests as oceanography, environmental chemistry, biogeochemistry, paleontology, and archaeology, to name a few. With recent developments in technology and a healthy stream of new ideas for its application, the field continues to evolve.

We are indebted to all of the authors for their cooperation and patience in preparing this text and to the reviewers who performed the task of challenging the authors to present more material in clearer fashion as concisely as possible. We wish to thank Diana Diez de Medina, Lisa Plaster-Kirk, and Kendra Brisman for their help with the initial editing process, and the editorial staff at Plenum—Amelia McNamara, John Matzka, and their assistants—for helping us to complete this formidable undertaking. We are extremely grateful to Joseph Senftle and ARCO Oil and Exploration for financial support for the color reproductions that appear in Chapter 15. S. A. Macko thanks Drs. Hillaire-Marcel at UQAM and Charles Prewitt at the Geophysical Laboratory for providing him space to continue his work on the text during his sabbatical leave. M. H. Engel acknowledges Dr. Frank Stehli for his continued encouragement to undertake this project.

Finally, we thank our advisors B. Nagy and P. L.

Parker for giving us the opportunity to work in this field, and the Geophysical Laboratory (Drs. P. E. Hare, T. C. Hoering, M. Fogel, and H. Yoder) for creating the scientific environment for nurturing the collaboration of the editors to eventually mix isotope research with amino acid analysis. Last, we would like to thank our spouses for their patience.

Stephen A. Macko
Michael H. Engel

Charlottesville, Virginia and Norman, Oklahoma

Contents

Part I • Introduction

Part II • Early Diagenesis of Organic Matter

Chapter 12 • The Early Diagenesis of Organic Matter: Reaction at the Air–Sea Interface

David J. Carlson

Part III • Kerogen and Related Materials

Chapter 13 • Determination of Structural Components of Kerogens by the Use of Analytical Pyrolysis Methods

S. R. Larter and B. Horsfield

Chapter 14 • Chemical Methods for Assessing Kerogen and Protokerogen Types and Maturity

Jean K. Whelan and Carolyn L. Thompson-Rizer

Chapter 15 • Organic Petrographic Approach to Kerogen Characterization

Joseph T. Senftle, Charles R. Landis, and Renee McLaughlin

Part IV • Thermal Alteration of Organic Matter and the Formation of Fossil Fuels

Part V • Applications of Organic Geochemical Research for Hydrocarbon Exploration

Chapter 24 • The Origin and Distribution of Gas Hydrates in Marine Sediments

Mahlon C. Kennicutt II, James M. Brooks, and H. Benjamin Cox

Part VI • Organic Matter and Metalliferous Deposits

Chapter 25 • The Role of Organic Matter in Ore Transport Processes

David A. C. Manning and Andrew P. Gize

Chapter 26 • Aspects of the Organic Geochemistry and Petrology of Metalliferous Ores

Andrew P. Gize and David A. C. Manning

Chapter 27 • Metals in Black Shales

Joel Leventhal

Chapter 28 • Trace and Minor Elements in Coal

Robert B. Finkelman

Part VII • The Precambrian

Chapter 29 • Evolution of the Earth's Atmosphere and Hydrosphere: Hadean to Recent

James F. Kasting

Chapter 30 • The Prebiotic Synthesis of Organic Compounds on the Early Earth

Stanley L. Miller

I

Introduction

Chapter 1

Biogeochemical Cycles

A Review of Fundamental Aspects of Organic Matter Formation, Preservation, and Composition

ROGER E. SUMMONS

1. Introduction

Carbon, hydrogen, nitrogen, oxygen, phosphorus, sulfur, calcium, and iron are the principal chemical elements that living organisms utilize in structural tissues, for replication, and for energy-harvesting activities. These same elements are also important components of the oceans, atmosphere, and crustal rocks. The physiologies of living organisms combine with chemical, physical, and geological forces to continually redistribute these elements between living and nonliving reservoirs in processes known as biogeochemical cycles. The word biogeochemistry implies separate contributions from biology, chemistry, and geology in this science. However, these facets cannot and should not be viewed in isolation. A rigorous appreciation of the interaction between geosphere and biosphere is reliant upon our awareness of the complexity and interdependence of the physical and biological worlds and an appreciation of the overwhelming importance of microbial life.

Organic matter, a fundamental constituent of the biosphere, is also the most important renewable chemical reductant on Earth. Organic carbon compounds are ubiquitous, abundant, and sometimes overlooked components of the oceans, lakes, and sedimentary rocks, and directly, or indirectly, they fuel all biogeochemical processes. Understanding the factors which control the formation, composition, and preservation of organic matter is central to comprehending the development and continuance of biogeochemical cycles of the life-forming elements, to understanding the formation of petroleum, coal, and certain mineral deposits, and to elucidating the beginnings and history of life.

Exploration for petroleum and minerals has traditionally induced and stimulated organic geochemical research for its potential economic benefits. Now, with the recognition of man as an active geological agent, a wider appreciation of the subtleties and interactions of the various elemental cycles is of growing importance. We presently live in an age where there is great concern about environmental deterioration and possible climatic stress caused by man's agricultural, mining, and lifestyle activities. In this regard, we need to know much more about the functioning and history of the biogeochemical cycles. Fossil organic matter comprises one of the more important and

ROGER E. SUMMONS • Australian Geological Survey Organisation, Canberra, ACT 2601, Australia. R. Summons publishes with permission from the Director, Australian Geological Survey Organisation.
Organic Geochemistry, edited by Michael H. Engel and Stephen A. Macko. Plenum Press, New York, 1993.

comprehensive records of environmental change on both local and global scales, extending over a time period from the present to at least three billion years ago. Accurate deciphering and interpretation of this record assumes a new and critical significance if man is to continue to exploit resources and yet retain a compatible position in the biosphere.

This chapter briefly summarizes some of the modes in which organic matter, by virtue of its composition, may store historical information. There is a short discussion of the pathways and budgets for the principal chemical elements which cycle between living and nonliving organic pools and inorganic reservoirs in the Earth's crust. Then follows a synopsis of how biological, chemical, and geological factors interact to effect environmental change and evolution on a global scale. Books, review articles and some of the key, primary research papers in biogeochemistry are cited, and one intended purpose is to introduce the reader to the more detailed primary literature about this extensive, important, and fascinating field of science.

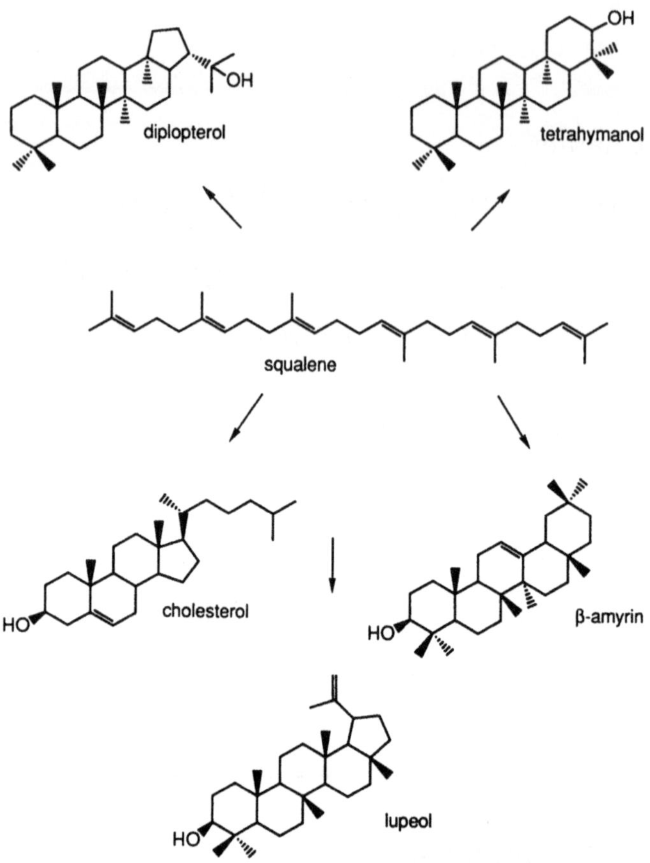

Figure 1. An illustration of the varied lipid structures which can be produced when the ubiquitous precursor hydrocarbon squalene is cyclized by different classes of organism. Diplopterol is synthesized by Eubacteria, cholesterol by eukaryotes, and tetrahymanol by some protozoans and bacteria. Lupeol and β-amyrin are higher plant products. Cyclizations to diplopterol and tetrahymanol take place anaerobically. The other transformations require molecular oxygen and are presumed to have appeared after the advent of oxygenic photosynthesis.

2. Organic Matter as an Information Source

We can study organic matter at its various structural levels and from diverse perspectives. At the *atomic* level, elemental and isotopic compositions are the essential parameters which are used to determine mass transport, elucidate pathways, and construct inventories. At the *molecular* level, the structures and stereochemical variations of individual organic compounds, *biological markers* or *biomarkers*, are important means of tracing source organisms, subsequent diagenetic affects, and thermal history. General and specific features operate simultaneously in the biosynthetic pathways of living organisms, and these are individually encoded into biomarker structures. For example, the C_{30} hydrocarbon squalene is a universal precursor molecule but, depending on the class of organism, can be transformed into a great variety of acyclic, tetracyclic, and pentacyclic product molecules (Fig. 1). These lipids and their fossil hydrocarbon analogs provide one of our chemical means for comparing and contrasting modern and ancient environments.

At the *macromolecular* level, genetic information is inherent in protein and nucleic acid biopolymers. These molecules are not refractory to the same degree as lipids and structural polymers and hence do not survive long periods of burial in sediments. Nevertheless, such information, particularly that which is encoded into ribosomal RNA sequences, is apparently highly conserved in extant organisms and can be retrieved using the techniques of modern molecular biology. The RNA and DNA sequence data, now being gathered from diverse living source organisms, very effectively inform us of the detailed phylogenetic relationships of early counterparts. At a still higher level of organization, the habit, morphology, and comparative biochemistry and physiology of whole *organisms* and *communities* of interdependent organisms constitute our principal means of inferring the behavior of ancient analogs. Finally, and most significantly, the preserved *geological record* of bacterial and protistan microfossils, microbial mat structures and textures, animal and plant macrofossils, isotopic anomalies and biomarkers tells us about the timing of events and how to construct evolutionary trees. The organic matter, continuously deposited and preserved in ancient sediments over significant periods of time, encapsulates and thus informs us of the evolutionary history of natural systems in response to tectonic, environmental, and biological change. The organic matter present in contemporary sedimentary environments assists in distinguishing between natural and anthropogenic phenomena.

3. Classification of Living Organisms; the Importance and Diversity of Microbes

Organisms have traditionally been divided into major taxonomic groups, defined by their physiological and biochemical capacities, their morphology, and their habit (i.e., *phenotypic* features). Genetic affinities, using data derived from *genotypic* analyses, now constitute the main means of assigning relationships (*phylogeny*). The new information derived from nucleic acid sequences is forcing a reappraisal of many phylogenetic relationships and a revision of classical taxonomy, particularly with respect to microbes. In contrast to the earlier five-kingdom classification of Whittaker (1969), it is now widely accepted that there are three primary kingdoms of organisms (Fig. 2), of roughly equal antiquity and which stem, in ways presently unknown, from a universal ancestral line (Woese and Fox, 1977; Woese, 1981; Woese and Wolfe, 1985; Olsen *et al.*, 1986). The simplest cellular organisms, the Prokaryotes, constitute two of the three primary kingdoms, that is, the Eubacteria and the Archaebacteria. Prokaryotes are generally very small objects (e.g., typical spheroids are about 1 μm in diameter or less), unicellular or living in simple colonies, and with little subcellular structure and minimal organization. The other primary kingdom, the Eukaryota, is a grouping of larger (e.g., a typical unicell would be 10 μm in diameter) and more complex organisms, many of them multicellular. Eukaryotes possess a membrane-bound cell nucleus and discrete intracellular organelles such as mitochondria and chloroplasts and have other morphological characteristics which distinguish them from the prokaryotes. While nucleic acid sequence data confirm the distinctiveness of the genetic material in the eukaryote nucleus, the mitochondria and chloroplasts have separate genetic material with close affinities to that of prokaryote groups, adding strong evidence to support the hypothesis that these organelles arose through evolution of earlier symbiotic associations between prokaryotes and the ancestral eukaryote as host (e.g., Margulis, 1981). For example, it has recently been established that there is a close phylogenetic relationship between the cyanobacteria and the chloroplasts of photosynthetic eukaryotes (e.g., Bonen *et al.*, 1979; Giovannoni *et al.*, 1988). Unicellular or simply organized colonies of eukaryotes are termed *protists* while complex and differentiated plants and animals are termed *metaphytes* or *metazoans*, respectively.

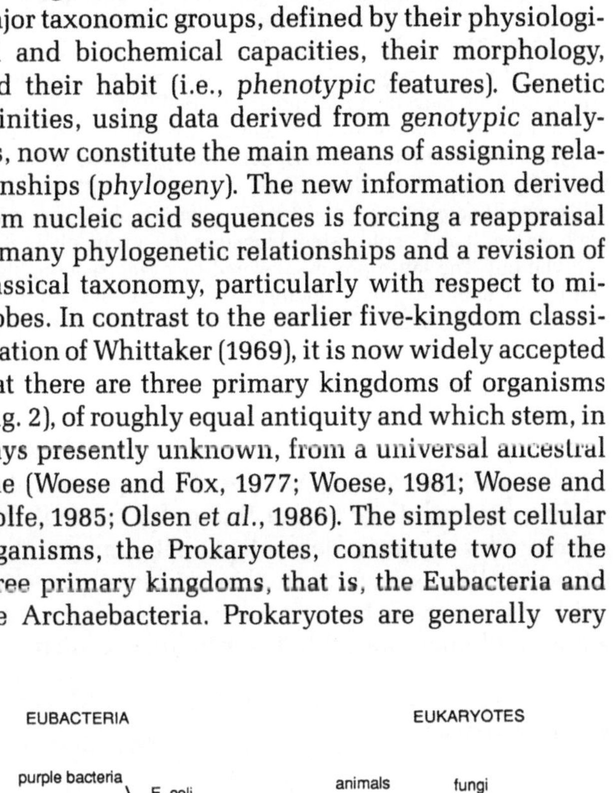

Figure 2. A diagram illustrating phylogenetic relationships between the three kingdoms of organisms. The data used to determine these comes from the base sequences of nucleic acids. The tree presently has no root because we have no unambiguous knowledge of the ancestral forms.

Microorganisms, that is, the prokaryotes and protists, comprise the bulk of living biomass; they generally grow and multiply faster than higher organisms and inhabit a wider range of environments. Microbial microfossils and microbially constructed stromatolites have an unambiguous record extending to at least 3.5 billion years of the 4.6 billion years of earth history (Awramik *et al.*, 1983; Walter, 1983; Schopf and Walter, 1983), although the earliest part of this record is poorly preserved and, accordingly, is somewhat cryptic. The oldest preserved microfossils (3.5 Ga) resemble modern Eubacteria while the oldest stromatolites were probably constructed by anoxygenic photosynthetic bacteria (for recent discussions, see Walter, 1983; Schopf and Walter, 1983; Knoll and Bauld, 1989; Pierson and Olson, 1989). The oldest visibly recognizable eukaryotes are in sediments dated at approximately 2.1 Ga (for discussions, see Schopf, 1983; Knoll, 1983; Knoll, 1989; Han and Runnegar, 1992; Schopf and Klein, 1992). By comparison, the metazoans and metaphytes have restricted distributions in space and have an undoubted fossil record of about 0.6 billion years (e.g., Glaessner, 1984). There is, however, evidence for a prehistory of organisms such as seaweeds extending to about 2.1 billion years (e.g., Walter *et al.*, 1990; Han and Runnegar, 1992). Microbes, in contrast to metazoans and metaphytes, show immense biochemical diversity (e.g., Starr *et al.*, 1981; Brock and Madigan, 1988). The chemical conversions effected by them are quantitatively significant on a global scale and, over geological time, have gradually transformed the surface of the planet from a hostile landscape to an environment supportive of complex life (e.g., Holland, 1978, 1984; Hayes *et al.*, 1983; Chapman and Schopf, 1983; Gregor *et al.*, 1988; Schopf and Klein, 1992). In the transfer of organic carbon through modern food chains, the final conversions are the milieu of bacteria. As outlined above, microbes, and principally the prokaryotes, were the only biological agents throughout much of Earth history. Thus, microbes and their metabolic products should be, and indeed are, the primary focus of organic geochemical interest.

Specific biochemical capacities, preferences, and tolerances generally mean that individual microbial types will have favored habitats, or ecological niches, defined by the availability of specific carbon and energy resources and by the absence of toxic substances (e.g., molecular oxygen or hydrogen sulfide). A suitable habitat may involve living in a symbiotic, or a close-community, relationship with interdependent organisms (consortium or syntropic system). *Autotrophic* organisms utilize inorganic carbon (CO_2 or HCO_3^-) as a starting material. To chemically reduce this carbon dioxide, they require energy harvested from sunlight (*photoautotrophs*) or energy from specific inorganic chemical reactions (*chemoautotrophs*). Another approach employed by some microbes (*photoheterotrophs*) is to use energy from sunlight to assimilate simple reduced carbon compounds. *Photosynthesis* is the process whereby energy from sunlight is utilized to fuel the conversion of carbon dioxide to simple organic intermediates and thence to carbohydrate (Fig. 3). Thus, photoautotrophs are the major "primary" producers of organic matter. They include prokaryotes such as the cyanobacteria, purple and green photosynthetic bacteria, and prochlorophytes, as well as the eukaryotic algae and the higher plants. *Heterotrophic* organisms obtain both their carbon and energy requirements from the reduced carbon compounds which were originally made by primary producers. They return inorganic carbon to the geosphere (*mineralization*). Heterotrophic organisms include heterotrophic bacteria, protists such as protozoans and fungi, and metazoans. Within the heterotrophs, there are a variety of different processes for obtaining energy through the oxidation of organic carbon, that is, *respiration* (Fig. 4). Some, including most eukaryotes and many bacteria, use molecular oxygen as an oxidant and are termed *aerobes*, qualified by the terms obligate or facultative depending on whether the requirement for oxygen is absolute or not, or the prefix micro if their tolerance for oxygen is dependent on its concentration. Others utilize substances such as nitrate or sulfate as oxidants, generally have no tolerance for oxygen, and are termed *anaerobes*. Photosynthetic organisms ob-

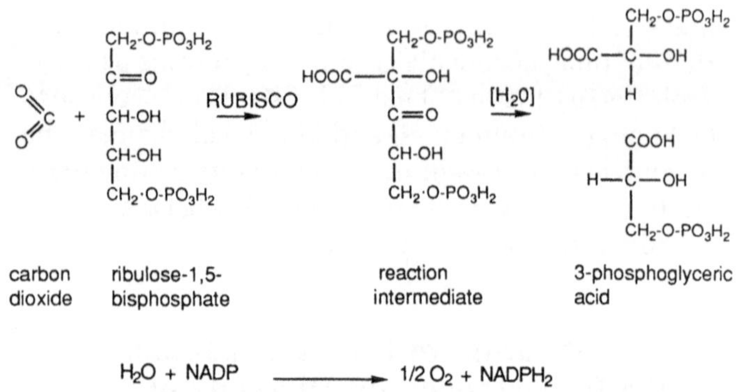

Figure 3. The critical steps of the Benson–Calvin Cycle, the predominant mode of carbon dioxide fixation in autotrophy. In the "dark" reaction, one molecule of carbon dioxide reacts with one molecule of ribulose bisphosphate (RUBP) in a reaction catalyzed by the enzyme ribulose bisphosphate carboxylase oxygenase (RUBISCO). The resulting adduct decomposes to yield two molecules of 3-phosphoglyceric acid (3-PGA). Some molecules of 3-PGA (5/6) are recycled to produce more RUBP while others (1/6) constitute the net gain of the cycle and are diverted to carbohydrate biosynthesis via fructose 6-phosphate. Energy for the dark reactions of photosynthesis is obtained from the "light" reactions, utilizing either photosystem I (cyclic photophosphorylation) or photosystem II and photosystem I together (noncyclic photophosphorylation). The latter process, favored by cyanobacteria, algae, and higher plants, involves photolysis of water with evolution of oxygen and is the source for the oxygen in the Earth's atmosphere and surface waters.

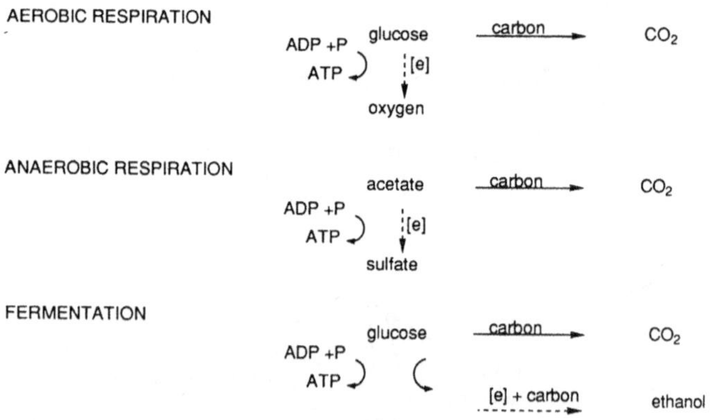

Figure 4. Modes of releasing energy by respiration or fermentation of simple organic carbon compounds. This energy is trapped and stored in the high-energy phosphate bonds of adenosine triphosphate (ATP) and rereleased when the ATP is converted to adenosine diphosphate (ADP). Full arrows show carbon flow while dashed arrows indicate the direction of the flow of electrons.

tain energy via conventional respiration when light is unavailable. In the process known as (anaerobic) *fermentation*, there is no net oxidation, and an organic carbon compound of intermediate oxidation state reacts in an internal disproportionation. Fermentation yields products which are both reduced and oxidized in comparison to the substrate, together with some energy which is harnessed for use in other cellular processes (Lehninger, 1982).

Archaebacteria, the third and most recently recognized primary kingdom of organisms, preferentially inhabit extreme environments and comprise taxa with a variety of quite distinctive biochemistries. They have no presently recognized morphological fossil record and only an uncertain detection through carbon isotopic anomalies in some Archaen and Early Proterozoic sediments (Schoell and Wellmer, 1981; Hayes, 1983). Nucleic acid sequence information and difference in membrane composition indicate that the kingdom is at least as old and as fundamentally distinct as the Eubacteria and the Eukaryota (Woese and Fox, 1977). Characteristic and relatively inert, ether-linked, isoprenoid membrane lipids (e.g., Langworthy, 1985), and the techniques of organic geochemistry, render Archaebacteria recognizable through their chemical fossils in some of the oldest, presently known, organic-rich sediments (Summons *et al.*, 1988). Archaebacteria include the anaerobic methanogens (Jones *et al.*, 1987) which have the capacity to produce methane from carbon dioxide (or simple carbon compounds) and a reductant such as hydrogen. Halophilic Archaebacteria are aerobic heterotrophs which exist only in environments with high salt concentration. The thermophilic sulfur-dependent Archaebacteria comprise both aerobic and anaerobic orders and, as their name implies, flourish at high temperatures and/or in sulfide-rich environments such as hot springs (Woese, 1981; Woese and Wolfe, 1985). Sulfide is unstable in the presence of oxygen but would have been omnipresent during the early stages of Earth history, and this, in combination with distinctive molecular biology and sulfur-dependent metabolism of these Archaebacteria, has led to the suggestion that they may be phylogenetically close to the last common ancestor of all extant life (e.g., Woese and Olsen, 1986; Woese, 1987; Lake, 1988).

For our understanding of the types of organisms which are geochemically and geologically significant, we rely heavily on knowledge and analogy generated from studies of modern microbes and microbial ecosystems. However, we should not lose sight of the fact that these organisms and communities are not the same as the ones which inhabited the Earth in times past. Modern organisms are just that and, in the course of their continuing evolution, have possibly lost and/or gained biochemical capacities and tolerances along with the morphological and structural changes we can recognize from the fossil record. We should never automatically assume that a "simple" contemporary microbe is identical to its ancestors.

4. Biogeochemical Cycles and Inventories of Major Bioactive Elements

The major chemical elements of living systems, C, H, N, O, P, and S, are in dynamic flux between their living and dead "organic" forms and one or more "inorganic" or nonbiological reservoirs. The transfer of each element between pools proceeds in cyclical fashion by spontaneous chemical reactions as well as by biological intervention. These conversions are therefore known as biogeochemical cycles (e.g., Odum, 1971; Garrels and Perry, 1974; Garrels *et al.*, 1975; Trudinger *et al.*, 1979; Ivanov and Freney, 1983). Traditionally, spatial separations of the cycles have been identified in the *endogenic*, or lithospheric or "rock," cycle and the *exogenic cycle*, or cycle in the sphere of living organisms (e.g., Golubic *et al.*, 1979).

The major biogeochemical cycles are intimately interconnected and are ultimately powered by energy from the sun via photosynthetic carbon fixation. Localized exceptions occur such as, for example, hydrothermal environments where the energy for processes such as carbon fixation may be provided (*lithotrophy*) by inorganic reducing agents emanating from molten rocks. The transformations within each biogeochemical cycle are oxidation and reduction reactions, each providing a basis for connection between cycles. In the exogenic carbon cycle (Fig. 5), for example, the inorganic pools consist of carbon in its oxidized states of CO_2 gas, in the atmosphere and dissolved in water, and HCO_3^- and CO_3^{2-}, mainly dissolved in the oceans, and as carbonate minerals in sediments. The organic pools comprise carbon in reduced states mainly confined in living biomass, in recently dead organic matter dissolved or suspended in the oceans and in surficial sediments, and in ancient, deeply buried sedimentary organic matter (kerogen, coal, and petroleum). The lithosphere is the reservoir for the bulk of global carbon, with about one-fifth of this being fossil organic carbon. This buried organic carbon far exceeds the amount of carbon confined in living biomass (Garrels and Mackenzie, 1971; Garrels *et al.*, 1975, 1976; Trudinger *et al.*, 1979; Berner, 1987, 1989). Table I shows current estimates of the most

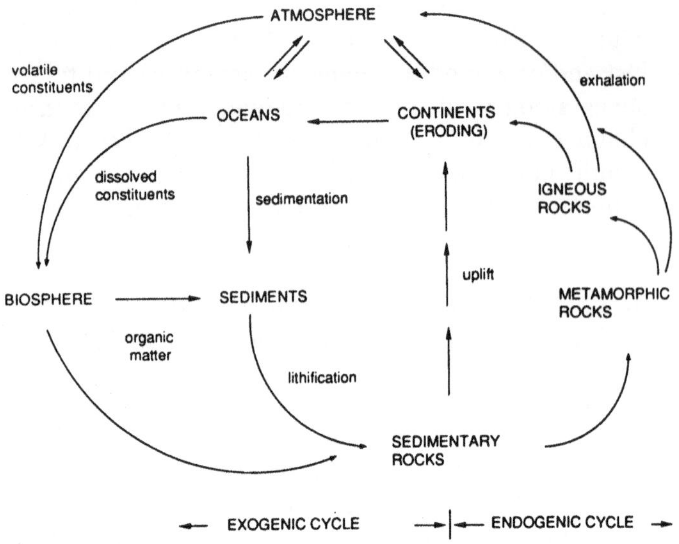

Figure 5. The exogenic and endogenic cycles, showing interchange of matter between the biosphere, oceans, atmosphere and "rocks." These features are common to the biogeochemical cycles of different elements.

TABLE I. Carbon Reservoirs: Residence Times and Isotopic Composition[a]

Species	Amount (10^{18} g C)	Residence time[b] (yr)	$\delta^{13}C$[c] (‰ PDB)
Sedimentary carbonate C	62,400	342,000,000	~0
Sedimentary organic C	15,600	342,000,000	~−24
Oceanic inorganic C	42	385	~+0.46
Necrotic C	4.0	20–40	~−27
Atmospheric CO_2	0.72	4	~−7.5
Living terrestrial biomass	0.56	16	~−27
Living marine biomass	0.007	0.1	~−22

Other carbon cycle parameters	Estimated value
Terrestrial net productivity	48×10^{15} g C yr^{-1}
Marine net productivity	35×10^{15} g C yr^{-1}
Burial of organic C in marine sediments	0.13×10^{15} g C yr^{-1}[d]
Isotopic fractionation of carbon assimilation	0–36‰

[a]Data extracted from Garrels *et al.* (1975), Deines (1980), Garrels and Lerman (1984), Dean *et al.* (1986), Berner (1987, 1989), and references therein.
[b]Residence times are "mean" times; e.g., some carbon in crust for $>10^9$ yr.
[c]Isotopic compositions are "average" values. There are significant variations with age and environment, viz. Chapter 3. Isotopic compositions are expressed relative to the international standard Pee Dee belemnite (PDB) where $\delta^{13}C$ (‰ PDB) = $[(^{13}C/^{12}C)_{sample}/(^{13}C/^{12}C)_{standard} - 1] \times 10^3$.
[d]The ultimate fractionation depends on assimilation pathway and will decrease as carbon becomes limiting.

important carbon reservoirs and mean residence times for carbon in these pools. Aspects of the carbon cycle are illustrated in Fig. 6.

Reduction of one mole of carbon dioxide in oxygenic photosynthetic formation of carbohydrate, and the concomitant splitting of a mole of water, releases one mole of oxygen. This, in turn, will be available for respiratory (oxidation) processes. Maintenance of an atmospheric oxygen concentration of 21%, therefore, requires rapid cycling of carbon in the atmospheric, hydrospheric, and biospheric pools, with the bulk of organic carbon remaining in the massive, and only slowly cycling, pool in the lithosphere (Garrels and Lerman, 1984; Berner, 1989). Oxygen-dependent respiration processes are the most efficient and quantitatively significant means of organic matter mineralization. Accordingly, the preservation and accumulation of organic matter in sediments depend strongly on the site of its production as well as the effectiveness of its removal and protection from the oxygen simultaneously produced at that same site. Quantitative data about the sites of production and reservoirs of living biomass are important, yet notoriously difficult, parameters to determine accurately (e.g., Duursma and Dawson, 1981; Romankevich, 1984). It is generally agreed that continental ecosystems produce and conserve more organic matter than do marine ecosystems, despite the area of the oceans being more than double that of the landmass (see Table I; Stumm, 1977; Holland, 1978).

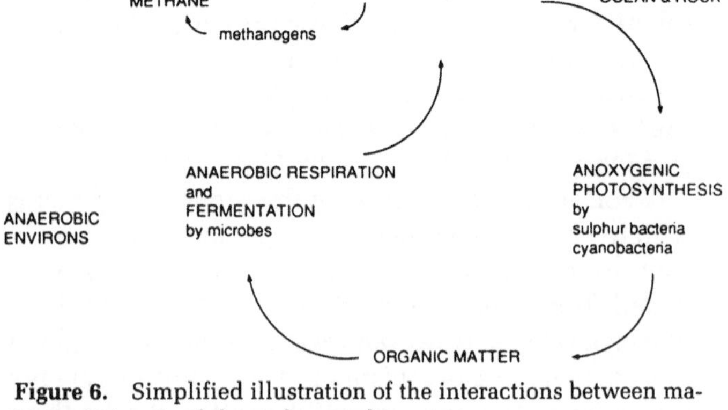

Figure 6. Simplified illustration of the interactions between major participants of the carbon cycle.

Organisms fixing carbon by photosynthesis also require a continual supply of nutrients such as nitrogen and phosphorus, thus forming a link between the cycles of these elements and the carbon cycle and influencing the spatial aspects of organic matter production. The nitrogen cycle (e.g., Trudinger *et al.*, 1979; Sweeney *et al.*, 1978; Schidlowski *et al.*, 1983), like the carbon cycle, involves rapid recycling within relatively small pools of organic nitrogen compounds in living organic matter and available inorganic nitrogen compounds such as ammonia, nitrate, and nitrite. Much larger and more slowly cycling nitrogen pools are confined as dinitrogen (N_2) in the atmosphere and as sedimentary inorganic and organic nitrogen. The large and small pools are linked by weathering processes, by biological oxidation of ammonium to nitrate (*nitrification*) and reduction of nitrate to N_2 (*denitrification*), and by the actions of nitrogen-fixing organisms converting atmospheric N_2 to ammonium iron (*nitrogen fixation*). The capacity for biological nitrogen fixation is a property of many prokaryote taxa (Gordon, 1981) and is probably quantitatively most important in cyanobacteria, phototrophic bacteria, and symbiotic nitrogen-fixing bacteria such as Rhizobiaceae. The atomic carbon-to-nitrogen ratio in living biomass is in the range of 100 (woody terrigenous materials) to about 6.6 (marine microorganisms). In dead organic matter, this ratio gradually increases during burial and diagenesis, and nitrogen is released and/or recycled. Data on inventories and fluxes for the global nitrogen cycle have been compiled by Sweeney *et al.* (1978) and reviewed by Schidlowski *et al.* (1983).

Phosphorus is less abundant than nitrogen in living organisms (for marine organic matter, atomic C:P $\approx$ 106:1; Redfield *et al.*, 1963) and even less abundant in fossil organic matter, consistent with rapid and efficient recycling of organic phosphorus. The slowly cycling lithospheric phosphate reservoir is massive (1.1×10^{25} g) in relation to the combined marine and terrestrial inorganic (2.9×10^{17} g) and marine and terrestrial organic (2×10^{15} g) pools (Pierrou, 1979). The availability of phosphorus is considered more likely to affect the fertility of a particular ecosystem than is that of nitrogen (Hayes *et al.*, 1983). A detailed account of the marine phosphorus cycle has been made by Froelich *et al.* (1982).

The biological sulfur cycle (Fig. 7; e.g., Ivanov and Freney, 1983) is one of the most important biogeochemical phenomena despite the fact that sulfur compounds themselves are a quantitatively minor, albeit essential, component of living biomass. Certain classes of bacteria, the sulfide-oxidizing bacteria and

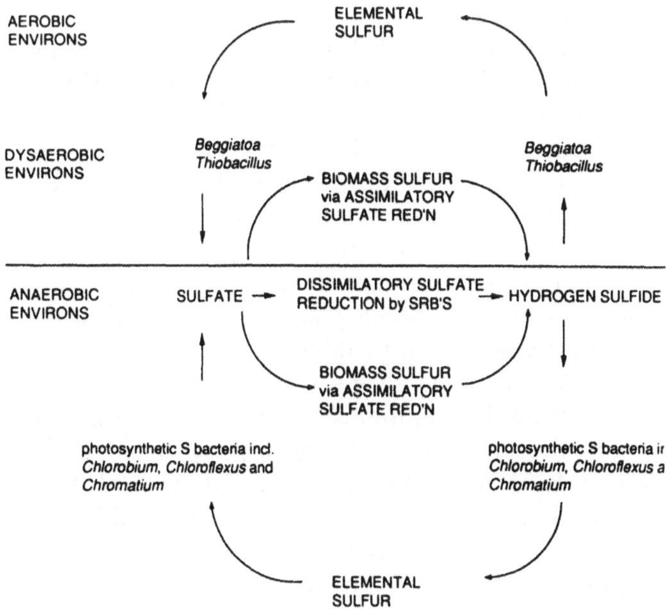

Figure 7. Simplified illustration of the interactions between major participants of the sulfur cycles. SRB = sulfate-reducing bacteria.

phototrophic sulfur bacteria, continually transform vast quantities of sulfur from sulfide to sulfate in energy transfer processes. Other groups of bacteria utilize sulfate as an oxidant for respiration. The majority of microbes and plants acquire sulfate for the production of biomass and reduce it in processes known as *assimilatory sulfate reduction*. Animals generally acquire sulfur as reduced sulfur compounds, mainly amino acids.

The most quantitatively important groups with respect to global sulfur cycling are the sulfate-reducing bacteria (SRB) and sulfide oxidizers. The former utilize sulfate as an oxidant to obtain energy from preformed organic carbon. Hydrogen sulfide is liberated as a by-product in the process known as *dissimilatory sulfate reduction* (e.g., Trudinger, 1979; Postgate, 1984; Skyring, 1987). The other geochemically significant group, living syntropically with SRB or close to magmatic sulfide sources, have the capacity to oxidize sulfide in chemolithotrophic or phototrophic processes leading to the primary fixation of CO_2. Thus, the biogeochemical carbon and sulfur cycles are tightly coupled. Fossil organic matter, particularly that deposited in certain marine environments, may contain vast amounts of organic sulfur (up to 10% w/w and atomic C:S = 1:1) and far in excess of the normal sulfur content of living biomass (0.5 to 1.5% dry weight or atomic C:S = 200:1). This observation and the specific structures of organic compounds in this organic matter (e.g., Tissot and Welte, 1984;

Brassell *et al.*, 1986; Sinninghe Damsté, 1988; Kohnen *et al.*, 1989) indicate that this sulfur is incorporated from bacterially generated sulfide and polysulfide in diagenetic processes during the early stages of sediment burial rather than from original organic sulfur bound in biomass. Iron compounds constitute another major group of redox partners in the biogeochemical sulfur cycle, and pyrite iron together with sedimentary organic matter are major sedimentary sinks for sulfide. Some data on the global sulfur inventories are given in Table II.

Many elements, in addition to C, H, N, S, O, and P, are minor, in quantitative terms, but nevertheless essential constituents of living organisms. Na, K, Ca, Mg, Si, F, Cl, Br, I, and certain transition metals (e.g., Fe, Co, Cu, Mo, Mn, Zn, Cr, and V) are notable, and hence their distribution in nature is affected, to varying degrees, by life processes (e.g., Trudinger *et al.*, 1979). Of these elements, iron is a particularly significant one. In living organisms, it is a constituent of a number of cellular enzymes, notably those associated with respiration processes. *Thiobacillus ferrooxidans* derives energy from the oxidation of ferrous iron, and some other organisms can use ferric iron as an electron acceptor in the oxidation of organic matter (e.g., Brock and Madigan, 1988). However, it is in the form of pyrite that iron is quantitatively important. Sedimentary pyrite and organic matter are the two principal reduced partners which balance atmospheric oxygen (e.g., see Berner, 1989, and references therein).

TABLE II. Sulfur Reservoirs and Isotopic Compositions

Species	Amount (10^{18} g S)[a]	$\delta^{34}S$[b] (‰ CDT)
Sedimentary sulfate	5400	~+17 ± 2
Sedimentary sulfide	6700	~−18 ± 6
Oceanic sulfate S	1300	~+20.0
Atmosphere (SO_2 + SO_4^{2-})	2.24×10^{-6}	
Atmosphere (H_2S)	0.96×10^{-6}	
Living biosphere	8×10^{-3}	~0
Dead biosphere	5×10^{-3}	~0

Metabolic process	Isotopic fractionation[c]
Assimilatory sulfate reduction	~0‰
Dissimilatory sulfate reduction	+5−−46‰
Oxidations of sulfur species[d]	+2−−18‰

[a]Inventories are based on data extracted from Holser and Kaplan (1966), Garrels *et al.* (1975), Schidlowski *et al.* (1983), Garrels and Lerman (1984), and Berner (1987, 1989).
[b]Isotopic compositions are "average" values and are expressed relative to the international standard troilite from the Canyon Diablo meteorite (CDT), where $\delta^{34}S$ (‰ CDT) = $[({}^{34}S/{}^{32}S)_{sample}/({}^{34}S/{}^{32}S)_{standard} - 1] \times 10^3$.
[c]Isotopic fractionations are relative to a substrate with $\delta^{34}S$ = 0‰ (Schidlowski *et al.*, 1983).
[d]These data on isotopic fractionations associated with oxidations of sulfur species are summarized by Fry *et al.* (1986).

In summary, biogeochemical cycles are redox processes whereby chemical elements in the Earth's crust are transferred between various living and nonliving reservoirs. Exchanges between reservoirs are mediated by living organisms, principally microbes, and by inorganic processes such as precipitation of minerals, chemical weathering, physical erosion, and oceanic and atmospheric circulation. Energy, originally from sunlight and trapped as a consequence of photosynthetic carbon fixation, is stored in organic carbon. It is then distributed and consumed by various living participants in the cycles. The energy becomes available when the oxidation state of the carbon compounds changes during conversion back to CO_2. The same biogeochemical cycles result in transfer of matter from place to place on a global scale and with average residence times ranging from weeks (carbon in living marine organisms) to hundreds of millions of years (organic and inorganic carbon in sedimentary rocks).

5. Biochemistry and Isotopic Consequences of Incorporation of Bioelements into Biomass; Diagnostic Isotopic Fossils

Table III summarizes the principal, primary biochemical mechanisms for incorporation of carbon, nitrogen, and sulfur into biomass and the classes of organisms in which the reactions operate. The listing is not exhaustive but covers those reactions where important metabolic sequences have identifiable biogeochemical consequences. In most cases, these are recognized through an isotopic fractionation or anomaly. When an element exists as a mixture of (stable) isotopes, as is the case with carbon, hydrogen, sulfur, nitrogen, and oxygen, biochemical and chemical transformations usually take place with a slight preference for one isotope. Most biochemical changes show preference for the light isotope, and this shows up as a measurable isotopic ratio difference between substrate and product. This is usually expressed in relative terms using the per mil (‰) notation and based on a system of international standards. The standards are generally chosen so as to be close to the isotopic composition of the elements in the major reservoir (e.g., the ocean) or as they were at the time of the formation of the Earth. For example, troilite, an iron sulfide mineral from a meteorite (the Canyon Diablo troilite; CDT), is used as the international

TABLE III. Major Biochemical Pathways for Carbon, Nitrogen, and Sulfur Incorporation into Biomass[a]

Element source	Substrate/product[b]	Organism class	Electron donor
CO_2	RuBP/3-PGA	Cyanobacteria,[c] algae,[c] C_3 plants,[c] CAM plants[c]	H_2O[d]
		Purple photosynthetic bacteria[c]	H_2, H_2S, S, $S_2O_3^{2-}$, organic C
		Chemoautotrophic bacteria	H_2, H_2S, S, $S_2O_3^{2-}$, NH_3, NO_2^-, Fe^{2+}
CO_2	PEP/OAA	C_3 plants (dark), C_4 plants,[c] CAM plants (dark)	H_2O[e]
		Anaerobic bacteria	Organic C
CO_2	AcCoA/pyruvate[f]	Green photosynthetic bacteria[c]	H_2, H_2S, S, $S_2O_3^{2-}$
		Methanogens	H_2
CO_2	CO_2/AcCOA[f]	Green photosynthetic bacteria,[c] anaerobic bacteria	H_2, H_2S, S, $S_2O_3^{2-}$
CH_4	$O_2 \rightarrow HCHO$	Methanotrophs[g]	
NH_4^+	Glu/Gln	Photosynthetic organisms and bacteria[h,i]	
NH_4^+	OAA/Glu	Photosynthetic organisms[j]	
SO_4^{2-}	ATP/Cys	Bacteria, fungi, photosynthetic organisms[k]	
SO_4^{2-}	Org-C$\rightarrow S^{2-}$ + CO_2	Sulfate-reducing bacteria[l]	

[a]Data extracted from reviews by Miflin and Lea (1976), Lea and Miflin (1979), Trudinger (1979), Schidlowski *et al.* (1983), and Brock and Madigan (1988), and references therein.
[b]Abbreviations: RuBP, Ribulose 1,5-bisphosphate; 3-PGA, 3-phosphoglyceric acid; PEP, phosphoenolpyruvate; OAA, oxaloacetic acid; AcCoA, acetyl coenzyme A; Glu, glutamic acid; Gln, glutamine; ATP, adenosine triphosphate; Cys, cysteine.
[c]Photosynthetically driven reactions.
[d]Calvin–Benson or three-carbon pathway catalyzed by ribulose bisphosphate carboxylase oxygenase (Rubisco).
[e]Hatch–Slack or four-carbon pathway catalyzed by phosphoenolpyruvate (PEP) carboxylase.
[f]Reverse or reductive tricarboxylic acid (TCA) cycle.
[g]Initial oxidation step followed by incorporation of formaldehyde.
[h]Glutamine synthase (GS-GOGAT) pathway.
[i]Assimilation of NH_3 may be preceded by reduction of N_2 catalyzed by nitrogenase or of NO_3/NO_2 catalyzed by nitrate and nitrite reductases.
[j]Glutamate dehydrogenase (GDH) pathway.
[k]Assimilatory sulfate reduction via adenosine phosphosulfate (and phosphoadenylyl sulfate in prokaryotes and fungi), thiosulfate, and sulfide.
[l]Dissimilatory sulfate reduction coupled to oxidation of organic acid, and particularly acetate and lactate in sulfate-reducing bacteria (SRBs); this S^{2-} utilized in further reactions but not fixed into biomass.

standard, and all sulfur isotope compositions are expressed relative to this, where:

$$\delta^{34}S\ (‰\ CDT) = \left(\frac{{}^{34}S/{}^{32}S_{\text{sample}}}{{}^{34}S/{}^{32}S_{\text{standard}}} - 1\right) \times 10^3$$

Similarly, carbon isotope compositions ($^{13}C/^{12}C$) are expressed relative to that of the Pee Dee belemnite (PDB) standard.

The assimilation of phosphorus from dissolved inorganic phosphate and organophosphorus compounds is absolutely essential for algal fertility. However, the nonexistence of a stable isotope (i.e., isotopic tracer) for this element and its rapid recycling in organisms and ecosystems largely inhibit the gathering of biogeochemical information about the phosphorus cycle from the nature of inorganic phosphate or organic phosphorus compounds in the sedimentary record. However, it should be noted that aspects of phosphate distributions in space and time may be informative (Cook and Shergold, 1986). The biogeochemical importance of phosphate stems from its capacity to form high-energy phosphate ester bonds. These provide a system of energy storage, transport, and release, exemplified in ATP (adenosine triphosphate), which is common to all living cells (Fig. 4).

Oxygen is assimilated from water, molecular oxygen, inorganic forms such as nitrate and sulfate, and organic compounds. The complexity of these multiple sources and processes has inhibited the use of oxygen-18 as a tracer for oxygen cycling between organic matter and the atmospheric and oceanic pools, although there are renewed efforts to address this problem (e.g., Guy *et al.*, 1987). The oxygen (and hydrogen) of the oceanic and atmospheric water, however, shows isotopic fractionations which are largely related to the temperature at which certain processes take place. Precipitation of carbonate minerals, for example, is subject to an equilibrium isotope effect which is informative and often preserved in the isotopic composition of oxygen in carbonates deposited simultaneously with organic matter.

Isotopic fractionations in nitrogen metabolism are associated with nitrogen fixation ($N_2 \rightarrow NH_3$; 0–3‰), nitrification ($NH_4^+ \rightarrow NO_2^-$ ~ 35‰), and denitrification ($NO_3^- \rightarrow N_2$; ~20‰ (Sweeney *et al.*, 1978; Mariotti *et al.*, 1981), and there is a major equilibrium isotope effect when aqueous ammonium ion is volatilized to gaseous ammonia. Nitrogen in nature and in sedimentary organic compounds has a wide dispersion in isotopic compositions ($\delta^{15}N$ from −40 to +45‰; e.g., Macko *et al.*, 1987; Owens, 1987; de Niro and Weiner, 1988), but progress in assignment of specific biogeochemical processes based on this information has been limited. The complexities inherent in isolating and measuring the biological, physical, and diagenetic controls on nitrogen isotope composition and the small number of $\delta^{15}N$ values accrued from the study of fossil organic matter are major factors in restricting our understanding of this area (e.g., Schidlowski *et al.*, 1983).

Multiple processes must be active in the assimilation of hydrogen into organic matter as is indicated by the considerable isotopic heterogeneity within components of living and dead biomass (e.g., Estep and Hoering, 1980). The hydrogen isotopic composition (δD) of organically bound hydrogen in plant cellulose (e.g., Epstein *et al.*, 1976, 1977) shows a strong correlation with the δD value for the meteoric water from which it is derived. δD of plant cellulose also correlates strongly with $\delta^{18}O$ of the same material (de Niro *et al.*, 1988). Lipidic material is more depleted in the heavy isotope deuterium than is cellulose from the same organism. Sapropelic marine phytoplankton debris is therefore isotopically lighter with regard to hydrogen than land plant debris (Estep and Hoering, 1980). The H/D composition of intact thermophilic algae and bacteria shows a strong correlation with environmental water temperature and with the pH as an additional complicating factor (Estep, 1984). In studies of fossil hydrogen, Hoering (1977), Smith *et al.* (1981), and Hayes *et al.* (1983) have established that the δD of organically bound hydrogen in large sample suites of Phanerozoic and Proterozoic kerogens and coals varies widely and is in the range −70 to −160‰ SMOW (the international **s**tandard **m**ean **o**cean **w**ater). The deuterium content of fossil organic matter appears to depend on the fraction of organic matter analyzed, its thermal history, and possibly its contact with groundwater or hydrothermal water. The weight of evidence tells us that the δD in sedimentary organic matter is comparable to that in living and recently deceased biomass, is affected by organism type, and shows regional differences which may relate to the isotopic content of the source water. However, the causes of primary fractionations are poorly understood (de Niro *et al.*, 1988), and it is likely that there are also significant and complex diagenetic effects on this parameter (e.g., Peters *et al.*, 1981). Extreme care must be exercised in interpretation of organic matter δD and $\delta^{18}O$ data until there is improved understanding of controlling factors.

As discussed above, the source of much sedimentary organic sulfur is H_2S originally produced by SRB during early diagenesis. This sulfur (sulfide, polysulfide, and possibly elemental sulfur) is trapped by receptive organic compounds, and this, as a polymerization mechanism, contributes to kerogen formation. We would therefore expect to find the same wide dispersion of negative sulfur isotope values in organically bound sulfur as is found in sedimentary pyrite of biogenic (as opposed to hydrothermal) origin, and the limited range of experimental data available does support this point. Assimilatory sulfate reduction is accompanied by a small discrimination against the heavy (^{34}S) isotope while the dissimilatory process leads to much larger discriminations, ranging as high as 60‰. Since the latter process is quantitatively more significant, and because the product pyrite is ubiquitous and tractable, most research work on sedimentary sulfur has concentrated on its inorganic forms of sulfide and sulfate. Biological mediation of the abundance and isotopic composition of this "inorganic" sulfur causes it to be a most useful and important biogeochemical fossil.

Maximal sulfur isotopic fractionations, leading to the most negative $\delta^{34}S$ values for pyrite sulfur (and organically bound sulfur), would normally be expected where there is an unlimited supply of sulfate and adequate mixing. Here, kinetic controls can exert the maximum discrimination in favor of the light (^{32}S) isotope over the heavy (^{34}S) isotope. Alternatively, reduced fractionations are usually considered to be evidence for closed or "euxinic" systems because, as more sulfate becomes consumed, there is progressive reduction in ^{32}S availability and hence the preference for ^{32}S over ^{34}S (e.g., Ohomoto and Rye, 1979). In nature, the sulfate-reducing/sulfide-oxidizing system is considerably more complex because of the effects of diffusion gradients, supply and nature of reductant, and reversibility of the reduction process as well as sedimentological variations peculiar to the environment (e.g., Chambers and Trudinger, 1979; Goldhaber and Kaplan, 1980; Fry *et al.*, 1986). In general, the features which reflect the operation of bacteriogenic dissimilatory sulfate reduction are a wide dispersion of negative values for pyrite S and organic S when examined in relation to the sulfur isotopic composition of the starting sulfate. This is currently about +20‰ CDT for "normal" marine sulfate, but the actual value has varied considerably over geological time (e.g., Holser and Kaplan, 1966; Holser, 1977; Claypool *et al.*, 1980; Schidlowski *et al.*, 1983; Hayes *et al.*, 1992).

In some special geological circumstances involving high temperature alteration, it is thought that organic matter in contact with sulfate from subsurface sediments or brines can participate in a thermochemical sulfate reduction process. Subsequent incorporation of this thermochemically produced sulfide, which is isotopically close to the starting sulfate, into the organic matter can lead to very heavy $\delta^{34}S$ values for sulfide and organic sulfur (e.g., Orr, 1974; Powell and McQueen, 1984). This, again, reinforces the need for great care in the interpretation of isotopic data and the importance of examining each situation in the light of accompanying geological information.

There is voluminous literature relating the carbon isotopic composition of living organic matter to the physical and biochemical processes of carbon assimilation. Sedimentary organic matter shows features that can generally be interpreted in the light of this data, and this includes knowledge about the $\delta^{13}C$ signatures of specific compounds and even individual carbon atoms within molecules. Recent findings in this area (Freeman *et al.*, 1989; Kohnen *et al.*, 1992) emphasize, yet again, the dominance of microbes as sources of sedimentary organic matter. This progress is clearly one of the greatest success stories of all biogeochemical research and is treated as a separate topic in Chapter 3.

6. Diagnostic Molecular Structures Encoded in Fossil Organic Matter

Identification of the structures and sources of hydrocarbons and a few other categories of biomarker has essentially dominated organic geochemical research direction because these molecules are small, ubiquitous, and stable and have proved tractable in the hands of analysts (e.g., Peters and Moldowan, 1993). However, since polymeric substances, such as kerogen, constitute the bulk of preserved sedimentary organic carbon, there is renewed appreciation of the need for improved methods for isolating and dismantling this material in a controlled and selective way.

Kerogen is a complex and generally heterogeneous substance formed from the most robust remains of dead organisms (e.g., Durand, 1980; Tegelaar *et al.*, 1989). The precursors include structural components such as polysaccharides, lignins, sporopollinen, and algal biopolymers. The storage and protective parts of cells and whole organisms, including lipids, resins, and waxes, are also important. Removal of the mineral matrix from a sediment, and microscopic examination of the organic residue, generally reveals a complex mixture comprising particles with morphology which can be recognized in terms of the original organisms (i.e., microfossils, pollen, reproductive cysts, etc.). There will also be particles of amorphous organic matter which are not readily assigned to specific origins. The different components in a kerogen complex undergo a well-documented color change from amber to black during thermal maturation. Accordingly, information from microscopic inspection may be a very valuable addition to chemical analysis, providing data concerning the age, community structure, paleoenvironment of deposition, degree of reworking, and thermal history. It also provides salutary reminders of the heterogeneous and complex characteristics of the material.

Successful approaches to kerogen degradation include oxidations and other reactions which attack specific sorts of chemical bonds, hydrogenolysis, hydrogenation, desulfurization, and controlled pyrolyses. Some examples of progress in this area, which should initiate further developments, are the release and analysis of specific archaebacterial and eubacterial lipid components from kerogen (e.g., Chappe *et al.*, 1980, 1982; Mycke *et al.*, 1987) and the recognition of the resistant algal biopolymers and their likely quantitative importance as readily preservable organic materials with isotopic and structural integrity (Chalansonnet *et al.*, 1988; Kadouri *et al.*, 1988; Tegelaar *et al.*, 1989).

Moving from a microscopic to molecular scale, hydrocarbons and porphyrins, about which we have considerably more structural knowledge, can be divided according to their origins from prokaryotic or eukaryotic sources (Fig. 8). Hopanoids and simple branched alkanes are principally derived from eubacterial lipids, while extended, regular, and irregular ($>C_{20}$) acyclic isoprenoids are derived predominantly from archaebacteria. The biochemistry of prokaryotic organisms is heavily dependent on the absence or presence of specific chemical substrates or energy sources. Accordingly, prokaryotic markers are good indicators of paleoenvironment.

An example of this type of correlation is the recognition, in sediments and oils, of aryl isoprenoids (Fig. 9) thought to be derived from carotenoids specific to the *Chlorobiaceae* family of green sulfur bacteria (Summons and Powell, 1986, 1987). These organisms are anaerobic and photosynthetic and so require light and H_2S for growth. In modern environments, they appear in sulfate-containing water bodies which are sufficiently quiescent and organic rich to enable sulfide production close to the photic zone. Thermal or salinity stratification is usually involved, such as may be found in hypersaline or fjord environments. With all caution, we can invoke these conditions in our assessment of depositional environments where *Chlorobium* biomarkers are abundant and, at the same time, look for independent geological support.

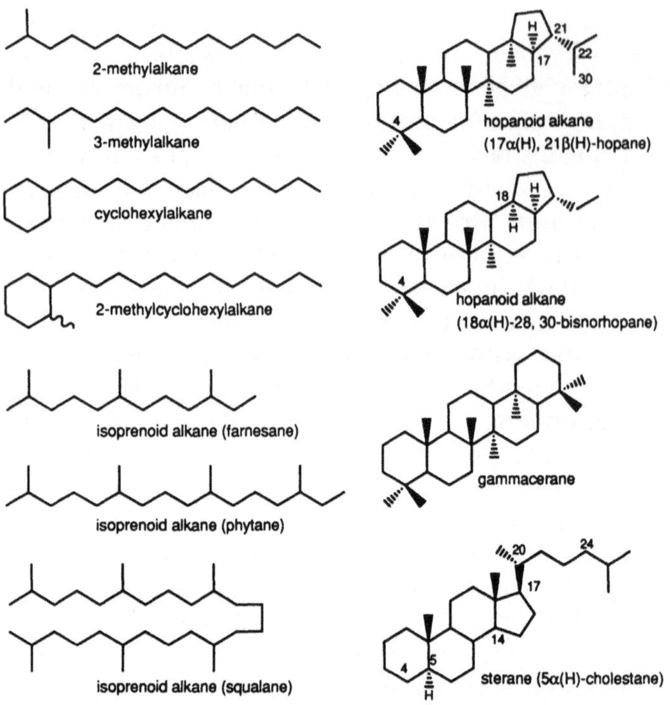

Figure 8. Illustration of the types of biogenic hydrocarbons routinely found in sedimentary bitumen. These molecules are typical of those which can be diagnostic for the major categories of organisms.

Hopanoid hydrocarbons with 28 carbon atoms are another class of biomarkers which evidently have a distinctive bacterial source and which display very selective distributions in sediments and petroleum. Known occurrences include the source rocks and oils of the Monterey Formation in California and the Kimmeridge oils of the North Sea. Organic matter in these deposits is considered to have been formed in high-productivity, open marine environments and subsequently buried below water which was anoxic and sulfide rich. The unusual 28,30-bisnorhopane biomarker (see Fig. 8), which is prominent in these strata (Grantham *et al.*, 1980; Moldowan *et al.*, 1984) and, in the case of the Monterey, clearly associated with a particular phosphatic carbonate lithofacies (Curiale and Odermatt, 1989), can be attributed to presently unknown bacteria which thrived somewhere in this system, most probably during the early diagenesis. Because of its infrequency, a high abundance of 28,30-bisnorhopane in sediments or oils strongly implies deposition under a specific set of circumstances influenced by a particular set of biogeochemical processes, and hence in a specific sort of paleoenvironment. A modern analog is not yet recognized because we have no knowledge of the connection between the fossil biomarker and the biochemistry of any extant organisms.

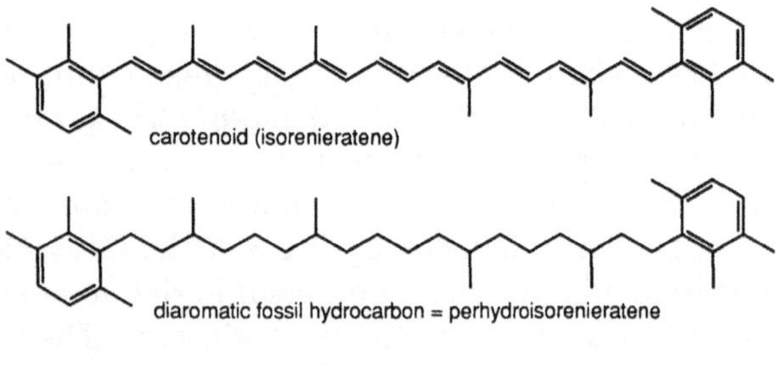

Figure 9. Structure of a principal and very specific carotenoid for green sulfur bacteria of the genus *Chlorobium* and sedimentary hydrocarbons derived from it.

Steroidal hydrocarbons are also prominent in sedimentary organic matter and display a great deal of structural variation. Steroidal lipids, from which these hydrocarbons are derived, are important building blocks in the membranes of eukaryotes. Only a few prokaryotes are known to have the capacity to biosynthesize sterols (Ourisson *et al.*, 1987), and the requirement for the presence of molecular oxygen is another constraint on their production. Consequently, steranes can be informative about the eukaryote composition of a microbial community and especially the nature of algal primary producers. Paleoenvironment, evolution, and changes in the dominance of algal lines over Proterozoic and Phanerozoic time appear to affect the steroid composition of preserved organic matter (e.g., Volkman, 1986; Grantham and Wakefield, 1988; Summons and Walter, 1990). Some steranes, derived from the sterols of specific classes or families of algae, are known to have unusual and diagnostic structural features and restricted distributions in space and/or time. Included in this category are several C_{30} compounds (Fig. 10) such as dinosterane (Summons *et al.*, 1987), 4-methyl-24-ethylcholestane (Goodwin *et al.*, 1988), and 24-*n*-propylcholestane (Moldowan *et al.*, 1990). These convey environmental information as well as imposing constraints on the geological ages of the materials in which they occur. The oldest steranes, hopanes and extended acyclic isoprenoids so far recognized are in sediments dated at approximately 1.7 Ga (Summons and Walter, 1990).

Vascular plants synthesize a wide variety of terpenoid natural products. These compounds may be transformed into stable hydrocarbons during burial while retaining their diagnostic structural features. In the case of some plant resins, the products do not

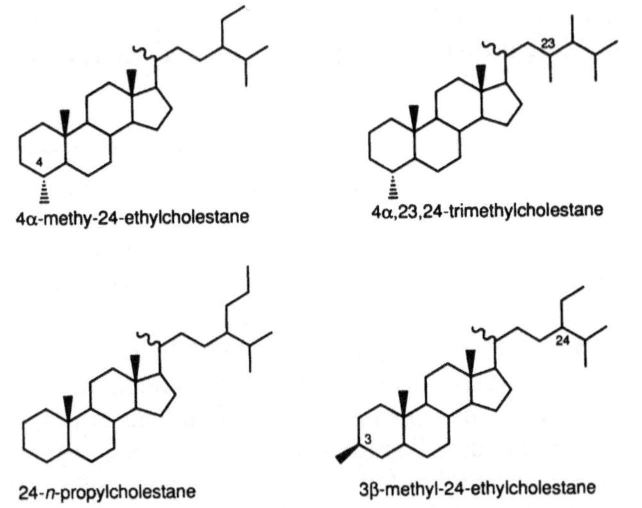

Figure 10. Different structural isomers found in sedimentary C_{30} steranes. Various combinations of these molecules are diagnostic for different sedimentary environments. The geologic age of sediments also exerts some control on their distributions.

outwardly resemble the precursors, but the connections can be made by artificial maturation experiments (e.g., van Aarssen *et al.*, 1990). As well as disclosing their terrestrial origins, plant terpenoid distributions also express evolutionary and paleobotanical information (e.g., Alexander *et al.*, 1988).

Making correct and definitive connections between biogenic organics and their fossil products in the sediments is never easy or complete. The above examples illustrate just a few of the ways in which molecular fossils can be informative. There are many more discussed in the following chapters of this book, but, for every answer, a new series of questions instantly arises. This, and the need to be always conscious of chemical, physical, biological, and geological considerations, is what makes organic geochemistry such an intriguing, intellectually challenging, and rewarding area for research.

7. Geological History of the Biogeochemical Cycles

The passage of the life-forming elements through the biogeochemical cycles often results in stable isotopic fractionations due to equilibrium isotope effects (e.g., CO_2/HCO_3^- equilibrium) and kinetic isotope effects (e.g., enzymic discriminations). Carbon, sulfur, oxygen, hydrogen, and nitrogen all exhibit significant isotopic heterogeneity within and between organic and inorganic reservoirs, and specific details of important fractionation processes are discussed in later chapters. Isotopic analyses constitute an informative and effective means of following pathways, constructing mass-balance equations, and timing events. For example, the significance and intensity of coupling of the global carbon and sulfur cycles during the Phanerozoic has become evident through recognition of an inverse relationship between the isotopic composition of marine carbonates and sulfates (Holser, 1977; Claypool *et al.*, 1980; Veizer *et al.*, 1980; Garrels and Lerman, 1984; Berner, 1987, 1989). Isotope stratigraphy, such as the investigation of fine-scale carbon isotopic variation in carbonates and organic matter deposited in former oceans, has shown that major faunal extinction events, such as those just prior to the Cambrian, the Permian Terminal Event and the Cretaceous–Tertiary Boundary Event, seem to be associated with rapid and major perturbations in the functioning of the carbon cycle (e.g., Tucker, 1986; Knoll *et al.*, 1986; Zachos *et al.*, 1989; Holser *et al.*, 1989). Since the carbon dioxide present in the ocean–atmosphere system is a major moderator of surface temperature through its capacity to act as a "greenhouse" gas, it is probable that the carbon cycle and global climate are tightly coupled (e.g., Broecker and Denton, 1990). The ultimate "holistic" view of a self-regulating Earth has been proposed in the Gaia metaphor (e.g., see Lovelock, 1990), and, in this context, other means of climate regulation have been suggested but are not proven.

Tectonism, climate, and oceanic circulation patterns all exert a profound influence on the spatial and temporal aspects of organic matter formation and deposition. Determination of the precise relationships between geochemical phenomena, such as sedimentary isotope fluctuations and global "anoxic events," and apparently simultaneous biological events signifying changed circumstances is a major challenge for geologists, climatologists, biogeochemists, and paleontologists to solve together.

Biogeochemical cycles had their origins in the early stages of biological evolution and were intimately connected with the progress and direction of that evolution (Fig. 11). Stromatolites, laminated sedimentary structures formed by microbes which trap, bind, and cement sediment, comprise ubiquitous and unambiguous pieces of evidence for the nature of early life. The morphology of the oldest known fossil stromatolites, approximately 3.5 Ga old (Lowe, 1983; Walter, 1983), resembles that of modern ones constructed by photosynthetic microbes. Isotopic differences between organic and inorganic carbon phases of the oldest sedimentary rocks are of "similar" mag-

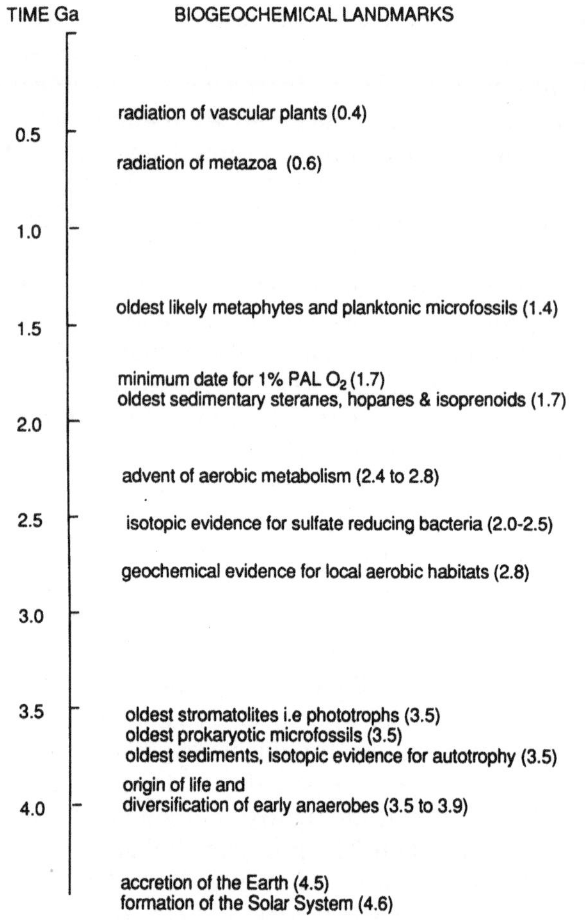

Figure 11. Time scale of Earth history, showing events of biogeochemical significance based on current assessments of fossil evidence. The data were extracted from Schopf (1983), Knoll and Bauld (1989), Hayes *et al.* (1990), and references therein.

nitude to those found today [approx −35‰ PDB; see Strauss *et al.* (1990) and Des Marais *et al.* (1992) for new information on this topic]. These and other observations suggest that photoautotrophy is as old, or almost as old, as life itself, that is, at least 3.5 billion years (e.g., Schidlowski *et al.*, 1983; Walter, 1983). Truly convincing evidence has not yet been found to establish whether this photosynthesis was based on hydrogen sulfide (anoxygenic; photosystem I), water (oxygenic; photosystems I + II), or some other substance as an electron donor, although there are biological grounds for the belief that anoxygenic photosynthesis was the primeval form (Pierson and Olson, 1989). Geological data suggests that photosynthesis operated as long as 2.7 billion years ago (Buick, 1992). However, there is general agreement that oceanic and atmospheric oxygen levels in the Archean and Early Proterozoic were very much lower than present-day atmospheric levels (PAL) (Cloud, 1976; Holland, 1978, 1984; Walker *et al.*, 1983; Kasting, 1987), and it is thought that oxygen production and accumulation from oxygenic photosynthesis eventually resulted in atmospheric levels capable of sustaining metazoan respiration (>0.1 PAL) by the end of the Proterozoic (Knoll, 1989). The availability of molecular oxygen is an essential factor in numerous biochemical transformations, and particularly for the biosynthesis of unsaturated fatty acids and sterols in eukaryotes. Molecular oxygen is also important as the precursor of ultraviolet (UV)-protective ozone in the Earth's atmosphere. Therefore, a certain level of molecular oxygen must have been present before organisms could survive damaging UV radiation in the surface waters of the oceans and on land, before eukaryote biochemistry could become a significant component of the carbon cycle, and before metazoans could become widespread.

Mass-balance calculations (Veizer *et al.*, 1980; Garrels and Lerman, 1984; Berner, 1987, 1989), based on isotopic compositions and the respective sizes of the various pools of carbon and sulfur, demonstrate the coupling of the cycles of these elements throughout the Phanerozoic and contribute to estimations of other parameters such as variations in atmospheric oxygen content. The record of sulfate-sulfur isotopes becomes sparse in rocks older than 1.2 billion years. Nevertheless, the characteristics of sulfide-sulfur isotopes, the record of which extends to about 3.8 billion years, allow geochemists to identify the advent of biogenic sulfide production by SRB at about 2.5 billion years ago. The sulfur isotopic composition of diagenetic pyrite is controlled by a number of parameters, including the availability of and isotopic composition of source sulfate and the rate at which it is utilized by the SRB (e.g., see Chambers and Trudinger, 1979). In the record of sedimentary systems, recent to about 2.5 billion years ago, it is usual to observe a wide spread and predominance of negative $\delta^{34}S$ values in sedimentary (diagenetic) sulfide (0 to −50‰ CDT), reckoned to typify the pattern imposed by bacterial sulfate reduction. Sulfates, on the other hand, show correspondingly positive values (+10 to +30‰ CDT). In strata older than 2.5 Ga, isotopic compositions of sulfides and sulfates (where they exist) cluster fairly tightly around 0.0‰ CDT, a value thought to be close to that of Earth's primordial sulfur, and reflecting a dominantly magmatic source for all sulfide. The main evolutionary radiation of terrestrial vascular plants took place during Devonian and Carboniferous times. This event flowed from the development of roots, stems, leaves, and reproductive apparatus enabling colonization of a new ecological niche (e.g., see Stewart, 1983; Raven *et al.*, 1986) and significantly altered the balance of the carbon cycle. The burial of massive amounts of nonmarine organic carbon is a consequence of this biological innovation and can be independently recognized in the sedimentary records of both sulfur and carbon isotopes (Berner, 1989).

The carbon cycle and its coupling to the sulfur cycle and concentration of atmospheric oxygen must have been operating with relative stability since the Latest Proterozoic. Evidence for this coupling is contained in systematic covariations in the isotopic records of sedimentary carbon and sulfur over Phanerozoic time (e.g., Veizer *et al.*, 1980), and there is some evidence to suggest that different circumstances may have existed in the Proterozoic (Lambert and Donnelly, 1988). During the Archean and much of the Proterozoic, ferric iron, hydrogen, and sulfur were each likely to have been significant redox partners with carbon, and the carbon cycle must have undergone dramatic evolutionary change in concert with biological innovations. The appearance of photosystem II biochemistry and physiology and the consequent rise in atmospheric oxygen concentration is particularly significant in this regard. Understanding the timing and course of atmospheric oxygenation is an important and formidable biogeochemical problem and topic of lively debate, detailed discussion having been made by Cloud (1968, 1983), Holland (1978, 1984), Towe (1981, 1990), Hayes *et al.* (1983), Schidlowski *et al.* (1983), Walker *et al.* (1983), Kasting (1987), Knoll (1989), Knoll and Bauld (1989), Holland and Beukes (1990), Hayes *et al.* (1992), and Des Marais *et al.* (1992).

8. Future Research Directions

Elucidating the history of planet Earth over geological time scales has always been a challenging and demanding task. It relies on the collection and assimilation of information from disparate disciplines and working with astonishingly cryptic clues. The complexity of biogeochemical processes and their networking provide especially formidable barriers to our understanding of biosphere–geosphere functioning. However, advances generally flow from the development of new analytical techniques, the discovery of new sources of data, and the occasional sagacious insight.

Important new data about the biogeochemical cycles and global change will come from fine-scale atomic, molecular, and isotopic analysis of sediments (including the polar ice caps). Isotope analysis at the molecular level will give more reliable clues about the sources and sinks for sedimentary organic carbon. Molecular biology is presently offering new ways to study the population structure and behavior of microbes living in their natural environments, and this will lead to enhanced appreciation of how entire communities function. Major changes to our views of the history of Earth and the history of life will inevitably follow.

ACKNOWLEDGMENTS. I am indebted to Marilyn Fogel, Glenn Hieshima, Phil Trudinger, and Malcolm Walter, who provided numerous helpful comments, corrections and suggestions on early versions of this manuscript. Errors, omissions, and oversights which remain are of my own making.

References

Alexander, R., Larcher, A. V., Kagi, R. I., and Price, P. L., 1988, The use of plant-derived biomarkers for correlation of oils with source rocks in the Cooper/Eromanga Basin system, Australia, *APEA J.* **28**:310–324.

Awramik, S. M., Schopf, J. W., and Walter, M. R., 1983, Filamentous fossil bacteria 3.5 × 10^9 years old from the Archean of Western Australia, *Precamb. Res.* **20**:357–374.

Berner, R. A., 1987, Models for carbon and sulfur cycles and atmospheric oxygen: Application to Paleozoic history, *Am. J. Sci.* **287**:177–196.

Berner, R. A., 1989, Biogeochemical cycles of carbon and sulfur and their effect on atmospheric oxygen over Phanerozoic time, *Palaeogeogr. Palaeoclimatol. Palaeoecol. (Global and Planetary Change Section)* **75**:97–122.

Bonen, L., Doolittle, W. F., and Fox, G. E., 1979, Cyanobacterial evolution: Results of 16S ribosomal RNA sequence analysis, *Can. J. Biochem.* **57**:879–888.

Brassell, S. C., Lewis, C. A., de Leeuw, J. W., de Lange, F., and Sinninghe Damsté, J. S., 1986, Isoprenoid thiophenes: Novel diagenetic products in sediments, *Nature* **320**:160–162.

Brock, T. D., and Madigan, M. T., 1988, *Biology of Microorganisms*, 5th ed., Prentice-Hall, Englewood Cliffs, New Jersey.

Broecker, W. S., and Denton, G. H., 1990, What drives glacial cycles? *Sci. Am.* **262**:43–50.

Buick, R., 1992, The antiquity of oxygenic photosynthesis: Evidence from stromatolites in sulfate-deficient Archaean lakes, *Science* **255**:74–77.

Chalansonnet, S., Largeau, C., Casadevall, E., Berkalof, C., Peniguel, G., and Couderc, R., 1988, Cyanobacterial resistant biopolymers. Chemical implications of the properties of *Schizothrix* sp. resistant material, in: *Advances in Organic Geochemistry 1987* (L. Mattavelli and L. Novelli, eds.), *Org. Geochem.* **13**: 1003–1010.

Chambers, L. A., and Trudinger, P. A., 1979, Microbiological fractionation of stable sulfur isotopes: A review and critique, *Geomicrobiol. J.* **1**:249–293.

Chapman, D. J., and Schopf, J. W., 1983, Biological and biochemical effects of the development of an aerobic environment, in:

Earth's Earliest Biosphere: Its Origin and Evolution (J. W. Schopf, ed.), Princeton University Press, Princeton, New Jersey, pp. 302–320.

Chappe, B., Michaelis, W., and Albrecht, P., 1980, Molecular fossils of Archaebacteria as selective degradation products of kerogen, in: *Advances in Organic Geochemistry 1979* (A. G. Douglas and J. R. Maxwell, eds.), Pergamon Press, Oxford, pp. 265–274.

Chappe, B., Albrecht, P., and Michaelis, W., 1982, Polar lipids of Archaebacteria in sediments and petroleums, *Science* **217:** 65–66.

Claypool, G. E., Holser, W. T., Kaplan, I. R., Sakai, H., and Zak, I., 1980, The age curves of sulfur and oxygen isotopes in marine sulfate and their mutual interpretation, *Chem. Geol.* **28:** 199–260.

Cloud, P., 1968, Atmospheric and hydrospheric evolution on the primitive earth, *Science* **160:**729–736.

Cloud, P., 1976, Beginnings of biospheric evolution and their biogeochemical consequences, *Paleobiology* **2:**351–387.

Cloud, P. 1983, Aspects of Proterozoic biogeology, *Geol. Soc. Am. Mem.* **161:**245–251.

Cook, P. J., and Shergold, J. H., 1986, Proterozoic and Cambrian phosphorites—nature and origin, in: *Proterozoic and Cambrian Phosphorites* (P. J. Cook and J. H. Shergold, eds.), Cambridge University Press, Cambridge, pp. 369–384.

Curiale, J. A., and Odermatt, J. R., 1989, Short-term biomarker variability in the Monterey Formation, Santa Maria Basin, *Org. Geochem.* **14:**1–13.

Dean, W. E., Arthur, M. A., and Claypool, G. E., 1986, Depletion of ^{13}C in Cretaceous marine organic matter: Source, diagenetic, or environmental signal? *Mar. Geol.* **70:**119–157.

Deines, P., 1980, The isotopic composition of reduced organic carbon, in: *Handbook of Environmental Isotope Geochemistry, Vol. 1, The Terrestrial Environment* (P. Fritz and J. P. Fontes, eds.), Elsevier, Amsterdam, pp. 329–406.

de Niro, M. J., and Weiner S., 1988, Chemical, enzymic and spectroscopic characterization of collagen and other organic fractions from prehistoric bones, *Geochim. Cosmochim. Acta* **52:** 2197–2206.

de Niro, M. J., Sternberg, L. D., Marino, B. D., and Druzik, J. R., 1988, Relation between D/H ratios and $^{18}O/^{16}O$ ratios in cellulose from linen and maize—implications for paleoclimatology and for sindonology, *Geochim. Cosmochim. Acta* **52:**2189–2196.

Des Marais, D. J., Strauss, H., Summons, R. E., and Hayes, J. M., 1992, Carbon isotopic evidence for the stepwise oxidation of the Proterozoic environment, *Nature* **359:**605–609.

Durand, B., 1980, *Kerogen—Insoluble Organic Matter from Sedimentary Rocks*, Editions Technip, Paris.

Duursma, E. K., and Dawson, R. (eds.), 1981, *Marine Organic Chemistry. Evolution, Composition, Interactions and Chemistry of Organic Matter in Seawater*, Elsevier, Amsterdam.

Epstein, S., Yapp, B. J., and Hall, J. H., 1976, The determination of the D/H ratio of non-exchangeable hydrogen in cellulose extracted from aquatic and land plants, *Earth Planet. Sci. Lett.* **30:**241–251.

Epstein, S., Thompson, P., and Yapp, B. J., 1977, Oxygen and hydrogen isotopic ratios in plant cellulose, *Science* **198:**1209–1214.

Estep, M. L. F., 1984, Carbon and hydrogen isotopic compositions of algae and bacteria from hydrothermal environments, Yellowstone National Park, *Geochim. Cosmochim. Acta* **48:** 591–599.

Estep, M. L. F., and Hoering, T. C., 1980, Biogeochemistry of stable hydrogen isotopes, *Geochim. Cosmochim. Acta* **44:**1197–1206.

Freeman, K. H., Hayes, J. M., Trendel, J.-M., and Albrecht, P., 1989, Evidence from carbon isotope measurements for diverse origins of sedimentary hydrocarbons, *Nature* **343:**254–256.

Froelich, P. N., Bender, M. L., Luedtke, N. A., Heath, G. R., and DeVries, T., 1982, The marine phosphorus cycle, *Am. J. Sci.* **282:**474–511.

Fry, B., Cox, J., Gest, H., and Hayes, J. M., 1986, Discrimination between ^{34}S and ^{32}S during bacterial metabolism of inorganic sulfur compounds, *J. Bacteriol.* **165:**328–330.

Garrels, R. M., and Lerman, A., 1984, Coupling of the sedimentary sulfur and carbon cycles—an improved model, *Am. J. Sci.* **284:** 989–1007.

Garrels, R. M., and Mackenzie, F. T., 1971, *Evolution of Sedimentary Rocks*, Norton, New York.

Garrels, R. M., and Perry, E. A., 1974, Cycling of carbon, sulfur and oxygen through geologic time, in: *The Sea*, Vol. 5 (E. D. Golgberg, ed.), John Wiley & Sons, New York, pp. 303–336.

Garrels, R. M., Mackenzie, F. T., and Hunt, C., 1975, *Chemical Cycles and the Global Environment*, Kaufmann, Los Altos, California.

Garrels, R. M., Lerman, A., and Mackenzie, F. T., 1976, Controls on atmospheric O_2 and CO_2: Past, present and future, *Am. Sci.* **64:** 306–315.

Giovannoni, S. J., Turner, S., Olsen, G. J., Barns, S. Lane, D. J., and Pace, N. R., 1988, Evolutionary relationships among cyanobacteria and green chloroplasts, *J. Bacteriol.* **170:**3584–3592.

Glaessner, M. F., 1984, *The Dawn of Animal Life: A Biohistorical Study*, Cambridge University Press, Cambridge.

Goldhaber, M. B., and Kaplan, I. R., 1980, Mechanisms of sulfur incorporation and isotope fractionation during early diagenesis in sediments of the Gulf of California, *Mar. Chem.* **9:** 95–143.

Golubic, S., Krumbein, W., and Schneider, J., 1979, The carbon cycle, in: *Biogeochemical Cycling of Mineral-Forming Elements* (P. A. Trudinger and D. J. Swaine, eds.), Elsevier, Amsterdam, pp. 293–314.

Goodwin, N. S., Mann, A. L., and Patience, R. L., 1988, Structure and significance of C_{30} 4-methylsteranes in lacustrine shales and oils, *Org. Geochem.* **12:**495–506.

Gordon, J. K., 1981, Introduction to the nitrogen fixing prokaryotes, in: *The Prokaryotes: A Handbook on Habitats, Isolation and Identification of Bacteria* (M. P. Starr, H. Stolp, H. G. Trüper, A. Balows, and H. G. Schlegel, eds.), Springer-Verlag, Berlin, pp. 781–794.

Grantham, P. J., and Wakefield, L. L., 1988, Variations in the sterane carbon number distributions of marine source rock derived oils through geological time, *Org. Geochem.* **12:**61–73.

Grantham, P. J., Posthuma, J., and De Groot, K., 1980, Variation and significance of the C_{27} and C_{28} triterpane content of a north Sea core and various North Sea crude oils in: *Advances in Organic Geochemistry 1979* (J. R. Maxwell and A. G. Douglas, eds.), Pergamon Press, Oxford, pp. 29–38.

Gregor, C. B., Garrels, R. M., Mackenzie, F. T., and Maynard, J. B. (eds.), 1988, *Chemical Cycles in the Evolution of the Earth*, John Wiley & Sons, New York.

Guy, R. D., Fogel, M. F., Berry, J. A., and Hoering, T. C., 1987, Isotope fractionation during oxygen production and consumption by plants, in: *Progress in Photosynthesis Research*, Vol. 3 (J. Biggens, ed.), Nijhoff Publishers, Boston, pp. 597–600.

Han, T., and Runnegar, B., 1992, Megascopic eukaryotic algae from the 2.1-billion year old Negaunee Iron-Formation, Michigan, *Science* **257:**232–235.

Hayes, J. M., 1983, Geochemical evidence bearing on the origin of aerobiosis, a speculative hypothesis, in: *Earth's Earliest Biosphere: Its Origin and Evolution* (J. W. Schopf, ed.), Princeton University Press, Princeton, New Jersey, pp. 291–301.

Hayes, J. M., Kaplan, I. R., and Wedeking, K. W., 1983, Precambrian organic geochemistry, in: *Earth's Earliest Biosphere: Its Origin and Evolution* (J. W. Schopf, ed.), Princeton University Press, Princeton, New Jersey, pp. 93–132.

Hayes, J. M., Des Marais, D. J., Lambert, I. B., Strauss, H., and Summons, R. E., 1990, Proterozoic biogeochemistry, in: *The Proterozoic Biosphere: A Multidisciplinary Study* (J. W. Schopf and C. Klein, eds.), Cambridge University Press, Cambridge, pp. 81–134.

Hoering, T. C., 1977, The stable isotopes of hydrogen in Precambrian organic matter, in: *Chemical Evolution of the Early Precambrian* (C. Ponnamperuma, ed.), Academic Press, New York, pp. 81–86.

Holland, H. D., 1978, *The Chemistry of the Atmosphere and Oceans*, John Wiley & Sons, New York.

Holland, H. D., 1984, *The Chemical Evolution of the Atmosphere and Oceans*, Princeton University Press, Princeton, New Jersey.

Holland, H. D., and Beukes, N. J., 1990, A paleoweathering profile from Griqualand West, South Africa: Evidence for a dramatic rise in atmospheric oxygen between 2.2 and 1.9 bypy, *Am. J. Sci.* **290A:**1–34.

Holser, W. T., 1977, Catastrophic chemical events in history of the ocean, *Nature* **276:**403–408.

Holser, W. T., and Kaplan, I. R., 1966, Isotope geochemistry of sedimentary sulfate, *Chem. Geol.* **1:**93–135.

Holser, W. T., Schönlaub, H.-P., Attrep, M., Boeckelmann, K., Klein, P., Magaritz, M., Orth, C. J., Fenninger, A., Jenny, C., Kralik, M., Mauritsch, H., Pak, E., Schramm, J.-M., Stattegger, K., and Schmöller, R., 1989, A unique geochemical record at the Permian/Triassic boundary, *Nature* **337:**39–44.

Ivanov, M. V., and Freney, J. R. (eds.), 1983, *The Global Biogeochemical Sulfur Cycle*, Scope 19, Wiley, Chichester, England.

Jones, W. J., Nagle, D. P., Jr., and Whitman, W. B., 1987, Methanogens and the diversity of Archaebacteria, *Microbiol. Rev.* **51:** 137–177.

Kadouri, A., Derenne, S., Largeau, C., Casadevall, E., and Berkaloff, C., 1988, Resistant biopolymer in the outer walls of *Botryococcus braunii*, *B* race, *Phytochemistry* **27:**551–557.

Kasting, J. F., 1987, Theoretical constraints on oxygen and carbon dioxide concentrations in the Precambrian atmosphere, *Precamb. Res.* **34:**205–229.

Knoll, A. H., 1983, Biological interactions and Precambrian eukaryotes, in: *Biotic Interactions in Recent and Fossil Benthic Communities* (J. S. Trevesz and P. L. McCall, eds.), Plenum Press, New York, pp. 251–283.

Knoll, A. H., 1989, Biological and biogeochemical preludes to the Ediacaran radiation, in: *Origins and Early Evolutionary History of the Metazoa* (J. H. Lipps and P. W. Signor, eds.), Plenum Press, New York, pp. 53–84.

Knoll, A. H., and Bauld, J., 1989, The evolution of ecological tolerance in prokaryotes, *Trans. R. Soc. Edinburgh: Earth Sci.* **80:**209–223.

Knoll, A. H., Hayes, J. M., Kaufman, A. J., Swett, K., and Lambert, I. B., 1986, Secular variation in carbon isotope ratios from Upper Proterozoic successions of Svalbad and East Greenland, *Nature* **321:**832–838.

Kohnen, M. E. L., Schouten, S., Sinninghe Damsté, J. S., de Leeuw, J. W., Merritt, D. A., and Hayes, J. M., 1992, Recognition of paleobiochemicals by a combined molecular sulfur and isotope geochemical approach, *Science* **256:**358–362.

Kohnen, M. E. L., Sinninghe Damsté, J. S., ten Haven, H. L., and de Leeuw, J. W., 1989, Early incorporation of polysulfides in sedimentary organic matter, *Nature* **341:**640–641.

Lake, J. A., 1988, Origin of the eukaryotic nucleus determined by rate-invariant analysis of r-RNA sequences, *Nature* **331:** 184–186.

Lambert, I. B., and Donnelly, T. H., 1988, The palaeoenvironmental significance of trends in sulfur isotope compositions in the Precambrian: A critical review, in: *Stable Isotopes and Fluid Processes in Mineralisation* (H. K. Herbert, ed.), Spec. Publ., Geological Society of Australia, Canberra.

Langworthy, T. A., 1985, The lipids of Archaebacteria, in: *The Bacteria, Vol. 8. Archaebacteria* (C. R. Woese and R. S. Wolfe, eds.), Academic Press, New York, pp. 459–497.

de Leeuw, J. L., van Bergen, P. F., van Aarssen, B. G. K., Gatellier J.-P. L. A., Sinninghe Damsté, J. S., and Collinson, M. E., 1991, Resistant biomacromolecules as major contributors to kerogen, *Phil. Trans. R. Soc. Lond. B.* **333:**329–337.

Lea, P. J., and Miflin, B. J., 1979, Photosynthetic ammonia assimilation, in: *Encyclopedia of Plant Physiology* (M. Gibbs and E. Latzko, eds.), Springer-Verlag, Berlin, Vol. 6, pp. 445–456.

Lehninger, A. L., 1982, *Principles of Biochemistry*, Worth Publishers, New York.

Lovelock, J. E., 1990, Hands up for the Gaia hypothesis, *Nature* **344:** 100–102.

Lowe, D. R., 1983, Restricted shallow-water sedimentation of early Archean evaporitic and stromatolitic strata of the Strelly Pool Chert, Pilbara Block, Western Australia, *Precamb. Res.* **19:** 239–283.

Macko, S. A., Fogel, M. L., Hare, P. E., and Hoering, T. C., 1987, Isotopic fractionations of nitrogen and carbon in the synthesis of amino acids by microorganisms, *Chem. Geol.* **65:**79–92.

Margulis, L., 1981, *Symbiosis and Cell Evolution*, W. H. Freeman and Co., San Francisco.

Mariotti, A., Germon, J. C., Hubert, P., Kaiser, P., Letolle, R., Tardieux, A., and Tardieux, P., 1981, Experimental determination of nitrogen kinetic isotope fractionation: Some principles; illustration for the denitrification and nitrification process, *Plant Soil* **62:**413–430.

Miflin, B. J., and Lea, P. J., 1976, The pathway of nitrogen assimilation in plants, *Phytochemistry* **15:**873–885.

Moldowan, J. M., Seifert, W. K., Arnold, E., and Clardy, J., 1984, Structure proof and significance of stereoisomeric 28,30-bisnorhopanes in petroleum source rocks, *Geochim. Cosmochim. Acta* **48:**1651–1661.

Moldowan, J. M., Fago, F. J., Lee, C. Y., Jacobson, S. R., Watt, D. S., Slougui, N.-E., Jeganathan, A., and Young, D. C., 1990, Sedimentary 24-n-propylcholestanes, molecular fossils diagnostic of marine algae, *Science* **204:**309–312.

Mycke, B., Narjes, F., and Michaelis, W., 1987, Bacteriohopanetetrol from chemical degradation of an oil shale kerogen, *Nature* **326:**179–181.

Odum, E. P., 1971, *Fundamentals of Ecology*, 3rd ed., W. B. Saunders, Philadelphia.

Ohomoto, H., and Rye, R. O., 1979, Isotopes of carbon and sulfur, in: *Geochemistry of Hydrothermal Ore Deposits*, 2nd ed. (H. L. Barnes, ed.), John Wiley & Sons, New York, pp. 509–567.

Olsen, G. J., Lane, D. L., Giovannoni, S. J., and Pace, N. R., 1986, Microbial ecology and evolution: A ribosomal RNA approach, *Annu. Rev. Microbiol.* **40:**337–365.

Orr, W. L., 1974, Changes in sulfur content and isotopic ratios of sulfur during petroleum maturation—study of Big Horn Basin Paleozoic oils, *Am. Assoc. Petrol. Geol. Bull.* **58:**2295–2318.

Ourisson, G., Rohmer, M., and Poralla, K., 1987, Prokaryotic hopanoids and other polyterpenoid sterol surrogates, *Annu. Rev. Microbiol.* **41:**301–333.

Owens, N. J. P., 1987, Natural variations in ^{15}N in the marine environment, *Adv. Mar. Biol.* **24:**389–451.

Peters, K. E., and Moldowan, J. M., 1993, *The Biomarker Guide: Interpreting Molecular Fossils in Petroleum and Ancient Sediments*, Prentice Hall, Englewood Cliffs, New Jersey.

Peters, K. E., Rohrback, B. G., and Kaplan, I. R., 1981, Carbon and hydrogen stable isotope variations in kerogen during laboratory simulated thermal maturation, *Am. Assoc. Petrol. Geol. Bull.* **65:**501–508.

Pierrou, U., 1979, The phosphorus cycle: Quantitative aspects and the role of man, in: *Biogeochemical Cycling of Mineral-Forming Elements* (P. A. Trudinger and D. J. Swaine, eds.), Elsevier, Amsterdam, pp. 205–210.

Pierson, B. K., and Olson, J. M., 1989, Evolution of photosynthesis in anoxygenic photosynthetic procaryotes, in: *Microbial Mats: Physiological Ecology of Benthic Microbial Communities* (Y. Cohen and E. Rosenberg, eds.), pp. 402–427.

Postgate, J. R., 1984, *The Sulfate-Reducing Bacteria*, 2nd ed., Cambridge University Press, Cambridge.

Powell, T. G., and McQueen, R. W., 1984, Precipitation of sulfide ores and organic matter: Sulfate reactions at Pine Point, Canada, *Science* **224:**63–66.

Raven, P. H., Evert, R. F., and Eichorn, S. H., 1986, *Biology of Plants*, 4th ed., Worth Publishers, New York.

Redfield, A. C., Ketchum, B. H., and Richards, F. A., 1963, The influence of organisms on the composition of seawater, in: *The Sea*, Vol. 2 (M. N. Hill, ed.), John Wiley & Sons, New York, pp. 26–87.

Romankevich, E. A., 1984, *Geochemistry of Organic Matter in the Ocean*, Springer-Verlag, Berlin.

Schidlowski, M., Hayes, J. M., and Kaplan, I. R., 1983, Isotopic inferences of ancient biochemistries: Carbon, sulfur, hydrogen and nitrogen, in: *Earth's Earliest Biosphere: Its Origin and Evolution* (J. W. Schopf, ed.), Princeton University Press, Princeton, New Jersey, pp. 147–186.

Schoell, M., and Wellmer, F. W., 1981, Anomalous 13C depletion in early Precambrian graphites from Superior Province, Canada, *Nature* **290:**696–699.

Schopf, J. W. (ed.), 1983, *Earth's Earliest Biosphere: Its Origin and Evolution*, Princeton University Press, Princeton, New Jersey.

Schopf, J. W., and Klein, C. (eds.), 1992, *The Proterozoic Biosphere: A Multidisciplinary Study*, Cambridge University Press, Cambridge.

Schopf, J. W., and Walter, M. R., 1983, Archean microfossils: New evidence of ancient microbes, in: *Earth's Earliest Biosphere: Its Origin and Evolution* (J. W. Schopf, ed.), Princeton University Press, Princeton, New Jersey, pp. 187–213.

Sinninghe Damsté, J. S., 1988, *Organically-bound sulfur in the geosphere: A molecular approach*, Ph.D. Thesis, Delft University of Technology.

Skyring, G. W., 1987, Sulfate reduction in coastal ecosystems, *Geomicrobiol. J.* **5:**295–374.

Smith, J. W., Gould, K. W., and Rigby, D., 1981, The stable isotopic geochemistry of Australian coals, in: Australian C.S.I.R.O. Fuel Geoscience Unit, Unpublished report.

Starr, M. P., Stolp, H., Trüper, H. G., Balows, A., and Schlegel, H. G. (eds), 1981, *The Prokaryotes: A Handbook on Habitats, Isolation and Identification of Bacteria*, 2 volumes, Springer-Verlag, Berlin.

Stewart, W. N., 1983, *Paleobotany and the Evolution of Plants*, Cambridge University Press, Cambridge.

Strauss, H., Des Marais, D. J., Hayes, J. M., and Summons, R. E., 1992, Proterozoic organic carbon—its preservation and isotopic record, in: *Early Evolution: Implications for Mineral and Energy Resources* (M. Schidlowski, D. M. McKirdy, and P. A. Trudinger, eds.), Springer-Verlag, Berlin, pp. 203–211.

Stumm, W. (ed.), 1977, *Global Chemical Cycles and Their Alteration by Man*, Dahlem Konferenzen, Akabon, Berlin.

Summons, R. E., and Powell, T. G., 1986, Chlorobiaceae in Palaeozoic Seas revealed by biological markers, isotopes and geology, *Nature* **319:**763–765.

Summons, R. E., and Powell, T. G., 1987, Identification of aryl isoprenoids in source rocks and crude oils: Biological markers for the green sulphur bacteria, *Geochim. Cosmochim. Acta* **51:** 557–566.

Summons, R. E., and Walter, M. R., 1990, Molecular fossils and microfossils of prokaryotes and protists from Proteozoic sediments, *Am. J. Sci.* **290A:**212–244.

Summons, R. E., Volkman, J. K., and Boreham, C. J., 1987, Dinosterane and other steroidal hydrocarbons of dinoflagellate origin in sediments and petroleum, *Geochim. Cosmochim. Acta* **51:**3075–3082.

Summons, R. E., Powell T. G., and Boreham, C. J., 1988, Petroleum geology and geochemistry of the Middle Proterozoic McArthur Basin, Northern Australia. III. Composition of extractable hydrocarbons, *Geochim. Cosmochim. Acta* **52:**1747–1763.

Sweeney, R. E., Liu, K. K., and Kaplan, I. R., 1978, Oceanic nitrogen isotopes and their uses in determining the source of sedimentary nitrogen, in: *Stable Isotopes in the Earth Sciences, N.Z. Dep. Sci. Ind. Res. Bull.***220:**9–26.

Tegelaar, E. W., de Leeuw, J. W., Derenne, S., and Largeau, C., 1989, A reappraisal of kerogen formation, *Geochim. Cosmochim. Acta* **53:**1303–1306.

Tissot, B. P., and Welte, D. H., 1984, *Petroleum Formation and Occurrence*, 2nd ed., Springer-Verlag, Berlin.

Towe, K. M., 1981, Environmental conditions surrounding the origin and early evolution of life, *Precamb. Res.* **16:**1–10.

Towe, K. M., 1990, Aerobic respiration in the Archean? *Nature* **348:** 54–56.

Trudinger, P. A., 1979, The biological sulfur cycle, in: *Biogeochemical Cycling of Mineral-Forming Elements* (P. A. Trudinger and D. J. Swaine, eds.), Elsevier, Amsterdam, pp. 293–314.

Trudinger, P. A., Swaine, D. J., and Skyring, G. W., 1979, Biogeochemical cycling of elements—general considerations, in: *Biogeochemical Cycling of Mineral-Forming Elements* (P. A. Trudinger and D. J. Swaine, eds.), Elsevier, Amsterdam, pp. 1–27.

Tucker, M. E., 1986, Carbon isotope excursions in Precambrian/Cambrian boundary beds, Morocco, *Nature* **319:**48–50.

van Aarssen, B. G. K., Cox, H. C., Hoorgendoorn, P., and de Leeuw, J. W., 1990, A cadinene biopolymer in fossil and extant dammar resins as a source for cadinanes and bicadinanes in crude oils from South East Asia, *Geochim. Cosmochim. Acta* **54:** 3021–3031.

Veizer, J., Holser, W. T., and Wilgus, C. K., 1980, Correlation of $^{13}C/^{12}C$ and $^{34}S/^{32}S$ secular variations, *Geochim. Cosmochim. Acta* **44:**579–587.

Volkman, J. K., 1986, A review of sterol markers for marine and terrigenous organic matter, Org. *Geochem.* **9:**83–99.

Walker, J. C. G., Klein, C., Schidlowski, M., Schopf, J. W., Stevenson, D. J., and Walter, M. R., 1983, Environmental evolution of the

Archean–Early Proterozoic Earth, in: *Earth's Earliest Biosphere: Its Origin and Evolution* (J. W. Schopf, ed.), Princeton University Press, Princeton, New Jersey, pp. 260–290.

Walter, M. R., 1983, Archean stromatolites: Evidence of the Earth's earliest benthos, in: *Earth's Earliest Biosphere: Its Origin and Evolution* (J. W. Schopf, ed.), Princeton University Press, Princeton, New Jersey, pp. 187–212.

Walter, M. R., Du, R., and Horodyski, R. J., 1990, Coiled carbonaceous megafossils from the Middle Proterozoic of Jixian (Tianjin) and Montana, *Am. J. Sci.* **290A:**133–148.

Whittaker, R. H., 1969, New concepts of kingdoms and organisms, *Science* **163:**150–160.

Woese, C. R., 1981, Archaebacteria, *Sci. Am.* **244:**98–106.

Woese, C. R., 1987, Bacterial evolution, *Microbiol. Rev.* **51:**221–271.

Woese, C. R., and Fox, G. E., 1977, Phylogenetic structure of the prokaryotic domain, *Proc. Natl. Acad. Sci. USA* **74:**5088–5090.

Woese, C. R., and Olsen, G. J., 1986, Archaebacterial phylogeny: Perspectives on the urkingdoms, *Syst. Appl. Microbiol.* **7:** 161–171.

Woese, C. R., and Wolfe, R. S., 1985, *The Bacteria, Vol. 8. Archaebacteria*, Academic Press, New York.

Zachos, J. C., Arthur, M. A., and Dean, W. E., 1989, Geochemical evidence for suppression of pelagic marine productivity at the Cretaceous/Tertiary boundary, *Nature* **337:**61–64.

Chapter 2

A Review of Macromolecular Organic Compounds That Comprise Living Organisms and Their Role in Kerogen, Coal, and Petroleum Formation

J. W. DE LEEUW and C. LARGEAU

1. Introduction

It is generally accepted that the formation of kerogens is the result of the so-called depolymerization–recondensation pathway (Tissot and Welte, 1984; Durand, 1980), Thus, naturally occurring macromolecular substances such as polysaccharides and proteins are enzymatically depolymerized to oligo- and monomers, which for the most part are mineralized. However, a small part of them are thought to condense with other substances such as low-molecular-weight lipids in a random way (Fig. 1). During diagenesis, the "geopolymers" thus formed continuously undergo chemical transformations by which they become more and more insoluble and resistant. Kerogens are being formed, which—depending on the nature of the original organic matter contributions—can generate various amounts and sorts of oil under thermal stress.

Recently, however, alternatives for the formation of kerogens—or parts thereof—have been proposed (Tegelaar *et al.*, 1989a). Based on the principle of selective preservation, it is assumed that specific organic substances, especially macromolecular substances, are selectively preserved during the processes of sedimentation and during diagenesis. In other words, biomolecules highly resistant to biodegradation and major chemical transformations such as hydrolysis and oxidation are thought to survive selectively and may become quantitatively important in kerogens, even when they constitute a minor part of their source organisms. Seen from this point of view kerogens can be thought of as mixtures of—sometimes partly changed—resistant macromolecular biomolecules. A more detailed discussion concerning this matter is given at the end of this chapter (see Section 3.1 and Fig. 62). Although this concept of kerogen formation by selective preservation is usually not considered in petroleum geochemistry, it should be mentioned that the reverse is true in coal geochemistry. The concept of coal macerals, although originally established on morphological links, sometimes strengthened chemically, is based on the relationship between fossil remains of higher plants, lower plants, and algae and their modern counterparts (Stach *et al.*, 1982).

J. W. DE LEEUW • Organic Geochemistry Unit, Faculty of Chemical Engineering and Materials Science, Delft University of Technology, 2628 RZ Delft, The Netherlands. C. LARGEAU • Laboratoire de Chimie Bioorganique et Organique Physique, UA CNRS 456, École Nationale Supérieure de Chimie de Paris, 75231 Paris Cedex 05, France. *Present address for J.W.D.*: Division of Marine Biogeochemistry, Netherlands Institute of Sea Research (NIOZ), Texel, The Netherlands.

Organic Geochemistry, edited by Michael H. Engel and Stephen A. Macko. Plenum Press, New York, 1993.

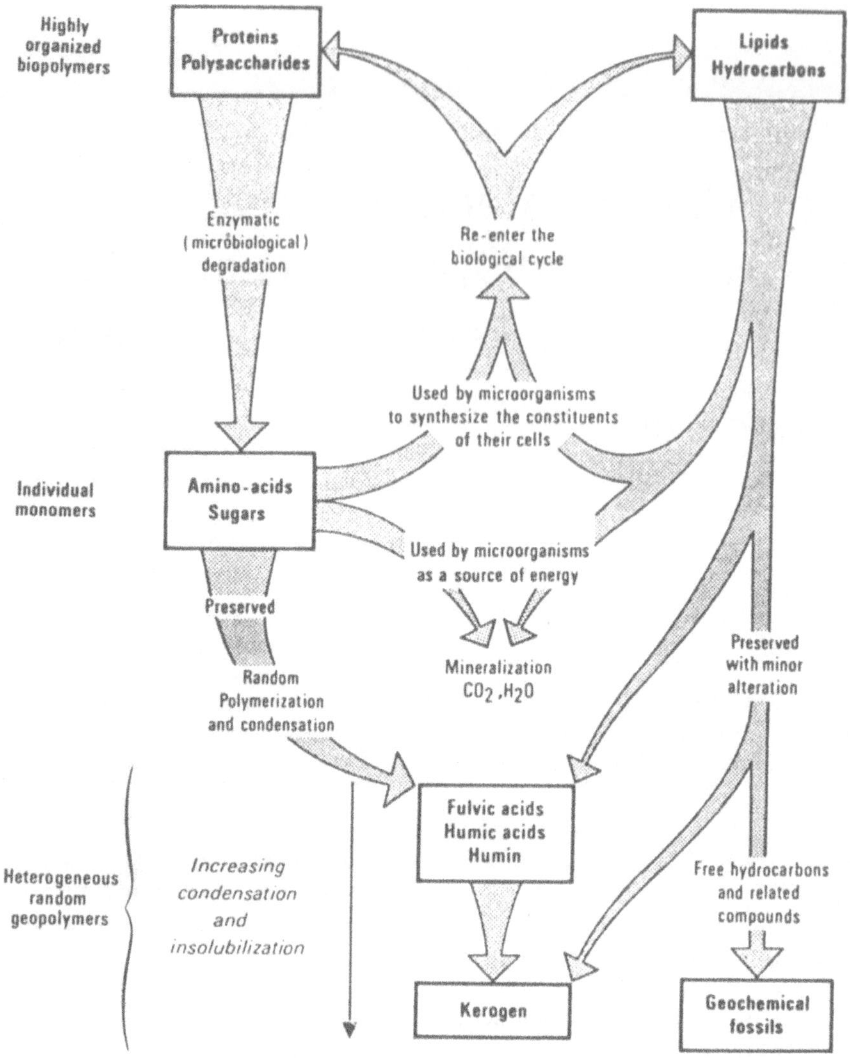

Figure 1. "Classical view" of the formation of kerogens. [After Tissot and Welte (1984).]

In this chapter, an attempt is made to discuss in some detail the above-mentioned alternative concept of kerogen formation. First of all, an overview is given of macromolecular organic substances occurring in nature. Chemical structures, natural occurrences, (bio)degradability, and survival potential will be discussed for individual macromolecules or groups of highly similar macromolecules. In this way, insight is obtained into the composition of resistant biomacromolecules present in recent sediments depending on the ecosystem, the source organisms, and depositional conditions. To complete the picture of kerogen formation, one has to take into account the formation of high-molecular-weight sedimentary organic matter resulting form incorporation of sulfur into low-molecular-weight lipids in specific depositional environments (Sinninghe Damsté *et al.*, 1988, 1989, 1990) (see Section 3.2 and Fig. 61). A short description of the "classical" kerogen types I, II, II-S, and III in a molecular fashion is given thereafter. Finally, the mechanisms of oil, gas, and coal genesis are briefly discussed on a molecular level, assuming thermal depolymerization of the resistant biomacromolecules and high-molecular-weight organosulfur substances.

2. Inventory of Macromolecular Substances, Their Occurrence in Organisms, and Their Potential for Survival in Recent Sediments

Although the inventory presented here attempts to cover the most important macromolecules presently known to occur in nature, emphasis is laid on those substances which are thought to be important from a geochemical point of view. As a result, the attention and space given to different groups of naturally occurring substances are heavily biased and not representative of the importance of their roles within the biosphere. As mentioned above, this inventory will be given mainly on a chemical basis. Thus, macromolecules with similar chemical structures will be treated together in the same section. A major distinction, however, is made first between the so-called storage materials, which are intracellular, and structural components occurring mainly in membranes, in cell walls, or extracellularly.

2.1. Storage Materials

The organic storage materials all serve as carbon and energy sources.

2.1.1. Starch

Starch is a major intracellular storage polysaccharide in vascular plant cells, though it occurs also in algae and some bacteria (Kennedy and White, 1983). Starch consists of two types of polymeric glucose: amylose and amylopectin [for recent reviews, see Zobel (1988) and Kainuma (1988)]. On average, amylose represents ca. 25% of starch and consists of long, unbranched chains of α-D-glucose units linked via (1→4)-glycosidic bonds (Fig. 2). The molecular mass varies from a few to 500 kilodaltons (kDa). As a consequence of the α-(1→4) linkage, amylose has a helically shaped tertiary structure.

Amylopectin, the other component of starch, is also of high molecular weight and built up from glucose units as well. After every 24–30 (1→4)-linked α-D-glucose units, a (1→6) linkage occurs, resulting in a highly branched structure (Fig. 3). The amylopectin contains some phosphate residues and Mg^{2+} and Ca^{2+} ions.

α-Amylase, widespread in plants, animals, and aerobic as well as anaerobic microorganisms, cleaves the (1→4)-glycosidic bonds in amylose and in the external amylose moieties of amylopectin. Another, also widely occurring, enzyme, amylo-1,6-glucosidase, cleaves the (1→6)-branches in amylopectin. As a result, mainly maltoses (4-*O*-α-D-glucopyranosyl-D-glucose) are produced from starch. A suite of other enzymes further biodegrade the maltose units. The products are finally used as carbon or energy sources.

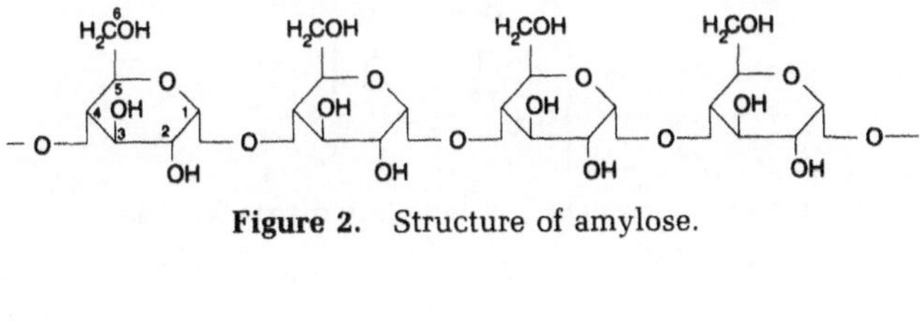

Figure 2. Structure of amylose.

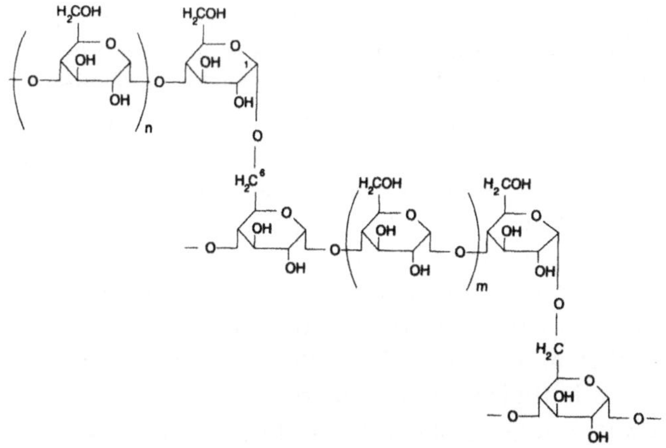

Figure 3. Structure of amylopectin.

Starches in algae and bacteria normally have much lower average molecular weights. They are more susceptible to enzymatic degradation. A striking example is known from the Solar Lake ecosystem in Egypt, where "cyanobacterial starch" disappears in the first few millimeters of this microbial mat sediment (Klok *et al.*, 1984b). Ittekkot *et al.* (1982) noticed the same in a study of particulate matter during a plankton bloom.

2.1.2. Glycogen

Glycogen is the main storage polysaccharide occurring in animal cells. This high-molecular-weight (up to several thousand kilodaltons) substance has a structure very similar to that of amylopectin, but is even more branched, with a (1→6) linkage for every 8–12 (1→4)-linked α-D-glucose units. Glycogen can be degraded by the combined action of α- and β-amylases. However, in animal cells a different enzyme, glycogen phosphorylase, transforms glycogen to glucose 1-phosphate.

2.1.3. Fructans

Fructans are water-soluble polymers of fructose with α-D-glucose usually as an end group. They occur mainly in vascular plants, especially in grasses, often together with starch. They are also found in algae and bacteria (Lewis, 1984). Apart from storage, they may be involved in other functions, such as osmoregulation and freezing-point lowering of tissue water by depolymerization (Nelson and Spollen, 1987, and references therein). The fructans seem to accumulate in periods of high irradiance or of low temperature. Two major groups of fructans occur. The inulins are β-(2→1)-linked fructose units, and the levans (in plants also called phleins) consist of β-(2→6)-linked fructose units (see Fig. 4). In plants, branches at some of the free hydroxyl groups of the fructose units occur.

Levans from bacteria such as those occurring in dental plaque are highly branched and have much higher molecular weights than the plant levans. In bacteria their formation by the reaction *n* sucrose → levan + *n* glucose is catalyzed by extracellular levan sucrase (Schlegel, 1986). Hydrolytic enzymes, converting inulins and levans to sucroses and fructoses, are omnipresent and widely distributed among bacteria (Pollock, 1986).

2.1.4. Laminarans

Laminarans occur in several brown algae. These storage polysaccharides consist mainly of (1→3)-linked β-D-glucose units with occasional β-(1→6) linkages (Fig. 5). D-Glucose or D-mannitol groups are present as branches via (1→6) linkages. The ratio of glucose and mannitol chains is about 1:1.

Laminaran type polysaccharides have also been encountered in other algae (Chrysophyceae and Bacillariophyceae) and in fungi (Saito *et al.*, 1987).

2.1.5. Poly-β-hydroxyalkanoates (PHA)

Poly-β-hydroxyalkanoates (PHA) occur in Eubacteria (Dawes and Senior, 1973). These linear polyesters are usually homogeneous and composed of β-hydroxybutanoate units (Fig. 6; R = H). Copolymers of β-hydroxybutanoate and pentanoate (R = CH_3) have also been identified in several species. In the latter copolymers, ratios of C_4 to C_5 units ranging from 0.5 to 2 were noted (Capon *et al.*, 1983). Regarding the degree of polymerization, an average molecular mass of ca. 135 kDa was reported for the copolymer of the freshwater cyanobacterium *Aphanothece* sp. (Capon *et al.*, 1983).

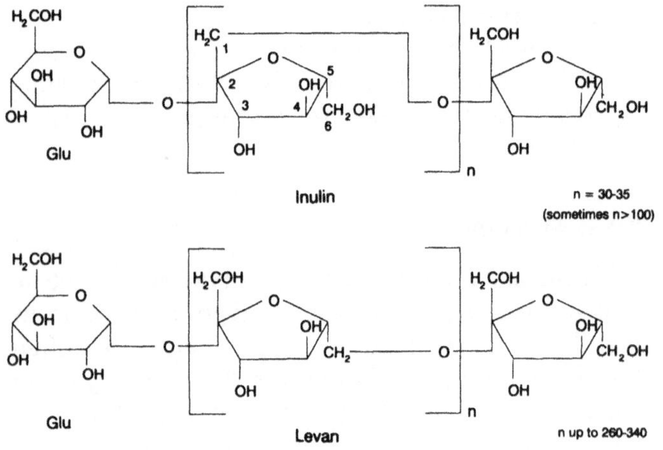

Figure 4. Structure of inulin and levan.

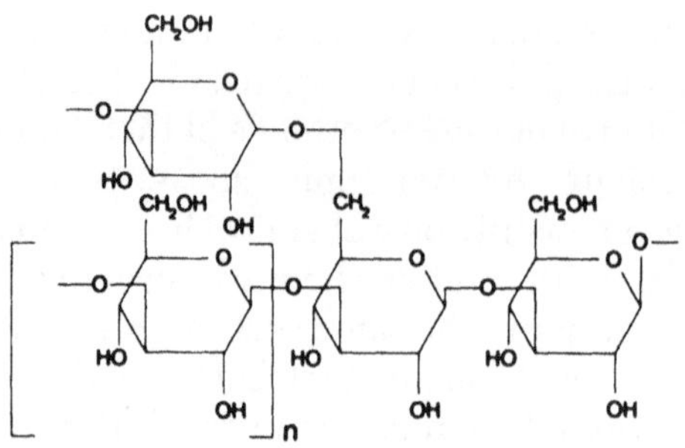

Figure 5. Structure of laminarans.

PHA are intracellular substances widely distributed in Eubacteria (Dawes and Senior, 1973). The kingdom Procaryotae (bacteria) comprises two phylogenetically distinct groups: the Archaebacteria and the Eubacteria (Woese *et al.*, 1978). The second group includes classical bacteria and cyanobacteria; the latter are still sometimes termed blue-green algae, since these are the sole prokaryotic organisms with an O_2-evolving photosynthetic system. Regarding cyanobacteria, PHA were identified in freshwater species, marine species, and mat-building halophilic species of cyanobacteria (Capon *et al.*, 1983). In all species of cyanobacteria tested, both marine and freshwater, only very low contents of PHA (0.02 to 0.04% of total dry weight), were observed. PHA have not been detected in eukaryotic algae (Lewin, 1974).

In numerous Eubacteria, PHA are important carbon reservoir and energy storage products as are polysaccharides (Smith, 1982). Considerable amounts of PHA can be accumulated by aerobic and facultative species when deprived of oxygen and used as energy and carbon sources on return to aerobic conditions (Schlegel, 1986). PHA are also abundant in muds containing sewage debris (Boon *et al.*, 1986).

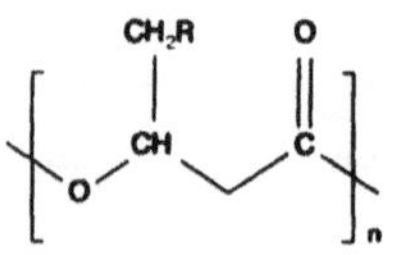

Figure 6. Structure of poly-β-hydroxyalkanoates.

PHA are based on readily cleaved ester bonds; furthermore, linkage of bifunctional units, like hydroxyalkanoates, cannot result in the formation of a two- or three-dimensional polymeric network. Accordingly, PHA are easily degraded via enzymatic and/or chemical hydrolysis, and their potential for survival in sediments should be very weak. This was demonstrated to be the case in cyanobacterial mats, where rapid metabolism of PHA occurs in the upper centimeters. In fact, PHA were only abundant in the purple bacterium-containing layers of the living part of the mats (Boon *et al.*, 1985; Boon and de Leeuw, 1987).

2.1.6. Triacylglycerols (Triglycerides)

Triacylglycerols are esters of one molecule of glycerol and three molecules of fatty acids (Fig. 7). Although they cannot be considered as real macromolecular substances, they are mentioned here because they represent major components of storage in plant and animal cells. The wide variety of the triacylglycerols is due to the variations in chain length and in the degree of unsaturation of the fatty acid moieties. Animal triacylglycerols are predominantly saturated, whereas plant triacylglycerols contain relatively high amounts of (poly)unsaturated fatty acids.

Triacylglycerols are very easily decomposed and are not thought to be important constituents of Recent sediments. They have, however, been reported in a 1700-year-old sediment of the Namibian Shelf (Boon *et al.*, 1980) and in surface sediments off the coast of Peru (Volkman and Wakeham, 1989). It is difficult to say whether the triglycerides encountered in these sediments are remains from organisms living in the water column or represent components of senescent microbial biomass living within the sediment. The latter explanation is favored by Volkman and Wakeham (1989).

Although the storage materials mentioned above can make up a very important part of the biomass in the biosphere, it is nevertheless assumed that their contribution to the geosphere is minor. Apart from their incidental presence in very recent sediments, they cannot be considered geochemically important. They are biosynthesized to be decomposed, and the enzymes to decompose or mineralize them, both inside and outside organisms, are very widespread and omnipresent. It is therefore believed that these substances are not, or virtually not, preserved in sediments either as such or altered.

2.2. Structural Components of Organisms

Structural components play many roles in organisms; they can protect and support cells, tissues, and organs.

2.2.1. Cellulose

Cellulose represents the most abundant biomolecule in nature and occurs in protective cell walls of higher and lower plants, especially in stalks, stems, trunks, and all woody parts of plant tissues. It is also present in cell walls of many algae (about 10% by weight) and in oomycetes but usually not in bacterial cell walls. It has been estimated that 50 kg of cellulose is produced per day per human being (Lehninger, 1982). Cellulose is a widely used substance as well: wood, cotton, paper, and cardboard all contain relatively large amounts of cellulose.

Cellulose is a linear homopolysaccharide consisting of 10,000 or more (1→4)-linked β-D-glucose units (Fig. 8). In contrast to the helical-type structure of amylose, the β-(1→4) linkages of cellulose imply an extended linear conformation of the chains. The chains are cross-linked to each other via numerous hydrogen bonds, thus building insoluble fibrils (Lehninger, 1982).

Cellulose is biodegraded by cellulase. The cellulase consists of three enzymes. Endo-β-1,4-glucanase cleaves β-(1→4) bonds in the center of the macromolecule, producing long-chain fragments. Exo-β-

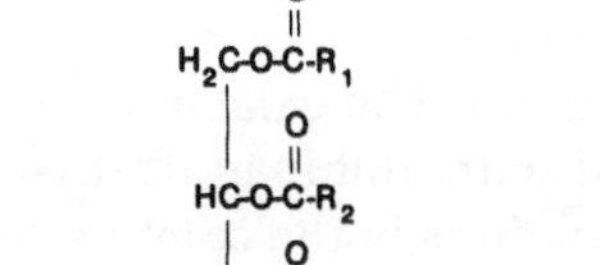

Figure 7. General structure of triacylglygerols.

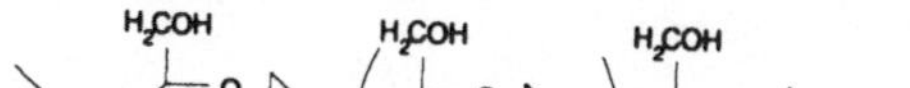

Figure 8. Structure of cellulose.

1,4-glucanase cleaves off the disaccharide cellobiose from the ends of the chains, and β-glucosidase converts cellobiose to glucose (Kluepfel, 1988).

Especially fungi, but also Eubacteria, degrade cellulose under aerobic conditions. Cellulose is degraded anaerobically by specific groups of bacteria, e.g., *Clostridia*. The end products are low-molecular-weight fatty acids such as acetic acid, propionic acid, and butyric acid.

In many Recent sediments, especially in peat environments, cellulose is preserved to some extent (e.g., Given *et al.*, 1984; Moers *et al.*, 1989). The relatively high amounts of glucose obtained after acid hydrolysis of peats (up to 10% dry weight organic matter) are for the most part thought to originate from cellulose of vascular plant tissues. In ancient sediments, the acid-released glucose is also ascribed mainly to remains of cellulose (Hedges *et al.*, 1985; Michaelis *et al.*, 1986). In vascular plants a considerable part of the cellulose is linked to lignin, forming the so-called lignin–cellulose complex (see Section 2.2.27). This probably accounts for a partial preservation of cellulose during diagenesis.

In many cell walls of vascular plants, cellulose fibrils are interwoven or cemented by other polysaccharides, previously called collectively "hemicelluloses." Since a number of polysaccharides occurring in vascular plant cell walls have also been encountered in algae, fungi, and bacteria, these polysaccharides will be discussed individually. However, the term "hemicellulose" is used throughout this chapter to indicate vascular plant cell wall polysaccharides closely associated with cellulose.

2.2.2. Xylans

Xylans are very widespread and abundant in vascular plants, where they are closely associated with cellulose (Kennedy and White, 1983). They also occur in certain algae. In woody tissues they might represent 10–25% of the carbohydrates.

In vascular plants, xylans consist of (1→4) linked β-D-xylose units (Fig. 9). Plant xylans also contain other sugars such as (1→3)-linked α-L-arabinose or (1→2)- and (1→3)-linked 4-*O*-methyl-D-glucuronic acid in side chains (Puls *et al.*, 1988). Especially, xylans of hardwoods are partly acetylated at C_2 or C_3. The molecular weight is much less than that of cellulose, and xylans are more rapidly biodegraded by xylanase. So-called L-arabino-D-xylans occur in cereal gums in association with acidic polysaccharides. These xylans represent highly, irregularly branched (1→4)-linked β-D-xylans attached via (1→3) links with α-L-arabinose.

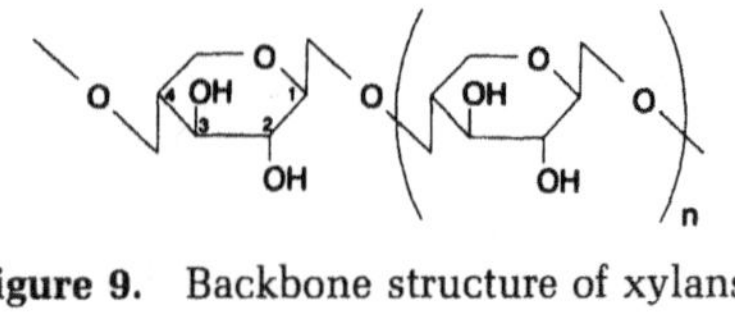

Figure 9. Backbone structure of xylans.

In algae the xylans are built up from D-xylose units linked (1→3) and/or (1→4). Contrary to plant xylans, the algal xylans contain hardly any other sugar units.

Many bacteria and fungi are able to decompose xylans (Kubicek, 1984; Biely, 1985; Kluepfel, 1988). Three different types of enzymes are involved; endo-D-xylanase and exo-D-xylanase produce oligosaccharides, which are further decomposed with β-D-xylosidase to xylose units. Other enzymes able to cleave the side groups, e.g., α-L-arabinofuranosidase, significantly enhance the decomposition.

In some cases xylose represents relatively high amounts among mixtures of sugars obtained after acid hydrolysis of Recent sediments (Liebezeit *et al.*, 1983; Moers *et al.*, 1989).

It should be noted, however, that relatively high amounts of specific sugars released by acid treatment does not necessarily imply the original presence of their homopolymeric counterparts. For example, the relatively high amounts of xylose observed in hydrolysates of Solar Lake cyanobacterial mats (Klok *et al.*, 1984b) are derived from the sheaths but cannot be ascribed to xylans straightforwardly.

2.2.3. Pectins

Pectins occur in primary cell walls and as intracellular substances in flowering plants and fruits. They are less abundantly present in woody tissues and in grasses (Kennedy and White, 1983; Jarvis, 1984).

Pectins can be considered as so-called block polysaccharides consisting of blocks of largely methyl-esterified galacturonan alternating with blocks of less esterified or nonesterified galacturonan. The methyl-esterified blocks contain rhamnose units, many of which carry side chains of galactans and branched arabinans (Fig. 10). The galactan side chains are further branched with arabinan chains.

Many variations in the composition of the pectin structures are known (BeMiller, 1986). Sometimes, relatively small amounts of other sugars such as xylose and fucose can be present. Acetyl groups have also been reported. The nonesterified or less es-

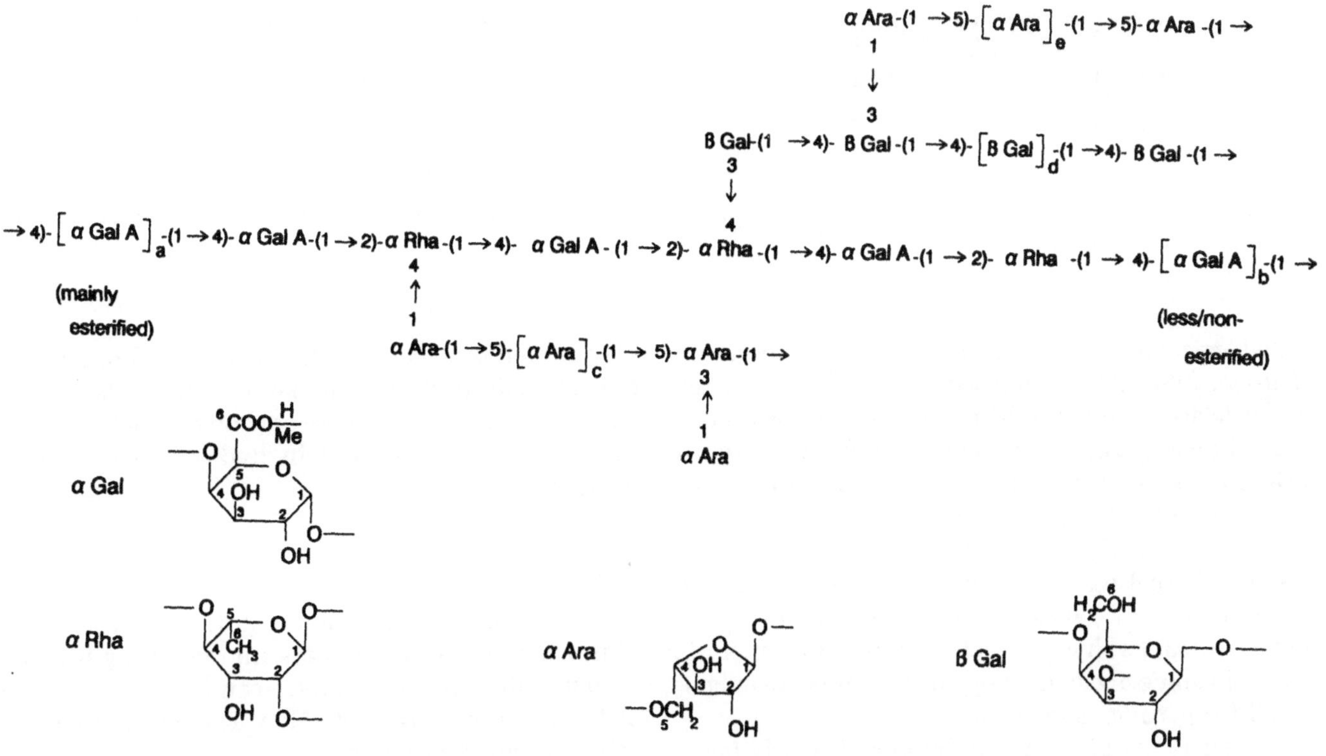

Figure 10. Partial structure of a pectin. [After Jarvis (1984).]

terified galacturonic acid chains are complexed with Ca^{2+} ions. Many aerobic and anaerobic bacteria and fungi are able to metabolize pectins enzymatically.

2.2.4. Mannans

Mannans occur in a number of vascular plants, algae, and fungi as homopolysaccharides or in heteropolysaccharides (Kennedy and White, 1983). The linear chains of (1→4)-linked β-D-mannose form the skeleton of these polysaccharides in vascular plants (Fig. 11). α-D-Galactose side chains are more or less frequently present via (1→6) linkages depending on the species. D-Gluco-D-mannans are present, among other mannans, in hardwood, the ratio of glucose to mannose being 1:2. These "hemicelluloses" consist of β-D-glucose and β-D-mannose units (1→4)-linked in linear chains. In coniferous woods, D-gluco-D-mannans occur with (1→6)-linked α-D-galactose units as side chain.

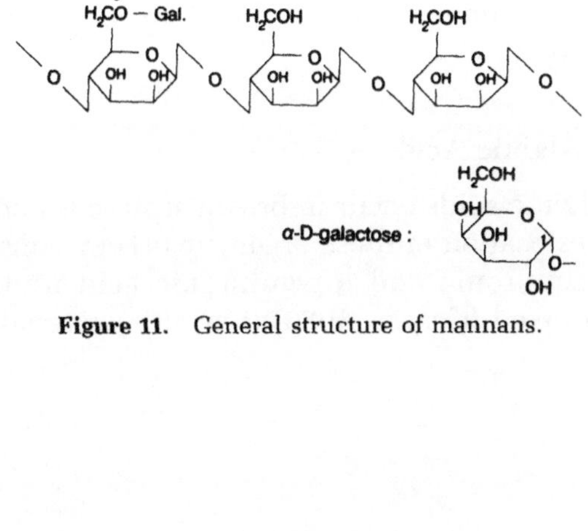

Figure 11. General structure of mannans.

In fungi, mannans are characterized by (1→6)- and (1→2)-linked α-D-mannose units. In some cases, side chains consisting of D-galactose and D-xylose are present. Enzymes capable of depolymerizing mannans are omnipresent in nature. As a consequence, a selective preservation of mannans can hardly be expected. However, the relatively high percentages of mannose and glucose encountered in hydrolysates of Mahakam Delta sediments and in the Brandon lignite might reflect preserved mannans (Moers, 1989).

2.2.5. Galactans

Vascular plant galactans are water-soluble, highly branched polysaccharides consisting of (1→3/6)-linked β-D-galactose units in the backbone chain. In these "hemicelluloses," a variety of side chains containing, for example, L-arabinose and L-rhamnose are present via (1→6) linkages (Fig. 12).

The carrageenans form a diverse group of sulfo-

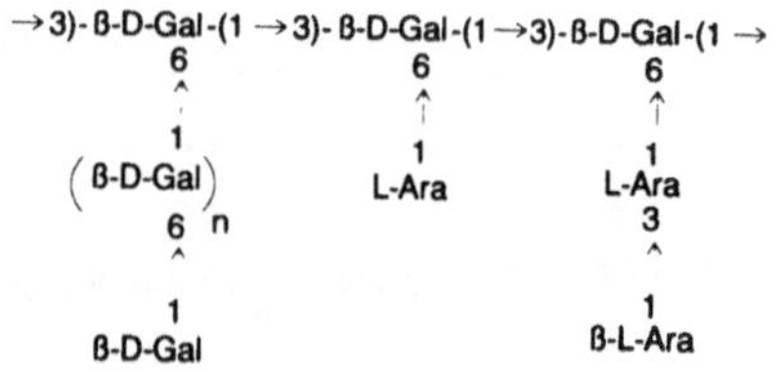

Figure 12. Partial structure of a vascular plant galactan. [After Kennedy and White (1983).]

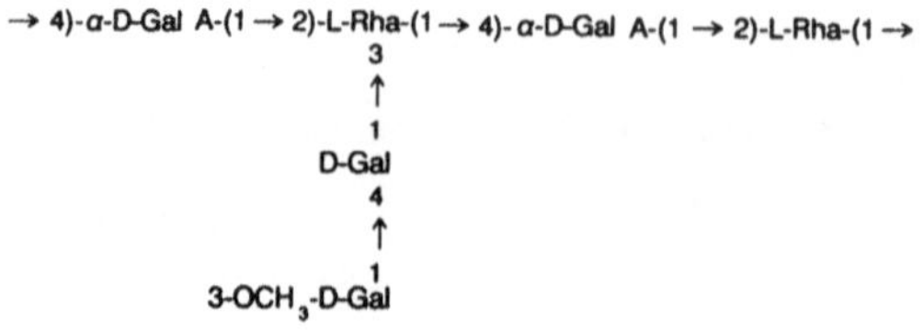

Figure 14. Example of a mucilage structure. [After Kennedy and White (1983).]

nated galactans occurring mainly in red seaweeds (Hirono, 1987). Alternating sequences of (1→3)-linked β-D-galactoses and (1→4)-linked α-D-galactoses bearing sulfonate groups at various positions (2,4,6) occur (Fig. 13). In many cases, the β-D-galactose is replaced by 3,6-anhydro-β-D-galactose.

The agars, also occurring in red seaweeds, consist of alternating (1→3)-linked β-D-galactoses and (1→4)-linked α-L-galactoses, sometimes sulfonated at O-6 or α-L-3,6-anhydrogalactose units (Brasch *et al.*, 1981). In some algae, the D-galactose units are substituted for pyruvic acid acetals.

Agar is not biodegraded by most bacteria (*sic!*). However, especially in tidal zones, agar decomposers can be abundantly present. It is therefore believed that agar is not preserved during deposition in Recent sediments (Schlegel, 1986).

2.2.6. Mucilages

D-Galacturonic acid is a characteristic component of many mucilages, substances which frequently occur in seeds, where they function as water reservoirs. An example of the structure of a mucilage (slippery elmbark) is given in Fig. 14. Many structural variations are known, however (Kennedy and White, 1983).

Since these polysaccharides occur in seeds, they may be well protected mechanically and thus survive biodegradation in very immature sediments. To the best of our knowledge, there are however no reports of galacturonic acid present in hydrolysates of fossil fruits.

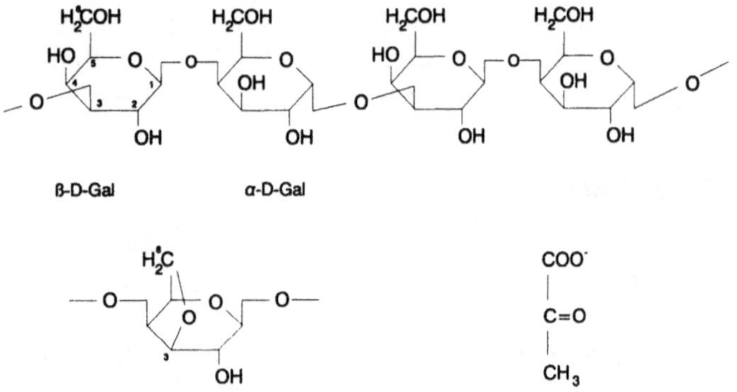

Figure 13. Structural elements of carrageenans.

2.2.7. Gums

Gums are produced and exuded as viscous liquids when barks or fruits are injured. They become hard nodules upon dehydration and thus seal the site of injury, providing protection against microorganisms. Several types exist.

- Gum arabic (e.g., *Acacia*) consists of (1→3)-linked β-D-galactose chains to which L-arabinose, L-rhamnose, and a D-galactosyl-L-arabinose are attached, mostly via (1→6) linkages. Also, D-glucuronic acids can be present in the side chain via (1→4) linkages (Kennedy and White, 1983; Glicksman, 1983a).
- Gum ghatti consists of chains of alternating (1→2)- and (1→4)-linked D-glucuronic acid and D-mannose units and numerous side chains containing L-arabinose and D-galactose (Kennedy and White, 1983).
- Gum tragacanth (e.g., *Astragalus*) consists of main chains of (1→4)-linked α-D-galacturonic acid units with (1→3)-linked branches containing among others xylose, galactose, and fucose. The structures are similar to those of pectic acids (Glicksman, 1983b).

Because of their natural function, it can be speculated that gums do have some resistance against microbial attack in sediments and may therefore be preserved to some extent.

2.2.8. Alginic Acids

Alginic acids occur in brown seaweeds and prevent desiccation of these organisms. They consist of β-D-mannuronic and α-L-guluronic acid units, all (1→4)-linked (Gacesa, 1988) (Fig. 15). Bacterial algi-

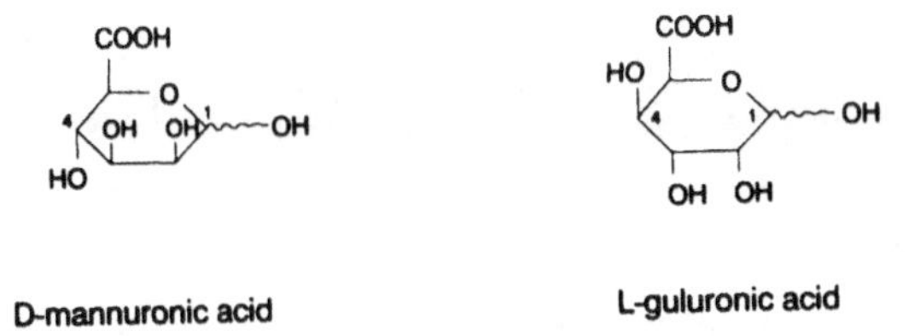

Figure 15. Building blocks of alginic acids.

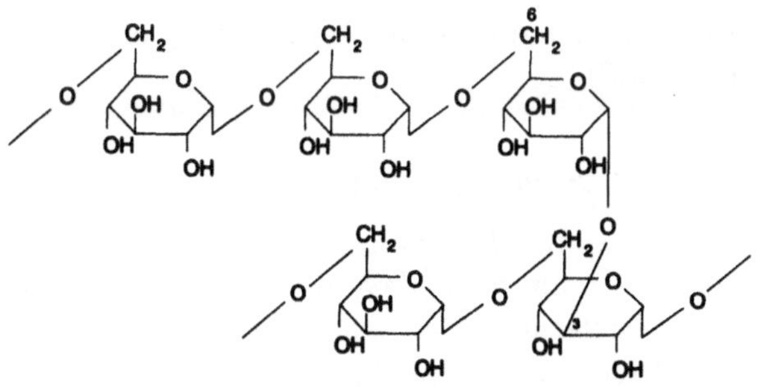

Figure 16. Partial structure of a dextran.

nates also occur but are invariably *O*-acylated in the mannuronic uits.

Alginic acids are well equipped to form complexes with metal ions such as Ca^{2+}. Many bacteria, but also fungi and even higher organisms, contain alginases. Therefore, preservation in Recent sediments is not expected.

2.2.9. Fungal Glucans

Fungi biosynthesize a variety of glucans. Nigeran is a polysaccharide present in plants and mycelia of some fungi and consists of an unbranched chain of alternating (1→3)- and (1→4)-linked α-D-glucose units (Kennedy and White, 1983; Schlegel, 1986). The yeast cell wall contains (1→6)-linked β-D-glucose units cross-linked with (1→3)-linked β-D-glucose units. Scleroglucans are secreted by several fungi and are built from tetrasaccharide units. This unit consists of three β-(1→3)-linked glucoses with a β-(1→6)-linked glucose attached to the second glucose (Holzwarth, 1985; Paul *et al.*, 1986).

Pullulans are exocellular α-glucans consisting of maltotriose units with (1→4) and (1→6) linkages (Paul *et al.*, 1986).

Presently, nothing is known about a preservation of fungal glucans in Recent sediments.

2.2.10. Dextrans

Dextrans are extracellular polysaccharides occurring in bacteria. The main chain consists of (1→6)-linked α-D-glucose units. Branching via (1→3) and (1→4) linkages is very general (Kennedy and White, 1983; Paul *et al.*, 1986) (Fig. 16).

2.2.11. Xanthans

Xanthans are extracellular polysaccharides biosynthesized by various *Xanthomonas* bacteria. They consist of a chain built up from (1→4)-linked β-D-glucose units. To this chain, trisaccharide side chains of α-D-mannose and D-glucuronic acid (ratio 2:1) are attached via (1→3) linkages. The mannoses are acetylated at C_6, and a part of the terminal mannose groups are linked to a pyruvic acid acetal group via O-4 and O-6 (Holzwarth, 1985; Paul *et al.*, 1986). These extracellular bacterial slimes, like the dextrans, may survive early diagenesis to some extent since bacteria represent the last stage of life in sediments. Nothing, however, is known about the fate of these substances in sediments.

2.2.12. Chitin

Chitin is a polysaccharide occurring in supporting skeletons and cell walls of arthropods, copepods, crustacea, many fungi, and some green algae. It is a homopolymer of (1→4)-linked 2-acetamido-2-deoxy-β-D-glucose (Fig. 17) and can be considered as an analog of cellulose (Kennedy and White, 1983).

Chitin is insoluble in water and many organic solvents and occurs in nature complexed or covalently bound to other substances such as proteins. Nevertheless, a variety of bacteria can decompose chitin using chitinase and chitiobiase, leading to the formation of the monomer. The possible occurrence of chitin in Recent sediments and soils has been concluded from flash pyrolysis studies (e.g., Saiz-Jimenez and de Leeuw, 1986b). Although chitin might be one of the contributors of glucosamines frequently found in immature sediments, the preservation of

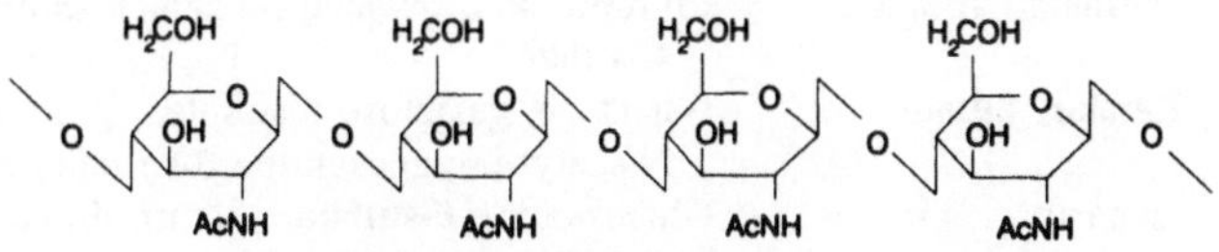

Figure 17. Structure of chitin.

chitin is expected to be small (Steinberg *et al.*, 1987; moers, 1989).

2.2.13. Glycosaminoglycans

Glycosaminoglycans are widespread in the cell envelope of mammal cells and occur occasionally in fish and bacteria and are part of proteoglycans. Because animals are not thought to be important contributors of organic matter to sediments, substances belonging to this group are only briefly considered here; names and structural elements are listed in Table I (Lehninger, 1976; Kennedy and White, 1983).

The structure of the polysaccharide moiety of the most abundant representative of this group, hyaluronic acid, is given in Fig. 18. In chondroitin, the N-acetyl-β-D-glucosamine (β-D-GluNAc) units are replaced by N-acetyl-β-D-galactosamine (β-D-GalNAc) units. Chondroitins with sulfate groups often present in the 4 or 6 position are encountered in vertebrate bones.

The glycosaminoglycans are linked to proteins via ether bonds and usually not via amide bonds as in glycoproteins (Kennedy and White, 1983) (see Fig. 19).

Many enzymes able to decompose these substances exist in nature. Moreover, as already mentioned, their main occurrence is in mammals, which makes their significant contribution to sediments rather unlikely.

2.2.14. Proteins

Proteins are the most abundant substances in cells. They consist of polypeptides, which are long chains of many amino acids. In most cases the pro-

TABLE I. Glycosaminoglycans

Name	Constituents
Chondroitin	Glucuronic acid, N-acetyl-D-galactosamine
Hyaluronic acid	Glucuronic acid, N-acetyl-D-glucosamine
Chondroitin 4-sulfate	Glucuronic acid, N-acetyl-D-galactosamine 4-sulfate
Chondroitin 6-sulfate	Glucuronic acid, N-acetyl-D-galactosamine 6-sulfate
Dermatan sulfate	Iduronic acid, N-acetyl-D-galactosamine 4-sulfate
Keratan sulfate	Galactose, galactose 6-sulfate, N-acetyl-D-glucosamine 6-sulfate
Heparin	Glucosamine 6-sulfate, glucuronic acid 2-sulfate, iduronic acid

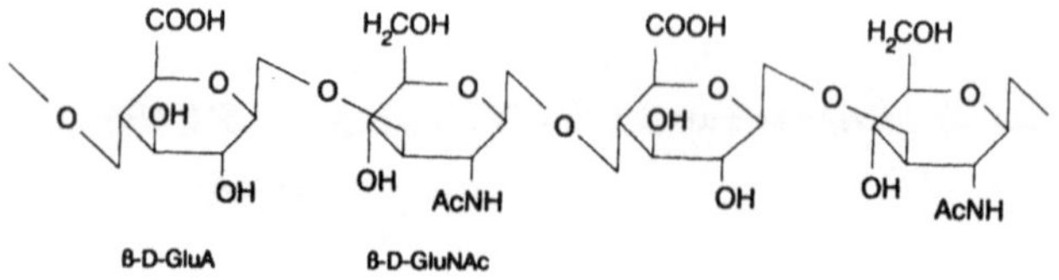

Figure 18. Structure of hyaluronic acid.

teins are built from a basic set of 20 amino acids. The huge variety in protein structures is a consequence of the immense number of possible sequences of amino acids.

The proteins serve many biological functions in nature. They can act as enzymes, as transporters (e.g., certain lipoproteins in blood plasma which carry lipids), and as regulators (e.g., insulin, which regulates sugar metabolism). They can also serve as storage substances (e.g., ovalbumin in egg white), as structural proteins (e.g., collagen, keratin), or as defense proteins (e.g., immunoglobulins, snake venoms, bacterial toxins). The proteins can be divided into the globular and the fibrous proteins. Globular proteins contain tightly folded chains and are usually water-soluble (enzymes, antibodies, nutrient storage proteins). The hydrophobic amino acid units are inside the macromolecule, whereas the more hydrophilic amino acid moieties are on the outside. The stability is mainly due to hydrogen bonding and sometimes cystine cross-linking. Fibrous proteins are water-insoluble and represent long stringy molecules. Most of the structural and protective proteins are fibrous ones.

Although proteins represent a considerable amount of the biomass on Earth, their significance for organic geochemistry is rather limited. Most proteins are very easily depolymerized by proteolytic exoenzymes and peptidases to the amino acid monomers. The amino acids are then metabolized further within the cells. Only the water-insoluble, structural proteins may survive biodegradation to some extent, especially when they are protected. Therefore, a few

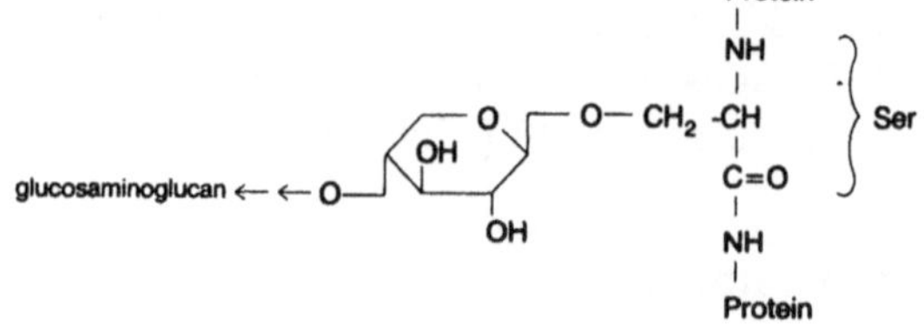

Figure 19. Partial structure of a proteoglycan. [After Kennedy and White (1983).]

representative examples of this group are briefly discussed here.

α-Keratins, found among others in hair, wool, scales, and hooves, consist of α-helically coiled polypeptide chains. The stability is due to intrachain hydrogen bonding of the peptide bonds and relatively many cystine cross-links. The α-keratins are rich in hydrophobic, helix-coil-favoring amino acid units: alanine, valine, methionine, isoleucine, and phenylalanine. In β-keratins (e.g., silk fibroin), the polypeptides are arranged in a so-called zigzag fashion and are hydrogen-bonded to each other. They are rich in glycine and alanine units.

Collagens are very abundant in vertebrates (e.g., in tendons, blood vessels, bones, and cartilages). They consist of three coiled polypeptide chains containing many proline and hydroxyproline units.

Elastin, a typical protein in elastic connective tissue, consists of many polypeptide chains cross-linked via desmosine, a special amino acid derived from four lysine molecules (Lehninger, 1982) (Fig. 20). Some proteins serve as matrix molecules for the formation of skeletons in shells and bones in a process called biomineralization (Omori and Watabe, 1980). These proteins are thus mechanically protected and are preserved quite well, as geoimmunological experiments have shown (Lowenstein, 1981; Muyzer et al., 1984). Flash pyrolysis studies of organic matter in Recent sediments also revealed the presence of hydrophobic proteins (e.g., Saiz-Jimenez and de Leeuw, 1986b). Further information on the fate of proteins is reported elsewhere in this volume.

2.2.15. Extensin

Extensin is a complex insoluble glycoprotein in plant cell walls covalently bound to cellulose. To some extent, it is comparable to collagen in animal tissue. L-arabinose and D-galactose units are frequently present as well as hydroxy-L-proline (Tierney and Varner, 1987; Kennedy and White, 1983).

Many other glycoproteins occur in vascular plants and algae, but their structures are far from understood. The algal ones are rich in hydroxy-L-proline and serine and in galactose and arabinose (Kennedy and White, 1983). The potential for survival of these glycoproteins in sediments is regarded as low.

2.2.16. Mureins

Mureins, sometimes referred to as peptidoglycans, constitute the rigid "skeleton" of the eubacterial cell walls (Schlegel, 1986). They consist of polysaccharide chains to which short peptides are attached. Cross-linking of the chains occurs via peptide bridges. The chain is built up from alternating 2-acetamido-2-deoxy-β-D-glucose (N-acetylglucosamine) and 2-acetamido-3-O-(1-carboxyethyl)-2-deoxy-β-D-glucose units via (1→4) linkages (Fig. 21). The latter unit is also referred to as N-acetyl muramic acid (MurNAc).

Penta- or tetrapeptides containing common and specific amino acids such as L and D-alanine (L and D-Ala), diaminopimelic acid (Dpm), L-lysine (L-Lys), and D-glutamic acid (D-Glu) are attached to the polysaccharide backbone via an amide linkage (see Fig.

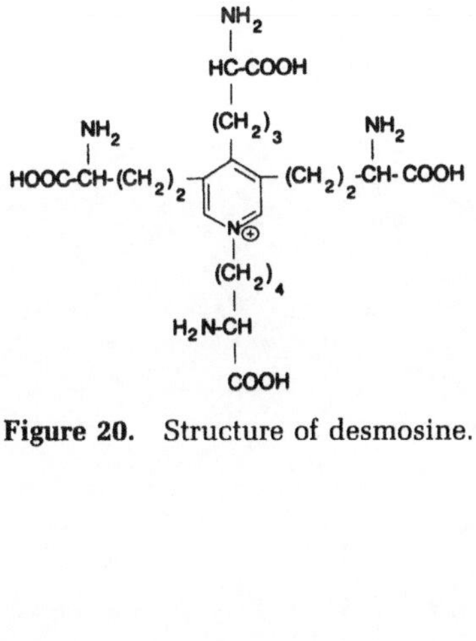

Figure 20. Structure of desmosine.

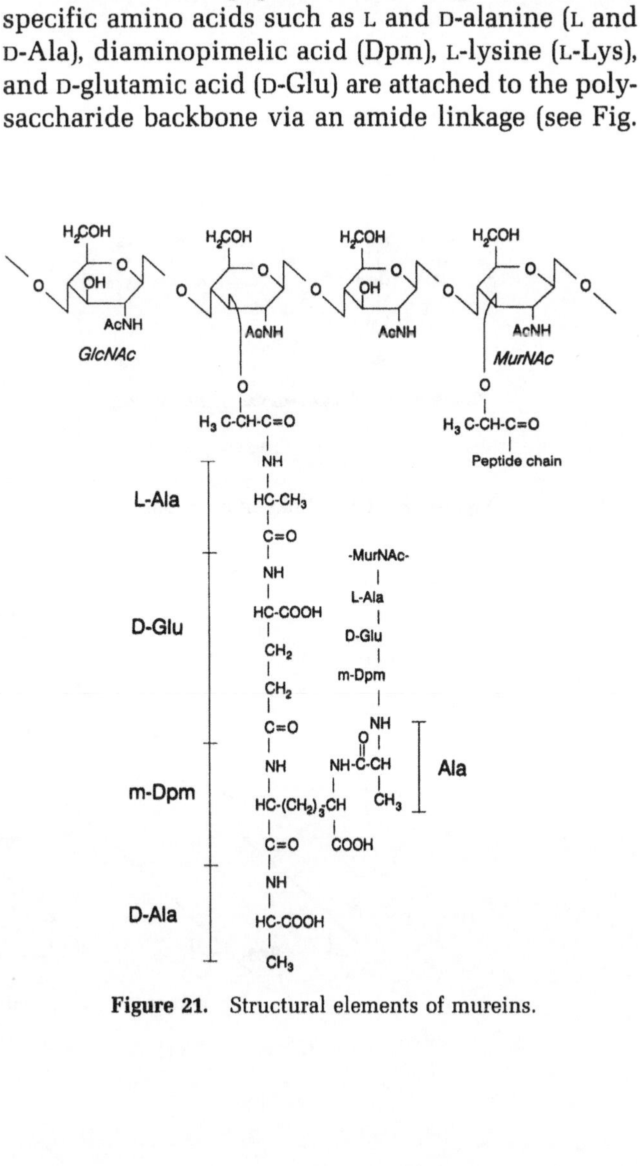

Figure 21. Structural elements of mureins.

21). The composition of these peptides varies considerably between gram-positive and gram-negative bacteria, but also between individual species. For example, L-lysine and LL-diaminopimelic acid (m-Dpm) occur only in gram-negative bacteria, and *meso*-diaminopimelic acid occurs only in gram-positive bacteria. The chains are cross-linked via amino acid moieties, such as alanine and diaminopimelic acid (see Fig. 22), in gram-negative bacteria or via interpeptide chains, such as pentaglycine between L-alanine and L-lysine, in gram-positive bacteria. As a consequence of this cross-linking, a very rigid and stable three-dimensional macromolecule is built up, which encases the cytoplasm membrane (see Fig. 23).

In gram-positive bacteria, the murein network represents on average 50% of the dry-weight cell wall material; in gram-negative bacteria the murein skeleton is much thinner and represents generally less than 10% of the cell wall dry weight.

Mureins are resistant to peptidases, but more specific enzymes such as lysozyme [cleavage of the (1→4) linkages of the polysaccharide backbone] and muroendopeptidase (cleavage of the cross-linking peptide bonds) degrade mureins to small fragments.

Specific moieties of mureins have not been reported from Recent sediments. Muramic acid or diaminopimelic acid may serve as a potential marker for the murein skeleton in sediments. However, procedures to isolate them intact and to identify them properly from sediments have not been explored. The role of microorganisms in a peat environment has been discussed based on muramic acid levels (Casagrande and Park, 1978). Due to the deviating stereochemistry of the amino acid moieties occurring in mureins, caution should be taken in the application of D/L ratios of amino acids occurring in sediments as maturity indicators (Bada, 1985). *N*-Acetylglucose moieties have been shown to be present in several Recent sediments (Moers, 1989; Steinberg *et al.*, 1987). Their occurrence is, however, not specific for mureins, since this amino sugar occurs in many natural substances.

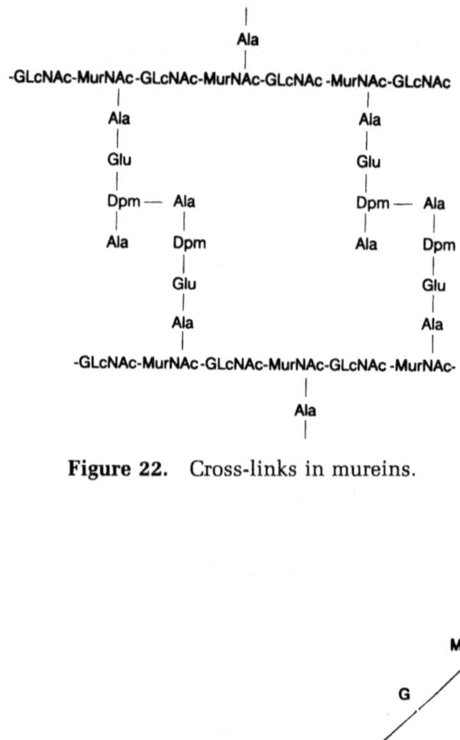

Figure 22. Cross-links in mureins.

2.2.17. Teichoic Acids

Teichoic acids consist of polyols (e.g., glycerol, ribitol, mannitol), in most cases linked via phosphate bridges. They are present exclusively in gram-positive bacteria covalently bound to the murein cell wall. Several types of teichoic acids are presently known (Kennedy and White, 1983; Hancock and Baddiley, 1985; Klosinski and Penczek, 1986):

- The 1,5-polyribitolphosphates often contain D-alanine esterified via O-2 of the D-ribitol, whereas sugar units such as glucoses or 2-acetamido-2-deoxyglucoses are attached glycosidically via O-4 (Fig. 24).
- The 1,3-polyglycerolphosphates are also widespread. D-Alanine, D-glucose, 2-acetamido-2-

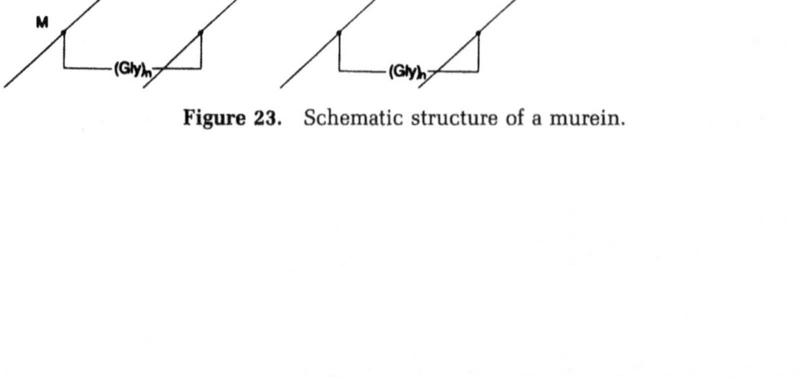

Figure 23. Schematic structure of a murein.

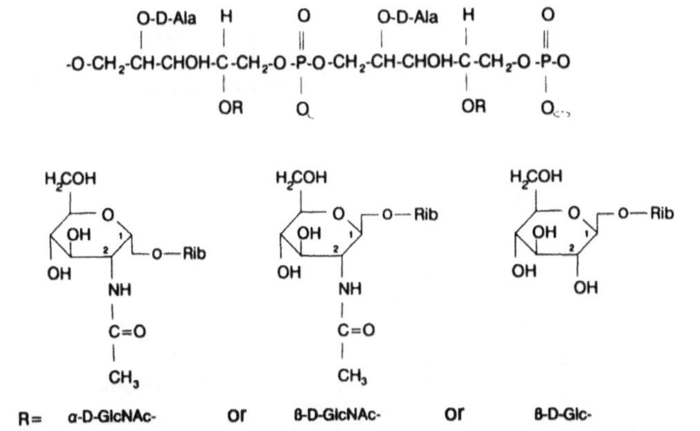

Figure 24. Structural elements of 1,5-polyribitolphosphates.

deoxy-D-glucose, or 2-acetamido-2-deoxy-D-galactose can be linked to the main chain via O-2 (Fig. 25).

- In some bacterial cell walls, polymannitolphosphates occur.
- In several cases, sugar units such as glucose or galactose are part of the chain (Fig. 26).
- Sugar 1-phosphate polymers also occur as a group of teichoic acids (Fig. 27).

The teichoic acids are covalently attached to the murein cell wall via a phosphodiester linkage (Fig. 28).

The teichoic acids are rather susceptible to depolymerization due to the labile phosphate bridges. As a consequence, only low-molecular-weight units, sugars as such, or sugars bound to glycerol, ribitol, or mannitol or D-alanylglycerol may survive in very recent sediments to some extent. The presence of ribitol and mannitol in sediments was investigated

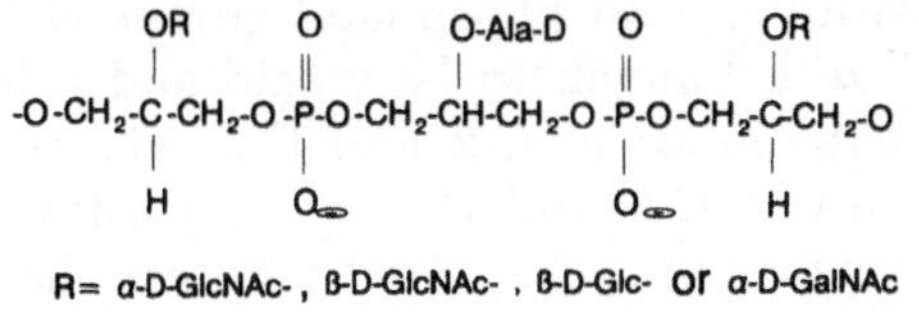

Figure 25. Structural elements of 1,3-polyglycerolphosphates.

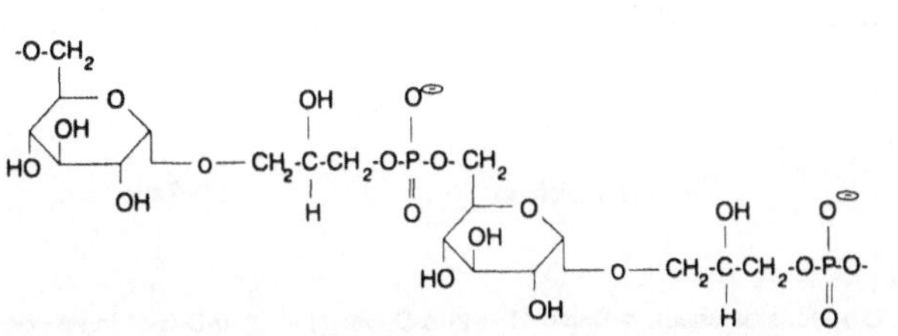

Figure 26. Partial structure of a teichoic acid with glucose in the chain.

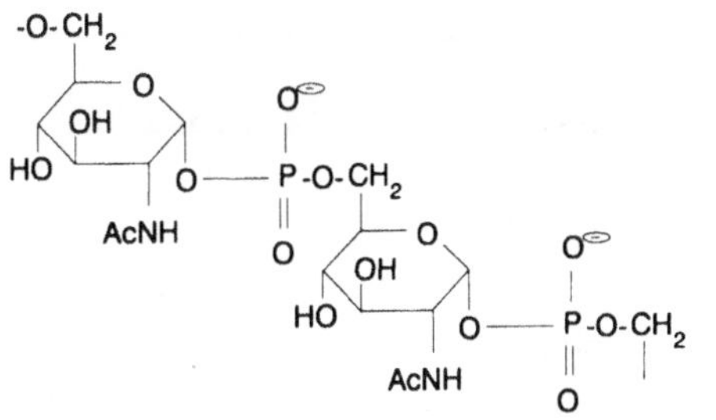

Figure 27. Sugar 1-phosphate polymers as teichoic acids.

after hydrolysis and $NaBD_4$ reduction. No undeuterated alditols were found, however (Moers, 1989).

2.2.18. Teichuronic Acids

If gram-positive bacteria are grown in phosphate- or glucose-poor media, teichoic acids are wholly or partly replaced by teichuronic acids, polymers mainly consisting of D-glucuronic acid and 2-acetamido-2-deoxygalactose and occasionally neutral sugar units such as rhamnose and glucose (Hancock and Baddiley, 1985; Kennedy and White, 1983). An example is given in Fig. 29.

In several cases molecular masses up to 500 kDa have been reported. It may be speculated that both teichoic and teichuronic acids can partly survive biodegradation in Recent sediments since gram-positive bacteria such as *Bacillus* spp. may be part of the bacterial community representing the last stages of life in sediments and soils. Aminosugar moieties do occur in many sediments, but they might reflect many precursors apart from these teichuronic acids (Moers, 1989; Moers *et al.*, 1989).

2.2.19. Lipoteichoic Acids (LTA)

Lipoteichoic acids (LTA) are widespread among the gram-positive bacteria and can be considered as

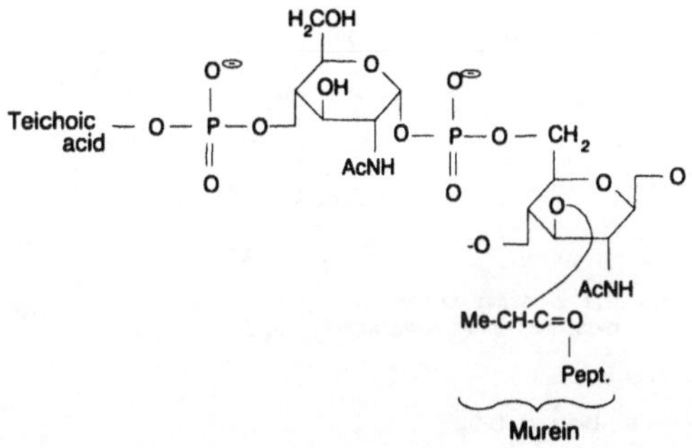

Figure 28. Example of a linkage of teichoic acids to a murein.

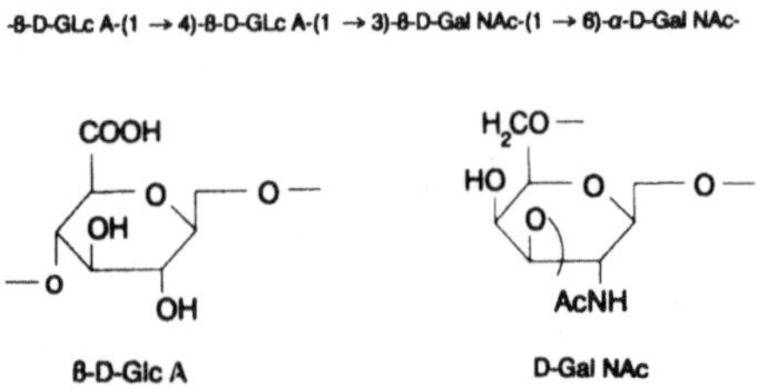

Figure 29. Structural elements of a teichuronic acid.

membrane-bound teichoic acids. They are basically polyglycerol phosphates anchored in the membrane via glycerophosphoglycolipid units. An example is given in Fig. 30 (Fischer *et al.*, 1978). It is thought that lipoteichoic acids undergoing enzymatic depolymerization give rise to the glycerophosphoglycolipids themselves.

It is likely that the corresponding esterified fatty acids might become incorporated to some extent in Recent sediments. However, these types of fatty acids are present in many other bacterial substances as well.

2.2.20. Bacterial Lipopolysaccharides (LPS)

The lipopolysaccharides (LPS) occur in gram-negative bacteria, including most cyanobacteria (Lüderitz *et al.*, 1982). They are not covalently bound to the murein cell wall; their lipid moieties form a double layer with a phospholipid layer directly located outside the murein (Fig. 31). Several moieties can be discriminated in the LPS structure: lipid A, an inner-core oligosaccharide, an outer-core oligosaccharide, and an O-specific side chain consisting of a polysaccharide (Brade *et al.*, 1988). The lipid A moiety is responsible for the toxic nature of LPS whereas the O-specific side chain is highly specific and contains the O-antigenic determinants. The O-specific side

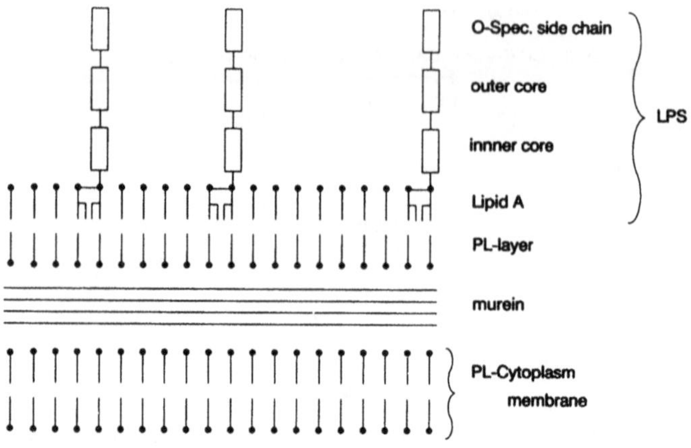

Figure 31. Location of LPS in bacterial cell wall material. PL, Phospholipid.

chain is a heteropolysaccharide which consists of repeating oligosaccharide units having short side chains. The variety of monosaccharides in this oligosaccharide unit is large between LPS of different species or even strains of bacteria. Neutral, amino, deoxy, and methoxy sugars as well as sugar acids both as pyranosides and furanosides in the α or β configuration are present. The so-called outer core consists of an oligosaccharide in many bacteria containing D-glucose, D-galactose, and *N*-actetyl-D-glucosamine. An example is given in Fig. 32. The inner core consists of an oligosaccharide containing a heptose, mainly L-glycero-D-manno- or D-glycero-D-manno-heptose and 3-deoxy-β-D-mannooctulosonic acid (KDO) (Fig. 33). Phosphate and ethanolamine groups are frequently encountered in the inner core.

The lipid A moiety is a disaccharide of 2-acetamido-2-deoxy-D-glycopyranosyl units attached to one of the KDO units of the inner core at C-4 or C-6′ and substituted with phosphatyl groups at C-1 and C-4′ and with long-chain fatty acid and β-hydroxy fatty acid moieties at C-2 and C-2′ via an amide linkage and at C-3 and C-3′ via an ester linkage (Wollenweber *et al.*, 1982). An example of the lipid A moiety of *Escherichia coli* is given in Fig. 34 (Imoto *et al.*, 1983).

The β-hydroxy fatty acids have the *R* stereochemistry unlike the S-β-hydroxy fatty acids which

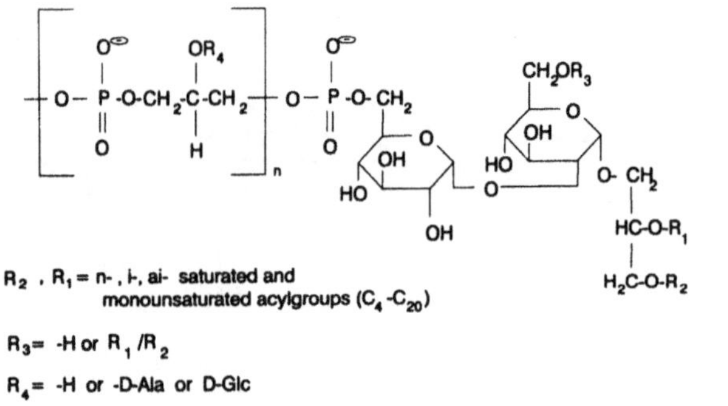

Figure 30. Example of a structure of lipoteichoic acid.

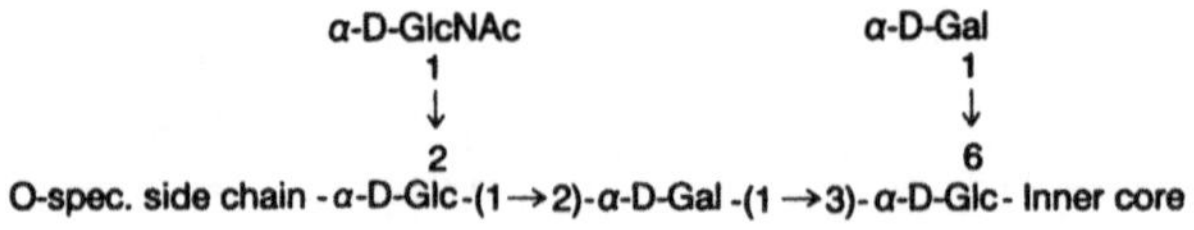

Figure 32. Example of the outer core of LPS. [After Kennedy and White (1983).]

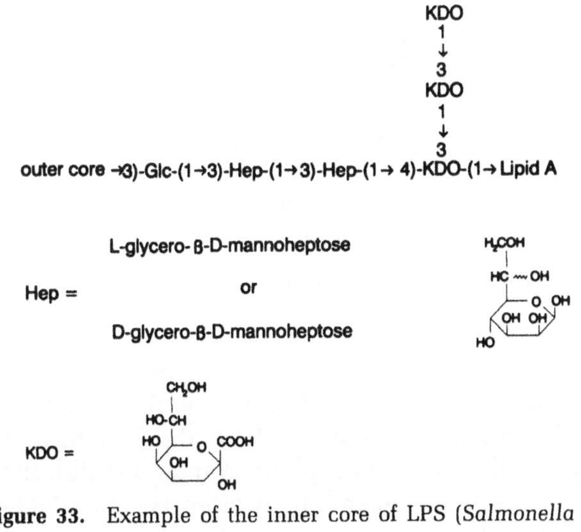

Figure 33. Example of the inner core of LPS (*Salmonella* sp.). [After Brade *et al.* (1988).]

are common intermediates in fatty acid biosynthesis or degradation. The β-hydroxy fatty acid moieties of lipid A can vary considerably in different bacteria. The *n*-β-hydroxy fatty acids mainly occur as C_{12}, C_{14}, or C_{16} homologs, the *iso*-β-hydroxy fatty acids have 12, 14, and 16 carbon atoms, and the *anteiso*-β-hydroxy fatty acid moieties mainly have 15 and 17 carbon atoms (Galanos et *al.*, 1977). In some specific cases, an esterified *iso*-C_{14}-α-hydroxy fatty acid moiety is also present in the lipid A structure (Imoto *et al.*, 1983; Galanos *et al.*, 1977).

Several components of LPS have been encountered in Recent sediments. Amide-bound β-hydroxy fatty acids released upon acid hydrolysis of extracted and saponified sediments are thought to be highly specific markers of lipid A constituents preserved in sediments. These amid-bound β-hydroxy fatty acids have been found in sediments up to 10^6 years old

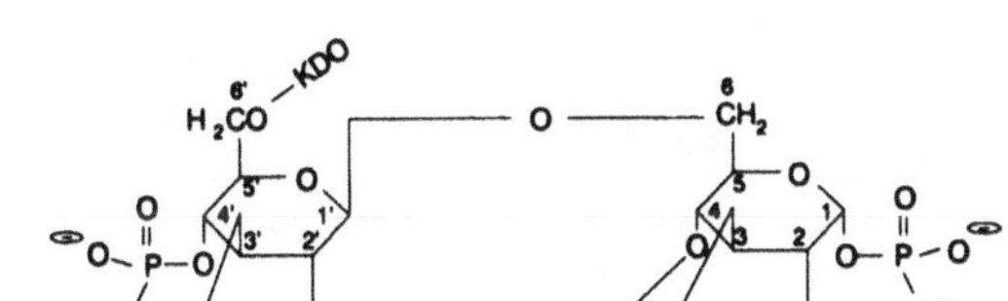

Figure 34. Lipid A part of LPS (*Escherichia coli*).

(Goossens *et al.*, 1986, 1989; Mendoza *et al.*, 1986; ten Haven *et al.*, 1987). Methylated and deoxy sugars as well as heptoses isolated after acid hydrolysis of recent and ancient sediments very likely originate form the O-specific side chain and inner core of LPS (Klok *et al.*, 1984a, b; Moers, 1989; Moers *et al.*, 1989). The conditions applied to isolate sugars from sediments are destructive for KDO. This might be the reason why KDO has not (yet) been shown to occur in sediments.

The amino sugars frequently identified in acid hydrolysates of Recent sediments might in part originate from LPS. However, the bacterial cell wall component murein as well as other naturally occurring amino sugars can act as precursors also. In summary, it can be concluded that LPS or remains thereof are selectively preserved in sediments. Although the macromolecules can be considered relatively unstable from a chemical point of view, their preservation might be due to the fact that omnipresent sedimentary gram-negative bacteria such as sulfate reducers and other heterotrophic bacteria represent the last stage of life and as a consequence are not "predated" extensively.

2.2.21. DNA and RNA

The biopolymers DNA and RNA are present in any living cell and contain or transfer genetic information. They consist of purine (adenine, guanine) and pyrimidine bases (cytosine, uracil, thymine) linked to either 2-deoxy-D-ribose (DNA) or D-ribose (RNA). The sugar units are linked together via phosphates (Fig. 35).

Outside living cells, RNA and DNA are biodegraded very easily. The relatively stable bases have, however, been reported in Recent sediments (Van der Velden and Schwartz, 1974). Obviously, minor amounts of DNA, RNA, or their bases can survive to some extent in the geosphere.

2.2.22. Glycolipids

2.2.22a. Glycosphingolipids. These animal glycolipids are generally derivatives of sphingoid bases (see Fig. 36). Because these types of glycolipids occur in animals only, they are not considered important within an organic geochemical context. In specific cases, the lipid moiety might survive biodegradation to some extent. No reports of sphingoid compounds or their partially altered counterparts in sediments exist, however.

A number of plant, algal, and microbial glyco-

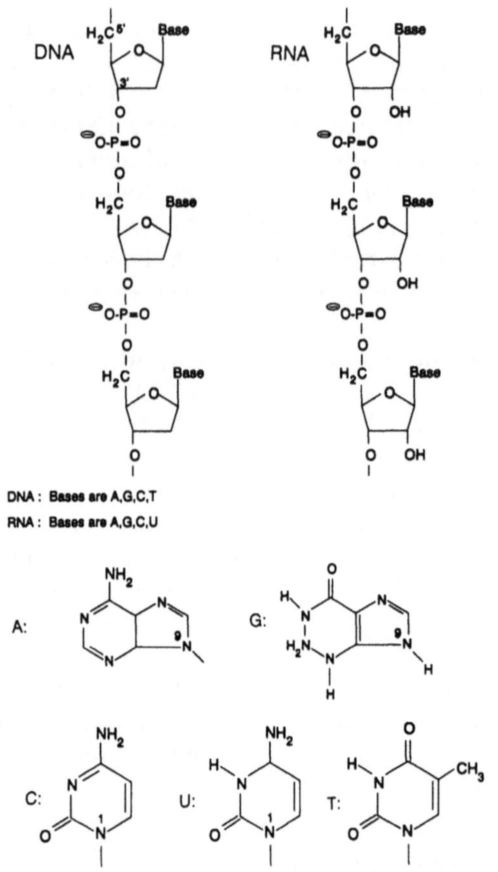

Figure 35. Structural units of DNA and RNA.

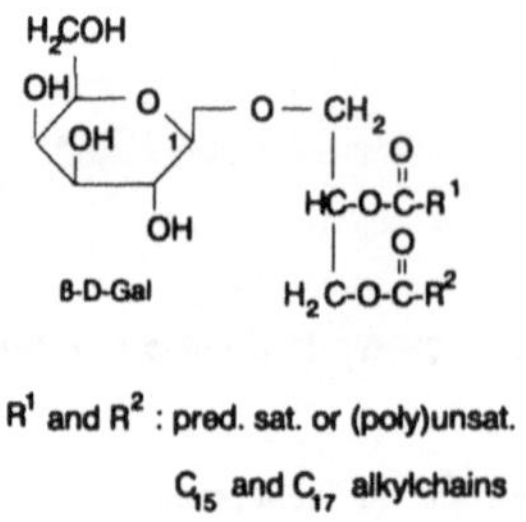

Figure 37. Structure of a glycosyldiacylglycerol.

lipids are discussed below, although they are not "real" macromolecular substances (Fuller and Nes, 1987).

2.2.22b. Glycosyldiacylglycerols. In chloroplasts of vascular plants and algae, galactosyl-, digalactosyl-, and sulfoquinovosyldiacylglycerols are the main representatives of this group (Kennedy and White, 1983) (Fig. 37). The fatty acid moieties mainly consist of C_{16} and C_{18} saturated, monounsaturated, and polyunsaturated fatty acids.

In bacteria, D-mannose and D-glucose occur more frequently than D-galactose. Moreover, the fatty acid moieties are often branched (*iso* and *anteiso*) and mostly saturated. In some bacteria, the diacylglycerol unit is linked via a phosphate to *myo*-inositol, inositol, or another glycerol with other sugars attached (see Fig. 38).

2.2.22c. Phytoglycolipids. Although minor compounds in plants, the so-called phytoglycolipids may be of interest to geochemists. In these compounds, a sphingoid base is linked via phosphate to a *myo*-inositol unit. Fatty acid moieties—frequently hydroxylated—are attached via an amide link to the sphingoid base (Fig. 39).

2.2.22d. Sterylglycosides. In many vascular plants and algae, sterylglycosides occur as minor components. In several cases, the sugar unit is substituted at position 6 with C_{16} and C_{18} saturated or unsaturated acids (Fig. 40).

2.2.22e. Mycobacterial Glycolipids. Mycobacteria, corynebacteria, and nocardia contain two specific groups of glycolipids, the so-called cord factor and the mycosides. The cord factor, present in the capsular material, consists of α,α-trehalose esterified with mycolic fatty acids (Brennan, 1988). The double bonds in the acyl chains are sometimes replaced by cyclopropyl or methoxy groups (see Fig. 41). The mycosides (e.g., phenolphthiocerol glycosides) consist of a long functionalized aliphatic chain to which

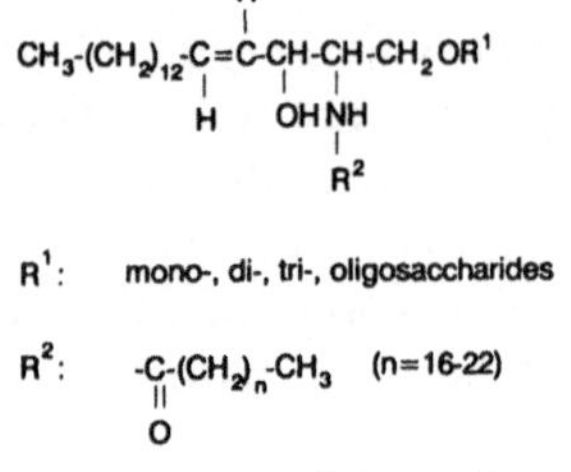

Figure 36. Structure of glycosphingolipids.

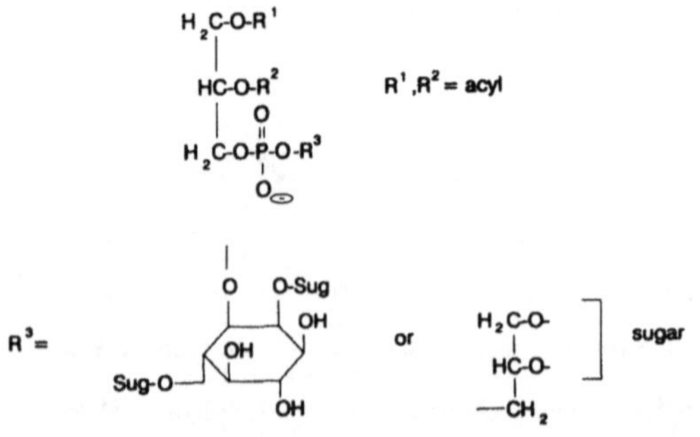

Figure 38. Structure of a diacylglycerol unit linked to sugars via a phosphate group.

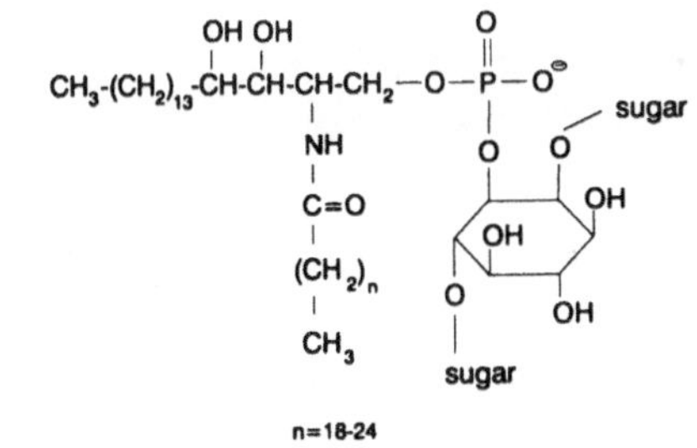

Figure 39. Structure of a phytoglycolipid.

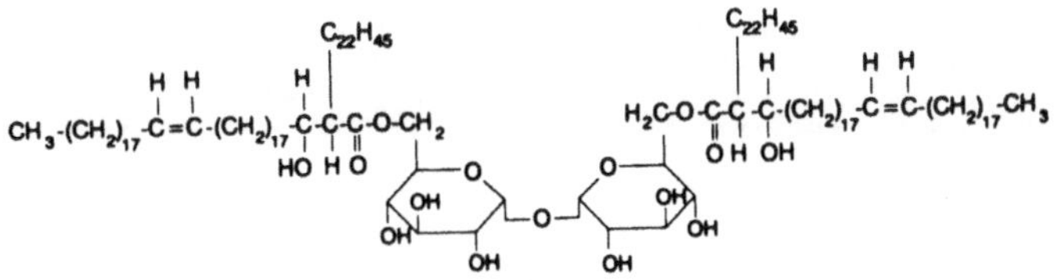

Figure 41. Structure of a mycobacterium glycoplipid. [After Kennedy and White (1983).]

straight-chain or branched acyl groups and a phenol group are attached. Glycosidically linked to the phenol is a di- or trisaccharide (Brennan, 1988) (Fig. 42).

The glycolipids of plants, algae, and bacteria described above, as well as others that have not been mentioned, are of interest from an organic geochemical perspective. If they are not fully decomposed, they may, after condensation, represent a minor part of the insoluble macromolecular organic material in Recent sediments and soils. Application of specific procedures similar to those applied to identify LPS hydroxy fatty acids (Goossens *et al.*, 1989) could lead to their discovery in sediments. Within this context, we might also wonder whether steranes, generated from kerogens during catagenesis, were partly originally present as sterylglycosides.

On the other hand, selective removal of the saccharide moieties and phosphate groups may have resulted in the release of less polar lipid moieties (sterols, fatty acids) which become part of an apolar low-molecular-weight fraction directly extractable from Recent sediments.

2.2.23. Polyisoprenes (Rubber and Gutta)

Apart from the well-known polyisoprene families (triterpenes and tetraterpenes), which will not be discussed here, high-molecular-weight polyisoprenes occur in some higher plants. These polymers result from the 1,4 (head-to-tail) linkage of isoprenoid units. High-molecular-weight polyisoprenes exclusively composed of *cis* units, i.e., rubber, and of *trans* units, i.e., gutta, have both been identified (Fig. 43) (Golub *et al.*, 1962; Chen, 1962; Schaefer, 1972; Buchanan *et al.*, 1978a; Uzabakiliho *et al.*, 1987). The degree of polymerization of rubber markedly depends on the species considered. The highest values are noted for *Hevea brasiliensis* (average molecular mass of ca. 500 kDa) (Archer *et al.*, 1963) while, in most other rubber-bearing species, average molecular masses range from 20 to 200 kDa (Buchanan *et al.*, 1978a, b; Nielsen *et al.*, 1977). In the *trans* isomer, the chains tend to be shorter, and average molecular masses from 20 to 150 kDa are noted for gutta (Duch and Grant, 1970).

Rubber is considered as an end product of metabolism and accumulates in latex. Rubber occurs in about 0.5% of higher plants, and over 7,000 rubber-producing species, belonging to different families, have been identified (Archer *et al.*, 1963). Among these families, the Euphorbiaceae, which comprise the genus *Hevea* and numerous rubber-producing plants of the genus *Euphorbia*, are however the most important in this respect. *Euphorbiaceae* are widespread both in tropical and temperate areas. Gutta is less widely distributed but, nevertheless, is observed in numerous plants (Roth *et al.*, 1985). The concomitant presence of rubber and gutta was noted in a limited number of species (Leeper and Schlesinger, 1954).

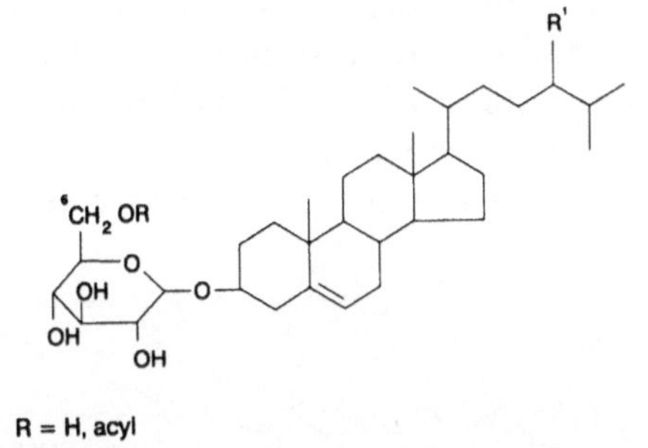

Figure 40. Structure of a sterylglycoside.

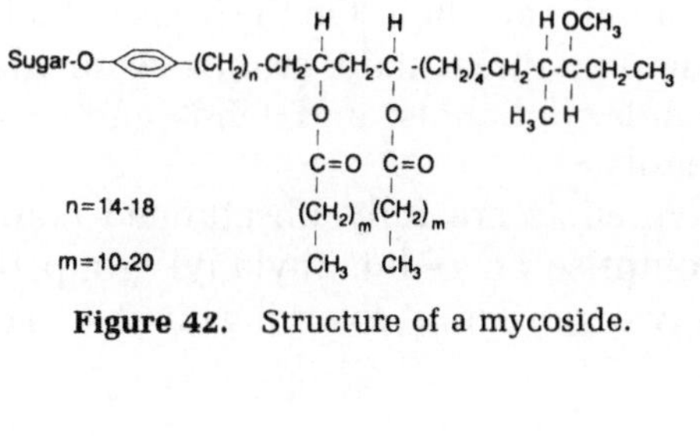

Figure 42. Structure of a mycoside.

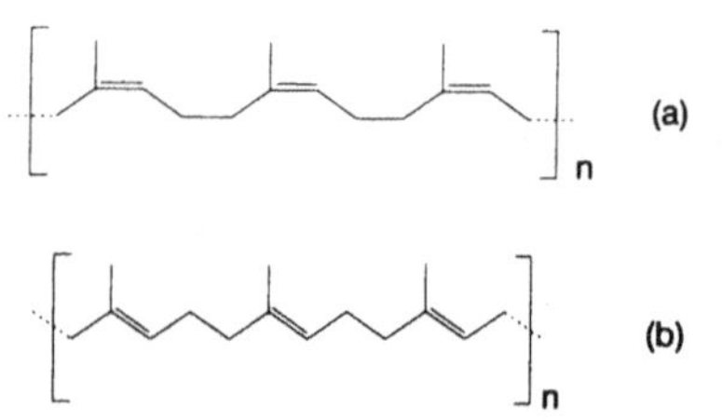

Figure 43. Structure of rubber (a) and gutta (b).

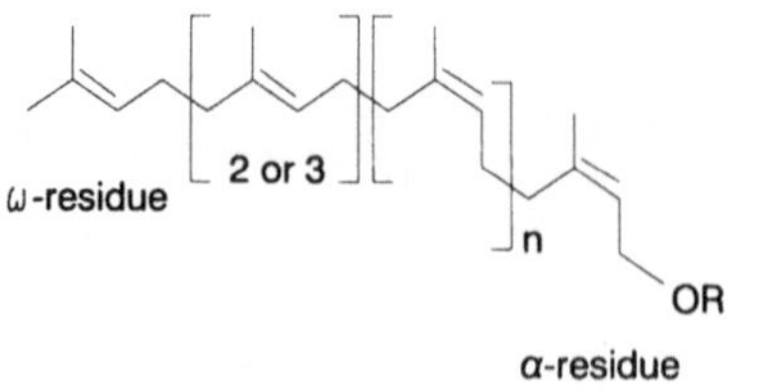

Figure 44. Structure of polyprenols and dolichols.

Rubber accounts for over 90% dry weight of latex in *H. brasiliensis*. In most other species, including *Euphorbia* spp., latex constituents are however dominated by polycyclic triterpenoids, and the rubber level is usually below 25% (Nielsen *et al.*, 1977; Nwadinigwe, 1981; Uzabakiliho *et al.*, 1987; Nwadinigwe, 1988). With respect to the total plant biomass, rubber usually appears as a minor constituent. Few plants show a rubber content of ca. 10%, but values around 0.5% are typical (Buchanan *et al.*, 1978a, b). Gutta levels ranging from 0.4 to 1.2% of total plant dry weight were noted in various plants (Roth *et al.*, 1985).

Rubber particles are supposed to be stabilized by a proteinaceous envelope and, also, by natural antioxidants contained in latices. A fast coagulation occurs on exposure of latex to the atmosphere, and enzymes, such as phenol oxidases, were shown to be involved in this process (Brzozowska-Hanower *et al.*, 1978). It seems that no information is available on the biodegradability of rubber and gutta and on their behavior during diagenesis.

2.2.24. Polyprenols and Dolichols

Polyprenols and dolichols, primary alcohols containing a very long unsaturated head-to-tail polyisoprenoid chain, are widespread in living organisms. A large variety of structures are observed, with respect to chain length, the number of *cis* and *trans* units, and the number and location of saturated units (Bu'lock, 1972; Carroll, 1987; Chojnacki *et al.*, 1987; Chojnacki and Dallner, 1988). Contrary to the chains of ubiquinones and of their precursor alcohols, characterized by usual all-*trans* stereochemistry of the isoprene residues, *cis* units always dominate in these polyisoprenoid alcohols. On the basis of their chemical structure, three main groups were identified: polyprenols, dolichols, and tetra- and hexahydropolyprenols.

Polyprenols are fully unsaturated compounds. They comprise an ω-dimethylallyl group linked to three or two *trans* units, a varying number of *cis* units, and a terminal α-*cis* unit with the alcohol group (Fig. 44). The total number of isoprene residues ranges from 6 to ca. 30. Both in shorter (6–8 units) and longer polyprenols (15–30 units), only two *trans* isoprene residues occur at the ω-end. In medium-length-polyprenols (9–14 units), three *trans* residues are observed. Polyprenols occur as free alcohols, acetates, esters of various fatty acids, and polyprenyl phosphates.

Dolichols only differ form polyprenols by the reduction of the terminal unit (α-dihydropolyprenols). Dolichols from animal tissues are characterized by an S configuration of carbon atom 3 and contain only two *trans* residues at the ω-end. The total number of C_5 units ranges from 12 to 24. In animal tissues the distribution of dolichols is species specific. Dolichols occur as free alcohols, esters of fatty acids, and dolichol phosphates; dolichyl pyrophosphates bound to oligosaccharides are also observed in animals.

Tetra and hexahydropolyprenols comprise, in addition to the α-unit, a varying number of other saturated residues. The latter are often located at the ω and/or ω-1 position. The presence of a methylene substituent on the ω-unit was sometimes noted (Fig. 45). The number of units in this group ranges form 18 to 24.

Very long chain unsaturated polyisoprenoid alcohols seem to be widespread in living organisms. Polyprenols have been identified in leaves, needles, or woody tissues of hundreds of higher plant species (Suga *et al.*, 1980; Ibata *et al.*, 1984a, b; Chojnacki *et al.*, 1987); they have also been detected in aquatic plants (Lanzetta *et al.*, 1988). Polyprenols with two *trans* residues at the ω-end mainly occur in gymnosperms, and those with three *trans* residues in an-

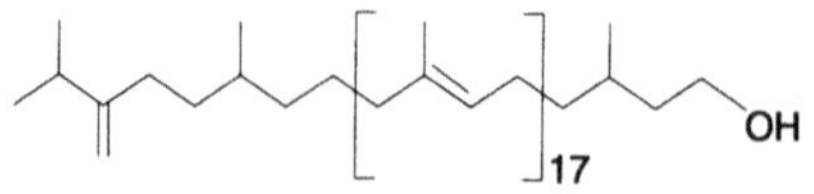

Figure 45. Hexahydropolyprenol with a methylene substituent.

giosperms (Ibata *et al.*, 1984b). Polyprenols are, by far, the most abundant polyisoprenoid alcohols in higher plants. Relatively high levels of polyprenols, up to 2.3% of dry weight, were observed in needles and leaves of a number of conifers (Ibata *et al.*, 1984a, b). Seeds of monocotyledons contain similar amounts of dolichols and polyprenols,and the seeds of dicotyledons contain dolichols exclusively (Carroll, 1987). Polyprenols also occur in fungi (Bu'lock, 1972), bacteria (e.g., 0.004% of dry weight in *Lactobacillus* sp.) (Bu'lock, 1972), and in minor quantities in animal tissues (Carroll, 1987; Chojnacki and Dallner, 1988), where they are dominated by dolichols and account for only ca. 1% of the total polyisoprenoid alcohols (Chojnacki and Dallner, 1988). Polyprenyl phosphates play an important role as acceptors and donors of sugar residues (Suga *et al.*, 1980; Chojnacki *et al.*, 1987). They are thus involved in the biosynthesis of cell wall constituents of bacteria (teichoic acids, polysaccharides) and of glycoproteins in eukaryotes.

Dolichols are widespread in animal tissues (Bu'lock, 1972; Carroll, 1987; Chojnacki and Dallner, 1988) but also in higher plants, yeasts, fungi, and microalgae (Bu'lock, 1972). Dolichyl phosphates are obligatory intermediates of glycoprotein biosynthesis in mammalians (Carroll, 1987; Chojnacki *et al.*, 1987; Chojnacki and Dallner, 1988). In animal cells, dolichols also play an important structural role in membranes (Chojnacki and Dallner, 1988).

The presence of tetra- and hexahydropolyprenols was reported in various fungi (Bu'lock, 1972).

No literature data exist about the diagenetic fate of very long chain unsaturated polyisoprenoid alcohols. Owing to their branched nature, some resistance to biodegradation can be anticipated. Head-to-tail, C_{21} to C_{42} isoprenoid hydrocarbons have been identified in various oils from terrigenous sources. It was suggested that these compounds are possibly derived from higher plant polyprenols via cracking and diagenetic reduction (Albaiges *et al.*, 1985; Philp and Gilbert, 1986).

2.2.25. Resinous Polysesquiterpenoids and Polyditerpenoids

Dammar, a well-known angiosperm resin, contains chloroform-soluble, methanol-insoluble polymers, sometimes called β-resenes (Tschirch and Glimmann, 1896; Shigemoto *et al.*, 1987). Shigemoto *et al.* (1987) studied the methanol-insoluble fraction of dammar by ^1H and ^{13}C nuclear magnetic resonance (NMR), and pyrolysis. These authors came to the conclusion that the monomeric unit was a cadinene

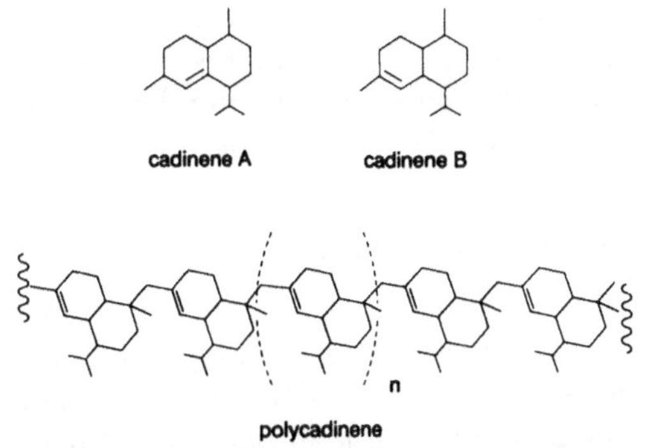

Figure 46. Proposed structure of polycadinene.

with structure A (Fig. 46). However, recent work by Van Aarssen *et al.* (1990) has indicated that the monomeric unit is cadinene with the double bond in the Δ^5 position (structure B, Fig. 46) and that the polymer is likely to have the polycadinene structure depicted in Fig. 46. These authors also made clear that the structure for the dammar polymer proposed by Brackman *et al.* (1984) is highly unlikely. The data presented by the latter authors can be interpreted better assuming the polycadinene structure.

Gymnosperm resins as well as some angiosperm resins contain soluble polymers produced from diterpenoid labdatriene monomers such as communic acid via 1,4 or 3,4 polymerizations (Carman *et al.*, 1970; Cunningham *et al.*, 1983; Mills *et al.*, 1984/85) (Fig. 47). In contrast to the formation of the polycadinenes, it is thought that this polymerization takes place after exudation of these diterpenoids under the influence of air and/or light.

The preservation of resinous polymeric substances seems to be relatively good, though structural changes such as ring condensation and aromatization take place during diagenesis (Langenheim, 1969; Van Aarssen *et al.*, 1992). The coal maceral resinite may consist predominantly of these types of structurally changed (bio)polymers (Wilson *et al.*, 1984; Tegelaar *et al.*, 1989a). There is good evidence that a considerable amount of saturated and aromatic sesquiterpenoids, diterpenoids, triterpenoids, and oligomeric

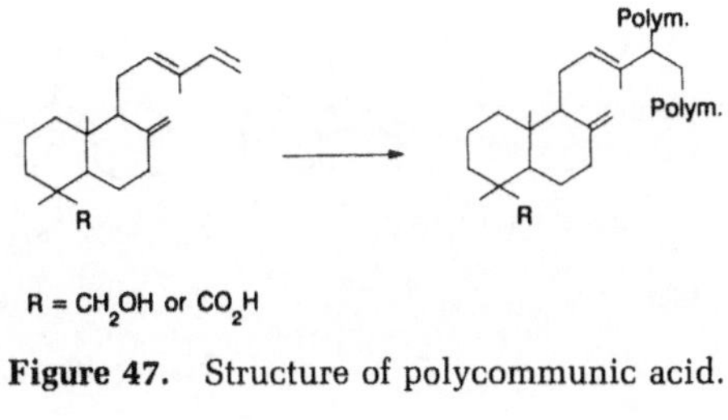

Figure 47. Structure of polycommunic acid.

terpenoids such as bi- and tricadinanes in coals and terrestrially derived crude oils results form thermodissociation of these altered biopolymers (Fig. 48) (Cox *et al.*, 1986; Van Aarssen *et al.*, 1992).

2.2.26. Cutins and Suberins

Cutins and suberins represent macromolecular polyesters occurring in vascular plants. Cutin is a characteristic component of the cuticular membrane, a thin continuous extracellular layer covering the aerial parts of all plants where no secondary thickening occurs (e.g., leaves, fruits, and nonwoody stems; Hallam, 1982; Holloway, 1982a). However, some internal tissues, such as the inner seed coats and juice sacs of citrus, contain cutin as well (Espelie *et al.*, 1980). Suberins are mainly present as wall components of cork cells, comprising the periderm layers of both aerial and subterraneous parts of woody plants. They also occur in the cell walls of specialized internal tissues such as in the endodermis (Casparian bands) (Robards *et al.*, 1973), bundle sheaths of grasses (O'Brian and Carr, 1970), and cell walls of calcium oxalate crystals isolated from leaf homogenates (idioblasts) (Espelie *et al.*, 1982). In cork, suberins can account for 30–35% of the dry weight. Cutins play a major role in the protection of vascular plants against infections of microorganisms. They are also important in preventing dessication. Suberins play a similar protective role in tissues where secondary growth occurs.

The characteristic monomeric units of cutins consist essentially of ω-hydroxy C_{16} and C_{18} fatty acids with mid-chain functionalities such as hydroxyl or epoxy groups (Fig. 49). It should be noted that, so far, epoxy groups have only been detected in angiosperm cutins (Hunneman and Eglinton, 1972; Holloway, 1982b). The composition of these fatty acids varies considerably among different taxa, but also within species depending on the stage of development and environmental conditions. The chemical analysis of suberins has been more limited. Nevertheless, nearly all of the C_{16} and C_{18} monomers identified in suberin have been reported in cutin as well, confirming their close affinity. The major difference is the occurrence in suberins of monomers with chain lengths longer than C_{18} having a lesser degree of mid-chain substitution (Holloway, 1984). These longer chain monomers consist predominantly of 1-alkanols and corresponding fatty acids (C_{20}–C_{30}) and ω-hydroxy

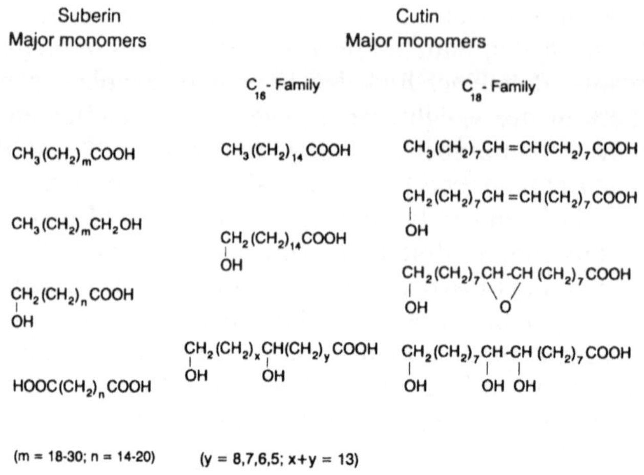

Figure 49. Characteristic monomers of cutins and suberins.

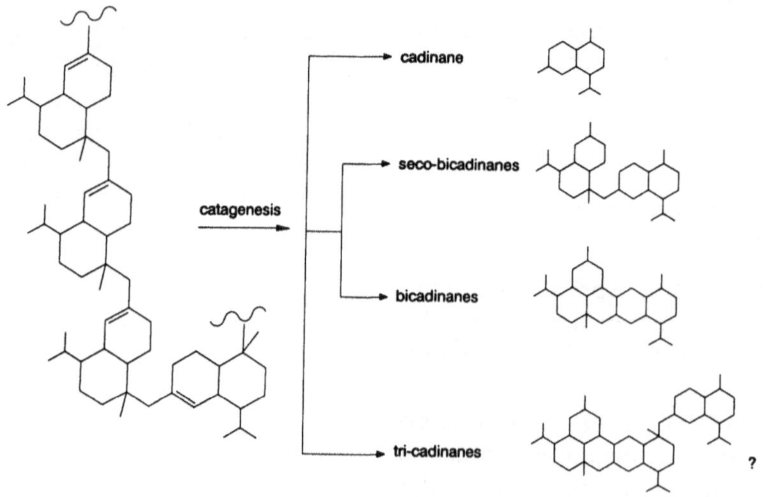

Figure 48. Catagenic products of polycadinene.

fatty acids (C_{16}–C_{24}) with minor contributions of α,ω-diols. Apart from aliphatic building blocks, it is suggested that suberins contain considerable amounts and cutins low amounts of esterified phenolic acids (Kolattukudy, 1980). However, according to Holloway (1982b), the contribution of phenolic acids esterified to carbohydrates, originating from cell wall material, must first be eliminated before the direct involvement of phenolic acids with cutin can be definitely established. The intramolecular structure of these types of polyesters is only poorly understood. For cutins, the results of several studies indicate that the primary hydroxyl groups in the polyester are almost completely esterified, whereas a substantial proportion of hydroxyl groups in mid-chain positions are free. These free hydroxyl groups and other mid-chain functional groups (e.g., epoxy) which cannot participate in cross-linking give rise to an essentially linear polyester (Agullo *et al.*, 1984; Tegelaar *et al.*, 1989d) (Fig. 50). The structure of suberins is even less well understood, and only a tentative working hypothesis has been proposed (Fig. 51) (Kolattukudy, 1980).

Cutins are easily depolymerized with cutinases, extracellular enzymes present in fungi. For example, yeasts can actually grow slowly on cutin as sole carbon source. Although less well studied, suberins are also depolymerized by extracellular enzymes from fungi (Kolattukudy, 1981).

To some extent, cutins and suberins may, however, be preserved in Recent sediments. Base hydrolysis of a number of very recent sediments and peats has released series of compounds similar to the building blocks of cutins and suberins (Cardoso *et al.*, 1977; Cardoso and Eglinton, 1983; ten Haven *et al.*, 1987; Tegelaar *et al.*, 1993a). Very recently, indirect evidence was obtained that epoxy group-containing cutins from angiosperms can be preserved via ring opening of the epoxy group and reactions of the intermediate oxygen radical with a macromolecular matrix (Tegelaar *et al.*, 1991b). Flash pyrolysis studies of very well preserved 50 × 10^6-years-old angiosperm cuticles from Geisltal (Germany) revealed series of unsaturated fatty acids with carbon chain lengths less than C_{16}. These fatty acids were thought to originate from cutin-derived substances as mentioned above.

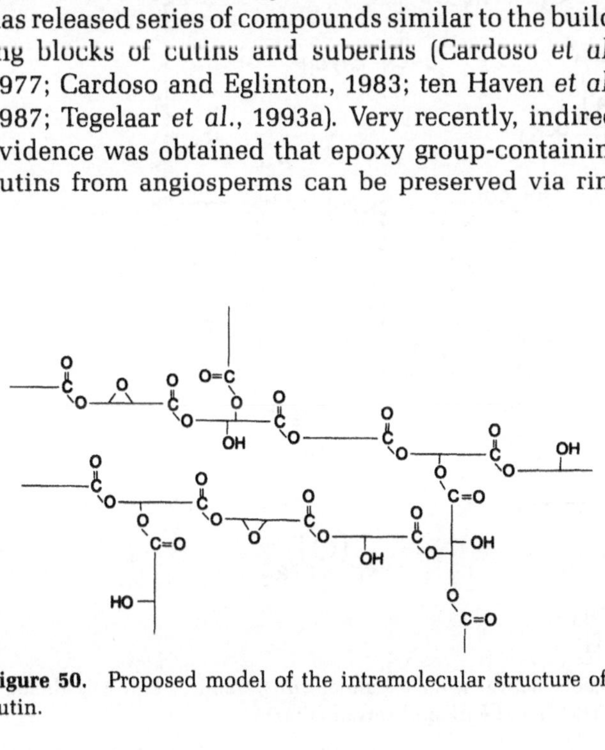

Figure 50. Proposed model of the intramolecular structure of a cutin.

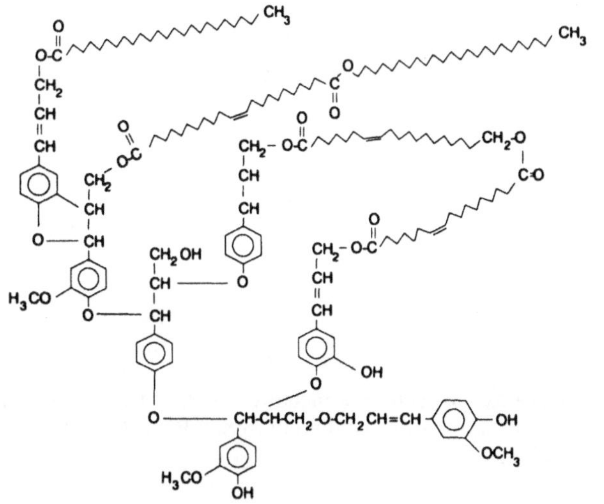

Figure 51. Proposed model of the intramolecular structure of suberins. [After Kolattukudy (1980).]

2.2.27. Lignins

Lignins are a group of wood constituents comprising high-molecular-mass polymers (600–1000 kDa) based on phenylpropanoid units. Lignins originate from the dehydrogenative condensation of three primary building blocks: p-coumaryl alcohol, coniferyl alcohol, and sinapyl alcohol (Goodwin and Mercer, 1972). The "monolignols" show a *trans* unsaturation and differ in the substitution pattern of their phenyl ring (Fig. 52). The three monolignols originate from the reduction of the corresponding acids—p-coumaric, ferulic, and sinapic acid. The latter acids derive from cinnamic acid via different oxidation patterns in the phenyl ring. For a given plant, the lignin structure is determined by the relative abundance of the monolignols. The availability of these primary building units is controlled at two levels: the formation of the precursor acids from cinnamic acid and their reduction to monolignols (Kutsuki *et al.*, 1982).

Most gymnosperm lignins are mainly derived from coniferyl alcohol while dicotyledonous angiosperm lignins are built up from both coniferyl and

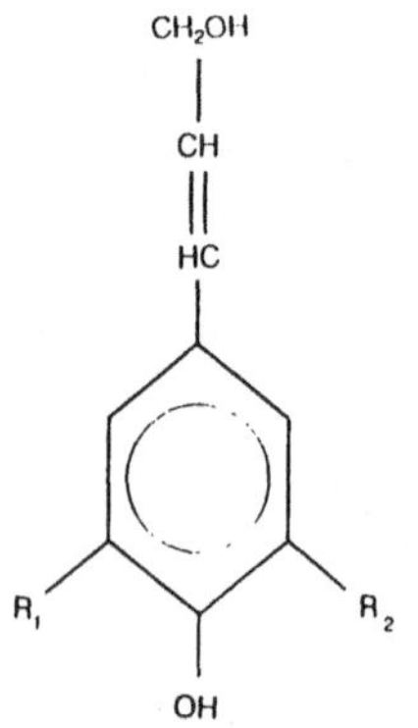

Figure 52. Structure of monolignols $R_1 = R_2 = H$: p-coumaryl; $R_1 = H$, $R_2 = OCH_3$: coniferyl alcohol; $R_1 = R_2 = OCH_3$: sinapyl alcohol).

sinapyl alcohol (Kirk and Farrell, 1987; Stafford, 1988). In these two types of lignins, a low contribution of p-coumaryl alcohol is generally noted as well. In monocotyledonous angiosperms, coniferyl and p-coumaryl alcohol contributions usually predominate (Kirk and Farrell, 1987; Stafford, 1988). Lignin from *Sphagnum* moss shows an unusual composition in that it is chiefly derived form p-coumaryl alcohol (Isherwood, 1965). Lignins arise from an enzyme-initiated dehydrogenative polymerization. A few dimers are formed within cells and excreted; radical polymerization of dimers and monomers then occurs extracellularly. This results in the random coupling of the mesomeric phenoxy free radicals derived from the monolignols (Isherwood, 1965; Goodwin and Mercer, 1972; Adler, 1977; Saiz-Jimenez and de Leeuw, 1984; Hatcher, 1987; Kirk and Farrell, 1987; Stafford, 1988). Over ten phenylpropanoid linkages have been identified in lignins. Alkylaryl ether bonds dominate, especially the β-O-4 type (Fig. 53), which accounts for 50–60% of interunit links in most lignins. A relatively large contribution of aryl–aryl and alkyl–aryl carbon–carbon bonds is also noted. Although head-to-tail linkages of the basic units dominate in lignins, some head-to-head and tail-to-tail linkages are also detected. In addition, condensed units exhibiting several linkages with adjoining units are commonly observed. Finally, most of the terminal

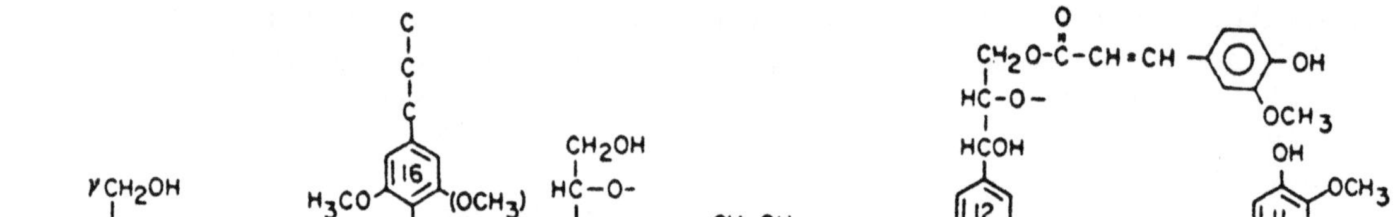

Figure 53. Schematic structure for lignins. A β-O-4 type linkage occurs between, for example, units 1 and 2; unit 5 is linked to "hemicellulose," and unit 12 is esterified to ferulic acid. [After Adler (1977) and Kirk and Farrell (1987).]

groups of the C_3 side chains in lignins are free alcohols, though a few aldehydes have been identified as well; in grasses and in some woods, some of the terminal hydroxy groups are also esterified with cinnamic acids (Kirk and Farrell, 1987). Due to the above features and to their formation via random free-radical copolymerization, lignins are highly heterogeneous, cross-linked, and optically inactive biopolymers.

Lignins are covalently linked to noncellulosic wall polysaccharides, especially "hemicelluloses" (Isherwood, 1965; Kirk and Farrell, 1987; Stafford, 1988). Various acids, including *p*-coumaric, ferulic, and diferulic acid, are linked to lignins via ester or ether bonds. Such bifunctional compounds, capable of forming both ester and ether linkages, act as bridging units between lignins and "hemicelluloses" (Tanner and Morrison, 1983; Scalbert *et al.*, 1985; Bolwell, 1988; Hartley *et al.*, 1988). Lignins are also covalently linked to some proteins of cell walls (Whitmore, 1982). Due to the occurrence of lignin–carbohydrate complexes, the isolation of undegraded lignins in a pure state free of sugars is practically impossible.

Lignins are universally distributed in the woody tissues of gymnosperms and angiosperms. Indeed, lignins occur in nearly all vascular plants, including herbaceous species and primitive groups such as ferns (Isherwood, 1965; Goodwin and Mercer, 1972; Ragan, 1984; Stafford, 1988), except in a few species that have secondarily evolved into an aquatic environment (Stafford, 1988). Lignins are not detected in most mosses and are also lacking in plants of lower taxonomic rank than mosses (Ragan, 1984; Kirk and Farrell, 1987).

Some recent reports suggested the presence of lignins in brown macroalgae; however, ^{13}C NMR spectra indicated that such products are not lignins (Ragan, 1984) but substances probably related to phlorotannins. "Lignin-like" compounds were isolated in some fungi; they actually contain units derived form *p*-coumaryl and sinapyl alcohol and may be similar to higher plant lignins (Towers, 1969; Turner, 1971).

Lignins exclusively occur in cell walls; they are major constituents of the walls of old xylem vessel cells of wood but also occur in young, growing cells in buds and seedlings (Goodwin and Mercer, 1972). The functions of lignins are both structural and protective (Isherwood, 1965; Goodwin and Mercer, 1972; Kirk and Farrell, 1987; Stafford, 1988). Most of the structural rigidity of vascular plants is due to the lignin–hemicellulose complex associated with the cellulose microfibrils. Lignins protect wall polysaccharides form enzymatic hydrolysis and chemical degradation (Isherwood, 1965).

Lignins are end products of metabolism and do not undergo any significant turnover. After cellulose, they are the most abundant organic compounds in nature. They account for 12–25% of the total dry biomass of vascular plants. In woody tissues, the average level of lignin is about 25%, and values up to 36% were observed in some arborescent gymnosperms and angiosperms (Odier and Rouau, 1985; Kirk and Farrell, 1987; Stafford, 1988).

Stable, nonhydrolyzable ether and carbon–carbon bonds are generated during the final steps of lignin formation. These insoluble polymers can be degraded by oxidation reactions, but complete degradation of lignin can hardly be achieved (Isherwood, 1965).

Lignins are relatively resistant to microbial degradations, and they are affected by a narrower array of microorganisms than the other major biopolymers, such as polysaccharides (Kirk *et al.*, 1980a, b; Odier and Rouau, 1985; William *et al.*, 1986; Kirk and Farrell, 1987; Haider, 1988; Stafford, 1988). The most efficient lignolytic organisms are fungi. Some species induce substantial changes in the chemical structure of lignin but are unable to cleave the interunit linkages; others can perform a complete degradation of lignins (Odier and Rouau, 1985; Kirk and Farrell, 1987). The highest lignolytic activity of the latter species commonly occurs during the stationary growth phases as a result of nitrogen starvation; however, few fungi also promote an efficient degradation during active growth phases (Odier and Rouau, 1985). A pronounced depolymerase activity is not sufficient to achieve a complete lignin degradation, since concomitant, fast repolymerization reactions then take place. Accordingly, extensive lignin degradation requires the cooperation of depolymerase and oxidase activities in fungi (Szklarz and Leonowicz, 1986). In addition, a high oxygen partial pressure is a prerequisite in fungal lignolysis. In fact, no degradations occur when the oxygen partial pressure decreases below 25% of its normal value in air (Odier and Rouau, 1985). Thus, rapid and extensive fungal lignolysis can occur only in highly aerobic environments; lignins seem not to be affected under anaerobic conditions (Kirk and Farrell, 1987). Recent studies also revealed some lignolytic activities by various bacteria. However, up till now, no extensive and rapid degradation of lignins was reported with any species of bacteria (Odier and Rouau, 1985; Kirk and Farrell, 1987).

Due to their relatively high chemical stability and resistance to microbial degradations, lignins can survive early diagenesis and have good chances to be preserved (Hatcher *et al.*, 1982; Saiz-Jimenez and de Leeuw, 1986a; Orem and Hatcher, 1987; Hatcher and Spiker, 1988). The presence of lignin derivatives was thus demonstrated in soils, river humic substances, Recent sediments, buried woods, coals, and kerogens (Given *et al.*, 1984; Mycke and Michaelis, 1986; Attalla *et al.*, 1988). In some of these samples, the structure of lignins was shown to be somewhat altered; a selective degradation of the sinapyl units was noted in lignins from buried woods (Hedges *et al.*, 1985; Attalla *et al.*, 1988). However, it is well documented that integral parts of lignin macromolecules can be incorporated into fossil organic matter. According to chemical degradation studies, a substantial amount of intact lignins occurs in some kerogens and brown coals comprising a large contribution from higher plants (Mycke and Michaelis, 1986).

In fact, lignins are often considered as the major input in vitrinite formation (Stach *et al.*, 1982). Pyrolysis–gas chromatography/mass spectrometry (Py-GC/MS) experiments (Nip *et al.*, 1987a) revealed that vitrinite macerals are enriched in lignin derivatives, relative to precursor woody tissues, as a result of the selective removal of the associated polysaccharides. Because lignins appear to be important source macromolecules of sedimentary organic matter, other chapters in this volume deal with the geological fate of these cell wall materials (see Chapters 4 and 24).

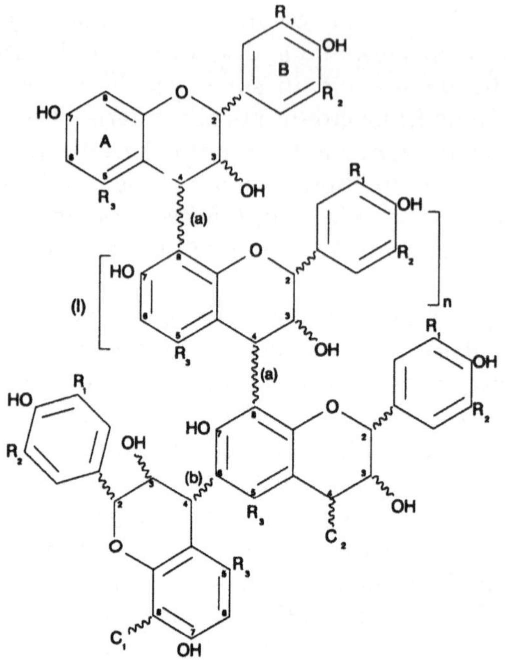

Figure 54. Structure of proanthocyanidin polymers (PAC). Procyanidin units: R_1 = OH, R_2 = H; prodelphinidin units: R_1 = R_2 = OH; prolargonidin units: R_1 = R_2 = H. (a) 4→8 interflavan bond; (b) 4→6 interflavan bond. (I) and C_1: Linear chains of flavan-3-ol units; when C_2 = C_1, a trilinked unit occurs and a branch point is obtained; when C_2 = H, an "angular" structure is obtained. Phloroglucinol-type A ring: R_3 = OH; resorcinol-type A ring: R_3 = H.

2.2.28. Tannins

Tannins are polyphenolic compounds that occur in higher plants and in some algae and are capable of precipitating proteins from aqueous solutions. These macromolecules are divided into three groups: the condensed tannins or proanthocyanidin polymers (PAC), the phlorotannins (PT), and the hydrolyzable tannins (HT).

2.2.28a. Proanthocyanidin Polymers (PAC). These are nonlignin polyphenols, resistant to hydrolysis by acids and bases and composed of chains of polyhydroxy flavan-3-ol units linked via C–C bonds (Fig. 54). The structures of PAC are further determined by the following features (Haslam, 1977; Foo and Porter, 1980; Foo, 1982; Mattice and Porter, 1984; Sun *et al.*, 1987):

- The hydroxylation pattern of the B-ring. The procyanidin units (R_1 = OH, R_2 = H) and the prodelphinidin units (R_1 = R_2 = OH) are widespread while the prolargonidin units (R_1 = R_2 = H) are only common in monocotyledons. Some tannins are exclusively composed of one of the above flavan units, but usually there is only a predominance of one type.
- The stereochemistry of the heterocyclic ring. The substituents at carbon atoms 2 and 3 show a *cis* or *trans* stereochemistry. In most cases, the absolute configuration of carbon atom 2 is the *R* configuration.
- The α or β stereochemistry of the interflavan bonds. In natural PAC these bonds are always *trans* relative to the C_3 substituent.
- The location of the interflavan bonds. Generally, carbon atom 4 of the heterocyclic ring is linked to carbon atom 8 of the adjacent unit. In some polymers, however, 4→6 links are observed, resulting in "angular" structures.
- The occurrence of branch points. A limited number of the flavan units (1 to 4%) are linked via carbon atoms 4, 6, and 8.
- The hydroxylation pattern of the A-ring. In most cases, phloroglucinol-type rings occur (R_3 = OH); in a few PAC, resorcinol-type rings (R_3 = H) are also noted.

- The average degree of polymerization. Most of the identified tannins consist of less than 10 flavan units. Some contain up to 40 units, and, in very limited cases, chains comprising hundreds of flavan units have been described. Even higher molecular weight tannins might exist in nature, but such substances will be difficult to analyze.

Due to the different features, a very large variety in PAC structures is observed. Furthermore, in some units, the hydroxyl group of the heterocyclic ring is sometimes esterified with gallic acid (Nonaka *et al.*, 1982; Sun *et al.*, 1987, 1988). Some PAC glycosides were also identified with a sugar residue linked either to a phenolic hydroxyl group in the A and B rings or to the hydroxyl group of the heterocyclic ring (Porter *et al.*, 1985). Finally, in all the PAC-containing tissues, the bulk of these polymers are glycosidically bound to a carbohydrate matrix, e.g., cell wall "hemicellulose" (Porter *et al.*, 1985; Shen *et al.*, 1986).

The biosynthesis of PAC occurs via the formation of a carbocation at carbon atom 4 of a flavan unit, followed by coupling of a second unit at carbon atoms 6 or 8 of the A ring. This polymerization is not enzymatically controlled, and PAC structures are therefore mainly determined by the nature and the relative abundance of the available flavan units and by thermodynamic considerations (Haslam, 1977; Hillis, 1985).

2.2.28b. Phlorotannins (PT). These macromolecules consist of phloroglucinol units linked via diaryl ether bridges (Ragan and Glombitza, 1986), and they are derived from the dehydropolymerization of phloroglucinol (Fig. 55). Monohalogenated PT, as well as polymers containing additional hydroxyl groups, have been identified (Koch and Gregson, 1984; Ragan and Glombitza, 1986). In some PT a part of the phloroglucinol units are linked directly via diaryl C–C bonds. The number of units ranges from 3 to 13 in most of the identified PT, but high-molecular-weight PT, with an average molecular mass of 10 kDa and values up to 400 kDa, have been observed (Grosse-Damhues and Glombitza, 1984; Koch and Gregson, 1984; Ragan and Glombitza, 1986). In addition to linear chains due to para linkages, various structures arising from other linkages (ortho, meta, trilinked units) have been reported (Grosse-Damhues and Glombitza, 1984; Ragan and Glombitza, 1986) (Fig. 55).

2.2.28c. Hydrolyzable Tannins (HT). This group comprises a polyhydroxy monomeric unit, or "core," esterified with a varying number of gallic acids, gallotannins, or with gallic and hexahydroxydiphenic acids, ellagitannins (Fig. 56). D-Glucopyranose is the most common core (Goodwin and Mercer, 1972; Nishizawa *et al.*, 1985; Tanaka *et al.*, 1985). However, some HT with an open-chain glucose core have also been identified (Sun *et al.*, 1988). Other monosaccharide (Dfructose and D-xylose) and polyalcohol (glycerol, quinic acid, quercitol, shikimic acid; Fig. 57) cores occur as well (Nishimura *et al.*, 1984, 1986;

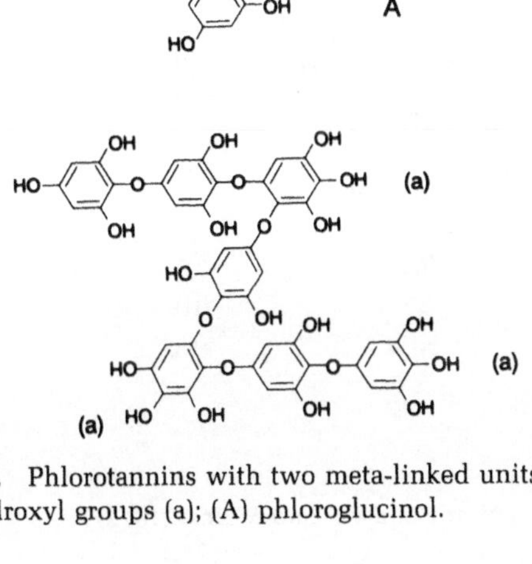

Figure 55. Phlorotannins with two meta-linked units and additional hydroxyl groups (a); (A) phloroglucinol.

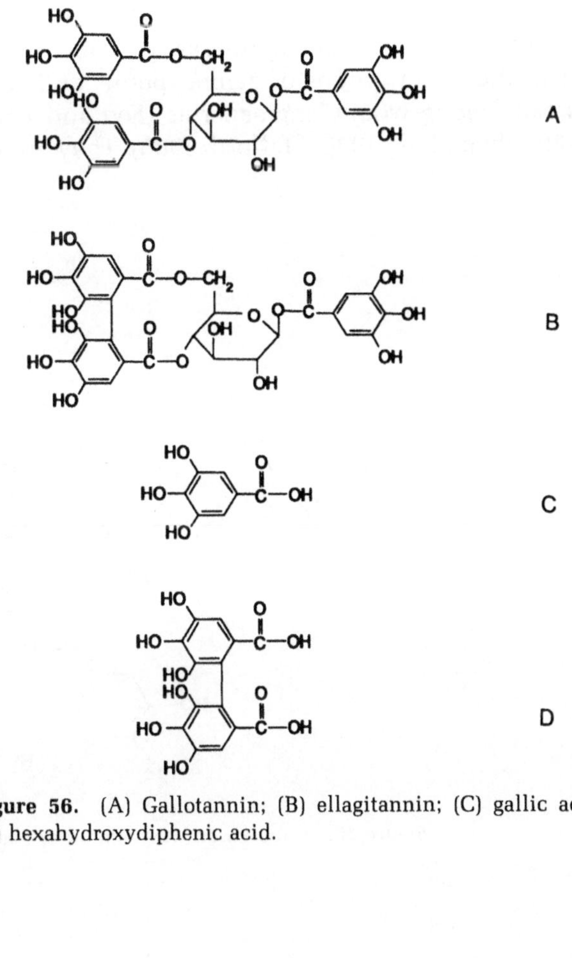

Figure 56. (A) Gallotannin; (B) ellagitannin; (C) gallic acid; (D) hexahydroxydiphenic acid.

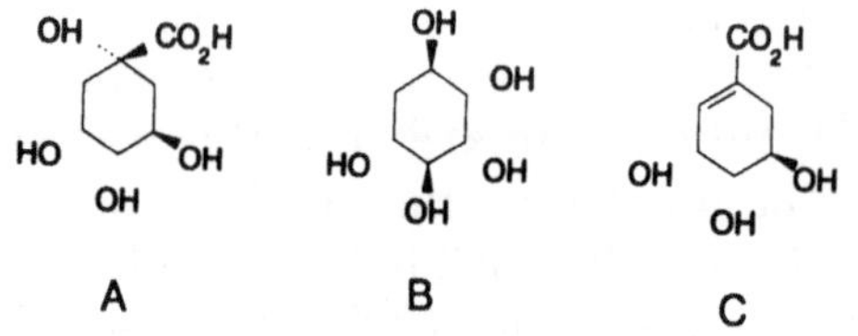

Figure 57. Some polyhydroxy cores occurring in hydrolyzable tannins: (A) Quinic acid; (B) quercitol; (C) shikimic acid.

Ishimaru *et al.*, 1987). The presence of two glucose molecules linked by a digalloyl bridge was observed as the core structure in a few cases (Yoshida *et al.*, 1985). In addition to monoesters, di- and trigalloyl units are commonly observed in HT (Fig. 58). Due to the above features, a large variety in HT structures is noted.

2.2.28d. Occurrence and Fate of Tannins. The occurrence of PT seems to be restricted to brown macroalgae (Grosse-Damhues and Glombitza, 1984; Koch and Gregson, 1984; Ragan and Glombitza, 1986). With levels up to 20% of dry weight, and an average content of 4% for a large number of species, phlorotannins are important constituents of these brown algae; PT accumulate intracellularly in vesicles, and a large fraction is also excreted (Ragan and Glombitza, 1986). PAC and HT are largely distributed in the plant kingdom (Goodwin and Mercer, 1972). Thus, PAC, while less common than lignins, are probably universal in the major groups of gymnosperms and widespread among woody angiosperms (Foo and Porter, 1980; Shen *et al.*, 1986; Stafford, 1988). They are rare or even lacking in nonwoody angiosperm families representing both aquatic and herbaceous plants. PAC were not identified in primitive vascular plants. Gallotannins are known in 21 families of higher plants, but they are absent in primitive vascular groups such as ferns, lycopods, liverworts, and mosses (Bate-Smith, 1984).

PAC and HT occur in cell walls of almost any plant organ. The three groups of tannins are supposed to play an important protective role as antifeedant (Lane and Schuster, 1981; Klocke *et al.*, 1986). PAC usually occur together with lignins in woody plants, and both types of polymers are considered to be important for structural rigidity (Stafford, 1988). High contents of PAC and HT are quite often observed. Thus, PAC are major costituents of barks in numerous plants such as pine (Foo, 1982; Sun *et al.*, 1987, 1988), and levels up to 45% occur in mangrove and eucalyptus bark (Goodwin and Mercer, 1972). PAC also predominate (ca. 20% of dry weight) in leaves of some cotton species (Lane and Schuster, 1981). Furthermore, the level of PAC might be underestimated in many cases, due to the presence of insoluble products covalently linked to carbohydrates. The highest concentration of HT (45% of dry weight) was reported in seed pods (Goodwin and Mercer, 1972).

Tannins exhibit antimicrobial properties because they are able to interact with proteins; thus, various PAC show a broad antibacterial spectrum (Porter and Woodruffe, 1984). In addition, it was noted that tannins tend to accumulate in dead or dying cells (Goodwin and Mercer, 1972). Due to these properties, it

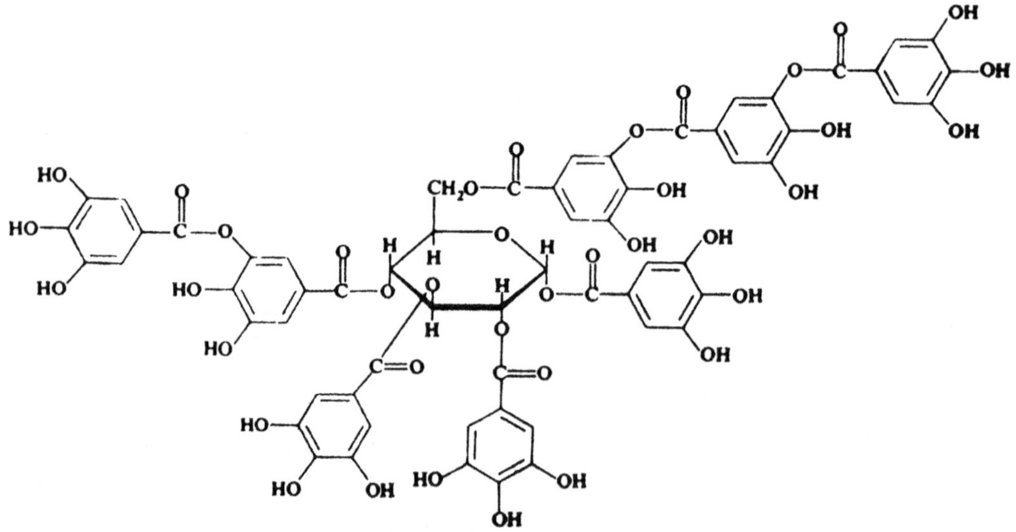

Figure 58. Gallotannin with mono-, di-, and trigalloyl units. [After Goodwin and Mercer (1972).]

might be speculated that tannins can make a substantial contribution to the formation of certain types of fossil organic matter (Wilson and Hatcher, 1988).

Regarding PT, it seems that no study was carried out on their biodegradability; based on their chemical structure, one could expect a relatively high chemical stability. As stressed above, the level of PT in brown algae is high. Furthermore, a continuous release of PT occurs in the surrounding water. Thus, the large amounts of yellow-colored products present in inshore seawater are considered as possibly derived from tannins of brown algae (Hellebust, 1974). Accordingly, PT might play a significant role in kerogen formation in such depositional environments.

HT can be degraded by a large number of esterases, and their chemical stability seems to be poor as well. Therefore, their preservation potential is thought to be lower.

Enzymatic hydrolysis of PAC via tannase activity was observed in a few cases only (Nonaka *et al.*, 1982; William *et al.*, 1986; Vennat *et al.*, 1987). Due to the presence of the labile trihydroxy B ring, prodelphinidin-rich PAC may be more sensitive to enzymatic or aerial oxidation than procyanidin- or propelargonidin-rich PAC (Foo and Porter, 1980). Attempts to hydrolyze soluble PAC resulted in the formation of an insoluble material via further polymerization (Goodwin and Mercer, 1972). As expected, the chemical stability of PAC to drastic hydrolysis conditions increases with the degree of polymerization (Foo, 1982). Owing to these features, a high preservation potential can be anticipated for PAC, especially for those based on procyanidin and propelargonidin units. Because large levels of PAC are commonly observed in higher plants, PAC should therefore provide a substantial contribution to land-derived sedimentary organic matter. Indeed, it has been suggested that PAC could be considered, in addition to cellulose and lignin, as a major source of the maceral vitrinite (Stach *et al.*, 1982; Given, 1984). A recent study using solid-state ^{13}C NMR (Wilson and Hatcher, 1988) indicated an important level of tannins in barks isolated from brown coals. The relative abundance of such compounds increases with coal rank. The presence of tannins was similarly detected in peats and buried woods.

2.2.29. Sporopollenins

Originally, sporopollenins were defined operationally as structural constituents of the outer walls, exines, of spores and pollen grains from vascular plants, surviving drastic nonoxidative chemical treatments such as acetolysis (Zetsche and Vicari, 1931). On the basis of the first studies, these macromolecules were assumed to be formed via an oxidative radical copolymerization of carotenoids and/or carotenoid esters (Brooks and Shaw, 1968a, b; Shaw, 1971; Brooks and Shaw, 1978). Later on, the presence of "sporopollenin-like" constituents was assumed in the walls of various microorganisms: algal vegetative cells and algal cysts,* bacterial vegetative cells and spores, and fungal spores. Recently, however, the relationship of certain chemically resistant cell wall macromolecules in microorganisms with sporopollenins was questioned (Berkaloff *et al.*, 1983; Brunner and Honegger, 1985; Puel *et al.*, 1987; Burczyk and Dworzanski, 1988).

In this section, sporopollenins are therefore considered to exclusively comprise nonhydrolyzable outer-wall macromolecules occurring in vascular plant pollen and spores. Furthermore, as explained below, most sporopollenins are unlikely to be derived from carotenoid-type precursors. Accordingly, the possible chemical structures, precursors, and biosynthesis of carotenoid-derived materials shall not be associated with the term sporopollenin.

With regard to the origin of sporopollenins of vascular plant spores and pollen, it has to be noted that the precursor–product relationship with carotenoids was established from investigations of a very limited number of samples, mainly the sporopollenins from *Lycopodium clavatum* spores and pollen from *Pinus species* and *Lilium henryii*. This precursor–product relationship was mainly concluded from indirect evidence. In fact, recent studies based on Fourier transform infrared (FTIR) and solid-state ^{13}C NMR spectroscopy, combined GM/MS analyses of oxidation and pyrolysis products, inhibition of carotenoid formation, and tracer studies (Schenck *et al.*, 1980; Prahl *et al.*, 1985, 1986; Schulze Osthoff and Wiermann, 1987; Guilford *et al.*, 1988) have indicated that carotenoids or carotenoid derivatives are hardly, or not at all, involved in the formation of most of the tested vascular plant sporopollenins. On the basis of these observations, the sporopollenins of various pollen are thought to be derived mainly either from phenylpropanoid units (Prahl *et al.*, 1986; Schulze Osthoff and Wiermann, 1987) or from long *n*-alkyl chains possibly derived from fatty acids (Guilford *et al.*, 1988). In the particular case of *L. Clavatum* spores, a small part of the sporopollenins might originate from carotenoids (Guilford *et al.*, 1988), although in

*Algal spores implicated in reproduction or present in the resting stages under adverse conditions.

this case, too, analyses by flash-pyrolysis GC/MS have indicated a predominant phenylpropanoid nature as well (Schenck *et al.*, 1980).

2.2.29a. What Are "Sporopollenin-Like" Macromolecules in Microorganisms? As noted above, "sporopollenin-like" constituents were also assumed to occur in various algal, bacterial, and fungal outer walls. The identification of these materials as sporopollenins was usually based on three main features: their resistance to acetolysis, their IR spectra (obtained after acetolysis), and their concomitant occurrence with oxygenated carotenoids. However, recently it was demonstrated that several drawbacks are associated with the use of direct acetolysis as a test for the occurrence of sporopollenins:

- Acetolysis results in a partial acetylation of hydroxyl groups. Therefore, the strong absorption bands associated with acetyl groups dominate the IR spectra, hampering severely a further interpretation of these spectra (Brunner and Honegger, 1985).
- Even more seriously, direct acetolysis of the unextracted biomass can produce artifacts due to condensation reactions of carotenoids and other compounds, thus yielding resistant insoluble clumps; moreover, the IR spectra of such condensates resemble those of acetolyzed sporopollenins (Brunner and Honegger, 1985).
- The concomitant presence, or absence, of resistant wall macromolecules and of oxygenated carotenoids was noted in numerous algae, bacteria, and fungi (Atkinson *et al.*, 1972; Gooday *et al.*, 1973; Strohl *et al.*, 1977; Burczyk and Hesse, 1981; Burczyk, 1987). However, the presence of resistant wall material was also demonstrated both in carotenoid-deficient mutants of fungal spores (Gooday *et al.*, 1974) and in the apochlorotic alga *Prototheca wickerhamii*, in which not even traces of carotenoids or carotenoid precursors occur (Puel *et al.*, 1987). Therefore, the co-occurrence of a resistant cell wall material and of oxygenated carotenoids in numerous microorganisms might be fortuitous, and this feature cannot be considered as evidence for a biosynthetic relationship. Indeed, a complete lack of relationship was demonstrated, in several microalgae, between the resistant cell wall macromolecules and carotenoids or carotenoid derivatives. We therefore consider that the term sporopollenins has to be *strictly limited*, as it was initially, to resistant materials isolated from the walls of vascular plant spores and pollens. As a consequence, the term "sporopollenin-like" must be banned to avoid further confusion.

2.2.29b. Occurrence and Fate of Sporopollenins. Sporopollenins are widespread substances since they build up the exines of spores in lower plants and of pollen in angiosperms and gymnosperms. The ornamentation pattern of the outermost layer of exines is typical of a given plant species and is determined by deposition on a matrix, probably polysaccharidic in nature (Heslop-Harrison, 1971). Sporopollenin formation always results from an extracellular polymerization occurring at specific surfaces. In the internal part of exines, sporopollenin deposition occurs on the two sides of lamellae of unit membrane dimension (Dickinson, 1971). Due to the accumulation of large amounts of sporopollenin on such lamellae, the resulting trilaminar structure is obscured in mature exines (Sengupta and Rowley, 1974). The various macromolecules implicated as deposit surfaces are embedded in sporopollenin and thus protected from external influences. Sporopollenins play a key role in the protection of spores and pollen grains.

Sporopollenin levels ranging from 1.4% in pollen of *Populus species* to 28% in pollen of *Pinus species* were observed (Brooks, 1971; Brooks and Shaw, 1972). In numerous angiosperms, sporopollenin accounts for ca. 5% of the pollen total dry weight. *L. clavatum* spores contain 23% of sporopollenin (Brooks, 1971), and even higher levels, up to 45%, occur in spores of ferns (Toia *et al.*, 1985). In addition to exine, the spore and pollen walls contain an inner polysaccharidic layer; the sporopollenin-to-polysaccharide ratios in these walls range from 0.6 to 9 and tend to be higher in the more primitive species (Brooks, 1971).

The chemical stability of sporopollenins is high except for oxidations. However, the rate of oxidative degradation strongly depends on the species considered. Thus, some exines completely disappear in a few minutes in concentrated nitric acid at room temperature while others withstand this treatment for weeks (Zetsche and Vicari, 1931). Under oxic conditions, in soils, several spores and pollen entirely disappear within years, while other exines, e.g., in *L. clavatum*, are unaffected by bacterial and fungal attacks and by chemical oxidations (Havinga, 1971). In Recent sediments, very large differences in degradation rates are also noted for various species of spores and pollens. In addition, the pH of the sediment is an important parameter, and a better preservation of sensitive exines was observed under acidic conditions

(Faegri, 1971). This differential preservation behavior may cause problems in the reconstruction of paleoenvironments based on spore and pollen analysis. As discussed above, sporopollenins show, at least in some species, a very high resistance to chemical and microbial degradations. The highly resistant exines survive morphologically intact in sediments. The initial morphological features of exines can be fully retained in some samples (Shaw, 1970; Brooks and Shaw, 1972). Thus, a similar ultrastructure was noted, via transmission electron microscopic (TEM) observations, in modern and fossil exines in *Lycopod* spores (Kovach and Dilcher, 1985). Spore and pollen exines appear as the most widely occurring plant fossils (Brooks, 1971).

Due to this high preservation potential, sporopollenins can be considered as the precursors of a typical maceral, sporinite, comprising fossil exines of both spores and pollens (Brooks, 1971). Sporinite is widespread in coals and in a few cases accounts for up to 50% of coal total mass (Cooper and Murchison, 1971). Sporinite is also widespread in kerogens with contribution of land-derived organic matter (Johnson, 1985). In fact, as discussed above, sporopollenin degradation is mainly associated with oxic and basic conditions of deposition. Fast deposition in coal-forming swamps, under stagnant conditions and with limited pH variations, should therefore allow for the preservation of the exines from a large array of spores and pollen.

Sporopollenins were often used as model substances in simulation experiments concerning thermal evolution (Monin *et al.*, 1980; Villey *et al.*, 1981). Mild treatments of *L. clavatum* sporopollenin result in the formation of a "geopolymer-like" material (Hayatsu *et al.*, 1987) similar, in several respects, to immature sporinites isolated from lignites. Further heat treatments of this material result in progressive transformations along the natural evolution path of sporinites. According to these observations, some variations in the chemical structure of sporopollenins take place during diagenesis and could be associated with the elimination of functional groups labile under mild thermal conditions. To test the above-mentioned assumptions, it is necessary, firstly, to gain a better knowledge of the chemical structure of modern sporopollenins and of their possible variations in connection with spore and pollen species and, secondly, to perform precise comparisons between modern sporopollenins and homogeneous concentrates of immature fossil exines of the same species.

In spite of the lack of detailed knowledge of the sporopollenin structure and of its diagenetic evolution, it is clear that sporopollenins often exhibit a high potential of preservation and can afford a substantial contribution in the genesis of land-derived fossil organic matter.

2.2.30. Algaenans

Nonhydrolyzable, insoluble, highly aliphatic substances have been detected in the cell walls of a number of species of microalgae since the pioneering work of Atkinson *et al.* (1972). Up to now, several terms have been used to indicate these substances. Recently, the term "algaenan" was introduced to refer to these biomacromolecules (Tegelaar *et al.*, 1989a). A polymethylenic-type algaenan was isolated for the first time from the cell walls of *Botryococcus braunii* *A* race. At the time this substance was referred to as "polymère résistant de *Botryococcus*," abbreviated "PRB *A*" (Berkaloff *et al.*, 1983). Later on (Kadouri *et al.*, 1988), a similar substance ("PRB *B*") was observed in *B. braunii B* race. Three different races have been identified for the green, colonial, freshwater alga *B. braunii*. These races show the same typical morphology but sharply differ by the chemical structure (Fig. 59) of the hydrocarbons they produce in large amounts (Maxwell *et al.*, 1968; Brown *et al.*, 1969; Metzger and Casadevall, 1983; Metzger *et al.*, 1986; Metzger and Casadevall, 1987): odd-carbon-numbered C_{23} to C_{31}, dienic and trienic, unbranched hydrocarbons in the *A* race; isoprenoid C_{30} to C_{37}, highly unsaturated hydrocarbons, termed botryococcenes in the *B* race; and a C_{40} isoprenoid, dienic hydrocarbon, lycopadiene, in the *L* race. The algaenans of the *A* and *B* races were studied by both pyrolytic (Py-GC/MS, "off line" pyrolysis) and spectroscopic [FTIR, cross polarization/magic angle spinning (CP/MAS) ^{13}C NMR] methods (Largeau *et al.*, 1984, 1986; Kadouri *et al.*, 1988). These two macromolecules were shown to exhibit similar chemical structures. These are based on a tridimensional net-

$$CH_3-(CH_2)_7-CH\overset{z}{=}CH-(CH_2)_x-CH=CH_2 \quad 1$$

x = 11, 13, 15 and 17

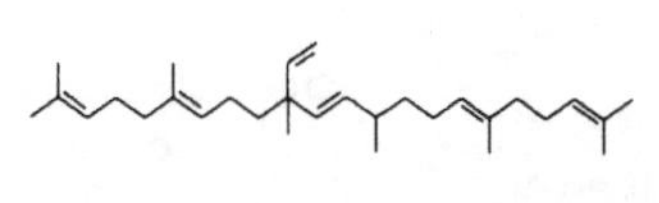

2

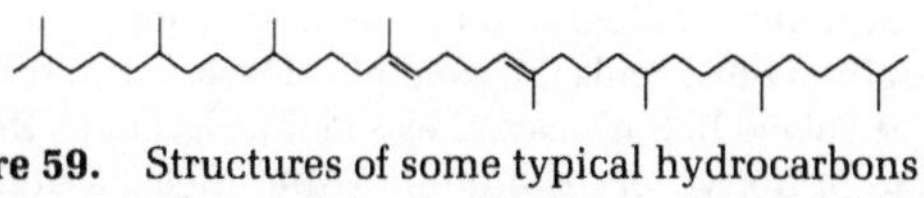

3

Figure 59. Structures of some typical hydrocarbons produced by *Botryococcus braunii* *A* race (1), *B* race (2), and *L* race (3).

work composed of ether-linked, long $(CH_2)_n$ polymethylenic chains (n up to 30). Few hydroxyl and ester functions are present within this network. Because of the resulting steric protection, they are not affected by chemical reagents; the esters can then survive drastic acid and basic treatments and will be cleaved only by pyrolysis. Saturated and unsaturated, even-carbon-numbered, linear, C_{12} to C_{18} fatty acids are linked to the macromolecular network via such protected ester functions. The algaenan isolated from *B. braunii L* race ("PRB *L*") exhibits the same major chemical features as those from the *A* and *B* races. In sharp contrast, however, the macromolecular network of the former is composed of long, saturated, isoprenoid chains instead of polymethylenic chains (Derenne *et al.*, 1988c, 1989, 1990a). This seems to be the only example of a polyisoprenoid-type polyalkylether reported so far. In addition to the linear fatty acids already observed in the algaenans of the *A* and *B* races, the protected esters of PRB *L* comprise a substantial contribution of C_{14} to C_{19} isoprenoid acids (Derenne *et al.*, 1991a).

Biosynthetic studies (Laureillard *et al.*, 1986, 1988) indicated that oleic acid is the precursor of the polymethylenic chains of the algaenans in *B. braunii* races *A* and *B*. Analysis of high-molecular-weight complex lipids, present in the *A* race and thought to be intermediates in the biosynthesis of algaenans, suggested the following steps (Fig. 60):

- elongation of oleic acid into long-chain C_{28}, C_{30}, and C_{32} fatty acid derivatives
- reduction to aldehydes
- formation of the basic building units by dimerization and epoxidation of these unsaturated aldehydes

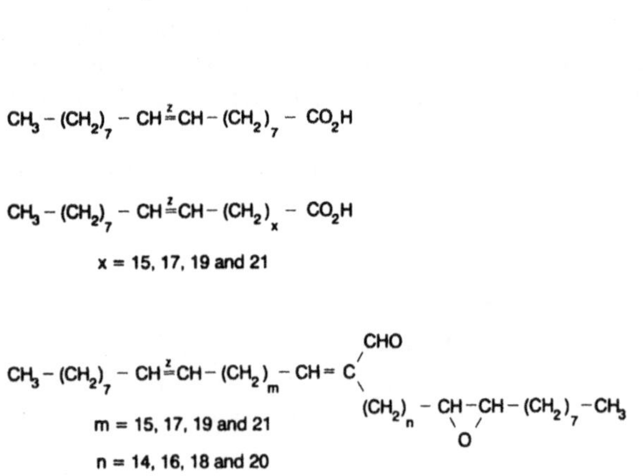
1
2
3

Figure 60. Structure of oleic acid (1) and of very long chain fatty acids derived from oleic acids (2); example of a basic building unit (3) probably involved in the biosynthesis of the algaenans from the *A* and *B* races of *Botryococcus braunii*. [After Metzger and Casadevall (1989).]

- polymerization and formation of the ether bridges via epoxide ring opening

The algaenans isolated from some other species of microalgae were examined by IR (Brunner and Honegger, 1985; Burczyk, 1987) and CP/MAS ^{13}C NMR spectroscopy (Guilford *et al.*, 1988), Py-GC (Burczyk and Dworzanski, 1988), and Py-GC/MS (Goth *et al.*, 1988). The structural information so obtained points to a major contribution of polymethylenic chains and to a close similarity with the algaenans of *B. braunii* races *A* and *B*. The ^{13}C NMR observations on "hydrolyzed alghumins," isolated from several green microalgae in a somewhat different fashion via successive extractions and acid hydrolysis, also suggest the presence of polymethylenic-type algaenans (Zelibor *et al.*, 1988).

The presence of algaenans was noted in a number of species of microalgae belonging to 26 different genera. Most of these are Chlorophyceae (green algae), but nonhydrolyzable macromolecules were also observed in Dynophyceae (dinoflagellates) and Prasinophyceae. Algaenans were detected in vegetative cells and/or cysts, and freshwater species account for most of the reported occurrences (Table II).

Various wall structures and nonhydrolyzable macromolecule locations within the wall were observed in algaenan-containing microalgae. In most green algae, algaenans are located in thin, 10- to 20-nm-thick, trilaminar outer walls, often termed "trilaminar sheaths" (TLS), comprising two electron-dense layers sandwiching an electron-lucent one (Brunner and Honegger, 1985). TLS organization is often retained following algaenan isolation, suggesting they built up the two dense peripheral layers. However, in various lichen phycobionts, the nonhydrolyzable material seems to be located in the central layer of the TLS (Honegger and Brunner, 1981). In other species, the trilaminar structure is no longer visible following the drastic chemical treatments required for algaenan isolation, and the precise location of the TLS within the outer wall cannot be established (Atkinson *et al.*, 1972). In the aerial alga *Phycopeltis* sp., the thick outer wall results from the stacking of numerous TLS (Good and Chapman, 1978). While the presence of algaenans is often associated with the occurrence of a trilaminar outer wall, there is not a strict relationship between these two features. Thus, the lack of algaenans was noted in TLS-containing species and vice versa (Morrill and Loeblich, 1981; Brunner and Honegger, 1985). *B. braunii* is characterized by thick (ca. 1 μm) basal outer walls which build up the matrix of the colonies. Perhaps due to a

TABLE II. Occurrence of Algaenans

Species	Method(s)	Reference(s)
Botryococcus braunii		
Race *A*—PRB *A*	Py-GC(-MS), NMR, FTIR	Largeau *et al.*, 1986
Race *B*—PRB *B*		Derenne *et al.*, 1988c
Race *L*—PRB *L*		Zelibor *et al.*, 1988
Tetraedron minimum	Py-GC(-MS), NMR, FTIR	Goth *et al.*, 1988; Tegelaar *et al.*, 1989e
Scenedesmus obliquus	Py-GC(-MS), NMR, FTIR	Zelibor *et al.*, 1988; Tegelaar, 1989
Scenedesmus quadricauda	Py-GC(-MS), NMR, FTIR	Derenne *et al.*, 1990d; Tegelaar *et al.*, 1989e
Pediastrum duplex	Py-GC(-MS)	Tegelaar, 1989
Dunaliella tertiolecta	NMR	Zelibor *et al.*, 1988
Chlorella pyronenosidosa	NMR	Zelibor *et al.*, 1988

very large accumulation of algaenans, no structures are clearly visible in such basal walls (Berkaloff *et al.*, 1983; Kadouri *et al.*, 1988). Finally, in some lichen phycobionts, trilayered walls are observed, and the nonhydrolyzable material is located in the thick central layer (König and Peveling, 1980). In the few algal cysts examined for the occurrence of algaenans, multilayered thick outer walls occur with various locations of the algaenan-containing layer (Devries *et al.*, 1983; Aken and Pienaar, 1985).

Various high-molecular-weight compounds, including enzymes, cannot enter algal cells surrounded by an algaenan-containing trilaminar wall. Such cells are therefore unaffected by extracellular enzymes, e.g., cellulases (Atkinson *et al.*, 1972). A close relationship was also noted between the ability of different strains of *Chlorella* and *Scenedesmus* to survive high concentrations of detergents and the presence of an algaenan-containing outer wall (Biedlingmaier *et al.*, 1987). The presence of algaenans can also provide a protection against fungal parasitism in symbiotic systems such as lichens (Honegger and Brunner, 1981).

As expected form the thickness of the basal outer walls in *B. braunii*, algaenans are major constituents (10 to 33% of the total biomass) of this colonial alga (Berkaloff *et al.*, 1983; Kadouri *et al.*, 1988; Derenne *et al.*, 1989). Most other algaenan-forming species contain a substantially lower level. Species surrounded by a thin TLS usually afford low amounts of algaenans, e.g., 0.3% of the total biomass in *Prototheca wickerhamii* and 0.6% in *Chlorella fusca* (Puel *et al.*, 1987; Atkinson *et al.*, 1972). It should however be kept in mind that upon release of daughter cells, following cell divisions, fragments of mother cell TLS are freed and accumulate in the growth medium (Atkinson *et al.*, 1972). In spite of a low content of algaenans in the biomass, such thin TLS-containing algae can therefore produce substantial amounts of nonhydrolyzable material.

A high resistance of the cell walls of some species and its relationship with the presence of algaenans was suggested both by "*in vitro*" laboratory experiments in which the degradation of various microalgae by bacteria was investigated (Gunnison and Alexander, 1975) and by observations on cores from lacustrine recent sediments (Stiller *et al.*, 1983). Thus, large amounts of well-preserved cell walls of green algae, retaining their typical size and shape, were observed in ca. 5.5×10^3-year-old samples. Moreover, most of these cell walls correspond to species now identified as algaenan producers. The occurrence of microfossils with a morphology close to that of algaenan-containing algae, such as *Pediastrum* sp., is also well documented in a number of kerogens, especially of type I.

Direct evidence for the selective preservation of some algaenans and their major contribution to the formation of kerogens was recently reported (Tegelaar *et al.*, 1989a). In fact, most of the type I kerogens, such as torbanites and Messel Oil Shale, seem to consist almost exclusively of selectively preserved algaenans (Largeau *et al.*, 1984, 1986; Derenne *et al.*, 1988c; Goth *et al.*, 1988). Marine kerogens of Ordovician age containing relatively high amounts of microscopically recognizable cell wall material of *Gleocapsomorpha prisca* yield abundant homologous series of alkenes and alkanes upon pyrolysis. Although the distribution patterns of the alkenes and alkanes are significantly different when compared to the extant algaenans so far studied, it has been concluded that *G. prisca* fossilization also likely occurred via algaenan selective preservation (Douglas *et al.*, 1992). Recent TEM observations, showing that cell contents are entirely eliminated in the fossil colonies of *G. prisca* while the morphology of the thick outer walls is perfectly retained (Largeau *et al.*, 1990b), are also consistent with the above conclusion.

2.2.31. Cutans

The heterogenous character of cuticles is well documented, and it has long been established (Holloway, 1982b) that they comprise a soluble wax fraction (mixture of long-chain lipids mainly composed of hydrocarbons, ketones, alcohols, fatty acids, and fatty esters) and an insoluble matrix. Recent studies have demonstrated that in addition to these waxes and the insoluble but hydrolyzable polyester cutin, another nonhydrolyzable insoluble macromolecular material called cutin is present in many plant cuticles (Nip *et al.*, 1986a, b, 1987b; Tegelaar *et al.*, 1989b).

Pyrolytic studies, using both "on-line" and "off-line" Py-GC and Py-GC/MS methods, were carried out on the isolated cutan of *Agave americana* cuticle (Nip *et al.*, 1986a, b; Tegelaar et al., 1989b). The pyrolysates are almost exclusively composed of homologous series of normal hydrocarbons (alkanes, alk-1-enes, α,ω-alkadienes) with carbon numbers ranging from C_7 to C_{35}, thus suggesting a polymethylenic-type structure.

Further examination with solid-state ^{13}C NMR spectroscopy, additional pyrolysis methods, differential scanning calorimetry, FTIR spectroscopy, and elemental analysis (Nip *et al.*, 1987b; Tegelaar *et al.*, 1989b) revealed the presence of a polysaccharide moiety, possibly covalently bound to the polymethylenic one. It was suggested that the polysaccharidic moiety consists of cellulose fibers surrounded by a sheath of covalently bound polymethylenic material. Owing to the resulting protection against hydrolysis reagents, such fibers can survive drastic acid treatments. In the nonhydrolyzable insoluble material isolated form *A. americana* cuticle, it was thus estimated (Tegelaar *et al.*, 1989b) that ca. 63% of aliphatic carbons corresponded to the polymethylenic-type moiety and 37% of carbons corresponded to the polysaccharide moiety. Nevertheless, polysaccharide-derived compounds only provide a minor contribution to the pyrolysis products identified on Py-GC and Py-GC/MS.

Several higher plants were tested for the presence of cuticular nonhydrolyzable insoluble macromolecules (Table III). The polymethylenic-type cutan was thus identified in the leaf cuticles of nine species. Taking into account the ubiquitous occurrence of cuticles, the presence of polymethylenic-type cutans can be anticipated in a very large number of higher plants. It is well documented that cuticles play a fundamental role as a barrier between the plant and its environment; cutans are probably involved in this protective function.

TABLE III. Chemical Characterization of Cuticular Matrices (CM) of Extant Higher Vascular Plants

Taxon (affinity)	CM type
Gymnosperms	
Ginkgo biloba (Ginkgoaceae)	Mixed[a]
Sciadopitys verticillata (Taxodiaceae)	Mixed
Amentotataxus argotaenia (Taxaceae)	Mixed
Monocotyledonous angiosperms	
Agave americana (Agavaceae)	Mixed
Clivia miniata (Amaryllidaceae)	Mixed
Zostera marina (Zosteraceae)	Mixed
Dicotyledonous angiosperms	
Lycopersicon esculentum (Solanaceae)	Cutin
Citrus limon (Rutaceae)	Cutin
Salicornia europaea (Chenopodiaceae)	Cutin
Erica carnea (Ericaceae)	Cutin
Symplocos paniculata (Symplocaceae)	Mixed
Hydrangea macrophylla (Saxifragaceae)	Mixed
Gossypium sp. (Malvaceae)	Mixed
Prunus laurocerasus (Rosaceae)	Mixed
Quercus robur (Fagaceae)	Mixed
Acer platanoides (Aceraceae)	Mixed
Malus pumila (Rosaceae)	Mixed
Limonium vulgare (Plumbaginaceae)	Mixed
Beta vulgaris ssp. *maritima* (Chenopodiaceae)	Cutan

[a]Mixed refers to the presence of both cutin and cutan.

Substantial amounts of cutans are present in *A. americana* and *Clivia miniata* cuticles (Nip *et al.*, 1986a, b). In *Beta vulgaris*, no cutin was detected, and the very thin leaf cuticle is exclusively composed of waxes and of a polymethylenic-type cutan (Nip *et al.*, 1986b).

Owing to the presence of cutans, cuticles appear as one of the most chemically resistant plant tissue. A high resistance to biodegradation can also be anticipated. Accordingly, polymethylenic-type cutans have a high fossilization potential and provide a substantial contribution to fossil organic matter of terrestrial origin.

Cutans, like algaenans, should exhibit a very high potential of preservation. The presence of cutans accounts for the commonly observed preservation of cuticles, as cutinite maceral, in peats, coals, and kerogens (Hunneman and Eglinton, 1968; Cardoso *et al.*, 1977; Stach *et al.*, 1982; Cardoso and Eglinton, 1983; Crelling and Bensley, 1984; Dimichele *et al.*, 1984). Indeed, a direct precursor–product relationship was demonstrated between the polymethylenic-type cutans isolated from modern plant cuticles and the major constituents of fossil cuticles (see Table IV) (Nip *et al.*, 1986a, b; Wilson *et al.*, 1987; Tegelaar *et al.*, 1991). Simulation experiments in which cutan was heated during several weeks indicated that the *n*-alkanes in high-wax crude oils originate from this biopolymer (Tegelaar *et al.*, 1989c).

2.2.32. Suberans

Recent investigations of periderm layers of extant trees indicate that a nonhydrolyzable polymethylenic type macromolecule very similar to cutan also occurs in suberized plant tissue (Table V) (Tegelaar *et al.*, 1993b).

This observation was based on spectroscopic and pyrolysis data. The degree of cross-linking in these so-called suberans is thought to be less than that in cutan since the suberans investigated are partly soluble in organic solvents. It has been speculated that the coal maceral suberinite consists mainly of this suberan biomacromolecule (Tegelaar *et al.*, 1993b).

TABLE IV. Chemical Characterization of Cuticular Matrices (CM) of Fossil Higher Plant Leaf Remains

Taxon (affinity)	Age	CM type
Gymnosperms		
Walchia speciosa (Walchiaceae)	L. Permian	Cutan
Autunia conferta (Peltaspermaceae)	L. Permian	Cutan
Ullmannia sp. (Voltziaceae)	U. Permian	Cutan
Ginkgo huttonii (Ginkgoaceae)	Jurassic	Cutan
Anomazomites nilsonii (Cycadophytae)	Jurassic	Cutan
Solenites murrayana (Czekanowskiales)	Jurassic	Cutan
Ptilophyllum pecten (Cycadophytae)	Jurassic	Cutan
Pachypteris papillosa (Corystospermaceae)	Jurassic	Cutan
Abietites linkii (Coniferales)	L. Cretaceous	Cutan
Amentotaxus gladifolia (Taxaceae)	M. Miocene	Cutan
Sciadopitys tertiaria (Taxodiaceae)	U. Miocene	Mixed[a]
Dicotyledonous angiosperms		
Quercus sp. (Fagaceae)	Miocene	Cutan
Ocotea obtusifolia (Lauraceae)	Eocene	Cutan
Rhodomyrtophyllum tristanioides (Myrtaceae)	U. Eocene	Cutan
Berryophyllum saffordii (Fagaceae)	Eocene	Cutan
Symplocos hallensis (Symplocaceae)	U. Eocene	Modified mixed[b]
cf. *Sapindus fructiferus* (Sapindaceae)	L. Eocene	Modified mixed
Protohedycarya ilicoides (Monimiaceae)	U. Cretaceous	Modified mixed

[a]Mixed refers to the presence of both cutin and cutan.
[b]Modified mixed means cutan and partly transferred cutin.

TABLE V. Occurrence of Suberan

Species	Methods	Reference
Betula japonica	Py-GG/MS	Tegelaar et *al.*, 1990
Betula alba	Py-GG/MS	Tegelaar et *al.*, 1990
Fagus sylvatica	Py-GG/MS	Tegelaar et *al.*, 1990
Malus pumila	Py-GG/MS	Tegelaar et *al.*, 1990
Ribes nigrum	Py-GG/MS	Tegelaar et *al.*, 1990
Erica carnea	Py-GG/MS	Tegelaar et *al.*, 1990

2.2.33. Miscellaneous

2.2.33a. Nonhydrolyzable, Insoluble Cell Wall Macromolecules in Bacteria and Fungi. The presence of nonhydrolyzable, insoluble macromolecular cell wall constituents (NIM) has been tested in a very limited number of Eubacteria, including cyanobacteria, and fungi. Few cyanobacterial NIM were studied by spectroscopic and pyrolytic methods (Chalansonnet *et al.*, 1988). Their precise structures are far from understood; it is clear, however, that they are not based, like the algaenans discussed above, on a polyalkyl network, in that the hydrocarbon chains only provide a minor contribution to the macromolecular structure. These cyanobacterial NIM constituents are all characterized by the presence of protected fatty esters and by high levels of methyl groups and of carbon–carbon double bonds.

Six species, representative of the main groups of cyanobacteria, were tested for the presence of NIM (Table VI). A nonhydrolyzable insoluble material was isolated from five of them (Chalansonnet et *al.*, 1988). In sharp contrast to the algaenans, where the initial

TABLE VI. Cyanobacteria Tested for the Presence of NIM[a]

Species	Occurrence[b]	Level[c]
Anacystis montana	+	2.1
Calothrix anomala	+	0.9
Fischerella muscicola	+	1.2
Gloeocapsa alpicola	−	—
Oscillatoria rubescens	+	2.9
Schizothrix sp.	+	5.6

[a]Chalansonnet et *al.*, 1988.
[b]Isolation via successive extractions by organic solvents and basic and acid hydrolysis; +, presence; −, absence.
[c]Percent of the dry weight of the initial biomass.

morphology of the walls is at least partly retained, such materials show an *amorphous* nature when observed by electron microscopy; accordingly, their location within cyanobacterial walls cannot be established. An insoluble organic residue was also isolated from recent cyanobacterial mats after prolonged acid hydrolysis (Philp and Calvin, 1976).

NIM were shown to occur in the walls of some bacterial and fungal spores and also in a few bacterial vegetative cells (Gooday *et al.*, 1973, 1974; Beckett, 1976; Strohl *et al.*, 1977; Furch and Gooday, 1978; Grippiolo and Bonfante-Fasolo, 1984). The typical shape and ornamentations of the wall are retained in several fungal spores, following NIM isolation (Gooday *et al.*, 1974; Beckett, 1976; Furch and Gooday, 1978). However, amorphous fungal NIM were also reported (Gooday *et al.*, 1973).

In bacteria the presence of NIM seems to be related to a greater than normal resistance to heat, dessication, and UV irradiation (Strohl *et al.*, 1977). Very little information is available on the chemical structure of fungal and bacterial NIM.

Various levels of NIM were noted in cyanobacteria (1–5%; Chalansonnet *et al.*, 1988) and in fungal spores (1–12%; Gooday *et al.*, 1973, 1974; Furch and Gooday, 1978). Nonhydrolyzable, insoluble cell wall macromolecules can therefore account for a substantial fraction of the biomass in a large array of microorganisms. All the cyanobacterial NIM isolated thus far, via drastic chemical treatments of modern organisms, appear as amorphous materials. This is consistent with the observations concerning cyanobacterium fossilization; a substantial contribution of cyanobacterial NIM can thus be expected in various amorphous kerogens. However, no direct chemical evidence has been obtained about the role of the selective preservation of such materials in kerogen formation.

2.2.33b. Polysaccharides of Cyanobacterial Sheaths. A separate section is devoted to these polysaccharides in view of the behavior of cyanobacterial sheaths during fossilization and of the major contribution of polysaccharides in such sheaths. Sheaths are the structured tubes surrounding filamentous cyanobacteria; they correspond to external layers of cell envelopes (Drews and Weckesser, 1982). Ultrastructural studies indicated that cyanobacterial sheaths are composed of a complex microfibrillar network (Tuffery, 1969).

Cyanobacterial sheaths are mainly polysaccharidic (Tuffery, 1969). In *Anabaena cylindrica*, a polysaccharide content of 66% was noted (Dunn and Wolk, 1970). Proteinaceous substances also occur in cyanobacterial sheaths (Schrader *et al.*, 1981; Boon, 1984), and specific staining pointed to the presence of acid mucopolysaccharides (Tuffery, 1969). A number of species were examined for the sugar composition of their sheaths (Dunn and Wolk, 1970; Drews and Weckesser, 1982; Boon *et al.*, 1983; Boon, 1984; Klok *et al.*, 1984a; Boon and de Leeuw, 1987). Eight neutral sugars were identified form *Microcoleus chtonoplastes*, including pentoses, hexoses, deoxyhexoses, and xylose as the dominant component (Boon *et al.*, 1983; Boon, 1984; Klok *et al.*, 1984b; Boon and de Leeuw, 1987). In *Chlorogloeopsis*, the sheaths contain a specific polysaccharide not found in the cell walls (Schrader *et al.*, 1981).

Sheaths can enclose a single chain of cells, or trichome, in some species of cyanobacteria such as *Lyngbya* sp. or *Nostoc* sp. (Tuffery, 1969). On the other hand, bundles of trichomes can be associated in a common sheath; the number of trichomes so enclosed can change with the growth conditions, e.g., from two to ten in *Microcoleus chtonoplastes* (Boon *et al.*, 1983).

Continuous secretion of sheath material takes place through cell walls while the outer layers of the sheath imbibe water and expand so that portions slough off (Tuffery, 1969). For a given species, sheath thickness sharply depends on growth conditions (Tuffery, 1969; Boon *et al.*, 1983). Sheaths can considerably increase the thickness of the cell envelope. Thus, in *Nostoc* sp., sheaths up to 30 μm thick were observed whereas the cells are only 4 μm in diameter (Tuffery, 1969). Accordingly, the sheath material can account for a major part of cyanobacterium total biomass.

Sheaths play a supportive role in filamentous cyanobacteria and also allow, as discussed below, for binding and trapping of mineral particles. An additional important function of sheaths is trichome protection against unfavorable environmental conditions; thus, thick sheaths provide protection against dessication during the subaerial exposures sometimes encountered by cyanobacteria in cyanobacterial mats (Boon and de Leeuw, 1983, 1987).

Studies on modern cyanobacterial mats, and on the derived laminated accumulations, indicated that sheaths are the most resistant parts of filamentous cyanobacteria. These recent laminated sediments, in the formation of which cyanobacteria play a major role, were largely studied since they are commonly considered as the precursors of ancient stromatolites.

Cyanobacterial mats are layered microbial ecosystems composed of phototrophic and heterotrophic microorganisms. The photic zone is generally a few millimeters thick; its surface is dominated by cyanobacteria, most generally filamentous. A second thin layer is mainly composed of photosynthetic bacteria. Below is an anoxic zone where anaerobic bacteria degrade the primary biomass produced in the photic zone (Schopf and Walter, 1982; Whitton and Potts, 1982; Boon *et al.*, 1983; Boon, 1984).

Modern cyanobacterial mats are generally confined to settings where the populations of grazing metazoans are strongly reduced by harsh conditions. Accordingly, they usually occur in hypersaline lakes, intermittently exposed marine settings, and hot springs (Gebelein, 1969; Gebelein and Hoffman, 1973; Horodyski and Vonder Haar, 1975; Horodyski *et al.*, 1977; Schopf and Walter, 1982; Whitton and Potts, 1982; Boon *et al.*, 1983; de Leeuw *et al.*, 1985). Mat accumulations commonly appear as laminated structures. The laminations generally reflect variations in sedimentation rate and/or in cyanobacterial population. Thus, a rapid influx of sediments will be followed by an overgrowth of the photic microbial community so as to recolonize the newly formed surface (Gebelein, 1969). Regarding the surface microbial population, changes can occur both on a daily basis (slow growth rate of cyanobacteria at night) (Gebelein and Hoffman, 1973) and on a yearly basis (variations in growth rate or in the species building up the surface layer) (Boon *et al.*, 1983). All the above changes will result in alternation of mineral- and organic-rich laminae. This laminated structure seems to survive reasonably well bacterial degradations occurring in the anoxic zone.

Observations of various mat accumulations revealed a rapid lysis—in a few years—of the cell contents and the cell walls of cyanobacteria (Boon *et al.*, 1983; Boon, 1984). In the deepest part of such recent deposits, the organic matter is almost exclusively composed of empty sheaths (Horodyski *et al.*, 1977; Boon *et al.*, 1983; Klok *et al.*, 1984b; Boon and de Leeuw, 1987). Sheaths are therefore quite resistant to microbial degradations. However, under given conditions of deposition, sheath preservation strongly depends on the species of filamentous cyanobacteria (Horodyski *et al.*, 1977). Analysis of a sheath fraction isolated from the deepest parts of the Solar Lake accumulation (ca. 2400 years old) revealed the presence of peptides and polysaccharides. It was suggested (Boon, 1984) that a part of the initial polysaccharides might be degraded in these sheaths and that only a core protein–polysaccharide fraction was retained.

Carbonate precipitation is a very common phenomenon in these mat systems. This process often starts just below the photic top layer of the mat and proceeds with depth (Horodyski *et al.*, 1977; Klok *et al.*, 1984b). The ability of cyanobacterial sheaths to trap inorganic particles and to bind inorganic products is well documented, and the sheaths are assumed to play a role in the lithification of mat accumulations (Gebelein and Hoffman, 1973; Boon *et al.*, 1983). The resulting encapsulation of cyanobacterial sheaths inhibits further degradations (Boon *et al.*, 1983; Boon and de Leeuw, 1983). The presence of large concentrations of polysulfides was also noted in the pore water of the anoxic zone of mat accumulations (Boon, 1984). This could contribute, in addition to lithification, to the preservation of sheath material. Cyanobacterial mats with a low mineral contribution were observed in alkaline hot springs (Doemel and Brock, 1977). The absence of lithification then results in a complete degradation of the primary organic matter in the anoxic zone. A complete degradation of cyanobacteria was also noted in subtidal and intertidal mats where no lithification occurs (Golubic and Focke, 1978).

In conclusion, cyanobacterial sheaths are more resistant than intracellular constituents; however, their preservation is a consequence of the conditions occurring in mat accumulations (lithification and high polysulfide level), but it does not reflect intrinsic resistance properties *per se*. The implications of this with regard to the fossilization of cyanobacteria in relation to stromatolite formation has been discussed by several authors (e.g., Golubic and Campbell, 1979; Knoll, 1985).

3. Structural Relationships between Kerogen/Coal and Selectively Preserved Resistant Biopolymers

The inventory of presently known biomacromolecules presented in the previous sections is a rather unbalanced one in terms of relative occurrences and biochemical importance. However, the inventory was made within a geochemical context, thus emphasizing those biomacromolecules thought to be resistant in the bio- and geosphere. Those biomacromolecules are important because they will be selectively enriched during biodegradation and early diagenesis. Even if they contribute less than 1% or 0.1% to the biomass of organisms, they can nevertheless become important geochemically because it is estimated that—in general—under natural conditions over 99% of the biomass is mineralized (see Hedges and Prahl, this volume, Chapter 11). As a consequence, it is worthwhile to estimate the so-called preservation potential of the known biomacromolecules in order to select the possible components of humic substances and kerogens.

3.1. Selective Preservation of Biomacromolecules

Table VII summarizes all the biomacromolecules discussed, their major occurrences, and their "preservation potential" (Tegelaar *et al.*, 1989a). It should be stressed that educated guesses concerning preservation potential are mainly made on the basis of anticipated chemical stabilities based on structures, on reported ease of biodegradability, and on reported presence in the geosphere. Other mechanisms for preservation such as mechanical protection or bacteriostatic activities of certain chemicals in the (paleo)-environment are not taken into account at this stage. In a recent paper (Tegelaar *et al.*, 1989a), it was stated that kerogens can be considered as physical mixtures mainly composed of selectively preserved and to

TABLE VII. Inventory of Presently Known Biomacromolecules, Their Occurrence in Extant Organisms, and Their Potential for Survival during Sedimentation and Diagenesis[a]

Biomacromolecules	Occurrence	"Preservation potential"[b]
Starch	Vascular plants; some algae; bacteria	−
Glycogen	Animals	−
Fructans	Vascular plants; algae; bacteria	−
Laminarans	Mainly brown algae; some other algae and fungi	−
Poly-β-hydroxyalkanoates (PHA)	Eubacteria	−
Cellulose	Vascular plants; some fungi	−/+
Xylans	Vascular plants; some algae	−/+
Pectins	Vascular plants	−/+
Mannans	Vascular plants; fungi; algae	−/+
Galactans	Vascular plants; algae	−/+
Mucilages	Vascular plants (seeds)	+
Gums	Vascular plants	+
Alginic acids	Brown algae	−/+
Fungal glucans	Fungi	+
Dextrans	Eubacteria; fungi	+
Xanthans	Eubacteria	+
Chitin	Arthropods; copepods; crustacea; fungi; algae	+
Glycosaminoglycans	Mammals; some fish; Eubacteria	−/+
Proteins	All organisms	−/+
Extensin	Vascular plants; algae	−/+
Mureins	Eubacteria	+
Teichoic acids	Gram-positive Eubacteria	+
Teichuronic acids	Gram-positive Eubacteria	+
Lipoteichoic acids (LTA)	Gram-positive Eubacteria	+
Bacterial lipopolysaccharides (LPS)	Gram-negative Eubacteria	++
DNA, RNA	All organisms	−
Glycolipids	Plants; algae; Eubacteria	+/++
Polyisoprenols (rubber and gutta)	Vascular plants	+
Polyprenols and dolichols	Vascular plants; bacteria; animals	+
Resinous polyterpenoids	Vascular plants	+/++
Cutins, suberins	Vascular plants	+/++
Lignins	Vascular plants	++++
Tannins	Vascular plants; algae	+++/++++
Sporopollenins	Vascular plants	+++
Algaenans	Algae	++++
Cutans	Vascular plants	++++
Suberans	Vascular plants	++++
Cyanobacterial sheaths	Cyanobacteria	+

[a]After Tegelaar *et al.* (1989a).
[b]The "preservation potential" ranges from − (extensive degradation under depositional conditions) to ++++ (no degradation under any depositional conditions).

some extent partly altered resistant biopolymers. Such a concept of kerogen seems to be in agreement with that of coal petrology (Stach *et al.*, 1982). The concept on coal macerals is based on morphologic similarities, sometimes strengthened chemically, between fossil remains of defined parts of higher plants, lower plants, and algae and their actuo-equivalents (Nip *et al.*, 1989; Tegelaar *et al.*, 1993). An indication that kerogens can indeed be considered as a complex mixture of resistant biomacromolecules and their slightly altered counterparts came from density-gradient centrifugation fractionation experiments performed with kerogens (Sentfle *et al.*, 1987). Flash-pyrolysis–gas chromatography of the fractions obtained showed differences in distribution patterns of alkanes, alkenes, and aromatic hydrocarbons. The kinds of differences observed are very similar to those observed in pyrolysates of macerals separated by the same method from a coal (Nip *et al.*, 1991). By assuming the concept of selective preservation of resistant biopolymers, it is possible to associate macerals with their chemical origin. Table VIII shows these relationships.

3.2. Sulfurization of Low-Molecular-Weight Functionalized Lipids

To complete the picture of kerogen structure, two other components of kerogens have to be mentioned. Marine kerogens and especially type II-S kerogens contain high-molecular-weight organic sulfur substances which are—in part—thought to be derived from intermolecular incorporation of inorganic sulfur species with functionalized lipids during early diagenesis (Sinninghe Damsté *et al.*, 1989, 1990; de Leeuw and Sinninghe Damsté, 1990). It should be noted that this process of "natural vulcanization" is restricted to specific depositional environments and results in the formation of structureless organic matter. As a consequence, this kerogen component is only revealed by chemical identification methods. Although of minor importance, the incorporation of low-molecular-weight compounds into kerogen has to be considered also. It has been demonstrated elegantly that relatively low amounts of hopanoid and archaebacterial isoprenoid ether moieties are preserved in kerogens (Michaelis and Albrecht, 1979; Mycke *et al.*, 1987; Chappe *et al.*, 1982). Because these biomolecules in general contain polar groups (De Rosa *et al.*, 1986), it is likely that incorporation of these complex lipids takes place via these polar groups during the very early stages of diagenesis. On the other hand, it cannot be fully excluded at this moment that biomacromolecules containing hopanoid or isoprenoid ether lipid moieties are absent in bacteria.

In summary, the formation of kerogen is thought to be the result of selective preservation and hence enrichment of resistant biomacromolecules, neogenesis via sulfur cross-linking, and, to some extent, incorporation of specific lipids into the macromolecular matrix. This concept of kerogen formation is shown in Fig. 61.

TABLE VIII. Resistant Biomacromolecules as Constituents of Kerogen, Their Corresponding Macerals, and Expected Thermal Degradation Products[a]

Resistant biomacromolecules	Macerals	Expected major catagenic products	Reference(s)
Algaenans	Alginite	Predominantly *n*-alkanes and some aromatics	Largeau *et al.*, 1984, 1986; Derenne *et al.*, 1989; Goth *et al.*, 1988; Burczyk and Dworzanski, 1988; Zelibor *et al.*, 1988
Cutans	Cutinite	Predominantly *n*-alkanes	Nip *et al.*, 1986a, b, 1989; Tegelaar *et al.*, 1989c
Suberans	Suberinite	Predominantly *n*-alkanes and aromatics	Tegelaar *et al.*, 1993b
Lignins	Vitrinite/ fusinite	Predominantly condensed aromatics and CH_4	Stach *et al.*, 1982
Polysesqui- and polyditerpenoids	Resinite	Predominantly bicadinanes, sesquiterpanes, and (condensed) aromatics	Mukhopadhyay and Gormly, 1984; Lewan and Williams, 1987; Van Aarssen *et al.*, 1990
Tannins	Vitrinite/ Fluorinite?	Predominantly condensed aromatics	Stach *et al.* 1982; Wilson and Hatcher, 1988
Sporopollenins	Sporinite	Predominantly *n*-alkanes or (condensed) aromatics	Prahl *et al.*, 1985; Schulze Osthoff and Wiermann, 1987; Guilford *et al.*, 1988

[a]After Tegelaar *et al.* (1989a).

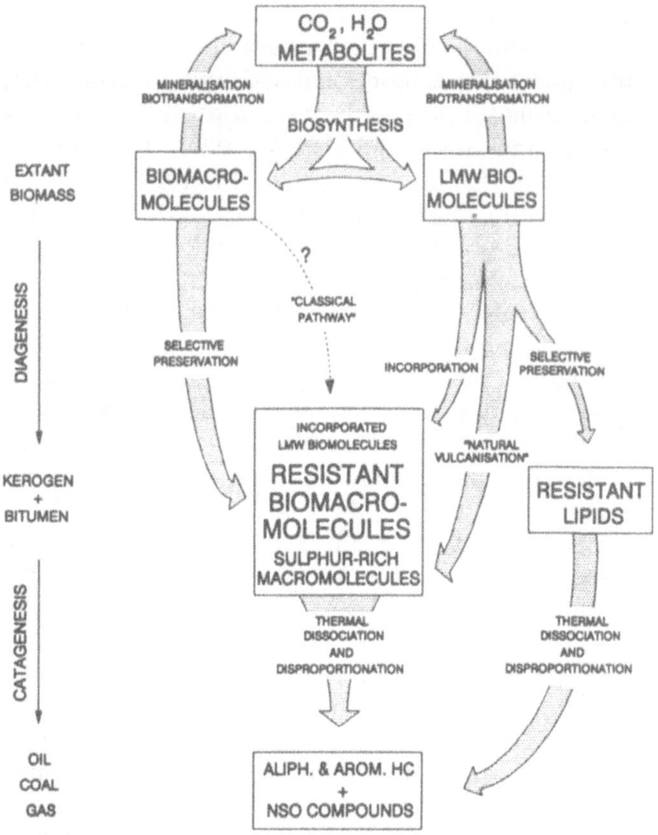

Figure 61. The selective preservation pathway model of kerogen. [After Tegelaar *et al.*, (1989a).]

3.3. The Relationship between Algaenans of *Botryococcus braunii* and Their Fossil Counterparts, Torbanites and Coorongites

To emphasize the importance of selective preservation of resistant biopolymers resulting in major contributions of these materials to kerogens, one example will be presented in more detail. It concerns the chemical relationship between algal cell wall components of *B. braunii* and torbanites and coorongites as representatives of type I kerogens. Torbanites, or boghead coals, are characterized by a very high kerogen level, up to 98%. Furthermore, these organic-rich deposits are homogeneous, and their kerogen is almost entirely composed of a single species of colonial microalgae with a typical morphology (Correia and Connan, 1974; Hutton *et al.*, 1980). It was noted, in early studies (Betrand, 1927; Temperley, 1936), that such a morphology is close to that observed in the extant, colonial, green microalga *Botryococcus braunii*. In addition, recent results about the occurrence of algaenans in *B. braunii*, their abundance, and their location in outer walls (i.e., in the matrix giving the colonies their typical organization) suggested that torbanites are indeed composed of fossil *Botryococcus* and that algaenans played a major role in torbanite formation.

The above assumptions were fully confirmed by a comparative study of immature torbanites and of the algaenans isolated from the different races of *B. braunii*. These modern and fossil materials were examined by a combination of spectroscopic and pyrolytic methods (Largeau *et al.*, 1984, 1986; Derenne *et al.*, 1988a, b). The chemical structures of the algaenans from the *A* and *B* races, on the one hand, and of the immature torbanites, on the other hand, were thus shown to be similar. *Botryococcus* therefore provided the first example of a very close structural relationship between biomacromolecules produced in large amounts by an extant microalga and the "geopolymer" representing an immature kerogen.

On the basis of these observations, it was therefore concluded that:

- Torbanites consist of fossil colonies of *Botryococcus* (*A* or *B* race).
- The algaenans of these algae made a major contribution to torbanite formation and did not undergo significant changes in chemical structure during diagenesis.
- The mechanism of "selective preservation" played a major role in torbanite formation. During fossilization, the algaenans originally present in *B. braunii* (*A* or *B* race) remained unaffected and were entirely retained as kerogen.
- The presence of high levels of algaenans in *B. braunii* and the involvement of a selective preservation mechanism in fossilization account for the occurrence of kerogens corresponding to massive accumulation of fossil *Botryococcus* and for the very high oil potential of such kerogens (the structure of races *A* and *B* algaenans is based on hydrocarbon chains).

Observations in recent sediments from an East Africa lake were also consistent with *B. braunii* fossilization via a selective preservation of algaenan. A large number of *B. braunii* colonies occurred in these sediments, and scanning electron microscopic examinations revealed a similar picture from the water–sediment interface down to the deepest part of the core (ca. 40,000 years old). Thus, whatever the age of the sample considered, the cell contents were entirely eliminated while the outer walls were still present and had fully retained their original morphology, even when their fine structure was observed at a high magnification.

The ability of algaenans to survive diagenesis even under highly oxic conditions was demonstrated by studies on coorongite. This rubbery material is formed on drying of large algal masses stranded on lake shores following profuse blooms of *B. braunii* (Cane, 1969; Douglas *et al.*, 1969; Cane, 1977). The coorongite sample examined was several years old and derived form a bloom of the *B* race. This coorongite was shown (Derenne *et al.*, 1988a; Dubreuil *et al.*, 1989) to be mainly composed (ca. 75%) of selectively preserved *B* race algaenan. In spite of coorongite formation under subaerial conditions, the algaenan only underwent minor alterations in chemical structure. In sharp contrast, the unsaturated isoprenoid hydrocarbons, botryococcenes, originally present in large amounts in the algal biomass were entirely degraded. Botryococcenes probably exhibit a quite low resistance to diagenesis since a complete lack of botryococcane was noted in the bitumen of 12 torbanites, even when deposited under partly oxic conditions (Derenne *et al.*, 1988b). This is also consistent with the occurrence of botryococcane in a very limited number of oils and bitumens while *Botryococcus* remains are widespread in numerous kerogens (Moldowan and Seifert, 1980; Seifert and Moldowan, 1981; Brassell *et al.*, 1986; McKirdy *et al.*, 1986). In addition to selectively preserved algaenan, the coorongite is composed of chemically labile constituents (ca. 25%), the bulk of which is likely of bacterial origin (Derenne *et al.*, 1988a; see also Glikson, 1983, 1984). The extensive bacterial reworking occurring during coorongite formation results not only in the elimination of cell contents but also in the coalescence of *B. braunii* outer walls and colonies. Coorongite appears therefore as a chiefly amorphous material where only concentric lamellae corresponding to outer wall limits are clearly visible (Dubreuil *et al.*, 1989). Accordingly, subsequent coorongite burial should yield kerogens with a chemical structure and an oil potential close to those of torbanites (major contribution of the algaenans from *B* race in coorongite formation) but distinct in morphological features (no outer wall and colony coalescence in torbanites) due to the different conditions during early diagenesis.

Kerogens derived from the *L* race of *B. braunii* have not yet been identified. Such kerogens should also originate from selective preservation of an algaenan. As a result, they would show the same morphology as torbanites (the three races of *B. braunii* exhibit an identical colony organization) but would markedly differ in chemical structure (the algaenans of the *L* race being based on lycopadiene units instead of polymethylenic ones) (Derenne *et al.*, 1989, 1990a).

Recent observations on the pronounced changes in colony morphology and in chemical composition taking place when *B. braunii* is grown in media containing increasing concentrations of NaCl (Derenne *et al.*, 1992) and parallel studies on the morphology and the pyrolysis products of *G. prisca* from Ordovician deposits of various origins (Derenne *et al.*, 1990b; Derenne *et al.*, 1991b, 1992) strongly suggest that fossil *G. prisca* in kukersite and in other Ordovician kerogens merely corresponds to *B. braunii* adapted to brackish or marine environments.

As previously studied, contrary to *B. braunii*, where the algaenan is located in thick outer walls, most of the other algaenan-containing green microalgae, including several species of the cosmopolitan genus *Scenedesmus*, show thin trilaminar resistant outer walls (TLS). As indicated before, these algae can produce large amounts of TLS fragments as a result of mother cell wall release following cell divisions. Transmission electron microscopic observations suggested that such TLS might have contributed, via a selective preservation process, to the formation of various kerogens (Raynaud *et al.*, 1988; Largeau *et al.*, 1990a, b). These observations were made for 40 low-maturity kerogens, Cambrian to Miocene in age, isolated from marine and lacustrine source rocks and oil shales. When previously examined by light microscopy, these kerogens looked amorphous. However, in addition to an actually amorphous fraction, this study revealed different structures in most of the samples examined. These structures are dominated by long "ultralaminae," 10 to 60 nm thick. The relative abundances of the amorphous and "ultralaminar" fractions strongly depend on the origin of the kerogen. Ultralaminae dominate most Mesozoic samples while amorphous materials account for the bulk of Cambrian kerogens. Similar morphological features of thickness and internal organization were noted between the resistant TLS isolated from extant green algae and the ultralaminar fraction of some kerogens (Largeau *et al.*, 1990a). The relationship between such extant and fossil structures was recently confirmed by spectroscopic (FTIR, CP/MAS ^{13}C NMR) and pyrolytic ("off line" pyrolysis) studies of the algaenan building up the TLS of *Scenedesmus quadricauda* (Derenne *et al.*, 1991c). This algaenan was shown to comprise a macromolecular network of ether-linked polymethylenic chains, similar to that of PRB A and B. However the pyrolysis products of *S. quadricauda* algaenan contain a substantial contribution, ca. 10% of the total pyrolysate, of C_{12} to C_{32} *n*-alkylnitriles with a bimodal distribution (maxima at C_{16} and C_{28}) whereas no traces of this type of compounds were detected in the case of *B. braunii* algaenan. Moreover, the same *n*-alkylnitriles with a similar bimodal distribution were observed both in the pyrolysate of *S. quadricauda* algaenan and the shale oils obtained from the Rundle Oil Shale (ROS) and the Green River Shale (GRS). The former is an important Eocene deposit of Australia in which the kerogen is nearly exclusively composed of accumulations of ultralaminae showing the same morphological features as isolated TLS (Largeau *et al.*, 1990a). On the contrary, the GRS only contains a low amount of ultralaminae embedded within an amorphous matrix. Interestingly, the level of *n*-alkylnitriles in shale oil from the GRS is markedly lower than in that from the ROS. According to the above observations, ultralaminae in these two oil shales originate from the selective preservation of the algaenan(s) building up TLS of *Scenedesmus* and/or of related green microalgae with similar outer walls. The formation of *n*-alkylnitriles with the same typical distribution as above was in fact reported from a number of oil shales (Iida *et al.*, 1966; Regtop *et al.*, 1982; Ingram *et al.*, 1983; Evans *et al.*, 1985). This feature, added to the widespread occurrence of ultralaminae in source rocks and oil shales, suggests (i) that the bulk of such ultralaminae likely originate from the selective preservation of thin, algaenan-containing outer walls (TLS) of microalgae and (ii) that the formation of *n*-alkylnitriles on pyrolysis might be a specific chemical signature of such outer walls and of derived fossil ultralaminae.

4. Consequences for Formation of Oil, Coal, and Gas

It is generally accepted that oil and gas are generated from kerogens by thermodissociation reactions. Coal is also a product of temperature-controlled reactions within a macromolecular matrix. The above-mentioned selective preservation pathway model of kerogen formation implies a much better understanding of the genesis of fossil fuels on a molecular level. By knowing the (bio)macromolecular structures of resistant biopolymers and of high-molecular-weight organic sulfur substances, it is much easier to understand the thermodissociation of these relatively well-defined materials. By direct comparisons of distribution patterns of low-molecular-weight compounds obtained by controlled heating of resistant biomacromolecules with naturally occurring distribution patterns of the same compounds in crude oils and gases, the gap between precursors (resistant biopolymers) and products (fossil fuel components) can be closed (Fig. 62).

Simulation experiments at elevated temperatures with a cutan isolated from *Agave americana* generated series of *n*-alkanes and branched alkanes indistinguishable from these series present in high-wax crude oils (Fig. 63) (Tegelaar *et al.*, 1989c). From this study it was anticipated that algaenans, the highly aliphatic biomacromolecules occurring in some extant and fossil microalgae, are major precursors of the *n*-alkanes in nonwaxy crude oils. Heating experiments performed with polycadinene, a macromolecular component isolated from dammar resin, strongly indicated that aliphatic and aromatic sesquiterpenoid hydrocarbons as well as the so-called bi- and tricadinanes, components all present in crude oils from Southeast Asia, are generated from this biomacromolecule during diagenesis and catagenesis (Van Aarssen *et al.*, 1992). Many more experiments of this type have to be performed before the above-mentioned concept can be fully proved. However, based on these successful experiments, major catagenic products derived from selectively preserved biomacromolecules or their slightly altered counterparts present in kerogens can be predicted (see Table VIII). Apart from these simulation experiments, specific chemical degradation reactions tuned to cleave well-defined functional groups in these biopolymers will also contribute considerably to our knowledge of oil and gas genesis on a molecular level. Based on the already existing knowledge of macromolecular organic sulfur substances, a number of specific products from these substances generated during late diagenesis and catagenesis were predicted (de Leeuw

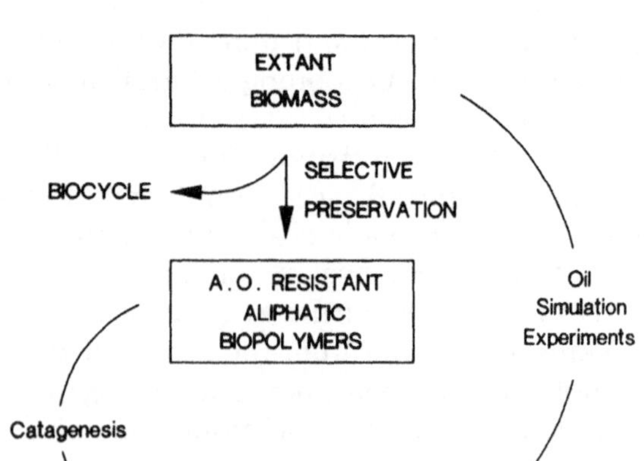

Figure 62. Relationship between extant biomass, kerogen, and fossil fuels.

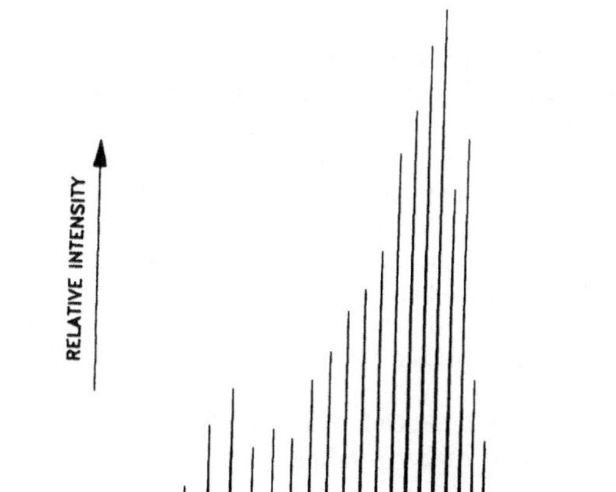

Figure 63. Alkane distribution patterns of a heated extant cutan (top) and an Indonesian crude oil (bottom).

and Sinninghe Damsté, 1990). Preliminary heating experiments with type II-S kerogen confirmed the presence of the predicted products in the product mixtures (M. E. L. Kohnen *et al.*, unpublished results).

It is to be expected that new discoveries will increase our knowledge of the occurrence of resistant biomacromolecules and consequently our knowledge of kerogen formation and hence genesis of fossil fuels.

The inventory of resistant biomacromolecules presently known and their occurrences in the biosphere might already indicate that the contribution of bacterial membrane lipids to fossil fuels is less important than postulated previously (Ourisson *et al.*, 1984).

ACKNOWLEDGMENTS. The authors are very grateful for the skillful and elaborate contributions of W. Pool and M. Baas (both TUD); W. Pool converted the many (hand-drawn) sketches into real figures, and M. Baas word processed and edited the text.

References

Adler, E., 1977, Lignin chemistry. Past, present and future, *Wood Sci. Technol.* **11:**169–218.

Agullo, C., Collar, C., and Seoane, E., 1984, Free and bound hydroxyl and carboxyl groups in the cutin of *Quercus robur* leaves, *Phytochemistry* **23:**2059–2060.

Aken, M. E., and Pienaar, R. N., 1985, Preliminary investigations on the chemical composition of the scale-boundary and cyst wall of *Pyramimonas pseudoparkeae* (Prasinophyceae), *S. Afr. J. Bot.* **51:**408–416.

Albaiges, J., Borbon, J., and Walker, W., II, 1985, Petroleum isoprenoid hydrocarbons derived from catagenetic degradation of archaebacterial lipids, *Org. Geochem.* **8:**293–297.

Archer, B. L., Audley, B. G., Cockbain, E. G., and McSweeney, G. P., 1963, The biosynthesis of rubber, of mevalonate and isopentenyl pyrophosphate into rubber, by *Hevea brasiliensis*-latex fractions, *Biochem. J.* **89:**565–585.

Atkinson, A. W., Gunning, B. E. S., and John, P. C. L., 1972, Sporopollenin in the cell wall of *Chlorella* and other algae: Ultrastructure, chemistry and incorporation of ^{14}C-acetate, studied in synchronous culture, *Planta (Berlin)* **107:**1–32.

Attalla, M. I., Serra, R. G., Vassallo, A. M., and Wilson, M. A., 1988, Structure of ancient buried wood from *Phyllocladus trichomanoides*, *Org. Geochem.* **12:**235–244.

Bada, J. L., 1985, Amino acid racemization dating of fossil bones, *Annu. Rev. Earth Planet. Sci.* **13:**241–268.

Bate-Smith, E. C., 1984, Age and distribution of galloyl esters, iridoids and certain other repellents in plants, *Phytochemistry* **23:**945–950.

Beckett, A., 1976, Ultrastructural studies of exogenously dormant ascospores of *Daldinia concentrica*, *Can. J. Bot.* **54:**689–697.

BeMiller, J. N., 1986, *An Introduction to Pectins: Structure and Properties*, ACS Symposium Series No. 310, American Chemical Society, Washington D.C., pp. 2–12.

Berkaloff, C., Casadevall, E., Largeau, C., Metzger, P., Peracca, S., and Virlet, J., 1983, The resistant polymer of the walls of the hydrocarbon-rich alga *Botryococcus braunii*, *Phytochemistry* **22:**389–397.

Bertrand, P., 1927, Les Botryococcacées actuelles et fossiles et les conséquences de leur activité biologique, *C. R. Mem. Soc. Biol. Paris* **96:**695–696.

Biedlingmaier, S., Wanner, G., and Schmidt, A., 1987, A correlation between detergent tolerance and cell wall structure in green algae, *Z. Naturforsch.*, C **42:**245–250.

Biely, P., 1985, Microbial xylanolytic systems, *Trends Biotechnol.* **3**(11)**:**286–290.

Bolwell, G. P., 1988, Synthesis of cell wall components: Aspects of control, *Phytochemistry* **27:**1235–1253.

Boon, J. J., 1984, Tracing the origin of chemical fossils in microbial mats: Biogeochemical investigations of Solar Lake cyanobacterial mats using analytical pyrolysis methods, in: *Microbial Mats: Stromatolites* (Y. Cohen, R. W. Castenholz, and H. O. Halvorson, eds.), Alan R. Liss, New York, pp. 313–342.

Boon, J. J., and de Leeuw, J. W., 1983, Early stromatolite lithification—organic geochemical aspects, in: *Biomineralization and Biological Metal Accumulation* (P. Westbroek and E. W. de Jong, eds.), D. Reidel, Dordrecht, pp. 327–334.

Boon, J. J., and de Leeuw, J. W., 1987, Organic geochemical aspects of cyanobacterial mats, in: *Cyanobacteria: Current Research* (P. Fay and C. van Baalen, eds.), Elsevier, Amsterdam, pp. 471–492.

Boon, J. J., Rijpstra, W. I. C., de Leeuw, J. W., and Burlingame, A. L., 1980, The occurrence of triglycerides in Namibian Shelf diatomaceous ooze, *Geochim. Cosmochim. Acta* 44:131–134.

Boon, J. J., Hines, H., Burlingame, A. L., Klok, J., Rijpstra, W. I. C., de Leeuw, J. W., Edmunds, K. E., and Eglinton, G., 1983, Organic geochemical studies of Solar Lake laminated cyanobacterial sediments, in: *Advances in Organic Geochemistry 1981* (M. Bjory *et al.*, eds.), Wiley, Chichester, England, pp. 207–221.

Boon, J. J., de Leeuw, J. W., and Krumbein, W. E., 1985, Biogeochemistry of Gavish Sabkha sediments, II—Pyrolysis mass spectrometry of the laminated microbial mat in the permanently water-covered zone before and after the desert sheetflood of 1979, *Ecological Studies 53–Hypersaline Ecosystems* (G. M. Friedman and W. E. Krumbein, eds.), Springer-Verlag, Berlin, pp. 368–380.

Boon, J. J., Brandt-de-Boer, B., Genuit, W., Dallinga, J., and Tukstra, E., 1986, in: *Organic Marine Geochemistry* (M. L. Sohm, ed.), ACS Symposium Series No. 305, American Chemical Society, Washington, D.C., pp. 76–90.

Brackman, W., Spaargaren, K., Van Dongen, J. P. C. M., Couperus, P. A., and Bakker, F., 1984, Origin and structure of the fossil resin from an Indonesian Miocene coal, *Geochim. Cosmochim. Acta* **48**:2485–2487.

Brade, H., Brade, L., and Rietschel, E. Th., 1988, Structure–activity relationships of bacterial lipopolysaccharides (endotoxins). Current and future aspects, *Zentralbl. Bakteriol. Hyg. A* **268:** 151–179.

Brasch, D. J., Chuah, Ch.-T., and Melton, L. D., 1981, A ^{13}C N.M.R. study of some agar-related polysaccharides from New Zealand seaweeds, *Aust. J. Chem.* **34**:1095–1105.

Brassell, S. C., Eglinton, G., and Fu Jia Mo, 1986, Biological marker compounds as indicators of the depositional history of the Maoming Oil Shale, in: *Advances in Organic Geochemistry 1985* (D. Leythaeuser and J. Rullkötter, eds.), Pergamon Press, Oxford, pp.927–941.

Brennan, P. J., 1988, Mycobacteria and other actinomycetes, in: *Microbial Lipids*, Vol. I (C. Ratledge and S. G. Wilkinson, eds.), Academic Press, London, pp. 203–298.

Britton, G., 1983, Carotenoids and polyterpenoids, *Terpenoids Steroids* **12**:235–266.

Brooks, J., 1971, Some chemical and geochemical studies on sporopollenin, in: *Sporopollenin* (J. Brooks, P. Grant, M. D., Muir, G. Shaw, and P. Van Gijzel, eds.), Academic Press, London, pp. 351–407.

Brooks, J., and Shaw, G., 1968a, Chemical structure of the exine of pollen walls and a new function for carotenoids in nature, *Nature* **219**:532–533.

Brooks, J., and Shaw, G., 1968b, The post-tetradontogeny of the pollen wall and the chemical structure of the sporopollenin of *Lilium henryii*, *Grana Palynol.* **8**:227–234.

Brooks, J., and Shaw, G., 1972, Geochemistry of sporopollenin, *Chem. Geol.* **10**:69–87.

Brooks, J., and Shaw, G., 1978, Sporopollenin: A review of its chemistry, palaeobiochemistry and geochemistry, *Grana* **17:** 91–97.

Brown, A. C., Knights, B. A., and Conway, E., 1969, Hydrocarbon content and its relationship to physiological state in the green alga *Botryococcus braunii*, *Phytochemistry* **8**:543–547.

Brunner, U., and Honegger, R., 1985, Chemical and ultra-structural studies on the distribution of sporopollenin-like biopolymers in six genera of lichen phycobionts, *Can. J. Bot.* **63**:2221–2230.

Brzozowska-Hanower, J., Hanower, P., and Lioret, C., 1978, Etude du mécanisme de la coagulation du latex d'*Hevea brasiliensis*, *Physiol. Veg.* **16**:231–236.

Buchanan, R. A., Cull, I. M., Otey, F. H., and Russell, C. R., 1978a, Hydrocarbon and rubber producing crops 1, *Econ. Bot.* **32:** 131–145.

Buchanan, R. A., Cull, I. M., Otey, F. H., and Russell, C. R., 1978b, Hydrocarbon and rubber producing crops 2, *Econ. Bot.* **32:** 146–153.

Bu'lock, J. D., 1972, Comparative and functional aspects of the isoprene pathway in fungi, in: *Chemistry in Evolution and Systematics* (T. Swain, ed.), Butterworths, London, pp. 435–461.

Burczyk, J., 1987, Biogenetic relationships between ketocarotenoids and sporopollenins in green algae, *Phytochemistry* **26**:113–119.

Burczyk, J., and Dworzanski, J., 1988, Comparison of sporopollenin-like algal resistant polymer from cell wall of *Botryococcus*, *Scenedesmus* and *Lycopodium clavatum* by GC-pyrolysis, *Phytochemistry* **27**:2151–2153.

Burczyk, J., and Hesse, M., 1981, The ultrastructure of the outer cell wall-layer of *Chlorella* mutants with and without sporopollenin, *Plant Syst. Evol.* **138**:121–137.

Button, D., and Hemmings, N. L., 1976, Teichoic acids and lipids associated with the membrane of a *Bacillus licheniformis* mutant and the membrane lipids of the parental strain, *J. Bacteriology* **128**(1):149–156.

Cane, R. F., 1969, Coorongite and the genesis of oil shale, *Geochim. Cosmochim. Acta* **33**:257–265.

Cane, R. F., 1977, Coorongite, Balkashite and related substances. An annotated bibliography. *Trans. R. Soc. South Aust.* **101:** 153–164.

Capon, R. J., Dunlop, R. W., Ghisalberti, E. L., and Jefferies, P. R., 1983, Poly-3-hydroxyalkanoates from marine and fresh water cyanobacteria, *Phytochemistry* **22:** 1181–1184.

Cardoso, J. N., and Eglinton, G., 1983, The use of hydroxyacids as geochemical indicators, *Geochim. Cosmochim. Acta* **47:** 723–730.

Cardoso, J. N., Eglinton, G., and Holloway, P. J., 1977, The use of cutin acids in the recognition of higher plant contribution to recent sediments, in: *Advances in Organic Geochemistry 1975* (R. Campos and J. Goni, eds.), ENADIMSA, Madrid, pp. 273–287.

Carman, R. M., Cowley, D. E., and Marty, R. A., 1970, Diterpenoids. XXV. Dundathic acid and polycommunic acid, *Aust. J. Chem.* **23**:1655–1665.

Carroll, K. K., 1987, Studies on the distribution and metabolism of dolichol and related compounds, *Chem. Scr.* **27**(1):73–77.

Casagrande, D. J., and Park, K., 1978, Muramic acid levels in Okefenokee peat: The role of microorganisms in the peat-forming system, *Soil Sci.* **125**:181–183.

Chalansonnet, S., Largeau, C., Casadevall, E., Berkaloff, C., Peniguel, G., and Couderc, R., 1988, Cyanobacterial resistant biopolymers. Geochemical implications of the properties of *Schizothrix* sp. resistant material, in: *Advances in Organic Geochemistry 1987* (L. Mattavelli and L. Novelli, eds.), Pergamon Press, Oxford, pp. 1003–1010.

Chappe, B., Albrecht, P., and Michaelis, W., 1982, Polar lipids of

Archaebacteria in sediments and petroleums, *Science* **217:** 65–66.

Chen, H. Y., 1962, Determination of cis-1,4 and trans-1,4 contents of polyisoprenes by high resolution nuclear magnetic resonance, *Anal. Chem.* **34:**1793–1795.

Chojnacki, T., and Dallner, G., 1988, The biological role of dolichol, *Biochem. J.* **251:**1–9.

Chojnacki, T., Swiezewska, E., and Vogtman, T., 1987, Polyprenols from plants—Structural analogues of mammalian dolichols, *Chem. Scr.* **27**(1):209–214.

Cooper, B. S., and Murchison, D. G., 1971, The petrology and geochemistry of sporinite, in: *Sporopollenin* (J. Brooks, P. Grant, M D. Muir, G. Shaw, and P. Van Gijzel, eds.), Academic Press, London, pp. 545–568.

Correia, M., and Connan, J., 1974, Analyses physico-chimiques et observations microscopiques de la matière organique de schistes bitumineux, in: *Advances in Organic Geochemistry 1973* (B. Tissot and F. Bienner, eds.), Editions Technip, Paris, pp. 153–161.

Cox, H. C., de Leeuw, J. W., Schenck, P. A., Van Koningsveld, H., Jansen, J. C., Van de Graaf, B., Van Geerestein, V. J., Kanters, J. A., Kruk, C., and Jans, A. W. H., 1986, Bicadinane—the first representative of a novel family of C_{30} pentacyclic isoprenoid hydrocarbons in crude oils, *Nature* **319:**316–318.

Crelling, J. C., and Bensley, D. E., 1984, Characterization of coal macerals by fluorescence microscopy, in: *Chemistry and Characterization of Coal Macerals* (R. E. Winans and J. C. Crelling, eds.), ASC Symp. Ser. 252, American Chemical Society, Washington, D.C., pp. 33–46.

Cunningham, A., Gay, I. D., Oehlschlager, A. C., and Langenheim, J. H., 1983, ^{13}C-NMR and IR analyses of structure, aging and botanical origin of Dominican and Mexican ambers, *Phytochemistry* **22:**965–968.

Dawes, E. A., and Senior, P. J., 1973, The role and regulation ot energy reserve polymers in microorganisms, *Adv. Microb. Physiol.* **10:**135–266.

De Leeuw, J. W., Sinninghe Damsté, J. S., Klok, J., Schenck, P. A., and Boon, J. J., 1985, Biogeochemistry of Gavish Sabkha sediments.—I. Studies on neutral reducing sugars and lipid moieties by gas chromatography–mass spectrometry, in: *Ecological Studies. Vol. 53; Hypersaline Ecosystems* (G. M. Friedman and W. E. Krumbein, eds.), Springer-Verlag, Berlin, pp. 350–367.

De Leeuw, J. W., and Sinninghe Damsté, J. S., 1990, Organic sulphur compounds and other biomarkers as indicators of paleosalinity: A critical evaluation, in: *Geochemistry of Sulphur in Fossil Fuels* (W. L. Orr and C. M. White, eds.), ACS Symposium Series, American Chemical Society, Washington, D.C., pp. 417–443.

Derenne, S., Largeau, C., Casadevall, E., and Connan, J., 1988a, Mechanism of formation and chemical structure of Coorongite. II—Structure and origin of the labile fraction. Fate of botryococcenes during early diagenesis, in: *Advances in Organic Geochemistry 1987* (L. Mattavelli and L. Novelli, eds.), Pergamon Press, Oxford, pp. 965–971.

Derenne, S., Largeau, C., Casadevall, E., and Connan, J., 1988b, Comparison of Torbanites of various origins and evolutionary stages. Bacterial contribution to their formation. Cause of the lack of botryococcane in bitumens, *Org. Geochem.* **12:**43–59.

Derenne, S., Largeau, C., Casadevall, E., Tegelaar, E. W., and de Leeuw, J. W., 1988c, Relationships between algal coals and resistant cell wall biopolymers of extant algae as revealed by PY-GC-MS, *Fuel Process. Technol.* **20:**93–101.

Derenne, S., Largeau, C., Casadevall, E., and Berkaloff, C., 1989, Occurrence of a resistant biopolymer in the L race of *Botryococcus braunii*, *Phytochemistry* **28:**1137–1142.

Derenne, S., Largeau, C., Casadevall, E., and Sellier, N., 1990a, Direct relationship between the resistant biopolymer and the tetraterpenic hydrocarbon in the lycopadiene race of *Botryococcus braunii*, *Phytochemistry* **29:**2187–2192.

Derenne S., and Largeau, C., 1991b, Morphological and chemical features of *Gloeocapsomorpha prisca* rich deposits from the Canning Basin, Australia. Comparison with other Ordovician *G. prisca*–derived kerogens, in: *Organic Geochemistry, Advances and Applications in Energy and the Natural Environment* (D. A. C. Manning, ed.), Manchester University, Manchester University Press, pp. 405–408.

Derenne, S., Largeau, C., Casadevall, E., 1990, Occurrence of tightly bound isoprenoid acids in an algal resistant biomacromolecule: Possible geochemical implications, *Org. Geochem.* **17:** 597–602.

Derenne, S., Largeau, C., Casadevall, E., Sinninghe Damsté, J. S., Tegelaar, E. W., and de Leeuw, J. W., 1990b, Characterization of Estonian Kukersite by spectroscopy and pyrolysis: Evidence for abundant alkyl phenolic moieties in an Ordovician, marine, type II/I kerogen, in: *Advances in Organic Geochemistry 1989* (B. Durand and F. Behar, eds.), *Org. Geochem.* **16:** 873–888.

Derenne, S., Largeau, C., Casadevall, E., and Berkaloff, C., 1990c, Chemical evidence of kerogen formation in source rocks and oil shales *via* selective preservation of thin resistant outer walls of microalgae. Origin of ultralaminae, *Geochim. Cosmochim. Acta* **55:**1041–1051.

Derenne, S., Metzger, P., Largeau, C., van Bergen, P. F., Gatellier, J. P., Sinninghe Damsté, J. S., de Leeuw, J. W., and Berkaloff, C., 1992, Similar morphological and chemical variations of *Gloeocapsomorpha prisca* in Ordovician sediments and cultured *Botryococcus braunii* as a response to changes in salinity. *Org. Geochem.* **19:**299–313.

De Rosa, M., Gambacorta, A., and Gliozzi, A., 1986, Structure, biosynthesis, and physicochemical properties of archaebacterial lipids, *Microbiol. Rev.* **50**(1):70–80.

Devries, P. J. R., Simmons, J., and Van Beern, A. P., 1983, Sporopollenin in the spore wall of Spirogyra (Zygnemataceae, Chlorophyceae), *Acta Bot. Neerl.* **32:**25–28.

Dickinson, H. G., 1971, The role played by sporopollenin in the development of pollen in *Pinus banksiana*, in: *Sporopollenin* (J. Brooks, P. Grant, M. D. Muir, G. Shaw, and P. van Gijzel, eds.), Academic Press, London, pp. 31–67.

Dimichele, W. A., Rischbieter, M. A., Eggert, D. L., and Gastaldo, R. A., 1984, Stem and leaf cuticles of *Karinopteris*: Source of cuticles from the Indiana "paper" coal, *Am. J. Bot.* **71:**626–637.

Doemel, W. N., and Brock, T. D., 1977, Structure, growth and decomposition of laminated algal–bacterial mats in alkaline hot springs, *Appl. Environ. Microbiol.* **34:**433–452.

Douglas, A. G., Eglinton, G., and Maxwell, J. R., 1969, The hydrocarbons of Coorongite, *Geochim, Cosmochim. Acta* **33:**569–577.

Douglas, A. G., Sinninghe Damsté, J. S., Eglinton, T. I., de Leeuw, J. W., and Fowler, M. G., 1990, Distribution and structure of hydrocarbons and heterocyclic sulphur compounds released from four kerogens of Ordovician age by means of flash pyrolysis, in: *Early Organic Evolution: Implications for Mineral and Energy Resources* (M. Schidlowski, ed.), Springer, Berlin, pp. 267–278.

Drews, G., and Weckesser, J., 1982, Function, structure and composition of cell walls and external layers, in: *Biology of Cyano-*

bacteria (N. G. Carr and B. A. Whitton, eds.), Blackwell Science Publishers, Oxford, pp. 333–357.

Dubreuil, C., Derenne, S., Largeau, C., Berkaloff, C., and Rousseau, B., 1989, Mechanism of formation and chemical structure of Coorongite—I. Role of the resistant biopolymer and of the hydrocarbons of *Botryococcus braunii*. Ultrastructure of Coorongite and its relationship with Torbanite, *Org. Geochem.* **14:**543–553.

Duch, M. V., and Grant, D. M., 1970, Carbon-13 chemical shift studies of the 1,4-polybutadienes and the 1,4-polyisoprenes, *Macromolecules* **3:**165–174.

Dunn, J. H., and Wolk, C. P., 1970, Composition of the cellular envelopes of *Anabaena cylindrica*, *J. Bacteriol.* **103:**153–158.

Durand, B. (ed.), 1980, *Kerogen—Insoluble Organic Matter from Sedimentary Rocks* Editions Technip, Paris.

Espelie, K. E., Davis, R. W., and Kolattukudy, P. E., 1980, Composition, ultrastructure and function of the cutin- and suberin-containing layers in the leaf, fruit peel, juice-sac and inner seed coat of grapefruit (*Citrus paradisi* Macfed.), *Planta* **149:** 498–511.

Espelie, K. E., Wattendorf, J., and Kolattukudy, P. E., 1982, Composition and ultrastructure of the suberized cell wall of isolated crystal idioblasts from *Agave americana* L. leaves, *Planta* **155:** 166–175.

Evans, E. J., Batts, B. D., Cant, N. W., and Smith, J. W., 1985, The origin of nitriles in shale oils, *Org. Geochem.* **8:**367–374.

Faegri, K., 1971, The preservation of sporopollenin membranes under natural conditions, in: *Sporopollenin* (J. Brooks, P. Grant, M. D. Muir, G. Shaw, and P. van Gijzel, eds.), Academic Press, London, pp. 256–272.

Fischer, W., Laine, R. A., and Nakano, M., 1978, On the relationship between glycerophosphoglycolipids and lipoteichoic acids in gram-positive bacteria. II. Structures of glycerophosphoglycolipids, *Biochim. Biophys. Acta* **528:**298–308.

Foo, L. Y., 1982, Polymeric proanthocyanidins of *Photinia glabrescens*, modification of molecular weight and nature of products from hydrogenolysis, *Phytochemistry* **21:**1741–1746.

Foo, L. Y., and Porter, L. J., 1980, The phytochemistry of proanthocyanidin polymers, *Phytochemistry* **19:**1747–1754.

Fuller, G., and Nes, W. D., 1987, Plant lipids and their interactions, in: *Ecology and Metabolism of Plant Lipids* (G. Fuller and W. D. Nes, eds.), ACS Symposium Series No. 325, American Chemical Society, Washington, D.C., pp. 2–8.

Furch, B., and Gooday, G. W., 1978, Sporopollenin in *Phycomyces blakesleeanus*, *Trans. Br. Mycol. Soc.* **70:**307–309.

Gacesa, P., 1988, Alginates, *Carbohydr. Polym.* **8:**161–182.

Galanos, C., Lüderitz, O., Rietschel, E. T., and Westphal, O., 1977, Newer aspects of the chemistry and biology of bacterial lipopolysaccharides, with special reference to their lipid A component, in: *Biochemistry of Lipids II*, Vol. 14 (T. W. Goodwin, ed.), University Park Press, Baltimore, pp. 288–335.

Gebelein, C. D., 1969, Distribution, morphology and accretion rate of recent subtidal algal stromatolites, Bermuda, *J. Sediment Petrol.* **39:**49–69.

Gebelein, C. D., and Hoffman, P., 1973, Algal origin of dolomite laminations in stromatolitic limestone, *J. Sediment Petrol.* **43:** 603–613.

Geisert, M., Rose, T., Bauer, W., and Zahn, R. K., 1987, Occurrence of carotenoids and sporopollenin in *Nanochlorum eucaryotum*, a novel marine alga with unusual characteristics, *Biosystems* **20:**133–142.

Given, P. H., 1984, An essay on the organic geochemistry of coal, in: *Coal Science*, Vol. 3 (M. L. Gorbaty, J. W. Larsen, and I. Wender, eds.), Academic Press, London, pp. 63–252.

Given, P. H., Spackman, W., Painter, P. C., Rhoads, C. A., Ryan, N. J., Alemany, L., and Pugmuire, R. J., 1984, The fate of cellulose and lignin in peats: An exploratory study of the input to coalification, in: *Advances in Organic Geochemistry 1983* (P. A. Schenck, J. W. de Leeuw, and G. W. M. Lijmbach, eds.), Pergamon Press, Oxford, pp. 399–407.

Glicksman, M. (ed.), 1983a, Gum arabic (gum acacia), in: *Food Hydrocolloids*, Vol. 2, CRC Press, Boca Raton, Florida, pp. 7–29.

Glicksman, M. (ed.), 1983b, Gum tragacanth, in: *Food Hydrocolloids*, Vol. 2, CRC Press, Boca Raton, Florida, pp. 49–60.

Glikson, M., 1983, Microbiological precursors of Coorongite and Torbanite and the role of microbial degradation in the formation of kerogen, *Org. Geochem.* **4:**161–172.

Glikson, M., 1984, Further studies on Torbanites and Coorongite using transmission electron microscopy and C-isotope analysis, *Org. Geochem.* **7:**151–160.

Golub, M. A., Fuqua, S. A., and Bhacca, N. S., 1962, High resolution nuclear magnetic resonance spectra of various polyisoprenes, *J. Am. Chem. Soc.* **84:**4981–4982.

Golubic, S., and Campbell, S. E., 1979, Analogous microbial forms in recent subaerial habitats and in Precambrian cherts, *Precamb. Res.* **8:**201–217.

Golubic, S., and Focke, J. W., 1978, *Phormidium hendersonii Howe*: Identity and significance of a modern stromatolitic building microorganism, *J. Sediment Petrol.* **48:**751–764.

Good, B. H., and Chapman, R. L., 1978, The ultrastructure of *Phycopeltis* (Chroolepidaceae: Chlorophyta), I. Sporopollenin in cell walls, *Am. J. Bot.* **65:**27–33.

Gooday, G. W., Fawcett, P., Green, D., and Shaw, G., 1973, The formation of fungal sporopollenin in the zygospore wall of *Mucor mucedo*: A role for the sexual carotenogenesis in the Mucorales, *J. Gen. Microbiol.* **74:**233–239.

Gooday, G. W., Green, D., Fawcett, P., and Shaw, G., 1974, Sporopollenin formation in the ascospore wall of *Neurospora crassa*, *Arch. Microbiol.* **101:**145–151.

Goodwin, T. W., and Mercer, E. I., 1972, *Introduction to Plant Biochemistry*, Pergamon Press, Oxford.

Goossens, H., Rijpstra, W. I. C., Düren, R. R., de Leeuw, J. W., and Schenck, P. A., 1986, Bacterial contribution to sedimentary organic matter; a comparative study of lipid moieties in bacteria and Recent sediments, in: *Advances in Organic Geochemistry 1985* (D. Leythaeuser and J. Rullkötter, eds.), Pergamon Journals, Oxford; *Org. Geochem.* **10:**683–696.

Goossens, H., de Leeuw, J. W., Rijpstra, W. I. C., Meyburg, G. J., and Schenck, P. A., 1989, Lipids and their mode of occurrence in bacteria and sediments I: A methodological study of the lipid composition of *Acinetobacter calcoacetices* LMD 79–41, *Org. Geochem.* **14:**15–25.

Goth, K., de Leeuw, J. W., Püttmann, W., and Tegelaar, E. W., 1988, Origin of Messel Oil Shale kerogen, *Nature* **336:**759–761.

Grippiolo, R., and Bonfante-Fasolo, P., 1984, Sporopollenin and melanin-like pigments in the wall of a *Glomus* spore, *G. Bot. Ital.* **118:**88–90.

Grosse-Damhues, J., and Glombitza, K. W., 1984, Isofuhalols, a type of phlorotannins from the brown alga *Chorda filum*, *Phytochemistry* **23:**2639–2642.

Guilford, W. J., Schneider, D. M., Labovitz, J., and Opella, S. J., 1988, High resolution solid state ^{13}C NMR investigation of sporopollenin from different plant taxa, *Plant Physiol.* **86:**134–136.

Gunnison, D., and Alexander, M., 1975, Basis for the resistance of several algae to microbial decomposition, *Appl. Microbiol.* **29:** 729–738.

Haider, K., 1988, Der Mikrobielle Abbau des Lignins und seine

Bedeutung für den Kreislauf des Kohlenstoffs, *Naturwissenschaftliche Mikrobiologie*, Forum Mikrobiologie 11/88, pp. 477–481.

Hallam, N. D., 1982, Fine structure of the leaf cuticle and origin of leaf waxes, in: *The Plant Cuticle* (D. F. Cutler, K. Alvin, and C. E. Price, eds.), Academic Press, London, pp. 197–214.

Hancock, I. C., and Baddiley, J., 1985, Biosynthesis of the bacterial envelope polymers teichoic acid and teichuronic acid, in: *The Enzymes of Biological Membranes*, Vol. 2 (A. N. Martonosi, ed.), 2nd ed., Plenum Press, New York, pp. 279–307.

Hartley, R. D., Whatley, F. R., and Harris, P., 1988, 4,4′-Dihydroxytruxillic acid as component of cell walls of *Lolium multiflorum*, *Phytochemistry* **27**:349–351.

Haslam, E., 1977, Symmetry and promiscuity in procyanidin biochemistry, *Phytochemistry* **16**:1625–1640.

Hatcher, P. G., 1987, Chemical structure of natural lignin by dipolar dephasing solid-state ^{13}C nuclear magnetic resonance, *Org. Geochem.* **11**:31–39.

Hatcher, P. G., and Spiker, E. C., 1988, Selective degradation of plant biomolecules, in: *Humic Substances and Their Role in the Environment* (F. H. Frimmel and R. F. Christman, eds.), John Wiley & Sons, New York, pp. 59–74.

Hatcher, P. G., Breger, I. A., Dennis, L. W., and Maciel, G. E., 1982, Chemical structure in coal: NMR studies and a geochemical approach, *Prepr. Pap.—Am. Chem. Soc., Div. Fuel Chem.* **27**: 172–183.

Havinga, A. J., 1971, An experimental investigation into the decay of pollen and spores in various soil types, in: *Sporopollenin* (J. Brooks, P. Grant, M. D. Muir, G. Shaw, and P. van Gijzel, eds.), Academic Press, London, pp. 446–479.

Hayatsu, R., Botto, R. E., McBeth, R. L., Scott, R. G., and Winans, R. E., 1987, Thermal reactions of sporopollenin and sporinite, *Prepr. Pap.—Am. Chem. Soc., Div. Fuel Chem.* **32**:1–8.

Hedges, J. I., Cowie, G. L., Ertel, J. R., Barbour, R. J., and Hatcher, P. G., 1985, Degradation of carbohydrates and lignins in buried woods, *Geochim. Cosmochim. Acta* **49**:701–711.

Hellebust, J. A., 1974, Extracellular products, in: *Algal Physiology and Biochemistry*, Botanical Monographs, Vol. 10 (W. D. P. Stewart, ed.), Blackwell, Oxford, pp. 838–863.

Heslop-Harrison, J., 1971, Sporopollenin in the biological context, in: *Sporopollenin* (J. Brooks, P. Grant, M. D. Muir, G. Shaw, and P. van Gijzel, eds.), Academic Press, London, pp. 1–30.

Hillis, W. E., 1985, Biosynthesis of tannins, in: *Biosynthesis and Biodegradation of Wood Compounds* (T. Higuchi, and T. Ayashodi, eds.), Academic Press, Orlando, pp. 325–347.

Hirono, I., 1987, Carrageenan, *Bioact. Mol.* **2**:121–126.

Holloway, P. J., 1982a, Structure and histochemistry of plant cuticular membranes: An overview, in: *The Plant Cuticle* (D. F. Cutler, K. L. Alvin, and C. E. Price, eds.), Linnean Society Symposium Series No. 10, Academic Press, London, pp. 1–32.

Holloway, P. J., 1982b, The chemical constitution of plant cutins, in: *The Plant Cuticle* (D. F. Cutler, K. L. Alvin, and C. E. Price, eds.), Linnean Society Symposium Series No. 10, Academic Press, London, pp. 45–85.

Holloway, P. J., 1984, Cutins and suberins, the polymeric plant lipids, in: *CRC Handbook of Chromatography, Lipids*, Vol. 1 (H. K. Mangold, ed.), CRC Press, Boca Raton, Florida, pp. 321–346.

Holzwarth, G., 1985, Xanthan and scleroglucan: Structure and use in enhanced oil recovery, *Dev. Ind. Microbiol.* **1985**:271–280.

Honegger, R., and Brunner, U., 1981, Sporopollenin in the cell walls of *Coccomyxa* and *Myrmecia* phycobionts of various lichens: An ultrastructural and chemical investigation, *Can. J. Bot.* **59**: 2713–2734.

Horodyski, R. J., and Vonder Haar, S. P., 1975, Recent calcareous stromatolites from Laguna Mormona, Baja California, Mexico, *J. Sediment Petrol.* **45**:894–906.

Horodyski, R. J., Bloeser, B., and Vonder Haar, S., 1977, Laminated algal mats from a coastal lagoon, Laguna Mormona, Baja California, Mexico, *J. Sediment Petrol.* **47**:680–696.

Hunneman, D. H., and Eglinton, G., 1968, Gas chromatographic–mass spectrometric identification of long chain hydroxy acids in plants and sediments, in: *Advances in Organic Geochemistry 1968* (P. A. Schenck and I. Havenaar, eds.), Pergamon Press, London, pp. 157–165.

Hunneman, D. H., and Eglinton, G., 1972, The constituent acids of gymnosperm cutins, *Phytochemistry* **11**:1989–2001.

Hutton, A. C., Kanstler, A. J., Cook, A. C., and McKirdy, D. M., 1980, Organic matter in oil shales, *J. Aust. Petrol. Explor. Assoc.* **20**: 44–65.

Ibata, K., Mizuno, M., Tanaka, Y., and Kageyu, A., 1984a, Long-chain polyprenols in the family Pinaceae, *Phytochemistry* **23**: 783–786.

Ibata, K., Kageyu, A., Takigawa, T., Okada, M., Nishida, T., Mizuno, M., and Tanaka, Y., 1984b, Polyprenols from conifers: Multiplicity in chain length distribution, *Phytochemistry* **23**:2517–2521.

Iida, T., Yoshii, E., and Kitatsuji, E., 1966, Identification of normal paraffins, olefins, ketones and nitriles from Colorado Shale oil, *Anal. Chem.* **38**:1224–1227.

Imoto, M., Kusumoto, S., Shiba, T., Naoki, H., Iwashita, T., Rietschel, E. Th., Wollenweber, H.-W., Galanos, C., and Lüderitz, O., 1983, Chemical structure of *E. coli* lipid A: Linkage site of acyl groups in the disaccharide backbone, *Tetrahedron Lett.* **24**:4017–4020.

Ingram, L. L., Ellis, J., Crisp, P. T., and Cook, A. C., 1983, Comparative study of oil shales and shale oils from the Mahogany zone, Green River formation (USA) and Kerosene Creek seam, Rundle formation (Australia), *Chem. Geol.* **38**:185–212.

Isherwood, F. A., 1965, Biosynthesis of lignin, in: *Biosynthetic Pathways in Higher Plants* (J. B. Pridham and T. Swain, eds.), Academic Press, London, pp. 133–146.

Ishimaru, K., Nonaka, G. I., and Nishioka, I., 1987, Gallic acid esters of proto-quercitol, quinic acid and (−)shikimic acid from *Quercus mongolica* and *Q. myrsinaefolia*, *Phytochemistry* **26**:1501–1504.

Ittekkot, V., Degens, E. T., and Brockman, U., 1982, Monosaccharide composition of acid-hydrolyzable carbohydrates in particulate matter during a plankton bloom, *Limnol. Oceanogr.* **27**:770–776.

Jarvis, M. C., 1984, Structure and properties of pectin gels in plant cell walls, *Plant Cell Environ.* **7**:152–164.

Johnson, N. G., 1985, Early Silurian palynomorphs from the Tuscarora formation in central Pennsylvania and their paleobotanical and geological significance, *Rev. Palaeobot. Palynol.* **45**:307–360.

Kadouri, A., Derenne, S., Largeau, C., Casadevall, E., and Berkaloff, C., 1988, Resistant biopolymer in the outer walls of *Botryococcus braunii B* race, *Phytochemistry* **27**:551–557.

Kainuma, K., 1988, Structure and chemistry of the starch granule, in: *The Biochemistry of Plants*, Vol. 14 (Preiss, ed.), Academic Press, San Diego, California, pp. 141–180.

Kennedy, J. F., and White, C. A., 1983, Bioactive carbohydrates, in: *Chemistry, Biochemistry and Biology*, 1st ed., Ellis Horwood, Chichester.

Kirk, T. K., and Farrell, R. L., 1987, Enzymatic "combustion": The microbial degradation of lignin, *Annu. Rev. Microbiol.* **41**: 465–505.

Kirk, T. K., Higuchi, T., and Chanc, H. M., 1980a, *Lignin Biodegradation, Microbiology, Chemistry and Potential Applications*, Vol. 1, CRC Press, Boca Raton, Florida.

Kirk, T. K., Higuchi, T., and Chanc, H. M., 1980b, *Lignin Biodegradation, Microbiology, Chemistry and Potential Applications*, Vol. 2, CRC Press, Boca Raton, Florida.

Klocke, J. A., Van Wagenen, B., and Balandrin, M. F., 1986, The ellagitannin geraniin and its hydrolysis products isolated as insect growth inhibitors from semi-arid land plants, *Phytochemistry* **25**:85–91.

Klok, J., Cox, H. C., Baas, M., Schuyl, P. J. W., de Leeuw, J. W., and Schenck, P. A., 1984a, Carbohydrates in recent marine sediments—I. Origin and significance of deoxy- and O-methyl-monosaccharides, *Org. Geochem.* **7**:73–84.

Klok, J., Cox, H. C., Baas, M., de Leeuw, J. W., and Schenck, P. A., 1984b, Carbohydrates in recent marine sediments—II. Occurrence and fate of carbohydrates in a recent stromatolitic deposit: Solar Lake, Sinai, *Org. Geochem.* **7**:101–109.

Klosinski, P., and Penczek, S., 1986, Teichoic acids and their models: Membrane biopolymers with polyphosphate backbones. Synthesis, structure and properties, *Adv. Polym. Sci.* **79**:140–308.

Kluepfel, D., 1988, Screening of prokaryotes for cellulose- and hemicellulose-degrading enzymes, *Methods Enzymol.* **160**: 180–186.

Knoll, A. H., 1985, A palaeobiological perspective on Sabkhas, *Ecological Studies 53—Hypersaline Ecosystems* (G. M. Friedman, and W. E. Krumbein, eds.), Springer-Verlag, Berlin, pp. 407–425.

Koch, M., and Gregson, R. P., 1984, Brominated phloroethols and nonhalogenated phlorotannins from the brown alga *Cystophora congesta*, *Phytochemistry* **23**:2633–2637.

Kolattukudy, P. E., 1980, Biopolyester membranes of plants: Cutin and suberin, *Science* **208**:990–1000.

Kolattukudy, P. E., 1981, Structure, biosynthesis, and biodegradation of cutin and suberin, *Annu. Rev. Plant Physiol.* **32**: 539–567.

König, J., and Peveling, E., 1980, Sporopollenin in the cell wall of the phycobiont *Trebouxia*, *Z. Pflanzenphysiol. Bot.* **98S**: 459–464.

Kovach, W. L., and Dilcher, D. L., 1985, Morphology, ultrastructure and paleoecology of *Paxillitriletes vittatus*, new species from the mid-Cretaceous cenomanian of Kansas USA, *Palynology* **9**: 85–94.

Kubicek, C. P., 1984, Bildung und Eigenschaften von Hemicellulasen aus Microorganismen: Xylanabbauende Enzyme, *Holzforschung Holzverwert.* **36**:1–5.

Kutsuki, H., Shimada, M., and Higuchi, T., 1982, Regulatory role of cinnamyl alcohol dehydrogenase in the formation of guaiacyl and syringyl lignins, *Phytochemistry* **21**:19–23.

Lane, H. C., and Schuster, M. F., 1981, Condensed tannins of cotton leaves, *Phytochemistry* **20**:425–427.

Langenheim, J. H., 1969, Amber: A botanical inquiry, *Science* **163**: 1157–1169.

Lanzetta, R., Monaco, P., Previtera, L., and Simaldone, A., 1988, Polyprenols and hydroxylated lycopersenes from *Myriophyllum verticillatum*, *Phytochemistry* **27**:887–890.

Lanzotti, V., De Rosa, M., Trincone, A., Basso, A. L., Gambacorta, A., and Zillig, W., 1987, Complex lipids from *Desulfurococcus mobilis*, a sulfur-reducing archaebacterium, *Biochim. Biophys. Acta* **922**:95–102.

Lanzotti, V., Nicolaus, B., Trincone, A., De Rosa, M., Grant, W. D., and Gambacorta, A., 1989, A complex lipid with a cyclic phosphate from the archaebacterium *Natronococcus occultus*, *Biochim. Biophys. Acta* **1001**:31–34.

Largeau, C., Casadevall, E., Kadouri, A., and Metzger, P., 1984, Formation of *Botryococcus*-derived kerogens. Comparative study of immature Torbanites and of the extant alga *Botryococcus braunii*, in: *Advances in Organic Geochemistry 1983* (P. A. Schenck, J. W. de Leeuw, and G. W. M. Lijmbach, eds.), Pergamon Press, Oxford, pp. 327–332.

Largeau, C., Derenne, S., Casadevall, E., Kadouri, A., and Sellier, N., 1986, Pyrolysis of immature Torbanite and of the resistant biopolymer (PRB *A*) isolated from extant alga *Botryococcus braunii*. Mechanism of formation and structure of Torbanite, in: *Advances in Organic Geochemistry 1985* (D. Leythaeuser and J. Rullkötter, eds.), Pergamon Press, Oxford, pp. 1023–1032.

Largeau, C., Derenne, S., Casadevall, E., Berkaloff, C., Corolleur, M., Lugardon, B., Raynaud, J. F., and Connan, J., 1990a, Occurrence and origin of "ultralaminar" structures in "amorphous" kerogens of various source rocks and oil shales, *Org. Geochem.* **16**: 889–895.

Largeau, C., Derenne, S., Clairay, C., Casadevall, E., Raynaud, J. F., Lugardon, B., Berkaloff, C., Corolleur, M., and Rousseau, B., 1990b, Characterization of various kerogens by scanning electron microscopy (SEM) and transmission electron microscopy (TEM). Morphological relationships with resistant outer walls in extant microorganisms, in: *Proceeding of the International Symposium on Organic Petrology, Zeist 1990* Vol. 45 (W. J. J. Fermont and J. W. Weegink, eds.), *Geological Survey of the Netherlands* special issue, pp. 91–101.

Laureillard, J., Largeau, C., Waeghemaker, F., and Casadevall, E., 1986, Biosynthesis of the resistant polymer in the alga *Botryococcus braunii*. Studies on the possible direct precursors, *J. Nat. Prod.* **49**:791–799.

Laureillard, J., Largeau, C., and Casadevall, E., 1988, Oleic acid in the biosynthesis of the resistant biopolymers of *Botryococcus braunii*, *Phytochemistry* **27**:2095–2098.

Leeper, H. M., and Schlesinger, W., 1954, Occurrence of both caoutchouc and gutta in additional plants, *Science* **120**: 185–188.

Lehninger, A. L., 1976, *Biochemistry*, Worth Publishers, New York.

Lehninger, A. L., 1982, *Principles of Biochemistry*, 1st ed., Worth Publishers, New York.

Lewan, M. D., and Williams, J. A., 1987, Evaluation of petroleum generation from resins by hydrous pyrolysis, *Am. Assoc. Petrol. Geol. Bull.* **71**:207–214.

Lewin, R. A., 1974, Biochemical taxonomy, in: *Algal Physiology and Biochemistry, Botanical Monographs, Vol. 10* (W. D. P. Stewart, ed.), Blackwell Scientific, Oxford, pp. 2–39.

Lewis, D. H., 1984, Occurrence and distribution of storage carbohydrates in vascular plants, in: *Storage Carbohydrates in Vascular Plants* (D. H. Lewis, ed.), Cambridge University Press, Cambridge, pp. 1–52.

Liebezeit, G., Böhm, L., Dawson, R., and Wefer, G., 1983, Estimation of algal carbonate input to marine aragonitic sediments on the basis of xylose contents, *Mar. Geol.* **54**:249–262.

Lowenstein, J. M., 1981, Immunological reactions from fossil material, *Philos. Trans. R. Soc. London B* **292**:143–149.

Lüderitz, O., Freudenberg, M. A., Galanos, C., Lehmann, V., Rietschel, E. Th., and Shaw, D. H., 1982, Lipopolysaccharides of gram-negative bacteria, in: *Current Topics in Membranes and Transport*, Vol. 17 (S. Razin and S. Rottem, eds.), Academic Press, New York, pp. 79–150.

Mattice, W. L., and Porter, L. J., 1984, Molecular weight averages and ^{13}C NMR intensities provide evidence for branching in proanthocyanidin polymers, *Phytochemistry* **23**:1309–1311.

Maxwell, J. R., Douglas, A. G., Eglinton, G., and McCormick, A., 1968, The botryococcenes—Hydrocarbons of novel structure from the alga *Botryococcus braunii*, Kützing, *Phytochemistry* **7**:2157–2171.

McKirdy, D. M., Cox, R. E., Volkman, J. K., and Howell, V. J., 1986, Botryococcane in a new class of Australian non-marine crude oils, *Nature* **320**:57–59.

Mendoza, Y. A., Gülacar, F. O., and Buchs, A., 1986, Comparison of extraction techniques for bound carboxylic acids in recent sediments Part II. β-Hydroxyacids, *Chem. Geol.* **62**:321–330.

Metzger, P., and Casadevall, E., 1983, Structure de trois nouveaux botryococcènes synthétisés par une souche de *Botryococcus braunii* cultivée en laboratoire, *Tetrahedron Lett.* **24**:4013–4016.

Metzger, P., and Casadevall, E., 1987, Lycopadiene, a tetraterpenoid hydrocarbon from new strains of the green alga *Botryococcus braunii*, *Tetrahedron Lett.* **28**:3931–3934.

Metzger, P., and Casadevall, E., 1989, Aldehydes, very long chain alkenylphenols, epoxides and other lipids from an alkadiene-producing strain of *Botryococcus braunii*, *Phytochemistry* **28**: 2097–2104.

Metzger, P., Templier, J., Largeau, C., and Casadevall, E., 1986, A *n*-alkatriene and some *n*-alkadienes from the *A* race of the green alga *Botryococcus braunii*, *Phytochemistry* **25**:1869–1872.

Michaelis, W., and Albrecht, P., 1979, Molecular fossils of archaebacteria in kerogen, *Naturwissenschaften* **66**:420–422.

Michaelis, W., Mycke, B., and Richnow, H. H., 1986, Organic chemical indicators for reconstructions of Angola Basin sedimentation processes, *Mitt. Geol.—Paläont. Inst. Univ. Hamburg* **60**:90–113.

Mills, J. S., White, R., and Gough, L. J., 1984/85, The chemical composition of Baltic amber, *Chem. Geol.* **47**:15–39.

Moers, M. E. C., 1989, Occurrence and fate of carbohydrates in recent and ancient sediments from different environments of deposition, Thesis, Delft University.

Moers, M. E. C., Boon, J. J., and de Leeuw, J. W., 1989, Carbohydrate speciation and PY-MS mapping of peat samples from a subtropical open marsh environment, *Geochim. Cosmochim. Acta* **53**:2011–2021.

Moldowan, J. M., and Seifert, W. K., 1980, First discovery of botryococcane in petroleum, *J. Chem. Soc., Chem. Commun.* **19**: 912–914.

Monin, J. C., Durand, B., Vandenbroucke, M., and Huc, A. Y., 1980, Experimental simulation of the natural transformation of kerogens, in: *Advances in Organic Geochemistry 1979* (A. G. Douglas and J. R. Maxwell, eds.), Pergamon Press, London, pp. 517–530.

Morrill, L. C., and Loeblich, A. R., 1981, The dinoflagellate pellicular wall layer and its occurrence in the division *Pyrrhophyta*, *J. Phycol.* **17**:315–323.

Mukhopadhyay, P. K., and Gormly, J. R., 1984, Hydrocarbon potential of two types of resinite. In: *Advances in Organic Geochemistry 1983* (P. A. Schenck, J. W. de Leeuw, and G. W. M. Lijmbach, eds.), *Org. Geochem.* **6**:439–454.

Mukhopadhyay, P. K., Gormly, J. R., and Zumberge, J. E., 1989, Generation of hydrocarbons from the Tertiary coals of Texas—Coal as potential source rock for liquid hydrocarbons in a deltaic basin? *Org. Geochem.* **14**:351–352.

Muyzer, G., Westbroek, P., De Vrind, J. P. M., Tanke, J., Vrijheid, T., De Jong, E. W., Bruning, J. W., and Wehmiller, J. F., 1984, Immunology and organic geochemistry, in: *Advances in Organic Geochemistry 1983* (P. A. Schenck, J. W. de Leeuw, and G. W. M. Lijmbach, eds.), Pergamon Press, Oxford, pp. 847–855.

Mycke, B., and Michaelis, W., 1986, Molecular fossils from chemical degradation of macromolecular organic matter, in: *Advances in Organic Geochemistry 1985* (D. Leythaeuser and J. Rullkötter, eds.), Pergamon Press, Oxford, pp. 847–858.

Mycke, B., Narjes, F., and Michaelis, W., 1987, Bacteriohopanetetrol from chemical degradation of an oil shale kerogen, *Nature* **326**:179–181.

Nelson, C. J., and Spollen, W. G., 1987, What's new in plant physiology. Fructans, *Physiol. Plant.* **71**:512–516.

Nes, W. R., 1987a, Structure–function relationships for sterols in *Saccharomyces cerevisiae*, in: *Ecology and Metabolism of Plant Lipids* (G. Fuller and W. D. Nes, eds.), ACS Symposium Series No. 325, American Chemical Society, Washington, D.C., pp. 252–267.

Nes, W. D., 1987b, Biosynthesis and requirement for sterols in the growth and reproduction of Oomycetes, in: *Ecology and Metabolism of Plant Lipids* (G. Fuller and W. D. Nes, eds.), ACS Symposium Series No. 325, American Chemical Society, Washington, D.C., pp. 304–328.

Nielsen, P. E., Nishimura, H., Otvos, J. W., and Calvin, M., 1977, Plant crops as a source of fuel and hydrocarbon-like material, *Science* **198**:942–947.

Nip, M., Tegelaar, E. W., Brinkhuis, H., de Leeuw, J. W., Schenck, P. A., and Holloway, P. J., 1986a, Analysis of modern and fossil plant cuticles by Curie point Py-GC and Curie point Py-GC-MS: Recognition of a new highly aliphatic and resistant biopolymer, in: *Advances in Organic Geochemistry 1985* (D. Leythaeuser and J. Rullkötter, eds.), *Org. Geochem.* **10**:769–778.

Nip, M., Tegelaar, E. W., de Leeuw, J. W., Schenck, P. A., and Holloway, P. J., 1986b, A new non-saponifiable highly aliphatic and resistant biopolymer in plant cuticles. Evidence from pyrolysis and ^{13}C-NMR analysis of present-day and fossil plants, *Naturwissenschaften* **73**:579–585.

Nip, M., de Leeuw, J. W., and Schenck, P. A., 1987a, Structural characterization of coals, coal macerals and their precursors by pyrolysis gas chromatography and pyrolysis gas chromatography–mass spectrometry, *Coal Sci. Technol.* **11**:89–92.

Nip, M., de Leeuw, J. W., Holloway, P. J., Jensen, J. P. T., Sprenkels, J. C. M., de Pooter, M., and Sleeckx, J. J. M., 1987b, Comparison of flash pyrolysis, differential scanning calorimetry, ^{13}C NMR and IR spectroscopy in the analysis of a highly aliphatic biopolymer from plant cuticles, *J. Anal. Appl. Pyrolysis* **11**: 287–295.

Nip, M., de Leeuw, J. W., Schenck, P. A., Windig, W., Meuzelaar, H. L. C., and Crelling, J. C., 1989, A flash pyrolysis and petrographic study of cutinite from the Indiana paper coal, *Geochim. Cosmochim. Acta* **53**:671–683.

Nip, M., de Leeuw, J. W., and Crelling, J. C., 1992, The chemical structure of a bituminous coal and its constituting maceral fractions as revealed by flash pyrolysis, *Energy Fuels* **6**: 125–136.

Nishimura, H., Nonaka, G. I., and Nishioka, I., 1984, Seven quinic acid gallates from *Quercus stenophylla*, *Phytochemistry* **23**: 2621–2623.

Nishimura, H., Nonaka, G. I., and Nishioka, I., 1986, Scyllo-quercitol gallates and hexahydroxydiphenoates from *Quercus stenophylla*, *Phytochemistry* **25**:2599–2604.

Nishizawa, M., Yamagishi, T., Nonaka, G. I., Nishioka, I., and Ragan, M. A., 1985, Gallotannins of the freshwater green alga *Spirogyra* sp., *Phytochemistry* **24**:2411–2413.

Nonaka, G. I., Miwa, N., and Nishioka, I., 1982, Stilbene glycoside gallates and proanthocyanidins from *Polygonum multiflorum*, *Phytochemistry* **21**:429–432.

Nwadinigwe, C. A., 1981, Comparative yield of cis-1,4-polyisoprene from stem latex of four *Landolphia* species, *Phytochemistry* **20**:2301–2302.

Nwadinigwe, C. A., 1988, Some hitherto uncharacterized latex polyisoprenes, *Phytochemistry* **27**:2135–2136.

O'Brian, T. P., and Carr, D. J., 1970, A suberized layer in the cell walls of the bundle sheath of grasses, *Aust. J. Biol. Sci.* **23**: 275–287.

Odier, E., and Rouau, X., 1985, Les cellulases et les enzymes de dépolymérisation de la lignine, in: *Hydrolases et Dépolymérases: Enzymes d'Intérêt Industriel* (A. Mouranches and C. Costes, eds.), Bordas, Paris, pp. 199–238.

Omori, M., Watabe, N., 1980, *The Mechanisms of Biomineralization in Animals and Plants*, Tokai University Press, Tokyo.

Orem, W. H., and Hatcher, P. G., 1987, Early diagenesis of organic matter in a sawgrass peat from the Everglades, Florida, *Int. J. Coal Geol.* **8**:33–54.

Ourisson, G., Albrecht, P., and Rohmer, M., 1984, The microbial origin of fossil fuels, *Sci. Amer.* **251**:34–41.

Paul, F., Morin, A., and Monsan, P., 1986, Microbial polysaccharides with actual potential industrial applications, *Biotech. Adv.* **4**:245–259.

Philp, R. P., and Calvin, M., 1976, Possible origin for insoluble organic (kerogen) debris in sediments from insoluble cell wall material of algae and bacteria, *Nature* **262**:134–136.

Philp, R. P., and Gilbert, T. D., 1986, Biomarker distribution in Australian oils predominantly derived from terrigenous source material, in: *Advances in Organic Geochemistry 1985* (D. Leythaeuser and J. Rullkötter, eds.), Pergamon Press, Oxford, pp. 73–84.

Pollock, C. J., 1986, Fructans and the metabolism of sucrose in vascular plants, *New Phytol.* **104**:1–24.

Poralla, K., and Kannenberg, E., 1987, Hopanoids: Sterol equivalents in bacteria, in: *Ecology and Metabolism of Plant Lipids* (G. Fuller and W. D. Nes, eds.), ACS Symposium Series No. 325, American Chemical Society, Washington, D.C., pp. 239–251.

Porter, L. J., and Woodruffe, J., 1984, Haemanalysis: The relative astringency of proanthocyanidin polymers, *Phytochemistry* **23**:1255–1256.

Porter, L. J., Foo, L. Y., and Furneaux, R. H., 1985, Isolation of three naturally occurring O-β-glucopyranosides of procyanidin polymers, *Phytochemistry* **24**:567–569.

Prahl, A. K., Springstubbe, H., Grumbach, K., and Wiermann, R., 1985, Studies on sporopollenin biosynthesis: The effect of inhibitors of carotenoid biosynthesis on sporopollenin accumulation, *Z. Naturforsch., C* **40**:621–626.

Prahl, A. K., Rittscher, M., and Wiermann, R., 1986, New aspects of sporopollenin biosynthesis, in: *Biotechnology and Biology of Pollen* (D. L. Mulcahy, G. Bergami, and E. Ottaviani, eds.), Springer-Verlag, Berlin, pp. 313–318.

Puel, F., Largeau, C., and Giraud, G., 1987, Occurrence of a resistant biopolymer in the outer walls of the parasitic alga *Prototheca wickerhamii* (Chlorococcales): Ultrastructural and chemical studies, *J. Phycol.* **23**:649–656.

Puls, J., Borchmann, A., Gottschalk, D., and Wiegel, J., 1988, Xylobiose and xylooligomers, *Methods Enzymol.* **160**:528–563.

Ragan, M. A., 1984, Fucus "Lignin": A reassessment, *Phytochemistry* **23**:2029–2032.

Ragan, M. A., and Glombitza, K. W., 1986, Phlorotannins, brown alga polyphenols, in: *Progress in Phycological Research* (F. E. Round and D. J. Chapman, eds.), Biopress Ltd., Bristol, pp. 129–253.

Raynaud, J. F., Lugardon, B., and Lacrampe-Couloume, G., 1988, Observation de membranes fossiles dans la matière organique "amorphe" de roches mères de pétrole, *C. R. Acad. Sci. Paris* **307-II**:1703–1709.

Regtop, R. A., Crisp, P. T., and Ellis, J., 1982, Chemical characterization of shale oil from Rundle, Queensland, *Fuel* **61**:185–192.

Robards, A. W., Jackson, S. M., Clarkson, D. T., and Sanderson, J., 1973, The structure of barley roots in relation to the transport into the stele, *Protoplasma* **77**:291–311.

Rogers, H. J., Perkins, H. R., and Ward, J. B., 1980, Lipopolysaccharides, in: *Microbial Cell Walls and Membranes*, 1st ed., Chapman and Hall, London, pp. 229–235.

Rohmer, M., and Ourisson, G., 1986, Unsaturated bacteriohopanepolyols from *Acetobacter aceti* ssp. *xylinum*, *J. Chem. Res. (S)* **1986**:356–357.

Rohmer, M., Bouvier, P., and Ourisson, G., 1979, Molecular evolution of biomembranes: Structural equivalents and phylogenetic precursors of sterols, *Proc. Natl. Acad. Sci. USA* **76**:847–851.

Roth, W. B., Carr, M. E., Davis, E. A., and Bagby, M. O., 1985, New sources of gutta-percha in *Garrya flavenscens* and *G. wrightii*, *Phytochemistry* **24**:183–184.

Saito, H., Tabeta, R., Yoshioka, Y., Hara, C., Kiho, T., and Ukai, S., 1987, A high-resolution solid state ^{13}C NMR study of the secondary structure of branched (1T)-β-D-glucans from fungi: Evidence of two kinds of conformers, curdlan-type single helix and laminaran-type triple helix forms, as manifested from the conformation-dependent ^{13}C chemical shifts, *Bull. Chem. Soc. Jpn.* **60**:4267–4272.

Saiz-Jimenez, C. and de Leeuw, J. W., 1984, Pyrolysis–gas chromatography–mass spectrometry of isolated, synthetic and degraded lignins, in: *Advances in Organic Geochemistry 1983* (P. A. Schenck, J. W. de Leeuw, and G. W. M. Lijmbach, eds.), Pergamon Press, Oxford, pp. 417–422.

Saiz-Jimenez, C., and de Leeuw, J. W., 1986a, Lignin pyrolysis products: Their structures and their significances as biomarkers, in: *Advances in Organic Geochemistry 1985* (D. Leythaeuser and J. Rullkötter, eds.), Pergamon Press, Oxford, pp. 869–876.

Saiz-Jimenez, C., and de Leeuw, J. W., 1986b, Chemical characterization of soil organic matter fractions by analytical pyrolysis–gas chromatography–mass spectrometry, *J. Anal. Appl. Pyrolysis* **9**:99–119.

Scalbert, E., Monties, B., Lallemand, J. Y., Guittet, E., and Rolando, C., 1985, Ether linkages between phenolic acids and lignin fractions from wheat straw, *Phytochemistry* **24**:1359–1362.

Schaefer, J., 1972, Comparison of the carbon-13 nuclear magnetic resonance of some solid cis- and trans-polyisoprenes, *Macromolecules* **5**:427–432.

Schenck, P. A., de Leeuw, J. W., Van Graas, G. Haverkamp, J., and Bouman, M., 1980, Analysis of recent spores and pollen and of thermally altered sporopollenin by flash pyrolysis–mass spectrometry and flash pyrolysis–gas chromatography–mass spectrometry, in: *Organic Maturation Studies in Fossil Fuel Exploration* (J. Brooks, ed.), Academic Press, London, pp. 225–237.

Schlegel, H. G., 1986, *General Microbiology* 6th ed., Cambridge University Press, Cambridge.

Schopf, J. W., and Walter, M. R., 1982, Origin and early evolution of cyanobacteria: The geological evidence, in: *The Biology of Cyanobacteria* (N. G. Carr, ed.), Blackwell, Oxford, pp. 543–564.

Schrader, M., Drews, G., Weckesser, J., and Meyer, H., 1981, Isolation and characterization of the sheath of the cyanobacterium *Chlorogloeopsis* PCC 6912, *J. Gen. Microbiol.* **128:**267–272.

Schulze Osthoff, S., and Wiermann, R., 1987, Phenols as integrated compounds of sporopollenin from *Pinus* pollen, *J. Plant Physiol.* **131:**5–15.

Seifert, W. K., and Moldowan, J. M., 1981, Paleoreconstruction by biological markers, *Geochim. Cosmochim. Acta* **45:**783–794.

Senftle, J. T., Yordy, K. L., Barron, L. S., and Crelling, J. C., 1987, Analysis of mixed kerogens from Upper Devonian New Albany Shale I. Evaluation of kerogen components derived from density separation, Presented at 1987 Eastern Oil Shale Symposium Lexington, Kentucky, November 18–20, 1987, Technical Paper, pp. 155–167.

Sengupta, S., and Rowley, J. R., 1974, Re-exposure of tapes at high temperatures and pressure in *Lycopodium clavatum* spore exine, *Grana* **14:**143–151.

Seydel, U., Lindner, B., Wollenweber, H.-W., and Rietschel, E. T., 1984, Structural studies on the lipid A component of enterobacterial lipopolysaccharides by laser desorption mass spectrometry, *Eur. J. Biochem.* **145:**505–509.

Shaw, G., 1970, Sporopollenin, in: *Phytochemical Phylogeny* (J. Harborne, ed.), Academic Press, London, pp. 31–58.

Shaw, G., 1971, The chemistry of sporopollenin, in: *Sporopollenin* (J. Brooks, P. Grant, M. D. Muir, G. Shaw, and P. van Gijzel, eds.), Academic Press, London, pp. 305–350.

Shen, Z., Haslam, E., Falshaw, C. P., and Begley, M. J., 1986, Procyanidins and polyphenols of *Larix gmelini* bark, *Phytochemistry* **25:**2629–2635.

Shigemoto, T., Ohtani, Y., Okagawa, A., and Summoto, M., 1987, NMR study of beta-resene, *Cell. Chem. Technol.* **21:**249–254.

Sinninghe Damsté, J. S., Rijpstra, W. I. C., de Leeuw, J. W., and Schenck, P. A., 1988, Origin of organic sulphur compounds and sulphur-containing high molecular weight substances in sediments and immature crude oils, in: *Advances in Organic Geochemistry 1987* (L. Mattavelli and L. Novelli, eds.), Org. *Geochim.* **13:**593–606.

Sinninghe Damsté, J. S.,Eglinton, T. I., de Leeuw, J. W., and Schenck, P. A., 1989, Organic sulphur in macromolecular sedimentary organic matter I. Structure and origin of sulphur-containing moieties in kerogen, asphaltene and coal as revealed by flash pyrolysis, *Geochim. Cosmochim. Acta* **53:** 873–889.

Sinninghe Damsté, J. S., Eglinton, T. I., Rijpstra, W.I. C., and de Leeuw, J. W., 1990, Characterisation of organically-bound sulphur in high-molecular-weight sedimentary organic matter using flash-pyrolysis and Raney Ni desulphurisation, in: *Geochemistry of Sulphur in Fossil Fuels* (W. L. Orr and C. M. White, eds.), ACS Symposium Series 429, American Chemical Society, Washington, D.C., in press, pp. 486–528.

Smith, A. J., 1982, Modes of cyanobacterial carbon metabolism, in: *The Biology of Cyanobacteria* (N. G. Carr, ed.), Blackwell Scientific, Oxford, pp. 47–85.

Stach, E., Mackowsky, M. T., Teichmüller, M., Taylor, G. H., Chandra, D., and Teichmüller, R., 1982, *Stach's Textbook of Coal Petrology* Gebruder Borntraeger, Berlin.

Stafford, H. A., 1988, Proanthocyanidins and the lignin connection, *Phytochemistry* **27:**1–6.

Steinberg, S. M., Venkaetsan, M. I., and Kaplan, I. R., 1987, Organic geochemistry of sediments from the continental margin off Southern New England, U.S.A.—Part I. Amino acids, carbohydrates and lignin, *Mar. Chem.* **21:**249–265.

Stiller, M., Ehrlich, A., Pollingher, U., Baruch, U., and Kaufman, A., 1983, The late Holocene sediments of Lake Kinneret (Israel). Multidisciplinary study of a five meter core, *Geological Survey of Israel Current Research*, pp. 83–88.

Strohl, W. R., Larkin, J. M., Good, B. H., and Chapman, R. L., 1977, Isolation of sporopollenin from four myxobacteria, *Can. J. Microbiol.* **23:**1080–1083.

Suga, T., Shishibori, T., and Nakaya, K., 1980, Structure and biosynthesis of malloprenols from *Mallotus japonicus*, *Phytochemistry* **19:**2327–2330.

Sun, D., Wong, H., and Foo, L. Y., 1987, Proanthocyanidin dimers and polymers from *Quercus dentata*, *Phytochemistry* **26:** 1825–1829.

Sun, D., Zhao, Z., Wong, H., and Foo, L. Y., 1988, Tannins and other phenolics from *Myrica esculenta* bark, *Phytochemistry* **27:** 579–583.

Szklarz, G., and Leonowicz, A., 1986, Cooperation between fungal laccase and glucose oxidase in the degradation of lignin derivatives, *Phytochemistry* **25:**2537–2539.

Tanaka, T., Nonaka, G. I., and Nishioka, I., 1985, Punicafolin and ellagitannin from the leaves of *Punica granatum*, *Phytochemistry* **24:**2075–2078.

Tanner, G. R., and Morrison, I. M., 1983, Phenolic–carbohydrate complexes in the cell walls of *Lolium perenne*, *Phytochemistry* **22:**1433–1439.

Tegelaar, E. W., Derenne, S., Largeau, C., and de Leeuw, J. W., 1989a, A reappraisal of kerogen formation, *Geochim. Cosmochim. Acta* **3:**3103–3107.

Tegelaar, E. W., de Leeuw, J. W., Largeau, C., Derenne, S., Schulten, H. R., Müller, R., Boon, J. J., Nip, M., and Sprenkels, J. C. M., 1989b, Scope and limitations of several pyrolysis methods in the structural elucidation of a macromolecular plant constituent in the leaf cuticle of *Agave americana L.*, *J. Anal. Appl. Pyrolysis* **15:**29–54.

Tegelaar, E. W., Matthezing, R. M., Jansen, B. H., Horsfield, B., and de Leeuw, J. W., 1989c, Origin of *n*-alkanes in high-wax crude oils, *Nature* **342:**529–531.

Tegelaar, E. W., de Leeuw, J. W., and Holloway, P. J., 1989d, Some mechanisms of flash pyrolysis of naturally occurring higher plant polyesters, *J. Anal. Appl. Pyrolysis* **15:**289–295.

Tegelaar, E. W., de Leeuw, J. W., and Saiz-Jimenez, C., 1989e, Possible origin of aliphatic moieties in humic substances, *Sci. Tot. Environ.* **81/82:**1–17.

Tegelaar, E. W., 1990, Resistant biomacromolecules in morphologically characterized constituents in kerogen: the key for the relationship between biopolymers and fossil fuels, Ph.D. Thesis, Delft, 191 pp..

Tegelaar, E. W., Wattendorf, J., and de Leeuw, J. W., 1993a, Possible effects of chemical heterogeneity in higher plant cuticles on the preservation on its ultrastructure upon fossilization, *Rev. Palaeobot. Palynol.*, in press.

Tegelaar, E.W., Kerp, J. H. F., Visscher, H., Schenck, P. A., and de Leeuw, J. W., 1991, Bias of the palaeobotanical records as a consequence of variations in the biochemical composition of vascular plant cuticles, *Paleobiology* **17:**133–144.

Tegelaar, E. W., Hollman, G., Vegt, P., van de Leeuw, J. W., and Holloway, P. J., 1993b, Chemical characterization of the outer bark tissue of some angiosperm species: Recognition of an insoluble, non-hydrolyzable highly aliphatic biopolymer (suberan), *Org. Geochem.*, submitted.

Temperley, B. N., 1936, The boghead controversy and the morphology of the boghead algae, *Trans. R. Soc. Edinburgh* **58:** 855–868.

Ten Haven, H. L., Baas, M., de Leeuw, J. W., and Schenck, P. A.,

1987, Late Quaternary Mediterranean sapropels I. On the origin of organic matter in sapropel S_7, *Mar. Geol.* **75**:137–156.

Thomas, B. R., 1969, Kauri resins—Modern and fossil, in: *Organic Geochemistry Methods and Results* (G. Eglinton and M. T. J. Murphy, eds.), Springer-Verlag, Berlin, pp. 599–618.

Tierney, M. L., and Varner, J. E., 1987, The extensins, *Plant Physiol.* **84**:1–2.

Tissot, B. P., and Welte, D. H., 1984, *Petroleum Formation and Occurrence*, 2nd ed., Springer-Verlag, Heidelberg.

Toia, R. E., Marsh, B. H., Perkins, S. K., McDonald, J. W., and Peters, G. A., 1985, Sporopollenin content of the spore apparatus of *Azolla*, *Am. Fern J.* **75**:38–43.

Towers, G. H. N., 1969, Metabolism of cinnamic acid and its derivatives in Basidiomycetes, in: *Perspectives in Phytochemistry* (J. B. Harborne and T. Swain, eds.), Academic Press, London, pp. 179–191.

Tschirch, A., and Glimmann, G., 1896, Untersuchungen über die Sekrete. 21. Über das Dammarharz, *Arch. Pharm.* **234**:585–589.

Tuffery, A. A., 1969, Light and electron microscopy of the sheath of a blue-green alga, *J. Gen. Microbiol.* **57**:41–50.

Turner, W. B., 1971, *Fungal Metabolites*, Academic Press, London.

Uzabakiliho, B., Largeau, C., and Casadevall, E., 1987, Latex constituents of *Euphorbia candelabrum*, *E. grantii*, *E. tirucalli*, and *Synadenium grantii*, *Phytochemistry* **26**:3041–3045.

Van Aarssen, B. G. K., Cox, H. C., Hoogendoorn, P., and de Leeuw, J. W., 1990, A cadinene biopolymer in fossil and extant dammar resins as a source for cadinanes and bicadinanes in crude oils from South East Asia, *Geochim. Cosmochim. Acta* **54**:3021–3031.

Van Aarssen, B. G. K., Hessels, J. K. C., Abbink, O. A., and de Leeuw, J. W., 1990, The occurrence of polycyclic sesqui-, tri- and oligoterpanes derived from a resinous polymeric cadinene in crude oils from South East Asia, *Geochim. Cosmochim. Acta* **56**:1231–1246.

Van der Velden, W., and Schwartz, A. L., 1974, Purines and pyrimidines in sediments from Lake Erie, *Science* **185**:691–693.

Vennat, B., Pourrat, A., Texier, O., Pourrat, H., and Gaillard, J., 1987, Proanthocyanidins from the roots of *Fragaria vesca*, *Phytochemistry* **26**:261–263.

Villey, M., Estrade-Szwarckopf, H., and Conard, J., 1981, Simulation thermique de l'évolution diagénétique des kérogènes: Étude par résonance paramagnétique électronique, *Rev. Inst. Fr. Pet.* **36**:3–15.

Volkman, J. K., and Wakeham, S. G., 1989, Capillary GC and GC-MS analysis of wax esters and triacylglycerols in marine sediments, in: *Fats for the Future* (R. C. Cambie, ed.), Ellis Horwood, Chichester, England, pp. 193–197.

Whitmore, F. W., 1982, Lignin–protein complex in cell walls of *Pinus elliottii*: Amino acid constituents, *Phytochemistry* **21**: 315–318.

Whitton, B. A., and Potts, M., 1982, Marine littoral, in: *The Biology of Cyanobacteria* (N. G. Carr, ed.), Blackwell Scientific, Oxford, pp. 515–542.

Wilkinson, S. G., 1977, Composition and structure of bacterial lipopolysaccharides, in: *Surface Carbohydrates of the Prokaryotic Cell* (J. Sutherland, ed.), Academic Press, London, pp. 97–175.

William, F., Boominathan, K., Vasudevan, N., Gurujeyalakshmi, G., and Mahadevan, A., 1986, Microbial degradation of lignin and tannins, *J. Sci. Ind. Res.* **45**:232–243.

Wilson, M. A., and Hatcher, P. G., 1988, Detection of tannins in modern and fossil barks and in plant residues by high-resolution solid-state ^{13}C nuclear magnetic resonance, *Org. Geochem.* **12**:539–546.

Wilson, M. A., Collin, P. J., Vassallo, A. M., and Russell, N. J., 1984, The nature of olefins and carboxyl groups in an Australian brown coal resin, *Org. Geochem.* **7**:161–168.

Wilson, M. A., Verheyen, T. V., Vassallo, M. A., Hill, R. S., and Perry, G. J., 1987, Selective loss of carbohydrates from plant remains during coalification, *Org. Geochem.* **11**:265–271.

Woese, C. R., Magrum, L. J., and Fox, G. E., 1978, Archaebacteria, *J. Mol. Evol.* **11**:245–252.

Wollenweber, H.-W., Broady, K. W., Lüderitz, O., and Rietschel, E. Th., 1982, The chemical structure of lipid A. Demonstration of amide-linked 3-acyloxyacyl residues in *Salmonella minnesota* Re lipopolysaccharide, *Eur. J. Biochem.* **124**:191–198.

Yoshida, T., Maruyama, Y., Memon, M. U., Shingu, T., and Okuda, T., 1985, Gemins D, E and F, ellagitannins from *Geum japonicum*, *Phytochemistry* **24**:1041–1046.

Zelibor, J. L., Romankiw, L., Hatcher, P. G., and Colwell, R. R., 1988, Comparative analysis of the chemical composition of mixed and pure cultures of green algae and their decomposed residues by ^{13}C nuclear magnetic resonance spectroscopy, *Appl. Environ. Microbiol.* **54**:1051–1060.

Zetsche, F., and Vicari, H., 1931, Untersuchugen über die membran der sporen und pollen II—*Lycopodium clavatum*, *Helv. Chim. Acta* **14**:58–62.

Zobel, H. F., 1988, Molecules to granules: A comprehensive starch review, *Starch/Stärke* **2**:44–50.

Chapter 3

Isotope Fractionation during Primary Production

MARILYN L. FOGEL and LUIS A. CIFUENTES

1. Introduction

The biological source of sedimentary organic matter can be inferred from detailed chemical studies on the structure of individual molecules extracted from sediments. These inferences are drawn from established relationships between biological source materials and diagenetically altered compounds (de Leeuw and Largeau, this volume, Chapter 2). A second method for inferring the biological source and the ecological setting in which organisms existed is with isotopic tracers of carbon, nitrogen, oxygen, hydrogen, and sulfur in organic matter and inorganic substances that have been processed by living organisms. Stable isotope compositions of bulk organic matter have integrative signals from processes that have occurred over the life of the organism, whereas those of individual compounds can record specific events in life. In this chapter, we will review recent advances in biological isotope fractionations of the light stable isotopes and attempt to offer an interpretation of processes that affected the isotopic signature of organic substances in living organisms that were preserved in the fossil record.

Since the first edition of *Organic Geochemistry* was published in 1969, the field of isotope biogeochemistry has been influenced by a number of discoveries in the biochemistry of carbon fixation during photosynthesis, which can be used to reinterpret isotopic ratios of both modern and fossil material. The first is the discovery of a widespread membrane phenomena in aquatic organisms in which plants and photosynthetic bacteria actively transport either HCO_3^- or CO_2 (see Lucas and Berry, 1985). As a significant proportion of the world's photosynthesis occurs in the ocean, dissolved inorganic carbon-concentrating mechanisms have an effect on the carbon isotope ratio of organic matter. The second discovery occurred in the biological sciences during the 1960s, yet it was not until the 1970s that C_4 photosynthesis was conclusively shown to produce a distinct carbon isotopic composition in terrestrial higher plants. Recent models of isotopic fractionation have incorporated some of these biochemical advances and have resulted in a greater understanding of the processes that actually control the isotope ratio of photosynthetically produced organic matter. The models rely on isotope fractionations that have been determined with key enzymes in photosynthesis. Of principal importance is the carbon isotope fractionation by ribulose 1,5-bisphosphate carboxylase, which is responsible for all carbon dioxide fixation in green plants, algae, and most autotrophic bacteria.

MARILYN L. FOGEL • Geophysical Laboratory, Carnegie Institution of Washington, Washington, D.C. 20015. LUIS A. CIFUENTES • Department of Oceanography, Texas A&M University, College Station, Texas 77843.

Organic Geochemistry, edited by Michael H. Engel and Stephen A. Macko. Plenum Press, New York, 1993.

With regard to the isotopes of nitrogen, oxygen, sulfur, and hydrogen, much less has been published, perhaps because of the difficulty of the analytical procedures used for making isotope measurements. In recent years, studies on nitrogen isotope fractionation have increased dramatically as simplified techniques for combustion of organic matter have become routine [see review by Owens (1987)]. In the case of nitrogen, recent technical advances have included the determination of nitrogen isotope ratios of ambient dissolved inorganic nitrogen, such that isotope mass balances can be understood. In the case of hydrogen, photosynthetic growth rather than heterotrophy imparts distinct isotope fractionations on organic matter, and different photosynthetic modes (C_3 vs. C_4) result in differing hydrogen isotope ratios. Oxygen isotope studies are currently limited to the analysis of polysaccharides. Oxygen isotopes in these compounds in plants are determined by isotopic exchange reactions with the cellular water [see review by Sternberg (1988a)].

The focus of this chapter will be on isotopic fractionation of carbon and nitrogen and the implications for modern and ancient sediments. A review of recent models for carbon isotope fractionation will be followed by a proposed biological model for the fractionation of nitrogen isotopes in aquatic photosynthetic organisms. Finally, we will discuss briefly the fractionation of oxygen, hydrogen, and sulfur isotopes during photosynthesis.

2. Definition of Isotope Terms

The isotope ratio of a sample or standard is defined as

$$R = \frac{X_h}{X_l} \tag{1}$$

where X_h and X_l refer to the heavier (e.g., ^2H, ^{13}C, ^{15}N, ^{18}O, ^{34}S) and lighter (e.g., ^1H, ^{12}C, ^{14}N, ^{16}O, ^{32}S) isotope, respectively. In most isotope-ratio determinations, the absolute abundance of a stable isotope is not measured precisely. Minute differences between a sample and standard (Table I), however, can be determined precisely with isotope-ratio mass spectrometers (e.g., Nier, 1947). The delta (δ) notation is most commonly used in reporting isotopic compositions of biological and geological materials and relates the isotopic ratio of a sample to that of a standard:

$$\delta X\ (‰) = \left(\frac{R_{sample} - R_{standard}}{R_{standard}}\right) \times 10^3 \tag{2}$$

TABLE I. Atom Percent Abundance (A) and Internationally Accepted Standards for the Stable Isotopes of Hydrogen, Carbon, Nitrogen, and Oxygen

Isotope	*A*	Reference standard
^{13}C	1.1100%	Pee Dee belemnite fossil, carbonate (PDB)
^{15}N	0.3663%	N_2, atmospheric nitrogen
^{18}O	0.2040%	Standard mean ocean water (SMOW)
^2H	0.0156%	Standard mean ocean water (SMOW)
^{34}S	4.2200%	Canyon Diablo troilite (CDT)

Many of the enzymatic reactions in photosynthetic and biosynthetic pathways have isotope effects associated with them, and, during the course of the reaction, there is a differential partitioning of isotopes or an isotopic fractionation between the substrate and product. Various conventions are used to describe isotope effects (e.g., O'Leary, 1981; Mariotti *et al.*, 1981). The kinetic fractionation factor, without making assumptions about the order of the reaction, is as follows:

$$\alpha = \frac{dX_{h,p}/X_{h,s}}{dX_{l,p}/X_{l,s}} \tag{3}$$

where X_h is the heavy isotope, X_l is the light isotope, *s* is the substrate, and *p* is the product. In this chapter, the isotope calculations will be calculated from the product relative to substrate reference (e.g., Mariotti *et al.*, 1981). Thus, negative isotope fractionations indicate a depletion of the heavier isotope in the product, and positive isotope fractionations indicate an enrichment of the heavier isotope. Care must be taken in the reading of these fractionation factors because other authors (e.g., O'Leary, 1981) have used the inverse of Eq. (3), where the reference is that of substrate relative to product. In the case that either the substrate concentration is sufficiently large so that the reservoir is not depleted significantly by the reaction (e.g., O'Leary, 1981) or the isotope ratio of the product is measured within an infinitely short period of time (e.g., Mariotti *et al.*, 1981), the fractionation factor can be defined as

$$\alpha = R_p/R_s \tag{4}$$

where R_p and R_s are the isotope ratios of the product and substrate, respectively. For a unidirectional reaction, the change in the isotopic ratio of the substrate is related to the fraction of unreacted substrate according to the "Rayleigh" equation (e.g., Mariotti *et al.*, 1981):

$$R_s/R_{s0} = f^{(\alpha-1)} \tag{5}$$

in which R_{s0} refers to the isotope ratio of the substrate at time zero, and f is the fraction of the substrate that is unreacted.

As the fractionation factor rarely differs from unity by more than 5%, it can be restated in terms of the enrichment factor, ϵ (e.g., Mariotti *et al.*, 1981), as

$$\epsilon = (\alpha - 1) \times 10^3 \quad (6)$$

In the case of a large substrate reservoir, the enrichment factor is related to measurements in δ notation by (e.g., Mariotti *et al.*, 1981)

$$\epsilon = \left(\frac{\delta_p - \delta_s}{\delta_s + 1000}\right) \times 10^3 \quad (7)$$

This equation is similar to that for the discrimination, D (O'Leary, 1981):

$$D = \frac{\delta_p - \delta_s}{1 + \delta_s/1000} \quad (8)$$

If the value of $\delta_s/1000$ differs from unity by less than 5%, as in most isotopic measurements, then Eq. (8) is similar to the commonly used

$$\Delta = \delta_p - \delta_s \quad (9)$$

where Δ is also the isotopic discrimination.

In some cases, equilibrium isotope effects are observed. Equilibrium isotope effects are simply related to kinetic isotope effects by

$$\alpha_{eq} = k_2/k_1 \quad (10)$$

where k_2 is the kinetic fractionation for the forward reaction, and k_1 is the kinetic fractionation for the reverse reaction (e.g., O'Leary, 1981). When equilibrium isotope effects are observed, Eq. (10) is equivalent to Eq. (4). Isotopic equilibrium occurs independently of chemical equilibrium and may often be reached slowly, particularly in the case of CO_2–HCO_3^- equilibrium. Equilibrium fractionation factors (Table II) should only be used when conditions are truly at equilibrium and not steady state.

TABLE II. Typical Equilibrium Fractionation Factors, σ_{eq}[a]

Reaction	Isotope	α_{eq}	Reference
$CO_2(g) \leftrightarrow CO_2(aq)$	^{13}C	0.9991	Mook *et al.*, 1974
$CO_2(g) \leftrightarrow CO_2(aq)$	^{18}O	0.9989	Vogel *et al.*, 1970
$CO_2 + H_2O \leftrightarrow HCO_3^- + H^+$	^{13}C	0.9921	Mook *et al.*, 1974
$O_2\ (g) \leftrightarrow O_2\ (aq)$	^{18}O	1.000	Kroopnick and Craig, 1972
$H_2O(s) \leftrightarrow H_2O(l)$	^{18}O	1.003	O'Neil, 1968
$H_2O(s) \leftrightarrow H_2O(l)$	D	1.019	O'Neil, 1968
$NH_4^+ \leftrightarrow NH_3 + H^+$	^{15}N	1.019–1.021	Hermes *et al.*, 1985

[a]With the exception of those for solid-to-liquid phase transitions, the equilibrium fractionation factors were measured at 20–25°C.

3. Carbon Isotopes

3.1. Fractionation during Photosynthesis

The enzyme ribulose 1,5-bisphosphate carboxylase (RuBP carboxylase) is responsible for the primary fractionation of carbon isotopes in photosynthetic plants [Fig. 1; see review by O'Leary (1988)]. Pioneering experiments of isotopic fractionation during photosynthesis by Park and Epstein in 1963 were repeated in the 1970s and 1980s with high-purity enzymes assayed under physiological conditions (see O'Leary, 1981, for review). The determination of an isotope fractionation requires the separation of the product of the reaction, phosphoglyceric acid (PGA), from the residual substrate of the reaction, CO_2, followed by measurement of their isotopic ratios. Early attempts to determine the fractionation accomplished this goal by purifying PGA and calculating the isotope composition of the fixed CO_2 with the following isotopic mass-balance equation:

$$\tfrac{5}{6}(\delta^{13}C\,RuBP) + \tfrac{1}{6}(\delta^{13}C \text{ fixed } CO_2) = \delta^{13}C \text{ PGA} \quad (11)$$

in which the $\delta^{13}C$ of the substrate RuBP and the product PGA were determined by direct analysis. The isotope fractionation was then calculated with the assumptions that CO_2 was an infinite pool and that the isotope ratio of the CO_2 in the reaction solution was unchanged.

Problems with this technique occurred in the isolation of pure PGA and in the accurate determination of the $\delta^{13}C$ of pure RuBP, which is very unstable (Roeske and O'Leary, 1984). Roeske and O'Leary (1984) measured the fractionation using the $\delta^{13}C$ of

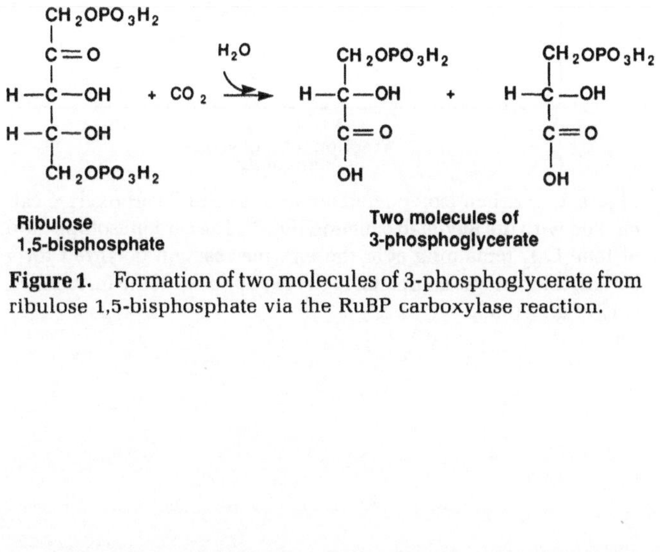

Figure 1. Formation of two molecules of 3-phosphoglycerate from ribulose 1,5-bisphosphate via the RuBP carboxylase reaction.

the carboxyl group of PGA, which is derived directly from CO_2 fixation. Guy *et al.* (1987) used an opposite approach of collecting sequential samples of residual CO_2 and calculating an isotope fractionation based on the "Rayleigh" fractionation (Eq. 5; Fig. 2). Both methods yielded fractionation values of −29.4‰ for enzyme extracted from higher terrestrial plants. Bacterial carboxylases, which have different molecular structures and thus different reaction mechanisms, produced smaller fractionations of −20‰ (O'Leary, 1988).

Once isotope fractionations began to be measured in various laboratories, it became clear that the $\delta^{13}C$ of photosynthetic organic carbon from both field samples and culture experiments could not be explained solely by enzymatic fixation of RuBP carboxylase. In most cases, the isotope fractionation between photosynthetic carbon ($\delta^{13}C$ = −27‰) and atmospheric or dissolved CO_2 ($\delta^{13}C$ = −7‰; e.g., Stuiver, 1978) is on average 20‰, a value 10‰ more positive than that for the enzymatic fractionation alone. With the fractionation factor for RuBP carboxylase established, models describing isotope fractionation were then developed (Farquhar, 1980; Farquhar *et al.*, 1982; Guy *et al.*, 1986). The model assumes that the amount of carbon isotope fractionation actually expressed in the tissues of plants is dependent on the ratio of the concentration of CO_2 inside of the plant to that in the external environment. The following simplified equation describes the model:

$$\Delta = a + (c_i/c_a)(b - a) \tag{12}$$

where Δ is the isotopic fractionation, a is the isotope effect occurring during diffusion of CO_2 into the plant (−4.4‰; O'Leary, 1988), b is the combined isotope effect during photosynthetic fixation by RuBP and phosphoenolpyruvate (PEP) carboxylases (−27‰), and c_i/c_a is the ratio of $[CO_2]$ internal to the plant to $[CO_2]$ in the atmosphere. With an unlimited amount of CO_2 available to the plant (i.e., c_i/c_a = 1), the enzymatic fractionation will determine the $\delta^{13}C$ of photosynthetic carbon, which could be as negative as −36‰. Conversely, when the concentration of CO_2 is limiting (i.e., $c_i/c_a \ll 1$), and the diffusion of CO_2 into the plant is the slow step of the reaction, $\delta^{13}C$ values will be determined primarily by the smaller isotope effect in diffusion. Therefore, when diffusion of CO_2 into the plant is a limiting process, the isotope fractionation of the plant decreases. Typically, the $\delta^{13}C$ of most plants (−20 to −30‰) shows that both diffusion and carboxylation are limiting processes (Fig. 3).

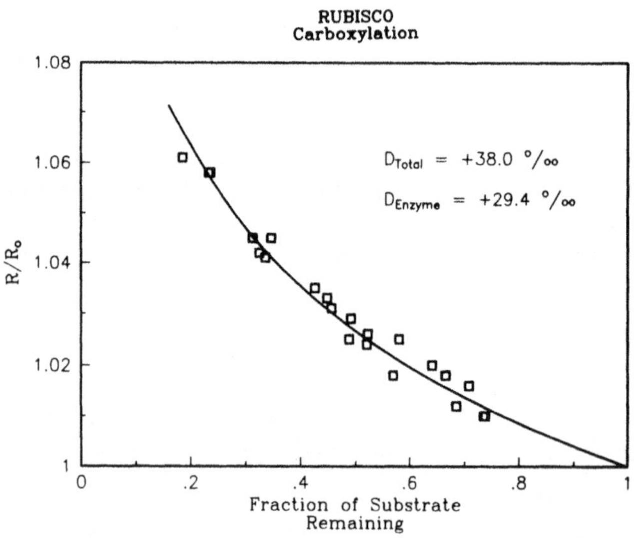

Figure 2. Carbon isotope fractionation by RuBP carboxylase calculated with the Rayleigh equation (Eq. 5). The carbon isotope ratio of total CO_2 remaining after the enzyme reaction occurred for a certain interval was compared to the isotope ratio of the initial CO_2. Guy *et al.*, (1987).

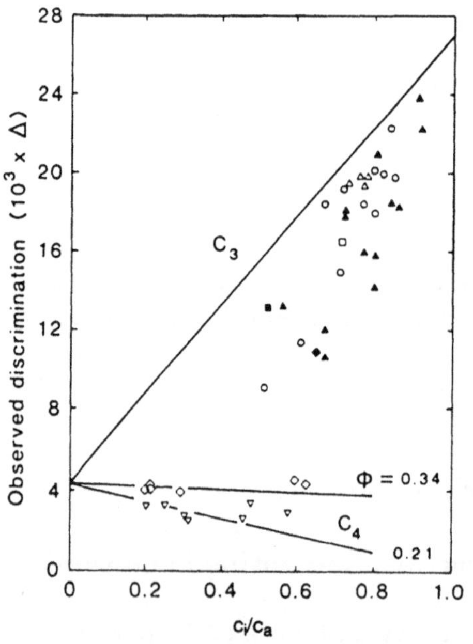

Figure 3. Relationship between observed fractionation of CO_2 and measured c_i/c_a (the ratio of the concentration of CO_2 internal to that in the atmosphere). The simplified theoretical lines for C_3 and C_4 plants are shown. Note that the isotope fractionation in C_4 plants shows no relationship to c_i/c_a (Farquhar, 1980). ○, △, ▲, □ refer to individual experiments.

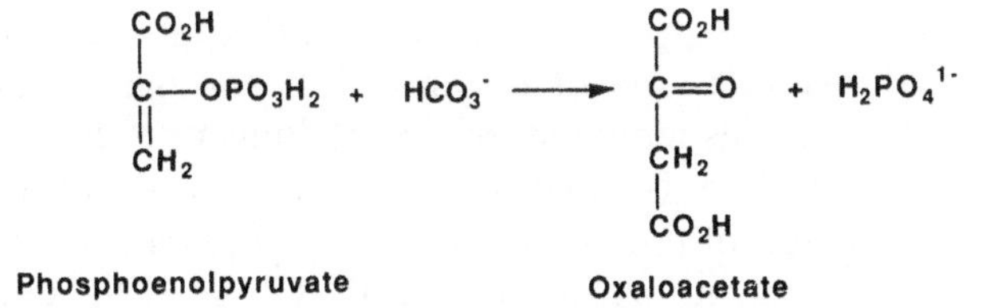

Figure 4. Formation of oxaloacetate from phosphoenolpyruvate by the PEP carboxylase reaction.

Plants that utilize the C_4 pathway for photosynthesis have an alternate enzyme, phosphoenolpyruvate carboxylase (PEP carboxylase; Fig. 4), for the first committed step of carbon dioxide fixation. The isotope effect associated with this enzyme (−2.2‰; O'Leary, 1988) is much smaller than that for RuBP carboxylase, such that C_4 plants have more positive isotope compositions in the range of −8 to −18‰ (Smith and Epstein, 1971; O'Leary, 1981; Deleens *et al.*, 1983). Essentially, the CO_2 fixed by PEP carboxylase is carried as part of a C_4 acid into internal bundle sheath cells of vascular plants (Fig. 5). The C_4 compound, either aspartate or malate, is then decarboxylated in close proximity to RuBP carboxylase. As the environment in the bundle sheath cells is almost a closed system, most of the CO_2 entering the cell is fixed back into organic matter. Thus, little isotope fractionation by the RuBP carboxylase is expressed in the whole plant tissue. Farquhar (1983) has described the fractionation (Δ) in C_4 plants in the following simplified equation:

$$\Delta = a + (b_4 + b_3\phi - a)\cdot c_i/c_a \qquad (13)$$

where a is the isotope effect during diffusion of CO_2, b_4 is the isotope effect in the diffusion of CO_2 in the bundle sheath cells, b_3 is the isotope fractionation during carboxylation (−2 to −4‰), and ϕ is the leakiness of the plant to CO_2. O'Leary (1988) described simplified versions of both models in a recent review.

Isotope fractionation in aquatic plants is more complex. Because CO_2 diffuses more slowly in water than air, diffusion to the plant or cell is often the limiting step (O'Leary, 1981; Raven *et al.*, 1987). Most aquatic plants have some membrane-bound mechanism that actively transports dissolved inorganic carbon (DIC) into the photosynthesizing cells (Lucas, 1983; Lucas and Berry, 1985). Studies have shown that when DIC (mostly CO_2 and HCO_3^-) concentrations in aquatic environments are low, the plants will "turn on" an HCO_3^- or CO_2 "pump," which will accumulate a higher concentration of DIC in the cell for photosynthesis (Fig. 6). The exact mechanism of the transport differs from organism to organism, but the final effects are the same: in spite of CO_2-limiting environments, aquatic plants are able to carry on high rates of photosynthesis.

In an early study on the effects of the DIC-concentrating mechanism on the isotope composition of algae, Sharkey and Berry (1985) showed that when cells are growing in high concentrations of CO_2 (5%), the carbon isotope fractionations between cells and DIC are large and similar to those seen in higher terrestrial plants. In low concentrations of CO_2 (0.03%), however, the concentrating mechanism is activated. Active transport of DIC into the cell results in a large internal pool of substrate available for photosynthesis. Much of this accumulated CO_2 does not leave the cell before it is fixed by RuBP carboxylase. Carbon isotope fractionations associated with algae can thus be quite small (~5‰), even though the primary enzyme for CO_2 fixation is RuBP carboxylase. The model describing the fractionation is a modification of that developed for C_4 photosynthesis:

$$\Delta = d + b_3\cdot(F_3/F_1) \qquad (14)$$

where d is the equilibrium isotope effect between CO_2 and HCO_3^-, b_3 is the isotope fractionation associ-

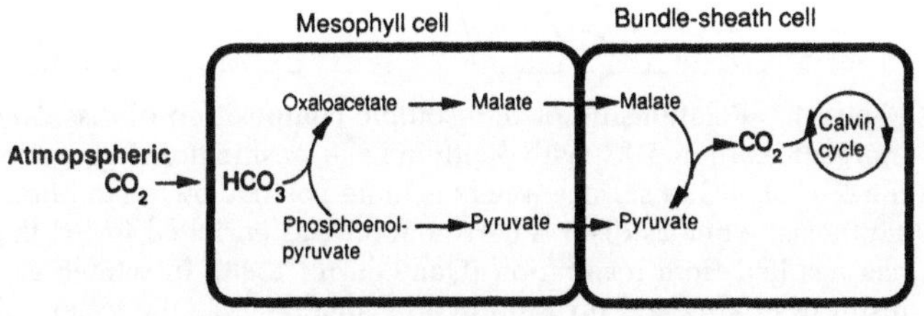

Figure 5. Simplified diagram of C_4 photosynthesis.

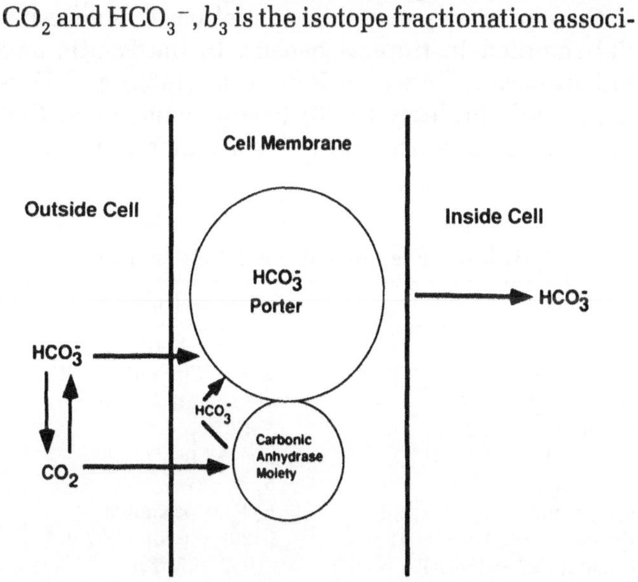

Figure 6. Model for dissolved inorganic carbon transport.

ated with carboxylation, and F_3/F_1 is the ratio of CO_2 leaking out of the cell to the amount inside the cell (Fig. 6). Sample calculations with data typically available are summarized in Table III. The model for C_3 plants is most useful in determining the gas exchange state of the plant cells. The ratio c_i/c_a can also be related to the water use efficiency of the plant. The model for aquatic plants could be used to determine the efficiency of an energy-dependent DIC-transport system based on the value obtained for F_3/F_1.

3.2. Interpretation of Carbon Isotope Compositions of Organic Matter in the Environment

Carbon isotope compositions of organic matter in aquatic systems or in sediments are commonly used for tracing the source of the organic matter to the sedimentary basin (Peters *et al.*, 1978; Fry and Sherr, 1984; Macko and Engel, this volume, Chapter 9). Suspended particulate carbon, including detritus, phytoplankton, microheterotrophs, and zooplankton, has also been analyzed in many areas of the world's oceans. The $\delta^{13}C$ of marine particulate material falls in the range of −20 to −28‰. A number of interpretations have been offered for these isotopic compositions. For example, Sackett *et al.* (1964) and Fontugne and Duplessy (1978, 1981) have linked variable isotope ratios of marine organic matter to direct or indirect effects of temperature on the isotope fractionation. In the Atlantic Ocean, a linear relationship exists between water temperature and the $\delta^{13}C$ of organic carbon. At higher latitudes and colder temperatures, the $\delta^{13}C$ are more negative than that of particulate material sampled in tropical oceans. In the Pacific and Indian oceans, however, Rau *et al.* (1982) and Fontugne and Duplessy (1978) have demonstrated that the $\delta^{13}C$ is associated with a particular water mass rather than a temperature. We conclude that the traditional explanations of oceanic $\delta^{13}C$ for particulate carbon, such as temperature, are no longer applicable.

What, then, is causing the change in the isotopic composition of the organic carbon in seawater? Mixing of terrestrial organic matter ($\delta^{13}C$ = −27‰) with that of *in situ* derived carbon can influence the isotope ratio of suspended material in plumes from major river basins such as the Amazon and the Mississippi (Hedges and Parker, 1976; Cai *et al.*, 1988). The biomass away from continental margins, however, is derived primarily from *in situ* production and therefore has $\delta^{13}C$ values that are independent of terrestrial sources.

A second possible modulator of the isotope ratio of particulate organic carbon is the isotopic composition of DIC available for the growth of the organisms. In excellent papers by Quay *et al.* (1986) and Herczeg and Fairbanks (1987), isotopic and chemical mass balances were calculated for carbon species in small lakes. These studies demonstrated that the $\delta^{13}C$ of DIC in surface waters was determined by a balance between photosynthetic uptake and respiratory production (Fig. 7). During summer and spring, when rates of photosynthesis were high, the isotope ratio of DIC was enriched in ^{13}C, whereas in fall, when respiration was the dominant process, the $\delta^{13}C$ of DIC became more negative as the organic carbon was remineralized. Dynamic fluctuations in the $\delta^{13}C$ of DIC were also seen in lakes (Oana and Deevey, 1960; Rau, 1978)

TABLE III. Examples of Model Calculations

C_3 plants	Aquatic plants
$\Delta = a + c_i/c_o \cdot (b - a)$	$\Delta = d + b_3 \cdot (F_3/F_1)$
$\Delta \approx \delta^{13}C(\text{plants}) - \delta^{13}C(CO_2)$	$\Delta \approx \delta^{13}C(\text{plant}) - \delta^{13}C(CO_2)$
$\delta^{13}C(\text{plant}) = -26$	$\delta^{13}C(\text{plant}) = -20$
$\delta^{13}C(CO_2) = -7.5$	$\delta^{13}C(CO_2) = -8$
$a = -4.4$	$d = -7.9$
$b = -27$	$b_3 = -27$
c_i/c_o = unknown	F_3/F_1 = unknown
$(-26) - (-7.5) = (-4.4) +$	$(-20) - (-8) = +7.9 +$
$c_i/c_o \cdot [-27 - (-4.4)]$	$(-27) \cdot (F_3/F_1)$
$c_i/c_o = 0.62$	$F_3/F_1 = 0.74$

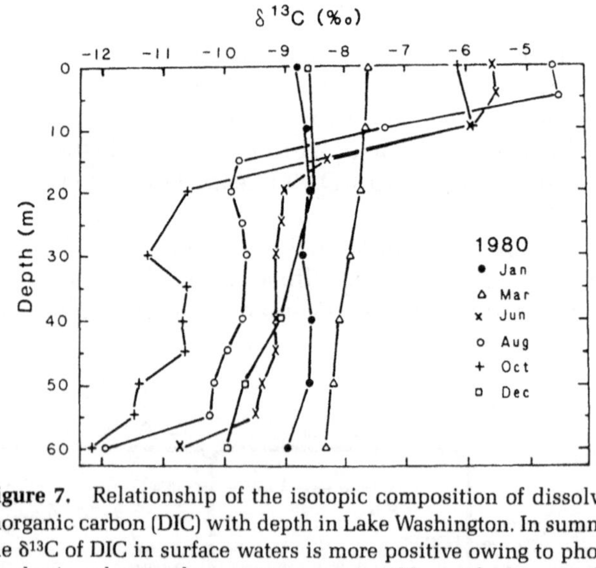

Figure 7. Relationship of the isotopic composition of dissolved inorganic carbon (DIC) with depth in Lake Washington. In summer the $\delta^{13}C$ of DIC in surface waters is more positive owing to photosynthesis, whereas deep waters contain DIC enriched in ^{12}C that has resulted from respiration (Quay *et al.*, 1986). In winter, as a result of mixing and invasion of atmospheric CO_2, the water column has a constant $\delta^{13}C$ of DIC.

and estuaries (Spiker and Schemel, 1979; Sherr, 1982; Fig. 8). In seawater the $\delta^{13}C$ of surface waters is enriched in ^{13}C relative to that of bottom waters by about 2‰ (Deuser and Hunt, 1969; Kroopnick *et al.*, 1970). This enrichment in ^{13}C has been attributed to the fractionation between ^{13}C and ^{12}C during photosynthesis by phytoplankton. The $\delta^{13}C$ of DIC in surface waters, however, does not vary sufficiently to explain the larger variability in $\delta^{13}C$ measured in particulate material.

In a long-term study of the biogeochemistry of carbon in the Delaware Estuary, the isotope ratios of particulate carbon cannot be related to simple mixing of two distinct organic matter sources, to temperature "isotope effects," or to the isotopic composition of DIC (Cifuentes *et al.*, 1988; Fogel *et al.*, 1988). In terms of source, only during winter when *in situ* production rates were very low was there evidence of mixing from terrestrial and marine organic matter sources (Fig. 9). As for temperature, in spring phytoplankton blooms first occurred when water temperatures were in the range of 5–10°C. The most positive isotope ratios of particulate material were measured at this time. In summer, when water temperatures were close to 25°C, some of the most negative isotopic ratios were recorded (Fig. 9). Last, although fluctuations in the $\delta^{13}C$ of DIC could be related to some of the changes in the isotope ratios of the particulate material, the magnitude and the direction of the shift in $\delta^{13}C$ of particulate and dissolved inorganic carbon phases were not always matched.

As discussed earlier, isotope fractionation during photosynthesis is influenced by the concentration of

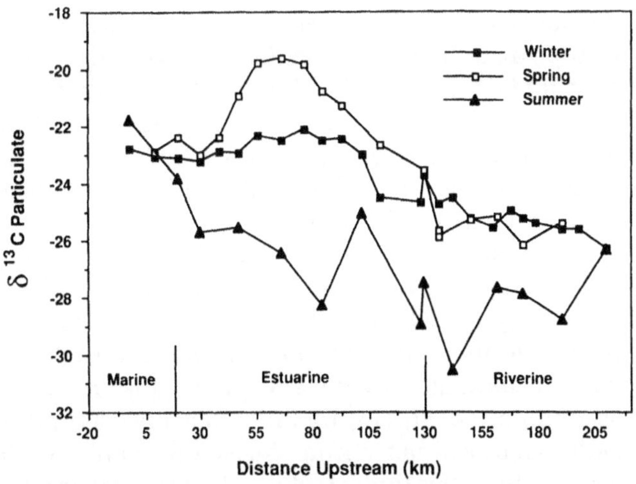

Figure 9. The variability in the $^{13}\delta C$ of particulate carbon (seston) in three different cruises along the Delaware Estuary in 1987. [Adapted from Fogel *et al.* (1988).]

CO_2 in the cell that is available for fixation (Eqs. 12–14; Rau *et al.*, 1989). In aquatic plants, diffusion of CO_2 limits photosynthesis (Raven *et al.*, 1987). As the rate of photosynthesis increases and the demand for CO_2 increases, algae become increasingly more stressed in terms of CO_2 availability. In order for many algae to sustain high rates of primary productivity, the organisms must actively transport DIC into the cell. Because the pH of seawater is basic (8.2), the majority of DIC (≈2mM) is in the form of HCO_3^- rather than CO_2 (≈12μM), which is the form of DIC actually used by RuBP carboxylase. At low concentrations of DIC, phytoplankton will begin to actively transport bicarbonate (Burns and Beardall, 1987). Internal pools of DIC will thus have enriched $\delta^{13}C$ values, as the isotope ratio of HCO_3^- is ~8‰ more positive than that of dissolved CO_2. In turn, the carbon isotope fractionation will decrease (Deuser, 1970; LaZerte and Szalados, 1982; Raven *et al.*, 1987). In support of this contention, Cifuentes *et al.* (1988) demonstrated an inverse relationship between the rate of primary production and the $\delta^{13}C$ of particulate carbon in an estuary.

Culture experiments with phytoplankton species reproduce the full range of $\delta^{13}C$ values that have been measured in particulate matter from oceanic waters (Deuser *et al.*, 1968; Pardue *et al.*, 1976; Estep *et al.*, 1978; Vogel, 1980; Descolas-Gros and Fontugne, 1985). Estep *et al.* (1978) proposed originally that diatoms fixed a portion of the internal CO_2 by PEP carboxylase in the C_4 pathway. Beardall *et al.* (1982) reinterpreted the data and suggested that active transport of DIC occurred in response to increased CO_2

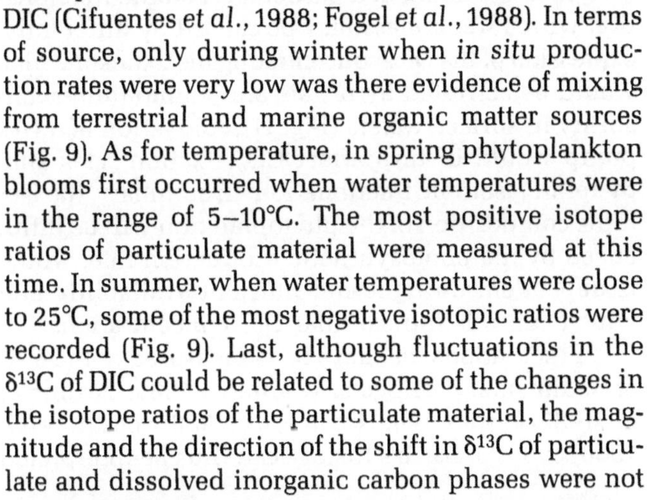

Figure 8. $\delta^{13}C$ of DIC as a function of season and location in the Delaware Estuary. Note that the variability of DIC cannot account fully for the shifts in $\delta^{13}C$ of particulate carbon in Fig. 9.

demand as cell cultures became more dense. In turn, isotope fractionation would decrease, with the $\delta^{13}C$ of phytoplankton becoming intermediate to that of C_3 and C_4 plants. The range of $\delta^{13}C$ values for culture experiments was most likely controlled by combinations of isotope effects from RuBP carboxylase, diffusion, and active transport.

The model for active transport (i.e., Eq. 14) predicts isotope fractionation should become smaller as the ability of CO_2 to leak out of the cell decreases relative to internal, elevated DIC concentrations that have accumulated by the DIC pump (Fig. 10). If the assumptions of Eq. (14) are valid, then the leakiness of cells can be estimated from the isotopic values of the algae. Thus, for typical phytoplankton growing in high-latitude, colder waters with a $\delta^{13}C$ of -28, the DIC-concentrating mechanism is probably not active, as all of the HCO_3^- pumped into the cell would theoretically leak out before it had been fixed. In lower, warmer latitudes, however, where phytoplankton may have a $\delta^{13}C$ of -20, the ratio of CO_2 leaking out of the cell relative to internal CO_2 would be 0.7 (see Table III). This means that approximately 70% of the DIC pumped into the cells would diffuse out before it is fixed. Why then should algae invest in the cellular energy to concentrate inorganic C in the cell?

Nitrogen limitation has been shown to activate HCO_3^- concentrating mechanisms in phytoplankton (Beardall *et al.*, 1982). As nitrogen limitation increases, the total amounts of carboxylating enzymes decrease. Therefore, increased CO_2 concentrations for the limited RuBP carboxylase activation presumably outweigh the energy needed to pump in HCO_3^- (see Burns and Beardall, 1987). Temperature isotope effects that have been used to explain a generally linear relationship between latitude and $\delta^{13}C$ values might alternatively be explained by the activation of a DIC-concentrating mechanism in certain regions. Polar waters in both the Atlantic and the Pacific Ocean generally have elevated concentrations of dissolved nitrate ($\approx 20\ \mu M$) in surface waters, whereas in lower latitudes, and especially in warmer months, nitrogen in the photic zone is barely detectable. We suggest that the association of a particular isotopic fractionation with a water mass may be caused by differential expression of the DIC-concentrating mechanism which would be activated by nitrate and ammonium availability in surface waters (Fig. 11). Activation of a DIC pump would in turn affect CO_2 fixation rates and, therefore, isotopic fractionation. In summary, the isotopic composition of the phytoplankton will be influenced by the photosynthetic production rate, which is dependent on light and nutrient availability and which controls intracellular DIC concentrations.

Fluctuations that have been measured in the $\delta^{13}C$ of sedimentary kerogens (Galimov, 1980; Arthur *et al.*, 1985; Lewan, 1986; Dean *et al.*, 1986; Schidlowski, 1988) over the Earth's history can thus be interpreted

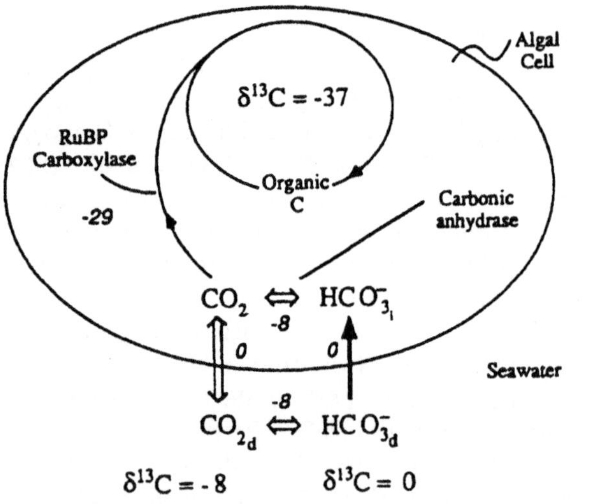

Figure 10. Carbon isotope discrimination model. Discrimination values (‰) are indicated for the reaction by RuBP carboxylase and equilibrium isotope effects for CO_2 and HCO_3^-. [Adapted from Fogel *et al.* (1988).]

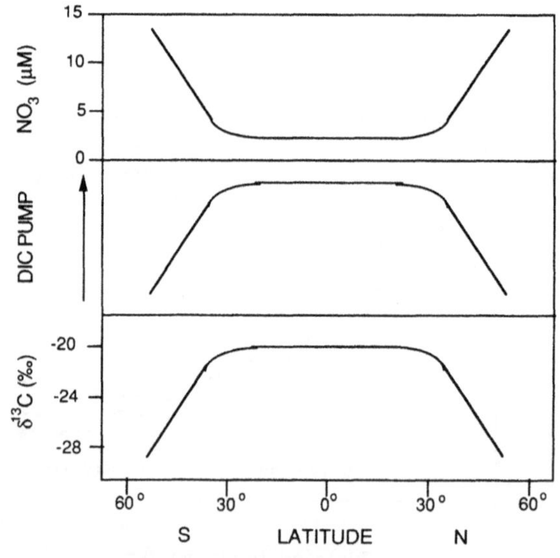

Figure 11. Hypothetical relationship between nitrate concentration, the establishment of a DIC-concentrating mechanism, and the $\delta^{13}C$ of particulate matter in the oceans.

in terms of the productivity in the water column and the availability of DIC at a particular geological time period. Part of the variability in $\delta^{13}C$ of sedimentary material may be caused by the species of DIC used by algae in photosynthesis. For example, at more acidic pH values, where CO_2 is the primary DIC species, the $\delta^{13}C$ of algae were more negative than at alkaline pH (Fig. 12).

Experiments conducted during isotopic disequilibrium of CO_2 and HCO_3^- have been used to show that some aquatic organisms preferentially transport CO_2 rather than HCO_3^- (Miller, 1985). More negative isotope ratios can be interpreted in terms of greater CO_2, rather than HCO_3^-, availability to the cells. In the Delaware Estuary, we have indirect evidence of the preferential uptake of CO_2 from the DIC pool (Fig. 9; Fogel *et al.*, 1988). In summer, both productivity and remineralization rates are high in the estuary. Remineralized CO_2 is isotopically light and will have $\delta^{13}C$ values (−25‰) similar to those of the organic carbon in the water column (Jacobsen *et al.*, 1970; Peterson and Fry, 1989). At that time, the isotope ratio of particulate material is the most negative. As the isotopic equilibrium between CO_2 and HCO_3^- is slow (hours) relative to the chemical equilibrium (seconds) (Deuser and Degens, 1967), uptake of CO_2 during photosynthesis occurs faster than isotopic equilibrium. If we assume that phytoplankton in an environment with low CO_2 concentrations available for growth will actively transport DIC, then in order

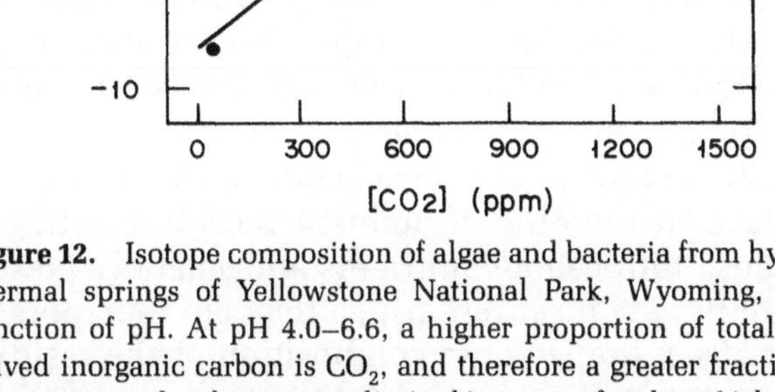

Figure 12. Isotope composition of algae and bacteria from hydrothermal springs of Yellowstone National Park, Wyoming, as a function of pH. At pH 4.0–6.6, a higher proportion of total dissolved inorganic carbon is CO_2, and therefore a greater fractionation occurs as the algae are not limited in terms of carbon. [Adapted from Estep (1984).]

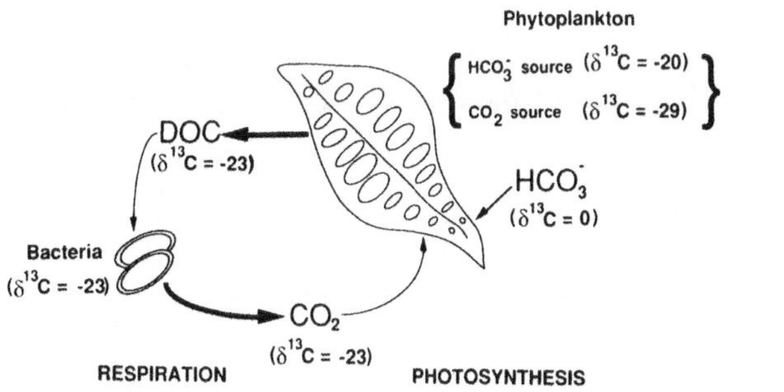

Figure 13. Carbon isotopic discrimination model of recycled CO_2 entering the photosynthetic pathway.

to produce the very negative $\delta^{13}C$ values *and* have a high photosynthetic rate, the cells are most likely transporting CO_2 at −25‰ rather than HCO_3^- at −3‰. The isotopic fractionation between particulate carbon ($\delta^{13}C$ = −28‰) and dissolved CO_2 would still be characteristically small (Fig. 13). Accumulation of isotopically light CO_2 in the algal cell may explain the more negative isotope ratios of amorphous kerogen in black shales (Lewan, 1986).

4. Nitrogen Isotopes

4.1. Fractionation during Photosynthesis

Isotopic discriminations between inorganic nitrogen sources and organic nitrogen in photosynthetic organisms are the result of fractionation during the primary assimilation of a particular nitrogen source (e.g., N_2, NO_3^-, NO_2^-, NH_4^+). There are various enzymatic pathways by which inorganic nitrogen can be fixed into organic matter during photosynthesis; however, in comparison to isotope measurements on CO_2-fixing enzymes, the isotopic studies on nitrogen assimilation are rudimentary. In addition to kinetic isotope effects by assimilatory enzymes, equilibrium isotope effects between NH_4^+ and NH_3, for example, may be important in the overall discrimination.

Ammonium can be incorporated into glutamate via the enzyme glutamate dehydrogenase (Fig. 14), which is a reversible reaction with a half-saturation constant (K_m) of 28m*M* (Falkowski and Rivkin, 1976). The other major enzyme responsible for assimilation of ammonia is glutamine synthetase (Fig. 15), which is a unidirectional reaction that requires cellular energy for catalysis. In diatoms, Falkowski and Rivkin (1976) measured a K_m of 29μ*M* for glutamine synthetase. As most freshwater and oceanic waters have

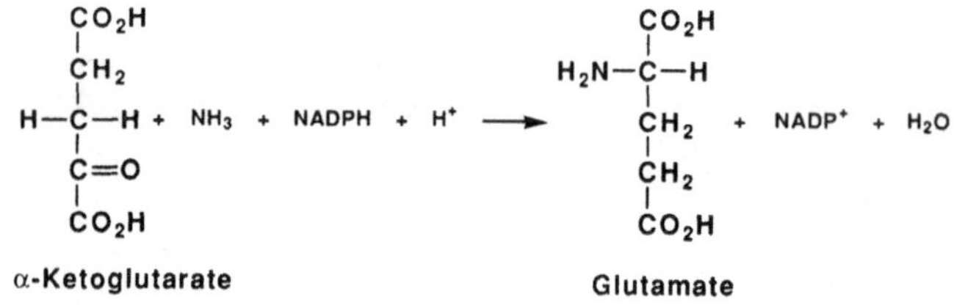

Figure 14. Formation of glutamate from α-ketoglutarate via the glutamate dehydrogenase reaction.

TABLE IV. Isotopic Fractionation for Nitrogen Assimilation[a]

Reaction	Fractionation[a]	References[b]
NO_3^- assimilation		
Cultures		
Millimolar concentrations	0–−24‰	1–7
Micromolar concentrations	−10‰	8
Field observations		
Micromolar concentrations	−4–−5‰	9, 10
NH_4^+ assimilation		
Cultures		
Millimolar concentrations	0–−15‰	7, 9, 11
Micromolar concentrations	−3–−27‰	8, 9, 11
Field observations		
Micromolar concentrations	−10‰	12, 13
N_2 fixation	−3–+1‰	5, 7, 14–17

[a]Negative values indicate that the product becomes enriched in ^{14}N as the reaction proceeds; positive values indicate that the product becomes depleted in ^{14}N as the reaction proceeds.

[b]References: 1, Miyake and Wada, 1971; 2, Wada and Hattori, 1976; 3, Sweeney *et al.*, 1978; 4, Wada and Hattori, 1978; 5, Kohl and Shearer, 1980; 6, Mariotti *et al.*, 1982; 7, Macko *et al.*, 1987; 8, Pennock *et al.*, 1988; 9, Wada, 1980; 10, Altabet and McCarthy, 1985; 11, Hoch *et al.*, 1989a; 12, Estep and Vigg, 1985; 13, Cifuentes *et al.*, 1988; 14, Hoering and Ford, 1960; 15, Mariotti *et al.*, 1980; 16, Minagawa and Wada, 1986; 17, Estep and Macko, 1984.

ambient NH_4^+ concentrations in the micromolar range, glutamine synthetase is the enzyme that is responsible for ammonium assimilation by phytoplankton and bacteria.

The oxidized forms of inorganic nitrogen, NO_3^- and NO_2^-, are reduced initially by nitrate or nitrite reductases to ammonia, which is eventually fixed into organic matter (Summons, this volume, Chapter 1). Both of these reductase enzymes have complex reaction mechanisms and have, therefore, not been studied in terms of their isotopic fractionations. Similarly, nitrogenase, the enzyme responsible for fixing molecular N_2 directly, has not been studied *in vitro*. Thus, the isotopic models of nitrogen isotope fractionation are more primitive in concept and preliminary in form than those for carbon.

Measurements of isotopic fractionation of dissolved inorganic nitrogen (DIN) during photosynthetic assimilation come mostly from laboratory culture studies, although there are a few estimates derived from natural samples (Table IV). The sum of all isotope fractionation experiments, however, should be interpreted carefully, particularly in the comparison of the culture studies to those conducted in the natural environment. One of the major differences between the two is that the concentrations of inorganic nitrogen varied by a factor of 500 among the various culture and field studies. In addition to the large range in substrate concentration, culture conditions differed in many parameters including light intensity, nutrient availability, and, consequently, growth rate. The algal cells from cultures were harvested during different growth phases, thus complicating interpretation of fractionations further. Needless to say, these culture conditions rarely matched those in the environment.

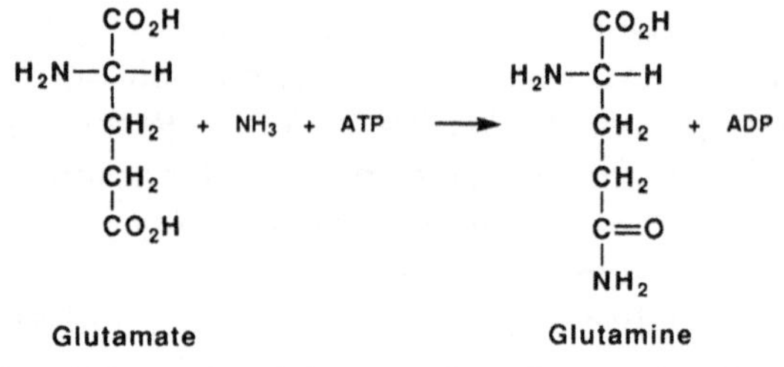

Figure 15. Formation of glutamine from glutamate via the enzyme glutamine synthetase.

Measurements of isotopic fractionation in the natural environment can be calculated with either pairs of organic or inorganic nitrogen isotope ratios (Eq. 9) or isotope ratios of inorganic nitrogen sources determined as a function of time (Eq. 6). Problems with the field-generated estimates that utilize $\delta^{15}N$ of organic matter are the following. Particulate nitrogen, presumably phytoplankton or algae, could be contaminated by variable mixtures of photosynthetic and nonphotosynthetic organisms in addition to detrital material. Estimates of isotopic fractionation could therefore be biased if the isotopic ratios of the different components differed greatly. Moreover, the interpretation of the field data is hampered by the paucity of data on the $\delta^{15}N$ of ambient inorganic nitrogen. Because nitrogenous nutrients are generally present in limiting concentrations, as they are incorporated into algae, the isotopic composition of the residual substrates is often altered (Mariotti *et al.*, 1982; Cifuentes *et al.*, 1989). In that situation, a simple pair of inorganic and organic measurements of $\delta^{15}N$ at any one time may be misleading.

4.2. Ammonium Assimilation and Fractionation

Isotopic fractionations during assimilation of NH_4^+ by algae varied extensively from 0 to −27‰ (Table IV). Some of the smallest values, 0 to −13‰ (Wada, 1980; Macko *et al.*, 1987), were observed when NH_4^+ levels in the growth media were at millimolar concentrations. In contrast, the largest fractionations, −19 to −27‰, have been measured in diatom cultures, where concentrations of NH_4^+ were in the micromolar range (Pennock *et al.*, 1988). The few models that have been developed to describe fractionation of dissolved inorganic nitrogen during assimilation by plants are based on diffusional transport across the cell membrane and enzyme kinetics (Wada and Hattori, 1978; Wada, 1980; Mariotti *et al.*, 1982). In recent years, active transport of NH_4^+ in both prokaryotic and eukaryotic organisms has been demonstrated with [^{14}C]methylamine as an analog for NH_4^+ (reviewed in Kleiner 1981, 1985). Ammonium transport occurs when concentrations in the water are below approximately 100μ*M*, and half-saturation constants were in the range of 10 to 100μ*M* (Kleiner 1981, 1985). Because natural waters, both freshwater and marine, have concentrations of NH_4^+ below 100μ*M*, models of isotopic fractionation during assimilation need to incorporate active transport mechanisms. Below, we describe models of nitrogen isotopic fractionation during the primary assimilation of ammonium for conditions of passive and active transport.

A simplified equation, which is analogous to that for carbon isotope fractionation in C_3 plants (Eq. 12), describes the dependence of the isotopic fractionation on the partitioning of ammonium inside and outside the cell:

$$\Delta = E_q + D + (C_i/C_o)[E_{enz} - D] \quad (15)$$

where Δ is the isotopic fractionation, E_q is the equilibrium isotope effect between NH_3 and NH_4^+ (Table II; −19 to −21‰; Hermes *et al.*, 1985), *D* is the isotope effect for diffusion of NH_3 in and out of the cell, C_i/C_o is the ratio of the concentration of NH_3 inside the cell to that outside the cell, and E_{enz} is the fractionation for the enzymatic fixation of NH_3 by either glutamine synthetase or glutamate dehydrogenase. This particular model applies to conditions such as cultures or sewage outfalls, where ammonium is present at millimolar concentrations, and assumes that NH_3 diffuses into the cell passively. It also assumes that the substrate concentration is sufficiently high such that both its concentration and isotopic value are not altered significantly by incorporation into organic matter. In this case, the discrimination (Δ) between the algal nitrogen and the ammonium in the medium should be equal to or greater than −19‰ (Fig. 16A). The diffusion of NH_3 through a semipermeable membrane is also associated with a considerable isotope effect (O'Leary, 1978). Taken together, the isotope fractionation from both equilibrium and diffusion is about −39‰.

Lower fractionation factors of 0 to −13‰, however, have been measured at these high concentrations. It is possible that, in spite of such high concentrations, the passive diffusion of NH_3 across the cell membrane in the rate-limiting step and controls the measured isotopic fractionation. Conversely, at higher ammonium concentrations, glutamate dehydrogenase may be the active enzyme in certain species (Ahmed *et al.*, 1977; Dortch *et al.*, 1979; Hoch *et al.*, 1989a). A positive isotope effect of about 2 to 4‰ has been measured with glutamate dehydrogenase (Weiss

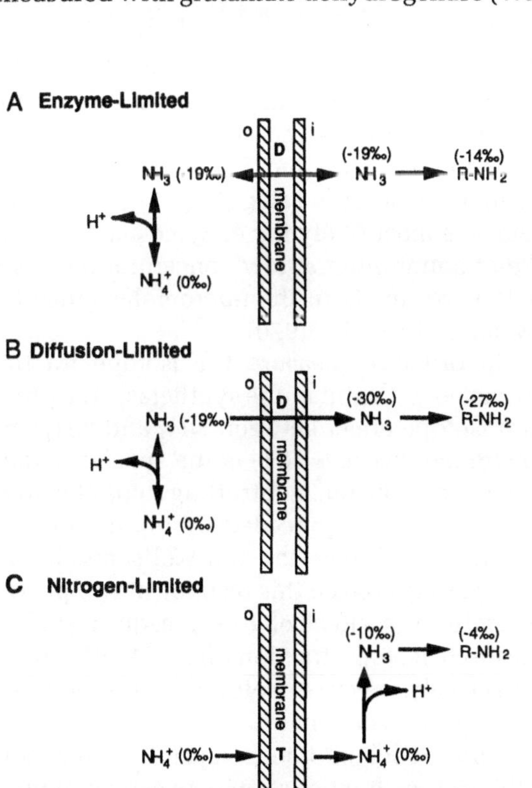

Figure 16. Maximal isotopic discrimination during ammonium assimilation when cells are enzyme limited (A), diffusion limited (B), and nitrogen limited (C). In all cases, the estimated isotopic fractionations for the enzymes, glutamine synthetase or glutamate dehydrogenase, are slightly positive.

et al., 1988). Positive isotope effects are normal in reversible reactions, because the heavy isotope of nitrogen is concentrated in the molecule with the strongest bond, glutamate.

Because the pK_a of ammonium is 9.25, the predominant chemical species in aquatic environments would be NH_4^+. At pH 7, for example, only 0.6% of the total nitrogen pool would be in the form of NH_3, the remainder being in the form of NH_4^+. If the transport of inorganic nitrogen across the cell membrane is driven by passive diffusion, then the value of C_i/C_o cannot reach values greater than 1. At external micromolar concentrations of NH_4^+, active transport of NH_4^+ is highly probable. Under these conditions, C_i/C_o in a variety of prokaryotic and eukaryotic organisms can reach values of at least 100 (Kleiner, 1981) owing to the establishment of energy-dependent NH_4^+-concentrating mechanisms. In these cases, the model of isotopic fractionation would be more analogous to the carbon isotope model for C_4 plants or aquatic organisms with CO_2-concentrating mechanisms, i.e.:

$$\Delta = (F_3/F_1)(E_q + E_{enz}) \tag{16}$$

where Δ is the isotopic fractionation, F_3/F_1 is the ratio of NH_4^+ leaking out of the cell to the concentration of $NH_3 + NH_4$ inside the cell, E_q is the equilibrium fractionation factor, and E_{enz} is the fractionation for the enzyme-catalyzed step (Fig. 16B). Glutamine synthetase is most likely the enzyme catalyzing the uptake of ammonium at low concentrations, as the K_m for the enzyme is in the micromolar range (see Falkowski and Rivkin, 1976).

In order to measure the isotope fractionation associated with glutamine synthetase, the preequilibrium isotope effect between NH_3 and NH_4^+ must be considered, because NH_3 is the species taken up by the enzyme. The rate-controlling steps of many chemical reactions are preceded by rapid and reversible preequilibria (Bigeleison and Wolfsberg, 1958). Hoch *et al.* (1989a) assayed this reaction at two pH values to determine the effect of this preequilibrium on the measured isotope fractionation. At pH 7.0 and 8.6, respectively, inverse isotope effects of +10.8 and +3.0‰ were calculated.

This model predicts a concentration dependency in the isotope fractionation between ammonium and organic nitrogen in that at slightly higher external concentrations (i.e., 100μ*M*) the fractionation would be the greatest, whereas at the lowest concentrations (i.e., < 10μ*M*) the amount able to leak out of the cell before assimilation would be essentially zero. At these low concentrations, the isotope fractionation between inorganic nitrogen and particulate nitrogen will be extremely small. In a lab culture experiment with an estuarine diatom, Pennock *et al.* (1988) found exactly that (Fig. 17). The largest fractionations (up to −27‰) measured during ammonium assimilation have been observed when the concentration of NH_4^+ in the medium was less than 100μ*M*, yet greater than 10μ*M* (Hoch *et al.*, 1989b; Fig. 17). Active transport of NH_4^+ could have been induced resulting in C_i/C_o greater than one, or essentially an excess of nitrogen for assimilation.

Hoch *et al.* (1989b) confirmed that the induction and operation of active NH_4^+ transport correlated with decreased isotope fractionation. From the model we would expect that as NH_3 uptake changed from passive to active transport, the internal concentrations of NH_3 were sufficient to support growth. Glutamine synthetase with its net positive isotope effect would be the limiting step in determining isotopic fractionation. At concentrations of extracellular ammonium less than 20μ*M*, total ammonium concentrations would be limiting. Isotopic fractionation decreased (Fig. 17) to approximately −2‰.

4.3. Nitrate Assimilation and Fractionation

A similar range of isotope fractionation has been measured with algae grown on nitrate as the source of nitrogen (Table IV). The major difference between the

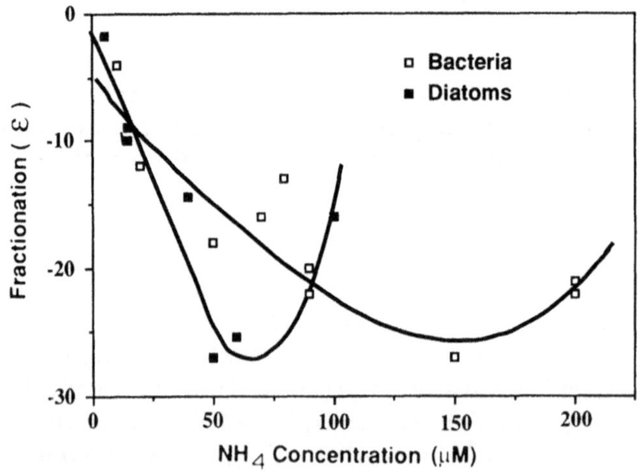

Figure 17. Nitrogen isotope fractionation by the diatom *Skeletonema costatum* (Pennock *et al.*, 1988) and a heterotrophic marine bacterium, *Vibrio harveyii* (Hoch *et al.*, 1989 and 1992b) as a function of the concentration of ammonium in the culture medium. Fractionation is greatest at around 50μ*M* NH_4^+ for phytoplankton and 150μ*M* for bacteria.

results with ammonium and those with nitrate is the dependence of isotopic fractionation on the concentration of nitrate at millimolar concentrations. In both diatoms (Wada and Hattori, 1978) and pearl millet (Mariotti *et al.*, 1982), the isotopic fractionation increased with increasing concentration of nitrate in the growth medium. No clear relationship between substrate concentration and isotopic fractionation of nitrate, however, is evident from the available data measured with algal cells cultured at lower concentrations (Pennock *et al.*, 1988).

Active transport of nitrate has been reported by Packard (1979) and Falkowski (1975) in marine phytoplankton. In a marine diatom, *Skeletonema costatum*, a membrane-bound enzyme appeared to be activated by nitrate. In studies with higher plants, Mariotti *et al.* (1982) demonstrated that there was no isotope effect for the transport of NO_3^- across the cell membrane. If an active transport mechanism for nitrate were to operate in natural conditions, nitrate may be in great enough supply to saturate the nitrate reductase such that fractionation would be expressed.

4.4. Interpretation of Nitrogen Isotope Compositions of Organic Matter in the Environment

Nitrogen isotope studies at the natural abundance level have provided critical information on phototrophic assimilation of nitrogen and on nutrient dynamics in aquatic ecosystems. Traditionally, oceanographers and limnologists centered their work on the analysis of nutrient concentrations to trace nitrogen through production and remineralization pathways. Commonly, ^{15}N-addition experiments with isotopically enriched inorganic nitrogen species were used for tracing biogeochemical processes. These additions of very small amount of either nitrate or ammonium often perturbed the system studied, such that erroneous conclusions were reached (e.g., Glibert *et al.*, 1982). Thus, a critical need developed for a tracer at the natural abundance level that could utilize the dissolved inorganic nitrogen already in the environment. The $\delta^{15}N$ of particulate nitrogen in relation to $\delta^{15}N$ of dissolved inorganic nitrogen has been useful in tracing nitrogen fixation, denitrification, new algal production from regenerated nitrogen, and trophic transfers in marine and estuarine ecosystems.

Isotopic compositions for particulate nitrogen, and non-nitrogen-fixing plankton, or macroalgae are typically -3 to $+18$‰ (Schoeninger and DeNiro, 1984, and references therein; Owens, 1987, and references therein; Cifuentes *et al.*, 1988). Terrestrial plants have a similar $\delta^{15}N$ range, -5 to $+18$, but are isotopically lighter on the average (see Schoeninger and DeNiro, 1984). In contrast, marine blue-green algae range from -2 to $+4$‰, and nitrogen-fixing terrestrial plants have a slightly larger range of -6 to $+6$‰. Some of the first papers to emerge in the early 1970s used nitrogen isotopes to trace the source of terrestrial organic matter to oceanic or estuarine surface waters and sediments. Simple mixing calculations with mass-balance equations were then used to describe the contribution of terrestrial and marine organic matter to the ecosystem (Sweeney and Kaplan, 1980; Sweeney *et al.*, 1980). More recently, biogeochemical processes affecting *in situ* produced material have been shown to control the isotope ratios of nitrogen species.

More negative isotopic compositions ($\delta^{15}N = -2$ to $+4$) may result from either microbial nitrogen fixation or the inclusion of terrestrial plant debris in the suspended particular material. Minagawa and Wada (1986) and Macko *et al.* (1984) have shown that in areas of the Pacific Ocean and the Gulf of Mexico, where the nitrogen-fixing, blue-green alga *Trichodesmium spp.* is the dominant species of plankter, the $\delta^{15}N$ of particulate nitrogen ($\delta^{15}N = -2$ to $+1$‰) is close to the value of atmospheric nitrogen (0‰). Rau *et al.* (1987) concluded that nitrogen fixation thus caused the isotopically light values measured in black shales from Cretaceous marine sediments. Without knowledge of the paleonutrient dynamics and possible diagenetic effects in the area at the time of the rock formation, other interpretations may be feasible.

Isotopic fractionation during primary assimilation of inorganic nitrogen will occur when the inorganic nitrogen source is present in excess (see Table IV). Saino and Hattori (1980) first postulated that fractionation of nitrate during algal assimilation explained the ^{15}N-depleted particulate organic nitrogen in the euphotic zone of North Pacific waters (Fig. 18). Altabet and McCarthy (1985, 1986) also demonstrated that isotopic fractionation of nitrate occurred in the water column with a fractionation value of -4‰. Subsequently, isotopic fractionation of ammonium in the urbanized Delaware Estuary (Cifuentes *et al.*, 1988) has also been shown to result in isotopically more negative organic matter (Fig. 19). Calculations with the isotopic composition of the residual substrate (Cifuentes *et al.*, 1989) resulted in a fractionation value of -9‰, which is comparable to the culture studies at low substrate concentrations (see Fig. 17).

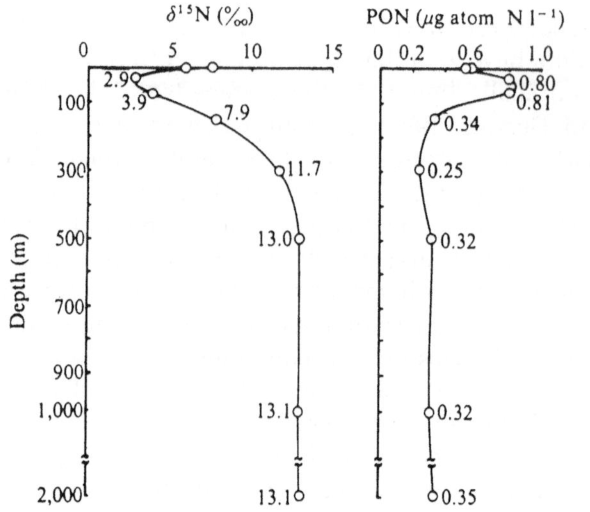

Figure 18. Profile of particulate organic nitrogen (PON) vs. its isotopic composition in the Pacific Ocean. Saino and Hattori (1980) proposed that the isotopically heavy values in subsurface waters were the result of material undergoing decomposition. The PON maximum corresponds with the most negative $\delta^{15}N$ values, possibly indicating isotopic fractionation (Saino and Hattori, 1980).

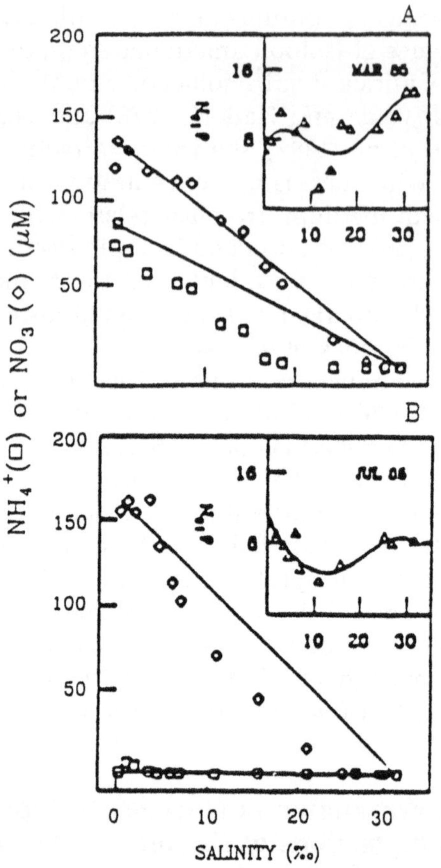

Figure 19. Relationship between $\delta^{15}N$ of suspended particulate matter and inorganic nitrogen distributions. NH_4^+ and NO_3^- concentrations are plotted vs. salinity (‰) for March (A) and July (B) 1985; the corresponding plots of $\delta^{15}N$ vs. salinity are shown in the insets. All concentrations are micromolar; $\delta^{15}N$ values are per mil. Included are conservative mixing lines for both nutrients (Cifuentes *et al.*, 1988).

In aquatic environments where combined nitrogen is limiting, Altabet and McCarthy (1985, 1986) and Altabet (1988) discounted isotopic fractionation as a likely mechanism of generating isotopically depleted nitrogen, because the substrate, nitrate, would be completely converted to product. Obviously, no isotope fractionation can occur under these conditions. These authors postulated that smaller photosynthetic organisms utilized isotopically light NH_4^+ regenerated from zooplankton (Checkley and Miller, 1989), which resulted in more negative $\delta^{15}N$. Moreover, as phytoplankton were consumed by zooplankton of a larger size and were repackaged as fecal pellets, particle size segregation occurred in the water column (Fig. 20).

Terrestrial sources of organic nitrogen range in $\delta^{15}N$ from −6 to +18‰ and result from complex nitrogen cycles that occur in microzones of soils. The average value for terrestrially derived organic nitrogen, +3‰ (Schoeninger and DeNiro, 1984, and references therein) is more negative than that for marine organic matter (~8‰). Today, fertilizer synthesized by the Haber process has flooded many populated coastal regions with inorganic nitrogen sources having isotopic compositions around 0‰ (Black and Waring, 1977; Kohl *et al.*, 1971). Thus, the presence of terrestrial organic nitrogen and anthropogenically derived inorganic nitrogen can be detected in estaurine and coastal transects, particularly during periods of high flow. For example, in early spring when flow increases and material from the drainage basins flows into estuaries, a stronger terrestrial signal can be observed (Cifuentes, 1991). Sweeney *et al.* (1980) were able to trace sewage-derived organic matter into the California Bight off of Los Angeles.

Particulate nitrogen with the most positive $\delta^{15}N$ values (+10 to +33‰) has been collected either below the photic zone in the open ocean or along salinity transects in estuaries. In the ocean, isotopic fractionation during decomposition is typically thought to result in residual material having an isotopically heavy composition (Saino and Hattori, 1980; Libes and Deuser, 1988; Altabet and McCarthy, 1985; Saino and Hattori, 1985; Altabet and McCarthy, 1986; Saino and Hattori, 1987). Altabet and McCarthy (1986) argued for a linear relationship between the $\delta^{15}N$ of particulate matter and the fraction of nitrogen that had been removed at depth in warm-core rings [Fig. 21;

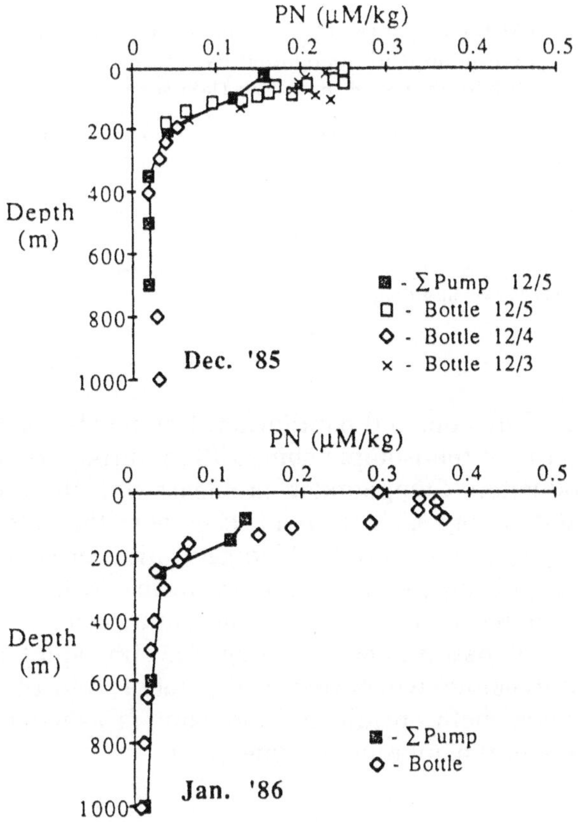

Figure 20. $\delta^{15}N$ of different size particles collected on glass fiber filters (GF/D or GF/F) demonstrating the enrichment of ^{15}N as material travels up the food chain (Altabet, 1988). The bottom figure indicates the concentration of particulate nitrogen collected on glass fiber filters at the same location.

see Eq. (5)], indicating that fractionation was occurring during degradation. The scatter in the data, however, suggests that the situation is more complex. Particle segregation can alter isotope ratios in the water column (Altabet, 1988) and could result in the type of scatter observed in the data. If decomposition were the mechanism by which the $\delta^{15}N$ became more positive, then ammonium with more negative $\delta^{15}N$ would necessarily be released. Unfortunately, no corresponding isotopic data for ammonium are available for these data sets. Sweeney and Kaplan (1980) did observe a close relationship between the isotopic ratio of pore-water ammonia and that of planktonic material in the Santa Barbara Basin. Similarly, in sediments from a salt marsh and a marine anoxic fjord, the isotopic composition of NH_4^+ was within 1‰ of that of the total particulate matter (Velinsky *et al.*, 1991; Fig. 22).

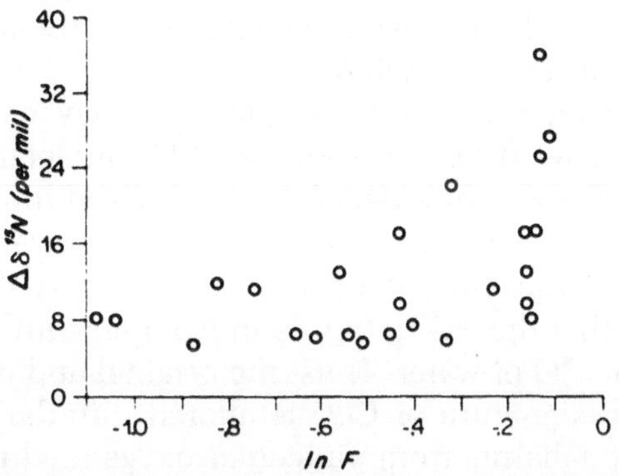

Figure 21. Change in $\delta^{15}N$ with degradation in the ocean (Altabet and McCarthy, 1986). F is the fraction of particulate material remaining.

Fractionation by bacteria during nitrification or uptake for biosynthesis may be additionally important in causing more positive $\delta^{15}N$ in oceanic depth profiles. In these instances, the residual, isotopically heavy ammonium may be incorporated into microbial biomass. This transfer of isotopically fractionated dissolved inorganic nitrogen into the particulate phase has been illustrated along salinity transects in estuaries. Both Mariotti *et al.* (1984) and Cifuentes *et al.* (1988, 1989) have measured $\delta^{15}N$ from particulate nitrogen, primarily fresh material, of +12 to +18‰. In both estuaries studied, the isotopic composition of the ammonium used by phytoplankton dur-

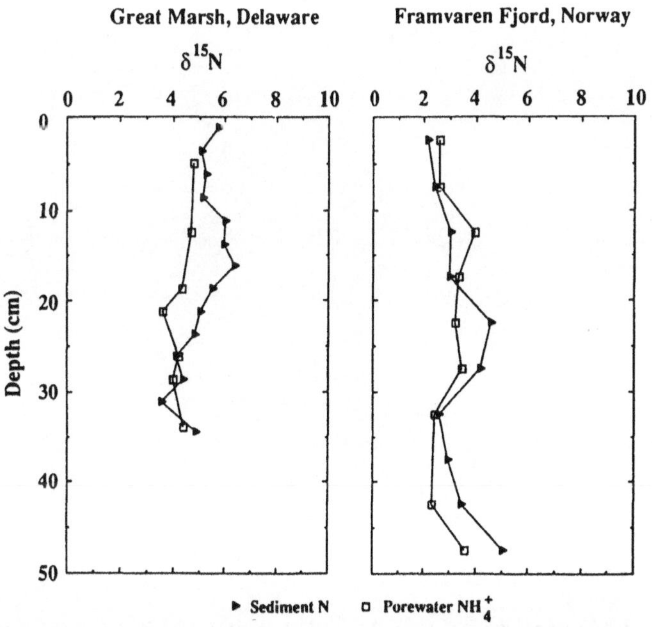

Figure 22. Depth profiles of the isotopic composition of pore water ammonium and particulate nitrogen from sediments sampled in the Great Marsh, Lewes, Delaware, and the Framvaren fjord, Norway (Velinsky *et al.*, 1991). Pore waters were obtained with sediment squeezers, and ammonium samples were processed by the method of Velinsky *et al.* (1989).

ing growth was shown to be enriched in ^{15}N. In the Scheldt Estuary, nitrification was shown to be the dominant process by which residual ammonium became enriched in the heavy isotope. In the Delaware Estuary, both algal assimilation and nitrification were responsible for altering the isotopic composition of dissolved inorganic nitrogen sources (Fig. 23). Accordingly, denitrification would increase the $\delta^{15}N$ of residual nitrate and would result in particulate nitrogen having more positive $\delta^{15}N$ values (Sigleo and Macko, 1985).

The three-dimensional plot of ammonium isotopic ratios generated from the Delaware Estuary is but one example of the complexity in the isotopic cycling of nitrogen in the field. The translation of surface water variability in dissolved inorganic $\delta^{15}N$ to suspended sediments and, finally, to bottom sediments has not been well documented. In the Delaware Estuary, the extreme $\delta^{15}N$ values in suspended organic nitrogen were not found in sedimentary nitrogen (Table V). Cifuentes *et al.* (1988) postulated that preferential remineralization of new surface organic matter reduced the amplitude of variation in bottom sediments. The isotopic composition of phototrophic organic matter is influenced by enzymatic fractionation, diffusion, and equilibrium isotope effects that determine the isotopic composition during primary production. Compounded problems with the interpretations of $\delta^{15}N$ in the field occur when the isotopic compositions of dissolved inorganic nitrogen sources vary with time or when the organisms switch from one source of nitrogen to another. Future use of nitrogen isotopes to interpret nitrogen cycling and food-web dynamics will have to incorporate a protocol that includes both organic and inorganic measurements made in the appropriate time frame.

TABLE V. Ranges of $\delta^{15}N$ for Surficial Bottom Sediments and Suspended Particulate Matter in Different Regions of the Delaware Estuary

Region[a]	Bottom sediments	Suspended particulates
0 to 40 km	+5.0–+9.8	+3.6–+11.5
40 to 80 km	+5.8–+13.1	+2.3–+17.5
80 to 140 km	+5.0–+8.0	+6.7–+18.7

[a]The mouth of the estuary is at 0 km, and Wilmington, Delaware, is approximately 130 km upstream.

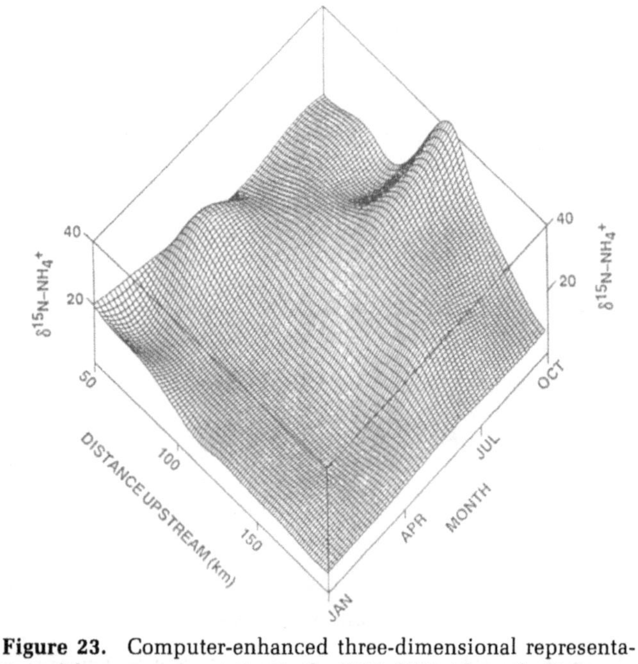

Figure 23. Computer-enhanced three-dimensional representation of the seasonal variation in the $\delta^{15}N$ of NH_4^+ from the Delaware Estuary. A cubic spline routine was used to interpolate between data points. The tidal river extends from 130 to 200 km upstream from the mouth of the estuary. The salinity gradient of the estuary is located downstream of 130 km. $\delta^{15}N$ minima in the Delaware Estuary. (Cifuentes *et al.*, 1989).

5. Oxygen Isotopes

5.1. Fractionation during Photosynthesis

Information concerning fractionation during photosynthesis of organic oxygen is limited to studies of the $\delta^{18}O$ of cellulose and other carbohydrates. Techniques for these analyses which are currently available involve heating organic matter with mercuric chloride or pyrolyzing the sample in a nickel reaction vessel (Thompson and Gray, 1977; Burk and Stuiver, 1981). Although the mercuric chloride technique has been adapted for the analysis of lipids and proteinaceous material, no results from biological work with this technique have been published (Schimmelmann and DeNiro, 1985).

During photosynthesis, oxygen may enter organic matter from three sources: CO_2, molecular O_2, or H_2O. Carbon dioxide, of course, is fixed into phosphoglyceric acid by RuBP carboxylase; however, DeNiro and Epstein (1979) have conclusively demonstrated that the ^{18}O in CO_2 is in isotopic equilibrium with the ^{18}O of water. Thus, the original and distinct isotopic signature of CO_2 is altered. Similarly, any isotopic labeling from molecular oxygen, which can also be fixed into organic matter, is lost by exchange with water (Berry *et al.*, 1978).

Therefore, it is the isotopic composition of plant

water that will determine the $\delta^{18}O$ of organically bonded oxygen in cellulose. Experiments conducted by Sternberg *et al.* (1986), who used plant or seed cultures grown heterotrophically, produced convincing evidence that the oxygen isotopic composition of cellulose is determined during isotopic exchange between carbonyl oxygens and water:

$$R_2C{=}^{16}O + H_2{}^{18}O \leftrightarrow R_2C(^{18}OH)(^{16}OH) \quad (17)$$

Fractionations of 16 to 27‰, in which $\delta^{18}O$ of carbonyl oxygen became isotopically heavier, were measured in these growth experiments and were dependent upon the particular organic substrate that was subjected to exchange (Table VI). A similar fractionation (27‰) was determined in the exchange of water with the carbonyl oxygen of acetone (Sternberg and DeNiro, 1983b). These values with both the model compound acetone and actual plant tissues are very similar to the overall fractionation measured in cellulose from most plants, $+27 \pm 3$‰. At some point in the Calvin cycle, all of the oxygen atoms in cellulose will be in a carbonyl form, and thus the exchange of oxygens is complete. Careful inspection of the data, however, does reveal some intriguing variations.

While there is no difference in $\delta^{18}O$ of cellulose that can be related to a photosynthetic pathway, such as the C_3 or C_4 carbon fixation pathways, real differences in the $\delta^{18}O$ of cellulose between species exist. One possible mechanism for these differences is differential enrichment of leaf water with ^{18}O during evaporative transpiration (Gonfiantini *et al.*, 1965; Dongmann *et al.*, 1974). As a leaf opens its stomates for gas exchange with the atmosphere, H_2O enriched in ^{16}O evaporates preferentially. The water remaining in the leaf, which is available for photosynthesis, can be enriched by several per mil relative to groundwater. Sternberg *et al.* (1984), however, grew submerged aquatic plants under controlled conditions, in which the isotopic composition of the leaf water was kept constant. Variations in the $\delta^{18}O$ were still recorded.

Another possibility for the observed variability might be the overall photosynthetic rate and productivity of the plant. Oxygen atoms will exchange at different rates in different molecules; thus, the residence time of oxygen atoms in particular carbonyl positions may be important in contributing to the total $\delta^{18}O$ value. A plant that is photosynthesizing rapidly, for example, may express less fractionation in cellulose, as the Calvin cycle intermediates will reside for only minutes or seconds before being synthesized into cellulose (Berry *et al.*, 1978). Sternberg *et al.* (1986) measured smaller fractionations than they had expected in their culture experiments (Table VI), which lends some credence to this idea.

TABLE VI. Predicted and Observed Percentage of Oxygens of Different Substrates That Exchanged with Water during Cellulose Synthesis and Observed Fractionation Factors for Different Substrates[a]

Species	Substrate	Exchangeable oxygen (%) Observed[b]	Exchangeable oxygen (%) Predicted[c]	Fractionation factor[d]
Daucus carota	Sucrose	47	45	16.3‰
	Glycerol	77	67	27.3‰
Acetobacter xylinum	Glucose	29	33	15.4‰
Ricinus communis	Unknown	76	—[e]	—[e]

[a]Taken from Sternberg (1988a).
[b]Slope of the linear regression line between oxygen isotope ratios of cellulose and water for each respective substrate.
[c]Calculated by counting carbonyl oxygens of the substrate and subsequent intermediates on the pathway of the substrate to cellulose.
[d]Calculated from the intercept of the linear regression line mentioned in footnote b.
[e]Could not be calculated since the actual substrate was not known.

5.2. Interpretations of Field Data

The isotopic composition of meteoric water varies as a function of latitude; seawater has an isotopic composition of 0‰, whereas Arctic precipitation has a $\delta^{18}O$ of approximately -55‰. In general, groundwater in warmer climatic regions has a more positive $\delta^{18}O$ value, and, conversely, the $\delta^{18}O$ of groundwater is more negative in the cooler climates. As the $\delta^{18}O$ of cellulose from plants is 27‰ more positive than that of the water in the leaf, both the $\delta^{18}O$ of cellulose and that of meteoric water can be related to climate. Thus, the oxygen isotopic composition in cellulose from modern and fossil wood has been used to determine paleoclimate (e.g., Epstein *et al.*, 1977; Burk and Stuiver, 1981).

This simple conceptual model can become complicated in large trees because the isotopic composition of water in different parts of the tree is not uniform. For example, leaf water is often enriched in ^{18}O relative to trunk or stem water (Dongmann *et al.*, 1974). Photosynthate formed in the leaves will be in equilibrium with water of a different isotopic composition than cellulose that is synthesized from sucrose in the wood. Sucrose is the principal photosynthetic product that is exported from the leaf into the tree trunk, and up to 45% of the oxygens in sucrose may still exchange with tissue water (Sternberg *et al.*, 1986). The relationship between climate, leaf water, and $\delta^{18}O$ of cellulose may thus be substantially modified by subsequent exchange events.

6. Hydrogen Isotopes

6.1. Fractionation during Photosynthesis

Isotope fractionation of hydrogen occurs in photosynthetic organisms such that the light isotope of hydrogen (H) is enriched in the organically bonded hydrogen. Quantitative modeling of individual and specific steps involved in this fractionation has not been accomplished to date owing to the complexity of the reactions of hydrogen in cellular metabolism. Hydrogen enters the plant as water either from roots in the case of terrestrial plants or via diffusion in the case of aquatic plants or algae. The water enters the plant without any apparent fractionation (Leaney *et al.*, 1985) and travels to the choloroplasts that are the sites of photosynthesis. In higher terrestrial plants, as the stomates in the leaves of plants open for gas exchange with the atmosphere, water transpires from the leaf due to evaporation. During this process, the remaining water in the leaf becomes enriched in the heavy isotope of hydrogen (D) by about 40 to 50‰, as isotopically lighter water tends to evaporate more readily (Estep and Hoering, 1980; Leaney *et al.*, 1985; White, 1988; see Fig. 24).

Research in stable hydrogen isotope fractionation in photosynthetic organisms has proceeded in two general directions: (1) investigations on the isotopic composition of the nonexchangeable hydrogen in cellulose and (2) investigations of the isotopic compositions of total organic hydrogen, lipids, or carbohydrates (Table VII).

Controversy surrounds the hypothesis that organically bonded hydrogen exchanges with the water in the surrounding tissue or environment. Estep and Hoering (1980) contended that the hydrogen in quaternary-bonded positions in macromolecules is not readily exchanged with free cellular water. Exchange experiments with both intact algal cells over a month's time and sonicated suspensions of algal organic matter showed that only with denaturation and heat do the hydrogen atoms exchange with the surrounding water. Thus, isotopic measurements on carefully dried whole plant tissue or algal material can give a consistent, meaningful, although complex, result (Ziegler *et al.*, 1976; Stiller and Nissenbaum, 1980).

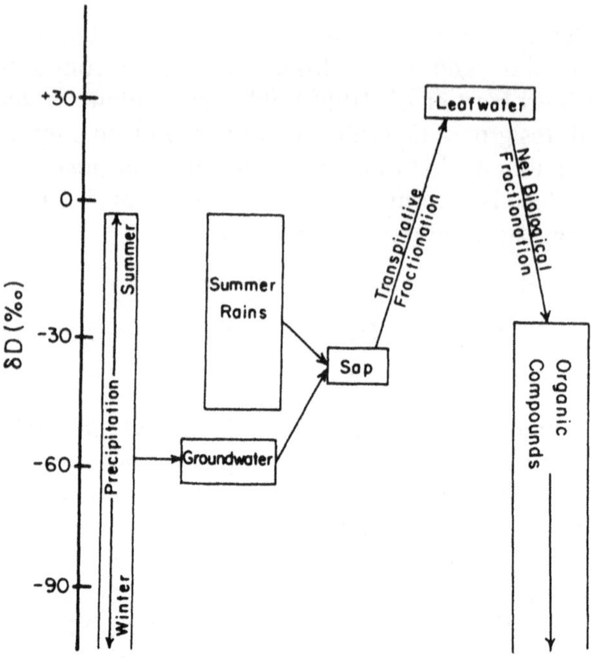

Figure 24. A generalized schematic of the changes in the hydrogen isotopic ratios of water and organic matter in the pathway of hydrogen into plants. Deuterium is enriched in the water in leaves; deuterium is depleted in the organic matter fixed in plants (White, 1988).

Nonexchangeable hydrogen in cellulose was thought to be an applicable material for paleoenvironmental studies; however, DeNiro (1981) has shown that the techniques used for isolating cellulose from hemicelluloses and lignin in some of the earlier studies (e.g., Epstein *et al.*, 1976) were not effective. In more recent years, Sternberg and DeNiro (1983a) have shown that plants growing within one area, where the isotopic composition of the water is the same, have highly variable hydrogen isotope ratios in the nonexchangeable hydrogens of cellulose (Fig. 25).

Enzymatic studies with specific reactions need to be attempted in order to elucidate the possible and myriad mechanisms of fractionation. As a start, Estep and Hoering (1980, 1981) cultured microalgae under controlled conditions, including light intensity and wavelength, nutrient concentrations, and temperature, to begin to elucidate the physiological pathways involved in hydrogen isotope fractionation. They cultured a species of green algae, *Chlorella sorokiniana*, under photosynthetic and heterotrophic conditions with either glucose or acetate as the substrate. When the algae grew photosynthetically, the organically bonded hydrogen was depleted in the heavy isotope of hydrogen relative to the composition of the water in the growth medium by approximately 100 to 150‰ (Fig. 26). When the algae were grown in the dark,

TABLE VII. Hydrogen Isotope Exchange in Isolated Compounds

	Type of hydrogen exchanged	Relative exchange rate
Acid–base-catalyzed exchange reactions		
$R{-}C(=O){-}OH + HDO \rightleftharpoons R{-}C(=O){-}OD + H_2O$	Acid	Extremely fast
$R{-}OH + HDO \rightleftharpoons R{-}OD + H_2O$	Alcohol	Very fast
$R{-}NH_2 + HDO \rightleftharpoons R{-}NHD + H_2O$	Amine	Very fast
$R{-}SH + HDO \rightleftharpoons R{-}SD + H_2O$	Thiol	Very fast
$C_6H_6 + HDO \rightleftharpoons C_6H_5D + H_2O$	Aromatic hydrogen	Intermediate
$R{-}C(=O){-}CH_3 + HDO \rightleftharpoons R{-}C(=O){-}CH_2D + H_2O$	Ketone	Intermediate
Hydrogen atom transfer exchange reactions		
$R{-}CH_3 + HDO \rightleftharpoons R{-}CH_2D + H_2O$	Methyl hydrogen	Slow
$R{-}C(H){=}C(H){-}R + HDO \rightleftharpoons R{-}C(H){=}C(D){-}R + H_2O$	Olefinic hydrogen	Slow
$R{-}C(H)(R){-}R + HDO \rightleftharpoons R{-}C(D)(R){-}R + H_2O$	Tertiary hydrogen	Slow
$R{-}CH_2{-}R + HDO \rightleftharpoons R{-}CHD{-}R + H_2O$	Secondary hydrogen	Very slow

however, with glucose as a substrate, the isotopic composition of the whole organism was almost identical with that of the glucose. Evidently, the isotopic fractionation occurs somewhere in the light reactions of photosynthesis. From experiments in which *Chorella sorokiniana* was grown under different light wavelengths and with inhibitors, Estep and Hoering hypothesized that the biochemical reducing power (NADPH) generated during photosynthesis controlled the isotopic composition of the whole cell. This reducing power, NAPDH, is used in a variety of biosynthetic reactions as the specific source of hydrogen in cellular metabolism.

Isotopic discrimination in higher terrestrial plants is of a similar magnitude to that measured in algae, $\Delta = -86$ to -120‰. Further research revealed a difference in the isotopic composition of organic hydrogen in the whole tissue and proteins from C_3 ($\Delta = -117$ to -121‰) and C_4 ($\Delta = -86$ to -109‰) plants (Ziegler *et al.*, 1976; Leaney *et al.*, 1985; Ziegler, 1979, 1988). Studies limited to measurement of nonexchangeable cellulosic hydrogen, however, show no consistent difference between these two photosynthetic pathways (Sternberg *et al.*, 1984), although a consistent difference with respect to a third photosynthetic pathway, crassulacean acid metabolism (CAM), was measured. It is possible that those very specific hydrogen atoms in cellulose may be labeled before any variations owing to photosynthetic pathway are expressed.

Leaney *et al.* (1985) have shown that differences in the δD of plants are not easily related to differences in the isotopic compositions of leaf water. They cultured both C_3 and C_4 plants together and monitored the isotopic ratio of leaf water throughout the photosynthetic period. Variations in leaf water did occur that were not mirrored by the organic hydrogen, most likely because the turnover in total cellular organic hydrogen is not attained over a daily cycle. Estep and Hoering (1981) noted that even when the isotopic ratio of the medium water was changed during the algal culture and growth did proceed, a lag occurred in the expression of the new isotopic ratio of hydrogen into all the cellular organic matter pools. They postulated that NADPH generated in previous light reactions was conserved for future biosynthetic reactions.

A consistency that does emerge from the two parallel studies on cellulose and whole tissues is that hydrogen isotope fractionation occurs in different reactions during the photosynthetic production of carbohydrates. For example, in both C_3 and C_4 plants, NADPH generated in photosynthesis results from the action of ferredoxin-$NADP^+$ reductase. In certain C_4 plants, an additional amount of NADPH is generated by NADPH-linked malic enzyme in the bundle sheath cells as malate, the C_4 acid, is decarboxylated. Other C_4 plants that produce aspartate as a first product decarboxylate with an NAD^+-linked malic enzyme. A third type of decarboxylating enzyme, PEP carboxykinase, which is found in the remainder of C_4 plants,

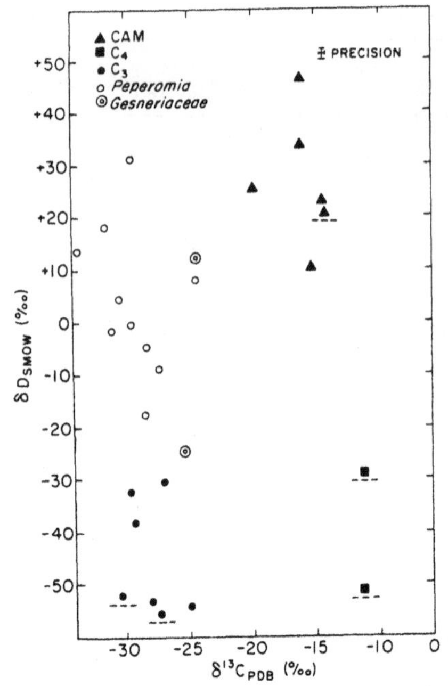

Figure 25. Hydrogen isotope ratios vs. carbon isotope ratios of cellulose nitrate from epiphytic plants collected and grown in La Selva, Costa Rica. The δD of cellulose has some relationship to the photosynthetic pathway of the plant, although even within one location the δD can be quite variable (Sternberg 1988a)

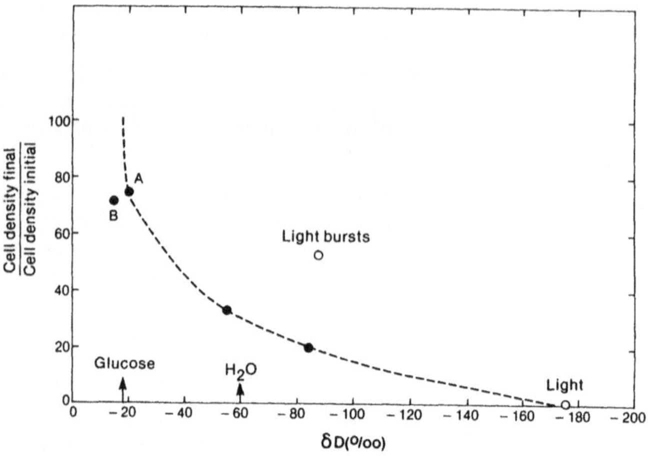

Figure 26. Hydrogen isotopic effects in *Chlorella sorokiniana*, a green alga, grown heterotrophically on glucose. Closed circles represent the average δD of a set of experiments where the algae were incubated in total darkness. The open circles represent the average δD of a set of experiments performed in the light (Estep and Hoering, 1980).

does not produce any associated reducing power. The isotopic compositions of these three types of C_4 plants are very distinct from one another (Ziegler, 1988; see Table VIII). It is interesting to note that plants with an alternate source of either NADPH or NADH are isotopically heavier than those which have NADPH derived only from ferredoxin-linked $NADP^+$ dehydrogenase.

6.2. Interpretation of Geochemical Material

The most promising material for geochemical studies is lipids, which have a majority of their hydrogen atoms bonded to carbon, bonds that break with difficulty at ambient temperatures. Total lipids that were isolated from photosynthetically and heterotrophically grown cells had isotope ratios that were always depleted in the lighter isotope, which indicated that the isotope fractionation associated with lipid metabolism was independent of photosynthesis (Estep and Hoering, 1980). Similarly, carbon isotope fractionation during lipid metabolism is dependent on an enzyme reaction in intermediary metabolism, that catalyzed by pyruvate dehydrogenase (DeNiro and Epstein, 1979), rather than RuBP carboxylase. When algal cells were cultured on acetate as substrate, however, the hydrogen isotope ratio of the extracted lipids was nearly identical with that of the whole cells (Estep and Hoering, 1980). This result is consistent with a fractionation occurring prior to the formation of acetyl-coenzyme A, the first substrate in fatty acid synthesis.

From plants collected in the field, the isotope fractionation between lipids and meteoric waters was about −120 ± 10‰ (Estep and Hoering, 1980; Sternberg, 1988a,b). When lipids were separated further

TABLE VIII. δD of Total Organic Matter Synthesized in C_4 Plant Types with Different Enzyme Pathways[a]

Main decarboxylating enzyme in bundle sheaths	Balance of decarboxylation in bundle sheaths	Main transport substance, mesophyll cells to bundle sheaths	δD total organic H (‰)
$NADP^+$-malic enzyme	Production of 1 NADPH per CO_2	Malate	−43.4 ± 9.2
NAD^+-malic enzyme	Production of 1 $NADH_2$ per CO_2	Aspartate	−72.8 ± 5.9
PEP carboxykinase	Consumption of 1 ATP per CO_2	PEP/alanine	−67.5 ± 10.4

[a]Modified from Ziegler (1988).

into individual classes (Estep and Hoering, 1980), an additional fractionation in the synthesis of nonsaponifiable lipids was detected (Fig. 27). Hoering (1977) found a similar difference in these two classes of lipids extracted from recent coastal sediments. Northfeldt *et al.* (1981) used lipids extracted from fossil woods to predict paleoclimates. Geochemical substances, petroleum, coal, and kerogen, will have formed from a mixture of plant and algal substances. Redding *et al.* (1980) have shown that isotopic ratios of the original plant source may be preserved in fossil materials, which typically have δD values of −90 to −130‰.

7. Sulfur Isotopes: Fractionation during Photosynthesis

The assimilation of sulfur by photosynthetic organisms primarily involves the reduction of sulfate to sulfide and incorporation into cysteine. It is not likely that either the transport of sulfate across the cell membrane or the activation reactions with ATP have an isotope effect; however, large fractionations should occur with the reduction steps (Rees, 1973; Winner *et al.*, 1981). In fact, the observed fractionation between sulfate and the cell organic sulfur is small. Values between 0.5 and −4.5‰ have been measured in marine and freshwater plants and microorganisms (Ishii, 1953; Kaplan *et al.*, 1963; Kaplan and Rittenberg, 1964; Mekhtieva and Kondrat'eva, 1966). Larger fractionations were documented in laboratory studies of *Chlorella* plants grown on concentrations of sulfate greater than 260μ*M* (Nriagu and Wong, 1975), but these results have not been duplicated. Thus, the consistently small fractionation during assimilation of sulfate suggests that the transport or activation steps are rate-limiting.

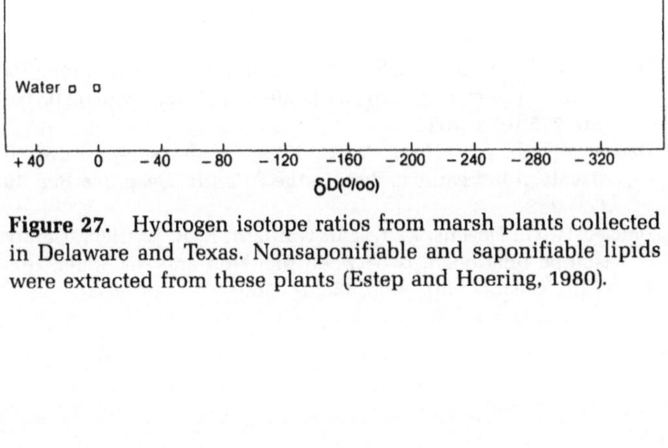

Figure 27. Hydrogen isotope ratios from marsh plants collected in Delaware and Texas. Nonsaponifiable and saponifiable lipids were extracted from these plants (Estep and Hoering, 1980).

8. Implications for Organic Geochemistry

What are the implications for the organic geochemist who is interested in tracing sources of organic matter in petroleum and kerogen with carbon isotope ratios? Basically, the δ¹³C of sedimentary carbon sources (marine and terrestrial) can be influenced by a number of environmental and biological parameters prior to deposition. Biomarkers are probably more specific indicators of the source of material to a kerogen or sediment, but isotope tracers in concert with this information are useful in describing paleobiochemistry of photosynthesis. Sharkey and Berry (1985) hypothesized that the negative δ¹³C values in Precambrian stromatolites and kerogens may be related to the fact that the active transport system for DIC had not evolved at that time (Raven and Sprent, 1989). The organic geochemist must be more discriminating in interpreting bulk δ¹³C values or even δ¹³C values for classes of compounds, as the variations are primarily controlled biologically by enzyme pathways in addition to the sources and availability of CO_2.

The δ¹⁵N of organic matter in sediments is influenced equally, or to a greater extent, by processes occurring after photosynthetic production. Although temporal variations have been observed in water column particulate matter, the data in sediments appear more conservative (Sweeney and Kaplan, 1980; Sweeney *et al.*, 1980, Cifuentes *et al.*, 1988). The discrepancy between water column and sedimentary data indicates that diagenetic changes occurring prior to deposition erased some of these significant variations. Gross isotopic differences between marine and terrestrial organic matter, however, can be maintained in refractory components that remain in sediments. When specific biomarkers (e.g., pigments; see Baker and Louda, 1986) are measured isotopically, it appears that some of the nitrogen and carbon isotopic imprints resulting from photosynthetic processes are preserved (Bidigare *et al.*, 1991; Hayes *et al.*, 1987; Takigiku, 1987; Popp *et al.*, 1989). This type of approach has important implications for research on source materials for petroleum. In addition, this approach can offer information about the paleobiochemistry of nitrogen assimilation during important geological periods such as the Cretaceous–Tertiary boundary.

The most recent developments in isotope geochemistry include the measurement of isotopic ratios in individual compounds separated by gas chromatography (Freeman *et al.*, 1990; Silfer *et al.*, 1991). It should be kept in mind, however, that all organic molecules that are preserved in ancient materials, such as chlorophyll, have been subjected to destruction by biological, chemical, and photooxidative processes. Therefore, it is imperative to establish whether these processes introduce fractionation, as typically 99 wt. % of the original molecules in living organisms are destroyed in recent sediments or the water column. Moreover, the internal fractionation of individual compounds within a living organism holds a complexity not yet thoroughly investigated (Macko *et al.*, 1987). The challenge for the future is to understand the detailed isotopic fractionations that lead to the distributions of isotopes within individual biochemical compounds and to relate this knowledge to the interpretation of isotopic compositions measured on similar molecules found in ancient rocks.

ACKNOWLEDGMENTS. We thank the following for critical reviews: Matthew P. Hoch, Thomas C. Hoering, Michael D. Lewan, Leo Sternberg, and David J. Velinsky. Unpublished data collected in collaboration with the following individuals have been used in some of the figures and interpretations in this chapter: Jonathan Sharp, Jonathan Pennock, David Velinsky, and John Ludlam (Delaware Estuary and phytoplankton cultures); Matthew Hoch and David Kirchman (bacterial ammonium uptake).

References

Ahmed, S. I., Kenner, R. A., and Packard, T. T., 1977, A comparative study of glutamate dehydrogenase activity in several species of marine phytoplankton, *Mar. Biol.* **39**:93–101.

Altabet, M. A., 1988, Variations in nitrogen isotopic compositions among particle classes: Implication for particle transformation and flux in the open ocean. *Deep-Sea Res.* **35**:535–554.

Altabet, M. A., and McCarthy, J. J., 1985, Temporal and spatial variations in the natural abundance of ^{15}N in PON from a warm core ring, *Deep-Sea Res.* **7**:755–772.

Altabet, M. A., and McCarthy, J. J., 1986, Vertical patterns in ^{15}N natural abundance in PON from the surface waters of several warm-core rings in the Sargasso Sea, *J. Mar. Res.* **44**:185–201.

Arthur, M. A., Dean, W. E., and Claypool, G. E., 1985, Anomalous ^{13}C-enrichment in modern marine organic carbon, *Nature* **315**: 216–218.

Baker, E. W., and Louda, J. W., 1986, Porphyrins in the geologic record, in: *Biological Markers in the Sedimentary Record* (R. B. Johns, ed.), Elsevier, Amsterdam, pp. 125–225.

Beardall, J., Griffiths, H., and Raven, J. A., 1982, Carbon isotopic discrimination and the CO_2 accumulating mechanisms in *Chlorella emersonii*, *J. Exp. Bot.* **33**:729–737.

Berry, J. A., Osmond, C. B., and Lorimer, G. H., 1978, Fixation of ^{18}O during photorespiration, *Plant Physiol.* **62**:739–742.

Bidigare, R. B., Kennicutt, M. C., Keeney-Kennicutt, W., and Macko, S. A., 1991, Isolation and purification of chlorophylls a and b for the determination of stable carbon and nitrogen isotope compositions, *Anal. Chem.* **63**:130–133.

Bigeleison, J., and Wolfsberg, M., 1958, Theoretical and experimental aspects of isotope effects in chemical kinetics, *Adv. Phys.* **1**: 15–76.

Black, A. S., and Waring, S. A., 1977, The natural abundance of ^{15}N in the soil–water system of a small catchment area, *Aust. J. Soil Res.* **15**:51–57.

Burk, R. L., and Stuiver, M., 1981, Oxygen isotope ratios in trees reflect mean annual temperature and humidity, *Science* **211**: 1417–1419.

Burns, B. D., and Beardall, J., 1987, Utilization of inorganic carbon by marine microalgae, *J. Exp. Mar. Biol. Ecol.* **107**:75–86.

Cai, D.-L., Tan, F. C., and Edmond, J. M., 1988, Sources and transport of particulate organic carbon in the Amazon River and Estuary, *Estuarine Coastal Shelf Sci.* **26**:1–14.

Checkley, D. M., and Miller, C. A., 1989, Nitrogen isotope fractionation by oceanic zooplankton, *Deep-Sea Res.* **36**:1449–1456.

Cifuentes, L. A., 1991, Spatial and temporal variations in terrestrially-derived organic matter from sediments of the Delaware Estuary, *Estuaries* **14**:414–429.

Cifuentes, L. A., Sharp, J. H., and Fogel, M. L., 1988, Stable carbon and nitrogen isotope biogeochemistry in the Delaware estuary, *Limnol. Oceanogr.* **33**:1102–1115.

Cifuentes, L. A., Fogel, M. L., Pennock, J. R., and Sharp, J. H., 1989, Seasonal variations in the stable nitrogen isotope ratio of ammonium in the Delaware estuary, *Geochim. Cosmochim. Acta* **53**:2713–2721.

Dean, W. E., Arthur, M. A., and Claypool, G. E., 1986, Depletion of ^{13}C in Cretaceous marine organic matter: Source, diagenetic, or environmental signal? *Mar. Geol.* **70**:119–157.

Deleens, E., Ferhi, A., and Queiroz, O., 1983, Carbon isotope fractionation by plants using the C4 pathway, *Physiol. Veg.* **21**: 897–905.

DeNiro, M. J., 1981, The effects of different methods of preparing cellulose nitrate on the determination of the D/H ratios of non-exchangeable hydrogen of cellulose, *Earth Planet. Sci. Lett.* **54**:177–185.

DeNiro, M. J., and Epstein, S., 1979, Relationship between the oxygen isotope ratios of terrestrial plant cellulose, carbon dioxide and water, *Science* **204**:51–53.

Descolas-Gros, C., and Fontugne, M. R., 1985, Carbon fixation in marine phytoplankton: Carboxylase activities and stable carbon isotope ratios; physiological and paleoclimatological aspects, *Mar. Biol.* **87**:1–6.

Deuser, W. G., 1970, Isotopic evidence for diminishing supply of available carbon during diatom bloom in the Black Sea, *Nature* **225**:1069–1071.

Deuser, W. G., and Degens, E. T., 1967, Carbon isotope fractionation in the system CO_2(gas)-CO_2(aqueous)-HCO_3^-(aqueous), *Nature* **215**:1033–1035.

Deuser, W. G., and Hunt, J. M., 1969, Stable isotope ratios of dissolved inorganic carbon in the Atlantic, *Deep-Sea Res.* **16**: 221–225.

Deuser, W. G., Degens, E. T., and Guillard, R. R. L., 1968, Carbon isotope relationships between plankton and sea water, *Geochim. Cosmochim. Acta* **32**:657–660.

Dongmann, G., Nurnberg, H. W., Forstel, H., and Wagener, K., 1974, On the enrichment of $H_2{}^{18}O$ in the leaves of transpiring plants, *Radiat. Environ. Biophys.* **11**:41–52.

Dortch, Q., Ahmed, S. I., and Packard, T. T., 1979, Nitrate reductase and glutamate dehydrogenase activities in *Skeletonema costatum* as measures of nitrogen assimilation rates, *J. Plank. Res.* **1**: 169–186.

Epstein, S., Yapp, C. J., and Hall, J. H., 1976, The determination of the D/H ratio of nonexchangeable hydrogen in cellulose extracted from aquatic and land plants, *Earth Planet. Sci. Lett.* **30**:241–251.

Epstein, S., Thompson, P., and Yapp, C. J., 1977, Oxygen and hydrogen isotope ratios in plant cellulose, *Science* **198**:1209–1215.

Estep, M. L. F., 1984, Carbon and hydrogen isotopic compositions of algae and bacteria from hydrothermal environments, Yellowstone National Park, *Geochim. Cosmochim. Acta* **48**:591–599.

Estep, M. F., and Hoering, T. C., 1980, Biogeochemistry of the stable hydrogen isotopes, *Geochim. Cosmochim. Acta* **44**:1197–1206.

Estep, M. F., and Hoering, T. C., 1981, Stable hydrogen isotope fractionations during autotrophic and mixotrophic growth of microalgae, *Plant Physiol.* **67**:474–477.

Estep, M. L. F., and Macko, S. A., 1984, Nitrogen isotope biogeochemistry of thermal springs, *Org. Geochem.* **6**:779–785.

Estep, M. L. F., and Vigg, S., 1985, Stable carbon and nitrogen isotope tracers of trophic dynamics in natural populations and fisheries of the Lahontan Lake System, Nevada, *Can. J. Fish. Aquat. Sci.* **42**:1712–1719.

Estep, M. F., Tabita, F. R., and Van Baalen, C., 1978, Purification of ribulose 1,5-bisphosphate carboxylase and carbon isotope fractionation by whole cells and carboxylase from *Cylindrotheca* sp. (Bacillariophyceae), *J. Phycol.* **14**:183–188.

Falkowski, P. G., 1975, Nitrate uptake in marine phytoplankton: (Nitrate, chloride)-activated adenosine triphosphatase from *Skeletonema costatum* (Bacillariophyceae), *J. Phycol.* **11**: 323–326.

Falkowski, P. G., and Rivkin, R. B., 1976, The role of glutamine synthetase in the incorporation of ammonium in *Skeletonema costatum* (Bacillariophyceae), *J. Phycol.* **12**:448–450.

Farquhar, G. D., 1980, Carbon isotope discrimination by plants: Effects of carbon dioxide concentration and temperature via the ratio of intercellular and atmospheric CO_2 concentrations, in: *Carbon Dioxide and Climate: Australian Research* (G. I. Pearman, ed.), Australian Academy of Science, pp. 105–110.

Farquhar, G. D., 1983, On the nature of carbon isotope discrimination in C_4 species, *Aust. J. Plant Physiol.* **10**:205–226.

Farquhar, G. D., O'Leary, M. H., and Berry, J. A., 1982, On the relationship between carbon isotope discrimination and the intracellular carbon dioxide concentration in leaves, *Aust. J. Plant Physiol.* **9**:121–137.

Fogel, M. L., Velinsky, D. J., Cifuentes, L. A., Pennock, J. R., and Sharp, J. H., 1988, Biogeochemical processes affecting the stable carbon isotopic composition of particulate carbon in the Delaware Estuary, *Carnegie Inst. Washington Annu. Rep. Director* **1988**:107–113.

Fontugne, M. R., and Duplessy, J.-C., 1978, Carbon isotope ratio of marine plankton related to surface water masses, *Earth Planet. Sci. Lett.* **41**:365–371.

Fontugne, M. R., and Duplessy, J.-C., 1981, Organic carbon isotopic fractionation by marine plankton in the temperature range −1 to 31°C, *Oceanol. Acta* **4**:85–90.

Freeman, K. H., Hayes, J. M., Trendel, J.-M., and Albrecht, P., 1990, Evidence from carbon isotope measurements for diverse origins of sedimentary hydrocarbons, *Nature* **343**:254–256.

Fry, B., and Sherr, E. B., 1984, $\delta^{13}C$ measurements as indicators of carbon flow in marine and freshwater ecosystems, *Contrib. Mar. Sci.* **27**:13–47.

Galimov, E. M., 1980, C13/C12 in kerogen, in: *Kerogen—Insoluble Organic Matter from Sedimentary Rocks* (B. Durand, ed.), Editions Technip., Paris, pp. 271–299.

Glibert, P. M., Lipshultz, F., McCarthy, J. J., and Altabet, M. A., 1982, Isotope dilution models and remineralization of ammonium by marine plankton, *Limnol. Oceanogr.* **27**:639–650.

Gonfiantini, R., Gratziu, S., and Tongoriori, E., 1965, Oxygen isotopic composition of water in leaves, in: *Isotope Radiation in Soil-Plant-Nutrition Studies*, International Atomic Energy Commission, Vienna, pp. 405–410.

Guy, R. D., Reid, D. M., and Krouse, H. R., 1986, Factors affecting $^{13}C/^{12}C$ ratios of inland halophytes. I. Controlled studies on growth and isotopic composition of *Puccinella nuttalliana*, *Can. J. Bot.* **64**:2693–2699.

Guy, R. D., Fogel, M. L., Berry, J. A., and Hoering, T. C., 1987, Isotope fractionation during oxygen production and consumption by plants, *Prog. Photosyn. Res. III.* **9**:597–600.

Hayes, J. M., Takigiku, R., Ocampo, R., Callot, H. J., and Albrecht, P., 1987, Isotopic compositions and probable origins of organic molecules in the Eocene Messel shale, *Nature* **329**:48–51.

Hedges, J. I., and Parker, P. L., 1976, Land-derived organic matter in surface sediments from the Gulf of Mexico, *Geochim. Cosmochim. Acta.* **40**:1019–1029.

Herczeg, A. L., and Fairbanks, R. G., 1987, Anomalous carbon isotope fractionation between atmospheric CO_2 and dissolved inorganic carbon induced by intense photosynthesis, *Geochim. Cosmochim. Acta* **51**:895–899.

Hermes, J. D., Weiss, P. M., and Cleland, W. W., 1985, Use of nitrogen-15 and deuterium isotope effects to determine the chemical mechanism of phenylalanine ammonia-lyase, *Biochemistry* **24**:2959–2967.

Hoch, M. P., Kirchman, D. L., and Fogel, M. L., 1989a, Nitrogen isotope fractionation in the uptake of ammonium by a marine bacterium, *Carnegie Inst. Washington Annu. Rep. Director* **1989**:117–123.

Hoch, M. P., Fogel, M. L., and Kirchman, D. L., 1992, The Isotope fractionation Uptake by *Vibrio harvey*, Limnol. Oceanogr., **37**: 1447–1459.

Hoering, T. C., 1977, The stable isotopes of hydrogen in Precambrian organic matter, in: *Chemical Evolution of the Early Precambrian* (C. Ponamperuma, ed.), Academic Press, New York, pp. 81–86.

Hoering, T. C., and Ford, H. T., 1960, Isotope effect in the fixation of nitrogen by *Azotobacter*, *J. Am. Chem. Soc.* **82**:376–378.

Ishii, M. M., 1953, A study of the fractionation of sulphur isotopes in the plant metabolism of sulphates, M.S. thesis, McMaster University.

Jacobsen, B. S., Smith, B. N., Epstein, S., and Laties, G. G., 1970, The prevalence of carbon-13 in respiratory carbon dioxide as an indicator of the type of endogenous substrate, *J. Gen. Physiol.* **55**:1–17.

Kaplan, I. R., and Rittenberg, S. C., 1964, Microbiological fractionation of sulfur isotopes, *J. Gen. Microbiol.* **34**:195–212.

Kaplan, I. R., Emery, K. O., and Rittenberg, S. C., 1963, The distribution and isotopic abundance of sulfur in recent marine sediments of Southern California, *Geochim. Cosmochim. Acta* **27**: 297–331.

Kleiner, D., 1981, The transport of NH_3 and $NH_4{}^+$ across biological membranes, *Biochim. Biophys. Acta* **639**:41–52.

Kleiner, D., 1985, Bacterial ammonium transport, *FEMS Microbiol. Rev.* **32**:87–100.

Kohl, D. H., and Shearer, G. B., 1980, Isotopic fractionation associated with symbiotic N_2 fixation and uptake of NO_3^- by plants, *Plant Physiol.* **66:**51–56.

Kohl, D. H., Shearer, G. B., and Commoner, B., 1971, Fertilizer nitrogen: Contribution to nitrate in surface water in a corn belt watershed, *Science* **174:**1331–1334.

Kroopnick, P., and Craig, H., 1972, Atmospheric oxygen—isotopic composition and solubility fractionation, *Science* **175:**54–55.

Kroopnick, P., Deuser, W. G., and Craig, H., 1970, Carbon 13 measurements on dissolved inorganic carbon at the North Pacific (1969) Geosecs Station, J. Geophys. Res. **75:**7668–7671.

LaZerte, B. D., and Szalados, J. E., 1982, Stable carbon isotope ratio of submerged freshwater macrophytes, *Limnol. Oceanogr.* **27:** 413–418.

Leaney, F. W., Osmond, C. B., Allison, G. B., and Ziegler, H., 1985, Hydrogen-isotope composition of leaf water in C3 and C4 plants: Its relationship to the hydrogen-isotope composition of dry matter, *Planta* **164:**215–220.

Lewan, M. D., 1986, Stable carbon isotopes of amorphous kerogens from Phanerozoic sedimentary rocks, *Geochim. Cosmochim. Acta* **50:**1583–1591.

Libes, S. M., and Deuser, W. G., 1988, The isotope geochemistry of particulate nitrogen in the Peru Upwelling area and the Gulf of Maine, *Deep-Sea Res.* **35:**517–533.

Lucas, W. J., 1983, Photosynthetic assimilation of exogenous HCO_3 by aquatic plants, *Annu. Rev. Plant Physiol.* **34:**71–104.

Lucas, W. J., and Berry, J. A., 1985, *Inorganic Carbon Uptake by Aquatic Photosynthetic Organisms*, American Society of Plant Physiologists, Rockville, Maryland.

Macko, S. A., Entzeroth, L., and Parker, P. L., 1984, Regional differences in nitrogen and carbon isotopes on the continental shelf of the Gulf of Mexico, *Naturwissen. Schaften* **71:**374–375.

Macko, S. A., Fogel, M. L., Hare, P. E., and Hoering, T. C., 1987, Isotopic fractionation of nitrogen and carbon in the synthesis of amino acids by microorganisms, *Chem. Geol.* **65:**79–92.

Mariotti, A., Mariotti, F., Amarger, N., Pizelle, G., Ngambi, J., Champigny, M., and Moyse, A., 1980, Fractionnements isotopiques de l'azote lors des processus d'absorption des nitrates et de fixation de l'azote atmosphérique par les plantes, *Physiol. Veg.* **18:**163–181.

Mariotti, A., Germon, J. C., Hubert, P., Kaiser, P., Letolle, R., Tardieux, A., and Tardieux, P., 1981, Experimental determination of nitrogen kinetic isotope fractionation, some principles; illustration for the denitrification and nitrification principles, *Plant Soil* **62:**413–430.

Mariotti, A., Mariotti, F., Champigny, M.-L., Amarger, N., and Moyse, A., 1982, Nitrogen isotope fractionation associated with nitrate reductase activity and uptake of NO_3^- by pearl millet, *Plant Physiol.* **69:**880–884.

Mariotti, A., Lancelot, C., and Billen, G., 1984, Natural isotopic composition of nitrogen as a tracer of origin for suspended organic matter in the Scheldt estuary, *Geochim. Cosmochim. Acta* **48:**549–555.

Mekhtieva, V. L., and Kondrat'eva, E. N., 1966, Isotope composition of sulfur of plants and animals from reservoirs of different salinity, *Geokhimiya* **6:**725–730.

Miller, A. G., 1985, Study of inorganic carbon transport: The kinetic approach, in: *Inorganic Carbon Uptake by Aquatic Photosynthetic Organisms*, (W. J. Lucas, and J. A., Berry, eds.), American Society of Plant Physiology, Rockville, Maryland.

Minagawa, M., and Wada, E., 1986, Nitrogen isotope ratios of red tide organisms in the East China Sea: A characterization of biological nitrogen fixation, *Mar. Chem.* **19:**245–259.

Miyake, Y., and Wada, E., 1971, The isotope effect on the nitrogen in biochemical, oxidation–reduction reactions, *Rec. Oceanogr. Works Jpn.* **11:**1–6.

Mook, W. G., Bommerson, J. C., and Staverman, W. H., 1974, Carbon isotope fractionation between dissolved bicarbonate and gaseous carbon dioxide, *Earth Planet. Sci. Lett.* **51:**64–68.

Nier, A. O., 1947, A mass spectrometer for isotope and gas analysis, *Rev. Sci. Instrum.* **18:**398–411.

Northfeldt, D. W., DeNiro, M. J., and Epstein, S., 1981, Hydrogen and carbon isotopic ratios of the cellulose nitrate and saponifiable lipid fraction prepared from annual growth rings of a California redwood, *Geochim. Cosmochim. Acta* **45:**1895–1898.

Nriagu, J. O., and Wong, P. T. S., 1975, Fractionation of sulfur isotopes by algae, in: *Proceedings of the Second International Conference on Stable Isotopes* (E. R. Klein and P. D. Klein, eds.), Springfield, Virginia, National Technical Information Service, pp. 734–738.

Oana, S., and Deevey, E. S., 1960, Carbon-13 in lake waters and its possible bearing on paleolimnology, *Am. J. Sci.* **258A:**811–814.

O'Leary, M. H., 1978, in: *Transition States of Biochemical Processes* (R. D. Gandour and R. L. Schowen, eds.), Plenum Press, New York, p. 292.

O'Leary, M. H., 1981, Carbon isotope fractionation in plants, *Phytochemistry* **20:**553–567.

O'Leary, M. H., 1988, Carbon isotopes in photosynthesis, *Bioscience* **38:**328–336.

O'Neil, A., 1968, Hydrogen and oxygen isotope fractionation between ice and water, *J. Phys. Chem.* **72:**3683–3684.

Owens, N. J. P., 1987, Natural variations in ^{15}N in the marine environment, *Adv. Mar. Biol.* **24:**411–451.

Packard, T. T., 1979, Half-saturation constants for nitrate reductase and nitrate translocation in marine phytoplankton, *Deep-Sea Res.* **26:**321–326.

Pardue, J. W., Scalan, R. S., Van Baalen, C., and Parker, P. L., 1976, Maximum carbon isotope fractionation in photosynthesis by blue-green algae and a green alga, *Geochim. Cosmochim. Acta* **40:**309–312.

Pennock, J. R., Sharp, J. H., Ludlam, J., Velinsky, D. J., and Fogel, M. L., 1988, Isotopic fractionation of nitrogen during the uptake of NH_4^+ and NO_3^- by *Skeletonema costatum*, *Eos* **69:** 1098.

Peters, K. E., Sweeney, R. E., and Kaplan, I. R., 1978, Corrolation of carbon and nitrogen stable isotope ratios in sedimentary organic matter, *Limnol. Oceanogr.* **23:**598–604.

Peterson, B. J., and Fry, B., 1989, Stable isotopes in ecosystems studies, *Annu. Rev. Ecol. Syst.* **18:**293–320.

Popp, B. N., Takigiku, R., Hayes, J. M., Louda, J. W., and Baker, E. W., 1989, The postpaleozoic chronology and mechanism of ^{13}C depletion in primary marine organic matter, *Am. J. Sci.* **289:** 436–454.

Quay, P. D., Emerson, S. R., Quay, B. M., and Devol, A. H., 1986, The carbon cycle for Lake Washington—A stable isotope study, *Limnol. Oceanogr.* **31:**596–611.

Rau, G., 1978, Carbon-13 depletion in a subalpine lake: Carbon flow implications, *Science* **201:**901–902.

Rau, G. H., Sweeney, R. E., and Kaplan, I. R., 1982, Plankton 13C:12C ratio changes with latitude: Differences between northern and southern oceans, *Deep-Sea Res.* **29:**1035–1039.

Rau, G. H., Arthur, M. A., and Dean, W. E., 1987, $^{15}N/^{14}N$ variations in Cretaceous Atlantic sedimentary sequences: Implications for past changes in marine nitrogen biogeochemistry, *Earth Planet. Sci. Lett.* **82:**269–279.

Rau, G. H., Takahashi, T., and DesMarais, D. J., 1989, Latitudinal variations in plankton $\delta^{13}C$: Implication for CO_2 and productivity in past oceans, *Nature* **341**:516–518.

Raven, J. A., and Sprent, J. I., 1989, Phototrophy, diazotrophy and paleoatmospheres: Biological catalysis and the H, C, N, and O cycles, *J. Geol. Soc.*, London **146**:167–170.

Raven, J. A., Macfarlane, J. J., and Griffiths, H., 1987, The application of carbon isotope discrimination techniques, in: *Plant Life in Aquatic and Amphibious Habitats*. B. E. S. Special Symposium (R. M. M. Crawford, ed.), Blackwell Scientific Publications, Oxford.

Redding, C. E., Schoell, M., Monin, J. C., and Durand, B., 1980, Hydrogen and carbon isotopic composition of coals and kerogens, in: *Advances in Organic Geochemistry 1979* (A. G. Douglas and J. R. Maxwell, eds.), Pergamon, London, pp. 711–723.

Rees, C. E., 1973, A steady-state model for sulfur isotope fractionation in bacterial reduction processes, *Geochim. Cosmochim. Acta* **35**:625–628.

Roeske, C. A., and O'Leary, M. H., 1984, Carbon isotope effects on the enzyme-catalyzed carboxylation of ribulose bisphosphate, *Biochemistry* **23**:6275–6284.

Sackett, W. M., Eckelamnn, W. R., Bender, M. L., and Be, A. W. H., 1964, Temperature dependence of carbon isotope composition in marine plankton and sediments, *Science* **148**:235–237.

Saino, T., and Hattori, A., 1980, ^{15}N natural abundance in oceanic suspended particulate matter, *Nature* **283**:752–754.

Saino, T., and Hattori, A., 1985, Variation in ^{15}N natural abundance of suspended organic matter in shallow oceanic waters, in: *Marine and Estuarine Geochemistry* (A. C. Sigleo and A. Hattori, eds.), Lewis Publishers, Chelsea, Michigan, pp. 1–13.

Saino, T., and Hattori, A., 1987, Geographical variation of the water column distribution of suspended particulate nitrogen and its ^{15}N abundance in the Pacific and its marginal seas, *Deep-Sea Res.* **28**:901–909.

Scheigl, W. E., and Vogel, J. C., 1970, Deuterium content of organic matter, *Earth Planet, Sci. Lett.* **7**:307–313.

Schidlowski, M., 1988, A 3,800-million-year isotopic record of life from carbon in sedimentary rocks, *Nature* **333**:313–318.

Schimmelmann, A., and DeNiro, M. J., 1985, Determination of oxygen stable isotope ratios in organic matter containing carbon, hydrogen, oxygen, and nitrogen, *Anal. Chem.* **57**:2644–2646.

Schoeninger, M. J., and DeNiro, M. J., 1984, Nitrogen and carbon isotope composition of bone collagen from marine and terrestrial animals, *Geochim. Cosmochim. Acta* **48**:625–639.

Sharkey, T. D., and Berry, J. A., 1985, Carbon isotope fractionation of algae as influenced by an inducible CO_2 concentrating mechanism, in: *Inorganic Carbon Uptake by Aquatic Photosynthetic Organisms* (W. J. Lucas and J. A. Berry, eds.), American Society of Plant Physiology, Rockville, Maryland.

Sherr, E. B., 1982, Carbon isotope composition of organic seston and sediments in a Georgia salt marsh, *Geochim. Cosmochim. Acta* **46**:1227–1232.

Sigleo, A. C., and Macko, S. A., 1985, Stable isotope and amino acid composition of estuarine dissolved colloidal material, in: *Marine and Estuarine Geochemistry* (A. C. Sigleo and A. Hattori, eds.), Lewis Publishers, Chelsea, Michigan, pp. 29–46.

Silfer, J. A., Engel, M. H., Macko, S. A., and Jumeau, E. J., 1991, Stable carbon isotope analysis of amino acid enantiomers by conventional isotope ratio mass spectrometry and combined gas-chromatography–isotope ratio mass spectrometry, *Anal. Chem.* **63**:370–374.

Smith, B. N., and Epstein, S., 1971, Two categories of $^{13}C/^{12}C$ ratios for higher plants, *Plant Physiol.* **47**:380–384.

Spiker, E. C., and Schemel, L. E., 1979, Distribution and stable-isotope composition of carbon in San Francisco Bay, in: *San Francisco Bay: The Urbanized Estuary*, Pacific Division of the American Association for the Advancement of Science, San Francisco.

Sternberg, L. S. L., 1988a, Oxygen and hydrogen isotope ratios in plant cellulose: Mechanisms and applications, in: *Stable Isotopes in Ecological Research* (P. W. Rundel, J. R. Ehleringer, and K. A. Nagy, eds.), Springer-Verlag, New York, pp. 124–141.

Sternberg, L. S. L., 1988b, D/H ratios of environmental water recorded by D/H ratios of plant lipids, *Nature* **333**:59–61.

Sternberg, L. S. L., and DeNiro, M. J., 1983a, Isotopic composition of cellulose from C_3, C_4, and CAM plants growing in the vicinity of one another, *Science* **220**:947–948.

Sternberg, L. S. L., and DeNiro, M. J., 1983b, Biogeochemical implications of the isotopic equilibrium fractionation factor between the oxygen atoms of acetone and water, *Geochim. Cosmochim. Acta* **47**:2271–2274.

Sternberg, L., DeNiro, M. J., and Keeley, J. E., 1984, Carbon, hydrogen and oxygen isotope ratios of cellulose from plants having intermediate photosynthetic modes, *Plant Physiol.* **74**:104–107.

Sternberg, L. S. L., DeNiro, M. J., and Savidge, R. A., 1986, Oxygen isotope exchange between metabolites and water during biochemical reactions leading to cellulose synthesis, *Plant Physiol.* **82**:423–427.

Stiller, M., and Nissenbaum, A., 1980, Variations of stable hydrogen isotopes in plankton from a freshwater lake, *Geochim. Cosmochim. Acta* **44**:1099–1101.

Stuiver, M., 1978, Atmospheric carbon dioxide and carbon reservoir changes, *Science* **199**:253–258.

Sweeney, R. E., and Kaplan, I. R., 1980, Natural abundance of ^{15}N as a source indicator for near-shore marine sedimentary and dissolved nitrogen, *Mar. Chem.* **9**:81–94.

Sweeney, R. E., Liu, K. K., and Kaplan, I. R., 1978, Oceanic nitrogen isotopes and their uses in determining the source of sedimentary nitrogen, in: Department of Scientific and Industrial Research Bulletin 220 (B. Robinson, ed.), Science Information Division, DSIR, Wellington, New Zealand, pp. 9–26.

Sweeney, R. E., Kahil, E. K., and Kaplan, I. R., 1980, Characterization of domestic and industrial sewage in Southern California coastal sediments using nitrogen, carbon, sulfur, and uranium tracers, *Mar. Environ. Res.* **3**:225–243.

Takigiku, R., 1987, Isotopic and molecular indicators of origins of organic compounds in sediments, Ph.D. thesis, University of Indiana.

Thompson, P., and Gray, J., 1977, Determination of $^{18}O/^{16}O$ ratios in compounds containing C,H and O, *Intr. J. Appl. Radiat. Isot.* **28**:411–415.

Velinsky, D. J., Sharp, J. H., Pennock, J. R., and Fogel, M. L., 1988, Isotopic variability of nitrate and ammonium in the lower Delaware Bay and coastal waters, *Eos* **69**:1103.

Velinsky, D. J., Pennock, J. R., Sharp, J. H., Cifuentes, L. A., and Fogel, M. L., 1989, Determination of the isotopic composition of ammonium-nitrogen at the natural abundance level from estuarine waters, *Mar. Chem.* **26**:351–361.

Velinsky, D. J., Burdige, D. J., and Fogel, M. L., 1991, Nitrogen diagenesis in anoxic marine sediments: Isotope effects, *Carnegie Inst. Washington Annu. Rep. Director* **1991**:154–162.

Vogel, J. C., 1980, Fractionation of the carbon isotopes during photosynthesis, *Sitzungsber, Heidelb. Akad. Wiss. Math.-Naturwiss. Kl* **3**:111–135.

Vogel, J. C., Grootes, P. M., and Mook, W. G., 1970, Isotopic fractionation between gaseous and dissolved carbon dioxide, *Z. Phys.* **230**:225–238.

Wada, E., 1980, Nitrogen isotope fractionation and its significance in biogeochemical processes occurring in marine environments, in: *Isotope Marine Chemistry* (E. D. Goldberg, Y. Horibe, and K. Saruhashi, eds.), Uchida Rokakudo Pub. Co., Tokyo, pp. 375–398.

Wada, E., and Hattori, A., 1976, Natural abundance of ^{15}N in particulate organic matter in the North Pacific Ocean, *Geochim. Cosmochim. Acta* **40**:249–251.

Wada, E., and Hattori, A., 1978, Nitrogen isotope effects in the assimilation of inorganic nitrogenous compounds by marine diatoms, *Geomicrobiology* **1**:85–101.

Weiss, P. M., Chen, C.-Y., Cleland, W. W., and Cook, P. F., 1988, Use of primary deuterium and ^{15}N isotope effects to deduce the relative rates of steps in the mechanisms of alanine and glutamate dehydrogenases, *Biochemistry* **27**:4814–4822.

White, J. W. C., 1988, Stable hydrogen isotope ratios in plants: A review of current theory and some potential applications, in: *Stable Isotopes in Ecological Research* (P. W. Rundel, J. R. Ehleringer, and K. A. Nagy, eds.), Springer-Verlag, New York, pp. 142–162.

Winner, W. E., Smith, C. L., Koch, G. W., Mooney, H. A., Bewley, J. D., and Krouse, H. R., 1981, Rates of emission of H_2S from plants and patterns of stable sulphur isotope fractionation, *Nature* **289**:672–673.

Ziegler, H., 1979, Diskriminierung von Kohlenstoff-und Wasserstoff isotopen: Zusammenhange mit dem Photosynthesemechanismus und den Standortbedingungen, *Ber. Deutsch. Bot. Ges. Bd.* **92**:169–184.

Ziegler, H., 1988, Hydrogen isotope fractionation in plant tissues, in: *Stable Isotopes in Ecological Research* (P. W. Rundel, J. R. Ehleringer, and K. A. Nagy, eds.), Springer-Verlag, New York, pp. 105–123.

Ziegler, H., Osmond, C. B., Stichler, W., and Trimborn, P., 1976, Hydrogen isotope discrimination in higher plants: Correlations with photosynthetic pathway and environment, *Planta* **128**:85–92.

II

Early Diagenesis of Organic Matter

Chapter 4

Early Diagenesis of Organic Matter: The Dynamics (Rates) of Cycling of Organic Compounds

SUSAN M. HENRICHS

1. Introduction

Major changes in the concentration and composition of organic matter occur in the upper layers of aquatic sediments [see chapters by Hedges and Prahl (Chapter 11), Macko, Engel, and Parker (Chapter 9), and Meyers and Ishiwatari (Chapter 8) in this volume]. The remineralization and most of the transformations of organic matter occur because of the energy these processes yield to sediment organisms. Certain transformations yield specific organic compounds required for growth and reproduction. Others occur because of changes in sediment chemistry resulting from organic matter oxidation. Much of the organic matter diagenesis in sediments is rapid compared to geologic time scales, occurring in less than 1000 years. Study of rates of organic matter decomposition can help to identify which characteristics of the environment, the sediment, or the deposited organic matter are important in determining whether an organic substance is remineralized, altered, or preserved unchanged during early diagenesis. Results of such studies have numerous broader implications and potential applications.

For example, information on the factors controlling the rate of decomposition of organic matter in different sedimentary environments leads to better understanding of differences in organic carbon burial rates. Because the composition of the atmosphere is linked to the amount of organic matter, carbonates, and sulfur buried in sediments, this understanding is necessary for identifying the processes responsible for changes in the atmosphere over geologic time (Berner, 1989).

Biomarkers (organic molecules which have a specific source organism or group of organisms) have widespread applications in identifying the sources of organic matter in both recent and ancient sediments and in fossil fuels [see chapters by Hedges and Prahl (Chapter 11), Philp (Chapter 19), Zumberge (Chapter 20), and Curiale (Chapter 21) in this volume]. Yet most biomarkers are subject to alteration or decomposition in the near-surface zone of sediments. If we understood why a fraction of such molecules are preserved, we would be in a much better position to quantitatively interpret their occurrence in sediments and sedimentary rocks.

In this chapter, both the available methods for

SUSAN M. HENRICHS • Institute of Marine Science, University of Alaska, Fairbanks, Alaska 99775-1080.
Organic Geochemistry, edited by Michael H. Engel and Stephen A. Macko. Plenum Press, New York, 1993.

rate measurements and selected results of rate measurements between the sediment–water interface and about 1-m depth in the sediment will be discussed. The discussion will be limited to naturally occurring organic compounds. Much of the discussion will apply to both freshwater and marine sediments, but there are significant differences in anaerobic organic matter decomposition processes (Capone and Kiene, 1988). The most significant are that little sulfate is present in freshwater sediments, whereas it is a major oxidant in marine sediments, and that lignin and structural polysaccharides such as cellulose are important constituents of the organic matter supplied to most freshwater sediments, but are minor constituents in nearly all marine sediments.

2. Methods of Measuring Decomposition and Transformation Rates during Early Diagenesis

Methods of biogeochemical rate measurement are summarized in Table I. Although efforts to measure the rates of early diagenesis began only about 25 years ago, the available techniques are too numerous to discuss in detail. Instead, the methods will be described in general categories, and discussion of the common advantages and disadvantages will be included. Examples which illustrate particular methods are described in Sections 4.1–4.4.

2.1. Direct Methods

Direct methods are those in which decomposition or alteration of a sediment organic substance is observed over time. Alternatively, consumption of the organic matter oxidant (e.g., oxygen or sulfate) or production of the reduced form of the oxidant is assayed. The time scales of experiments range from a few minutes to several years. Changes in either a natural sediment constituent or an added material can be followed. Reeburgh (1983) reviewed direct rate measurements in anoxic sediments.

An important category within this group of methods is the radiotracer techniques, in which a radioactively labeled form of the substance being studied is added to the sediment. After incubation, labeled products are collected. Tracer approaches usually improve sensitivity, allowing measurements of slower rates or use of shorter incubation times than would

TABLE I. Methods of Direct and Indirect Rate Measurements for Early Diagenetic Processes

Process measured	Brief description	Reference(s)
A. *Direct methods*		
Electron-acceptor consumption or reduced product formation		
Oxygen consumption	Decrease of oxygen concentration in seawater overlying incubated cores	Jørgensen, 1983[a]
	Decrease of oxygen concentration in water contained in *in situ* benthic flux chamber	Smith and Hinga, 1983[a]
Denitrification	N_2 production by sediments incubated in atmospheric-nitrogen-free chambers	Seitzinger, 1988[a]
	N_2O production by sediment cores after addition of acetylene, which inhibits reduction of N_2O, an intermediate in denitrification	
	$^{15}N_2$ production from $^{15}NO^-{}_3$ and $^{15}NH_4{}^+$ added to water overlying incubated cores	
Iron reduction	Increase in Fe(II) concentration in pore waters of incubated sediments	Lovely and Phillips, 1986
Sulfate reduction	Various	Skyring, 1987[a]
	Decrease of sulfate concentration in pore waters of sediments incubated in sealed tubes	Crill and Martens, 1987
	Reduction of $^{35}SO_4{}^{2-}$ to ^{35}S-labeled reduced sulfur compounds[b]	Jørgensen, 1978; Howarth and Jørgensen, 1984; Edenborn *et al.*, 1987[c]
Methane production	$^{14}CH_4$ production from ^{14}C-labeled acetate and $HCO_3{}^{-}$[b]	Crill and Martens, 1986
Hydrogen production	Increase in sediment H_2 concentration after $SO_4{}^{2-}$ reduction and CH_4 production inhibited by $MoO_4{}^{2-}$ and BES[d]	Novelli *et al.*, 1987

TABLE I. *(Continued)*

Process measured	Brief description	Reference(s)
Remineralization product production		
Ammonium production	$^{15}NH_4^+$ isotope dilution	Blackburn, 1979
Remineralization of organic substances		
Acetate oxidation	$^{14}CO_2$ production from ^{14}C-labeled acetate[b]	Christensen and Blackburn, 1982
Short-chain organic acid oxidation or fermentation	$^{14}CO_2$ and $^{14}CH_4$ production from ^{14}C-labeled acetate[b]	Shaw *et al.*, 1984; Sansone, 1986
Acetate oxidation by sulfate reduction	As above, except sulfate reduction inhibited by MoO_4^{2-}; decrease in rate gives oxidation by sulfate	Shaw *et al.*, 1984
Amino acid oxidation	$^{14}CO_2$ and ^{14}C-biomass production from ^{14}C-labeled amino acids[b]	Christensen and Blackburn, 1980; Sugai and Henrichs, 1992
Glucose uptake by bacteria	Uptake of ^{14}C-labeled glucose from pore water	King and Klug, 1982
Methane oxidation	$^{14}CO_2$ production from $^{14}CH_4$, anaerobic	Alperin and Reeburgh, 1985
Remineralization of algal detritus, bacterial polymer, melanoidins	$^{14}CO_2$ production from ^{14}C-labeled organic matter incubated in jars containing sediment and overlying seawater	Henrichs and Doyle, 1986
Remineralization of sediment particulate organic carbon	Decrease in POC concentration in incubated cores; oxic, bioturbated, and anoxic cores compared	Kristensen and Blackburn, 1987
B. *Indirect methods*		
Electron-acceptor consumption		
Oxygen consumption	Sediment pore-water oxygen concentration gradient measured using microelectrode; consumption calculated from gradient by one-dimensional diffusion model	Reimers *et al.*, 1986
Sulfate reduction	Model-calculated reduction rate from sediment pore-water sulfate concentration vs. depth[e]	Berner, 1980b; Martens and Klump, 1984[f]
Remineralization product production		
Total carbon dioxide production (and ammonium production)	Model of dissolved total carbon dioxide (and ammonium) concentrations in pore water vs. depth[e]	Henrichs and Farrington, 1987
Decomposition of organic substances		
Total organic matter decomposition	Model of sediment total organic carbon concentration vs. depth; illustrates large effect of bioturbation on calculated rates[e]	Grundmanis and Murray, 1982; Berner and Westrich, 1985
Organic nitrogen and phosphorus remineralization	Model of sediment total nitrogen and phosphorus concentrations vs. depth[e]	Klump and Martens, 1987
Total hydrolyzable amino acid decomposition	Model of total hydrolyzable amino acid concentrations vs. depth[e]	Henrichs and Farrington, 1987; Burdige and Martens, 1988

[a]Review.
[b]A small volume (100 μl or less) of a solution of the specified radiotracer was injected into sediment, and the products were collected after incubation.
[c]This reference is one of few where sulfate reduction rates in offshore sediments were measured.
[d]BES, 2-Bromoethanesulfonic acid. Sediment H_2 production was calculated from the H_2 concentration in a headspace above the sediment.
[e]The model used to calculate the specified rate was a version of the one-dimensional, steady-state sediment model described in detail by Berner (1980a). Particular adaptations of the model are noted in the table. Unless stated otherwise, the model used did not include terms for adsorption of dissolved solutes, bioirrigation, or bioturbation.
[f]Model result was compared to direct sulfate reduction rate measurements.

otherwise be possible. Tracer methods are also particularly useful when the product(s) of the process being studied could have other sources. For example, CO_2 produced in sediments is the product of the oxidation of many organic compounds, but use of [^{14}C]acetate makes it possible to measure the rate of $^{14}CO_2$ production from acetate alone. If the substance under study is both produced and consumed in sediments, radiotracer techniques usually measure gross consumption rates, since the precursors do not become radiolabeled during short-term experiments.

A variation on the direct rate measurement approach involves addition of a metabolic inhibitor affecting one process, so that rates of other processes can be measured. This approach is particularly useful when the concentration of the sediment constituent being studied is normally invariant, because production and consumption rates are equal. If production or consumption can be stopped, then the rate of concentration decrease or increase will give the rate of the inhibited reaction. However, nonspecific effects of inhibitors can occur, and effects on rates other than the "target" rate can invalidate the results.

Another important direct approach to measuring rates of sediment diagenesis is the measurement of fluxes of reactants or products across the sediment–water interface. If steady state is assumed, the influx of reactant is the sum of the depth-integrated rate of the consuming reaction(s) and the burial rate of reactant. If the system is in steady state and if the reaction product remains in solution, the efflux of product is the depth-integrated reaction rate minus the product burial rate.

An advantage of direct methods is that the duration of the experiment is known and data can be collected at many time points. The time over which decomposition has occurred is often a major uncertainty in indirect rate estimates (see Section 2.2). Another advantage is that the conditions of the experiment can be controlled and manipulated. A serious disadvantage is that very slow processes cannot be studied, since incubations longer than a year are usually impractical. A second concern is that the experimental manipulations or incubation conditions will change the rates being measured. Sediment biology and chemistry can change, especially during long incubations, altering rates.

2.2. Indirect Methods

Indirect methods use as their data base sediment depth profiles of organic substances or inorganic molecules which are produced or consumed during organic matter decomposition. Rate information is obtained by means of mathematical models (Berner, 1980a). The models require knowledge of or assumptions about the processes responsible for producing a given depth distribution. For a solid-phase constituent, at least the following questions must be answered. Is the composition of the material deposited to the sediment–water interface constant, or have changes occurred over time? What is the sediment accumulation rate, and has it remained constant over time? Does bioturbation affect the depth distribution of sediment particles, and, if so, what is the bioturbation rate? What chemical and biological processes affect the substance of interest, and what kinetic model is appropriate for these processes? For a dissolved constituent, the following questions must also be answered. What is the sediment diffusion coefficient for this substance? Is it reversibly or irreversibly adsorbed by the solid phase of the sediment? Does bioirrigation affect the depth profile?

Obviously, considerable information about the sediment and the processes affecting organic matter are needed to apply this approach, and this is the major disadvantage of indirect techniques. Certain kinds of required information can be difficult to obtain. In particular, sediment composition is usually studied at only one time. It is often not possible to prove that the depth profiles have remained constant over time. Another limitation is that only net rates can be estimated. If a species is both produced and consumed within the sediment, concentration changes with depth will depend on the production rate minus the consumption rate. The kinetics of important biological and geochemical reactions are unknown. The remineralization of reactive organic matter is usually assumed to be first-order with respect to organic matter concentration, because concentration versus depth profiles of sediment organic matter and porewater sulfate, total CO_2, and ammonium often can be fit to exponential equations. Models can include multiple pools of organic matter decomposing with several different rate constants (Westrich and Berner, 1984) or a rate parameter which decreases with time (Middelburg, 1989).

Despite these problems, the indirect approach has been used successfully. It is particularly useful for estimating rates which are too slow to measure directly, on time scales longer than a few years. Another advantage of this approach is that the decomposition of organic matter occurred under *in situ* conditions; there is no concern about artifacts introduced by artificial incubation conditions.

3. Key Processes and Rates

It might seem that the best way to measure rates of organic matter decomposition in sediments would either be to directly observe the rate of decrease of organic content of incubated sediments or to model total organic carbon (TOC) concentrations versus depth in sediments. These approaches are rarely very useful, however. Sediment organic matter decomposes so slowly that TOC concentrations may not change over months or even years. Models can give inaccurate results when applied to TOC distributions because rapid decomposition near the sediment–water interface results in steep gradients in concentration which are difficult to measure (e.g., Martens and Klump, 1984) and because bioturbation homogenizes surface sediments, leading to large underestimates of decomposition rates (Berner and Westrich, 1985).

Oxygen consumption and sulfate reduction are the two rate measurements done most often in sediments (Jørgensen, 1983; Smith and Hinga, 1983; Skyring, 1987). Oxygen and sulfate are the major organic matter oxidants in oxic and anoxic marine sediments, respectively, and thus their consumption rates are closely related to total rates of organic matter remineralization in many sediments. Because of the presence of sulfate, the oxidizing capacity of marine sediments is greater than that of freshwater sediments. In freshwater anoxic sediments, fermentation and methanogenesis, which do not result in net oxidation of carbon, are dominant (Capone and Kiene, 1988).

Denitrification rates are measured because of the importance of the loss of fixed nitrogen to ecosystem nitrogen budgets (Seitzinger, 1988). However, denitrification is not responsible for the remineralization of a significant fraction of the organic matter in sediments (Sørensen *et al.*, 1979; Bender and Heggie, 1984). Iron and manganese also appear to be relatively minor oxidants in most sediments (Bender and Heggie, 1984), probably because most of the iron and manganese is not available for reduction (Lovely, 1987). However, the microbial reduction of iron and manganese has a major effect on the geochemistry of suboxic sediments and overlying oxidized layers.

Remineralization rates of specific organic compounds have also been measured. The short-chain organic acids have been studied more often than all other compounds combined. These acids, particularly acetate, are key intermediates in the remineralization of organic matter in anoxic sediments. That is, much of the organic matter remineralized is decomposed to acetate before being oxidized to CO_2. However, there are some puzzling inconsistencies in acetate decomposition studies (see Section 4.4).

Oxygen consumption, sulfate reduction, methanogenesis (in freshwater sediments), and acetate oxidation have so far been identified as key processes in early diagenesis, because they are closely related to the major pathways of carbon oxidation. The following discussion will consider these processes and related rate measurements in more detail.

4. Accuracy of Rate Measurements

A major difficulty in assessing the accuracy of rate measurements is determining the "right" answer. Two approaches have been used: comparison to another method of measuring the same rate, and comparison to other rates measured in the same sediment. The rationale for the second approach is given in Table II. Certain rates have expected quantitative relationships to other rates; if the expectation is met, then confidence in the accuracy of both rate measurements is increased.

4.1. Oxygen Consumption Rates

Two potential sources of error in the measurement of oxygen consumption rates are artifacts introduced by enclosing the sediment for incubation and the difficulty in distinguishing oxygen consumption due to oxygen respiration from that due to oxidation of H_2S and other reduced substances produced in anoxic sediments (Smith and Hinga, 1983). *In situ* respirometer measurements give the most accurate results for the deep sea (Smith and Hinga, 1983), because of temperature and pressure effects on benthic metabolism. In deep-sea sediments, oxygen fluxes predicted by a diffusive model of pore-water oxygen concentration gradients measured using a microelectrode and fluxes measured using an *in situ* benthic respirometer sometimes agreed within the errors of the methods (Reimers and Smith, 1986). Discrepancies found in some sediments were attributed to nondiffusive exchange processes, i.e., bioirrigation (Reimers *et al.*, 1986).

Another way to assess the accuracy of oxygen consumption rate measurements is to compare them to the flux of organic matter to the sediment surface measured using sediment traps. In the Atlantic, the supply of carbon to the seafloor is approximately equal to the consumption by oxidation (Smith and

TABLE II. Expected Quantitative Relationships between Rates of Different Biogeochemical Processes in Sediments

Type of sediments	Hypothesized decomposition pathway	Expected rate relationships[a]
Oxic sediments	Organic molecules are broken down into monomers such as sugars and amino acids, which are oxidized directly by bacteria using oxygen as the electron acceptor. Organic matter is also ingested and oxygen respired by higher organisms.	Sediment oxygen uptake rate is equal to total organic matter oxidation rate (moles of CO_2 produced per unit time). Oxidation rate of any single organic molecule (e.g., monosaccharide, amino acid, short-chain organic acid), in terms of the CO_2 production rate, is much less than the oxygen consumption rate.
Anoxic marine sediments (sulfate present at non-limiting concentrations)	Organic molecules are broken down into monomers which are fermented to short-chain organic acids including acetate; acetate and other acids are oxidized to CO_2 by sulfate-reducing bacteria.	Total organic matter oxidation rate (in terms of the CO_2 production rate) is equal to twice the sulfate reduction rate. Acetate oxidation rate is less than or equal to sulfate reduction rate.
	Oxidation of amino acids and other organic nitrogen compounds releases ammonium.	Amino acid oxidation rate (in terms of the CO_2 production rate) is much less than twice the sulfate reduction rate. Oxidation rate of all amino acids is less than or equal to ammonium production rate.
Methanogenic, low-sulfate sediments	As for sulfate-containing anoxic sediments, except that, in the absence of sulfate, organic acids are fermented to methane, other acids, CO_2, and H_2.[b]	Methane efflux rate from the methanogenic zone is equal to the production rate (rate of CO_2 reduction plus acid fermentation) within the zone.
Oxic sediments overlying anoxic sediments	See above descriptions of oxic and anoxic sediments. In addition, soluble reduced substances produced in the anoxic zone (e.g., Fe(II), S^{2-}, CH_4) diffuse upward[c] and are oxidized in the oxic zone.[d]	Oxygen consumption rate at the sediment–water interface is less than the total organic matter oxidation rate (in terms of CO_2 produced) within the sediment. However, the two rates can be similar if retention of reduced products of organic matter oxidation in the anoxic zone is small and these products are oxidized near the sediment–water interface. The flux of a reduced product (e.g., methane) out of the sediment is equal to the production rate in the anoxic zone minus the oxidation rate in overlying sediment.

[a]All rates expressed in units of moles consumed per unit time, unless otherwise noted.
[b]Some of these fermentation reactions also occur in the presence of sulfate, particularly with substrates not used by sulfate-reducing bacteria.
[c]Reduced substances can also be transported into the oxic zone by bioturbation and bioirrigation, and by bubble ebullition in the case of methane.
[d]Methane is also oxidized in sulfate-containing zones of anoxic sediments.

Hinga, 1983). However, five eastern and central North Pacific stations showed a deficiency in carbon input to sediments compared to oxygen uptake rates; the supply/consumption ratio ranged as low as 0.03 and was greater than 1 only once in 17 measurements (Smith, 1987). Further, input and consumption rates were weakly correlated ($r^2 = 0.59$). Smith (1987) discussed how sources of organic matter other than the sinking particles collected by sediment traps and temporal variability could have accounted for some of the excess oxygen consumption. However, the apparent deficiencies in supply are so large in many cases that either the sediment-trap carbon flux or the sediment oxygen consumption rate data may be inaccurate.

4.2. Sulfate Reduction Rates

Cape Lookout Bight (CLB), North Carolina, sediments are organic-rich and anoxic, with unusually high rates of organic matter degradation by sulfate reduction and methanogenesis. Table III gives the results of the measurement of sulfate reduction rates in CLB sediments by several techniques. Table IV compares these rates to other measures of sediment organic matter mineralization. Results obtained using different methods agree reasonably well (relative standard deviation = 18%). Further, these rates are consistent with the total carbon dioxide flux out of the sediment, corrected for CO_2 production due to methanogenesis (compare items 4 and 5 in Table IV). The directly measured sulfate reduction rate, integrated over sediment depth, is also equal to the sum of sulfide burial and efflux across the sediment–water interface (compare items 5 and 9). The sum of sulfate reduction, methane production, and dissolved organic carbon (DOC) efflux plus burial rates is consistent with the rate of disappearance of organic carbon (compare the sum of items 2, 3, and 5 to items 10 and 11). In general, the rates of carbon remineralization measured by several independent techniques

TABLE III. Comparison of Sulfate Reduction Rates Obtained by Different Methods

Sulfate reduction rate (mol m^{-2} yr^{-1})[a]	Method	Reference
13	Berner type kinetic model of SO_4^{2-} concentration profile	Martens and Klump, 1984
16	SO_4^{2-} flux across the sediment–water interface, calculated from the interfacial concentration gradient[b]	Martens and Klump, 1984
15 ± 7[c]	Decrease in SO_4^{2-} concentration during incubation of sediments in sealed tubes	Crill and Martens, 1987
20 ± 11[c]	$^{35}SO_4^{2-}$ reduction rate	Crill and Martens, 1987

[a]Rates are integrated over 30-cm depth in the sediment.
[b]An empirical sediment diffusion coefficient was used which included the effects of bubble tubes generated during methane ebullition.
[c]The error limits represent one standard deviation of data collected over several years. Winter rates were consistently lower than summer rates.

agree well in CLB sediments. In particular, direct measurements of sulfate reduction rate and methane and ΣCO_2 fluxes at the sediment–water interface are reasonably accurate and are consistent with less precise estimates obtained using Berner type kinetic models.

Alperin (1988) and Alperin *et al.* (1992) compared the results of several approaches to estimating rates of biogeochemical processes in Skan Bay sediments and found some discrepancies (Table V). They assumed that the process of decomposition followed a pathway where sediment particulate organic carbon (POC) was first converted to DOC in sediment pore water, then remineralized to dissolved inorganic carbon (DIC equivalent to ΣCO_2 in Table IV). There was no methanogenesis in the upper 39 cm of sediment, but methane diffusing from underlying sediments was oxidized anaerobically. Sulfate was the only significant electron acceptor for organic matter oxidation.

Under such conditions, the DOC production rate in the sediment must be equal to the POC consumption rate, and the DIC production rate equal to the DOC consumption rate plus the methane oxidation rate. Alperin (1988) found that the DIC production rate calculated from a Berner type diagenetic model of DIC vs. depth was much less than POC deposition minus burial, implying that the DOC production rate was much greater than the consumption rate within the sediment. One possible explanation was that most of the DOC produced in the sediment diffused into the overlying water column before it was completely oxidized to CO_2, but the calculated DOC diffusive flux was too small to account for the discrepancy. Another inconsistency was that the sulfate reduction rate measured using $^{35}SO_4^{2-}$ was five times the DIC production rate estimated using the diagenetic model,

TABLE IV. Comparison of Rates Measured in Cape Lookout Bight Sediments

Process measured	Rate (mol C m^{-2} yr^{-1})[a]	Reference
1. ΣCO_2 efflux from the sediment + ΣCO_2 burial	36 ± 5	Martens *et al.*, 1992
2. CH_4 efflux from the sediment + CH_4 burial	9 ± 3	Martens *et al.*, 1992
3. DOC efflux from the sediment + DOC burial	2.6 ± 1.7	Martens *et al.*, 1992
4. ΣCO_2 production due to SO_4^{2-} reduction (1. − 2.)	27 ± 5	Martens *et al.*, 1992
5. SO_4^{2-} reduction[b]	41	Crill and Martens, 1987
6. O_2 uptake at sediment surface	25	Chanton *et al.*, 1987
7. ΣS^{2-} efflux from sediment	9.2	Chanton *et al.*, 1987
8. ΣS^{2-} burial[c]	31	Chanton *et al.*, 1987
9. SO_4^{2-} reduction (7. + 8.)	41	Chanton *et al.*, 1987
10. Total organic carbon loss (from sediment TOC vs. depth data)	22–50[d]	Martens and Klump, 1984
11. Total organic carbon remineralization + solubilization (1. + 2. + 3.)	48 ± 6	Martens *et al.*, 1992
12. Oxygen uptake *not* explained by dissolved ΣS^{2-} oxidation[e]	15	Chanton *et al.*, 1987
13 CH_4 production[f]	10	Crill and Martens, 1986
14. Acetate oxidation[g]	30	Crill and Martens, 1986

[a]The reported rates have been converted to units of equivalent moles of carbon remineralized.
[b]$^{35}SO_4^{2-}$ reduction rate from Table III.
[c]Includes acid-volatile sulfide, organic sulfur, elemental sulfur, and pyrite.
[d]The large range of values reflects the large uncertainty in this estimate caused by variability in the organic carbon content of surface sediments.
[e]Oxygen uptake rate minus the ΣS^{2-} efflux rate, in units of equivalent moles of carbon remineralized.
[f]Methane production rates were measured directly using radiolabeled bicarbonate and acetate. The annual rate was extrapolated from rates measured during July and August, and so probably is an overestimate.
[g]The rates of acetate remineralization to $^{14}CO_2$ were measured directly using [U-^{14}C]acetate. Rates were measured during the summer only, so the extrapolated annual rate is probably an overestimate.

TABLE V. Comparison of Rates Measured in Skan Bay Sediments

Process measured[a]	Rate (mol C m^{-2} yr^{-1})	Reference
POC deposition	22 ± 5	Alperin *et al.*, 1992
POC burial	3.9 ± 0.9	Alperin *et al.*, 1992
POC deposition–burial	18 ± 5	Alperin *et al.*, 1992
Efflux + burial rates[b]		Alperin *et al.*, 1992
DOC	3.6 ± 1.5	
DIC	18 ± 2	
CH_4	0.10 ± 0.02	
Total	22 ± 3	
Oxidant consumption rates[c]		Alperin *et al.*, 1992
NO_3^-	0.63 ± 0.01	
SO_4^{2-}	18 ± 1	
CH_4 production	0.6 ± 0.1	
Total	19 ± 1	
DIC production[d]	3.3 ± 0.8	Alperin, 1988
Sulfate reduction[e]	7.3	Alperin, 1988
Acetate oxidation[f]	52	Shaw *et al.*, 1984

[a]POC, particulate organic carbon; DOC, dissolved organic carbon; DIC, dissolved inorganic carbon or ΣCO_2.
[b]Efflux from the sediment was measured using a benthic chamber. Burial was calculated from the porewater concentration at 40 cm and the sediment accumulation rate.
[c]The nitrate oxidation rate was determined from the flux of nitrate into the sediment measured using a benthic chamber. Sulfate reduction rates were measured using a $^{35}SO_4^{2-}$ tracer technique. The methane production rate was assumed equal to the methane oxidation rate, which was measured using a $^{14}CH_4$ tracer technique. The integrated values are for the upper 39 cm of sediment. All oxidant consumption rates are expressed in terms of equivalent C oxidized.
[d]This rate was calculated from the porewater DIC distribution using a steady-state model including terms for advection (sediment accumulation), diffusion of dissolved substances, and consumption or production due to decomposition and remineralization of sediment organic matter.
[e]This rate is the diffusive flux of sulfate into the sediment, calculated from the vertical porewater dissolved sulfate concentration gradient. The units are the equivalent carbon oxidized by the sulfate flux.
[f]Acetate oxidation rates were measured using a radiotracer. Production of $^{14}CH_4$ was not significant in the upper 40 cm of sediment. The acetate oxidation rate was multiplied by 2 to give the equivalent carbon remineralization.

although it was equal to the POC consumption rate. The radiotracer sulfate reduction rates were also too large to be consistent with the diffusive flux of sulfate into the sediment calculated from the sulfate concentration gradient in the pore water. However, the sum of DIC, DOC, and CH_4 efflux plus burial was approximately equal to POC deposition minus POC burial and to the sum of nitrate and sulfate reduction and methanogenesis. The most likely reason for the discrepancies is a process, not considered by the diagenetic model, which decreases DIC and increases sulfate concentrations in pore waters. As there is no evidence of bioirrigation, the nature of this process is unknown.

4.3. Methane Oxidation and Production Rates

Methane is the major end product of carbon metabolism in anoxic freshwater sediments (Capone and Kiene, 1988). However, quantitative studies of methane cycling in freshwater sediments are few. Kuivila *et al.* (1988) measured rates of methane production and oxidation in Lake Washington sediments. They found good agreement between the difference of the methane fluxes into and out of the surficial oxic sediment zone (250–350 μmol m^{-2} h^{-1}) and methane oxidation rates measured using a radiotracer. Methane production was equivalent to about 20% of the total organic matter decomposition in the sediment.

Crill and Martens (1986) measured methane production rates in CLB sediments directly by measuring $^{14}CH_4$ production from radiolabeled acetate and bicarbonate (Table IV), and also by measuring the increase in methane concentrations in sediments incubated in sealed tubes. The two methods gave very similar results. The methane production rates also were consistent with the measured flux of methane across the sediment–water interface (Table IV), when it is noted that production rates were measured only during the summer whereas fluxes were measured year-round.

Methane oxidation rates in anoxic Skan Bay sediments have been measured directly using radiotracer techniques (Alperin and Reeburgh, 1985; Alperin, 1988). The direct measurements were in reasonable agreement with rates estimated from a diagenetic model when errors in both are considered.

4.4. Acetate Oxidation Rates

Tables IV and V show acetate oxidation rates in CLB and Skan Bay sediments. In CLB sediments, the measured acetate oxidation rate is similar to the sulfate reduction rate and the CO_2 production rate; thus, the data are internally consistent if acetate is the major energy source for sulfate-reducing bacteria. However, in Skan Bay sediments, the acetate oxidation rate is three times the directly measured sulfate reduction rate and 20 times the model-estimated DOC consumption rate. Acetate oxidation rates have also been measured in Limfjord sediments. The acetate oxidation rate measured was again too large relative to other measures of the rate of sediment metabolism (Ansbaek and Blackburn, 1980; Christensen and Blackburn, 1982).

Ansbaek and Blackburn (1980) proposed that the

acetate oxidation rate was incorrect because the measured acetate concentration in pore water was greater than the concentration available to microorganisms. Several studies have shown that the acetate in pore water can be separated into several fractions by ether extraction or gel-permeation chromatography (Christensen and Blackburn, 1982; Parkes *et al.*, 1984; Thompson and Nedwell, 1985; Michelson *et al.*, 1989). King (1991) reported an enzymatic technique for measuring bioavailable acetate in pore waters.

Acetate and other carboxylic acids are adsorbed by sediment particles (Shaw *et al.*, 1984; Sansone *et al.*, 1987). Adsorption to particles, however, does not seem to be responsible for the rate overestimates cited. Shaw and McIntosh (1990) found acetate concentrations varying over a factor of 10 when different methods, including squeezing, centrifuging, and dialysis, were used to extract pore water from the same Skan Bay sediment depth interval. They attributed the high acetate oxidation rates reported earlier (Shaw *et al.*, 1984) to artifactually high acetate concentrations. Adsorption–desorption reactions and release or uptake of acetate by cells during pore-water extraction could be responsible for the concentration variability.

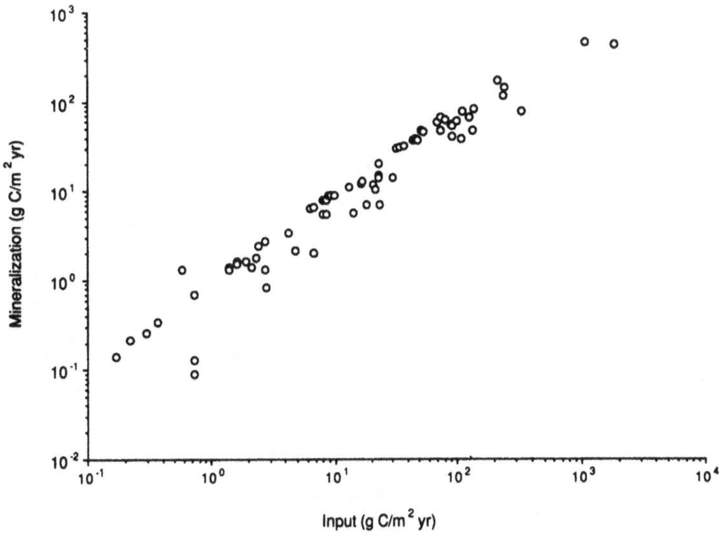

Figure 1. Relationship between the flux of organic carbon to the sediment surface and organic carbon remineralization rate in marine sediments. A few points have been added to the data set used by Henrichs (1992). Data were taken from the following sources: Jørgensen (1977), Murray *et al.* (1978), Crisp *et al.* (1979), Hinga *et al.* (1979), Prahl and Carpenter (1979), Rowe and Gardner (1979), Deuser and Ross (1980), Prahl *et al.* (1980), Lorenzen *et al.* (1981, 1983), Reimers and Suess (1983), Smith and Hinga (1983), Bender and Heggie (1984), Henrichs and Farrington (1984), Howarth and Jørgensen (1984), Martens and Klump (1984), Wassmann (1984), Berner and Westrich (1985), Emerson (1985), Glenn and Arthur (1985), Craven *et al.* (1986), Hong (1986), Vaynshteyn *et al.* (1986), Berelson *et al.* (1987), Heggie *et al.* (1987), Henrichs and Farrington (1987), Silverberg *et al.* (1987), Smith (1987), Alperin (1988), Hedges *et al.* (1988), Rowe *et al.* (1988), Canfield (1989), Bender *et al.* (1989), Fossing (1990), Jahnke (1990). Berelson *et al.* (1990), Jørgensen *et al.* (1990), Reimers *et al.* (1992).

5. Patterns of Variation in Rates of Early Diagenesis

5.1. Effects of Sedimentary Environment

The depth-integrated rate of organic matter remineralization in marine sediment correlates very closely with carbon flux to the sediment surface (Fig. 1). In the oceans, the carbon flux to the sediment surface varies over about four orders of magnitude and, in general, decreases with increasing water column depth and decreasing total sediment accumulation rate (Heath *et al.*, 1977; Suess, 1980; Betzer *et al.*, 1984). On a per-unit-volume, rather than a per-unit-area, basis, decomposition rates in deep-sea sediments are likewise slow relative to rates in continental margin sediments. The deep-sea sites in Fig. 1 have organic matter oxidation rates of about 0.005 to 0.2 mg C cm^{-3} yr^{-1} in the top few centimeters of sediments. The oxygen uptake rate of Danish coastal sediments indicates a carbon oxidation rate of about 16 to 36 mg C cm^{-3} yr^{-1} in the surface centimeter (Sørensen *et al.*, 1979; Jørgensen and Sørensen, 1985). The sulfate reduction rate at 4–6-cm depth in Skan Bay sediments is equivalent to oxidation of 1.7 mg C cm^{-3} yr^{-1} (Alperin, 1988), while the rate at this depth in CLB sediments ranged between 25 (average winter) and 50 (average summer) mg C cm^{-3} yr^{-1} (Crill and Martens, 1987).

Paradoxically, the fraction of incoming carbon not remineralized in surface sediments also correlates with sediment accumulation rate (Henrichs and Reeburgh, 1987) and with carbon flux to the sediment surface (Fig. 2). That is, although the rate of remineralization and the total amount of organic matter remineralized are much greater in rapidly accumulating sediments, the efficiency of remineralization is less than in slowly deposited, deep-sea sediments. The fraction of deposited carbon buried ranges from less than 1% in some deep-sea sediments to more than 50% at coastal sites such as CLB and the "Black Hole" in Long Island Sound.

What factors are responsible for differences in the rate and extent of organic matter remineralization? Several recent papers have analyzed available

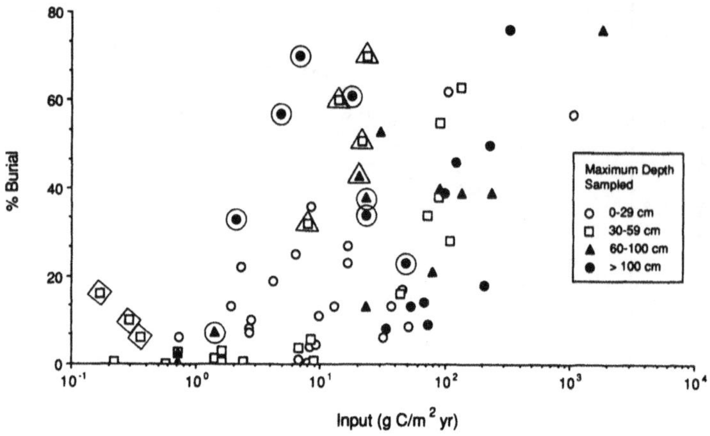

Figure 2. Flux of organic carbon to the sediment surface compared to the percentage of deposited organic carbon buried below the maximum depth sampled, as indicated. The data sources are given in the caption to Fig. 1. The circled points are from the Black Sea (Vaynshteyn *et al.*, 1986). The points enclosed by squares are from southwest Pacific sediments for which no organic matter input data were available (Reimers and Suess, 1983); input was estimated from the primary productivity and water depth after Suess (1980). Points inside triangles are from the California continental borderland (Jahnke, 1990).

data from ancient (Stein *et al.*, 1986; Stein, 1990) and modern (Müller and Suess, 1979; Emerson, 1985; Henrichs and Reeburgh, 1987; Emerson and Hedges, 1988; Mayer *et al.*, 1988; Berner, 1989; Canfield, 1989; Calvert and Pedersen, 1992) sediments and have identified factors which correlate with organic matter preservation: the flux of terrestrial and marine-derived organic matter to the sediment surface (see also Fig. 2); the bulk or total sediment accumulation rate; sediment grain size and surface area; and the availability of oxygen. These factors are similar to those identified by Aller and Mackin (1984) as controlling the extent of preservation predicted by a diagenetic model (Berner, 1980a). Tegelaar *et al.* (1989) proposed that refractory macromolecules synthesized by plants, such as lignins, tannins, algaenans, cutans, and suberans, make up the bulk of organic matter preserved in sediments. In their view, differences in preservation between sediments would result from variations in the proportion of these refractory substances in the total organic matter deposited.

Somewhat conflicting evidence on the effects of bottom-water oxygenation has been found. Figure 2 and Canfield (1989) show that several sites in the Black Sea have greater organic matter burial than sediments overlain by oxic water. However, Calvert *et al.* (1991) presented evidence that carbon accumulation rates in the Black Sea are not greater than those for similar oxygenated environments, and Calvert and Pedersen (1992) concluded that high organic matter input rates, not anoxia, are primarily responsible for enhanced organic matter preservation in rapidly accumulating sediments. Emerson (1985) concluded that low-oxygen bottom waters decreased remineralization of organic matter in California continental borderland sediments, but Jahnke (1990) found no evidence that bottom-water oxygen concentration was negatively correlated with sediment organic matter preservation there. Perhaps the strongest evidence for higher rates of organic matter oxidation in the presence of oxygen is the case of northeast Atlantic turbidites. Organic matter in these sediments does not oxidize appreciably in the presence of pore-water sulfate but is remineralized by oxygen and nitrate diffusing from above (Emerson and Hedges, 1988).

Figure 2 shows enhanced preservation of organic matter for some Black Sea sediments [data from Vaynshteyn *et al.* (1986)], California continental borderland sediments (Jahnke, 1990), and three sites from the southwest Pacific (Reimers and Suess, 1983). Differences in percent burial at these sites are not an artifact of differences in the maximum sediment depth sampled; a large fraction of total remineralization takes place in the upper 10 cm of most sediments. The southwest Pacific points are possibly inaccurate because organic matter influx was estimated from primary productivity according to Suess (1980), not directly measured.

Of course, the time interval represented by surface sediments is much different for sites with high and low sediment accumulation rates. Figure 3 shows a negative correlation between percent burial and time since the organic matter was deposited. However, this relationship should not lead to the conclusion that time is the sole factor controlling preservation. Marine sediments from 0.01 to 1 km below the sediment surface, with ages up to 10 million years, have TOC contents and accumulation rates similar to those found at 1-m depth in comparable sedimentary environments (e.g., Suess and von Huene, 1988; Wolf and Thiede, 1991). Also, ancient marine shales have organic contents in a range similar to that of modern marine sediments at 1–2-m depth (Berner, 1982). Thus, although slow remineralization and changes in composition continue, the bulk of organic matter buried below 1 m is preserved long-term. The trend in Fig. 3 largely reflects the relationship between or-

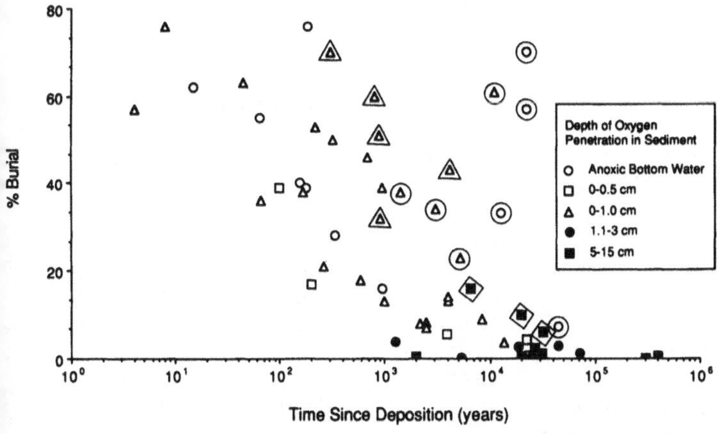

Figure 3. The percentage of deposited organic carbon buried below the maximum depth sampled vs. the time since organic matter at that depth was deposited, i.e., the depth in sediment divided by the sediment accumulation rate. Points representing data from the Black Sea, the southwest Pacific, and the California continental borderland are indicated as in Fig. 2. Data sources are given in the caption to Fig. 1.

ganic matter preservation and sediment accumulation rate (Müller and Suess, 1979; Henrichs and Reeburgh, 1987; Stein, 1990). Possibly, the time that organic matter resides in the surface sediment zone of rapid diagenesis, rather than the total time since deposition, strongly influences preservation.

The availability of oxygen is also indicated in Fig. 3. The depth of oxygen penetration was estimated from the depth of the pore-water nitrate maximum in most cases; for a few sites, oxygen data were available. The Black Sea and the California continental borderland show unusually high percent burial, but overall there is no consistent relationship between oxygen availability and organic matter preservation at a given time.

The underlying reason why more organic matter is preserved in rapidly accumulating sediments is unknown. All of these sediments are anoxic below the upper few centimeters, and it may be that a fraction of organic matter is refractory in the absence of oxygen. However, experimental studies of the decomposition of algae (Otsuki and Hanya, 1972a, b; Foree and McCarty, 1970; Jewell and McCarty, 1971) and sediment organic matter (Andersen and Hargrave, 1984; Kristensen and Blackburn, 1987) have found no or small differences in oxic and anoxic decomposition rates. Fewson (1988) pointed out that nearly all molecules studied to date that decompose aerobically also decompose anaerobically, although sometimes at a slower rate. Possible exceptions are large, heterogeneous, nonhydrolyzable substances such as lignins (Zeikus *et al.*, 1982; Young and Frazer, 1987), some petroleum hydrocarbons (Atlas, 1981), and humic substances. Because time scales of laboratory experiments are very short compared to the time available for decomposition in sediment, even a "slow" decomposition rate would not necessarily mean that the substance would survive early diagenesis.

Another factor may be the presence of macrofaunal organisms, bioturbation, or bioirrigation, which could enhance organic matter decomposition by direct digestion, by breaking up or redistributing sediment particles, or by flushing metabolic inhibitors out of the sediment. Bioturbation was found to increase organic matter decomposition rates in oxic sediments (Kristensen and Blackburn, 1987). Organic matter spends a much longer time in oxygenated, bioturbated surficial sediments in the deep sea than in shallow water, because the bulk sedimentation rate is so much lower. Another alternative is that the decomposition of refractory organic molecules is enhanced by the presence of labile cometabolites (Brock, 1988). Near the sediment–water interface these are continuously added, but there is probably little readily metabolized organic matter present once sediment is buried below the zone of bioturbation.

5.2. Variation with Organic Matter Composition

Table VI summarizes first-order decomposition rate constants (k) which have been measured for organic substances in sediments. One pattern which emerges is that k is smaller in deep-sea sediments than in coastal sediments. A related pattern has been noted in sulfate-reducing sediments by Toth and Lerman (1977): $k = (0.04)\omega^2$, where ω is the sediment accumulation rate. The explanation proposed for these trends is that the more rapidly decomposed fractions of sedimenting organic matter are more likely to reach shallow-water sediments, since less time is required for sinking to the sediment–water interface, and are more likely to be buried to the depth of the sulfate-reducing zone in more rapidly accumulating sediments.

Different constituents of sediment organic matter also have very different decomposition rates. The most rapidly decomposed substances in Table VI are

TABLE VI. First-Order Decomposition Rate Constants for Organic Substances in Sediments

Sediment location/type	Sediment accumulation rate	Organic substance	Rate constant (yr^{-1})[a]	Reference
Cape Lookout Bight, anoxic, 3–5% TOC	10 cm yr^{-1}	Acetate	300 ± 6	Crill and Martens, 1986
		Free amino acids	7500	Burdige and Martens, 1990
		TOC	0.7–11	Martens and Klump, 1984
		Total nitrogen	0.8–5.4	Klump and Martens, 1987
		Total hydrolyzable amino acids	1.4	Burdige and Martens, 1988
		Fatty acids	0.76	Haddad, 1989
		Carbohydrates	0	Haddad, 1989
		Lignin	0	Haddad and Martens, 1987
Resurrection Bay, oxic surface sediment, 0.7% TOC	0.4 cm yr^{-1}	Alanine	18,000	Sugai and Henrichs, 1992
		Glutamic acid	19,000	
		Lysine	45,000	
		Soluble fraction of algae	>50	Henrichs and Doyle, 1986
		Particulate fraction of algae	1–3	
		Arthrobacter and associated polysaccharide	0.8–6	
		Glutamic acid melanoidin	0.2–0.7	
		Alanine melanoidin	<0.09	
		Natural sediment TOC	<0.03	
Long Island Sound, anoxic sediment with high sulfate concentration, from 1-m depth, 1–2% TOC	0.06 cm yr^{-1}	Fresh plankton	8.8	Westrich and Berner, 1984
		Plankton which had previously decomposed for 72 days in oxic seawater	0.84	
Same, except 0–6-cm depth	0.06 cm yr^{-1}	Natural sediment TOC, more reactive fraction	7.2	
		Natural sediment TOC, less reactive fraction	1.0	
Buzzards Bay, upper 70 cm, oxic 0–2 cm, anoxic below, 1.1–2% TOC	0.3 cm yr^{-1}	Natural sediment TOC	0.02	Henrichs and Farrington, 1987
		Total hydrolyzable amino acids	0.02	
		Fatty acids[b]	0.01	Farrington *et al.*, 1977a
		Sterols[c]	0.015	Lee *et al.*, 1977
		Hydrocarbon[d]	<0.001	Farrington *et al.*, 1977b
Central equatorial Pacific, red clay and carbonate ooze, oxic, 0.05–0.7% TOC	1–10 mm (1000 yr)$^{-1}$	Natural sediment TOC	0.0003–0.01[e]	Grundmanis and Murray, 1982
Northeast Pacific, red clay, oxic, 0.1–0.4% TOC	1 mm (1000 yr)$^{-1}$	Natural sediment TOC	0.000083	Murray and Kuivila, 1990

[a]The rate constant k was calculated by fitting concentration vs. time (or depth divided by sediment accumulation rate) data to the following equation: $G_t = (G_0 - G_i)e^{-kt} + G_i$, where G_t is the concentration of the organic substance at time t, G_0 is the concentration of decomposable organic matter at $t = 0$, and G_i is the concentration of inert organic matter.
[b]Both solvent-extractable and saponifiable fatty acids decomposed at the same rate.
[c]Solvent-extractable free sterols.
[d]Nonacosane (n-$C_{29}H_{60}$).
[e]Mean = 0.004.

the amino acids; other simple organic molecules, including short-chain carboxylic acids (Sansone, 1986) and glucose (King and Klug, 1982; Meyer-Reil, 1986), have similarly rapid decomposition rates. Experiments have shown that phytoplankton appears to consist of three fractions of differing reactivity in decomposition experiments conducted over periods of several months: a very reactive or soluble fraction which decomposes on time scales of a few days to weeks; a less reactive fraction which decomposes over months to a few years; and a fraction which appears inert. Decomposition rates of a few other organic compounds have been tested; a bacterially produced polysaccharide decomposed over a few months to a year, while melanoidins (sugar–amino acid condensation products) decomposed on time scales longer than one year. Certain structural characteristics are known to decrease biodegradability of organic molecules (Fewson, 1988). These include polymerization, branching, bonds between subunits which are not readily hydrolyzed, heterocyclic, polycyclic, and aromatic components, and heteroatoms such as chlorine. Present data on the relative rates of decomposition of organic compounds in sediments are broadly consistent with this pattern.

Changes in concentration of several organic compound classes have been measured versus depth in Buzzards Bay (Station P) sediments. Little difference was found in the decomposition rate of the reactive fraction of hydrolyzable amino acids (THAA), sterols, fatty acids, and TOC, but these substances did differ in the fraction of the total which was reactive: 100% for the solvent-extractable free sterols, but only 40% for the fatty acids, TOC, and THAA. Long-chain normal alkanes did not decompose detectably over the 200-year time span represented by sediment cores.

It is not clear why 40% of the TOC, THAA, and fatty acids decomposed within a few hundred years, while the remaining 60% did not. The composition of the individual compounds making up the THAA and fatty acids did not change significantly over time, for example. One hypothesis is that the physical or chemical environment of a fraction of the organic matter prevents its rapid decomposition. For example, decomposition may be slowed because of protection by biogenic mineral matrices (Schroeder and Bada, 1976), adsorption by sediment particles (Mayer *et al.*, 1988; Sugai and Henrichs, 1992), incorporation by refractory macromolecules (i.e., humic substances) (Harvey *et al.*, 1986), or complexation by metals (Degens and Mopper, 1976). Sugai and Henrichs (1992) found that remineralization of amino acids was slowed by adsorption, particularly in the case of the strongly adsorbed basic amino acid lysine. Henrichs and Sugai (1993) found evidence that a melanoidin-type condensation reaction may be partly responsible for lysine adsorption by sediments. If the humification process requires intermolecular reactions such as the melanoidin reaction between sugars and amino acids (Nissenbaum and Kaplan, 1972), the rate of formation of refractory molecules could increase with the organic matter content of sediment, consistent with the variation in carbon preservation illustrated in Fig. 2.

6. Goals for the Future

In reviewing the available information on rates of organic matter diagenesis in marine sediments, a number of areas where more work is clearly needed have been identified.

6.1. Improving the Accuracy of Rate Measurements

Comparisons of independent measurements of the rates of the same or related processes in sediments have sometimes shown encouraging consistency, but not always. The inaccuracy of some acetate oxidation rate measurements is particularly troublesome, especially because it raises questions about other direct rate measurements carried out by similar methods. Accurate pore-water concentration measurements, essential for calculating accurate remineralization rates, are problematic for substances such as acetate and amino acids which have large adsorbed or intracellular concentrations, have very short turnover times, and possibly form biologically refractory dissolved "complexes." Too few detailed studies of single sites, with the aim of intercalibrating rate measurements, have been done; these constitute one of the best available approaches for evaluating techniques of measurement.

6.2. Reducing the "Hot Spot" Bias

Direct measurements of the rates of early diagenetic processes have been concentrated at sites where the rates are extremely high. This may have led to exaggerated estimates of the global importance of anaerobic decomposition processes (Henrichs and Reeburgh, 1987). It also is not clear that techniques developed and verified in sediments with extremely high rates will work equally well in sediments with lower activity.

6.3. Improving Techniques for the Study of the Sediment–Water Interface

Much of the early diagenesis of organic matter takes place within a few centimeters of the sediment–water interface. Rates in this zone are difficult to measure, for several reasons: the interfacial sediment is often lost or disturbed during sampling; vertical gradients in concentrations and rates are large, so that millimeter-scale measurement intervals are often needed; horizontal variability, particularly that associated with bioturbation and bioirrigation, is often substantial; and temporal (e.g., seasonal) variability is greater than in subsurface sediments. The recent development of oxygen microelectrodes (Revsbech and Jørgensen, 1986; Reimers *et al.*, 1986) represents a significant advance in our ability to study processes in this interfacial zone, as are the variety of "free vehicle" benthic samplers being used to measure fluxes across the sediment–water interface (e.g., Smith and Baldwin, 1983; Devol, 1987). More tools, capable of a greater variety of rate measurements, are needed.

6.4. Investigating the Physical and Chemical Environment of Organic Molecules

It remains uncertain why some fractions of organic matter are rapidly decomposed, while others are preserved on geologic time scales. Although sometimes differences in preservation can be related to composition, in other cases the preserved and decomposed fractions cannot be distinguished on that basis. The effects of the physical and chemical environment of organic molecules on their rate of diagenesis require more study.

7. Conclusions

The rates of several processes important in early diagenesis, oxygen consumption, sulfate reduction, and TOC consumption, have been measured in a large number of marine sediments. The available evidence indicates that most of these rate measurements are reasonably accurate, but more evaluation is needed. Rates of total organic matter oxidation in the upper meter of sediments range from about 0.1 to 500 g C m^{-2} yr^{-1} and are positively correlated with the flux of organic matter to the sediment surface and bulk sediment accumulation rate. Rates are generally higher in coastal, shallow-water sediments than in the deep sea because of the greater rate of supply of labile organic material.

On the other hand, remineralization of organic matter is less efficient in rapidly accumulating sediments. The paradox of greater carbon preservation in sediments where much of the organic matter decomposes very rapidly can be explained if a property of high-accumulation-rate sediments inhibits decomposition of only some of the many compounds making up sediment organic matter. In general, the longer organic matter resides in the upper meter of sediments, the less is preserved. This upper layer differs from deeper sediments in several ways: oxygen availability (usually); the presence of macrofauna, bioturbation, and bioirrigation (usually); and continuing inputs of "fresh" organic material.

The underlying reasons why certain organic molecules are rapidly decomposed, while others persist unchanged, are only beginning to be understood. In some cases, slow decomposition is clearly related to the refractory nature of the molecule itself. In other cases, the physical or chemical environment of identical molecules causes them to decompose at different rates.

ACKNOWLEDGMENTS. I thank W. S. Reeburgh, S. F. Sugai, and L. Mayer for reviewing this chapter. I also thank the National Science Foundation for support of my research on organic matter decomposition in sediments under grants OCE-8214537, 8415557, 8516178, 8900362, and 9102995. This is contribution number 729 from the Institute of Marine Science, University of Alaska, Fairbanks.

References

Aller, R. C., and Mackin, J. E., 1984, Preservation of reactive organic matter in marine sediments, *Earth Planet. Sci. Lett.* **70**:260–266.

Alperin, M. J., 1988, The carbon cycle in an anoxic marine sediment: Concentrations, rates, isotope ratios, and diagenetic models, Ph.D. Thesis, University of Alaska, Fairbanks.

Alperin, M. J., and Reeburgh, W. S., 1985, Inhibition experiments on anaerobic methane oxidation, *Appl. Environ. Microbiol.* **50**:940–945.

Alperin, M. J., Reeburgh, W. S., and Devol, A. H., 1992, Organic carbon remineralization and preservation in sediments of Skan Bay, Alaska, in: *Organic Matter: Productivity, Accumulation, and Preservation in Recent and Ancient Sediments* (J. Whelan and J. W. Farrington, eds.), Columbia University Press, New York, pp. 99–122.

Andersen, F. Ø., and Hargrave, B. T., 1984, Effects of *Spartina* detritus enrichment on aerobic/anaerobic benthic metabolism in an intertidal sediment, *Mar. Ecol. Prog. Ser.* **15**:161–171.

Ansbaek, J., and Blackburn, T. H., 1980, A method for the analysis of acetate turnover in a coastal marine sediment, *Microb. Ecol.* **5**:253–264.

Atlas, R. M., 1981, Microbial degradation of petroleum hydrocarbons: An environmental perspective, *Microbiol. Rev.* **45**: 180–209.

Bender, M. L., and Heggie, D. T., 1984, Fate of organic carbon reaching the deep-sea floor: A status report, *Geochim. Cosmochim. Acta* **48**:977–986.

Bender, M., Jahnke, R., Weiss, R., Martin, W., Heggie, D., Orchado, J., and Sowers, T., 1989, Organic carbon oxidation and benthic nitrogen and silica dynamics in San Clemente Basin, a continental borderland site, *Geochim. Cosmochim. Acta* **53**: 685–697.

Berelson, W. M., Hammond, D. E., and Johnson, K. S., 1987, Benthic fluxes and the cycling of biogenic silica and carbon in two southern California borderland basins, *Geochim. Cosmochim. Acta* **51**:1345–1363.

Berelson, W. M., Hammond, D. E., O'Neill, D. E., Xu, X.-M., Chin, C., and Zukin, J., 1990, Benthic fluxes and porewater studies from sediment of the central equatorial north Pacific: Nutrient diagenesis, *Geochim. Cosmochim. Acta* **54**:3001–3012.

Berner, R. A., 1980a, *Early Diagenesis: A Theoretical Approach*, Princeton University Press, Princeton, New Jersey.

Berner, R. A., 1980b, A rate model for organic matter decomposition during bacterial sulfate reduction in marine sediments, in: *Biogéochimie de la Matière Organique à l'Interface Eau–Sédiment Marin, Colloques Internationaux du CNRS*, No. 293, pp. 35–44.

Berner, R. A., 1982, Burial of organic carbon and pyrite in the modern ocean: Its geochemical and environmental significance. *Am. J. Sci.* **282**:451–475.

Berner, R. A., 1989, Biogeochemical cycles of carbon and sulfur and their effect on atmospheric oxygen over Phanerozoic time, *Palaeogeogr. Palaeoclimatol. Palaeoecol. (Global and Planetary Change Section)* **75**:97–122.

Berner, R. A., and Westrich, J. T., 1985, Bioturbation and the early diagenesis of carbon and sulfur, *Am. J. Sci.* **285**:193–206.

Betzer, P. R., Showers, W. J., Laws, E. A., Winn, C. D., DiTullio, G. R., and Kroopnick, P. M., 1984, Primary productivity and particle fluxes on a transect of the equator at 153° W in the Pacific Ocean, *Deep-Sea Res.* **31**:1–12.

Blackburn, T. H., 1979, Method for measuring rates of NH_4^+ turnover in anoxic marine sediments, using a $^{15}NH_4^+$ dilution technique, *Appl. Environ. Microbiol.* **37**:760–765.

Brock, T. D., 1988, *Biology of Microorganisms*, Prentice-Hall, Englewood Cliffs, New Jersey, p. 640.

Burdige, D. J., and Martens, C. S., 1988, Biogeochemical cycling in an organic-rich marine basin: 10. The role of amino acids in sedimentary carbon and nitrogen cycling, *Geochim. Cosmochim. Acta* **52**:1571–1584.

Burdige, D. J., and Martens, C. S., 1990, Biogeochemical cycling in an organic-rich coastal marine basin: 11. The sedimentary cycling of dissolved, free amino acids, *Geochim. Cosmochim. Acta* **54**:3033–3052.

Calvert, S. E., and Pedersen, T. F., 1992, Organic carbon accumulation and preservation in marine sediments: How important is anoxia?, in: *Organic Matter: Productivity, Accumulation, and Preservation in Recent and Ancient Sediments* (J. Whelan and J. W. Farrington, eds.), Columbia University Press, New York, pp. 231–263.

Calvert, S. E., Karlin, R. E., Toolin, L. J., Donahue, D. J., Southon, J. R., and Vogel, J. S., 1991, Low organic carbon accumulation rates in Black Sea sediments, *Nature* **350**:692–695.

Canfield, D. E., 1989, Sulfate reduction and oxic respiration in marine sediments: Implications for organic carbon preservation in euxinic environments, *Deep-Sea Res.* **36**:121–138.

Capone, D. G., and Kiene, R. P., 1988, Comparison of microbial dynamics in marine and freshwater sediments: Contrasts in anaerobic carbon metabolism, *Limnol. Oceanogr.* **33**:725–749.

Chanton, J. P., Martens, C. S., and Goldhaber, M. B., 1987, Biogeochemical cycling in an organic-rich coastal marine basin. 7. Sulfur mass balance, oxygen uptake, and sulfide retention, *Geochim. Cosmochim. Acta* **51**:1187–1199.

Christensen, D., and Blackburn, T. H., 1980, Turnover of tracer (^{14}C, ^{3}H labeled) alanine in inshore marine sediments, *Mar. Biol.* **58**: 97–103.

Christensen, D., and Blackburn, T. H., 1982, Turnover of ^{14}C-labeled acetate in marine sediment, *Mar. Biol.* **71**:113–119.

Craven, D. B., Jahnke, R. A., and Carlucci, A. F., 1986, Fine-scale vertical distributions of microbial biomass and activity in California borderland sediments, *Deep-Sea Res.* **33**:379–390.

Crill, P. M., and Martens, C. S., 1986, Methane production from bicarbonate and acetate in an anoxic marine sediment, *Geochim. Cosmochim. Acta* **50**:2089–2097.

Crill, P. M., and Martens, C. S., 1987, Biogeochemical cycling in an organic-rich coastal marine basin. 6. Temporal and spatial variations in sulfate reduction rates, *Geochim. Cosmochim. Acta* **51**:1175–1186.

Crisp, P. T., Brenner, S. Ventkatesan, M. J., Ruth, E., and Kaplan, I. R., 1979, Organic geochemical characterization of sediment-trap particulates from San Nicholas, Santa Barbara, Santa Monica, and San Pedro Basins, California, *Geochim. Cosmochim. Acta* **43**:1791–1801.

Degens, E. T., and Mopper, K., 1976, Factors controlling the distribution and early diagenesis of organic material in marine sediments, in: *Chemical Oceanography*, 2nd ed., Vol. 6 (J. P. Riley and R. Chester, eds.), Academic Press, New York, pp. 59–113.

Deuser, W. G., and Ross, E. H., 1980, Seasonal change in the flux of organic carbon to the deep Sargasso Sea, *Nature* **283**:364–365.

Devol, A. H., 1987, Verification of flux measurements made with *in situ* benthic chambers, *Deep-Sea Res.* **34**:1007–1026.

Doyle, A. P., 1988, Respiration and adsorption of alanine, glutamic acid, lysine, and glucose in sediments from Resurrection Bay, Alaska, M.S. Thesis, University of Alaska, Fairbanks.

Edenborn, H. M., Silverberg, N., Mucci, A., and Sundby, B., 1987, Sulfate reduction in deep coastal marine sediments, *Mar. Chem.* **21**:329–345.

Emerson, S., 1985, Organic carbon preservation in marine sediments, in: *The Carbon Cycle and Atmospheric CO_2: Natural Variations Archaean to Present* (E. Sundquist and W. S. Broecker, eds.), Geophysical Monograph 32, American Geophysical Union, Washington, D.C., pp. 78–87.

Emerson, S. E., and Hedges, J. I., 1988, Processes controlling the organic carbon content of open ocean sediments, *Paleoceanography* **3**:621–634.

Farrington, J. W., Henrichs, S. M., and Anderson, R. F., 1977a, Fatty acids and Pb-210 geochronology of a sediment core from Buzzards Bay, Massachusetts, *Geochim. Cosmochim. Acta* **41**: 289–296.

Farrington, J. W., Frew, N. M., Gschwend, P. M., and Tripp, B. W., 1977b, Hydrocarbons in cores of northwestern Atlantic coastal and continental margin sediments, *Estuarine Coastal Mar. Sci.* **5**:793–808.

Fewson, C. A., 1988, Biodegradation of xenobiotic and other persistent compounds: The causes of recalcitrance, *TIBTECH* **6**: 148–153.

Foree, E. G., and McCarty, P. L., 1970, Anaerobic decomposition of algae, *Environ. Sci. Technol.* **4**:842–849.

Fossing, H., 1990, Sulfate reduction in shelf sediments in the upwelling region off central Peru, *Cont. Shelf Res.* **10**:355–367.

Glenn, C. R., and Arthur, M. E., 1985, Sedimentary and geochemical indicators of productivity and oxygen contents in modern and ancient basins: The Holocene Black Sea as the "type" anoxic basin, *Chem. Geol.* **48**:325–354.

Grundmanis, V., and Murray, J. W., 1982, Aerobic respiration in pelagic marine sediments, *Geochim. Cosmochim. Acta* **46**: 1101–1120.

Haddad, R. I., 1989, Sources and reactivity of organic matter accumulation in a rapidly depositing, coastal marine sediment, Ph.D. Thesis, University of North Carolina at Chapel Hill.

Haddad, R. I., and Martens, C. S., 1987, Biogeochemical cycling in an organic-rich coastal marine basin: 9. Sources and accumulation rates of vascular plant-derived organic material, *Geochim. Cosmochim. Acta* **51**:2991–3001.

Harvey, H. R., Fallon, R. D., and Patton, J. S., 1986, The effect of organic matter and oxygen on the degradation of bacterial membrane lipids in marine sediments, *Geochim. Cosmochim. Acta* **50**:795–804.

Heath, G. R., Moore, T. C., Jr., and Dauphin, J. P., 1977, Organic carbon in deep-sea sediments, in: *The Fate of Fossil Fuel CO_2 in the Oceans* (N. R. Andersen and A. Malahoff, eds.), Plenum Press, New York, pp. 605–625.

Hedges, J. I., Clark, W. A., and Cowie, G. L., 1988, Fluxes and reactivities of organic matter in a coastal marine bay, *Limnol. Oceanogr.* **33**:1137–1152.

Heggie, D., Maris, C., Hudson, A., Dymond, J., Beach, R., and Cullen, J., 1987, Organic carbon oxidation and preservation in NW Atlantic continental margin sediments, in: *Geology and Geochemistry of Abyssal Plains* (P. P. E. Weaver and J. Thomson, eds.), Geological Society Special Publication 31, pp. 215–236.

Henrichs, S. M., 1992, Early diagenesis of organic matter in marine sediments: Progress and perplexity, *Mar. Chem.* **39**:119–149.

Henrichs, S. M., and Doyle, A. P., 1986, Decomposition of ^{14}C-labeled organic substances in marine sediments, *Limnol. Oceanogr.* **31**:765–778.

Henrichs, S. M., and Farrington, J. W., 1984, Peru upwelling region sediments near 15° S. 1. Remineralization and accumulation of organic matter, *Limnol. Oceanogr.* **29**:1–19.

Henrichs, S. M., and Farrington, J. W., 1987, Early diagenesis of amino acids and organic matter in two coastal marine sediments, *Geochim. Cosmochim. Acta* **51**:1–15.

Henrichs, S. M., and Reeburgh, W. S., 1987, Anaerobic mineralization of marine sediment organic matter: Rates and the role of anaerobic processes in the oceanic carbon economy, *Geomicrobiol. J.* **5**:191–237.

Henrichs, S. M., and Sugai, S. F., 1993, Adsorption of amino acids and glucose by sediments of Resurrection Bay, Alaska, USA: Functional group effects, *Geochim. Cosmochim. Acta.* **57**: 823–835.

Hinga, K. K., Sieburth, J. M., and Heath, G. R., 1979, The supply and use of organic material at the deep-sea floor, *J. Mar. Res.* **37**: 557–579.

Hong, G. H., 1986, Fluxes, dynamics, and chemistry of suspended particulate matter in a southeast Alaska fjord, Ph.D. Thesis, University of Alaska, Fairbanks.

Howarth, R. W., and Jørgensen, B. B., 1984, Formation of ^{35}S-labeled elemental sulfur and pyrite in coastal marine sediments (Limfjorden and Kysing Fjord, Denmark) during short-term $^{35}SO_4^{2-}$ reduction measurements, *Geochim. Cosmochim. Acta* **48**:1807–1818.

Jahnke, R. A., 1990, Early diagenesis and recycling of biogenic debris at the sea floor, Santa Monica Basin, California, *J. Mar. Res.* **48**:413–436.

Jewell, W. J., and McCarty, P. L., 1971, Aerobic decomposition of algae, *Environ. Sci. Technol.* **5**:1023–1031.

Jørgensen, B. B., 1977, The sulfur cycle of a coastal marine sediment (Limfjorden, Denmark), *Limnol. Oceanogr.* **22**:814–832.

Jørgensen, B. B., 1978, A comparison of methods for the quantification of bacterial sulfate reduction in coastal marine sediments I. Measurement with radiotracer techniques, *Geomicrobiol. J.* **1**:11–27.

Jørgensen, B. B., 1983, Processes at the sediment–water interface, in: *The Major Biogeochemical Cycles and Their Interactions*, SCOPE 21 (B. Bolin and R. Cook, eds.), John Wiley & Sons, New York, pp. 477–509.

Jørgensen, B. B., and Sørensen, J., 1985, Seasonal cycles of O_2, NO_3^-, and SO_4^{2-} reduction in estuarine sediments: The significance of a NO_3^- reduction maximum in spring, *Mar. Ecol. Prog. Ser.* **24**:65–74.

Jørgensen, B. B., Bang, M., and Blackburn, T. H., 1990, Anaerobic mineralization in marine sediments from the Baltic–North Sea transition, *Mar. Ecol. Prog. Ser.* **59**:39–54.

King, G. M., 1991, Measurement of acetate concentrations in marine pore waters by using an enzymatic approach, *Appl. Environ. Microbiol.* **57**:3476–3481.

King, G. M., and Klug, M. J., 1982, Glucose metabolism in sediments of a eutrophic lake: Tracer analysis of uptake and product formation, *Appl. Environ. Microbiol.* **44**:1308–1317.

Klump, J. V., and Martens, C. S., 1987, Biogeochemical cycling in an organic-rich marine basin. 5. Sedimentary nitrogen and phosphorus budgets based upon kinetic models, mass balances, and the stoichiometry of nutrient regeneration, *Geochim. Cosmochim. Acta* **51**:1161–1173.

Kristensen, E., and Blackburn, T. H., 1987, The fate of organic carbon and nitrogen in experimental marine sediment systems: Influence of bioturbation and anoxia, *J. Mar. Res.* **45**: 231–257.

Kuivila, K. M., Murray, J. W., Devol, A. H., Lidstrom, M. E., and Reimers, C. E., 1988, Methane cycling in the sediments of Lake Washington, *Limnol. Oceanogr.* **33**:571–581.

Lee, C., Gagosian, R. B., and Farrington, J. W., 1977, Sterol diagenesis in recent sediments from Buzzards Bay, Massachusetts, *Geochim. Cosmochim. Acta* **41**:985–992.

Lorenzen, C. J., Schuman, F. R., and Bennett, J. T., 1981, *In situ* calibration of a sediment trap, *Limnol. Oceanogr.* **26**:580–585.

Lorenzen, C. J., Welschmeyer, N. A., and Copping, A. E., 1983, Particulate organic flux in the subarctic Pacific, *Deep-Sea Res.* **30**:639–643.

Lovely, D. R., 1987, Organic matter remineralization with the reduction of ferric iron: A review, *Geomicrobiol. J.* **5**:375–399.

Lovely, D. R., and Phillips, E. J. P., 1986, Organic matter mineralization with reduction of ferric iron in anaerobic sediments, *Appl. Environ. Microbiol.* **51**:683–689.

Martens, C. S., and Klump, J. V., 1984, Biogeochemical cycling in an organic-rich coastal marine basin. 4. An organic carbon budget for sediments dominated by sulfate reduction and methanogenesis, *Geochim. Cosmochim. Acta* **48**:1987–2004.

Martens, C. S., Haddad, R. I., and Chanton, J. P., 1992, Organic matter accumulation, remineralization, and burial in an anoxic coastal sediment, in: *Organic Matter: Productivity, Accumulation, and Preservation in Recent and Ancient Sediments* (J. Whelan and J. W. Farrington, eds.), Columbia University Press, New York, pp. 82–98.

Mayer, L. M., Macko, S. A., and Cammen, L., 1988, Provenance, concentrations, and nature of sedimentary organic nitrogen in the Gulf of Maine, *Mar. Chem.* **25**:291–304.

Meyer-Reil, L.-A., 1986, Measurement of hydrolytic activity and incorporation of dissolved organic substrates by microorganisms in marine sediments, *Mar. Ecol. Prog. Ser.* **31**:143–149.

Michelson, A. R., Jacobsen, M. E., Scranton, M. I., and Mackin, J. E., 1989, Modeling the distribution of acetate in anoxic estuarine sediments, *Limnol. Oceanogr.* **34**:747–757.

Middelburg, J. J., 1989, A simple rate model for organic matter decomposition in marine sediments, *Geochim. Cosmochim. Acta* **53**:1577–1581.

Müller, P. J., and Suess, E., 1979, Productivity, sedimentation, and sedimentary organic carbon in the oceans. I. Organic carbon preservation, *Deep-Sea Res.* **26**:1347–1362.

Murray, J. W., and Kuivila, K. M., 1990, Organic matter diagenesis in the northeast Pacific: Transition from aerobic red clay to suboxic pelagic sediments, *Deep-Sea Res.* **37**:59–80.

Murray, J. W., Grundmanis, V., and Smethie, W. M., Jr., 1978, Interstitial water chemistry in the sediments of Saanich Inlet, *Geochim. Cosmochim. Acta* **42**:1011–1026.

Nissenbaum, A., and Kaplan, I. R., 1972, Chemical and isotopic evidence for the *in situ* origin of marine humic substances, *Limnol. Oceanogr.* **17**:570–581.

Novelli, P. C., Scranton, M. I., and Michener, R. H., 1987, Hydrogen distributions in marine sediments, *Limnol. Oceanogr.* **32**: 565–576.

Otsuki, A., and Hanya, T., 1972a, Production of dissolved organic matter from dead green algal cells. I. Aerobic microbial decomposition, *Limnol. Oceanogr.* **17**:248–257.

Otsuki, A., and Hanya, T., 1972b, Production of dissolved organic matter from dead green algal cells. II. Anaerobic microbial decomposition, *Limnol. Oceanogr.* **17**:258–264.

Parkes, R. J., Taylor, J., and Jorck-Ramberg, D., 1984, Demonstration, using *Desulfobacter* sp., of two pools of acetate with different biological availabilities in marine pore water, *Mar. Biol.* **83**: 271–276.

Prahl, F. G., and Carpenter, R., 1979, The role of zooplankton fecal pellets in the sedimentation of polycyclic aromatic hydrocarbons in Dabob Bay, Washington, *Geochim. Cosmochim. Acta* **43**:1959–1972.

Prahl, F. G., Bennett, J. T., and Carpenter, R., 1980, The early diagenesis of aliphatic hydrocarbons and organic matter in sedimentary particulates from Dabob Bay, Washington, *Geochim. Cosmochim. Acta* **44**:1967–1976.

Reeburgh, W. S., 1983, Rates of biogeochemical processes in anoxic sediments, *Annu. Rev. Earth Planet. Sci.* **11**:269–298.

Reimers, C. E., and Smith, K. L., Jr., 1986, Reconciling measured and predicted fluxes of carbon across the sediment–water interface, *Limnol. Oceanogr.* **31**:305–318.

Reimers, C. E., and Suess, E., 1983, The partitioning of organic carbon fluxes and sedimentary organic matter decomposition rates in the ocean, *Mar. Chem.* **13**:141–168.

Reimers, C. E., Fisher, K. M., Merewether, R., Smith, K. L., Jr., and Jahnke, R. A., 1986, Oxygen profiles measured *in situ* in deep ocean sediments, *Nature* **320**:741–744.

Reimers, C. E., Jahnke, R. A., and McCorkle, D. C., 1992, Carbon fluxes and burial rates over the continental slope and rise off central California with implications for the global carbon cycle, *Glob. Biogeochem. Cycles* **6**:199–224.

Revsbech, N. P., and Jørgensen, B. B., 1986, Microelectrodes: Their use in microbial ecology, *Adv. Microb. Ecol.* **9**:293–352.

Rowe, G. T., and Gardner, W. D., 1979, Sedimentation rates in the slope water of the northwest Atlantic Ocean measured directly with sediment traps, *J. Mar. Res.* **37**:581–600.

Rowe, G. T., Theroux, R., Phoel, W., Quimby, H., Wilke, R., Koschoreck, D., Whitledge, T. E., Falkowski, P. G., and Fray, C., 1988, Benthic carbon budgets for the continental slope south of New England, *Cont. Shelf Res.* **8**:511–527.

Sansone, F. J., 1986, Depth distribution of short-chain organic acid turnover in Cape Lookout Bight sediments, *Geochim. Cosmochim. Acta* **50**:99–105.

Sansone, F. J., Andrews, C. C., and Okamoto, M. Y., 1987, Adsorption of short-chain organic acids onto nearshore marine sediments, *Geochim. Cosmochim. Acta* **51**:1889–1896.

Schroeder, R., and Bada, J. L., 1976, A review of the geochemical applications of the amino acid racemization reaction, *Earth Sci. Rev.* **12**:347–391.

Seitzinger, S. P., 1988, Denitrification in freshwater and coastal marine ecosystems: Ecological and geochemical significance, *Limnol. Oceanogr.* **33**:702–724.

Shaw, D. G., and McIntosh, D. J., 1990, Acetate in recent anoxic sediments: Direct and indirect measurements of concentration and turnover rates, *Estuarine Coastal Shelf Sci.* **31**:775–788.

Shaw, D. G., Alperin, M. J., Reeburgh, W. S., and McIntosh, D. J., 1984, Biogeochemistry of acetate in the anoxic sediments of Skan Bay, Alaska, *Geochim. Cosmochim. Acta* **48**:1819–1825.

Silverberg, N., Bakker, J., Edenborn, H. M., and Sundby, B., 1987, Oxygen profiles and organic carbon fluxes in Laurentian trough sediments, *Neth. J. Sea Res.* **21**:95–105.

Skyring, G., 1987, Sulfate reduction in coastal ecosystems, *Geomicrobiol. J.* **5**:295–374.

Smith, K. L., Jr., 1987, Food energy supply and demand: A discrepancy between particulate organic carbon flux and sediment community oxygen consumption in the deep ocean, *Limnol. Oceanogr.* **32**:201–220.

Smith, K. L., Jr., and Baldwin, R. J., 1983, Deep-sea respirometry, in: *Polarographic oxygen sensors* (E. Gnaiger and H. Forstner, eds.), Springer-Verlag, New York, pp. 298–319.

Smith, K. L., Jr., and Hinga, K. R., 1983, Sediment community respiration in the deep sea, in: *The Sea*, Vol. 8 (G. T. Rowe, ed.), John Wiley & Sons, New York, pp. 331–370.

Sørensen, J., Jørgensen, B. B., and Revsbech, N. P., 1979, A comparison of oxygen, nitrate, and sulfate respiration in coastal marine sediments, *Microb. Ecol.* **5**:105–115.

Stein, R., 1990, Organic carbon sedimentation rate relationship and its paleoenvironmental significance for marine sediments, *Geo-Mar. Lett.* **10**:37–44.

Stein, R., Rullkötter, J., and Welte, D., 1986, Accumulation of organic-carbon-rich sediments in the late Jurassic and Cretaceous Atlantic Ocean—a synthesis, *Chem. Geol.* **56**:1–32.

Suess, E., 1980, Particulate organic flux in the oceans—surface productivity and oxygen utilization, *Nature* **288**:260–263.

Suess, E., and von Huene, R., 1988, Ocean drilling program Leg 112, Peru continental margin: Part 2, sedimentary history and diagenesis in a coastal upwelling environment, *Geology* **16**: 939–943.

Sugai, S. F., and Henrichs, S. M., 1992, Rates of amino acid uptake and mineralization in Resurrection Bay (Alaska) sediments, *Mar. Ecol. Prog. Ser.* **88**:129–141.

Tegelaar, E. W., de Leeuw, J. W., Derenne, S., and Largeau, C., 1989, A reappraisal of kerogen formation, *Geochim. Cosmochim. Acta* **53**:3103–3106.

Thompson, L. A., and Nedwell, D. B., 1985, Existence of different pools of fatty acids in anaerobic model ecosystems and their availability to microbial metabolism, *FEMS Microbiol. Ecol.* **31**:141–146.

Toth, D. J., and Lerman, A., 1977, Organic matter reactivity and sedimentation rates in the ocean, *Am. J. Sci.* **277**:465–485.

Vaynshteyn, M. B., Tokarev, V. G., Shakola, V. A., Lein, A., and Ivanov, M. V., 1986, The geochemical activity of sulfate-reducing bacteria in sediments in the western part of the Black Sea, *Geochem. Int.* **23**:110–122.

Wassmann, P., 1984, Sedimentation and benthic mineralization of organic detritus in a Norwegian fjord, *Mar. Biol.* **83**:83–94.

Westrich, J. T., and Berner, R. A., 1984, The role of sedimentary organic matter in bacterial sulfate reduction: The G model tested, *Limnol. Oceanogr.* **29**:236–249.

Wolf, T. C., and Thiede, J., 1991, History of terrigenous sedimentation during the past 10 m.y. in the North Atlantic (ODP Legs 104 and 105 and DSDP Leg 81), *Mar. Geol.* **101**:83–102.

Young, L. Y., and Frazer, A. C., 1987, The fate of lignin and lignin-derived compounds in anaerobic environments, *Geomicrobiol. J.* **5**:261–293.

Zeikus, J. G., Wellstein, A. L., and Kirk, T. K., 1982, Molecular basis for biodegradation recalcitrance of lignin in anaerobic environments, *FEMS Microbiol. Lett.* **15**:193–197.

Chapter 5

The Early Diagenesis of Organic Matter: Bacterial Activity

JODY W. DEMING and JOHN A. BAROSS

1. Introduction to Bacteria as Primary Agents

Bacteria are the primary agents of the early diagenesis of organic matter (OM) in marine sediments. The reasons for their predominant roles are straightforward: (1) they occur abundantly and universally throughout all marine sediments; (2) they can respire, reproduce, and, therefore, use OM more rapidly than any other organisms; (3) they possess enzymes and enzyme systems (in many cases, unique to the prokaryotic kingdom) that make them extraordinarily versatile in their nutritional requirements and abilities to alter a wide variety of particulate and dissolved organic (and inorganic) materials; and (4) they readily enter into complex associations with each other and with higher organisms in ways that produce powerful degradative capabilities beyond those of a single organism in isolation.

Perhaps the most fundamental of these characteristics, when considering the fate of OM in marine sediments, is the possession by bacteria of unique catalytic properties—specifically, enzymes that act on highly refractory organic carbon compounds (cellulose, chitin, lignin, agar) and complex enzyme systems that fix and recycle inorganic carbon, nitrogen, sulfur, and other elemental species in ways that contribute to the further mobilization and degradation of organic matter. In the well-studied, organic-rich sediments of estuaries, bacterial degradation of refractory OM proceeds most efficiently and effectively through the actions of microbial consortia that develop and persist in anaerobic (or microaerophilic) microenvironments. However, the vast portion of the global seafloor can be characterized as well oxygenated and organically depauperate. How bacteria work to prevent the accumulation of undegraded OM in these deeper environments is the focus of this chapter. We draw upon the greater body of literature from shallow environments, and an increasing number of studies from the deep sea, to develop rationales for understanding the microbial fate of OM in oceanic sediments and ways to test hypotheses in future studies. Central to our arguments are the concepts of bacteria acting in consortia with other bacteria and microorganisms (the microbial loop), especially in anaerobic microenvironments, and bacteria acting through a variety of mutualistic and other symbiotic associations with marine invertebrates.

2. Current Conceptual Models and Experimental Approaches

The ultimate model of early diagenesis of organic matter in marine sediments will predict pathways

JODY W. DEMING and JOHN A. BAROSS • School of Oceanography, University of Washington, Seattle, Washington 98195.
Organic Geochemistry, edited by Michael H. Engel and Stephen A. Macko. Plenum Press, New York, 1993.

and reaction rates for the wide diversity of organic compounds commonly found in sediments and almost continuously attacked by living organisms and their extracellular enzymes. Its construction and validation will require, ideally, direct measurements of organisms and enzymes working to degrade complex particulate and dissolved OM in undisturbed sediments and pore waters under *in situ* conditions. Though modelers and measurers are approaching this ideal, we remain ignorant of some basic aspects of OM diagenesis (for example, the types of organic compounds degraded in sediment layers dominated by nitrate-reducing bacteria are not known) and dependent on indirect or oversimplified experimental and modeling approaches. Four current approaches to understanding OM diagenesis, and clarifying questions yet to be resolved, are described in the following sections.

2.1. Biogeochemical Pore-Water Models

The classic biogeochemical pore-water model for interpreting early diagenesis of OM in marine sediments (*sensu* Berner, 1974) orders horizontal sediment strata according to the oxidants available to bacteria for their use in degrading organic compounds (Fig. 1). Bacterial oxidant uptake is deduced from the shape of an oxidant depth profile (or the profile of some product of the oxidation pathway), after accounting for passive diffusion. Carbon degradation is then inferred (calculated) from oxidant uptake, assuming traditional Redfield stoichiometry (Berner, 1977). The standard approach requires no measurements or knowledge *per se* of the biota acting within each stratum, only measured depth profiles of chemical reactants or products assumed to be indicative of bacterial action (Fig. 1). However, since differ-

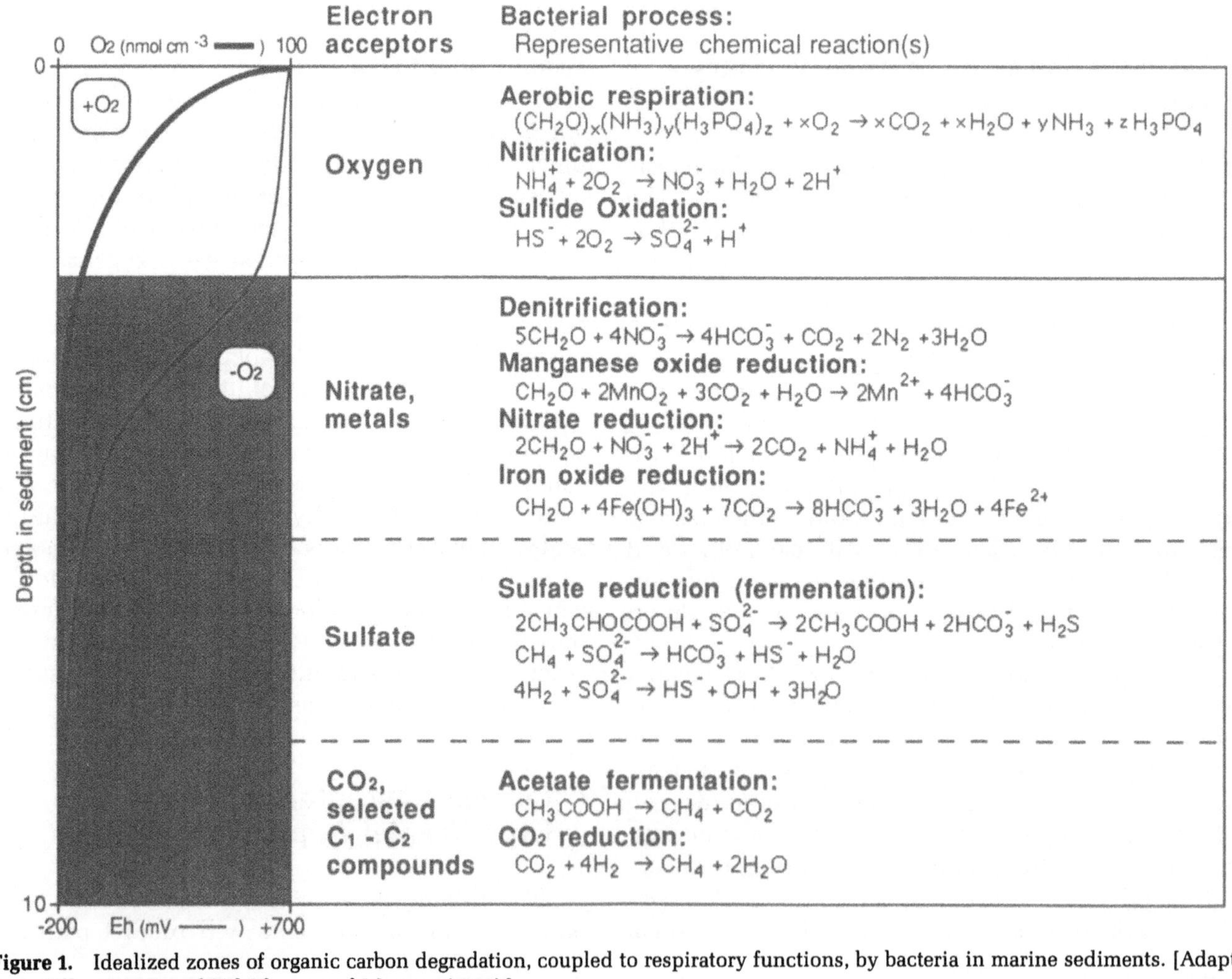

Figure 1. Idealized zones of organic carbon degradation, coupled to respiratory functions, by bacteria in marine sediments. [Adapted from Billen (1982) and Val Klump and Martens (1983).]

ent physiological types of bacteria degrade different classes of organic compounds using specific oxidants, profiling a single oxidant or by-product is insufficient to estimate total OM degradation. Oxygen profiles, for example, fail to account for anaerobic OM degradation, which can be substantial in shallow water and other organic-rich sediments (Henrichs and Reeburgh, 1987; Capone and Kiene, 1988).

The history and general predictive capabilities of pore-water models, as well as their added power if actual rate measurements (using radiolabeled tracers) of specific bacterial processes are made in each stratum, were reviewed by Reeburgh (1983). He considered data obtained primarily from shallow coastal environments at a time when little information was available on the pelagic supply of OM to the sediments. Since then, progress has been made on at least four fronts. First, efforts to model pore-water processes in deeper oceanic sediments of more global importance from a mass-balance standpoint have increased (e.g., Bender *et al.*, 1977; Froelich *et al.*, 1979; Jahnke *et al.*, 1982, 1986, 1990; Grundmanis and Murray, 1982; Bender and Heggie, 1984; Christensen and Rowe, 1984; McCorkle *et al.*, 1985; Emerson *et al.*, 1980, 1985, 1987; Sayles and Curry, 1988; Canfield, 1989). Second, microelectrodes that provide more refined pore-water profiles, especially at the sediment–water interface, have been developed and deployed at both shallow and deep stations (e.g., Revsbech *et al.*, 1979; Revsbech and Jorgensen, 1986; Reimers, 1987; Reimers *et al.*, 1984, 1986; Reimers and Smith, 1986; Archer *et al.*, 1989; Gundersen and Jorgensen, 1990). Third, ways to measure the net flux of a critical gas or solute across the sediment–water interface under *in situ* conditions, using submersible-deployed or remote-controlled "landers," have become well established (e.g., Smith and Teal, 1973; Rowe *et al.*, 1975; Smith *et al.*, 1976, 1979, 1987; Berelson and Hammond, 1986; Devol, 1987; Jahnke and Jackson, 1987; Jahnke *et al.*, 1990). Fourth, sediment-trap studies, providing direct measurements of the vertical supply of OM to the seafloor (though sometimes of arguable accuracy) against which pore-water predictions can be tested, have entered the literature in great numbers (e.g., Honjo, 1978; Rowe and Gardner, 1979; Hinga *et al.*, 1979; Suess, 1980; Wakeham *et al.*, 1980; Deuser *et al.*, 1981; Ittekkot *et al.*, 1984; Martin *et al.*, 1987; Biscaye *et al.*, 1988; Walsh *et al.*, 1988).

This burst of research activity has yielded better techniques for measuring and assessing specific pore-water processes, but it has also left exposed the conceptual weaknesses of pore-water models as predictors of OM diagenesis. For example, two elements basic to the approach, reliance on Redfield stoichiometry and assumption of steady state at the sediment–water interface, are difficult to justify for the vast majority of marine sediments [see Jahnke and Jackson (1987) for a slightly different perspective on this issue]. Sayles and Curry (1988) recently identified significant stoichiometric departures from Redfield convention in deep-sea sediments by measuring $\delta^{13}C$ of pore-water CO_2, a more direct tracer of degrading OM than the oxidants typically measured (see also McCorkle and Emerson, 1988). Others have detected substantial seasonal variations in the pelagic supply of OM (Deuser and Ross, 1980; Deuser *et al.*, 1981; Ittekkot *et al.*, 1984) and corresponding biological responses (Smith and Baldwin, 1984; Lochte and Turley, 1988; Graf, 1989) in the very types of environments (again, the deep sea) where steady-state assumptions have been made with most confidence in the past. As long as pore-water models predict rates that conflict with measured OM fluxes to the seafloor (and vice versa), OM degradation rates calculated from them can only be considered approximations (Smith *et al.*, 1987; Sayles and Curry, 1988; McCorkle and Emerson, 1988).

Other persisting limitations of pore-water models derive from the natural biological and physical complexities of sediments. Protozoa, meiofauna, macrofauna, and megafauna regularly produce heterogeneities and disturbances beyond the perceived boundaries of a bacterial oxidant layer (Aller, 1982; Revsbech and Jorgensen, 1986), as do erosional events (Aller and Aller, 1986; Aller, 1989). Recent studies in the Santa Catalina Basin (depths of about 1200 m) suggest that near-surface horizontal mixing rates of sediment particles by animals far exceed vertical mixing rates in the same stratum, introducing error to one-dimensional (vertical) pore-water models (Wheatcroft *et al.*, 1990). Recognizing and accounting for these natural, multidimensional heterogeneities and disturbances in sediments are not trivial exercises (Gundersen and Jorgensen, 1990). Even within an undisturbed sediment layer, microscale environments can support bacterial reactions in conflict with the larger scale oxidizing or reducing state of the sediment (Jorgensen, 1977; see later sections). The modeling efforts of Jahnke (1985) demonstrate that if the magnitudes of the conflicting reaction rates are significant, they can cause shifts in expected vertical profiles of the relevant solutes and gases.

An improved understanding of OM diagenesis can be expected as the classic biogeochemical pore-water models, based on specific gases or inorganic oxidants, merge with newer models of the organic chemistry of marine sediments. The latter are perhaps best represented by the "multiple-G" model of Berner (1980) and Westrich and Berner (1984). Originally based on simple bulk measurements of total OM in sediments, it describes the decomposition of different "pools" of OM at different rates, depending on the relative lability of the organic compounds involved. Some experimental systems and procedures for testing the model have been reported (Battersby and Brown, 1982; Westrich and Berner, 1984; Henrichs and Doyle, 1986), but initial experiments examined primarily the "end point" pools of OM—dissolved labile versus particulate refractory forms of carbon. Future tests of the multiple-G model can be expected to include a broader range of the complex intermediate compounds generated during early diagenesis, as well as parallel measurements of bacterial and extracellular enzyme activities associated with them.

2.2. Carbon Budget Models

In contrast to pore-water models, the carbon budget approach to estimating rates of OM degradation in sediments focuses directly on particulate organic carbon (POC), tracking it from sources to sediment biota and final burial. An example for a coastal upwelling environment is depicted in Fig. 2 (Rowe, 1981). Input terms on the left include POC from the pelagic environment, suspended and bedload transport from nearby and more distant (turbidite) sources, and *in situ* chemoautotrophic bacterial conversion of CO_2 to POC. Loss terms on the right include similar sediment transport processes plus respiratory losses in the form of CO_2. The latter are subdivided according to the typical aerobic (both animal and bacterial) and anaerobic bacterial oxidation pathways shown in Fig. 1. Theoretically, rate of total input minus rate of total loss equals the accumulation rate of POC in sediments, below the reach of further significant biological action.

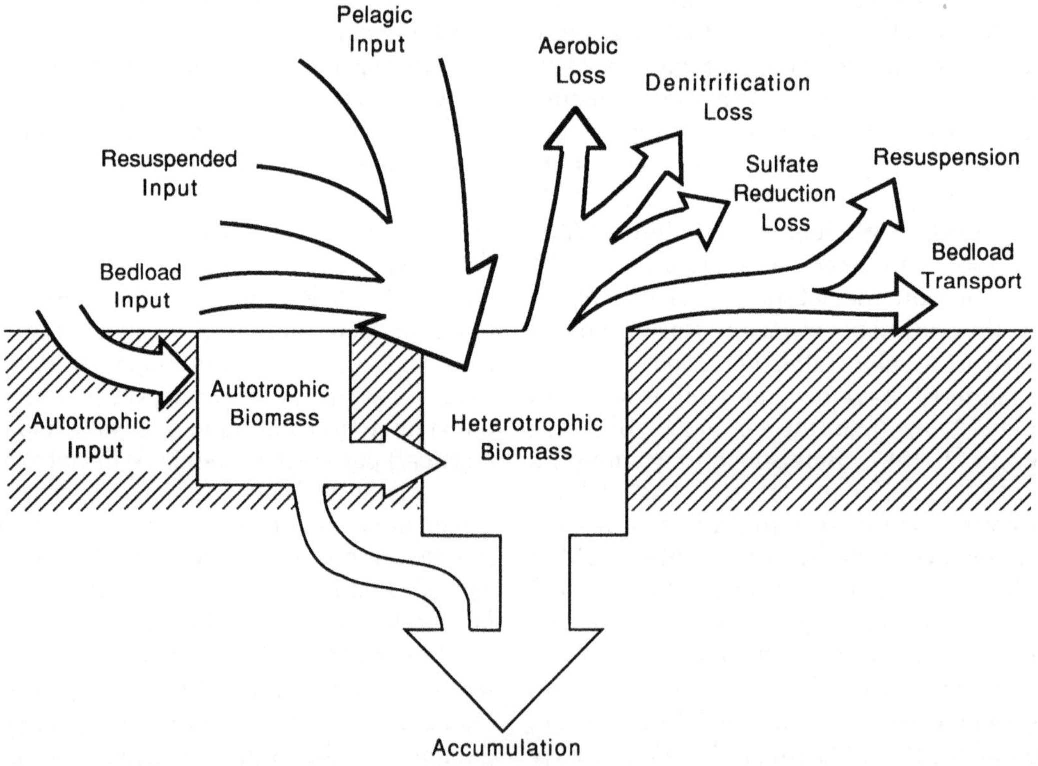

Figure 2. General conceptual approach to modeling sources and sinks of organic matter (OM) in a coastal (upwelling) benthic ecosystem, where the difference between biological and physical input terms on the left (*in situ* autotrophic conversion of CO_2 to OM, OM transport) and biological and physical loss terms on the right (respiratory production of CO_2 from OM, OM transport) equals the accumulation rate in sediments below the reach of further significant biological action (arrows not indicative of term magnitudes). Significant *in situ* autotrophic input is restricted to shallow-water sediments that receive light (photoautotrophic input) or unusually organic-rich, anaerobic sediments that release reduced chemical energy sources, e.g., sulfides, metals, and ammonia (chemoautotrophic input). [Adapted from Rowe (1981).]

Few benthic carbon budgets have been constructed in practice. Inferring OM degradation rates within sediments from measurements made with an oxygen electrode is much easier, for example, than obtaining the full suite of multidisciplinary measurements required to balance a carbon budget. Many of the measurements are either difficult to achieve technically or yield results with multiple or ambiguous interpretations [see report of U.S. GOFS Working Group (1989) on the problems with sediment-trap measurements]. Early POC budgets (e.g., Rowe and Gardner, 1979) were constructed from simplest terms, inferring biological utilization of carbon in sediments (essentially synonymous with OM diagenesis) from two basic measurements: POC flux to the seafloor, using sediment traps, and final burial rate, deduced from profiles of total organic carbon (TOC) in sediment cores. Budgets have become more sophisticated recently, as measurements of additional terms (Fig. 2) have been attempted. Though these more complex efforts seldom yield balanced budgets (Smith, 1987; Smith *et al.*, 1987), they have at least illuminated terms most critical to OM diagenesis and new questions and directions for research. Of particular relevance to this chapter are the following examples: (1) Rowe *et al.* (1986) were able to refute the hypothesis that continental shelves export significant amounts of POC (Walsh *et al.*, 1981) by including measurements of microbial POC consumption in the budgeting process and considering non-steady-state conditions (a lag in coupling between pelagic POC supply and benthic demand); and (2) Rowe and Deming (1985), Smith *et al.* (1987), and Rowe *et al.* (1991) obtained quantitative evidence for the long-held belief that heterotrophic bacteria, rather than other sediment biota, account for the major fraction of OM degradation in the deep sea.

The carbon budgeting exercise, as approached to date, is at once limited by multidisciplinary complexities and conceptual oversimplification. As with the classic pore-water modeling approach, neither the quality of POC entering and buried within the sediments (the multiple-G model) nor the energetics of POC degradation apart from living organisms (mediated by extracellular enzymes) has yet been incorporated into the conceptual framework of the model.

2.3. Bacteriological Laboratory Models

A third approach to understanding OM diagenesis in sediments is the historical hallmark of the bacteriologist: to characterize the physiological potential of bacterial isolates from natural sediments and the environmental parameters that constrain their potential through experimentation in the laboratory, either in pure or mixed culture. This approach has been essential to confirming conclusions based on isotopic tracer experiments or chemical measurements in natural sediments and pore waters. To a significant degree, experimental evidence for a new or unusual microbial process remains suspect until an individual microorganism or consortium of isolates is demonstrated in the laboratory to achieve the same results. More importantly, significant progress toward understanding environmental constraints in nature frequently is not achieved until appropriate isolates are available in culture.

For example, the idea that indigenous barophiles (pressure-loving bacteria) might be the primary agents of OM degradation in oxygenated abyssal sediments was first suggested by ZoBell and Morita (1957). However, no further support for this concept was obtained until after the first barophile was isolated in pure culture two decades later (Yayanos *et al.*, 1979). Soon thereafter, experiments examining the bacterial fate of isotopically labeled carbon compounds in deep-sea sediments under *in situ* conditions demonstrated, without exception, that more rapid bacterial carbon utilization occurred under elevated pressures than in controls at atmospheric pressure (Deming and Colwell, 1982, 1985; Rowe and Deming, 1985; Cahet and Sibuet, 1986). The conclusion that barophiles were responsible for this activity was considered firm when numerous strains of barophiles were also isolated readily from the sediments (Deming, 1985, 1986; Yayanos, 1986). As the role of heterotrophic barophiles in deep-sea carbon cycles has become increasingly evident through field experiments and carbon budgeting efforts (Rowe and Deming, 1985; Cole *et al.*, 1987; Lochte and Turley, 1988), the positive influences of deep-sea pressures on rates of other biogeochemical cycles driven by bacteria have also been discovered: for example, barophilic manganese oxidation (Cowen, 1989) and barophilic (aerobic) methane oxidation (deAngelis *et al.*, 1991).

Considerable geochemical and experimental tracer evidence also exists for the occurrence of an unusual bacterial process, anaerobic methane oxidation (Reeburgh, 1976; Alperin and Reeburgh, 1984; Alperin *et al.*, 1988), yet no bacterium capable of oxidizing methane anaerobically has been cultured. In this case, the process has been explored and documented so thoroughly via tracer experiments across a wide spectrum of anaerobic marine environments (Reeburgh, 1989), including deep basins (Black Sea;

Reeburgh *et al.*, 1991), that little skepticism remains about the validity of the process. Nevertheless, until an appropriate isolate, or (more likely) consortium of microorganisms (see later sections), is obtained in culture, progress in understanding the physiological requirements of and environmental constraints on anaerobic methane oxidation will be limited.

Only a fraction of a percentage of the bacteria known to be present and active by nonculturing techniques in natural sediments have been isolated in pure culture. The laboratory conditions required to obtain isolates, and our imaginations and incentives for choosing the conditions, are simply too selective. Perhaps more importantly, many bacteria may survive and function in nature only as members of microbial consortia; laboratory isolation of individual strains in pure culture may not be achievable (or relevant to nature, if achieved). Realistically, models of bacterial action based on laboratory culture work cannot be expected to predict rates of OM diagenesis in natural marine sediments. Their power lies more in a conceptual understanding of the ecological niche occupied by specific microbes, environmental factors that might constrain their activities, and pathways provided for the degradation of specific organic compounds.

The continuing utility of laboratory models is well illustrated by recent discoveries of entirely novel modes of anaerobic microbial energy metabolism in sediments. Whereas only abiotic mechanisms for the oxidation of aromatic carbon compounds with iron had been known previously, Lovley and his colleagues isolated the first organism of any type (an unnamed bacterial strain, designated GS-15) that can oxidize aromatic hydrocarbons anaerobically (Lovley *et al.*, 1989b). They demonstrated further that anaerobic bacterial iron-reducing activity is required for that degradative process to occur in natural sediments (Lovley *et al.*, 1989a). Isolate GS-15 can also couple hydrogen and formate oxidation to the reduction of iron and manganese (Lovley and Phillips, 1988). By considering how this unique (among heretofore cultured strains) bacterium might interact with other known strains of the sediment bacteria *Alteromonas*, *Clostridium*, and *Bacillus*, these authors have developed a conceptual model that explains how an anaerobic microbial consortium might oxidize fermentable OM completely to carbon dioxide using Fe(III) or Mn(IV) as the sole electron acceptor (Lovley *et al.*, 1989a).

The identification of iron as an important electron acceptor in OM degradation, especially in reducing sediments, and the still mysterious nature of the electron acceptors involved in anaerobic methane oxidation underscore the limitations of pore-water and carbon-budget modeling approaches (Sections 2.1 and 2.2) that address only one or more of the conventional electron acceptors (oxygen, nitrate, and sulfate). Laboratory studies of the physiology of new bacterial isolates hold much promise for a more sophisticated and accurate view of OM diagenesis in marine sediments.

2.4. Microbial Loop Models

Underlying each of the above efforts to model the process of OM diagenesis in marine sediments, whether quantitatively or conceptually, is the need for more information on the magnitude and quality of OM entering the sediment ecosystem. The direct approach of analyzing particulate materials recovered in bottom-moored sediment traps is arguably compromised by problems associated with the design and deployment of the traps themselves (U.S. GOFS Working Group, 1989; Butman, 1986; Butman *et al.*, 1986). An evolving alternative approach to the same issue relies on the use of pelagic productivity models, including new production estimates (Eppley and Peterson, 1979) and knowledge of community structure, to predict both the amount and quality of OM exported from the euphotic ecosystem to the seafloor.

Current understanding of pelagic ecosystems is based on three relatively new discoveries: (1) a significant portion of the CO_2 converted to OM by phytoplankton is channeled into bacteria (in pelagic terminology, picoheterotrophs): (2) small cyanobacteria (picoautotrophs) frequently constitute a large fraction of the primary producers; and (3) both types of picoplankton are consumed by microflagellates, ciliates, and other Protozoa. The development of this "microbial loop" portrait of pelagic ecosystems (Fig. 3; Pomeroy, 1974, 1979; Azam *et al.*, 1983) has altered the long-held view that the principal consumers of phototrophs were larger zooplankton. Knowledge of the dominant grazers in a pelagic ecosystem carries with it important implications for the flux of OM from the euphotic zone, since different species and size classes of planktonic grazers excrete OM of varying chemical quality and physical form, only some of which escapes *in situ* decomposition by sinking into the ocean interior (e.g., copepod fecal pellets). With enough information on species composition and prey size specificity of the predator community, the expectation is that "microbial loop" models can help to elucidate the chemical nature and thus the nutri-

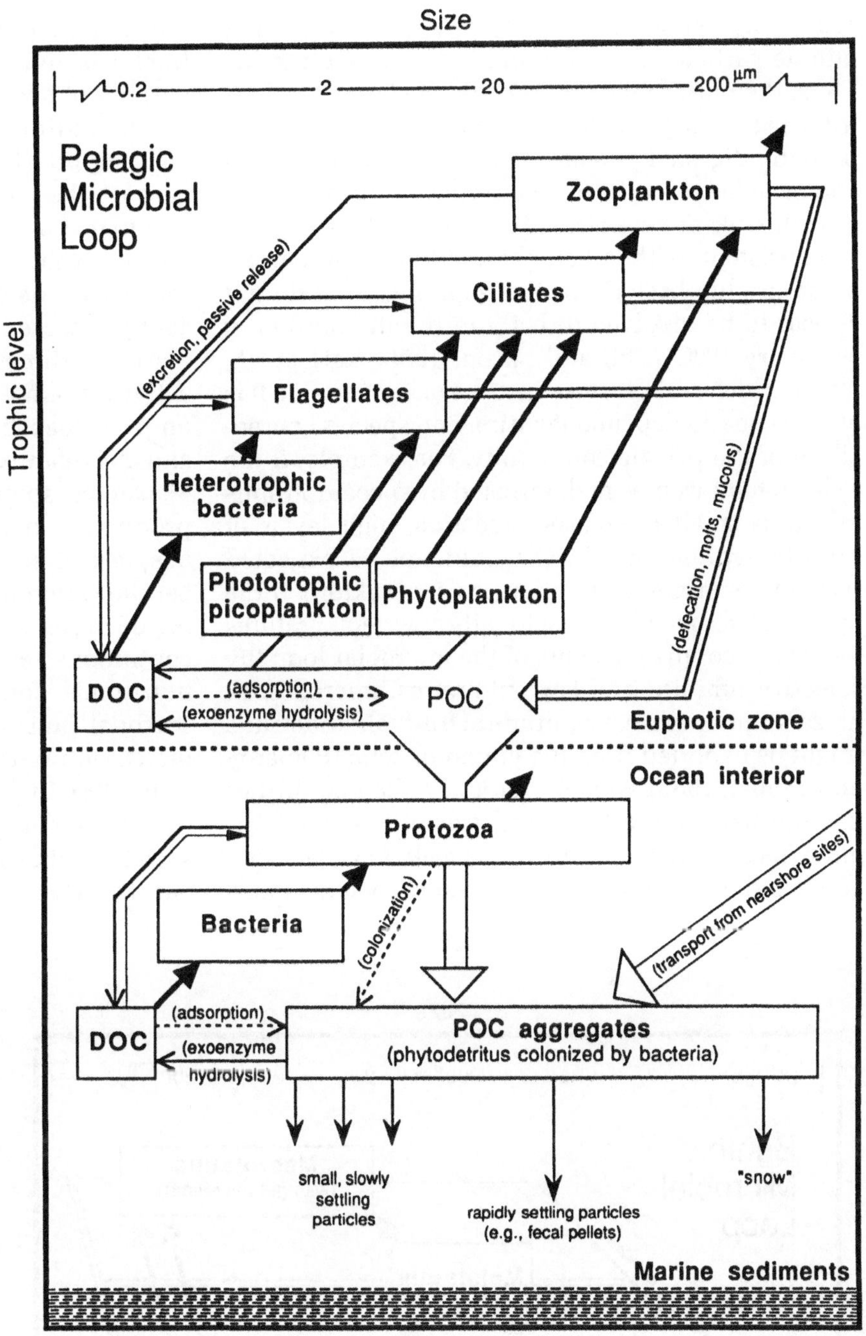

Figure 3. Sources and general description of particulate organic carbon (POC) that reaches the benthic environment. Excess POC produced but not recycled within the pelagic microbial loop of the euphotic zone [upper panel, taken from Azam *et al.* (1983)], as well as POC (including terrestrial phytodetritus) transported laterally from nearshore environments, forms aggregates that sink into the ocean interior, where they may be altered further, depending on sinking rate, by passage through additional (nonphotosynthetic) microbial loops (an example is shown in the lower panel) before reaching the seafloor. Solid triangular arrowheads indicate energy flow to higher trophic levels; open arrowheads, loss of nonliving POC; line arrowheads, flow of dissolved organic carbon (DOC).

tional value of OM exported from the pelagic environment (Michaels and Silver, 1988).

There is much to learn, however, before this goal can be achieved. Bacterial growth rates (based on nucleotide uptake experiments and protozoan grazing rates) clearly indicate that heterotrophic bacteria must be consuming 20 to 40% of the carbon fixed by phototrophs (Azam *et al.*, 1983; Jumars *et al.*, 1989), yet the major pathway of carbon flow between the two groups is still widely debated. Evidence for the popular explanation that healthy, or not so healthy, phytoplankton directly leak the necessary dissolved organic carbon (DOC) to the bacteria, as proposed by Azam *et al.* (1983), is not compelling. More recent data (Isao *et al.*, 1990) indicate that much of the DOC in the euphotic zone may be in the form of submicron-sized, highly labile, nonliving organic particles, believed to derive from leaking or dissolving (by physical or hydrolytic action) fecal excretions of zooplankton and other types of detrital particles (Jumars *et al.*,

1989; Toggweiler, 1990; Smith *et al.*, 1992). Similar minute particles are not found in the ocean interior. Thus, the most labile and nutritious elements of detrital particles produced in the euphotic zone (e.g., partially digested phototrophs) may be both released and recycled there, fully accounting for the measured growth rates of heterotrophic bacteria. This scenario is consistent with other observations that pelagic heterotrophic bacteria in the euphotic zone do not appear to be involved directly in the dissolution of refractory POC (Cho and Azam, 1988; Karl *et al.*, 1988), which sinks instead into the ocean interior. It is also a scenario dependent entirely on species composition of the pelagic community. For example, if the pelagic food web were dominated by picoautotrophs and thus small protozoan predators, high levels of rapidly sedimenting POC would not be expected (since they originate from larger zooplankton). If the effects of seasonally or regionally low temperatures were to decouple elements of the microbial loop (by selective inhibition of key autotrophs, microbes, or grazers), an unusual magnitude of fresh phytodetritus might be exported from the euphotic zone (Pomeroy and Deibel, 1986). Model predictions can be further complicated by the biological exceptions: large diatom flocs that sink rapidly (Smetacek, 1985; Jackson, 1990) and dominant predator species that consume a broad, rather than narrow, size range of prey [see Michaels and Silver (1988) for a more complete discussion of species composition and sinking fluxes].

In addition to a better understanding of the pelagic supply of OM to the seafloor, the increasing data base on the concept of a pelagic microbial loop has stimulated new thinking about the potential for similar microbial pathways and interactions to occur in the seabed itself. Over 25 years ago, Fenchel (1967, 1969) observed that detrital food webs in shallow marine sediments were dependent on the catalytic activities of both aerobic and anaerobic bacteria and on protozoan links between bacteria and meiofauna and macrofauna. Such detrital "microbial loops" (Fig. 4) can be spatially and temporally continuous in productive neritic and shelf environments. That deeper regions of the seafloor might support similar benthic microbial loops was never imagined, since the deep sea was believed to receive only a slow and continuous rain of low levels of OM, very poor in nutritional content (Jannasch, 1979). As reports of seasonal pulses of more labile OM to the deep sea increased in number (Deuser and Ross, 1980; Deuser *et al.*, 1981; Ittekkot *et al.*, 1984), the potential for more active and complex microbial communities to be associated with sinking detritus was also demonstrated, particularly for marine snow transiting the ocean

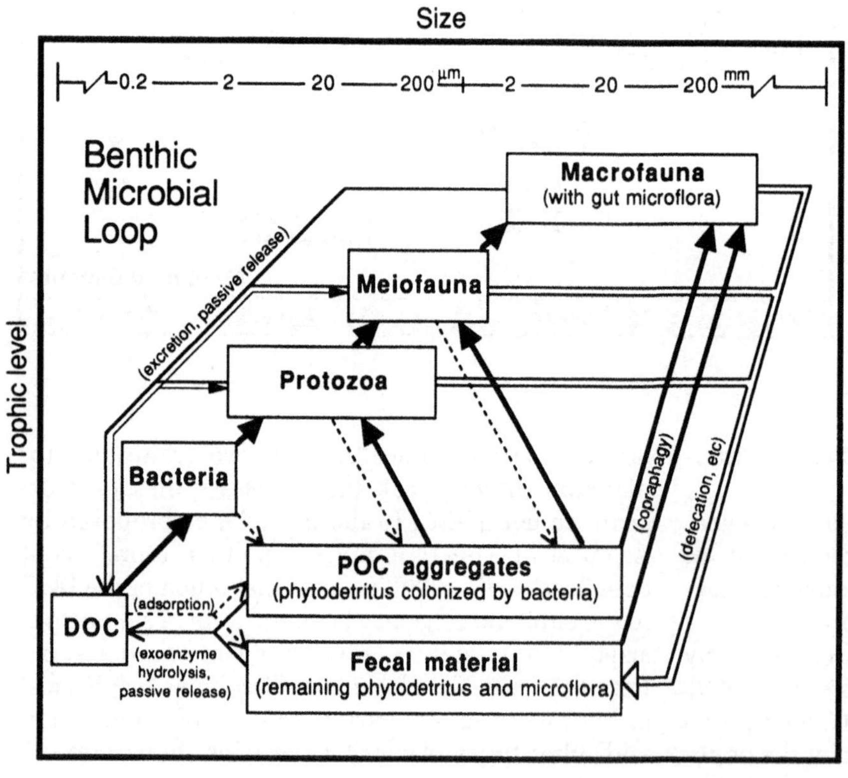

Figure 4. Benthic microbial loop in marine sediments, based on the supply of POC aggregates escaping pelagic loops (Fig. 3), adapted from the detrital food chain model of Fenchel and Jørgensen (1977) and the pelagic microbial loop described by Azam *et al.* (1983). Solid triangular arrowheads indicate energy flow to higher trophic levels; open arrowheads, loss of POC; line arrowheads, flow of dissolved organic carbon (DOC); dotted arrows, colonization. Since virtually all POC in surface marine sediments passes through animal guts before final burial, the distinction between POC aggregates and fecal material is conceptually ambiguous: both are given equal status in providing base support for the microbial loop [according to Pace *et al.* (1984)]. For the same reason, the importance of gut microflora in sediment diagenesis cannot be overemphasized.

interior (Alldredge and Cohen, 1987; Alldredge and Silver, 1988) and fecal pellets reaching abyssal depths (Deming, 1985; Deming and Colwell, 1985). Lochte and Turley (1988) and Turley *et al.* (1988) forever altered the conventional view of the deep sea by recovering from the surface of an abyssal sediment core a sample of unusually fresh phytodetritus, heavily colonized by pelagic picoplankton from the euphotic zone but also by a complex community of barophilic bacteria and Protozoa, including a novel, barophilic bacterivorous flagellate unique to the deep sea. This abyssobenthic microbial loop was further demonstrated to decompose the phytodetritus at rates more rapid than previously thought possible under deep-sea conditions. Future research is needed to determine how widespread and important such benthic microbial loops may be to rates and pathways of OM diagenesis across the global ocean seafloor. In later sections, we examine specific degradative pathways inherent to microbial loops and implications for anaerobic processes when microbial activity on particles reduces oxygen concentrations on a scale of micrometers.

3. Bacterial Abundance and Global Distribution in Sediments

Until recently, perceptions of the number of bacteria residing in marine sediments were limited by available methods. Culturing efforts clearly underestimated the total population by orders of magnitude, while biochemical and molecular approaches missed the answer by being too specific (muramic acid assays, genetic probes) or too broad (ATP assays). [For a complete review of methods for measuring bacterial and general microbial abundance, biomass, and various activities, see Karl (1986).] Only through the introduction of epifluorescence microscopy as a convenient tool for examining sediments (Dale, 1974) have the great densities of benthic bacteria been appreciated. The current maximum density, reported for Australian mangrove environments (Alongi, 1989), exceeds 5×10^{11} bacteria per gram dry wt. of sediment.

While technical details of the ideal microscopic method are still debated (Newell and Fallon, 1982; Dye, 1983; Ellery and Schleyer, 1984; Velji and Albright, 1986), the data base on direct counts of bacteria from a wide variety of marine sediments has grown large. Partial tabulations of these data from nearshore environments are available (Meyer-Reil, 1987b; Alongi, 1989; Kemp, 1990), but analyses of bacterial abundance in light of specific environmental factors, or comparable information from deeper regions of the ocean, are rare. An important early analytical effort along these lines resulted in the paradigm that mean grain size and TOC content determine bacterial densities in marine sediments (Dale, 1974). (Dale was careful to explain that a measure of organic quality more refined than bulk TOC would provide a better determinant of bacterial abundance, but the paradigm is frequently generalized to total organic content.) Thus, fine-grained muds rich in organic carbon should harbor the largest bacterial populations, and coarse-grained beach sands of low organic content, the smallest. For the most part, such generalizations persist in the literature and planning of related studies. However, closer statistical analyses of old and new data sets (Bird and Duarte, 1989), as well as overt exceptions to the rule (Thistle *et al.*, 1985; Graf, 1986), have left the grain-size/TOC paradigm less well grounded (soiled). Increasingly, other factors, such as animal grazing pressure (Morrison and White, 1980; Kemp, 1987, 1988), physical disturbance of the sediment (Findlay *et al.*, 1985; Miller, 1989), and timing and quality of the OM supply to the seafloor (Lochte and Turley, 1988; Graf, 1989), have been seen to override grain size or TOC as more important determinants of bacterial abundance in marine sediments. Of relevance to our purposes here is the fact that many findings forcing new questions to be asked about sediment bacteria have derived from work in deep-sea environments.

In an attempt to provide a more global picture of bacterial densities in marine sediments than offered by other reviews (Meyer-Reil, 1987b; Alongi, 1989; Kemp, 1990), we examined direct-count data from the wide variety of sedimentary environments that have been studied over the last 15 years, including as much comparable data from the deep sea as we could locate. In the process, we also reviewed the limited data available on more dynamic features of sediment bacteria, specifically bacterial population growth and carbon production rates [see data in Deming and Yager (1992), which build even further on those presented here]. The results of this literature survey were condensed to produce Table I and Fig. 5, in which the range of densities (and growth and carbon production rates) observed in the surface layer of sediments are indicated according to conventional depth zones of the ocean (Sverdrup *et al.*, 1942): <10 m (intertidal and subtidal zones); 10 to <200 m (bays, harbors, and coastal areas); 200 to <2000 m (offshore basins and continental shelf and slope regions); 2000 to <6000 m (continental rise areas and abyssal plains); and

TABLE I. Ranges of Bacterial Densities and Population Growth and Carbon Production Rates in the Surface Layer of Marine Sediments

	Range (n)		
Depth zone (m)	Density[a] (No. $\times 10^9$ g^{-1} dry wt.)	Growth rate[b] (day^{-1})	Carbon production rate[c] (mg C m^{-2} day^{-1})
<10[d]	0.10[e]–560[f] (345)	0.006[g]–5.5[h] (82)	1.0[i]–4035[j] (73)
10–<200[k]	0.18[l]–20[m] (186)	0.019[g]–1.7[n] (13)	28.8[e]–466[l] (14)
200–<2000[o]	0.01[p]–7.1[p] (37)	0.013[q]–1.0[r] (25)	9.36[q]–156[s] (6)
2000–<6000[t]	0.20[u]–21.6[v] (83)	0.001[w]–0.17[x] (11)	0.068[w]–3.48[w] (11)
≥6000[y]	0.32–1.2 (3)	<0.03 (1)	0.013–0.014 (2)

[a]Measured by epifluorescence microscopy in 0–1-cm layer.
[b]Measured by variety of methods (as footnoted) in 0–1-cm layer (except as footnoted); generation times (G) converted to growth rate (u) using $u = 0.693/G$.
[c]Measured by variety of methods (as footnoted) in 0–1-cm layer or else corrected to that depth (as footnoted) for comparative purposes.
[d]Includes intertidal and subtidal environments (salt marshes, mud and sand flats, beaches, mangrove, and coral reef environments); draws upon information in Meyer-Reil *et al.* (1980), Moriarty and Pollard (1981, 1982), Rublee (1982), Fallon *et al.* (1983), Crraven and Karl (1984), Moriarty *et al.* (1985), Meyer-Reil (1987b), Hansen *et al.* (1987), Karl *et al.* (1987), Alongi (1987a, 1989), Karl and Novitsky (1988), and Boto *et al.* (1989).
[e]Hansen *et al.*, 1987; min. density, Australian coral reef sediment, 5-m depth; min. production, by tritiated thymidine method (TTM), coral reef sediment, 12-m depth.
[f]Alongi, 1989; Australian mangrove sediment.
[g]Fallon *et al.*, 1983; min. growth, by TTM, Georgia salt marsh sediment (0.006 day^{-1}); offshore sediment, 16-m depth (0.019).
[h]Alongi, 1987a; by TTM, Australian mangrove sediment, 0–2 cm.
[i]Moriarty *et al.*, 1985; by TTM, Australian beach sand.
[j]Craven and Karl, 1984; by tritiated adenine method (TAM), Hawaiian intertidal sediment, 0–2 cm (0.5 × reported value); higher production rates, calculated from frequency of dividing cells (FDC) in other intertidal sediments, considered suspect (Fallon *et al.*, 1983).
[k]Includes bays, harbors, and coastal (including Arctic) environments; draws upon information in Aller and Yingst (1980), Fallon *et al.* (1983), Novitsky (1983a, 1983b, 1987), Atlas and Griffiths (1984), Yingst and Rhoads (1985), Meyer-Reil (1987b), Karl *et al.* (1987), and Montagna *et al.* (1989).
[l]Montagna *et al.*, 1989; min. density, California hydrocarbon seep sediment, 18-m depth; max. production, by FDC, same site.
[m]Meyer-Reil, 1987b; Kiel Bight muddy sediment, 28-m depth.
[n]Karl *et al.*, 1987; by TAM, Hawaiian Sand Island, 75-m depth.
[o]Includes offshore basins and continental shelf and slope environments; draws upon information in Craven and Karl (1984), Deming (1985), Crave *et al.* (1986), Smith *et al.* (1987), and Alongi (1987b, 1990).
[p]Alongi, 1987b, Australian reef slope sediment, 695-m depth.
[q]Alongi, 1990, min. growth, by TTM, 0–0.5 cm, Australian trough sediment, 695-m depth; min. production, by TTM, 0–0.5 cm (2 × reported value), same trough, 1131-m depth.
[r]Smith *et al.*, 1987, by TAM, 0–1 mm, Santa Catalina Basin, 1300-m depth.
[s]Craven *et al.*, 1986; by TAM, 0–4 cm (calculated for 0–1-cm layer from data in their Fig. 3), San Pedro Basin sediment, 850-m depth.
[t]Includes continental rise and abyssal plain environments; draws upon information in Deming and Colwell (1982, 1985), Harvey *et al.* (1984), Deming (1985), Rowe and Deming (1985), Wilke *et al.* (1985), Thistle *et al.* (1985), Aller and Aller (1986), Aller (1989), and Alongi (1990).
[u]Rowe and Deming, 1985, French Bay of Biscay sediment, 4100-m depth.
[v]Thistle *et al.*, 1985; Nova Scotian continental rise sediment (HEBBLE site), 4626-m depth.
[w]Alongi, 1990; min. growth, by TTM, 0–0.5 cm, Australian Woodlark Basin sediments, 2395-m and 2800-m depths; min. production, by TTM, 0–0.5 cm (2 × reported values), trough sediment, 2426-m depth; max. production, same except Coral Sea Basin sediment, 4008-m depth.
[x]Wilke *et al.*, 1985; by density increase at *in situ* temperature and pressure (DI), Hatteras Abyssal Plain sediment, 5411-m depth.
[y]Includes sites along axis of Puerto Rico Trench; draws upon information in Tietjen *et al.* (1989) and from Deming (unpublished); max. density, 1–2 cm, 7460-m depth; growth, by DI, 7460-m depth; min. production, using $u = 0.001$ day^{-1} and bacterial density (0–1 cm), 7460-m depth; max. production, same except 8189-m depth.

≥6000 m (hadal trenches). Considering the somewhat arbitrary (from a biological perspective) breakpoints of these depth zones, the diversity of study sites and investigators, and our general trend-seeking purpose, we have presented only ranges of values and not calculations of mean or median values.

The results of reviewing available bacterial density measurements are instructive from several perspectives. First, on the basis of number of data points alone, it is clear that the most understudied regions of the seafloor include important continental shelf and slope environments in the depth range of 200 to <2000 m and, perhaps not as surprisingly, many deeper regions of the seafloor. General perceptions of how bacteria contribute to OM diagenesis are no doubt colored by the preponderance of research in nearshore sediments, where levels of OM, flora, and fauna are higher, and oxygen lower, than in offshore sediments. Second, efforts to improve counting techniques and study new environments have expanded, not tightened, the range of bacterial densities known for surface sediments from each of the depth zones. Without exception, the end points of each range in Table I derive from studies published since 1985. This trend for range expansion can be expected to continue, as evidenced by new data in Fig. 6. Direct counts in surface sediments from depths of 5840 and 5845 m in the Nares Abyssal Plain, underlying the oligotrophic Sargasso Sea, are even lower then the published deep-sea values shown in Table I. Third,

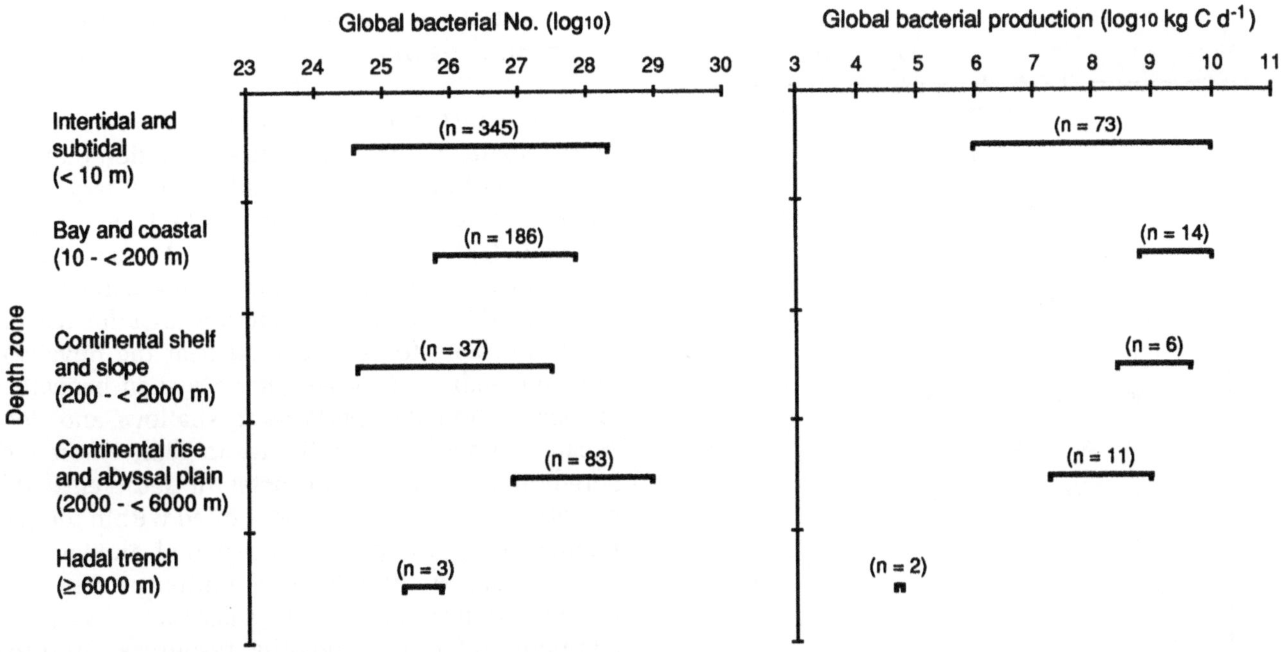

Figure 5. Global estimates of bacterial numbers and carbon production rates in the surface layer (0–1 cm) of major marine sedimentary environments [calculated from ranges listed in Table I, assuming 1.5 g dry wt. of sediment per cubic centimeter when necessary and using depth zone areal coverages reported by Sverdrup *et al.* (1942).] The global totals fall in the range of 1×10^{27}–1×10^{29} bacteria and 1×10^{9}–3×10^{10} kg C day^{-1}.

the widespread perception that bacterial abundances (and many other organic features of marine sediments; see Muller and Suess, 1979; Suess, 1980) generally decrease with increasing ocean depth is verified only shakily. The anomalies and scatter seem more compelling than the overall trend. The maximal density for the deep sea, observed near the *HEBBLE* study site in sediments of the Nova Scotian continental rise at a depth of 4626 m (Thistle *et al.*, 1985), is actually somewhat higher than the maximum observed for the much shallower depth zone of 10 to <200 m, in muddy sediments of Kiel Bight at a depth of 28 m (Meyer-Reil, 1987b). The lowest density of the global data set derives not from the greatest depth examined (8189 m in the Puerto Rico Trench; Tietjen *et al.*, 1989), but from sediment on the Australian continental slope at the relatively modest depth of 695 m (Alongi, 1987b). Fourth, when scaled to areal coverage of the global ocean seafloor (Fig. 5), the largest number of bacteria are seen to reside in the surface layer of deep-sea sediments under a water column of 2000 m or greater [maximal estimate of 10^{29} bacteria or about three times the maximal estimate for the global intertidal and subtidal depth zones of <10 m, where specific densities can be extremely high (Table I)].

For over a century, the vast portion of these deep-sea bacteria was believed to be operating suboptimally, if at all, as a result of restrictive temperatures and pressures (Certes, 1884; Jannasch and Wirsen, 1973). The global trend of maximal bacterial productivity in sediments diminishing with increased ocean depth, evident in both Table I and Fig. 5, would appear to confirm this notion. However, it is important to realize that the available information on dynamic features of bacteria in deep-sea sediments is limited, and maxima presented for deep-sea growth and carbon production rates in Table I and Fig. 5 are probably underestimates. For example, no growth or carbon production estimates have been made at the deep-sea sites where remarkably high bacterial densities were observed (Thistle *et al.*, 1985; Aller and Aller, 1986; Aller, 1989), and the very rapid rates of deep-sea bacterial activity observed by Lochte and Turley (1988) were not reported in terms that allowed their inclusion in Table I or Fig. 5 [see Deming and Yager (1992) and Lochte (1992) for an updated treatment of the data]. In the next section, we examine why so great a reservoir of diagenetic agents in the deep sea appears to be held in check and what events or conditions might trigger a different response from them.

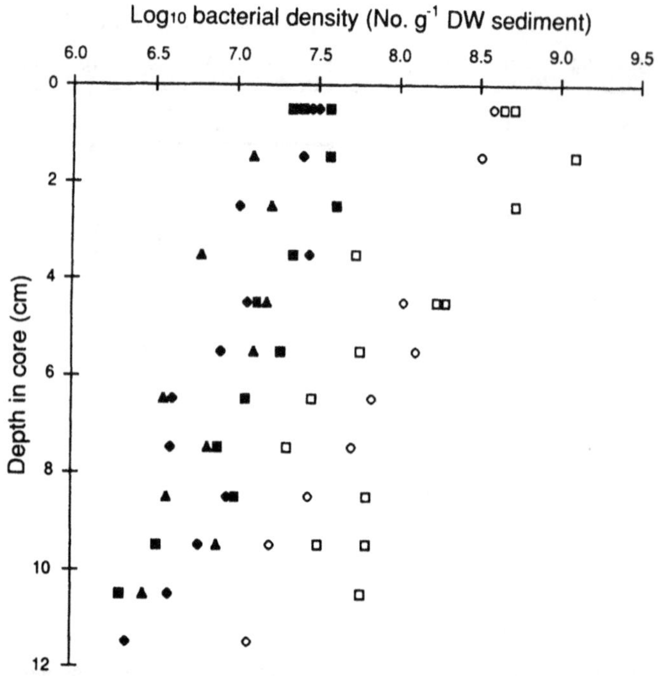

Figure 6. Bacterial abundance in sediments from the Nares Abyssal Plain at depths of 5840 m (■, ▲) and 5845 m (●) and from the Puerto Rico Trench at depths of 7460 m (□) and 8189 m (○), determined by acridine orange staining and epifluorescence microscopy [from Tietjen *et al.* (1989) and Deming (unpublished data)]. Unusually low densities in the Nares sediments (underlying the oligotrophic Sargasso Sea) and significantly higher densities in the deeper Trench sediments (proximal to land) belie the paradigm that benthic bacterial abundance decreases simply as a function of increasing ocean depth, supporting instead the hypothesis that flux of organic materials to the seafloor provides the primary constraint on benthic productivity.

4. Constraints on Bacterial Activities in Sediments

From research on shallow sediments (particularly Novitsky, 1987) has come the notion that most bacteria within an undisturbed sedimentary matrix use energy from the oxidation of OM (or inorganic compounds) for cell maintenance and survival, rather than for the explicit production of new cells. Significant growth or carbon production rates for bacteria are more likely to be measured near the physically dynamic sediment–water interface than beneath it, an observation that pertains to shallow- and deep-ocean sediments alike (all rates in Table I were measured in the upper centimeter of sediment). Why bacterial activities should be limited within the sedimentary matrix is poorly understood, though some combination of the effects of limited access to required dissolved and gaseous substrates, interspecific competition for these and other resources, buildup of inhibitory metabolites, and, possibly, absence of grazing pressure seems likely. A discussion of factors restricting specific metabolic processes on a microscale within the sedimentary matrix is beyond the scope of this section; we examine instead some of the factors operating on an oceanic scale to constrain overall benthic bacterial activity and diagenetic rates, especially in the deep sea (Table II).

"The initial approach to understanding a complex ecological system is often, sensibly, to look for general quantitative relationships between two or

TABLE II. Combination of General Parameters Believed to Regulate Microbial Activity in Sedimentary Environments

	Potential regulatory parameters					
Type of sedimentary environment	Input of organic materials	General availability of oxygen	Physical flushing	Temperature	Pressure	Predation and complexity of trophic levels
Estuarine, nearshore	High	Seasonably variable	Periodic (tidal)	Seasonably variable	Negligible	High, complex[a] (well-known microbial loop[b])
Deep-sea	Low[a]	High[a]	Constant and slow[a]	Low[a]	High	Low, loop exists[a] (data limited)
Polar (shallow to deep)	Seasonally variable	High	Usually dynamic	Low	Low to high	Variable, loop probable (data limited)
Anoxic (Black Sea, Cariaco Trench)	Intermediate to high	Negligible	Low to negligible	Low	Intermediate to high	Low to negligible, loop possible (data limited)
Hydrothermal (Guaymas Basin)	High	Spatially variable	Complex convection	Spatially variable (low to high)	Intermediate	Variable?, loop possible (no data)

[a]Evidence exists for important exceptions to the generalized paradigm: see Rowe *et al.* (1988, 1990, 1991) for evidence that the microbial loop[b] may be less significant in nearshore sandy sediments; see text for exceptions to deep-sea generalizations.
[b]Microbial loop refers to the cycling of organic materials through microscale flora (bacteria) and fauna (protozoa); see Figs. 3 and 4.

more variables describing the system (Rigler, 1982). The search is not done blindly, but proceeds on the basis of previous successes in other systems" (from Bird and Duarte, 1989). This is how Dale (1974), in his consideration of intertidal sediments, launched the study of benthic bacterial ecosystems, which over time has become more experimental in nature. The understanding of bacterial ecosystems in deep-sea sediments, however, has been approached in the opposite manner. Initial studies were by individual investigators on isolated cruises addressing specific questions with the methods of their choice, the result being a set of unique and frequently enlightening experiments but no broad set of like variables from which to make oceanwide comparisons or extrapolations. Only recently have enough data been accumulated from a range of offshore sites in one ocean to enable a search for general quantitative relationships between a sediment bacterial parameter and other deep-sea variables. Rowe *et al.* (1991) considered, among other things, the relationships of total bacterial abundance in sediment to ocean depth (continental shelf to abyssal plains of the North Atlantic), TOC content of the sediments, and supply of POC to the seafloor (measured via bottom-moored sediment-trap collections). We broadened their data set by adding unpublished data, obtained by the same team of investigators at new sites in the North Atlantic, and then calculated new correlation coefficients, the results of which are shown in Fig. 7 [see Deming and Yager (1992) for an expansion of these data to include abyssal Arctic basins]. In spite of the simplicity and limits of the approach, particularly that total bacterial abundance is not necessarily synonymous with activity, the results confirm findings from more experimental studies of deep-sea bacteria and provide a framework for discussing potential constraints on their activities.

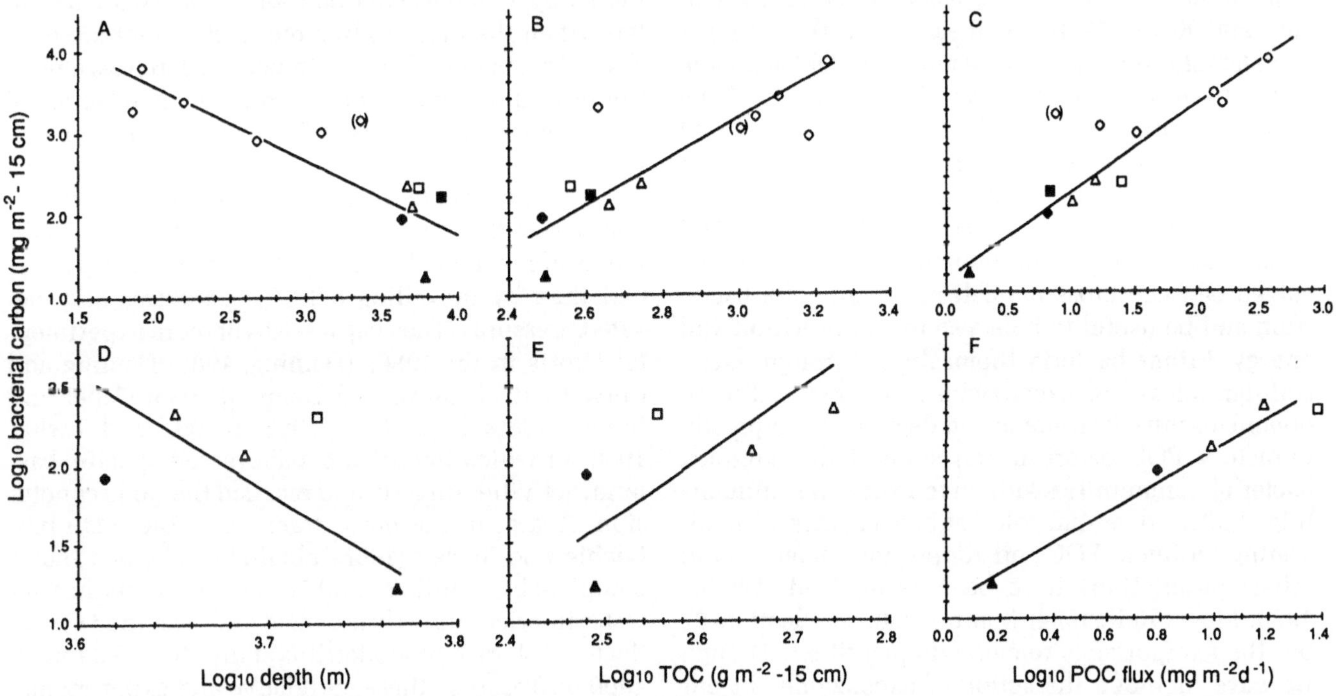

Figure 7. Correlations between total bacterial carbon content of oceanic sediments (determined by epifluorescence microscopy, using a conversion factor of 10^{-11} mg C bacterium^{-1}) and total organic carbon (TOC) content, water depth, and particulate organic matter (POC) flux (based on sediment-trap collections) based on data from the following sites: continental margin off U.S. east coast (○), from Rowe *et al.* (1991); Demerara Abyssal Plain (△), from Rowe and Deming (1985); Hatteras Abyssal Plain (□), from Wilke *et al.* (1985) and Tietjen *et al.* (1989); Bay of Biscay (●), from Rowe and Deming (1985); Nares Abyssal Plain (▲), from Rowe and Deming (unpublished data); and Puerto Rico Trench (■), from Wilke *et al.* (1985), Tietjen *et al.* (1989), and Deming (unpublished data). All bacterial carbon and TOC values are based on integrations to a depth of 15 cm (corrected to that depth, when necessary for comparative purposes, using data in published depth profiles). Graphs in the upper panel (A–C) include all known data from a depth range of 70 to 7460 m ($n = 12$). Open circles shown parenthetically represent data from the one station located on a steep slope (and, therefore, anomalously influenced by turbidity flows), which were omitted from statistical analyses. Graphs in the lower panel (D–F) focus on data from abyssal plains only ($n = 5$). In either case (broad depth range or deep sea only), bacterial carbon correlated most strongly with the supply of POC to the seafloor: (A) $n = 11$, $r^2 = 0.78$, $P < 0.01$; (B) $n = 11$, $r^2 = 0.63$, $P < 0.01$; (C) $n = 11$, $r^2 = 0.91$, $P < 0.01$; (D) $n = 5$, $r^2 = 0.28$, $P > 0.05$; (E) $n = 5$, $r^2 = 0.42$, $P > 0.05$; (F) $n = 5$, $r^2 = 0.96$, $P < 0.01$.

The expected inverse relationship between bacterial abundance and station depth, when examined across the entire depth range of 70 to 7460 m (Fig. 7A), was present though unimpressive ($r^2 = 0.78$, $P < 0.01$, $n = 11$). It was surprisingly weak when a subset of the abyssal data, from the narrower depth range of 4100 to 5840 m (Fig. 7D), was considered ($r^2 = 0.28$, $P > 0.05$, $n = 5$). That some factor(s) other than depth—and the low temperatures and elevated hydrostatic pressures inherent to ocean depth—must be more critical to the constraint of deep-sea bacteria is also made clear in Fig. 6, where bacterial abundances at the bottom of the Puerto Rico Trench (7460 and 8189 m) were found to be greater by an order of magnitude that those on the Nares Abyssal Plain at shallower depths (5840 and 5845 m).

Building upon the findings of Dale (1974) for bacteria, and upon conclusions reached for other populations of marine organisms [Banse (1964) and Vinogradov (1968) for zooplankton; Thiel (1983) for meiofauna; Rowe (1983) for macrofauna; and Haedrich and Rowe (1978) for megafauna], the stronger correlate should be some measure of the OM available to the bacteria as food. Herein lies the heart of the problem in understanding the early diagenesis of OM mediated by sediment bacteria: how to measure or classify OM in natural sediments according to relative availability as food for bacteria. The bulk of OM in sediments exists in particulate form, yet only dissolved OM can cross the cell membrane of a bacterium and be useful to it as a source of nutrition and energy. Either bacteria themselves, through extracellular release of exoenzymes, or other sediment organisms must first attach and degrade this typically refractory POC before it can support the complex bacterial communities known to reside in sediments (Fig. 1). The potential roles of exoenzymes in modulating sediment POC and supporting bacterial (and other) populations have been recognized (Wiebe, 1979; Meyer-Reil, 1984; Mann, 1988; see Section 5), but their importance remains unquantified. Though we have included the action of exoenzymes in our renditions of pelagic and benthic microbial loops (Figs. 3 and 4; see also Smith *et al.*, 1992), exoenzymes do not appear, even conceptually, in the more classic models of OM diagenesis (Figs. 1 and 2). As organic geochemists and microbiologists chip away at the problem in accessible sedimentary environments (Westrich and Berner, 1984; King, 1986; Meyer-Reil 1986, 1987a, 1987b; Henrichs and Doyle, 1986; Mayer *et al.*, 1988; Mayer, 1989), the deep sea remains unexplored in this context.

Thus, the only correlations that can be attempted at this time between a bacterial parameter of deep-sea sediments and potential organic resources are those between total microscopic counts and bulk measurements of TOC in the sediments or POC flux to the sediment–water interface. To the extent that recently settled particles represent a more labile, and therefore more immediately attacked, source of OM than "aged" particles already incorporated into the sedimentary matrix (Wakeham *et al.*, 1980; Khripounoff and Rowe, 1985; Lochte and Turley, 1988), POC flux measurements can be viewed as better gross indices to available bacterial "food" than TOC measurements on whole-sediment cores. The correlations between bacterial abundance and TOC for both the broad and narrow depth ranges considered in Fig. 7 were weakly positive at best ($r^2 = 0.63$, $P < 0.01$, $n = 11$, Fig. 7B; $r^2 = 0.42$, $P > 0.05$, $n = 5$, Fig. 7E). A much stronger relationship was apparent between bacterial abundance and POC flux to the seafloor ($r^2 = 0.91$, $P < 0.01$, $n = 11$, Fig. 7C). The tightest correlation of the study was observed between these two parameters when the data set was reduced to include only abyssal sediments, thus minimizing differences among sampling sites due to depth and the related factors of temperature and pressure ($r^2 = 0.96$, $P < 0.01$, $n = 5$, Fig. 7F).

The results of these first simple correlation studies emphasize that bacterial activities in deep-sea sediments are regulated more by the incoming supply of OM than by prevailing cold temperatures and elevated pressures. They cap a series of more experimental efforts in the 1980s (Deming, 1985; Deming and Colwell, 1985; Rowe and Deming, 1985; Cahet and Sibuet, 1986; Cole *et al.*, 1987; Lochte and Turley, 1988) in which investigators addressed specific bacterial activities directly and reached the same conclusion. Again, in the most dramatic of these studies, Lochte and Turley (1988) obtained evidence that a complete barophilic microbial loop at the sediment–water interface quickly attached and degraded OM in the form of fresh phytodetritus at the abyssal depth of 4500 m. Together, these correlative and experimental studies overturn the conventional wisdom of the 1970s (and earlier), which held that bacteria were incapable of utilizing fresh OM at detectable rates in the deep sea because of severe restrictions placed on their metabolic systems by *in situ* temperatures and pressures.

The question of temperature–pressure regulation of OM diagenesis in the deep sea can now be framed somewhat differently. If OM were nonlimiting in the deep sea (as occurs on the microscale in association with animal activities, seasonally via pulses of phytodetritus, and perhaps globally during glacial changes), what absolute constraints might cold temperatures and elevated pressures place on rates of bacterial OM degradation? This question, which has been addressed theoretically (see Fig. 8 in Jannasch and Wirsen, 1983), can now be approached in practice, by examining measured growth rates of natural assemblages of deep-sea bacteria under *in situ* temperatures and pressures in samples supplemented with readily utilizable forms of OM. If extreme physical conditions played no regulatory roles, then bacterial populations from hadal sediments (under such nonlimiting nutrient conditions) should grow as rapidly as those from shallow sediments. The pattern of decreasing maximal growth rate with increasing station depth (and therefore increasing pressure of incubation), shown in Fig. 8, suggests that hydrostatic pressures do place some absolute limits on rates of bacterial activity in the deep sea. [The issue of temperature regulation is not addressed by these data (but see Deming and Yager, 1992), since incubation temperatures were uniformly cold at <3°C. However, studies of barophiles in pure culture have indicated that modest temperature increases (of 5 to 7°C) under deep-sea pressures would stimulate OM degradation (Deming *et al.*, 1984; Yayanos and DeLong, 1987).]

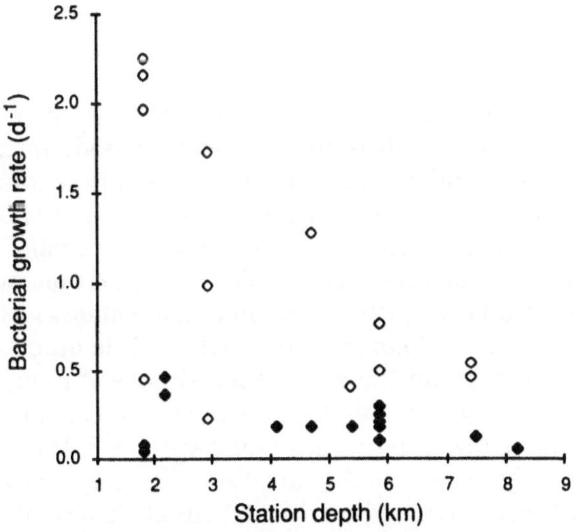

Figure 8. Comparative growth rates of bacterial populations present in samples of recently settled particulates (sediment traps deployed 10–92 meters off bottom; ○) and sediment slurries (boxcores; ●) recovered from depths of 1850 to 8189 m in the North Atlantic (locations described in caption to Fig. 7). Growth rates were determined under nonlimiting nutrient conditions (samples supplemented with ⩾0.01% yeast extract, amino acids, or chitin) by bacterial density increase, using epifluorescence microscopy, in well-mixed (oxygenated) samples incubated shipboard at *in situ* temperature and pressure (Deming, 1985). Published generation times (G) were converted to growth rates (u) for comparative purposes according to the equation: $u = \ln 2/G$. Data from Wilke *et al.* (1985) and Deming (1985, 1986, and unpublished data).

The observed pressure effect is more pronounced for bacteria associated with recent accumulations of sediment (represented by bottom-moored sediment-trap collections; open symbols, Fig. 8) than for those associated with long-buried sediments (sediment core slurries; closed symbols, Fig. 8). In fact, the minimal effect of pressure on sediment bacteria, observed across this broad sampling depth range of greater than 6000 m, and their low maximal growth rates (0.05 to 0.5 divisions day^{-1}), suggest that some factor(s) other than pressure or available OM limits their absolute rates of activity. Since potential inhibitory factors inherent to undisturbed sediments (discussed earlier) cannot easily be invoked in this case (the growth rates were measured in samples of sediments resuspended in seawater), that factor may be genetic rather than environmental. Most long-term bacterial residents of deep-sea sediments may have evolved substrate uptake kinetics that do not allow them to take advantage of elevated levels of OM: they may be oligotrophic microorganisms. Some evidence supports this idea (Deming, 1986), but no oligotrophic barophile has been obtained in culture. All known barophiles have been obtained with organic-rich media. Their maximal growth rates under *in situ* temperatures and pressures (Fig. 9) reveal a degree of absolute pressure constraint similar to that observed with natural bacterial populations (Fig. 8). However, differentiation between sediment-trap and sediment-core bacteria is no longer apparent (perhaps because the sediment strains were isolated in rich media). The only barophilic isolates that seem notable as a group are those derived from animal-associated samples (solid symbols, Fig. 9). At abyssal pressures, they tend to grow faster than their counterparts recovered from the same depths but different types of samples (Deming, 1985).

It is in the study of animal-associated microflora that the potential for most rapid and unconstrained rates of bacterial activity has been recognized. Bacteria in the guts of abyssal invertebrates (Hessler *et al.*, 1978; Deming *et al.*, 1981; Deming and Colwell, 1982) appear to utilize OM (reproduce) *in situ* almost as rapidly as those within the digestive tracts of the shallow benthic polychaete *Abarenicola vagabunda* (Plante *et al.*, 1989). By focusing more on such animal-associated microenvironments in progressively deeper marine sediments (especially anaerobic niches, as discussed below), more may be learned about constraints on deep-sea microbial activities

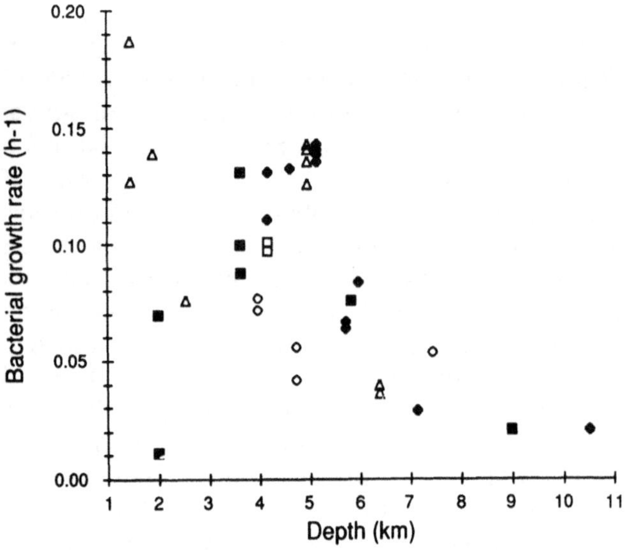

Figure 9. Comparative growth rates under *in situ* temperature and pressure for all known deep-sea bacteria isolated in pure culture under nonlimiting nutrient conditions from animal guts (●), seawater surrounding amphipods in pressurized traps (■), seawater (△), sediment-trap collections (○), and sediment cores (□) recovered from depths of 1410 to 10,476 m in the Atlantic and Pacific oceans. Published generation times (G) were converted to growth rates (u) for comparative purposes according to the equation: $u = \ln 2/G$. Data from Yayanos *et al.* (1982), Deming *et al.* (1984, 1988), Jannasch and Wirsen (1984), Deming (1985), Yayanos (1986), and Yayanos and DeLong (1987).

and about why bulk chemical and microbial measurements may have more limited meaning in the deep sea than in organic-rich, shallow-water sediments.

5. Unique Catalytic Properties and Microbial Consortia

Particulate OM reaching deep sediments will consist primarily of refractory polymeric carbohydrates, particularly cellulose and chitin but also dextrins, alginates, lignins, and other complex carbohydrates from neritic and terrestrial sources. Depending on the "freshness" of the flux (e.g., Lochte and Turley, 1988), the particulates will also contain a smaller protein and lipid fraction derived from living components of the detritus. This more labile material includes not only attached organisms but also the bacterial exoenzymes and polymers (glycocalyx and other "adhesins") involved in the attachment process and potential formation of more complex biofilms. Extracellular enzymes are required to break down both the living and nonliving components of this detrital carbon, and, as discussed in Section 4, none of the arriving particulate material can be utilized by indigenous bacteria in the absence of such enzymatic activity. It should come as no surprise that within the broad class of microorganisms that includes aerobic and anaerobic bacteria, all of the catalytic properties required to degrade particulate OM reaching the seafloor have evolved (Table III). Some marine animals are also believed to produce as digestive enzymes one or more of the carbohydrases essential to the breakdown of complex OM in sediments (Table III), but in many cases the proof (clear separation from bacterial enzymes) is equivocal (Foulds and Mann, 1978; Fong and Mann, 1980; Vonk and Western, 1984).

Understanding decomposition rates, efficiencies of conversion to soluble carbon, and environmental constraints on the diagenesis of particulate OM in marine sediments is made difficult by the fact that the breakdown of most polymeric carbohydrates requires multiple enzymes and, thus, the activities of multiple species of bacteria. When such an interactive microbial community or consortium is involved, conventional analytical approaches (for example, first-order Michaelis–Menten kinetics) are no longer applicable. As members of a consortium, heterotrophic bacteria can compete for products of enzyme decomposition while producing proteases that decompose other exoenzymes. They may also saturate an attachment glycocalyx with enzyme, thus ensuring maximal yield of product and perhaps endurance of the enzyme beyond the active life of the microbes. Indeed, the most complex of microbial consortia, known from studies of macrophyte detritus in nearshore sediments (Fenchel, 1987), are enzyme-rich biofilms of aerobic and anaerobic bacteria, diatoms, Protozoa, and meiofauna (Costerton *et al.*, 1987) functioning together as a benthic microbial loop (Fig. 4). Finally, while much may be known about the general physiological groups of microorganisms operating together as a consortium, little is known about individual species within each group. In marine sediments, the best studied species are those involved in the terminal stages of OM diagenesis—sulfate reduction, acetogenesis, and methanogenesis (Jorgensen, 1977, 1982; Ljungdahl, 1986; Capone and Kiene, 1988; Oremland, 1988); the least understood are those involved in earlier stages of OM attack—the aerobic and anaerobic, heterotrophic bacteria.

In freshwater and marine sediments that receive a sufficient input of OM (as well as in some soils and terrestrial animals with fermentative endosymbionts), oxygen disappears rapidly, due to the activities of aerobic organisms, and fermentative processes of the

TABLE III. Common Organic Substrates in Marine Sediments and the Enzymes Required for Their Degradation

Substrate(s)	Sources	Enzymes	Known enzyme sources[a] Animal	Microbial	Microbial consortia[b]	Reference(s)[c]
Lignin	Vascular plants	Ligninases	–	+ (fungi in soils)	+ (in wood-boring marine bivalves, sediments, termite guts, cow rumen)	1–5
Agarose, agaropectin, "agar–agar"	Marine macrophytes	Agarase	–[d]	+ (specific genera)	+	6
Cellulose (β-1,4-glucose)	Marine and terrestrial plants	Cellulases	–, +[d] (specific genera)	+ (common)	+	3, 6–12
Chitin (N-acetylglucosamine)	Crustaceans, mollusks, fungi, and fecal pellets	Chitinase	–, +[d] (whales, fish, crustaceans)	+ (common)	+ (sediments)	6, 13–17
Starch (α-1,4-glucose)	Marine and terrestrial plants	Amylases	+	+ (common)	+	6, 18, 19
Reserve sugars (β-1,3-glucans)	Algae, fungi, yeast, lichens, and protozoa	Laminarases	+ (pelagic copepods)	+	+	6
Alginic acid	Brown algae	Alginase	+	?	?	6
Other polysaccharides (uronic acid, dextran, exopolysaccharide)	Bacteria, algae, and animals	Carbohydrases	?	+	?	6, 18–20
Murein	Eubacterial cell walls	Muramidase, lysozyme	+[d] (fish and crustaceans)	+ (specific genera)	+ (parasitic bacteria)	6
Protein	Bacteria, plants, and animals	Proteases	+	+ (common)	+	6, 20, 21
Lipids (fatty acids, phospholipids, esters, etc.)	Bacteria, plants, and animals	Lipases	+	+ (common)	+	6
Nucleic acids (RNA, DNA)	Bacteria, plants, and animals	Nucleases	+	+ (common)	+	6

[a]Minus sign indicates source(s) examined but no enzymes found; question mark, sources not yet examined.
[b]The best studied consortia derive from specific animal guts, sediments, and soils, in which they mediate anaerobic decomposition of complex polysaccharides. However, it is probably safe to conclude that in nature the decomposition of all types of particulate organic substrates involves multiple microbial species.
[c]References: 1, Benner *et al.*, 1984; 2, Breznak, 1982; 3, Butler and Buckerfield, 1979; 4, Colberg and Young, 1982; 5, Crawford, 1981; 6, Vonk and Western, 1984; 7, Foulds and Mann, 1978; 8, Friesan *et al.*, 1986; 9, Hungate, 1975; 10, Sinsabaugh *et al.*, 1985; 11, Taylor, 1982; 12, Yokoe and Yasumasu, 1964; 13, Boyer, 1986; 14, Danulat, 1986; 15, Goodrich and Morita, 1977; 16, Herwig and Staley, 1986; 17, Herwig *et al.*, 1984; 18, Prim and Lawrence, 1975; 19, Vitalis *et al.*, 1988; 20, Martin *et al.*, 1980; 21, Mayer, 1989.
[d]Evidence for occurrence of animal enzymes is frequently equivocal, since the separation of enzymes produced by bacteria (either ingested or carried as gut flora) from those produced by the animal is difficult.

anaerobic heterotrophs become dominant. In such anaerobic systems, consortia of bacteria are known to decompose cellulose and chitin to organic acids, which are fermented further by sulfate reducers, and hydrogen and carbon dioxide, then consumed by methanogens (Fig. 10; Boyer, 1986; Ljungdahl, 1986; Dwyer *et al.*, 1988). Similar consortia have been demonstrated for the degradation of lignin compounds (Colberg and Young, 1982; Benner *et al.*, 1984, 1986, 1988). A prevailing misconception about anaerobic consortia, however, is that their effective operation requires anaerobic areas that measure from tens of centimeters to meters (anaerobic sediments or animal guts). Nothing could be further from reality. Both indirect and direct evidence exist for anaerobic processes occurring on the micron scale in otherwise oxygenated environments (Jorgensen, 1977; Paerl and Carlton, 1988). For example, it has been known for years that gaseous end products of anaerobic bacterial activities (e.g., methanogenesis) appear in aerobic

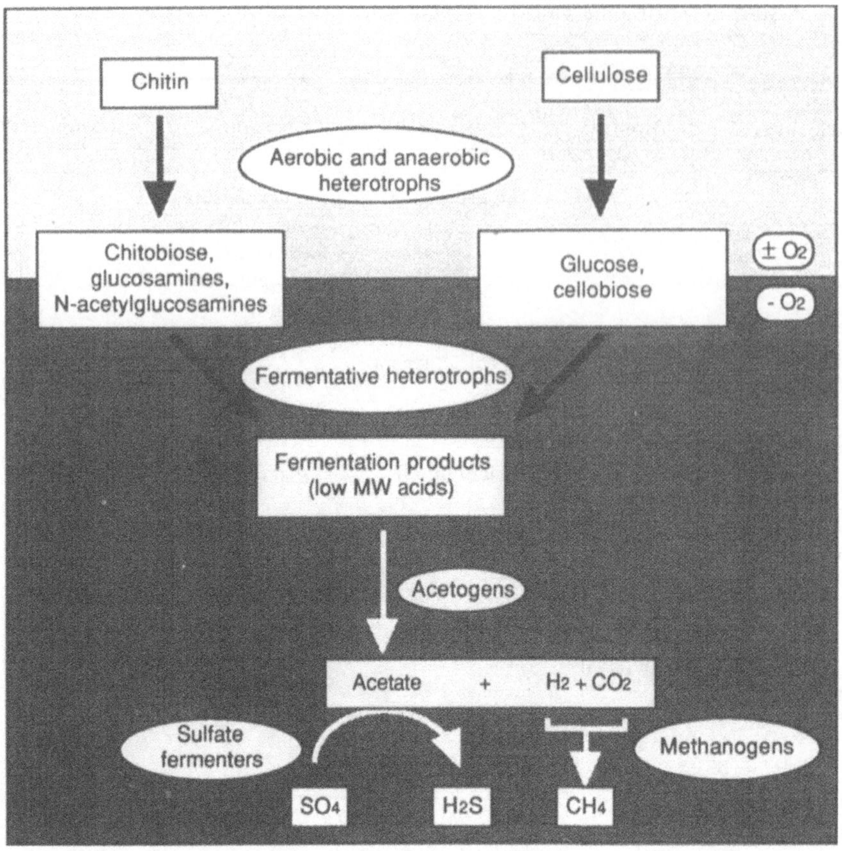

Figure 10. Types of bacteria involved in the degradation of refractory carbon (e.g., chitin, cellulose) in marine sediments, with emphasis on the predominant role of anaerobic processes. Black arrows indicate degradative steps dependent on bacterial production of extracellular enzymes.

environments, particularly open-ocean seawater (Lilley *et al.*, 1982; Sieburth, 1987). That suitable microenvironments existed within fecal pellets and animal guts seemed the most likely explanation for the repeated detection of anaerobic end products in oxygenated seawater. Unequivocal proof of this hypothesis was obtained when microelectrodes were applied to detrital particles, fecal pellets, and biofilms suspended in aerated seawater (Alldredge and Cohen, 1987) and to animal guts (Plante and Jumars, 1992). Furthermore, results from laboratory studies of mixed cultures in an oxygen-limiting chemostat have demonstrated that obligately aerobic heterotrophs and obligately anaerobic methanogens can grow together (Gerritse *et al.*, 1990).

Recognizing that aerobic and anaerobic consortia of decomposers and fermenters can coexist on the microscale (Fig. 11) has important implications for the oxygenated deep sea. Aerobic heterotrophs are more efficient in deriving energy from the oxidation of OM than anaerobic fermenters, but they utilize only limited components of complex detrital compounds and leave the bulk refractory carbon unaffected. However, as soon as aerobes reduce oxygen levels on the micron scale (as in biofilm formation), anaerobes can colonize a detrital particle and begin their degradative work. As organic intermediates and metabolites are released from bulk OM and consumed by aerobes and other anaerobes, an interdependent and stable consortium capable of fermenting all of the carbon ultimately to methane becomes established. The quantitative importance of such consortia in anaerobic microenvironments of the deep sea is not known. However, the microscopic observation of microbial consortia colonizing millimeter slivers of cellulose and chitin, implanted in artificial sediments on the abyssal seafloor and recovered after periods of 5 days to 11 months (Fig. 12), suggests that the issue should be investigated further, both on natural detrital particles in the seabed and in the guts of animals that feed on them.

6. Concluding Remarks

In the past, microbial ecologists have approached the study of diagenetic processes in sediments much like organic chemists—dissecting out a key process

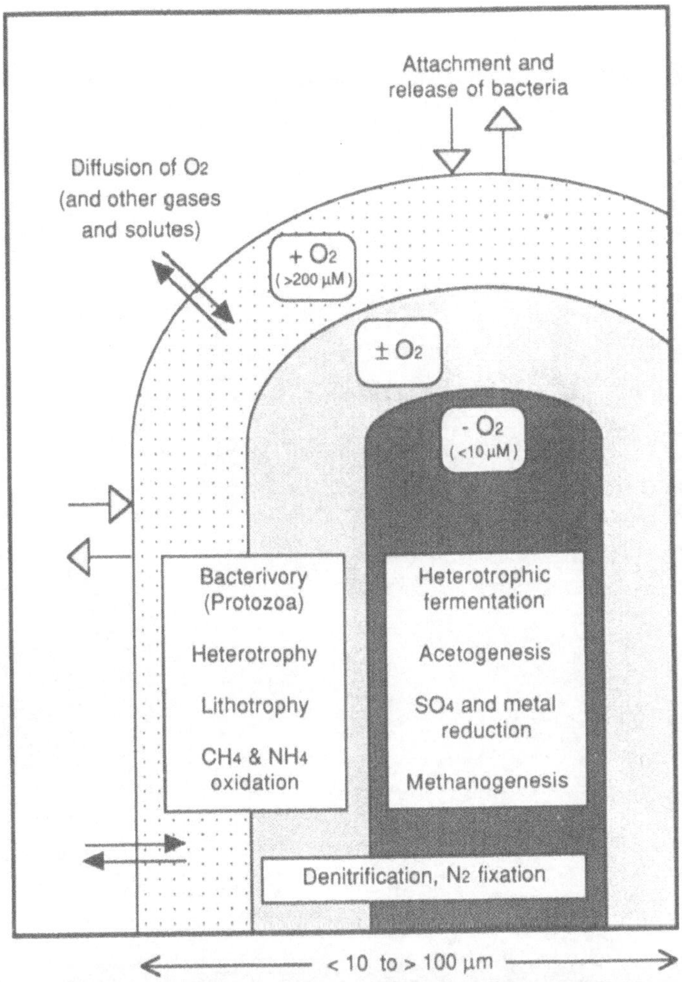

Figure 11. Microbial processes proposed to occur on a microscale in association with small organic particles (e.g., fecal pellets, POC aggregates in Fig. 4), even in well-oxygenated sediment strata. [Adapted, in part, from Aller (1982).]

from the complexity of activities ongoing in a sedimentary matrix in hopes that intense study of that one process will help to elucidate the overall rate of diagenesis. Simplifying the system has its advantages, as illustrated by laboratory work with newly cultured isolates, but a complete and coherent picture of OM diagenesis in marine sediments will emerge only as information from each of the four current modeling approaches (and new ones) is assimilated across disciplines. For example, knowing the concentrations of electron acceptors for specific aerobic and anaerobic pathways of OM decomposition is vital, since they will set constraints on the spatial distribution and magnitude of the corresponding bacterial activity, but the full range of potential electron acceptors (and bacterial players and degradative pathways) is not yet appreciated. The primary electron acceptor(s) driving the globally significant process of anaerobic methane oxidation is not even known. Carbon budget models have helped to identify important biological variables and organisms associated with the supply of organic particles to the seabed, especially the dominant role of bacteria in the deep sea, but they have not yet accounted for differences in the chemical nature of the particles or the role of extracellular enzymes in decomposing them. It is virtually impossible for sediment traps even to collect the particles that may be most important to the overall rate of OM diagenesis in sediments—the fragile, highly labile, and heavily colonized particles (Smith *et al.*, 1992) like those recovered opportunistically by Lochte and Turley (1988) on the surface of some of their sediment cores. Ultimately, it may be the microbial consortia and benthic loops associated with such detrital particles on the microscale that determine the overall rate and progress of OM diagenesis in oceanic sediments. Increased knowledge of the members of these associations, the sum of their activities, and the environmental and genetic constraints that act upon them will contribute to a better predictive understanding of the fate of OM reaching the seafloor.

Bacterial activities associated with OM diagenesis are better understood in estuarine and nearshore environments than in the deep sea, not simply because the sites are more accessible to study but because bacterial densities and processes occur on larger scales when organic resources are not limiting. When a process occurs homogeneously on a scale of centimeters to meters, bulk chemical or microbial measurements can provide meaningful information. This is especially true for anaerobic bacterial processes essential to the degradation of refractory organic compounds—they are well known only from studies of shallow-water environments. In deeper oceanic sediments, where refractory carbon compounds dominate the OM supply but do not accumulate, it seems likely that effective degradation by anaerobic microbial consortia must be occurring on the microscale within organic-rich microenvironments of heterogeneous occurrence. Indeed, the same reducing conditions grossly evident in horizontal strata of estuarine sediments (Fig. 1) occur on the microscale scale in association with detrital particles (Fig. 11) and in animal digestive tracts (Plante *et al.*, 1990; Plante and Jumars, 1992), through which the vast majority of sediments pass repeatedly before final burial. Since the activities of benthic animals (from locomotion, burrowing, and tube building to ingestion, digestion, and egestion of sedimentary materials) are essential to the creation of microenviron-

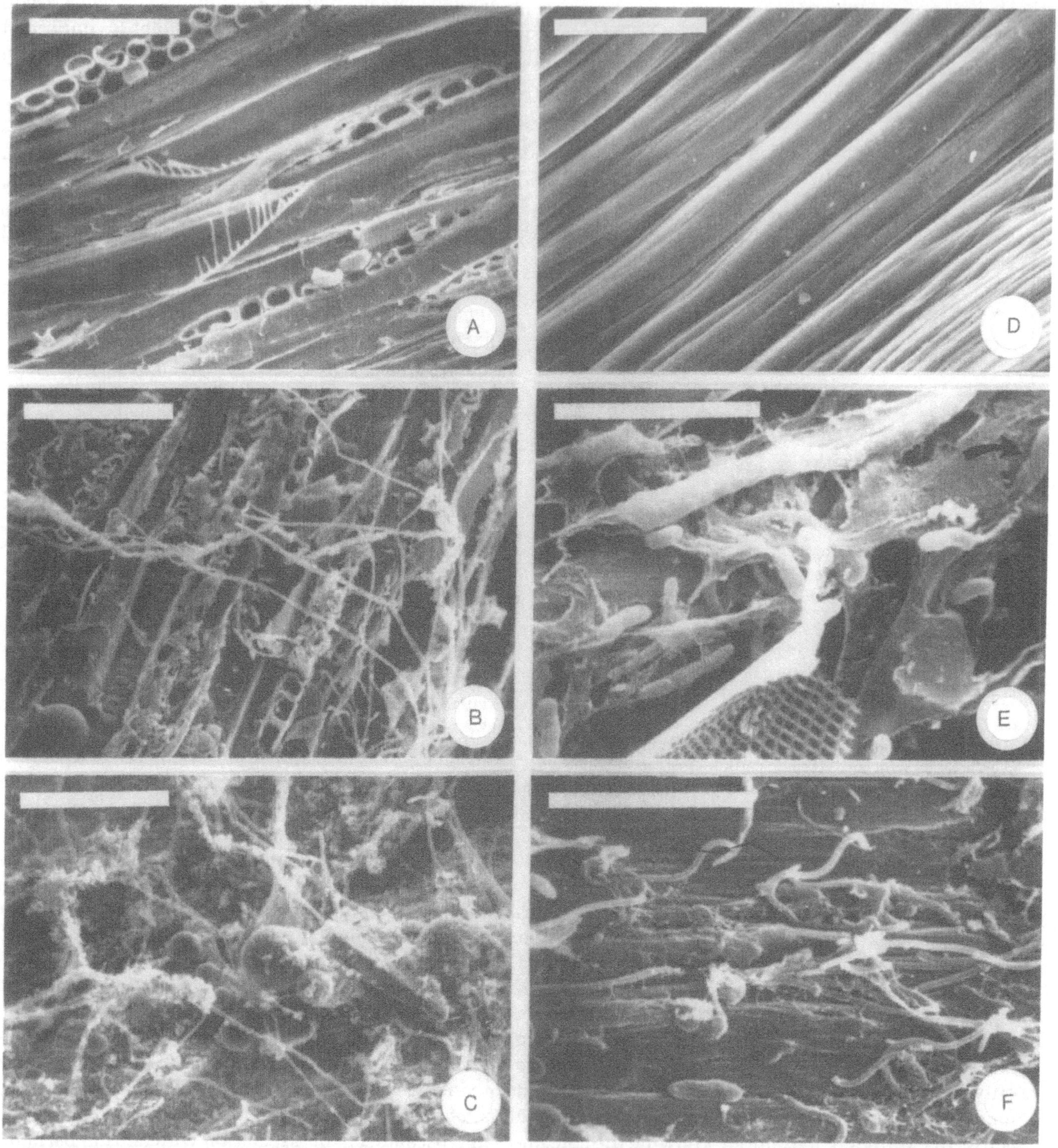

Figure 12. Progressive microbial colonization of pieces of cellulose (A–C) and chitin (D–F) buried in the surface layer of artificial sediments (glass beads) incubated on the seafloor at 2°C and 4100 m in the Bay of Biscay for periods of 5 days (A, D; control pieces exposed to mooring deployment and recovery procedures), 6 mos. (B, E), and 11 mos. (C, F). Note the Actinomycete-like filaments on the cellulose (pieces of wood), eventually obscuring it (B, C; the spheres are glass beads from the artificial sediment) and typical vibrio- and spirillum-shaped bacteria on the pieces of chitin which were derived from squid pens (E, F). Bars = 100 μm (A–D) and 2 μm (E, F). [Adapted from Desbruyeres *et al.* (1985).]

ments with characteristics contrary to those of the surrounding milieu, their importance to early diagenesis of OM in the deep sea cannot be overemphasized (Jumars *et al.*, 1990; Plante *et al.*, 1990; Deming and Yager, 1992; Lee, 1992). Future studies that adopt a microscale approach to the role of bacteria, microbial consortia, and benthic animals in OM diagenesis across vast but understudied regions of the deep sea should be enlightening.

ACKNOWLEDGMENTS. Support for preparation of this chapter was provided by the National Science Foundation through grants DPP-8800401 and OCE-8917822. We thank our colleagues Peter Jumars, Stephen Macko, and Gilbert Rowe for access to unpublished data and their insights on many of the issues we have tried to address. We also greatly appreciate the assistance of Patricia Yager in the analysis of data and preparation of the figures.

References

Alldredge, A. L., and Cohen, Y., 1987, Can microscale chemical patches persist in the sea? Microelectrode study of marine snow, fecal pellets, *Science* **235**:689–691.

Alldredge, A. L., and Silver, M., 1988, Characteristics, dynamics, and significance of marine snow, *Prog. Oceanogr.* **20**:41–82.

Aller, J. Y., 1989, Quantifying sediment disturbance by bottom currents and its effect on benthic communities in a deep-sea western boundary zone, *Deep-Sea Res.* **36**:901–934.

Aller, R. C., 1982, The effects of macrobenthos on chemical properties of marine sediment and overlying water, in: *Animal–Sediment Interactions* (P. L. McCall and M. J. S. Tevesz, eds.), Plenum Press, New York, pp. 53–102.

Aller, R. C., and Aller, J. Y., 1986, Evidence for localized enhancement of biological activity associated with tube and burrow structures in deep-sea sediments at the HEBBLE site, western North Atlantic, *Deep-Sea Res.* **33**:755–790.

Aller, R. C., and Yingst, J. Y., 1980, Relationships between microbial distributions and the anaerobic decomposition of organic matter in surface sediments of Long Island Sound, USA, *Mar. Biol.* **56**:29–42.

Alongi, D., 1987a, Bacterial productivity and microbial biomass in tropical mangrove sediments, *Microb. Ecol.* **15**:59–78.

Alongi, D., 1987b, The distribution and composition of deep-sea microbenthos in a bathyal region of the western Coral Sea, *Deep-Sea Res.* **34**:1245–1254.

Alongi, D., 1989, The role of soft-bottom benthic communities in tropical mangrove and coral reef ecosystems, *Rev. Aquat. Sci.* **1**(2):243–280.

Alongi, D. M., 1990, Bacterial growth rates, production and estimates of detrital carbon utilization in deep-sea sediments of the Solomon and Coral Seas, *Deep-Sea Res.* **37**:731–746.

Alperin, M. J., and Reeburgh, W. S., 1984, Geochemical observations supporting anaerobic methane oxidation, in: *Microbial Growth on C-1 Compounds* (R. L. Crawford and R. S. Hanson, eds.), American Society for Microbiology, Washington, D.C., pp. 282–289.

Alperin, M. J., Reeburgh, W. S., and Whiticar, M. J., 1988, Carbon and hydrogen isotope fractionation resulting from anaerobic methane oxidation, *Glob. Biogeochem. Cycles* **2**:279–288.

Archer, D., Emerson, S., and Smith, C. R., 1989, Direct measurements of the diffusive sublayer at the deep sea floor using oxygen microelectrodes, *Nature* **340**:623–626.

Atlas, R. M., and Griffiths, R. P., 1984, Bacterial populations of the Beaufort Sea, in: *The Alaskan Beaufort Sea: Ecosystems and Environments* (P. W. Barnes, D. M. Schell, and E. Reimnitz, eds.), Academic Press, New York, pp. 327–345.

Azam, F., Fenchel, T., Field, J. G., Gray, J. S., Meyer-Reil, L. A., and Thingstad, F., 1983, The ecological role of water-column microbes in the sea, *Mar. Ecol. Prog. Ser.* **10**:257–263.

Banse, K., 1964, On the vertical distribution of zooplankton in the sea, *Prog. Oceanogr.* **2**:53–125.

Battersby, N. S., and Brown, C. M., 1982, Microbial activity in organically enriched marine sediments, in: *Sediment Microbiology* (D. B. Nedwell and C. M. Brown, eds.), Academic Press, New York, pp. 147–170.

Bender, M. L., and Heggie, D. T., 1984, Fate of organic carbon reaching the deep sea floor: A status report, *Geochim. Cosmochim. Acta* **48**:977–986.

Bender, M. L., Fanning, K. A., Froelich, P. N., Heath, G. R., and Maynard, V., 1977, Interstitial nitrate profiles and oxidation of sedimentary organic matter in the Eastern Equatorial Atlantic, *Science* **198**:605–609.

Benner, R., Maccubbin, A. E., and Hodson, R. E., 1984, Anaerobic biodegradation of the lignin and polysaccharide components of lignocellulosic and synthetic lignin by sediment microflora, *Appl. Environ. Microbiol.* **47**:998–1004.

Benner, R., Moran, M. A., and Hodson, R. E., 1986, Biogeochemical cycling of lignocellulosic carbon in marine and freshwater ecosystems: Relative contributions of procaryotes and eucaryotes, *Limnol. Oceanogr.* **31**:89–100.

Benner, R., Lay, J., K'nees, E., and Hodson, R. E., 1988, Carbon conversion efficiency for bacterial growth on lignocellulose: Implications for detritus-based food webs, *Limnol. Oceanogr.* **33**:1514–1526.

Berelson, W. M., and Hammond, D. E., 1986, The calibration of a new free vehicle benthic flux chamber for use in the deep sea, *Deep-Sea Res.* **33**:1439–1454.

Berner, R. A., 1974, Kinetic models for the early diagenesis of nitrogen, sulfur, phosphorus and silicon in anoxic marine sediments, in: *The Sea*, Vol. 5 (E. D. Goldberg, ed.), Wiley-Interscience, New York, pp. 427–449.

Berner, R. A., 1977, Stoichiometric models for nutrient regeneration in anoxic sediments, *Limnol. Oceanogr.* **22**:781–786.

Berner, R. A., 1980, *Early Diagenesis: A Theoretical Approach*, Princeton University Press, Princeton, New Jersey.

Billen, G., 1982, Modelling the processes of organic matter degradation and nutrients in sedimentary systems, in: *Sediment Microbiology* (D. B. Nedwell and C. M. Brown, eds.), Academic Press, New York, pp. 15–52.

Bird, D. F., and Duarte, C. M., 1989, Bacteria–organic matter relationship in sediments: A case of spurious correlation, *Can. J. Fish. Aquat. Sci.* **46**:904–908.

Biscaye, P., Anderson, R. F., and Deck, B. L., 1988, Fluxes of particles and constituents to the eastern United States continental slope and rise: SEEP I, *Cont. Shelf Res.* **8**:855–904.

Boto, K. G., Alongi, D. M., and Nott, A. L. J., 1989, Dissolved organic carbon–bacteria interactions at sediment–water interface in a tropical mangrove system, *Mar. Ecol. Prog. Ser.* **51**:243–251.

Boyer, J., 1986, End products of anaerobic chitin degradation by salt marsh bacteria as substrates for dissimilatory sulfate reduction and methanogenesis, *Appl. Environ. Microbiol.* **52:**1415–1418.

Breznak, J. A., 1982, Intestinal microflora of termites and other xylophagous insects, *Annu. Rev. Microbiol.* **36:**323–343.

Butler, J. H. A., and Buckerfield, J. C., 1979, Digestion of lignin by termites, *Soil Biol. Biochem.* **11:**507–513.

Butman, C. A., 1986, Sediment trap biases in turbulent flow: Results from a laboratory flume study, *J. Mar. Res.* **44:**645–693.

Butman, C. A., Grant, W. D., and Stolzenbach, K. D., 1986, Predictions of sediment trap biases in turbulent flow: A theoretical analysis based on observations from the literature, *J. Mar. Res.* **44:**601–644.

Cahet, G., and Sibuet, M., 1986, Activité biologique en domaine profond: Transformations biochimiques *in situ* de composés organiques marqués au carbone-14 à l'interface eau–sediment par 2000 m de profondeur dans le golfe de Gascogne, *Mar. Biol.* **90:**307–315.

Canfield, D. E., 1989, Sulfate reduction and oxic respiration in marine sediments: Implications for organic carbon preservation in euxinic environments, *Deep-Sea Res.* **36:**121–138.

Capone, D. G., and Kiene, R. P., 1988, Comparison of microbial dynamics in marine and freshwater sediments: Contrasts in anaerobic carbon catabolism, *Limnol. Oceanogr.* **33:**725–749.

Certes, A., 1884, Sur la culture, à l'abri des germes atmosphériques, des eaux et des sédiments rapportés par les expéditions du Travailleur et du Talisman, 1882–1883, *C. R. Acad. Sci.* **98:**690–693.

Cho, B. C., and Azam, F., 1988, Major role of bacteria in biogeochemical fluxes in the ocean's interior, *Nature* **332:**441–443.

Christensen, J. P., and Rowe, G. T., 1984, Nitrification and oxygen consumption in Northwest Atlantic deep-sea sediments, *J. Mar. Res.* **42:**1099–1116.

Colberg, P. J., and Young, L. Y., 1982, Biodegradation of lignin-derived molecules under anaerobic conditions, *Can. J. Microbiol.* **28:**886–889.

Cole, J. J., Honjo, S., and Erez, J., 1987, Benthic decomposition of organic matter at a deep-water site in the Panama Basin, *Nature* **327:**703–704.

Costerton, J. W., Cheng, K.-J., Geesey, G. G., Ladd, T. I., Nickel, J. C., Dasgupta, M., and Marrie, T. J., 1987, Bacterial biofilms in nature and disease, *Annu. Rev. Microbiol.* **41:**435–464.

Cowen, J. P., 1989, Positive pressure effect on manganese binding by bacteria in deep-sea hydrothermal plumes, *Appl. Environ. Microbiol.* **55:**764–766.

Craven, D. B., and Karl, D. A., 1984, Microbial RNA and DNA synthesis in marine sediments, *Mar. Biol.* **83:**129–139.

Craven, D. B., Jahnke, R. A., and Carlucci, A. F., 1986, Fine-scale vertical distributions of microbial biomass and activity in California Borderland sediments, *Deep-Sea Res.* **33:**379–390.

Crawford, R. L., 1981, *Lignin Biodegradation and Transformation*, Wiley-Interscience, New York.

Dale, N. G., 1974, Bacteria in intertidal sediments: Factors related to their distribution, *Limnol. Oceanogr.* **19:**509–518.

Danulat, E., 1986, Role of bacteria with regard to chitin degradation in the digestive tract of the cod *Gadus morhua*, *Mar. Biol.* **90:** 335–343.

deAngelis, M., Baross, J. A., and Lilley, M. D., 1991, Enhanced microbial methane oxidation in water from a deep-sea hydrothermal vent field at simulated *in situ* hydrostatic pressures, *Limnol. Oceanogr.* **36:**565–569.

Deming, J. W., 1985, Bacterial growth in deep-sea sediment trap and boxcore samples, *Mar. Ecol. Prog. Ser.* **25:**305–312.

Deming, J. W., 1986, Ecological strategies of barophilic bacteria in the deep ocean, *Microbiol. Sci.* **3:**205–211.

Deming, J. W., and Colwell, R. R., 1982, Barophilic bacteria associated with digestive tracts of abyssal holothurians, *Appl. Environ. Microbiol.* **44:**1222–1230.

Deming, J. W., and Colwell, R. R., 1985, Observations of barophilic microbial activity in samples of sediments and intercepted particulates from the Demerara Abyssal Plain, *Appl. Environ. Microbiol.* **50:**1002–1006.

Deming, J. W., and Yager, P. L., 1992, Natural bacterial assemblages in deep-sea sediments: Towards a global view, in: *Deep-Sea Food Chains—Their Relation to the Global Carbon Cycles* (G. T. Rowe and V. Pariente, eds.), Kluwer Academic Publishers, Dordrecht, The Netherlands, pp. 11–27.

Deming, J. W., Tabor, P. S., and Colwell, R. R., 1981, Barophilic growth of bacteria from intestinal tracts of deep-sea invertebrates, *Microb. Ecol.* **7:**85–94.

Deming, J. W., Hada, H., Colwell, R. R., Luehrsen, K. R., and Fox, G. E., 1984, The ribonucleotide sequence of 5S rRNA from two strains of deep-sea barophilic bacteria, *J. Gen. Microbiol.* **130:** 1911–1920.

Deming, J. W., Somers, L. K., Straube, W. L., Swartz, D. G., and MacDonell, M. T., 1988, Isolation of an obligately barophilic bacterium and description of a new genus, *Colwellia* gen. nov., *Syst. Appl. Microbiol.* **10:**152–160.

Desbruyeres, D., Deming, J. W., Dinet, A., and Khripounoff, A., 1985, Réactions de l'écosystème benthique profond aux perturbations: Nouveaux résultats expérimentaux, in: *Peuplements Profonds du Golfe de Gascogne* (L. Laubier and C. Monniot, eds.), IFREMER (Institute Français de Recherche pour l'Exploitation de la Mer), Brest, France, pp. 193–208.

Deuser, W. G., and Ross, E. H., 1980, Seasonal change in the flux of organic carbon to the deep Sargasso Sea, *Nature* **283:**364–365.

Deuser, W. G., Ross, E. H., and Anderson, R. F., 1981, Seasonality in the supply of sediment to the deep Sargasso Sea and implications for the rapid transfer of matter to the deep ocean, *Deep-Sea Res.* **28A:**495–505.

Devol, A. H., 1987, Verification of flux measurements made with *in situ* benthic chambers, *Deep-Sea Res.* **34:**1007–1026.

Dwyer, D. F., Weeg-Aerssens, E., Shelton, D. R., and Tiedje, J. M., 1988, Bioenergetic conditions of butyrate metabolism by a syntrophic, anaerobic bacterium in co-culture with hydrogen-oxidizing methanogenic and sulfidogenic bacteria, *Appl. Environ. Microbiol.* **54:**1354–1359.

Dye, A. H., 1983, A method for the quantitative estimation of bacteria from mangrove sediments, *Estuarine Coastal Shelf Sci.* **17:**207–212.

Ellery, W. N., and Schleyer, M. H., 1984, Comparison of homogenization and ultrasonication as techniques in extracting attached sedimentary bacteria, *Mar. Ecol. Prog. Ser.* **15:**247–250.

Emerson, S. Jahnke, R., Bender, M., Froelich, P., Klinkhammer, G., Bowser, C., and Setlock, G., 1980, Early diagenesis in sediments from the eastern equatorial Pacific: 1. Pore water nutrient and carbonate results, *Earth Planet. Sci. Lett.* **49:**57–80.

Emerson, S., Reimers, C., Fischer, K., and Heggie, D., 1985, Organic carbon dynamics and preservation in deep-sea sediments, *Deep-Sea Res.* **32:**1–22.

Emerson, S., Stump, C., Grootes, P. M., Stuiver, M., Farwell, G. W., and Schmidt, F. H., 1987, Estimates of degradable organic carbon in deep-sea surface sediments from ^{14}C concentrations, *Nature* **329:**51–53.

Eppley, R. W., and Peterson, B. J., 1979, Particulate organic matter flux and planktonic new production in the deep ocean, *Nature* **282:**677–680.

Fallon, R. D., Newell, S. Y., and Hopkinson, C. S., 1983, Bacterial production in marine sediments: Will cell-specific measures agree with whole-system metabolism?, *Mar. Ecol. Prog. Ser.* **11:** 119–127.

Fenchel, T. M., 1967, The ecology of marine microbenthos I. The quantitative importance of ciliates as compared with metazoans in various types of sediments, *Ophelia* **4:**121–137.

Fenchel, T. M., 1969, The ecology of marine microbenthos IV. Structure and function of the benthic ecosystem, its chemical and physical factors and the microfauna communities with special reference to the ciliated protozoa, *Ophelia* **6:**1–182.

Fenchel, T. M., 1987, *Ecology of Protozoa, The Biology of Free-Living Phagotrophic Protists*, Brock-Springer Series in Contemporary BioScience, Science Tech. Publishers, Madison, Wisconsin.

Fenchel, T. M., and Jorgensen, B. B., 1977, Detritus food chains of aquatic ecosystems: The role of bacteria, in: *Advances in Microbial Ecology*, Vol. 1 (M. Alexander, ed.), Plenum Press, New York, pp. 1–58.

Findlay, R. H., Pollard, P. C., Moriarty, D. J. W., and White, D. C., 1985, Quantitative determination of microbial activity and community nutritional status in estuarine sediments: Evidence for a disturbance artifact, *Can. J. Microbiol.* **31:**493–498.

Fong, J. B., and Mann, K. H., 1980, Role of gut flora in the transfer of amino acids through a marine food chain, *Can. J. Fish. Aquat. Sci.* **37:**88–96.

Foulds, J. B., and Mann, K. H., 1978, Cellulose digestion in *Mysis stenolepsis* and its ecological implications, *Limnol. Oceanogr.* **23:**760–766.

Friesan, J. A., Mann, H. K., and Novitsky, K. H., 1986, *Mysis* digests cellulose in the absence of a gut microflora, *Can. J. Zool.* **64:** 431–441.

Froelich, P. N., Klinkhammer, G. P., Bender, M. L., Luedtke, N. A., Heath, G. R., Cullen, D., Dauphin, P., Hammond, D., Hartman, B., and Maynard, V., 1979, Early oxidation of organic matter in pelagic sediments of the eastern equatorial Atlantic: Suboxic diagenesis, *Geochim. Cosmochim. Acta* **43:**1075–1090.

Gerritse, J., Schut, F., and Gottschal, J. C., 1990, Mixed chemostat cultures of obligately aerobic and fermentative or methanogenic bacteria grown under oxygen-limiting conditions, *FEMS Microbiol. Lett.* **66:**87–94.

Goodrich, T. D., and Morita, R. Y., 1977, Bacterial chitinase in the stomachs of marine fishes from Yaquina Bay, Oregon, USA, *Mar. Biol.* **41:**355–360.

Graf, G., 1986, Winter inversion of biomass and activity profile in a marine sediment, *Mar. Ecol. Prog. Ser.* **33:**231–235.

Graf, G., 1989, Benthic–pelagic coupling in a deep-sea benthic community, *Nature* **341:**437–439.

Grundmanis, V., and Murray, J. W., 1982, Aerobic respiration in pelagic marine sediments, *Geochim. Cosmochim. Acta* **46:**1101–1120.

Gundersen, J. K., and Jorgensen, B. B., 1990, Microstructure of diffusive boundary layers and the oxygen uptake of the sea floor, *Nature* **345:**604–607.

Haedrich, R. L., and Rowe, G. T., 1978, Megafaunal biomass in the deep sea, *Nature* **269:**141–142.

Hansen, J. A., Alongi, D. M., Moriarty, D. J. W., and Pollard, P. C., 1987, The dynamics of benthic microbial communities at Davies Reef, central Great Barrier Reef, *Coral Reefs* **6:** 63–70.

Harvey, H. R., Richardson, M. D., and Patton, J. S., 1984, Lipid composition and vertical distribution of bacteria in anaerobic sediments of the Venezuela Basin, *Deep-Sea Res.* **31:**403–413.

Henrichs, S. M., and Doyle, A. P., 1986, Decomposition of ^{14}C-labeled organic substances in marine sediments, *Limnol. Oceanogr.* **31:**765–778.

Henrichs, S. M., and Reeburgh, W. S., 1987, Anaerobic mineralization of marine sediment organic matter: Rates and the role of anaerobic processes in the oceanic carbon economy, *Geomicrobiol. J.* **5:**191–237.

Herwig, R. P., and Staley, J. T., 1986, Anaerobic bacteria from the digestive tract of North Atlantic fin whales (*Balaenoptera physalus*), *FEMS Microbiol. Lett.* **38:**361–371.

Herwig, R. P., Staley, J. T., Nerini, M. K., and Braham, H. W., 1984, Baleen whales: Preliminary evidence for forestomach microbial fermentation, *Appl. Environ. Microbiol.* **47:**421–423.

Hessler, R. R., Ingram, C. L., Yayanos, A. A., and Burnett, B. R., 1978, Scavenging amphipods from the floor of the Philippine Trench, *Deep-Sea Res.* **25:**1029–1048.

Hinga, K., Sieburth, J. McN., and Heath, G. R., 1979, The supply and use of organic material at the deep sea floor, *J. Mar. Res.* **37:** 557–579.

Honjo, S., 1978, Sedimentation of materials in the Sargasso Sea at 5,367 m deep station, *J. Mar. Res.* **36:**469–492.

Hungate, R. E., 1975, The rumen microbial ecosystem, *Annu. Rev. Ecol. Syst.* **6:**39–66.

Isao, K., Hara, S., Terauchi, K., and Kogure, K., 1990, Role of submicrometre particles in the ocean, *Nature* **345:**242–244.

Ittekkot, V., Deuser, W. G., and Degens, E. T., 1984, Seasonality in the fluxes of sugars, amino acids, and amino sugars to the deep ocean: Panama Basin, *Deep-Sea Res.* **31:**1057–1069.

Jackson, G. A., 1990, A model of the formation of marine algal flocs by physical coagulation processes, *Deep-Sea Res.* **37:**1197–1211.

Jahnke, R., 1985, A model of microenvironments in deep-sea sediments: Formation and effects on porewater profiles, *Limnol. Oceanogr.* **30:**956–965.

Jahnke, R. A., and Jackson, G. A., 1987, Role of sea floor organisms in oxygen consumption in the deep North Pacific Ocean, *Nature* **329:**621–623.

Jahnke, R., Emerson, S., and Murray, J. W., 1982, A model of oxygen reduction, denitrification and organic matter mineralization in marine sediments, *Limnol. Oceanogr.* **27:**610–623.

Jahnke, R., Emerson, S. R., Cochran, J. K., and Hirshberg, D. J., 1986, Fine scale distributions of porosity and particulate excess ^{210}Pb, organic carbon and $CaCO_3$ in surface sediments of the deep equatorial Pacific, *Earth Planet. Sci. Lett.* **77:**59–69.

Jahnke, R., A., Reimers, C. E., and Craven, D. B., 1990, Intensification of recycling of organic matter at the sea floor near ocean margins, *Nature* **348:**50–54.

Jannasch, H. W., 1979, Microbial turnover of organic matter in the deep sea, *BioScience* **29:**228–232.

Jannasch, H. W., and Wirsen, C. O., 1973, Deep-sea microorganisms: *In situ* response to nutrient enrichment, *Science* **180:** 641–643.

Jannasch, H. W., and Wirsen, C. O., 1983, Microbiology of the deep sea, in: *The Sea, Vol. 8, Deep-Sea Biology* (G. T. Rowe, ed.), John Wiley & Sons, New York, pp. 231–259.

Jannasch, H. W., and Wirsen, C. O., 1984, Variability of pressure adaptation in deep sea bacteria, *Arch. Microbiol.* **139:**281–288.

Jorgensen, B. B., 1977, The sulfur cycle of a coastal marine sediment (Limfjorden, Denmark), *Limnol. Oceanogr.* **5:**814–832.

Jorgensen, B. B., 1982, Mineralization of organic matter in the seabed—the role of sulfate reduction, *Nature* **296:**643–645.

Jumars, P. A., Penry, D. L., Baross, J. A., Perry, M. J., and Frost, B. W., 1989, Closing the microbial loop: Dissolved carbon pathway to heterotrophic bacteria from incomplete ingestion, digestion and absorption in animals, *Deep-Sea Res.* **36:**483–495.

Jumars, P. A., Mayer, L. M., Deming, J. W., Baross, J. A., and Wheatcroft, R. A., 1990, Deep-sea deposit-feeding strategies suggested by environmental and feeding constraints, *Philos. Trans. R. Soc. London Ser. A* **331**:85–101.

Karl, D. M., 1986, Determination of *in situ* microbial biomass, viability, metabolism, and growth, in: *Bacteria in Nature*, Vol. 2, *Methods and Special Applications in Bacterial Ecology* (J. S. Poindexter and E. R. Leadbetter, eds.), Plenum Press, New York, pp. 85–176.

Karl, D. M., and Novitsky, J. A., 1988, Dynamics of microbial growth in surface layers of a coastal marine sediment ecosystem, *Mar. Ecol. Prog. Ser.* **50**:169–176.

Karl, D. M., Jones, D. R., Novitsky, J. A., Winn, C. D., and Bossard, P., 1987, Specific growth rates of natural microbial communities measured by adenine nucleotide pool turnover, *J. Microbiol. Methods* **6**:221–235.

Karl, D. M., Knauer, G. A., and Martin, J. H., 1988, Downward flux of particulate organic matter in the ocean: A particle decomposition paradox, *Nature* **332**:438–441.

Kemp, P. F., 1987, Potential impact on bacteria of grazing by a macrofaunal deposit-feeder, and the fate of bacterial production, *Mar. Ecol. Prog. Ser.* **36**:151–161.

Kemp, P. F., 1988, Bacterivory by benthic ciliates: Significance as a carbon source and impact on sediment bacteria, *Mar. Ecol. Prog. Ser.* **49**:163–169.

Kemp, P. F., 1990, The fate of benthic bacterial production, *Rev. Aquat. Sci.* **2**:109–123.

Khripounoff, A., and Rowe, G. T., 1985, Les apports organiques et leur transformation en milieu abyssal à l'interface eau–sédiment dans l'ocean Atlantique tropical, *Oceanol. Acta* **8**: 293–301.

King, G. M., 1986, Characterization of β-glucosidase activity in intertidal marine sediments, *Appl. Environ. Microbiol.* **51**: 373–380.

Lee, C., 1992, Controls on organic carbon preservation: The use of stratified water bodies to compare intrinsic rates of decomposition in oxic and anoxic systems, *Geochim. Cosmochim. Acta* **56**:3323–3335.

Lilley, M. D., Baross, J. A., and Gordon, L. I., 1982, Dissolved hydrogen and methane in Saanich Inlet, British Columbia, *Deep-Sea Res.* **29**:1471–1487.

Ljungdahl, L. G., 1986, The autotrophic pathway of acetate synthesis in acetogenic bacteria, *Annu. Rev. Microbiol.* **40**:415–450.

Lochte, K., 1992, Bacterial standing stock and consumption of organic carbon in the benthic boundary layer of the abyssal North Atlantic, in: *Deep-Sea Food Chains—Their Relation to the Global Carbon Cycles* (G. T. Rowe and V. Pariente, eds.), Kluwer Academic Publishers, Dordrecht, The Netherlands, pp. 1–10.

Lochte, K., and Turley, C. M., 1988, Bacteria and cyanobacteria associated with phytodetritus in the deep sea, *Nature* **333**: 67–69.

Lovley, D. R., and Phillips, E. J. P., 1988, Novel mode of microbial energy metabolism: Organic carbon oxidation coupled to dissimilatory reduction of iron or manganese, *Appl. Environ. Microbiol.* **54**:1472–1480.

Lovley, D. R., Phillips, E. J. P., and Lonergan, D. J., 1989a, Hydrogen and formate oxidation coupled to dissimilatory reduction of iron or manganese by *Alteromonas putrefaciens*, *Appl. Environ. Microbiol.* **55**:700–706.

Lovley, D. R., Baedecker, M. J., Lonergan, D. J., Cozzarelli, I. M., Phillips, E. J. P., and Siegel, D. I., 1989b, Oxidation of aromatic contaminants coupled to microbial iron reduction, *Nature* **339**:297–299.

Mann, K. H., 1988, Production and use of detritus in various freshwater, estuarine, and coastal marine ecosystems, *Limnol. Oceanogr.* **33**:910–930.

Martin, J. H., Knauer, G. A., Karl, D. M., and Broenkow, W. M., 1987, VERTEX: Carbon cycling in the northeast Pacific, *Deep-Sea Res.* **34**:267–285.

Martin, M. M., Martin, J. S., Kukor, J. J., and Merritt, R. W., 1980, The digestion of protein and carbohydrate by the stream detritivore, *Tipula abdominalis* (Diptera, Tipulidae), *Oecologia* **46**: 360–364.

Mayer, L. M., 1989, Extracellular proteolytic enzyme activity in sediments of an intertidal mudflat, *Limnol. Oceanogr.* **34**: 973–981.

Mayer, L. M., Macko, S. A., and Cammen, L., 1988, Provenance, concentrations and nature of sedimentary organic nitrogen in the Gulf of Maine, *Mar. Chem.* **25**:291–304.

McCorkle, D. C., and Emerson, S. R., 1988, The relationship between pore water carbon isotopic composition and bottom water oxygen concentration, *Geochim. Cosmochim. Acta* **52**: 1169–1178.

McCorkle, D. C., Emerson, S. R., and Quay, P. D., 1985, Stable carbon isotopes in marine porewaters, *Earth Planet. Sci. Lett.* **74**:13–26.

Meyer-Reil, L.-A., 1984, Bacterial biomass and heterotrophic activity in sediments and overlying waters, in: *Heterotrophic Activity in the Sea* (J. E. Hobbie and P. J. leB. Williams, eds.), Plenum Press, New York, pp. 523–546.

Meyer-Reil, L.-A., 1986, Measurement of hydrolytic activity and incorporation of dissolved organic substrates by microorganisms in marine sediments, *Mar. Ecol. Prog. Ser.* **31**:143– 149.

Meyer-Reil, L.-A., 1987a, Seasonal and spatial distribution of extracellular enzymatic activities and microbial incorporation of dissolved organic substrates in marine sediments, *Appl. Environ. Microbiol.* **53**:1748–1755.

Meyer-Reil, L.-A., 1987b, Biomass and activity of benthic bacteria, in: *Lecture Notes on Coastal and Estuarine Studies*, Volume XIII (M. J. Bowman, R. T. Barber, C. N. K. Mooers, and J. A. Raven, eds.), Springer-Verlag, New York, pp. 93–110.

Meyer, Reil, L.-A., Bolter, M., Dawson, R., Liebezeit, G., Szwerinski, H., and Wolter, K., 1980, Interrelationships between microbiological and chemical parameters of sandy beach sediments, a summer aspect, *Appl. Environ. Microbiol.* **39**:797–802.

Michaels, A. F., and Silver, M. W., 1988, Primary production, sinking fluxes and the microbial food web, *Deep-Sea Res.* **35**: 473–490.

Miller, D. C., 1989, Abrasion effects on microbes in sandy sediments, *Mar. Ecol. Prog. Ser.* **55**:73–82.

Montagna, P. A., Bauer, J. E., Hardin, D., and Spies, R. B., 1989, Vertical distribution of microbial and meiofaunal populations in sediments of a natural coastal hydrocarbon seep, *J. Mar. Res.* **47**:657–680.

Moriarty, D. J. W., and Pollard, P. C., 1981, DNA synthesis as a measure of bacterial productivity in seagrass sediments, *Mar. Ecol. Prog. Ser.* **5**:151–156.

Moriarty, D. J. W., and Pollard, P. C., 1982, Diel variation of bacterial productivity in seagrass (*Zostera capricorni*) beds measured by rate of thymidine incorporation into DNA, *Mar. Biol.* **72**: 165–173.

Moriarty, D. J. W., Pollard, P. C., Hunt, W. G., Moriarty, C. M., and Wassenberg, T. J., 1985, Productivity of bacteria and microalgae and the effect of grazing by holothurians in sediments on a coral reef flat, *Mar. Biol.* **85**:293–300.

Morrison, S. J., and White, D. C., 1980, Effects of grazing by estuarine gammaridean amphipods on the microbiota of allochthonous detritus, *Appl. Environ. Microbiol.* **40**:659–671.

Muller, P. J., and Suess, E. 1979, Productivity, sedimentation rate, and sedimentary organic matter in the oceans—I. Organic carbon preservation, *Deep-Sea Res.* **26**:1347–1362.

Newell, S. Y., and Fallon, R. D., 1982, Bacterial productivity in the water column and sediments of the Georgia (USA) coastal zone: Estimates via direct counting and parallel measurement of thymidine incorporation, *Microb. Ecol.* **8**:33–46.

Novitsky, J. A., 1983a, Heterotrophic activity throughout a vertical profile of seawater and sediment in Halifax Harbor, Canada, *Appl. Environ. Microbiol.* **45**:1753–1760.

Novitsky, J. A., 1983b, Microbial activity at the sediment–water interface in Halifax Harbor, Canada, *Appl. Environ. Microbiol.* **45**:1761–1766.

Novitsky, J. A., 1987, Microbial growth rates and biomass production in a marine sediment: Evidence for a very active but mostly nongrowing community, *Appl. Environ. Microbiol.* **53**: 2368–2372.

Oremland, R. S., 1988, The biogeochemistry of methanogenic bacteria, in: *The Biology of Anaerobic Microorganisms* (A. Zehnder, ed.), John Wiley & Sons, New York, pp. 405–447.

Pace, M. L., Glasser, J. E., and Pomeroy, L. R., 1984, A simulation analysis of continental shelf food webs, *Mar. Biol.* **82**:47–63.

Paerl, H. W., and Carlton, R. G., 1988, Control of nitrogen fixation by oxygen depletion in surface-associated microzones, *Nature* **332**:260–262.

Plante, C. J., and Jumars, P. A., 1992, The microbial environment of marine deposit-feeder guts characterized via microelectrodes, *Microb. Ecol.* **23**:257–277.

Plante, C. J., Jumars, P. A., and Baross, J. A., 1989, Rapid bacterial growth in the hindgut of a marine deposit feeder, *Microb. Ecol.* **18**:29–44.

Plante, C. J., Jumars, P. A., and Baross, J. A., 1990, Digestive associations between marine detritivores and bacteria, *Annu. Rev. Ecol. Syst.* **21**:93–127.

Pomeroy, L. R., 1974, The ocean's food web, a changing paradigm, *BioScience* **24**:499–504.

Pomeroy, L. R., 1979, Secondary production mechanisms of continental shelf communities, in: *Ecological Processes in Coastal and Marine Systems* (R. J. Livingston, ed.), Plenum Press, New York, pp. 163–186.

Pomeroy, L. R., and Deibel, D., 1986, Temperature regulation of bacterial activity during the spring bloom in Newfoundland coastal waters, *Science* **18**:359–361.

Prim, P., and Lawrence, J. M., 1975, Utilization of marine plants and their constituents by bacteria isolated from the gut of echinoids (Echinodermata), *Mar. Biol.* **33**:167–173.

Reeburgh, W. S., 1976, Methane consumption in Cariaco Trench waters and sediments, *Earth Planet. Sci. Lett.* **28**:337–344.

Reeburgh, W. S., 1983, Rates of biogeochemical processes in anoxic sediments, *Annu. Rev. Earth Planet. Sci.* **11**:269–298.

Reeburgh, W. S., 1989, Interaction of sulfur and carbon cycles in marine sediments, in: *Evolution of the Global Biogeochemical Sulfur Cycle SCOPE 39* (P. Brimblecombe and A. Yu Lein, eds.), John Wiley & Sons, New York, pp. 125–159.

Reeburgh, W. S., Ward, B. B., Whalen, S. C., Sandbeck, K. A., Kilpatrick, K. A., and Kerkhof, L. J., 1991, Black Sea methane geochemistry, *Deep-Sea Res.* **38**:S1189–S1210.

Reimers, C. E., 1987, An *in situ* microprofiling instrument for measuring interfacial pore water gradients: Methods and oxygen profiles from the North Pacific Ocean, *Deep-Sea Res.* **34**: 2019–2035.

Reimers, C. E., and Smith, K. L., Jr., 1986, Reconciling measured and predicted fluxes of oxygen across the deep-sea sediment–water interface, *Limnol. Oceanogr.* **31**:305– 318.

Reimers, C. E., Kalhorn, S., Emerson, S. R., and Nealson, K. H., 1984, Oxygen consumption rates in pelagic sediments from the Central Pacific: First estimates from microelectrode profiles, *Geochim. Cosmochim. Acta* **48**:903–910.

Reimers, C. E., Fischer, K. M., Merewether, R., Smith, K. L., Jr., and Jahnke, R. J., 1986, Oxygen microprofiles measured *in situ* in deep ocean sediments, *Nature* **320**:741–744.

Revsbech, N. P., and Jorgensen, B. B., 1986, Microelectrodes: Their use in microbial ecology, in: *Advances in Microbial Ecology*, Vol. 9 (K. C. Marshall, ed.), Plenum Press, New York, pp. 293–352.

Revsbech, N. P., Jorgensen, B. B., and Blackburn, T. H., 1979, Oxygen in the sea bottom measured with a microelectrode, *Science* **207**:1355–1356.

Rigler, F. H., 1982, Recognition of the possible: An advantage of empiricism in ecology, *Can. J. Fish. Aquat. Sci.* **39**:1323–1331.

Rowe, G. T., 1981, The benthic processes of coastal upwelling ecosystems, in: *Coastal Upwelling* (F. A. Richards, ed.), American Geophysical Union, Washington, D.C., pp. 464–471.

Rowe, G. T., 1983, Biomass and production of the deep-sea macrobenthos, in: *The Sea, Vol. 8, Deep-Sea Biology* (G. T. Rowe, ed.), Wiley-Interscience, New York, pp. 97–122.

Rowe, G. T., and Deming, J. W., 1985, The role of bacteria in the turnover of organic carbon in deep-sea sediments, *J. Mar. Res.* **43**:925–950.

Rowe, G. T., and Gardner, W., 1979, Sedimentation rates in the slope water of the northwest Atlantic Ocean measured directly with sediment traps, *J. Mar. Res.* **37**:581–600.

Rowe, G. T., Clifford, C. H., Smith, K. L., and Hamilton, P. L., 1975, Benthic nutrient regeneration and its coupling to primary productivity in coastal waters, *Nature* **255**:215–217.

Rowe, G. T., Smith, S., Falkowski, P. G., Whitledge, T. E., Theroux, R., Phoel, W., and Ducklow, H., 1986, Do continental shelves export organic matter?, *Nature* **324**:559–561.

Rowe, G. T., Theroux, R., Phoel, W., Quinby, H., Wilke, R., Koschoreck, D., Whitledge, T. E., Falkowski, P. G., and Fray, C., 1988, Benthic carbon budgets for the continental shelf south of New England, *Cont. Shelf Res.* **8**:511–527.

Rowe, G. T., Sibuet, M., Deming, J. W., Khripounoff, A., and Tietjen, J., 1990, Organic carbon residence time in the deep-sea benthos, *Prog. Oceanogr.* **24**:141–160.

Rowe, G. T., Sibuet, M., Deming, J. W., Khripounoff, A., Tietjen, J., Macko, S., and Theroux, R., 1991, "Total" sediment biomass and preliminary estimates of organic carbon residence time in deep-sea benthos, *Mar. Ecol. Prog. Ser.* **79**:99–114.

Rublee, P. A., 1982, Bacterial and microbial distribution in estuarine sediments, in: *Estuarine Comparisons* (V. S. Kennedy, ed.), Academic Press, New York, pp. 159–182.

Sayles, F. L., and Curry, W. B., 1988, Delta ^{13}C, TCO_2, and the metabolism of organic carbon in deep sea sediments, *Geochim. Cosmochim. Acta* **52**:2963–2978.

Sieburth, J. M., 1987, Contrary habitats for redox-specific processes: Methanogenesis in oxic waters and oxidation in anoxic waters, in: *Microbes in the Sea* (M. A. Sleigh, ed.), Ellis Horwood, Chichester, and John Wiley & Sons, New York, pp. 11–38.

Sinsabaugh, R. L., Linkins, A. E., and Benfield, E. F., 1985, Cellulose digestion and assimilation by three leaf-shredding aquatic insects, *Ecology* **66**:1464–1471.

Smetacek, V. S., 1985, Role of sinking in diatom life-history cycles: Ecological, evolutionary and geological significance, *Mar. Biol.* **84**:239–251.

Smith, D. C., Simon, M., Alldredge, A. L., and Azam, F., 1992, Intense hydrolytic enzyme activity on marine aggregates and implications for rapid particle dissolution, *Nature* **359**:139–142.

Smith, K. L., Jr., 1987, Food energy supply and demand: A discrepancy between particulate organic carbon flux and sediment community oxygen consumption in the deep sea, *Limnol. Oceanogr.* **32**:201–220.

Smith, K. L., Jr., and Baldwin, R. J., 1984, Seasonal fluctuations in deep-sea sediment community oxygen consumption: Central and eastern Pacific, *Nature* **307**:624–626.

Smith, K. L., Jr., and Teal, J. M., 1973, Deep-sea benthic community respiration—an *in situ* study to 1850 meters, *Science* **179**: 282–283.

Smith, K. L., Jr., Clifford, C., Eliason, A., Walden, B., Rowe, G., and Teal, J. 1976, A free vehicle for measuring benthic community respiration, *Limnol. Oceanogr.* **21**:164–170.

Smith, K. L., Jr., White, G. A., and Laver, M. B., 1979, Oxygen uptake and nutrient exchange of sediments measured *in situ* using a free vehicle grab respirometer, *Deep-Sea Res.* **16A**:337–346.

Smith, K. L., Jr., Carlucci, A. F., Jahnke, R. A., and Craven, D. B., 1987, Organic carbon mineralization in the Santa Catalina Basin: Benthic boundary layer metabolism, *Deep-Sea Res.* **34**: 185–211.

Suess, E., 1980, Particulate organic carbon flux in the oceans—surface productivity and oxygen utilization, *Nature* **288**: 260–263.

Sverdrup, H. U., Johnson, M. W., and Fleming, R. H., 1942, *The Oceans: Their Physics, Chemistry, and General Biology*, Prentice-Hall, New York, 1060 pp.

Taylor, E. C., 1982, Role of aerobic microbial populations in cellulose digestion by desert millipedes, *Appl. Environ. Microbiol.* **44**:281–291.

Thiel, H., 1983, Meiobenthos and nanobenthos of the deep sea, in: *The Sea*, Vol. 8, *Deep-Sea Biology* (G. T. Rowe, ed.), Wiley-Interscience, New York, pp. 167–230.

Thistle, D., Yingst, J. Y., and Fauchald, K., 1985, A deep-sea benthic community exposed to strong near-bottom currents on the Scotian Rise (Western Atlantic), *Mar. Geol.* **66**:91–112.

Tietjen, J. H., Deming, J. W., Rowe, G. T., Macko, S., and Wilke, R. J., 1989, Meiobenthos of the Hatteras Abyssal Plain and Puerto Rico Trench: Abundance and associations with bacteria and particulate fluxes, *Deep-Sea Res.* **36**:1567–1577.

Toggweiler, J. R., 1990, Diving into the organic soup, *Nature* **345**: 203–204.

Turley, C. M., Lochte, K., and Patterson, D. J., 1988, A barophilic flagellate isolated from 4500 m in the mid-North Atlantic, *Deep-Sea Res.* **35**:1079–1092.

U.S. GOFS Working Group, 1989, Sediment trap technology and sampling, U.S. Global Ocean Flux Study Planning Report Number 10 (G. Knauer and V. Asper, co-chairs), Woods Hole, Massachusetts, 94 pp.

Val Klump, J., and Martens, C. S., 1983, Benthic nitrogen regeneration, in: *Nitrogen in the Marine Environment* (E. J. Carpenter and D. G. Capone, eds.), Academic Press, New York, pp. 411–457.

Velji, M. I., and Albright, L. J., 1986, Microscopic enumeration of attached bacteria of seawater, marine sediment, fecal matter, and kelp blade samples following pyrophosphate and ultrasound treatments, *Can. J. Microbiol.* **32**:121–126.

Vinogradov, M. E., 1968, *Vertical Distribution of the Oceanic Zooplankton*, Nauka, Moscow.

Vitalis, T. Z., Spence, M. J., and Carefoot, T. H., 1988, The possible role of gut bacteria in nutrition and growth of the sea hare *Aplysia*, *Veliger* **30**:333–341.

Vonk, H. J., and Western, J. R. H., 1984, *Comparative Biochemistry and Physiology of Enzymatic Digestion*, Academic Press, London.

Wakeham, S. G., Farrington, J., Gagosian, R. B., Lee, C., DeBaar, H., Nigerelli, G., Tripp, B., Smith, S., and Frew, N., 1980, Fluxes of organic matter from a sediment trap experiment in the equatorial Atlantic Ocean, *Nature* **286**:798–800.

Walsh, J. J., Rowe, G. T., Iverson, R. L., and McRoy, C. P., 1981, Biological export of shelf carbon is a sink of the global CO_2 cycle, *Nature* **291**:196–201.

Walsh, I., Dymond, J., and Collier, R., 1988, Rates of recycling of biogenic components of settling particles in the ocean derived from sediment trap experiments, *Deep-Sea Res.* **35**:43–58.

Westrich, J. T., and Berner, R. A., 1984, The role of sedimentary organic matter in bacterial sulfate reduction: The G model tested, *Limnol. Oceanogr.* **29**:236–249.

Wheatcroft, R. A., Jumars, P. A., Smith, C. R., and Nowell, A. R. M., 1990, A mechanistic view of the particulate biodiffusion coefficient: Step lengths, rest periods and transport directions, *J. Mar. Res.* **48**:177–207.

Wiebe, W. J., 1979, Anaerobic benthic microbial processes: Changes from the estuary to the continental shelf, in: *Ecological Processes in Coastal and Marine Systems* (R. J. Livingston, ed.), Plenum Press, New York, pp. 469–485.

Wilke, R. J., Deming, J. W., Macko, S., Tietjen, J., Rowe, G., Stein, D., Fray, C., Khripounoff, A., Koshorek, D., Stepien, J., and Voras, B., 1985, Low-Level-Waste Ocean Disposal Project: R/V Iselin Cruise 19 June–10 July 1984, Data Report, Brookhaven National Laboratory Associated Universities, Inc., Upton, New York.

Yayanos, A. A., 1986, Evolutionary and ecological implications of the properties of deep-sea barophilic bacteria, *Proc. Natl. Acad. Sci. U.S.A.* **83**:9542–9546.

Yayanos, A. A., and DeLong, E. F., 1987, Deep-sea bacterial fitness to environmental temperatures and pressures, in: *Current Perspectives in High Pressure Biology* (H. W. Jannasch, R. E. Marquis, and A. M. Zimmerman, eds.), Academic Press, London, pp. 17–32.

Yayanos, A. A., Dietz, A. S., and Van Boxtel, R., 1979, Isolation of a deep-sea barophilic bacterium and some of its growth characteristics, *Science* **205**:808–810.

Yayanos, A. A., Dietz, A. S., and Van Boxtel, R., 1982, Dependence of reproduction rate on pressure as a hallmark of deep-sea bacteria, *Appl. Environ. Microbiol.* **44**:1356–1361.

Yingst, J. Y., and Rhoads, D. C., 1985, The structure of soft-bottom benthic communities in the vicinity of the Texas Flower Garden Banks, Gulf of Mexico, *Estuarine Coastal Shelf Sci.* **20**: 569–592.

Yokoe, Y., and Yasumasu, I., 1964, The distribution of cellulase in invertebrates, *Comp. Biochem. Physiol.* **13**:323–338.

ZoBell, C. E., and Morita, R. Y., 1957, Barophilic bacteria in some deep-sea sediments, *J. Bacteriol.* **73**:563–568.

Chapter 6

Production, Transport, and Alteration of Particulate Organic Matter in the Marine Water Column

STUART G. WAKEHAM and CINDY LEE

1. Introduction

Most of the organic matter sequestered in Recent and ancient marine sediments is ultimately derived from organic matter biosynthesized by marine organisms inhabiting the surface waters of the oceans and transported to the seafloor as particulate organic matter (POM). However, most of the organic matter produced in the upper ocean is recycled in the upper few hundred meters of the water column (the epipelagic zone). Only a small fraction of the particulate material produced in the euphotic zone sinks into deeper waters (the mesopelagic and bathypelagic zones). In turn, only a small fraction of this sinking material survives transport to the seafloor to be preserved in the sediments. Extensive alteration of organic matter in the water column and at the sediment–water interface can yield sedimentary organic matter having a chemical composition markedly different from that of the material originally biosynthesized. A major goal of marine organic geochemists is to understand the qualitative and quantitative changes which occur in the water column and at the sediment–water interface (Wakeham and Lee, 1989). In this chapter we will discuss recent advances in our understanding of water column processes which influence the cycling of organic matter in the ocean.

A variety of biological, chemical, and physical processes work in concert to alter the organic and inorganic chemical composition of particulate material as it is transported through the water column (Fig. 1). Present estimates of the primary production of particulate organic carbon (POC) by photosynthetic organisms (photoautotrophs) in surface waters are on the order of 10^{16} g C yr^{-1} (assuming a mean carbon fixation rate of 200 mg C m^{-2} day^{-1} over a range of for oligotrophic 100 and for upwelling areas 2000 mg C m^{-2} day^{-1}; (Jenkins, 1982; Platt and Harrison, 1985; Martin *et al.*, 1987). Rivers and rainfall may contribute an additional 10^{14} g C yr^{-1} to the ocean (SCOPE, 1979; Mackenzie, 1981; Meybeck, 1982). By comparison, the reservoir (standing stock) of POC is about 10^{16} g C, while the dissolved organic carbon (DOC) pool in seawater is about 10^{18} g C. Consumption of POC by bacteria and zooplankton (heterotrophs) in the epipelagic ocean recycles >90% of the organic matter produced by photosynthesis, converting some of it to new biomass, some to degraded particulate matter

STUART G. WAKEHAM • Skidaway Institute of Oceanography, Savannah, Georgia 31416. CINDY LEE • Marine Sciences Research Center, State University of New York, Stony Brook, New York 11794.

Organic Geochemistry, edited by Michael H. Engel and Stephen A. Macko. Plenum Press, New York, 1993.

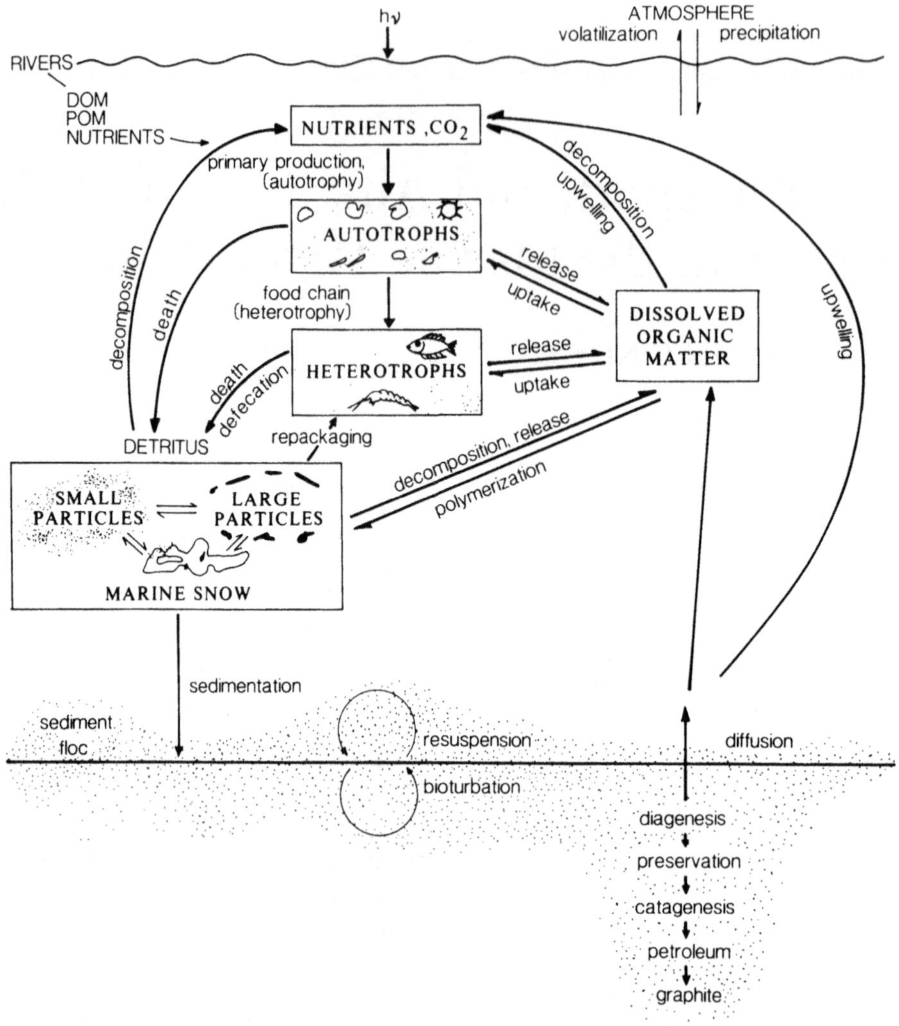

Figure 1. Schematic of the organic matter cycle in the ocean. DOM, Dissolved organic matter; POM, particulate organic matter.

and regenerated nutrients, and some to DOC. This recycled material is continuously replaced by newly photosynthesized material that is produced using nutrients recycled in surface waters or upwelled from deep waters. Thus, the oceanic carbon pools remain at steady-state concentrations.

Less than 5–10% of the primary production (100 in oligotrophic 2000 mg C m^{-2} day^{-1} in upwelling areas, respectively) passively sinks out of the epipelagic zone as particulate matter (Eppley and Peterson, 1979; Martin *et al.*, 1987). Vertically migrating zooplankton may actively transport additional organic matter to depth (Vinogradov, 1955; Urrerre and Knauer, 1981; Angel, 1984; Bishop *et al.*, 1986; Lee *et al.*, 1988), but the quantitative significance of this process is poorly known. As particles sink through the oceanic water column, they are continually altered by mesopelagic and bathypelagic heterotrophs. The result is that more than 90% of the POC which exits the epipelagic zone is recycled in the deep ocean (Suess, 1980; Honjo, 1980; Wakeham *et al.*, 1984; Lee and Cronin, 1984), leaving less than a few percent (1–5 mg C m^{-2} day^{-1}) of the material produced in surface waters to reach the sediments. This material may be highly altered compared to its original source, the more labile components having been preferentially removed and the more refractory components concentrated. Material which survives transport through the water column is then subjected to further alteration and decomposition by benthic heterotrophs; only a fraction of a percent of the original material may be buried in the sediment and contribute to the sedimentary organic geochemical record (Romankevich, 1984; Emerson and Hedges, 1988).

Several factors act to control both the vertical flux and chemical composition of particulate material: (1) the amount and composition of organic material biosynthesized in surface waters; (2) the transport mechanisms involved; (3) the biological community structure mediating the transformation reactions; and (4) the physical character of the water column (water column depth, redox state, temperature, etc.). It is important to recognize that these processes are important in deep waters as well as at the sea surface. The discussion which follows will consider how these several processes influence the flux and composition of material being delivered to the sediment–water interface.

2. Vertical Flux of Organic Matter Sinking through the Water Column

2.1. Organic Matter Production in Surface Waters

Primary photosynthetic production by phytoplankton in surface seawater is the largest source of organic carbon in marine systems. This is an autotrophic process using dissolved inorganic carbon and nutrients with light as the energy source (Fig. 1). Chemosynthesis is also an autotrophic process but uses reduced inorganic compounds for energy. Chemosynthesis can be an important production process in localized areas, such as hydrothermal vents or hot springs, but accounts for only a small percentage of the net global production of organic matter in the sea. Photosynthetic processes produce POC in the form of algal cells and DOC as excretion products of living algae. Particulate carbon can sink directly out of the euphotic zone under some conditions, but both dissolved and particulate organic matter produced during photosynthesis usually enter the marine food chain in the surface waters through the activity of heterotrophs.

Heterotrophs, or secondary producers, derive their energy from oxidizing organic matter rather than from light or from reduced inorganic compounds. The heterotrophic zooplankton, Protozoa, and bacteria acting together can use both the dissolved and particulate products of photosynthesis. The zooplankton can directly consume both living and nonliving (detrital) particles, thus removing much of the phytoplankton (and bacteria and Protozoa as well) from the surface waters. Bacteria can take up dissolved organic matter naturally excreted by phytoplankton during their growth or released as a result of cell rupture during feeding processes or after death. Bacteria can also colonize fecal pellets and other detrital particles. Protozoa graze on the bacteria or may also take up dissolved organic compounds directly. Nearly all of the phytoplankton production in the surface waters passes through either the herbivorous zooplankton or the microbial community, thus allowing only a small fraction to pass unaltered to deeper waters.

In spite of the great consumption of organic matter by heterotrophs in surface waters, the amount of material leaving the euphotic zone on particles appears for the most part to be directly related to the level of primary production. Results of sediment-trap studies have shown a correlation between the flux of bulk organic carbon and primary production (Suess, 1980; Betzer *et al.*, 1984; Pace *et al.*, 1987). A similar correlation exists between primary production and the fluxes of various classes of organic compounds (Fig. 2). It is interesting to note that the flux of organic matter varies by at least an order of magnitude more than the corresponding production values in these measurements. Several explanations could account for this. This relative difference between flux and production appears to be different for different compound classes. One possible explanation for this would be if biosynthesis of certain compounds were dependent on growth rate of the organisms. At higher productivity rates, for example, more of the carbon could be produced as amino or fatty acids. Concomitantly, less of these compounds could be produced at slower growth rates. This, however, seems unlikely. We know that areas of higher primary productivity frequently are dominated by different organisms than areas of lower productivity. Thus, the organisms commonly found in different productivity regimes might have greater ranges of cellular carbon divided between different compound classes. Another possible explanation is that the proportion of organic matter which sinks out of the euphotic zone on particles may vary as a function of decomposition rate, which in turn depends on productivity. Lee and Cronin (1984) compared decomposition rates for amino acids in areas of very different productivities, ranging from the highly productive Peru upwelling area to the oligotrophic central Pacific gyre. They found that the loss rate was higher when the productivity was higher, but that a low percentage of the amino acids produced were destroyed. If particles with a higher percentage of amino acids relative to total carbon exit the euphotic zone in high-productivity areas and particles with a lower percentage exit in less productive areas, then the amino acid flux would show a larger range than that seen in productivity in different areas, as shown in Fig. 2. This effect would be most

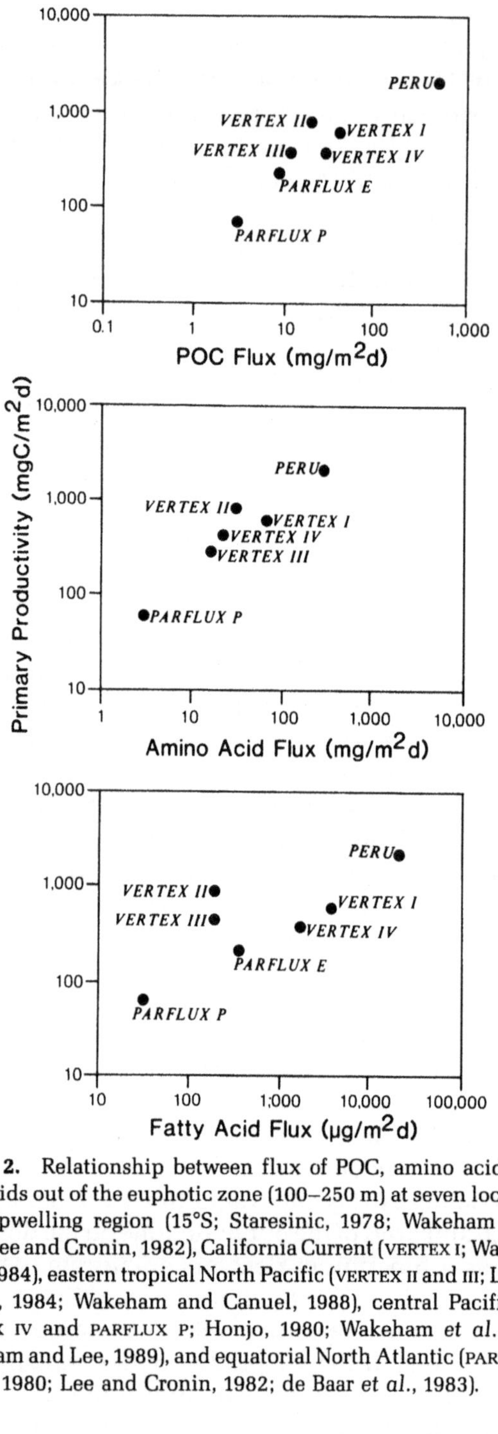

Figure 2. Relationship between flux of POC, amino acids, and fatty acids out of the euphotic zone (100–250 m) at seven locations: Peru upwelling region (15°S; Staresinic, 1978; Wakeham *et al.*, 1983; Lee and Cronin, 1982), California Current (VERTEX I; Wakeham *et al.*, 1984), eastern tropical North Pacific (VERTEX II and III; Lee and Cronin, 1984; Wakeham and Canuel, 1988), central Pacific gyre (VERTEX IV and PARFLUX P; Honjo, 1980; Wakeham *et al.*, 1984; Wakeham and Lee, 1989), and equatorial North Atlantic (PARFLUX E; Honjo, 1980; Lee and Cronin, 1982; de Baar *et al.*, 1983).

pronounced for the most labile classes of organic compounds.

Although there may be a good general correlation between primary productivity and POC flux, the analogous productivity–organic compound flux relationship does not always hold. At the VERTEX II and III sites off the western coast of Mexico, the fatty acid flux out of the euphotic zone (100 m) was considerably less than that found at other sites of comparable primary productivity (Fig. 2). POC and amino acid fluxes at VERTEX II/III were not comparably low. One explanation for the depletion of fatty acids in the particles at 100 m is that biosynthesis of storage lipids, and hence fatty acid production, by organisms in surface waters was inhibited in the warm subtropical waters at this site compared to colder waters at the other study sites. On the other hand, preferential decomposition of fatty acids may have been enhanced by the biogeochemical environment in the upper hundred meters of the water column, where a very intense oxygen minimum exists at 120–140 m.

2.2. Temporal Variations in Vertical Flux

Since the flux of particulate organic matter appears to be correlated with surface production, it might be expected that flux would vary with seasonal and diel changes in productivity. This has been clearly documented in several sediment-trap studies. Bimonthly collections of sediment-trap material taken over several years from 3200 m at a station in the Sargasso Sea have shown variations in flux which correspond to annual cycles of primary production at the sea surface (Deuser *et al.*, 1981; Ittekkot *et al.*, 1984a, b; Deuser, 1986). Figure 3 shows the variation in flux of <37-μm particles, POC, carbohydrates, and amino acids observed at this site over a four-year period. Three successive flux maxima correspond with spring blooms. In 1981, the spring bloom resulted in an unusually high peak in flux, showing that large interannual variations can occur and that long-term sampling efforts are necessary for a complete picture of oceanic productivity and carbon flux.

Ittekkot *et al.* (1984a) suggested that certain organic nitrogen compounds present in particles can be used as source markers, specifically the ratio of glutamic acid to γ-aminobutyric acid (glu/γ-aba) as an indicator of microbial degradation and that of amino acid to hexosamine (AA/HA) as an indicator of zooplankton biomass. Thus, the positive correlation between these ratios and POC flux during the first three years at the Sargasso Sea station (Fig. 3) indicates that material sinking during the spring blooms is relatively fresh, that is, relatively low in γ-aba and hexosamines. It is therefore interesting to note that during the anomalous year of 1981, hexosamine and γ-aba contents were relatively higher, suggesting that this sinking material was more degraded compared to that in previous years.

Hedges *et al.* (1988a, b) also found a strong sea-

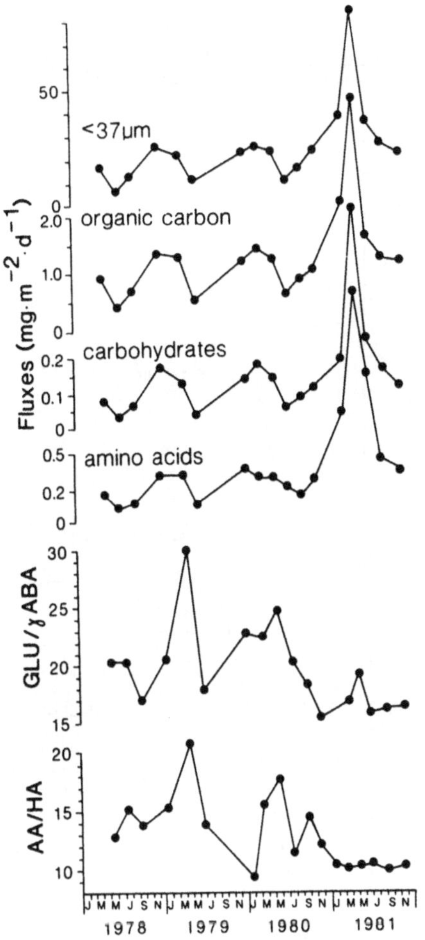

Figure 3. Seasonal fluxes of <37-μm particles, organic carbon, carbohydrates, and amino acids and the ratios of glutamic acid to γ-aminobutyric acid (glu/γ-aba) and of amino acids to hexosamines (AA/HA) for four years at a 3200-m sediment-trap site off Bermuda. [Data replotted from Ittekkot *et al.* (1984a).]

sonal trend in sediment-trap samples collected monthly over a one-year period in Dabob Bay, an arm of Puget Sound near Seattle, Washington. They saw major changes in particle composition and flux, based on their analyses of carbohydrates and lignin. Although bulk flux of particulate material did not vary greatly in their shallow traps, they observed about a fivefold change in the percent of organic carbon between the winter and nonwinter months. Changes in the flux of lignin and carbohydrates showed a small but continual input of terrestrial material throughout the year and a larger input of planktonic material during nonwinter months.

On much shorter time scales, measurable diel variations in flux and organic matter composition can occur in highly dynamic areas, such as where coastal upwelling occurs. Staresinic (1978), Lee and Cronin (1982), Wakeham *et al.* (1983), Gagosian *et al.* (1983), and Wakeham (1985) observed major diel changes in the flux of POC, amino acids, fatty acids, wax esters, triacylglycerols, and sterols in waters of the Peru upwelling zone. The exact nature of the diel patterns found in these studies was greatly dependent on the compound class and reflected the complex character of the biological processes producing the particles. However, in most instances, nocturnal fluxes were significantly greater than daytime fluxes, presumably as a consequence of increased grazing activity by herbivores and thus greater production of fecal pellets at night (Boyd *et al.*, 1980).

2.3. Vertical Transport of Particulate Organic Compounds

Organic matter is deposited at the seafloor in the form of particles sinking from above. As recently as about a decade ago, the most common conceptual model of particle dynamics was that slow sinking of fine suspended particles, which constitute most of the standing stock of particles in seawater, resulted in the vertical transport of material to the deep ocean (e.g., Parsons, 1975; Lal, 1977). Recycling of organic matter occurred primarily in the upper 400 m (Menzel, 1974) with the material at depth being, in general, "refractory." Recent observations, however, have led to revisions in our understanding of particle distributions and dynamics (McCave, 1975, 1984). For example, the finding of intact calcareous tests at depths below the carbonate compensation depth (Honjo, 1976) and "labile" organic compounds in the bathypelagic ocean (Wakeham *et al.*, 1980) suggests the presence of a second, geochemically distinct, pool of particles. These large and fast-sinking aggregates are now recognized as transporting material rapidly to depth.

Today, particulate matter in the ocean is operationally divided (based primarily on sampling strategy) into small and large particles. In reality, of course, a continuum of particle sizes exists (from colloids to whales). The bulk (>95%) of the POC pool comprises small (<20 μm) particles which sink very slowly (<1 m day^{-1}) and contribute little to the vertical flux of material toward the seafloor. Small, suspended particles are comprised of bacteria and small algal cells, clay minerals, and detrital fragments of large particles. They are readily collected by conventional water bottles and *in situ* filtration but are inefficiently sampled by sediment traps. Large particles (>20 μm) are relatively rare, but their large size and rapid sinking rates (hundreds to thousands of meters per day) make them the dominant mechanism transporting material vertically through the ocean. Sediment traps provide the primary means of collecting the large algal cells, fecal pellets, zooplankton carcasses and molts, and amorphous aggregates known as "marine snow" which constitute the large, sinking particle pool. Physical and biological disaggregation of large particles renews the small-particle pool, while aggregation, thought to be primarily biologically mediated, produces large particles *in situ*. Interactions between large and small particles are currently the topic of considerable interest (Simpson, 1982; McCave, 1975, 1984; Bacon *et al.*, 1985; Wakeham and Canuel, 1988).

The flux of individual organic compounds associated with vertically sinking large particles in the water column reflects the dominant primary and secondary sources of organic matter in surface waters. As depth increases, the particles disaggregate and decompose, and the flux of particulate organic carbon and organic compounds generally decreases with depth. Figure 4 illustrates these trends, showing fluxes for POC (Staresinic, 1978; Honjo, 1980; Martin *et al.*, 1987), amino acids (Lee and Cronin, 1982, 1984), and fatty acids (de Baar *et al.*, 1983; Wakeham *et al.*, 1984; Wakeham and Canuel, 1988) in a series of sediment-trap experiments in the Atlantic and Pacific Oceans. Other measurements of carbohydrates (Ittekkot *et al.*, 1984a, b; Tanoue and Handa, 1987), amino sugars (Ittekkot *et al.*, 1984a, b), pigments (Repeta and Gagosian, 1984; Welschmeyer and Lorenzen, 1985), wax esters and triacylglycerols (Wakeham, 1982; Wakeham *et al.*, 1984), and sterols (Gagosian *et al.*, 1982; Wakeham and Canuel, 1988) show comparable trends.

The flux of material measured at any given depth in the water column is the sum of the flux of material from above and *in situ* biosynthesis minus the loss due to decomposition and transformation. Decomposition and transformation rates vary depending on the molecular structure of individual compounds and their availability as substrates for heterotrophic metabolism (i.e., their physical packaging within the particle matrix). Preferential removal of certain organic components can be inferred from measurements of bulk compositional parameters. As particles sink, organic carbon is enriched relative to organic nitrogen and phosphorus with increased depth (C/N and C/P ratios increase; Gordon, 1970, 1971; Knauer *et al.*, 1979; Wefer *et al.*, 1982) due to preferential decomposition of nitrogen- and phosphorus-containing organic compounds (e.g., amino acids and phospholipids).

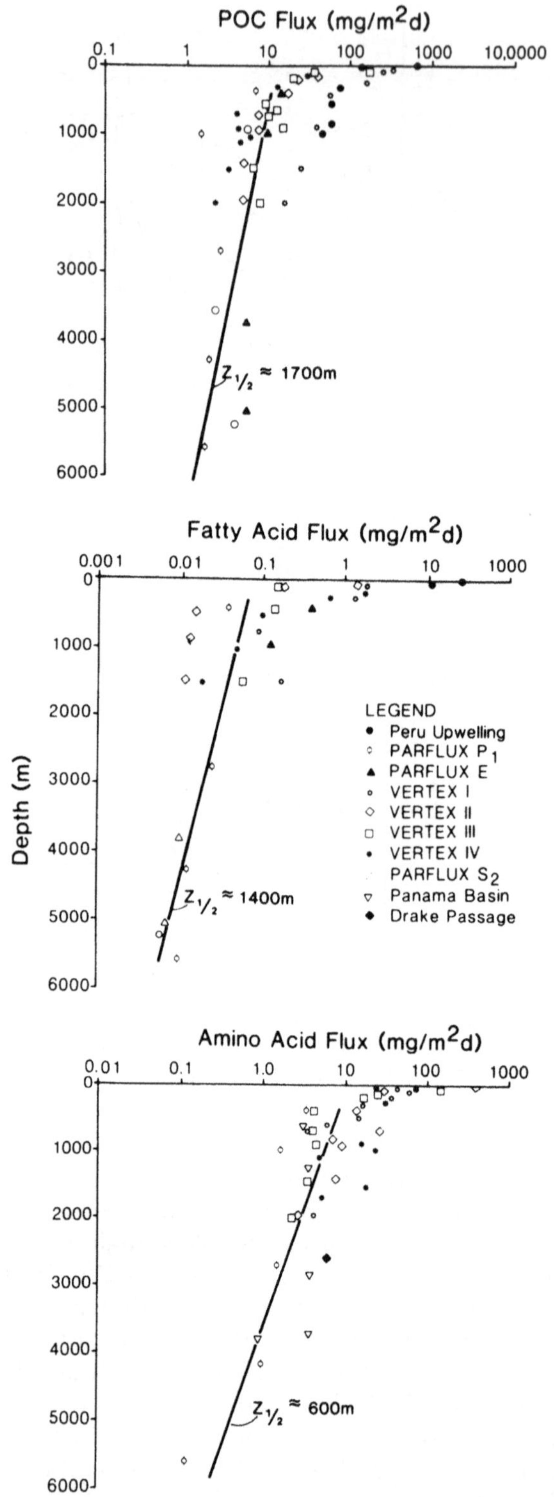

Figure 4. Vertical fluxes of POC, fatty acids, and amino acids at several locations in the ocean. Study sites as listed for Fig. 2, and including PARFLUX S_2 (Sargasso Sea; Honjo, 1980; de Baar *et al.*, 1983), Panama Basin (Lee and Cronin, 1984), and Drake Passage (Wefer *et al.*, 1982). Lines are regressions for flux data at depths greater than 400 m; $z_{1/2}$ are calculated half-depths (see text).

2.4. What Percent of the Various Classes of Organic Compounds Produced in Surface Waters Reach the Deep Sea?

After undergoing alteration in the water column during transit to the seafloor, a small proportion of the organic matter reaches the seafloor. Gagosian *et al.* (1982) estimated that less than 1% of the sterols produced in surface waters at the PARFLUX E site in the North Atlantic Ocean survived transport to 5068 m. An even smaller proportion (0.01–0.02%) of the surface-produced fatty acids reached the bottom waters at this site and at PARFLUX P in the Pacific (Wakeham *et al.*, 1984). This proportion may be higher in productive coastal waters. Lee and Cronin (1982) estimated that about 10% of the surface-produced amino acids reached the bottom at a site in the coastal Peru upwelling zone.

Because there are few studies comparing sediment accumulation rates with sinking particle fluxes of organic compounds, it is difficult to assess the proportion of the surface production of various classes of organic compounds reaching the deep sea. However, by looking at the alteration of the compound classes on sinking particles with depth, we can determine their relative reactivities in a qualitative fashion. Differential behavior of some specific organic compound classes has been observed in particles obtained in sediment-trap experiments in the deep ocean. For example, at the 5200 m-deep PARFLUX E site in the equatorial North Atlantic Ocean, lipid and amino acid fluxes were measured (Wakeham *et al.*, 1980; Lee and Cronin, 1982; Wakeham, 1982; Gagosian *et al.*, 1982; de Baar *et al.*, 1983; Wakeham *et al.*, 1984). The flux of particulate material decreased by a factor of about 0.3 and that of POC by a factor of 5 between the shallowest trap at 389 m and the deepest at 5068 m. Specific organic compounds, however, showed variable but generally greater losses due to their greater lability relative to bulk organic carbon. Hydrocarbon flux decreased by a factor of 14, amino acids by 25, sterols by 45, and fatty acids by 60.

Once particles arrive at the seafloor, they undergo further decomposition at the sediment–water interface. Hedges *et al.* (1988b) calculated relative reactivity relationships for a large variety of organic compounds and tied these reactivities to alteration in the water column. They found that the organic compounds produced in the shallow coastal waters of Dabob Bay varied greatly in reactivity. Some compounds, such as the vanillyl and cinnamyl structural units derived from lignin in vascular plant debris, were not appreciably degraded. However, syringyl phenols, also derived from vascular plants, were measurably degraded (30–40%). Approximately 70% of the neutral sugars and total organic nitrogen was lost at the sediment–water interface. Plankton-derived lipids (>90% loss; Prahl *et al.*, 1980) and plant pigments (>99%; Furlong and Carpenter, 1988) were even more reactive.

Flux data in Fig. 4 can be used to estimate net decomposition rate constants for different classes of organic matter in the mesopelagic and bathypelagic zones. The lines shown on the figure were obtained from a first-order exponential regression ($\ln F = \ln F_0 - kz$) of flux versus depth data for all samples below 400-m depth. Half-depths ($z_{1/2}$) are the depth range over which 50% of the material has been lost ($z_{1/2} = 0.693/k$). Compounds or classes with smaller $z_{1/2}$ would be more rapidly degraded (i.e., more labile) than materials having larger $z_{1/2}$ values. This exercise suggests that amino acids, with $z_{1/2} \approx 600$ m, are more labile than fatty acids ($z_{1/2} \approx 1400$ m), and both amino acids and fatty acids are lost more quickly than bulk POC.

A more qualitative comparison between organic matter flux through the water column and sediment accumulation rates can be made with data from the Peru upwelling area. For example, fluxes of fatty acids and sterols measured in free-drifting sediment traps from this very productive coastal region were compared to accumulation rates in the underlying surface sediments (Fig. 5; Volkman *et al.*, 1983; Gagosian *et al.*, 1983; Wakeham *et al.*, 1983). For certain predominantly marine-derived components (e.g., palmitic acid and 24-methylcholesta-5,22*E*-dien-3β-ol), there was a substantial decrease in flux measured in traps at the base of the euphotic zone (14 m) and at the thermocline (50 m) compared with fluxes estimated to arrive at the surface sediment (400 m). On the other hand, the fluxes of tetracosanoic acid and 24-ethylcholest-5-en-3β-ol, compounds usually (but not exclusively) attributed to terrigenous sources, and the marine-derived C_{37}/C_{38} methyl alkenones were more similar at both sediment-trap depths and in surface sediment. In the unique case of the highly unsaturated alkenones (di-, tri-, and tetraenones), which normally would be expected to be highly labile, their resistance to degradation apparently results from their unusual *trans*, rather than *cis*, configurations (Rechka and Maxwell, 1988) and the inability of heterotrophs to metabolize them.

In the case of the two fatty acids, differing chain length may be a factor in relative resistance, since a C_{24} compound can be degraded less readily than a C_{16} compound (Van Vleet and Quinn, 1979; de Baar *et al.*,

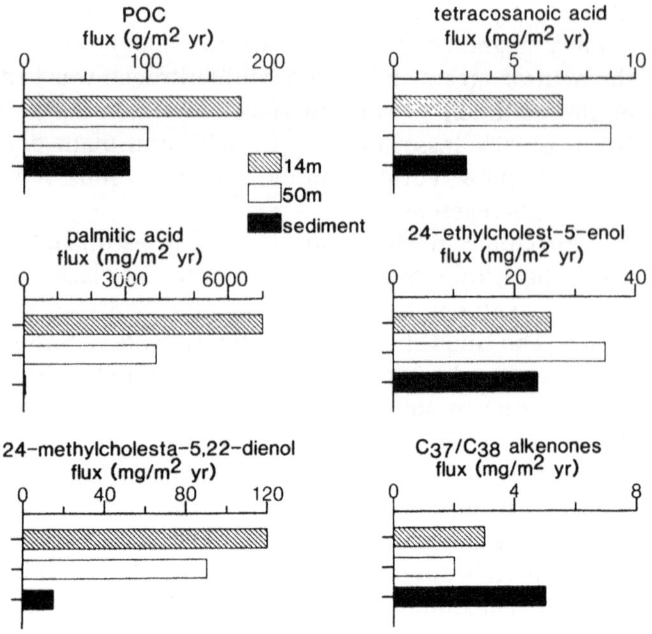

Figure 5. Fluxes (means of day and night sediment traps) of material at 14 m (base of euphotic zone), at 50 m (depth of thermocline), and at the sediment–water interface for Peru upwelling. POC data from Staresinic (1978); fatty acid and alkenone data from Wakeham *et al.* (1983) and Volkman *et al.* (1983); sterol data from Gagosian *et al.* (1983).

1983). This explanation is, however, unsatisfactory for the sterols, as the two described differ by only one carbon atom. Alternatively, the degree of decomposition may depend on the matrix in which the organic compounds are packaged within the particles; the terrigenous compounds may simply be better protected against decomposition. A third possibility is that some of the terrestrially derived material may reach the sediment without passing through the water column. For example, lateral transport by bottom currents would bring material to the sediments, bypassing the sediment traps deployed at shallower depths in the water column. It must also be remembered that short-term sediment-trap experiments, e.g., with 12-h time scales off Peru, may poorly mimic the longer time periods during which sediments accumulate.

Amino acid data from the Peru upwelling region also allow a comparison of water column fluxes (Lee and Cronin, 1982) and sediment accumulation rates [calculated from Henrichs and Farrington (1984) and Henrichs *et al.* (1984)]. Although the water column flux of every amino acid decreased between 14 and 52 m, the fluxes of serine, glycine, and threonine decreased more slowly than those of the others (Fig. 6). Comparing water column fluxes with sediment accumulation rates shows that amino acid sedimentation rates are generally lower, as might be expected due to continuing decomposition processes in the water column as particles sink to the sediment. However, the sedimentation rate of serine is actually higher than its flux out of the euphotic zone, and glycine and threonine sedimentation rates are higher than the flux at 52 m.

Previous studies of suspended and sinking matter and sediments have also observed relative increases in serine, threonine, and/or glycine with water or sediment depth (Degens, 1970; Siezen and Mague, 1978; Lee and Cronin, 1982, 1984; Lee *et al.*, 1987; Burdige and Martens, 1988) although this predominance is not a characteristic of all sediments. This increase has generally been attributed to the persistence of a protein–silica complex found in diatom cell walls which is enriched in serine, threonine, and glycine (Hecky *et al.*, 1973). However, resistance alone cannot account for sediment accumulation rates of serine and, to a lesser extent, glycine and threonine being higher than the flux out of the water column. One possible explanation would be lateral transport of material enriched in serine, glycine, and threonine, either below the 52-m trap or along the sediment bottom, as mentioned above as a mechanism for transporting terrestrially derived lipids. Since the predominance of serine, glycine, and threonine occurs in open ocean environments as well, lateral transport probably does not account for all the increase in sedimentation rate. Another possible explanation based on this evidence is that chemical or more likely biological production of these three amino acids occurs with depth, either *de novo* or from other amino acids, during particle or sediment diagenesis.

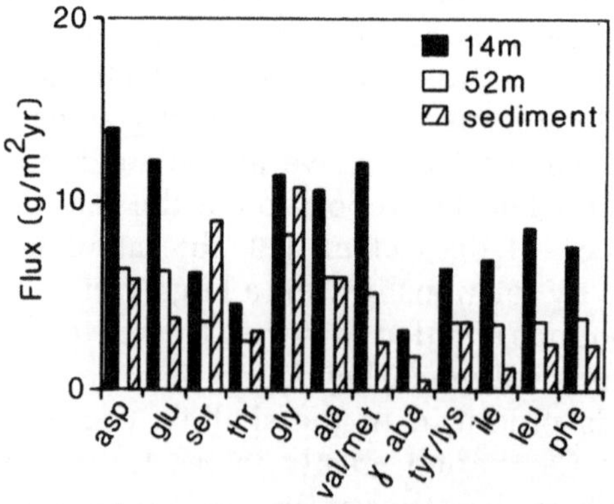

Figure 6. Water column fluxes of amino acids at 14 m and 52 m (Lee and Cronin, 1982) and amino acid sedimentation rates [calculated from data in Henrichs and Farrington (1984) and Henrichs et *al.* (1984)] for the Peru upwelling area.

3. Influences on the Quality of Particulate Organic Matter Sinking through the Water Column

3.1. Biological Community in Surface Waters

Particulate matter in the epipelagic zone reflects its planktonic source [see review by Lee and Wakeham (1988)]. Small, suspended particles collected by water bottles or *in situ* filtration are similar to phytoplankton in composition, while the larger particles reflect a mixed phytoplankton and zooplankton origin. Details of the organic composition of particles can vary significantly depending on the biological community producing the particles; these differences form the basis of using biomarkers as indicators of the source of sedimentary organic matter.

Lipids have been used extensively as source markers for sediment-trap material. For example, the fatty acid and sterol compositions of fine particles in such diverse oceanic regimes as the highly productive Peru upwelling area and the oligotrophic north central Pacific gyre have been compared (Fig. 7 and Table I). Fatty acid and sterol distributions are markedly different in the two samples, reflecting the different phytoplankton communities inhabiting the areas. Dinoflagellates, and to a lesser extent diatoms and coccolithophores, were prevalent in surface waters off the Peruvian coast (Staresinic, 1978) and resulted in the presence of 16:0, 16:1Δ^9, and 20:5 fatty acids (Orcutt and Patterson, 1975; Joseph, 1975; Volkman *et al*, 1980b; Nichols *et al*., 1984) and the sterols cholesta-5,22*E*-dien-3β-ol, 24-methylcholesta-5,22*E*-dien-3β-ol, 24-methylcholest-22*E*-en-3β-ol, 23,24-dimethylcholest-22*E*-en-3β-ol, and 4α,23,24-trimethylcholest-22-en-3β-ol (Rubenstein and Goad, 1974; Orcutt and Patterson, 1975; Alam *et al*., 1979; Boon *et al*., 1979; Volkman *et al*., 1980b, 1981; Volkman, 1986). In contrast, small flagellates (prymnesiophytes and prasinophytes) and cyanobacteria dominated the oligo-

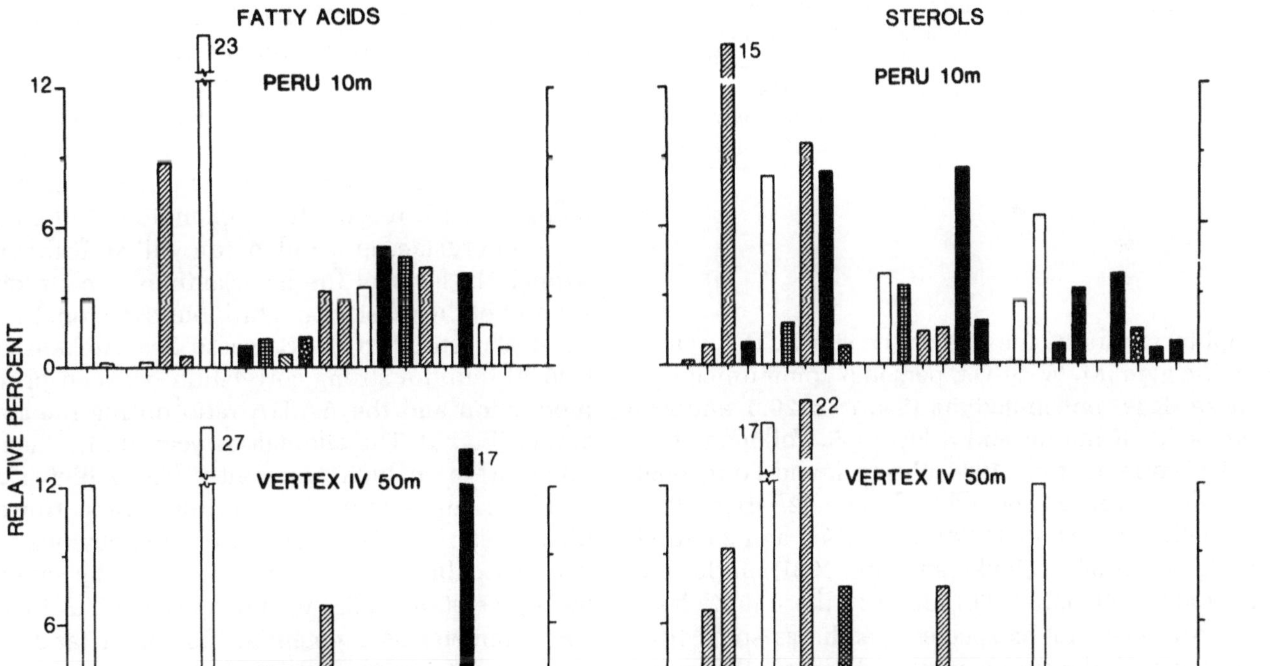

Figure 7. Fatty acid and sterol distributions for suspended particles (*in situ* filter samples) collected at 10 m in the Peru upwelling and at 50 m at the oligotrophic VERTEX IV site (Wakeham and Lee, 1989). Bar designations refer to Table I.

TABLE I. Fatty Acid and Sterol Assignments in Peru and VERTEX IV Particulate Matter[a]

Bar assignment	Fatty acid	Sterol
1	14:0	24-Norcholesta-5,22*E*-dien-3β-ol
2	15:0	27-Nor-24-methylcholesta-5,22*E*-dien-3β-ol
3	15:0 *iso*	Cholesta-5,22*E*-dien-3β-ol
4	15:0 *anteiso*	5α(H)-Cholest-22*E*-en-3β-ol
5	16:1Δ^9	Cholest-5-en-3β-ol
6	16:1Δ^{11}	5α(H)-Cholestan-3β-ol
7	16:0	24-Methylcholesta-5,22*E*-dien-3β-ol
8	17:0	24-Methyl-5α(H)-cholest-22*E*-en-3β-ol
9	18:5+6	24-Methylcholesta-5,24(28)-dien-3β-ol
10	18:4	24-Methylcholest-24(28)-en-3β-ol
11	18:3	24-Methylcholest-5-en-3β-ol
12	18:2	24-Methyl-5α(H)-cholestan-3β-ol
13	18:1Δ^9	23,24-Dimethylcholesta-5,22*E*-dien-3β-ol
14	18:1Δ^{11}	24-Ethylcholesta-5,22*E*-dien-3β-ol
15	18:0	23,24-Dimethylcholest-22*E*-en-3β-ol
16	20:5	4α,24-Dimethyl-5α(H)-cholest-22*E*-en-3β-ol
17	20:4	24-Ethyl-5α(H)-cholest-22*E*-en-3β-ol
18	20:1Δ^{11}	23,24-Dimethylcholest-5-en-3β-ol
19	20:0	24-Ethylcholest-5-en-3β-ol
20	22:6	24-Ethyl-5α(H)-cholestan-3β-ol
21	22:0	4α,24-Dimethyl-5α(H)-cholestan-3β-ol
22	24:0	24-Ethylcholesta-5,24(28)-dien-3β-ol
23	26:0	4α,23,24-Trimethylcholest-22*E*-en-3β-ol
24		4α,23,24-Trimethylcholest-8(14)-en-3β-ol
25		4α,23*S*,24*R*-Trimethyl-5α(H)-cholestan-3β-ol
26		4α,23*R*,24*R*-Trimethyl-5α(H)-cholestan-3β-ol

[a]See Fig. 7.

trophic waters at the VERTEX IV site in the central Pacific gyre (M. W. Silver, personal communication). These algae contributed the 18:4, 18:2, 20:5, and 22:6 fatty acids (Chuecas and Riley, 1969; Volkman *et al.*, 1981; Piorreck *et al.*, 1984; Piorreck and Pohl, 1984) and the sterols 24-methylcholesta-5,22*E*-dien-3β-ol, 24-methylcholesta-5,24(28)-dien-3β-ol, and 24-ethylcholest-5-en-3β-ol (Volkman *et al.*, 1981; Marlowe *et al.*, 1984; Volkman, 1986) found in these particles.

Although not as species-specific as some lipid components, the amino acids and carbohydrates can also reflect the biological community in the surface waters. These two compound classes account for the major part of the organic carbon in particles. They are largely structural components in marine organisms and do not vary as greatly in composition between organisms as do the lipids. However, subtle variations do occur. Inputs from algal producers of siliceous versus carbonate skeletons can be distinguished by relative abundances of amino acids and sugars (Ittekkot *et al.*, 1984b). Fucose is a major constituent of the siliceous cell walls of diatoms (Hecky *et al.*, 1973), while arabinose is commonly found in algae producing carbonate skeletons like the coccolithophores (Ittekkot *et al.*, 1982). Thus, the ratio of arabinose to fucose reflects the proportion of carbonate- to silicate-producing organisms. Similarly, siliceous diatoms appear to be enriched in glycine relative to aspartic acid (Hecky *et al.*, 1973), while carbonate organisms like coccolithophores and foraminifera appear to be relatively richer in aspartic acid (King, 1974).

Ittekkot *et al.* (1984a, b) used this information in their investigation of sinking particles in the Sargasso Sea and Panama Basin. Figure 8 shows the fluxes of sugars and amino acids at three depths for a time-series sediment-trap experiment in the Panama Basin. The large peak observed during early summer coincided with production by a single species of coccolithophore. Carbonate accounted for 60–90% of the total mass flux in these samples. Higher arabinose:fucose and aspartic acid:glycine ratios clearly reflect the prevalence of carbonate-producing organisms at this time (Fig. 8). In the Sargasso Sea, arabinose:fucose ratios were close to unity over the three-year period studied, reflecting a more balanced input of both carbonate and silicate producers.

In these same sediment-trap experiments from the Sargasso Sea and Panama Basin, Ittekkot *et al.* (1984a, b) used the ratio of hexosamines to amino acids as an indicator of the prevalence of phytoplankton with respect to zooplankton. A dominant input of crustacean zooplankton will shift this ratio strongly in favor of the hexosamines, the structural subunit of the crustacean chitinous skeleton. In their three-year study in the Sargasso Sea, Ittekkot *et al.* (1984a) found a strong correlation between primary production and the AA/HA ratio during the spring bloom (Fig. 3). The anomalous year, 1981, when the fluxes were so high apparently had a higher zooplankton input than the previous years. However, Ittekkot *et al.* (1984a) pointed out the obvious problem that collection of active swimmers by sediment traps (Lee *et al.*, 1988) would seriously bias the AA/HA parameter as a potential biomarker (and lipid-based indicators as well).

3.2. Heterotrophic Alteration of Particulate Organic Matter

3.2.1. Bacterial Transformations

The composition of sinking particulate organic matter, and thus the organic matter reaching the seafloor, is influenced by the alteration reactions which

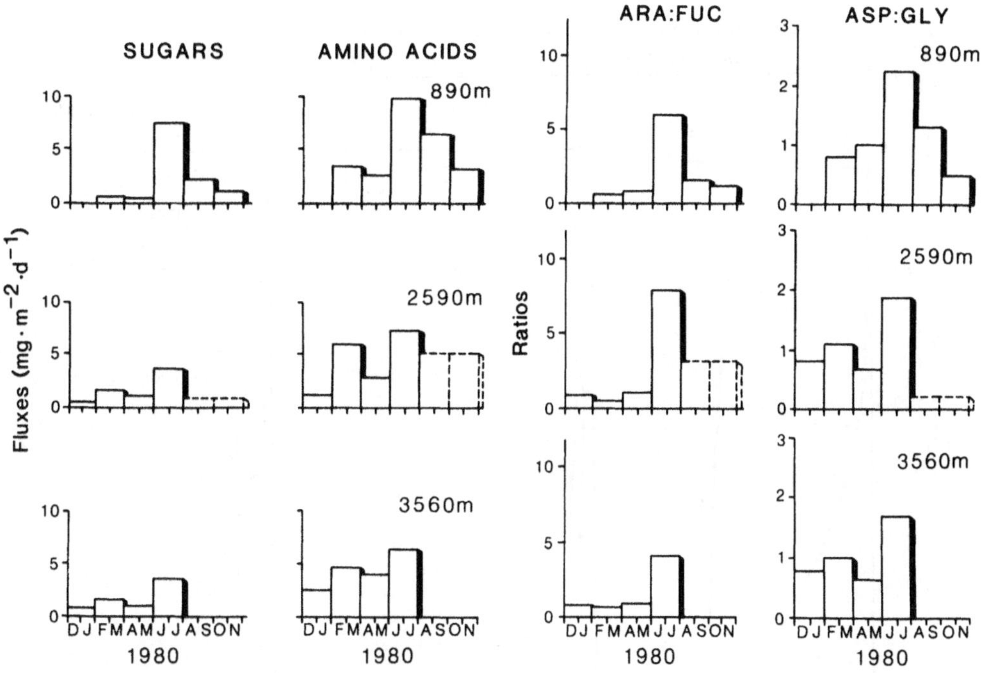

Figure 8. Bimonthly fluxes of sugars and amino acids and arabinose:fucose (ara:fuc) and aspartic acid:glycine (asp:gly) ratios for a sediment-trap experiment in the Panama Basin. [Data replotted from Ittekkot *et al.* (1984b).]

can occur during particle transit. Both zooplankton and bacteria consume organic matter, changing particle composition by preferential degradation and transformation reactions. Organic compounds present on particles can be consumed, and waste products which are a structural portion of the original compound can be left behind, or bacteria colonizing a particle can metabolize organic matter and in the process synthesize new bacterial biomass. For example, de Baar *et al.* (1983) found specific branched-chain fatty acids (*iso-* and *anteiso-*C_{15} and C_{17}) in deep traps in the North Atlantic ocean. These compounds are produced mainly by bacteria, and their persistence with depth probably indicates a viable bacterial community colonizing the particles.

The organic nitrogen compounds are particularly useful as markers of bacterial alterations. Lee and Cronin (1982) found increasing concentrations with depth of the nonprotein amino acid β-alanine on particles sinking in the Peru upwelling area. They attributed this increase to bacterial decomposition, possibly of aspartic acid, on the sinking particles. Ittekkot *et al.* (1984a) also used the ratios of aspartic acid to β-alanine (asp/β-ala) and glutamic acid to γ-aminobutyric acid (glu/γ-aba) (Fig. 3) as indicators of microbial decomposition on sinking particles in the Sargasso Sea. They found that these ratios varied seasonally, with minimal microbial alteration manifested at times of highest productivity [i.e., lower efficiency of decomposition associated with higher productivity as also reported by Lee and Cronin (1984)]. γ-Aba and β-ala are sometimes found as the predominant amino acids in deep-sea sediments (Schroeder, 1975; Whelan, 1977; Henrichs, 1980). It is unclear whether these sedimentary amino acids are transported on particles to the seafloor or are produced *in situ* in the sediments. Cole and Lee (1986) found that γ-aba and β-ala could be quickly degraded by marine bacteria and suggested that a difference in relative degradation rates between these two nonprotein amino acids and other amino acids was not the most likely mechanism for enrichment on the seafloor. *In situ* production of γ-aba and β-ala at the sediment–water interface and preferential adsorption of these compounds by clay minerals (Friebele *et al.*, 1980; Weliky, 1983) were thought to be more likely mechanisms to account for the presence and enrichment of these nonprotein amino acids in the sediment.

The amino acid ornithine can also be a marker for bacterial decomposition on particles. Degens (1970) suggested that the presence of ornithine in particulate matter indicated that dead plankton cells in the particles had begun to decompose. Ornithine is a decomposition product of the protein amino acid arginine. Lee and Cronin (1984) observed increased amounts of this compound in particles from the strong oxygen minimum zone off the coast of Mexico (Fig. 9). The presence of muramic acid, a component of bacterial cell membranes, is a rather specific bacterial marker as it is rarely found in any other capacity (Salton, 1960). Concentrations of muramic acid in

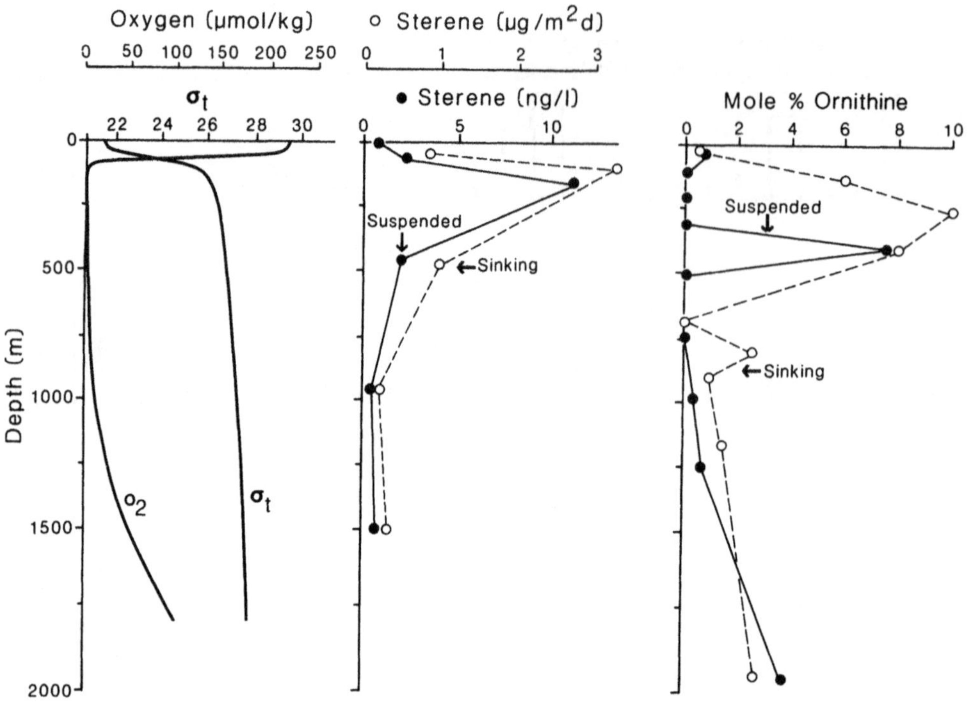

Figure 9. Dissolved oxygen and density σ_t at the VERTEX II/III site off western Mexico, and sterene flux and concentration and mole % ornithine depth profiles for sinking and suspended particles sampled across the oxygen minimum zone. [Data replotted from Lee and Cronin (1984) and Wakeham (1987).]

sediment have been used to estimate bacterial biomass (King and White, 1977; Moriarty, 1977). Lee *et al.* (1983) and Ittekkot *et al.* (1984b) both found muramic acid in sinking particles, suggesting the presence of bacterial colonization on the particles.

3.2.2. Zooplankton Alteration of Particulate Organic Matter

A major fate of phytoplankton cells growing in the upper ocean is to be consumed by zooplankton and fish. Experiments by Copping and Lorenzen (1980) showed that some 45% of the phytoplankton carbon ingested by herbivorous zooplankton goes into growth of the animal, while about 50% is respired or lost as DOC. The remaining 3–4% is egested as fecal pellets. Fecal pellets can have very different organic compositions compared to those of the animal's original diet of living or detrital particles (e.g., Corner *et al.*, 1986). The alteration of dietary organic compounds may occur within the animal's gut via enzymatic reactions or by the action of enteric microflora; alteration can also occur within fecal pellets if enzymes and microorganisms remain active after the pellet is defecated. Fecal matter comprises a significant portion of the particulate material sinking through the water column (reviewed in Fowler and Knauer, 1986), although recent studies indicate that intact fecal pellets (as opposed to amorphous fecal matter) may contribute only about 10% of the vertical flux of organic carbon in the deep sea (Pilskaln and Honjo, 1987).

The importance of modification of the organic composition of particulate matter by zooplankton has led to a series of laboratory studies which evaluated the qualitative and quantitative effects of these modifications. An early study by Cowey and Corner (1966) focused on changes in amino acid composition as unicellular algae were fed to *Calanus finmarchicus* and found that the fecal pellets produced were not very different in composition from the algal diet consumed. Similar qualitative results were reported by Tanoue *et al.* (1982), who fed *Dunaliella tertiolecta* to the euphausiid *Euphausia superba*; however, they found that algal amino acids and fatty acids were preferentially depleted compared to sugars during metabolism and fecal pellet production. Studies by Wakeham and Canuel (1986) and Harvey *et al.* (1987) have shown that dietary fatty acids are more readily assimilated by zooplankton than are dietary sterols (see Fig. 10).

Marked changes in composition can occur within specific compound classes; lipids have been used to document these changes in a series of studies examining herbivorous, carnivorous, and coprophagic feeding. In one such study, Harvey *et al.* (1987) investigated changes in fatty acid and sterol composition in experiments with *Calanus helgolandicus* feeding

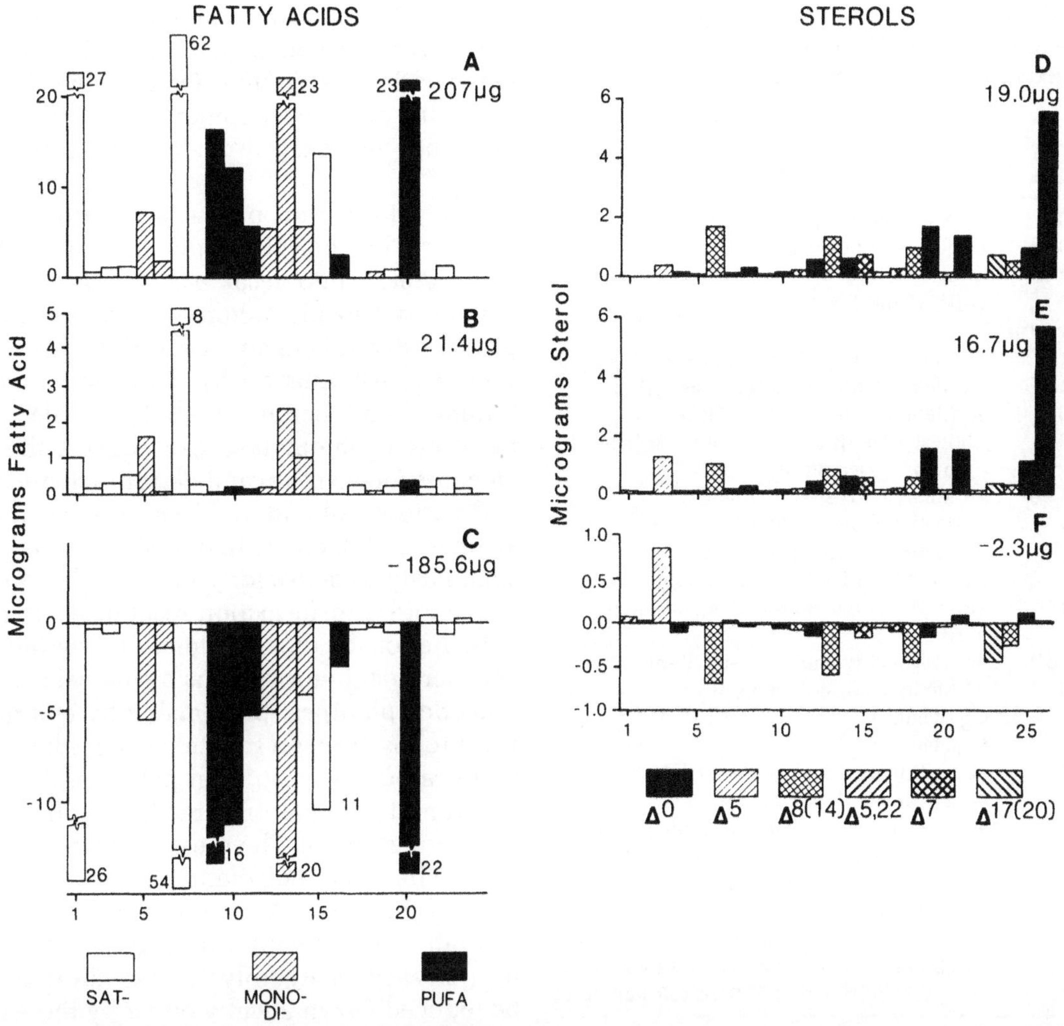

Figure 10. Distributions and concentrations (micrograms/experiment) of fatty acids and sterols in a feeding experiment involving *Calanus helgolandicus* feeding on *Scrippsiella trochoidea*. (A) and (D): *S. trochoidea* lipids ingested; (B) and (E): lipids in fecal pellets of *C. helgolandicus*; (C) and (F): difference plots for fecal pellet lipid minus lipid ingested. The large numbers at the upper right of each panel are the total or difference of fatty acid or sterol for that sample. Bar designations refer to Table II. [Data plotted from Harvey *et al.* (1987).]

on the dinoflagellate *Scrippsiella trochoidea* (Fig. 10 and Table II). In these experiments, up to 97% of ingested fatty acids were assimilated by the copepod, with a marked preference for polyunsaturated components. Algal sterols were removed to a lesser extent, but significant amounts of 4-methyl- and 4-desmethyl sterols with 8(14) and 17(20) unsaturation, abundant in the dinoflagellate, were removed by the copepod. Ring-saturated stanols passed through the animal's gut quantitatively. The selective removal of polyunsaturated fatty acids and sterols having certain unsaturation patterns is consistent with results reported by other investigators (Prahl et *al.*, 1984a; Wakeham and Canuel, 1986). In the case of sterols, for example, C_{28} and C_{29} sterols with Δ^5 and $\Delta^{5,7}$ unsaturation from the green alga *Dunaliella primolecta* were selectively removed compared to Δ^7 components, and the Δ^7 sterols were thus enriched in fecal pellets of *C. helgolandicus*.

In addition to preferential depletion of certain algal dietary components, fecal pellets often are enriched in zooplankton-derived compounds. For example, cholesterol and wax esters, absent in algal dietary lipids, are found in fecal pallets produced by herbivores (Volkman et *al.*, 1980a; Prahl et *al.*, 1984a). Carnivorous and coprophagic feeding results in fecal material enriched in components endogenous to the consumer (Prahl et *al.*, 1985; Wakeham and Canuel, 1986; Neal et *al.*, 1986). Neal et *al.* (1986) also showed that feeding mechanism plays a role in determining

TABLE II. Fatty Acid and Sterol Designations for the *S. trochoidea–C. helgolandicus* Feeding Experiment of Harvey *et al.* (1987)[a]

Bar assign-ment	Fatty acid	Sterol
1	14:0	24-Norcholesta-5,22*E*-dien-3β-ol
2	15:0	Cholesta-5,22*E*-dien-3β-ol
3	15:0 *iso*	Cholest-5en-3β-ol
4	15:0 *anteiso*	5α(H)-Cholestan-3β-ol
5	16:1Δ^9	27-Nor-24-methyl-5α(H)-cholestan-3β-ol
6	16:1Δ^{11}	4α-Methylcholesta-8(14),22-dien-3β-ol
7	16:0	4α-Methylcholesta-22*E*-en-3β-ol
8	17:0	Cholest-7-en-3β-ol + 24-methyl-5α(H)-cholest-22-en-3β-ol
9	18:5+6	4α-Methylcholest-8(14)-en-3β-ol
10	18:4	4α-Methylcholestan-3β-ol
11	18:3	4α-Methylcholest-8(14)-en-3β-ol
12	18:2	24-Methyl-5α(H)-cholestan-3β-ol
13	18:1Δ^9	4α,24-Dimethylcholesta-8(14),22*E*-dien-3β-ol
14	18:1Δ^{11}	4α,24-Dimethylcholest-22-en-3β-ol
15	18:0	24-Methylcholest-7-en-3β-ol
16	20:5	C_{29}-stenol
17	20:4	C_{29}-stenol
18	20:1Δ^{11}	4α,24-Dimethylcholest-8(14)-en-3β-ol
19	20:0	4α,24-Dimethylcholestan-3β-ol
20	22:6	4α,23,24-Trimethylcholest-5,22-dien-3β-ol
21	22:0	4α23,24-Dimethylcholest-22-en-3β-ol
22	24:0	4α,24-dimethylcholest-7-en-3β-ol
23	26:0	4α,23,24-Trimethylcholest-17(20)-en-3β-ol
24		4α,23,24-Trimethylcholest-8(14)-en-3β-ol
25		4α,23*S*,24*R*-Trimethyl-5α(H)-cholestan-3β-ol
26		4α,23*R*,24*R*-Trimethyl-5α(H)-cholestan-3β-ol

[a]See Fig. 10.

the composition of fecal matter. In a planktonic feeding sequence, the lipid composition of fecal pellets produced by barnacle nauplii (*Elminius modestus*) feeding herbivorously on unicellular algae (*Hymenomonas carterae* and *Dunaliella primolecta*) was compared with the composition of pellets produced by adult copepods (*C. helgolandicus*) feeding coprophagously on the barnacle nauplii fecal pellets. The major changes in dietary lipid occurred when the plant cells were initially processed by the barnacle, with subsequent coprophagy causing further, more minor alterations. Characteristic algal hydrocarbons and polyunsaturated fatty acids were completely removed by the barnacle nauplii, whereas sterols were less affected. Fecal pellets of both *Elminius* and *Calanus* were enriched in C_{26} and C_{27} sterols characteristic of the zooplankton, along with small amounts of algal C_{28} and C_{29} sterols. Thus, feeding involving coprophagy is in some ways similar to herbivory in that labile hydrocarbons and fatty acids are completely removed in fecal pellets. On the other hand, pellets from direct herbivory contain a wider range of sterols than the "repackaged" pellets produced by coprophagy.

Transformation products formed by enzymatic reactions in guts of zooplankton and fish can be incorporated into fecal pellets. The conversion of chlorophyll to pheophorbide via hydrolysis of the phytyl ester side-chain is a well-known example of such a transformation by herbivorous zooplankton (Shuman and Lorenzen, 1977). Similar hydrolysis reactions by herbivores can convert the algal carotenoids fucoxanthin and peridinin to fucoxanthinol and peridininol and yield particles enriched in the alcohols and depleted in the esters (Repeta and Gagosian, 1984; see below for a more detailed discussion of carotenoid transformation reactions). Oxidation and dehydration of dietary sterols to steroidal ketones and steroidal hydrocarbons by the pelagic crab *Pleuroncodes planipes*, presumably by enteric bacteria, was proposed as the source for steroidal ketones and hydrocarbons in crab fecal pellets and in sediment-trap samples containing crab fecal pellets (Wakeham and Canuel, 1986). The presence of dihydrophytol in fecal pellets of *C. helgolandicus* fed a diet containing none of this saturated alcohol has been documented by Prahl *et al.* (1984b), who suggested that algal phytol (released by hydrolysis from chlorophyll) might be reduced enzymatically either by the animal itself or by gut microflora.

3.3. Production of Organic Matter at Depth

While most of the particulate organic matter in the water column originates in the euphotic zone, there is in fact production of material in deep waters as well. Bacteria, Protozoa, and numerous species of meso- and bathypelagic zooplankton and fish inhabit deep waters and produce organic compounds as part of their own living biomass. Heterotrophic (secondary) production of organic matter may be responsible for part of the increased organic carbon fluxes sometimes observed at depth. Karl *et al.* (1984) and Karl and Knauer (1984), for example, reported that bacterial chemolithotrophy at depth (700–900 m at the VERTEX II/III site off Mexico) may produce "new bacterial carbon" at a rate equivalent to about 1% of the integrated autotrophic primary production in surface waters. This production may result in slightly increased "fluxes" of organic carbon to sediment traps deployed deeper in the water column. Bacteria may be the source of branched C_{15} and C_{17} fatty acids (de Baar *et al.*, 1983; Wakeham and Canuel, 1988) and steroid ketones (Gagosian *et al.*, 1982) detected in deep particle samples. Deep-sea barophilic bacteria often contain elevated levels of mono- and polyunsaturated fatty acids (e.g., 16:1, 18:1, 20:5, and 22:6) in order to maintain membrane fluidity in the high-pressure/low-temperature environment of the deep sea (Russell, 1983; DeLong and Yayanos, 1985; Wirsen *et al.*, 1987). Perhaps such a source could account for the high levels of polyunsaturated fatty acids observed in suspended particles at depths up to 1500 m (Wakeham and Canuel, 1988). Slight increases in protein amino acid fluxes were observed below the oxygen minimum at VERTEX II and III (Lee and Cronin, 1984; Lee *et al.*, 1988). This increase could be due to bacterial production or to mesopelagic inputs of "minipellets" at this location by deep-dwelling Protozoa and small invertebrates (Gowing and Silver, 1985).

Large maxima in fluxes of various lipid classes which cannot be attributed to microbial production have been observed in several deep-water sediment-trap experiments: wax esters, steryl esters, and sterols at 988 m at PARFLUX E (equatorial North Atlantic; Wakeham, 1982; Gagosian *et al.*, 1982); fatty acids and wax esters at 1500 m at VERTEX I (California Current site; Wakeham *et al.*, 1984); and fatty acids and wax esters at 2778 m at PARFLUX P (north central Pacific; Wakeham *et al.*, 1984). The increased fluxes of lipids in these samples do not appear to be from surface-dwelling zooplankton which sink from the surface waters. This is seen from detailed compositional analyses conducted on the lipid classes. For example, the wax ester compound distributions for PARFLUX P samples are shown in Fig. 11. The distribution of wax esters at 2778 m is markedly different from that in both shallower and deeper traps. Three compounds, 34:1, 34:2, and 36:2, comprise about 80% of total wax esters in the 2778-m sample whereas compound abundances in the other samples are more evenly distributed. It seems unlikely, therefore, that the wax esters in the 2778-m sample could have originated at a shallower depth; rather the data are interpreted as resulting from inputs of particulate matter rich in wax esters biosynthesized by mesopelagic zooplankton (Wakeham *et al.*, 1984). A similar explanation is consistent with the observed deep-water flux maxima observed for this and other lipids at the several sites listed above. Whether these flux increases result from passive input of truly sinking particulate material, such as carcasses of mesopelagic organisms, or from

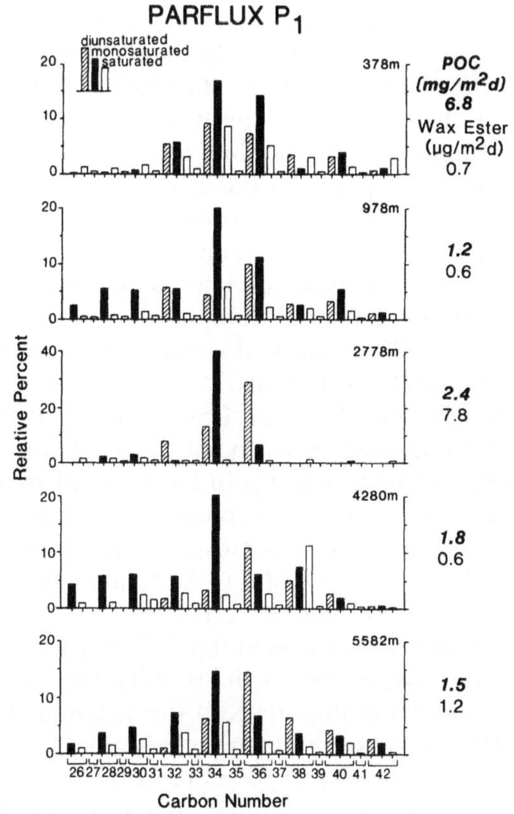

Figure 11. Wax ester distributions at the central Pacific PARFLUX P site. [Data plotted from Wakeham *et al.* (1984).]

active input of mesopelagic "swimmers" (Lee *et al.*, 1988) remains unknown.

3.4. Particle Size—Sinking versus Suspended Particles

Particulate matter in the ocean exists in a variety of size classes, and, although there must be exchange between various size classes, there are significant differences in organic composition of the different sizes of particles. This has been documented in comparisons of amino acids, sugars, pigments, and lipids in sediment-trap (large, sinking particles) and suspended (fine, suspended particles) samples. These differences are partly due to differences in source and partly due to the aggregation–disaggregation processes which occur within the water column.

Suspended particles were relatively enriched in the amino acids serine and glycine compared to sinking particles collected in field experiments in the eastern equatorial North Pacific by Lee and Cronin (1984). As mentioned earlier, this prevalence of glycine and serine on particles found deeper in the water column may be due to the persistence of these amino acids in diatom cell walls or to chemical or biological production during particle diagenesis. This would suggest that suspended particles have undergone more extensive alteration than the larger sinking particles since they are enriched in the more resistant amino acids, serine and glycine.

Lipid distributions are significantly different in sinking particles compared to suspended particles (Saliot *et al.*, 1982; Wakeham *et al.*, 1983; Gagosian *et al.*, 1983; Wakeham and Canuel, 1988; Wakeham and Lee, 1989). At the VERTEX IV site in the central Pacific gyre, Wakeham and Lee (1989) found sinking particles collected in sediment traps [PITs (particle interceptor traps); Fig. 12] to be enriched in sterols and triacylglycerols and depleted in fatty acids compared to suspended particles collected at the same depths by *in situ* filtration. At 1500-m depth, the lowest depth sampled, steroidal ketones were a significant part of the lipids in the sinking particles, but not in the suspended particles, whereas wax esters were much more important in the suspended particles. The increased abundance of steroidal ketones on sinking particles may be due to *in situ* conversion of steroidal alcohols to ketones; the wax esters were thought to be biosynthesized by mesopelagic microzooplankton readily sampled by *in situ* filtration.

Wakeham and Canuel (1988) and Wakeham and Lee (1989) have reported on comparative details of the lipid compound distribution in several sets of sinking and suspended particles. The strong signature of phytoplankton lipids in epipelagic suspended particles can be seen in the dominance of 14:0, 16:0, and 22:6 fatty acids and 24-methylcholesta-5,22*E*-dien-3β-ol, and 24-ethylcholest-5-en-3β-ol among the sterols (e.g., Fig. 7). This contrasts with the composition of compounds in large, sinking particles at the same locations, with zooplankton fatty acids (e.g., $18{:}1\Delta^9$) and sterols (cholesta-5,22*E*-dien-3β-ol and cholest-5-en-3β-ol) most abundant. Thus, the suspended particles tend to reflect a more phytoplankton-dominated source of lipids while the sinking particles more clearly reflect a zooplankton influence. This may, in part, reflect the sampling bias of *in situ* filtration versus sediment trapping. *In situ* filtration is more efficient for suspended particles, such as algal cells, while live zooplankton may be more able to avoid capture in the pump intake. Zooplankton fecal pellets, carcasses, and molts (as well as some "swimmers") may contribute a greater proportion of animal-derived lipids to trap samples.

Further evidence of differences in organic composition between sinking and suspended particles is seen in Fig. 13. Fatty acids in particles from the VERTEX II, III, and IV studies are plotted on a ternary diagram with saturated, mono- and diunsaturated, and polyunsaturated components as the apices. Sterols are plotted with C_{27}, C_{28}, and C_{29} components as the apices. Sinking particles are generally richer in mono- and diunsaturated fatty acids and C_{27} sterols than suspended particles. The suspended particles tend to contain more saturated and polyunsaturated fatty acid relative to sinking particles and more C_{28} and C_{29} sterols.

Recent advances in sampling strategy, due in part to the recognition that particle size classes can have distinctly different compositions, have led to further fractionation of the suspended particle pool during *in situ* filtration. At present, results from these samples are preliminary, but significant compositional differences between "small" suspended and "large" suspended particles are now also apparent. During a recent field program in the anoxic Cariaco Trench (Wakeham, 1990; see discussion in following sections), suspended particles were fractionated *in situ* into <53-μm and >53-μm classes, and separate analyses conducted on the two sets of samples. That the two particle size classes have different compositions, and thus perhaps different sources, is immediately apparent from gas chromatograms shown in Fig. 14 for the hydrocarbons obtained at 750 m in the Cariaco Trench. The <53-μm particles are dominated by a variety of branched hydrocarbons, presumably mostly of microbial origin (see below) while in the >53-μm particles, the same branched hydrocarbons are relatively minor components compared to the homologous series of *n*-alkanes over the C_{17}–C_{40} range.

3.5. Water Column Character

The characteristics of the water column through which particles pass can greatly affect the processes of degradation which particles undergo during transit to the seafloor. This is most clear for water columns with little or no oxygen, where microbial reactions typical of suboxic or anoxic environments occur. One example of this is found in analyses of sediment-trap and suspended particles from the eastern tropical North Pacific Ocean. This area off the west coast of Mexico is characterized by a shallow but strong oxygen minimum (Fig. 9); this O_2 minimum is known to be an area of enhanced microbial activity (Garfield *et al.*, 1983). As mentioned earlier, Lee and Cronin (1984) found an inverse correlation between particulate ornithine concentrations and concentra-

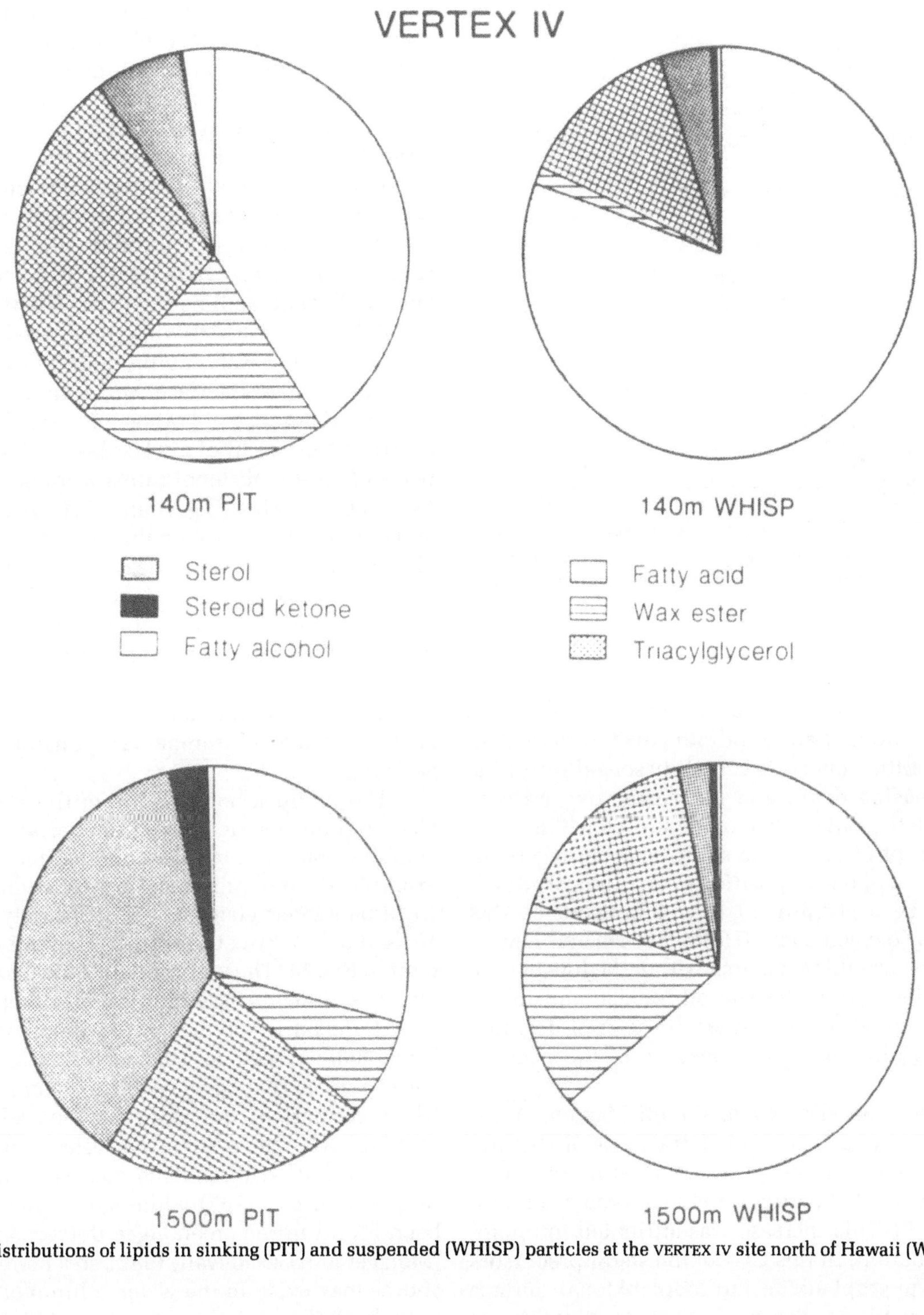

Figure 12. Distributions of lipids in sinking (PIT) and suspended (WHISP) particles at the VERTEX IV site north of Hawaii (Wakeham and Lee, 1989).

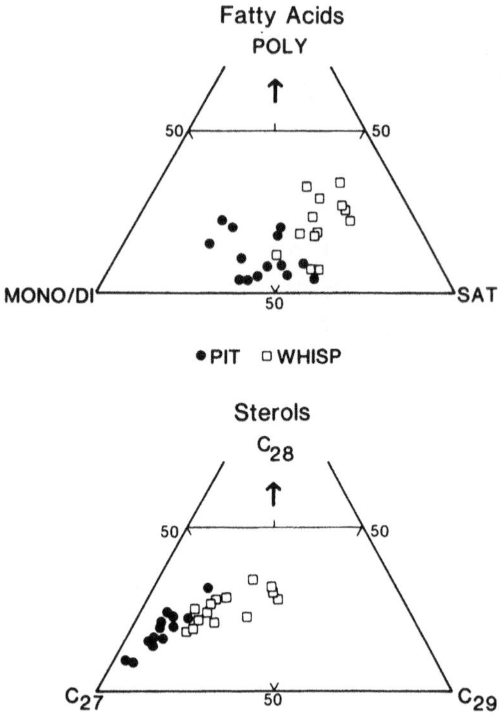

Figure 13. Ternary diagrams of saturated vs. mono/diunsaturated vs. polyunsaturated fatty acids and C_{27} vs. C_{28} vs. C_{29} sterols for sinking (●) and suspended (□) particles at the VERTEX II, III, and IV sites.

tions of dissolved oxygen at this site (Fig. 9). In the Baltic Sea, Mopper and Lindroth (1982) observed an increase in the concentration of dissolved ornithine with increasing depth and decreasing oxygen content, which they attributed to a depth-related increase in heterotrophic activity. In addition to being a product of microbial decomposition of arginine, ornithine may also be a constituent of microorganisms that inhabit low-oxygen zones (Holm-Hansen and Lewin, 1965). Thus, ornithine maxima within the low-oxygen zones at the VERTEX II/III site off Mexico and in the Baltic Sea could be due either to bacterial degradation of arginine or to the presence of the bacteria themselves.

At the same VERTEX II/III site off Mexico, Wakeham (1987) found increased abundances of sterenes (steroidal hydrocarbons) associated with both sinking and suspended particles in the oxygen minimum zone (Fig. 9). This increase was attributed to microbial degradation, in this case of the sterol precursors formed by phytoplankton and zooplankton in surface waters. Formation of sterenes in Recent sediments has been well established (Dastillung and Albrecht, 1977; Gagosian and Farrington, 1978), but the finding of such high abundances of sterenes in the water column is apparently unique to this oceanographic environment.

Another example where the extent of oxygenation influences the composition of particulate organic matter is in the Cariaco Trench, an anoxic marine basin in the Caribbean Sea north of Venezuela. In a study of the organic chemical composition of suspended particulate matter in the water column of the Cariaco Trench, Wakeham and co-workers (Wakeham and Ertel, 1988; Wakeham, 1990) detected a variety of microbial products immediately below the oxic–anoxic interface. For example, the microbial conversion of ring-unsaturated sterols (stenols) to ring-saturated sterols (stanols) is well documented in anoxic Recent sediments (e.g., Nishimura, 1977, 1978). In the Cariaco Trench, this transformation occurs within the water column, and primarily below the oxic–anoxic interface at 275 m as indicated by increased stanol:stenol ratios, illustrated in Fig. 15 for 5α(H)-cholestan-3β-ol:cholest-5-en-3β-ol. Such high particulate stanol:stenol ratios were not observed in oxic oceanic sites (Gagosian *et al.*, 1982; Wakeham and Canuel, 1988); even in the oxygen minimum zone of the VERTEX II/III site, ratios were markedly lower. The high stanol:stenol ratios were found predominately in the <53-μm particle size class (particles were fractionated *in situ* into <53-μm and >53-μm fractions; Fig. 15), suggesting that the smaller particles contain the microorganisms and/or the biogeochemical microenvironment responsible for this conversion.

The hydrocarbon fraction of the <53-μm particles from the anoxic zone of the Cariaco Trench also contains abundant markers of microbial activity. For example, the most abundant hydrocarbons are 2,6,10, 15,19-pentamethyleicosane (PME) and lycopane (2,6, 10,14,19,23,27,31-octamethyldotriacontane) as is evident in Fig. 14. These compounds have been found in anoxic sediments (Dastillung and Corbet, 1975; Brassell *et al.*. 1981; Volkman *et al.*, 1986; Wakeham and Ertel, 1988; Wakeham, 1990). PME is produced by a variety of methanogens (Holzer *et al.*, 1979), most likely as part of the isoprenoid-based membrane structure for which Archaebacteria are now noted. A carotenoid source has been suggested for lycopane (Murphy *et al.*, 1967), although a lycopodiene has been found in the green algae *Botryococcus braunii* (Metzger and Casadevall, 1987), so a more direct algal source may exist. In the water column of the Cariaco Trench, PME and lycopane were most abundant immediately below the oxic–anoxic boundary, where they constituted up to 30% of total hydrocarbons on <53-μm particles. Neither was abundant in the >53-

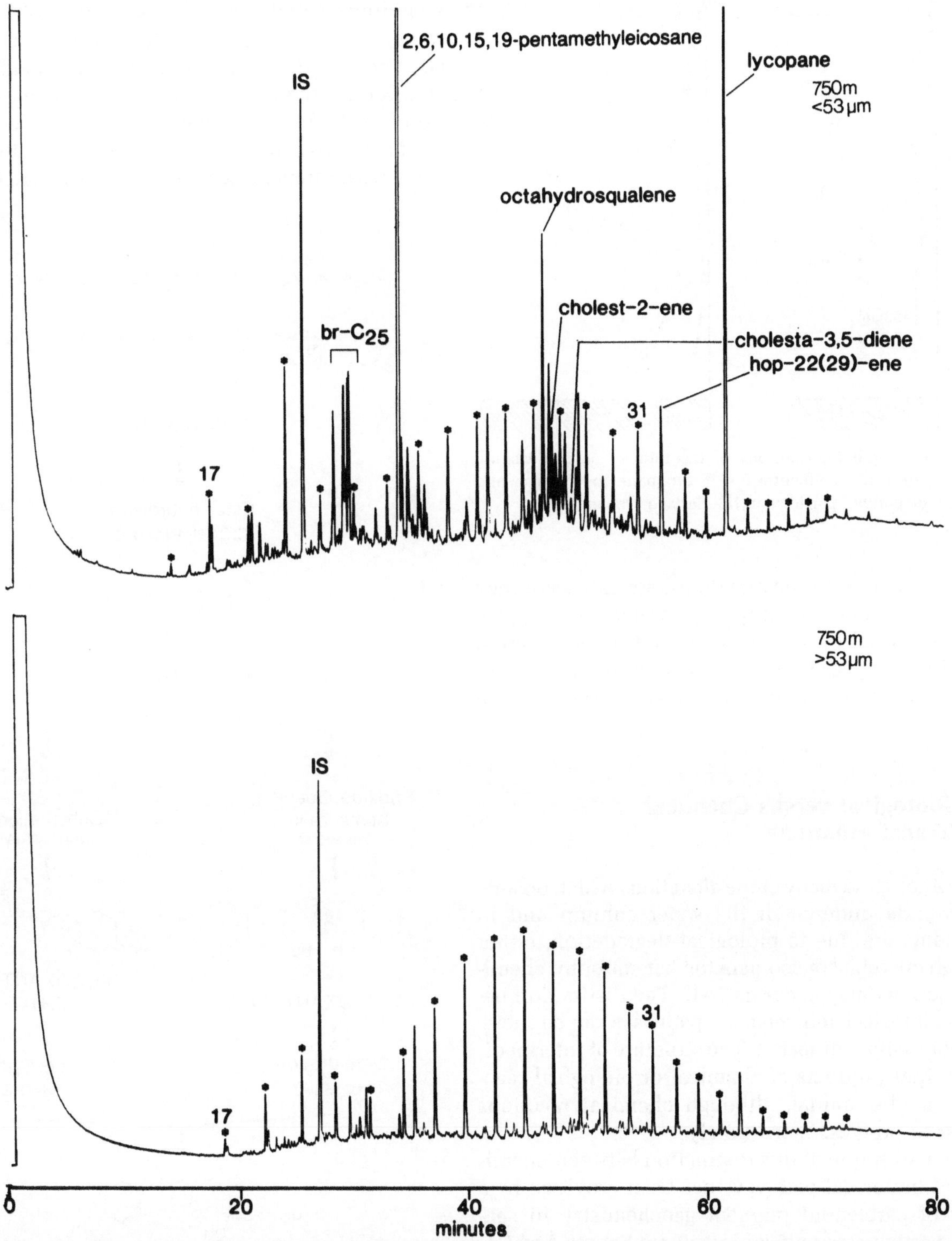

Figure 14. Gas chromatograms (DB-5, 100–320°C at 3° /min^{-1}; hydrogen carrier at 0.5 kg /cm^{-2}; Carlo Erba 4160, FID, on-column injection) of hydrocarbons isolated from <53-μm and >53-μm suspended particles from 750 m in the anoxic bottom waters of the Cariaco Trench.

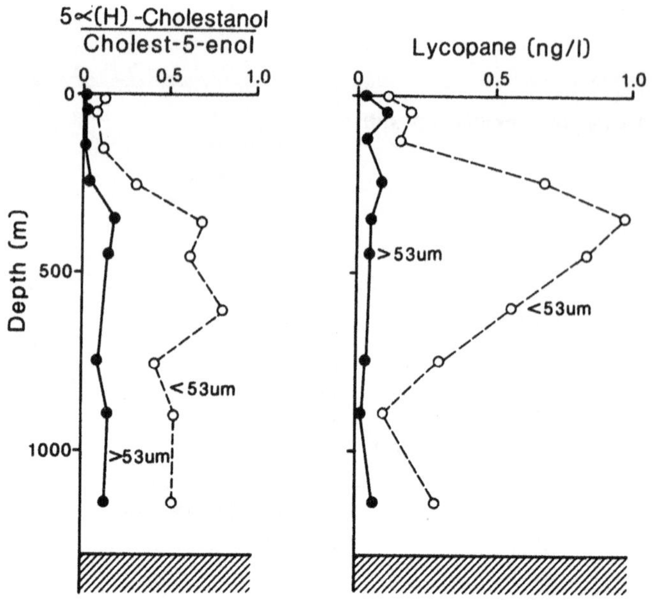

Figure 15. Depth distributions of the ratio of 5α-cholestanol/cholest-5-enol and concentration of lycopane in <53-μm and >53-μm suspended particles in the Cariaco Trench.

μm size fraction in the anoxic waters, nor were they abundant in the oxic epipelagic waters. The appropriate bacterial or algal sources and transformation conditions appear to be associated with the <53-μm particles uniquely found in the anoxic Cariaco Trench waters.

3.6. Biological versus Chemical Transformations

Although so many of the alterations which organic compounds undergo in the water column and in sediments are due to biological degradation, either through microbial or zooplankton heterotrophy, chemical reactions may occur as well. The distinction between biological and chemical pathways can be subtle and not easily delineated. The structure of intermediate or final products of chemical or biological reactions can be similar, although chemical reactions appear to progress more slowly.

One example of this distinction between chemical and biological pathways has been provided by a study of carotenoid pigment geochemistry in seawater particulates and sediments by Repeta and Gagosian (1982, 1984). Their studies strongly suggest that the carotenoid esters fucoxanthin and peridinin, which are produced by phytoplankton, are hydrolyzed in the guts of zooplankton and other herbivores to the alcohols fucoxanthinol (Fig. 16) and peridininol. These four compounds, the esters and corresponding alcohols, are the major carotenoid pigments transported by particles to the seafloor. Repeta and Gagosian further suggested that microbially mediated, rather than acid-catalyzed chemical, dehydration occurs. Microbial dehydration of fucoxanthinol yields fucoxanthinol-5′-dehydrate (Fig. 16), which was present in particles they analyzed from Buzzards

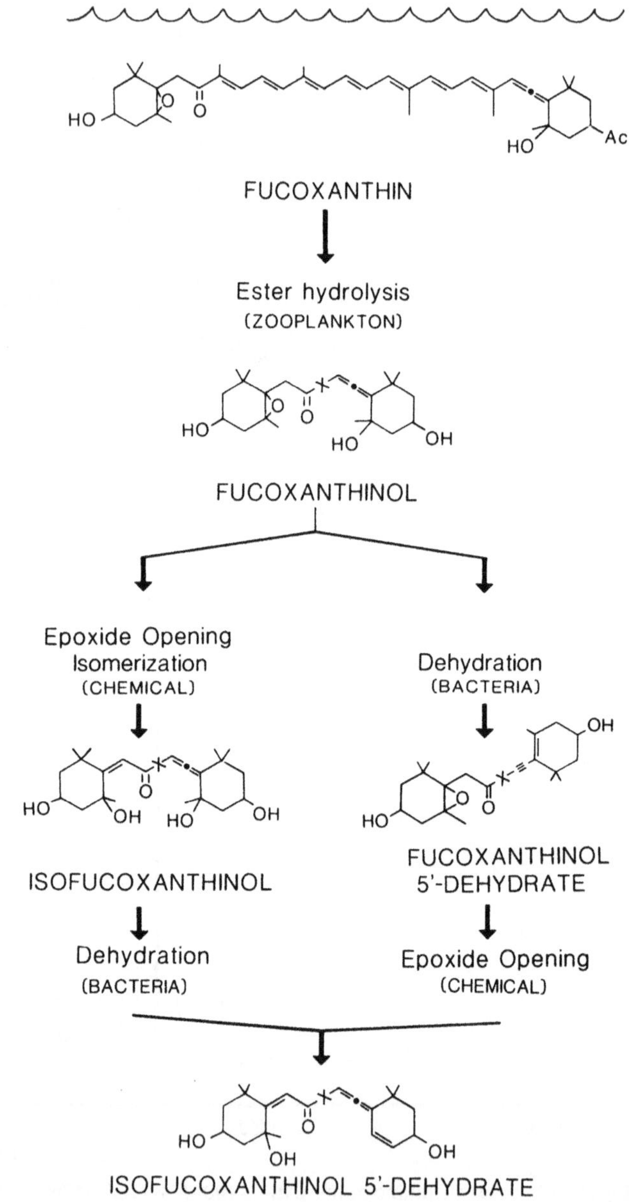

Figure 16. Transformation scheme proposed by Repeta and Gagosian (1984) and Repeta and Frew (1988) for conversion of algal fucoxanthin to isofucoxanthinol 5′-dehydrate in the water column and surface sediments.

Bay, Massachusetts. Acid-catalyzed chemical dehydration would yield a mixture of acetylenic and allenic isomers of fucoxanthinol-5′-dehydrate, but only the acetylenic isomer was observed (Repeta and Frew, 1988). In an alternate pathway, chemical epoxide opening and isomerization of fucoxanthinol would form isofucoxanthinol. However, no isofucoxanthinol was observed in any samples analyzed. Subsequent chemical epoxide opening and isomerization of fucoxanthinol-5′-dehydrate (or microbial dehydration of isofucoxanthinol) would yield the final product isofucoxanthinol-5′-dehydrate. Very small amounts of the dehydrates were found in sinking particles in the Peru upwelling region, and none in suspended particles. Since the dehydrates were present in zooplankton and anchovy fecal pellets, microbial transformations within the fecal pellets were presumed to be the source of dehydrates in sediment-trap samples, in which fecal pellets were important components. Opened epoxides found in near-bottom sediment traps in Buzzards Bay (Repeta and Gagosian, 1982) were probably associated with resuspended sediments, so the chemical reaction pathway seems limited to the sediments. The dehydration and epoxide-opening reactions were thought to be much slower, with time scales of years in sediments, while the more rapid ester hydrolysis by zooplankton occurs in surface waters on a scale of days.

A similar distinction between chemical and biological pathways can be made for the diagenetic formation of sterenes in the water column (Wakeham, 1987) and in Recent sediments (Dastillung and Albrecht, 1977; Gagosian and Farrington, 1978). Acid-catalyzed chemical (and thermal) reactions of sterols produce abundant rearranged Δ^4 and Δ^5-sterenes in laboratory experiments (Rhead *et al.*, 1971). Compounds with this unsaturation pattern (Δ^4, Δ^5, $\Delta^{4,22}$, and $\Delta^{5,22}$) are major components of the sterenes in immature Monterey shale (Giger and Schaffner, 1981). However, in both water column and sediment samples, the dominant sterenes are Δ^2, $\Delta^{3,5}$, and $\Delta^{3,5,22}$ compounds. No Δ^4- and Δ^5-sterenes have been found in water column particles or Recent sediments, and Δ^2-, $\Delta^{3,5}$-, and $\Delta^{3,5,22}$-sterenes are not major components in laboratory experiments or immature sediments. Thus, it is generally accepted that the Δ^2-, $\Delta^{3,5}$-, and $\Delta^{3,5,22}$-sterenes are products of microbial transformation of precursor sterols. These sterenes, in turn, are intermediates for chemical reactions which yield the rearranged Δ^4, Δ^5, and $\Delta^{13(17)}$-sterenes and the diasterenes found in older sediments (Dastillung and Albrecht, 1977).

4. Concluding Remarks

A number of biological, chemical, and physical processes alter the organic chemical composition of particulate matter as it is transported through the water column. These diagenetic processes can be observed in the water column by following changes in the particle composition with depth. Because only a small percentage of the organic matter biosynthesized in surface waters reaches the seafloor, the composition of particulate organic matter in the sediments can be very different from that which was initially biosynthesized. The alteration processes which change this composition can be complex and include metabolic transformations, and to a lesser extent chemical reactions, which depend on seasonal and spatial variations in the character of the water column.

During the past decade, there have been many studies examining the source, transport, and alteration processes affecting particulate matter in the ocean, and we have summarized many of the results from these studies in this chapter. However, as we have shown, much of the work has focused on particulate matter and its behavior. There is still much to be done in characterizing the organic composition and dynamics of the more abundant dissolved organic matter (DOM) pool. Without information on the sources and behavior of DOM, we cannot hope to fully understand the cycling of organic matter—both particulate and dissolved—in the marine water column.

ACKNOWLEDGMENTS. This research and the preparation of this chapter were supported by grants from the National Science Foundation and the Office of Naval Research.

References

Alam, M., Sansing, T. B., Busby, E. L., Martinez, D. R., and Ray, S. M., 1979, Dinoflagellate sterols I: Sterol composition of the dinoflagellates of *Gonyaulax* species, *Steroids* **33**:197–201.

Angel, M. V., 1984, Detrital organic fluxes through pelagic ecosystems, in: *Flows of Energy and Material in Marine Ecosystems* (M. J. R. Fasham, ed.), Plenum Press, New York, pp. 475–516.

de Baar, H. J. W., Farrington, J. W., and Wakeham, S. G., 1983, Vertical flux of fatty acids in the North Atlantic Ocean, *J. Mar. Res.* **41**:19–41.

Bacon, M. P., Huh, C. A., Fleer, A. P., and Deuser, W. G., 1985, Seasonality in the flux of natural radionuclides and plutonium in the deep Sargasso Sea, *Deep-Sea Res.* **32**:273–286.

Betzer, P. R., Showers, W. J., Laws, E. A., Winn, C. D., DiTullio, G. R., and Kroopnik, P. M., 1984, Primary productivity and particle

fluxes on a transect of the equator at 153°W in the Pacific Ocean, *Deep-Sea Res.* **31**:1–11.
Bishop, J. K. B., Stepien, J. C., and Wiebe, P. H., 1986, Particulate matter distributions, chemistry and flux in the Panama Basin: Response to environmental forcing, *Prog. Oceanogr.* **17**:1–59.
Boon, J. J., Rijpstra, W. I. C., de Lang, F., and de Leeuw, J. W., 1979, Black Sea sterol—a molecular fossil for dinoflagellate blooms, *Nature* **277**:125–127.
Boyd, C. M., Smith, S. L., and Cowles, T. J., 1980, Grazing patterns of copepods in the upwelling system off Peru, *Limnol. Oceanogr.* **25**:583–596.
Brassell, S. C., Wardroper, A. M. K., Thomson, I. D., Maxwell, J. R., and Eglinton, G., 1981, Specific acyclic isoprenoids as biological markers of methanogenic bacteria in marine sediments, *Nature* **290**:693–696.
Burdige, D. J., and Martens, C. S., 1988, Biogeochemical cycling in an organic-rich coastal marine basin: 10. The role of amino acids in sedimentary carbon and nitrogen cycling, *Geochim. Cosmochim. Acta* **52**:1571–1584.
Chuecas, L., and Riley, J. P., 1969, Component fatty acids of the total lipids of some marine phytoplankton, *J. Mar. Biol. Assoc. U.K.* **49**:97–116.
Cole, J., and Lee, C., 1986, Rapid metabolism of non-protein amino acids in seawater, *Biogeochemistry* **2**:299–312.
Copping, A. E., and Lorenzen, C. J., 1980, Carbon budget of a marine phytoplankton–herbivore system with carbon-14 as a tracer, *Limnol. Oceanogr.* **25**:873–882.
Corner, E. D. S., O'Hara, S. C. M., Neal, A. C., and Eglinton, G., 1986, Copepod faecal pellets and the vertical flux of biolipids, in: *The Biological Chemistry of Marine Copepods* (E. D. S. Corner and S. C. M. O'Hara, eds.), Oxford University Press, Oxford, pp. 260–322.
Cowey, C. B., and Corner, E. D. S., 1966, The amino acid composition of certain unicellular algae, and of the faecal pellets produced by *Calanus finmarchicus* when feeding on them, in: *Some Contemporary Studies in Marine Science* (H. Barnes, ed.), Allen and Unwin, London, pp. 225–231.
Dastillung, M., and Albrecht, P., 1977, Δ^2-Sterenes as diagenetic intermediates in sediments, *Nature* **269**:678–679.
Dastillung, M., and Corbet, B., 1975, La géochimie organique des sédiments marins profond, in: *Orgon II, Atlantique-N.E. Brésil*, (A. Combaz and R. Pelet, eds.), CNRS, Paris, pp. 295–326.
Degens, E. T., 1970, Molecular nature of nitrogenous compounds in seawater and recent marine sediments, in: *Organic Matter in Natural Waters*, (D. W. Hood, ed.), University of Alaska Press, Fairbanks, pp. 77–106.
DeLong, E. F., and Yayanos, A. A., 1985, Adaptation of the membrane lipids of a deep-sea bacterium to changes in hydrostatic pressure, *Science* **228**:1101–1103.
Deuser, W. G., 1986, Seasonal and interannual variations in deep-water particle fluxes in the Sargasso Sea and their relation to surface hydrography, *Deep-Sea Res.* **33**:225–246.
Deuser, W. G., Ross, E. H., and Anderson, R. F., 1981, Seasonality in the supply of sediment to the deep Sargasso Sea and implications for the rapid transfer of matter to the deep sea, *Deep-Sea Res.* **28A**:495–505.
Emerson, S., and Hedges, J. I., 1988, Processes controlling the organic carbon content of open ocean sediments, *Paleoceanography* **3**:621–634.
Eppley, R. W., and Peterson, B. J., 1979, Particulate organic matter flux and planktonic new production in the deep ocean, *Nature* **282**:677–680.
Fowler, S. W., and Knauer, G. A., 1986, Role of large particles in the transport of elements and organic compounds through the oceanic water column, *Prog. Oceanogr.* **16**:147–194.
Friebele, E., Shimoyama, A., and Ponnamperuma, C., 1980, Adsorption of protein and non-protein amino acids on a clay mineral: A possible role of selection in chemical evolution, *Mol. Evol.* **16**:269–278.
Furlong, E. T., and Carpenter, R., 1988, Pigment preservation and remineralization in oxic coastal marine sediments, *Geochim. Cosmochim. Acta* **52**:87–99.
Gagosian, R. B., and Farrington, J. W., 1978, Sterenes in surface sediments from the southwest African shelf and slope, *Geochim. Cosmochim. Acta* **42**:1091–1101.
Gagosian, R. B., Smith, S. O., and Nigrelli, G. E., 1982, Vertical transport of steroid alcohols and ketones measured in a sediment trap experiment in the equatorial Atlantic Ocean, *Geochim. Cosmochim. Acta* **46**:1163–1172.
Gagosian, R. B., Volkman, J. K., and Nigrelli, G. E., 1983, The use of sediment traps to determine sterol sources in coastal sediments off Peru, in: *Advances in Organic Geochemistry 1981* (M. Bjørøy *et al.*, eds.), John Wiley & Sons, New York, pp. 369–379.
Garfield, P. C., Packard, T. T., Friederich, G. E., and Codispoti, L. A., 1983, The occurrence of a subsurface particle maximum layer and associated increase in microbial activity in the secondary nitrite maximum of the northeastern tropical Pacific Ocean, *J. Mar. Res.* **41**:747–768.
Giger, W., and Schaffner, C., 1981, Unsaturated steroid hydrocarbons as indicators of diagenesis in immature Monterey shales, *Naturwissenschaften* **68**:37–39.
Gordon, D. C. J., 1970, Some studies on the distribution and composition of particulate organic carbon in the North Atlantic Ocean, *Deep-Sea Res.* **17**:233–243.
Gordon, D. C. J., 1971, Distribution of particulate organic carbon and nitrogen at an oceanic station in the central Pacific, *Deep-Sea Rea.* **18**:1127–1134.
Gowing, M. M., and Silver, M. W., 1985, Minipellets: A new and abundant size class of marine fecal pellets, *J. Mar. Res.* **43**: 395–418.
Harvey, H. R., Eglinton, G., O'Hara, S. C. M., and Corner, E. D. S., 1987, Biotransformation and assimilation of dietary lipids by *Calanus* feeding on a dinoflagellate, *Geochim. Cosmochim. Acta* **51**:3031–3040.
Hecky, R. E., Mopper, K., Kilham, P., and Degens, E. T., 1973, The amino acid and sugar composition of diatom cell-walls, *Mar. Biol.* **19**:323–331.
Hedges, J. I., Clark, W. A., and Cowie, G. L., 1988a, Organic matter sources to the water column and surficial sediments of a marine bay, *Limnol. Oceanogr.* **33**:1116–1136.
Hedges, J. I., Clark, W. A., and Cowie, G. L., 1988b, Fluxes and reactivities of organic matter in a marine bay, *Limnol. Oceanogr.* **33**:1137–1152.
Henrichs, S. M., 1980, Biogeochemistry of dissolved free amino acids in marine sediments, Ph.D. Thesis, Woods Hole Oceanographic Institution/Massachusetts Institute of Technology.
Henrichs, S. M., and Farrington, J. W., 1984, Peru upwelling region sediments near 15°S. 1. Remineralization and accumulation of organic matter. *Limnol. Oceanogr.* **29**:1–19.
Henrichs, S. M., Farrington, J. W., and Lee, C., 1984, Peru upwelling region sediments near 15°S. 2. Dissolved free and total hydrolyzable amino acids. *Limnol. Oceanogr.* **29**:20–34.
Holm-Hansen, O., and Lewin, R. A., 1965, Bound ornithine in certain flexibacteria and algae, *Physiol. Plant.* **18**:418–423.
Holzer, G., Oro, J., and Tornabene, T. G., 1979, Gas chromato-

graphic–mass spectrometric analysis of neutral lipids from methanogenic and thermoacidophilic bacteria, *J. Chromatogr.* **186:**795–809.

Honjo, S., 1976, Coccoliths: Production, transportation and sedimentation, *Mar. Micropaleo.* **1:**65–79.

Honjo, S., 1980, Material fluxes and modes of sedimentation in the mesopelagic and bathypelagic zones, *J. Mar. Res.* **38:**53–97.

Ittekkot, V., Degens, E. T., and Brockmann, U., 1982, Monosaccharide spectra of acid-hydrolyzable carbohydrates in particulate matter during a plankton bloom, *Limnol. Oceanogr.* **27:** 711–716.

Ittekkot, V., Deuser, W. G., and Degens, E. T., 1984a, Seasonality in the fluxes of sugars, amino acids, and amino sugars to the deep ocean: Sargasso Sea, *Deep-Sea Res.* **31:**1057–1069.

Ittekkot, V., Degens, E. T., and Honjo, S., 1984b, Seasonality in the fluxes of sugars, amino acids, and amino sugars to the deep ocean: Panama Basin, *Deep-Sea Res.* **31:**1071–1083.

Jenkins, W., 1982, Oxygen utilization rates in the North Atlantic subtropical gyre and primary production in oligotrophic systems, *Nature* **300:**246–248.

Joseph, J. D., 1975, Identification of 3,6,9,12,15-octadecapentaenoic acid in laboratory-cultured photosynthetic dinoflagellates, *Lipids* **10:**395–403.

Karl, D. M., and Knauer, G. A., 1984, Vertical distribution, transport, and exchange of carbon in the northeast Pacific Ocean: Evidence for multiple zones of biological activity, *Deep-Sea Res.* **31:**221–243.

Karl, D. M., Knauer, G. A., Martin, J. H., and Ward, B. B., 1984, Bacterial chemolithotrophy in the ocean is associated with sinking particles, *Nature* **309:**54–56.

King, J. D., and White, D. C., 1977, Muramic acid as a measure of microbial biomass in estuarine and marine samples, *Appl. Environ. Microbiol.* **33:**777–783.

King, K., Jr., 1974, Preserved amino acids from silicified protein in fossil radiolaria, *Nature* **252:**690–692.

Knauer, G. A., Martin, J. H., and Bruland, K. W., 1979, Fluxes of particulate carbon, nitrogen, and phosphorus in the upper water column of the northeast Pacific, *Deep-Sea Res.* **26A:** 97–108.

Lal, D., 1977, The oceanic microcosm of particles, *Science* **198:** 997–1009.

Lee, C., and Cronin, C., 1982, The vertical flux of particulate organic nitrogen in the sea: Decomposition of amino acids in the Peru upwelling area and the equatorial Atlantic, *J. Mar. Res.* **40:**227–251.

Lee, C., and Cronin, C., 1984, Particulate amino acids in the sea: Effects of primary productivity and biological decomposition, *J. Mar. Res.* **42:**1075–1097.

Lee, C., and Wakeham, S. G., 1988, Organic matter in seawater: Biogeochemical processes, in: *Chemical Oceanography*, Vol. 9 (J. P. Riley, ed.), Academic Press, New York, pp. 1–51.

Lee, C., Wakeham, S. G., and Farrington, J. W., 1983, Variations in the composition of particulate organic matter in a time-series sediment trap, *Mar. Chem.* **13:**181–194.

Lee, C., McKenzie, J. A., and Sturm, M., 1987, Carbon isotope fractionation and changes in the flux and composition of particulate matter resulting from biological activity during a sediment trap experiment in Lake Greifen, Switzerland, *Limnol. Oceanogr.* **32:**83–96.

Lee, C., Wakeham, S. G., and Hedges, J. I., 1988, The measurement of oceanic particle flux—are "swimmers" a problem?, *Oceanography* **1:**34–36.

Mackenzie, F. T., 1981, Global carbon cycle: Some minor sinks for CO_2, in: *Flux of Organic Carbon by Rivers to the Ocean* (G. E. Likens, F. T. Mackenzie, J. E. Richey, J. R. Sedell, and K. K. Turekian, eds.), U.S. Department of Energy, Washington, D.C., pp. 360–384.

Marlowe, I. T., Green, J. C., Neal, A. C., Brassell, S. C., Eglinton, G., and Course, P. A., 1984, Long-chain (n-C_{37}–C_{39}) alkenones in the Prymnesiophyceae. Distribution of alkenones and other lipids and their taxonomic significance, *Br. Phycol. J.* **19:** 203–216.

Martin, J. H., Knauer, G. A., Karl, D. M., and Broenkow, W. W., 1987, VERTEX: Carbon cycling in the northeast Pacific, *Deep-Sea Res.* **34:**267–285.

McCave, I. N., 1975, Vertical flux of particles in the ocean, *Deep-Sea Res.* **22:**491–502.

McCave, I. N., 1984, Size spectra and aggregation of suspended particles in the ocean, *Deep-Sea Res.* **31:**329–352.

Menzel, D. W., 1974, Primary productivity, dissolved and particulate organic matter and the sites of oxidation of organic matter, in: *The Sea*, Vol. 5 (E. D. Goldberg, ed.), John Wiley & Sons, New York, pp. 657–678.

Metzger, P., and Casadevall, E., 1987, Lycopadiene, a tetraterpenoid hydrocarbon from new strains of the green alga *Botryococcus braunii*, *Tetrahedron Lett.* **28:**3931–3934.

Meybeck, M., 1982, Carbon, nitrogen and phosphorus transport by world rivers, *Am. J. Sci.* **282:**401–450.

Mopper, K., and Lindroth, P., 1982, Diel and depth variations in dissolved free amino acids and ammonium in the Baltic Sea determined by shipboard HPLC analyses, *Limnol. Oceanogr.* **27:**336–347.

Moriarty, D. J. W., 1977, Improved method using muramic acid to estimate biomass of bacteria in sediments, *Oecologia* **26:** 317–323.

Murphy, M. T. J., McCormick, A., and Eglinton, G., 1967, Perhydro-β-carotene in Green River Shale, *Science* **157:**1040–1042.

Neal, A. C., Prahl, F. G., Eglinton, G., O'Hara, S. C. M., and Corner, E. D. S., 1986, Lipid changes during a planktonic feeding sequence involving unicellular algae, *Elminius* nauplii, and adult *Calanus*, *J. Mar. Biol. Assoc. U.K.* **66:**1–13.

Nichols, P. D., Jones, G. J., de Leeuw, J., and Johns, R. B., 1984, The fatty acid and sterol composition of two marine dinoflagellates, *Phytochemistry* **23:**1043–1047.

Nishimura, M., 1977, Origin of stanols in young lacustrine sediments, *Nature* **270:**711–712.

Nishimura, M., 1978, Geochemical characteristics of the high reduction zone of stenols in Suwa sediments and the environmental factors controlling the conversion of stenols to stanols, *Geochim. Cosmochim. Acta* **42:**349–357.

Orcutt, D. M., and Patterson, G. W., 1975, Sterol, fatty acid, and elemental composition of diatoms grown in chemically defined media, *Comp. Biochem. Physiol.* **50B:**579–583.

Pace, M. L., Knauer, G. A., Karl, D. M., and Martin, J. H., 1987, Primary production, new production and vertical flux in the eastern Pacific Ocean, *Nature* **325:**803–805.

Parsons, T. R., 1975, Particulate organic carbon in the sea, in: *Chemical Oceanography*, Vol. 2 (J. P. Riley and G. Skirrow, eds.), Academic Press, Oxford, pp. 365–383.

Pilskaln, C. H., and Honjo, S., 1987, The fecal pellet fraction of biogeochemical particle fluxes to the deep sea, *Glob. Biogeochem. Cycles* **1:**31–48.

Piorreck, M., and Pohl, P., 1984, Formation of biomass, total protein, chlorophylls, lipids and fatty acids in green and blue-green algae during one growth phase, *Phytochemistry* **23:** 217–223.

Piorreck, M., Baasch, K. H., and Pohl, P., 1984, Biomass production, total protein, chlorophylls, lipids and fatty acids of freshwater green and blue-green algae under different nitrogen regimes, *Phytochemistry* **23**:207–216.

Platt, T., and Harrison, W. G., 1985, Biogenic fluxes of carbon and oxygen in the ocean, *Nature* **318**:55–58.

Prahl, F. G., Bennett, J. T., and Carpenter, R., 1980, The early diagenesis of aliphatic hydrocarbons and organic matter in sedimentary particulates from Dabob Bay, Washington, *Geochim. Cosmochim. Acta* **44**:1967–1976.

Prahl, F. G., Eglinton, G., Corner, E. D. S., O'Hara, S. C. M., and Forsberg, T. E. V., 1984a, Changes in plant lipids during passage through the gut of *Calanus*, *J. Mar. Biol. Assoc. U.K.* **64**: 317–334.

Prahl, F. G., Eglinton, G., Corner, E. D. S., and O'Hara, S. C. M., 1984b, Copepod fecal pellets as a source of dihydrophytol in marine sediments, *Science* **224**:1235–1237.

Prahl, F. G., Eglinton, G., Corner, E. D. S., and O'Hara, S. C. M., 1985, Faecal lipids released by fish feeding on zooplankton, *J. Mar. Biol. Assoc. U.K.* **65**:547–560.

Rechka, J. A., and Maxwell, J. R., 1988, Characterization of alkenone temperature indicators in sediments and organisms, in: *Advances in Organic Geochemistry 1987* (L. Mattavelli and L. Novelli, eds.), Pergamon Press, New York, *Org. Geochem.* **13**: 727–734.

Repeta, D. J., and Frew, N. M., 1988, Carotenoid dehydrates in recent marine sediments. The structure and synthesis of fucoxanthin dehydrate, *Org. Geochem.* **12**:469–477.

Repeta, D. J., and Gagosian, R. B., 1982, Carotenoid transformations in coastal marine waters, *Nature* **295**:51–54.

Repeta, D. J., and Gagosian, R. B., 1984, Transformation reactions and recycling of carotenoids and chlorins in the Peru upwelling region (15°S, 75°W), *Geochim. Cosmochim. Acta* **48**:1265–1277.

Rhead, M. M., Eglinton, G., and Draffan, G. H., 1971, Hydrocarbons produced by the thermal alteration of cholesterol under conditions simulating maturation of sediments, *Chem. Geol.* **8**: 277–297.

Romankevich, E. A., 1984, *Geochemistry of Organic Matter in the Ocean*, Springer-Verlag, New York.

Rubenstein, I., and Goad, L. J., 1974, Occurrence of (24S)-24-methylcholesta-5,22*E*-dien-3β-ol in the diatom *Phaeodactylum tricornutum*, *Phytochemistry* **13**:485–487.

Russell, N. J., 1983, Adaption to temperature in bacterial membranes, *Biochem. Soc. Trans.* **11**:333–335.

Saliot, A., Goutx, M., Fevrier, A., Tusseau, D. and Andrie, D., 1982, Organic sedimentation in the water column in the Arabian Sea: Relationship between the lipid composition of small and large-size, surface and deep particles, *Mar. Chem.* **11**:257–278.

Salton, M., 1960, *Microbial Cell Walls*, John Wiley & Sons, New York.

Schroeder, R. A., 1975, Absence of beta-alanine and gamma-aminobutyric acid in cleaned foraminiferal shells: Implication for use as a chemical criterion to indicate removal of nonindigenous amino acid contaminants, *Earth Plant. Sci. Lett.* **25**: 274–278.

SCOPE, 1979, *The Global Carbon Cycle* (B. Bolin, E. T. Degens, S. Kempe, and P. Ketner, eds.), John Wiley & Sons, New York.

Shuman, F. R., and Lorenzen, C. J., 1977, Quantitative degradation of chlorophyll by a marine herbivore, *Limnol. Oceanogr.* **20**: 580–586.

Siezen, R. J., and Mague, T. H., 1978, Amino acids in suspended particulate matter from oceanic and coastal waters of the Pacific, *Mar. Chem.* **6**:215–231.

Simpson, W. R., 1982, Particulate matter in the oceans—sampling methods, concentration, size distribution and particle dynamics, *Oceanogr. Mar. Biol. Annu. Rev.* **20**:119–172.

Staresinic, N., 1978, The vertical flux of particulate organic matter in the Peru upwelling as measured with a free-drifting sediment trap, Ph.D. Thesis, Woods Hole Oceanographic Institution/Massachusetts Institute of Technology.

Suess, E., 1980, Particulate organic carbon flux in the oceans—surface productivity and oxygen utilization, *Nature* **288**: 260–263.

Tanoue, E., and Handa, N., 1987, Monosaccharide composition of marine particles and sediments from the Bering Sea and northern North Pacific, *Oceanol. Acta* **10**:91–99.

Tanoue, E., Handa, N., and Sakugawa, H., 1982, Difference of the chemical composition of organic matter between fecal pellet of *Euphausia superba* and its feed, *Dunaliella tertiolecta*, *Trans. Tokyo Univ. Fish.* **5**:189–196.

Urrerre, M. A., and Knauer, G. A., 1981, Zooplankton fecal pellet fluxes and vertical transport of particulate organic material in the pelagic environment, *J. Plank. Res.* **3**:369–387.

Van Vleet, E. S., and Quinn, J. G., 1979, Diagenesis of marine lipids in ocean sediments, *Deep-Sea Res.* **26A**:1225–1236.

Vinogradov, M. E., 1955, Vertical migrations of zooplankton and their importance for the nutrition of abyssal pelagic fauna, *Tru. Inst. Oceanol.* **13**:71–76.

Volkman, J. K., 1986, A review of sterol markers for marine and terrigenous organic matter, *Org. Geochem.* **9**:83–99.

Volkman, J. K., Corner, E. D. S., and Eglinton, G., 1980a, Transformations of biolipids in the marine food web and in underlying bottom sediments, in: *Colloques Internationaux du C.N.R.S. No. 293—Biogéochimie de la Matière Organique à l'Interface Eau–Sédiment Marin*, Editions CNRS Paris, pp. 185–197.

Volkman, J. K., Eglinton, G. and Corner, E. D. S., 1980b, Sterols and fatty acids of the marine diatom *Biddulphia sinensis*, *Phytochemistry* **19**:1809–1813.

Volkman, J. K., Smith, D. J., Eglinton, G., Forsberg, T. E. V., and Corner, E. D. S., 1981, Sterol and fatty acid composition of four marine Haptophycean algae, *J. Mar. Biol. Assoc. U.K.* **61**: 509–527.

Volkman, J. K., Farrington, J. W., Gagosian, R. B., and Wakeham, S. G., 1983, Lipid composition of coastal marine sediments from the Peru upwelling region, in: *Advances in Organic Geochemistry 1981* (M. Bjøroy et al., eds.), John Wiley & Sons, New York, pp. 228–240.

Volkman, J. K., Allen, D. I., Stevenson, P. L., and Burton, H. R., 1986, Bacterial and algal hydrocarbons from a saline Antarctic lake, Ace Lake. *Org. Geochem.* **10**:671–681.

Wakeham, S. G., 1982, Organic matter from a sediment trap experiment in the equatorial North Atlantic: Wax esters, steryl esters, triacylglycerols and alkyldiacylglycerols, *Geochim. Cosmochim. Acta* **46**:2239–2257.

Wakeham, S. G., 1985, Wax esters and triacylglycerols in sinking particulate matter in the Peru upwelling area (15°S, 75°W), *Mar. Chem.* **17**:213–235.

Wakeham, S. G., 1987, Steroid geochemistry in the oxygen minimum zone of the eastern tropical North Pacific Ocean, *Geochim. Cosmochim. Acta* **51**:3051–3069.

Wakeham, S. G., 1990, Algal and bacterial hydrocarbons in particulate matter and interfacial sediment of the Cariaco Trench, *Geochim. Cosmochim. Acta* **54**:1325–1336.

Wakeham, S. G., and Canuel, E. A., 1986, Lipid composition of the pelagic crab *Pleuroncodes planipes*, its feces, and sinking particulate organic matter in the equatorial North Pacific Ocean, *Org. Geochem.* **9**:331–343.

Wakeham, S. G., and Canuel, E. A., 1988, Organic geochemistry of particulate matter in the eastern tropical North Pacific Ocean: Implications for particle dynamics, *J. Mar. Res.* **46:**183–213.

Wakeham, S. G., and Ertel, J. R., 1988, Diagenesis of organic matter in suspended particles and sediments in the Cariaco Trench, in: *Advances in Organic Geochemistry 1987* (L. Mattavelli and L. Novelli, eds.), *Org. Geochem.* **13:**815–822.

Wakeham, S. G., and Lee, C., 1989, Organic geochemistry of particulate matter in the ocean: The role of particles in oceanic sedimentary cycles, *Org. Geochem.* **14:**83–96.

Wakeham, S. G., Farrington, J. W., Gagosian, R. B., Lee, C., de Baar, H., Nigrelli, G. E., Tripp, B. W., Smith, S. O., and Frew, N. M., 1980, Organic matter fluxes from sediment traps in the equatorial Atlantic Ocean, *Nature* **286:**789–800.

Wakeham, S. G., Farrington, J. W., and Volkman, J. K., 1983, Fatty acids, wax esters, triacylglycerols and alkyldiacylglycerols associated with particles collected in sediment traps in the Peru upwelling, in: *Advances in Organic Geochemistry 1981* (M. Bjørøy *et al.*, eds.), John Wiley & Sons, New York, pp. 185–197.

Wakeham, S. G., Lee, C., Farrington, J. W., and Gagosian, R. B., 1984, Biogeochemistry of particulate organic matter in the oceans: Results from sediment trap experiments, *Deep-Sea Res.* **31:** 509–528.

Wefer, G., Suess, E., Balzer, W., Liebezeit, G., Müller, P. J., Ungerer, C. A., and Zenk, W., 1982, Fluxes of biogenic components from sediment trap deployment in circumpolar waters of the Drake Passage, *Nature* **299:**145–147.

Weliky, K., 1983, Clay organic associations in marine sediments: Carbon, nitrogen, and amino acids in the fine grained fractions, M.S. Thesis, Oregon State University.

Welschmeyer, N. A., and Lorenzen, C. J., 1985, Chlorophyll budgets: Zooplankton grazing and phytoplankton growth in a temperate fjord and the Central Pacific gyres, *Limnol. Oceanogr.* **30:**1–21.

Whelan, J. K., 1977, Amino acids in a surface sediment core of the Atlantic abyssal plain, *Geochim. Cosmochim. Acta* **41:**803–810.

Wirsen, C. O., Jannasch, H. W., Wakeham, S. G., and Canuel, E. A., 1987, Membrane lipids of a psychrophilic and barophilic deep-sea bacterium, *Curr. Microbiol.* **14:**319–322.

Chapter 7

Organic Matter at the Sediment–Water Interface

LAWRENCE M. MAYER

1. Introduction

This chapter will consider the term "sediment–water interface" as a double entendre and discuss the interface both as the horizontal zone where the water column meets the sediment column and as the wetted surface of mineral grains which occur in this horizontal zone. Hereafter, the first definition will be termed the "sediment–water interface," and the second the "mineral–water interface." The interface in both of these senses is a zone of concentration of organic matter, albeit for quite different reasons.

2. Organic Material at the Mineral–Water Interface

2.1. Evidence for Mineral–Organic Matter Association

In most sediments, organic matter makes up a small fraction (<10% w/w) of the material present. The mineral grains that make up the bulk of the sediment are associated in various ways with organic matter. The modes of these associations have received far less attention than have the chemical structures of the organic material itself. Is organic matter present mainly as organic particulates diluted by mineral grains, or is it present in a more dispersed form?

Most organic matter in the oceans originates as tissue material, which in some organisms is associated with a mineral skeleton. Only in a small fraction of marine sediments (e.g., those receiving large inputs of vascular plant matter) is this tissue form preserved. Microscopic observations of the sediment–water interface indicate numerous organic materials related to the high levels of biological activity in this zone (e.g., Johnson, 1974). Detrital fragments of organisms and their excreta abound. Histochemical observations show a marked increase at the interface of organic material sufficiently concentrated to stain; this pattern is particularly marked for lipid and protein stains in shallow-water sediments (Fig. 1; data from Johnson, 1977). Periodic acid–Schiff reagent (PAS) staining, indicative of carbohydrates, shows less intense enrichments at the interface. These data imply that substantial amounts of organic matter occur as discrete organic particulates enmeshed in a matrix of minerals. The quantitative determination of the fraction of total organic matter contained in these recognizable detritus fractions has received little attention. The most suggestive data addressing this question are those deriving from heavy-liquid separations, in which discrete organic particulates of density much lower than that of mineral grains are centrifugally separated in liquids of density 1.5–2.1 g cm^{-3} (e.g., Prahl and Carpenter, 1983; Ertel and Hedges, 1985; Murdoch *et al.*, 1986; Gershanovich and Zaslavskiy, 1983). The few studies of this type indicate that a minor or insignificant fraction of the total organic matter partitions into the low-density phase, or discrete organic particulates—never more

LAWRENCE M. MAYER • Department of Oceanography, Darling Marine Center, University of Maine, Walpole, Maine 04573.
Organic Geochemistry, edited by Michael H. Engel and Stephen A. Macko. Plenum Press, New York, 1993.

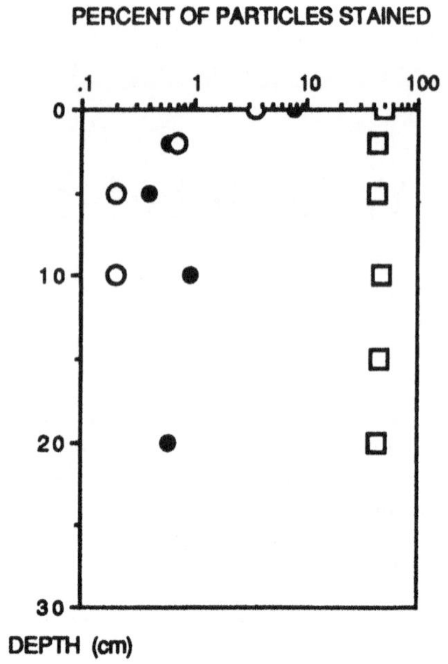

Figure 1. Histochemical staining of particles from a nearshore sediment (Barnstable Harbor, U.S.A.), as percent of total particles. □, Carbohydrate; •, protein; ○, lipid. [Data taken from Johnson (1977).]

than a few tens of percent. Instead, most of the organic matter appears to be tightly associated with the higher density material—that is, the mineral phase.

There is a very strong relationship between total organic matter concentrations and sediment grain size, particularly in shallower marine environments (Premuzic *et al.*, 1982). This relationship is well known from spatial surveys but is also seen with time-series sampling at a single site (van Es, 1982; Sargent *et al.*, 1983; Mayer *et al.*, 1985). Grain size control is much stronger in marine sediment systems than in soil environments, which is perhaps due to the fact that marine sediments are exposed to a narrower range of environmental conditions which affect organic matter accumulation and decay (e.g., temperature, water content) than are soils. The grain size relationship in marine sediments is likely due to a combination of two factors. First, discrete organic particulates should behave hydrodynamically like finer grained minerals and hence accumulate in the same places. Second, the greater specific surface area of finer grained minerals should allow greater adsorptive accumulation by these fractions. Measurements of the relationship between mineral specific surface area and organic matter content have led to the hypothesis that a significant fraction of the organic matter is present in the form of adsorbed monolayers of moderate-size organic molecules (Suess, 1973; Weiler and Mills, 1965; Tanoue and Handa, 1979; Mayer *et al.*, 1985, 1988). These studies have shown that the organic carbon (OC) content of sediments with surface areas of 1–40 $m^2\ g^{-1}$ is 1–40 mg OC g^{-1}. The resultant ratios of organic carbon to surface area—typically about 1 mg OC m^{-2}—are consistent with monolayer thicknesses of organic molecules at other interfaces (e.g., air–water). The low concentrations of discrete organic particulates measured to date are consistent with these ratios and suggest that the adsorptive hypothesis is likely more important than the hydrodynamic equivalence hypothesis as an explanation for grain size influence on total organic matter concentrations.

There is some evidence to indicate that grain size does not have nearly as much control on the labile fraction of sedimentary organic matter, which probably makes up a minor fraction of the total organic matter. Mayer *et al.* (1988) found much poorer correlations between labile proteins or detrital phytopigments and the sediment grain size than were observed for total organic carbon or nitrogen. These data imply that the labile fraction of organic matter may be concentrated in the discrete organic particulate fraction rather than in the adsorbed fraction (Mayer, 1989).

This apparent homeostasis of total organic matter concentrations at roughly monolayer coverage in most shallow-water sediments has not been explained. One possible explanation lies in the nature of the surface of mineral grains. These surfaces are generally quite rough, and the resultant surface areas exceed those predicted from grain size and shape by one to two orders of magnitude (Slabaugh and Stump, 1964; Weiler and Mills, 1965; Mayer and Rossi, 1982). The cause of most of this roughness is tiny pores with diameters typically <10 nm (e.g., Weiler and Mills, 1965). Thus, if all grain surfaces are covered by organic matter in a monolayer fashion and if most surface area lies inside these small pores, then most organic matter will lie in a microenvironment that is partially unavailable, for stereochemical reasons, to the action of many large hydrolytic exoenzymes. This unavailability ought to result in at least slower kinetics of organic matter degradation, which may explain why so many sediments are apparently buffered at this surface area-normalized concentration. This hypothesis is a variant of the argument used to suggest protection of amino acids in the interlamellar spaces of clay minerals (e.g., Müller, 1977).

Grain size does not seem to exert a strong influence on bulk organic matter concentrations at the sediment–water interface in the deep sea (Bezrukov *et al.*, 1977) and in some shallow-water environments where there is extremely high primary production.

2.2. Chemistry of Organic Matter Adsorption onto Mineral Surfaces

The mechanism(s) of attachment of most organic matter to most sedimentary minerals is very incompletely understood. There are certain organic matter–mineral interactions that have been identified as being important in some systems, such as basic amino acid adsorption into the interlamellar spaces of expandable clay minerals (Hedges and Hare, 1987) and the attachment of material relatively rich in acidic amino acids to carbonate mineral surfaces (Mitterer, 1972; Carter, 1978). Both of these interactions are based on ionic interactions between charged organic functional groups and surface atoms of the respective minerals. However, neither of these two direct ionic interactions explains the composition of most organic matter associated with the surfaces of minerals in the ocean, which are dominated by neither carbonate nor expandable clay minerals. Even the nitrogenous material adsorbed to carbonate surfaces is only somewhat enriched in acidic amino acids; the rest of its amino acid composition is much like that of average dissolved peptide materials (Müller and Suess, 1977).

Most attempts to unravel these adsorptive mechanisms have involved either extractions of natural organic material by a variety of solvents or adsorption experiments with model or "humic" compounds. The latter approaches have shown strong enhancement of adsorption by relatively high proton or cationic metal concentrations (both alkali/alkaline-earth and transition metals) (e.g., Preston and Riley, 1982). These results have implied some importance of the carboxyl functionality of natural organic matter. The carboxyl group is likely important for two reasons. First, its role as the major charge-bearing functionality causes partitioning toward the aqueous phase in the absence of metal cations, because of its hydrophilic nature. Second, it may serve as a binding moiety to surfaces, either directly to surface metal ions such as Al or to other otherwise negatively charged surfaces via cationic "water-bridging" connections (Greenland, 1971). Extraction work corroborates this interpretation, with enhanced removal of natural sedimentary organic matter by alkali, sodium pyrophosphate, and metal chelator (e.g., EDTA) solutions, each of which removes charge-reducing and water-bridging cations from carboxyl groups.

Amine–silicate interactions may also play a role in adsorption. The chemistry of these interactions has been well studied (Bull, 1956; Weldes, 1967). Evidence for their potential importance in sediments comes from the melanoidin adsorption experiments of Hedges (1978), in which lysine-rich melanoidins adsorbed onto clays more strongly than those of nonbasic amino acids. The common inclusion of a significant amount of nitrogen in marine sediments, exhibited as C/N ratios of 8–12, may indicate the importance of this interaction, although amino acid spectra do not show basic amino acids to be remarkably enriched in sediments relative to their occurrence in planktonic material.

It is clear, however, that these mechanisms based directly or indirectly on ionic interactions alone cannot provide the explanation for much of the organic matter in sediments. Both adsorption and extraction experiments also indicate a role for weaker interactions, such as hydrogen bonding, the hydrophobic effect, and van der Waals bonding. While these mechanisms are weak relative to ionic or covalent attachments, they are additive forces which can cause a significant free energy of association for larger molecules involved in a multipoint attachment to a surface. Extracted organic compounds generally exhibit relatively high molecular weights (e.g., Hayase and Tsubota, 1985), which may be indicative of either heterocondensates (i.e., humified substances) or original detrital compounds with a high molecular weight. Selective adsorption of the higher molecular weight fractions of natural organic matter has also been observed (Davis and Gloor, 1981; Zutic and Tomaic, 1988; Preston and Riley, 1982). Homologous series of compounds usually show increased adsorption with increasing chain length of a hydrocarbon moiety (e.g., Zullig and Morse, 1988), and their adsorption behavior implies the importance of adsorbate–adsorbate interactions caused by forces such as the hydrophobic effect. A variety of evidence indicates a strong aliphaticity for bulk organic matter in marine sediments (Hatcher and Orem, 1985; Poutanen, 1986; Klok *et al.*, 1984a), also consistent with some role for the hydrophobic effect.

It is also likely that adsorption of many natural organic materials is a heterogeneous process, with attachment to organic molecules already adsorbed on the surface. This feature is well appreciated in the literature on hydrophobic organic pollutant adsorption, wherein adsorption onto sediments is largely a function of the preexisting organic carbon concentrations of those sediments (Karickhoff *et al.*, 1979; Brown and Flagg, 1981). What is quite unclear at this time is the "fouling sequence" of natural organic matter on mineral surfaces—that is, what is the sequence of compound types and attachment mechanisms involved in coating a surface?

3. Organic Matter at the Sediment–Water Interface

3.1. The Nature of the Sediment–Water Interface

The "sediment–water interface" is not as simple a descriptor as the phrase perhaps implies. The term nominally implies no more than the meeting zone of the overlying water column and the underlying deposited sediments—that is, a geometric surface. However, this term often has other connotations. For example, it may refer to a section of finite thickness, rather than a surface. To the organic geochemist, it often connotes that zone in which organic matter is first accumulated from the water column and is initially metabolized by the sediment heterotrophic community (and hence undergoes initial sedimentary diagenesis). In this latter case, there is no convention that defines the boundary of this initial metabolism. From the perspective of an organic geochemist considering diagenesis over sediment depths of tens or hundreds of meters, this zone may be a meter in thickness. From the perspective of more surface-oriented geochemists, the zone may be on the order of a millimeter thick.

The physical characteristics of the top several centimeters of sediment are dictated by a combination of physical and biological processes. These characteristics are reviewed well by Rhoads (1974). In summary, this zone is characterized by high water content, a variety of forms of biological aggregation of mineral grains (e.g., fecal pellets, tube structures), vertical and horizontal mixing, and frequent resuspension.

The initial metabolism and diagenesis of organic matter after deposition can occur in any of several zones located near the physical sediment–water interface. Resuspension of this organic matter may cause as much as a major fraction of this metabolism to occur in the water column directly overlying the sediment (Smith *et al.*, 1987)—the benthic nepheloid layer. Likewise, emplacement of initial deposits into burrow structures (e.g., Aller and Aller, 1986) can cause injection of this material into the subsurface sediments. Aller (1978, 1980) has extensively explored the implications of this "invagination" of the sediment–water interface in terms of its effects on exchange between the sediments and the overlying water column. The impact of this invagination on downcore distributions of materials initially deposited from the water column, such as labile organic compounds or radionuclides, has not received as much attention as is warranted.

From a practical standpoint, the sediment–water interface has tended to be defined by the sampling method used. Most commonly, a core is carefully retrieved and a thin surficial plane section is removed. While surface planes of 1–2 cm in thickness are most commonly reported in the literature, there has been a recent trend to sample on the millimeter scale. This sampling scale has led to greater care in the coring process, as surficial material of this thickness can be easily disrupted or even lost by resuspension during core emplacement or retrieval. As little organic geochemistry has been done on the top few millimeters, most of the work reported here will be based on centimeter-scale sampling.

The plane sections obtained during most core dissections provide only an approximation of a surface that is nonplanar on millimeter to centimeter scales. The roughness of this surface is reflected in distributions of the initial metabolic heterotrophic activity, as indicated in microelectrode studies (Fig. 2; Jorgensen and Revsbech, 1985). This roughness, in addition to the invagination by deeper burrows noted above, adds uncertainty to downcore interpretations of organic matter distributions that assume the geological law of superposition. This variability can be seen as depth-dependent horizontal variability in small-scale (several centimeters) core replication studies; for example, Christensen and Kanneworff (1986) found horizontal variability of pigment concentrations to decrease sharply with depth over a 2-cm surface interval. Concentrations of organic compounds typical of the aerobic surficial few millimeters can therefore be elevated in subsurface burrow linings. For example, Dobbs and Guckert (1988) found fatty acids and pigments typical of aerobic surface sediment to be concentrated along burrow linings. It is indeed possible that depth-dependent concentration changes observed with successive plane sections reflect nothing more than the depth dependence of the density of animal burrows and their accompanying "sediment–water interfaces." This possibility ap-

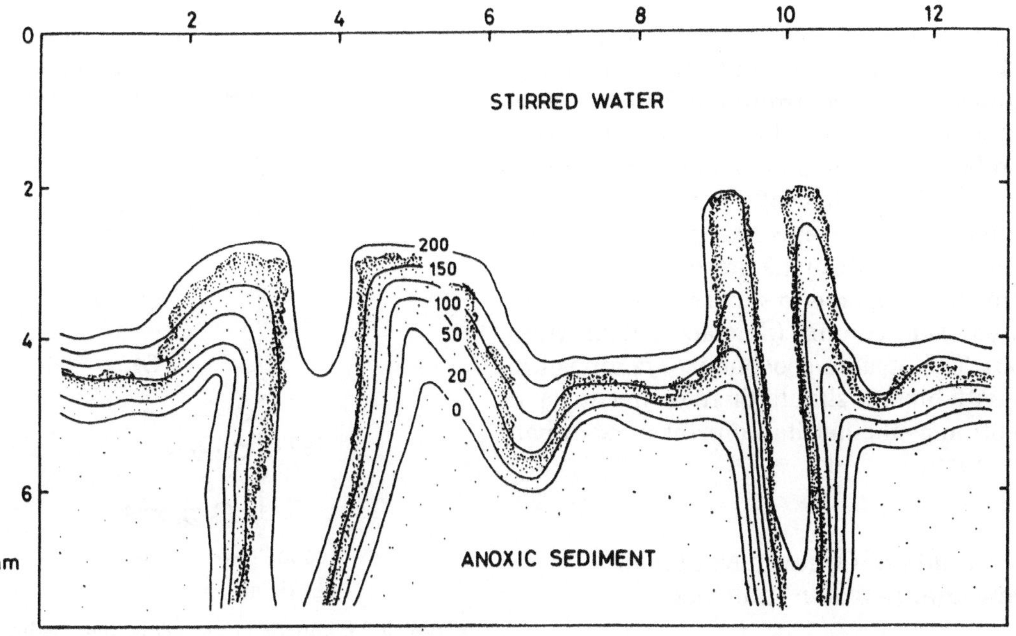

Figure 2. Vertical section of sediment topography and oxygen concentration isopleths (μM), showing the microstructure of the sediment–water interface in a nearshore sediment. [Modified from Fig. 7 in Jorgensen and Revsbech (1985).]

plies to a number of water column-derived radionuclides as well as organic compounds.

Microspatial variability at the sediment–water interface is matched by temporal variability. Because the interface is subject to temporally varying movement of the overlying water column in most environments, sedimentological characteristics of the interface (e.g., topography, grain size) can change on a variety of time scales. Biological processes at the interface provide another source of temporal variability, which also ranges over many time scales. This source of variability results from temporally changing food inputs from the water column, varying temperature, and various biotic rhythms. As a result, caution must be used in interpreting surficial compositions of organic matter as representative starting compositions for the deeper material.

The concentration of living organisms at this interface, normalized to volume, is perhaps greater than in any other marine zone. Bacterial populations alone typically show 10^3–10^4 times the population numbers of the overlying water column (Karl and Novitsky, 1988). In shallow, photic zone sediments these sediment–water interface organisms may be dominated by photosynthetic autotrophs. Highly anoxic sediments can be overlain by dense populations of chemoautotrophic organisms. In the majority of marine sediments, however, these organisms are virtually all heterotrophs that respond to the collection of settling food particles at the interface. These biotic enrichments at the interface are seen clearly in fine-interval measurements of ATP (Novitsky and Karl, 1986; Craven *et al.*, 1986).

A special feature occasionally observed at the sediment–water interface is the so-called "fluff layer," which is a millimeter to several centimeters thick layer of flocculent material of markedly higher water content than the more consolidated sediment below. The most common literature reports of fluff layers are in connection with water column bloom events, which can result in a sudden deposition of a large amount of algal detritus. Deposition of the algal detritus can occur with such rapidity that significant amounts of the detritus can arrive even at the deep-sea floor (Lampitt, 1985). Bottom photograph sequences indicate a short lifetime for these fluff layers; their labile and relatively erodible nature leaves them subject to biological consumption or rapid resuspension (and export). These layers are easily resuspendible by coring devices, and it is possible that such layers are more widespread—in space and in time—than is commonly appreciated.

Little organic geochemistry has been performed on these fluff layers. This lack of attention results from their ephemeral nature and the difficulty of their sampling. However, the existing literature reporting visual and microscopic examinations (e.g., Lochte and Turley, 1988), pigment and total organic matter analyses (e.g., Christensen and Kanneworff, 1986), and fatty acid analyses (Rice *et al.*, 1986) indicates that these layers include labile organic material of phytoplanktonic origin. To emphasize the importance of topography in controlling downcore distributions in sediments, this fluff material has been found to preferentially accumulate in empty burrow structures in sedimentologically active areas (Aller and Aller, 1986). Bloom events likely result in a relatively high loss of organic matter from the photic zone to the sediments compared to slower algal dynamics in a stratified water column (Smetacek, 1980; Wassmann, 1983). The organic geochemical composition associated with blooms should therefore receive particular attention as the detrital imprint for sedimentary organic matter.

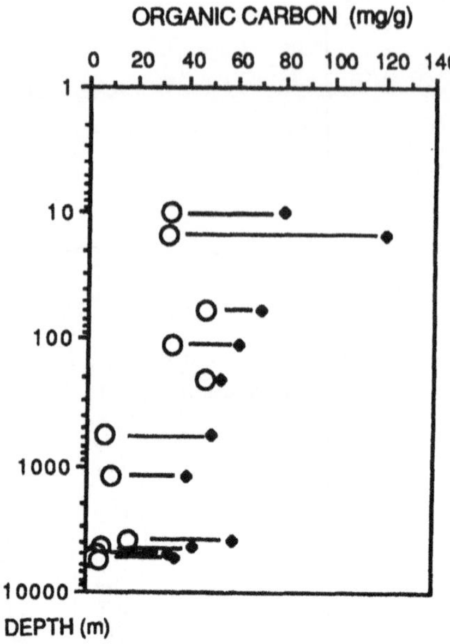

Figure 3. Comparison of near-bottom, sediment-trap TOC (♦) with sediment–water interface TOC (○) at various depths in the oceans. [Data taken from Webster *et al.* (1975), Prahl *et al.* (1980), Honjo *et al.* (1982), Taguchi (1982), Sibuet *et al.* (1984), Khripounoff and Rowe (1985), Anderson *et al.* (1988), Biscaye *et al.* (1988), Hamilton and Hedges (1988), Venkatesan *et al.* (1988), and Faganeli (1989).]

3.2. Enrichments of Organic Matter at the Sediment–Water Interface

The nature of organic matter enriched at the sediment–water interface, rather than the characteristics of the overall sedimentary organic matter found at the interface will be addressed here. It should be reemphasized that the following discussion is based on data from plane sections of sediment taken near the interface.

The major potential source of enrichments at this interface in most marine sediments beneath the photic zone is the rain of organically enriched particles from the overlying water column. The chemistry of these particulates is reviewed in detail by Wakeham and Lee (this volume, Chapter 6) and will not be repeated here. It is worth examining, however, the degree of enrichment of organic matter in sedimenting particulates relative to that found in the receiving sediments. Some representative data of this type (Fig. 3) show that organic carbon concentrations in near-bottom sediment traps are usually several tens of milligrams per gram higher than those of bottom sediments; indeed, severalfold enrichments over the sediments are often seen. These sediment-trap values may not be truly representative of organic carbon enrichment processes. For example, relatively few data are available for annually averaged particle compositions, so seasonal bias can be expected. Also, near-bottom sediment traps usually exhibit dilution of organic matter by resuspended bottom sediments, which are poor in organic material relative to the detrital rain (e.g., Gardner *et al.*, 1985). If organic-poor bottom sediments are "refluxed" into the apparent sedimentation flux from the water column, then the time-integrated organic enrichments predicted from sediment-trap compositions will be underestimates. Thus, the enrichments seen in Fig. 3 represent minimal amounts of carbon (on a wt:wt basis) that can be expected to decay in the surface mixed layer of the sediment column.

The intensity of enrichment at the sediment–water interface varies considerably and is summarized here as total organic carbon (TOC) enrichments in the top 10 cm of sediment from various depth zones of the ocean (Fig. 4). These enrichments are calculated as the difference between the TOC concentration in the top 1–2 cm and that at a depth of 10±2 cm. In some publications only a graph of the data has been provided, and I have estimated the difference from the plot. I have also ignored those (infrequent) data sets in which a negative difference was found. Assuming TOC to constitute one-half of organic matter, these numbers can be simply doubled to obtain organic matter enrichments. This data review reflects a bias toward sediments around North America, particularly for shallower environments.

Shallow-water environments generally show relatively large enrichments, with a broad mode of the distribution from 1 to 5 mg C g^{-1}. Enrichments in sediments from shelf and slope environments (50–2000m water depth) are generally smaller, with the mode below 1 mg C g^{-1}. There are surprisingly many essentially unchanging distributions with core depth

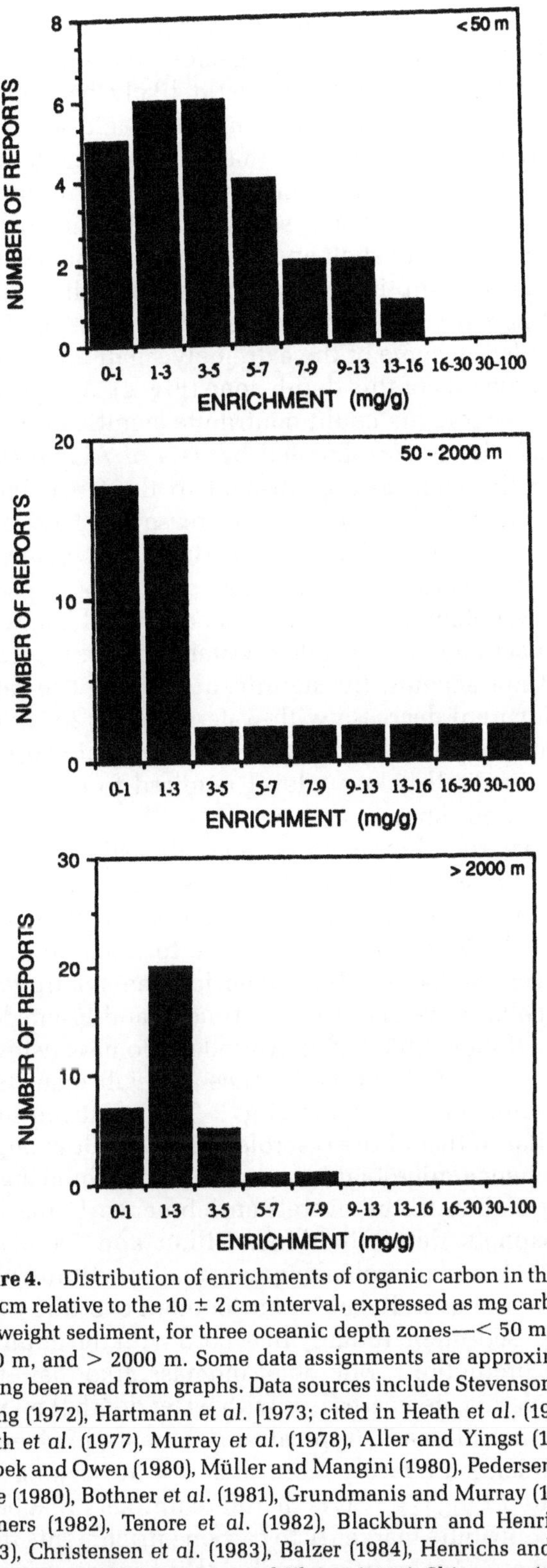

Figure 4. Distribution of enrichments of organic carbon in the top 1–2 cm relative to the 10 ± 2 cm interval, expressed as mg carbon/g dry weight sediment, for three oceanic depth zones—< 50 m, 50–2000 m, and > 2000 m. Some data assignments are approximate, having been read from graphs. Data sources include Stevenson and Cheng (1972), Hartmann *et al.* [1973; cited in Heath *et al.* (1977)], Heath *et al.* (1977), Murray *et al.* (1978), Aller and Yingst (1980), Filipek and Owen (1980), Müller and Mangini (1980), Pedersen and Price (1980), Bothner *et al.* (1981), Grundmanis and Murray (1982), Reimers (1982), Tenore *et al.* (1982), Blackburn and Henriksen (1983), Christensen *et al.* (1983), Balzer (1984), Henrichs and Farrington (1984, 1987), Martens and Klump (1984), Shirayama (1984), Briggs *et al.* (1985), Gillan and Sandstrom (1985), Khripounoff and Rowe (1985), Walsh *et al.* (1985), Alongi (1987, 1989), Carpenter (1987, personal communication), Emerson *et al.* (1987), Heggie *et al.* (1987), Piper *et al.* (1987), Sheu (1987), Volkman *et al.* (1987), Anderson *et al.* (1988), Froelich *et al.* (1988), Furlong and Carpenter (1988), Hamilton and Hedges (1988), McNichol *et al.* (1988), Venkatesan (1988), Bernhard (1989), and Mayer (unpublished data).

reported in the literature. This 50–2000 m water depth zone, however, also has the greatest number of very large enrichments reported. The deeper parts of the ocean show a mode of 1–3 mg C g^{-1}, with a more peaked distribution than the shallow-water sediments and no large enrichments (>9 mg C g^{-1}) reported. It should be reemphasized that these enrichments would be larger if finer intervals at the sediment–water interface were used, as TOC concentrations often increase markedly upward in the top few millimeters (e.g., Emerson *et al.*, 1985; Silverberg *et al.*, 1987). The negative enrichments found in this literature search (data not shown) may have been due to changing sedimentation patterns during the time interval represented by the top 10 cm, or perhaps to enrichments of organic matter at depth by biological processes.

These positive enrichments likely can be explained by one or both of the following arguments. First, the most common interpretation of such profiles assumes a simple model in which organic-rich material is deposited at the sediment–water interface and then is either buried or mixed downward while undergoing decomposition. Numerous studies have modeled this process, using variants of Berner's (1974) general diagenetic equation in conjunction with sedimentation rate or bioturbation data, in order to obtain the decomposition rate constants. These models commonly assume that the organic matter consists of two pools—a labile fraction which accounts for the material lost downcore, and a refractory fraction which remains essentially unchanged. The intensity of enrichment under this scenario will result from competition between the deposition–decomposition processes, which cause enrichments, and mixing–sedimentation processes, which vertically homogenize the sediment and hence eliminate the enrichment. The importance of mixing in this scenario is evident in the marked enrichments found in Saanich Inlet (Hamilton and Hedges, 1988) and the Peru shelf (Henrichs and Farrington, 1984; Volkman *et al.*, 1987; Froelich *et al.*, 1988) sediments—which account for all of the data greater than 7 mg C g^{-1} of the 50–2000 m zone in Fig. 4. Both of these areas are characterized by lack of bioturbation. Notably, the sediment-trap particulates examined by Hamilton and Hedges (1988) in Saanich Inlet also showed minimal difference from the material at the sediment–water interface (Fig. 3), suggesting this case to be one in which the material at the interface may indeed represent the initial material undergoing decay. Conversely, the larger differences observed between sediment-trap particulates and surficial sediments in the case of most of the data in Fig. 3 may thus be due to mixing of the sedimented particulates downward (e.g., by bioturbation), diluting the interfacial material with less organic-rich sediment.

A second interpretation, not so commonly discussed, is that macrofaunal deposit feeding and defecation activities can lead to an enrichment of organic-rich particles at the sediment–water interface (Jahnke *et al.*, 1986). Many animals selectively ingest organically enriched particles, using mechanical attributes such as size or density (Self and Jumars, 1978), and subsequently defecate the material at the sediment–water interface. Although organic material is removed from the particles during gut passage, the residual fecal material is still often enriched in organic matter relative to bulk sediment (Hylleberg and Gallucci, 1975; Brown, 1986). Jumars *et al.* (1981) modeled this process and predicted an increased residence time of these selected particles at the sediment–water interface; however, the organic geochemical implications of this process remain to be investigated. A variant of this scenario may explain some of the negative enrichments. Some subsurface-dwelling animals may remove deposited material from the sediment–water interface and transport it to depth (Graf, 1989; Jumars *et al.*, 1990).

3.3. Composition of Enriched Organic Matter

The nature of the enriched material at the sediment–water interface ought to reflect a combination of the inherited, labile material from the water column and living and dead biomass-derived materials associated with the communities at the interface.

The contribution of live biomass to the organic material at the interface has not been often determined, in part because it is difficult to convert many measures of sediment organisms to carbon equivalents. We will consider here the potential biomass contributions from the macrofauna, meiofauna, and microbial components. Summaries of macrofaunal biomass as a function of depth in the ocean have been provided by Menzies *et al.* (1973), Whittle (1977), and Rowe (1983); the distributions are characterized as an exponentially decreasing function with depth. In shallow waters, corresponding to the 0–50 m interval of Fig. 4, estimates range from 0.5 to about 20 g C m^{-2}. Assuming that up to 100% of this biomass is concentrated in the upper 2 cm, with an average sediment porosity of 70%, the macrofaunal biomass will contribute 0.03–1.4 mg C g^{-1}. The upper end of this range, therefore, approaches the mode of the carbon enrichments for this depth zone. It is likely, however, that the upper end of this range applies to sediments with abundant epibenthic macrofauna (e.g., surface-dwelling mollusks) that would likely be excluded from an organic carbon determination, and so it seems reasonable to conclude that macrofauna biomass forms but a minor fraction of the carbon enrichments for most shallow-water sediments. The macrofaunal contributions in shelf and slope environments, using similar assumptions of sediment depth distribution and porosity, contribute a maximum of 0.1 mg C g^{-1}. Coupled to some of the extremely small total carbon enrichments of this depth zone (Fig. 4), it is possible that macrofauna could contribute significantly. This possibility is corroborated by data of Alongi (1989) from the shelf sands adjacent to the Great Barrier Reef. In the deep sea, the decrease in macrofaunal biomass causes this pool to drop well below the observed carbon enrichments.

Meiofaunal biomass is generally well below that of macrofauna in shallow-water sediments, and so will not account for significant carbon. The rate of meiofaunal decrease with water depth is lower than for macrofauna (Pfannkuche, 1985), but biomass still remains well below a level required to explain the carbon enrichments.

Microbial biomass is generally estimated either by epifluorescence counting of cells or measurement of lipid phosphate. Each method has weaknesses in the calculation of biomass, due to uncertainties in conversion factors. Typical enrichments of up to 5×10^9 cells between the top centimeter and 10-cm depth (e.g., Rublee, 1982), if converted to biomass using the factor 2×10^{-13} g C $cell^{-1}$ (Lee and Fuhrman, 1987), are equivalent to 1 mg C g^{-1}. This contribution is similar to that of the macrofauna for shallow depths, and hence microbial biomass can contribute significant carbon to the overall enrichments. Using lipid phosphate determinations, Gillan and Sandstrom (1985) found microbial biomass that constitutes about 6% of the overall carbon enrichment over the top 10 cm of sediment, roughly in accord with the enumeration predictions. Microbial biomass, like that of the meiofauna, decreases less with water depth than macrofaunal biomass. Typical enrichments in deep-sea sediments are rarely more than 5×10^8 cells g^{-1}, or about 0.1 mg C g^{-1}. It is therefore unlikely that bacteria contribute markedly to carbon enrichments in the deep sea. Again, it is possible that bacteria can contribute markedly to the very small carbon enrichments often found in shelf depths.

In summary, it appears that carbon enrichments at the sediment–water interface are not usually dominated by living biomass, although they may be in certain situations. The techniques of biomass measurement need considerable improvement and need to be applied to the same cores as organic geochemical measurements, before good answers to this question can be obtained.

To what extent do different compound classes account for the carbon enrichments? Generally, the fraction of organic matter than can be identified decreases with depth in the sediment, as shown by decreasing compound class:total carbon ratios. Among the major compound classes that can be identified, lipid concentrations often show the greatest relative enrichments near the sediment–water interface. However, they do not account for more than a few percent of the total carbon enrichments (e.g., van Vleet and Quinn, 1979; Harvey *et al.*, 1984; Gillan and Sandstrom, 1985). Acid-hydrolyzable amino acids often show mildly greater enrichments than total carbon in the top several centimeters, in terms of amino acid C:total organic C ratios. If their contribution to the enriched fraction of carbon is calculated, the values are usually about twice that of the ratios found for bulk organic matter at 10–20-cm depth (e.g., Henrichs and Farrington, 1984, 1987; Burdige and Martens, 1988), indicating that the organic enrichments are of relatively labile material. However, even this class of compounds accounts for no more than 45% of the organic carbon enrichments, and usually considerably less. These data are corroborated by the fact that the stoichiometric changes in carbon and nitrogen downcore are usually in the range of 6–8 (C:N, as wt:wt). These changes indicate a significant but not major contribution of nitrogenous compounds to the enrichments—also on the order of tens of percent. Downcore data for carbohydrates are particularly scarce. Hamilton and Hedges (1988) found total hydrolyzable carbohydrates to account for about 15% of TOC loss over the top 14 cm of anoxic Saanich Inlet sediments. The combined data of Steinberg *et al.* (1987) and Anderson *et al.* (1988) indicate that carbohydrate loss plays an even smaller role in total carbon loss in open shelf and slope sediments of the Northwest Atlantic. This finding is corroborated by constant carbohydrate:TOC ratios in the top 10 cm of Dabob Bay sediments (Cowie and Hedges, 1984). This constancy of carbohydrate concentrations agrees with the PAS-stain observations (Fig. 1; Johnson, 1977) discussed above.

Thus, the carbon enrichments of the top few centimeters remain largely unidentified at this time, as is the bulk organic matter. Nitrogenous compounds make up the most important identified component of this enriched material, with lesser amounts of carbohydrates and lipids. Because much of the organic material found in sediment traps is also unidentified (e.g., Liebezeit and von Bodungen, 1987), these compositional trends do not help to differentiate between the two hypothesized mechanisms for carbon enrichments discussed above.

One important question that is poorly answered in the literature is the degree to which either the bulk organic matter or even the enriched fraction at the sediment–water interface is primarily of detrital or reworked composition. That is, do the compounds represent residues of material rained from the water column above or transformation products of heterotrophic communities in the sediments? The answer(s) to this question cannot, of course, be universal; certainly, the biomarker literature indicates the presence of both types of organic matter in different relative amounts in different sediments. In nearshore sediments in particular, there are clear indications of refractory organic matter deposited to the sediments, such as terrestrial organic matter and phenolic compounds derived from nearshore vascular plants (Prahl, 1985). Focusing our attention on systems dominated by planktonic organic matter inputs, however, we can make a prediction by coupling estimates of the fraction of settled organic matter respired at the sediment–water interface with estimates of the assimilation efficiency of this organic matter by heterotrophic populations. Studies of organic matter diagenesis (Reimers and Suess, 1983; Silverberg *et al.*, 1985; McNichol *et al.*, 1988; Hedges *et al.*, 1988) indicate that a significant if not overwhelming fraction of settled organic matter is respired at the interface. The efficiency with which organic matter consumed by benthic heterotrophs is resynthesized into organic materials, as opposed to remineralized product, is poorly known, but values of as much as several tens of percent, found for marine bacterial decomposition of algal material (e.g., Robinson *et al.*, 1982; Linley and Newell, 1984; Goldman *et al.*, 1987), would be reasonable. Thus, if 70% of the deposited material is respired at the interface, then the remaining 30% of organic matter might well be composed of primarily new biomass and its detrital products.

Biomarkers have been used extensively to seek an answer to this question (cf. Hedges and Prahl, this volume, Chapter 11). Carbohydrate studies provide equivocal help in distinguishing between algal detritus and, say, microbial products, due to the similarity in sugar compositions among these biota. Klok *et al.* (1984b) used ratios of minor to major sugars to argue that the bulk of hydrolyzable carbohydrate in a variety of sedimentary environments was bacterially derived rather than a residuum of incomplete algal decay. Cowie and Hedges (1984) were unable to distinguish between these two sources on the basis of major sugar types. Similar problems confront the use of nitrogenous materials in making this distinction. The presence of nonprotein and microbial cell wall amino acids (e.g., β-alanine, ornithine) have indicated the presence and activity of bacteria (e.g., Henrichs and Farrington, 1987; Burdige and Martens, 1988). However, bulk amino acid spectra of bacteria are quite similar to those of marine algae, and both are similar to those of marine sediments. Algal fatty acids are rapidly replaced with microbial ones (e.g., Boon *et al.*, 1978; Perry *et al.*, 1979; Parkes and Taylor, 1983; Gillan and Sandstrom, 1985; Baird and White, 1985) while other lipids indicative of planktonic material often remain evident to depths of several centimeters or deeper (e.g., steroids and pigments; Volkman *et al.*, 1983; Furlong and Carpenter, 1988). It is important to note, also, that few of these biomarker studies have been done with careful dissection of cores at the sediment–water interface. Certainly, an enrichment of algal material can be found with close-interval sampling—on the order of millimeters (Christensen and Kanneworff, 1986). Under a highly productive water column, the top 1–2 mm of sediment can exhibit a lipid composition very representative of the algal producers (Smith and Eglinton, 1983; Smith *et al.*, 1983 a, b). However, Sargent *et al.* (1983) concluded that the lipid compositions in the top 5 mm of a fjord sediment implied an extensive reworking of the algal detritus by microbes. In summary, the data provide equivocal answers, because (1) the biomarkers are not quantitative indicators of bulk organic matter composition, (2) the original material and transformation products often have similar compositions, and (3) most of the organic matter is still unidentified.

It is quite possible, of course, that evidence for biological transformations should point toward water column microbial transformations rather than those occurring at the sediment–water interface. Wakeham and Lee (1989) have pointed out the dearth of coordinated sediment-trap–sediment studies that would allow an answer to this question. Wakeham and Ertel (1987), however, make a convincing case for this alternative in the Cariaco Trench, and it seems reasonable to suppose that the importance of this alternative increases with increasing water column depth.

In summary, techniques that measure the truly labile fractions of sedimentary organic matter show this material to be particularly enriched near the sediment–water interface. However, the bulk of the organic matter is still dominated by material of indeterminate composition and origin. While it seems likely that even this unidentified material is of greater lability than organic matter in deeper sediments, at least some fraction of the enrichment may be due to no more than a concentration of refractory material due to some process such as size grading.

4. Synthesis and Future Directions

This chapter has considered two size scales of marine interfaces at which organic matter is concentrated. At the mineral–water interface in most of the ocean's sediments, there is an enrichment of on the order of 1 mg C per square meter of mineral surface area, which likely accounts for the bulk of organic matter found in this environment. At the sediment–water interface are found both positive and negative enrichments of carbon; the positive enrichments range from 0 to 10^2 mg C g^{-1} over the carbon content at 10-cm depth. This enrichment translates to as much as 10^5 mg C per square meter of bottom area. The enriched organic matter of the sediment–water interface is likely composed of relatively labile organic matter, while the adsorbed phase is likely quite inert.

Our understanding of the manner in which heterotrophs process organic matter deposited from the water column is minimal and is critical to determining the nature and causes of both of these enrichments. Metabolism of this deposited organic matter by both metazoans and microbial communities clearly has strong influences on both the nature and vertical distribution of organic matter in the top several centimeters of the sediment column. Likewise, the interplay between digestive mechanisms and the physical nature of organic matter in the sediments needs to be addressed before an understanding of the evolution of the inert residue is achieved. The possibility that this inert residue is largely associated with mineral surfaces (see above and also Suess and Müller, 1980) points toward work addressing the interaction of digestive enzymes with adsorbed organic materials.

Considerable effort has been devoted by organic geochemists to determining the differences between the detrital input and sediment processing signals. This work needs to continue, with emphases on finer scale sampling (controlled by the three-dimensional microgeography of sediments rather than simply a series of horizontal plane sections), identification of the functional role of compounds in various forms (e.g., bound versus unbound lipids, enzyme-hydrolyzable versus acid-hydrolyzable carbohydrates and amino acids), greater attention to the nature and dynamics of the heretofore unidentified organic matter, and combined analytical and experimental approaches.

Organic geochemists are well accustomed to the value of separations as a tool of analytical chemistry. Study of these two interfaces will require application of new types of separations, based on spatial, physical, and chemical properties of the sediment. The occurrences of various compounds in sediments need to be associated with other specific chemical, physical, and biological features. Histochemical approaches may prove especially fruitful.

ACKNOWLEDGMENTS. This work was supported by NSF OCE 87-00358 and OCE 89-12433. Reviews by S. Henrichs and J. Hedges were very helpful. This paper represents contribution number 214 from the Darling Marine Center.

References

Aller, J. Y., and Aller, R. C., 1986, Evidence for localized enhancement of biological activity associated with tube and burrow structures in deep-sea sediments at the HEBBLE site, western North Atlantic, *Deep-Sea Res.* **33:**755–790.

Aller, R. C., 1978, The effects of animal–sediment interactions on geochemical processes near the sediment–water interface, in: *Estuarine Interactions* (M. L. Wiley, ed.), Academic Press, New York, pp. 157–172.

Aller, R. C., 1980, Relationships of tube-dwelling benthos with sediment and overlying water chemistry, in: *Marine Benthic Dynamics* (K. R. Tenore and B. C. Coull, eds.), University of South Carolina Press, Georgetown S.C., pp. 285–310.

Aller, R. C., and Yingst, J. Y., 1980, Relationships between microbial distributions and the anaerobic decomposition of organic matter in surface sediments of Long Island Sound, USA, *Mar. Biol.* **56:**29–42.

Alongi, D. M., 1987, The distribution and composition of deep-sea microbenthos in a bathyal region of the western Coral Sea, *Deep-Sea Res.* **34:**1245–1254.

Alongi, D. M., 1989, Benthic processes across mixed terrigenous-carbonate sedimentary facies on the central Great Barrier Reef continental shelf, *Cont. Shelf Res.* **9:**629–663.

Anderson, R. F., Bopp, R. F., Buesseler, K. O., and Biscaye, P. E., 1988, Mixing of particles and organic constituents in sediments from the continental shelf and slope off Cape Cod: SEEP-I results, *Cont. Shelf Res.* **8:** 925–946.

Baird, B. H., and White, D. C., 1985, Biomass and community structure of the abyssal microbiota determined from the ester-linked phospholipids recovered from Venezuela Basin and Puerto Rico Trench sediment, *Mar. Geol.* **68:**217–231.

Balzer, W., 1984, Organic matter degradation and biogenic element cycling in a nearshore sediment (Kiel Bight), *Limnol. Oceanogr.* **29:**1231–1246.

Berner, R. A., 1974, Kinetic models for the early diagenesis of nitrogen, sulfur, phosphorus, and silicon in anoxic marine sediments, in: *The Sea*, Vol. 5 (E. D. Goldberg, ed.), John Wiley & Sons, New York, pp. 427–450.

Bernhard, J. M., 1989, The distribution of benthic Foraminifera with respect to oxygen concentration and organic carbon levels in shallow-water Antarctic sediments, *Limnol. Oceanogr.* **34:**1131–1141.

Bezrukov, P. L., Yemel'yanov, Ye. M., Lisitzyn, A. P., and Romankevich, Ye. A., 1977, Organic carbon in the upper sediment layer of the world ocean, *Oceanology* **17:**561–564.

Biscaye, P. E., Anderson, R. F., and Deck, B. L., 1988, Fluxes and constituents to the eastern United States continental slope and rise: SEEP-1, *Cont. Shelf Res.* **8:**855–904.

Blackburn, T. H., and Henriksen, K., 1983, Nitrogen cycling in different types of sediments from Danish waters, *Limnol. Oceanogr.* **28:**477–493.

Boon, J. J., de Leeuw, J. W., and Burlingame, A. L., 1978, Organic geochemistry of Walvis Bay diatomaceous ooze—III. Structural analysis of the monoenoic and polycyclic fatty acids, *Geochim. Cosmochim. Acta* **42:**631–644.

Bothner, M. H., Spiker, E. C., Johnson, P. P., Rendigs, R. R., and Aruscavage, P. J., 1981, Geochemical evidence for modern sediment accumulation on the continental shelf off southern New England, *J. Sediment. Petrol.* **51:**281–292.

Briggs, K. B., Richardson, M. D., and Young, D. K., 1985, Variability in geoacoustic and related properties of surface sediments from the Venezuela basin, Caribbean Sea, *Mar. Geol.* **68:** 73–106.

Brown, D. S., and Flagg, E. W., 1981, Empirical prediction of organic pollutant sorption in natural sediments, *J. Environ. Qual.* **10:**382–386.

Brown, S. L., 1986, Feces of intertidal benthic invertebrates: Influence of particle selection in feeding on trace element concentration, *Mar. Ecol. Prog. Ser.* **28:**219–231.

Bull, H. B., 1956, Adsorption of bovine serum albumin on glass, *Biochim. Biophys. Acta* **19:**464–471.

Burdige, D. J., and Martens, C. S., 1988, Biogeochemical cycling in an organic-rich coastal marine basin: 10. The role of amino acids in sedimentary carbon and nitrogen cycling, *Geochim. Cosmochim. Acta* **52:**1571–1584.

Carpenter, R., 1987, Has man altered the cycling of nutrients and organic C on the Washington continental shelf and slope?, *Deep-Sea Res.* **34:**881–896.

Carter, P. W., 1978, Adsorption of amino acid-containing organic matter by calcite and quartz, *Geochim. Cosmochim. Acta* **42:** 1239–1242.

Christensen, H., and Kanneworff, E., 1986, Sedimentation of phytoplankton during a spring bloom in the Oresund, *Ophelia* **26:** 109–122.

Christensen, J. P., Rowe, G. T., and Clifford, C. H., 1983, The possible importance of primary amino nitrogen in nitrogen regeneration by coastal marine sediments in Buzzards Bay, Massachusetts, *Int. Rev. Gesamten Hydrobiol.* **68:**501–512.

Cowie, G. L., and Hedges, J. I., 1984, Carbohydrate sources in a coastal marine environment, *Geochim. Cosmochim. Acta* **48:** 2075–2087.

Craven, D. B., Jahnke, R. A., and Carlucci, A. F., 1986, Fine-scale vertical distributions of microbial biomass and activity in California Borderland sediments, *Deep-Sea Res.* **33:**379–390.

Davis, J. A., and Gloor, R., 1981, Adsorption of dissolved organics in lake water by aluminum oxide. Effect of molecular weight, *Environ. Sci. Technol.* **15:**1223–1229.

Dobbs, F. C., and Guckert, J. B., 1988, *Callianassa trilobata* (Crustacea: Thalassinidea) influences abundance of meiofauna and biomass, composition, and physiologic state of microbial communities within its burrow, *Mar. Ecol. Prog. Ser.* **45:**69–79.

Emerson, S., Fischer, K., Reimers, C., and Heggie, D., 1985, Organic carbon dynamics and preservation in deep-sea sediments, *Deep-Sea Res.* **32**:1–21.

Emerson, S., Stump, S., Grootes, P. M., Stuiver, M., and Farwell, G. W., 1987, Estimates of degradable organic carbon in deep-sea surface sediments from ^{14}C concentrations, *Nature* **329**:51–53.

Ertel, J. R., and Hedges, J. I., 1985, Sources of sedimentary humic substances: Vascular plant debris, *Geochim. Cosmochim. Acta* **49**:2097–2107.

Faganeli, J., 1989, Sedimentation of particulate nitrogen and amino acids in shallow coastal waters (Gulf of Trieste, Northern Adriatic), *Mar. Chem.* **26**:67–80.

Filipek, L. H., and Owen, R. M., 1980, Early diagenesis of organic carbon and sulfur in outer shelf sediments from the Gulf of Mexico, *Am. J. Sci.* **280**:1097–1112.

Froelich, P. N., Arthur, M. A., Burnett, W. C., Deakin, M., Hensley, V., Jahnke, R., Kaul, L., Kim, K.-H., Roe, K., Soutar, A., and Vathakanon, C., 1988, Early diagenesis of organic matter in Peru continental shelf sediments: Phosphorite precipitation, *Mar. Geol.* **80**:309–343.

Furlong, E. T., and Carpenter, R., 1988, Pigment preservation and remineralization in oxic coastal marine sediments, *Geochim. Cosmochim. Acta* **52**:87–99.

Gardner, W. D., Southard, J. B., and Hollister, C. D., 1985, Sedimentation, resuspension and chemistry of particles in the Northwest Atlantic, *Mar. Geol.* **65**:199–242.

Gershanovich, D. Ye., and Zaslavskiy, Ye. N., 1983, Geochemical characterization of the organic matter in bottom sediments in the upwelling zone in the Southeast Pacific, *Geochem. Int.* **20**: 88–96.

Gillan, F. T., and Sandstrom, M. W., 1985, Microbial lipids from a nearshore sediment from Bowling Green Bay, North Queensland: The fatty acid composition of intact lipid fractions, *Org. Geochem.* **8**:321–328.

Goldman, J. C., Caron, D. A., and Dennett, M. R., 1987, Regulation of gross growth efficiency and ammonium regeneration in bacteria by substrate C:N ratio. *Limnol. Oceanogr.* **32**:1239–1252.

Graf, G., 1989, Benthic–pelagic coupling in a deep-sea benthic community, *Nature* **341**:437–439.

Greenland, D. J., 1971, Interactions between humic and fulvic acids and clays, *Soil Sci.* **111**:34–41.

Grundmanis, V., and Murray, J. W., 1982, Aerobic respiration in pelagic marine sediments, *Geochim. Cosmochim. Acta* **46**: 1101–1120.

Hamilton, S. E., and Hedges, J. I., 1988, The comparative geochemistries of lignins and carbohydrates in an anoxic fjord, *Geochim. Cosmochim. Acta* **52**:129–142.

Hartmann, M., Kogler, F.-D., Müller, P., and Suess, E., 1973, Preliminary results of geochemical and soil mechanical investigations on Pacific Ocean sediments, *Papers on the Origin and Distribution of Manganese Nodules in the Pacific and Prospects for Exploration*, University of Hawaii, Honolulu, pp. 71–76.

Harvey, H. R., Richardson, M. D., and Patton, J. S., 1984, Lipid composition and vertical distribution of bacteria in aerobic sediments of the Venezuela Basin, *Deep-Sea Res.* **31**:403–413.

Hatcher, P. G., and Orem, W. H., 1985, Structural interrelationships among humic substances in marine and estuarine sediments as delineated by cross-polarization/magic angle spinning ^{13}C NMR, in: *Organic Marine Geochemistry* (M. Sohn, ed.), ACS Symp. Series No. 305, American Chemical Society Washington, D.C., pp. 142–157.

Hayase, K., and Tsubota, H., 1985, Sedimentary humic acid and fulvic acid as fluorescent organic materials, *Geochim. Cosmochim. Acta* **49**:159–163.

Heath, G. R., Moore, T. C., Jr., and Dauphin, J. P., 1977, Organic carbon in deep-sea sediments, in: *The Fate of Fossil Fuel CO_2 in the Oceans* (N. R. Andersen and A. Malahoff, eds.), Plenum Press, New York, pp. 605–625.

Hedges, J. I., 1978, The formation and clay mineral reactions of melanoidins, *Geochim. Cosmochim. Acta* **42**:69–76.

Hedges, J. I., and Hare, P. E., 1987, Amino acid adsorption by clay minerals in distilled water, *Geochim. Cosmochim. Acta* **51**: 255–259.

Hedges, J. I., Clark, W. A., and Cowie, G. L., 1988, Fluxes and reactivities of organic matter in a coastal marine bay, *Limnol. Oceanogr.* **33**:1137–1152.

Heggie, D., Maris, C., Hudson, A., Dymond, J., Beach, R., and Cullen, J., 1987, Organic carbon oxidation and preservation in NW Atlantic continental margin sediments, in: *Geology and Geochemistry of Abyssal Plains* (P. P. E. Weaver and J. Thomson, eds.), Geological Society of America Spec. Publ. No. 31, pp. 215–236.

Henrichs, S. M., and Farrington, J. W., 1984, Peru upwelling region sediments near 15°S. 1. Remineralization and accumulation of organic matter, *Limnol. Oceanogr.* **29**:1–19.

Henrichs, S. M., and Farrington, J. W., 1987, Early diagenesis of amino acids and organic matter in two coastal marine sediments, *Geochim. Cosmochim. Acta* **51**:1–15.

Honjo, S., Manganini, S. J., and Cole, J. J., 1982, Sedimentation of biogenic matter in the deep ocean, *Deep-Sea Res.* **29**:609–625.

Hylleberg, J., and Gallucci, V. F., 1975, Selectivity in feeding by the deposit-feeding bivalve *Macoma nasuta*, *Mar. Biol.* **32**:167–178.

Jahnke, R. A., Emerson, S. R., Cochran, J. K., and Hirschberg, D. J., 1986, Fine scale distributions of porosity and particulate excess ^{210}Pb, organic carbon and $CaCO_3$ in surface sediments of the deep equatorial Pacific, *Earth Planet. Sci. Lett.* **77**:59–69.

Johnson, R. G., 1974, Particulate matter at the sediment–water interface in coastal environments, *J. Mar. Res.* **32**:313–330.

Johnson, R. G., 1977, Vertical variation in particulate matter in the upper twenty centimeters of marine sediments, *J. Mar. Res.* **35**: 273–282.

Jorgensen, B. B., and Revsbech, N. P., 1985, Diffusive boundary layers and the oxygen uptake of sediments and detritus, *Limnol. Oceanogr.* **30**:111–122.

Jumars, P. A., Nowell, A. R. M., and Self, R. F. L., 1981, A simple model of flow–sediment–organism interaction, *Mar. Geol.* **42**: 155–172.

Jumars, P. A., Mayer, L. M., Deming, J. W., Baross, J. A., and Wheatcroft, R. A., 1990, Deep-sea deposit-feeding strategies suggested by environmental and feeding constraints, *Philos. Trans. R. Soc. London, Ser. A.* **331**:85–101.

Karickhoff, S. W., Brown, D. S., and Scott, T. A., 1979, Sorption of hydrophobic pollutants on natural sediments, *Water Res.* **13**: 241–248.

Karl, D. M., and Novitsky, J. A., 1988, Dynamics of microbial growth in surface layers of a coastal marine sediment ecosystem, *Mar. Ecol. Prog. Ser.* **50**:169–176.

Khripounoff, A., and Rowe, G. T., 1985, Les apports organiques et leur transformation en milieu abyssal à l'interface eau–sédiment dans l'océan Atlantique tropical, *Oceanol. Acta* **8**: 293–301.

Klok, J., Baas, M., Cox, H. C., de Leeuw, J. W., Rijpstra, W. I. C., and Schenck, P. A., 1984a, Qualitative and quantitative characterization of the total organic matter in a recent marine sediment (Part II), *Org. Geochem.* **6**:265–278.

Klok, J., Cox, H. C., Baas, M., Schuyl, P. J. W., de Leeuw, J. W., and Schenck, P. A., 1984b, Carbohydrates in recent marine sediments I. Origin and significance of deoxy- and O-methyl-monosaccharides, *Org. Geochem.* **7**:73–84.

Lampitt, R. S., 1985, Evidence for the seasonal deposition of detritus to the deep-sea floor and its subsequent resuspension, *Deep-Sea Res.* **32**:885–897.

Lee, S., and Fuhrman, J. A., 1987, Relationships between biovolume and biomass of naturally derived marine bacterioplankton, *Appl. Environ. Microbiol.* **53**:1298–1303.

Liebezeit, G., and von Bodungen, B., 1987, Biogenic fluxes in the Bransfield Strait: Planktonic versus macroalgal sources, *Mar. Ecol. Prog. Ser.* **36**:23–32.

Linley, E. A. S., and Newell, R. C., 1984, Estimates of bacterial growth yields based on plant detritus, *Bull. Mar. Sci.* **35**: 409–425.

Lochte, K., and Turley, C. M., 1988, Bacteria and cyanobacteria associated with phytodetritus in the deep sea, *Nature* **333**: 67–69.

Martens, C. S., and Klump, J. V., 1984, Biogeochemical cycling in organic-rich coastal marine basin 4. An organic carbon budget for sediments dominated by sulfate reduction and methanogenesis, *Geochim. Cosmochim. Acta* **48**:1987–2004.

Mayer, L. M., 1989, The nature and determination of non-living sedimentary organic matter as a food source for deposit-feeders, in: *Ecology of Marine Deposit Feeders* (G. Lopez, G. Taghon, and J. Levinton, eds.), Springer-Verlag, New York, pp. 98–113.

Mayer, L. M., and Rossi, P. M., 1982, Specific surface areas in marine sediments: Relationships with other textural factors, *Mar. Geol.* **45**:241–252.

Mayer, L. M., Rahaim, P. T., Guerin, W., Macko, S. A., Watling, L., and Andersen, F. E., 1985, Biological and granulometric controls on sedimentary organic matter of an intertidal mudflat, *Estuarine Coastal Shelf Sci.* **20**:491–504.

Mayer, L. M., Macko, S. A., and Cammen, L., 1988, Provenance, concentrations and nature of sedimentary organic nitrogen in the Gulf of Maine, *Mar. Chem.* **25**:291–304.

McNichol, A. P., Lee, C., and Druffel, E. R. M., 1988, Carbon cycling in coastal sediments: 1. A quantitative estimate of the remineralization of organic carbon in the sediments of Buzzards Bay, MA, *Geochim. Cosmochim. Acta* **52**:1531–1543.

Menzies, R. J., George, R. Y., and Rowe, G. T., 1973, *Abyssal Environment and Ecology of the World Oceans*, John Wiley & Sons, New York.

Mitterer, R. M., 1972, Biogeochemistry of aragonite mud and oolites, *Geochim. Cosmochim. Acta* **36**:1407–1422.

Müller, P. J., 1977, C/N ratios in Pacific deep-sea sediments: Effects of inorganic ammonium and organic nitrogen compounds sorbed by clays, *Geochim. Cosmochim. Acta* **41**:765–776.

Müller, P. J., and Mangini, A., 1980, Organic carbon decomposition rates in sediments of the Pacific manganese nodule belt dated by ^{230}Th and ^{231}Pa, *Earth Planet. Sci. Lett.* **51**:94–114.

Müller, P. J., and Suess, E., 1977, Interaction of organic compounds with calcium carbonate III. Amino acid composition of sorbed layers, *Geochim. Cosmochim. Acta* **41**:941–949.

Murdoch, M. H., Barlocher, F., and Laltoo, M. L., 1986, Population dynamics and nutrition of *Corophium volutator* (Pallas) in the Cumberland Basin (Bay of Fundy), *J. Exp. Mar. Biol. Ecol.* **103**: 235–249.

Murray, J. W., Grundmanis, V., and Smethie, W. M., Jr., 1978, Interstitial water chemistry in the sediments of Saanich Inlet, *Geochim. Cosmochim. Acta* **42**:1011–1026.

Novitsky, J. A., and Karl, D. M., 1986, Characterization of microbial activity in the surface layers of a coastal sub-tropical sediment, *Mar. Ecol. Prog. Ser.* **28**:49–55.

Parkes, R. J., and Taylor, J., 1983, The relationship between fatty acid distributions and bacterial respiratory types in contemporary marine sediments, *Estuarine Coastal Shelf Sci.* **16**:173–189.

Pedersen, T. F., and Price, N. B., 1980, The geochemistry of iodine and bromine in sediments of the Panama Basin, *J. Mar. Res.* **38**: 397–411.

Perry, G. J., Volkman, J. K., and Johns, R. B., 1979, Fatty acids of bacterial origin in contemporary marine sediments, *Geochim. Cosmochim. Acta* **43**:1715–1725.

Pfannkuche, O., 1985, The deep-sea meiofauna of the Porcupine Seabight and abyssal plain (NE Atlantic): Population structure, distribution, standing stocks, *Oceanol. Acta* **8**:343–353.

Piper, D. Z., Rude, P. D., and Monteith, S., 1987, The chemistry and mineralogy of haloed burrows in pelagic sediment at domes site A: The equatorial North Pacific, *Mar. Geol.* **74**:41–55.

Poutanen, E.-L., 1986, Characterization of humic and fulvic acid isolated from Baltic Sea sediments using ^{13}C and ^{1}H nuclear magnetic resonance spectra, *Org. Geochem.* **9**:163–170.

Prahl, F. G., 1985, Chemical evidence of differential particle dispersal in the southern Washington coastal environment, *Geochim. Cosmochim. Acta* **49**:2533–2539.

Prahl, F. G., and Carpenter, R., 1983, Polycyclic aromatic hydrocarbon (PAH)-phase associations in Washington coastal sediment, *Geochim. Cosmochim. Acta* **47**:1013–1023.

Prahl, F. G., Bennett, J. T., and Carpenter, R., 1980, The early diagenesis of aliphatic hydrocarbons and organic matter in sedimentary particulates from Dabob Bay, Washington, *Geochim. Cosmochim. Acta* **44**:1967–1976.

Premuzic, E. T., Benkovitz, C. M., Gaffney, J. S., and Walsh, J. J., 1982, The nature and distribution of organic matter in the surface sediments of world oceans and seas, *Org. Geochem.* **4**:63–77.

Preston, M. R., and Riley, J. P., 1982, The interactions of humic compounds with electrolytes and three clay minerals under simulated estuarine conditions, *Estuarine Coastal Shelf Sci.* **14**:567–576.

Reimers, C. E., 1982, Organic matter in anoxic sediments off central Peru: Relations of porosity, microbial decomposition and deformation properties, *Mar. Geol.* **46**:175–197.

Reimers, C. E., and Suess, E., 1983, The partitioning of organic carbon fluxes and sedimentary organic matter decomposition rates in the ocean, *Mar. Chem.* **13**:141–168.

Rhoads, D. C., 1974, Organism–sediment relations on the muddy sea floor, in: *Annual Reviews of Oceanography and Marine Biology* Vol. 12 (H. Barnes, ed.), George Allen & Unwin, London, pp. 263–300.

Rice, A. L., Billett, D. S. M., Fry, J., John, A. W. G., Lampitt, R. S., Mantoura, R. F. C., and Morris, R. J., 1986, Seasonal deposition of phytodetritus to the deep-sea floor, *Proc. R. Soc. Edinburgh* **88B**:265–279.

Robinson, J. D., Mann, K. H., and Novitsky, J. A., 1982, Conversion of the particulate fraction of seaweed detritus to bacterial biomass, *Limnol. Oceanogr.* **27**:1072–1079.

Rowe, G. T., 1983, Biomass and production of the deep-sea macrobenthos, in *The Sea*, Vol. 8 (G. T. Rowe, ed.), John Wiley & Sons, New York, pp. 97–122.

Rublee, P. A., 1982, Seasonal distribution of bacteria in salt marsh sediments in North Carolina, *Estuarine Coastal Shelf Sci.* **15**: 67–74.

Sargent, J. R., Hopkins, C. C. E., Seiring, J. V., and Youngson, A.,

1983, Partial characterization of organic material in surface sediments from Balsfjorden, northern Norway, in relation to its origin and nutritional value for sediment-ingesting animals, *Mar. Biol.* **76**:87–94.

Self, R. F. L., and Jumars, P. A., 1978, New resource axes for deposit feeders?, *J. Mar. Res.* **36**:627–641.

Sheu, D.-D., 1987, Sulfur and organic carbon contents in sediment cores from the Tyro and Orca basins, *Mar. Geol.* **75**:157–164.

Shirayama, Y., 1984, The abundance of deep sea meiobenthos in the western Pacific in relation to environmental factors, *Oceanol. Acta* **7**:113–121.

Sibuet, M., Monniot, C., Desbruyeres, D., Dinet, A., Khripounoff, A., Rowe, G., and Segonzac, M., 1984, Peuplements benthiques et caractéristiques trophiques du milieu dans la plaine abyssale de Demerara, *Oceanol. Acta* **7**:345–358.

Silverberg, N., Edenborn, H. M., and Belzile, N., 1985, Sediment response to seasonal variations in organic matter input, in: *Marine and Estuarine Geochemistry* (A. C. Sigleo and A. Hattori, eds.), Lewis Publishers, Chelsea, Michigan, pp. 69–80.

Silverberg, N., Bakker, J., Edenborn, H. M., and Sundby, B., 1987, Oxygen profiles and organic carbon fluxes in Laurentian Trough sediments, *Neth. J. Sea Res.* **21**:95–105.

Slabaugh, W. H., and Stump, A. D., 1964, Surface areas and porosity of marine sediments, *J. Geophys. Res.* **69**:4773–4778.

Smetacek, V., 1980, Annual cycle of sedimentation in relation to plankton ecology in Western Kiel Bight, *Ophelia* **1**:65–76.

Smith, D. J., and Eglinton, G., 1983, Interfacial sediment and assessment of organic input from a highly productive water column, *Nature* **304**:259–262.

Smith, D. J., Eglinton, G., and Morris, R. J., 1983a, The lipid chemistry of an interfacial sediment from the Peru continental shelf: Fatty acids, alcohols, aliphatic ketones and hydrocarbons, *Geochim. Cosmochim. Acta* **47**:2225–2232.

Smith, D. J., Eglinton, G., Morris, R. J., and Poutanen, E. L., 1983b, Aspects of the steroid geochemistry of an interfacial sediment from the Peruvian upwelling, *Oceanol. Acta* **6**:211–219.

Smith, K. L., Carlucci, A. F., Jahnke, R. A., and Craven, D. B., 1987, Organic carbon mineralization in the Santa Catalina Basin: Benthic boundary layer metabolism, *Deep-Sea Res.* **34**:185–211.

Steinberg, S. M., Venkatesan, M. I., and Kaplan, I. R., 1987, Organic geochemistry of sediments from the continental margin off southern New England, U.S.A.—Part I. Amino acids, carbohydrates and lignin, *Mar. Chem.* **21**:249–265.

Stevenson, F. J., and Cheng, C.-N., 1972, Organic geochemistry of the Argentine Basin sediments: Carbon–nitrogen relationships and Quaternary correlations, *Geochim. Cosmochim. Acta* **36**:653–671.

Suess, E., 1973, Interaction of organic compounds with calcium carbonate—II. Organo-carbonate association in Recent sediments, *Geochim. Cosmochim. Acta* **37**:2435–2447.

Suess, E., and Müller, P. J., 1980, Productivity, sedimentation rate and sedimentary organic matter in the oceans II.—Elemental fractionation, in: *Biogéochimie de la Matière Organique à l'Interface Eau–Sédiment Marin*, Colloques Internationaux du CNRS, pp. 17–26.

Taguchi, S., 1982, Sedimentation of newly produced particulate organic matter in a subtropical inlet, Kaneohe Bay, Hawaii, *Estuarine Coastal Shelf Sci.* **15**:533–544.

Tanoue, E., and Handa, N., 1979, Differential sorption of organic matter by various sized sediment particles in recent sediment from the Bering Sea, *J. Oceanogr. Soc. J.* **35**:199–208.

Tenore, K. R., Boyer, L. F., Cal, R. M., Corral, J., Garcia-Fernandez, C., Gonzalez, N., Gonzalez-Gurriaran, E., Hanson, R. B., Iglesias, J., Krom, M., Lopez-Jamar, E., McClain, J., Pamatmat, M. M., Perez, A., Rhoads, D. C., de Santiago, G., Tietjen, J., Westrich, J., and Windom, H. L., 1982, Coastal upwelling in the Rias Bajas, NW Spain: Contrasting the benthic regimes of the Rias de Arosa and de Muros, *J. Mar. Res.* **40**:701–772.

van Es, F. B., 1982, Community metabolism of intertidal flats in the Ems-Dollard estuary, *Mar. Biol.* **66**:95–108.

van Vleet, E. S., and Quinn, J. G., 1979, Diagenesis of marine lipids in ocean sediments, *Deep-Sea Res.* **26A**:1225–1236.

Venkatesan, M. I., 1988, Organic geochemistry of marine sediments in Antarctic region: Marine lipids in McMurdo Sound, *Org. Geochem.* **12**:13–27.

Venkatesan, M. I., Steinberg, S., and Kaplan, I. R., 1988, Organic geochemical characterization of sediments from the continental shelf south of New England as an indicator of shelf edge processes, *Cont. Shelf Res.* **8**:905–924.

Volkman, J. K., Farrington, J. W., Gagosian, R. B., and Wakeham, S. G., 1983, Lipid composition of coastal marine sediments from the Peru upwelling region, in: *Advances in Organic Geochemistry*, M. Bjorøy, ed.), John Wiley & Sons, New York, pp. 228–240.

Volkman, J. K., Farrington, J. W., and Gagosian, R. B., 1987, Marine and terrigenous lipids in coastal sediments from the Peru upwelling region at 15°S: Sterols and triterpene alcohols, *Org. Geochem.* **11**:463–477.

Wakeham, S. G., and Ertel, J. R., 1987, Diagenesis of organic matter in suspended particles and sediments in the Cariaco Trench, *Org. Geochem.* **13**:815–822.

Wakeham, S. G., and Lee., C., 1989, Organic geochemistry of particulate matter in the ocean: The role of particles in oceanic sedimentary cycles, *Org. Geochem.* **14**:83–96.

Walsh, J. J., Premuzic, E. T., Gaffney, J. S., Rowe, G. T., Harbottle, G., Stoenner, R. W., Balsam, W. L., Betzer, P. R., and Macko, S. A., 1985, Organic storage of CO_2 on the continental slope off the mid-Atlantic bight, the southeastern Bering Sea, and the Peru coast, *Deep-Sea Res.* **32**:853–883.

Wassmann, P., 1983, Sedimentation of organic and inorganic particulate material in Lindaspollene, a stratified, land-locked fjord in western Norway, *Mar. Ecol. Prog. Ser.* **13**:237–248.

Webster, T. J. M., Paranjape, M. A., and Mann, K. H., 1975, Sedimentation of organic matter in St. Margaret's Bay, Nova Scotia, *J. Fish. Res. Board Can.* **32**:1399–1407.

Weiler, R. R., and Mills, A. A., 1965, Surface properties and pore structure of marine sediments, *Deep-Sea Res.* **12**:511–529.

Weldes, H. H., 1967, Interaction of alkali metal silicates with amino acids and soybean protein, *Adhes. Age* **10**:32–35.

Whittle, K. J., 1977, Marine organisms and their contribution to organic matter in the ocean, *Mar. Chem.* **5**:381–411.

Zullig, J. J., and Morse, J. W., 1988, Interactions of organic acids with carbonate mineral surfaces in seawater and related solutions: I. Fatty acids adsorption, *Geochim. Cosmochim. Acta* **52**:1667–1678.

Zutic, V., and Tomaic, J., 1988, On the formation of organic coatings on marine particles: Interactions of organic matter at hydrous alumina/seawater interfaces, *Mar. Chem.* **23**:51–67.

Chapter 8

The Early Diagenesis of Organic Matter in Lacustrine Sediments

PHILIP A. MEYERS and RYOSHI ISHIWATARI

1. Sources and Sedimentation of Organic Matter in Lakes

1.1. Introduction

Lakes are diverse in their size, biological communities, watershed types, bottom morphology, and other important limnological factors which impact the accumulation and composition of their sediments. The diversity of lacustrine depositional systems is useful to organic geochemical studies. Different aspects of the early, as well as the more advanced, stages of diagenesis can be investigated, using lakes that have sedimentary characteristics which highlight the aspects of interest. Although sediment deposition in an individual lake is usually unique, such studies have led to generalizations about the processes affecting the organic matter contents of lacustrine sediments.

Organic matter constitutes a minor but important fraction of lake sediments. It is a mixture of the myriad lipids, carbohydrates, proteins, and other biochemicals contained in the tissues of living benthic microorganisms and contributed from the detritus of organisms formerly living in the lake and its watershed, plus humic substances diagenetically formed from these biochemical starting materials. Because organic matter is particularly subject to geochemical oxidation and to biological utilization, it is an especially dynamic component of sediments.

1.2. Sources of Organic Matter Incorporated into Lacustrine Sediments

The sources of organic matter incorporated into lacustrine sediments and the processes modifying it as incorporation into sediments proceeds are summarized in Fig. 1. The two primary sources of organic matter are the particulate detritus of aquatic organisms and the residues of biota from the land surrounding a lake. Nearly all of this detritus originates from plants; less than 10% comes from animals. Plants can be divided into two geochemically significant groups according to their biochemical compositions: those that lack woody and cellulosic tissues,

PHILIP A. MEYERS • Department of Geological Sciences and Center for Great Lakes and Aquatic Science, The University of Michigan, Ann Arbor, Michigan 48109. RYOSHI ISHIWATARI • Department of Chemistry, Tokyo Metropolitan University, Tokyo 192-03, Japan.

Organic Geochemistry, edited by Michael H. Engel and Stephen A. Macko. Plenum Press, New York, 1993.

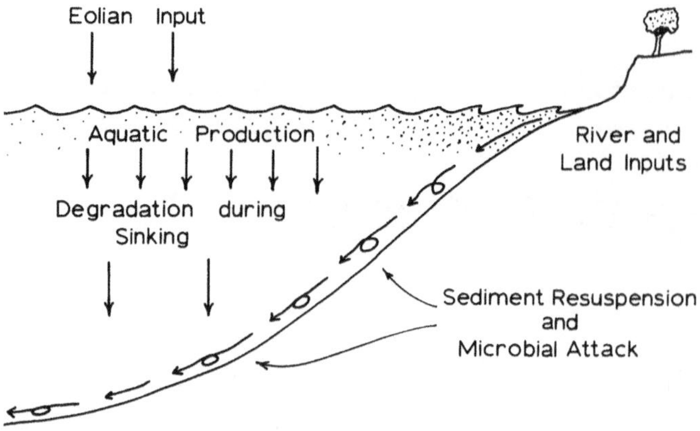

Figure 1. Schematic representation of the sources of organic matter to lake sediments and of the processes acting to alter and degrade organic matter during its sedimentation. The major sources are organic matter from algal production within a lake and from plants within the watershed. Airborne material can originate from outside the watershed and is normally a minor contribution. Resuspension and down-slope transport move sediments from shallow locations to basinal depths. Microbial reprocessing of organic matter occurs in the water and in the lake bottom, diminishing the total amount and substituting microbial components.

and those that have these tissues. Aquatic plants that are simple in form, such as algae, lack cellulosic supportive structures and therefore have elemental compositions that typically have C/N ratios in the range of 4 to 10. Plants having cellulose, such as grasses, shrubs, and trees, contain proportionally more carbon. Their C/N ratios range from 20 to 80 or so. These types of plants exist on land and in the shallow parts of lakes as bottom-rooted, emergent vegetation.

A third important type of organic matter components in lacustrine sediments originates from the bacteria and other microbes that inhabit the water and sediment of lakes and the soil of the surrounding watershed. Although important in identifying organic geochemical processes, these components are not an addition to the pool of organic carbon; they are derived at the expense of primary materials. Microbes biosynthesize their characteristic biomarker compounds while breaking down the residua of the organic matter created by the plants and animals of the lake and watershed.

1.3. Changes to Organic Matter during Sediment Deposition

Organic matter in a lake is subject to a variety of processes which can alter its character in the relatively short time between its synthesis and its long-term burial in the lake bottom. Foremost among these processes is the biochemical oxidation of organic matter within the water column. Vallentyne (1962) presented the simple concept that the water-soluble fractions of biological organic matter are easily remineralized by bacteria in lake waters; these forms of organic matter are not available for incorporation into the bottom sediments. Only the hydrophobic, poorly soluble types of organic components reach the water–sediment interface. The earliest stages of diagenesis occur prior to sedimentation, and they can modify greatly the character of organic matter. Subsequent studies have verified this concept.

The results of sediment-trap studies in Lake Michigan provide an example of the magnitude of oxidative destruction that can occur before organic matter reaches the sediments at the bottom of a lake. Eadie *et al.* (1984) estimated that 83% of the organic carbon that is photosynthetically formed in the photic zone is remineralized before leaving the epilimnion, the part of the water column above the thermocline (Table I). Ninety-four percent of the primary aquatic production of this lake is recycled before reaching the sediment surface, only 100 m below. For shallower lakes, organic matter would have shorter sinking times and consequently shorter exposures to oxidation in the water column, but the concept of preferential losses of water-soluble components would still apply.

Once reaching the bottom, organic matter is subject to additional processes leading to its oxidative alteration and destruction. As noted by Eadie *et al.* (1984) and Meyers *et al.* (1984a) in their sediment-trap studies in Lake Michigan, extensive resuspension of bottom sediments occurs. In one location, the amount of this resuspension is estimated to total 75 g C_{org} m^{-2} yr^{-1} at a depth of 100 m (Table I), or nearly ten

TABLE I. Organic Carbon Fluxes in Lake Michigan[a]

Element of carbon cycle	Flux (gC_{org} m^{-2} yr^{-1})	Percentage
Primary production	139	100
Oxidation in epilimnion	−115	−83
Sinking into hypolimnion	24	17
Oxidation in hypolimnion	−16	−11
Sinking to sediment surface	8	6
Resuspension into hypolimnion	75	54

[a]Data from Eadie *et al.* (1984); a 100-m-deep water column is assumed.

times the annual flux of organic carbon to the bottom. Organic matter so resuspended is exposed to renewed oxidation in the water column. In addition to resuspension, biological mixing, or bioturbation, of surface sediments prolongs the exposure of organic matter to oxidation. The benthic animals mixing the sediments impose a further degradative demand on the organic matter through their nutritional use of it.

Once buried below the zone of bioturbation, organic matter is much more slowly altered. Sulfate-reducing bacteria are less common in freshwater than in marine sediments because of the lesser supply of dissolved sulfate in most lakes. Methanogenic bacteria, however, are active below the zone of bioturbation in lacustrine sediments. As a consequence, the postdepositional degradation of organic matter continues. For example, Johnson *et al.* (1982) calculated a half-life of about 100 yr for organic carbon in the sediments of Lake Superior.

2. Indicators of Early Diagenesis of Organic Matter in Lacustrine Sediments

2.1. Introduction

Alteration and degradation of organic matter components during their sinking to lake bottoms modify the overall character of sedimented organic matter. Because organic matter is a mixture of many different molecules and combinations of molecules, degradative processes selectively alter its geochemical character. Less reactive forms of organic matter become more dominant as the more reactive and water-soluble forms are utilized by microbes and by deposit-feeding animals for food. As an example, Kemp and Johnston (1979) noted that humic substances and lipids survive postdepositional degradative processes better than do proteinaceous compounds and carbohydrates and consequently become larger proportions of the total organic matter in deeper sediments of lakes Erie, Huron, and Ontario.

Throughout this reprocessing, residues of microbially synthesized organic matter are added to the mixture, some to be in turn modified or destroyed, but others to be preserved. Ultimately, the sediment organic matter is a small fraction of the starting material. Nonetheless, many characteristics of the initial materials are preserved, and they provide evidence of the effects of early diagenesis in reshaping the geochemical character of sediment organic matter.

2.2. Effects of Early Diagenesis on the Carbon Stable Isotope Ratios of Organic Matter

The dominant pathway used by photosynthetic plants to incorporate carbon into organic matter is the C_3 Calvin process, which preferentially incorporates ^{12}C into organic matter, producing an average shift of −20‰ from the carbon isotope ratio of the inorganic carbon source. Some plants use the C_4 Hatch–Slack pathway, and others, mostly succulents, utilize the crassulacean acid metabolism (CAM) pathway. The isotope shift for the C_4 pathway is −8 to −12‰; that of the CAM pathway can vary from −20 to −10‰.

The existence of the characteristic C_3 and C_4 isotopic signatures of photosynthetically fixed organic carbon offers opportunities to trace sources of organic matter in lacustrine food webs and incorporated into lake sediments. In most lakes, however, the differences in plant isotope contents evidently are erased in sedimented organic matter by food web integration of the different carbon sources. For example, photosynthesizers living in Fayetteville Green Lake, New York, have carbon isotope values ranging from −15‰ to −41‰. Despite this range among available food sources, most animals in the lake have isotope signatures between −25‰ and −30‰, and bottom sediments have a value of −28‰ (Fry, 1986). The sediment $\delta^{13}C$ value is typical of C_3 land plants, in which a biological shift of −20‰ is superimposed on the −7‰ value of atmospheric CO_2.

Microbial reworking of organic matter during early diagenesis has the potential to modify the original isotopic content of organic matter. Organic matter is a mixture of different types of compounds which deviate differently in isotopic content from the inorganic carbon pool. Amino acids, for example, are depleted in ^{13}C by −17‰ on average, whereas lignin has a value of −23‰ (Degens, 1969). Selective loss of more reactive fractions of the total organic matter may create a diagenetic shift in the isotope ratio of bulk organic matter which will mask the original source signature. The magnitude of this possible effect of diagenesis is illustrated by a study in which the C_4 marsh grass *Spartina alterniflora* was decomposed under controlled laboratory conditions. Progressive losses of nonlignin material created an isotope shift which was estimated to change the original ratio from −13‰ to −17‰ (Benner *et al.*, 1987). A similar isotopic shift of −4‰ is found in a sediment core from Mangrove Lake, Bermuda (Hatcher *et al.*, 1983). The isotope ratio of total organic carbon changes from

−17‰ in surface sediments to −21‰ in sediments below 5 m in this core, most likely because of selective loss of the isotopically heavy carbohydrate fraction of total organic matter (Spiker and Hatcher, 1984). In the organic-carbon-rich sediment of this 2-m-deep lake, the diagenetic isotope shift is complete in less than 4000 yr (Hatcher *et al.*, 1983).

Diagenetic isotopic equilibrium appears to be achieved faster in lakes where low organic carbon concentrations indicate more extensive microbial reworking, with the result that a downcore isotope shift is not found in their sediment organic matter. In a core from Lake Michigan, for example, $\delta^{13}C$ values remain at −26‰ in sediments spanning modern deposition to 3500 yr B.P. in which organic carbon concentrations were between 1 and 3% (Rea *et al.*, 1980), indicating that no diagenetic isotope shift took place in these sediments. Similarly, Jasper and Gagosian (1989) reported that organic matter source controls the $\delta^{13}C$ values of a 100-ka sediment core from the Pygmy Basin in the Gulf of Mexico. Fluctuations from −22.5‰ to −26.5‰ are caused by shifts in terrigenous/marine inputs, and diagenetic isotope shifts appear not to be significant. Finally, Nakai (1986) presented $\delta^{13}C$ data from a core of sediment from Lake Biwa, Japan, that preserve evidence of changes in proportions of land-derived and aquatic-sourced organic matter in deposits laid down as long ago as the Brunhes–Matayama paleomagnetic reversal 730,000 yr B.P. It seems that in sediments having organic carbon concentrations below a few percent, no diagenetic isotope effect occurs. These are conditions typical of oligotrophic and mesotrophic lakes.

2.3. Effects of Early Diagenesis on Organic Matter C/N Ratios

The elemental composition of organic matter can be modified by microbial reworking during early diagenesis. Comparison of the C/N ratios of a sample of living spruce wood with that of wood that had been buried in a lake bottom for 10,000 yr shows a diagenetic decrease in the C/N ratio from 46 to 20 (Bourbonniere, 1979). A similar change, although less dramatic, occurs during early deposition of organic matter in Lake Michigan. Settling particles have a C/N value of 9; the ratio in resuspended bottom sediments is 8 (Meyers *et al.*, 1984a). The lowering of C/N ratios has also been observed in soils (e.g., Sollins *et al.*, 1984), where it involves the microbial immobilization of nitrogenous material accompanied by the remineralization of carbon. The magnitude of these changes in the elemental composition of organic matter is not large enough, however, to erase completely the difference between vascular and nonvascular plants.

After permanent burial in lake bottoms, C/N ratios appear to be insensitive to further diagenetic alteration. The C/N values of Quaternary sediment samples from the upper 10 m of a core from Mangrove Lake vary irregularly between 11 and 17 (Fig. 2). This variation resembles a similar downcore C/N profile

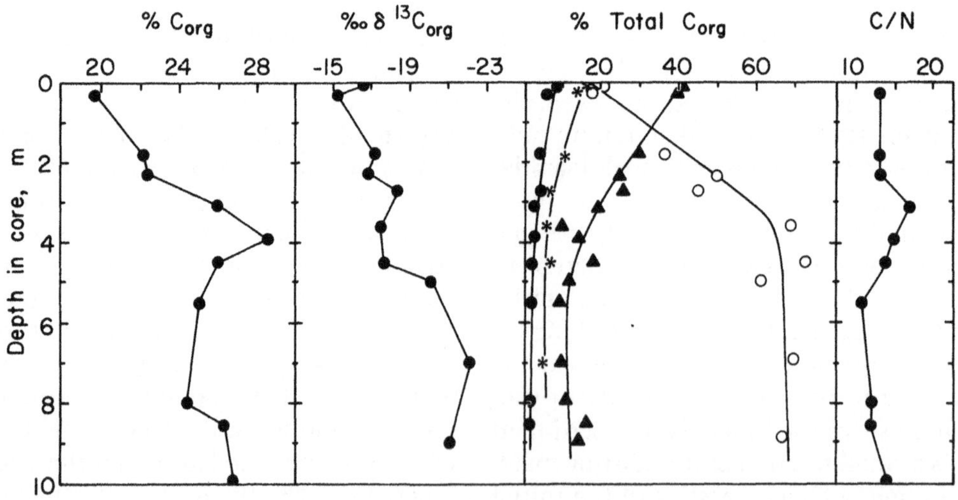

Figure 2. Organic carbon concentrations, isotope values, and compositions at different depths in a sediment core from Mangrove Lake, Bermuda. Isotope ratios are relative to the PDB standard. Organic carbon fractions are proteins (●), lipids (*), carbohydrates (▲), and humin (○). Concentrations of these fractions are expressed as percentage contribution of carbon to the total sediment organic carbon. The downcore increase in humin contribution reflects decreases in the contributions of the other fractions with progressively greater sediment age, which is less than 4000 yr. Data are from Hatcher *et al.* (1982, 1983).

from Lake Yunoko, Japan, which Ishiwatari *et al.* (1977) interpreted to reflect changes in the source mixture. In contrast to the apparent preservation of C/N values in the sediments of Mangrove Lake, significant diagenetic alteration trends are seen in the carbon isotope contents and the organic matter composition (Fig. 2). Preservation of the C/N source signal was also found by Ertel and Hedges (1985), who demonstrated that vascular plant debris isolated from the organic matter of a sediment sample having an averaged C/N ratio of 15 retained the elevated C/N values characteristic of cellulosic plants. In their study of the carbon isotope ratios and C/N values of organic matter in Quaternary sediments from the northern Gulf of Mexico, Jasper and Gagosian (1989) found that C/N ratios reliably recorded changes in the proportional contributions of land-derived and marine materials over a time span of 100,000 yr. The record of changes in land-derived and aquatic dominance of organic matter supply is preserved in the organic C/N ratios of Lake Biwa sediments deposited over 730 ka (Nakai, 1986). Diagenesis evidently modifies, but does not drastically change, the source signal contained in C/N ratios of organic matter dispersed in sediments.

Sources of organic matter incorporated into lake sediments deposited in the past can be inferred from C/N ratios. In one example, Ishiwatari and Uzaki (1987) reported a sharp transition in atomic C/N values from an average of 9 to an average of 13 at the 250-m level in a drilled core of sediment from Lake Biwa in Japan. They interpreted this change to record a shift in dominance of organic matter source from aquatic plants in younger sediments to land plants in sediments deposited more than 630 ka. This interpretation is consistent with a change in sediment types from fine-grained, deep-water sediment to sands and gravels at the 250-m depth. Another example is from the sediments of Pleistocene Lake Karewa in India. Several marked increases to values of 10 or more occur in C/N ratios that typically remain below 4 in sediments deposited between 2.4 Ma and 400 ka (Krishnamurthy *et al.*, 1986). Lighter $\delta^{13}C$ values in this core interval indicate that land plant contributions to sediment organic matter increased at these times.

2.4. Early Diagenesis of Humic Substances

Humic substances are diagenetically produced from biological organic matter by the alteration and recombination of precursor compounds into macromolecular mixtures. Humic substances are operationally divided into fulvic acid (soluble in base and acid), humic acid (soluble in base but not acid), and humin (insoluble in base and acid). Humin is sometimes called "protokerogen" because its definition is the same as that of kerogen, which is present in lithified sediments. Humic substances typically constitute between 60 and 70% of the organic matter in young lacustrine sediments and more than 90% in older sediments (Ishiwatari, 1985).

The formation of humic substances evidently involves incorporation of both altered and intact biochemical compounds into the geochemical organic matrix. The process is probably microbially mediated, but the specifics of the process are not known. The fact that humic substances constitute the majority of organic matter in surficial lacustrine sediments indicates rapid formation. Microbial alteration and recombination of biochemical materials is probably one element in the formation of humic substances. Selective preservation of high-molecular-weight biochemical compounds is another aspect that may be involved.

Because the chemical structures grouped as humic substances are complex and diverse, studies of this organic matter fraction have traditionally employed general characterizations, such as elemental and spectroscopic analyses. These studies show that humic substances, the products of rapid diagenetic alteration of biogenic organic matter, undergo further slow diagenesis after burial in lake sediments. Bourbonniere and Meyers (1978) used infrared and visible spectroscopy to detect evidence for the condensation of fulvic acid into humic acid in a core of Lake Huron sediment. The effects of this diagenetic conversion are minor, inasmuch as H/C ratios of the fulvic acid (ca. 2.2) and the humic acid (ca. 1.6) do not change downcore. The oberservation of H/C values greater than one indicates the aliphatic nature and probable aquatic origin of these humic substances in Lake Huron. The lack of substantial change in the H/C values indicates that the 550-year period represented in this core is inadequate for major diagenetic effects.

Over multi-thousand-year periods of time, the effects of humic substance diagenesis in lake sediments become more evident. The fulvic acid, humic acid, and humin fractions isolated from a core of Lake Michigan sediment illustrate some of these effects (Fig. 3). The humin fraction increases from 65% of the total humics at the sediment surface to 85% of the total at the bottom of this 1-m core, which is dated to be ca. 3500 years old (Rea *et al.*, 1980). Some of this increase may arise from conversion of fulvic and

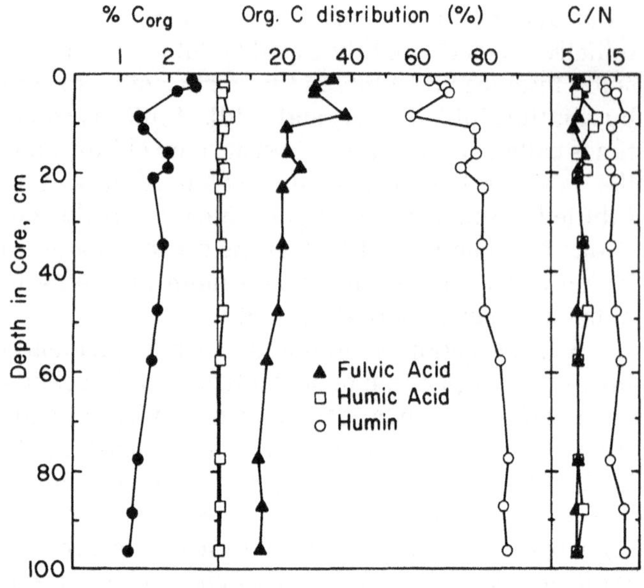

Figure 3. Fulvic acid, humic acid, and humin C/N values and proportional contributions to total organic carbon in a core of Lake Michigan sediment. Sediment age is from modern to 3500 yr B.P. The downcore decrease in organic carbon concentration is mostly from decreases in fulvic and humic acid concentrations, producing a proportional increase in the relative contribution of humin. Atomic C/N ratios are not affected by the diagenetic changes. Core location, sedimentology, and dating are described by Rea *et al.* (1980) and Bourbonniere (1979).

humic acids to humin, but the overall decrease in total sediment organic carbon indicates that diagenetic loss of organic matter continues throughout the core and preferentially degrades fulvic and humic acids. Despite this evidence of continued degradation of organic matter in this core, the $\delta^{13}C$ ratio of total organic carbon does not deviate from −26.3‰ (Rea *et al.*, 1980). The C/N ratios of both fulvic and humic acids remain ca. 7 (Fig. 3) and suggest a common origin from aquatic sources. Although the higher C/N ratio of humin suggests an important component of cellulosic land plant material, humin typically has an atomic C/N value of 14 to 16 (Ishiwatari, 1985). The difference between the C/N ratios of fulvic and humic acids and humin may result from diagenetic loss of nitrogen instead of different starting materials.

Comparisons of samples of wood from modern trees and of wood that has been buried for thousands of years in sediments have been used to investigate the early diagenesis of humic substances (e.g., Meyers *et al.*, 1980b; Hedges *et al.*, 1985; Spiker and Hatcher, 1987). A consistent observation of these studies is that cellulose, a polymer made up of aliphatic carbohydrate monomers, decreases in relation to lignin, which is a polyphenolic aromatic macromolecule.

Lignin components constitute up to one-third of sediment organic matter and therefore are major constituents of humic matter. Diagenetic changes in lignin composition are evident in lake sediments. The ratio of vanillic acid to the related aldehyde, vanillin, increases with depth in sediments of Lake Biwa (Ishiwatari and Uzaki, 1987). Vanillic acid is the more oxidized component in this ratio, suggesting continued oxidative degradation of lignins with time in this core. Furthermore, the ratio is 0.5 in surface sediments, which is substantially higher than found in fresh plant material (0.15, from Hedges *et al.*, 1982), indicating that the lignins delivered to Lake Biwa have been extensively oxidized during transport to the lake. Further evidence of lignin degradation comes from studies of wood buried in sediments for up to 25,000 yr, which shows that as much as 30% of the original amount of syringyl components can be diagenetically lost from lignins, whereas vanillyl components seem to be preserved (Hedges *et al.*, 1985).

3. Molecular Indicators of Early Diagenesis of Organic Matter

3.1. Effects of Early Diagenesis on the Geolipid Contents of Lacustrine Sediments

Lipids are operationally defined as the fraction of organic matter which is isolated from biological material by extraction with an organic solvent, such as methanol, dichloromethane, toluene, or hexane. This extraction yields hydrocarbons and the hydrocarbon-like carboxylic acids, alcohols, ketones, aldehydes, and related compounds. These latter compounds are often combined as esters and triacylglycerols. Most lipids have molecular structures which reflect their enzymatically controlled biosynthesis, being built of either acetyl units or branched isoprenoid units. These two main pathways yield, respectively, straight-chain compounds and branched and cyclic compounds.

The lipid content of lacustrine sediments commonly differs from the lipids in biological matter, although the operational definition remains the same. Many of the original compounds having oxygen-containing functional groups and carbon–carbon double bonds are absent. Because of these early diagenetic modifications of molecular compositions, the term "geolipid" is often applied to the solvent-extractable fraction of sedimentary organic matter.

The original carbon skeleton of many molecules, even diagenetically altered ones, is often preserved. The retention of their biological heritage earns the name "biological marker," commonly abbreviated to "biomarker," for many of these compounds. This characteristic of the geolipid fraction makes it important to geochemical studies, even though it typically constitutes only a few percent of the total sediment organic matter.

3.1.1. Alteration of Lipids during Sedimentation

Studies of settling particles intercepted by sediment traps indicate that lipids undergo degradative alterations during sinking to the bottoms of lakes. Proportional contributions of alkanoic acids and of alkanols become smaller in settling sediment collected at progressively deeper depths in Lake Michigan (Meyers *et al.*, 1980b, 1984a). As the contributions diminish, the mixture of molecular components comprising each of these lipid fractions undergoes change. The n-C_{16} and n-C_{18} acids and alcohols which distinguish algal material become less dominant, whereas the n-C_{24}, n-C_{26}, n-C_{28}, and n-C_{30} components contributed from the waxy coatings of land plants become proportionally more important. Similarly, Kawamura *et al.* (1987) found that distributions of n-alkanes in surficial sediments from Lake Haruna, Japan, were depleted in n-C_{17}, an algal biomarker, relative to the hydrocarbon contents of sediment-trap material in this lake. Aquatic lipid matter is evidently preferentially degraded during the sinking of particulate material. The apparent greater susceptibility of the aquatic fraction to alteration reflects the relative freshness of this material. In contrast, land-derived organic matter has been microbially reworked before arrival in the lake; only the least degradable components have survived transport from their land sources. These components also survive well during incorporation into lake sediments.

Comparison of the lipid compositions of aquatic organisms with those of underlying sediments shows degradation trends similar to those found in sediment-trap studies. Robinson *et al.* (1984a) analyzed the lipid contents of bacteria, rotifers, and protozoans from Priest Pot, a eutrophic lake in the English Lake District. They found that shorter-carbon-chain n-alkanols degrade faster than their longer homologs and that sterols having two double bonds do not survive sinking to the lake bottom. Land plant components are preferentially preserved in surface sediments. Alteration of geolipid compositions continues in the sediments of this lake, most intensely in the surface layer. In particular, easily extractable geolipids become less abundant, and "bound" geolipids, those that can be isolated only after acid or base hydrolysis of sediment samples, increase. Also, evidence of microbial oxidation of n-alkanes, the least reactive type of geolipid, is found in the appearance of n-alkan-2-ones having chain-length distributions similar to those of the hydrocarbons (Cranwell *et al.*, 1987). As seen in sediment-trap studies, the lipid constituents of aquatic biota experience greater amounts of degradation and alteration than do those of plant biota before becoming incorporated in bottom sediments. Goossens *et al.* (1989) show that land plant lipids dominate the biomarker content of modern sediment from Lake Vechten, The Netherlands, although carbon budgets indicate that land contributions comprise a small part of the input to the lake system. Preferential loss of many of the algal components during early diagenesis biases the eventual record of geolipid sources contained in lacustrine sediments.

3.1.2. Selective Diagenesis of Geolipids in Sediments

The geolipid content of sediments is influenced both by the original sources of organic matter and by diagenesis during and after sedimentation. Some types of geolipid compounds are less likely to be altered by diagenesis than are others, and comparisons between easily altered and less easily altered molecules help delineate diagenetic changes from source changes. Hydrocarbons, for example, are not as subject to postdepositional alteration as are lipids having oxygen-containing functional groups. Within the n-alkane fraction of hydrocarbons, longer-chain-length molecules are less degradable than are their shorter-chain-length counterparts. As an illustration, the longer carbon-chain-length components of the extractable n-alkane distribution of sediments of Priest Pot retain a record of a change in watershed plant cover which has occurred since 1900. The dominant n-alkane has shifted from being n-C_{27} in older sediments to n-C_{31} in modern sediment as swamp and shore grasses replaced trees as the major source of land-derived organic matter (Cranwell *et al.*, 1987). In contrast to evident preservation of the long-chain n-alkanes, the short-chain components, indicative of aquatic contributions, which are large in this productive lake, display substantial diagenetic loss.

A study of surficial sediments from four lakes in Japan having different rates of aquatic production provides a comparison of the effects of selective microbial reworking of geolipid distributions and trophic status (Kawamura and Ishiwatari, 1985). In the oligotrophic lakes Motosu and Biwa, the *n*-alkane distributions of sediments are dominated by the C_{29}, C_{31}, and C_{33} components diagnostic of land-plant waxes (Fig. 4). The *n*-alkanol distributions from these two low-productivity lakes contain dominant-to-large proportions of long-chain, plant-wax components as well. In the *n*-alkanoic distributions, however, the C_{16} component is dominant, although their bimodal distributions reveal important amounts of land-plant-derived fatty acids. The C_{16} *n*-alkanoic acid is ubiquitous in the biosphere; it can be found in land plants, in aquatic algae, and in bacteria and other microbes. Its dominance in these sediments, in which both the *n*-alkanes and the *n*-alkanols record major contributions of land-plant lipids, indicates that microbial utilization and resynthesis of the more reactive fatty acids has occurred. The sediment from eutrophic Lake Suwa contains a notable amount of the C_{17} *n*-alkane diagnostic of algal production, and the *n*-alkanol and *n*-alkanoic acid fractions are dominated by C_{16} components. The relatively small contributions of long-chain acids and alcohols indicate that contributions of land-plant lipids to the sediment of this lake are small compared to the aquatic contributions. The significant difference in chain-length distributions between the *n*-alkanes and the *n*-alkanoic acids of the Lake Suwa sediments again illustrates the selective loss of aquatic components of organic matter relative to land-derived material.

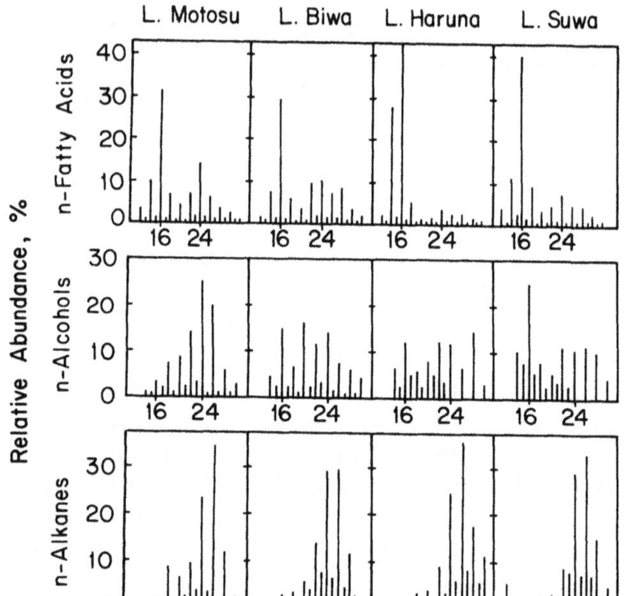

Figure 4. Chain-length distributions of *n*-alkanes, *n*-alkanols, and *n*-alkanoic acids in surficial sediments from four Japanese lakes. Proportions of the components in each lipid fraction are given as percentage of the total fraction. Long-chain-length components originate from land plants, whereas short components are produced by aquatic organisms. The productivities of the lakes increase progressively from oligotrophic Lake Motosu to eutrophic Lake Suwa, and the relative proportions of long to short components decrease as lake productivities increase. Data are from Kawamura and Ishiwatari (1985).

A type of geolipid material which is relatively resistant to diagenetic alteration is the fraction comprised of intact, high-molecular-weight wax esters. Fukushima and Ishiwatari (1984) compared the molecular distributions of the component fatty acids and alcohols comprising the wax esters in surface sediments of three lakes in Japan. Those from oligotrophic Lake Motosu are dominated by the C_{24} *n*-alkanoic acids and *n*-alkanols characteristic of land-plant epicuticular waxes, whereas those from eutrophic Lake Shoji contain a predominance of the *n*-C_{16} components expected from aquatic plants and few of the long-chain, land-plant materials. The sediment from mesotrophic Lake Haruna displays an intermediate wax ester composition, having the *n*-C_{16} acid and *n*-C_{24} alkanol as the major contributions to their respective distributions. The wax esters, having low solubilities and presumably low susceptibilities to microbial utilization, record the relative proportions of aquatic and watershed plant origins of the geolipid fraction of sediment organic matter.

3.1.3. Early Diagenesis of Fatty Acids in Lacustrine Sediments

Some types of fatty acids are more susceptible to early diagenesis than are others, so that comparisons between different components can distinguish diagenetic effects from source changes. Furthermore, variations in preservational conditions can be reflected by the extent of diagenetic alterations of different components of the fatty acid contents of sediments.

The results of a study by Matsuda and Koyama (1977a, b) of a core from Lake Suwa, Japan, illustrate some of the effects of diagenesis and of source change on the fatty acid contents of sediments. As seen in Fig. 5, the concentrations of total fatty acids and of total organic carbon decrease rapidly over the top 25 cm of the 150-cm-long core. The decrease in the ratio of fatty acid carbon to total organic carbon shows that this lipid fraction is degraded more readily than is total organic matter in the upper sediments of this core, presumably by microbial activity. The magni-

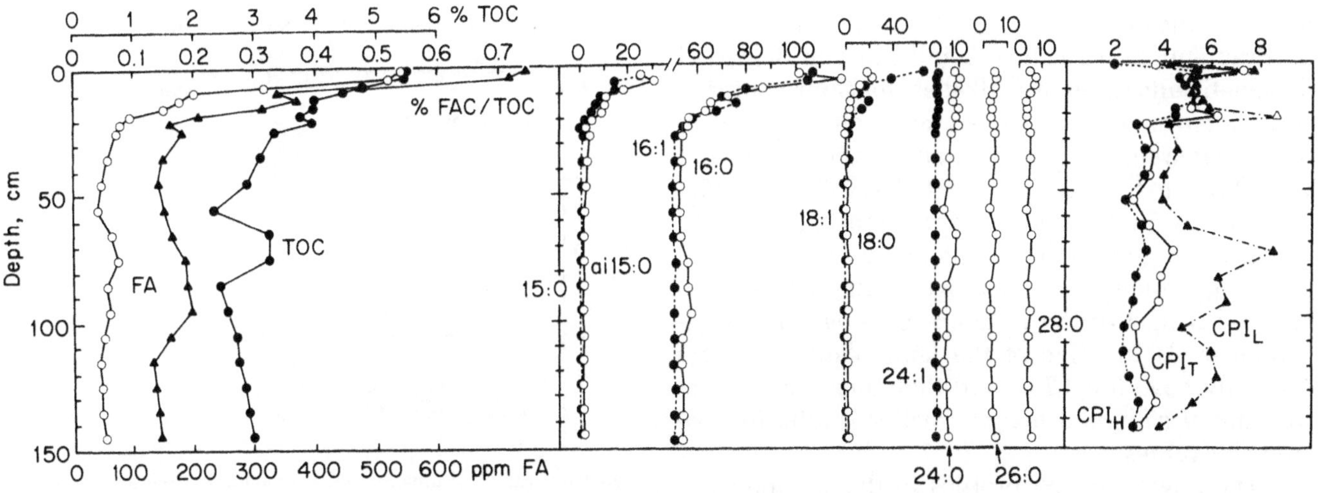

$$CPI_T = \frac{1}{2}\,\frac{(\Sigma C_{12}\text{-}C_{16} + \Sigma C_{22}\text{-}C_{30}) + (\Sigma C_{14}\text{-}C_{18} + \Sigma C_{24}\text{-}C_{32})}{\Sigma C_{13}\text{-}C_{17} + \Sigma C_{23}\text{-}C_{31}}$$

$$CPI_L = \frac{1}{2}\,\frac{\Sigma C_{12}\text{-}C_{16} + \Sigma C_{14}\text{-}C_{18}}{\Sigma C_{13}\text{-}C_{17}}$$

$$CPI_H = \frac{1}{2}\,\frac{\Sigma C_{22}\text{-}C_{30} + \Sigma C_{24}\text{-}C_{32}}{\Sigma C_{23}\text{-}C_{31}}$$

Figure 5. Downcore concentrations of total fatty acids and of representative components of the total acids in parts per million in sediments of Lake Suwa, Japan. Ratios of total fatty acid carbon to total organic matter are given as FAC/TOC. Carbon preference indices of total alkanoic acids (CPI_T), of higher molecular weight acids (CPI_H), and of lower molecular weight acids (CPI_L) are shown. Major shifts in all parameters at a sediment depth of 20 cm record modern eutrophication of Lake Suwa. Microbial reprocessing of fatty acids in surface sediments is also evident from the distribution of C_{15} acids. Data are from Matsuda and Koyama (1977a, b).

tude of the decrease with depth is different for separate components of the fatty acid fraction. Long-chain components, represented by 24:0, 26:0, and 28:0 (number of carbon atoms:number of double bonds) in Fig. 5, change little with depth. Short-chain acids, such as 15:0, 16:0, and 18:0, decrease markedly in the upper part of the core. Shorter chain unsaturated acids, typified by 16:1 and 18:1, have higher concentrations than saturated acids of the same carbon chain length in the topmost sediment sample but disappear faster than their saturated analogs. In contrast, the concentration of the long-chain 24:1 alkenoic acid appears to experience little change with depth. Shorter chain molecules appear to be more susceptible to early diagenesis than long-chain fatty acids.

The fatty acid components in Lake Suwa sediments may have different sources, as suggested by their chain lengths, which influence their diagenetic susceptibility. The long-chain components probably originate from land plants, and they are the less reactive survivors of the transport of land-derived debris to the lake bottom. The shorter chain components, particularly the alkenoic acids, are products of algal and microbial liposynthesis in the lake. These lipids are easily metabolized by sediment microbes. The *n*-15:0 and *anteiso*-15:0 are indicators of microbial fatty acids (cf. Cranwell, 1973). They represent *in situ* production of secondary lipids at the expense of original forms of organic matter. Most of the concentration changes occur in the top 25 cm of these sediments, corresponding to 60 yr, from an estimated sedimentation rate of 4 mm yr^{-1} (Matsuda and Koyama, 1977a). Although the decrease in the contribution of fatty acids to total organic matter (FAC/TOC in Fig. 5) is compelling evidence for diagenetic losses of these lipids, it is probable that there have been source changes as a result of the documented increasing algal productivity of Lake Suwa over this time. The lower concentrations of the C_{16} and C_{18} acids in deeper sediments may arise jointly from formerly lower production and from diagenetic removal. The similar pattern for the C_{15} acids may reflect the presence of larger microbial populations in response to larger amounts of easily metabolized organic matter in younger sediments than in older deposits.

Matsuda and Koyama (1977b) examined the possibility that source changes as well as diagenetic changes impact downcore fatty acid compositions of Lake Suwa sediments by comparing sediment contents to those of potential source materials. They modified the carbon preference index (CPI), devised by Bray and Evans (1961) to indicate the degree of diagenetic alteration of *n*-alkanes, for their comparison. Moreover, they separated fatty acids into lower molecular weight (C_{12} to C_{18}) and higher molecular weight (C_{22} to C_{32}) groups, denoted as CPI_L and CPI_H, respectively. In the lipid content of most plants, CPI_L values are high (Fig. 5), ranging between 12 and ca. 100; CPI_H values are not as high, being between 0.9 and 8, because of the presence of long-chain odd-carbon acids in some of the land plants. Bacteria generally have low CPIs, and therefore bacterial reworking of sediment organic matter results in lowered CPI values.

The downcore CPI profiles in the sediments of Lake Suwa are variable, and they are lower than CPI values of potential source materials (Fig. 5). The variability indicates that a combination of source changes and of changes in degree of microbial reworking are involved. The index for all *n*-alkanoic acids, CPI_T, is similar to CPI_L in the upper ca. 20 cm of sediment and becomes more like CPI_H in sediments lower than 25 cm in the core. The generally lower CPI values imply that microbial resynthesis of fatty acids has been important in erasing the original compositions. Much of this resynthesis has evidently occurred in the upper part of the core as part of the early diagenesis of this geolipid fraction.

3.1.3a. Effects of Carbon–Carbon Double Bonds. Unsaturated fatty acids are more susceptible to alteration in sediments than are saturated acids. Unsaturated C_{16} and C_{18} acids are major constituents of the lipids of freshwater algae, and they are rapidly degraded by microbial attack (Cranwell, 1974). In a dated core of sediment from Lake Haruna, Japan, ratios of $C_{16:1}$, $C_{18:1}$, $C_{18:2}$, and $C_{18:3}$ to their corresponding alkanoic acids decreased by a factor of ten in the upper 8 cm of sediment (Fig. 6). Below this depth, corresponding to 120 yr of sediment accumulation, little unsaturated acid remained, and consequently the ratios decreased much more slowly (Kawamura *et al.*, 1980). The change with depth found in Lake Haruna sediments is similar to the unsaturated-to-saturated change also observed in Lake Suwa sediments (Fig. 5) and in sediments of Lake Huron (Meyers *et al.*, 1980c). The existence of carbon–carbon double bonds evidently facilitates microbial utilization of these lipids and enhances their rapid degradation.

3.1.3b. Effects of Depositional Conditions. Factors in addition to the molecular compositions of fatty acids participate in affecting the amount and type of these lipids in lake sediments. Subtle differences in depositional parameters, such as sedimentation rates and degree of lake water oxygenation, impact the preservation of the relatively reactive fatty acids incorporated in sediments laid down at different times.

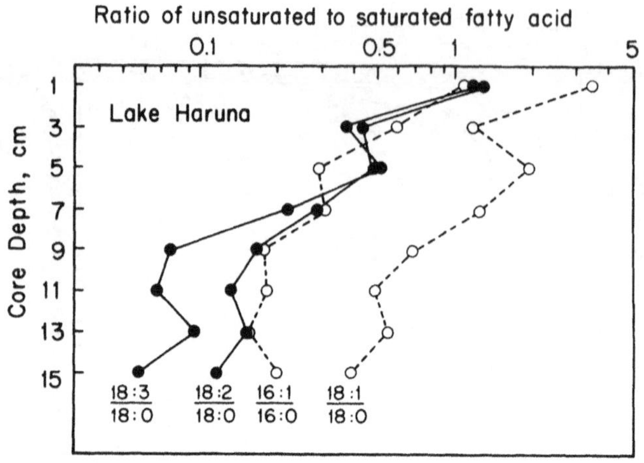

Figure 6. Ratios of unsaturated fatty acids to their saturated analogs at different core depths in sediments of Lake Haruna, Japan. The downcore decreases in these ratios result from the greater susceptibility of unsaturated compounds to microbial degradation. Data are from Kawamura *et al.* (1980).

Two examples illustrate the effects of changing sedimentation rates. Mendoza *et al.* (1987) reported fatty acid compositions of a core of sediment from Lake Leman, Switzerland, in which the expected downcore decreases in concentration and in proportion of unsaturated acids are not found (Fig. 7). Concentrations of lower-molecular-weight and higher-molecular-weight acids begin to decrease over the top meter of this core, but they increase in sediments between 2.5 and 3.5 m deep, presumably because of an interlude of higher sedimentation rate. Unsaturated C_{16} and C_{18} acids show a similar pattern, indicating better preservation during the time of faster burial.

A dramatic example of enhanced preservation of fatty acids through faster burial in the sediments of Pyramid Lake, Nevada, is given by Meyers *et al.* (1980d). In this setting, the concentrations of total fatty acids increase nearly tenfold with depth, and the proportions of unsaturated acids more than double over the top 25 cm of the sediment core (Fig. 8). The cause of these atypical patterns is postulated to be slumping of deltaic sediments. This episode of rapid burial removed surface sediments from prolonged exposure to the aerobic bottom waters of this lake and from continued microbial reprocessing in the bioturbated zone. Source changes can be ruled

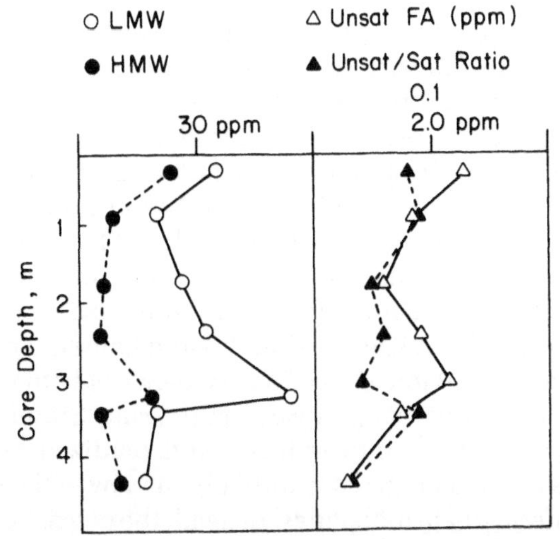

Figure 7. Concentrations of total lower molecular weight fatty acids (LMW), total higher molecular weight fatty acids (HMW), and total unsaturated fatty acids at different core depths in sediments of Lake Leman, Switzerland. The ratio of total unsaturated acids to total saturated acids is also shown. Downcore degradation of fatty acids is interrupted between 2 and 4 m sub-bottom by a period of increased sedimentation rate, which improved preservation of these reactive lipids. Data are from Mendoza *et al.* (1987).

out, inasmuch as total hydrocarbon concentrations show no enhanced inputs of lipids over this sediment depth range (Fig. 8). A source change to more terrigenous lipid sources and to better lipid preservation is indicated, however, by *n*-alkane and fatty acid compositions near the bottom of this core, corresponding to 1200 to 1400 A.D., illustrating that source changes can be distinguished from preservational effects in the geolipid contents of sediments.

3.1.4. Early Diagenesis of Cyclic Triterpenoids in Lacustrine Sediments

Cyclic triterpenoids include the steroid and hopanoid compounds, both of which have special geochemical significance. These lipids are synthesized from six isoprenoid units; hence, their nominal number of carbon atoms is 30. Deletion and addition of carbon side chains to the basic ring structures of these two types of lipids expands the range in carbon numbers of their members to between 26 and 34. Early diagenesis typically modifies the side chains, the functional groups, and the number of C=C bonds while preserving the ring geometry of these molecules.

3.1.4a. Steroid Compounds. Sterols are important biomarker compounds. The molecular structures of several representatives of this family of tetracyclic alcohols are shown in Fig. 9. The presence or absence of double bonds within the ring system, the positioning of methyl groups at various points on the ring system, the length of the branched side chain at the C_{17} position, and the stereochemistry of the substituent bonds are responsible for the many diverse sterols that have been identified (e.g., de Leeuw and Baass, 1986; Volkman, 1986). The structural diversity of sterols and their derivatives provides important

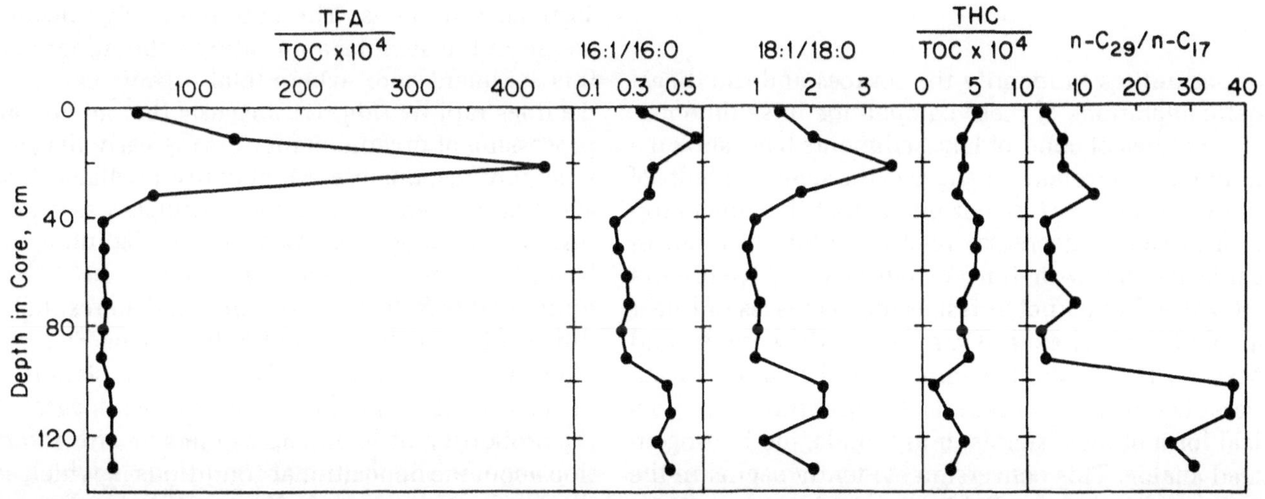

Figure 8. Evidence of changes in depositional conditions affecting lipid contents of sediments of Pyramid Lake, Nevada. Ratios of concentrations of total fatty acids (TFA) and total aliphatic hydrocarbons (THC) to total organic carbon (TOC), of monounsaturated C_{16} and C_{18} acids to their saturated analogs, and of C_{29}/C_{17} *n*-alkanes vary at different core depths. Enhanced preservation is indicated by maxima in the fatty acid parameters; enhanced contribution of land plant hydrocarbons is indicated by the increase in the C_{29}/C_{17} ratio. Data are from Meyers *et al.* (1980d).

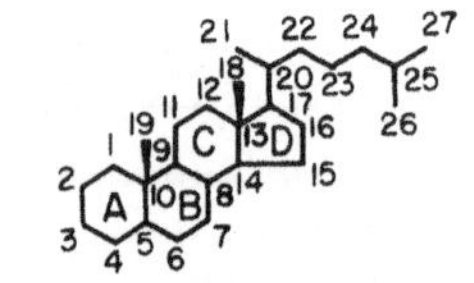

General steroid structure, with rings labelled and positions numbered

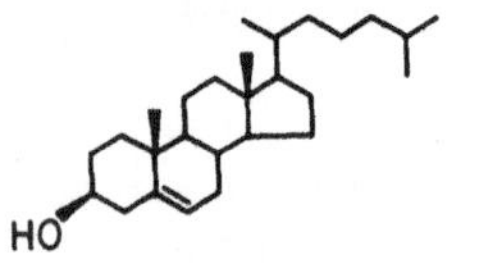

Cholest-5-en-3β-ol (Cholesterol)

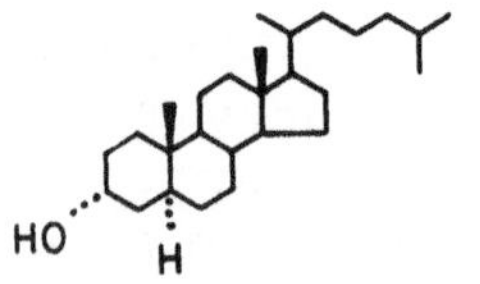

5α(H)-Cholestan-3α-ol (Epicholestanol)

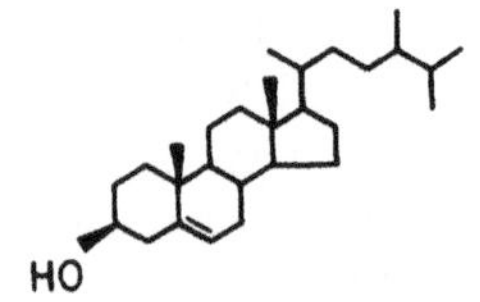

24-Methyl-cholest-5-en-3β-ol (Campesterol)

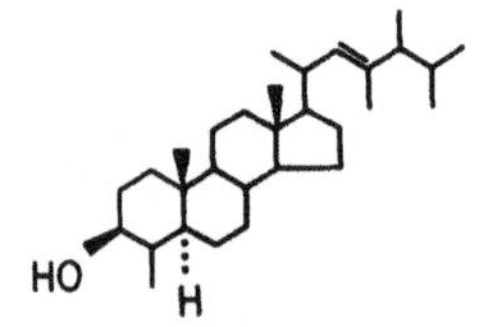

4,23,24-Trimethyl-5α(H)-cholest-22-en-3β-ol (Dinosterol)

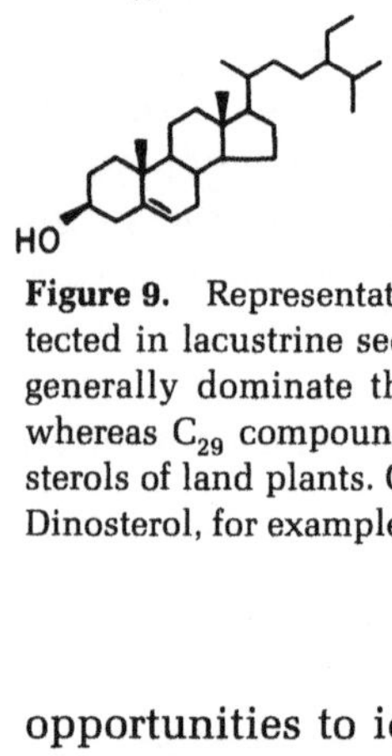

24-Ethyl-cholest-5-en-3β-ol (β-Sitosterol)

Figure 9. Representative sterol compounds which have been detected in lacustrine sediments. C_{27} compounds (e.g., cholesterol) generally dominate the sterol compositions of aquatic plants, whereas C_{29} compounds (e.g., sitosterols) constitute most of the sterols of land plants. Other sterols give greater source specificity. Dinosterol, for example, is indicative of dinoflagellate production.

opportunities to identify the sources and the diagenetic alterations of these compounds in sediments.

Studies of cores of lake sediments have shown a number of diagenetic changes in the sterol contents of these sediments. A major one is that the concentration of sterols decreases relative to that of total organic carbon with time of burial (e.g., Nishimura 1977a), although not as fast as the decreases of linear alcohols or fatty acids (Cranwell, 1981; Leenheer and Meyers, 1983). Another important change is the apparent conversion of stenols, the unsaturated biological form of most sterols, into stanols, the hydrogenated analog. This conversion evidently begins in the water column (e.g., Robinson *et al.*, 1984a, 1986; Wünsche *et al.*, 1987), with the consequence that both stenols and derived stanols are common in surficial lacustrine sediments.

Diagenetic alteration of the biological forms of sterols into their geochemical derivatives proceeds rapidly in the surface sediments of lakes, implying that the microorganisms that are abundant in these sediments are responsible. A variety of steroid compounds have been tentatively identified in the surficial mixed layer of sediments in Lake Haruna, Japan (R. Ishiwatari, unpublished results). Many of these compounds are intermediates in alteration pathways that lead to either saturation or aromatization of the steroid carbon skeleton (Fig. 10) and consequently to greater geochemical stability. Evidence of both reductive and oxidative pathways is present. Simple disproportionation into reduced and oxidized derivatives of the precursors is unlikely in view of the types of intermediate molecules. Instead, the coexistence of both pathways indicates a mixture of different types of microbes and microbial processes. Nearly all of the Lake Haruna steroids are steps in diagenetic pathways that have been described by Mackenzie *et al.*, (1982) and Brassell *et al.* (1983), except that the diketone (Fig. 10) is a newly identified compound. Also, steranes, the hydrocarbons derived from sterols, do not appear in lacustrine sediments until much later in diagenetic pathways.

Conversion of stenols to stanols and the resulting stabilization of the sterol carbon skeleton continues deeper into lake sediments. Evidence of this process is exemplified in Fig. 11, which shows the ratio of total unsaturated sterols (stenols) to total saturated sterols (stanols) in sediments of Lake Suwa, Japan. The stenol/stanol ratio is ca. 50 in the diatom type that dominates aquatic productivity in Lake Suwa (Nishimura and Koyama, 1976), yet it is only 5 in the surficial sediments. The continued, significant decrease in the stenol/stanol ratio in the upper part of this sediment core, where total organic carbon also declines rapidly (Fig. 11), suggests that microbial reprocessing of organic matter during early diagenesis is largely responsible. Microbially mediated hydrogenation of stenols has been postulated as a part of this reprocessing (e.g., Gaskell and Eglinton, 1976; Nishimura and Koyama, 1977; Reed, 1977; Nishimura, 1977b; R. Ishiwatari, unpublished results). Hydrogenation of the Δ^5 double bond (see Fig. 9) can yield two isomers, the 5α(*H*) and 5β(*H*) stanols. Evidence from field and laboratory studies suggests that the proportion of 5α and 5β isomers records information about the depositional conditions in which early diagenesis proceeded. Hydrogenation in oxygenated sedimentary environments yields predominantly the 5α form (Reed, 1977), whereas in sediments made anoxic by enhanced aquatic production of organic matter the 5β form dominates (Reed, 1977; Nishimura,

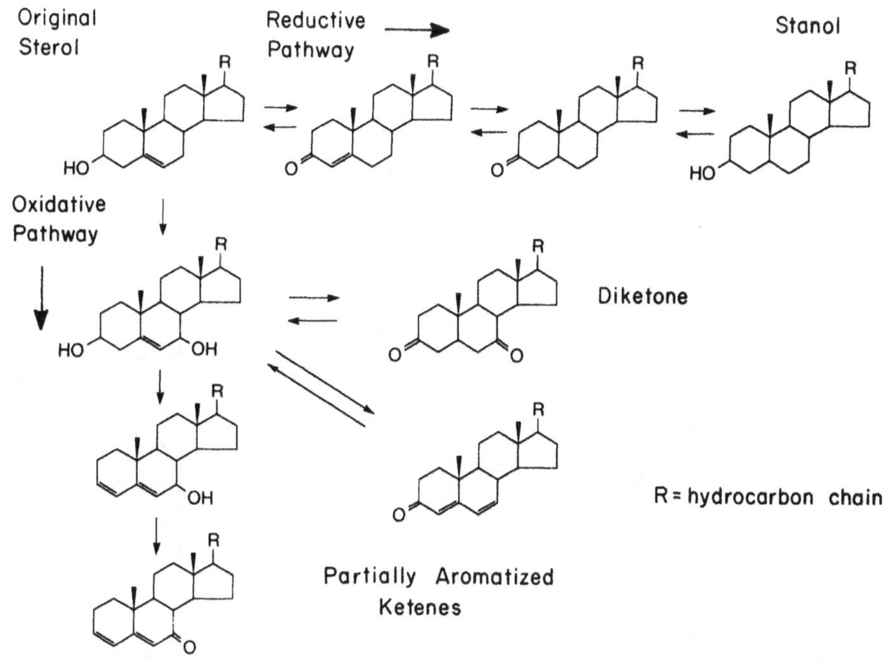

Figure 10. Preliminary identifications of products of diagenetic alteration of sterol precursors found in the surface sediments of Lake Haruna, Japan. These derived sterols illustrate the variety of possible diagenetic transformations and also demonstrate the retention of the source-identifying molecular carbon skeleton. Postulated diagenetic routes are indicated. (R. Ishiwatari, unpublished results.)

1982). An additional factor contributing to the shift away from stenol predominance in sediments is the preferential preservation of the small fractions of stanols originally synthesized by living organisms (Nishimura and Koyama, 1977; Nishimura, 1977b,

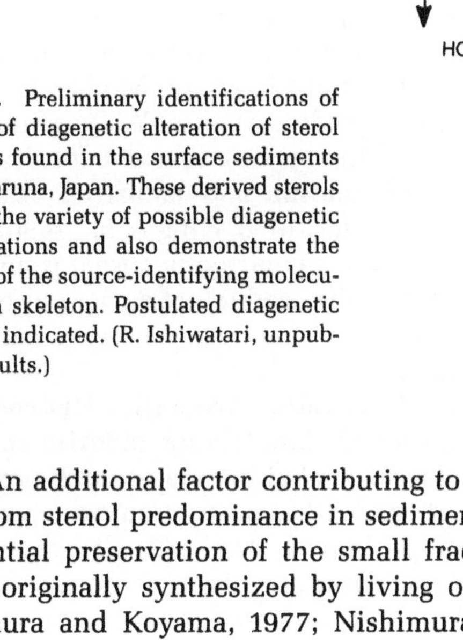

Figure 11. Downcore changes in the concentration of total organic carbon (TOC) and in ratio of total unsaturated sterols to total saturated sterols (stenol/stanol) in sediments of Lake Suwa, Japan. Similarity of the two depth profiles suggests that much of the change in the stenol/stanol ratio arises from selective preservation of stanols. Data adapted from Nishimura and Koyama (1976).

1978; Robinson *et al.*, 1984b). Preferential preservation may be involved in explaining why the stenol/stanol ratios of aquatic C_{27} sterols decrease more rapidly with sediment depth than do those of land-derived C_{29} sterols (Nishimura, 1978; Leenheer and Meyers, 1983).

In their comparison of sterol contents of sediments from three depths in Rostherne Mere, England, Gaskell and Eglinton (1976) found that cholesterol, campesterol, and β-sitosterol (see Fig. 9) dominated the composition of surface sediments. In deeper sections, cholesterol became a lesser component (Fig. 12), indicating that the change to the presently enhanced aquatic productivity of Rostherne Mere is recorded in the sterol contents of the sediments. The relative contributions of C_{27}, C_{28}, and C_{29} stenols and stanols are consistently similar in the three sediment sections studied (Fig. 12), although concentrations diminish by a factor of ten and the stenol/stanol ratios change from 3.1 to 0.2 over the total depth interval. Stenol/stanol ratios below unity indicate substantial diagenetic modification of the biochemical precursor compounds. An overwhelming predominance of 5α(*H*)-stanols suggests that sterol reduction had occurred in oxygenated bottom-water conditions. Retention of the relative molecular proportions indicates that the source information present in the carbon skeletons is preserved despite the effects of early diagenesis.

3.1.4b. Hopanoid Compounds. Hopanoid compounds are pentacyclic triterpenoids produced by

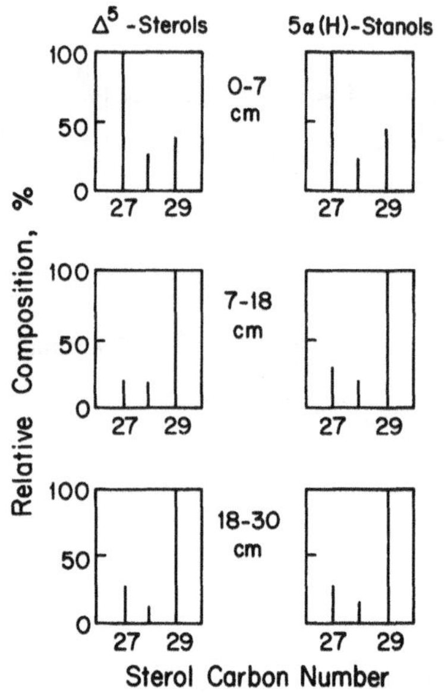

Figure 12. Relative contributions of C_{27}, C_{28}, and C_{29} unsaturated (Δ^5) and saturated [5α(*H*)] sterols to the sterol contents of sediments from Rostherne Mere, England. Contributions are shown relative to the major component of each sterol fraction. Dominance of C_{27} components in the surface sediment indicates modern eutrophication of this lake. Similarity of stenol and stanol distributions at each depth suggests preservation of source information despite changes in stenol/stanol ratios. Data are from Gaskell and Eglinton (1976).

bacteria for use in their cell membranes. The presence of these biomarker lipids in sedimentary organic matter is an indication of bacterial contributions of organic compounds produced by reworking of primary organic matter. As with steroid compounds, the diversity of side chains, the presence of stereospecific bonds in the isoprenoid ring structures, and the various types of functional groups endow hopanoids with a large amount of geochemical information.

Evidence of early diagenetic transformations of hopanoid compounds has been found in the upper 15 cm of sediments of Lake Haruna, Japan, by Uemura and Ishiwatari (unpublished). As shown in Fig. 13, both oxidized and reduced forms of the presumed hopanol precursors are present in these young sediments. These compounds illustrate several of the stages of early diagenesis which are active in lacustrine sediments: (1) microbial reprocessing of primary organic matter, (2) creation of microbial biomarkers, and (3) diagenetic modification of the secondary organic matter, the microbial biomarkers. In addition, the presence of the 17α hopanone suggests contribution of older organic matter, perhaps from soils eroded from land areas around Lake Haruna. The detection of friedelin, which is derived from vascular plants, is consistent with the postulated contribution of soil organic matter to these lake sediments.

3.1.5. Early Diagenesis of Hydrocarbons in Lacustrine Sediments

Because of their relatively low susceptibility to microbial degradation compared to other types of organic matter, saturated and aromatic hydrocarbons can record the depositional history of organic matter sources in lake sediments. *n*-Alkane biomarkers are synthesized by many biota (cf. Muller, 1987). Polycyclic aromatic hydrocarbons (PAH) are ubiquitous components of the organic matter contents of sediments in both marine and lacustrine areas (e.g., Laflamme and Hites, 1978; Hites *et al.*, 1980), yet this class of compounds is virtually absent from the biosphere. PAHs are the products of diagenesis instead of biosynthesis, unlike most other geolipid components of sediments.

3.1.5a. Polycyclic Aromatic Hydrocarbons. Within the general classification of PAHs, individual compounds are created by one of three types of diagenesis: (1) early diagenesis from reasonably well defined biogenic precursors, (2) diagenesis acting slowly on precursor materials over longer periods of time, and (3) high-temperature processes which fragment and restructure kerogen into characteristic PAH distributions. Nearly all PAH molecules fit into one of these three categories (e.g., Wakeham *et al.*, 1980; Tan and Heit, 1981; Furlong *et al.*, 1987), making them useful indicators of geochemical transport and transformation processes.

Downcore profiles of individual PAH concentrations illustrate patterns typical of components of the separate categories. Studies of the sediments of Mountain Pond, Coburn Mountain, Maine, have shown variations in the formation of a PAH representative of the early diagenesis group and in the inputs of compounds indicative of the high-temperature, or pyrolysis, group (Gschwend and Hites, 1981; Gschwend *et al.*, 1983). The concentration of perylene increases regularly with depth in a half-meter core which records about two centuries of sediment accumulation, although the concentration of total organic matter remains essentially constant (Fig. 14). In contrast, the concentrations of pyrolytically derived PAHs decrease with depth after peaking at a depositional horizon corresponding to ca. 1958. As postulated by Gschwend *et al.* (1983), the perylene pattern reflects *in situ* formation of this compound from some biogenic precursor or precursors via either first-order or second-order reaction kinetics. The patterns of benzo-

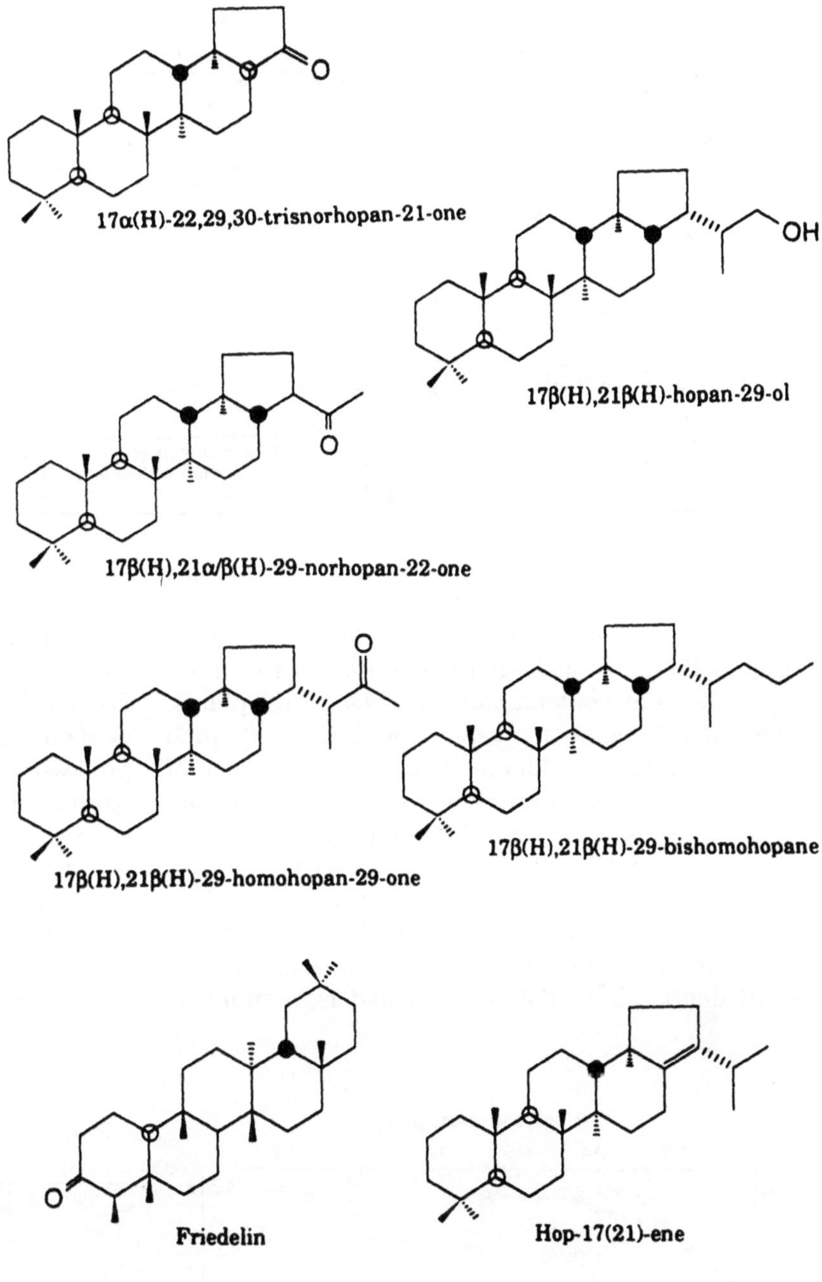

Figure 13. Types of hopanoid compounds produced by early diagenesis in the surface sediments of Lake Haruna, Japan (K. Uemura and R. Ishiwatari, unpublished results). The presumed precursors are hopanols produced by sediment bacteria, such as the hopanol reported in these sediments, and both oxidized and reduced by-products are formed. Friedelin is derived from vascular plants.

(*a*)pyrene, chrysene, and pyrene record the history of coal combustion in the northern United States and airborne transport of some of the combustion products to isolated depositional sites, such as Mountain Pond (Gschwend and Hites, 1981; Furlong *et al.*, 1987). The preservation of these two types of records attests to the low reactivity of these molecules once they have been formed.

3.1.5b. Hydrocarbon Contents of Greifensee Sediments. A core from Greifensee, Switzerland, consists of three main stratigraphic units in which concentrations of organic carbon generally decrease with depth (Fig. 15). Significant downcore changes occur in the compositions of the hydrocarbons in each of the three layers. In the upper sapropelic muds, n-C_{17} is the dominant alkane, a product of the presently high rates of algal production in this eutrophic lake. Concentrations of this n-alkane decrease in deeper sediments and record times when the lake evidently was oligotrophic. Part of the decrease in contributions of n-C_{17}, however, is due to diagenetic loss of this algal biomarker relative to the long-chain plant-wax hydrocarbons (Giger *et al.*, 1980). Important contributions of long-chain n-alkanes, represented by n-C_{29} in Fig. 15, are also present in the upper unit, as well as in the deeper units. High CPI

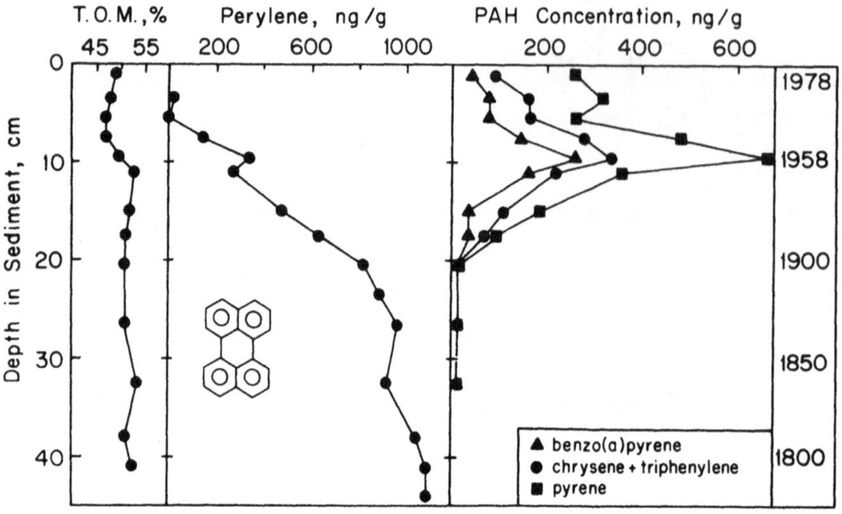

Figure 14. Concentrations of total organic matter (T.O.M.), perylene, and three pyrolytic polycyclic aromatic hydrocarbons (PAH) in a dated sediment core from Mountain Pond, Maine. The progressive increase in perylene concentration as sediment age increases indicates diagenetic formation of this PAH. The concentration maxima of the three pyrolytically formed PAHs coincides with the time of peak coal combustion. Data are compiled from Norton *et al.* (1981), Gschwend and Hites (1981), and Gschwend *et al.* (1983).

values (Bray and Evans, 1961) of these long-chain hydrocarbons throughout the core verify that their origin is from contemporaneous watershed plants, inasmuch as ancient *n*-alkanes would have lost much of their odd/even predominance. The dominance of *n*-C_{29} in all samples except the topmost suggests that deciduous trees have been the major watershed plant type (cf. Cranwell, 1973), which is consistent with what is known of the historical record.

The isoprenoid hydrocarbon phytane, a diagenetic product of chlorophyll *a*, is found at all sediment depths, but pristane, a related isoprenoid, is present only in the postglacial clays and in small amounts in some samples of the sapropelic mud. Pristane is common in modern marine sediments and evidently is a by-product of zooplanktonic digestive processing of chlorophyll *a* (Blumer *et al.*, 1971), a specialized type of early diagenesis! Its absence in some modern lacustrine sediments reflects differences between freshwater and marine zooplankton assemblages, specifically the absence or abundance of calanoid copepods, respectively (Ho and Meyers, 1993). Both phytane and pristane are also produced from slow-acting diagenetic reworking of phytol and

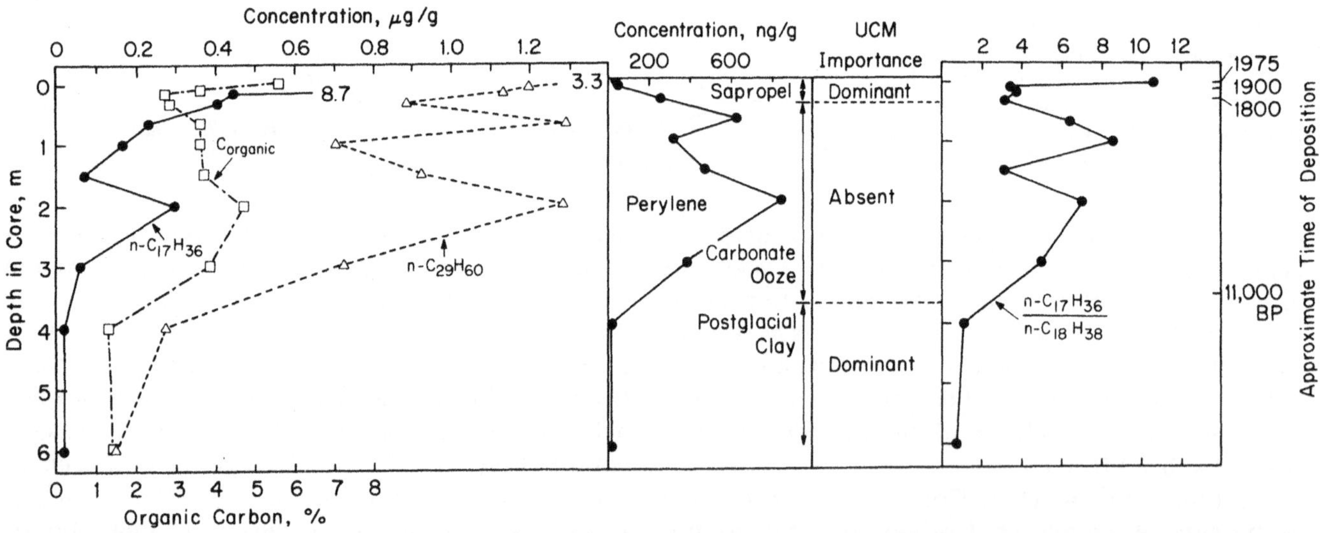

Figure 15. Downcore concentrations of organic carbon and of two biomarked *n*-alkanes in sediments of Greifensee, Switzerland. Perylene concentrations and unresolved mixture of aliphatic hydrocarbons (UCM) are shown at different core depths. Changes in the whole lake environment, recorded in changes in the sediment types, appear also in the hydrocarbon molecular compositions. Data are from Giger *et al.* (1980) and Wakeham *et al.* (1980).

consequently are common constituents of the hydrocarbon contents of ancient sedimentary rocks. Giger *et al.* (1980) interpreted the presence of pristane in Greifensee postglacial sediments to indicate that hydrocarbons from erosion of Miocene rocks in the surrounding land area were incorporated in the lake bottom at this time. With the development of forests in the watershed after about 11,000 yr, erosion became negligible, and pristane inputs essentially disappeared.

The downcore profile of the diagenetically formed polycyclic aromatic hydrocarbon perylene is very similar to that of the n-C_{29} alkane (Fig. 15). The only difference is in the topmost part of the core, where perylene concentrations are low and increase with depth, whereas the alkane concentrations decrease from their maximum value. The difference in the profiles in the upper three samples in Fig. 15 is probably a consequence of the time lag in the diagenetic appearance of perylene (cf. Gschwend *et al.*, 1983). The similarities deeper in the core, however, point to a land-derived biogenic precursor for this polycyclic aromatic hydrocarbon.

An unresolved complex mixture of hydrocarbons (UCM) is a major component of the total hydrocarbon contents of the sapropelic muds and the postglacial muds, but not of the carbonate oozes. UCMs are common in geologically old hydrocarbon distributions, such as found in petroleum and in ancient sedimentary rocks. The UCM in the postglacial mud arises from deposition of hydrocarbon mixtures obtained from erosion of watershed rock formations, and it accompanies the presence of pristane in these sediments (Giger *et al.*, 1980). Further evidence of the deposition of ancient hydrocarbons is given by the lack of odd/even predominance of the C_{17} to C_{21} n-alkanes in the postglacial muds, as represented by the C_{17}/C_{18} ratio shown in Fig. 15. This ratio is high where algal inputs are large, such as in the surface sediments, and low where geologically old n-alkanes are major geolipid components.

3.1.5c. Hydrocarbon Contents of Sediments of Lake Washington. The early diagenetic formation of several polycyclic aromatic hydrocarbons in the sediments of Lake Washington is related to deforestation of the watershed during historical times (Fig. 16). Retene is a diagenetic PAH which is believed to be derived from abietic acid, a common constituent of conifer sap (Simoneit, 1977). Wakeham *et al.* (1980) postulated that accelerated erosion of soil from the conifer-covered watershed of Lake Washington since the early 1900s has added large amounts of abietic acid present in forest litter to the lake sediments. The core profile in Fig. 16a reflects the delay between addition of the precursor material and diagenetic production of the more stable aromatic hydrocarbon.

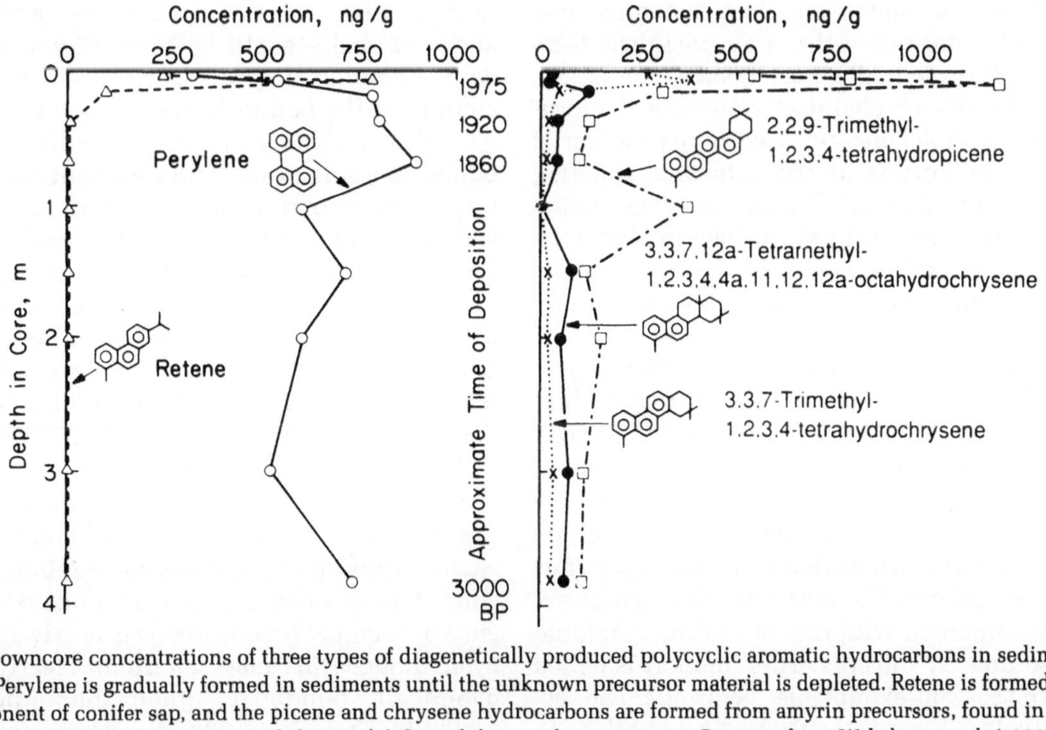

Figure 16. Downcore concentrations of three types of diagenetically produced polycyclic aromatic hydrocarbons in sediments of Lake Washington. Perylene is gradually formed in sediments until the unknown precursor material is depleted. Retene is formed from abietic acid, a component of conifer sap, and the picene and chrysene hydrocarbons are formed from amyrin precursors, found in conifer bark. Widespread logging temporarily enhanced the availability of the conifer precursors. Data are from Wakeham *et al.* (1980).

The profiles of downcore concentrations of the pentacyclic and two tetracyclic triterpenoids shown in Fig. 16b are similar to the retene profile. Probable precursors for these triterpenoid hydrocarbons are α- and β-amyrin, which are found in conifer bark. These precursors evidently share the same source enhancement from watershed erosion that abietic acid is believed to have had.

Comparison of the contents of biological precursor PAHs of sediments from Greifensee and Lake Washington shows differences in source inputs. Wakeham *et al.* (1980) noted that the sediments of Greifensee contain negligible amounts of retene, whereas those of Lake Washington are rich in this diagenetic PAH. The probable precursor of retene is abietic acid, a component of conifer resin, and conifers are not abundant around Greifensee. Similarly, pimanthrene, for which resin-derived pimaric acid is the probable precursor, and a wide range of triterpenoid hydrocarbons derived from amyrin precursors are abundant in the sediments of Lake Washington but not in those of Greifensee. Differences in the watershed plant types clearly affect the types of hydrocarbons produced by early diagenesis of the precursor material available in the sediments of these two lakes.

3.1.6. Free and Bound Geolipids in Lacustrine Sediments

Lipid components of living biological matter are typically chemically bound into the biochemical matrix of the cells and cell walls of an organism. Geochemical studies generally ignore such bonds and instead concentrate on characterizations of the individual types of molecules released by extraction of samples and hydrolysis of the extracted material, which severs the chemical linkages between molecules. Some investigators, however, have explored the possibilities of obtaining geochemical information by separating geolipid fractions while preserving any retained bonds.

The most common approach to the study of geolipid materials in which some of their original biochemical associations may be preserved is to employ a two-step sediment extraction procedure. In the first step, organic solvents are used to remove the soluble geolipid components from a sample. These include truly free molecules which have no residual bonds with other compounds and also molecules which are chemically combined with others, yet the combinations are soluble in solvents. This fraction is called the "free" or "solvent-extractable" geolipids. The second step involves hydrolysis of chemical bonds with hot acid or base, followed by a second extraction with organic solvents. These lipids are normally called the "bound" or "non-solvent-extractable" fraction. The additional geolipids released by the hydrolysis procedure probably include materials incorporated into humic substances, retained in biological detritus, and associated with minerals. The actual nature of any chemical bonds cannot be absolutely ascertained, yet inferences are possible. For example, acidic hydrolysis cleaves amide linkages, whereas alkaline hydrolysis severs ester linkages.

Comparisons of the compositions of free and bound geolipid fractions of sediment organic matter show that the contents of the two compartments appear to have different sources and appear to respond differently to early diagenesis. Based on carbon-chain-length distributions of *n*-alkanoic acids and *n*-alkanols and C_{29}/C_{27} sterol ratios, the free geolipid fraction contains a large proportion of land-plant material, whereas the bound fraction is constituted mostly of algal and bacterial components (Cranwell, 1978, 1984; Ishiwatari *et al.*, 1980; Cranwell *et al.*, 1987; Wünsche *et al.*, 1988). An example of the difference in *n*-alkanoic acid and *n*-alkane contents of the free and bound geolipids is illustrated in Fig. 17 (P. A. Meyers, unpublished results). This sediment sample, from Heart Lake, New York, contains free geolipids dominated by the long-chain components diagnostic of land-plant epicuticular waxes. The bound geolipids contain primarily short-chain, algal biomarkers. In particular, in this example, few of the branched and odd-chain-length bacterial biomarkers are evident, although they do constitute a slightly larger amount of the bound fraction than of the free one.

Observations from some studies that free-to-bound ratios decrease with sediment age suggest that free geolipid components become diagenetically converted into bound forms (e.g., Cranwell *et al.*, 1987). Such conversion is not universally true in lacustrine sediments. Comparisons of concentrations and molecular compositions of the free and bound geolipid fractions of sediment cores from lakes in which pollen data record watershed vegetation histories show that a complex mix of source and preservation factors influence these geolipid fractions (Meyers *et al.*, 1984b; Wünsche *et al.*, 1988). As illustrated in Fig. 18, the concentrations of the free and bound portions of each geolipid fraction share some relationship, but a consistent downcore pattern does not exist. The presence of "bound" hydrocarbons is nearly always found, even though this class of lipids clearly cannot be chemically bonded to other compounds. As suggested by Cranwell (1978), the bound materials may

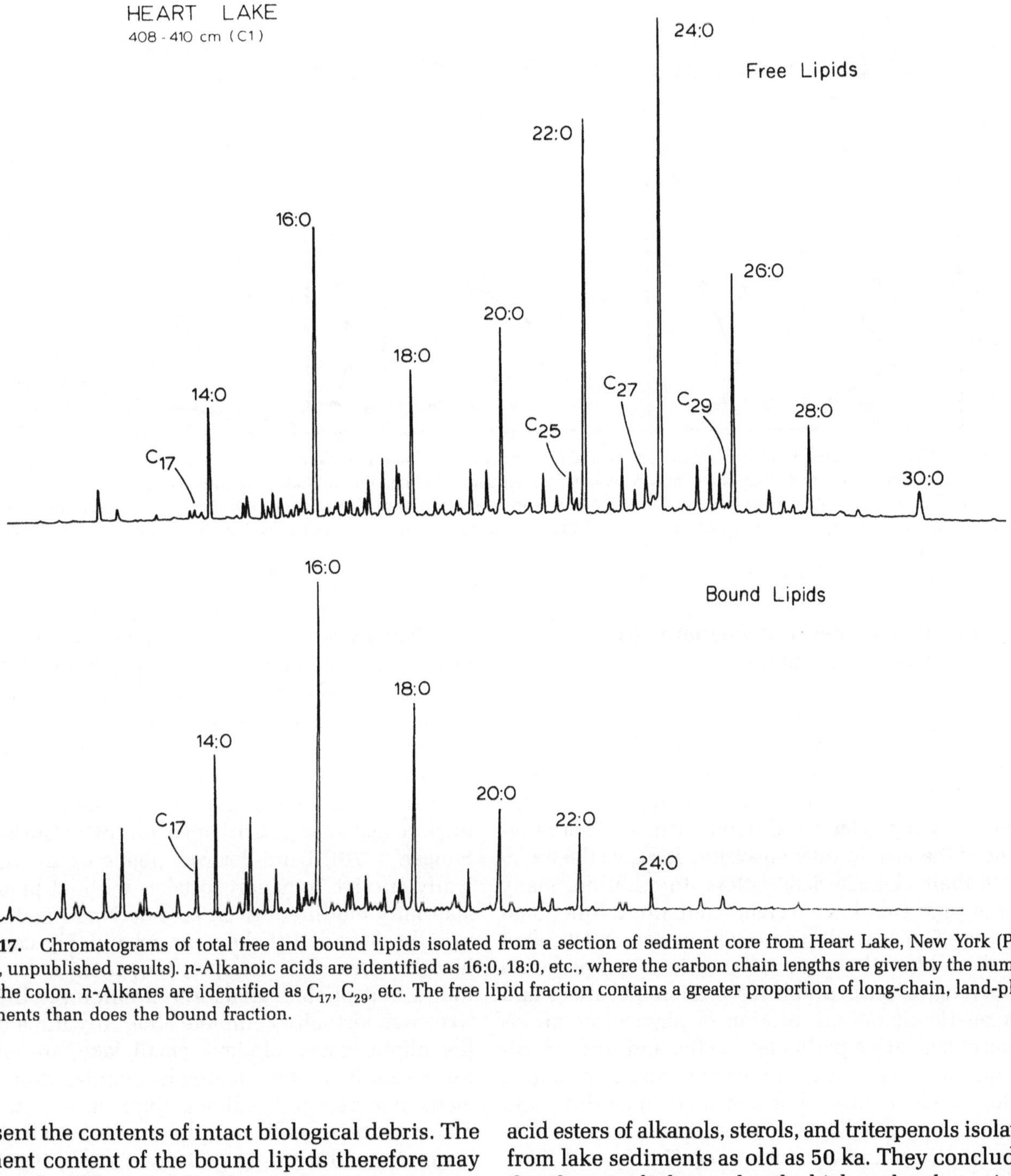

Figure 17. Chromatograms of total free and bound lipids isolated from a section of sediment core from Heart Lake, New York (P. A. Meyers, unpublished results). *n*-Alkanoic acids are identified as 16:0, 18:0, etc., where the carbon chain lengths are given by the number before the colon. *n*-Alkanes are identified as C_{17}, C_{29}, etc. The free lipid fraction contains a greater proportion of long-chain, land-plant components than does the bound fraction.

represent the contents of intact biological debris. The sediment content of the bound lipids therefore may be strongly influenced by fluctuations in preservational conditions.

3.1.7. Intact Geolipid Esters in Lacustrine Sediments

Another approach that has been used to investigate the nature of chemical linkages in lipid fractions has been to extract and to isolate intact esterified compounds from sediment samples. Cranwell (1986) and Cranwell and Volkman (1981) have analyzed the fatty acid esters of alkanols, sterols, and triterpenols isolated from lake sediments as old as 50 ka. They concluded that the ester linkage, plus the high molecular weights of the esters, enhances the resistance of these lipid forms to microbial attack. As a consequence, these biomarker compounds retain evidence of their respective algal or vascular plant origins. Bacterial biomarkers constitute little of the ester fraction. Despite the enhanced preservation displayed by these compounds, losses of the lower-molecular-weight components have been inferred from lower than expected contributions of algal ester biomarkers in sediments of eutrophic Tokyo Bay (Fukushima and Ishiwatari, 1984).

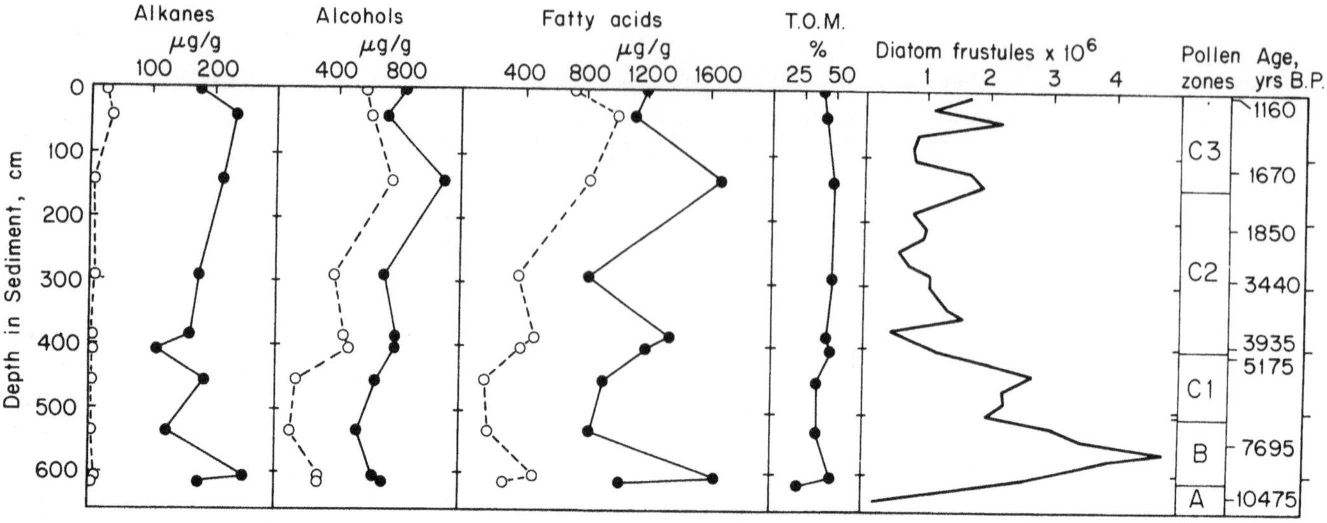

Figure 18. Concentrations of free (solid lines) and bound (dashed lines) *n*-alkanes, *n*-alkanols, *n*-alkanoic acids, and total organic matter (T.O.M.) in sediments of Heart Lake, New York. Postglacial pollen zones and changes in diatom abundance are shown with radiocarbon ages. The lack of correlation between the lipid fractions and the microfossil components indicates that complicated source and preservation changes impact the lipid compositions of the organic-carbon-rich sediments. Data are from Meyers *et al.* (1984b).

3.2. Early Diagenesis of Pigments in Lacustrine Sediments

Photosynthetic plants synthesize a variety of pigmented organic compounds. Some of these molecules serve as biomarkers specific to various types of plants. All of them contain chromophore groups, typically conjugated C═C bonds, which absorb portions of the visible solar spectrum and give the molecules their characteristic colors. In addition, many of the pigments have oxygen-containing functional groups. The various combinations of unsaturated bonds and functional groups give these compounds a wide range of susceptibilities to microbial attack and diagenetic alterations. Studies of pigment compositions of lacustrine particulate matter and surface sediments have shown that some pigments, for example, chlorophyll *a*, pheophorbide *a*, fucoxanthin, and peridinin, degrade much faster than others, such as pheophytin *a* and β-carotene (cf. Hurley and Armstrong, 1990; Leavitt and Carpenter, 1990). Nonetheless, a portion of the pigment input to lakes can escape degradation and be made more resistant to diagenetic losses. The major alterations are stabilization of the chlorophyll tetrapyrrole ring by hydrolysis and hydrogenation of side chains and by incorporation of nickel or vanadium to form porphyrins and hydrogenation of carotenoid chains to form isoprenoid alkanes. These diagenetic products can be found in ancient lake sediments.

Downcore changes in pigment contents of lake sediments record both source changes and the effects of diagenetic alterations of pigment molecules. Sanger (1988) reviews the factors important to the accumulation of pigments in lacustrine sediments. In general, lakes having higher aquatic productivities contain greater proportions of pigments in their sedimented organic matter (e.g., Gorham *et al.*, 1974; Gorham and Sanger, 1976). Land-derived pigments do not typically survive transport to lakes without being substantially modified. Even lake-derived pigments are rapidly degraded if they are not quickly carried to lake bottoms. For example, Carpenter *et al.* (1986) found that photodegradation of chlorophyll derivatives was virtually complete after only three days in the photic zones of three small lakes in southern Michigan. Increased grazing by zooplankton, which leads to more rapid sedimentation, appears to be the major factor in enhancing pigment preservation in lacustrine sediments (e.g., Leavitt *et al.*, 1989).

After incorporation into lake sediments, pigments are subject to rapid diagenetic modification. Keely and Brereton (1986) documented the progressive conversion of chlorophyll *a* into pheophytin *a* in a core representing 100 yr of sediment accumulation in Priest Pot, England (Fig. 19). They noted that the major chlorophyll degradation product is pheophytin, whereas in marine sediments pheophorbide commonly is the major product. The dominance of pheophytin *a*, the demetallated form of chlorophyll

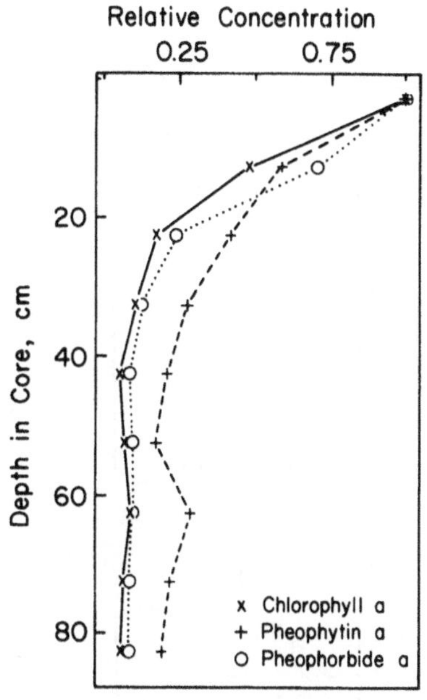

Figure 19. Downcore changes in relative concentrations of chlorophyll *a*, pheophytin *a*, and pheophorbide *a* in sediments of Priest Pot, England. Concentrations are normalized to surface sediment (1). The downcore decrease in all three compounds underscores the degradability of pigments, yet pheophytin survives best. Data are from Keely and Brereton (1986).

a, rather than pheophorbide *a*, in which both magnesium and ester side chains are lost, in these sediments probably results from the absence in this eutrophic lake of the calanoid copepod zooplankton that are the dominant grazers in marine food chains. This food chain effect is reflected similarly in the absence of pristane in the sediments of some other lakes (cf. Giger *et al.*, 1980; Ho and Meyers, 1993) and is another factor influencing the early diagenesis of organic matter prior to sedimentation in lake bottoms. The presence of a suite of pyrochlorins in the sediments of Priest Pot proves that decarbomethoxylation of chlorophyll *a* and its derivatives can occur at the earliest stages of diagenesis, perhaps starting in the water column (Keely *et al.*, 1988).

Diagenesis of chlorophyll derivatives continues deeper into lacustrine sediments. A progressive change in the relative proportions of degradation products of chlorophyll *a* has been described in the 200-m sediment core from Lake Biwa, Japan, and indicates the succession of alterations this pigment molecule experiences (Hata *et al.*, 1980). Intact chlorophyll *a* is produced within the waters of the lake by phytoplankton; this compound is quickly demetallized to pheophytin *a*, which then proceeds by both the chlorin and the pyrochlorin degradative pathways to arrive at a more reduced degradation product (Fig. 20). An intermediate degradation product is delivered to the lake from land runoff. This compound is the pyrochlorin degradation product derived from the chlorophyll *a* of land plants. It, too, continues to be altered to the more reduced degradation product found in sediments several hundreds of thousands of years old.

4. Summary and Conclusion

Lake systems are highly diverse, and their sedimentary deposits have several differences from marine sediments. Lacustrine sediments generally accumulate faster and have higher concentrations of organic matter than do typical marine deposits. Land-derived materials often constitute a major fraction of lake sediments. Differences between marine and lacustrine food chains influence diagenesis during sinking of organic matter to bottom sediments. The enhanced accumulation rates found in the sediments of many lakes allow the effects of early diagenesis to be more readily observed. Organic matter is rapidly degraded and altered during early diagenesis.

Humic substances dominate the compositions of lacustrine sedimentary organic matter. Diagenesis does not modify the $\delta^{13}C$ or C/N values of this material over time spans of hundreds and even thousands of years in most lake sediments.

Microbial reworking of organic matter is active in lake sediments. Aquatic components are rapidly degraded, and microbial components partially replace them. Molecular components of land plants are less susceptible to reworking in lake sediments, and biomarker distributions can become biased toward land contributions. Molecules containing oxygenated functional groups or unsaturated carbon–carbon bonds are easily degraded and disappear rapidly with time of burial in lake bottoms. Straight-chain and isoprenoid hydrocarbons are more resistant to degradation.

Conversion of precursor molecules into diagenetic products occurs in lake sediments at different rates for different types of compounds. Polycyclic aromatic hydrocarbons are produced in concentrations that evidently reflect availability of starting materials. Production of chlorophyll by-products, such as pristane and pheophorbide, is influenced by food chain preprocessing before organic matter is sedimented. Generalized patterns of early diagenetic al-

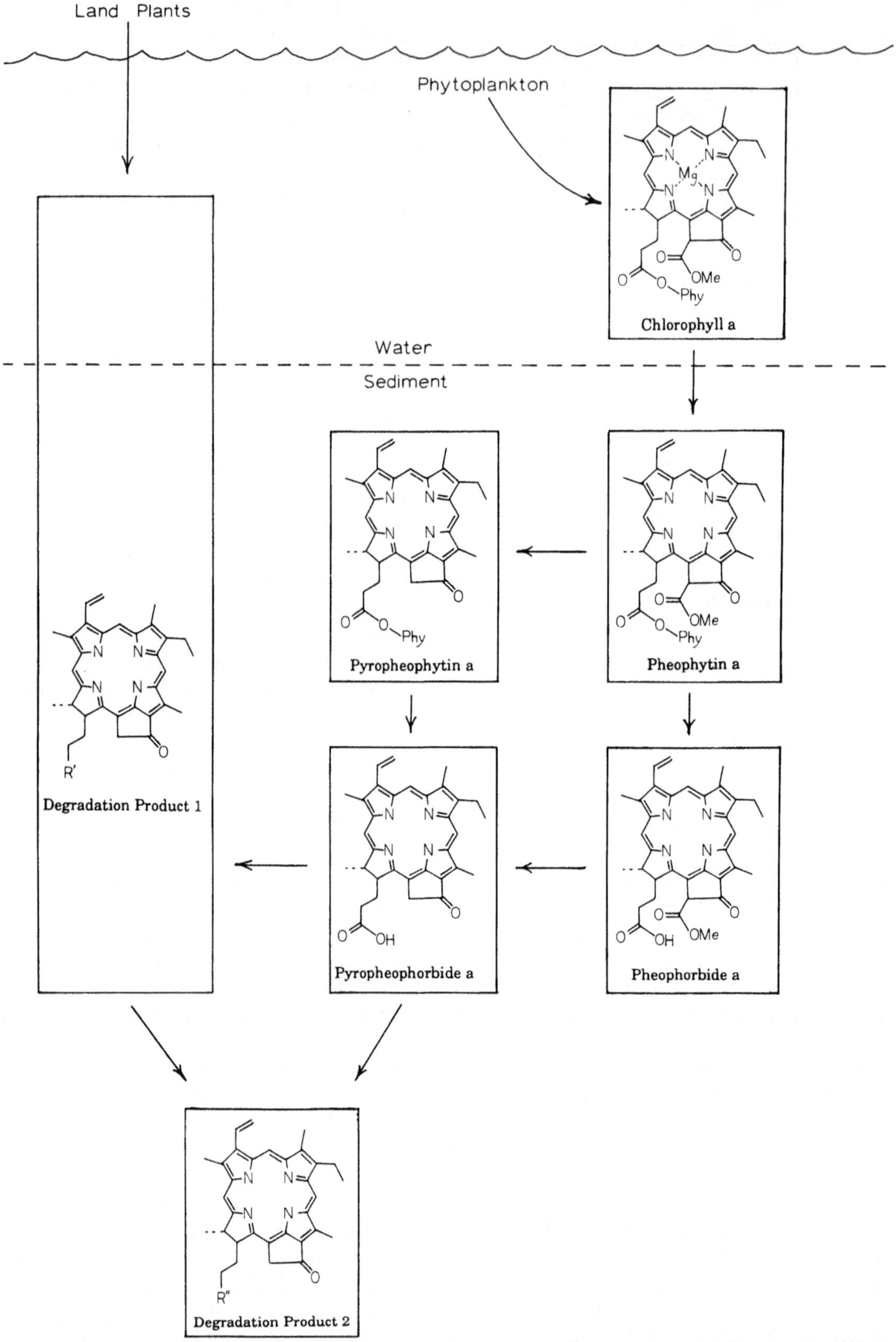

Figure 20. A degradation scheme for chlorophyll *a* proposed by Hata *et al.* (1980) and based on amounts of the various diagenetic products found in the 200-m sediment core from Lake Biwa, Japan. Aquatic production supplies unaltered chlorophyll *a*, whereas the land-plant supply of this pigment has been degraded during transport to the lake.

terations of organic matter are partially overprinted by local factors of individual lakes.

References

Benner, R., Fogel, M. L., Sprague, E. K., and Hodson, R. E., 1987, Depletion of ^{13}C in lignin and its implications for stable carbon isotope studies, *Nature* **329**:708–710.

Blumer, M., Guillard, R. R. L., and Chase, T., 1971, Hydrocarbons of marine plankton, *Mar. Biol.* **8**:183–189.

Bourbonniere, R. A., 1979, Geochemistry of organic matter in Holocene Great Lakes sediments, Thesis, The University of Michigan, Ann Arbor.

Bourbonniere, R. A., and Meyers, P. A., 1978, Characterization of sedimentary humic matter by elemental and spectroscopic methods, *Can. J. Spectrosc.* **23**:35–41.

Brassell, S. C., Eglinton, G., and Maxwell, J. R., 1983, The geochemistry of terpenoids and steroids, *Biochem. Soc. (U.K.) Trans.* **11**:575–586.

Bray, E. E., and Evans, E. D., 1961, Distribution of *n*-paraffins as a clue to the recognition of source beds, *Geochim. Cosmochim. Acta* **27**:1113–1127.

Carpenter, S. R., Elser, M. M., and Elser, J. J., 1986, Chlorophyll production, degradation, and sedimentation: Implications for paleolimnology, *Limnol. Oceanogr.* **31**:112–124.

Cranwell, P. A., 1973, Branched-chain and cyclopropanoid acids in a recent sediment, *Chem. Geol.* **11**:307–313.

Cranwell, P. A., 1974, Monocarboxylic acids in lake sediments: Indicators, derived from terrestrial and aquatic biota, of paleoenvironmental trophic levels, *Chem. Geol.* **14**:1–14.

Cranwell, P. A., 1978, Extractable and bound lipid components in a freshwater sediment, *Geochim. Cosmochim. Acta* **42**:1523–1532.

Cranwell, P. A., 1981, Diagenesis of free and bound lipid components in terrestrial detritus deposited in a lacustrine sediment, *Org. Geochem.* **3**:79–89.

Cranwell, P. A., 1984, Lipid geochemistry of sediments from Upton Broad, a small productive lake, *Org. Geochem.* **7**:25–37.

Cranwell, P. A., 1986, Esters of acyclic and polycyclic isoprenoid alcohols: Biochemical markers in lacustrine sediments, *Org. Geochem.* **10**:891–896.

Cranwell, P. A., and Volkman, J. K., 1981, Alkyl and steryl esters in a recent lacustrine sediment, *Chem. Geol.* **32**:29–43.

Cranwell, P. A., Eglinton, G., and Robinson, N., 1987, Lipids of aquatic organisms as potential contributors to lacustrine sediments—II, *Org. Geochem.* **11**:513–527.

Degens, E. T., 1969, Biogeochemistry of stable carbon isotopes, in: *Organic Geochemistry—Methods and Results* (G. Eglinton and M. T. J. Murphy, eds.), Springer-Verlag, New York, pp. 304–329.

de Leeuw, J., and Baass, M., 1986, Early-stage diagenesis of sterols, in: *Biological Markers in the Sedimentary Record* (R. B. Johns, ed.), Elsevier, Amsterdam, pp. 101–123.

Eadie, B. J., Chambers, R. L., Gardner, W. S., and Bell, G. E., 1984, Sediment trap studies in Lake Michigan: Resuspension and chemical fluxes in the southern basin, *J. Great Lakes Res.* **10**:307–321.

Ertel, J. R., Hedges, J. I., 1985, Sources of sedimentary humic substances: Vascular plant debris, *Geochim. Cosmochim. Acta* **49**:2097–2107.

Fry, B., 1986, Sources of carbon and sulfur nutrition for consumers in three meromictic lakes of New York State, *Limnol. Oceanogr.* **31**:79–88.

Fukushima, K., and Ishiwatari, R., 1984, Acid and alcohol compositions of wax esters in sediments from different environments, *Chem. Geol.* **47**:41–56.

Frulong, E. T., Cessar, L. R., and Hites, R. A., 1987, Accumulation of polycyclic aromatic hydrocarbons in acid sensitive lakes, *Geochim. Cosmochim. Acta* **51**:2965–2975.

Gaskell, S. J., and Eglinton, G., 1976, Sterols of a contemporary lacustrine sediment, *Geochim. Cosmochim. Acta* **40**:1221–1228.

Giger, W., Schaffner, C., and Wakeham, S. G., 1980, Aliphatic and olefinic hydrocarbons in recent sediments of Greifensee, Switzerland, *Geochim. Cosmochim. Acta* **44**:119–129.

Goossens, H., Duren, R. R., de Leeuw, J. W., and Schenck, P. A., 1989, Lipids and their mode of occurrence in bacteria and sediments—II. Lipids in the sediment of a stratified, freshwater lake, *Org. Geochem.* **14**:27–41.

Gorham, E., and Sanger, J. E., 1976, Fossilized pigments as stratigraphic indicators of cultural eutrophication in Shagawa Lake, northeastern Minnesota, *Geol. Soc. Am. Bull.* **87**:1638–1642.

Gorham, E., Lund, J. W. G., Sanger, J., and Dean, W. E., Jr., 1974, Some relationships between algal standing crop, water chemistry, and sediment chemistry in the English Lakes, *Limnol. Oceanogr.* **19**:601–617.

Gschwend, P. M., and Hites, R. A., 1981, Fluxes of polycyclic aromatic hydrocarbons to marine and lacustrine sediments in the northern United States, *Geochim. Cosmochim. Acta* **45**: 2359–2367.

Gschwend, P. M., Chen, P. H., and Hites, R. A., 1983, On the formation of perylene in recent sediments: Kinetic models, *Geochim. Cosmochim. Acta* **47**:2115–2119.

Hata, K., Handa, N., and Ohta, K., 1980, Early diagenetic change in chlorophyll *a* in sediment collected from Lake Biwa, *Geochemistry* **14**:23–29 [in Japanese].

Hatcher, P. E., Simoneit, B. R. T., Mackenzie, F. T., Neumann, A. C., Thorstenson, D. C., and Gerchakov, S. M., 1982, Organic geochemistry and pore water chemistry of sediments from Mangrove Lake, Bermuda, *Org. Geochem.* **4**:93–112.

Hatcher, P. E., Spiker, E. C., Szeverenyi, N. M., and Maciel, G. E., 1983, Selective preservation and origin of petroleum-forming aquatic kerogen, *Nature* **305**:498–501.

Hedges, J. I., Ertel, J. R., and Leopold, E. B., 1982, Lignin geochemistry of a Late Quaternary core from Lake Washington, *Geochim. Cosmochim. Acta* **46**:1869–1877.

Hedges, J. I., Cowie, G. L., Ertel, J. R., Barbour, R. J., and Hatcher, P. L., 1985, Degradation of carbohydrates and lignins in buried woods, *Geochim. Cosmochim. Acta* **49**:701–711.

Hites, R. A., Laflamme, R. E., and Windsor, J. G., Jr., 1980, Polycyclic aromatic hydrocarbons in marine/aquatic sediments: Their ubiquity, L. Petrakis and F. T. Weiss, Eds., *Petroleum in the Marine Environment* Adv. Chem. Series No. 185, American Chemical Society, Washington, D.C., pp.289–311.

Ho, E. S., and Meyers, P. A., 1993. Variability of early diagenesis in lake sediments: Evidence from the sedimentary geolipid record in an isolated tarn. *Chem. Geol.* In press.

Hurley, J. P., and Armstrong, D. E., 1990, Fluxes and transformations of aquatic pigments in Lake Mendota, Wisconsin, *Limnol. Oceangr.* **35**:384–389.

Ishiwatari, R., 1985, Geochemistry of humic substances, in: *Humic Substances in Soil, Sediment, and Water: Geochemistry, Isolation, and Characterization* (D. M. McKnight, ed.), John Wiley & Sons, New York, pp. 147–180.

Ishiwatari, R., and Uzaki, M., 1987, Diagenetic changes of lignin compounds in a more than 0.6 million-year-old lacustrine sediment (Lake Biwa, Japan), *Geochim. Cosmochim. Acta* **51:** 321–328.

Ishiwatari, R., Takamatsu, N., and Ishibshi, T., 1977, Separation of autochthonous and allochthonous materials in lacustrine sediments by density differences, *Jpn. J. Limnol.* **38:**94–99.

Ishiwatari, R., Ogura, K., and Horie, S., 1980, Organic geochemistry of a lacustrine sediment (Lake Haruna, Japan), *Chem. Geol.* **29:** 261–280.

Jasper, J. P., and Gagosian, R. B., 1989, Glacial–interglacial climatically-forced sources of sedimentary organic matter to the late Quaternary northern Gulf of Mexico, *Nature* **342:**60–62.

Johnson, T. C., Evans, J. E., and Eisenreich, S. J., 1982, Total organic carbon in Lake Superior sediments: Comparisons with hemipelagic and pelagic marine environments, *Limnol. Oceanogr.* **27:**481–491.

Kawamura, K., and Ishiwatari, R., 1985, Distribution of lipid-class compounds in bottom sediments of freshwater lakes with different trophic status, in Japan, *Chem. Geol.* **51:**123–133.

Kawamura, K., Ishiwatari, R., and Yamazaki, M., 1980, Identification of polyunsaturated fatty acids in surface lacustrine sediments, *Chem. Geol.* **28:**31–39.

Kawamura, K., Ishiwatari, R., and Ogura, K., 1987, Early diagenesis of organic matter in the water column and sediments: Microbial degradation and resynthesis of lipids in Lake Haruna. *Org. Geochem.* **11:**251–264.

Keely, B. J., and Brereton, R. G., 1986, Early chlorin diagenesis in a Recent aquatic sediment, *Org. Geochem.* **10:**975–980.

Keely, B. J., Brereton, R. G., and Maxwell, J. R., 1988, Occurrence and significance of pyrochlorins in a lake sediment, *Org. Geochem.* **13:**801–805.

Kemp, A. L. W., and Johnston, L. M., 1979, Diagenesis of organic matter in the sediments of Lakes Ontario, Erie, and Huron, *J. Great Lakes Res.* **5:**1–10.

Krishnamurthy, R. V., Bhattacharya, S. K., and Kusumgar, S., 1986, Palaeoclimatic changes deduced from $^{13}C/^{12}C$ and C/N ratios of Karewa lake sediments, India, *Nature* **323:**150–152.

Laflamme, R. E., and Hites, R. A., 1978, The global distribution of polycyclic aromatic hydrocarbons in recent sediments, *Geochim. Cosmochim. Acta* **42:**289–303.

Leavitt, P. R., and Carpenter, S. R., 1990, Aphotic pigment degradation in the hypolimnion: Implications for sedimentation studies and paleolimnology, *Limnol. Oceanogr.* **35:**520–534.

Leavitt, P. R., Carpenter, S. R., and Kitchell, J. F., 1989, Whole-lake experiments: The annual record of fossil pigments and zooplankton, *Limnol. Oceanogr.* **34:**700–717.

Leenheer, M. J., and Meyers, P. A., 1983, Comparison of lipid compositions in marine and lacustrine sediments, in: *Advances in Organic Geochemistry 1981* (M. Bjoroy, ed.), John Wiley, Chichester, pp. 309–316.

MacKenzie, A. S., Brassell, S. C., Eglinton, G., and Maxwell, J. R., 1982, Chemical fossils: The geologic fate of steroids, *Science* **217:**491–504.

Matsuda, H., and Koyama, T., 1977a, Early diagenesis of fatty acids in lacustrine sediments—I. Identification and distribution of fatty acids in recent sediment from a freshwater lake, *Geochim. Cosmochim. Acta* **41:**777–783.

Matsuda, H., and Koyama, T., 1977b, Early diagenesis of fatty acids in lacustrine sediments—II. A statistical approach to changes in fatty acid composition from recent sediments and source materials, *Geochim. Cosmochim. Acta* **41:**1825–1834.

Mendoza, Y. A., Gulacar, F. O., Hu, Z.-L., and Buchs, A., 1987, Unsubstituted and hydroxy substituted fatty acids in a recent lacustrine sediment, *Int. J. Environ. Anal. Chem.* **31:**107–127.

Meyers, P. A., Edwards, S. J., and Eadie, B. J., 1980a, Fatty acid and hydrocarbon content of settling sediments in Lake Michigan, *J. Great Lakes Res.* **6:**331–337.

Meyers, P. A., Leenheer, M. J., Erstfeld, K. M., and Bourbonniere, R. A., 1980b, Changes in spruce composition following burial in lake sediments for 10,000 yr, *Nature* **287:**534–536.

Meyers, P. A., Bourbonniere, R. A., and Takeuchi, N., 1980c, Hydrocarbons and fatty acids in two cores of Lake Huron sediments, *Geochim. Cosmochim. Acta* **44:**1215–1221.

Meyers, P. A., Maring, H. B., and Bourbonniere, R. A., 1980d, Alkane and alkanoic acid variations with depth in modern sediments of Pyramid Lake, in: *Advances in Organic Geochemistry 1979* (A. G. Douglas and J. R. Maxwell, eds.), Pergamon Press, Oxford, pp. 365–374.

Meyers, P. A., Leenheer, M. J., Eadie, B. J., and Maule, S. J., 1984a, Organic geochemistry of suspended and settling particulate matter in Lake Michigan, *Geochim. Cosmochim. Acta* **48:** 443–452.

Meyers, P. A., Kawka, O. E., and Whitehead, D. R., 1984b, Geolipid, pollen, and diatom stratigraphy in postglacial lacustrine sediments, *Org. Geochem.* **6:**727–732.

Muller, H., 1987, Hydrocarbons in the freshwater environment, *Advances in Limnology, Arch. Hydrobiol.*, Stuttgart.

Nakai, N., 1986, Paleoenvironmental features of Lake Biwa deduced from carbon isotope compositions and organic C/N ratios of the upper 800m sample of 1,400m cored column, *Proc. Jpn. Acad.* **62B:**279–282.

Nishimura, M., 1977a, The geochemical significance in early sedimentation of geolipids obtained by saponification of lacustrine sediments, *Geochim. Cosmochim. Acta* 41:1817–1823.

Nishimura, M., 1977b, Origin of stanols in young lacustrine sediments, *Nature* **270:**711–712.

Nishimura, M., 1978, Geochemical characteristics of the high reduction zone of stenols in Suwa sediments and the environmental factors controlling the conversion of stenols into stanols, *Geochim. Cosmochim. Acta* **42:**349–357.

Nishimura, M., 1982, 5β-Isomers of stanols and stanones as potential markers of sedimentary organic quality and depositional paleoenvironments, *Geochim. Cosmochim. Acta* **46:**423–432.

Nishimura, M., and Koyama, T., 1976, Stenols and stanols in lake sediments and diatoms, *Chem. Geol.* **17:**229–239.

Nishimura, M., and Koyama, T., 1977, The occurrence of stanols in various living organisms and the behavior of sterols in contemporary sediments, *Geochim. Cosmochim. Acta* **41:**379–385.

Norton, S. A., Hess, C. T., and Davis, R. B., 1981, Rates of accumulation of heavy metals in pre- and post-European sediments in New England lakes, in: *Pollutants in Natural Waters* (S. J. Eisenreich, ed.), Ann Arbor Science, Ann Arbor, Michigan, pp. 409–421.

Rea, D. K., Bourbonniere, R. A., and Meyers, P. A., 1980, Southern Lake Michigan sediments: Changes in accumulation rate, mineralogy, and organic content, *J. Great Lakes Res.* **6:**321–330.

Reed, W. E., 1977, Biogeochemistry of Mono Lake, California, *Geochim. Cosmochim. Acta* **41:**1231–1245.

Robinson, N., Cranwell, P. A., Finlay, B. J., and Eglinton, G., 1984a, Lipids of aquatic organisms as potential contributors to lacustrine sediments, *Org. Geochem.* **6:**143–152.

Robinson, N., Eglinton, G., and Brassell, S. C., 1984b, Dinoflagellate origin for sedimentary 4α-methylsteroids and 5α(*H*)-stanols, *Nature* **308:**439–441.

Robinson, N., Cranwell, P. A., Eglinton, G., Brassell, S. C., Sharp, C. L., Gophen, M., and Pollinger, U., 1986, Lipid geochemistry of Lake Kinneret, Org. *Geochem.* **10:**733–742.

Sanger, J. E., 1988, Fossil pigments in paleoecology and paleolimnology, *Palaeogeogr. Palaeoclimatol. Palaeoecol.* **62:**343–359.

Simoneit, B. R. T., 1977, The Black Sea, a sink for terrigenous lipids, *Deep-Sea Res.* **24:**813–830.

Sollins, P., Spycher, G., and Glassman, C. A., 1984, Net nitrogen mineralization from light-fraction and heavy-fraction forest soil organic matter, *Soil Biol. Biochem.* **16:**31–37.

Spiker, E. C., and Hatcher, P. G., 1984, Carbon isotope fractionation of sapropelic organic matter during early diagenesis. Org. *Geochem.* **5:**283–290.

Spiker, E. C., and Hatcher, P. G., 1987, The effects of early diagenesis on the chemical and stable carbon isotopic composition of wood, *Geochim. Cosmochim. Acta* **51:**1385–1391.

Tan, Y. L., and Heit, M., 1981, Biogenic and abiogenic polynuclear aromatic hydrocarbons in sediments from two remote Adirondack lakes, *Geochim. Cosmochim. Acta* **45:**2267–2279.

Vallentyne, J. R., 1962, Solubility and the decomposition of organic matter in nature, *Arch. Hydrobiol.* **58:**423–434.

Volkman, J. K., 1986, A review of sterol markers for marine and terrigenous organic matter, Org. *Geochem.* **9:**83–99.

Wakeham, S. G., Schaffner, C., and Giger, W., 1980, Polycyclic aromatic hydrocarbons in Recent lake sediments—II. Compounds derived from biogenic precursors during early diagenesis, *Geochim. Cosmochim. Acta* **44:**415–429.

Wünsche, L., Gulacar, F. O., and Buchs, A., 1987, Several unexpected marine sterols in a freshwater environment. Org. *Geochem.* **11:**215–219.

Wünsche, L., Mendoza, Y. A., and Gulacar, F. O., 1988, Lipid geochemistry of a post-glacial lacustrine sediment, Org. *Geochem.* **13:**1131–1143.

Chapter 9

Early Diagenesis of Organic Matter in Sediments

Assessment of Mechanisms and Preservation by the Use of Isotopic Molecular Approaches

STEPHEN A. MACKO, MICHAEL H. ENGEL, and PATRICK L. PARKER

1. Introduction

Substantial information regarding the history and source of organic matter and transfer of both carbon nitrogen to and within marine environments lies in the chemical and isotopic signals of organic materials. Origins and diagenetic histories of organic materials can be assessed on a bulk basis, using molar ratios of carbon to nitrogen and stable isotope compositions, or on a molecular level, using characteristic molecules or compositional changes to indicate the major processes occurring in the sedimentary deposit. More recently, clear identification of the preservation of particular components has been achieved, and the processes involved with the formation of an organic deposit have been established through the isotopic characterization of the compounds themselves. This chapter endeavors to lay a foundation for such molecular isotopic approaches in studies of the early diagenesis of sedimentary organic material. Specific details on molecular carbon approaches, especially those having to do with lipid characterization, are discussed in other chapters of this book [see chapters by Brassell (Chapter 34), Hedges and Prahl (Chapter 11), and Meyers and Ishiwatari (Chapter 8)] or larger compilations related to the field of marine organic geochemistry (Eglinton and Murphy, 1969; Duursma and Dawson, 1981; Brooks and Welte, 1984; Tissot and Welte, 1984; Sigleo and Hattori, 1985; Sohn, 1986).

2. Amino Acid Nitrogen in Sediments

Amino acids are present in sediments, bound in peptides, proteins, humic acids, and proteinaceous fragments and also in the free state, likely a product of decomposition. The sediment–water interface contains the highest concentrations of bound amino acids, which decrease rapidly with depth in the sediment (Hare, 1969; Whelan, 1977; Montani *et al.*, 1980). Much of the nitrogen that can be characterized in marine sediments is in the form of these labile compounds. Amino acids play an important role in a number of chemical and biochemical processes which take place in the water column and sediments (Starikova and Korzhikova, 1969; Bada and Mann, 1980; Wakeham and Lee, this volume, Chapter 6). Thus, the presence or absence of amino acids may be due to certain reactions taking place. For instance, an increase in glycine concentrations and a corresponding decrease in serine and threonine concentrations may

STEPHEN A. MACKO • Department of Environmental Sciences, The University of Virginia, Charlottesville, Virginia 22903. MICHAEL H. ENGEL • School of Geology and Geophysics, The University of Oklahoma, Norman, Oklahoma 73019. PATRICK L. PARKER • Department of Marine Sciences, The University of Texas at Austin, Austin, Texas 78712.

Organic Geochemistry, edited by Michael H. Engel and Stephen A. Macko. Plenum Press, New York, 1993.

be related to a specific reaction pathway of decomposition (Bada and Mann, 1980). Alternatively, the differences in amino acid compositions can be explained in terms of depositional environments and different stabilities through microbial or other geochemical reactions (Degens, 1970; Morris, 1975; Degens and Mopper, 1976; Dungworth *et al.*, 1977; Whelan, 1977; Carter and Mitterer, 1978; Gonzalez, 1983). For example, analyses on two cores from the Ebro delta (Spain) revealed that the branched-chain amino acids (valine, leucine, and isoleucine) are the least stable geochemically (Gonzalez, 1983). Black Sea sediments, deposited under varying oxidizing and reducing conditions, reflect variations of amino acid compositions (Degens and Mopper, 1976). Also, threonine and serine are not expected to be observed in older sediments due to their thermal instabilities (Wehmiller and Hare, 1972). The overall relative contributions of acidic, basic, and neutral amino acids to the total appear to remain fairly constant (Macko *et al.*, 1987b).

Large changes occur in the amino acid composition of material found in the water column and that which is observed in the sediments. In many instances, nearly 100% of the nitrogen present in sediment-trap materials can be characterized as amino acids (Biggs *et al.*, 1988, 1989), whereas only approximately 2 to 12% of the nitrogen in upper sediments can be identified as being in amino acids (Rosenfeld, 1979; Mayer *et al.*, 1988). Much higher proportions have been reported for upwelling environments (Henrichs and Farrington, 1984; Henrichs *et al.*, 1984; Patience *et al.*, 1990). In some environments, amino acids have been useful indicators of organisms or depositional environments (Mopper and Degens, 1972; Morris, 1975; Dungworth *et al.*, 1977; Gonzalez, 1983; Henrichs and Farrington, 1984; Macko *et al.*, 1987b).

The amino acid composition may reflect the source of the sedimentary organic matter, and variations of this may then be related to the processes associated with sedimentation. It may be an indicator of marine or terrigenous debris or, more specifically, of a particular plant type. Relatively large amounts of acidic amino acids are present in the organic materials of terrigenous sediments (Akiyama and Johns, 1972). Within the acidic fraction of amino acids, aspartic acid appears to be more abundant than glutamic acid in marine-derived organics as compared to terrestrial organics (Macko *et al.*, 1987b). In the neutral fraction, a predominance of threonine over serine could indicate a marine planktonic source (Gonzalez, 1983). Furthermore, specific plant types have unusual amino acid relationships when compared to marine sedimentary organics. For example, red mangroves from Florida contain high levels of glycine, and the relative abundance of tyrosine with respect to phenylalanine is about 0.3 (Casagrande and Given, 1980). In seagrasses, this ratio has been observed to be about 1, whereas in brown algae it appears to be near 0.6 (for *Ascophyllum*, *Fucus*, and *Sargassum*; S. A. Macko and K. Pulchan, unpublished results). In surficial sediments, phenylalanine can be oxidized to tyrosine, resulting in tyrosine/phenylalanine values greater than 1, a ratio observed in sediments associated with macrophytic alga (S. A. Macko and K. Pulchan, unpublished results). Basic amino acids were in greater relative abundance in surficial sediments from deeper water depths of the Gulf of Maine, whereas aromatic amino acids had heightened abundances in estuarine Maine sediments (Mayer *et al.*, 1988). To a large extent, however, these labile amino acid components are rapidly altered and lost from most sedimentary environments as a result of diagenesis (Wakeham *et al.*, 1984; Morris, 1975; Mayer *et al.*, 1988; Burdige and Martens, 1988). In many surficial sediments, the overall amino acid distributions are remarkably similar (e.g., Degens and Mopper, 1976; Rosenfeld, 1979; Sargent *et al.*, 1983; Henrichs and Farrington, 1987; Mayer *et al.*, 1988) although in abyssal sediments, unusually low levels, consisting predominantly of uncommon amino acids, have been reported (Whelan, 1977). Nonprotein amino acids, possibly associated with bacterial cell walls, have been found in coastal deposits (Casagrande and Given, 1980; Burdige and Martens, 1988).

Sedimentary dissolved free amino acids (DFAA) and other organic components of pore waters have been employed in the estimation of organic transformations within marine sediments (Henrichs and Farrington, 1979, 1980; Henrichs *et al.*, 1984; Macko, 1992). Amino acids from sediment pore waters have typically been observed in concentrations on the order of 1 to 300μM whereas DFAA in overlying waters are orders of magnitude lower in concentration. The amino acid distribution in sediment pore waters is also chemically distinct from both hydrolyzable sedimentary amino acids and the DFAA of overlying waters, being dominated by a few compounds including glutamic acid, aspartic acid, glycine, and alanine. Usually, pore-water amino acids have been observed to decrease rapidly with depth in the sediment, virtually vanishing below about 10 cm. The decline in pore-water amino acids correlates closely, on a molar basis, with the increase in the pore-water ammonium over that depth range.

3. Bulk Characterization of Sediments: C/N

Bulk parameters, including the molar ratio of organic carbon to nitrogen (C/N) and stable isotopes, have been useful in distinguishing sources of organic matter, and, when compared downcore, the variations have been useful in indicating diagenetic activity (Muller and Suess, 1979). Sedimentary increases in C/N are usually suggestive of preferential loss of nitrogen during diagenesis (Bordovskiy, 1965; Muller, 1977; Rosenfeld, 1979; Patience *et al.*, 1990). Little change in C/N downcore has suggested similar mineralization of carbon and nitrogen or preservation (Dungworth *et al.*, 1977; Henrichs and Farrington, 1987). Decreases in C/N have also been observed and may result from absorption of organic materials or inorganic forms of nitrogen onto clay or other silicate surfaces (Degens and Mopper, 1976; Muller, 1977; Rosenfeld, 1979; Macko and Pereira, 1990).

3.1. Bulk Characterization of Sediments: Stable Carbon Isotopes

Stable isotope compositions may serve as signatures of both the origin of the organic materials and the specific factors influencing that origin, such as temperature, nutrient levels, and productivity. In numerous marine environments, relative contributions of terrigenous and marine inputs have been documented through the use of stable carbon isotopes. For example, in the deltas of the Pedernales (Venezuela) (Eckelmann *et al.*, 1962), Niger (Gearing *et al.*, 1977), Mississippi (Hedges and Parker, 1976), Amazon (Cai *et al.*, 1988), Orinoco, Nile, and Changjiang rivers (Kennicutt *et al.*, 1987), woody fragments and more finely disseminated terrestrial plant debris give a clear terrigenous isotopic signature to deltaic sediments. Generally, surficial sediments contain increasing amounts of the heavier isotope of carbon (^{13}C) with increasing proximity to the sea, evidence for decreasing influence of land-derived detritus (Sackett and Thompson, 1963; Gearing *et al.*, 1977). In more northern environments, the lateral transport of terrigenous debris may be enhanced by ice rafting (Gearing *et al.*, 1977). Variations in the $\delta^{13}C$ of organic matter thus record the sedimentary history of an area. In the Gulf of Mexico, variations downcore have been correlated with glacial and interglacial periods which are related to sea level lowering and extent of influence of the Mississippi River (Parker *et al.*, 1972; Newman *et al.*, 1973; Jasper and Gagosian, 1990). In systems which are exclusively or heavily dominated by marine productivity, other environmental parameters such as temperature, growth rate, species distribution, and CO_2 availability may affect carbon isotopic composition (Sackett *et al.*, 1965; Degens *et al.*, 1968a, b; Degens, 1969; Gearing *et al.*, 1977; Fontugne and Duplessy, 1978, 1981; Rau *et al.*, 1982, 1987, 1989; Jasper and Gagosian, 1989; Fogel and Cifuentes, this volume, Chapter 3). Large temperature differences in water masses have been correlated with planktonic organic material which is depleted in $\delta^{13}C$. The interpretation of this observation is based on the premise that the isotopic fractionations by phytoplankton growing in the open ocean will be affected by temperature. These temperature-induced signatures have been interpreted to be the cause for carbon isotopic variations observed in high-latitude deep-sea sediments (Rogers *et al.*, 1972; Sackett *et al.*, 1974; Sackett, 1986; Rau *et al.*, 1991). However, carbon isotopic variations may also be associated with changing populations of phytoplankton. For example, in the Antarctic, the transition of diatom ooze to a dinoflagellate/coccolithophore-dominated sediment at the Antarctic Convergence is associated with a change in carbon isotopic signature (Sackett *et al.*, 1974). Isotopic differences exist among phytoplankton of different size classes (hence different species) from the same population of a bloom (Gearing *et al.*, 1984). Such variability would certainly have an impact on the organic isotopic record preserved in a sediment.

3.2. Bulk Characterization of Sediments: Stable Nitrogen Isotopes

In similar fashion, stable nitrogen isotopes may be useful indicators of the source of organic material entering the marine environment. Because the major source of nitrogen utilized in terrestrial systems (N_2 via nitrogen fixation) is isotopically distinguishable from that utilized in marine systems (oceanic nitrate through nitrate reduction), these sources may be resolved in many instances (Wada *et al.*, 1975; Kaplan, 1975). Furthermore, in purely aqueous environments, the processes through which phytoplankton (or bacteria) incorporate nitrogen may be resolved if one can eliminate sources of nitrogen from land. The process of nitrogen fixation has a small isotopic fractionation associated with the utilization of molecular nitrogen (Hoering and Ford, 1960; Macko *et al.*, 1987a). Such a process is easily distinguished from nitrate reduction in which phytoplankton fully utilize dissolved nitrate and reflect its isotopic signature (Macko *et al.*,

1984; Fogel and Cifuentes, this volume, Chapter 3). When algae are only able to use a small portion of the dissolved nitrate, more depleted $\delta^{15}N$ values may be evident, reflecting the isotopic fractionation associated with that incorporation (Macko *et al.*, 1987a). Marine organic nitrogen, predominantly from nitrate reduction, could then be distinguished from terrigenously sourced organic materials which derive their nitrogen from nitrogen fixation mechanisms. Peters *et al.* (1978) correlated nitrogen isotope signatures with relative contributions of terrigenous and marine inputs into a sediment. Such contributions of organics can be quantified by mass-balance calculations in which source estimates are made (Macko, 1981, 1983). With a knowledge of the isotopic composition of the sources of organic matter influencing an environment, it is possible to quantitatively assess their relative contributions. This calculation is based on the premise that the isotopic composition of a sediment is the sum for all sources of the fraction contributed from a source (F_i) multiplied by its isotopic ratio (δ_i) (isotopic mass balance):

$$\delta_{\text{sample}} = \sum_{i=1}^{n} F_i \delta_i$$

Using a single isotope, this equation may be solved to estimate the relative contribution from two sources:

$$F_1 = (\delta_s - \delta_2)/(\delta_1 - \delta_2)$$

where δ_s is the isotopic composition of the sample mixture. In offshore environments, variations in the type of primary production (from nitrate reduction to nitrogen fixation) or increases in the nitrate concentrations will be reflected in the $\delta^{15}N$ of the sedimentary organic matter (Macko *et al.*, 1984). Such changes may then be useful in the interpretation of the paleoceanographic record of an area (Rau *et al.*, 1987, 1989; Macko, 1989; Muzuka *et al.*, 1991). Sediments derived from periods of higher productivity, in which phytoplankton fully utilized the oceanic nitrate, may be more enriched in ^{15}N than sediments associated with lower productivity, in which larger fractionations are possible (Macko and Pereira, 1990).

4. Isotopes and Diagenetic Fractionation

The aforementioned situation is complicated, however, by factors which include microbial action, diagenesis, and recycling of organics. The use of stable isotopic signatures as indicators of the provenance of sedimentary organic matter is based on the premise that no change occurs in the isotopic compositions during the process of early diagenesis. Generally, it is assumed that changes in isotopic abundances are small or negligible during diagenesis (Sackett, 1964; Sweeney *et al.*, 1978; Myers, 1974; Sweeney and Kaplan, 1980a, b; Arthur *et al.*, 1985; Dean *et al.*, 1986; Schidlowski *et al.*, 1983). However, in some organic-rich sediments, isotopic shifts as a consequence of diagenesis have been observed (Behrens and Frishman, 1971; Macko, 1981; Ostrom and Macko, 1991a, b). Some variation may occur as a consequence of the preferential loss of isotopically enriched (i.e., amino acids) or depleted (i.e., lipids) components. Other variations in isotopic signature may occur as a result of deamination or decarboxylation reactions associated with the decomposition of organic-rich materials (Wada, 1980; Zieman *et al.*, 1984; Sigleo and Macko, 1985). Such processes of alteration may be especially important in changing isotopic signatures below the euphotic zone (Altabet and McCarthy, 1986; Altabet, 1988). In the transition from labile, protein-rich organic matter to more refractory materials found in sediments, peptide bond rupture (Silfer *et al.*, 1992) may also be important in determining the final isotopic compositions of the organic nitrogen and carbon. An enrichment in the $\delta^{13}C$ of algal mat organic matter was reported to be associated with the loss of depleted lipid carbon (Behrens and Frishman, 1971; Macko, 1981; Fig. 1). Spiker and Hatcher (1984, 1987) concluded that the decreases in $\delta^{13}C$ they observed in a marine sapropel were the result of diagenetic losses of carbohydrates.

The diagenesis of organic matter in sediments is thought to proceed via a series of chemical and microbially mediated reactions that results in the formation of geopolymers which in turn are transformed to kerogen (Nissenbaum and Schallinger, 1974; Welte, 1974). In general, these geopolymers tend to be depleted in ^{13}C relative to the organic matter from which it forms (Nissenbaum *et al.*, 1972; Nissenbaum, 1974; Galimov, 1980). Abelson and Hoering (1961) reported that the carboxyl groups of amino acids are more enriched in ^{13}C relative to the remaining carbon constituents in these molecules. It has been speculated that the loss of ^{13}C-enriched carboxyl groups during early diagenesis could in part account for the depletion in ^{13}C observed in residual organic materials (Nissenbaum *et al.*, 1972; Nissenbaum, 1974; Galimov, 1980). The carbon derived from the decarboxylation of amino acids is dispersed in the pore waters of sediments and is mixed with inorganic CO_2. The Maillard reaction (Maillard, 1912) represents one possible pathway for the formation of humic materials

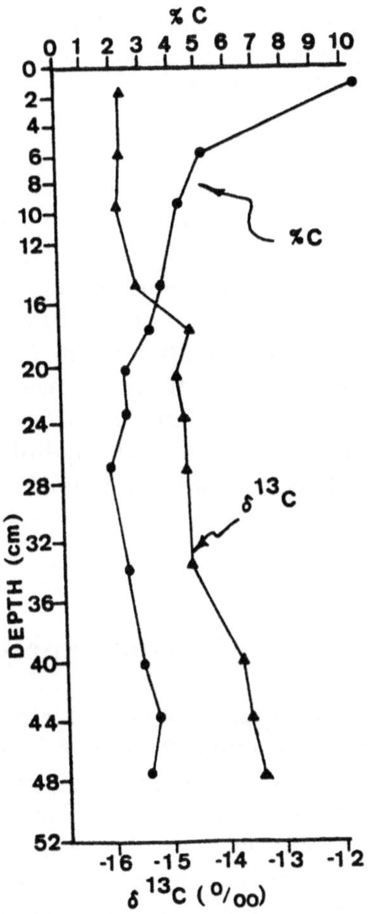

Figure 1. Variations in $\delta^{13}C$ and %C with depth in a sedimentary core from the Baffin Bay, Texas, algal mat laminations.

via the condensation of amino acids and sugars (Hedges, 1978; Larter and Douglas, 1980; Taguchi and Sampei, 1986). Detailed investigations of the structure and composition of melanoidins, the products of the Maillard reaction, indicate that they are similar to natural humic substances (Hoering, 1973; Ertel and Hedges, 1983; Rubinsztain *et al.*, 1984, 1986a, b; Ishiwatari *et al.*, 1986; Aizenshtat *et al.*, 1987; Ikan *et al.*, 1988). Hence, the Maillard reaction has been proposed as a possible model for the diagenesis of organic matter in the water column and in sediments (Abelson and Hare, 1971; Ertel and Hedges, 1983; Engel *et al.*, 1986; Yamamoto and Ishiwatari, 1989; Rafalska *et al.*, 1991). The products of a series of Maillard reactions have been observed to be depleted in $\delta^{13}C$ and $\delta^{15}N$ compositions relative to the reactants (Qian *et al.*, 1992). This depletion can in part be explained by kinetic isotope fractionation effects during the formation of melanoidins. Thus, variations seen in carbon isotopes during diagenesis may be influenced by a number of effects, including breakdown, polymerization, and the increase or decrease of compound classes.

Nitrogen isotopes may also be affected by diagenetic processes involving bond rupture. Decomposing plankton have been observed to become enriched in ^{15}N (Wada, 1980). Downcore enrichments in ^{15}N of the organic matter in a seagrass bed (Fig. 2) have been suggested to be the result of diagenesis, possibly arising from deamination or hydrolysis reactions (Macko, 1981), with preferential bond rupture or loss of ^{14}N amines. Organic matter enriched in ^{15}N has also been observed in oceanic core rings, upwelling areas, and open-ocean particulate materials (Saino and Hattori, 1980, 1985, 1987; Altabet and McCarthy, 1985, 1986; Altabet, 1988; Libes and Deuser, 1988); all of these observations were thought to be associated with residual organic materials following diagenetic loss of ^{14}N-enriched materials. An enrichment in nitrogen isotopes with a decline in the amount of nitrogen characterizable as amino acids is indicative of the diagenetic pathway; it is possibly the result of peptide bond hydrolysis, followed by bacterial utilization of the free or shorter chain peptides. Peptide bond hydrolysis is the fundamental diagenetic process responsible for the degradation of native pro-

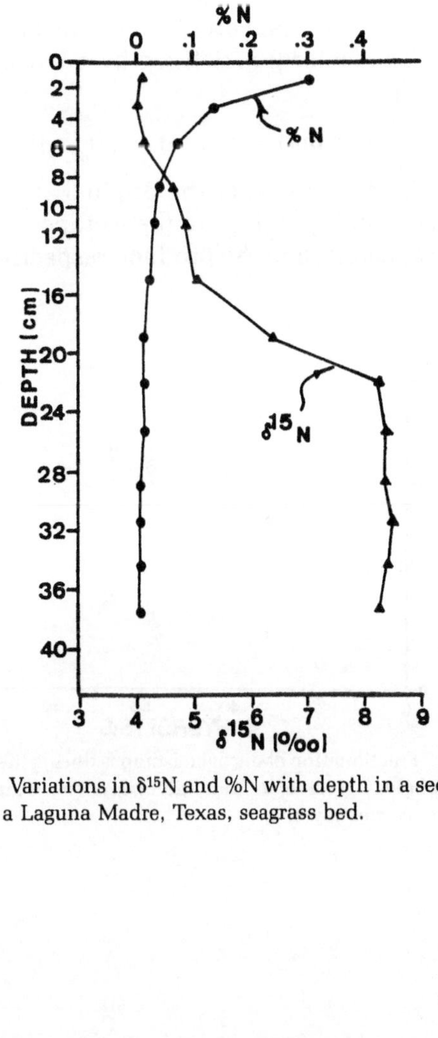

Figure 2. Variations in $\delta^{15}N$ and %N with depth in a sedimentary core from a Laguna Madre, Texas, seagrass bed.

teins. Serban *et al.* (1988) proposed that the fractionation of carbon and nitrogen isotopes during peptide bond hydrolysis may influence molecular isotope compositions in fossil "proteins." Isotope fractionation was invoked to explain observed differences between the $\delta^{13}C$ and $\delta^{15}N$ compositions of D- and L-glutamic acid isolated from a Pleistocene mollusk shell (Serban *et al.*, 1988). Kinetic isotope effects have been associated with the hydrolysis of esters (O'Leary and Marlier, 1979) and amides (Stacey *et al.*, 1952; O'Leary and Kluetz, 1972). Carbon and nitrogen isotope effects during peptide bond cleavage clearly would be expected (Bada *et al.*, 1989). In the hydrolysis of glycylglycine (Silfer *et al.*, 1992), distinct fractionations of carbon and nitrogen isotopes occur (Fig. 3). The free glycine was significantly depleted in ^{13}C and ^{15}N relative to the $\delta^{13}C$ and $\delta^{15}N$ compositions of the original peptide and the residual glycylglycine. The isotope compositions of the glycine and residual peptide reflect the irreversible, bond rupture process under laboratory conditions.

The nitrogen kinetic isotope effects associated with the hydrolysis of glycylglycine and for the decomposition of the seagrass bed organic matter can be evaluated from knowledge of the isotopic compositions of the residual organic materials and the initial substrates, together with the amount of substrate converted or lost. The fractionation factor, β, corresponds to the ratio of the apparent rate constants, for the process. The factor β is defined by a Rayleigh equation:

$$\beta = \ln(1 - f)/\ln[1 - f(r_p/r_s)]$$

where f is the fraction of the original substrate that has reacted and r_s and r_p are the isotope ratios of the original substrate and the product, respectively (Hoering, 1957; Mariotti *et al.*, 1980, 1981, 1982; Macko *et al.*, 1986). The equation can be rewritten in terms of the more familiar δ notation:

$$\delta_p = [1 - (1 - f)^{1/\beta}](1000 + \delta_s) - 10^3$$

where δ_p and δ_s represent the $\delta^{15}N$ compositions of the product and the initial starting material, respectively. The nitrogen isotope fractionation factors are 1.004 (at 100°C) for the peptide bond rupture and 1.006 (ambient temperature range, $10°C < T < 40°C$) for the seagrass bed. Extrapolation of the peptide bond rupture fractionation to the lower temperature regime indicates nearly identical fractionation factors for the two processes. Potentially, hydrolytic cleavage of amino acids is the controlling mechanism for the isotopic compositions of residual organic-rich materials.

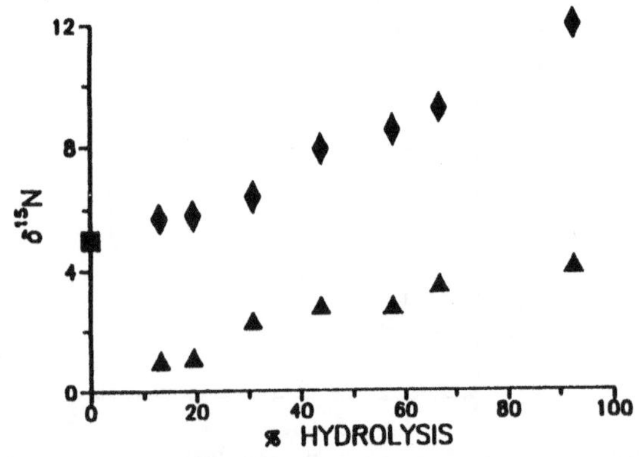

Figure 3. Fractionation of nitrogen isotopes during the hydrolytic cleavage of glycylglycine. ■, Initial glycylglycine; ◆, residual glycylglycine; ▲, free glycine.

5. Molecular-Level Isotope Analyses

Sedimentary organic materials are mixtures of hundreds to thousands of chemical components, each having its own isotopic composition. Stable isotopic determinations made on bulk materials are, in reality, weighted averages of the isotopic compositions of this variety of compounds. The relative contribution of each of these isotopic signatures to that of the bulk material can, in theory, be quantified through mass-balance or isotopic proportionation equations. The analysis of individual molecular components for their respective stable isotopic compositions holds the potential for authenticating the sources of the molecules and offers a unique way of determining the fate, be it destruction or preservation, of the organic molecules in a sedimentary environment. Over the years, a limited number of attempts have been made to isolate individual molecular components for stable isotope analysis, using liquid- or gas-chromatographic techniques, in order to better interpret or trace the source or history of an organic material. The initial efforts typically collected components as they eluted from a chromatograph and combusted the isolated compounds to CO_2 and N_2, which were then cryogenically purified and analyzed in a mass spectrometer. The major limitation of this approach was that of sample size. Because the technique is extremely labor-intensive, only a restricted number of additional reports utilizing molecular/stable isotope characterizations have appeared in the open literature. In addition to establishing separation techniques to provide sufficient material for stable isotope analysis (usually milligram quantities), the analytical scheme must produce little or no isotopic fractionation. Certain compounds may require unique separation schemes owing to close similarities in chemical structure. However, the studies to date have yielded important information regarding the source and history of the compounds characterized. Most of the earlier studies emphasized hydrocarbons (Welte, 1969; Des Marais *et al.*, 1981; Vogler *et al.*, 1981; Gilmour *et al.*, 1984) owing to the ease of separation and potential economic benefits, although some application was made to natural products (Des Marais *et al.*, 1980), with primary production sources of petroleum-derived or related materials or food chain origins of the hydrocarbons able to be suggested. Through recent technological advancements, gas-chromatographic (GC) effluents can be pyrolyzed and the resulting carbon dioxide directly introduced into the stable-isotope ratio mass spectrometer (IRMS). This modification, GC/IRMS, allows for rapid analysis of the carbon isotopes of components in a mixture, and with increased sensitivity, on the order of 0.5nM of each compound (Freedman *et al.*, 1988). Gas-chromatographic-based systems are presently constrained by chromatographic considerations and the volatility of the components investigated. For this reason, hydrocarbons appear to be ideal components for direct GC/IRMS analyses, owing to their excellent resolution using capillary gas chromatography and little need for derivatization to increase volatility. As a result of these developments, the evaluation of components in low abundances, i.e., hydrocarbons (Freeman *et al.*, 1990; Hayes *et al.*, 1990; Rieley *et al.*, 1991), or amino acids (Engel *et al.*, 1990; Silfer *et al.*, 1991) can now be attempted in the reconstruction of organic deposition in the sedimentary record. The union of chemical analysis with isotopic compositions now offers a unique, extremely powerful tool in the delineation of sources, production, and utilization of organic compounds.

The foundation for the recent advancements in the use of a combination of molecular techniques in conjunction with stable carbon isotope compositions may be ascribed to the initial work by Abelson and Hoering (1961). Using a liquid-chromatographic separation procedure for the isolation of sufficient material for stable carbon isotope analyses of individual amino acids and their carboxyl groups, this keynote study strengthened our understanding of natural biosynthetic pathways and the effects of decarboxylation during diagenesis. Studies on nitrogen metabolism and biosynthesis have also been attempted using stable isotopic compositions of individual amino acids (Schoenheimer and Rittenberg, 1939; Gaebler *et al.*, 1963, 1966) based on similar separation approaches and subsequent isotopic analysis. The possibility of comparative biochemistry in modern or fossil organisms has been suggested through the assessment of the isotopic differences between compounds of a family of components (Macko *et al.*, 1987a; Hare *et al.*, 1991). Such differences are the result of enzymatic fractionation effects during synthesis or metabolism of the compound; an example of such an effect has been clearly seen using the enzyme transaminase (Macko *et al.*, 1986), with nitrogen isotope fractionations being observed in acetylglucosamine (Macko *et al.*, 1990) and in the amino acids aspartic acid and glutamic acid (and others) in both cultured and natural populations of organisms (Macko *et al.*, 1986, 1987a, 1991).

Amino acid isotopic compositions have also continued to be investigated, with pathways for nitrogen metabolism (Macko *et al.*, 1987a) and comparative biochemistry in fossil materials (Hare *et al.*, 1991) being elucidated. Isotopic compositions of amino acids preserved in sediments or isolated from pore waters will enable the resolution of the sources of these components and the assessment of the kinds of labile components that are being preserved in sediments. Nitrogen isotope abundances in the individual amino acids appear to be related to kinetic isotope effects associated with transamination reactions during synthesis (Macko *et al.*, 1986). With the observation that the process of amino acid racemization has little apparent effect on the isotopic compositions of stereoisomers (Engel and Macko, 1986), the possibility of an absolute criterion for determining the indigeneity of amino acids in fossil or extraterrestrial materials has been suggested (Engel and Macko, 1984; Serban *et al.*, 1988). Similarly, the isotopic analysis of the stereoisomers of amino acids will allow for the delineation and quantification of the various sources of organic inputs (bacterial, phytoplankton) to a sedimentary environment. Amino acid analysis, using GC/IRMS techniques, is complicated by the fact that the amino acids are nonvolatile, multifunctional molecules that require derivatization prior to gas-chromatographic analysis. Through the derivatization, additional carbon is added to the parent compound. Further, a fractionation of ^{13}C occurs during the esterification and acylation (Silfer *et al.*, 1991). The original carbon isotopic compositions of individual amino acids and/or their stereoisomers can be computed, however, through analysis of standards prepared in a similar fashion. Extreme enrichments in amino acids and similar enrichments in both stereoisomers of the same amino acid from a meteorite, analyzed using GC/IRMS, have confirmed the extraterrestrial origin of those components and supported the lack of contamination by terrestrial compounds in the absolute concentrations and stereoisomer relationships (Engel *et al.*, 1990). Such a study required that the technology for the isotope analysis have the sensitivity for determinations on only fractions of nanomoles of the individual stereoisomers. Applications of this technology (GC/IRMS) to interpret fossil organic matter, to establish its indigeneity, or to suggest modern biosynthetic relationships have recently been explored (Macko and Engel, 1991). Few studies have analyzed the isotopic compositions of chemical components other than hydrocarbons or amino acids.

Isotopic compositions of individual hydrocarbons have the potential for establishing sources for the materials, bacterial or otherwise, and have been useful in correlation techniques both in the petroleum industry and in pollution assessment. Petroleum-derived or related components appear to show a relationship to original biological metabolism prior to deposition (Summons and Powell, 1986; Hayes *et al.*, 1989; Kennicutt and Brooks, 1990; Freeman *et al.*, 1990; Hayes *et al.*, 1990; Bjorøy *et al.*, 1991). Extraterrestrial origins of certain hydrocarbon components which were extracted from a meteorite were analyzed using this new technology (Franchi *et al.*, 1989). Origins of sedimentary lipids from tree waxes were also documented using GC/IRMS (Rieley *et al.*, 1991).

Diagenetically altered tetrapyrroles were among the first sedimentary compounds isolated and suggested to be directly related to components found in living organisms (Treibs, 1935). A recent development, in the area of compound-specific isotope analysis of lipid-related materials, has been the utilization of a separate isolation scheme using liquid chromatography for the purification of chlorophylls, chlorophyll-derived pigments, and other tetrapyrroles, followed by isotopic characterization (Hayes *et al.*, 1987; Ocampo *et al.*, 1989; Boreham *et al.*, 1989; Bidigare *et al.*, 1991; Kennicutt *et al.*, 1992). This separation allows for the determination of both carbon and nitrogen isotopic compositions of the pigments. It appears that the isotopic signatures associated with the original pigments are preserved following deposition (Takigiku, 1987; Popp *et al.*, 1989). Analyses of this sort enable a strictly biochemical basis for inputs and preservation of primary production in sediments to be addressed in both modern and ancient depositional environments. Isotope studies of chlorophylls (Bidigare *et al.*, 1991; Kennicutt *et al.*, 1992) or their breakdown products (Hayes *et al.*, 1987) have the potential for the evaluation of the paleo-oceanographic character and quantity of productivity (Kennicutt *et al.*, 1991). To our knowledge, few attempts at the assessment of the isotopic compositions of long-chain fatty acids have been made (Parker, 1962, 1964; Vogler and Hayes, 1980). Depletions observed in the individual fatty acids were consistent with fractionations associated with lipid synthesis and could yield clues as to the origins of long-chain hydrocarbons in sediments and fossil fuels, as well as the sources of the fatty acids themselves. Fatty acids, as methyl ester derivatives, are also amenable to isotope analysis by GC/IRMS (Abrajano *et al.*, 1992).

Carbohydrates represent the largest reservoir of carbon which can be chemically characterized. Through diagenesis, this reservoir is altered, with the more labile portions being utilized or degraded while the more refractory materials are preserved. Individual carbohydrate isotope compositions have shown great potential in enhancing the understanding of diagenetic processes and resolving sources of carbohydrates bound in sedimentary materials (Macko *et al.*, 1990, 1991). Depletions in the carbon isotopes of the products of reactions allow for calculations to be done which quantify use and production of new organic materials and resolve them from native materials, even though the chemical compositions of the substances are identical. Degens *et al.* (1968a), through isolation and isotopic analysis of major biochemical fractions of phytoplankton, which included different groupings of carbohydrates, concluded that the most labile materials were carbohydrates and proteins. Utilizing an ion-exchange technique, Shimmelmann and DeNiro (1986) isolated *N*-acetyl-D-glucosamine, the monomeric unit of chitin. The study of this monosaccharide has suggested its use in paleoenvironmental and paleoclimatic reconstruction (Shimmelmann *et al.*, 1986). Recent developments in high-pressure liquid-chromatographic (HPLC) techniques and stable isotope analysis enabled Macko *et al.* (1990) to study the isotopic compositions of carbohydrates isolated from individual organisms. In general, the carbohydrates from a plant are isotopically similar to the plant bulk carbon. Individual sugar isolates from organisms or polymers, including glucans, galactans, and chitins, all have carbon isotopic compositions similar to, but typically depleted by approximately 1 to 2‰ with respect to, those of the organism from which they were isolated. The monosaccharide essentially maintains the isotopic signal of the organism and reflects the primary source of carbon entering the organism, i.e., Calvin cycle or Hatch–Slack cycle metabolism or marine bicarbonate. Carbohydrates may be derivatized to acetates and analyzed for their carbon stable isotopic compositions using GC/IRMS (Macko *et al.*, 1992).

The similarity in isotopic composition of the monosaccharide to the organism is not maintained in the nitrogen isotopic compositions of the N-acetylglucosamine isolated from chitin. An approximately 9‰ depletion is observed in samples of the polymer and pure compound relative to the $\delta^{15}N$ composition of the organism (Shimmelmann and DeNiro, 1986; Macko *et al.*, 1990). As sources of nitrogen for growth and metabolism, amino acids donate nitrogen for cellular biosynthesis. The isotope ratio of the carbohydrate which receives a donated nitrogen reflects the action of the enzymatic isotopic fractionation which occurs during the transfer of nitrogen. Aspartic acid has been observed to be depleted by up to 9‰ in ^{15}N relative to glutamic acid in the transamination of glutamic acid to aspartic acid (Macko *et al.*, 1986). Transaminase enzymes are likely influential in the transfer of nitrogen from glutamine to fuctose in the initial synthesis of glucosamine, the precursor of N-acetylglucosamine. Marine environments receive large amounts of chitin from the exoskeletons of marine invertebrates. The consequences of not considering the above depletions in the nitrogen isotopic signature, in environments impacted by inputs of chitin, could include inappropriate assignment of sources of organic nitrogen and incorrect assessment of the amount of contributions from potential sources.

In an initial investigation on a commercial peat, it was observed that the hexose sugar mannose was isotopically similar to the whole peat and the carbohydrate extract (Macko *et al.*, 1991). Xylose showed a distinct depletion in ^{13}C by over 7‰, when compared with other carbohydrates or the bulk carbon of the peat. This fractionation is consistent with a direction which indicated production of new material (Macko *et al.*, 1986, 1987a). In a *Sphagnum* peat, similar isotopic trends are noted, with certain of the sugars remaining constant (rhamnose, arabinose) whereas other sugars (xylose, galactose, and glucose) become increasingly depleted in ^{13}C with depth. Such a trend likely indicates that these sugars are being produced and, when compared with the changes in the chemical compositions, gives an indication of the lability and rates of utilization and turnover. The constancy of the arabinose and rhamnose isotopic compositions with the relative increases in the molar concentrations of these materials (three- to fourfold) may indicate the amount of the plant debris which has been lost during decomposition reactions.

The application of GC/IRMS analysis to a multitude of environmental, ecological, or biochemical research areas holds great potential, which is only beginning to be realized. The extension of compound-specific isotopic analytical data, which have been derived from modern organisms and settings, to yield interpretations of ancient depositional environments, past climates, and insights into the mechanisms of preservation certainly appears possible. It should be noted, however, that all organic materials deposited in a sedimentary environment, as a result of only a small portion surviving diagenesis, may undergo some isotopic fractionation during preservation. The same caution must be exercised in employing molecular isotopic approaches as in studies using bulk measurements in the interpretation of these new measurements. Further applications of GC/IRMS techniques to the understanding of the cycling of carbon and nitrogen or to the identification and of pollutants and their alterations or to the elucidation of metabolic relationships between compounds in living or extinct organisms are all within the scope of future research, given the proper discretion and comprehension of the measurements.

ACKNOWLEDGMENTS. We would like to express our appreciation to our collaborators, students, and advisors, P. L. Parker (S.A.M.) and B. Nagy (M.H.E.), for many stimulating discussions and the opportunities to address the questions which always arose. We also thank the Geophysical Laboratory of the Carnegie Institution for support of our initial endeavors. We would like to acknowledge NASA, NSF, the Petroleum Research Fund, and NSERC (Canada) for support.

References

Abelson, P. H., and Hare, P. E., 1971, Reactions of amino acids with natural and artificial humus and kerogens, *Carnegie Inst. Washington Yearb.* **69**:327–334.

Abelson, P. H., and Hoering, T. C., 1961, Carbon isotope fractionation in formation of amino acids by photosynthetic organisms, *Proc. Natl. Acad. Sci. U.S.A.* **47**:623–632.

Abrajano, T. A., Fang, J., Comet, P. A., and Brooks, J., 1992, Compound-specific carbon analysis of fatty acids, 203rd National Meeting of the American Chemical Society, San Francisco, Abstract #104.

Aizenshtat, Z., Rubinsztain, Y., Ioselis, P., Miloslavski, I., and Ikan, R., 1987, Long-living free radicals study of stepwise pyrolyzed melanoidins and humic substances, *Org. Geochem.* **11**:65–71.

Akiyama, M., and Johns, W. D., 1972, Amino acids in the Cretaceous Pierre Shale of eastern Wyoming, *Pac. Geol.* **4**:79–89.

Altabet, M. A., 1988, Variations in nitrogen isotopic composition between sinking and suspended particles: Implications for nitrogen cycling and particle transformation in the open ocean, *Deep-Sea Res.* **35**:535–554.

Altabet, M. A., and McCarthy, J. J., 1985, Temporal and spatial variations in the natural abundance of ^{15}N in PON from a warm core ring, *Deep-Sea Res.* **7**:755–772.

Altabet, M. A., and McCarthy, J. J., 1986, Vertical patterns in ^{15}N natural abundance in PON from the surface waters of warm-core rings, *J. Mar. Res.* **44**:185–201.

Arthur, M. A., Dean, W. E., and Claypool, G. E., 1985, Anomalous ^{13}C enrichment in modern marine organic carbon, *Nature* **315:** 216–218.

Bada, J. L., and Mann, E. H., 1980, Amino acid diagenesis in DSDP cores: Kinetics and mechanisms of some reactions and their applications in geochronology and in paleotemperature and heat flow determinations, *Earth Sci. Rev.* **16**:21–57.

Bada, J. L., Schoeninger, M. J., and Shimmelmann, A., 1989, Isotopic fractionation during peptide bond hydrolysis, *Geochim. Cosmochim. Acta* **53**:3337–3341.

Behrens, E. W., and Frishman, S. A., 1971, Stable carbon isotopes in blue-green algal mats, *J. Geol.* **79**:95–100.

Bidigare, R. R., Kennicutt, M. C., Keeney-Kennicutt, W. L., and Macko, S. A., 1991, Isolation and purification of chlorophylls *a* and *b* for the determination of stable carbon and nitrogen isotope compositions, *Anal. Chem.* **63**:130–133.

Biggs, D. C., Berkowitz, S. P., Altabet, M. A., Bidigare, R. R., DeMaster, D. J., Dunbar, R. B., Leventer, A., Macko, S. A., Nittrouer, C. A., and Ondrusek, M. E., 1988, A cooperative study of upper ocean particulate fluxes in the Weddell Sea, in: *Initial Reports of the Ocean Drilling Program Leg 113*, Part A (P. F. Barker, J. P. Kennett, *et al.*, eds.), College Station, TX pp. 77–85.

Biggs, D. C., Berkowitz, S. P., Altabet, M. A., Bidigare, R. R., DeMaster, D. J., Macko, S. A., Ondrusek, M. E., and Noh, I., 1989, A cooperative study of upper ocean particulate fluxes, in: *Initial Reports of the Ocean Drilling Program Leg 119*, Part A (B. Larren, and J. Barron, *et al.*, eds.), College Station, TX, pp. 109–119.

Bjorøy, M., Hall, K., Gillyon, P., and Jumeau, J., 1991, Carbon isotope variations in *n*-alkanes and isoprenoids of whole oils, *Chem. Geol.* **93**:13–20.

Bordovskiy, O. K., 1965, Accumulation and transformation of organic substances in marine sediments, *Mar. Geol.* **3**:3–114.

Boreham, C. J., Fookes, C. J. R., Popp, B. N., and Hayes, J. M., 1989, Origins of etioporphyrins in sediments: Evidence from stable carbon isotopes, *Geochim. Cosmochim. Acta* **53**:2451–2455.

Brooks, J., and Welte, D. (eds.), 1984, *Advances in Petroleum Geochemistry*, Academic Press, New York.

Burdige, D. J., and Martens, C. S., 1988, Biogeochemical cycling in an organic-rich coastal marine basin: 10. The role of amino acids in sedimentary carbon cycling and nitrogen cycling, *Geochim. Cosmochim. Acta* **52**:1571–1584.

Cai, D. L., Tan, F. C., and Edmond, J. M., 1988, Sources and transport of particulate organic carbon in the Amazon River and Estuary, *Estuarine Coastal Shelf Sci.* **26**:1–14.

Carter, P. W., and Mitterer, R. M., 1978, Amino acid composition of organic matter associated with carbonate and non-carbonate sediments, *Geochim. Cosmochim. Acta* **38**:341–364.

Casagrande, D. J., and Given, P. H., 1980, Geochemistry of amino acids in some Florida peat accumulations—II. Amino acid distributions, *Geochim. Cosmochim. Acta* **44**:1493–1507.

Dean, W. E., Arthur, M. A., and Claypool, G. E., 1986, Depletion of ^{13}C in Cretaceous marine organic matter: Source, diagenetic, or environmental signal? *Mar. Geol.* **70**:119–157.

Degens, E. T., 1969, Biogeochemistry of stable carbon isotopes, in: *Organic Geochemistry*, (G. Eglinton and M. T. J. Murphy, eds.), Springer-Verlag, New York, pp. 304–329.

Degens, E. T., 1970, Molecular nature of nitrogenous compounds in seawater and recent sediments, in: *Organic Matter in Natural Waters* (D. W. Wood, ed.), Marine Sciences Institute, University of Alaska, pp. 77–106.

Degens, E. T., and Mopper, K., 1976, Factors controlling the distribution and early diagenesis of organic material in marine sediment, in: *Chemical Oceanography* (J. P. Riley and R. Chester, eds.), Academic Press, New York, pp. 60–112.

Degens, E. T., Behrend, M., Gottharl, B., and Reppmann, E., 1968a, Metabolic fractionation of carbon isotopes in marine plankton—II. Data on samples collected off the coasts of Peru and Ecuador, *Deep-Sea Res.* **15**:11–20.

Degens, E. T., Gaillard, R. R. L., Sackett, W. M., and Hellebust, J. A., 1968b, Metabolic fractionation of carbon isotopes in marine plankton—I. Temperature and respiration experiments, *Deep-Sea Res.* **15**:1–9.

Des Marais, D. J., Mitchell, J. M., Meinschein, W. G., and Hayes, J. M., 1980, The carbon isotopic biogeochemistry of individual hydrocarbons in bat guano and the ecology of insectivorous bats in the region of Carlsbad, New Mexico, *Geochim. Cosmochim. Acta* **44**:2075–2086.

Des Marais, D. J., Donchin, J. H., Nehring, N. L., and Truesdell, A. H., 1981, Molecular carbon evidence for the origin of geothermal hydrocarbons, *Nature* **292**:826–828.

Dungworth, G., Thijssen, M., Zuusveld, J., Van der Velden, W., and Schwartz, A. W., 1977, Distribution of amino acids, amino sugars, purines and pyrimidines in a Lake Ontario sediment core, *Chem. Geol.* **19**:295–308.

Duursma, E. K., and Dawson, R. (eds.), 1981, *Marine Organic Chemistry*, Elsevier, Amsterdam.

Eckelmann, W. R., Broecker, W. S., Whitlock, D. W., and Allsup, J. R., 1962, Implications of carbon isotopic composition of total organic carbon of some recent sediments and ancient oils, *Am. Assoc. Petrol. Geol. Bull.* **46**:699–704.

Eglinton, G., and Murphy, M. T. J. (eds.), 1969, *Organic Geochemistry*, Springer-Verlag, New York.

Engel, M. H., and Macko, S. A., 1984, Separation of amino acid enantiomers for stable nitrogen and carbon isotopic analyses, *Anal. Chem.* **56**:2598–2600.

Engel, M. H., and Macko, S. A., 1986, Stable isotope evaluation of the origins of amino acids in fossils, *Nature* **323**:531–533.

Engel, M. H., Rafalska-Bloch, J., Schiefelbein, C. F., Zumberge, J. E., and Serban, A., 1986, Simulated diagenesis and catagenesis of marine kerogen precursors: melanoidins as model systems for light hydrocarbon generation, in: *Advances in Organic Geochemistry 1985* (D. Leythaeuser and J. Rullkötter, eds.), Org. *Geochem.* **10**:1073–1079.

Engel, M. H., Macko, S. A., and Silfer, J. A., 1990, Carbon isotope composition of individual amino acids in the Murchison meteorite, *Nature* **348**:47–49.

Ertel, J. R., and Hedges, J. I., 1983, Bulk chemical and spectroscopic properties of marine and terrestrial humic acids, melanoidins and catechol-based synthetic polymers, in: *Aquatic and Terrestrial Humic Substances* (R. F. Christman and E. T. Gjessing, eds.), Ann Arbor Science, Ann Arbor, Michigan, pp. 143–163.

Fontugne, M. R., and Duplessy, J. C., 1978, Carbon isotope ratio of marine plankton related to surface water masses, *Earth Planet. Sci. Lett.* **41**:365–371.

Fontugne, M. R., and Duplessy, J. C., 1981, Organic carbon isotopic fractionation by marine plankton in the temperature range −1 to 31°C, *Oceanol. Acta* **4**:85–90.

Franchi, I. A., Exley, R. A., Gilmour, I., and Pillinger, C. T., 1989, Stable isotope and abundance measurements of solvent extractable compounds in Murchison, Extended Abstract, 14th

Symposium on Antarctic Meteorites, June 1989, National Institute for Polar Research, Tokyo.

Freedman, P. A., Gillyon, E. C. P., and Jumeau, E. J., 1988, Design and application of a new instrument for GC-isotope ratio MS, *Am. Lab.* **1988**(June):114–119.

Freeman, K. H., Hayes, J. M., Trendel, J.-M., and Albrecht, P., 1990, Evidence from carbon isotope measurements for diverse origins of sedimentary hydrocarbons, *Nature* **343**:254–256.

Gaebler, O. H., Choitz, H. C., Vitti, T. G., and Vukmirovich, R., 1963, Significance of ^{15}N excess in nitrogenous compounds of biological origin, *Can. J. Biochem.* **41**:1089–1097.

Gaebler, O. H., Vitt, T. G., and Vukmirovich, R., 1966, Isotope effects in metabolism of ^{14}N and ^{15}N from unlabeled dietary proteins, *Can. J. Biochem.* **44**:1249–1257.

Galimov, E. M., 1980, C^{13}/C^{12} in kerogen, in: *Kerogen—Insoluble Organic Matter from Sedimentary Rocks* (B. Durand, ed.), Editions Technip, Paris, pp. 271–299.

Gearing, J. N., Gearing, P. J., Rudnick, D. T., Requejo, A. G., and Hutchings, M. J., 1984, Isotopic variability of organic carbon in a phytoplankton-based, temperate estuary, *Geochim. Cosmochim. Acta* **48**:1089–1098.

Gearing, P., Plucker, F. E., and Parker, P. L., 1977, Land-derived organic matter in surface sediments from the Gulf of Mexico, *Geochim. Cosmochim. Acta* **40**:1019–1029.

Gilmour, I., Swart, P. K., and Pillinger, C. T., 1984, The isotopic composition of individual petroleum lipids, *Org. Geochem.* **6**: 665–670.

Gonzalez, J. M., 1983, Amino acid composition of sediments from a deltaic environment, *Mar. Chem.* **14**:61–71.

Hare, P. E., 1969, Geochemistry of proteins, peptides and amino acids, in: *Organic Geochemistry* (G. Eglinton and M. T. J. acids, in: *Organic Geochemistry* (G. Eglinton and M. T. J. Murphy, eds.), Springer-Verlag, New York, pp. 438–463.

Hare, P. E., Fogel, M. L., Stafford, T. W., Mitchell, A. D., and T. C. Hoering, 1991, The isotopic composition of carbon and nitrogen in individual amino acids isolated from modern and fossil proteins, *J. Arch. Sci.* **18**:277–292.

Hayes, J. M., Takigiku, R., Ocampo, R., Callot, H. J., and Albrecht, P., 1987, Isotopic compositions and probable origins of organic molecules in the Eocene Messel shale, *Nature* **329**:48–51.

Hayes, J. M., Popp, B. N., Takigiku, R., and Johnson, M. W., 1989, An isotopic study of biogeochemical relationships between carbonates and organic carbon in the Greenhorn Formation, *Geochim. Cosmochim. Acta* **53**:2961–2972.

Hayes, J. M., Freeman, K. H., Popp, B. N., and Hoham, C. H., 1990, Compound-specific isotope analysis: A novel tool for reconstruction of ancient biogeochemical processes, *Org. Geochem.* **16**:1115–1128.

Hedges, J. I., 1978, The formation and clay mineral reactions of melanoidins, *Geochim. Cosmochim. Acta* **42**:69–76.

Hedges, J. I., and Parker, P. L., 1976, Land-derived organic matter in surface sediments of the Gulf of Mexico, *Geochim. Cosmochim. Acta* **40**:1019–1029.

Henrichs, S. M., and Farrington, J. W., 1979, Amino acids in interstitial waters of marine sediments, *Nature* **279**:319–322.

Henrichs, S. M., and Farrington, J. W., 1980, Amino acids in interstitial waters of marine sediments: A comparison of results from different sedimentary environments, in: *Advances in Organic Geochemistry 1979* (A. G. Douglas and J. R. Maxwell, eds.), Pergamon Press, New York, pp. 435–443.

Henrichs, S. M., and Farrington, J. W., 1984, Peru upwelling region sediments near 15°S. 1. Remineralization and accumulation of organic matter, *Limnol. Oceanogr.* **29**:1–19.

Henrichs, S. M., and Farrington, J. W., 1987, Early diagenesis of amino acids and organic matter in two coastal marine sediments, *Geochim. Cosmochim. Acta* **51**:1–15.

Henrichs, S. M., and Farrington, J. W., and Lee, C., 1984, Peru upwelling region sediments near 15°S. 2. Dissolved free and total hydrolyzable amino acids, *Limnol. Oceanogr.* **29**:20–34.

Hoering, T. C., 1957, The isotopic composition of ammonia and the nitrate ion in rain, *Geochim. Cosmochim. Acta* **12**:97–102.

Hoering, T. C. 1973, A comparison of melanoidin and humic acid, *Carnegie Inst. Washington Yearb.* **72**:682–690.

Hoering, T., and Ford, H. T., 1960, The isotope effect in the fixation of nitrogen by *Azotobacter*, *J. Am. Chem. Soc.* **82**:376–378.

Ikan, R., Ioselis, P., Rubinsztain, Y., Aizenshtat, Z., Muller-Vonmoos, M., and Rub, A., 1988, Light hydrocarbons and volatile compounds produced during the thermal treatment of melanoidins and humic substances, *Org. Geochem.* **12**:273–279.

Ishiwatari, R., Morinaga, S., Yamamoto, S., Machihara, T., Rubinsztain, Y., Ioselis, P., Aizenshtat, Z., and Ikan, R., 1986, A study of formation mechanism of sedimentary humic substances—I. Characterization of synthetic humic substances (melanoidins) by alkaline potassium permanganate oxidation, *Org. Geochem.* **9**:11–23.

Jasper, J. P., and Gagosian, R. B., 1989, Glacial–interglacial climatically forced ^{13}C variations in sedimentary organic matter, *Nature* **342**:60–62.

Jasper, J. P., and Gagosian, R. B., 1990, The sources and deposition of organic matter in the Late Quaternary Pigmy Basin, Gulf of Mexico, *Geochim. Cosmochim. Acta* **54**:1117–1132.

Kaplan, I. R., 1975, Stable isotopes as a guide to biogeochemical processes, *Proc. R. Soc. London, Ser. B* **189**:183–211.

Kennicutt, M. C., and Brooks, J. M., 1990, Unusual normal alkane distributions in offshore New Zealand sediments, *Org. Geochem.* **15**:193–197.

Kennicutt, M. C., Barker, C., Brooks, J. M., DeFreitas, D. A., and Zhu, G. H., 1987, Selected organic matter indicators in the Orinoco, Nile and Changjiang deltas, *Org. Geochem.* **11**:41–51.

Kennicutt, M. C., Macko, S. A., Harvey, H. R., and Bidigare, R. R., 1991, Preservation of *Sargassum* under anoxic conditions: Isotopic and molecular evidence, in: *Productivity Accumulation and Preservation of Organic Matter in Recent and Ancient Sediments* (J. K. Whelan and J. W. Farrington, eds.) Columbia University Press, New York, pp. 123–141.

Kennicutt, M. C., Bidigare, R. R., Macko, S. A., and Keeney-Kennicutt, W. L., 1992, The stable isotopic composition of photosynthetic pigments and related biochemicals, *Isot. Geosci.* **15**:235–246.

Larter, S. R., and Douglas, A. G., 1980, Melanoidins—kerogen precursors and geochemical lipid sinks: A study using pyrolysis gas chromatography (PGC), *Geochim. Cosmochim. Acta* **44**:2087–2095.

Libes, S. M., and Deuser, W. G., 1988, The isotope geochemistry of particulate nitrogen in the Peru Upwelling Area and the Gulf of Maine, *Deep-Sea Res.* **35**:517–533.

Macko, S. A., 1981, Stable nitrogen isotope ratios as tracers of organic geochemical processes, Ph.D. Thesis, University of Texas, Austin.

Macko, S. A., 1983, Sources of organic nitrogen in mid-Atlantic coastal bays and continential shelf sediments of the United States: Isotopic evidence, *Carnegie Inst. Washington Yearb.* **82**:390–394.

Macko, S. A., 1989, Stable isotope organic geochemistry of sediments from the Labrador Sea (Sites 646, 647) and Baffin Bay (Site 645) ODP Leg 105, in: *Initial Reports of the ODP Leg 105*,

Part B (M. Arthur, S. Srivastava, *et al.*, eds.), College Station, TX, pp. 209–232.

Macko, S. A., 1992, The characterization of organic matter in abyssal sediments, pore waters and sediment traps, in: *Deep-Sea Food Chains and the Global Carbon Cycle* (G. Rowe, ed.), NATO Advanced Research Workshop, Kluwer Academic Publ., Dordrecht, pp. 325–338.

Macko, S. A., and Engel, M. H., 1991, Assessment of indigeneity in fossil organic matter: Amino acids and stable isotopes, *Philos. Trans. R. Soc. London* **333**:367–374.

Macko, S. A., and Pereira, C. P. G., 1990, Neogene paleoclimate development of the Antarctic Weddell Sea Region: Organic geochemistry, in: *Initial Reports of the Ocean Drilling Program Leg 113*, College Station, TX, Part B (P. F. Barker, J. P. Kennett, *et al.*, eds.), pp. 881–897.

Macko, S. A., Entzeroth, L., and Parker, P. L., 1984, Regional differences in nitrogen and carbon isotopes on the continental shelf of the Gulf of Mexico, *Naturwissenschaften* **71**:374–375.

Macko, S. A., Estep, M. F., Engel, M. H., and Hare, P. E., 1986, Kinetic fractionation of stable nitrogen isotopes during amino acid transamination, *Geochim. Cosmochim. Acta* **50**:2143–2146.

Macko, S. A., Estep, M. L. F., Hare, P. E., and Hoering, T. C., 1987a, Isotopic fractionation of nitrogen and carbon in the synthesis of amino acids by microorganisms. *Isot. Geosci.* **65**:79–92.

Macko, S. A., Pulchan, K., and Ivany, D. E., 1987b, Organic Geochemistry of Baffin Island Fjords. Sedimentology of Arctic Fjords Experiment, Geological Survey of Canada Open File Report 1589-13, pp. 1–34.

Macko, S. A., Helleur, R., Hartley, G., and Jackman, P., 1990, Diagenesis of organic matter—a study using stable isotopes of individual carbohydrates, *Org. Geochem.* **16**:1129–1137.

Macko, S. A., Engel, M. H., Hartley, G., Hatcher, P., Helleur, R., Jackman, P., and Silfer, J., 1991, Isotopic compositions of individual carbohydrates as indicators of early diagenesis of organic matter, *Chem. Geol.* **93**:147–161.

Macko, S. A., Leskey, T., Ryan, M., and Engel, M. H., 1992, Carbon isotopic analysis of individual carbohydrates by GC/IRMS, 203rd National Meeting of the American Chemical Society, San Francisco, Abstract No. 107.

Maillard, L. C., 1912, Action des acides amines sur les sucres: Formation des mélanoidines par voie méthodiques, *C. R. Acad. Sci.* **154**:66–68.

Mariotti, A., Mariotti, F., Amarger, N., Pizelle, G., Ngambi, J.-M., Champigny, M.-L., and Moyse, A., 1980, Fractionnements isotopiques de l'azote lors des processus d'absorption des nitrates et de fixation de l'azote atmosphérique par les plantes, *Physiol. Veg.* **18**:163–181.

Mariotti, A., Germon, J. C., Hubert, P., Kaiser, P., Letolle, R., Tardieux, A., and Tardieux, P., 1981, Experimental determination of nitrogen kinetic isotope fractionation: Some principles: Illustration for the denitrification and nitrification processes, *Plant Soil* **62**:423–430.

Mariotti, A., Germon, J. C., Leclerc, A., Catroux, G., and Letolle, R., 1982, Experimental determination of kinetic isotope fractionation of nitrogen isotopes during denitrification, in: *Stable Isotopes* (H. L. Schmidt, *et al.*, eds.), Elsevier, Amsterdam, pp. 459–464.

Mayer, L. M., Macko, S. A., and Cammen, L., 1988, Provenance, concentrations and nature of sedimentary organic nitrogen in the Gulf of Maine, *Mar. Chem.* **25**:291–304.

Montani, S., Yoshaki, M., and Fukase, S., 1980, Flux of nitrogen compounds in coastal sediments and pore water, *Chem. Geol.* **30**:35–45.

Mopper, K., and Degens, E. T., 1972, Aspects of biogeochemistry of carbohydrates and proteins in aquatic environments, Technical Reports, Woods Hole Oceanographic Institution, Ref. # 72-68. Woods Hole, Massachusetts.

Morris, R. J., 1975, The amino acid composition of a deep water sediment from the upwelling region northwest of Africa, *Geochim. Cosmochim. Acta* **39**:381–388.

Muller, P. J., 1977, C/N ratios in Pacific deep sea sediments: Effect of inorganic ammonium and organic nitrogen compounds adsorbed to clays, *Geochim. Cosmochim Acta* **41**:765–776.

Muller, P. J., and Suess, E., 1979, Productivity, sedimentation rate, and sedimentary carbon content in the oceans. 1. Organic carbon preservation, *Deep-Sea Res.* **26A**:1347–1362.

Muzuka, A. N. N., Macko, S. A., and Pedersen, T. F., 1991, Stable carbon and nitrogen isotope compositions of organic matter from ODP Sites 724 and 725, Oman Margin, in: *Initial Reports of the Ocean Drilling Program Leg 117*, College Station, TX, Part B (W. L. Prell, N. Niitsuma, *et al.*, eds.), pp. 571–586.

Myers, E. P., 1974, The concentration and isotopic composition of carbon in marine sediments affected by sewage discharge, Ph.D. Thesis, California Institute of Technology.

Newman, J. W., Parker, P. L., and Behrens, E. W., 1973, Organic carbon isotope ratios in Quaternary cores from the Gulf of Mexico, *Geochim. Cosmochim. Acta* **37**:225–238.

Nissenbaum, A., 1974, The organic geochemistry of marine and terrestrial humic substances: Implications of carbon and hydrogen isotope studies, in: *Advances in Organic Geochemistry 1973* (B. Tissot and F. Bienner, eds.), Editions Technip, Paris, pp. 39–52.

Nissenbaum, A., and Schallinger, K. M., 1974, The distribution of the stable carbon isotope ($^{13}C/^{12}C$) in fractions of soil organic matter, *Geoderma* **11**:137–145.

Nissenbaum, A., Presley, B. J., and Kaplan, I. R., 1972, Early diagenesis in a reducing fjord, Saanich Inlet, British Columbia—I. Chemical and isotopic changes in major components of interstitial water, *Geochim. Cosmochim. Acta* **36**:1007–1027.

Ocampo, R., Callot, H. J., Albrecht, P., Popp, B. N., Horowitz, M. R., and Hayes, J. M., 1989, Different isotopic compositions of C_{32} etioporphyrin II in oil shale. Origin of etioporphyrin II from heme? *Naturwissenschaften* **76**:419–421.

O'Leary, M. H., and Kluetz, M. D., 1972, Nitrogen isotope effects on the chymotrypsin-catalyzed hydrolysis of *N*-acetyl-L-tryptophanamide, *J. Am. Chem. Soc.* **94**:3585–3589.

O'Leary, M. H., and Marlier, J. F., 1979, Heavy-atom isotope effects on the alkaline hydrolysis and hydrazinolysis of methyl benzoate, *J. Am. Chem. Soc.* **101**:3300–3306.

Ostrom, N. E., and Macko, S. A., 1991a, Sources, cycling and distribution of water column particulate and sedimentary organic matter in northern Newfoundland fjords and bays: A stable isotope study, in: *Productivity, Accumulation and Preservation of Organic Matter in Recent and Ancient Sediments* (J. K. Whelan and J. W. Farrington, eds.), Columbia University Press, New York, pp. 55–81.

Ostrom, N. E., and Macko, S. A., 1991b, Late Wisconsinan to present sedimentation of organic matter off northern Newfoundland in response to climatological events, *Cont. Shelf Res.* **11**:1285–1296.

Parker, P. L., 1962, The isotopic composition of the carbon of fatty acids, *Carnegie Inst. Washington Ann. Rep.* **1961–1962**:187–190.

Parker, P. L., 1964, The biogeochemistry of the stable isotopes of carbon in a marine bay, *Geochim. Cosmochim. Acta* **28:**1155–1164.

Parker, P. L., Behrens, E. W., Calder, J. A., and Shultz, D., 1972, Stable carbon isotope ratio variations in the organic carbon from Gulf of Mexico sediments, *Contrib. Mar. Sci.* **16:**139–147.

Patience, R. L., Clayton, C. J., Kearsley, A. T., Rowland, S. J., Bishop, A. N., Rees, A. W. G., Bibby, K. G., and Hopper, A. C., 1990, An integrated biochemical, geochemical and sedimentological study of organic diagenesis in sediments from ODP, Leg 112, in: *Initial Reports of the Ocean Drilling Program Leg 112*, College Station TX, Part B (E. Suess, R. von Huene, *et al.*, eds.), pp. 135–153.

Peters, K. E., Sweeney, R. E., and Kaplan, I. R., 1978, Correlation of carbon and nitrogen stable isotope ratios in sedimentary organic matter, *Limnol. Oceanogr.* **23:**598–604.

Popp, B. N., Takigiku, R., Hayes, J. M., Louda, J. W., and Baker, E. W., 1989, The post-paleozoic chronology and mechanism of ^{13}C depletion in primary marine organic matter, *Am. J. Sci.* **289:** 436–454.

Qian, Y., Engel, M. H., and Macko, S. A., 1992, Stable isotope fractionation of biomonomers during protokerogen formation, *Isotope Geosci.* **15:**201–210.

Rafalska, J. K., Engel, M. H., and Lanier, W. P., 1991, Retardation of racemization rates of amino acids incorporated into melanoidins, *Geochim. Cosmochim. Acta* **55:**3669–3676.

Rau, G. H., Sweeney, R. E., and Kaplan, I. R., 1982, Plankton $^{13}C/^{12}C$ ratio changes with latitude: Differences between northern and southern oceans, *Deep-Sea Res.* **29:**1035–1039.

Rau, G. H., Arthur, M. A., and Dean, W. E., 1987, $^{15}N/^{14}N$ variations in Cretaceous Atlantic sedimentary sequences: Implications for past changes in marine nitrogen biogeochemistry, *Earth Planet. Sci. Lett.* **82:**269–279.

Rau, G. H., Takahashi, T., and Des Marais, D. J., 1989, Latitudinal variations in plankton $\delta^{13}C$: Implications for CO_2 and productivity in past oceans, *Nature* **341:**516–518.

Rau, G. H., Takahashi, T., Des Marais, D. J., and Sullivan, C. W., 1991, Particulate organic matter $\delta^{13}C$ variations across the Drake Passage, *J. Geophys. Res.* **96:**15131–15135.

Rieley, G., Collier, R. J., Jones, D. M., Eglinton, G., Eakin, P. A., and Fallick, A. E., 1991, Sources of sedimentary lipids deduced from stable carbon isotope analyses of individual compounds, *Nature* **352:**425–427.

Rogers, M. A., van Hinte, J., and Sugden J. G., 1972, Organic carbon ^{13}C values from Cretaceous, Tertiary and Quaternary marine sequences in the North Atlantic, in: *Initial Reports of the Deep Sea Drilling Project, Vol. XII* (T. A. Davies, ed.), U.S. Government Printing Office, Washington D.C., pp. 1115–1126.

Rosenfeld, J. K., 1979, Amino acid diagenesis and absorption in nearshore anoxic sediments, *Limnol. Oceanogr.* **24:**1014–1021.

Rubinsztain, Y., Ioselis, P., Ikan, R., and Aizenshtat, Z., 1984, Investigations on the structural units on melanoidins, *Org. Geochem.* **6:**791–804.

Rubinsztain, Y., Yariv, S., Ioselis, P., Aizenshtat, Z., and Ikan, R., 1986a, Characterization of melanoidins by IR spectroscopy—I. Galactose-glycine melanoidins, *Org. Geochem.* **9:**117–125.

Rubinsztain, Y., Yariv, S., Ioselis, P., Aizenshtat, Z., and Ikan, R., 1986b, Characterization of melanoidins by IR spectroscopy—II. Melanoidins of galactose with arginine, isoleucine, lysine and valine, *Org. Geochem.* **9:**371–374.

Sackett, W. M., 1964, The depositional history and isotopic organic carbon composition of marine sediments, *Mar. Geol.* **2:**173–185.

Sackett, W. M., 1986, $\delta^{13}C$ signatures of organic carbon in southern high latitude deep sea sediments; paleotemperature implications, *Org. Geochem.* **9:**63–68.

Sackett, W. M., and Thompson, R. R., 1963, Isotope organic carbon composition of Recent continental derived clastic sediments of Eastern Gulf Coast, Gulf of Mexico, *Am. Assoc. Petrol. Geol.* **147:**535–531.

Sackett, W. M., Eckelmann, W. R., Bender, M. L., and Be, A. W. H., 1965, Temperature dependence of carbon isotope compositions in marine plankton and sediments, *Science* **148:**235–237.

Sackett, W. M., Eadie, B. J., and Exner, M. E., 1974, Stable isotopic composition of organic carbon in recent Antarctic sediments, in: *Advances in Organic Geochemistry 1973* (B. Tissot, and F. Bienner, eds.), Editions Technip, Paris, pp. 661–671.

Saino, T., and Hattori, A., 1980, ^{15}N natural abundance in oceanic suspended particulate matter, *Nature* **283:**752–754.

Saino, T., and Hattori, A., 1985, Variation in ^{15}N natural abundance of suspended organic matter in shallow oceanic waters, in: *Marine and Estuarine Geochemistry* (A. C. Sigleo and A. Hattori, eds.), Lewis Publishers, Chelsea, Michigan, pp. 1–14.

Saino, T., and Hattori, A., 1987, Geographical variation of the water column distribution of suspended particulate nitrogen and its ^{15}N abundance in the Pacific and its marginal seas, *Deep-Sea Res.* **28:**901–909.

Sargent, J. R., Hopkins, C. C. E., Seiring, J. V., and Youngson, A., 1983, Partial characterization of organic material in surface sediments from Balsfjorden, northern Norway, in relation to its origin and nutritional value for sediment-ingesting animals, *Mar. Biol.* **76:**87–94.

Schidlowski, M., Hayes, J. M., and Kaplan, I. R., 1983, Isotopic inferences of ancient biochemistries: Carbon, sulfur, hydrogen and nitrogen, in: *Earth's Earliest Biosphere: Its Origin and Evolution* (J. W. Schopf, ed.), Princeton University Press, Princeton, New Jersey, pp. 149–186.

Schoenheimer, R., and Rittenberg, D., 1939, Studies in protein metabolism, *J. Biol. Chem.* **127:**285–344.

Serban, A., Engel, M. H., and Macko, S. A., 1988, The distribution, stereochemistry and stable isotopic composition of amino acid constituents of fossil and modern mollusk shells, *Org. Geochem.* **13:**1123–1129.

Shimmelmann, A., and DeNiro, M. J., 1986, Stable isotopic studies on chitin. II. The $^{13}C/^{12}C$ and $^{15}N/^{14}N$ ratios in arthropod chitin, *Contrib. Mar. Sci.* **29:**113–130.

Shimmelmann, A., DeNiro, M. J., Poulicek, M., Voss-Foucart, A., Goffinet, G., and Jeuniaux, C., 1986, Stable isotopic composition of chitin from arthropods recovered in archaeological contexts as palaeoenvironmental indicators, *J. Arch. Sci.* **13:** 553–566.

Sigleo, A. C., and Hattori, A. (eds.), 1985, *Marine and Estuarine Geochemistry*, Lewis Publishers, Chelsea, Michigan.

Sigleo, A. C., and Macko, S. A., 1985, Stable isotope and amino acid composition of estuarine dissolved colloidal material, in: *Marine and Estuarine Geochemistry* (A. C. Sigleo and A. Hattori, eds.), Lewis Publishers, Chelsea, Michigan, pp. 30–46.

Silfer, J. A., Engel, M. H., Macko, S. A., and Jumeau, E. J., 1991, Stable carbon isotope analysis of amino acid enantiomers by conventional isotope ratio mass spectrometry and combined gas chromatography–isotope ratio mass spectrometry, *Anal. Chem.* **63:**370–374.

Silfer, J. A., Engel, M. H., and Macko, S. A., 1992, Kinetic fractionation of stable carbon and nitrogen isotopes during peptide bond hydrolysis: Experimental evidence and geochemical implications, *Isotope Geosci.* **15:**211–222.

Sohn, M. L. (ed.), 1986, *Organic Marine Chemistry*, American Chemical Society, Washington, D.C.

Spiker, E. C., and Hatcher, P. G., 1984, Carbon isotope fractionation of sapropelic organic matter during early diagenesis, *Org. Geochem.* **5**:283–290.

Spiker, E. C., and Hatcher, P. G., 1987, The effects of early diagenesis on the chemical and stable carbon isotopic composition of wood, *Geochim. Cosmochim. Acta* **51**:1385–1391.

Stacey, F. W., Lindsay, J. G., and Bourns, A. N., 1952, Isotope effects in the thermal deammonation of phthalamide, *Can. J. Chem.* **30**:135–145.

Starikova, N. D., and Korzhikova, R. I., 1969, Amino acids in the Black Sea, *Oceanology* **9**:509–518.

Summons, R., and Powell, T. G., 1986, Chlorobiaceae in Paleozoic seas revealed by biological markers, isotopes and geology, *Nature* **319**:763–765.

Sweeney, R. E., and Kaplan, I. R., 1980a, Tracing flocculent industrial and domestic sewage transport on San Pedro shelf, southern California by nitrogen and sulfur isotope ratios, *Mar. Environ. Res.* **3**:214–224.

Sweeney, R. E., and Kaplan, I. R., 1980b, Natural abundances of ^{15}N as a source indicator for near-shore marine sedimentary and dissolved nitrogen, Mar. Chem. **9**:81–94.

Sweeney, R. E., Liu, K. K., and Kaplan, I. R., 1978, Oceanic nitrogen isotopes and their uses in determining the source of sedimentary nitrogen, in: *Stable Isotopes in the Earth Sciences* (B. W. Robinson, ed.), New Zealand Department of Scientific and Industrial Research Bulletin 220, Wellington, New Zealand, pp. 9–26.

Taguchi, K., and Sampei, Y., 1986, The formation, and clay mineral and $CaCO_3$ association reactions of melanoidins, in: *Advances in Organic Geochemistry 1985* (D. Leythaeuser and J. Rullkötter, eds.), *Org. Geochem.* **10**:1081–1089.

Takigiku, R., 1987, Isotopic and molecular indicators of origins of organic compounds in sediments, Ph.D. Thesis, University of Indiana.

Tissot, B. P., and Welte, D. H., 1984, *Petroleum Formation and Occurrence*, Springer-Verlag, New York.

Treibs, A., 1935, Chlorophyll- und Haminderivivate in bituminosen Gesteinen, Erdolen, Erdwachsen und Asphalten, *Ann. Chem.* **510**:42–62.

Vogler, E. A., and Hayes, J. M., 1980, Carbon isotopic compositions of carboxyl groups of biosynthesized fatty acids, in: *Advances in Organic Geochemistry 1979* (A. G. Douglas and J. R. Maxwell, eds.), Pergamon Press, Oxford, pp. 697–704.

Vogler, E. A., Meyers, P. A., and Moore, W. A., 1981, Comparison of Michigan basin crude oils, *Geochim. Cosmochim. Acta* **45**: 2287–2293.

Wada, E., 1980, Nitrogen isotope fractionation and its significance in biogeochemical processes occurring in marine environments, in: *Isotope Marine Chemistry* (E. D. Goldberg, and Y. Horibe, eds.), Uchida Rokakuho Publ., Tokyo, pp. 375–398.

Wada, E., Kadonaga, T., and Matsuo, S., 1975, ^{15}N abundance in nitrogen of naturally occurring substances and global assessment of denitrification from isotopic viewpoint, *Geochem. J.* **9**:139–148.

Wakeham, S. G., Lee, C., Farrington, J. W., and Gagosian, R. B., 1984, Biochemistry of particulate organic matter in the oceans: Results from sediment trap experiments, *Deep-Sea Res., Part A* **31**:509–528.

Wehmiller, J. F., and Hare, P. E., 1972, Amino acid content of some samples from the Deep Sea Drilling Project, in: *Initial Reports of the Deep Sea Drilling Project*, Vol. IX, U.S. Government Printing Office, Washington, D.C., pp. 903–905.

Welte, D. H., 1969, Determination of C^{13}/C^{12} isotope ratios of individual higher *n*-paraffins from different petroleums, in: *Advances in Organic Geochemistry 1968* (P. A. Schenck and I. Havenaar, eds.), Pergamon Press, Oxford, pp. 269–277.

Welte, D. H., 1974, Recent advances in organic geochemistry of humic substances and kerogen. A review, in: *Advances in Organic Geochemistry 1973* (B. Tissot and F. Bienner, eds.), Editions Technip, Paris, pp. 39–52.

Whelan, J. K., 1977, Amino acids in a surface sediment core of the Atlantic abyssal plain, *Geochim. Cosmochim. Acta* **41**:803–810.

Yamamoto, S., and Ishiwatari, R., 1989, A study of the formation mechanism of sedimentary humic substances—II. Protein-based melanoidin model, *Org. Geochem.* **14**:479–489.

Zieman, J. C., Macko, S. A., and Mills, A. L., 1984, Role of seagrasses and mangroves in estuarine food webs: Temporal and spatial changes in stable isotope composition and amino acid content during decomposition, *Bull. Mar. Sci.* **35**:380–392.

Chapter 10

Marine Invertebrate Feeding and the Sedimentary Lipid Record

STUART A. BRADSHAW and GEOFFREY EGLINTON

1. Introduction

Although it is recognized that many different marine invertebrates affect the flux of lipids in the water column, studies have considered invertebrates in total, that is, as a single factor (Wakeham, 1982; Wakeham *et al.*, 1984; Wakeham and Canuel, 1986). These studies have provided gross information on how invertebrate feeding is important in the processes which determine the eventual lipid distributions of interfacial sediments. In the consideration of invertebrates, it is usually the crustaceans that have been cited as the important mediators in the flux of organic C (Corner *et al.*, 1986, and references therein). Whereas this invertebrate group is without doubt important in the marine environment, its prominence does not preclude the possibility that in some situations other invertebrate groups may be dominant. Furthermore, little consideration has been given to the effects other groups may have on specific compounds. For example, although crustaceans may be responsible for the major part of the invertebrate effect on sedimentary lipid distributions, a single, not necessarily dominant, species of invertebrates may have a significant effect on a specific lipid. In situations where interpretation is dependent on such a single lipid component, for example, the C_{15} and C_{17} branched (*iso* and *anteiso*) fatty acids as biomarkers for bacterial activity (Perry *et al.*, 1979; Volkman *et al.*, 1980b), then clearly the reconstruction of the processes determining sedimentary lipids might be flawed unless the effects of feeding by all invertebrate groups are considered.

To provide more detailed information concerning invertebrate processing of dietary lipids, a series of laboratory experiments has been carried out. These experiments have concentrated on a limited number of species, chosen as representatives of various animal groups. However, they indicate the manner in which the effects of feeding on dietary lipids can vary from animal to animal, species to species, and group to group.

Most of the experiments have concentrated on pelagic crustaceans, particularly the copepod *Calanus helgolandicus*, modeling the feeding processes which might occur in the water column. Some experiments have studied benthic animals, extending the investigation to groups such as mollusks and annelids (Fig. 1).

These experiments fall into two main categories. The majority of experiments sought to model natural feeding conditions. These have investigated herbivory (with a variety of cultured algal species fed to the animals), that is, the situation where an animal feeds upon an algal bloom or intact algal cells in the feces of other animals (Volkman *et al.*, 1980a; Tanoue *et al.*, 1982; Prahl *et al.*, 1984a, b; Neal *et al.*, 1986; Harvey *et al.*, 1987; Bradshaw *et al.*, 1990a, b, 1991), and copro-

STUART A. BRADSHAW and GEOFFREY EGLINTON • Organic Geochemistry Unit, School of Chemistry, University of Bristol, Bristol BS8 1TS, England. *Present address of S.A.B.*: Department of Retail Automation, Mobil Oil Company, Ltd., London SW1E GQB, England.

Organic Geochemistry, edited by Michael H. Engel and Stephen A. Macko. Plenum Press, New York, 1993.

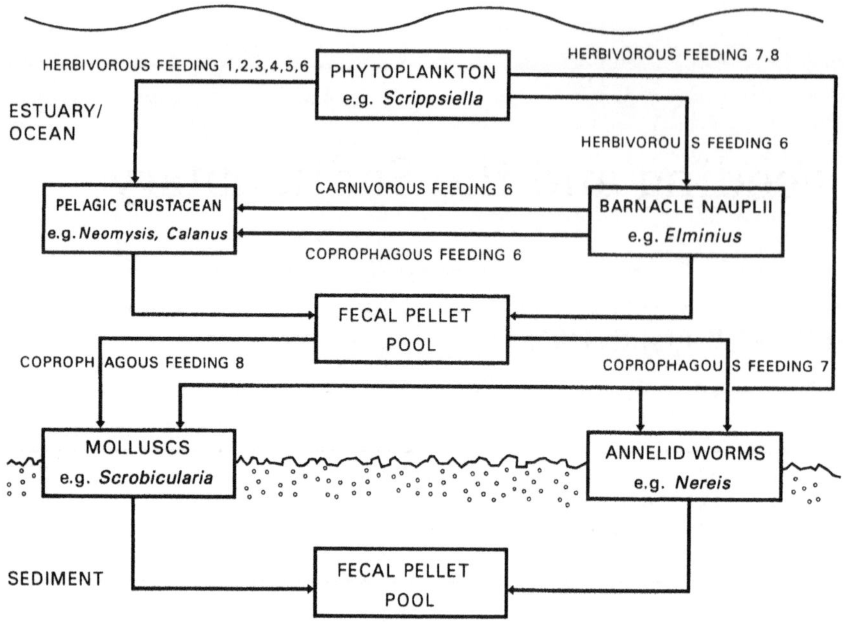

Figure 1. Summary of the previously conducted laboratory feeding experiments: 1, Prahl *et al.*, 1984 a, b; 2, Tanoue *et al.*, 1982; 3, Volkman *et al.*, 1980 d; 4, Harvey *et al.*, 1987; 5, Bradshaw *et al.*, 1990a; 6, Neal *et al.*, 1986; 7, Bradshaw *et al.*, 1990b; 8, Bradshaw *et al.*, 1991.

phagy, where the diet is feces derived from algal material which has already passed through the gut of another animal (Neal *et al.*, 1986; Bradshaw *et al.*, 1990b, 1991).

A few experiments have concentrated on more specific aspects of dietary lipid changes during feeding. These have included the effects of feeding on single dietary sterols (Bradshaw *et al.*, 1989; Harvey *et al.*, 1991), the contribution of animal lipids to the digested material (Bradshaw *et al.*, 1989), the microbially mediated changes in fecal lipids after egestion (Harvey *et al.*, 1989), and the effects of different food levels on dietary lipids (Harvey *et al.*, 1987). The laboratory experiments have provided basic information regarding the possible effects invertebrates may have upon the distribution of sedimentary lipids in the marine environment, and the means by which to interpret those natural distributions in terms of such feeding. Usually, the experiments have considered the changes in more than one lipid class at a time. In addition, lipid samples have generally been analyzed after saponification to break down lipids such as wax esters and triglycerides. Techniques of analysis have included high-pressure liquid chromatography, gas chromatography, and gas chromatography/mass spectrometry. The following discussion details the main conclusions from the feeding experiments concerning the effects of marine invertebrates on the passage of total fatty acids, fatty alcohols, and sterols through the food chains.

2. Fatty Acids

Algal species grown as monocultures representing chlorophytes, prymnesiophytes, and dinoflagellates have been fed to the crustaceans *C. helgolandicus*, the barnacle *Elminius modestus*, and the estuarine shrimp *Neomysis integer*. In one study with *Euphausia superba*, little difference was observed between the fatty acid distributions of the food alga and the feces (Tanoue *et al.*, 1982). However, crustacean herbivory was generally characterized by preferential assimilation of dietary polyunsaturated fatty acids (PUFAs) over other saturation types (>90% and 60–70% assimilated, respectively). The extent to which crustaceans such as *C. helgolandicus* assimilate unsaturated fatty acids has also been shown to depend upon dietary levels (Harvey *et al.*, 1987). Whereas PUFAs appeared to be assimilated to a high degree at all food levels, including those representing algal bloom conditions, the extent of removal of unsaturated and saturated fatty acids was dependent upon the total quantity of algal material ingested, with greater amounts of these fatty acids being assimilated at lower food levels. Although the proportion of unsaturated fatty acids removed from the material passing through the food chains may vary, it is clear that crustaceans are, in part, responsible for a decrease in the flux of PUFAs with depth in the water column (Wakeham *et al.*, 1984; Wakeham and Canuel, 1986).

Such differences in the assimilation of dietary PUFAs during feeding have also been demonstrated in herbivory experiments with the benthic mollusk *Scrobicularia plana* and annelid worm *Nereis diversicolor* (Bradshaw *et al.*, 1990b, 1991). The mollusk acted in a similar way to *C. helgolandicus* while the annelid assimilated all fatty acids to a high degree. Although an explanation for this variation between species might be responses to different food levels, evidently species from different groups can have different effects upon dietary fatty acids. The low abundances of PUFAs observed in some sediments can be due to the feeding activities of pelagic and benthic animals as well as microbial degradation.

Experiments simulating coprophagy have also been conducted with *C. helgolandicus*, *S. plana*, and *N. diversicolor* in order to evaluate the importance of invertebrate reworking of particulate material in the water column and in the sediment (Neal *et al.*, 1986; Bradshaw *et al.*, 1990b, 1991). Although there are situations where invertebrates feed on relatively unaltered algal material (feeding on algal blooms or fecal material containing relatively unchanged algal material due to superfluous feeding), for many species, coprophagy is the normal mode of feeding. Because a significant proportion of the organic material produced in the euphotic zone eventually reaches the sediment as zooplankton fecal material, the experiments have involved the feeding of feces from pelagic species of crustaceans to the species under study. Although these experiments do not simulate the ingestion of material which has undergone microbial attack as might occur with sinking fecal material or in the sediment, they have addressed the changes in dietary lipids during successive feeding by invertebrates.

The effects of coprophagy on the unsaturated fatty acids of the digested material seem to differ with the species under investigation. Two experiments with *C. helgolandicus* feeding coprophagously on the feces from the barnacle *E. modestus*, which had been fed *Hymenomonas carterae* and *Dunaliella primolecta*, apparently resulted in little change in the fatty acid distributions of the diet. Most of the PUFAs were removed during the herbivory by *E. modestus*, suggesting that primary crustacean feeding in a water column will rapidly remove PUFAs from the material passing through the food chains. Reworking of organic material by pelagic crustaceans is likely to remove residual PUFAs and lead to the sedimentation of material which is greatly depleted in these fatty acids. However, it has to be remembered that in conditions of superfluous feeding, PUFAs may survive herbivory. Here, coprophagy will be of more importance in determining sedimentary fatty acid distributions and will probably remove most of these fatty acids from the food material.

The diet used in the coprophagy experiments with *S. plana* and *N. diversicolor* did not contain PUFAs (feces from *N. integer* feeding on *Scripsiella trochoidea*). Whereas *N. diversicolor* was seen to remove all monounsaturated fatty acids from the food, *S. plana* nevertheless appeared to increase the relative abundance of these acids. The increased quantities of these fatty acids seemed to originate from the animal itself, and, indeed, one study has shown that the mollusk is capable of contributing some unsaturated fatty acids such as 20:1 to its feces (Bradshaw *et al.*, 1989).

Benthic animals are capable of both enriching and depleting the unsaturated fatty acid content of sediments, the effect being partly species dependent. However, invertebrate feeding as a whole generally decreases the flux of unsaturated fatty acids through the food chains, and it is only during reworking of sedimentary material that the reverse may occur.

The capacity for preferential assimilation of unsaturated fatty acids has, to some extent, been demonstrated in experiments where animals have been fed lipid-free diets. The three species *N. integer*, *S. plana*, and *N. diversicolor* have all been shown to contribute fatty acids to their feces (Bradshaw *et al.*, 1989). However, these fatty acids are more saturated than those in the body tissues. They probably have an origin in gut cells, which have, at least in the case of crustaceans, been shown to disintegrate during the feeding process (Nott *et al.*, 1985; Al-Mohanna and Nott, 1986). Although the different fatty acid distributions may reflect differences in the composition of the gut cells and body tissues as a whole, it seems likely that the lysed gut cells are subject to the same digestive processes as ingested material. Preferential assimilation of PUFAs should therefore occur, and the observed contribution of animal fatty acids shows a depletion in these components.

Furthermore, many of the fatty acids which are contributed to the animal feces (whether they are of animal or gut microbial origin), are of the same identity as those in common phytoplanktonic algae, e.g., 16:0 and 18:0. Although it is not known how important these contributed fatty acids are in relation to those fatty acids ingested as dietary lipids, they should be noted in consideration of the distributions of fatty acids in fecal pellets.

Some sedimentary fatty acids are considered characteristic of bacterial activity. These include the normal and particularly the branched *iso-* and *anteiso-*

C_{15} and C_{17} fatty acids, and the 18:1 Δ^{11} fatty acid (Perry *et al.*, 1979; Volkman *et al.*, 1980b). However, experiments have shown that invertebrate feeding is also likely to affect the abundances of these compounds in sediments.

C_{15} and C_{17} fatty acids were quantitatively contributed to the digested material during feeding by *N. integer*, *S. plana*, and *N. diversicolor*. (Bradshaw *et al.*, 1989). Furthermore, the 18:1 isomers were contributed to the feces in such a way that there was a predominance of the 18:1 Δ^{11} isomer over the 18:1 Δ^{9} isomer. Although these fatty acids could have an origin in animal tissues, reports of microbial populations in the guts of many marine animals (Sochard *et al.*, 1979; Nagasawa and Nemoto, 1988) suggest an enteric bacterial origin. Sedimentary distributions of these fatty acids are likely to have contributions not only from free-living bacteria but also from the enteric bacteria of invertebrates. There has also been a report of C_{15} and C_{17} fatty acids occurring in significant abundances in a phytoplankton species (Harvey *et al.*, 1988), indicating that these organisms are a direct source of these compounds in sediments.

The herbivory experiments with various species have shown that feeding leads to greater relative abundances of the C_{15} and C_{17} bacterial fatty acids in the digested material than in the food. Although these greater abundances may be due to relative decreases in other fatty acids (e.g., the unsaturated fatty acids), experiments have demonstrated that the C_{15} and C_{17} fatty acids are in fact quantitatively increased during feeding (Bradshaw *et al.*, 1990a, b, 1991) and suggest that the enteric microbial contribution of these fatty acids is significant relative to dietary fatty acids. The C_{15} and C_{17} fatty acids were increased further in relative abundance during coprophagy. However, the relative increase was shown to be associated with a quantitative decrease, at least with *S. plana* and *N. diversicolor* (Bradshaw *et al.*, 1990b, 1991). This apparent contradiction may be due to a lower degree of assimilation relative to other fatty acids. The bacterial C_{15} and C_{17} fatty acids observed in sediments originate with free-living microbes and invertebrate feeding. Whereas reprocessing of material within the sediments by invertebrates is likely to lead to the greatest abundances of these fatty acids, in cases of coprophagy there can be quantitative decreases in these fatty acids.

The situation with the 18:1 isomers was different. Herbivory did not generally lead to increases in the Δ^{11} isomer relative to the Δ^{9} isomer as might be expected if there is a contribution from enteric microbes. The feces contained a predominance of the Δ^{9} isomer even where the diet contained the reverse. This can be explained by an increased animal contribution of the Δ^{9} isomer (which is usually the dominant isomer in animal tissues), owing to greater turnover of the gut epithelium with an algal diet than with a lipid-free starch diet. Generally, animal feeding seems unlikely to contribute to the high abundances of the 18:1Δ^{11} fatty acid in sediments. However, one invertebrate species, *N. diversicolor*, affected these isomers in a different way from the majority of invertebrate species studied to date. It has been shown that, during herbivory, feces are produced with a predominance of the 18:1Δ^{11} isomer (Bradshaw *et al.*, 1991). Furthermore, coprophagy resulted in feces with an absence of 18:1 fatty acid altogether.

Annelids clearly have some effect upon the abundance of the 18:1 Δ^{11} bacterial marker in sediments, but the effects seem to depend upon the mode of feeding. The results from the feeding experiments suggest that the use of this fatty acid as a quantitative and qualitative bacterial marker should be treated with caution.

Invertebrate feeding then, appears to influence two aspects of sedimentary fatty acid distributions. Firstly, unsaturated fatty acids, particularly PUFAs, are rapidly removed from the material passing through the food chains by herbivory or, if present in the diet, by coprophagy. Although pelagic herbivory by crustaceans may remove substantial quantities of unsaturated fatty acids from the particulate material in the water column, benthic invertebrates can also play an important role in determining the quantities of unsaturated fatty acids in sediments.

Secondly, the abundances of bacterial fatty acids in sediments are also likely to be affected. There is evidence that invertebrate feeding can alter the abundance of the bacterial marker 18:1 Δ^{11}; whether this results in an increase or decrease depends on the species involved. The branched C_{15} and C_{17} fatty acids are generally increased in relative abundance during feeding, with coprophagy producing the highest abundances. Assessment of bacterial activity in a sediment using these compounds should take account of invertebrate feeding.

3. Phytol and Fatty Alcohols

Phytol (3,7,11,15-tetramethylhexadec-2-enol) is usually the most abundant fatty alcohol in the total-saponifiable lipid extracts of phytoplankton, being derived from chlorophylls *a* and *b* (Ragan and Chapman, 1978).

Pelagic crustaceans decrease the abundance of this compound when feeding herbivorously (Prahl *et al.*, 1984a, b; Neal *et al.*, 1986; Harvey *et al.*, 1987; Bradshaw *et al.*, 1990a). Experiments with benthic invertebrates have shown that mollusks and annelids are also capable of removing this compound from food while feeding both herbivorously and coprophagously (Bradshaw *et al.*, 1990b, 1991). Coprophagy was found to decrease the abundances of phytol still further from those produced after the initial crustacean herbivory. Phytol is rapidly removed from the material passing through the food chains in a similar way as has been described for PUFAs. Any phytol remaining in the feces of pelagic zooplankton or benthic invertebrates after herbivorous feeding (on phytoplankton in the water column or settled on the sea bottom) will be quickly removed on subsequent reprocessing of the fecal material. High abundances of phytol in sediments are probably due to the deposition of large quantities of phytoplanktonic material during bloom conditions when the invertebrate communities are unable to process all available food material.

The hydrocarbon pristane (2,6,10,14-tetramethylpentadecane), found in many marine organisms and particularly calanoid copepods (Blumer *et al.*, 1963, 1964), is known to be derived from dietary phytol (Avigan and Blumer, 1968). An experiment with *C. helgolandicus* has shown that feces from the animal contain an intermediate, dihydrophytol (3,7,11,15-tetramethylhexadecanol; DHP), in this conversion (Prahl *et al.*, 1984b). Although neither DHP nor pristane has been detected in the feces of other marine invertebrates (Bradshaw *et al.*, 1990a, b, 1991), the decreases in phytol observed during feeding experiments suggest that many marine invertebrates are capable of removing significant quantities of this compound from the material passing through the food chains. It seems probable that phytol is converted to DHP or another intermediate in the formation of pristane.

The DHP detected in marine and lacustrine sediments is proposed to be microbially formed by hydrogenation from phytol (Sever and Parker, 1969; Brooks and Maxwell, 1974; Van Vleet and Quinn, 1979; Cranwell, 1980). However, with the confirmation that at least one species of invertebrate, *C. helgolandicus*, produced DHP in its feces and the observed decreases in phytol during feeding in a number of other studies, it is probable that invertebrates can directly input DHP to the sediments. It is, of course, possible that the conversion of phytol into DHP might be mediated by the enteric microbial population which is present in many marine invertebrates (Sochard *et al.*, 1979) rather than the animals themselves.

Fatty alcohols are generally only a minor class of lipids in the material analyzed in feeding experiments. These compounds are present as components of wax esters in some species of cryptomonads (Antia *et al.*, 1974) and dinoflagellates (Withers and Nevenzel, 1977) and have been identified in other phytoplankton species (Volkman *et al.*, 1981; Nichols *et al.*, 1983; Harvey *et al.*, 1988). However, fatty alcohols are not commonly abundant in phytoplankton, and the diets used in feeding experiments also have been low in these components.

These compounds are, however, frequently present in many crustaceans as constituents of wax ester food reserves (Lee *et al.*, 1971; Sargent *et al.*, 1978, 1981). Nevertheless, although *C. helgolandicus* contains greater quantities of fatty alcohols than fatty acids (Prahl *et al.*, 1984b), the crustacean *N. integer*, which apparently has no wax ester reserves (Morris *et al.*, 1977), has few fatty alcohols (Bradshaw *et al.*, 1990a). Furthermore, the other invertebrates which have been used in feeding studies, *S. plana* and *N. diversicolor*, also contain quantitatively little fatty alcohols compared with the amounts of fatty acids (Bradshaw *et al.*, 1990b, 1991), and wax esters are not major lipids in *S. plana* (Payne, 1978). Generally then, feeding experiments have been conducted with food algae and animals which do not contain great quantities of fatty alcohols. The information gained on the changes in these compounds in the diet during feeding is therefore limited.

With *C. helgolandicus* feeding on the green alga *D. primolecta*, it was found that the fecal total fatty alcohol distribution comprised both algal- and animal-derived compounds, indicating a contribution by the copepod to the feces. However, the underlying dietary fatty alcohol distribution appeared to be little changed after feeding. This is consistent with the proposal that copepod wax esters are formed from *de novo* synthesized fatty alcohols and dietary fatty acids (Sargent *et al.*, 1981). In this instance, the dietary fatty alcohols would not be required for wax ester production. However, an experiment with *C. helgolandicus* feeding on different dietary levels of the dinoflagellate *S. trochoidea* (Harvey *et al.*, 1987) found somewhat different results. The fecal fatty alcohol distribution contained additional components not present in alga or animal. It was suggested that active biotransformation of lipids was occurring in the gut. Furthermore, the quantities of fatty alcohols in animals after feeding for 20 h were greatest in those with the highest dietary level of *S. trochoidea*, and it was suggested that some dietary fatty alcohols were being incorporated into copepod wax esters.

Similarly, the feces produced by *N. integer* feeding herbivorously also do not show the simple combination of dietary and animal distributions. The major fatty alcohols in a feed alga, *S. trochoidea*, are 16:0 and 18:0. After feeding, the 18:0 was quantitatively increased and the 16:0 was quantitatively decreased. Although the animal has been shown to contribute some of its own fatty alcohols to the feces (Bradshaw *et al.*, 1989), it also appears to assimilate some of the dietary fatty alcohols.

Both herbivory and coprophagy with *S. plana* and *N. diversicolor* have been shown to lead to quantitative increases in fatty alcohols in the digested material (Bradshaw *et al.*, 1990b, 1991). Feeding with both species also increases the range of fatty alcohols present in the feces compared with in the food, as was observed with *C. helgolandicus* (Harvey *et al.*, 1987). Again, these invertebrates have been shown to contribute fatty alcohols to feces when feeding on lipid-free diets (Bradshaw *et al.*, 1989). Evidently, although the animals themselves do not contain great quantities of fatty alcohols or wax esters, they are capable of contributing fatty alcohols to the material passing through the food chains. Distributions of fatty alcohols in sedimenting particles, especially zooplankton fecal pellets, and in surface sediments are likely to be affected by invertebrate feeding. The effects will depend upon the species involved, but it is clear that feeding activity can lead to the removal or input of fatty alcohols.

4. Sterols

Sterols have attracted a great deal of attention in terms of invertebrate feeding because some of these compounds are commonly used as biomarkers.

One striking feature of sterol distributions observed in feeding experiments modeling herbivory is the presence of animal-derived sterols in the feces. Early experiments demonstrated that typical C_{26} and C_{27} animal 4-desmethyl sterols were contributed to the original algal distributions (often containing greater abundances of C_{28} and C_{29} sterols) in the feces (Volkman *et al.*, 1980a; Prahl *et al.*, 1984a). Analyses of natural feces from a crab (Wakeham and Canuel, 1986) and a bivalve mollusk (Yasuda, 1978) also suggested that sterols are contributed to the feces in food chains. Recently, it has been confirmed that a number of invertebrates are capable of contributing a range of their own sterols to the digested material (Bradshaw *et al.*, 1989; Harvey *et al.*, 1989). As with the fatty acids, the source of these sterols is probably the disintegrating gut cells associated with the feeding process, and the distribution is not necessarily the same as that of the animal as a whole.

In experiments where animals are fed sterol-free diets, both *N. diversicolor* and especially *N. integer* have been found to produce feces with greater abundances of 24-ethyl-4-desmethyl sterols but lower abundances of cholest-5-en-3β-ol (cholesterol) than in their tissues (Bradshaw *et al.*, 1989). *S. plana*, however, produces feces with a very similar distribution. The digestive processes may have some affect upon the distributions. Crustaceans are incapable of *de novo* sterol synthesis (Goad, 1978) but can dealkylate C-24 of 24-methyl and 24-ethyl sterols (Teshima, 1971; Teshima and Kanazawa, 1971, 1972). Both uptake of cholesterol and a preferential dealkylation of 24-methyl over 24-ethyl sterols from lysed gut cells by *N. integer* were invoked to explain the fecal distributions with this invertebrate, which cannot synthesize its own cholesterol (Bradshaw *et al.*, 1989). *N. diversicolor* is apparently capable of *de novo* sterol synthesis (Wooten and Wright, 1962), suggesting that the processes of cholesterol uptake and sterol C-24 dealkylation are not as important in this species. The distribution of sterols in the feces from *S. plana* suggested that the mollusk, in contrast, probably takes up dietary sterols with little or no modification and/or is capable of *de novo* sterol synthesis.

Feeding experiments using diets containing [^{14}C] cholesterol as the only sterol have confirmed that *C. helgolandicus*, *N. integer*, *S. plana*, and *N. diversicolor* can indeed assimilate dietary cholesterol (Bradshaw *et al.*, 1989; Harvey *et al.*, 1989).

The abundance of cholesterol in a diet is likely to affect the contribution of animal sterols to the feces. Generally, herbivory experiments with crustaceans have found that cholesterol is the most abundant sterol which is contributed to the feces (Volkman *et al.*, 1980a; Prahl *et al.*, 1984a; Neal *et al.*, 1986; Harvey *et al.*, 1987; Bradshaw *et al.*, 1990a). These experiments have been carried out with cholesterol-poor algal diets. The animals appear to lose more cholesterol in the disintegrating gut epithelium than they take up from the diet. This has also been demonstrated with *S. plana*, and *N. diversicolor* feeding herbivorously (Bradshaw *et al.*, 1990b, 1991). However, in the experiments where animals are fed material which has already been processed by one animal (simulating coprophagy), the diet is generally cholesterol-rich (Neal *et al.*, 1986; Bradshaw *et al.*, 1990b, 1991). When *C. helgolandicus* was fed on diets of feces produced by *E. modestus* feeding on the green algae *H. carterae* and *D. primolecta*, the relative abundance of cholesterol was increased in the feces from the copepod after coprophagy. The crustacean was probably dealkylating dietary 4-desmethyl sterols at position C-24 to form cholesterol and/or contributing further animal cholesterol to the digested material during feeding. In contrast, the reverse was noted with *S. plana* and *N. diversicolor* (Bradshaw *et al.*, 1990b, 1991). Cholesterol was little changed quantitatively in the experiment with *S. plana*, although the relative abundances of other 4-desmethyl sterols were increased. It appeared that the mollusk was taking up a similar quantity of cholesterol from the diet to the amount that it was losing to its feces. *N. diversicolor*, however, quantitatively removed cholesterol from the diet. The diet used in these coprophagy experiments was derived from the dinoflagellate *S. trochoidea* (after feeding by *N. diversicolor*. Although some C_{28} and C_{29} 4-desmethyl sterols were present in the diet from the first feeding, the majority of the sterols were 4-methyl sterols. In the absence of high abundances of C_{28} and C_{29} 4-desmethyl sterols in the diet, these two benthic species apparently assimilate the dietary cholesterol.

The level of cholesterol in the diet also probably affects assimilation of other dietary sterols. *C. helgolandicus* has been noted to remove $\Delta^{5,7}$ sterols from a diet during herbivory, and it was suggested that a conversion to cholesterol was taking place (Prahl *et al.*, 1984a). Recently, it has become clear that other saturated sterols may be susceptible to assimilation. With the crustaceans *C. helgolandicus* and *N. integer*, in particular, it has been observed that dietary $\Delta^{8(14)}$ and $\Delta^{17(20)}$ sterols are assimilated during herbivorous feeding (Harvey *et al.*, 1987; Bradshaw *et al.*, 1989). What is more, some of these sterols are 4-methyl substituted. Again, it was suggested that the animals utilize these sterol saturation types for conversion to cholesterol. The $\Delta^{8(14)}$ and $\Delta^{5,7}$ sterols apparently fit into the scheme for conversion of $\Delta^{8(14)}$ sterols to Δ^5 sterols proposed to operate in mammals (Schroepfer, 1982). With a cholesterol-poor diet, such as *S. trachoidea*, such a conversion of dietary sterols might occur in these marine invertebrates. Indirect support for such a conversion comes from the experiments conducted with *S. plana* and *N. diversicolor* (Bradshaw *et al.*, 1990b, 1991). Similar, though lesser, decreases in $\Delta^{8(14)}$ sterols were observed during herbivory by these two species. However, during coprophagy, such sterols are little changed. The availability of cholesterol in the latter mode of feeding might make a $\Delta^{8(14)}$ to Δ^5 conversion unnecessary.

Similar decreases in dietary sterols during feeding have been noted with the $\Delta^{17(20)}$ sterol 4α,23,24-trimethylcholest-17(20)-en-3β-ol (peridinosterol), particularly in the crustaceans *C. helgolandicus* and *N. integer* (Harvey *et al.*, 1987; Bradshaw *et al.*, 1990a). This, together with the removal of $\Delta^{8(14)}$ sterols, may explain the absence of these compounds from many sediments even when there is a source present (e.g., Nishimura and Koyama, 1976; Nishimura, 1978; Huang and Meinschein, 1978; Smith *et al.*, 1982; de Leeuw *et al.*, 1983; Volkman, 1986; Volkman *et al.*, 1987). One particularly striking example is at Lake Kinneret in Israel where surface sediments contain low abundances of peridinosterol even though the dominant phytoplankton species in the lake, *Peridinium gatunense*, contains an abundance of this sterol (Robinson *et al.*, 1986). Analyses of the feces of four species of fish from the lake, including the principal grazer of the alga, *Sarotherodon galilaeus*, show that dietary peridinosterol is decreased after feeding (Bradshaw, unpublished results).

The apparent removal of these dietary sterols raises important questions, not just regarding the possible conversion routes of the saturation types, but also the fate of those sterols which are 4-methyl substituted (originating with the dinoflagellate *S. trochoidea*). Many of the $\Delta^{8(14)}$ sterols and, of course, peridinosterol, are 4-methyl sterols. These sterols have traditionally been used as biomarkers for dinoflagellates since, apart from a few instances in bacteria (Bird *et al.*, 1971; Bouvier *et al.*, 1976) and yeasts (Barton *et al.*, 1970), they were only recognized in this group of algae (Withers, 1983). Recently, there has been a report of 4-methyl sterols occurring in prymnesiophyte microalgae (Volkman *et al.*, 1990). The 4-methyl sterol 4α,23,24-trimethylcholest-22-en-3β-ol (dinosterol), in particular, is often used as a dinoflagellate marker (Boon *et al.*, 1979). The 4-methyl sterols have been important in the assessment of dinoflagellate input to many sediments (e.g., Boon *et al.*, 1979; Robinson *et al.*, 1984) and of the composition of the material passing through the water column (Gagosian *et al.*, 1982). However, apart from the question of the sources of these compounds, such assessment depends on a knowledge of any processes which might affect them between biosynthesis and incorporation into the sediment. Obviously, invertebrate feeding is one such process.

The feeding experiments have indicated that at least some dietary 4-methyl sterols are affected by the feeding process, particularly with crustaceans. However, quantitative data indicate that although 4-methyl $\Delta^{8(14)}$ and $\Delta^{17(20)}$ sterols are removed relative to other sterols, the total quantity of 4-methyl sterols remains unchanged after feeding (Bradshaw *et al.*, 1990a). The possible conversion products were not identified, but it seems probable that the 4-methyl sterols, as such, altered in quantity during invertebrate feeding.

More specific evidence comes from a series of experiments in which [U-^{14}C] dinosterol was fed as a single sterol to a number of marine invertebrate species (Bradshaw *et al.*, 1989; Harvey *et al.*, 1989). It has been established that this dinoflagellate marker sterol is not assimilated or bioconverted in any way during the feeding process and passes through experimental animals quantitatively. Furthermore, it has been established with fecal material from *C. helgolandicus* that dinosterol does not undergo any structural alteration during a period of 19 days simulating sedimentation after egestion (Harvey *et al.*, 1989). Dinosterol can be considered as a conservative, quantitative biomarker for dinoflagellates in recent sediments. Results from experiments such as that reported for *C. helgolandicus* feeding on the dinoflagellate *Gonyaulax tamarensis*, where dinosterol was enriched in the feces compared with the diet (Neal, 1984), undoubtedly relate to relative changes in 4-desmethyl sterols rather than 4-methyl sterols.

The fecal contribution and assimilation of dietary 4-desmethyl sterols will thus affect the 4-methyl: 4-desmethyl (4m:4d) sterol ratio. With cholesterol-poor diets, the contribution of 4-desmethyl sterols to the feces will decrease the 4m:4d ratio of the digested material. The reverse seems to occur in coprophagy experiments where a cholesterol-rich diet is used. Care is evidently needed when assessing the importance of dinoflagellate input to sediments based on relative abundance of 4-methyl sterols. As mentioned above, quantitative data on abundances of these sterols are probably a better indicator of such input.

The fecal contribution or assimilation of cholesterol will also affect 5α-stanol:stenol ratios in sediments. Once again, the mode of feeding could lead to increases or decreases. Nuclear saturated sterols are thought to be formed by microbial reduction of the ring of the unsaturated precursor stenols, either in the sediment (Gaskell and Eglinton, 1975, 1976; Nishimura and Koyama, 1977), in seawater (Gagosian and Heinzer, 1979), or in sedimenting material (Wakeham, 1987). However, various studies have shown that there are other sources of stanols in sediments. A dinoflagellate origin has been proposed for the 4α-methyl 5α-stanols in Black Sea sediments (de Leeuw *et al.*, 1983) and 4-desmethyl 5α-stanols in a lacustrine sediment (Robinson *et al.*, 1984). Indeed, many dinoflagellate species have been found to contain stanols (e.g., Alam *et al.*, 1978, 1984; Jones *et al.*, 1983; Volkman *et al.*, 1984), and these sterols are also common in many invertebrate species (Goad, 1978). Sediments may have a direct input of stanols from phytoplankton or invertebrates in addition to *in situ* microbial formation. Pertinent to the observation that invertebrates are a potential source for these sterols are the findings that they can be formed in the gut systems of some animals (Björkhem and Gustafsson, 1971; Parmentier and Eyssen, 1974). Feeding experiments with *C. helgolandicus* and *E. modestus* have shown that small abundances of 5α-cholestan-3β-ol are present in the feces from these animals after feeding on diets lacking this stanol (Neal *et al.*, 1986). Other experiments have shown that, in some cases, stanols are indeed quantitatively contributed to the digested material during feeding (Bradshaw *et al.*, 1990a, 1991). Invertebrate feeding may thus affect sedimentary 5α-stanol:stenol ratios in two ways—by quantitative input of stanols and by input or removal of stenols from the sediments.

Ultimately, the effect an animal has upon sedimentary sterol distributions will depend upon the sterol metabolism of the species involved and the sterol composition of the dietary material. Cholesterol appears to be of considerable importance in both respects. What is clear is that invertebrates may substantially alter sedimentary sterol distributions and should be considered as an important factor in the reconstruction of the processes which give rise to those distributions.

5. Hydrocarbons and Long-Chain Lipids

Reports of the effects of invertebrate feeding on hydrocarbons and long-chain lipids are somewhat more limited than for those compounds already discussed. *C. helgolandicus* completely removes the algal *n*-alkane C_{17} and *n*-alkenes $C_{17:1}$, $C_{17:2}$, and $C_{21:6}$ (heneicosahexaene) during feeding. Studies of sediment-trap material have noted a range of aliphatic hydrocarbons including long-chain components, C_{25}–C_{33}, probably indicative of higher plant material (Eglinton and Hamilton, 1967). Pristane, which, as mentioned previously, probably originates from crustaceans, and C_{15}–C_{17} alkanes have a similar origin. Crustacean feeding in the water column seems unlikely to greatly affect the distribution of aliphatic hydrocarbons in material reaching the sediments, particularly the higher plant *n*-alkanes. At present, no further information is available regarding other species of invertebrates.

Long-chain alkenes ($C_{31:2}$ and $C_{33:4}$) and di- and

triunsaturated alkenones (C_{37}–C_{39} methyl and ethyl ketones) have only been recognized, to date, in a few species of prymnesiophyte algae (Volkman *et al.*, 1980c, 1990; Marlow *et al.*, 1984) and as such have been used as biomarkers for this group. The long-chain alkenones have been recognized in ancient sediments (Marlowe *et al.*, 1984; Farrimond *et al.*, 1986), illustrating their excellent degree of preservation in the sedimentary record. Furthermore, the degree of unsaturation in the alkenones, as described by the Uk_{37} index, has been shown to vary with the temperature at which the biosynthesizing algae grow (Marlow, 1984; Prahl and Wakeham, 1987). This has enabled the alkenones present in sediments to be used as a basis for the evaluation of sea surface temperature variations in the contemporary and ancient oceans (Poynter *et al.*, 1989; Prahl and Wakeham, 1987). These compounds, then, are important biomarkers. However, little attention has been given to the possibility that processes occurring in the water column and sediments might have some effect on these compounds.

A few experiments have been conducted in which algae containing the alkenones were fed to invertebrates. When the coccolithophorid prymnesiophyte alga *Emiliania huxleyi* was fed to *C. helgolandicus*, no alkenones were subsequently found (on analysis of unsaponified lipid extracts) in the animals, although fecal pellets contained high concentrations of the alkenones (Volkman *et al.*, 1980d). The copepod apparently does not assimilate any of these compounds or, if it does, destroys them in some way. However, in an experiment where *Isochrysis galbana* was fed to the salp *Salpa fusiformis*, individual animals were found to affect the lipid composition of the digested material in different ways. In one case, the distribution altered little during the course of feeding. In others, however, the alkenones were very much decreased in abundance (Bradshaw, unpublished results) relative to other lipid components. It appears that salps might have some effect upon the passage of these compounds to the sediments through their feeding activities. These experiments were, however, qualitative, and lower abundances of ketones in the feces compared with the alga may simply reflect relative increases in other components.

6. Conclusions

Overall, the laboratory feeding experiments indicate that invertebrate feeding substantially affects lipids in marine food chains, particularly the fatty acids and sterols, and thus determines in part the distributions of these compounds in sediments. The results of invertebrate feeding are likely to affect the estimations of microbial activity in sediments because may of the features of lipid distributions which are associated with bacterial activity may also be brought about by such feeding.

ACKNOWLEDGMENT. Dr. Q. Bone of Plymouth Marine Laboratories is thanked for conducting the experiments with *Salpa* at Villefranche-sur-Mer, France.

References

Alam, M., Schram, K. H., and Ray, S. M., 1978, 24-Demethyldinosterol: An unusual sterol from the dinoflagellate, *Gonyaulax diagensis*, *Tetrahedron Lett.* **38:**3517–3518.

Alam, M., Sanduja, R., Watson, D. A., and Loeblich, A. R., 1984, Sterol distribution in the genus *Heterocapsa* (Pyrrhophyta), *J. Phycol.* **20:**331–335.

Al-Mohanna, F. Y., and Nott, J. A., 1986, B-Cells and diagestion in the hepatopancreas of *Penaeus semisulcatus* (Crustacea: Decapoda), *J. Mar. Biol. Assoc. U.K.* **66:**403–414.

Antia, N. J., Lee, R. F., Nevenzel, J. C., and Cheng, J. Y., 1974, Wax ester production by the marine cryptomonad *Chroomonas salina* grown photoheterotrophically on glycerol, *J. Protozool.* **21:**768–771.

Avigan, J., and Blumer, M., 1968, On the origin of pristane in marine organisms, *J. Lipid Res.* **9:**34–36.

Barton, D. H. R., Harrison, D. M., Moss, G. P., and Widdowson, D. A., 1970, Investigations on the biosynthesis of steroids and terpenoids. Part II. Role of 24-methylene derivatives in the biosynthesis of steroids and terpenoids. *J. Chem. Soc. (C)* **1970:** 775–785.

Bird, C. W., Lynch, J. M., Pirt, F. J., Reid, W. W., Brooks, C. J. W., and Middleditch, B. S., 1971, Steroids and squalene in *Methylococcus capsulatus* grown on methane, *Nature* **230:**473–474.

Björkhem, I., and Gustafsson, J. -A., 1971, Mechanism of microbial transformation of cholesterol into coprostanol, *Eur. J. Biochem.* **21:**428–432.

Blumer, M., Mullin, M. M., and Thomas, D. W., 1963, Pristane in zooplankton, *Science* **140:**974.

Blumer, M., Mullin, M. M., and Thomas, D. W., 1964, Pristane in the marine environment, *Helgol. Wiss. Meeresunters* **10:**187–201.

Boon, J. J., Rijpstra, W. I. C., De Lange, F., de Leeuw, J. W., Yoshioka, M., and Shimizu, Y., 1979, Black Sea sterol—a molecular fossil for dinoflagellate blooms, *Nature* **227:**125–127.

Bouvier, P., Rohmer, M., Benveniste, P., and Ourisson, G., 1976, $\Delta^{8(14)}$-Steroids in the bacterium *Methylococcus capsulatus*, *Biochem. J.* **159:**267–271.

Bradshaw, S. A., O'Hara, S. C. M., Corner, E. D. S., and Eglinton, G., 1989, Assimilation of dietary sterols and faecal contribution of lipids by the marine invertebrates *Neomysis integer*, *Scrobicularia plana* and *Nereis diversicolor*, *J. Mar. Biol. Assoc. U.K.* **69:**891–911.

Bradshaw, S. A., O'Hara, S. C. M., Corner, E. D. S., and Eglinton, G., 1990a, Changes in lipids during simulated herbivorous feeding by the marine crustacean *Neomysis integer*, *J. Mar. Biol. Assoc. U.K.* **70:**225–243.

Bradshaw, S. A., O'Hara, S. C. M., Corner, E. D. S., and Eglinton, G., 1990b, Dietary lipid changes during herbivory and coprophagy by the marine invertebrate *Nereis diversicolor*, *J. Mar. Biol. Assoc. U.K.* **70**:771–787.

Bradshaw, S. A., O'Hara, S. C. M., Corner, E. D. S., and Eglinton, G., 1991, Effects on dietary lipids of the marine bivalve *Scrobicularia plana* feeding in different modes. *J. Mar. Biol. Assoc. U.K.* **71**:635–653.

Brooks, C. J. W., and Maxwell, J. R., 1974, Early stage fate of phytol in a recently-deposited lacustrine sediment, in: *Advances in Organic Geochemistry 1973* (B. Tissot and F. Bienner, eds.), Editions Technip, Paris, pp. 977–991.

Corner, E. D. S., O'Hara, S. C. M., Neal, A. C., and Eglinton, G., 1986, Copepod faecal pellets and the vertical flux of biolipids, in: *The Biological Chemistry of Marine Copepods* (E. D. S. Corner and S. C. M. O'Hara, eds.), Oxford University Press, Oxford, pp. 261–321.

Cranwell, P. A., 1980, Branched/cyclic alkanols in lacustrine sediments (Great Britain): Recognition of *iso*- and *anteiso*-branching and stereochemical analysis of homologous alkan-2-ols, *Chem. Geol.* **30**:15–26.

de Leeuw, J. W., Rijpstra, W. I. C., Schenck, P. A., and Volkman, J. K., 1983, Free, esterified and residual bound sterols in Black Sea Unit I sediments, *Geochim. Cosmochim. Acta* **47**:455–465.

Eglinton, G., and Hamilton, R. J., 1967, Leaf epicuticular waxes, *Science* **156**:1322–1335.

Farrimond, P., Eglinton, G., and Brassell, S. C., 1986, Alkenones in Cretaceous Black Shales, Blake—Bahama Basin, Western North Atlantic, Org. *Geochem.* **10**:897–903.

Gagosian, R. B., and Heinzer, F., 1979, Stanols and stenols in the oxic and anoxic waters of the Black Sea, *Geochim. Cosmochim. Acta* **43**:471–486.

Gagosian, R. B., Smith, S. O., and Nigrelli, G. E., 1982, Vertical transport of steroid alcohols and ketones measured in a sediment trap experiment in the equatorial Atlantic Ocean, *Geochim. Cosmochim. Acta* **46**:1163–1172.

Gaskell, S. J., and Eglinton, G., 1975, Rapid hydrogenation of sterols in a contemporary lacustrine environment, *Nature* **254**:209–211.

Gaskell, S. J., and Eglinton, G., 1976, Sterols of a contemporary lacustrine environment, *Geochim. Cosmochim Acta* **40**:1221–1228.

Goad, L. J., 1978, The sterols of marine invertebrates: Composition, biosynthesis and metabolites, in: *Marine Natural Products: Chemical and Biological Perspectives. II* (P. J. Scheuer, ed.), Academic Press, New York, pp. 75–172.

Harvey, H. R., Eglinton, G., O'Hara, S. C. M., and Corner, E. D. S., 1987, Biotransformation and assimilation of dietary lipids by *Calanus* feeding on a dinoflagellate, *Geochim. Cosmochim. Acta* **51**:3031–3040.

Harvey, H. R., Bradshaw, S. A., O'Hara, S. C. M., Eglinton, G., and Corner, E. D. S., 1988, Lipid composition of the marine dinoflagellate *Scrippsiella trochoidea*, *Phytochemistry* **27**:1723–1729.

Harvey, H. R., O'Hara, S. C. M., Eglinton, G., and Corner, E. D. S., 1989, The comparative fate of dinosterol and cholesterol in copepod feeding: Implications for a conservative molecular biomarker in the marine water column, Org. *Geochem.* **14**: 635–641.

Harvey, H. R., Eglinton, G., O'Hara, S. C. M., and Corner, E. D. S., 1991, Microbial transformation of faecal pellet lipids during sedimentation, in: *Diversity of Environmental Biogeochemistry* (J. Berthelin, ed.), Elsevier, Amsterdam, pp. 25–36.

Huang, W. -Y., and Meinschein, W. G., 1978, Sterols in sediments from Baffin Bay, Texas, *Geochim. Cosmochim. Acta* **42**:1391–1396.

Jones, G. J., Nichols, P. D., and Johns, R. B., 1983, The lipid composition of *Thoracosphaera heimii*: Evidence for inclusion in the Dinophyceae, *J. Phycol.* **19**:416– 420.

Lee, R. F., Nevenzel, J. C., and Paffenhofer, G. -A., 1971, Importance of wax esters and other lipids in the marine food chain: Phytoplankton and copepods, *Mar. Biol.* **9**:99–108.

Marlowe, I. T., 1984, Lipids as palaeoclimatic indicators, Ph.D. Thesis, University of Bristol, England.

Marlowe, I. T., Green, J. C., Neal, A. C., Brassell, S. C., Eglinton, G., and Course, P. A., 1984, Long chain (n-C_{37}–C_{39}) alkenones in the Prymnesiophyceae. Distribution of alkenones and other lipids and their taxonomic significance, *Br. Phycol. J.* **19**: 203–216.

Morris, R. J., Armitage, M. E., Raymont, J. E. G., Ferguson, C. F., and Raymont, J. K. B., 1977, Effects of a starch diet on the lipid chemistry of *Neomysis integer* (Leach), *J. Mar. Biol. Assoc. U.K.* **57**:181–189.

Nagasawa, S., and Nemoto, T., 1988, Presence of bacteria in guts of marine crustaceans and on their fecal pellets, *J. Plank. Res.* **10**: 559–564.

Neal, A. C., 1984, The biogeochemical significance of copepod feeding, Ph.D. Thesis, University of Bristol, England.

Neal, A. C., Prahl, F. G., Eglinton, G., O'Hara, S. C. M., and Corner, E. D. S., 1986, Lipid changes during a planktonic feeding sequence involving unicellular algae, *Elminius* nauplii and adult *Calanus*, *J. Mar. Biol. Assoc. U.K.* **66**:1–13.

Nichols, P. D., Volkman, J. K., and Johns, R. B., 1983, Sterols and fatty acids of the marine unicellular alga FCRG 51. *Phytochemistry* **22**:1447–1452.

Nishimura, M., 1978, Geochemical characteristics of the high reduction zone of stenols in Suwa sediments and the environmental factors controlling the conversion of stenols into stanols, *Geochim. Cosmochim. Acta* **42**:349–357.

Nishimura, M., and Koyama, T., 1976, Stenols and stanols in lake sediments and diatoms, *Chem. Geol.* **17**:229–239.

Nishimura, M., and Koyama, T., 1977, The occurrence of stanols in various living organisms and the behaviour of sterols in contemporary sediments, *Geochim. Cosmochim. Acta* **41**:379–385.

Nott, J. A., Corner, E. D. S., Mavin, L. J., and O'Hara, S. C. M., 1985, Cyclical contributions of the digestive epithelium to faecal pellet formation by the copepod *Calanus helgolandicus*, *Mar. Biol.* **89**:271–279.

Parmentier, G., and Eyssen, H., 1974, Mechanism of biohydrogenation of cholesterol to coprostanol by *Eubacterium* ATCC 2108, *Biochim. Biophys. Acta* **348**:279–284.

Payne, D. W., 1978, Lipid digestion and storage in the littoral bivalve *Scrobicularia plana* (da Costa), *J. Molluscan Stud.* **44**: 295–304.

Perry, G. J., Volkman, J. K., Johns, R. B., and Bavor, H. J., 1979, Fatty acids of bacterial origin in contemporary marine sediments, *Geochim. Cosmochim. Acta* **43**:1715–1725.

Poynter, J. G., Farrimond, P., Brassell, S. C., and Eglinton, G., 1989, Molecular stratigraphic study of sediments from Holes 658A and 660A, Leg 108, *Proceedings of the Ocean Drilling Program, Scientific Results*, Vol. 108 (W. Ruddiman and M. Sarnthein, eds.), Ocean Drilling Program, Texas A&M University, College Station, Texas, pp. 387–394.

Prahl, F. G., and Wakeham, S. G., 1987, Calibration of unsaturation patterns in long-chain ketone compositions for palaeotemperature assessment, *Nature* **330**:367–368.

Prahl, F. G., Eglinton, G., Corner, E. D. S., and O'Hara, S. C. M.,

1984a, Copoepod fecal pellets as a source of dihydrophytol in marine sediments, *Science* **224**:1235–1237.

Prahl, F. G., Eglinton, G., Corner, E. D. S., O'Hara, S. C. M., and Forsberg, T. E. V., 1984b, Changes in plant lipids during passage through the gut of *Calanus*, *J. Mar. Biol. Assoc. U.K.* **64**: 317–334.

Ragan, M. A., and Chapman, D. J., 1978, *A Biochemical Phylogeny of the Protists*, Academic Press, New York.

Robinson, N., Eglinton, G., Brassell, S. C., and Cranwell, P. A., 1984, Dinoflagellate origin for sedimentary 4α-methylsteroids and 5α(*H*)-stanols, *Nature* **308**:439–441.

Robinson, N., Cranwell, P. A., Eglinton, G., Brassell, S. C., Sharp, C. L., Gophen, M., and Pollingher, U., 1986, Lipid geochemistry of Lake Kinneret. *Org. Geochem.* **10**:733–742.

Sargent, J. R., Morris, R. J., and McIntosh, R., 1978, Biosynthesis of wax esters in oceanic crustaceans, *Mar. Biol.* **46**:315–320.

Sargent, J. R., Gatten, R. R., and Henderson, R. J., 1981, Marine wax esters, *Pure Appl. Chem.* **53**:867–871.

Schroepfer, G. J., 1982, Sterol biosynthesis, *Annu. Rev. Biochem.* **51**:555–585.

Sever, J., and Parker, P. L., 1969, Fatty alcohols (normal and isoprenoids) in sediments, *Science* **164**:1052–1054.

Smith, D. J., Eglinton, G., Morris, R. J., and Poutanen, E. -L., 1982, Aspects of the steroid geochemistry of a recent diatomaceous sediment from the Namibian shelf, *Oceanolo. Acta* **5**:365–378.

Sochard, M. R., Wilson, D. F., Austin, B., and Colwell, R. R., 1979, Bacteria associated with the surface and gut of marine copepods, *Appl. Environ. Microbiol.* **37**:750–759.

Tanoue, E., Handa, N., and Sakugawa, H., 1982, Difference of the chemical composition of organic matter between fecal pellet of *Euphausia superba* and its feed, *Dunaliella tertiolecta*, *Trans. Tokyo Univ. Fish.* **5**:189–196.

Teshima, S., 1971, Bioconversion of β-sitosterol and 24-methylcholesterol to cholesterol in marine Crustacea, *Comp. Biochem. Physiol.* **39B**:815–822.

Teshima, S., and Kanazawa, A., 1971, Utilization and biosynthesis of sterols in *Artemia salina*, *Bull. Jpn. Soc. Sci. Fish.* **37**: 720–723.

Teshima, S., and Kanazawa, A., 1972, Bioconversion of the dietary brassicasterol to cholesterol in *Artemia salina*, *Bull, Jn. Soc. Sci. Fish.* **38**:1305–1310.

Van Vleet, E. S., and Quinn, J. G., 1979, Diagenesis of marine lipids in ocean sediments, *Deep-Sea Res.* **26**:1225– 1236.

Volkman, J. K., 1986, A review of sterol markers for marine and terrigenous organic matter, *Org. Geochem.* **9**:83–99.

Volkman, J. K., Corner, E. D. S., and Eglinton, G., 1980a, Transformations of biolipids in the marine food web and in underlying bottom sediments, *Colloq. In. CNRS* **293**:185–197.

Volkman, J. K., Johns, R. B., Gillan, F. T., Perry, G. J., and Bavor, H. J., Jr., 1980b, Microbial lipids of an intertidal sediment—I. Fatty acids and hydrocarbons, *Geochim. Cosmochim. Acta* **44**: 1133–1143.

Volkman, J. K., Eglinton, G., Corner, E. D. S., and Forsberg, T. E. V., 1980c, Long chain alkenes and alkenones in the marine coccolithophorid *Emiliania huxleyi*, *Phytochemistry* **19**:2619–2622.

Volkman, J. K., Eglinton, G., Corner, E. D. S., and Sargent, J. R., 1980d, Novel unsaturated straight-chain C_{37}–C_{39} methyl and ethyl ketones in marine sediments and in a coccolithophore, *Emiliania huxleyi*, in: *Advances in Organic Geochemistry 1979* (A. G. Douglas and J. R. Maxwell, eds.), Pergamon Press, Oxford, pp. 219–227.

Volkman, J. K., Smith, D. J., Eglinton, G., Forsberg, T. E. V., and Corner, E. D. S., 1981, Sterol and fatty acid composition of four Haptophycean algae, *J. Mar. Biol. Assoc. U.K.* **61**:509–527.

Volkman, J. K., Gagosian, R. B., and Wakeham, S. G., 1984, Free and esterified sterols of the marine dinoflagellate *Gonyaulax polygramma*, *Lipids* **19**:457–465.

Volkman, J. K., Farrington, J. W., and Gagosian R. B., 1987, Marine and terrigenous lipids in coastal sediments from the Peru upwelling region at 15°S: Sterols and triterpene alcohols, *Org. Geochem.* **11**:463–477.

Volkman, J. K., Kearney, P. S., and Jeffrey, S. W., 1990, A new source of 4-methyl sterols and 5⁻(*H*)-stanols in sediments: Prymnesiophyte microalgae of the genus *Pavlova*, *Org. Geochem.* **15**:489–497.

Wakeham, S. G., 1982, Organic matter from a sediment trap experiment in the equatorial North Atlantic: Wax esters, steryl esters, triacylglycerols, and alkyldiacylglycerols, *Geochim. Cosmochim. Acta* **46**:2239–2257.

Wakeham, S. G., 1987, Steroid geochemistry in the oxygen minimum zone of the eastern tropical North Pacific Ocean, *Geochim. Cosmochim. Acta* **51**:3051–3069.

Wakeham, S. G., and Canuel, E. A., 1986, Lipid composition of the pelagic crab *Pleuroncodes planipes*, its feces, and sinking particulate organic matter in the Equatorial North Pacific Ocean, *Org. Geochem.* **9**:331–343.

Wakeham, S. G., Lee, C., Farrington, J. W., and Gagosian, R. B., 1984, Biogeochemistry of particulate organic matter in the oceans: Results from sediment trap experiments, *Deep-Sea Res.* **31**: 509–528.

Withers, N. W., 1983, Dinoflagellate sterols, in: *Marine Natural Products: Chemical and Biochemical Perspectives*, Vol. 5 (P. J. Scheuer, ed.), Plenum Press, New York, pp. 87–130.

Withers, N. W., and Nevenzel, J. C., 1977, Phytl esters in marine dinoflagellate, *Lipids* **12**:989–993.

Wooten, J. A. M., and Wright, L. D., 1962, A comparative study of sterol biosynthesis in Annelida, *Comp. Biochem. Physiol.* **5**: 235–264.

Yasuda, S., 1978, Steroids in the faeces of a marine bivalve, *Bull. Jpn. Soc. Sci. Fish.* **44**:525–528.

Chapter 11

Early Diagenesis: Consequences for Applications of Molecular Biomarkers

JOHN I. HEDGES and FREDRICK G. PRAHL

1. Introduction

Detailed analyses of organic materials in soils, recent sediments, and natural waters invariably indicate the presence of complex mixtures of organic molecules (Eglinton and Murphy, 1969; Thurman, 1985). Typically, a major fraction of the total organic matter is highly degraded and occurs in structurally complex polymers such as humic substances (Christman and Frimmel, 1988). In spite of these complexities, organic materials in modern natural environments are of interest as sources of energy and nutrition, as recorders of past environmental conditions, and as precursors for the formation of fossil resources. The amounts and distributions of organic materials from different biological or geographic sources often are a fundamental consideration in studies of processes, such as production, transport, and degradation, that affect organic remains in natural settings. Thus, the development and use of dependable source indicators has been a major thrust in organic geochemistry.

Treibs's (1934) demonstration that petroleums contain porphyrins derived from the tetrapyrrole ring system of chlorophyll pigments was the first application of specific molecules as indicators of the origin of organic matter in environmental samples. Since this early work, many different types of organic molecules which can be structurally related to a specific biological source have been identified (Eglinton, 1969). Collectively, these source-specific molecules have become known as "biomarkers." For example, certain aromatic and open-chain carotenoid lipids are presently recognized as biomarkers for green and purple photosynthetic bacteria, respectively (Gillan and Johns, 1986). In addition, 4-methyl sterols such as dinosterol, C_{37}–C_{39} *n*-alkenones, and the xanthophyll, fucoxanthin, are now viewed as indicators of dinoflagellates, prymnesiophytes, and diatoms, respectively (Boon *et al.*, 1979; Marlowe *et al.*, 1984; Stauber and Jeffrey, 1988). Long-chain *n*-alkyl lipids present in simple waxes (Prahl and Pinto, 1987) or as components of cutin and suberin polymers (Kolattukudy, 1980; Holloway, 1982; Kolattukudy and Espelie, 1985), diterpenoid resin acids (Simoneit, 1977), and triterpenoids such as α- and β-amyrins (Simoneit, 1986), as well as lignins (Haddad and Martens, 1987; Hedges *et al.*, 1988b, c), are all considered to be useful tracers for organic matter derived from vascular land plants.

JOHN I. HEDGES • School of Oceanography, University of Washington, Seattle, Washington 98195. FREDRICK G. PRAHL • College of Oceanography, Oregon State University, Corvallis, Oregon 97331.

Organic Geochemistry, edited by Michael H. Engel and Stephen A. Macko. Plenum Press, New York, 1993.

For source analysis, biomarkers afford several advantages over measurements of bulk chemical properties such as elemental and stable isotope composition. First, the diversity and great number of different organic molecules in natural samples allow finer resolution of source contributions. Second, biomarkers often occur in suites within which the relative abundances of the congeners sometimes carry additional environmental or diagenetic information. For example, changes in unsaturation within the suite of long-chain ketones biosynthesized by prymnesiophyte algae correspond to the water temperature at which these plants grow (Brassell *et al.*, 1986a; Prahl and Wakeham, 1987), and variations in chain length within series of vascular plant-derived *n*-alkanes, *n*-alcohols, and *n*-acids from atmospheric dusts reflect temporal changes in regional wind patterns (Gagosian *et al.*, 1987). Finally, source distinctions based on molecular-level analyses often afford orders-of-magnitude greater sensitivity than is possible with measurements of bulk chemical properties because biomarker molecules often can be quantified against an analytical background of essentially zero.

Biomarker methods, however, pose significant disadvantages as well. One of these is that information on the distributions of organic molecules in living organisms is incomplete and spotty. As more comprehensive surveys are conducted, presumably unique biomarker–source relationships, such as that between β-sitosterol and vascular land plants, are often later found to be equivocal (Volkman, 1986). Notably, many natural products such as the extended ($>C_{35}$) hopanoids (Ourisson *et al.*, 1984), the multibranched acyclic C_{20}, C_{25}, and C_{30} isoprenoid hydrocarbons (Rowland *et al.*, 1985), and the long-chain alkane-1,5-diols (de Leeuw *et al.*, 1981) were identified in environmental samples well before their biological sources were discovered. A second common drawback is that many biomarkers occur only as trace constituents of living organisms and geochemical mixtures, with the result that these molecules typically are not important nutrient sources or contributors to bulk chemical properties. Finally, biomarkers, like bulk organic materials, are sensitive to diagenetic effects. Often, only a minor structural alteration, such as migration of a double bond, a stereochemical rearrangement, or a change in one functional group within the molecule, is sufficient to remove a biomarker from its analytical window or to confuse an initially straightforward biomarker–source relationship. Thus, a fundamental understanding of how diagenesis can affect various biomarker applications is an important consideration if biomarkers are to be used effectively to investigate environmental processes.

The focus of this chapter will be a critical examination of the consequences of early diagenesis on different molecular biomarker applications. For simplicity, we will limit the discussion to biomarkers from modern terrestrial and aquatic plants. Early diagenesis will be taken to represent decomposition of such plant materials by microbial metabolism, as opposed to abiotic degradation by such reactions as thermal alteration, humification, or photolysis. Although most of the examples given in this chapter are for aquatic environments, they should provide useful analogies for many other natural settings (previous chapters).

2. The Nature of Early Diagenesis

Before examining the effects of early diagenesis on compositional trends within organic mixtures, it is useful to review the nature of the process. One important characteristic of early diagenesis is that it is extensive. Global estimates indicate that only about 0.1% of total primary production is ultimately preserved, primarily in fine-grained coastal marine sediments (Berner, 1982; Romankevich, 1984). On a smaller scale, rivers are estimated to export 1% or less of the primary production within their drainage basins (Meybeck, 1982), which typically accumulate essentially no organic matter on a long-term basis due to physical erosion. In the open ocean, sediment-trap studies indicate that less than 5% of the total organic carbon flux from surface waters sinks through an average water column (~4 km) to the ocean floor (Suess, 1980; Martin *et al.*, 1987). Rapid degradation also occurs at the immediate water–sediment interface (Emerson and Dymond, 1984) and within surface sediment deposits (Emerson *et al.*, 1985). Slower degradation can continue for thousands of years within bioturbated, oxygen-rich open-ocean sediments (Emerson *et al.*, 1987; Wilson *et al.*, 1985). In addition, numerous laboratory studies have demonstrated extensive microbial remineralization of a variety of plant materials, ranging from algae (e.g., Newell *et al.*, 1981; Westrich and Berner, 1984) to wood (Otjen *et al.*, 1987; Hedges *et al.*, 1988a).

Early diagenesis, although extensive, is typically selective in its initial and intermediate stages. This fact is well documented in laboratory experiments involving the utilization of a variety of plant materials by heterotrophic microorganisms and metazoans. For example, white-rot and brown-rot fungi are capable of selectively decaying the lignin and polysaccharide components, respectively, of wood (Kirk and Highley, 1973) and preferentially degrading syringyl versus guaiacyl structural units within the lignin component (Kirk and Farrell, 1987; Hedges *et al.*, 1988a). Bacteria also remineralize carbohydrates more rapidly than lignin under both aerobic and anaerobic conditions (Benner *et al.*, 1984a, b). Microbial communities also have been shown in laboratory simulations to degrade biopolymers in preference to synthetic humic-like polymers such as melanoidins (Henrichs and Doyle, 1986) and to selectively attack molecules with natural, rather than rearranged, stereochemistries (Alexander, 1973; Wardroper *et al.*, 1985). Laboratory experiments with zooplankton fed various types of phytoplankton have demonstrated a high degree of selectivity when lipid, protein, and carbohydrate components of algal diets are altered in animal guts (Corner *et al.*, 1986; Cowie, 1990).

Under field conditions, even within a chemically similar mixture such as petroleum hydrocarbons, weathering is well documented to be more rapid for low- versus high-molecular-weight *n*-alkanes, linear versus branched-chain structures, and cyclic versus acyclic molecules (Atlas *et al.*, 1981; Alexander *et al.*, 1983). Preferential losses of polysaccharides versus lignin also are observed in peats (Stout *et al.*, 1988) and buried woods (Hatcher and Breger, 1981; Spiker and Hatcher, 1987), where the individual polysaccharide components appear to degrade at different rates (Hedges *et al.*, 1985). In a comparison of the compositions of sedimentary versus fresh conifer needles, Hedges and Weliky (1989) found patterns of preferential degradation among and within the lignin, aldose, and cyclitol constituents of the tissues. Hatcher *et al.* (1983) also have reported selective degradation of carbohydrates versus paraffinic polymers in freshwater peats. In addition, comparisons of the settling and burial fluxes of particulate organic materials at various ocean sites clearly indicate contrasting rates of degradation among different lipid classes within the water column (Wakeham *et al.*, 1980, 1984; Lee and Wakeham, 1988) and preferential degradation of marine versus terrestrially derived organic materials at the water–sediment interface (Prahl *et al.*, 1980; Hedges *et al.*, 1988c).

In spite of the previous trends, it is clear that susceptibility to diagenetic alteration is not related to molecular composition alone. Both laboratory simulations and studies in natural settings such as aerobic and anaerobic sediments have shown that concentrations of individual molecule types almost never drop to zero during diagenesis (Barrick *et al.*, 1980; Hamilton and Hedges, 1988; Burdige and Martens, 1988). Organic molecules of identical structure often occur together in both labile and relatively unreactive forms. In addition, intrinsically unstable molecules such as cellulose (Brasch and Jones, 1959) and bacteriohopanetetrols (Mycke *et al.*, 1987) have been recovered from ancient sediments.

The most probable explanation for these apparent inconsistencies is that molecules from some sources are either naturally contained, or become incorporated, within microenvironments that physically shield them from microbial enzymes (Alexander, 1973). Examples of such protective matrices include proteinaceous material in carbonate shells (King and Hare, 1972; Robbins and Brew, 1990), hydrocarbons within reweathered shales and coals (Rowland and Maxwell, 1984; Barrick *et al.*, 1984), and molecules adsorbed to mineral surfaces (Theng, 1974; Weliky, 1983). Although such physically protected molecules may be initially rare, they would become concentrated in geochemical samples as the bulk organic matter is extensively degraded.

A third characteristic of early diagenesis is that the rate of degradation typically is first-order with respect to the concentration of labile organic matter (Berner, 1980a). This relationship has been demonstrated by numerous laboratory studies in which plankton were microbially degraded within aerobic water (e.g., Grill and Richards, 1964; Jewell and McCarty, 1968; Westrich and Berner, 1984) and with the fungal degradation of wood (Hiroi and Tamai, 1983; Hedges *et al.*, 1988a). In addition, Westrich and Berner (1984) have shown in the laboratory that the rate of microbial sulfate reduction in anaerobic sediments is first-order with respect to the amount of available labile organic matter (plankton remains). First-order kinetics are also used almost exclusively to model the degradation of organic matter in reducing (Berner, 1980a) and aerobic (Grundmanis and Murray, 1982; Emerson *et al.*, 1985) sediments.

The simple first-order relationship describing the rate of degradation of a given concentration of labile organic matter (G) can be expressed as

$$dG/dt = -kG \tag{1}$$

where t is time and k is the first-order decay constant (Berner, 1980a). Equation (1) can be integrated between the boundary conditions $t = 0$, $G = G_0$ and $t \to \infty$, $G \to 0$, to obtain

$$G_t = G_0 [\exp(-kt)] \tag{2}$$

The recognition that organic matter in natural environments is comprised of many components that fall into different classes of substrate quality, with characteristically different decay kinetics, led to the formulation of the "multi-G" model (Berner, 1980b). In this model, total organic matter (G_t) is treated as having multiple components (G_i) with different decay constants (k_i). Thus, the form of Eq. (2) for such a mixture would be

$$G_t = \Sigma G_{0i} [\exp(-k_i t)] \quad (3)$$

This formulation is functionally equivalent to assigning higher order kinetics to the decomposition of the total organic matter pool (Emerson and Hedges, 1988). The demonstration by Westrich and Berner (1984) that the rate constant for degradation of plankton remains in reducing marine sediment is dependent upon the freshness of the plankton supports the multi-G model, which will be used in the following exercise.

3. Compositional Trends in Degrading Organic Matter—a Simple Model

A biomarker is initially only one component of a complex biochemical mixture comprising a living organism. Upon the death of the organism, the bulk organic matter and any component biomarkers can be expected to undergo extensive and selective degradation following first-order decay kinetics (see previous discussion). In the following hypothetical model, we will assume that the previous conditions hold during the early diagenesis of simple (3–5 component) plant tissues and examine the trends among the masses of the degrading components. Obviously, this treatment represents an oversimplification of natural diagenetic processes, which involve many more components. Furthermore, some of these components are likely to be inhomogeneous, to occur within protective matrices, or to recombine during diagenesis to form byproducts, such as humic substances, that have very different reactivities. Nonetheless, the model sufficiently demonstrates compositional trends that can occur during early diagenesis and sets the stage for the following discussion of the effects of these trends on different basic biomarker applications.

The diagenesis model will first be applied using a hypothetical plant tissue that contains three components (C_1, C_2, and C_3) of initially equal concentration. Any one of these constituents could represent a biomarker. The three components will be assigned different reactivities, as represented by first-order degradation rate constants that vary with respect to each other by a factor of five (k_1, k_2, and k_3 = 0.2, 1.0, and 5.0, respectively). Wood, which is comprised almost exclusively of comparable amounts of cellulose, hemicellulose, and lignin (Fengel and Wegener, 1984), would be a compositionally similar starting material.

The calculated exponential decreases in the initial masses of the three tissue components over a five-year period are illustrated in Fig. 1A. Although the individual components exhibit simple first-order decay curves, the total mass of the tissue decreases over time with a complex functionality. The curvature of the total-mass line is fit well by first-order equations only during the very first and last stages of degradation, where the decay kinetics closely match those of the least stable (C_3) and most stable (C_1) components, respectively. As would be expected, the three components drop to trace levels (<1% of initial concentrations) at different times (about 1, 5, and 25 years for C_3, C_2, and C_1). After five years, the degraded plant tissue consists almost exclusively of the most stable component (C_1).

Figure 1B illustrates the percent of the total tissue weight that each of the previously discussed three components represents over the course of the simulated degradation. Each percentage corresponds to a point on a decay curve (Fig. 1A) divided by the corresponding total tissue mass remaining at that time. The weight percentages of all three mixture components vary widely with degradation. Notably, regardless of relative reactivity, no component remains a constant "average" fraction of the remaining tissue throughout the degradation period. None of the three weight percentage curves in Fig. 1B represents a pure first-order trend because each is generated from division by the complex curve for total remaining mass.

The weight percent of the least stable component (C_3) in Fig. 1B decreases rapidly and resembles the form of the corresponding mass trend in Fig. 1A. This, however, is the only mixture component that exhibits parallel trends over time in weight percentage and mass. For example, the weight percentage of the component with intermediate stability (C_2) initially rises and then decreases, whereas the weight percent of the most stable of the three mixture components (C_1) rises asymptotically toward 100% (Fig. 1B). This factor of three increase over the original concentration corresponds to the inverse of the fraction of C_1 in the initial plant material.

A plot of the same compositional trends in Fig. 1B versus the percent of total organic matter degradation (Fig. 1C) tends to compress elapsed time and thereby linearizes early changes in the weight per-

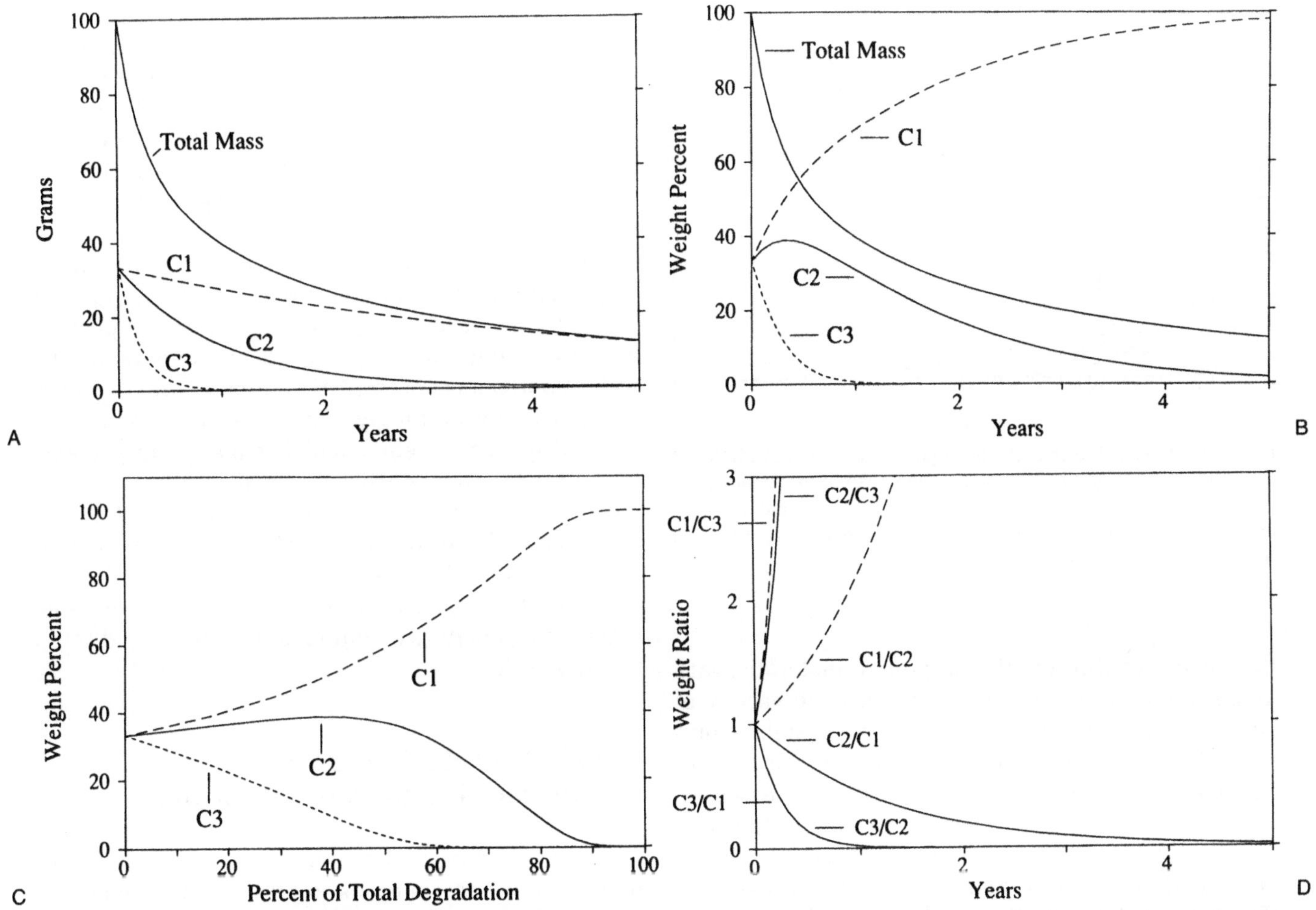

Figure 1. Changes in initial mass vs. time (A), weight percentages of individual components vs. time (B), and percent degradation (C), as well as weight ratios of individual components vs. time (D) which occur during the degradation of a hypothetical plant material. The plant material initially contains equal amounts of three components (C_1, C_2, and C_3) which have first-order decay constants (k_i) of 0.2, 1.0, and 5.0, respectively.

centages of the three degrading components. This format illustrates the important point that compositional changes *per unit of organic matter lost* are most extreme in the latter stages of degradation. Thus, the mixtures of highly degraded materials characteristic of natural environments can be expected to exhibit pronounced compositional changes in response to any further *in situ* diagenesis.

The six ratios that can be generated by all possible combinations of the three masses in simple one-to-one normalizations are plotted in Fig. 1D. All of these ratios are diagenetically sensitive and follow nonlinear trends over time. Any ratio with a concentration term for the more reactive component in the denominator eventually "explodes" to infinity whereas ratios with the more reactive component in the numerator asymptotically approach zero. When relatively reactive components (e.g., C_3) occur in the numerator, the percent change in the ratio closely parallels the percent mass loss of that component. The closer the reactivities of the constituents, the less diagenetically sensitive are their ratios. Changes in the relative reactivities (*k* values) of the various components alter the abruptness, but not the form, of this and the other trends in Fig. 1.

In order to examine the effects of the number of constituents in the degrading material, a second hypothetical model was considered in which the tissue was treated as having five initially equal constituents (C_1–C_5) with *k* values of 0.2, 0.4, 0.8, 1.6, and 3.2, respectively. Temporal trends in the weight percentages of the five individual components (Fig. 2) are similar in general form to those discussed previously for the three-component system (Fig. 1). It is more evident, however, that the weight percentages of intermediately reactive constituents, which would pre-

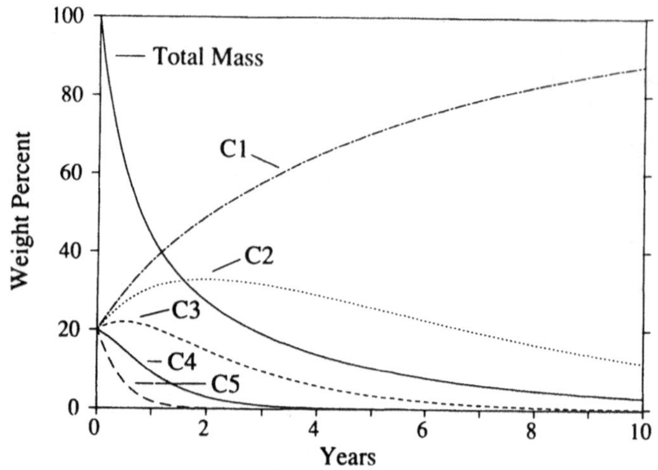

Figure 2. Plots of the changes in weight percentages of individual components vs. degradation time for a more complex hypothetical plant tissue. This plant material initially contains equal amounts of five constituents (C_1–C_5) which have first-order decay constants (k_i) of 0.2, 0.4, 0.8, 1.6, and 3.2, respectively.

dominate in natural multicomponent materials, vary along different curves that exhibit a wide variety of shapes (Fig. 2). In this example, the most stable component (C_1), which originally accounted for 20% of the mass of the tissue, asymptotically increases toward a final value of five times this concentration.

Finally, a third three-component model was investigated which was identical to the first example (Fig. 1), except the most, second most, and least stable components constituted 0.9, 9.0 and 90.1 wt. % of the degrading tissue, respectively. This composition is probably more representative of natural organic materials, such as plankton remains, that are comprised predominantly of reactive constituents (Westrich and Berner, 1984). Trends versus time (Fig. 3A) and percentage degradation (Fig. 3B) in the weight percentages of the three components are similar to those for the previous three-component model (Figs. 1B and 1C). However, the weight percentage of the intermediately reactive component (C_2) exhibits an extremely pronounced bidirectional trend, with an inflection point near the beginning of the temporal plot (Fig. 3A) and near the end of the plot versus percent overall degradation (Fig. 3B). The increase in the weight percentage of C_2 results from a rapid decrease in C_3. This loss strongly concentrates the other two remaining constituents, which subsequently decay away. Eventually, the tissue consists almost exclusively of C_1 (initially 1%), whose concentration increases by a factor of 100 overall. Biomarkers, which typically are minor plant tissue components, can generally be expected to exhibit the type of extreme diagenetic sensitivity demonstrated for C_1 and C_2 in this example.

4. Effects of Early Diagenesis on Different Biomarker Applications

Given that organic materials from essentially all plant sources are complex mixtures of biochemicals with different reactivities toward microbial degradation, it is prudent to be concerned about the possible effects of early diagenesis on biomarker applications. In this section, the diagenetic sensitivities of five

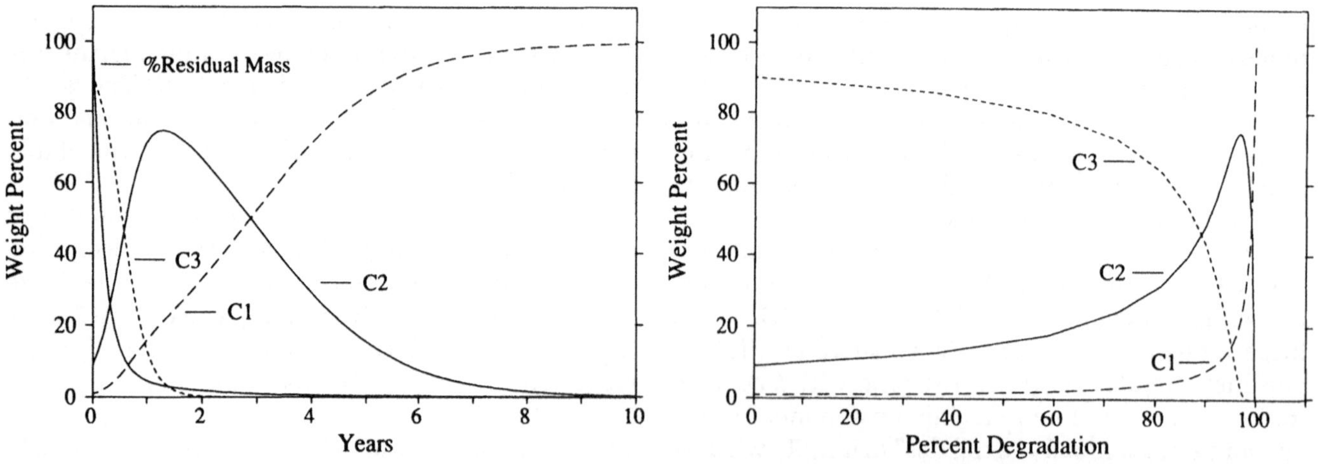

A B

Figure 3. Changes in weight percentages of individual components vs. time (A) and percent degradation (B) during the degradation of a hypothetical plant tissue. This plant material initially contains three constituents (C_1–C_3) which have first-order decay constants (k_i) of 0.2, 1.0, and 5.0 and initially account for 0.9, 9.0, and 90.1 wt. %, respectively, of the fresh tissue mass.

different categories of biomarker applications will be discussed. The individual application types (Table I) are arranged in order of increasing uncertainty (tenuousness); these range from simple detection of the biomarker in a field sample (conservative) to estimates of the relative productivity of a particular source organism within a given environment (highly speculative). The possible influences of early diagenesis on each type of application will be presented in light of the previous compositional models (Figs. 1–3), and ways of minimizing some of the associated interpretive complications will be discussed.

4.1. Application #1: Biomarker Detection

The most straightforward application of a molecular biomarker is a qualitative test for its presence in a natural organic mixture. The procedure is simply to determine whether a specific biomarker is detectable by a given analytical method. If so, then input of organic matter from the biomarker's characteristic source is indicated. This is the approach that Treibs (1934) used to establish that porphyrins in petroleum are ultimately derived from chlorophylls and thus that petroleums are at least in part biogenic. Importantly, this application necessitates no assumption about the amounts of other organic matter from the same source that might be present in the sample. The major assumptions involved are only that the biomarker is (a) properly identified and (b) unique to its assigned source (Table I). Neither of these assumptions is directly affected by diagenesis.

TABLE I. Assumptions Involved in Five Basic Applications with Biomarkers (B) and Associated Organic Matter (AOM)

Application type	Assumption(s)
I. Detection of source	1. B properly identified 2. B source(s) known and unique
II. Determination of relative biomarker levels	1. All the above 2. B concentrations quantified with proportional accuracy 3. Relative concentrations of B diagenetically unchanged
III. Determination of relative concentration of AOM	1. All the above 2. B/AOM is constant in source (or mixed to uniformity) 3. B/AOM is not diagenetically altered *in situ*
IV. Quantification of absolute concentration of AOM	1. All the above 2. B/AOM in natural sample(s) is known
V. Quantification of the paleoproduction of AOM	1. All the above 2. B/AOM is constant in source over a long time period 3. Efficiency of physical input of B and AOM is unchanged 4. B/AOM of source is not diagenetically altered

The simplicity of this application, however, does not mean that conclusions drawn from spot tests for the presence of a biomarker are entirely free from the consequences of diagenesis. The primary complication which arises is that the biomarker may have been initially present in the sampled material, but subsequently was affected by *in situ* diagenesis to the extent that it is no longer detectable. This could happen either because the biomarker was completely remineralized or because it was masked by chemical changes which move it outside the initial analytical window (see above). The biomarker also could become bound to humic polymers (Furlong and Carpenter, 1988), included in diagenetically formed sulfur compounds (Brassell *et al.*, 1986b; Moers *et al.*, 1988; Sinninghe Damsté *et al.*, 1989), or adsorbed to mineral surfaces (Theng, 1974) in such a way as to not be extractable. The consequence of these various diagenetic reactions is that a negative test for the presence of a biomarker does not provide conclusive evidence that organic matter of the characteristic source was initially absent from the sample. A definitive interpretation of a negative test can only be made if the biomarker is known to be more stable than all the other constituents from the source and their possible reaction products. This condition is almost never met.

The key to increasing the power of qualitative biomarker assays, therefore, lies in maximizing the potential for a positive test. There are several ways in which this can be done. First, plants which are quantitatively important sources of organic matter to natural environments can be systematically surveyed for potential biomarkers. Ideally, these biomarkers would be unusually stable and/or occur in homologous suites that provide a multicomponent source signature, as well as "nested" diagenetic information (see below). In particular, the identification of multiple biomarkers from common sources would allow for tests of concordance as are used in radiochemical dating methods. Second, the diagenetic pathways of the more promising biomarkers should be defined so that their degradative intermediates can be recognized. Broader analytical methods can then be developed to test more sensitively for molecules carrying the source-specific structural features. Third, laboratory and field studies are sorely needed to establish the relative reactivities of different biomarkers characteristic of a specific source. From such characterizations, it should be possible to identify the most stable molecules that have the highest likelihood of persisting in the environment as definitive tracers. Although little can be done in the eventuality of the total destruction of a biomarker, at least the analytical sensitivity and blank characteristics of the analytical procedure can be reported so that effective thresholds for negative tests can be meaningfully compared.

4.2. Application #2: Measurements of Biomarker Ratios

Measurements of biomarker ratios are similar to simple detection tests (application #1) in that no assumption beyond the presence of the biomarker(s) is necessarily involved (Table I). Ratio measurements, however, fall into two separate subcategories. The simpler of these is the use of ratios as qualitative source indicators, with no attempt to deconvolute mixture compositions. This approach often is taken when individual molecules from a given source are not unique to that material, but occur at characteristic ratios within it. Examples include the high odd-to-even predominance of long-chain (C_{25}–C_{35}) *n*-alkanes which characterize hydrocarbons from vascular land plants versus those in petroleum and bacteria (Cranwell, 1982), the high ratios of glucose to ribose (or fucose) in vascular plants versus plankton (Degens and Mopper, 1975; Cowie and Hedges, 1984), and the higher ratios of glycine to aspartic acid which distinguish siliceous from calcareous phytoplankton (King, 1977; Ittekkot *et al.*, 1984). This application is similar in assumptions and liabilities to that described in Section 4.1; i.e., the ratios characteristic of the various contributing sources must be known. An additional assumption involved, however, is that the characteristic ratio is not diagenetically sensitive (see below).

A second related use of ratio measurements is to assess the comparative amounts of biomarkers that come from multiple sources. This simple form of mixture analysis provides only a measure of relative biomarker amounts because absolute quantities are lost in normalization. This apparent limitation, however, has associated advantages. Biomarker ratios can be precisely determined from a single chromatographic analysis without the need for rigorous control of extensive variables such as sample size, injection volume, or overall recovery. A common example of this type of application would be measurement of the carbon preference index (CPI or "odd-to-even predominance") in a homologous series of *n*-alkanes in order to determine the relative amounts of vascular plant (CPI = 5–10) and mature fossil hydrocarbons (CPI = ~1) within an environmental sample (Cranwell, 1982). Similarly, ratios of vanillyl, syringyl, and cinnamyl phenols obtained from the CuO oxidation of lignin have been used to estimate relative proportions of different classes of vascular plant tissues in sedimentary mixtures (Hedges *et al.*, 1984, 1988b). This application, like the previous one, also involves the assumption that early diagenesis does not affect the value of the ratio characterizing any of the pure end members contributing to the natural mixture. Although the previous examples have been presented in terms of the biomarker ratios, applications of weight percentages within a biomarker suite (e.g., Huang and Meinschein, 1976; Pauly and Van Vleet, 1986) involve similar assumptions and pitfalls, including the fact that weight percentages are sensitive to changes in *all* measured components of the end member suites.

The basic assumption that ratios among biomarkers from a common origin are diagenetically invariant is troublesome for several reasons. For example, as illustrated in Fig. 1D, unless components of a common origin have identical diagenetic reactivities and histories, their ratios will be changed during degradation. These changes will be nonlinear and can be extremely large per unit degradation (Figs. 1C and 3B). The most effective way of minimizing diagenetic effects on biomarker ratios is to match the reactivities of the individual molecule types as closely as possible. The CPI, for example, is a diagenetically insensitive ratio because it is based on the relative abundances of chemically similar compounds within a limited molecular weight range. In addition, *n*-alkanes are relatively unreactive and unlikely to be affected by any mechanism that would be selective for alternating homologs in the sequence. In comparison, the ratio of cinnamyl to vanillyl phenols derived from the CuO oxidation of lignin has the potential to be diagenetically sensitive because the two reaction products derive from structural units which differ from each other in both their intermolecular bonding and their distribution within nonwoody plant tissues (Whitehead *et al.*, 1981).

Relative reactivity should also be a consideration in selecting the form of a biomarker ratio. Under circumstances where a ratio is to be used as a source indicator, the concentration of the more reactive compound(s) should be placed in the numerator so that the value of the parameter does not increase wildly during diagenesis (Fig. 1D). Unfortunately, insufficient information presently exists about the relative reactivities of most biomarkers to confidently formulate the least diagenetically sensitive parameters.

An additional strategy to minimize uncertainty in source assessments based on ratio analyses is to use multiple biomarkers that occur together in suites that bear independent information about the diagenetic state of the component molecules. An example of such an internal indicator is the ratio of acidic to aldehydic phenols within individual families of CuO oxidation products obtained from lignin. For example, essentially all fresh hardwoods yield a suite of six vanillyl and syringyl phenols within which the ratio of vanillic acid to vanillin increases progressively with the extent of degradation by most white-rot fungi (Hedges *et al.*, 1988a). The same fungi preferentially degrade syringyl versus vanillyl structural units within the lignin polymer, decreasing the initially high ratio of these two phenol types (S/V) that is characteristic of angiosperm tissues. The result of progressive fungal degradation is that the lignin phenol composition of the angiosperm wood becomes more and more like that of fresh gymnosperm tissue (e.g., softwood), which produces only vanillyl phenols upon CuO oxidation (Hedges and Mann, 1979; Mycke and Michaelis, 1986).

In this case, an elevated vanillic acid/vanillin ratio for angiosperm wood debris in an environmental sample is evidence that the material has been degraded by white-rot fungi and therefore likely has also suffered a decrease in its initial S/V. Different white-rot fungi, however, exhibit a range of decreases in S/V per unit increase in acid/aldehyde ratio, and all brown-rot fungi tested to date are capable of decreasing S/V without causing changes in the relative yields of phenols within a structural family. Thus, although acid/aldehyde ratios can serve as useful danger signals for diagenetic alteration by white-rot fungi, they do not provide proof of good preservation nor a means of correcting for diagenetic alteration in natural lignin mixtures. Similarly, apparent depletions of more highly unsaturated C_{37} alkenones in marine sediments may be useful qualitative indicators for the diagenetic alteration of lipids derived from prymnesiophyte algae (ten Haven *et al.*, 1987; Prahl *et al.*, 1988). Unfortunately, few other examples of built-in diagenetic indicators are presently known.

A final point should be made before leaving the topic of ratio parameters. When two (or more) components characterized by different values of a biomarker ratio are mixed, the value of the ratio *for the mixture* usually varies nonlinearly (and disproportionately) with the weight percentages of the end members. This complexity is illustrated in Fig. 4, which shows compositional trends for *n*-alkane mixtures comprised of varying fractions of fossil (CPI = 1) and modern vascular plant (CPI = 7) hydrocarbons. The three lines were generated by mixing end members having different relative concentrations (per unit total mass) of the hydrocarbons which determine the CPI ratio. As can be seen from the middle curve, mixing of two materials containing equal total concentrations of the ratioed hydrocarbons does not necessarily result in a linear CPI trend. In this and other such mixtures, a linear relationship between the value of the ratio and the fractional weights of the end members will only occur when the concentrations of the component(s) comprising the denominator of the ratio are equal in the two materials being mixed—in this case when the concentration of vascular plant hydrocarbons is four times that of fossil hydrocarbons. Even under such a highly unusual circumstance of initially equal concentrations, selective degradation of molecules within either one of the two suites will result in a nonlinear trend in the CPI of the resulting mixtures.

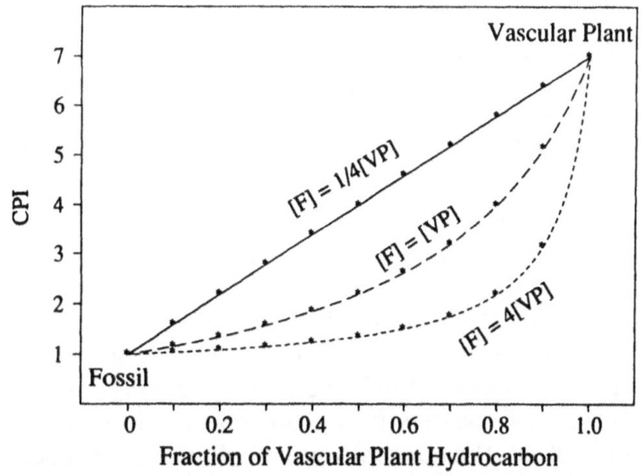

Figure 4. Calculated trends in carbon preference index (CPI) vs. the weight fraction of total vascular plant material in hypothetical mixtures of fossil (CPI = 1) and higher plant (CPI = 7) *n*-alkanes. The upper, middle, and bottom mixing curves correspond to binary mixtures where the concentrations of hydrocarbon in the fossil vs. vascular plant end members occur in ratios of 1:4, 1:1, and 4:1, respectively.

4.3. Application #3: Relative Extrapolations from Biomarker Concentrations to Concentrations of Associated Organic Matter

This and the following two applications have the common characteristic that all three include some form of quantitative extrapolation from the immediate biomarker(s) to bulk organic material of a common origin. Such extrapolations are made either to the coexisting amount of residual organic material from that source or to the amount of total organic material that was once produced by the identified source organisms. To simplify discussions of these relationships, the biomarker and its associated (same source) organic material will be abbreviated as B and AOM, where brackets [ ] about these symbols indicate the corresponding concentrations or amounts.

The least uncertain of the last three applications involves the use of a biomarker as an indicator of the relative amount of total organic matter from the same source that coexists in a given sample. In a typical application, the concentration of biomarker in a suite of samples is measured (in terms of either total mass or total organic carbon content) within each sample. A difference in biomarker concentration between any two samples is assumed to indicate a proportional (or at least parallel) change in the associated bulk organic matter. In this case, no attempt is made to quantify the absolute amount of associated bulk organic matter. This method requires that a reproducible fraction, but not necessarily all, of the total biomarker be measured. Jasper (1988) has published measurements of this type for long-chain (C_{37}–C_{39}) alkenones and for individual (C_{27}–C_{29}) desmethyl sterols, which were applied as respective indicators of marine and terrestrial organic materials preserved in Late Quaternary sediments from the northern Gulf of Mexico.

The major new assumptions involved in this application are (a) that the quantification of [B] is accurate and precise and (b) that [B] and [AOM] occur at a constant ratio within all samples of the set. The latter condition necessitates either that the source is compositionally uniform with respect to B or that the samples are well mixed on a time scale that is short with respect to compositional changes in the source. The value of [B]/[AOM] within a sample set, however, is sensitive to *in situ* diagenesis. As illustrated in Figs. 1B, 2, and 3A, selective degradation within a multicomponent source material always leads to changes in the weight percentage of each constituent, and therefore to variations in [B]/[AOM]. Under such conditions, [B]/[AOM] changes throughout the diagenetic process, regardless of the relative reactivity of the biomarker. Furthermore, [B]/[AOM] can vary bidirectionally at different stages of *in situ* diagenesis (Figs. 1C, 2, and 3B), leading either to underestimates or overestimates of the associated total plant tissue (see below). A constant, direct proportionality will only exist in the absence of *in situ* degradation or when B (or material of identical reactivity) is the only remaining component.

There is no direct way out of this dilemma other than to analyze sample sets for which *in situ* diagenesis is minimal. It is possible, however, to test for the effects of *in situ* diagenesis by looking for nonlinear trends between B and a characteristic bulk physical property of the source that is relatively insensitive to diagenesis (Jasper, 1988). Stable carbon isotope compositions often are relatively insensitive to diagenesis (Fry and Sherr, 1984) and can be a useful bulk property in such applications. However, both Benner *et al.* (1987) and Spiker and Hatcher (1987) have reported significant shifts in $\delta^{13}C$ during early diagenesis of vascular plant tissues that could be confusing. Although bulk physical properties have the advantage that they more comprehensively represent total organic matter than most molecular tracers, all exhibit the associated drawback that they are affected by diagenetic change in any of the mixture components.

Under conditions where *in situ* degradation of a biomarker or its congeners is known to occur, it is useful to bear in mind that a relatively reactive biomarker will trace AOM better during the earlier stages of diagenesis (e.g., C_3 in Fig. 1A) whereas a more stable one (e.g., C_1) will work better during latter stages. All other considerations being equal, an initially abundant biomarker will be more representative of bulk AOM than a trace component (e.g., compare the variability of C_1 in Figs. 1B and 3A). Importantly, this application is not affected by diagenesis which may have occurred before the sample suite was emplaced.

4.4. Application #4: Use of a Biomarker to Estimate the Absolute Amount of Associated Organic Matter in a Sample

This application goes beyond the previous premise (that a fixed [B]/[AOM] exists in a sample suite) to further assume that the value of this ratio can be reliably determined. Once established, the value of [B]/[AOM] can be used to calculate the absolute quantity of total organic matter from that source which is contained in a sample. For example, a remarkably linear relationship exists between plant wax *n*-alkanes (predominantly C_{25}, C_{27}, C_{29}, and C_{31}) and total organic carbon (OC) concentrations in bottom sediments through a 1000-km section of the lower Columbia River drainage basin (Fig. 5). Prahl and Muehlhausen (1989) assumed that the slope of the regression of these data ([B]/[AOM] = 280 mg/g OC) is characteristic of particulate organic material exported by the Columbia River. They then used this value and the measured plant wax hydrocarbon content of Wash-

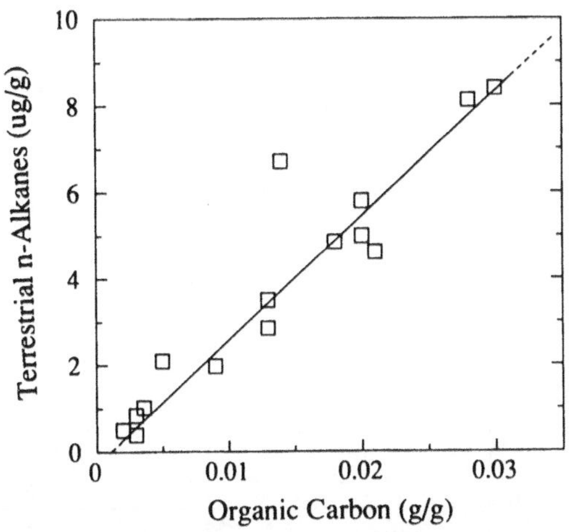

Figure 5. A correlation plot between concentrations of plant wax n-alkanes (C_{25}, C_{27}, C_{29}, and C_{31}) and total organic carbon in sediments from the Columbia River drainage basin. [Data from Prahl (1982).]

ington coastal sediments to calculate bulk terrestrial organic carbon content.

The key difficulty in such applications is that [B]/[AOM] may not be accurately known for the organic end member to be quantified, often because the organic matter in this fraction is degraded. One approach is to establish a finite [B]/[AOM] for a geochemical sample by estimating the ratio based on the composition of the original plant source(s)—as has been done for lignin-derived phenol produced from different types of fresh vascular plant tissues (Hedges *et al.*, 1984; Haddad and Martens, 1987). This choice is understandable because fresh plant tissues are readily available and are most commonly used for natural product analyses.

Nevertheless, any method which relates the [B]/[AOM] ratio of environmental materials to fresh plant material is presumptive. One difficulty, discussed previously under application #2, is that "representative" biomarker concentrations in the source material must be assigned under conditions where these values may change with growth conditions or over the life history of the plant. Such compositional changes are known to occur in the surface waxes (Caldicott and Eglinton, 1973; Prahl and Pinto, 1987) and lignins (Sarkanen and Ludwig, 1971; Haddad *et al.*, 1992) of vascular plants and among the sterol components of living phytoplankton (Ballantine *et al.*, 1979). In the case where the biomarker has multiple origins within a characteristic source category (e.g., among nonwoody tissues of angiosperms or gymnosperms), it also is necessary to make some assumption about the relative contributions made by locally abundant species to the environmental mixtures. The use of "average" or "universal" literature values for plant sources is particularly dangerous unless it is known that the actual sources are in fact the same and that identical analytical methods have been employed. Unfortunately, in many studies, fresh plant materials provide the only feasible reference for interpreting bulk organic matter sources to soils and sediments. An even greater danger of using fresh plant tissues as models for quantitative analyses of geochemical mixtures is that the original organic matter likely has suffered degradation and an almost unavoidable resultant shift in [B]/[AOM] (e.g., Figs. 1, 2, and 3). This is particularly likely in this case because AOM represents the entire fresh plant material, which almost invariably includes high concentrations of reactive biochemicals. The potential for error is great, especially for trace biomarker constituents (e.g., C_1 in Fig. 3).

One interesting variant on plant tissue end members is the use of reactive biochemicals, such as ATP or phospholipids, as biomarkers. Since these molecules persist for short periods of time after the death of the source organism, they potentially can be used to quantify living biomass (Guckert *et al.*, 1985; Smith *et al.*, 1986). In this application, the [B]/[AOM] ratio may be maintained within a narrow range by the living organism, which serves in effect as a protective matrix. In addition, some labile biochemicals such as the phospholipids have remarkable source specificity (Guckert *et al.*, 1985). Potential problems with this method are that [B]/[AOM] may change with the growth stage (or environment) of the organism or that small amounts of biomarker may persist after the source organism dies. This is one application where the definitiveness of the technique deceases with increasing biomarker stability. On the other hand, the possibility of a living source is sometimes overlooked when interpretations are made of labile biochemical distributions in samples such as sediment cores where profiles of degrading substrates and microbial biomass may both decrease exponentially with depth and be confused.

A second means of addressing the plant end member problem is to separate the organic fraction of interest from selected samples so that its biomarker content can be directly measured. This direct approach avoids assumptions about the original biological sources and automatically incorporates the effects of prior diagenesis. Vascular plant tissues, for example, can be isolated from soils or sediments by size or density separations (Thompson and Eglinton, 1978; Prahl and Carpenter, 1983; Ertel and Hedges, 1985), and in some instances, such as conifer needles (Hedges and Weliky, 1989) or wood fragments (Hedges *et al.*, 1985), it is possible to identify the specific plant source. The isolated plant material, however, may represent only a minor fraction of the total biomarker in the bulk soil or sediment sample, and methods for physically isolating the remains of small organisms such as phytoplankton and bacteria are not well developed at present.

A third solution is to sample a pure end member at the latest possible stage of introduction to the environment under study. Thus, for an assessment of terrestrial organic material in coastal marine sediments, erodible bottom sediment along a river drainage (Fig. 5) or suspended particles from the lower river (e.g., Pempkowiak and Pocklington, 1983) can be collected and analyzed. Likewise, sediment traps can be used to sample particulate plankton remains as they sink through the water column to the ocean floor. Although such approaches account for previous diagenesis, they do not rule out subsequent degradation or differential physical sorting and consequent chemical fractionation (Prahl, 1985; Jasper, 1988). In addition, time-averaged samples often are required due to subannual fluctuations in source composition.

A final, direct means of establishing [B]/[AOM] within a given fraction of an environmental mixture is to independently measure an additional property (such as stable isotope composition) among a suite of related samples. Observation of a direct correlation between biomarker concentration and another property of the samples (e.g., Figs. 5 and 6) is evidence that material carrying the biomarker has mixed conservatively with a second organic fraction of different origin (see discussion under application #2). If such a direct relationship is obtained and the value of the additional property is well known for organic matter from the origin of interest, then the correlation line can be extrapolated to this value to establish the biomarker concentration in the pure end member. If the biomarker is unique to only this one mixture component, then the average value of the additional property in the remaining bulk organic matter can be estimated by extrapolation of the correlation line to a biomarker concentration of zero. If the data give a good fit to a straight line and this second extrapolation is not extensive, then the value of the complementary property in the remaining organic matter can be determined with relatively high confidence because a biomarker concentration of zero involves no measurement error. Once the properties of both organic fractions have been established, their weight percentages can be determined from the location of the data point along the mixing line between the two end members. If the two parameters do not mix linearly, then their weight percentages must be calculated algebraically.

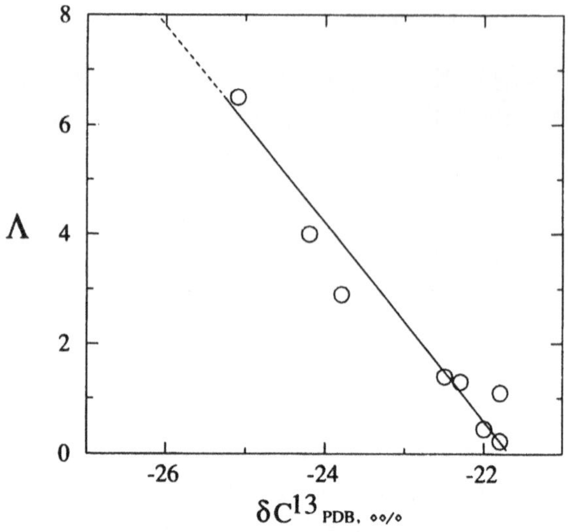

Figure 6. Average total yields of lignin-derived phenols, Λ (mg/100 mg organic carbon), plotted vs. $\delta^{13}C$ values for surface sediments from the southern Washington coast. The vertical mark at −25.5‰ on the correlation line indicates the estimated $\delta^{13}C$ value of organic material introduced from land by the Columbia River. [From Hedges and Mann (1979).]

An example of this overall approach is given in Fig. 6, where lignin phenol yield is used as an unambiguous biomarker for land-derived organic matter in coastal marine sediments. In this case, stable carbon isotope composition ($\delta^{13}C$) constituted the additional parameter and was assumed, based on analyses of local river particulates, to have a value of −25.5‰. The $\delta^{13}C$ of marine plankton remains in the sample was then estimated by extrapolating the correlation line to a lignin phenol yield of zero (plankton contain no lignin), yielding a value near −21.5‰. In this case the two properties mix linearly, and samples near the middle of the line contain approximately equal amounts of marine and terrigenous organic matter. Although informative, this method can be affected by *in situ* diagenesis and requires that a representative value for at least one property be independently measured or assumed.

4.5. Application #5: Use of a Biomarker to Estimate Changes in the Production of a Specific Type of Organic Matter within an Environment

The key characteristic of this type of application is that a biomarker is used to estimate the relative amounts of fresh plant material that were produced some time in the past within a given source environment. For example, vertical changes in the sedimentary concentration (or flux) of a characteristic biomarker might be used to estimate changes in the paleoproduction of its specific plankton source. An example of this type of application would be the use of long-chain alkenone concentration profiles in marine sediment cores to reconstruct the productivity history of prymnesiophytes at a given ocean site (Farrington *et al.*, 1988).

In addition to most of the assumptions involved with quantitative biomarker applications in general (Table I), this method depends on a proportional relationship between B (in concentration or flux units) and the amount of organic matter produced by the source organism(s). This direct relationship back to initial productivity in turn necessitates other conditions. First, the concentration of B in the source organism(s) must remain constant over the time period represented by the sample suite. Second, the efficiency of physical transport from the biomarker source region to the repository must be unchanged. In the case of a marine sedimentary record, this condition might be violated if the direction of the locally prevailing current changed or if turbidity currents introduced organic matter from a distant source. Finally, the extent of dilution of B by sedimentary particles or bulk organic matter (depending on how concentrations of B are normalized) must remain steady or be known by independent means, such as radiochemical measurements of net sediment accumulation rates. In general, absolute (sediment mass-normalized) concentrations are dangerous because they are strongly affected by physical processes involving the inorganic phases, such as particle sorting, mineral dissolution, and long-range horizontal advection.

Biomarker-based production estimates are also sensitive to diagenetic effects. The preserved mass of B can directly reflect its production (or input) rate only if B does not degrade or degrades to the same extent under all circumstances represented by the sample suite. The importance of this condition is illustrated in Fig. 7 by two hypothetical paleoreconstructions of relative changes in organic carbon production by prymnesiophytes, based on a simple analysis of the stratigraphic record of long-chain alkenone biomarkers. In this example the behavior of the alkenones during diagenesis was assumed to follow the trend depicted for C_2 in the 0.9:9.0:90.1 model illustrated in Fig. 3B. This representation is not necessarily unreasonable because alkenones have been shown to constitute about 10% of the total organic carbon biosynthesized by the prymnesiophyte *Emiliania huxleyi* (Prahl *et al.*, 1988). The two reconstructions demonstrate that small changes (1%) in the extent of *in situ* diagenesis can lead either to significant overestimates or underestimates of prymnesiophyte production, depending on the stage of diagenesis. This variability in the direction of the effect occurs because the ratio of alkenone to remaining prymnesiophyte carbon, [B]/[AOM], will vary bidirectionally over the last stages of diagenesis (Fig. 3B). If past environmental conditions favored 3%, rather than 2%, preservation of organic carbon from this source (Fig. 7, case 1), the biomarker record would *over*estimate the changes in prymnesiophyte production by 26%. On the other hand, if conditions favored 5%, versus 4%, preservation of organic carbon from the same source, the biomarker record would *under*estimate the changes in prymnesiophyte production by 28%. Such extreme sensitivity poses a significant limitation to the use of biomarkers as reliable tools for quantitative assessments of paleoproductivity until methods are devel-

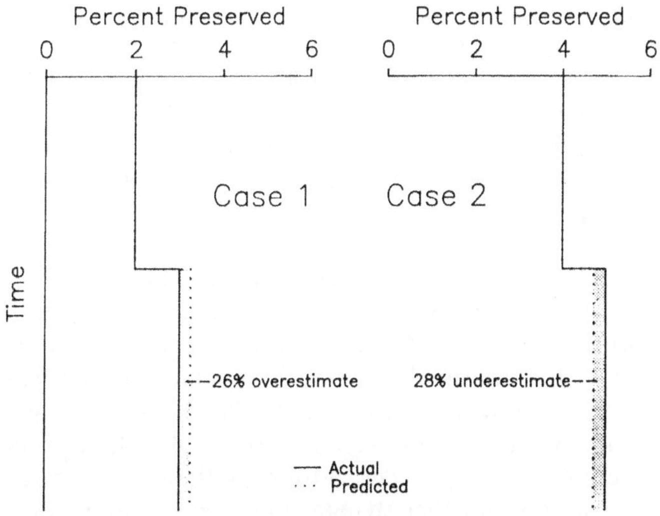

Figure 7. Estimations of relative changes in paleoproductivity for a given source based on quantitative analyses of biomarker information contained in a hypothetical sediment core. The modeled source contains three components (C_1, C_2, and C_3) of organic material that initially occur in the relative proportions of 90.1:9.0:0.9 and have first-order decay constants of 0.2, 1.0, and 5.0, respectively. In the present case, calculated changes (Fig. 3) of the weight percentage of the intermediately reactive component (C_2) were used to make the paleoproductivity reconstructions. The dotted curves are predicted values of the percent organic carbon from this source that is preserved in the record. These predictions are calculated from values for B/AOM (Fig. 3) corresponding to 3% versus 2% (case 1), and 5% versus 4% (case 2), organic carbon actually preserved in the hypothetical mixtures.

oped that distinguish between diagenetic alteration and changes in source strength.

In all paleoproduction estimates, the assumption of constant proportionality is particularly tenuous because reference is back to the original organism, thereby allowing for maximal diagenetic alterations (see previous discussion for vascular plant sources). Completely stable biomarkers are as yet unknown and, because of the overall extent of remineralization, would necessarily be trace components (<1%) in the original organism. Also, there are many plant sources for which intermediately stable biomarkers are presently unknown (e.g., diatoms). Thus, accurate estimates of total productivity based on molecular tracers are not yet feasible. For sediments, there is the complicating factor that changes in organic matter input typically alter preservation efficiencies (Emerson and Hedges, 1988) so that proportional responses cannot be confidently assumed. This uncertainty is further magnified by the circumstance that compositional changes per unit degradation are most pronounced for heavily altered organic mixtures (Figs. 1C, 3B, and 7), which are characteristic of sediments in general.

5. Conclusions

Due to the specificity of their origin and the sensitivity with which they can be measured, biomarkers afford a powerful and often unique means of investigating the extent to which various organisms contribute organic materials to complex environmental mixtures. However, early diagenesis affects almost all organic remains and essentially all biomarker applications. In order to more accurately and effectively employ biomarkers as geochemical tools for source assessments, a number of advancements will be necessary. These include the need to:

1. comprehensively define the distributions of potential biomarkers in living organisms; in particular, it will be necessary to search for (a) relatively stable biomarkers, (b) biomarkers that occur within homologous suites, (c) biomarkers which in their structures or associations carry information about their diagenetic histories, and (d) biomarkers from quantitatively important organisms for which good tracers do not presently exist;
2. determine the relative reactivities of different biomarkers under various natural conditions so that their diagenetic sensitivities can be predicted and taken into account in formulating parameters and interpreting results;
3. better elucidate biomarker diagenesis pathways and products and develop more comprehensive analytical procedures so that molecules carrying a characteristic structure can be traced further; and
4. seek greater uniformity in analytical techniques, commonly available reference materials, and modes of reporting data so that both qualitative and quantitative biomarker information from different laboratories can be confidently compared.

References

Alexander, M., 1973, Nonbiodegradable and other recalcitrant molecules, *Biotechnol. Bioeng.* **15:**611–647.

Alexander, R., Kagi, R. I., Woodhouse, G. W., and Volkman, J. K., 1983, The geochemistry of some biodegraded Australian oils, *APEA J.* **23:**53–63.

Atlas, R., Boehm, P. D., and Calder, J. A., 1981, Chemical and biological weathering of oil, from the Amoco Cadiz Spillage, within the littoral zone, *Estuarine Coastal Shelf Sci.* **12:**589–608.

Ballantine, J. A., Lavis, A., and Morris, R. J., 1979, Sterols of the phytoplankton—effects of illumination and growth stage, *Phytochemistry* **18:**1459–1466.

Barrick, R. C., Hedges, J. I., and Peterson, M. L., 1980, Hydrocarbon geochemistry of the Puget Sound region—I. Sedimentary acyclic hydrocarbons, *Geochim. Cosmochim. Acta* **44:**1349–1362.

Barrick, R. C., Furlong, E. T., and Carpenter, R., 1984, Hydrocarbon and azaarene markers of coal transport to aquatic sediments, *Environ. Sci. Technol.* **18:**846–854.

Benner, R., Maccubbin, A. E., and Hodson, R. E., 1984a, Anaerobic biodegradation of the lignin and polysaccharide components of lignocellulose and synthetic lignin by sediment micro-flora, *Appl. Environ. Microbiol.* **47:**998–1004.

Benner, R., Newell, S. Y., Maccubbin, A. E., and Hodson, R. E., 1984b, Relative contributions of bacteria and fungi to rates of degradation of lignocellulosic detritus in salt-marsh sediments, *Appl. Environ. Microbiol.* **48:**36–40.

Benner, R., Fogel, M. L., Sprague, E. K., and Hodson, R. E., 1987, Depletion of ^{13}C in lignin and its implications for stable carbon isotope studies, *Nature* **329:**708–710.

Berner, R. A., 1980a, *Early Diagenesis: A Theoretical Approach*, Princeton University Press, Princeton, New Jersey.

Berner, R. A., 1980b, A rate model for organic matter decomposition during bacterial sulfate reduction in marine sediments, in: *Biogeochimie de la Matière Organique à l'Interface Eau–Sediment Marin* (R. Dumas, ed.), Editions du CNRS, Marseilles, France, pp. 35–44.

Berner, R. A., 1982, Burial of organic carbon and pyrite sulfur in the modern ocean: Its geochemical and environmental significance, *Am. J. Sci.* **282:**451–473.

Boon, J. J., Rijpstra, W. I., de Lange, F., and de Leeuw, J. W., 1979, Black Sea sterol—a molecular fossil of dinoflagellate blooms, *Nature* **277:**125–127.

Brasch, D. J., and Jones, J. K. N., 1959, Investigation of some ancient woods, *Tappi* **42:**913–920.

Brassell, S. C., Eglinton, G., Marlowe, I. T., Pflaumann, U., and Sarnthein, M., 1986a, Molecular stratigraphy: A new tool for climatic assessment, *Nature* **320:**129–133.

Brassell, S. C., Lewis, C. A., de Leeuw, J. W., de Lange, F., and Sinninghe Damsté, J. S., 1986b, Isoprenoid thiophenes: Novel products of sediment diagenesis?, *Nature* **320:**160–162.

Burdige, D. J., and Martens, C. S., 1988, Biogeochemical cycling in an organic-rich coastal marine basin: 10. The role of amino acids in sedimentary carbon and nitrogen cycling, *Geochim. Cosmochim. Acta* **52:**1571–1584.

Caldicott, A. B., and Eglinton, G., 1973, Surface waxes, in: *Phytochemistry III* (L. P. Miller, ed.), Van Nostrand Reinhold, New York, pp. 162–194.

Christman, R. F., and Frimmel, F. H. (eds.), 1988, *Humic Substances and Their Role in the Environment*, John Wiley & Sons, New York.

Corner, E. D. S., O'Hara, S. C. M., Neal, A. C., and Eglinton, G., 1986, Copepod faecal pellets and the vertical flux of biolipids, in: *The Biological Chemistry of Marine Copepods* (E. D. S. Corner and S. C. M. O'Hara, eds.), Oxford Science Publications, Oxford, pp. 260–321.

Cowie, G. L., 1990, Marine organic diagenesis: A comparative study of amino acids, neutral sugars and lignin, Ph.D. Thesis, University of Washington.

Cowie, G. L., and Hedges, J. I., 1984, Carbohydrate sources in a coastal marine environment, *Geochim. Cosmochim. Acta* **48:** 2075–2087.

Cranwell, P. A., 1982, Lipids of aquatic sediments and sedimenting particles, Prog. *Lipid Res.* **21:**271–308.

Degens, E. T., and Mopper, K., 1975, Early diagenesis of organic matter in marine soils, *Soil Sci.* **119:**65–72.

de Leeuw, J. W., Rijpstra, W. I. C., and Schenck, P. A., 1981, The occurrence and identification of C_{30}, C_{31} and C_{32} alkan-1,15-diols and alkan-15-one-1-ols in Unit I and Unit II Black Sea sediments, *Geochim. Cosmochim. Acta* **45:**2281–2285.

Eglinton, G., 1969, Organic chemistry: The organic chemist's approach, in: *Organic Geochemistry* (G. Eglinton and M. T. J. Murphy, eds.) Springer-Verlag, New York, pp. 20–71.

Eglinton, G., and Murphy, M. T. J., 1969, *Organic Geochemistry*, Springer-Verlag, New York.

Emerson, S., and Dymond, J., 1984, Benthic organic carbon cycles: Toward a balance of fluxes from particle settling and pore water gradients, in: *Workshop on Global Ocean Flux Study*, National Academy Press, Washington, D.C., pp. 284–304.

Emerson, S., and Hedges, J. I., 1988, Processes controlling the organic carbon content of open ocean sediments, *Paleoceanography* **3:**621–634.

Emerson, S., Fischer, K., Reimers, C., and Heggie, D., 1985, Organic carbon dynamics and preservation in deep-sea sediments, *Deep-Sea Res.* **32:**1–21.

Emerson, S., Stump, C., Grootes, P. M., Stuiver, M., Farwell, G. W., and Schmidt, F. H., 1987, Organic carbon in surface deep-sea sediments: C-14 concentration, *Nature* **329:**51–54.

Ertel, J. R., and Hedges, J. I., 1985, Sources of sedimentary humic substances: Vascular plant debris, *Geochim. Cosmochim. Acta* **49:**2097–2107.

Farrington, J. W., Davis, A. C., Sulanowski, J., McCaffrey, M. A., McCarthy, M., Clifford, C. H., Dickinson, P., and Volkman, J. K., 1988, Biogeochemistry of lipids in surface sediments of the Peru upwelling area at 15°S, Org. *Geochem.* **12:**607–617.

Fengel, D., and Wegener, G., 1984, *Wood: Chemistry, Ultrastructure, Reactions*, de Gruyter, Berlin.

Fry, B., and Sherr, E. B., 1984, $\delta^{13}C$ measurements as indicators of carbon flow in marine and freshwater ecosystems, *Contrib. Mar. Sci.* **27:**13–47.

Furlong, E. T., and Carpenter, R., 1988, Pigment preservation and remineralization in oxic coastal marine sediments, *Geochim. Cosmochim. Acta* **52:**87–99.

Gagosian, R. B., Peltzer, E. T., and Merrill, J. T., 1987, Long-range transport of terrestrially derived lipids in aerosols from the south Pacific, *Nature* **325:**800–803.

Gillan, F. T., and Johns, R. B., 1986, Chemical markers for marine bacteria: Fatty acids and pigments, in: *Biological Markers in the Sedimentary Record* (R. B. Johns, ed.), Elsevier, Amsterdam, pp. 291–309.

Grill, E. V., and Richards, F. A., 1964, Nutrient regeneration from phytoplankton decomposing in seawater, *J. Mar. Res.* **22:** 51–59.

Grundmanis, V., and Murray, J. W., 1982, Stoichiometry of decomposing organic matter in aerobic marine sediments, *Geochim. Cosmochim. Acta* **46:**1101–1121.

Guckert, J. B., Antworth, C. P., Nichols, P. D., and White, D. C., 1985, Phospholipid, ester-linked fatty acid profiles as reproducible assays for change in prokaryotic community structure of estuarine sediments, *FEMS Microbiol. Ecol.* **31:**147–158.

Haddad, R. I., and Martens, C. S., 1987, Biogeochemical cycling in an organic-rich coastal marine basin: 9. Sources and accumulation rates of vascular plant-derived organic material, *Geochim. Cosmochim. Acta* **51:**2991–3001.

Haddad, R. I., Newell, S. Y., Martens, C. S., and Fallon, R. D, 1992, Lignin diagenesis of lignin-associated phenolics in the salt marsh grass *Spartina alterniflora*, *Geochim. Cosmochim. Acta* **56:**3751–3764.

Hamilton, S. E., and Hedges, J. I., 1988, The comparative geochemistries of lignins and carbohydrates in an anoxic fjord, *Geochim. Cosmochim. Acta* **52:**129–142.

Hatcher, P. G., and Breger, I. A., 1981, Nuclear magnetic resonance studies of ancient buried wood—I. Observations on the origins of coal to the brown coal stage, Org. *Geochem.* **3:**49–55.

Hatcher, P. G., Spiker, E. C., Szeverenyi, N. M., and Maciel, G. E., 1983, Selective preservation and origin of petroleum-forming aquatic kerogen, *Nature* **305:**498–501.

Hedges, J. I., and Mann, D. C., 1979, The lignin geochemistry of marine sediments from the southern Washington coast, *Geochim. Cosmochim. Acta* **43:**1809–1818.

Hedges, J. I., and Weliky, K., 1989, The diagenesis of conifer needles in a coastal marine environment, *Geochim. Cosmochim. Acta* **53:**2659–2673.

Hedges, J. I., Turin, H. J., and Ertel, J. R., 1984, Sources and distributions of sedimentary organic matter in the Columbia River drainage basin, *Limnol. Oceanogr.* **29:**35–46.

Hedges, J. I., Cowie, G. L, Ertel, J. R., Barbour, R. J., and Hatcher, P. G., 1985, Degradation of carbohydrates and lignins in buried woods, *Geochim. Cosmochim. Acta* **49:**701–711.

Hedges, J. I., Blanchette, R. A., Weliky, K., and Devol, A. H., 1988a, Effects of fungal degradation on the CuO oxidation products of lignin: A controlled laboratory study, *Geochim. Cosmochim. Acta* **52:**2717–2726.

Hedges, J. I., Clark, W. A., and Cowie, G. L., 1988b, Organic matter sources to the water column and surficial sediments of a marine bay, *Limnol. Oceanogr.* **33:**1116–1136.

Hedges, J. I., Clark, W. A., and Cowie, G. L., 1988c, Fluxes and reactivities of organic matter in a coastal marine bay, *Limnol. Oceanogr.* **33**:1137–1152.

Henrichs, S. M., and Doyle, A. P., 1986, Decomposition of ^{14}C-labeled organic substances in marine sediments, *Limnol. Oceanogr.* **31**:765–778.

Hiroi, T., and Tamai, A., 1983, Degradation of beech wood components by a white-rot fungus, *Grifora frondosa*, in: *Recent Advances in Lignin Biodegradation Research* (T. Higuchi, H.-M., Chang, and T. K. Kirk, eds.), UNI Publishers, Tokyo, pp. 34–43.

Holloway, P. J., 1982, The chemical constitution of plant cutins, in: *The Plant Cuticle* (D. F. Cutler, K. L. Alvin, and C. E. Price, eds.), Linnean Society Symposium Series, No. 10, Academic Press, London, pp. 45–85.

Huang, W-Y., and Meinschein, W. G., 1976, Sterols as source indicators of organic materials in sediments, *Geochim. Cosmochim. Acta* **40**:323–330.

Ittekkot, V., Deuser, W. G., and Degens, E. T., 1984, Seasonality in the fluxes of sugars, amino acids and amino sugars to the deep ocean: Sargasso Sea, *Deep-Sea Res.* **31**:1057–1069.

Jasper, J. P., 1988, An organic geochemical approach to problems of glacial–interglacial climatic variability, Ph.D. Thesis, Woods Hole Oceanographic Institution.

Jewell, W. J., and McCarty, P. L., 1968, Aerobic decomposition of algae, Technical Report 91, Department of Civil Engineering, Stanford University.

King, K., Jr., 1977, Amino acid survey of Recent calcareous and siliceous deep-sea microfossils, *Micropaleontology* **23**:180–193.

King, K., Jr., and Hare, P. C., 1972, Amino acid composition of planktonic foraminifera: A paleobiochemical approach to evolution, *Science* **175**:1461–1463.

Kirk, T. K., and Farrell, R. L., 1987, Enzymatic "combustion": The microbial degradation of lignin, *Annu. Rev. Microbiol.* **41**: 465–505.

Kirk, T. K., and Highley, T. L., 1973, Quantitative changes in structural components of conifer woods during decay by white and brown-rot fungi, *Phytopathology* **63**:1338–1342.

Kolattukudy, P. E., 1980, Biopolyester membranes of plants: Cutin and suberin, *Science* **208**:990–1000.

Kolattukudy, P. E., and Espelie, K. E., 1985, Biosynthesis of cutin, suberin and associated waxes, in: *Biosynthesis and Biodegradation of Wood Components* (T. Huguchi, ed.), Academic Press, London, pp. 161–207.

Lee, C., and Wakehan, S. G., 1988, Organic matter in seawater: Biogeochemical processes, in: *Chemical Oceanography*, Vol. 9 (J. P. Riley and G. Skirrow, eds.), Academic Press, London, pp. 1–51.

Marlowe, I. T., Green, J. C., Neal, A. C., Brassell, S. C., Eglinton, G., and Course, P. A., 1984, Long chain alkenones in the Prymnesiophyceae. Distribution of alkenones and other lipids and their taxonomic significance, *Br. Phycol. J.* **19**:203–216.

Martin, J. H., Knauer, G. A., Karl, D. M., and Broenkow, W. W., 1987, VERTEX: Carbon cycling in the northeast Pacific, *Deep-Sea Res.* **34**:267–285.

Meybeck, M., 1982, Carbon, nitrogen, and phosphorus transport by world rivers, *Am. J. Sci.* **282**:401–450.

Moers, M. E. C., de Leeuw, J. W., Cox, H. C., and Schenck, P. A., 1988, Interaction of glucose and cellulose with hydrogen sulphide and polysulphides, *Org. Geochem.* **13**:1087–1091.

Mycke, B., and Michaelis, W., 1986, Lignin-derived molecular fossils from geological materials, *Naturwissenschaften* **73**: 731–734.

Mycke, B., Narjes, F., and Michaelis, W., 1987, Bacteriohopane-tetrol from chemical degradation of an oil shale kerogen, *Nature* **326**:179–181.

Newell, R. C., Lucas, M. I., and Linley, E. A. S., 1981, Rate of degradation and efficiency of conversion of phytoplankton debris by marine microorganisms, *Mar. Ecol. Prog. Ser.* **6**: 123–136.

Otjen, L., Blanchette, R., Effland, M., and Leatham, G., 1987, Assessment of 30 white rot basidiomycetes for selective lignin degradation, *Holzforschung* **41**:343–349.

Ourisson, G., Albrecht, P., and Rohmer, M., 1984, The microbial origin of fossil fuels, *Sci. Am.* **251**:44–51.

Pauly, G. G., and Van Vleet, E. S., 1986, Archaebacterial ether lipids: Natural tracers of biogeochemical processes, *Org. Geochem.* **10**:859–867.

Pempkowiak, J., and Pocklington, R., 1983, Phenolic aldehydes as indicators of the origin of humic substances in marine environments, in: *Aquatic and Terrestrial Humic Materials* (R. F. Christman and E. T. Gjessing, eds.), Ann Arbor Science, Ann Arbor, Michigan, pp. 371–386.

Prahl, F. G., 1982, The geochemistry of polycyclic aromatic hydrocarbons in Columbia River and Washington coastal sediments, Ph.D. Thesis, University of Washington, Seattle.

Prahl, F. G., 1985, Chemical evidence of differential particle dispersal in the southern Washington coastal environment, *Geochim. Cosmochim. Acta* **49**:2533–2539.

Prahl, F. G., and Carpenter, R., 1983, Polycyclic aromatic hydrocarbon (PAH)-phase associations in Washington coastal sediment, *Geochim. Cosmochim. Acta* **47**:1013–1024.

Prahl, F. G., and Muehlhausen, L. A., 1989, Lipid biomarkers as geochemical tools for paleoceanographic study, in: *Productivity of the Ocean: Present and Past*, Dahlem Conference, John Wiley & Sons, New York, pp. 271–289.

Prahl, F. G., and Pinto, L. A., 1987, A geochemical study of long-chain *n*-aldehydes in Washington coastal sediments, *Geochim. Cosmochim. Acta* **51**:1573–1582.

Prahl, F. G., and Wakeham, S. G., 1987, Calibration of unsaturation patterns in long-chain ketone compositions for palaeotemperature assessment, *Nature* **330**:367–369.

Prahl, F. G., Bennett, J. T., and Carpenter, R., 1980, The early diagenesis of aliphatic hydrocarbons and organic matter in sedimentary particulates from Dabob Bay, Washington, *Geochim. Cosmochim. Acta* **44**:1967–1976.

Prahl, F. G., Muehlhausen, L. A., and Zahnle, D. A., 1988, Further evaluation of long-chain alkenones as indicators of paleoceanographic conditions, *Geochim. Cosmochim. Acta* **52**:2303–2310.

Robbins, L. L., and Brew, K., 1990, Proteins from the organic matrix of core-top and fossil planktonic foraminifera, *Geochim. Cosmochim. Acta* **54**:2285–2292.

Romankevich, E. A., 1984, *Geochemistry of Organic Matter in the Ocean*, Springer-Verlag, New York.

Rowland, S. J., and Maxwell, J. R., 1984, Reworked triterpenoid and steroid hydrocarbons in a recent sediment, *Geochim. Cosmochim. Acta* **48**:617–624.

Rowland, S. J., Yon, D. A., Lewis, C. A., and Maxwell, J. R., 1985, Occurrence of 2,6,10-trimethyl-7-(3-methylbutyl)dodecane and related hydrocarbons in the green alga *Enteromorpha prolifera* and sediments, *Org. Geochem.* **8**:207–213.

Sarkanen, K. V., and Ludwig, C. H., 1971, *Lignins*, Wiley-Interscience, New York.

Simoneit, B. R. T., 1977, Diterpenoid compounds and other lipids in deep-sea sediments and their geochemical significance, *Geochim. Cosmochim. Acta* **41**:463–476.

Simoneit, B. R. T., 1986, Cyclic terpenoids of the geosphere, in: *Biological Markers in the Sedimentary Record* (R. B. Johns, ed.), Elsevier, New York, pp. 43–99.

Sinninghe Damsté, J. S., Rijpstra, W. I. C., Kock-van Dalen, A. C., de Leeuw, J. W., and Schenck, P. A., 1989, Quenching of labile functionalised lipids by inorganic sulphur species: Evidence for the formation of sedimentary organic sulphur compounds at the early stages of diagenesis, *Geochim. Cosmochim. Acta* **53**:1343–1355.

Smith, G. A., Nichols, P. D., and White, D. C., 1986, Fatty acid composition and microbial activity of benthic marine sediment from McMurdo Sound, Antarctica, *FEMS Microbiol. Ecol.* **38**:219–231.

Spiker, E. C., and Hatcher, P. G., 1987, The effects of early diagenesis on the chemical and stable carbon isotopic composition of wood, *Geochim. Cosmochim. Acta* **51**:1385–1391.

Stauber, J. L., and Jeffrey, S. W., 1988, Photosynthetic pigments in fifty-one species of marine diatoms, *J. Phycol.* **24**:158–172.

Stout, S. A., Boon, J. J., and Spackman, W., 1988, Molecular aspects of the peatification and early coalification of angiosperm and gymnosperm woods, *Geochim. Cosmochim. Acta* **52**:405–414.

Suess, E., 1980, Particulate organic carbon flux in the oceans—surface productivity and oxygen utilization, *Nature* **288**: 260–263.

ten Haven, H. L., Baas, M., Kroot, M., de Leeuw, J. W., Schenck, P. A., and Ebbing, J., 1987, Late Quaternary Mediterranean Sapropels. III. Assessment of source of input and palaeotemperature as derived from biological markers, *Geochim. Cosmochim. Acta* **51**:803–810.

Theng, B. K. G., 1974, *The Chemistry of Clay–Organic Reactions*, John Wiley & Sons, New York.

Thompson, S., and Eglinton, G., 1978, The fractionation of a recent sediment for organic geochemical analysis, *Geochim. Cosmochim. Acta* **42**:199–207.

Thurman, E. M., 1985, *Organic Geochemistry of Natural Waters*, W. Junk, Boston.

Treibs, A., 1934, Chlorophyll und Häminderivate in bituminösen, Gesteinen, Erdölen, Erdwachsen und Asphalten, *Ann. Chem.* **510**:42–62.

Volkman, J. K., 1986, A review of sterol markers for marine and terrigenous organic matter, *Org. Geochem.* **9**:83–99.

Wakeham, S. G., Farrington, J. W., Gagosian, R. B., De Baar, H., Nigrelli, G. E., Tripp, B. W., Smith, S. O., and Frew, N. M., 1980, Organic matter fluxes from sediment traps in the Equatorial Atlantic Ocean, *Nature* **286**:798–800.

Wakeman, S. G., Farrington, J. W., and Gagosian, R. B., 1984, Variability in lipid flux and composition of particulate matter in the Peru upwelling region, *Org. Geochem.* **6**:203–215.

Wardroper, A. M. K., Hoffman, C. F., Maxwell, J. R., Barwise, A. J. G., Goodwin, N. S., and Park, P. J. D., 1985, Crude oil biodegradation under simulated and natural conditions, Part II—Aromatic steroid hydrocarbons, *Org. Geochem.* **6**:605–618.

Weliky, K., 1983, Clay–organic associations in marine sediment: Carbon, nitrogen, and amino acids in fine grained fractions, Master's Thesis, Oregon State University.

Westrich, J. T., and Berner, R. A., 1984, The role of sedimentary organic matter in bacterial sulfate reduction: The G model tested, *Limnol. Oceanogr.* **29**:236–249.

Whitehead, D. C., Dibb, H., and Hartley, R. D., 1981, Extractant pH and the release of phenolic compounds from soils, plant roots and leaf litter, *Soil Biol. Biochem.* **13**:343–348.

Wilson, T. R. S., Thomson, J., Colley, S., Hydes, D. J., Higgs, N. C., and Soresen, J., 1985, Early organic diagenesis: The significance of progressive subsurface oxidation fronts in pelagic sediments, *Geochim. Cosmochim. Acta* **49**:811–822.

Chapter 12

The Early Diagenesis of Organic Matter: Reaction at the Air–Sea Interface

DAVID J. CARLSON

1. Introduction

The central question of this chapter is to what extent are oceanic or atmospheric organic materials altered as a result of having spent time at or passed through the air–sea interface. Such alterations are not, strictly speaking, diagenetic, because they occur before the organic materials are deposited in sediments, but they are worthy of consideration if they affect what is eventually preserved relative to what was produced in the ocean or supplied from the atmosphere. In addition to products destined for ocean sediments, alterations at the air–sea interface may result in products which leave the ocean for the atmosphere. One can imagine, for example, photochemical processes at the interface which produce volatile products "lost" to the atmosphere and residual materials which eventually settle to the seafloor. There may be more evidence for interfacial processes leading to atmospheric products than for processes leading to sedimentary products, but the discussion in this chapter will be confined to the latter.

Three potentially important interfacial processes will be discussed: conversions of dissolved organic materials to particulate form, conversions of low-molecular-weight materials into materials of higher molecular weight, and photodegradation. Phase transformations of dissolved materials to particulate form may affect the downward flux of organic carbon, while chemical or photochemical alterations may alter the molecular composition of oceanic or atmospheric organic materials, perhaps obscuring original source information and replacing that information with unique interfacial signals. Some of these conversion and alteration processes are unique to the air–sea interface. Others may occur in bulk solution as well as at the air–sea interface but may be uniquely effective at the interface. Reverse reactions are also possible (particulate organic materials could dissolve or complex materials could break apart), but the products of the reverse reactions seem less likely to reach the sediments than the products of the forward reactions.

Evidence for any of these "diagenetic" processes at the ocean surface is very scarce—the study of oceanic interfacial chemistry has not gotten much beyond the descriptive phase. In some cases, it is possible to extrapolate from laboratory experiments. When, as is more often the case, laboratory conditions are not representative of ocean surfaces, some reasonable estimates will be given of the extent and effectiveness of natural interfacial processes. There will be

DAVID J. CARLSON • TOGA/COARE International Project Office, UCAR, Boulder, Colorado 80307.
Organic Geochemistry, edited by Michael H. Engel and Stephen A. Macko. Plenum Press, New York, 1993.

a few putative processes for which there is no reliable evidence. These will be discussed on the basis of their potential effect and in an effort to instigate some research.

At the end of the chapter, brief mention will be made of the special case in intertidal regions, where the air–sea interface periodically comes in direct contact with marsh, mudflat, or beach sediments.

2. The Air–Sea Interface: Description and Composition

This chapter will describe an air–sea interfacial region, defined as a region at the ocean surface in which solution properties differ from those in the underlying bulk solution. The dimensions of this interfacial region, appropriately termed the "surface microlayer," vary depending on which physical or chemical property is being considered. Temperature profiles near the ocean surface made with infrared sensors (McAlister and McLeish, 1969, 1970) or fine-scale thermocouples (Khundzhua *et al.*, 1977) indicate a surface thermal boundary region several hundred micrometers to a few millimeters thick in which temperatures are different from (generally cooler than) that of the subsurface water. Likewise, near-surface velocity profiles obtained by observing the movements of microscopic bubbles indicate viscous boundary layers several hundred micrometers thick (McAlister and McLeish, 1969; McLeish and Putland, 1975). For dissolved organic materials, surface microlayers seem to be on the order of 100–200 μm because per volume dissolved organic material concentrations do not continue to increase as the sample depth gets shallower than about 100–200 mm (Daumas *et al.*, 1976; Carlson, 1982a, 1983). Particulate materials with dimensions of tens to hundreds of micrometers also accumulate near the ocean surface. Although there are uncertainties in each of these estimates, having to do with the accuracy and response of the profiling techniques and the efficiencies of microlayer sampling techniques, the combined thermal, viscous, and chemical evidence indicates a relatively thick interfacial region with distinct physical properties containing a mixture of dissolved and particulate materials.

There are some important differences between these relatively thick oceanic surface microlayers and typical monolayers formed in laboratory troughs, differences which arise from the chemical complexity of the ocean microlayers, from the fact that organic materials accumulate into oceanic surface microlayers from below but are generally spread on a surface from above in laboratory experiments, and from the thickness of the oceanic microlayers. There are also important differences in atmospheric conditions between laboratory surfaces and ocean surfaces. Relative humidity is nearly always kept near 100% over laboratory surfaces but varies widely over the ocean surface. Properties of laboratory monolayers are especially sensitive to humidity and surface temperature (Bilkadi and Neuman, 1979). Because of the differences between laboratory and oceanic surface conditions, much of the laboratory literature about rates of exchange between bulk and surface regions, about surface conformation, and about surface reactions is of limited use in describing oceanic surface conditions.

There have been a few compounds added to the list of microlayer components since Hunter and Liss (1981) reviewed microlayer composition, but there is still little evidence that surface microlayers contain any material that is not also in bulkwaters or that microlayers are devoid of any bulkwater material excepting perhaps organisms which may actively avoid microlayer regions. Thus, the starting materials for any diagenetic alterations are similar in microlayers and in bulkwaters. It is also true, as in bulkwaters, that large portions of microlayer dissolved organic carbon (DOC) remain uncharacterized (Williams *et al.*, 1986). This general similarity in the composition of microlayer and bulkwater organic materials indicates that bulkwaters are the predominant source of microlayer materials. The similarity may also reflect the fact that few studies have simultaneously measured atmospheric materials and microlayer and bulkwater materials in order to distinguish atmospheric or bulkwater inputs to surface microlayers. Lipid materials were measured in aerosols and microlayers and bulkwaters by Marty *et al.* (1979), but the *n*-alkane, hydrocarbon, and fatty acid fractions in the microlayer samples showed no indication of having come from other than a bulkwater source.

Major stable elements such as calcium, a variety of particulate and dissolved trace metals, nutrients, and several radionuclides have all been found enriched in surface microlayers on occasion (Barker and Zeitlin, 1972; Duce *et al.*, 1972; Elzerman and Armstrong, 1979; Bacon and Elzerman, 1980; Hunter, 1980; Lyons *et al.*, 1980; Pattenden *et al.*, 1981; Harvey *et al.*, 1982; Heyraud and Cherry, 1983; Hardy *et al.*, 1985a; Williams *et al.*, 1986). Of these, only the trace metals might be directly involved in microlayer diagenetic reactions, but as a group these data indicate that the microlayer inorganic matrix may be considerably different from the bulkwater matrix.

Under certain conditions, the abundance of organic materials in surface microlayers is sufficient to alter light reflection from the surface, giving rise to surface slicks, regions of surface which appear glassy or smooth relative to adjacent regions. It is thought that the organic materials alter capillary wave generation or propagation with consequent changes in reflectivity, although much of the evidence for capillary wave alteration comes from experiments with monolayer films rather than natural oceanic microlayers (Hunter and Liss, 1981; Carlson, 1987). The association between slicks and surface microlayer organic materials seems clear: microlayer enrichments (ratio of microlayer to subsurface concentrations) of dissolved UV-absorbing materials and particulate chlorophyll-containing materials are strongly correlated with the occurrence of slicks (Carlson, 1982b, c; Carlson *et al.*, 1988; Cullen *et al.*, 1989). These slicks present an unusual challenge to marine chemists—a visible accumulation of organic materials which should also be measurable. Chemical studies of oceanic surface microlayers have not always distinguished slicked surfaces from clean (unslicked) surfaces. The importance of making such distinctions will become evident in the ensuing descriptions and discussions.

In the discussion that follows, concentrations of microlayer materials will be described as microlayer excesses or microlayer enrichments. Microlayer excesses are the amounts of individual or total materials in microlayer samples in excess of the amounts of the same materials in bulkwaters, simply the difference between microlayer and bulkwater concentrations. Microlayer enrichments are the ratios of microlayer to bulkwater concentrations. Bulkwater samples typically come from 10–100-cm depth below the microlayer samples. Because of the apparent thickness of surface microlayers, all microlayer concentrations will be per volume concentrations rather than per area concentrations, used to describe monolayers.

3. Transformation of Dissolved to Particulate Carbon in Surface Microlayers

Dissolved organic materials in surface microlayers may be converted to particulate organic materials by direct precipitation due to aggregation or polymerization, by coagulation or compression associated with wave motions or bubble collapse, or by adsorption to other atmospheric or oceanic particles in the interface. There is no doubt that some phase transformations take place: intense slicks often contain bands or lines oriented across the direction of waves or currents in which it is visually obvious that dissolved slick materials have accumulated into a particulate, usually brown, matrix. Tar balls are another visual example of accumulation of interfacial materials into particulate form. This section will describe the processes by which such phase transformations occur and estimate the quantitative importance of these interfacial processes. The discussion will center on the relative enrichments of dissolved and particulate materials and on presumed and observed transformation processes.

3.1. Quantities of Dissolved and Particulate Organic Materials in Surface Microlayers

On the whole, there is not much more dissolved organic material in surface microlayers than there is in an equivalent volume of bulkwater, at least not when the dissolved organic matter is measured as DOC by persulfate oxidation to CO_2. The average microlayer DOC excess (microlayer concentration minus bulkwater concentration) in data compiled from coastal and oceanic waters is around 1 mg liter^{-1} (Carlson, 1983; Williams *et al.*, 1986), with the result that most microlayer enrichment factors are small (<2) except in some oceanic samples where bulkwater DOC concentrations are less than 1 mg liter^{-1}. There are also reports of instances in which microlayer DOC concentrations were lower than corresponding bulkwater concentrations (Dietz *et al.*, 1976; Carlson, 1983). Microlayer DOC concentrations may turn out to be higher than previously reported when microlayer samples are analyzed by new high-temperature oxidation techniques (Sugimura and Suzuki, 1988), but because there are no *a priori* reasons to expect that microlayer DOC will be more or less susceptible to high-temperature oxidation than bulkwater DOC, microlayer DOC enrichments may stay the same, namely, small. Thus, based only on the amounts of dissolved organic materials present, diagenetic reactions in oceanic surface microlayers should not be much different from those in oceanic bulkwaters, except in surface slicks. Some slicks contain much more DOC than adjacent unslicked microlayers and associated bulkwater (Carlson, 1983).

In contrast to the small average enrichments of DOC in oceanic surface microlayers, microlayer enrichments and excesses of particulate organic carbon (POC) can be very large (see compilation in Williams *et al.*, 1986). Bits of leaves, pine needles, or detached floating macroalgae may have biased POC measurements in coastal waters, but even oceanic samples away from obvious terrestrial influence have POC enrichments of 5 or greater (Carlson, 1983; Williams *et al.*, 1986). As a result, the proportions of particulate and dissolved organic carbon are different in microlayers than they are in bulkwaters. This relative abundance of particulate materials may be an important factor in microlayer diagenetic reactions, and it may be a consequence of unique microlayer processes. As much as 20% f the POC may represent biomass (Williams *et al.*, 1986), although the enhanced microlayer biomass is not necessarily an indication of enhanced biological activity. As in the case of dissolved materials, the highest microlayer POC excesses and enrichments are found in slicks (Carlson, 1983).

3.2. Conversion of Dissolved Materials to Particulate Materials

Although measuring these disproportionately high total POC concentrations in microlayers is relatively easy, determining which fractions of the POC have been produced from dissolved organic materials in the surface microlayer by microlayer diagenetic processes is difficult. Some fractions may not come from microlayer DOC but from direct inputs of organic particulates, by buoyant rise of subsurface organic particles or by deposition of organic particles from the atmosphere.Those fractions that are the result of microlayer transformation processes, by adsorption of microlayer dissolved components onto inorganic particles or by association and interaction among the microlayer dissolved components themselves, may subsequently interact with the preformed POC. Thus, POC in surface microlayers, and POC which eventually sinks from surface microlayers, may include components from two inputs and several transformation processes, making it difficult to find definitive indications of microlayer diagenetic alterations in either the amount or the composition of such a mixture.

One pathway for conversion of microlayer dissolved organic materials to particulate form may be via adsorption to other oceanic or atmospheric particles. Particles rising to the surface presumably carry a coating of adsorbed organic materials and may only exchange these adsorbed materials with microlayer components. Terrestrial particles deposited from the atmosphere, on the other hand, may include aluminosilicate or metallic dusts or ashes which are not fully coated with organic materials. These uncoated particles may then adsorb microlayer dissolved organic components, most of which have, by definition, some tendency to leave seawater for surfaces. This adsorption of microlayer materials is a special case of the whole set of seawater adsorption processes, special because the adsorbing molecules in the microlayer may be somewhat more hydrophobic than those available in bulkwater and because the contact time between the particle and the dissolved organics while the particle resides in the surface microlayer may be greater than contact times between bulkwater dissolved organics and a sinking particle. There is considerable evidence that adsorption increases as hydrophobicity of the adsorbate increases (e.g., Elliot and Huang, 1980; Sullivan *et al.*, 1982), indicating that hydrophobic molecules in surface microlayers should adsorb to particles. However, hydrophobicity, expressed as octanol–water partition coefficients, does not account for interactions among a mixture of potential adsorbates or between adsorbates and other organic components (Boehm and Quinn, 1973; Hassett and Anderson, 1979; Means and Wijayaratne, 1982; Brownawell and Farrington, 1985; Whitehouse, 1985) in a heterogeneous organic system such as surface microlayers. For example, while Rice *et al.* (1983) found that the proportion of total polychlorinated biphenyls (PCBs) in particulate phases was much higher in surface microlayers of Lake Michigan than in bulkwaters, indicative of enhanced microlayer adsorption, the distribution of those PCBs was not what one would predict from bulk partition coefficients: there were higher proportions of more soluble PCBs on microlayer particles than of less soluble PCBs. The important terms in estimating contact times between microlayer organics and particles are adsorption rates and particle residence times. Kinetic studies of the adsorption of phthalate esters from seawater onto clay minerals indicate that initial adsorption is relatively fast, most of the adsorption occurring within the first 10 or 15 minutes (Sullivan *et al.*, 1982). Residence times for particulate metals carried by dry deposition to surface microlayers may be minutes to hours (Hunter, 1980; Eisenreich, 1982; Hardy *et al.*, 1985b), long enough for considerable adsorption (or exchange) to take place. Separate from their role in adsorption, atmospheric particles may serve as condensation nuclei for aggregation or polymerization while they reside in microlayers, in dilute imitation of the way particles stabilize emulsions and foams (Adamson, 1982).

Another pathway from microlayer DOC to POC

may occur when dissolved organic molecules associate or interact directly to form particulate (filterable) materials. These transformations, a sort of heterogeneous precipitation, may be favored in surface microlayers because, in slicks at least, the microlayers contain relatively high concentrations of dissolved organic compounds and because those microlayer compounds are subjected to unique interfacial stresses. There is laboratory and field evidence that such transformations occur.

Sutcliffe *et al.* (1971) measured particle abundance in the upper ($\sim$ 1 m) ocean at times when the ocean surface had parallel bands of streaks and floating *Sargassum* debris of a type and scale ascribed to Langmuir circulation. They found that particles were more abundant in convergent zones beneath the streaks than in water from an equal depth between the streaks. They also found that, in summer at least, the relative concentrations of POC near the surface, collected without reference to convergences, increased as the wind speed increased from 3 to about 13 knots (1.5 to 6.5 m s^{-1}). These observations demonstrate only that there was an accumulation of particulate materials near, but not necessarily from, surface microlayers, but they could indicate that particle production, presumably at the surface, increased due to wave motions and larger scale convergences. Wheeler (1975) actually produced particulate material by compressing natural slicks on Nova Scotian lakes and in Nova Scotia coastal seawaters. He was also able to produce particles from Sargasso Sea seawater by compression in laboratory tanks, but only after he supplemented that seawater with protein (ovalbumin) at 1.7 mg $liter^{-1}$. Assuming the ovalbumin was 50% carbon by weight, the ovalbumin supplement represents a carbon concentration of 0.8 mg $liter^{-1}$, well within the range of values for microlayer DOC excess. However, the ovalbumin would also represent a protein carbon concentration of about 67 mmol $liter^{-1}$, a higher value than reported in most screen-collected microlayer samples (e.g., Williams *et al.*, 1986). Protein carbon concentrations in excess of 400 mmol $liter^{-1}$ have been measured in glass plate and glass cylinder (Carlson *et al.*, 1988) samples of slicks in coastal Maine seawater (D. J. Carlson and J. E. Brophy, unpublished data). Thus, the transformations that Wheeler induced by mechanically compressing surface microlayers could occur under natural convergent pressures on the sea surface. As mentioned, transformation processes may affect contaminants as well as natural components of surface microlayers. Gordon *et al.* (1976) followed the fate of crude oil spread on tanks of natural seawater. After about three days, the oil slicks began to break up and tar balls began to form, often around a nucleus of particulate debris. Eventually, the only oil visible or measurable on the surface was in the form of tar balls. Although the tanks minimized dispersive forces which occur on natural ocean surfaces, these experiments again illustrate the surface conversion of dissolved (or in this case, liquid) components into particulate components.

Laboratory experiments provide some information about possible mechanisms for the DOC to POC transformation. In the absence of wind and waves, i.e., on a calm surface, microlayer organic molecules may interact due to a combination of factors including proximity, hydrophobic effects, and surface configuration. If the interactions result in stable products sufficiently large (or sufficiently "sticky") to be filterable, than phase transformation from dissolved to particulate has in effect occurred. Such transformations may account for the frequent observations of hysteresis in laboratory film balance experiments (e.g., Jarvis *et al.*, 1967; Dragcevic *et al.*, 1979): dissolved materials originally spread on a surface may irreversibly change to particulate form when the molecules are crowded together. (It is also possible that materials are forced away from the surface and do not easily or quickly return.) There is no documentation of the complete phase transformation on natural ocean surfaces, but evidence will be presented later in this chapter that a part of the process, namely, the association of low-molecular-weight organic compounds to form higher molecular weight products, does occur in surface microlayers.

The surface of the ocean is not usually calm: convergent and divergent motions occur at the surface due to waves and to larger scale advection. Laboratory experiments suggest that both compression and dilation can result in phase changes of surface molecules. Monolayers of many proteins coagulate under compression (MacRitchie, 1963). Under extreme compression, the coagulation may be irreversible (MacRitchie, 1963), leaving a stable particulate phase. More interesting, when considering potential diagenetic changes, is the fact that the coagulated proteins are often insoluble, and their configurations are not the same as the native configurations in bulk solution. This alteration occurs because the protein has been denatured at the surface and because there are energy barriers to desorption which prevent the molecule from reassuming its original configuration (MacRitchie, 1986). Dilation of a surface may also induce coagulation. Peters and Heller (1970) have shown that the increase in coagulation of α-FeOOH sols upon stirring is actually enhanced surface coagulation due to increased surface-to-volume ratios rather than mechanical coagulation due to turbulence. Local surface dilation as a wave passes on the ocean may provide similar surface extensions and transient changes in local surface-to-volume ratios. Finally, Ferguson *et al.* (1987) have shown that some very high molecular weight polymer solutions form particulate fibers as thin streams of the solution are stretched, in the same way that freshly extruded silk proteins form a fibrous thread almost instantaneously to support a spider. None of these laboratory polymer solutions, sols, or monolayers is analogous in composition, concentration, or arrangement to ocean microlayers, but the laboratory evidence is nonetheless useful because it illustrates processes in simple systems which may occur in modified form on the more complex ocean surface. One important feature of the ocean surface, namely, that compression and dilation occur repetitively, has never been duplicated in laboratory experiments.

Bubbles are another important feature of the ocean surface rarely included in laboratory surface chemistry experiments. Bubbles, which may have carried organic materials to surface microlayers, can cause the transformation of microlayer dissolved materials to particulates as they burst or as they are forced into solution at depth. The production of aerosol particles from bursting bubbles is well known (Blanchard and Woodcock, 1957). Most of those aerosol particles carry organic material (Bezdek and Carlucci, 1974; Hoffman and Duce, 1976; Graham et al., 1979; Gershey, 1983a), and the bubbling process (rising and then bursting) can be used to produce particulate organic carbon from seawater solutions (Wallace and Duce, 1975; Ichikawa, 1980; Gershey, 1983b). It is not known what fraction of the organic matter in these bubble-generated particulates derives from the surface microlayer and what fraction derives from adsorption to the bubble surface during its rise, what fraction of the oceanic aerosols are transported from their production site, or whether the bubble-generated particulate materials redissolve when they eventually settle again on the sea surface. Particulate materials are also produced when microbubbles collapse under pressure as they are mixed in the upper ocean (Johnson and Cooke, 1981; Johnson and Wangersky, 1987), but again it is not clear whether the organic materials that stabilize the bubbles and aggregate when the bubbles collapse or dissolve derive from surface microlayers or from bulk solution.

The difficulty in quantifying bubble effects extends to the whole process of transforming microlayer DOC to POC: too many of the transformation pathways are known only from laboratory experiments, and too few rates are known for oceanic conditions. It is instructive, however, to gather a few terms and estimate some rates to see whether these transformations have any quantitative significance. If we assume that 10% of a region of the ocean surface is covered with slicks (not an unreasonable estimate for calm conditions or coastal areas), that the DOC excess in the slicks is 3 mg liter^{-1}, and that all of that excess is converted to POC during a diel cycle of slick accumulation and dispersal, then POC production by microlayer transformation processes is 2.5 μmol C m^{-2} day^{-1} for a 100 μm-thick microlayer, almost 1 mmol C m^{-2} y^{-1}. Compared to primary biological production of particulate carbon, which can exceed 8 mol C m^{-2} y^{-1}, microlayer processes are insignificant. Very little of primary production reaches the sediments, however, so a more appropriate comparison is between microlayer transformation to POC and the flux of POC to the sediments. Deep fluxes of POC are usually tens to hundreds of mmoles of C per square meter per year (Bishop *et al.*, 1980; Brewer *et al.*, 1980; Honjo *et al.*, 1982), still considerably larger than production of POC estimated for microlayer processes. These estimates are too uncertain to warrant much discussion, except for three points. The first is that particulate products of some microlayer transformations, such as insoluble proteins produced by surface coagulation, may be sufficiently altered so that they are less susceptible to enzymatic degradation. In that way, a predeposition diagenetic process in surface microlayers may affect what is eventually deposited and preserved. The second related point is that many toxic organic materials are highly enriched in surface microlayers, especially in slicks (Seba and Corcoran, 1969; Biddleman and Olney, 1974; Williams and Robertson, 1975; Wu *et al.*, 1980; Rice *et al.*, 1983), and very likely to be transformed, especially by adsorption, to particulate form in surface microlayers. In this case, microlayer diagenesis of those contaminants, once they are deposited on the ocean surface, may determine whether they pass quickly and inertly through the ocean on sinking particles or become immediately available for biological ingestion as surface or near-surface aggregates. Finally, although microlayer transformations from DOC to POC may be difficult to quantify, they may be diagenetically important if they add microlayer and oceanic organic signals to terrestrial inorganic substrates delivered by the atmosphere.

4. Transformation from Low to High Molecular Weight

The second microlayer diagenetic process to be discussed is conversion of low-molecular-weight materials into higher molecular weight materials. These transformations are distinct from DOC to POC transformations described in the previous section because the products remain in dissolved form. It will be argued that organic materials in surface slicks are sufficiently abundant to undergo such transformations, but it should be noted that the presence of slicks may indicate that interactions among small molecules have already occurred to a sufficient extent to affect microlayer rheology and the propagation of small waves.

Several mechanisms have been proposed by which small organic compounds in seawater might interact or combine to produce higher molecular weight heteropolycondensates, which might then be components of the portion of oceanic DOC that resists characterization and biodegradation. These mechanisms include Maillard reactions, metal bridging, hydrophobic interactions, and others (Degens, 1970; Hedges, 1978; Dawson and Duursma, 1981; Gagosian and Lee, 1981; Harvey *et al.*, 1983). It has been difficult to demonstrate that any of these mechanisms are effective in seawater because the potential reactants are so dilute. Transformation of carbon from monomers such as leucine and glucose to molecules of higher molecular weight has been shown in surface and deep waters of the Atlantic and Pacific oceans (Carlson *et al.*, 1985; Brophy and Carlson, 1989). This section proposes that a general transformation from low to high molecular weight is favored and that certain interactions occur uniquely in surface microlayers, due to three factors: the relative abundance of potential organic reactants in surface microlayers, the relative immobility of microlayer molecules, and the unusual conformations of molecules near the interface. An additional factor, the abundance of light, especially low-wavelength light, will be mentioned briefly and then discussed in the final section of the chapter.

4.1. Abundance of Microlayer Molecules

In the previous section, dissolved organic materials in unslicked microlayers at the air–sea interface were shown not to be much more abundant than dissolved organic materials in bulkwaters, or much different in composition. Those dissolved organic materials in unslicked microlayers may, however, be less mobile and have different configurations than similar molecules in bulkwaters. Surface slicks, on the other hand, are visible evidence that microlayer organic molecules can occasionally be abundant enough, immobile enough, and in such a configuration to affect the physical structure of the interface. Under these conditions, chemical interactions between individual molecules to form larger molecules may be (or may have been) especially effective.

Some slick DOC enrichments are as large as tenfold (Carlson, 1982b). Components of DOC in slicks can in some instances have even larger enrichment ratios: more than 25 for UV-absorbing components (Carlson, 1982b), more than 12 for protein (Williams *et al.*, 1986; D. J. Carlson and J. E. Brophy, unpublished data), etc. These high enrichments represent exceptional conditions, but they may be the conditions under which reaction pathways hypothesized for bulk solution are actually effective. For example, Harvey *et al.* (1983) suggested that polyunsaturated lipids in seawater might undergo photoinduced, trace metal-catalyzed cross-linking and oxidations to form substances fitting the description of marine humic materials. The polyunsaturated lipids are rare and insoluble in bulk seawater, but they could be sufficiently abundant in slicks (e.g., Lee and Williams, 1974) so that, under high surface UV irradiation, such transformations might occur.

4.2. Immobility of Microlayer Molecules

In addition to being abundant, organic molecules in slicks may be relatively immobile: unable to rotate or bend freely or to depart from the interface region into bulk solution. This immobility may have large consequences for intermolecular interactions. A molecule whose ability to rotate or move is hindered may undergo internal alteration or orient itself toward a second molecule differently in an interfacial region than it would in isotropic bulk solution. Depending on the degree to which molecules are pushed together by convergence, close-range interactions such as hydrophobic effects, hydrogen bonding, or even covalent bonding may be more effective between the interfacial molecules. The interfacial region is thus a potential catalyst, lowering the energy barriers for reaction by inducing reactive orientations.

Evidence for restricted movement of organic materials at interfaces comes largely from single-component monolayer films spread on simple laboratory bulk solutions, leaving a large gap between the experimental data and actual oceanic conditions. There are, however, some experimental data in which molecules reached the interfacial region from below, as in the ocean, and some general observations from the laboratory work which can serve as hints to the oceanic situation.

It is generally assumed in laboratory surface chemistry work that some portion of the "surface-active" molecule actually penetrates the air–water interface. This assumption is illustrated in familiar form in virtually every interfacial chemistry textbook and review: the zigzag hydrocarbon "tail" of a long-chain fatty acid is shown on the gaseous side of a planar interface, while the hydrophilic carboxyl "head" is shown immersed in liquid below the interface. One challenge to marine chemists is to extrapolate from that idealized situation to the ocean, where the interface itself may be diffuse and dynamic rather than planar and quiescent, where the accumulation zone may extend many molecular dimensions beneath the interfacial region, and where there are complex mixtures of dissolved and particulate organic materials present in place of pure laboratory compounds.

If even a small part of a molecule penetrates the interface, the entire molecule will have limited freedom to rotate or move. If surface molecules do not actually penetrate the interface, there may still be viscous constraints on their movements in the microlayer region: Carlson (1987) showed that the rotation of fluorescent probes was hindered in samples from oceanic surface slicks. This diminished ability to rotate or move in the surface microlayer region may subject microlayer molecules to shear or compressive forces not experienced by similar molecules in bulk solution and lead to interactions among small molecules which might not be possible or favored in bulk solution. These interactions, if durable, may constitute an effective transformation from small to large molecular weight.

Measurements of surface potential on the ocean surface (Williams *et al.*, 1980) indicate the presence of oriented dipoles at the interface. These data and the viscosity data of Carlson (1987) indicate that molecules in oceanic surface microlayers are likely to have their movements restricted in a manner similar to molecules on the surface of laboratory tanks. Direct confirming evidence from the ocean surface is not available and is unlikely to become available soon. It should be noted that wave damping in surface slicks can be modeled in terms of surface dilational elasticity, which implies that the two-dimensional mobility of surface molecules is limited relative to the three-dimensional mobility of the same molecules below the surface.

Immobility within the surface microlayer is accompanied by inability to leave: there is no evidence that molecules attracted to the ocean interface region redisperse into bulkwater despite the large microlayer-to-bulkwater concentration gradients, at least not until breaking waves forcibly remove them from the interface. Diffusion or weak small-scale turbulence which disperses bulkwater molecules will be less effective in dispersing microlayer molecules. There are suggestions (Carlson, 1983) that wave and water motions that lack sufficient energy to erode microlayers may reorder the microlayer molecules into thinner layers, layers in which the molecules are perhaps less mobile.

In laboratory surface chemistry literature, the inability of surface molecules to redissolve in bulkwater once they are spread on a surface is taken as an indication of insolubility of the surface molecules—spread films of such molecules are said to be "stable." In fact, for monolayer-forming molecules such as palmitic acid (a fatty acid) or dimyristoylphosphatidylcholine (a membrane phospholipid), the films are stable not because of insolubility of their components but because of energy barriers to the movement of the molecules from the interface into solution (Gershfeld, 1982). Similar energy barriers have been shown for the desorption of proteins applied onto a surface from above (Gonzalez and MacRitchie, 1970). As a result, the per area concentrations of laboratory film-forming molecules spread onto a surface are likely to exceed the per area concentrations which occur when molecules reach the interface from bulk solution. In the ocean, most organic molecules reach the surface from below and may not actually have to "cross" the interface to leave the surface region. Nonetheless, once an organic molecule reaches the surface microlayer, it may be delayed in leaving it by attraction of all or part of the molecule to the interface itself or to other molecules already in the interfacial region. Attraction to the interface itself is of course surface activity in the conventional sense. Attraction to neighboring molecules in the surface microlayer rather than to the interface itself expands the concept of surface enrichment to include not only those molecules with classically defined hydrophobicity (as, for example, a high octanol–water partition coefficient) but also molecules that are more hydrophilic but which interact with surface-active molecules. Such interactions may be responsible for commonly observed surface microlayer enrichments of monomeric and polymeric saccharides (Dietz *et al.*, 1976; Sieburth *et al.*, 1976; Henrichs and Williams, 1985; Williams *et al.*, 1986) which by themselves are not generally considered surface-active.

Although there is no evidence that oceanic microlayer molecules leave the microlayer region, there is also no evidence that they do not: fluxes to or from oceanic microlayers are unknown. The predominant condition of the ocean surface is nonetheless one of organic enrichment, indicating that some fraction of oceanic DOC consistently prefers, and may be constrained within, surface microlayers.

4.3. Conformation of Microlayer Molecules

In addition to having their movements constrained, molecules in surface microlayers may assume unusual configurations. It is known from laboratory data that protein molecules unfold and assume linear configurations if spread on an air–water interface (e.g., MacRitchie, 1986). Similar changes in configuration, which result in the hydrophobic segments of the proteins being directed toward the interface while the hydrophilic portions remain in bulk solution, presumably occur when proteins and other amphiphilic biopolymeric materials reach the oceanic surface microlayer. Once in the surface microlayer, these denatured molecules may expose reactive regions which were inaccessible in the native solution configuration and thus undergo interactions with neighboring molecules which were precluded in bulk solution. The surface molecular configuration may also have different light absorption spectra or cross sections and be more susceptible to coagulation processes mentioned above.

4.4. Evidence for Microlayer Transformation from Low to High Molecular Weight

There are several hints from laboratory experiments that small molecules may interact to form larger molecules in interfacial regions and one report that such transformations may occur in oceanic surface microlayers. Some laboratory evidence has to do with the binding of hydrocarbons and fatty acids to natured and denatured proteins (Ray *et al.*, 1966; Reynolds *et al.*, 1968). Although not at an interface, these protein data are significant because they demonstrate that unfolding of proteins during denaturation exposes additional binding sites not available in the native conformation. It is likely that unfolding as proteins come near the ocean surface will likewise expose reactive sites which are not accessible in bulk solution. Other laboratory evidence showed that stable associations were formed between tannic acids in bulk solution and proteins in monolayers (Schulman and Rideal, 1937a), due evidently to interactions between the submerged polar portions of the proteins and polar components of the tannins. The protein monolayers alone could be dispersed by addition of soap, but the protein monolayers stabilized by tannins could not. Phenolic monomers also interacted with the monolayer proteins but did not affect surface pressure or surface potential to the degree that the polymeric tannin molecules did, indicating that interactions between macromolecules, one in solution and one at an interface, are possible in interfacial regions, with potentially important rheological consequences. The same authors also showed that lipid components introduced beneath a monolayer of other lipids could actually penetrate the monolayer film and form stable complexes with the monolayer molecules (Schulman and Rideal, 1937b). One might predict that lipophilic compounds, especially lipophilic contaminants, in seawater would undergo similar reactions in ocean surface microlayers. Platford (1982) showed that partition coefficients between octanol and water for two PCB isomers and the pesticides DDT, Lindane, and hexachlorobenzene were ten to a thousand times greater if the octanol was in a film on water rather than in its bulk form. In the only oceanographic data, Carlson *et al.* (1985) showed that leucine, glucose, and especially palmitic acid were incorporated into higher molecular weight fractions when added to a sample from a surface slick. The amount of palmitic acid incorporated into high-molecular-weight materials in the slick sample was greater than incorporation of any of the compounds in any other seawater or freshwater sample.

Despite these several strong hints and the available evidence that transformation from low to higher molecular weight occurs among dissolved components in surface microlayers, especially in surface slicks, there is not sufficient evidence to attempt to quantify the process. It is clear that the unique conditions in oceanic surface microlayers may favor or permit molecular interactions that occur slowly or not at all in the bulk ocean. If the higher molecular weight products of the transformations are resistant to biodegradation, then the processes may be qualitatively and diagenetically important even if they are relatively infrequent or slow. The higher molecular weight materials may be an intermediate product in the overall transformation from DOC to POC and may not accumulate in the ocean or in sediments. They do however have potential importance in determining surface rheological properties, and they may represent an important pathway for the incorporation of atmospheric organic contaminants into oceanic organic pools.

This discussion has focused on interaction among microlayer organic materials. As mentioned earlier, oceanic surface microlayers also include dissolved inorganic components, some of which may be highly enriched and diagenetically important. Trace metals and other inorganic cations can serve as bridges or otherwise facilitate interactions between organic molecules, as in the formation of metal–organic complexes (e.g., Mills *et al.*, 1982) or the involvement of magnesium in the gelation of ionomeric polymers (Broze *et al.*, 1981). Considerable attention has been given to the scavenging of trace metals by surface-active organic materials in the oceans (e.g., Wallace, 1982), but little attention has been directed toward the associated possibilities that metals induce interactions among surface-active materials. Laboratory studies indicate that silicic acid polymers in solution can affect the surface properties of protein monolayers (Minones *et al.*, 1973), but there have not been similar investigations of interactions among inorganic and organic components at the ocean surface. There are also organisms at the ocean surface, especially in slicks. Some of these organisms can move into and out of surface microlayers. Others may arrive passively and be trapped until they decompose, are eaten, or are incorporated into sinking particles. These organisms may be important sources of reactive organic materials, and they may play roles in various transformations. Just as at the sediment–water interface, diagenetic processes at the air–water interface include important biological factors.

5. Photochemical Alterations

There are several good recent reviews of marine photochemistry. Photochemical alterations are mentioned here only to point out the obvious but as yet little explored role of photochemistry at the air–sea interface and to present some laboratory evidence suggesting that photochemical alterations in surface microlayers may be very different from photochemical processes in the upper ocean.

A book edited by Zika and Cooper (1987) and a review resulting from the 1987 CHEMRAWN Workshop (Waite *et al.*, 1988) provide clear up-to-date evidence of the importance of photochemical processes in the upper ocean. Much effort has been directed toward detecting low-molecular-weight products such as gases or aldehydes in bulkwater, while the overall DOC degradation rates and any residual organic materials from such photodegradations remain largely unknown. It seems that investigators have not yet explored ocean surface microlayers, where there is high light intensity, especially at high-energy low wavelengths, and a relative abundance of potential absorbers and reactants. If photochemical processes are important in global nitrogen, sulfur, or iodide cycles, in air–sea exchange of reactive species, or in degrading oceanic organic carbon, then it might be informative to look for evidence of such processes in what are likely to be the most reactive sites. Laboratory evidence suggests that interfacial photochemical reactions may be much different from reactions in bulk solution, for the reasons already mentioned: abundance of organic materials, constraints on movement, and unusual surface configurations.

Wheeler (1972) noted that unconjugated fatty acids on the surface of artificial seawater developed conjugation when they were exposed to solar levels or UV irradiation. That work and the laboratory investigations of photoinduced lipid oxidation and cross-linking by Harvey *et al.* (1983), along with studies of photodegradation of films of contaminants such as petroleum (Hansen, 1975) or alkylbenzenes (Ehrhardt and Petrick, 1984), seem to be the only work related to oceanic surface microlayers. It is not clear from any of those works that the photochemical reactions described would only occur in surface microlayers or would be especially effective in surface microlayers. Laboratory evidence from non-seawater systems suggests that photochemical reactions that occur slowly or rarely in bulk solution can occur quickly and effectively at interfaces. Whitten *et al.* (1977) demonstrated several photochemical reactions within and between organic molecules, including *cis–trans* isomerization, ligand photoejection, and dimerization, that were enhanced in monolayer films because of unique orientation and close packing of the substrates at the interface. In related work, Horsey and Whitten (1978) observed unique photodegradation products from protoporphyrin when the protoporphyrin was spread in a monolayer film as the dihydrocholesterol ester. The degradation products from the monolayer molecules were related to the unique geometry of the protoporphyrin molecules at the interface.

As mentioned before, the types and abundances of molecules in laboratory films are not likely to be representative of the organic chemical composition of the ocean surface. However, many of the same interfacial effects obtain: molecules may be relatively more abundant, may assume surface configurations, and may be uniquely reactive toward other surface molecules or toward molecules in bulk solution. Our understanding of marine photochemistry would probably be different, and might be improved, if we investigated photochemical processes in oceanic surface microlayers. We at least need to test the assumptions that rates of photodegradation of natural compounds or contaminants which accumulate in slicks at the ocean surface can be calculated by extrapolating bulk solution reaction rates to surface light intensities (Zepp *et al.*, 1975).

6. Summary and a Special Case

The surface microlayer occupies the upper 100 μm of the ocean, an average of 3800 meters above the ocean sediments. That is a long distance, vertically and conceptually, between the surface and the sediments, and it is not obvious that organic reactions and interactions at the surface have any importance with regard to the type, amount, or information content of organic carbon buried in the sediments. This chapter has suggested three processes—transformation from dissolved to particulate carbon, transformation from small to larger molecules, and photoalteration—which might affect the organic chemical composition of the ocean surface. If the products of any of these processes sink, and survive the descent, they could contribute to the sedimentary organic reservoir and record. Quantitatively, the cumulative effect of surface interfacial processes on sedimentary organic carbon composition or flux seems likely to be small. However, sedimentary organic geochemists are expert at finding large amounts of information in a few organic molecules, and it is not inconceivable that the products of reactions unique to organically enriched, compressed, highly irradiated surface slicks could be found and interpreted in sediment records. The largest barrier to obtaining such information seems to be knowing what products to look for, that is, knowing the reactions at the ocean surface. Because oceanic surface microlayers contain complex mixtures of dissolved and particulate organic and inorganic materials, and biota, understanding ocean surface reactions will require simultaneous attention to dissolved organic chemistry, heterogeneous interactions, and biological processes.

In intertidal zones over beaches, mudflats, or marshes, surface microlayers periodically come in direct contact with sediments. Slicks are perhaps more frequent in these relatively calm regions, so that the interaction between sediment and water surface could often include a highly enriched surface microlayer. Foam, drift algae, and other less pleasant debris on beaches or at the high-tide swash line in marshes and mudflats are visible evidence of deposition directly from the ocean surface. Storm events may erode any surface microlayer materials deposited, but deposition must on the average exceed erosion or else the sedimentary environments would not persist. Most of the surface microlayer carbon might be welcome fodder for sediment organisms, but recalcitrant products of some of the unique microlayer diagenetic processes described above might accumulate and leave a recognizable signal. The distance between surface microlayers and intertidal sediments seems, physically at least, a much easier gap to close than the distance between surface microlayers and deep ocean sediments. Finally, although attention has recently been diverted to hydrothermal systems, it is still true that many of the elements thought necessary for the development of life—thin membrane-like layers of oriented organic molecules, biopolymers in reactive configurations, clay mineral templates, and high UV irradiation—can be present when surface microlayers interact with intertidal sediments. It may be that current surface microlayer diagenetic processes are the faint descendants of momentous ancient reactions.

ACKNOWLEDGMENTS. Many of my investigations into surface microlayer chemistry have been supported by the Office of Naval Research. The ideas in this chapter were influenced by discussion with Pete Williams and Nelson Frew, even though they may not concur with all that I have written.

References

Adamson, A. W., 1982, *Physical Chemistry of Surfaces*, 4th ed., John Wiley & Sons, New York, p. 470.

Bacon, M. P., and Elzerman, A. W., 1980, Enrichment of ^{210}Pb and ^{210}Po in the sea-surface microlayer, *Nature* **284**:332–334.

Barker, D. R., and Zeitlin, H., 1972, Metal-ion concentrations in sea-surface microlayer and size-separated aerosol samples in Hawaii, *J. Geophys. Res.* **77**:5076–5086.

Bezdek, H. F., and Carlucci, A. F., 1974, Concentration and removal of liquid microlayers from a seawater surface by bursting bubbles, *Limnol. Oceanogr.* **19**:126–132.

Biddleman, T. G., and Olney, C. E., 1974, Chlorinated hydrocarbons in the Sargasso-Sea atmosphere and surface water, *Science* **183**:516–518.

Bilkadi, Z., and Neuman, R. D., 1979, Effect of humidity of monolayer desorption at the air–water interface, *Nature* **278**:842.

Bishop, J. K. B., Collier, R. W., Kettens, D. R., and Edmond, J. M., 1980, The chemistry, biology, and vertical flux of particulate matter from the upper 1500 m of the Panama Basin, *Deep-Sea Res.* **27**:615–640.

Blanchard, D. C., and Woodcock, A. H., 1957, Bubble formation and modification in the sea and its meteorological significance, *Tellus* **9**:145–158.

Boehm, P. D., and Quinn, J. G., 1973, Solubilization of hydrocarbons by the dissolved organic matter in sea water, *Geochim. Cosmochim. Acta* **37**:2459–2477.

Brewer, P. G., Nozaki, Y., Spencer, D. W., and Fleer, A. P., 1980, Sediment trap experiments in the deep North Atlantic: Isotopic and elemental abundances, *J. Mar. Res.* **38**:703–728.

Brophy, J. E., and Carlson, D. J., 1989, Production of biologically-refractive dissolved organic carbon by natural seawater microbial populations, *Deep-Sea Res.* **36**:497–507.

Brownawell, B. J., and Farrington, J. W., 1985, Partitioning of PCB's in marine sediments, in: *Marine and Estuarine Geochemistry* (A. C. Sigleo and A. Hattori, eds.), Lewis Publishers, Chelsea, Michigan, pp. 97–120.

Broze, G., Jerome, R., and Teyssie, Ph., 1981, Halato-telechelic polymers. 1. Gel formation and its dependence on the ionic content, *Macromolecules* **14**:224–225.

Carlson, D. J., 1982a, A field evaluation of plate and screen microlayer sampling techniques, *Mar. Chem.* **11**:189–208.

Carlson, D. J., 1982b, Surface microlayer phenolic enrichments indicate sea surface slicks, *Nature* **296**:426–429.

Carlson, D. J., 1982c, Phytoplankton in marine surface microlayers, *Can. J. Microbiol.* **28**:1226–1234.

Carlson, D. J., 1983, Dissolved organic materials in surface microlayers: Temporal variability and relation to sea state, *Limnol. Oceanogr.* **28**:415–431.

Carlson, D. J., 1987, Viscosity of sea-surface slicks, *Nature* **329**: 823–825.

Carlson, D. J., Mayer, L. M., Brann, M. L., and Mague, T. H., 1985, Binding of monomeric organic compounds to macromolecular dissolved organic matter in seawater, *Mar. Chem.* **16**:141–153.

Carlson, D. J., Cantey, J. L., and Cullen, J. J., 1988, Description of and results from a new surface microlayer sampling device, *Deep-Sea Res.* **35**:1205–1213.

Cullen, J. J., McIntyre, H. L., and Carlson, D. J., 1989, Distributions and photosynthesis of phototrophs in sea-surface films. *Mar. Ecol. Prog. Ser.* **55**:271–278.

Daumas, R. A., Laborde, P. L., Marty, J. C., and Saliot, A., 1976, Influence of sampling method on the chemical composition of the water surface film, *Limnol. Oceanogr.* **21**:319–326.

Dawson, R., and Duursma, E. K., 1981, State of the art, in: *Marine Organic Chemistry* (E. K. Duursma and R. Dawson, eds.), Elsevier, Amsterdam, pp. 497–512.

Degens, E. T., 1970, Molecular nature of organic nitrogen compounds in seawater and recent marine sediments, in: *Organic Matter in Natural Waters* (D. W. Hood, ed.), University of Alaska, Fairbanks, pp. 77–106.

Dietz, A. S., Albright, L. J., and Tuominen, T., 1976, Heterotrophic activities of bacterioneuston and bacterioplankton, *Can. J. Microbiol.* **22**:1699–1709.

Dragcevic, D., Vukovic, M., Cukman, D., and Pravdic, V., 1979, Properties of the seawater–air interface. Dynamic surface tension studies, *Limnol. Oceanogr.* **24**:1022–1030.

Duce, R. A., Quinn, J. G., Olney, C. E., Piotrowicz, S. R., Ray, B. J., and Wade, T. L., 1972, Enrichment of heavy metals and organic compounds in the surface microlayer of Narragansett Bay, Rhode Island, *Science* **176**:161–164.

Ehrhardt, M., and Petrick, G., 1984, On the sensitized photo-oxidation of alkylbenzenes in seawater, *Mar. Chem.* **15**:47–58.

Eisenreich, S. J., 1982, Overview of atmospheric inputs and losses from films, *J. Great Lakes Res.* **8**:241–242.

Elliot, H. A., and Huang, C., 1980, Adsorption of some copper (II)–amino acid complexes at the solid–solution interface. Effect of ligand and surface hydrophobicity, *Environ. Sci. Technol.* **14**: 87–93.

Elzerman, A. W., and Armstrong, D. E., 1979, Enrichment of Zn, Cd, Pb, and Cu in the surface microlayer of Lakes Michigan, Ontario, and Mendota, *Limnol. Oceanogr.* **24**:133–144.

Ferguson, J., Hudson, N. E., Warren, B. C. H., and Tomatarian, A., 1987, Phase changes during elongational flow of polymer solutions, *Nature* **325**:234–236.

Gagosian, R. B., and Lee, C., 1981, Processes controlling the distribution of biogenic organic compounds in seawater, in: *Marine Organic Chemistry* (E. K. Duursma and R. Dawson, eds.), Elsevier, Amsterdam, pp. 91–123.

Gershey, R. M., 1983a, Characterization of seawater organic matter carried by bubble-generated aerosols, *Limnol. Oceanogr.* **28**: 309–319.

Gershey, R. M., 1983b, A bubble adsorption device for the isolation of surface-active organic matter in seawater, *Limnol. Oceanogr.* **28**:395–400.

Gershfeld, N. L., 1982, The liquid condensed/liquid expanded transition in lipid films: A critical analysis of the film balance experiment, *J. Colloid Interface Sci.* **85**:28–40.

Gonzalez, G., and MacRitchie, F., 1970, Equilibrium adsorption of proteins, *J. Colloid. Interface Sci.* **32**:55–61.

Gordon, D. C., Keizer, P. D., and Aldous, D. G., 1976, Fate of crude oil spilled on seawater contained in outdoor tanks, *Environ. Sci. Technol.* **10**:580–585.

Graham, W. F., Piotrowicz, S. R., and Duce, R. A., 1979, The sea as a source of atmospheric phosphorus, *Mar. Chem.* **7**:325–342.

Hansen, H. P., 1975, Photochemical degradation of petroleum hydrocarbon surface films on seawater, *Mar. Chem.* **3**:183–195.

Hardy, J. T., Apts, C. W., Crecelius, E. A., and Bloom, N. S., 1985a, Sea-surface microlayer metals enrichments in an urban and rural bay, *Estuarine Coastal Shelf Sci.* **20**:299–312.

Hardy, J. T., Apts, C. W., Crecelius, E. A., and Fellingham, G. W., 1985b. The sea-surface microlayer: Fate and residence times of atmospheric metals, *Limnol. Oceanogr.* **30**:93–101.

Harvey, G. R., Boran, D. A., Chesal, L. A., and Tokar, J. M., 1983, The structure of marine humic acids, *Mar. Chem.* **12**:119–132.

Harvey, R. W., Lion, L. W., Young, L. Y., and Leckie, J. O., 1982, Enrichment and association of lead and bacteria at particulate surfaces in a salt-marsh surface layer, *J. Mar. Res.* **40**:1201–1211.

Hassett, J. P., and Anderson, M. A., 1979, Association of hydrophobic organic compounds with dissolved organic matter in aquatic systems, *Environ. Sci. Technol.* **13**:1526–1529.

Hedges, J. I., 1978, The formation and clay mineral reactions of melanoidins, *Geochim. Cosmochim. Acta* **42**:69–86.

Henrichs, S. M., and Williams, P. M., 1985, Dissolved and particulate amino acids and carbohydrates in the sea surface microlayer, *Mar. Chem.* **17**:141–163.

Heyraud, M., and Cherry, R. D., 1983, Correlation of ^{210}Pb and ^{210}Po enrichments in the sea-surface microlayer with neuston biomass, *Cont. Shelf Res.* **1**:283–293.

Hoffman, E. J., and Duce, R. A., 1976, Factors influencing the organic carbon content of marine aerosols: A laboratory study, *J. Geophys. Res.* **81**:3667–3670.

Honjo, S., Manganini, S. J., and Cole, J. J., 1982, Sedimentation of biogenic matter in the deep ocean, *Deep-Sea Res.* **29**:609–625.

Horsey, B. E., and Whitten, D. G., 1978, Environmental effects on photochemical reactions: Contrasts in the photooxidation behavior of protoporphyrin IX in solution, monolayer films, organized monolayer assemblies, and micelles, *J. Am. Chem. Soc.* **100**:1293–1295.

Hunter, K. A., 1980, Processes affecting particulate trace metals in the sea surface microlayer, *Mar. Chem.* **9**:49–70.

Hunter, K. A., and Liss, P. S., 1981, Organic sea surface films, in: *Marine Organic Chemistry* (E. K. Duursma and R. Dawson, eds.), Elsevier, Amsterdam, pp. 259–298.

Ichikawa, T., 1980, Particle formation from dissolved material by the use of a continuous harvesting system, *Rep. Fac. Sci., Kagoshima Univ.* **13**:131–144.

Jarvis, N. L., Garrett, W. D., Scheiman, M. A., and Timmons, C. O., 1967, Surface chemical characterization of surface-active material in seawater, *Limnol. Oceanogr.* **12**:88–96.

Johnson, B. D., and Cooke, R. C., 1981, Generation of stabilized microbubbles in seawater, *Science* **213**:209–211.

Johnson, B. D., and Wangersky, P. J., 1987, Microbubbles: Stabilization by monolayers of adsorbed particles, *J. Geophys. Res.* **92**: 14641–14647.

Khundzhua, G. G., Gusev, A. M., Andreyev, Y. G., Gurov, V. V., and Skorohkvatov, N. A., 1977, Structure of the cold surface film of the ocean and heat transfer between the ocean and the atmosphere, Izv. *Atmos. Ocean. Phys.* **13**(7):506–509.

Lee, R. F., and Williams, P. M., 1974, Copepod "slick" in the northwest Pacific Ocean, *Naturwissenschaften* **61**:505–506.

Lyons, W. B., Pybus, M. J. S., and Coyne, J., 1980, The seasonal variation in the nutrient chemistry of the surface microlayer of Galway Bay, Ireland, *Oceanol. Acta* **3**:151–155.

MacRitchie, F., 1963, Collapse of protein monolayers, *J. Colloid Sci.* **18**:555–561.

MacRitchie, F., 1986, Spread monolayers of proteins, *Adv. Colloid Interface Sci.* **25**:341–385.

Marty, J. C., Saliot, A., Buat-Menard, P., Chesselet, R., and Hunter, K. A., 1979, Relationship between the lipid compositions of marine aerosols, the sea surface microlayer, and subsurface water, *J. Geophys Res.* **84**:5707–5716.

McAlister, E. D., and McLeish, W., 1969, Heat transfer in the top millimeter of the ocean, *J. Geophys. Res.* **74**:3408–3414.

McAlister, E. D., and McLeish, W., 1970, A radiometer system for measurement of total heat flow from the sea surface, *Appl. Opt.* **9**:2697–2705.

McLeish, W., and Putland, G. E., 1975, Measurements of wind-driven flow profiles in the top millimeter of water, *J. Phys. Oceanogr.* **5**:516–518.

Means, J. C., and Wijayaratne, R., 1982, Role of natural colloids in the transport of hydrophobic pollutants, *Science* **215**: 968–970.

Mills, G. L., Hanson, A. K., Quinn, J. G., Lammela, W. R., and Chasteen, N. D., 1982, Chemical studies of copper–organic complexes isolated from estuarine waters using C-18 reverse-phase liquid chromatography, *Mar. Chem.* **11**:355–377.

Minones, J., Fernandez, S. G., Irigarnegaray, E., and Pedrero, P. S., 1973, The interaction of silicic acid with protein monolayers: Effect of pH and ionic strength of substrate, *J. Colloid. Interface Sci.* **42**:503–515.

Pattenden, N. J., Cambray, R. S., and Playford, K., 1981, Trace and major elements in the sea-surface microlayer, *Geochim. Cosmochim. Acta* **45**:93–100.

Peters, J., and Heller, W., 1970, Mechanical and surface coagulation. II. Coagulation by stirring of α-FeOOH sols, *J. Colloid Interface Sci.* **33**:578–585.

Platford, R. F., 1982, Pesticide partitioning in artificial surface films, *J. Great Lakes Res.* **8**:307–309.

Ray, A., Reynolds, J. A., Polet, H., and Steinhardt, J., 1966, Binding of large organic anions and neutral molecules by native bovine serum albumin, *Biochemistry* **5**:2606–2616.

Reynolds, J., Herbert, S., and Steinhardt, J., 1968, The binding of some long-chain fatty acid anions and alcohols by bovine serum albumin, *Biochemistry* **7**:1357–1361.

Rice, C. P., Meyers, P. A., and Brown, G. S., 1983, Role of surface microlayers in the air–water exchange of PCBs, in: *Physical Behavior of PCBs in the Great Lakes* (D. Mackay, S. Paterson, S. J. Eisenreich, and S. S. Milagros, eds.), Ann Arbor Science, Ann Arbor, Michigan, pp. 157–179.

Schulman, J. H., and Rideal, E. K., 1937a, Molecular interactions in monolayers: II—The action of haemolytic and agglutinizing agents on lipo-protein monolayers, *Proc. R. Soc. London, Ser. B* **122**:46–57.

Schulman, J. H., and Rideal, E. K., 1937b, Molecular interactions in monolayers: I—Complexes between large molecules, *Proc. R. Soc. London, Ser. B* **122**:29–45.

Seba, D. B., and Corcoran, E. F., 1969, Surface slicks as concentrators of pesticides in the marine environment, *Pestic. Monit. J.* **3**:190–193.

Sieburth, J. McN., Willis, P. J., Johnson, K. M., Burney, C. M., Lavoie, D. M., Hinga, K. R., Caron, D. A., French, F. W., Johnson, P. W., and Davis, P. G., 1976, Dissolved organic matter and heterotrophic microneuston in surface microlayers of the North Atlantic, *Science* **194**:1415–1418.

Sugimura, Y., and Suzuki, Y., 1988, A high temperature catalytic oxidation method of non-volatile organic carbon in seawater by direct injection of liquid sample, *Mar. Chem.* **24**:105–131.

Sullivan, K. F., Atlas, E. L., and Giam, C.-S., 1982, Adsorption of phthalic acid esters from seawater, *Environ. Sci. Technol.* **16**: 428–432.

Sutcliffe, W. H., Sheldon, R. W., Prakash, A., and Gordon, D. C., 1971, Relations between wind speed, Langmuir circulation, and particle concentration in the ocean, *Deep-Sea Res.* **18**: 639–643.

Waite, T. D., Sawyer, D. T., and Zafiriou, O. C., 1988, Oceanic reactive chemical transients, *Appl. Geochem.* **3**:9–17.

Wallace, G. T., 1982, The association of copper, mercury and lead with surface-active organic matter in coastal seawater, *Mar. Chem.* **11**:379–394.

Wallace, G. T., and Duce, R. A., 1975, Concentration of particulate trace metals and particulate organic carbon in marine surface waters by a bubble flotation mechanism, *Mar. Chem.* **3**: 157–181.

Wheeler, J. R., 1972, Some effects of solar levels of ultraviolet radiation on lipids in artificial seawater, *J. Geophys. Res.* **77**: 5302–5306.

Wheeler, J. R., 1975, Formation and collapse of surface films, *Limnol. Oceanogr.* **20**:338–342.

Whitehouse, B., 1985, The effects of dissolved organic matter on the aqueous partitioning of polynuclear aromatic hydrocarbons, *Estuarine Coastal Shelf Sci.* **20**:393–402.

Whitten, D. G., Hopf, F. R., Quina, F. H., Sprintschnik, G., and Sprintschnik, H. W., 1977, Photochemistry of organic chromophores incorporated into fatty acid monolayers, *Pure Appl. Chem.* **49**:379–388.

Williams, P. M., and Robertson, K. J., 1975, Chlorinated hydrocarbons in sea-surface films and subsurface waters at nearshore stations and in the north central Pacific gyre, *Fish Bull.* **73**: 445–447.

Williams, P. M., Van Vleet, E. S., and Booth, C. R., 1980, *In situ* measurements of sea-surface film potentials, *J. Mar. Res.* **38:** 193–204.

Williams, P. M., Carlucci, A. F., Henrichs, S. M., Van Vleet, E. S., Horrigan, S. G., Reid, F. M. H., and Robertson, K. J., 1986, Chemical and microbiological studies of sea-surface films in the southern Gulf of California and off the west coast of Baja California, *Mar. Chem.* **19:**17–98.

Wu, T. L., Lambert, L., Hastings, D., and Banning, D., 1980, Enrichment of the agricultural herbicide atrazine in the microsurface water of an estuary, *Bull. Environ. Contam. Toxicol.* **24:**411–414.

Zepp, R. G., Wolfe, N. L., Gordon, J. A., and Baughman, G. L., 1975, Dynamics of 2,4-D esters in surface water, *Environ. Sci. Technol.* **9:**1144–1150.

Zika, R. G., and Cooper, W. J. (eds.), 1987, *Photochemistry of Environmental Aquatic Systems*, ACS Symposium Series No. 327, American Chemical Society, Washington, D. C.

III

Kerogen and Related Materials

Chapter 13

Determination of Structural Components of Kerogens by the Use of Analytical Pyrolysis Methods

S. R. LARTER and B. HORSFIELD

1. Introduction

Kerogen, defined as organic material in sedimentary rocks that is insoluble in common organic solvents, alkali, and nonoxidizing acids, represents the major organic carbon reservoir in the Earth's crust. The chemical composition of kerogen is a complex function of its biochemical source-related mechanical composition (maceral composition) and diagenetic modification (both facies related) and its degree of thermal evolution (maturity level). Bulk chemical analysis has been most useful in providing generalized chemical descriptions of kerogens. However, the detailed molecular configuration of a kerogen cannot be obtained from its bulk chemistry due to the isomeric complexity possible with high-molecular-weight macromolecules. Many sophisticated chemical degradation (including pyrolysis) and spectroscopic techniques have therefore been applied to the structural characterization of kerogen (see Durand, 1980). Some of the techniques most commonly applied today are those involving analytical pyrolysis methods, and it is those we discuss here.

Analytical pyrolysis, for the purposes of this chapter, involves the on-line coupling of an inert atmosphere (He or N_2) pyrolyzer with a gas chromatograph (Py-GC) or with a mass spectrometer either directly (Py-MS) or via a GC column (Py-GC/MS) (Larter, 1984; Horsfield, 1984). This chapter attempts to draw together the core literature relevant to the utilization of analytical pyrolysis as a kerogen structural analysis tool, that is, as a tool suitable for the description of a kerogen in terms of the quantitative distribution of various molecular types in the kerogen macromolecule. The reader is referred to earlier review articles for a discussion of qualitative analytical pyrolysis studies of kerogens (Meuzelaar *et al.*, 1982; Philip, 1982; Larter and Douglas, 1982; Horsfield, 1984; Larter, 1984). Throughout the chapter, we stress the fact that kerogen is a polymaceralic material of variable mechanical as well as chemical composition, and therefore our kerogen structural models need to take this factor into account. We also attempt to apply our understanding of kerogen chemical composition to assess the relative contributions made to specific kerogens by direct selective preservation of biopolymers, as opposed to kerogen formation through chemical condensation processes.

S. R. LARTER • Newcastle Research Group in Fossil Fuels & Environmental Geochemistry, University of Newcastle, Newcastle upon Tyne NE1 7RU, England. B. HORSFIELD • Institute of Petroleum and Organic Geochemistry (ICG-4), KFA-Jülich GmbH, D-5170 Jülich, Germany.
Organic Geochemistry, edited by Michael H. Engel and Stephen A. Macko. Plenum Press, New York, 1993.

2. Elucidation of Kerogen Composition by Pyrolysis

Concerns regarding the reproducibility of Py-GC have not been substantiated, and it appears from review of well over a decade of pyrolytic data that kerogens, already substantially heroroatom depleted during diagenesis and maturation, are much less sensitive to pyrolyzer and sample environment than heteroatom-rich biopolymers. Providing adequate care is given to rapid product removal, a wide range of pyrolyzers will provide very similar and reproducible qualitative results (Larter, 1984; Horsfield and Larter, 1981). In the evolution of this area of study, coupled use of Py-GC/MS and Py-MS techniques has proven most beneficial (Maters *et al.*, 1977; Larter, 1984, and references therein), but to date most quantitative data are associated with Py-GC studies. As structural determination requires both molecular identification of pyrolysis products and knowledge of their relationship to precursor structures, plus determination of the representativity of structures with respect to a kerogen compositional model, we stress quantitative rather than qualitative observations. The vast bulk of modern, kerogen concentrate, analytical pyrolyses have been performed using two main types of pyrolyzer:

(a) Filament pyrolyzers, either resistively or inductively heated, with isothermal pyrolyses of submilligram samples in the temperature range 600–800°C for several seconds. Quantitation has, to date, used polymer-based internal quantitation methodologies (Larter and Senftle, 1985).

(b) Furnace pyrolyzers which can pyrolyze larger samples either isothermally or, more typically, under programmed temperature conditions over the range 300–800°C at heating rates usually similar to those of a Rock-Eval instrument (ca. 25°C min^{-1}). Quantitation is somewhat easier as the milligram size samples used in this approach can be readily weighed, and both internal and external quantitation have been used (Horsfield *et al.*, 1989a).

Results from both systems appear to be qualitatively and quantitatively quite comparable.

Py-GC-MS has enabled the identification of a great many components in kerogen pyrolysates including light hydrocarbon gases (Giraud, 1970), inorganic gases (Larter, 1984; Eglinton *et al.*, 1988), alkylaromatic and aliphatic hydrocarbons, phenols, acids, and oxygen-, nitrogen-, and sulfur-bearing heterocycles (Maters *et al.*, 1977; Van Graas *et al.*, 1980, 1981; Larter and Douglas, 1978; Van de Meent *et al.*, 1980a; Philp *et al.*, 1982a, b; Meuzelaar *et al.*, 1982, and references therein; Sinninghe-Damsté *et al.*, 1989).

Depending on the aromaticity of the kerogens (a function of source type and maturity level), the proportions of these different components in the pyrolysates change systematically, with highly aromatic kerogens (vitrinite- and inertinite-rich kerogens) having pyrolysates relatively rich in aromatic components such as phenols and alkyaromatic hydrocarbons, and aliphatic carbon- and hydrogen-rich kerogens (liptinite-rich kerogens) having relatively aliphatic hydrocarbon-rich pyrolysates (Horsfield, 1984; Larter, 1984) in terms of the GC-amenable fraction. The absolute yields of kerogen pyrolysis products of all types, however, are simply a function of the overall aliphatic carbon and hydrogen content of the kerogen. Thus, type I and type II kerogens, rich in carbon-bound hydrogen, produce higher absolute yields on pyrolysis of both aliphatic and aromatic hydrocarbons than type III kerogens, despite the higher aromaticity of the latter (Zilm *et al.*, 1981; Miknis *et al.*, 1982; Larter and Senftle, 1985; see Fig. 1 and discussion below). However, only general relationships between kerogen structure and pyrolysis products are known for most of these species, and the quantitative relevance of many of these components to kerogen structural determination has not been determined. Based on the results of many workers, we consider that the major identified (GC-resolvable) and readily quantifiable pyrolysis products of kerogens under pyrolysis–gas-chromatographic (Py-GC) conditions are saturated and unsaturated normal, branched, and cyclic aliphatic hydrocarbons in the carbon number range C_1–C_{35}; alkylated one- to three-ring aromatic and naphthenoaromatic hydrocarbons with side chains from C_1 to C_{30+}; and alkylphenols with various alkyl, methoxyl, and vinyl substituents [many references referred to in Larter (1984) and Horsfield (1984)]. [The large amounts of polar, relatively high molecular weight material that are also produced on analytical pyrolysis, the major pyrolysis products of kerogens (see below), have not yet been characterized systematically and will be discussed elsewhere.]

2.1. Aliphatic Hydrocarbons

Many studies have indicated that the dominant source of aliphatic hydrocarbons in kerogen pyrolysates is from scission of lipid-derived elements in the kerogen macromolecule. The dominant species iden-

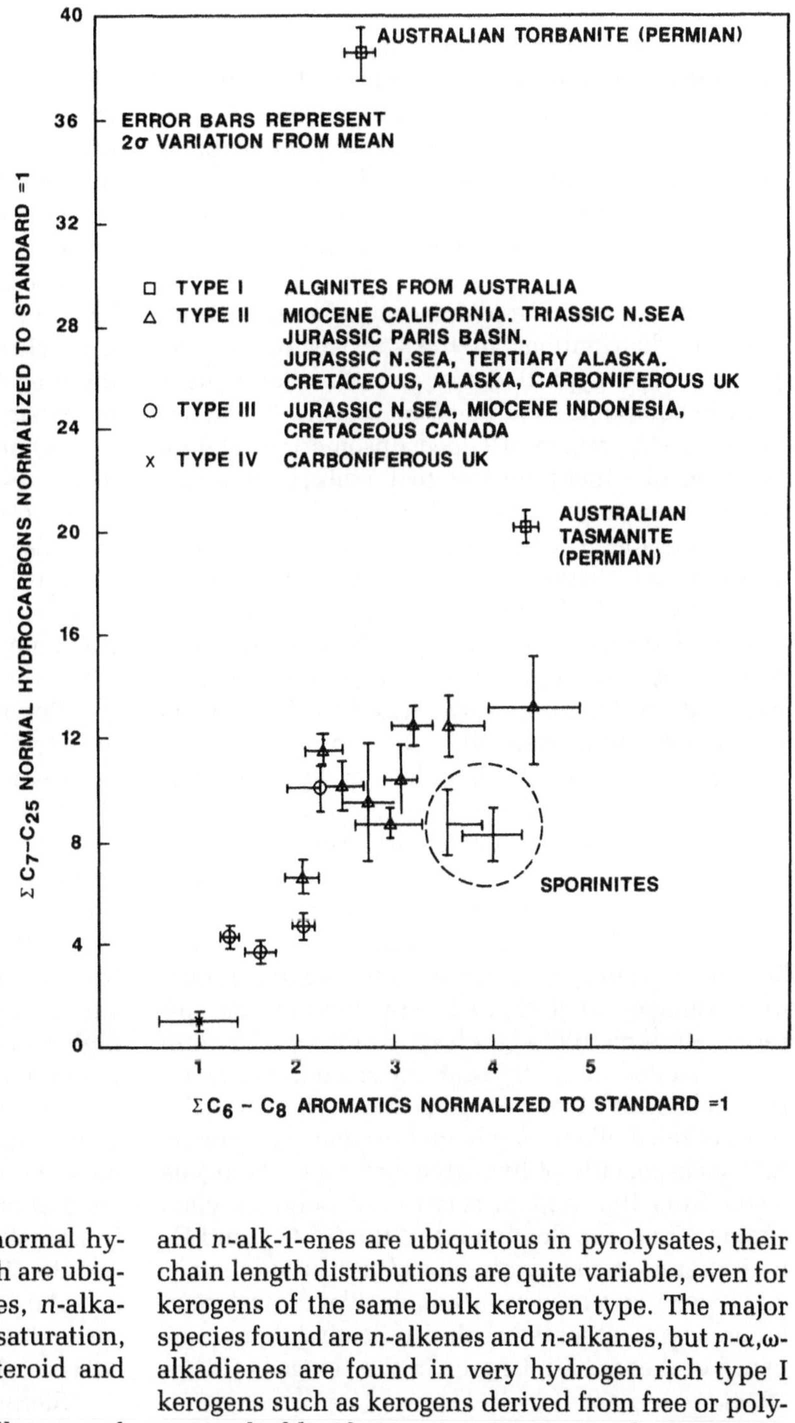

Figure 1. Yields of normal hydrocarbons and alkylaromatic hydrocarbons as a function of kerogen type for 800°C/20-s flash pyrolysates of kerogens. One standard unit approximately equals 0.25% by weight of kerogen. Quantitative data obtained by automatic peak recognition/peak area determination. [After Larter and Senftle (1985).]

tified to date in kerogen pyrolysates are normal hydrocarbons (*n*-alkanes, *n*-alk-1-enes), which are ubiquitous components of kerogen pyrolysates, *n*-alkadienes with terminal and mid-chain unsaturation, acyclic isoprenoid alkenes, and cyclic steroid and terpenoid alkanes and alkenes.

Normal hydrocarbons, particularly *n*-alkanes and alk-1-enes, represent the visually dominant species in pyrograms of all kerogens with the exception of very hydrogen deficient inertinitic or coal vitrain vitrinite kerogens, resinite kerogens (Snowdon, 1980; Horsfield *et al.*, 1983; Dembicki *et al.*, 1983; Larter and Senftle, 1985), or in unusual, possibly radiation-damaged, kerogens from the Cambrian Alum Shale, Scandinavia (Bharati *et al.*, 1992), where these species are in very low concentrations. Although *n*-alkanes and *n*-alk-1-enes are ubiquitous in pyrolysates, their chain length distributions are quite variable, even for kerogens of the same bulk kerogen type. The major species found are *n*-alkenes and *n*-alkanes, but *n*-α,ω-alkadienes are found in very hydrogen rich type I kerogens such as kerogens derived from free or polymeric algal lipids (Larter, 1984; Nip *et al.*, 1986; Goth *et al.*, 1988). Terminal unsaturation in *n*-hydrocarbons is produced by β scission of radicals during pyrolysis; nonterminal unsaturation is probably indicative of structures within the kerogens containing relict olefinic sites. The general resistance of kerogens to saponification and the presence of long-chain *n*-hydrocarbon-producing structures in even overmature kerogens (see Section 4.) or bomb pyrolysis residues (Eglinton, 1988) indicates to us that most alkyl fragments in kerogens are strongly bonded through C—C or C—O bonds. Further, the strong correlations between generation on pyrolysis of *n*-hydrocarbons of all chain lengths in a given kerogen [that is, long- and short-chain *n*-hydrocarbon are produced on pyrolysis by related processes (Øygard *et al.*, 1989)] strongly indicate a common source of these species, most probably linear, lipid-derived elements in the kerogens.

Odd-dominated *n*-alkane and even-dominated *n*-alkene distributions in the C_{22+} region of flash pyrolysates of type II kerogens with a land plant-derived component (Van de Meent *et al.*, 1980a) and in the C_9–C_{18} region of *Gloeocapsomorpha* alginite (Reed *et al.*, 1986) suggest that *n*-alkyl pyrolysate distributions are not controlled entirely by random chain scission. Further, model experiments (Larter and Douglas, 1980b; Larter *et al.*, 1983) suggest that proximity of alkyl groups in kerogen molecules to inherited or diagenetically related functional groups or aromatic rings may give rise on pyrolysis to alkyl fragments of length closely related to those of the incorporated (diagenetically or biosynthetically) parent lipid fragments. As bimolecular recombination reactions are not a major process in analytical pyrolysis, pyrolysate chain lengths represent minimum values for the alkyl chain lengths actually present in the kerogen. It is concluded that in most kerogens the chain length distributions seen in normal hydrocarbons in kerogen pyrolysates are to some extent related to in-kerogen alkyl chain length distributions preserved through to the pyrolysates or even into crude oils generated naturally from the same kerogens (cf. Horsfield, 1989), It must be remembered however, that some of the *n*-alkenes/*n*-alkanes produced during pyrolysis, especially of hydrogen-rich type I kerogens, result from the random scission of polymethylene chains. Thus, the dominance of the C_6, C_{10}, and C_{14} *n*-alk-1-enes in the pyrolysates of torbanite (cf. polyethylene) is governed by intramolecular radical transfer processes rather than being indicative of predominance of these chain lengths in the kerogens (Larter, 1978).

Acyclic isoprenoid hydrocarbons have been identified in many high-temperature kerogen pyrolysates (Maters *et al.*, 1977; Larter *et al.*, 1979), the principal isoprenoid hydrocarbon found being the C_{19} component prist-1-ene (Fig. 2) (Larter *et al.*, 1979), with a C_{14} regular isoprenoid alkene often being present (Larter, 1984). The origin of prist-1-ene is not completely defined, but model compound pyrolysis studies (Larter *et al.*, 1983, and references therein) indicated that the kerogen-bound precursor was not ester bound and was most probably bonded by a strong C—C or C—O bond. Goossens *et al.* (1984) subsequently showed that likely candidates for the prist-1-ene precursor in kerogens were kerogen-bound tocopherols (Fig. 2), although kinetic studies (Burnham, 1989; S. R. Larter and T. I. Eglinton, unpublished results) showed a spread of activation energies for catagenic pristene precursor loss, suggesting that a single kerogen-bound precursor is unlikely. Recently, we have found that phenol–phytol condensation products (chromans) bound into kerogens, possibly during diagenesis, may also be a source of pristene on pyrolysis (M. Li and S. R. Larter, unpublished results).

As prist-1-ene abundance in kerogen pyrolyzates decreases as pristane is generated in the subsurface (Van Graas *et al.*, 1981; Goossens *et al.*, 1988) or in artificial oils during artificial maturation (Burnham *et al.*, 1982; Larter, 1985; Eglinton, 1988), it is probably that the prist-1-ene precursors in flash pyrolysis are the same as those producing pristane in the subsurface.

Phytane precursors in kerogens are less well understood. No single abundant C_{20} hydrocarbon fragment is seen in high-temperature kerogen pyrolysates, although several phytenes have been identified in hydrous kerogen pyrolysates (Eglinton, 1988) and phytane is produced by hydrogenation of collected high-temperature kerogen pyrolysates (Behar and Pelet, 1985), It may also be possible that, if archaebacterial biphytanyl ether lipid structures are the subsurface catagenetic phytane source, these produce higher molecular weight ($\gg C_{20}$) pyrolysis products under analytical pyrolysis conditions.

Isoprenoid and other biomarker structures are substantially more labile during maturation than the bulk of the kerogen (Øygard *et al.*, 1989), the absolute yield of prist-1-ene decreasing from initial high values in immature kerogens to values near zero near peak oil retention (about 0.8% R_0; see Fig. 2 and Curry and Simpler, 1988). This quantifies observations made earlier by Van Graas *et al.* (1980) and Burnham *et al.*, (1982).

Steroid and terpenoid hydrocarbons have been observed in high-temperature pyrolysates of oil shale and coal kerogens by many authors (Gallegos, 1975, 1978; Siefert, 1978; Philp, 1982, Philp *et al.*, 1982a; Larter, 1984; Philp and Gilbert, 1985). Both classes of compounds in kerogen pyrolysates showed changes in isomerization at chiral centers in the hopanoid and steroid skeletons produced on pyrolysis that could be related to the maturity level of the kerogens (Philp and Gilbert, 1985; Behar and Pelet, 1985; Eglinton, 1988). Unfortunately, we do not know the quantitative

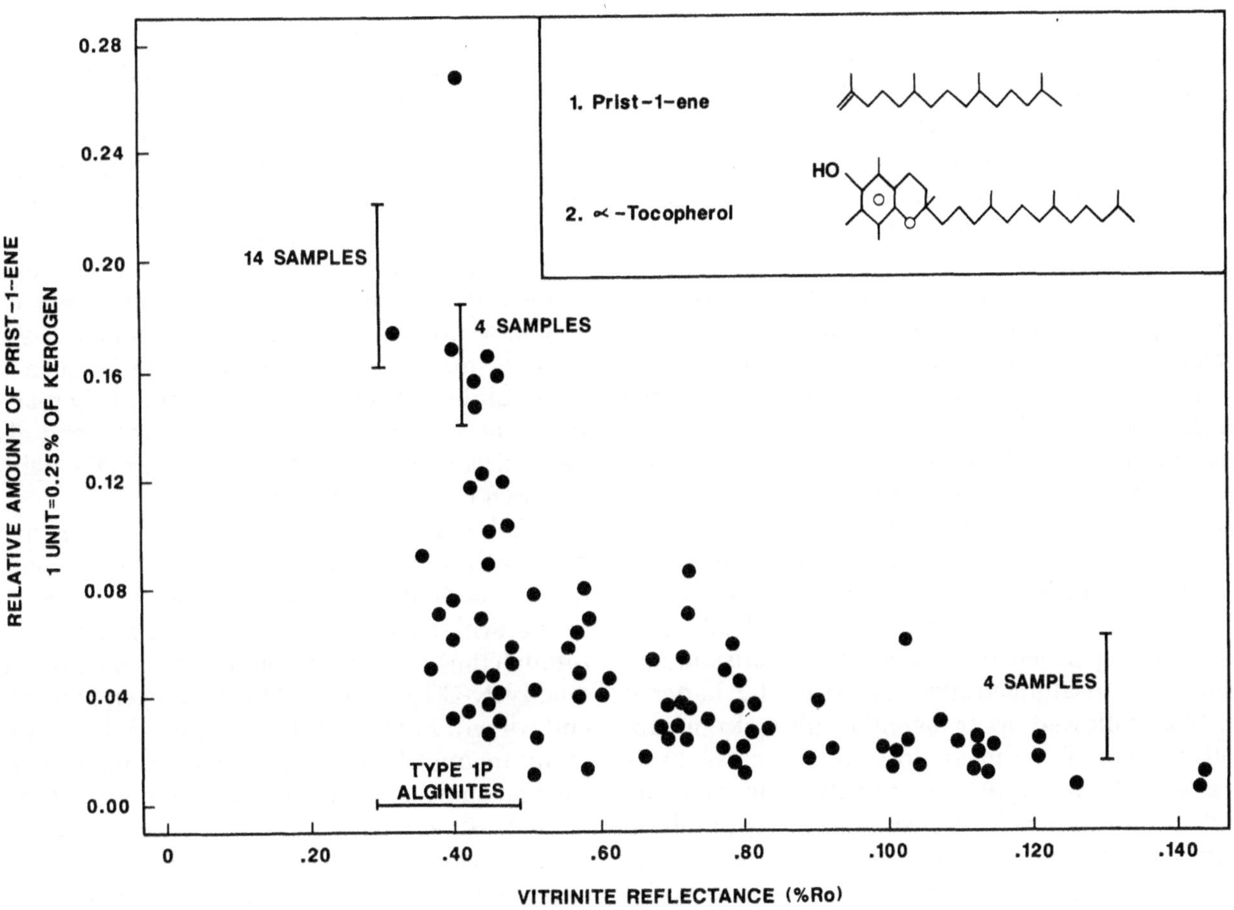

Figure 2. Prist-1-ene yields as a function of maturity for 800°C/2-s flash pyrolysates of kerogens [original data from Øygard *et al.* (1980)]. The structures of prist-1-ene (1) and α-tocopherol (2) are shown.

contribution to kerogens made by such biological marker structures, but analytical pyrolysis clearly indicates that these are present in kerogen macromolecules at some, probably low, level. This is also suggested by the work of Eglinton (1988), who obtained hopanoid and steroid hydrocarbon yields (on bomb pyrolysis at temperatures up to 375°C) on the order of 0.01–0.1% by weight of kerogen for type I and type II kerogens.

2.2. Aromatic Hydrocarbons

The common presence of extended-chain alkylaromatic hydrocarbons in kerogen pyrolysates (Larter *et al.*, 1978; Solli *et al.*, 1980a,b; Allan *et al.*, 1980; Baset *et al.*, 1980), especially *n*-alkylbenzenes with one long *n*-alkyl chain and two or more methyl groups on the aromatic ring and with similar carbon number distributions to the *n*-hydrocarbon distributions, suggests that many long alkyl groups are associated with aromatic species in kerogens. Cleavage of bonds β to the aromatic rings in long-chain alkylbenzene structures bonded to kerogens could readily produce *n*-hydrocarbons on catagenesis or pyrolysis. Cleavage of other alkyl bonds connecting the aromatic ring to the kerogen could produce methylalkylbenzenes. The similarities between alkylbenzene distributions in source rock flash kerogen pyrolysates and in associated crude oils (Solli *et al.*, 1980a) suggest that the isomeric distribution of these extended-chain alkylbenzenes (produced during pyrolysis or catagenesis) is controlled in part by kerogen structure.

Two- and three-ring fully aromatic hydrocarbons (naphthalenes and phenanthrene/anthracenes) in kerogen pyrolysates have been also been reported by Van Graas *et al.* (1980), Solli *et al.* (1980b), and others (see Larter, 1984, and references therein), the main species being the parent nucleus and the mono- and dimethyl homologs. These species are probably derived from β cleavage of alkylaromatic structures in kerogens derived, in part, from catagenic aromatization of naphthenoaromatic structures (Van Graas *et al.*, 1980; Allan and Larter, 1983). Naphthenoaromatic hydrocarbons, including alkylindanes and alkylindenes, have been reported, as have heterocyclic aromatic compounds including benzofurans (Van Graas *et al.*, 1981) and alkylthiophenes and benzothiophenes (Sinninghe-Damsté *et al.*, 1989; Eglinton *et al.*, 1988, and references therein). The precursors in the kerogens for these species are presumably closely related aromatic structures, those containing sulfur being derived from reaction of unsaturated sites in the kerogens during diagenesis with reactive species such as hydrogen sulfide followed by later aromatization of the cyclic moieties produced (Sinninghe-Damsté *et al.*, 1989). While some alkylbenzenes are derived ultimately through preservation or alkylation of inherited benzenoid structures from lignins or amino acids, some aromatic structures are probably produced during diagenetic condensation of unsaturated lipids with unsaturated sites in kerogens (cf. Larter *et al.*, 1983) followed by, or merely resulting from, aromatization and hydrogen loss during catagenesis (Tissot and Welte, 1978). Aromatization of some aliphatic and alicyclic structures during pyrolysis cannot be completely ruled out. For instance, rare aliphatic kerogens that are strongly cross-linked by source-related controls (e.g., *Botryococcus* rubber; Larter, 1978) or irradiation (Horsfield *et al.*, 1992) are certainly prone to generate aromatic hydrocarbons on pyrolysis. Nuclear magnetic resonance (NMR) studies on commonly occurring kerogens and pyrolysates suggest that kerogen pyrolysates accurately reflect the aromaticity of the parent kerogens (see Fig. 3 below; Larter, 1984; Horsfield, 1989), the hydrogen sources for radical-terminating reactions being predominantly the kerogen rather than the pyrolysates.

2.3. Heterocompounds

While many oxygen-bearing pyrolysis products have been identified in flash kerogen pyrolysates, the most abundant are the alylated phenols derived from lignin input into the kerogen mixture. These are discussed in detail below in Section 3. Other heteroatom-containing kerogen pyrolysate components chiefly present in very immature kerogens or humic substances are those ultimately derived from carbohydrate and peptide sources. Carbohydrate-related fragments have been found in very immature kerogen pyrolysates and humic materials, these being chiefly furan derivatives and short-chain aliphatic alcohols, aldehydes, and ketones (Maters *et al.*, 1977; Van de Meent *et al.*, 1980b; Meuzelaar *et al.*, 1982). The main processes involved in pyrolytic conversion of bound carbohydrate fragments are dehydration, retroaldolization, and decarboxylation and are discussed by Schulten and Gortz (1978). Peptide and protein fragments in geopolymers have been identified in pyrolysates as nitriles, dimethyl disulfide, pyrroles, and indoles (Maters *et al.*, 1977; Van de Meent *et al.*, 1980b; Meuzelaar *et al.*, 1982). In general, these are visually minor components of most kerogen pyrograms and have not yet been quantitatively related to part structures in kerogens. Alkylpyrroles may also be derived from pyrolysis of bile pigments (T. I. Eglinton, personal communication).

Following on from the discussion above, we now explore the contribution made by analytical pyrolysis to the quantitative study of kerogen structure. Much of the evolution of the subject area has been made through the use of analytical pyrolysis methods, principally Py-CG and Py-GC/MS, as kerogen typing tools, and we will follow the development of the approach in this manner. In particular, a distinct theme emerges from an analysis of this type in that most of the work of the 1970s and early 1980s was aimed at assessment of parameters that are now known to be clearly related to qualitative estimates of kerogen aromaticity.

3. Pyrolysis as a Structural Tool

In view of the inherent uncertainties concerning the origin of the individual pyrolysis products discussed earlier, for structural considerations, pyrolysis should always be used in conjunction with established chemical degradation methods (Stefanovic and Vitorovic, 1959; Burlingame and Simoneit, 1969; Deno *et al.*, 1978; Allan *et al.*, 1980) or spectroscopic techniques, especially infrared and NMR approaches (Miknis *et al.*, 1982; Wilson, 1987; Rouxhet and Robin, 1978). It is also noted that compositional inferences from pyrolysis are based on only a rather small mass fraction of the resolved pyrolysates and an even smaller and usually uncalculated mass fraction of the total pyrolysates ("humps" in pyrograms) plus undetermined condensed tarry materials in the pyrolyzer interface. Further, they take no direct account of the involatile kerogen component at all. While the pyrolytically volatile portion of kerogens can be broadly considered to be the same portion that generates petroleum in the subsurface (Cooles *et al.*, 1986), the pyrolytic inaccessibility of large fractions of all kerogens potentially renders analytical pyrolysis less applicable than bulk spectroscopic methods to chemical structural definition problems.

Recent observations, however, concerning ^{13}C NMR- and Py-GC-determined aromaticities suggest that the picture described above is overly pessimistic (Horsfield, 1989). Shown in Fig. 3 is the relationship between atomic H/C and the ratio of the peak abundances of major GC-resolvable aromatic (hydrocarbon and alkylphenol) and *n*-alkyl moieties (*n*-alkanes, *n*-alk-1-enes) in C_{5+} programmed temperature pyrolysates. The choice of these moieties was based upon earlier approaches which utilized aromatic and aliphatic hydrocarbon ratios in pyrolysates to infer kerogen aromaticity (Leplat, 1967; Romovacek and Kubat, 1968; Giraud, 1970; Larter and Douglas, 1978, 1980; Larter, 1984). Also shown on the figure is the relationship between aromaticity (f_a) determined using solid-state ^{13}C NMR methods and atomic H/C ratio from published sources, including studies on coal macerals (Zilm *et al.*, 1981; Dereppe *et al.*, 1983; Barwise *et al.*, 1984; Saxby and Stephenson, 1987). These NMR estimates of aromaticity are based on the whole kerogen and therefore take into account the aromatic and aliphatic portions of both inert and reactive kerogen components. It should be noted that Py-GC aromaticities are base on ratios of peak areas (Ar/Al)

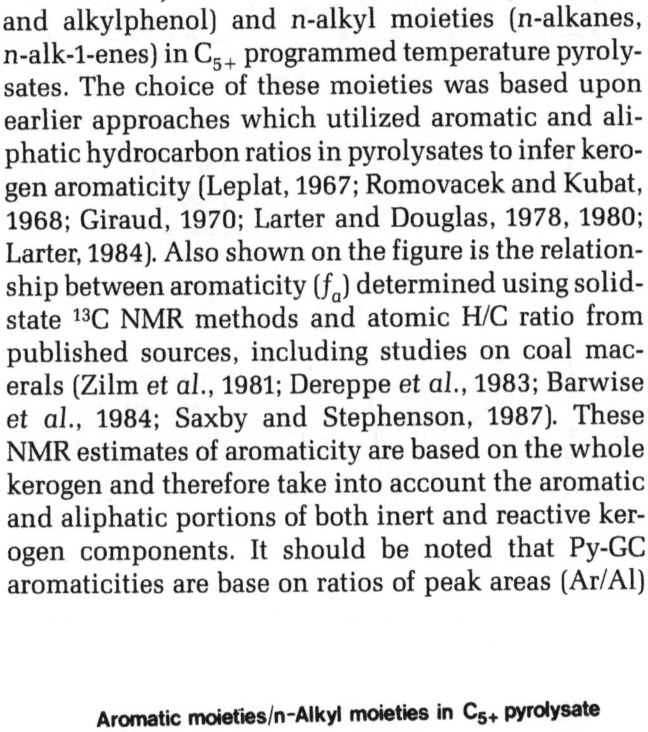

Figure 3. The aromaticity of kerogens indicated by both Py-GC ratios and NMR measurements. See text for discussion. [After Horsfield (1989).]

and ^{13}C NMR aromaticities are expressed as proportions (in keeping with tradition) but that the scales are directly numerically equivalent. The remarkable overlap of the H/C versus aromaticity relationships using the two methods is readily discernible from Fig. 3 and suggests that (1) Py-GC-determined aromaticities are representative of the aromaticity of kerogen as a whole and not just the volatile portion and (2) the aromatic structures in pyrolysates are quantitatively proportionally representative of the major aromatic moieties in kerogens. Thus, one- to three-ring alkylated aromatic groups are the dominant species in kerogens, even those of coals, at least up to low-volatile bituminous rank levels, consistent with published coal kerogen models (Given, 1984).

The agreement between ^{13}C NMR and Py-GC aromaticities is particularly good for kerogens with H/C ratios less than 1.1 (Fig. 3). Note in particular the excellent agreement for the Silkstone vitrinite (#11) and Silkstone sporinite (#13) in the inset portion of Fig. 3, these samples having been analyzed by both Zilm *et al.* (1981) and Horsfield (1989). The divergence between Py-GC and NMR aromaticities above atomic H/C ratios of 1.1 is probably related to the NMR aromaticies being too high as olefinic sites are present in type I kerogens, for example, the case of the torbanite kerogen (#4) (cf. Allan *et al.*, 1980; Derenne *et al.*, 1987). In such cases, Py-GC aromaticities may be more reliable as they can differentiate aromatic from unsaturated olefinic species.

In addition to providing reliable estimates of kerogen aromaticity, Py-GC provides quantitatively relevant structural information at the molecular level. We illustrate this with an examination of phenolic and thiophenic species in kerogen pyrolysates. Immature humic kerogens of rank levels up to and including sub-bituminous produce pyrolysates relatively rich in methoxylated and vinyl-substituted phenolic components derived from diagenetically modified lignin moieties (Van de Meent *et al.*, 1980b; Saiz-Jimenez and de Leeuw, 1984; Philp *et al.*, 1982b; Hatcher *et al.*, 1988). In contrast, coal vitrain kerogens of high-medium-volatile bituminous rank level ("oil window maturity") produce pyrolysates rich in simple alkylphenols with no significant extent of vinyl or methoxyl substitution on the phenol nucleus. Figure 4 shows pyrolysis mass spectra of a sub-bituminous rank woody kerogen and a high-volatile bituminous ranked vitrinite kerogen that illustrate this point.

Alkylphenols in type III kerogen pyrolysates represent one of the most abundant individual species found in kerogen pyrolysates (Senftle *et al.*, 1986). Figure 5 shows the absolute yields of phenol plus C_1-

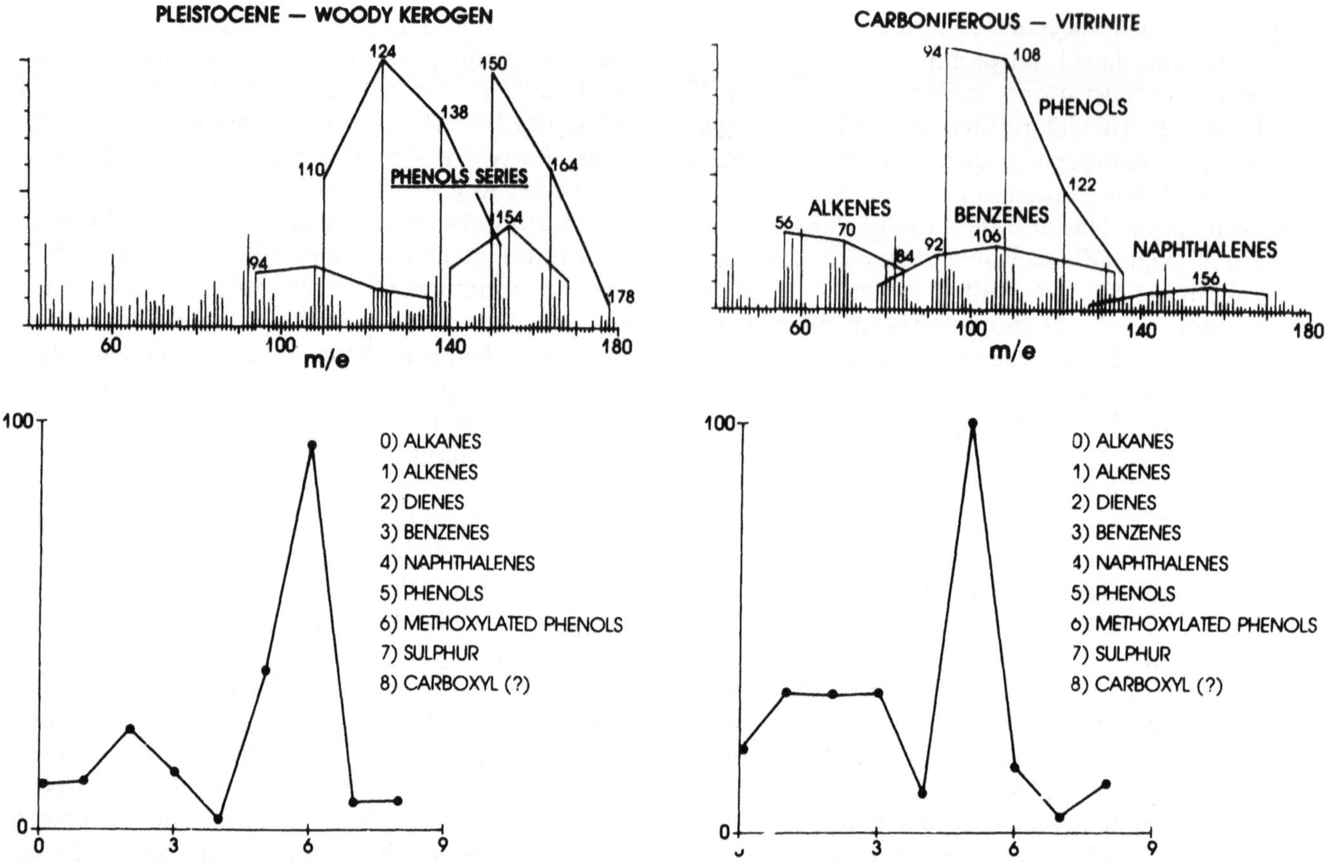

Figure 4. Pyrolysis mass spectra (610°C/10 s) of a sub-bituminous woody kerogen and a high-volatile bituminous ranked vitrinite kerogen. The histograms show the relative distributions of different pyrolysis products produced by summing the appropriate molecular-ion series in the spectra.

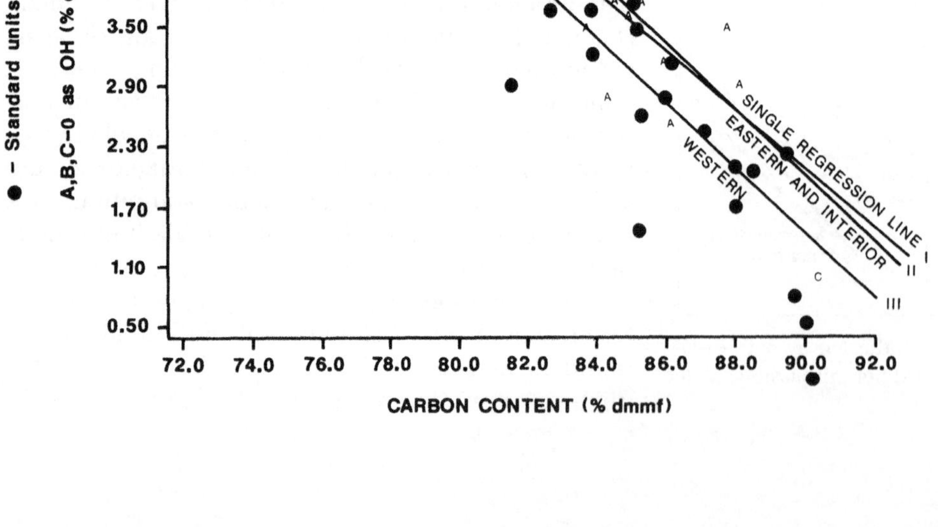

Figure 5. Hydroxyl oxygen content for U.S. coals of eastern (A), interior (B), and western provinces (C) [after Yarzab *et al.* (1979)] and pyrolytic phenol yields for 800°C/20-s flash pyrolysates of eastern U.S. vitrinite kerogens (●) [after Senftle *et al.* (1986)] as a function of maturity.

and C_2-substituted alkylphenols determined by these authors in flash pyrolysates of a series of vitrinite kerogens from the eastern United States. The phenol data are plotted together with the chemical hydroxyl oxygen content data of Yarzab *et al.* (1979) and Given (1984). The yields of phenols in vitrinite pyrolysates decrease dramatically over the rank range 81–91% C (ca. approx. 0.5–1.7% R_0), in agreement with accepted views of coal diagenesis. Although the absolute yield of phenols is much lower than the concentration of phenolic hydroxyl oxygen in related kerogens, as determined by wet chemical methods (see Fig. 5), the overall trends are remarkably concordant (the common *y*-axis scale is serendipitous). This again confirms the proposition made above that, although GC-amenable analytical pyrolysates are not quantitatively representative of kerogens, they are proportionally structurally representative. Thus, even though the actual mechanisms of alkylphenol production from kerogens may be complex (Larter, 1989), phenolic structures in pyrolysates appear proportionally quantitatively representative of oxygen-substituted benzenoid structures in kerogens.

Similar observations have been made by Eglinton *et al.* (1990), who have observed a close relationship between atomic S/C ratios in kerogens and asphaltenes and Py-GC ratios such as 2,3-dimethylthiophene/(1,2-dimethylbenzene + n-nonene ratios (cf. Fig. 6). These authors have also demonstrated a quantitative dependence of absolute pyrolysate alkylthiophene yield on kerogen S/C atomic ratio, suggesting again that pyrolytic sulfur species are proportionally representative of kerogen-bound sulfur species.

The major point to be made is that Py-GC data appear to provide compositional information about gross kerogen structure as a whole and not about atypical, readily volatilizable part structures. Py-GC therefore lends itself admirably to the rapid qualitative and quantitative characterization of gross kerogen structure and type as part, of course, of an integrated multiparameter characterization approach.

4. Analytical Pyrolysis and Implications for Kerogen Formation and Petroleum Generation

Analytical pyrolysis has had only minor impact upon the way in which we routinely and globally define kerogen types, despite known serious shortcomings with established elemental atomic ratio-based schemes, especially in provinces dominated by terrigenous organic matter (Snowdon and Powell, 1982; Horsfield, 1984; Larter, 1985). Several analytical pyrolysis-based schemes have been proposed to remedy this situation (cf. Larter, 1985; Larter and Senftle, 1985, and references therein), but for one reason or another these did not provide generally accepted practical alternatives to elementally based classifications. We do not fully understand this in view of the fact that nonmarine petroleum constitutes a large fraction of the world's reserves!

Renewed motivation for changes in the way in which we type kerogens has emerged as a result of increased application of reservoir engineering principles to petroleum geochemistry (England *et al.*, 1987). Here, the composition and molecular weight distributions of petroleums (particularly the mass fraction of gas) produced on catagenesis, as determined by analytical pyrolysis methods, are most relevant to problems associated with secondary migration and petroleum volumetrics (England and Mackenzie, 1989; Horsfield and Larter, 1989). In view of the fact that elemental-based kerogen typing discriminates solely according to *genetic potential* ([mg hydrocarbon (HC)/g total organic carbon (TOC)], no basis exists for such estimations from these methods. Analytical pyrolysis fills this gap by providing two additional criteria, the *kerogen quality*, which describes the composition and molecular weight distribution of petroleum precursors in the kerogen, and the *maturation characteristics* of the kerogen, which describe the manner in which these distributions change with maturity.

With regard to kerogen quality assessment, careful examination of a wealth of Py-GC data (Larter and Douglas, 1980; Larter, 1978, 1985; Schenck *et al.*, 1981; Horsfield, 1984; Larter and Senftle, 1985) indicates that immature to early mature kerogens are most readily discriminated from one another by (1) the relative abundances of selected aromatic and aliphatic pyrolysate components and (2) the relative abundances of short-chain versus long-chain doublets of *n*-alkanes and *n*-alk-1-enes ("chain length distribution"). Furthermore, it has been shown by correlation studies that these compositional attributes in pyrolysates are manifested accordingly in related natural petroleums (Horsfield, 1989), and thus pyrolysates can be described by analogy to their natural petroleum analogs (cf. zones on Fig. 7, defined by reference to related petroleums).

Defining kerogen quality on the basis of the molecular distribution of petroleum precursors, in addition to genetic potential considerations (hydrogen index), provides a step toward the unification of petroleum and kerogen geochemistry through genetic

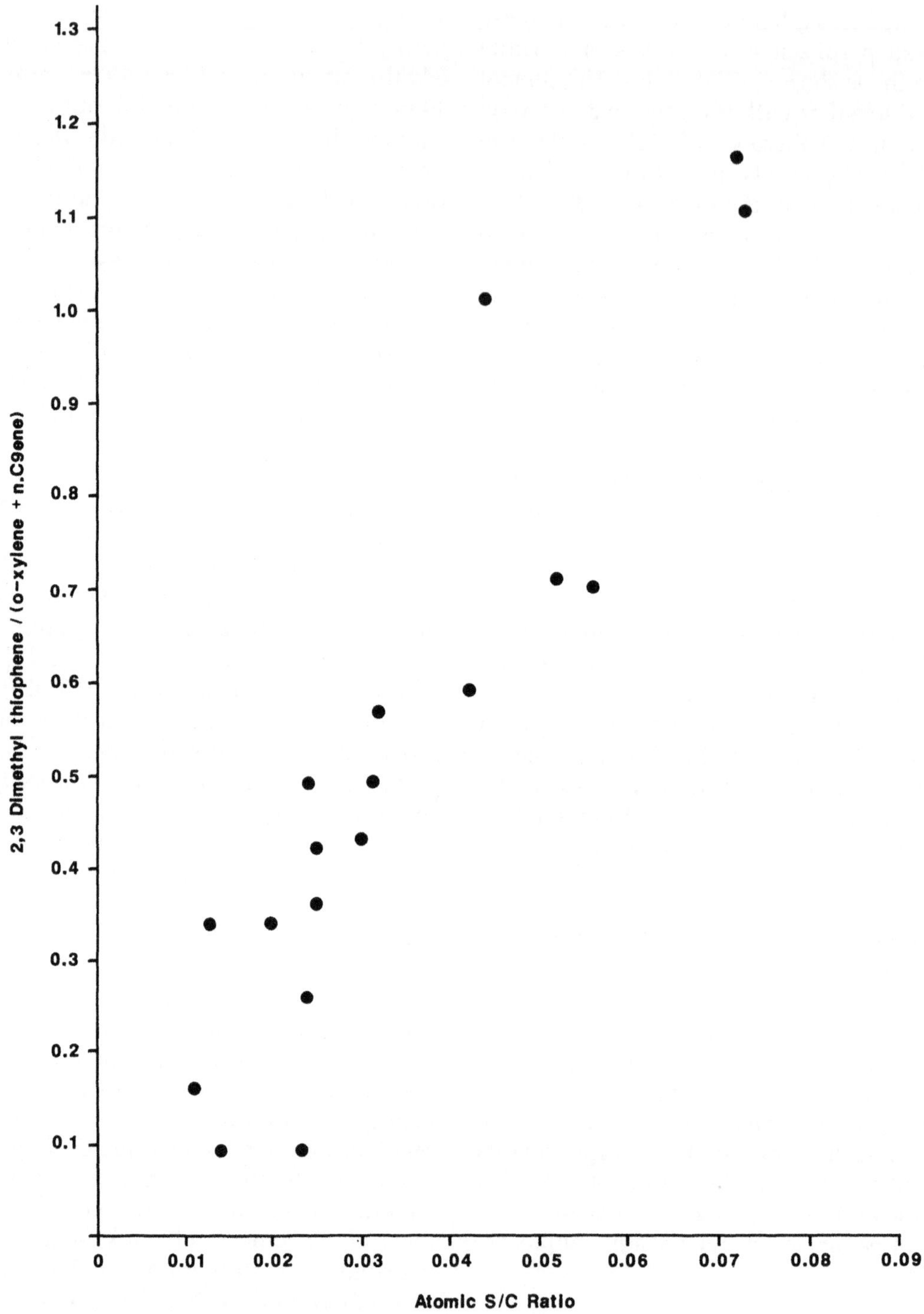

Figure 6. Plot of the 2,3-dimethylthiophene/(1,2-dimethylbenzene + *n*-nonene) ratio from 800°C flash pyrolysis of oil asphaltenes and coil kerogens vs. chemically determined S/C atomic ratio. [After Eglinton *et al.* (1990).]

correlations between kerogen structure and petroleum molecular compositions. It is worth noting that a great deal of progress has been made in this and the broader area of geopolymer characterization, in the investigations summarized by Pelet *et al.* (1986) on the compositional similarities of kerogens and related asphaltenes, as revealed by, *inter alia*, Py-GC.

The second attribute of kerogen composition addressed here is the way in which the molecular composition changes with increasing maturity. Two-step

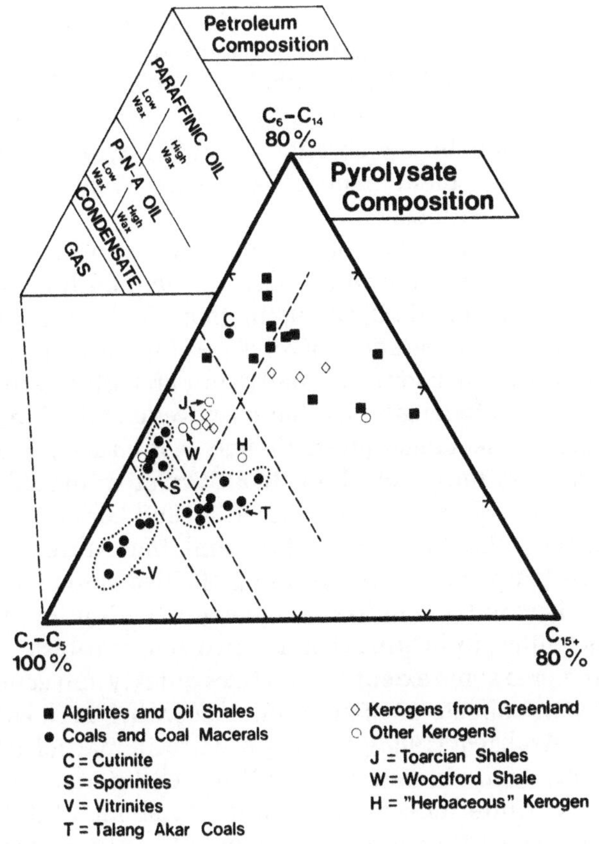

Figure 7. The *n*-hydrocarbon chain length distributions of programmed temperature kerogen pyrolysates (320–600°C/40°C/min^{-1}).

programmed temperature Py-GC, in which kerogens are pyrolyzed first from 320 to 450°C and then from 450 to 600°C, both pyrolysates being analyzed by GC, reveals that immature and mature kerogens can reveal quite different behavior on thermal decomposition. As shown in Fig. 8, *n*-hydrocarbons are released from the structure of a Talang Akar coal (Indonesia) under the first mild pyrolysis step (320–450°C), in addition to a complex spectrum of aromatic hydrocarbons (A), phenols (P), and prist-1-ene (I). In contrast, the second pyrolysis of the coal produced only aromatic compounds in abundance (Fig. 8), the latter fingerprint resembling coal vitrain kerogen pyrograms. The chain length distributions of the C_1–C_5, C_6–C_{14}, and C_{15+} *n*-hydrocarbons in the pyrolysates change dramatically between the two pyrolysates. The changes in chain length distribution, particularly the more gas-prone nature of the second pyrolysate, are broadly consistent with the chemical concepts of oil-prone labile and gas-prone refractory kerogen components as defined by Cooles *et al.* (1986). In contrast, and in line with pyrolysates for natural maturation series of Permo-Carboniferous age coals (Allan, 1975; McHugh *et al.*, 1976; Larter and Douglas, 1980b), an early mature sporinite kerogen generated low- and high-temperature pyrolysates with similar chain length distributions (clearly, the similarity does not extend to other compound classes; cf. also Van Graas *et al.*, 1981). Similar observations apply to hydrogen-rich kerogens (Fig. 9). The sample from Greenland in Fig. 9 appears to be thermally very labile, with most C_{5+} normal hydrocarbons being produced during the first pyrolysis step, and a significant difference in chain length distribution between the first and second pyrolysates is noted. This is similar to the case of the kerogens analyzed by Øygard *et al.* (1989), where progressive chain shortening was observed with increasing maturity for a natural series of kerogens consisting mainly of amorphous marine (type II) and paludal coal (type III) kerogens. In contrast, the torbanite (type I) pro-

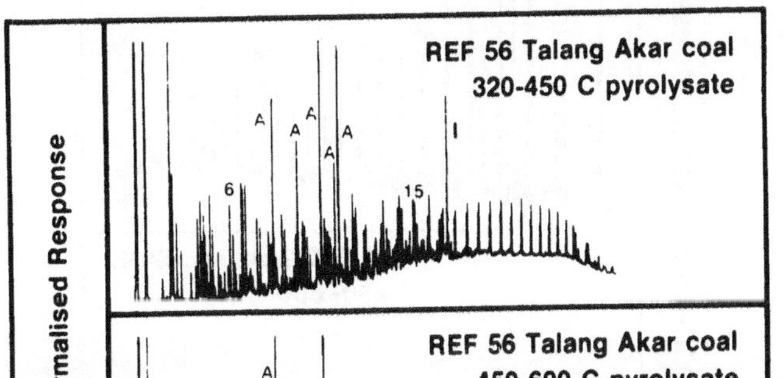

Figure 8. Two-step Py-GC traces of kerogens from Talang Akar coal and sporinite. For conditions, see text. The response scale of each step in respective experiments is the same.

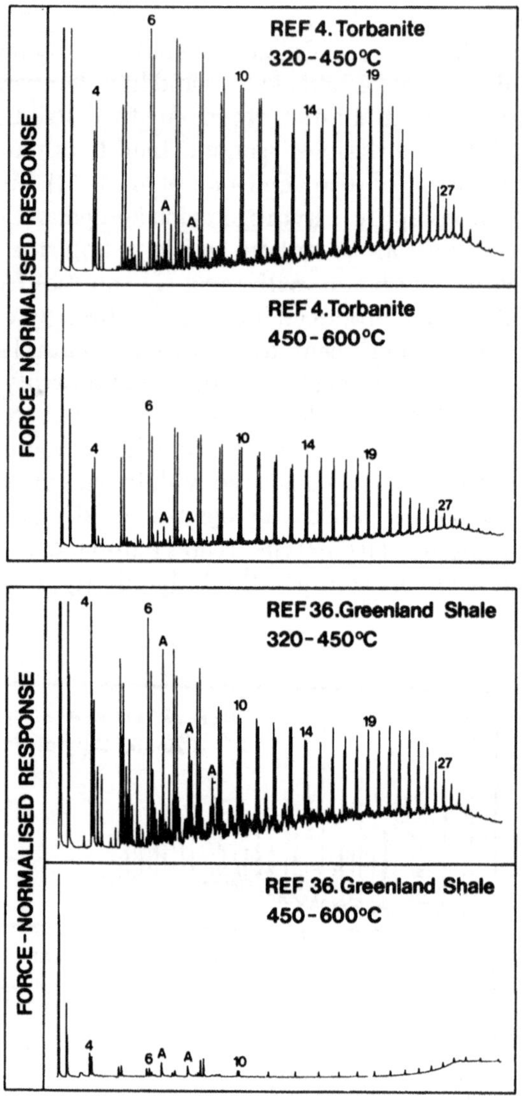

Figure 9. Two-step Py-GC traces of kerogens from torbanite and a Greenland shale. For conditions, see text. The response scale of each step in respective experiments is the same.

duced similar yields and distributions of *n*-hydrocarbons on pyrolysis at both temperatures, reflecting clearly different chemical environments for the alkyl chains in this kerogen type. In the case of the torbanite, a largely C—C-bonded polymethylene structure for the kerogen and low concentrations of structurally weakening functional groups or aromatic rings (Allan *et al.*, 1980; Zilm *et al.*, 1981) confer great stability on the kerogen. Interestingly, Dungworth and Schwartz (1972), McKirdy *et al.* (1980), McKirdy and Hahn (1982), and Jackson *et al.* (1984) have found highly overmature kerogens which generate long *n*-hydrocarbon chains in high relative abundance relative to total pyrolysates. This suggests that with oxygen-deficient, highly polymethylenic kerogens, the base of liquid-hydrocarbon generation in the subsurface is controlled by oil-to-gas cracking kinetics rather than by an absence of liquid-hydrocarbon-generating potential in the kerogens, even in overmature situations.

The distribution of *n*-alkyl moieties in sporinite and torbanite (*inter alia*) has been termed homogeneous by Horsfield (1989) in that single-step programmed temperature pyrolysates (up to 600°C) of sporinites are compositionally similar, with regard to *n*-alkyl chain length distributions, over an oil window maturity sequence (even though pyrolysate phenol concentrations drop dramatically, relative to hydrocarbons, over the same range; Allan and Larter, 1983). Further, the low-temperature and the subsequent high-temperature pyrolysis step of a two-step pyrolysis also produce similar chain length distributions for sporinites, indicating that this two-step pyrolysis approach to some extent summarizes quickly and generally the effects of maturation. According to Takeda and Asakawa (1988), heating of vitrinite particles to 450°C under standard Rock-Eval conditions raised the vitrinite reflectance from 0.4 to about 1.5% R_0. We suspect therefore that the cutoff between step 1 and step 2 corresponds approximately to an equivalent natural maturity of about 1.5–1.6% R_0. This is very broadly equivalent to the maturity at which refractory kerogens start to undergo major gas generation (Cooles *et al.*, 1986), though clearly the torbanite discussion above shows that even some Type 1 kerogens preserve (even liquid) petroleum potential be at this maturity level. Note, however, that under the very open system conditions employed in analytical pyrolysis, no "oil cracking" occurs in the first pyrolysis step, unlike the situation in equivalent natural systems. This has been discussed fully by Horsfield and Larter (1989).

Homogeneous chain length distributions in kerogens, as visualized by Py-GC and indicative of chemically simple uniform kerogen structure, may not be consistent with that view of kerogen formation involving condensation of lipid moieties on to a functionalized and condensed nucleus (Tissot and Welte, 1978) but is suggestive of the purity of biosynthesis (cf. de Leeuw and Largeau, this volume, Chapter 2; Goth *et al.*, 1988). However, it has been suggested that Diels–Alder type reactions may allow incorporation of olefinic lipids into highly aliphatic, rare protokerogens of the coorongite type which may produce relatively homogeneous kerogens (Larter, 1984). That is, chemical homogeneity may not always reflect biosynthesis, but it appears that, in most cases we know of, pyrolytically homogeneous kerogens may be associated with selectively preserved, microscopically structured, lipid-rich biopolymeric material. For instance, sporinite, *Botryococcus* derived alginite A (cf. Hutton *et al.*, 1980), *Gleocapsomorpha* alginite A, some vitrinites, and Indiana cutinite are "homogeneous."

In this regard, nonsaponifiable, highly aliphatic biopolymeric substances have been found in extant leaf cuticle (Nip *et al.*, 1986) in Messel Shale kerogen (Goth *et al.*, 1988) and in *Botryococcus braunii* (Largeau *et al.*, 1984; de Leeuw and Largeau, this volume, Chapter 2). In the latter case, the polymer was viewed as the major organic contributor to torbanite (Largeau *et al.*, 1986; Derenne *et al.*, 1988; see also Philp and Calvin, 1976) though alternate mechanisms for coorongite kerogen formation have been suggested (cf. above). While the true extent of the contribution of aliphatic biopolymers (de Leeuw and Largeau, this volume, Chapter 2) to hydrogen-rich kerogens is equivocal at this point, and more preserved biopolymeric materials need to be investigated, it does seem that homogeneous kerogens have thermally resistant structures which preserve their compositional characteristics to high levels of maturity. The significant concentrations of aromatic carbon in all crude oils, and immature oil-prone kerogens of all source facies, including those without land plant contributions, does suggest however that selective preservation of aliphatic biopolymers cannot be the dominant contributor to most source rock kerogen systems—though aliphatic biopolymers are certainly locally important (cf. de Leeuw and Largeau, this volume, Chapter 2).

In contrast, the kerogens from east Greenland and the Talang Akar coal (Figs. 8 and 9) may be described as being "heterogeneous" with regard to alkyl moiety distribution in the kerogen, possibly consistent with a condensation-related kerogen genesis, i.e., a lipid envelope on an initially functionalized, now refractory core. Interestingly, the precursors of the *n*-hydrocarbons in the pyrolysates of these Talang Akar kerogens are associated with amorphous and finely comminuted liptinite (Horsfield *et al.*, 1988), optically structureless alginite B being the main maceral in the Greenland type II kerogen. Clearly, consideration of the relative proportions of preserved biopolymeric kerogen and diagenetically produced kerogens must be incorporated into any structural model of kerogen components. With this apparent relationship between optical structure and chemical character, we stress again that physical examination of the kerogen by optical methods is a powerful and necessary adjunct to chemical characterization (Hutton *et al.*, 1980; Senftle *et al.*, 1987; Senftle *et al.*, this volume, Chapter 15).

5. Summary

The main quantified components of kerogen pyrolysates, normal and branched/cyclic hydrocarbons, alkylaromatic hydrocarbons, sulfur-bearing alkylated aromatic compounds, and alkylphenols derived principally from kerogen-bound lipid, lignin, and diagenetic modification of structures of uncertain origin, show pyrogram distributions that can be related to source facies and maturity. Many of the detailed mechanisms relating specific kerogen pyrolysis products to precursor elements are not completely understood. Further, it is known that instrumentally resolvable kerogen pyrolysates are quantitatively minor portions of kerogens. However, it can be shown that analytical pyrolysates are quantitatively proportionally representative of gross structural parameters such as total kerogen aromaticity and molecular parameters such as absolute concentrations of aryl–oxygen bonds or organic sulfur in kerogens.

This is illustrated by a schematic model of kerogen composition as shown in Fig. 10 (S. R. Larter, B. Horsfield, and T. I. Eglinton, manuscript in preparation). Under geological heating rates, the component structural elements of reactive kerogen are to a greater extent thermally disassembled to give large hydrocarbon concentrations. Under open-system pyrolysis, more nonhydrocarbons (many probably non-GC-amenable) are formed, with only relatively small hydrocarbon concentrations. The proportions of aliphatic (R.H) and aromatic hydrocarbons Ar.H, R.Ar) and GC-amenable nonhydrocarbons such as alkylphenols (R.As.H, Rs.H) are, however, proportionally representative of the parent structural element concentrations in the kerogen. Thus, Py-GC provides a powerful tool for rapid kerogen structural elucidation.

The distributions of structural elements such as alkyl chain distributions in kerogens as a function of facies and maturity can therefore be followed by examination of analytical pyrolysates, and these distributions not only can be used to provide information about kerogen structure and facies, but also allow inferences about kerogen genesis mechanisms.

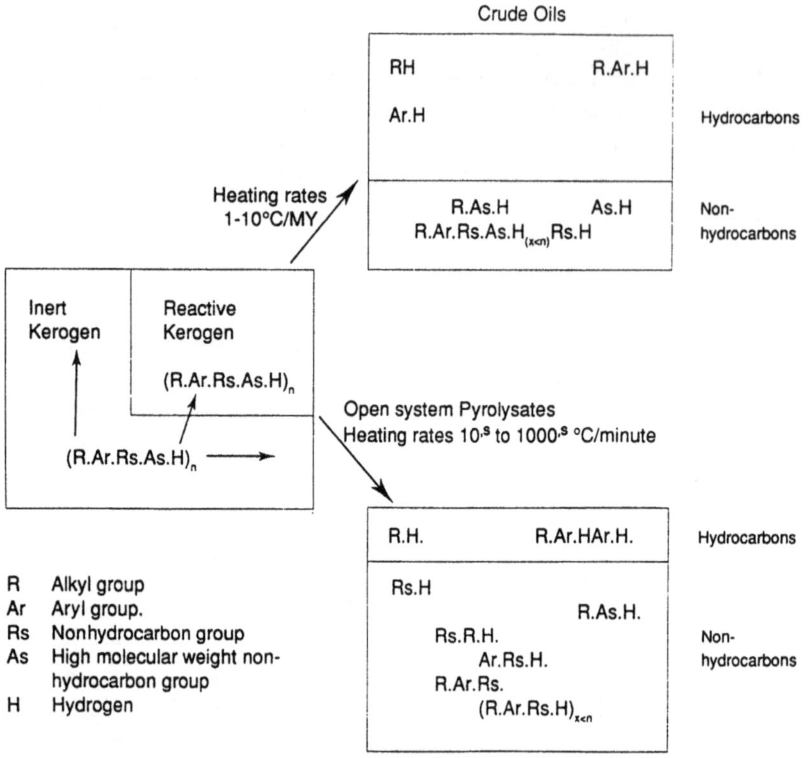

Figure 10. A schematic kerogen decomposition model suggesting a conceptual explanation of the utility of open-system pyrolysis (Py-GC) as a proportionally representative structural tool for characterizing kerogens. See text for explanation.

ACKNOWLEDGMENTS. Archie Douglas was instrumental in setting both of us (and many others) in pursuit of kerogens and their pyrolysates, and we acknowledge his contributions in these areas. We are grateful to Tim Eglinton for permission to use his data and for useful discussions and review. S. R. L. acknowledges financial and drafting support from BP and financial support from the Nordisk Ministerad Energy Research program. B. H. acknowledges Professor D. H. Welte's support of pyrolysis-related research at ICH5. Yvonne Hall prepared the manuscript.

References

Allan, J., 1975, Natural and artificial diagenesis of coal macerals, Ph.D. Thesis, University of Newcastle upon Tyne.

Allan, J., and Larter, S. R., 1983, Aromatic structures in coal maceral extract and kerogens, in: *Advances in Organic Geochemistry 1981* (M. Bjorøy *et al.*, eds.), Wiley Heyden, London, pp. 534–545.

Allan, J., Bjorøy, M., and Douglas, A. G., 1980, A geochemical study of the exinite group maceral alginite, selected from three Permo-Carboniferous torbanites, in: *Advances in Organic Geochemistry 1979* (A. G. Douglas and J. R. Maxwell, eds.), Pergamon Press, Oxford, pp. 239–248.

Barwise, A. J. G., Mann, A. L., Eglinton, G., Goward, A. P., Wardroper, A. M. K., and Gutteridge, C. S., 1984, Kerogen characterisation by ^{13}C NMR spectroscopy and pyrolysis mass-spectrometry, in: *Advances in Organic Geochemistry 1983* (P. A. Schenck, *et al.*, eds.), Pergamon Press, Oxford, pp. 343–349.

Baset, Z. H., Pancirov, R. J., and Ashe, T. R., 1980, Organic compounds in coal: Structure and origin, in: *Advances in Organic Geochemistry 1979* (A. G. Douglas and J. R. Maxwell, eds.), Pergamon Press, Oxford, pp. 619–630.

Behar, F., and Pelet, R., 1985, Pyrolysis gas chromatography applied to organic geochemistry. Structural similarities between kerogens and asphaltenes from related source rocks extracts and oils, *J. Anal. Appl. Pyrolysis* **8**:173–187.

Bharati, S., Larter, S. R., and Horsfield, B., 1992, The unusual source potential of the Cambrian Alum Shale in Scandinavia as determined by quantitative pyrolysis methods, in: *Generation, Accumulation and Production of Europe's Hydrocarbons II*, (A. M. Spencer, ed.), Special Publication of the European Association of Petroleum Geochemists No. 2, Springer-Verlag Berlin, Heidelberg, pp. 103–110.

Burlingame, A. L., and Simoneit, B. R., 1969, High resolution mass spectometry of Green River Formation kerogen oxidations, *Nature* **222**:741–747.

Burnham, A. K., 199, On the validity of the pristane formation index, *Geochim. Cosmochim. Acta* **53**:1693–1697.

Burnham, A. K., Clarkson, J. E., Singleton, M. F., Wong, C. M., and Crawford, R. W., 1982, Biological markers from Green River Shale kerogen decomposition, *Geochim. Cosmochim. Acta* **46**:1243–1251.

Cooles, G. P., Mackenzie, A. S., and Quigley, T. M., 1986, Calculation of petroleum masses generated and expelled from source rocks, in: *Advances in Organic Geochemistry 1985* (D. Leythaeuser and J. Rullkotter, eds.), Pergamon Press, Oxford, pp. 235–245.

Curry, D. J., and Simpler, T. K., 1988, Isoprenoid constituents in kerogens as a function of depositional environment and catagenesis, in: *Advances in Organic Geochemistry 1987* (L. Mattavelli and L. Novelli, eds.), Pergamon Press, Oxford, pp. 995–1001.

Dembicki, H., Horsfield, B., and Ho, T. T. Y., 1983, Sources rock evaluation by pyrolysis gas chromatography, *Bull. Am. Assoc. Petrol. Geol.* **67**:1094–1103.

Deno, N. C., Greigger, B. A., and Stroud, S. G., 1978, New method for elucidating the structures of coal, *Fuel* **57**:455–459.

Derenne, S., Largeau, C., Casadevall, E., and Laupretre, F., 1987, Structural analysis of two torbanites at different evolutionary stages. Investigation of the quantitative reliability of f_a determination by ^{13}C CP/MAS n.m.r., *Fuel* **66**:1084–1090.

Derenne, S., Largeau, C., Casadevall, E., and Connan, J., 1988, Comparison of torbanites of various origins and evolutionary stages. Bacterial contribution to their formation. Cause of the lack of botryococcane in bitumens, *Org. Geochem.* **12**:43–59.

Dereppe, J. M., Boudon, J. P., Moreaux, C., and Durand, B., 1983, Structural evolution of a sedimentologically homogeneous coal series as a function of carbon content by solid state 13-C n.m.r., *Fuel* **62**:575–579.

Dungworth, G., and Schwartz, A. W., 1972, Kerogen isolates from the Precambrian of South Africa and Australia, in: *Advances in Organic Geochemistry 1971* (H. R. V. Gaertner and H. Wehner, eds.), Pergamon Press, Oxford, pp. 699–706.

Durand, B., 1980, *Kerogen—Insoluble Organic Matter from Sedimentary Rocks*, Editions Technip, Paris.

Eglinton, T. I., 1988, An investigation of kerogens using pyrolysis methods, Ph.D. Thesis, University of Newcastle Upon Tyne.

Eglinton, T. I., Philp, R. P., and Rowland, S. J., 1988, Flash pyrolysis of artificially matured kerogens from the Kimmeridge Clay Fm., UK, *Org. Geochem.* **12**:33–41.

Eglinton, T. I., Sinninghe Damste, J. S., Kohnen, M. E. L., de Leeuw, J. W., Larter, S. R., Thatcher, M., and Patience, R. S., 1990, Analysis of maturity related trends in the organic sulphur composition of kerogens by flash pyrolysis gas-chromatography, in: *Geochemistry of Sulphur in Fossil Fuels* (W. L. Orr and C. M. White, eds.), ACS Symposium Series, American Chemical Society, Washington, D.C., pp. 471–493.

England, W. A., and Mackenzie, A. S., 1989, Some aspects of the organic geochemistry of petroleum fluids, *Geol. Rundsch.* **78/1**:291–303.

England, W. A., Mackenzie, A. S., Mann, D. M., and Quigley, T. M., 1987, The movement and entrapment of petroleum fluids in the subsurface, *J. Geol. Soc.* **144**:327–347.

Gallegos, E. J., 1975, Terpane and sterane release from kerogen by pyrolysis gas chromatography mass spectrometry, *Anal. Chem.* **47**:1524–1528.

Gallegos, E. J., 1978, Analysis of five US coals. Pyrolysis gas chromatography mass spectrometry computer methods, in: ACS Advances in Chemistry Series No. 170 (P. C. Uden, S. Siggia, and H. G. Jensen, eds.), American Chemical Society, Washington, D.C., p. 236.

Giraud, A., 1970, Application of pyrolysis and pyrolysis gas chromatography to the geochemical characterisation of kerogen in sedimentary rocks, *Bull. Am. Assoc. Petrol. Geol.* **54**:439–455.

Given, P. H., 1984, An essay on the organic geochemistry of coal, in: *Coal Science* (M. L. Gorbaty *et al.*, eds.), Vol. 3, Academic Press, New York, pp. 63–251.

Goossens, H., de Leeuw, J. W., Schenck, P. A., and Brassell, S. C., 1984, Tocopherols as likely precursors of pristane in ancient sediments, *Nature* **312**:440–442.

Goossens, H., de Lange, F., de Leeuw, T. W., and Schenck, P. A., 1988, The Pristane Formation Index, a molecular maturity parameter. Confirmation in samples from the Paris Basin, *Geochim. Cosmochim. Acta* **52**:2439–2444.

Goth, K., de Leeuw, J. W., Püttmann, W., and Tegelaar, E. W., 1988, Origin of Messel Oil Shale, *Nature* **336**:759–761.

Hatcher, P. G., Lerch, H. E., Kotra, R. K., and Verheyen, T. V., 1988, Pyrolysis-gc-ms of degraded woods and coalified logs that increase in rank from peat to subbituminous coal, *Fuel* **67**: 1069–1075.

Horsfield, B., 1984, Pyrolysis studies and petroleum exploration, in: *Advances in Petroleum Geochemistry* (J. Brooks and D. H. Welte, eds.), Vol. I, Academic Press, London, pp. 247–292.

Horsfield, B., 1989, Practical criteria for classifying kerogens: Some observations from pyrolysis-gas chromatography, *Geochim. Cosmochim. Acta* **53**:891–901.

Horsfield, B., and Larter, S. R., 1981, Unpublished round robin analytical pyrolysis study.

Horsfield, B., and Larter, S. R., 1989, Practical kerogen typing for petroleum exploration, American Association of Petroleum Geologists Annual Meeting **73**:3, 151.

Horsfield, B., Dembicki, H., and Ho, T. T. Y., 1983, Some potential applications of pyrolysis to basin studies, *J. Geol Soc. London* **140**:431–443.

Horsfield, B., Yordy, K. L., and Crelling, J. C., 1988, Determining the petroleum generating potential of coal using organic geochemistry and organic petrology, in: *Advances in Organic Geochemistry 1987* (L. Mattavelli, and L. Novelli, eds.), Pergamon Press, Oxford, pp. 121–131.

Horsfield, B., Disko, U., and Leistner, F., 1989, The microscale simulation of maturation: Outline of a new technique and its potential applications, *Geol. Rundsch.* **78**:361–373.

Horsfield, B., Bharati, S., Larter, S. R., Leistner, F., Littke, R., Mann, U., Schenk, H. J., and Dypvik, H., 1992, On the atpyical petroleum generating characteristics of alginite in the Cambrian Alum Shale, in: *Early Organic Evolution: Implications for Mineral and Energy Resources* (M. Schidlowski *et al.*, ed.), Springer-Verlag, Berlin, Heidelberg, pp. 257–266.

Hutton, A. C., Kantsler, A. J., Cook, A. C., and McKirdy, D. M., 1980, Organic matter in oil shales, *J. Aust. Petrol. Explor. Assoc.* **20**: 68–86.

Jackson, K. S., McKirdy, D. M., and Deckelmann, J. A., 1984, Hydrocarbon generation in the Amadeus Basin, Central Australia, *APEA J.* **24**:43–65.

Largeau, C., Casadevall, E., Kadouri, A., and Metzger, P., 1984, Formation of *Botryococcus* derived kerogens: Comparative study of immature torbanites and of the extant alga *Botryococcus braunii*, in: *Advances in Organic Geochemistry 1983* (P. A. Schenck, J. W. de Leeuw, and G. W. M. Lijmbach, eds.), Pergamon Press, Oxford, pp. 327–332.

Largeau, C., Derenne, S., Casadevall, E., Kadouri, A., and Sellier, N., 1986, Pyrolysis of immature torbanite and of the resistant biopolymer (PRBA) isolated from extant alga *Botryococcus braunii*. Mechanism of formation and structure of torbanite, in: *Advances in Organic Geochemistry 1985* (D. Leythaeuser and J. Rullkötter, eds.), Pergamon Press, Oxford, pp. 1023–1032.

Larter, S. R., 1978, A geochemical study of kerogens and related material, Ph.D. Thesis, University of Newcastle upon Tyne.

Larter, S. R., 1984, Application of analytical pyrolysis techniques to kerogen characterisation and fossil fuel exploration/exploitation, in: *Analytical Pyrolysis Methods and Applications* (K. Voorhees, ed.), Butterworths, London, pp. 212–275.

Larter, S. R., 1985, Integrated kerogen typing in the recognition and quantitative assessment of petroleum source rocks, in: *Petroleum Geochemistry in Exploration of the Norwegian Shelf* (B. M. Thomas *et al.*, eds.), Graham and Trotman, London, pp. 269–286.

Larter, S. R., 1989, Chemical models of vitrinite reflectance evolution, *Geol. Rundsch.* **78/1:**349–359.

Larter, S. R., and Douglas, A. G., 1978, Low molecular weight aromatic hydrocarbons in coal maceral pyrolysates as indicators of diagenesis and organic matter type, in: *Environmental Biogeochemistry and Geomicrobiology*, Vol. 1 (W. Krumbein, ed.), Ann Arbor Science Publishers Inc., Ann Arbor, Michigan, pp. 373–386.

Larter, S. R., and Douglas, A. G., 1980a, Melanoidins—kerogen precursors and geochemical lipid sinks: A study using pyrolysis–gas chromatography (Py-GC), *Geochim. Cosmochim. Acta* **44:**2087–2095.

Larter, S. R., and Douglas, A. G., 1980b, A pyrolysis–gas chromatographic method for kerogen typing, in: *Advances in Organic Geochemistry 1979* (A. G. Douglas and J. R. Maxwell, eds.), Pergamon Press, Oxford, pp. 579–584.

Larter, S. R., and Douglas, A. G., 1982, Pyrolysis methods in organic geochemistry. An overview, *J. Anal. Appl. Pyrolysis* **4:**1–19.

Larter, S. R., and Senftle, J., 1985, Improved kerogen typing for petroleum source rock analysis, *Nature* **318:**277–280.

Larter, S. R., Solli, H., and Douglas, A. G., 1978, Analysis of kerogens by pyrolysis–gas chromatography mass spectrometry using selective ion detection, *J. Chromatogr.* **167:**421–431.

Larter, S. R., Solli, H., Douglas, A. G., De Lange, F., and de Leeuw, J. W., 1979, The occurrence and significance of prist-1-ene in kerogen pyrolysates, *Nature* **279:**405–408.

Larter, S. R., Solli, H., and Douglas, A. G., 1983, Phytol containing melanoidins and their bearing on the fate of isoprenoid structures in sediments, in: *Advances in Organic Geochemistry 1981* (M. Bjorøy *et al.*, eds.), John Wiley, Chichester, England, pp. 513–521.

Leplat, P., 1967, Application of pyrolysis gas chromatography to the study of non-volatile petroleum fractions, *J. G.C.* **1967**(March):128–135.

Maters, W. L., Van De Meent, D., Schuyl, P. J. W., de Leeuw, J. W., and Schenck, P. A., 1977, Curie-point pyrolysis in organic geochemistry, in: *Analytical Pyrolysis* (Jones and Cramer, eds.), Elsevier, Amsterdam, pp. 203–216.

McHugh, D. J., Saxby, J. D., and Tardif, J. W., 1976, Pyrolysis hydrogenation gas chromatography of carbonaceous material from Australian sediments. Part I. Some Australian coals, *Chem. Geol.* **17:**243–259.

McKirdy, D. M., and Hahn, J. H., 1982, The composition and kerogen and hydrocarbons in Precambrian rocks, in: *Mineral Deposits and the Evolution of the Biosphere* (H. D. Holland and M. Sckidlowski, eds.), Springer-Verlag, pp. 123–154.

McKirdy, D. M., McHugh, D. J., and Tardif, J. W., 1980, Comparative analysis of stromatolitic and other microbial kerogens by pyrolysis–hydrogenation gas chromatography (PHGC), in: *Biogeochemistry of Ancient and Modern Environments* (P. A. Trudinger, M. R. Walter, and B. J. Ralph, eds.), Australian Academy of Sciences and Springer-Verlag, Berlin, Heidelberg, pp. 187–200.

Meuzelaar, H. L. C., Haverkamp, J., and Hileman, F. D. (eds.), 1982, *Py-MS of Recent and Fossil Biomaterials Compendium and Atlas*, Elsevier, Amsterdam.

Miknis, F. P., Smith, J. W., Maughan, E. K., and Maciel, G. E., 1982, Nuclear magnetic resonance: A technique for direct nondestructive evaluation of source rock potential, *Am. Assoc. Petrol. Geol. Bull.* **66:**1396–1401.

Nip, M., Tegelaar, E. W., Brinkhuis, H., de Leeuw, J. W., Schenck, P. A., and Halloway, P. J., 1986, Analysis of modern and fossil plant cuticles by Curie-point Py-GC and Curie-point Py-GC-MS: Recognition of a new, highly aliphatic and resistant biopolymer, in: *Advances in Organic Geochemistry 1985* (D. Leythaeuser and J. Rullkötter, eds.), Pergamon Press, Oxford, pp. 769–778.

Øygard, K., Larter, S. R., and Senftle, J. T., 1989, The control of maturity and kerogen type on analytical pyrolysis data, in: *Advances in Organic Geochemistry 1987* (L. Mattavelli and M. Novelli, eds.), Pergamon Press, Oxford, pp. 1153–1162.

Pelet, R., Behar, F., and Monin, J. C., 1986, Resins and asphaltenes in the generation and migration of petroleum, in: *Advances in Organic Geochemistry 1985* (D. Leythaeuser and J. Rullkötter, eds.), Pergamon Press, Oxford, pp. 481–498.

Philp, R. P., 1982, Application of pyrolysis-GC and pyrolysis-GC-MS to fossil fuel research, *Trends Anal. Chem. 1* **10:**237–241.

Philp, R. P., and Calvin, M., 1976, Possible origin for insoluble organic kerogen debris in sediments from insoluble wall materials of algae and bacteria, *Nature* **262:**134–136.

Philp, R. P., and Gilbert, T. D., 1985, Source rock and asphaltene biomarker characterisation by pyrolysis–gas chromatography–mass spectrometry multiple ion detection, *Geochim. Cosmochim. Acta* **49:**1421–1432.

Philp, R. P., Gilbert, T. D., and Russell, N. J., 1982a, Characterisation by pyrolysis gas chromatography mass spectrometry of the insoluble organic residues derived from hydrogenation of *Tasmanites* sp. oil shale, *Fuel* **61:**221–237.

Philp, R. P., Russell, N. J., Gilbert, T. D., and Friedrich, J. M. 1982b, Characterisation of Victoria Brown Coal wood by microscopic techniques and Curie-point pyrolysis combined with gas chromatography mass spectrometry, *J. Anal. Appl. Pyrolysis* **4:** 143–161.

Reed, J. D., Illich, H. A., and Horsfield, B., 1986, Biochemical evolutionary significance of Ordovician oils and their sources, in: *Advances in Organic Geochemistry 1985* (D. Leythaeuser and J. Rullkötter, eds.), Pergamon Press, Oxford, pp. 347–358.

Romovacek, J., and Kubat, J., 1968, Characterisation of coal substance by pyrolysis–gas chromatography, *Anal. Chem.* **40:** 1119–1127.

Rouxhet, P. G., and Robin, P. L., 1978, Infrared study of the evolution of kerogens of different origins during catagenesis and pyrolysis, *Fuel* **57:**533–540.

Saiz-Jimenez, C., and de Leeuw, J. W., 1984, Pyrolysis gas chromatography mass spectrometry of isolated, synthetic and degraded lignins, in: *Advances in Organic Geochemistry 1983* (P. A. Schenck *et al.*, eds.), Pergamon Press, Oxford, pp. 417–422.

Saxby, J. D., and Stephenson, L. C., 1987, Effect on an igneous intrusion on oil shale at Rundle (Australia), *Chem. Geol.* **63:** 103–116.

Schenck, P. A., de Leeuw, J., van Graas, J. W., Haverkamp, J., and Bouman, M., 1981, Analysis of Recent spores and pollen and of thermally altered sporopollenin by flash pyrolysis–mass spectrometry and flash pyrolysis–gas chromatography–mass spectrometry, in: *Organic Maturation Studies and Fossil Fuel Exploration* (J. Brooks, ed.), Academic Press, New York, pp. 225–237.

Schulten, H. R., and Gortz, W., 1978, Curie-point pyrolysis and field ionisation mass spectrometry of polysaccharides, *Anal. Chem.* **50:**428–433.

Senftle, J. T., Larter, S. R., Bromley, B. W., and Brown, J. B., 1986, Quantitative chemical characterization of vitrinite concentrates using pyrolysis–gas chromatography. Rank variation of pyrolysis products, Org. *Geochem.* **9**:345–350.

Senftle, J. T., Brown, J. B., and Larter, S. R., 1987, Refinement of organic petrographic methods for kerogen characterisation, *Int. J. Coal Geol.* **7**:105–117.

Siefert, W. K., 1978, Steranes and terpanes in kerogen pyrolysis for correlation of oils and source rocks, *Geochim. Cosmochim. Acta* **42**:473–484.

Sinninghe-Damsté, J. S., Eglinton, T. I., de Leeuw, J. W., and Schenck, P. A., 1989, Organic sulphur in macromolecular sedimentary organic matter. I. Structure and origin of sulphur containing moieties in kerogen, asphaltenes and coal as revealed by flash pyrolysis, *Geochim. Cosmochim. Acta* **53**: 873–889.

Snowdon, L. R., 1980, Resinite: A potential petroleum source in the Upper Cretaceous/Tertiary of the Beaufort Mackenzie Basin, *Can. Soc. Petrol. Geol. Mem.* **6**:509–521.

Snowdon, L. R., and Powell, J. G., 1982, Immature oil and condensate: Modification of hydrocarbon generation model for terrestrial organic matter, *Am. Assoc. Petrol. Geol. Bull.* **66**:775–788.

Solli, H., Larter, S. R., and Douglas, A. G., 1980a, Analysis of kerogens by pyrolysis–gas chromatography mass spectrometry using selective ion detection. Part II. Alkylnaphthalenes, *J. Anal. Appl. Pyrolysis* **1**:231–241.

Solli, H., Larter, S. R., and Douglas, A. G., 1980b, Analysis of kerogens by pyrolysis–gas chromatography mass spectrometry using selective ion detection. Part III. Long chain alkylbenzenes, in: *Advances in Organic Geochemistry 1979* (A. G. Douglas and J. R. Maxwell, eds.), Pergamon Press, Oxford, pp. 591–597.

Stefanovic, G., and Vitorovic, D., 1959, Nature of oil shale kerogen, *J. Chem. Eng. Data* **4**:162–167.

Takeda, N., and Asakawa, T., 1988, Study of petroleum generation by pyrolysis I. Pyrolysis experiments by Rock Eval and assumption of molecular structural change of kerogen using 13-C NMR, *Appl. Geochem.* **3**:441–453.

Tissot, B. P., and Welte, D. H., 1978, *Petroleum Formation and Occurrence*, Springer-Verlag, Berlin.

Van de Meent, D., Brown, S. C., Philp, R. P., and Simoneit, B. R. T., 1980a, Pyrolysis high resolution gas chromatography and pyrolysis–gas chromatography mass spectrometry of kerogen precursors, *Geochim. Cosmochim. Acta* **44**:999–1014.

van de Meent, D., de Leeuw, J. W., and Schenck, P. A., 1980b, Chemical characterisation of non-volatile organics in suspended matter and sediments of the Rhine River delta, *J. Anal. Appl. Pyrolysis* **2**:249–263.

Van Graas, G., de Leeuw, J. W., and Schenck, P. A., 1980, Analysis of coals of different rank by Curie point pyrolysis–gas chromatography and Curie point pyrolysis–mass spectrometry, in: *Advances in Organic Geochemistry 1979* (A. G. Douglas and J. R. Maxwell, eds.), Pergamon Press, Oxford, pp. 485–497.

Van Graas, G., de Leeuw, J. W., Schenck, P. A., and Haverkamp, J., 1981, Kerogen of Toarcian shales of the Paris Basin. A study of its maturation by flash pyrolysis techniques, *Geochim. Cosmochim. Acta* **45**:2465–2474.

Wilson, M. A., 1987, *NMR Techniques and Applications in Geochemistry and Soil Chemistry*, Pergamon Press, Oxford.

Yarzab, R. F., Abdel-Basset, Z., and Given, P. H., 1979, Hydroxyl contents of coals, *Geochim. Cosmochim. Acta* **43**:281–287.

Zilm, K. W., Pugmire, R. J., Larter, S. R., Allan, J., and Grant, D. M., 1981, Carbon-13 CP/MAS spectroscopy of coal macerals, *Fuel* **60**:717–722.

Chapter 14

Chemical Methods for Assessing Kerogen and Protokerogen Types and Maturity

JEAN K. WHELAN and CAROLYN L. THOMPSON-RIZER

1. Introduction and Scope

Kerogen is the complex, high-molecular-weight, disseminated organic matter (OM) in sediments. It is operationally defined as OM that is insoluble in nonpolar organic solvents and in nonoxidizing mineral acids (HCl and HF). It is generally believed to be the major starting material for most oil and gas generation as sediments are subjected to geothermal heating in the subsurface. It is the most abundant form of organic carbon on Earth—about 1000 times more abundant than coal, which forms primarily from terrigenous higher plant remains. Kerogen consists of the altered remains of marine and lacustrine microorganisms, plants, and animals, with variable amounts of terrigenous debris in sediments. It represents about 1% of the organic matter which originates from biological sources and undergoes extensive degradation and alteration within the water column and at the sediment–water interface prior to burial in sediments (Hedges and Prahl, this volume, Chapter 11). The structured terrestrial (e.g., woody) portions of kerogen have elemental compositions similar to that of coal (Hunt, 1979, pp. 279–280; Hunt, 1991). In the past, kerogen was often depicted as a complex high-molecular-weight material formed from random condensation of monomers generated by the initial breakdown of polymeric biological precursor molecules following sediment burial. However, kerogen may contain significant contributions from biopolymers which are altered to varying degrees by biogenic and diagenetic processes during deposition and early stages of sediment burial (e.g., Zumberge *et al.*, 1978; Curry, 1981; Nip *et al.*, 1986). Microscopic and, increasingly, chemical evidence is more in accord with substantial incorporation of biological macromolecules that have undergone varying degrees of alteration prior to and after burial (Goth *et al.*, 1988; Tegelaar *et al.*, 1989). Because kerogen is formed from the altered remains of marine, lacustrine, and terrigenous microorganisms, plants, and animals, it contains information about the depositional, geological, and geothermal history of sediments.

During the last 15 years, chemical and optical methods for examining kerogen in ancient sediments have "come of age" and are currently widely used in oil exploration. This chapter summarizes what is known at the present time about changes which occur

JEAN K. WHELAN • Department of Marine Chemistry and Geochemistry, Woods Hole Oceanographic Institution, Woods Hole, Massachusetts 02543. CAROLYN L. THOMPSON-RIZER • Conoco, Inc., Ponca City, Oklahoma 74602-1267.

Organic Geochemistry, edited by Michael H. Engel and Stephen A. Macko. Plenum Press, New York, 1993.

in kerogen structure as a result of biological, geological, and geothermal processes. These chemical–physical changes in kerogen structure, which will be considered here very broadly as changes in kerogen type and maturity, can be examined at the present time by a variety of chemical and spectroscopic methods, which will be the focus of this chapter. Microscopic methods, which are presented by Sentfle *et al.* (this volume, Chapter 15), will be summarized here as they pertain to kerogen chemical structure. Petrographic techniques, as they relate to bulk organic phases, provide important complementary information about the chemistry of the kerogen.

Kerogen analysis methods have the advantage over those for bitumen in that kerogen cannot migrate and, therefore, kerogen and the surrounding sediments (except for recycled materials) have been subjected to the same depositional and thermal histories. Microscopic methods work very well for determination of kerogen type and maturity of that portion of kerogen which is microscopically recognizable, or structured. However, the amorphous organic matter in sediments is often present in much greater abundance than the structured kerogen and is very important with respect to bulk organic geochemical properties, especially petroleum-generating potential (e.g., Peters *et al.*, 1977; Batten, 1983; Tissot, 1984; Thompson and Dembicki, 1986). Therefore, this chapter will be particularly concerned with those methods with the potential to obtain better quantitation of proportions of all constituents that make up the high-molecular-weight organic matter in a particular sediment. Several of the topics in this chapter overlap with others in this volume so that we will refer to background material provided by those authors and will limit discussion here to an overview of the strengths and weaknesses of those methods for characterization of kerogen structure, type, and degree of maturation at the present time.

All scientists who have developed and thought carefully about chemical kerogen classification schemes are in agreement that there are no "magic bullets"—no one method currently exists which will rapidly provide qualitative and quantitative information on kerogen type and maturity. All of the existing chemical methods should be used together with as much ancillary geological, geochemical, and microscopic information as possible.

Chemical techniques of kerogen typing will be discussed in terms of their usefulness with respect to: (1) routine analyses in oil and gas exploration and (2) providing information about the origin and subsequent geological history of kerogen in a particular sediment interval. A survey of the recent literature rapidly leads to the conclusion that methodology suitable for the former often does not provide a very complete picture of the latter. This is because determination of the total oil and gas generation potential of kerogens is directly related to the availability of various types of hydrogen-rich linkages in the kerogen. Thus, in theory, any combination of measurements which provide information on the amount, abundance, and thermal conditions under which various CH-containing moieties in the kerogen can be broken off and expelled will be suitable for oil and gas exploration. The most widely used "screening" methods for this purpose at the present time are a combination of pyrolysis and microscopic methods to measure both kerogen type and maturity. Rock-Eval pyrolysis is a popular technique used to measure residual oil- and, to some extent, gas-generating potential and thermal maturation via T_{max} (the temperature at which maximum pyrolyzable organic matter evolves). Microscopic characterization includes qualitative determinations of proportions of woody OM, amorphous OM, etc., and, for maturation, measurement of thermal alteration index (TAI), fluorescence, and vitrinite reflectance (% R_0).

"Routine" chemical kerogen typing methods now commonly in use (Tissot, 1984), while extremely useful in predicting bulk source rock potential, are very limited in their ability to relate specific kerogen properties to specific hydrocarbon generation potential as well as specific paleoproductivity and paleodepositional environments. This is partly because the latter tend to be very complex and can show both temporal and spatial variabilities on time scales which are geologically short (1 day, 1 year, and tens of years) while it is generally difficult to carry out closely spaced sampling on ancient cores on time scales of less than about 100 to 1000 years. The "end member" depositional and chemical characteristics of standards from specific areas for which the various kerogen typing schemes were developed are fairly clear (Fig. 1). These standards include algal-rich kerogens, particularly those containing lacustrine *Botryococcus* spp. (such as are typically present in kerogens in some layers or horizons of the Green River Shale) or their marine equivalent, tasmanite (type I), the Lower Toarcian Shale from the Paris Basin (type II), and the Upper Cretaceous sediments from the Douala Basin (type III) (Tissot *et al.*, 1974). However, sediments from other areas tend to consist of mixed types which are difficult to quantitate. There is the additional problem that many types of depositional environments exist which are not included in the stan-

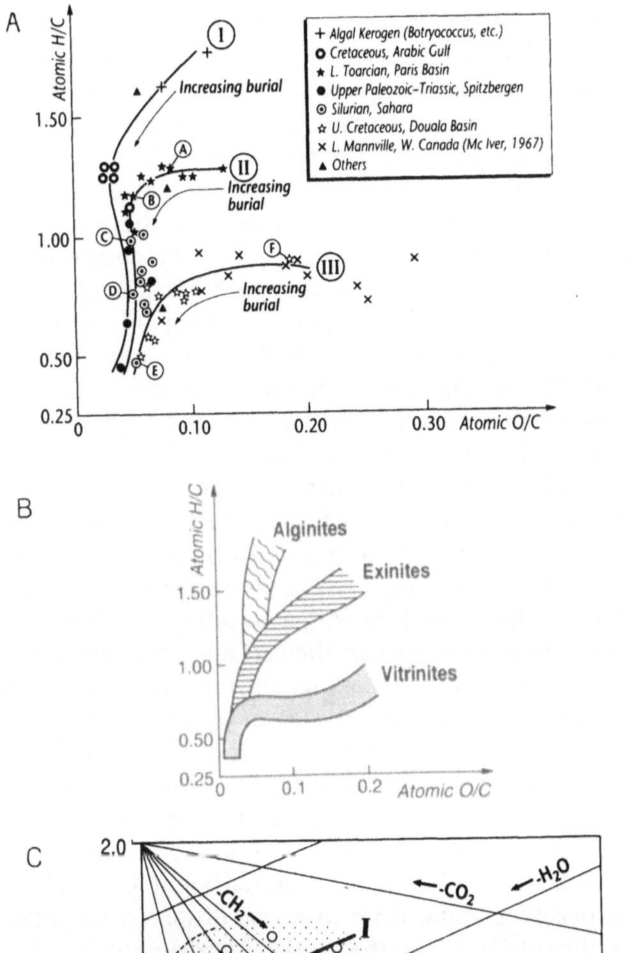

Figure 1. (A) Van Krevelen diagram of elemental H/C and O/C as applied to coal and kerogen typing. [From Tissot *et al.* (1974).] (B) Principal coal types, based on van Krevelen (1961). (C) Principal evolution pathways of kerogen maturation from original data set for kerogen types I, II, and III. [From Tissot and Welte (1984).]

dard set and that almost all depositional settings which are examined in detail tend to have a variety of kerogen types, as currently defined by the standard Tissot (1984) classifications (discussed below). For example, the complexity and diversity of lacustrine depositional environments is well known to limnologists (Fleet *et al.*, 1988), and the influence of this diversity on petroleum generation and geochemical indicators is beginning to be recognized (e.g., Talbot, 1988; Bjorøy *et al.*, 1988). Similar arguments apply to "terrigenous" sediments (e.g., Huc *et al.*, 1986) and to marine sediments (e.g., Mitterer, 1988).

Kerogen typing becomes even more problematic in examining ancient sediments, where the various depositional environments may not have good modern analogs. For example, the present-day ocean circulation patterns are apparently much different today than in past periods when much of the world's current oil appears to have been generated (Demaison and Moore, 1980). Over half of the world's current conventional petroleum reserves appear to have been generated in the Jurassic and Cretaceous (Klemme and Ulmishek, 1989), in which epicontinental seas and transgressive cycles were important. Modern analogs to these seas are very difficult to find. The problem of relating recent to ancient kerogens becomes more difficult as sediments are exposed to increasing thermal alteration during burial because, even for the simplest kerogen classification schemes, chemical and optical properties tend to converge at higher maturities (e.g., Durand and Monin, 1980). Thus, in the case of higher maturity ancient sediments, only a geological and stratigraphic or sedimentological reconstruction is likely to give a reasonable estimate of past generation potential, since immature samples or analogs of the source kerogen in question are often unavailable. In addition, the complexity of kerogen demands that a variety of chemical and microscopic techniques be applied together if "origin" versus "maturation" versus "biodegradation" effects on kerogen alteration (directly related to oil and gas generation potential) are to be successfully distinguished.

A problem with most of the microscopic kerogen classification schemes now in common use is that they are derived from the coal literature, where most identifiable organic matter is derived from higher plants and generally shows distinctive structures which can often be related back to the biological precursor, or at least can be interrelated. In contrast, the majority of visual kerogen data reported in the literature for petroleum source rocks show more than 50% amorphous kerogen, with little or no information on the biological origin of this important fraction. Bulk chemical methods provide data on the overall generating potential of a sample, while optical methods show the physical makeup of the sample, such as a dominant type of amorphous fluorescent kerogen and/or varying amounts of structured kerogens (oil-prone or gas-prone). Thus, microscopic and chemical kerogen characterization should be combined, whenever possible, as proposed by Thompson and Dembicki (1986) and Jones (1987).

2. Definitions and Isolation Procedures

Kerogen is defined here as the insoluble organic matter in sediment and rocks which remains after sequential treatment with common organic solvents (such as benzene/methanol, toluene, and methylene chloride) followed by dissolution of the rock matrix with cold or warm hydrochloric and hydrofluoric acids via methods described more fully by Durand *et al.* (1972) and Durand and Nicaise (1980). Various earlier operational definitions of the term "kerogen" were reviewed by Durand (1980). The definition given above is somewhat narrower and conforms with general usage at the present time: kerogen is a material derived from a widely used extraction procedure (an operational definition). This definition reflects the complex chemical nature of kerogen, which is not a well-defined chemical substance.

Rock samples are coarse ground (20–40 mesh) to minimize mechanical damage to the OM, then mixed with various acid solutions and water rinses over a period of several days. Usually, the ground sample is not solvent extracted; therefore, some of the more bitumen-like OM may be preserved. Very low heat, never more than 50°C, is occasionally necessary to dissolve dolomite in HCl (Bostick and Alpern, 1977).

Extraction procedures used in kerogen isolation can alter chemical, physical, and microscopic structure, thereby giving erroneous information about both kerogen type and maturity. Descriptions of nonoxidative acid (HCl and HF) isolation of kerogen and preparation of strewn slides for microscopic work have been given by Correia (1969, 1971), Bostick and Alpern (1977), Thompson and Dembicki (1986), and several others.

Organic matter initially deposited with unconsolidated sediments has generally not been considered to be kerogen, but a "protokerogen" (sometimes also called humin, although this term is more commonly reserved for the insoluble material which remains after extraction of soils or sediments with 3*N* to 6*N* sodium hydroxide; for a recent review, see Francois, 1990) which is converted to kerogen during early stages of sediment burial and diagenesis by biological, chemical, and physical processes (Connan, 1967; Huc, 1980; summarized in Hunt, 1979, pp. 118 and 550). The distinction of kerogen from protokerogen was based on observations in many laboratories that a significant fraction of surface sediment organic matter can often be removed by hydrolysis with refluxing 6*N* hydrochloric acid for periods of typically several hours or days. For example, 50% of the organic matter in a Black Sea surface sediment could be removed by a 16-h hydrolysis with refluxing 6*N* hydrochloric acid, which hydrolyzes protein and carbohydrates and causes a number of other functional-group changes in the sediment organic matrix (Huc, 1980). This percentage dropped gradually with depth to 10% by about 700 m and then remained constant to about 1100 m, by which time the sediment could more properly be termed as lithified. However, Connan (1967) found that the percentage of amino acids recoverable in this way was also highly dependent on the geologic age of the sediment. On the basis of pyrolysis data, little loss was observed for recent sediment OM when the mild acid treatment (i.e., acid isolation, no heat) was used to isolate protokerogens from a variety of contemporary depositional environments (see microscopic slides, Fig. 13, C. L. Thompson-Rizer and J. K. Whelan, unpublished data).

Although protokerogen is rather loosely defined in the literature, it is presumed in the examples above that the acid-hydrolyzable material would be called "protokerogen" while the nonhydrolyzable residue would be "kerogen." Chemically, the acid hydrolysis removes the more acid-labile portions of the biological macromolecules making up protokerogen (such as carbohydrate, protein, and ester bonds and alcohol functionalities via acid-catalyzed elimination reactions). However, treatment with refluxing 6*N* HCl is not generally part of the normal kerogen isolation scheme, so that there is no quantitative general method for making the distinction between protokerogen and kerogen in the definition of kerogen given above and as commonly applied in most laboratories. The currently common kerogen isolation schemes do use 4*N* HCl at 60–70°C for significant periods of time (Durand and Nicaise, 1980); this treatment undoubtedly also causes some hydrolysis of acid-labile groups in unconsolidated sediments. Therefore, it should be kept in mind that the discussion below includes some fraction of protokerogen within the term "kerogen" for thermally immature sediments, although we will use the term protokerogen occasionally to refer to the disseminated high-molecular-weight organic matter in recent unconsolidated sediments. This nomenclature conforms to that used by Curry (1981) in a study of kerogens, humic acids, and fulvic acids extracted from recent sediments from the Gulf of Mexico and Baffin Bay. That study also concluded, on the basis of isotopic, spectral, and chemical evidence, that humic acids and fulvic acids, extracted by base from surface sediments, have different precursors and follow completely different evolutionary paths from those of kerogen.

Protokerogens have been shown to be much less predictable than kerogen with regard to the changes which occur in their chemical and spectral properties during early stages of maturation (Durand, 1980; Curry, 1981; Peters *et al.*, 1977, 1980, 1981a). More care must be taken in isolation procedures to ensure that protokerogens are not altered, particularly by reaction with oxygen or heat generated by acid treatment, which can occur during the isolation procedure (e.g., Haddad *et al.*, 1989). The justifications for discussing protokerogen and kerogen together in this chapter are: (1) there is no clear method of experimentally separating the two using currently common isolation procedures; (2) even though protokerogens do not undergo chemical changes as predictably as kerogens, they do have many characteristics in common—particularly with respect to their microscopic properties and the nature of most of their C—H bonds (discussed further in a later section); and (3) understanding the features which protokerogens and kerogens have in common is the key to a better understanding of the depositional environments from which ancient kerogen types were derived.

It is nearly impossible to obtain a mineral-free kerogen concentrate due to acid-resistant minerals, neoformed minerals from HF complex fluorides (ralstonite), and minerals protected by OM coating (Durand and Nicaise, 1980). *Complete removal of pyrite, which is intimately mixed with many kerogens, is a particular problem, and all procedures currently in use for its removal also cause some alteration of the kerogen structure* (Durand and Monin, 1980). Robert (1974) developed a "froth flotation" technique to isolate kerogen, using a water–oil–soap–air mixture without the use of acids, so that large fragments which maintain part of the organic-mineral matrix are isolated. Some laboratories use heavy liquids (water-soluble $ZnBr_2$ adjusted to a specific gravity of 2.0; Staplin, 1969) to separate the pyrite from the isolated kerogen (Robinson, 1969), as the pyrite can be troublesome for both optical and chemical analyses. Lithium aluminum hydride has been used (Saxby, 1970; Yurum *et al.*, 1985) to dissolve pyrite; however, the OM is chemically changed. Some work has been done on electrolytically removing pyrite from kerogen (Wen *et al.*, 1976); however, no optical descriptions have been published. Kinghorn and Rahman (1983) used a series of heavy liquids to segregate different kinds of kerogen, including oil-prone from gas-prone amorphous kerogens. Visual kerogen strewn slides can be observed during application of any of these techniques to monitor qualitative changes made in the OM distribution, particularly of particle size fractionations caused by density differences.

3. Kerogen Type

3.1. Bulk Chemical Methods—Provision of the Chemical Framework

Many different chemical methods have been used with varying degrees of success to determine kerogen type and maturity. Material balance is always important in deducing both chemical and geochemical processes so that, ideally, these methods should always be applied within the framework provided by the chemist's old classical standby of elemental analysis. Some elements, particularly oxygen and sometimes hydrogen, are difficult to determine reliably, so that pyrolysis techniques have largely replaced elemental analysis, particularly for petroleum source rock analysis. Because of the complexity of kerogen, many other techniques have also been explored or utilized over the years, including various spectral techniques [infrared (IR) spectroscopy, nuclear magnetic resonance (NMR) spectroscopy, and electron spin resonance (ESR) spectroscopy], chemical degradation techniques, and isotopic methods. Most of these methods are restricted to kerogen typing in source rocks of low to moderate maturity and are not applicable to high-maturity kerogens, unless an immature analog is available. That is, there are almost no generally applicable chemical methods currently available for telling what a kerogen was like prior to thermal maturation—information needed to determine the composition and amount of oil and gas the "thermally spent kerogen" might have generated in the past. One approach to solving this problem might be to apply hydrous pyrolysis, a laboratory procedure which mimics the chemistry and physical properties of the geological environment in many respects (see Lewan, this volume, Chapter 18), to a variety of low-maturity sediments where the depositional environment has been well defined.

An excellent general review of kerogen analyses methods and results as applied to Precambrian rocks (Hayes *et al.*, 1983) points out that the complexity of kerogen is far beyond the capabilities of any single technique and that a variety of analytical tools must be applied to any successful classification scheme. Three groups of chemical tools are defined:

- (i) "overall" techniques that provide information about the average-bulk structure and composition of all organic matter in a given sediment (such as elemental and isotopic analysis);
- (ii) structural techniques that allow qualitative recognition or semiquantitative (and sometimes quantitative) measurement of some of the chemical structural features of organic matter (e.g., spectral properties); and
- (iii) degradative techniques that allow isolation and detailed characterization of some more or less well-defined subunits [e.g., pyrolysis–gas chromatography/mass spectrometry (Py-GC/MS), chemical degradative schemes].

The authors pointed out that application of this sequence is to know progressively more and more about less and less. The degradation techniques can provide very elegant structural information about specific portions of the kerogen. However, these products may be very unrepresentative of the kerogen as a whole and these techniques should always be applied within the framework of overall characteristics of the kerogen or rock, which can only be crudely determined. The "overall" methods, such as elemental analysis or measurement of any other chemical parameter proportional to elemental analysis, can only provide average (bulk) data for the entire sample and cannot provide information about detailed components of the mixture. Durand (1980) also emphasized this point in his thoughtful discussion of the limitations of elemental analysis as applied to kerogens, particularly those which have undergone extensive burial and maturation.

No one type of kerogen chemical classification method is intrinsically superior to another. The choice of whether the "bulk" or "detailed" methods are most suitable must be made by considering the purpose of the classification. For example, hydrogen transport is the crucial overall chemical process for oil and gas formation, for which kerogen can be considered to be simply the transport agent. Therefore, the most basic chemical parameter required for kerogen classification with respect to oil and gas formation is the amount of elemental hydrogen bonded to the kerogen. However, if the purpose of kerogen classification is to gain information about detailed geological processes, then detailed chemical methods used in conjunction with good microscopic information are more suitable.

3.1.1. Elemental Analysis

One of the earliest and still most unambiguous chemical methods of kerogen characterization is based on a technique widely and successfully applied to coal (van Krevelen, 1961, Chapter 6): determination of the elemental hydrogen, carbon, and oxygen composition of the organic material and plotting of the H/C and O/C ratios on a van Krevelen diagram, as shown in Fig. 1 (Tissot *et al.*, 1974; Durand and Monin, 1980). The kerogens showing the highest H/C ratio are classified as type I and the lowest as type III, with type II being intermediate. Note that kerogen types I, II, and III as originally defined by Tissot *et al.* (1974) follow the same H/C versus O/C lines as the coal macerals alginite, exinite, and vitrinite, respectively (compare Figs. 1A and 1B). During thermal maturation or catagenesis, all kerogen types lose hydrogen- as well as oxygen-containing functional groups and generally progress along the solid lines in the direction of the arrows toward the lower left-hand corner. The superimposed diagonal lines (Fig. 1C) show the effects on the elemental composition of "formal" elimination of water, carbon dioxide, or paraffins from the initial kerogen. Some representative carbon skeletons which would correspond to various H/C ratios are shown in Fig. 2.

It has been shown experimentally that all kerogen types initially expel hydrogen and oxygen predominantly as water and carbon dioxide during the lower temperature (diagenetic) stage of maturation and via hydrocarbon (HC) loss (oil and gas generation) during the higher temperature catagenetic maturation stages (e.g., Durand and Monin, 1980; Curry, 1981; Tissot and Welte, 1984; Hunt, 1979). Interestingly, when laboratory sealed tube heating experiments were carried out, both types of elimination (CO_2 and H_2O together with hydrocarbons) occurred simultaneously from the protokerogen (Peters *et al.*, 1980, 1981a), which suggests that there may also be a heating rate effect on the apparent maturation-

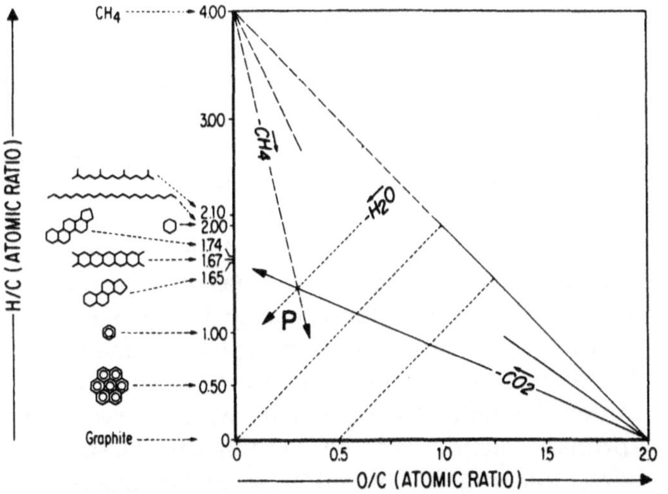

Figure 2. Typical kerogen elemental loss patterns and representative organic structures corresponding to various H/C and O/C atomic ratios.

dependent "selectivity" of evolution of the CO_2 and H_2O versus HCs.

The scheme outlined above, as originally proposed by Tissot *et al.* (1974), was the first purely *chemical* (elemental) kerogen classification to appear and be widely applied. This original scheme, which followed the coal classification scheme of van Krevelen (1961), was *not* based on either microscopic or pyrolysis data, and none are presented in the initial papers. In the original kerogens to which the Tissot classification scheme was applied (Fig. 1A), a type I classification was given to paraffinic kerogens where the H/C ratio was greater than 1.25 and the O/C ratio was low (less than about 0.15; Tissot *et al.*, 1974). Typical type I kerogens are found in boghead coals and shales containing abundant *Botryococcus* algae derived from lacustrine sedimentation or tasmanite, a marine equivalent. Using the criteria of high H/C ratios, some carbonate kerogens also fall into the type I category, such as some of the Persian Gulf Cretaceous limestones (Tissot *et al.*, 1974).

True type I source rocks, which tend to be primarily oil prone upon maturation, are relatively rare. One reason may be an artifact of the definition scheme shown in Fig. 1: the type I curve tends to merge with that of type II during the maturation process so that type I can only be recognized at fairly low (below about 0.8% R_0) maturation levels. At all higher maturation stages, types I and II kerogens would be typed together as type II.

The original "reference" type II kerogens shown on Fig. 1 came from the Lower Toarcian Shale of the Paris Basin (Tissot *et al.*, 1971). In comparison to type I kerogens, immature type II kerogens had lower H/C ratios (less than about 1.3) and equivalent or somewhat higher O/C ratios (in the range of about 0.03–0.18). Many organic-rich ancient and recent low-maturity marine sediments have predominantly type II kerogen. Tissot *et al.* (1974) also described elemental analyses of other type II kerogens including Silurian shales from the Sahara, Algeria, and Libya, one Cretaceous Persian Gulf sample, and several Upper Paleozoic–Triassic shales from Spitsbergen. The reference type II kerogens typically tend to generate a mixture of oil and gas upon maturation and are believed to be the immature analogs of the major kerogen type in many highly productive oil and gas fields. However, this nomenclature is confusing, since the "type" infers a *chemical* rather than a microscopic or depositional group. This confusion in nomenclature pervades the kerogen literature. For example, type II kerogens are sometimes referred to broadly as "liptinitic" or more narrowly as "algal," even though these terms are also often applied to type I kerogens, which have an even higher H/C ratio. Other examples are given later in this chapter, where "typical" lacustrine, marine, and terrigenous sediments are characterized by any of the four chemical kerogen types, or as mixtures (e.g., Huc *et al.*, 1986; Mitterer, 1988; Bjorøy *et al.*, 1988). Tissot (1984) observed similar problems in chemical and optical relationships: "Systematic elemental analysis performed on a set of amorphous kerogens from various origins has shown that, although some of them belong to type II, the chemical composition of the amorphous kerogen may spread over the entire van Krevelen diagram" (i.e., Fig. 1).

The type III kerogens for the samples shown in Fig. 1 have relatively low H/C ratios (generally less than 1) and higher O/C ratios (in the range of about 0.3 to 0.03). The specific reference type III samples in Fig. 1A include Upper Cretaceous sediments from the Douala Basin, Cameroon, in which planktonic remains are virtually absent (Albrecht *et al.*, 1976; Durand and Espitalié, 1976), and the Lower Mannville shales of the Western Canada Basin (McIver, 1967). Note that it is generally assumed that there is a significant higher plant and "woody" fragment contribution to the type III kerogens, so that they are often referred to as "woody, coaly, vitrinitic, or humic" in reference to the relatively low hydrogen and high oxygen content generally found in these materials. However, microscopic results for the "standard" Douala Basin type III kerogen show 0.4–2.7% exinite, 5.5–11.7% vitrinite, 0.4–2% inertinite, and 78–93% "amorphous cement" (Durand and Espitalié, 1976). No microscopic data are reported by McIver (1967) for the Mannville shales. Therefore, it is not obvious that the higher O/C ratios of type III kerogens come *predominantly* from higher plant fragments, since the amorphous fraction from an unknown precursor is by far the largest fraction in the Douala Basin kerogens. The type III kerogens, which tend to be more gas prone than either of the other two types, are chemically broadly analogous to the coal maceral vitrinite, although wide variations occur (e.g., Huc *et al.*, 1986). The amorphous material may be similar to the gas-prone amorphous types documented by Thompson and Dembicki (1986).

A fourth coal maceral type, inertinite, has a H/C ratio always less than 0.5 and which follows its own maturation line, below that for the other macerals. The corresponding kerogen constitutes a fourth kerogen type, called a "residual type" by Tissot and Welte (1984, p. 155). Other scientists often designate this kerogen as type IV or inertinite (e.g., Waples, 1981; Orr, 1983), with a maturation line near the bottom axis of the van Krevelen diagram.

More recently, Orr (1986) has defined an additional unusual high-sulfur (8–14%) type II kerogen, designated type II-S, which is the source for heavy high-sulfur oils from the onshore and offshore Monterey Formation in California (Fig. 3). These heavy sulfur-rich oils are apparently generated at much lower maturities than observed for other kerogen types. The distinction between types II and II-S is based on the high kerogen S/C ratios (Fig. 3); the visual characteristics of type II-S are not different from those of type II, with both being microscopically classifiable as amorphous type A as defined in Section 3.2 (Thompson and Dembicki, 1986).

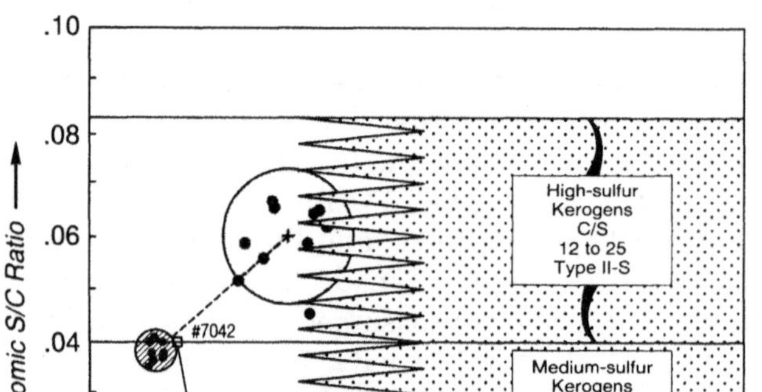

Figure 3. Use of O/C and S/C ratios in classification of type II-S kerogen. [From Orr (1986).]

Because of the time and technical difficulties of carrying out kerogen isolation and elemental analyses reproducibly, other faster techniques have been explored and developed in recent times. Many of these "secondary" classification parameters serve as a proxy for elemental data so that development of a new kerogen typing parameter generally involves a demonstration of proportionality to the elemental data for a variety of kerogen types and maturities. Therefore, it is important to know the reliability of the initial elemental measurements which serve as the "primary" standards for these comparisons. High ash or residual mineral content in isolated kerogens often causes difficulties in obtaining reproducible elemental analyses (Durand and Monin, 1980; Hayes *et al.*, 1983; Thompson and Dembicki, 1986).

The reliability and precision of carrying out kerogen elemental analysis of C, H, O, N, S, and Fe have been carefully reviewed by Durand and Monin (1980), who applied these measurements to 440 kerogens. The experimental procedures used by these workers are fairly typical: C, H, and N were measured on a single aliquot by sample oxidation at about 1000–1100°C under a stream of oxygen to produce CO_2, H_2O, and a mixture of nitrogen oxides. The latter were reduced to nitrogen by passage over copper at 500°C. The resultant CO_2, H_2O, and N_2—generally separated and measured by GC—gave a measure of kerogen C, H, and N. Oxygen was measured on a separate aliquot by pyrolysis in inert atmosphere at about 1000°C, followed by passing the free gas through pure carbon at 1100–1500°C, then over copper at 900°C in order to convert all oxygenated molecules into CO and at the same time to remove sulfur. The CO was either measured directly by IR (the best procedure for samples containing very low oxygen contents of <0.1%) or was oxidized to CO_2 and then measured by either GC or coulometry. Sulfur was oxidized to SO_2 at 1300°C and then measured coulometrically.

Measurements of carbon and hydrogen were found to be the most precise (1.25 and 2.65% mean standard deviation, respectively) with oxygen, nitrogen, and sulfur showing considerably greater scatter (4.2, 4.6, and 10.1% mean relative deviation). The scatter was attributed predominantly to natural intrasample variation and problems caused by the kerogen isolation rather than the elemental analysis itself. Utilizing ratios of H/C and O/C for a specific sample gave better precision (1.65 and 3.5% mean standard deviation, respectively) than the individual analyses for C, H, and O.

Residual minerals were found to systematically increase the H and O values so that it was found to be important to eliminate as much as possible the neoformed mineral fluorides that resulted from dissolution of the silicate mineral matrix with HF during the kerogen isolation procedure. This was accomplished by conducting the HF dissolution in 4*N* HCl at somewhat elevated temperature (70–90°C), which was thought to be responsible for much of the variability observed in the kerogen nitrogen analyses.

Kerogen elemental analysis of sulfur gave the greatest problem and the most scatter (10.1% mean deviation) because of the difficulty of completely separating organic and inorganic species, particularly pyrite, prior to analysis. This problem was attributed to the close association of sedimentary organic matter with pyrite, so that organic matter apparently coats the pyrite, as shown by the detection of small amounts of carbon and hydrogen in isolated pyrites (Durand and Monin, 1980, p. 116). Thus, oxidative procedures for removing pyrite can cause errors in kerogen sulfur as well as carbon, hydrogen, and oxygen analysis.

Representation of elemental data as ratios, as in Fig. 1, is helpful in viewing C, H, and O data because (1) no primary standardization of the C, H, and O is required for a specific sample; (2) the experimental uncertainty in measuring H/C and O/C via kerogen elemental analysis is less than that for measuring individual C, H, and O values; and (3) the diagram provides information about allowable structures for the carbon skeleton [Fig. 2, from Durand and Monin (1980)]. For example, loss of hydrogen can correspond to cyclization of straight-chain hydrocarbons or formation of double bonds or aromatic rings. Other spectral and chemical information would be needed to determine exactly which of these changes had actually occurred.

As is the case with coal, the van Krevelen diagram provides an initial idea of allowable chemical changes which may be occurring in the complex mixture of substances called kerogen. An example is shown in Fig. 2, which shows representative degrees of condensation of aromatic rings for various H/C ratios. This figure emphasizes the important point that the reliability of other more sophisticated kerogen analysis techniques for characterizing detailed structures should be tested for their potential generality within this H/C framework. For example, at higher maturities involving kerogens with H/C of less than 0.5, any technique looking only at hydrogen can give very little information on the kerogen as a whole and its probable source, because all carbon skeletal structure is rapidly approaching the graphitic state.

Elemental analysis of sulfur as well as C, H, and O has been used by Orr (1986) in deducing the mechanism of thermal evolution of type II-S (Monterey) kerogens to heavy sulfur-rich oils. In contrast to the common Type II kerogens, which generate oil largely by cleavage at carbon–oxygen and carbon–carbon bonds [eliminating CO, CO_2, H_2O, and hydrocarbon fragments, as summarized by Durand (1980)], the type II-S kerogens have additional weak bonds associated with the bound sulfur. Orr postulated that preferential cleavage at the weak sulfur linkages produces larger fragments, leading to high initial amounts of asphaltenes, resins, and sulfur-rich aromatics together with the smaller amounts of saturated hydrocarbons during the maturation process. Elemental analyses of Monterey kerogens and asphaltenes have led to the concept that the asphaltenes may be smaller fragments of the kerogen, consistent with the elemental, isotopic, and thermal cracking properties (Orr, 1986). The chemical similarity of asphaltene/kerogen pairs for other kerogen types has also been demonstrated based on elemental composition, spectroscopic (NMR and ESR) data, and chemical degradative (pyrolysis-GC and sealed-tube pyrolysis) analyses (Bandurski, 1982; Horsfield *et al.*, 1983; Behar and Pelet, 1985; Aizenshtat *et al.*, 1986a).

Because considerable overlap in kerogen types occurs within the simple classification scheme based strictly on elemental analysis (Fig. 1), Jones (1987) has proposed a modified scheme, which is summarized in Table I. A series of "organic facies" are defined

TABLE I. Generalized Geochemical and Microscopic Characteristics of Organic Facies A–D via Kerogen Classification Scheme of Jones (1987)

Organic facies	H/C at $\%R_0$ about 0.5	Pyrolysis yield HI	Pyrolysis yield OI	Dominant organic matter
A	≧1.45	>850	10–30	Algal; amorphous
AB	1.35–1.45	650–850	20–50	Amorphous; minor terrestrial
B	1.15–1.35	400–650	30–80	Amorphous; common terrestrial
BC	0.95–1.15	250–400	40–80	Mixed; some oxidation
C	0.75–0.95	125–250	50–150	Terrestrial; some oxidation
CD	0.60–0.75	50–125	40–150+	Oxidized; reworked
D	≦0.6	<50	20–200+	Highly oxidized; reworked

based on H/C values at low maturities $R_0 < 0.5\%$), and the general elemental and microscopic classification of each is summarized. However, because low-maturity samples are not always available for any given kerogen, considerable geological and geochemical interpretation was needed in making the assignments to each facies. The advantages of this scheme over that of Tissot and Durand and co-workers are not completely clear, since the initial Tissot scheme is often broadened by defining "mixed" kerogen types such as the commonly occurring mixed type II/III. The Jones "organic facies" classification scheme provides somewhat different divisions between depositional and maturation facies than the original Tissot scheme. Both schemes are limited in descriptions of individual samples in which several kerogen types coexist. For example, it is difficult to differentiate oil- from gas-prone kerogen at intermediate H/C ratios (about 0.8) by either the Tissot or Jones classification scheme. The use of microscopic fluorescence methods for low-maturity kerogens can resolve this problem in many cases, because oil-prone kerogens generally fluoresce, while gas-prone kerogens do not (see Section 4.1.2).

The kerogen classification schemes based primarily on elemental analysis, described above, have proved to be very valuable in petroleum source rock classification where the main objective is to obtain an estimate of the amount of oil plus gas potentially available from a particular rock. However, a number of authors have pointed out that strict adherence to these criteria can cause problems in prediction of oil versus gas generation potential (Horsfield *et al.*, 1988, and references therein), particularly for terrigenous organic matter and fluviodeltaic–lacustrine systems. Ambiguities can also arise in applying these schemes to deducing depositional environments. For example, maturation lines for types I, II, and III of Tissot and facies A through D of Jones (1987) all tend to merge as they approach the origin at the lower left. Therefore, at higher maturities, it is not possible to deduce the kerogen type prior to maturation by elemental analysis alone. In addition, biodegradation of either type I or type II kerogens (facies A and B) prior to or just after deposition tends to decrease the H/C ratio and increase the O/C ratio. Therefore, a biodegraded marine kerogen would be classified as type III (or facies C) and might be incorrectly classified as having a terrigenous source. This is a common problem for deep-sea sediments so that the organic source (in comparison to oil/gas potential) must be determined by other means. For example, Fig. 4 shows kerogen elemental analysis data for immature kerogens isolated from South Atlantic Cretaceous Black Shales (Tissot *et al.*, 1980). Based only on elemental analysis, the observed differences in H/C and O/C ratios between samples can be attributed either to a greater contribution of terrigenous organic matter to the type III kerogens in groups C and D or to biodegradation of the hydrogen-rich portions of initially type I or II marine kerogens during or shortly after deposition. Microscopic examination showed the presence of significant amounts of coaly or more oxidized organic matter in both groups C and D in Fig. 4 with both reflected and transmitted light. Infrared analysis also showed a greater abundance of more oxygen-rich and aromatic functional groups in the group C and D samples in comparison to the group A and B samples in Fig. 4. Therefore, the greater contribution of terrigenous material appears to be responsible for the greater type III character of some of these samples. This interpretation is consistent with microscopic evidence (i.e., more woody kerogen, typical of terrigenous plants, rather than biodegraded phytoplankton) and IR characteristics (relatively high amount of aromatic C—H, characteristic of woody material in immature kerogens). However, note that it is not possible to determine by either of these techniques whether the terrigenous fraction is quantitatively sufficient to account for the observed differences in elemental ratios.

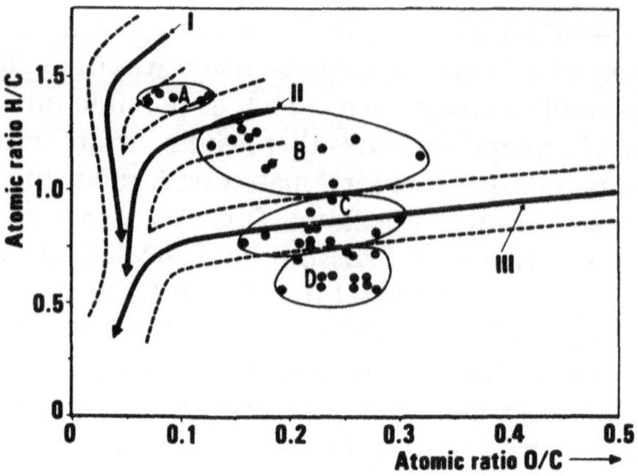

Figure 4. Van Krevelen diagram for immature Cretaceous Black Shales from the South Atlantic. All samples are immature. Group A: marine organic matter; Group B: mixture of marine and terrestrial organic matter; Group C: terrestrial, moderately degraded; and Group D: residual organic matter. [Figure and classifications from Tissot (1984); data from Tissot *et al.* (1980).]

Because of limitations of both the Tissot and Jones kerogen classification schemes described above, recent modifications have been to include more detailed microscopic and chemical structural information, such as that available from pyrolysis-GC and -GC/MS, spectral (infrared and NMR) analyses, and various chemical degradation schemes, as described later in this chapter.

Alternatively, in the future, because of the high abundance of amorphous kerogen in many sedimentary rocks, improved classification of this material via automated microscopic fluorescence measurements (see Section 4.1.2) could provide a semiquantitative approach to kerogen classification that would relate better to chemical classification parameters, i.e., observed oil/gas generation potential, elemental analysis, and spectral properties. For example, in the Conoco Geochemical Data Base, approximately 60% of the samples with visual kerogen data contain 50% or more amorphous kerogen (Thompson and Dembicki, 1986). One scheme for classification of these predominant kerogens into four categories, amorphous types A through D, together with a description of their relative oil to gas generation potentials, was recently described by Thompson and Dembicki (1986) and Thompson-Rizer (1987; Fig. 12, Table II and Section 3.2). The authors did not feel that a genetic classification of the four amorphous types was possible without further work, although comparison with kerogens isolated from well-characterized contemporary surface sediments, such as shown for some examples in Fig. 13, may make this possible in the future.

3.1.2. Pyrolysis Methods for Bulk Chemical Kerogen Typing—Hydrogen Index (HI) and Oxygen Index (OI)

Because kerogen isolation and elemental analysis are time-consuming and often leave behind residues which can interfere with both hydrogen and oxygen elemental analyses (Durand and Monin, 1980; Hayes *et al.*, 1983), efforts have been made to determine a parameter which will correlate with the H/C and O/C ratios by other techniques which can be applied more rapidly and routinely.

Early thermal methods for kerogen classification were summarized by Hunt (1962), with the first being that of von Gaertner and Kroepelin (1956), where two types of organic matter were distinguished by differential thermal analysis (DTA), one with combustion characteristics similar to those of coal and the other to those of oil. Forsman and Hunt (1958), who measured the thermal characteristics of isolated kerogen from several marine sediments, also found two basic types—one resembling coal, which gave mostly gas and very little liquid upon heating, and a second "oil shale type," which yielded a large proportion of liquid hydrocarbons. This study was followed by that of Radchenko (1960) in the USSR, who showed similar differences in sediment organic matter using DTA. Erdman (1961) found that the rates at which organic matter from sediments could be oxidized in the laboratory differed greatly, with that from nonmarine shales having the same oxidation rate as coals and that from marine limestones having the same oxidation rate as petroleum asphalts. Yen *et al.* (1961) used X-ray diffraction techniques to suggest that marine organic matter may have a structure similar to asphalt but of higher molecular weight. Gransch and Eisma (1970) demonstrated a good inverse correlation between the kerogen H/C ratio and C_R/C_T where C_T is the weight of total organic carbon and C_R is the weight of residual carbon left after heating the kerogen at 900°C for 2 hours in a nitrogen stream (Fig. 5).

The modern pyrolysis procedures generally measure the total hydrocarbons and other gases evolved during pyrolysis via a GC detector as the sample is heated at a fixed rate in a stream of inert gas. Early work was carried out primarily on whole rocks, rather than isolated kerogens (Barker, 1974; Claypool and Reed, 1976). A variation of this technique, where only the total evolved pyrolyzed gases are measured using a flame ionization detector (FID) for hydrocarbons and a thermal conductivity detector (TCD) for total pyrolyzable gases, became widely utilized for petroleum exploration after the group at the Institut Français du Pétrole applied the technique to the same sample set as in Fig. 1 and demonstrated a response which correlated with the elemental analysis of the various kerogen types (Fig. 6; Espitalié *et al.*, 1977, 1984). The kerogen or whole-rock sample is gradually heated in a helium stream (rate about 30°C min^{-1}), and pyrolyzed hydrocarbons and carbon dioxide are measured as a function of temperature, to give a pyrogram (Fig. 7A), where S_1 represents free hydrocarbons in the kerogen, related to oil already generated, and S_2 represents pyrolyzable hydrocarbons generated by pyrolysis, related to the petroleum-generating potential of the sediment. S_2 is normalized to total organic carbon (TOC) to give the hydrogen index, which is proportional to the kerogen elemental H/C ratio (Figs. 6, 10, and 11). In a similar way, the pyrolyzable CO_2 is measured and normalized to TOC to give the oxygen index, which is sometimes proportional to the kerogen elemental O/C ratio (Fig. 10). Note that the two ratios obtained will be called the "hydrogen index" (HI) and "oxygen index" (OI), respectively, in this chapter, in order to avoid confusion with data obtained by elemental analysis, which

TABLE II. Some Proposed Optical Kerogen Type Classifications[a]

Staplin (1969)[b]	Alpern (1970)[b]	Bostick (1970)[b]	Correia (1971)	Burgess (1974)	Combaz (1975)	Bujak *et al.* (1977)	Alpern (1980)[b]	Masran & Pocock (1981)[b]
Primary materials	Textites	Liptinite	Lignitic	Amorphous	Figured fraction	Amorphogen	Phytoclasts	Structured terrestrial
Terrestrial source	Xylotextite	Low-gray	Tracheids, sharp edges	Finely disseminated	Algae	Fine or fluffy, structureless	Xylites	Leaf, root, stem
Plant cuticles	Exotextite	High-gray	Sapropelic	Herbaceous plant	Microfossils	Phyrogen	Gelites	Spores & pollen
Plant spores, pollen	Fungitextite	Fusinite	Algae, tissues, degraded	Woody plant	Spore-pollen	Cuticle, spore, dinoflagellate, cysts, etc.	Detrites	Biodegraded terrestrial
Lignified wood	Gelites	Flocules	Smooth membranes	Coaly	Cuticle	Hylogen	Epiderms	Sheets of cells with or without walls
Mineral charcoal	Detrites		Fine particles	Algal	Tissue	Woody, fibrous, nonopaque	Fungi	Charcoal
Resins	Bitumites				Fusinite	Melanogen	Spores	Fungi
Plankton, freshwater algae	Spherolites				Amorphous fraction	Opaque	Organisms	Resins
Marine source	Sporites				Clots		Phytoplankton	Amorphous
Organisms—phytoplankton, benthos (bacterial, algal, fungal)	Organites				Coproliths		Zooplankton	Yellow, gray, granular, structureless, spherulitic
Modified materials					Subcolloidal		Bacteria	Biodegraded aqueous
Sapropelic					Pellicle		Bitumens	Structured aqueous
Fluffy mass—extremely fine					Waxy			Algal, dinoflagellates, acritarchs
Platy, translucent					Resinous globules			
Very resistant elements					Gellified fraction			
Platy, inert, cuticular reworked					Vitrinite			
Thermal transformation products					Crushed fraction			
Heat-blackened macerals from above					Inertinite			
Pyrobitumens & sooty structureless					Bituminites			
					Metamorphosed OM			

Venkatachala (1981)[b]	van Gijzel (1981)[b]	Dow and O'Connor (1982)	Batten (1983)[b]	Mukhopadhyay *et al.* (1985)[b]	Massoud and Kinghorn (1985)[b]	Robert (1985)[b]	Thompson and Dembicki (1986)[b]	Sentfle *et al.* (1987)[b]
Biodegraded terrestrial	Identifiable primary	Vitrinite	Amorphous	Primary	Keroginite	Coaly—primary	Amorphous A	Vitrinite
Biodegraded aqueous	Inertinite	Exinite	Fibrous	Liptinite	Liptikeroginite	Inertinite	Oil-prone	Liptinite
Phytoplankton, thalloid algae, filamentous algae	Vitrinite	Inertinite	Membranous	Alginite	Resikeroginite	Vitrinite	Amorphous B	Inertinite
Amorphous	Exinite	Amorphous	Pellicular	Unfigured amorphous	Vitrikeroginite	Exinite	Gas-prone	Amorphinite
Bacterial origin	"Incertae Sedis"		Spongy	Algo-detrinite	Inertikeroginite	Structureless—primary	Amorphous C	Fluoramorphinite
Gray amorphous	Zooorganism		Flakey	Sapropelinite I	Liptinite	Sapropelic groundmass	Gas-prone	Hebamorphinite
	Unstructured primary		Aggregated	Sapropelinite II	Autochthonous	Humic groundmass	Amorphous D	Faunal relics
	Detrinite		Finely disseminated	Liptodetrinite	Detrital	Bituminite	Oil- or gas-prone	
	Amorphinite		Subcolloidal	Particulate liptinite A	Allochthonous	Secondary products		
	Secondary products			Particulate liptinite B	Vitrinite	Oil		
	Bitumen			Resinite (A+B) + fluorinite	Autochthonous	Exsudatinite		
	Micrinite-carbon residues			Vitrinite	Parauchthonous	Micrinite		
	Gelites			Autochthonous	Allochthonous	Bitumens (soluble & insoluble)		
	Inclusions			Humosapropelinite	Inertinite			
				Vitrinite	Structured			
				Recycled	Detrital—unstructured			
				Vitrodetrinite & oxidized				
				Inertinite				
				Micrinite				
				Inertodetrinite				
				Scloerotinite				
				Fusinite/semifusinite				
				Secondary				
				Granular vitrinite				
				Rank inertinite (meta-alginite, -sapropelinite)				
				Micrinite (clustered)				
				Solid & liquid bitumen				
				Oil droplets				
				Exsudatinite				

[a]No attempt has been made to put equivalent terms in order.
[b]Photomicrographs have been published with classification.

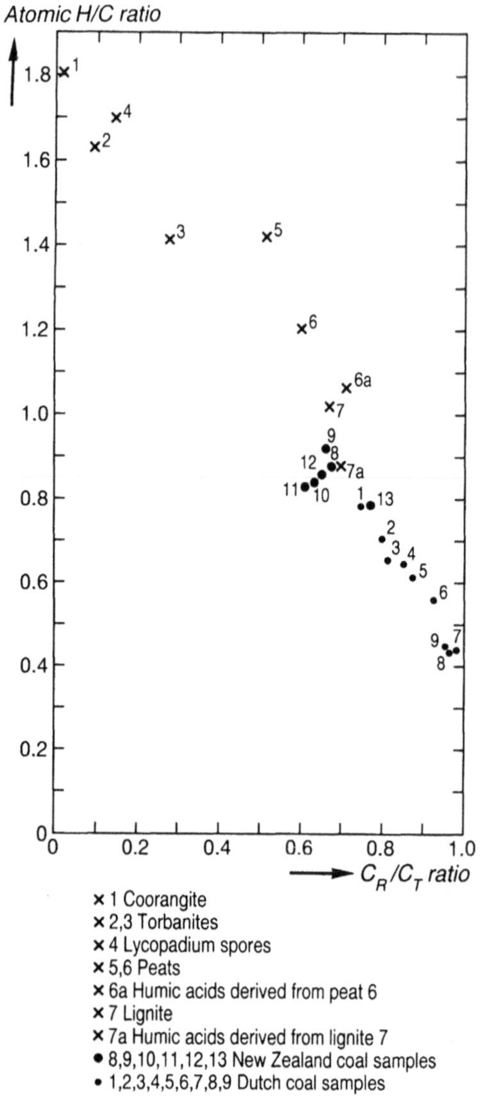

Figure 5. Atomic H/C vs. C_R/C_T, where C_T is total organic carbon and C_R is residual carbon remaining after heating kerogen or coal at 900°C for 2 h. [Adapted from Gransch and Eisma (1970).]

gives somewhat different curves (compare Figs. 1 and 6). The ratios obtained by elemental analysis will be referred to as H/C and O/C ratios throughout this chapter.

In rocks with a high carbonate content and low (<2%) TOC, interference of carbonate mineral pyrolysis can elevate the oxygen index when whole rocks are pyrolyzed. Therefore, oxygen indices are now generally measured on isolated kerogens rather than on whole rocks. Alternatively, pyrolysis kerogen typing is also commonly carried out by plotting the hydrogen index versus T_{max} [Fig. 8, from Espitalié *et al.*, 1984]. T_{max} is the temperature at which the S_2 peak reaches its maximum evolution. T_{max} is a pyrolysis maturation indicator which is discussed in detail in Section 4.2.1.

Excellent discussions of kerogen typing techniques via "Rock-Eval" type pyrolysis are included in two widely used texts (Hunt, 1979; Tissot and Welte, 1984) as well as in several review articles on use of the technique in petroleum exploration (Philp *et al.*, 1982; Philp and Gilbert, 1984, 1985; Dembicki *et al.*, 1983; Dembicki, 1984; Katz, 1983; Orr, 1983; Espitalié *et al.*, 1984; Larter, 1984, 1985, 1989; Tissot, 1984; Peters, 1986). A recent bibliography by Barker and Wang (1988) demonstrates the exponential growth rate of publications on pyrolysis as applied to petroleum geochemistry.

Because pyrolysis has so widely replaced elemental hydrogen analysis in the kerogen classification schemes described above, it is important to compare hydrogen determinations obtained by the two methods. A number of workers have now found that certain minerals, particularly smectite and illite, often cause catalytic effects with whole rock pyrolysis and give an artificially low hydrogen index when compared to isolated kerogen elemental analysis data

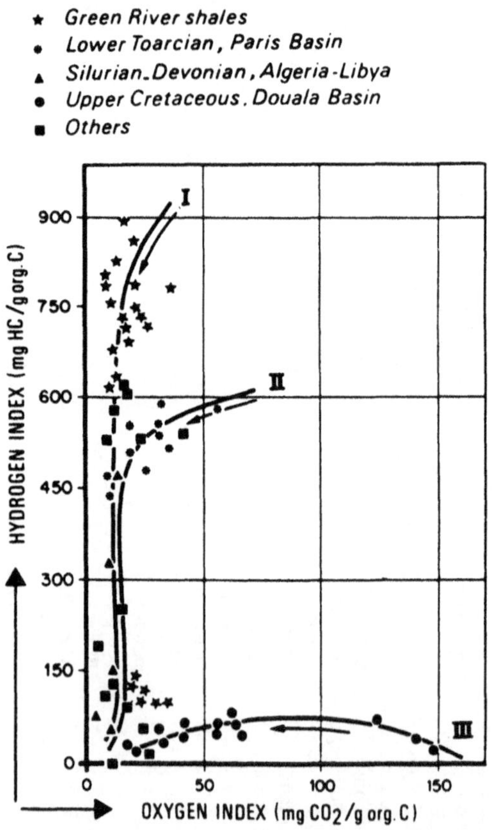

Figure 6. Classification of kerogen types using hydrogen and oxygen indices obtained from Rock-Eval pyrolysis. [From Tissot and Welte (1984); data from Espitalié *et al.* (1977).]

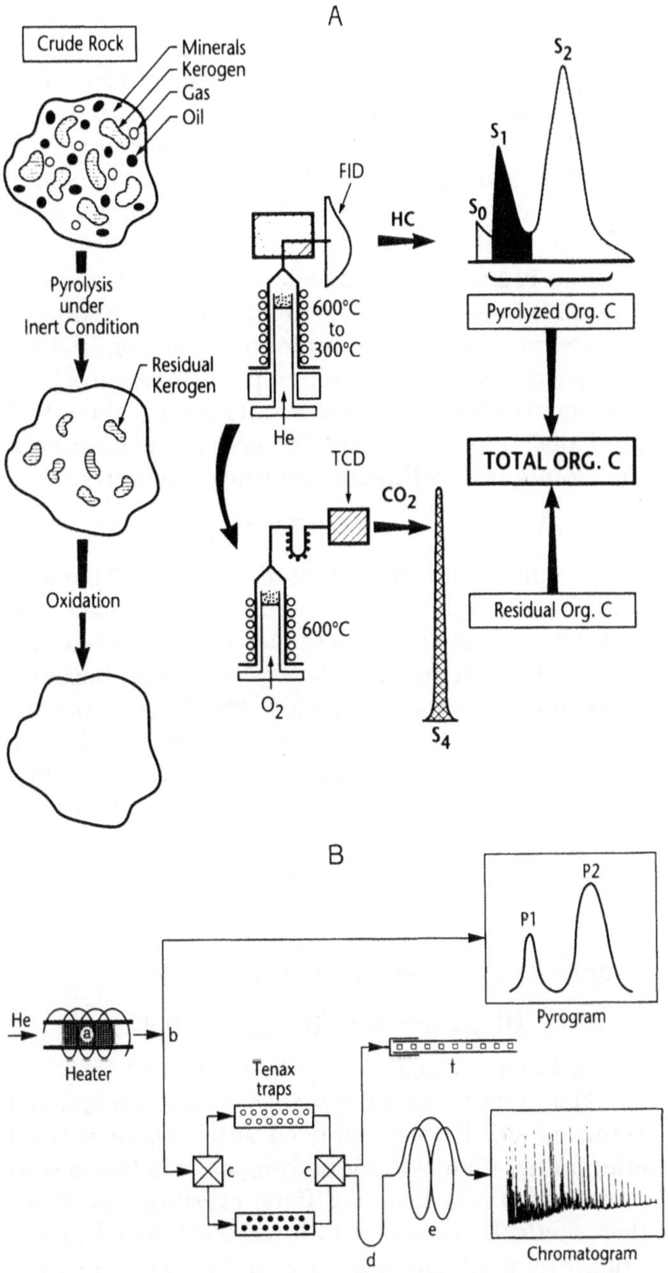

Figure 7. Schematic representation of pyrolysis apparatus and typical kerogen pyrolysis data: (A) Rock-Eval apparatus [from Espitalié *et al.* (1984)]; (B) typical pyrolysis–GC apparatus [adapted from Whelan *et al.* (1980)]. a) Sample being heated; b) helium stream split; c) multiport valves; d) chill loop; e) capillary GC column; f) Tenax trap for MS sample.

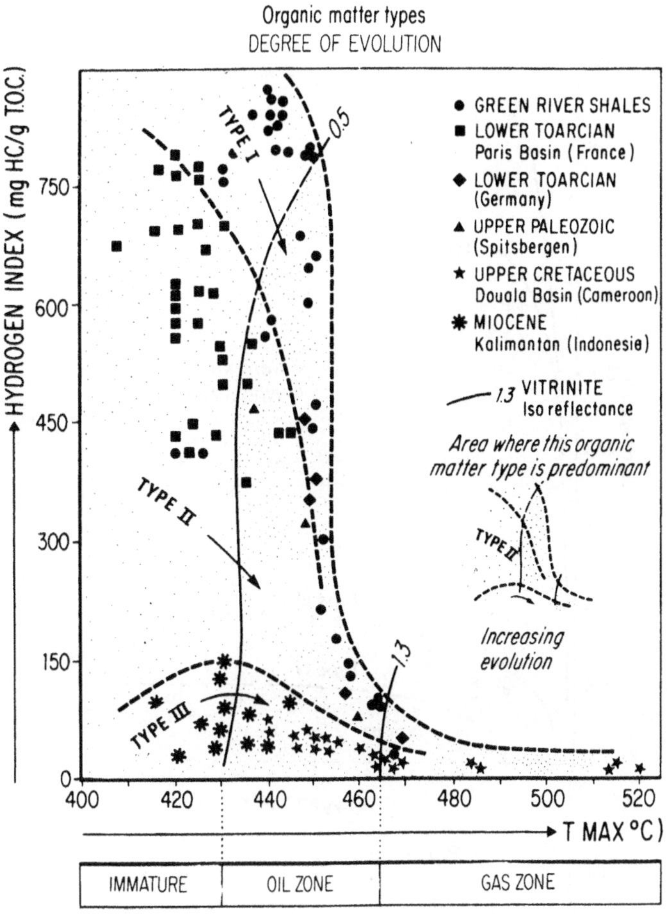

Figure 8. Kerogen typing via plot of hydrogen index vs. T_{max}. [From Espitalié *et al.* (1984).]

[as summarized by Orr (1983), Katz (1983), Peters (1986) Espitalié (1986), and Larter (1989), among others]. The problem becomes particularly acute for rocks containing low amounts of carbon. Typical examples are shown in Fig. 9 for miscellaneous Gulf Coast samples (Fig. 9A) and a set of down-hole samples from a well in the East Cameron Field off the Louisiana Coast in the Gulf of Mexico (Fig. 9B; J. K. Whelan and W. G. Dow, unpublished data). The kerogens are type II/III based on the isolated kerogen data but would be mistakenly classified as type IV based on the whole-rock data. These sediments do not contain calcium carbonate. Gulf sediments tend to contain high amounts of smectite, which is probably responsible for the artificially low HI and high OI values obtained for the whole-rock pyrolyses.

A second problem can arise if either the whole rock or the kerogens have not been solvent extracted and contain significant amount of heavy hydrogen-rich generated or migrated bitumens. In this case, the hydrogen index will be overestimated. These potential problems need to be kept in mind when pyrolysis hydrogen indices are substituted for kerogen elemental hydrogen analyses. To complicate matters further, in the case of high-maturity sediments containing pyrobitumens, it is often not possible to remove this nonindigenous material via solvent washing (e.g., Tarafa *et al.*, 1988).

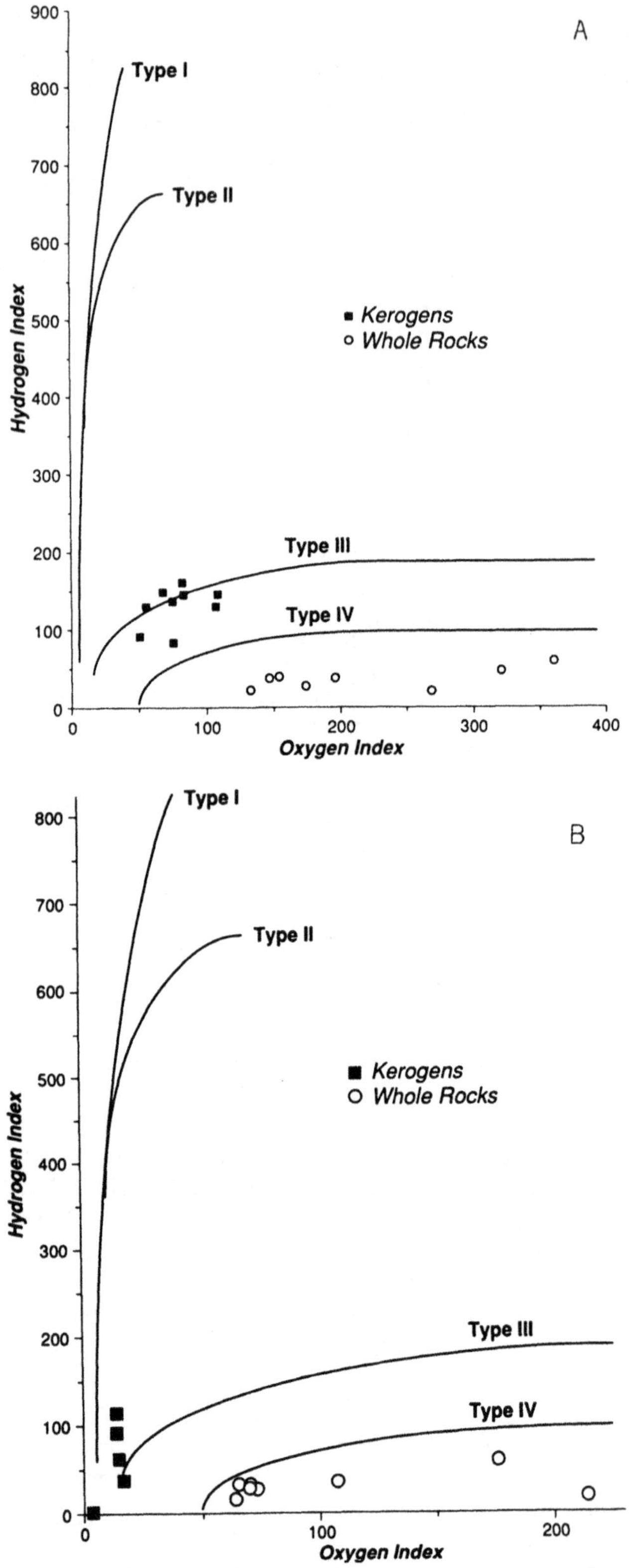

Figure 9. Hydrogen versus oxygen indices for Gulf Coast kerogens versus whole rocks: (A) Miscellaneous samples (W. G. Dow, unpublished data); (B) East Cameron well, Tertiary section, Louisiana Gulf Coast (J. K. Whelan and W. G. Dow, unpublished data).

How well do kerogen H/C and O/C ratios, as determined by elemental analysis, correlate with the pyrolysis hydrogen and oxygen indices? Good correlations were found for whole rocks for both hydrogen index versus H/C and oxygen index versus O/C for high-TOC samples analyzed by Espitalié *et al.* (1977) (Fig. 10), so that mineral matrix effects and pyrolysis artifacts do not appear to have been a problem in these early analyses. An excellent paper by Crossey *et al.* (1984) compares H/C elemental ratios to hydrogen indices for a suite of 70 samples representing all three kerogen types, a range of minerals, and a range of TOC contents (<1%–5%). These results clearly show that the isolated kerogens give the best correlation with the atomic H/C, following the linear relation:

$$\text{Atomic H/C} = 0.00073(\text{HI}_{\text{ext}}) + 0.541$$

with a correlation coefficient of 0.92 (Fig. 11), where "HI_{ext}" represents the solvent-extracted kerogen. The correlation is also good for the 18 whole-rock samples with TOC contents of >5% ($r = 0.96$) and becomes worse for whole rocks having 1–5% TOC ($r = 0.62\%$) and <1% TOC ($r = 0.64$). The measurement of hydrogen index on whole rocks gives a line parallel to that for isolated kerogens when the pyrolysis hydrogen indices for each are plotted against the elemental analysis data. Thus, the whole-rock hydrogen index (HI_{bulk}) was found to underestimate the isolated and solvent-extracted kerogen value $\text{HI}_{\text{(extract)}}$ for the whole sample set according to the relationship:

$$\text{HI}_{\text{(extract)}} = 1.077(\text{HI}_{\text{bulk}}) + 76.95$$

which has a correlation coefficient of 0.87.

Note that these analyses were done on isolated kerogens which were extracted with organic solvent after (rather than before) the kerogen isolation procedure, which may lead to different results from those that would be obtained following the usual procedure, where solvent extraction and kerogen isolation are carried out in the reverse sequence. Durand and Monin (1980) found that solvent extraction after kerogen isolation tends to remove significantly more bitumen in some cases, in comparison to cases where solvent extraction is carried out on the whole rocks before kerogen isolation. It has been found that refluxing smectite with benzene or toluene and an alcohol can cause Friedel-Crafts condensation of the alcohol alkyl group with the aromatic rings of the kerogen (Riolo and P. Albrecht, unpublished data). Therefore, solvent extraction after kerogen isolation may be the preferred method of removing generated or migrated bitumens from kerogens, particularly for low-maturity samples where functional groups are still present on the bitumen and kerogen.

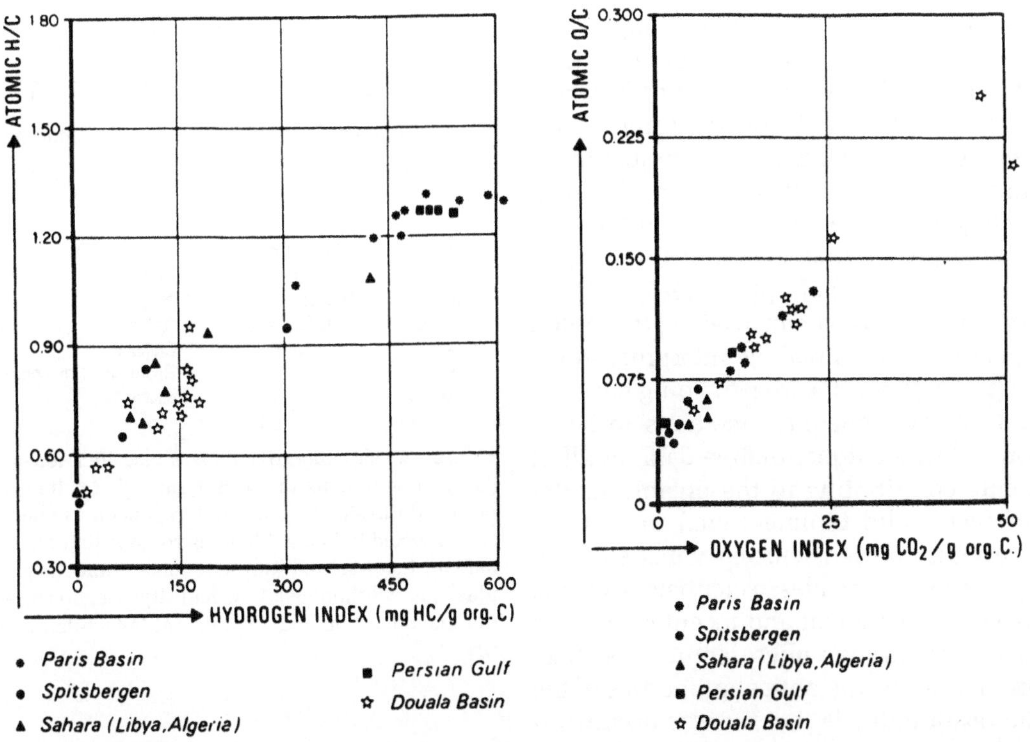

Figure 10. Linearity of elemental analysis H/C vs. pyrolysis HI and O/C vs. OI. [Data from Espitalié *et al.* (1977); from Tissot and Welte (1984).]

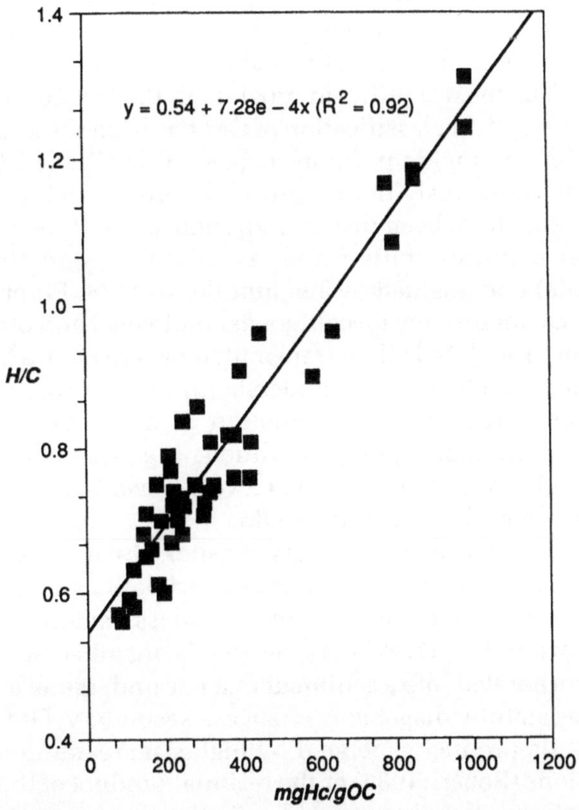

Figure 11. Elemental H/C vs. pyrolysis hydrogen index, HI for isolated and solvent-extracted kerogens. [Data from Crossey *et al.* (1984).]

If the precautions outlined above are observed, the pyrolysis hydrogen index can give a good estimate of elemental H/C, even for very organic lean and reworked kerogens. In an extreme case, Hayes *et al.* (1983) noted that a kerogen with an H/C ratio of 0.6 will fairly reliably give a hydrogen index of about 40 mg HC/g organic carbon, while those with an H/C ratio of 0.5 or less will yield essentially no S_2 peak. The absence of an S_2 peak has been reported to be indicative of an inability to generate further gas or oil (Espitalié *et al.*, 1977), although small amounts of gas generation may continue to lower H/C ratios.

It is generally assumed that pyrolysis oxygen index data are proportional to elemental O/C and can be fairly reliably obtained from isolated kerogens, in which carbonate-rich minerals, which can cause artifacts, have been removed during the acid isolation procedure. However, to our knowledge, the only experimental information available on this point relates to the data set reported in the initial paper on the pyrolysis technique, by Espitalié *et al.* (1977) (Fig. 10).

3.2. Optical Microscope Classification

Optical microscope techniques, particularly as applied to the structured components of kerogen, are covered by Sentfle *et al.* (this volume, Chapter 15). These techniques are also discussed here as they relate to bulk chemical characterization of kerogen, particularly for the abundant nonstructured amorphous phases. Even though changes in optical properties are undoubtedly closely related to chemical changes in kerogen, the chemical aspects of the relationship are very poorly understood at the present time. Various combinations of optical kerogen types may result in similar bulk chemistries (Larter, 1985) so caution must be used in making interpretations.

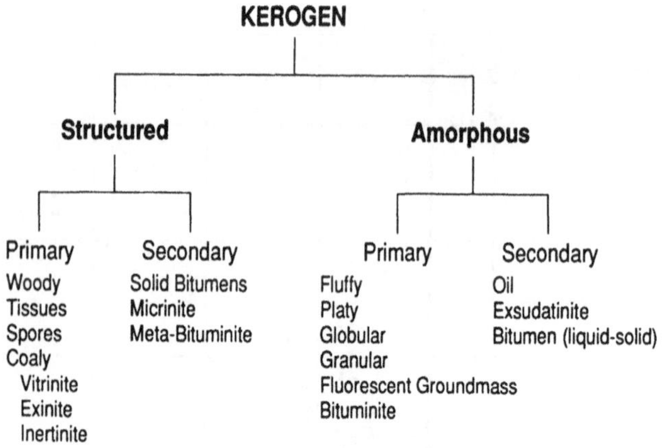

Figure 12. Possible divisions in classification of organic matter (OM) in sediments or rocks from either whole rock or isolates prepared physically or by acid treatment, as viewed via transmitted, reflected (white or blue) microscope lighting. Primary = original OM incorporated into sediments, some diagenetic changes in place (autochthonous), low maturity; secondary = products from thermal alteration, mobile, migrant (allochthonous), higher maturity.

Microscopic methods of kerogen typing have potential advantages over chemical methods in being capable of providing semiquantitative data on all of the components contributing to the organic matter (OM, i.e., kerogen, solid bitumen, etc.) in a single sample. However, little use is currently made of that capability. For several decades, scientists working with OM dispersed in ancient and recent sediments have struggled with various microscopic classification schemes to describe the optical properties of the material. The result today is that there is no universally accepted nomenclature for the optical description of kerogen, even though the International Committee for Coal Petrology (ICCP) has established detailed nomenclature for the organic components of coal, with published descriptions and photomicrographs in their handbooks (1963, 1971, 1975). The ICCP is currently working on the problem of classifying dispersed OM in sedimentary rocks. In the meantime, nearly two dozen classifications for optical kerogen types have been proposed, some of which are listed in Table II and Fig. 12. Many of those classifications combine terms from coal petrology, palynology, and/or organic geochemistry. The subjective nature of many of the terms makes it nearly impossible to compare optical OM descriptions between analysts or among laboratories. An example of the confusion can be seen in the classification scheme proposed by Massoud and Kinghorn (1985) and the criticism (Mukhopadhyay, 1986) and reply (Massoud and Kinghorn, 1986). The two groups of researchers propose different terminology, definitions, and thoughts about the origin of various kerogens, which need to be verified and standardized for others to use.

Perhaps one of the reasons for a lack of standardized nomenclature is the fact that a variety of sample preparations and microscope lighting conditions are being used for the optical study of kerogen. Often, workers are trying to describe the same material, which looks vastly different in thin section in transmitted light compared to the concentrated form in reflected white light. At the present time, it is essential that photomicrographs be published with OM descriptions, so workers can begin to agree that they are indeed seeing the same types of kerogen. Two excellent textbooks (Durand, 1980; Robert, 1985) with many color photomicrographs summarize many of the current ideas about visual kerogen types.

The most significant progress in the last 20 years in the optical classification of OM has been the application of incident fluorescence (near-UV or blue light). Previously unrecognized, hydrogen-rich components have been discovered, such as the three new coal macerals (bituminite, exsudatinite, and fluorinite) documented by Teichmuller in 1974. Fluorescence microscopy reveals greater biological and other structural details than transmitted or reflected white light, which helps in understanding the origins of some kerogens. A combination of analyses using three different lighting conditions (i.e., transmitted, reflected white, and blue) is a very powerful tool for complete identification of OM.

Kerogen or phytoclasts (Bostick, 1970) in sediments can be either recognizable primary or secondary structured particles or shapeless material, as shown in Fig. 12. Primary kerogen is organic material incorporated into a sediment that has undergone low-temperature diagenetic changes; secondary OM is mobile product generated by higher temperature reactions (Robert, 1985) or the residual product of those reactions. The biggest problem in the classification of OM is the shapeless or amorphous fraction, which is

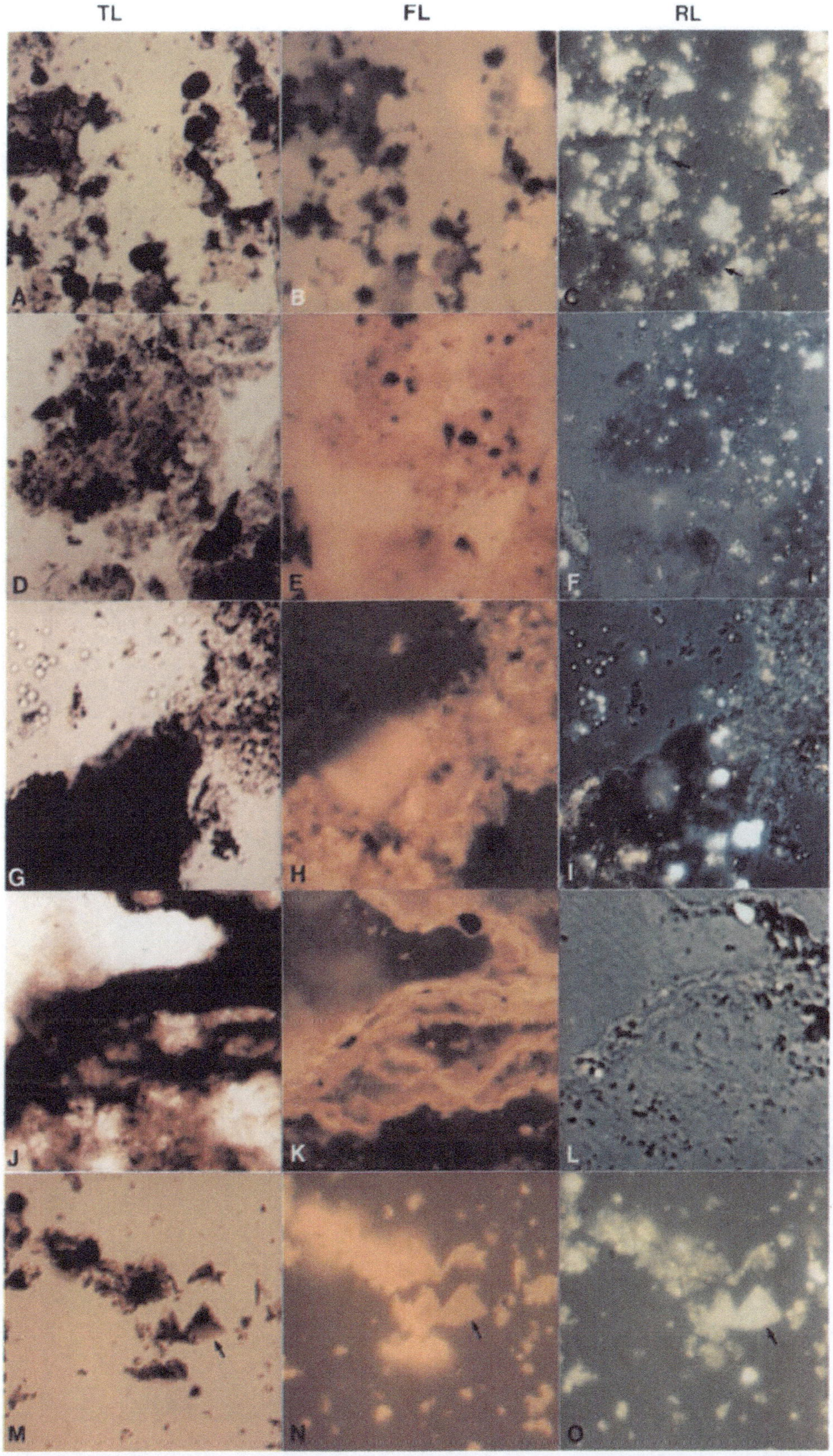

Figure 13. Photomicrographs of textural differences among ancient and recent amorphous kerogens [nomenclature according to Thompson and Dembicki (1986)] and solid bitumen. Five sets of photographs show five fields of view in different microscope lighting conditions: transmitted (TL), incident blue or "fluorescence" (FL), and incident white or reflected (RL). Fields of view are approximately 150 μm wide. A, B, and C show recent sediment from oxic bottom waters of the Peru upwelling area including amorphous type B (arrows). Vitrinitic particles are abundant with some structured kerogens, such as cuticular and fungal grains (acid isolation, no heat). D, E, and F illustrate recent sediment from an anoxic silled basin on the Pettaqusmscutt River estuary adjacent to a bay in Rhode Island; dominated by amorphous type A kerogen (acid isolation, no heat). G, H, and I show ancient kerogen from carbonate potential source rocks of Dubai; mostly amorphous type A, acid isolation. J, K, and L are same as G, H, and I; however, the amorphous type A kerogen is shown *in situ* in a doubly polished thin section. M, N, and O show solid bitumen (arrow) from ancient sediments of the Norwegian North Sea (acid isolation).

A

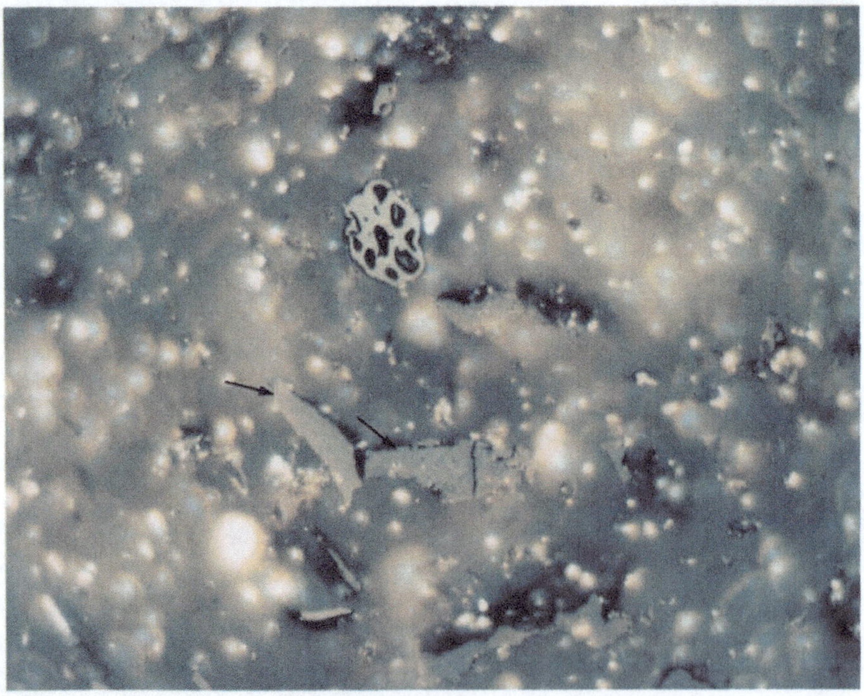

B

Figure 15. Photomicrographs showing vitrinite (arrows) in coal (A) and an acid-isolated kerogen (B). The coal is from Utah; the shale is from Indonesia. (Incident white light, oil immersion, field of view approximately 300 μm.)

the dominant type of kerogen in most petroleum source rocks, but is relatively minor in coals. Examples of amorphous kerogens are shown in Fig. 13. The amorphous material constitutes both degraded OM and bituminous substances, at various levels of maturity. The main difference among the classification schemes shown in Table II lies in the attempts at describing amorphous kerogens. Some workers describe it genetically, linking specific types to biological precursors (Mukhopadhyay *et al.*, 1985); others describe its appearance (Staplin, 1969; Combaz, 1975), and some describe appearance in combination with its relationship to organic geochemical traits (infrared, pyrolysis-GC, etc.) and generating potential (Thompson and Dembicki, 1986).

The chemical kerogen classifications devised by Tissot *et al.* (1974; types I, II, III, and residual) for both ancient and recent sediments are derived from bulk compositions (elemental ratios of hydrogen, oxygen, and carbon) representing all kinds of combinations of optical kerogen types. The idea that type I is composed predominantly of algae (Tissot *et al.*, 1974) does not always hold true when the kerogen is examined under the microscope. Samples from the Green River Formation contain amorphous kerogen and "have not been reported as including algae" (Robert, 1985). Similarly, the other chemical kerogen types contain a variety of optical kerogen types. Both the chemical and optical attributes of kerogen are necessary for a complete description; care must be taken not to overinterpret the results from either technique alone. The OM in recent sediments, "protokerogen," optically appears to be very much like kerogen of ancient rocks (Staplin, 1969) with an abundance of amorphous material, as illustrated in Fig. 13. *The observation that much protokerogen is amorphous suggests that preservation of structured kerogen is not common and that structured kerogen is not the major input to sediment. As a result, many present-day microscopic (and chemical) classification schemes may be overinterpreting a minor input, rather than focusing on the major (amorphous) kerogen input as the major contributor to oil and gas formation.*

Some Optical Classification Schemes

Two distinct methods of identifying kerogen have developed. The variety of terms in Table II reflect these methods as well as the subjective nature of kerogen description. However, it should be noted that in many cases the authors have published photomicrographs with their classifications, and, in those cases, the different kinds of OM being described are very similar among authors. One of the identification methods stems from palynology and is based upon the study in transmitted light (white, polarized, generally utilizing a daylight filter) of isolated kerogen particles mounted on a cover slip attached to a glass slide (strewn slide). Strewn slides are examined with relatively high magnification (400–500×) with oil immersion. Reflected white and blue light can also be used to examine strewn slides (Thompson and Dembicki, 1986). Classifications based on this type of preparation and examination tend to emphasize plant anatomy or biological origin, such as those proposed by Masran and Pocock (1981) and Venkatachala (1981). The other identification method comes from coal petrography. Samples (either isolated kerogen or whole rock) are polished and viewed with reflected white and blue light. The terms applied to OM identified in this preparation are coal petrographic (vitrinite, exinite, inertinite, and submacerals) as well as some terms derived to describe amorphous or bituminous OM in petroleum source rocks. Van Gijzel's (1981) classification system is typical of combined coal petrographic and other terms. Frequently, it is difficult to distinguish kerogen types microscopically because some of them (e.g., huminite to vitrinite and the various bituminites) change with increasing maturity.

Structured Kerogens. Structured kerogens are the primary focus of Sentfle *et al.* (this volume, Chapter 15). The pieces of recognizable plant and animal remains in kerogen were first noticed by observers (palynologists) using transmitted light microscopy. Often, workers describing structured kerogen types provide both a biological term and a coal maceral equivalent. Masran and Pocock (1981) recognized plant material such as "leaf, root, stem" which they related to "telinite" of coal maceral nomenclature and put in parentheses in their classification. Similarly, Massoud and Kinghorn (1985) provided coal maceral equivalencies for their proposed kerogen types. In general, "woody" is the palynological or transmitted light term for material which is vitrinitic in reflected light. Herbaceous is the transmitted light term for tissues which are not usually visible in reflected light, some of which may be equivalent to cutinite. The transmitted light term "coaly" (Burgess, 1974) can include vitrinities (with or without other macerals such as oil-prone liptinites), inertinites, pyrite, dead oil, and other unrelated material; the term and the material are better defined in reflected light.

Most structured kerogen is generally recognized by petrographers, even though the exact names may differ. Quantitative fluorescence may help to subdivide some classes of liptinites such as resinites (Crelling, 1982). Occasionally, the scanning electron microscope (SEM) has ben used to study kerogen (Masran and Pocock, 1981) or the mineral inclusions of kerogen (Combaz, 1980). Transmission electron microscopy (TEM) (Taylor *et al.*, 1988) has been used to show tiny inclusions in inertinite grains. SEM and TEM analyses are very time-consuming and represent only small grains or parts of the sample. Therefore, such techniques are impractical for routine kerogen typing work.

Some remarkable structured kerogens have been documented in both ancient and recent sediments by very careful and detailed study. Algae in oil shales (Hutton, 1987) and in Ordovician rocks (Foster *et al.*, 1986), bacteria in recent mats and the Monterey Formation (Williams and Reimers, 1983), and chitinous fragments in coals (Goodarzi, 1984) are some good examples of the specific biological origins of OM, all of which would be expected to produce very different detailed chemical signals, if the various structures could be separated and analyzed individually.

Amorphous Kerogen. From the transmitted light (palynological) description of OM, the term "amorphous" kerogen was developed to categorize the material with no structural relation to floral or faunal remains. It has become a "catchall" term, lumping together all kinds of material which have different origins, chemistries, and optical properties (Figs. 12 and 13). As shown in Fig. 12, amorphous kerogen can be primary (deposited or formed *in situ*) or secondary (formed by increased thermal reaction or migration). For example, Bujak *et al.* (1977) described their "amorphogen" category as "unorganized, structureless OM which may be finely disseminated or coagulated into fluffy masses" which, unfortunately may include both primary and secondary OM. Some authors, including Robert (1985), believe that much amorphous material is produced during sample preparation (grinding and acid removal of minerals); however, a great deal of structureless OM can be seen in whole-rock preparations (especially thin sections) and in recent sediments (e.g., Fig. 13 and Reimers, 1982).

Coal petrographers have noted amorphous bituminite (ICCP, 1975), "mineral bituminous groundmass" (Teichmuller and Ottenjann, 1977), fluorescent groundmass, or organomineral association (Robert, 1985) in polished source rock samples viewed with incident white and blue light. In fact, there is considerable overlap in the chemical and microscopic properties of hydrogen-rich coals and kerogens; both are known to be important oil and gas source rocks, as recently summarized by Hunt (1991).

Studies of recent sediments and modern aquatic environments (Reimers, 1982) help us to understand that much OM is amorphous at early stages of incorporation into sediments (Fig. 13). The heating experiments of Peters *et al.* (1977) showed that the heating of protokerogen may produce amorphous kerogen, so increasing maturity may increase the amount of such material in source rocks. It has been suggested by Suzuki (1984) that amorphous kerogens can be mixtures of different amorphous materials, making it nearly impossible to determine exact origins by microscopic means alone. Amorphous kerogen is often linked to an algal or marine origin (Combaz, 1974) and to sapropel. Robert (1985) stated that in 1920 Potonie originally used the term "sapropel" to mean "algal ooze"; however, Durand (1980) defined sapropel as a "gel derived from organic debris which has putrified in an anaerobic medium." The latter definition may be more suitable to amorphous kerogens, since, in most cases, their biological origin cannot be established. Staplin (1969) described the complex physical transformation of OM by ingestion and processing of organisms, coagulation, precipitation, and reworking, resulting in fluffy or platy sapropels. It would be expected that these processes would lead to distinct chemical changes in these kerogens—both with respect to indicators of source and degree of reworking—which have not been examined in detail. There is currently a great need for better interpretation of existing bulk chemical kerogen characteristics by relating geological and chemical data with more complete and objective optical information.

The experiments and optical descriptions of Masran and Pocock (1981; Table II) show the importance of the Eh along with original plant input of the environment in which different kinds of amorphous (yellow-amber versus gray) kerogens are likely to be preserved. There has been a tendency by most authors to assume that the amorphous material is derived from breakdown of the few structured kerogens which are rarely observed or are closely associated, or juxtaposed, in the same sample. For example, the occasionally well-preserved algae in the Posidonia Shale is assumed to give rise to amorphous bituminite (Teichmuller, 1986). Two cases listed in Table II which appear to depend upon the associated structured kerogen are the different "keroginite" categories of Massoud and Kinghorn (1985) and the marine and lacustrine sapropelic groundmass categories of Robert (1985). The latter depend on the association of "various unicellular algae" and *Botryococcus* algae for the marine and lacustrine classifications, respectively.

Fluorescence and textural properties of amorphous kerogen, used together with organic geochemical analyses and geology, hold promise in solving the age-old frustration of organic petrographers, organic geochemists, and geologists who have seen the failed attempts of relating optical kerogen properties to specific geochemical traits (e.g., oil-versus gas-generating potential). These failures result from the difficulty of relating existing chemical and optical kerogen typing techniques to the bulk of the organic matter which is actually found in various "organic facies" in geological systems. The majority of visual kerogen data in the literature report most samples with more than 50% amorphous kerogen, with no clue as to specific biological origin, which has been assumed to be necessary to interpret generating potential. For this reason, Powell *et al.* (1982) and Price and Barker (1985) concluded that optical techniques cannot be used to distinguish hydrogen-rich from hydrogen-poor amorphous kerogens. Jones (1987) referred to this problem as the "cloud" over organic petrography; however, he stated that fluorescence can be used to help distinguish oil-prone from gas-prone amorphous kerogens. Van Gijzel (1981) classified three "amorphinites" on the basis of fluorescence; similarly, Mukhopadhyay *et al.* (1985) documented three "sapropelinites" (Table II). A more complicated approach has been attempted by Smith (1984) using a library of structured kerogen fluorescence spectra to resolve the components making the amorphous fluorescence spectrum. Combaz (1980) distinguished several amorphous kerogen types on the basis of textures such as flakes, clots, dusts, etc. Batten (1983) also tried using morphologies similar to those of Combaz; however, most of his descriptions tend to rely upon associated structured kerogen and depositional environment. Textural differences in all three microscope lighting conditions were used by Thompson and Dembicki (1986) to distinguish four kinds of amorphous kerogens which are related to the hydrocarbon-generating potential of source rocks. They attributed the textural differences to preservation and/or environment of deposition. Therefore, amorphous kerogens do have optical differences which can be related to hydrocarbon-generating potential and, perhaps, unique depositional environments which are probably important in discerning organic facies. These differences were also discernible from distinct chemical characteristics.

The optical recognition of secondary bitumens (at all maturity levels) and their distinction from primary kerogen is important to the organic geochemist's interpretation of bulk geochemistry and to the geologist's generation–migration conceptual reconstruction of the past, such as demarcation of paleoreservoir facies or carrier beds and migration routes. Secondary amorphous OM (Fig. 12) is defined as generated product, usually moved from some external source into its present location. These substances appear structureless when viewed in whole-rock or isolated preparations and need to be described and distinguished from the primary amorphous kerogen described above. Oils are frequently seen in whole-rock preparations (van Gijzel, 1981; Robert, 1985; Teichmuller, 1986). More insoluble secondary OM, such as exsudatinite or solid bitumens, are detectable by their reflectance and fluorescence (Jacob, 1966, 1967; Robert, 1985) and their ability to flow into voids, cracks, and spaces between mineral grains. This OM is the solid form of liquid hydrocarbons (bitumens) which may develop from the hardening of liquids due to thermal alteration, oxidation, water contact, or the loss of the more mobile phases, leaving behind a residue which has the structure or shape of the spaces between the mineral grains which were filled (Robert, 1985). The material is called solid bitumen or "migrabitumen" (Jacob, 1985) and has been documented in polished samples. Metabituminite (Teichmuller, 1971) is fine-grained micrinite, which is distinctive in reflected light appearance (white granules) and represents an early transformation of bituminous OM. Thompson-Rizer (1987) illustrated how solid bitumen can be identified in visual kerogen strewn slides by its translucency and high reflectivity and geometrical shape, as shown in Fig. 13. This material displays a range of properties, depending on the maturity level of the rocks in which it occurs, and has been characterized by chemical means to be greater than 70% carbon (Jacob, 1985) with less than 20% H and less than 10% S, N, and O. It is best defined operationally by its solubility in carbon disulfide. Names such as ozocerite, gilsonite, wurtzilite, and impsonites have been applied to this material.

4. Principal Kerogen Maturation Techniques Now in Use

A summary of correlations between organic and inorganic indicators of thermal alteration in sediments is shown in Fig. 14 (Hayes *et al.*, 1983). Excellent summaries have appeared which compare a number of organic geochemical maturity indicators for kerogens (i.e., pyrolysis hydrogen/oxygen index data and microscopic data) with those for oil and gas in source rocks (i.e., biomarkers and other bitumen

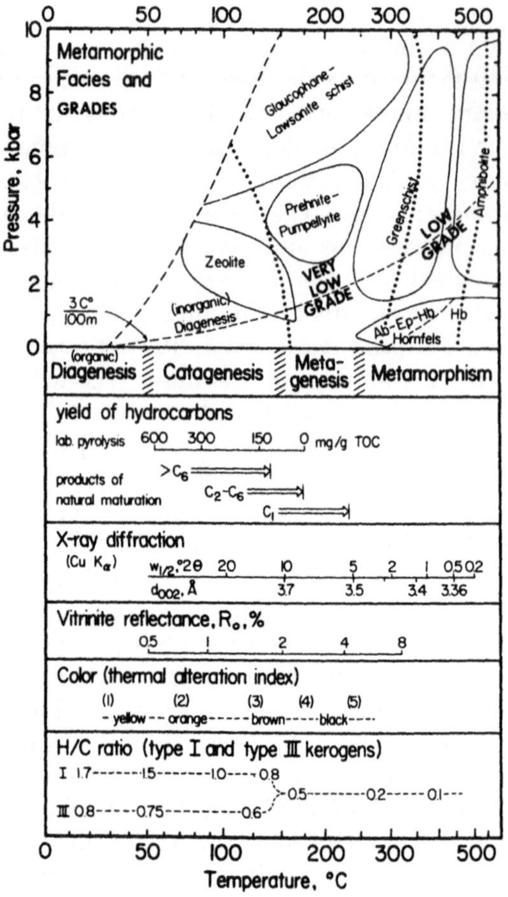

Figure 14. Relationship between organic and inorganic maturation scales. The uppermost frame refers to inorganic minerals while all lower frames refer to coexisting organic materials. The indicated geothermal gradient of 3°C/100 m is shown only for purposes of reference. Ab, Albite; Ep, epidote; Hb, hornblende. [From Hayes *et al.* (1983).]

indicators; Tissot *et al.*, 1987; Curiale *et al.*, 1989). The optical technique of vitrinite reflectance and the chemical technique of determination of pyrolysis T_{max} are currently those most frequently used for quantitative determinations of kerogen maturity. The techniques of fluorescence and spore color, summarized below and by Sentfle *et al.*, this volume, Chapter 15), are also important. Recent developments promise to make both of these latter types of measurements more quantitative and less operator dependent in the future. Today, much effort is being placed on modeling the thermal histories of sedimentary basins with a key building block to the models being the calibration of measured vitrinite reflectance (R_0) data to a calculated time–depth model such as those of Lopatin (1971), Bostick (1974), Waples (1980), and many others. Recent comparisons of the activation energies and time–temperature scales, including the dependence on kerogen type needed to carry out reliable maturation reconstructions in basins, can be found in Wood (1988) and Hunt *et al.* (1991).

Vitrinite reflectance measurements are currently accepted as the standard against which organic geochemists calibrate all other maturation indicators, both microscopic and chemical. However, the validity of this practice has recently been questioned (McCulloh and Naeser, 1989). Some of these problems arise because vitrinite reflectance measurements are subject to a number of limitations which are not widely understood by geochemists, including the existence of at least two distinct types of vitrinite which respond very differently to the maturation process. Therefore, a brief discussion of the "state of the art" for several widely used microscopic techniques and their dependence on kerogen type is given below to allow a more objective comparison to "bulk phase" chemical methods of determining kerogen maturity.

4.1. Kerogen Maturation via Microscopic Techniques

Microscopic techniques are discussed here as they relate to bulk kerogen chemistry or where kerogen typing is critical to correct maturity determinations, as is the case for vitrinite reflectance.

Measurements of maturity from three different kinds of optical measuring techniques (transmittance, reflectivity, fluorescence) are becoming more quantitative, more objective, and less operator dependent. Vitrinite reflectance photometric measurements were proven to be very reliable rank (maturity) indicators for coals in the 1950s and 1960s. In the 1970s the same techniques were applied to the measurement of organic particles in sedimentary rocks (Teichmuller, 1971; Bostick, 1974, 1979). Some attempts have been made to photometrically measure the translucency or color of spores or pollen in transmitted light (Gutjahr, 1960; Grayson, 1975; Smith, 1983); however, most laboratories use subjectively determined maturity scales. Tremendous progress has been made in the use of qualitative organic matter (OM) fluorescence colors and quantitative fluorescence color and intensity measurement by photometry for application to maturity study (Ottenjann *et al.*, 1974).

There is no one infallible analytical technique for maturity determination. The exact correlation or equivalent numerical values of parameters generated by the different techniques is uncertain (Heroux *et al.*, 1979). Optical maturation changes involve very complex chemical and physical processes which depend on time, thermal energy, kerogen type, and temperature (Heroux *et al.*, 1979) as well as the stage of diagenesis, facies, and tectonic conditions (Teichmuller, 1971). The optical changes which occur in OM with increasing maturity include darkening; decrease of light transmission; increase of index of refraction; reflection and luster in reflected light; a loss of the finer structural details (Staplin, 1969); and a shift in fluorescence colors from bright yellows to dull reds. Hilt's law of 1873 refers to the increase of coalification (maturity) with increase in depth of burial (Robert, 1980). White (1915, 1935) noted the relation between the degree of coal metamorphism and the occurrence of oil in Pennsylvania. Coalification is equivalent to kerogen maturation in that it is irreversible and is the result of concentrating carbon via loss of volatile breakdown products (H_2O, CO_2, CH_4, N_2, and hydrocarbons; Robert, 1980).

Naeser and McCulloh (1989) have recently summarized inorganic geothermal and maturation indicators and have concluded that some are more reliable than those currently available for kerogen. However, direct maturity information from OM, rather than from the associated minerals, is necessary because of the higher sensitivity of the OM to smaller temperature changes and its lower sensitivity to pressure changes (Teichmuller and Teichmuller, 1966). Most mineral paleogeothermometers do not cover the entire temperature range (25–400°C) affecting OM as does vitrinite reflectance (Hayes *et al.*, 1983; Price and Barker, 1985). Optical methods require only a small amount of sample (important in some wells where few cuttings are available) and are relatively inexpensive and rapid compared to some chemical or mineral maturity analyses. Among the three major optical methods, various combinations can be used for complementary information; for example, transmitted light characteristics may be more useful for low-maturity samples and vitrinite reflectance more reliable for higher-maturity samples where translucency and fluorescence cannot be measured (Grayson, 1975).

The petroleum generation limits should be considered gradational and may vary with OM type or analytical technique (Heroux *et al.*, 1979). In addition, Price and Barker (1985) noted that all optical and chemical parameters can be affected by OM associations (such as abundant exinite). Figure 3 in Chapter 15 (of this volume) shows some of the relations among the optical maturity indicators, petroleum generation, and coal rank.

4.1.1. Vitrinite Reflectance

The use of vitrinite reflectance data has become routine as the primary maturation parameter in the coal and petroleum industries for determining coal quality or rank and for locating the depth of the petroleum "liquid window" (Pusey, 1973; Dow and O'Connor, 1982). Often problems arise because those making geological interpretations from reflectance data are unaware of what vitrinite is or how reflectance is measured. Potential problems with vitrinite reflectance measurements can be produced either from kerogen isolation or from microscopic interpretation, summarized below and discussed by Sentfle *et al.* (this volume, Chapter 15).

4.1.1a. Experimental Considerations. Vitrinite reflectance can be measured on either whole rocks or kerogen isolates. However, in whole-rock mounts, the vitrinite often becomes "lost" among mineral grains and occurs in less abundance than in the acid-concentrated kerogen and is usually of poorer quality due to scratches from mineral grains falling out of the sample during polishing. Therefore, acid isolation of kerogen (see Section 2) is often carried out for source rocks which contain low abundances of small-size vitrinite particles (Castano and Sparks, 1974). Even after kerogen isolation, vitrinite characterization is not trivial. For example, Fig. 15B is typical of kerogens isolated from many sediments, containing a low abundance of small-size grains assumed to be vitrinite, surrounded by the epoxy mounting medium. Such grains are difficult to identify because of their small size and because of the absence of adjacent macerals for comparison of shades of gray colors, as is the case in coal (Fig. 15A).

At least 30 reflectance measurements should be made on the OM from rocks (Robert, 1985), with less than 10 measurements per sample being unacceptable (Castano and Sparks, 1974). An average % R_0 value is reported for the sample, but this is not a mathematical average. Instead, the data are plotted in histograms, as shown in Fig. 16, and "representative" populations are chosen by the analyst or the geologist (as described below). Both histogram picking and difficulties in recognizing true vitrinite particles probably contribute to the general lack of agreement among laboratories in comparisons of R_0 data (Dembicki, 1984). Generally, reflectance values for a specific population of kerogen (Fig. 16) are reliable to one decimal place, with the standard deviation of the picked population being less than 0.1% R_0 (Robert, 1985; discussed in more detail in Section 4.1.1b).

Kerogen preparation can also affect the vitrinite

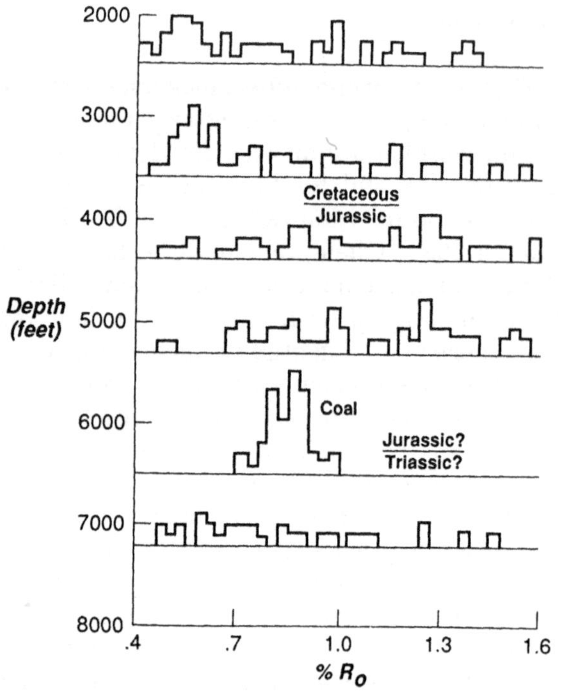

Figure 16. Histograms of vitrinite reflectance measurements from rocks and coal. [Adapted from Jones and Edison (1978, p. 8).]

reflectance. Castano and Sparks (1974) demonstrated that acid treatment has little effect on vitrinite reflectance of coals. Oxidation can increase reflectivity although this is not always the case, as shown by Kalkreuth (1982), who found a decrease in reflectivity for weathered coals. For this reason, Teichmuller (1971) advised that sandstones should not be used for reflectance studies as the vitrinite may be oxidized by groundwater. Sandstones also often contain large coaly fragments which are oxidized (Castano and Sparks, 1974), as well oxidized OM from redeposition or reworking (Durand *et al.*, 1986).

The ideal materials for vitrinite characterization are coal layers or coal particles dispersed in a sediment. If there are no coal layers available in the area of interest, Teichmuller (1971) showed that small coal inclusions in sedimentary rocks (cores and drill cuttings) can be measured for reflectance. Vitrinite in shales, and sedimentary rocks other than coal, is measured for reflectance in the same way as in coal. Low-rank vitrinite (huminite, less than 0.5% R_0), usually has a lower reflectance when dispersed in shale than in coal of the same level of maturity (Teichmuller, 1971; Goodarzi, 1987). Possibly, bitumen in the shale is adsorbed into the vitrinite, lowering the reflectivity (Teichmuller, 1971). Coal and shale pairs of R_0 values were shown to be in close agreement by Castano and Sparks (1974). However, detailed work by Bostick and Foster (1975) on a 66-m core from the Illinois Basin shows the variation in vitrinite reflectance data for coals and adjacent sandstones, argillites, and/or limestones. In almost all cases, the R_0 value from the rock samples was lower than for the coal, with the greatest difference being 0.17% R_0. However, the coals themselves showed similar variability and ranged in reflectance from 0.51 to 0.70% R_0 (a difference of 0.19% R_0). The difference in the coal layers may be caused by variations in the depositional environment of the coal layers, particularly oxidizing versus reducing conditions.

Much has been written about the problems associated with vitrinite reflectance measurements in low-rank coals and petroleum source rocks of maturities less than 1.3% R_0. In many cases, optical studies can help to distinguish indigenous vitrinite from other organic components which may confuse the bulk organic geochemical maturity measurements. However, the problem of specific vitrinite maceral identification in low-rank sediments persists. Huminite (Stach *et al.*, 1982) is the name given to low-rank vitrinite, and humotelinite is the submaceral variety that should be measured for reflectance (Goodarzi, 1987), but it is not easy to identify when dispersed in rocks. Cuttings samples from deep wells are also plagued with a number of other problems such as mixtures of lithologies, cavings falling down the hole, drilling mud additives (lignites, etc.), pre-Devonian sections with no wood, carbonate sections with no wood, liquid hydrocarbon impregnation (stain), and small sample size.

Two different types of vitrinite have been recognized in samples, and it is often difficult to identify a particular vitrinite consistently and unambiguously in either whole-rock preparation or the acid-isolated kerogen. The most recently discussed problem in measuring vitrinite reflectance, in both low-rank coals and source rocks, is the recognition and/or choice of measuring high or low gray vitrinites, which have different chemical compositions and coalification pathways (mature at different rates). Band vitrinite in coals (high gray) usually consists of telinite, collinite, or telocollinite and is sometimes called vitrinite "A" or "1," which is usually measured for reflectance. The more hydrogen-rich, lower reflecting variety (low gray) consists of desmocollinite, heterocollinite, and degradinite and may be referred to as vitrinite "B" or "2." Price and Barker (1985) pointed out that there is a tendency for many analysts to measure one kind exclusively while other analysts prefers to make a total scan of the sample, measuring both kinds of vitrinite, introducing considerable subjectivity into the technique. It was first suggested (Bostick, 1974; Castano and Sparks, 1974; Dow, 1977) that the lower reflecting vitrinitic particles be measured in source rocks to avoid the reworked and oxidized OM. With the recognition that hydrogen-rich, low reflecting vitrinites produce unreliable maturity trends, Toxopeus (1983) recommends that the higher gray reflecting vitrinite be measured. It is often not possible to select specific grains in source rock preparations because there are so few available to measure.

Besides the different kinds of vitrinite, it has also been found that the association of vitrinite with other macerals, in particular, the hydrogen-rich liptinites (such as alginite, sporinite, and cutinite) tends to lower reflectivity. Hutton and Cook (1980) showed that Australian coals containing more than 50% alginite can have the reflectance of the band vitrinite decreased by 0.24% R_0 and that of the desmocollinite by 0.31% R_0. Similarly, Kalkreuth (1982) found that different liptinites lower reflectance by as much as 0.27% R_0, although some pure vitrinites without associated liptinites also had low R_0 values, possibly due to original plant type or depositional environment. The differences in vitrinite reflectance among plant types, such as the Carboniferous lycopod and the Cretaceous–Tertiary conifer, have been determined to be about 0.15% R_0 by Sitler (1979), with the conifer being lower. Depositional environmental differences explain vitrinite reflectance variations of 0.28% R_0 between two New Zealand coal seams separated by only 100 m. According to Newman and Newman (1982), the higher reflecting coals are from a well-drained swamp facies, whereas the lower reflecting coals are from swamps with little drainage where bacterial activity produced a hydrogen-rich vitrinite (perhydrous). Hydrogen-rich vitrinites can be detected in low-rank samples because they fluoresce dull yellow to orange-brown color (when excited with a mercury vapor lamp, blue light). Vitrinite maturation measurements should not be carried out on fluorescing vitrinites. Alternatively, fluorescence should be noted so the vitrinite data can be interpreted with caution (Toxopeus, 1983).

Samples of potential source rocks are often very difficult subjects for vitrinite reflectance measurement. Robert (1980) suggested that other optical techniques, for example, spore color or fluorescence, be used to help pick the correct reflectance value. Alpern (1980) believes that the fluorescence of algae may be more syngenetic with the sediments and therefore a better maturity parameter. Durand *et al.* (1986) acknowledged that vitrinite reflectance is a "precious tool" for determining the maturity of terrestrial sequences, as long as the limitations of the technique are known; *they feel that it is dangerous to use* R_0 *in marine, lacustrine, and pre-Devonian sediments because of the confusion with OM of different origins and compositions.* Basically, "coals are safer than dispersed organic matter" for establishing vitrinite reflectance trends (Durand *et al.*, 1986). Thus, reliable determination of maturity from vitrinite may require knowledge about kerogen type as related to depositional environment as well as about maceral associations.

4.1.1b. Data Interpretation. Once the "proper" kind of vitrinite has been properly polished and measured for reflectance, the problem of data interpretation begins, especially for source rocks. Figure 16 shows some typical histograms for OM reflectance measurements. Only the coal interval at 6,500 ft provides a single vitrinite population (although not a tight histogram), showing that the bottom of the oil window (approximately 1.2 or 1.3% R_0) has not yet been reached in this well. By focusing on the coal, the two sets of data at 2000 and 3000 ft can be interpreted as "immature" (about 0.5% R_0). This sort of reasoning, employing all available geological and geochemical information, is what typically goes into picking representative vitrinite reflectance populations for a sample. The broad histograms may be due to bioturbated, winnowed, weathered, caved, or contaminated (drilling mud) OM in the samples. The taller histogram plots or those where less than 10 measurements are recorded for an interval do not always represent the "true" vitrinite. The lowest reflecting population has been suggested to be more indigenous than the populations with higher values (Bostick, 1974; Castano and Sparks, 1974). Therefore, it is very important to examine histograms before blindly accepting an interpreted R_0 value.

The exact chemical changes which occur as vitrinite reflectance increases are currently unknown. The change in vitrinite reflectivity with increasing maturity correlate with the loss of hydrogen (Durance *et al.*, 1986) which accompanies increasing temperature. For many years, workers have tried to establish exact correlations of R_0 to temperature or paleotemperature; however, it is now generally accepted that R_0 responds to maturation (time and temperature) rather than to just temperature. The complexities of geological time, the poorly defined chemistry of the process leading to increased R_0, and the susceptibility of R_0 to heating rate and vitrinite type, as well as maximum paleotemperature, currently place limitations on understanding of the relationship between

R_0 and time and temperature (Curiale *et al.*, 1989; Naeser and McCulloh, 1989; Wood, 1988). Therefore, many publications (e.g., Vassoyevitch *et al.*, 1969; Pusey, 1973) show different temperature limits for the oil- and gas-generating windows. Teichmuller (1971) noted that oil generation occurs in the Paris Basin at a cooler temperature than in the Los Angeles–Ventura basins because more time was available for generation in the former. In general, vitrinite reflectance measurements help to establish maturity trends in sedimentary sequences (Robert, 1980); however, many problems are associated with source rock reflectance measurements. Most difficulties occur in the low-rank (less than 1% R_0) coals and source rocks. The best way to be confident about maturity level is to measure several optical and chemical parameters of the kerogen. Information on sample lithology, age, geographic location, and unconformities are required to establish a logical maturity trend. Chemical or optical maturity measurements on the extractable bitumen fraction will be unreliable unless the bitumen can be unambiguously related to the rock that it presently occupies (Robert, 1980).

In the oil industry, vitrinite reflectance data are most used for establishing well maturity profiles. From linear trends on a log scale, Dow (1977) found that it is possible to interpret a variety of geological events from vitrinite reflectance data, such as erosional unconformities, igneous intrusions, and faults (Fig. 5 in Chapter 15 of this volume). Once the linear trend is established, there is a tendency to use the calculated values (the line) rather than the actual measured data and to extrapolate or project both ends of the line to predict the maturity at depths which were not sampled. Both of these practices can introduce errors into the understanding of the geothermal history of an area.

Robert (1985) preferred the linear scale curves which help to depict time as well as temperature histories and compiled enormous quantities of worldwide data which statistically show about 0.65% R_0 at 4000 m (subsea) for rocks younger than Carboniferous; those older at the same depth show about 0.85% R_0. Neither the curves nor the linear well profile plots can be accounted for by changes in lithology, which Price and Barker (1985) suggested should be plotted separately. Generally, vitrinite is measured from shales in most wells; shale values have lower reflectances than found for siltstones or sandstones (Heroux *et al.*, 1979).

There is some question as to whether thick coals reflect the geothermal gradient of the rest of the sediment column; for example, Heroux *et al.* (1979) have found that R_0 increases in thick coal beds. In other cases, a plot of R_0 versus depth gives no increase in reflectivity. This lack of increase is attributed to the associated liptinite macerals, which depress vitrinite reflectance (Kalkreuth, 1982). A shift from higher to lower reflectance with depth can also accompany a change in OM type; for example, a shift from 1.2 to 0.4% R_0 accompanied a shift to more liptinitic OM (Price and Barker, 1985).

Reflectance can be measured on other kinds of organic matter in rocks, such as solid bitumen. Measurements on solid bitumen become particularly important when vitrinite is absent (Robert, 1980, 1985), especially in carbonate rocks. Jacob (1985) showed the linear relationship between bitumen reflectance and vitrinite reflectance from about 0.5 to 2% R_0.

In conclusion, reversals or insignificant increases in R_0 with increasing depth and maturity are very common for all of the reasons discussed above and can cause very large problems in geological interpretation. Therefore, it must be remembered that vitrinite reflectance data are not "magic bullets"—they need to be interpreted along with other geological and geochemical data to give an overall picture which is reasonable.

4.1.2. Fluorescence

The fluorescent properties of organic matter have been observed since the 1930s and 1940s, but not in a systematic way (Robert, 1985). Not until the late 1960s and early 1970s was the relationship between fluorescence colors and intensities and coalification established (van Gijzel, 1967). In the 1970s and 1980s, many improvements occurred in fluorescence microscope equipment.

The molecular features of kerogen causing fluorescence are not well understood even though it is commonly observed and measured. Teichmuller and Durand (1983) explained that there is a close relationship between liptinite fluorescence and the formation of petroleum. Aromatic molecular structures fluoresce with great intensity which is diluted or quenched as aliphatic or saturated hydrocarbons are mixed in, as may occur during oil generation (Bertrand *et al.*, 1986; Lin *et al.*, 1987; Khavari Khorasani and Murchison, 1988). Among the many fluorescence parameters measured or calculated, the one dealing with alteration may be most closely linked to chemical reactions in organic matter. Fluorescence spectral fading (van Gijzel, 1971) or alteration (Ottenjann *et al.*, 1975) is the change of color and intensity with prolonged exposure to near-UV light. The phenomenon appears to be connected to the maturity and overall organic geochemical properties of the sample via factors which are still under investigation. Spectra can be measured at various time intervals during continuous excitation of the sample (30 minutes to several hours), resulting in spectral alteration whereby the color changes irreversibly from the original spectrum to more red (positive alteration) or more green (negative alteration) due to photochemical transformations (van Gijzel, 1975). The decrease in the measured fluorescence intensity is directly related to the increase in rank of coalification (van Gijzel, 1967).

Alteration of fluorescence intensity with time can be determined in relation to a standard or in relation to the original measurement of the object. Ottenjann (1985) described intensity alteration measurements which show increases (positive) or decreases (negative) relative to the initial measurement. He suggested that positive intensity alteration is related to the formation of new chemical bonds in the particle being excited; negative alteration appears to be caused by the partial evaporation of fluorescent volatile components from the object. Figure 17 illustrates the use of fluorescence intensity alteration as an additional maturity indicator for samples with different bulk chemical kerogen types. Positive alteration corresponds to immaturity, negative is mature, and

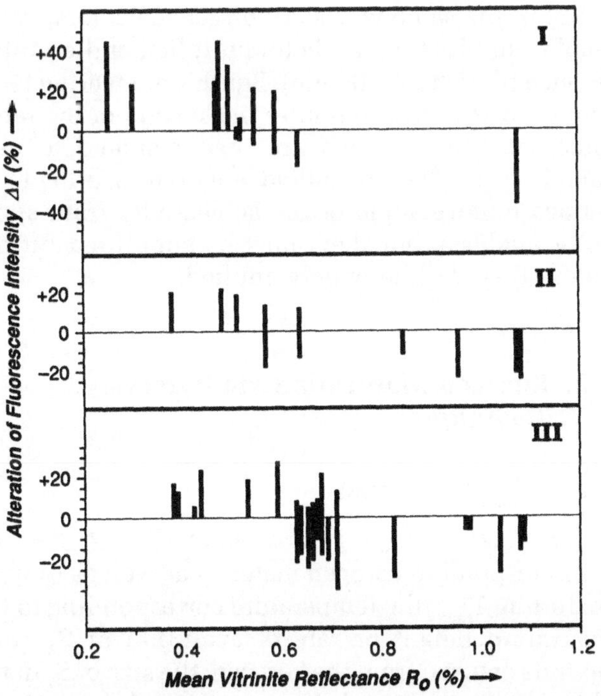

Figure 17. An illustration of fluorescence intensity alteration trends with maturity and kerogen type. [Adapted from Leythaeuser *et al.* (1980, p. 35).]

the transition (both positive and negative) is related to the onset of hydrocarbon generation (Leythaeuser *et al.*, 1980).

The different fluorescence parameters, such as peak of maximum intensity or the red/green ratio (Sentfle *et al.*, this volume, Chapter 15), can be very confusing to geologists and geochemists trying to establish maturity trends. Conversion into vitrinite reflectance equivalents has been suggested as a solution to those problems of usage and reliability (Robert, 1985). Therefore, Thompson-Rizer and Woods (1987) have proposed the term "R_F" (reflectance determined from fluorescence) to convey spectral fluorescence information.

Although much progress has been made in our ability to measure fluorescence, such measurements are not commonly performed by most laboratories, because of instrumentation and other experimental problems. For example, problems with standardization are illustrated by a comparison of quantitative fluorescence data from laboratories all over the world, in which Thompson-Rizer *et al.* (1988) found generally good agreement on samples of Plexiglas with strong florescence intensities, but lack of agreement on organic particles, especially from shale samples.

4.1.3. Transmittance or Spore-Pollen Color

The relation between the color of spore tissues and degree of metamorphism has been known from the work of Schopf (1948). The art of describing transmitted light colors of microscopic organic grains is highly subjective, and when arbitrary numerical values are assigned to those colors, there is much confusion, as shown by the color scales in Table III. To aid in reproducibility, several workers have created numerical scales, reference sets of samples, and photomicrographs which an operator can compare to that of the unknown sample. These scales are referred to as spore coloration index (SCI), thermal alteration index (TAI), or kerogen alteration index (KAI). It has been suggested by Staplin (1969) that because different reference sets may be needed for different basins (due to changing OM types, thermal regimes, etc.), these sets should be calibrated to various geochemical properties spanning the entire petroleum generation range. Many of the numerical scales are further subdivided to indicate gradational colors. Decimal places are often used with the Staplin (1969) and the Robertson Research International Ltd. (Barnard *et al.*, 1976) scales which make these color descriptions look more precise than they really are. The Chevron thermal alteration scale (Jones and Edison, 1978) arbi-

TABLE III. Transmitted Light Color Scales[a,b]

Pollen and spores[c]	Organic matter[d]	Palynomorph[e]	Palynomorph[f]
1. Transparent yellow	1. Fresh yellow	1. Colorless pale yellow	1. Colorless to pale yellow
2. Orange to reddish	2. Brownish yellow	2. Pale yellow to lemon yellow	2. Yellow
3. Brown	3. Brown	3. Lemon yellow	3. Light brownish yellow, yellowish orange
4. Opaque (black)	4. Black	4. Golden yellow	4. Light-medium brown
5. Brown spherules	5. Black, with additional evidence of rock metamorphism	5. Yellow orange	5. Dark brown
6. Indeterminable		6. Orange	6. Very dark brown-black
		7. Orange brown	7. Black (opaque)
		8. Dark brown	
		9. Dark brown-black	
		10. Black	

[a]Modified from Smith (1983).
[b]No attempt has been made to put equivalent colors in order.
[c]Correia, 1967, 1971.
[d]Staplin, 1969.
[e]Barnard *et al.* (Robertson Research International), 1976.
[f]Batten, 1981.

trarily assigned 0.1 changes in the TAI scale to correspond with 0.1% R_0 changes in the oil window. Other scales indicate intermediate colors by using plus and minus signs (Bujak *et al.*, 1977) or fractions (Batten, 1981). Some lab standard reference sets have been made by artificially heating a single sample and relating OM color to measured fixed carbon content (Burgess, 1974).

Saxby (1982) suggested that spore color changes are due to chemical decomposition reactions which produce, by way of free radicals, gas or oil from polymerized lipid-rich components of the spore: yellow to orange was proposed to represent the breakdown of carboxyl (CO_2) groups in acids and esters, brown to correspond to oil evolution at a maturity level of 1.6% R_0 (where the reflectance of spores and that of vitrinite are the same), and black to occur at the point where aromatic and aliphatic carbon bonds break to form methane. Detailed chemical analyses (perhaps on single grains) would be required to verify these hypotheses. Teichmuller (1974a) stated that black spore color appears when the reflectance of sporinite and that of vitrinite are the same. Gutjahr (1960) noted that some pollen or spore grains can become a lighter color by autoxidation whereby the outer walls take up oxygen until the grain is completely destroyed.

An effort has been made to quantitatively measure the transmittance of OM using photometric equipment. Gutjahr (1966) measured the light absorption of spores and related it to increases in carbonization and depth for both laboratory-heated samples and well samples. These measurements are very time-consuming requiring the averaging of a large number of measurements. Staplin (1969) mentioned tissue thickness as a cause of differences in light absorption and felt that there is still a lot of subjectivity in Gutjahr's method for picking grains to measure and in interpreting the results. In 1975, Grayson used photometry to measure transmittance or translucency (the reciprocal of absorption) of a sample's first 30 spores or pollen of the same taxon (if possible) at exactly the same midpoint in each. He compared his data to elemental analyses to determine the petroleum generation potential relationship. The color of OM in transmitted light can be measured (Smith, 1983) in the same way as fluorescence colors, using stable illumination, a photomultiplier, and an interference filter (400–700 nm). Smith's pioneering technique saves in the computer the spectra of the reference set slides, which are then matched to new sample data. These quantitative light-measuring techniques remove some of the subjectivity from spore color analysis, but they must be simplified further before they will be widely applied.

4.2. Kerogen Maturation via Pyrolysis Technique

4.2.1. T_{max} (Pyrolysis)

Several pyrolysis parameters (Fig. 7; Section 3.1.2) respond to kerogen maturity as well as to type, including T_{max}, the temperature corresponding to the maximum height or rate of evolution of S_2 compounds cracked from the kerogen; the size of S_1 in the absence of migrational effects; and, to a lesser extend, the size of S_2, which decreases relative to TOC as kerogen cracking progresses to produce petroleum through the oil and gas windows. Pyrolytic determination of kerogen maturation parameters related to oil and gas exploration has been recently reviewed by several authors (Peters, 1986; Orr, 1983; Katz, 1983; Tissot, 1984; Tissot *et al.*, 1987; Espitalié *et al.*, 1984). Maturity is difficult to determine using elemental ratios or HI–OI plots alone (e.g., Fig. 1C), particularly in the absence of ancillary information on kerogen type and richness prior to maturation. Pyrolysis T_{max} values provide a much more reliable measure of maturity if the limitations discussed below are kept in mind.

Espitalié *et al.* (1984), using a large group of relatively homogeneous type III isolated kerogens and coals, have shown that T_{max} correlates with vitrinite reflectance, but that the correlation is not linear (Fig. 18A). Nonlinearity between the two maturation parameters is not unexpected if T_{max} and R_0 respond to different chemical reactions which do not have the same time and temperature response. Although the exact chemistry involved in causing R_0 to increase with increasing thermal maturation is unknown, the

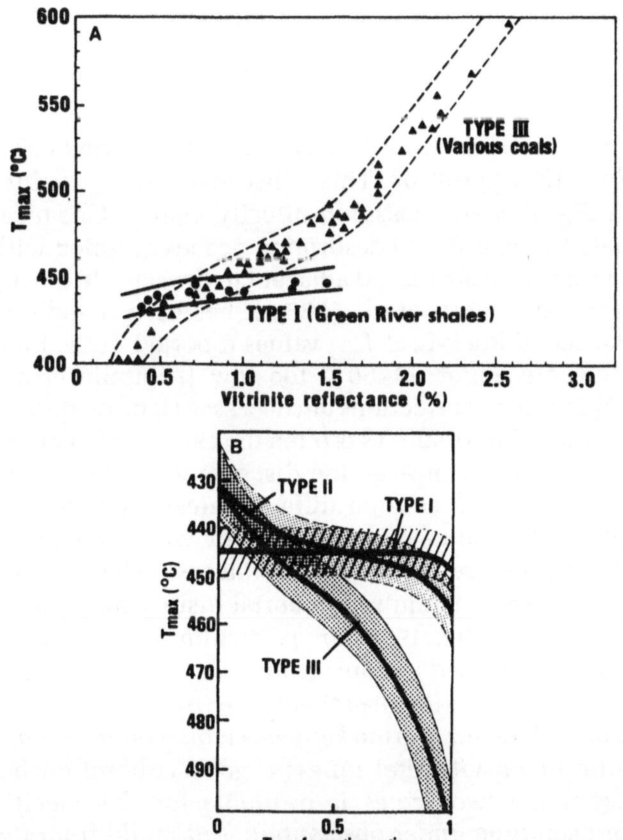

Figure 18. (A) T_{max} vs. R_0, for kerogen types III and I. [From Espitalié *et al.* (1986).] (B) Change in T_{max} and transformation ratio with kerogen type. [From Tissot *et al.* (1987).]

corresponding decrease in kerogen H/C ratios suggest that reactions leading to increased aromatization within the kerogen matrix are involved. A number of papers have pointed out that, if done carefully on the well-characterized macerals via procedures described earlier, R_0 is a true "maturation" indicator, in that it responds to increases in both time and temperature. T_{max}, on the other hand, results from a variety of hydrocarbon fragments being broken off of the kerogen matrix in a large variety of parallel reactions which occur progressively throughout the pyrolysis heating process (Waples, 1980; Wood, 1988; Sweeney *et al.*, 1986; Tissot *et al.*, 1987). As a result, the relative importance of time and temperature to overall thermal cracking and, therefore, to T_{max} can vary considerably depending on heating rate (both in the geological system and in the pyrolysis system) as well as on the initial types of chemical bonds which are present within the kerogen. The observed increase in T_{max} of S_2 with increasing maturation is a result of the weakest chemical bonds breaking first during geological thermal maturation so that only the residual stronger bonds are left when the kerogen is analyzed in the laboratory by pyrolysis. Thus, the temperature corresponding to the S_2 peak maximum or the highest rate of kerogen cracking, T_{max}, is observed to shift to progressively higher temperature as the "envelope" of residual kerogen thermal cracking reactions progressively shifts to the stronger bonds at higher maturities.

Because the chemical nature of a particular kerogen is intimately related to the observed T_{max}, it would be expected that different kerogen types would show different responses of T_{max} to the maturation process. T_{max} data have recently been summarized for kerogen types I, II, and III by Tissot *et al.* (1987), who concluded that T_{max} is a good maturation indicator for kerogen types II and III, but not for type I because the T_{max} values remain very constant as a function of maturity for type I (Fig. 18). In fact, kerogen maturation, as reflected by T_{max}, is observed not to be very sensitive to time (unlike vitrinite reflectance), so that temperature is more important than time in the overall kinetics of kerogen breakdown and the associated oil and gas generation (i.e., Wood, 1988; Hunt *et al.*, 1991).

Some of the factors which can complicate interpretation of T_{max} as a maturation indicator are illustrated in Fig. 19, which shows some typical examples of pyrolyses of kerogens within whole rocks recovered from two Alaskan North Slope wells (Whelan *et al.*, 1986). Whole-rock T_{max} measurements are much more common than isolated kerogen measurements,

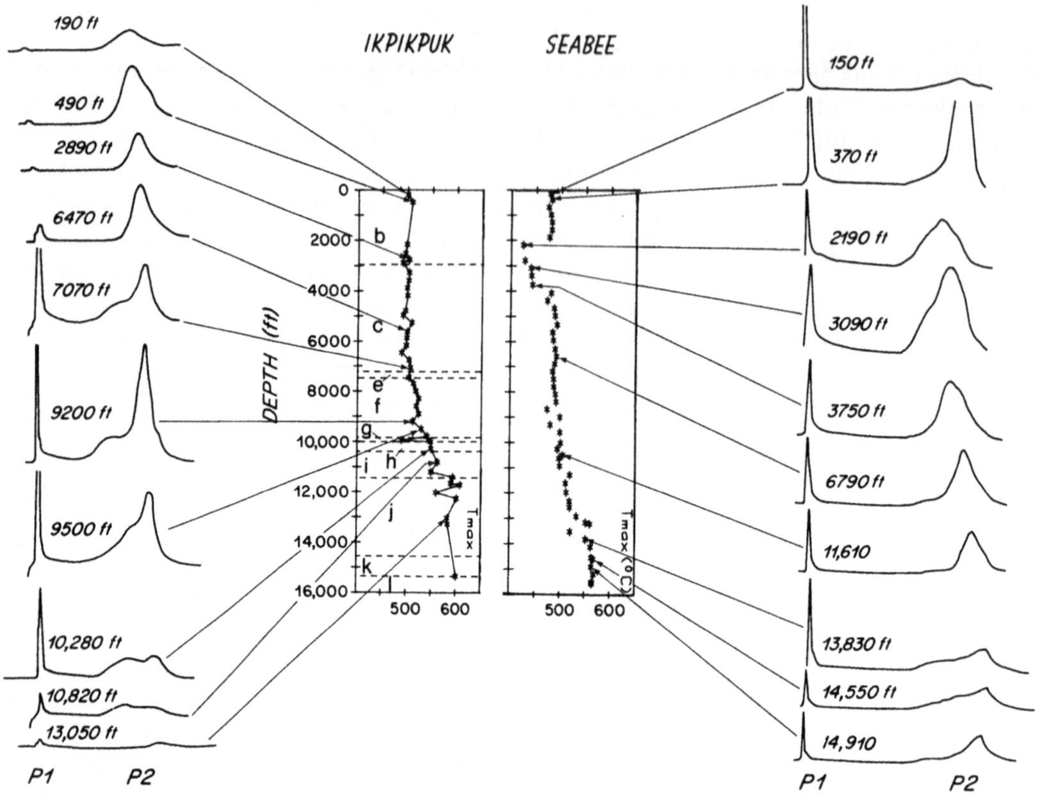

Figure 19. Typical down-hole pyrolysis data, Alaskan North Slope. [From Whelan *et al.* (1986).]

particularly in oil exploration studies, so that potential problems caused by the rock matrix discussed above need to be kept in mind. These two Alaskan wells, the Ikpikpuk and Seabee wells, penetrated all of the oil generation and a significant portion of the gas generation zones with the down-hole vitrinite reflectance curves for the two holes being almost superimposable (R_0 = 0.5–2.6%). With increasing depth and maturity, a number of changes are observed: (a) T_{max} increases with increasing depth and maturity in both wells; (b) P_1 (equivalent to S_1, migrated or generated bitumen) first increases as oil generation begins and subsequently decreases as oil is expelled with increasing depth in the Ikpikpuk well; and (c) P_2 (S_2) decreases at the bottom of both wells below the oil window. The pyrolysis data shown in Fig. 19 for the Ikpikpuk well are very typical of the normal changes in T_{max}, P_1, and P_2 which occur as a function of increasing kerogen maturation. No oil shows were observed in the Ikpikpuk well, so that the pyrolysis data can be interpreted in terms of increasing kerogen maturation with increasing depth, and effects of migrated bitumens can be neglected (Whelan *et al.*, 1986).

The pyrolysis T_{max} values shown in Fig. 19 are about 50°C higher than those normally reported using Rock-Eval pyrolysis. This is because the T_{max} values in Fig. 19 were measured directly using a Chemical Data Systems (CDS) desorption probe equipped with a thermocouple placed adjacent to the sample during pyrolysis (Tarafa *et al.*, 1988). It has been noted that the actual Rock-Eval T_{max} values reported in the literature are about 50–60°C too low (Espitalié *et al.*, 1977), but no corrections are necessary if only the T_{max} *profile* is important, as is often the case in oil exploration studies. However, the discrepancy is important when T_{max} values from different areas, laboratories, and instruments are compared, or when the actual T_{max} values are required, as is the case when kinetic parameters are required in burial history reconstructions (Espitalié, 1986). *In particular, great caution needs to be used in converting pyrolysis T_{max} values to "equivalent vitrinite reflectance values" (as is commonly done for marine kerogens which often contain little or no vitrinite) unless a good calibration between the two scales is available for the specific kerogen type under observation, optimally from the well or formation under investigation.*

The data for the Seabee well demonstrate how migrated bitumen can alter the relationship between

T_{max} and R_0. The Seabee well shows a trend in T_{max} similar to that of Ikpikpuk, except that the T_{max} values are shifted about 30°C to lower temperatures throughout the well (Fig. 19) in spite of the similar vitrinite reflectance profiles for the two wells. Such a lowering of T_{max} is consistent with the presence of migrated bitumens (Vandenbrourke *et al.*, 1981; Orr, 1983; Whelan *et al.*, 1986; Peters, 1986). For example, the shallow Seabee samples between 2000 and 4000 ft (Fig. 19) give T_{max} values of <450°C, which are anomalously low in comparison to the values for the rest of the well, characteristic of an interval containing significant migrated bitumen. Abnormally high P_1 values (in comparison to the values for the Ikpikpuk well samples of comparable depth) accompany the anomalously low T_{max} values through the 2000- to 4000-ft section of the well, the interval which penetrates the producing Umiat reservoir facies (Bird, 1985). T_{max} values throughout deeper sections of the well are somewhat lower than those of other rocks of similar maturity, so that the entire profile was initially thought to be significantly influenced by migrated bitumen. However, biomarker maturation data, particularly the regular increases of ratios of methyl phenanthrenes with increasing depth below 4000 ft (Farrington *et al.*, 1988b), suggest that migrated bitumens are present only in the shallower sections of the well.

The shoulder on the front of the Alaskan North Slope P_2 peak in Fig. 19 represents unexpelled bitumens which remain in the fine-grained rocks as they pass through the generation zone (Tarafa *et al.*, 1988). This phenomenon can be seen for a number of samples from both wells, particularly within the oil window. This shoulder becomes the predominant P_2 fraction in some deeper rocks which are either within or have passed through the oil window (e.g., samples from 10,280 to 13,050 ft in the Ikpikpuk well). It is not possible to remove this material via solvent extraction, so that it probably represents a pyrobitumen (i.e., unexpelled source rock bitumen which has passed through the oil window), an identification which has been confirmed optically (L. Eglinton and J. K. Whelan, unpublished results).

An advantage of the CDS instrumentation used for the Alaskan North Slope study (Whelan *et al.*, 1986; Peters, 1986) as well as a variety of custom-designed pyrolysis instruments in use in many other laboratories (Barker, 1974; Claypool and Reed, 1976; Horsfield, 1984; Larter, 1984; Tarafa *et al.*, 1988) in comparison to the more widely used Rock-Eval unit is the frequently better separation of the P_1 and P_2 peaks as well as the ability to observe multiple lobes of the P_2 peaks, which can provide information on oil generation and expulsion, as discussed above for Fig. 19. It is often difficult to pick the exact position of T_{max} in Rock-Eval data, particularly via automated procedures utilizing computerized data collection, because of the "overlapping" and "multiple lobe" peak problem. In other types of apparatus where the P_1 and multiple lobes of the P_2 peaks are better separated, as is the case with the CDS unit, consistent identification of which peak to use for T_{max} identification is much less of a problem. In the Alaskan North Slope case, the T_{max} measurements shown in Fig. 19 were all made on the second, larger P_2 lobe, rather than on the first bitumen shoulder peak, which would have given an anomalously low T_{max} reading. Note that if these T_{max} values had been measured on an instrument where the two lobes were not as well separated, an anomalously low P_2 reading would have been obtained. This "multiple lobe" P_2 behavior is ubiquitous in a number of wells throughout the North Slope for all samples recovered either in or below the oil generation zone (Tarafa *et al.*, 1988; Whelan *et al.*, 1986).

T_{max} can be changed by factors other than thermal maturation, including heating rate during the pyrolysis experiment, as discussed above. T_{max} also changes as a function of maceral type [Fig. 20, from Dembicki *et al.* (1983)] and has also been found to be affected by air oxidation in coals, as shown in Fig. 21 (Landais *et al.*, 1984), where T_{max} of kerogen increased by up to 20°C when the humic coal was exposed to air at 200°C for short time periods. An exponential rise in T_{max} occurred during the first 5 h followed by a leveling off as heating was continued up to the maximum of 48 h. T_{max} values can vary considerably between whole rocks and their corresponding isolated

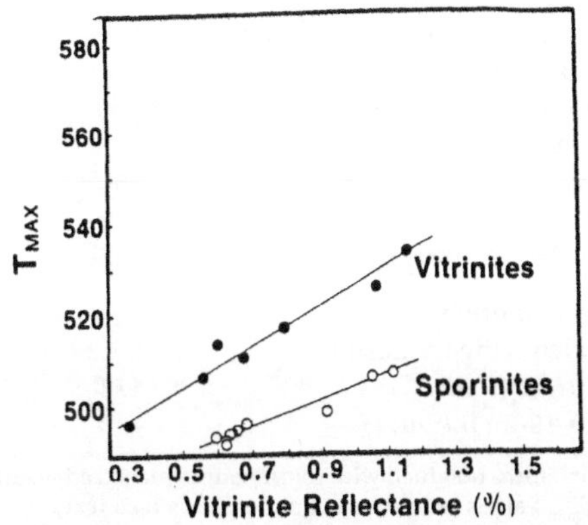

Figure 20. Changes in T_{max} as a function of maceral type. [From Dembicki *et al.* (1983).]

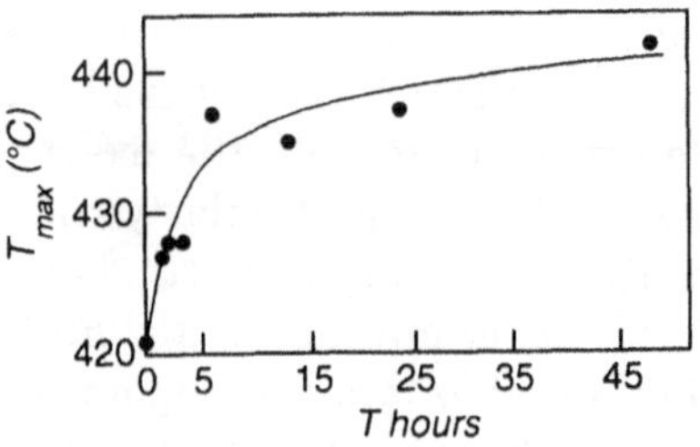

Figure 21. Change in T_{max} with air oxidation for T hours. [From Landais et al. (1984).]

kerogens or protokerogens, as shown in Table IV for a series of recent sediments. All of these samples are immature with respect to petroleum generation.

4.2.2. Thermogravimetric (TG) Techniques

In thermogravimetric (TG) analysis, the weight loss of a sample is measured continuously as the temperature is increased linearly. The TG analysis of a few samples of kerogen types I, II, and III has been described by Durand-Souron (1980), who measured the nature of the gases evolved during analysis using a mass spectrometer. Behar and Vandenbroucke (1986, 1987) combined the quantitative data available from TG analysis with the chemical structural information available from elemental analysis, electron microscopy, ^{13}C NMR, chemical degradation analysis, and pyrolysis in order to present average structures consistent with representative chemical properties of kerogen types I, II, and III at various stages of maturation. These authors pointed out that their postulated structures cannot and are not meant to reflect the heterogeneous nature of the various kerogen structures. However, the structures probably do represent a good "average" composite structure consistent with the extensive kerogen data currently available within the Institut Français du Pétrole.

TABLE IV. T_{max} Values for Kerogen Isolates and Whole Rocks from Immature Recent Sediments

Source of sample	T_{max}[a] Kerogen	T_{max}[a] Whole rock
Surface sediments		
Cape Lookout Bight	470	465
Peru upwelling[b]		
0.4 to 7 m, water depth = 150 m		457–466
1 m, water depth = 3000 m		462–472
Surface sediments, 0–70 cm, water depths 85–500 m		445–465
Surface core from anoxic bottom water zone (0.3 mbsf[c])	484 ± 1	468 ± 3
Georgia Bay	450	438
Buzzards Bay	489	490
Guaymas surface sediment	455	470
Smectite-rich samples		
Gulf of Mexico—Pigmy Basin[d]		
DSDP 619-1-2 (5 m)	473	495
DSDP 619-19-2 (160 m)	433	478

[a]Actual temperatures obtained with thermcouple placed adjacent to sample; Rock-Eval T_{max} values would be about 50°C less (see text).
[b]Whelan et al., 1990.
[c]mbsf, Meters below seafloor.
[d]Whelan et al., 1986b.

Preliminary work has also been carried out on analysis of kerogens and sedimentary rocks using TG coupled with Fourier transform infrared analysis (TG-FTIR) to continuously monitor the gaseous mixture evolved during pyrolysis (Whelan et al., 1988b, Whelan et al., 1990a). Gas-phase infrared spectra typically give very high resolution so that even gases in mixtures can be identified and quantitated unambiguously on the basis of the fine-line structure. Therefore, TG-FTIR shows promise of extending the kerogen T_{max} scale into the higher maturation ranges typical of gas generation using the T_{max} of methane (Fig. 22A). The values shown in Fig. 22A, which include whole rocks and isolated kerogens from a variety of areas and depositional environments, show a fairly good correlation with vitrinite reflectance to very high reflectance values. T_{max} of ammonia derived from clay minerals has also been measured via the TG-FTIR technique (Fig. 22B; Whelan et al., 1988b, 1990a). The high T_{max} values for ammonia above 13,000 ft in Fig. 22B are typical of those associated with clay mineral phases. Isolated kerogens show much lower values which do not increase with increasing maturation (Whelan, unpublished results). This mineral ammonia signal remains, even after oxidants such as ferric oxide, ferric chloride, or manganese dioxide are added to the sample (J. K. Whelan, unpublished results). The anomalously low ammonia T_{max} values below 13,000 ft in Fig. 22B are attributed to the presence of asphaltenes or pyrobitumens, which were identified microscopically and chemically throughout deeper sections of this well (Whelan et al., 1988a).

TG-FTIR also provides continuous monitoring of total weight loss from the sample, the time, and the temperature and a continuous record of the amount of each gas lost during the pyrolysis process, the data required to determine the kinetic parameters for breakage of specific types of bonds within the kerogen matrix during pyrolysis, as recently demonstrated for coals (Solomon et al., 1990). Thus, measurements via TG techniques utilizing specific gas

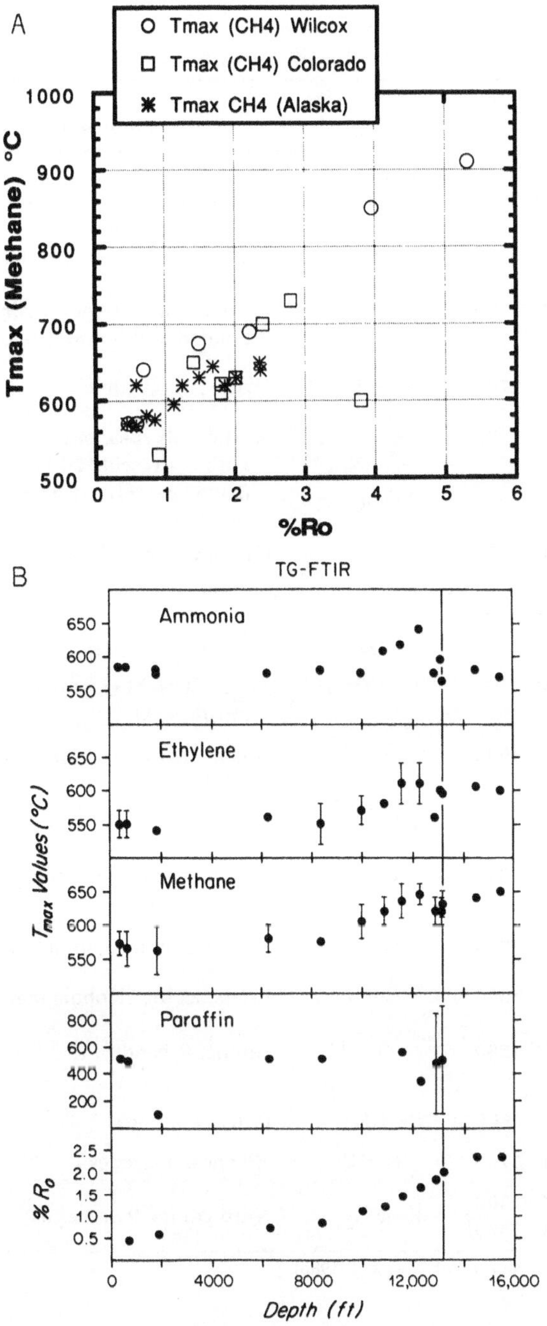

Figure 22. (A) T_{max} of CH_4 evolution from isolated kerogens and whole rocks. [From Whelan *et al.* (1990a)]. (B) T_{max} of NH_3, C_2H_4, CH_4, and paraffin from whole rocks, Alaskan North Slope. [Adapted from Whelan *et al.* (1988b).]

monitoring can be related to cleavage of specific (rather than very broad) groups of bonds in the kerogen structure. The potential exists for more accurately determining the kinetic parameters for specific kerogen reactions whose time–temperature response can be more easily extrapolated to reconstruction of well burial histories than can be done with current techniques utilizing a series of parallel reactions. Using kinetics of specific reactions would overcome some of the problems of current procedures which arise from the necessity of having to simultaneously follow the total product evolution from the many parallel reactions involved in S_2 evolution, all having their own frequency factors and activation energies (Tissot *et al.*, 1987; Sweeney *et al.*, 1986; Waples, 1980, 1981; Wood, 1988; Yukler and Kokesh, 1984; Hunt *et al.*, 1991).

5. Spectral Methods of Determining Kerogen Type and Maturation

Spectral techniques fall somewhere between bulk and very detailed structural degradation methods in their potential to provide information about molecular structures of kerogen and protokerogen. Spectral techniques are capable of providing considerable structural information about pure organic compounds. This description obviously does not apply to kerogens which are much more complex and of higher molecular weight.

5.1. Infrared Spectroscopy

Infrared (IR) spectroscopy as applied to kerogen was the subject of an excellent review by Rouxhet *et al.* (1980), which includes considerable detail on the experimental procedure. If the atoms in molecules are considered to be tiny balls on the ends of springs which represent chemical bonds, then absorption of infrared radiation occurs as a result of the discrete amounts of energy (corresponding to specific frequencies of light) required to stretch or bend these bonds. Therefore, the absorption frequencies for specific molecules obtainable from IR spectroscopy provide organic structural information about the presence of specific bond types and functional groups. Some examples of useful absorptions for kerogens (i.e., those which do not significantly overlap with frequencies typical of other groups; see Fig. 23 and Tables V and VI) are those of aliphatic and aromatic C—H bonds (in the range of 3100–2900 cm^{-1}), C═O groups (1800–1650 cm^{-1}), and O—H and N— H (3600–3200 cm^{-1}). There are a number of other lower frequency bands which have been used, as discussed in detail by Rouxhet *et al.* (1980) and more recently by Monthioux *et al.* (1985) and Monthioux and Landais (1988) (Fig. 23), although these are not as reliable as those listed above because of ambiguities arising be-

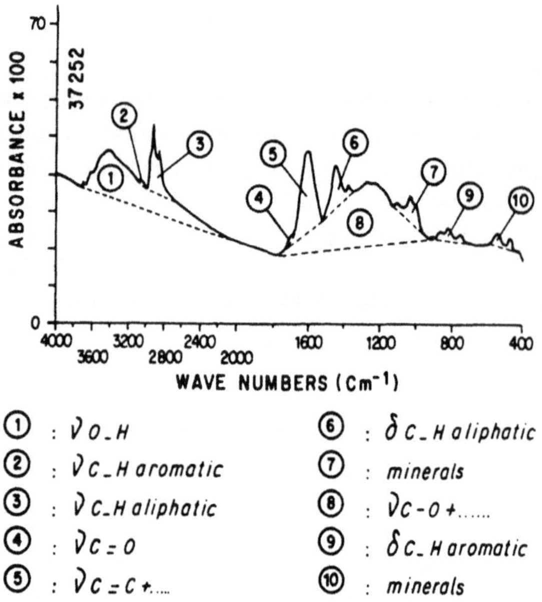

Figure 23. Positions of infrared absorption bands used to quantitate alteration of macerals of Mahakam Delta coal. [From Monthioux and Landais (1988).]

cause of potential overlap with absorptions of other groups.

Even though IR spectroscopy of kerogen is not used as much today as it was in the past, the technique gives valuable information about both kerogen type and maturity when used together with other data. For example, Thompson and Dembicki (1986) have shown that typical amorphous kerogen types A and D (described in Section 3.2) give substantially stronger C—H absorption than either amorphous types B or C. In addition, IR spectroscopy can provide a quantitative measure of specific bond types and functional groups in favorable cases (Fig. 24). Also, Fig. 25A shows an excellent correlation between the Rock-Eval S_2 yield and the amount of aliphatic C—H as measured by IR spectroscopy (Schenk *et al.*, 1986). This impressive correlation was obtained using several types of kerogens isolated from a number of different geographic regions, although some dependence on maturation was also detected. Schenk *et al.* (1986) also found an excellent correlation between carbonate carbon as determined via IR spectroscopy (C_{min}, Fig. 25B) and as measured from total carbon minus TOC ($C_{tot} - C_{org}$). These studies demonstrate that good quantitation is possible with IR spectroscopy when the bands utilized are well resolved and well separated from other absorptions.

Typical infrared spectra for kerogen types I through IV are shown in Fig. 26 (Rouxhet *et al.*, 1980). The most obvious difference is the much stronger

TABLE V. Summary of Primary and Confirming IR Absorption Bands[a]

Wave number, cm⁻¹	Functional group	Comment
Primary bands (bond stretching)		
3600–3500	—OH	Alcohol or phenol (non-hydrogen-bonded)
3500–3200	—OH	Alcohol or phenol (hydrogen-bonded)
3500–3300	—N—H (with bond below N)	Amine or amide (non-hydrogen-bonded and hydrogen-bonded)
3200–2700	—C—H (with bonds above and below C)	Hydrocarbon stretch
2300–2100	—C≡C—C, —C≡N, >C=C=C<	Nitriles, alkynes, other groups with *sp*-hybridized carbon (bands usually weak; may be missing)
1800–1600	>C=O	Carbonyl derivatives
Confirming bands		
3300–2500	R—C(=O)O—H	Acids (stretching, single, very broad peak)
1690–1600	>C=C<, >C=N—	Alkene, benzene, imine (often weak)
1450–1350	—C—H (with bonds above and below C)	Alkane groups (bending)
1250–1000	—C—O— (with bonds above and below C)	Alcohols, ethers (stretching)
1400–1300	—O—H	Esters, acids, alcohols (bending)
1650–1500	O=C(—)—N(H)—	Amide, II band
1650–1600	R—NH_2	Primary amines
1000–670	>C=C<	Alkenes, benzenes
1570–1500, 1390–1330	R—NO_2	Nitro groups (two bands)

[a]From Kemp and Vellaccio (1980).

C—H stretching band (at about 3000 cm⁻¹) for kerogen types I and II than for types III and IV, which is consistent with results from elemental analysis, pyrolysis, and NMR spectroscopy. The differences in oxygen functionality are not as obvious. For example, C=O bands at about 1750–1700 cm⁻¹ are apparent for all kerogen types. However, the broad band from 3400 to 3200 cm⁻¹ is typical of carboxyl groups and is considerably stronger for kerogen types III and IV than for types I and II. Care needs has to be taken in using this band, since water gives strong absorption in the same region.

TABLE VI. Summary of IR Band Positions for Specific Organic Functional Groups[a]

Group	Band position, cm^{-1}	Functional group	Comment
O—H, N—H	3700–3500	—OH (non-hydrogen-bonded)	Sharp peak
	3500–3200	—OH (hydrogen-bonded)	Broad
	3500–3300	—N—H (with a single bond down from the N)	Primary, two bands; secondary, one band
C—H	3200–3000	Vinyl or phenyl C—H	
	3000–2800	Alkane C—H	
	2750	Aldehyde C—H	
—C≡X, =C=	2300–2000	—C≡C—, —C≡N, >C=C=C<, >C=C=O, etc.	Usually weak band
>C=X	1840–1800	Anhydride	Two bands usually present
	1800	Acid chloride	
	1770	Cyclobutanone or phenol or enol ester	
	1740	Cyclopentanone	
	1740–1725	Acyclic ester	
	1735–1720	Aldehyde	C—H at 2750 cm^{-2} confirms
	1725–1700	Carboxylic acid	O—H at 3300–2500 cm^{-1} (broad) confirms
	1710	Acyclic ketone; cyclohexanone	
	1680	α,β-Unsaturated or phenyl ketone	
	1690–1640	Amide	
	1690–1640	Imine	Usually weak
	1670–1600	Alkene	Usually weak
Confirmatory bands	1300–1100	C—O	Esters in range 1250–1190 cm^{-1}
	1560–1540 1390–1360	Nitro—NO_2	Strong bands
	970–960	*Trans* alkene	RCH=CHR
	700–650	*Cis* alkene	RCH=CHR
	1000–990 920–900	Monosubstituted alkene	
	890–860	2,2-Disubstituted alkene	
	800–700	Alkyl chlorides	
	600–500	Alkyl bromides	

[a]From Kemp and Vellaccio (1980).

Use of these and other less well resolved IR bands both for kerogen and coal, shown in Fig. 26, for typing and maturation was discussed by Rouxhet *et al.* (1980) and more recently for Mahakam Delta coals by Monthioux and Landais (1988), who carried out extensive comparisons of natural and artificial maturation series (Fig. 24). Both the aliphatic C—H and carbonyl absorption intensities decrease with increasing maturation.

5.2. Nuclear Magnetic Resonance (NMR) Spectroscopy

Very simplistically, NMR spectroscopy is a nondestructive technique which provides information about the number and types of the nearest neighbor atoms in particular organic structures. The type of information potentially available about kerogen structure is illustrated in Fig. 27, which shows the diagnostic resonances for proton NMR and ^{13}C NMR [from Wilson (1987)].

Any atomic nucleus which possesses either an odd mass or an odd atomic number, or both, has a quantized spin angular moment and a magnetic moment, the basic atomic property required to obtain an NMR signal. The more common nuclei which possess "spin" are ^{1}H, ^{13}C, ^{14}N, ^{15}N, ^{17}O, and ^{19}F. Those used for kerogen characterization to date are ^{1}H and ^{13}C.

The basic theory of NMR spectroscopy is presented clearly in most modern organic textbooks [e.g., Kemp and Vellaccio, 1980; Morrison and Boyd, 1973; for a more complete introductory discussion, see Pavia *et al.* (1979)]. A brief summary is also given here in order to demonstrate some of the limitations of the NMR techniques as applied to quantitative

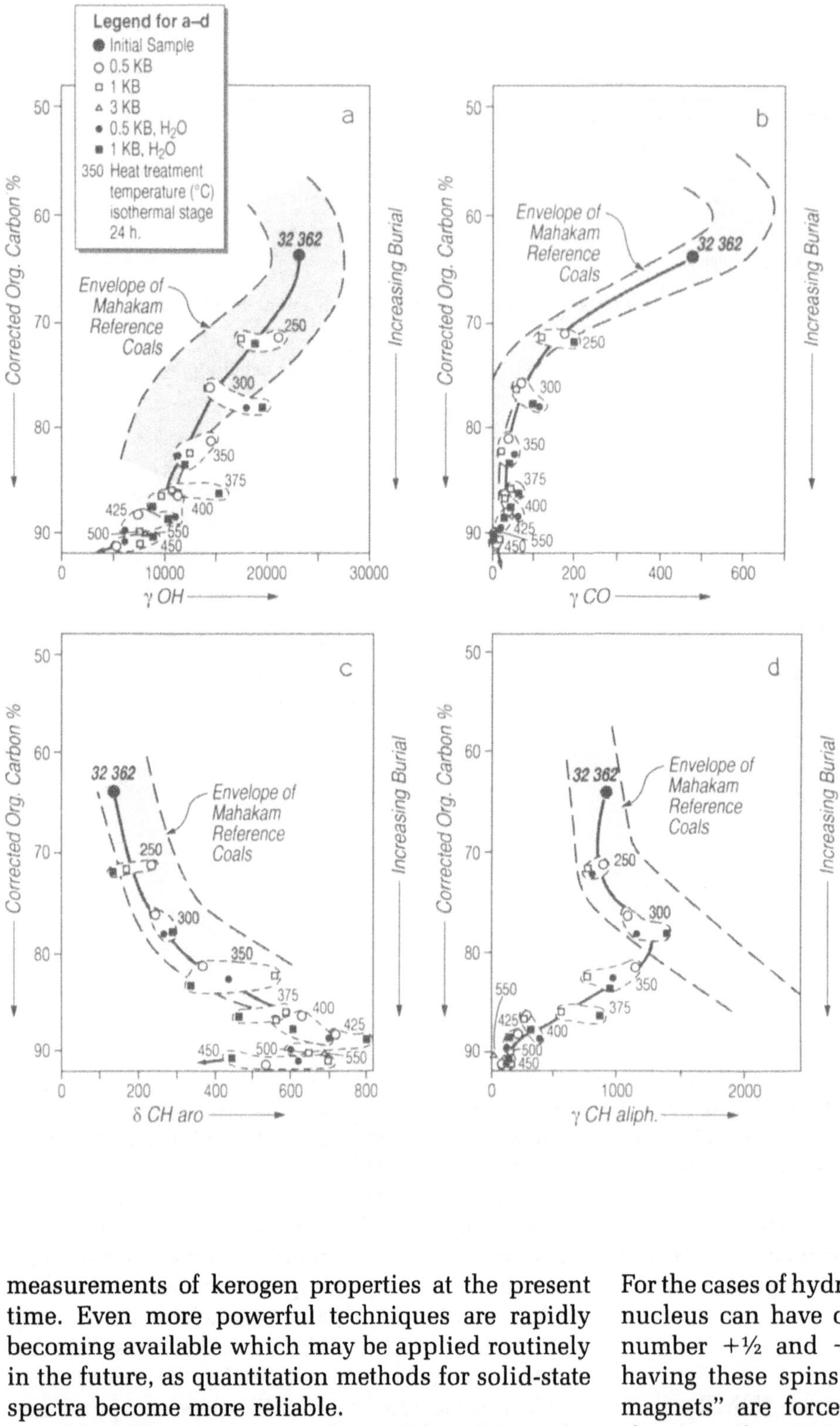

Figure 24. Quantitative loss of various IR bands in spectra of Mahakam Delta reference coals (envelope between dashed lines) compared to that observed with maturation by heat treatment in confined pyrolysis system. [From Monthioux and Landais (1988).]

measurements of kerogen properties at the present time. Even more powerful techniques are rapidly becoming available which may be applied routinely in the future, as quantitation methods for solid-state spectra become more reliable.

An NMR spectrum is obtained by placing the sample in a strong magnetic field and noting the absorption of very low frequency radio frequency radiation by the nuclei in the sample. The atomic nucleus can be considered to be a spinning nuclear charge which behaves like a tiny bar magnet (Fig. 28). For the cases of hydrogen (protons) and ^{13}C nuclei, the nucleus can have only two spin states of quantum number $+\frac{1}{2}$ and $-\frac{1}{2}$. When two identical nuclei having these spins are present, these nuclear "bar magnets" are forced by the laws of quantum mechanics to face in opposite directions and have equal energies. When placed in an external magnetic field, the two nuclei acquire two different energy states: one aligned with and one against the external magnetic field (Fig. 28A) and which depend on the magnetic quantum number, m, for the particular nucleus.

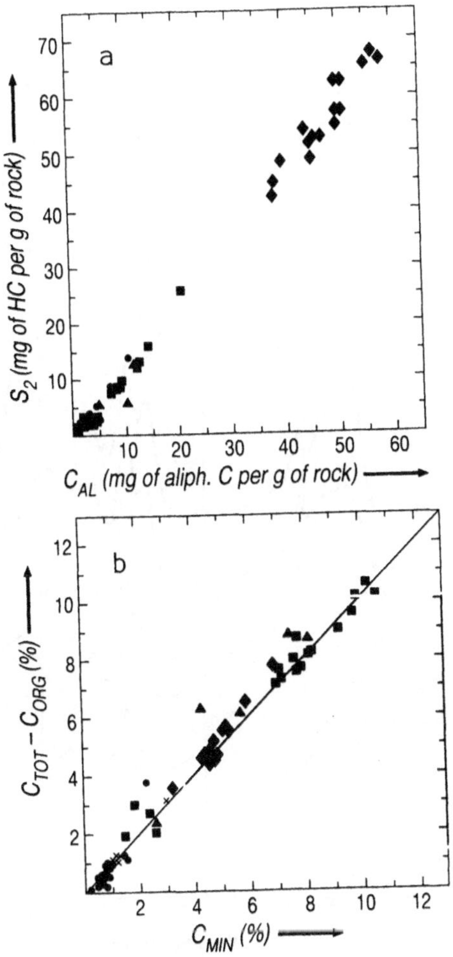

Figure 25. Quantitative IR data, showing correlation between intensity of IR C—H absorption band (C_{al}) and Rock-Eval pyrolysis yields of S_2 (A) and between carbonate carbon determined by IR spectroscopy (C_{min}) and carbonate carbon calculated as total carbon minus TOC (B). [From Schenk *et al.* (1986).]

The magnitude of the difference between the two energy states depends on the strength of the external magnetic field (Fig. 28B) as well as on the nature of the molecular electronic environment surrounding the nuclei. If radio frequency energy, ν, of frequency $\Delta E/h$, where ΔE is the energy difference between these two energy states and h is the Boltzmann constant, is applied to a population of such nuclei within a strong external magnetic field, that energy will be absorbed as nuclei in the lower state are elevated to the higher state, and the resulting absorption spectra will be observed as a single line characteristic of that particular nucleus. However, the other spinning electrons and nuclei within the molecule also create their own magnetic fields and are said to cause magnetic shielding of the nuclei whose NMR frequencies are being measured. This shielding must also be added to or subtracted from the external magnetic field experienced by each nucleus in production of the resonance effect. The exact radio frequencies absorbed by each nucleus depend both on the intrinsic electronic and nuclear structure of each atom and on the surrounding electronic and magnetic molecular structure. These properties all influence the primary measured parameter, called the chemical shift (Fig. 27), which is related to the frequency at which nuclei resonate in an external magnetic field.

For practical reasons, an NMR spectrum is usually obtained by keeping the radio frequency radiation at constant frequency and varying the external magnetic field over a small range. Because the frequency differences are very small for all nuclei of a specific type and would thus be hard to measure on an absolute scale, the positions of the peaks are reported as chemical shifts from the peak position of a standard reference compound—usually tetramethylsilane for protons and ^{13}C (Fig. 27).

The chemical shift is often experimentally observed to be split into several lines, which are caused by the interaction of the magnetic fields of the surrounding nuclei with those of the nucleus under observation. For example, if a carbon-13 atom is attached to a proton, the proton magnetic field can be aligned either with or against the field of the carbon magnet, causing the observed ^{13}C signal to be "split" into two lines of about equal intensity. The ^{13}C signal is said to be "coupled" to the proton signal, with the distance between the two "split" ^{13}C lines being called the "coupling constant," a characteristic of the two interacting nuclei. In the simplest case of n equivalent coupling protons, the ^{13}C signal will be split into $n + 1$ lines. For example, Fig. 28C shows splitting caused by two equivalent methylene protons adjacent to a ^{13}C nucleus. In the proton NMR spectrum, the methylene protons of an ethyl group would cause the same splitting to be observed for the methyl protons (i.e., X = CH_3). Normally, these splitting effects fall off rapidly with distance, so that only those nuclei directly attached to the nucleus under observation cause strong splitting. If two different nuclei have identical chemical shifts, no splitting will be observed. More generally, the nuclei will have at least small chemical shift differences, and complex splitting patterns may be observed. When chemical shifts of the two interacting nuclei are close, splitting generally does not occur into lines of equal intensity. In fact, the exact splitting patterns can become more complex as the chemical shift difference between the two interacting nuclei decreases. These splitting effects can interact in complex ways, either giving rise

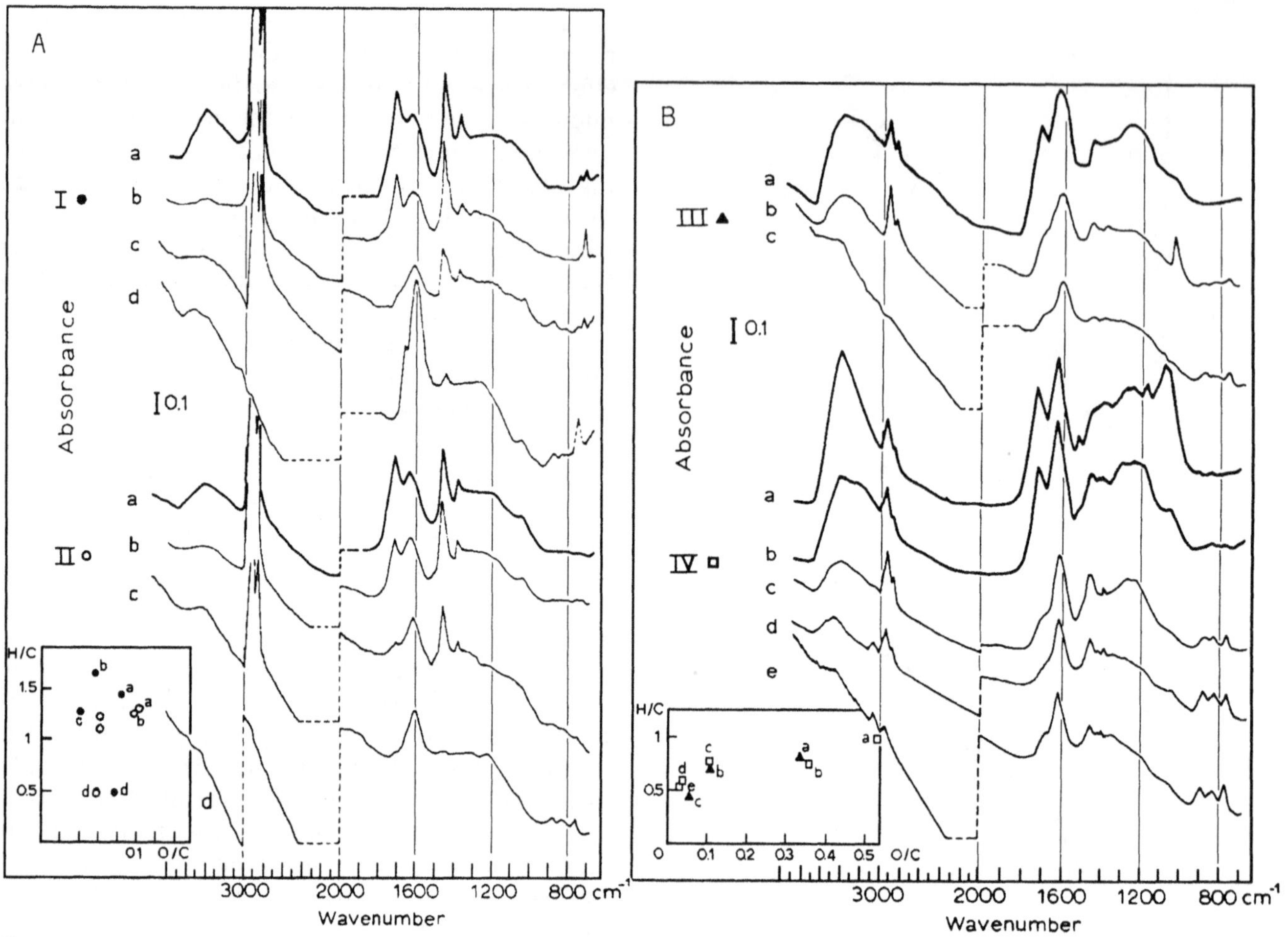

Figure 26. Typical infrared spectral absorption spectra for kerogen types I and II (A) and III and IV (B). Elemental H/C versus O/C data for the same samples are shown in insets. [From Rouxhet *et al.* (1980).]

to a series of lines at a particular chemical shift or, if several nuclei of very similar chemical shift and slightly different "coupling" constants are present, the observed NMR signal will not be resolved into separate lines, but will be observed to broaden.

Very sophisticated methods have been developed to analyze the details of this coupling and broadening which can give a great deal of molecular information about complex molecular structures [for recent reviews, see Wilson (1987), Wershaw and Mikita (1987), and Axelson (1985)]. In modern NMR instruments, very high sensitivities and signal-to-noise ratios are obtained by using Fourier transform NMR (FT-NMR), in which many spectra are collected, processed, and averaged via computer during a typical experiment. These techniques allow NMR spectra to be obtained on isotopes of low abundance, such as ^{13}C, which makes up only 1.1% of the carbon in normal samples. This low abundance makes it unlikely that two ^{13}C nuclei will be bonded in the same molecule, so that splitting in ^{13}C spectra is due primarily to interactions with protons.

Because protons and ^{13}C absorb at different radio frequencies, the carbon–hydrogen splitting can be removed by decoupling, which involves the use of powerful radio frequency radiation to cause rapid exchange of hydrogen spins, in effect eliminating the influence of the protons on the ^{13}C NMR spectrum. Thus, in the decoupled spectra, only the unsplit chemical shifts of the ^{13}C nuclei in the structure are observed, as shown in Fig. 27A.

For proton NMR spectra measured in solution, the relative intensity of signals at particular chemical shifts can be integrated to give a measure of the relative number of nuclei of a particular type. The ability to carry out such measurements reliably on other nuclei depends on the time that a particular nucleus spends in a specific magnetic environment (related to both molecular motion, which will differ for liquids and solids, and to the rates of "relaxation"

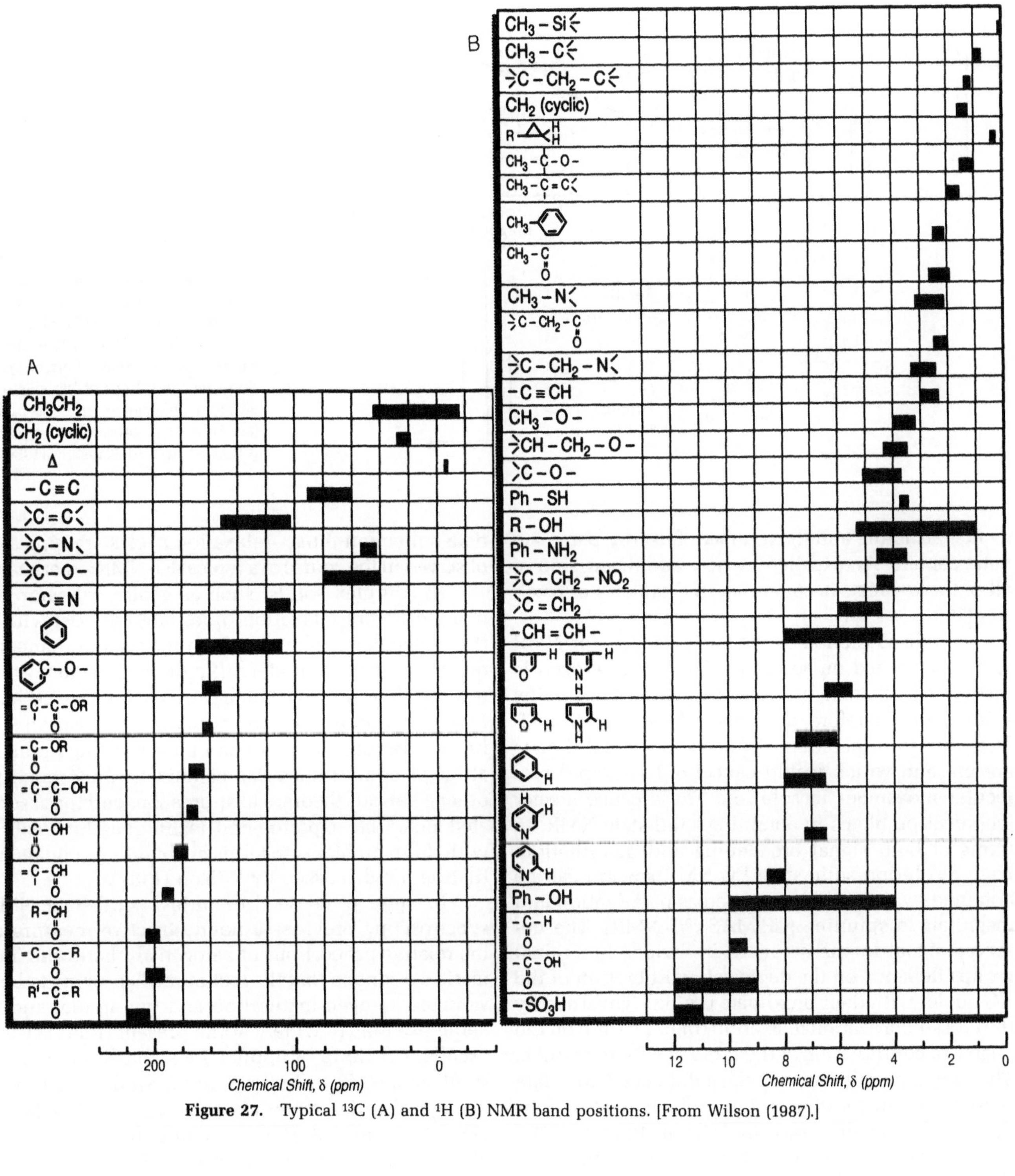

Figure 27. Typical ^{13}C (A) and ^{1}H (B) NMR band positions. [From Wilson (1987).]

of the nuclei from the higher to the lower energy states) with respect to the experimental observation time. Nuclei can lose energy in two ways: (1) spin–spin relaxation, in which nuclei exchange energy with other nuclei, characterized by a time constant, T_2, and (2) spin–lattice relaxation, in which nuclei lose energy to the surroundings and realign themselves back to the ground state, characterized by a spin–lattice relaxation time, T_1. In quantitation, the time for each NMR pulse-decay sequence, or contact time, is critical in any attempt to directly compare numbers of different types of ^{13}C bonds, since the signal strength is strongly affected by the relaxations listed above, which vary in complex ways with mo-

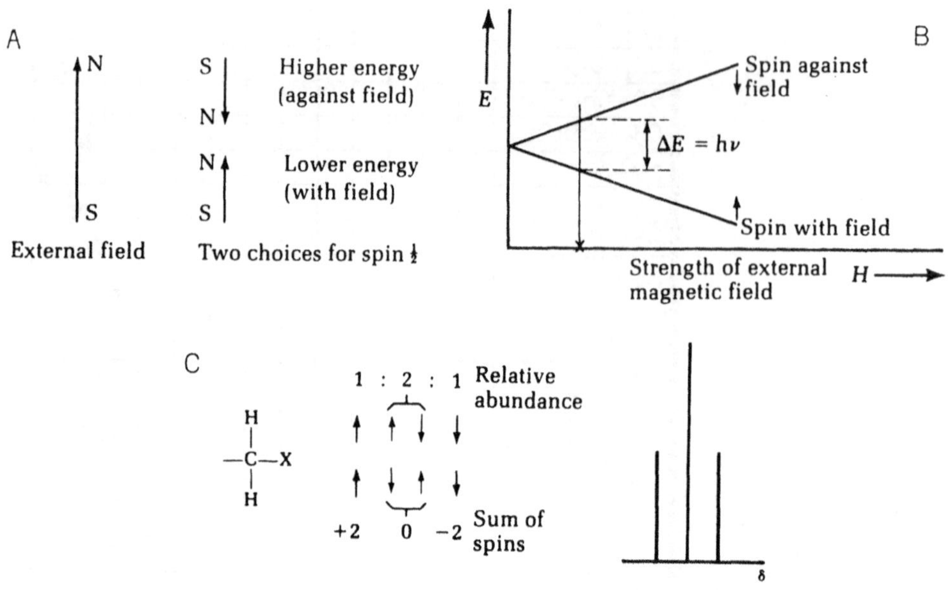

Figure 28. (A) Two possible orientations of protons in a magnetic field, with and against the external magnetic field. (B) Energy difference (ΔE) of two possible proton orientations in an increasing magnetic field. At constant field strength, energy difference between the two spin states is shown by the dashed lines, where ν is frequency, and h is Planck's constant. (C) Spin effects of two identical methylene protons causing splitting into three lines of the signal of either an adjacent ^{13}C nucleus or adjacent methyl protons (X = CH_3). [From Kemp and Vellaccio (1980).]

lecular structure and the number of nearby paramagnetic centers. An excellent review and discussion of all of these effects is contained in Chapter 3 of Axelson (1985).

The discussion above is fairly general for NMR spectra obtained on solutions. However, kerogen is insoluble so that analyses must be carried out in the solid state. Therefore, a number of complicating effects due to orientation of nuclei in the solid state are present, which would cancel out via rapid molecular movement in solution. These cause greater technical problems in obtaining solid-state NMR, in terms of both signal broadening and quantitation. High-resolution solid-state ^{13}C NMR spectra can be obtained by the techniques of cross polarization and magic angle spinning (CP/MAS ^{13}C NMR). The observed signal intensity for ^{13}C in various functional groups depends on the degree of equilibration of the ^{13}C nuclei with their proximate protons (governed by the cross-polarization relaxation time, T_{cp}) and on the degree of equilibration of the protons with the lattice. Therefore, quantitation of such solid-state NMR data presents a number of potential problems not present in solutions, as discussed by Wilson (1987, pp. 75–77). The most important is that the observed signal does not necessarily quantitatively represent the number of nuclei of a specific type in the sample because of the different rates at which various nuclei relax after each pulse. Nuclei with very short relaxation times may relax before they are observed. Alternatively, very slowly relaxing nuclei must be allowed to relax after each pulse or else they will not be observed with full signal intensity because several data collection pulse–relaxation cycles are typically observed in the course of a typical FT-NMR experiment.

In complex solids, such as geological samples, many different relaxation times often coexist within the molecular structure, which makes simultaneous quantitation of all nuclei difficult. Paramagnetic centers (i.e., free radicals which can exist for long periods of time within the kerogen structure, see Section 5.3) further complicate the situation by assisting in relaxation of some molecular centers in preference to others. Various theoretical approaches and the associated potential experimental techniques for dealing with these problems in terms of obtaining good quantitation are discussed by Wilson (1987).

In spite of these difficulties, solid-state NMR spectroscopy provides a nondestructive measure of the relative proportions of kerogen aliphatic and aromatic carbons so that the technique is proving to be a valuable kerogen maturation and typing tool and is being used fairly widely at the present time. Several excellent reviews of applications and limitations of solid-state NMR techniques in the study of geo- and macromolecular materials have appeared (Wilson, 1987; Axelson, 1985; Wershaw and Mikita, 1987). Use of NMR for direct and nondestructive evaluation of source rock potential of kerogens in organic-rich whole rocks in the Permian Phosphoria Formation has been described by Miknis *et al.* (1982). An excellent review of applications of the technique to characterization of very organic rich coals and oil shales has recently appeared (Axelson, 1985, Chapter 7).

Caution must be used in measuring the NMR signals of aromatic and aliphatic ^{13}C signals in complex heterogeneous materials such as kerogens and coals because (1) paramagnetic centers (free radicals) can cause major chemical shift displacements of carbons in close proximity, as well as increased tailing, making precise integration difficult; and (2) inadvertent transformations of carbonyl, carboxyl, and alkoxyl (e.g., carbohydrate) moieties during kerogen or protokerogen isolation can lead to additional intensity in the aromatic and olefinic portions of the NMR spectra (Maciel and Dennis, 1982).

An example of the application of ^{13}C NMR to determination of kerogen maturation is shown in Fig. 29, where a Cretaceous Black Shale from DSDP Leg 41 in the eastern Atlantic was thermally altered during the Miocene by an intrusive basalt (Dennis *et al.*, 1982). The kerogen becomes progressively hotter and more aromatic in moving toward the sill contacts

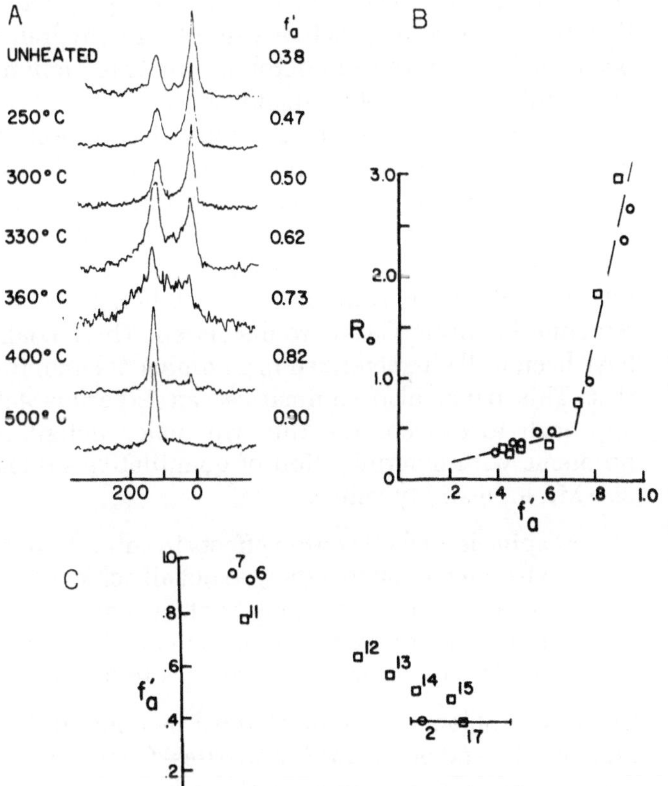

Figure 29. Quantitative NMR data of kerogens obtained from above and below an intrusive sill. (A) Solid-state ^{13}C NMR spectra of samples heated to various temperatures. Temperatures and apparent aromaticity, f'_a, are indicated. (B) Vitrinite reflectance (R_0) vs. apparent aromaticity, f'_a; ○, *in situ* samples; □, laboratory-heated samples. (C) Plot of apparent aromaticity against distance from sill contact; ○, samples above sill; □, samples below sill. [From Dennis *et al.* (1982).]

where the heating was the most intense, as shown by the increase in the lower field aromatic NMR peak and a decrease in the area of the higher field aliphatic peak. The ratio of the integrated intensities of the aromatic/total NMR signals is called the "apparent aromaticity," f'_a, and shows a positive correlation with increasing maturation as measured by increasing vitrinite reflectance (Fig. 29B) and with decreasing distance from the sill, or decreasing kerogen H/C ratios (Fig. 29C). Similar changes in f'_a were produced by laboratory heating of isolated immature kerogen from these samples (Fig. 29A, B). Note that f'_a is much more sensitive to increased heating than R_0 at low maturations (less than R_0 = 0.5%).

A complex NMR signal was noted for the 330°C and 360°C laboratory heating experiments (Fig. 29A), where very broad resonances were produced along with a low maturation range (R_0 = 0.62–0.73) within the range where catagenetic oil production would be expected to begin. The authors postulated that both the signal broadening and increased intensity in the 100-ppm range may be due to paramagnetic centers in the samples, possibly induced by kerogen free-radical production during the heating. The latter suggestion is in accord with results of Baker *et al.* (1977), who found that the ESR signal (see below) of samples taken from the same strata and subjected to artificial laboratory heating showed significant changes in just 2 h at 320°C.

Figure 30 shows how the percentages of aromatic and aliphatic carbon atoms for kerogen types I, II, and III for a series of isolated immature kerogens as determined by solid-state ^{13}C NMR (Barwise *et al.*, 1984). The kerogen types show some overlap, although there is a general progression from more aliphatic carbons in type I to a higher proportion of aromatic carbons in type III with type II being intermediate. There is no clear division between the three types, which the authors attributed to the continuous spectrum of mixed inputs into the potential source rocks. Similar results were reported by Vucelic *et al.* (1979), who also found no sharp breaks between kerogen types.

Carbon-13 NMR also provides molecular information about kerogen structure, as shown by examples in Fig. 31A for a type II kerogen (Solli *et al.*, 1985). Sample K7 shows a weak signal from 50 to 100 ppm, just below the aliphatic carbon signal which has tentatively been assigned to aliphatic carbons bound to oxygen and nitrogen. This signal is not observed in the slightly more mature sample, K5, consistent with early loss of functional groups during the early stages of kerogen maturation. A second example showing the results of laboratory heating of a

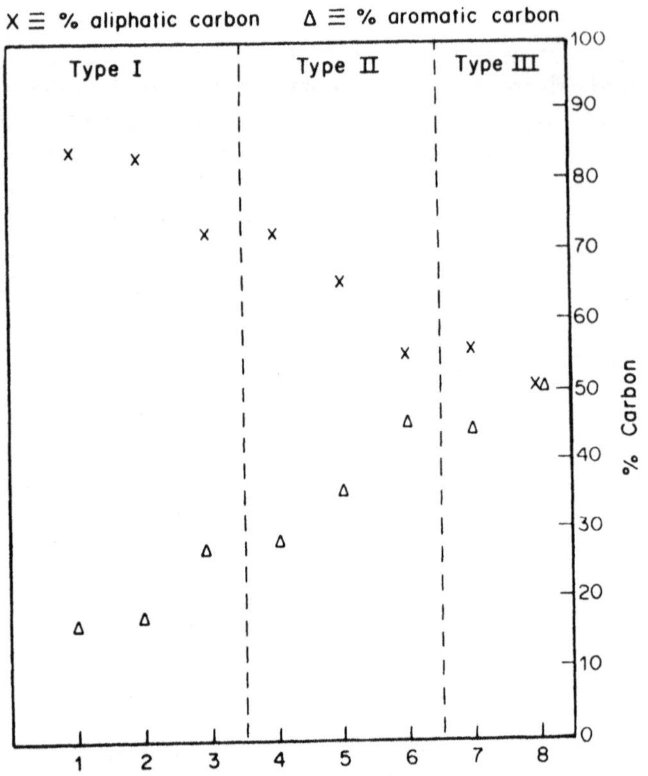

Figure 30. Percent of aliphatic vs. aromatic carbons as determined by solid-state ^{13}C NMR spectroscopy for kerogen types I, II, and III. [From Barwise *et al.* (1984).]

type III kerogen is shown in Fig. 31B, where an even stronger signal is observed from 50 to 100 ppm, consistent with more carbons bonded to oxygen and nitrogen for an immature type III than for a type II kerogen. After heating at 350°C for 18 h, the aromatic protons from 100 to 230 ppm are the predominant peaks. These are of approximately the same intensity as those observed for the unheated sample, suggesting that kerogen maturation and oil generation primarily affect the aliphatic and hetero linkages in the kerogen, leaving the aromatic part of the structure relatively unaffected.

Solli *et al.* (1985) also pointed out that the aliphatic C signal for sample K2 (Fig. 31A) is slightly narrower than that for sample K7 and suggested that the difference is due to the presence of more methylene groups in aliphatic chains in the former and more cycloalkane groups, which show a broader aliphatic signal, in the latter case. These features are consistent with the pyrolysis–GC data, where sample K2 shows a higher proportion of the homologous *n*-alkane/1-alkene chains in the pyrolysate, while K7 gives a larger proportion of a broad unresolved hump characteristic of a naphthenic envelope.

Horsfield (1989) has recently presented data showing the correlation between published NMR f'_a and H/C values for a variety of kerogens and coals and kerogen aromaticity as determined by pyrolysis-GC (see Larter and Horsfield, this volume, Chapter 13). There is an excellent overlap of the H/C versus aromaticity relationships between the two methods, even though aromaticity is expressed somewhat differently in each case [i.e., as the ratio of pyrolyzable aromatic to aliphatic moieties in Py-GC and as the proportion of aromatic carbon in NMR, where f'_a = aromatic C/(aromatic + aliphatic C)].

Solid-state ^{13}C NMR spectra were used to follow the loss of carbohydrate carbon (72 ppm) accompanied by the enhancement of aliphatic hydrocarbon carbon (0–50 ppm) from 0 to 290 cm in a surface sediment recovered from an anoxic coastal marine lake in Bermuda (Orem *et al.*, 1986). A gradual decrease in the atomic O/C ratio was also observed with depth. Aromatic peaks (110–160 ppm) were very small throughout the core. Pore-water analyses show that this early and selective loss of carbohydrates occurs below the zone of microbial sulfate reduction, but within the zone of methanogenesis.

An excellent and very complete description of quantitative information potentially available on kerogen molecular structure, related to both source (i.e., depositional environment) and degree of maturation, has recently appeared (Solum *et al.*, 1989) in which analyses of solid-state ^{13}C NMR spectra of Argonne Premium Coals are discussed. These coals have been well characterized in a number of laboratories. This paper also summarizes an experimental approach to overcoming the two major potential problems with determination of quantitative values of NMR aromaticity values:

1. spin–lattice relaxation effects in solids which alter signals so that they do not all achieve full polarization at the same contact time
2. presence of paramagnetic centers which may render some carbon resonances invisible

The first problem can be minimized by determining aromaticity (and other NMR properties) from variable contact angle time experiments, which were described in detail as applied to these coals. The second problem can cause erratic results, with as little as 26% of carbons in one maceral being seen while others are unaffected. Paramagnetic centers tend to concentrate on aromatic carbons but can potentially obscure all carbons within 8–11 Å. Therefore, an open question of all solid-state NMR results must always be whether or not the observed carbons are representative of the whole sample. The intimate association of

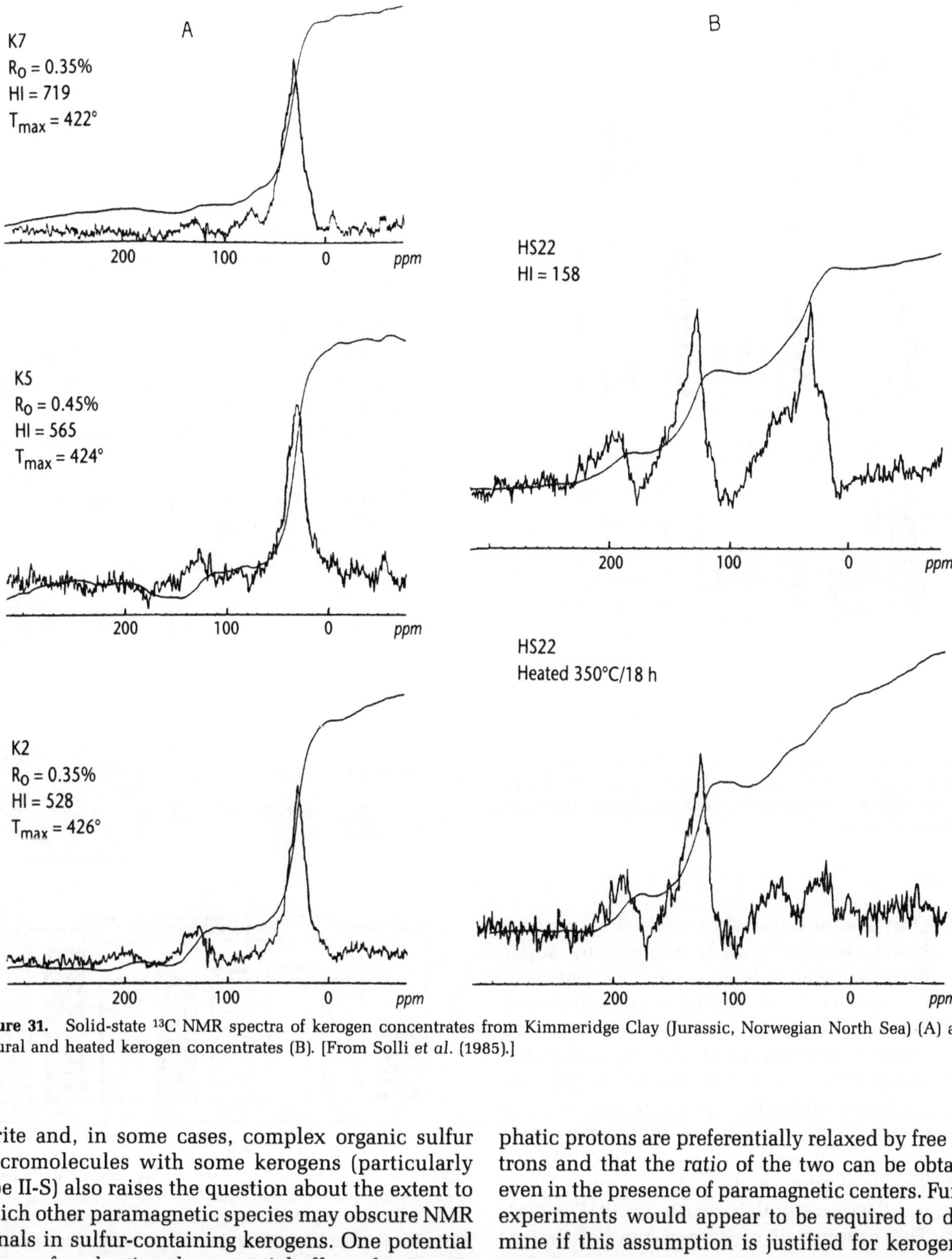

Figure 31. Solid-state ^{13}C NMR spectra of kerogen concentrates from Kimmeridge Clay (Jurassic, Norwegian North Sea) (A) and of natural and heated kerogen concentrates (B). [From Solli *et al.* (1985).]

pyrite and, in some cases, complex organic sulfur macromolecules with some kerogens (particularly type II-S) also raises the question about the extent to which other paramagnetic species may obscure NMR signals in sulfur-containing kerogens. One potential means of evaluating the potential effect of paramagnetic species on NMR signals might be to obtain ESR data on the same samples. Another approach is that of Landais *et al.* (1988), where it is assumed, based on studies of five coals, that neither aromatic nor aliphatic protons are preferentially relaxed by free electrons and that the *ratio* of the two can be obtained even in the presence of paramagnetic centers. Further experiments would appear to be required to determine if this assumption is justified for kerogens of various types.

The dependence of various solid-state ^{13}C NMR molecular features on increasing coal rank (comparable to kerogen degree of maturation), shown in Fig. 32, can be summarized as follows:

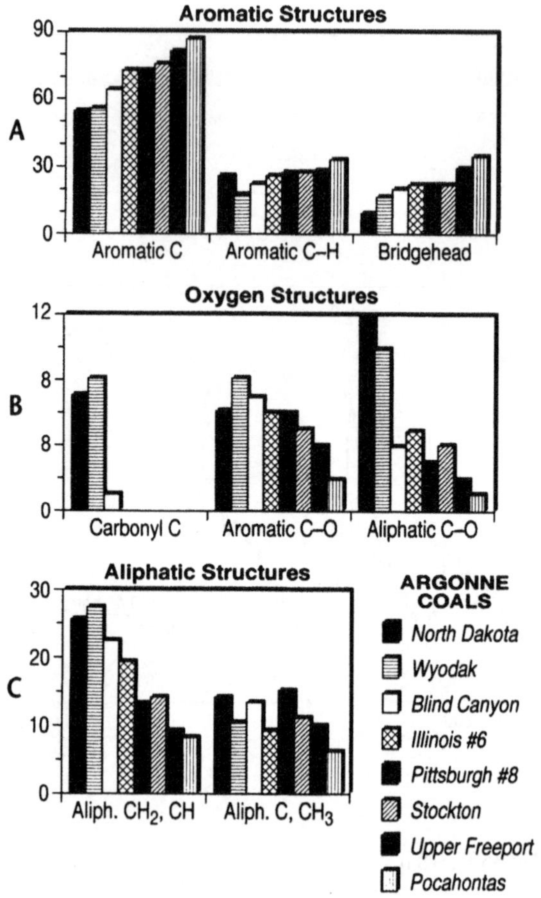

Figure 32. Structural distribution of carbons as a function of increasing rank (degree of maturation) from left to right, for Argonne Premium Coals as determined by solid-state ^{13}C NMR [integration of signals via dipolar-dephasing techniques (Solum *et al.*, 1989)]: (A) Aromatic carbons; (B) carbons associated with oxygen; (C) aliphatic carbons.

1. Only the most immature coals show any carbonyl carbon (Fig. 32B). The amount of carbonyl oxygen is highest in three immature oxidized coals. This trend is consistent with elemental analysis, which also indicates a decrease in oxygen content with increasing rank.
2. The proportions of both protonated and nonprotonated aromatic carbons increase with increasing rank (maturity), with the effect being most pronounced for the latter (Fig. 32A).
3. The proportion of aliphatic CH_2 and CH carbon decreases with increasing rank (Fig. 32C).
4. The proportion of carbon-bonded methyl groups depends on the specific coal and is not strongly rank dependent (Fig. 32C).
5. The proportion of aromatic bridgehead carbons (i.e., unprotonated) increases with increasing rank (Fig. 32A). This parameter was used for this set of coals in calculating the average size of aromatic clusters within the coals, as shown in Fig. 33. The aromatic cluster sizes reported here are similar to those reported for other coals of similar carbon content (Van Krevelen, 1961, p. 116; 1984)

Three low-rank but more oxidized coals which were examined in the same study might be comparable to kerogens deposited in more oxic environments. These coals were fairly similar to the other low-rank coals except for the following:

1. The fraction of carbon attached to carbonyl carbon tended to be somewhat higher.
2. The fraction of nonprotonated and aromatic carbons and aromatic bridgehead carbons were somewhat higher at comparable rank.

Surprisingly, the proportions of aliphatic carbon and phenolic ether carbon were about equal for the "oxidized" and "normal" coals.

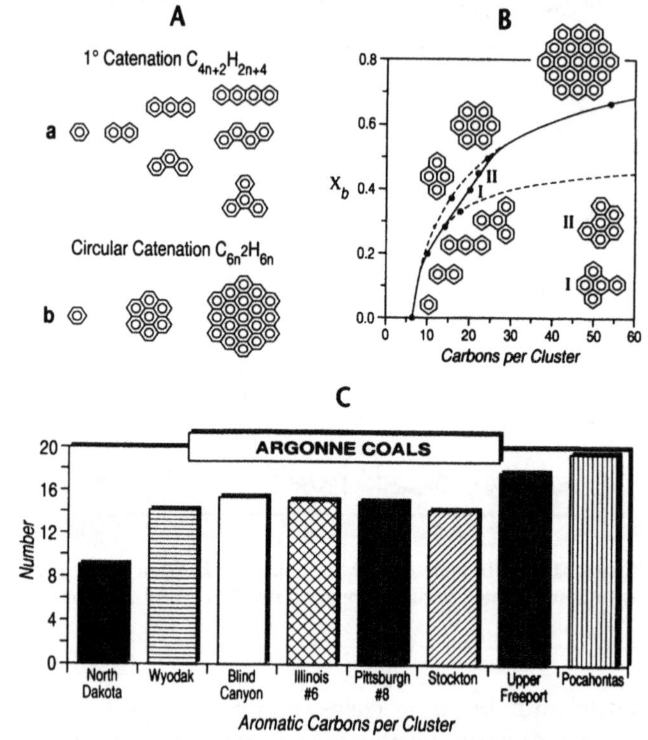

Figure 33. (A) Two limiting cases of polyaromatic hydrocarbon condensation: primary catenation and circular catenation. (B) Plot of mole fraction of bridgehead carbons, X_b, vs. number of carbon atoms per atomic cluster. Also shown are potential representative aromatic clusters in the kerogen structures corresponding to these X values. The upper dashed curve is for the circular catenation model, and the lower dashed curve for the primary catenation model. (C) Atomic cluster size determined for Argonne coals using a model combining both types of catenation. [From Solum *et al.* (1989).]

These properties are consistent with the solid-state ^{13}C NMR data currently available for kerogens, even though the accuracy of the integration for the latter is more open to question. For example, Solli *et al.* (1985) noted that the apparent aromaticity decrease during maturation was reflected more strongly by a decrease of aliphatic carbons rather than by creation of new aromatic centers during the early stages of maturation. Elemental, infrared, and gas evolution and model experimental data are all consistent with loss of carbonyl and some carboxyl groups very early in the kerogen diagenesis process, as also observed by Durand (1980). The oil generation process which occurs subsequently causes preferential loss of aliphatic groups (preferentially methylene), followed during gas generation by loss of methyls bound to aromatic rings.

5.3. Electron Spin Resonance (ESR) Spectroscopy

The technique of electron spin resonance [ESR, also called electron paramagnetic resonance (EPR)] is a method of measuring the concentration and molecular environment of unpaired electrons in much the same way that NMR is used to examine the molecular surroundings of nuclei. In fact, the techniques of ESR and NMR are closely analogous. In the ESR experiment, the energy difference between the two spin states of an unpaired electron at a fixed frequency is detected by measuring the magnetic field strength that causes the odd unpaired electron to change from the low- to the high-energy spin state. The frequency and splitting pattern are diagnostic of the molecular environment of the free radical being examined. The free radicals detected by ESR can be short-lived intermediates in many organic reactions, but often become stable over geologic times within the complex solid matrices of both coals and kerogens.

The three diagnostic ESR parameters commonly reported for kerogens are: (1) the position of the ESR signal, diagnostic of the electronic environment of the free radical, which is reported as a "g value" at the position of maximum absorption; (2) the spin concentration (also referred to sometimes as spin number or spin density), or X_p per mass unit, which is the measured intensity of the ESR signal and is proportional to the number of paramagnetic centers or free radicals in the sample; and (3) the peak width, Ho or Hms, or G, which is influenced by a number of factors including kerogen homogeneity, relaxation times, interactions between paramagnetic centers, and hyperfine coupling with neighboring nuclei.

Electron spin resonance has ben tested as a kerogen maturation and typing method by a number of laboratories. The early work was reviewed by Marchand and Conard (1980) and summarized by Horsfield (1984, p. 258). Free-radical concentrations, even in complex kerogen matrices, are strongly influenced by changes in kerogen molecular structure. Because most kerogen isolates are a complex mixture of macerals and molecular structures, a great deal of scatter typically occurs in all of the kerogen ESR parameters. Therefore, ESR has not found wide acceptance as a kerogen maturation or typing technique, probably largely because of the difficulty of isolating homogeneous maceral fractions needed for reproducible measurements (Morishima and Matsubayashi, 1978; Aizenshtat *et al.*, 1986b). For example, Morishima and Matsubayashi (1978) found that even though a fairly wide distribution of radical concentrations was typical for isolated kerogens and coals (Fig. 34A), there was a good correlation of radical concentration with geothermal history in samples from more homogeneous maceral concentrates (Fig. 34B). Aizenshtat *et al.* (1986b) concluded that ESR could be used as a kerogen typing and maturation technique, but that other data on the origin and chemical composition was required to interpret the g values, line widths, and spin concentrations. Zumberge *et al.* (1978) reported an unusually high concentration of radicals in Precambrian Vaal Reef kerogens which had been exposed to radiation from uranium. The observed g value of 2.0067 is close to the free-electron spin value of 2.0023, consistent with delocalized unpaired electron spins in aromatic organic systems and inconsistent with the unpaired spin being present in the associated minerals. Vacuum pyrolysis–GC of the Vaal Reef kerogens gave predominantly alkyl aromatics and very little aliphatic pyrolysis product yield.

Purification of kerogens for ESR studies is probably made even more difficult by their intimate association with pyrite, iron sulfides, and probably other transition-metal and organic sulfides, which often possess paramagnetic atoms which could easily obscure the kerogen ESR signal. Even in cases where cleaning might effectively remove these interfering substances, care would have to be taken to avoid alteration of the kerogen structure and associated ESR signal.

During maturation or laboratory pyrolysis under nitrogen (to prevent oxygen from reacting with the kerogen, which can generate spurious free radicals

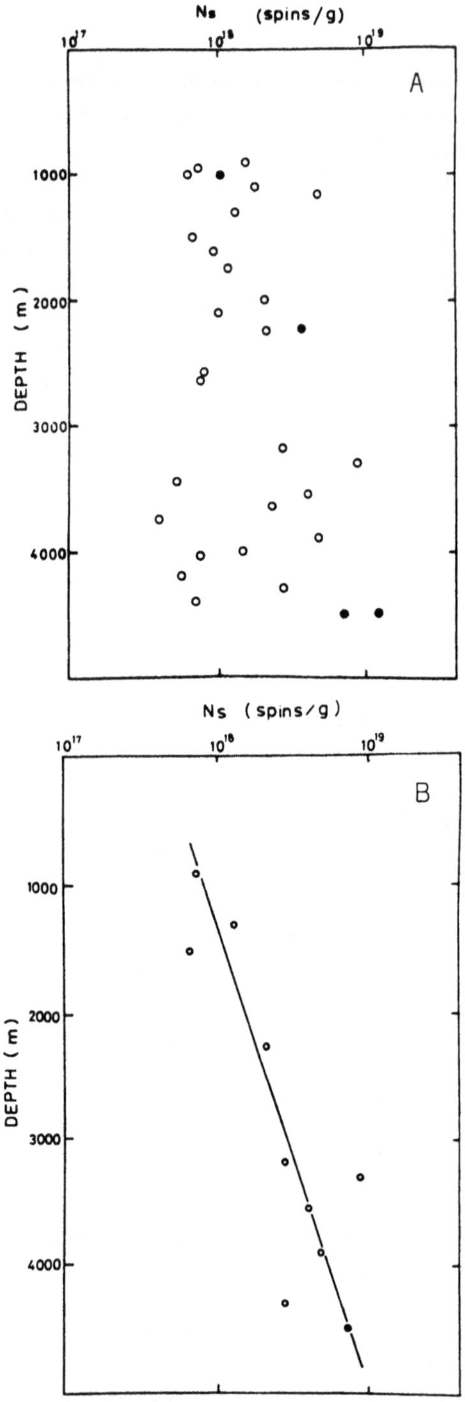

Figure 34. ESR spin number (N_s) versus depth of mixed macerals from coal particles (A) and of vitrinite-rich homogeneous coal particles (B). [From Morishima and Matsubayashi (1978).]

and cause artifacts in the ESR signal), the free-radical concentration typically increases with maturation up to a maximum value and then decreases [Fig. 35A, from Marchand and Conard (1980); Fig. 35B, from Horsfield (1984); also see Peters *et al.* (1980 and 1981a), Simoneit *et al.* (1981), and Aizenshtat *et al.* (1986b)]. The maximum in the plot of temperature versus spin concentration, which can vary considerably because of production and destruction of free-radical sites, is highly dependent on the specific molecular features of the kerogen. Thus, it is generally difficult to determine the meaning of the maxima in these curves because the positions are strongly influenced by extraneous factors, such as degree of biodegradation, etc., which have much less influence on other maturation parameters, such as vitrinite reflectance (Marchand and Conard, 1980).

The g-value is related to the molecular environment of the radical and, therefore, changes position with kerogen type and degree of maturation. Higher g-values are characteristic of immature kerogens, with a rapid decrease occurring during the initial stages of maturation followed by a second slow increase during the very late stages of maturation. The g-value increases as larger and larger condensed aromatic structures form with increasing maturation, up to a maximum value corresponding to that of pure graphite, $g = 2.018$, at room temperature (Marchand and Conard, 1980). The reasons for the initial rapid *decrease* in g values during the early stages of maturation are not well understood, although they may be related to initial radical stabilization by kerogen functional groups and double bonds which are lost as CO_2 and water and other gases during early stages of thermal maturation prior to oil and gas generation.

The response of both spin concentration and g values to increasing maturation for a few examples of kerogen types I, II, and III was discussed by Marchand and Conard (1980). Studies have been carried out on the response of g value, spin concentration, and ESR line shape during laboratory heating experiments of lignite, sporopollenin, and kerogen types I, II, and III (Villey *et al.*, 1981). However, some of these patterns may not be very general because of potential problems associated with analysis of maceral mixtures, described above.

Suzuki and Taguchi (1983) found a linear increase in the log of the radical concentration with increasing burial depth by using kerogen isolated from only the clay-size fraction (Fig. 36). They postulated that the large amount of scatter found in this parameter by previous workers may be caused by analyzing a mixture of more "coaly" and highly aromatic kerogen from the coarser-sediment fraction along with the more amorphous kerogen from the fine-grained sediment. Therefore, the simple expedient of fractionating sediment according to grain size prior to kerogen isolation may provide a means of

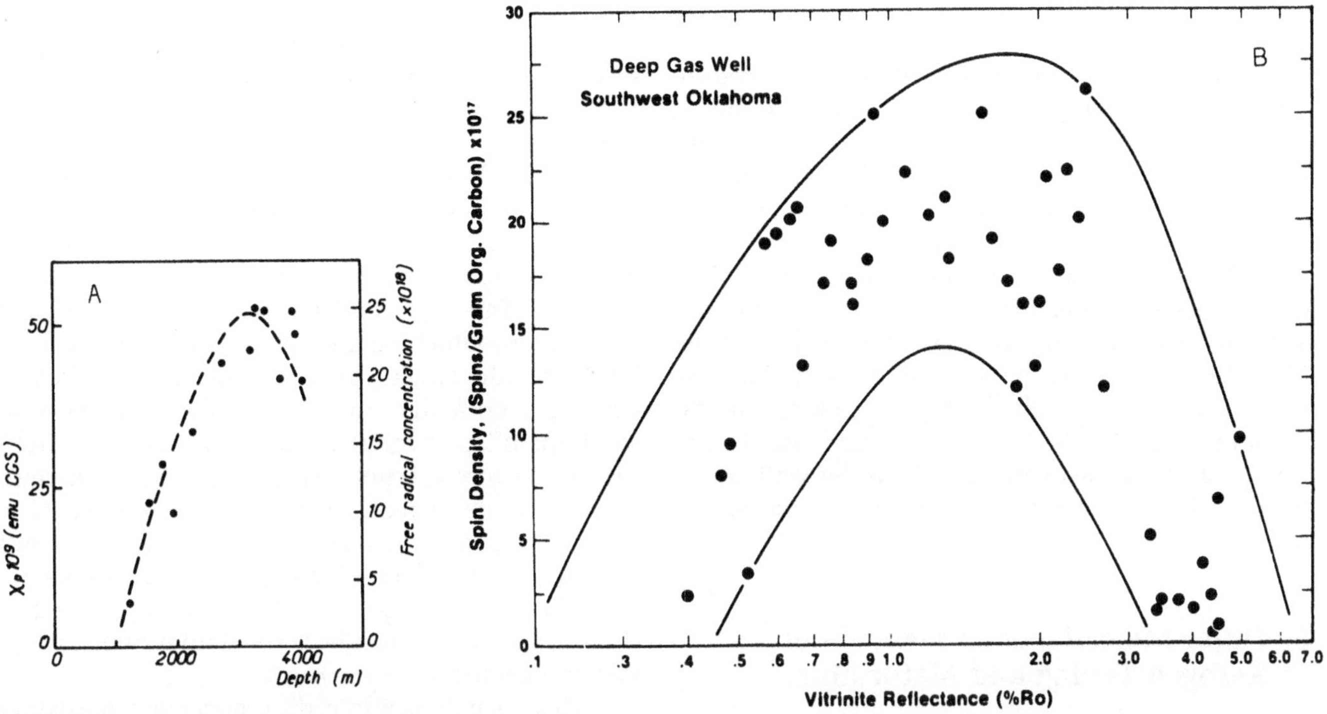

Figure 35. (A) ESR spin concentration of unpurified kerogen (from the Douala Basin) at room temperature as a function of depth of burial. [From Marchand and Conard (1980).] (B) Similar data for a deep gas well in southwest Oklahoma (Horsfield, 1984).

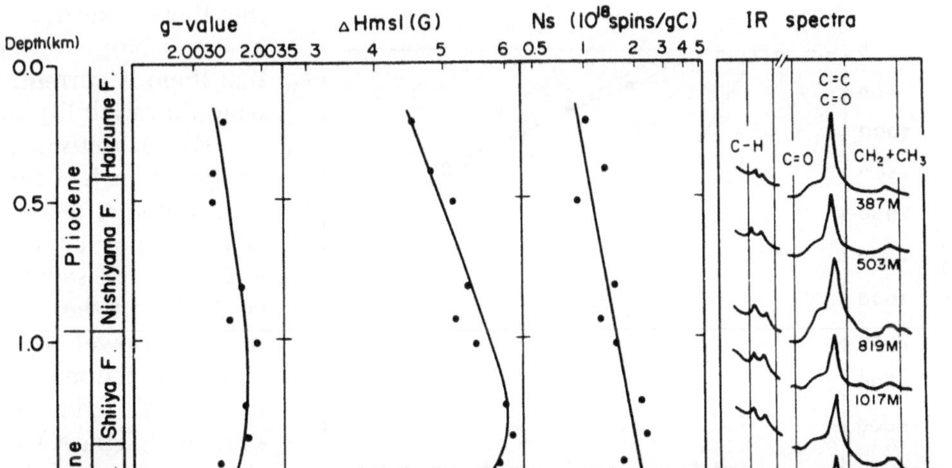

Figure 36. Vertical variation in ESR properties and IR spectra of kerogens in clay-size fraction from a well penetrating the Pliocene in Japan. [From Suzuki and Taguchi (1983).]

concentrating specific maceral fractions in favorable cases.

Bakr *et al.* (1988) reported a different procedure for obtaining a more homogeneous kerogen fraction, in which the isolated kerogen was extracted with pyridine or dioxane (which, in this case, solubilized a large fraction of the macromolecular organic matrix). The ESR experiment was then performed on the residue. The resulting log of the magnetic susceptibility, or spin concentration, for type II/III kerogens showed a linear increase with increasing burial depth (Fig. 37). The less steeply inclined regression line was found for the well with the higher geothermal gradient (estimated to be 70°C km^{-1}), while the steeper line (open triangles) corresponds to the well calculated to have a maximum geothermal gradient of about 24°C km^{-1}.

6. Other Miscellaneous Methods of Kerogen Typing and Maturation

6.1. Isotopic Techniques

Different isotopes of the same element have the same electronic configuration and similar chemical properties. Their rates of reaction, however, are not identical due to their different masses. For example, the activation energy for making or breaking bonds with the heavier isotope of an element is slightly increased so that the rate of the resulting reaction decreases by a small amount. Similarly, the higher reaction rate observed for molecules containing the lighter isotope produces an isotopic discrimination in which the heavier isotope is relatively depleted in the product in comparison to the source material.

As a result, different biological and geological sources for specific molecules can show small differences in isotopic abundances for a particular chemical structure. This is particularly true for some biological reactions in which smaller molecular structures are built up into larger structures. Because geological processes normally involve complex sequences of reactions often involving precursor molecules from diverse sources, it is not surprising to find that certain specific kinds of reaction sequences show specific isotopic fractionations. The accumulated source- and reaction-process-derived isotopic differences are reflected in small changes in relative isotope concentrations in specific molecules derived from some biological and geological processes.

These small isotopic differences are measured mass spectrometrically. Because these differences are small, values are normally reported relative to a standard in order to minimize variations between laboratories, as parts per thousand (parts per mil, or ‰) from a widely accepted standard, the Pee Dee belemnite (PDB) standard, via the following formula:

$$\delta^{13}C = \left(\frac{{}^{13}C/{}^{12}C_{\text{sample}} - {}^{13}C/{}^{12}C_{\text{standard}}}{{}^{13}C/{}^{12}C_{\text{standard}}} \right) \times 10^3$$

In reviews on the use of stable isotopes in petroleum geochemistry, Schoell (1984a, b) pointed out that there is surprisingly little systematic data available on isotopic variations in coals and kerogens and that there is currently no overall concept to explain isotopic variability in kerogens.

For kerogens in their immature stage, the isotopic composition is a result of the various contributing source constituents (Deines, 1980). Hatcher *et al.* (1983) and Spiker and Hatcher (1984) found that the $\delta^{13}C$ of recent sediments varied as the carbohydrate content, with the carbohydrate carbon being enriched in ^{13}C. However, the $\delta^{13}C$ of pore-water carbon of the 0–55-cm section of an anoxic surface core was found to be 2.7‰ lighter than the total sediment carbon, even though the NMR spectrum showed that the pore-water carbon consisted primarily of carbohydrates (Orem *et al.*, 1986). The authors postulated that the residual heavy carbohydrate carbon in the sediment must be balanced by an isotopically lighter fraction, most likely lipids, to account for the high sediment H/C ratios, determined by elemental analysis, and the

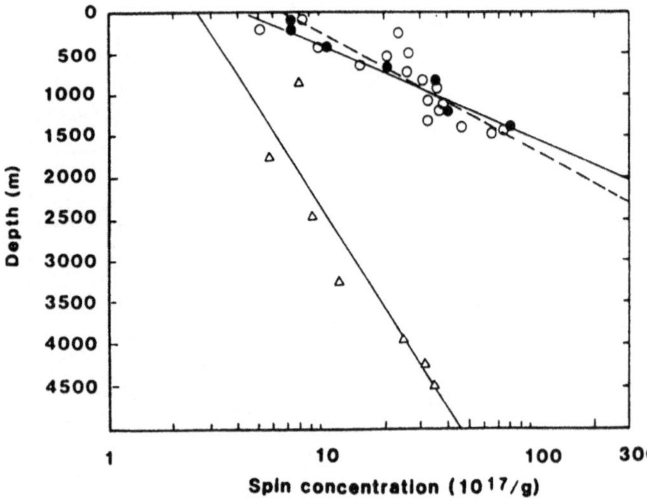

Figure 37. ESR data relating to maturation, after extraction of isolated kerogen with organic solvent. ○ and ---, Extraction with dioxane; ● and ——, extraction with pyridine; Δ, sample from a second well extracted with pyridine. [From Bakr *et al.* (1988).]

high aliphatic and low aromatic carbon content shown by NMR.

Systematic changes in C, H, and N have been noted by Peters *et al.* (1980, 1981b) during laboratory thermal maturation of protokerogens (Fig. 38). These authors found that the maximum changes in $\delta^{13}C$ caused by maturation (up to 4‰ for any single protokerogen) were considerably less than the approximately 10‰ difference observed between protokerogens from different depositional environments (about −22‰ for the sapropelic compared to −32‰ for the more terrigenous and humic protokerogens). Thus, the isotopic differences for these low-maturity kerogens reflect depositional environment (source) more strongly than maturation. Similar conclusions have been reached with regard to $\delta^{13}C$ isotopic compositions of oils, coals, and kerogens in ancient sediments, as reviewed by Fuex (1977), Deines (1980), and, more recently, Schoell (1984a, b). The change in both carbon and sulfur isotopes as a function of maturation, as measured by vitrinite reflectance and H/C ratios, has been discussed for intrusion of a diabase sill into an organic-rich Cretaceous Black Shale in the eastern Atlantic (Simoneit *et al.*, 1981).

A homogenization of carbon isotopic values occurs with maturation of coals (Rigby and Smith, 1981) and for isotopic differences between kerogens and their associated bitumens (Schoell, 1984b) within a particular formation. Deines (1980) concluded that carbon isotopic composition changes during thermal alteration are small (1–2‰), leading to $\delta^{13}C$ enrichment in the residual petroleum. Most of the petroleum samples analyzed to date fall in the range of $\delta^{13}C$ = −21 to −32‰ and generally show $\delta^{13}C$ depletion in comparison to the organic matter incorporated into recent sediments and the kerogen of older sediments. These generally light isotopic values for petroleum (oil) are consistent with oil generation occurring primarily from kerogen enriched in the isotopically lighter lipid fraction. Oil generation predominantly

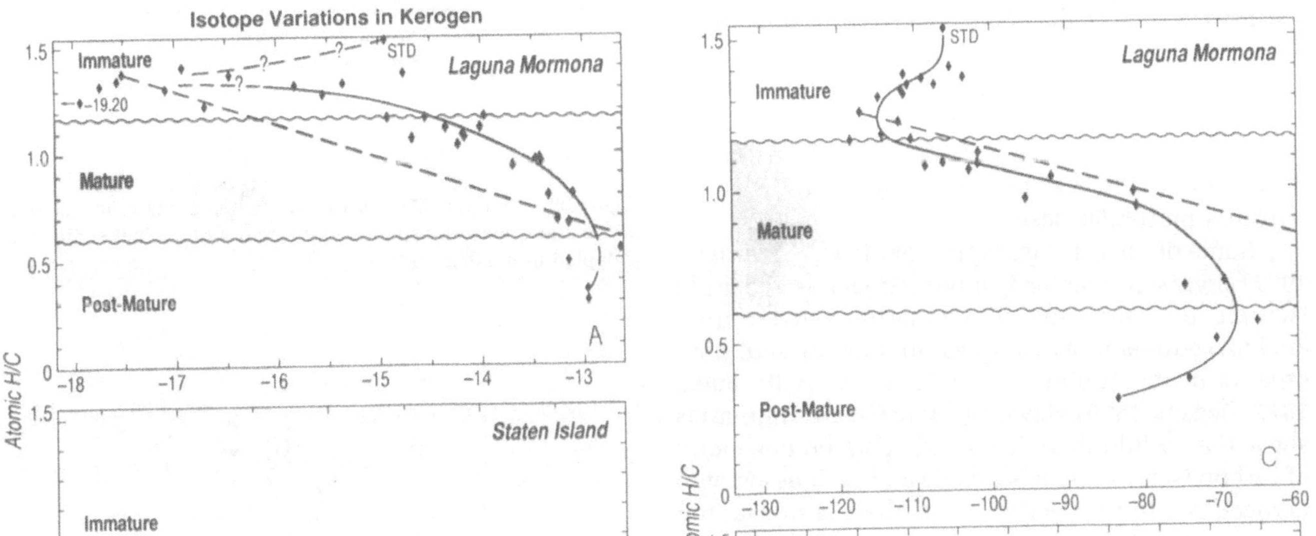

Figure 38. Effect of artificial heating on atomic H/C ratio and isotopic composition [$\delta^{13}C$(ppt) and δD(ppt)] of protokerogen in recent sediments for Laguna Mormona algal mat (A and C) and Staten Island peat (C and D). [From Peters *et al.* (1981b).]

from lipid-rich kerogen is also consistent with the elemental, pyrolysis, spectroscopic, and chemical evidence discussed previously. For example, Dean *et al.* (1986) have pointed out that the consistently even lighter isotopic values observed for Cretaceous oils (about 3‰ less than for other oils) may be caused by isotopically lighter algal (possibly more lacustrine?) sources during that geologic period rather than to a more terrigenous source, as had been postulated previously. However, recent work by Hayes *et al.* (1987) suggests that such light isotopic values may also be typical of biogenic turnover in highly organic rich and anoxic depositional environments. A depletion of kerogen $\delta^{13}C$ by about 3‰ has been noted by several researchers as being typical of kerogen samples with high amounts of hydrogen-rich kerogen (e.g., Dean *et al.*, 1986; Schoell, 1984a, b; Hayes *et al.*, 1987). It has been suggested that in the case of the Messel Shale, this signature may be indicative of organic-rich and highly oxygen depleted environments where incorporation of microbial methane into the biomass has occurred (Hayes *et al.*, 1987). Such an explanation may also be applicable to other highly organic rich sediments, such as the organic-rich Cretaceous Black Shales in the South Atlantic, where $\delta^{13}C$ depletion is also observed (Dean *et al.*, 1986). Schoell (1984b) proposed that, from the point of view of petroleum geology, such isotopically depleted kerogens might be useful in tracing the most prolific oil sources within a particular basin.

Some of the early isotopic work (e.g., Silverman, 1967) suggested that terrigenous kerogen is generally isotopically lighter than more marine kerogen from ancient sediments, as is the case for modern sediment organic matter (Deines, 1980; Galimov, 1980; Fuex, 1977; Degans, 1969). However, more recent summaries show that, while there is considerable homogeneity of carbon isotopic signals for oils and their associated kerogens within a specific region, there is substantial overlap of $\delta^{13}C$ ranges for marine and terrigenous kerogens when distributions over many areas are combined (Deines, 1980; Schoell, 1984a; Fuex, 1977) as summarized in Fig. 39.

The relation of carbon isotopes to kerogen type, as defined on the basis of hydrogen and oxygen indices is not clear. The additional isotopic changes that occur upon further maturation of specific kerogen types also do not appear to be consistent or even very linear for the limited number of samples examined to date, as shown by examples in Figs. 40 and 41. Type I kerogens from the Uinta Basin show wide variations with increasing maturation (up to 10‰), and types II and III show smaller changes (about 4‰).

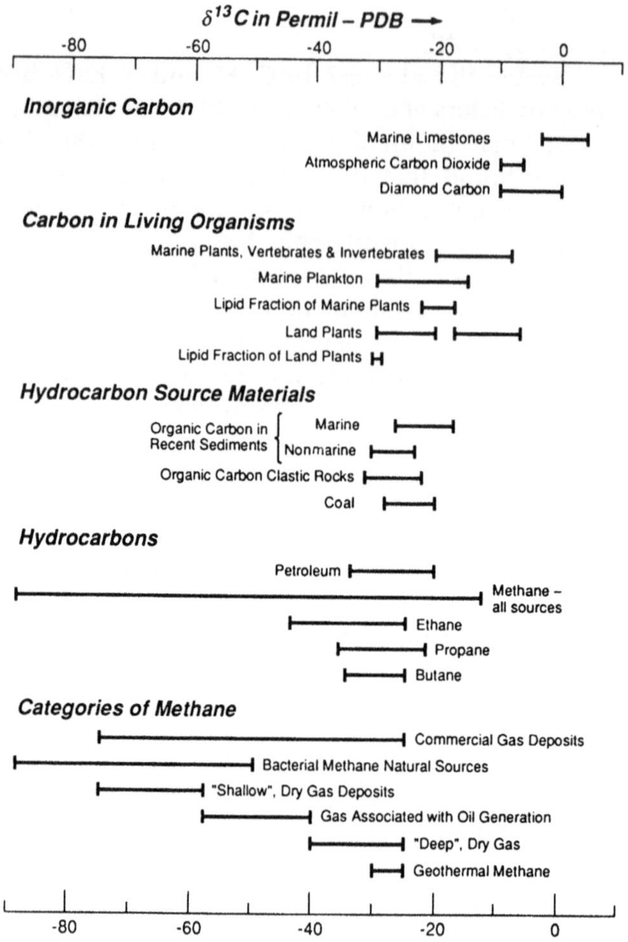

Figure 39. Values of $\delta^{13}C$ for various carbon-containing materials in the atmosphere, biosphere, and recent and ancient sediments. [Adapted from Fuex (1977).]

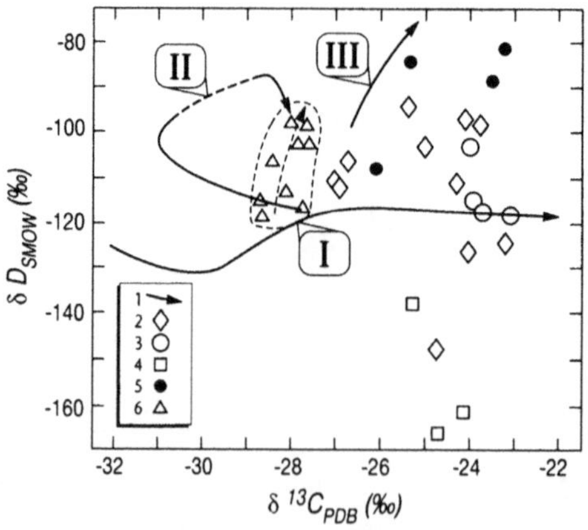

Figure 40. Isotopic changes upon maturation of different kerogen types. Arrows point to increasing maturity of samples. Samples are coals from Europe and North America (◇); coals from the Ruhr area, Germany (○); Permian coals from Australia (□); various macrolithotypes in Australian coals (●); and Mahakam Delta type III kerogens (△). [From Schoell (1984b).]

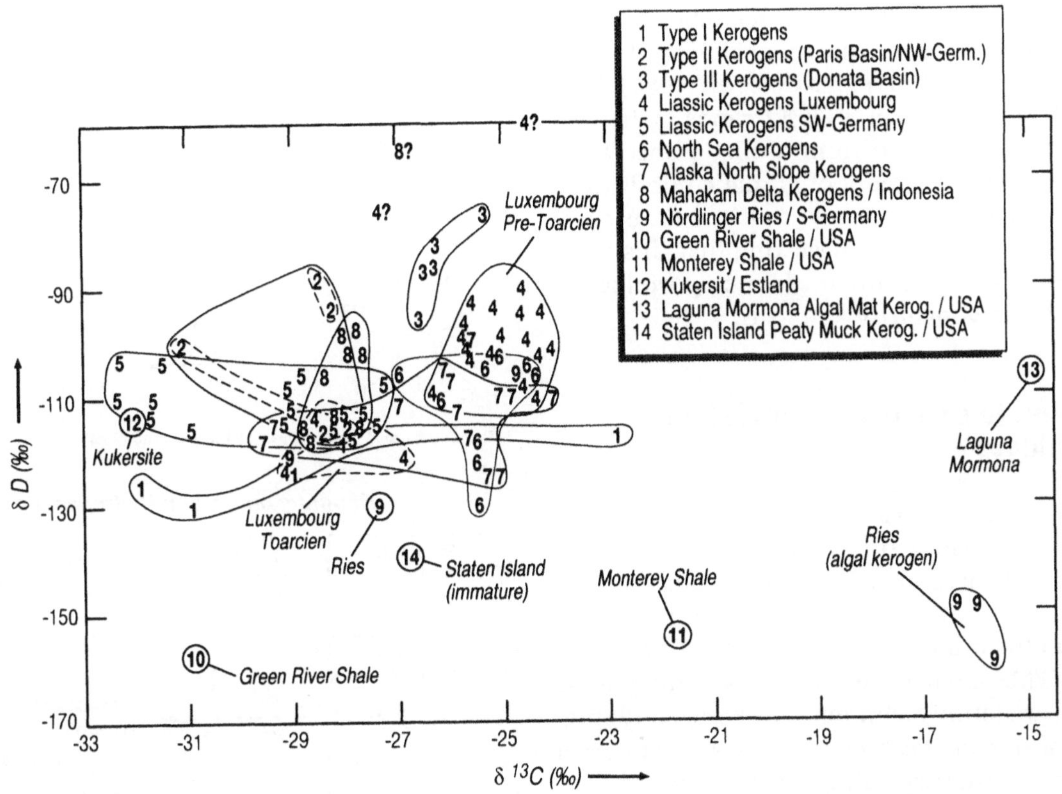

Figure 41. Summary of carbon and hydrogen isotopic data for kerogens of various origins. [From Schoell (1984a).]

If ancient kerogens are comprised of mixtures which have not been well characterized petrographically or chemically, the problem of interpreting isotopic values becomes very difficult, considering the wide source-related variability (Fig. 39).

Galimov (1980) summarized carbon isotopic data from kerogens and their associated bitumens for a number of Paleozoic deposits in the Volga–Ural area of the Russian Platform which show a consistent depletion of $\delta^{13}C$ in the bitumen of 2‰ in comparison to the associated kerogen. All of these were considerable leaner (0.2 to 3% TOC) than the Messel Shale discussed above and also showed somewhat heavier $\delta^{13}C$ values (in the range of about -22 to -28 for the kerogens as compared to -31‰ for the Messel Shale). An excellent correlation between $\delta^{13}C$ of the kerogens and that of their associated bitumens was also observed for the Permian–Ural kerogens (Galimov, 1980).

A summary of literature isotopic data on kerogens utilizing both ^{13}C and deuterium isotopes is shown in Figs. 40 and 41 (Schoell, 1984a, b). The effect of source organic matter is clear for the Laguna Mormona and Ries algal material and the Monterey kerogen, which are rich in marine organic matter (Fig. 41). All show enrichment of $\delta^{13}C$, but variable deuterium concentrations. However, substantial overlap in isotope values occurs for the other kerogens including the type II kerogens from the Mahakam Delta and Toarcian of the Paris Basin and the type I kerogens from the Green River Shale, although values within one geographic region appear to fall in a specific area of the graph.

It can be concluded from the above discussion that carbon isotopic variations of kerogens are highly dependent on depositional environment and that the final isotopic compositions generally strongly reflect the contributions of the various contributing organic inputs. Kerogen isotopic data, therefore, would appear to have the potential to provide a good "typing" tool—i.e., a bulk parameter strongly reflecting depositional environment with smaller or minimal effects from degree of maturation and other processes (e.g., deasphalting, migration, and biodegradation).

The limited evidence in the literature to date suggests that other isotopes may also be useful as source indicators. For example, Estep and Hoering (1980, 1981) discussed fractionations of hydrogen isotopes as a function of source in biogeochemical systems, although caution must be used in interpreting results because extensive hydrogen scrambling within the hydrocarbon skeleton can occur during thermal maturation (Hoering, 1984). DeNiro *et al.* (1988) demonstrated a correlation of $^{18}O/^{16}O$ ratios with D/H ratios for textiles derived from a single species of plant grown under a variety of climatic conditions. Separate correlation lines were produced for each of three single plant species, suggesting that this technique might also be useful in deducing biological precursors for low-maturity kerogens.

6.2. Relation to Chemical Degradation Techniques

A vast literature exists on chemical reactions which have, at various times, been applied to kerogens. Many of these alter the kerogen structure so profoundly that little useful information can be gained about the initial kerogen structure. Pyrolysis–GC and –GC/MS are currently the most widely used chemical degradation techniques for kerogen typing. Examples of use of these techniques as qualitative kerogen typing (fingerprinting) techniques are described in Section 6.3. Some general comments and examples of requirements for a reaction to be useful in discerning chemical structural features related to kerogen type are given here. In a few cases, results from chemical degradation methods (other than pyrolysis) have been compared to kerogen structural features as discerned by spectral methods.

The "ideal" chemical reagents for use in kerogen typing would be capable of reacting with specific chemical bond types in high yield and would cause minimal rearrangement of the basic carbon skeleton in the reaction products. Examples of two such reactions being used at the present time which appear to be suitable for kerogens are Raney nickel desulfurization, which specifically attacks carbon–sulfur bonds, leaving the carbon skeleton intact (see examples in Fig. 42; Pettit and van Tamelen, 1962; Sinninghe Damsté *et al.*, 1990, and references cited therein), and hydrogenation via rhodium on carbon, which cleaves ether linkages as well as reducing various functional groups, such as C=C and C=O double bonds (Fig. 42; Mycke and Michaelis, 1986). Raney nickel desulfurization has been extensively used in determining structures of carbon skeletons in high-sulfur oils and kerogens (Sinninghe Damsté *et al.*, 1990, and references cited therein). The rhodium on carbon hydrogenation has been used on coals and kerogens by Michaelis and co-workers. When hydrogenation is carried out in the presence of deuterium instead of

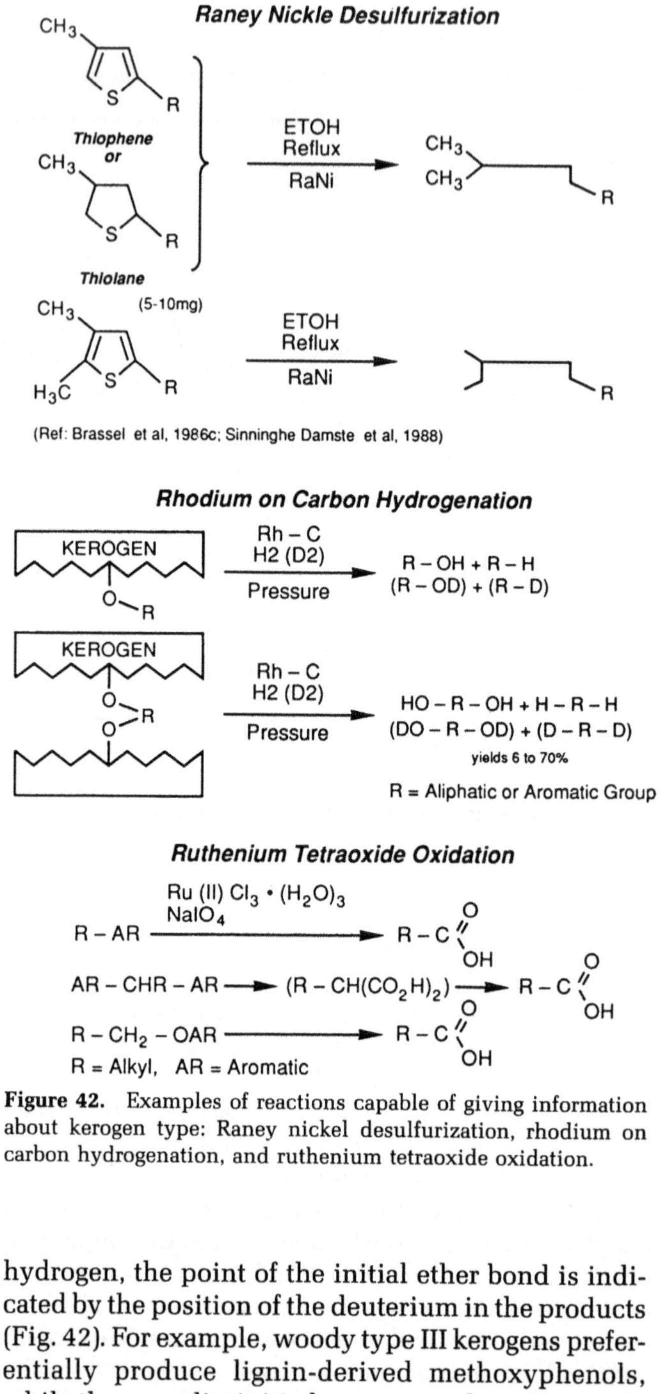

Figure 42. Examples of reactions capable of giving information about kerogen type: Raney nickel desulfurization, rhodium on carbon hydrogenation, and ruthenium tetraoxide oxidation.

hydrogen, the point of the initial ether bond is indicated by the position of the deuterium in the products (Fig. 42). For example, woody type III kerogens preferentially produce lignin-derived methoxyphenols, while the more liptinitic kerogens such as the type II Messel Shale tend to give predominantly hydrocarbons, including *n*-alkanes, steranes, and hopanes. Even though these reactions are not adaptable to routine measurements, they can provide detailed information about incorporation of biologically derived carbon skeletons into the initial kerogen structure. For example, Mycke and Michaelis (1986) concluded that the Messel type II kerogen contains about 10–15% terrestrial contribution based on phenol yields of this material in comparison to a coal which was assumed to be derived from approximately 100% terrigenous matter. These authors pointed out the potential for application of this method for gaining a semiquantitative measure of terrigenous input to pre-Tertiary marine kerogens, including those which have undergone some degree of alteration.

Many different oxidation procedures have been used on kerogens. This can give useful structural information if the specificity of the reaction is well understood and if the stability of the oxidation products to the reaction conditions can be demonstrated. For example, Barakat and Yen (1988) compared a stepwise oxidation of a Green River Shale (type I) kerogen with that of a Monterey Shale (type II-S) kerogen using sodium dichromate in glacial acetic acid. Overall yields were greater than 50% in both cases, with the major products being a series of mono- and dicarboxylic acids. The reagent is known to preferentially attack alkane CH groups, particularly those adjacent to phenyl groups. The approximate relative oxidation rates are 1:114:7000–18,000 for methyl: methylene:methine (i.e., CH_3:CH_2:CH), so that methyl groups are relatively inert. The reagent also oxidizes other vulnerable organic groups [i.e., alcohols, aldehydes, etc; for a recent review, see Cainelli and Cardillo (1984)] and has been applied successfully to polymers in coal and coal products (Hayatsu *et al.*, 1978). In the work of Baraket and Yen (1988), both the type I and type II-S kerogens produced significant amounts of C_{10} to about C_{30} mono and dicarboxylic acids, consistent with a high aliphatic content in both. The major difference was that the ratio of di- to monocarboxylic acids was lower for the Monterey kerogen compared to the Green River kerogen, suggesting a higher degree of cross-linking in the latter. Greater cross-linking in the Green River kerogen would provide a molecular explanation for the high pyrolysis T_{max} values and activation energies observed for the Green River kerogen, in comparison to any other kerogen types of comparable maturity.

A fairly clean and specific oxidative procedure utilizing ruthenium tetraoxide has also recently been described for the purified coal macerals liptinite, vitrinite, and inertinite (Choi *et al.*, 1988). This use of the reagent, which specifically oxidizes aromatic rings and leaves carboxyl groups on aliphatic chains at the point of attachment (Fig. 42), would appear to be suitable for providing similar information about kerogens. Quantitative analysis of the volatile acids produced, indicative of the lengths of the side chains bonded to aromatic rings in the initial coal structure, was carried out by an isotope dilution technique. About 50% of the methyl groups measurable by FTIR and NMR spectroscopy were oxidizable by this procedure. An estimation of aromatic to aliphatic carbons and hydrogens was made by a combination of FTIR and NMR procedures. For this particular coal, the spectral methods show that the aliphatic hydrogens are more concentrated in the liptinite than in the vitrinite or inertinite, while the inertinite contains the highest proportion of aromatic hydrogens. However, the ratios of aromatic H/C are about equal for the three maceral types, being slightly higher in the inertinite while vitrinite shows the highest ratio of aliphatic H/C, suggesting a higher proportion of methyl groups in this maceral. This conclusion is borne out by the RuO_4 oxidation, which shows yields of ethanoic (acetic) acid which are about two times higher for vitrinite than for the other two macerals, indicative of a higher proportion of methyls bonded to aromatic rings in vitrinite than in liptinite or inertinite. In this particular case, none of the macerals gave significant amounts of acids longer than C_3 even though it is known that longer carboxylic acids are stable to the reaction conditions. Thus, it would appear that the original OM also did not generally contain alkyl groups longer that C_3, suggesting that this particular coal is more gas- than oil-prone, a conclusion which is not immediately obvious from the spectral and elemental characteristics alone (Table VII).

6.3. Kerogen Typing via Pyrolysis–GC and –GC/MS Fingerprints

Pyrolysis–GC and –GC/MS techniques are very powerful for detailed typing of kerogens, although their full potential is only now beginning to be realized (Larter and Horsfield, this volume, Chapter 13). For example, it has been noted by Horsfield (1984), Horsfield *et al.* (1983), and Larter and Douglas (1980) that the four kerogen types can be readily distinguished by pyrolysis techniques up to about 1.2% vitrinite reflectance. At higher maturities, their compositions tend to converge (Allan and Douglas, 1974).

Pyrolysis coupled to detailed analysis of products via GC (Py-GC) was applied fairly early as a means of characterization of kerogen with respect to petroleum source rock potential by Giraud (1970), who described pyrolysis of whole-rock samples at 280°C for 20–30 min. It was found that continental sediments produced predominantly methane, ethane, and benzene, whereas more marine sediments gave higher proportions of paraffinic hydrocarbons. As this early pyrolysis work was extended and the

TABLE VII. Spectral and RuO_4-Catalyzed Oxidation Data for Macerals Isolated from a Coal[a]

	Coal	Liptinite	Vitrinite	Inertinite
Chemical data				
Percent carbon	82.4	84.2	82.1	85
Acid yield (μmoles of acid/g maceral)				
Ethanoic	676	578	1003	627
Propanoic	161	173	195	99
Butanoic	18	20	27	3
Spectral data				
^{13}C NMR data				
Aromaticity, f_a	0.78	0.59	0.79	0.87
IR data (hydrogens)				
Aliphatic/total	0.55	0.76	0.61	0.38
Aromatic/total	0.39	0.21	0.33	0.58
OH/total	0.06	0.03	0.06	0.04
H(aliphatic)/C (aliphatic)	1.67	1.71	1.98	1.61
H(aromatic)/ C(aromatic)	0.33	0.33	0.28	0.37

[a]Data from Choi *et al.* (1988).

hydrogen and oxygen indices began to be widely applied for petroleum source rock screening, detailed analyses of pyrolysis products from coals and kerogens by GC and GC/MS (Allan and Douglas, 1974; Claypool and Reed, 1976; Larter, 1978; Whelan *et al.*, 1980; Philp *et al.*, 1982; Larter and Douglas, 1982) and by MS (Maters *et al.*, 1977; Meuzelaar *et al.*, 1982; Van Graas *et al.*, 1981) were also being carried out in a number of laboratories as a means of further "fingerprinting" and trying to gain information about sources of kerogens and coals. Oygard *et al.* (1988) and Horsfield (1989) described techniques for combining Py-GC with elemental and optical data in kerogen typing for use in petroleum exploration. Py-GC and –GC/MS techniques have only recently begun to be applied routinely to kerogen typing in the petroleum industry, probably because of the difficulties in quantitatively interpreting the voluminous data output from these techniques. Automated statistical and computerized data handling techniques such as those developed in the pioneering work of Meuzelaar and co-workers (e.g., Meuzelaar *et al.*, 1982) have the potential to make these measurements and data reduction faster in the future. If such procedures could be used to better relate detailed structures of Py-GC and –GC/MS products to complex features in the initial kerogen structure, they would provide a detailed chemical window to kerogen sources and subsequent alterations which is not vulnerable to expulsion of the altered products out of the system—always a potential problem with organic-solvent-extractable biomarkers.

Py-GC/MS was used to characterize Precambrian kerogens in order to minimize the contamination problems which often occur for solvent extracts from these very old samples (Philp and Van de Meent, 1983). The general Py-GC/MS characteristics of kerogen types I, II, and III and their related depositional environments are discussed by Larter and Horsfield (this volume, Chapter 13). In addition, qualitative Py-GC and -GC/MS fingerprints can also be used to rapidly determine changes in kerogen and associated minerals, as shown by several examples below.

"Open-system" pyrolyses of both kerogens and whole rocks can be carried out with excellent reproducibility (±5% is typical for resolved GC peaks; Whelan *et al.*, 1986) even on sediments containing only small amounts of TOC (<0.5%; Tarafa *et al.*, 1987). In fact, observation of the reproducibility of the Py-GC pattern has been used in a number of studies to determine where and when changes in the detailed chemistry of the kerogen (and in some cases, the associated rock matrix) occur. If analysis is carried out on isolated kerogens, structural information about the initial kerogen matrix can also be obtained, as discussed later in this section. In the presence of either extensive secondary reactions or mineral matrix effects during pyrolysis, as sometimes occurs for whole rocks, it is difficult or impossible to relate the structures and quantities of pyrolysis products to structural features in the initial kerogen prior to pyrolysis. However, even in extreme cases where these problems are known to occur, it is still often possible to use Py-GC to "fingerprint" changes in kerogens and/or their associated rocks. Such a Py-GC pattern can be broadly considered to be a "kerogen typing" tool, because it shows where depositional conditions changed in a much more detailed way than is possible with any other "bulk" chemical technique at the present time. An example of this technique is given in Fig. 43, which shows the similarity of the Py-GC patterns for Cariaco Trench kerogens from core A1160-1798G, a surface sediment, and DSDP core 15-147C-3-1, a deeper and older sediment (van de Meent *et al.*, 1980). This similarity strongly suggests that conditions for the deposition of the bulk organic matter were comparable in both cases, even though deposition took place at different geologic times, separated by about 10^3 to 10^4 years, and in slightly different locations.

An example of the geological usefulness of the ability of "fingerprint" the kerogen in whole-rock samples via such Py-GC fingerprints is shown by the

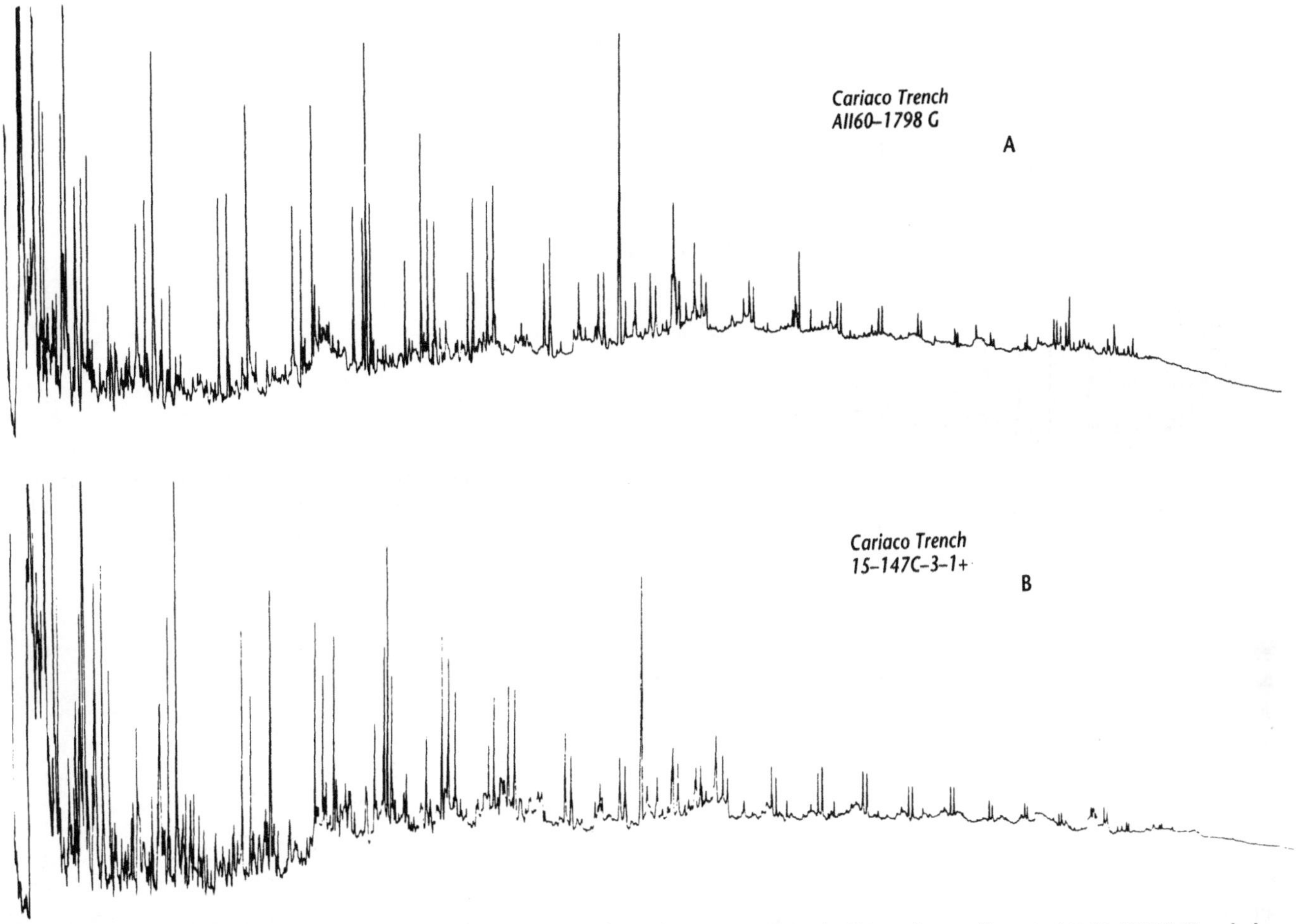

Figure 43. Similarity in Py-GC patterns for isolated kerogens from the Cariaco Trench: (A) surface sediment, A1160-1798G [0 m below seafloor (mbsf.)]; (B) DSDP sediment, 15-147C-3-1, 138 mbsf. [From van de Meent *et al.* (1980).]

detailed reproducibility of the P_2 Py-GC pattern for a series of immature gray and black Cretaceous Black Shales from the South Atlantic (Fig. 44) containing a range of TOC values (2–17%). These two samples differ in depth by 86 m, corresponding roughly to a difference of 10^6 to 10^7 years between the times of their deposition. However, the remarkable similarity of the Py-GC patterns suggest very similar depositional conditions in spite of the episodic nature of the deposition of these organic-rich layers (Jasper *et al.*, 1984; Whelan *et al.*, 1984). Note that the kerogens could be examined in the whole rock in this study because mineral matrix effects tend to be minimal in organic-rich marine sediments, such as these (as established by comparing Py-GC patterns of the isolated kerogens with those of the whole rocks; J. K. Whelan, unpublished results).

Dembicki *et al.* (1983) have pointed out the usefulness of Py-GC fingerprinting in avoiding the embarrassing difficulty of accidentally identifying a well contaminant as a new kerogen facies.

A case where whole-rock Py-GC fingerprinting was found to be useful in spite of mineral matrix effects is shown in Fig. 45, where two different kerogen/mineral assemblages could be distinguished. This information was then used to distinguish "*in situ*" generation from a migration source for the occurrence of a C_8 alkane, 1,1-dimethylcyclohexane (11DMCH), in sediments from the Gulf Coast East Cameron well (Whelan *et al.*, 1984). 11DMCH was found to be associated only with the sediment showing whole-rock Py-GC pattern A and not with that showing pattern B [Fig. 45], so that vertical migration of this C_8 alkane did not occur in this well. It was concluded that 11DMCH is being generated by sediment type A but not by B [Fig. 45]. Note that the kerogens associated with both sediments "A" and "B" are type II/III—no distinction was possible on the

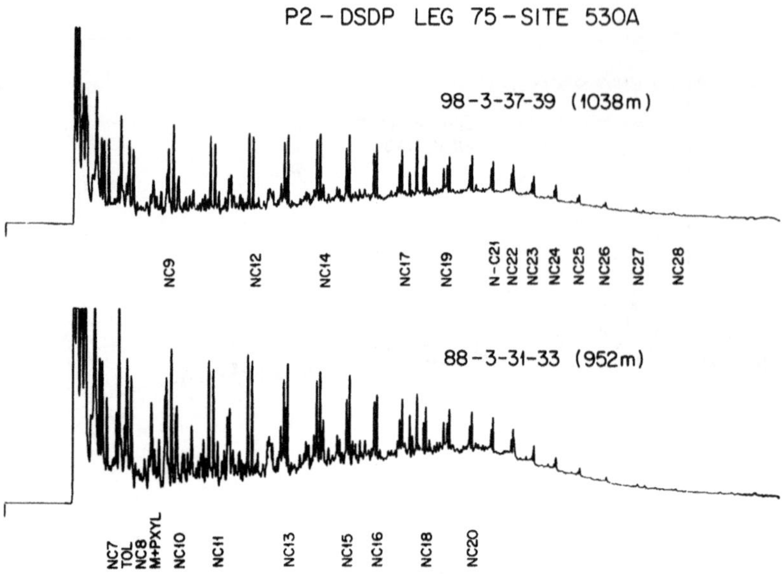

Figure 44. Similarity of Py-GC patterns for Cretaceous black and gray shales from the South Atlantic. [Adapted from Jasper *et al.* (1984).]

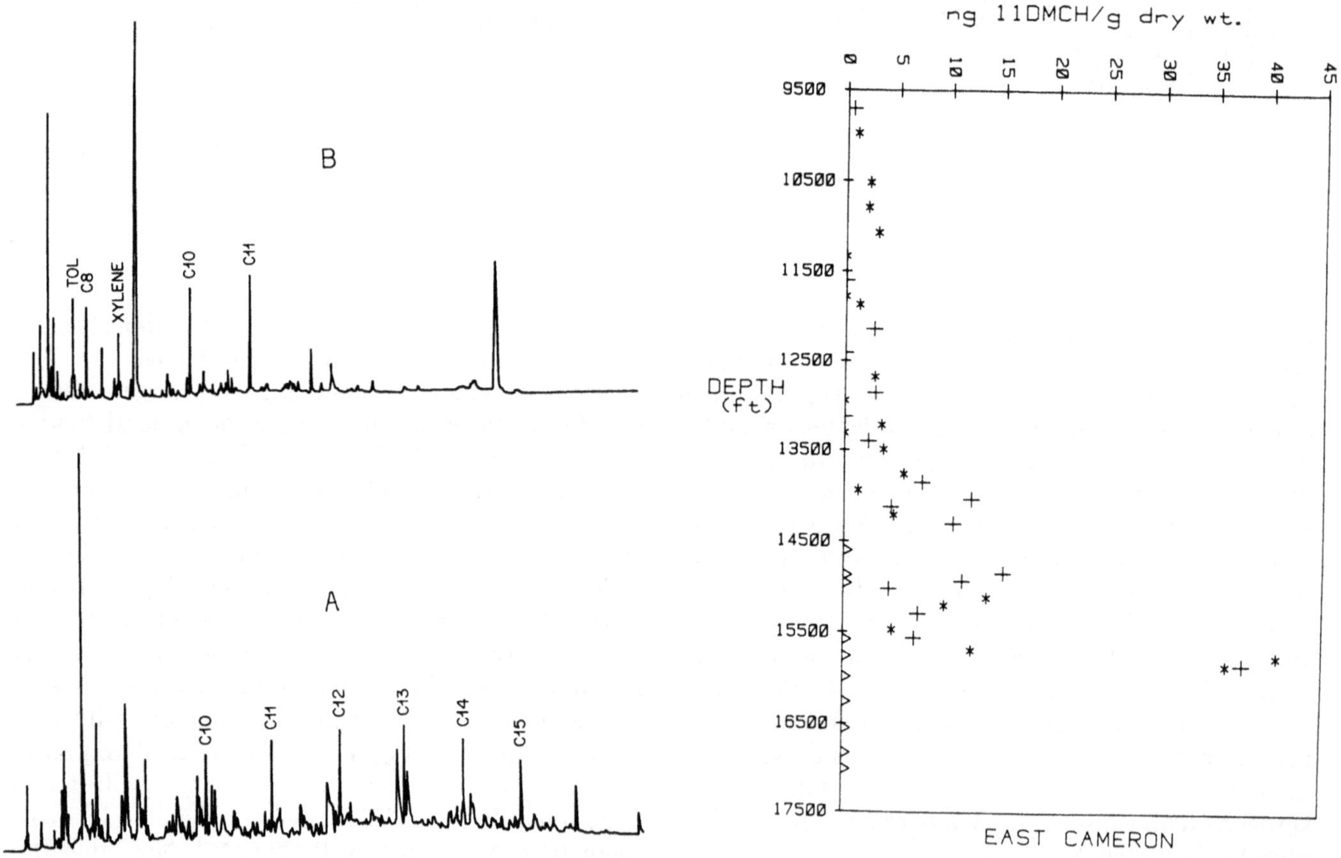

Figure 45. (i) Py-GC patterns A and B for whole rocks from the East Cameron well, Louisiana Gulf Coast. (ii) 1,1-Dimethylcyclohexane (11DMCH) is associated with kerogen/rock type A (+ and *) but not with B (△). [From Whelan *et al.* (1984).]

basis of either Rock-Eval or microscopic data (Whelan *et al.*, 1988a). However, the Py-GC patterns for the two rock types are completely different [Fig. 45(i)], implying at least two distinct depositional styles for this well.

Pyrolysis–GC and –GC/MS fingerprinting has also been used to demonstrate similarities between a number of asphaltenes and kerogens (Bandurski, 1982; Behar *et al.*, 1984; Behar and Pelet, 1985; Macko *et al.*, 1988). These and many other similar studies have led to the conclusion that asphaltenes are produced early in the oil formation process (Rubinstein *et al.*, 1979). On the basis of these data, it has been proposed that asphaltenes may represent small pieces of kerogen which can migrate with the oil while the kerogen remains in the source rock. If this finding proves to be general, then comparison of asphaltene to kerogen Py-GC/MS pairs could prove to be not only a powerful oil source–reservoir correlation tool (Behar *et al.*, 1984) but might provide a means of detailed "typing of kerogen" from which a specific oil was derived. For example, such a correlation was recently carried out for an interbedded sand–shale sequence from the Upper Jurassic Hareelve Formation in East Greenland (Requejo *et al.*, 1989). Py-GC/MS patterns of isolated kerogens from the shales matched very closely those of asphaltenes derived from the oil in the interbedded sandstones. Solvent-extractable biomarker patterns for steranes and triterpanes from the same intervals also matched for the shales and oils.

Py-GC/MS patterns of kerogens are also often reproducible over wide maturation ranges, providing a means of "typing" kerogen by tying it to an immature analog in a much more detailed manner than is currently possible by relying on elemental analysis-related methods. For example, Larter (1989) demonstrated that the Py-GC/MS pattern was relatively rank independent, particularly in the high molecular weight range, over much of the oil window for kerogens isolated from the Lower Kittaning Formation in the North Sea (R_0 = 0.59–1.51%), where the detailed pyrolysis peak composition remains fairly constant.

6.4. Electron Microscopy (Diffraction)

Electron microscopy has been used to study the molecular structure of kerogen (Boulmier *et al.*, 1975; Oberlin *et al.*, 1980; Wedeking and Hayes, 1983). Boulmier *et al.* (1975) demonstrated a narrowing of interlayer spacings and a gradual ordering of aromatic clusters in the kerogen matrix during heating up to 1000°C. Oberlin *et al.* (1980) demonstrated that the technique distinguished a small set of initially oxygen-rich from oxygen-poor kerogens, including both natural and artificially heated types I, II, and III maturation series. The authors suggested that the extent of molecular orientation can be used as an absolute criterion for distinguishing between kerogen types, even for high-maturity samples (R_0 = 2% and greater) where other techniques are not applicable. Any factor causing a greater degree of oxidation of the initial kerogen was found to cause a lowering in the extent of molecular orientation during the maturation process. Therefore, molecular orientation as determined by this technique was postulated to be closely related to past oil potential of a kerogen, although distinction between the various natural factors which might have caused greater kerogen oxidation, such as source variations, bacterial influences, and detrital additions, was not possible without further research.

Wedeking and Hayes (1983) modified the technique somewhat and analyzed a wide variety of Precambrian kerogens, including cherts, shales, diamonds, carbonates, and other mixtures. This varied sample mixture produced a remarkably consistent curve showing the variations of diffraction peak widths and kerogen interplanar spacing as a function of decreasing H/C ratios in the range of 0 to 1.2. They interpreted their results as demonstrating that the increased stacking of kerogen aromatic layers, corresponding to increasing graphite crystallinity, occurs during catagenesis down to H/C ratios of about 0.5, after which a rapid apparent increase occurred. The authors concluded that, because of these ambiguities in response, the kerogen elemental H/C ratios were a more reliable absolute indicator of kerogen maturity for this sample set. However, the technique would appear to have value in learning more about "spent" kerogens, a common problem encountered in the oil industry, where the rocks that have generated a particular gas and oil have already "lost" much of the kerogen chemical structural information which could be deduced by one or more of the techniques discussed earlier in this chapter.

Zumberge *et al.* (1978) have described the use of a combined scanning electron microscope–electron microprobe analysis on a series of Precambrian Vaal Reef kerogens. This analysis was useful in delineating two finely admixed substances in these kerogens—one consisting mainly of silicon, aluminum, and iron together with subordinant amounts of sulfur (associated with the iron) and gold. The other component consisted of kerogen containing nonpyrite sulfur and occasionally nitrogen. Even though the technique was not useful in looking at anything other than small areas of the kerogen, it did show that gold frequently occurred adjacent to the clay–kerogen boundaries in both the clay and organic phases. Thus, this might be a useful technique in some situations for examining the nature of the inorganic phase which normally accompanies kerogen isolates (see Section 3.2) and the nature of its binding (if any) to the kerogen phase.

ACKNOWLEDGMENTS. We wish to thank Mary Zawoysky and Lorraine Eglinton for help in editing this chapter. Helpful comments from Tom Hoering and Mike Engel are gratefully acknowledged.

Financial support to JKW was provided from the Department of Energy, Grant DE-FG02-86ER13466. Woods Hole Contribution No. 7238.

References

Aizenshtat, Z., Miloslovski, I., and Tannenbaum, E., 1986a, Thermal behavior of immature asphalts and related kerogens, *Org. Geochem.* **10:**537–546.

Aizenshtat, Z., Pinsky, I., and Baruch, B., 1986b, Electron spin resonance of stabilized free radicals in sedimentary organic matter, *Org. Geochem.* **9:**321–329.

Albrecht, P., Vandenbroucke, M., and Mandengue, M., 1976, Geochemical studies on the organic matter from the Douala Basin (Cameroon)—I. Evolution of the extractable organic matter and the formation of petroleum, *Geochim. Cosmochim. Acta* **40:**791–799.

Allan, J., and Douglas, A. G., 1974, Alkanes from pyrolytic degradation of bitumenous vitrinites and sporinites. In *Advances in Organic Geochemistry 1973* (B. Tissot and G. Bienner, eds.), Editions Technip, Paris, pp. 203–206.

Alpern, B., 1970, Classification pétrographique des constituants organiques fossiles des roches sédimentaries, *Rev. Insti. Fr. Pet.* **25:**1233–1267.

Alpern, B., 1980, Pétrographie du kérogène, in: *Kerogen–Insoluble Organic Matter from Sedimentary Rocks* (B. Durand, ed.), Editions Technip, Paris, pp. 339–371.

Axelson, D. E., 1985, *Solid State NMR of Fossil Fuels: An Experimental Approach*, Multiscience Publications Limited, Montreal, 226 pp.

Baker, E. W., Huang, W. Y., Rankin, J. A., Castano, J. R., Guinn, J. R., and Fuex, A. N., 1977, Electron paramagnetic resonance study of thermal alteration of kerogen in deep sea sediments by basaltic sill intrusions. *Initial Reports of DSDP* (Y. Lancelot *et al.*, eds.), Vol. 41, U.S. Government Printing Office, Washington, D.C., pp. 839–847.

Bakr, M., Akiyama, M., Sanada, Y., and Yokono, T., 1988, Radical concentration of kerogen as a maturation parameter, *Org. Geochem.* **12:**29–32.

Bandurski, E., 1982, Structural similarities between oil-generating kerogens and petroleum asphaltenes, *Energy Sources* **6:**47–66.

Barakat, A. O., and Yen, T. F., 1988, Size distribution of the straight-chain structures in type I and II kerogens, *Energy and Fuels* **2:**181–185.

Barker, C., 1974, Pyrolysis techniques for source rock evaluation, *Am. Assoc. Petrol. Geol. Bull.* **58:**2349–2361.

Barker, C., and Wang, L., 1988, Applications of pyrolysis in petroleum geochemistry: A bibliography, *J. Anal. Appl. Pyrolysis* **13:**9–61.

Barnard, P. C., Cooper, B. S., and Fisher, M. J., 1976, Organic maturation and hydrocarbon generation in the Mesozoic sediments of the Sverdrup basin Arctic Canada, 4th International Palynology Conference, Lucknow, pp. 581–588.

Barwise, A. J. G., Mann, A. L., Eglinton, G., Gowar, A. P., Wardroper, A. M. K., and Gutteridge, C. S., 1984, Kerogen characterization by ^{13}C NMR spectroscopy and pyrolysis–mass spectrometry, *Org. Geochem.* **6:**343–349.

Batten, D. J., 1981, Palynofacies, organic maturation and source potential for petroleum, in: *Organic Maturation Studies and Fossil Fuel Exploration* (J. Brooks, ed.), Academic Press, London, pp. 201–223.

Batten, D. J., 1983, Identification of amorphous sedimentary organic matter by transmitted light microscopy, in: *Petroleum Geochemistry and Exploration in Europe* (J. Brooks, ed.), Blackwell Scientific, London, pp. 275–287.

Behar, F., and Pelet, R., 1985, Pyrolysis–gas chromatography applied to organic geochemistry. Structural similarities between kerogens and asphaltenes from related rock extracts and oils, *J. Anal. Appl. Pyrolysis* **8:**173–187.

Behar, F., and Vandenbroucke, M., 1986, Representation chemique de la structure des kérogènes et des asphaltènes en fonction de leur origine et de leur degré d'évolution, *Rev. Inst. Fr. Pet.* **41:** 173–188.

Behar, F., and Vandenbroucke, M., 1987, Chemical modelling of kerogens, *Org. Geochem.* **11:**15–24.

Behar, F., Pelet, R., and Roucache, J., 1984, Geochemistry of asphaltenes, *Org. Geochem.* **6:**587–595.

Bertrand, P., Pittion, J.-L., and Bernaud, C., 1986, Fluorescence of sedimentary organic matter in relation to its chemical composition, *Org. Geochem.* **10:**641–647.

Bird, K. J., 1985, The framework geology of the North Slope of Alaska as related to oil-source rock correlations, in: *Alaskan North Slope Oil/Rock Correlation Study* (L. B. Magoon and G. E. Claypool, eds.), Am. Assoc. Petrol. Geol. Spec. Stud. Geol., No. 20, pp. 3–29.

Bjorøy, M., Hall, P. B., Loberg, R., McDermott, J. A., and Mills, N., 1988, Hydrocarbons from non-marine source rocks, *Org. Geochem.* **13:**221–244.

Bostick, N. H., 1970, Thermal alteration of clastic organic particles (phytoclasts) as an indicator of contact and burial metamorphism in sedimentary rocks, Ph.D. Thesis, Stanford University.

Bostick, N. H., 1974, Phytoclasts as indicators of thermal metamorphism, Franciscan Assemblage and Great Valley Sequence (Upper Mesozoic), California, Geological Society of America Special Paper 153, pp. 1–17.

Bostick, N. H., 1979, Microscopic measurement of the level of catagenesis of solid organic matter in sedimentary rocks to aid exploration for petroleum and to determine former burial temperatures—a review, *Soc. Econ. Paleontol. Mineral. Spec. Publ.* **26:**17–43.

Bostick, N. H., and Alpern, B., 1977, Principles of sampling, preparation and constituent selection for microphotometry in measurement of maturation of sedimentary organic matter, *J. Microsc.* **109:**41–47.

Bostick, N. H., and Foster, J. N., 1975, Comparison of vitrinite reflectance in coal seams and in kerogen of sandstones, shales, and limestones in the same part of a sedimentary section, in:

Pétrographie de la Matière Organique des Sédiments, Relations avec la Paléotempérature et le Potentiel Pétrolier (B. Alpern, ed.), Editions du Centre National de la Recherche Scientifique, Anatole, pp. 13–25.

Boulmier, J. L., Oberlin, A., and Durand, B., 1975, Structural study of some series of kerogens and relation to carbonization process (high resolution electron microscopy), in: *Advances in Organic Geochemistry* (R. Campos, ed.), Wiley Interscience, New York, pp. 781–796.

Bujak, J. P., Barss, M. S., and Williams, G. L., 1977, Offshore East Canada's organic type and color and hydrocarbon potential, Parts I and II, *Oil Gas J.* **75**(14):198–202; **75**(15):96–100.

Burgess, J. D., 1974. Microscopic examination of kerogen (dispersed organic matter) in petroleum exploration, Geological Society of America Special Paper 153, pp. 19–30.

Cainelli, G., and Cardillo, G., 1984, *Chromium Oxidations in Organic Chemistry*, Springer-Verlag, Berlin.

Castano, J. R., and Sparks, D., 1974, Interpretation of vitrinite reflectance measurements in sedimentary rocks and determination of burial history using vitrinite reflectance and authigenic minerals, Geological Society of America Special Paper 153, pp. 31–52.

Choi, Chol-yoo, Wang, Shih-Hsien, and Stock, L. M., 1988, Ruthenium tetraoxide catalized oxidation of maceral groups, *Energy Fuels* **2**:37–48.

Claypool, G. E., and Reed, P. R., 1976, Thermal analysis technique for source-rock evaluation: Quantitative estimate of organic richness and effects of lithologic variation, *Am. Assoc. Petrol. Geol. Bull.* **60**:608–612.

Combaz, A., 1974, La matière algaire et l'origine du pétrole, in: *Advances in Organic Geochemistry 1973* (B. Tissot and F. Bienner, eds.), Editions Technip, Paris, pp. 423–438.

Combaz, A., 1975, Essai de classification des roches carbonées et des constituants organiques des roches sédimentaries, in: *Pétrographie de la Matière Organique des Sédiments, Relations avec la Paléotempérature et le Potentiel Pétrolier* (B. Alpern, ed.), Editions du Centre National de La Recherche Scientifique, Anatole, pp. 93–101.

Combaz, A., 1980, Les Kérgènes vus au microscope, in: *Kerogen—Insoluble Organic Matter from Sedimentary Rocks* (B. Durand, ed.), Editions Technip, Paris, pp. 55–112.

Connan, J., 1967, Geochemical significance of the extraction of amino acids from sediments, *Bull. Cent. Rech. Pau-SNPA* **1**: 165–171.

Correia, M., 1969, Contribution à la recherche de zones favorables à la genése du pétrole par l'observation microscopique de la matière organique figurée, *Rev. Inst. Fr. Pet.* **24**:1417–1454.

Correia, M., 1971, Diagenesis of sporopollenin and other comparable organic substances: Application to hydrocarbon research, in: *Sporopollenin* (J. Brooks *et al.*, eds.), Academic Press, London, pp. 569–620.

Crelling, J. C., 1982, Current uses of fluorescence microscopy in coal petrology, *J. Microsc.* **132**:251–266.

Crossey, L. J., Hagen, E. S., and Surdam, R. C., 1984, Correlation of organic parameters derived from elemental analysis and programmed pyrolysis of kerogen, in: *Roles of Organic Matter in Sediment Diagenesis* (D. L. Gautier, ed.), *Soc. Econ. Paleontol. Mineral. Spec. Special Publ.* **38**:35–45.

Curiale, J. A., Larter, S. R., Sweeney, R. E., and Bromley, B. W., 1989, Molecular thermal maturity indicators in oil and gas source rocks, in: *Thermal History of Sedimentary Basins, Methods and Case Histories* (N. D. Naeser and T. H. McCulloh, eds.), Springer-Verlag, New York, pp. 53–72.

Curry, D. J., 1981, *The organic geochemistry of kerogen and humic acids in recent sediments from the Gulf of Mexico*, Ph.D. Thesis, The University of Texas at Austin.

Dean, W. E., Arthur, M. A., and Claypool, G. E., 1986, Depletion of ^{13}C in Cretaceous marine organic matter: Source, diagenetic, or environmental signal? *Mar. Geol.* **70**:119–157.

Degans, E. T., 1969, Biochemistry of stable carbon isotopes, in, *Organic Geochemistry Methods and Results* (G. Eglinton and M. T. J. Murphy, eds.), Springer-Verlag, New York, pp. 304–329.

Demaison, G. J., and Moore, G. T., 1980, Anoxic environments and oil source bed genesis, *Am. Assoc. Petrol. Geol. Bull.* **64**:1179–1209.

Dembicki, H., Jr., 1984, An interlaboratory comparison of source rock data, *Geochim. Cosmochim. Acta* **48**:2641–2649.

Dembicki, H., Jr., Horsfield, B., and Ho, T. T. Y., 1983, Source rock evaluation by pyrolysis–gas chromatography, *Am. Assoc. Petrol. Geol. Bull.* **67**:1094–1103.

DeNiro, M. J., Sternberg, L. D., Marino, B. D., and Druzik, J. R., 1988, *Geochim. Cosmochim. Acta* **52**:2189–2196.

Dennis, L. W., Maciel, G. E., Hatcher, P. G., and Simoneit, B. R. T., (1982), ^{13}C nuclear magnetic resonance studies of kerogen from Cretaceous black shales thermally altered by basaltic intrusions and laboratory simulations, *Geochim. Cosmochim. Acta* **46**:901–907.

Deines, P., 1980, The isotopic composition of reduced organic carbon, in: *Handbook of Environmental Isotope Geochemistry*, Vol. 1 (P. Friz and J. Ch. Fantes, eds.), Elsevier, Amsterdam, pp. 329–406.

Dow, W. G., 1977, Kerogen studies and geological interpretations, *J. Geochem. Explor.* **7**:79–99.

Dow, W. G., and O'Connor, D. I., 1982, Kerogen maturity and type by reflected light microscopy applied to petroleum exploration, in: *How to Assess Maturation and Paleotemperature*, Society of Economic Paleontologists and Mineralogists, Short Course Number Seven, Tulsa, Oklahoma, pp. 133–157.

Durand, B., 1980, Sedimentary organic matter and kerogen. Definition and quantitative importance of kerogen, in: *Kerogen—Insoluble Organic Matter from Sedimentary Rocks* (B. Durand, ed.), Editions Technip, Paris, pp. 13–34.

Durand, B., and Espitalié, J., 1976, Geochemical studies on the organic matter from the Douala Basin (Cameroon)—II. Evolution of kerogen, *Geochim. Cosmochim. Acta* **40**:801–808.

Durand, B., and Monin, J. C., 1980, Elemental analysis of kerogens (C, H, O, N, S, Fe), in: *Kerogen—Insoluble Organic Matter from Sedimentary Rocks* (B. Durand, ed.), Editions Technip, Paris, pp. 113–142.

Durand, B., and Nicaise, G., 1980, Procedures for kerogen isolation, in: *Kerogen—Insoluble Organic Matter from Sedimentary Rocks* (B. Durand, ed.), Editions Technip, Paris, pp. 35–53.

Durand, B., Espitalié, J., Nicaise, G., and Combaz, A., 1972, Etude de la maitère organique insoluble (kérogène) des argiles du Toarcien du Bassin de Paris: I, Etude par les procédés optiques, analyse élémentaire, étude en microscopie et diffraction électroniques, *Rev. Inst. Fr. Pet.* **27**:865–884.

Durand, B., Alpern, B., Pittion, J. L., and Pradier, B., 1986, Reflectance of vitrinite as a control of thermal history of sediments, in: *Thermal Modeling in Sedimentary Basins* (J. Burrus, ed.), Editions Technip, Paris, pp. 441–474.

Durand-Souron, C., 1980, Thermogravimetric analysis and associated techniques applied to kerogen, in: *Kerogen—Insoluble Organic Matter from Sedimentary Rocks* (B. Durand, ed.), Editions Technip, Paris, pp. 143–161.

Erdman, J. G., 1961, Some chemical aspects of petroleum genesis as related to the problem of source bed recognition, *Geochim. Cosmochim. Acta* **22**:16–36.

Espitalié, J., 1986, Use of T_{max} as a maturation index for different types of organic matter. Comparison with vitrinite reflectance, in: *Thermal Modeling in Sedimentary Basins* (J. Burrus, ed.), Editions Technip, Paris, pp. 475–496.

Espitalié, J., LaPorte, J. L., Madec, M., Marquis, F., Leplat, P, Paulet, J., and Boutefeu, A., 1977, Méthode rapide de caractériasation des roches mères de leur potential pétrolier et de leur degré d'évolution, *Rev. Inst. Fr. Pet.* **32**:23–42.

Espitalié, J., Marquis, F., and Barsony, I., 1984, Geochemical logging, in: *Analytical Pyrolysis* (K. J. Voorhees, ed.), Butterworths, Boston.

Estep, M. F., and Hoering, T. C., 1980, Biogeochemistry of the stable hydrogen isotopes, *Geochim. Cosmochim. Acta* **44**:1197–1206.

Estep, M. F., and Hoering, T. C., 1981, Stable hydrogen isotope fractionations during autotrophic and mixotrophic growth of microalgae, *Plant Physiol.* **67**:474–477.

Farrington, J. W., Davis, A. C., Sulanowski, J., McCaffrey, M. A., McCarthy, M., Clifford, C. H., Dickinson, P., and Volkman, J. K., 1988a, Biogeochemistry of lipids in surface sediments of the Peru upwelling area at 15°S, *Org. Geochem.* **13**:607–617.

Farrington, J. W., Davis, A. C., Tarafa, M. E., McCaffrey, M. A., Whelan, J. K., and Hunt, J. M., 1988b, Bitumen molecular maturity parameters in the Ikpikpuk well, Alaskan North Slope, *Org. Geochem.* **13**:303–310.

Fleet, A. J., Kelts, K., and Talbot, M. R., 1988, *Lacustrine Petroleum Source Rocks*, Blackwell Scientific, Oxford.

Forsman, J. P., and Hunt, J. M., 1958, Insoluble organic matter (kerogen) in sedimentary rocks, *Geochim. Cosmochim. Acta* **15**:170–182.

Foster, C. B., O'Brien, G. W., and Watson, S. T., 1986, Hydrocarbon source potential of the Goldwyer formation, Barbwire Terrace, Canning Basin, Western Australia, *APEA J.* **26** (Part 1):142–155.

Francois, R., 1991, Marine sedimentary humic substances: structure, genesis, and properties, *Reviews in Aquatic Sciences* **3**: 41–80.

Fuex, A. N., 1977, The use of stable carbon isotopes in hydrocarbon exploration, *J. Geochem. Explor.* **7**:155–188.

Galimov, E. M., 1980, C^{13}/C^{12} in kerogen, in: *Kerogen—Insoluble Organic Matter from Sedimentary Rocks* (B. Durand, ed.), Editions Technip, Paris, pp. 271–299.

Giraud, A., 1970, Applications of pyrolysis and gas chromatography to geochemical characterization of kerogen in sedimentary rock, *Am. Assoc. Petrol. Geol. Bull.* **54**:439–455.

Goodarzi, F., 1984, Chitinous fragments in coal, *Fuel* **63**:1504–1507.

Goodarzi, F., 1987, Comparison of reflectance data from various macerals from sub-bituminous coals, *J. Petrol. Geol.* **10**:219–226.

Goth, K., de Leeuw, J. W., Puttmann, W., and Tegelaar, E. W., 1988, Origin of Messel Oil Shale kerogen, *Nature* **336**:759–761.

Gransch, J. A., and Eisma, E., 1970, Characterization of the insoluble organic matter of sediments by pyrolysis, in: *Advances in Organic Geochemistry 1966* (G. D. Hobson and G. C. Speers, eds.), Pergamon Press, Oxford, pp. 407–426.

Grayson, J. F., 1975, Relationship of palynomorph translucency to carbon and hydrocarbons in clastic sediments, in *Pétrographie de la Matière Organique des Sédiments, Relations avec la Paléotempérature et le Potentiel Pétrolier* (B. Alpern, ed.), Editions du Centre National de la Recherche Scientifique, Anatole, pp. 261–273.

Gutjahr, C. C. M., 1960, Palynology and its application in petroleum exploration, *Trans. Gulf Coast Assoc. Geol. Soc.* **10**:175–187.

Gutjahr, C. C. M., 1966, Carbonization measurements of pollen-grains and spores and their application, *Leidse Geol. Meded.* **38**:1–29.

Haddad, R. I., Rohrback, B. G., and Kaplan, I. R., 1990, Hydrofluoric acid induced alterations of sedimentary humic acids, in: *Facets of Modern Biogeochemistry* (Z. Ittekot, F. Kemp, W. Michaelis and A. Spitzy, eds.), Springer-Verlag, Berlin, Heidelberg, pp. 415–425.

Hatcher, P. G., Spiker, E. C., Szeverenyi, N. M., and Maciel, G E., 1983, Selective preservation and origin of petroleum-forming aquatic kerogen, *Nature* **305**:498–501.

Hayatsu, R., Winans, R. E., Scott, R. G., Moore, L. P., and Sudier, M. H., 1978, *Fuel* **57**:541–548.

Hayes, J. M., Kaplan, I. R., and Wedeking, K. W., 1983, Precambrian organic geochemistry, preservation of the record, in: *Earth's Earliest Biosphere: Its Origin and Evolution* (J. W. Schopf, ed.), Princeton University Press, Princeton, New Jersey, pp. 93–134.

Hayes, J. M., Takigiku, R., Ocampo, R., Callot, H. J., and Albracht, P., 1987, Isotopic compositions and probable origins of organic molecules in Eocene Messel shale, *Nature* **329**:48–51.

Henry, G., and Perrodon, A., 1988, Recherche scientifique et technique en exploration, *Bull. Cent. Rech. Explor.–Prod. Elf Aquitaine* **12**:1–27.

Heroux, Y., Chagnon, A., and Bertrand, R., 1979, Compilation and correlation of major thermal maturation indicators, *Am. Assoc. Petrol. Geol. Bull.* **63**:2128–2144.

Hoering, T. C., 1984, Thermal reactions of kerogen with added water, heavy water and pure organic substances, *Org. Geochem.* **5**:267–278.

Horsfield, B., 1984, Pyrolysis studies and petroleum exploration, in: *Advances in Petroleum Geochemistry*, Vol. 1 (J. Brooks and D. Welte, eds.), Academic Press, London, pp. 247–298.

Horsfield, B., 1989, Practical criteria for classifying kerogens: Some observations from pyrolysis–gas chromatography, *Geochim. Cosmochim. Acta* **53**:891–901.

Horsfield, B., and Woods, R. A., 1984, The preparation of drill cuttings for organic geochemical analysis using a water-flow device, *J. Sediment. Petrol.* **54**:637–638.

Horsfield, B., Dembicki, H., and Ho, T. T. Y., 1983, Some potential applications of pyrolysis to basin studies, *J. Geol. Soc. London* **140**:431–443.

Horsfield, B., Yordy, K. L., and Crelling, J. C., 1988, Determining the petroleum-generating potential of coal using organic geochemistry and organic petrology, *Org. Geochem.* **13**:121–129.

Huc, A. Y., 1980, Origin and formation of organic matter in recent sediments and its relation to kerogen, in: *Kerogen—Insoluble Organic Matter from Sedimentary Rocks* (B. Durand, ed.), Editions Technip, Paris, pp. 445–474.

Huc, A. Y., Durand, B., Roucachet, J., Vandenbroucke, M., and Pittion, J. L., 1986, Comparison of three series of organic matter of continental origin, *Org. Geochem.* **10**:65–72.

Hunt, J. M., 1962, Geochemical data on organic matter in sediments, in: *Proceedings of the International Scientific Oil Conference*, Budapest, Hungary, Hungarian Trading Company for Books and Newspapers KULTUA, Budapest, pp. 394–412.

Hunt, J. M., 1979, *Petroleum Geochemistry and Geology*, W. H. Freeman and Co., San Francisco.

Hunt, J. M., 1991, Generation of gas and oil from coal and terrestrial organic matter, *Org. Geochem.* **17**:673–680.

Hunt, J. M., Lewan, M. K., and Hennet, R., 1991, Modeling oil generation with Arrhenius time–temperature index graphs, *Am. Assoc. Petrol. Geol. Bull.* **75**:795–807.

Hutton, A. C., 1987, Petrographic classification of oil shales, *Int J. Coal Geol.* **8**:203–231.

Hutton, A. C., and Cook, A. C., 1980, Influence of alginite on the reflectance of vitrinite from Joadja, NSW, and some other coals and oil shales containing alginite, *Fuel* **59**:711–714.

International Committee for Coal Petrology, 1963, 1971, 1975, *International Handbook of Coal Petrography*, CNRS, Paris.

Jacob, H., 1966, Uber Beziehungen zwischen Kohle und Erdöl, *Erdöl Kohle* **19**:5–11.

Jacob, H., 1967, Petrologie von Asphaltiten und asphaltischen Pyrobitumina, *Erdöl Kohle* **20**:393–400.

Jacob, H., 1985, Dispersed solid bitumens as an indicator for migration and maturity in prospecting for oil and gas, *Erdöl Kohle, Erdgas, Petrochem.* **38**:365–392.

Jasper, J. P., Whelan, J. K., and Hunt, J. M., 1984, Migration of C_1–C_8 volatile organic compounds in sediments from the Deep Sea Drilling Project, Leg 75, Hole 530A, *Initial Reports of DSDP*, Vol. 75 (W. W. Hay, J. C. Sibuet, *et al.*, eds.), U.S. Government Printing Office, Washington, D.C., pp. 1001–1008.

Jones, R. W., 1987, Organic facies, in: *Advances in Petroleum Geochemistry*, Vol. 2 (J. Brooks and D. Welte, eds.), Academic Press, London, pp. 1–90.

Jones, R. W., and Edison, T. A., 1978, Microscopic observations of kerogen related to geochemical parameters with emphasis on thermal maturation, in: *Symposium in Geochemistry: Low Temperature Metamorphism of Kerogen and Clay Minerals* (D. F. Oltz, ed.), Society of Economic Paleontologists and Mineralogists, Los Angeles, pp. 1–12.

Kalkreuth, W. D., 1982, Rank and petrographic composition of selected Jurassic–Lower Cretaceous coals of British Columbia, Canada, *Bull. Can. Petrol. Geol.* **30**:112–139.

Katz, B. J., 1983, Limitations of "Rock-Eval" pyrolysis for typing organic matter, *Org. Geochem.* **4**:195–199.

Kemp, D. S., and Vellaccio, F., 1980, *Organic Chemistry*, Worth, New York, Chapter 12.

Khavari Khorasani, G., and Murchison, D. G., 1988, Order of generation of petroleum hydrocarbons from liptinitic macerals with increasing thermal maturity, *Fuel* **67**:1160–1162.

Kinghorn, R. R. F., and Rahman, M., 1983, Specific gravity as a kerogen type and maturation indicator with special reference to amorphous kerogens, *J. Petrol. Geol.* **6**:179–194.

Klemme, H. D., and Ulmishek, G. F., 1989, Depositional controls, distribution and effectiveness of world's petroleum source rocks, *Am. Assoc. Petrol. Geol. Bull.* **73**:372–373.

Landais, P., Monthioux, M., and Meunier, J. D., 1984, Importance of the oxidation/maturation pair in the evolution of humic coals, *Org. Geochem.* **7**:249–260.

Landais, P., Monthioux, M., Dereppe, J. M., and Moreaux, C., 1988, Analyses of insoluble residues of altered organic matter by ^{13}C CP/MAS nuclear magnetic resonance, *Org. Geochem.* **13**:1061–1065.

Larter, S., 1978, A geochemical study of kerogen and related materials, Ph.D. Dissertation, University of Newcastle upon Tyne.

Larter, S. R., 1984, Application of analytical pyrolysis techniques to kerogen characterization and fossil fuel exploration/exploitation, in: *Analytical Pyrolysis Techniques and Exploration* (K. J. Voorhees, ed.), Butterworths, Boston, pp. 212–275.

Larter, S., 1985, Integrated kerogen typing in the recognition and quantitative assessment of petroleum source rocks, in: *Petroleum Geochemistry in Exploration of the Norwegian Shelf* (B. M. Thomas *et al.*, eds.), The Norwegian Petroleum Society, Graham and Trotman, London, pp. 269–286.

Larter, S., 1988, Pragmatic perspective in petroleum geochemistry, Invited contribution, *Marine and Petroleum Geology* **5**: 194–204.

Larter, S., and Douglas, A. G., 1980, in: *Advances in Organic Geochemistry 1979* (A. G. Douglas and J. R. Maxwell, eds.), Pergamon Press, Oxford, pp. 579–584.

Larter, S., and Douglas, A. G., 1982, Pyrolysis methods in organic chemistry: An overview, *J. Anal. Appl. Pyrolysis* **4**:1–19.

Leythaeuser, D., Hagemann, H. W., Hollerbach, A., and Schaefer, R. G., 1980, Hydrocarbon generation in source beds as a function of type and maturation of their organic matter: A mass balance approach, *Proceedings of the 10th World Petroleum Congress*, Vol. 2, John Wiley & Sons, Chichester, pp. 31–41.

Lin, R., Davis, A., Bensley, D. F., and Derbyshire, F. J., 1987, The chemistry of vitrinite fluorescence, *Org. Geochem.* **11**:393–399.

Lopatin, N. V., 1971, Temperature and geologic time as factors in coalification, *Izv. Akad. Nauk SSR, Ser. Geol.* **3**:95–106 [in Russian].

Maciel, G. E., and Dennis, L. W., 1982, Comparison of oil shales and kerogen concentrates by ^{13}C nuclear magnetic resonance, *Org. Geochem.* **3**:105–109.

Macko, S. A., Curry, D. J., Kiceniuk, J. W., and Simpler, K. T., 1988, Variations in geochemical character of asphaltenes from a source rock in contact with an oil reservoir, *Org. Geochem.* **13**: 273–281.

Marchand, A., and Conard, J., 1980, Electron paramagnetic resonance in kerogen studies, in: *Kerogen—Insoluble Organic Matter from Sedimentary Rocks* (B. Durand, ed.), Editions Technip, Paris, pp. 243–270.

Masran, Th. C., and Pocock, S. J., 1981, The classification of plant-derived particulate organic matter in sedimentary rocks, in: *Organic Maturation Studies and Fossil Fuel Exploration* (J. Brooks, ed.), Academic Press, London, pp. 145–175.

Massoud, M. S., and Kinghorn, R. R. F., 1985, A new classification for the organic components of kerogen, *J. Petrol. Geol.* **8**: 85–100.

Massoud, M. S., and Kinghorn, R. R. F., 1986, A new classification of the organic components of kerogen: Reply, *J. Petrol. Geol.* **9**: 227–231.

Maters, W. L., van de Meent, D., Schuyl, P. J. W., de Leeuw, J. W., Schenck, P. A., and Meuzelaar, H. L. C., 1977, Curie point pyrolysis in organic geochemistry, in: *Analytical Pyrolysis* (C. E. Roland Jones and C. A. Cramers, eds.), Elsevier, Amsterdam, pp. 203–216.

McCulloh, T. H., and Naeser, N. D., 1989, Thermal history of sedimentary basins: Introduction and overview, in: *Thermal History of Sedimentary Basins, Methods and Case Histories* (N. D. Naeser and T. H. McCulloh, eds.), Springer-Verlag, New York, pp. 1–12.

McIver, R. D., 1967, Composition of kerogen—clue to its role in the origin of petroleum, in: *Proceedings of the 7th World Petroleum Congress*, Mexico City, Vol. 2, Elsevier, London, pp. 26–36.

Meuzelaar, H. L. C., Haverkamp, J., and Hileman, F. D., 1982, *Pyrolysis Mass Spectrometry of Recent and Fossil Biomaterials: Compendium and Atlas*. Techniques and Instrumentation in Analytical Chemistry, Vol. 3, Elsevier, Amsterdam.

Miknis, F. P., Smith, J. W., Maughan, E. K., and Maciel, G. E., 1982, Nuclear magnetic resonance: A technique for direct nondestructive evaluation of source-rock potential, *Am. Assoc. Petrol. Geol. Bull.* **66**:1396–1401.

Mitterer, R. M., 1988, Extractable and pyrolyzable hydrocarbons in shallow-water carbonate sediments, Florida Bay, Florida, *Org. Geochem.* **13**:283–294.

Monthioux, M., and Landais, P., 1988, Natural and artificial maturations of a coal series: Infrared spectrometry study, *Energy Fuels* **2**:794–801.

Monthioux, M., Landais, P., and Monin, J. C., 1985, Comparison between natural and artificial maturation series of humic coals from the Mahakam delta, Indonesia, *Org. Geochem.* **8**:275–292.

Morishima, H., and Matsubayashi, H., 1978, ESR diagram: A method to distinguish vitrinite macerals, *Geochim. Cosmochim. Acta* **42**:537–540.

Morrison, R. T., and Boyd, R. N., 1973, *Organic Chemistry*, 3rd ed., Allyn and Bacon, Boston, Chapter 13.

Mukhopadhyay, P. K., 1986, A new classification for the organic components of kerogen: Discussion, *J. Petrol. Geol.* **9**:111–114.

Mukhopadhyay, P. K., Hagemann, W. H., and Gormly, J. R., 1985, Characterization of kerogens as seen under the aspect of maturation and hydrocarbon generation, *Erdöl Kohle, Erdgas, Petrochem.* **38**:7–18.

Mycke, B., and Michaelis, W., 1986, Molecular fossils from chemical degradation of macromolecular organic matter, *Org. Geochem.* **10**:846–858.

Naeser, N. D., and McCulloh, T. H., 1989, *Thermal History of Sedimentary Basins, Methods and Case Histories*, Springer-Verlag, New York.

Newman, J., and Newman, N. A., 1982, Reflectance anomalies in Pike River coals: Evidence of variability in vitrinite type, with implications for maturation studies and "Suggate rank," *N. Z. J. Geol. Geophys.* **25**:233–243.

Nip, M., Tegelaar, E. W., Brinkhuis, H., de Leeuw, J. W., Schenk, P. A., and Holloway, P. J., 1986, Analysis of modern and fossil plant cuticles by Curie point Py-GC and Curie point Py-GC-MS: Recognition of a new, highly aliphatic and resistant biopolymer, *Org. Geochem.* **10**:769–778.

Oberlin, A., Boulmier, J. L., and Villey, M., 1980, Electron microscopic study of kerogen microtexture. Selected criteria for determining the evolution path and evolution stage of kerogen, in: *Kerogen–Insoluble Organic Matter from Sedimentary Rocks* (B. Durand, ed.), Editions Technip, Paris, pp. 191–24.

Orem, W. H., Hatcher, P. G., Spiker, E. C., Szeverenyi, N. M., and Maciel, G. E., 1986, Dissolved organic matter in anoxic pore waters from Mangrove Lake, Bermuda, *Geochim. Cosmochim. Acta* **50**:609–618.

Orr, W. L., 1983, Comments on pyrolytic hydrocarbon yields in source-rock evaluation, *Advances in Organic Geochemistry* (M. Bjorøy *et al.*, eds.), Wiley, New York, pp. 775–787.

Orr, W. L., 1986, Kerogen/asphaltene/sulfur relationships in sulfur-rich Monterey oils, *Org. Geochem.* **10**:499–516.

Ottenjann, K., 1985, Fluoreszenzeigenschaften von Flozporben aus stratigraphisch jungen Schichten des Oberkarbons, *Fortschr. Geol. Rheinl. Westfalen* **33**:169–175.

Ottenjann, K., Teichmuller, M., and Wolf, M., 1974, Spektrale fluorezenzmessungen an sporiniten mit auflicht-anregung, eine mikroskopische methode zur bestimmung des inkohlungsgrades geringinkohlter kohlen, *Fortschr. Geol. Rheinl. Westfalen* **24**:1–36.

Ottenjann, K., Teichmuller, M., and Wolf, M., 1975, Spectral fluorescence measurements of sporinities in reflected light and their applicability for coalification studies, in *Petrographie de la Matiere Organique des Sediments, Relations avec la Paleotemperature et le Potentiel Petrolier* (B. Alpern, ed.), Editions du Centre National de la Recherche Scientifique, Anatole, pp. 49–65.

Oygard, K., Larter, S., and Sentfle, J., 1988, The control of maturity and kerogen type on quantitative analytical pyrolysis data, *Org. Geochem.* **13**:1153–1162.

Pavia, D. L., Lampman, G. M., and Kriz, G. S., Jr., *Introduction to Spectroscopy*, Saunders, Philadelphia, 1979, pp. 81–182.

Peters, K. E,., 1986, Guidelines for evaluating petroleum source rock using programmed pyrolysis, *Am. Assoc. Petrol. Geol. Bull.* **70**:318–329.

Peters, K. E., Ishiwatari, R., and Kaplan, I. R., 1977, Color of kerogen as index of organic maturity, *Am. Assoc. Petrol. Geol. Bull.* **61**: 504–510.

Peters, K. E., Rohrback, B. G., and Kaplan, I. R., 1980, Laboratory-simulated thermal maturation of recent sediments, *9th International Congress on Organic Geochemistry* (A. G. Douglas and J. R. Maxwell, eds.), Pergamon Press, New York, pp. 547–557.

Peters, K. E., Rohrback, B. G., and Kaplan, I. R., 1981a, Geochemistry of artificially heated humic and sapropelic sediments—I: Protokerogen, *Am. Assoc. Petrol. Geol. Bull.* **65**:688–705.

Peters, K. E., Rohrback, B. G., and Kaplan, I. R., 1981b, Carbon and hydrogen stable isotope variations in kerogen during laboratory-simulated thermal maturation, *Am. Assoc. Petrol. Geol. Bull.* **65**:501–508.

Pettit, G. R., and van Tamelen, E. E., 1962, Raney nickel desulfurization, in: *Organic Reactions*, Vol. 12, John Wiley & Sons, pp. 356–380.

Philp, R. P., and Gilbert, T. D., 1984, Characterization of petroleum source rocks and shales by pyrolysis–gas chromatography–mass spectrometry–multiple ion detection, *Org. Geochem.* **6**: 489–501.

Philp, R. P., and Gilbert, T. D., 1985, Source rock and asphaltene biomarker characterization by pyrolysis–gas chromatography–mass spectrometry–multiple ion detection, *Geochim. Cosmochim. Acta* **49**:1421–1432.

Philp, R. P., and Van de Meent, D., 1983, Characterization of Precambrian kerogens by analytical pyrolysis, *Precamb. Res.* **20**: 3–16.

Philp, R. P., Gilbert, T. D., and Russell, N. J., 1982, Characterization of Victorian brown coals by pyrolysis techniques coupled with gas chromatography and gas chromatography–mass spectrometry, *Aust. Coal Geol.* **4**:228–243.

Powell, T. G., Creaney, S., and Snowdon, L. R., 1982, Limitations of use of organic petrographic techniques for identification of petroleum source rocks, *Am. Assoc. Petrol. Geol. Bull.* **66**: 430–435.

Price, L. C., and Barker, C. E., 1985, Suppression of vitrinite reflectance in amorphous rich kerogen—a major unrecognized problem, *J. Petrol. Geol.* **8**:59–84.

Pusey, W. C., 1973, How to evaluate potential gas and oil source rocks, *World Oil* **4**:71–75.

Radchenko, O. A., 1960, Differential thermal analysis in the study of dispersed organic matter in rocks, *Dokl. Akad. Nauk SSSR* **135**:713–716.

Reimers, C. E., 1982, Organic matter in anoxic sediments off central Peru: Relations of porosity, microbial decomposition and deformation properties, *Mar. Geol.* **46**:175–197.

Requejo, A. G., Hollywood, J., and Halpern, H. I., 1989, Recognition and source correlation of migrated hydrocarbons in upper Jurassic Hareelv Formation, Jameson Land, East Greenland, *Am. Assoc. Petrol. Geol. Bull.* **73**:1065–1088.

Rigby, D., and Smith, J. W., 1981, *Anisotrophic study of gases in hydrocarbons in the Cooper Basin, APEA J.* **21**:222–229.

Robert, P., 1974, Analyse microscopique des charbons et des bitumes dispersés dans les roches et mesure de leur pouvoir reflecteur: Application à l'etude de la paléogéothermie des bassins sédimentaries et de la genèse des hydrocarbures, in: *Advances in Organic Geochemistry 1973* (B. Tissot and F. Bienner, eds.), Editions Technip, Paris, pp. 549–569.

Robert, P., 1980, The optical evolution of kerogen and geothermal histories applied to oil and gas exploration, in: *Kerogen—Insoluble Organic Matter from Sedimentary Rocks* (B. Durand, ed.), Editions Technip, Paris, pp. 385–414.

Robert, P., 1985, *Organic Metamorphism and Geothermal History: Microscopic Study of Organic Matter and Thermal Evolution of Sedimentary Basins*, D. Reidel, Dordrecht [English ed., 1988].

Robinson, W. E., 1969, Isolation procedures for kerogens and associated soluble organic materials, in: *Organic Geochemistry Methods and Results* (G. Eglinton and M. Murphy, eds.), Springer-Verlag, New York, pp. 181–195.

Rouxhet, P. G., Robin, P. L., and Nicaise, G., 1980, Characterization of kerogens and of their evolution by infrared spectroscopy, in: *Kerogen—Insoluble Organic Matter from Sedimentary Rocks* (B. Durand, ed.), Editions Technip, Paris, pp. 163–190.

Rubinstein, I., Spyckerelle, C., and Strausz, O. P., 1979, Pyrolysis of asphaltenes, *Geochim. Cosmochim. Acta* **43**:1–6.

Saxby, J. D., 1970, Isolation of kerogen in sediments by chemical methods, *Chem. Geol.* **6**:173–184.

Saxby, J. D., 1982, A reassessment of the range of kerogen maturities in which hydrocarbons are generated, *J. Petrol. Geol.* **5**: 117–128.

Schenk, H. J., Witte, E. G., Muller, P. J., and Schowochau, K., 1986, Infrared estimates of aliphatic kerogen carbon in sedimentary rocks, *Org. Geochem.* **10**:1099–1104.

Schoell, M., 1984a, Recent advances in petroleum isotope geochemistry, *Org. Geochem.* **6**:645–663.

Schoell, M., 1984b, Stable isotopes in petroleum research, in: *Advances in Petroleum Geochemistry*, Vol. 1 (J. Brooks and D. Welte, eds.), Academic Press, New York, pp. 215–245.

Schopf, J M., 1948, Variable coalification: The processes involved in coal formation, *Econ. Geol.* **43**:207–225.

Senftle, J. T., Brown, J. H., and Larter, S. R., 1987, Refinement of organic petrographic methods for kerogen characterization, *Int. J. Coal Geology* **7**:105–117.

Silverman, S. R., 1967, Carbon isotopic evidence for the role of lipids in petroleum geochemistry, *J. Am. Oil Chem. Soc.* **44**: 691–695.

Simoneit, B. R. T., Brenner, S., Peters, K. E., and Kaplan, I. R., 1981, Thermal alteration of Cretaceous black shale by diabase intrusions in the Eastern Atlantic—II. Effects of bitumen and kerogen, *Geochim. Cosmochim. Acta* **45**:1581–1602.

Sinninghe Damsté, J. S., Eglington, T. I., Rijpstra, W. I. C., and de Leeuw, J. W., 1990, Characterization of sulfur-rich high molecular weight substances by flash pyrolysis and Raney Ni desulfurisation, in: *Geochemistry of Sulfur in Fossil Fuels* (W. L. Orr and C. M. White, eds.), ACS Symposium Series, American Chemical Society, Washington, D.C., pp. 486–528.

Sitler, J. A., 1979, Effects of source material on vitrinite reflectance, M.S. Thesis, West Virginia University, p. 103.

Smith, P. M. R., 1983, Spectral correlation of spore coloration standards, in: *Petroleum Geochemistry and Exploration in Europe* (J. Brooks, ed.), Blackwell Scientific, London, pp. 289–294.

Smith, P. M. R., 1984, The use of fluorescence microscopy in the characterisation of amorphous organic matter, *Org. Geochem.* **6**:839–845.

Solli, H., Schou, L., Krane, J., Sjketne, T., and Leplat, P., 1985, Characterization of sedimentary organic matter using nuclear magnetic resonance and pyrolysis techniques, in: *Petroleum Geochemistry in Exploration of the Norwegian Shelf* (B. M. Thomas et *al.*, eds.) Graham and Trotman, London, pp. 309–317.

Solomon, P. R., Serio, M. A., Carangelo, R. B., Bassilakis, R., Gravel, D., Baillargeon, M., Baudais, F., and Vail, G., 1990, Analysis of the Argonne Premium Coal samples by TG-FTIR, *Energy Fuels* **4**:319–333.

Solum, M. S., Pugmire, R. J., and Grant, D. M., 1989, ^{13}C solid-state NMR of Argonne Premium Coals, *Energy Fuels* **3**:187–193.

Spiker, E. C., Hatcher, P. G., 1984, Carbon isotope fractionation of sapropelic organic matter during early diagenesis, *Org. Geochem.* **5**:283–290.

Stach, E., Mackowsky, M.-Th., Teichmuller, M., Taylor, G. H., Chandra, D., and Teichmuller, R., 1982, *Stach's Textbook of Coal Petrology*, Gebruder Borntraeger, Berlin.

Staplin, F. L., 1969, Sedimentary organic matter, organic metamorphism, and oil and gas occurrence, *Bull. Can. Petrol. Geol.* **17**: 47–66.

Suzuki, N., 1984, Characteristics of amorphous kerogen fractionated from terrigenous sedimentary rocks, *Geochim. Cosmochim. Acta* **48**:243–249.

Suzuki, N., and Taguchi, K., 1983, Characteristics and diagenesis of kerogens associated with clay fraction of mudstone, in: *Advances in Organic Geochemistry 1981* (M. Bjorøy et *al.*, eds.), John Wiley & Sons, New York, pp. 607–612.

Sweeney, J. J., Burnham, A. K., and Braun, R. L., 1986, A model of hydrocarbon maturation in the Uinta Basin, Utah, U.S.A., in: *Thermal Modeling in Sedimentary Basins* (J. Burrus, ed.), Editions Technip, Paris, pp. 547–561.

Talbot, M. R., 1988, The origins of lacustrine oil source rocks: Evidence from the lakes of tropical Africa, in: *Lacustrine Petroleum Source Rocks* (A. J. Fleet, K. Kelts, and M. R. Talbot, eds.), Geological Society Special Publication No. 40, pp. 29–43.

Tarafa, M. E., Whelan, J. K., and Mountain, G. S., 1987, Sediment slumps in the middle and lower Eocene of Deep Sea Drilling Project Holes 605 and 613: Chemical detection by pyrolysis techniques, in: *Initial Reports of DSDP*, Vol. 95 (C. W. Poag, A. B. Watts et *al.*, eds.), U.S. Government Printing Office, Washington, D. C., pp. 661–669.

Tarafa, M. E., Whelan, J. K., and Farrington, J. W., 1988, Investigation on the effects of organic solvent extraction on whole-rock pyrolysis: Multiple-lobed and symmetrical P_2 peaks, *Org. Geochem.* **12**:137–149.

Taylor, G. H., Liu, S. Y., and Smyth, M., 1988, New light on the origin of Cooper Basin oil, *APEA J.* **28**(Part 1):303–309.

Tegelaar, E. W., de Leeuw, J. W., Dereene, S., and Largeau, C., 1989, A reappraisal of kerogen formation, *Geochim. Cosmochim. Acta* **53**:3103–3106.

Teichmuller, M., 1971, Anwendung kohlenpetrographischer Methoden bei der Erdol und Erdgasprospektion, *Erdöl Kohle, Erdgas, Petrochem.* **24**:69–76.

Teichmuller, M., 1974a, Generation of petroleum-like substances in coal seams as seen under the microscope, in: *Advances in Organic Geochemistry* 1973 (B. Tissot and F. Bienner, eds.), Editions Technip, Paris, pp. 379–407.

Teichmuller, M., 1974b, Uber neue Macerale der Liptinit-Gruppe und die Entstehung von Micrinit, *Fortschr. Geol. Rheinl. Westfalen* **24**:37–74.

Teichmuller, M., 1986, Organic petrology of source rocks, history and state of the art, *Org. Geochem.* **10**:581–599.

Teichmuller, M., and Durand, B., 1983, Fluorescence microscopical rank studies on liptinites and vitrinites in peat and coals, and comparison with results of the Rock-Eval pyrolysis, *Int. J. Coal Geol.* **2**:197–230.

Teichmuller, M., and Ottenjann, K., 1977, Liptinite und lipoide stoffe in einem Erdolmuttergestein, *Erdöl Kohle, Erdgas, Petrochem.* **30:**387–398.

Teichmuller, M., and Teichmuller, R., 1966, Geological causes of coalification, in: ACS Advances in Chemistry Series No. 155, American Chemical Society, Washington, D, C,., pp. 133–155.

Thompson, C. L., and Dembicki, H., Jr., 1986, Optical characteristics of amorphous kerogens and the hydrocarbon-generating potential of source rocks, *Int. J. Coal Geol.* **6:**229–249.

Thompson-Rizer, C. L., 1987, Some optical characteristics of solid bitumen in visual kerogen preparations, *Org. Geochem.* **11:** 385–392.

Thompson-Rizer, C. L., and Woods, R. A., 1987, Microspectrofluorescence measurements of coals and petroleum source rocks, *Int. J. Coal Geol.* **7:**85–104.

Thompson-Rizer, C. L., Woods, R. A., and Ottenjann, K., 1988, Quantitative fluorescence results from sample exchange studies, *Org. Geochem.* **12:**323–332.

Tissot, B. P., 1984, Recent advances in petroleum geochemistry applied to hydrocarbon exploration, *Am. Assoc. Petrol. Geol. Bull.* **68:**545–563.

Tissot, B. P., and Welte, D. H., 1984, *Petroleum Formation and Occurrence*, Springer-Verlag, Berlin.

Tissot, B., Califet-Debyser, Y., Deroo, G., and Oudin, J. L., 1971, Origin and evolution of hydrocarbons in Early Toarcian shales, Paris Basin, France, *Am. Assoc. Petrol. Geol. Bull.* **55:** 2177–2193.

Tissot, B. P., Durand, B., Espitalié, J., and Combaz, A., 1974, Influence of nature and diagenesis of organic matter in formation of petroleum, *Am. Assoc. Petrol. Geol. Bull.* **58:**499–506.

Tissot, B. P., Demaison, G., Masson, P., Delteil, J. R., and Combaz, A., 1980, Paleoenvironment and petroleum potential of middle Cretaceous black shales in Atlantic basins, *Am. Assoc. Petrol. Geol. Bull.* **64:**2051–2063.

Tissot, B. P., Pelet, R., and Ungerer, P. H., 1987, Thermal history of sedimentary basins, maturation indices, and kinetics of oil and gas generation, *Am. Assoc. Petrol. Geol. Bull.* **71:**1445–1466.

Toxopeus, Buiskool, J. M. A., 1983, Selection criteria for the use of vitrininte reflectance as a maturity tool, in: *Petroleum Geochemistry and Exploration in Europe* (J. Brooks, ed.), Blackwell Scientific, London, pp. 295–307.

van de Meent, D., Brown, S. C., Philp, R. P., and Simoneit, B. R. T., 1980, Pyrolysis–high resolution gas chromatography and pyrolysis gas chromatography–mass spectrometry of kerogens and kerogen precursors, *Geochim. Cosmochim. Acta* **44:**999–1013.

Vandenbourke, M., Durand, B., and Oudin, J. L., 1981, Detecting migration phenomena in a geological series by means of C_1–C_{35} hydrocarbon amounts and distributions, in: *Advances in Organic Geochemistry 1981* (M. Bjorøy *et al.*, eds.), Wiley, New York, pp. 147–155.

van Gijzel, P., 1967, Autofluorescence of fossil pollen and spores with special reference to age determination and coalification, *Leidse Geol. Meded.* **40:**263–317.

van Gijzel, P., 1971, Review of the UV-fluorescence microphotometry with fresh and fossil exines and exosporia, in: *Sporopollenin* (J. Brooks *et al.*, eds.), Academic Press, London, pp. 659–682.

van Gijzel, P., 1975, Polychromatic UV-fluorescence microphotometry of fresh and fossil plant substances, with special reference to the location and identification of dispersed organic material in rocks, in: *Pétrographie de la Matière Organique des Sédiments, Relations avec la Paléotempérature et le Potentiel Pétrolier* (B. Alpern, ed.), Editions du Centre National de la Recherche Scientifique, Anatole, pp. 67–91.

van Gijzel, P., 1981, Application of the geomicrophotometry of kerogen, solid hydrocarbons and crude oils to petroleum exploration, in: *Organic Maturation Studies and Fossil Fuel Exploration* (J. Brooks, ed.) Academic Press, London, pp. 351–377.

Van Graas, G., de Leeuw, J. W., Schenck, P. A., and Haverkamp, J., 1981, Kerogen of Toarcian shales of the Paris Basin. A study of its maturation by flash pyrolysis techniques, *Geochim. Cosmochim. Acta* **45:**2465–2474.

van Krevelen, D. W., 1961, *Coal*, Elsevier, Amsterdam.

van Krevelen, D. W., 1984, Organic geochemistry—old and new, *Org. Geochem.* **6:**1–10.

Vassoyevitch, N. B., Korchagina, J. I., Lopatin, N. V., and Tchernitchev, V. V., 1969, Die Hauptphase der Erdolbildung, *Z. Angew. Geol.* **15:**611–622.

Venkatachala, B. S., 1981, Differentiation of amorphous organic matter types in sediments, in: *Organic Maturation Studies and Fossil Fuel Exploration* (J. Brooks, ed.), Academic Press, London, pp. 177–200.

Villey, M., Estrade-Szwarckopf, H., and Conard, J., 1981, Thermal simulation of the diagenetic evolution of kerogens: Electron paramagnetic resonance analysis, *Rev. Inst. Fr. Pet.* **36:**3–16.

von Gaertner, H. R., and Kroepelin, H., 1956, Petrographic and chemical investigations in the Posidonian Shales of Northwestern Germany, *Erdöl. Kohle 1956* **9:**588–592.

Vucelic, D., Juranic, N., and Vitorović, D., 1979, Potential of proton-enhanced ^{13}C NMR for classification of kerogens, *Fuel* **58:** 759–764.

Waples, D. W., 1980, Time and temperature in petroleum formation: Application of Lopatin's method to petroleum exploration, *Am. Assoc. Petrol. Geol. Bull.* **64:**916–926.

Waples, D. W., 1981, *Organic Geochemistry for Exploration Geologists*, Burgess, Minneapolis.

Wedeking, E. W., and Hayes, J. M., 1983, Carbonization of Precambrian kerogens, in: *Advances in Organic Geochemistry* (M. Bjorøy *et al.*, eds.), Wiley, New York, pp. 546–553.

Wen, C.-S., Kwan, J., and Yen, T. F., 1976, Exploratory experiments on pyrite removal from oil shale by an electrolytic process, *Fuel* **55:**75–78.

Wershaw, R. L., and Mikita, M. A., 1987, *NMR of Humic Substances and Coal*, Lewis Publishers, Chelsea, Michigan.

Whelan, J. K., Hunt, J. M., and Huc, A. Y., 1980, Applications of thermal distillation–pyrolysis to petroleum source rock studies and marine pollution, *J. Anal. Appl. Pyrolysis* **2:**79–96.

Whelan, J. K., Hunt, J. M., Jasper, J., and Huc, A., 1984, Migration of C_1–C_8 hydrocarbons in marine sediments, *Org. Geochem.* **6:** 683–694.

Whelan, J. K., Farrington, J. W., and Tarafa, M. E., 1986, Maturity of organic matter and migration of hydrocarbons in two Alaskan North Slope wells, *Org. Geochem.* **10:**207–219.

Whelan, J. K., Hunt, J. M., Hennet, R., Tarafa, M. E., Dickinson, P. A., and Johnson, C., 1988a, Organic geochemistry of continental margin and deep ocean sediments. A progress report to the U.S. Department of Energy, Department of Energy Report No. COO-13466-3, U.S. Government Printing Office, Washington, D.C., pp. 1–11.

Whelan, J. K., Solomon, P. R., Deshpande, G. V., and Carangelo, R. M., 1988b, Thermogravimetric Fourier transform infrared spectroscopy (TG-FTIR) of petroleum source rocks. Initial results, *Energy Fuels* **2:**65–73.

Whelan, J. K., Carangelo, R. M., and Solomon, P. R., 1990a, TG/

plus—A pyrolysis method for following maturation of oil and gas generation zones using Tmax of methane, *Org. Geochem.* **16**:1187–1201.

Whelan, J. K., Kanyo, Z., Tarafa, M., and McCaffrey, M. A., 1990b, Organic matter in Peru upwelling sediments—analysis by pyrolysis, pyrolysis-gas chromatography (PY-GC) and PY-GC mass spectrometry (PY-GCMS), Vol. 112 (M. Kaster, E. Suess *et al.*, eds.), Ocean Drilling Program, College Station, TX, pp. 573–590.

White, D., 1915, Some relations in origin between coal and petroleum, *J. Washington Acad. Sci.* **5**:189–212.

White, D., 1935, Metamorphism of organic sediments and derived oils, *Am. Assoc. Petrol. Geol. Bull.* **18**:589–617.

Williams, L. A., and Reimers, C., 1983, Role of bacterial mats in oxygen-deficient marine basins and coastal upwelling regimes: Preliminary report, *Geology* **11**:267–269.

Wilson, M. A., 1987, *N.M.R. Techniques and Applications in Geochemistry and Soil Chemistry*, Pergamon Press, Oxford.

Wood, D. A., 1988, Relationships between thermal maturity indices calculated using Arrhenius equation and Lopatin method, *Am. Assoc. Petrol. Geol. Bull.* **72**:115–134.

Yen, T. F., Erdman, J. G., and Pollack, S. S., 1961, Investigation of the structure of petroleum asphaltenes by X-ray diffraction, *Anal. Chem.* **33**:1587–1594.

Yukler, M. A., and Kokesh, F., 1984, A review of models used in petroleum resource estimation and organic geochemistry, in: *Advances in Petroleum Geochemistry*, Vol. 1 (J. Brooks and D. Welte, eds.), Academic Press, London, pp. 69–114.

Yurum, Y., Dror, Y., and Levy, M., 1985, Effect of acid dissolution on the mineral matrix and organic matter of Zefa Efe oil shale, *Fuel Process. Technol.* **11**:71–86.

Zumberge, J. E., Sigleo, A. C., and Nagy, B., 1978, Molecular and elemental analyses of the carbonaceous matter in the gold and uranium bearing Vaal Reef carbon seams, Witwatersrand sequence, *Miner. Sci. Eng.* **10**:223–246.

Chapter 15

Organic Petrographic Approach to Kerogen Characterization

JOSEPH T. SENFTLE, CHARLES R. LANDIS, and RENEE L. McLAUGHLIN

1. Introduction

Organic matter in sedimentary sequences ranges from finely disseminated occurrences of kerogen to concentrated organic matter in coals. Kerogen, the organic fraction which is insoluble in organic solvents, comprises up to 90% or more of the organic matter in sedimentary rocks and is 1000 times more abundant than the sum of coal and reservoired oil (Hunt, 1979). Evaluation of the role of kerogen in oil and gas formation is a major concern in oil exploration. The organic facies of a sedimentary section is a composite function of the origin and nature of the organic matter (kerogen type) and its state of thermal evolution (maturity).

The first consideration in evaluating the source potential of a kerogen is to accurately classify the organic material. The traditional geochemical approach classifies kerogen by bulk elemental composition into types I, II, and III (Tissot *et al.*, 1974) and type IV (Harwood, 1977). Although geochemical methods are analytically accurate and objective, kerogen is not a discrete, uniform component but rather a complex, diverse mixture of organic materials which make up a chemical and physical continuum. Further, the different kerogen components undergo diagenetic, catagenetic, and metagenetic changes at different rates. Thus, the maturation process nonuniformly influences the chemical composition of diverse kerogen mixtures. Furthermore, chemical data measures only the present condition of organic matter. Oil source beds at advanced levels of maturation, for example, may exhibit the same chemical characteristics as less mature gas source beds (Dow, 1977). Thus, organic petrography is a means to characterize complex kerogen assemblages simultaneously with respect to organic constituent distribution and thermal maturity.

Visual kerogen type, a first approximation of the hydrocarbon generative capacity, can be determined by means of transmitted, reflected, and fluorescence light microscopy. Visual kerogen analysis (VKA) has the advantage of identifying various components of the kerogen mixture which cannot be distinguished by chemical methods alone. However, the visual method does not provide a direct assessment of kerogen convertibility to hydrocarbons, and the visual data can be misinterpreted, especially when the organic material is finely dispersed. Although both the chemical and physical approaches to kerogen classification have their strengths and weaknesses, the objective chemical and subjective optical methods are complementary. These techniques should be used together to minimize the possibilities of incorrect evaluations of the organic facies and source potential of a sedimentary section.

JOSEPH T. SENFTLE and CHARLES R. LANDIS • ARCO Exploration Research and Technical Services, Plano, Texas 75075. RENEE L. McLAUGHLIN • Core Laboratories, Carrollton, Texas 75006.

Organic Geochemistry, edited by Michael H. Engel and Stephen A. Macko. Plenum Press, New York, 1993.

Thermal evolution of sediments during diagenesis, catagenesis, and metagenesis systematically changes many of the chemical and physical properties of the organic matter. These systematic changes can be used as indicators of thermal maturity. Clearly, integrated geochemical and microscopic analysis of organic matter provides the most accurate assessment of thermal maturity. This is particularly true since the bulk of the geochemical methods are applied to whole-rock and/or kerogen samples and do not resolve discrete component (maceral) compositions. Conversely, microscopic methods contain a subjective element, relying on judgments made by the petrographer. There are several optical methods of maturation determination. Vitrinite reflectance and spore coloration are the most widely applied while liptinite fluorescence and conodont color are also utilized as maturity indicators (especially when well-preserved vitrinite and spore materials are not present).

2. Optical Methods of Kerogen Characterization

2.1. Preparation

Optical characterization of sedimentary organic matter utilizes many of the methods developed in coal petrology. Optical analysis offers several unique advantages, including (1) preserved plant parts and organisms from ancient environments are observed, (2) the sample is preserved during the analysis, (3) sample preparation cost is minimal, and (4) microscopic inspection reveals general fabric relationships not directly obtained from bulk rock chemistry.

Two methods are used routinely to prepare samples for optical examination. The first method is the analysis of whole-rock (WR) samples. Drill core or well cuttings are mounted in epoxy binder, polished with aluminum hydroxide slurries, and examined under incident white and near-UV/visible light. The second method involves the use of concentrated kerogen, which may be observed as a strewn slide in transmitted or reflected and/or fluorescent light. This chapter will emphasize concentrated kerogen microscopy due to its widespread use by contract service and petroleum industry companies, although whole-rock analysis is still commonly used.

Concentrated strewn mounts have been long used for the study of kerogen composition. A common kerogen isolation method involves hydrochloric and hydrofluoric acid extractions, which remove carbonate and silicate mineral phases, respectively. The goal of acid maceration is to remove the mineral phase without significantly altering the organic matter composition. Unfortunately, the resultant acid-insoluble residual often contains pyrite and other minerals, in addition to the organic matter. Heavy-liquid density separation will fractionate pyrite from kerogen, at the expense of associated amorphous organic matter components. Thus, this final removal treatment can alter kerogen compositions of samples.

Other problems associated with kerogen isolation include chemical alteration of the original sample. By-products of acid demineralization are complex neoformed fluorides, mainly CaF_2. These fluorides can interfere with the identification of organic matter. These salts are formed during the hydrofluoric acid maceration step and can be minimized by ensuring that the sample has been rinsed well with distilled water after the hydrochloric acid treatment and brought to a neutral pH. Also, a recent International Committee of Coal Petrology (ICCP) study of an acid-macerated sample revealed that the fluorescence properties of kerogen were altered by the isolation procedure (Baranger *et al.*, 1989), while the reflectance properties of the vitrinite components were not measurably affected (Kalkreuth, 1989). Durand and Nicaise (1980) presented an excellent review of the problems associated with the isolation of kerogen from sediment and rock.

Many problems with kerogen isolation relate to the lack of a standard procedure throughout the industry; the methods employed by contract service laboratories vary significantly. For example, a recent round robin exercise sponsored by The Society for Organic Petrology (TSOP) indicates that significant sample preparation variation exists between laboratories. In this exercise, a sample of the Devonian Woodford Shale, circulated to eight laboratories, was prepared as strewn slide mounts. Examination of these samples showed that the kerogen varied significantly, ranging from finely comminuted amorphous organic matter to large organic and mineral matter clumps. In order to accurately study kerogen using optical methods, care must be taken to prepare concentrates with as little mineral matter present as possible and containing the most representative assemblage of organic matter present within the original sediment.

Traditionally, strewn mounts on glass slides are used in transmitted light to examine concentrated kerogen samples from petroleum source rocks. Kerogen remains translucent through the bulk of the oil window (up to a vitrinite reflectance of 1.3%). However, application of this method can be limited since thermally mature kerogen is opaque in transmitted light due to increased absorption properties produced by the maturation process.

2.2. Classification

Potonie (1908) first distinguished between sapropelic and humic organic matter in sediments. The term sapropelic was used to group lipid-rich materials which were preserved under anoxic conditions in aquatic muds occurring in marine and lacustrine environments. The term humic material describes products of peatification and coalification including land plant matter preserved in swamps in the presence of some oxygen. These terms are still in use but are applied in a wider context than originally defined.

The organic matter in sedimentary rocks, mainly kerogen and to a lesser extent coal, is a heterogeneous assemblage of petrographically unique substances. Using transmitted light microscopy, Staplin (1969) initially classified kerogen into (1) primary materials including cuticle, pollen, spores, wood fragments, charcoal, algae, and phytoplankton, (2) modified materials including sapropelic organic matter and cuticular remains, and (3) thermal-transformation products including pyrobitumen and sooty amorphous organic matter particles. Several workers have expanded and modified this original classification system (Burgess, 1974; Combaz, 1975; Teichmuller and Ottenjann, 1977; Bostick, 1979; Alpern, 1980; van Gijzel, 1982; Massoud and Kinghorn, 1985; Mukhopadhyay *et al.*, 1985; Senftle *et al.*, 1987). These additional classification schemes utilize not only transmitted light but also incident white-light and fluorescence light microscopy methods (Fig. 1).

The details of the above kerogen classifications will not be expanded upon here; rather, an overview of the common critical elements and the importance of sample type upon the classification of the organic matter will be discussed. [Two main categories of fossil organic material are generally distinguished in many laboratories, structured and unstructured (amorphous) debris.] This general review and description of kerogen components is presented solely as a guide to understanding the complex assemblages of organic matter observed in sedimentary kerogen assemblages.

2.2.1. Structured Organic Matter

The petrographic nomenclature which is governed by the ICCP defines three maceral (microscopically identifiable constituents in coal) groups that occur in varying amounts in most coals and sedimentary rocks and commonly exhibit morphological integrity. Structured organic matter includes vitrinite (wood remains), inertinite (carbonized remains), and liptinites (spores, pollen grains, cuticles, and algae), in addition to a suite of faunal remains.

Vitrinite Group. Vitrinite is preserved woody tissue. The term vitrinite was originally used by coal petrographers to describe a maceral group observed in coal (ICCP 1963, 1971, 1975). These materials are believed to be derived predominantly from lignified primary xylem of woody and cortical tissues. Wood remains are observed within kerogen with transmitted light as translucent to opaque, blocky material and as dark to light gray matter using reflected light microscopy (Fig. 2a–c). Cell structures are often preserved in vitrinites and can distinguish these macerals. These cell structures include fossilized tracheid, fiber-tracheid, and vessel structural plant elements. In general, tracheid elements are common in conifers, other gymnosperms, and ferns. They are generally thick-walled with imperforated cells of fairly uniform size that are tapered at each end. In contrast, angiosperm wood is characterized by fiber-tracheid and vessel elements.

Subdivision of vitrinites is based upon visual structure (Figs. 1 and 2b). Telinite refers to vitrinite with defined cell structure; collinite is vitrinite derived from solidified humic gels, often colloidal with no discernible cell structure. Telocollinite is composed of collinite intermixed with telinite (ICCP, 1975). The term vitrodetrinite is applied to describe vitrinite of fine particle size (<10 μm in diameter) (ICCP, 1971, 1975). Gutjahr (1983) has further distinguished the vitrinite macerals based upon fluorescence properties. His vitrinite-1 is nonfluorescing vitrinite, and vitrinite-2 fluoresces weakly. He suggested that vitrinite-2 is a hydrogen-enriched variety of vitrinite due to finely disseminated liptinites in the vitrinite emplaced by syndepositional bacterial transformation of organic matter or by hydrocarbon impregnation. Price and Barker (1985) summarized control of these processes on vitrinite reflectivity and fluorescence. Recent work by Taylor and Liu (1987) supports the inclusion of submicroscopic liptinite constituents within vitrinite-2.

Vitrinite is known to occur as thick layers to small discrete fragments within coals and other sedimentary rocks. Degradation often occurs, whereby these materials are broken down by biological processes such as fungal and bacterial attack, chemical degradation including oxidation, and mechanical

TYPICAL SCHEMES FOR ORGANIC PETROGRAPHIC CHARACTERIZATION OF SEDIMENTARY ORGANIC MATTER

Reflected Light	***Transmitted Light***	***Multimode Illumination***
huminite/vitrinite telinite collinite vitrodetrinite	woody	vitrinite
inertinite fusinite semifusinite sclerotinite macrinite inertodetrinite micrinite	coaly	inertinite
liptinite sporinite cutinite suberinite fluorinite resinite chlorophyllinite liptodeterinite	herbaceous	liptinite
alginite	algal	
bituminite exsudatinite oil expulsions	amorphous	amorphinite
faunal relics		faunal relics
mineral-bituminous groundmass	amorphous	bituminous mineral groundmass sold bitumen
Teichmuller and Ottenjann (1977)	Burgess (1974)	Modified from Van Gijzel (1979), Senftle et al., (1987)

Figure 1. Compilation of kerogen classification schemes based upon sample type and illumination mode.

processes associated with sedimentation which fragment and rework the vitrinite. Degradation results generally in the deterioration of structured appearance, development of ragged, indistinct, and discolored edges, and/or the development of a granular or pitted appearance. Successive, transitional stages of degradation can yield partially altered woody tissue to highly altered and amorphous kerogen, producing changes in chemical composition as well (Tissot and Pelet, 1971).

Inertinite Group. The inertinite maceral group consists of organic material that has been altered by rapid oxidation, moldering, and biological attack. This term is used to describe remains of structured debris, predominantly woody tissues and fungal bodies but also subordinate amounts of carbonized spores, algae, cuticles, and resins. Inertinite is thought to be produced during forest fires in the ancient peat swamp. These materials are relatively hydrogen poor and chemically inert. In general, inertinite is the result of a "fusinization" process which produces altered debris with relatively enriched carbon contents and low hydrogen contents via aromatization and condensation processes. Chemical processes, including dehydration and oxidation in addition to fungal attack, also contribute to the fusinization process.

Inertinite can be classified into subgroups (including fusinite, semifusinite, macrinite, sclerotinite, and inertodetrinite) based upon preservation of cell structure, intensity of carbonization, derivation, and

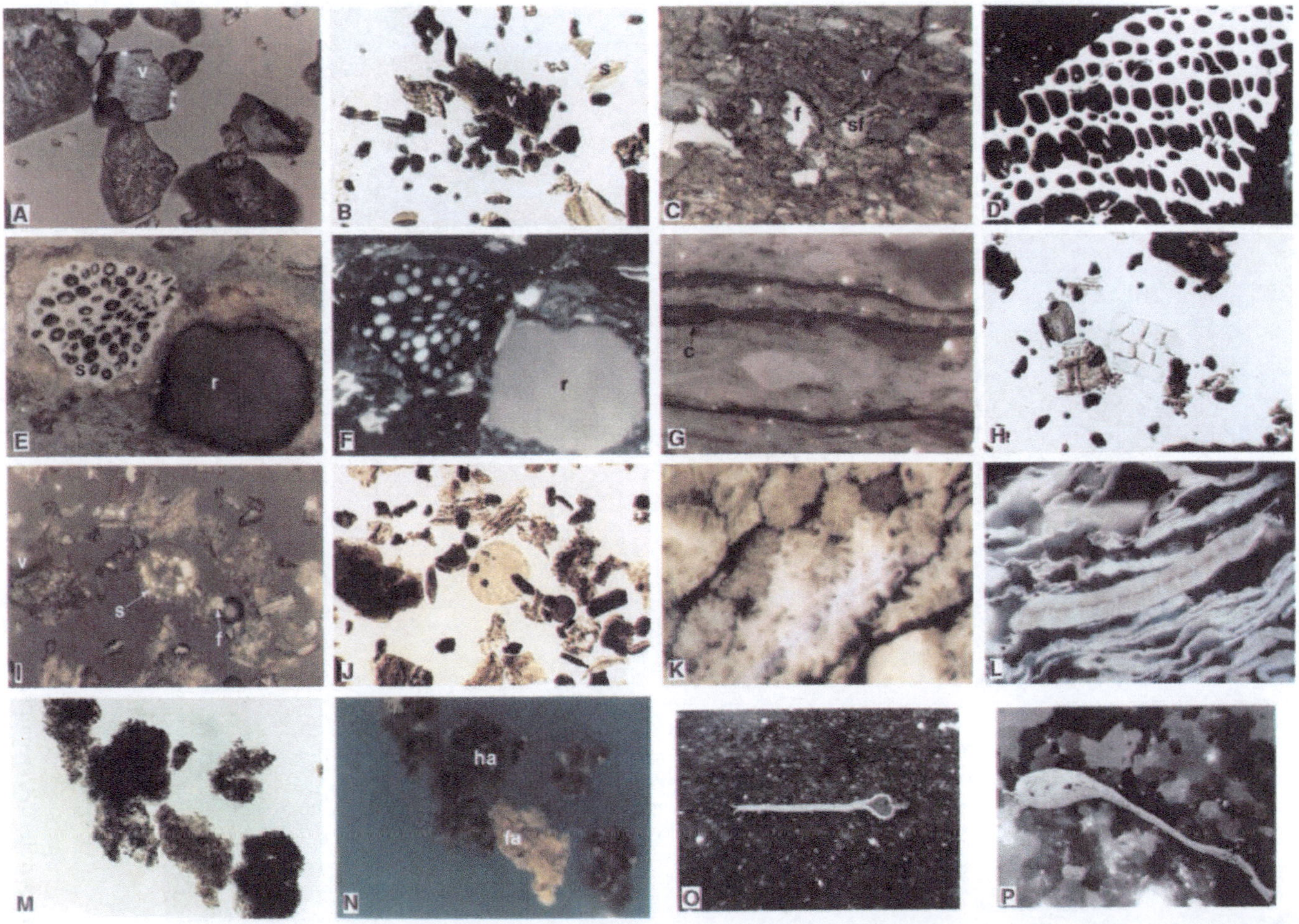

Figure 2. Photomicrographs of petrographic constituents of kerogen. (A) Concentrated kerogen slide mount of (collinite) vitrinite from Tertiary shale in reflected white light exhibiting characteristic gray color (field length = 200 μm). (B) Concentrated kerogen slide mount of vitrinite (v) from Tertiary shale in transmitted white light exhibiting reddish-brown color (field length = 200 μm). (C) Crushed-particle mount (in Pennsylvanian coal) of fusinite (f) and semifusinite (sf) in reflected white light (> 20 mesh). Note the reflectivity and morphological heterogeneity characteristic of the inertinites (field length = 200 μm). (D) Crushed-particle mount (<20 mesh) of Pennsylvanian coal illustrating characteristic cellular structure of fusinite in reflected white light (field length = 200 μm). (E) Crushed-particle mount of Tertiary coal illustrating sclerotinite (s) and resinite in reflected white light (field length = 200 μm). (F) Same field as in (E) in reflected blue light illustrating exsudatinite in cell lumens of sclerotinite and moderately intense, yellowish-orange resinite fluorescence (field length = 200 μm). (G) Crushed-particle mount (>20 mesh) illustrating cutinite in reflected white light as dark linear features in desmocollinite matrix (field length = 200 μm). (H) Concentrated kerogen slide mount from Tertiary (?) shale illustrating cutinite in plan view in transmitted white light (field length = 200 μm). (I) Concentrated kerogen slide mount from Tertiary (?) coal of sporinite in reflected white light exhibiting prominent internal reflections (s). Note also associated fusinite (f) and vitrinite (v) (field length = 200 μm). (J) Same field as in (I) in transmitted white light illustrating yellowish-brown sporinite in plan view (field length = 200 μm). (K) Whole-rock mount of *Botryoccoccus* alginite of torbanite exhibiting strong yellow fluorescence in reflected blue light (field length = 200 μm). (L) Whole-rock mount of *Tasmanites* alginite in reflected blue light exhibiting strong fluorescence intensity from Quamby mudstone in Tasmania, Australia (field length = 200 μm). (M) Concentrated kerogen slide mount of amorphous kerogen in transmitted white light (field length = 200 μm). (N) Same field as in (M) in reflected blue light of hebamorphinite (nonfluorescent) and fluoramorphinite (fluorescent) varieties of amorphous kerogen, ha and fa, respectively (field length = 200 μm). (O) Chitinozoan showing long oral tube and small chamber, middle Devonian (field length = 250 μm) (courtesy of F. Goodarzi). (P) Graptolite fragment, argillaceous limestone, early Early Silurian (field length = 250 μm) (courtesy of F. Goodarzi).

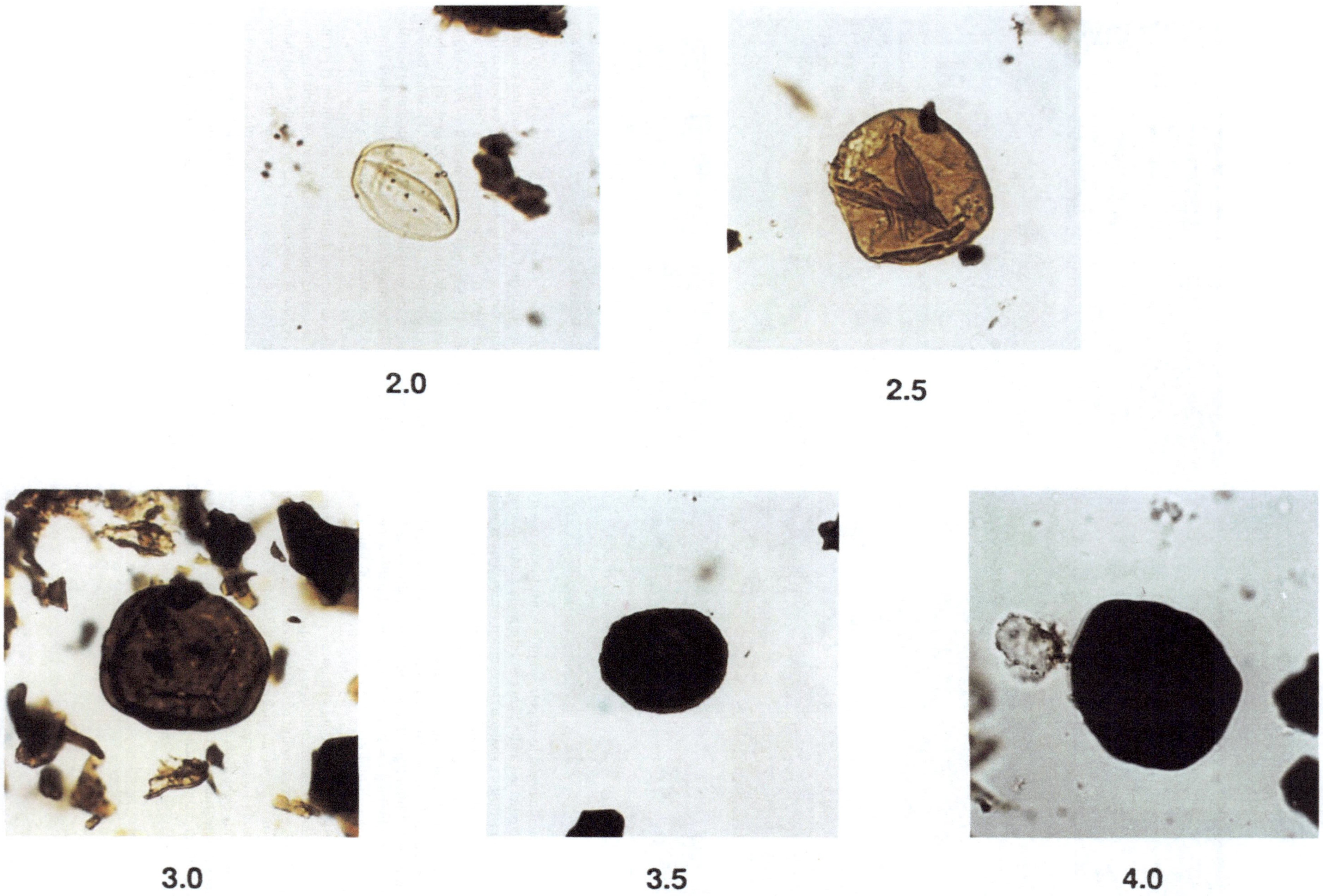

Figure 8. Thermal alteration index (TAI) for a series of sporinite in transmitted white light, exhibiting the shift from translucent yellow to opaque with increasing TAI.

size (Figs. 2c,d,e,i). Stach *et al.* (1982) reviewed the origin of these materials in detail. In transmitted light, these constituents occur as solid opaque, angular fragments or large black angular fragments with open cell lumens. In reflected light, inertinite material is highly reflective gray to white in color, and structural elements can be observed. Fusinite and semifusinite often exhibit some degree of cell structure. Fusinite has undergone a higher degree of carbonization and correspondingly is more conspicuous in reflected light as a white to yellow-white maceral. Semifusinite, in contrast, represents partially or incompletely altered plant tissue that exhibits reflectance and morphological properties intermediate to those of fusinite and vitrinite. Numerous studies have indicated that a fraction of the semifusinite maceral is not completely inert (Nandi and Montgomery, 1975; Nandi *et al.*, 1977; Steyn and Smith, 1977; Chadhuri and Ghose, 1978). Sclerotinite is derived from fungal mycelia. These dark pigmented (melanin) fungal remains, including sclertia, hyphae, and spores, exhibit characteristic structure and have a strong reflectance (Fig. 2e). Finally, physically degraded and redeposited debris of fusinite, semifusinite, and sclerotinite, often angular and less than 30 μm in diameter, is referred to as inertodetrinite. The distinction among inertinites can be difficult to make using transmitted light since it is difficult to distinguish between thick organic debris particles, pyrite, and opaque organic debris.

Liptinite Group. The liptinite group is derived from leaf cuticle, spores, pollen, plant waxes, fats, oils, and resins. Major liptinite maceral subgroups include sporinite, cutinite, resinite, alginite, and subernite. Chemically, the liptinite group is a suite of aliphatic, hydrogen-rich macerals. These materials characteristically exhibit a relatively low reflectance compared to that of vitrinite and inertinite (Fig. 2e–l). In contrast to vitrinite and inertinite debris, they are more translucent and often fluoresce upon excitation by UV/blue light, commonly up to a maturity level equivalent to a vitrinite reflectance of 1.1%.

Sporinite is the term used to classify the resistant outer walls of pollen and spores. Pollen and spores are part of the reproductive system of terrestrial plants. Spores are produced by lower vascular plants (i.e., psilopsids, lycopsids, sphenopsids, and ferns) as well as the nonvascular bryophytes (mosses and liverworts). The outer walls of spores are commonly resistant to degradation and can exhibit ornamentation which permits characterization on the basis of morphology. The occurrence of spores ranges from Silurian to Recent. Since the parent plants may be restricted to particular geologic times, spores represent useful stratigraphic markers (i.e., distinction of Carboniferous and Mesozoic vegetation). However, they may be shed from plants in large numbers and transported by wind into nonterrestrial depositional systems.

Pollen grains are derived from immature male gametophytes that are shed from either the pollen cones of gymnosperms or the anthers of angiosperms. Gymnosperms range from Carboniferous to Recent whereas angiosperms, which evolved in the late Neocomian, are found from the Albian to Recent. The outer walls of pollen have structures similar to those observed in spores and are also valuable indicators of the original plant community. The color of sporinite in transmitted light changes progressively from a translucent light yellow-green in immature sediments, to medium yellow-brown in mature sedimentary rocks, to dark brown to black in overmature sequences. This systematic change in spore color is the basis of a thermal indicator technique (Staplin, 1969; Bayliss, 1980) which will be discussed later in this chapter.

Cutinite is a liptinite maceral derived from the cuticular layers of some terrestrial plants ranging in age from the Silurian to Recent, although not generally found in abundance before the Devonian. These macerals represent the waxy protective coatings on the outer cell walls of leaves and other exposed nonwoody parts of plants such as stems, needles, stalks, or shoots. The cuticular layers protect land plants from desiccation. These layers are generally composed of hydrogen-rich pure cutin, saturated hydroxy-fatty acids, and wax alcohols.

The cuticular layers typically form homogeneous, translucent laminated sheets which express the interface between the outer cell wall and the coating. A cerated edge often then characterizes cutinite. These cell structures range from irregular polygonal to nearly square to generally rectangular with either straight or undulatory walls. Cutinite associated with the woody portion of leaves tends to exhibit polygonal-shaped cell wall structure (Fig. 2h). In contrast, cutinite derived from elongated parts of plants such as stems, stalks, or ribs tends to exhibit elongated cell structures. Much like that of sporinite, the color of cutinite in transmitted light changes progressively from a transparent light yellow-green in immature sediments, to medium yellow to yellow-brown to brown in mature sediments, to very dark brown to black in overmature sedimentary rocks.

Resinite is derived from plant resins and secretions such as waxes, oils, and fats. Since resins are derived from a number of different sources, they can consist of a variety of substances including terpenes, phenols, alcohols, and acids. Their exact composition and mode of occurrence is dependent upon the type of plant from which they are derived and their mode of deposition (e.g., cell filling, exudate, etc.). Resinite often occurs as cell fillings (Fig. 2e,f), appearing in strewn slide preparations as rounded to oval-shaped bodies or as elongated or stringy blebs within vitrinite. They are yellowish orange to amber color in transmitted light and gray in reflected light. Due to their origin, resinites exhibit a variety of fluorescence colors and alteration properties (Teichmuller and Durand, 1983; Senftle *et al.*, 1987). Consequently, the optical properties of resinites are not a good indicator of thermal maturity.

Alginite macerals are fossil algae. Generally, alginites are chemically notable by their high content of fatty and proteinaceous substances (Stach *et al.*, 1982). In recent algae, normal fatty acids (n-C_{16} and n-C_{18}) are abundant. Similar abundances of corresponding olefinic acids are often observed in alginite, while acids of odd carbon numbers are rarely found (Stach *et al.*, 1982).

Alginite represents preserved chlorophytes (green algae) of predominantly marine environments, fresh and brackish water chlorophytes, dinoflagellates, and acritarchs. Alginite derived from chlorophytes ranging in age from Precambrian to Recent can be relatively abundant in sediments. Alginite may be either colonial or unicellular. *Botryococcus* is a colonial green alga which occurs today in freshwater lakes or in brackish to saline pools and lagoons (Fig. 2k). These algae secrete oil from their cell membranes. In areas such as Lake Balkhash in Kazakhstan and lakes in the Coorong region of southern Australia, *Botryococcus braunii* is so abundant that combustible sediments form (Stach *et al.*, 1982). Marine unicellular *Tasmanites* is considered to be a fossil Leiosphere (cyst) derived from planktonic, marine algae bearing an affinity to the present-day spherical alga *Packysphaera peolgica* (Cane, 1968). *Tasmanites* was abundant during the Silurian, Devonian, Permian, and Jurassic periods and was generally deposited in shallow marine depositional environments (Cook et *al.*, 1981). Among the unicellular Tasmanales alginite, *Tasmanites* has been identified in North American black shales stratigraphically equivalent to the Woodford Shale of the Permian Basin (Landis *et al.*, 1992). *Nostocopsis* is a small, spherical unicellular alga belonging to Schizophyceae. This alga has been documented in Liassic source rocks (Madler, 1964). Finally, *Pediastrum* is a relatively abundant colonial green alga in certain Cretaceous and Tertiary units (Hutton, 1987). This alga is common among plankton in recent freshwater lakes and ponds (Cookson, 1953) and has been reported in brackish environments of deposition.

In blue light, *Tasmanites* exhibits a strong fluorescence, particularly in immature samples (Fig. 2l). Commonly, the fluorescence properties of alginite permit the observation of these macerals in strewn slide mounts. Unlike sporinite, alginite macerals do not commonly exhibit a systematic and unique change in color with maturation, and thus are less useful as indicators of thermal alteration. Alginites generally are fluorescent up to a maturity level equivalent to a vitrinite reflectance of 1.1%.

Two morphological types of alginite are readily identified (Hutton *et al.*, 1980; Sherwood *et al.*, 1984; Cook, 1987) in whole-rock preparations. Telalginite describes discrete, well-preserved oval or ellipsoidal algal bodies. Lamalginite is defined as thin, crenulated, "filamentous" bodies that accommodate lithic fabric along the extent of the maceral. Both have been identified in oil shales (Sherwood *et al.*, 1984; Cook, 1987).

Suberinite is a maceral which has been observed in Tertiary sediments and in a few Mesozoic coals. This maceral is derived from suberin which is found in corkified cell walls occurring mainly in barks and is considered to be chemically similar to cutin but less polymerized (Stach *et al.*, 1982). The occurrence of suberinite depends to some extent on the orientation of the section, ranging in appearance from sheets with irregular polygonal four- to six-sided cells to laminar masses. The color ranges from yellow to brown depending upon the thickness and degree of maturation. Suberinite fluorescence is yellow at low levels of maturity and orange at higher levels of maturity.

Dinoflagellates occur as cysts which represent one stage in the life cycle of the planktonic unicellular algae. These cysts are composed of sporopollenin, complex polymers resulting from oxidative polymerization of carotenoids or carotenoid esters (Brooks and Shaw, 1973). Occurrence of dinoflagellates ranges from Silurian to Recent, although their period of maximum diversity was in the Mesozoic. Acritarchs are small cysts known from the Cambrian to Recent which lack paratabulation and/or archeopyle characteristics of the otherwise morphologically similar dinoflagellates (Evitt, 1963). Acritarchs are abundant especially in Paleozoic rocks but are much more limited in the Mesozoic. Their cell wall composition is organic, likely sporopollenin.

Faunal Relics. In addition to the three maceral groups discussed above, kerogen samples from many sedimentary rocks contain microfossils from both animal and plant remains which differ in composition from the other kerogen components. This group contains scolecodonts, chitinozoans, conodonts, and graptolites. Scolecodonts are the organic fossil remains of worm jaw elements of eucinid polychaete annelid worms. They occur most commonly as lamellar to tooth-like structures, in Ordovician to Recent marine rocks. They are particularly abundant in Paleozoic rocks. Spectrographic and X-ray diffraction analysis of Ordovician and Devonian scolecodonts indicate that they are composed of fluorapatite with minor amounts of calcium, copper, silica, and magnesium (Schwab, 1966). Goodarzi *et al.* (1985) and Goodarzi and Higgins (1987) have discussed the morphology of scolecodonts and their optical properties. They reported that the reflectance of scolecodonts systematically increases with rank and thus can be used to assess thermal maturity of sediments.

Chitinozoans comprise a group of vase-shaped tests of uncertain protozoan affinity, ranging from amber to dark brown or black in color in transmitted light, depending on maturity. They may be either ornamented or unornamented with a wide variety of spines. Chitinozoans often do not transmit light well due to their thick walls (Traverse, 1974). In reflected light, they are highly reflective gray to white in color (Fig. 2o). The composition of their walls is similar to that of pseudochitin (Tschudy and Scott, 1969). Chitinizoans are known to range from Ordovician to Devonian (Jansonius and Jenkins, 1978).

Conodonts are phosphatic marine microfossils, commonly 0.1 to 1 mm in size, that range in age from Cambrian to the Triassic. Conodonts are composed predominantly of apatite (Pietzner *et al.*, 1968) with subordinate amounts of organic matter. They are most common in carbonates but also occur in calcareous and dolomitic shale. Conodont morphology is variable, but often well-defined denticles and blades are preserved. In transmitted light, conodonts range from pale yellow to light brown to dark brown to black (Epstein *et al.*, 1977). In blue light, conodonts which have experienced minor thermal alteration exhibit a yellow to orange fluorescence (Barrick *et al.*, 1990). Commonly, denticles exhibit a yellow to orange fluorescence while blades are characteristically orange. Mastalerz *et al.* (1992) reported a loss of fluorescence intensity as CAI increases to 3.

Graptolites occur mainly in Lower Paleozoic rocks as thin, elongated bodies, showing complex skeletal morphology, and are highly reflective gray to white in color (Fig. 2p). Graptolites are composed of pseudochitin and thus are similar chemically to chitinozoans. Since pre-Silurian kerogen consists of mainly aquatic constituents, graptolites have been used as a means to evaluate thermal maturity in lieu of vitrinite. These studies (Clausen and Teichmuller, 1982; Kurylowicz *et al.*, 1976; Goodarzi, 1982, 1984, 1985) have shown that the morphology and reflectance of graptolite fragments can be compared to vitrinite reflectance in younger rocks.

2.2.2. Unstructured Organic Matter

Kerogen from many sedimentary rocks contains a significant portion of organic matter which does not exhibit morphological integrity. These materials include a broad range of constituents which can have both primary and secondary origins. Welte (1974) has summarized the origins of these materials as random condensation and polymerization products and other monomeric materials such as lipids.

The formation and preservation of amorphous kerogen is considered to be a complex process that involves the degradation of organic matter under a combination of aerobic and anaerobic conditions. A primary source of aquatic organic matter is phytoplankton (Bordovsky, 1965) and, to a lesser extent, zooplankton and fecal material from larger invertebrates. Terrestrial organic matter deposited in aquatic settings includes woody material, pollen, spores, and plant cuticles. Together, these materials can decompose due mainly to aerobic bacterial activity in oxygen-rich water columns, in interstitial sediment waters, and, to a lesser extent, in the digestive tracts of scavengers. If left unimpeded, oxidation will proceed to completion and essentially convert the material into carbon dioxide and water. Anaerobic conditions, however, can arrest the oxidation process due to nitrate-reducing and sulfate-reducing bacteria. When the oxygen concentration falls well below 0.1 ml liter^{-1}, benthic metazoans (bioturbation) are eliminated. At this point, bacterial sulfate reduction is slowed or ceases, thereby increasing the probability that the amorphous and other organic matter will be preserved (Demaison and Moore, 1980).

There have been many authors who have included amorphous organic matter in classification systems of sedimentary organic matter (Combaz, 1964, 1980; Staplin, 1969; Burgess, 1974; Robert, 1979, 1980; Alpern, 1980; Mukhopadhyay *et al.*, 1985; Thompson-Rizer and Dembicki, 1986; Senftle *et al.*, 1987). Due to the complex origins of these materials, the chemical composition of amorphous organic material varies considerably, ranging from hydrogen-rich to hydrogen-poor material. Powell *et al.* (1982) related bulk chemical parameters and the occurrence of unstructured organic matters in sediments. They noted that some kerogens rich in amorphous organic material exhibited high H/C ratios while others did not. They were unable to distinguish hydrogen-rich types microscopically. Senftle *et al.* (1987) discussed a refinement of the classification of unstructured organic matter. Up to a maturation level equivalent to a vitrinite reflectance level of 1.1%, they were able to distinguish fluorescent (fluoramorphinite) from nonfluorescent (hebamorphinite) amorphous kerogen (Fig. 2m,n). These authors observed a poor correlation of atomic H/C ratio with unstructured organic matter content, as well as a poor correlation of atomic H/C ratio with liptinite and with amorphous kerogen types. They did however correlate atomic H/C ratio with total liptinite content plus fluoramorphinite content (%AMEX). They stated that the %AMEX parameter can be used as an estimate of the hydrogen content of the kerogen assemblage and thus could be used as a rough estimate of oil/gas potential.

Thompson-Rizer and Dembicki (1986) have differentiated amorphous kerogen types based upon petrographic texture. They reported that four types of amorphous kerogen occur in sediments: chunky compact masses with mottled network or weak polygonal textures (type A); very small, dense, elongated, oval, or rounded individual grains (type B); clumps with granular, fragmented, or globular textures (type C); and thin platy or rectangular individual grains (type D). Pyrolysis, infrared, and atomic-ratio data correlate with their optically defined types.

The petrographic characterization of amorphous organic material is important for many reasons. In addition to correlation of visual and chemical kerogen types, recognition of various degradational states of amorphous organic matter can contribute to biostratigraphic studies in conjunction with palynomorphs. Amorphous organic matter also shows promise as a tool for paleoenvironmental evaluation. Finally, some workers routinely make source rock assessments based on the presence or absence of particular amorphous kerogen matter. Masran and Pocock (1981) related not only the varieties of amorphous matter but also their colors to hydrocarbon source potential.

3. Optical Methods to Evaluate Thermal Maturity

In addition to the measure of kerogen quality and quantity, the degree of thermal maturation of organic matter is essential for the evaluation of kerogen source potential. Petrographic methods such as vitrinite reflectance, spore coloration, fluorescence color, and conodont coloration are the most common petrographic means of assessing thermal maturity. Any of these methods should be integrated with geochemical data due to their analytical subjectivity and the complexity of most source rocks. Despite the limitations of these approaches, petrographic evaluation of organic maturity is a fundamental method routinely applied throughout the petroleum and coal industries.

3.1. Vitrinite Reflectance

Thermal maturation of kerogen and coalification are somewhat analogous terms which describe the stage of chemical and physical changes of organic matter as it is altered during burial. Accompanying compaction and dewatering, organic matter undergoes thermal alteration. This irreversible alteration includes a gradual increase in carbon content, a corresponding reduction in hydrogen, and the decomposition of various chemical compounds. The stage of coalification, ranging from peat and passing from lignite to bituminous coal and then to anthracite, has traditionally been evaluated by coal petrographers using the reflectance properties of the vitrinite maceral. Coal rank has been evaluated in the coal industry using the mean maximum vitrinite reflectance while average vitrinite reflectance is commonly used to evaluate shales and other sedimentary rocks. The use of vitrinite reflectance as an indicator of organic matter thermal maturity is now widespread in the petroleum industry for the evaluation of kerogen and dispersed organic matter in sedimentary rocks (Castano and Sparks, 1974).

Reflectance is defined as the proportion of normally incident light reflected by a polished surface. The measurement of vitrinite reflectance is expressed as the percentage of the incident light that is reflected from the vitrinite. Typically, it is only a small fraction of the incident light, less than 4% in most cases. The complete procedure for reflectance analysis of coals is outlined by the American Society for Testing and Materials (1976). As outlined in this method description, a stabilized light source, oil immersion of known refractive index, and a microscope photometer are used to compare the reflected intensity of macerals to that of a glass or synthetic mineral standard. Reflective analysis can be performed on any of the macerals, but vitrinite has been used as the standard maceral for comparison. Vitrinite is best suited for this purpose because it is often the predominant maceral in terrestrially derived sediments and occurs in minor amounts in various sediments. It often occurs as a homogeneous constituent, and the particle size is large enough to permit measurment. As can be seen in Fig. 3, vitrinite can be used to assess a broad range of thermal maturity, ranging from early diagenesis through catagenesis to low grades of metamorphism. This range of thermal maturity includes the sequence of generation, preservation, and destruction of hydrocarbons in rocks. Consequently, the use of vitrinite reflectance has gained widespread usage by coal petrographers in the coal and steel industry and by kerogen petrographers in the petroleum industry.

While vitrinite reflectance has the advantage of providing a thermal indicator for rocks which have experienced a wide range of maturation levels, there are several technical limitations associated with this method. Early work described by Stach *et al.* (1975) suggested that the reflectance of vitrinite would provide a maturity indicator which was not sensitive to variations in depositional processes. Later work by several authors (Jones and Edison, 1978; Bostick, 1979; Price and Barker, 1985) has shown that this

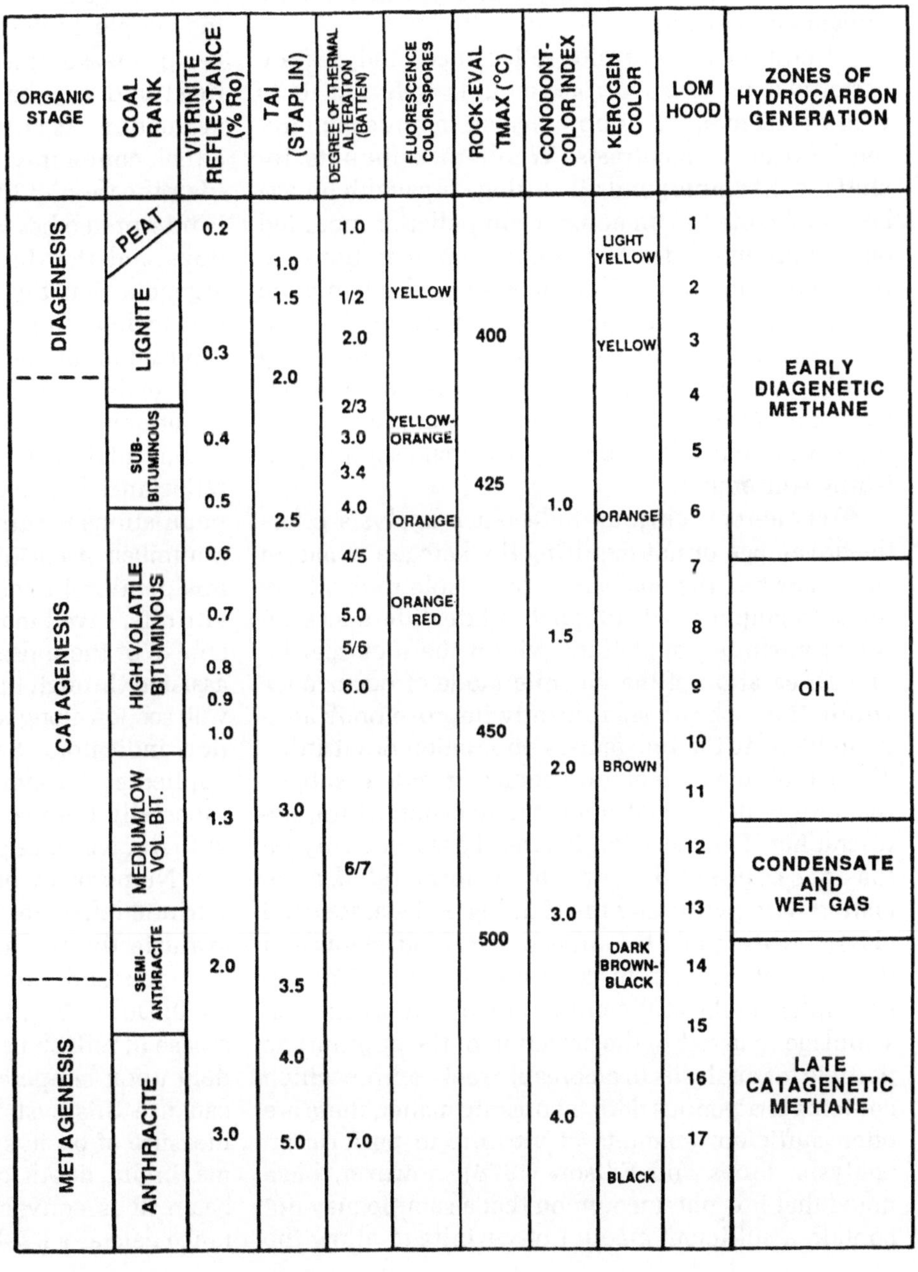

Figure 3. Correlation of vitrinite reflectance (% R_0) to indicators of thermal alteration in sediments. [Reprinted by permission from Senftle and Landis (1991).]

assumption does not always apply; thus, the identification of vitrinite in shale and other sedimentary rocks needs to be made cautiously. Further, sampling methods can influence the quality of the analysis. Specifically, oxidation of outcrop samples or alternation of ditch cutting samples through the use of high-temperature drying ovens has been observed to significantly alter the reflectance properties of vitrinite. Finally, the use of vitrinite reflectance is limited by the occurrence of vitrinite in rock. It is not uncommon for the petrographer to encounter fragments of semi-inertinite macerals, reworked vitrinite (allochthonous), or solid hydrocarbons which can be mistakenly identified as vitrinite. Thus, care and experience are needed to correctly identify and measure indigenous vitrinite.

Traditionally, vitrinite reflectance analyses are performed on crushed rock mounts (referred to as whole-rock mounts) or on polished mounts containing kerogen concentrates. Whole-rock mounts include rock samples or drill cutting chips which have been embedded in an epoxy resin pellet or mounted on microscope slides. These sample preparations are polished using conventional methods, which include the use of silicon carbide or aluminum oxide powders. Kerogen mounts are commonly made using smaller mounts which can be embedded within conventional sized whole-rock mount sample holders, thereby allowing the use of conventional sample polishing equipment.

Whole-rock vitrinite reflectance analysis offers the advantage of not requiring the kerogen isolation steps. Further, the analysis of the whole rock allows the petrographer to distinguish subtle differences between vitrinite populations within the rock matrix. The observation of the vitrinite mode of occurrence within the rock can significantly improve one's ability to identify the indigenous population of vitrinite. The actual selection of indigenous vitrinite is subjective and is dependent upon the decision of the petrographer. This decision is often based not only on the appearance of the vitrinite, its form, relief, association with pyrite, and the staining of the associated clays, but also upon the appearance of sediments and the color of bitumen in the mineral matrix. The petrographer looks at the entire organic–inorganic assemblage to assist in the selection of the appropriate particle for analysis. In an organic rock sample which contains terrigenous derived organic matter, there are often sufficient amounts of vitrinite to perform an analysis. Jones and Edison (1978), however, have noted that it is not uncommon that a sample may not contain a sufficient amount of vitrinite to allow the determination of a statistically significant average of the measurements. It is a general rule that approximately 20 measurements from indigenous vitrinite are needed to evaluate a sample. It is useful for the petrographer in all cases to report the number of particles analyzed along with the standard deviation and average vitrinite reflectance value. Clearly, any description of the mode of occurrence of the vitrinite should be noted, thereby giving the interpreter some measure to evaluate the criteria used to identify the indigenous vitrinite population.

Kerogen concentrate preparations are widely used within the petroleum industry. The use of this type of preparation allows the petrographer to rapidly view a large portion of the organic matter occurring in the sample. In the course of this analysis, a large number of measurements are often taken (50–100 readings) and can be summarized into histograms for evaluation. As noted, the organic matter derived from a rock commonly occurs as a complex mixture. The identification of the indigenous vitrinite population is interpreted based upon the petrographer's observations and the distribution of reflectance data. It is arguable that the use of an additional interpretive step to determine the indigenous vitrinite population can lead to different results when compared to whole-rock analyses. This problem is further complicated by variations in kerogen preparation procedures used among analytical labs. Figure 4 presents examples of reflectance histograms that can be observed. Multiple populations within the reflectance data can be due to a number of reasons, including the measurement of misidentified constituents, multiple populations of vitrinite, caved material, and reworked material. The choice of the indigenous vitrinite population can be assisted through the study of samples throughout the well section along with the use of other thermal maturation indicators. Several of these indicators can be applied at the same time as the reflectance analysis, especially the use of spore and kerogen color along with fluorescence properties of the liptinite macerals.

Numerous workers have documented the use of vitrinite reflectance versus depth plots as a means to evaluate the thermal history of the section studied (Castano and Sparks, 1974; Bostick, 1979; Robert, 1980; Dow, 1977; Dow and O'Connor, 1982). The increase in reflectance of vitrinite is systematic, dependent upon temperature and time. Consequently, one can use this systematic change in reflectance as a measure of basin subsidence and the corresponding maximum depth of burial and heat flow within a basin. It is conventional to present vitrinite reflectance data as a log-linear plot, thereby permitting the

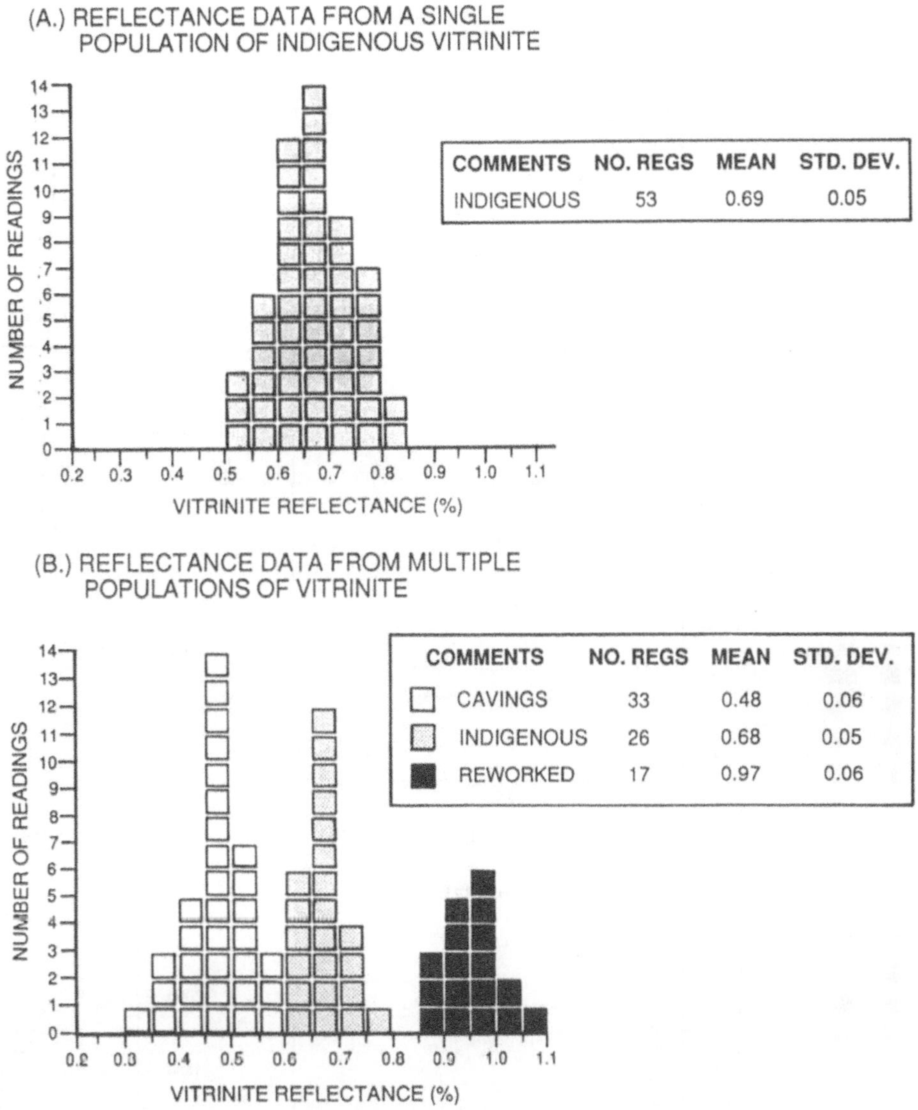

Figure 4. Vitrinite reflectance histograms indicating a normal distribution of data (A) and relative reflectances of indigenous, reworked, and caved vitrinite (B). [Reprinted by permission from Senftle and Landis (1991).]

geologist to observe changes in the reflectance profile with increasing depth over a wide range of maturities. The shape of the reflectance profiles (Fig. 5) can be used, at times, to interpret geologic processes including continuous sedimentation, tectonic activity, interruption of sedimentation, erosional events, and localized thermal events (such as intrusions).

A final note regarding vitrinite reflectance concerns the suppression of reflectance due to poorly understood processes. Figure 6 illustrates that a downward drift of reflectance data can be observed superimposed upon the general reflectance versus depth relationship. Many studies have linked this suppression of vitrinite reflectance to significant concentrations of liptinite or amorphous kerogens in coals and rocks (Ting, 1977; Jones and Edison, 1978; Hutton and Cook, 1980; Kalkreuth, 1982) as would seem to be suggested in Fig. 6. Other workers have suggested that the suppression of vitrinite reflectance may be related to differences in depositional environments, plant communities, and the occurrence of hydrogen-enriched saprovitrinites in some coals and oil shales (Sitler, 1979; Newman and Newman, 1982; Toxepeus, 1982; Teichmuller, 1987). It has been noted that low-reflecting, hydrogen-enriched vitrinites are common in oil-prone kerogen assemblages (Price and Barker, 1985; Teichmuller, 1987) and that this hydrogen enrichment of vitrinite occurs in source rocks during deposition in anoxic environments and early diagenesis. Other workers have concluded that vitrinite is impregnated by bitumen or mobile hydrocarbons (Jones and Edison, 1978; Hutton and Cook, 1980; Hutton *et al.*, 1980; Kalkreuth, 1982; Teichmuller and Durand, 1983).

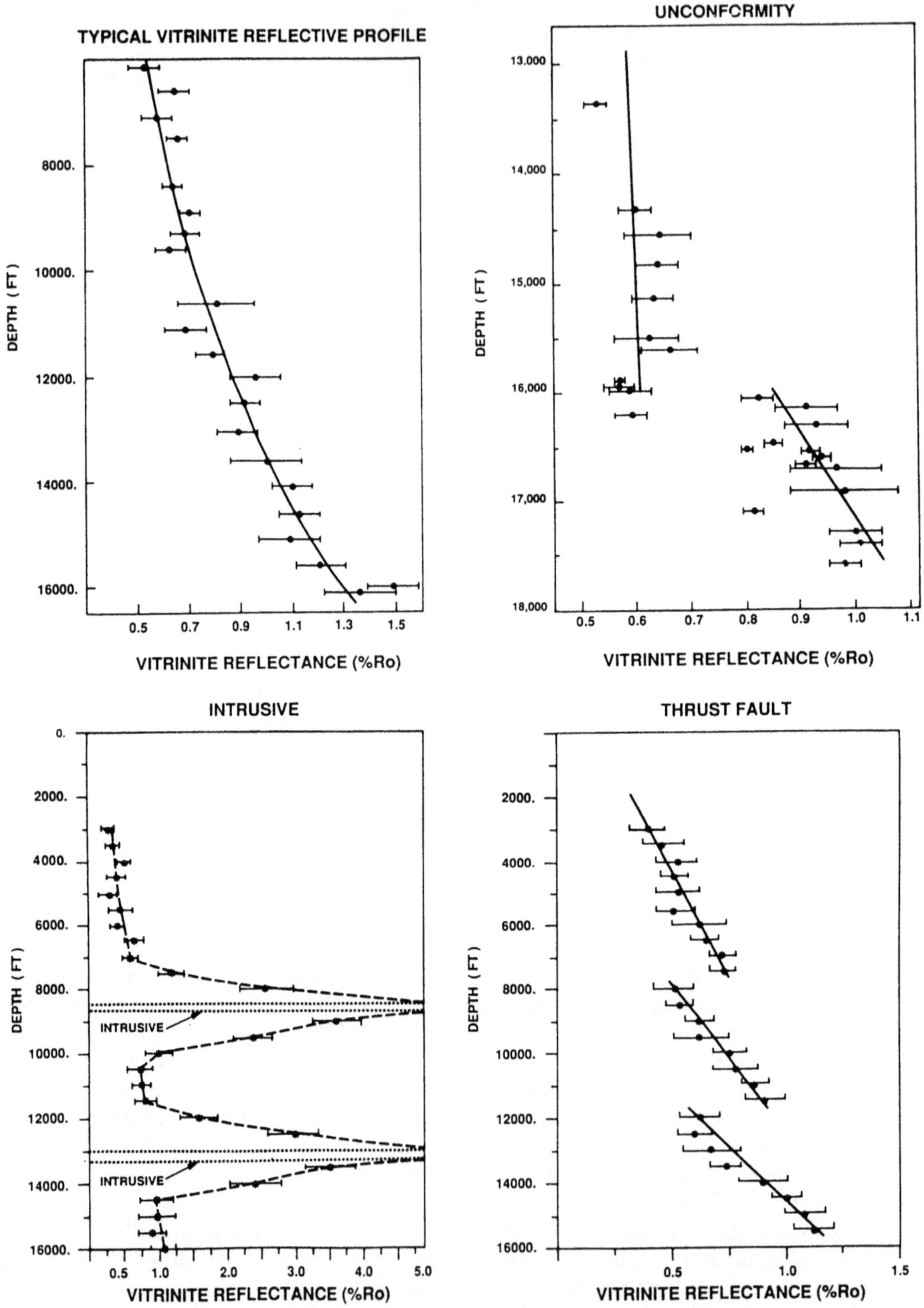

Figure 5. Vitrinite reflectance profiles exhibiting gradients for several geologic scenarios. [Reprinted by permission from Senftle and Landis (1991).]

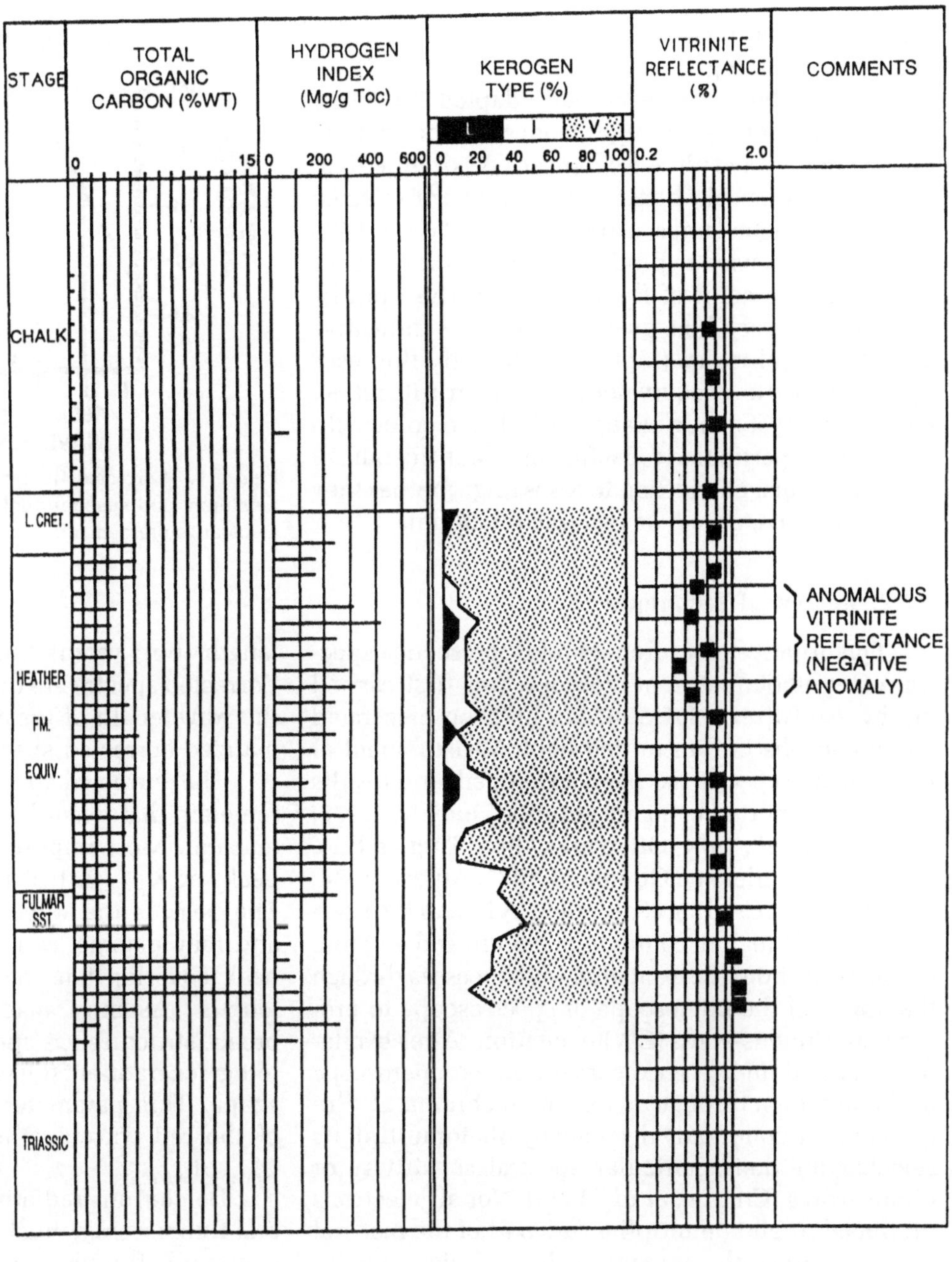

Figure 6. Vitrinite reflectance profile exhibiting suppressed reflectance data at depth. [Reprinted by permission from Senftle and Landis (1991).]

3.2. Fluorescence Microscopy

3.2.1. Introduction

The application of fluorescence microscopic techniques to the study of dispersed sedimentary organic material is appealing for many reasons. Fluorescence microscopy is a useful tool for the direct examination of those macerals prone to the generation of liquid hydrocarbons and of the generated products themselves. Type I and type II kerogen heavily favor the liptinites (Tissot and Welte, 1978). The liptinites are a major source of the deleterious bitumens (Teichmuller and Durand, 1983) that impregnate vitrinite during bituminous coalification or catagenesis and are the earliest hydrocarbons generated from the kerogen.

Fluorescence microscopy has been used as a coal petrographic technique for over 50 years. Schochardt (1943) probably first initiated the study of polished coal surfaces under UV radiation. Since then, numerous authors have studied the fluorescence of coal macerals or kerogens (Ammosov, 1956; Jacob, 1952, 1964; Stach, 1969; van Gijzel 1966, 1967a,b, 1971; Ottenjann *et al.*, 1974, 1975; Robert, 1980; Lin *et al.*, 1986). Crelling *et al.* (1989) compiled an excellent review of the utilization of fluorescence microscopy in organic petrology.

Visible fluorescence is well adapted for either transmitted- or reflected-light modes of illumination. To date, long-wave ultraviolet (near-UV) or blue-light (VIS) emission peaks from mercury lamps or selected bands from xenon lamps, usually between 365 and 435 nm, have been used to produce kerogen fluorescence emission. The color of fluorescence of the liptinite macerals can vary from green to red and is indicative of maturation history. (Liptinites generally fluoresce up to maturity level equivalent to 1.1% vitrinite reflectance.) The systematic change of liptinite color with maturity is particularly useful for identification of these macerals, in addition to assessing whether they are indigenous, caved, or reworked materials.

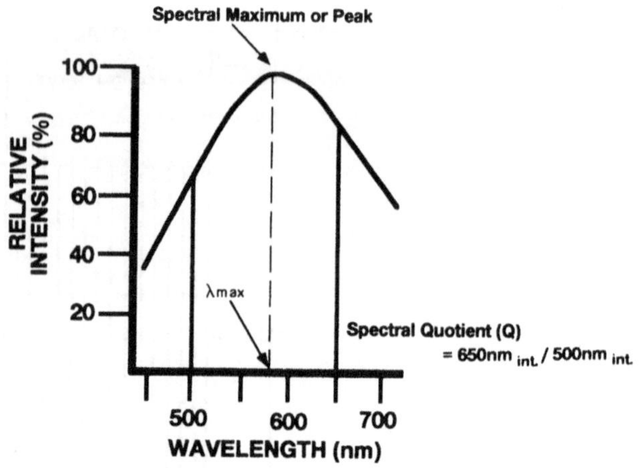

Figure 7. Corrected fluorescence spectrum illustrating spectral maximum and spectral quotient parameters used to characterize maceral fluorescence.

3.2.2. Review of Technique

Quantitative measurement of fluorescence spectra, using photometric microscopy, was first carried out by van Gijzel (1961, 1967a,b, 1971) on macerated palynomorphs in transmitted light. Measurement of fluorescence spectra has been developed, and results have been reported by many authors (Homann, 1972; Ottenjann *et al.*, 1974; Ottenjann, 1982; Ting and Lo, 1975; Teichmuller and Ottenjann, 1977; Alpern, 1980; Crelling, 1983; Crelling *et al.*, 1989). A line in the near-UV or visible region, commonly 365 nm and 450 nm, respectively, from the excitation source passes through the optics of the microscope or spectroscope to provide continuous-wave (CW) excitation. After excitation, maceral fluorescence passes through microscope optics and then through an exit monochromator. Fluorescence is commonly detected by photomultipliers selected for their particular spectral sensitivity or diode arrays (Crelling *et al.*, 1989). Not all electrons produced in photomultipliers are part of the maceral intensity. Since the amount of thermal electrons increases as temperature increases, the relative contribution of maceral intensity to the total signal may be enhanced by as much as an order of magnitude by simply cooling the detector to −25°C for mature shales (Landis *et al.*, 1992). As a result, the capability of measuring very weakly emitting macerals is greatly improved.

Uncorrected spectra are a convolution of the system response and the maceral signal. Correction for the inherent system response over the entire scanning range must be completed in order to produce the corrected fluorescence spectrum (Fig. 7). The corrected spectrum which is indicative of the fluorescence color of the maceral is a binary plot of wavelength (nm) versus the intensity of emission (mV). Corrected spectra are then normalized with respect to the wavelength of greatest intensity to produce normalized corrected spectra.

The corrected spectrum yields several useful parameters that describe the fluorescence color of the maceral. Most important is the spectral maximum, peak, or λ_{max}. All of these terms are synonymous. The peak is the wavelength of maximum intensity and may emerge as a spectral line or from a broad peak covering tens of competitively intense wavelengths. As most macerals characteristically yield broad, structureless spectra, the spectral quotient or red/green quotient (Q) is used to convey gross spectral shape. This parameter compares the intensities (I) of the red wavelengths to those of the green, i.e., $I_{650\ nm}/I_{500\ nm}$.

Deleterious radiative species not related to maceral fluorescence must be minimized. Immersion oil commonly fluoresces or induces fluorescence by rapidly acting as a solvent. Other potentially troublesome constituents include polishing slurry residuals, epoxy binder, and mineral matter. Polishing slurry is easily recognized in blue light by its very strong, blue, fibrous nature. Epoxy binder is particularly troublesome for macerals located near the edge of the sample, and mineral matter adjacent to the liptinite in question should be avoided. These problems are minimized in strewn mounts of kerogen concentrates.

A problem common to all sample mounting techniques is photochemical alteration or "fading." Alteration is the photochemical process whereby the maceral fluorescence is affected during the period of irradiation. Alteration has been reported for some time (van Gijzel, 1967a,b; Ottenjann *et al.*, 1975; Teichmuller and Wolf, 1977; Senftle and Larter, 1987; Pradier *et al.*, 1990). Very little is known regarding the extent and magnitude of alteration and the photochemical processes associated with it.

3.2.3. Maturation Effects

Nearly 30 years of research indicate that fluorescence intensity and vitrinite reflectance are inversely proportional. Thus, fluorescence spectral analysis and vitrinite reflectance analyses can be used in a complementary manner. Perhaps the most studied maceral has been sporinite. Sporinite spectra have been shown to shift to the red or longer visible wavelengths as a function of organic carbon (Ottenjann *et al.*, 1975), during artificial maturation (Ting and Lo, 1975), and in sampled Carboniferous coal seams (Teichmuller and Durand, 1983). The latter workers illustrated that sporinites from a series of coals extending from peat to high-volatile bituminous A rank exhibit a shift toward the red, as expressed by the shift of the peak and the change in the red/green quotient (Q), with increasing maturity. Radke *et al.* (1980) reported that the spectral quotient increases rapidly near a vitrinite reflectance of 1.0% R_0. A similar red shift has been reported for alginite in coal (Teichmuller and Wolf, 1977) and in Toarcian oil shale in the Paris Basin (Alpern and Cheymol, 1978).

In a comprehensive fluorescence spectral study, Teichmuller and Durand (1983) reported the fluorescence spectral variation of the common liptinites in high-volatile bituminous coals. Their findings indicate that each liptinite varies uniquely at coal ranks equivalent to catagenesis. For example, resinite exhibited a spectral continuum, as measured by spectral maximum and quotient, perhaps due to the diverse origin of resinite macerals. Thus, in contrast to sporinite spectra, the usefulness of resinite spectra as a maturation parameter is not readily apparent.

Fluorescence properties can be influenced by maceral degradation. Spiro and Mukhopadhyay (1983) illustrated that the relationship between fluorescence maturity parameters and vitrinite reflectance is inconsistent with a strong shift toward the red when degradation has occurred. Another cause for a shift in fluorescence parameters is the severe desiccation and heating of ditch samples using heat sources. This is more a problem with well ditch cutting samples which have been stored for many years or dried using a heat source.

Exposure of a sample to UV or blue-light excitation can result in an alteration of fluorescence spectra. Teichmuller and Ottenjann (1977) considered the alteration of fluorescence measured on the mineral-bituminous groundmass an important parameter for distinguishing immature, mature, and overmature stages of source rock evolution. Measuring fluorescence intensity at 546 nm, they reported that a positive alteration (increase of fluorescence intensity at 546 nm with time of irradiation) occurs in immature source rocks, a negative alteration (decrease in fluorescence intensity at 546 nm) occurs in mature source rocks, and organic fluorescence is lost in overmature source rocks.

Kerogen fluorescence is a complex phenomenon. Results to date have largely been empirical. Some progress in understanding the relationships between fluorescence and the chemistry controlling this phenomenon is being made, however. Bertrand *et al.* (1986) discussed factors which they believed control the fluorescence properties of kerogen. They discussed (a) the structure of chromophores which are generally associated with aromatic moieties, (b) the concentration of chromophores, and (c) the role of matrix–solvent effects which allow energy exchanges through inter- or intramolecular interactions, with the absorbed energy diffusing outside of the visible spectrum or in some other form of energy. Lin *et al.* (1986) have studied solvent extracts of vitrinites and have suggested that secondary fluorescence in these vitrinites is due to a mobile phase containing fluorophores. This mobile phase includes smaller molecules trapped within the vitrinite structure during condensation reactions and absorption of hydrocarbons after their expulsion from the liptinite macerals. Clearly, further understanding of kerogen fluorescence phenomena is developing, and this technique holds great promise as a tool for evaluating maturity.

3.3. Thermal Alteration Index

Examination of chemical changes of organic matter using transmitted light microscopy is an effective means to evaluate changes in maturation. Maturation produces a gradual coloring and darkening, culminating in opaque palynodebris. These palynomorphic coloration changes can be used to evaluate states of thermal alteration on a scale referred to as the thermal alteration index (TAI). The TAI is based upon color changes of one variety of pollen or spore. Therefore, TAI is more useful for stratigraphically marked beds within a basin than a cored interval at one or even a few localities. A main advantage of this method is the ability to study palynological residues in order to distinguish between indigenous and reworked material or between indigenous material and contaminants from drilling mud or from overlying sediments.

Spore color standards are gradational (Fig. 8). An arbitrary numerical value is used to delineate the color change sequence and to relate the numerical values with other maturity indicators. The TAI scale used here is from 1 (unaltered) to 2 (yellow) to 3 (brown) to 4 (black) to 5 (severely altered). TAI is generally related to vitrinite reflectance, coal rank, and hydrocarbon quality potential (Fig. 3). Limitations of this method include the lack of standardization, subjectivity of the measurement, and a limited ability to relate appropriate geochemical parameters to TAI.

As discussed above, the preparation of isolated kerogen and palynological residue samples is not a standardized technique. For determination of TAI, only HCl and HF treatment should be used since oxidizing agents can significantly alter kerogen color. The mounting media should not react with the organic matter and should be clear, preferably with a refractive index near 1.54. A stable light source equivalent to that used to generate the standard reference should be used; polarization filters should be avoided as they can affect the image contrast and inhibit color discrimination (Staplin *et al.*, 1982).

With regard to the practical aspects of carrying out a TAI determination, several suggestions are made:

1. Make the determination on relatively unornamented grains when possible to avoid variations in shading caused by ornamentation. When it is not feasible to make determinations on such grains, use grains with fine ornamentation.
2. Avoid making determinations on very thin grains or thick grains, on the saccus of saccate grains, or on highly folded grains.
3. Avoid etched grains. If this is not feasible, make determinations on the thicker, unetched parts.
4. Cuticles can be used for determinations by using the standard reference plate as a color standard. Make determinations on lumina rather than ridges. These data should be recorded separately from spore color data. The TAI determination should be an average of the spore color and the cuticle data.

In addition to thermal maturation, the environment of deposition and its effect upon the preservation of the organic residue influences the color of spores and pollen. This problem can be significant and needs to be considered when interpreting data. It is this problem along with the subjective nature of the analysis which has disenchanted workers with this technique. Regardless, the TAI approach has been developed and applied widely since it offers a means to determine maturity levels within the range of oil generation, it is relatively inexpensive, and it complements additional petrographic methods such as vitrinite reflectance.

Gutjahr (1966) and Grayson (1975) have quantified the amount of light transmitted by specific spore and pollen types at different levels of thermal maturity. Gutjahr (1966) developed a method to measure light absorption of spores and related it to oil and gas occurrences along the Gulf Coast of Texas. He showed that the maximum rate of change and the maximum total change in spore absorption began within the high-volatile bituminous rank coals and continued into medium-volatile bituminous coals. This corresponds to the maturation range of maximum hydrocarbon generation. Grayson (1975) reached similar conclusions by measuring the translucency of Carya pollen in relation to oil maturity in the Tertiary of southern Louisiana.

Jones and Edison (1978) discussed the correlation between TAI determinations and vitrinite reflectance and geochemical measurements on kerogen and bitumens from the Toarcian of the Paris Basin. Using published data (Correia, 1969; Tissot and Pelet, 1971; Durand *et al.*, 1972; Huc and Durand, 1977), they reported a strong correlation with the increase in hydrocarbon content and TAI. Perhaps this correlation best illustrates the relationship between hydrocarbon generation in spores, and pollens and the evaluation of maturity using spores and pollen themselves.

3.4. Conodont Color Index

Epstein *et al.* (1977) established the conodont color alteration index (CAI) from experimental and field data. They establish thermal maturity indices, corresponding to experimental and field data for temperatures ranging from 50 to 300°C. Change in conodont color is attributed to the irreversible and systematic thermal alteration of trace amounts of organic material in the conodont (Harris, 1979). The color changes are assigned numerical color indices (CAI) from petrographic observation which range from CAI 1 through 5. Subsequent work by Harris (1979) and

Rejebian *et al.* (1987) established CAI values of 6 through 8, corresponding to higher thermal regimes. Conodont color alteration ranges from pale yellow (CAI 1) to brown (CAI 2) to black (CAI 5). At temperatures ranging from 300 to 550°C, conodonts vary from black to gray to white to transparent (Harris, 1979). These changes are due to the loss of carbon, release of water of crystallization, and recrystallization (Rejebian *et al.*, 1987).

Consistency in determination of CAI requires the visual comparison of a standard reference set, analogous to the procedure used to evaluate palynomorph thermal alteration index. Further, conodont elements can exhibit variability in size, shape, and thickness, and, consequently, color. At low degrees of alteration, the distinctions between conodont coloration are difficult and are subjective. For example, CAI 1 to 2 covers the temperature range of hydrocarbon generation (approximately 50–120°C). Epstein *et al.* (1977) recommended the determination of the alteration index from the lightest colored elements or parts of elements.

Epstein *et al.* (1977) were able to correlate conodont color alteration to vitrinite reflectance and TAI for samples from the Appalachian Basin. The correlation of CAI to TAI was based upon 66 limestone samples ranging in age from Ordovician to Mississippian. Based upon this study, it was shown that conodont color alteration does not begin until later stages of spore thermal alteration. Thus, CAI is considered an indicator well suited for later stages of catagenesis and organic metamorphism. Conodonts are useful for the evaluation of regional thermal maturity trends of pre-Devonian and/or marine carbonate rocks where palynomorphs and vitrinite are absent (Epstein *et al.*, 1977; Harris, 1979; Rejebian *et al.*, 1987). Harris (1979) delineated geographic limits of gas generation and structural events by using CAI.

4. Conclusions

Sampling methods and optical kerogen classification schemes are not standardized worldwide. This chapter presents a broad description of the optical petrographic methods and terminology currently used to evaluate kerogen composition. A uniform classification of kerogen is necessary to understand the influence of sedimentology on the composition of organic matter. Progress in this area is necessary to better predict the hydrocarbon generative capacity of source rocks. Linking visual kerogen composition to source rock sedimentology is a necessary first step to permit the prediction of lateral and vertical variation in the distribution of organic matter in sedimentary sequences.

Petrographic indices used to evaluate the organic thermal maturity of sediments have been better standardized throughout the world. Four common indices have been presented. Clearly, these parameters are reinforced when they are used in a complementary fashion. It is best to use more than one approach, when possible, to ensure the most reliable interpretation of thermal maturity. Remaining calibration research is needed to correlate subsurface temperatures, sedimentation history, and inorganic indices to organic thermal indicators.

References

Alpern, B., 1980, Pétrographie du kérogène, in: *Kerogen—Insoluble Organic Matter from Sedimentary Rocks* (B. Durand, ed.), Editions Technip, Paris, pp. 339–371.

Alpern, B., and Cheymol, D., 1978, Réflectance et fluorescence des organoclastes du Toarcian du Bassin de Paris en fonction de la profondeur et de la température, *Rev. Inst. Fr. Pet.* **33**:515–535.

American Society for Testing and Materials (ASTM), 1976, *Annual Book of ASTM Standards*, ASTM, Philadelphia, 2798–72.

Ammosov, I. I., 1956, New Methods for Coal Petrography—Studies of the Laboratory of Coal Geology 6, Moscow [in Russian].

Baranger, R., Faggionato, J. L., and Nicolas, G., 1989, Round Robin Exercise, International Committee for Coal Petrology, Annual Meeting, 1990, Commission II, Minutes.

Barrick, J. E., Over, D. J., Landis, C. R., and Borst, W. L., 1990, Conodont fluorescence emission spectra: Potential for high resolution evaluation of organic maturation in hydrocarbon-bearing basins, Geological Society of America, South-Central Section Meeting, Abstracts with Programs, Stillwater, Oklahoma, Vol 22(1), pp. 1–2.

Bayliss, G. S., 1980, *Source Rock Evaluation Reference Manual*, Geochem Laboratories, Houston, Texas.

Bertrand, P., Pittion, J., and Bernaud, C., 1986, Fluorescence of sedimentary organic matter in relation to its chemical composition, *Org. Geochem.* **10**:641–647.

Bordovsky, O. K., 1965, Accumulation and transformation of organic substances in marine sediments, *Mar. Geol.* **3**:3–114.

Bostick, N. H., 1979, Microscopic measurement of the level of catagenesis of solid organic matter in sedimentary rocks to aid exploration for petroleum and to determine former burial temperatures—a review, *Soc. Econ. Paleontol. Mineral Spec. Publ.* **26**:17–23.

Brooks, J., and Shaw, G., 1973, Geochemistry of sporopollenin, *Chem. Geol.* **10**:69–87.

Burgess, J. D., 1974, Microscopic examination of kerogen (dispersed organic matter) in petroleum exploration, *Geological Society of America Special Paper 153*, pp. 19–30.

Cane, R. F., 1968, *Proceedings of Symposium on the Development and Utilization of Oil Shale Resources*, Tallin, Valgus, U.S.S.R.

Castano, J. R., and Sparks, D. M., 1974, Interpretation of vitrinite reflectance measurements in sedimentary rocks and determination of burial history using vitrinite reflectance and authigenic minerals, Geological Society of America Special Paper 153, pp. 31–52.

Chadhuri, S. G., and Ghose, S., 1978, A preliminary study of reactive semifusinites in Indian coking coals, *J. Mines, Met. Fuels* **1978**:137–141.

Clausen, C. D., and Teichmüller, M., 1982, Die Bedeutung der Graptolithenfragmente in Palaozoikum von Soest—Erwitte für Stratigraphie und Inkohlung, *Fortschr. Geol. Rhein. Westfalen* **30**:145–167.

Combaz, A., 1964, Les palynofacies, *Rev. Micropaleo.* **7**:205–218.

Combaz, A., 1975, Essai de classification, des roches carbonées et des constituants organiques des roches sédimentaires, in: *Pétrographie de la Matière Organique des Sédiments, Relations avec la Paléotempérature et le Potentiel Pétrolier* (B. Alpern, ed.), CNRS, Paris, pp. 93–102.

Combaz, A., 1980, Les kérogènes vus au microscope, in: *Kerogen—Insoluble Organic Matter from Sedimentary Rocks* (B. Durand, ed.), Editions Technip, Paris, pp. 55–112.

Cook, A. C., 1987, Organic petrological studies of oil shales, Abstracts and Program, 4th Annual Meeting of The Society for Organic Petrology, pp. 15–25.

Cook, A. C., Hutton, A. C., and Sherwood, N. R., 1981, Classification of oil shales, *Bull. Cent. Rech. Expl.-Prod. Elf Aquitaine* **5**: 353–381.

Cookson, I., 1953, Records of the occurrence of *Botryococcus braunii*, *Pediastrum* and the Hystrichosphaerideae in Cainozoic deposits of Australia, *Mem. Nat. Mus. Melbourne* **18**: 107–125.

Correia, M., 1969, Contribution à la recherche de zones favorables à la genèse du pétrole par l'observation microscopique de la matière organique figurée, *Rev. Inst. Fr. Pet.* **24**:1417–1454.

Crelling, J. C., 1983, Current uses of fluorescence microscopy in coal petrology, *J. Microsc.* **132**(3):151–266.

Crelling, J. C., Bensley, D. F., Landis, C. R., and Rimmer, S. M., 1989, Fluorescence Microscopy: Workshop Lecture Notes, Annual Meeting Workshop sponsored by the Society for Organic Petrology, Carbondale, Illinois (TSOP).

Demaison, G. J., and Moore, G. T., 1980, Anoxic environments and oil source bed genesis, *Am. Assoc. Petrol. Geol. Bull.* **64**:1179–1209.

Dow, W. G., 1977, Kerogen studies and geological interpretations, *J. Geochem. Explor.* **7**:79–99.

Dow, W. G., and O'Connor, D. E., 1982, Kerogen maturity and type by reflected light microscopy, in: *How to Assess Maturation and Paleotemperatures*, Society of Economic Paleontologists and Mineralogists, Short Course Number 7, Tulsa, Oklahoma, pp. 133–158.

Durand, B., and Nicaise, G., 1980, Procedures for kerogen isolation, in: *Kerogen—Insoluble Organic Matter from Sedimentary Rocks* (B. Durand, ed.), Editions Technip, Paris, pp. 35–53.

Durand, B., Espitalié, J., Nicaise, G., and Combaz, A., 1972, Etude de la matière organique insoluble (kérogène) des argiles du Toarcien du Bassin de Paris. Première partie: Etude par les procèdès optiques, analyse elementaire, étude en microscopie et diffraction eletroniques, *Rev. Inst. Fr. Pet.* **27**:865–884.

Epstein, A. G., Epstein, J. B., and Harris, L. D., 1977, Conodont Color Alteration—An Index to Organic Metamorphism, U.S. Geological Survey Professional Paper 995.

Evitt, W. R., 1963, A discussion and proposals concerning fossil dinoflagellates, hystrichospheres and acritarchs, I and II, *Proc. Natl. Acad. Sci. U.S.A.* **49**:158–164.

Goodarzi, F., 1982, A Brief Hydrocarbon Potential Study of Southeast Turkey Using Organic Petrography, Report to Turkish Petroleum Research Centre, Ankara, Turkey.

Goodarzi, F., 1984, Organic petrology of graptolite fragments from Turkey, *Mar. Petrol. Geol.* **1**:202–210.

Goodarzi, F., 1985, Dispersion of optical properties of graptolite epiderms with increased maturity in early Paleozoic organic sediments, *Fuel* **64**:1735–1740.

Goodarzi, F., and Higgins, A. C., 1987, Optical properties of scolecodonts and their use as indicators of thermal maturity, *Mar. Petrol. Geol.* **4**:353–359.

Goodarzi, F., Snowdon, L. R., Gunther, P. R., and Jenkins, W. A. M., 1985, Preliminary organic petrography of Paleozoic rocks from Grand Banks, Newfoundland, *Mar. Petrol. Geol.* **2**:145–166.

Grayson, J. F., 1975, Relationship of palynomorph translucency to carbon and hydrocarbons in clastic sediments, in: *Pétrographie de la Matière Organique des Sédiments, Relations avec la Paléotempérature et le Potentiel Pétrolier* (B. Alpern, ed.), CNRS, Paris, pp. 261–273.

Gutjahr, C. C. M., 1966, Carbonization measurements of pollen grains and spores and their application, *Leidse Geol. Meded.* **38**:1–30.

Gutjahr, C. C. M., 1983, Introduction to incident-light microscopy of oil and gas source rocks, *Geol. Mijnbouw* **62**:417–425.

Harris, A. G., 1979, Conodont color alteration, an organo-mineral metamorphic index, and its application to Appalachian Basin Geology, *Soc. Econ. Paleontol. Mineral Spec. Publ.* **26**:3–16.

Harwood, R., 1977, Oil an gas generation by laboratory pyrolysis of kerogen, *Am. Assoc. Petrol. Geol. Bull.* **61**:2082–2102.

Homann, W., 1972, Zum spektralen Fluoreszenz-Verhalten des Sporinits in Kohlen-Anschliffen und die Bedeutung fur die Inkohlungsgrad-Bestimmung, Presented at the 25th Meeting of the International Committee for Coal Petrology, Belgrade, Yugoslavia.

Huc, A. Y., and Durand, B., 1977, Occurrence and significance of humic acids in ancient sediments, *Fuel* **56**:73–80.

Hunt, J. M., 1979, *Petroleum Geochemistry and Geology*, W. H. Freeman and Co., San Francisco, pp. 44–67.

Hutton, A. C., 1987, Petrographic classification of oil shales, *Int. J. Coal Geol.* **8**:203–231.

Hutton, A. C., and Cook, A. C., 1980, Influence of alginite on the reflectance of vitrinite from Joadja, NSW, and some other coals and oil shales containing alginite,, *Fuel* **59**:711–714.

Hutton, A. C., Kantsler, A. J., Cook, A. J., and McKindy, D. M., 1980, Organic matter in oil shales, *APEA J.* **20**(1):44–67.

International Committee for Coal Petrology, 1963, *International Handbook of Coal Petrology*, 2nd ed., CNRS, Paris.

International Committee for Coal Petrology, 1971, *International Handbook of Coal Petrology*, 2nd ed., 1st Supplement, CNRS, Paris.

International Committee for Coal Petrology, 1975, *International Handbook of Coal Petrology*, 2nd ed., 2nd Supplement, CNRS, Paris.

Jacob, H., 1952, *Forschritte auf dem Gebier der Braunkohlem-Lumineszenz-Microskopie-Bergakademic*, Vol. 4, Leipzig, Bergakud., Freiberg, pp. 337–347.

Jacob, H., 1964, Neue Erkenntinusse auf dem Gebiet der Lumineszenzmikroskopie fossiler Brennstoffe, *Fortschr. Geol. Rheinl. Westfalen.* **12**:569–588.

Jansonius, J., and Jenkins, W. A. M., 1978, Chitinozoan, in: *Introduction to Marine Micropaleontology* (B. V. Haq and A. Boersma, eds.), Elsevier, North Holland, New York, pp. 341–357.

Jones, R., and Edison, T., 1978, Microscopic observations of kerogen related to geochemical parameters with emphasis on thermal maturation, in: *Low-Temperature Metamorphism of*

Kerogen and Clay Minerals (D. F Oltz, ed.), Society of Economic Paleontologists and Mineralogists, Los Angeles, pp. 1–12.

Kalkreuth, W. D., 1982, Rank and petrographic composition of selected Jurassic–Lower Cretaceous coals of British Columbia, Canada, *Can. Petrol. Geol. Bull.* **30**:112–139.

Kalkreuth, W. D., 1989, Round Robin Exercise, International Committee for Coal Petrology, 1990, Annual Meeting, Commission II, Minutes.

Kurylowicz, L. E., Ozimic, S., McKirdy, D. M., Kantsler, A. J., and Cook, A. C., 1976, Reservoir and source rock potential of the Larapinta Group, Amadeus Basin, Central Australia, *J. Aust. Petrol. Explor. Assoc.* **16**:49–65.

Landis, C. R., Trabelsi, A., and Strathearn, G., 1992, Hydrocarbon Potential of Selected Permian Basin Shales as Classified within the Organic Facies Concept, in: *Source Rocks, Generation, and Migration of Hydrocarbons and Other Fluids in the Southern Mid-Continent*, Oklahoma Geological Survey, Circ. 93, pp. 229–247.

Lin, R., Davis, A., Bensley, D. F., and Derbyshire, F. J., 1986, Vitrinite secondary fluorescence, its chemistry and relation to the development of a mobile phase and thermoplasticity in coal, *Int. J. Coal Geol.* **6**:215–288.

Madler, K., 1964, Die geologische Verbreitung von Sporen und Pollen in der Deutschen Trias, *Beih. Geol. Jahrb.* **65**:1–147.

Masran, T. C., and Pocock, S. A. J., 1981, The classification of plant-derived particular organic matter in sedimentary rocks, in: *Organic Maturation Studies and Fossil Fuel Exploration* (J. Brooks, ed.), Academic Press, London, pp. 145–175.

Massoud, M. S., and Kinghorn, R. R. F., 1985, A new classification for the organic components of kerogen, *J. Petrol. Geol.* **8**(1): 85–100.

Mastalerz, M., Bustin, R. M., Orchard, M., and Forster, P. J. L., 1992, Fluorescence of conodonts: Implications for organic maturation analysis, *Org. Geochem.* **18(1)**, pp. 93–101.

Mukhopadhyay, P. K., Hagemann, H. W., and Gormly, J. R., 1985, Characterization of kerogens as seen under the aspects of maturation and hydrocarbon generation, *Erdöl Kohle, Erdgas, Petrochem.* **38**:7 18.

Nandi, B. N., and Montgomery, D. S., 1975, Nature and thermal behavior of semifusinite in Cretaceous coal from western Canada, *Fuel* **54**:193–196.

Nandi, B. N., Brown, T. D., and Lee, G. K., 1977, Inert coal macerals in combustion, *Fuel* **56**:125–130.

Newman, J., and Newman, N. A., 1982, Reflectance anomalies in Pike River coals, evidence of variability in vitrinite type, with implications for maturation studies and "Suggate rank," *N. Z. J. Geol. Geophys.* **25**:233–243.

Ottenjann, K., 1982, Fluoreszenzeigenschaften von Flozproben aus stratigraphisch jungen Schichten des Oberkarbons, *Fortschr. Geol. Rheinl. Westfalen* **33**:169–195.

Ottenjann, K., Teichmuller, M., and Wolf, M., 1974, Spektrale fluoreszenzmessungen an sporiniten mit auflicht-anregung, eine mikroscopische method zur Bestimmung des inkohlungsgrades gering inkohlter kohlen, in: *Inkohlung und Erdöl* (E. Wiegel, ed.), *Fortsch. Geol. Rheinl. Westfalen* **24**:1–35.

Ottenjann, K., Teichmuller, M., and Wolf, M., 1975, Spectral fluorescence measurements of sporinites in reflected light and their applicability for coalification studies, in: *Pétrographie de la Matière Organique des Sédiments, Relations avec la Paléotempérature et le Potentiel Pétrolier* (B. Alpern, ed.), CNRS, Paris, pp. 67–91.

Pietzner, H., Vahl, J., Werner, H., and Ziegler, W., 1968, Zur Chemischen Zusammensetzung und Micromorphologie der Conodoten, *Paleontographica* **128**:115–152.

Potonie, H., 1908, Die rezenten Kaustobiolithe und ihre Lagerstatten: Die Sapropeliten, *Abh. Kgl. Preuss. Geol. Landesanstalt*, New Series, **1(55)**:361–392.

Powell, T. G., Creaney, S., and Snowdon, L. R., 1982, Limitations of use of organic petrographic techniques for identification of petroleum source rocks, *Am. Assoc. Petrol. Geol. Bull.* **66**: 430–435.

Pradier, B., Larqeau, C., Derenne, S., Martinez, L., Bertrand, P., and Pouet, Y., 1990, Chemical basis of fluorescence alteration of crude oils and kerogens—I. Microfluorimetry of an oil and its isolated fractions; relationships with chemical structures, in: *Advances in Organic Geochemistry 1989*, pp. 457–460.

Price, L. C., and Barker, C. E., 1985, Suppression of vitrinite reflectance in amorphous rich kerogen—A major unrecognized problem, *J. Petrol. Geol.* **8**:59–84.

Radke, M., Schaefer, R. G., Leythaeuser, D., and Teichmuller, M., 1980, Composition of soluble organic matter in coals: Relation to rank and liptinite fluorescence, *Geochim. Cosmochim. Acta* **44**:1787–1800.

Rejebian, V. A., Harris, A. G., and Huebner, J. S., 1987, Conodont color and textural alteration, An index to regional metamorphism, contact metamorphism, and hydrothermal alteration, *Geol. Soc. Am. Bull.* **99**:471–479.

Robert, P., 1979, Classification des matières organiques en fluorescence, Applications aux roches-mères petrolières, *Bull. Cent. Rech. Explor.–Prod., Elf Aquitaine* **3**:223–263.

Robert, P., 1980, The optical evolution of kerogen and geothermal histories applied to oil and gas exploration, in: *Kerogen—Insoluble Organic Matter from Sedimentary Rocks* (B. Durand, ed.), Editions Technip, Paris, pp. 385–414.

Schochardt, M., 1943, *Grundlagen und neuere Erkenntnisse der angewandten Braunkohlenpetrographic*, Knapp, Halle, Germany.

Schwab, K. W., 1966, Microstructure of some fossil and Recent scolecodonts, *J. Paleont.* **40**:416–423.

Senftle, J. T., and Landis, C. R., 1991, Vitrinite reflectance as a tool to assess thermal maturity, in: *Source Migration Processes and Evaluation Techniques*, (R. K. Merrill, ed.), AAPG Treatise of Petroleum Geology, AAPG, Tulsa, OK, pp. 119–126.

Senftle, J. T., and Larter, S. R., 1987, The geochemistry of exinites—1. Evaluation of spectral fluorescence of a series of modern resins and fossil resinites, *Org. Geochem.* **13**:973–980.

Senftle, J. T., Brown, J. H., and Larter, S. R., 1987, Refinement of organic petrographic methods for kerogen characterization, *Int. J. Coal Geol.* **7**:105–117.

Sherwood, N. R., Cook, A. C., Gibling, M., and Tantisukrit, C., 1984, Petrology of a suite of sedimentary rocks associated with some coal-bearing basins in NW Thailand, *Int. J. Coal Geol.* **4**(1): 45–71.

Sitler, J. A., 1979, Effects of source material on vitrinite reflectance, Master's Thesis, West Virginia University.

Spiro, B., and Mukhopadhyay, K., 1983, Effects of degradation on spectral fluorescence of alginite, *Erdöl Kohle* **36**(7):297–299.

Stach, E., 1969, Fortschritte der Auflicht-Fluoreszenz-Mikroskopie in der Kohlenpetrographie, *Frieberg Forschungs.* **242**:35–55.

Stach, E., Mackowsky, M., Teichmuller, M., Taylor, G., Chandra, D., and Teichmuller, R., 1975, *Stach's Textbook of Coal Petrology*, Gebruder Borntraeger, Berlin, Stuttgart, 428 pp.

Stach, E., Chandra, D., Mackowsky, M., Taylor, G., Teichmuller, M., and Teichmuller, R., 1982, *Stach's Textbook of Coal Petrology*, Gebruder Borntraeger, Berlin.

Staplin, F. L., 1969, Sedimentary organic matter, organic metamorphism, and oil and gas occurrences, *Bull. Can. Petrol. Geol.* **17:** 47–66.

Staplin, F. L., Down, W. G., Milner, C. W. D., O'Connor, D. I., Pocock, S. A. J., van Gijzel, P., Welte, D. H., and Yukler, M. A., 1982, *How to Assess Maturation and Paleotemperatures*, Society of Economic Paleontologists and Mineralogists, Short Course Number 7, Tulsa, Oklahoma.

Steyn, J. G. D., and Smith, W. H., 1977, Coal petrography in the evaluation of South Africa coals, *Coal, Gold, and Base Minerals of Southern Africa* **25(9):**107–117.

Taylor, G. H., and Liu, S. Y., 1987, Biodegradation in coals and other organic-rich rocks, *Fuel* **66:**1269–1273.

Teichmuller, M., 1982, *Fluoreszenzmikroskopische Anderungen von Liptiniten und Vitriniten mit zunehmenden Inkohlungsgrad und ihre Beziehungen zu Bitumenbildung und Verkokungsverhalten*, Geologische Landesamt Nordrhein-Westfalen, Krefeld, Germany.

Teichmuller, M., 1987, Organic petrology of source rocks, history and state of art, *Organic Geochem.* **10(1–3):**581–599.

Teichmuller, M., and Durand, B., 1983, Fluorescence microscopical rank studies on liptinites and vitrinites in peat and coals, and comparisons with results of the Rock-Eval pyrolysis, *Int. J. Coal. Geol.* **2:**197–230.

Teichmuller, M., and Ottenjann, K., 1977, Art und Diagenese von Liptiniten und lipoiden Stoffen in einem Erdolmuttergestein aufgrund fluoreszenmikroskopischer Untersuchungen, *Erdöl, Kohle, Petrochem.* **30:**387–398.

Teichmuller, M., and Wolf, M., 1977, Application of fluorescence microscopy in coal petrology and oil exploration, *J. Microsc.* **109:**49–73.

Thompson-Rizer, C. L., and Dembicki, H., Jr., 1986, Optical characteristics of amorphous kerogens and the hydrocarbon generating potential of source rocks, *Int. J. Coal Geol.* **6:**229–249.

Ting, F. T. C., 1977, Microscopial investigation of the transformation (diagenesis) from peat to lignite, *J. Microsc.* **109:**75–83.

Ting, F. T. C., and Lo, H. B., 1975, Fluorescence characteristics of thermo-altered exinites (sporinites), *Fuel* **54:**201–204.

Tissot, B. P., and Pelet, R., 1971, Nouvelles donnees sur les mécanismes de genèse et de migration du pétrole: Simulation mathémratique et application à la prospection, in: *Proceedings of the 8th World Petroleum Congress*, Vol. 2, Applied Science Publishers, London, pp. 35–46.

Tissot, B. P., and Welte, D. H., 1978, *Petroleum Formation and Occurrence*, Springer-Verlag, Berlin.

Tissot, B. P., Durand, B., Espitalié, J., and Combax, A., 1974, Influence of the nature of diagenesis of organic matter in formation of petroleum, *Bull. Am. Assoc. Petrol. Geol.* **58:**499–506.

Toxepeus, Buiskool, J. M. A., 1982, Selection criteria for the use of vitrinite reflectance as a maturity tool in petroleum exploration, *J. Petrol. Geol.* **5:**295–307.

Traverse, A. F., 1974, Paleopalynology (1947–1972), *Ann. Mi. Bot. Gard.* **61:**203–206.

Tschudy, R. H., and Scott, R. A., 1969, *Aspects of Palynology*, John Wiley & Sons, London, pp. 469–474.

van Gijzel, P., 1961, Autofluorescence and age of some fossil pollen and spores, *Proc. K. Ned. Akad. Wet.*, 61(1), pp. 56–63.

van Gijzel, P., 1966, Die Fluoreszenz-Mikrophotometrie von Mikrofossilien mit dem Zweistrahl-Mikroskopphotometer von Berek, *Leitz-Mitt. Wiss. Tech.* **3:**206–214.

van Gijzel, P., 1967a, Autofluorescence of fossil pollen and spores with special reference to age determination and coalification, *Leidse Geol. Meded*, **40:**263–317.

van Gijzel, P., 1967b, Palynology and fluorescence microscopy, *Rev. Palaeobot. Palynol.* **2:**49–79.

van Gijzel, P., 1971, Review of the UV-fluorescence microphotometry of fresh and fossil exines and exosporia, in: *Sporopollenin* (J. Brooks *et al.*, eds.), Academic Press, London, pp. 659–682.

van Gizjel, P., 1979, Manual of the technique and some geological applications of fluorescence microscopy; workshop sponsored by American Association of Stratigraphic Palynologist Core Laboratories Inc., Dallax, Texas.

van Gijzel, P., 1982, Characterization and identification of kerogen and bitumen and determination of thermal maturation by means of qualitative and quantitative microscopical techniques, in: *How to Assess Maturation and Paleotemperatures*, Society of Economic Paleontologists and Mineralogists, Short Course Number 7, Tulsa, Oklahoma, pp. 159–207.

Welte, D., 1974, Recent advances in organic geochemistry of humic substances and kerogen, in: *Advances in Organic Geochemistry 1973* (B. Tissot and F. Bienner, eds.), Editions Technip, Paris, pp. 3–14.

IV

Thermal Alteration of Organic Matter and the Formation of Fossil Fuels

Chapter 16

The Thermal Alteration of Kerogen and the Formation of Oil

JÜRGEN RULLKÖTTER

1. Introduction

Oil and gas found today as accumulations of fossil hydrocarbons in porous reservoir rocks were originally generated from fine-grained source rocks enriched in organic matter and later migrated out to their place of entrapment. Hydrocarbon generation in sediments involves the thermal decomposition of macromolecular organic matter of biogenic origin, so-called kerogen (Durand, 1980), over periods of geologic time [see Hunt (1979) and Tissot and Welte (1984) for more detailed overviews]. Organic geochemical research over the last three decades has provided overwhelming evidence for the precursor/product relationship between organic matter of decayed organisms incorporated into sediments and petroleum in reservoir rocks. Nevertheless, the origin of petroleum from abiogenic hydrocarbon formation in the Earth's mantle or deeper crust still has at least one popular proponent (Gold, 1987).

The quantities of hydrocarbons that can be generated from a given source rock depend mainly on the concentration and initial composition of the kerogen and the extent to which the hydrocarbon potential has been realized as a consequence of the geothermal evolution during the burial history of this rock. The product distribution (gas versus oil) is determined by the kerogen concentration and composition as well as the geothermal heating rate. A detailed understanding of the hydrocarbon generation process in sedimentary basins thus will largely depend on a knowledge of the chemical structure of kerogen and the kinetics of the chemical reactions involved.

2. Thermal Alteration of Kerogen

2.1. Diagenesis, Catagenesis, Metagenesis, and the Oil Window

In petroleum geochemistry, the evolution of sedimentary organic matter under the influence of increasing burial depth and related temperature rise (maturation) is commonly subdivided into three stages, i.e., diagenesis, catagenesis, and metagenesis (Fig. 1). Diagenesis begins in recently deposited sediments and comprises microbial and chemical transformations of sedimentary organic matter at low tem-

JÜRGEN RULLKÖTTER • Institut für Chemie und Biologie des Meeres (ICBM), Carl von Ossietzky Universität Oldenburg, D-26111 Oldenburg, Germany.

Organic Geochemistry, edited by Michael H. Engel and Stephen A. Macko. Plenum Press, New York, 1993.

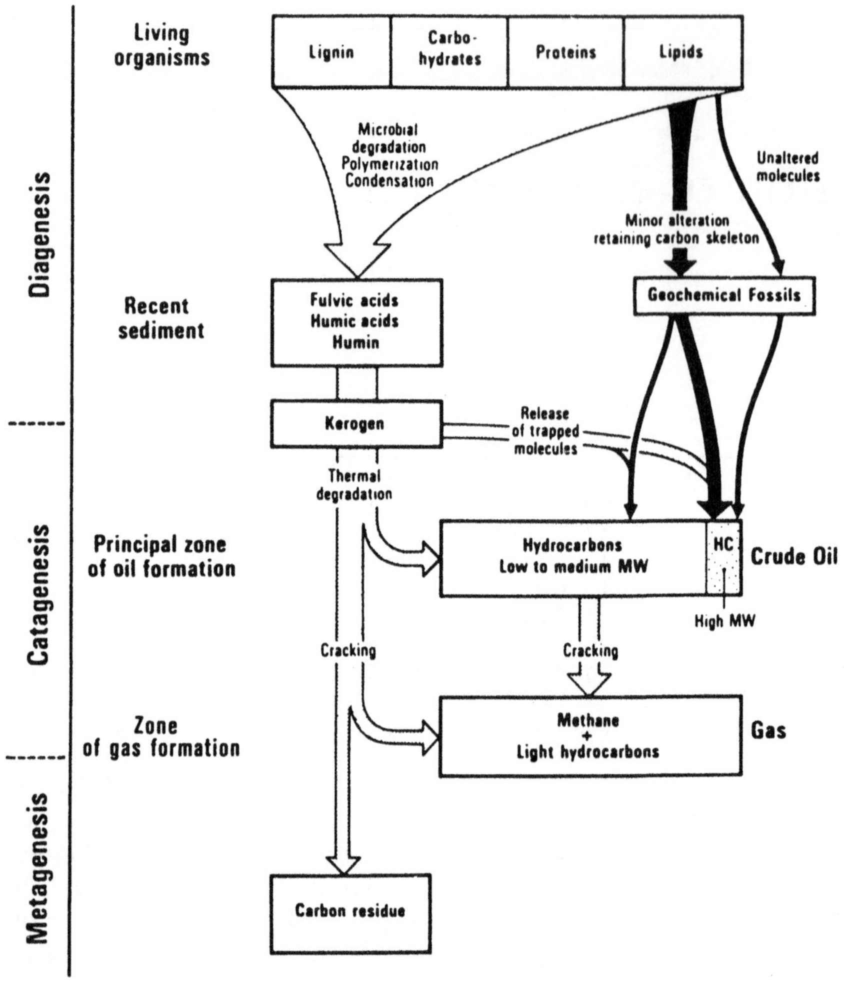

Figure 1. Stages of kerogen transformation and hydrocarbon formation pathways in geological situations. Direct inheritage of low-molecular-weight compounds is indicated by block arrows. Kerogen transformation is shown by gray arrows (from Tissot and Welte, 1984, p. 94).

peratures. Catagenesis results from an increase in temperature during burial of sediments and involves the formation of oil and gas from thermal breakdown of kerogen. The subsequent and last stage of the evolution of organic matter in sedimentary basins at high temperatures is called metagenesis.

Transformation of the initially deposited remains of decayed organisms starts in the water column and upper sediment layers with microbial action and low-temperature chemical reactions. Random polymerization and condensation reactions of degraded biopolymers are believed to form the initial "geopolymers," i.e., fulvic acids, humic acids, and humin. On a molecular level, these components are not well defined, but they are differentiated on the basis of their acid or base solubility. Several attempts were made to simulate this initial process by reacting sugars with amino acids in the laboratory to form so-called melanoidins (e.g., Hedges, 1978; Engel *et al.*, 1986), but the significance of the results is ambiguous.

Kerogen derives from preserved biogenic organic matter as well as humic materials and is an inhomogeneous macromolecular aggregate rather than a single polymer with well-defined monomer building blocks. It is insoluble in acid and base as well as in organic solvents. Its formation is complete by the end of the first transformation sequence, which is called diagenesis. The recent discovery of largely intact biogenic building blocks in the kerogen by mild chemical degradation (Michaelis and Albrecht, 1979) and of a new highly aliphatic polymer in plants as well as in fossil organic matter (e.g., Derenne *et al.*, 1989; Nip *et al.*, 1986; Tegelaar *et al.*, 1989a) raises doubt as to the validity of the concept that initially most of the biopolymers are hydrolyzed to their monomeric units in the uppermost sediment layers and then recombined by polycondensation to form kerogen (Tegelaar *et al.*, 1989b).

Thermal breakdown of kerogen, possibly supported by mineral catalysis, occurs in the catagenesis stage (Fig. 1). Under the influence of elevated temperatures (usually >50°C) and geologic time, liquid and gaseous hydrocarbons together with organic compounds containing heteroatoms such as oxygen, sulfur, and nitrogen are released from the macromolecular kerogen network, mainly by cleavage of carbon–carbon bonds. Oil not expelled from a source rock will be cracked to gas if subsidence continues and higher temperatures are reached.

The ultimate stage of hydrocarbon generation is called metagenesis (Fig. 1). Here, only dry gas (mainly methane) is stable. It derives from previously formed liquid hydrocarbons and to a certain extent is generated in the course of the high-temperature decomposition of kerogen.

Petroleum geologists commonly call the catagenesis range, in which oil is effectively produced from kerogen, the "oil window." This is usually illustrated with reference to a bell-shaped generation curve (Fig. 2) where the amount of total extract or extractable hydrocarbons (normalized to total rock or the organic carbon content) or the grossly equivalent S_1 peak from Rock-Eval pyrolysis (Espitalié *et al.*, 1985a, b, 1986) is shown as a function of burial depth or a maturation parameter (e.g., temperature, vitrinite reflectance). The geochemical processes within this oil window and their assessment by laboratory analysis and mathematical modeling will be described in more detail in the following sections.

2.2. Structural and Molecular Evolution of Kerogen

Formation of oil from kerogen in bulk chemical mass-balance terms is a disproportionation reaction, yielding a hydrogen-rich mobile phase and a hydrogen-depleted carbon residue. The latter may ultimately form graphite after crystalline reordering at the late metagenesis stage or beyond. This systematic elemental evolution of kerogen is commonly illustrated in a diagram of H/C versus O/C atomic ratios (van Krevelen diagram; Fig. 3), where different kerogen types are distinguished due to the different hydrogen and oxygen contents of the precursor materials (types I–III; Tissot *et al.*, 1974). Although to a variable extent for the different kerogen types, the main evolution pathways indicated in Fig. 3 involve an initial decrease of the O/C atomic ratio due to the loss of small oxygen-bearing molecules (CO_2, H_2O) during diagenesis and the release of bitumen enriched in heteroatomic substances during early catagenesis. During catagenesis and metagenesis, there is a drastic decrease of the H/C atomic ratio until the kerogen is transformed into an inert carbon residue. A diagram similar to that in Fig. 3 can be used to display kerogen evolution based on the results of Rock-Eval pyrolysis (Espitalié *et al.*, 1985a, 1986), which are more readily obtained. In this case, hydrogen indices (HI) and oxygen indices (OI) represent the pyrolysis yields of organic compounds

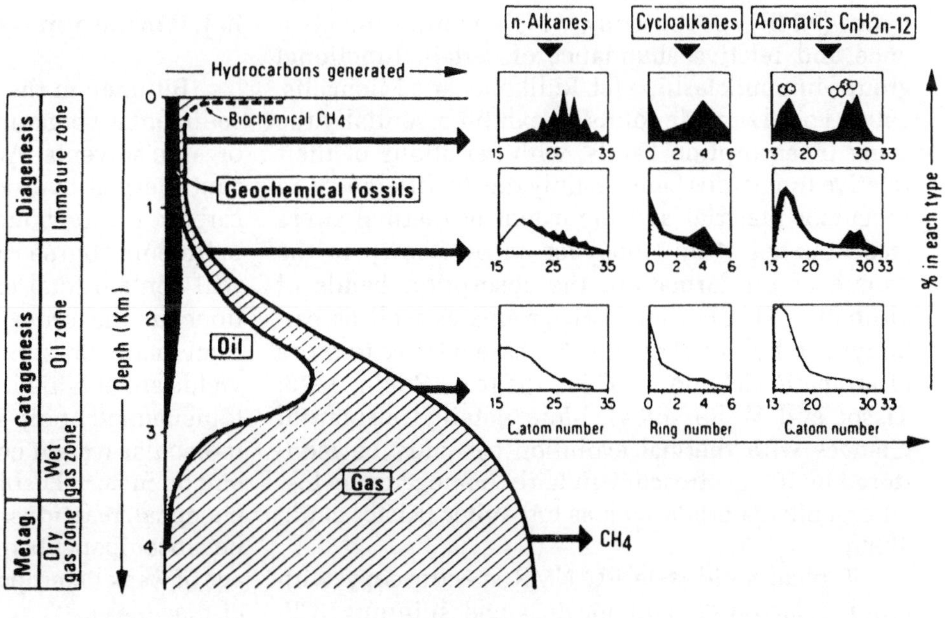

Figure 2. General scheme of hydrocarbon formation as a function of burial of the source rock. The evolution of the hydrocarbon composition for three compound classes is shown schematically in the insets. Depths are only indicative and may vary according to the actual geological situation. [From Tissot and Welte (1984, p. 215); modified after Tissot *et al.* (1974).]

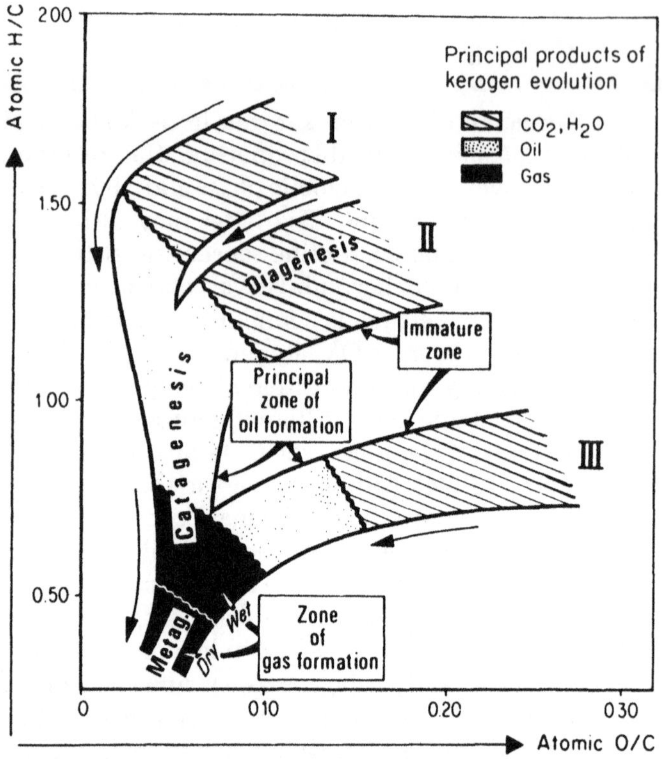

Figure 3. General scheme of kerogen evolution in a van Krevelen diagram. [From Tissot and Welte (1984, p. 216).]

and CO_2, respectively, both normalized to organic carbon content.

A step toward studying the effects of thermal alteration of kerogen on a more detailed structural level is the use of spectroscopic techniques, particularly infrared (IR) and ^{13}C nuclear magnetic resonance (NMR) spectroscopy, which monitor the presence and relative abundance of certain functional groups in a bulk fashion (cf. Rullkötter and Michaelis, 1990). Fossil organic solids all exhibit essentially the same IR absorption bands, with variations of their relative intensities being mainly due to differences in precursor material and the extent of thermal stress [see Rouxhet *et al.* (1980) for an overview]. In the course of oil formation, the absorption bands of aliphatic CH, CH_2, and CH_3 groups as well as carbonyl and hydroxyl groups decrease relative to those of aromatic CH and C=C groups (e.g., Robin, 1975; Tissot and Welte, 1984). More detailed structural changes with thermal evolution can only be monitored by IR spectroscopy in fairly homogeneous fossil organic materials such as torbanites (Kister *et al.*, 1990).

Typical solid-state ^{13}C NMR spectroscopy with cross polarization and magic angle spinning (CP/MAS) distinguishes two main broad signals in most fossil organic materials; these are assigned to an "aliphatic" and an "aromatic" region, respectively [see Wilson (1987) for an overview]. The ratio of these two broad signals is used to calculate the aromaticity factor (f_a) of the organic matter, which systematically increases during oil formation as a consequence of the release of aliphatic hydrocarbons (e.g., Witte *et al.*, 1988). As maturity increases, a methyl group signal is partially resolved, indicating the importance of small aliphatic substituents on an increasingly aromatic kerogen nucleus (Witte *et al.*, 1988).

In contrast to the bulk elemental and structural changes, the molecular evolution of kerogen is not easy to follow because of the heterogeneous macromolecular nature of this fossil organic matter. A detailed molecular investigation of kerogen is tedious and requires the careful selection of reagents for chemical degradation (Rullkötter and Michaelis, 1990). The application of analytical pyrolysis with subsequent gas chromatography or mass spectrometry (or a combination of both) is often used as a compromise, particularly because it is fast relative to chemical degradation (cf. Chapter 13). Certain structural units can be recognized in this way, but there is no information on how these are linked together. Both open-system and closed-system pyrolysis provide a reasonable estimate, however, of the remaining hydrocarbon potential of a specific rock and of the type of hydrocarbon compounds that can be expected from this rock under natural conditions of oil and gas formation.

2.3. Formation and Composition of Bitumen

Bitumen in the sense used here is the portion of sedimentary organic matter which is extractable with organic solvents. At the diagenesis level, no significant thermal hydrocarbon generation by cracking of carbon–carbon bonds has occurred, but good-quality petroleum source rocks, even in the immature stage, will contain total extractable material at concentrations in the order of 30–50 mg/g C_{org} (in dichloromethane or similar solvents; somewhat higher yields are obtained with more polar solvents such as toluene/methanol). Most of this is inherited as low-molecular-weight compounds from the biogenic precursor material, slightly altered by diagenetic geochemical reactions, but some also consist of polar heterocomponents released from the kerogen by cleavage of weak heteropolar bonds, particularly at the end of diagenesis.

Carbonate source rocks or other clay-mineral-poor sediments rich in organic matter (evaporites, diatomites, phosphorites) may yield higher amounts of extractable bitumen also at low levels of thermal stress, particularly when large amounts of sulfur were incorporated in the organic matter during early diagenesis (e.g., Hunt and McNichol, 1984). The extractable material of these rocks predominantly contains high-molecular-weight material (asphaltenes) which is considered to constitute smaller building blocks of kerogen (Palacas, 1983; Béhar *et al.*, 1984). Apparently, in carbonates (and the other types of rocks mentioned) the organic matter does not reach the same high degree of condensation as in clay-bearing rocks.

The increase in bitumen (oil) in source rocks during catagenesis shown as an idealized hydrocarbon generation curve (Fig. 2) was originally based on investigations in the Paris Basin and the Douala Basin (Tissot *et al.*, 1974). Actually, the curve is an envelope of a multitude of broadly scattered data points. Maximum values typically observed in source rock at peak generation range from 150 to an exceptional 250 mg hydrocarbon/g C_{org}. Attempts to construct similar generation curves in other prolific sedimentary basins often failed to show the bell-shaped feature (e.g., Larter, 1988). This is due to common vertical facies variations in many well profiles (even if several wells are combined to show mainly one particular source rock horizon), as well as to the fact that the concentration of (residual) bitumen in a source rock is the net result of hydrocarbon generation, hydrocarbon expulsion, and cracking of bitumen into gas. In contrast to the concept prevailing in petroleum geochemistry for a long time, recent investigations have shown that the latter two processes are far from being negligible or even only moderate compared to the first one (cf. Section 3). With respect to gas generation, the representation given in Fig. 2 is just a conceptual extension of the oil generation curve because measured gas concentrations in source rock are much lower than bitumen concentrations both for technical reasons (depressurizing during sampling) and due to the greater mobility of gases even in rocks with low permeabilities.

Figure 2 also illustrates some of the compositional changes occurring in source rock bitumen as a consequence of continuing oil formation. Molecular distributions typical of inherited biological material are progressively masked by the smoother hydrocarbon distributions derived from random thermocatalytic cracking of carbon–carbon bonds. In addition, there is a tendency for the conversion of heteroatom-bearing compounds into hydrocarbons (mainly an increase in the concentration of saturated hydrocarbons at the expense of heteroatom-bearing compounds with the relative proportion of aromatic hydrocarbons being less affected) and the thermal destruction of larger molecules into smaller ones as illustrated by the shift of carbon number maxima to lower values in Fig. 2.

2.4. Crude Oil from Coal

Coal has long been considered a source rock mainly for gas (Karweil, 1969; Stahl, 1978; Bartenstein and Teichmüller, 1974), although minor oil occurrences related to coal-bearing strata had been documented (e.g., Kent, 1954). There are, however, only a few larger oil accumulations where massive coal beds were identified as the source [e.g., Durand *et al.*, 1983; see Tissot and Welte (1984) for an overview]. This is in contrast to the reasonably high hydrogen indices and pyrolysis yields of many coals (Durand and Parratte, 1983; Durand *et al.*, 1983; Horsfield *et al.*, 1988; Littke and ten Haven, 1989). Durand *et al.* (1983) invoked a geological reason for the low frequency of oil accumulations associated with coals, i.e., a dominance of dissipative migration in most cases. With a better understanding of the relationship between oil generation and expulsion from the source rock, however, the high gas and low oil potential of many coals was related to the fact that, due to the occurrence of coals as massive beds and their internal microporous structure, initially formed liquid hydrocarbons are mostly retained in the coal and at a later maturation stage largely converted to gas (Tissot *et al.*, 1984; Monthioux and Landais, 1987; Hvoslef *et al.*, 1988; Littke *et al.*, 1990), although some doubts as to the validity of these ideas persist (Durand, 1988).

The Mahakam (Durand and Oudin, 1979; Huc *et al.*, 1986) and Niger deltas (Bustin, 1988), the Gippsland Basin in Australia (Thomas, 1982), and the Haltenbanken area in the Norwegian North Sea (Hvoslef *et al.*, 1988; Khorasani, 1989) are well-documented cases where oil pools are found together with gas condensate accumulations and the presence of humic coals as their (co)source can be invoked. Associations of coal-bearing nonmarine source rocks with (marginally) marine source rocks are particularly common in Tertiary basins in Indonesia and other Southeast Asian countries (e.g., Roe and Polito, 1977; Curry, 1985; Gordon, 1985; Noble *et al.*, 1991). These coals are often rich in liptinite macerals (e.g., resinite) with hydrogen indices of 200–400 mg HC/g C_{org} (Horsfield *et al.*, 1988). Expulsion of oil from these coaly layers into carrier beds may be facilitated by small-scale interbedding of coals and shales. The alternative possibility that the high conversion rates of liptinites to liquid hydrocarbons results in deactivation of adsorptive sites and provides the critical oil saturation required for primary migration by continuous monophasic flow has been suggested by Horsfield *et al.* (1988).

2.5. Maturation Parameters

The progress of the thermal alteration of kerogen and of oil formation can be monitored by a number of so-called maturation parameters. Systematic changes in elemental composition and spectroscopic properties of kerogen with increasing thermal stress have already been mentioned. The kerogen maturation parameters most widely applied at present, however, are vitrinite reflectance (Stach *et al.*, 1982) and the temperature of maximum pyrolysis yield (T_{max}) from Rock-Eval pyrolysis (Espitalié *et al.*, 1985a). A comprehensive review of bulk and optical maturity indicators and their relation to hydrocarbon generation was published by Héroux *et al.* (1979).

On the product side, molecular maturation parameters range from the relatively simple carbon preference index (CPI) of *n*-alkanes (Bray and Evans, 1961) to the use of sophisticated geochemical reactions of biological markers [see Mackenzie (1984) for an overview]. Recent studies have shown, however, that the understanding of the reaction mechanisms involved is far from being adequate and that the chemical basis for some of the systematic changes observed (e.g., sterane isomer ratios) is more complex than previously thought (Rullkötter and Marzi, 1989; Abbott *et al.*, 1990; Peters *et al.*, 1990; Marzi and Rullkötter, 1990). Thus, the related maturity parameters should be applied with the necessary caution. While biological marker parameters are most useful in the early-to-peak generation range, aromatic hydrocarbon parameters can also be applied at higher maturity levels [see Radke (1987) and Alexander *et al.* (1988) for overviews].

3. Quantitation of Hydrocarbon Generation

Although the consecutive processes of hydrocarbon generation, expulsion (primary migration), secondary migration, and accumulation have been reasonably well understood in qualitative terms for quite some time (cf. Tissot and Welte, 1984), there has only more recently been some significant improvement in the quantitative assessment of hydrocarbon generation and expulsion (Merewether and Claypool, 1980; Cooles *et al.*, 1986; Quigley *et al.*, 1987; Durand, 1988; Leythaeuser *et al.*, 1988; Rullkötter *et al.*, 1988; Ungerer *et al.*, 1990). This is considered an important progress toward a more successful application of petroleum geochemistry in exploration (Tissot *et al.*, 1987a; Mackenzie and Quigley, 1988; Tissot and Ungerer, 1990).

3.1. The Hils Syncline as a Natural Laboratory

In the assessment of a mass balance for petroleum generation in field studies, two major problems are usually faced. Firstly, the hydrocarbons formed in a source rock are mobile, and a certain portion may have left the system and thus escaped detection. The use of a hydrocarbon generation curve (Fig. 2) to follow the progress of petroleum formation based on the amount of extractable bitumen in the source rock, however, assumes this effect to be relatively small and thus negligible (expulsion efficiencies of up to 10% are cited in the earlier literature; e.g., Hunt, 1977; Barker, 1979). Secondly, organic matter is not homogeneously distributed within a source rock layer, and it is difficult to find representative samples at the initial immature stage and all levels of hydrocarbon generation without changes in organic carbon content or organic facies from the basin margin to the center.

A favorable geological situation for a mass-balance study of hydrocarbon generation and expulsion exists in the Hils syncline, northern Germany (Fig. 4; Rullkötter *et al.*, 1988), which is a topographic high located near the eastern flank of a deep-seated intrusive body (Vlotho Massif). Heat released from this intrusion during the Cretaceous has accelerated maturation of organic matter in most sediment layers in the vicinity. Among these rocks is the Lower Toarcian Posidonia Shale, a petroleum source rock in northern Germany and other parts of Europe. Since the Cretaceous this shale has been uplifted to the surface in the Hils syncline, and unweathered core material could be recovered in shallow boreholes at depths between about 30 and 80 m. The samples from the different drilling locations (Fig. 4) span the whole range of organic matter maturity, as measured by percent vitrinite reflectance (% R_0), from immature at Wenzen in the southeast (0.48% R_0) to overmature at

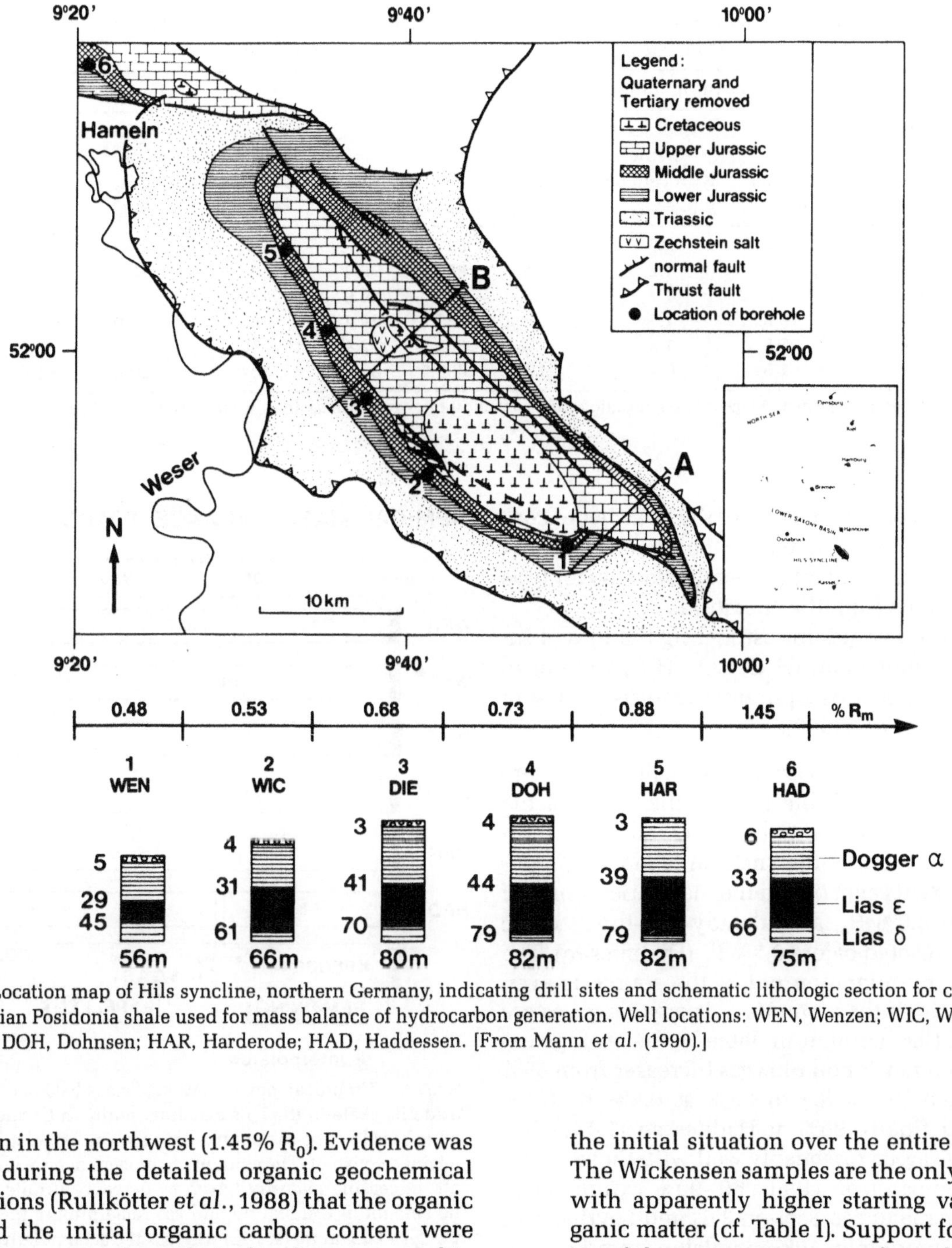

Figure 4. Location map of Hils syncline, northern Germany, indicating drill sites and schematic lithologic section for core samples of Lower Toarcian Posidonia shale used for mass balance of hydrocarbon generation. Well locations: WEN, Wenzen; WIC, Wickensen; DIE, Dielmissen; DOH, Dohnsen; HAR, Harderode; HAD, Haddessen. [From Mann *et al.* (1990).]

Haddessen in the northwest (1.45% R_0). Evidence was obtained during the detailed organic geochemical investigations (Rullkötter *et al.*, 1988) that the organic facies and the initial organic carbon content were very uniform over the relatively short geographic distance (a few tens of kilometers).

The mass-balance calculations for the amounts of hydrocarbons generated and expelled from the Lower Toarcian shales in the Hils syncline are based on average organic geochemical data of about 10–20 samples per well and the following major assumptions:

- The average organic carbon content of the Wenzen samples (10.7%) and their organic matter type were taken as being representative of the initial situation over the entire study area. The Wickensen samples are the only exception, with apparently higher starting values of organic matter (cf. Table I). Support for the validity of the assumption came from the results of sedimentological and organic petrographic investigations as well as dead carbon measurements (see Section 3.2) on a number of samples.
- The amount of mineral matrix was taken as being constant. All data refer to dry sediment, so that there is no effect of the loss of compaction water. In addition, rapid heating of the sediments apparently left the clay mineral composition largely unchanged.

TABLE I. Extent of Kerogen Conversion and Expulsion Efficiencies Calculated from Mass-Balance Results in the Hils Syncline

Borehole	Residues (mg/g initial rock)							Losses				
	Kerogen (mg)	Kerogen (%)	Extract (C_{15+})	Medium-volatile hydrocarbons (C_8–C_{14})	Light hydrocarbons (C_2–C_7)	Matrix		Total (mg)	Percent of transformed kerogen	H_2O + H_2S + CO_2 (mg)	Oil + gas (mg)	Expulsion efficiency (oil + gas) (%)
Wenzen	140.8	100	4.2	0.9	0.6	853.5		—	—	—	—	—
Wickensen	129.1	92	6.4	1.2	0.4	853.5		9.4	80	12.8	n.c.[a]	n.c.
Dielmissen	94.2	67	12.0	1.1	0.2	853.5		39.0	84	14.4	24.6	65
Dohnsen	94.9	67	10.5	1.5	0.2	853.5		39.4	86	16.0	23.4	66
Harderode	65.4	46	6.8	1.2	0.7	853.5		72.4	96	17.6	54.8	86
Haddessen	61.5	44	1.6	0.4	0.3	853.5		82.7	104	20.8	61.9	96

[a]n.c., Not calculated, because apparently higher initial organic matter content in the Wickensen samples would lead to unrealistic mass-balance data.

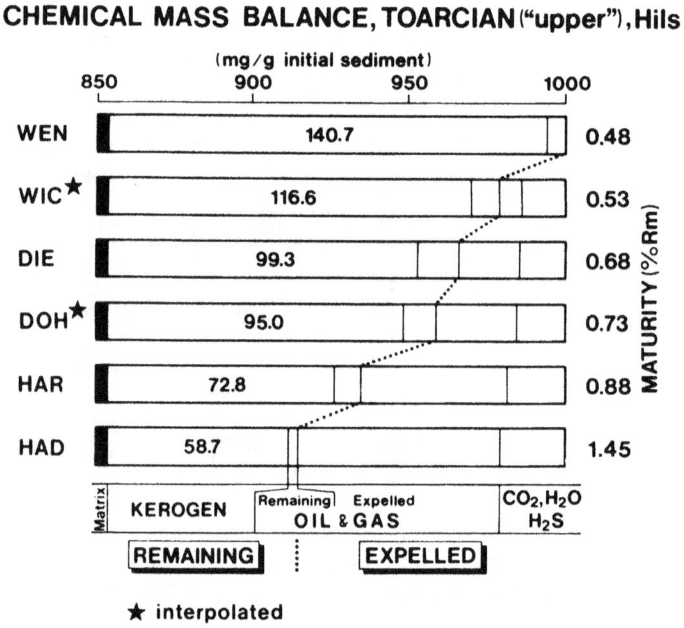

Figure 5. Graphical representation of mass-balance results for the Posidonia shale in the Hils syncline, northern Germany. Calculation is based on the assumption that the amount of mineral matrix remained constant. The amount of kerogen then steadily decreases with increasing maturity (given as percent vitrinite reflectance). The amount of residual oil and gas remaining in the rock passes through a maximum (at DIE) whereas the accumulative quantity of expelled oil and gases increases with increasing maturity as does the loss of small inorganic molecules (CO_2, H_2O, H_2S). The values for WIC and DOH are interpolated (compare with experimentally determined values in Table I). For method of determination, see Rullkötter *et al.* (1988). See caption to Fig. 4 for definitions of abbreviations of well locations.

- Organic carbon contents of the rocks were converted to organic matter contents by using maturity-dependent conversion factors for type II kerogens from the literature.
- Maximum values for estimating the loss of inorganic molecules (H_2O, CO_2, H_2S) were used based on spectroscopic and pyrolysis studies of Ungerer *et al.* (1983).

The results of the mass-balance calculations are shown in Fig. 5 and Table I. More than 50% of the kerogen is transformed into bitumen and small inorganic molecules under the "natural laboratory" conditions of the Hils syncline in the maturation range of about 0.5 to 0.9% R_0, and already one-third of the kerogen has reacted before 0.7% R_0 (Dielmissen). The amount lost by migration out of the source is even more dramatic. If the amounts of inorganic gases are disregarded (last column in Table I), the average expulsion efficiency for oil plus gas increases from 65% at Dielmissen (0.7% R_0) to 86% at 0.9% R_0 (Harderode) and finally 96% at Haddessen (1.45% R_0). These results agree reasonably well with those calculated by Cooles *et al.* (1986; 60–90% expulsion of total oil generated; see Section 3.2) and Leythaeuser *et al.* (1987; more than 50% expulsion from kerogen type II source rocks at about 0.8% R_0). This clearly shows that an attempt to quantitate hydrocarbon generation based on extract yields of mature petroleum source rocks will lead to a great underestimation of the hydrocarbon potential already realized. The cumulative amount of hydrocarbons generated up to 0.88% R_0 (Harderode) in the Hils syncline, based on the figures in Table I, would be about 600 mg per initial gram of organic carbon. This compares well with the approximately 500 mg HC/g C_{org} determined by Leythaeuser *et al.* (1987) for a similar source rock type. The corresponding measured value in the Harderode samples was less than 150 mg/g C_{org} (Rullkötter *et al.*, 1988). Expulsion apparently starts at a relatively early maturation stage when the capacity of the source rock pores for the storage of the newly generated hydrocarbons is exceeded.

3.2. The "Dead Carbon" Approach

Cooles *et al.* (1986) solved the problem of assessing the total hydrocarbon potential of a source rock from a mature sample which has already spent some of its potential and lost part (or most) of the hydrocarbons generated during primary migration, by measuring the amount of "dead carbon" in this particular rock. A review of the problems involved in this approach and of earlier attempts in this direction was given by Larter (1988).

The philosophy of the "dead carbon" approach is based on a subdivision of the kerogen in a source rock into several fractions (Fig. 6). Their relative abundances are predetermined by the initial composition of the organic matter before the onset of thermocatalytic hydrocarbon generation. The fraction of the kerogen which will be hydrocarbons when all of the generation potential is used up is called *reactive*; the remaining fraction is termed *inert*. The reactive portion is further subdivided into a *labile* component, which can generate both oil and gas at low to moderate maturity, and a *refractory* component, which mainly yields gas at higher maturities. The central assumption is that the concentration of the inert kerogen fraction remains constant during the main phase of petroleum generation; i.e., this fraction can be determined experimentally at each maturity level by heating the kerogen until all of the remaining generation potential is spent and by measuring the amount of residual organic matter ("dead carbon"). Multiplying factors are used to convert from mass of organic carbon (determined experimentally) to mass of organic matter in order not to experimentally deal with other elements leaving the system during heating.

Because the initial kerogen concentration within a source rock usually is not spatially constant, it is necessary to normalize all quantities to the amount of inert kerogen fraction for any given sample. On the other hand, it has to be assumed that the type of organic matter is homogeneous within reasonable limits for a particular source rock. The definition of a petroleum generation index (PGI),

$$\text{PGI} = \frac{\text{Petroleum generated} + \text{Initial petroleum}}{\text{Total petroleum potential}} \tag{1}$$

where petroleum means the sum of oil plus gas, uses the concentrations of the different kerogen fractions (Fig. 6) in the following way:

$$\text{PGI} = \frac{[(c_{KL}{}^{0'} + c_{KR}{}^{0'}) - (c_{KL} + c_{KR}) + c_{O}{}^{0'}]}{[c_{KL}{}^{0'} + c_{KR}{}^{0'} + c_{O}{}^{0'}]} \tag{2}$$

where the subscripts are as defined in the caption to Fig. 6 and the superscripts 0′ relate to the initial concentrations of kerogen fractions in the now mature rock sample when it was immature. Using an immature source rock sample for comparison and making the assumption that the relative concentrations of the kerogen fractions in the now mature rock and in the presently immature rock, normalized to

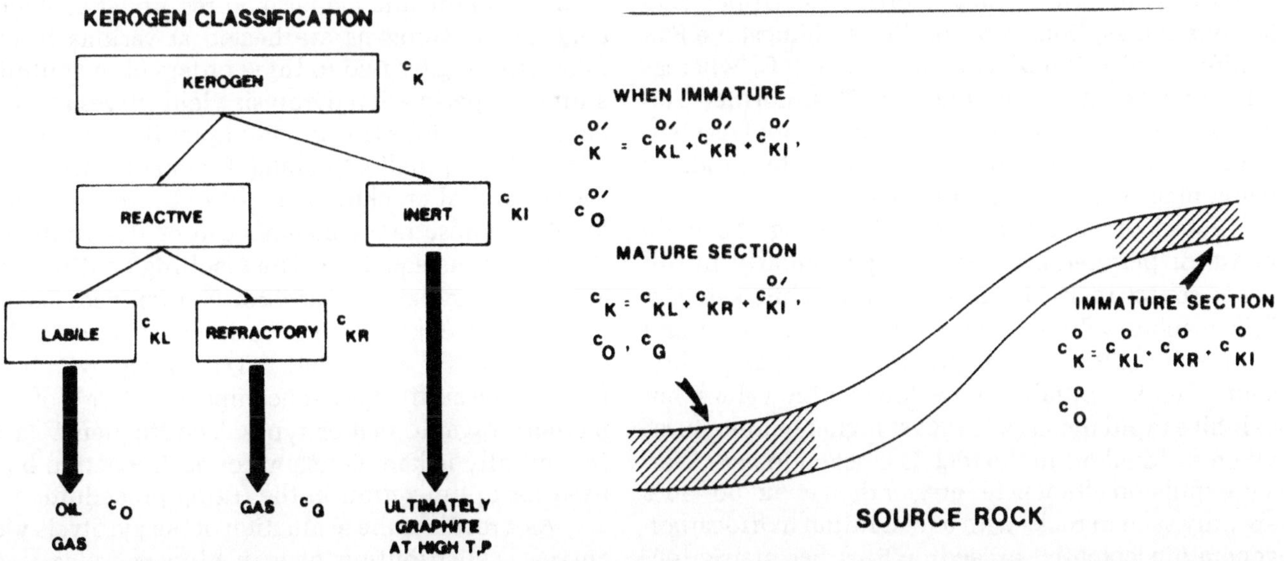

Figure 6. Schematic classification of kerogen according to its potential to generate petroleum in the subsurface, and definition of relative concentrations of total kerogen (c_K), inert kerogen (c_{KI}), labile kerogen (c_{KL}), and refractory kerogen (c_{KR}) for mature and immature source bed sections together with corresponding oil (c_O) and gas (c_G) concentrations. [After Cooles *et al.* (1986).]

dead carbon, are constant, all parameters in Eq. (2) can be determined experimentally. PGI may vary from close to zero in immature rocks to 1.0 for source rocks which have realized all their petroleum potential.

In a similar way, the petroleum expulsion efficiency (PEE) is defined as

$$\text{PEE} = \frac{\text{Petroleum expelled}}{\text{Petroleum generated} + \text{Initial petroleum}} \quad (3)$$

and, using the concentrations defined above,

$$\text{PEE} = \frac{[(c_{KL}{}^{0'} + c_{KR}{}^{0'}) - (c_{KL} + c_{KR}) + c_{O}{}^{0'}] - (c_O + c_G)}{[c_{KL}{}^{0'} + c_{KR}{}^{0'} - (c_{KL} + c_{KR}) + c_{O}{}^{0'}]} \quad (4)$$

PEE may vary from zero (no expulsion) to 1.0 (all hydrocarbons expelled).

Applying this approach to petroleum source rocks in a number of different sedimentary basins, Cooles *et al.* (1986) found that PGI increases steadily with increasing maturity (vitrinite reflectance) although the rates differ as a function of the type of organic matter. In a source rock containing mainly labile reactive kerogen (e.g., Kingak Shale, Alaska North Slope), PGI has reached its end value (1.0) at 1% vitrinite reflectance; PGI at the same level of vitrinite reflectance is only about 0.6 in a source rock containing a mixture of approximately equal amounts of labile and refractory reactive kerogen (e.g., Lias δ shales in northern Germany). The study of Cooles *et al.* (1986), in addition, showed that the main temperature zone of petroleum generation for rocks containing labile reactive kerogen (e.g., Kingak Shale, Alaska; Kimmeridge Clay, North Sea; Upper Cretaceous shales, Douala Basin; Brown Limestone Formation, Gulf of Suez) is about 120–150°C, whereas refractory kerogen (e.g., Lias δ shales, norther Germany; Westphalian coals, northwest Europe) reacts at temperatures up to about 250°C, with the products being mainly gas in the latter case.

Petroleum expulsion is high during the main phase of petroleum generation, particularly in oil-prone source rocks containing mainly labile kerogen. PEE reaches values of 0.6 to 0.9 in this range and approaches 1.0 when PGI tends to 1.0. Less prolific source rocks initially show lower PEE values but exhibit a rapid increase in PEE at higher temperatures when residual oil in the rock is cracked to gas. Average expulsion efficiencies greater than about 60–90% are only seen in rocks with a total initial hydrocarbon generation potential exceeding 5 kg per metric ton, corresponding to an initial total organic carbon content of about 1.5%. Below this level of organic matter richness, oil expulsion is considerably less efficient. This means that most of the oil generated in poor source rocks is never expelled as oil but rather is cracked to gas at higher temperatures associated with deeper burial [cf. Durand (1988) and discussion on oil generation from coal]. The results of Cooles *et al.* (1986) also indicate again that the geochemical assessment of source rocks heated naturally to temperatures above 120°C (or vitrinite reflectance greater than 0.6%) will seriously underestimate initial (total) petroleum potential.

3.3. Kinetics of Oil Formation

It has been shown empirically that kinetic models based on first-order reactions with rate constants following the Arrhenius law can satisfactorily describe the formation of oil (and gas) from kerogen in sediments in bulk terms ("macrokinetics"). This approach has been developed and refined over the last 20 years, initially starting from geological observations (Tissot and Espitalié, 1975; Hood *et al.*, 1975; Mackenzie and McKenzie, 1983; Ungerer *et al.*, 1986; Sweeney *et al.*, 1987) and pyrolysis of oil shales (e.g., Campbell *et al.*, 1980; Burnham and Braun, 1985) or coal (Hanbaba, 1967; Jüntgen and Klein, 1975).

Presently, the kinetic data for hydrocarbon generation from a given kerogen are commonly determined from microscale pyrolysis in the laboratory (Ungerer and Pelet, 1987; Burnham *et al.*, 1987; Schaefer *et al.*, 1990), which avoids the uncertainties involved in the reconstruction of time–temperature histories of sediments with organic matter matured under geological conditions. Kerogens are heated at various heating rates differing by two to three orders of magnitudes, and the experimental pyrolysis yield curves are modeled using various fitting techniques (Burnham *et al.*, 1988). Conceptually, the models consider an infinite set of parallel cracking reactions obeying first-order kinetics whose rate constants can be determined by the Arrhenius equation. The resulting continuum of activation energies is arbitrarily digitized, e.g., in intervals of 2 kcal mol^{-1} between 40 and 80 kcal mol^{-1} (Ungerer and Pelet, 1987), a range which was found to be sufficient to encompass all kinds of sedimentary organic matter types. The frequency factor (A) initially is kept constant for each reaction but is used for optimization in the fitting procedure.

As a result of the evaluation of the pyrolysis yield curves, a distribution of activation energies is obtained for a specific kerogen sample with a certain fraction of the initial hydrocarbon potential (X_{i0}, expressed in terms of milligrams of hydrocarbons per gram of organic carbon) assigned to each activation energy (Fig. 7). As is obvious from Fig. 7, the activation energy distributions for different kerogen types differ both in their shapes and in the position of their maxima. Type III organic matter of terrigenous origin (Tertiary coal, Mahakam Delta, Indonesia) has a broad

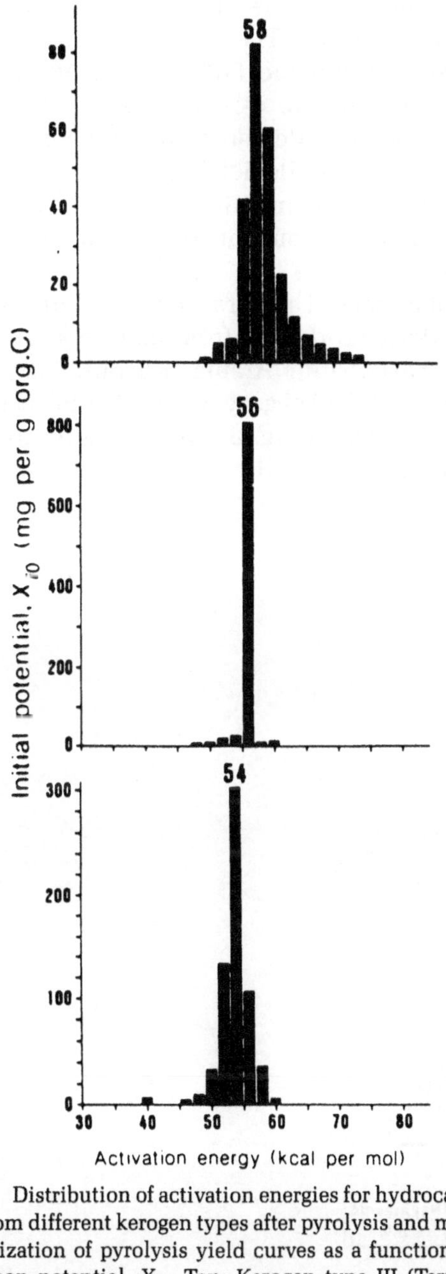

Figure 7. Distribution of activation energies for hydrocarbon generation from different kerogen types after pyrolysis and mathematical optimization of pyrolysis yield curves as a function of initial hydrocarbon potential. X_{i0}. Top: Kerogen type III (Tertiary coal, Mahakam Delta, Indonesia); middle: kerogen type I (Eocene Green River Shale, Uinta Basin, USA); bottom: kerogen type II (Lower Toarcian shales, Paris Basin, France). [From Ungerer and Pelet (1987); Copyright by Macmillan Journals, 1987).]

and asymmetric distribution with activation energies extending from 50 to more than 70 kcal mol^{-1} and the highest activation energy maximum of the three kerogen types shown in Fig. 7. The cumulative hydrocarbon potential (250 mg/g C_{org}), however, is relatively low. Lacustrine aquatic organic matter (type I; Eocene Green River Shale, Uinta Basin, Utah) exhibits an extremely narrow activation energy distribution with a maximum slightly lower than that of continental organic matter, discussed before. The total hydrocarbon potential (900 mg/g C_{org}) is the highest of the three samples shown. Finally, well-preserved marine type II organic matter from the Lower Toarcian shales of the Paris Basin (Jurassic) has a slightly broader activation energy distribution again and an intermediate total potential (630 mg/g C_{org}) but the lowest activation energy maximum.

With kinetic models of this type, the influence of time and temperature on the formation of oil and gas from kerogen can be determined more accurately than with the earlier, simpler models. Most importantly, it was found that reaction rates do not double for each temperature increase of 10°C as in the Lopatin/Waples model (Waples, 1980), but rather for a 3–5°C increase (Sweeney *et al.*, 1987; Tissot *et al.*, 1987b).

4. Composition of Crude Oils

4.1. Crude Oil Classification

Crude oils can be characterized by their bulk properties as well as their chemical compound class or molecular composition [see Tissot and Welte (1984) for a more detailed overview]. Distillation of crude oils provides fraction profiles over a certain boiling range. The crude oils as well as the distillation fractions can be described in term of density, viscosity, refractive index, sulfur content, or other bulk parameters. A widely used parameter in the oil industry is the API gravity, which is inversely proportional to the density. Most conventional crude oils range from about 25 to 45°C API gravity. Oils with API gravities less than about 20° are called heavy oils, and those with API gravities above 50°C can occur as condensates, although these are not strict boundaries (Waples, 1985).

In organic geochemical studies, crude oils are commonly separated into fractions of different polarities using column or thin-layer liquid chromatography (e.g., Radke *et al.*, 1980). Usually, after removal of the low-boiling components at a certain temperature and under reduced pressure ("topping"), asphaltenes are precipitated by the addition of a nonpolar solvent (e.g., *n*-hexane). The soluble (maltene) portion is separated into saturated hydrocarbons, aromatic hydrocarbons, and a fraction containing the polar, heteroatomic compounds (NSO compounds or resins).

Further subfractionation facilitates subsequent studies on a molecular level using less complex mixtures. For example, the saturated hydrocarbon fraction can be treated with 5-Å molecular sieves or urea for the removal of the *n*-alkanes, leaving behind a fraction of branched and cyclic alkanes. Aromatic hydrocarbons are often further fractionated according to the number of aromatic rings into mono-, di-, tri-, and polyaromatic hydrocarbons (Radke *et al.*, 1984).

The ternary diagram in Fig. 8 shows the composition of more than 500 crude oils based on the content of normal plus isoalkanes (paraffins), cycloalkanes (naphthenes), and aromatic hydrocarbons plus NSO compounds. This compound class subdivision provides an optimal spread of the data over the diagram, thus allowing a classification of the oils into six groups. Crude oils with more than 50% saturated hydrocarbons are called "paraffinic" or "naphthenic" depending on their relative contents of normal and branched alkanes and of cycloalkanes. Paraffinic oils are light, but many have a high viscosity at room temperature due to their high content of long-chain wax alkanes. A high wax content is often an indication of a strong terrigenous contribution to the organic matter in the source rock, and thus such oils are commonly generated from deltaic or lacustrine sediments as in Indonesia and West Africa. However, type I kerogen may also generate highly paraffinic crude oils as is the case in the Green River Formation of the Uinta Basin, Utah. The most common oil type is paraffinic–naphthenic (40% of the oils in Fig. 8), whereas naphthenic oils are scarce and often the result of microbial degradation in the reservoir (Connan, 1984). All the oils rich in saturated hydrocarbons are commonly low in sulfur content (<1%).

Aromatic intermediate oils contain 40–70% aromatic hydrocarbons and a high proportion of resins and asphaltenes. They are rich in sulfur and mostly heavy. Many crude oils from marine sources in the Middle East (Saudi Arabia, Kuwait, Iraq, etc.) are included in this category, which is the second most important class in Fig. 8. The aromatic–naphthenic and aromatic–asphaltic classes are mostly repre-

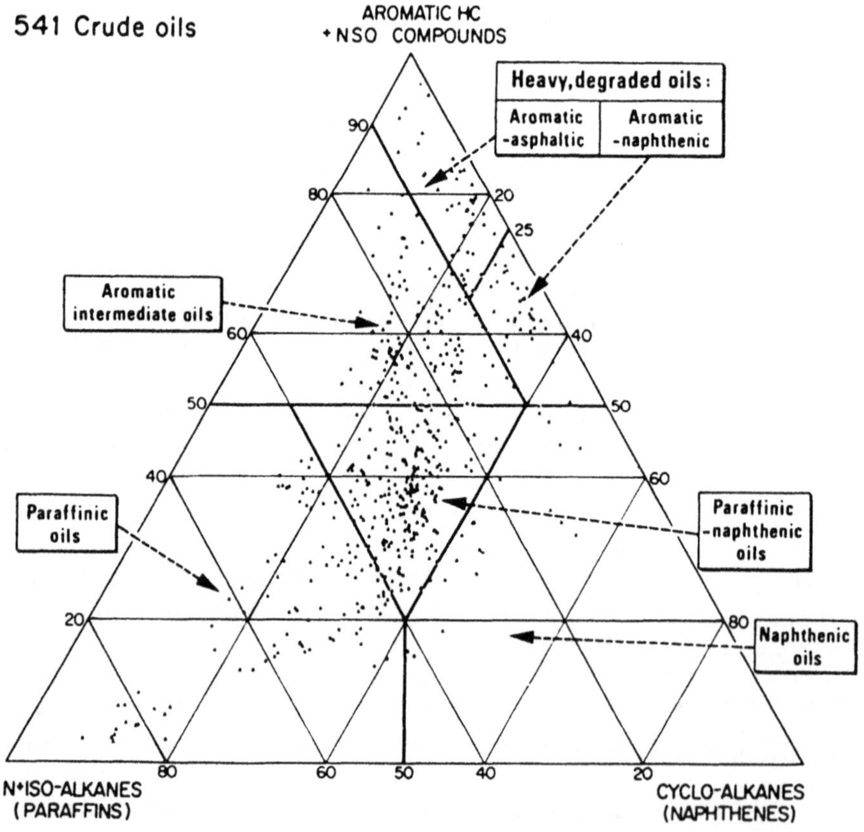

Figure 8. Ternary diagram showing the composition of six classes of crude oils based on the analysis of 541 oils. [From Tissot and Welte (1984, p. 419).]

sented by altered crude oils. They are heavy but the sulfur content may vary according to the original type of the crude oil. The Cretaceous heavy oils of Athabasca are typical examples of these last two classes.

4.2. Molecular Constituents of Crude Oils

Crude oils not affected by secondary alteration commonly contain saturated hydrocarbons as the most abundant group of compounds. A considerable proportion of these may be in the low-molecular-weight range. Identification of individual molecules in this range started with API Research Project No. 6 and led to the recognition of several hundred different structures with up to 15 carbon atoms (Martin *et al.*, 1963; Poulet and Roucaché, 1970; Erdman and Morris, 1974). Although this is far from a comprehensive elucidation of all the molecular structures in crude oils, these low-molecular-weight compounds may represent more than 50% of a crude oil. This is often not considered when geochemical work is concentrated on the C_{15+} saturated hydrocarbon fraction obtained from liquid-chromatographic separation.

n-Alkanes are the most significant compounds in almost all unaltered oils, followed by the isoprenoid hydrocarbons pristane and phytane in the C_{15+} carbon number range. Biological markers with polycyclic saturated hydrocarbon structures, particularly steranes and triterpanes, although present in relatively low concentrations in most crude oils, have received a lot of attention because of the high information content in their specific structures [see Mackenzie (1984), Moldowan *et al.* (1991), and Peters and Moldowan (1993) for overviews]. They have been extensively used for the assessment of the source rock type and its depositional environment, oil–oil and oil–source rock correlation, and maturity assessment. The biological marker compounds are efficiently resolved and identified by gas-chromatographic/mass-spectrometric analysis. A group of molecules which probably is always present but becomes particularly important in microbially altered (biodegraded) oils is not resolved even by high-resolution capillary columns and then forms the so-called "unresolved complex mixture" (UCM) or "hump." A recent attempt to obtain structural information on these molecules by chemical degradation provided some evidence that they may be "T"-shaped alkanes (Gough and Rowland, 1990).

Analysis of aromatic hydrocarbons, apart from low-molecular-weight constituents and aromatic steroid biological markers, concentrated on alkylated naphthalenes and phenanthrenes and their maturity aspects (Radke, 1987). In addition, aromatic sulfur-bearing compounds have been identified in particularly sulfur-rich oils (e.g., Orr and Sinninghe Damsté, 1990; Strausz *et al.*, 1990; Schmid *et al.*, 1987; Sinninghe Damsté *et al.*, 1988).

The heterocomponent fractions of crude oils have not been characterized in the same detail as the hydrocarbon fractions. There are some studies, however, on the molecular structure of carboxylic acids (e.g., Seifert and Howells, 1969; Seifert *et al.*, 1972) and nitrogen-bearing heterocyclic components (Schmitter and Arpino, 1985).

4.3. Geological Factors Influencing Crude Oil Composition

The depositional environment of the source rock, its thermal evolution, and secondary alteration processes are the most important factors which determine the composition of crude oils. Among the environmental factors, those that influence the nature of the organic matter in the source rock and its mineral composition are of primary significance.

Although all hydrocarbon source rocks are deposited under aquatic conditions, they may contain varying amounts of land-derived organic matter. The terrestrial contribution can be predominant particularly in intracontinental basins and in the deltas of large rivers which may extend far into the open sea. Continental organic matter is usually rich in cellulose and lignin, which, due to their oxidation state, are not considered to contribute significantly to oil formation. The subordinate lipid fraction together with the biomass of sedimentary microorganisms incorporated into the source rock yields crude oils which are rich in aliphatic units (from waxes, fats, aliphatic biopolymers, etc.), i.e., straight-chain and branched alkanes (paraffins). Polycyclic naphthenes, particularly steranes, are present in very low concentration. Total aromatic hydrocarbons are also significantly less abundant than in crude oil derived from marine organic matter, as is the sulfur content.

Marine organic matter (usually type II kerogen) produces oils of paraffinic–naphthenic or aromatic-intermediate type (Fig. 8). The amount of saturated hydrocarbons is moderate, but isoprenoid alkanes and polycyclic alkanes, like steranes (from algal steroids) and hopanes (from membranes of cyanobacteria and bacteria), are relatively more abundant than in oils from terrigenous organic matter. Kerogen derived from marine organic matter, particularly when it is very rich in sulfur, is particularly suited to release resin- and asphaltene-rich heavy crude oils at a very early stage of catagenesis. Type II kerogens are preferentially deposited where the environmental conditions are favorable for organic matter preservation (anoxic water column in silled basins or in areas of coastal upwelling) and where the continental runoff is limited due to physiographical or climatic reasons.

The sulfur content of crude oils shows a close relationship to the type of mineral matrix in the source rocks. Sediments consisting of calcareous (e.g., from dinoflagellates or foraminifera) or siliceous shell fragments (e.g., from diatoms or radiolaria) of decayed planktonic organisms and at the same time containing abundant organic matter are enriched in sulfur. The reason for this is that under the anoxic conditions which are required to preserve the organic matter, sulfate-reducing bacteria form hydrogen sulfide (H_2S) or other reactive inorganic sulfur species. These may react with the organic matter, and the sulfur will become incorporated into the kerogen (see Orr and White, 1990). Examples are the Monterey Formation and the related crude oils produced onshore and offshore in southern California and many of the carbonate source rocks of the Middle Eastern crude oils.

In clastic rocks containing an abundance of detrital clay minerals, the iron content usually is high enough to remove most of the H_2S generated by the sulfate-reducing bacteria through the formation of iron sulfides. Because terrigenous organic matter is commonly deposited together with detrital mineral matter (e.g., in deltas), waxy crude oils derived from type III kerogen usually are depleted in sulfur.

Oil generation from a source rock is a long-term dynamic process. During the many millions of years that hydrocarbon generation is active, the source rock may undergo further burial due to basin subsidence. The increasing temperatures to which the source rock is exposed will lead to changes in the chemical composition of the hydrocarbons generated. If migration in the subsurface then leads to accumulation of the hydrocarbons in different reservoirs at different times, the oils will differ in their chemical and physical properties (cf., e.g., Gussow, 1954, 1968; Mackenzie *et al.*, 1985).

Large sedimentary basins may have different geothermal heat flows in certain parts of the basin. Timing of oil generation and the composition of the products may then vary even if the source rock is of the same type all over the basin. An equally frequent reason for variations of the oil composition within a sedimentary basin is lateral or vertical changes of the organic facies of a source rock. This may happen due to water depth variations in the depositional environment associated with fluctuations of the anoxicity of the water column above the sediment surface.

Petroleum reservoir rocks within the drainage area of two or more hydrocarbon source rocks of different age or depth setting may receive a mixture of oils from different sources. If several reservoirs are present, the relative contributions of the source rocks may vary and lead to regional differences in crude oil composition (e.g., Michigan Basin: Vogler *et al.*, 1981; Illich and Grizzle, 1983; Pruitt, 1983; Rullkötter *et al.*, 1992).

5. Timing of Oil Generation

5.1. Time–Temperature Dependence

Oil formation from kerogen, being a function of chemical reaction kinetics, depends both on geologic time and geothermal heat flow (temperature). As pointed out before, the kinetics of oil generation follows the Arrhenius law. Thus, the dependence on geologic time is linear while that on temperature is exponential.

Because geological and geothermal histories of petroleum source rocks often are not known (exactly), the progress of hydrocarbon generation is often represented as a function of a maturity parameter, the most common one being vitrinite reflectance. The chemical (and physical) processes responsible for the increase of vitrinite reflectance with geothermal heating of organic matter are different from those of thermal cracking of kerogen during hydrocarbon generation. Thus, not surprisingly, the kinetic data for the increase of vitrinite reflectance differ from those of oil formation. This means that the relationship between vitrinite reflectance and hydrocarbon yield varies with geological heating rates as illustrated in Fig. 9. The "oil window" at high heating rates is relatively narrow on the vitrinite reflectance scale and shifted toward lower reflectance values, whereas at low heating rates hydrocarbon generation extends over a wide range of reflectance values and is relatively flat, compared to the average (medium heating rate) geological situation (Fig. 9).

High geothermal heating rates are typical of geologically active areas such as in the Los Angeles or Pannonian (Hungary) basins during the recent geological past. Here, the oil generation threshold is reached within only a few million years after deposition due to rapid and extensive basin subsidence and

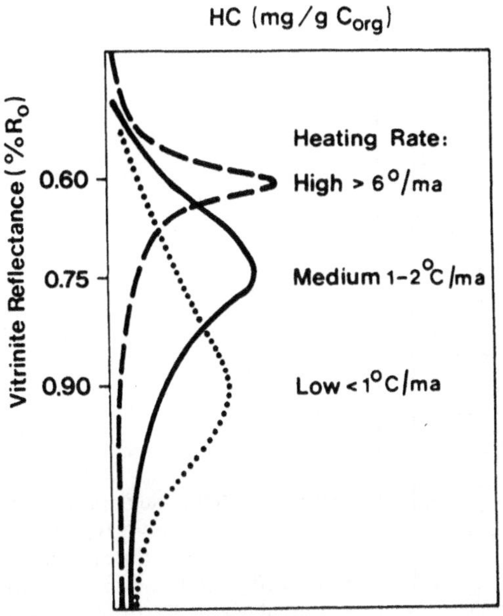

Figure 9. The influence of geological heating rates upon the position of the oil window relative to the vitrinite reflectance maturity parameter. HC (mg/g C_{org}) = yield of extractable hydrocarbons in milligrams per gram of C_{org}. [From Welte (1989).]

related higher sedimentation rates, higher geothermal heat flow, and thus rapid temperature increase with depth. Absolute temperatures necessary to generate hydrocarbons are considerably higher (150°C and more) than in "normal" geological situations. On the other hand, very low temperatures (e.g., <40°C) even over extremely long geologic times will not be able to initiate thermal breakdown of kerogen, i.e., hydrocarbon generation, for kinetic reasons.

What has been said before about the relationship between hydrocarbon yield and vitrinite reflectance is also true for other maturation parameters. *A priori*, there is no way of converting the data of one maturation parameter to those of other ones without taking into account the geological (heating) conditions of a particular sedimentary basin.

5.2. Basin Modeling

Integrated basin modeling is used to understand and reconstruct the geological and geothermal evolution of a sedimentary basin, to quantify this information numerically, and in this way to derive the history of generation, migration, and accumulation of petroleum (Welte and Yalcin, 1988). With the more sophisticated techniques of basin modeling, a three-dimensional view of the basin can be constructed and followed through geologic time. Overviews of the present state of the art of basin modeling and the different approaches to it have recently been published by Welte and Yalcin (1988), Lerche (1990), and Ungerer *et al.* (1990).

In basin modeling a sedimentary basin is taken as a rock volume with changing boundary conditions. Mathematical equations control the mass balance (e.g., sediment supply and water leaving the system) as well as the energy balance (e.g., heat flow from the basement, mainly as a function of plate tectonics, and energy loss to the atmosphere or as heated water). The model results are compared with the real system (calibration by well and surface information) and optimized by a series of iteration steps until a satisfactory match with the real system is obtained. Due to the close interrelationship between hydrocarbon generation and expulsion, not only the generation aspect is considered in modern modeling systems but migration is integrated as well.

One of the key types of information provided by basin modeling is the timing of oil generation and migration together with an impression of the geological situation at that time, which primarily relates to the presence or absence of suitable migration pathways and traps for hydrocarbon accumulation.

Figure 10 reproduces the curves for oil yield as a function of geologic time for a source rock at three locations in the San Joaquin Basin, California. The different timing of oil generation at the three locations is a function of source rock subsidence, being higher at gridpoint A than at B and higher at B than at C. The broken sections on the yield curves indicate time intervals where the pressure buildup in the source rock was considered sufficient to initiate hydrocarbon expulsion according to the migration model employed. The simulation results show that the events of oil expulsion do not only occur at different times during basin evolution at each point but that the expulsion can be discontinuous, i.e., it may occur in several successive phases. These phases at the three different locations differ with respect to not only the onset but also the individual timing and duration (Welte and Yalcin, 1988).

In another case history, the numerical simulation of hydrocarbon expulsion was coupled with an integrated model of the geological and geochemical history of the Lower Jurassic Posidonia Shale in the Lower Saxony Basin in northern Germany (Düppenbecker *et al.*, 1991). This mathematical approach considered facies variations, compositional changes of the petroleum generated, pore size distributions, and pressure-induced microfracturing in order to deter-

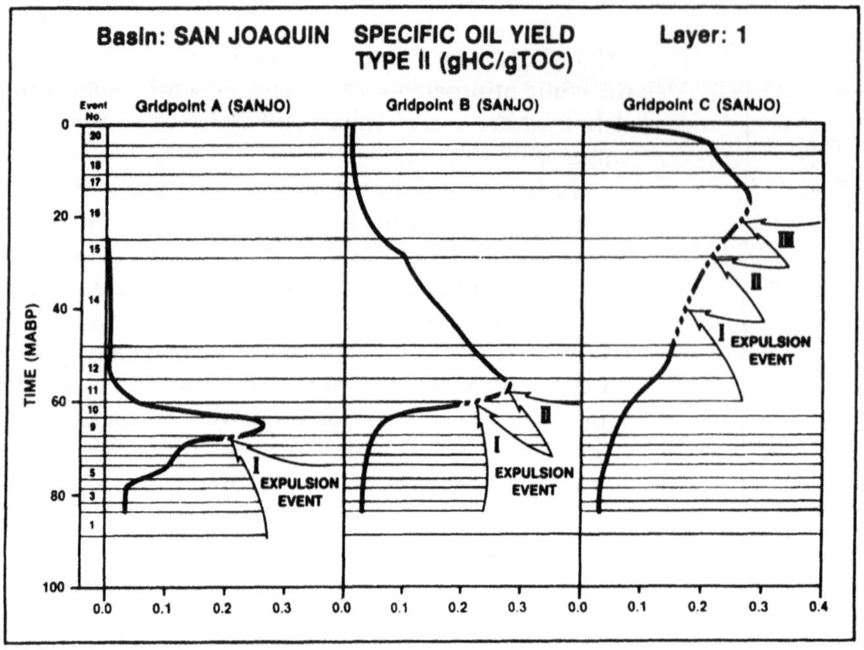

Figure 10. Temporal development of hydrocarbon formation from type II kerogen in a source rock layer in the San Joaquin basin at three different locations and timing and phases of expulsion. [From Welte and Yalcin (1988).]

mine the timing, intensity, and mechanism of petroleum expulsion. It was found that the influence of a deep-seated igneous intrusion on the petroleum generation and expulsion processes in the eastern part of the basin was quantitatively in good agreement with present-day geological and geochemical observations (cf. Rullkötter *et al.*, 1988).

Basin modeling and the consequent utilization of numerical simulation techniques force geologists to think quantitatively and in terms of geological *processes* (Welte and Yalcin, 1988). This approach also provides a better means to interrelate systematically the many factors influencing geological processes (Welte and Yalcin, 1988; Ungerer *et al.*, 1990). It also will have a significant economic effect by increasing exploration efficiency, particularly in frontier areas.

6. Conclusions

The thermal alteration of kerogen leading to the formation of oil has been one of the key objectives of organic geochemical research ever since this scientific discipline emerged from the interests and needs of the petroleum industry. Although many of the largely empirical concepts that were developed early still hold and are successfully applied in exploration, truly interdisciplinary fundamental research has nevertheless had a strong impact on the progressively changing view of the processes involved.

The detection of nonsaponifiable, highly aliphatic biopolymers both in extant plants and in macromolecular fossil materials is one of the most important recent achievements in this area of research and significantly modified the ideas about the chemical nature of kerogen. Further efforts toward a better understanding of the chemical structure of the starting material of petroleum generation are presently being made and involve the concerted application of (1) consecutive specific chemical degradation reactions, (2) the detailed quantitative molecular analysis of small degradation products, (3) the study of the degradation residues by spectroscopy and pyrolysis, and (4) refined concepts of the preservation of biological macromolecules during kerogen formation (Rullkötter and Michaelis, 1990).

On the other hand, integrated systems for routine analysis and computer modeling will be a tool to lead applied petroleum research toward better integration of geology, geochemistry, and geophysics, considering parameters such as structural evolution, heat transfer, chemistry and kinetics of kerogen transformation, rock matrix/organic matter interactions, and hydrocarbon expulsion and migration (Tissot and Ungerer, 1990).

Acknowledgments. The author is grateful to Dr. B. Horsfield (KFA Jülich) and John Zumberge (Core Lab, Houston) for their critical reviews of the manuscript and for their helpful comments. The manuscript was typed by Mrs. M. Sostmann.

References

Abbott, G. D., Wang, G. Y., Eglinton, T. I., Home, A. K., and Petch, G. S., 1990, The kinetics of sterane biological marker release and degradation processes during the hydrous pyrolysis of vitrinite kerogen, *Geochim. Cosmochim. Acta* **54**:2451–2461.

Alexander, R., Kagi, R. I., Tok, E., and van Bronswijk, W., 1988, The use of aromatic hydrocarbons for assessment of thermal histories in sediments, in: *Proceedings of the North West Shelf Symposium* (P. G. Purcell and R. R. Purcell, eds.), Petroleum Exploration Society of Australia, Perth, pp. 559–562.

Barker, C., 1979, *Organic Geochemistry in Petroleum Exploration*, Education Course Note Series 10, American Association of Petroleum Geologists, Tulsa, Oklahoma.

Bartenstein, H., and Teichmüller, R., 1974, Inkohlungsuntersuchungen; ein Schlüssel zur Prospektierung von paläozoischen Kohlenwasserstoff-Lagerstätten? *Fortschr. Geol. Rheinl. Westfalen* **24**:129–160.

Béhar, F., Pelet, R., and Roucachet, J., 1984, Geochemistry of asphaltenes, in: *Advances in Organic Geochemistry 1983* (P. A. Schenck, J. W. de Leeuw, and G. W. M. Lijmbach, eds.), Pergamon Press, Oxford, pp. 587–595.

Bray, E. E., and Evans, E. D., 1961, Distribution of *n*-paraffins as a clue to recognition of source beds, *Geochim. Cosmochim. Acta* **22**:2–15.

Burnham, A. K., and Braun, R. L., 1985, General kinetic model of oil shale pyrolysis, *In Situ* **9**:1–23.

Burnham, A. K., Braun, R. L., Gregg, H. R., and Samoun, A. M., 1987, Comparison of methods for measuring kerogen pyrolysis rates and fitting kinetic parameters, *Energy Fuels* **1**:452–458.

Burnham, A. K., Braun, R. L., and Samoun, A. M., 1988, Further comparison of methods for measuring kerogen pyrolysis rates and fitting kinetic parameters, in: *Advances in Organic Geochemistry 1987* (L. Mattavelli and L. Novelli, eds.), Pergamon Press, Oxford, pp. 839–845.

Bustin, R. M., 1988, Sedimentology and characteristics of dispersed organic matter in Tertiary Niger Delta: Origin of source rocks in a deltaic environment, *Am. Assoc. Petrol. Geol. Bull.* **72**:277–298.

Campbell, J. H., Gallegos, G., and Gregg, M., 1980, Gas evolution during oil shale pyrolysis. 2. Kinetic and stoichiometric analysis, *Fuel* **59**:727–732.

Connan, J., 1984, Biodegradation of crude oils in reservoirs, in: *Advances in Petroleum Geochemistry*, Vol. 1 (J. Brooks and D. Welte, eds.), Academic Press, London, pp. 299–335.

Cooles, G. P., Mackenzie, A. S., and Quigley, T. M., 1986, Calculation of petroleum masses generated and expelled from source rocks, in: *Advances in Organic Geochemistry 1985* (D. Leythaeuser and J. Rullkötter, eds.), Pergamon Press, Oxford, pp. 235–245.

Curry, D. J., 1985, Organic geochemistry and oil generation potential of Tertiary age coals from the West Natuna basin, South China Sea, Book of Abstracts, 12th International Meeting on Organic Geochemistry, Jülich, Germany, p. 171.

Derenne, S., Largeau, C., Casadevall, E., and Berkaloff, C., 1989, Occurrence of a resistant biopolymer in the *L* race of *Botryococcus braunii*, *Phytochemistry* **28**:1137–1142.

Düppenbecker S. J., Dohmen, L., and Welte, D. H., 1991, Numerical modelling of petroleum expulsion in two areas of the Lower Saxony Basin, Northern Germany, in: *Petroleum Migration* (A. Fleet and W. England, eds.), Geological Society Special Publication, Blackwell Scientific, Oxford, pp. 47–64.

Durand, B. (ed.), 1980, *Kerogen—Insoluble Organic Matter from Sedimentary Rocks*, Editions Technip, Paris.

Durand, B., 1988, Understanding of HC migration in sedimentary basins (present state of knowledge), in: *Advances in Organic Geochemistry 1987* (L. Mattavelli and L. Novelli, eds.), Pergamon Press, Oxford, pp. 445–459.

Durand, B., and Oudin, J. L., 1979, Exemple de migration des hydrocarbures dans une série deltaique: La Delta de la Mahakam, Kalimantan, Indonésia, *Proceedings of the 10th World Petroleum Congress*, Vol. 2, Heyden & Sons, London, pp. 1–9.

Durand, B., and Parratte, M., 1983, Oil potential of coals, a geochemical approach, in: *Petroleum Geochemistry and Exploration of Europe* (J. Brooks, ed.), Blackwell Scientific, Oxford, pp. 255–265.

Durand, B., Parratte, M., and Bertrand Ph., 1983, Le potentiel en huile des charbons: Une approche géochimique, *Rev. Inst. Fr. Pét.* **38**:709–721.

Engel, M. H., Rafalska-Bloch, J., Schiefelbein, C. F., Zumberge, J. E., and Serban, A., 1986, Simulated diagenesis and catagenesis of marine kerogen precursors: Melanoidins as model system for light hydrocarbon generation, in: *Advances in Organic Geochemistry 1985* (D. Leythaeuser and J. Rullkötter, eds.), Pergamon Press, Oxford, pp. 1073–1079.

Erdman, J. G., and Morris, D. A., 1974, Geochemical correlation of petroleum, *Am. Assoc. Petrol. Geol. Bull.* **58**:2326–2337.

Espitalié, J., Deroo, G., and Marquis, F., 1985a, La pyrolyse Rock-Eval et ses applications. Première partie, *Rev. Inst. Fr. Pét.* **40**:563–579.

Espitalié, J., Deroo, G., and Marquis, F., 1985b, La pyrolyse Rock-Eval et ses applications, Deuxième partie, *Rev. Inst. Fr. Pét.* **40**:755–784.

Espitalié, J., Deroo, G., and Marquis, F., 1986, La pyrolyse Rock-Eval et ses applications. Troisième partie, *Rev. Inst. Fr. Pét.* **41**: 73–89.

Gold, T., 1987, *Power from the Earth*, J. M. Dent and Sons, London.

Gordon, T. L., 1985, Talang Akar coals—Ardjuna sub-basin oil source, *Proceedings of the Indonesian Petroleum Association, 14th Annual Convention*, Indonesian Petroleum Association, Djakarta, pp. 91–120.

Gough, M. A., and Rowland, S. J., 1990, Characterization of unresolved complex mixtures of hydrocarbons in petroleum, *Nature* **344**:648–650.

Gussow, W. C., 1954, Differential entrapment of oil and gas—a fundamental principle, *Am. Assoc. Petrol. Geol. Bull.* **38**:816–853.

Gussow, W. C., 1968, Migration of reservoir fluids, *J. Petrol. Technol.* **1968**:353–363.

Hanbaba, P., 1967, Reaktionskinetische Untersuchungen zur Kohlenwasserstoffentbindung aus Steinkohlen bei niedrigen Aufheizgeschwindigkeiten (10^{-3} bis 1 grad/min), Ph.D. Thesis, RWTH Aachen.

Hedges, J. I., 1978, The formation and clay mineral reactions of melanoidins, *Geochim. Cosmochim. Acta* **42**:69–76.

Héroux, Y., Chagnon, A., and Bertrand, R., 1979, Compilation and correlation of major thermal maturation indicators, *Am. Assoc. Petrol. Geol. Bull.* **63**:2128–2144.

Hood, A., Gutjahr, C. C., and Heacock, R. L., 1975, Organic metamorphism and the generation of petroleum, *Am. Assoc. Petrol. Geol. Bull.* **59**:986–996.

Horsfield, B., Yordy, K. L., and Crelling, J. R., 1988, Determining the petroleum-generating potential of coal using organic geochemistry and organic petrology, in: *Advances in Organic Geochemistry 1987* (L. Mattavelli and L. Novelli, eds.), Pergamon Press, Oxford, pp. 121–129.

Huc, A. Y., Durand, B., Roucachet, J., Vandenbroucke, M., and Pittion, L., 1986, Comparison of three series of organic matter of continental origin, in: *Advances in Organic Geochemistry 1985* (D. Leythaeuser and J. Rullkötter, eds.), Pergamon Press, Oxford, pp. 65–72.

Hunt, J. M., 1977, Ratio of petroleum to water during primary migration in western Canada basin, *Am. Assoc. Petrol. Geol. Bull.* **61:**434–435.

Hunt, J. M., 1979, *Petroleum Geochemistry and Geology*, W. H. Freeman and Co., San Francisco.

Hunt, J. M., and McNichol, A. P., 1984, The Cretaceous Austin Chalk of South Texas—a petroleum source rock, in: *Petroleum Geochemistry and Source Rock Potential of Carbonate Rocks* (J. G. Palacas, ed.), American Association of Petroleum Geologists, Tulsa, Oklahoma, pp. 117–125.

Hvoslef, S., Larter, S. R., and Leythaeuser, D., 1988, Aspects of generation and migration of hydrocarbons from coal-bearing strata of the Hitra formation, Haltenbanken area, offshore Norway, in: *Advances in Organic Geochemistry 1987* (L. Mattavelli and L. Novelli, eds.), Pergamon Press, Oxford, pp. 525–536.

Illich, H. A., and Grizzle, P. L., 1983, Comment on "Comparison of Michigan Basin crude oils" by Vogler *et al.*, *Geochim. Cosmochim. Acta* **47:**1151–1155.

Jüntgen, H., and Klein, J., 1975, Entstehung von Erdgas aus kohligen Sedimenten, *Erdöl Kohle, Erdgas, Petrochem.* **28:**65–73.

Karweil, J., 1969, Aktuelle Probleme der Geochemie der Kohle, in: *Advances in Organic Geochemistry 1968* (P. A. Schenck and I. Havenaar, eds.), Pergamon Press, Oxford, pp. 59–84.

Kent, P. E., 1954, Oil occurrences in coal measures of England, *Am. Assoc. Petrol. Geol. Bull.* **38:**1699–1712.

Khorasani, G. K., 1989, Factors controlling source rock potential of the Mesozoic coal-bearing strata from offshore central Norway: Application to petroleum exploration, *Bull. Can. Soc. Petrol. Geol.* **37:**417–427.

Kister, J., Guiliano, M., Largeau, C., Derenne, S., and Casadevall, E., 1990, Characterization of chemical structure, degree of maturation and oil potential of torbanites (type I kerogens) by quantitative FT-i.r. spectroscopy, *Fuel* **69:**1356–1361.

Larter, S. R., 1988, Some pragmatic perspectives in source rock geochemistry, *Mar. Petrol. Geol.* **5:**194–204.

Lerche, I., 1990, *Basin Analysis: Quantitative Methods*, Vol. 1, Academic Press, New York.

Leythaeuser, D., Schaefer, R. G., and Radke, M., 1987, On the primary migration of petroleum, in: *Proceedings of the 12th World Petroleum Congress*, Vol. 2, Heyden, London, pp. 31–41.

Leythaeuser, D., Schaefer, R. G., and Radke, M., 1988, Geochemical effects of primary migration of petroleum in Kimmeridge source rocks from Brae field area, North Sea: I. Gross composition of C_{15+}-soluble organic matter and molecular composition of C_{15+}-saturated hydrocarbons, *Geochim. Cosmochim. Acta* **52:**701–713.

Littke, R., and ten Haven, H. L., 1989, Palaeoecologic trends and petroleum potential of Upper Carboniferous coal seams of western Germany as revealed by their petrographic and organic geochemical characteristics, *Int. J. Coal Geol.* **13:**529–574.

Littke, R., Leythaeuser, D., Radke, M., and Schaefer, R. G., 1990, Petroleum generation and migration in coal seams of the Carboniferous Ruhr Basin, northwest Germany, in: *Advances in Organic Geochemistry 1989* (B. Durand and F. Béhar, eds.), Pergamon Press, Oxford, pp. 247–258.

Mackenzie, A. S., 1984, Applications of biological markers in petroleum geochemistry, in: *Advances in Petroleum Geochemistry*, Vol. 1 (J. Brooks and D. Welte, eds.), Academic Press, London, pp. 115–214.

Mackenzie, A. S., and McKenzie, D. P., 1983, Isomerization and aromatization of hydrocarbons in sedimentary basins formed by extension, *Geol. Mag.* **120:**417–470.

Mackenzie, A. S., and Quigley, T. M., 1988, Essentials of geochemical prospect appraisal, *Am. Assoc. Petrol. Geol. Bull.* **72:** 399–415.

Mackenzie, A. S., Rullkötter, J., Welte, D. H., and Mankiewicz, P., 1985, Reconstruction of oil formation and accumulation in North Slope, Alaska, using quantitative gas chromatography–mass spectrometry, in: *Alaska North Slope Oil/Rock Correlation Study* (L. B. Magoon and G. E. Claypool, eds.), AAPG Studies in Geology No. 20, American Association of Petroleum Geologists, Tulsa, Oklahoma, pp. 319–377.

Mann, U., Düppenbecker, S., Langen, A., Ropertz, B., and Welte, D.H., 1990, Pore network evolution of the Lower Toarcian Posidonia shale during petroleum generation and expulsion—a multidisciplinary approach, *Zentralbl. Geol. Paläontol., Teil 1*, 1051–1071.

Martin, R. L., Winters, J. C., and Williams, J. A., 1963, Distribution of *n*-paraffins in crude oils and their implications to origin of petroleum, *Nature* **199:**1190–1193.

Marzi, R., and Rullkötter, J., 1992, Qualitative and quantitative evolution and kinetics of biological marker transformations–laboratory experiments and application to the Michigan Basin, in: *Biological Markers in Sediments and Petroleum* (J. M. Moldowan, P. Albrecht, and R. P. Philp, eds.), Prentice-Hall, Englewood Cliffs, New Jersey, pp. 18–41.

Merewether, E. A., and Claypool, G. E., 1980, Organic composition of some Upper Cretaceous shale, Powder River Basin, Wyoming, *Am. Assoc. Petrol. Geol. Bull.* **64:**488–500.

Michaelis, W., and Albrecht, P., 1979, Molecular fossils of Archaebacteria in kerogen, *Naturwissenschaften* **66:**420–421.

Moldowan, J. M., Albrecht, P., and Philp, R. P. (eds.), 1991, *Biological Markers in Sediments and Petroleum*, Prentice-Hall, Englewood Cliffs, New Jersey.

Monthioux, M., and Landais, P., 1987, Evidence for free but trapped hydrocarbons in coal, *Fuel* **66:**1703–1708.

Nip, M., Tegelaar, E. W., de Leeuw, J. W., Schenck, P. A., and Holloway, P. J., 1986, A new non-saponifiable highly aliphatic and resistant biopolymer in plant cuticles: Evidence from pyrolysis and ^{13}C-NMR analysis of present day and fossil plants, *Naturwissenschaften* **73:**579–585.

Noble, R. A., Wu, C. H., and Atkinson, C. D., 1991, Petroleum generation and migration from Talang Akar coals and shales offshore N.W. Java, Indonesia, *Org. Geochem.* **17:**363–374.

Orr, W. L., and Sinninghe Damsté, J. S., 1990, Geochemistry of sulfur in petroleum systems, in: *Geochemistry of Sulfur in Fossil Fuels* (W. L. Orr and C. M. White, eds.), ACS Symposium Series No. 429, American Chemical Society, Washington, D.C., pp. 2–29.

Orr, W. L., and White, C. M. (eds), 1990, *Geochemistry of Sulfur in Fossil Fuels*, ACS Symposium Series No. 429, American Chemical Society, Washington, D.C.

Palacas, J. G., 1983, Carbonate rocks as sources of petroleum: Geological and chemical characteristics and oil-source correlations, in: *Proceedings of the 11th World Petroleum Congress*, Vol. 2., Wiley, Chichester, England, pp. 31–43.

Peters, K. E., and Moldowan, J. M. (1993). *The Biomarker Guide*, Prentice Hall, Englewood Cliffs, New Jersey.

Peters, K. E., Moldowan, J. M., and Sundararaman, P., 1990, Effects of hydrous pyrolysis on biomarker thermal maturity parameters: Monterey phosphatic and siliceous members, *Org. Geochem.* **15**:249–265.

Poulet, M., and Roucaché, J., 1970, Influence du mode d'échantillonage sur la composition chimique des fractions légères d'une huile brute, in: *Advances in Organic Geochemistry 1966* (G. D. Hobson and G. C. Speers, eds.), Pergamon Press, London, pp. 155–179.

Pruitt, J. D., 1983, Comment on "Comparison of Michigan Basin crude oils" by Vogler *et al.*, *Geochim. Cosmochim. Acta* **47**: 1157–1159.

Quigley, T. M., Mackenzie, A. S., and Gray, J. R., 1987, Kinetic theory of petroleum generation, in: *Migration of Hydrocarbons in Sedimentary Basins* (B. Doligez, ed.), Editions Technip, Paris, pp. 649–665.

Radke, M., 1987, Organic geochemistry of aromatic hydrocarbons, in: *Advances in Petroleum Geochemistry*, Vol. 2 (J. Brooks and D. H. Welte, eds.), Academic Press, London, pp. 141–207.

Radke, M., Willsch, H., and Welte, D. H., 1980, Preparative hydrocarbon group type determination by automated medium pressure liquid chromatography, *Anal. Chem.* **52**:406–411.

Radke, M., Willsch, H., and Welte, D. H., 1984, Class separation of aromatic compounds in rock extracts and fossil fuels by liquid chromatography, *Anal. Chem.* **56**:2538–2546.

Robin, P. L., 1975, Caractérisation des kérogènes et de leur évolution par spectroscopie infrarouge, Ph.D. Dissertation, Université de Louvain.

Roe, G. D., and Polito, L. J., 1977, Source rocks for oils in the Ardjuna sub-basin of the Northwest Java basin, Indonesia, *UN/ESCAP, CCOP Tech. Publ.* **6**:180–194.

Rouxhet, P. G., Robin, P. L., and Nicaise, G., 1980, Characterization of kerogens and of their evolution by infrared spectroscopy, in: *Kerogen—Insoluble Organic Matter in Sedimentary Rocks* (B. Durand, ed.), Editions Technip, Paris, pp. 163–190.

Rullkötter, J., and Marzi, R., 1989, New aspects of the application of sterane isomerisation and steroid aromatisation to petroleum exploration and the reconstruction of geothermal histories of sedimentary basins, *Prepr., Am. Chem. Soc., Div. Petrol. Chem.* **34**:126–131.

Rullkötter, J., and Michaelis, W., 1990, The structure of kerogen and related materials—a review of recent progress and future trends, in: *Advances in Organic Geochemistry 1989* (B. Durand and F. Béhar, eds.), Pergamon Press, Oxford, pp. 829–852.

Rullkötter, J., Leythaeuser, D., Horsfield, B., Littke, R., Mann, U., Müller, P. J., Radke, M., Schaefer, R. G., Schenk, H.-J., Schwochau, K., Witte, E. G., and Welte, D. H., 1988, Organic matter maturation under the influence of a deep intrusive heat source: A natural experiment for quantitation of hydrocarbon generation and expulsion from a petroleum source rock (Toarcian shale, northern Germany), in: *Advances in Organic Geochemistry 1987* (L. Mattavelli and L. Novelli, eds.), Pergamon Press, Oxford, pp. 847–856.

Rullkötter, J., Marzi, R., and Meyers, P. A., 1992, Biological markers in Paleozoic sedimentary rocks and crude oils from the Michigan Basin: Reassessment of sources and thermal history of organic matter, in: *Early Organic Evolution: Implications for Mineral and Energy Resources* (M. Schidlowski, M. M. Kimberley, D. M. McKirdy, P. A. Trudinger, and S. Golubic, eds.), Springer-Verlag, Heidelberg, pp. 324–335.

Schaefer, R. G., Schenk, H. J., Hardelauf, H., and Harms, R., 1990, Determination of gross kinetic parameters for petroleum formation from Jurassic source rocks of different maturity levels by means of laboratory pyrolysis, in: *Advances in Organic Geochemistry 1989* (B. Durand and F. Béhar, eds.), Pergamon Press, Oxford, pp. 115–120.

Schmid, J. C., Connan, J., and Albrecht, P., 1987, Occurrence and geochemical significance of long-chain dialkylthiacyclopentanes, *Nature* **329**:54–56.

Schmitter, J. M., and Arpino, P. J., 1985, Azaarenes in fuels, *Mass Spectrom. Rev.* **4**:87–121.

Seifert, W. K., and Howells, W. G., 1969, Interfacially active acids in a California crude oil. Isolation of carboxylic acids and phenols, *Anal. Chem.* **41**:554–562.

Seifert, W. K., Gallegos, E. J., and Teeter, R. M., 1972, Proof of structure of steroid carboxylic acids in a California petroleum by deuterium labelling, synthesis and mass spectrometry, *J. Am. Chem. Soc.* **94**:5880–5887.

Sinninghe Damsté, J. S., Rijpstra, W. I. C., de Leeuw, J. W., and Schenck, P. A., 1988, Origin of organic sulphur compounds and sulphur-containing high molecular weight substances in sediments and immature crude oils, in: *Advances in Organic Geochemistry 1987* (L. Mattavelli and L. Novelli, eds.), Pergamon Press, Oxford, pp. 593–606.

Stach, E., Mackowsky, M. Th., Teichmüller, M., Taylor, G. H., Chandra, D., and Teichmüller, R., 1982, *Textbook of Coal Petrology*, 3rd ed., Gebrüder, Borntraeger, Stuttgart.

Stahl, W., 1978, Reifeabhängigkeit der Kohlenstoff-Isotopenverhältnisse des Methans von Erdgasen aus Norddeutschland, *Erdöl Kohle, Erdgas, Petrochem.* **31**:516–518.

Strausz, O. P., Lown, E. M., and Payzant, J. D., 1990, Nature and geochemistry of sulfur-containing compounds in Alberta petroleums, in: *Geochemistry of Sulfur in Fossil Fuels* (W. L. Orr and C. M. White, eds.), ACS Symposium Series No. 429, American Chemical Society, Washington, D.C., pp. 366–396.

Sweeney, J. J., Burnham, A. K., and Braun R. L., 1987, A model of hydrocarbon maturation from type I kerogen: Application to Uinta basin, Utah, *Am. Assoc. Petrol. Geol. Bull.* **71**:967–985.

Tegelaar, E. W., de Leeuw, J. W., and Saiz-Jimenez, C., 1989a, Possible origin of aliphatic moieties in humic substances, *Sci. Total Environ.* **81/82**:1–17.

Tegelaar, E. W., de Leeuw, J. W., Derenne S., and Largeau, C., 1989b, A reappraisal of kerogen formation, *Geochim. Cosmochim. Acta* **53**:3103–3106.

Thomas, B. M., 1982, Land-plant source rocks for oil and their significance in Australian basins, *APEA J.* **22**:164–178.

Tissot, B. P., and Espitalié, J., 1975, L'évolution thermique de la matière organique des sédiments: Applications d'une simulation mathématique, *Rev. Inst. Fr. Pét.* **30**:743–777.

Tissot, B. P., and Ungerer, Ph., 1990, Advances in organic geochemistry, a narrow path between fundamental interest and industrial needs, in: *Advances in Organic Geochemistry 1989* (B. Durand and F. Béhar, eds.), Pergamon Press, Oxford, pp. XXV–XLV.

Tissot, B. P., and Welte, D. H., 1984, *Petroleum Formation and Occurrence*, 2nd ed., Springer-Verlag, Heidelberg.

Tissot, B. P., Durand, B., Espitalié, J., and Combaz, A., 1974, Influence of nature and diagenesis of organic matter in the formation of petroleum, *Am. Assoc. Petrol. Geol. Bull.* **58**:499–506.

Tissot, B. P., Welte, D. H., and Durand, B., 1987a, The role of geochemistry in exploration risk evaluation and decision making, *Proceedings of the 12th World Petroleum Congress*, Vol. 2, Heyden, London, pp. 99–112.

Tissot, B. P., Pelet, R., and Ungerer, P., 1987b, Thermal history of sedimentary basins, maturation indices, and kinetics of oil and gas generation, *Am. Assoc. Petrol. Geol. Bull.* **71**:1145–1466.

Ungerer, P., and Pelet, R., 1987, Extrapolation of the kinetics of oil and gas formation from laboratory experiments to sedimentary basins, *Nature* **327**:52–54.

Ungerer, Ph., Béhar, F., and Discamps, D., 1983, Tentative calculation of the overall volume expansion of organic matter during hydrocarbon genesis from geochemistry data. Implications for primary migration, in: *Advances in Organic Geochemistry 1981* (M. Bjorøy *et al.*, eds.), Wiley, Chichester, England, pp. 129–135.

Ungerer, P., Espitalié, J., Marquis, F., and Durand, B., 1986, Use of kinetic models of organic matter evolution for the reconstruction of paleotemperatures, in: *Thermal Modeling of Sedimentary Basins* (J. Burrus, ed.), Editions Technip, Paris, pp. 531–546.

Ungerer, P., Burrus, J., Doligez, B., Chénet, P. Y., and Bessis, F., 1990, Basin evaluation by integrated two-dimensional modeling of heat transfer, fluid flow, hydrocarbon generation, and migration, *Am. Assoc. Petrol. Geol. Bull.* **74**:309–335.

Vogler, E. A., Meyers, P. A., and Moore, W. E., 1981, Comparison of Michigan Basin crude oils, *Geochim. Cosmochim. Acta* **45**: 2287–2293.

Waples, D. W., 1980, Time and temperature in petroleum formation: Application of Lopatin's method to petroleum exploration, *Am. Assoc. Petrol. Geol. Bull.* **64**:916–926.

Waples, D. W., 1985, *Geochemistry in Petroleum Exploration*, International Human Resources Development Corp., Boston.

Welte, D. H., 1989, The changing face of geology and future needs, *Geol. Rundsch.* **78**:7–20.

Welte, D. H., and Yalcin, M. N., 1988, Basin modelling—A new comprehensive method in petroleum geology, in: *Advances in Organic Geochemistry 1987* (L. Mattavelli and L. Novelli, eds.), Pergamon Press, Oxford, pp. 141–151.

Wilson, M. A., 1987, *NMR Techniques and Applications in Geochemistry and Soil Chemistry*, Pergamon Press, Oxford.

Witte, E. G., Schenk, H. J., Müller, P. J., and Schwochau, K., 1988, Structural modifications of kerogen during natural evolution as derived from ^{13}C CP/MAS NMR, IR spectroscopy and Rock-Eval pyrolysis of Toarcian shales, in: *Advances in Organic Geochemistry 1987* (L. Mattavelli and L. Novelli, eds.), Pergamon Press, Oxford, pp. 1039–1044.

Chapter 17

Hydrothermal Alteration of Organic Matter in Marine and Terrestrial Systems

BERND R. T. SIMONEIT

1. Introduction

The discovery in 1977 of active hydrothermal venting and associated mineralization at the seafloor spreading center of the Galápagos (Corliss *et al.*, 1979) initiated extensive surveys of global ridge crests. Hydrothermal systems associated with ocean spreading centers are now recognized as a relatively common phenomenon and are thus one of the most actively researched topics in geology, biology, and geochemistry. The morphology, mineralization, and fluid chemistry of the active systems in spreading centers have been reviewed (e.g., Barrett and Jambor, 1988; Rona, 1988; Rona *et al.*, 1983; Von Damm, 1990) and related to ancient deposits (e.g., Franklin *et al.*, 1981; Scott, 1985). Other nonrift areas with analogous hydrothermal systems are tectonically active subduction zones, fracture zones, and back-arc basins. The active seafloor systems are a major niche for fauna carrying out chemosynthetic biochemistry. This and the characterization of new species is now a main topic in biology (e.g., Jones, 1985; Childress, 1988; Grassle, 1985). The geochemistry of vent fluids, interstitial fluids in sediments, and organic matter associated with hydrothermal systems is also a major area of research effort (e.g., Edmond and Von Damm, 1983; Gieskes *et al.*, 1982a, b, 1988; Simoneit, 1990; Simoneit *et al.*, 1984; Von Damm, 1990).

Because there is extensive geological evidence for the occurrence of hydrothermal systems back to Archean time (e.g., Haymon *et al.*, 1984; de Wit *et al.*, 1982), the hypothesis of a relationship between such systems and the origin of life has merit (Corliss *et al.*, 1981). Some authors propose that life could not have originated in high-temperature hydrothermal systems, but only that the latter provided reduced compounds and chemical recycling in the Archean (Miller and Bada, 1988; Oró *et al.*, 1990). Nevertheless, this exciting topic of the origin of life is still under extensive investigation (e.g., Corliss, 1989; Yanagawa and Kobayashi, 1989).

It is the purpose of this chapter to provide a description of the fate and chemical alterations of organic matter under hydrothermal conditions, with a description of the localities where these processes have been studied to date. Two major but overlapping aspects will be considered: (a) hydrothermal petroleum and gas generation with concomitant expulsion and migration, and (b) hydrothermal or geothermal remobilization and migration of conventional petroleum.

Petroleums generated in high temperature and with high-fluid-flow regimes are defined here as hy-

BERND R. T. SIMONEIT • Petroleum Research Group, College of Oceanic and Atmospheric Sciences, Oregon State University, Corvallis, Oregon 97331.

Organic Geochemistry, edited by Michael H. Engel and Stephen A. Macko. Plenum Press, New York, 1993.

drothermal because the agent of thermal alteration and mass transfer, hot circulating water, is responsible for petroleum generation and migration from the source rocks or unconsolidated sediments (Didyk and Simoneit, 1989, 1990). In contrast, conventional oils are natural products of basin evolution and are generated contemporaneously with sediment compaction and heating. Generation of hydrothermal oils and gases is a relatively rapid process, whereas geothermal oils are generated at a rate that is tied to basin subsidence (Tissot and Welte, 1984; Hunt, 1979).

2. Geological Locales with Hydrothermal Petroleum

The locations with known hydrothermal activity and associated mineralization at seafloor spreading centers (divergent plate boundaries) currently number about 100 and are cataloged in the reviews by Rona (1984, 1988). Those with associated organic matter alteration are indicated on the tectonic sketch map in Fig. 1. Both of these lists are expected to expand as exploration continues. Two continental systems are also shown in Fig. 1.

Hydrothermal systems operate under a pressure head of circulating water and yield fluids with varying temperatures depending on the depth (below sea or land surface) of venting and the heat dissipation. Thus, shallow vents generally have exit temperatures less than 100°C (e.g., Lake Tanganyika, 20 m below lake level; Tiercelin *et al.*, 1989), and deep vents have exit temperatures ranging from warm to >300°C (e.g., Guaymas Basin, Gulf of California, 2000-m water depth). The physical aspects of the two most extensively studied examples will be discussed briefly.

Guaymas Basin (Fig. 1) is an actively spreading oceanic basin (2000-m water depth in the rifts) where sedimentation is rapid (> 2 m/1000 yr), covering the rift floors to a depth of ~300–500 m (Curray *et al.*, 1982; Einsele, 1985; Lonsdale, 1985; Lonsdale and Becker, 1985). The organic matter of these recent sediments is derived primarily from diatomaceous and microbial detritus and averages about 2% organic carbon. Influx of terrigenous organic matter is low because deserts border the Gulf of California (Simoneit and Philp, 1982). Numerous hydrothermal mounds rise to 20–30 m about the south rift floor, and most are actively discharging vent fluids with water temperatures ranging from warm up to 315°C at ~200 bar (Lonsdale, 1985; Lonsdale and Becker, 1985; Merewether *et al.*, 1985). Some photographs of an example of the vent/mound systems and fluid discharges are given in Fig. 2. The mounds are composed of complex deposits of sulfide, sulfate, silicate, and carbonate minerals and colonies of tube worms, bacterial mats, and other chemosynthetic organisms (Peter, 1986; Koski *et al.*, 1985; Jones, 1985).

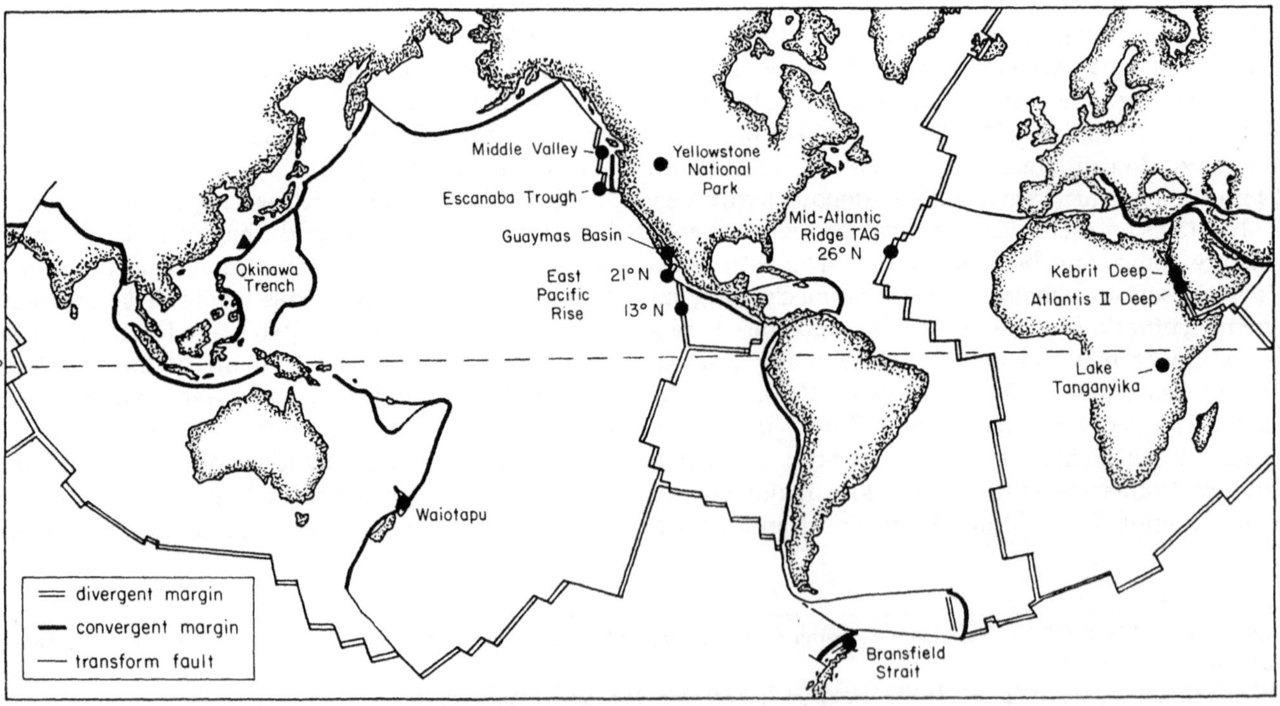

Figure 1. General location map of the hydrothermal vent fields discussed in this chapter with the sketched global tectonics.

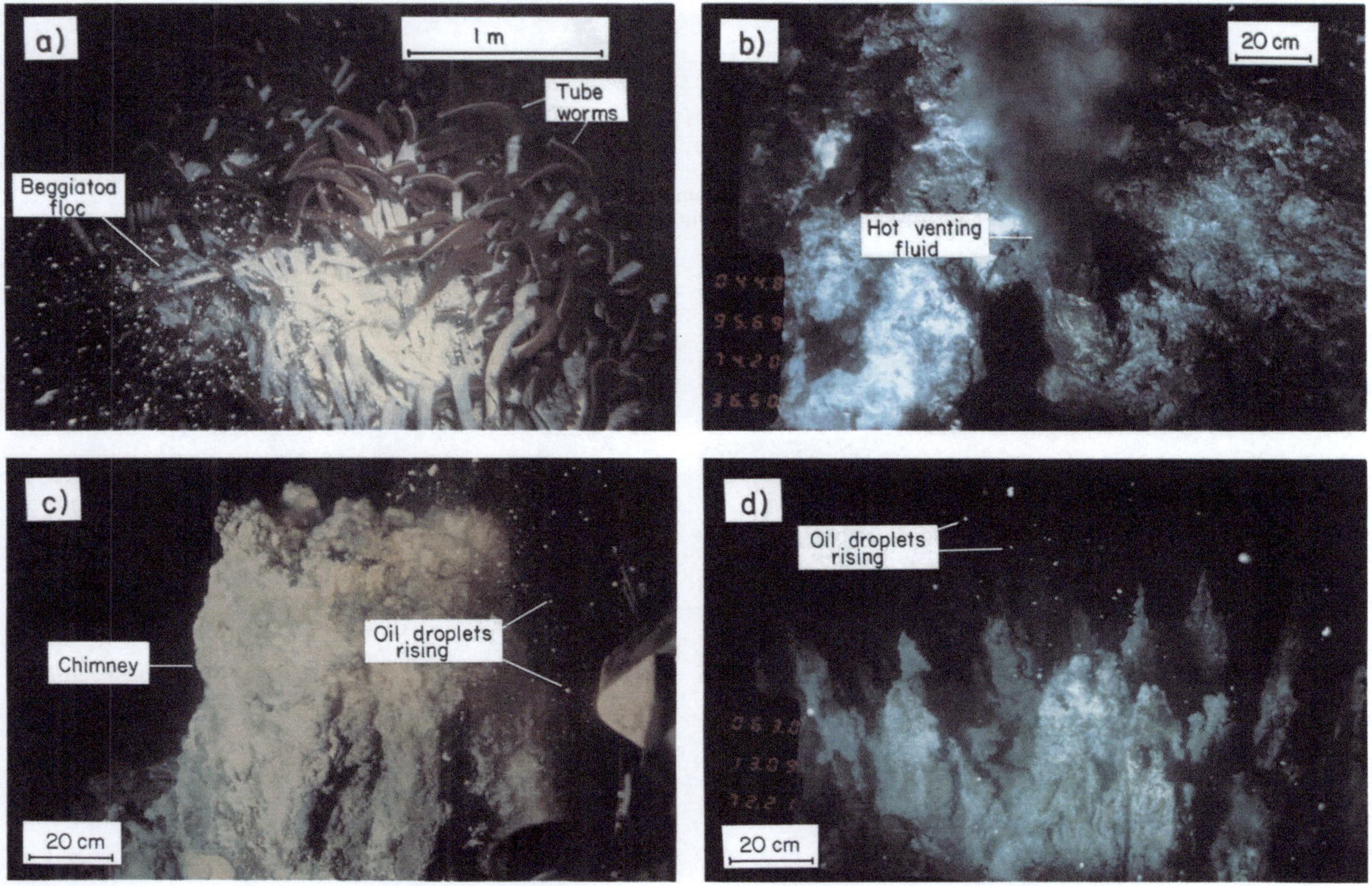

Figure 2. Photographs of examples of Guaymas Basin vent systems: (a) mound macrofauna (tube worms and beggiatoa floc); (b) vent discharging hot fluid, (c) petroleum rising from base of chimney, (d) petroleum rising from top of mound with small chimneys.

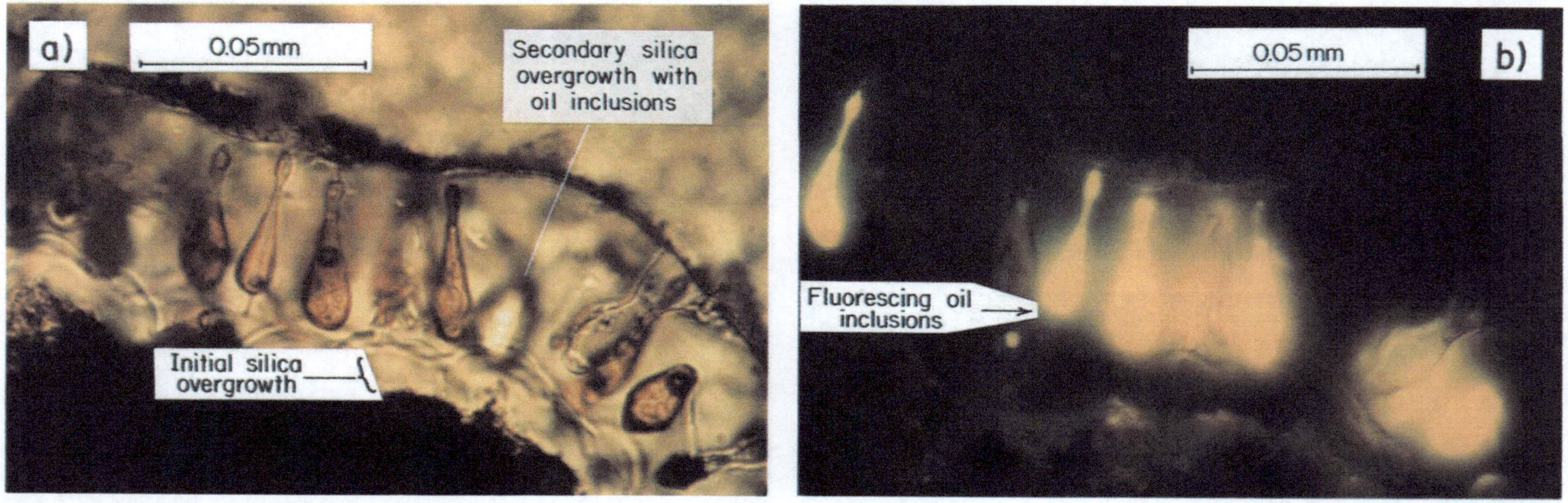

Figure 8. Photomicrographs of oil fluid inclusions in amorphous silica of vent/mount material from Guaymas Basin [sample 1175-8, as described by Peter *et al.* (1990)]: (a) transmitted light; (b) UV fluorescence light.

Guaymas Basin is to date (1991) the only marine hydrothermal system with associated petroleum that has been drilled. Leg 64 of the Deep Sea Drilling Project (DSDP) encountered intrusives and hydrothermal alteration at depth in all holes drilled in the basin (Curray *et al.*, 1982). Thermogenic hydrocarbon gas, H_2S, and CO_2 were identified for all sites in Guaymas Basin, and the lipids (bitumen) were thermally altered close to and below intrusive sills, especially for Site 477 in the southern rift (Simoneit and Philp, 1982).

The Escanaba Trough in the northeastern Pacific (Fig. 1) is the southern extension of the Gorda Ridge, an active oceanic spreading center about 300 km long and bounded on the north and south by the Blanco and Mendocino fracture zones, respectively. It is filled with up to 500 m of Quaternary turbidite sediments (Kvenvolden *et al.*, 1986). The petroleum which saturates the sediments and mineral ores that blanket the ridge axis is derived from hydrothermal alteration of sedimentary organic matter (Kvenvolden *et al.*, 1986, 1990). Middle Valley is another actively venting and sediment-covered hydrothermal system in the northeastern Pacific (Fig. 1). Preliminary analyses of shallow core material and one vent chimney from the area revealed only traces of organic thermal alteration products derived from low-temperature stress in the sediments, but the vent/mound material contained hydrothermal tar (Simoneit *et al.*, unpublished data).

The Bransfield Strait, Antarctica (Fig. 1), is a typical example of a back-arc rift, which is tectonically active with extensional features such as dip-slip faults and intrusives, and is also heavily sedimented (Whiticar *et al.*, 1985; Suess *et al.*, 1993). Hydrothermal activity is evidenced by mineral alteration and a slight petroliferous odor of the sediments.

The examples of sediment-starved hydrothermal systems are located on basaltic rift areas as, for example, 26°N on the Mid-Atlantic Ridge and 13° and 21°N on the East Pacific Rise (Fig. 1). Continental systems are in volcanic or failed rift terranes as, for example, Yellowstone National Park, Lake Tanganyika, and Waiotapu (Fig. 1).

3. Interactions of Hydrothermal Activity with Organic Matter

3.1. Source Organic Matter

The nature or constitution of the organic matter being altered determines the types of petroleum products that form under both kinds of generation regimes, i.e., normal slow thermal maturation (termed geothermal herein) and hydrothermal. In both regimes the sedimentary lipid matter initially undergoes alteration due to mainly diagenetic and early catagenetic processes, thus changing the hydrocarbon signature. For example, in Guaymas Basin and the Escanaba Trough the two end-member lipid signatures (marine and terrigenous, respectively) can be observed. The lipid hydrocarbons of the seabed sediment in Guaymas Basin (Fig. 3a) exhibit *n*-alkanes primarily shorter than C_{21}, and only a minor series of homologs longer than C_{23} with an odd-carbon-number predominance. This composition is typical of a predominantly microbial origin from marine productivity and a minor influx of terrigenous vascular plant wax (> C_{25}) (Simoneit, 1978). The kerogen composition also indicates such an origin. The other example, the lipid hydrocarbons of the seabed sediment in the Escanaba Trough (Fig. 3b), shows *n*-alkanes primarily longer

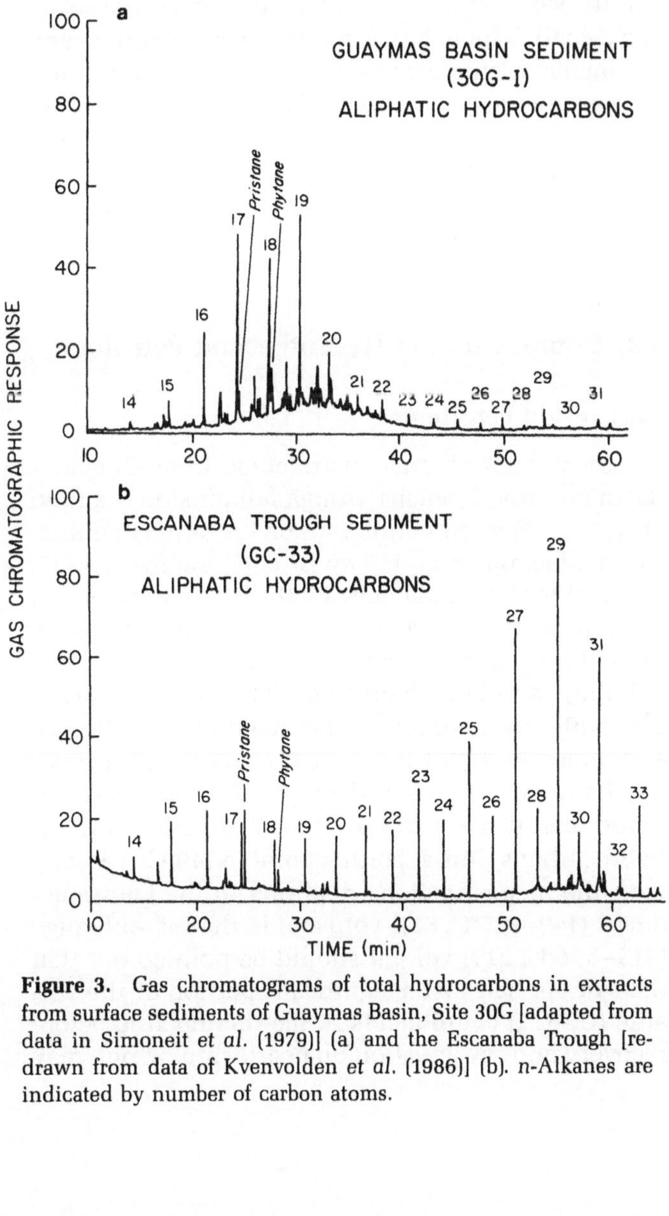

Figure 3. Gas chromatograms of total hydrocarbons in extracts from surface sediments of Guaymas Basin, Site 30G [adapted from data in Simoneit *et al.* (1979)] (a) and the Escanaba Trough [redrawn from data of Kvenvolden *et al.* (1986)] (b). *n*-Alkanes are indicated by number of carbon atoms.

than C_{23} with a strong odd-carbon-number predominance and lesser amounts at lower carbon numbers. This signature is derived mainly from terrigenous vascular plant wax, with a minor microbial component (Kvenvolden *et al.*, 1986). The kerogens of similar samples from that region also reflect a primarily terrigenous composition (Simoneit, 1978).

During conventional and hydrothermal oil generation, large amounts of additional hydrocarbons are superimposed on the syngenetic lipids, thus diluting such signatures (e.g., loss or reduction of the odd-carbon-number predominance for *n*-alkanes longer than C_{25}; compare Fig. 3 to Fig. 9 in later section). The major source of petroleum compounds is from the sedimentary macromolecular organic detritus, called kerogen, which generally constitutes the bulk of the total organic carbon content (Tissot and Welte, 1984). The constitution and source of the kerogen partially determines the nature of the petroleum generated. In general, terrestrial organic detritus from mainly vascular plants yields an aromatic kerogen (e.g., coal) which has a natural gas potential, and marine/lacustrine organic matter from primarily microbial residues yields an aliphatic kerogen (e.g., sapropel) which has a paraffinic petroleum potential (Tissot and Welte, 1984; Hunt, 1979). Kerogens in sedimentary basins are generally mixtures of these inferred end members.

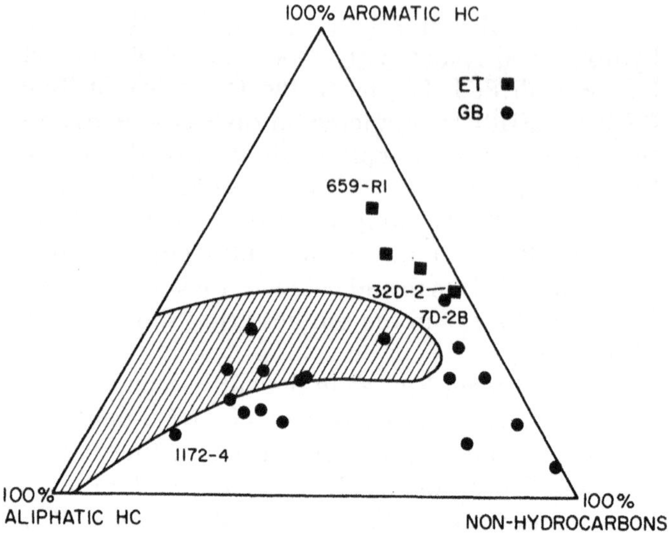

Figure 4. Ternary diagram of saturated hydrocarbons, aromatic hydrocarbons, and NSO plus asphaltic components (Kvenvolden and Simoneit, 1990; Kawka and Simoneit, 1987; Didyk and Simoneit, 1990). Typical crude oils fall within the hachured area (Tissot and Welte, 1984). ●, Guaymas Basin oils; ■, Escanaba Trough oils.

3.2. Composition of Hydrothermal Petroleum

3.2.1. Fluid Petroleum

Most hydrothermal petroleums from Guaymas Basin and the Escanaba Trough fall outside the field of typical reservoir petroleums on the ternary composition diagram (Fig. 4; Kawka and Simoneit, 1987; Kvenvolden and Simoneit, 1990). This indicates that they are of diverse compositions and generally more polar than conventional petroleums. One typical chimney sample with oil from Guaymas Basin (sample 1630) was compared to normal crude oils (Didyk and Simoneit, 1990). Its total extractable organic matter distilled over the full range, similar to normal crude oils. It has a lower content of gasoline range hydrocarbons (initial boiling point to 149°C; 3% vol.), with the major fractions distilling in the kerosene–diesel (149–315°C; 61% vol.) and in the gas–oil range (315–538°C; 21% vol.). It should be pointed out that the sample had lost a significant portion of the volatile (C_1–C_{10}) components in the thermocline before recovery and preservation on board ship, which may in part explain the low gasoline content. The relatively high content of distillation residue (> 538°C; 15% vol.) is due to a significant content of asphaltene and polar compounds, which are somewhat high for an oil of an intermediate gravity (sample 1630, API = 27.6). These results indicate that the bulk of this oil is compositionally the same as conventional crude oil, except its content of polar material is enhanced.

A comparison of the stable carbon isotope compositions of hydrothermal petroleum samples from Guaymas Basin with data from DSDP Holes 477 and 477A (Jenden *et al.*, 1982; Simoneit *et al.*, 1984), indicates that the oils are very similar to each other and confirms that they are derived from similar sources (Schoell *et al.*, 1990). The most reasonable source areas for the hydrothermal petroleums are the surrounding sediments of the vent area. The comparison of the carbon isotope variations in the hydrothermal petroleums with those of the sediment kerogens (Fig. 5) indeed shows the same range of isotope values. The low atomic H/C ratios of the kerogens from Guaymas Basin sediments approaching intrusive sills (Fig. 5, samples 4–6 and 8–10; Jenden *et al.*, 1982) were obviously caused by thermal stress. The considerable decrease of the H/C ratios suggests that some of the kerogen in those samples has already been converted to bitumen or oil.

Typical oils from Guaymas Basin (e.g., sample 1630; Didyk and Simoneit, 1990) have an intermedi-

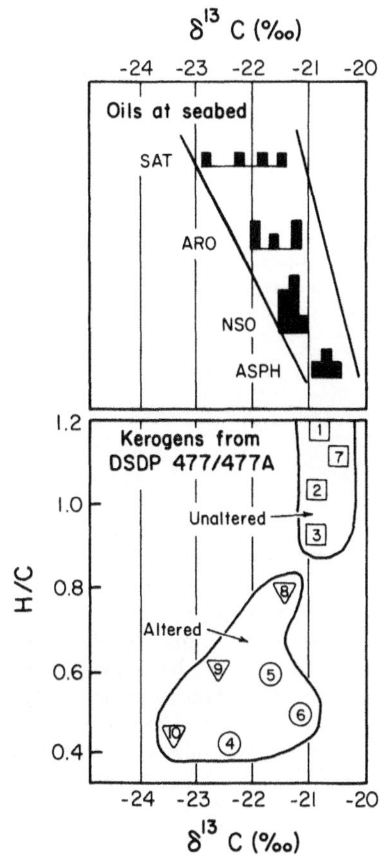

Figure 5. Carbon isotopic oil-to-source correlation of hydrothermal petroleum fractions with sedimentary organic matter from the vicinity in Guaymas Basin. Data for the sedimentary kerogens are from Jenden *et al.* (1982) [the histograms represent the data of Schoell *et al.* (1990).]

ate content of *n*-alkanes (18.4%) and a relatively normal content of *iso*, *anteiso*, isoprenoid, and naphthenic hydrocarbons (81.6%) (Fig. 6a and b), both percentages being comparable to those in normal crude oils. The *n*-alkane range for the stabilized oil (C_{10}–C_{35}, max. C_{15}, Fig. 6a) was a monomodal and wide carbon number distribution with no carbon number preference [$CPI_{10-32} = 0.99$, where CPI is the carbon preference index (modified); for hydrocarbons, it is expressed as a summation of the odd-carbon-number homologs over the range present (in this case ~C_{10} to C_{32}), divided by a summation of the even-carbon-number homologs over the same range (Cooper and Bray, 1963; Simoneit, 1978)]. A suite of isoprenoid hydrocarbon biomarkers is present, ranging from C_{15} to C_{20} (no C_{17}), with pristane as the dominant homolog (Fig. 6b) and a pristane/phytane ratio of 1.85. The aromatic hydrocarbons are comprised of *n*-alkylbenzenes and *n*-alkyltoluenes with significant amounts of polycyclic aromatic hydrocarbons (PAHs) and alkyl PAHs (Fig. 6c). This hydrocarbon distribution skewed to low carbon numbers is compatible with a predominantly bacterial–algal origin for the source organic matter. The CPI of ~1 indicates complete maturation. The typical diagnostic biomarkers consist of the triterpenoid, steroid, and tricyclic terpane hydrocarbons as are generally found in crude oils.

3.2.2. Volatile Components

Hydrothermal fluids in Guaymas Basin convert immature organic matter in the overlying sedimentary cover to petroleum, which is also scavenged by these fluids and migrates upward. Upon exiting at the seabed, the fluids are often saturated with a broad range of volatile hydrocarbons (C_1 to *n*-C_{10}) as well as lower concentrations of heavy ends (>C_{15}) (Simoneit *et al.*, 1988). These volatile components are missing in most of the mound samples (i.e., old, established systems). Interstitial gas in sediments of DSDP cores consists of biogenic methane (C_1) overprinted by thermogenic C_1 to C_5 hydrocarbons near the sills and, to a lesser extent, at increasing sub-bottom depths. These are of a similar composition to that of the venting volatile hydrocarbons (Simoneit *et al.*, 1988; Whelan *et al.*, 1988).

Many vent water samples contain high amounts of light hydrocarbons, with C_1 at corrected concentrations of about 150 cm^3 (STP) kg^{-1} (Welhan and Lupton, 1987). For comparison, the C_1 concentrations in vent fluids from the East Pacific Rise at 21°N, a sediment-starved rift system, have been reported to be 1–2 cm^3 (STP) kg^{-1} (Welhan and Lupton, 1987). Thus, sedimented hydrothermal systems generate more natural gas. Typical condensate hydrocarbon distributions (C_1 to C_{10}) are present in the Guaymas Basin vent fluids, as, for example, for sample 1620-2C, where the exit temperature was 308°C (Fig. 7b; Simoneit *et al.*, 1988). Compared with the headspace gases of the mound sample 1629-A3 (Fig. 7a), the hydrocarbon content of the 308°C vent water is highly enriched in the lower alkanes (<C_7). Quantitatively, the compositions of normal, branched, and cyclic alkanes shorter than C_8 are the same for the high-temperature water and the mound samples. However, the hot water has an enhanced content of aromatic (benzene, toluene, ethylbenzene, and xylenes) versus aliphatic hydrocarbons, which is indicated by ratios of *n*-heptane to toluene, generally ≤1.0 for vent fluids (Simoneit *et al.*, 1988).

Acetate and propionate ions are highly enriched

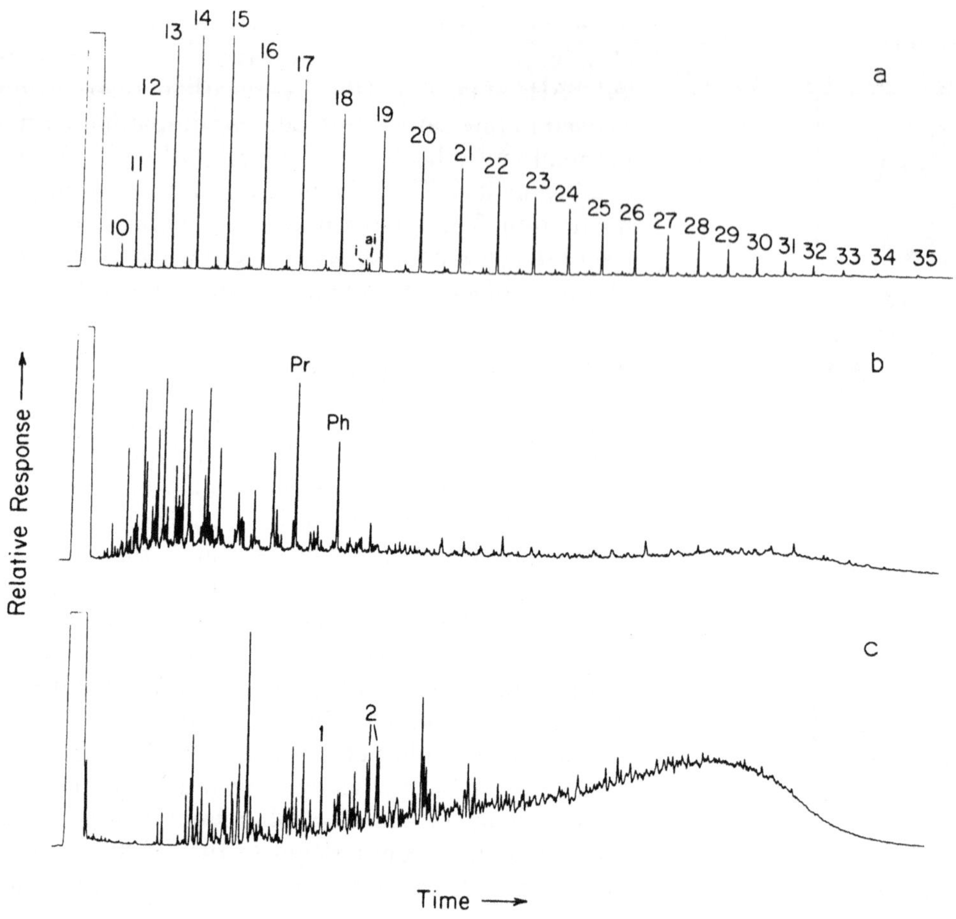

Figure 6. Gas chromatograms of the hydrocarbon fractions of the Guaymas Basin oil from chimney sample 1630: (a) normal alkanes; (b) branched and cyclic hydrocarbons; (c) aromatic hydrocarbons (F2). Numerals refer to carbon number of *n*-alkanes. Pr, Pristane; Ph, phytane. In (a), minor peaks are *iso*- (*i*) and *anteiso*-alkanes (*ai*); in (c), 1 = phenanthrene, 2 = methylphenanthrenes. [Adapted from Didyk and Simoneit (1990).]

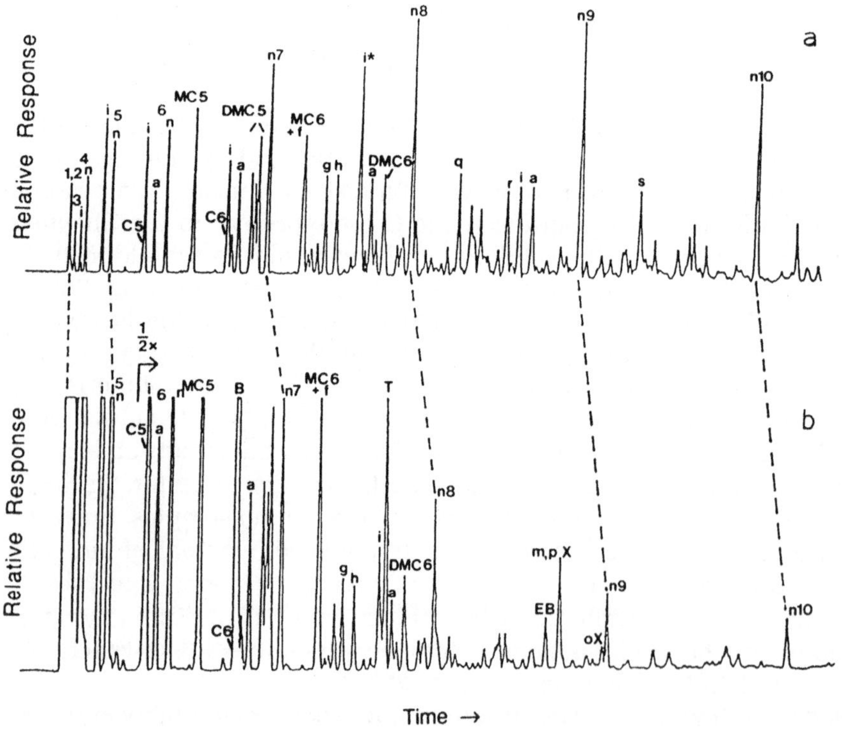

Figure 7. Gas chromatograms of headspace analyses for comparison of the volatile hydrocarbons from: (a) hydrothermal crust, 1629-A3; (b) hot venting water, 1620-2C (T = 308°C) (Simoneit *et al.*, 1988). Numbers refer to carbon chain length with n = normal, i = *iso* (2-methyl-), and a = *anteiso* (3-methyl-) compounds of corresponding chain length. Other acyclic compounds are 2,6-dimethylheptane (q), 2,3-dimethylheptane (r), and 2,6-dimethyloctane (s). Cyclic compounds are the cyclo- (C), methylcyclo- (MC), and dimethylcycloalkanes (DMC). The DMC5 triplet contains the *cis*-1,3, *trans*-1,3, and *trans*-1,2 isomers. Other individual alkylcyclopentanes are 1,1,3-trimethyl- (f), 1,2,4-trimethyl- (g), and 1,2,3-trimethylcyclopentane (h). The aromatics are benzene (B), ethylbenzene (EB), toluene (T), and xylenes [X, with ortho (o), meta (m), and para isomers (p)]. The asterisk indicates a coeluting unknown, and the symbol ½× reflects a signal attenuation by a factor of 2.

in the hydrothermally altered sediments of Guaymas Basin relative to normal marine sediments (Martens, 1990). These volatile and water-soluble acids are important as complexing agents for metals, substrates for heterotrophic microbial processes, and buffers, as well as possibly in CO_2, C_1, and C_2 generation.

Generally, the volatile hydrocarbons exhibit large variations in character in terms of carbon number range (C_1–C_{10+}), structural diversity (relative contents of the normal, branched, and cyclic components), and polarity (aliphatic versus aromatic components). This character is controlled by a number of factors, primarily temperature, aqueous solubility, and biodegradation/water washing. Migration of these volatile hydrocarbons occurs through dispersion in vent fluids and as bulk phase in the sediments and vein systems. The more soluble and volatile hydrocarbons are released into the water column by rapidly venting fluids, rising in some cases as large plumes (Merewether *et al.*, 1985; Simoneit *et al.*, 1990a), and by aqueous remobilization from some exposed hydrothermal mounds.

Although direct measurement of the oil flow rate at the vent sites of Guaymas Basin has so far not been feasible, the collection of oil globules under "*in situ*" conditions showed that the gas/oil ratio is approximately 5 at standard temperature and pressure, with possible minor gas loss due to dissolution in hydrothermal water and seawater by exposure of the oil prior to and during expulsion from the vent system (Simoneit *et al.*, 1988). Low bottom temperatures (2–3°C) contribute to condensation and retention of some of the oil, which generates oil impregnations in (Peter *et al.*, 1990) and on (Carranza-Edwards *et al.*, 1990) inorganic substrates of the hydrothermal vent system. In general, the low pour point (<18°C; Didyk and Simoneit, 1990) of the hydrothermal oil allows the oil to remain fluid at these bottom temperatures. Volatile hydrothermal oil is also trapped in fluid inclusions in the amorphous silica of veins and conduits of the vent/mound systems (Fig. 8; Peter *et al.*, 1990).

3.3. Interactions and Rates

The thermal alteration products of organic matter in hydrothermal systems can be considered to be in a metastable equilibrium state (e.g., Shock, 1988, 1989). In this state, not all of the stable equilibrium species are present due to kinetic constraints, but theoretical evaluations of the distributions of species at metastable equilibrium are analogous to those for stable equilibrium. Thus, Guaymas Basin vent fluids, for example, contain hydrogen, hydrogen sulfide, and C_1 (reduced species), CO_2 and acetate (oxidized species) (Martens, 1990; Welhan and Lupton, 1987).

The interactions of the fluid medium in terms of chemistry and solvent properties are not well understood. The dominant fluid is water, and in the examples of Guaymas Basin and the Escanaba Trough it is under temperatures approaching 350 and 400°C, respectively, with pressures exceeding 200 and 300 bar (2×10^7 and 3×10^7 Pa), respectively. These are in the near-critical domain of water (Chen, 1981; Bischoff and Rosenbauer, 1984, 1988; Pitzer, 1986). Supercritical water has enhanced solvent capacity for organic compounds and reduced solvation properties for ionic species due to its loss of aqueous hydrogen bonding (Tödheide, 1982). It is also a reactive medium for either reductive or oxidative reactions (Ross, 1984). Thus, the near-critical domain of water in hydrothermal systems is expected to aid reaction rates and solvation capacity for organic matter.

Fluids in hydrothermal systems also contain large concentrations of C_1 and CO_2 (Sakai *et al.*, 1990; Simoneit and Galimov, 1984; Simoneit *et al.*, 1988; Welhan and Lupton, 1987). These gases are supercritical under the temperature and pressure conditions of the hydrothermal systems, and their effects on the critical point of seawater are not known. Phase separation of CO_2 from water at reduced temperatures has been proposed for liquid CO_2 vents in a back-arc hydrothermal system (Sakai *et al.*, 1990). Carbon dioxide liquid is also an excellent solvent for organic compounds. Thus, hydothermal fluids are efficient solvents for scavenging hydrothermal petroleum from the source and migrating it away from the hot zone.

The reaction rates of organic matter conversion to petroleum in hydrothermal systems are also efficient, and the process occurs over a geologically brief time. For example, carbon-14 dates have been obtained from hydrothermally derived petroleum and calcite from the southern trough of Guaymas Basin (Peter *et al.*, 1991). The ages for petroleum range from 4240 to 5705 yr B.P. whereas those for calcite are from 7340 to 31,450 yr B.P. These are not true ages, but rather they reflect the age of carbon within these materials. The data confirm the rapid conversion of sedimentary organic matter to hydrothermal petroleum and its mobilization from the upper 30 m of the sediment column (≤500 m thick) underlying the seafloor deposits. The large difference in ages between the petroleum and most of the calcites indicates that a major portion of the carbon in the latter is scoured (dissolved) from much greater depths.

4. Hydrothermal Petroleum Generation and Migration

Organic matter of sedimentary sinks (basins), usually marine and of either recent or geologically old origin, is derived from the residues of biogenic debris (Simoneit, 1978, 1982a, 1983). This detritus is composed of both autochthonous (locally derived) detritus and allochthonous residues originating from continental sources (Simoneit, 1982a). Aquatic sediments receive allochthonous organic detritus primarily by river wash-in and eolian transport, with ice rafting and sediment recycling as minor contributing processes (Simoneit, 1978).

The preservation of organic matter in sediments depends on the initial diagenetic processes involving microbial degradation and chemical conversion, coupled with the environmental conditions of acidity and redox potential (e.g., Didyk *et al.*, 1978; Demaison and Moore, 1980). Then, during normal sediment maturation and lithification, the organic matter is modified by the metamorphic effects of the geothermal gradient (temperature and pressure) and petrology (e.g., Hunt, 1979; Kartsev *et al.*, 1971; Simoneit, 1978; Tissot and Welte, 1984), ultimately yielding petroleum products over long geologic time periods (millions of years). The late stages of catagenesis and the subsequent high-temperature phase called metagenesis (very deep burial) generate primarily methane from both bitumen and kerogen (Hunt, 1979; Tissot and Welte, 1984).

Hydrothermal processes also act on sedimentary organic matter and on entrained ambient organic detritus (e.g., in water), resulting in "instantaneous" diagenesis and catagenesis of this biogenic material and thus producing analogous petroleum products (Simoneit and Lonsdale, 1982; Simoneit 1983, 1984a, b, 1985a, 1988). Gas (CH_4–C_{8+}, CO_2, and H_2S) and bitumen [n-C_8–n-C_{40+}, with a large unresolved complex mixture (UCM) of branched and cyclic hydrocarbons] are cracked from the pseudokerogen and biopolymers, becoming superimposed on the endogenous gas and lipids. Additionally, products characteristic of high-temperature processes (e.g., PAHs, stabilized molecular markers, etc.) are also found in the petroleums (e.g., Kawka and Simoneit, 1987, 1990; Simoneit, 1984a). The similarities and differences between hydrothermal petroleums and conventional reservoir petroleums are summarized in Table I. The spent kerogen remains as amorphous "activated" carbon (Simoneit, 1982b).

TABLE I. Hydrothermal Petroleum Compared to Reservoir Petroleum

Similarities to reservoir petroleum
1. Natural gas and gasoline-range hydrocarbons
2. Full range of *n*-alkanes, no carbon number predominance (CPI = 1.0–1.2)
3. Naphthenic components (major hump, UCM)
4. Isoprenoid hydrocarbons (including significant but variable pristane and phytane)
5. Biomarkers [e.g., mature 17α(*H*)-hopanes and steranes]
6. Alkylaromatic hydrocarbons and asphaltenes
Differences from reservoir petroleum
1. High concentrations of polynuclear aromatic hydrocarbons (PAHs)
2. Residual immature biomarkers and intermediates [e.g., 17β(*H*)-hopanes, hopenes, sterenes]
3. Degraded biomarkers (e.g., Diels' hydrocarbon, porphyrins with C_{27} max.)
4. Significant heteroaromatic compounds
5. High sulfur content
6. Alkene content in bitumen near "source rock"

The organic matter associated with deeper hydrothermal systems (e.g., epithermal ores in volcanic terranes) is usually more asphaltic with a high PAH content. Such organic matter is widely distributed and has, for example, been studied from the California mercury deposits (idrialite; Blumer, 1975; Geissman *et al.*, 1967) and other hydrothermal sulfide deposits (Germanov and Bannikova, 1972). The advent of deep-well drilling (>7000 m) has yielded core materials (e.g., Cretaceous shales) which were at *in situ* temperatures of about 260–300°C (e.g., Price, 1982; Price *et al.*, 1981). These samples had high concentrations of bitumen components, and the kerogens still had a significant hydrocarbon generation potential. This result has been suggested to indicate that *in situ* petroleum can be stable at high temperatures and pressures and over long geologic time periods (Price, 1982; Price *et al.*, 1981). Metagenesis also appears to occur over much wider and higher temperature conditions than normally believed.

Migration processes are understood least. In sedimentary basins, migration appears to occur by diffusion and in solution ($CH_4/CO_2/H_2O$ solvent; Hunt, 1979), whereas, in hydrothermal areas, migration proceeds by the previous processes and also by thermally driven diffusion, advection, and mass transport as oils and/or emulsions (e.g., Whelan *et al.*, 1984, 1988; Kawka and Simoneit, 1987; Simoneit *et al.*, 1988). Although the overall result may be the same for both regimes, migration of petroleum in hydrothermal systems is often more rapid and efficient due to the inherent high fluid flow. For example, in Guaymas Basin the petroleum products appear to migrate by advection, diffusion, and hydrothermal circulation as fluids away from the heat sources upward to the seabed and laterally into porous sections. There the petroleum condenses (solidifies) according to the ambient temperatures in the conduits (veins) and vugs of the hydrothermal mineral mounds. PAHs condense in the hot vents, waxes crystallize, and oil solidifies in intermediate-temperature regions (~20–80°C), and the volatile petroleum partially collects in cold areas (~3°C) and emanates mainly into the ambient seawater as plumes (Simoneit, 1984a, 1985a; Simoneit *et al.*, 1990a; Merewether *et al.*, 1985). Volatile oil is found in fluid inclusions in silica overgrowths of the mound minerals (Fig. 8). The trapping temperatures of these hydrocarbons were estimated to be within the range of 116 to 226°C (Peter *et al.*, 1990).

4.1. Sedimented Hydrothermal Systems

4.1.1. Guaymas Basin, Gulf of California

Petroleum products have been characterized in samples obtained by shallow gravity coring (Simoneit *et al.*, 1979), piston coring (Simoneit, 1983), and deep coring by Leg 64 of the DSDP (Curray *et al.*, 1982; Simoneit *et al.*, 1984). Petroleum-bearing samples have also been recovered from the seafloor by dredging operations (Simoneit and Lonsdale, 1982) and submersible sampling with the *Alvin* in 1982, 1985, and 1988 (Simoneit, 1984a, b, 1985a; Kawka and Simoneit, 1987; Simoneit and Kawka, 1987; Simoneit *et al.*, 1988, 1990a, 1992). The samples have very diverse petroleum contents and hydrocarbon distributions [e.g., Fig. 6 and Fig. 9a; other examples are presented by Simoneit (1984a, b, 1985a), Kawka and Simoneit (1987), and Simoneit and Kawka (1987)] and are analogous to those described for bitumens at depth in the DSDP holes (Simoneit, 1983, 1984b; Simoneit *et al.*, 1984; Kawka, 1990). The *n*-alkanes range from C_1 to *n*-C_{40+}, with usual maxima in the mid-C_{20} region and no carbon number predominance (CPI = 1.0).

The biomarkers, mainly the steranes and triterpanes, of the hydrothermal petroleums are generally mature. The steranes are present as complex mixtures ranging from C_{27} to C_{29} (e.g., Fig. 9b), and the dominant sterane in all samples is 5α(*H*),14α(*H*),17α(*H*)-cholestane (20*R*), the immature isomer. Diasteranes are also present, with the 13β(*H*),17α(*H*)-diacholestanes (20*S* and 20*R*) most abundant. The triterpanes consist primarily of the 17α(*H*),21β(*H*)-hopanes, with minor amounts of 17β(*H*), 21α(*H*)-hopanes (moretanes) and 17β(*H*),21β(*H*)-hopanes (biological configuration), and range from C_{27} to C_{34} (C_{28} absent) (Fig. 9c). The various biomarker ratios confirm their high degree of maturity (Kawka and Simoneit, 1987; Kawka, 1990) and, along with the previous data, indicate that the petroleums were generated by rapid and intense heating.

An example of gas-chromatographic (GC) trace of an aromatic/naphthenic fraction of an oil sample (GB-7D-2B) is shown in Fig. 10a. The major resolved peaks are unsubstituted PAHs, a group of compounds uncommon in petroleums but ubiquitous in higher temperature pyrolysates (Geissman *et al.*, 1967; Blumer, 1975; Hunt, 1979). The dominant analogs are the pericondensed aromatic series, for example, phenanthrene, pyrene, chrysene, benzopyrenes, perylene, benzoperylene, and coronene. A pyrolytic origin is also supported by the presence of five-membered alicyclic rings (e.g., fluorene, methylenephenanthrene, fluoranthene, benzofluoranthene, and indenopyrene). These compounds are found in all pyrolysates from organic matter, and, once formed, they do not easily revert to periocondensed aromatic hydrocarbons (Blumer, 1975, 1976; Scott, 1982). The aromatic/naphthenic fractions of the Guaymas oils also contain significant amounts of NSO (nitrogen-, sulfur-, oxygen-containing) hetero-PAHs (e.g., Gieskes *et al.*, 1988), Diels' hydrocarbon (Simoneit *et al.*, 1992), and toxic PAHs, e.g., the benzopyrenes (Kawka and Simoneit, 1990). In addition, perylene is present; it is the predominant PAH of unaltered lipids in sediments deposited under oxygen-minimum environments in the Gulf (Simoneit and Philp, 1982; Simoneit *et al.*, 1979, 1984). The chemical composition of the aromatic fractions suggests derivation from a combination of high-temperature pyrolysis and admixture of less mature bitumen.

4.1.2. Escanaba Trough and Middle Valley, Northeastern Pacific

An example of a GC trace of a hydrocarbon fraction of hydrothermal petroleum from the Escanaba Trough is shown in Fig. 9d. The organic source material for these hydrocarbons appears to be terrigenous, based on the CPI, carbon number range (especially >*n*-C_{25}), biomarker composition, and sedimentological considerations (Kvenvolden *et al.*, 1990; Kvenvolden and Simoneit, 1990). The *n*-alkanes of the petroleum range from C_{14} to C_{40}, with a carbon number maximum at *n*-C_{27} and still a significant odd-carbon-number predominance for alkanes longer than *n*-C_{25} (CPI = 1.25, Fig. 9d), typical of a terrestrial,

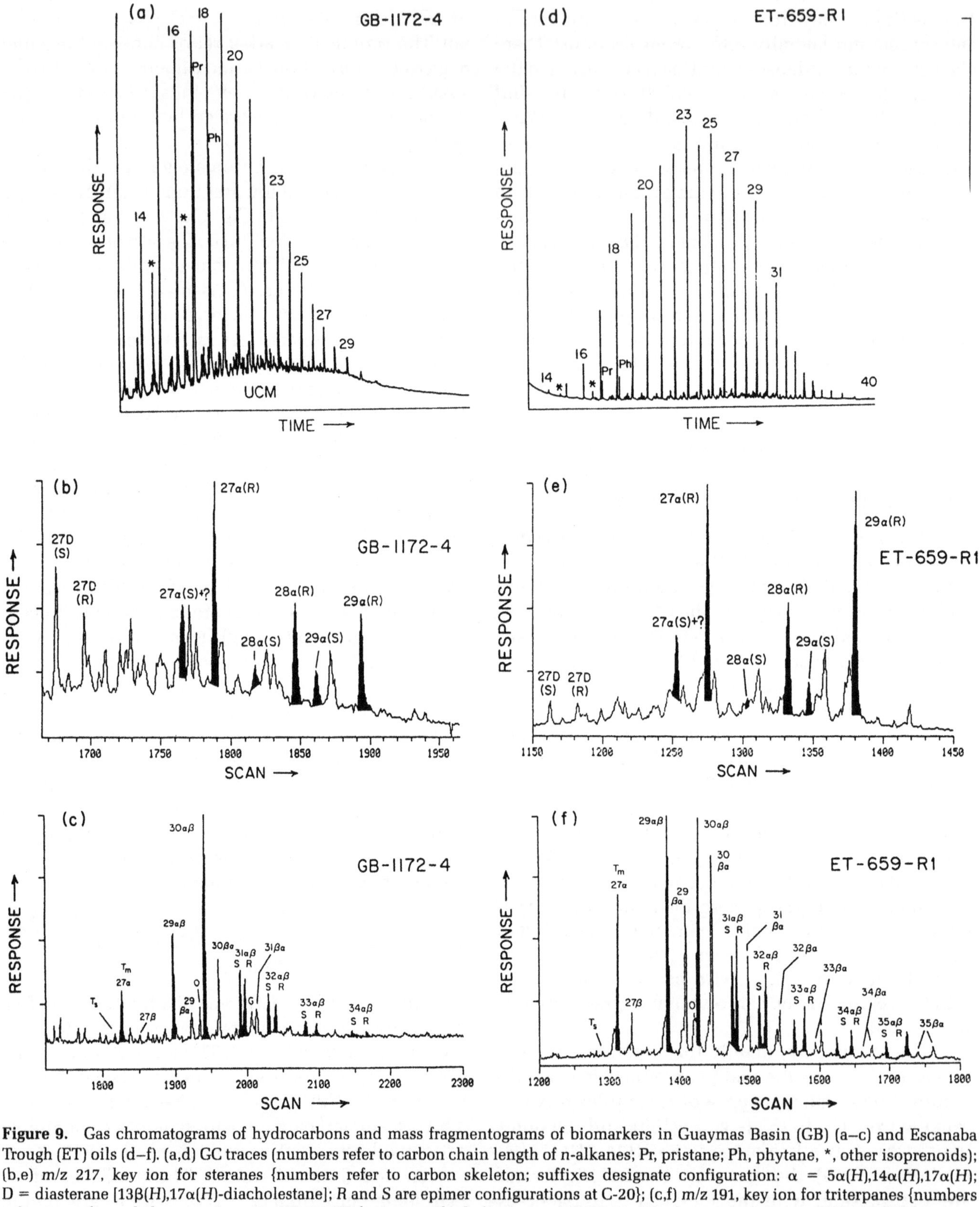

Figure 9. Gas chromatograms of hydrocarbons and mass fragmentograms of biomarkers in Guaymas Basin (GB) (a–c) and Escanaba Trough (ET) oils (d–f). (a,d) GC traces (numbers refer to carbon chain length of *n*-alkanes; Pr, pristane; Ph, phytane, *, other isoprenoids); (b,e) *m/z* 217, key ion for steranes {numbers refer to carbon skeleton; suffixes designate configuration: α = 5α(*H*),14α(*H*),17α(*H*); D = diasterane [13β(*H*),17α(*H*)-diacholestane]; *R* and S are epimer configurations at C-20}; (c,f) *m/z* 191, key ion for triterpanes {numbers refer to carbon skeleton; αβ = 17α(*H*),21β(*H*)-hopanes (shaded); βα = 17β(*H*),21α(*H*)- hopanes; ββ = 17β(*H*),21β(*H*)-hopanes; T_s = 18α(*H*)-22,29,30-trisnorneohopane; T_m = 17α(*H*)-22,29,30-trisnorhopane; O = oleanane; G = gammacerane; in the pairs of shaded α-hopanes, the 22*S* epimer is followed by the 22*R* epimer}. [Adapted from Kvenvolden and Simoneit (1990).]

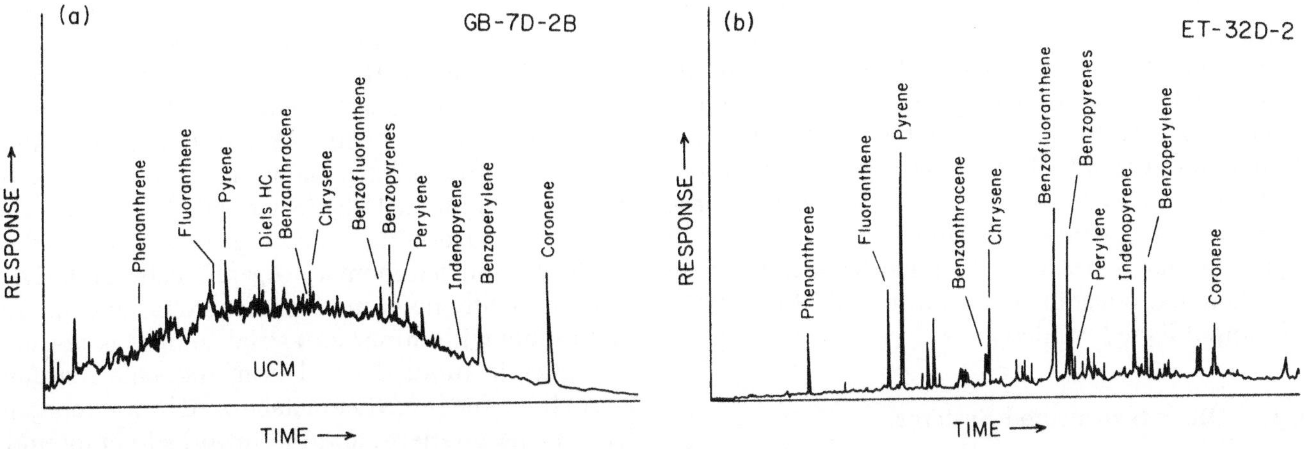

Figure 10. Gas chromatograms showing the distributions of aromatic hydrocarbons in the oil samples from Guaymas Basin (a) and Escanaba Trough (b) (the PAHs are labeled). [Adapted from Kvenvolden and Simoneit (1990).]

higher plant origin. The CPI may indicate admixture of bitumens with various maturities. Homologs of a marine origin ($<n$-C_{25}) are less concentrated. In general, the biomarkers of these petroleums, i.e., steranes and triterpanes, are less mature than in the case of the Guaymas petroleums. The steranes range from C_{27} to C_{29}, with $5\alpha(H),14\alpha(H),17\alpha(H)$-cholestane (20*R*) slightly less concentrated than the C_{29} homolog (Fig. 9e). Diasteranes are minor components. The triterpanes consist of the $17\alpha(H),21\beta(H)$-hopanes with major amounts of the $17\beta(H),21\alpha(H)$-hopanes (Fig. 9f) and ranging from C_{27} to C_{35} (C_{28} absent). The PAHs of these oils are also dominated by the unsubstituted analogs (Fig. 10b; Kvenvolden and Simoneit, 1990). The generation of this petroleum was probably by intense heating of short duration, as indicated by the biomarker distributions and the high concentrations of unsubstituted PAH.

Solid petroleum/tar has also been recovered from Middle Valley (Fig. 1) in an extinct hydrothermal chimney consisting primarily of barite. The GC traces of the total saturated and aromatic hydrocarbons are shown in Fig. 11. The saturated hydrocarbons are comprised primarily of an UCM, without *n*-alkanes

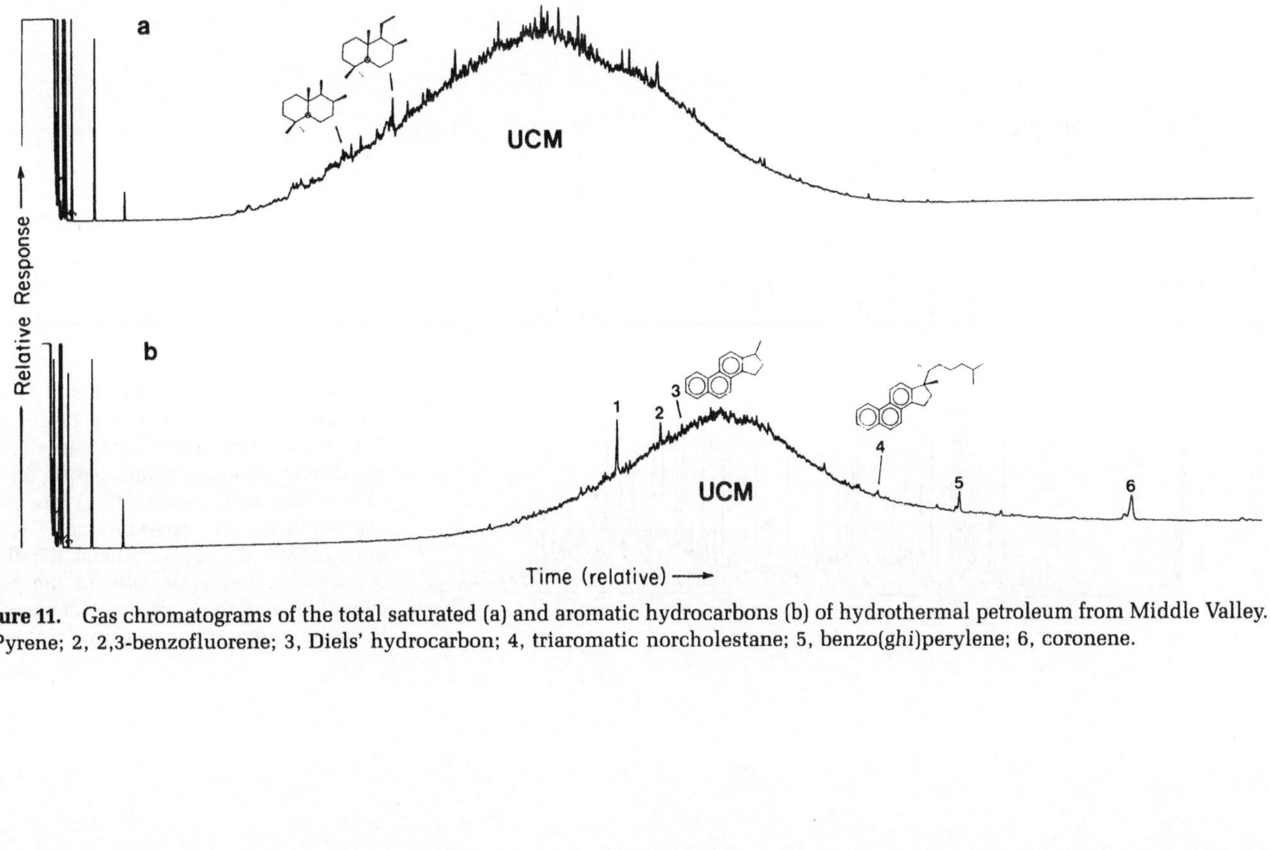

Figure 11. Gas chromatograms of the total saturated (a) and aromatic hydrocarbons (b) of hydrothermal petroleum from Middle Valley. 1, Pyrene; 2, 2,3-benzofluorene; 3, Diels' hydrocarbon; 4, triaromatic norcholestane; 5, benzo(*ghi*)perylene; 6, coronene.

and with minor amounts of various mature biomarkers. The aromatic hydrocarbons are also composed of an UCM and major PAHs, consisting of the heavy analogs ranging from pyrene (*m/z* 202) to coronene (*m/z* 300) with additional PAHs to *m/z* 352. Minor amounts of Diels' hydrocarbon and the triaromatic steroid hydrocarbons are also present. These compound signatures indicate that this sample is a hydrothermal petroleum which has been extensively weathered, biodegraded, and reworked (Simoneit *et al.*, unpublished results).

4.1.3. Other Sedimented Systems

Gravity cores from the eastern part of the Bransfield Strait, Antarctica (Fig. 1), have a weak petroliferous odor in the hydrothermally altered zones analogous to that described for the Guaymas Basin (Simoneit and Kawka, 1987). The lipid/bitumen compositions of two piston cores have been analyzed (Brault and Simoneit, 1988, 1990). The unaltered surface samples exhibit compound distributions (e.g., Fig. 12a) that can be correlated with their marine biogenic origin, where the bulk of the *n*-alkanes and additional biomarkers [e.g., hop-22(29)-ene, C_{28}-steradienes, and C_{25}-polyalkenes] are derived from autochthonous marine microbial sources, as was also reported by Venkatesan and Kaplan (1987). The hydrocarbon patterns in the hydrothermally altered zones are dramatically different (e.g., Fig. 12b), with a superposition of complex resolved and unresolved (UCM) thermal products on the *n*-alkane pattern. These patterns indicate only mild and localized heating, which resulted in accelerated diagenesis and limited migration (Brault and Simoneit, 1988).

The Atlantis II Deep (Fig. 1) contains stratified brine layers, the deepest of which is at a temperature of 62°C (Hartmann, 1980, 1985). Bulk organic matter and hydrocarbons have been analyzed in two sediment cores from the Deep (Simoneit *et al.*, 1987). The dense brine overlying the coring areas is reported to be sterile, and sedimentary organic material derived from autochthonous marine planktonic and microbial inputs with minor terrestrial sources is present. The organic input derived from the water column above the brine is further metabolized by microorganisms at the interface, and the reworked compounds with organic detritus are apparently then incorporated into the sediments under the brine by sinking while adsorbed or bound together with particles of metallic oxide precipitates.

Low-temperature maturation in the sediments results in petroleum generation, even from low amounts of organic matter (average 0.05%). Both steroid and triterpenoid hydrocarbons (biomarkers) show that extensive acid-catalyzed reactions occur in the sediments. In comparison with other hydrothermal systems (e.g., Guaymas Basin), these sediments exhibit a lower degree of thermal maturation, based on the elemental composition of the kerogens and the absence of pyrolytic PAHs in the bitumen. The lack of a carbon number preference for the *n*-alkanes (CPI = 1.0), especially in the case of the long-chain homologs (e.g., Fig. 13), indicates catagenetic alteration of the organic matter. Related data on hydrothermal petroleum from the Kebrit and Shaban Deeps of the Red

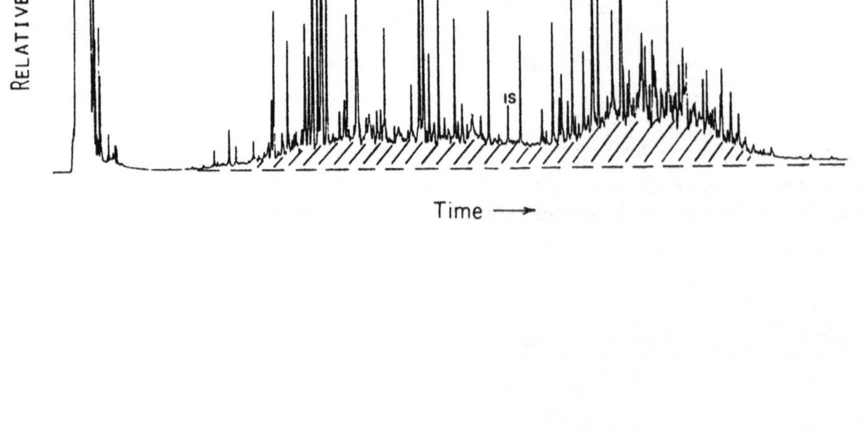

Figure 12. Gas chromatograms of total hydrocarbon fractions from core 1341-1 in the Bransfield Strait [edited from data shown in Brault and Simoneit (1988, 1990)]: (a) 500–510 cm, thermally unaltered; (b) core catcher 750 cm, thermally altered products (IS, internal standard; DS, C_{28} steradiene; DP, diploptene; arabic numerals refer to the carbon chain length of the *n*-alkanes; hachured area is the UCM of a thermogenic origin).

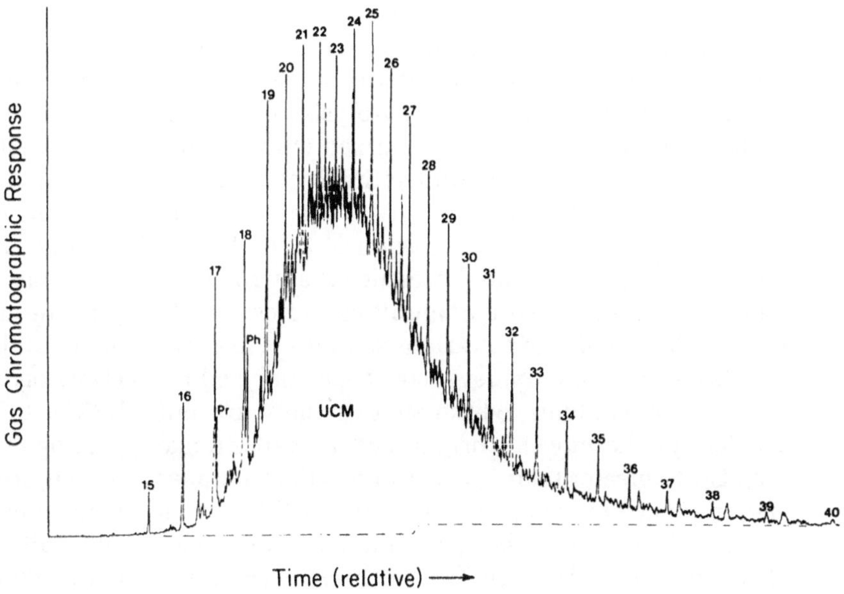

Figure 13. Gas chromatogram of the total hydrocarbon fraction extracted from Atlantis II Deep core sample 84, 443–453 cm. *n*-Alkanes are indicated by number of carbon atoms. [Adapted from Simoneit *et al.* (1987).]

Sea have also been reported; however, these systems appear to be at higher temperatures (Michaelis *et al.*, 1990).

4.2. Sediment-Starved Hydrothermal Systems

Hydrothermal activity and associated massive sulfide deposits, with abundant faunal communities, are found on the unsedimented axis of the East Pacific Rise (EPR) in the region of 13°N (Fig. 1; Hékinian *et al.*, 1983). Aliphatic hydrocarbons have been analyzed in hydrothermal plumes and in metalliferous sediments near the active vents and at the base of an inactive chimney (Brault *et al.*, 1985, 1988; Simoneit *et al.*, 1990b). Hydrocarbons from metalliferous sediments have distributions characteristic of immature organic matter which has recently been biosynthesized and microbiologically degraded, as indicated by the abundance of low-molecular-weight ($<C_{25}$) *n*-alkanes and phytane (Fig. 14a). A minor contribution of continental higher plant material is indicated by the presence of high-molecular-weight *n*-alkanes with an odd-carbon-number predominance. The immature character of the organic matter is also suggested by the presence of biomarker hydrocarbons derived from steroids and triterpenoids, which are the result of low temperatures, as might be expected in the surrounding talus of a vent system. The contents of a sediment trap deployed in the area within ~20 m of the vents are characterized by biologically derived material and also by thermally altered biomarker compounds (Brault *et al.*, 1985). This result indicates that higher temperature alteration of entrained organic detritus is a significant process near hydrothermal discharge sites. Thermally mature com-

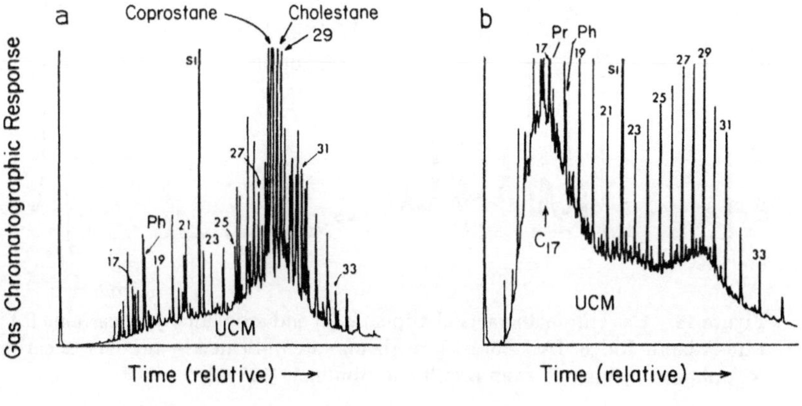

Figure 14. Gas chromatograms of aliphatic hydrocarbons from extracts of hydrothermal metalliferous sediment (a) and surrounding ambient water (b) from the East Pacific Rise at 13°N. *n*-Alkanes are indicated by number of carbon atoms. SI, Internal standard (*n*-C_{22}); Pr, pristane; Ph, phytane. [Adapted from Brault *et al.* (1985).]

pounds (Fig. 14b) are also present at trace levels in waters collected within ~1 km above the hydrothermal vents. The hydrocarbon patterns of these waters are indicative in many cases of pyrolysis of bacterial matter entrained in ocean water mixed with the discharging high-temperature fluids (Brault *et al.*, 1988).

Extensive hydrothermal activity has also been described for the EPR in the region of 21°N (Fig. 1), occurring on unsedimented oceanic crust with associated abundant faunal communities (Spiess *et al.*, 1980; Ballard *et al.*, 1981). Various samples of massive sulfides from vent chimneys have been analyzed for hydrocarbon contents, which are extremely low but of a definite thermogenic origin. The *n*-alkanes range from C_{14} to greater than C_{40}, with no carbon number predominance for samples of massive sulfides and a slight odd-carbon-number predominance for a sample of pyritized tube worm from a chimney (Brault *et al.*, 1989). All samples contain PAHs, supporting evidence for an origin of the hydrocarbons from hydrothermal activity. This observation coupled with the carbon number maxima at n-C_{27} or higher indicates that these hydrocarbons were entrapped/condensed in a high-temperature regime such as an active chimney. The sample with the pyritized tube worm residues also contains hydrothermally altered derivatives (e.g., cholestenes, hopenes) of biomarkers from the vent biota, mainly tube worms and bacteria.

The Trans-Atlantic Geotraverse (TAG) hydrothermal field on the Mid-Atlantic Ridge crest at 26°N (Fig. 1) is an active vent system on a slow-spreading mid-oceanic ridge (Rona *et al.*, 1984). Various hydrothermal ores deposited directly on oceanic crust have been dredged from the area (TAG 1985-1), and four types of samples have been examined for lipid/bitumen content (Brault and Simoneit, 1989). A sample consisting of predominantly ferric oxide contained no hydrocarbons attributable to hydrothermal alteration of associated organic detritus. However, three other samples (consisting of mainly anhydrite, sphalerite, and chalcopyrite, respectively) did contain minor amounts of lower molecular weight (C_{10}–C_{22}) hydrothermal petroleums. The saturated and aromatic hydrocarbon fractions separated from the extract of the sphalerite sample are shown as an example in Fig. 15. The *n*-alkanes range from C_{11} to C_{22} with a CPI of 1.0, pristane and phytane are present, and the UCM maximizes at the GC retention time for n-C_{17}. This pattern is analogous to that observed for the samples from the EPR at 13°N and from the Atlantis II Deep. The supporting evidence for a hydrothermal origin is found in the aromatic fraction, which contains naphthalene, phenanthrene, pyrene, their alkyl homologs, and sulfur aromatic compounds (Brault and Simoneit, 1989).

4.3. Continental Systems

In the Waiotapu geothermal region of New Zealand (Fig. 1), small amounts of oil are presently being generated from volcanic sedimentary rocks of Lower Pleistocene age (Czochanska *et al.*, 1986). The source material is terrigenous organic matter present in a vitric tuff, sandwiched between layers of volcanic breccia which have been rapidly buried by volcanic overburden. The breccias serve as regional aquifers and surround the tuff with high-temperature water. The generated oil, however, lacks the disequilibrium reaction products seen in the hydrothermal petroleums. Other than a few subtle geochemical indicators, the seep oils are remarkably similar to conventional petroleums.

Petroleums from Yellowstone National Park (Fig.

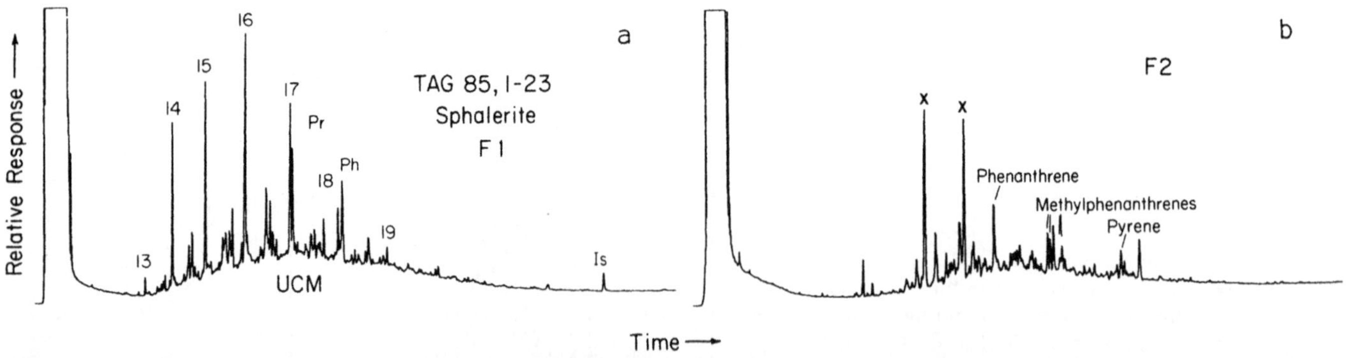

Figure 15. Gas chromatograms of aliphatic (a) and aromatic hydrocarbons (b) from the extract of a massive sphalerite sample from the Mid-Atlantic Ridge, TAG '85, 1-23. *n*-Alkanes are indicated by number of carbon atoms; Is, internal standard; Pr, pristane; Ph, phytane, ×, unknown. [Adapted from Brault and Simoneit (1989).]

1) are the first major petroleum seeps directly associated with an active continental hydrothermal system that have been studied (Love and Good, 1970; Clifton et al., 1990). The vent condensates at Calcite Springs are dominated by aromatic and polar compounds which are similar, but not identical, in composition to bitumens emanating from high-temperature deep-sea hydrothermal vents (Fig. 16a). Clifton et al. (1990) concluded that two hydrothermal sources have contributed to the fluids. A deep source contributed PAHs, thio-PAHs, and other compounds formed under extreme temperature conditions. This deep source unit is highly altered by ~250°C water, which is the agent for mobilization. A second source in contact with a shallow hydrothermal reservoir (~200-m depth) has been thermally altered. Generation and/or remobilization of a bitumen, also rich in aromatic and polar compounds with traces of aliphatic and biomarker hydrocarbons, has occurred in this shallow reservoir following short exposure (~62 ka) at lower temperatures (~215°C). Permian Phosphoria Formation sedimentary rocks are suggested as the source for this second component, which constitutes the bulk of the total hydrothermal petroleum. Organic matter

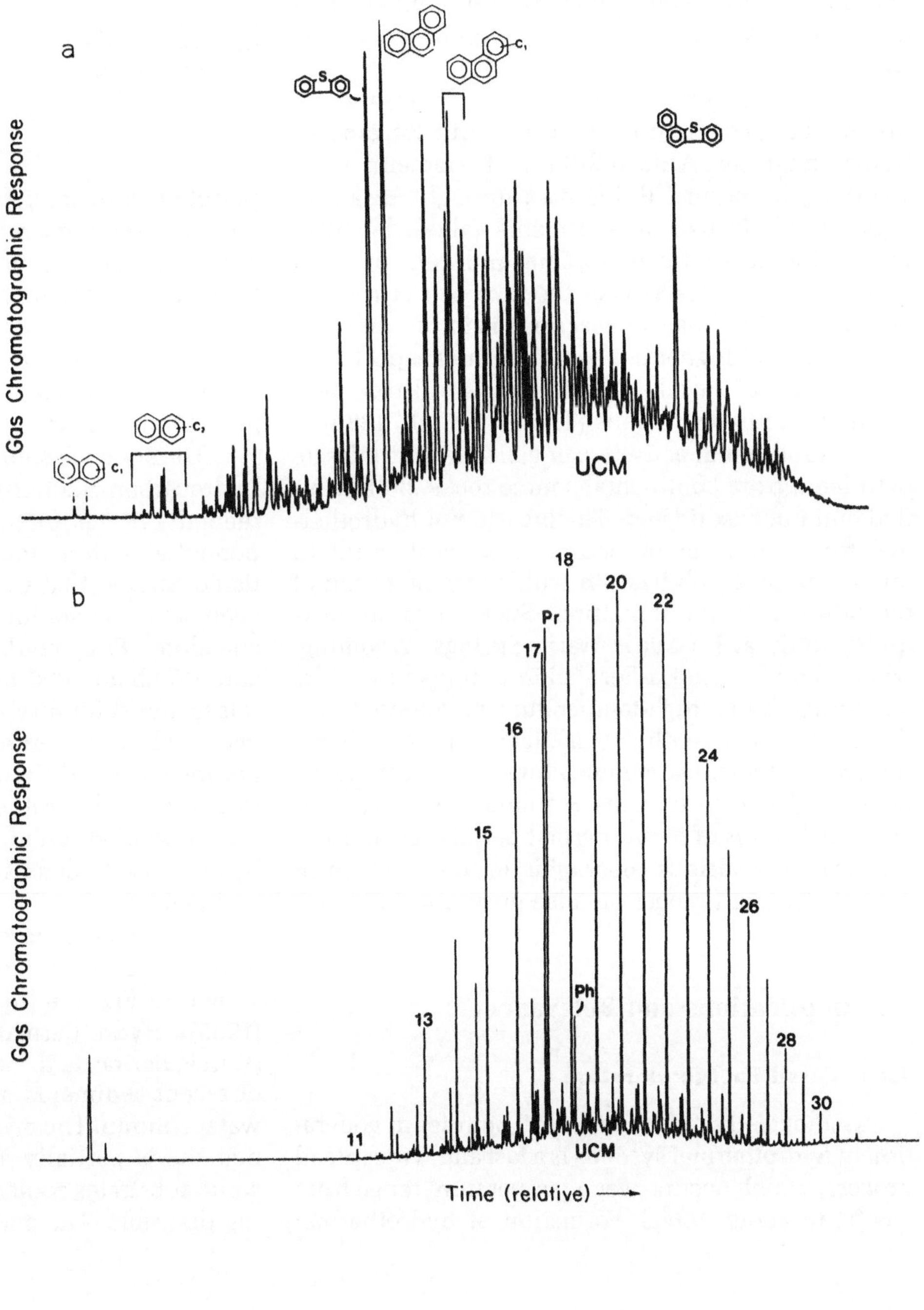

Figure 16. (a) Reconstructed ion chromatogram of the C_{15+} aromatic fraction from a Calcite Springs vent petroleum sample (YCS-1); (b) whole oil chromatogram of a petroleum from Rainbow Springs (YRS-1) (adapted from data in Clifton et al. 1990). n-Alkanes are indicated by number of carbon atoms; Pr, pristane; Ph, phytane.

from near-surface terrestrial plants and bacterial microorganisms have also contributed trace amounts of hydrocarbons.

In contrast, the petroleum at Rainbow Springs is dominated by paraffinic hydrocarbons (Clifton *et al.*, 1990; Fig. 16b). Unusual distributions of tricyclic diterpanes and nonhopanoid terpanes suggest that this oil was generated from Tertiary coaly or lacustrine sediments dominated by terrestrial higher plant input. Both gymnosperm and angiosperm markers are present. Geological and temperature considerations indicate that such sediments must be interspersed among the volcanics which blanket the area. Petroleum generation occurred at lower temperatures than at Calcite Springs (~ 150°C), requiring longer periods of time (<7.5 Ma), and an intravolcanic equivalent of the Eocene Aycross Formation is proposed as the source.

Another occurrence of massive sulfides and petroleum has been described for the north Tanganyika trough in the East African Rift (Fig. 1; Tiercelin *et al.*, 1989). Hydrothermal fluids pass through ~2 km of organic-rich lacustrine sediments (algal detritus), mobilizing asphaltic petroleum and venting with temperatures of 65–80°C at the lake bed at a water depth of ~20 m (the site described is in proximity to shore; vents at higher temperatures are suspected in deeper water of the lake). The vent waters also contain thermogenic hydrocarbons (Tiercelin *et al.*, 1989).

Hydrothermal activity can generate and migrate petroleum from continental source rocks, both lithified and unconsolidated. The invasion of hydrothermal fluids into mature source rocks will result in migration by remobilization with some alteration of the bitumen in the formation. Such an example is under study at De Maris Warm Springs, Wyoming, where the Phosphoria Formation is transgressed by hot fluids. Thus, high-temperature petroleums, i.e., those associated with magmatically driven hydrothermal systems, can be encountered in both continental and marine domains. A summary of the ranges of typical yields of hydrothermal petroleums in samples from the various geographic locales is given in Table II. The CPI ranges are also given for reference.

TABLE II. Summary of Hydrothermal Petroleum from Various Geographic Areas

Location	TOC (%)	Total hydrocarbons ($\mu g\ g^{-1}$)[a]	CPI
Guaymas Basin, Gulf of California	2	1000–550,000	1.02
Escanaba Trough, Gorda Ridge[b]	0.4–5.6	500–55,000	1.25
Kebrit Deep, Red Sea[c]	0.3–2.3	250–3800	1.5–4.2
Shaban Deep, Red Sea[c]	0.6–1.5	240–2700	1.6–3.9
Bransfield Strait, Antarctica	0.8	0.5–1.2	1.6–2.0
Atlantis II Deep, Red Sea	0.14	0.23	1.1
East Pacific Rise, 13°N	0.4	1	1.1
East Pacific Rise, 21°N	n.d.[d]	0.0002–0.006	1.01
Mid-Atlantic Ridge, 26°N	n.d.	n.d.	1.01

[a]Based on dry weight.
[b]Kvenvolden *et al.*, 1986.
[c]Michaelis *et al.*, 1990.
[d]n.d., Not determined

4.4. Implications and Relevance

4.4.1. Fossil Fuel Exploration

Geologically "instantaneous" petroleum generation in hydrothermal systems is a fast and widespread process, which occurs over a temperature range from ~60°C to about 400°C. Formation of hydrothermal petroleums seems to commence in low-temperature regions, generating products from weaker bonds, and, as the temperature regime rises, products are derived from more refractory organic matter and are even "reformed" (e.g., PAHs). At the high temperatures, organic matter is only partly destroyed, probably because the thermogenic products are rapidly removed from the hot zone.

The aqueous solubility of petroleum and various hydrocarbon fractions has been determined experimentally (Price, 1976). Their solubility increased exponentially from 100°C to 180°C. It has also been demonstrated that C_1 in the presence of water is an even better carrier for petroleum than water or methane alone (Price *et al.*, 1983). Both increases in pressure (to about 1800 bar) and temperature (to 250°C) raised the solubility of petroleum, and cosolubility was found at rather moderate conditions. These experimental conditions generally approximate those found in the hydrothermal environments and therefore support efficient migration by bulk oil flow and by aqueous/gaseous solubilization of the hydrothermal petroleums.

A schematic representation of the prevailing conditions existing in the Guaymas Basin vent systems is shown in Fig. 17a [redrawn and adapted from Scott (1985)]. Hydrothermal fluids, driven by a deep heat source, permeate through an open, fine-grained body of recent sediments and discharge directly into the water column. The oil discharged with the hydrothermal fluids partially adsorbs or condenses on inorganic substrates cooled by seawater (~3°C) surrounding the vents. The major part of the oil plume above

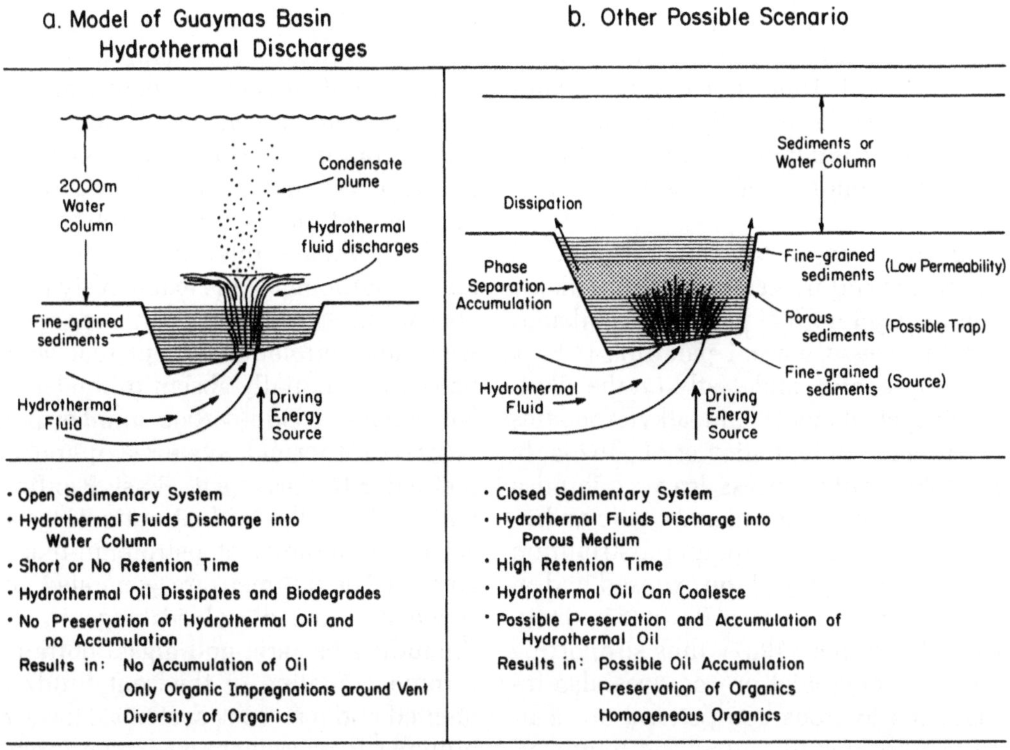

Figure 17. Schematic models for hydrothermal petroleum generation and migration scenarios: (a) Guaymas Basin open system; (b) hypothetical closed system. [Adapted from Didyk and Simoneit (1989, 1990).]

the vent area dissipates into the water column mainly by dispersion, dissolution, and eventual biodegradation. Another scenario could be postulated, as for example in Fig. 17b, where a similar hydrothermally generated oil is discharged into a porous sediment body with a finite retention time for the fluids (Didyk and Simoneit, 1989, 1990). There, the hydrothermal oil–water mixtures can undergo phase separation, and petroleum can eventually accumulate if adequate sedimentary and tectonic features are available to constitute a reservoir. Such a scenario could possibly lead to a hydrothermal oil accumulation which could have a potential for exploration.

Hydrothermal systems may be particularly active during the early rifting of ocean basins along continental margins (Lonsdale, 1985). Thus, geological locales where this process should be considered in exploration are, for example, split rift basins, failed rifts with hemipelagic or lacustrine sediments, and pull-apart basins and rifts overridden by continental drift. Remobilization of petroleum by hydrothermal fluids from magmatic activity affecting conventional sedimentary basins is another aspect for consideration in exploration.

The Canning Basin, Australia, can serve to illustrate this point (Reeckmann and Mebberson, 1984; Peter *et al.*, 1988). It covers the offshore extension of the Lennard Shelf and northern Fitzroy Graben in northern Australia. Drilled wells and seismic lines indicate the presence of dolerite sills and dikes, and vitrinite reflectance and fission track analyses indicate that the intrusives had a significant thermal effect on the sedimentary host rocks. Mineral alteration effects around the sills and dikes extend to > 3 km, indicative of extensive hydrothermal circulation concomitant with intrusions into the sediments. Thus, these intrusions have briefly placed a considerable section of regionally immature sediments into the "oil window." Whole-rock geochemistry indicates that these sills and dikes are of a continental origin and that they may have been emplaced during initial rifting of the Western Australian margin. Hydrothermal petroleum may have been produced and subsequently migrated away during the time these sedimentary rocks were within the oil window.

4.4.2. Biodegradation and Carbon Source for Ecosystem

Exposed hydrothermal petroleum, i.e., on exterior surfaces of seafloor mounds and accessible in, for example, unconsolidated surface sediments, is rapidly degraded to a residue of unresolved complex hydrocarbons (UCM or "hump"; Simoneit, 1985b).

These residues do not result from water leaching alone, because that would leave the heavy components (> n-C_{20}) unaltered. They appear to be a result of microbial alteration; however, the possibility of some water leaching cannot be excluded (Kawka and Simoneit, 1987; Simoneit *et al.*, 1988; Kawka, 1990).

Microbial degradation of petroleum, as studied both in the laboratory and in reservoirs, proceeds by initial attack on the light normal paraffins (n-alkanes $<C_{25}$), then on the heavy normal paraffins ($>C_{25}$), followed by the isoprenoids, and, lastly, on the alkylnaphthenes (polycycloalkanes) and alkylaromatics (Winters and Williams, 1969; Bailey *et al.*, 1973a, b; Connan *et al.*, 1980). This process leaves behind a nondegradable residue of branched and cyclic hydrocarbons, the UCM. The same compound distribution patterns are observed for many of the exposed hydrothermal petroleums in the Guaymas Basin (Simoneit, 1985b; Kawka and Simoneit, 1987), thus supporting biodegradation. This microbial process was also invoked to explain the hydrocarbon patterns in near-surface cores (Bazylinski *et al.*, 1988), and hydrocarbon utilization by microbes has been demonstrated (Bazylinski *et al.*, 1989).

Thus, microbiological alteration and utilization of the organic carbon from the petroliferous matter accessible in hydrothermal vent systems constitute a major biogenic carbon source in these environments of the ocean. It may be equivalent in magnitude to CO_2 fixation by vent fauna in sedimented rift basins. Therefore, a carbon source from hydrothermal petroleum should be incorporated in the ecological studies of the biota of such areas.

5. Summary

Conventional petroleum formation is a geologically slow process tied to basin subsidence and organic matter maturation. In the case of hydrothermal systems, organic matter maturation, petroleum generation, expulsion, and migration are compressed into an "instantaneous" geologic time frame. At seafloor spreading axes, hydrothermal systems active under sedimentary cover (e.g., Guaymas Basin and Escanaba Trough) generate petroleum from generally immature organic matter in the sediments. The hydrothermal petroleum migrates rapidly away from the high-temperature zone, usually upward, and leaves behind a spent carbonaceous residue. The Guaymas Basin petroleum is composed of: (1) gasoline-range hydrocarbons (C_1–C_{12}); (2) a broad distribution of n-alkanes (C_{12}–C_{40+}) with essentially no carbon number predominance; (3) a naphthenic hump, UCM; (4) pristane and phytane at significant concentrations; (5) mature biomarkers (e.g., α-hopanes); and (6) major concentrations of PAHs and thio-PAHs (Table I). Exposed petroleum and petroleums in unconsolidated surface sediments are microbially degraded and leached, whereas interior samples are essentially unaltered, although some extensively reworked oils can also occur. Similar compositions are observed for the Escanaba petroleum, except that volatile hydrocarbons are essentially absent in the latter and the n-alkanes have an odd-carbon-number predominance.

Hydrothermal systems operating in unsedimented rift areas (e.g., East Pacific Rise at 13°N and 21°N and the Mid-Atlantic Ridge at 26°N) generate trace amounts of petroleum-like material. Low amounts of bitumen are generated by pyrolysis of suspended and dissolved biogenic organic detritus (including bacteria and algae) entrained during the turbulent cooling of the vent fluids for both sedimented and bare rock systems. However, this type of bitumen is overwhelmed in the former case by the large quantity of petroleum generated from the sedimentary organic matter. In addition, low-level maturation is observed in the surrounding areas at vent sites, probably due to warming of ambient detritus in hydrothermal talus. Hydrothermal processes also generate and migrate, as well as remobilize, petroleum in continental systems (e.g., Wyoming).

In general, hydrothermal oil generation processes differ significantly from the conventionally accepted scenario for petroleum formation in sedimentary basins, where organic matter input, subsidence, geothermal maturation, oil generation, and oil migration are discrete successive steps that occur over a long period of geologic time. In hydrothermal petroleum formation, several of the steps of oil generation occur almost simultaneously, and the oil-generating process has been shown to be completed in short periods of geologic time.

It is not known to what extent hydrothermal petroleum generation processes have contributed to the formation of crude oil accumulations, but they appear to be highly efficient for organic matter maturation and oil generation and migration. Hydrothermal oil generation should be evaluated thoroughly to document its occurrence and implications, and to obtain a better understanding of the geological constraints for the process. This approach would open potentially new and, up to now, unconventional oil exploration targets. Also, hydrothermal alteration of organic matter is probably a ubiquitous process along the global rift systems and is a phenomenon that has been active over most of geologic time.

ACKNOWLEDGMENTS. I thank the Deep Sea Drilling Project (DSDP) and the National Science Foundation for access to DSDP samples and participation on the D.S.V. *Alvin* cruises and Dr. J. Baross, Dr. M. Brault, Mr. C. G. Clifton, Dr. E. M. Galimov, Dr. J. M. Gieskes, Dr. W. D. Goodfellow, Dr. P. D. Jenden, Dr. O. E. Kawka, Dr. K. A. Kvenvolden, Mr. R. N. Leif, Dr. P. F. Lonsdale, Dr. A. Lorre, Dr. J. D. Love, Dr. M. A. Mazurek, Dr. J. M. Peter, Dr. R. P. Philp, Dr. P. A. Rona, Mr. E. Ruth, Dr. M. Schoell, Dr. K. A. Sundell, Dr. J. Tiercelin, and Dr. C. C. Walters for samples, data, and assistance. I am grateful to Drs. O. E. Kawka and K. A. Kvenvolden for comments and suggestions that greatly improved this chapter. Funding from the National Science Foundation, Division of Ocean Sciences (grants OCE81-18897, OCE-8312036, OCE-8512832, OCE-8601316, and OCE-9002366), is gratefully acknowledged.

References

Bailey, N. J. L., Jobson, A. M., and Rogers, M. A., 1973a, Bacterial degradation of crude oil: Comparison of field and experimental data, *Chem. Geol.* **11**:203–221.

Bailey, N. J. L., Krouse, H. R., Evans, C. R., and Rogers, M. A., 1973b, Alteration of crude oils by waters and bacteria—evidence from geochemical and isotope studies, *Am. Assoc. Petrol. Geol. Bull.* **57**:1276–1290.

Ballard, R. D., Francheteau, J., Juteau, T., Rangan, C., and Normark, W., 1981, East Pacific Rise at 21°N: The volcanic, tectonic and hydrothermal processes of the central axis, *Earth Planet. Sci. Lett.* **55**:1–10.

Barrett, T. J., and Jambor, J. L. (eds), 1988, Seafloor hydrothermal mineralization, *Can. Mineral.* **26**:429–888.

Bazylinski, D. A., Farrington, J. W., and Jannasch, H. W., 1988, Hydrocarbons in surface sediments from a Guaymas Basin hydrothermal vent site, *Org. Geochem.* **12**:547–558.

Bazylinski, D. A., Wirsen, C. O., and Jannasch, H. W., 1989, Microbial utilization of naturally occurring hydrocarbons at the Guaymas Basin hydrothermal vent site, *Appl. Environ. Microbiol.* **55**:2832–2836.

Bischoff, J. L., and Rosenbauer, R. J., 1984, The critical point and two-phase boundary of sea water, 200–500°C, *Earth Planet. Sci. Lett.* **68**:172–180.

Bischoff, J. L., and Rosenbauer, R. J., 1988, Liquid–vapor relations in the critical region of the system NaCl–H_2O from 380 to 415°C: A refined determination of the critical point and two-phase boundary of seawater, *Geochim. Cosmochim. Acta* **52**: 2121–2126.

Blumer, M., 1975, Curtisite, idrialite and pendletonite, polycyclic aromatic hydrocarbon minerals: Their composition and origin, *Chem. Geol.* **16**:245–256.

Blumer, M., 1976, Polycyclic aromatic compounds in nature, *Sci. Am.* **234**:34–45.

Brault, M., and Simoneit, B. R. T., 1988, Steroid and triterpenoid distributions in Bransfield Strait sediments: Hydrothermally-enhanced diagenetic transformations, in: *Advances in Organic Geochemistry 1987* (L. Matavelli and L. Novelli, eds.), *Org. Geochem.* **13**:697–705.

Brault, M., and Simoneit, B. R. T., 1989, Trace petroliferous organic matter associated with hydrothermal minerals from the Mid-Atlantic Ridge at the Trans-Atlantic Geotraverse 26°N site, *J. Geophys. Res.* **94**:9791–9798.

Brault, M., and Simoneit, B. R. T., 1990, Mild hydrothermal alteration of immature organic matter in sediments from the Bransfield Strait, Antarctica, in: *Organic Matter Alteration in Hydrothermal Systems—Petroleum Generation, Migration and Biogeochemistry* (B. R. T. Simoneit, ed.), *Appl. Geochem.* **5**:149–158.

Brault, M., Simoneit, B. R. T., Marty, J. C., and Saliot, A., 1985, Les hydrocarbures dans le système hydrothermal de la ride Est-Pacifique, à 13°N, *C. R. Acad. Sci. Paris* **301**(II):807–812.

Brault, M., Simoneit, B. R. T., Marty, J. C., and Saliot, A., 1988, Hydrocarbons in waters and particulate material from hydrothermal environments at the East Pacific Rise, 13°N, *Org. Geochem.* **12**:209–219.

Brault, M., Simoneit, B. R. T., and Saliot, A., 1989, Trace petroliferous organic matter associated with massive hydrothermal sulfides from the East Pacific Rise at 13°N and 21°N, *Oceanol. Acta* **12**:405–415.

Carranza-Edwards, A., Rosales-Hoz, L., Aguayo-Camargo, J. E., Lozano-Santa Cruz, R., and Hornelas-Orozco, Y., 1990, Geochemical study of hydrothermal core sediments and rocks from the Guaymas Basin, Gulf of California, in: *Organic Matter Alteration in Hydrothermal Systems—Petroleum Generation, Migration and Biogeochemistry* (B. R. T. Simoneit, ed.), *Appl. Geochem.* **5**:77–82.

Chen, C.-T. A., 1981, Geothermal systems at 21°N, *Science* **211**:298.

Childress, J. J. (ed.), 1988, Hydrothermal vents, a case study of the biology and chemistry of a deep-sea hydrothermal vent of the Galapagos Rift, the Rose Garden in 1985, *Deep-Sea Res.* **35**: 1677–1849.

Clifton, C. G., Walters, C. C., and Simoneit, B. R. T., 1990, Hydrothermal petroleums from Yellowstone National Park, Wyoming, U.S.A., in: *Organic Matter in Hydrothermal Systems—Petroleum Generation, Migration and Biogeochemistry* (B. R. T. Simoneit, ed.), *Appl. Geochem.* **5**:169–191.

Connan, J., Restle, A., and Albrecht, P., 1980, Biodegradation of crude oil in the Aquitaine basin, in: *Advances in Organic Geochemistry 1979* (A. G. Douglas and J. R. Maxwell, eds.), Pergamon Press, Oxford, pp. 1–17.

Cooper, J. E., and Bray, E. E., 1963, A postulated role of fatty acids in petroleum formation, *Geochim. Cosmochim. Acta* **29**: 1113–1127.

Corliss, J. B., 1989, Submarine hot springs again, International Society for Study of the Origin of Life, Prague, July 3–8, Abstract.

Corliss, J. B., Dymond, J., Gordon, L. I., Edmond, J. M., von Herzen, R. P., Ballard, R. D., Green, K., Williams, D., Bainbridge, A., Crane, K., and van Andel, T. H., 1979, Submarine thermal springs on the Galapagos Rift, *Science* **203**:1073–1083.

Corliss, J. B., Baross, J. A., and Hoffman, S. E., 1981, An hypothesis concerning the relationship between submarine hot springs and the origin of life on Earth, in: Proceedings of the 26th International Geological Congress, *Oceanol. Acta*, 59–69.

Curray, J. R., Moore, D. G., Aguayo, J. E., Aubry, M. P., Einsele, G., Fornari, D. J., Gieskes, J., Guerrero, J. C., Kastner, M., Kelts, K., Lyle, M., Matoba, Y., Molina-Cruz, A., Niemitz, J., Rueda, J.,

Saunders, A. D., Schrader, H., Simoneit, B. R. T., and Vacquier, V. (eds.), 1982, *Initial Reports of the Deep Sea Drilling Project*, Vol. 64, Parts I and II, U.S. Government Printing Office, Washington, D.C.

Czochanska, Z., Sheppard, C. M., Weston, R. J., Woolhouse, A. D., and Cook, R. A., 1986, Organic geochemistry of sediments in New Zealand, Part I. A biomarker study of the petroleum seepage at the geothermal region of Waiotapu, *Geochim. Cosmochim. Acta* **50**:507–515.

Demaison, G. J., and Moore, G. T., 1980, Anoxic environments and oil source bed genesis, *Org. Geochem.* **2**:9–31.

de Wit, M. J., Hart, R., Martin, A., and Abbott, P., 1982, Archean abiogenic and probable biogenic structures associated with mineralized hydrothermal vent systems and regional metasomatism, with implications for greenstone belt studies, *Econ. Geol.* **77**:1783–1802.

Didyk, B. M., and Simoneit, B. R. T., 1989, Hydrothermal oil of Guaymas Basin and implications for petroleum formation mechanisms, *Nature* **342**:65–69.

Didyk, B. M., and Simoneit, B. R. T., 1990, Petroleum characteristics of the oil in a Guaymas Basin hydrothermal chimney, in: *Organic Matter Alteration in Hydrothermal Systems—Petroleum Generation, Migration and Biogeochemistry* (B. R. T. Simoneit, ed.), *Appl. Geochem.* **5**:29–40.

Didyk, B. M., Simoneit, B R. T., Brassell, S. C., and Eglinton, G., 1978, Organic geochemical indicators of paleoenvironmental conditions of sedimentation, *Nature* **272**:216–222.

Edmond, J. M., and Von Damm, K., 1983, Hot springs on the ocean floor, *Sci. Am.* **248**(4):78–93.

Einsele, G., 1985, Basaltic sill–sediment complexes in young spreading centers: Genesis and significance, *Geology* **13**: 249–252.

Franklin, J. M., Lydon, J. W., and Sangster, D. F., 1981, Volcanic-associated massive sulfide deposits, *Econ. Geol. 75th Anniv. Vol.*, 485–627.

Geissman, T. A., Sim, K. Y., and Murdoch, J., 1967, Organic minerals. Picene and chrysene as constituents of the mineral curtisite (idrialite), *Experientia* **23**:793–794.

Germanov, A. I., and Bannikova, L. A., 1972, Alternation of organic matter of sedimentary rocks during hydrothermal sulfide concentration, *Dokl, Akad, Nauk SSSR* **203**:1180–1182.

Gieskes, J. M., Elderfield, H., Lawrence, J. R., Johnson, J., Meyers, B., and Campbell, A. C., 1982a, Geochemistry of interstitial waters and sediments, Leg 64, Gulf of California, in: *Initial Reports of the Deep Sea Drilling Project*, Vol. 64, Part II (J. R. Curray, D. G. Moore *et al.*, eds.), U.S. Government Printing Office, Washington, D.C., pp. 675–694.

Gieskes, J. M., Kastner, M., Einsele, G., Kelts, K., and Niemitz, J., 1982b, Hydrothermal activity in the Guaymas Basin, Gulf of California: A synthesis, in: *Initial Reports of the Deep Sea Drilling Project*, Vol. 64, Part II (J. R. Curray, D. G. Moore *et al.*, eds.), U.S. Government Printing Office, Washington, D.C., pp. 1159–1168.

Gieskes, J. M., Simoneit, B. R. T., Brown, T., Shaw, T., Wang, Y.-C., and Magenheim, A., 1988, Hydrothermal fluids and petroleum in surface sediments of Guaymas Basin, Gulf of California: A case study, *Can. Mineral.* **26**:589–602.

Grassle, J. F., 1985, Hydrothermal vent animals: Distribution and biology, *Science* **229**:713–717.

Hartmann, M., 1980, Atlantis II Deep geothermal brine system. Hydrographic situation in 1977 and changes since 1965, *Deep-Sea Res.* **27**:161–171.

Hartmann, M., 1985, Atlantis II Deep geothermal brine system. Chemical processes between hydrothermal brines and Red Sea deep water, *Mar. Geol.* **64**:157–177.

Haymon, R. M., Koski, R., and Sinclair, C., 1984, Fossils of hydrothermal vent worms from Cretaceous sulfide ores of the Somail ophiolite, Oman, *Science* **223**:1407–1409.

Hékinian, R., Fevrier, M., Avedik, F., Cambon, P., Charlou, J. L., Needham, H. D., Raillard, J., Boulègue, J., Merlivat, L., Moinet, A., Manganini, S., and Lange, J., 1983, East Pacific Rise near 13°N: Geology of new hydrothermal fields, *Science* **219**:1321–1324.

Hunt, J. M., 1979, *Petroleum Geochemistry and Geology*, W. H. Freeman and Company, San Francisco.

Jenden, P. D., Simoneit, B. R. T., and Philp, R. P., 1982, Hydrothermal effects on protokerogen of unconsolidated sediments from Guaymas Basin, Gulf of California, elemental compositions, stable carbon isotope ratios and electron spin resonance spectra, in *Initial Reports of the Deep Sea Drilling Project*, Vol. 64 (J. R. Curray, D. G. Moore, *et al.*, eds.), U.S. Government Printing Office, Washington, D.C., pp. 905–912.

Jones, M. L. (ed), 1985, Hydrothermal vents of the Eastern Pacific: An overview, *Bull. Biol. Soc. Washington* **6**:1–566.

Kartsev, A. A., Vassoevich, N. B., Geodekian, A. A., Neruchev, S. G., and Sokolov, V. A., 1971, The principal stage in formation of petroleum, *Proceedings of the 8th World Petroleum Congress*, Institute of Petroleum, London, Vol. 2, pp. 3–11.

Kawka, O. E. M., 1990, Hydrothermal alteration of sedimentary organic matter in Guaymas Basin, Gulf of California, Ph.D. Thesis, Oregon State University.

Kawka, O. E., and Simoneit, B. R. T., 1987, Survey of hydrothermally-generated petroleums from the Guaymas Basin spreading center, *Org. Geochem.* **11**:311–328.

Kawka, O E., and Simoneit, B. R. T., 1990, Polycyclic aromatic hydrocarbons in hydrothermal petroleums from the Guaymas Basin spreading center, in: *Organic Matter Alteration in Hydrothermal Systems—Petroleum Generation, Migration and Biogeochemistry* (B. R. T. Simoneit, ed.), *Appl. Geochem.* **5**: 17–27.

Koski, R. A., Lonsdale, P. F., Shanks, W. C., Berndt, M. E., and Howe, S. S., 1985, Mineralogy and geochemistry of a sediment-hosted hydrothermal sulfide deposit from the southern trough of Guaymas Basin, Gulf of California, *J. Geophys. Res.* **90**: 6695–6707.

Kvenvolden, K. A., and Simoneit, B. R. T., 1990, Hydrothermally derived petroleum: Examples from Guaymas Basin, Gulf of California and Escanaba Trough, Northeast Pacific, *Am. Assoc. Petrol. Geol. Bull.* **74**:223–237.

Kvenvolden, K. A., Rapp, J. B., Hostettler, F. D., Morton, J. L., King, J. D., and Claypool, G. E., 1986, Petroleum associated with polymetallic sulfide in sediment from Gorda Ridge, *Science* **234**: 1231–1234.

Kvenvolden, K. A., Rapp, J. B., and Hostettler, F. D., 1990, Hydrocarbon geochemistry of hydrothermally-generated petroleum from Escanaba Trough, offshore California, in: *Organic Matter Alteration in Hydrothermal Systems—Petroleum Generation, Migration and Biogeochemistry* (B. R. T. Simoneit, ed.), *Appl. Geochem.* **5**:83–91.

Lonsdale, P., 1985, A transform continental margin rich in hydrocarbons, Gulf of California, *Am. Assoc. Petrol. Geol. Bull.* **69**: 1160–1180.

Lonsdale, P., and Becker, K., 1985, Hydrothermal plumes, hot springs, and conductive heat flow in the Southern Trough of Guaymas Basin, *Earth Planet. Sci. Lett.* **73**:211–225.

Love, J. D., and Good, J. M., 1970, Hydrocarbons in thermal areas, northwestern Yellowstone National Park, Wyoming, U.S. Geological Survey Professional Paper 7644-B.

Martens, C. S., 1990, Generation of short chain organic acid anions in hydrothermally altered sediments of the Guaymas Basin, Gulf of California, in: *Organic Matter Alteration in Hydrothermal Systems—Petroleum Generation, Migration and Biogeochemistry* (B. R. T. Simoneit, ed.), *Appl. Geochem.* **5**:71–76.

Merewether, R., Olsson, M. S., and Lonsdale, P., 1985, Acoustically detected hydrocarbon plumes rising from 2-km depths in Guaymas Basin, Gulf of California, *J. Geophys. Res.* **90**:3075–3085.

Michaelis, W., Jenisch, A., and Richnow, H. H., 1990, Hydrothermal petroleum generation in Red Sea sediments from the Kebrit and Shaban Deeps, in: *Organic Matter Alteration in Hydrothermal Systems—Petroleum Generation, Migration and Biogeochemistry* (B. R. T. Simoneit, ed.), *Appl. Geochem.* **5**:103–114.

Miller, S. L., and Bada, J. L., 1988, Submarine hot springs and the origin of life, *Nature* **334**:609–611.

Oró, J., Miller, S. L., and Lazcano, A., 1990, The origin and early evolution of life on earth, *Annu. Rev. Earth Planet. Sci.* **18**: 317–356.

Peter, J. M., 1986, Genesis of hydrothermal vent deposits in the southern trough of Guaymas Basin, Gulf of California: A mineralogical and geochemical study, M.Sc. thesis, University of Toronto.

Peter, J. M., Simoneit, B. R. T., and Scott, S. D., 1988, Geological controls on thermogenic oil production in Guaymas Basin, Gulf of California, a modern sedimented submarine rift, in: *Proceedings of the 26th Annual Conference of the Ontario Petroleum Institute*, Oct. 1987, Lambeth, Ont., Canada, Paper 14, pp. 1–16.

Peter, J. M., Simoneit, B. R. T., Kawka, O. E., and Scott, S. D., 1990, Liquid hydrocarbon-bearing inclusions in modern hydrothermal chimneys and mounds from the southern trough of Guaymas Basin, in: *Organic Matter Alteration in Hydrothermal Systems—Petroleum Generation, Migration and Biogeochemistry* (B. R. T. Simoneit, ed.), *Appl. Geochem.* **5**:51–63.

Peter, J. M., Peltonen, P., Scott, S. D., Simoneit, B. R. T., and Kawka, O. E., 1991, ^{14}C ages of hydrothermal petroleum and carbonate in Guaymas Basin, Gulf of California: Implications for oil generation, expulsion and migration, *Geology* **19**:253–256.

Pitzer, K. S., 1986, Large-scale fluctuations and the critical behavior of dilute NaCl in H_2O, *J. Phys. Chem.* **90**:1502–1504.

Price, L. C., 1976, Aqueous solubility of petroleum as applied to its origin and primary migration, *Am. Assoc. Petrol. Geol. Bull.* **60**:213–244.

Price, L. C., 1982, Organic geochemistry of core samples from an ultra-deep hot well (300°C, 7 km), *Chem. Geol.* **37**:215–228.

Price, L. C., Clayton, J. L., and Rumen, L. L., 1981, Organic geochemistry of the 9.6 km Bertha Rogers No. 1 well, Oklahoma, *Org. Geochem.* **3**:59–77.

Price, L. C., Wenger, L. M., Ging, T., and Blount, C. W., 1983, Solubility of crude oil in methane as a function of pressure and temperature, *Org. Geochem.* **4**:201–221.

Reeckmann, S. A., and Mebberson, A. J., 1984, Igneous intrusions in the northwest Canning Basin and their impact on oil exploration, in: *The Canning Basin, W.A.* (P. G. Purcell, ed.), Proceedings of the Geological Society of Australia, Perth, pp. 42–52.

Rona, P. A., 1984, Hydrothermal mineralization at seafloor spreading centers, *Earth Sci. Rev.* **20**:1–104.

Rona, P. A., 1988, Hydrothermal mineralization at oceanic ridges, *Can. Mineral.* **26**:431–465.

Rona, P. A., Boström, K., Laubier, L., and Smith, K. L., Jr. (eds), 1983, *Hydrothermal Processes at Seafloor Spreading Centers*, NATO Conference Series, Plenum Press, New York.

Rona, P. A., Thompson, G., Mottl, M. J., Karson, J. A., Jenkins, W. J., Graham, D., Mallette, M., Von Damm, K., and Edmond, J. M., 1984, Hydrothermal activity at the Trans-Atlantic Geotraverse hydrothermal field, Mid-Atlantic Ridge Crest at 26°N, *J. Geophys. Res.* **89**:11365–11377.

Ross, D. S., 1984, Coal conversion in carbon monoxide–water systems, *Coal Sci.* **3**:301–337.

Sakai, H., Gamo, T., Kim, E.-S., Tsutsumi, M., Tanaka, T., Ishibashi, J., Wakita, H., Yamano, M., and Oomori, T., 1990, Venting of carbon dioxide-rich fluid and hydrate formation in mid-Okinawa Trough backarc basin, *Science* **248**:1093–1096.

Schoell, M., Hwang, R. J., and Simoneit, B. R. T., 1990, Carbon isotope composition of hydrothermal petroleums from Guaymas Basin, Gulf of California, in: *Organic Matter Alteration in Hydrothermal Systems—Petroleum Generation, Migration and Biogeochemistry* (B. R. T. Simoneit, ed.), *Appl. Geochem.* **5**:65–69.

Scott, L. T., 1982, Thermal rearrangements of aromatic compounds, *Acc. Chem. Res.* **15**:52–58.

Scott, S. D., 1985, Seafloor polymetallic sulfide deposits: Modern and ancient, *Mar. Min.* **5**:191–212.

Shock, E. L., 1988, Organic acid metastability in sedimentary basins, *Geology* **16**:886–890.

Shock, E. L., 1989, Corrections to "Organic acid metastability in sedimentary basins," *Geology* **17**:572–573.

Simoneit, B. R. T., 1978, The organic chemistry of marine sediments, in: *Chemical Oceanography, 7* (J. P. Riley and R. Chester, eds.), Academic Press, London, New York, pp. 233–311.

Simoneit, B. R. T., 1982a, The composition, sources and transport of organic matter to marine sediments—the organic geochemical approach, in: *Proceedings of the Symposium on Marine Chemistry into the Eighties* (J. A. J. Thompson and W. D. Jamieson, eds.), National Research Council of Canada, Ottawa, pp. 82–112.

Simoneit, B. R. T., 1982b, Shipboard organic geochemistry and safety monitoring, Leg. 64, Gulf of California, in: *Initial Reports of the Deep Sea Drilling Project*, Vol. 64 (J. R. Curray, D. G. Moore *et al.*, eds.), U.S. Government Printing Office, Washington, D.C., pp. 723–728.

Simoneit, B. R. T., 1983, Organic matter maturation and petroleum genesis: Geothermal versus hydrothermal, in: *The Role of Heat in the Development of Energy and Mineral Resources in the Northern Basin and Range Province*, Geothermal Research Council, Special Report No. 13, Davis, California, pp. 215–241.

Simoneit, B. R. T., 1984a, Hydrothermal effects on organic matter-high versus low temperature components, in: *Advances in Organic Geochemistry 1983* (P. A. Schenck, J. W. de Leeuw, and G. W. M. Lijmbach, eds.), *Org. Geochem.* **6**:857–864.

Simoneit, B. R. T., 1984b, Effects of hydrothermal activity on sedimentary organic matter: Guaymas Basin, Gulf of California—petroleum genesis and protokerogen degradation, in: *Hydrothermal Processes at Seafloor Spreading Centers* (P. A. Rona, K. Boström, L. Laubier, and K. L. Smith, Jr., eds.), Proc. NATO Advanced Research Institute Series, Plenum Press, New York, pp. 453–474.

Simoneit, B. R. T., 1985a, Hydrothermal petroleum: Genesis, migration and deposition in Guaymas Basin, Gulf of California, *Can. J. Earth Sci.* **22**:1919–1929.

Simoneit, B. R. T., 1985b, Hydrothermal petroleum: Composition and utility as a biogenic carbon source, in: *Hydrothermal*

Vents of the Eastern Pacific: An Overview (M. L. Jones, ed.), *Bull. Biol. Soc. Washington* **6**:49–56.

Simoneit, B. R. T., 1988, Petroleum generation in submarine hydrothermal systems: An update, in: *Recent Hydrothermal Mineralization at Seafloor Spreading Centers* (T. J. Barrett and J. S. Fox, eds.), *Can. Mineral.* **26**:827–840.

Simoneit, B. R. T. (ed), 1990, *Organic Matter in Hydrothermal Systems—Petroleum Generation, Migration and Biogeochemistry*, *Appl. Geochem.* **5**:1–248.

Simoneit, B. R. T., and Galimov, E. M., 1984, Geochemistry of interstitial gases in Quaternary sediments of the Gulf of California, *Chem. Geol.* **43**:151–166.

Simoneit, B. R. T., and Kawka, O. E., 1987, Hydrothermal petroleum from diatomites in the Gulf of California, in: *Marine Petroleum Source Rocks* (J. Brooks and A. J. Fleet, eds.), Geological Society of London, Special Publication No. 26, pp. 217–228.

Simoneit, B. R. T., and Lonsdale, P. F., 1982, Hydrothermal petroleum in mineralized mounds at the seabed of Guaymas Basin, *Nature* **295**:198–202.

Simoneit, B. R. T., and Philp, R. P., 1982, Organic geochemistry of lipids and kerogen and the effects of basalt intrusions on unconsolidated oceanic sediments: Sites 477, 478 and 481, Guaymas Basin, Gulf of California, in: *Initial Reports of the Deep Sea Drilling Project*, Vol. 64 (J. R. Curray, D. G. Moore et al., eds.), U.S. Government Printing Office, Washington, D.C., pp. 883–904.

Simoneit, B. R. T., Mazurek, M. A., Brenner, S., Crisp, P. T., and Kaplan, I. R., 1979, Organic geochemistry of recent sediments from Guaymas Basin, Gulf of California, *Deep-Sea Res.* **26A**: 879–891.

Simoneit, B. R. T., Philp, R. P., Jenden, P. D., and Galimov, E. M., 1984, Organic geochemistry of Deep Sea Drilling Project sediments from the Gulf of California—hydrothermal effects on unconsolidated diatom ooze, *Org. Geochem.* **7**:173–205.

Simoneit, B. R. T., Grimalt, J. O., Hayes, J. M., and Hartman, H., 1987, Low temperature hydrothermal maturation of organic matter in sediments from the Atlantis II Deep, Red Sea, *Geochim. Cosmochim. Acta* **51**:879–894.

Simoneit, B. R. T., Kawka, O. E., and Brault, M., 1988, Origin of gases and condensates in the Guaymas Basin hydrothermal system, in: *Origins of Methane in the Earth* (M. Schoell, ed.), *Chem. Geol.* **71**:169–182.

Simoneit, B. R. T., Lonsdale, P. F., Edmond, J. M., and Shanks, W. C., III, 1990a, Deep-water hydrocarbon seeps in Guaymas Basin, Gulf of California, in: *Organic Matter Alteration in Hydrothermal Systems—Petroleum Generation, Migration and Biogeochemistry* (B. R. T. Simoneit, ed.), *Appl. Geochem.* **5**:41–49.

Simoneit, B. R. T., Brault, M., and Saliot, A., 1990b, Hydrocarbons associated with hydrothermal minerals, vent waters and talus on the East Pacific Rise and Mid-Atlantic Ridge, in: *Organic Matter Alteration in Hydrothermal Systems—Petroleum Generation, Migration and Biogeochemistry* (B. R. T. Simoneit, ed.), *Appl. Geochem.* **5**:115–124.

Simoneit, B. R. T., Kawka, O. E., and Wang, G.-M., 1992, Biomarker maturation in contemporary hydrothermal systems, alteration of immature organic matter in zero geological time, in: *Biological Markers in Sediments and Petroleum* (J. Moldowan, P. Albrecht, and R. P. Philp, eds.), Prentice Hall, Englewood Cliffs, N.J., pp. 124–141.

Spiess, F. N., Macdonald, K. C., Atwater, T., Ballard, R., Carranza, A., Cordoba, D., Cox, C., DiazGarcia, V. M., Francheteau, J., Guerrero, J., Hawkins, J., Haymon, R., Hessler, R., Juteau, T., Kastner, M., Larson, R., Luyendyk, B., Macdougall, J. D., Miller, S., Normark, W., Orcutt, J., and Rangin, C., 1980, East Pacific Rise; hot springs and geophysical experiments, *Science* **207**: 1421–1433.

Suess, E., Simoneit, B. R. T., Wefer, G., Whiticar, M. J., Fisk, M., von Breymann, M., Han, M. W., Wittstock, R., and Laban, C., 1993, Hydrothermalism in the Bransfied Strait, Antarctica, *Geol. Rundsch.*, in preparation.

Tiercelin, J.-J., Thourin, C., Kalala, T., and Mondegeur, A., 1989, Discovery of sublacustrine hydrothermal activity and associated massive sulfides and hydrocarbons in the north Tanganyika trough, East African Rift, *Geology* **17**:1053–1056.

Tissot, B. P., and Welte, D. H., 1984, *Petroleum Formation and Occurrence: A New Approach to Oil and Gas Exploration*, 2nd ed., Springer-Verlag, Berlin.

Tödheide, K., 1982, Hydrothermal solutions, *Ber. Bunsenges. Phys. Chem.* **86**:1005–1016.

Venkatesan, M. I., and Kaplan, I. R., 1987, The lipid geochemistry of Antarctic marine sediments: Bransfield Strait, *Mar. Chem.* **21**: 347–375.

Von Damm, K. L., 1990, Seafloor hydrothermal activity: Black smoker chemistry and chimneys, *Annu. Rev. Earth Planet. Sci.* **18**:173–204.

Welhan, J. A., and Lupton, J. E., 1987, Light hydrocarbon gases in Guaymas Basin hydrothermal fluids: Thermogenic versus abiogenic origin, *Am. Assoc. Petrol. Geol. Bull.* **71**:215–223.

Whelan, J. K., Hunt, J. M., Jasper, J., and Huc. A., 1984, Migration of C_1–C_8 hydrocarbons in marine sediments, in: *Advances in Organic Geochemistry 1983* (P. A. Schenck, J. W. de Leeuw, and G. W. M. Lijmbach, eds.), *Org. Geochem.* **6**:683–694.

Whelan, J. K., Simoneit, B. R. T., and Tarafa, M., 1988, C_1–C_8 hydrocarbons in sediments from Guaymas Basin, Gulf of California—comparison to Peru Margin, Japan Trench and California Borderlands, *Org. Geochem.* **12**:171–194.

Whiticar, M. J., Suess, E., and Wehner, H., 1985, Thermogenic hydrocarbons in surface sediments of the Bransfield Strait, Antarctic Peninsula, *Nature* **314**:87–90.

Winters, J. C., and Williams, J. A., 1969, Microbiological alteration of crude oil in the reservoir, in: *Symposium on Petroleum Transformation in Geologic Environments*, Am. Chem. Soc., Div. Petrol. Chem., Prepr. **14**(4):E22–E31.

Yanagawa, H., and Kobayashi, K., 1989, Formation of amino acids, peptide-like polymers, and microspheres in superheated hydrothermal environments, *International Society for Study of Origin of Life*, Prague, July 3–8, Abstract.

Chapter 18

Laboratory Simulation of Petroleum Formation

Hydrous Pyrolysis

M. D. LEWAN

1. Introduction

The importance of water in laboratory experiments designed to simulate natural processes is well documented in the studies of granite melts (Goranson, 1931, 1932; Tuttle and Bowen, 1958), metamorphic reactions (Winkler, 1974, p. 15; Rumble *et al.*, 1982; Ferry, 1983), and coal formation (Berl and Schmidt, 1932; Schuhmacher *et al.*, 1960). Industrial processes also benefit from the presence of water as demonstrated in oil shale retorting (Gavin, 1922, p. 181), conversion of coal to oil (Fischer, 1925, p. 180), heavy oil upgrading (McCollum and Quick, 1976a, b), and conversion of organic refuse to oil (Appell *et al.*, 1971, 1975). Prior to 1979, organic geochemists inadvertently ignored these observations and the ubiquity of water in sedimentary basins when considering the natural process of petroleum generation. A notable exception is the work reported by Jurg and Eisma in 1964. Noting differences in the thermal decomposition of behenic acid in the presence and absence of liquid water, these investigators suggested that water played an important role in petroleum generation. Although a subsequent study in 1969 by Brooks and Smith employed water in laboratory simulations of petroleum generation from coals, laboratory experiments over the next decade did not consider the role of water in petroleum generation (e.g., Tissot *et al.*, 1974; Larter *et al.*, 1977; Harwood, 1977).

In 1979, Lewan *et al.* reported that heating organic-rich rocks in the presence of liquid water resulted in the generation and expulsion of an oil-like pyrolysate. The expelled pyrolysate in this type of experiment was a free-flowing oil that accumulated on the water surface above the submerged rock. Physically, chemically, and isotopically, the expelled oil pyrolysate was similar to natural crude oils. This experimental approach was referred to as hydrous pyrolysis (Lewan *et al.*, 1979) and has been shown to provide useful information on primary migration, stages and kinetics of petroleum generation, and thermal maturity indices (Lewan, 1983, 1985, 1987; Winters *et al.*, 1983; Lewan *et al.*, 1986). The following discussion elaborates on these attributes within the context of methodology employed in hydrous pyrolysis experimentation.

2. Definition of Hydrous Pyrolysis

Hydrothermal experiments involve the heating of samples in the presence of water irrespective of whether it occurs in the form of a vapor, liquid, or fluid phase. Hydrous pyrolysis is a hydrothermal experiment, but more specifically its definition requires that the heated samples be in contact with liquid water (Lewan *et al.*, 1979) and not in contact with water vapor or supercritical water fluid. Exclu-

M. D. LEWAN • Amoco Production Company, Research Center, Tulsa, Oklahoma 74102. *Present address*: U.S. Geological Survey, Denver, Colorado 80225.

Organic Geochemistry, edited by Michael H. Engel and Stephen A. Macko. Plenum Press, New York, 1993.

sion of these other water phases from the definition is based on their absence under normal burial diagenesis conditions and their different chemical reactivities relative to liquid water. Water vapor in the subsurface is limited to shallow geothermal vent areas, and its reactivity with organic and inorganic components in rocks is less than that of liquid water. Supercritical water fluid is limited to subsurface regimes in excess of greenschist metamorphic conditions, and its reactivity with organic and inorganic components in rocks may be significantly greater than that of liquid water.

Water in hydrous pyrolysis experiments typically occurs in the liquid and vapor phases (e.g., Lewan, 1985), unless pressure regulators are employed to maintain only a liquid phase (e.g., Monthioux *et al.*, 1985). It is important in two-phase water systems to optimize the amount of water used for a specific reactor volume to ensure the sample is submerged in the liquid phase. The volume of liquid water in a reactor at an experimental temperature depends on the amount of thermal expansion of the liquid phase and the amount of water needed to generate the required vapor pressure in the remaining gas space of the reactor. Specific volumes of water in the liquid (γ_l^T) and vapor (γ_v^T) phases at experimental temperatures (T) are critical to these volume determinations and are available in the *ASME Steam Tables* (ASME, 1979). Using these specific volumes, the volume of liquid water (V_l^T) at an experimental temperature in a given reactor volume (V_r) may be determined by the expression

$$V_l^T = \frac{(M_w^0 \gamma_v^T - V_r)\gamma_l^T}{\gamma_v{}^T - \gamma_l^T}$$

where M_w^0 is the mass of liquid water added to the reactor at room temperature.

This volume of liquid water may be expressed as a percentage of the reactor volume [i.e., $(V_l^T \times 10^2)/V_r$]. Several of these percentages are shown graphically as a series of lines in Fig. 1 for an experimental temperature of 330°C. Experimental conditions plotting on or above the zero-volume percentage line represent experiments with only water vapor in the reactor, which does not constitute hydrous pyrolysis conditions. Below this line, hydrous pyrolysis conditions exist provided the amount of liquid water present is sufficient to cover the heated sample. It is helpful in this determination to have an estimate of the sample density from which its volume may be approximated. The optimum condition at room temperature for a loaded reactor is a water surface several centimeters above the submerged sample. This makes collection of expelled oil from the water surface, after an experiment has been cooled to room temperature, more manageable without the interference of emergent sample in the expelled-oil layer.

Another consideration in loading a reactor is not to add too much water. If thermal expansion of the added water exceeds the available volume of the reactor at an experimental temperature, a dangerous condition will occur in which the reactor may violently rupture. Although commercially available reactors are usually equipped with a safety rupture disk, it is a good practice never to allow the total volume of sample plus expanded water to exceed 85 vol. % of the reactor at an experimental temperature. Pressures that develop within the remaining gas space are predominantly the result of the water vapor, which increases with temperature as shown by the vapor–liquid line in Fig. 2. The pressures in most hydrous pyrolysis experiments usually exceed these vapor pressures because of additional gases generated from the pyrolyzed sample. If a constant gas pressure is desired over a range of experimental temperatures, an inert gas such as helium may be added to the gas space in the appropriate amounts to compensate for lower vapor pressures at lower temperatures.

The definition of hydrous pyrolysis also specifies that water at experimental temperatures within a reactor is not in the supercritical fluid phase. This prerequisite is based on the exceptionally high solubilities organic substances have in supercritical fluid water (Deshpande *et al.*, 1984; Houser *et al.*, 1986; Amestica and Wolf, 1986) and the absence of this phase in the subsurface under the conditions in which petroleum generation occurs. Experiments using distilled water are limited to temperatures below the critical point of 374°C. As shown in Fig. 2, this critical-point temperature may be raised with the addition of NaCl to the water. While NaCl solutions provide subcritical temperatures as high as 700°C (Sourirajan and Kennedy, 1962), they are also highly corrosive, and care must be taken in selecting an appropriate reactor-wall composition.

Figure 2 also shows that the critical-point temperature of water may be lowered with the addition of CO_2. Supercritical fluid water may occur at temperatures below 300°C in experiments that generate high CO_2 concentrations relative to the amount of water present in the reactor. An example of this is the experiments conducted by Monthioux *et al.* (1985), in which high CO_2 concentrations were generated from pyrolyzed coals at high pressures in the presence of small quantities of added water. Calculations based

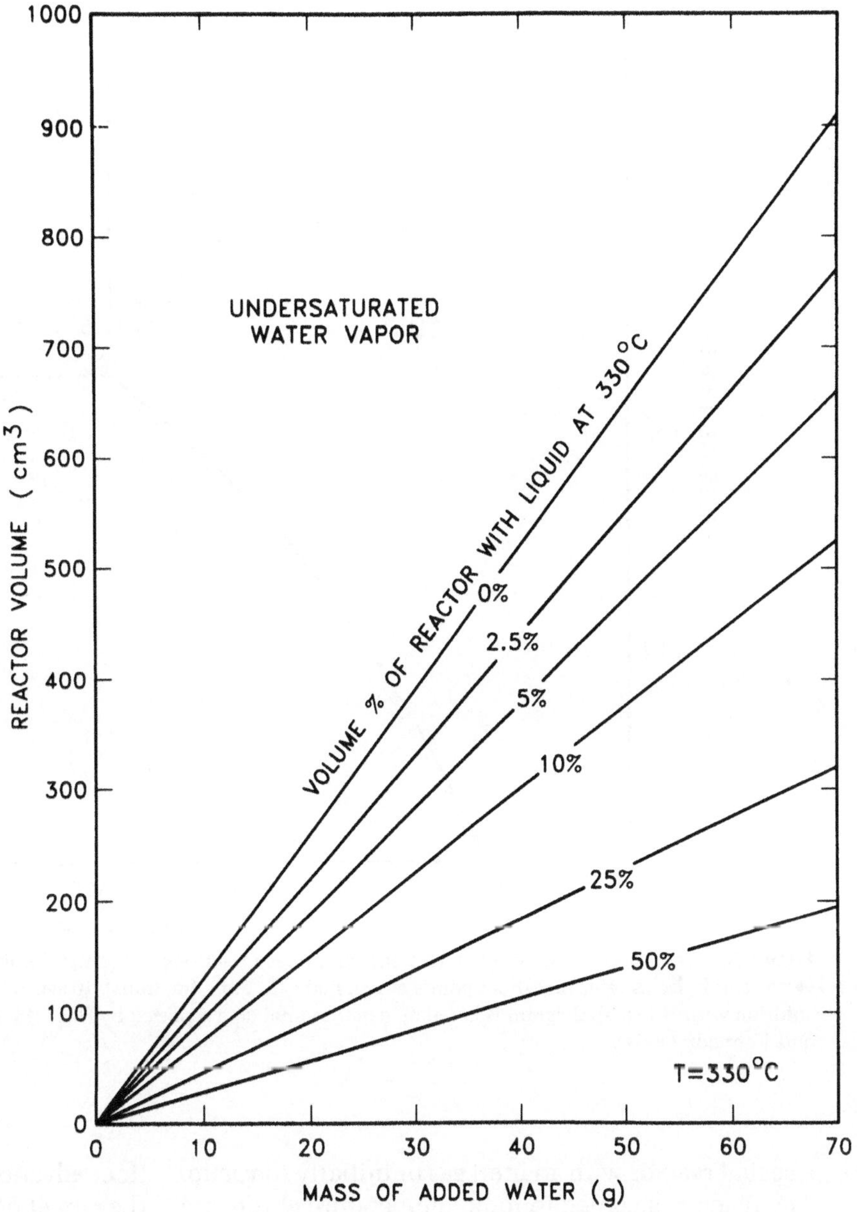

Figure 1. Plot showing volume percent of reactor filled with liquid water at 330°C for varying amounts of added water and reactor volumes.

on their reported experimental data indicate supercritical water fluid and not liquid water existed in their experiments at temperatures as low as 300°C. Supercritical hydrothermal experiments such as these are not hydrous pyrolysis experiments as originally defined.

A salient point of this discussion is that sample size, reactor volume, and amount of added water are key variables in determining whether or not a hydrous pyrolysis experiment has been conducted. Neglecting to report any one of these variables makes it impossible to determine the water phase or phases in contact with a sample during an experiment. If appropriate comparisons of experimental data from different laboratories are to be made, it is vital that experimental conditions be thoroughly described when reporting results in the literature.

3. Significance of Expelled Oil

Petroleum formation may be defined as hydrocarbon generation within and expulsion from a source rock. The major problem with anhydrous pyrolysis experiments (i.e., no added water) is their inability to generate and expel an oil-like pyrolysate in a manner analogous to that in the natural system. Closed anhydrous experiments involve pyrolyzing a sample in

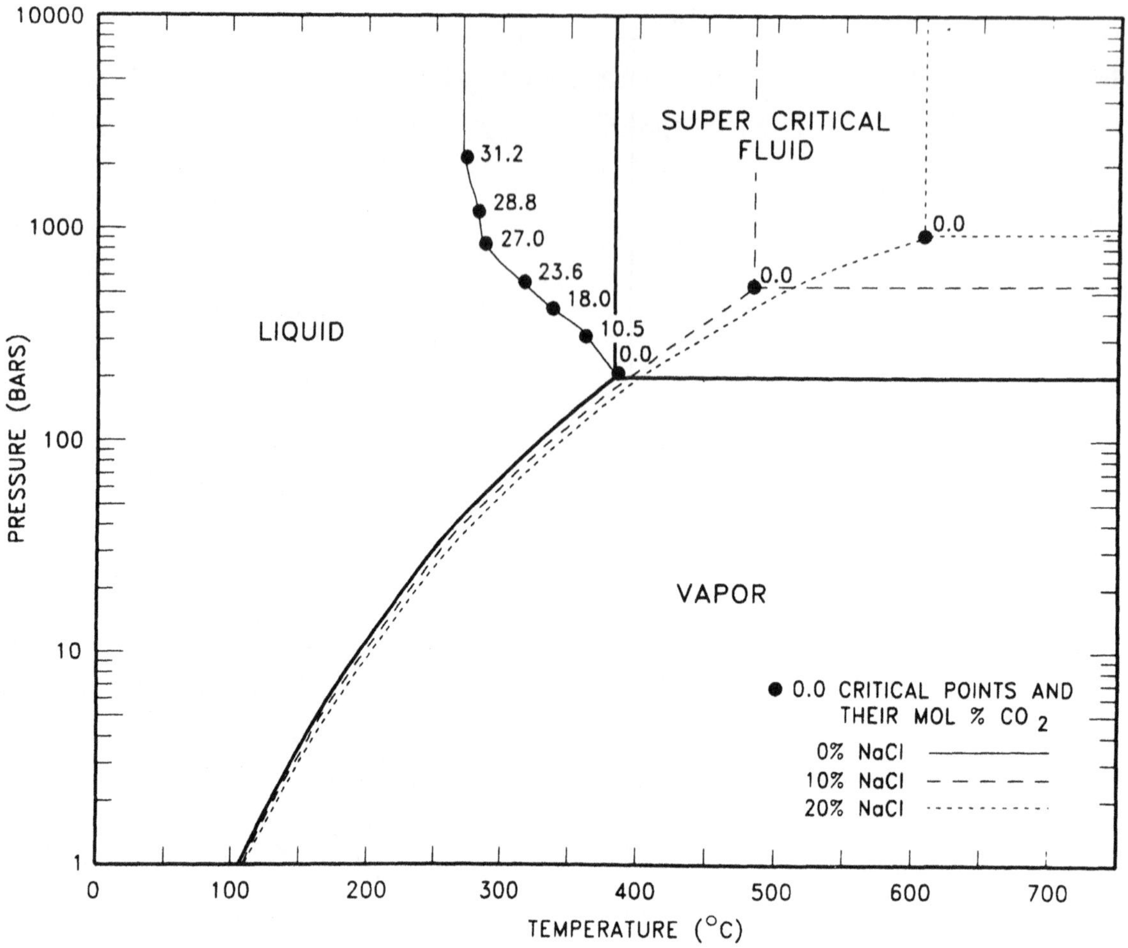

Figure 2. Pressure–temperature diagram showing phase relations and critical points for pure H_2O (——), CO_2 in solution with H_2O (●—●; solid circles refer to critical points at denoted CO_2 mole fractions), 10 wt. % NaCl solution with H_2O (– – –), and 20 wt. % NaCl solution with H_2O (---). Diagram is based on experimental data reported by Haas (1976), Sourirajan and Kennedy (1962), and Takenouchi and Kennedy (1964).

a sealed reactor with an inert gas or initially *in vacuo*. The reactor is cooled to room temperature at the end of the experiments, and the pyrolyzed sample is extracted with an organic solvent to obtain a pyrolysate product. Open anhydrous experiments allow the escape of components volatilized from a sample while it is being pyrolyzed. The volatile products are removed rapidly from the reactor by sweeping the sample with a carrier gas or by maintaining a vacuum on the system through a cold trap. Although these experiments have provided structural information on kerogens and relative assessments of petroleum potential of rocks, they have not provided information on generation and expulsion of oil from source rocks in the natural system. The ability of hydrous pyrolysis experiments to generate an oil compositionally similar to crude oil and to expel it in a manner similar to that in naturally maturing source rocks gives them a distinct advantage over anhydrous pyrolysis experiments (Lewan *et al.*, 1979). Some studies have concentrated on compositional similarities between pyrolysates from hydrous and anhydrous pyrolysis experiments (Saxby *et al.*, 1986; Comet *et al.*, 1986) but have not considered the advantage of expelling an oil-like pyrolysate from a rock in hydrous pyrolysis experiments.

Monitoring expelled oil, retained bitumen, and kerogen of a potential source rock through a series of hydrous pyrolysis experiments at different temperatures has shown that petroleum formation may be described by four stages: (1) pre-oil generation, (2) incipient-oil generation, (3) primary-oil generation, and (4) post-oil generation (Lewan, 1985). An example of this is shown in Fig. 3 for aliquots of a sample of Woodford Shale (WD-5) heated isothermally at different temperatures under hydrous py-

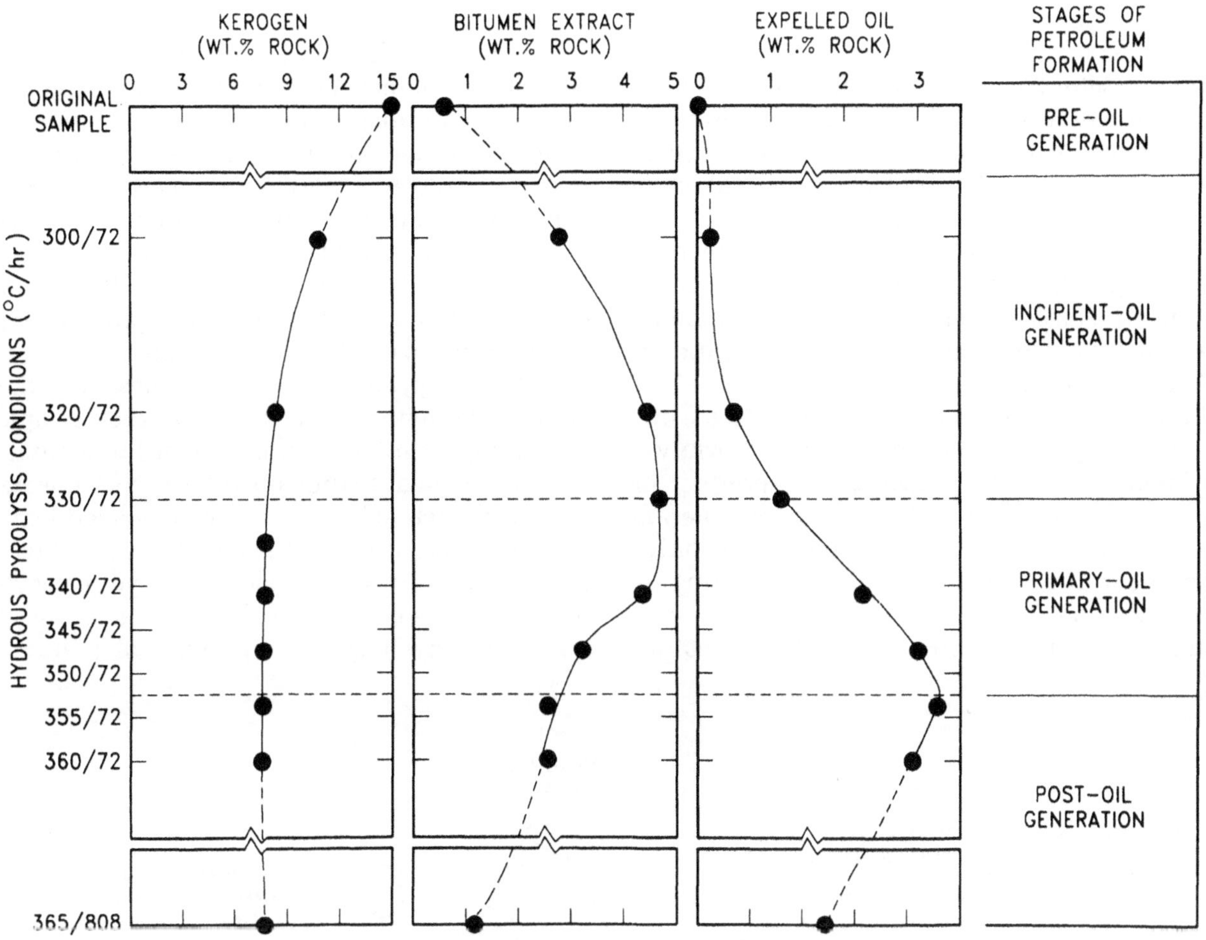

Figure 3. Variations in the amounts of kerogen, bitumen, and expelled oil from aliquots of a sample of Woodford Shale (WD-5) subjected to different temperatures for 72 to 808 h under hydrous conditions.

rolysis conditions. Pre-oil generation represents a thermally immature stage in which the organic matter occurs predominantly as a solid insoluble kerogen. Incipient-oil generation begins when thermal stress is sufficient to initiate decomposition of the kerogen into a tarry-soluble bitumen, which is enriched in high-molecular-weight hydrocarbons and heteroatom compounds. The amount of kerogen remains constant through the primary-oil generation stage, while thermal decomposition of bitumen results in the generation of an expelled oil, which is enriched in saturated hydrocarbons. Thermal decomposition of the expelled oil denotes the start of post-oil generation, in which the expelled oil and bitumen thermally decompose into gas and pyrobitumen.

Thermal decomposition of kerogen to bitumen and subsequent thermal decomposition of bitumen to oil have long been recognized in studies on oil shale retorting (Engler, 1913; McKee and Lyder, 1921; Franks and Goodier, 1922). A bitumen intermediate has also been suggested by Louis and Tissot (1967) for petroleum formation in the natural system. Hydrous pyrolysis experiments define this series of overall reactions and demonstrate that maximum bitumen generation does not correspond to maximum oil generation (Fig. 3). These two organic phases are distinguished in this approach by whether they are expelled from or retained in a source rock, which is the same operational distinction usually made in the natural system. Crude oils are expelled products typically reservoired in nonsource rocks, and bitumen extracts are the soluble products retained within a source rock. This experimental and natural distinction is supported by differences in their physical and chemical character. Physical differences are most apparent, with bitumens being viscous tars and oils being free-flowing liquids. This physical difference is also reflected in their chemical composition. Figure 4 shows the differences in saturated, aromatic, and polar constituents of bitumen extracts and expelled oils from hydrous pyrolysis experiments on the Woodford Shale (Lewan, 1983, 1985). Expelled oils plot within the compositional range of normal crude oils and have less compositional variability than the bitumen extracts. Despite the compositional variations through the different stages of petroleum formation, the expelled oils show no compositional similarities or parallel trends to the bitumen extracts.

Collection of an oil from the water surface with a pipet or syringe at the end of a hydrous pyrolysis experiment provides a full-range C_{5+} product for study. Conversely, solvent extraction of a sample subjected to closed anhydrous pyrolysis provides only a C_{15+} product due to solvent evaporation during the concentration procedure. Low-molecular-weight compounds make up a significant portion of natural crude oils, and, in addition to C_5–C_{15} *n*-alkanes, this boiling range includes thiophenes, benzothiophenes, benzenes, naphthalenes, tetralins, indanes, biphenyls, phenols, and adamatanes. Inclusion of these low-molecular-weight compounds in expelled oils from hydrous pyrolysis experiments is a useful attribute in studying natural petroleum generation. This has been demonstrated in evaluating the types of oil generated from resinites (Lewan and Williams, 1987) and the effects of radiation damage on organic matter in uranium-rich black shales (Lewan and Buchardt, 1989).

Hydrous pyrolysis as originally described (Lewan *et al.*, 1979) emphasized the use of rock samples in simulating petroleum formation in the natural system. Subsequent hydrous pyrolysis experimentation has been extended to the study of isolated kerogens (Comet *et al.*, 1986). Although studies of this type provide information on compositional aspects and hydrocarbon potential of isolated kerogens, they have not taken advantage of the ability of hydrous py-

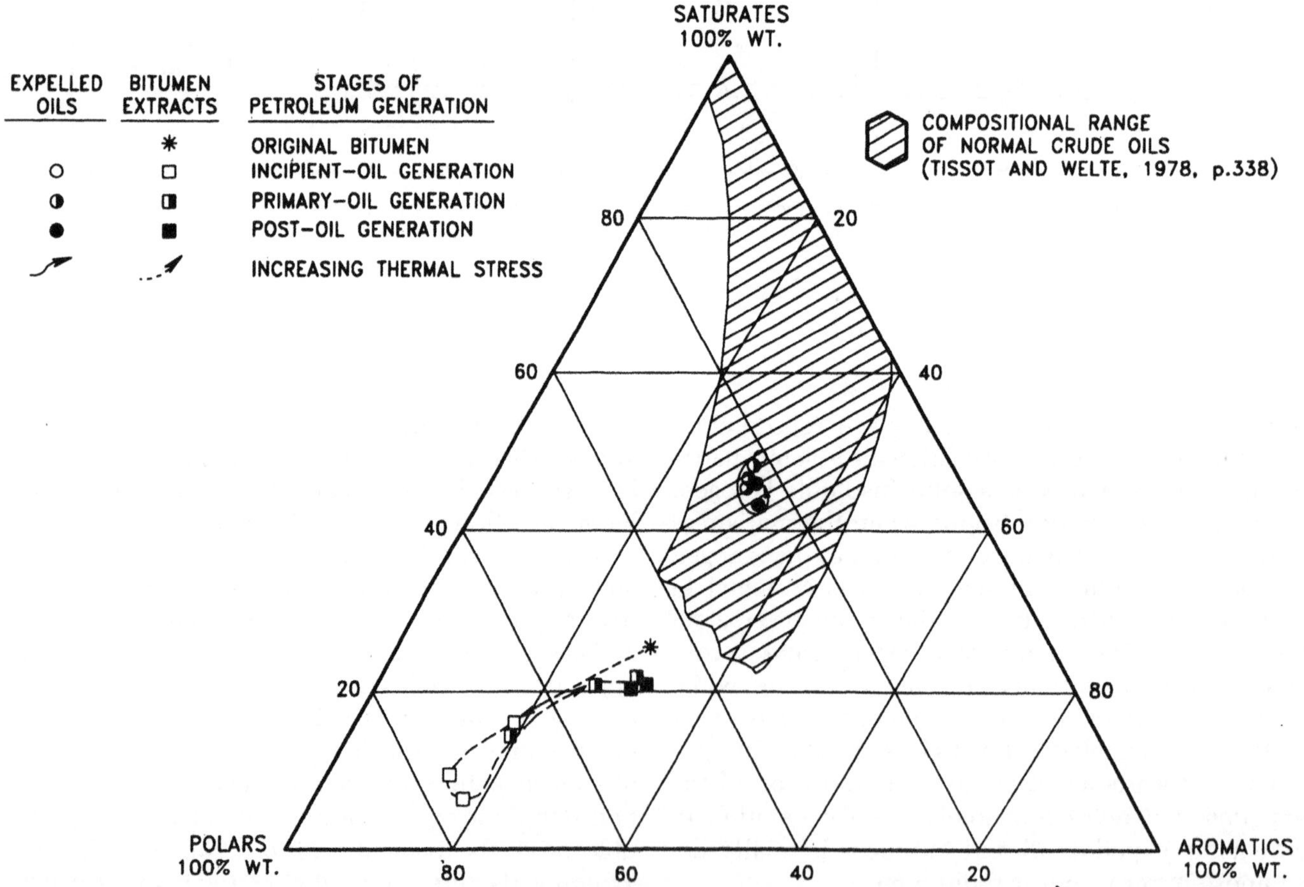

Figure 4. Ternary diagram showing the proportionality of saturate, aromatic, and polar fractions of expelled oils and bitumen extracts from Woodford Shale (WD-5) subjected to hydrous pyrolysis at the temperatures shown in Fig. 3 for 72-h durations. Fractions are defined by solid–liquid chromatography using a 40-cm alumina (MCB Alcoa F-20) column with saturates eluting with *n*-heptane, aromatics with benzene, and polars with a 40:60 (w/w) methanol:benzene mixture.

rolysis to differentiate between expelled oil and retained bitumen. Isolated kerogen that is oil-prone does generate an oil that accumulates on the water surface under hydrous conditions. Distinguishing oil from bitumen in these hydrous pyrolysis experiments is based solely on density differences relative to water. Therefore, the term floating oil rather than expelled oil is more appropriate in these types of experiments (e.g., Lewan and Williams, 1987).

4. Comparison of Pyrolysate Yields

In addition to generating an expelled oil, the total pyrolysate yields (bitumen + oil + gas) from hydrous pyrolysis are significantly greater than those from closed anhydrous pyrolysis. Comet *et al.* (1986) reported hydrous pyrolysis yields 1.2 to 1.7 times greater than closed anhydrous pyrolysis yields from isolated kerogens at 330°C after 72 h. Similarly, data reported by Tannebaum *et al.* (1986) on kerogen–mineral mixtures indicate hydrous pyrolysis yields 1.1 to 1.8 times greater than closed anhydrous pyrolysis yields at 300°C after 10 and 100 h, respectively. Neither of these comparative studies made a distinction between expelled oil and retained bitumen in their hydrous pyrolysis experiments. In order to make a more complete comparison between hydrous and anhydrous conditions, a series of 72-h pyrolysis experiments at 300, 330, and 350°C were conducted on aliquots of an immature sample of Woodford Shale (WD-26). These experimental conditions were used because they include stages of petroleum formation for the Woodford Shale from incipient-oil generation through primary-oil generation (Fig. 3). All of the experiments were conducted in 1-liter stainless steel-316 reactors with 400 g of crushed rock ranging in size from 0.5 to 2.0 cm. Hydrous pyrolysis experiments were conducted with 320 g of deionized (ASTM type I) water and an initial helium gas-space pressure of 241 kPa. Expelled oil and bitumen were collected by the procedures described by Lewan (1983). Anhydrous pyrolysis experiments at the three different temperatures were conducted in the same manner with the exclusion of added water and an initially evacuated gas space. The total amount of gas generated in each of the experiments was calculated on the basis of the ideal gas law at room temperature using gas-space volumes and mass-spectrometric analyses of the gases. Dissolved gases in the water at the conclusion of the hydrous pyrolysis experiments were not included in the total amount of generated gases at room temperature.

Pyrolysate yields from these comparative experiments are given in Table I in terms of generated gas, expelled oil, bitumen extract, and total pyrolysate (i.e., gas + oil + bitumen). Greater amounts of generated gas, and lack of an expelled oil in anhydrous pyrolysis compared to hydrous pyrolysis are the most obvious differences. The total pyrolysate for hydrous pyrolysis and anhydrous pyrolysis is essentially the same at 300°C after 72 h. This condition represents the incipient-oil generation stage, in which a portion of the kerogen decomposes into bitumen and only a small amount of expelled oil is generated from the bitumen. As the amount of expelled oil generated under hydrous conditions increases at the higher temperatures (i.e., 330 and 350°C), the difference in total pyrolysate between hydrous and anhydrous conditions becomes greater. Total pyrolysate under hydrous conditions is 1.4 and 2.0 times greater than under anhydrous conditions after 72 h at 330 and 350°C, respectively. This suggests that under closed anhydrous conditions the inability to generate an expelled oil results in a reduction in the total pyrolysate yield.

Table I shows the final pressures at experimental temperatures to be significantly higher in hydrous pyrolysis than in anhydrous pyrolysis. The higher pressures in hydrous pyrolysis are a result of water vapor, generated gas, and helium, and the lower pressures in anhydrous pyrolysis are only a result of generated gas. The significance of these pressure differences was evaluated in another series of experiments at 350°C for 72 h using aliquots of the same sample of Woodford Shale (WD-26). The first experiment in this series was under anhydrous conditions with an initial helium pressure of 7.72 MPa at room temperature. After 72 h at 350°C, the final pressure was 23.48 MPa of helium and generated gas. Table I shows that this higher pressure has no significant effect on the amount of total pyrolysate compared to that of the low-pressure anhydrous experiment. Once again, the inability of closed anhydrous pyrolysis to generate an expelled oil reduces its total pyrolysate relative to that generated in hydrous pyrolysis by a factor greater than 2 (Table I).

The second experiment in this series considers the conclusion by Monthioux and co-workers that confinement of generated gas in closed anhydrous pyrolysis produces results similar to those of hydrous pyrolysis (Monthioux *et al.*, 1985; Monthioux, 1988). This experiment involved mechanically pressing 500 g of crushed rock (0.5–1.0 cm) into a 500-ml stainless steel reactor, which was then evacuated. After 72 h at 350°C, a final pressure of 21.76 MPa was reached by

TABLE I. Comparison of Experimental Conditions and Pyrolysate Yields from Hydrous, Anhydrous (No Added Water), and Steam Pyrolysis of Crushed Aliquots (0.5–2.0 cm) of an Immature Sample of Woodford Shale (WD-26)[a]

Experimental conditions	Final pressure at experimental temperature (MPa)	Generated gas (wt. % of rock)	Expelled oil (wt. % of rock)	Bitumen extract (wt. % of rock)	Total pyrolysate[b] (wt. % of rock)
300°C/72 h					
Hydrous[c]	9.76	0.43	0.65	8.35	9.43
Anhydrous[d]	2.17	0.78	0.00	8.66	9.44
Δ%[e]	+77.8	−81.4	+100.0	−3.7	−0.1
330°C/72 h					
Hydrous[c]	15.89	1.05	2.79	8.19	12.03
Anhydrous[d]	4.24	1.74	0.00	6.62	8.36
Δ%	+73.3	−65.7	+100.0	+19.2	+30.5
350°C/72 h					
Hydrous[c]	21.13	1.68	4.15	5.71	11.54
Anhydrous[d]	5.69	2.40	0.00	3.30	5.70
Δ%	+73.1	−42.9	+100.0	+42.2	+50.6
Helium–anhydrous[f]	23.48	2.27	0.00	2.99	5.26
Δ%	−11.1	−35.1	+100.0	+47.6	+54.4
Confined gas–anhydrous[g]	21.76	2.60	0.00	2.27	4.87
Δ%	−3.0	−54.8	+100.0	+60.2	+57.8
Steam pyrolysis[h]	18.31	2.08	0.00	6.01	8.09
Δ%	+13.3	−23.8	+100.0	−5.3	+29.9

[a]All of the experiments were conducted with 400 g of rock in 1-liter stainless steel-316 reactors, except for the confined-gas anhydrous experiment, in which 500 g of rock were pyrolyzed in a 500-ml stainless steel-316 reactor.
[b]Total pyrolyzate = generated gas + expelled oil + bitumen extract.
[c]Crushed rock with 320 g of deionized (ASTM type I) water under an initial He pressure of 241 kPa.
[d]Crushed rock initially in an evacuated reactor.
[e]Δ% = (Hydrous − Anhydrous)100/Hydrous.
[f]Crushed rock under an initial He pressure of 7.72 MPa at room temperature.
[g]500g of crushed rock mechanically pressed at 5000 psi into a 500-ml reactor, which was initially evacuated
[h]Crushed rock with 75 g of dionized (ASTM type I) water under an initial He pressure of 241 kPa.

the generated gas as a result of the reduced gas space. As shown in Table I, this confined gas pressure is similar to the pressure obtained in hydrous pyrolysis at 350°C after 72 h. However, similar to the He-pressured anhydrous pyrolysis, the total pyrolysate is less than that from hydrous pyrolysis by a factor greater than 2 (Table I). Therefore, these experiments do not support the conclusion that confined gas under anhydrous conditions gives the same results as hydrous pyrolysis.

Although Monthioux *et al.* (1985) used supercritical-water conditions rather than hydrous conditions, this is not likely to be the major cause of the discrepancy in conclusions. More likely, this discrepancy may be explained by the inability of the coals used in their study to generate an expelled oil under hydrous conditions. Pyrolysate yields in Table I show that the inability of anhydrous conditions to generate an expelled oil reduces the total pyrolysate compared to that obtained under hydrous conditions. Under experimental conditions where little or no expelled oil is generated (e.g., 300°C for 72 h), the difference between hydrous and anhydrous pyrolysis is negligible. Similarly, if the type, amount, or distribution of organic matter in a rock is not conducive for generation of an expelled oil by hydrous pyrolysis, no significant differences in total pyrolysate yields are expected between hydrous and anhydrous pyrolysis. Most of the comparable Rock-Eval data presented by Monthioux *et al.* (1985) are on a sample (no. 32362) that has an initial hydrogen index of only 135 mg HC/g organic C. Coals with hydrogen indices this low do not generate an expelled oil under hydrous conditions, and, therefore, their total pyrolysate yield is expected to be essentially the same as that obtained under anhydrous conditions.

The significance of water is also demonstrated in the third experiment of this series. An aliquot of the same sample of Woodford Shale (WD-26) with sufficient water to provide only water vapor in the reactor was heated at 350°C for 72 h. Results of this steam pyrolysis are given in Table I. Although no expelled oil was generated in this experiment, the total pyrolysate yield is intermediate between the total pyrolysate yields for hydrous pyrolysis and anhydrous pyrolysis under the same time and temperature conditions. It becomes apparent from this series of experiments that the presence of liquid water is important in generating an expelled oil and that it does not merely act as a confining medium.

Comparisons with hydrous pyrolysis in this discussion have thus far concentrated on closed anhydrous pyrolysis. Another consideration is open anhydrous pyrolysis, which has been used successfully to characterize kerogen (Larter and Douglas, 1980; Larter and Senftle, 1985) and to provide rapid source rock evaluations (Espitalié *et al.*, 1977; Claypool and Reed, 1976). This type of pyrolysis typically uses a carrier gas to sweep volatiles generated from a rock or kerogen as it is heated from room temperature to temperatures in excess of 500°C. Open anhydrous pyrolysis makes no pretense to simulate natural petroleum formation, which is evident by the high concentrations of alkenes in its pyrolysates. Table II compares open anhydrous yields (i.e., Rock-Eval pyrolysis) from thermally immature samples of Woodford Shale and Phosphoria Retort Shale with their hydrous pyrolysis yields at the end of primary-oil generation. The total pyrolysates for both methods are similar, but the measurement of volatile hydrocarbons by a flame ionization detector used in Rock-Eval pyrolysis makes no distinction between gas, oil, or bitumen. Although this composite pyrolysate from Rock-Eval pyrolysis provides a relative assessment of hydrocarbon potential of a rock, it does not provide an assessment of the amounts of generated oil expelled from a rock. As shown in Table II, the hydrocarbons generated by Rock-Eval pyrolysis (i.e., S_2) are greater than the expelled oils from hydrous pyrolysis by a factor of 2.

TABLE II. Comparison of Rock-Eval and Hydrous Pyrolysis Yields from Woodford Shale and Phosphoria Retort Shale[a]

	Yield (wt. % of rock)	
Type of pyrolysis	Woodford Shale WD-5	Phosphoria Retort Shale P-64
Rock-Eval pyrolysis[b] of original sample		
Volatile HC (S_1)	0.25	0.50
Generated HC (S_2)	7.11	13.45
Total pyrolysate	7.36	13.95
Hydrous pyrolysis[c] at the end of primary-oil generation		
Bitumen extract	2.56	6.52
Expelled oil	3.14	6.52
Generated HC gases	0.63	0.94
Total pyrolysate	6.33	13.98

[a]Lewan, 1985.
[b]Rock-Eval II, cycle 1.
[c]Experimental conditions: 72 h at 355°C for WD-5 and at 350°C for P-64.

Neglecting to distinguish oil from bitumen in open anhydrous pyrolysis limits its use in determining kinetic parameters for modeling natural petroleum formation. Determining kinetic parameters on a composite product of bitumen, oil, and gas results in the need to use activation energy distributions that predict petroleum formation at lower temperatures and over broader temperature ranges than if only generated oil expelled from a rock is considered. Lewan (1985) has shown that the overall reaction of oil generation in hydrous pyrolysis experiments is adequately described by a single set of kinetic parameters, which are also applicable to predicting oil generation in the natural system. A comparative study by Burnham *et al.* (1987a, b) on kinetic parameters determined by Rock-Eval pyrolysis and hydrous pyrolysis shows the consequences of not making a distinction between oil and bitumen. Figure 5 compares the oil generation curves for Phosphoria Retort Shale determined by Rock-Eval-derived kinetics using activation energy distributions and by hydrous pyrolysis-derived kinetics using one activation energy. It is apparent for laboratory (Fig. 5a) and geological (Fig. 5b) heating rates that the inclusion of bitumen formation in Rock-Eval kinetics results in oil generation at significantly lower temperatures and over a broader temperature range than for kinetics determined by hydrous pyrolysis. Although Rock-Eval kinetics may adequately describe *hydrocarbon* generation (i.e., bitumen + oil + gas), hydrous pyrolysis kinetics more specifically describe *oil* generation.

A bitumen and oil may be distinguished in isothermal experiments using open anhydrous pyrolysis on the basis of pyrolysate volatility (e.g., Hubbard and Robinson, 1950). These experiments define oil as the condensable organic pyrolysate that is subsequently condensed to a liquid, and bitumen as the solvent-extractable organic pyrolysate that does not vaporize at experimental conditions and remains within the rock. Braun and Rothman (1975) have shown that by making this distinction the generation of vaporized pyrolysate oil may be described with a single set of kinetic parameters without the use of an activation energy distribution. The major concern with this type of pyrolysis is that the oil is obtained by vaporization and condensation, which is not operative in the natu-

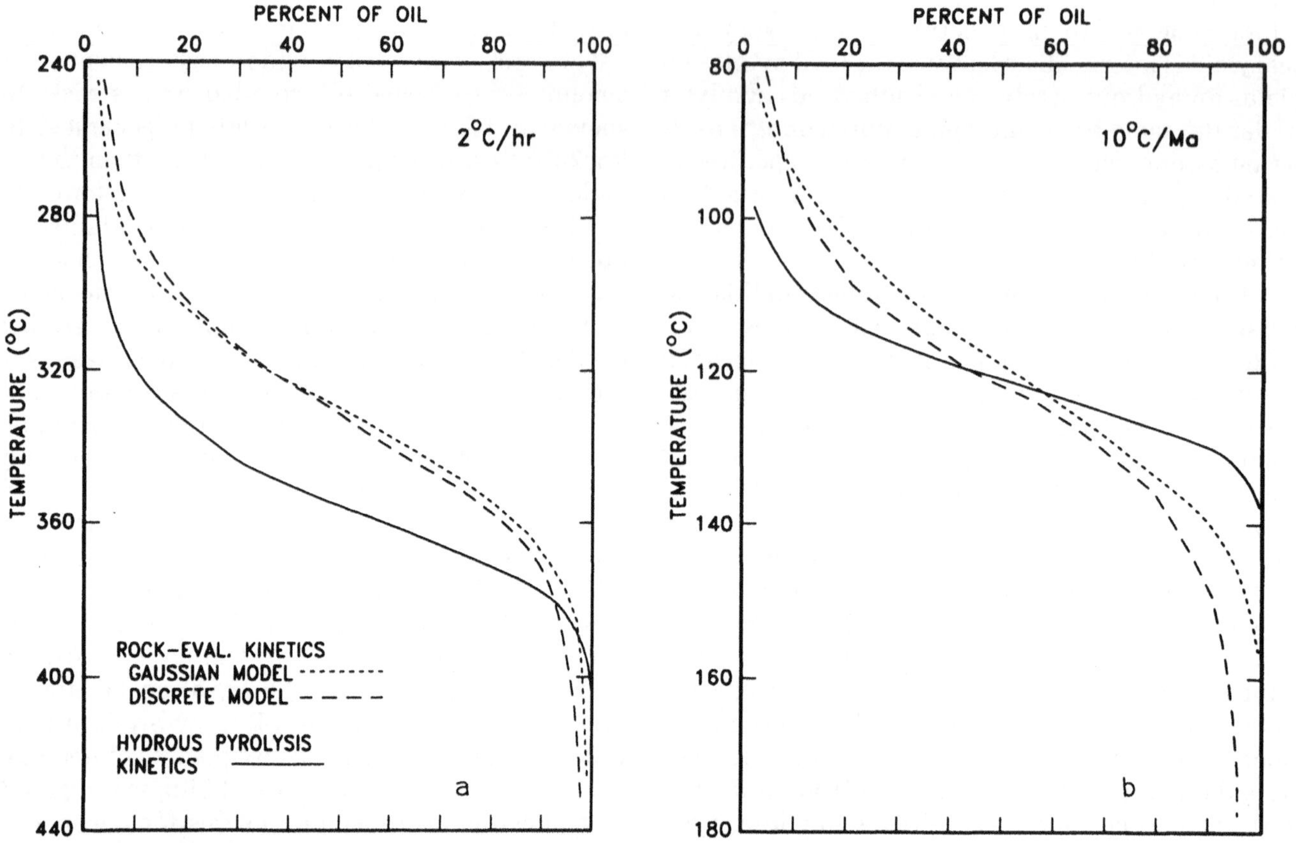

Figure 5. Oil generation curves determined from Rock-Eval kinetics and hydrous pyrolysis kinetics for Phosphoria Retort Shale at two different heating rates: (a) 2°C h^{-1} (Burnham *et al*, 1987b); (b) 10°C Ma^{-1} (Burnham *et al.*, 1987a).

ral system. As a result of vaporization playing a key role in obtaining an oil by open anhydrous pyrolysis, experimental factors including heating rate, rock sample size, and gas pressure have a significant effect on the results.

5. Utility of Rock Samples

Hydrous pyrolysis as described by Lewan *et al.* (1979) used rock samples rather than isolated kerogens to simulate natural petroleum formation. Unlike kerogens isolated by harsh acid treatments involving HCl and HF, organic matter in unpulverized rock samples occurs in its natural embedded state. Tannenbaum *et al.* (1986) have tried an alternative by subjecting artificial mixtures of isolated kerogens and minerals to closed anhydrous and hydrous pyrolysis. They observed that under hydrous conditions, water prevented kerogen–mineral interactions and concluded that these experiments are not representative of the natural system. The problem with simulating natural petroleum formation in these experiments is not the use of hydrous conditions but rather the use of unconsolidated powdered mixtures instead of indurate rock samples.

Rock samples in the form of cores or crushed chips allow the embedded organic matter to mature in contact with natural mineral fabrics and interstitial waters. The hydrous conditions allow the interstitial waters to be maintained and behave as liquids within the rock during heating as they would in the natural system. As demonstrated in Table I, the added water surrounding the rock chips or cores allows oil expulsion to occur. This added water is considered analogous to water in joints (i.e., regional fractures), which are ubiquitous in the natural system (Pollard and Aydin, 1988). Cores that fill most of the subaqueous portion of a reactor are ideal, but preparing for a series of experiments requiring cores of one rock sample is difficult and tedious. A more practical approach is to use crushed rock in the form of chips ranging in size from 0.5 to 2.0 cm. Provided the rock sample has not been pulverized to a fine powder (>62 μm) that destroys the organic and inorganic matrix-relationship, the size of the rock chips does not have a significant effect on the amounts of expelled oil and generated gas. Table III gives the yields from three

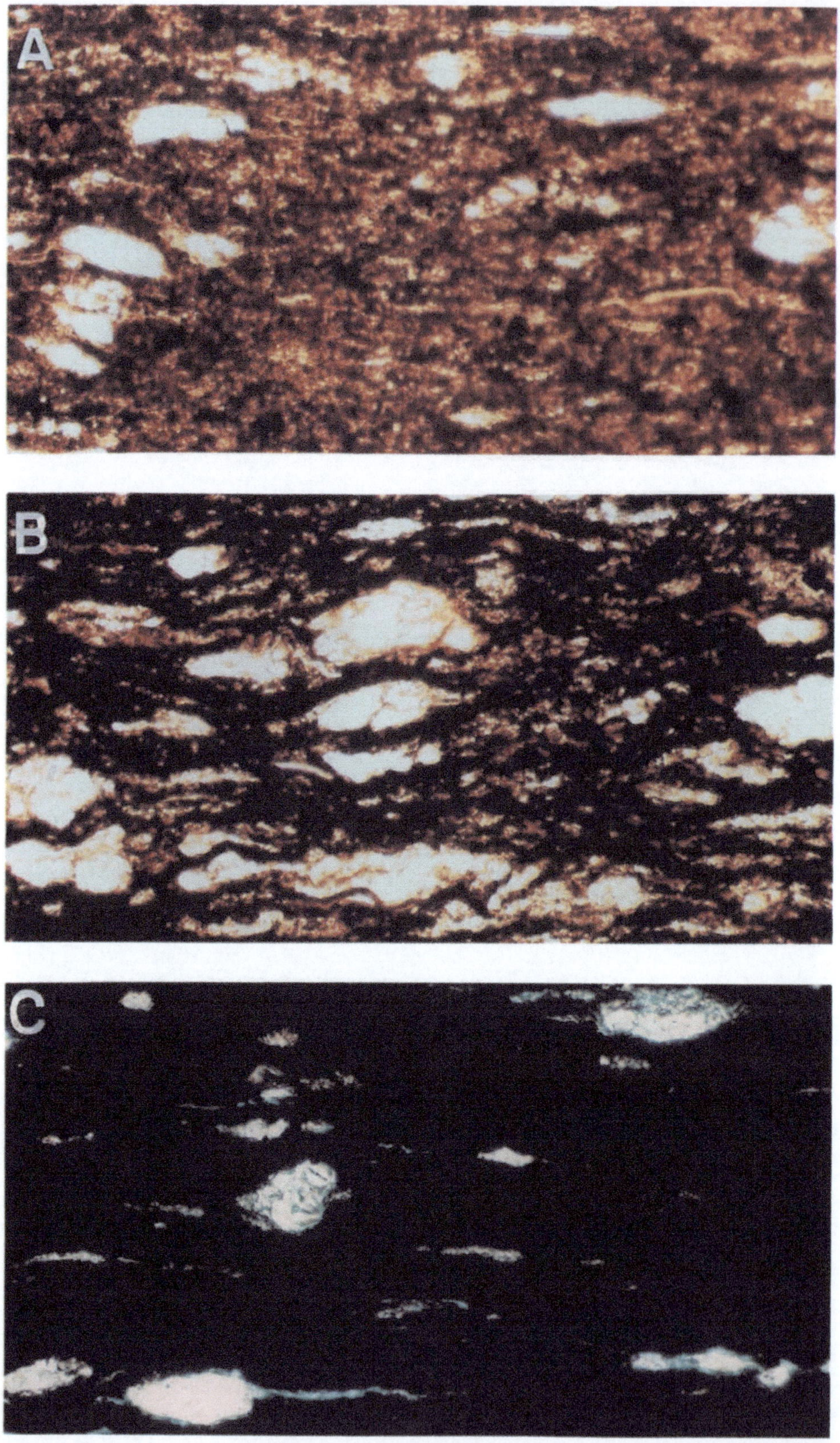

Figure 6. Photomicrographs of thin sections of cores taken from the same sample of Woodford Shale (WD-7) before hydrous pyrolysis (A), after hydrous pyrolysis for 72 h at 300°C (b), and after hydrous pyrolysis for 72 h at 352°C (C). Basal lengths of photomicrographs are 0.52 mm.

different size ranges of a rock sample from the Monterey Shale after 72 h at 330°C. As shown, there are no significant differences in amounts of expelled oil or generated gas. The recommended 0.5 to 2.0-cm size range can be easily obtained by sieving crushed rock through standard sieves. This size range packs reasonably well into a reactor and is large enough for petrographic thin sections and rock property determinations following the experiments.

Rocks that have generated an expelled oil have a swollen appearance as a results of *en echelon* parting separations that develop parallel to the bedding fabric. These parting separations pinch out laterally with maximum separations less than 150 μm thick (Lewan, 1987). Although these parting separations render the rocks more fissile, their discontinuous and *en echelon* character allows the rocks to remain coherent if handled carefully. This ability of rocks to remain coherent after hydrous pyrolysis has been observed for a variety of source-rock lithologies. Based on the laboratory classification by Lewan (1978), these lithologies include quartzose claystones, argillaceous claystones and mudstones, argillaceous and calcareous marlstones, and calcitic micstones and mudstones (i.e., limestones). A notable exception is argillaceous or siallitic claystones with high concentrations (>25 wt. %) of smectite. Hydration of this expandable clay mineral during hydrous pyrolysis results in a mud slurry at the end of an experiment, which makes recovery of products more difficult. The occurrence of shales rich in smectite is not common during natural petroleum generation because the apparent conversion of smectite to illite in the subsurface usually occurs at vitrinite reflectance values less than 0.7% R_0 (Kisch, 1987). Soldan and Cerqueira (1986) have simulated this clay–mineral conversion in hydrous pyrolysis experiments using potassium-rich formation waters and demonstrated that such a conversion improved the total pyrolysate yield of the rock.

TABLE III. Weight Percent of Expelled Oil and Generated Gas from Hydrous Pyrolysis of Different Size Fractions of a Rock Sample (MR-83) from the Monterey Formation[a]

Size range (cm)	Expelled oil (wt. % of rock)	Generated gas (wt. % of rock)
0.5–2.0	2.61	2.53
1.0–1.5	2.47	2.56
0.1–0.5	2.50	2.56
Mean ± std. dev.	2.52 ± 0.06	2.55 ± 0.01

[a]Experiments involved heating 300 g of rock with 365 g of deionized water in a 1-liter stainless steel-316 reactor at 330°C for 72 h.

The use of a rock in hydrous pyrolysis experiments is also a more relevant approach to studying primary oil migration and expulsion than the use of isolated kerogens. Studies on primary oil migration have benefited from petrographic thin-section examination of rocks that have been thermally matured in nature and in hydrous pyrolysis experiments (Bertrand *et al.*, 1987; Lewan, 1987; Talukdar *et al.*, 1987). Results from these studies suggest expulsion of oils from rocks in hydrous pyrolysis experiments is mechanistically similar to that in the natural system. This is exemplified by the petrographic study by Lewan (1987) on Devonian–Mississippian black shales of the midcontinent region in the United States. These shales in the pre-oil generation stage contain amorphous masses of kerogen dispersed within a translucent orange-brown groundmass (Fig. 6A). As thermal stress increases, the viscous bitumen generated from the decomposition of the kerogen impregnates the planar bedding fabric and micropores of the rock. The continuous bitumen network that develops during this incipient-oil generation stage results in a dark brown groundmass (Fig. 6B). Increasing thermal stress results in partial decomposition of the bitumen network into a liquid oil enriched in saturated hydrocarbons. The net volume increase generated by this overall reaction within the confined inorganic rock matrix results in expulsion of the oil. Following primary-oil generation, the bitumen network is carbonized to pyrobitumen with increasing thermal stress, leaving a black opaque groundmass (Fig. 6C). The net volume increase responsible for bitumen migration within and oil expulsion from the rocks is the result of an overall decrease in density of reaction products and thermal expansion of the generated organic liquids (Lewan, 1987).

The significance of these observations is that petroleum formation takes place within the rock chips or cores and not within the added water which surrounds the rock. Once petroleum formation commences, the rock matrix becomes oil wet as observed in the natural system (Meissner, 1978; Lewan, 1987). This bituminous system replaces the preexisting aqueous system, and, as a result, mineral transformations and dissolutions are not controlled by aqueous solubilities. The inability of closed anhydrous pyrolysis to generate an expelled oil and the excess water present in hydrous pyrolysis of isolated kerogens emphasizes the importance of using rock samples in hydrous pyrolysis to simulate natural petroleum formation. Water surrounding the rock maintains the liquid water phases within the rock prior to petroleum formation and provides an accommodating medium in which generated oil may be expelled during petroleum formation.

6. Effects of Reactor Wall Composition

Alteration of experimental results by the interaction of reactants or products with reactor walls is a common concern. Hydrous pyrolysis experiments using isolated kerogens (e.g., Comet *et al.*, 1986; Eglinton and Douglas, 1988) or model compounds (e.g., Palmer and Drummond, 1986) allow both the reactants and products to be in direct contact with the reactor walls. Conversely, the use of rock chips or cores in hydrous pyrolysis minimizes the contact of reactor walls with reactants (i.e., kerogen, bitumen, and rock matrix), but not with expelled products (i.e., oil and gas). In order to evaluate this latter interaction, a series of hydrous pyrolysis experiments using a crushed rock sample was conducted with reactor walls composed of stainless steel-316, Hastelloy C-276, plated gold, and borosilicate glass.

Reactors composed of stainless steel-316 have proven to be reliable for hydrous pyrolysis experiments using deionized or distilled water (e.g., Lewan, 1983; Lundegard and Senftle, 1987; Eglinton *et al.*, 1988). This austenitic iron-base Cr–Ni–Mo alloy is one of the more corrosion-resistant commercial stainless steels and also has high creep and rupture strengths (King, 1983, p. 614). Another austenitic alloy that has a better corrosion resistance to chloride-bearing solutions such as seawater and brines is Hastelloy C-276, which is a nickel-base Mo–Cr–Fe–W alloy (Haynes International, Inc., Kokomo, Indiana). Expelled products from hydrous pyrolysis experiments conducted in reactors made of these two alloys were compared with those from a gold-plated stainless steel reactor. All of the reactors had fresh metal walls that had not been subjected to prior hydrous pyrolysis experiments or other heat treatments. The experiments consisted of heating 300 g of crushed (0.5–2.0 cm) Monterey Shale (MR-216) with 365 g of deionized (ASTM type I) water for 72 h at 330°C in 1-liter reactors with the different wall compositions.

In Table IV, expelled products generated in these

TABLE IV. Amount and Composition of Expelled Oils from 300-g Aliquots of a Sample of Monterey Shale (MR-216) Heated with 365 g of Deionized Water for 72 h at 330°C in 1-Liter Reactors with Different Wall Compositions

	Reactor					
	Fresh metal walls			Carburized metal walls		
	Gold-plated SS-316	Stainless steel-316	Hastelloy C-276	Stainless steel-316	Hastelloy C-276	Borosilicate glass liner in SS-316
Amount of expelled oil (g/300 g rock)	4.38	4.10	4.52	4.53	4.71	4.19
GC/MS ratios[a]						
$\frac{n\text{-C}_{10}}{n\text{-C}_{10} + n\text{-C}_{20}}$	0.33	0.32	0.33	0.34	0.35	0.34
$\frac{n\text{-C}_{30}}{n\text{-C}_{30} + n\text{-C}_{20}}$	0.21	0.20	0.22	0.27	0.24	0.23
$\frac{\text{Pristane}}{\text{Pristane} + n\text{-C}_{20}}$	0.56	0.58	0.55	0.55	0.55	0.56
$\frac{\text{Thiophenes}^b}{\text{Thiophenes} + n\text{-C}_{10}}$	0.74	0.72	0.73	0.74	0.74	0.74
$\frac{\text{Hopanes}^c}{\text{Hopanes} + n\text{-C}_{30}}$	0.09	0.11	0.09	0.09	0.10	0.08
$\frac{\text{Steranes}^d}{\text{Steranes} + n\text{-C}_{30}}$	0.10	0.13	0.11	0.10	0.12	0.08
$\frac{\text{TA steroids}^e}{\text{TA steroids} + n\text{-C}_{30}}$	0.33	0.39	0.33	0.33	0.34	0.30

[a]These ratios are based on intensities of specific peaks from the *m/z* 57 ion chromatograms, except where otherwise noted.
[b]Thiophenes include peak intensities of C_1- and C_2-substituted thiophenes from the *m/z* 45 ion chromatograms.
[c]Hopanes include peak intensities of C_{29}, C_{30}, and C_{31} hopanes from the *m/z* 191 ion chomatograms.
[d]Steranes include peak intensities of C_{26}, C_{27}, and C_{29} ααS steranes from the *m/z* 217 ion chromatograms.
[e]Triaromatic steroid hydrocarbons (TA) include peak intensities of C_{27}, C_{28}, and C_{30} steroids from the *m/z* 231 ion chromatograms.

experiments from the original 300-g aliquots of rock sample are tabulated. The amount of expelled oil for the three fresh-metal walls is highest for Hastelloy C-276, lowest for stainless steel-316, and intermediate for gold plating. Comparison of the resulting expelled oils on the basis of gas chromatography (Fig. 7) and gas-chromatographic/mass-spectrometric analyses (Table IV and Figs. 8–10) show no significant differences in the relative concentrations or distributions of *n*-alkanes, acyclic isoprenoids, thiophenes, terpanes, steranes, and triaromatic steroids. Therefore, it is apparent that these compound classes are not involved in the reactions responsible for the differences in amounts of expelled oil from the fresh-metal reactors.

Amounts of generated gases from these experiments are less variable, with stainless steel-316 having the higher yield and the other two reactor types having similar lower yields (Table V). Compositions of these generated gases presented in Table V show that H_2S and CO_2 have the most variability and an inverse relationship with one another. The variation in the amount of H_2S is related to the amount of iron sulfide formed on the reactor walls. In the gas space of the fresh stainless steel-316 reactor, the walls are coated with a thin (<3 μm) plating of iron sulfide at the end of the experiment. This is a result of H_2S gas reacting with iron from the reactor walls. X-ray diffraction patterns and scanning electron microscopic (SEM) examinations clearly identify this iron sulfide coating as hexagonal pyrrhotite ($Fe_{1-x}S$ with $x < 0.1$). In contrast, the low-iron-bearing, fresh Hastelloy C-276 walls had only a light dusting of pyrrhotite in the gas space of the reactor. The composition of generated gas from this experiment (Table V) has considerably more H_2S than that from the experiment with fresh stainless steel-316 walls. The gold-plated walls had a coating of pyrrhotite powder in the gas space of the reactor where the gold plating had deteriorated and exposed the underlying stainless steel-316 surface. This intermediate amount of pyrrhotite corresponds to an intermediate amount of H_2S relative to that for the other two fresh metal walls (Table V). Qualitatively, these results indicate that the amount of H_2S gas recovered in a hydrous pyrolysis experiment is inversely proportional to the amount of available iron in the reactor walls.

Although pyrrhotite in coals has been reported to catalyze coal-liquefaction reaction through the enhancement of free-radical generation (Srinivasan and Seehra, 1982, 1983), the amount of pyrrhotite on the gas-space reactor walls is inversely proportional to the amounts of expelled oil. This lack of a catalytic effect in hydrous pyrolysis may be explained by the limited exposure expelled oils have with pyrrhotite in the gas space of the reactors. Despite variations in the amount of pyrrhotite in the gas space of the reactors, there is an equal amount of pyrrhotite occurring as a light dusting in the liquid space of all three fresh-metal reactors. In addition, significant increases in free-radical generation with pyrrhotite in coals only occurs at temperatures in excess of 400°C (Srinivasan and Seehra, 1983; Murakami *et al.*, 1986). It is also worth noting that while pyrrhotite is forming on the reactor walls, the pyrite (FeS_2) within the rock sample remains unchanged. This apparent disequilibrium is explained by the bitumen-impregnated matrix of the rock (Fig. 6B and C) preventing interaction with the surrounding water.

In addition to pyrrhotite formation, the bright silvery finish of the fresh-metal reactors dull and darken with successive experiments in spite of wire-brush cleanings after each experiment. X-ray photoelectron spectroscopic analyses of these darkened but smooth surfaces indicate that graphitic carbon is being introduced into the reactor walls. Auger electron spectroscopic analysis on a piece of reactor wall (stainless steel-316) subjected to more than 50 hydrous pyrolysis experiments showed that this introduced carbon extended only 3.9 μm into the reactor wall (C. Willey and T. Fleisch, 1985, personal communication). This process by which carbon is introduced into steel or other alloys is referred to as carburization. The source of carbon that diffuses into alloys during this process may be hydrocarbons or carbon dioxide (Stanley, 1970). Consecutive hydrous pyrolysis experiments show that the gas space of the reactors carburizes more readily than the liquid portion, but eventually the entire wall surface is carburized. It has also been observed in hydrous pyrolysis experiments that stainless steel-316 carburizes more readily than Hastelloy C-276.

The degree of carburization is variable in the gas space of the three fresh metal reactors. A dark carburized surface was apparent under the pyrrhotite plating on the stainless steel-316 reactor, but only a dull silvery surface was observed under the light pyrrhotite dusting on the Hastelloy C-276 reactor. Under the pyrrhotite powder coating on the gold-plated reactor, the exposed stainless steel surface was dull and slightly darkened. These different degrees of carburization correspond inversely with the amounts of expelled oil. This suggests that expelled oil is the source of carbon for carburization of the fresh metal walls. Additional support of this comes from two hydrous pyrolysis experiments conducted with the

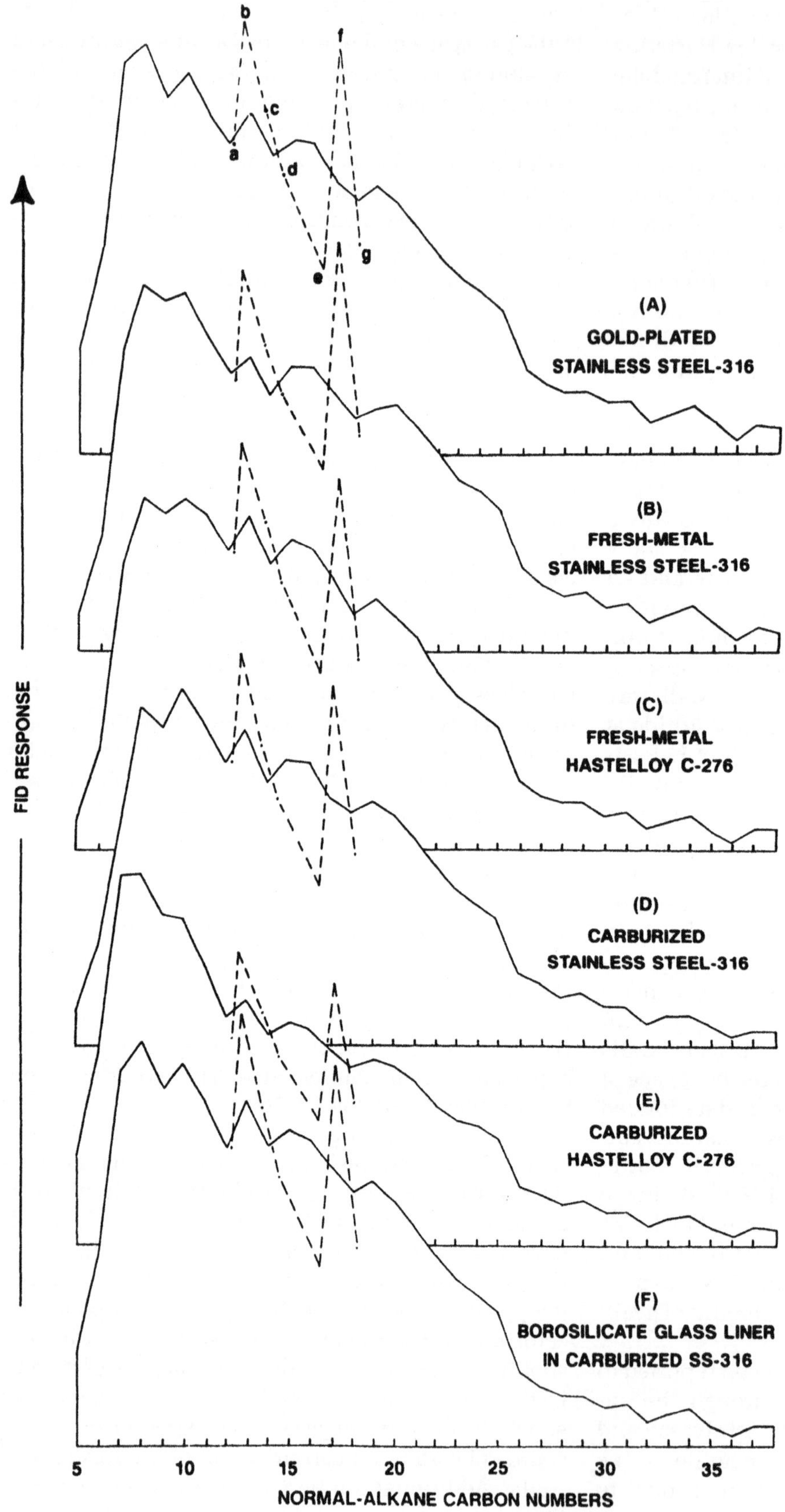

Figure 7. Comparison of *n*-alkane (——) and acyclic isoprenoid (– – –) distributions from gas chromatograms of expelled oils generated from aliquots of a sample of Monterey Shale (MR-216) subjected to hydrous pyrolysis at 330°C for 72 h in 1-liter reactors with different wall compositions. The solid lines represent *n*-alkane distributions and dashed lines represent acyclic isoprenoid distributions with a, C_{13}; b, C_{14}; c, C_{15}; d, C_{16}; e, C_{18} (norpristane); f, C_{19} (pristane); g, C_{20} (phytane).

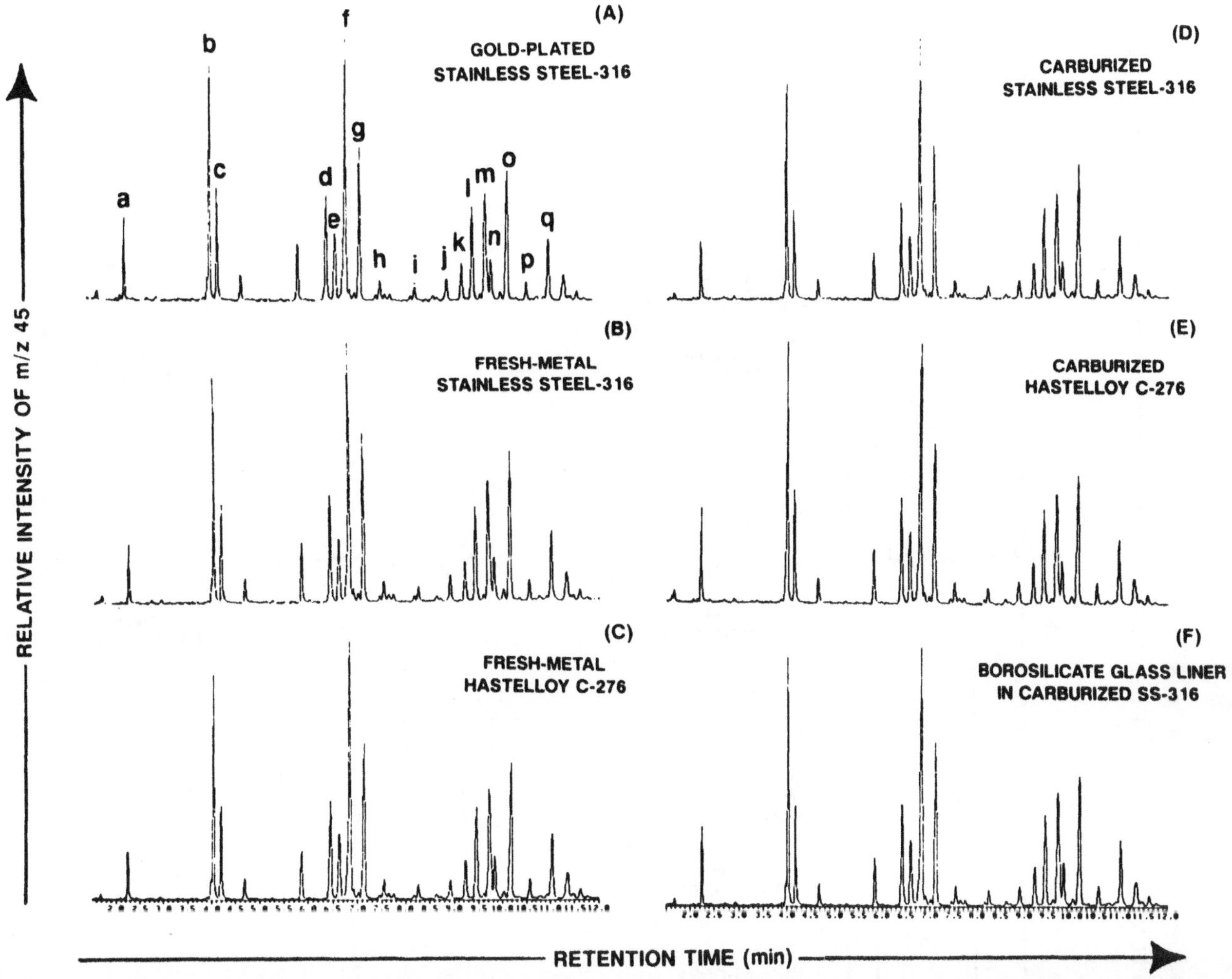

Figure 8. Comparison of m/z 45 ion chromatograms of expelled oils generated from aliquots of a sample of Monterey Shale (MR-216) subjected to hydrous pyrolysis in reactors with different wall compositions at 330°C for 72 h. The thiophenes identified in these chromatograms are as follows: a, thiophene; b, 2-methylthiophene; c, 3-methylthiophene; d, 2-ethylthiophene; e, 2,5-dimethylthiophene; f, 2,4-dimethylthiophene + 3-ethylthiophene; g, 2,3-dimethylthiophene; h, 3,4-dimethylthiophene; i, 2-isopropylthiophene; j, thiolane; k, 2-n-propylthiophene; l, 2-methyl-5-ethylthiophene; m, 3-methyl-2-ethylthiophene; n, 2-methyl-4-ethylthiophene; o, 2,3,5-trimethylthiophene; p, 3-methyl-4-ethylthiophene; q, 2,3,4-trimethylthiophene.

same sample (MR-216) under the same conditions (330°C/72 h) in stainless steel-316 and Hastelloy C-276 reactors that were completely carburized in prior experiments. Table IV shows that the amounts of expelled oils in these experiments are higher than in those with their fresh-metal equivalents.

Tannenbaum and Kaplan (1985a) stated that borosilicate-glass reactors minimize catalytic effects in hydrous pyrolysis experiments compared with metal reactors, but no supportive data or references were presented. An actual comparison of hydrous pyrolysis in glass and metal reactors was conducted with isolated kerogens by Comet *et al.* (1986). Their results showed that steranes and hopanes in solvent extracts of isolated kerogens after hydrous pyrolysis were present in stainless steel reactors but absent in silica-glass reactors. In order to provide an assessment of the effects of glass reactors on hydrous pyrolysis of rock samples, an additional experiment was conducted on an aliquot of sample MR-216 in a borosilicate-glass liner within a carburized 1-liter stainless steel-316 reactor. The closed base and full-reactor length of the glass liner prevented rocks chips, expelled oil, and liquid water from making contact with the metal walls, but the open top of the glass liner did not prevent generated gas and water vapor from making contact with the metal walls.

Similar to results reported by Tannenbaum and Kaplan (1985b) and Comet *et al.* (1986), corrosion of the originally transparent glass linear resulted in a

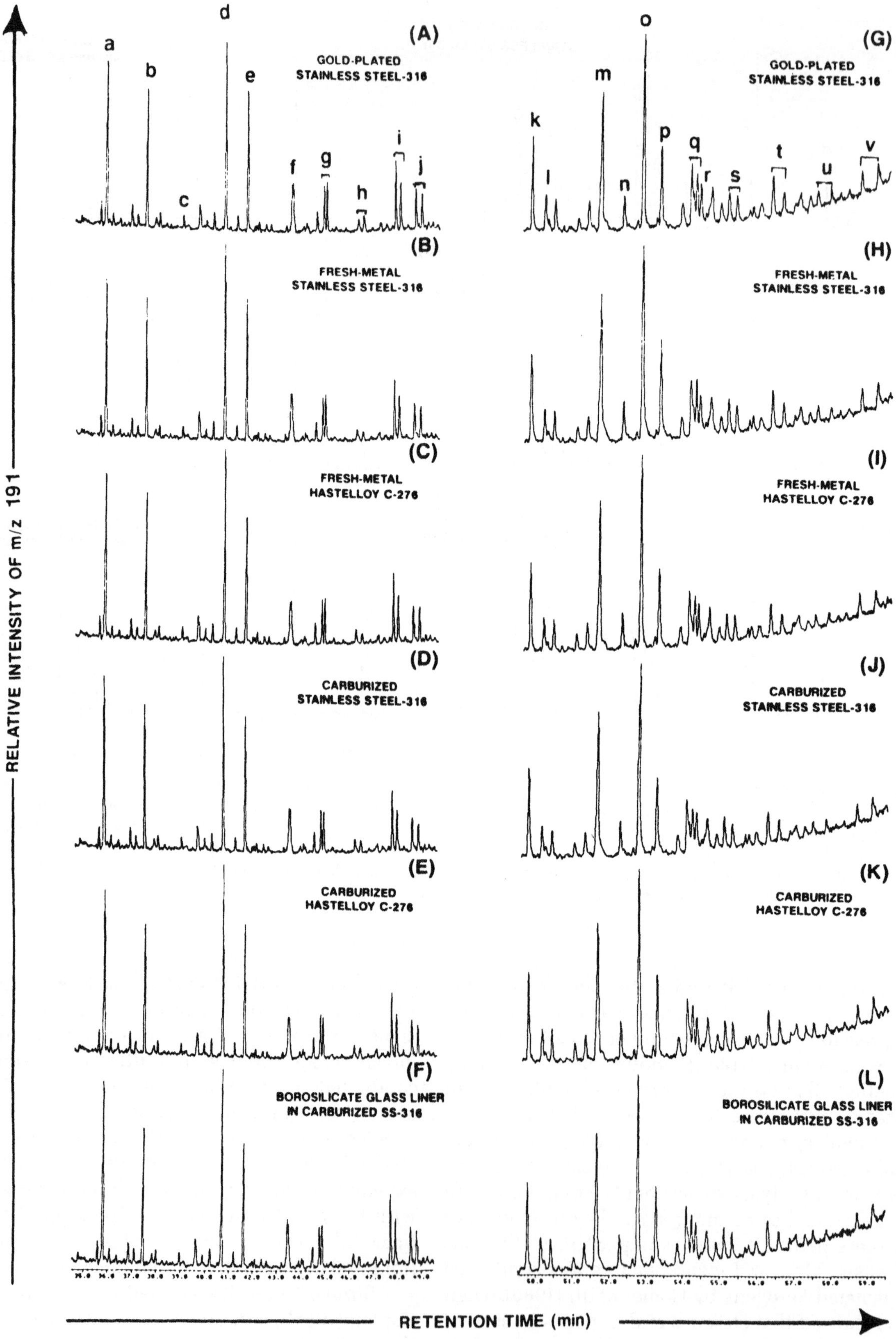
RELATIVE INTENSITY OF m/z 191
a
b
c
d
e
f
g
h
i
j
(A)
GOLD-PLATED
STAINLESS STEEL-316
(B)
FRESH-METAL
STAINLESS STEEL-316
(C)
FRESH-METAL
HASTELLOY C-276
(D)
CARBURIZED
STAINLESS STEEL-316
(E)
CARBURIZED
HASTELLOY C-276
(F)
BOROSILICATE GLASS LINER
IN CARBURIZED SS-316
k
l
m
n
o
p
q
r
s
t
u
v
(G)
GOLD-PLATED
STAINLESS STEEL-316
(H)
FRESH-METAL
STAINLESS STEEL-316
(I)
FRESH-METAL
HASTELLOY C-276
(J)
CARBURIZED
STAINLESS STEEL-316
(K)
CARBURIZED
HASTELLOY C-276
(L)
BOROSILICATE GLASS LINER
IN CARBURIZED SS-316
RETENTION TIME (min)

white frosted surface. The expelled oil layer at the experimental temperature (i.e., 330°C) was marked by a 2.8-cm interval of black to medium gray frosted glass at the expanded water level within the liner. Analysis of silica scraped from this interval indicated the darkening resulted from introduction of carbon into the glass. Carbon deposits on glasses during pyrolysis experiments are enhanced by the presence of boron (Dawidowicz *et al.*, 1984, 1986). The lower yield of expelled oil in the glass liner experiment compared to that in the carburized metal reactors (Table IV) is attributed to carbon deposits resulting from diffusion of oil into the glass. Similar to carburization of metal reactor walls, this process does not appear to be selective of particular compound classes as shown by the analyses on the gases (Table V) and expelled oils (Table IV and Figs. 7–10). This is contrary to observations made by Comet *et al.* (1986), but their comparative experiments involved isolated kerogens in direct contact with silica glass.

Overall, these results indicate that the effects of reactor wall composition on expelled oils generated by hydrous pyrolysis of rock samples do not significantly alter molecular compositions but do significantly alter recoverable yields. Alterations in the amount of expelled oil are minimized by using well-carburized metal reactors. This pretreatment is especially important for stainless steel-316 reactors and less critical for Hastelloy C-276 reactors. Carburization also reduces the interaction of generated H_2S gas with iron in the reactor walls (Table V). New and remachined metal reactors should be subjected to several preliminary hydrous pyrolysis experiments with oil-prone rock samples to carburize the walls before conducting quantitative experiments. This is a common practice that is an important prerequisite for experimental studies involving organic compounds. Palmer and Drummond (1986) reported in their study on thermal decarboxylation of acetate that several preliminary experiments were necessary to render the titanium alloy walls of their reactors inert to the reaction.

7. Consideration of Experimental Artifacts

One of the most noticeable artifacts of open anhydrous pyrolysis is the generation of alkenes, which rarely occur in natural crude oils (Hoering, 1977). Lewan *et al.* (1979) reported no detectable alkenes in oil-like pyrolysates generated from rock samples by closed pyrolysis under hydrous and helium-pressured anhydrous conditions. Similar results were also reported in subsequent studies using isolated kerogens (Hoering, 1985; Saxby *et al.*, 1986; Comet *et al.*, 1986). Contrary to these observations, Huizinga *et al.* (1987) reported three normal alkenes (C_{21}, C_{22}, and C_{24}), prist-1-ene, and phytene in closed hydrous and anhydrous pyrolysis of isolated kerogens. Generation of pristenes and phytenes by hydrous and anhydrous pyrolysis has also been reported by Eglinton *et al.* (1988) for isolated kerogen from the Kimmeridge Blackstone Band and by Jones *et al.* (1987) for isolated asphaltenes from Douk Daka and Boscan crude oils. Alkene concentrations in all of these studies are most apparent at temperatures less than 300°C for 72-h durations and are negligible at higher thermal stress levels (i.e., ≥300°C for ≥72 hr). At the lower thermal stress levels, bitumen generation dominates (Lewan, 1985), and this does not represent generation of expelled oils, which do not contain these alkenes when derived from rock samples (Lewan *et al.*, 1979) rather than isolated kerogens.

A study by Tannenbaum and Kaplan (1985b) of gases from the same experiments reported by Huizinga *et al.* (1987) also showed the presence of significant quantities of low-molecular-weight alkenes under hydrous and anhydrous conditions. Analyses of their gases generated at 300°C under hydrous conditions by kerogens isolated from the Monterey Shale and Green River Shale are compared in Table VI with analyses of gases generated under hydrous conditions by rock samples from the same rock units. The only gaseous alkenes generated from the rock samples are a trace of ethene from the Monterey Shale and

Figure 9. Comparison of *m/z* 191 ion chromatograms of expelled oils generated from aliquots of a sample of Monterey Shale (MR-216) subjected to hydrous pyrolysis in reactors with different wall compositions at 330°C for 72 h. Tricyclic terpanes are shown in chromatograms A through F with the following peak identifications: a, C_{20}; b, C_{21}; c, C_{22}; d, C_{23}; e, C_{24}; f, extended C_{25} (C-22S and R); g, extended C_{26} (C-22S and R); h, extended C_{27} (C-22S and R); i, extended C_{28} (C-22S and R); j, extended C_{29} (C-22S and R). Pentacyclic terpanes are shown in chromatograms G through L with the following peak identifications: k, 18α(*H*)-22,29,30-trisnorneohopane (T_s), l, 17α(*H*)-22,29,30-trisnorhopane (T_m); m, 17α(*H*),21β(*H*)-30-norhopane; n, 17β(*H*),21α(*H*)-30-normoretane; o, 17α(*H*),21β(*H*)-hopane; p, 17β(*H*),21α(*H*)-moretane; q, 17α(*H*),21β(*H*)-30-homohopane (C-22S and R); r, gammacerane; s, 17α(*H*),21β(*H*)-30,31-bishomohopane (C-22S and R); t, 17α(*H*),21β(*H*)-30,31,32-trishomohopane (C-22S and R); u, 17α(*H*),21β(*H*)-30,31,32,33-tetrakishomohopane (C-22S and R); v, 17α(*H*),21β(*H*)-pentakishomohopane (C-22S and R).

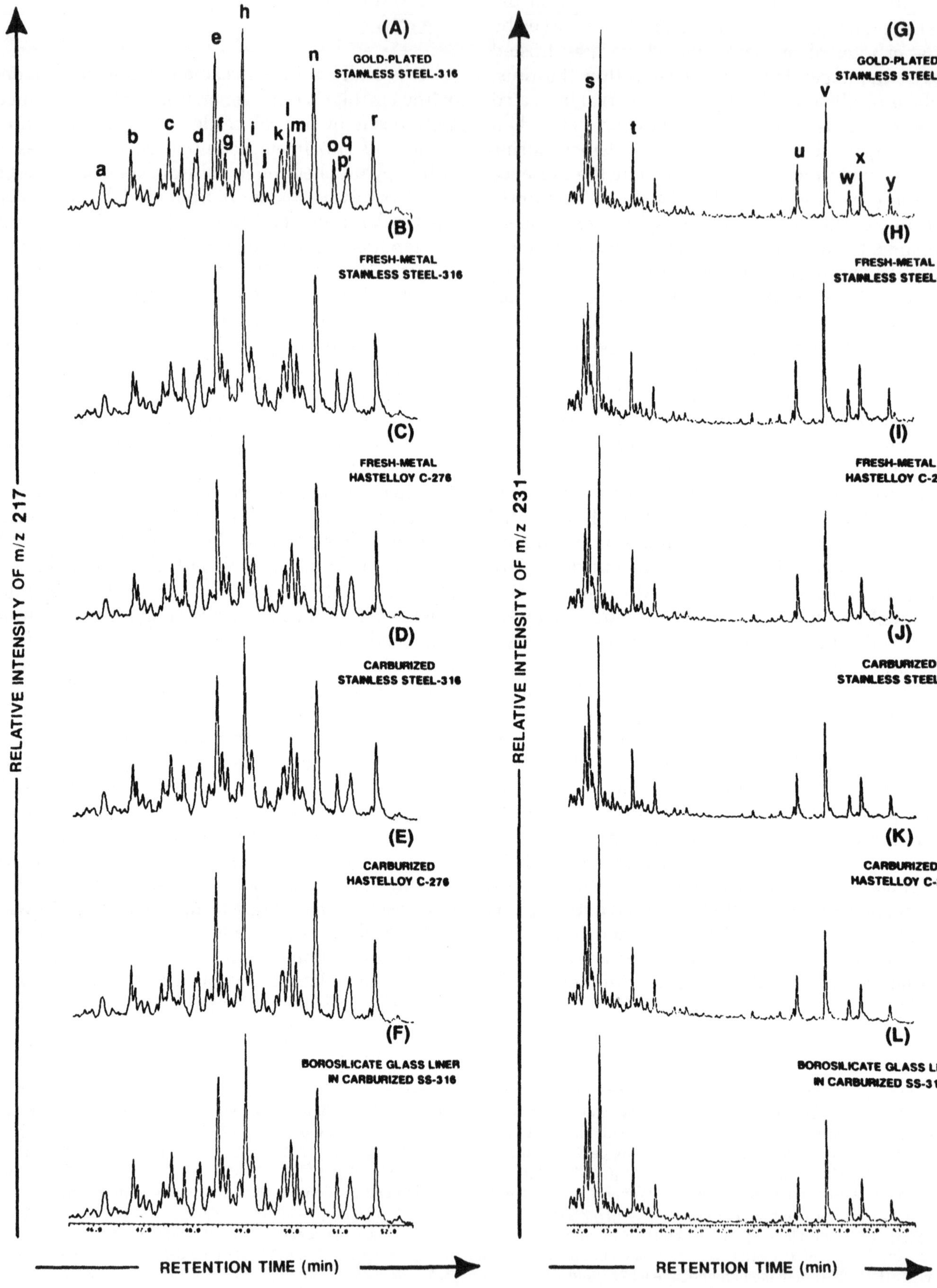
(A)
GOLD-PLATED STAINLESS STEEL-316
(B)
FRESH-METAL STAINLESS STEEL-316
(C)
FRESH-METAL HASTELLOY C-276
(D)
CARBURIZED STAINLESS STEEL-316
(E)
CARBURIZED HASTELLOY C-276
(F)
BOROSILICATE GLASS LINER IN CARBURIZED SS-316
(G)
GOLD-PLATED STAINLESS STEEL-316
(H)
FRESH-METAL STAINLESS STEEL-316
(I)
FRESH-METAL HASTELLOY C-276
(J)
CARBURIZED STAINLESS STEEL-316
(K)
CARBURIZED HASTELLOY C-276
(L)
BOROSILICATE GLASS LINER IN CARBURIZED SS-316
RELATIVE INTENSITY OF m/z 217
RELATIVE INTENSITY OF m/z 231
RETENTION TIME (min)
RETENTION TIME (min)

TABLE V. Amount and Composition of Gases Generated from 300-g Aliquots of a Sample of Monterey Shale (MR-216) Subjected to Hydrous Pyrolysis for 72 h at 330°C in 1-Liter Reactors with Different Wall Compositions

	Amount of gas (mg/300 g rock) collected from reactor[a]					
	Fresh metal walls			Carburized metal walls		
Generated gases	Gold-plated SS-316	Stainless steel-316	Hastelloy C-276	Stainless steel-316	Hastelloy C-276	Borosilicate glass liner in SS-316
Methane	172	179	180	194	189	174
Ethane	137	143	135	146	140	146
Propane	101	101	75	103	83	101
Isobutane	51	46	23	45	23	46
n-Butane	48	45	88	52	84	50
Pentanes	4	20	24	25	33	30
Hexanes	0	0	0	0	0	0
C_7+[b]	0	16	0	0	0	0
Ethene	1	2	2	2	3	1
Propene	3	4	1	1	0	1
Butenes	0	0	0	0	0	0
Butadienes	0	1	1	0	2	0
C_5H_{10}[c]	28	24	33	37	32	41
C_6H_{12}[c]	34	15	34	31	0	39
CO_2	4284	4489	4198	4668	4306	4573
CO	0	0	0	0	0	5
H_2	14	13	10	9	11	10
H_2S	577	504	649	706	604	644
Total	5454	5602	5453	6019	5510	5856

[a]Generated gases were collected from reactors at room temperature and analyzed on a mass spectrometer.
[b]Hydrocarbon gases with 7 or more carbon atoms.
[c]Analysis does not differentiate between cyclic alkanes and acyclic alkenes.

a trace of propene from the Green River Shale. As shown in Table VI, these trace amounts of alkenes are significantly less than those from isolated kerogens reported by Tannenbaum and Kaplan (1985b). In addition, the gas analyses for isolated kerogens show an increase in the ratio of alkenes to alkanes plus alkenes with increasing carbon number, which is not observed in the gases generated from rock samples. Although an explanation for this discrepancy requires additional experimentation, it is apparent that conclusions made by Tannenbaum and Kaplan (1985a) may not be extended to hydrous pyrolysis of rock samples.

The most apparent artifact of hydrous pyrolysis is the large polar fraction that is not recovered from solid–liquid column chromatography of the expelled oils (Lewan *et al.*, 1979). Figure 4 shows that the proportionality of the eluted saturate, aromatic, and polar fractions of expelled oils is similar to that for natural crude oils. However, Table VII shows that the fraction not recovered from activated alumina columns typically comprises 31 to 40 wt. % of these expelled oils. The majority of this unrecovered fraction consists of highly polar organics that remain on the column and, to a lesser extent, low-molecular-weight organics that are lost during solvent evapora-

←

Figure 10. Comparison of *m/z* 217 and 231 ion chromatograms of expelled oils generated from aliquots of a sample of Monterey Shale (MR-216) subjected to hydrous pyrolysis in reactors with different wall compositions at 330°C for 72 h. Ion chromatograms A through F show the steranes (*m/z* 217) in saturate column cuts of the expelled oils. Identifications of the steranes are as follows: a, 13β(*H*),17α(*H*)-diacholestane (20*S*); b, 13β(*H*),17α(*H*)-diacholestane (20*R*); c, 24-methyl-13β(*H*),17α(*H*)-diacholestane (20*S*); d, 24-methyl-13β(*H*),17α(*H*)-diacholestane (20*R*); e, 14α(*H*),17α(*H*)-cholestane (20*S*) + 17α(*H*)-diacholestane (20*S*); f, 14β(*H*),17β(*H*)-cholestane (20*R*) + 24-ethyl-13α(*H*),17α(*H*)-diacholestane (20*S*); g, 14β(*H*),17β(*H*)-cholestane (20*S*) + 24-methyl-13β(*H*),17β(*H*)-diacholestane (20*R*); h, 14α(*H*)17α(*H*)-cholestane (20*R*); i, 24-ethyl-13β(*H*),17α-diacholestane (20*R*); j, 24-ethyl-14α(*H*),17α(*H*)-diacholestane (20*S*); k, 24-methyl-14α,17α-cholestane (20*S*); l, 24-methyl-14β(*H*),17β(*H*)-cholestane (20*R*) + 24-ethyl-13α(*H*),17β(*H*)-diacholestane (20*R*); m, 24-methyl- 24β(*H*),-17β(*H*)-cholestane (20*S*); n, 24-methyl-14σ(*H*),17α(*H*)-cholestane (20*R*); o, 24-ethyl-14α(*H*),17α(*H*-cholestane (20*S*); p, 24-ethyl-14β(*H*),17β(*H*)-cholestane (20*R*); q, 24-ethyl-14β(*H*),17β(*H*)-cholestane (20*S*); r, 24-ethyl-14α(*H*),17α(*H*)-cholestane (20*R*). Ion chromatograms G through L show the triaromatic steroid hydrocarbons (*m/z* 231) in aromatic column cuts of the expelled oils. Identification of these steroids is as follows: s, C_{20}; t, C_{21}; u, C_{26} (20*S*); v, C_{26} (20*R*) + C_{27} (20*S*); w, C_{28} (20*S*); x, C_{27} (20*R*); y, C_{28} (20*R*).

TABLE VI. Comparison of Gases Generated by Hydrous Pyrolysis of Isolated Kerogens and Rock Samples from the Monterey Shale and Green River Shale[a]

	Monterey Shale		Green River Shale	
	Kerogen[b]	Rock[c]	Kerogen	Rock[d]
Generated gases	300 °C/100-h pyrolysis	300 °C/72-h pyrolysis	300 °C/100-h pyrolysis	300 °C/144-h pyrolysis
Methane	3.7	3.02	2.2	3.15
Ethane	2.2	2.08	0.80	1.08
Ethene	0.39	0.03	0.14	0.00
Propane	1.8	1.50	0.87	1.09
Propene	0.53	0.00	0.27	0.07
n-Butane	0.86	0.69	0.32	0.00
Isobutane	0.47	0.48	0.48	1.37
Butenes	0.93	0.00	0.50	0.00
Butadienes	—	0.00	—	0.00
Pentanes	0.87	0.66	0.59	0.00
C_5H_{10}[e]	0.49	0.42	0.56	0.94
Hexanes	0.86	0.00	1.02	0.00
C_6H_{12}[e]	0.80	0.11	0.61	0.00
Total	13.90	8.99	8.36	7.70
$\frac{\text{Ethene}}{\text{Ethane + ethene}}$	0.15	0.01	0.15	0.00
$\frac{\text{Propene}}{\text{Propane + propene}}$	0.23	0.00	0.24	0.06
$\frac{\text{Butenes}}{\text{Butanes + butenes}}$	0.41	0.00	0.39	0.00

[a]Data for isolated kerogens are from Tannenbaum and Kaplan (1985a, b); data for rock samples are from the author's laboratory. Concentrations are expressed as mg/g of kerogen for rock samples and isolated kerogens.
[b]This analysis is actually from an anhydrous experiment, but is described by Tannenbaum and Kaplan (1985a, b) as being comparable to the hydrous experiment under the same time–temperature conditions.
[c]Experiment involved heating 300 g of crushed (0.5–2.0 cm) Monterey Shale (MR-178) with 365 g of deionized water in a 1-liter carburized stainless steel-316 reactor. The reactor gas space was purged three times with 8–14 MPa of helium and then filled with 240 kPa of helium at the start of the experiment.
[d]Experiment involved heating 400 g of crushed (0.5–2.0 cm) Green River Shale (GR-49) with 325 g of deionized water in a 1-liter stainless steel-316 reactor. Procedure for purging and filling with helium is the same as described in footnote *c*.
[e]No distinction is made between cyclic alkanes and acyclic alkenes.

TABLE VII. Compositional Fractions[a] of Expelled Oils Generated from Aliquots of a Sample of Woodford Shale (WD-5) Subjected to Hydrous Pyrolysis under Different Temperature–Time Conditions as Described by Lewan (1983)

Stage of petroleum formation and experimental conditions	Saturate fraction (wt. %)	Aromatic fraction (wt. %)	Polar fraction (wt. %)	Unrecovered fraction (wt. %)
Incipient-oil stage				
330°C/72 h	29.8	17.6	12.2	40.4
Primary-oil stage				
340°C/72 h	31.6	19.0	14.0	35.4
345°C/72 h	32.1	20.2	16.2	31.5
350°C/72 h	30.9	20.9	16.1	32.1
Post-oil stage				
355°C/72 h	26.1	19.7	14.4	39.8
360°C/72 h	29.6	19.7	14.7	36.0
365°C/808 h	12.0	20.3	6.6	61.1

[a]Fractions were separated on a 40-cm alumina (MCB Alcoa F-20) column with eluting solvents being *n*-heptane for saturates, benzene for aromatics, and a 40:60 (w/w) methanol:benzene mixture for polars. Unrecovered fraction includes highly polar organics that remained on the column, and to a lesser extent, low-molecular weight components lost during evaporation of eluting solvents.

tion in the isolation of the eluted fractions. One explanation for this highly polar fraction is that it remains within the rock as bitumen and is not expelled with the oil at the lower temperatures under which the natural system operates. This is supported in part by the observation that polyaromatic hydrocarbons such as perylene and triaromatic steroids are preferentially concentrated in the bitumens of rocks subjected to hydrous pyrolysis versus the expelled oils (Lewan *et al.*, 1986). Another possible explanation is that the expelled oils from hydrous pyrolysis migrate only short distances in the rock samples without being subjected to secondary migration. The highly polar organics in the natural system may be chromatographically removed from crude oils during secondary migration. As discussed by Hunt (1979, pp. 498–513), it is not uncommon for the highly polar fraction of natural crude oils to decrease with increasing migration distance from their source rocks.

Although these explanations may account for the large polar fractions in expelled oils generated in the incipient- and primary-oil generation stages, it is doubtful that they account for the even larger polar fractions of expelled oils generated in the post-oil generation stage (Table VII). Formation of these polars at the higher experimental temperatures may be attributed to aromatization of saturated cyclic hydrocarbons and their condensation with preexisting aromatic hydrocarbons to form polycyclic aromatic compounds. Compared with the overall thermal decomposition of bitumen to oil, these aromatization and condensation reactions appear to be more temperature dependent. This results in larger aromatic and polar fractions in expelled oils generated by hydrous pyrolysis than in crude oils generated by the natural system. Although this degree of aromatization and condensation is not likely during natural oil generation, it is likely in high-thermal-stress regimes where reservoired crude oils develop into pyrobitumens.

8. Conclusions

Hydrous pyrolysis is a laboratory technique used to simulate petroleum formation. The technique maintains a liquid water phase in contact with potential source rocks while they are heated at subcritical water temperatures. If the proper experimental time and temperature conditions are employed, a free-flowing liquid oil similar in composition to natural crude oil is generated and expelled from the rock. The expelled oil accumulates on the surface of the water, where it may be quantitatively collected at the end of an experiment. Petrographic studies on naturally and experimentally matured source rocks indicate that a net volume increase of the organic matter within the confining rock matrix is the driving force for primary migration of bitumen and expulsion of oil. The ability of hydrous pyrolysis to generate an oil compositionally similar to natural crude oil and to expel it in a manner similar to that operative in naturally maturing source rocks gives it a distinct advantage over anhydrous pyrolysis experiments. Depending on whether an open or closed system is employed, anhydrous pyrolysis experiments either generate products compositionally different from natural crude oil or obtain a pyrolysis product by processes that are not operative in the natural system (e.g., vacuum cryogenic trapping, carrier gas flushing, and organic solvent refluxing).

Simulating petroleum formation by hydrous pyrolysis has shown that four stages may be defined: (1) pre-oil generation, (2) incipient-oil generation, (3) primary-oil generation, and (4) post-oil generation. Pre-oil generation represents a thermally immature stage in which the organic matter occurs predominantly as a solid insoluble kerogen dispersed within a rock matrix. Incipient-oil generation begins when thermal stress is sufficient to initiate the decomposition of kerogen into a tarry-soluble bitumen, which is enriched in high-molecular-weight hydrocarbons and heteroatom compounds. The net volume increase generated by this overall reaction results in bitumen expansion along the planar bedding fabric of the rock matrix to form a continuous bitumen network. As thermal stress increases, primary-oil generation begins, with the bitumen network partially decomposing into a liquid oil, which is enriched in saturated hydrocarbons. The net volume increase generated by this overall reaction results in expulsion of the generated oil from the bitumen-impregnated rock. This suggests that oil expulsion from a rock is a consequence of its generation within a rock, and, therefore, generation and expulsion of oil may be considered collectively as one overall process. Post-oil generation begins when oil generation ceases. As thermal stress continues to increase, the bitumen generates gas and condenses into an insoluble pyrobitumen.

Hydrous pyrolysis of rock samples in the form of chips (0.5 to 0.2 cm) or cores provides a better simulation of natural petroleum generation than that of isolated kerogens or unconsolidated mixtures of kerogen and mineral powders. Organic matter in a rock sample is embedded naturally within its inorganic matrix. This condition more accurately simulates the natural system than the use of kerogens, which are isolated from their host rock with HCl, HF, and $LiAlH_4$. Although the use of the term "excess" water in describing hydrous pyrolysis of isolated kerogens or their mixture with unconsolidated mineral powder is appropriate, it is not appropriate in describing hydrous pyrolysis of rock samples. The pore system of source rocks during thermal maturation in the natural system and in hydrous pyrolysis experiments changes from an aqueous system to a bituminous system. Once the bituminous system has developed within a rock, only the water surrounding a rock in joints, fractures, or faults is important in facilitating oil expulsion. As shown in anhydrous pyrolysis experiments with and without the confinement of generated gas, the lack of liquid water prevents oil expulsion from a rock and significantly reduces the total pyrolysate yield (i.e., gas + oil + bitumen). Although specifics on the role of water remain to be elucidated, the presence of water appears to be an important factor in simulating petroleum formation in the laboratory.

ACKNOWLEDGMENTS. The author thanks the management of Amoco Production Company for support of this research and permission to publish. John C. Winters (Amoco Production Co.) initiated the experimental concept of hydrous pyrolysis in 1977, which was fostered by his earlier association with research on upgrading heavy oils by Dr. John D. McCollum (Amoco Research). The author is indebted to John C. Winters for introducing him to this type of experimentation in 1977 and for supporting the author's research and development of hydrous pyrolysis in the subsequent years. Drs. Attila Kilinc and John Grover (University of Cincinnati) are gratefully acknowledged for their lectures and discussions, which instilled in the author an interest in hydrothermal experimentation and a realization of the importance of water in natural processes. Drs. Barry Maynard and Wayne A. Pryor (University of Cincinnati) are acknowledged for indoctrinating the author into the tenet that one should look at the rocks one studies.

References

Amestica, L. A., and Wolf, E. E., 1986, Catalytic liquefaction of coal with supercritical water/CO/solvent media, *Fuel* **65**:1226.

Appell, H. R., Fu, Y. C., Friedman, S., Yavorsky, P. M., and Wender, I., 1971, Converting Organic Wastes to Oil: A Replenishable Energy Source, U.S. Department of the Interior, Bureau of Mines Rept. Invest. 7560.

Appell, H. R., Fu, Y. C., Illig, E. G., Steffgen, F. W., and Miller, R. D., 1975, Conversion of Cellulosic Wastes to Oil, U.S. Department of the Interior, Bureau of Mines Rept. Invest. 8013.

ASME, 1979, *ASME Steam Tables—Thermodynamic and Transport Properties of Steam*, American Society of Mechanical Engineers, New York.

Berl, E., and Schmidt, A., 1932, Über die Entstehung der Kohlen, II. Die Inkohlung von Cellulose and Lignin in neutralem Medium, *Ann. Chemie* **493**:97.

Bertrand, P., Martinez, L., and Pracher, B., 1987, Petrology Study of Primary Migration by Hydrous Pyrolysis, in: *Migration of Hydrocarbons in Sedimentary Basins* (B. Doligez, ed.), Editions Technip, Paris, pp. 633–647.

Brooks, J. D., and Smith, J. W., 1969, The diagenesis of plant lipids during the formation of coal, petroleum, and natural gas—II. Coalification and the formation of oil and gas in the Gippsland Basin, *Geochim. Cosmochim. Acta* **33**:1183.

Braun, R. L., and Rothman, A. J., 1975, Oil-shale pyrolysis: Kinetics and mechanisms of oil production, *Fuel* **54**:129.

Burnham, A. K., Braum, R. L., Gregg, H. R., and Samoun, A. M., 1987a, Comparison of methods for measuring kerogen pyrolysis rates and fitting kinetic parameters, *Energy Fuels* **1**:452.

Burnham, A. K., Braum, R. L., and Samoun, A., 1987b, Further Comparison of Methods for Measuring Kerogen Pyrolysis Rates and Fitting Kinetic Parameters, Lawrence Livermore National Laboratory, UCRL-97352.

Claypool, G. E., and Reed, R. P., 1976, Thermal-analysis technique for source-rock evaluation: Quantitative estimate of organic richness and effects of lithology, *Am. Assoc. Petrol. Geol. Bull.* **60**:608.

Comet, P. A., McEvoy, J., Giger, W., and Douglas, A. G., 1986, Hydrous and anhydrous pyrolysis of DSDP Leg 75 kerogens—A comparative study using a biological marker approach, *Org. Geochem.* **9**:171.

Dawidowicz, A. L., Nazimek, D., Pikus, S., and Skubiszewska, J., 1984, The influence of boron atoms on the surface of controlled porous glasses on the properties of the carbon deposit obtained by pyrolysis of alcohol, *J. Anal. Appl. Pyrolysis* **7**:53.

Dawidowicz, A. L., Pikus, S., and Nazimek, D., 1986, Properties of the material surfaces obtained by pyrolysis of alkanols on boron-enriched controlled porous glasses, *J. Anal. Appl. Pyrolysis* **10**:59.

Deshpande, G. V., Holder, G. D., Bishop, A. A., Gopal, J., and Wender, I., 1984, Extraction of coal using supercritical water, *Fuel* **63**:956.

Eglinton, T. I., and Douglas, A. G., 1988, Quantitative study of biomarker hydrocarbons released from kerogens during hydrous pyrolysis, *Energy Fuels* **2**:81.

Eglinton, T. I., Curtis, C. D., and Rowland, S. J., 1987, Generation of water-soluble organic acids from kerogen during hydrous pyrolysis: Implications for porosity development, *Miner. Mag.* **51**:495.

Eglinton, T. I., Douglas, A. G., and Rowland, S. J., 1988, Release of aliphatic, aromatic, and sulfur compounds from Kimmeridge kerogen by hydrous pyrolysis: A quantitative study, *Org. Geochem.* **13**:655.

Engler, K. O. V., 1913, *Die Chemie und Physik des Erdöls*, Vol. 1, S. Hirzel, Leipzig.

Espitalié, J., Laporte, J. L., Madec, M., Marquis, F., Leplat, P., Paulet, J., and Boutefeu, A., 1977, Méthode rapide de caractérisation des roches mères de leur potential pétrolier et de leur degré d'évolution, *Rev. Inst. Fr. Pet.* **32**:23.

Ferry, J. M., 1983, Regional metamorphism of the Vassalboro Formation, south-central Maine, U.S.A.: A case study of the role of fluid in metamorphic petrogenesis, *J. Geol. Soc. London* **140:**551.

Fischer, F., 1925, *The Conversion of Coal into Oils*, Ernest Benn Ltd., London.

Franks, A. J., and Goodier, B. D., 1922, Preliminary study of the organic matter of Colorado Oil Shales, *Quart. Colo. Sch. Mines* **17:**3.

Gavin, M. J., 1922, Oil-shale: An historical, technical, and economic study, U.S. Department of the Interior, Bureau of Mines Bulletin 210, Bradford-Robinson, Denver.

Goranson, R. W., 1931, The solubility of water in granite magmas, *Am. J. Sci.* **22:**481.

Goranson, R. W., 1932, Some notes on the melting of granite, *Am. J. Sci.* **23:**227.

Haas, J. L., Jr., 1976, Thermodynamical properties of the NaCl component in boiling NaCl solutions, U.S. Geological Survey Bulletin 1421-B.

Harwood, R. J., 1977, Oil and gas generation by laboratory pyrolysis of kerogen, *Am. Assoc. Petrol. Geol. Bull.* **61:**2082.

Hoering, T. C., 1977, Olefinic hydrocarbons in Bradford, Pennsylvania, crude oil, *Chem. Geol.* **20:**1.

Hoering, T. C., 1985, Thermal reactions of kerogen with added water, heavy water, and pure organic substances, *Org. Geochem.* **5:**267.

Houser, T. J., Tiffany, D. M., Li, Z., McCarville, M. E., and Houghton, M. E., 1986, Reactivity of some organic compounds with supercritical water, *Fuel* **65:**827.

Hubbard, A. B., and Robinson, W. E., 1950, A Thermal Decomposition Study of Colorado Oil Shale, U.S. Department of the Interior, Bureau of Mines Rept. Invest. 4744.

Huizinga, B. J., Tannenbaum, E., and Kaplan, I. R., 1987, The role of minerals in the thermal alteration of organic matter—IV. Generation of *n*-alkanes, acyclic isoprenoids, and alkenes in laboratory experiments, *Geochim. Cosmochim. Acta* **51:**1083.

Hunt, J. M., 1979, *Petroleum Geochemistry and Geology*, W. H. Freeman and Co., San Francisco.

Jones, M., Douglas, A. G., and Connan, J., 1987, Hydrocarbon distributions in crude oil asphaltene pyrolyzates. 1. Aliphatic compounds, *Energy Fuels* **1:**468.

Jurg, J. W., and Eisma, E., 1964, Petroleum hydrocarbons: Generation from fatty acid, *Science* **144:**1451.

King, R. J., 1983, Steel, in: *Kirk-Othmer Encyclopedia of Chemical Technology*, Vol. 21, John Wiley & Sons, New York, pp. 552–625.

Kisch, H. J., 1987, Correlation between indicators of very low-grade metamorphism, in: *Low Temperature Metamorphism* (M. Frey, ed.), Blackie, Glasgow, pp. 227–300.

Larter, S. R., and Douglas, A. G., 1980, A pyrolysis–gas chromatographic method for kerogen typing, in: *Advances in Organic Geochemistry 1979* (A. G. Douglas and J. R. Maxwell, eds.), Pergamon Press, Oxford, pp. 579–584.

Larter, S. R., and Senftle, J. T., 1985, Improved kerogen typing for petroleum source rock analysis, *Nature* **318:**277.

Larter, S. R., Horsfield, B., and Douglas, A. G., 1977, Pyrolysis as a possible means of determining petroleum generating potential of sedimentary organic matter, in: *Analytical Pyrolysis* (C. E. R. Jones and C. A. Cramers, eds.), Elsevier, Amsterdam, pp. 189–202.

Lewan, M. D., 1978, Laboratory classification of very fine-grained sedimentary rocks, *Geology* **6:**745.

Lewan, M. D., 1983, Effects of thermal maturation on stable organic carbon isotopes as determined by hydrous pyrolysis of Woodford Shale, *Geochim. Cosmochim. Acta* **47:**1471.

Lewan, M. D., 1985, Evaluation of petroleum generation by hydrous pyrolysis experimentation, *Philos. Trans. R. Soc. London Ser. A* **315:**123.

Lewan, M. D., 1987, Petrographic study of primary petroleum migration in the Woodford Shale and related rock units, in: *Migration of Hydrocarbons in Sedimentary Basins* (B. Doligez, ed.), Editions Technip, Paris, pp. 113–130.

Lewan, M. D., and Buchardt, B., 1989, Irradiation of organic matter by uranium decay in the Alum Shale, Sweden, *Geochim. Cosmochim. Acta* **53:**1307.

Lewan, M. D., and Williams, M. D., 1987, Evaluation of petroleum generation from resinites by hydrous pyrolysis, *Am. Assoc. Petrol. Geol. Bull.* **71:**207.

Lewan, M. D., Winters, J. C., and McDonald, J. H., 1979, Generation of oil-like pyrolyzates from organic-rich shales, *Science* **203:**897.

Lewan, M. D., Bjorøy, M., and Dolcater, D. L., 1986, Effects of thermal maturation on steroid hydrocarbons as determined by hydrous pyrolysis of Phosphoria Retort Shale, *Geochim. Cosmochim. Acta* **50:**1977.

Louis, M. C., and Tissot, B. P., 1967, Influence de la température et de la pression sur la formation des hydrocarbures dans les argiles à kérogen, *Proceedings of the 7th World Petroleum Congress*, Vol. 2, Elsevier, Amsterdam, p. 47.

Lundegard, P. D., and Senftle, J. T., 1987, Hydrous pyrolysis: A tool for the study of organic acid synthesis, *Appl. Geochem.* **2:**605.

McCollum, J. D., and Quick, L. M., 1976a, Process for upgrading a hydrocarbon fraction, U.S. Patent 3,960,708.

McCollum, J. D., and Quick, L. M., 1976b, Process for upgrading a hydrocarbon fraction, U.S. Patent 3,989,618.

McKee, R. H., and Lyder, E. E., 1921, The thermal decomposition of shales. I—Heat effects, *J. Ind. Eng. Chem.* **13:**613.

Meissner, F. F., 1978, Petroleum geology of the Bakken Formation, Williston Basin, in: *Williston Basin Symposium*, Montana Geological Society 24th Annual Conference, pp. 207–227.

Monthioux, M., 1988, Expected mechanisms in nature and in confined-system pyrolysis, *Fuel* **67:**843.

Monthioux, M., Landais, P., and Monin, J.-C., 1985, Comparison between natural and artificial maturation series of humic coals from the Mahakam delta, Indonesia, *Org. Geochem.* **8:**275.

Murakami, K., Yokono, T., and Sanada, Y., 1986, An investigation of the role of hydrogen sulfide in coal liquefaction catalysis by high-temperature and high-pressure e.s.r., *Fuel* **65:**1079.

Palmer, D. A., and Drummond, S. E., 1986, Thermal decarboxylation of acetate. Part I. The kinetics and mechanism of reaction in aqueous solution, *Geochim. Cosmochim. Acta* **50:**813.

Pollard, D. D., and Aydin, A., 1988, Progress in understanding jointing over the past century, *Geol. Soc. Am. Bull.* **100:**1181.

Rullkötter, J., Aizenshtat, Z., and Spiro, B., 1984, Biological markers in bitumens and pyrolzates of Upper Cretaceous bituminous chalks from the Ghareb Formation (Israel), *Geochim. Cosmochim. Acta* **48:**151.

Rumble, D., III, Ferry, J. M., Hoering, T. C., and Boucot, A. J., 1982, Fluid flow during metamorphism at the Beaver Brook fossil locality, New Hampshire, *Am. J. Sci.* **282:**886.

Saxby, J. D., and Riley, K. W., 1984, Petroleum generation by laboratory scale pyrolysis over six years simulating conditions in a subsiding basin, *Nature* **308:**177.

Saxby, J. D., Bennett, A. J. R.,Corcoran, J. F., Lambert, D. E., and Riley, K. W., 1986, Petroleum generation: Simulation over six years of hydrocarbon formation from torbanite and brown coal in a subsiding basin, *Org. Geochem.* **9:**69.

Schuhmacher, J. P., Huntjens, F. J., and van Krevelen, D. W., 1960,

Chemical structure and properties of coal XXVI. Studies on artificial coalification, *Fuel* **39:**223.

Soldan, A. L., and Cerqueira, J. R., 1986, Effects of thermal maturation on geochemical parameters obtained by simulated generation of hydrocarbons, *Org. Geochem.* **10:**339.

Sourirajan, S., and Kennedy, G. C., 1962, The system H_2O–NaCl at elevated temperatures and pressures, *Am. J. Sci.* **260:**115.

Srinivasan, G., and Seehra, M. S., 1982, Changes in free radicals in coal-derived pyrites upon heating in N_2, H_2, and vacuum: Role of pyrite–pyrrhotite conversion, *Fuel* **61:**1249.

Srinivasan, G., and Seehra, M. S., 1983, Effects of pyrite and pyrrhotite on free radical formation in coal, *Fuel* **62:**792.

Stanley, J. K., 1970, The carburization of four austenitic stainless steels, *J. Matter* **5:**957.

Takenouchi, S., and Kennedy, G. C., 1964, The binary system H_2O–CO_2 at high temperatures and pressures, *Am. J. Sci.* **262:**1055,

Talukdar, S., Gallango, O., Vallejas, C., and Ruggiero, A., 1987, Observations on the primary migration of oil in the LaLuna Source rocks of the Marracaibo Basin, Venezuela, in: *Migration of Hydrocarbons in Sedimentary Basis* (B. Doligez, ed.), Editions Technip, Paris, pp. 59–77.

Tannenbaum, E., and Kaplan, I. R., 1985a, Role of minerals in the thermal alteration of organic matter—I: Generation of gases and condensates under dry conditions, *Geochim. Cosmochim. Acta* **49:**2589.

Tannenbaum, E., and Kaplan, I. R., 1985b, Low-M_r hydrocarbons generated during hydrous and dry pyrolysis of Kerogen, *Nature* **317:**708

Tannenbaum, E., Huizinga, B. J., and Kaplan, I. R., 1986 Role of minerals in thermal alteration of organic matter—II: A material balance, *Am. Assoc. Petrol. Geol. Bull.* **70:**1156.

Tissot, B. P., and Welte, D. H., 1978, *Petroleum Formation and Occurrence*, Springer-Verlag, Berlin.

Tissot, B., Durand, B., Espitalié, J., and Combaz, A., 1974, Influence of nature and diagenesis of organic matter in formation of petroleum, *Am. Assoc. Petrol. Geol. Bull.* **58:**499.

Tuttle, O. F., and Bowen, N. L., 1958, Origin of granite in the light of experimental studies in the system NaAl Si_3O_8–$KAlSi_3O_8$–SiO_2–H_2O, *Geol. Soc. Am. Mem.* **74.**

Winkler, H. G. F., 1974, *Petrogenesis of Metamorphic Rocks*, 3rd ed., Springer-Verlag, New York.

Winters, J. C., Williams, J. A., and Lewan, M. D., 1983, A laboratory study of petroleum generation by hydrous pyrolysis, in: *Advances in Organic Geochemistry 1981* (M. Bjorøy, ed.), John Wiley & Sons, New York, pp. 524–533.

V

Applications of Organic Geochemical Research for Hydrocarbon Exploration

Chapter 19

Oil–Oil and Oil–Source Rock Correlations: Techniques

R. P. PHILP

1. Introduction

It is commonly accepted, with few exceptions, that crude oils have a biogenic origin and are formed as a result of the diagenetic and thermal conversion of organic matter deposited and preserved in various types of aquatic sedimentary environments (Hunt, 1979). A large amount of organic matter from living systems is incorporated into macromolecular structures (kerogen) prior to thermal degradation and formation of the mobile liquid products (Rullkötter and Michaelis, 1990). The organic matter responsible for the kerogen formation is predominantly derived from algal, microbial, and higher plant sources. The amount of the organic matter actually preserved, converted into kerogen, and ultimately available for conversion into fossil fuels will depend on a variety of conditions including various chemical and microbial aspects of the depositional environments.

Geochemical correlations are based on the ability to recognize distinct physical and chemical similarities, or differences, between the hydrocarbons or oil in a reservoir and the extractable bitumen plus the residium in the original source rock. A lack of correlation is equally important in that it permits one to say that samples are not genetically related, and hence the search for related oils or their source rocks should continue in other directions. The ability to establish oil–oil, oil–source, gas–gas, gas–oil, and gas–source relationships is based on the assumption that the processes acting on the separated hydrocarbons such as migration, thermal maturation, biodegradation, and water washing do not adversely affect the parameters necessary for correlation. Furthermore it is assumed that since the organic matter in the source rock is responsible for producing the oil, the "fingerprints" of organic compounds in both the oil and source rock extract will be very similar. In addition to establishing correlations between genetically related pools of oil, or respective source rocks, a similar approach can be used to prove communication between fault blocks and different producing horizons.

Correlation techniques based on geochemical properties can be divided into two main groups depending on whether they measure (1) *bulk parameters*, which describe properties of the whole sample, either whole oil or total extract; or (2) *specific properties*, generally based on molecular properties, which describe detailed chemical characteristics of specific fractions.

Analytical techniques available for geochemical correlation purposes over the past two decades have increased dramatically both in number and in level of sophistication. Examination of the early literature will show that correlation techniques commonly used 20 or 30 years ago were very crude by current standards and certainly lacked the specificity available today. For example, API gravity has been, for a

R. P. PHILP • School of Geology and Geophysics, The University of Oklahoma, Norman, Oklahoma 73019.
Organic Geochemistry, edited by Michael H. Engel and Stephen A. Macko. Plenum Press, New York, 1993.

long time, a popular technique for correlation of oils. However, oils from many different regions of the world have similar API gravities and are obviously not related, limiting the value of this parameter for detailed correlations. Some early workers also used the color of an oil as a possible means of correlation, but many factors can affect the color of a crude oil, including water washing, thermal alteration, and biodegradation. Despite the lack of specificity associated with techniques such as color and API gravity for correlation purposes, an even greater drawback is the fact that it is impossible to undertake oil–source rock correlations based on these bulk property parameters. The added specificity and sensitivity available from the techniques that will be described below have played a major role in leading to the enhanced importance of geochemistry in correlation studies.

This chapter is concerned primarily with reviewing correlation techniques, and therefore it is not proposed to spend a great deal of space discussing related topics such as evaluation of source materials, levels of thermal maturity, extent of biodegradation, relative migration distances, or nature of depositional environments. Many aspects of these applications are covered in other chapters of this volume (Larter and Horsfield, this volume, Chapter 13; Zumberge, this volume, Chapter 20). There are also many papers which review the literature describing geochemical correlation techniques and discuss specific examples of these applications in the form of case histories. The best collections of such papers for the reader not familiar with the topic but wanting a comprehensive introduction can be found in the series of books entitled *Advances in Organic Geochemistry*. These are the proceedings of the International Meeting of Organic Geochemistry held biennially in Europe and provide a good compilation of up-to-date references and many case histories. There are, of course, the more specialized journals for the interested reader such as *Geochimica et Cosmochimica Acta*, *Chemical Geology*, *Organic Geochemistry*, *Applied Geochemistry*, and the *American Association of Petroleum Geologists Bulletin*.

Why has geochemistry become such an important and integral part of petroleum exploration studies? There are many reasons for this which will be discussed throughout this chapter, but a great deal of the credit must be given to the pioneering work of Wolf Seifert and his group. Seifert, who worked with Chevron until his untimely death in the early 1980s, was the person in the industry who used the ideas developed in the laboratories of Eglinton, Maxwell, Ourisson, and Albrecht and applied them to industrial problems. However, he took this one step further, despite working in an industrial laboratory, and was successful in publishing many of his results (e.g., Seifert and Moldowan, 1978). The rest as they say is history since once one of the major oil companies publishes some new ideas, the other major companies will not be far behind in applying the same approach. Another major factor in the use of geochemical techniques has been the rapid development of a wide variety of sophisticated and sensitive analytical techniques capable of analyzing the very complex mixtures of organic compounds present in oils and source rocks extracts, often in relatively low concentrations. A number of these techniques and recent developments will be described in the following sections.

2. Gas Chromatography and Mass Spectrometry

The trend toward the use of sophisticated and sensitive instrumentation in petroleum geochemistry became apparent in the early 1960s with the development of commercially available gas chromatographs and the appearance of several articles in the literature illustrating the possibility of obtaining *n*-alkane distributions from crude oils using this approach (e.g., Schenck and Eisma, 1964). Compared with data obtained today, these chromatograms were relatively crude, with the low resolution of the columns available at that time producing very broad chromatographic peaks. In the past 20 years, there have been many advances in gas chromatography, particularly in the development of different types of columns and column coatings and in the extension of the temperature ranges at which these columns can be operated. Wide-bore packed columns are rarely, if ever used in geochemical correlation studies today. Instead, there has been a steady transition to the use of capillary columns, which were initially made of stainless steel and then glass, culminating in the development, and widespread use, of the fused-silica capillary columns. More recently, fused-silica columns have been coated with aluminum to provide greater stability at high temperatures so that it is now possible to operate them at oven temperatures up to 500°C, permitting analysis of relatively high molecular weight components, up to at least C_{100}. The flexibility and durability of the fused-silica columns has made them extremely easy to use and has eliminated many of the mechanical problems associated with the less sturdy glass capillary columns. Surprisingly enough, it has been observed recently that there are a number of stability problems associated with the aluminum-coated columns. The latest development in terms of high-temperature gas chromatography (HTGC) has involved development of deactivated stainless steel columns that can be used up to 480°C. The use of HTGC in this type of work will be discussed below.

At the same time as developments were being made in the technology associated with gas chromatography (GC), the combination of gas chromatography and mass spectrometry (GC/MS) was rapidly developing to take advantage of the higher resolution capillary columns available. The availability of mass spectrometers with faster pumping capacities started to eliminate the need for any type of separator between the gas chromatograph and the mass spectrometer to reduce the relative concentration of the carrier gas. Faster analyzer scanning speeds made it possible to obtain data even on the relatively small components present in complex mixtures. The recognition in the early 1970s by Hites and Biemann (1970) that mass spectrometers could be used in the single-ion monitoring mode (SIM) or the multiple-ion detection mode (MID) certainly provided a major impetus for the development of methods applicable to determination of trace components in complex mixtures.

Changes and developments in the design of mass spectrometers have also proliferated. Whereas most of the mass spectrometers in use two decades ago were slow-scanning magnetic sector instruments, of either high- or low-resolution capability, the emergence of commercially available quadrupole instruments heralded a major development in available analytical capabilities. The faster scanning speeds, although with lower resolution, available with the quadrupoles as compared to the magnetic sector instruments of the time made them more acceptable for the detection of the components rapidly eluting from the higher resolution capillary columns. As time passed, laminated magnets were developed which enabled magnetic sector instruments to scan almost as rapidly as quadrupoles, eliminating the earlier advantage held by the quadrupoles for this type of application.

More recently, we have seen the development of a wide variety of hybrid instruments combining quadrupoles and magnetic and electric sector analyzers, all of which can operate in the so-called MS/MS mode. One of the main uses of these hybrid instruments, particularly for geochemical purposes, is to provide an additional degree of separation, spectrometrically rather than chromatographically, and to provide information on the structure of unknown components through collision-activated decomposition (CAD) reactions of specific fragment ions. The addition of the GC to the MS/MS system further enhances the analytical power of the system. It would appear, from papers published in the recent literature, that many of the major geochemical laboratories are now routinely utilizing triple-stage quadrupole instruments or other hybrid systems, for the analyses of geochemical samples, namely, oils and source rock extracts (Moldowan *et al.*, 1983, 1985; Philp and Oung, 1988; Summons *et al.*, 1988).

Since one of the many roles of this book is obviously to serve as a reference work, it is appropriate in this chapter to provide a few examples of the types of data produced by GC, GC/MS, and GC/MS/MS. Although there is a trend toward the analysis of whole oils, many laboratories still prefer to fractionate their samples, oils and rock extracts, into saturates, aromatics, and polar fractions prior to GC, GC/MS and GC/MS/MS analyses—as well as isolating the asphaltenes by precipitation with pentane. For GC alone, analysis of whole oils is not a major problem since many oils are dominated by saturated hydrocarbons unless weathering, biodegradation, water washing, or thermal alteration has reduced the concentrations of the *n*-alkanes. A comparison between the chromatograms of a whole oil and the corresponding saturate fraction isolated from the same oil is shown in Figs. 1a and b, and the similarities between these chromatograms are very clear. One advantage of fractionating the oils is that chromatograms can also be obtained for the aromatic fractions of the oils, as shown in Fig. 1c.

Gas chromatography is important for the preliminary screening of oils and extracts, but in many cases it is impossible to unambiguously identify components from this type of chromatogram alone. For the saturate hydrocarbon fractions, the problem is somewhat less complex since it is possible to obtain carbon numbers for the major components by comparison of relative retention times and knowing that the major isoprenoids, pristane and phytane, elute with very similar retention times to those of the C_{17} and C_{18} *n*-alkanes (Fig. 1b). The use of chromatograms such as these for correlation purposes is fairly limited in view of the similarities in many samples worldwide. Saturate hydrocarbon fractions contain a great deal of information not clearly visible in the chromatograms but "hidden" in the baseline, and it is these trace components that are often the most useful for correlation purposes (Mackenzie, 1984; Philp, 1985).

At the next higher level of analytical sophistication for obtaining information on the identity and distribution of these trace components, it is necessary

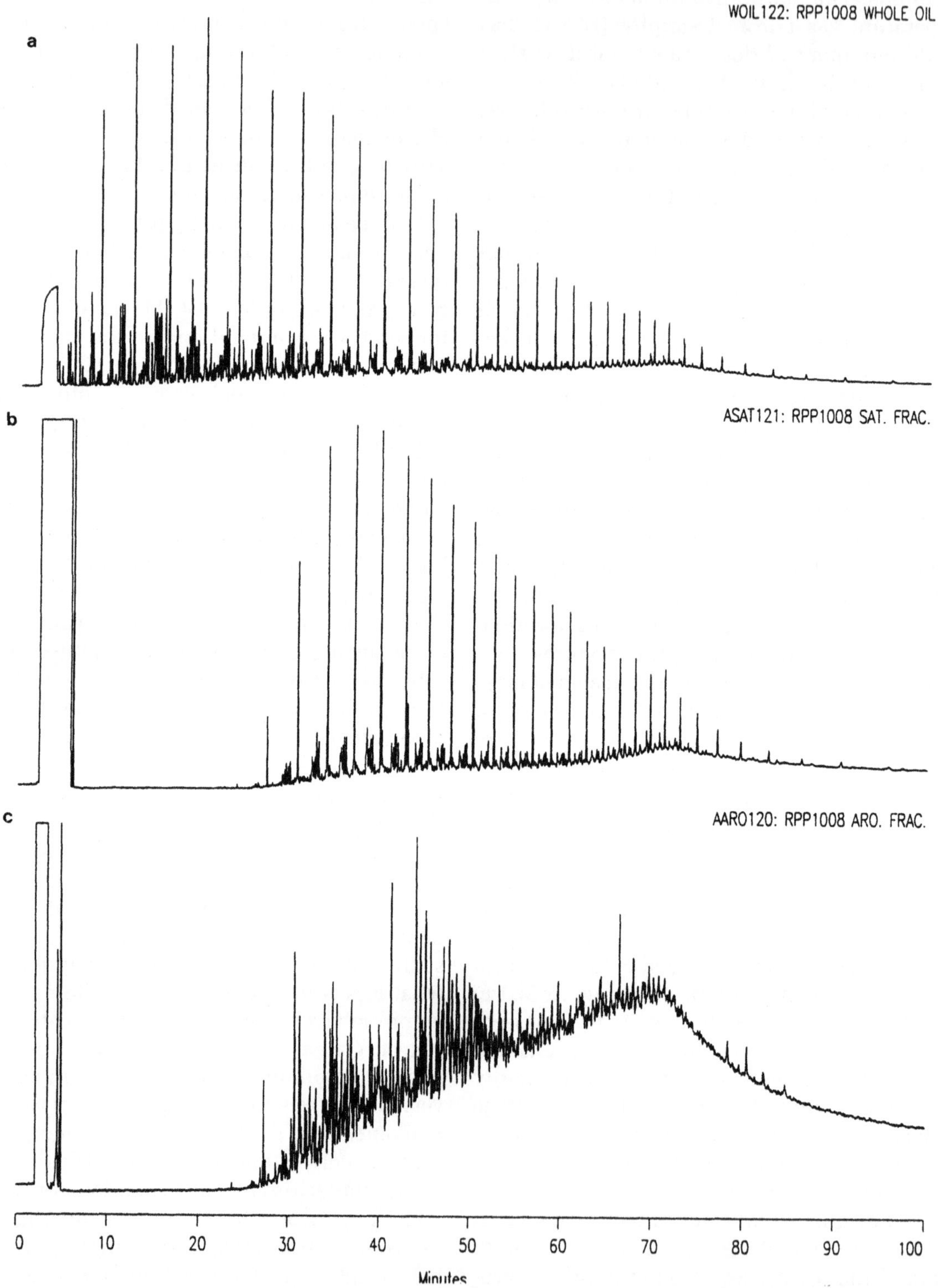

Figure 1. Examples of the gas chromatograms obtained from a whole oil (a), the saturate hydrocarbon fraction from the same oil (b), and the aromatic fraction from the same oil (c). All of these chromatograms were obtained under the same conditions.

to undertake analysis of the samples by GC/MS and related ancillary techniques such as MID and SIM. GC/MS has been utilized for many years for the characterization of all types of fossil fuels and is the first choice of analytical techniques for many organic geochemistry laboratories. There are large amounts of GC/MS data from fossil fuel analyses recorded in the literature, and many reviews which summarize these data have been published (Mackenzie, 1984; Philp, 1985; Burlingame *et al.*, 1988). Bench-top mass spectrometers are available which are capable of providing results similar to those obtained with more expensive, low-resolution mass spectrometers and are extremely useful for screening or routine analysis of fossil fuel samples (Philp *et al.*, 1989). Many of the GC/MS reports in the literature are concerned with operation of the MS in SIM or MID mode, to obtain fingerprints of specific classes of biomarkers. For most applications in fossil fuel analyses based on GC/MS, ionization of the molecules is typically performed in electron-ionization (EI) mode at an electron energy of −70 eV. Alternative methods of ionization, such as chemical ionization (CI), have been used to enhance the intensity of the molecular-ion peaks and to assist in the identification of unknown components (Buchanan, 1982; Dzidic *et al.*, 1988; Philp and Johnston, 1988).

For the majority of correlation studies, a fairly standard set of ions can be monitored when analyzing either whole oils or saturated hydrocarbon fractions for geochemical purposes. Ions commonly monitored during the routine analyses of oils and source rock extracts are summarized in Table I, along with an indication of the compound classes that these ions are used to detect. Molecular ions of many compound classes may also be included as an additional parameter for tentatively confirming the identity of the compounds.

TABLE I. Ions Monitored in Routine GC/MS Analysis of Oils and Source Rock Extracts

Ion (m/z)	Compound class
95, 98	Dinosteranes/methylsteranes
99	Saturated hydrocarbons
123	Bicyclic sesquiterpanes
177	Demethylated terpanes
183	Isoprenoids
191	Terpanes
205	Methylated terpanes
217	Steranes (αα-stereochemistry)
218	Steranes (ββ-stereochemistry)
231	Methylsteranes or triaromatic steroid H/C
253	Monoaromatic steroid H/C
259	Rearranged steranes

The result of analyzing samples by the SIM or the MID approach is a number of single ion chromatograms which alone do not permit the unequivocal identification of individual components. Generally, it is possible to compare distributions of various components in the chromatograms with previously published data and assign identifications to these components with a fairly high degree of confidence. Novel compounds apparent in the chromatograms can be subject to further, more detailed studies. For correlation purposes, it should be remembered that it is not always necessary to identify all of the components in the chromatograms since, for the most part, correlations consist of comparing and matching fingerprints of as many classes of compounds as possible. Typically, such correlations have only been undertaken qualitatively; however, quantitation of the distributions of individual components provides a powerful and additional means of confirming any correlations being made.

In order to illustrate typical biomarker distributions obtained by GC/MS and MID, chromatograms showing the distributions of terpanes and steranes for three oils are shown in Fig. 2. The major classes of compounds and their carbon numbers are shown on these chromatograms. These oils were collected from different fields and, as clearly illustrated here, have quite different biomarker fingerprints. The biomarker distributions will vary between oils from different sources and depend on such factors as source, maturity, biodegradation, and migration mechanisms.

Finally, it is informative to undertake these analyses at an even higher level of sophistication and illustrate, with examples, how additional information can be obtained from the analysis of these samples in the GC/MS/MS mode. In recent years, tandem mass spectrometry (MS/MS) with or without a gas chromatograph has played an increasingly important role in the analysis of fossil fuels. Tandem mass spectrometers may operate in a number of ways. In high-resolution magnetic instruments, metastable-ion monitoring (MIM) detects the spontaneous fragmentation of parent ions which occurs in the first field-free region of a double-focusing mass spectrometer. These ions can be observed separately by using a programmable power supply to vary the accelerating voltage while holding magnetic and electrostatic fields at constant values. One of the first examples of the application of MS/MS to fossil fuel analysis was from the work of Gallegos (1976), who

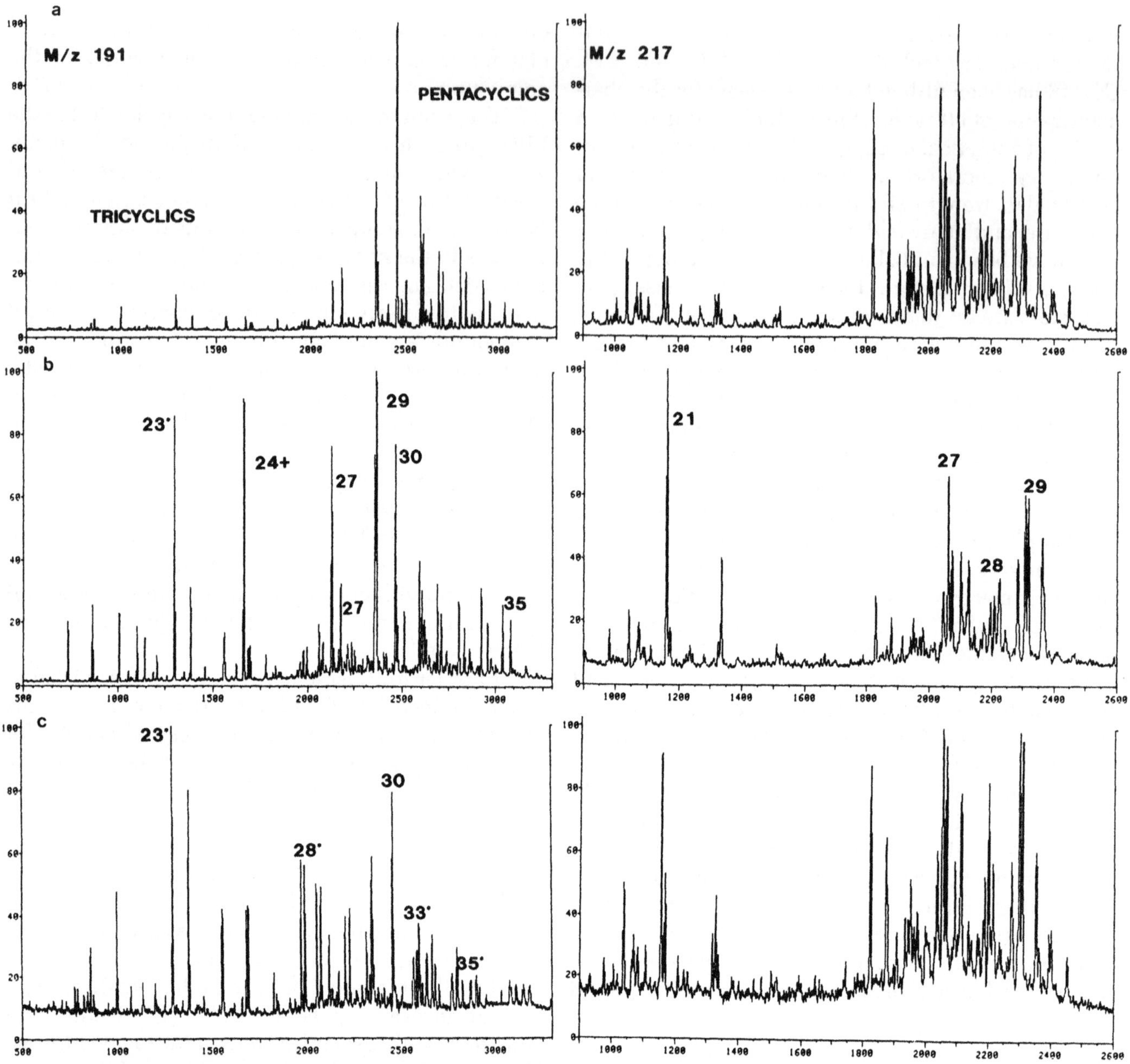

Figure 2. Single-ion chromatograms showing the distributions of steranes (*m/z* 217) and terpanes (*m/z* 191) for oils from three different sources (a–c). Peaks labeled 27–35 in the *m/z* 191 chromatogram are regular hopanes; 23* are tricyclic terpanes and 24+ is a tetracyclic terpane. In the *m/z* 217 chromatogram, the labels refer to the carbon numbers of the individual steranes.

used MIM to analyze terpanes and steranes present in crude oils. MS/MS with MIM has also been used to characterize various steranes in fossil fuels (Warburton and Zumberge, 1983; Fowler and Brooks, 1990). Tricyclic terpanes from C_{19} to C_{45} have been identified by Moldowan *et al.* (1983) in various crude oils by the MIM approach. Summons and co-workers used this approach to characterize and identify a number of branched and isoprenoid alkanes in various Proterozoic oils and source rock samples (Summons, 1987; Summons *et al.*, 1988). Brooks *et al.* (1984) and Telnaes and Dahl (1986) used high-resolution selected MIM to determine the distribution of various biomarkers in oils and source rocks.

The second major type of tandem mass spectrometer consists of three quadrupoles, coupled in tandem. The middle quadrupole is a collision cell, operated only in a radio frequency mode. Ions formed in the ion source of the mass spectrometer are separated in the first quadrupole, and selected ions pass into the collision cell. The ions formed as a result of the collisions of these ions with the inert gas, such as argon, in the collision cell pass into the third quadrupole for subsequent separation and analysis. The system can be operated in three different scan modes—parent, daughter, and neutral— and all three of these have proved to be extremely helpful for the analysis of the complex fossil fuel type samples and the determination of compounds, such as steranes, hopanes, and bicyclic terpanes, in oils and source rock extracts (Philp and Oung, 1988; Philp *et al.*, 1990).

Selected-reaction monitoring (SRM) can also be used in the MS/MS scan modes and is analogous to SIM in the MS scan modes. In SRM, a particular reaction or set of reactions, such as the fragmentation of an ion or the loss of a neutral moiety, is monitored. SRM can be employed in any of the commonly used MS/MS scan modes. Like SIM, SRM allows very rapid analysis of trace components in complex mixtures. However, because two sets of ions are being selected, the specificity obtained in SRM can be much greater than that obtained in SIM. Any interfering compound would not only have to form a parent ion of the same mass-to-charge ratio as the selected parent ion from the target compound, but that parent ion would also have to fragment to form a daughter ion of the same mass-to-charge ratio as the selected daughter ion from the target compound. Many different variables can be utilized in the analysis, such as collision gas pressure and collision energy. Changing any of these parameters will affect the extent of ionization resulting from the collision process, and, as an additional variable, the system can be operated in alternative modes of ionization (Philp *et al.*, 1990).

An illustration of the additional resolving power can be seen using the steranes, which produce a complex *m/z* 217 chromatogram consisting of many overlapping homologs and isomers. To simplify this chromatogram, the MS/MS parent mode can be used to monitor various parent/daughter ion transitions, particularly those from parent ions of the C_{27}–C_{30} steranes to the *m/z* 217 daughter ion. After deconvoluting the data, it becomes possible to obtain individual chromatograms for the C_{27}, C_{28}, C_{29}, and C_{30} homologs, hence removing the problem of co-eluting components, as shown in Fig. 3. In addition to reducing the complexity of the fingerprints, this process also makes it much easier to quantitate relative proportions of the individual steranes and obtain an accurate assessment of the relative proportions of the C_{27}, C_{28}, C_{29}, and C_{30} steranes, which is extremely important information for both correlation and source determinations.

A development that will lead to the introduction of some new correlation parameters is the availability of high-temperature GC columns. These columns, which, as mentioned above, are typically coated with aluminum, has led to the possibility of analyzing components up to C_{100} and probably higher. Analyses of components above C_{40} will lead to a reevaluation of certain basic geochemical concepts. For instance it can be shown that certain marine oils, typically thought to be dominated by *n*-alkanes maximizing around C_{20}, actually have another maximum in the C_{40} region. Relatively few papers have been published on the use of HTGC. At the Organic Geochemistry Meeting in Manchester in September 1991, two papers were presented which demonstrated the

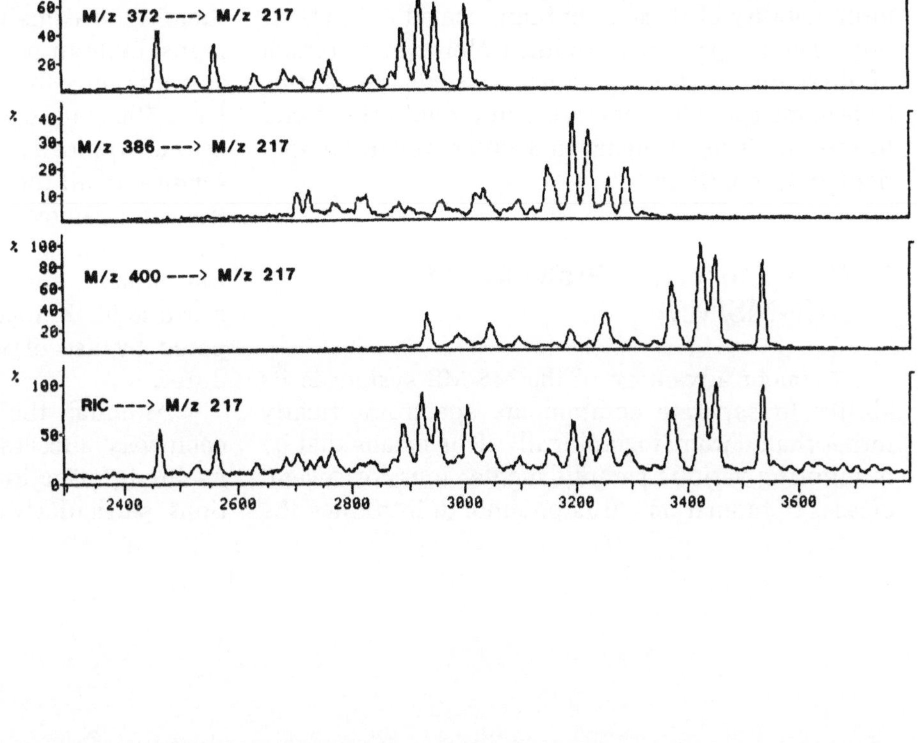

Figure 3. MS/MS combined with GC provides a powerful method for the analysis of complex mixtures of the type encountered in geological samples. In this example, GC/MS/MS is used to separate a complex mixture of steranes into its individual homologs.

potential use of HTGC in geochemical studies both for correlation purposes and for determination of source materials (Carlson *et al.*, 1991; Del Rio and Philp, 1991).

In a related area, a few papers have also started to appear on the use of HTGC columns for analyses of porphyrins, both free base and metalloporphyrins. Porphyrins (Treibs, 1934, 1936; Hajibrahim *et al.*, 1981; Gill *et al.*, 1985, 1986) have not been widely used for correlation purposes because of their insufficient volatility for analysis by GC/MS and the need to use high-pressure liquid chromatography (HPLC) and HPLC/MS. However, it has been shown that it is possible to analyze porphyrins by GC on fused-silica capillary columns, following the preparation of various derivatives (Alexander *et al.*, 1980; Marriott and Eglinton, 1982; Marriott *et al.*, 1984; Gill *et al.*, 1986). The continued development of GC and GC/MS methods for porphyrin analysis complements the HPLC/MS methods developed for petroporphyrins (Eckardt *et al.*, 1988). Recent papers by Blum *et al.* (1989; 1990) have demonstrated the use of high-temperature capillary columns in the GC/MS analysis of free-base porphyrins and metalloporphyrins without the need for prior derivatization. Apart from describing the analytical results, these papers also describe the preparation of the high-temperature GC/MS interface. This, of course, is an essential prerequisite for the use of the high-temperature columns since cold spots and condensation of components in the interface need to be avoided at all costs. Gallegos *et al.* (1991a, b) in recent papers also demonstrated the use of HTGC columns combined with MS for the analysis of porphyrins. The results shown in these papers clearly demonstrate the viability of using this approach for the analysis of metalloporphyrins. As mentioned previously, it would appear that the long-term stability of these aluminum-coated columns is not as good as previously thought. However, a number of alternative high-temperature columns are starting to become available, and without a doubt this trend toward analysis of higher molecular weight components will continue.

3. Direct Insertion Probe MS/MS (DIP-MS/MS)

A major advantage of the MS/MS system is its ability to separate components spectrometrically rather than chromatographically. This means that by monitoring various parent/daughter ions for several classes of biomarkers, it is possible to introduce the sample via the direct insertion probe (DIP) while monitoring these transitions. This relatively rapid analysis (5 min) can be performed on whole oils to produce a composite spectrum showing parent-ion distributions for all classes of compounds monitored (Fig. 4). These distributions, or fingerprints, can then be further resolved into individual compound classes or compared directly with those of other oils and source rocks for purposes of correlation (Philp *et al.*, 1990).

The major disadvantage of this approach is that it does not separate isomers with the same carbon number. Hence, for instance, all the C_{27} steranes (MW 372) will show up as one parent-ion peak which cannot be resolved. However, the rearranged steranes have an intense daughter-ion peak at *m/z* 189, not *m/z* 217. Hence, the two classes of compounds can be resolved by monitoring the two transitions to the daughter ions at *m/z* 189 and *m/z* 217. In many cases, the fingerprints obtained by the DIP-MS/MS approach provide sufficient information for a rapid screening process from which individual samples can be selected for more detailed analyses by GC/MS or GC/MS/MS.

4. Pyrolysis

From the discussions above and reference to the literature, it is clear that extensive use has been made of GC/MS, GC/MS/MS, and MS/MS for undertaking oil–oil and oil–source rock correlations. Equally important, but relatively overlooked for correlation purposes, are the insoluble organic fractions. In potential source rocks, the insoluble organic fraction (kerogen) can make up 90%, or more, of the total organic carbon in the rock (Hunt, 1979). In oils and rock extracts, there is the asphaltene fraction, which, by definition, is insoluble in pentane, but is soluble in more polar solvents and can be characterized by techniques similar to those used for kerogen characterization. Kerogens are typically isolated, or concentrated, following removal of all the soluble material by solvent extraction and removal of carbonates by treatment with hydrochloric acid, silicates by treatment with hydrofluoric acid, and, in certain cases, pyrite by use of nitric acid. Kerogen concentrates can also be prepared by use of various density separation procedures.

Although the analysis of soluble fractions has been very successful and provides a great deal of useful information for many exploration applications, particularly correlation studies, the potential is

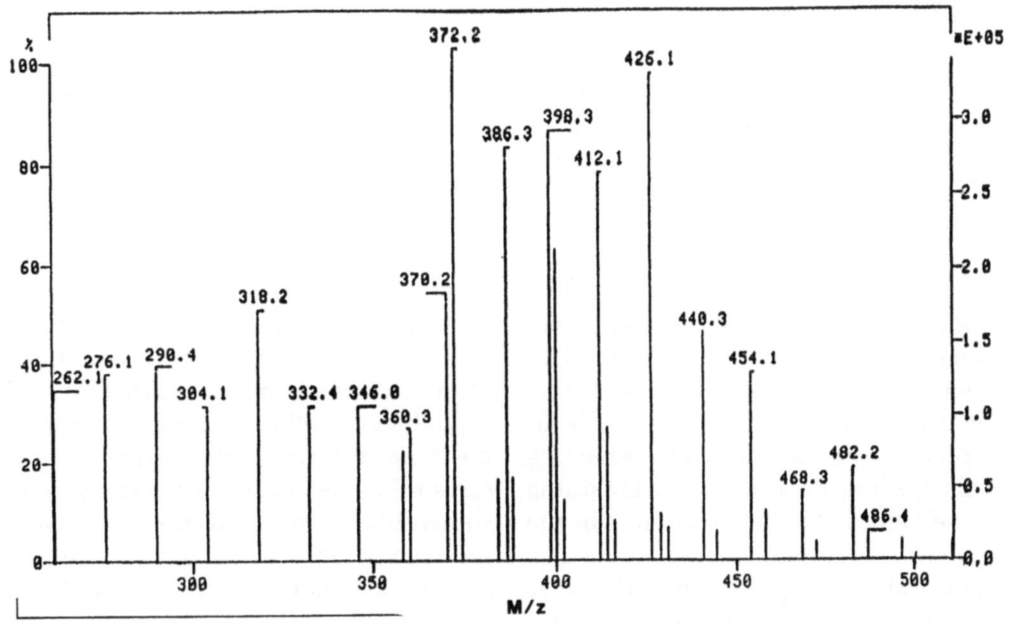

Figure 4. Example of the analysis of whole oils by use of the direct insertion probe and MS/MS, a very powerful method for the fingerprinting or rapid characterization of crude oils for the purposes of correlation.

always there for contamination to occur either from natural sources, i.e., migration, or during sample collection and laboratory handling. The advantage of utilizing the insoluble organic fractions is that virtually all potential contaminants are removed during the actual isolation as a result of the extraction process. The problem is how to utilize the insoluble fractions for correlation purposes. The published methods for kerogen characterization consist of a wide array of chemical and physical methods, many of which are described in review articles and books. A particularly useful reference book for this topic is entitled *Kerogen*, edited by Durand (1980). For correlation purposes, it is commonly accepted that the most appropriate method for degrading the kerogens or asphaltenes into more volatile fragments prior to analysis is pyrolysis.

A wide variety of methods are available for the pyrolytic degradation of the insoluble organic fractions, and the results obtained can be used in a variety of ways (Philp and Gilbert, 1987). Prior to discussing results from such pyrolysis methods, a brief summary of the major pyrolysis techniques will be given along with an indication of which methods are the most useful, or appropriate, for correlation studies. A more detailed discussion on the use of pyrolysis in geochemistry can be found in Chapter 13 of this volume, by Larter and Horsfield.

One of the most rapid forms of kerogen characterization, without the actual isolation of kerogen or any chromatographic separation of the pyrolysis products, permits passage of the pyrolysis products directly into a flame ionization detector (FID). Such an approach is well suited for the rapid screening of large numbers of samples and is commonly referred to as the Rock-Eval technique (Rock Eval = Rock Evaluation) (Espitalié *et al.*, 1985). The Rock-Eval approach is very limited in its application for correlation purposes since it basically provides information on the maturity of the organic matter and the petroleum potential of a suspected source rock. None of the pyrolysis products is resolved chromatographically, and only two major peaks are produced—S_1, corresponding to the components normally extracted with organic solvents, and S_2, corresponding to the products produced from the pyrolytic degradation of the kerogen or asphaltenes. More detailed fingerprinting of the insoluble organic material can be undertaken by using various types of flash pyrolysis methods in conjunction with GC or GC/MS.

"Off-line" pyrolysis reactions can be performed in a variety of ways, but each has the same goal, namely, thermal degradation of the insoluble residue. In the "off-line" process, the products from pyrolysis are trapped and separated into various fractions prior to analysis by GC, GC/MS, and GC/MS/MS. The products generated from the pyrolysis are presumed to resemble those that will be generated naturally from the rock as it matures. A statement such as this is always subject to some debate since the process of anhydrous artificial maturation will generate alkenes, which are not generated naturally, as well as alkanes. As will be seen below, "off-line" anhydrous pyrolysis reactions designed to simulate maturation have been superseded by hydrous pyrolysis methods. An advantage of the anhydrous pyrolysis approach is that it provides the opportunity to pyrolyze relatively large sample sizes and obtain significant quantities of products, including biomarkers, that can be used for correlation purposes. The products can be separated into compound classes and analyzed by GC and GC/MS. A disadvantage of this approach is that, during the fractionation process, many of the lower molecular weight and volatile components are lost, reducing the utility of this method as a means of obtaining information on the nature of source materials in the various samples.

"On-line" pyrolysis of kerogens or asphaltenes is one of the more powerful pyrolysis techniques available for oil–source rock or oil–oil correlation purposes. In this approach, the pyrolyzer, of which there are many designs, is attached to the injector of the gas chromatograph, and the pyrolysis products are transmitted directly to the front end of the GC column by the use of a carrier gas, such as He, flowing rapidly over the sample being pyrolyzed. Once the pyrolysis is completed (typical pyrolysis times are 20 s to 2 min at temperatures around 600°C), the experiment becomes a regular GC/MS or GC/MS/MS analysis, with the GC oven temperature being programmed from the temperature at which the products are trapped (typically −25°C) to the upper temperature limit. Pyrolysis–GC (Py-GC) experiments are performed with extracted rocks or isolated kerogens unless a stepwise pyrolysis reaction is undertaken. If the rocks are not preextracted, pyrolysates will represent a mixture of products released from the kerogen plus those present in the extractable fraction that will be thermally distilled at the higher temperatures. The pyrograms that are obtained by Py-GC can be used for purposes of obtaining information on the types of source material in a rock and estimating the relative maturity level of the samples. By undertaking Py-GC/MS analyses, biomarkers can be monitored and analyzed in the same manner as in a regular GC/MS analysis. The biomarkers released from the kerogens in this way are truly indigenous to the kerogen. The distributions can be used in the same way as the "free" biomarkers for source determinations. However, maturity measurements based on biomarkers released by pyrolysis will always yield lower maturities than those based on the "free" biomarkers. This is generally considered to be due to the fact that these biomarkers attached to the kerogens are sterically hindered, making it more difficult for any conformational change to occur (Philp and Gilbert, 1985).

An important use of Py-GC characterization of asphaltenes has been demonstrated in pyrolysis of asphaltenes from heavily biodegraded oils. Py-GC or Py-GC/MS of such materials produces a pyrogram that closely resembles the chromatogram of the oil prior to biodegradation, which in turn permits correlation to be made between degraded and nondegraded oil samples (Behar and Pelet, 1985). Whether this is due to products actually being released by breakdown of the asphaltene or simply products trapped in the asphaltene released at the higher temperatures of pyrolysis has yet to be resolved. A recent paper has shown that pyrolysis of the NSO fractions will also produce pyrograms that are very similar to the original chromatograms of the nondegraded oil samples (Philp *et al.*, 1988).

One of the most recent areas of interest is the characterization of kerogens with high sulfur content due to their capability for early oil generation. An updated compilation of papers concerned with this topic can be found in the recent Proceedings of the American Chemical Society Dallas Congress (Orr and White, 1990). Many of these papers describe some parameters related to sulfur compounds that can be used to assess crude oil maturation and for correlation purposes. Eglinton *et al.* (1988) used Py-GC to study various samples that had been artificially matured by pyrolysis. The sulfur compounds produced in this way appeared to show some systematic variations with maturity and nature of source material. Philp and Bakel (1988) noticed a number of variations in both the organosulfur and organonitrogen compounds produced by pyrolysis of various source rocks, with respect to depositional environment of the source material and maturity. A disadvantage of the Py-GC and Py-GC/MS approach is the fact that the sample size has to be somewhat smaller than those used for off-line pyrolysis studies. If the samples are lean, particularly for extracted source rocks, the concentrations of the biomarkers produced in this way will also be rather low.

Anhydrous pyrolysis methods are generally used for the characterization or fingerprinting of kerogens (Philp and Gilbert, 1985), whereas hydrous pyrolysis is a valuable approach for simulating processes involved in petroleum generation, or source rock maturation, on a relatively short time scale in the laboratory (Winters *et al.*, 1983; Lewan, 1985). A combination of hydrous and anhydrous pyrolysis techniques can provide an extremely powerful laboratory method for monitoring changes resulting from maturation of kerogens and the nature of the products produced at various levels of maturity (Eglinton *et al.*, 1988). This information can be used subsequently for correlation purposes.

Hydrous pyrolysis is a widely used technique for a variety of geochemical purposes including correlation studies. As described in the papers of Winters *et al.* (1983) and Lewan (1985), who were responsible for developing the hydrous pyrolysis approach, the method is used primarily for simulating natural maturation in the laboratory. The maturation reactions are undertaken in the presence of water, and the products generated closely resemble those of naturally generated crude oils. By performing the experiments over a range of times and temperatures, it is possible to mimic natural maturation processes if relatively immature rocks are initially used in the experiments. Hydrous pyrolysis also provides an extremely powerful method for undertaking oil–source rock correlation without necessarily obtaining a sample of the mature source rock from the field, eliminating the need for costly drilling operations. Although relatively expensive to initially set up, hydrous pyrolysis allows a number of potential source rocks to be characterized very rapidly. The expenses saved as a result of the information obtained from the hydrous pyrolysis experiments and by not drilling dry holes permits relatively rapid recovery of the expenses involved in setting up the hydrous pyrolysis systems. After the experiments have been performed and the oils and extracts obtained, the analytical routines undertaken are then the standard type of GC and GC/MS analyses. The biomarker distributions obtained by this process are virtually identical to those in naturally occurring samples. Hence, oil–source rock studies can be undertaken using the samples obtained from these experiments.

5. High-Pressure Liquid Chromatography and Mass Spectrometry

The use of high-pressure liquid chromatography (HPLC) to fractionate fossil fuels into their various aromatic fractions has become particularly commonplace in the laboratory. In order to reduce the time-consuming column-chromatographic processes, conventional methods of hydrocarbon group type separation have been replaced by HPLC in many laboratories. Radke *et al.* (1988, and references therein) described an automated medium-pressure liquid chromatograph, for the fractionation of aromatic hydrocarbons according to their degree of aromatization. This technique has been widely used to isolate aromatic fractions. A combination of normal- and reverse-phase LC has been used by Garrigues *et al.* (1988, and references therein) to discriminate between different aromatic ring systems and degrees of methylation in order to characterize maturity.

HPLC has been used in a number of geochemical correlation studies, most of which involve determination of porphyrin distributions. For example, HPLC/MS was used to study demetallated porphyrins in shales and oils from the Shengli oil field in China as part of an attempt to correlate the oils with suspected source rocks (Jiyang *et al.*, 1982). The fingerprints obtained by HPLC are useful for correlation purposes although the complexity of the porphyrin mixtures is such that total resolution of the individual porphyrins is still not possible by HPLC.

Eckardt *et al.* (1990) have discussed recent developments in the use of LC/MS for the analysis of tetrapyrroles from various geological sources. They used a reverse-phase column for the separation of both free-base and metallated porphyrins. The data obtained show major improvements over that in early publications and clearly demonstrate the value of the LC/MS approach for the analysis of these more polar tetrapyrrole-type compounds. The authors also provided a brief review of the other major developments in the use of LC/MS for tetrapyrrole analyses, and hence these references will not be repeated herein.

6. Supercritical Fluid Chromatography and Mass Spectrometry

Supercritical fluid chromatography/mass spectrometry (SFC/MS) is a relatively new approach for geochemical correlation studies, and few papers have appeared in the literature on this topic. SFC has been used in many other disciplines for some time and has a strong potential in many areas of geochemistry. One reason why it has yet to gain widespread acceptance in geochemistry is that for compounds in the C_{15}–C_{40} range there is little difference in results obtained by GC/MS versus those obtained by SFC/MS. A major advantage of SFC/MS will probably be in the analysis of higher molecular weight hydrocarbons (C_{40}–C_{100}). Analysis of compounds in this carbon number range can also be approached by HTGC. However, the use of SFC avoids the decomposition of components that may result from exposures to such high temperatures.

Monin *et al.* (1988) coupled the extraction capacity of supercritical fluid extraction (SFE) methods with SFC for a few samples of crude oils, source rock extracts, and oil shales in order to compare the efficiency of this method with the classical organic solvent extraction and GC.

Hopfgartner *et al.* (1990) recently compared the results of extracting various rock samples using both SFE methods with CO_2 and conventional solvent extraction methods. In this process, liquid CO_2 (possibly with additional small amounts of polar solvents such as methanol) can be used to extract source rocks relatively rapidly, with the advantage of not extracting the more polar NSO and asphaltene fractions. The extracts obtained with CO_2 are predominantly aliphatic and aromatic components and can be analyzed directly by GC and GC/MS without purification. Biomarker distributions were virtually identical using either procedure. However, the advantages of the SFE method were its rapidity, absence of thermal degradation, less contamination, and greater selectivity. Smaller sample sizes also introduce the possibility of a much larger number of replicate analyses, leading to better reproducibility of these methods.

There have not been any papers published in the literature where SFC has been used directly for correlation purposes. However, the work of Hawthorne and others has demonstrated that it is possible to utilize this method for the analysis of high-molecular-weight hydrocarbons of the type appearing in waxes (Hawthorne and Miller, 1987). With the development and availability of SFE methods and the desire to develop methods suited for analysis of high-molecular-weight biomarkers, it is inevitable that SFC/MS methods will ultimately become an important analytical technique in future geochemical analyses.

7. Gas Chromatography and Carbon and Deuterium Isotopic Compositions

Variations in isotopic compositions have been used for many years as a powerful screening and correlation tool for oils and suspected source rocks, and a number of comprehensive reviews have been published on this approach (Stahl, 1978; Schoell, 1984).

$\delta^{13}C$ isotopic values have been used far more extensively than δD values for geochemical purposes. Their uses include providing information on relative maturity levels of rocks and oils, extent of biodegradation, depositional environments, and possible migration distances. Since source rocks are mixtures of large amounts of many different types of organic materials, the isotopic composition reflects the average of all the isotopic compositions of these materials and cannot be used to directly infer the specific types of source material in a sample (Schoell, 1984). However, when the oil is generated from a particular source rock, the $\delta^{13}C$ value of the oil produced will resemble that of the source rock from which it was generated. In this way it is possible to use these $\delta^{13}C$ values for either oil–oil or oil–source rock correlation studies. It should be remembered that the $\delta^{13}C$ value of the oils will vary as a result of factors such as maturity, biodegradation, and migration to some extent. However, the extent to which these factors change the $\delta^{13}C$ value is predictable to a large degree, and appropriate compensation can be made during the correlation.

A number of approaches utilizing $\delta^{13}C$ isotopic compositions of crude oils and source rock extracts have been used for correlation purposes. The two most widely cited approaches are those based either on the so-called Sofer plot (Sofer, 1984) or the Stahl curve (Stahl, 1978). In the former method, the $\delta^{13}C$ compositions of the saturate and the aromatic fractions are plotted against each other. In this situation, oils and suspected source rocks extracts that are derived from similar types of source materials will plot in the same region whereas those derived from different types of source material will plot in other regions of the graph. Stahl type curves are somewhat different in that isotopic compositions of the saturates, aromatics, NSO, asphaltenes, whole oil, and source rock extracts are plotted as shown in Fig. 5. In this situation, if the oil is derived from the suspected source rock extract used in the analyses, then a relatively

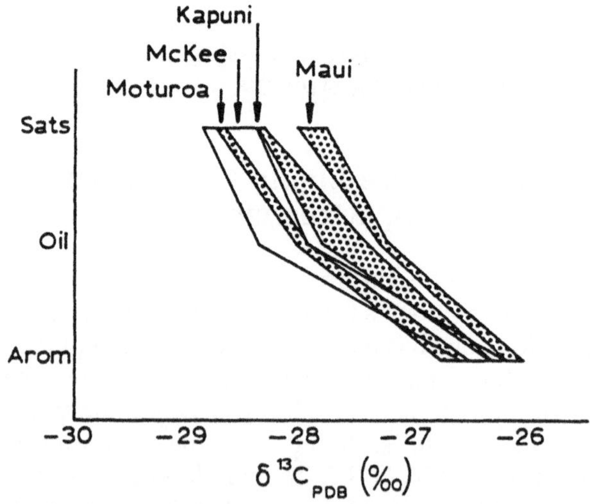

Figure 5. Stahl type curves for four oil fields within the Taranaki Basin, New Zealand. (Stippled areas used to separate samples from individual fields.)

straight-line plot will be obtained. If not, the extract will not plot on that line. Naturally, oils that are closely related to each other will plot in similar positions on this graph. Biodegradation, migration, and other effects will change the shape of the curve in a fairly predictable manner.

One of the major analytical advances that has occurred in recent years that is applicable to geochemistry has been the development of the combination of gas chromatography and isotope-ratio mass spectrometry (GC/IRMS). Much of the initial work on this combination was published from the laboratories of John Hayes at Indiana University. The early work has concentrated to a large extent on the measurement of the $\delta^{13}C$ compositions of individual biomarkers. These results have gone a long way toward changing some of our ideas about the origins of many of these compounds and about the processes involved in diagenesis and maturation.

There are many ways in which GC/IRMS will aid in geochemical research, and some of the basic ideas were presented by the Hayes group at the recent international geochemistry meetings in Paris (1989) and Manchester (1991) and the Geological Society of America meeting in St. Louis (Hayes *et al.*, 1989a, b; Freeman *et al.*, 1989; Hayes *et al.*, 1990). For instance, it has been commonly assumed that the isoprenoids, pristane and phytane, had a common origin, namely, the phytol side chain of chlorophyll. It was assumed that the formation of pristane was favored and under oxidizing conditions phytane formation was favored under reducing conditions. This in turn led to the extensive use of the pristane/phytane ratio as an indicator of the nature of the depositional environment (Didyk et *al.*, 1978). Recently, the validity of this ratio has been subject to some debate since a number of alternative sources for pristane (α-tocopherol) (Goossens *et al.*, 1984) and phytane (Archaebacteria) have been proposed (ten Haven *et al.*, 1987). The results obtained by the Indiana group using GC/IRMS have shown that the isotopic composition of pristane and phytane in some of the samples examined differed by several per mil, suggesting that these two compounds did indeed have different origins (Freeman, 1989, 1990). Several other examples have also been presented to show that other classes of biomarkers, such as the hopanes, are not always derived from a common precursor, as previously proposed. These results indicate that the origin and fate of organic matter in the sedimentary record are far more complicated than was previously assumed. The ability to obtain isotopic data on individual biomarkers will provide us with a powerful tool to differentiate between the origins of different biomarkers and also another useful tool for correlation purposes.

A more practical example that demonstrates the value of this technique in the research in our laboratory was based on the analysis of a suspected oil seep from Tanzania. An evaluation was being made of various oil seeps to determine their origin and to possibly correlate them with oils in the region. Initial analysis of the total hydrocarbons produced a very unusual hopane distribution for a seep sample in terms of the total absence of all components apart from hopanes and the fact that these hopanes had the stereochemistries associated with immaturity. Analysis of this sample in the laboratory of Dr. Hayes revealed that the major hopanes had isotopic compositions in the range of −62 to −65‰ (Fig. 6). Once these isotopic values were obtained for the individual hopanes, it was obvious that this was not a normal seep. Light isotopic compositions for individual components suggested that they had a microbial and not a thermal origin. Further evaluation of the situation has led us to the conclusion that in the area where the so-called seep was found, natural gas known to be seeping to the surface is acting as a substrate for the bacteria. The bacteria in turn are biosynthesizing various hopanoids, including hopanoic acids, which are also present in high abundance in these samples. Hence, the hopanes observed in the hydrocarbons of this supposed seep have isotopic compositions consistent with a bacterial origin.

To date, there are no published reports of the use

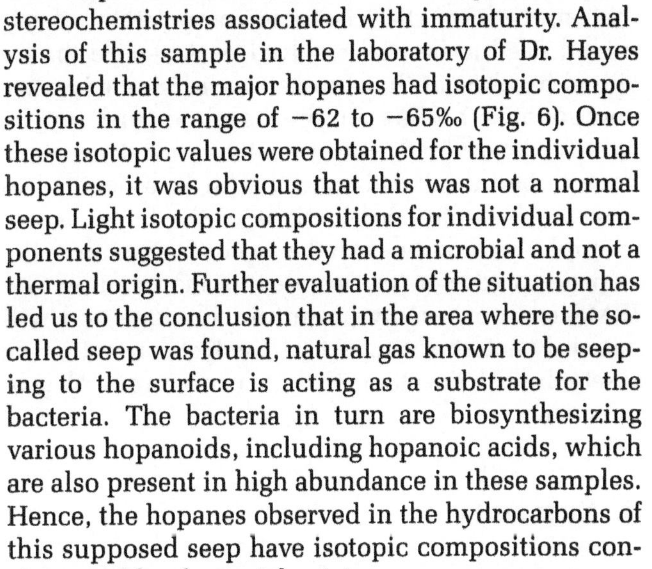

Figure 6. Example of the use of gas chromatography–isotope-ratio mass spectrometry (GC/IRMS) for the analysis of a seep sample from Tanzania to determine the isotopic compositions for the individual biomarkers.

of GC/IRMS as a correlation tool. However, this approach could have a major impact on correlation work. At this time, many of our correlation studies are based on qualitative and quantitative comparisons of biomarker fingerprints. By analyzing the samples in the GC/IRMS mode, isotopic compositions of individual components can be obtained, and, clearly, if the samples are similar, they will have similar isotopic compositions.

One of the problems currently associated with this approach is that there is no equivalent MID approach to GC/IRMS. In other words, components that are to have their isotopic compositions determined need to be well-resolved peaks above the residual baseline of the chromatogram. Minor or trace components hidden in the baseline will not be readily determined. Additional fractionation steps have to be introduced such that, if it is decided to measure isotopic compositions of the hopanes, for example, a fraction will be needed in which the hopanes predominate. Despite this shortcoming, GC/IRMS is an extremely powerful technique and will certainly act as an important catalyst for furthering research and developments in many areas of geochemistry.

8. Summary

One of the most important factors in the rapid development of organic geochemistry and its application to petroleum exploration problems has been the continued development of sophisticated analytical techniques. Advances in the design and technology of gas chromatographs, mass spectrometers, computers, and chromatographic columns have been uppermost in providing the capability to detect and determine ever-decreasing amounts of complex organic molecules in mixtures of ever-increasing complexity. These developments are continuing; indeed, the past year or so has seen the commercial availability of GC/IRMS systems and development of high-temperature GC columns. Both of these advances are going to lead to improvement in our correlation methods and provide a greater degree of confidence in any correlation that is being undertaken.

Where the future lies in correlation studies is difficult to predict. Undoubtedly, methods that provide the ability to analyze much larger and more polar molecules will become available. At the present time, most correlation work is undertaken using the hydrocarbon fractions so the investigation of the polar fractions will become particularly important. One only has to look at the major advances made in the past two decades to imagine the level of sophistication we may expect to appear in the next two decades.

References

Alexander, R., Eglinton, G., Gill, J. P., and Volkman, J. K., 1980, Capillary GC and GCMS of bis(trimethylsiloxy) silicon (IV) derivatives of alkyl porphyrins, *J. High Resolut. Chromatogr. Chromatogr. Commun.* **3**:521–522.

Behar, F., and Pelet, R., 1985, Pyrolysis–gas chromatography applied to organic geochemistry. Structural similarities between kerogens and asphaltenes from related source rocks, extracts and oils, *J. Anal. Appl. Pyrolysis* **8**:173–187.

Blum, W., and Eglinton, G., 1989, Glass capillary gas chromatography–alkali flame ionization detection at high temperatures. Direct analysis of free base petroporphyrins, *J. High Resolut. Chromatogr.* **12**:621–623.

Blum, W., Ramstein, P., and Eglinton, G., 1990, Coupling of high temperature glass capillary columns to a mass spectrometer. GC/MS analysis of metalloporphyrins from Julia Creek oil shale samples, *J. High Resolut. Chromatogr.* **13**:85–93.

Brooks, P. W., Meyer, T., and Christie, O. H. J., 1984, A comparison of high resolution GCMS and selected metastable ion monitoring of polycyclic steranes and triterpanes, *Org. Geochem.* **6**: 813–816.

Buchanan, M. V., 1982, Mass spectral characterization of nitrogen containing compounds with ammonia chemical ionization, *Anal. Chem.* **54**:570–574.

Burlingame, A. L., Maltby, D., Rusell, D. H., and Holland, P. T., 1988, Mass spectrometry, *Anal. Chem.* **60**:294R–342R.

Carlson, R. M. K., Moldowan, J. M., Gallegos, E. J., Peters, K. E., Smith, K. S., and Seetoo, W C., 1991, Biological markers in the C_{40} to C_{60} range. New marine/lacustrine source indicators, Presented at 15th Meeting of European Association of Organic Geochemists, Manchester, England, 1991.

Del Rio, J. C., and Philp, R. P., 1991, Occurrence of high molecular weight hydrocarbons (above C_{40}) in geological materials, in: *Organic Geochemistry* (D. A. C. Manning, ed.), Manchester University Press, Manchester, England, pp. 183–185.

Didyk, B. M., Simoneit, B. R., Brassell, S. C., and Eglinton, G., 1978, Organic geochemical indicators of palaeoenvironmental conditions of sedimentation, *Nature* **272**:216–222.

Durand, B. (ed.), 1980, *Kerogen—Insoluble Organic Matter from Sedimentary Rocks*, Editions Technip, Paris.

Dzidic, I., Balicki, M. D., and Hart, H. V., 1988, Identification and quantifications of carbazole and benzocarbazole homologues in clarified slurry oils by ammonia chemical ionization, *Fuel* **67**:1155–1159.

Eckardt, C. B., Dyas, L., Yendle, P. W., and Eglinton, G., 1988, Multimolecular data processing and display in organic geochemistry: The evaluation of petroporphyrin GCMS data, *Org. Geochem.* **13**:573–582.

Eckardt, C. B., Carter, J. F., and Maxwell, J. R., 1990, Combined liquid chromatography/mass spectrometry of tetrapyrroles of sedimentary significance, *Energy Fuels* **4**:741–747.

Eglinton, T., Philp, R. P., and Rowland, S. J., 1988, Flash pyrolysis of artificially matured kerogens from the Kimmeridge Clay Fm. U.K., *Org. Geochem.* **12**:33–41.

Espitalié, J., Deroo, G., and Marquis, F., 1985, La pyrolyse Rock-Eval et ses applications, *Rev. Inst. Fr. Pet.* **40**:755–784.

Fowler, M. G., and Brooks, P., 1990, Organic geochemistry as an aid in the interpretation of the history of oil migration into different reservoirs at the Hibernia K-18 and Ben Nevis I-45 wells, Jeanne d'Arc Basin, offshore eastern Canada, *Org. Geochem.* **16:**461–475.

Freeman, K. H., 1989, Distinct origins of pristane and phytane: Isotopic evidence, Abstract, 1989 Geological Society of America Annual Meeting, St. Louis.

Freeman, K. H., 1990, Evidence from carbon isotope measurements for diverse origins of sedimentary hydrocarbons, *Nature* **343:** 254–256.

Freeman, K. H., Hayes, J. M., Trendel, J.-M., and Albrecht, P., 1989, Evidence from GC/MS carbon isotopic measurements for multiple origins of sedimentary hydrocarbons, *Nature* **353:** 254–256.

Gallegos, E. J., 1976, Analysis of organic mixtures using metastable transition spectra, *Anal. Chem.* **48:**1348–1351.

Gallegos, E. J., Fetzer, J. C., Carlson, R. M.,and Pena, M. M., 1991a, High temperature GCMS characterization of porphyrins and high molecular weight saturated hydrocarbons, *Energy Fuels* **5:**376–381.

Gallegos, E. J., Fetzer, J. C., Carlson, R. N. K., and Pena, M. M., 1991b, High temperature GCMS of porphyrins, high molecular weight polynuclears and saturated hydrocarbons, Paper presented at 15th Meeting of European Association of Organic Geochemists, Manchester, England, 1991.

Garrigues, P., de Sury, R., Angelin, M. L., Bellocq, J., Oudin, J. L., and Ewald, M., 1988, Relation of the methylated aromatic hydrocarbon distribution pattern to the maturity of organic matter in ancient sediments from the Mahakam delta, *Geochim. Cosmochim. Acta* **52:**375–384.

Gill, J. P., Evershed, R. P., Chicarelli, M. I., Wolff, G. A., Maxwell, J. R., and Eglinton, G., 1985, Computerized capillary gas chromatography–mass spectrometric studies of the petroporphyrins of the Gilsonite bitumen (Eocene, U.S.A.), *J. Chromatogr.* **350:**37–62.

Gill, J. P., Evershed, R. P., and Eglinton, G., 1986, Comparative computerized gas chromatographic–mass spectrometric analysis of petroporphyrins, *J. Chromatogr.* **369:**281–312.

Goossens, H., de Leeuw, J. W., Schenck, P. A., and Brassell, S. C., 1984, Tocopherols as likely precursors of pristane in ancient sediments and crude oils, *Nature* **312:**440–442.

Hajibrahim, S. K., Quirke, J. M. E., and Eglinton, G., 1981, Petroporphyrins V. Structurally-related porphyrin series in bitumens, shales and petroleums—evidence from HPLC and mass spectrometry, *Chem. Geol.* **32:**173–188.

Hawthorne, S. B., and Miller, D. J., 1987, Analysis of commercial waxes using capillary supercritical fluid chromatography mass spectrometry, *J. Chromatog.* **388:**397–409.

Hayes, J. M., Freeman, K. H., and Popp, B. N., 1989a, Carbon isotopic information of individual organic compounds: New information for organic geochemistry, Abstract, *International Meeting of Organic Geochemistry*, Paris, September 1989.

Hayes, J. M., Takigiku, R., Ocampo, R., Callot, H. J., and Albrecht, P., 1989b, Isotopic compositions and probable origins of organic molecules in the Eocene Messel shale, *Nature* **329:**48–51.

Hayes, J. M., Freeman, K. H., Popp, B. N., and Hokam, C. H., 1990, Compound specific isotopic analyses: A novel tool for reconstruction of ancient biogeochemical processes, *Org. Geochem.* **16:**1115–1128.

Hites, R. H., and Biemann, K., 1970, Computer evaluation of continuously scanned mass spectra of gas chromatographic effluents, *Anal. Chem.* **42:**855–860.

Hoffmann, C. F., Mackenzie, A. S., Lewis, C. A., Maxwell, J. R., Oudin, J. C., Durand, B., and Vandenbroucke, M., 1984, A biological marker study of coals, shales, and oils from the Mahakam Delta, Kalimantan, Indonesia, *Chem. Geol.* **42:**1–23.

Hopfgartner, G., Veuthey, J.-L., Gulacar, O., and Buchs, A., 1990, Extraction of biomarkers from sediments with supercritical carbon dioxide: A comparative study with solvent extraction and thermodesorption methods, *Org. Geochem.* **15:**397–401.

Hunt, J. M., 1979, *Petroleum Geochemistry and Geology*, W. H. Freeman and Co., San Francisco.

Jiyang, S., MacKenzie, A. S., Alexander, R., Eglinton, G., Gowar, A. P., Wolff, G. A., and Maxwell, J. R., 1982, A biological marker investigation of petroleums and shales from the Shengli Oilfield, The People's Republic of China, *Chem. Geol.* **35:**1–31.

Lewan, M. D., 1985, Evaluation of petroleum generation by hydrous pyrolysis experimentation, Phil. Trans. Roy. Soc., London, **A315:**124–134.

Mackenzie, A. S., 1984, Application of biological markers in petroleum geochemistry, in: *Advances in petroleum Geochemistry*, Vol. 1 (J. Brooks and D. Welte, eds.), Academic Press, London, pp. 115–215.

Marriott, P. J., and Eglinton, G., 1982, Capillary gas chromatography and gas chromatography–mass spectrometry of silicon (IV) derivatives of porphyrins with polar substituents, *J. Chromatogr.* **249:**311–321.

Marriott, P. J., Gill, J. P., Evershed, R. P., Hein, C. S., and Eglinton, G., 1984, Computerized gas chromatographic–mass spectrometric analysis of complex mixtures of alkyl porphyrins, *J. Chromatogr.* **301:**107–128.

Moldowan, J. M., Seifert, W. K., and Gallegos, E. J., 1983, Relationship between composition and depositional environment of petroleum, *Am. Assoc. Petrol. Geol. Bull.* **69:**1255–1268.

Moldowan, J. M., Seifert, W. K., and Gallegos, E. J., 1985, Identification of an extended series of tricyclic terpanes in petroleum, *Geochim. Cosmochim. Acta* **47:**1531–1534.

Monin, J. C., Barth, D., Perrot, M., Espitalié, M., and Durand, B., 1988, Extraction of hydrocarbons from sedimentary rocks by supercritical carbon dioxide, *Org. Geochem.* **13:**1079–1086.

Orr, W., and White, C. M. (eds.), 1990, *Geochemistry of Fossil Fuels*, ACS Symposium Series No. 429, American Chemical Society, Washington, D.C.

Philp, R. P., 1985, *Fossil Fuel Biomarkers: Applications and Spectra*, Elsevier, Amsterdam.

Philp, R. P., and Bakel, A., 1988, Production of heteroatomic compounds by pyrolysis of asphaltenes, coals and source rocks, *Energy Fuels* **2:**59–64.

Philp, R. P., and Gilbert, T. D., 1985, Source rock and asphaltene biomarker characterization by pyrolysis–gas chromatography–mass spectrometry–multiple ion detection, *Geochim. Cosmochim. Acta* **49:**1421–1432.

Philp, R. P., and Gilbert, T. D., 1987, Biomarkers in kerogens as determined by pyrolysis–gas chromatography and pyrolysis–gas chromatography–mass spectrometry, *Anal. Pyrolysis* **11:** 93–108.

Philp, R. P., and Johnston, M., 1988, Determination of biomarkers in geological samples by tandem mass spectrometry, in: *Geochemical Markers* (T. F. Yen and J. M. Moldowan, eds.), Harwood Academic Publishers, New York, pp. 253–274.

Philp, R. P., and Oung, J.-N., 1988, Biomarkers, occurrence, utility and detection, *Anal. Chem.* **60:**887A–896A.

Philp, R. P., Bakel, A. J., Galvez-Sinibaldi, A., and Lin, L. H., 1988, A comparison of organosulphur compounds produced by pyrolysis of asphaltenes and those present in related crude oils and tar sands, in: *Advances in Organic Geochemistry 1987* (L. Mattavelli and L. Novelli, eds.), Pergamon Press, Oxford, pp. 915–926.

Philp, R. P., Lewis, C. A., Campbell, C., and Johnson, E., 1989, The use of the ion trap detector in petroleum geochemistry, *Org. Geochem.* **14**:183–187.

Philp, R. P., Oung, J.-N., Yu, C. P., and Lewis, S., 1990, The determination of biomarker distributions by tandem mass spectrometry, *Org. Geochem.* **16**:1211–1220.

Radke, M., Willsch, H., and Ramanampisoa, L., 1988, Isolation of aromatic compounds from a coal extract by semipreparative HPLC, *Chromatographia* **16**:259–266.

Rinaldi, G. G. L., Leopold, V. M., and Koons, C. B., 1988, Presence of benzohopanes, nonaromatic secohopanes, and saturate hexacyclic hydrocarbons in petroleums from carbonate environments, in: *Geochemical Markers* (T. F. Yen and J. M. Moldowan, eds.), Harwood Academic Publishers, New York, pp. 331–355.

Rullkötter, J., and Michaelis, W., 1990, The structure of kerogen and related materials. A review of recent progress and future trends, *Org. Geochem.* **16**:829–852.

Schenck, P. A., and Eisma, E., 1964, Quantitative determination of *n*-alkanes in crude oils and rock extracts by gas chromatography, in: *Advances in Organic Geochemistry 1962* (V. Colombo and G. D. Hobson, eds.), Pergamon Press, Oxford, pp. 403–416.

Schoell, M., 1984, Stable isotope studies in petroleum exploration, in: *Advances in petroleum Geochemistry* (J. Brooks and D. H. Welte, eds.), Academic press, London, pp. 215–225.

Seifert, W. K., and Moldowan, J. M., 1978, Application of steranes, terpanes and monoaromatics to the maturation, migration and source of crude oils, *Geochim. Cosmochim. Acta* **42**:77–95.

Snowdon, L. R., Brooks, P. W., Williams, G. K., and Goodarzi, F., 1987, Correlation of the Canol Formation source rock with oil from Norman Wells, *Org. Geochem.* **11**:529–548.

Sofer, Z., 1984, Stable carbon isotope compositions of crude oils: Application to source depositional environments and petroleum alteration, *Am. Assoc. Petrol. Geol. Bull.* **68**:31–49.

Stahl, H., 1978, Source rock/crude oil correlation by isotopic type curves, *Geochim. Cosmochim. Acta* **42**:1573–1577.

Summons, R. E., 1987, Branched alkanes from ancient and modern sediments: Isomer discrimination by GCMS with multiple reaction monitoring, *Org. Geochem.* **11**:281–289.

Summons, R. E., Powell, T. G., and Boreham, C. J., 1988, Petroleum geology and geochemistry of the Middle Proterozoic McArthur Basin, Northern Australia: III. Composition of extractable hydrocarbons, *Geochim. Cosmochim. Acta* **52**:1747–1764.

Telnaes, N., and Dahl, B., 1986, Oil–oil correlation using multivariate techniques, *Org. Geochem.* **10**:425–432.

ten Haven, H. L., de Leeuw, J. W., Rullkötter, J., and Sinninghe Damsté, J. S., 1987, Restricted utility of pristane/phytane ratio as a palaeoenvironmental indicator, *Nature* **330**:641–643.

Treibs, A., 1934, Uber das Vorkommen von Chlorophyllderivatin in einem Olschiefer aus der oberen Trias, *Ann. Chem.* **509**:103–114.

Treibs, A., 1936, Chlorophyll und Haminderivate in organischen Mineralstoffen, *Angew. Chem.* **49**:682–686.

Warburton, G. A., and Zumberge, J. E., 1983, Determination of petroleum steranes and triterpanes distributions by mass spectrometry with selective metastable ion monitoring, *Anal. Chem.* **55**:123–125.

Winters, J. C., Williams, J. A., and Lewan, M. D., 1983, A laboratory study of petroleum generation by hydrous pyrolysis, in: *Advances in Organic Geochemistry 1981* (M. Bjorøy *et al.*, eds.), Wiley, Chichester, pp. 524–533.

Chapter 20

Organic Geochemistry of Estancia Vieja Oils, Rio Negro Norte Block

Correlation with Other Neuquen Basin, Argentina, Oils

JOHN E. ZUMBERGE

1. Introduction

Previous organic geochemical studies of crude oils and possible source sediments from the Neuquen Basin, Argentina (e.g., Leenheer, 1982, 1984) have shown that crude oils produced in this basin were generated from source rocks deposited in a characteristically marine depositional environment with one exception, the Puesto Flores Oeste (PFO) X-1 oil. Furthermore, the Upper Jurassic/Lower Cretaceous Vaca Muerta Formation was determined to be the source of these marine oils. However, certain facies of the Lower Jurassic Los Molles Formation were rated as good to excellent possible source rocks (generally below 2500 m in the PFO X-1 well). In fact, oil recovered from PFO X-1 appeared to have been generated from the Los Molles Formation, which exhibited a relatively large lateral variability in organofacies. Also, the Los Molles Formation contains abundant organic matter of terrigenous origin, allowing for a relatively straightforward differentiation between oils generated from these disparate source rocks.

In the present study, four crude oils recovered from three drill stem tests (DST) and one repeat formation test (RFT) from the Estancia Vieja X-1 well drilled by Occidental Petroleum in the Rio Negro Norte Block in the Neuquen Basin were geochemically characterized using gravimetric, chromatographic, stable carbon isotopic, and mass spectrometric techniques (e.g., Sofer *et al.*, 1986; Vlierboom *et al.*, 1986; Zumberge, 1987b). Sample identifications, producing formations, and depths are listed in Table I. A location map is given in Fig. 1. Any "new" oil (i.e., from a source other than Vaca Muerta) discovered in this region could result in more favorable pricing structures.

2. Results and Discussion

Table I lists the gross compositional data and certain gas chromatographic ratios of the oils recovered from the Estancia Vieja X-1 well. In addition, compositional data from the PFO X-1 well from the Rio Norte Block and two representative Neuquen Basin oils sourced from the marine Vaca Muerta Formation are also given.

Figure 2 is a plot of API gravity versus percent sulfur; with increasing thermal maturity and API gravity, sulfur decreases. Also, oils from marine sources tend to contain more sulfur than oils generated from source rocks with abundant terrestrial organic matter. Little information regarding the source of the Estancia Vieja (EV) oils is discernible from Fig. 2. Figure 3 illustrates the C_{15+} gross hydrocarbon composition of Neuquen Basin oils on a ternary diagram. The PFO, EV DST #2 (EV2), and #4 (EV4), and RFT #4 (EVR) oils all plot within the terrigenous source region while the Vaca Muerta oils and the EV DST #8 (EV8) oil plot within the marine origin area.

JOHN E. ZUMBERGE • GeoMark Research, Inc., Houston, Texas 77095.

Organic Geochemistry, edited by Michael H. Engel and Stephen A. Macko. Plenum Press, New York, 1993.

TABLE I. Gross Compositional and Gas-Chromatographic Data

	Sample name						
	EV RFT#4 D5401	EV DST#2 D5493	EV DST#4 D5494	EV DST#8 D5497	PFO DST#1	CG-9	PH250
Depth (m)	2336	2264	2042	1785	2168	1363	—
Formation	Lajas	Lajas	Punta Rosada	Sierras Blancas	—	Lotena	Agrio
Gross composition							
$<C_{15+}$	20.6	23.2	19.5	33.7	15.1	31.5	24.1
API gravity	35.2	34.3	34.7	42.3	23.1	30.4	24.3
% S	0.47	0.42	0.08	0.07	0.04	0.52	0.25
% Saturates	67	68	68	73	48	47	61
% Aromatics	20	23	25	12	19	23	25
% NSO	9	5	4	13	17	22	10
% Asphaltenes	5	4	3	2	17	8	5
C_{15+} *Hydrocarbon composition*							
% Paraffins	36	38	44	39	40	20	11
% Naphthenes	41	37	29	47	32	47	60
% Aromatics	23	25	27	14	28	33	29
Stable carbon isotope composition							
$\delta^{13}C$ (Saturates)	−31.92	−31.20	−31.70	−28.70	−33.94	−29.08	−28.02
$\delta^{13}C$ (Aromatics)	−30.88	−30.10	−31.50	−28.70	−32.79	−27.60	−26.79
Gas-chromatographic data							
Pr/Ph[a]	1.02	0.93	1.99	1.25	1.52	0.79	0.90
Pr/C_{17}	0.50	0.57	0.39	0.32	0.43	0.63	1.02

[a]Pr/Ph, Pristane/phytane ratio.

Figure 4 shows the C_{15+} saturate gas chromatograms of the oils from the EV well. From these chromatograms, *n*-paraffin distributions (Fig. 5) and isoprenoid cross plots (Fig. 6) can be constructed. The *n*-paraffin distribution of EV8 is significantly different from those of the other EV oils (Fig. 5) in that it is less waxy with a greater abundance of lighter paraffins. Either EV8 is significantly more mature than EVR, EV2, or EV4 or it has another source. In comparison with other Neuquen Basin oils, the PFO oil is waxier than EV2 with a more distinct odd-carbon *n*-paraffin preference; Vaca Muerta oils tend to have a slight even-carbon preference (Leenheer, 1984).

Regarding the C_{27} to C_{34} *n*-paraffin region of the gas chromatograms in Fig. 4, it can be seen that the unresolved "hump" and certain biomarkers (e.g., the peak eluting immediately after *n*-C_{31}) decrease in relative abundance in the order EV2 > EVR >> EV4. This suggests that EV2 and EV4 are less thermally mature than EV4, assuming an equivalent source. Also, pristane/phytane ratios should increase with maturity while pristane/*n*-C_{17} values should decrease. In the isoprenoid plot shown in Fig. 6, EV4 could be a more mature oil than EV2 and EVR, again assuming the same source for these three oils. The source/maturity relationships among the Vaca Muerta, PFO, and EV oils are not clear from Fig. 6, although EV8 could be a more mature Vaca Muerta oil.

The stable carbon isotope diagram (Sofer, 1984) in Fig. 7 greatly clarifies the source relationship of the Neuquen Basin oils. It appears that EVR, EV2, and EV4 are genetically related while EV8 has an isotopic composition similar to that of the Vaca Muerta oils (CG-9 and PH250). The PFO oil is isotopically lighter than the EV oils, suggesting yet another source, perhaps a different facies of the same formation that generated the EV oils.

In order to confirm and expand upon the source/maturity relationships suggested from the carbon isotopic data, terpane and sterane biomarker distributions were obtained. The tricyclic terpane, pentacyclic triterpane, and sterane distributions are shown in Figs. 8A, 8B, and 8C, respectively. Table II identifies the peaks labeled in Fig. 8. The close similarity between the EV2 and EVR tricyclic terpanes (Fig. 8A) again suggests equivalent sources, while the greater background noise and relative abundance of the C_{19} relative to, for example, the C_{25} component for the EV4 tricyclic terpanes suggests greater cracking and increased thermal maturity (Leenheer and Zumberge, 1987), consistent with the gas-chromatographic results. Because of the small amounts of tricyclic ter-

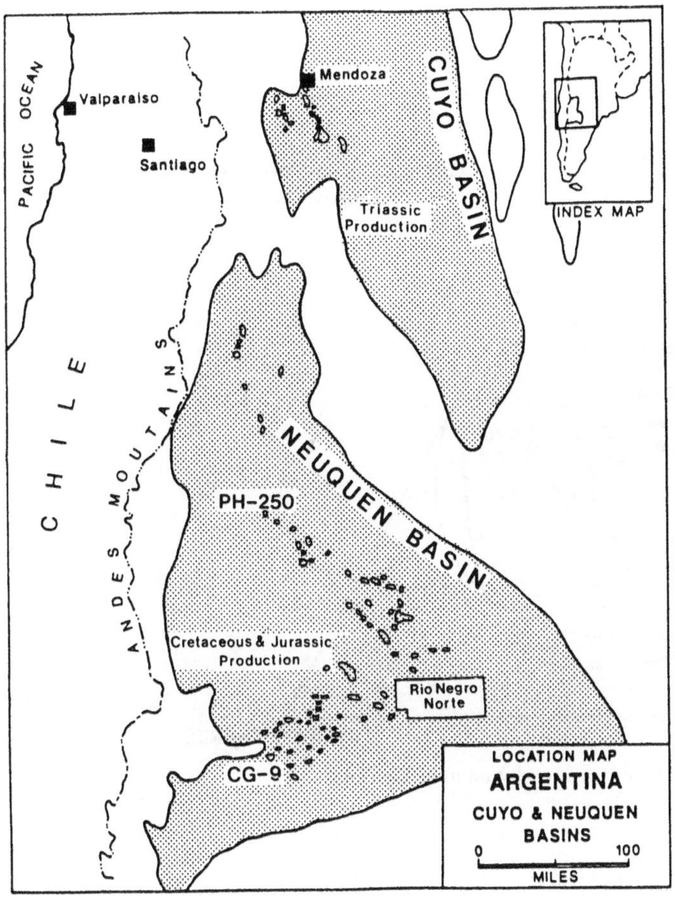

Figure 1. Location map.

panes in the EV8 oil, it is difficult to use these compounds as correlation parameters, although the Vaca Muerta oils also contain relatively small quantities of tricyclics (Leenheer, 1984). The PFO oil contains relatively more of the C_{21} component, suggesting some difference in source facies.

The triterpane distributions (Fig. 8B) of EV2 and EVR are unique in that a large gammacerane (shaded peak I) component is present. Gammacerane is abundant in many crude oils generated from lacustrine source rocks (Zumberge, 1987a) such as the Colorado Green River marl, many China source rocks, and pre-salt Cretaceous sediments from West Africa. However, it is now generally accepted that a large gammacerane component is an indicator of hypersaline/evaporite depositional environments and is not necessarily a lacustrine marker (ten Haven *et al.*, 1988). Although not shown in Fig. 8B, the C_{35} extended hopanes are more abundant than the C_{34} hopanes, another indication of hypersaline conditions. Gammacerane is much less abundant in EV4, PFO, and EV8, which may be due to maturity differences in EV4 and the absence of this compound in Vaca Muerta oils (Leenheer, 1984) for EV8. In fact, the C_{27} Ts/Tm (A/B) and C_{29*}/C_{29} (Cl/C) triterpane ratios are known to increase with increasing thermal maturity (Seifert and Moldowan, 1978; Leenheer and Zumberge, 1987) for oils with equivalent sources. This is exactly the case for EV4

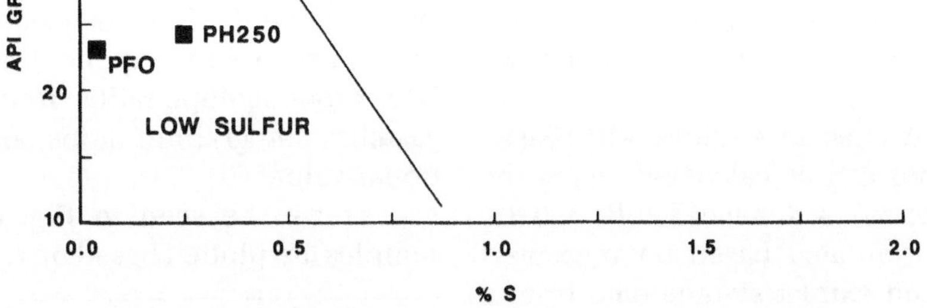

Figure 2. API gravity vs. percent sulfur for Neuquen Basin oils.

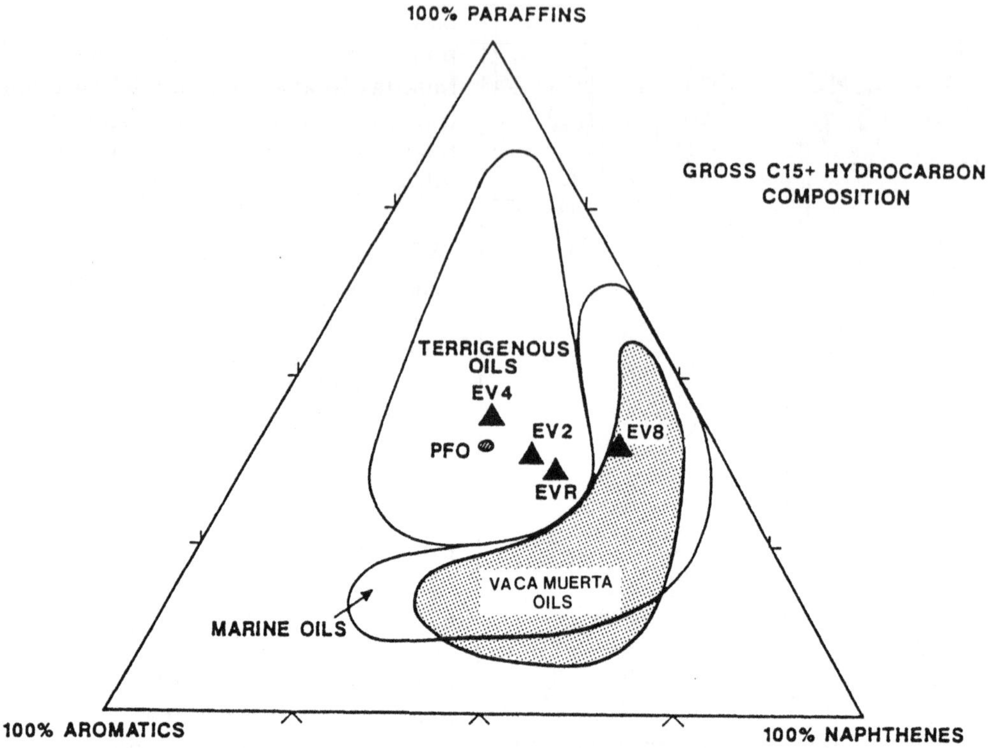

Figure 3. Gross C_{15+} hydrocarbon composition of Neuquen Basin oils.

relative to its lower maturity counterparts, EVR and EV2. Also, these maturity indicators suggest that EV8 is at least as mature as the Vaca Muerta oils CG-9 and PH250. Selected biomarker ratios for all oils are summarized in Table III.

Sterane distributions are shown in Fig. 8C. The relative distributions of peaks 11, 18, and 22 (C_{27}, C_{28}, and C_{29} 20*R* configuration, respectively) are a function of source; the ratio of peak 19 to peak 22 (C_{29} 20*R*/20S) is a function of thermal maturity; and the ratio of peak 1 to peak 11 (C_{27} diasterane/C_{27} regular sterane) is dependent upon both source and maturity (e.g., Seifert and Moldowan, 1978; Zumberge, 1987b). Again, the sterane distributions are the same for EVR and EV2, suggesting the same source and similar thermal histories. Oil EV4 steranes are consistent with a more mature counterpart of EV2, while EV8 appears to be from another source of about the same maturity as EV4. EV8 steranes are similar to Vaca Muerta steranes, and PFO steranes are poorly resolved (Leenheer, 1984).

Equivalent vitrinite reflectance values (VRE) based upon C_{29} 20S/20*R* ratios can be calculated using the following relationship: $\%R_0 = 0.49(20S/20R) + 0.33$. This equation was formulated based on measured vitrinite reflectance and extract sterane data from a number of sediment samples of varying thermal maturities and depositional environments (Zumberge, 1984). EV2 and EVR oils were generated from sediments at R_0 values equal to about 0.73% while EV4 and EV8 have VRE values of approximately 0.86%.

3. Conclusions

In order to coherently tie together all of the disparate geochemical data, multivariate statistical techniques can be used (e.g., Zumberge, 1987a; Engel *et al.*, 1988). In this manner, variables based on gas chromatography, stable carbon isotopes, and biomarker distributions can all be employed to form new variables (factor or principal component analysis) which best describe the variation in the data. The pattern recognition computer program EinSight (Infometrix, Inc., Seattle) for the personal computer was applied to this correlation problem in the Neuquen Basin. Table IV lists the geochemical variables used in the principal component analysis; these include two carbon isotope ratios, four biomarker ratios, two gas-chromatographic ratios, and one gross compositional value.

As can be seen in Fig. 9A, in which the oil samples are plotted based on the two new variables or

Figure 4. C_{15+} saturate gas chromatograms for oils from the Estancia Vieja (EV) well.

EVR

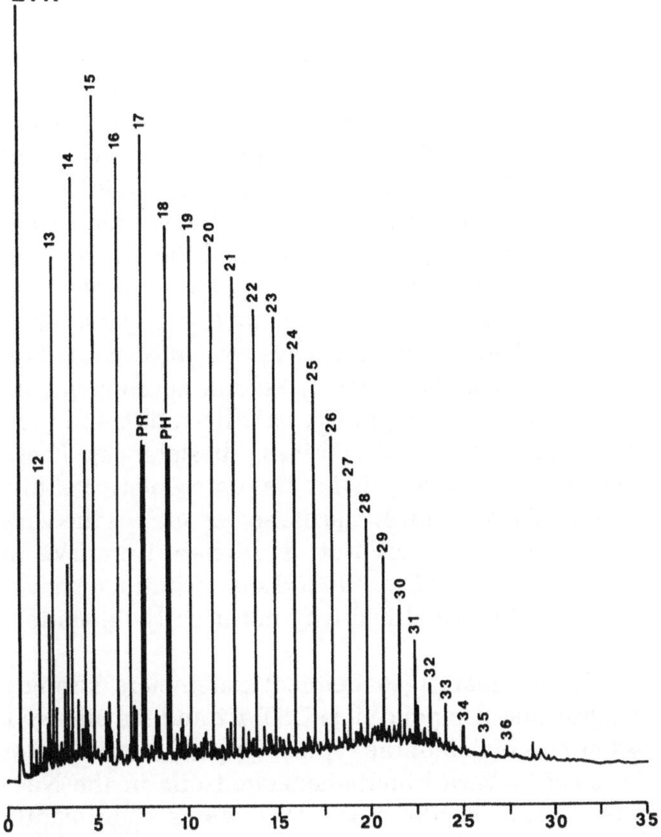

Sample: ESTANCIA X1
Raw: SD5401 Proc: PD5401
Enlargement= 570

EV2

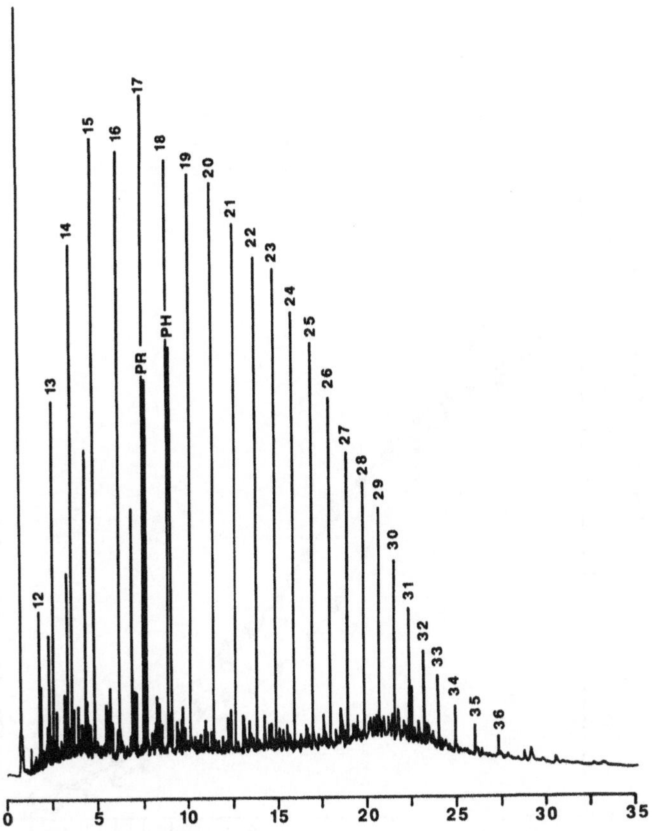

Sample: ESTANV DST2
Raw: SD5493 Proc: PD5493
Enlargement= 310

EV4

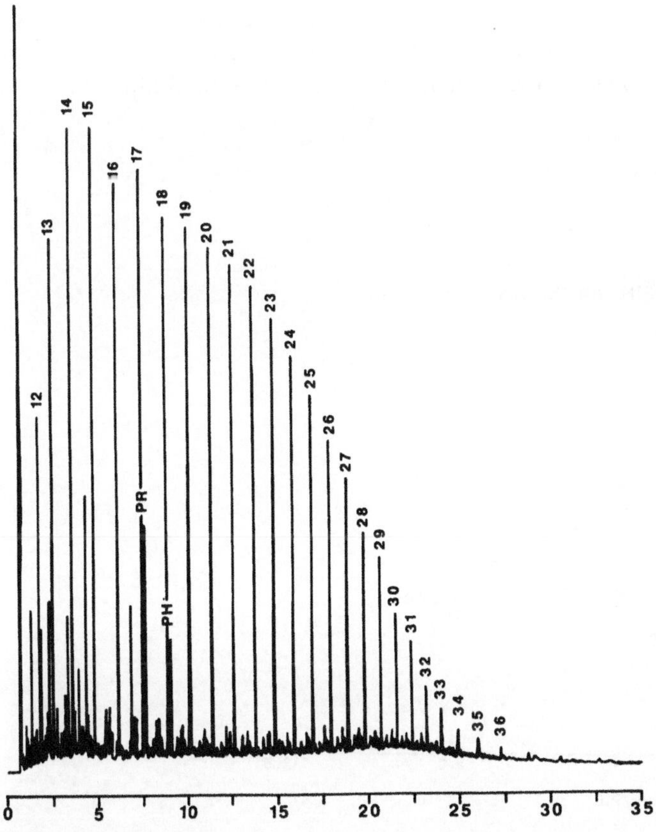

Sample: ESTANV DST4
Raw: SD5494 Proc: PD5494
Enlargement= 130

EV8

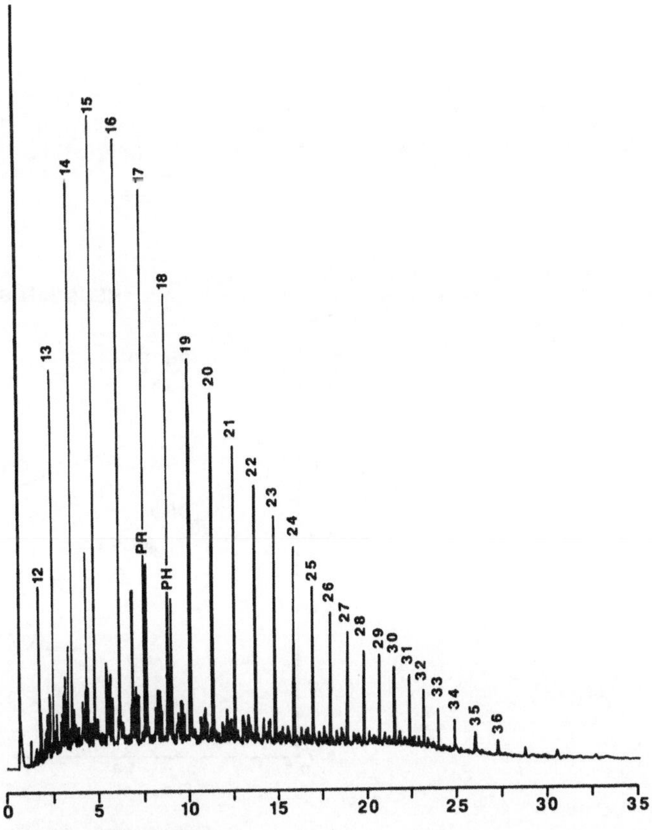

Sample: ESTANV DST8
Raw: SD5497 Proc: PD5497
Enlargement= 190

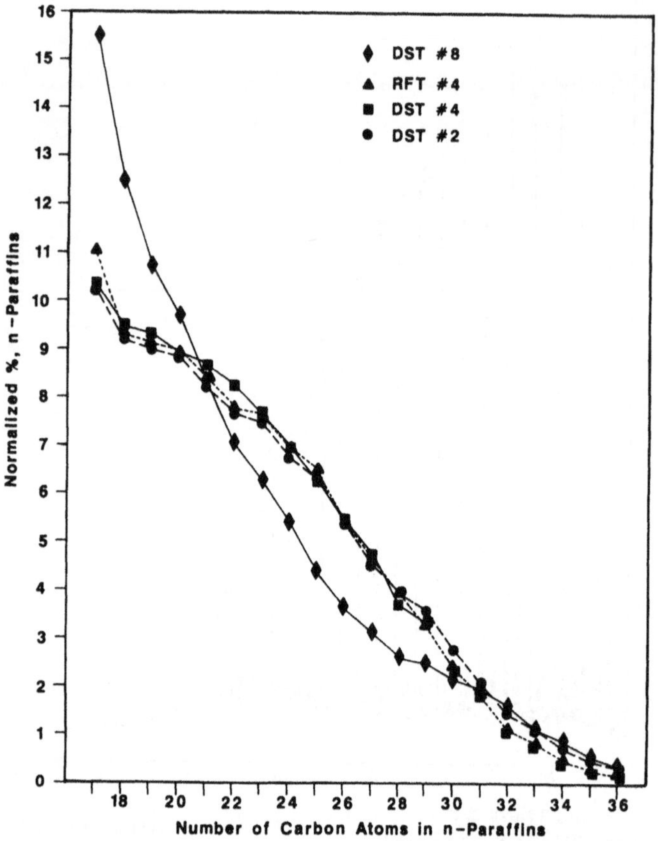

Figure 5. n-Paraffin distributions for the EV oils.

factors constructed by the program (and accounting for over 85% of the total variation in the data), EV2 and EVR are highly correlated, while EV8 appears to be associated with the Vaca Muerta oils (CG-9 and PH250). Oil EV4 is closer in composition to PFO than the other oil groups. Figure 9B illustrates which original geochemical parameters caused the sample separations or similarities in Fig. 9A. For example, EVR and EV2 have high gammacerane/C_{31} hopane ratios while the Vaca Muerta oils have more positive carbon isotope vales and a larger C_{27} sterane component, are less waxy (high n-C_{17}/n-C_{27} paraffin ratio), and contain more rearranged steranes (diasteranes). These geochemical characteristics are representative of oils generated from typical marine shales such as the Vaca Muerta Formation. EV4 oil is different from EV2 in that it has a larger C_{29} 20R/20S sterane ratio, a known measure of thermal maturity not usually dependent on source.

In summary, geochemical parameters strongly suggest that Estancia Vieja DST #2 and #4 and RFT #4 are unrelated to the typical Upper Jurassic/Lower Cretaceous Vaca Muerta-generated oils in the Neuquen Basin. Furthermore, DST #4 is significantly more mature than DST #2. These oils were likely generated from source rocks deposited in a hypersaline environment, perhaps a facies of the Lower Jurassic Los Molles Formation. However, DST #8 from the Sierras Blancas Formation appears to have been sourced from the Vaca Muerta shale.

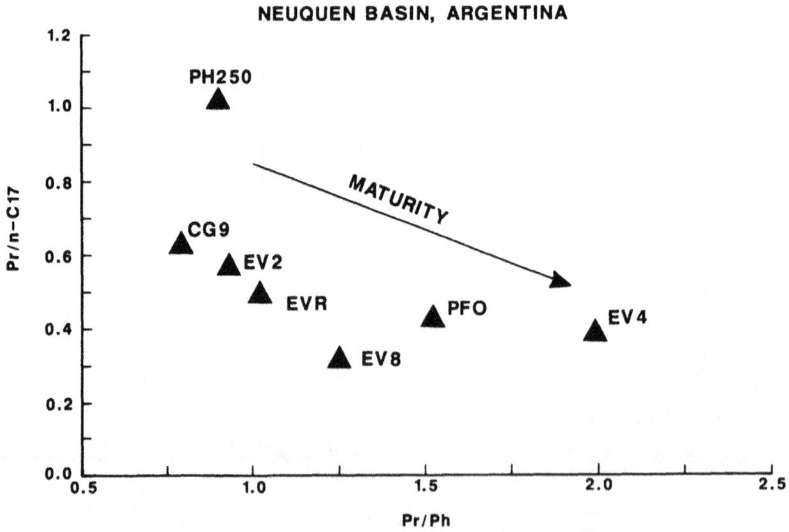

Figure 6. Isoprenoid ratios for Neuquen Basin oils: Pristane/phytane ratio (Pr/Ph) vs. Pr/n-C_{17}.

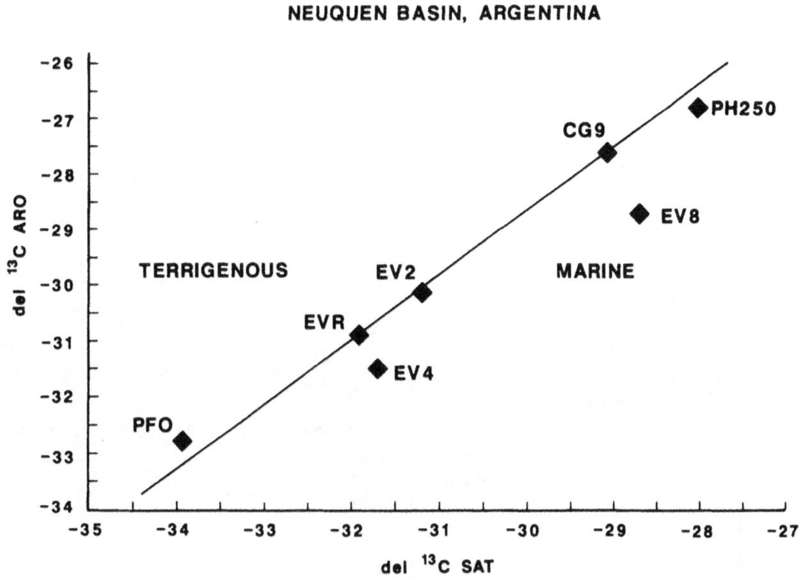

Figure 7. Stable carbon isotope ratios for Neuquen Basin oils: Saturates vs. aromatics.

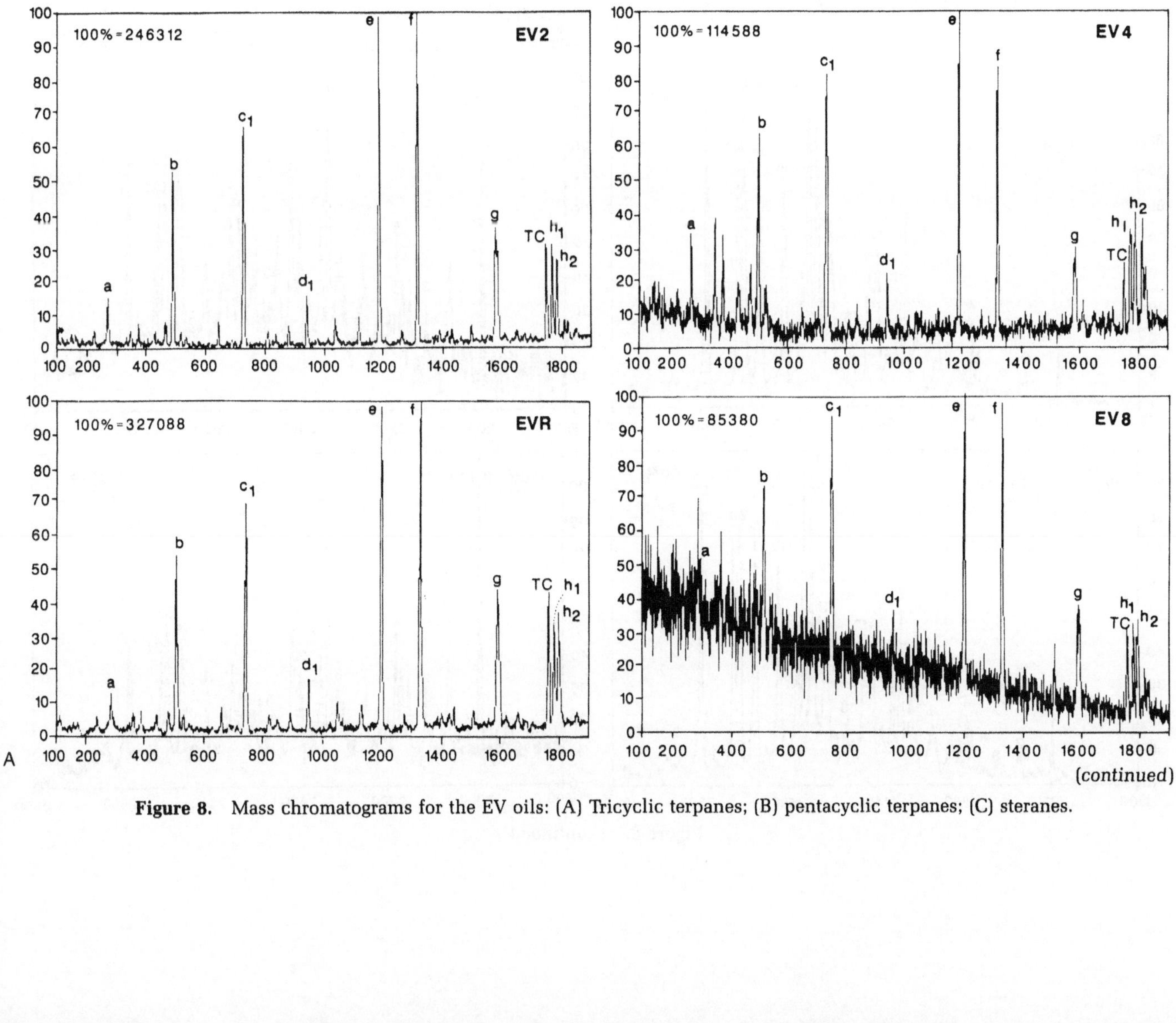

(continued)

Figure 8. Mass chromatograms for the EV oils: (A) Tricyclic terpanes; (B) pentacyclic terpanes; (C) steranes.

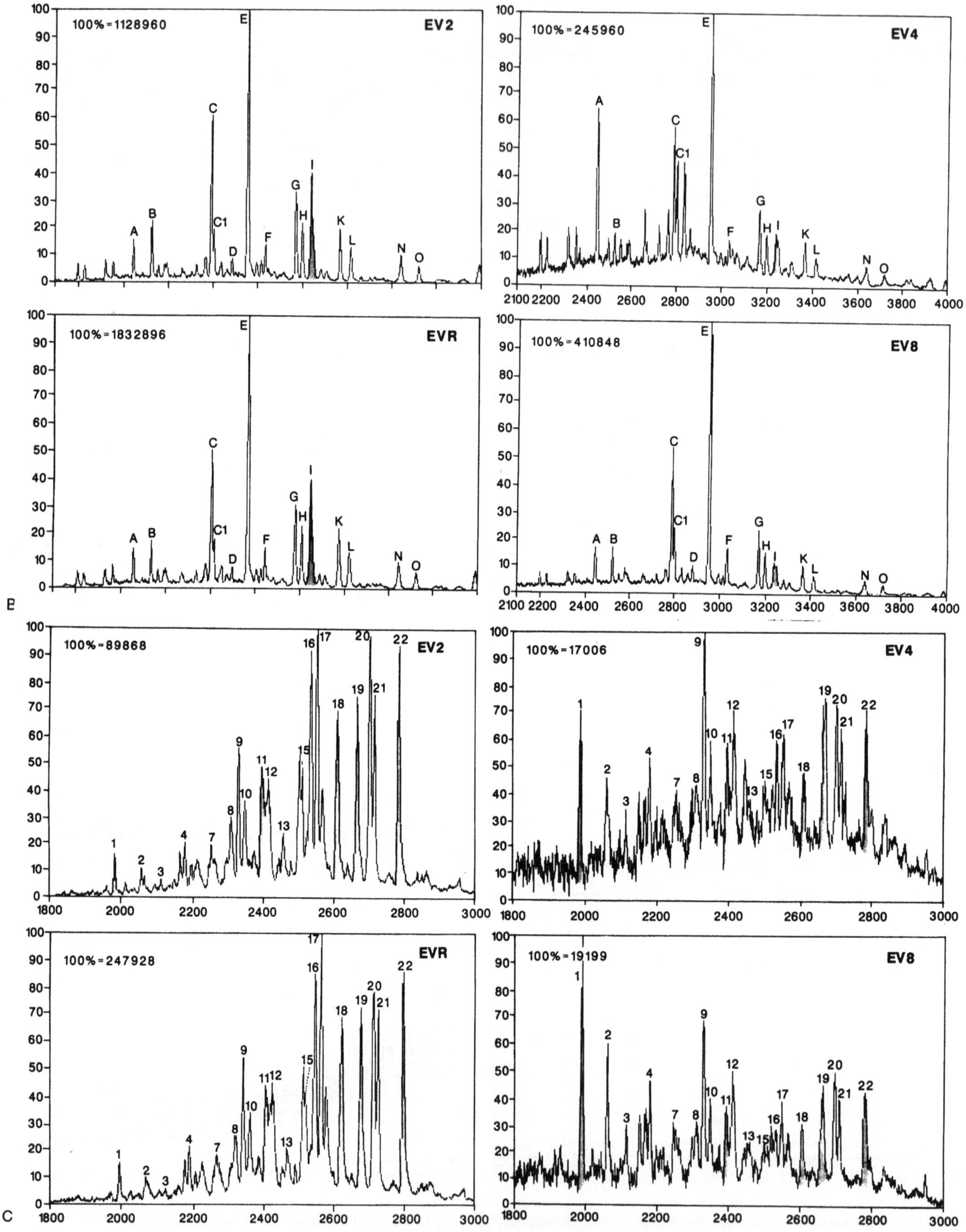

Figure 8. *(Continued)*

TABLE II. List of Identified Peaks in Fig. 8

Peak	Formula	Weight	Identification	Peak	Formula	Weight	Identification
			Steranes[a]				Terpanes
1	$C_{27}H_{48}$	372	13β,17α-Diacholestane (20S)	a	$C_{19}H_{34}$	262	Tricyclic diterpane
2	$C_{27}H_{48}$	372	13β,17α-Diacholestane (20R)	b	$C_{20}H_{36}$	276	Tricyclic diterpane
8	$C_{27}H_{48}$	372	5α-Cholestane (20S) + 5β-cholestane (20R)	c_1	$C_{21}H_{38}$	290	Tricyclic diterpane
				d_1	$C_{22}H_{40}$	304	Tricyclic diterpane
9	$C_{27}H_{48}$	372	5α,14β,17β-Cholestane (20R) + 13β,17α-diastigmastane (20S)	e	$C_{23}H_{42}$	318	Tricyclic terpane
				f	$C_{24}H_{44}$	332	Tricyclic terpane
11	$C_{27}H_{48}$	372	5α-Cholestane (20R)	g	$C_{25}H_{46}$	346	Tricyclic terpane
12	$C_{29}H_{52}$	400	Rearranged C_{29} sterane	h_1, h_2	$C_{26}H_{48}$	360	Tricyclic terpane
15	$C_{28}H_{50}$	386	5α-Ergostane (20S)	TC	$C_{24}H_{42}$	330	Tetracyclic terpane
16	$C_{28}H_{50}$	386	5α,14β,17β-Ergostane (20R) + 5β-ergostane (20R)	A	$C_{27}H_{46}$	370	18α,21β-22,29,30-Trisnorhopane
18	$C_{28}H_{50}$	386	5α-Ergostane (20R)	B	$C_{27}H_{46}$	370	17α,21β-22,29,30-Trisnorhopane
19	$C_{29}H_{52}$	400	5α-Stigmastane (20S)	C	$C_{29}H_{50}$	398	17α,21β-30-Norhopane
20	$C_{29}H_{52}$	400	5α,14β,17β-Stigmastane (20R)	C1	$C_{29}H_{50}$	398	18α,17α-Methyl-28,30-dinorhopane
21	$C_{29}H_{52}$	400	5α,14β,17β-Stigmastane (20S) + 5β-stigmastane (20R)	D	$C_{29}H_{50}$	398	17β,21α-30-Normortane
				E	$C_{30}H_{52}$	412	17α,21β-Hopane
22	$C_{29}H_{52}$	400	5α-Stigmastane (20R)	F	$C_{30}H_{52}$	412	17β,21α-Moretane
				G, H	$C_{31}H_{54}$	426	17α,21β-30-Homohopane (22S + 22R)
				I	$C_{30}H_{52}$	412	Gammacerane
				K, L	$C_{32}H_{56}$	440	17α,21β-Bishomohopane (22S + 22R)
				N, O	$C_{33}H_{58}$	454	17α, 21β-Trishomohopane (22S + 22R)

[a]Assumes 8β,9α,14α,17α unless otherwise stated; dia = rearranged.

TABLE III. Selected Biomarker Ratios

	Triterpanes			Steranes			
Sample Name	GA/C_{31}[a]	Ts/Tm[b]	$C_{29}*/C_{29}$	Peak 1/11	C_{27}/C_{29}	20S/20R	%VRE[c]
EV RFT #4 D5401	1.35	0.80	0.33	0.35	0.50	0.84	0.74
EV DST #2 D5493	1.25	0.68	0.30	0.36	0.49	0.79	0.72
EV DST #4 D5494	0.71	4.30	0.76	1.50	0.80	1.08	0.86
EV DST #8 D5497	0.42	1.00	0.40	3.10	0.84	1.06	0.85
PFO DST #1	0.60	0.64	0.45	0.43	0.60	1.10	0.87
CC-9	0.14	1.23	0.89	3.45	1.60	1.00	0.82
PH250	0.15	1.73	0.91	2.61	1.41	0.94	0.79

[a]GA, gammacerane; C_{31}, homohopane.
[b]Ts, 18α, 21β-22,29,30-trishorhopane; Tm, 17α, 21β-22,29,30-trisnorhopane.
[c]$\%VRE(R_o) = 0.49\ (20S/20R) + 0.33$.

TABLE IV. Variables Used in Principal Component Analysis

Sample name	Para[a] Naph	n-C_{17} / n-C_{27}	Pr/Ph[b]	$\delta^{13}C$[c] Sat.	Arom.	Terpane GA/C_{31}[d]	Sterane C_{27}/C_{29}[e]	Sterane Peak 1/11[f]	Sterane 20S/20R[g]
EV RFT #4 D5401	0.88	2.4	1.02	−31.9	−30.9	1.35	0.50	0.35	0.84
EV DST #2 D5493	1.03	2.2	0.93	−31.2	−30.1	1.25	0.49	0.36	0.79
EV DST #4 D5494	1.52	2.1	1.99	−31.7	−31.5	0.71	0.80	1.50	1.08
EV DST #8 D5497	0.83	5.0	1.25	−28.7	−28.7	0.42	0.84	3.10	1.06
PFO DST #1	1.25	1.4	1.52	−33.9	−32.8	0.60	0.60	0.43	1.10
CG-9	0.43	2.6	0.79	−29.1	−27.6	0.14	1.60	3.45	1.00
PH250	0.18	3.1	0.90	−28.0	−26.8	0.15	1.41	2.61	0.94

[a]Para, paraffins; Naph, naphthenes.
[b]Pr/Ph, pristane/phytane ratio.
[c]Sat., saturate; Arom., aromatic.
[d]GA, gammacerane; C_{31}, homohopane.
[e]C_{27}, 5α-cholestane; C_{29}, 5α-stigmastane.
[f]11,13β,17α-diacholestane; 11,5α-cholestane.
[g]20S, 5α-stigmastane; 20R, 5α-stigmastane.

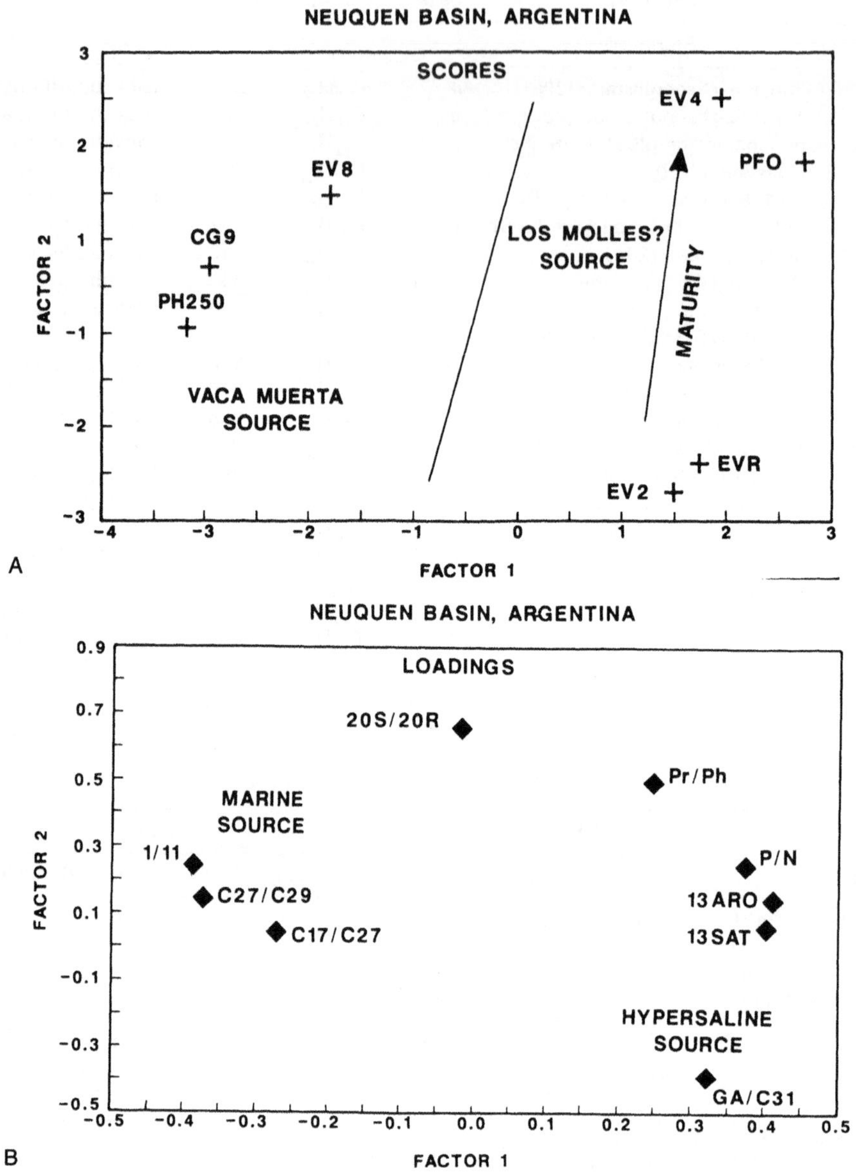

Figure 9. Principal component analysis: (A) Oil samples; (B) geochemical ratios.

Acknowledgments. I thank Janet Williamson and Shirley Sellers for performing the analytical analyses and Occidental International Exploration and Production Company for permission to publish.

References

Engel, M. H., Imbus, S. W., and Zumberge, J. E., 1988, Organic geochemical correlation of Oklahoma crude oils using R- and Q-mode factor analysis, *Org. Geochem.* **12**:157–170.

Leenheer, M., 1982, Organic Geochemistry of Oils, Cores, Cuttings, and Outcrop Samples from the Neuquen Basin, Argentina, Cities Service Research Technical Report #104, G82-09, Tulsa, Oklahoma.

Leenheer, M., 1984, Oil–Source Rock Correlation in the Neuquen Basin, Argentina, Cities Service Research Technical Report #126, G84-14, Tulsa, Oklahoma.

Leenheer, M., and Zumberge, J., 1987, Correlation and thermal maturity of Williston Basin crude oils and Bakken source rocks using terpane biomarkers, in: *Williston Basin: Anatomy of a Cratonic Oil Province* (M. Longman, ed.), Rocky Mountain Association of Geologists, Denver, pp. 287–301.

Seifert, W., and Moldowan, M., 1978, Application of steranes, terpanes, and monoaromatics to the maturation, migration, and source of crude oils, *Geochim. Cosmochim. Acta* **42:** 77–95.

Sofer, Z., 1984, Stable carbon isotope composition of crude oils: Application to source depositional environments and petroleum alterations, *Am. Assoc. Petrol. Geol. Bull.* **68**:31–49.

Sofer, Z., Zumberge, J. E., and Lay, V., 1986, Stable carbon isotopes and biomarkers as tools in understanding genetic relationship, maturation, biodegradation, and migration of crude oils in the Northern Peruvian Oriente (Maranon) Basin, *Org. Geochem.* **10**:377–389.

ten Haven, H., de Leeuw, J., Sinninghe Damsté, J., Schenck, P., Palmer, S., and Zumberge, J., 1988, Application of biological markers in the recognition of paleo-hypersaline environments, in: *Lacustrine Petroleum Source Rocks* (A. J. Fleet, *et al.*, eds.), Geological Society Special Publication No. 40, Blackwell Scientific, Oxford, pp. 123–130.

Vlierboom, F. W., Collini, B., and Zumberge, J. E., 1986, The occurrence of petroleum in sedimentary rocks of the meteor impact crater at Lake Siljan, Sweden, *Org. Geochem.* **10**:153–161.

Zumberge, J. E., 1984, Terpenoid Biomarker Distributions in Low Maturity Crude Oils, Cities Service Research Report #140, G84-18, Tulsa, Oklahoma.

Zumberge, J. E., 1987a, Prediction of source rock characteristics based on terpane biomarkers in crude oils: A multivariate statistical approach, *Geochim. Cosmochim. Acta* **51**:1625–1637.

Zumberge, J. E., 1987b, Terpenoid biomarker distributions in low maturity crude oils, *Org. Geochem.* **11**:479–496.

Chapter 21

Oil to Source Rock Correlation

Concepts and Case Studies

JOSEPH A. CURIALE

1. Introduction

The relationship between a crude oil and its source rock is of interest for scientific and economic reasons, and understanding this relationship is a major goal of organic geochemistry. Successful oil–source rock correlations require knowledge of petroleum generation, expulsion, migration, and entrapment, along with a detailed molecular understanding of how petroleum composition alters during these processes. Pairing oil with its source rock represents the culmination of the various technical aspects of organic geochemistry as applied in the exploration for, and prediction of, oil and gas accumulations.

An oil–source rock correlation is the discovery of a genetic relationship between a petroleum source rock and a crude oil that is consistent with all known geological and geochemical facts. The interdisciplinary character of oil–source rock correlation efforts requires close cooperation among geochemists and geologists before a successful solution can be proposed. Because correlation questions are basically geological problems addressed with geochemical methods, an understanding of basinal history and structural setting is critical. Consequently, efforts at geologically "blind" correlations, such as isolated molecular-matching schemes, are likely to be unsuccessful. It has become increasingly clear that a simple set of chemical correlations between an oil and the extractable organic matter in a proposed source rock is not sufficient to establish an oil–source correlation. Considerable supporting geologic data are also necessary. When such supporting data are unavailable, the limitations of solely chemically based correlations must be clearly stated, and conclusions tempered with experienced judgment.

The economic importance of correctly assigning a crude oil to its source rock is one aspect of the industrial applications of organic geochemistry. The primary goal of the petroleum explorationist is the discovery of commercial accumulations of hydrocarbons, not the discovery of reservoir, structure, or trap; a prerequisite for the existence of such accumulations is an effective petroleum source rock. Without such a rock, reservoir porosity, structural closure, and trap integrity become problems of academic interest. Conversely, with the discovery of an effective source rock, determination of expulsion volumes and pathways can be pursued by geologic methods.

The current interest in fluid volumetric estimations in sedimentary basins (Mackenzie and Quigley, 1988) emphasizes the major economic application of oil–source rock correlations. Mass-balance calculations now make possible the estimation of undis-

JOSEPH A. CURIALE • Unocal Energy Resources, Brea, California 92621.
Organic Geochemistry, edited by Michael H. Engel and Stephen A. Macko. Plenum Press, New York, 1993.

covered oil quantities, given known oil reserves and estimates of hydrocarbon fluid expulsion factors for petroleum source rocks (England *et al.*, 1987). Because such a mass-balance effort depends upon a particular source unit giving rise to a particular oil, it is clear that the first required criterion for establishing the validity of such estimations is the existence of a geochemically verifiable, geologically reasonable oil–source rock correlation.

The purpose of this chapter is to explore documented, successful cases of oil–source rock correlation, with an emphasis on how the search for petroleum is assisted by correlation concepts and applications. A conceptual framework for oil–source rock correlations is presented, stressing the understanding that is required to generate geologically acceptable results. The strong interaction between geologic knowledge and chemical conclusions will be stressed, along with the potential pitfalls encountered when data are misinterpreted. A generalized design for oil–source rock correlations is also presented, to provide the reader with a working format for future such efforts. Once the conceptual groundwork is laid, a discussion of typical case studies is pursued, with emphasis on the interpretation of geochemical source rock and petroleum data within a geologic framework. Case studies addressing specific exploration situations are stressed. The chapter concludes with an analysis of the future directions of oil–source rock correlation efforts, from both a conceptual and an analytical point of view.

2. Oil–Source Rock Correlations—A Conceptual Framework

As we have noted earlier, oil–source rock correlations are based on chemical relationships supported by geologic data. Numerous correlation techniques have been proposed, using both bulk geochemical data and detailed molecular distributions (See Philp, this volume, Chapter 19). Whereas differences of opinion exist among geochemists concerning the utility of these techniques, it is critical that we recognize the importance of using a diversified set of methods in oil–source rock correlation. As is the case with most attempts to understand geologic data, no single method can provide definitive proof of the existence of a genetic relationship between oil and rock; supportive evidence from numerous correlation techniques is necessary. This need becomes even more critical when the importance of distinguishing genetic and nongenetic correlation parameters is considered.

2.1. Genetic and Nongenetic Correlation Parameters

All oil–source rock correlation techniques rely upon a positive comparison between crude oil and the organic matter of a prospective source rock (either solvent-extractable *bitumen* or insoluble *kerogen*). Further, although the organic geochemist may utilize several techniques for correlation purposes, he relies upon the validity of a single concept, namely, that the particular parameter under study is invariant under all nongenetic processes. In other words, for each specific method used, it must be clear that the observed parameter variation is due solely to variability in the *origin* of the organic matter involved. This concept governs the interpretation of all oil–source rock correlation data and, where neglected, can lead to gross errors concerning the origin of oil in sedimentary basins. Successful correlation efforts therefore rely upon distinguishing genetic and nongenetic influences on crude oil chemistry, which in turn requires an understanding of the origin of oil. For the purposes of the following discussion, the objective of oil–source rock correlation efforts will be to resolve the effects of nongenetic overprinting on original organic matter characteristics.

Four basic processes operate on the chemical composition of a crude oil. Only the first of these, *the influence of original organic matter (source) input*, is considered genetic in character. The second, *thermal maturation of organic matter*, involves kinetic considerations governing compositional alteration. The third and fourth processes, *migration* and *postmigration alteration*, cause the most severe nongenetic compositional changes. All chemical characteristics of all oils derive from some combination of these four processes.

At the instant of separation of an oil from its source rock, the differences in the chemical composition of the oil and the soluble organic matter retained in the source rock are a function of migration fractionation (Leythaeuser *et al.*, 1987). Following departure of the oil from its sourcing location, several processes alter both the oil and the organic matter remaining in the source. In the case of the source rock, maturation processes continue, further altering the chemical composition of both the kerogen and the bitumen. The oil itself is affected by further migration fractionation and postmigrational alteration, the latter occurring mostly after the oil has reached a reservoir. Each of these factors will change the chemical character of the organic matter in the source rock or the oil, resulting in extensive overprinting of the original genetic characteristics linking the oil to its source, and thus complicating the correlation attempt.

2.2. The Occurrence of Multiple Source Intervals

Our discussion to this point has assumed that there is a one-to-one correspondence between an oil and its source rock; early oil–source rock correlation studies were conducted under this assumption. It was initially convenient to consider that a single source rock had sourced a single chemically distinctive crude oil, which had migrated in a single pulse to a nearby trap. As exploration concepts grew more sophisticated, it became evident that the situation was considerably more complex and that such one-to-one correspondences should not be assumed. In fact, given the multiple sourcing horizons now known to exist in most sedimentary basins, it is reasonable to expect that the contents of a subsurface trap represent a mixture of fluids from more than one source.

The possibility that a single oil may in fact represent the chemical summation of fluids from multiple sources demonstrates the need for geologic input to any oil–source correlation effort. Specifically, basin-wide understandings of paleogeologic structural trends and paleohydrology become critical factors in assessing past fluid avenues (England *et al.*, 1987). Further, the chemical overprinting effects of migration through multiple pathways, and the consequences of mixing of fluids having differing initial source maturities, must be understood. At the least, it is apparent that the maturities and source input characteristics of a crude oil could represent weighted averages of fluids expelled from the various responsible source units.

The problem of multiple source contributions to a single oil also raises a fundamental question concerning the nature of an oil–source rock correlation. Whereas the relationship between an oil and its specific source rock has been pursued for economic reasons (as noted earlier), it seems clear that from a practical perspective such a correlation is actually established, not between oil and rock, but between oil and organic facies (Jones, 1987). This conceptual distinction is critical, because it implies that two sedimentary units containing the identical indigenous organic matter would be indistinguishable using the genetic correlation criteria discussed earlier. Again, this stresses the importance of applying geologic input to the problem. Only a strong geologic framework for the basin in question would allow us to choose the correct oil–source rock relationship from among the possible oil–organic facies relationships. The recognition of such problems also raises questions about published conclusions of source rock character based upon crude oil chemistry. Such problems are obviously real, and their solutions represent challenges to those attempting oil–source rock correlations in frontier basins.

2.3. Design Concepts for Basin-Wide Oil–Source Rock Correlation Projects

Many of the concepts discussed in this section are summarized in the basin-wide oil–source rock correlation project design chart shown in Fig. 1. This overview is intended to provide an ideal schema containing the major steps and objectives of an integrated geologic and geochemical oil–source rock correlation effort. It is stressed that this comprehensive approach is a model only; complete details of all inputs will rarely be available in any single instance.

The correlation chart begins with the factor most critical to a successful final conclusion: accurate knowledge of the geologic history of the sedimentary basin. This input parameter consists mainly of a geologic basin model and includes information about tectonic history, depositional environment, stratigraphic succession, and subsequent structural deformation. A knowledge of paleohydrodynamics is important as input for assessing the reliability of subsequent correlations, particularly those requiring long-distance migration in carrier beds. Once the geologic history is reasonably understood, sample selection may begin. This aspect of the project design is often the most limiting: sample availability usually cannot be controlled by the analyst. Subsurface rock samples and unaltered oils are only available where previous drilling has provided them. In addition, oil seeps and outcrop rock samples, while perhaps more accessible, are available only where they are adequately exposed. These problems are further exacerbated in cases where the sedimentary basin is predominantly offshore.

When an adequate sample suite is assembled, oils are grouped into genetic families and rock samples are evaluated for source potential, using techniques presented elsewhere in this volume. At this point, a suite of samples is chosen for use in the geochemical aspect of the oil–source rock correlation. This suite will include representative (and, if available, unaltered) oils from each genetic family

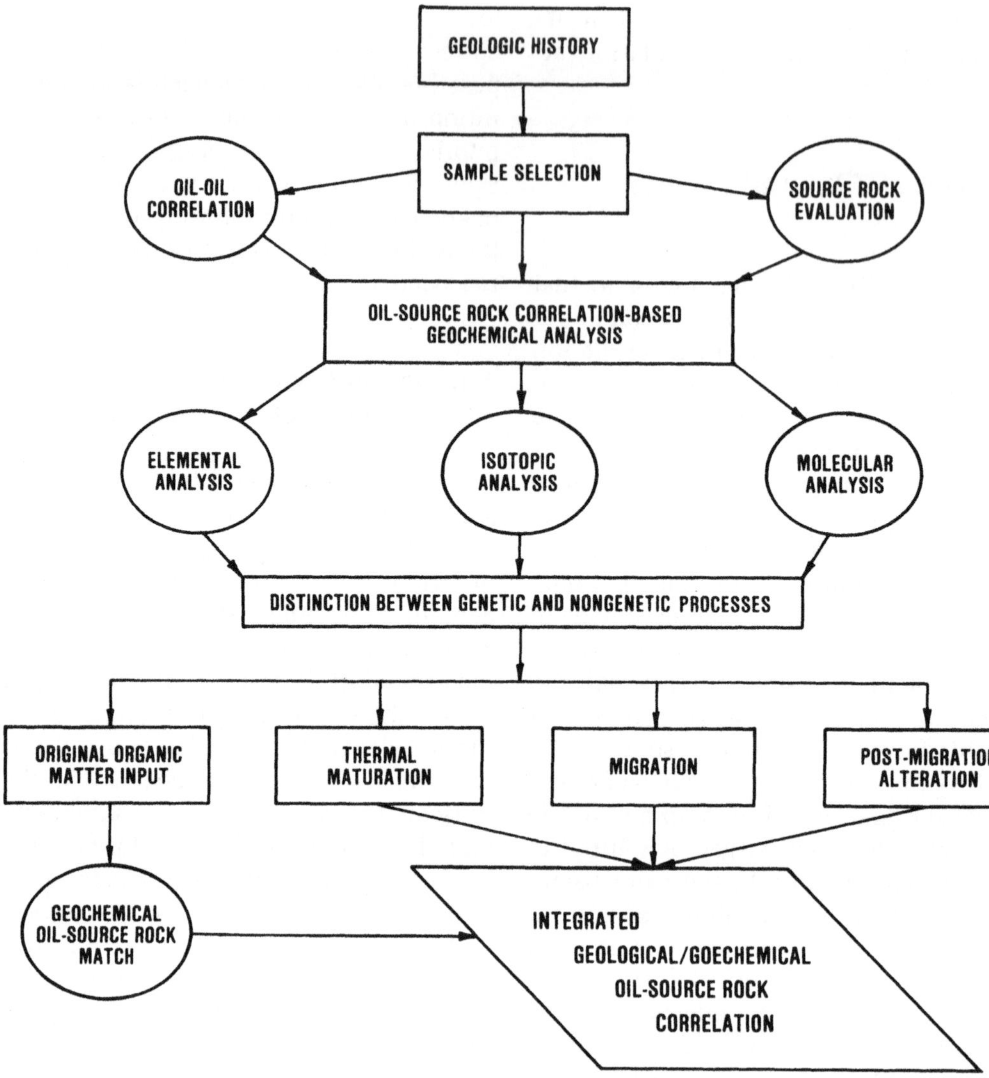

Figure 1. Conceptual framework for the project design of a typical integrated geological/geochemical oil–source rock correlation project. See the text for a conceptual discussion of the various stages of this design. For oil–oil correlation and source rock evaluation details, see Chapters 19 and 20 in this volume.

and rock samples representative of each potential petroleum source rock unit in the basin.

Subsequent analyses of source rock organic matter are then chosen to complement those already available as a result of the previous oil–oil correlation. These are summarized as elemental, isotopic, and molecular analyses (examples of which are presented later with the case studies). Analytical results must then be used to distinguish among the effects of the four processes responsible for the chemical composition of the organic matter in the source rock and oil. The accuracy of the geochemical result is most dependent upon the correct solution to this problem, as the case studies below demonstrate.

When the effects of original organic matter input are differentiated from those of nongenetic processes, the geochemical oil–source rock match is complete. Combining this result with the influences of thermal alteration, migration, and postmigration alteration will finally provide an integrated geological/geochemical oil–source rock correlation. The purpose of this important last step is to mesh the geochemical relationship, based on an analytical match, with the geologic realities as initially understood. The process is an iterative and synergistic one, whereby analytical correlation knowledge and geologic understanding are simultaneously enhanced. In the following section, several case studies will be cited and discussed, in an effort to show how the application of this basic project design can lead to increased understanding of real basins.

3. Case Studies

3.1. Review of Early Oil–Source Rock Correlation Efforts

Whereas the genetic link between crude oil and organic-rich sedimentary rocks has served as a working hypothesis within organic geochemistry for most of this century, prior to the mid-1950s no geologic studies reliably documented such a link. As late as 25 years ago, most oil–source rock "correlations" were based entirely upon geologic inference, without any support from geochemical data (e.g., Barbat, 1967). However, with the advent and widespread availability of gas-chromatographic (GC) and mass-spectrometric (MS) techniques in the late 1960s, detailed molecular correlations began to appear in oil company reports and eventually in the open literature (Barker, 1981). Technological advances and geologic acceptance have now progressed to the point where even the most sophisticated correlation techniques are commonplace.

Excellent reviews of published oil–source rock correlations are available (Barker, 1981; Hunt, 1979, pp. 497–513; Tissot and Welte, 1984, pp. 561–570), and the reader is directed to these for a detailed synopsis of early efforts. This section will briefly cite those studies which document technical or conceptual breakthroughs in the field. Following this section, case studies involving common geologic situations are discussed in some detail.

John Hunt and colleagues carried out the earliest well-documented effort at oil–source rock correlation, which involved noncommercial liquid and solid oil deposits in the Uinta Basin of Utah (Hunt *et al.*, 1954). These workers used infrared spectra, refractive indices, solvent-extractable organic matter yields, compound class distributions, and elemental data (hydrogen and carbon) to relate these deposits to adjacent lacustrine source beds. This classic study was the first attempt of its kind, as well as the first successful integration of oil–oil/oil–source rock correlations with field-checked geologic data. Several important concepts, including migration fractionation and mixing from multiple sources, were introduced to the literature in the context of the genetic relationships between an oil and its source rock. This work became a conceptual model after which future correlation studies could be patterned.

The Uinta Basin was an ideal location for developing oil–source rock correlation methods, in that most of the potential source rocks are still thermally immature, and migration distances are generally short. It remained for later efforts to investigate petroleum generation in more typical areas. Baker (1962) studied the complicated oil–source rock relationships of the Cherokee Group rocks and oils of Kansas and Oklahoma. Although detailed oil–source rock correlations, using today's technical standards, were not possible, Baker recognized the overprinting effects of migration and maturation processes on correlation efforts. We now know that the migrational complexity and potential source mixing present in the marine and nonmarine rocks of these cyclothems make such correlations a formidable problem, even with present-day technology.

As noted earlier, the emergence of gas chromatography and mass spectrometry in the late 1960s led to the application of new methods for oil–source rock correlations. Detailed molecular correlations involving both light (Young and McIver, 1977) and heavy (Welte *et al.*, 1975; Seifert, 1977; Leythaeuser *et al.*, 1977) hydrocarbons became possible, allowing the direct compositional comparison of crude oils with source rock extracts. Subsequently, Seifert (1978) introduced the concept of molecular correlation involving kerogen pyrolysates and crude oils, successfully showing that the insoluble fraction of the source organic matter can be used as a matching parameter. Whereas each of these workers successfully developed methods for oil–source rock correlation, none fit their geochemical conclusions into a strong geologic framework. Such a combination of modern molecular and isotopic methods with robust geologic support was first successfully accomplished by Williams (1974) and Dow (1974), in their classic study of oils and source rocks in the Williston Basin.

Just as the Uinta Basin provided Hunt and coworkers with a conceptually straightforward location for the first oil–source rock correlations, so did the Williston Basin provide Williams and Dow with an ideal site for applying the geochemical correlation methods available in the 1970s. Williams (1974) used carbon isotope ratios, gas-chromatographic analyses of light and heavy hydrocarbons, infrared spectroscopy, optical rotation, and distillation curves to group 125 unaltered oils into three families, which were postulated to be source-distinctive. Three separate source units were then identified in the basin, using geochemical and geological criteria, and were subsequently matched to the three oil types using the same analytical techniques. The successful assignment of one-to-one oil–source rock matches in the study was accommodated by the occurrence of strong seals (salt beds) in the stratigraphic section. These evaporite barriers were subsequently invoked to explain the lack of considerable source mixing in the center of the basin (Dow, 1974).

The Williston Basin effort was the first comprehensive oil–source rock correlation study to involve molecular geochemistry as a correlative parameter. At about the same time, Welte *et al.* (1975), Seifert (1977), and others introduced the use of a particular subdiscipline of molecular geochemistry to the correlation field, namely, biological marker analysis. The application of biomarkers began to revolutionize oil–source rock correlation efforts in the late-1970s, and biomarker analysis still remains the correlation method of choice. In the generalized case studies that follow, biomarker geochemistry plays a key role in deciphering the processes of source organic matter input and crude oil maturation, migration, and alteration.

3.2. Northern Alaska—Multiple Oil Families/ Multiple Source Input

The North Slope of Alaska lies between the Brooks Range on the south and the Beaufort Sea on the north (Fig. 2). The largest producing oil field in North America, Prudhoe Bay, was discovered on the North Slope in 1968, and over 50 billion barrels of oil in place have been discovered throughout the entire region. In this section we discuss the sources of oil on the North Slope of Alaska, as a representative case of multiple oil families derived from multiple source rocks.

3.2.1. Geologic Framework

The petroliferous sedimentary rocks of the North Slope consist of Mississippian through Tertiary sandstones, conglomerates, shales, and carbonates (Fig. 3). Two major rock sequences comprise the sedimentary succession. The older Ellesmerian sequence, deposited through the Early Cretaceous, is derived from a postulated northern landmass, whereas the younger Brookian sequence consists of sediments shed from the southerly Brooks Range since the Early Cretaceous (Bird, 1985). Present-day structure is dominated by (from south to north) the Brooks Range thrust belt rocks, the broad Colville trough, and the eastward-plunging Barrow arch (Fig. 4, top panel). Almost all of the known commercial oil on the Slope is reservoired in close proximity to the Barrow arch and is presumed to be derived from deeper source rocks to the south.

The great volumes of oil in traps along the Barrow arch suggest that significant petroleum source rock potential is present on the North Slope. In addition, present-day subsurface structures suggest that fluid movement northward from the trough to the arch occurs along unconformities and coarse-grained carrier beds; several such conduits are juxtaposed against organic-rich fine-grained rocks (compare Fig. 3 with the bottom panel of Fig. 4). Excellent source rocks, focused migration pathways, good reservoirs, and adequately sealed traps have combined to create a world-class petroleum province.

3.2.2. Identification of Oil Families and Petroleum Source Rocks

Two basic oil types are recognized on the North Slope, defined initially on the basis of family characteristics (Magoon and Claypool, 1981) and more recently on the basis of source rock organic facies (Curiale, 1987a, b). Type A oils (e.g., Prudhoe Bay) have, relative to type B oils (e.g., Umiat Field), high heteroatom contents, high V/(V + Ni) ratios, and high tricyclic/pentacyclic terpane and 17α(*H*)-trisnorhopane/18α(*H*)-trisnorneohopane ratios. A set of chemical parameters used to distinguish the two oil types is shown in Table I. The distinctions between the oil types have been interpreted to reflect different proportions of marine- and terrigenous-derived organic matter in the primary source rock(s). Specifically, type A oils are considered to be derived from source rocks containing organic matter from the marine environment, whereas type B oils contain some organic remains derived from a terrigenous environment.

The numerous fine-grained rock sequences in the sedimentary section of the North Slope (including Cretaceous, Jurassic, Triassic, and Mississippian rocks) suggest that several candidates are available as possible petroleum source rocks. Previous work (Bird, 1985; Claypool and Magoon, 1985) indicates that the Torok Formation and pebble shale unit (Early Cretaceous), the Kingak Formation (Jurassic), and the Shublik Formation (Triassic) are organic-rich over widespread areas of the Slope, whereas Late Cretaceous shales (Colville Group) and Pennsylvanian/Mississippian carbonates (Lisburne Formation) may contain local source potential. Thus, there is a temporal coincidence of potential source rocks and reservoir rocks (Fig. 3) in the sedimentary section.

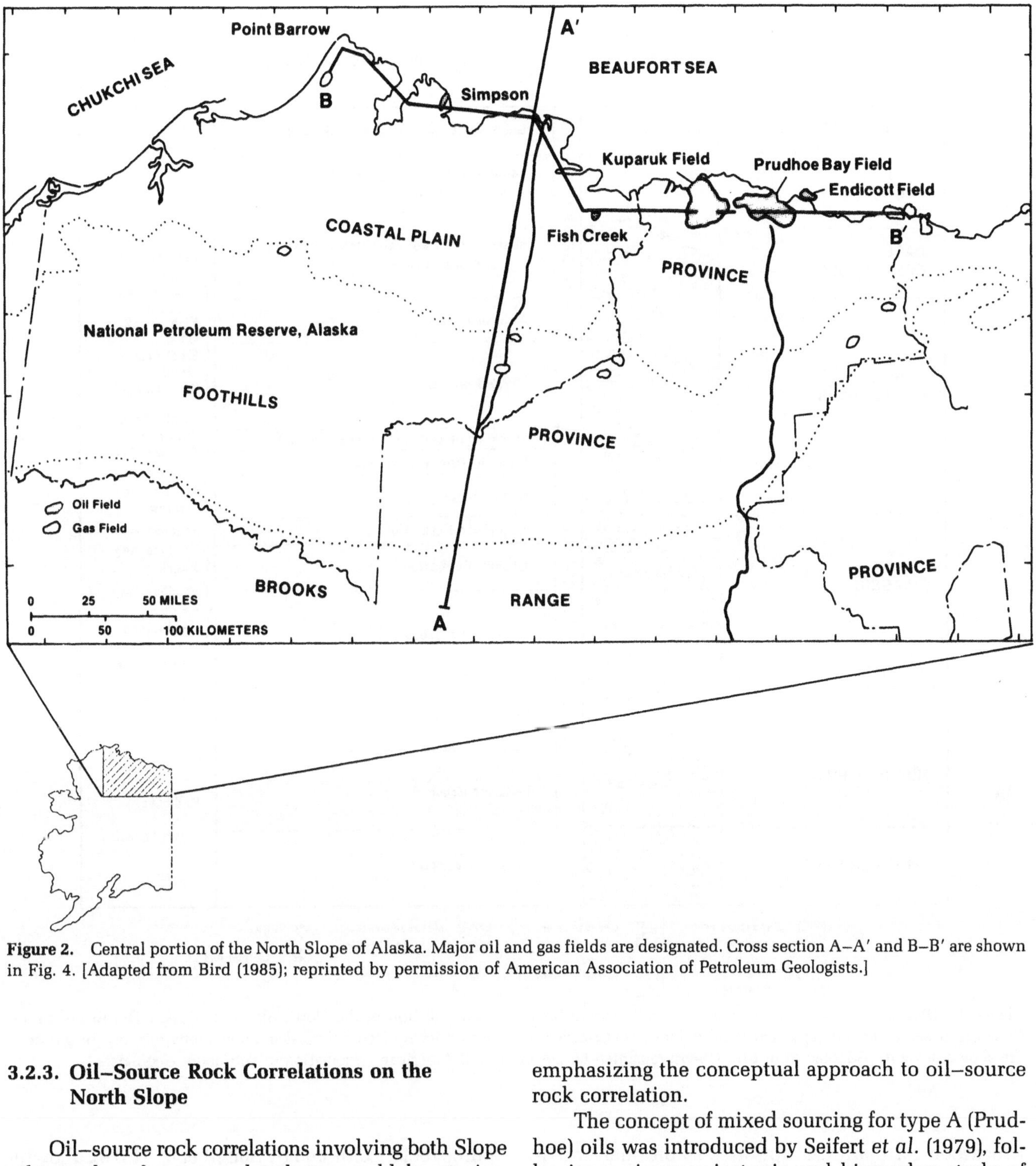

Figure 2. Central portion of the North Slope of Alaska. Major oil and gas fields are designated. Cross section A–A′ and B–B′ are shown in Fig. 4. [Adapted from Bird (1985); reprinted by permission of American Association of Petroleum Geologists.]

3.2.3. Oil–Source Rock Correlations on the North Slope

Oil–source rock correlations involving both Slope oil types have been completed at several laboratories, mostly as part of a study organized by the United States Geological Survey (USGS) in the early eighties (Claypool and Magoon, 1985). A wide assortment of correlation methods was utilized, and results varied significantly among the participating institutions. This section presents the highlights of these efforts, emphasizing the conceptual approach to oil–source rock correlation.

The concept of mixed sourcing for type A (Prudhoe) oils was introduced by Seifert *et al.* (1979), following a rigorous isotopic and biomarker study of selected oil and rock samples from the Prudhoe Bay area. The organic matter in Mississippian through Late Cretaceous shales was compared to that of type A oils in the major reservoir sands, and biological marker data were relied upon heavily for the final correlation conclusions. Monoaromatic steroid hy-

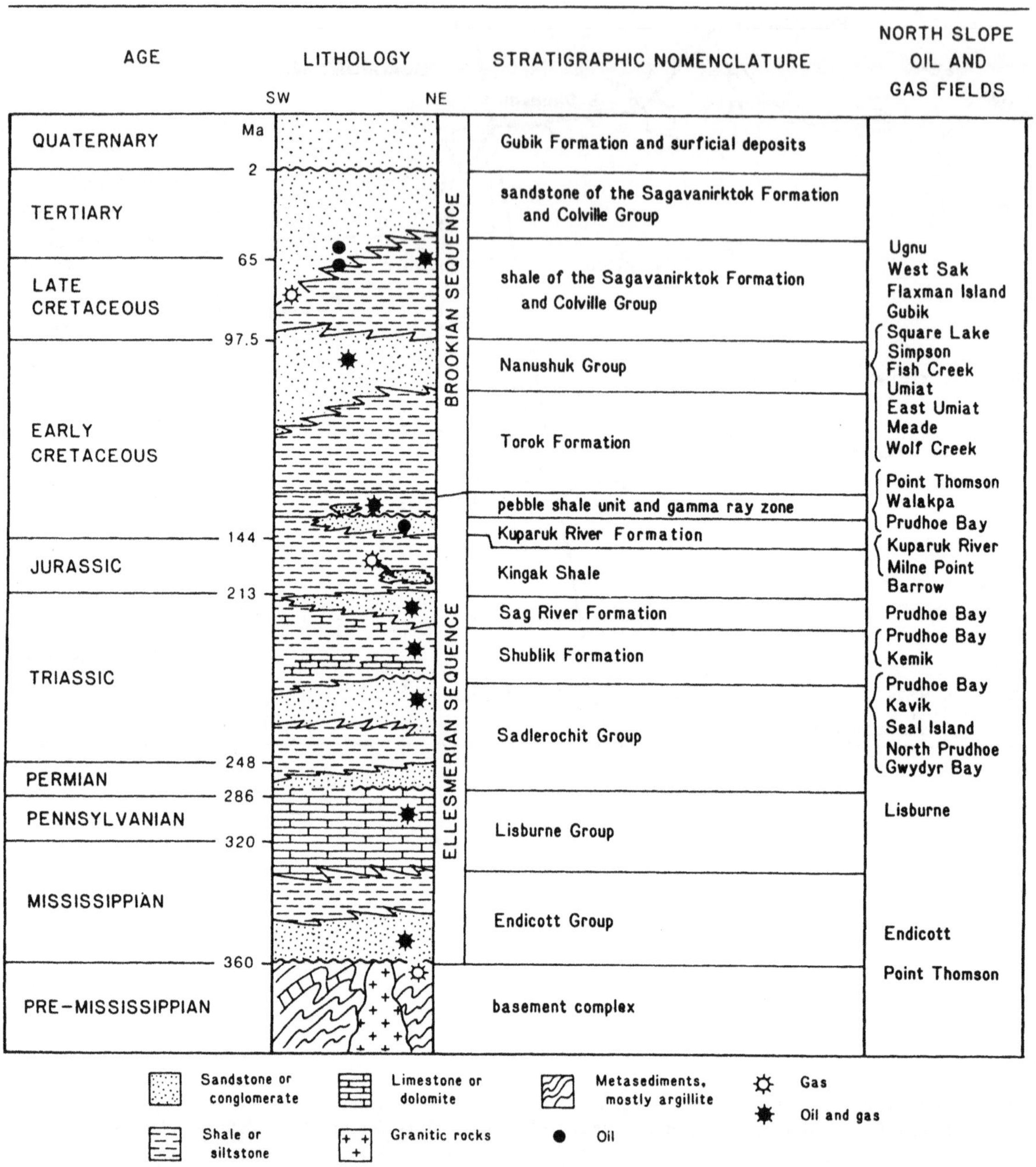

Figure 3. Stratigraphic section above economic basement for the central portion of the North Slope of Alaska. The Brookian and Ellesmerian sequences are separated at the major Lower Cretaceous unconformity. Both commercial and noncommercial oil and gas fields are shown at right. [Adapted from Bird (1985); reprinted by permission of American Association of Petroleum Geologists.]

drocarbon correlations, supported by bitumen and oil carbon isotope ratios that differ by only 0.3‰, successfully demonstrated that an oil recovered from a sand stringer in the Kingak Formation (not type A) is locally derived from the Kingak. The case of the oil from the Permo-Triassic Sadlerochit Formation (type A) however, is more involved.

The complete biomarker patterns of the Sadlerochit oil were not observed in any of the prospective source rocks examined by Seifert *et al.* (1979), allowing the authors to summarily eliminate certain candidates and pursue the concept of multiple source rock contributions from others. The Shublik Formation, Kingak Formation, and deep (off-structure) "post-Neocomian" shales were proposed as co-sources for oil of the Sadlerochit Formation, based upon several biomarker indicators, and supported by carbon isotope ratios of the oil and source rock bitumens and kerogens.

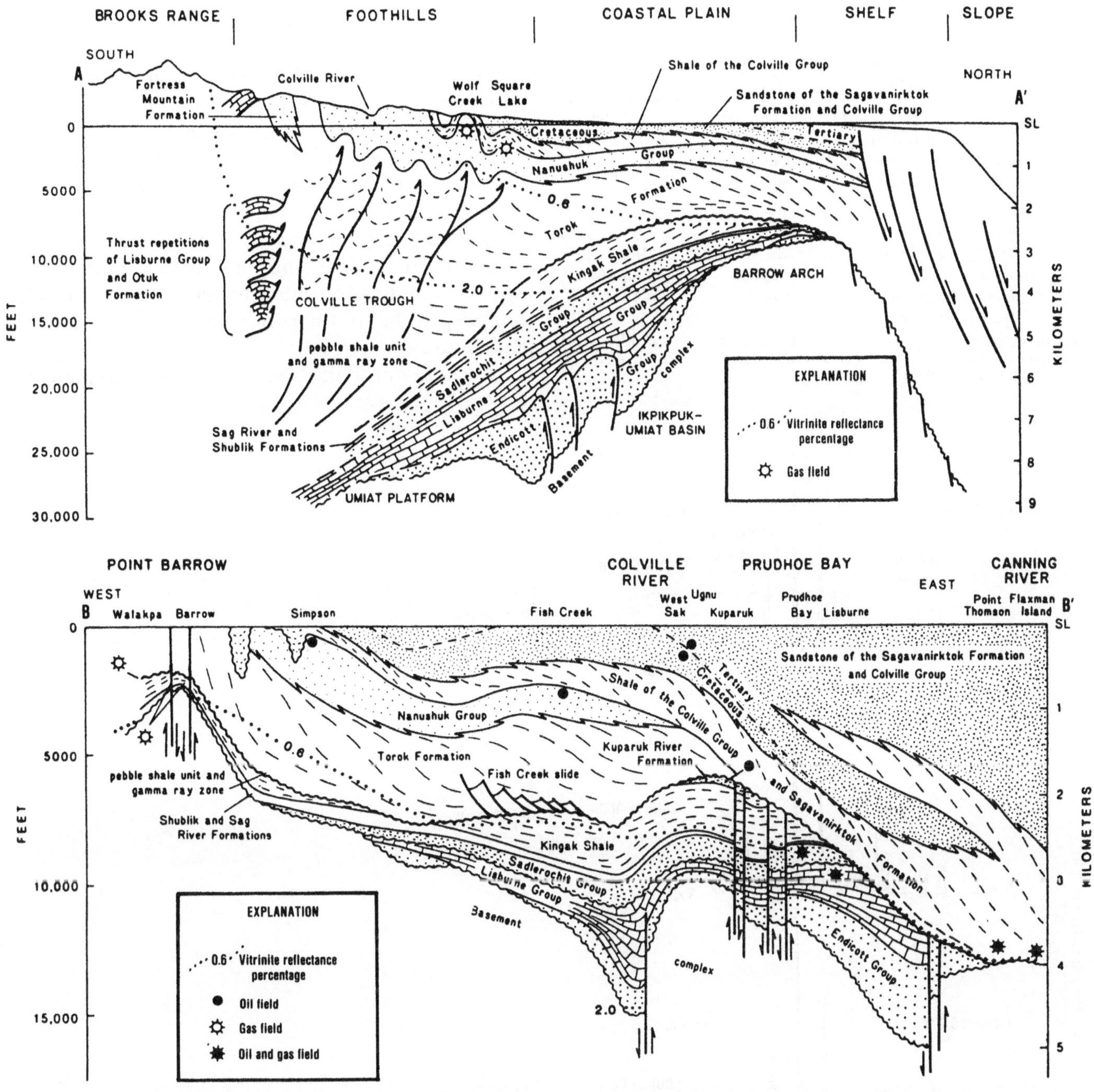

Figure 4. Schematic cross sections for the North Slope, showing major stratigraphic units, approximate thermal maturity (vitrinite reflectance) contours, and location of oil and gas accumulations. Lines of section are shown in Fig. 2, and lithologic symbolism as in Fig. 3. [From Bird (1985); reprinted by permission of American Association of Petroleum Geologists.]

Figure 5, from Seifert *et al.* (1979), shows the distribution of tricyclic and pentacyclic terpanes (m/z 191 mass chromatograms) of the oil (top) and the three source rock candidates, along with accompanying carbon isotope data. The occurrence of relatively high amounts of tricyclic terpanes in the Shublik Formation and the presence of compounds 6 and 7 in the Shublik and Lower Cretaceous samples (see Fig. 5) support these units as co-sources of the Sadlerochit oil. The inclusion of the Kingak Formation as a co-source derives from monoaromatic steroid hydrocarbon correlations. Finally, each of the proposed co-sources is geologically proximate to a major migratory conduit, namely, the Lower Cretaceous unconformity.

This early effort introduced several important concepts to oil–source rock correlation. Seifert *et al.*

TABLE I. Oil Type Characteristics, North Slope of Alaska[a]

Compositional parameter	Type A	Type B
Vanadium in whole oil (ppm)	11–169	0.2–68
Nickel in whole oil (ppm)	6.1–65	1.0–3.1
V/(V + Ni) in whole oil	0.44–0.72	0.11–0.46
Carbon isotope ratios (‰)		
Whole oil	−29.6–−29.1	−28.7–−25.3
Aliphatic hydrocarbons	−29.6–−29.0	−28.7–−26.2
Aromatic hydrocarbons	−28.7–−28.2	−26.9–−25.9
%C_{27} steranes	55–58	44–46
Methyldibenzothiophene/ methylphenanthrene	0.46–0.92	0.14–0.23
Sulfur in whole oil (%)	0.86–1.85	0.01–0.27

[a]Data from Curiale (1985).

(1979) were faced with some prospective source rock samples that were thermally immature and others that contained nonindigenous (migrated) organic matter. These complications were resolved using pyrolysis and biomarker interpretations based upon prior experience, and, in this way, the influence of such nongenetic processes on the correlation effort could be recognized during the search for genetic parameters. Further Slope correlation efforts stemming from the USGS study addressed another nongenetic complication often encountered in correlation efforts, that of the comparison of a crude oil to postmature source rocks.

Participants in the USGS effort [see Claypool and

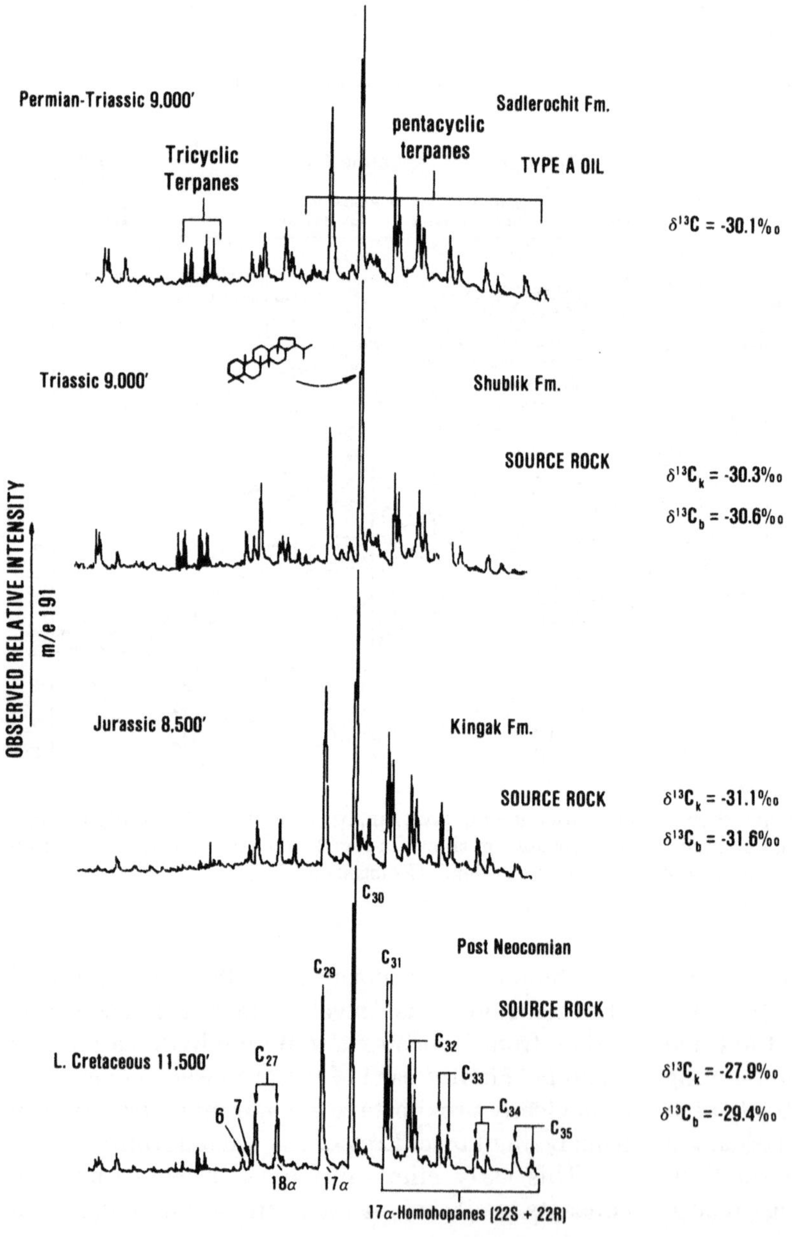

Figure 5. *M/z* 191 mass chromatograms, showing distribution of higher tricyclic terpanes and C_{27}–C_{35} pentacyclic terpanes (hopanes) for a typical type A oil (Prudhoe Bay; top) and three co-sources. Compounds 6 and 7 (bottom trace) are noted in the text. Note that each co-source makes a distinct contribution, although carbon isotope ratios for kerogen ($\delta^{13}C_k$) and bitumen ($\delta^{13}C_b$) suggest that the post-Neocomian contribution is minor. [From Seifert *et al.* (1979); reprinted by permission of John Wiley & Sons, Inc., copyright 1979.]

Magoon (1985) for a summary of this study] examined 9 oils and 15 rock samples from across the Slope. The oils included one condensate (Seabee well), two samples each from the Barrow and Simpson areas, and one sample each from Prudhoe Bay and Fish Creek (Fig. 2); types A and B were both represented. The rock samples studied included four each from the Torok Formation and the pebble shale unit, three each from the Kingak and Shublik Formations, and one from the Echooka unit (Sadlerochit Group; Fig. 3). Of these 15 possible source rocks, seven (three Shublik, two Kingak, one Torok, and one Echooka) are thermally postmature, having vitrinite reflectance values greater than 1.4%; only four are at or near peak generation (Curiale, 1985). Further, only two samples would be considered petroleum source rocks, using conventional criteria. These complications made oil–source rock correlations particularly difficult, although it should be noted that such problems are not uncommon in correlation projects.

Most participants in this multilaboratory study agreed with Seifert *et al.* (1979) that the Shublik Formation is an important source for type A oil. This is interesting because the Shublik samples of the USGS group study are thermally mature or overmature, thus limiting the applications of biomarker results. Nevertheless, the Shublik Formation was suggested as a major type A source, based upon either carbon isotope data [less than 1.0‰ difference between oil and source rock kerogen; see Fig. 6, from Burkley and Castano (1985)] or simply the absence of other reasonable possibilities (Elrod *et al.*, 1985; Anders *et al.*, 1985).

Exceptions to the Shublik Formation as the predominant type A source were also proposed, most notably by Mackenzie *et al.* (1985), who used quantitative molecular geochemistry and integrated geological concepts to propose a significant Kingak Formation source component for both oil types. These authors used the ratio of 17α(*H*),21β(*H*)-hopane to the sum of the major regular ethylcholestanes (which is relatively low in the oils and much higher in the Shublik extracts) to infer that the Shublik Formation could not have been the major source rock involved. It should be noted that the effect of postmaturity on rock extract hopane/sterane ratios was not discussed. This effect could be critical to the interpretation, because both Shublik extracts examined by Mackenzie *et al.* (1985) were overmature ($VR_0 > 1.5\%\ R_0$). Indeed, in previous studies, an increase in hopane/sterane ratio was observed following laboratory pyrolysis (see Mackenzie, 1984).

Uncertainties also exist concerning the source rock for type B oils. Several authors reported a Kingak Formation origin for type B oils (Mackenzie *et al.*, 1985; Hughes *et al.*, 1985). Hughes *et al.* (1985) suggested that admixtures of two distinct organic facies within the Kingak Formation could give rise to the Umiat type oil. However, retene and cadalene (observed in mass chromatograms *m/z* 234 and *m/z* 183, respectively), both terrigenous molecular markers, are present in the oil and absent in the Kingak samples. Hughes *et al.* (1985) attributed this to either (a) thermal destruction of these compounds in the mature Kingak samples or (b) a contribution from another source unit containing high concentrations of terrigenous organic matter.

The majority of workers assign type B oils to a pebble shale source (Early Cretaceous, Fig. 3; Claypool and Magoon, 1985), although support for this assignment is often not based on correlational data, but rather on geologic inference and source quality (e.g., Walters *et al.*, 1985). Williams *et al.* (1985), using tri- and tetramethylnaphthalene distributions and geologic inference, proposed a dominant pebble shale source for type B, whereas Kennicutt *et al.* (1985) came to the same conclusion using three-dimensional scanning fluorescence.

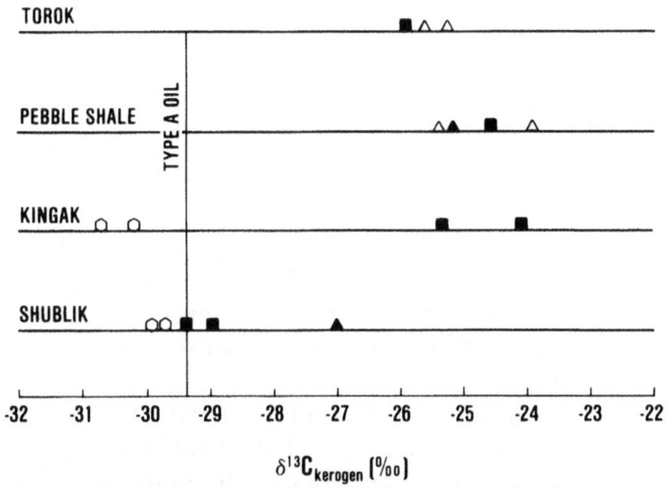

Figure 6. Distribution of kerogen carbon isotope ratios for selected subsurface samples from four potential North Slope source rock units. ■, Source rocks; ▲, marginal source rocks; △, nonsource rocks [data of Burkley and Castano (1985]. Remaining data are from Seifert *et al.* (1979). The carbon isotope ratio for Sadlerochit oil (type A; Prudhoe Bay Field) is shown as a vertical line; note that, on this basis, only the Shublik samples correlate. Note also that the isotope ratio for the Sadlerochit oil in this study is significantly different from that determined by Seifert *et al.* (1979), as indicated in Fig. 5. [Adapted from Burkley and Castano (1985); reprinted by permission of American Association of Petroleum Geologists.]

This brief case study of the search for source rocks of oils on the North Slope of Alaska reflects the many subtleties involved in all oil–source rock correlation efforts. Interpretational difficulties were presented by each of the three nongenetic process parameters:

1. *Thermal Maturation.* Both immature and overmature source rock organic matter was involved in the correlation effort; in addition, one of the oils was actually a thermal condensate.
2. *Migration.* The oils are recovered from an area encompassing tens of thousands of square miles (Figs. 3 and 4); paleohydrodynamic intricacies are poorly understood and difficult to integrate into a regional oil–source rock correlation effort (Mackenzie *et al.*, 1985).
3. *Postmigration alteration.* Of the nine oils analyzed in the USGS study, at least four are biodegraded (Curiale, 1985); several are waterwashed; and at least three are the product of source mixing.

Geochemists who have studied the Slope problem have used numerous methods to address the interfering effects of these nongenetic processes. In addition to the now conventional carbon isotope and biological marker analyses, other oil–source rock correlation parameters measured and techniques employed included sulfur and hydrogen isotope ratios, naphthalene distributions, transition-metal concentrations and ratios, fluorescence intensity and distribution, pyrolysis–GC and pyrolysis–GC/MS, low-voltage mass spectrometry for aromatic distributions, pyrolysis–carbon isotope ratios, and optical rotation measurements (Claypool and Magoon, 1985). It is noteworthy that, even with such a battery of technologies, significant and consistent differences exist among the final interpretations, reflecting the complexity of the geologic problem involved.

3.3. Source Correlations Involving Biodegraded Oils

Oil–source rock correlations involving biodegraded oils have been attempted in many sedimentary basins. Exploration efforts over the next several years will undoubtedly require reliable correlations involving such oils. In this section, the question of oil–source rock correlations involving heavily biodegraded oils will be discussed, with emphasis on petroleum exploration applications.

Biodegraded oil–source rock correlation attempts can be subdivided into those that are indirect and those that are direct. Prior to and in the early developmental stages of biological marker applications, most such correlations were indirect: a biodegraded oil was found to be part of a genetic oil family having known nonbiodegraded members, and these nonbiodegraded members were correlated to prospective source rocks. That is, the biodegraded oil–source rock correlation was accomplished by inference. Examples include studies by Welte *et al.* (1982) and Wehner and Teschner (1981) in the Vienna and Molasse basins, respectively. Because such indirect correlation efforts fit comfortably under the heading of oil–oil correlation, they will not be discussed further here (see Palmer, this volume, Chapter 23).

A biodegraded oil may be directly correlated to prospective source rocks using either biomarker and supporting isotope methods or pyrolysis, commonly followed by molecular analysis. Even in heavily biodegraded oils, certain genetic molecular parameters often remain intact, and the use of biomarkers for such problems is now routine. Recent published examples include the use of aromatic hydrocarbon carbon isotope ratios and aromatic steroid hydrocarbons (which are often unchanged by biodegradation) in oils and source rocks from California (Curiale *et al.*, 1985), Alaska (Anders and Magoon, 1986), Oklahoma (Curiale, 1986), and Peru (Sofer *et al.*, 1986). Such correlations represent conventional molecular and isotopic comparisons, given prior understanding of the effects of microbial alteration on chemical composition.

The application of molecular marker technology to biodegraded oil–source rock correlations, while successful in many instances, suffers from the same disadvantages as all petroleum biomarker efforts: in most cases, fewer than 1% of the compounds present in an oil are involved in the study. This limitation becomes a significant problem if the provenance of such compounds is not the same as that of the remainder of the oil, as in instances of migrational contamination (Philp and Gilbert, 1982). Indeed, any comparison of a source rock extract with a crude oil must consider the possibility of nonindigenous material in the extract. This problem can often be resolved by comparison of oils with pyrolysates of high-molecular-weight organic matter fractions (e.g., asphaltenes and kerogens) separated from prospective source rocks. This approach is particularly suitable in biodegraded oil–source rock correlations, where pyrolysates of crude oil asphaltenes can be directly compared with those of source rock asphaltenes and kerogens. In such cases, the complete chemical characteristics of the material (not just the cyclic biomarkers) can be used as correlation tools.

The concept of using pyrolysis for biodegraded oil–source rock correlations derives from studies suggesting that pyrolysates of asphaltenes isolated from biodegraded oils resemble compositionally the hydrocarbons contained in the oil prior to biodegradation. Thus, asphaltene pyrolysates from a single genetic oil family will be similar compositionally, even though individual oils may be heavily biodegraded. This has been shown by Rubinstein *et al.* (1979), Curiale *et al.* (1983), and Cassani and Eglinton (1986) for extremely biodegraded oils and by Wehner *et al.* (1986) in controlled laboratory biodegradation experiments. Consequently, biodegraded oil–source rock correlations are feasible, where hydrocarbons pyrolyzed from biodegraded oil asphaltenes are compared to those pyrolyzed from source rock kerogens and asphaltenes.

Pyrolytic oil–source rock correlations involving nonbiodegraded oils have been successfully carried out recently by Behar and Pelet (1988) for source rocks and oils from Venezuela. Figure 7 shows pyrolysis–gas chromatograms for two oil–source rock pairs, using kerogen and source rock asphaltenes as comparison parameters. In addition to whole pyrolysis–GC efforts, Behar and Pelet (1988) also compared

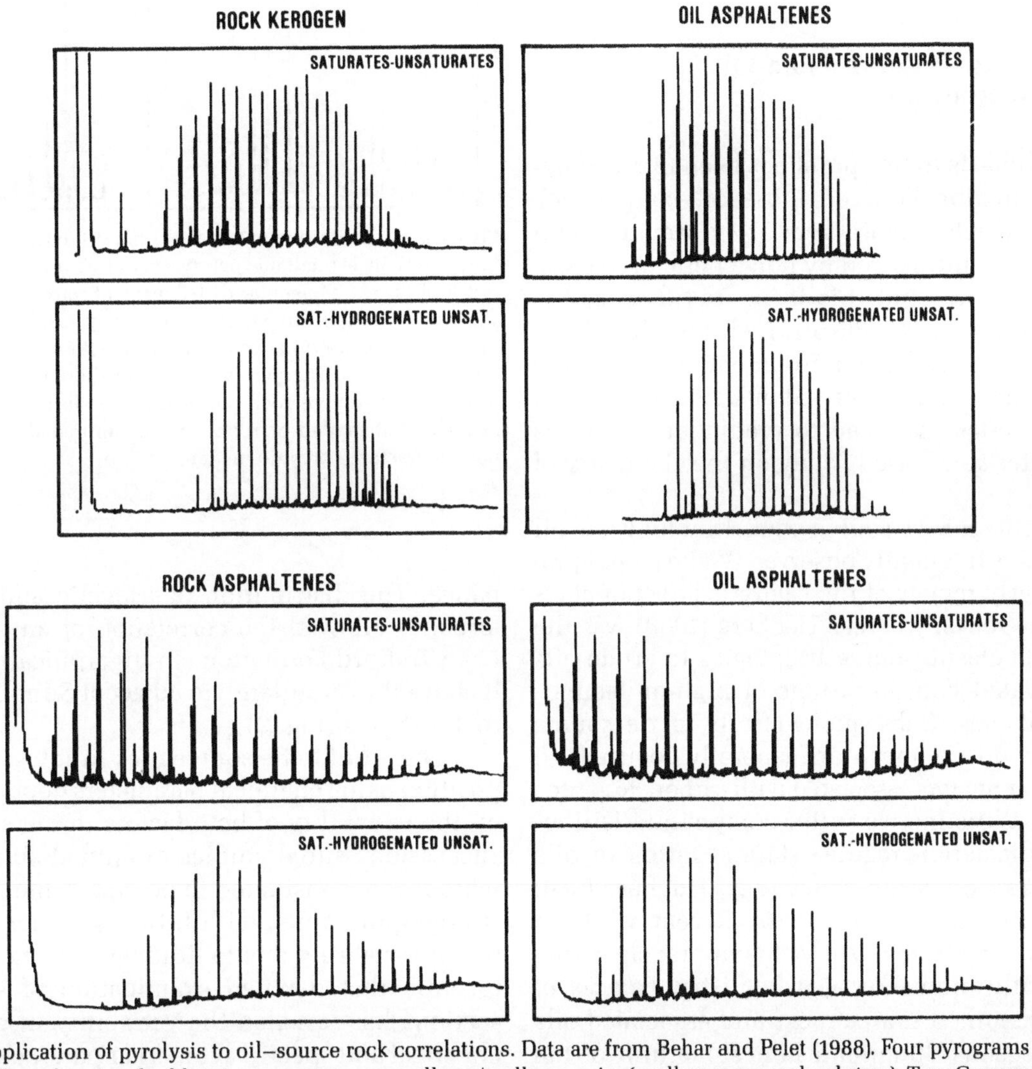

Figure 7. Application of pyrolysis to oil–source rock correlations. Data are from Behar and Pelet (1988). Four pyrograms are shown for each of two oil–rock pairs; doublets in pyrograms are *n*-alkene/*n*-alkane pairs (*n*-alkenes are early eluters). Top: Comparison of source rock kerogen pyrolysates (left) with oil asphaltene pyrolysates (right), for oil–rock set A. Total pyrogram (above) and hydrogenated pyrolysate (below) are shown. Note similar distribution of *n*-alkanes in kerogen and asphaltene hydrogenated pyrolysates. Bottom: Comparison of source rock asphaltene pyrolysates (left) with oil asphaltene pyrolysates (right) for oil–rock set B. Total pyrogram (above) and hydrogenated pyrolysate (below) are shown.

biomarker distributions in oil and source rock asphaltene pyrolysates of one of their oil–rock sets (Fig. 8). Note the remarkable similarity in terpane distributions of the oil and rock asphaltene pyrolysates in Fig. 8.

Efforts which specifically utilize biomarkers in pyrolysis-based correlations have been reviewed in some detail by Philp and Gilbert (1987). Such pyrolytic methods, involving as they do both whole-rock thermal extraction and compositional analysis of the entire molecular suite amenable to GC analysis, represent the ideal approach to holistic correlations. It is expected that as such methods develop even further, oil–source rock correlation involving biodegraded oils, using high-resolution on-line pyrolysis–GC/MS techniques, will become routine.

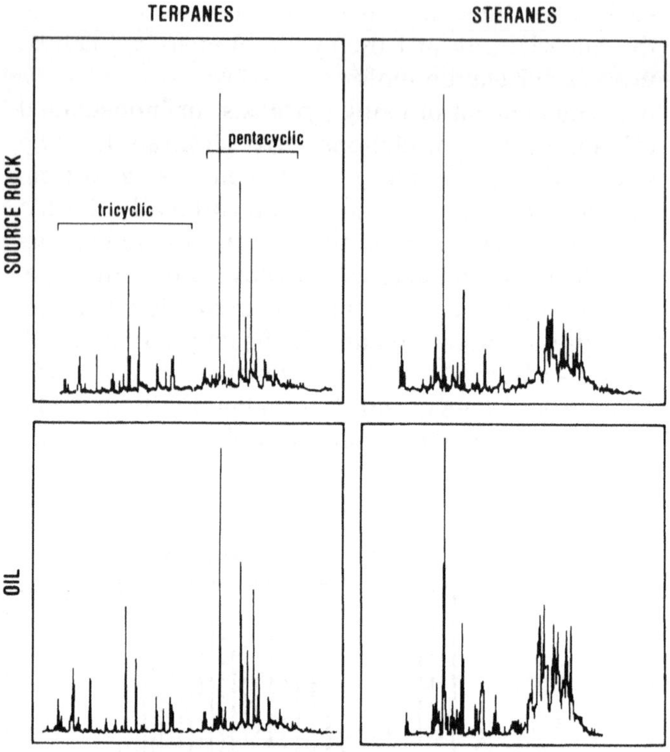

Figure 8. Distribution of terpanes (*m/z* 191, left) and steranes (*m/z* 217, right) in asphaltene pyrolysates of an oil–source rock pair (set B of Fig. 7). Molecular distribution of rock asphaltene pyrolysate (top) is genetically similar to that of oil asphaltene pyrolysate (bottom). Data are from Behar and Pelet (1988); all pyrolyses conducted at 550°C. The pyrolytic method shown here and in Fig. 7 provides oil–source rock correlation possibilities when oils are heavily biodegraded or rocks are contaminated with nonindigenous extractable organic matter.

3.4. Source Rock Controls on Oil Composition

Our emphasis to this point has been in recognizing the distinction between those oil–source rock parameters which contain genetic information and those which are influenced by nongenetic processes. In this section we shift this focus and discuss the effect that source rock controls, including lithologic input and depositional environment, have on the chemical composition of derived oils. Detailed knowledge of this effect can lead to prediction of source rock characteristics based solely on the chemistry of the oil.

The influence of source rock lithology on oil composition is frequently observed. Welte (1965) presented an early review of the catalytic effect of clays in oil source rocks, whereas Hedberg (1968) was the first to relate clastic source lithologies to crude oils having elevated concentrations of high-molecular-weight *n*-alkanes. Subsequent efforts to tie source lithology to oil composition generally focused on biomarker parameters associated with carbonate source rocks, such as distinctive sulfur compound distributions and diasterane/regular sterane ratios in oils sourced from nonclastic facies (e.g., Hughes, 1984; Philp, 1985; Connan *et al.*, 1986). Recent work on source rocks in Florida typify this approach. Using predominantly biomarker geochemistry, Palacas *et al.* (1984) examined source rocks and associated oils in the Cretaceous Sunniland Limestone unit of the South Florida Basin. Unusual hydrocarbon distributions were observed in this carbonate unit, including high C_{34}/C_{33} extended hopane ratios and relatively high concentrations of C_{24} and C_{26} tetracyclic terpanes. The distribution of tricyclic and tetracyclic terpanes allowed the correlation of an oil from the Lake Trafford Formation (stratigraphically about 100 ft above the Sunniland) to adjacent Sunniland source rock extracts (Fig. 9).

The effort to tie source rock lithology to oil composition using chemical parameters depends strongly on the variability of both factors in the stratigraphic succession. Initial studies examined single samples which were considered to be representative of entire stratigraphic units, with little regard for intraformational variability due to lithologic changes. It is only recently that detailed examination of source rock sections has revealed the variability in general geochemical and specific biomarker parameters. Considerable elemental and molecular variations are observed over intervals as small as 10 cm in source rock cores from Germany (Moldowan *et al.*, 1986), Kansas (Wenger and Baker, 1986), and California (Curiale and Odermatt,

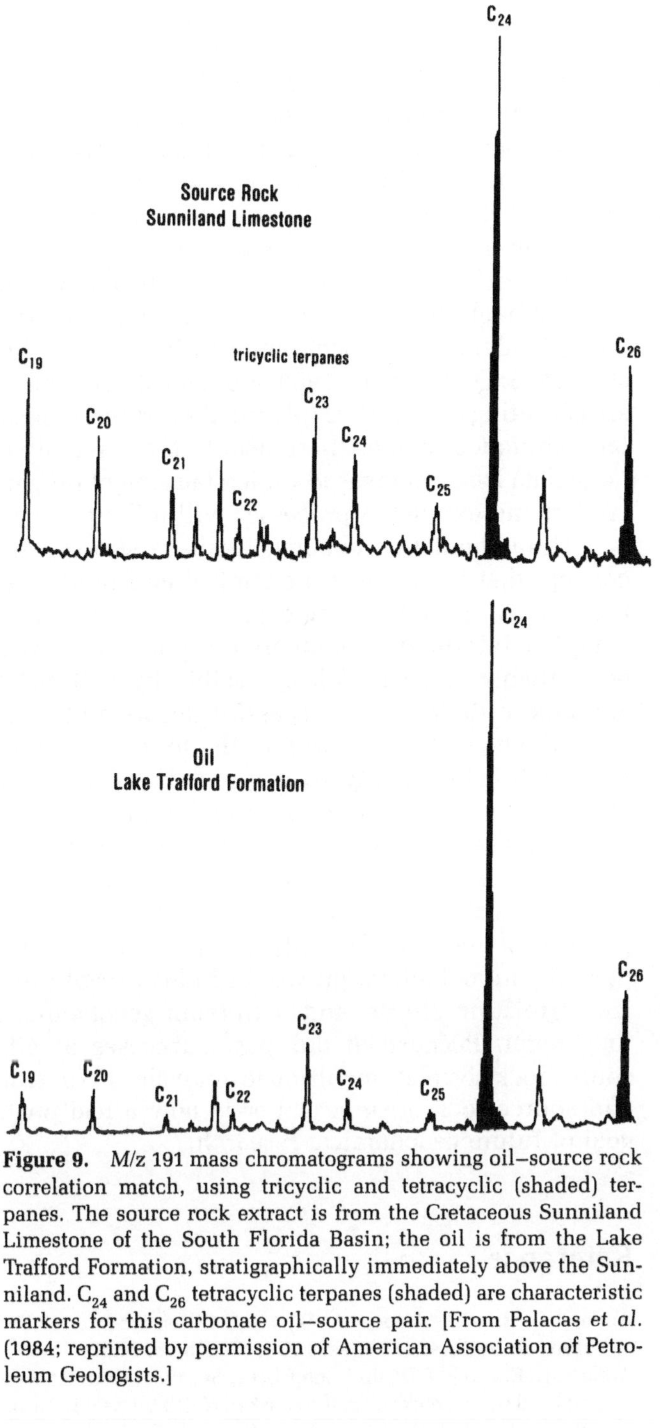

Figure 9. *M/z* 191 mass chromatograms showing oil–source rock correlation match, using tricyclic and tetracyclic (shaded) terpanes. The source rock extract is from the Cretaceous Sunniland Limestone of the South Florida Basin; the oil is from the Lake Trafford Formation, stratigraphically immediately above the Sunniland. C_{24} and C_{26} tetracyclic terpanes (shaded) are characteristic markers for this carbonate oil–source pair. [From Palacas *et al.* (1984; reprinted by permission of American Association of Petroleum Geologists.]

1989). This suggests that source-related petroleum chemistry could vary substantially within a source rock from a single stratigraphic unit. Such difficulties have precluded serious efforts at correlation-motivated source rock character predictions until very recently.

Initial efforts at source rock character prediction based upon oil chemistry stressed rudimentary distinctions involving depositional environment (marine versus nonmarine) and lithology (shale versus carbonate). Moldowan *et al.* (1985) found that distinctive carbon number distributions of monoaromatic steroid hydrocarbons and steranes possessing a 30th carbon atom located on their side chains (i.e., C_{30} desmethylsteranes) often characterized oils derived from marine source rocks. Adding carbon and hydrogen isotope data to the oil parameters used, Peters *et al.* (1986) extended this work and successfully distinguished marine shale, nonmarine shale, and marine carbonate sources using discriminant analysis on a set of 28 oils.

More recently, Mello *et al.* (1988, 1989) confirmed and extended previous studies and successfully identified five distinct source rock depositional environments using the compositional characteristics of a set of offshore Brazilian oils. Using 16 oils chosen to minimize postsourcing compositional alteration, the authors "correlated" the oils to lacustrine freshwater, lacustrine saline water, marine evaporitic, marine carbonate, and marine deltaic depositional regimes. Depositional settings were deduced using a combination of (a) elemental (S, V, Ni), compositional, isotopic, and quantitative biomarker data and (b) geologic information derived from tectonic and environmental reconstruction. Thus, without prior knowledge about actual source rock units, oil–source rock predictions were possible. It is anticipated that future efforts in this direction, assisted by analytical advances which will make "ultimate" oil analyses feasible (see below), will eventually allow such predictions to be routine.

4. Future Directions

The review of the case studies and project design concepts in the previous sections illustrates the widespread geologic application of oil–source rock correlations. These applications will undoubtedly expand in both academic and industrial environments, as geochemical exploration in mature basins becomes commonplace. This increase will inevitably expose certain present-day limitations of oil–source rock correlation techniques and interpretations. Future directions in these areas are discussed in this section, with an emphasis on the development of correlation methods for problematic cases.

Current oil–source rock correlations rely heavily on comparison of crude oil with organic matter that is solvent-extracted from rocks. The comparison usually involves some common feature of the oil and the extractable organic matter (EOM), such as a carbon isotope ratio or a molecular distribution, and the suitability of comparing these two organic mixtures is rarely questioned. Yet it is obvious that the methods by which we derive these mixtures are radically different: the oil represents part of a reservoir pore-filling fluid, transported through pipe to the surface, whereas the EOM represents the organic matter which we extract from the rock with (relatively) strong organic solvents using widely varying extraction techniques. The different chemical characteristics of the mixtures derived from these two processes are obvious in some cases (e.g., variable percentages of polar compounds in the EOM) and subtle in other cases (e.g., slight variation in EOM molecular ratios due to laboratory preparation steps). Such contrasts should not be surprising, considering the disparity inherent in the two separation methods. Yet these differences represent a distinct limitation of our current methods for oil–source rock correlation and indicate the importance of development of more "natural" methods of source rock organic matter extraction. Possibilities for more accurate extraction of the oil-like organic matter from source rocks include supercritical fluid extraction (using, e.g., carbon dioxide) and high-pressure thermal distillation, analogous to natural subsurface conditions, followed by off-line chemical analysis. Alternatively, on-line thermal microextraction methods, typified by low-temperature "pyrolysis"–gas chromatography/mass spectrometry using whole source rock, could be developed further as a quantitative extraction technology capable of rapid generation of correlation data. Such a method would eliminate the biasing effects of solvent extraction, while reducing the time required to execute molecular oil–source rock correlations.

The difficulty of accurately comparing an oil with a solvent extract is further aggravated in the case of postmature source rocks. As discussed earlier in this chapter, sample limitations often require that oils be compared to possible source rocks which are no longer at the maturity levels they were at when the oil left the rock. The problem is most acute when the only available source rock samples are thermally postmature. It then becomes necessary to find remnant genetic similarities between the postmature rock and the oil generated from it at some time in the geologic past. Although this case is certainly nonideal, it is often encountered (cf., the Alaskan studies discussed earlier), and methods must be developed to deal with the situation. Perhaps the most promise lies in those source rock parameters that are least sensitive to thermal alteration, such as kerogen carbon isotope ratios and nonhydrocarbon molecular markers. Obviously, a clearer understanding of organic matter transformations with increased temperature and pressure would be of great benefit in this regard.

Finally, because oil–source rock correlation has become more accurate as our analytical methods have become more sophisticated, it is evident that new analytical techniques, many not yet conceived, will play a significant role in the development of the field. Recent geochemical developments in the analysis of conventional biomarkers and geometalloporphyrins using tandem mass spectrometry (Philp and Johnston, 1988; Quirke *et al.*, 1988) are typical of the many areas waiting to be fully explored. As further molecular techniques such as these lead to the assembly of large data bases, increased understanding of oil generation and expulsion processes will follow.

Throughout this chapter, we have stressed the concept that crude oil is a chemical end product of the effects of four geologic processes. Ultimately, complete information about all four phenomena will be available in every oil, accessible by "ultimate" analysis of the oil and successful deconvolution of the oil's chemistry into each of the four responsible processes. This ability to completely analyze the chemical composition of an oil and simultaneously assess (a) the character of the organic matter input to the source rock, (b) the extent of maturation of the source rock at the time of sourcing, (c) the extent and pathway of migration, and (d) the cause and extent of postmigrational alteration would indeed revolutionize correlation efforts, and petroleum geochemistry in general. Because of our past successes at oil–source rock correlation, ultimate analysis of crude oil and source rock organic matter is now a legitimate goal of future geochemical research.

References

Anders, D. E., and Magoon, L. B., 1986, Oil–source correlation study in northeastern Alaska, *Org. Geochem.* **10**:407–416.

Anders, D. E., King, J. D., and Lubeck, C., Sr., 1985, Correlation of oils and source rocks from the Alaskan North Slope, in: *Alaska North Slope Oil–Rock Correlation Study* (L. B. Magoon and G. E. Claypool, eds.), AAPG Studies in Geology No. 20, American Association of Petroleum Geologists, Tulsa, Oklahoma, pp. 281–304.

Baker, D., 1962, Organic geochemistry of Cherokee Group in southeastern Kansas and northeastern Oklahoma, *Am. Assoc. Petrol. Geol. Bull.* **46**:1621–1642.

Barbat, W. N., 1967, Crude-oil correlations and their role in exploration, *Am. Assoc. Petrol. Geol. Bull.* **51**:1255–1292.

Barker, C., 1981, Oil source rock correlation aids drilling site selection, *World Oil* **181**(5):121–126.

Behar, F., and Pelet, R., 1988, A new technique for oil/source rock and oil/oil correlation, in: *Geochemical Biomarkers* (T. F. Yen and J. M. Moldowan, eds.), Harwood Academic, London, pp. 51–70.

Bird, K. J., 1985, The framework geology of the North Slope of Alaska as related to oil–source rock correlations, in: *Alaska North Slope Oil–Rock Correlation Study* (L. B. Magoon and G. E. Claypool, eds.), AAPG Studies in Geology No. 20, American Association of Petroleum Geologists, Tulsa, Oklahoma, pp. 3–30.

Burkley, L. A., and Castano, J. R., 1985, Alaska North Slope oil–source rock correlation study, in: *Alaska North Slope Oil–Rock Correlation Study* (L. B. Magoon and G. E. Claypool, eds.), AAPG Studies in Geology No. 20, American Association of Petroleum Geologists, Tulsa, Oklahoma, pp. 95–122.

Cassani, F., and Eglinton, G., 1986, Organic geochemistry of Venezuelan extra-heavy oils: Pyrolysis of asphaltenes, *Chem. Geol.* **46**:167–183.

Claypool, G. E., and Magoon, L. B., 1985, Comparison of oil–source rock correlation data for Alaskan North Slope: Techniques, results and conclusions, in: *Alaska North Slope Oil–Rock Correlation Study* (L. B. Magoon and G. E. Claypool, eds.), AAPG Studies in Geology No. 20, American Association of Petroleum Geologists, Tulsa, Oklahoma, pp. 49–84.

Connan, J., Bouroullec, J., Dessort, D., and Albrecht, P., 1986, The microbial input in carbonate-anhydrite facies of a sabkha palaeoenvironment from Guatemala: A molecular approach, *Org. Geochem.* **10**:29–50.

Curiale, J. A., 1985, Oil types and source rock–oil correlation on the North Slope, Alaska—A cooperative USGS–industry study, in: *Alaska North Slope Oil–Rock Correlation Study* (L. B. Magoon and G. E. Claypool, eds.), AAPG Studies in Geology No. 20, American Association of Petroleum Geologists, Tulsa, Oklahoma, pp. 203–232.

Curiale, J. A., 1986, Origin of solid bitumens, with emphasis on biological marker results, *Org. Geochem.* **6**:559–580.

Curiale, J. A., 1987a, Crude oil chemistry and classification, Alaska North Slope, in: *Alaskan North Slope Geology*, Vol. 1 (I. Tailleur and P. Weimer, eds.), Society of Economic Paleontologists and Mineralogists, Bakersfield, California, pp. 161–167.

Curiale, J. A., 1987b, Distribution of transition metals in north Alaskan oils, in: *Metal Complexes in Fossil Fuels: Geochemistry, Characterization and Processing* (R. H. Filby and J. F. Branthaver, eds.), American Chemical Society, Washington, D.C., pp. 135–145.

Curiale, J. A., and Odermatt, J. R., 1989, Short-term biomarker variability in the Monterey Formation, Santa Maria Basin, Org. *Geochem.* **14**:1–13.

Curiale, J. A., Harrison, W. E., and Smith, G., 1983, Sterane distribution of solid bitumen pyrolyzates. Changes with biodegradation of crude oil in the Ouachita Mountains, Oklahoma, *Geochim. Cosmochim. Acta* **47**:517–523.

Curiale, J. A., Cameron, D., and Davis, D. V., 1985, Biological marker distribution and significance of oils and rocks of the Monterey Formation, California, *Geochim. Cosmochim. Acta* **49**:271–288.

Dow, W. G., 1974, Application of oil-correlation and source-rock data to exploration in Williston Basin, *Am. Assoc. Petrol. Geol. Bull.* **58**:1253–1262.

Elrod, L. W., Bissada, K. K., and Colling, E. L., 1985, Alaskan North Slope oil–rock correlation, in: *Alaska North Slope Oil–Rock Correlation Study* (L. B. Magoon and G. E. Claypool, eds.), AAPG Studies in Geology No. 20, American Association of Petroleum Geologists, Tulsa, Oklahoma, pp. 233–242.

England, W. A., Mackenzie, A. S., Mann, D. M., and Quigley, T. M., 1987, The movement and entrapment of petroleum fluids in the subsurface, *J. Geol. Soc.* **144**:327–347.

Hedberg, H. D., 1968, Significance of high-wax oils with respect to genesis of petroleum, *Am. Assoc. Petrol. Geol. Bull.* **42**:736–750.

Hughes, W. B., 1984, Use of thiophenic organosulfur compounds in characterizing crude oils derived from carbonate versus siliciclastic sources, in: *Petroleum Geochemistry and Source Rock Potential of Carbonate Rocks* (J. G. Palacas, ed.), American Association of Petroleum Geologists, Tulsa, Oklahoma, pp. 181–196.

Hughes, W. B., Holba, A. G., and Miller, D. E., 1985, North Slope Alaska oil–rock correlation study, in: *Alaska North Slope Oil–Rock Correlation Study* (L. B. Magoon and G. E. Claypool, eds.), AAPG Studies in Geology No. 20, American Association of Petroleum Geologists, Tulsa, Oklahoma, pp. 379–402.

Hunt, J. M., 1979, *Petroleum Geochemistry and Geology*, W. H. Freeman and Co., San Francisco.

Hunt, J. M., Stewart, F., and Dickey, P. A., 1954, Origin of hydrocarbons of Uinta Basin, Utah, *Am. Assoc. Petrol. Geol. Bull.* **38**: 1671–1698.

Jones, R. W., 1987, Organic facies, in: *Advances in Petroleum Geochemistry*, Vol. 2 (J. Brooks and D. Welte, eds.), Academic Press, London, pp. 1–90.

Kennicutt, M. C., II, Brooks, J. M., and Denoux, G. J., 1985, Carbon isotope, gas chromatography and fluorescence techniques applied to the North Slope of Alaska correlation, in: *Alaska North Slope Oil–Rock Correlation Study* (L. B. Magoon and G. E. Claypool, eds.), AAPG Studies in Geology No. 20, American Association of Petroleum Geologists, Tulsa, Oklahoma, pp. 639–650.

Leythaeuser, D., Hollerbach, A., and Hagemann, H. W., 1977, Source rock/crude oil correlation based on distribution of C_{27+} cyclic hydrocarbons, in: *Advances in Organic Geochemistry 1975* (R. Campos and J. Goni, eds.), ENADIMSA, Madrid, pp. 3–20.

Leythaeuser, D., Schaefer, R. G., and Radke, M., 1987, On the primary migration of petroleum, in: *Proceedings of the 12th World Petroleum Congress*, Vol. 2, John Wiley & Sons, London, pp. 227–236.

Mackenzie, A. S., 1984, Applications of biological markers in petroleum geochemistry, in: *Advances in Petroleum Geochemistry*, Vol. 1 (J. Brooks and D. Welte, eds.), Academic Press, London, pp. 115–214.

Mackenzie, A. S., and Quigley, T. M., 1988, Principles of geochemical prospect appraisal, *Am. Assoc. Petrol. Geol. Bull.* **72**:399–415.

Mackenzie, A. S., Rullkötter, J., Welte, D. H., and Mankiewicz, P., 1985, Reconstruction of oil formation and accumulation in North Slope, Alaska, using quantitative gas chromatography–mass spectrometry, in: *Alaska North Slope Oil–Rock Correlation Study* (L. B. Magoon and G. E. Claypool, eds.), AAPG Studies in Geology No. 20, American Association of Petroleum Geologists, Tulsa, Oklahoma, pp. 319–378.

Magoon, L. B., and Claypool, G. E., 1981, Two oil types on North Slope of Alaska—Implications for exploration, *Am. Assoc. Petrol. Geol. Bull.* **65**:644–652.

Mello, M. R., Telnaes, N., Gaglianone, P. C., Chicarelli, M. I., Brassell, S. C., and Maxwell, J. R., 1988, Organic geochemical characterization of depositional palaeoenvironments of source rocks and oils in Brazilian marginal basins, *Org. Geochem.* **13**: 31–46.

Mello, M. R., Doutsoukos, E. A. M., Hart, M. B., Brassell, S. C., and Maxwell, J. R., 1989, Late Cretaceous anoxic events in the Brazilian continental margins, *Org. Geochem.* **14**:529–543.

Moldowan, J. M., Seifert, W. K., and Gallegos, E. J., 1985, The relationship between petroleum composition and the environment of deposition of petroleum source rocks, *Am. Assoc. Petrol. Geol. Bull.* **69**:1255–1268.

Moldowan, J. M., Sundararaman, P., and Schoell, M., 1986, Sensitivity of biomarker properties to depositional environment and/or source input in the Lower Toarcian of SW-Germany, *Org. Geochem.* **10**:915–926.

Palacas, J. G., Anders, D. E., and King, J. D., 1984, South Florida Basin—prime example of carbonate source rocks of petroleum, in: *Petroleum Geochemistry and Source Rock Potential of Carbonate Rocks* (J. G. Palacas, ed.), American Association of Petroleum Geologists, Tulsa, Oklahoma, pp. 71–96.

Peters, K. E., Moldowan, J. M., Schoell, M., and Hempkins, W. B., 1986, Petroleum isotopic and biomarker composition related to source rock organic matter and depositional environment, *Org. Geochem.* **10**:17–27.

Philp, R. P., 1985, Biological markers in fossil fuel production, *Mass Spectrom. Rev.* **4**:1–54.

Philp, R. P., and Gilbert, T. D., 1982, Unusual distribution of biological markers in an Australian crude oil, *Nature* **299**: 245–247.

Philp, R. P., and Gilbert, T. D., 1987, A review of biomarkers in kerogen as determined by pyrolysis–gas chromatography and pyrolysis–gas chromatography–mass spectrometry, *J. Anal. Appl. Pyrolysis* **11**:93–108.

Philp, R. P., and Johnston, M., 1988, Determination of biomarkers in geological samples by tandem mass spectrometry, in: *Geochemical Biomarkers* (T. F. Yen and J. M. Moldowan, eds.), Harwood Academic, London, pp. 253–273.

Quirke, J. M. E., Perez, M., Britton, E. D., and Yost, R. A., 1988, Tandem mass spectrometric analyses of geoporphyrins, in: *Geochemical Biomarkers* (T. F. Yen and J. M. Moldowan, eds.), Harwood Academic, London, pp. 355–372.

Rubinstein, I., Spyckerelle, C., and Strausz, O. P., 1979, Pyrolysis of asphaltenes—a source of geochemical information, *Geochim. Cosmochim. Acta* **43**:1–6.

Seifert, W. K., 1977, Source rock–oil correlations by C_{27}–C_{30} biological marker hydrocarbons, in: *Advances in Organic Geochemistry 1975* (R. Campos and J. Goni, eds.), ENADIMSA, Madrid, pp. 21–44.

Seifert, W. K., 1978, Steranes and terpanes in kerogen pyrolysis for correlation of oils and source rocks, *Geochim. Cosmochim. Acta* **42**:473–484.

Seifert, W. K., Moldowan, J. M., and Jones, R. W., 1979, Application of biological marker chemistry to petroleum exploration, in: *Proceedings of the Tenth World Petroleum Congress*, Heyden and Son, London, pp. 425–440.

Sofer, Z., Zumberge, J. E., and Lay, V., 1986, Stable carbon isotopes and biomarkers as tools in understanding genetic relationship, maturation, biodegradation and migration of crude oils in the Northern Peruvian Oriente (Maranon) Basin, *Org. Geochem.* **10**:377–390.

Tissot, B. P., and Welte, D. H., 1984, *Petroleum Formation and Occurrence*, 2nd ed., Springer-Verlag, Berlin.

Walters, C. C., Burruss, R. C., and Ashbaugh, T. C., 1985, Biodegradation and other complications in the organic geochemistry of oils and rocks from the NPRA–USGS interlaboratory study, in: *Alaska North Slope Oil–Rock Correlation Study* (L. B. Magoon and G. E. Claypool, eds.), AAPG Studies in Geology No. 20, American Association of Petroleum Geologists, Tulsa, Oklahoma, pp. 139–164.

Wehner, H., and Teschner, M., 1981, Correlation of crude oils and source rocks in the German Molasse Basin, *J. Chromatogr.* **204**: 481–490.

Wehner, H., Teschner, M., and Bosecker, K., 1986, Chemical reactions and stability of biomarkers and stable isotope ratios during *in vitro* biodegradation of petroleum, *Org. Geochem.* **10**:463–471.

Welte, D. H., 1965, Relationship between oil and source rock, *Am. Assoc. Petrol. Geol. Bull.* **49**:2246–2268.

Welte, D. H., Hagemann, H. W., Hollerbach, A., Leythaeuser, D., and Stahl, W., 1975, Correlation between petroleum and source rock, in: *Proceedings of the Ninth World Petroleum Congress*, Vol. 2, Applied Science Publishers, London, pp. 179–191.

Welte, D. H., Kratochvil, H., Rullkötter, J., Ladwein, H., and Schaefer, R. G., 1982, Organic geochemistry of crude oils from the Vienna Basin and an assessment of their origin, *Chem. Geol.* **35**:33–68.

Wenger, L. M., and Baker, D. R., 1986, Variations in organic geochemistry of anoxic–oxic black shale–carbonate sequences in the Pennsylvanian of the Midcontinent, USA, *Org. Geochem.* **10**:85–92.

Williams. J. A., 1974, Characterization of oil types in Williston Basin, *Am. Assoc. Petrol. Geol. Bull.* **58**:1243–1252.

Williams, J. A., Dolcater, D. L., and Olson, R. K., 1985, Oil-source rock correlations, Alaskan North Slope, in: *Alaska North Slope Oil–Rock Correlation Study* (L. B. Magoon and G. E. Claypool, eds.), AAPG Studies in Geology No. 20, American Association of Petroleum Geologists, Tulsa, Oklahoma, pp. 165–184.

Young, A., and McIver, R. D., 1977, Distribution of hydrocarbons between oils and associated fine-grained sedimentary rocks—physical chemistry applied to petroleum geochemistry, II, *Am. Assoc. Petrol. Geol. Bull.* **61**:1407–1436.

Chapter 22

The Kinetics of Biomarker Reactions

Implications for the Assessment of the Thermal Maturity of Organic Matter in Sedimentary Basins

C. ANTHONY LEWIS

1. Introduction

1.1. General Remarks

The application of biomarkers, often qualitatively, to problem solving in organic geochemistry is discussed in other chapters; however, in this chapter their use in attempting to quantify the thermal history of sedimentary basins is described. Prediction of gross maturity [e.g., time–temperature index (TTI) or vitrinite reflectance (%R_0)] or hydrocarbon generation from a known or modeled burial history has attracted considerable interest (e.g., Turcotte and McAdoo, 1979; McKenzie, 1981; Welte and Yukler, 1981; Angevine and Turcotte, 1983; De Bremaecker, 1983; Middleton and Falvey, 1983; Ungerer *et al.*, 1984); however, these advances will not be discussed.

1.2. Biomarkers

The term "biological marker" introduced by Eglinton *et al.* (1964), and more recently abbreviated to "biomarker" (e.g., Philp and Lewis, 1987; Philp and Oung, 1988), has eclipsed other terms such as "biochemical fossil" (Fox, 1944), "chemical fossil" (Eglinton and Calvin, 1967), and "molecular fossil" (Calvin, 1969) and has been defined as "any organic compound detected in the geosphere whose basic skeleton suggests an unambiguous link with a known, contemporary natural product" (Mackenzie, 1984a). As well as the traditional biomarkers (e.g., steroids and triterpenoids), some aromatic hydrocarbons, most of which are not generally thought of as biomarkers, are discussed in this chapter.

Although many types of biomarkers have been

C. ANTHONY LEWIS • Centre for Petroleum and Environmental Organic Geochemistry, School of Applied Chemistry, Curtin University of Technology, Perth 6001, Western Australia, Australia. *Present address:* Department of Environmental Sciences, University of Plymouth, Plymouth, Devon PL4 8AA, England.

Organic Geochemistry, edited by Michael H. Engel and Stephen A. Macko. Plenum Press, New York, 1993.

reported in the literature (e.g., Brassell *et al.*, 1983; Philp, 1985), only a relatively small number have been studied quantitatively or semiquantitatively, especially with respect to basin modeling. The reactions and biomarkers discussed herein are shown in Fig. 1. These are (i) the aromatization of two C_{29} monoaromatic steroid hydrocarbons (5α and 5β) to a C_{28} triaromatic steroid hydrocarbon (Fig. 1a), (ii) the configurational isomerization at C-20 of 5α-24-ethylcholestane (Fig. 1b), (iii) the configurational isomerization at C-22 of 17α,21β-bishomohopane (Fig. 1c), (iv) the cyclization of 2,3-dimethylbiphenyl to 1-methylfluorene (Fig. 1d), (v) the "rearrangement" of 1,8-dimethylnaphthalene to 2,7-dimethylnaphthalene (Fig. 1e), (vi) the "rearrangement" of 1,4,6- and 1,3,5-trimethylnaphthalene to 2,3,6-trimethylnaphthalene (Fig. 1f), (vii) the "rearrangement" of 1-methylphenanthrene to 2- and 3-methylphenanthrene (Fig. 1g), and finally (viii) the pristane formation index (PFI; Fig. 1h).

Although reaction (iv) has been observed in the laboratory and is believed to occur in the geosphere, the decrease in the relative amount of 2,3-dimethylbiphenyl observed in sediments has been normalized to that of 3,5-dimethylbiphenyl (Alexander *et al.*, 1988a, b, 1989). Alexander *et al.* (1986) termed examples (v), (vi), and (vii) pseudoreactions since each "rearrangement" is not a single-step reaction, if it occurs at all. Thus, in these latter three cases the amount of the component(s) in the denominator is assumed to remain approximately constant while that of the numerator changes (decreases). In other words, the numerator is normalized relative to the denominator. Similarly, Goossens *et al.* (1988a, b) considered the PFI as quantifying a pseudoreaction since the same precursor(s) were assumed to yield pristane or pristenes under natural or analytical conditions, respectively. Burnham (1989) has subsequently corroborated the validity of using the PFI as a maturity parameter.

Considerable interest has been shown in these biomarkers and the various parameters derived from them, not only because they cover a wide range of maturity (e.g., Mackenzie *et al.*, 1980; Alexander *et al.*, 1985, 1988b), but also because their relative rates of change may be dependent on the thermal history of the sediment in which they occur (e.g., Mackenzie *et al.*, 1982; Mackenzie and McKenzie, 1983; McKenzie *et al.*, 1983) and/or differences in the sediment matrices (Alexander *et al.*, 1986; Strachan *et al.*, 1989a, b).

2. Chemical Kinetics

2.1. Isothermal Analysis

Before proceeding further with the application of biomarker reactions to the elucidation of the thermal history of sedimentary basins, it will be constructive to review some aspects of chemical kinetics. The discussion will be limited to first-order reactions, since, in the geosphere, the above reactions (Fig. 1) are believed to follow this type of rate equation (e.g., Mackenzie and McKenzie, 1983; Abbott *et al.*, 1984, 1985a, b; Alexander *et al.*, 1986, 1988a; Strachan *et al.*, 1989a, b; Kagi *et al.*, 1990). However, it should be mentioned that higher order reaction kinetics have been suggested for certain geochemical processes, for example, hydrocarbon generation reactions (Price, 1983, 1985), the formation of perylene in a lacustrine sediment (Gschwend *et al.*, 1983), and the release of hydrocarbons during pyrolysis of shales (Butler and Barker, 1986).

Chemical kinetics is the study of how fast a chemical reaction proceeds from the substrate(s) to the product(s) (e.g., Moore, 1972). Typically, this is accomplished by following the change in concentration of the substrate(s) and/or product(s). For a simple, first-order reaction scheme, e.g., Eq. (1),

$$\text{Substrate} \underset{k_b}{\overset{k_f}{\rightleftharpoons}} \text{Product} \tag{1}$$

the following rate law applies:

$$\frac{-d[S]_t}{dt} = k_f[S]_t - k_b[P]_t \tag{2a}$$

Figure 1. Reactions of biomarkers discussed and the ratios typically employed to follow their progress in the subsurface. (a) The aromatization of two C_{29} monoaromatic steroid hydrocarbons (5α and 5β) to a C_{28} triaromatic steroid hydrocarbon, measured as tri/(mono + tri); (b) the configurational isomerization at C-20 of 5α-24-ethylcholestane, measured as 20S/(20R + 20S); (c) the configurational isomerization at C-22 of 17α,21β-bishomohopane, measured as 22S/(22R + 22S); (d) the cyclization of 2,3-dimethylbiphenyl to 1-methylfluorene, measured as 3,5-DMB/2,3-DMB; (e) the amount of 1,8-dimethylnaphthalene relative to that of 2,7-dimethylnaphthalene, measured as 1,8-DMN/2,7-DMN; (f) the amount of 1,4,6- and 1,3,5-trimethylnaphthalene relative to that of 2,3,6-trimethylnaphthalene, measured as (1,4,6-TMN + 1,3,5-TMN)/2,3,6-TMN; (g) the amount of 1-methylphenanthrene relative to that of 2- and 3-methylphenanthrene, measured as 1-MP/(2-MP + 3-MP); (h) the amount of pristane relative to that of pristane and pristenes [pristane formation index (PFI)], measured as PR/(PR + PR-1-ENE + PR-2-ENE). DMB, Dimethylbiphenyl; DMN, dimethylnaphthalene; TMN, trimethylnaphthalene; MP, methylphenanthrene; PR, prist(ane).

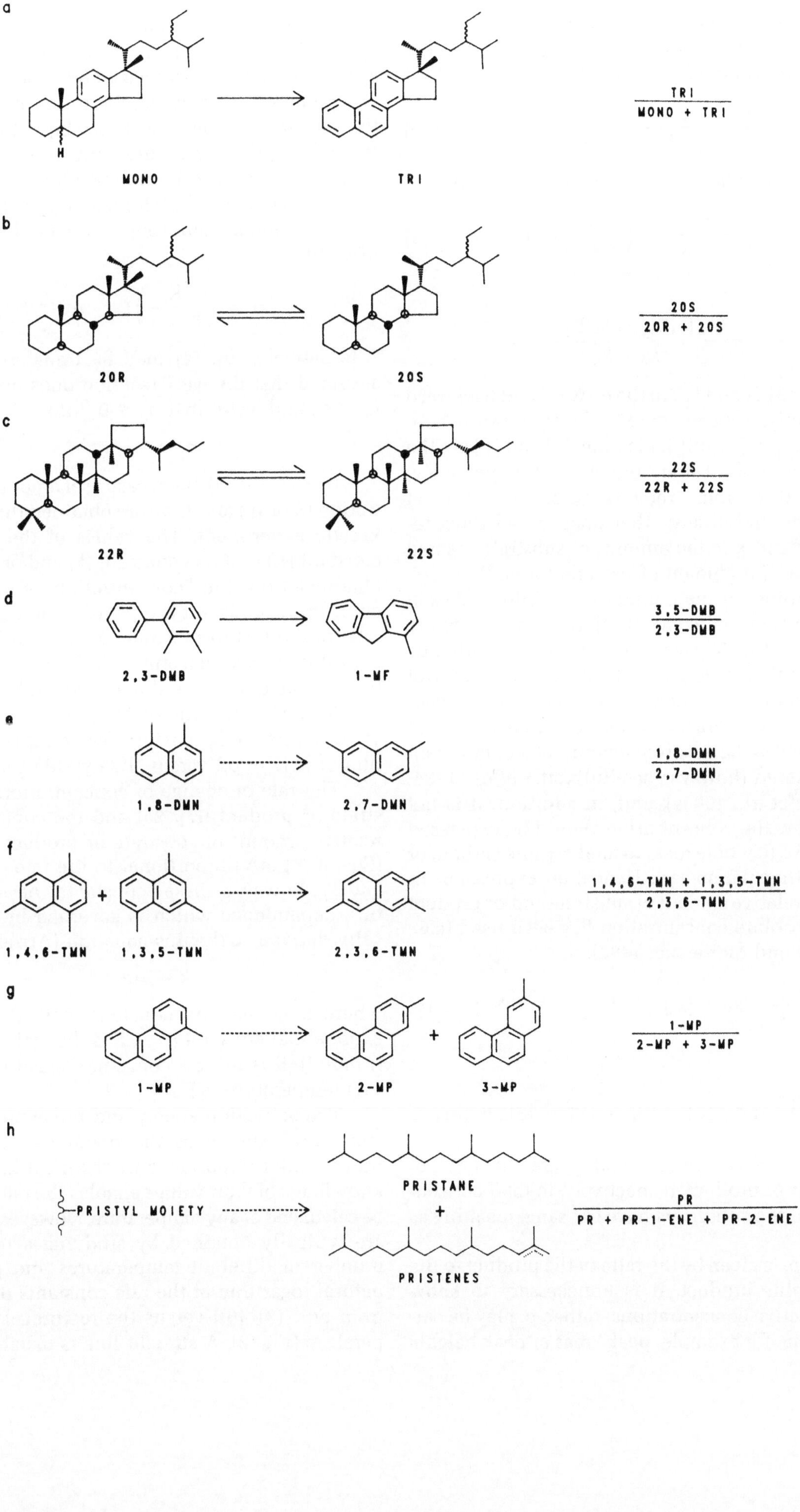
a
MONO
TRI
TRI / MONO + TRI
b
20R
20S
20S / 20R + 20S
c
22R
22S
22S / 22R + 22S
d
2,3-DMB
1-MF
3,5-DMB / 2,3-DMB
e
1,8-DMN
2,7-DMN
1,8-DMN / 2,7-DMN
f
1,4,6-TMN + 1,3,5-TMN
2,3,6-TMN
1,4,6-TMN + 1,3,5-TMN / 2,3,6-TMN
g
1-MP
2-MP + 3-MP
1-MP / 2-MP + 3-MP
h
PRISTYL MOIETY
PRISTANE
+
PRISTENES
PR / PR + PR-1-ENE + PR-2-ENE

Alternatively, the rate law may be expressed in terms of an amount, x, that has been transformed from substrate into product:

$$\frac{dx}{dt} = \left(\frac{K+1}{K}\right)k_f(m - x) \tag{2b}$$

from which may be derived (e.g., Moore, 1972):

$$\ln\left(\frac{m}{m - x}\right) = \left(\frac{K+1}{K}\right)k_f t \tag{3}$$

with

$$m = \frac{k_f[S]_0 - k_b[P]_0}{k_f + k_b}$$

In Eqs. (1)–(3), k_f and k_b are the forward and backward rate constants, respectively; K is the equilibrium constant ($= k_f/k_b = [P]_e/[S]_e$); t is time; [S] and [P] are the concentrations of substrate and product, respectively; subscripts 0, e, and t refer to values at $t = 0$, at equilibrium, and at any other unspecified time, respectively; and x is the amount of substrate reacted after a time t (or amount of product formed).

Depending on the boundary conditions chosen (e.g., value of initial concentration), it is possible to simplify Eq. (3); however, one difficulty with this equation is that it employs the absolute amount of substrate and/or product. Typically, organic geochemical analyses are not performed in a manner that readily enables the absolute amount of a component to be measured (however, see Rullkötter *et al.*, 1984; Mackenzie *et al.*, 1985a), and, in addition, it is not obvious how the concentration should be expressed (e.g., relative to whole rock, to total organic carbon, or to some other parameter). Hence, an expression in which the relative amount of substrate and/or product replaces absolute concentration has been used (e.g., Mackenzie and McKenzie, 1983):

$$p_t = \left(\frac{1}{k_f + k_b}\right)[k_b + (k_b p_0 - k_f s_0)e^{-(k_f + k_b)t}] \tag{4}$$

with

$$p_t = \frac{[P]_t}{[C]} \quad \text{and} \quad [C] = [S]_t + [P]_t = [S]_0 + [P]_0$$

where [C] is a constant, s and p are the ratio of substrate or of product, respectively, to total concentration, and the subscripts have the same meaning as before.

Since p_t is given by the ratio of the product to the substrate plus product, it is unnecessary to know their respective concentrations; rather, p_t may be calculated from, for example, peak areas or peak heights furnished by chromatographic analysis (assuming response factors for substrate and product are either known or may be assumed to be equal).

Equation (4) may be simplified by introducing the equilibrium constant [K; see Eq. (3)], which, over the restricted temperature range typically studied, may be assumed to be independent of temperature (e.g., Mackenzie and McKenzie, 1983) and by assuming the boundary condition $p_0 = 0$ (and hence $s_0 = 1$), to yield

$$p_t = \left(\frac{K}{K+1}\right)(1 - e^{-(K+1)k_b t}) \tag{5}$$

Alternatively, Eq. (4) may be transformed, if it is assumed that the back reaction does not occur (i.e., $k_b = 0$) and again that $p_0 = 0$, into

$$p_t = 1 - e^{-k_f t} \tag{6}$$

In Eqs. (3)–(6), the measure of amount or concentration (x or p_t) and time are obtained directly from a kinetic experiment. The values of the equilibrium constant (K) and rate constant (k_f and/or k_b) are then obtained from the "concentration"–time data. The equilibrium constant is typically calculated from the ratio of product to substrate for an experiment where equilibrium has been attained. The rate constant may be calculated from the slope obtained by plotting the applicable "concentration" function versus time that yields a linear relationship. For example, for Eq. (6) a plot of $\ln(1 - p_t)$ versus time yields a slope of $-k_f$.

The rate of change of concentration of the substrate or product (Eq. 2a) and the concentration or relative amount of substrate or product at any time (Eqs. 3–6) are proportional to the rate constants, k_f and k_b; however, these "constants" have a temperature dependence which is governed by the empirically derived Arrhenius equation (Arrhenius, 1889):

$$k = Ae^{-E/RT} \tag{7}$$

where k is the rate constant (s^{-1}), A is the pre-exponential constant (s^{-1}), E is the activation energy ($J\ mol^{-1}$), R is the gas constant ($8.3143\ J\ K^{-1}\ mol^{-1}$), T is temperature (K).

The activation energy and pre-exponential constant (also known as the frequency factor) are of fundamental importance in chemical kinetics since knowledge of their values enables the rate constant to be calculated at any temperature. However, their values are typically obtained by studying a reaction at a number of different temperatures and plotting the natural logarithm of the rate constants derived [e.g., from Eqs. (3)–(6)] versus the reciprocal of the temperature (Fig. 2). A straight line is usually obtained,

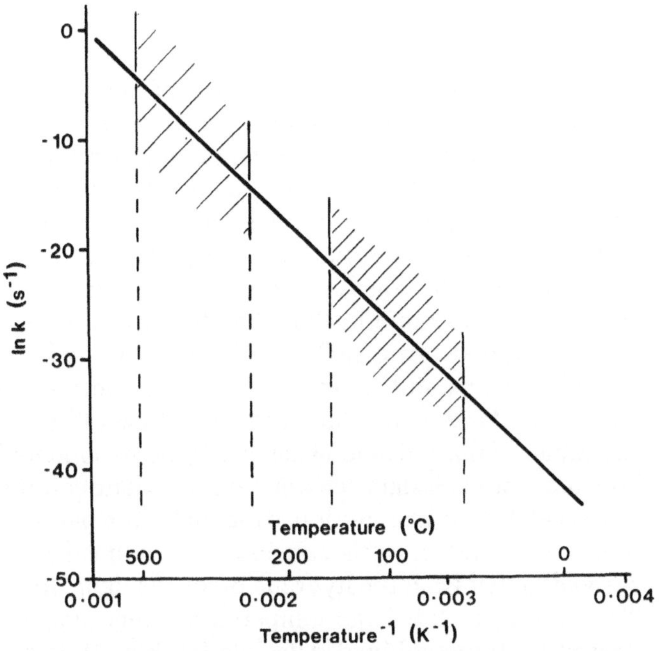

Figure 2. Plot of the natural logarithm of the rate constant (ln k) vs. reciprocal temperature (1/T). This type of graph is often described as an Arrhenius plot, and the values of the activation energy (E) and pre-exponential constant (A) may be obtained from the slope and the y-axis intercept, respectively. Note that the region typically studied in the laboratory is well separated along the x axis from that typically encountered in the subsurface where the reactions of biomarkers proceed.

and the activation energy and pre-exponential constant are calculated from the slope and y-axis intercept, respectively. Since each value of the rate constant is determined at a constant temperature, this treatment is termed an isothermal kinetic analysis.

2.2. Nonisothermal Analysis

Although isothermal kinetic analysis is presented in many physical chemistry textbooks (e.g., Moore, 1972; Atkins, 1986) as the only way to perform chemical kinetics, an alternative approach exists as there is no *a priori* reason why the temperature cannot be increased during a kinetic experiment. This approach is called nonisothermal kinetics (e.g., Wold and Ahlberg, 1970; Bunce *et al.*, 1988). Thus, substituting Eq. (7) into a generalized rate equation [i.e., $dc/dt = k{\cdot}f(c)$; cf. Eq. (2b)] yields

$$\frac{dc}{dt} = Ae^{-E/RT}f(c) \tag{8}$$

where $f(c)$ is a function, describing the applicable reaction scheme involving some measure of a change in the concentration of substrate and/or product. Since the temperature is changed during the course of the kinetic experiment, the rate of change of temperature may be given by

$$\frac{dT}{dt} = q \tag{9}$$

where q is the heating rate (i.e., the temperature increase per unit time). Substitution of Eq. (9) into Eq. (8) and integration gives

$$g(c) = \frac{A}{q}\int_{T_0}^{T} e^{-E/RT}\,dT \tag{10}$$

where $g(c)$ is the integral $1/f(c)$.

The integral in Eq. (10) has no exact analytical solution (e.g., Zsakó, 1968); however, it may be approximated by a truncated series expansion (e.g., Doyle, 1961; Abramowitz and Stegun, 1965), yielding

$$g(c) = \left(\frac{ART^2}{qE}\right)e^{-E/RT}\left[1 - \left(\frac{2RT}{E}\right) + 6\left(\frac{RT}{E}\right)^2\right] \tag{11}$$

Assuming that the heating rate is known, then Eq. (11) may be fitted, using an iterative least-squares procedure, to the experimental concentration data to yield values of the activation energy and pre-exponential constant.

3. Elucidation of the Activation Energy and Pre-Exponential Constant

As already stated, the activation energy and pre-exponential constant are of fundamental importance to the study of chemical kinetics. However, these parameters have to be determined by experimental measurement, and, for reactions believed to occur in the sedimentary column, this has been accomplished in a number of ways.

3.1. Laboratory Studies

One way to derive the activation energy and pre-exponential constant is to perform laboratory experiments where either a pure substrate (e.g., Abbott *et al.*, 1984, 1985a, b; Abbott and Maxwell, 1988; Alexander *et al.*, 1988a, 1989) or a sediment (e.g., Mackenzie *et al.*, 1981; Mackenzie and McKenzie, 1983; Suzuki, 1984; Lewan *et al.*, 1986; Rullkötter and Marzi, 1988, 1989; Marzi and Rullkötter, 1992), which may or may not have been extracted to remove soluble organic matter, is heated at a number of different temperatures under conditions attempting to model those in the subsurface. Hence, rate constants may be calculated (e.g., Eqs. 3–6) which may then be used to calculate the activation energy and pre-exponential constant (Eq. 7). Table I summarizes the values derived in the laboratory for the activation energy and pre-exponential constant of biomarker reactions.

Abbott *et al.* (1984, 1985a, b) investigated the aromatization of monoaromatic to triaromatic steroids (Fig. 1a) by heating the pure substrate (as a mixture of 5α and 5β stereoisomers) between 147 and 187°C on a sedimentary carbonate in the presence of elemental sulfur. Although the concentration of the triaromatic steroid product was also found to decrease, they were able to derive the activation energy and pre-exponential constant for the aromatization of the monoaromatic species (Abbott *et al.*, 1985a, b; Table I). The presence of sulfur was believed to promote the reaction via a radical mechanism.

Alexander *et al.* (1988a, 1989) heated pure 2,3-dimethylbiphenyl between 460 and 500°C under conditions chosen to minimize catalytic effects and determined the activation energy and pre-exponential constant for its cyclization (Fig. 1d) to 1-methylfluorene (Table I). The reaction is believed to proceed via a purely thermally induced radical mechanism due to the lack of catalytically active sites under the conditions chosen.

Suzuki (1984) heated a natural mudstone (Nishiyama Formation, Japan) between 330 and 400°C under anhydrous conditions and investigated the configurational isomerization at C-20 of 5α-24-ethylcholestane (Fig. 1b) and at C-22 of 17α,21β-bishomohopane (Fig. 1c). He derived values for the activation energy and pre-exponential constant for these reactions (Table I), and, although concern was expressed as to whether the same mechanism was operating in the laboratory as in nature, he was encouraged that the hopane reaction was faster below 320°C in the laboratory, as is observed in the sedimentary column (e.g., Mackenzie *et al.*, 1980).

Hydrous pyrolysis of a shale (Lias ϵ, Germany) between 210 and 350°C enabled Rullkötter and Marzi (1988, 1989) and Marzi and Rullkötter (1992) to derive the activation energy and pre-exponential constant for the aromatization of monoaromatic steroids and the configurational isomerization at C-20 of 5α-24-ethylcholestane and at C-22 of 17α,21β-bishomohopane (Table I). These authors also calculated the kinetic parameters by combining data from hydrous pyrolysis experiments with those from natural samples. They found that there was good agreement with the kinetic parameters for the sterane isomerization reaction calculated using solely hydrous pyrolysis data and concluded that the reaction was probably proceeding via the same mechanism in the laboratory as in nature. However, in the case of the hopane isomerization and the steroid aromatization reactions, the agreement between the kinetic parameters calculated from the combined data sets and those calculated from each individual data set (i.e., laboratory and natural) was found to be poor and led to speculation as to whether the mechanism for these reactions is the same in the laboratory as in nature.

A detailed error analysis of the combined hydrous pyrolysis and sediment sterane isomerization data sets (Marzi *et al.*, 1990) indicated that, although the values of the activation energy and pre-exponential constant for each data set were almost identical, the elliptical 95% joint confidence region for the parameters (i.e., E and $\ln A$) is considerably reduced with respect to that from the hydrous pyrolysis data alone. This is due to the large temperature range encompassed by the combined data sets (cf. Fig. 2) as opposed to the restricted range studied in the laboratory. As a result of the reduced confidence region for the kinetic parameters, the uncertainty in variables derived from basin model simulations (e.g., temperature and burial depth) is also reduced. However, Marzi *et al.* (1990) drew attention to the fact that the confidence intervals of any predictions made remain significantly large and, therefore, that results should not be overinterpreted.

Although laboratory studies enable the experimental conditions to be precisely controlled, there are disadvantages to using this approach to calculate the activation energy and pre-exponential constant. These are (i) the high temperature, relative to that in the geosphere (Fig. 2), which must typically be used to bring about the reaction in a convenient laboratory time scale, and (ii) the fact that the reaction studied in the laboratory, irrespective of the temperature, may not actually be the one that occurs in the geosphere.

The first problem may manifest itself in the reaction mechanism in the laboratory being different from that occurring in the geosphere, which may in turn cause a completely different reaction to occur in the laboratory. Alternatively, with respect to the second problem mentioned above, the other experimental conditions in the laboratory simulation (e.g., matrix and/or catalytic effects, pressure regime) may be sufficiently different from those in the geosphere so that a different reaction becomes favorable.

For the values of activation energy and pre-exponential constant derived in the laboratory to have any relevance to the geosphere, the same reaction, proceeding via the same mechanism, must be

TABLE I. Summary of Activation Energies and Pre-Exponential Constants Derived for Biomarker Reactions

Reaction	*A* (s^{-1})	E (kJ mol^{-1})	Notes	Reference(s)[a]
Aromatization				
Monoaromatic steroid	1.8×10^{14}	200	Isothermal kinetic–thermomechanical model, shale matrix	1
	6.7×10^{12}	145	Laboratory thermal simulation, anhydrous, radical, pure substrate	2
	4.85×10^{10}	181.4	Laboratory thermal simulation, hydrous, shale matrix, selected natural samples	3, 4
Isomerization				
C-20 5α-24-ethyl-cholestane	6.0×10^{-3}	91	Isothermal kinetic–thermomechanical model, shale matrix	1
	6.6×10^{-3}	84	Nonisothermal kinetic model, natural samples, shale matrix	5, 6
	2.4×10^{-3}	91.6	Nonisothermal kinetic model, natural samples, shale matrix	7
	—[b]	26	Nonisothermal kinetic model, natural samples, coal matrix	6
	—	59	Nonisothermal kinetic model, natural samples, intermediate shale/coal matrix	6
	—	34	Laboratory thermal simulation, hydrogen donor solvent, coal matrix	6
	6.5×10^{7}	147	Laboratory thermal simulation, anhydrous, shale matrix	8
	4.86×10^{8}	169.0	Laboratory thermal simulation, hydrous, shale matrix, selected natural samples	3, 4
C-22 17α, 21β-bishomohopane	1.6×10^{-3}	91	Isothermal kinetic–thermomechanical model, natural samples, shale matrix	1
	3.5×10^{-2}[c]	87.8[c]	Nonisothermal kinetic model, natural samples, shale matrix	7
	2.0×10^{4}	98	Laboratory thermal simulation, anhydrous, shale matrix	8
	8.1×10^{8}	168	Laboratory thermal simulation, hydrous, shale matrix, selected natural samples	4
Cyclization				
2,3-Dimethylbiphenyl	8.2×10^{7}	180	Laboratory thermal simulation, noncatalytic, pure substrate	9, 10
2,2′-Dimethylbiphenyl	0.1×10^{7}	153	Laboratory thermal simulation, noncatalytic, pure substrate	10
Pseudoreactions				
DP-1[d]	3.6×10^{-14}	14	Nonisothermal kinetic model, natural samples, coal matrix	5
	1.4×10^{-5}	73	Nonisothermal kinetic model, natural samples, shale matrix	5
TP-1[e]	1.8×10^{-14}	29	Nonisothermal kinetic model, natural samples, coal matrix	5
	3.0×10^{-13}	19	Nonisothermal kinetic model, natural samples, shale matrix	5
PP-1[f]	1.0×10^{-14}	14	Nonisothermal kinetic model, natural samples, coal matrix	5
	7.7×10^{-13}	22	Nonisothermal kinetic model, natural samples, shale matrix	5
PFI[g]	2.2×10^{-7}	59	Isothermal kinetic–thermomechanical model	11
	1.1×10^{13}	198.7[h]	Rock-Eval pyrolysis, nonisothermal kinetic model	12

[a]References: 1, Mackenzie and McKenzie, 1983; 2, Abbott *et al.*, 1985a,b; 3, Rullkötter and Marzi, 1989; 4, Marzi and Rullkötter, 1992; 5, Alexander *et al.*, 1986; 6, Strachan *et al.*, 1989a,b; 7, Sajgó and Lefler, 1986; 8, Suzuki, 1984; 9, Alexander *et al.*, 1988a; 10, Alexander *et al.*, 1989; 11, Goossens *et al.*, 1988a; 12, Burnham, 1989.
[b]—, Value not given.
[c]Derived from the average composition data of C_{31} and C_{32} 17α,21β-hopanes.
[d]1,8-DMN/2,7-DMN (DMN = dimethylnaphthalene).
[e](1,4,6-TMN + 1,3,5-TMN)/2,3,6-TMN (TMN = trimethylnaphthalene).
[f]1-MP/(2-MP + 3-MP) (MP = methylphenanthrene).
[g]Pristane Formation Index, pristane/(pristane + prist-1-ene + prist-2-ene).
[h]Distribution of activation energies assumed (σ = 11.6 kJ mol^{-1}).

occurring in both environments. If this criterion is not met, then, whatever the cause, the laboratory-derived activation energy and pre-exponential constant cannot be applied to the geosphere (e.g., Snowdon, 1979; Quigley and Mackenzie, 1988).

Thermal simulation experiments, on pure substrates, have demonstrated that reactions thought to occur in the geosphere can indeed be brought about in the laboratory under conditions that attempt to model those found in sediments (e.g., Abbott *et al.*, 1984, 1985a, b; Abbott and Maxwell, 1988; Alexander *et al.*, 1988a, 1989). However, Abbott *et al.* (1990), in their continuing studies on the kinetics of hydrous pyrolysis, have demonstrated a minimal extent of isomerization of a deuterated 5α-cholestane. This indicates that direct isomerization of free extractable hydrocarbons may not occur, and, therefore, it is possible that the apparent isomerization observed in the geosphere is intimately associated with the generation of hydrocarbons from kerogen. Although observation of the same reaction in the laboratory and the geosphere does not prove that it occurs by the same mechanism, evidence in support of this conclusion may be obtained if values of the activation energy and pre-exponential constant, derived in the laboratory, satisfactorily predict the extent of reaction in the geosphere.

3.2. Basin Studies

Another method of calculating values of the activation energy and pre-exponential constant employs geochemical data obtained directly from sediments. To achieve this, two different kinetic approaches have been investigated. The first employs isothermal and the second nonisothermal chemical kinetic equations.

3.2.1. Isothermal Kinetic Approach

Mackenzie *et al.* (1982) proposed that the relative extent to which the aromatization of monoaromatic steroids and the configurational isomerization at C-20 of 5α-24-ethylcholestane (Fig. 1a and 1b, respectively) had occurred appeared to be different in basins that had experienced different thermal histories. Thus, young basins having a high average heating rate showed an enhanced extent of aromatization over that of isomerization, while older basins having a low average heating rate showed an opposite trend. To explain these observations, Mackenzie *et al.* (1982) proposed that the temperature dependence of the aromatization reaction was greater than that for the isomerization reaction.

Mackenzie and McKenzie (1983), in a classic paper, developed this idea and derived values of the activation energy and pre-exponential constant for these isomerization and aromatization reactions. They used samples from the Pannonian Basin (Hungary) and North Sea, which are each believed to have experienced a different but relatively simple thermal history with little or no uplift (e.g., Mackenzie, 1984b, and references therein), to obtain a single set of values for the activation energy and pre-exponential constant of each reaction that could adequately describe their extent in both sedimentary sequences (Table I).

The approach of Mackenzie and McKenzie (1983) utilized isothermal chemical kinetic equations (Eq. 4) to calculate an approximate value of the rate constant (k_f or k_b) for each sediment sample, from which rough estimates of the activation energy and pre-exponential constant were then obtained by applying the Arrhenius equation (Eq. 7). Since Eq. (4) is only true under conditions of constant temperature, its application to geological samples, where the temperature of a particular rock particle changes with time, can only yield approximate values of these parameters. However, these initial estimates were used in conjunction with the temperature history obtained from a basin model (McKenzie, 1978, 1981; Mackenzie and McKenzie, 1983) to numerically integrate the rate law (Eq. 2) over short time intervals during which the temperature was assumed to be approximately constant. In effect, the value of p_t (Eqs. 4–6) was used to calculate another value a short time later (i.e., $p_{t+\delta t}$), which in turn was used to calculate the next value, after a further time δt.

Mackenzie and McKenzie (1983) assumed, for the short time interval δt, that the temperature remained constant and, hence, for *two different reactions* that Eqs. (5) and (6) could be equated, eliminating the time, *t*. Thus, the data, calculated from the numerical integration procedure, for the aromatization of monoaromatic steroids and the configurational isomerization at C-20 of 5α-24-ethylcholestane were plotted against each other on an aromatization–isomerization diagram ("A–I" diagram). The shape of the curve on this type of plot depends on the difference between the activation energies and the ratio of the pre-exponential constants of the respective reactions as well as the temperature history (i.e., heating rate). For a given sedimentary column, the temperature history is fixed (although not necessarily accurately known), and, hence, the values of the activation energy and pre-exponential constant, for each reaction, were adjusted until the calculated curve satisfactorily matched the observed data. Mackenzie and McKenzie (1983) found that the same parameters could be used to obtain reasonably good agreement between the calculated and observed A–I curves for the Pannonian Basin and the North Sea, which have each experienced a different temperature history.

The values of the activation energy and pre-exponential constant that Mackenzie and McKenzie (1983) obtained for the steroid aromatization and isomerization reactions were very different (Table I), and therefore they proposed that these two reactions could be used to distinguish basins having different thermal histories since the data would plot in different regions of an A–I diagram.

In addition to the relationship between these steroid aromatization and isomerization reactions, Mackenzie and McKenzie (1983) investigated the relative extent of (i) the aromatization of monoaromatic steroids and the configurational isomerization at C-22 of 17α,21β-bishomohopane, and (ii) the configurational isomerization at C-20 of 5α-24-ethylcholestane and at C-22 of 17α,21β-bishomohopane. Thus, data for the steroid aromatization were plotted against those for the hopane isomerization or data for the two isomerization reactions were plotted against each other on an isomerization–isomerization diagram ("I–I" diagram). In the latter case, both sets of data were obtained using Eq. (5), since both reactions were assumed to be reversible. Mackenzie and McKenzie (1983) proposed that the activation energy was the same for the two isomerization reactions and that the pre-exponential constants were similar (Table I). Hence, the shape of the curve on the I–I diagram is independent of the thermal history (and therefore geophysical model) and depends only on the kinetic assumptions made. Mackenzie and McKenzie (1983) found that there was good agreement between the observed and predicted shape of the curve on the I–I diagram for a number of basins, including those having experienced uplift, and cited this as evidence in favor of their assumption that these reactions are first-order and unimolecular.

The Pannonian Basin and North Sea were, therefore, used to "calibrate" these biomarker reactions. The activation energies and pre-exponential constants obtained (Table I) were subsequently used to study the timing and extent of uplift in more complex basins or, in frontier basins, the depth at which sediments might be expected to be sufficiently mature to have generated oil and/or gas.

3.2.2. Nonisothermal Kinetic Approach

Nonisothermal kinetic equations [for example, of the form of Eq. (11)] have been employed to derive values of the activation energy and pre-exponential constant directly from geochemical data (e.g., Alexander *et al.*, 1986; Sajgó and Lefler, 1986; Strachan *et al.*, 1989a, b).

Thus, Alexander *et al.* (1986) and Strachan *et al.* (1989a, b) fitted an equation similar to Eq. (11), using an iterative least-squares procedure, to data for the configurational isomerization at C-20 of 5α-24-ethylcholestane (Fig. 1b) in sediments from the Pannonian Basin (Fig. 3; Sajgó, 1980; Mackenzie and McKenzie, 1983) and, hence, derived values of the activation energy and pre-exponential constant (Table I). The values calculated using nonisothermal equations were found to be in good agreement with those previously reported by Mackenzie and McKenzie (1983).

Using a similar methodology, Alexander *et al.* (1986) investigated three pseudoreactions (Fig. 1e–g), formally involving methyl shifts of methylated naphthalenes and phenanthrenes, and derived values of the activation energy and pre-exponential constant for different sedimentary systems (see Section 5.).

Sajgó and Lefler (1986) also investigated sediments from the Pannonian Basin and applied non-

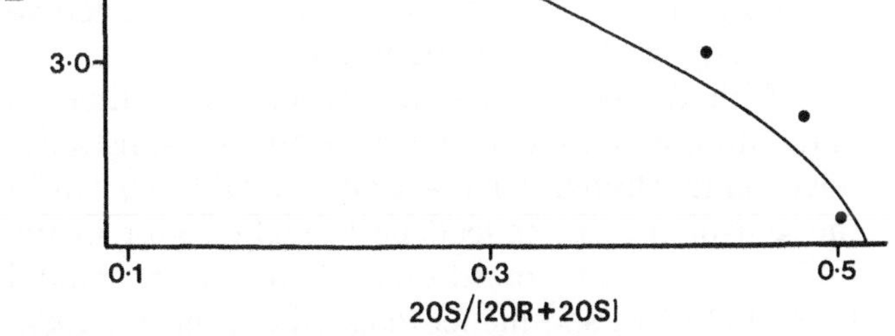

Figure 3. Plot of the observed extent of reaction for the configurational isomerization at C-20 of 5α-24-ethylcholestane (Fig. 1b) versus depth for the Hód 1 well of the Pannonian Basin, Hungary. The curve is the best-fit line to the observed data, obtained from an equation similar to Eq. (11) which describes the variation of the extent of a reaction under nonisothermal conditions. [Modified and reproduced with permission from Alexander *et al.* (1986).]

isothermal kinetic equations to obtain values of the activation energy and the pre-exponential constant for the configurational isomerization at C-20 of 5α-24-ethylcholestane and at C-22 of 17α,21β-hopanes (composition data for the C_{31} and C_{32} species were averaged; Table I). In addition, they attempted to derive values for the aromatization of monoaromatic steroids; however, different methods of calculation were found to yield substantially different values, and it was concluded that this reaction was not well enough understood to be used for reconstructing the geothermal history of sedimentary basins.

3.2.3. Problems Associated with Basin Studies

Although, in a strict kinetic sense, the approach of Alexander *et al.*, (1986), Sajgó and Lefler (1986), and Strachan *et al.* (1989a, b) is more correct than that of Mackenzie and McKenzie (1983) both suffer from similar problems. For example, it is typically assumed that (i) for reactions involving steroids and hopanoids, the compounds under investigation are actually the substrate and the product of a single reaction [i.e., a conversion of S to P occurs; Eq. (1)] whereas for reactions involving other aromatic hydrocarbons one member is normalized against another, that is, the "substrate" changes in amount while the "product" is (relatively) unaffected, (ii) the initial amount of substrate (and product) is the same for each sample studied down the sedimentary column, (iii) for reactions involving steroids and hopanoids, the total absolute amount of substrate plus product remains constant with increasing maturation whereas for reactions involving other aromatic hydrocarbons the component used for normalization (i.e., the denominator or "product") remains constant with increasing maturation, and (iv) the reactions proceed via the same mechanism in the different basins that are compared when determining values of the activation energy and pre-exponential constant.

With respect to the first assumption, thermal simulation experiments, on pure organic substrates, have demonstrated that reactions thought to occur in the geosphere can indeed be brought about in the laboratory under conditions attempting to model those found in sediments (e.g., Abbott *et al.*, 1984, 1985a, b; Abbott and Maxwell, 1988; Alexander *et al.*, 1988a, 1989), although Abbott *et al.* (1990) subsequently demonstrated a minimal extent of isomerization of a deuterated 5α-cholestane during hydrous pyrolysis. Other laboratory thermal simulation experiments, using extracted or unextracted sediments without addition of further organic substrate, have also demonstrated that changes similar to those occurring in sediments appear to occur (e.g., Mackenzie *et al.*, 1981; Mackenzie and McKenzie, 1983; Suzuki, 1984; Lewan *et al.*, 1986; Rullkötter and Marzi, 1988, 1989; Marzi and Rullkötter, 1992). However, the observation of a reaction in the laboratory does not prove that it occurs in the geosphere. For example, Alexander *et al.* (1988a, 1989) observed the cyclization of 2,3-dimethylbiphenyl to 1-methylfluorene in the laboratory but observed no systematic relationship between these compounds in sediments from the Carnarvon Basin (Australia) although 1-methylfluorene was present in all samples as the major methylfluorene isomer. Thus, 2,3-dimethylbiphenyl may or may not cyclize to 1-methylfluorene in these sediments and/or the latter may be affected by one or more other reactions.

The second assumption appears to be the least important, since the progress of the reactions is followed by the ratio of product to either total substrate plus product [*i.e.*, P/(S + P); Eq. (4)] or to another independent, unchanging species. Thus, in either case a small range of values may be tolerated as the "initial value," with the reaction assumed to have occurred to a significant extent when this range is exceeded.

The third assumption is potentially the most serious, since it has been observed that the absolute amount of many biomarkers decreases with increasing thermal stress, both in the laboratory (e.g., Mackenzie *et al.*, 1981; Abbott *et al.*, 1984, 1985a, b; Abbott and Maxwell, 1988; Marzi and Rullkötter, 1992) and in the geosphere (Rullkötter *et al.*, 1984; Mackenzie *et al.*, 1985a). This implies that there are one or more additional reactions, other than those typically used to quantify thermal maturation, affecting the substrate and product. Therefore, the assumption of a simple reaction scheme (e.g., Eq. 1) may be invalid.

The complexity of geochemical reactions is illustrated by the isomerization of 5α-cholestane. Abbott *et al.* (1984) heated this pure substrate on a shale matrix with added elemental sulfur at 259°C and observed the formation of four isomeric cholestanes in addition to the major product, the (20S)-5α-isomer. All these components have been observed in geological samples (e.g., Seifert and Moldowan, 1979). A more thorough study (Lewis, 1985) investigated the concentration–time profiles of 5α-cholestane and seven saturated as well as five triaromatic steroid products.

The final assumption is also important, since in the same way that the mechanism of a reaction observed in the laboratory and in the geosphere has to be the same if meaningful comparisons are to be obtained, then in different basins the same mechanism should also apply if the progress of a reaction in each sedimentary sequence is to be comparable. The effect of the organic matter type, amount of organic matter, and mineralogical composition on the activation energy and pre-exponential constant of biomarker reactions is discussed in Section 5.

Although these assumptions are potentially serious and, in a strict kinetic sense, question the verisimilitude of the models, the fact that the models can predict, to a first approximation, the progress of reactions in the geosphere appears to bear out their validity.

However, a problem that affects kinetic models based on geochemically derived data, whether isothermal or nonisothermal, is the accuracy with which the thermal history of the basin is known. Models based on geochemically derived data typically employ a basin having a simple thermal history (e.g., one having experienced only subsidence) to provide values of the activation energy and pre-exponential constant. These parameters may then be applied to other basins where the thermal history is less well understood, with a view to estimating, for example, the amount and timing of uplift or the depth at which sediments may be expected to be sufficiently mature to have generated oil and/or gas. An incorrect thermal model for the simple basin may lead to the derivation of incorrect values of the activation energy and pre-exponential constant, which in turn may lead to erroneous conclusions concerning other more complex basins. It is apparent, therefore, that a thorough understanding of the thermal history of the basin used to "calibrate" the activation energy and pre-exponential constant of a particular reaction is essential if accurate values are to be obtained. This has been discussed by Kagi *et al.* (1990), who emphasized the utility of laboratory experiments to derive values of the activation energy and pre-exponential constant.

3.3. Concluding Remarks Concerning the Activation Energy and Pre-Exponential Constant

A number of attempts have been made to calculate the activation energy and pre-exponential constant of biomarker reactions, and the values reported are given in Table I. Mackenzie and McKenzie (1983) determined the kinetic parameters for three biomarker reactions and then applied these values to other basins. Their methodology depends on the values of the activation energy and pre-exponential constant for two reactions being substantially different so that basins experiencing different heating rates may be differentiated. However, recent work (Rullkötter and Marzi, 1989; Marzi and Rullkötter, 1992) has shown that the kinetic parameters for the configurational isomerization at C-20 of 5α-24-ethylcholestane are, in fact, quite similar to those for the aromatization of monoaromatic steroids. Hence, it is not possible to differentiate basins that have experienced different heating rates using a plot of the extent of one reaction versus the extent of another (i.e., on an A–I diagram). However, Rullkötter and Marzi (1989) and Marzi and Rullkötter (1992) have demonstrated that a single reaction may be used to investigate the thermal history of sedimentary basins (see Section 4.4.).

Although it appears that the values of the activation energy and pre-exponential constant derived by Mackenzie and McKenzie (1983) are incorrect, this does not invalidate their methodology of comparing the relative extent of two reactions to assess the thermal history of a sedimentary basin. For this reason, a number of case histories that have relied on the values reported by Mackenzie and McKenzie (1983) are presented in the next section, even though some of the conclusions reached may be erroneous.

Recently, Marzi (1992) has illustrated how the activation energy and pre-exponential constant of any reaction must fall within a defined region on a plot of log A versus E if that reaction is to have any geological relevance. If values of the activation energy and pre-exponential constant for a reaction, however derived, do not fall within this region then their use to assess thermal or subsidence histories is inappropriate. Either the reaction will proceed too quickly and therefore reach completion or equilibrium in too short a time, or will proceed too slowly for a significant extent or reaction to have occurred within the lifetime of a sedimentary basin.

4. Case Studies

4.1. The Paris Basin, France, and Lower Saxony Basin, Germany

Values of the activation energy and pre-exponential constant derived from the Pannonian Basin and North Sea have been used to calculate the timing and the amount of uplift experienced by the Toarcian and Pliensbachian shales of the Paris Basin (France) and Lower Saxony Basin (Germany), respectively (Mackenzie and McKenzie, 1983; Mackenzie, 1984b; Mackenzie *et al.*, 1984, 1988).

The thermal history of the Paris Basin was modeled assuming a crustal stretching event ending 180 Ma B.P. (Mackenzie and McKenzie, 1983; Mackenzie, 1984b). Since that time, the basin has experienced both subsidence and uplift and erosion, with different regions of the basin having experienced differing amounts of these processes. Attempts by these authors to predict the progress of the sterane and hopane isomerization and monoaromatic steroid aromatization reactions using previously available estimates of the amount (<200 m in the center and 700 m on the eastern margin of the basin) and timing of uplift and erosion (e.g., Deroo, 1967) were unsuccessful. They concluded that these estimates of the amount of uplift and erosion were too small and that larger amounts were necessary to explain the relative extents of these reactions, particularly on the eastern margin of the basin. An A–I plot illustrating the relationship between the time since a stretching event and the expected sample depth (Fig. 4) proved useful in estimating the amount of uplift. Thus, the difference between the present-day depth and that of the isobath on which the data plots gives an estimate of the amount of uplift experienced by the sample. From this type of analysis, Mackenzie and McKenzie (1983) proposed that there was no detectable uplift in the center of the basin (within the total experimental error of the analysis), but toward the eastern margin of the basin the uplift increases to at least 2.0 km.

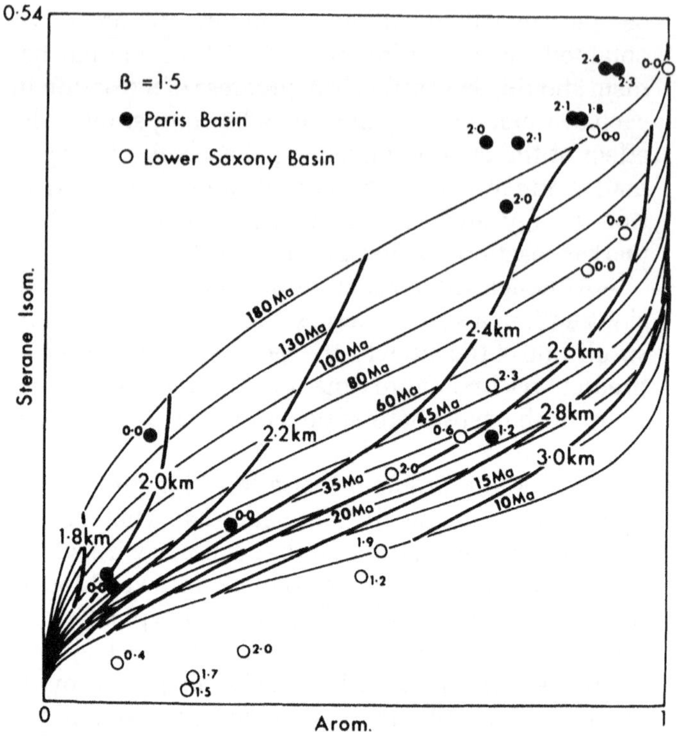

Figure 4. Diagram used to determine the amount and timing of uplift based on the steroid aromatization and sterane isomerization reactions (Fig. 1a and 1b, respectively). If both reactions are stopped by uplift, the position of the observed data yields the time elapsed between the stretching and uplift events (thin lines), and the difference between the observed depth and that calculated (thick lines) yields the amount of uplift. [Reproduced with permission from Mackenzie and McKenzie (1983).]

Although it should be possible to estimate the time elapsed between the stretching and uplift events from Fig. 4, this proved to be difficult for the Paris Basin because of the poor agreement between the geological and geochemical data for samples on the eastern margin of the basin. However, the extent of the reactions indicates that the shales spent only a brief period at depths sufficient to cause the reactions to proceed, and it was suggested that the uplift and erosion on the eastern margin had occurred within 40–50 Myr of the stretching event. Earlier reports that a widespread regression had occurred at about this time, which may have been associated with major uplift and erosion, was cited as corroborative evidence (Mackenzie and McKenzie, 1983; Mackenzie, 1984b).

Goossens *et al.* (1988b) modeled the thermal history of the Paris Basin sediments using the pristane formation index (PFI). They were able to corroborate the results of Mackenzie and McKenzie (1983) and Mackenzie (1984b) and demonstrated that the surface sediments at the eastern margin of the basin had experienced an elevated temperature for about 10–30 Myr, whereas those deeply buried at the center of the basin had not undergone significant temperature changes in the past.

The organic-rich Pliensbachian shales (Lias δ, Jurassic) were used to elucidate the timing and extent of uplift in the Lower Saxony Basin (Mackenzie and McKenzie, 1983; Mackenzie, 1984b; Mackenzie *et al.*, 1988). Major subsidence in the Lower Saxony Basin is believed to have occurred during the early to middle Cretaceous in response to lithospheric stretching; however, the basin subsequently experienced uplift and inversion as a result of compressional forces reactivating normal faults as thrust faults (Ziegler, 1980).

When the observed extent of reaction was plotted on Fig. 4, it was found that the uplift had occurred about 50 Ma after the extensional event (i.e., in the late Cretaceous to early Tertiary) and that the most deeply buried samples before uplift were now the shallowest (Mackenzie and McKenzie, 1983; Mackenzie, 1984b; Mackenzie *et al.*, 1988). Thus, samples from the southern margin have been uplifted 2.0–2.5 km. This is in agreement with the belief that the Lower Saxony Basin is a type example of an inverted basin (Ziegler, 1980).

4.2. The Alberta Basin, Canada

Unlike the basins discussed in Section 4.1, whose formation has been modeled using an extensional thinning of the lithosphere (e.g., McKenzie, 1978, 1981; Mackenzie and McKenzie, 1983), the Alberta Basin has been modeled using a lithospheric flexure model (Beaumont *et al.*, 1985).

A major difference between these two subsidence models is the temporal variation of the heat flow. The extensional model predicts that the heat flow of a subsiding basin decreases to the ambient value with a time constant proportional to the regional thickness of the lithosphere (Turcotte and Ahern, 1977; McKenzie, 1978, 1981; Turcotte and McAdoo, 1979). However, the flexural model predicts that the heat flow is time-invariant if the underlying lithosphere is thermally mature (Beaumont, 1981; Beaumont *et al.*, 1982a, 1985). It is apparent therefore that these models should predict different reaction profiles as a function of burial depth or time. Since the values of the activation energy and pre-exponential constant previously derived for the sterane and hopane isomerization and steroid aromatization reactions (e.g., Mackenzie and McKenzie, 1983) are independent of the subsidence model employed to determine the thermal history of a basin, the values derived from the Pannonian Basin and North Sea (subsidence modeled by an extension of the lithosphere) may be applied to the Alberta Basin (subsidence modeled by lithospheric flexure).

Beaumont *et al.* (1985) showed that the extent of the sterane and hopane isomerization and steroid aromatization reactions in the Alberta Basin could be adequately described by a lithospheric flexure model. Furthermore, they demonstrated that the paleogeothermal gradient increased with increasing distance from the edge of thrusting, a situation that still exists at the present time. At a distance of approximately 50 km from the edge of thrusting, the paleogeothermal gradient was proposed to be between 26 and 29°C km^{-1} depending on the surface temperature (10 and 0°C, respectively). Approximately 100 km from the edge of faulting, the paleogeothermal gradient was proposed to have increased to between 30 and 35°C km^{-1} (surface temperature 10 and 0°C, respectively). At a distance of approximately 200 km from the edge of faulting, the paleogeothermal gradient was proposed to have increased still further to between 43 and 49°C km^{-1} (surface temperature 10 and 0°C, respectively); however, the agreement between the predicted and the observed data was not as good at this distance as that for samples closer to the edge of thrusting. The fact that the present-day geothermal gradient behaves in a similar manner was used to corroborate these conclusions.

Although the extent of these reactions may be used to predict the value of the paleogeothermal gradient, they cannot yield information about the cause(s) of the perturbation. This information must be obtained from subsidence models and/or other geological information. Since neither the flexural nor the extensional model can account for this increase in heat flow in the updip, shallow region of the Alberta Basin, a different explanation was sought. It was suggested (Beaumont *et al.*, 1985) that water heated at depth in the faulted foothills area, and consequently depressing the temperature gradient in that region, migrates updip, enhancing the thermal gradient in the shallower areas to the east. This explanation has been previously discussed (see Beaumont *et al.*, 1985, and references therein).

4.3. The Nova Scotia Continental Margin, Offshore Canada

The formation and evolution of the continental margin off Nova Scotia (Canada) has been modeled on the basis of lithospheric extension during rifting (Beaumont *et al.*, 1982b; Mackenzie *et al.*, 1985b). Mackenzie *et al.* (1985b) employed the same basic thermomechanic model as Beaumont *et al.* (1982b); however, the values of the thermal conductivity of the sediments were modified to be more representative of the actual samples studied. This was done because values typically quoted in the literature resulted in the model predicting too low a geothermal gradient. When the new values were used, Mackenzie *et al.* (1985b) obtained good agreement between the predicted and observed present-day bottom-hole temperature for wells drilled on the continental slope and rise. Furthermore, there was good agreement between the predicted and observed extent of the sterane and hopane isomerization and steroid aromatization reactions for these wells. On the continental slope and rise, the isomerization and aromatization reactions are currently under way in an approximately

1-km zone that is independent of stratigraphic horizons, and the top of the oil generation zone was placed at a depth of approximately 4 km (Mackenzie *et al.*, 1985b).

Further offshore, there had been little or no drilling, and, hence, the maturity of the sediments in this area could not be obtained by measurement but had to be predicted. The agreement between predicted and observed maturity levels in the continental slope and rise area encouraged Mackenzie *et al.* (1985b) to predict the timing of oil generation in the Verrill Canyon Formation (Middle Jurassic to Lower Cretaceous), which had been previously proposed as the major source rock for oils and condensates found on the Scotian Shelf (Powell, 1982). Considering the large-scale structure of the basin (i.e., ignoring small-scale perturbations such as variations in lithology, sediment thickness, or salt-shale diapirism), Mackenzie *et al.* (1985b) proposed that oil generation has been occurring in the Verrill Canyon Formation beneath the slope and outer shelf area since the Barremian (Middle Cretaceous) and continues to the present; however, significant volumes of sediment only entered this zone during the Tertiary. Overlying formations (Logan Canyon and Wyandot) were not predicted to be regions of significant oil generation.

4.4. The Michigan Basin, United States

The Michigan Basin has caused considerable controversy in the literature because some authors maintain that significant oil generation has not occurred in Silurian and younger sediments (Nunn *et al.*, 1984, 1985a, b), while others believe that Silurian, Devonian, and even Mississippian source rocks have generated oil (Cercone, 1984; Daly and Lilly, 1985; Illich and Grizzle, 1985; Rullkötter *et al.*, 1986). Recently, the thermal history of the basin has been reappraised by employing values of the activation energy and pre-exponential constant derived for the configurational isomerization at C-20 of 5α-24-ethylcholestane (Rullkötter and Marzi, 1989; Marzi and Rullkötter, 1992; Rullkötter *et al.*, 1992; Table I; see Section 3.1).

The progress of the sterane isomerization reaction has been modeled (Rullkötter and Marzi, 1989; Marzi and Rullkötter, 1992; Rullkötter *et al.*, 1992) assuming the thermal history proposed by Nunn *et al.* (1984), and it was observed that a good match between the calculated and the observed extent of reaction was obtained for the base of the Silurian. In addition, the reaction was calculated to have reached equilibrium for the base of the Middle Ordovician ca. 200 Ma B.P., and samples of this age were indeed observed to be overmature. However, the calculated extent of reaction was significantly underestimated for the base of the Devonian.

A more severe thermal history than that of Nunn *et al.* (1984) was proposed by Cercone (1984). Using this thermal history, a considerably better fit between the calculated and the observed extent of reaction for the early Devonian, the Middle Devonian (Antrim Shale), and the early Carboniferous (Coldwater Shale) was obtained (Rullkötter and Marzi, 1989; Marzi and Rullkötter, 1992; Rullkötter *et al.*, 1992). However, the extent of reaction in an outcropping Pennsylvanian coal was significantly underestimated, and this was thought to imply that the coal had been more deeply buried in the past than predicted by the model.

Although there is no reported geological evidence to support further subsidence of the basin, Marzi and Rullkötter (1992) and Rullkötter *et al.* (1992) modified the thermal history presented by Cercone (1984) to include ca. 0.75 km extra burial, with subsequent continuous erosion between the late Carboniferous and early Jurassic. The geothermal gradient was assumed to have increased to 35°C km^{-1} during the period of subsidence, remained constant for ca. 60 Myr into the Permian and thereafter declined during the Permian and Jurassic to the present value (21°C km^{-1}; Nunn *et al.*, 1984). As a result, a better agreement was obtained between the calculated and the observed extent of reaction for all Devonian and younger sediments, including the Pennsylvanian coal, than was obtained with the other thermal histories (Marzi and Rullkötter, 1992; Rullkötter *et al.*, 1992). This finding, as well as the fact that the geothermal gradient is decreased, relative to that presented by Cercone (1984), was used by these authors to justify the more complex burial history proposed.

4.5. The Carnarvon Basin, Australia

The cyclization of 2,3-dimethylbiphenyl to 1-methylfluorene (Fig. 1d) has been studied in the laboratory under noncatalytic conditions, and values of the activation energy and pre-exponential constant calculated (Alexander *et al.*, 1988a, 1989; Table I).

The kinetic parameters obtained from the laboratory experiments were used to study the thermal history of sediments from three wells in the Carnarvon Basin (Alexander *et al.*, 1988a, b, 1989, 1990) by employing nonisothermal kinetic equations [e.g., of the form of Eq. (11)] to fit the predicted extent of reaction to that observed in the sediments. Although the reaction followed in the laboratory was the cyclization of 2,3-dimethylbiphenyl to 1-methylfluorene, in the sediments there appeared to be no systematic relationship between these two components. To overcome this problem, the decrease in the amount of 2,3-dimethylbiphenyl was measured relative to the 3,5-isomer, which had been shown to be unchanged in the laboratory under conditions causing complete cyclization of 2,3-dimethylbiphenyl (Alexander *et al.*, 1988a). Evidence for the validity of using the depletion of the substrate, rather than the formation of the product, has subsequently been presented by Kagi *et al.* (1990), who observed a kinetic isotope effect between the cyclization of deuterated (benzylic carbon) and nondeuterated 2-methylbiphenyl.

Alexander *et al.* (1988a, b, 1989) found that there was good agreement between the predicted and observed extent of reaction for sediments from two of the wells (Jupiter #1 and Madeleine #1) when the present-day geothermal gradient (23 and 34°C km^{-1}, respectively) was used. Since these authors assumed that the geothermal gradient had been constant throughout time, they therefore proposed that the present-day geothermal gradient is the maximum that these sediments have experienced. More recently, Alexander *et al.* (1990) refined the thermal history for Jupiter #1 to include a comparatively low heat flow (1.1 Heat Flow Unit, HFU) until ca. 88 Ma B.P. followed by a gradual increase to the calculated present-day value (1.3 HFU) in the Tertiary, after which the heat flow remained constant. To accomplish this, they used a combination of data from six independent maturity indicators, based on alkylbiphenyls, and a sophisticated computer modeling package.

In the case of sediments from the Barrow #1 well, Alexander *et al.* (1988a, b, 1989) found that a geothermal gradient of 42°C km^{-1}, slightly higher than that existing today (38°C km^{-1}), was found to yield the best fit between the predicted and observed data. Based on the relative amount of 2,3-dimethylbiphenyl, Alexander *et al.* (1988b) calculated that the Barrow Island oil reservoired in Jurassic sands (ca. 2040-m depth) was sourced from rocks at a depth of ca. 2800 m and a temperature of 140°C. This conclusion is in good agreement with that of Volkman *et al.* (1983), who made a detailed study of steranes, hopanes, and other hydrocarbons in oils and source rocks from the Barrow Sub-basin.

4.6. The Niigata Basin, Japan

The Yoshii gas and condensate reservoir is situated in the Nishiyama/Chuo oil field within the Niigata Basin, a tectonically active back-arc basin in which 5 km or more of Miocene–Pleistocene clastic sediments have accumulated (Suzuki, 1990). Since the Yoshii reservoir is one of the largest in Japan, it has attracted considerable geological and geochemical interest (e.g., Suzuki, 1990 and references therein).

In an attempt to elucidate the thermal and subsidence histories of the sediments in the Nishiyama/Chuo oil field, the configurational isomerization at C-20 of 5α-24-ethylcholestane (Fig. 1b) and at C-22 of 17α,21β-bishomohopane (Fig. 1c) have been studied by Suzuki (1990). In particular, he used published values of the activation energy and pre-exponential constant for the former reaction (Alexander *et al.*, 1986; Strachan *et al.*, 1989a, b; Table I).

Suzuki (1990) concluded that the Nishiyama/Chuo oil field region of the Niigata Basin has been subjected to extensive tectonism during the late Quaternary. Similar rates of both subsidence and uplift have occurred (2–4 km Myr^{-1}), concomitantly resulting in similar heating and cooling rates (80–160°C Myr^{-1}). The greatest amount of subsidence and uplift has occurred in the eastern part of the area. Accumulation of hydrocarbons in the Yoshii reservoir started in the late Pliocene and possibly continues to the present day; however, the major phase of hydrocarbon generation and accumulation, during the Quaternary, has resulted in the Yoshii reservoir being composed of gas and condensate.

5. The Effect of Mineral Matrix

Alexander *et al.* (1986) and Strachan *et al.* (1989a, b) have also applied nonisothermal kinetic equations [e.g., of the form of Eq. (11)] to sedimentary sequences having different mineralogy and/or types of organic matter.

Sediments from two wells in Western Australia were studied by Alexander *et al.* (1986). The first consists of mainly siltstone and claystones with type II organic matter (Dingo Claystone; Upper Jurassic; Madeleine #1), while the second is a fluvio-deltaic sequence comprising coal seams and dispersed coaly fragments of type II–III organic matter (Mungaroo Formation; Upper Triassic; Jupiter #1). As well as having different mineral matrices and organic matter types, these sediments have experienced different geothermal gradients (see Section 4.5). The progress of three aromatic maturity indicators (Fig. 1e–g) was investigated, and it was observed that for each indicator different values of the activation energy and pre-exponential constant (Table I) were needed to fit the predicted maturity values to those measured for the shaley and coaly sediments. Alexander *et al.* (1986) proposed that source and mineral matrix effects might be responsible for this effect.

Alexander *et al.* (1986) and Strachan *et al.* (1989a, b) compared the extent of configurational isomerization at C-20 of 5α-24-ethylcholestane (Fig. 1b) in sediments from the Hód #1 well (Pannonian Basin, Hungary) with that in sediments from the Jupiter #1 well. Again, it was found that different values of the activation energy and pre-exponential constant (Table I) were needed to fit the predicted extent of reaction to that observed for these sedimentary sequences.

These data appear to contradict the observations of Hoffmann *et al.* (1984), who studied three pairs of coal and shale sediments, each having a similar burial depth, from the HD1 well (Mahakam Delta, Indonesia) and found almost identical values for a number of parameters based on steroid and hopane biomarkers. However, Strachan *et al.* (1989a, b) calculated values of the activation energy and pre-exponential constant for the sterane isomerization reaction in the HD1 sediments and found values intermediate between those calculated for the Jupiter #1 and Hód #1 sediments (Table I). They suggested that this result may be due to the steranes being associated with both coal and shale matrices in sediments from the HD1 well, thus yielding values of the activation energy and pre-exponential constant intermediate between those for coaly and shaley sediments (Jupiter #1 and Hód #1, respectively). This does not seem unreasonable since the coals studied by Hoffmann *et al.* (1984) have total organic carbon values ranging 12 to 26% (Lewis, 1985), while Strachan *et al.* (1989a, b) applied the term coal to sediments with at least 30% total organic carbon.

Strachan *et al.* (1989b), in a detailed investigation of the effect of mineral matrix on the sterane isomerization reaction, reported a different extent of reaction in each member of nine pairs of adjacent coal and shale sediments. They also observed that in some cases the extent of reaction was greater in the coal than in the shale, while in others the shale exhibited the greater value. Since the activation energy and pre-exponential constant for the isomerization have been proposed to be different in coal and shale sediments (Table I; Alexander *et al.*, 1986; Strachan *et al.*, 1989a, b), the relative extent of reaction in each matrix pair might be expected to depend on the heating rate experienced (cf. Mackenzie and McKenzie, 1983). Thus, Strachan *et al.* (1989b) demonstrated that under a low-heating-rate regime (0.1°C Myr^{-1}) the extent of reaction, with respect to temperature (burial depth), was predicted to be greater in a coal than in a shale, while the opposite was predicted under a high-heating-rate regime (10°C Myr^{-1}). Hence, they proposed that the difference observed in the relative extent of reaction for these nine coal and shale sediment pairs was due to variation in the heating rate.

Further evidence for the applicability of different values of the activation energy and pre-exponential constant to the sterane isomerization was supplied by the Volador #1 (Gippsland Basin, Australia) and Rayeu #1 (North Sumatra Basin, Indonesia) wells (Strachan *et al.*, 1989b). In sediments from Volador #1, the progress of the sterane isomerization, with respect to the temperature (burial depth), could be explained either by values of the activation energy and pre-exponential constant derived for shales and an extremely high heating rate (40–300°C Myr^{-1}) or by values of the activation energy and pre-exponential constant derived for coals. The lack of evidence for such a high heating rate and the relatively high total organic carbon content of most samples prompted Strachan *et al.* (1989b) to conclude that the reaction had occurred in a "coal" matrix. In the Rayeu #1 sediments, values of the $20S/(20R + 20S)$ ratio appear to fall in two major groups, and this was interpreted to be due to the sterane isomerization obeying two different values of the activation energy and of the pre-exponential constant. Thus, samples having a higher amount of total organic carbon were associated with a predicted extent of reaction for a coal matrix, while those having a lower amount of total organic carbon were associated with a predicted extent of reaction for a shale matrix. In addition, for samples with a similar extent of reaction, an increasing total organic carbon content was found to correlate with an increasing present-day sediment temperature. This behavior is the same as that predicted for coal and shale sediments experiencing the heating rate observed for the Rayeu #1 well (ca. 14°C Myr^{-1}).

These examples demonstrate that the activation energy and pre-exponential constant for the configurational isomerization at C-20 of 5α-24-ethylcholestane appear to be matrix-dependent. Hence, values of these kinetic parameters derived from one type of sediment (i.e., coal or shale) may not be applicable to another type. However, Strachan *et al.* (1989b) suggested that the values of the activation energy and pre-exponential constant derived for shales and coals should only be regarded as a general indication of the range of possible values and that there may be instances when intermediate values are observed.

6. Conclusions

Over the past three decades, organic geochemistry has evolved from a purely descriptive science into one attempting to understand why and how the changes observed in the geosphere occur. Reactions of organic compounds in the geosphere obey the same chemical principles as those in the laboratory, although their study is hindered by the lack of "experimental" control available to the geochemist and the complexity of the sedimentary environment. However, advances have begun to be made in elucidating the kinetic parameters of some of the reactions believed to occur in sediments.

The reactions include configurational isomerization of a sterane and a hopane, aromatization of monoaromatic steroids, and cyclization of a dimethylbiphenyl. The values of the activation energy and pre-exponential constant for these reactions have been obtained from laboratory thermal simulation studies or from a combination of thermomechanical basin subsidence models with chemical kinetic models and in some cases from both methods. The studies have shown that for some reactions it is possible to derive values of the activation energy and pre-exponential constant that appear to be applicable to many different types of sedimentary system (i.e., different mineralogy and/or organic matter type), while for others these values appear to be dependent on the composition of the sediment.

These values of the activation energy and pre-exponential constant have been used to gain a better understanding of the subsidence history of sedimentary basins. Thus, attempts to derive the amount and timing of uplift in the Paris and the Lower Saxony Basins have been described. In addition, the thermal history of sediments from the Alberta, Carnarvon, and Michigan Basins and the Nova Scotia Continental Margin has been described.

At the present time, there remains considerable difficulty with both the derivation of the kinetic parameters of subsurface reactions and their application as a predictive tool. However, this area of geochemistry is the subject of continued investigation, since it is of academic as well as economic interest, and in the future a more thorough understanding of the kinetics of biomarker reactions in the subsurface will be obtained, which will in turn lead to an increase in the accuracy and reliability of their predictive ability.

ACKNOWLEDGMENTS. I would like to thank the many colleagues who have contributed to my understanding of the subject and especially Dr. R. Marzi for his comments and improvements to earlier versions of the manuscript. The editors are also thanked for the opportunity to contribute to the volume.

References

Abbott, G. D., and Maxwell, J. R., 1988, Kinetics of the aromatisation of rearranged ring-C monoaromatic steroid hydrocarbons, in: *Advances in Organic Geochemistry 1987* (L. Mattavelli and L. Novelli, eds.), Pergamon Press, Oxford, pp. 881–885.

Abbott, G. D., Lewis, C. A., and Maxwell, J. R., 1984, Laboratory simulation studies of steroid aromatisation and alkane isomerisation, in: *Advances in Organic Geochemistry 1983* (P. A. Schenck, J. W. de Leeuw, and G. W. M. Lijmbach, eds.), Pergamon Press, Oxford, pp. 31–38.

Abbott, G. D., Lewis, C. A., and Maxwell, J. R., 1985a, Laboratory models for aromatization and isomerization of hydrocarbons in sedimentary basins, *Nature* **318**:651–653.

Abbott, G. D., Lewis, C. A., and Maxwell, J. R., 1985b, The kinetics of specific organic reactions in the zone of catagenesis, *Philos. Trans. R. Soc. London, Ser. A* **315**:107–122.

Abbott, G. D., Wang, G. Y., Eglinton, T. I., Home, A. K., and Petch, G. S., 1990, The kinetics of sterane biological marker release and degradation processes during the hydrous pyrolysis of vitrinite kerogen, *Geochim. Cosmochim. Acta* **54**:2451–2461.

Abramowitz, M., and Stegun, I. A., 1965, *Handbook of Mathematical Functions with Formulas, Graphs, and Mathematical Tables*, 3rd printing, U.S. Government Printing Office, Washington, D.C.

Alexander, R., Kagi, R. I., Rowland, S. J., Sheppard, P. N., and Chirila, T. V., 1985, The effects of thermal maturity on distributions of dimethylnaphthalenes and trimethylnaphthalenes in some Ancient sediments and petroleums, *Geochim. Cosmochim. Acta* **49**:385–395.

Alexander, R., Strachan, M. G., Kagi, R. I., and van Bronswijk, W., 1986, Heating rate effects in aromatic maturity indicators, in: *Advances in Organic Geochemistry 1985* (D. Leythaeuser and J. Rullkötter, eds.), Pergamon Press, Oxford, pp. 997–1003.

Alexander, R., Fisher, S. J., and Kagi, R. I., 1988a, 2,3-Dimethylbiphenyl: Kinetics of its cyclisation reaction and effects of maturation upon its relative concentration in sediments, in: *Advances in Organic Geochemistry 1987* (L. Mattavelli and L. Novelli, eds.), Pergamon Press, Oxford, pp. 833–837.

Alexander, R., Kagi, R. I., Toh, E., and van Bronswijk, W., 1988b, The use of aromatic hydrocarbons for assessment of thermal histories of sediments, in: *The North West Shelf, Australia: Proceedings of Petroleum Exploration Society of Australia Symposium, Perth, 1988* (P. G. Purcell and R. R. Purcell, eds.), Petroleum Exploration Society of Australia, Perth, pp. 559–562.

Alexander, R., Kagi, R. I., and Toh, E., 1989, Organic reaction kinetics and the thermal histories of sediments, *Chem. Aust.* **56**:74–76.

Alexander, R., Marzi, R., and Kagi, R. I., 1990, A new method for assessing the thermal history of sediments: A case study from the Exmouth Plateau in northwestern Australia, *APEA J.* **30**(1): 364–372.

Angevine, C. L., and Turcotte, D. L., 1983, Oil generation in overthrust belts, *Am. Assoc. Petrol. Geol. Bull.* **67**:235–241.

Arrhenius, S., 1889, Über die Reaktionsgeschwindigkeit bei der Inversion von Rohrzucker durch Säuren, *Z. Phys. Chem. (Leipzig)* **4**:226–248.

Atkins, P. W., 1986, *Physical Chemistry*, 3rd ed., Oxford University Press, Oxford.

Beaumont, C., 1981, Foreland basins, *Geophys. J. R. Astron. Soc.* **65**: 291–329.

Beaumont, C., Keen, C. E., and Boutilier, R., 1982a, A comparison of foreland and rift margin sedimentary basins, *Philos. Trans. R. Soc. London, Ser. A* **305**:295–317.

Beaumont, C., Keen, C. E., and Boutilier, R., 1982b, On the evolution of rifted continental margins: Comparison of models and observations for the Nova Scotian margin, *Geophys. J. R. Astron. Soc.* **70**:667–715.

Beaumont, C., Boutilier, R., Mackenzie, A. S., and Rullkötter, J., 1985, Isomerization and aromatization of hydrocarbons and the paleothermometry and burial history of Alberta Foreland basin, *Am. Assoc. Petrol. Geol. Bull.* **69**:546–566.

Brassell, S. C., Eglinton, G., and Maxwell, J. R., 1983, The geochemistry of terpenoids and steroids, *Biochem. Soc. Trans.* **11**: 575–586.

Bunce, N. J., Forber, C. L., McInnes, C., and Hutson, J. M., 1988, Single-step methods for calculating activation parameters from raw kinetic data, *J. Chem. Soc., Perkin Trans. 2* **1988**: 363–368.

Burnham, A. K., 1989, On the validity of the Pristane Formation Index, *Geochim. Cosmochim. Acta* **53**:1693–1697.

Butler, E. B., and Barker, C., 1986, Kinetic study of bitumen release from heated shale, *Geochim. Cosmochim. Acta* **50**:2281–2288.

Calvin, M., 1969, *Chemical Evolution: Molecular Evolution Towards the Origin of Living Systems on the Earth and Elsewhere*, Oxford University Press, London.

Cercone, K. R., 1984, Thermal history of Michigan Basin, *Am. Assoc. Petrol. Geol. Bull.* **68**:130–136.

Daly, A. R., and Lilly, D. H., 1985, Thermal subsidence and generation of hydrocarbons in Michigan Basin: Discussion, *Am. Assoc. Petrol. Geol. Bull.* **69**:1181–1184.

De Bremaecker, J.-Cl., 1983, Temperature, subsidence, and hydrocarbon maturation in extensional basins: A finite element model, *Am. Assoc. Petrol. Geol. Bull.* **67**:1410–1414.

Deroo, G., 1967, Influence de la température et de la pression sur la genèse des hydrocarbures. Etude des argiles du Lias du bassin de Paris. Fascicule 1: Reconstitution de l'histoire de l'enfouissement du Toarcian dans le bassin de Paris, Institut Français du Pétrole Report 14427.

Doyle, C. D., 1961, Kinetic analysis of thermogravimetric data, *J. Appl. Polym. Sci.* **5**:285–292.

Eglinton, G., and Calvin, M., 1967, Chemical fossils, *Sci. Am.* **216** (1):32–43.

Eglinton, G., Scott, P. M., Belsky, T., Burlingame, A. L., and Calvin, M., 1964, Hydrocarbons of biological origin from a one-billion-year-old sediment, *Science* **145**:263–264.

Fox, D. L., 1944, Biochemical fossils, *Science* **100**:111–113.

Goossens, H., Due, A., de Leeuw, J. W., van de Graff, B., and Schenck, P. A., 1988a, The Pristane Formation Index, a new molecular maturity parameter. A simple method to assess maturity by pyrolysis/evaporation–gas chromatography of unextracted samples, *Geochim. Cosmochim. Acta* **52**:1189–1193.

Goossens, H., de Lange, F., de Leeuw, J. W., and Schenck, P. A., 1988b, The Pristane Formation Index, a molecular maturity parameter. Confirmation in samples from the Paris Basin, *Geochim. Cosmochim. Acta* **52**:2439–2444.

Gschwend, P. M., Chen, P. H., and Hites, R. A., 1983, On the formation of perylene in recent sediments: Kinetic models, *Geochim. Cosmochim. Acta* **47**:2115–2119.

Hoffmann, C. F., Mackenzie, A. S., Lewis, C. A., Maxwell, J. R., Oudin, J. L., Durand, B., and Vandenbroucke, M., 1984, A biological marker study of coals, shales and oils from the Mahakam delta, Kalimantan, Indonesia, *Chem. Geol.* **42**:1–23.

Illich, H. A., and Grizzle, P. L., 1985, Thermal subsidence and generation of hydrocarbons in Michigan Basin: Discussion, *Am. Assoc. Petrol. Geol. Bull.* **69**:1401–1403.

Kagi, R. I., Alexander, R., and Toh, E., 1990, Kinetics and mechanism of the cyclisation reaction of *ortho*-methylbiphenyls, in: *Advances in Organic Geochemistry 1989* (B. Durand and F. Behar, eds.), Pergamon Press, Oxford, pp. 161–166.

Lewan, M. D., Bjorøy, M., and Dolcater, D. L., 1986, Effects of thermal maturation on steroid hydrocarbons as determined by hydrous pyrolysis of Phosphoria Retort Shale, *Geochim. Cosmochim. Acta* **50**:1977–1987.

Lewis, C. A., 1985, Studies of pristane isomerisation under laboratory conditions, Ph.D. Thesis, University of Bristol.

Mackenzie, A. S., 1984a, Applications of biological markers in petroleum geochemistry, in: *Advances in Petroleum Geochemistry*, Vol. 1 (J. Brooks and D. Welte, eds.), Academic Press, London, pp. 115–214.

Mackenzie, A. S., 1984b, Organic reactions as indicators of the burial and temperature histories of sedimentary sequences, *Clay Miner.* **19**:271–286.

Mackenzie, A. S., and McKenzie, D., 1983, Isomerization and aromatization of hydrocarbons in sedimentary basins formed by extension, *Geol. Mag.* **120**:417–470.

Mackenzie, A. S., Patience, R. L., Maxwell, J. R., Vandenbroucke, M., and Durand, B., 1980, Molecular parameters of maturation in the Toarcian shales, Paris Basin, France—I. Changes in the configurations of acyclic isoprenoid alkanes, steranes and triterpanes, *Geochim. Cosmochim. Acta* **44**:1709–1721.

Mackenzie, A. S., Lewis, C. A., and Maxwell, J. R., 1981, Molecular parameters of maturation in the Toarcian shales, Paris Basin, France—IV. Laboratory thermal alteration studies, *Geochim. Cosmochim. Acta* **45**:2369–2376.

Mackenzie, A. S., Lamb, N. A., and Maxwell, J. R., 1982, Steroid hydrocarbons and the thermal history of sediments, *Nature* **295**:223–226.

Mackenzie, A. S., Beaumont, C., and McKenzie, D. P., 1984, Estimation of the kinetics of geochemical reactions with geophysical models of sedimentary basins and applications, in: *Advances in Organic Geochemistry 1983* (P. A. Schenck, J. W. de Leeuw, and G. W. M. Lijmbach, eds.), Pergamon Press, Oxford, pp. 875–884.

Mackenzie, A. S., Rullkötter, J., Welte, D. H., and Mankiewicz, P., 1985a, Reconstruction of oil formation and accumulation in North Slope, Alaska using quantitative gas chromatography–mass spectrometry, in: *Alaska North Slope Oil–Rock Correlation Study* (L. B. Mangoon and G. E. Claypool, eds.), AAPG Studies in Geology No. 20, American Association of Petroleum Geologists, Tulsa, Oklahoma, pp. 319–377.

Mackenzie, A. S., Beaumont, C., Boutilier, R., and Rullkötter, J., 1985b, The aromatization and isomerization of hydrocarbons and the thermal and subsidence history of the Nova Scotia margin, *Philos. Trans. R. Soc. London, Ser. A* **315**:203–232.

Mackenzie, A. S., Leythaeuser, D., Altebäumer, F.-J., Disko, U., and

Rullkötter, J., 1988, Molecular measurements of maturity for Lias δ shales, N.W. Germany, *Geochim. Cosmochim. Acta* **52:** 1145–1154.

Marzi, R., 1992, Comment on "The kinetics of sterane biological marker release and degradation processes during the hydrous pyrolysis of vitrinite kerogen" by G. D. Abbott, G. Y. Wang, T. I. Eglinton, A. K. Home, and G. S. Petch. *Geochim. Cosmochim. Acta* **56:**533–534.

Marzi, R., and Rullkötter, J., 1992, Qualitative and quantitative evolution and kinetics of biological marker transformations—laboratory experiments and application to the Michigan Basin, in: *Biological Markers in Sediments and Petroleum* (J. M. Moldowan, P. Albrecht, and R. P. Philp, eds.), Prentice-Hall, Englewood Cliffs, New Jersey, pp. 18–41.

Marzi, R., Rullkötter, J., and Perriman, W. S., 1990, Application of the change of sterane isomer ratios to the reconstruction of geothermal histories: Implications of the results of hydrous pyrolysis experiments, in: *Advances in Organic Geochemistry 1989* (B. Durand and F. Behar, eds.), Pergamon Press, Oxford, pp. 91–102.

McKenzie, D., 1978, Some remarks on the development of sedimentary basins, *Earth Planet. Sci. Lett.* **40:**25–32.

McKenzie, D., 1981, The variation of temperature with time and hydrocarbon maturation in sedimentary basins formed by extension, *Earth Planet. Sci. Lett.* **55:**87–98.

McKenzie, D., Mackenzie, A. S., Maxwell, J. R., and Sajgó, Cs., 1983, Isomerisation and aromatisation of hydrocarbons in stretched sedimentary basins, *Nature* **301:**504–506.

Middleton, M. F., and Falvey, D. A., 1983, Maturation modeling in Otway Basin, Australia, *Am. Assoc. Petrol. Geol. Bull.* **67:** 271–279.

Moore, W. J., 1972, *Physical Chemistry*, 5th ed., Longman, London.

Nunn, J. A., Sleep, N. H., and Moore, W. E., 1984, Thermal subsidence and generation of hydrocarbons in Michigan Basin, *Am. Assoc. Petrol. Geol. Bull.* **68:**296–315.

Nunn, J. A., Sleep, N. H., and Moore, W. E., 1985a, Thermal subsidence and generation of hydrocarbons in Michigan Basin: Reply, *Am. Assoc. Petrol. Geol. Bull.* **69:**1185–1187.

Nunn, J. A., Sleep, N. H., and Moore, W. E., 1985b, Thermal subsidence and generation of hydrocarbons in Michigan Basin: Reply, *Am. Assoc. Petrol. Geol. Bull.* **69:**1404.

Philp, R. P., 1985, Biological markers in fossil fuel production, *Mass Spectrom. Rev.* **4:**1–54.

Philp, R. P., and Lewis, C. A., 1987, Organic geochemistry of biomarkers, *Annu. Rev. Earth Planet. Sci.* **15:**363–395.

Philp, R. P., and Oung, Jung-Nan, 1988, Biomarkers—occurrence, utility and detection, *Anal. Chem.* **60:**887A–896A.

Powell, T. G., 1982, Petroleum geochemistry of the Verrill Canyon Formation: a source for Scotian Shelf hydrocarbons, *Bull. Can. Petrol. Geol.* **30:**167–179.

Price, L. C., 1983, Geologic time as a parameter in organic metamorphism and vitrinite reflectance as an absolute paleogeothermometer, *J. Petrol. Geol.* **6:**5–38.

Price, L. C., 1985, Geologic time as a parameter in organic metamorphism and vitrinite reflectance as an absolute paleogeothermometer: Reply, *J. Petrol. Geol.* **8:**233–240.

Quigley, T. M., and Mackenzie, A. S., 1988, The temperature of oil and gas formation in the sub-surface, *Nature* **333:**549–552.

Rullkötter, J., and Marzi, R., 1988, Natural and artificial maturation of biological markers in a Toarcian shale from northern Germany, in: *Advances in Organic Geochemistry 1987* (L. Mattavelli and L. Novelli, eds.), Pergamon Press, Oxford, pp. 639–645.

Rullkötter, J., and Marzi, R., 1989, New aspects of the application of sterane isomerisation and steroid aromatisation to petroleum exploration and the reconstruction of geothermal histories of sedimentary basins, *Prepr., Am. Chem. Soc., Div. Petrol. Chem.* **34**(1):126–131.

Rullkötter, J., Mackenzie, A. S., Welte, D. H., Leythaeuser, D., and Radke, M., 1984, Quantitative gas chromatography–mass spectrometry analysis of geological samples, in: *Advances in Organic Geochemistry 1983* (P. A. Schenck, J. W. de Leeuw, and G. W. M. Lijmbach, eds.), Pergamon Press, Oxford, pp. 817–827.

Rullkötter, J., Meyers, P. A., Schaefer, R. G., and Dunham, K. W., 1986, Oil generation in the Michigan Basin: A biological marker and carbon isotope approach, in: *Advances in Organic Geochemistry 1985* (D. Leythaeuser and J. Rullkötter, eds.), Pergamon Press, Oxford, pp. 359–375.

Rullkötter, J., Marzi, R., and Meyers, P. A., 1992, Biological markers in Paleozoic sedimentary rocks and crude oils from the Michigan Basin: Reassessment of sources and thermal history of organic matter, in: *Early Organic Evolution: Implications for Mineral and Energy Resources* (M. Schidlowski, S. Golubic, M. M. Kimberly, D. M. McKirdy, and P. A. Trudinger, eds.), Springer-Verlag, Heidelberg, pp. 324–335.

Sajgó, Cs., 1980, Hydrocarbon generation in a super-thick Neogene sequence in South-east Hungary. A study of the extractable organic matter, in: *Advances in Organic Geochemistry 1979* (A. G. Douglas and J. R. Maxwell, eds.), Pergamon Press, Oxford, pp. 103–113.

Sajgó, Cs., and Lefler, J., 1986, A reaction kinetic approach to the temperature–time history of sedimentary basins, in: *Paleogeothermics* (G. Buntebarth and L. Stegena, eds.), *Lecture Notes in Earth Sciences*, Vol. 5, Springer-Verlag, Berlin, pp. 119–151.

Seifert, W. K., and Moldowan, J. M., 1979, The effect of biodegradation on steranes and terpanes in crude oil, *Geochim. Cosmochim. Acta* **43:**111–126.

Snowdon, L. R., 1979, Errors in extrapolation of experimental kinetic parameters to organic geochemical systems, *Am. Assoc. Petrol. Geol. Bull.* **63:**1128–1138.

Strachan, M. G., Alexander, R., van Bronswijk, W., and Kagi, R. I., 1989a, Source and heating rate effects upon maturity parameters based on ratios of 24-ethylcholestane diastereomers, *J. Geochem. Explor.* **31:**285–294.

Strachan, M. G., Alexander, R., Subroto, E. A., and Kagi, R. I., 1989b, Constraints upon the use of 24-ethylcholestane diastereomer ratios as indicators of the maturity of petroleum, *Org. Geochem.* **14:**423–432.

Suzuki, N., 1984, Estimation of maximum temperature of mudstone by two kinetic parameters; epimerization of sterane and hopane, *Geochim. Cosmochim. Acta* **48:**2273–2282.

Suzuki, N., 1990, Application of sterane epimerization to evaluation of Yoshii gas and condensate reservoir, Niigata Basin, Japan, *Am. Assoc. Petrol. Geol. Bull.* **74:**1571–1589.

Turcotte, D. L., and Ahern, J. L., 1977, On the thermal and subsidence history of sedimentary basins, *J. Geophys. Res.* **82:** 3762–3766.

Turcotte, D. L., and McAdoo, D. C., 1979, Thermal subsidence and petroleum generation in the southwestern block of the Los Angeles Basin, California, *J. Geophys. Res., B* **84:**3460–3464.

Ungerer, P., Bessis, F., Chenet, P. Y., Durand, B., Nogaret, E., Chiarelli, A., Oudin, J. L., and Perrin, J. F., 1984, Geological and geochemical models in oil exploration; principles and practical examples, in: *Petroleum Geochemistry and Basin Evalua-*

tion (G. Demaison and R. J. Murris, eds.), *AAPG Memoir 35*, American Association of Petroleum Geologists, Tulsa, Oklahoma, pp. 53–77.

Volkman, J. K., Alexander, R., Kagi, R. I., Noble, R. A., and Woodhouse, G. W., 1983, A geochemical reconstruction of oil generation in the Barrow Sub-basin of Western Australia, *Geochim. Cosmochim. Acta* **47**:2091–2105.

Welte, D. H., and Yukler, M. A., 1981, Petroleum origin and accumulation in basin evolution—a quantitative model, *Am. Assoc. Petrol. Geol. Bull.* **65**:1387–1396.

Wold, S., and Ahlberg, P., 1970, Evaluation of activation parameters for a first order reaction from one kinetic experiment. Theory, numerical methods and computer program, *Acta Chem. Scand.* **24**:618–632.

Ziegler, P. A., 1980, Northwestern Europe: Geology and hydrocarbon provinces, in: *Facts and Principles of World Petroleum Occurrence* (A. D. Miall, ed.), Canadian Society of Petroleum Geologists Memoir 6, Calgary, Alberta, pp. 653–706.

Zsakó, J., 1968, Kinetic analysis of thermogravimetric data, *J. Phys. Chem.* **72**:2406–2411.

Chapter 23

Effect of Biodegradation and Water Washing on Crude Oil Composition

SUSAN E. PALMER

1. Introduction

The study of crude oil geochemistry becomes difficult when crude oils are altered by microbial action (biodegradation) and water washing. Parameters used for comparison of oils to determine genetic relationships (i.e., oil–oil correlation), depositional environments, and time of oil generation (i.e., thermal maturity of source rock at time of oil generation/expulsion) are altered by microbial degradation and/or water washing. Much of this knowledge comes from petroleum geochemists, who have been documenting case histories of such occurrences through the years (e.g., Winters and Williams, 1969; Bailey *et al.*, 1973a, b; Rubinstein *et al.*, 1977; Connan *et al.*, 1975, 1980; Seifert and Moldowan, 1979; Rullkötter and Wendisch, 1982; Volkman *et al.*, 1983; Momper and Williams, 1984; Palmer, 1984; Williams *et al.*, 1986).

In addition, microbiologists and petroleum geochemists have studied the action of bacteria on petroleum in the laboratory (e.g., McKenna, 1972; Horowitz *et al.*, 1975; Jobson *et al.*, 1979; Connan, 1981; Goodwin *et al.*, 1983). Some workers have isolated products of bacterial metabolism of crude oils or classes of hydrocarbons (e.g., Gibson, 1976; Higgins and Gilbert, 1978; Cripps and Watkinson, 1978; Cain, 1980; Mackenzie *et al.*, 1983). Connan (1984) illustrated the metabolic products recognized by some of these workers and others. Products of aerobic degradation are often organic acids and CO_2. Anaerobic bacteria can live on the metabolites of the aerobes but do not grow on hydrocarbons. Thus, the impact of anaerobes on oil biodegradation is only slight.

Alteration of crude oil by water washing has been indirectly studied in the laboratory through determination of water solubilities of individual hydrocarbons and of mixtures of several hydrocarbons and by studying compositional changes of whole crude oils (e.g., McAuliffe, 1966; Bailey *et al.*, 1973b; Price, 1976; Eganhouse and Calder, 1976; McAuliffe, 1980; May *et al.*, 1978a, b; Lafargue and Barker, 1988). Extensive review articles covering work done prior to 1985 on biodegradation and water washing have been prepared by Milner *et al.* (1977) and Connan (1984). These two reviews give an overview of the effects of biodegradation and water washing gleaned from the many exemplary papers found in the literature. Also, Lafargue and Barker (1988) reviewed and discussed data obtained from laboratory water washing experiments and presented their own observations and conclusions.

SUSAN E. PALMER • Amoco Production Company, Research Center, Tulsa, Oklahoma 74102.
Organic Geochemistry, edited by Michael H. Engel and Stephen A. Macko. Plenum Press, New York, 1993.

2. Geologic Constraints and Physicochemical Conditions for Biodegradation and Water Washing

The processes of microbial degradation and water washing of crude oils occur when certain conditions are met. Milner *et al.* (1977) and Connan (1984) outlined the requirements for both processes in their review articles. Their findings are summarized below.

Biodegradation occurs in surface seeps and in relatively shallow reservoirs, e.g., 6000 ft or less. Case histories and microbiological studies both show that aerobic bacteria are the major agents of crude oil degradation. Aerobic bacteria can grow in relatively cool reservoirs, i.e., below 80°C (176°F), which are invaded by oxygen-charged waters. In addition to dissolved oxygen, nutrients such as nitrate and phosphate must be present and the salinity of the water must be less than 100 to 150‰ (D. Munneke, personal communication). Also, as H_2S is toxic to aerobic bacteria, the amount of H_2S in an oil must be very low, unless aerobic bacteria are living in H_2S-free microenvironments, or in reservoirs which experience seasonal flushing be meteoric water or removal of H_2S by another mechanism. Thus, cool, shallow reservoirs that are flushed by oxygenated, nutrient-rich fresh water can be expected to contain oil that is being actively biodegraded.

However, biodegraded oils are also present in deeper reservoirs. In areas where tectonic activity causes subsidence of reservoirs, biodegraded oils are found preserved far below the arbitrary 6000-ft cutoff for bacterial activity. Thus, cases where biodegraded oils occur in reservoirs of, e.g., 8000 to 10000 ft have been reported (e.g., Palmer and Russell, 1988). Also, with regard to maximum depth of reservoirs and ongoing degradation, lower thermal gradients can permit deeper occurrences. Connan (1984) pointed out that aerobic bacteria degrade oil at the oil–water interface. In such a case, one would expect the lower part to be degraded rather than shallower portions of the same accumulation. However, some reservoirs have multiple oil–water contacts and different hydrologic regimes which could lead to a complex and perhaps confusing array of degraded and nondegraded oils showing no relationship to depth. Lafargue and Barker (1988) mentioned that hydrodynamically tilted oil–water contacts are indicators of actively flowing waters and delineate areas where degradation is occurring.

Physical and chemical processes other than biodegradation and water washing (e.g., in reservoir maturation, fractionation of light and heavy ends due to migration) can add to the difficulty of understanding transformation of oil from its initial state to the time when it is recovered in a discovery well or surface seep. Some of these other processes could be mistaken as biodegradation/water washing; thus, a good understanding of the postgeneration history of an oil is of major importance for establishing cause and effect relationships.

Because water is a necessary ingredient for biodegradation, the process of water washing generally accompanies biodegradation. In spite of many studies that have attempted to separate the effects of water washing from microbial alteration, questions concerning the effects of these processes on crude oil geochemistry are still open for discussion.

Conditions favorable for water washing exist during oil migration if oil is passing through a water–wet carrier bed–reservoir system. However, Lafargue and Barker (1988) suggested that water washing during migration must be minimal because highly water soluble molecules, namely, benzene and toluene, are present in most oils. They suggested that most water washing happens after accumulation, where, given the proper conditions, biodegradation can also occur. Water washing can take place outside the temperature, oxygen, and salinity constraints of biodegradation. Price (1976) and Lafargue and Barker (1988) noted that solubility of crude oil components increases at higher temperatures. These results demonstrate that water washing can occur in zones where microbial activity is precluded by high temperature. The work of Price (1976) shows that high salinities (over 270‰) cause exsolution of hydrocarbons; thus, salinity may control the occurrence of water washing. Price's work is supported by the results of Lafargue and Barker (1988).

3. Effects of Biodegradation and Water Washing on Crude Oil Composition

3.1. Biodegradation

Biodegradation produces heavy, low-API-gravity oils depleted in hydrocarbons and enriched in the nonhydrocarbon NSO compounds (nitrogen-, sulfur-, oxygen-bearing) and asphaltenes. Removal of hydrocarbon compound classes in order of increasing resistance to biodegradation and a scale of degrees of biodegradation have been presented by Volkman *et al.* (1984). Other workers have provided slightly different scales based on their own suite of samples (e.g., Williams *et al.*, 1986; Dajiang *et al.*, 1988). Although oil biodegradation in general follows the path outlined below, deviations are frequently observed, as illustrated by some of the examples given in Section 4. These deviations indicate that oil transformation is the result of a complex process and that some factors might not be known for a given case. Mild to moderate effects of biodegradation can be readily detected in gas chromatograms of whole oils or of saturate hydrocarbons, but more extensive degradation requires gas-chromatographic–mass-spectrometric analysis.

Volkman *et al.* (1984) indicated *initial or mild* biodegradation as the removal of low-molecular-weight *n*-paraffins (e.g., *n*-C_{10} to *n*-C_{14}), which is most readily observed on whole-oil gas chromatograms. *Moderate* biodegradation is marked by a nearly total loss of *n*-paraffins. At slightly higher levels of biodegradation (i.e., *moderate* to *extensive*), branched paraffins (pristane and phytane) and single-ring naphthenes are removed. Volkman *et al.* (1984) stated that alkylbenzenes are depleted and selective removal of dimethylnaphthalenes occurs during moderate biodegradation; however, the examples given below suggest that the dimethylnaphthalenes can be altered during initial biodegradation. *Extensive* biodegradation is indicated by removal of two-ring naphthenes (C_{14}–C_{16} bicyclics), detected by changes in mass chromatograms of the *m/z* 123 ion. *Very extensive* biodegradation is denoted as loss of the C_{27}–C_{29} "normal" steranes. Of particular importance is the selective removal of the 20*R* 5α(*H*) steranes, which are ratioed against the 20S 5α(*H*) steranes to assess maturity level of an oil (i.e., timing of oil generation and expulsion of an oil from its source rock). Depletion of 20*R* 5α(*H*) relative to 20S 5α(*H*) steranes would give an erroneously high 20S/20*R* ratio, indicating a higher maturity than the actual maturity of the oil. Cases where the steranes are degraded include La Salina seep oils (Palmer, unpublished results), Greek oil seeps (Seifert *et al.*, 1984), and Monterey oil seeps (Sofer, 1988). In the latter two cases, the hopane distributions are altered so that the C_{30+} components are diminished, but demethylated hopanes are absent.

Severe biodegradation is indicated by the initial demethylation of the C_{27}–C_{35} hopanes. A methyl group is removed from the A/B ring, producing a series of C-10 demethylated hopanes, the 25-norhopanes, detected by the *m/z* 177 ion (Seifert and Moldowan, 1979; Rullkötter and Wendisch, 1982). According to the scheme presented by Volkman *et al.* (1984), demethylated hopanes become abundant following degradation of C_{27}, C_{28}, and C_{29} steranes and indicate *extreme* biodegradation. To date, the occurrence of demethylated hopanes in source rocks has not been reported in the literature although Howell *et al.* (1984) mentioned some unpublished data showing that they do occur, along with demethylated tricyclic terpanes, in rock extracts. Philp (1985) added an additional biodegradation step, the alteration of the rearranged steranes; i.e., *very extreme* degradation.

An alternative degradation path for hopanes also appears to exist; a series of C_{27} to C_{30} (and possibly C_{31}) tetracyclic compounds, the 8,14-*seco*-hopanes, are formed by opening the C ring (Rullkötter and Wendisch, 1982; Schmitter *et al.*, 1982). In such cases demethylated hopanes can also be present, and the steranes may be only slightly altered. The presence of 9,10-*seco*-steranes as well as 8,14-*seco*-hopanes has been reported in a case of extreme biodegradation (Dajiang *et al.*, 1988). These examples suggest that various degradative processes can operate to produce severely biodegraded oils. Perhaps, certain environmental conditions are required to allow specific bacteria to grow on oils.

The C_{19}–C_{26} tricyclic terpanes survive extreme biodegradation, although demethylated tricyclic terpanes have been tentatively identified (e.g., Howell *et al.*, 1984; Philp, 1985). Because of their resistance to biodegradation, tricyclic terpanes have been used for oil–oil correlation in severely biodegraded oils. Their distributions also supply information concerning depositional environments (e.g., Zumberge, 1987).

The stable carbon isotopic composition of crude oils can also be altered by biodegradation, although not in a consistent manner. For example, in a 42-day simulated oil biodegradation study, Stahl (1980) observed that the saturate hydrocarbon fraction was enriched in ^{13}C (i.e., more positive $\delta^{13}C$ values), but the isotopic composition of the aromatic hydrocarbon fraction remained unchanged. Sofer (1984) and Momper and Williams (1984) showed that the saturate fraction of naturally biodegraded oils is also enriched in ^{13}C. However, field examples showing no or little change in isotopic composition or changes in both the saturate and aromatic fractions have also been reported (Sofer, 1984). Connan (1984) reviewed other studies where the isotopic composition of crude oil fractions other than the saturate hydrocarbons also become enriched in ^{13}C.

3.2. Water Washing

Water washing usually accompanies biodegradation, removing the more water-soluble hydrocarbons, and aids in concentration of the heavier molecules in the residual oil. This process is most readily recognized by changes in the composition of the gasoline-range hydrocarbons because these compounds are more water soluble than are the C_{15+} hydrocarbons (McAuliffe, 1966; Price, 1976). For a given carbon number, ring formation, unsaturation, and branching cause an increase in water solubility. Thus, one would expect that aromatic hydrocarbons of a given carbon number would decrease first when water washing occurs, followed by naphthenes, branched paraffins, and *n*-paraffins. Generally, the loss of benzene and toluene is a good indicator that water washing has occurred. These low-molecular-weight aromatics are also biodegradable; however, their high water solubilities make them a useful parameter of water washing. Other indicators of water washing are the loss of ethylnaphthalenes relative to dimethylnaphthalenes (Eganhouse and Calder, 1976) and possibly the loss of C_{20} and C_{21} relative to C_{26}–C_{28} triaromatic steranes (Wardroper *et al.*, 1984). In the latter study, monoaromatics were shown not to be susceptible to water washing/biodegradation. It is important to recognize a change in the triaromatic sterane composition because these compounds are useful indicators of oil maturity. Water washing has been invoked as an explanation for changes observed in the C_{15+} aromatic fraction (e.g., Wardroper *et al.*, 1984; Palmer, 1984); however, these ideas are not consistent with moderate solubility of C_{15+} aromatics. Connan (1984) showed that bacteria can degrade aromatics and suggested that the aromatics can be degraded with little effect on the saturate hydrocarbons. Alternatively, other factors not yet considered or known could be influencing the C_{15+} aromatic hydrocarbon composition.

Experimental water washing studies by Lafargue and Barker (1988) support the loss of gasoline-range (i.e., $<C_{15}$) aromatic hydrocarbons relative to naphthenes and paraffins, in line with the solubility studies noted above. However, this experimental work showed the loss of *n*-paraffins prior to naphthenes, which is not expected. These results may indicate loss due to differences in volatility (rather than solubility) of *n*-paraffins versus naphthenes and aromatics. These experiments precluded bacterial growth, which is the accepted cause of selective removal of *n*-paraffins, leaving the loss of *n*-paraffins prior to the more water-soluble naphthenes unexplained.

4. Examples of Occurrences of Biodegradation and Water Washing

Several examples of biodegradation and water washing are given below. In general, the degrees of degradation are similar to those outlined by Volkman *et al.* (1984), but deviations are also noted. These studies are by no means exhaustive, nor do the discussions contain comparisons with all studies currently under way. However, the discussions of the data show that alternative interpretations could apply. This is especially true when all of the facts concerning the postmigration history of an oil and the effects of agents other than bacteria and water are not considered or known.

4.1. Central and Southern Llanos Basin, Colombia, and Oriente Basin, Peru

Crude oils from the central and southern Llanos Basin, Colombia, and the Peruvian Oriente Basin provide useful examples of biodegraded oils because they contain both abundant *n*-paraffins and demethylated hopanes. Other examples of similar occurrences include oils from Argentina (Philp, 1983), Colombia (Howell *et al.*, 1984; Palmer and Russell, 1988), and Australia (Alexander *et al.*, 1983). If it is accepted that demethylation of hopanes at C-10 is a product of severe biodegradation, as proposed by Rullkötter and Wendisch (1982), then the presence of these compounds in an otherwise nondegraded oil suggests mixing of severely degraded and nondegraded oils. The implications concerning oil generation and accumulation in Oriente Basin, Peru, and in the southern and central Llanos Basin, Colombia, are described below.

4.1.1. Llanos Basin

Oils from northern, central, and southern Llanos Basin are designated as Families 1 through 5 based on their geochemical features (Palmer and Russell, 1988). Family 1 oils of northern Llanos Basin are not discussed here because they are not biodegraded and are not related to Families 2–5. Figures 1 and 2 show that waxy to semiwaxy oils in Families 2 and 3 have similar distributions of *n*-paraffins, isoprenoids, tricyclic and pentacyclic terpanes, and C_{27}–C_{29} steranes, which suggests that they are genetically related (Palmer and Russell, 1988). Oils in Families 2 and 3 contain only 0.03–0.5% sulfur, in line with their proposed nonmarine clastic origin. As shown in Fig. 3, most of these oils plot on the nonmarine side of the best separation line between nonmarine (waxy) and marine (nonwaxy) oils (Sofer, 1984). The proximity of the isotopic composition of Family 3 to that of Family 2 also shows that oils in these families are genetically

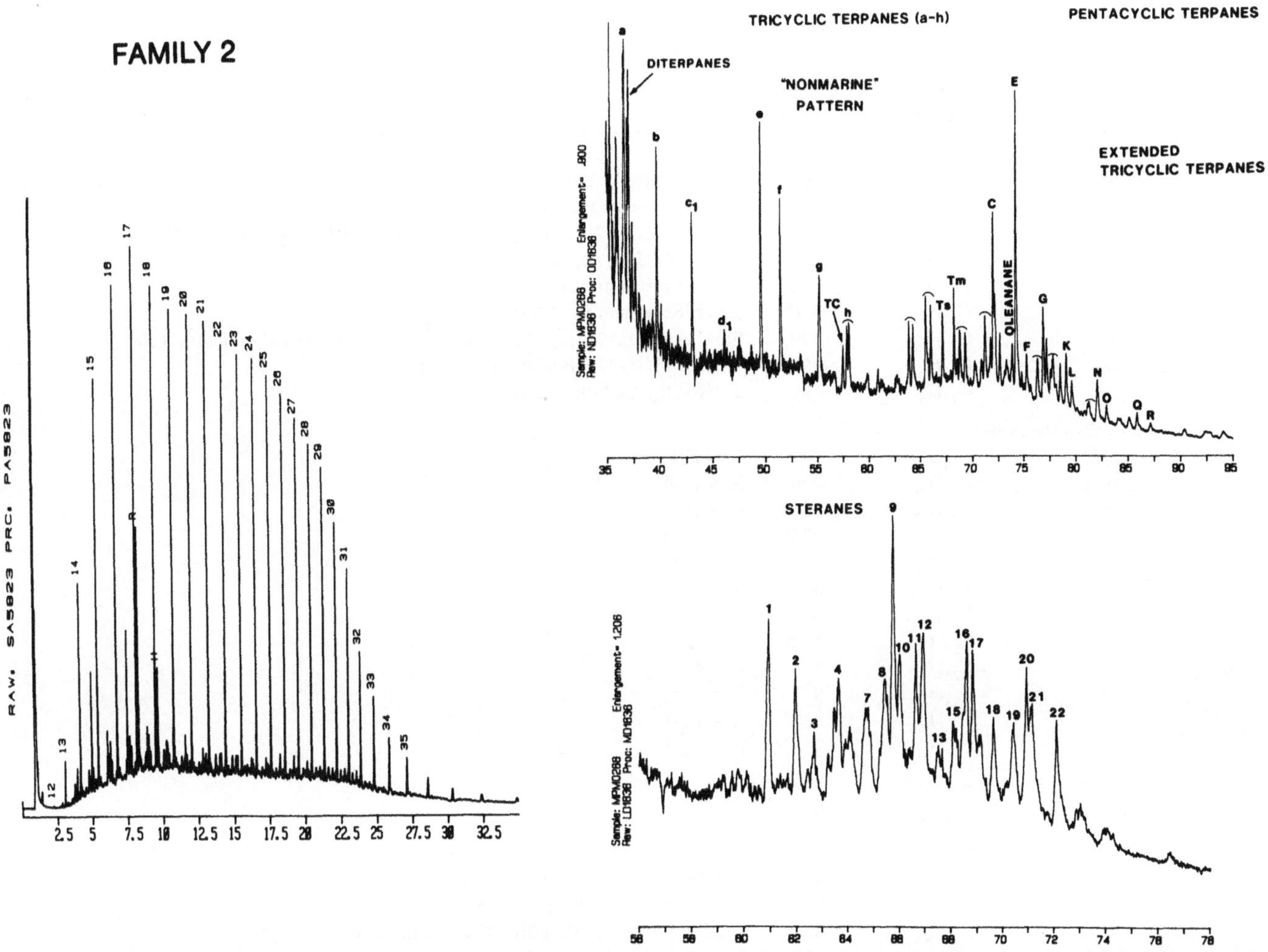

Figure 1. Gas chromatograms of C_{15+} saturate hydrocarbons and distributions of tricyclic and pentacyclic terpanes (*m/z* 191) and steranes (*m/z* 217) for Family 2 oils from central Llanos Basin, Colombia. Labeled terpanes and steranes are identified in Table I. [From Palmer and Russell (1988).]

TABLE I. Identification of Steranes and Terpanes

Peak designation	Molecular formula	Molecular weight	Identification
Steranes			
1	$C_{27}H_{48}$	372	13β(*H*),17α(*H*)-Diacholestane (20*S*)
2	$C_{27}H_{48}$	372	13β(*H*),17α(*H*)-Diacholestane (20*R*)
3	$C_{27}H_{48}$	372	13α(*H*),17β(*H*)-Diacholestane (20*S*)
4	$C_{27}H_{48}$	372	13α(*H*),17β(*H*)-Diacholestane (20*R*)
	$C_{28}H_{50}$	386	13β(*H*),17α(*H*)-Diaergostane (20*S*)
5	$C_{28}H_{50}$	386	Rearranged C_{28} sterane
6	$C_{28}H_{50}$	386	Rearranged C_{28} sterane
7	$C_{28}H_{50}$	386	13β(*H*),17α(*H*)-Diaergostane (20*R*)
8	$C_{27}H_{48}$	372	5α(*H*)-Cholestane (20*S*) + 5β(*H*)-cholestane (20*R*)
9	$C_{27}H_{48}$	372	5α(*H*),14β(*H*),17β(*H*)-Cholestane (20*R*)
	$C_{29}H_{52}$	400	13β(*H*),17α(*H*)-Diastigmastane (20*S*)
10	$C_{27}H_{48}$	372	5α(*H*),14β(*H*),17β(*H*)-Cholestane (20*S*)
11	$C_{27}H_{48}$	372	5α(*H*)-Cholestane (20*R*)
12	$C_{29}H_{52}$	400	Rearranged C_{29} sterane
13	$C_{29}H_{52}$	400	Rearranged C_{29} sterane

(*continued*)

TABLE I. *(Continued)*

Peak designation	Molecular formula	Molecular weight	Identification
Steranes			
14	$C_{29}H_{52}$	400	Rearranged C_{29} sterane
15	$C_{28}H_{50}$	386	5α(*H*)-Ergostane (20*S*)
16	$C_{28}H_{50}$	386	5α(*H*),14β(*H*),17β(*H*)-Ergostane (20*R*) + 5β(*H*)-ergostane (20*R*)
17	$C_{28}H_{50}$	386	5α(*H*),14β(*H*),17β(*H*)-Ergostane (20*S*)
18	$C_{28}H_{50}$	386	5α(*H*)-Ergostane (20*R*)
19	$C_{29}H_{52}$	400	5α(*H*)-Stigmastane (20*S*)
20	$C_{29}H_{52}$	400	5α(*H*),14β(*H*),17β(*H*)-Stigmastane (20*R*) + 5β(*H*)-stigmastane (20*R*)
21	$C_{29}H_{52}$	400	5α(*H*),14β(*H*),17β(*H*)-Stigmastane (20*S*)
22	$C_{29}H_{52}$	400	5α(*H*)-Stigmastane (20*R*)
Tricyclic terpanes			
a	$C_{19}H_{34}$	262	$R = CH_3$
b	$C_{20}H_{36}$	276	$R = C_2H_5$
c_1	$C_{21}H_{38}$	290	$R = C_3H_7$
c_2	$C_{21}H_{38}$	290	$R = C_3H_7$
c_3	$C_{21}H_{38}$	290	$R = C_3H_7$
d_1	$C_{22}H_{40}$	304	$R = C_4H_9$
d_2	$C_{22}H_{40}$	304	$R = C_4H_9$
e	$C_{23}H_{42}$	318	$R = C_5H_{11}$
f	$C_{24}H_{44}$	332	$R = C_6H_{13}$
g	$C_{25}H_{46}$	346	$R = C_7H_{15}$
h_1	$C_{26}H_{48}$	360	$R = C_8H_{17}$
h_2	$C_{26}H_{48}$	360	$R = C_8H_{17}$
Tetracyclic terpane			
TC	$C_{24}H_{42}$	330	
Pentacyclic terpanes			
A	$C_{27}H_{46}$	370	18α(*H*)-22,29,30-Trisnorhopane
B	$C_{27}H_{46}$	370	17α(*H*)-22,29,30-Trisnorhopane
BP	$C_{28}H_{48}$	384	17α(*H*),18α(*H*),21β(*H*)-28,30-Bisnorhopane
DM	$C_{29}H_{50}$	398	Demethylated hopane at A/B ring
C	$C_{29}H_{50}$	398	17α(*H*),21β(*H*)-30-Norhopane
D	$C_{29}H_{50}$	398	17β(*H*),21α(*H*)-30-Normoretane
OL	$C_{30}H_{52}$	412	Oleanane
E	$C_{30}H_{52}$	412	17α(*H*),21β(*H*)-Hopane
F	$C_{30}H_{52}$	412	17β(*H*),21α(*H*)-Moretane
G	$C_{31}H_{54}$	426	17α(*H*),21β(*H*)-30-Homohopane (22*S*)
H	$C_{31}H_{54}$	426	17α(*H*),21β(*H*)-30-Homohopane (22*R*)
I	$C_{30}H_{52}$	412	Gammacerane
J	$C_{31}H_{54}$	426	17β(*H*),21α(*H*)-30-Homomoretane
K	$C_{32}H_{56}$	440	17α(*H*),21β(*H*)-30,31-Bishomohopane (22*S*)
L	$C_{32}H_{56}$	440	17α(*H*),21β(*H*)-30,31-Bishomohopane (22*R*)
M	$C_{32}H_{56}$	440	17β(*H*),21α(*H*)-30,31-Homomoretane
N	$C_{33}H_{58}$	454	17α(*H*),21β(*H*)-30,31,32-Trishomohopane (22*S*)
O	$C_{33}H_{58}$	454	17α(*H*),21β(*H*)-30,31,32-Trishomohopane (22*R*)
Q	$C_{34}H_{60}$	468	17α(*H*),21β(*H*)-30,31,32,33-Tetrakishomohopane (22*S*)
R	$C_{34}H_{60}$	468	17α(*H*),21β(*H*)-30,31,32,33-Tetrakishomohopane (22*R*)
S	$C_{35}H_{62}$	482	17α(*H*),21β(*H*)-30,31,32,33,34-Pentakishomohopane (22*S*)
T	$C_{35}H_{62}$	482	17α(*H*),21β(*H*)-30,31,32,33,34-Pentakishomohopane (22*R*)

related. However, the Family 3 oils have naphthenic–aromatic C_{15+} hydrocarbon compositions rather than paraffinic compositions, which seems unusual for waxy to semiwaxy oils. In contrast, Family 2 oils are paraffinic–naphthenic. The low amount of *n*-paraffins and the presence of demethylated hopanes suggest that Family 3 oils are a mixture of a severely degraded oil and a nondegraded waxy oil which is similar to Family 2 oils.

Family 5 oils of southern Llanos Basin appear to

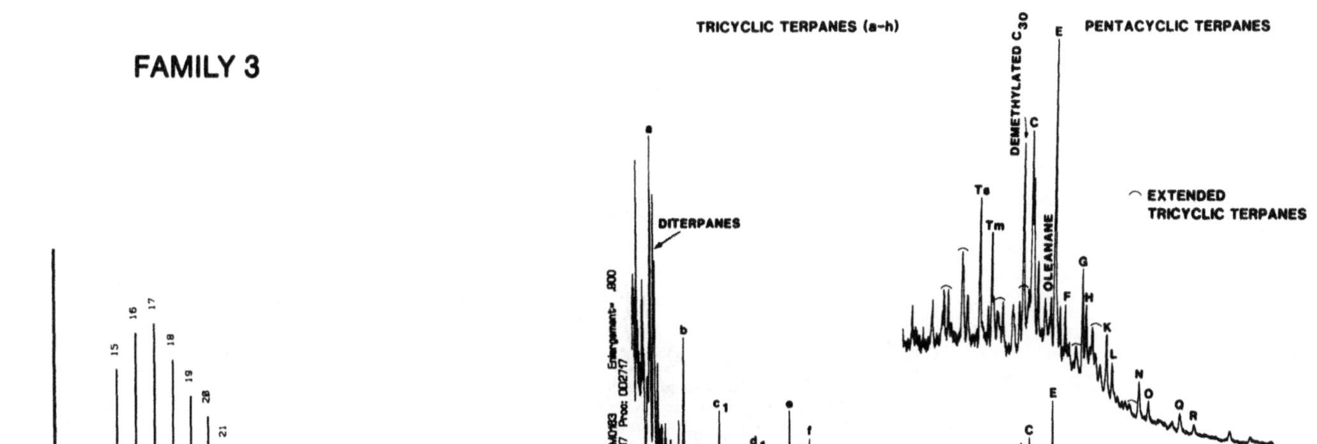

Figure 2. Gas chromatograms of C_{15+} saturate hydrocarbons and distributions of tricyclic and pentacyclic terpanes (*m/z* 191) and steranes (*m/z* 217) for Family 3 oils, from central and southern Llanos Basin. [From Palmer and Russell (1988).]

be genetically related to the oils in Families 2 and 3 based on hydrocarbon distributions shown in Figs. 1, 2, and 4. These waxy oils show the presence of demethylated hopanes, are naphthenic–aromatic in gross hydrocarbon composition, but, unlike the oils in Families 2 and 3, have sulfur contents of 1.1–1.3%. The higher sulfur content of Family 5 oils and marine isotopic composition (Fig. 3) suggest a marine carbonate source, which is inconsistent with the hydrocarbon distributions shown in Fig. 4. Perhaps, Family 5 oils are a mixture of a severely biodegraded marine oil and a nonmarine, nondegraded oil similar to those of Family 2.

Family 4 oils from central Llanos Basin are a loose collection of some other naphthenic–aromatic, waxy oils that contain demethylated hopanes. Family 4 has features that show that these oils are not totally related to those of Families 2, 3, and 5. These oils have low (0.3–0.5%) to high (1.2–1.7%) sulfur contents and have different tricyclic terpane distributions from those of the oils of Families 2, 3, and 5. Family 4 oils lack the interfering, unidentified compounds in the C_{19} tricyclic terpane region (Fig. 5). Also, the tricyclics have more of a marine pattern, i.e., lower amounts of C_{19} to C_{21} (peaks a, b, and c_1) relative to C_{23} (peak e) (Zumberge, 1987). Chromatographic patterns in Fig. 5 suggest that the Family 4 oils are mixtures of severely biodegraded and nondegraded waxy oils. The biodegraded oil may be of marine origin, as indicated by the higher sulfur contents in some of the

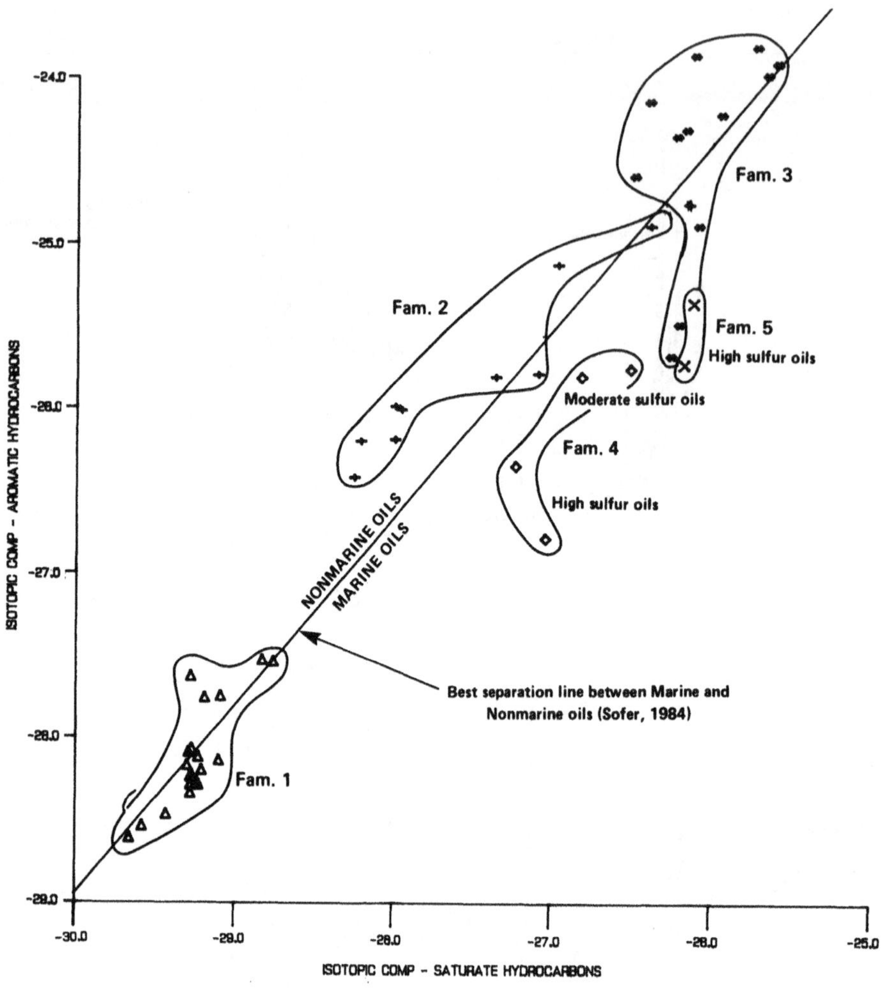

Figure 3. Stable carbon isotopic composition of C_{15+} saturate and aromatic hydrocarbons of Llanos Basin oil families 1–5. [From Palmer and Russell (1988).]

oils and the marine isotopic compositions shown in Fig. 4. The marine character of the tricyclic terpane distributions might also be attributed to a biodegraded marine oil.

Palmer and Russell (1988) described a tectonic history for the Llanos Basin which would allow two phases of oil generation and migration. The first phase of oil generation occurred during Late Cretaceous time; this process was interrupted by the Andean orogeny in Miocene time. Basin tilting allowed remigration of the oil and possibly also promoted a later phase of oil generation from nonmarine rocks during the Miocene to Pliocene. Constraints based on geologic history support the proposed ideas of migration, biodegradation, remigration, and a second phase of source rock maturation. Total understanding of Llanos Basin oil geochemistry requires the identification of the source beds for the high-sulfur marine oils and the waxy oils. At present, the best candidates are Early and Late Cretaceous rocks.

4.1.2. Oriente Basin

Sofer *et al.* (1986) described the geologic events that resulted in the mixtures of degraded plus nondegraded oils present today in Cretaceous reservoirs of the Oriente Basin of northern Peru. Severe biodegradation by invasion of meteoric waters from the north led to the formation of demethylated hopanes in the shallower reservoirs. Degradation was halted when the reservoirs were sealed off by deposition of impermeable red beds. Tilting of the basin to the west and southwest, associated with the Andean orogeny, allowed the oils to remigrate and produced mixtures of degraded and undegraded oils (Fig. 6). The proportion of degraded to nondegraded oil can be estimated based on the relative abundance of the demethylated hopanes. In line with the meteoric water invasion from the north, the demethylated hopanes are more prominent in the northeastern fields than in the southeastern fields. Sofer *et al.* (1986) offered geologi-

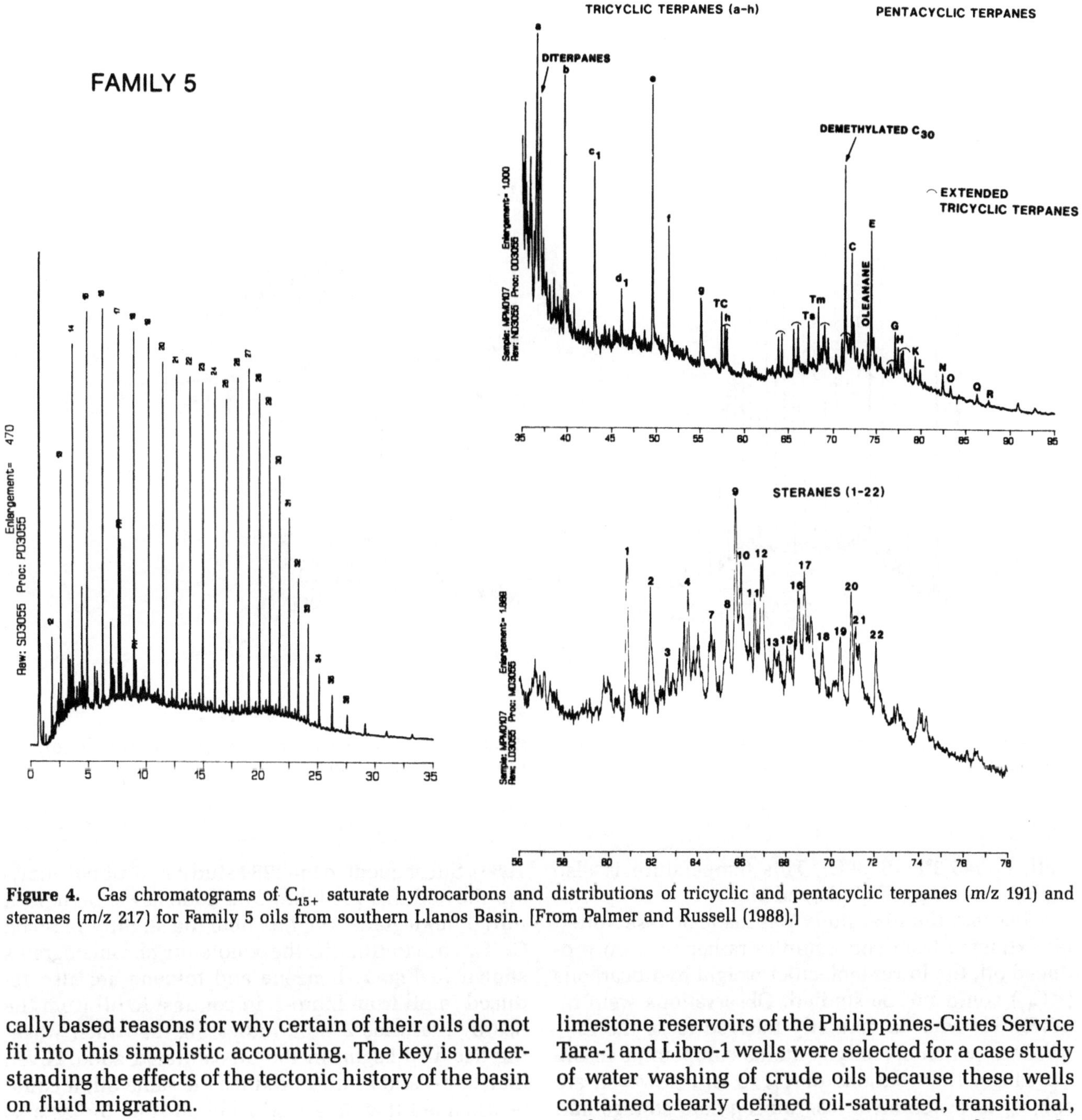

Figure 4. Gas chromatograms of C_{15+} saturate hydrocarbons and distributions of tricyclic and pentacyclic terpanes (*m/z* 191) and steranes (*m/z* 217) for Family 5 oils from southern Llanos Basin. [From Palmer and Russell (1988).]

cally based reasons for why certain of their oils do not fit into this simplistic accounting. The key is understanding the effects of the tectonic history of the basin on fluid migration.

4.2. Northwest Palawan, Philippines

4.2.1. Tara-1 and Libro-1

Changes in oil composition due to water washing/biodegradation of marine oils from northwest Palawan have been studied. Details concerning the geochemistry of these oils are given in Palmer (1984) and references therein. Cores of the Eocene–Miocene limestone reservoirs of the Philippines-Cities Service Tara-1 and Libro-1 wells were selected for a case study of water washing of crude oils because these wells contained clearly defined oil-saturated, transitional, and water-saturated zones. A petrographic study showed that the water-saturated lower part of the reservoir had been flushed with nonmeteoric water. The lack of meteoric diagenesis implies that the reservoir was not invaded by oxygenated fresh water. However, *nonmeteoric, slightly saline water* could also promote bacterial destruction of oil if it was oxygen-charged (i.e., seawater). In fact, present-day salinity of the formation water in Tara-1 is only 36‰, well within the range in which biodegradation occurs. The reservoir formation temperature, measured in the Libro-1

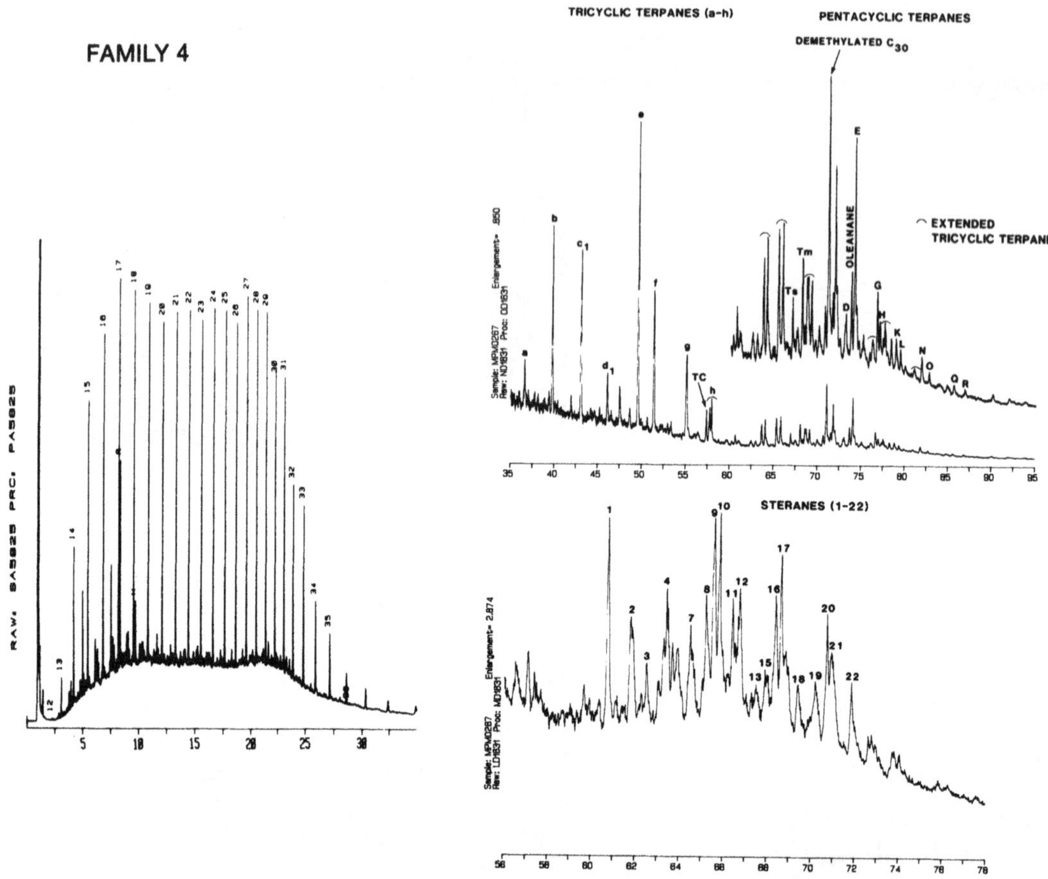

Figure 5. Gas chromatograms of C_{15+} saturate hydrocarbons and distributions of tricyclic and pentacyclic terpanes (m/z 191) and steranes (m/z 217) for Family 4 oils from central Llanos Basin. [From Palmer and Russell (1988).]

well, is 145.3°F (62.9°C). This temperature is also within the range of biodegradation.

Because the 1984 study was made on reservoired oil extracted from core samples rather than on produced oil, the lower molecular weight hydrocarbons ($<C_{12}$) could not be studied. Observations were reported on what is normally referred to as the C_{15+} fraction, which actually contains compounds in the C_{12} range. Distributions of C_{12} to C_{17}, however, are subject to evaporative loss during core storage and hydrocarbon analysis. In this regard, the Libro-1 cores were stored for two years prior to analysis while Tara-1 cores were analyzed five months after the well was drilled. In order to monitor any losses from the C_{15+} fractions due to evaporation, drill-stem test oil samples from both wells were analyzed and their composition compared with that of the oil extracted from the cores.

The C_{15+} fraction is useful to show the effects of *mild to severe biodegradation* but does not allow the detection of *initial biodegradation* (Volkman *et al.*, 1984). Subsequent to the 1984 study, additional analyses showed that the Libro-1 oil zone has experienced initial biodegradation; i.e., note the slightly reduced C_7–C_{13} *n*-paraffins in the whole-oil chromatograms shown in Fig. 7. Benzene and toluene are also reduced in oil from Libro-1, in contrast to oil from the nearby Matinloc-2 well, indicating that water washing has also begun in the oil-saturated zone. Given that Tara-1 and Libro-1 produce the same oil, it could be assumed that Tara-1 oil zone oil (Fig. 8) has also experienced initial degradation and some degree of water washing.

Geochemical data obtained on the reservoir cores from the oil-saturated, oil–water transition, and water-saturated zones and drill-stem test oils are given in Table II. The strong similarity of the data obtained for Tara-1 oil column core samples and the drill-stem test sample indicates that selective evaporative loss during core storage was negligible. The lengthy storage time of the Libro-1 cores, however, did result in loss of the more volatile compounds. Because the Libro-1

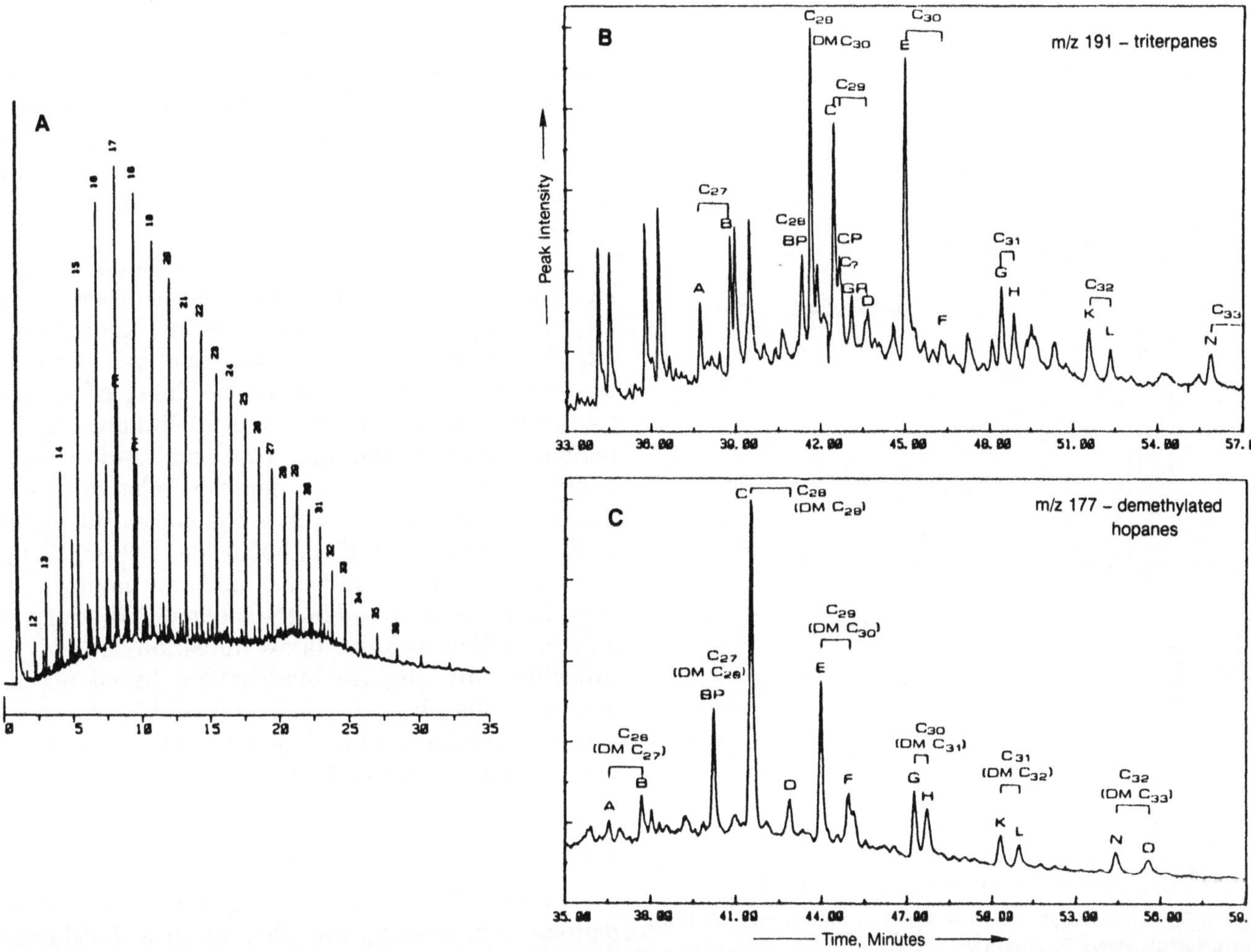

Figure 6. Gas chromatogram of C_{15+} saturate hydrocarbons and terpane distributions of a commingled oil from Oriente Basin, Peru. The chromatogram (A) shows *n*-paraffins of the nondegraded oil; (B) and (C) are distributions of the pentacyclic terpane fraction (m/z 191) and demethylated hopanes (*m*/*z* 177). [Reprinted by permission from Sofer *et al.* (1986).]

core data follow trends similar to that of Tara-1, it is believed that the data are reliable for the purpose of this study.

In the water-saturated zone, mild biodegradation of the Tara-1 and Libro-1 crude oil is demonstrated by the loss of *n*-paraffins relative to branched paraffins, naphthenes, and aromatics (Fig. 9). The gas chromatogram of the C_{15+} saturate fraction from the Tara-1 well shows a preferential loss of the lower molecular weight *n*-paraffins, up to *n*-C_{25}. Similar trends were observed in the Libro-1 well even though pristane and *n*-paraffins up to *n*-C_{22} experienced evaporative loss (Table II). In contrast to samples from the water-saturated zone, the saturate fractions of samples from the oil–water transition zone are compositionally similar to those from the oil column and are assumed to be only slightly degraded (Fig. 8).

Gas chromatograms of the aromatic hydrocarbons from both wells demonstrate a progressive loss of some of the aromatic components, beginning in the oil–water transition zone and continuing into the water-saturated zone until only an unresolved mixture remains (Fig. 10). The relative quantities of C_2- and C_3-naphthalenes, dibenzothiophene, and C_1-dibenzothiophenes are reduced in the oil–water transition zone relative to oil zone samples (see Table III for identification of peaks). In addition, the distributions of the C_2- and C_3-naphthalenes are altered in the oil–water transition zone (see discussion below).

Gas chromatograms for Tara-1 (Fig. 10) show that

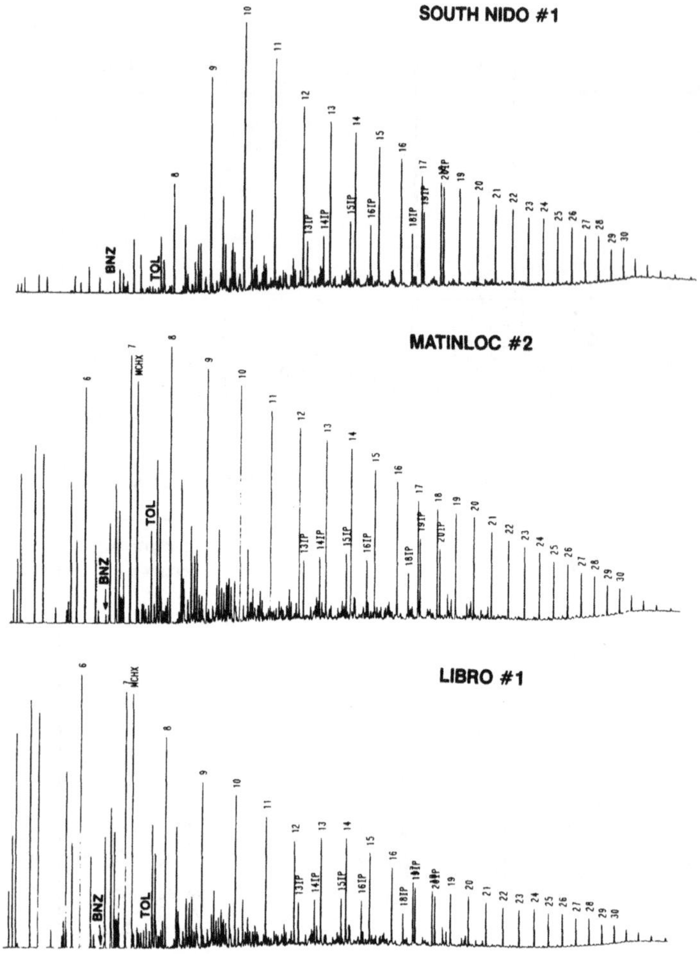

Figure 7. Whole-oil gas chromatograms of three crude oils from northwest Palawan, Philippines.

the distribution of alkylphenanthrenes remains relatively unchanged in the oil–water transition zone but that the alkyldibenzothiophene series is enriched in the C_2 and C_3 homologs. The aromatic fractions from the Libro-1 well follow the same trends as those in Tara-1, as indicated by the ratios dibenzothiophene/phenanthrene and C_1-dibenzothiophene/C_1-phenanthrene (i.e., 9-methylphenanthrene) in Table IV.

Because the relative amount of *n*-paraffins is not significantly depleted in the oil–water transition zone while dibenzothiophene and C_1-dibenzothiophenes are, it seems that the oils are water washed and only slightly biodegraded. Both water washing and, to a lesser extent, biodegradation occurred in the water-saturated zone. These data agree with experimental data of Price (1976) and Lafargue and Barker (1988), which showed that heterocompounds are more water-soluble than hydrocarbons. Heterocompounds such as alkyldibenzothiophenes are generally believed to be resistant to mild biodegradation; however, they have been degraded by cultured bacteria (Connan, 1984). Unless the Tara-1 and Libro-1 oils are being degraded by an atypical population of bacteria, the changes observed in the concentrations of alkyldibenzothiophenes appear to be due to water washing.

Changes in the distributions of the alkylnaphthalenes as shown in Fig. 10 have been observed in moderately (i.e., total loss of *n*-paraffins) biodegraded oils (Volkman *et al.*, 1984; Williams *et al.*, 1986). In both of these studies and in the Tara-1 oil–water transition zone oil, the 2,6- and 2,7-dimethylnaphthalenes (peak 1) and 1,3-, 1,7-, and 1,6-dimethylnaphthalenes (peak 2) are depleted relative to the more water-soluble ethylnaphthalenes (unlabeled peak prior to peak 2) (Eganhouse and Calder, 1976). Thus, this change suggests that moderate biodegradation has occurred, based on the scheme of Volkman *et al.* (1984). However, the Tara-1 transition zone oil is probably only slightly biodegraded based on its *n*-paraffin distribution. Also, one would not expect loss of alkyldibenzothiophenes during either slight or moderate biodegradation.

4.2.2. Linapacan-1A and South Nido-1

In order to demonstrate the effect of biodegradation/water washing on other northwest Palawan crude oils, chromatograms of the saturate and aromatic fraction of oils from the Linapacan-1A well and the South Nido-1 well are given in Fig. 11. The production of oil from the Nido fields was facilitated by an active aquifer and resulting bottom-water drive (Harry, 1979; Withjack, 1983). Consistent with the carbonate nature of the reservoir, oils from the South Nido fields contain moderate amounts of H_2S, e.g., 1800–3300 ppm, which would inhibit growth of aerobic bacteria.

The South Nido-1 C_{15+} saturate fraction remains unaltered, but the whole-oil chromatogram shows the loss of the low-molecular-weight hydrocarbons, including benzene and toluene (Fig. 7). Biodegradation of the saturate hydrocarbon fraction has not occurred, based on the full range of n-C_{10} to n-C_{30} paraffins and the presence of isoprenoids. The depleted aromatic hydrocarbon fraction shown in Fig. 11 may be partly the result of water washing. However, the overall depletion of the aromatics and resulting large hump, composed primarily of higher molecular weight unresolved compounds, may be the result of interaction

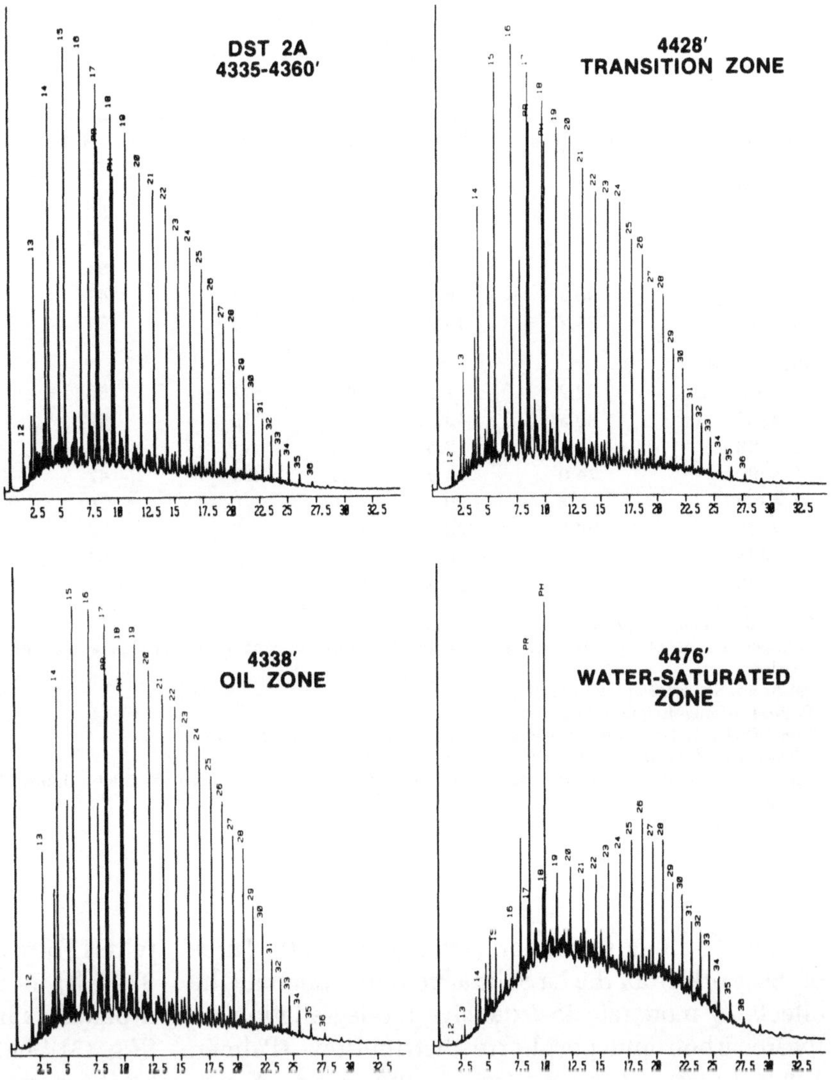

Figure 8. Gas chromatograms of C_{15+} saturate hydrocarbon fractions from Tara-1 well, northwest Palawan. [Reprinted by permission from Palmer (1984).]

with sulfur (J. A. Williams, personal communication). A similar chromatogram was reported by Volkman *et al.* (1984), for Mardie oil from Western Australia, although the Mardie oil is more severely degraded than is the South Nido oil. All of the C_2-naphthalenes are reduced in the South Nido oil, with the 2,6- and 2,7-dimethylnaphthalenes (peak 1) being the most significantly depleted (Fig. 12). The alteration of the C_2-naphthalenes, therefore, is similar to that observed for the Tara-1 oil–water transition zone (Fig. 10) and contrasts with that of the nondegraded Libro-1 and Matinloc-2 oils shown in Fig. 12.

In contrast to the Nido Field oils, the saturate and aromatic fractions are both altered in the shallow Linapacan-1A oil (i.e., DST 1 from 2354 ft). The saturate fraction shows mild to moderate biodegradation, having lost most of its *n*-paraffins, irrespective of chain length. The aromatic fraction is depleted in dibenzothiophene relative to phenanthrene, but the methyldibenzothiophenes are similar in abundance to those of nearby nondegraded oils and therefore do not appear to have been altered. The C_2-naphthalenes show alteration similar to that observed for the South Nido-1 oil; i.e., peak 1 is reduced; however, peak 2b is also lower, indicative of initial biodegradation of the aromatic fraction (Fig. 11).

4.3. La Salina Field, Middle Magdalena Valley, Colombia

Crude oils from the Las Monas area of Middle Magdalena Valley, Colombia, are genetically related and are sourced from the carbonate-rich Upper Cre-

TABLE II. Organic Geochemical Data Obtained from Tara-1 and Libro-1 DST Oils and Extracted Reservoir Cores, Northwest Palawan

DST oil or core extract[a]	Gross C_{15}^{+} Composition (%)[b] Hydrocarbons	NSO	Asphalt	*n*-Para	Br-para + naph	Arom	Sat[c]/Arom	Pr[d]/Ph	Pr/*n*-C_{17}	Zone
Tara-1 well										
T-1 DST 2A[e]	92.5	7.2	0.3	19	50	31	2.2	1.10	0.83	Oil sample
T-4338	93.1	5.3	1.6	18	53	29	2.4	1.06	0.87	Oil
T-4355	91.4	6.9	1.7	19	51	30	2.3	1.10	0.87	Oil
T-4428	90.3	6.8	2.9	31	41	28	2.6	1.05	0.87	Transition
T-4476	74.2	15.9	9.9	4	51	45	1.2	0.88	4.50	Water
Libro-1 well										
L-1 DST 3[f]	92.2	7.3	0.5	17	50	33	2.0	1.10	0.88	Oil sample
L-4222	80.5[g]	13.7	5.8	16	52	32	2.1	0.83[g]	0.84[g]	Oil
L-4258	83.3	12.6	4.1	—[h]	—	—	2.4	0.74	0.90	Oil
L-4260	74.0	14.9	11.1	23	47	30	2.4	0.64	0.93	Oil
L-4398	76.2	20.2	3.6	18	50	32	2.1	0.77	1.50	Transition
L-4412	68.8	20.9	10.3	2	58	40	1.5	0.52	1.49	Water
L-4425	64.7	26.6	8.7	2	54	44	1.3	0.77	1.42	Water
L-4435	59.9	23.9	16.2	6	53	41	1.4	0.68	1.32	Water

[a]Core extracts are designated by their depth in feet.
[b]Relative amounts of *n*-paraffins (*n*-Para), branched paraffins plus naphthenes (Br-para + naph), and aromatic hydrocarbons (Arom) are based on weighed fractions.
[c]Sat/Arom, Saturates/aromatics ratio.
[d]Pr/Ph, Pristane/phytane ratio.
[e]Tara-1 DST 2A: 4335–4360 ft, API gravity (60°F) = 40.9°, wt. % S = 0.85.
[f]Libro-1 DST 3: 4400–4425 ft, API gravity (60°F) = 41.4°, wt. % S = 0.88.
[g]Values for core extracts L-4222, L-4258, and L-4260 do not agree with those obtained in Libro-1 DST 3 oil sample because of evaporative loss during core storage.
[h]Analyses not made.

taceous La Luna Formation (Zumberge, 1980). Three of these oils from the La Salina Field demonstrate the effects of moderate to extensive biodegradation on hydrocarbon and porphyrin distributions (Palmer, 1983). LSB11 oil, reservoired in the Eocene Esmeraldes Formation at a relatively shallow depth (3287–3744 ft), is nondegraded on the basis of abundant *n*-paraffins in the C_{15+} saturate hydrocarbon fraction (Fig. 13). Shallow oils (LS2), reservoired in the Eocene La Paz Formation (755–2008 ft), are moderately biodegraded to very extensively degraded. Seep oils (SO2) showing very extensive degradation are present where the La Paz crops out. The stages of biodegradation used here are based on the scheme of Volkman *et al.* (1984). Following this scheme, the moderately degraded LS2 oil is depleted in C_{15+} *n*-paraffins relative to naphthenes. However, the distribution of *n*-paraffins in LS2 is unusual because of the abundant low-molecular-weight components (*n*-C_{14} to *n*-C_{19}) relative to n-C_{20+}. It is generally believed that initial biodegradation removes lower molecular weight *n*-paraffins, i.e., *n*-C_{10} to *n*-C_{14}, prior to the higher molecular weight components. Perhaps, the LS2 reservoir received a secondary pulse of light oil following degradation, resulting in the pattern shown in Fig. 13. In contrast to LSB11 and LS2, the seep oil (SO2) is depleted in *n*-paraffins and isoprenoids (Fig. 13). Also, only traces of the regular C_{27}–C_{29} steranes remain in the seep oil while the rearranged steranes (i.e., diasteranes) remain intact (Fig. 14).

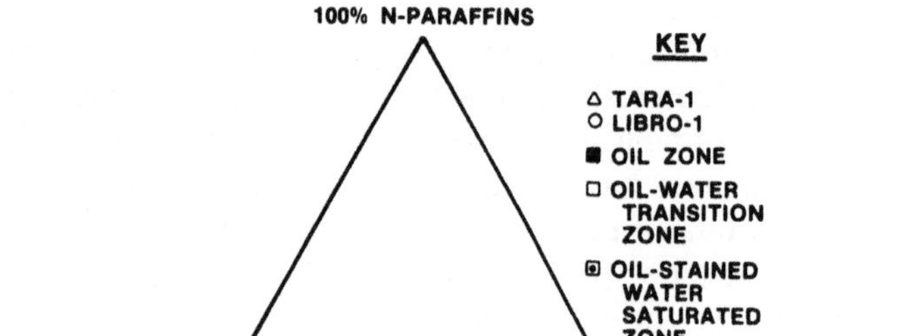

Figure 9. Gross C_{15+} hydrocarbon composition of Tara-1 and Libro-1 oil samples.

In addition to the obvious changes in *n*-paraffin,

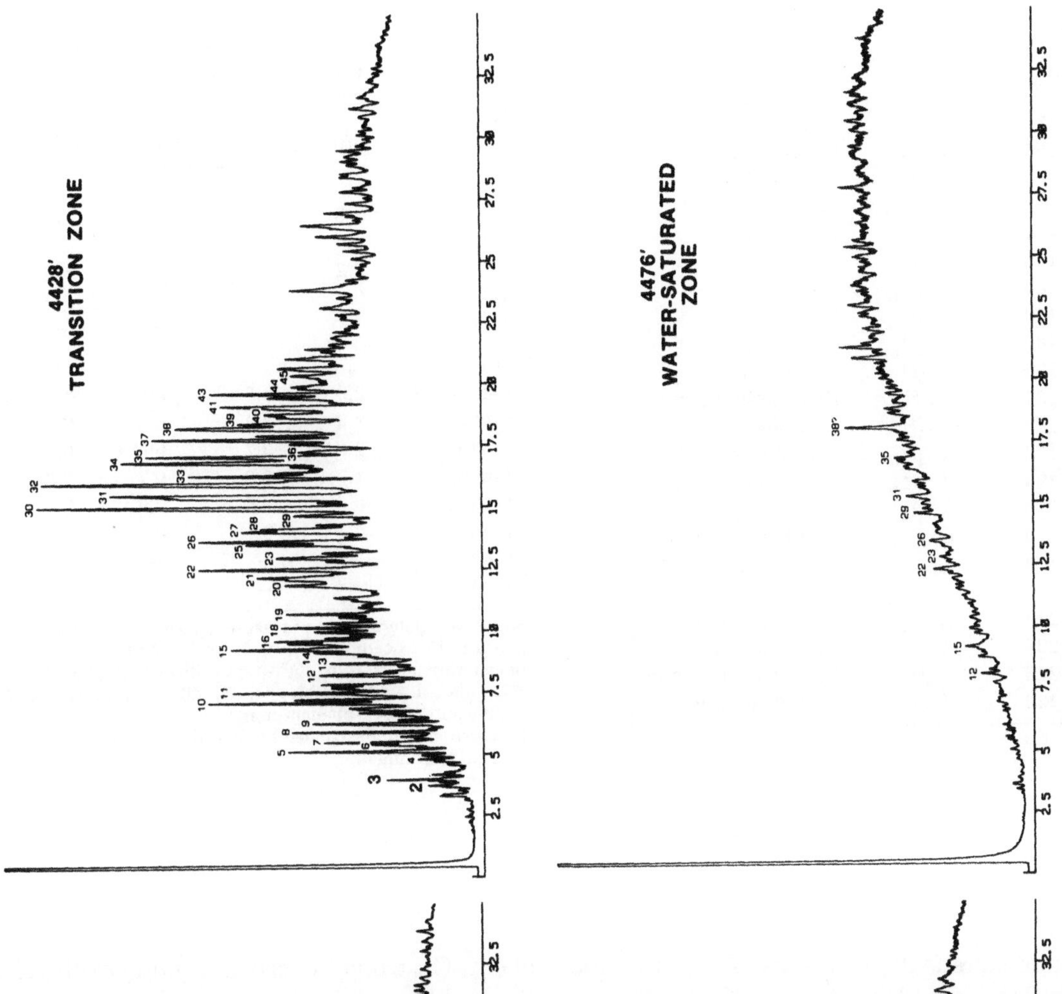

Figure 10. Gas chromatograms of C_{15+} aromatic hydrocarbons from Tara-1 oil. See Table III for identification of numbered peaks. [Reprinted with permission from Palmer (1984).]

TABLE III. Identification of Aromatic Hydrocarbons

Peak(s)	Molecular weight	Molecular formula	Compound (major component)
1–3	156	$C_{12}H_{12}$	C_2-naphthalene
4	168	$C_{13}H_{12}$	C_1-biphenyl
5–9	170	$C_{13}H_{14}$	C_3-naphthalene
7, 10–11	182	$C_{14}H_{14}$	C_2-biphenyl
12, 13	184	$C_{14}H_{16}$	C_4-naphthalene
14	196	$C_{15}H_{16}$	C_3-biphenyl
	180	$C_{14}H_{12}$	C_1-fluorene
15	180	$C_{14}H_{12}$	C_1-fluorene
16	180	$C_{14}H_{12}$	C_1-fluorene
	196	$C_{15}H_{16}$	C_3-biphenyl
17	184	$C_{12}H_8S$	Dibenzothiophene
18	196	$C_{15}H_{16}$	C_3-biphenyl
19	178	$C_{14}H_{10}$	Phenanthrene
20, 21	194	$C_{15}H_{14}$	C_2-fluorene
	210	$C_{16}H_{18}$	C_4-biphenyl
22–24	198	$C_{13}H_{10}S$	C_1-dibenzothiophene
25–28	192	$C_{15}H_{12}$	C_1-phenanthrene (3-,2-,9-,1-methyl)
29–33	212	$C_{14}H_{12}S$	C_2-dibenzothiophene
33–36	206	$C_{16}H_{14}$	C_2-phenanthrene
37–41	226	$C_{15}H_{14}S$	C_3-dibenzothiophene
42–45	220	$C_{17}H_{16}$	C_3-phenanthrene

TABLE IV. Relative Amounts of Phenanthrene and Dibenzothiophene in Tara-1 and Libro-1 Aromatic Fractions[a]

DST oil or core extract[b]	d/p	C_1-d/C_1-p	p/C_1-p	p/C_1-p	Zone
Tara-1 well					
T-1 DST 2A	1.8	3.6	0.5	1.0	Oil sample
T-4338	1.9	3.9	0.5	1.0	Oil
T-4355	2.2	4.2	0.5	1.0	Oil
T-4428	0	1.4	0	0.8	Transition
T-4476	—[c]	—	—	—	Water
Libro-1 well					
L-1 DST 3	2.0	3.8	0.6	1.1	Oil sample
L-4222	2.3	3.4	0.5	0.8	Oil
L-4258	2.7	3.7	0.4	0.6	Oil
L-4260	1.7	2.9	0.4	0.6	Oil
L-4398	0.5	0.4	0.9	0.8	Transition
L-4412	0.4	0.1	0.7	0.2	Water
L-4425	—	—	—	—	Water
L-4435	—	—	—	—	Water

[a]Ratios were determined by measuring the following ratios of peak heights of aromatic hydrocarbons in Fig. 7: peaks 17/19 (d/p = dibenzothiophene/phenanthrene), 22/27 (C_1-d/C_1-p = C_1-dibenzothiophene/C_1-phenanthrene), 17/22 (d/C_1-d = dibenzothiophene/C_1-dibenzothiophene, and 19/27 (p/C_1-p = phenanthrene/C_1-phenanthrene).

[b]Core extracts are designated by their depth in feet.

[c]—, Not determined.

isoprenoid, and sterane distributions of the saturated hydrocarbon fraction, degradative effects are observed in the aromatic hydrocarbon fraction (Fig. 13). The C_{15+} aromatic hydrocarbon fractions of the nondegraded LSB11 and degraded LS2 oils contain a full complement of C_2-naphthalenes. In contrast, both 1,6-dimethylnaphthalene (peak 2b) and the ethylnaphthalenes are depleted in the seep oil (SO2). Loss of peak 2b suggests that biodegradation of the aromatic fraction has just begun. However, SO2 has been depleted in regular steranes (Fig. 14), which suggests extensive biodegradation according to Volkman *et al.* (1984).

Absence of ethylnaphthalenes indicates that SO2 has been water washed. In addition, dibenzothiophene and methyl- and dimethyldibenzothiophenes are reduced relative to phenanthrenes in the seep oil, SO2 (Fig. 13). Loss of dibenzothiophene and its higher homologs could be due to water washing, as described for the Philippine oils. However, some bacteria can degrade benzothiophene and the dibenzothiophene series (Connan, 1984). Soil bacteria may be able to oxidize these compounds in seep oils.

Biodegradation has not altered the distributions of C_{19}–C_{26} tricyclic terpanes, pentacyclic terpanes, or the stable carbon isotopic composition of the C_{15+} saturate and aromatic hydrocarbon fractions (Figs. 15 and 13, respectively).

Vanadyl porphyrin distributions in these three oils are also not affected by degradation. As shown in Fig. 16, distributions of the DPEP and ETIO series and the ratio of the amount of DPEP to ETIO vanadyl porphyrins are similar among the three oils. Nickel porphyrins are low in concentration (in line with the marine origin of these oils), and reliable mass-spectrometric data were not obtained; thus, their distributions could not be studied. Although the distributions of vanadyl porphyrins did not change, the LS2 and SO2 oils are enriched in vanadyl relative to nickel porphyrins (Table V). The enrichment of the vanadyl fraction in SO2 shows that these compounds are not biodegraded; however, one would not expect the nickel porphyrins to be degraded, either. Thus, the cause of their depletion relative to vanadyl cannot be explained; however, fractionation during migration may be a factor. In this regard, LS2 and SO2 may be a more polar, late-migrating fraction than the nondegraded LSB11 oil.

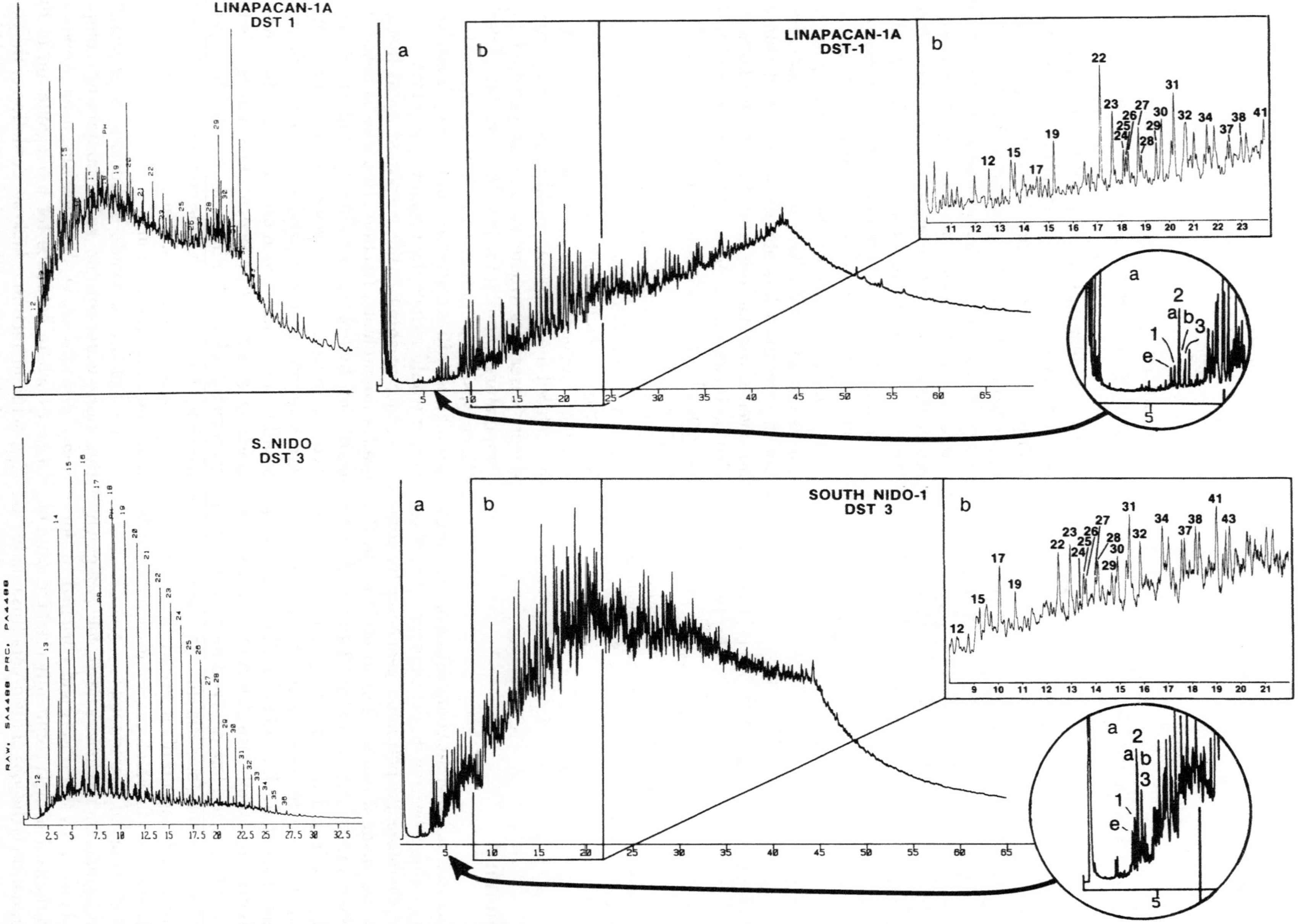

Figure 11. Gas chromatograms of C_{15+} saturate and aromatic fractions of Linapacan-1A and South Nido-1 oils. See Table III for identification of peaks. C_2-naphthalenes are labeled in inset (a): (e) 1- and 2-ethylnaphthalene, (1) 2,6- and 2,7-dimethylnaphthalene, (2a) 1,3- and 1,7-dimethylnaphthalene, (2b) 1,6-dimethylnaphthalene, (3) 1,4- 1,5-, and 2,3-dimethylnaphthalene, (4) 1,2-dimethylnaphthalene.

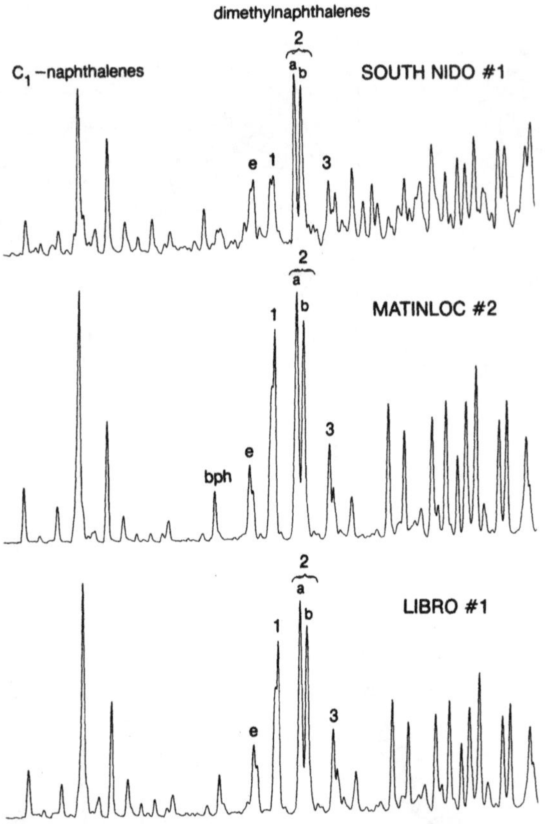

Figure 12. Partial gas chromatograms of aromatic hydrocarbon fraction in three crude oils from northwest Palawan, showing comparison of C_2-naphthalene distributions. See Fig. 11 for peak identification.

5. Summary

Processes that alter the composition of a crude oil can have a direct bearing on the commerciality of an oil field; therefore, the petroleum geochemistry community has made a major effort to understand the occurrence and causes of biodegradation and water washing. Biodegradation generally occurs in relatively shallow, cool reservoirs which are charged with oxygenated water. Such conditions allow the growth of aerobic bacteria. Both case history studies of biodegraded crude oils and laboratory-induced changes in oil composition using bacteria show that aerobic bacteria are the major agents of oil degradation. However, special cases exist in which anaerobic bacteria are the suspected agents of degradation.

Biodegradation produces a heavy, low-API-gravity oil depleted in hydrocarbons and enriched in nonhydrocarbons. The most commonly used indicator of biodegradation is the loss of the *n*-paraffins relative to branched paraffins, naphthenes, and aromatic hydrocarbons. This is because bacteria consume *n*-paraffins prior to branched paraffins, naphthenes, and aromatic hydrocarbons. The loss of *n*-paraffins is readily observed in whole-oil gas chromatograms and in chromatograms of the C_{15+} saturate hydrocarbon fraction.

Petroleum geochemists are also concerned with recognizing the effects of biodegradation on correlation parameters. Bacteria consume hydrocarbons used in correlation (e.g., *n*-paraffins, pristane, phytane, steranes, pentacyclic terpanes) and alter the isotopic composition of an oil. These changes must be recognized during oil geochemical studies where one is determining genetic relationships among oils, proposing source rock types and depositional environments based on oil composition, and determining the thermal maturity of the source rock at time of oil generation. Most correlation parameters are based on the saturate fraction; therefore, a greater emphasis has been placed on understanding how biodegradation affects the saturate fraction while the aromatic fraction has received less attention. Microbial studies, however, show that the aromatic hydrocarbons can also be biodegraded. More case studies need to be performed for an assessment of bacterial attack on these compounds.

Because water is present during biodegradation, water washing and biodegradation can occur simultaneously. In such cases it is difficult to determine whether biodegradation or water washing, or both, has altered the oil. Water washing can also occur in deeper reservoirs where lack of oxygen and high temperatures prohibit the growth of aerobic bacteria. Water washing is enhanced at higher temperatures, but high salinities cause exsolution of hydrocarbons, which would reduce their removal by water.

Water washing is the suspected agent that degrades the aromatic fraction. This idea is based primarily on hydrocarbon solubility studies; aromatic hydrocarbons are more water-soluble than saturate hydrocarbons (paraffins and naphthenes). Because smaller molecules are more water-soluble, water washing is most readily recognized by changes in the composition of the gasoline-range hydrocarbons ($<C_{15}$ fraction). The loss of benzene and toluene relative to $<C_{15}$ paraffins and naphthenes is the most useful parameter for this fraction.

Loss of C_{15+} aromatic hydrocarbons, especially the more water-soluble sulfur-containing components, due to water washing has also been proposed. In cases where evidence for biodegradation of the C_{15+} saturate fraction is lacking, depletion of the C_{15+} aromatic fraction is attributed to water washing. However, in cases of extensive biodegradation, as demon-

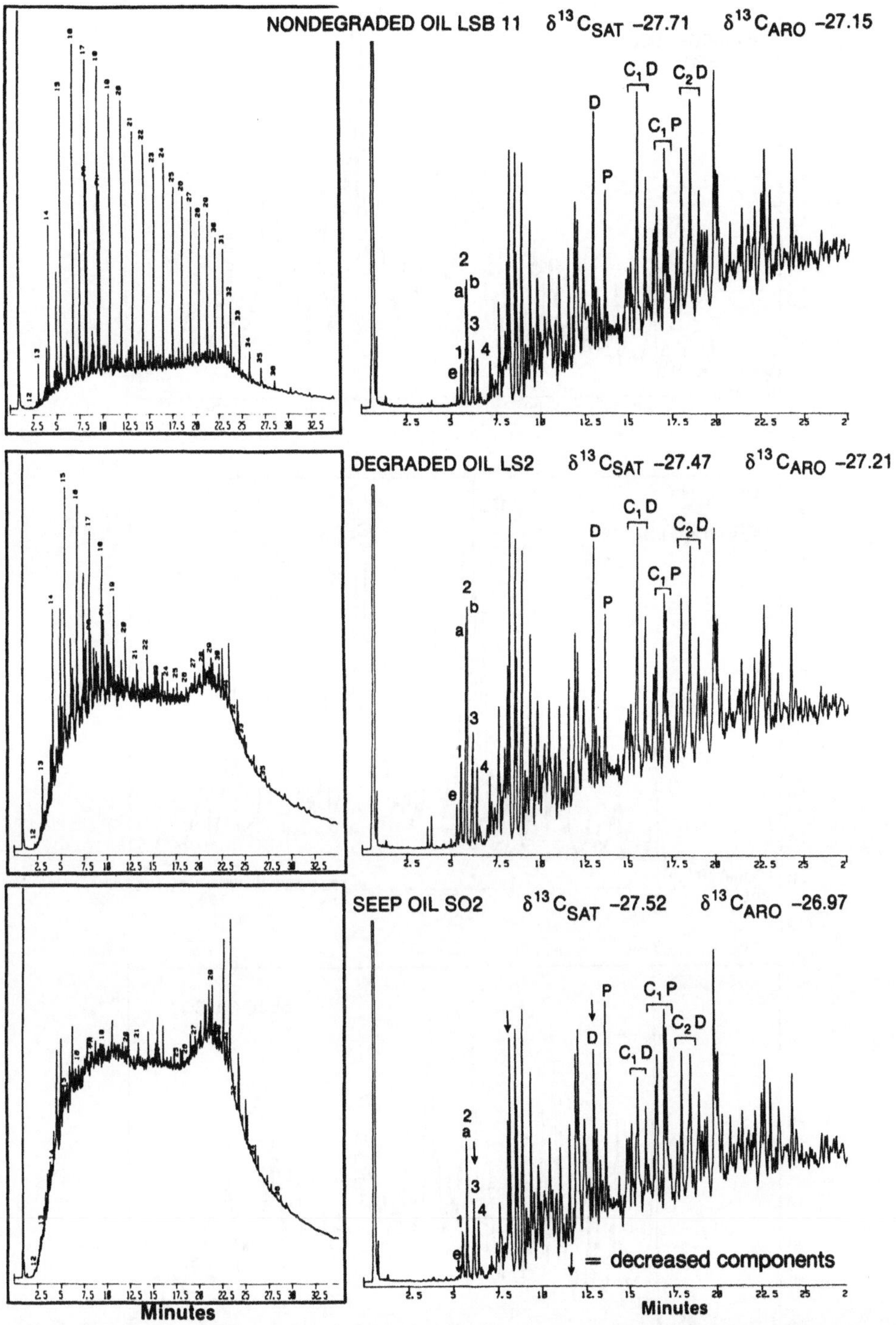

Figure 13. Gas chromatograms of C_{15+} saturate and aromatic hydrocarbons and stable carbon isotopic compositions of three La Salina Field oils, Colombia. Labeled peaks: (e) and (1)–(4), C_2-naphthalenes; D, C_1-D, and C_2-D, dibenzothiophenes; P and C_1-P, phenanthrenes.

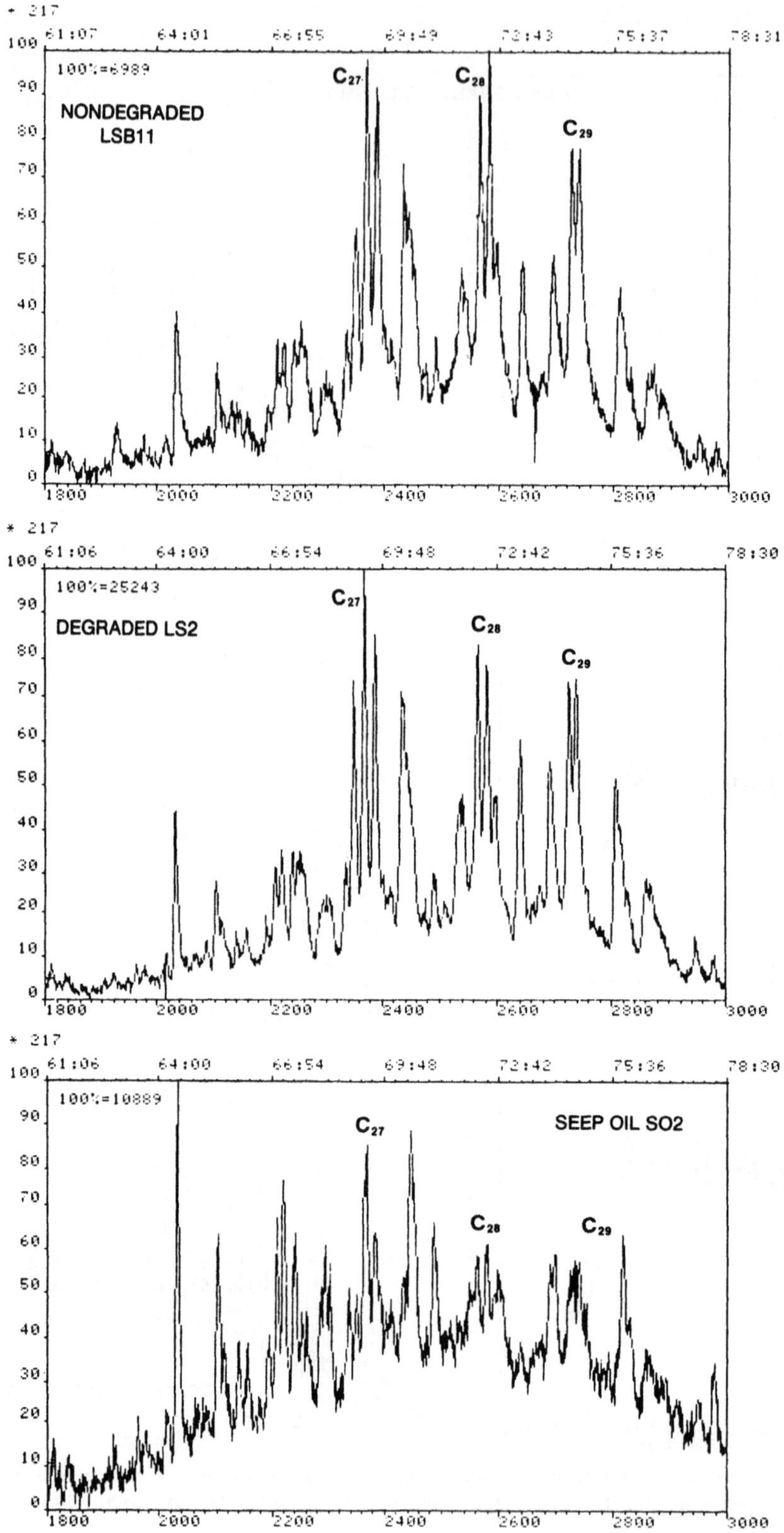

Figure 14. Mass chromatograms of steranes (*m/z* 217) for three La Salina Field oils.

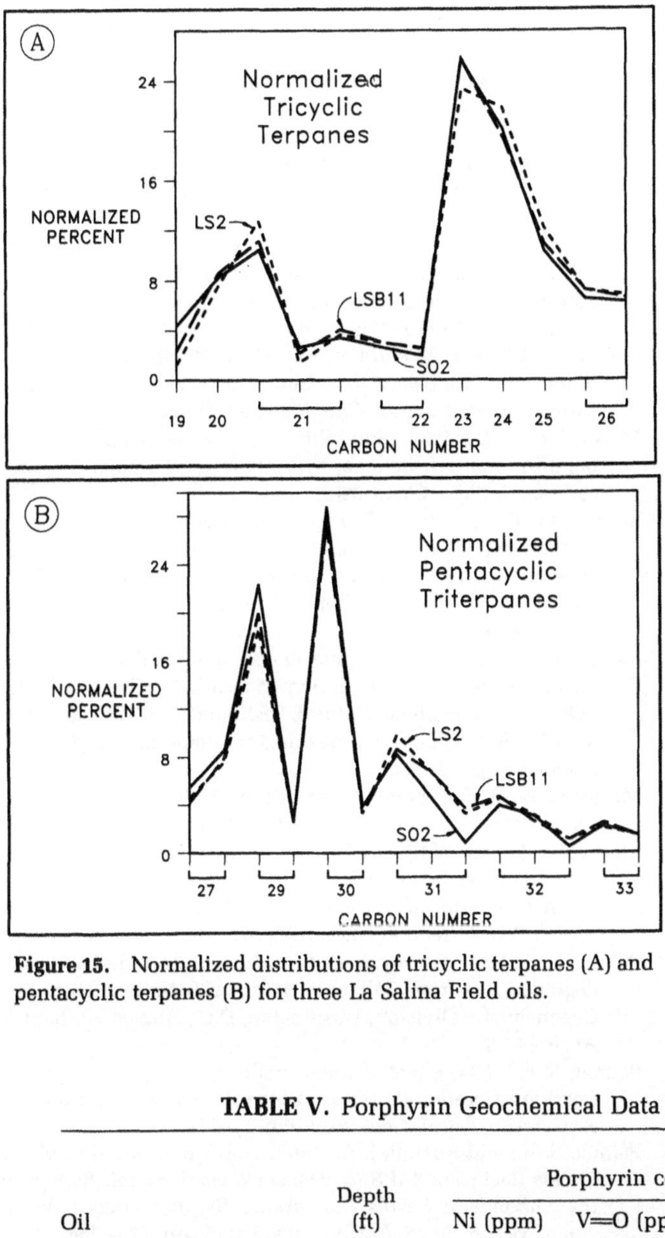

Figure 15. Normalized distributions of tricyclic terpanes (A) and pentacyclic terpanes (B) for three La Salina Field oils.

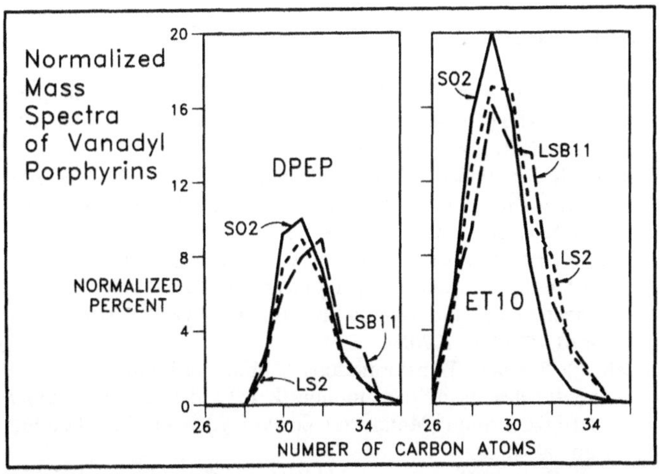

Figure 16. Normalized mass spectra of vanadyl porphyrins from three La Salina Field oils.

TABLE V. Porphyrin Geochemical Data for Three La Salina Field Oils, Colombia

Oil	Depth (ft)	Porphyrin concn.			DPEP/ETIO of V=O	Ni and V concn. (porphyrin complexed)		
		Ni (ppm)	V=O (ppm)	V=O/Ni	porphyrins[a]	Ni (ppm)	V (ppm)	V/Ni
Seep oil (SO2)	0	3.3	66.1	20.0	0.42	0.4	5.8	14.5
Degraded (LS2)	755–2088	3.8	40.5	10.7	0.37	0.4	3.8	9.5
Nondegraded (LSB11)	3287–3744	11.1	26.2	2.4	0.46	1.2	2.5	2.1

[a]DPEP, deoxophylloerythroetioporphyrin. ETIO, etioporphyrin.

strated by the destruction of the saturate hydrocarbons including naphthenic steranes and pentacyclic terpanes, degradation of the aromatic fraction may also be due to biodegradation. Alternatively, other processes, such as combining with sulfur, may remove aromatic hydrocarbons from crude oil.

Acknowledgments. The following colleagues offered many valuable suggestions and critically reviewed this chapter: Mike Lewan, Zvi Sofer, Jack Williams, and John Winters. I thank the technical staff members of the organic geochemistry group of Occidental Oil and Gas Corporation Technology Center, Tulsa, for oil analysis and the graphics staffs of Occidental Technology Center and Amoco Production Company Research Center for preparation of illustrations. Thanks go to Occidental Oil and Gas Corporation for permission to publish these data.

References

Alexander, R., Kagi, R. I., Woodhouse, G. W., and Volkman, J. K., 1983, The geochemistry of some biodegraded Australian oils, *Aust. Petrol. Explor. Assoc. J.* **23**:53–63.

Bailey, N. J. L., Jobson, A. H., and Rogers, M. A., 1973a, Bacterial degradation of crude oil: Comparison of field and experimental data, *Chem. Geol.* **11**:203–221.

Bailey, N. J. L., Krouse, H. R., Evans, C. R., and Rogers, M. A., 1973b, Alteration of crude oil by waters and bacteria—evidence from geochemical and isotope studies, *Am. Assoc. Petrol. Geol. Bull.* **57**:1276–1290.

Cain, R. B., 1980, Transformations of aromatic hydrocarbons: in: *Hydrocarbons in Biotechnology* (R. B. Cain, D. E. F. Harrison, I. J. Higgins, and R. Watkinson, eds.), Heyden and Son, London, pp. 99–132.

Connan, J., 1981, Un exemple de biodégradation préférentielle des hydrocarbonures aromatiques dans des asphaltes du bassin Sud-Aquitain (France), *Bull. Cent. Rech. Explor.-Prod., Elf-Aquitaine* **5**(1):151–171.

Connan, J., 1984, Biodegradation of crude oils in reservoirs, in: *Advances in Petroleum Geochemistry*, Vol. 1 (J. Brooks and D. H. Welte, eds.), Academic Press, London, pp. 300–330.

Connan, J., Le Tran, K., and van der Weide, B., 1975, Alteration of petroleum in reservoirs, in: *9th World Petroleum Congress, Tokyo, Proceedings*, Vol. 2, Applied Science Publishers, London, pp. 171–178.

Connan, J., Restle, A., and Albrecht, P., 1980, Biodegradation of crude oil in the Aquitaine basin, in: *Advances in Organic Geochemistry 1979* (A. A. Douglas and J. R. Maxwell, eds.), Pergamon Press, Oxford, pp. 1–17.

Cripps, R. E., and Watkinson, R. J., 1978, Polycyclic aromatic hydrocarbons: Metabolism aspects, in: *Developments in Biodegradation of Hydrocarbons*, Vol. 1, Applied Science Publishers, London, pp. 113–134.

Dajiang, Z., Difan, H., and Jinchao, L., 1988, Biodegraded sequence of Karamay oils and semi-quantitative estimation of their biodegraded degrees in Junggar Basin, China, Org. Geochem. **13**: 295–302.

Eganhouse, R. P., and Calder, J. A., 1976, The solubility of medium molecular weight and aromatic hydrocarbons and the effects of hydrocarbon co-solutes and salinity, *Geochim. Cosmochim. Acta* **40**:556–561.

Gibson, D. T., 1976, Microbial metabolism of polycyclic aromatic hydrocarbons, *Prepr., Am. Chem. Soc. Div. Petrol. Chem.* **21** (3):409.

Goodwin, N. S., Park, P. J. D., and Rawlinson, A. P., 1983, Crude oil biodegradation under simulated and natural conditions, in: *Advances in Organic Geochemistry 1981* (M. Bjorøy *et al.*, eds), John Wiley, Chichester, England, pp. 650–658.

Harry, R. Y., 1979, Cities Service develop offshore Philippines, *Oil Gas J.* **77**(18):180–195.

Higgins, I. J., and Gilbert, P. D., 1978, The biodegradation of hydrocarbons, in: *The Oil Industry and Microbial Ecosystems* (K. W. A. Chater and H. J. Somerville, eds.), Heyden and Son, London, pp. 80–117.

Horowitz, A., Gutnick, D., and Rosenberg, E., 1975, Sequential growth of bacteria on crude oil, *Appl. Microbiol.* **30**:10–19.

Howell, V. J., Connan, J., and Aldridge, A. K., 1984, Tentative identification of demethylated tricyclic terpanes in nonbiodegraded and slightly biodegraded crude oils from the Los Llanos Basin, Colombia, *Org. Geochem.* **6**:83–92.

Jobson, A. M., Cook, F. D., and Westlake, D. W. S., 1979, Interaction of aerobic and anaerobic bactéria in petroleum biodegradation, *Chem. Geol.* **24**:355–365.

Lafargue, E., and Barker, C., 1988, Effect of water washing on crude oil compositions, *Am. Assoc. Petrol. Geol. Bull.* **72**:263–276.

Mackenzie, A. S., Wolff, G. A., and Maxwell, J. R., 1983, Fatty acids in some biodegraded petroleums: Possible origins and significance, in: *Advances in Organic Geochemistry 1981* (M. Bjorøy *et al.*, eds.), John Wiley, Chichester, England, pp. 637–649.

May, W. E., Wasik, S. P., and Freeman, D. H., 1978a, Determination of the aqueous solubility of polynuclear aromatic hydrocarbons by a coupled column liquid chromatographic technique, *Anal. Chem.* **50**:175–179.

May, W. E., Wasik, S. P., and Freeman, D. H., 1978b, Determination of the solubility behavior of some polycyclic aromatic hydrocarbons in water, *Anal. Chem.* **50**:997–1000.

McAuliffe, C. D., 1966, Solubility in water of paraffin, cycloparaffin, olefin, acetylene, cycloolefin, and aromatic hydrocarbons, *J. Phys. Chem.* **70**:1267–1275.

McAuliffe, C. D., 1980, The multiple gas-phase equilibrium method and its application to environmental studies, in: *Petroleum in the Marine Environment* (L. Petrakis and F. T. Weiss, eds.), American Chemical Society, Washington, D.C., pp. 193–218.

McKenna, E. J., 1972, Microbial metabolism of normal and branched chain alkanes, in: *Degradation of Synthetic Organic Molecules in the Biosphere: Natural, Pesticidal, and Various Other Man-Made Compounds*, Washington, National Academy of Sciences, pp. 73–97.

Milner, C. W. D., Rogers, M. A., and Evans, C. R., 1977, Petroleum transformations in reservoirs, *J. Geochem. Explor.* **7**:101–153.

Momper, J. A., and Williams, J. A., 1984, Geochemical exploration in the Powder River basin, in: *Petroleum Geochemistry and Basin Evaluation* (G. Demaison and R. J. Murris, eds.), *Am. Assoc. Petrol. Geol. Mem.* **35**:181–191.

Palmer, S. E., 1983, Porphyrin distributions in degraded and nondegraded oils from Colombia, 186th ACS National Meeting, Geochemistry Division, Washington, D.C., August 25–Sept 2, Abstract 23.

Palmer, S. E., 1984, Effect of water washing on C_{15+} hydrocarbon fraction of crude oils from northwest Palawan, Philippines, *Am. Assoc. Petrol. Geol. Bull.* **68**:137–149.

Palmer, S. E., and Russell, J. A., 1988, The five oil families of the Llanos Basin, in: *3rd Soc. Venezuela Geol. Petrol. Explor. in the Subandean Basins Bolivariano Symp., Caracas, Venezuela, March 13–16, Proceedings*, Vol. 2, pp. 723–754.

Philp, R. P., 1983, Correlation of crude oils from the San Jorges Basin, Argentina, *Geochim. Cosmochim. Acta* **47**:267–275.

Philp, R. P., 1985, *Fossil Fuel Biomarkers: Applications and Spectra*, Methods in Geochemistry and Geophysics, Vol. 23, Elsevier, New York.

Price, L. C., 1976, Aqueous solubility of petroleum as applied to its origin and primary migration, *Am. Assoc. Petrol. Geol. Bull.* **60**:213–244.

Rubinstein, I., Strausz, O. P., Spyckerelle, C., Crawford, R. J., and Westlake, D. W. S., 1977, The origin of the oil sand bitumens of Alberta: A chemical and microbial simulation study, *Geochim. Cosmochim. Acta* **41**:1341–1353.

Rullkötter, J., and Wendisch, D., 1982, Microbial alteration of 17α(*H*)-hopanes in Madagascar asphalts: Removal of C-10 methyl group and ring opening, *Geochim. Cosmochim. Acta* **46**:1545–1553.

Schmitter, J. M., Sucrow, W., and Aprino, P. J., 1982, Occurrence of novel tetracyclic geochemical markers: 8,14-Seco-hopanes in a

Nigerian crude oil, *Geochim. Cosmochim. Acta* **46**:2345–2350.

Seifert, W. K., and Moldowan, J. M., 1979, The effect of biodegradation on steranes and terpanes in crude oils, *Geochim. Cosmochim. Acta* **43**:111–126.

Seifert, W. K., Moldowan, J. M., and Demaison, G. J., 1984, Source correlation of biodegraded oils, *Org. Geochem.* **6**:633–643.

Sofer, Z., 1984, Stable carbon isotope compositions of crude oils: Application to source depositional environments and petroleum alteration, *Am. Assoc. Petrol. Geol. Bull.* **68**:31–49.

Sofer, Z., 1988, Hydrous pyrolysis of Monterey asphaltenes, *Org. Geochem.* **13**:939–945.

Sofer, Z., Zumberge, J. E., and Lay, V., 1986, Stable carbon isotopes and biomarkers as tools in understanding genetic relationship, maturation, biodegradation, and migration of crude oils in the Northern Peruvian Oriente (Maranon) Basin, *Org. Geochem.* **10**:377–389.

Stahl, W. J., 1980, Compositional changes and $^{13}C/^{12}C$ fractionations during the degradation of hydrocarbons by bacteria, *Geochim. Cosmochim. Acta* **44**:1903–1907.

Volkman, J. K., Alexander, R., Kagi, R. I., and Woodhouse, G. W., 1983, Demethylated hopanes in crude oils and their applications in petroleum geochemistry, *Geochim. Cosmochim. Acta* **47**:1033–1040.

Volkman, J. K., Alexander, R., Kagi, R. I., Rowland, S. J., and Sheppard, P. N., 1984, Biodegradation of aromatic hydrocarbons in crude oils from the Barrow Subbasin of western Australia, *Org. Geochem.* **6**:619–632.

Wardroper, A. M. K., Hoffmann, C. F., Maxwell, J. R., Barwise, A. J. G., Goodwin, N. S., and Park, P. J. D., 1984, Crude oil biodegradation under simulated and natural conditions—II. Aromatic steroid hydrocarbons, *Org. Geochem.* **6**:605–617.

Williams, J. A., Bjorøy, M., Dolcater, D. L., and Winters, J. C., 1986, Biodegradation in South Texas Eocene oils–effects on aromatic and biomarkers, *Org. Geochem.* **10**:451–461.

Winters, J. C., and Williams, J. A., 1969, Microbial alteration of crude oil in the reservoir, in: *Symposium of Petroleum Transformation in Geologic Environments*, American Chemical Society, Division of Petroleum Chemistry, Paper PETR. 86, pp. E22–E31.

Withjack, E. M., 1983, Analysis of naturally fractured reservoirs with bottom-water drive: Nido A and B fields, Offshore N. W. Palawan, Philippines, 58th Annual Society of Petroleum Engineers of AIME Technical Conference, San Francisco, October 5–8, Preprint SPE-12019.

Zumberge, J. E., 1980, Source rocks of the La Luna Formation (Upper Cretaceous) in the Middle Magdalena Valley, Colombia, in: *Petroleum Geochemistry and Source Rock Potential of Carbonate Rocks* (J. C. Palacas, ed.), AAPG Studies in Geology No. 18, American Association of Petroleum Geologists, Tulsa, Oklahoma, pp. 127–133.

Zumberge, J. E., 1987, Prediction of source rock characteristics based on terpane biomarkers in crude oils: A multivariate statistical approach, *Geochim. Cosmochim. Acta* **51**:1625–1637.

Chapter 24

The Origin and Distribution of Gas Hydrates in Marine Sediments

MAHLON C. KENNICUTT II, JAMES M. BROOKS, and H. BENJAMIN COX

1. Introduction

At low temperatures and high pressures, icelike crystalline inclusion compounds, gas hydrates, may form when high concentrations of methane and water cooccur in subsurface strata. Sediments in large areas of the permafrost and on the outer continental margins of the oceans occur within the appropriate temperature and pressure regime for hydrates to be stable. Since Makogon (1965) first reported the occurrence of hydrates in permafrost regions, it has been determined that an even larger reservoir of hydrate gas is probably present in marine sediments (Fig. 1). Estimates of the amount of hydrate gas in permafrost and oceanic sediments vary from 1.4×10^{13} to 3.4×10^{16} and from 3.1×10^{15} to 7.6×10^{18} m^3, respectively (Kvenvolden, 1988). The reservoir of hydrates in oceanic sediments is believed to be at least two orders of magnitude greater than that in permafrost regions. Given these two repositories of hydrates, even the most conservative estimates indicate the enormity of the carbon reservoir. On a global scale, it has been estimated that the only pool of organic carbon larger than methane hydrates is disseminated carbon in sediments and rocks (Kvenvolden, 1988). It has been estimated that the natural gas reservoired in hydrates surpasses total fossil fuel reserves by as much as a factor of 2 (Kvenvolden, 1988). Hydrates represent a substantial energy resource of the future if the appropriate technologies can be developed for their recovery.

In order to form hydrates, sufficient gas molecules must be present to stabilize the hydrate cavities, water molecules must be present to form the hydrate lattice, and appropriate conditions of temperature and pressure must prevail (Sloan, 1989). These conditions are potentially present over a significant percentage of the Earth's surface, particularly under the oceans. The present-day inventory of known or inferred gas hydrates probably represents only a small portion of the actual occurrences (Fig. 2).

The methane gas sequestered in hydrates can be biogenic or thermogenic in origin, or both. Thermogenic methane is produced by catagenesis at high temperatures, which also produces ethane and higher hydrocarbons. Microbial alteration of organic matter produces methane in anaerobic sediments. Methane produced by biogenic processes ranges in stable isotopic composition ($\delta^{13}C$) from about −55 to −85‰ whereas thermogenic methane $\delta^{13}C$ values generally range from −35 to −60‰ (Hunt, 1979). A second characteristic of biogenic gas is that it is predominantly methane. Biogenic gas has $C_1/(C_2 + C_3)$ ratios greater than 1000 whereas thermogenic gas has ratios lower than 100 (Bernard *et al.*, 1977). Empirically, biogenic gas hydrates are by far predominant in permafrost regions (Kvenvolden, 1988).

The mechanism of hydrate formation, though not

MAHLON C. KENNICUTT II and JAMES M. BROOKS • Geochemical and Environmental Research Group (GERG), Texas A&M University, College Station, Texas 77845. H. BENJAMIN COX • Radian Corporation, Austin, Texas 78759.
Organic Geochemistry, edited by Michael H. Engel and Stephen A. Macko. Plenum Press, New York, 1993.

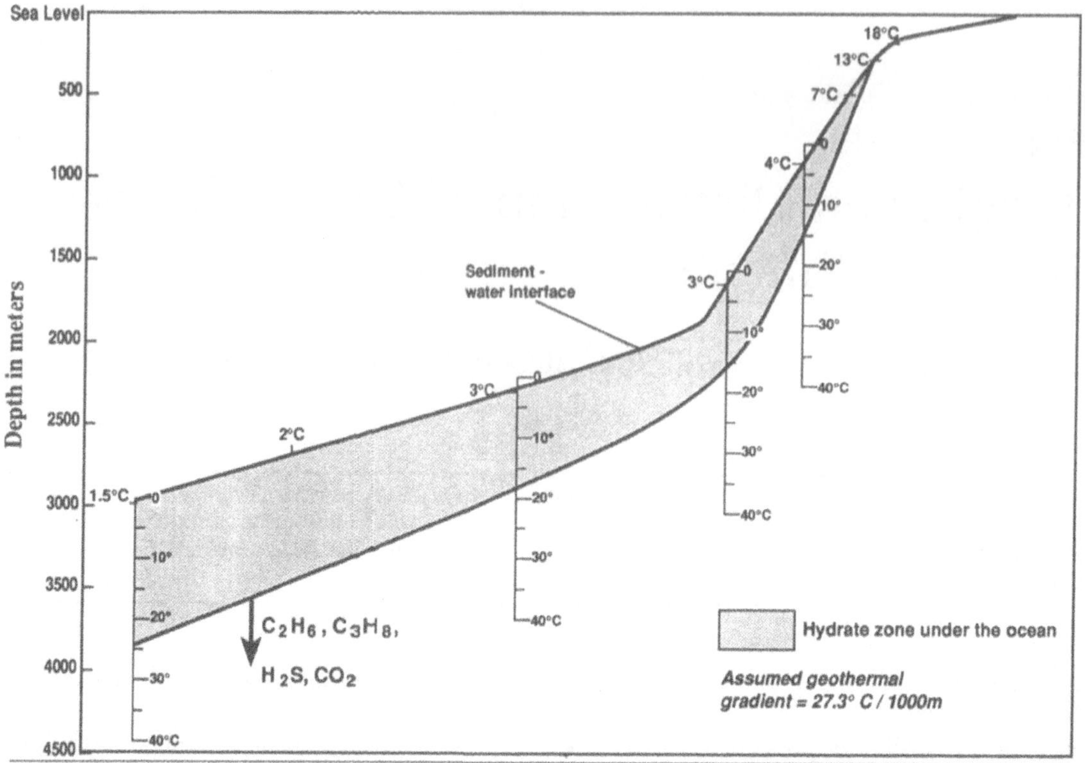

Figure 1. An example of the hydrate stability zone in an outer continental margin setting. [Modified from Kvenvolden (1988).]

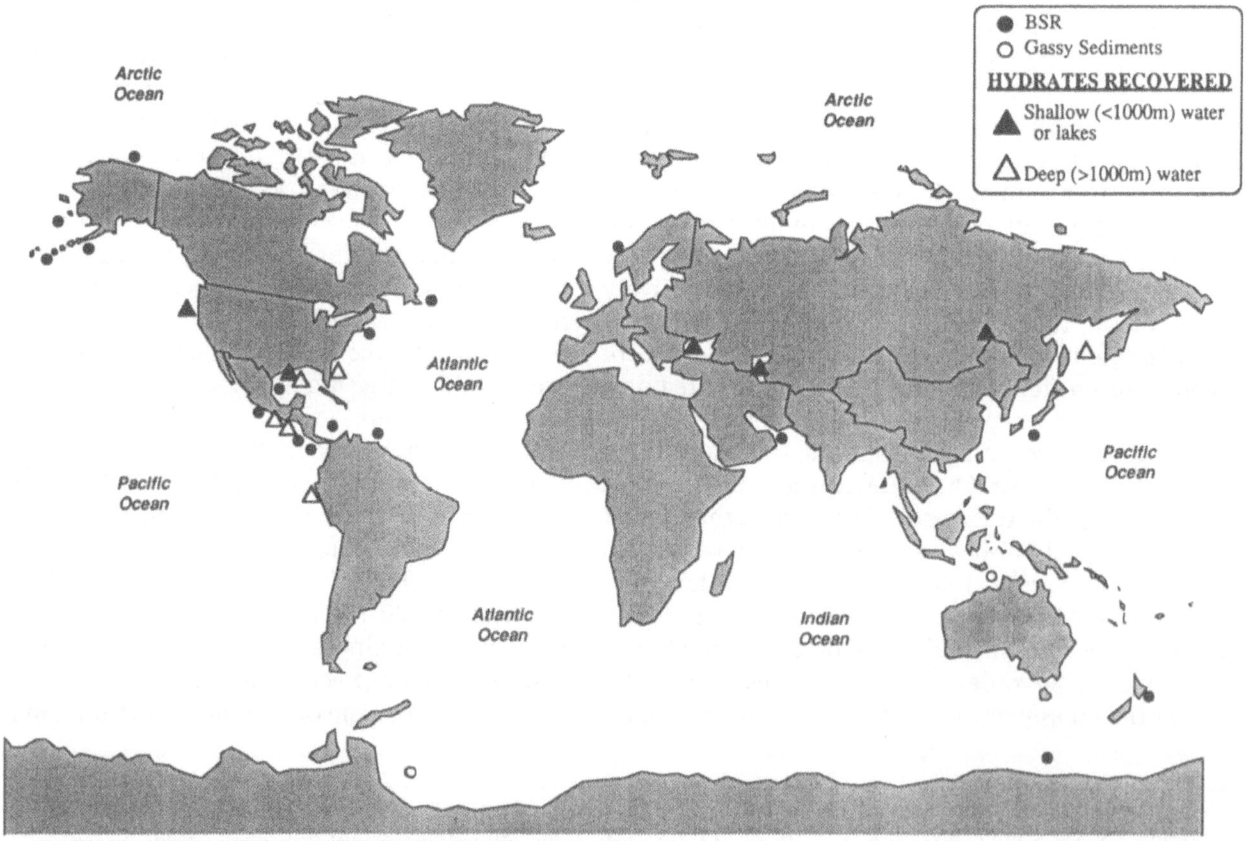

Figure 2. Summary of the worldwide occurrence of marine gas hydrates (including lakes) as inferred from a BSR, gassy sediments, and/ or actual recovery of hydrates. [After Kvenvolden (1988).]

completely understood, appears to depend on the source of the methane. Since thermogenic gas is produced at high temperatures, it must migrate from its place of origin to the hydrate stability zone. Hydrates most likely form at the two-phase (gas–liquid) interface; however it has also been suggested that hydrates can form from gases dissolved in the liquid phase (Sloan, 1989). Whatever the mechanism, it is clear that gas must be supplied to the zone of stability in quantities greatly in excess of the gas solubility in the liquid phase (supersaturation; Sloan, 1989). Hydrates occur as finely disseminated crystals, nodules, layers, and massive accumulations. It has been hypothesized that the evolution from smaller to larger accumulations represents an orderly progression in the formation of hydrates in deep-sea sediments (Fig. 3).

Certain sedimentary settings and features are believed to encourage the formation of hydrates. These include rapid sedimentation rates, total organic carbon content in excess of 0.5%, and methane interstitial concentrations in excess of 10 ml liter^{-1}. Hydrate lattices are stabilized or destabilized by co-occurring organic and inorganic pore-water constituents. The exclusion of ions from the hydrate lattice is a well-documented phenomenon which causes freshening of hydrate lattice water as well as a potential destabilization of the hydrate when formed in saltwater. Co-occurring higher hydrocarbon gases such as propane can cause changes in the crystal lattice. In the presence of propane, hydrates can be stable at much higher temperatures or lower pressures than pure methane hydrates. Certain solids such as clays also appear to stabilize hydrates. Hydrate formation is also influenced by sediment texture, authigenic carbonate rubble formation, and shallow faulting and fracturing of sediments (Brooks *et al.*, 1986).

The following sections summarize the geophysical and geochemical evidence for the presence of hydrates and describe locations where hydrates have been retrieved from marine sediments. Hydrates in permafrost regions have been extensively reviewed elsewhere (Kvenvolden and McMenamin, 1980; Makogon, 1981; Sloan, 1989).

2. Structure and Stability

Gas hydrates are icelike crystalline substances made up of two or more components. One component (the host molecule) forms an expanded framework with void spaces while the guest component(s) fills the void spaces (MacLeod, 1982). The interaction between the two components is not by chemical bonding, but by van der Waals forces similar to those found in a simple solution (Hand *et al.*, 1974). Natural gas hydrates have a water framework with void spaces filled by one or more gases. In marine sediments, gas hydrates are found in regions where high pressure, low temperature, and gas in excess of solubility are present. Low-molecular-weight hydrocarbons (LMWH), i.e., methane through butane, carbon dioxide, and possibly hydrogen sulfide are the only gases found in sufficient concentrations in marine sediments to form gas hydrates (Brooks *et al.*, 1984). The pressure–temperature stability boundary for a pure methane gas hydrate is accepted as at least 50 atm (500 m) and just above the freezing point of water, 4–6°C (Fig. 4; Kvenvolden and McMenamin, 1980). While this criterion is met in a vast portion of deep-sea sediments, only limited regions have gas concentrations sufficient to form gas hydrates.

Hydrates are stabilized during their formation by inclusion of gases within the structure. Stabilization depends on the pressure–temperature conditions and the types and amounts of gases present. The type

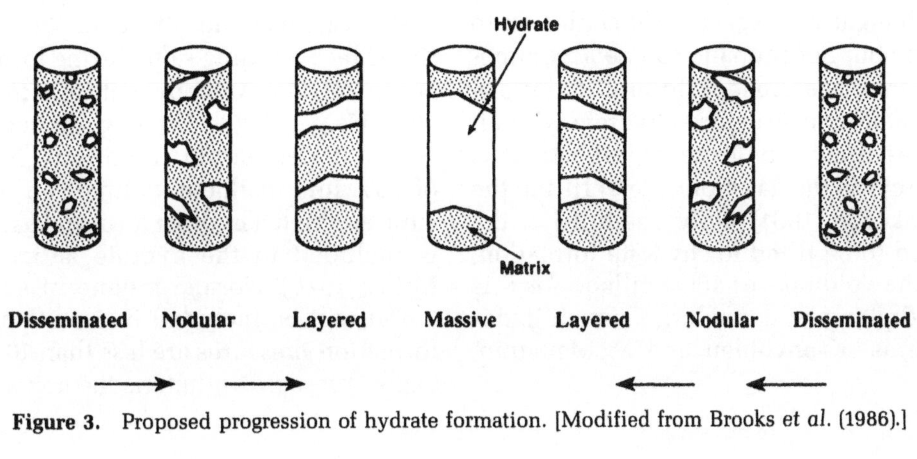

Figure 3. Proposed progression of hydrate formation. [Modified from Brooks *et al.* (1986).]

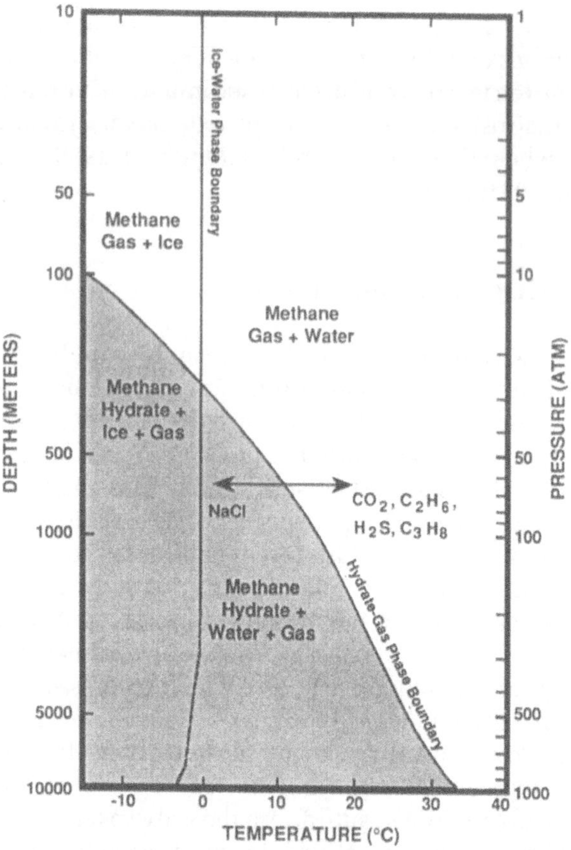

Figure 4. The distribution of the hydrate stability zone in marine sediments. [After Kvenvolden and McMenamin (1980).]

of gas also determines the structure of the gas hydrate. Inclusion of gases within the framework causes water to freeze in a cubic system rather than the hexagonal configuration of normal ice. The cubic arrangement can form two structures that are differentiated by the size of gas molecules that can be included within the respective void spaces (Claypool and Kvenvolden, 1983).

Structure I hydrates are a 2 pentagonal dodecahedra (12 sides) and a 6 tetrakaidecahedra (14 sided) polyhedron with eight void spaces. Molecules up to the size of ethane but not propane can be accommodated in a structure I cage. An ideal formula or ratio of hydrate gas to water for a structure I hydrate with all eight void spaces filled would be X(gas):5.75 H_2O molecules, where X is the gas molecule(s) filling the void spaces (Makogon, 1981). Only one-third of the void spaces need to be filled for hydrate formation, and rarely are the void spaces 100% filled. For this reason, gas hydrates are considered nonstoichiometric compounds (Kvenvolden and McMenamin, 1980).

Structure II hydrates have a 17-Å cell containing 136 water molecules arranged in a pentagonal dodecahedra and an 8 hexakaidecahedra (16-sided) polyhedron. This structure has 16 small and 8 large void spaces that can accommodate molecules up to 4.8 and 6Å, respectively (Makogon, 1981). The ideal formula or ratio of hydrate gas to water molecules for structure II is X(gas):17 H_2O. Only the eight large void spaces are filled when an individual gas large enough to form a structure II hydrate is present (Hand *et al.*, 1974). Thus, for a natural structure II hydrate the ideal formula will not occur. Structure II hydrates have void spaces in a diamond packing arrangement which can accommodate propane and isobutane (Miller, 1974; Claypool and Kvenvolden, 1983).

The theoretical conditions for gas hydrate formation have been derived from laboratory experiments (Katz *et al.*, 1959; Hand *et al.*, 1974). A pure methane–water hydrate will form at pressures of 48–50 atm and temperatures of 3–4°C (Fig. 4). The addition of higher molecular weight gases will allow the hydrate to form at a higher temperature or a lower pressure (Kvenvolden and McMenamin, 1980). The various higher molecular weight gases that can enter a methane–water system have different effects on hydrate formation. Tests with different natural gases reveal propane to have the greatest stabilizing effect followed by hydrogen sulfide, ethane, and carbon dioxide (Hitchon, 1974).

Several factors can inhibit gas hydrate formation. NaCl in an aqueous system functions as an "antifreeze," effectively lowering the temperature at which a hydrate will form (Fig. 4; Kvenvolden and McMenamin, 1980). A 3.5% NaCl solution will reduce the stabilization temperature of a methane–water hydrate by 2–3°C. The decrease in formation temperature caused by salinity is usually offset by the inclusion of heavier gases (Claypool and Kvenvolden, 1983). The effect of salinity has to be considered for hydrate formation at water depths approaching the pressure–temperature stability boundary. Conversely, gas hydrates can have an effect on the surrounding pore fluids by excluding salts during formation and freshening the pore-water by melting (Hesse and Harrison, 1981; Hesse, 1990). Certain gases can cause inhibitory effects under certain conditions. Nitrogen in excess of 50% in a methane hydrate can increase the pressure at which a hydrate forms by as much as 30% but is included in the hydrate lattice (Hitchon, 1974; Miller, 1974). Normal butane in small amounts (1–5.8%) will be included in a methane hydrate when formation pressures are less than 103 atm. At its own vapor pressure, *n*-butane is not a hydrate former, although it does not seem to inhibit the formation of methane hydrates (Ng and Robinson, 1976; John and Holder, 1982).

3. Bottom-Simulating Reflector

Gas hydrates have been predicted in deep-sea sediments using various seismic techniques. Markl *et al.* (1970) observed an anomalous seismic reflector on the Blake–Bahama Outer Ridge off the east coast of Florida. The authors speculated that the reflector could be either an old erosional surface or a diagenetic process. Subsequent drilling on the Deep Sea Drilling Project (DSDP) Leg 11 at and adjacent to the reflector site revealed gassy sediments with large amounts of methane (Hollister *et al.*, 1972). Additional core analyses led the authors to consider the reflector to be a diagenetic feature. Stoll *et al.* (1971) conducted laboratory experiments to determine acoustic velocities through gas hydrate layers and found that acoustic velocities increased from 1.85 to 2.69 km s^{-1} through hydrated sediments and decreased when the gas hydrates were allowed to decompose. The velocity of the hydrate layer corresponded with the 2.2 km $^{-1}$ velocity reported in gassy sediments from the Blake–Bahama Outer Ridge. These data suggested that the anomalous reflector reported by Markl *et al.* (1970) was probably sediment containing gas hydrates.

The velocity change from gas hydrate sediments to nongassy sediments is commonly called the bottom-simulating reflector (BSR) (Kvenvolden and McMenamin, 1980). The BSR is an anomalous reflector that parallels seafloor topography lying 100 to 1000 m below the sediment surface. BSRs are normally observed cutting across other reflectors and are found at water depths greater than 400 m. Shipley *et al.* (1979) reviewed some 30,000 km of multichannel seismic data and identified BSRs in the western Gulf of Mexico, the Middle American Trench off the west coast of Central America, and the western Atlantic (Blake–Bahama Outer Ridge). The strong BSR from the Blake–Bahama Outer Ridge indicated that the reflector first reported by Markl *et al.* (1970) could be a gas hydrate zone. The BSRs reported by Shipley *et al.* (1979) were confined to terrigenous continental margins within the pressure–temperature stability boundary for methane hydrates. Field and Kvenvolden (1985) reported the probable presence of gas hydrates off the coast of northern California. BSRs were observed at water depths of 800 to 1200 m and sub-bottom depths of 135 to 300 m, respectively. The BSRs were attributed to gas hydrated sediments because (1) they occurred at depths within the hydrate pressure–temperature stability boundary, and (2) the depth of the BSR in sediments increased with increasing water depth.

Bottom-simulating reflectors are not always indicative of gas hydrate layers. Reflections from slump surfaces, interbed seafloor multiples, and diagenetic boundaries can be confused with gas hydrate BSRs (MacLeod, 1982). An anomalous reflector in the Bering Sea was thought to be the BSR for a gas hydrate layer. DSDP Leg 19 found no evidence of gas hydrates at this site (Scholl and Creager, 1973). The evidence contrary to the presence of gas hydrates was the lack of gas in the cores, a large velocity increase below the reflector, and an increase in sub-bottom depth with decreasing water depth (Shipley *et al.*, 1979; MacLeod, 1982). Hein *et al.* (1978) concluded that the boundary represented a diatom dissolution layer; since the diagenetic change is temperature dependent, the reflection parallels the ocean bottom.

4. Geochemical Evidence and Hydrate Occurrence

Data presented by Shipley *et al.* (1979) led to investigations of several BSR horizons and the discovery of gas hydrates during DSDP/IPOD Legs 66, 67, 76, and 84 (Shipley and Didyk, 1982; Harrison and Curiale, 1982; Kvenvolden and Barnard, 1983; Kvenvolden and McMenamin, 1985). These hydrates, listed in Table I, were recovered from water more than 2000 m deep and at sub-bottom depths of 150–350 m. Only gas hydrate samples from Leg 84 were collected and preserved for shore-based analyses.

4.1. Deep Subsurface Occurrences

4.1.1. Blake–Bahama Outer Ridge

Leg 76 returned to the site of Leg 11 on the Blake–Bahama Outer Ridge. Small white crystals were recovered in a gassy sediment layer. The layer from a sub-bottom depth of 238 m had a gas:pore water ratio of 20:1 or four times the gas solubility in pore water at *in situ* pressure and temperature (Kvenvolden and Barnard, 1983). Molecular and isotopic values from the decomposed hydrate were 99% methane with a $\delta^{13}C$ of −70‰. Headspace gas analyses from adjacent sediment ranged from 76 to 97% methane with ppm levels of ethane, propane, and butane. Isotopic values of the

TABLE I. Gas Hydrate Data from DSDP Legs 66, 67, 76, and 84

DSDP cruise	Sub-bottom depth (m)	C_1 (%)	C_2 (%)	CO_2 (%)	C_1/C_2	Gas:fluid ratio	$\delta^{13}C(C_1)$ (‰)
Leg 66[a]							
Site 490	139	76.8	—[b]	23.2	—	0.9:1	—
Site 491	89	98.0	0.03	—	2809	7.2:1	—
Site 492	170	99.5	0.18	0.18	—	20:1	—
Leg 67[c]							
Site 497	360	85.0	ppm	0.2	2800	—	—
Site 498A	307	83.0	1.9	0.2	417	80:1	—
Leg 76,[d] Site 533	238	99.0	ppm	—	9000	20:1	−70.0
Leg 84[e]							
Site 565	319	89.0	ppm	0.03	2000	133:1	—
Site 568	404	64.0	ppm	0.1	530	30:1	—
Site 570	249	61.3	0.14	0.3	400	64:1	−44.0

[a]From Shipley and Didyk, 1982.
[b]—Data not available.
[c]From Harrison and Curiale, 1982.
[d]From Kvenvolden and Barnard, 1983.
[e]From Kvenvolden and McDonald, 1985.

methane ranged from −62 to −70‰ with methane: ethane (C_1/C_2) ratios of 4000 to 9000, indicating that the gas was biogenic in origin (Brooks *et al.*, 1983).

4.1.2 Middle America Trench

DSDP Leg 66 discovered gas hydrates at three drill sites off the Pacific coast of Mexico. These sites, on the landward side of the Middle America Trench, had a discernible BSR suggesting a gas hydrate layer. Shipley and Didyk (1982) recovered frozen sediments at three sites, one suspected and two confirmed gas hydrates. All three sites had gas of a biogenic origin based on methane:(ethane + propane) [$C_1/(C_2 + C_3)$] ratios of >1000.

Suspected gas hydrates were observed at Site 490 in frozen sediment at sub-bottom depths of 139 and 166 m. Frozen pieces, with impure ice up to 2 cm in diameter, were in void spaces or expansion zones within the unconsolidated mud. Only 0.91 ml of gas was evolved per milliliter of ice. This was not significantly greater than the gas released from interstitial water in nearby sections (0.47 ml gas per milliliter of water). Due to the small gas:pore fluid ratio, Shipley and Didyk (1982) were not certain whether the ice inclusions represented gas hydrates or water in expansion voids frozen by hydrate sublimation during core retrieval.

Gas hydrates were recovered from Site 491 at sub-bottom depths of 89 and 162 m in 2883 m of water. The estimated *in situ* pressures and temperatures were 295 atm, 4°C and 305 atm, 6°C, respectively. The hydrate at 89 m was 2 cm in diameter and had a gas: pore fluid ratio of 7.2:1, or two times the amount of gas that could be dissolved in water at that pressure. $C_1/(C_2 + C_3)$ ratios for the 89- and 162-m hydrates were 2809 and 105, respectively. Heavier hydrocarbons in the deeper hydrate suggested a thermogenic component. Gas hydrates at Site 492 were recovered in cores 16 (140 m) and 19 (170 m). The gas composition for both was nearly identical, with the sediments being predominantly volcanic ash and fine sands with a porosity of 46%. The 20:1 gas:fluid ratio for core 19 is roughly six times the volume of gas that can be dissolved in water at the estimated pressure. The estimated *in situ* pressures and temperatures of these three sediments are well within the stability zone for methane hydrates.

Gas hydrates were recovered unexpectedly on DSDP Leg 67 in the Middle America Trench off the Guatemala coast. Unlike Leg 66, there was no discernible BSR observed in seismic records at the discovery sites. Harrison and Curiale (1982) reported that drilling was prematurely terminated at two sites (Sites 497 and 498A) due to the excess pressures and low C_1/C_2 ratios associated with gas hydrates. Both sites contained high-porosity (20%) sediments that were retrieved frozen solid but collapsed when allowed to warm. The sites had estimated pressures (250–275 atm) and measured temperatures (8°C) that were well within the pressure–temperature stability zone for methane hydrates.

Gas hydrates from Site 497 (Leg 67) were recovered at a sub-bottom depth of 360 m with gas pressure great enough to eject sediment from the core barrel. Methane solubility for the hydrate was 4.08 mol % (50:1 gas:pore fluid) compared to the calculated 0.3 mol % for methane in pore water at *in situ* pressure. This is ten times the amount of methane that could be dissolved in the pore water. Declining C_1/C_2 ratios (6000 to 2800 in 25 m) in the hydrate zone caused drilling to be stopped.

Site 498A was similar to Site 497 in that a gas hydrate was recovered in the bottom core that terminated drilling. The gas hydrate was incorporated in cemented sands that collapsed into a bubbling mass when the hydrate decomposed. Methane solubilities were 6.18 mol % (80:1 gas:pore fluid) or approximately 15 times the solubility in pore water at *in situ* pressure. Owing to the low C_1/C_2 ratio (417) and the recovery of gas hydrates, drilling was terminated as a safety precaution (Harrison and Curiale, 1982).

Seismic data from Leg 67 were reevaluated, and a BSR was found along the landward slope of the Middle America Trench (von Huene *et al.*, 1982). DSDP Leg 84 returned to this area to recover gas hydrates and preserve them for laboratory analysis. At three sites, gas hydrates were recovered at sub-bottom depths greater than 256 m and at estimated *in situ* pressures ranging from 195 atm at Site 570 to 332 atm at Site 565. Gas hydrates at two sites (566 and 569) were inferred on the basis of pore-water and organic geochemical evidence (Kvenvolden and McDonald, 1985).

Site 565 was located in an area offshore Costa Rica with hydrates recovered at 319 m. Gas analysis revealed a gas:pore fluid ratio of 133:1 or 30 times the solubility of methane at *in situ* temperature and pressure. Site 568 was drilled adjacent to Site 496 (Leg 67) where a review of seismic data confirmed the presence of a BSR (von Huene *et al.*, 1982). Gas hydrates were observed at a depth of 404 m in a tuffaceous mudstone. Gas:pore fluid ratios were 30:1, corresponding to approximately seven times the dissolved gas found in water at *in situ* pressure and temperature (Kvenvolden and McDonald, 1985).

The largest gas hydrate was recovered from Site 570 at sub-bottom depths of 246–355 m. A massive hydrate, 1.05 m in length, was recovered at 250 m sub-bottom. Well logs indicated that this hydrate was 3–4 m thick (Kvenvolden *et al.*, 1984). This hydrate was preserved frozen (−20°C) in pressure vessels for laboratory analysis. Gas analysis showed 44–75% methane, 0.25–0.34% carbon dioxide, and up to 2000 ppm ethane (C_1/C_2 = 220–840; Brooks *et al.*, 1985). The massive hydrate had $\delta^{13}C$ values of −42.5‰ for methane and −26.8‰ for ethane. Heavy isotopic methane values suggested gas of a thermogenic source. However, this was discounted since the $\delta^{13}C$ of CO_2 (+2.9‰) in the core suggested that the methane was formed from biogenic reduction of heavy CO_2 (Jeffrey *et al.*, 1985; Kvenvolden *et al.*, 1984). Brooks *et al.* (1985) suggested the heavy methane was from low-temperature diagenesis in immature sediments or upward migration from a thermogenic source.

It was suggested that the formation of this unusually large accumulation of massive hydrate probably involved upward migration and accumulation of gas into a hydrate-stabilizing zone, unusually high microbial gas production, and a release of gas due to a pressure drop brought on by sea level lowering or tectonic uplift (Kvenvolden *et al.*, 1984).

4.1.3. Orca Basin

Gas hydrates were recovered at 20–40-m sub-bottom depths at Site 618 in the Orca Basin on DSDP Leg 96 (Brooks *et al.*, 1986; Pflaum *et al.*, 1986). The hydrate decomposition gas was primarily methane with trace amounts of ethane, propane, and CO_2. The methane $\delta^{13}C$ value was −71.3‰, indicating a biogenic origin. Ethane and propane concentrations occluded in the hydrate were much higher than those found in adjacent gas pockets [$C_1/(C_2 + C_3)$ = 159 and 18,400 for the hydrate and a gas pocket, respectively]. It was suggested that hydrate formation involves no stable isotopic fractionation of methane but may involve a preferential incorporation of ethane and propane.

4.1.4. Peruvian Slope

Gas hydrates were recovered on the Peruvian outer continental shelf during coring on Leg 112 (sites 685 and 688) of the Ocean Drilling Project (Kvenvolden and Kastner, 1990). Gas hydrates were recovered at 99, 141, and 166 m below the seafloor in water depths of 3820 and 5070 m. Nearby sites (682 and 683) exhibited bottom-simulating reflectors, and sediments contained large amounts of gas, suggesting that hydrates were widespread. The hydrate decomposition gas was 99% methane, and stable isotope compositions varied from −50 to −80‰, indicating that microbial processes were responsible for supplying most of the hydrate methane. Pore-water chemistry reflected sulfate depletion, methane production, hydrate decomposition, and salt enrichment that was associated with the formation of hydrates.

4.2. Shallow Subsurface Occurrences

4.2.1. Northern Gulf of Mexico

Prior to 1988, gas hydrate discoveries in marine sediments had been in deep water (>1000 m) at subbottom depths greater than 100 m. The exception was an icelike material recovered in 6.5 m of sediment in the Black Sea (Yefremova and Zhizhchencko, 1974). No gas hydrates had been recovered at or near the pressure–temperature stability boundary for a methane–water hydrate. In 1983, the first Gulf of Mexico hydrate discovery was reported in shallow water (<1000 m). These hydrates were recovered in three 5-m cores in 530 m of water on the continental slope offshore Louisiana (Brooks *et al.*, 1984). The hydrates were distributed throughout oil-stained cores and occurred at temperatures and pressures near the presumed limit of pure methane hydrate stability. Gas analysis indicated that the hydrates were thermogenic in origin. The decomposition gas of the hydrate had $C_1/(C_2 + C_3)$ ratios of 3.5 and a methane $\delta^{13}C$ value of −44.6‰. The high concentrations of propane and isobutane (14.4 and 4.4%) indicated the presence of structure II hydrates. The gas hydrates ranged in size from small crystals to nodules several centimeters in diameter dispersed in carbonate rubble. This discovery demonstrated that hydrates can exist (1) in shallow subsurface sediments (<10 m), (2) in shallow water, (3) commingled with oil seepage, and (4) as structure II hydrates.

Gas hydrates have been recovered at more than a dozen locations on the northern Gulf of Mexico continental slope in water depths ranging from 530 to 2400 m (Brooks *et al.*, 1986). The hydrates occur as fine disseminated crystals, nodules, interspersed layers, and solid masses (>150 mm thick). $C_1/(C_2 + C_3)$ ratios range from 1.9 to >1000, and methane $\delta^{13}C$ values vary from −43 to −71‰. Thermogenic gas hydrates are often associated with oil-stained cores containing up to 7% extractable oil exhibiting moderate to severe microbial degradation. Hydrate-associated sediments contain elevated levels of carbon isotopically light $CaCO_3$ (up to 65%). Hydrates in the northern Gulf of Mexico are preferentially associated with collapsed structures, diapiric crests, and deep faults on the flanks of diapirs.

4.2.2. Offshore Northern California

Methane hydrates were recovered in shallow cores (<6 m) from the Eel River Basin offshore northern California as dispersed crystals, small nodules, and layered bands (Brooks *et al.*, 1991). Hydrates occurred very close to the sediment–seawater interface, in some instances within 0.3 m. Hydrates were preferentially associated with coarser grained sediments in water depths between 510 and 672 m adjacent to an area exhibiting a BSR. Hydrates were present north of the Mendocine Fracture Zone but were probably absent south of the zone. This absence was attributed to lower heat flows in areas of active subduction. C_1/C_2 ratios ranged from 456 to 8030, and methane $\delta^{13}C$ values varied from −57.6 to −69.1‰, indicating a microbial biogenic gas source. One sample had sufficient ethane for a stable carbon isotopic measurement ($\delta^{13}C = -27.9‰$). All hydrate-containing cores also contained hydrogen sulfide, suggesting that the sulfate-free zone coincided with hydrate formation.

5. The Significance of Gas Hydrates

Gas hydrates represent an important potential energy resource of the future. The estimated energy value of methane hydrates greatly exceeds that reservoired in proved and recoverable reserves though many technological problems still remain to be solved before this potential can be realized. It has been suggested that hydrate zones are also viable exploration targets if they form impermeable seals under which oil and gas can accumulate (Hedberg, 1980; Dillon *et al.*, 1980). On the other hand, hydrates are seen as drilling hazards due to the potential for the catastrophic release of gases, resulting in the alteration of sedimentary geotechnical properties (Carpenter, 1981; Prior *et al.*, 1989). The formation/decomposition of hydrates can act to destabilize sediments, resulting in slumps, mass sediment movements, and the loss of bearing capacity.

In recent years, methane has been recognized as an important greenhouse gas. Gas hydrates, a major repository of methane, are subject to decomposition brought on by increases in water temperature or decreases in pressure. The discovery of significant amounts of hydrates in shallow waters at or near their stability limit suggests that subtle changes in global temperature could induce a redistribution of global methane from sedimentary hydrates to atmospheric gas. This is particularly critical in permafrost regions due to the shallow depth of hydrate occurrence and the effect of the transition of ice to water (Kvenvolden, 1988). The addition of hydrate-derived methane to the atmosphere is believed to be presently augmenting the global warming trend (MacDonald,

1982; Bell, 1983; Chamberlain *et al.*, 1983; Revelle, 1983). The arctic permafrost and gas hydrates of the Canadian Beaufort Shelf are degrading, but the contributions of gas hydrate methane to the global warming process are believed to be minimal (Kvenvolden, 1988).

As pointed out by many authors, hydrates are an important repository of carbon that should not be ignored in any evaluation or modeling of the global carbon cycle.

References

Bell, P. R., 1983, Methane hydrate and the carbon dioxide question, in: *Carbon Dioxide Review, 1982* (W. C. Clark, ed.), Oxford University Press, New York, pp. 401–406.

Bernard, B. B., Brooks, J. M., and Sackett, W. M., 1977, A geochemical model for characterizing hydrocarbon gas sources in marine sediments, in: *Proceedings of Offshore Technology Conference* OTC 2934, OTC, Houston, pp. 435–438.

Brooks, J. M., Barnard, L. A., Weisenberg, D. A., Kennicutt, M. C., II, and Kvenvolden, K. A., 1983, Molecular and isotopic compositions of hydrocarbons at Site 533, DSDP Leg 76, in *Initial Reports of Deep Sea Drilling Project*, Vol. 76 (R. E. Sheridan, F. M., Gradstein, *et al.*, eds.), U.S. Government Printing Office, Washington, D.C., pp. 337–390.

Brooks, J. M., Kennicutt, M. C., II, Fay, R. R., McDonald, T. J., and Sassen, R., 1984, Thermogenic gas hydrates in the Gulf of Mexico, *Science* **225**:409–411.

Brooks, J. M., Jeffrey, A. W. A., McDonald, T. J., Pflaum, R. C., and Kvenvolden, K. A., 1985, Analysis of hydrate gas and water from Site 570 DSDP Leg 84, in: *Initial Reports of Deep Sea Drilling Project*, Vol. 84 (R. von Huene, J. Aubouin *et al.*, eds), U.S. Government Printing Office, Washington, D.C., pp. 699–704.

Brooks, J. M., Cox, H. B., Bryan, W. R., Kennicutt, M. C., II, Mann, R. G., and McDonald, T. J., 1986, Association of gas hydrates and oil seepage in the Gulf of Mexico, *Org. Geochem.* **10**:221–234.

Brooks, J. M., Field, M. E., and Kennicutt, M. C., II, 1991, Observations of gas hydrates offshore northern California, *Mar. Geol.* **96**:103–109.

Carpenter, G., 1981, Coincident sediment slum/clathrate complexes on the U.S. Atlantic continental slope, *Geo-Mar. Lett.* **1**:29–32.

Chamberlain, J. W., Foley, H. M., MacDonald, G. J., and Ruderman, M. A., 1983, Climate effects of minor atmospheric constituents, in: *Carbon Dioxide Review, 1982* (W. C. Clark, ed.), Oxford University Press, New York, pp. 255–277.

Claypool, G. E., and Kvenvolden, K. A., 1983, Methane and other hydrocarbon gases in marine sediment, *Annu. Rev. Earth Planet. Sci.* **11**:294–297.

Dillon, W. P., Grow, J. A., and Paull, C. K., 1980, Unconventional gas hydrate seals may trap gas off the southeastern U.S., *Oil Gas J.* **1980**(January 7):124–130.

Field, M. E., and Kvenvolden, K. A., 1985, Gas hydrates on the northern California continental margin, *Geology* **13**:517–520.

Hand, J. H., Katz, D. L., and Verma, V. K., 1974, Review of gas hydrates with implication for ocean sediments, in: *Natural Gases in Marine Sediments* (I. R. Kaplan, ed.), Plenum Press, New York, pp. 179–194.

Harrison, W. E., and Curiale, J. A., 1982, Gas hydrates in sediments of holes 497 and 498A DSDP Leg 67, in: *Initial Reports of Deep Sea Drilling Project*, Vol. 67 (J. Aubouin, R. von Huene, *et al.*, eds.), U.S. Government Printing Office, Washington, D.C., pp. 591–594.

Hedberg, H. D., 1980, Methane generation and petroleum migration, in *Problems of Petroleum Migration* (W. H. Robers, III and R. J. Cordell, eds.), AAPG Studies in Geology No. 10, American Association of Petroleum Geologists, Tulsa, Oklahoma, pp. 179–206.

Hein, J. R., Scholl, D. W., Barron, J. A., Jones, M. G., and Miller, J., 1978, Diagenesis of later Cenozoic diatomaceous deposits and formation of the bottom simulating reflector in the southern Bering Sea, *Sedimentology* **25**:155–181.

Hesse, R., 1990, Pore-water anomalies in gas hydrate-bearing sediments of the deeper continental margins: Facts and problems, *J. Inclusion Phenom. Mol. Recognit. Chem.* **8**:117–138.

Hesse, R., and Harrison, W. E., 1981, Gas hydrates (clathrates) causing pore-water freshening and oxygen isotope fractionation in deep-water sedimentary sections of terrigenous continental margins, *Earth Planet. Sci. Lett.* **55**:453–462.

Hitchon, B., 1974, Occurrence of natural gas hydrates in sedimentary basins, in: *Natural Gases in Marine Sediments* (I. R. Kaplan, ed.), Plenum Press, New York, pp. 195–225.

Hollister, C. D., Ewing, J. I., Habib, D., Hathaway, J. C., Lancelot, Y., Luterbacher, H., Paulus, F. J., Poag, C. W., Wilcoxon, J. A., and Worstell, P., 1972, Sites 102–103–104 Blake–Bahama Outer Ridge (Northern End), in: *Initial Reports of Deep Sea Drilling Project*, Vol. 11 (C. D. Hollister, J. I. Ewing *et al.*, eds.), U.S. Government Printing Office, Washington, D.C., pp. 135–143.

Hunt, J. M., 1979, Methane hydrates, in: *Petroleum Geochemistry and Geology* (J. M. Hunt, ed.), W. H. Freeman and Co., San Francisco, pp. 156–162.

Jeffrey, A. W. A., Pflaum, R. C., McDonald, T. J., Brooks, J. M., and Kvenvolden, K. A., 1985, Isotopic analysis of core gases at site 565–570, in: *Initial Reports of Deep Sea Drilling Project*, Vol. 84 (R. von Huene, J. Aubouin *et al.*, eds.), U.S. Government Printing Office, Washington, D.C., pp. 719–726.

John, V. T., and Holder, G. D., 1982, Hydrates of methane + *n*-butane below the ice point, *J. Chem. Eng.* **27**:18–21.

Katz, D. L., Cornell, D., Kobayashi, R., Poettmann, F. H., Vary, J. A., Elenbaas, J. R., and Winaug, C. F., 1959, *Handbook of Natural Gas Engineering*, McGraw-Hill, New York.

Kvenvolden, K. A., 1988, Methane hydrate—a major reservoir of carbon in the shallow geosphere? *Chem. Geol.* **71**:41–51.

Kvenvolden, K. A., and Barnard, L.A., 1983, Gas hydrates of the Blake Outer Ridge, Site 533, DSDP Leg 76, in: *Initial Reports of Deep Sea Drilling Project*, Vol. 76 (R. E. Sheridan, F. M. Gradstein, *et al.*, eds.), U.S. Government Printing Office, Washington, D.C., pp. 353–365.

Kvenvolden, K. A., and Kastner, M., in press, Gas hydrates of the Peruvian continental margin, in: *Initial Reports ODP*, Vol. 112, U.S. Government Printing Office, Washington, D.C.

Kvenvolden, K. A., and McDonald, T. J., 1985, Gas hydrates of the Middle America Trench-DSDP/IPOD Leg 84, in: *Initial Reports of Deep Sea Drilling Project*, Vol. 84 (R. von Huene, J. Aubouin, *et al.*, eds.), U.S. Government Printing Office, Washington, D.C., pp. 667–682.

Kvenvolden, K. A., Claypool, G. E., Threlkeld, C. N., and Dendy, E., 1984, Geochemistry of a naturally occurring massive marine gas hydrate, *Org. Geochem.* **6**:703–713.

Kvenvolden, K. A., and McMenamin, M. A., 1980, Hydrates of

natural gas: A review of their geologic occurrence, *U.S. Geol. Surv. Circ.* **825**.

MacDonald, G. J., 1982, *The Long-Term Impacts of Increasing Atmospheric Carbon Dioxide Levels*, Ballinger, Cambridge, Massachusetts.

MacLeod, M. K., 1982, Gas hydrates in ocean bottom sediments, *Am. Assoc. Petrol. Geol. Bull.* **66**:2649–2662.

Makogon, Y. F., 1965, Hydrate formation in gas bearing strata in permafrost regions, *Gazov. Promst. Izd. Nedra* **5**.

Makogon, Y. F., 1981, *Hydrates of Natural Gas*, Pennwell Publishing Co., Tulsa, Oklahoma [translation by W. J. Cieslewicz].

Markl, R. G., Bryan, G. M., and Ewing, J. I., 1970, Structure of the Blake–Bahama Outer Ridge, *J. Geophys. Res.* **75**:4539–4555.

Miller, S. L., 1974, The nature and occurrence of clathrate hydrates, in: *Natural Gases in Marine Sediments* (I. R. Kaplan, ed.), Plenum Press, New York, pp. 151–178.

Ng, H. J., and Robinson, D. B., 1976, The role of *n*-butane in hydrate formation, *AIChE J.* **22**:656–661.

Pflaum, R. C., Brooks, J. M., Cox, H. B., Kennicutt, M. C., II, and Sheu, D. D., 1986, Molecular and isotopic analysis of core gases and gas hydrates from sites 618 and 619, DSDP Leg 96, in: *Initial Reports of Deep Sea Drilling Project*, Vol. 96 (A. H. Bouma, J. M. Coleman, *et al.*, eds.), U.S. Government Printing Office, Washington, D.C., pp. 781–784.

Prior, D. B., Doyle, E. H., and Kaluza, M. J., 1989, Evidence for sediment eruption on deep sea floor, Gulf of Mexico, *Science* **243**:517–519.

Revelle, R. R., 1983, Methane hydrates in continental slope sediments and increasing atmospheric carbon dioxide, in: *Changing Climates*, National Academy Press, Washington, D.C., pp. 252–261.

Scholl, D. W., and Creager, J. S., 1973, Geologic synthesis of Leg 19 (DSDP) results for North Pacific and Aleutian Ridge, and Bering Sea, in: *Initial Reports of Deep Sea Drilling Project*, Vol. 19 (J. S. Creager, D. W. Scholl, *et al.*, eds.), U.S. Government Printing Office, Washington, D.C., pp. 897–913.

Shipley, T. H., and Didyk, B. M., 1982, Occurrence of methane hydrates offshore southern Mexico, in: *Initial Reports of Deep Sea Drilling Project*, Vol. 66 (J. S. Watkins, J. C. Moore, *et al.*, eds.), U.S. Government Printing Office, Washington, D.C., pp. 547–555.

Shipley, T. H., Houston, M. H., Buffler, R. T., Shaub, F. J., McMillen, K. J., Ladd, J. W., and Worzel, J. L., 1979, Seismic evidence for widespread possible gas hydrate horizons on continental slopes and rises, *Am. Assoc. Petrol. Geol. Bull.* **63**:2204–2213.

Sloan, E. D., 1989, *Clathrate Hydrates of Natural Gases*, Marcel Dekker, New York.

Stoll, R. D., Ewing, J., and Bryan, G. M., 1971, Anomalous wave velocities in sediments containing gas hydrates, *J. Geophys. Res.* **76**:2090–2094.

von Huene, R., Ladd, J., and Norton, I., 1982, Geophysical observations of slope deposits, Middle America Trench off Guatemala, in: *Initial Reports of Deep Sea Drilling Project*, Vol. 67 (J. Aubouin, R. von Huene, *et al.*, eds.), U.S. Government Printing Office, Washington, D.C., pp. 719–732.

Yefremova, A. G., and Zhizhchencko, B. P., 1974, Occurrence of crystal hydrates of gases in the sediments of modern marine basins, *Dokl. Akad. Nauk SSSR*, Earth Science Section [Engl. translation 1975], **214**:219–220.

VI

Organic Matter and Metalliferous Deposits

Chapter 25

The Role of Organic Matter in Ore Transport Processes

DAVID A. C. MANNING and ANDREW P. GIZE

1. Introduction

Organic matter is a common constituent of low-temperature hydrothermal mineral deposits hosted by sedimentary rocks which are themselves organic-rich or part of a maturing sedimentary basin. Hydrothermal mineral deposits in basement rocks have also been reported to contain hydrocarbons, which may have migrated into their present sites from adjacent sedimentary basins. Of particular interest is the well-known association between organic matter and Pb–Zn mineralization of Mississippi Valley type (e.g., Barton, 1967; Macqueen and Powell, 1983; Marikos *et al.*, 1986; Gize and Barnes, 1987), where bitumens and petroleum liquids are reported to occur. In addition, a close relationship between organic matter and mineralization is known for Carlin type Au deposits (Hausen and Park, 1986) and for disseminated sandstone-hosted U mineralization (Vine, 1962; Leventhal, 1986). Organic matter may also be involved in the formation of sandstone-hosted Pb and Cu(–V–Co) deposits, which characteristically show many diagenetic features (Bjørlykke and Sangster, 1981; Rickard *et al.*, 1979; Sverjensky, 1987). The low temperatures involved in the formation of these types of mineral deposit (rarely in excess of 200°C) are not incompatible with the presence of a wide variety of organic compounds.

The role of organic constituents in the formation of the host mineralization has been subject to considerable debate in descriptions of the above mineralization types. Considerable emphasis has been placed on the role of organic matter in controlling the precipitation of ore mineral components, by reduction, chemical reaction, or adsorption (e.g., Saxby, 1976; Barton, 1967; Disnar, 1981; Nakashima *et al.*, 1984). More recently, the potential of organic components as agents for the transport of metals to the sites of deposition has been considered as a means of overcoming the problem of transporting elements which are normally poorly soluble in aqueous fluids under geological conditions (Gardner, 1974; Giordano, 1985; Manning, 1986; Hennet *et al.*, 1988). Organic matter has a role to play in both deposition and transport processes, which necessarily differ in their mechanisms. It is the purpose of this chapter to concentrate on transport processes in the presence of organic matter and to assess the role of petroleum liquids in ore transport as well as the role of organic components of the aqueous phase. Emphasis will be placed on the

DAVID A. C. MANNING and ANDREW P. GIZE • Department of Geology, University of Manchester, Manchester M13 9PL, England.
Organic Geochemistry, edited by Michael H. Engel and Stephen A. Macko. Plenum Press, New York, 1993.

behavior of Cu, Pb, and Zn: for more detailed consideration of V and Ni the recent review by Filby and Van Berkel (1987) should also be consulted.

2. Metals in Petroleum Liquids

Petroleum represents a complex mixture of organic compounds, mixed with "impurities" such as included water or salts. The organic constituents can be subdivided into several fractions, including saturated and aromatic hydrocarbons, resins, and asphaltenes (Speight, 1980; Tissot and Welte, 1984). Resins and asphaltenes are composed of relatively high molecular weight species, with complex structures which include the heteroatoms N, S, and O and which are host to numerous nonhydrocarbon functional groups.

Several studies have been carried out to determine the inorganic constituents of petroleum, principally from the point of view of environmental problems arising from combustion. Petroleum is particularly rich in certain metals, such as vanadium which reaches 2000 ppm in Venezuelan crude oil (Boscan; Tissot and Welte, 1984). Most alkalis and alkaline earths are present in small amounts (of the order of parts per million or less; Jones, 1975); the alkalis at least are probably present as extrinsic matter. Of greatest interest are the transition elements, which occur in significant quantities and which, for theoretical reasons (discussed below), are likely to dominate the intrinsic metallic component of the oil. Typical values for many elements are given in Table I (together with sources of data), and maximum reported values are illustrated in Fig. 1. Particular attention will be paid to the dominant metallic components, V and Ni, and selected minor components of ore genetic interest, Cu, Pb, and Zn.

TABLE I. Metals in Petroleum

	Concentration (ppm)										
	Western Canada[a]		Saudi Arabia[b]		Mollase Basin[c]		Italy[d]		Venezuela[e]		Jones review[f]
Metal	Avg	Max	Avg	Max	Avg	Max	Avg	Max	Avg	Max	Max
Ag			0.16	0.39		0.32					0.004
Al			1.00	1.40							7.8
As	0.11	1.99									2.4
Au	(0.44)[g]	(1.32)[g]									0.003
Ba											11.9
Ca			1.31	4.00					6.38	16.0	10
Cd			0.84	2.37							—
Co	0.05	2.00			0.13	0.64		0.003[h]			12.75
Cr	0.09	1.68	0.65	1.18	0.11	0.45	3.15	10.7			3.6
Cu			0.32	0.88		2.50		5.5[h]	0.105	0.170	11.7
Fe	10.8	254	2.17	6.64	10.0	82.0		0.9[h]	4.98	26.0	120
Ga		0.04									0.8
Hg	0.05	0.40				0.50					29.6
Mn	0.01	3.85	0.05	0.15			1.2	9.7			0.15
Mo							2.27	183	0.572	2.00	10.1
Ni	9.38	74.1	7.96	18.3	13.5	55.0	46.6	155	40.2	105	105
Pb			0.89	1.57							2.1
Sb	0.01	0.03						0.035[h]			11
Sn				1.74							2.2
U											0.43
V	13.6	177	25.5	54.6	1.01	11.3	64.5	248	147	433	1580
Zn	0.46	5.92	0.81	2.55	1.54	21.0		0.9[h]			160
Zr											2.7

[a]Data for up to 88 samples, from Hitchon *et al.* (1975).
[b]Data for 10 samples, from Ali *et al.* (1983).
[c]Data or 43 samples, from Ellrich *et al.* (1985).
[d]Data for 37 samples (except where otherwise noted), from Colombo and Sironi (1981) and Colombo *et al.* (1964).
[e]Data from Simoza *et al.* (1985).
[f]Data from Jones (1975).
[g]Concentrations in parentheses are ppb.
[h]Data for 4 samples.

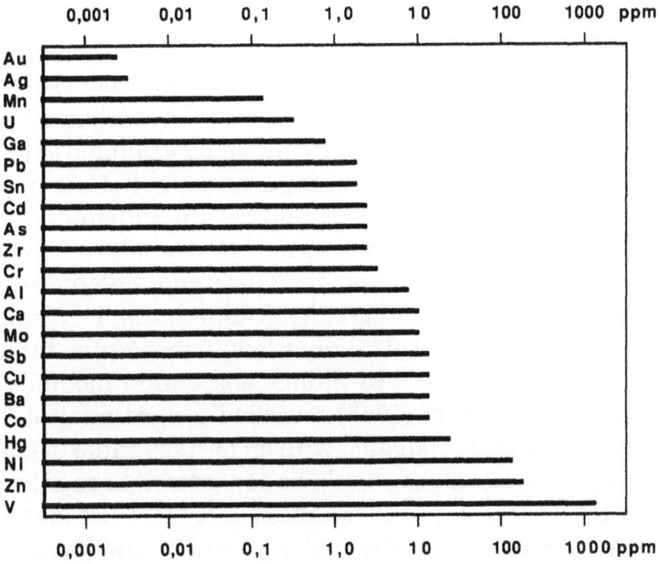

Figure 1. Maximum values reported for selected elements in petroleum. Data taken from Jones (1975) and tabulated in Table I.

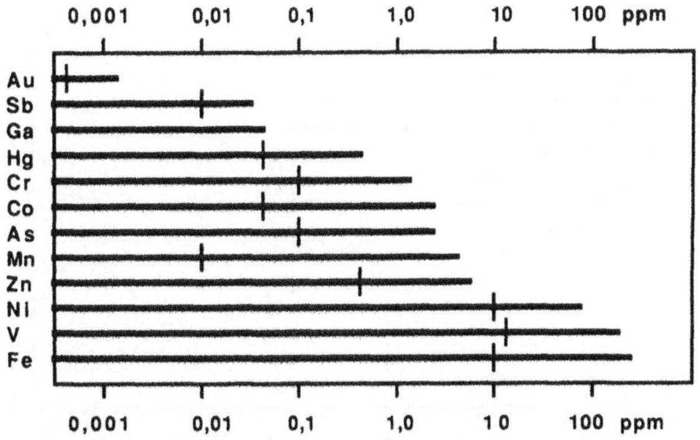

Figure 2. Maximum values of contents of oils from the Western Canada Basin (Hitchon *et al.*, 1975; Hitchon and Filby, 1983). The vertical bar on each line marks the mean value.

The data shown in Table I and Fig. 1 are drawn from several studies of the contents of metals within petroleum. The purposes of the source studies are varied and include characterization for tracing oil spills, investigation of environmental aspects of combustion, evaluation of potential of petroleum as a source of metals (especially V), and assessment of possible catalyst contamination. Several reviews of the metal contents of petroleum are available, of which the most exhaustive is that of Jones (1975). Others include those of Speight (1980), who provides similar data to Jones (1975), drawing largely from Shah *et al.* (1970a, b), and contributors to Yen's volume on trace metals in petroleum (Yen, 1975a; Filby, 1975; Filby and Shah, 1975; Hitchon *et al.*, 1975; Smith *et al.*, 1975; Gleim *et al.*, 1975). Studies of the geological significance of the metal contents of petroleum are relatively sparse. Some concentrate on the importance of V and Ni (e.g., Hodgson, 1954; Hodgson and Baker, 1957, 1959; Al-Shahristani and Al-Atyia, 1972), whereas important regional studies which also include data for other elements are provided, for example, by Hitchon *et al.* (1975) and Hitchon and Filby (1983) for the Western Canada Basin (Fig. 2), by Colombo and Sironi (1961) and Colombo *et al.* (1964) for Italian oils, by Ellrich *et al.* (1985) for the central European Molasse Basin, and by Simoza *et al.* (1985) for Venezuela (Fig. 3). Regional data are also available for Arabian oils (Al-Shahristani and Al-Atyia, 1972; Ali *et al.*, 1983) and are shown in Fig. 4. All of these studies are limited by the choice of elements considered (which depends to some extent on the analytical strategy used—often a combination of neutron activation and X-ray fluorescence techniques, which have their limitations) and the availability of uncontaminated samples. Thus, for geological purposes, the value of reviews such as those of Jones (1975), Smith *et al.* (1975), and Speight (1980) is limited to a sometimes tantalizing glimpse of potential metal contents without any geological context. Furthermore, high values for certain metal contents that have been reported in the literature are occasionally disputed or are for the asphaltene fraction, which, of the individual components of petroleum, is naturally richest in metals (e.g., Eldib et *al.*, 1960).

In detail, it can be seen from the data presented in Table I and Figs 1–4 that several metals are present within petroleum in amounts that are of major significance, especially when it is considered that many

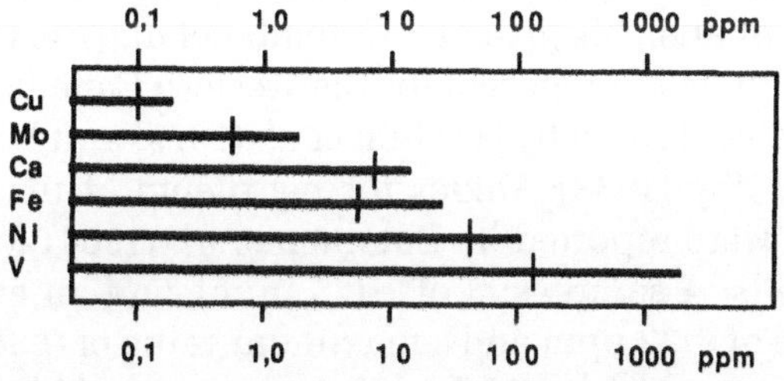

Figure 3. Maximum values of metal contents of oils from Venezuela (Simoza *et al.*, 1985). The vertical bar on each line marks the mean value.

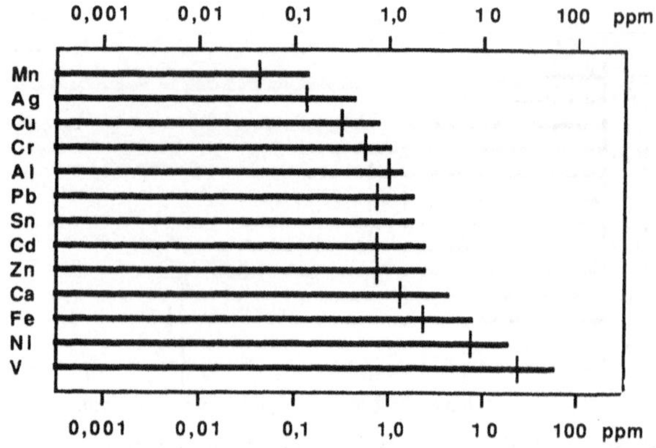

Figure 4. Maximum value of metal contents of oils from the Middle East (Al-Shahristani and Al-Atyia, 1972; Ali et *al.*, 1983). The vertical bar on each line marks the mean value.

postulated ore fluids have metal contents of the order of parts per million. Of particular interest are V and Ni, which frequently dominate and have received considerable attention already (e.g., Al-Shahristani and Al-Atyia, 1972; Hodgson, 1954; Hodgson and Baker, 1957; Quirke, 1987). The few available data suggest that Co, Cu, Mo, Pb, and Zn are also present in significant quantities. This superficial observation can be clarified by close examination of the sources of these data, concentrating on individual elements.

Co was reported by Jones (1975) to reach concentrations of 12.75 ppm for a single sample of California crude oil analyzed by Shah *et al.* (1970b); this sample also contained 85.8 ppm Zn and was rich in Sc (9.5 ppb; cf. ~1 ppb for other analyzed samples in the same study). In other studies lower Co concentrations have typically been reported: 53.7 ppb (average)–2000 ppb (max.) for Western Canada Basin oils (84 samples; Hitchon *et al.*, 1975), and 130 ppb (average)–640 ppb (max.) for oils from the Molasse Basin (Ellrich *et al.*, 1985). Co contents of oils should be regarded as typically of the order of 100 ppb, with exceptional examples at the parts per million level.

Cu occurs in amounts up to 11.7 ppm according to Jones (1975); however, the latter value refers to an asphalt analysis given by Colombo *et al.* (1964). This element is not reported for the Western Canada Basin samples studied by Hitchon *et al.* (1975) and Hitchon and Filby (1983). Values for petroleum of up to 5.5 ppm were reported by Colombo *et al.* (1964) for Italian oils (4 samples; 3 of ~0.5 ppm), and an average value of 0.32 ppm and a maximum value of 0.88 ppm were reported for 10 Saudi Arabian oils (Ali *et al.*, 1983). Simoza *et al.* (1985) gave values up to 0.17 ppm for Venezuelan oils, averaging 0.105 ppm. Typical values should be considered to be of the order of 0.1–0.5 ppm.

Mo has been reported for relatively few petroleum samples. Jones (1975) reported up to 10.1 ppm for Italian crude oil, as determined by Colombo *et al.* (1964); again this figure refers to an asphalt. Colombo *et al.* (1964) reported 8–53 ppb for 4 petroleum samples determined by neutron activation analysis, compared with up to 183 ppm for 32 samples determined by ultraviolet spectrography (Colombo and Sironi, 1961). Gleim *et al.* (1975) reported up to 7.3 ppm for 3 samples determined by atomic absorption methods, and Simoza *et al.* (1985) recorded a maximum of 2 ppm for Venezuelan oils, with an average value of 0.5 ppm. Typical values should be considered to be of the order of 0.5–1 ppm. Colombo and Sironi's study (1961) shows unusually high Mo values which correlate positively with V and Ni, which suggests that there may be spectral interference. Clearly, further study of the Mo content of petroleum is required.

Ni is reported for most analyzed petroleums, reaching concentrations of up to 155 ppm (Table I).

Pb is reported for relatively few petroleum analyses. Maximum contents of 2.1 ppm have been recorded for a Venezuelan crude oil (Jones, 1975). Ali *et al.* (1983) gave values of 0.89 ppm (average) and 1.57 ppm (Max.) for 10 samples from Saudi Arabia. Bratzel and Chakrabarti (1972) reported 0.31 and 0.17 ppm for two Venezuelan crude oils. Overall, Pb contents of the order of 0.2–0.5 ppm should be regarded as typical (Jones, 1975).

V is the dominant metallic component of petroleum, followed by Ni. Many determinations are available, showing contents of up to 1580 ppm for Boscan crude oil from Venezuela (Jones, 1975). A wide range of V contents has been observed (Table I).

Zn has been reported for many petroleum samples, with extreme high values of 160 ppm reported for a single sample by Jones (1975). There is some dispute over the high Zn contents reported by Shah *et al.* (1970b), which range from 3 to 86 ppm [see Jones (1975) for details], but values of 21 ppm are reported for the southern German Molasse Basin (Ellrich *et al.*, 1985), and 5.9 ppm for the Western Canada Basin (Hitchon *et al.*, 1975). The average Zn content for Venezuelan oils studied by Simoza *et al.* (1985) was 0.5 ppm, with a maximum of 1.2 ppm. Some of the high values reported in the literature could be due to contamination (Jones, 1975), but, based on studies where this factor has been considered and where large data sets are available (e.g., Hitchon *et al.* 1975; Hitchon and Filby, 1983), values of 0.5 ppm (average)–6 ppm (max.) can be considered as representative.

When considering the possible role of petroleum liquids as ore-transporting agents, it can be seen from the above discussion that Cu and Zn, in particular, are present in petroleum and that there are reports of elevated concentrations of these elements. V and Ni predominate, whereas Pb is present only in minor amounts. However, from Table II it can be seen that Pb, Cu, and Zn contents of postulated aqueous ore fluids appropriate for Mississippi Valley-type mineralization are similar to or less than those reported as maximum values for certain oils. Petroleum liquids therefore occasionally have the potential to contribute to the metals available for mineralization; however, in order to evaluate this possibility fully, it is necessary to understand the mechanisms by which metals are incorporated into petroleum.

2.1. The Incorporation of Metals within Petroleum

The dominant metals within petroleum are V and Ni, and most studies of the incorporation of these metals have concentrated on their occurrence within porphyrins. The porphyrins are extractable, organic chelating species, within which a central metal ion is coordinated with four nitrogen atoms, which form part of a porphin ring structure (Fig. 5), as a tetrapyrrole complex. The porphin ring is an essential characteristic of the porphyrins, of which at least 35 species, differing in peripheral substitutions, have been characterized (Filby and Van Berkel, 1987). Within petroleum, five types of porphyrins predominate, each type consisting of a homologous series with variations in peripheral substituent groups (e.g., Barwise and Whitehead, 1980). Although a number of possible biological precursors are recognized, the chlorophylls are considered to be the most significant. In these, the central metal is Mg^{2+}, which is replaced at an early stage in diagenesis by other, usually divalent, cations. The mechanisms of this replacement have been reviewed by many authors, including Filby and Van Berkel (1987), Buchler *et al.* (1973), and Buchler (1975, 1978). For the purposes of this chapter, the reaction pathways by which chlorophyll or other precursors are modified to become "geoporphyrins" will be neglected, as, while the details of these pathways are important, they do not affect the overall stability and properties of the porphin ring from the point of view of metal substitution. Instead, the substitution of metals within porphyrin structures will be considered as analogous to the substitution of metals within mineral structures, where substitution sites are clearly defined crystallographically. Finally, for the purposes of simplicity, it will be assumed that the behavior of the porphyrins with respect to metal complexation will be representative of metal complexation by analogous tetrapyrrole complexes within nonporphyrin species. For example, tetrapyrrole sites can be assumed to be present within asphaltenes (Yen, 1975b), but, as components of high-molecular-weight species, they are not easily isolated and are not necessarily porphyrins *sensu strictu*

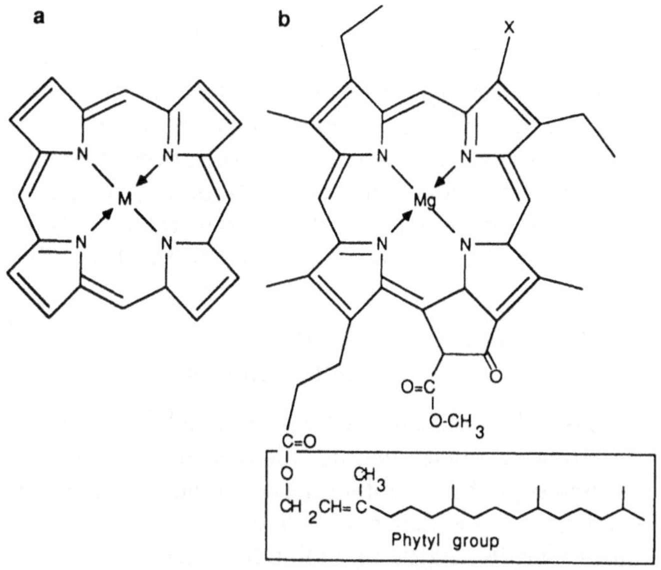

Figure 5. Structures of (a) the porphin ring and (b) chlorophyll. X denotes -CH_3 for chlorophyll *a* and -CHO for chlorophyll *b*.

TABLE II. Metal Contents of Petroleum and Mississippi Valley Type (MVT) Ore Fluids

	Concentration (ppm)		
		Petroleum	
Metal	MVT ore fluid[a]	Typical	Extreme
Zn	1.3	0.5	6
Pb	2.0	0.2–0.5	1.6
Cu	0.1	0.1–0.5	6

[a]Values for typical potential ore-forming fluid for MVT mineralization taken from Sverjensky (1984).

2.2. Theoretical Considerations

The considerable importance of porphyrins, particularly for biological processes, is reflected by a long and extensive history of study (Treibs, 1934; Falk, 1964; Smith, 1975; Dolphin, 1978). Their significance as complexing agents for metals was reviewed by Buchler (1978), who compared the relative stability of metalloporphyrins as a function of metal species and the ability of the metal to be "removed" from the porphyrin molecule. The substitution of metals within porphyrins can essentially be regarded as a process of ion exchange, where the metal substitutes for two hydrogen atoms according to the following scheme:

$$\underset{\text{porphyrin}}{P[H_2]} + M^{2+} = \underset{\text{metalloporphyrin}}{P[M]} + 2H^+ \quad (1)$$

This reaction will clearly be influenced by the pH of the exchange medium, and this has been recognized in the empirical establishment of metalloporphyrin stability classes according to the acid strength required for demetallation (Table III). Furthermore, the geometry of the porphin ring structure favors cations with an ionic radius of approximately 0.63Å (Buchler *et al.*, 1973; Buchler, 1978). This factor has been used in combination with ionic charge and electronegativity to establish a stability index S (Buchler *et al.*, 1973):

$$S = ZE_n/r$$

where Z is ionic charge, E_n is electronegativity, and r is ionic radius (Å). Stability indices for selected elements are shown in Table IV. Most stability data for metalloporphyrins are comparative in that they relate only to the porphyrin type of complex, and so comparison of the complexing behavior of porphyrins to that of other ligands within aqueous solutions cannot be made in detail; there is also a lack of pertinent data. However, a stability constant of log β = 29 has been obtained for a zinc porphyrin (Phillips, 1960), which illustrates the extremely high stability of porphyrin complexes of metals compared with, for example, those with chloride (log β = 0.4–0.6 for Zn; 1.4–1.8 for Pb) or acetate (log β = 1.36–1.57 for Zn; 2.9–4.08 for Pb [data from Martell and Smith (1977) and Smith and Martell (1978)].

TABLE III. Stability Classes for Metal–Porphyrin Complexes

Stability class	Reagent	Degree of demetallation
I	Pure H_2SO_4	Partial
II	Pure H_2SO_4	Complete
III	$HCl/H_2O-CH_2Cl_2$	Complete
IV	Pure CH_3COOH	Complete
V	$H_2O-CH_2Cl_2$	Complete

TABLE IV. Stability Indices for Selected Metal Ions[a]

Charge, Z	Cation	Stability index	Stability class
2	VO^{2+}	11.05	I
	Pd^{2+}	6.9	I
	Ni^{2+}	6.8	II
	Cu^{2+}	6.1	II
	Co^{2+}	5.8	II
	Pt^{2+}	5.5[b]	I
	Fe^{2+}	4.8	III
	Zn^{2+}	4.5	III
	Ag^{2+}	4.3[b]	II
	Pb^{2+}	3.9	V
	Hg^{2+}	3.9	V
	Mn^{2+}	3.8	IV
	Mg^{2+}	3.6	IV
	Cd^{2+}	3.6	IV
	Ca^{2+}	3.0	V
	Sr^{2+}	1.6	V
	Ba^{2+}	1.3	V
3	Co^{3+}	10.9	II
	Au^{3+}	10.3[b]	I
	Al^{3+}	9.1	I
	Ga^{3+}	8.8	II
	Fe^{3+}	8.6	II
	Cr^{3+}	8.0	I
	Mn^{3+}	7.2	II
4	Si^{4+}	19.0	I
	Pt^{4+}	14.0[b]	I
5	Mo^{5+}	17.2	II
	W^{5+}	18.8	I

[a]Values taken from Buchler *et al.* (1973) and Buchler (1978).
[b]Calculated using ionic radii from Shannon and Prewitt (1969) and Pauling electronegativities.

On the basis of the information given in Tables III and IV, it is possible to account for the predominance of vanadyl and nickel porphyrins (c.f. Quirke, 1978) and to speculate on the possible stabilities of other species which have not yet been described in detail for natural samples. VO^{2+} and Ni^{2+} both have high stability indices and belong to stability classes for which partial or complete demetallation involves conditions of extremely low pH. Mg^{2+} has a relatively low stability index and is removed by treatment with water. It is clear that natural porphyrins will effectively accumulate VO^{2+} and Ni^{2+} in preference to Mg^{2+}. Other metal ions with relatively high stability indices include Cu^{2+}, Co^{2+} and Co^{3+}, Fe^{2+} and Fe^{3+}, and Zn^{2+}. Porphyrin complexes of these ions belong to stability classes II and III and might be expected to occur in nature; the Cu porphyrin has already been reported (Baker and Louda, 1984; Palmer and Baker, 1978). It is also notable that Si^{4+} has a high stability index; its occurrence is precluded by the absence of ionic Si under normal geological conditions. However, the high stability constants for both Au^{3+} and Pt^{4+} suggest that these metal ions may occur within porphyrin complexes. The abundance of both elements is so low in most geological materials that they are unlikely to be detected in porphyrin-bearing samples. Both elements are, however, present in relatively elevated amounts in organic-rich rocks. Similarly, reported enrichments of Mo in petroleum and organic-rich shales may relate to the presence of porphyrin-type complexes of Mo.

The data discussed above can be used not only to identify metals prone to form porphyrin complexes, but also to indicate those which are unlikely to occur in nature. Metal–porphyrin complexes in stability classes IV and V are least stable. Mg, belonging to class IV, is removed during diagenesis, and so all other elements of classes IV and V can be regarded as equally unstable within a porphyrin host under diagenetic conditions. Thus, Pb in particular (class V) is unlikely to be present in geoporphyrins.

The relative stability of metalloporphyrins can be discussed in terms of an exchange equilibrium analogous to Eq. (1):

$$P[M] + M' = P[M'] + M \qquad (2)$$

where M and M′ refer to different cationic species. From this:

$$K_2 = \{P[M']\}\cdot\{M\}/\{P[M]\}\cdot\{M'\}$$

where { } designates concentration. For a natural mixture of metalloporphyrins, the ratios of different metal species will reflect the magnitude of K_2 for an appropriate exchange reaction. Thus, as equilibrium is approached during the geological history of a porphyrin-rich material, the proportion of porphyrin-hosted metals will tend to favor those metals whose porphyrin complexes have high stability indices. The predominance of vanadyl and nickel metalloporphyrins can again be explained in this way (cf. Quirke, 1987). Additionally, the possibility of incomplete exchange allows bulk chemical data for the metal contents of oils to be interpreted in terms of the feasibility of porphyrin complexing for minor components (see Section 2.3).

2.2. Trace Metals in Petroleum

The occurrence of V and Ni within petroleum has long been recognized, and at an early stage it was noted that the two elements vary sympathetically so that for a given suite of oils they maintain an approximately constant ratio. This ratio has been used in attempts to correlate oils and has been related to their geological age (e.g., Al-Shahristani and Al-Atyia, 1972; Hodgson, 1954; Hodgson and Baker, 1959). An example of the value of V and Ni for this purpose is shown in Fig. 6, for data from the Western Canada Basin (Hitchon and Filby, 1983). Three suites of petroleum have been distinguished on geological and organic geochemical grounds (Deroo *et al.*, 1977), and plot as three separate families with average V/Ni ratios of approximately 1.7 (group 1; Upper Cretaceous), 2.5 (group 2; Upper Cretaceous–Carboniferous), and 0.75 (group 3; Upper–Middle Devonian). However, Fig. 6 clearly shows one sample assigned to group 3 which plots with those of group 2, suggesting that the status of this sample may need to be reexamined. Similarly, V and Ni correlate for oils from Venezuela (Simoza *et al.*, 1985), which illustrate the relatively high proportions of metals in low-API-gravity (heavy) oils (Fig. 7). Other data which clearly show the positive correlation of V and Ni are summarized by Tissot and Welte (1984). To some extent, the variation in V and Ni reflects the proportion of asphaltenes within the petroleum (Hodgson and Baker, 1959), as this fraction is known to be particularly rich in metals and is most abundant in relatively heavy oils.

If it is assumed that V and Ni are present within petroleum principally as porphyrin complexes, it follows from the reaction in Eq. (2) that other metals present within porphyrins will similarly show a positive correlation with either element. The ratios of V or Ni to other elements will reflect the relative stabilities of their porphyrin complexes. In Fig. 6, Co is plotted against V for samples from the Western Canada Basin (Hitchon and Filby, 1983), permitting the three suites of petroleum to be distinguished. Approximate V/Co ratios are 0.1 for group 1, 2.7 for group 2, and 0.15 for group 3. It would thus appear reasonable to assume that Co is present within these oils as a porphyrin complex and is competing with V and Ni for substitution in the porphyrin structure. This is consistent with the stability index and stability class for Co (Tables III and IV). Similar plots have been made for Cu, Zn, and Mo, for oils from Venezuela (Figs. 7 and 8), but these show no clear correlation with V, apart from a general tendency for those oils richest in V to also be richest in other metals. Although a Cu porphyrin has been isolated and, on theoretical grounds, the behavior of Cu should resemble that of Co, there is still no firm evidence that Cu porphyrins form a significant component of the petroleum porphyrin fraction. A proportion of petroleum-hosted Cu, and other metals, may be present in a form which involves

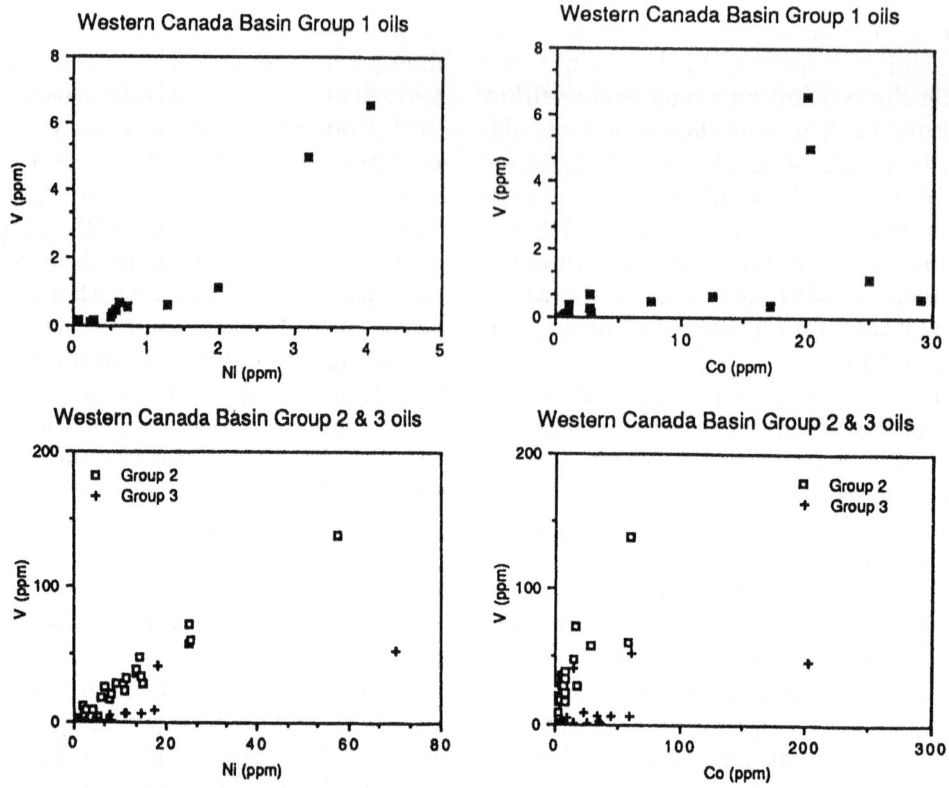

Figure 6. Ni and Co contents vs. V content for petroleum from the Western Canada Basin. [Data from Hitchon and Filby (1983).]

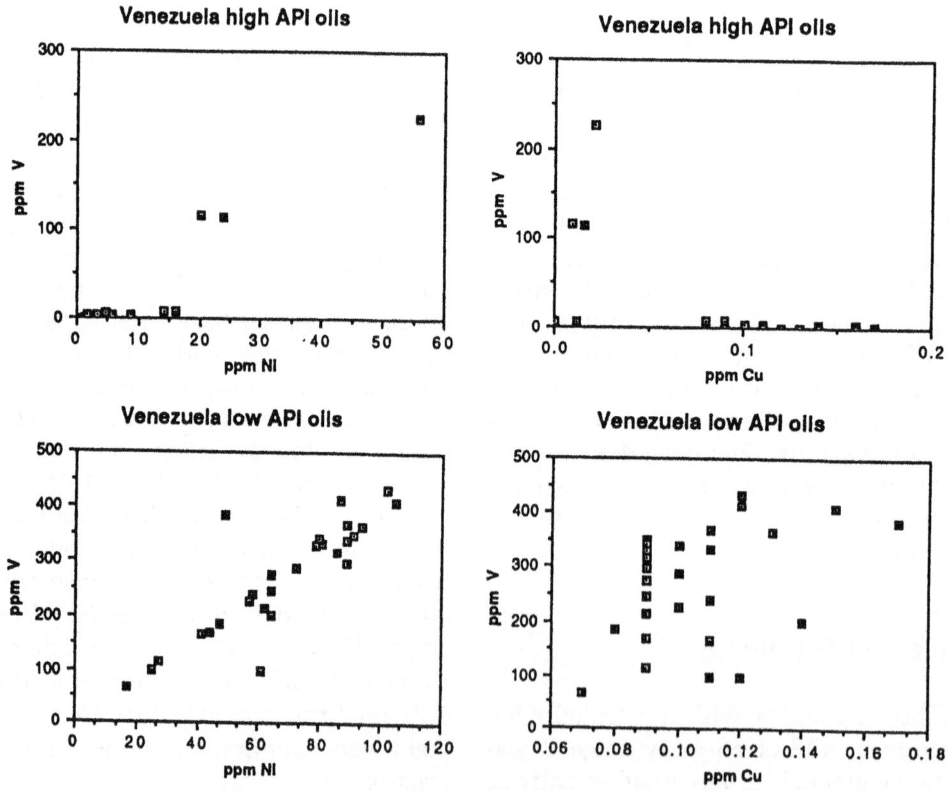

Figure 7. Ni and Cu contents vs. V content for petroleum from Venezuela. [Data from Simoza *et al.* (1985).]

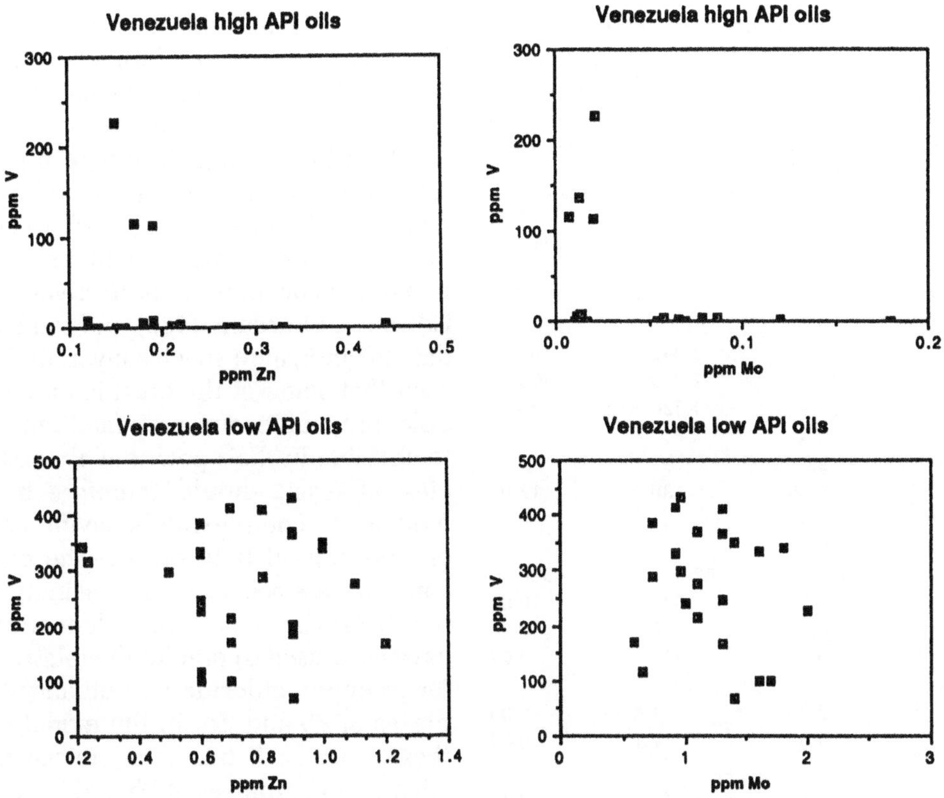

Figure 8. Zn and Mo contents vs. V content for petroleum from Venezuela. [Data from Simoza *et al.* (1985).]

complexing with other functional groups principally within high-molecular-weight components of petroleum. Direct bonding between metals and carbon is unlikely to be significant under geological conditions (Yen, 1975b).

3. Metals in Oil Field and Formation Waters

Oil field brines have been proposed as possible ore-forming fluids for Mississippi Valley-type mineral deposits by a number of authors (e.g., Billings *et al.*, 1969b; Carpenter *et al.*, 1974; Manning, 1986; Sverjensky, 1984, 1987). They are characteristically low-temperature brines, with a wide range in composition. Their development, as oil field brines, is considered to have taken place in the presence of organic matter, alongside the development of petroleum. There are, however, very few published accounts of coordinated studies of petroleum and brines from single geological systems. Without these, it is impossible to evaluate fully the extent to which petroleum and brine interact and exchange inorganic and organic constituents.

Oil field brines vary greatly in their composition, as shown by studies carried out by Billings *et al.* (1969a), Rittenhouse *et al.* (1969), Hitchon *et al.* (1972), Kharaka *et al.* (1987), and other authors. A wide range of cations is usually reported, including the base metals Cu^{2+}, Pb^{2+}, and Zn^{2+}, but anions are usually restricted to SO_4^{2-}, Cl^-, and other halides. Studies of the presence of organic anions have tended to be in isolation from inorganic components, and very few studies present data for coordinated brine and oil samples.

Overall, the contents of the base metals Cu, Pb, and Zn tend to be greatest, as would be expected, in those brines which have the greatest total dissolved solids. Examples of highly saline brines are given in Table V, which illustrates compositional variation as a function of dissolved species. The base metal contents of more diluted brines are correspondingly reduced (e.g., Kharaka *et al.*, 1987). Taken individually, the greatest Cu content for brines is 0.49 mg liter^{-1}, reported for the Western Canada Basin; Zn reaches 27.5 mg liter^{-1} (Billings *et al.*, 1969a; Hitchon *et al.*, 1971). Pb and Zn are reported to reach 111 and 575 mg liter^{-1}, respectively, in oil field brines from central Mississippi (Carpenter *et al.*, 1974), for which Cu data

TABLE V. Inorganic Components of Oil Field Brines

	Mississippi Salt Dome Basin[a]		Western Canada Basin[b]	
	mg liter^{-1}	Molarity $\times 10^3$	mg liter^{-1}	Molarity $\times 10^3$
Li	35	5.04	46	6.63
Na	61,700	2684	53,100	2310
K	990	25.4	4,400	112.8
Rb	4.6	0.05	5.2	0.06
Cs	1.5	0.01		
Mg	3,050	125.5	3,990	164.2
Ca	48,600	1215	32,000	800
Sr	1,920	21.9	1,320	15.1
Ba	60	0.44		
NH_3	34	2.0		
B	59	5.46	130	12.0
SiO_2	27.8	0.46		
F	1.5	0.08		
Cl	198,000	5580	162,000	4560
Br	2,020	25.3	848	10.6
I	17	0.13	26	0.20
SO_4	64	0.67	256	2.67
H_2S	<0.1	—		
Fe	465	8.3	1.8	0.03
Mn	212	3.9	1.1	0.02
Pb	70.2	0.34		
Zn	243	3.72	27.5	0.42
Al	(367)[c]	(13.6)[c]		
Cd	0.99	0.009		
Cu	(21)[c]	(0.33)[c]	(200)[c]	(3.15)[c]
TDS[d]	316,000		258,400	
pH	5.08		5.98	

[a]Data from Kharaka *et al.* (1987); sample no. 84-MS-11.
[b]Data from Hitchon *et al.* (1971); sample no. 3 (Upper Devonian).
[c]Values in parentheses have been multiplied by 1000.
[d]TDS, Total dissolved solids.

are not available. Typical values for brines thought to be appropriate for ore-forming solutions are of the order of 0.1 mg Cu liter^{-1}, 2.0 mg Pb liter^{-1} and 1.3 mg Zn liter^{-1} (Sverjensky, 1984).

In order to assess the role of organic matter in the transport of ore metals within the aqueous phase, it is necessary to consider the possible significance of the organic anions which might be present. Reported maximum concentrations are given in Table VI (MacGowan and Surdam, 1988), and can be compared with data for inorganic species given in Table V. It can be seen that some of the organic anions (notably ethanoic, propanoic, and propanedioic) are second only to chloride in their potential abundance. These species, at least, should therefore be considered in models for complexing behavior within postulated ore-forming oil field brines. The pertinent stability constants are becoming increasingly available (Gardner, 1974; Giordano, 1985; Hennet *et al.*, 1988), and have been used to predict the relative proportions of, for example, chloride and ethanoate (acetate) complexes of Pb and Zn. In the model systems used by these authors, it has been consistently shown that ethanoate complexes of Pb and Zn predominate over those with chloride for compositions rich in the ethanoate anion. However, very few studies provide compositions for brines which include data for both the inorganic and organic constituents. In the study by Kharaka *et al.* (1987) of Mississippi Salt Dome Basin fluids, ethanoate contents only exceptionally exceeded 100 mg liter^{-1} (0.002*M*), emphasizing the hypothetical nature of chemical modeling metal-

TABLE VI. Organic Acid Anions in Oil Field Brines[a]

Organic acid	Formula	Anion concentration: ppm	Anion concentration: Molality $\times 10^3$
Methanoic (formic)	CHOOH	62.6	1.36
Ethanoic (acetic)	CH_3COOH	10,000	166.5
Propanoic	C_2H_5COOH	4,400	59.4
Butanoic	C_3H_7COOH	44.0	0.499
Pentanoic	C_4H_9COOH	32.0	0.313
Ethanedioic (oxalic)	HOOC—COOH	494	5.49
Propanedioic (malonic)	HOOC—CH_2—COOH	2,540	24.4
Butanedioic (succinic)	HOOC—$(CH_2)_2$—COOH	63	0.529
Pentanedioic	HOOC—$(CH_2)_3$—COOH	36	0.272
Butenedioic (maleic)	(Z)-HOOC—CH═CH—COOH	26	0.224
Others		<10	

[a]Data from MacGowan and Surdam (1988).

organic complex species distribution involving maximum observed abundances of organic anions. Indeed, this point was acknowledged by Hennet *et al.* (1988), who suggested that significant complexing of Pb with ethanoate only takes place when the concentrations of this anion are in excess of 0.05*m*.

4. Discussion

The role of organic matter in processes of ore genesis can be subdivided into a number of contributing phenomena. First, organic materials may be relatively rich in metals of interest and act as sources of ore-forming components in their own right. Second, organic liquids may act as transporting agents for ore components, enabling them to be transported to sites where concentration may take place. Third, the evolution of organic matter as a diagenetic process may result in the modification of aqueous pore fluids and affect their efficiency as transporting agents. Finally, organic materials may influence the precipitation of ore minerals from aqueous fluids. It is assumed that ore minerals precipitate from an aqueous fluid phase and are not precipitated directly from hydrocarbon fluids. For the purposes of this chapter, discussion will concentrate on the role of organic matter in ore transport processes.

4.1. Base Metal Transport by Petroleum Liquids

It is quite clear from the compositional data presented in Tables I and II that petroleum liquids are potentially capable of transporting Cu and Zn in amounts which exceed or equal those thought to be appropriate for aqueous ore fluids. This is consistent with the possible presence in petroleum of at least Cu as porphyrin or analogous tetrapyrrole complexes. The Pb contents of petroleum are, however, low compared with those of aqueous fluids, reflecting the difficulty with which Pb is incorporated into the porphyrin structure. These observations can be combined with an empirical assessment of the relative ability of petroleum and aqueous fluids to transport base metals. In Table VII, the contents of metals within petroleum and brines are given, together with the partition coefficients calculated from these data to indicate the preference of each metal for either phase (Fig. 9; Manning, 1986).

Partition coefficients of this type have little geological significance as they are not derived from concentration data for corresponding oil and water samples, which may be impossible to obtain for technical reasons. However, they do illustrate the way in which those metals which are well known to be hosted by porphyrins (V, Ni, and Co) are clearly partitioned in favor of the petroleum component. On the other hand, Pb clearly prefers the aqueous phase, as does Zn. Cu apparently partitions in favor of the organic phase. For the latter three metals, the choice of concentration taken for the calculation of the partition coefficient will clearly have a major effect on the result, as the amounts of the metals present in the two phases are similar in magnitude. On the other hand, the V, Ni, and Co contents of petroleum and aqueous fluids differ by several orders of magnitude, and so the partition coefficients for these elements are unlikely to vary qualitatively as a function of their variable abundances in natural systems.

TABLE VII. Partition Coefficients of Selected Metals Calculated from Maximum Values for Their Concentrations within Petroleum and Brines[a]

Metal	Concentration (mg liter^{-1}) Petroleum	Brine	Partition coefficient
Cu	5.5	0.49	11.2
Pb	2.1	111	0.019
Zn	21	575	0.037
V	1580	<0.001	>10^6
Ni	155	<0.390	>400
Co	2.0	0.029	69.0
Fe	254	490	0.52

[a]Petroleum data from Table I; brine data from Carpenter *et al.* (1971), Hitchon *et al.* (1971), and Manning (1986).

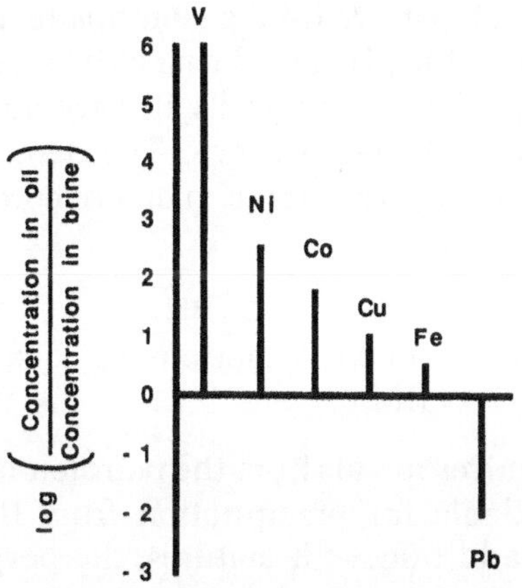

Figure 9. Schematic representation of partitioning of metals between petroleum and brine, deduced from maximum reported values.

It is more appropriate to determine partition coefficients in this way for corresponding petroleum and brine samples. Unfortunately, very few studies report the elements of interest for samples of petroleum and fluid obtained by separation of a composite wellhead sample, and in any case the assumption that the oil and water sampled in this way represent the assemblage present under reservoir conditions is unlikely to be strictly valid. However, data for corresponding samples are available for the Western Canada Basin (Hitchon *et al.*, 1971; Hitchon and Filby, 1983) and the southern German Molasse Basin (Ellrich *et al.*, 1985). The elements considered include Zn, Co, and Fe, as summarized in Table VIII.

In Table VIII, the data for the southern German Molasse basin illustrate the clear preference of Co for the petroleum phase, and this preference is shared by both Fe and Zn. The Western Canada Basin data emphasize, on the other hand, the potentially variable behavior of Zn. Overall, the consideration of partition coefficients is limited to demonstrating the clear preference for the petroleum phase shown by those metals which can be confidently associated with the porphyrins, namely, V, Ni, and Co. It is disappointing that there are insufficient published data to compare the partitioning behavior of Cu, as this element might be expected to show consistent behavior, similar to that of Co, in view of its similar stability within porphyrin complexes (see Table IV and associated discussion).

The partitioning of porphyrin-hosted cations between petroleum and a coexisting aqueous phase can be regarded as being analogous to the partitioning of metals between minerals and melt in view of the structural control on metal incorporation into the tetrapyrrole complex. An additional, more specific, control involves the role of cation exchange and the influence of pH. Within a composite petroleum–brine system, the pH of the brine will be controlled by mineralogical reactions and will vary depending on the lithology of the host rock. The demetallation of porphyrins may then take place according to the reaction

$$\underset{\text{petroleum}}{MP} + \underset{\text{aqueous fluid}}{2H^+} = \underset{\text{petroleum}}{PH_2} + \underset{\text{aqueous fluid}}{M^{2+}} \quad (3)$$

which removes metals from the petroleum and makes them available for precipitation from the aqueous phase. In addition, as it matures, the porphyrin will become metallated to an increasing extent by those metals which are stable to acids, such as V and Ni. This results in an overall exchange reaction of the type

TABLE VIII. Trace Element Partition Coefficients for Oil and Water Samples

Sample	Fe(oil/aq)	Co(oil/aq)	Zn(oil/aq)
Southern German Molasse Basin[a]			
Fr 12		8.33	48.4
Ill 1	15.0	6.06	147
M 13		61.5	4.62
A 14	30.0	26.2	81.5
A 25		23.3	19.6
Ai 5		70.0	14.3
Ass 2		108	17.7
Hof 1	66.7	457	21.6
Mu 4	50.0	920	163
Mu 5	40.0	307	87.5
Am 5	130	214	75.0
Am 9	95.0	175	292
Am 11	35.0	142	17.8
Am 25	70.0	221	33.6
Ve 6	13.3	50.0	69.6
Western Canada Basin[b]			
2/81			1.63
9/52			0.409
24/66			15.9
14/11			0.606
15/29			1.67
16/36			6.89
35/58			4.37
26/34			6.01
38/30			0.091
46/93			1.49
48/24			12.1
49/20			5.79
59/8			0.185
56/72			0.134
65/B			7.48
63/56			84.6
71/16			0.034
70/19			0.150
76/9			1.38
79/31			0.082
82/54			6.27

[a]Data from Ellrich *et al.* (1985).
[b]Sample numbers refer to corresponding oil and water analyses from Hitchon and Filby (1983) and Hitchon *et al.* (1971), respectively.

$$\underset{\text{petroleum}}{M'P} + \underset{\text{fluid}}{M} = \underset{\text{petroleum}}{MP} + \underset{\text{fluid}}{M'} \quad (4)$$

where M has a higher stability index than M′. In this way, the metal component of porphyrins will become dominated by those metals in stability classes I and II, and other metals (if originally present) will be progressively removed to enter the aqueous fluid phase.

In view of the observed stability of metalloporphyrins to acids, the partitioning behavior of metals between petroleum and brines can be predicted qualitatively. Those cations in stability classes I and II

TABLE IX. Stability Classes of Cations of Metallogenetic Interest

Stability class	Cations
I	VO^{2+}, Pd^{2+}, Pt^{2+}, Au^{3+}
II	Ni^{2+}, Cu^{2+}, Co^{2+}, Ag^{2+}
III	Fe^{2+}, Zn^{2+}
IV	Mn^{2+}, Mg^{2+}, Cd^{2+}
V	Pb^{2+}, Hg^{2+}, Ca^{2+}, Sr^{2+}, Ba^{2+}

(Table IX) are unlikely to be removed from porphyrin complexes under conditions of pH achieved under natural conditions; these elements include V, Ni, Cu, and Co. Conversely, those in stability classes V and IV are unlikely [in view of the known removal of Mg (class IV) from chlorophyll during diagenesis] to be present as a porphyrin complex under natural geological conditions. Elements in stability class III are likely to show behavior which is variable, depending on local geological environmental circumstances. This behavior is demonstrated by the available compositional data, and derived partition coefficients, for petroleum and brines.

4.2. The Role of Organic Anions in Brines

The second way in which organic matter can influence the transport of ore-forming components is by contributing to the transport potential of aqueous pore fluids, by modifying their composition. Oil field brines are known to be rich in dissolved organic anions, especially ethanoate (Carothers and Kharaka, 1978; Fisher, 1987; MacGowan and Surdam, 1988), and several authors have speculated on their potential as complexing agents for the transport of base metals (Gardner, 1974; Giordano, 1985; Hennet *et al.*, 1988). It is clear that the contribution of ethanoate to the transport potential of a brine may be significant compared to that of chloride and other inorganic anions, in view of the known potentially high abundance of ethanoate in oil field brines. However, it is not clear whether other organic anions (which may form complex species with favorable stability constants) are present in significant quantities in natural brines.

One of the problems associated with assessing the role of organic anions in aqueous transport processes is that there may be species generated during catagenesis whose stability is low over periods of time appropriate for geological processes. Complex organic anions and polar species are obtained as a result of experimental treatment of organic-rich materials under laboratory conditions (e.g., hydrous pyrolysis), which cannot claim to represent equilibrium geological conditions. These include principally mono- and dicarboxylic aliphatic acids with subsidiary aromatic acids (e.g., Cooles *et al.*, 1987; Eglinton *et al.*, 1987; Kawamura *et al.*, 1986; Kawamura and Kaplan, 1987). However, studies of natural oil field waters have shown that abundances decrease as aliphatic chain length increases beyond that of ethanoic acid and that the abundance of total acids decreases as a function of temperature. Furthermore, the decarboxylation of dissolved ethanoic acid is considered to take place as follows (Carothers and Kharaka, 1978):

$$CH_3COO^- + H_2O = CH_4 + HCO_3^- \qquad (5)$$

and it follows that higher molecular weight organic acid anions will similarly decompose according to

$$RCOO^- + H_2O = RH + HCO_3^- \qquad (6)$$

where R represents an aliphatic chain. Other more complex reactions may take place which involve the formation of ethanoic acid as an intermediate product.

In terms of transport processes which may take place over periods of geologic time, there is no doubt that the role of ethanoic acid anions should be considered in low-temperature solution modeling. However, the role of other organic anions is more difficult to assess, as it may involve short time periods prior to their decomposition. Two lines of indirect evidence suggest that the role of organic anions may be greater than that suggested by the presence of ethanoic acid anions alone in present-day brines. First, there is abundant evidence that kerogen loses oxygen-bearing functional groups as maturity increases, as shown by a van Krevelen type diagram (Fig. 10); at least a proportion of these may contribute to a water-soluble population of geologically unstable organic anions. Second, studies of the phenomenon of secondary porosity in sandstones (Surdam *et al.*, 1984; Surdam and MacGowan, 1987) provide abundant evidence for extensive dissolution of aluminosilicate minerals, for which organic anions are believed to be responsible. Neither the phenomenon of functional group removal from kerogen nor that of secondary porosity generation can be fully described at present, in the absence of pertinent experimental data, but their potential for ore transport processes must be regarded as significant.

The observed extent of secondary porosity generation within sandstones is potentially of considerable importance for studies of ore transport and deposition.

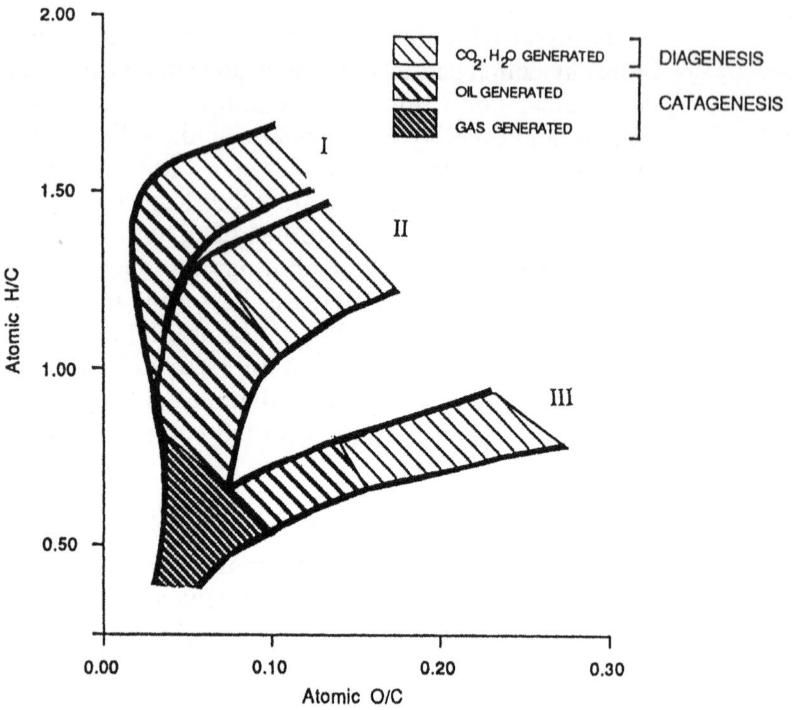

Figure 10. Van Krevelen diagram illustrating the initial reduction in oxygen-bearing functional groups during diagenesis for all three kerogen types. [Modified from Tissot and Welte (1984).]

It is apparent that significant quantities of aluminosilicate minerals may be dissolved as a consequence of normal diagenetic processes, thereby releasing trace elements as well as major components to the aqueous phase. Of particular interest is the potential of feldspars to liberate Pb and Ba, which are present as trace substituents for K (Manning *et al.*, 1991). This process is of particular importance when the known stability of Pb complexes with ethanoic acid anions is considered.

5. Behavior of Metals in a Composite Petroleum–Brine System

Using the above discussion, it is possible to describe the behavior of ore metals during transport within complex organic–aqueous fluid systems. Theoretical considerations suggest strongly that of the three base metals Cu, Pb, and Zn, Cu in particular may be transported by petroleum liquids in preference to an aqueous phase. Pb is unlikely to be present as a significant component of petroleum but clearly prefers the aqueous phase, where it may be complexed in a significant proportion by organic acid anions, especially ethanoic. Zn is less predictable but is unlikely to show a strong affinity for petroleum-hosted porphyrins. Other ore metals may be transported by petroleum, such as the precious metals and Mo, but insufficient abundance data are available to assess the extent to which this process might occur.

The complicated processes involved in the development of ore-forming aqueous fluids, and their precipitated ore minerals, are summarized in a simplified form as an adapted geochemical cycle in Fig. 11. This figure can be used to explain the preferential assemblages of minerals shown by certain low-temperature sediment-hosted ore deposits. For example, the predominance of Pb and Zn minerals in Mississippi Valley-type deposits may be due to the absence of Cu from the aqueous ore fluids, which may, in turn, be due to the absence of conditions appropriate for the removal of any Cu present within coexisting petroleum. The removal of Cu from petroleum-hosted tetrapyrrole complexes requires a relatively low pH, and this may not be achieved within the calcium carbonate-dominated host rocks typical of Mississippi Valley-type deposits. Alternatively, in an immature mineralizing system, exchange of porphyrin-hosted Cu for cations whose porphyrin complexes have higher stability indices may not be complete. On the other hand, low pH values are more likely to be achieved in sandstone-hosted ore fluids, with feldspar/mica/clay reactions controlling pH, and may account for some of the sandstone-hosted Cu deposits. However, it is most unlikely that natural pH values will ever be reduced to values appropriate for the removal of V from petroleum, and so alternative mechanisms are

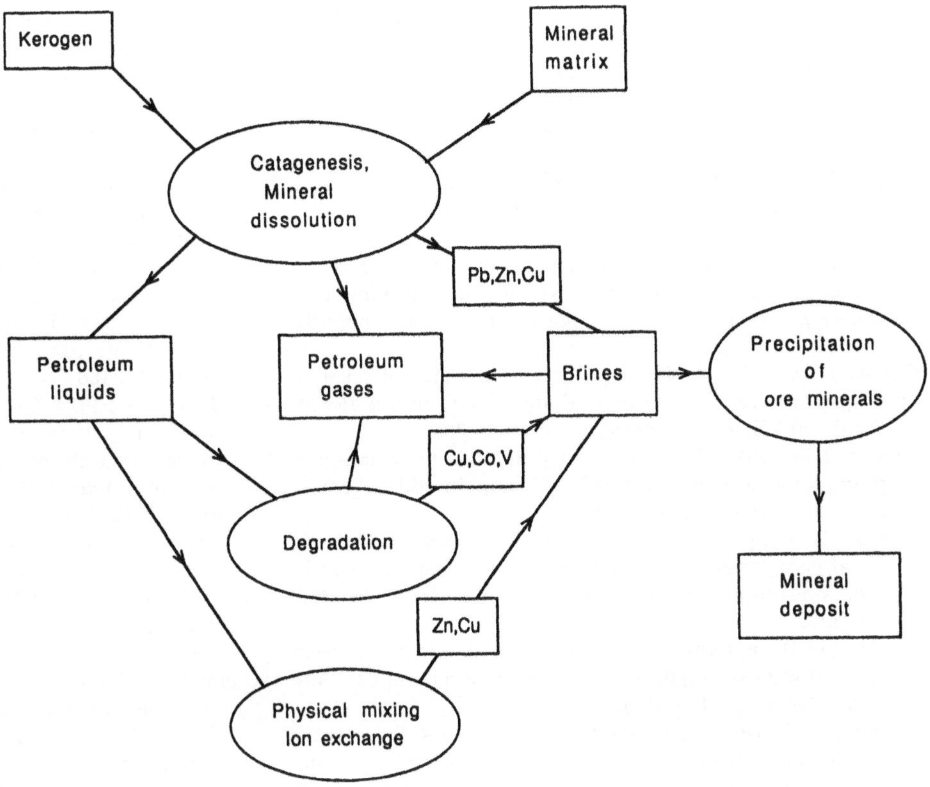

Figure 11. Simplified geochemical cycle to show the participation of organic matter in pathways leading to the formation of ore minerals.

required when considering the origin of those deposits which characteristically contain V or other metals whose porphyrin complexes have high stability indices, such as Co. It is most likely that destruction of the porphyrin complex, by biological or thermal degradation, is necessary to release these elements for mineralization.

In Fig. 11, there are essentially three pathways which contribute to the ore-forming potential of formation waters and brines. First, dissolution during catagenesis and diagenesis will enable trace components of minerals (such as Pb from feldspars) to enter solution. Organic anions, whether long-lived or transient, may augment this process, which includes the generation of secondary porosity. Second, petroleum liquids may lose base metals originally incorporated during their formation as they mature, as a consequence of exchange with coexisting aqueous pore fluids. They are unlikely to contribute Pb or, to a lesser extent, Zn but may act as a source of Cu. Third, biological or thermal processes of degradation of petroleum will ultimately involve the destruction of the porphyrin complex and the release of elements such as V, Ni, and Co as well as Cu and other constituents. These three pathways will contribute in varying proportions to the formation of ore fluids responsible for the precipitation of low-temperature base metal mineral deposits.

ACKNOWLEDGMENTS. D.A.C.M. acknowledges support in the form of a Nuffield Science Research Fellowship.

References

Ali, M. F ., Bukhari, A., and Saleem, M., 1983, Trace metals in crude oils from Saudi Arabia, *Ind. Eng. Chem. Prod. Res. Dev.* **22**:691–694.

Al-Shahristani, H., and Al-Atyia, M. J., 1972, Vertical migration in Iraqi oil fields: Evidence based on vanadium and nickel concentrations, *Geochim. Cosmochim. Acta* **36**:929–938.

Baker, E. W., and Louda, J. W., 1984, Highly dealkylated copper and nickel etioporphyrins in marine sediments, *Org. Geochem.* **6**: 183–192.

Barton, P. B., 1967, Possible role of organic matter in the precipitation of Mississippi Valley ores, in: *Genesis of Stratiform Lead–Zinc–Barite–Fluorite Deposits in Carbonate Rocks* (J. S. Brown, ed.), Econ. Geol. Monograph No. 3, pp. 371–378.

Barwise, A. J. G., and Whitehead, E. V., 1980, Separation and structure of petroporphyrins, in: *Advances in Organic Geochemistry 1979* (A. G. Douglas and J. R. Maxwell, eds.), Physics and Chemistry of the Earth, Pergamon Press, Oxford, Vol. 12, pp. 181–192.

Billings, G. K., Hitchon, B., and Shaw, D. R., 1969a, Geochemistry and origin of formation waters in the western Canada sedimentary basin, 2. Alkali metals, *Chem. Geol.* **4**:211–223.

Billings, G. K., Kesler, S. E., and Jackson, S. A., 1969b, Relation of zinc-rich formation waters, northern Alberta, to the Pine Point ore deposit, *Econ. Geol.* **64**:385–391.

Bjørlykke, A., and Sangster, D. F., 1981, An overview of sandstone lead deposits and their relation to red-bed copper and carbonate-hosted lead–zinc deposits, *Econ. Geol.*, 75th anniversary volume, pp. 179–213.

Bratzel, M. P., and Chakrabarti, C. L., 1972, Determination of lead in petroleum and petroleum products by atomic absorption spectrometry with a carbon rod atomizer, *Anal. Chim. Acta* **61**: 25–32.

Buchler, J. W., 1975, Static coordination chemistry of metalloporphyrins, in: *Porphyrins and Metalloporphyrins* (K. M. Smith, ed.), Elsevier, Amsterdam, pp. 157–231.

Buchler, J. W., 1978, Synthesis and properties of metalloporphyrins, in: *The Porphyrins*, Vol. 1 (D. Dolphin, ed.), Academic Press, New York, pp. 389–483.

Buchler, J. W., Puppe, L., Rohbock, K., and Schneehage, H. H., 1973, Metal complexes of octaethylporphin: Preparation, axial ligand substitution and reduction, *Ann. N.Y. Acad. Sci.* **206**: 116–137.

Carothers, W. W., and Kharaka, Y. K., 1978, Aliphatic acid anions in oil-field waters—implications for origin of natural gas, *Am. Assoc. Petrol. Geol. Bull.* **62**:2441–2453.

Carpenter, A. B., Trout, M. L., and Pickett, E. E., 1974, Preliminary report on the origin and chemical evolution of lead- and zinc-rich oil field brines in central Mississippi, *Econ. Geol.* **69**: 1191–1206.

Colombo, U., and Sironi, G., 1961, Geochemical analysis of Italian oils and asphalts, *Geochim. Cosmochim. Acta* **25**:24–51.

Colombo, U., Sironi, G., Fasolo, G. B., and Malvano, R., 1964, Systematic neutron activation technique for the determination of trace metals in petroleum, *Anal. Chem.* **36**:802–807.

Cooles, G. P., Mackenzie, A. S., and Parkes, R. J., 1987, Non-hydrocarbons of significance in petroleum exploration: Volatile fatty acids and non-hydrocarbon gases, *Mineral. Mag.* **51**: 483–493.

Deroo, G., Powell, T. G., Tissot, B., and McCrossan, R. G., 1977, The origin and migration of petroleum in the western Canada sedimentary basin, Alberta, *Bull. Geol. Surv. Can.* **262**.

Disnar, J. R., 1981, Etude expérimentale de la fixation de métaux par un matériau sédimentaire actuel d'origine algaire II. Fixation "in vitro" de UO_2^{2+}, Cu^{2+}, Ni^{2+}, Zn^{2+}, Pb^{2+}, Co^{2+}, Mn^{2+}, ainsi que VO_3^-, MoO_4^{2-} et GeO_3^{2-}, *Geochim. Cosmochim. Acta* **45**: 363–379.

Dolphin, D. (ed.), 1978, *The Porphyrins*, Academic Press, New York.

Eglinton, T., Curtis, C. D., and Rowland, S. J., 1987, Generation of water-soluble organic acids from kerogen during hydrous pyrolysis: Implications for porosity development, *Mineral. Mag.* **51**:495–503.

Eldib, I. A., Dunning, H. N., and Bolen, R. J., 1960, Nature of colloidal materials in petroleum, *J. Chem. Eng. Data* **5**:550–553.

Ellrich, J., Hirner, A., and Stark, H., 1985, Distribution of trace elements in crude oils from southern Germany, *Chem. Geol.* **48**:313–323.

Falk, J. E., 1964, *Porphyrins and Metalloporphyrins*, Elsevier, Amsterdam.

Filby, R. A., and Shah, K. R., 1975, Neutron activation methods for trace elements in crude oils, in *The Role of Trace Metals in Petroleum* (T. F. Yen, ed.), Ann Arbor Science Publishers, Ann Arbor, Michigan, pp. 89–110.

Filby, R. H., 1975, The nature of metals in petroleum, in: *The Role of Trace Metals in Petroleum* (T. F. Yen, ed.), Ann Arbor Science Publishers, Ann Arbor, Michigan, pp. 31–58.

Filby, R. H., and Van Berkel, G. J., 1987, Geochemistry of metal complexes in petroleum, source rocks and coal: An overview, in: *Metal Complexes in Fossil Fuels, Geochemistry, Characterization and Processing* (R. H. Filby and J. F. Branthaver, eds.), ACS Symposium Series No. 344, American Chemical Society, Washington, D.C., pp. 2–39.

Fisher, J. B., 1987, Distribution and occurrence of aliphatic acid anions in deep subsurface waters, *Geochim. Cosmochim. Acta* **51**:2459–2468.

Gardner, C. R., 1974, Organic versus inorganic trace metal complexes in sulfidic marine waters—some speculative calculations based on available stability constants, *Geochim. Cosmochim. Acta* **38**:1297–1302.

Giordano, T. H., 1985, A preliminary evaluation of organic ligands and metal–organic complexing in Mississippi Valley-type ore solutions, *Econ. Geol.* **80**:96–106.

Gize, A., and Barnes, H. L., 1987, The organic geochemistry of two Mississippi Valley-type lead–zinc deposits, *Econ. Geol.* **82**: 457–470.

Gleim, W. K. T., Gatsis, J. G., and Perry, C. J., 1975, The occurrence of molybdenum in petroleum, *The Role of Trace Metals in Petroleum* (T. F. Yen, ed.), Ann Arbor Science Publishers, Ann Arbor, Michigan, pp. 161–166.

Hausen, D. M., and Park, W. C., 1986, Observations on the association of gold mineralisation with organic matter in Carlin-type ores, in: *Organics and Ore Deposits* (W. E. Dean, ed.), Proceedings of the Denver Regional Exploration Geologists Society Symposium, pp. 119–136.

Hennet, R. J.-C., Crerar, D. A., and Schwartz, J., 1988, Organic complexes in hydrothermal systems, *Econ. Geol.* **83**:742–764.

Hitchon, B., and Filby, R. H., 1983, Geochemical Studies—1. Trace Elements in Alberta Crude Oils, Alberta Research Council Open-File Report 1983–02.

Hitchon, B., Billings, G. K., and Klovan, J. E., 1971, Geochemistry and origin of formation waters in the western Canada sedimentary basin—III. Factors controlling chemical composition, *Geochim. Cosmochim. Acta* **35**:567–598.

Hitchon, B., Filby, R. H., and Shah, K. R., 1975, Geochemistry of trace elements in crude oils, Alberta, Canada, in: *The Role of Trace Metals in Petroleum* (T. F. Yen, ed.), Ann Arbor Science Publishers, Ann Arbor, Michigan, pp. 111–122.

Hodgson, G. W., 1954, Vanadium, nickel, and iron trace metals in crude oils of western Canada, *Am. Assoc. Petrol. Geol. Bull.* **38**:2537–2554.

Hodgson, G. W., and Baker, B. L., 1957, Vanadium, nickel, and porphyrins in thermal geochemistry of petroleum, *Am. Assoc. Petrol. Geol. Bull.* **41**:2413–2426.

Hodgson, G. W., and Baker, B. L., 1959, Geochemical aspects of petroleum migration in Pembina, Redwater, Joffre, and Lloydminster oil fields of Alberta and Saskatchewan, Canada, *Am. Assoc. Petrol. Geol. Bull.* **43**:311–328.

Jones, P., 1975, Trace Elements and Other Elements in Crude Oil—a Literature Review, Report of British Petroleum Research Centre, Sunbury.

Kawamura, K., and Kaplan, I. R., 1987, Dicarboxylic acids generated by thermal alteration of kerogen and humic acids, *Geochim. Cosmochim. Acta* **51**:3201–3207.

Kawamura, K., Tannenbaum, E., Huizinga, B. J., and Kaplan, I. R., 1986, Long-chain carboxylic acids in pyrolysates of Green River kerogen, *Org. Geochem.* **10**:1059–1065.

Kharaka, Y. K., Maest, A. S., Carothers, W. W., Law, L. M., Lamothe, P. J., and Fries, T. L., 1987, Geochemistry of metal-rich brines from central Mississippi Salt Dome basin, U.S.A., *Appl. Geochem.* **2**:543–561.

Leventhal, J. S., 1986, Roles of organic matter in ore deposits, in: *Organics and Ore Deposits* (W. E. Dean, ed.), Proceedings of the Denver Regional Exploration Geologists Society Symposium, pp. 7–20.

MacGowan, D. B., and Surdam, R. C., 1988, Difunctional carboxylic acid anions in oilfield waters, *Org. Geochem.* **12**:245–259.

Macqueen, R. W., and Powell, T. G., 1983, Organic geochemistry of the Pine Point lead–zinc orefield and region, Northwest Territories, Canada, *Econ. Geol.* **78**:1–25.

Manning, D. A. C., 1986, Assessment of the role of organic matter in ore transport processes in low-temperature base-metal systems, *Trans. Inst. Min. Metall.* **95**:B195–B200.

Manning, D. A. C., Rae, E. I. C., and Small, J. S., 1991, An exploratory study of acetate decomposition and dissolution of quartz and Pb-rich potassium feldspar at 150°C, 50 MPa (500 bars), *Mineral. Mag.* **55**:183–195.

Marikos, M. A., Laudon, R. C., and Leventhal, J. S., 1986, Solid insoluble bitumen in the Magmont West orebody, southeast Missouri, *Econ. Geol.* **81**:1983–1988.

Martell, A. E., and Smith, R. M., 1977, *Critical Stability Constants, Volume 3: Other Organic Ligands*, Plenum Press, New York.

Nakashima, S., Disnar, J. R., Perruchot, A., and Trichet, J., 1984, Experimental study of mechanisms of fixation and reduction of uranium by sedimentary organic matter under diagenetic or hydrothermal conditions, *Geochim. Cosmochim. Acta* **48**: 2321–2329.

Palmer, S. E., and Baker, E. W., 1978, Copper porphyrins in deep-sea sediments: A possible indicator of oxidized terrestrial organic matter, *Science* **201**:49–51.

Phillips, J. N., 1960, The ionization and coordination behaviour of porphyrins, *Rev. Pure Appl. Chem.* **10**:35–60.

Quirke, J. M. E., 1987, Rationalization for the predominance of nickel and vanadium porphyrins in the geosphere, in: *Metal Complexes in Fossil Fuels. Geochemistry, Characterization and Processing* (R. H. Filby and J. F. Branthaver, eds.), ACS Symposium Series No. 344, American Chemical Society, Washington, D.C., pp. 74–83.

Rickard, D. T., Willden, M. Y., Marinder, N.-E., and Donnelly, T. H., 1979, Studies on the genesis of the Laisvall sandstone lead–zinc deposit, Sweden, *Econ. Geol.* **74**:1255–1285.

Rittenhouse, G., Fulton, R. B., Grabowski, R. J., and Bernard, J. L., 1969, Minor elements in oil-field waters, *Chem. Geol.* **4**:189–209.

Saxby, J. D., 1976, The significance of organic matter in ore genesis, in: *Handbook of Strata-Bound and Stratiform Ore Deposits*, Vol. 2 (K. H. Wolf, ed.), Elsevier, Amsterdam, pp. 111–134.

Shah, K. R., Filby, R. H., and Haller, W. A., 1970a, Determination of trace elements in petroleum by neutron activation analysis I. Determination of Na, S, Cl, K, Ca, V, Mn, Cu, Ga and Br, *J. Radioanal. Chem.* **6**:185–192.

Shah, K. R., Filby, R. H., and Haller, W. A., 1970b, Determination of trace elements in petroleum by neutron activation analysis II. Determination of Sc, Cr, Fe, Co, Ni, Zn, As, Se, Sb, Eu, Au, Hg and U, *J. Radioanal. Chem.* **6**:413–422.

Shannon, R. D., and Prewitt, C. T., 1969, Effective ionic radii in oxides and fluorides, *Acta Crystallogr. Sect. B* **25**:925–946.

Simoza, R., Carrión de Rosa-Brussin, N., Torres, R. J., and Eyzaguirre, C. L., 1985, Distributión de V, Ni, Zn, Mo, Ca, K, Fe y Cu en los crudos de la región sur de la cuenca oriental de Venezuela, *VI Congreso Geológico Venezolano, Memoria* [Proceedings of the 6th Venezuelan Geological Congress], Vol. 3, Sociedad Venezolana de Geologos, Caracas, pp. 2088–2116.

Smith, I. C., Ferguson, T. L., and Carson, B. L., 1975, Metals in new and used petroleum products and by-products—quantities and consequences, in: *The Role of Trace Metals in Petroleum* (T. F. Yen, ed.), Ann Arbor Science Publishers, Ann Arbor, Michigan, pp. 123–148.

Smith, K. M. (ed.), 1975, *Porphyrins and Metalloporphyrins*, Elsevier, Amsterdam.

Smith, R. M., and Martell, A. E., 1978, *Critical Stability Constants, Volume 4: Inorganic Complexes*, Plenum Press, New York.

Speight, J. G., 1980, *The Chemistry and Technology of Petroleum*, Marcel Dekker, New York.

Surdam, R. C., and MacGowan, D. B., 1987, Oilfield waters and sandstone diagenesis, *Appl. Geochem.* **2**:613–619.

Surdam, R. C., Boese, S. W., and Crossey, L. J., 1984, The chemistry of secondary porosity, in: *Clastic Diagenesis* (D. A. McDonald and R. C. Surdam, eds.), American Association of Petroleum Geologists Memoir 37, pp. 127–149.

Sverjensky, D. A., 1984, Oil field brines as ore-forming solutions, *Econ. Geol.* **79**:23–37.

Sverjensky, D. A., 1987, The role of migrating oil field brines in the formation of sediment-hosted Cu-rich deposits, *Econ. Geol.* **82**:1130–1141.

Tissot, B. P., and Welte, D. H., 1984, *Petroleum Formation and Occurrence*, Springer-Verlag, Berlin.

Treibs, A., 1934, The occurrence of chlorophyll derivatives in an oil shale of the upper Triassic, *Annalen Chem.* **517**:103–114.

Vine, J. D., 1962, Geology of uranium in coaly carbonaceous rocks, Professional Paper of the United States Geological Survey 356-D, pp. 113–170.

Yen, T. F. (ed.), 1975a, *The Role of Trace Metals in Petroleum*, Ann Arbor Science Publishers, Ann Arbor, Michigan.

Yen, T. F., 1975b, Chemical aspects of metals in native petroleum, in: *The Role of Trace Metals in Petroleum* (T. F. Yen, ed.), Ann Arbor Science Publishers, Ann Arbor, Michigan, pp. 1–30.

Chapter 26

Aspects of the Organic Geochemistry and Petrology of Metalliferous Ores

ANDREW P. GIZE and DAVID A. C. MANNING

1. Introduction

1.1. Organic–Ore Association

Reduced organic matter occurs in intimate association with many types of metalliferous ore deposits. The Kupferschiefer, an extensive Middle Permian Cu–Pb–Zn–S horizon in Europe, is a bituminous shale. Bitumens and liquid petroleum frequently occur in Mississippi Valley-type Pb–Zn–S deposits, or in the geologically similar Illinois Fluorite District. In the South African Witswatersrand, gold and uraninite (UO_2) occur within the "Carbon Leader." Pyrobitumens are present in hydrothermal vein systems, such as the Ag–Ni–Co deposits at Kongsberg, Norway. "Hydrocarbon fronts" are considered by some exploration geologists to be an essential feature of Carlin-type, "no see-um," disseminated gold deposits.

It should not be surprising that organic matter is associated with metalliferous ore deposits. Carbon is a relatively common element in the Earth's crust, with an average concentration of approximately 200 ppm (Mason, 1966). It has been estimated that 82% of the total carbon in sedimentary rocks is present as carbonates, the remaining 18% being organic carbon (Schidlowski *et al.*, 1974). This quantity of organic carbon is of approximately the same order of magnitude as the zinc content (70 ppm), an order of magnitude greater than that of lead and uranium (13 ppm and 1.8 ppm, respectively), and three orders of magnitude greater than that of silver (0.07 ppm). Given such a relatively large reservoir of organic carbon, it would be more surprising to find geochemical processes which eliminated organic carbon during ore deposition.

Simplistically, the steps leading to the formation of an orebody can be considered as (1) source, (2) migration, and (3) precipitation. This sequence of steps is similar to that involved in hydrocarbon migration. Considering the source, black shales are not only potential petroleum and gas sources, but can also be enriched in trace metals. Migration of both hydrocarbons and metals is a fluid-flow process, with the potential for the same permeable units to be used by both hydrocarbons and ore solutions. There is the

ANDREW P. GIZE and DAVID A. C. MANNING • Department of Geology, University of Manchester, Manchester M13 9PL, England.

Organic Geochemistry, edited by Michael H. Engel and Stephen A. Macko. Plenum Press, New York, 1993.

potential for metals and sulfur to be transported together during hydrocarbon migration, as metalloporphyrins, organosulfur compounds, and other species. Organic chelates or complexes can increase the solubility of metals in the mineralizing solutions. At the site of metal precipitation, organic matter may provide reduced sulfur or act as a reductant.

The broad similarities between organic processes in the Earth's crust and those leading to ore formation leads to the question of how the ore–organic association can be used. Simplistically, there are two approaches. First, the organic geochemistry and petrology can provide information about ore genesis. Alternatively, the ores can provide information on organic processes. This chapter is not intended to be a comprehensive review of all reported ore–organic associations. After a preliminary review of certain aspects of ore genesis, the objective is to highlight certain associations and to discuss the information available.

1.2. Ore Deposit Classifications

Ore deposits can be classified in several ways. One approach is to group ores which were formed under broadly similar geological conditions into "types" (e.g., Stanton, 1972). For example, the Mississippi Valley-type deposits are characterized by Pb–Zn–S ores, dolomitic host rocks, and low fluid inclusion homogenization temperatures (≤200°C) among other features. There will be differences between the deposits but shared gross characteristics. The recognition of these "types" indicates that ore deposits are not formed by random geochemical processes, but rather that ore genesis is controlled by specific events. In other words, there is a predictability in ore genesis, and organic studies in one ore deposit can, with care, be correlated with those in other deposits of the same "type."

There are two other approaches to ore deposit classification. The first disregards to some extent the geological processes and emphasizes the temperatures of the ore-forming processes. The temperature classification provides useful guides to the probable thermal maturity of the organic matter (Park and MacDiarmid, 1970). The ore deposits are classified as telethermal (low temperatures), epithermal (50–200°C), mesothermal (200–300°C), and hypothermal (300–500°C). The last classification emphasizes the timing of ore precipitation relative to host rock formation. In a syngenetic deposit, the ore is precipitated at approximately the same time as the host rock. Alternatively, in an epigenetic deposit, the ore is emplaced after host rock lithification. If the ore is epigenetic, then the potential of multiple generations of organic matter arises. For example, the situation can exist where the host rocks contain inherent hydrocarbons, and then the ore solution introduces an epigenetic organic generation.

There are several advantages to working within ore deposits, among which are ready access and the ability to sample on different lateral and vertical scales. Perhaps most important of all is the wealth of inorganic and physical geochemical information available to provide the background to the organic processes. In addition to fluid inclusion data, ore mineral assemblages can be used to constrain parameters such as fO_2 (oxygen fugacity) and sulfur speciation. The dynamic inorganic processes represented by ores lead to an important interpretation of the organic matter present: were the hydrocarbons chemically active or passive? If the hydrocarbons were chemically active, then organic geochemistry and petrology have an as yet unfulfilled contribution to make to ore genesis studies. If the ore–organic association is passive, then ore deposits provide exciting opportunities to examine organic geochemical processes under a variety of well-characterized, sometimes extreme, conditions.

2. Low-Temperature Deposits

2.1. Introduction

One of the geological environments of prime interest in organic studies is that of an evolving sedimentary basin. This geological situation is also considered to be responsible for the generation of Mississippi Valley-type deposits and the geologically similar Illinois Fluorite District.

The mineralogy of these ores is simple (Ohle, 1959). They consist predominantly of galena (PbS), sphalerite (ZnS), barite ($BaSO_4$), and fluorite (CaF_2). Subordinate pyrite (FeS_2) and marcasite (FeS_{2-x}) occur, with trace chalcopyrite ($CuFeS_2$), millerite (NiS), and siegenite $(NiCo)_3S_4$. The gangue (noneconomic minerals) are dolomite, calcite, and quartz. The host rock is typically dolomite, although some deposits are sandstone hosted (e.g., Laisvall, Sweden; Rickard *et al.*, 1979). Igneous rocks are usually absent. The ore tends to occur in structurally positive features around paleobasins, such as reefs. At the time of host rock formation, paleolatitudes have been suggested to be within 30° of the equator (Dunsmore, 1975; Dunsmore and Shearman, 1977). Precipitation temperatures range from approximately 220°C to 50°C for late-stage fluids (Upper Mississippi District; McLimans *et al.*, 1980).

Further reviews of these deposits include those of Ohle (1959, 1980) and Stanton (1972).

It is generally accepted that the ores are epigenetic, although syngenetic formation has been argued for some deposits (Amstutz, 1964). The geological and geochemical characteristics have led to the concept that the mineralizing fluids are formed during sedimentary basin evolution (Jackson and Beales, 1967; Hanor, 1979). The potential exists in Mississippi Valley-type ores that not only have migrating hydrocarbons been incorporated passively into the mineralizing fluids, but that there has been active ore–organic interaction. Genetic arguments invoking active organic interaction have included organometallic complexing (Giordano and Barnes, 1981), reduction (Barton, 1967), sulfate reduction (Dunsmore and Shearman, 1977), and sulfur being contributed by organic sulfur (Skinner, 1967).

There have been several organic geochemical studies of Mississippi Valley-type deposits. In the Derbyshire Pb–Zn–F District, the associated oil seepages were shown to have been derived from local shales and limestones (Pering and Ponnamperuma, 1969; Pering, 1973) and to have been extensively biodegraded. Rickard *et al.* (1975, 1979) isolated hydrocarbons from the Laisvall deposit and supported oil field brines as mineralizing fluids. The Pine Point (Northwest Territories, Canada) deposits have locally high quantities of bitumens. The bitumen source has been shown to be a type I kerogen, migration has been local, and extensive biodegradation is reported (Powell and Macqueen, 1980; Macqueen and Powell, 1983). On the basis of bitumen elemental compositions and sulfur isotopes, Powell and Macqueen (1984) argued in favor of active chemical involvement of the organic matter in sulfate reduction.

2.2. Role of Sulfur Bacteria

Considerable attention has been focused on the potential roles of sulfur bacteria, not only in Mississippi Valley-type ores, but also in other relatively low temperature deposits. Reviews of the roles of sulfur bacteria include those of Trudinger (1982) and Trudinger and Williams (1982). The latter authors have indicated that the best evidence for a microbial role in ore genesis is to be found in sediment-hosted copper deposits (White Pine, United States; Kupferschiefer, Europe). Such evidence includes relatively low sulfide concentrations, variable $\delta^{34}S$ compositions, framboidal pyrite textures, and positive correlations between sulfide and organic carbon contents.

However, when other studies on the Kupferschiefer are considered which invoke low-temperature mineralizing fluids ascending into the bituminous shales (Jowett, 1986; Speczik *et al.*, 1986; Speczik and Puttmann, 1989), it becomes evident that multiple synchronous processes may have contributed to mineralization. Similar multiple processes have occurred at McArthur River (Australia), where isotopic evidence exists for at least two sulfur sources (Smith and Croxford, 1973) or underlying processes (Williams and Rye, 1974). The sulfur isotopic compositions of the pyrites support bacterial sulfate reduction in a closed basin, with a systematic increase in $\delta^{34}S$ of the pyrites from the bottom to the top of the deposit. However, the sphalerite and galena sulfur isotopic ratios do not follow a systematic trend. Trudinger (1982) concluded that the evidence for microbial sulfate reduction indicates, at best, an indirect contribution to the formation of major ore deposits.

2.3. Biodegradation

In addition to studies largely dependent upon sulfur isotopes, evidence based on microbial degradation of hydrocarbons has been used to advocate the presence of sulfur bacteria. The formation of sulfides in sulfate-reducing bacterial cultures has been reported by Rubentschik (1928), Baars (1930), and Selwyn and Postgate (1959). In these studies acetate was used as the substrate, but unfortunately the bacterial cultures used were not single strains. More recently, Pfennig and Widdel (1982) reviewed the ability of single strains to reduce sulfate using various organic acids as substrates. Of note are the three strains *Desulfococcus multivorans*, *Desulfosarcina variablis*, and *Desulfonema magnum*, which have the ability to oxidize not only acetate and other aliphatic acids, but also benzoate and phenyl-substituted acids. The extensive biodegradation of bitumens associated with barite and sulfides at the Saint Privat deposits has been postulated to be indicative of the presence of sulfate-reducing bacteria (Connan and Orgeval, 1973; Connan, 1977).

Extensive biodegradation may be a feature of the postmineralization history of the deposit. Deposits with reported extensive biodegradation (Derbyshire, Saint Privat, Pine Point) are at, or near, the surface and are from open-pit operations. In contrast, deposits which are subsurface and mined underground (the Upper Mississippi Valley District and Gays River in Nova Scotia) show limited or no biodegradation (Hatch *et al.*, 1985; Gize and Barnes, 1987). At Gays River, limited biodegradation at the top of the deposit was related to a Cretaceous exhumation.

2.4. Thermal Anomalies

The possibility of local thermal alteration of the host rock organic matter by heated mineralizing fluids has been explored with the objective of using thermal anomalies as prospecting guides. In Derbyshire, thermal alteration was postulated by Mueller (1954) but later discounted by Pering (1973). At Pine Point, thermal alteration has been described by Powell and Macqueen (1980) and Macqueen and Powell (1983). By contrast, detailed attempts to use biomarker isomerization to detect thermal alteration approaching, and within, orebodies in the Upper Mississippi Valley District have been unsuccessful (Gize, 1984; Hatch *et al.*, 1985; Gize and Barnes, 1987). Methane anomalies, stemming from work in the North Pennine Orefield, have been applied as an exploration tool (Carter and Cazalet, 1984) with mixed success. The rationale for the anomalies has been thermal alteration by the hydrothermal fluids or generation of methane as a result of sulfate reduction (Dunsmore, 1975). Thermal degradation of organosulfur compounds, with the sulfur being incorporated in the ores (Skinner, 1967), seems quantitatively minor on the basis of mass-balance evidence (Gize and Barnes, 1987).

In the Upper Mississippi Valley District, although no evidence for thermal alteration has been observed, marked and contrasting compositional changes have been reported. Rock-Eval analyses indicated a decrease in the S_1 and S_2 yields as an ore body was approached, together with a decrease in the relative amounts of n-C_{25} and lower, saturated hydrocarbons (Hatch *et al.*, 1985). In the same district, and in Gays River, marked increases in the n-C_{20} and lower saturates were observed as a result of mineralization (Gize and Barnes, 1987; Fig. 1). In conclusion, the presence of thermal anomalies with Mississippi Valley-type deposits is currently equivocal.

One possible interpretation of the data is that the mineralizing fluids have acted as natural analogs of "hydrous pyrolysis" experiments. This interpretation opens up the possibility of δD shifts in the indigenous hydrocarbons if H is taken from the mineralizing fluids.

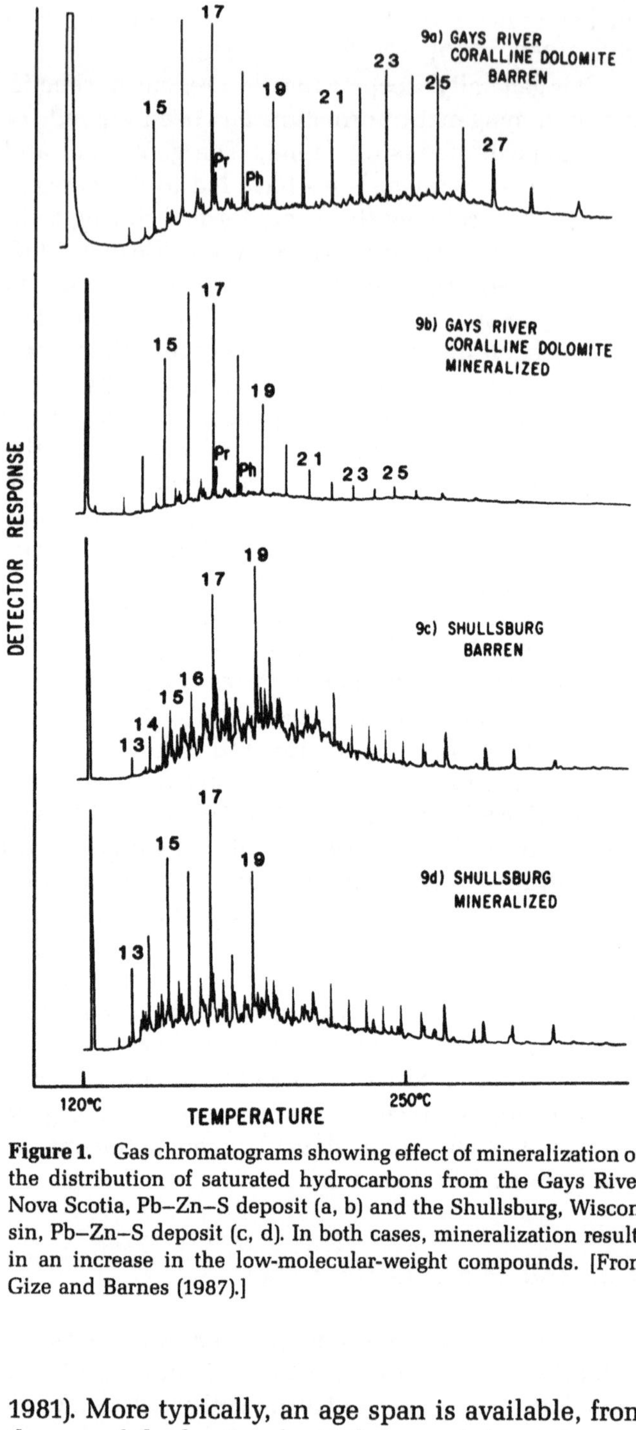

Figure 1. Gas chromatograms showing effect of mineralization on the distribution of saturated hydrocarbons from the Gays River, Nova Scotia, Pb–Zn–S deposit (a, b) and the Shullsburg, Wisconsin, Pb–Zn–S deposit (c, d). In both cases, mineralization results in an increase in the low-molecular-weight compounds. [From Gize and Barnes (1987).]

2.5. Timing of Mineralization

One vexing problem in Mississippi Valley-type genesis is estimating the time of mineralization. In exceptional circumstances, an age can be obtained, for example, by use of remnant magnetism (Krs and Stovickova, 1966; Beales *et al.*, 1974; Wu and Beales, 1981). More typically, an age span is available, from the age of the host rocks to the age of the sediments containing eroded ore.

Hydrocarbon maturation modeling can potentially be used to constrain the mineralization age. The prerequisites are that the hydrocarbons can be related temporally to all or part of the mineral paragenesis and that the hydrocarbons are epigenetic (not derived locally). If ore deposition and secondary hydrocarbon migration are synchronous, then ore precipitation can be related to the onset of catagenesis, providing an estimate of time.

The Illinois Fluorite District may offer such an opportunity. Liquid hydrocarbons occur as seeps in the mines, in vugs, or as petroleum inclusions in the ore (Fig. 2a). Solid bitumens, reported as the commonest form of organic matter (Brecke, 1962), appear as froth, droplets, or smears on the ore minerals. Petroleum inclusions occur throughout the fluorite paragenesis, being commonest toward the end of the paragenesis (Fig. 2a). If the petroleum source rock can be identified, then the fluorite deposition can be linked to the time of catagenesis. Another example is the Elmwood Mine (Tennessee), where rare petroleum inclusions occur on the margins of some sphalerites (Fig. 2b). In the latter case, the two bitumen reflectances ($R_0' = 0.2\%$ and $R_0'' = 0.87\%$) suggest hydrothermal activity after cessation of sphalerite precipitation.

3. High-Temperature Deposits: Optical Anisotropy

The optical properties of bitumens associated with ores are very variable. In Mississippi Valley-type ores, optically isotropic bitumens are most common. Rarely, the bitumens are optically anisotropic, as in the Illinois Fluorite District. In higher temperature deposits (≥200°C), optical anisotropy becomes relatively common as the organic matter is coked. Local intrusions can also produce optical anisotropy, as well as an increase in reflectance.

The development of optical anisotropy reflects some degree of molecular ordering. The end product, in appropriate material, will be graphite, which can be considered in terms of a thermodynamic phase. Higher temperature deposits therefore offer opportunities to pursue organic matter from systems which are interpreted by kinetics toward thermodynamic treatment as a phase in ore genesis.

The molecular route by which carbonaceous material becomes graphitized has been extensively studied by the coking industry. The term "graphitizing carbons" was used by Franklin (1951) for organic matter which, upon laboratory heating, produced ordered graphite-like layers. "Nongraphitizing carbons," upon heating, also contained graphite-like layers in parallel groups, but not orientated as in crystalline graphite. The nongraphitizing carbons tended to be hydrogen-poor or oxygen-rich relative to graphitizing carbons. Using reflected light microscopy, Brooks and Taylor (1965) observed graphitizing carbons to develop optically anisotropic spheres in an isotropic medium at temperatures of 400–500°C. It was concluded that the anisotropic spheres consisted of condensed, planar, polynuclear aromatics, with a preferred orientation toward the centers. The texture was considered to have affinity to the intermediary state between a true crystalline lattice and a true isotropic liquid referred to as "liquid crystals" or a "mesophase" [for a review, see Brown and Shaw (1957)].

Mesophases have been classified by Friedel (1922) as occurring in one of two structures. In the "smectic" structure, the ordering consists of the molecules being arranged in layers, but with their long axes normal to the layer planes. Alternatively, in the "nematic" structures, the only constraint on ordering is that the molecules have a parallel, or near-parallel, arrangement. A third structure, "cholesteric," is exhibited largely by cholesterol derivatives. Two different processes can produce a mesophase. A thermotropic mesophase is produced by heating, whereas a lyotropic mesophase is produced by the action of solvents (e.g., detergent in water).

The anisotropic spheres observed by Brooks and Taylor (1965) were considered as a thermotropic mesophase with a nematic structure. This interpretation has continued (Marsh, 1973; White, 1976; Davis *et al.*, 1983). Graphitizing carbons from geological sources have included coals (Taylor, 1961; Kisch, 1966; Cook *et al.*, 1972; Jones and Creaney, 1977) and gilsonite (Khavari-Khorasani *et al.*, 1978).

Optical anisotropy in bitumens associated with ores has been reported from the Kongsberg Ag–Co–Ni vein deposit in Norway (Lietz, 1939; Neumann, 1944; Dons, 1956) and from the Carlin disseminated Au deposit in Nevada (Hausen and Kerr, 1968). Interpretations of the textures as mesophases have been presented (Gize and Rimmer, 1983; Gize, 1985).

Experimentally, the formation of a thermotropic mesophase from natural organic materials starts at about 350°C (White, 1976) when anisotropic spheres form in the isotropic mixture. The spheres sink under gravity and coalesce into larger bodies. The surrounding isotropic liquid generates low-molecular-weight hydrocarbons and hydrogen, eventually forming a semi-coke. Experimentally, the conversion from an isotropic medium to a mesophase is completed within 10–24 h at temperatures of 350–400°C (Honda *et al.*, 1970). Once the bitumen has been converted to a mesophase, increasing the experimental temperature to 600°C only produces shrinkage cracks associated with increased density (Brooks and Taylor, 1966, 1968; White, 1974). In Fig. 2c–e, the sequence of sphere growth, coalescence, and development of long-range order are shown in bitumens from the Nanisivik Mississippi Valley-type deposit (Baffin Island), where coking has occurred in response to diabase dike in-

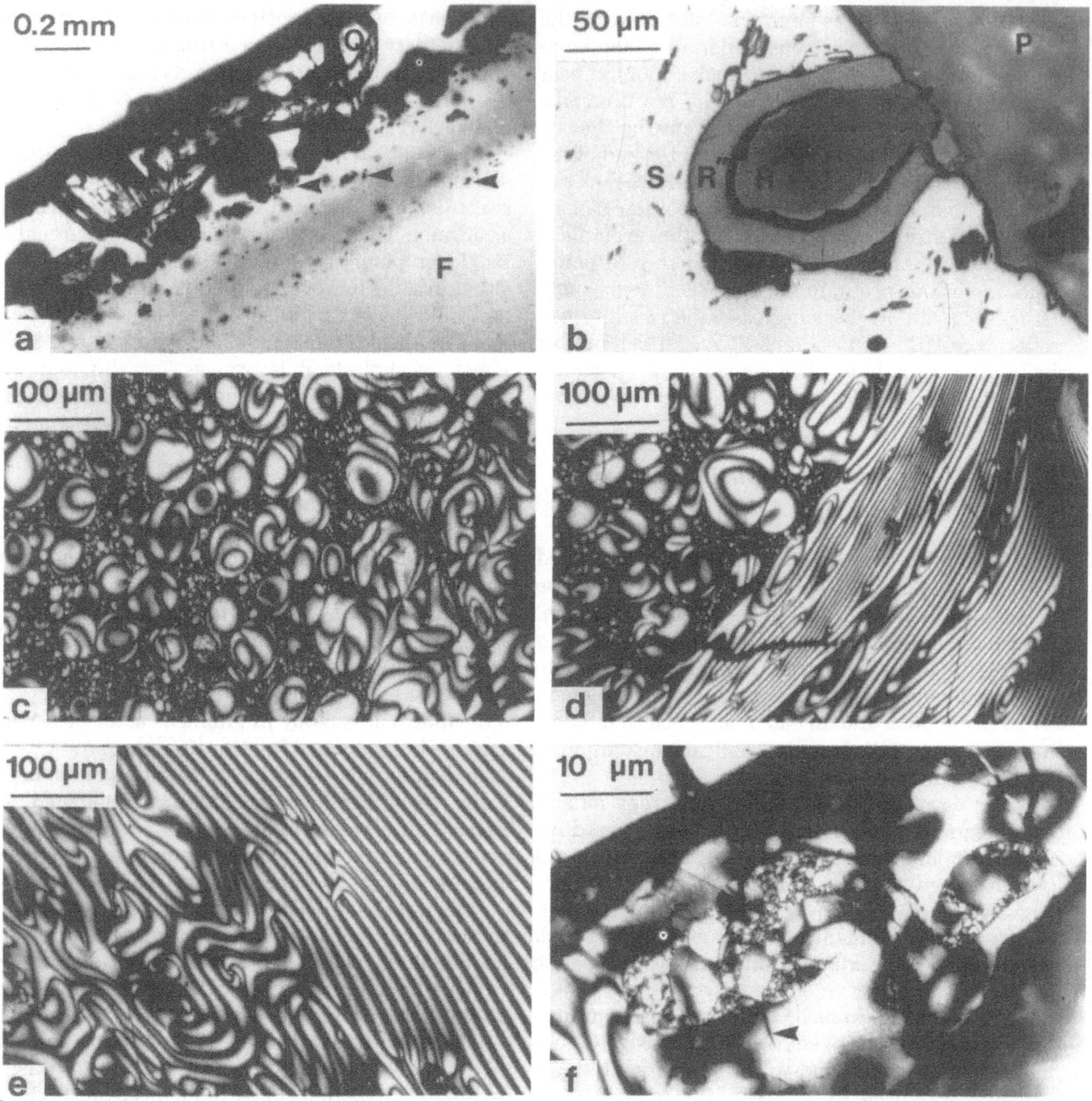

Figure 2. (a) Cross section of fluorite (F) with petroleum inclusions (arrows) in distinct planes. Note growth of quartz (Q) from the latest plane of petroleum inclusions. Transmitted light, polars uncrossed, in air. (b) Bitumen inclusion in sphalerite (S) from Elmwood, Tennessee. Note two different reflectances ($R_0'' = 0.87\%$, $R_0' = 0.2\%$), suggesting two generations of hydrocarbons. P, Mounting plastic. Reflected light, polars uncrossed, in air. (c) Growth of anisotropic spheres from an isotropic bitumen, from Nanisivik, Baffin Island. The spheres, constituting a mesophase, coalesce to form a larger anisotropic body to the right. Reflected light, crossed polars, in air. (d) Further coalescence of anisotropic spheres to form an anisotropic body with long-range optical ordering. The material is bitumen from Nanisivik, Baffin Island, Reflected light, crossed polars, in air. (e) Detail of long-range optical ordering. The material is bitumen from Nanisivik, Baffin Island. Reflected light, crossed polars, in air. (f) Residual anisotropic spheres enclosed within a larger anisotropic body. Features interpreted as shrinkage cracks are indicated by arrows. The material is "coalblende" from Kongsberg, Norway. Reflected light, crossed polars, in oil.

strusions. Scanning electron photomicrographs of similar material are shown in Figs. 3a and 3b. In Fig. 3a, a layer of anisotropic carbon is shown, with characteristic structureless textures. The earlier carbonaceous material (B′) is a "nongraphitizing" carbon, with the development of pores contrasting it from the "graphitizing" carbon. The lamellar material formed by coalescence of anisotropic spheres is shown in Fig. 3b, where the lamellae are clearly visible. Similar material, with shrinkage cracks, from the Kongsberg Ag–Ni–Co vein deposit is shown in Figs. 2f and 3f.

Experimentally, a nongraphitizing carbon produces optical domains on the micron scale (Khavari-Khorasani *et al.*, 1978), and these do not develop long-range order. Examples of such material are shown in Fig. 3c, from the Nanisivik deposit, and in Fig. 4d–f, from the Carlin disseminated Au deposit in Nevada.

The development of a thermotropic mesophase experimentally starts at about 350°C. Geologically, the mesophase textures form at significantly lower temperatures, based on fluid inclusion homogenization temperatures. For example at Kongsberg, 300°C is suggested to have been the highest temperature of ore precipitation (J. Johansen and T. V. Segalstad, personal communication). In the case of the Illinois Fluorite District, where some anisotropic bitumens occur, maximum temperatures were approximately 150°C (Richardson and Pinckney, 1984). Presumably, the longer periods of time during mineralization permit lower temperatures to be active.

Superimposed on the optical anisotropy are other effects. Marginal to the coked bitumens, there may be rims of increased reflectance, as found at Kongsberg (Fig. 3d). Such rims are suggestive of oxidation processes (cf. Stach *et al.*, 1982, p. 165, Fig. 63b).

Another type of rim texture is one in which the optical anisotropy becomes ordered perpendicular to the bitumen–aqueous interface. The scale of the ordering may extend from 1 to 15 μm (Fig. 3e). The cause of the ordering is unknown, but it may result from a heat flow gradient. The major use of this texture is to provide a ready means of separating different hydrocarbon generations in cokes. Lastly, the optical ordering provides ready evidence of stress, with deformation styles ranging from plastic to brittle (Fig. 3f).

4. Hydrocarbon Phases

The composition of the organic matter discussed in the preceding sections has ranged from complex hydrocarbon mixtures in Mississippi Valley-type deposits to optically ordered cokes in higher temperature deposits. This compositional simplification leads to the possibility that the organic matter can start to be considered in terms of thermodynamics, rather than kinetics, with the development of organic phases.

Organic materials can occur as phases in ore deposits, with the carbon being reduced, elemental, or oxidized. The presence of elemental carbon (graphite, diamond) and carbonate will not be dealt with in this chapter. The two types of organic phases that will be briefly discussed are polycyclic aromatic hydrocarbons and oxalates.

4.1. Polycyclic Aromatic Hydrocarbons

Polycyclic aromatic hydrocarbon minerals tend to be associated with hot-spring mercury deposits. Initially, distillates were reported during anoxic roasting of the ores (Blumer, 1975, and references therein), but it was unclear as to whether the hydrocarbons (such as fluoranthrene) were from the ores or pyrolysates. "Curtisite" was described by Wright and Allen (1930) from hot-spring surface vents at Skaggs Springs, California. The molecular composition of curtisite was deduced by Geissman *et al.* (1967) to consist predominantly of picene and chrysene with lower concentrations of C_2–C_3 homologs, and a structural similarity to steroids and triterpenoids was noted. A yellow prismatic mineral has been known from the New Idria mercury deposits in California since the early 1900s. Initially, the mineral was identified as the antimony oxide "valentinite," but Murdoch and Geissman (1967) deduced the mineral to be coronene and renamed the phase as "pendletonite."

The classic study of polycyclic aromatic hydrocarbons is that of Blumer (1975), involving column and gas chromatography, UV spectroscopy, and probe distillation mass spectrometry. Recently, Wise *et al.* (1986) have characterized curtisite, idrialite, and pendletonite by high-pressure liquid chromatography (HPLC), gas chromatography/mass spectrometry (GC/MS), and nuclear magnetic resonance (NMR), and their findings support the original interpretations of Blumer (1975). Pendletonite has been shown to be virtually pure coronene (Blumer, 1975; Wise *et al.*, 1986), whereas a number of polycyclic aromatic hydrocarbons are present in idrialite, and curtisite contain many of these same components, but in considerably different amounts. The major constituents of curtisite have been identified as picene, dibenzo[*a*,*h*]-fluorene, 11*H*-[2,1-a]phenanthrene, benzo[b]phen

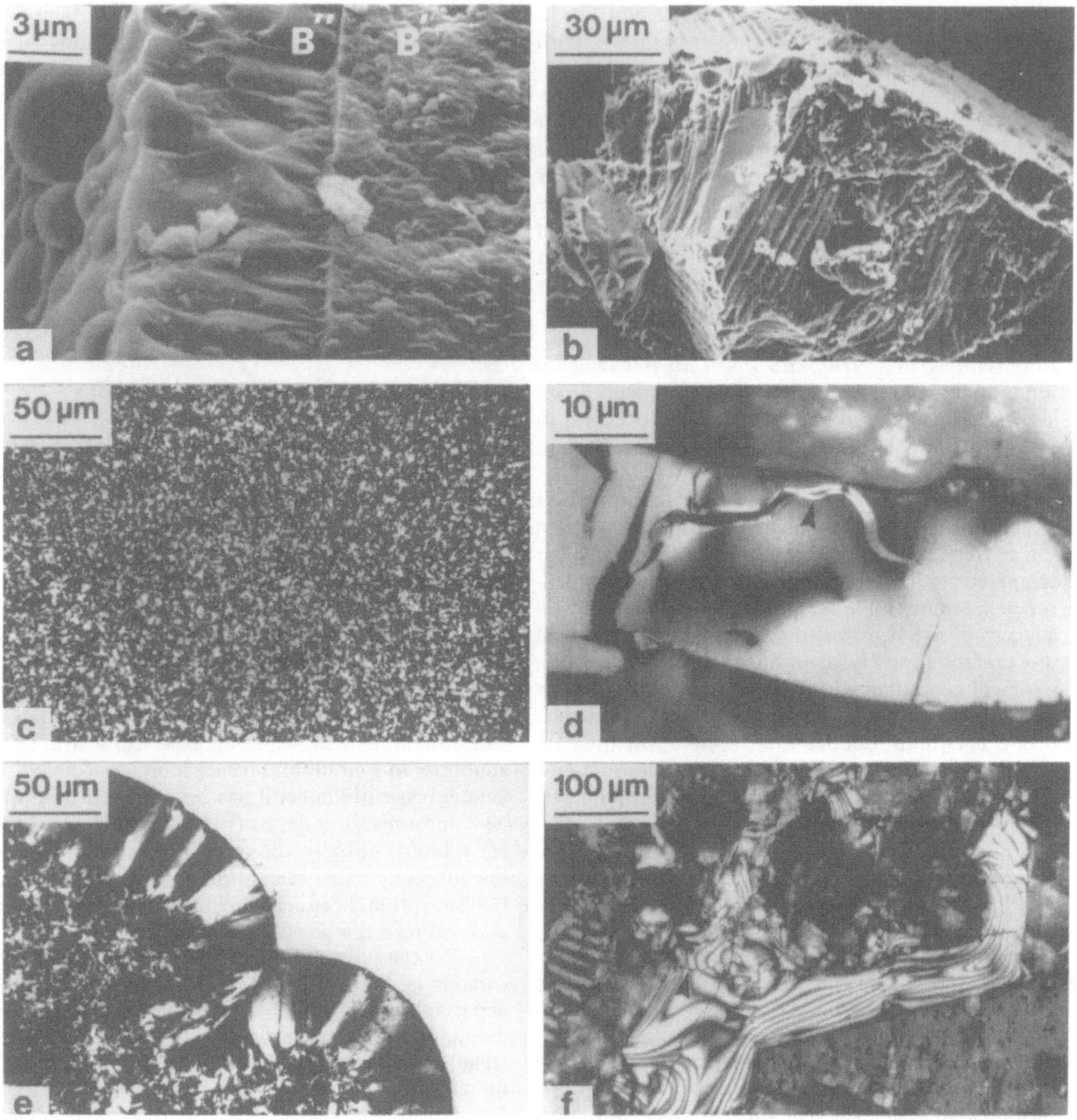

Figure 3. (a) Scanning electron photomicrograph of mesophase, from Nanisivik, Baffin Island. The material to the left (B″) is the last material in the hydrocarbon paragenesis and corresponds to the material shown in Fig. 2d. The paragenetically earlier material (B′) shows optical domains on the micron scale and corresponds to the material shown in Fig. 2c. Note the porosity present in B′, but not in B″. (b) Scanning electron photomicrograph of lamellar textures, corresponding to Fig. 2e. The long-range ordering visible indicates a "graphitizing" carbon. The material is from Nanisivik, Baffin Island. (c) "Nongraphitizing" carbon, from Kongsberg, Norway. The optical domains are small and will not develop the long-range order shown in (b) and in Figs. 2d and 2e. Reflected light, crossed polars, in oil. (d) Postulated oxidation along margins of a fracture, resulting in a reflectance increase (indicated by arrow). The material is "coalblende" from Kongsberg, Norway. Reflected light, partially crossed polars, in oil. (e) Development of optical ordering perpendicular to bitumen–mineralizing brine contact. This is a diagnostic feature, permitting resolution of multiple generations of hydrocarbons in some deposits. A similar features can be seen in Fig. 2a, B″. The material is "coalblende" from Kongsberg, Norway. Reflected light, crossed polars, in air. (f) Variable deformation styles in mesophases, including ductile and brittle (arrow). The material is "coalblende" from Kongsberg, Norway. Reflected light, crossed polars, in oil.

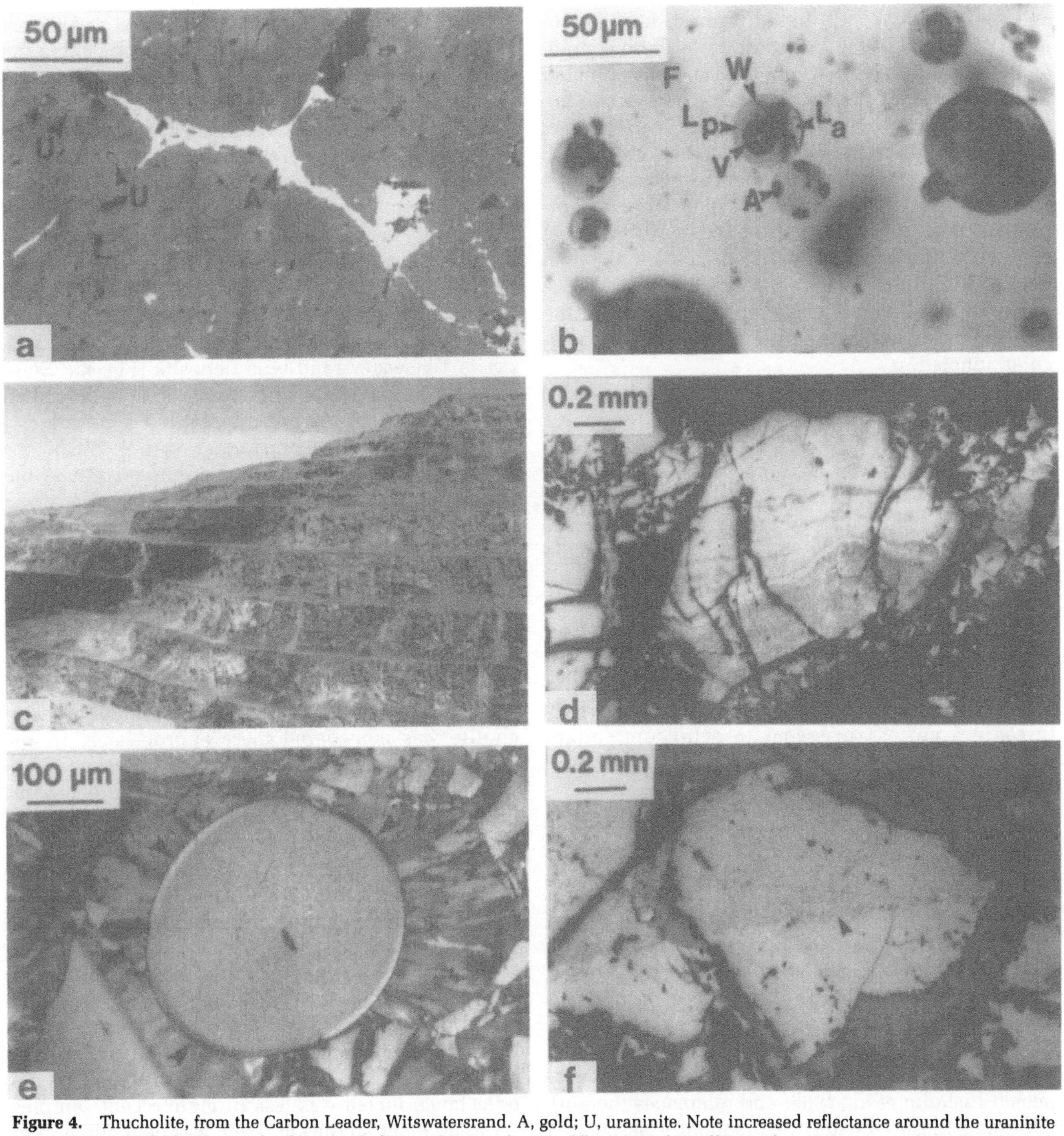

Figure 4. Thucholite, from the Carbon Leader, Witswatersrand. A, gold; U, uraninite. Note increased reflectance around the uraninite grains. Reflected light, uncrossed polars, in oil. (b) Petroleum inclusion of fluorite (F) from Illinois Fluorite District. V, Vapor; La, aqueous phase; Lp, liquid petroleum; W, wax; A, asphaltene (?). Note arrangement of asphaltenes in a tetrad symmetry, reflecting possibly that of fluorite. Transmitted light, uncrossed polars, in air. (c) Hydrocarbon "front," Carlin Mine, Nevada. For scale, the benches are approximately 60 m apart. The white body crossing the "front" is a hornblende–andesite dike. (d) Detail of a vein containing pyrobitumen, from Carlin, Nevada. Although fragmented by later processes, it is clear that the original hydrocarbons migrated as a single phase. Reflected light, partially crossed polars, in oil. (e) Sphere of pyrobitumen in a vein, from Carlin, Nevada. In contrast to (d), some hydrocarbons were originally transported as immiscible globules. Later stresses have caused some orientation of the optical domains around the rims (indicated by arrows). Reflected light, partially cross polars, in air. (f) Two pulses of hydrocarbons, recognizable by different coking behaviors (arrow indicates boundary). Carlin, Nevada. Reflected light, partially crossed polars, in oil.

anthro[2,1-d]thiophene, indenofluorenes, chrysene, and their methyl- and dimethyl-substituted homologs (Wise *et al.*, 1986). In addition to these components, idrialite has been reported to contain higher molecular weight species, including benzonaphthofluorenes, benzoindenofluorenes, and benzopicene.

The interpretation of Blumer (1975), corroborated by Wise *et al.* (1986), is that the minerals represent a "medium-temperature" pyrolysis of geochemical organic matter, followed by "extended equilibration at elevated temperatures in the subsurface, and the final fractionation during migration to the surface." The purity of pendletonite was ascribed, in part, to the pericondensed lattice rejecting components with angular annelation.

Apart from the polycyclic aromatic hydrocarbons, petroleum, asphalt, and hydrocarbon gases have been reported from all of the world's six largest mercury deposits. For the deposits in the California coast ranges, Peabody (1987) has suggested the Franciscan Complex to contain the hydrocarbon source rocks. Recently, Peabody (1989) concluded that the hydrocarbons and mercury were introduced simultaneously, together with CO_2 and H_2S, and speculated that the hydrocarbons and mercury were both derived from sedimentary sources. The ore deposits themselves represent the site where hydrocarbons and mercury were focused by large-scale anticlinal structures, in a process analogous to petroleum entrapment.

4.2. Oxalates

The conditions under which the polycyclic aromatic hydrocarbons concentrate are generally reducing. In marked contrast, the oxalates are characteristic of supergene conditions, where metal remobilization occurs under oxidizing conditions. Oxalates reported to date include moolooite ($CuCu_2O_4 \cdot nH_2O$), glushinskite ($MgC_2O_2 \cdot 2H_2O$), weddelite ($CaC_2O_4 \cdot 2H_2O$), and whewellite ($CaC_2O_4 \cdot 2H_2O$) together with iron oxalates (Chisholm *et al.*, 1987, and references therein). Although these oxalates are frequently associated with lichens, there is no evidence that they represent the extreme oxidation of reduced hydrocarbons.

5. Thucholite

A well-recognized metal–organic association is that of uranium. A complete summary of the research on this topic is beyond the scope of this chapter, and only the anomalous material thucholite will be discussed. The interested reader is referred to Landais (1986) for a comprehensive review emphasizing the organic geochemistry of the uranium–organic association. Thucholite is considered here because of the great age of some of the deposits, which consequently represent some of the earliest terrestrial reduced carbons available for study in organic geochemistry.

The term "thucholite" was derived by Ellsworth (1928) from the elemental symbols Th, U, C, H, and O to describe carbonaceous material from a pegmatite. Thucholite has been described from vein and pegmatite deposits, but has significance in the economically important quartz conglomerate–gold–uranium–pyrite Witswatersrand-type deposits (Fig. 4a; Stanton, 1972). Davidson and Bowie (1950) described thucholite as "not a true mineral but a complex of uraninite" and considered formation to be a result of a "hydrocarbon of hydrothermal origin, resulting from a polymerization or condensation of hydrocarbon fluids effected by radiations from uranium ore." In some early literature, "uraniferous anthraxolite" is probably thucholite (e.g., Obalski, 1904).

The organic composition of Witswatersrand thucholite was described by Zumberge *et al.* (1978) to be "a complex, solid, and solvent insoluble, polymer-like substance, containing mainly hydrocarbons and some organic sulfur and oxygen compounds." Thucholite was considered more appropriately in the Witswatersrand to be a kerogen. Pyrolysis–GC/MS yielded "alkyl substituted aromatic hydrocarbons, low molecular weight aliphatic hydrocarbons, with some aromatic sulfur and aliphatic oxygen compounds" [Zumberge *et al.* (1978)]. A later survey of European thucholites by Eakin (1989) has corroborated the predominance of aromatics together with oxygen and sulfur species.

Petrographically, thucholites show a nonlinear increase in reflectance, or a halo, around uraninite grains (Stach *et al.*, 1982; Eakin, 1989). Eakin (1989) has reported a reflectance increase of 2% (in oil) adjacent to the Witswatersrand uraninites. As the reflectance increase is less in more recent samples, it was argued that the alteration was dependent upon the amount of radiation received.

The original source of the organic matter in thucholites has been extensively debated. In the Witswatersrand, the carbon has been suggested to be a polymer formed by radiation-induced free radicals in liquid petroleum or gases (Davidson and Bowie, 1951) or to have been derived from Precambrian microorganisms (Eichmann and Schidlowski, 1974; Zumberge *et al.*, 1978). Although textural evidence for biological sources in the Witswatersrand has been presented (Prashnowsky and Schidlowski, 1967; Hallbauer, 1975), textural evidence from migrated thucholites (Eakin, 1989) indicates probable multiple sources for the hydrocarbons.

In addition to optical changes, the carbon isotopic composition is very variable between thucholites from different localities, and often extreme. In the majority of reports, the carbon isotope ratio tends to be lighter than the ranges reported for petroleums (−20 to −35‰ relative to the Pee Dee belemnite (PDB) standard; Deines, 1980). Examples include the thucholites from Witswatersrand (−22 to −33‰; Hoefs and Schidlowski, 1967), Blind River (−21.7 to −38.8‰; Eichmann and Schidlowski, 1974), Vastervik in Sweden (−32 to −46‰; Aberg *et al.*, 1985) and the Lake George, New Brunswick, antimony deposit thucholites (−42.1 to 45.9‰; Gize, unpublished data). However, Leventhal and Threlkeld (1978), noting a systematic increase in ^{13}C with uranium content in sandstone-hosted deposits from the Grants Uranium Region, New Mexico, advocated carbon isotope fractionation as a result of radiation. Similarly, Eakin (1989) has reported a tendency for a heavier carbon isotope ratio with increasing uranium content (from −29.3‰ with 0.15 wt. % uranium to −11.4‰ with 35.78 wt. % uranium) and suggested that the increase in the ^{13}C content is related to increased aromaticity. Combining the petrographic and carbon isotopic evidence, it is clear that the processes and sources of carbon in thucholites are still poorly understood.

6. Petroleum Inclusions

The majority of deposits discussed in this chapter are epigenetic, with the ores formed after the host rocks. As the ore solutions were migratory, the potential exists for ore deposits to provide information on petroleum migration.

6.1. Introduction

One technique which provides direct observations on secondary petroleum migration is that of inclusion analysis (Fig. 4b). At the time of ore mineral growth, aliquots of the surrounding fluids can be trapped within the minerals, for example, in lattice defects. Later fracturing of the mineral can also trap fluids during fracture healing. The inclusions can contain solids, fluids, or gas. For comprehensive reviews, the reader is referred to Roedder (1972, 1984), and for aspects related to hydrocarbons, to Burruss (1981) and Pagel *et al.* (1986).

6.2. Analysis

A variety of analytical techniques are available to study inclusions including heating and freezing techniques (e.g., Roedder, 1984) on intact inclusions and methods for opening inclusions in order to analyze the components. Kvenvolden and Roedder (1971) were able to extract hydrocarbons for GC/MS analysis by crushing quartz crystals with a mortar and pestle. A stainless steel crusher, with direct flow to a gas chromatograph, was developed by Andrawes and Gibson (1979). A similar technique, but with the crushing device in the gas chromatograph, has been reported by Burruss (1987). Another technique applied has been release of inclusion contents by thermal decrepitation. Baker and Smith (1986) have reported a vacuum thermal decrepitation–mass spectrometer configuration capable of analyzing up to 225 inclusions per hour in 10-mg mineral samples.

6.3. Observations

The Illinois Fluorite District is an example of a deposit in which plentiful petroleum inclusions occur (Fig. 4b) (Roedder, 1972). In common with several deposits of the Mississippi Valley-type, the paragenetic sequence can be correlated between deposits on the basis of color banding (McLimans *et al.*, 1980; Richardson and Pinckney, 1984), implicating district-scale fluid flow. Petroleum inclusions occur throughout the paragenetic sequence, but tend to be concentrated at three of the seven recognized color bands. Aqueous fluid inclusion data indicate that secondary petroleum migration occurred in pulses under predominantly isothermal (135–151°C) and isohaline (18–20 wt. % NaCl equivalent) conditions (Richardson and Pinckney, 1984). The simplest primary petroleum inclusions contain liquid + vapor phases (Fig. 4b); additional phases can include clear wax precipitates, black organic precipitates (asphaltenes?), aqueous fluid, and inorganic daughter phases (e.g., NaCl). Using Fourier transform infrared spectroscopy, CO_2 has been observed in the vapor phase (A. P. Gize and C. K. Richardson, in preparation).

In addition to providing information on secondary migration conditions and the petroleum composition, each inclusion represents a miniature petroleum reservoir. For example, in the case of the Illinois fluorite inclusions, the precipitation of waxes, asphaltenes, and inorganic phases indicates mechanisms by which the petroleum can impede migration by blockage of pores.

7. Geological Controls

In the introduction to this chapter, the similarities between ore and petroleum genesis were emphasized. Perhaps the major difference between the two processes lies in the formation of an economic concentration. The formation of a petroleum reservoir is the cumulation of several processes, including buoyancy, water potential, and capillary pressure (England *et al.*, 1987). A classical geological structure for petroleum entrapment is an anticline. In contrast, although mineralizing fluids can be driven by similar processes as petroleum (Bethke, 1986), the concentration of the elements into an ore deposit is a chemical process. Precipitation mechanisms can include temperature and pressure changes as well as oxidation or reduction mechanisms.

Nevertheless, the situation can be envisaged where, for example, in an anticline, ore precipitation can occur in a petroleum reservoir. Such a situation may be present in some examples of the Carlin-type disseminated gold deposits (Fig. 4c).

7.1. Carlin Gold Deposit

The Carlin-type deposits are a relatively new type of deposit, in which the gold is rarely visible, particles being <1–5 μm in size (Hausen and Park, 1986). Ore grades are typically 1–10 g of gold per tonne, with individual deposits yielding $>5 \times 10^6$ tonnes of gold (Thorpe, 1986). Among the characteristics of these deposits is a carbonaceous carbonate or carbonate-shale host rock. In the case of the classic deposits of Nevada, the ore deposits are located along "trends," which are anticlines (Hausen and Park, 1986).

To date, organic carbon has been reported to be virtually ubiquitous in the nonoxidized sediment-hosted disseminated gold deposits of the western United States (Bagby and Berger, 1986). Characteristically, the organic matter is present in higher concentrations than expected for the host rock (Joralemon, 1951; Hausen and Kerr, 1968; Wells *et al.*, 1969; Radtke *et al.*, 1980; Birak and Hawkins, 1985; Radtke, 1985). Although the carbonaceous matter is not always auriferous, Radtke and Scheiner (1970) argued that the organic matter was introduced with the gold. Other workers have argued in favor of concentration of the hydrocarbons during hydrothermal mineralization and alteration (Hausen and Kerr, 1968; Radtke *et al.*, 1980; Radtke, 1985; Hausen and Park, 1986).

In contrast, at the Alligator Ridge deposit, Ilchick *et al.* (1986) showed that the organic matter was indigenous to the host rocks and recorded (by use of Rock-Eval pyrolysis and reflectance) the later hydrothermal event responsible for mineralization. Similarly, at the Carlin deposit, Kuehn *et al.* (in press) have established that the hydrocarbons preceded gold mineralization.

7.2. Mechanisms of Secondary Migration

Petrographically, the organic matter at Carlin is a coke, possibly formed from a nongraphitizing precursor (Fig. 4d–f). Fluid inclusion data indicate that secondary migration occurred within a methane-saturated brine at temperatures of 155 ± 20°C and at pressures of 1.0 ± 0.2 kbar. Kuehn *et al.* (in press) have advocated that the carbonaceous matter at Carlin represents a coked petroleum reservoir in an anticlinal trap. Preservation of textures by hardening during coking has permitted observation of migration mechanisms. Initially, secondary migration occurred parallel to bedding to form stringers in low-pressure regions. Larger scale secondary migration occurred in extension fractures and veinlets, possibly developed by fluid overpressuring in more competent interbeds. Migration in veinlets involved either hydroca bon phases or immiscible droplets in the brines.

Although at Carlin and Alligator Ridge the hydrocarbons were in place prior to gold mineralization, the thermal maturity of the organic matter was high at the time of gold mineralization. For example, at Carlin, thermal modeling (Kuehn *et al.*, in press) has suggested that the organic matter had a reflectance of approximately 3%+ (in oil) at the time of mineralization. It is possible, therefore, that the coked organic matter provided a locally, strongly reducing ("active" carbon) environment which triggered gold precipitation.

8. Conclusion

In a sedimentary basin, the reservoir of reduced organic carbon may be significantly larger than that of certain metals. Consequently, it is not surprising that organic carbon can be intimately associated with some metalliferous ores. The question to be asked is how this association can be developed to promote understanding of organic geochemistry and of the processes leading to the formation of both commercially viable fossil fuel and metalliferous ore deposits.

The information available from the organic matter is dependent primarily upon the thermal maturity. In the case of relatively low temperature deposits, such as the Mississippi Valley-type, the organic matter can be investigated by the use of biomarkers and other petroleum research techniques. Problems of potentially mutual interest to both organic geochemists and ore geochemists involve oil–source correlations, hydrocarbon mixing, microbial activity, water washing, and hydrous pyrolysis. In the cases of the Illinois Fluorite District and the California mercury deposits, secondary hydrocarbon migration was synchronous with ore precipitation. Fluid inclusion studies of the ores can provide information on the physical and chemical conditions under which secondary petroleum migration can occur; biomarker studies of the included petroleum can help define source rocks and provide an estimate of timing.

Under certain circumstances, the organic matter in some ore deposits becomes close to being present as a phase, in a thermodynamic sense. In some mercury deposits, the polycyclic aromatic hydrocarbons "crystallize" as minerals. The high thermal effects associated with some types of mineralization cause coking in hydrocarbons and the development of a thermal mesophase, or "liquid crystals."

The complex association termed thucholite is of interest to the organic geochemist as it represents some of the earliest reduced carbon available for study. There remain problems in interpretation of the source of the reduced carbon and of the conflicting trends of the carbon isotopic data.

The three basic steps leading to the formation of a hydrocarbon reservoir or ore deposits are similar: source, migration, and development of the economically viable deposit. The last stage may be the most important step in segregating ore deposits and hydrocarbon reservoirs. In the case of hydrocarbons, important factors include buoyancy and the presence of suitable stratigraphic or sedimentological traps. In contrast, development of ore deposits requires processes such as a reduction in solubility as a result of pressure and temperature changes or reduction–oxidation processes. In spite of these different processes, structural geology can focus both the ore solutions and hydrocarbons into approximately the same final reservoir. In both the California mercury deposits and the Carlin gold deposit, anticlines focused both metals and hydrocarbons together. In the former example, the mercury and hydrocarbons are believed to have migrated together. At Carlin, the hydrocarbons were clearly emplaced and then coked prior to gold mineralization.

The objective of this brief review has been to focus on certain aspects of the organic–ore association and to indicate to the interested organic geochemist some of the processes of mutual interest to organic and inorganic geochemists. In our opinion, a clear causative relationship between organic matter and precipitation of a major ore deposit remains to be proven. In part, it may be that the organic matter is a "trigger" for precipitation. However, part of the lack of knowledge may also be due to differences in emphasis and terminology between organic and inorganic geochemists. To quote from Mark Twain's *A Connecticut Yankee in King Arthur's Court*, "that neat idea hit the boy in a blank place, for geology hadn't been invented yet. However, I made a note of the remark, and calculated to educate the commonwealth up to it if I pulled through. It is no use to throw a good thing away merely because the market isn't ripe yet."

References

Aberg, G., Lofvendahl, R., Nord, A. G., and Holm, E., 1985, Radionuclide mobility in thucholitic hydrocarbons in fractured quartzite, *Can. J. Earth Sci.* **22**:959–967.

Amstutz, G. C. (ed.), 1964, *Developments in Sedimentology*, Vol. 2, Elsevier, Amsterdam.

Andrawes, F. F., and Gibson, E. K., Jr., 1979, Release and analysis of gases from geological samples, *Am. Mineral.* **64**:453–463.

Baars, J. K., 1930, Over Sulfaatreductie door Bacterien, Proefschrift, Techn. Hoogeschool. Delft (Holland), Mienema (Thesis).

Bagby, W. C., and Berger, B. R., 1986, Geological characteristics of sediment-hosted disseminated precious-metal deposits in the western United States, in: *Geology and Geochemistry of Epithermal Systems* (B. R. Berger and P. M. Bethke, eds.), *Rev. Econ. Geol.* **2**:73–98.

Barker, C., and Smith, M. P., 1986, Mass spectrometric determination of gases in individual fluid inclusions in natural minerals, *Anal. Chem.* **58**:1330–1333.

Barton, P. B., 1967, Possible role of organic matter in the precipitation of the Mississippi Valley ores, in: *Genesis of Stratiform Lead–Zinc–Barite–Fluorite Deposits* (J. S. Brown, ed.), Econ. Geol Monograph 3, pp. 371–378.

Beales, F. W., Carracedo, J. W., and Strangway, D. W., 1974, Palaeomagnetism and the origin of Mississippi Valley-type ore deposits, *Can. J. Earth Sci.* **11**:211–223.

Bethke, C. M., 1986, Hydrologic constraints on the genesis of the Upper Mississippi Valley District from Illinois Basin brines, *Econ. Geol.* **81**:233–249.

Birak, D. J., and Hawkins, R. H., 1985, The geology of the Enfield Bell mine and Jerritt Canyon district, Elko County, Nevada, in: *Characteristics of Sediment and Volcanic-Hosted Dissemi-*

nated Gold Deposits—Search for an Occurrence Model (E. W. Tooker, ed.), U.S. Geological Survey Bulletin 1646, pp. 95–105.

Blumer, M., 1975, Curtisite, idrialite, and pendletonite, polycyclic aromatic hydrocarbon minerals: Their composition and origin, *Chem. Geol.* **16**:245–256.

Brecke, E. A., 1962, Ore genesis of the Cave-in-Rock fluorspar district, Hardin County, Illinois, *Econ. Geol.* **57**:499–535.

Brooks, J. D., and Taylor, G. H., 1965, Formation of graphitizing carbons from the liquid phase, *Nature* **206**:697–699.

Brooks, J. D., and Taylor, G. H., 1966, Development of order in the formation of coke, in: Advances in Chemistry Series No. 55, American Chemical Society, Washington, D.C., pp. 549–563.

Brooks, J. D., and Taylor, G. H., 1968, The formation of some graphitizing carbons, *Chem. Phys. Carbon* **4**:243–286.

Brown, G. H., and Shaw, W. G., 1957, The mesomorphic state, *Chem. Rev.* **57**:1049–1157.

Burruss, R. C., 1981, Hydrocarbon fluid inclusions in studies of sedimentary diagnesis, in: *Short Course in Fluid Inclusions, Applications to Petrology* (L. S. Hollister and M. L. Crawford, eds.), Mineralogical Association of Canada, pp. 138–156.

Burruss, R. C., 1987, Crushing-cell capillary column gas chromatography of petroleum fluid inclusions: Method and application to petroleum source rocks, reservoirs, and low temperature hydrothermal ores, American Current Research on Fluid Inclusions, Jan. 5–7, Program and Abstracts (unpaginated).

Carter, J. S., and Cazalet, P. C. D., 1984, Hydrocarbon gases in rocks as pathfinders for mineral exploration, in: *International Symposium on Prospecting in Areas of Glaciated Terrain*, Vol. 6, Institute of Mining and Metallurgy, London, pp. 11–20.

Chisholm, J. E., Jones, G. C., and Purvis, O. W., 1987, Hydrated copper oxalate, moolooite, in lichens, *Mineral Mag.* **51**:715–718.

Coonan, J., 1977, Relationship between oil and ore in the Saint Privat barite deposit, Lodeve Basin, France, in: *Proceedings of the Forum on Oil and Ore in Sediments*, Imperial College, London, pp. 167–188.

Connan, J., and Orgeval, J-J., 1973, Les bitumes des minéralisations barytiques et sulfurées de St. Privat (Bassin de Lodève, France), *Bull. Cent. Rech. Pau-SNPA* **6**:195–214.

Cook, A. C., Murchison, D. G., and Scott, E., 1972, A British meta-anthracitic coal of Devonian age, *Geol. J.* **8**:83–94.

Davidson, C. F., and Bowie, S. H. U., 1951, On Thucholite and Related Hydrocarbon Complexes, Geological Survey of Great Britain, Atomic Energy Division, Report No. 92.

Davis, A., Hoover, D. S., Wakely, L. D., and Mitchell G. D., 1983, The microscopy of mesophase formation and of anisotropic cokes produced from solvent-refined coals, *J. Microsc.* **132**:315–331.

Deines, P., 1980, The isotopic composition of reduced organic carbon, in: *Handbook of Environmental Isotope Geochemistry*, Vol. 1 (P. Fritz and J. Ch. Fontes, eds.), Elsevier, Amsterdam, pp. 329–406.

Dons, J. A., 1956, Coal blend and uraniferous hydrocarbon in Norway, *Nor. Geol. Tidsskr.* **36**:249–266.

Dunsmore, H. E., 1975, Origin of lead–zinc ores in carbonate rocks: A sedimentary-diagenetic model, Ph.D. Thesis, University of London.

Dunsmore, H. E., and Shearman, D. J., 1977, Mississippi Valley-Type lead–zinc orebodies: A sedimentary and diagenetic origin, in: *Proceedings of the Forum on Oil and Ore Sediments*, Imperial College, London, pp. 189–205.

Eakin, P. A., 1989, The organic petrography and geochemistry of uraniferous hydrocarbons, Ph.D. Thesis, Queen's University, Belfast.

Eichmann, R., and Schidlowski, M., 1974, Isotopic composition of carbonaceous matter from the Precambrian uranium deposits of the Blind River District, Canada, *Naturwissenschaften* **61**:449.

Ellsworth, H. V., 1928, Thucholite, a remarkable primary carbon mineral from the vicinity of Parry Sound, Ontario, *Am. Mineral.* **13**:419–441.

England, W. A., Mackenzie, A. S., Mann, D. M., and Quigley, T. M., 1987, The movement and entrapment of petroleum fluids in the subsurface, *J. Geol. Soc. London* **144**:327–347.

Franklin, R. E., 1951, Crystallite growth in graphitizing and non-graphitizing carbons, *Proc. R. Soc. London, Ser. A.* **209**:196–218.

Friedel, G., 1922, The mesomorphic state, *Ann. Phys.* **18**:273–474.

Geissman, T. A., Sim, K. Y., and Murdoch, J., 1967, Organic minerals, picene and chrysene as constituents of the mineral curtisite (idrialite), *Experientia* **23**:793–794.

Giordano, T. H., and Barnes, H. L., 1981, Lead transport in Mississippi Valley-type ore solutions, *Econ. Geol.* **76**:2200–2211.

Gize, A. P., 1984, The organic geochemistry of three Mississippi Valley-type ore deposits, Ph.D. Thesis, The Pennsylvania State University.

Gize, A. P., 1985, The development of a thermal mesophase in bitumens from high temperature ore deposits, in: *Organics and Ore Deposits* (W. E. Dean, ed.), Proceedings of the Denver Region Exploration Geologists Society Symposium, Denver Region Exploration Geologists Society, Denver, pp. 137–150.

Gize, A. P., and Barnes, H. L., 1987. The organic geochemistry of two Mississippi Valley-type lead–zinc deposits, *Econ. Geol.* **82**:457–470.

Gize, A. P., and Rimmer, S. M., 1983, Mesophase development in a bitumen from the Nanisivik Mississippi Valley-type deposit, *Carnegie Inst. Washington Yearb.* **82**:414–419.

Hallbauer, D. K., 1975, Geochemistry and morphology of the mineral components from the fossil gold and uranium placers of the Witswatersrand, in: *Genesis of Uranium- and Gold-Bearing Precambrian Quartz Pebble Conglomerates* (F. C. Armstrong, ed.), U.S. Geological Survey Professional Paper, 1161-A-BB, pp. M1–M18.

Hanor, J. S., 1979, The sedimentary genesis of hydrothermal fluids, in: *Geochemistry of Hydrothermal Ore Deposits* (H. L. Barnes, ed.), 2nd ed., Wiley-Interscience, New York, pp. 137–172.

Hatch, J. R., Heyl, A. V., and King, J. D., 1985, Organic geochemistry of hydrothermal alteration, basal shale and limestone beds, Middle Ordovician Quimbys Mill Member, Platteville Formation, Thompson–Temperly Zinc–Lead Mine, Lafayette County, Wisconsin, in: *Organics and Ore Deposits* (W. E. Dean, ed.), Proceedings of the Denver Region Exploration Geologists Society Symposium, Denver, Colorado, pp. 93–104.

Hausen, D. M., and Kerr, P. F., 1968, Fine gold occurrence at Carlin, Nevada, in: *Ore Deposits in the United States*, Vol. 1 (J. D. Ridge, ed.), American Institute of Mining, Metallurgical, and Petroleum Engineers, pp. 908–940.

Hausen, D. M., and Park, W. C., 1986, Observations on the association of gold mineralization with organic matter in Carlin-Type ores, in: *Organics and Ore Deposits* (W. E. Dean, ed.), Proceedings of the Denver Region Exploration Geologists Society Symposium, pp. 119–136.

Hoefs, J., and Schidlowski, M., 1967, Carbon isotope composition of carbonaceous matter from the Precambrian of the Witswatersrand series, *Science* **155**:1096.

Honda, H., Kimura, H., Sanada, Y., Sugawara, S., and Furota, T., 1970, Optical mesophase texture and X-ray diffraction pattern of the early-stage carbonization of pitches, *Carbon* **8**:181–189.

Ilchick, R. P., Brimhall, G. H., and Schull, H. W., 1986, Hydrothermal maturation of indigenous organic matter at the Alligator Ridge gold deposits, Nevada, *Econ. Geol.* **81**:113–130.

Jackson, S. A., and Beales, F. W., 1967, An aspect of sedimentary basin evolution: The concentration of Mississippi Valley-type ores during late stages of diagenesis, *Bull. Can. Petrol. Geol.* **15**: 383–433.

Jones, J. M., and Creaney, S., 1977, Optical character of thermally metamorphosed coals of northern England, *J. Microsc.* **109**: 105–118.

Joralemon, P., 1951, The occurrence of gold at the Getchell mine, Nevada, *Econ. Geol.* **46**:267–310.

Jowett, E. C., 1986, Genesis of Kupferschiefer Cu–Ag deposits by convective flow of Rotliegend brines during Triassic rifting, *Econ. Geol.* **81**:1823–1837.

Khavari-Khorasani, G., Blayden, H. E., and Murchison, D. G., 1978, Refractive index and absorption coefficients as measures of structural organization in carbonized bitumen, *J. Microsc.* **114**: 199–204.

Kisch, H. J., 1966, Carbonization of semi-anthracitic vitrinite by an analcime basanite sill, *Econ. Geol.* **61**:1043–1063.

Krs, M., and Stovickova, N., 1966, Palaeomagnetic investigation of hydrothermal deposits in the Jackymov (Joachimsthal) region, Western Bohemia, *Trans. Inst. Min. Metall.* **75**:B51–B57.

Kuehn, C. A., and Bodnar, R. J., 1984, P–T–X characteristics of fluids associated with the Carlin sediment-hosted gold deposit, *Geological Society of America Abstracts with Programs*, Vol. 16, pp. 566.

Kuehn, C. A., Gize, A. P., and Furlong, K. P., Temporal and P–T–X history of secondary migration and thermal maturation of organic matter at the Carlin gold deposit, Nevada, *Econ. Geol.*, in press.

Kvenvolden, K. A., and Roedder, E., 1971, Fluid inclusions in quartz crystals from South-West Africa, *Geochim. Cosmochim. Acta* **35**:1209–1229.

Landais, P., 1986, *Géologie et Géochimie de l'uranium*, Centre de Recherches sur la Géologie de l'Uranium, Mémoire 10.

Leventhal, J. S., and Threlkeld, C. N., 1978, Carbon-13/carbon-12 isotope fractionation of organic matter associated with uranium ores induced by alpha radiation, *Science* **207**:430–432.

Lietz, J., 1939, Mikroskopische und chemische Untersuchungen an Kongsberger Silberezen, *Z. Min.* **2**:65.

Macqueen, R. W., and Powell, T. G., 1983, Organic geochemistry of the Pine Point lead-zinc field and region. Northwest Territories, Canada, *Econ. Geol.* **78**:1–25.

Marsh,, H., 1973, Carbonization and liquid-crystal (mesophase) development: Part 1. The significance of the mesophase during carbonization of coking fuels. *Fuel* **52**:205–212.

Mason, B., 1966, *Principles of Geochemistry*, 3rd ed., John Wiley & Sons, New York.

McLimans, R. K., Barnes, H. L., and Ohmoto, H., 1980, Sphalerite stratigraphy of the Upper Mississippi Valley Zinc–Lead District, southwest Wisconsin, *Econ. Geol.* **75**:351–361.

Mueller, G., 1954, The theory of genesis of oil through hydrothermal alteration of coal type substances within certain Carboniferous strata of the British Isles, *International Geologica Congress, 19th Session, Algiers*, No. 12, pp. 279–328.

Murdoch, J., and Geissman, T. A., 1967, Pendletonite, a new hydrocarbon mineral from California, *Am. Mineral.* **52**:611–616.

Neumann, H., 1944, Silver deposits at Kongsberg, *Nor. Geol. Unders. Publ.* **162**.

Obalski, J., 1904, On a mineral containing "Radium" in the Province of Quebec, *J. Can. Min. Inst.* **7**:245.

Ohle, E. L., 1959, Some considerations in determining the origin of ore deposits of the Mississippi Valley-type, *Econ. Geol.* **54**: 769–789.

Ohle, E. L., 1980, Some considerations in determining the origin of ore deposits of the Mississippi Valley-type, Part 2, *Econ. Geol.* **75**:161–172.

Pagel, M., Walgenwitz, F., and Dubessy, J., 1986, Fluid inclusions in oil and gas-bearing sedimentary formations, in: *Thermal Modelling in Sedimentary Basins* (J. Burrus, ed.), Editions Technip, Paris, pp. 565–583.

Park, C. F., Jr., and MacDiarmid, R. A., 1970, *Ore Deposits*, 2nd ed., W. H. Freeman and Co., San Francisco.

Peabody, C. E., 1987, The association of petroleum and cinnabarat the Culver-Bear Mine, Sonoma County, California, *Geological Society of America, 1987 Annual Meeting, Abstracts with Programs*, p.801.

Peabody, C. E., 1989, Cinnabar–petroleum deposits: Nature and sources of mineralizing fluids, *28th International Geological Congress*, Abstracts, Vol. 2, pp. 582–583.

Pering, K. L., 1973, Bitumens associated with lead, zinc, and fluorite ore minerals in North Derbyshire, England, *Geochim. Cosmochim. Acta* **37**:401–417.

Pering, K. L., and Ponnamperuma, C., 1969, Alicyclic hydrocarbons from an unusual deposit in Derbyshire, England, *Geochim. Cosmochim. Acta* **33**:528–532.

Pfennig, N., and Widdel, F., 1982, The bacteria of the sulphur cycle, *Philos. Trans. R. Soc. London* **B298**:433–441.

Powell, T. G., and Macqueen, R. W., 1980, Organic geochemistry of the Pine Point region, NWT, Canada, in: *Extended Abstracts of the Conference on the Geochemistry of Organic Matter in Ore Deposits* (M. F. Estep, P. E. Hare, and T. C. Hoering, eds.), Carnegie Institution of Washington, Geophysical Laboratory, p. 113.

Powell, T. G., and Macqueen, R. G., 1984, Precipitation of sulfide ores and organic matter: Sulfate reactions at Pine Point, Canada, *Science* **224**:63–66.

Prashnowksy, A. A., and Schidlowski, M., 1967, Investigation of Pre-Cambrian thucholite, *Nature* **216**:560–563.

Radtke, A. S., 1985, Geology of the Carlin Gold Deposit, Nevada, U.S. Geological Survey Professional Paper 1267.

Radtke, A. S., and Scheiner, B. J., 1970, Studies of hydrothermal gold deposition (1). Carlin gold deposit, Nevada: The role of carbonaceous materials in gold deposition, *Econ. Geol.* **65**: 87–102.

Radtke, A. S., Rye, R. O., and Dickson, F. W., 1980, Geology and stable isotope studies of Carlin gold deposit, Nevada, *Econ. Geol.* **75**:641–672.

Richardson, C. K., and Pinckney, D., 1984, The chemical and thermal evolution of the fluids in the Cave-in-Rock fluorspar district, Illinois, *Econ. Geol.* **79**:1833–1856.

Rickard, D. T., Willden, M. Y., Marde, Y., and Ryhage, R., 1975, Hydrocarbons associated with lead–zinc ores at Laisvall Sweden, *Nature* **255**:131–133.

Rickard, D. T., Willden, M. Y., Marinder, N. E., and Donelly, T. H., 1979, Studies on the genesis of the Laisvall sandstone lead–zinc deposit, Sweden, U.S. Government Printing Office, Washington, D.C.

Roedder, E., 1972, Composition of Fluid Inclusions, U.S. Geological Survey, U.S. Government Printing Office, Washington, D.C., pp. 1–164.

Roedder, E., 1984, *Fluid Inclusions*, in: Min. Soc. Am. Reviews in Mineralogy, Vol. 12 (P. H. Ribbe, ed.), Minerological Society of America, pp. 1–644.

Rubentschik, L., 1928, Über sulfatreduktion durch Bakterien bei

Zellulosegarungsprodukten als Energiquelle, *Zentralbl. Bakteriol. Parasiten. II Abt* **73:**483–496.

Schidlowski, M., Eichmann, R., and Junge, C. E., 1974, Evolution des iridischen Sauerstoff-Budgets und Entwicklung der Erdatmosphäre, *Umschau* **22:**703–707.

Selwyn, S. C., and Postgate, J. R., 1959, A search for the Rubentschikii group of *Desulfovibrio*, *J. Microbial. Serol.* **25:**465–472.

Skinner, B. J., 1967, Precipitation of Mississippi Valley-type ores: A possible mechanism, in: *Genesis of Stratiform Lead–Zinc–Barite–Fluorite Deposits* (J. S. Brown, ed.), Econ. Geol. Monograph No. 3, pp. 363–370.

Smith, J. W., and Croxford, N. J. W., 1973, Sulphur isotope ratios in the McArthur River lead–zinc–silver deposit, *Nature* **245:** 10–12.

Speczik, S., and Puttmann, W., 1989, Oxidation of organic matter and its influence on Kupferschiefer Mineralization of south western Poland, *28th International Geological Congress*, Abstracts, Vol. 3, p. 160.

Speczik, S., Skowronek, G., Friedrich, R., and Diedel, G., 1986, The environment of generation of some base metal Zechstein occurrences in Central Europe, *Acta Geol. Polon.* **36:**1–35.

Stach, E., Mackowsky, M.-Th., Teichmuller, M., Taylor, G. H., Chandra, D., and Teichmuller, R., 1982, *Coal Petrology*, 3rd ed., Gebruder Borntraeger, Berlin/Stuttgart.

Stanton, R. L., 1972, *Ore Petrology*, McGraw-Hill, New York.

Taylor, G. H., 1961, Development of optical properties of cokes during carbonization, *Fuel* **40:**465–471.

Thorpe, R. I., 1986, Gisements d'or dans des sédiments clastiques, in: *Types de Gisements Minéraux du Canada: Un Bref Exposé Géologique* (O. R. Eckstrand, ed.), Rapport de Géologie Economique 36, Geological Survey of Canada, p. 30.

Trudinger, P. A., 1982, Geological significance of sulphur oxidoreduction by bacteria, *Philos. Trans. R. Soc. London, Ser. B* **298:**563–581.

Trudinger, P. A., and Williams, N. 1982, Stratified sulfide depositions in modern and ancient environments, in: *Mineral Deposits and the Evolution of the Biosphere* (H. D. Holland and M. Schidlowski, eds.), Springer-Verlag, Berlin, pp. 177–198.

Wells, J. D., Stoiser, L. R., and Elliot, J. E., 1969, Geology and geochemistry of the Cortez gold deposit, Nevada, *Econ. Geol.* **64:**526–537.

White, J. L., 1974, The formation of the microstructure in graphitizable materials, in: *Progress in Solid State Chemistry*, Vol. 9 (J. O. McCaldin and G. Smorjai, eds.), Pergamon Press, pp. 59–104.

White, J. L., 1976, Mesophase mechanisms in the formation of the microstructure of petroleum coke, in: *Petroleum Derived Carbons* (M. L. Deviney and T. M. O'Grady, eds.), ACS Symposium, Series No. 21, American Chemical Society, Washington, D.C., pp. 282–314.

Williams, N., and Rye, D. O., 1974, Alternative interpretation of sulphur isotope ratios in the McArthur lead–zinc–silver deposits, *Nature* **247:**535–537.

Wise, J. A., Campbell, R. M., West, W. R., Lee, M. L., and Bartle, K. D., 1986, Characterization of polycyclic aromatic hydrocarbon minerals curtisite, idrialite and pendletonite using high-performance liquid chromatography, gas chromatography, mass spectrometry and nuclear magnetic resonance spectroscopy, *Chem. Geol.* **54:**339–357.

Wright, F. E., and Allen, E. T., 1930, Curtisite. A new organic mineral from Skaggs Springs, Sonoma County, California, *Am. Mineral.* **15:**169–173.

Wu, Y., and Beales, F. W., 1981, A reconnaissance study by paleomagnetic methods of the age of mineralization along the Viburnam Trend, Southeast Missouri, *Econ. Geol.* **76:**1879–1894.

Zumberge, J. E., Sigleo, A. C., and Nagy, B., 1978, Molecular and elemental analyses of the carbonaceous matter in the gold and uranium bearing Vaal Reef carbon seams, Witswatersrand sequence, *Miner. Sci. Eng.* **10:**223–246.

Chapter 27

Metals in Black Shales

JOEL LEVENTHAL

1. Introduction

Shales that are rich in organic matter occur throughout the geologic record, but special conditions are responsible for their occurrence (Tourtelot, 1979). Before considering organic matter- and metal-rich shales, some background information is necessary. A shale is a sedimentary rock that is composed of small (mostly less than 0.1 mm) particles dominated by phyllosilicate (clay) minerals and containing subordinate amounts of quartz, carbonate, and phosphate minerals and, in some cases, organic matter and pyrite [for a discussion, see Spears (1980)]. Usually, the only detrital minerals are quartz and kaolinite; the others are authigenic. Organic-rich shales often show bedding laminations ranging from less than millimeter to centimeter thickness that are the result of size sorting, organic content, or pyrite layers, caused by changes in depositional environment. Shales can be formed in fresh, brackish, marine, or hypersaline water bodies, but most of the examples in this chapter are from the marine environment (Degens, 1965). The organic matter in shales is mainly a macromolecular "geopolymer" called kerogen (Hunt, 1979) that is insoluble (in normal organic solvents and nonoxidizing inorganic acids and bases). Kerogen forms from the residues of land and marine organisms (and plants) that have been degraded by microorganisms, buried, and sometimes heated. Only a very small fraction (usually less than 3%) of the organic matter is soluble in common solvents. This soluble material may be nonpolymerized original material, petroleum-like material that has been produced as a result of burial, or more recent in-migrated (contaminant) material. In addition to being organic-rich, as are petroleum source rocks, shales are also sometimes enriched in metals. However, unlike petroleum source rocks, which are not of interest after they have been heated and are through the oil and gas generation zone, metalliferous shales almost always seem to retain their metals and thus are interesting both as potential resources and for scientific study. Another difference from petroleum source rocks is that areally small occurrences of metal-enriched shales can have economic potential.

2. Organic Matter in Shales

2.1. Processes of Deposition and Accumulation

Organic carbon content of sediments is due to productivity, preservation, and sedimentation rate—these factors are not mutually exclusive. It is esti-

JOEL LEVENTHAL • U.S. Geological Survey, Federal Center MS 973, Denver, Colorado 80225.
Organic Geochemistry, edited by Michael H. Engel and Stephen A. Macko. Plenum Press, New York, 1993.

mated that only 1% organic C is needed in a sediment to make it anoxic, which allows sulfate-reducing microorganisms to operate and produce HS^-, which then reacts with Fe and other metals. An organic carbon content of 1% can be sufficient to coat all mineral surfaces and to deplete oxygen (bacterially or chemically) to zero. This amount of organic matter can set the stage for subsequent diagenesis, sulfide production, and preservation of organic matter and metals.

By the time the organic C content is 10% then, because of the associated O, H, and N and the 2.5 times lower density than that of the bulk sediment, the rock is about 30% organic matter by volume, coats everything, and fills in much of the porosity. When the organic carbon content is about 20%, the rock effectively has an organic matrix or groundmass. The organic matter occurs in a variety of forms and can be centimeter- to millimeter-sized remnants of structured land-plant woody tissues (vitrinite, etc), or it can be submillimeter-sized microorganisms that are still sufficiently structured to identify (e.g., the marine alga *Tasmanites*), or it can be amorphous and/or unstructured humic material and kerogen. Both humic material and kerogen can have either a land or a marine source; the classification is based on solubility, not source.

2.2. Organic Matter Type

The type of organic matter in metalliferous black shales can originate from both marine and terrestrial (land) sources, but before Silurian or Devonian time there were no (or few) land plants. Thus, the pre-Silurian organic matter came only from marine sources and consisted of the remains of marine organisms including plants, algae, bacteria, animals, and their decomposition products. For example, the Chattanooga Shale of Tennessee (Conant and Swanson, 1961; Leventhal *et al.*, 1981) contains mainly terrestrially derived organic matter, whereas the stratigraphically and age-equivalent Devonian shales from West Virginia and Ohio contain mainly marine-sourced organic matter (Leventhal, 1981). Marine organic matter is more readily metabolized by microorganisms and thus is more efficient in depleting oxygen and, as a result, more efficient in producing sulfide. However, sulfide is often produced in excess of the available and reactive Fe and other metals, and therefore some H_2S escapes from the sediment to the water column, where it is oxidized to elemental S and SO_4. However, organic matter of terrestrial origin in the Chattanooga Shale has a sulfur:carbon ratio as high as that of the marine organic matter in the equivalent shale units of West Virginia and Pennsylvania [see Raiswell and Berner (1986) for contrasting examples]. Terrestrial organic matter, with its more abundant oxygen functional groups from lignin, is probably more efficient at complexing or chelating metals than marine organic matter. In addition, the terrestrial organic matter may be pre-enriched in metals by complexing the metals in the rivers by the organic matter before the metals enter the sea. Thus, the type of organic matter is only one of several variables.

For sediments to accumulate organic matter, it is not necessary to have high productivity, but only to have high preservation of the organic matter. Preservation can be accomplished by a layer of water above the sediment that is anoxic that forms due to lack of mixing. That is, the organic matter has depleted the oxygen in the water faster than it is replenished by mixing. This stratification is usually caused by temperature or density differences but is also influenced by geologic and/or topographic features, such as a sill, ridge, or deep depression, that limit the flow and/or mixing of bottom water.

3. Metals in Shales

As well as being enriched in organic matter, black shales are commonly enriched in sulfides and metals relative to average shales (see Table I). However, organic-rich shales (Duncan and Swanson, 1965) are not always metal-rich because, in some cases, metals are either unavailable for concentration or the sedimentation rate is so high that clastic dilution keeps metal values low. Therefore, metal-lean black shales can occur despite high organic productivity and/or preservation. At the other extreme are metal-rich shales that form in a variety of geologic settings—some of which have modern analogs; however, many ancient black shale environments are not represented in modern settings.

3.1. Sedimentary Basins

Like petroleum source rocks, metal-rich black shales form in basins including rift zones and fjords, but these shales are often not preserved due to subsequent plate tectonic movements. It is also important to consider the locations in a basin where these sediments are deposited—such as basin catchment area, nearshore, offshore, or center of a restricted basin—because the basin geometry may control the depositional environment and determine the extent of metal enrichment. For example, in the Holocene Black Sea

TABLE I. Representative Element Values for Selected Shale Units

Shale	Al (%)	Fe (%)	S (%)	Org. C (%)	Parts per million U	Mo	V	Ni	Cu	Cr	Co	As	Pb	Ba	Zn	Ag	Cd	REE + Y	P	Mn	Ca (%)	Carbonate (%)	Reference(s)[a]
Average shale	8	4.7	0.2	2.1[b]	4	3	130	68	45	90	19	13	20	580	95	0.1	0.3	220	700	850	2.2	—	1
Average black shales	7	2	—	3.2	—	10	150	50	70	100	10	—	20	300	—	—	—	—	—	150	1.5	0.33	2
	8	3.7	—	—	15	—	500	—	—	111	17	29		1120	—	—	—	195[c]	—	383	1.7	—	3
Atlantic Cretaceous	—	3.5	1.3	6.2	—	60	820	190	160	200	30	—	16	530	830	4.5	0.9	—	—	910	—	—	4
Green River Fm., Eocene	—	2	1.6	30	10	30	150	30	70	40	10	35	40	300	—	—	1	—	—	220	—	—	5
Black Sea layer C	—	3.8	1.6	3.5	12	25	250	100	55	150	—	—	11	—	108	—	—	—	—	—	—	1.7	6
Appalachian Devonian	—	4	3	10	60	90	300	150	150	60	25	60	30	400	200	0.2	—	150	—	100	—	—	7, 8
Condor, Australia	—	4.6	0.7	11.7	1	30	110	25	25	55	15	9	—	200	65	—	—	—	—	350	0.4	0.6	9, 10
Alum, Cambrian	—	7.1	6.7	13.7	206	270	680	160	190	94	50	17	140	500	150	1	—	200	500	—	—	—	11
Kolm from Alum	6	6	7.7	50	4500	90	290	95	60	—	—	150	230	400	200	—	3	1300	500	240	—	—	12
Mecca Quarry, Pennsylvanian	—	3.6	3	30	130	1100	1800	400	100	400	—	30	30	300	1500	—	70	150	4000	250	1	0.1	13
New Albany, Devonian	7	2	1.6	8.4	11	15	200	50	150	70	15	17	15	700	<300	<0.2	<1	—	—	200	1.5	0.8	14
Falling run/ Henryville	—	4.5	1	17	180	70	1000	200	300	100	5	—	150	300	1000	—	—	6	2%	50	2.8[b]	0.4	15, 16
Associated phosphate	—	4	3	1.2	125	20	100	100	70	20	70	—	200	70	500	—	—	5000	35	100	47[b]	1.5	15, 16

[a]References: 1, Turekian and Wedepohl, 1961; 2, Vine and Tourtelot, 1970; 3, Quinby-Hunt *et al.*, 1989; 4, Brumsack, 1980; 5, Desborough *et al.*, 1976; 6, Hirst, 1974; 7, Leventhal *et al.*, 1978; 8, Leventhal *et al.*, 1983; 9, Patterson *et al.*, 1988; 10, Glikson *et al.*, 1985; 11, Armands, 1972; 12, Leventhal, 1990; 13, Coveney *et al.*, 1987; 14, Shaffer *et al.*, 1984; 15, Leventhal and Kepferle, 1982; 16, Leventhal, unpublished results.

[b]Degens, 1965.

[c]Not including Y.

(Hirst, 1974) the center deeps are enriched in metals because of the greater extent of euxinic water (thicker H_2S column) and lack of clastic dilution relative to the nearshore, oxic shelf environment. In the case of the nearshore enrichment of uranium and other metals found in the Devonian shales of the Appalachian basin (Leventhal, 1981; Leventhal *et al.*, 1981) its source was, in part, volcanic ash that was leached to provide the metals (Leventhal and Kepferle, 1982). Another case is the Middle and Upper Pennsylvanian thin black shales of the mid-continent United States (Coveney *et al.*, 1987), which are rich in terrestrial organic matter. Control of metals and organic matter enrichment in these Pennsylvanian shales was by the transgression and regression cycles at the basin edge as sea level varied with tectonic adjustment in response to plate collision.

3.2. Basin Environment

Rapid burial is the usual mode for organic preservation below normal oxic marine water, but this results in dilution of metals. Therefore, special circumstances are necessary to make shales that are both organic-rich and metal-rich. Organic matter can be preserved if there is an overlying anoxic water column, in which case the sedimentation rate and burial are not critical to the preservation of organic matter because oxidation is prevented by the anoxic water column. Therefore, metal enrichment and preservation are most favored by a euxinic (H_2S-containing) water column within a starved basin where there is high preservation of organic matter and low dilution of metals.

Once established, a euxinic water column will act to maintain itself (positive feedback) because the organic matter, which falls into the sulfidic water column, is largely preserved rather than partly utilized by the usual aerobic microorganisms, which cannot operate in the absence of O_2 and presence of H_2S (Goldhaber and Kaplan, 1973). The following equations show the metabolic reactions:

$$RCH_2O + O_2 \rightarrow R + H_2CO_3 \quad \text{(aerobic)}$$

$$R(CH_2O)_2 + SO_4^{2-} \rightarrow R + 2HCO_3^- + H_2S \quad \text{(anaerobic)}$$

where CH_2O is chemical shorthand for carbohydrate, and R is the remainder of the molecule. Therefore, the dissolved organic matter is metabolized anaerobically in the water column itself, with no limitation on available sulfate.

Euxinic conditions are more favorable to the concentration of metals than anoxic conditions because there is a direct combination of metals with sulfides in the water column (Hallberg, 1978). This reaction is as follows:

$$HS^- + Me^{2+} \rightarrow MeS + H^+$$

where Me is a metal.

The removal of metals from a euxinic water column is superimposed on the metal sulfides formed in the sediment. Euxinic conditions also can consume oxygen by reaction with aqueous sulfide and still remain anoxic, whereas strictly anoxic water can be oxygenated by addition of small amounts of oxygen. Euxinic conditions were probably more important in the past (Rhoades and Morse, 1971) when there were large inland marine seas such as the Devonian Chattanooga sea and the Cretaceous Pierre sea as well as narrow restricted ocean basins such as accompanied the opening of the Atlantic during the Cretaceous (Stein, 1986). A modern example is the Black Sea, which is euxinic because of a low-density cap of less saline water. Euxinic conditions are also favorable for an increased amount of sulfate reduction because sulfate reduction can occur in the water column and top few millimeters of the sediment as well as below the sediment surface. Other modern euxinic examples are nearshore basins such as offshore southern California and Saanich Inlet offshore Vancouver, Canada (Presley *et al.*, 1972), the Cariaco Trench offshore Venezuela (Richards, 1965), fjords off the coast of Norway (Jacobs and Emerson, 1982), and Canada (Loring, 1976), and parts of the Baltic Sea.

Metal-rich black shales are also formed near ocean floor vents (Koski *et al.*, 1984) and hot springs, producing "sedex" (sedimentary–exhalative) deposits. Modern examples of these metal enrichments are usually relatively small and local in extent and not of sufficient size to be considered as a petroleum resource. However, ancient analogs of these "black smokers" are often large and contain enough metal (grade and tonnage) to be an ore deposit. These black shales could be a host for sedimentary–exhalative metal sulfide deposits in Ireland (Mills *et al.*, 1987) and the Selwyn basin, Northwest Territories, Canada (Goodfellow and Jonasson, 1986) and may occur as stratiform deposits in Australia (Williams, 1978; Lambert, 1976). Other organic matter-associated metal enrichments are discussed in Leventhal (1986).

3.3. Differences from Petroleum Source Rocks

How do metal-rich black shales differ from petroleum source rocks? The occurrence of these shales can be more limited in extent. These shales need to have a sufficient metal availability, slow sedimentation rate (low clastic dilution), and abundant sulfate (marine environment) to be reduced to sulfide so as to form metal sulfides.

4. Metal Sources

The immediate source of moderate amounts of metals available for syngenetic incorporation into shales is seawater. Calculations show that normal seawater is only a good source in special cases, involving very slow sedimentation rates and efficient removal of metals (Holland, 1979). Unusual water circulation patterns, such as anti-estuarine circulation, have also been suggested (Brongersma-Sanders, 1965) as a mechanism to increase metal availability. Other mechanisms include localized input of extra metals from rapid leaching of volcanic ash (Leventhal and Kepferle, 1982) and hydrothermal activity from submarine vents.

Even in metal-rich shales, the only metal that is usually observable in hand specimens or by optical microscopy is iron as pyrite (FeS_2). Other metals that occur in concentrations of tens to hundreds of parts per million (ppm) (U, Cr, Mo, Ni, Cu, As, etc.) and thousands of parts per million (Zn and V) are usually not visible because they do not occur in a discrete mineral phase of the metal. Instead, these metals are dispersed in the organic matter, clay, or sulfide host and are not usually observed, even by electron microprobe (e.g., see Coveney *et al.*, 1987; Shaffer *et al.*, 1984; Leventhal and Knapp, 1988).

Colors of an organic-rich and metal-rich shale range from black to dark gray to brown. These colors are related to both organic carbon content and organic maturity. Immature samples that have been buried range from gray (about 0.5% organic C) to black (4% or more C). Very young sediments with high organic matter content that have not experienced burial or lithification may appear green, whereas ancient lithified shales are black. Samples that have been heated to more than 200°C will appear black even if they have only 0.5% C because the carbon has been partly converted to graphite.

4.1. Modes of Metal Occurrence

Metals can be associated with sulfides, organic matter, and/or clays (LeRiche, 1959). After syngenetic enrichment of metals, diagenetic and epigenetic mobilization of metals may translocate them to new sites. For example, vanadium originally chelated to organic matter may become a constituent of clay minerals after diagenesis and/or metamorphism. Nickel in organic matter may be translocated to a sulfide phase on the surface of pyrite. Leventhal and Knapp (1988) searched for metals in Devonian black shales using fission track methods. They showed that uranium was mainly in terrestrial vitrinite but not in pyrite or in marine *Tasmanites* microfossils. They looked for other trace elements (Ni, Cu, As, Mo, Co, and V) in the pyrite and pyrite rims, where concentrations of greater than 500 ppm could have been detected, but did not find any metals at this level.

4.2. Sites for Metals

Fe, Zn, and Mo can form intermetallic sulfides or pure sulfide phases. Other chalcophile elements (Cu, Ni) are not often observed as discrete phases but probably occur in pyrite. Elements such as U and V do not form sulfides under low-temperature conditions and seem to be associated with organic matter or perhaps clays. Although Mo can form sulfides, it is probably associated with organic matter because it mimics U in its enrichment (Hirst, 1974; Coveney *et al.*, 1987). Figure 1 shows a schematic of organic matter and trace elements (for example, U and a metalloid, As) as they go from the land by rivers to the

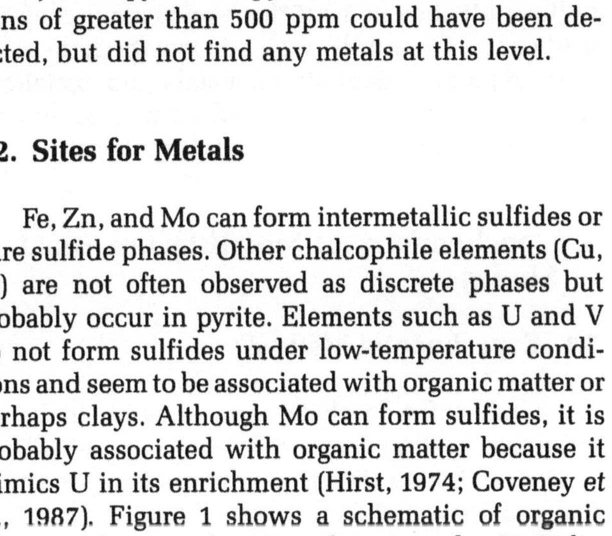

Figure 1. Pathways of organic matter and elements from terrestrial sources to the ocean. Associations of organic matter from terrestrial sources (in rectangles) and marine sources (in ellipses) with dissolved U and a metalloid (in circles) are shown. [From Leventhal (1983b).]

sea (Sholkovitz, 1978) and into the sediment and illustrates how the varying environment affects the metal and organic matter.

If a sediment has 10% organic C, it is accompanied by 1% organic oxygen in the organic matter. If each two oxygen atoms are chelating one metal atom, this sediment bears a metal (atomic weight 48) content of around 15,000 ppm and constitutes a metal-rich shale. Thus, even marine organic matter has the capacity to complex/fix a large amount of metals. The possibilities of forming metal sulfides containing metals other than Fe are also important. A marine sediment that contains 10% organic carbon may typically have 4% sulfide, which is mostly pyrite; however, important amounts of metals and metalloids such as As, Zn, Ni, Cu, Hg, and Ag may occur combined with the pyrite, or as separate phases.

5. Geochemistry of Metalliferous Black Shales

5.1. Experimental Methods

Total sulfur in shales was determined by the induction heating combustion and infrared (IR) detection method. Bulk organic carbon data (see Leventhal and Shaw, 1980) are obtained from Rock-Eval or LECO (Stanton *et al.*, 1983) analysis. Pyrolysis–gas chromatography is also used to characterize the kerogen (Leventhal, 1981; Leventhal *et al.*, 1986). Inductively coupled plasma (ICP) argon emission spectroscopy is used for multielement measurements (Lichte *et al.*, 1987).

5.2. Examples of Metalliferous Black Shales

Table I gives representative elemental values for a number of well-characterized organic-rich and/or metal-rich black shales. These shales represent a variety of geologic environments and depositional settings and range in age from Proterozoic to Quaternary. To set a reference frame, the average shales of Turekian and Wedepohl (1961) are not particularly high in organic C or metals and represent data available up to 1961. The average black shales of Vine and Tourtelot (1970) represent data based on 20 shale units, with a total of 779 samples. Vine and Tourtelot also gave 90th percentile values for the metals in their black shale set to define metal-rich shales. Relative to average shales, certain elements are greatly enriched (30-fold); this group comprises U, Mo, and sometimes V. Others are enriched by about the same amount as organic matter (4-fold); V, Ni, Cu, As, Pb, and Zn constitute this group. This enrichment appears to be a direct organic matter or nondilution effect for the latter group but a special organic or environmental effect for U, Mo, and sometimes V.

Metal enrichment is not due only to organic matter type, because the Scandinavian Alum Shale is marine organic matter, whereas the Mecca Quarry Shale has a substantial terrestrial component. Other enrichments that occur are rare earth elements (REE) with phosphate and Mn with calcite. Not listed in Table I are important examples of metal enrichment in freshwater sediments reported by Jenne (1977), Jackson *et al.* (1978), and Stevenson (1983) and in marine sediments from China (Fan, 1983).

5.3. Relationship between Organic Matter and Metals

The relationship between organic matter and trace elements/metals in shales can be investigated in a number of ways. It is useful to compare individual and mean values of trace elements/metals and to consider absolute ranges or mean values plus or minus one standard deviation. Another way is given in the compendium of Vine and Tourtelot (1970), where they report the value of the 90th percentile. The ratio of a metal to organic carbon can be used to normalize samples for purposes of comparison (Leventhal and Hosterman, 1982). More sophisticated statistical treatments such as linear least-squares regression analysis can be used. To obtain more detail for several elements, multiple regression analysis, factor analysis, and principal component analysis can be used (Davis, 1973; Tardy, 1975).

Graphical techniques can also be used. Trace elements can be plotted against each other and against major elements, such as U versus As, U versus organic carbon, and As versus S. Another graphical plot that is quite useful for recognizing multiple relationships is a "down-hole" plot. As a sample set, the core of Devonian Antrim Shale from the Michigan basin will be used below (Leventhal, 1980).

5.4. Antrim Shale

Table II gives data for shale samples from the Michigan Basin that include the Antrim Shale, Traverse limestone, and the overlying "false" Antrim that is sometimes mistaken in hand specimens. Ex-

TABLE II. Results of Analyses of Core Samples of Antrim Shale, Sanilac County, Michigan

Depth	Semiquantitative (%)						Quantitative														
							Parts per million											Percent		
(ft)	SiO_2	Al_2O_3	Fe_2O_3	CaO	K_2O	TiO_2	V	Mo	Co	Ni	Cu	Zn	Mn	Hg	As	U	Th	S	Organic C	Carbonate C
832	55	18	3.6	0.2	3.6	0.8	505	57	14	100	38	641	147	0.05	23	9.9	21	0.91	2.9	<0.01
871	54	21	4.8	0.1	3.8	0.8	185	<3	14	65	32	70	177	0.02	16	3.6	18	1.69	0.84	<0.01
913	56	16	3.6	0.2	3.4	0.7	195	19	14	55	30	114	117	0.03	18	7.6	16	1.44	2.4	<0.01
953	58	20	4.4	0.2	3.9	0.8	284	<3	17	65	30	55	141	0.03	25	5.0	14	1.28	0.96	<0.01
996	81	4	1.4	2.1	0.60	0.2	<25	<3	3	15	10	22	696	0.01	3.9	1.2	5	<0.01	0.22	0.95
1066	59	19	4.6	0.3	3.8	0.9	185	<3	35	75	25	69	364	0.12	46	4.0	19	0.09	0.42	<0.01
1136	62	18	5.1	0.3	3.3	0.9	160	<3	18	45	30	202	498	0.01	16	3.8	17	<0.01	0.31	0.41
1162	54	18	6.8	0.5	3.6	0.8	155	<3	40	60	42	1390	832	0.11	17	3.5	19	<0.01	0.41	0.87
1195	53	23	5.0	0.3	4.2	0.8	160	<3	20	65	35	78	379	0.02	5.2	4.5	19	<0.01	0.81	0.21
1216	51	20	6.8	0.4	3.7	0.7	140	<3	19	50	38	56	1180	0.02	5.1	3.5	16	<0.01	0.65	1.0
1221	51	19	4.2	0.2	3.5	0.7	1060	261	48	235	64	1720	114	0.08	50	36	—[a]	1.61	8.9	<0.01
1266	59	17	4.0	0.3	3.7	0.6	235	107	14	110	47	40	122	0.05	21	25	—	2.00	5.5	<0.01
1299	61	14	3.6	0.7	3.2	0.6	175	69	17	105	42	51	159	0.03	23	20	—	1.46	4.7	0.14
1320	60	12	3.1	0.3	3.0	0.5	155	83	21	100	44	186	110	0.03	15	23	—	1.51	6.8	0.07
1382	45	13	5.1	6.2	3.1	0.6	135	9	8	50	36	99	1190	0.03	27	18	19	1.19	0.96	2.97
1384	56	13	4.5	0.4	3.1	0.7	180	64	30	140	87	74	159	0.06	25	43	—	2.79	10.6	<0.01
1418	56	14	4.0	2.0	3.4	0.7	160	50	26	95	84	167	309	0.05	13	13	16	1.82	6.0	0.53
1431	43	12	15	0.4	2.8	0.5	200[b]	30[b]	7[b]	100[b]	100[b]	<300[b]	100[b]	—	—	23	—	15.2	6.3	0.07
1460	2.6	0.9	6.9	24	0.1	0.05	<25	<3	<1	25	7	18	3200	0.01	0.7	0.6	2	<0.01	0.02	12.0

[a]—, Not analyzed.
[b]Emission spectroscopy data.

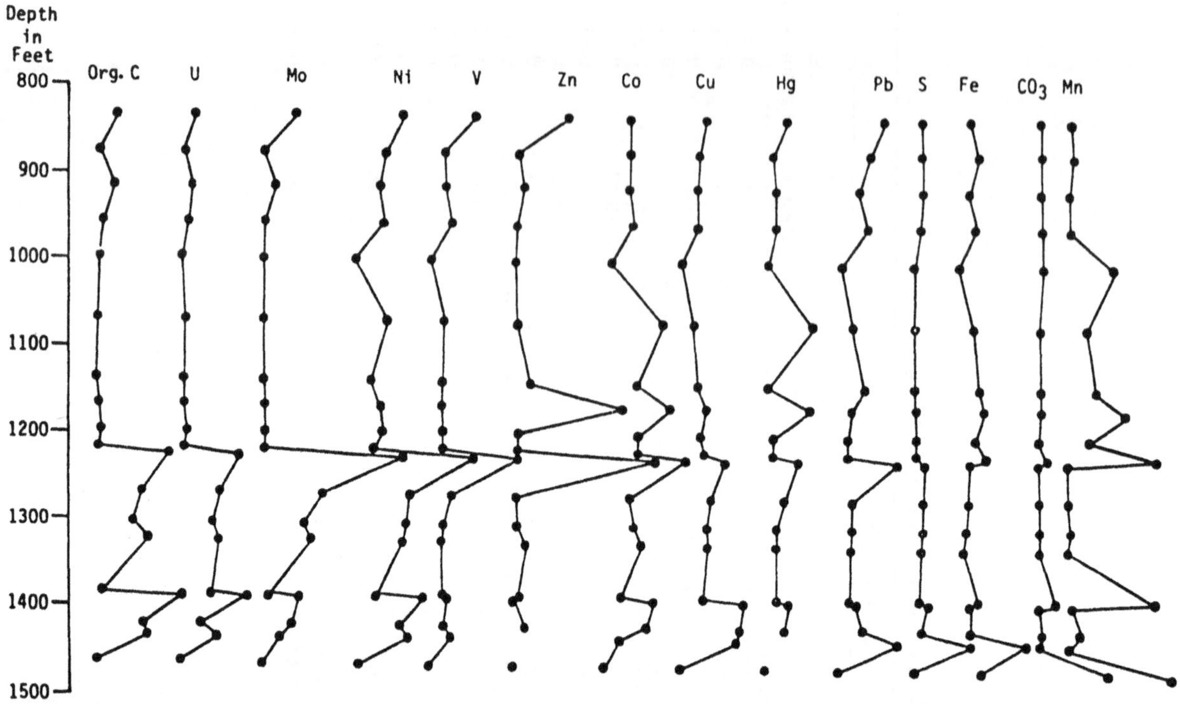

Figure 2. Down-hole plot of element concentrations for shale samples from Sanilac County, Michigan. Increasing element content (Table II) is to the right (different scales) and constituents showing similar patterns are nested together. [From Leventhal (1980).]

cept for the top sample of false Antrim, this unit is not enriched in metals, nor is the underlying Traverse limestone. Figure 2 shows the down-hole plot of the data, where the major constituents such as organic carbon and, to a lesser degree, sulfur control the enrichments of U, Mo, Ni, V, Zn, Co, Cu, As, Hg, and Pb. The abundance of sulfide commonly controls As, Hg, and Pb enrichment. Carbonate (CO_3) minerals often control Mn enrichments.

6. Carbon and Sulfur Relationships

The relationships between important redox-controlling inorganic and organic abundances can be used to define or constrain the environment of deposition.

The environment of deposition is usually an oxygenated water column overlying anoxic sediments (usually just below the water–sediment interface) and can be determined by several methods. Linear regression of a plot of S versus C gives an intercept on the S axis of 0 to a few tenths of a percent sulfur (Leventhal, 1987). In the case of dysaerobic (suboxic) water column (very low O_2, no H_2S), linear regression of the S versus C plot gives an intercept at the origin. For an euxinic environment (H_2S, no O_2), the linear regression of the S (total S or sulfide S) versus C (organic C) plot shows a positive intercept on the S axis that is greater than 1% (Leventhal, 1983a).

Figure 3 shows an idealized S versus C plot. The euxinic, normal marine, and nonmarine lines are shown. The data from real samples will not fall exactly on these lines. Data for samples from euxinic environments will fall in the region between A and B, and those for normal marine samples will fall in a region between lines X and Y. The region between the normal marine and euxinic lines represents the sulfide formed in the water column (Leventhal, 1983a, b, c).

Carbon–sulfur plots are used to infer the environment of deposition. Post-Devonian S/C values are usually around 0.4, but Ordovician and older host sediments (especially Proterozoic ones) often give S/C values that are higher, sometimes as high as 1. It is not clear whether these higher values are a result of lower oxygen values of the atmosphere and water column (especially in the Archean and Early Proterozoic) or are due to the more readily metabolized organic matter solely of marine origin (that is, pre-land plant, pre-Silurian age).

Problems of interpretation of S versus C plots are caused by lack of sulfate for reduction to sulfide or lack of reactive iron to combine with and preserve the sulfide. Usually, organic carbon, not iron or sulfate, is

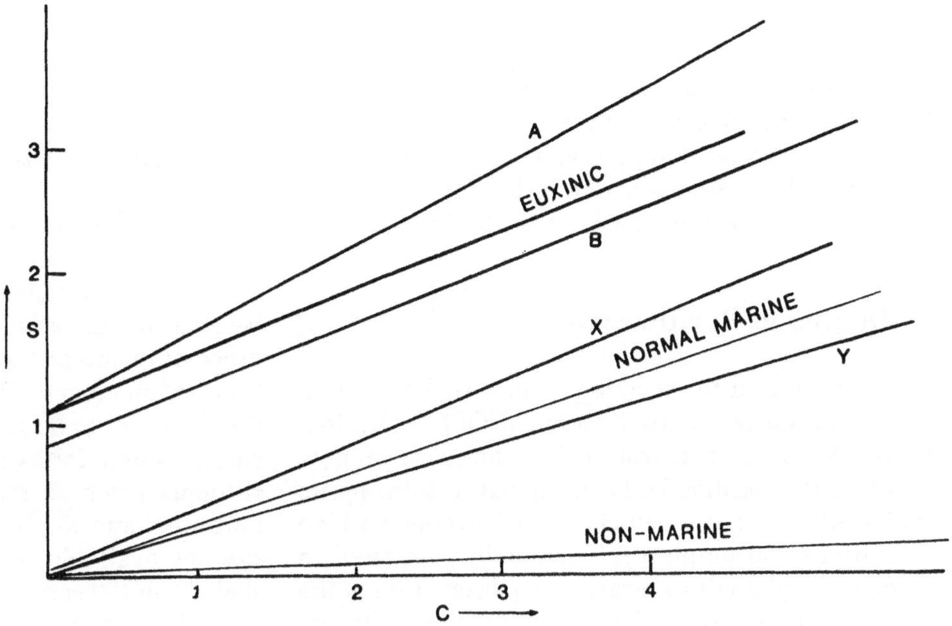

Figure 3. Schematic plot of sulfur organic carbon, showing the fields for euxinic environments (range from line A to line B), normal marine environments (range from line X to line Y), and nonmarine environments. [From Leventhal (1983a).]

the limiting factor (Fig. 3). For example, in rocks with very high organic carbon content (usually >10%), the Fe can be completely sulfidized (Fe limiting) and additional sulfide escapes because of lack of reactive iron. Another example of limiting Fe is in carbonate rocks that have low Fe contents (<1%) and moderate organic carbon; in this case, the Fe is all sulfidized. Thus, the S versus C plots only work well on shales with less than about 15% organic C and normal amounts of Fe (4–5%) because, again, iron becomes limiting to preserve the sulfide. Figure 4 shows an idealized S versus Fe plot. For a euxinic environment, if all the Fe is reactive, then the line will coincide with the pyrite line, corresponding to a stoichiometry of FeS_2. If some of the Fe is not reactive, the result will be a line that parallels the pyrite line and inter-

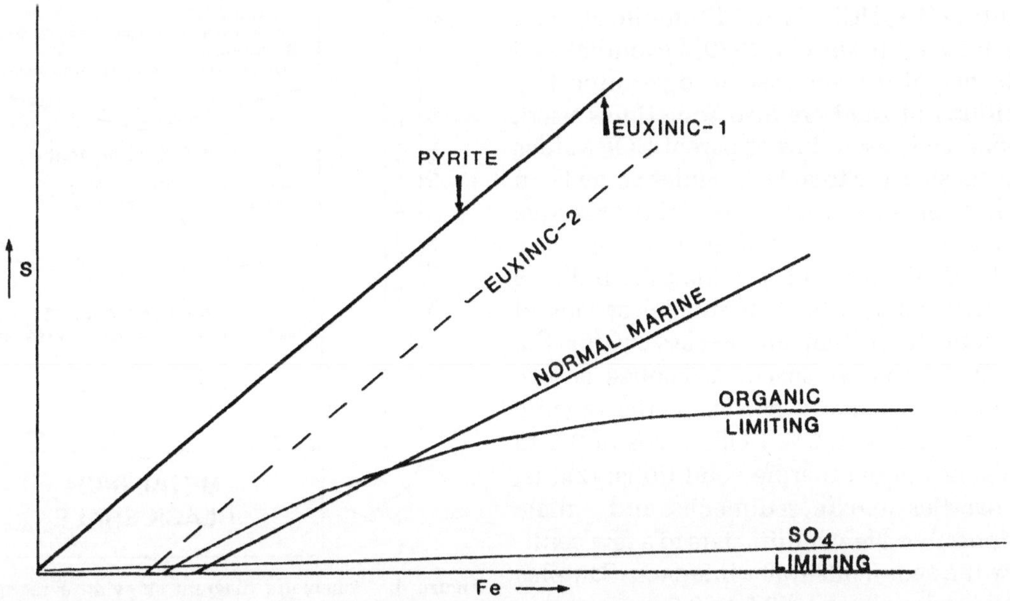

Figure 4. Schematic plot of sulfur vs. iron, showing fields for euxinic (euxinic-1 and -2), normal marine, and freshwater (SO_4 limiting) environments. The pyrite line corresponds to FeS_2 stoichiometry. The organic limiting line shows the effect of insufficient organic matter to provide enough sulfide to reduce the iron (Leventhal, 1983c).

cepts the Fe axis (in the euxinic-2 region). For a noneuxinic environment, the slope of this line is commonly not as steep as the pyrite stoichiometry line. If organic matter is limiting, the S values will have a lower slope at high Fe values. Sulfate-limiting (freshwater) conditions or the lack of anoxic conditions will give results that lie near the Fe axis.

7. Degree of Pyritization

Iron and sulfur reactions can also be demonstrated by degree of pyritization (DOP) determinations. Degree of pyritization is a measure of how much of the reactive Fe is sulfidized to form pyrite (Raiswell and Berner, 1985). It has several working definitions: the simplest is based on inspection of an S versus Fe plot with a pyrite line. Observation of the closeness (or lack thereof) of the data to the pyrite line or of the extent to which the data are distributed as an array, or cluster, parallel to the pyrite line (or randomly) provides a measure of sulfidation (Leventhal, 1980, 1983b, c). The S versus Fe plot is similar to a plot of pyrite iron (Fe_{py}) versus Fe. According to the original definition (Berner, 1970), DOP is the ratio of the pyrite iron to the sum of pyrite iron + leachable iron in the sample. The leachable iron is assumed to be reactive iron that would have reacted with sulfide, given enough time or sufficient sulfide concentration. By the Berner definition, DOP = $Fe_{py}/(Fe_{py} + Fe_x)$, where Fe_x is the Fe that is solubilized by a hot 12*N* HCl treatment of the sample for 1 min. Treatment with more dilute (1*N*) HCl or with dithionite at room temperature for 24 h (Canfield, 1989; Leventhal and Taylor, 1990) has also been used to determine Fe_x. Other definitions of DOP are also sometimes used; when Fe_{py}/total Fe is used, low apparent DOP values are obtained because the total Fe includes some Fe in silicate or other nonsulfidizable sites. This unreactive Fe may be as much as one-half of the total Fe.

Generally, DOP values of less than 0.1 indicate nonmarine environments (lack of sulfate) or lack of the anoxic conditions that are necessary for the sulfate-reducing microorganisms to metabolize; anoxic conditions are not generally present if the organic carbon content is below 0.2%. DOP values of 0.2 to 0.5 are typical of normal marine conditions; that is, oxic water overlies anoxic sediments, and sulfate reduction is possible a few millimeters to a few centimeters below the sediment–water interface. Samples with DOP values from around 0.5 to 0.7 were probably deposited below anoxic water column conditions. DOP values greater than about 0.7 indicate a euxinic, H_2S-containing, water column (Raiswell *et al.*, 1988).

8. Resources and Raw Materials

Future use of metal- and organic-rich shales, which often occur near industrial centers, as resources for raw materials is likely in the 21st century because of increased energy costs associated with processing and/or transporting higher grade ores in/from remote areas. The organic matter contained in the shale would be an inexpensive (readily available) energy source for extraction of the other useful constituents (such as metals, potash, phosphate, etc.). Leventhal and Kepferle (1982) estimated that the resources available from mining a 60-ft-thick black shale from 150 acres would yield V, Ti, Co, Sb, U, Mo, and Ni in amounts corresponding to the following percentages of annual U.S. production: V, 100%; Ti, 75%; Co, 47%; Sb, 30%; U, 14%; Mo, 4%; Ni, 3%.

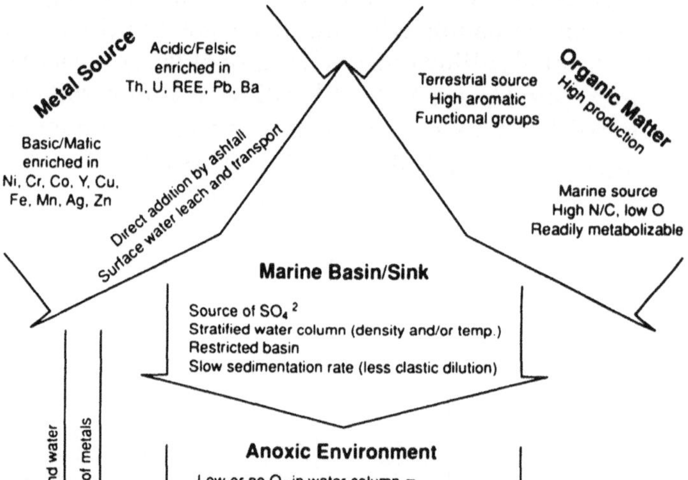

Figure 5. Schematic diagram showing a recipe for producing metalliferous black shales, starting with raw materials and suitable environments at the top. DOP is degree of pyritization (see text).

9. Summary

The possible roles of metal sources, organic matter in metal enrichment, and depositional environments of metalliferous black shales (Fig. 5) are summarized here:

(a) mobilization of metals from soil and source rock
(b) transportation of metals as soluble complexes
(c) concentration of metals directly by chelation (early in rivers) and by reduction and indirectly by facilitating sulfate reduction
(d) preservation of minerals by filling pores and keeping fO_2 low

ACKNOWLEDGMENTS. I thank the following colleagues for detailed and informative discussions: Joe Hatch, Marty Goldhaber, Roy Kepferle, Harry Tourtelot, George Desborough, and David G. Gee. I thank R. David Matthews and Dow Chemical for the Antrim core samples. I also thank Mark Stanton for lab support on carbon, sulfur, and pyrolysis–GC experiments. I acknowledge efforts of adapting ICP spectroscopy to geological samples by Fred Lichte. I thank Harry Tourtelot and Ken Watts for helpful reviews of the manuscript. This manuscript was written while I was an active member of the International Geological Correlation Program project 254 "Metalliferous Black Shales and Related Ore Deposits."

References

Armands, G., 1972, Geochemical studies of the Swedish Alum Shale, *Stockholm Contrib. Geol.* **27**:1–148.

Berner, R., 1970, Sedimentary pyrite formation, *Am. J. Sci.* **268**: 1–23.

Brongersma-Sanders, M., 1965, Metals of Kupferschiefer supplied by normal sea water, *Geol. Rundsch.* **55**:363–375.

Brumsack, H. J., 1980, Geochemistry of Cretaceous black shales from the Atlantic Ocean, *Chem. Geol.* **31**:1–25.

Canfield, D., 1989, Reactive iron in marine sediments, *Geochim. Cosmochim. Acta* **53**:619–632.

Conant, L., and Swanson, V., 1961, Chattanooga Shale and Related Rocks of Central Tennessee, U.S. Geological Survey Professional Paper 357.

Coveney, R., Leventhal, J., Glasscock, M., and Hatch, J., 1987, Origins of metals and organic matter in Mecca Quarry Member and stratigraphically equivalent beds in the midwest, *Econ. Geol.* **81**:915–933.

Davis, J. C., 1973, *Statistics and Data Analysis in Geology*, John Wiley & Sons, New York.

Degens, E., 1965, *Geochemistry of Sediments*, Prentice-Hall, Englewood Cliffs, New Jersey, p. 202.

Desborough, G., Pitman, J., and Huffman, C., 1976, Concentration and mineralogical residence of elements in the Green River shale, *Chem. Geol.* **17**:13–26.

Duncan, D., and Swanson, V., 1965, Organic rich shales of the U.S. and world land areas, *U.S. Geol. Surv. Circ.* **523**.

Fan Delian, 1983, Polyelements in the Lower Cambrian black shale series in southern China, in: *The Significance of Trace Elements in Solving Petrogenetic Problems*, Theophrastus Publications, Athens, pp. 447–474.

Glikson, M., Chappell, B., Freeman, R., and Webber, E., 1985, Trace elements in oil shales, source, organic association in Australian deposits, *Chem. Geol.* **53**:155–174.

Goldhaber, M., and Kaplan, I., 1973, The sulfur cycle, in: *The Sea*, Vol. 5 (E. Goldberg, ed.), John Wiley & Sons, New York, pp. 569–655.

Goodfellow, W., and Jonasson, I., 1986, Environment of formation of the Howards Pass Zn–Pb deposit, Canada, in: *Mineral Deposits of the Northern Cordillera* (J. A. Morin, ed.), *Can. Inst. Min. Metall. Spec. Vol.* **37**:19–50.

Hallberg, R., 1978, Metal–organic interaction at the redoxcline, in: *Environmental Biogeochemistry and Geomerobiology* (W. E. Krumbein, ed.), Vol. 3, pp. 947–953.

Hirst, D., 1974, Geochemistry of sediments from Black Sea cores, in: *The Black Sea* (D. Ross and E. Degens, eds.), American Association of Petroleum Geologists Memoir 20, pp. 430–455.

Holland, H. D., 1979, Metals in black shales: A reassessment, Econ. Geol. **74**:1676–1680.

Hunt, J. M., 1979, *Petroleum Geochemistry and Geology*, W. H. Freeman and Co., San Francisco.

Jackson, K., Jonasson, I., and Skippen, G., 1978, Nature of metals–sediment–water interactions in freshwater with the role of organic matter, *Earth Sci. Rev.* **14**:97–146.

Jacobs, L., and Emerson, S., 1982, Trace metal solubility in an anoxic fjord, *Earth Planet. Sci. Lett.* **60**:237–252.

Jenne, E., 1977, Trace element sorption by sediments and soils, in: *Molybdenum in the Environment* (W. Chappell and K. Peterson, eds.), Marcel Dekker, New York, pp. 425–553.

Koski, R., Clague, D., and Oudin, E., 1984, Mineralogy and chemistry of massive sulfide deposits from the Juan de Fuca Ridge, *Geol. Soc. Am. Bull.* **95**:930–945.

Lambert, I., 1976, The McArthur Zn–Pb–Ag deposit, in: *Handbook of Stratabound and Stratiform Ore Deposits* (K. H. Wolf, ed.), Vol. 6, Elsevier, Amsterdam, pp. 535–585.

LeRiche, H., 1959, Distribution of metals in Lower Lias of southern England, *Geochim. Cosmochim. Acta* **16**:101–122.

Leventhal, J., 1980, Comparative Geochemistry of Devonian Shale Cores from West Virginia, Illinois, Indiana and Michigan, U.S. Geological Survey Open-File Report 80-938, pp. 1–32.

Leventhal, J., 1981, Pyrolysis–gas chromatography–MS to characterize organic matter related to uranium from Devonian shales, *Geochim. Cosmochim. Acta* **45**:883–889.

Leventhal, J., 1983a, Interpretation of C and S relationships in Black Sea sediments, *Geochim. Cosmochim. Acta* **47**:133–137.

Leventhal, J., 1983b, Organic carbon, sulfur and iron relationships in ancient shales as indicators of environments of deposition, *U.S. Geol. Surv. Circ.* **822**:34–36.

Leventhal, J., 1983c, Organic carbon, sulfur and iron relationships in ancient shales as indicators of environment of deposition, EoS **64**:739.

Leventhal, J., 1986, Roles of organic matter in ore deposits, in: *Organics and Ore Deposits* (W. Dean, ed.), Denver Region Exploration Geologists Symposium, Wheat Ridge, Colorado, pp. 7–20.

Leventhal, J., 1987, Carbon and sulfur relationships in Appalachian Devonian shales as an indicator of environment of deposition, *Am. J. Sci.* **287**:23–40.

Leventhal, J., 1990, Comparative geochemistry of metals and trace elements from Cambrian Alum Shale of Sweden, in: *Sediment Hosted Mineral Deposits* (J. Parnell, ed.), International Association of Sedimentology Spec. Publ. No. 11, Blackwell Scientific, Oxford, pp. 203–215.

Leventhal, J., and Hosterman, J., 1982, Chemical and mineralogical analysis of Devonian black shale samples from Kentucky, Ohio, Virginia and Tennessee, *Chem. Geol.* **37**:239–264.

Leventhal, J., and Kepferle, R., 1982, Geochemistry and geology of strategic metals and fuels in Devonian shales of the eastern interior basins, U.S., in: *Synthetic Fuels from Oil Shale II* (R. D. Matthews, ed.), Institute of Gas Technology, Chicago, pp. 73–96.

Leventhal, J., and Knapp, T., 1988, Preliminary examination of Devonian Appalachian black shales by optical microscopy, electron microprobe and fission track methods, in: *Proceedings of IGCP Inaugural Meeting* (J. Pasava and Z. Gabriel, eds.), Czechoslovakia Geological Survey, Prague, pp. 57–65.

Leventhal, J., and Shaw, V., 1980, Organic matter in Devonian black shale: Comparison and short range variations, *J. Sediment. Petrol.* **50**:77–81.

Leventhal, J., and Taylor, C., 1990, Comparison of methods to determine degree of pyritization, *Geochim. Cosmochim. Acta* **54**:2621–2625.

Leventhal, J., Crock, J., Mountjoy, W., Thomas, J., Shaw, V. E., Briggs, P. H., Wahlberg, J. S., and Malcolm, M. J., 1978, Results of Analysis of USGS Black Shale Standard SDO-1, U.S. Geological Survey Open-File Report 78-447.

Leventhal, J., Crock, J., and Malcolm, M., 1981, Geochemistry of Trace Elements and Uranium in Devonian Shales of the Appalachian Basin, U.S. Geological Survey Open-File Report 81-778.

Leventhal, J., Briggs, P., and Baker, J. W., 1983, Geochemistry of the Chattanooga Shale, Tennessee, *Southeast. Geol.* **24**:101–116.

Leventhal, J., Daws, T., and Frye, J., 1986, Organic geochemical analysis of organic matter associated with uranium, *Appl. Geochem.* **1**:241–247.

Lichte, F., Golightly, D., and Lamothe, P., 1987, Inductively coupled plasma-atomic emission spectroscopy, in: *Methods for Geochemical Analysis* (P. Baedecker, ed.), U.S. Geological Survey Bulletin 1770, pp. B1–B10.

Loring, D., 1976, Distribution of metals in sediments of Saguenay fjord, *Can. J. Earth Sci.* **13**:1706–1718.

Mills, H., Halliday, A., Ashton, J., Anderson, I., and Russell, M., 1987, Origin of a giant orebody at Navan, Ireland, *Nature* **327**: 223–226.

Patterson, J., Ramsden, A., and Dale, L., 1988, Geochemistry and mineralogy residences of trace elements in oil shales from Australia, *Chem. Geol.* **67**:327–340.

Presley, B., Kolodny, Y., Nissenbaum, A., and Kaplan, I., 1972, Early diagenesis in a reducing fjord—trace element distribution in pore water and sediment, *Geochim. Cosmochim. Acta* **36**: 1073–1090.

Quinby-Hunt, M., Wilde, P., Orth, C., and Berry, W., 1989, Elemental geochemistry of black shales—statistical comparison, in: *Metalliferous Black Shales and Related Ore Deposits—1988 Working Group Meeting of IGCP 254* (R. Grauch and J. Leventhal, eds.), U.S. Geological Survey Circular 1037, pp. 10–15.

Raiswell, R., and Berner, R., 1985, Pyrite formation in euxinic and semi-euxinic sediments, *Am. J. Sci.* **285**:710–724.

Raiswell, R., and Berner, R., 1986, Pyrite and organic matter in Phanerozoic normal marine shales, *Geochim. Cosmochim. Acta* **50**:1967–1976.

Raiswell, R., Buckley, F., Berner, R., and Anderson, T., 1988, Degree of pyritization of iron as a paleoenvironmental indicator of bottom-water oxygenation, *J. Sediment. Petrol.* **58**:812–819.

Rhoades, D., and Morse, J., 1971, Evolutionary and ecologic significance of oxygen deficient basins, *Lethaia* **4**:413–428.

Richards, F. A., 1965, Anoxic basins and fjords, in: *Chemical Oceanography* (J. Riley and G. Skirrow, eds.), Vol. I, Academic Press, New York, pp. 611–645.

Shaffer, N., Leininger, R., and Gilstrap, M., 1984, Composition of New Albany Shale in southeastern Indiana, in: *1983 Eastern Oil Shale Symposium*, University of Kentucky, Institute for Mining and Minerals Research, Lexington, pp. 195–205.

Sholkovitz, E., 1978, Flocculation of dissolved Fe, Mn, Al, Cu, Ni and Cd during estuarine mixing, *Earth Planet. Sci. Lett.* **41**: 77–86.

Spears, D. A., 1980, Towards a classification of shales, *J. Geol. Soc. London* **137**:125–129.

Stanton, M., Leventhal, J., and Hatch, J., 1983, Short Range Vertical Variation in Organic C and S in Upper Pennsylvanian Stark Shale, Kansas, U.S. Geological Survey Open-File Report 83-315.

Stein, R., 1986, Surface water paleo-productivity as inferred from sediments in oxic and anoxic deepwater, in: *Biogeochemistry of Black Shales* (E. Degens, ed.), *Mitt. Geol.-Palaeont. Inst. Univ. Hamburg.* **60**:55–70.

Stevenson, F. J., 1983, Trace metal–organic interactions in geologic environments, in: *The Significance of Trace Elements in Solving Petrogenic Problems*, Theophrastus Publications, Athens, pp. 671–691.

Tardy, Y., 1975, Element partition ratios in sedimentary environment, *Sci. Geol. Bull. Strasbourg* **28**:75–95.

Tourtelot, H. A., 1979, Black Shale—its deposition and diagenesis, *Clays Clay Miner.* **27**:313–320.

Turekian, K., and Wedepohl, K., 1961, Distribution of elements in some major units of the earth's crust, *Geol. Soc. Am. Bull.* **72**: 175–192.

Vine, J., and Tourtelot, E., 1970, Geochemistry of black shale deposits, *Econ. Geol.* **65**:253–272.

Williams, N., 1978, Studies of base metal sulfide deposits at McArthur River, Northern Territory, Australia, *Econ. Geol.* **73**: 1005–1056.

Chapter 28

Trace and Minor Elements in Coal

ROBERT B. FINKELMAN

1. Introduction

Coal will be a major energy source in the United States and in many other countries well into the 21st century. Although coal is composed predominantly of organic matter, inorganic constituents in coal commonly attract more attention and can ultimately determine how the coal will be used.

Coal is defined as having more than 50% carbonaceous material (Wood *et al.*, 1983). The remainder of the coal consists of discrete mineral grains and organically bound or associated inorganic elements, together referred to as the inorganic constituents. Except for a few extremely rare elements (polonium, astatine, francium, actinium, protactinium), every element has been found in coal. In ash (the inorganic residue from the complete incineration of coal), elemental concentrations range from the parts per trillion level to more than 50 wt. %. Table I contains estimates of the arithmetic and geometric means for each of 79 elements reported in U.S. coal. Although few, if any, coal samples have compositions similar to that of the "average" U.S. coal, these values may be useful in establishing norms and for comparisons or other evaluations. Table I also contains information on the maximum concentrations encountered and the number of analyzed samples in the U.S. Geological Survey's National Coal Resources Data System (NCRDS). In most coal samples the contents of silicon, aluminum, sulfur, and iron may be as much as several percent by weight. For all other elements (excluding carbon, oxygen, hydrogen, and nitrogen), contents are usually less than 1 wt. %. For this review, we will consider all inorganic elements in coal to be trace or minor constituents.

The elements in coal can occur in a wide variety of chemical forms or modes of occurrence. It is important to emphasize that it is the mode of occurrence of an element that dictates its behavior and its technological, environmental, and economic impact. This aspect will be discussed more fully below. Because most inorganic elements in coal, especially those in high-rank (bituminous and anthracite) coal, occur in minerals (Finkelman, 1981a), the following discussion will deal not only with trace and minor elements in coal but also with the minerals in which they reside.

The inorganic constituents in coal can have profound effects on just about every aspect of coal utilization. They affect mining, cleaning, combustion, conversion, and disposal (see examples in Section 5 of this chapter). The magnitude of these effects is dependent, in part, on the element's concentration and mode of occurrence and on the presence of other elements. In the following section I shall describe the factors that cause the vertical and lateral variation of

ROBERT B. FINKELMAN • U.S. Geological Survey, Reston, Virginia 22092.
Organic Geochemistry, edited by Michael H. Engel and Stephen A. Macko. Plenum Press, New York, 1993.

TABLE I. Arithmetic and Geometric Means for Ash and Chemical Elements in U.S. Coal

Component	Arithmetic		Geometric		Max	No. of samples
	Mean	SD	Mean	SD		
Ash, %	13.1	8.3	10.9	1.9	50.0	7976
Aluminum (Al), %	1.5	1.1	1.1	2.1	10.6	7882
Antimony (Sb), ppm	1.2	1.6	0.61	3.6	35	7473
Arsenic (As), ppm	24	60	6.5	5.5	2200	7676
Barium (Ba), ppm	170	350	93	3.0	22000	7836
Beryllium (Be), ppm	2.2	4.1	1.3	3.5	330	7484
Bismuth (Bi), ppm	(<1.0)[b]	n.d.[c]	n.d.	n.d.	14	128
Boron (B), ppm	49	54	30	3.1	1700	7874
Bromine (Br), ppm	17	19	9.1	4.1	160	4999
Cadmium (Cd), ppm	0.47	4.6	0.02	18	170	6150
Calcium (Ca), %	0.46	1.0	0.23	3.3	72	7887
Carbon (C), %	63	15	62	1.3	90	7154
Cerium (Ce), ppm	21	28	5.1	7.1	700	5525
Cesium (Cs), ppm	1.1	1.1	0.70	3.2	15	4972
Chlorine (Cl), ppm	614	670	79	41	8800	4171
Chromium (Cr), ppm	15	15	10	2.7	250	7847
Cobalt (Co), ppm	6.1	10	3.7	2.9	500	7800
Copper (Cu), ppm	16	15	12	2.1	280	7911
Dysprosium (Dy), ppm	1.9	2.7	0.008	35	28	1510
Erbium (Er), ppm	1.0	1.1	0.002	73	11	1792
Europium (Eu), ppm	0.40	0.33	0.12	5.8	4.8	5268
Fluorine (F), ppm	98	160	35	15	4000	7376
Gadolinium (Gd), ppm	[1.8][d]	n.d.	n.d.	n.d.	39	2376
Gallium (Ga), ppm	5.7	4.2	4.5	2.1	45	7565
Germanium (Ge), ppm	5.7	14	.59	16	780	5689
Gold (Au), ppm	(<0.05)	n.d.	n.d.	n.d.	n.d.	n.d.
Hafnium (Hf), ppm	0.73	0.68	0.04	38	18	5120
Holmium (Ho), ppm	[0.35]	n.d.	n.d.	n.d.	4.5	1130
Hydrogen (H), %	5.2	0.9	5.2	1.2	9.5	7155
Indium (In), ppm	(<0.3)	n.d.	n.d.	n.d.	n.d.	n.d.
Iodine (I), ppm	(<1.0)	n.d.	n.d.	n.d.	n.d.	n.d.
Iridium (Ir), ppm	(<0.001)	n.d.	n.d.	n.d.	n.d.	n.d.
Iron (Fe), %	1.3	1.5	0.75	2.9	24	7882
Lanthanum (La), ppm	12	16	3.9	6.0	300	6235
Lead (Pb), ppm	11	37	5.0	3.7	1900	7469
Lithium (Li), ppm	16	20	9.2	3.3	370	7848
Lutetium (Lu), ppm	0.14	0.10	0.06	4.7	1.8	5008
Magnesium (Mg), %	0.11	0.12	0.07	2.7	1.5	7887
Manganese (Mn), ppm	43	84	19	3.9	2500	7796
Mercury (Hg), ppm	0.17	0.24	0.10	3.1	10	7649
Molybdenum (Mo), ppm	3.3	5.6	1.2	6.5	280	7107
Neodymium (Nd), ppm	[9.5]	n.d.	n.d.	n.d.	230	4749
Nickel (Ni), ppm	14	15	9.0	2.8	340	7900
Niobium (Nb), ppm	2.9	3.1	1.0	7.7	70	6843
Nitrogen (N), %	1.3	0.4	1.3	1.4	13	7153
Osmium (Os), ppm	(<0.001)	n.d.	n.d.	n.d.	n.d.	n.d.
Oxygen (O), %	16	12	12	2.0	60	7151
Palladium (Pd), ppm	(<0.001)	n.d.	n.d.	n.d.	n.d.	n.d.
Phosphorus (P), ppm	430	1500	20	20	58000	5079
Platinum (Pt), ppm	(<0.001)	n.d.	n.d.	n.d.	n.d.	n.d.
Potassium (K), %	0.18	0.21	0.10	3.5	2.0	7830
Praseodymium (Pr), ppm	(2.4)	n.d.	n.d.	n.d.	65	1533
Rhenium (Re), ppm	(<0.001)	n.d.	n.d.	n.d.	n.d.	n.d.
Rhodium (Rh), ppm	(<0.001)	n.d.	n.d.	n.d.	n.d.	n.d.
Rubidium (Rb), ppm	21	20	0.62	41	140	2648
Ruthenium (Ru), ppm	(<0.001)	n.d.	n.d.	n.d.	n.d.	n.d.

TABLE I. *(Continued)*

Component	Arithmetic Mean	Arithmetic SD	Geometric Mean	Geometric SD	Max	No. of samples
Samarium (Sm), ppm	1.7	1.4	0.35	13	18	5151
Scandium (Sc), ppm	4.2	4.4	3.0	2.3	100	7803
Selenium (Se), ppm	2.8	3.0	1.8	3.1	150	7563
Silicon (Si), %	2.7	2.4	1.9	2.4	20	7846
Silver (Ag), ppm	(<0.1)	0.35	0.01	9.1	19	5038
Sodium (Na), %	0.08	0.12	0.04	3.5	1.4	7784
Strontium (Sr), ppm	130	150	90	2.5	2800	7842
Sulfur (S), %	1.8	1.8	1.3	2.4	25	7214
Tantalum (Ta), ppm	0.22	0.19	0.02	13	1.7	4622
Tellurium (Te), ppm	(<0.1)	n.d.	n.d.	n.d.	n.d.	n.d.
Terbium (Tb), ppm	0.30	0.23	0.09	7.7	3.9	5024
Thallium (Tl), ppm	1.2	3.4	0.00004	205	52	1149
Thorium (Th), ppm	3.2	3.0	1.7	5.0	79	6866
Thulium (Tm), ppm	[0.15]	n.d.	n.d.	n.d.	1.9	365
Tin (Sn), ppm	1.3	4.3	0.001	54	140	3004
Titanium (Ti), (%)	0.08	0.07	0.06	2.2	0.74	7653
Tungsten (W), ppm	1.0	7.6	0.10	14	400	4714
Uranium (U), ppm	2.1	16	1.1	3.5	1300	6923
Vanadium (V), ppm	22	20	17	2.2	370	7924
Ytterbium (Yb), ppm	[0.95]	n.d.	n.d.	n.d.	20	7522
Yttrium (Y), ppm	8.5	6.7	6.6	2.2	170	7897
Zinc (Zn), ppm	53	440	13	3.4	19000	7908
Zirconium (Zr), ppm	27	32	19	2.4	700	7913

[a]All values are on a coal basis. Data are exclusively from the U.S. Geological Survey's National Coal Resources Data System (NCRDS), except where otherwise noted.
[b]Values in parentheses are estimates based on NCRDS and literature data.
[c]n.d., No data.
[d]Values in brackets are calculated from cerium and lanthanium data and assuming a chondrite normalized rare-earth element distribution pattern.

the inorganic constituents in coal; then I shall discuss the modes of occurrence and organic–inorganic interactions.

2. Source and Variation of Inorganic Constituents

The abundance of the inorganic constituents in coal varies at every level—between coal basins, between coal beds within a basin, and within coal beds—over distances from micrometers to kilometers. The short-range variations can be as significant as the long-range ones (Dulong *et al.*, 1986). Figure 1 illustrates the types of short-range variations that are common in many coals.

Due to the influence of inorganic constituents on coal utilization, they have received considerable attention (Gluskoter *et al.*, 1977; Valkovic, 1983; Glick and Davis, 1987; Swaine, 1990; Finkelman, 1990), yet we have not achieved widely applicable models to predict their distribution.

The difficulty in developing accurate predictive models is evident when one considers that biological, geological, hydrologic, and geochemical factors (Fig. 2) all can strongly influence the distribution of the inorganic constituents, causing substantial vertical and lateral variations. Current geochemical research is attempting to sort out the relative influences of these factors on coal quality.

2.1. Detrital Input

Detrital material (air- and water-borne particulates and dissolved species) is a major source of the inorganic constituents in many coals (Lindahl and Finkelman, 1986) as shown by the mineralogical composition and the textural relationships of the minerals in moderate- to high-ash coals. Many of the minerals in the moderate- to high-ash coals occur in bands intermixed with fragments of organic matter. The minerals in these bands are generally subangular to subrounded and consist largely of quartz (some

Figure 1. Scanning electron photomicrograph of a polished block of coal, depicting short-range heterogeneity of minerals and organic matter. Scale bar = 100 μm.

rutilated), illite, mixed-layer clays, rutile, rare-earth phosphates, and zircon (Finkelman, 1981b).

Although many of the chemical elements may have been originally associated with detrital particles, some elements can be substantially remobilized. Cecil *et al.* (1979) noted the strong correlation of copper, lead, and zinc with silicon, aluminum, and ash in an Appalachian bituminous coal. Yet mineralogical data indicate that the copper, lead, and zinc are not associated with the aluminosilicates but are found in micrometer-size accessory sulfide (Fig. 3) and selenide minerals in the organic matrix (Finkelman, 1985). The textural relationships indicate that the accessory sulfide minerals are authigenic. The chalcophile elements were probably introduced to the coal-forming basin along with other detrital components, accounting for the strong positive statistical correlation with aluminosilicates, then remobilized and precipitated as newly formed sulfides and selenides.

The positive correlation of the remobilized species with components with which they were formerly associated has led some researchers to postulate or imply chemical associations that no longer exist in the coal. For example, Roscoe and Hopke (1982), using factor analysis, showed that the rare-earth elements are predominantly associated with aluminosilicates. This type of approach cannot distinguish the physical association of rare-earth phosphates with clays and quartz, thus leaving the erroneous impression of a chemical association.

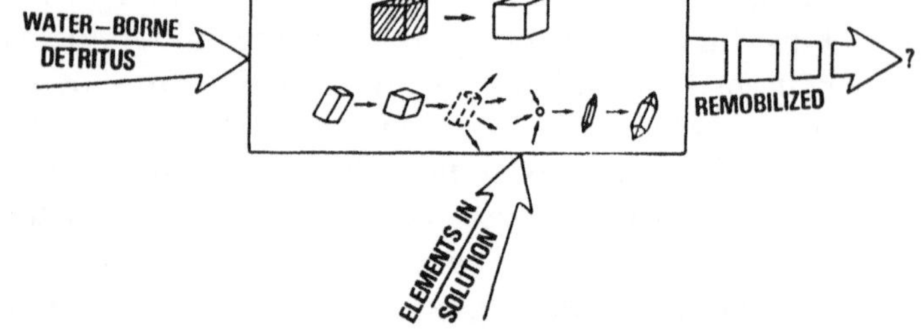

Figure 2. Diagram depicting the pathways available to inorganic constituents for entering and leaving the coal system. [From Finkelman (1981c).]

Figure 3. Scanning electron photomicrograph depicting sphalerite (ZnS) containing a clausthalite (PbSe) inclusion. Scale bar = 1 μm.

Identifying the source of the detrital components in coal may help to decipher the evolution of the depositional basin. In the Powder River Basin coals and overburden, selenium is primarily associated with organic matter; lesser amounts are found in pyrite, sphalerite, clausthalite, and other forms. The selenium content of the Powder River Basin coals decreases from south to north (from 1.7 ppm to 0.7 ppm) while maintaining a strong positive statistical correlation ($r = 0.6$) with sulfur, ash, and other ash-forming components in the coal. Oman *et al.* (1988) suggested that the inorganic components of the Powder River Basin coals were primarily derived from seleniferous Cretaceous sedimentary rocks that crop out to the south of the basin. Soluble selenates and sulfates were transported by a northward-flowing fluvial system, along with particulate detritus, into Paleocene swamps where the selenium and sulfur were fixed by the organic matter. The geochemical evidence from the coals is more consistent with the northward-flowing fluvial system proposed by Flores and Hanley (1984) than with the alternate concept that these coals formed in a lake with a source from the east (Ayers and Kaiser, 1984).

The inorganic constituents in coal have played a role in one of the most stimulating geological controversies, the extinction of the dinosaurs. Some of the earliest evidence cited for the impact event that purportedly caused the demise of much of the life in the Cretaceous period came from studies of iridium-rich partings in coal beds straddling the Cretaceous–Tertiary boundary (Gilmore *et al.*, 1984). The source of the anomalous trace-element concentrations in these coal partings is still hotly debated. In contrast, there is little doubt that windblown dust and "cosmic" dust occur in all coals, but these sources are generally of minor significance. Of far more importance is the airborne volcanic detritus that can be a major source of trace elements in some coals, especially Cretaceous western U.S. coal (Bohor *et al.*, 1979; Crowley *et al.*, 1989). It has been widely believed that Pennsylvanian coal from the eastern United States received little volcanic input. However, recent detailed mineralogical work by Lyons *et al.* (1990) indicates that volcanic input in Appalachian coals may be significant. They found evidence of substantial volcanic input in partings from at least nine Appalachian coals.

2.2. Biological Input

Detrital input plays a minor role in most low-ash (less than about 8 wt. %), high-rank coals and in many low-rank coals (Finkelman, 1981c). In these coals many inorganic constituents are associated with the organic fraction; they are primarily attached to carboxylic functional groups or occur as chelate complexes (Miller and Given, 1978; Benson *et al.*, 1992). In low-rank coal (lignite and sub-bituminous), most of the sodium, magnesium, calcium, strontium, and barium is generally associated with the organic matter, as are lesser amounts of many other elements (Finkelman, 1981c).

Most of the elements that are organically complexed in the low-rank coals are essential for plant nutrition. Major elements essential for plant metabolism include phosphorus, potassium, sulfur, calcium, and magnesium; minor elements include boron, chlorine, copper, iron, manganese, molybdenum, and zinc (Severson and Shacklette, 1988). Plant material accumulating in the peat bog may contribute a substantial proportion of these elements to the resulting coal.

Plants may contribute inorganic constituents to coal in other ways. Several authors have noted the presence of crystalline and noncrystalline biological material, such as phytoliths and sponge spicules in peats (Andrejko and Cohen, 1984; Raymond *et al.*, 1990). Some of this biogenically derived material may be preserved in coal.

Bacteria may play an important role in modifying concentrations by remobilizing or fixing elements in coal. Microbial degradation of plant material is a crucial step in the recycling of nutrients (Gorham and Santelmann, 1984). Recent work indicates that microbial precipitation of iron, manganese, selenium, sulfur, and uranium may be an important mechanism for concentrating these elements in coals (Robbins *et al.*, 1990a, b).

2.3. Effect of Climate

Climate can influence coal geochemistry through its effects on peat development. Not only can the climate dictate the types and amounts of plant material and detrital input, but it can significantly affect the chemistry of surface and ground waters. Cecil *et al.* (1985) have suggested that in dry-seasonal tropical climates, where evaporation periodically exceeds precipitation, flat or planar peat bodies are favored. These peats are primarily nurtured by the nutrients in near-neutral surface and ground waters. These conditions can result in peats and subsequent coals having relatively high ash and high sulfur contents. In an ever-wet tropical climate, where rainfall exceeds evaporation, the formation of domed or raised peat bogs is favored. These low-pH peats are nurtured primarily by rainwater with a low complement of dissolved solids. These conditions result in peats and subsequent coals that are relatively low in ash and low in sulfur.

2.4. Adsorption from Water

Various forms of carbon are widely used as filters to remove undesired chemicals from drinking water, laboratory water, and streams. Coals can behave as natural filters, effectively adsorbing ions from percolating groundwater. Enrichment of elements such as germanium, beryllium, vanadium, nickel, antimony, molybdenum, and uranium in the margins of coal beds is likely the result of this process (Zubovic, 1966; Karner *et al.*, 1984; Kuellmer *et al.*, 1987). In many locations, coals act as aquifers and are exposed to trace elements in groundwater throughout their history. The propensity of coal to act as a sink for certain inorganic constituents is being investigated as a mechanism for cleaning up degraded groundwater (Davis and Dodge, 1986).

Similar processes may have been in effect long before the formation of the coal. McCarthy *et al.* (1989) suggested that subaqueous decaying plant material can incorporate significant amounts of metals, which will then be incorporated into the resultant coal.

2.5. Diagenesis

In the swamp environment, significant changes occur that involve many of the inorganic constituents. As the decayed plant matter is coalified and upranked (altered from peat to bituminous coal and ultimately to anthracite), the organic functional groups are destroyed, thereby releasing the chemically combined trace metals. Some of these metals precipitate as sulfides, some may be adsorbed onto clays in the coal, and some are flushed from the system. Data from Swanson *et al.* (1976) show a progressive decline in the concentrations of calcium, magnesium, sodium, boron, barium, and strontium with increasing coal rank. Hatch (1983) illustrated the decrease of magnesium with increasing coal rank in Rocky Mountain and Northern Great Plains coal samples.

Ferromagnesian minerals (amphiboles, pyroxenes, micas, olivines) and volcanic glasses are particularly unstable in the acidic swamp environment. Most are destroyed, and their component elements incorporated into newly formed minerals (Cecil *et al.*, 1979; Pevear *et al.*, 1980). However, some of these phases undergo pseudomorphic replacement, such as the replacement of volcanic glass shards by calcium–barium–strontium aluminum phosphate (Triplehorn and Finkelman, 1989).

Altschuler (1988) suggested that during diagenesis (peat through sub-bituminous coal stages) some of the organically bound sulfur (preferentially oxygen-bonded rather than carbon-bonded) is released, allowing sulfides (primarily pyrite) to form in the coal's cleat and fractures.

2.6. Epigenetic Mineralization

The largest and most obvious minerals in coal are those that have precipitated in the cleat and fractures. These epigenetic minerals, including sulfides (pyrite, sphalerite, galena), carbonates (calcite, siderite), and kaolinite, are common, but volumetrically minor, constituents. Epigenetic mineralization is particularly notable in coals from the mid-continent region, which exhibit greatly enriched contents of chalcophile elements, such as zinc, lead, and cadmium (Gluskoter and Lindahl, 1973; Hatch *et al.*, 1976a, b; Cobb *et al.*, 1980). High contents of epigenetic analcime (a sodium aluminum silicate) in the coals of the Wasatch Plateau in Utah have profoundly affected their utilization because of concerns over their high sodium content (Finkelman *et al.*, 1987). Recent work by Daniels *et al.* (1990) on the anthracites of Pennsylvania has shown that the mineralogy and chemistry of epigenetic minerals varies with cleat direction. These differences were attributed to mineralization by chemically distinct fluids at different stages of coalification.

3. Modes of Occurrence

Information on variations in element content, however useful, provides only a partial picture of the potential behavior of the trace elements in coal. The picture can be brought into better focus if we know an element's mode of occurrence. It is the mode of occurrence of an element that will determine its technological behavior, environmental impact, and economic potential and will reveal its geologic significance.

The classic example is that of sulfur. It has long been known that sulfur in coal occurs in sulfides, sulfates, and organic association; it is now known that elemental sulfur may exist in coal (Duran *et al.*, 1986), that almost 30 different sulfide minerals have been identified as occurring in coal (Finkelman, 1985), and that coal contains a complex suite of organic sulfur compounds (Casagrande, 1987). Each form of sulfur responds differently to physical, chemical, and biological processes.

Many other elements exist in coal in a variety of forms. For example, Dreher and Finkelman (1992) found that selenium in Powder River Basin coal occurred in at least six different forms, including selenium-bearing pyrite and sphalerite, organically bound selenium, lead selenide, water-soluble selenium, and ion-exchangeable selenium. They concluded that the selenium associated with pyrite had the greatest potential for entering surface and ground waters.

There has been considerable work to determine the modes of occurrence of elements in coal. Table II contains a list of the more likely modes of occurrence

TABLE II. Probable Modes of Occurrence of Selected Elements in Coal[a]

Element	Probable modes of occurrence
Aluminum	Clays, feldspars, perhaps some organic association
Antimony	Accessory sulfide, some organic association
Arsenic	Solid solution in pyrite
Barium	Barite, crandallite, organic association in low-rank coal
Beryllium	Organic association, clay
Bismuth	Accessory sulfide
Boron	Organic association, illite
Bromine	Organic association
Cadmium	Sphalerite
Calcium	Calcite, organic association, sulfates, phosphates, silicates
Cesium	Clays, feldspar, mica
Chlorine	Organic association
Chromium	Clays (?)
Cobalt	Accessory sulfides, pyrite
Copper	Chalcopyrite
Fluorine	Perhaps apatite, clays, mica, amphiboles
Gallium	Clays, organics, sulfides
Germanium	Organic association
Gold	Native gold
Hafnium	Zircon
Indium	Probably sulfides
Iodine	Probably organic association
Iron	Pyrite, siderite, sulfates, oxides, some organic association
Lead	Galena, PbSe
Lithium	Clays
Magnesium	Clays
Manganese	Siderite, calcite
Mercury	Solid solution in pyrite
Molybdenum	Unclear; perhaps with sulfides or organics
Nickel	Unclear; perhaps sulfides, organics, or clay
Niobium	Oxides
Phosphorus	Phosphates
Platinum	Native alloys, perhaps some organic association
Rare earths	Phosphates, some organic association
Rubidium	Probably illite
Scandium	Unclear; clays, phosphates, or organics
Selenium	Organic association, pyrite, PbSe
Silicon	Quartz, clays, silicates
Silver	Perhaps silver sulfides
Sodium	Organic association, clays, zeolites, silicates
Strontium	Carbonates, phosphates, organic association
Tantalum	Oxides
Tellurium	Unclear
Thorium	Rare-earth phosphates
Tin	Inorganic: tin oxides or sulfides
Titanium	Oxides, clays, some organic association
Tungsten	Oxides, organic association
Uranium	Organic association, zircon
Vanadium	Clays, perhaps some organic association
Yttrium	Rare-earth phosphates
Zinc	Sphalerite
Zirconium	Zircon

[a]Modified from Finkelman (1982).

for selected elements in coal. It is apparent from the variety of mineral species and organic associations that determining the modes of occurrence can be a challenging task. The job is further complicated by the low concentration levels, the intimate intermixtures of organic and inorganic materials, and the very fine-grained sizes of the minerals (most are less than 2 μm in diameter). For these reasons, most efforts to determine the modes of occurrence have been qualitative. Future efforts will have to focus on ways to quantify the results.

4. Organic–Inorganic Interactions

Compared to most other geologic materials, coal represents the ultimate in organic–inorganic interactions. Coal provides an organic framework within which all the minerals and organically bound elements reside. For some minerals, the organic matrix is merely a physical framework, but other inorganic constituents can interact with the organic matter over geologic time.

For many elements, coal and the precursor organic matter in peats and plants provide a chemical framework for attachment. It is probably not too extreme to say that at some time in the coalification process most elements experience some degree of organic complexing. The role of growing and decaying plants in concentrating elements has already been discussed here. Acids generated from the organic matter can contribute to the breakdown of minerals brought into a peat swamp (Bennett and Siegel, 1987). Elements released from organic matter can precipitate in the coal as well as in adjacent clastic rocks. The reducing environment in peat, lignite, and coal fosters the formation of a wide range of sulfide and other minerals. The cleat and fracture system in coal provides an ideal network for the transport of fluids, allowing for continuous interaction between organic and inorganic matter.

5. Why Concern Ourselves with the Elements in Coal?

I mentioned earlier that the inorganic constituents in coal can have a profound effect on just about every aspect of coal utilization. Table III lists about 20 ways in which they influence coal use or can be used in coal science, and the following brief examples illustrate their influence.

TABLE III. Effects of the Inorganic Constituents on Coal Utilization and Evaluation

Effect	Reference(s)
Contributes to boiler fouling	Reid, 1981
Contributes to slagging properties	Reid, 1981
Contributes to corrosion	Reid, 1981
Contributes to erosion of combustors	Reid, 1981
Contributes to catalysis of conversion processes	Guin *et al.*, 1979
Poisons methanation catalysis	Jenkins and Walker, 1978
Contributes to abrasion of mining and grinding equipment	Callcott and Smith, 1981
Affects washability of coal	Falcon, 1978
Lessens tendency of coal to form dust	Falcon, 1978
Causal factor in pneumoconiosis	Falcon, 1978
Affects oxidizability of coal	Falcon, 1978
Affects tendency of coal to combust spontaneously	Falcon, 1978
Affects calorific value of coal	Falcon, 1978
Affects coke strength	Jenkins and Walker, 1978
Affects accuracy of ultimate analysis	Given and Yarzab, 1978
Useful in seam identification and correlation	Bouska, 1981; Falcon, 1978
Contributes to environmental pollution	U.S. National Committee for Geochemistry, 1980
Potential economic resource	Finkelman and Brown, 1989
Contains information on environment of deposition, diagenesis, source material	Finkelman, 1981a

5.1. Technological Behavior

One of the most costly problems of coal utilization is the buildup of sintered ash deposits on the heat-exchange surfaces of coal-fired boilers (Reid, 1981; Fig. 4). These deposits not only drastically reduce the efficiency of the boiler but also promote corrosion and erosion (Honea *et al.*, 1982). The size and strength of these deposits depend on the configuration of the boiler, the operating conditions, and the inorganic composition of the coal being combusted. Finkelman and Dulong (1989) found that ash yield and the concentrations of sodium, calcium, and magnesium were the most important factors influencing the fouling behavior of coal; a high silicon:aluminum ratio, a function of the coal's mineralogy, was also important. On the basis of these and other compositional factors, the investigators developed models that accurately predicted the weight of fouling deposits formed in a test combustion unit (Fig. 5).

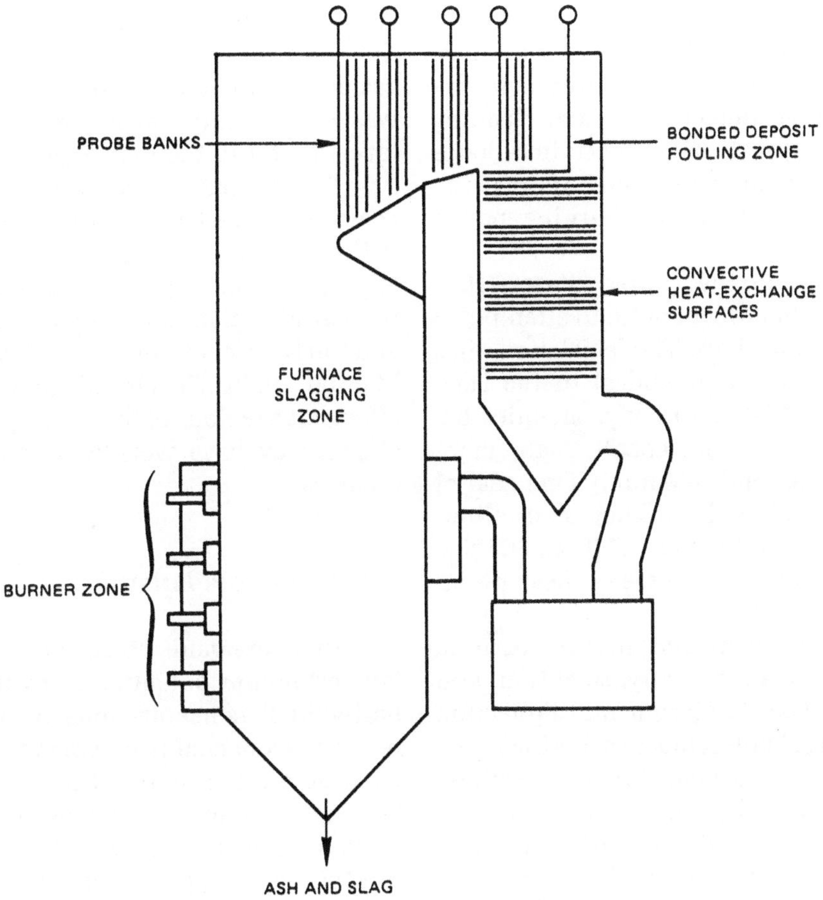

Figure 4. Cross section of a coal-fired boiler, showing fouling and slagging zones.

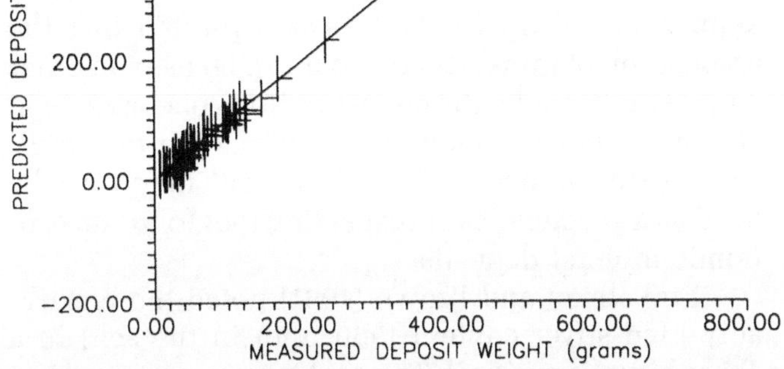

Figure 5. Plot of actual weight of fouling deposit in a test combustor vs. predicted weight based on a model developed from compositional factors.

The U.S. Department of Energy recently sponsored a program (see, e.g., Helble *et al.*, 1990) designed to form a better understanding of what transformations the various elements and minerals in coal undergo during the combustion process. This information will ultimately result in the development of methods to reduce the fouling and slagging of coal-fired boilers.

5.2. Environmental Impact

The environmental impact of sulfur released during coal mining and combustion has been of concern for more than one hundred years (Graham, 1907). The recent increased awareness of such environmental effects has focused attention on a number of other elements in coal. Information coming out of Eastern Europe may provide an insight into the full magnitude of the environmental impact of uncontrolled coal combustion. Recent data (Chandler *et al.*, 1990) indicate that lead and arsenic released from coal-burning power plants in Czechoslovakia may have retarded the bone growth of many children in the nearby communities. Fortunately, the coal cleaning and effluent treatment practices in the United States and most developed countries greatly reduce the level of toxic emissions from coal-burning power plants.

Recent legislation in the United States (H.R. 3030, Clean Air Act Amendment of 1990) attempts to further reduce these emissions. The legislation cites 11 elements (excluding radionuclides) in the Hazardous Air Pollutant List (antimony, arsenic, beryllium, cadmium, chromium, cobalt, lead, manganese, mercury, nickel, and selenium). One goal of the legislation is to "reduce the volume of, or eliminate emissions of, such pollutants. . ." (H.R. 3030, p. 295). A potentially significant source of these pollutants is the combustion of coal.

Many of the same elements (arsenic, barium, cadmium, chromium, lead, mercury, selenium, and silver) appear in the U.S. Environmental Protection Agency's list of potential pollutants in drinking water. Standards setting maximum concentrations for these elements have been published (U.S. EPA, 1976).

The U.S. EPA (1987) found that most coal wastes are not hazardous as defined by the Resources Conservation and Recovery Act. It was noted, however, that some groundwater contamination has occurred in the vicinity of waste disposal sites. At one site in Virginia, drinking-water wells were contaminated with vanadium and selenium from nearby coal combustion wastes (U.S. EPA, 1987).

Many state regulatory agencies now recognize the potentially harmful effects of trace elements released during coal mining and utilization. Boon *et al.* (1987) noted that most western U.S. coal-mining states now require baseline information on several trace elements, especially selenium, boron, and molybdenum. The release of selenium from coal combustion wastes has caused significant environmental damage (extensive fish kills) at two sites in North Carolina and one in Texas (Shepard, 1987). New Mexico and Colorado now require data on 10 to 20 trace elements (Boon *et al.*, 1987).

Knowing the concentration of the element is generally not sufficient to assess its potential impact. Dreher and Finkelman (1992) found that the mode of occurrence of selenium in coal and in overburden played a major role in its release during mining. Moreover, they found that the selenium in the coal occurred in at least six different forms. Although, selenium associated with pyrite accounted for only 10–15% of the approximately 1 ppm of selenium in their samples, Dreher and Finkelman (1992) suggested that the oxidation of selenium-bearing pyrite was the primary cause for the elevated selenium levels found in the recharged groundwater.

The environmental problems associated with coal are not restricted to the Appalachian Basin or to the industrialized Midwest. Coal deposits occur in at least 38 states (Fig. 6). Coal-burning utility and industrial boilers are operating in at least 45 states and the District of Columbia (Keystone Coal Industry Manual, 1989). The emissions from coal combustion affect large regions of the country, and the cumulative effects may have worldwide environmental consequences.

5.3. Economic Potential

Some inorganic constituents of coal have important economic potential as by-products. Extraordinarily high concentrations of elements have been reported from coal (some many time higher than the maximum values in the NCRDS data base; Table I). In fact, mineral resources have historically been extracted from coal. Jenny (1903) described several pre-1900 mining operations in which minerals were extracted from coal and coal-bearing rocks, including vanadium from lignite in Argentina, antimonial silver from coal in Peru, and sphalerite (ZnS) and galena (PbS) from Missouri coal. In the 1960s there was limited commercial production of uranium from North and South Dakota lignites (Noble, 1973; Schnable, 1975). In one area, molybdenum was enriched to such a degree that it was recovered along with the uranium. However, burning of the molybdenum-rich lignite allegedly caused molybdenosis in local cattle (E. A. Noble, personal communication, 1990). More recent efforts at by-product recovery from coal have been described by Burnet (1986) and Cobb *et al.* (1979).

Hatch *et al.* (1976b), Cobb *et al.* (1979), and Finkelman and Brown (1989) have suggested that the association of minerals and coal can be used in other ways. For example, the concentrations of elements in coal may not be economical to extract but may indicate nearby mineralization. The coal may thus be used as a geochemical prospecting tool to locate economic mineral deposits.

Finkelman and Brown (1991) noted the remarkably high silver content (500 ppm in the ash) in a Texas bituminous coal. The coal is from an area about 80 miles northwest of the Llano Uplift in west-central Texas (Fig. 7). Zinc, lead, and silver have been mined

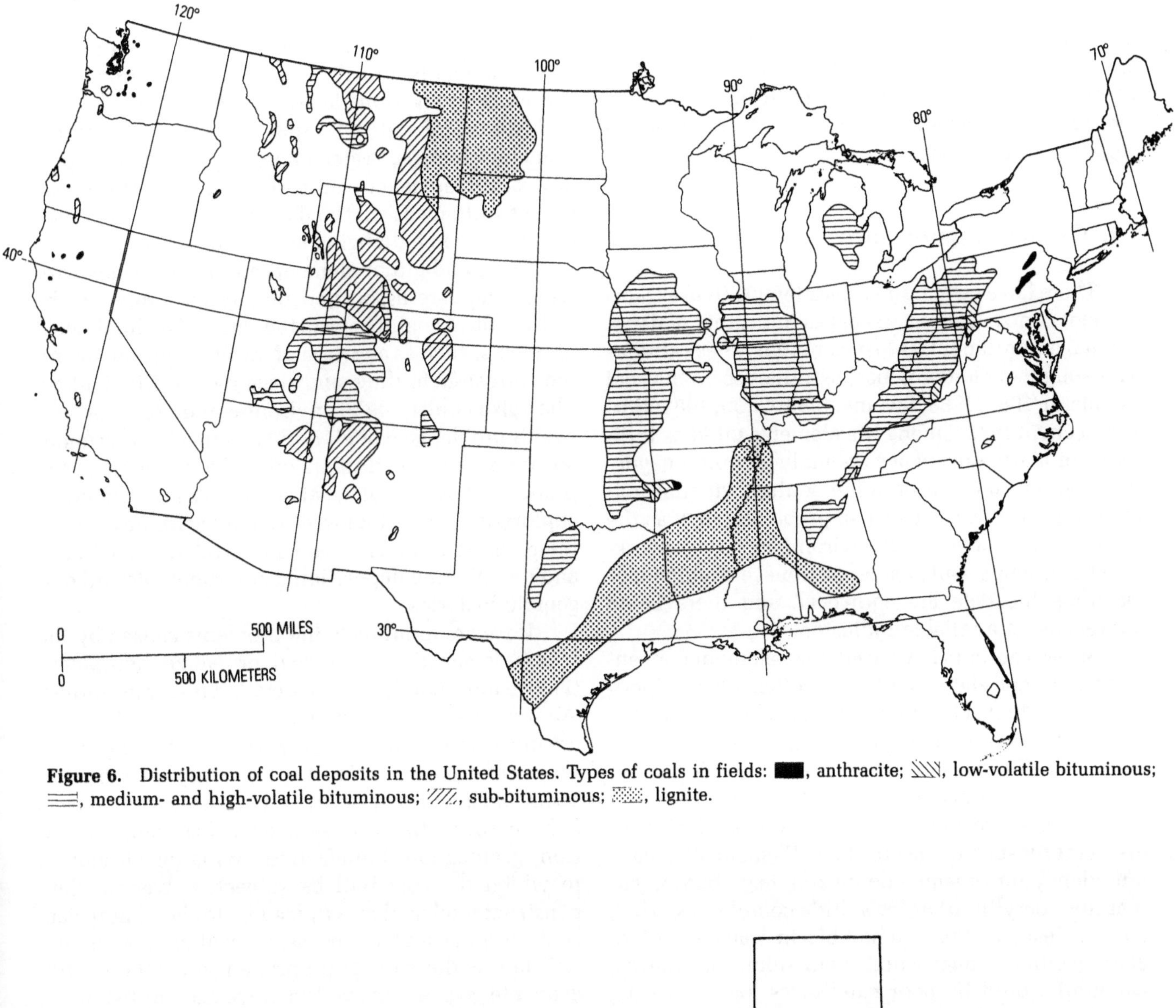

Figure 6. Distribution of coal deposits in the United States. Types of coals in fields: ■, anthracite; ⑊⑊⑊, low-volatile bituminous; ☰, medium- and high-volatile bituminous; ////, sub-bituminous; ░, lignite.

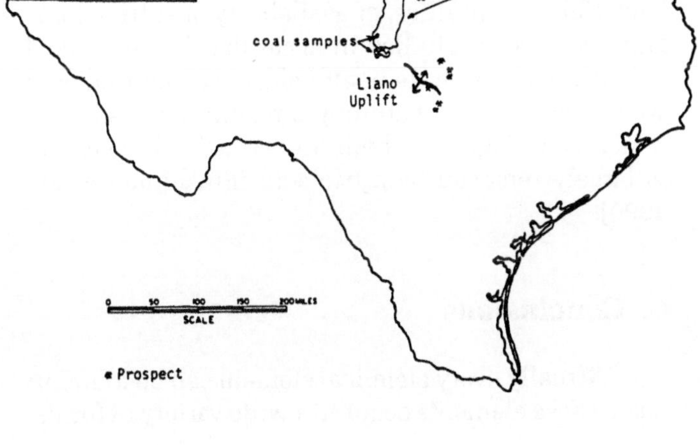

Figure 7. Map of Texas illustrating the proximity of mineralized coal to the Llano Uplift and nearby historic mining areas.

east and southeast of the uplift. Finkelman and Brown (1991) suggested that this anomalously high silver value in the coal may be indicative of nearby economic deposits and recommended a detailed exploration of the area.

5.4. Geologic Significance

There have been numerous attempts to use trace-element concentrations in coal as indicators of depositional environments. Most commonly these studies have sought evidence of marine influence on the coal (Goodarzi, 1987, 1988; Swaine, 1983; Chou, 1984; Hart and Leahy, 1983). Among the elements cited as indicators of marine influence are molybdenum, magnesium, boron, chlorine, bromine, sodium, yttrium, and uranium. However, because of problems, such as the blurring effect of brackish environments, the reworking of sediments, and postdepositional enrichment or leaching, the data are equivocal, and there is no consensus on a reliable indicator.

Several authors have used trace-element concentrations to correlate coal beds (Butler, 1953; Alpern and Morel, 1968; O'Gorman, 1971; Nichols and D'Auria, 1981; Swaine, 1983; Cebulak, 1983; O'Connor, 1988). Significant vertical and lateral variation of the trace-element concentrations within coal beds has limited the success of these attempts. The selection of elements for such usage is critical. Elements that have a tendency for organic complexing (e.g., boron, germanium, beryllium) or for sulfide complexing (zinc, copper, lead) or that are soluble in water or dilute acids (sodium, magnesium, potassium, manganese, calcium) would be poor candidates for correlating coal beds, but those that form relatively inert minerals (zirconium, hafnium, niobium, tantalum, cesium, scandium) should be more useful (Finkelman, 1981a). The concentrations or ratios of these elements would not be significantly affected by changes in Eh and pH, rate of detrital influx, changes in plant or bacterial communities, or availability of sulfide ions. However, recent studies indicate that in many coal samples even these elements behave in unanticipated ways; for example, hafnium is readily leached from many coal samples by hydrofluoric acid, and cesium is largely removed by nitric acid (Finkelman *et al.*, 1990).

6. Conclusions

Virtually every chemical element can be found in coal. These elements occur in a wide variety of forms, with most occurring in minerals. The concentrations of the elements and their modes of occurrence are influenced by various factors, including the amount and source of detrital input, organic complexing, climatic changes, water chemistry, diagenetic processes, and epigenetic mineralization. It is this diversity of influences that makes the study of coal geochemistry so challenging and so rewarding.

Trace and minor elements and the minerals with which they are associated play crucial roles in the utilization of coal. They are the cause of technological problems, such as abrasion of mining equipment and the corrosion, fouling, and slagging of utility boilers. They also contribute to environmental pollution and health problems. Not all of their impact is detrimental; inorganic constituents offer the potential of byproduct recovery from coal and use as a geochemical exploration tool. Moreover, we may learn how to use their distribution as an aid in correlating coal beds and deciphering depositional environments and diagenetic histories.

If we are to mitigate the problems caused by the inorganic constituents in coal, or realize their potential, we must conduct more detailed research studies. We must better understand the ways in which they are incorporated in coal and what factors dictate their modes of occurrence. We must quantify the modes of occurrence and determine how these various forms behave during the processes (e.g., cleaning, combustion, gasification, liquefaction, oxidation, leaching) to which the coal will be subjected. We can then construct models that will lead us to the appropriate coal for a particular process, as well as models that will tell us the most appropriate use for the readily available coal resources. When we accomplish these objectives, coal will be a more efficient, cost-effective, and environmentally compatible energy source.

ACKNOWLEDGMENTS. The incisive reviews of my U.S. Geological Survey colleagues Z. A. Altschuler, J. R. Hatch, and E. A. Noble greatly improved this paper. Linda J. Bragg of the U.S. Geological Survey compiled the data for Table I. I appreciate their time, effort, and patience.

References

Alpern, B., and Morel, P., 1968, Examen, dans le cadre du bassin houiller lorrain, des possibilités stratigraphiques de la géochimie, *Ann. Soc. Geol. Nord* **88**(4):185–202.

Altschuler, Z. S., 1988, Evolution in forms of sulfur from peat to coal, *Geological Society of America Abstracts with Programs*, Vol. 20, No. 7, p. A90.

Andrejko, M. J., and Cohen, A. D., 1984, Scanning electron micros copy of silicophytoliths from the Okefenokee swamp–marsh complex, in: *The Okefenokee Swamp* (A. D. Cohen, D. J. Casagrande, M. J. Andrejko, and G. R. Best, eds.), Wetland Survey, Los Alamos, New Mexico, pp. 468–491.

Ayers, W. B., Jr., and Kaiser, W. R., 1984, Lacustrine–interdeltaic coal in the Fort Union Formation (Paleocene), Powder River Basin, Wyoming and Montana, U.S.A., in: *Sedimentology of Coal and Coal-Bearing Sequences* (R. A. Rahmani and R. M. Flores, eds.), Special Publication No. 7, International Association of Sedimentologists, Blackwell Scientific, Oxford, pp. 61–84.

Bennett, P., and Siegel, D. I., 1987, Increased solubility of quartz in water due to complexing by organic compounds, *Nature* **326:** 684–686.

Benson, S. A., Zygarlicke, C. J., Steadman, E. N., and Karner, F. R., 1992, Geochemistry of Fort Union lignite, in: *Geology and Utilization of Fort Union Lignites* (R. B. Finkelman, S. J. Tewalt, and D. J. Daly, eds.), Environmental and Coal Associates, Reston, Virginia, pp. 111–120.

Bohor, B. F., Phillips, R. E., and Pollastro, R. M., 1979, Altered Volcanic Ash Partings in Wasatch Formation Coal Beds of the Northern Powder River Basin: Composition and Geologic Applications, U.S. Geological Survey Open-File Report 79-1203.

Boon, D. Y., Munshower, F. F., and Fisher, S. E., 1987, Overburden chemistry: A review and update, in: Billings Symposium on Surface Mining and Reclamation in the Great Plains, American Society for Surface Mining and Reclamation, Reclamation Research Unit Report No. 8704, pp. A-1-1–A-1-18.

Bouska, V., 1981, *Geochemistry of Coal, Coal Science and Technology 1*, Elsevier, Amsterdam.

Burnet, G., 1986, Newer technologies for resource recovery from coal combustion solid wastes, *Energy* **11:**1363–1375.

Butler, J. R., 1953, Geochemical Affinities of Some Coals from Svalbard, *Nor. Polarinst. Skr.* **96**.

Callcott, T. G., and Smith, G. B., 1981, Mechanical properties of coal, in: *Chemistry of Coal Utilization*, 2nd supplementary volume (M. A. Elliot, ed.), John Wiley & Sons, New York, pp. 285–315.

Casagrande, D. J., 1987, Sulphur in peat and coal, in: *Coal and Coal-Bearing Strata: Recent Advances* (A. C. Scott, ed.), Geol. Soc. Spec. Publ. No. 32, pp. 87–105.

Cebulak, S., 1983, Correlation of coal seams in the Central Coal Region of the Lublin Coal Basin, on the basis of geochemical data, Rozkowska Anna. *Kwart. Geol.* **27**(1):25–39 [in Polish].

Cecil, C. B., Stanton, R. W., Allshouse, S. D., and Finkelman, R. B., 1979, Interbed variation in and geologic controls on element concentrations in the Upper Freeport Coal, Abstracts of Papers, Fuel Chemistry Division, American Chemical Society and Chemical Society of Japan, Chemical Congress, Honolulu, Hawaii, Paper No. 48.

Cecil, C. B., Stanton, R. W., Neuzil, S. G., Dulong, F. T., Ruppert, L. F., and Pierce, B. S., 1985, Paleoclimate controls on Late Paleozoic sedimentation and peat formation in the central Appalachian Basin (U.S.A.), *Int. J. Coal Geol.* **5:**195–230.

Chandler, W. U., Makarov, A. A., and Dadi, Z., 1990, Energy for the Soviet Union, Eastern Europe and China, *Sci. Am.* **263**(3): 121–127.

Chou, C.-L., 1984, Relationship between geochemistry of coal and the nature of strata overlying the Herrin coal in the Illinois Basin, U.S.A., *Memoir of the Geological Society of China*, No. 6, pp. 269–280.

Cobb, J. C., Masters, J. M., Treworgy, C. G., and Helfinstine, R. J., 1979, Abundance and recovery of sphalerite and fine coal from mine waste in Illinois, *Ill. State Geol. Surv. Ill. Miner. Note* **71**.

Cobb, J. C., Steele, J. D., and Treworgy, C. G., 1980, The abundance of zinc and cadmium in sphalerite-bearing coals in Illinois, *Ill. State Geol. Surv. Ill. Miner. Note* **74**.

Crowley, S. S., Stanton, R. W., and Ryer, T. A., 1989, The effects of volcanic ash on the maceral and chemical composition of the C coal bed, Emery Coal Field, Utah, *Org. Geochem.* **14:**315–331.

Daniels, E. J., Altaner, S. P., Marshak, S., and Eggleston, J., 1990, Hydrothermal alteration in anthracite from eastern Pennsylvania: Implications for mechanisms of anthracite formation, *Geology* **18:**247–250.

Davis, R. E., and Dodge, K. A., 1986, Results of Experiments Related to Contact of Mine-Spoils Water with Coal, West Decker and Big Sky Mines, Southeastern Montana, U.S. Geological Survey Water-Resources Investigation Report 86-4002.

Dreher, G. B., and Finkelman, R. B., 1992, Selenium mobilization in a surface coal mine, Powder River Basin, Wyoming, U.S.A., *Environ. Geol. Water Sci.* **19**(3):115–167.

Dulong, F. T., Cecil, C. B., and Stanton, R. W., 1986, Regional and local variation of coal quality parameters in the Upper Freeport coal bed, western Pennsylvania, *Geological Society of America Abstracts with Programs*, Vol. 18, No. 6, p. 589.

Duran, J. E., Mahasay, S. R., and Stock, L. M., 1986, The occurrence of elemental sulfur in coals, *Fuel* **65:**1167–1168.

Falcon, R. M. S., 1978, Coal in South Africa, Part II: The application of petrography to the characterization of coal, *Miner. Sci. Eng.* **10**(1):28–52.

Finkelman, R. B., 1981a, Modes of Occurrence of Trace Elements in Coal, U.S. Geological Survey Open-File Report 81-99.

Finkelman, R. B., 1981b, Recognition of authigenic and detrital minerals in coal, *Geological Society of America Abstracts with Programs*, Vol. 13, No. 7, p. 451.

Finkelman, R. B., 1981c, The origin, occurrence, and distribution of the inorganic constituents in low-rank coals, in: *Proceedings of the Basic Coal Science Workshop* (H. H. Schobert, compiler), Grand Forks Energy Research Center, Grand Forks, North Dakota, pp. 70–90.

Finkelman, R. B., 1982, Modes of occurrence of trace elements and minerals in coal: An analytical approach, in: *Atomic and Nuclear Methods in Fossil Energy Research* (R. H. Filby, B. S. Carpenter, and R. C. Ragaini, eds.), Plenum Press, New York, pp. 141–149.

Finkelman, R. B., 1985, Mode of occurrence of accessory sulfide and selenide minerals in coal, in: *Neuvième Congrès International de Stratigraphie et de Géologie du Carbonifère, Compte Rendu*, Vol. 4 (A. T. Cross, ed.), Southern Illinois University Press, Carbondale, Illinois, pp. 407–412.

Finkelman, R. B., 1990, What we *don't* know about the occurrence and distribution of trace elements in coal, *J. Coal Qual.* **8**(3/4): 63–66.

Finkelman, R. B., and Brown, R. D., Jr., 1989, Mineral resources and geochemical exploration potential of coal that has anomalous metal concentrations, in: *USGS Research on Energy Resources—1989 Program and Abstracts* (K. S. Schindler, ed.), U.S. Geological Survey Circular 1035, pp. 18–19.

Finkelman, R. B., and Brown, R. D., Jr., 1991, Coal as a host and as an indicator of mineral resources, in: *Geology in Coal Resource Utilization* (D. C. Peters, ed.), TechBooks, Fairfax, Virginia, pp. 471–481.

Finkelman, R. B., and Dulong, F. T., 1989, Development and Evaluation of Deterministic Models for Predicting the Weights of Fouling Deposits from Coal Combustion, U.S. Geological Survey Open-File Report 89-208.

Finkelman, R. B., Yeakel, J. D., and Harrison, W. J., 1987, Sodium in the Upper Cretaceous coal beds of the Wasatch Plateau, Utah:

Mode of occurrence, geologic controls, possible source, and effects of coal utilization, *Geological Society of America Abstracts with Programs*, Vol. 19, No. 7, p. 663.

Finkelman, R. B., Palmer, C. A., Krasnow, M. R., Aruscavage, P. J., Sellers, G. A., and Dulong, F. T., 1990, Combustion and leaching behavior of elements in the Argonne Premium Coal Samples, *Energy Fuels* **4**:755–766.

Flores, R. M., and Hanley, J. H., 1984, Anastomosed and associated coal-bearing fluvial deposits: Upper Tongue River Member, Palaeocene Fort Union Formation, northern Powder River Basin, Wyoming, U.S.A., in: *Sedimentology of Coal and Coal-Bearing Sequences* (R. A. Rahmani and R. M. Flores, eds.), Int. Assoc. Sediment. Spec. Publ. No. 7, pp. 85–103.

Gilmore, J. S., Knight, J. D., Orth, C. J., Pillmore, C. L., and Tschudy, R. H., 1984, Trace element patterns at a nonmarine Cretaceous–Tertiary boundary, *Nature* **307**:224–228.

Given, P. H., and Yarzab, R. F., 1978, Analysis of the organic substance of coals: Problems posed by the presence of mineral matter, in: *Analytical Methods for Coal and Coal Products*, Vol. 2 (C. Karr, Jr., ed.), Academic Press, New York, pp. 3–41.

Glick, D. C., and Davis, A., 1987, Variability in the inorganic element content of U.S. coals including results of cluster analysis, *Org. Geochem.* **11**:331–342.

Gluskoter, H. J., and Lindahl, P. C., 1973, Cadmium: Mode of occurrence in Illinois coals, *Science* **181**:264–266.

Gluskoter, H. J., Ruch, R. R., Miller, W. G., Cahill, R. A., Dreher, G. B., and Kuhn, J. K., 1977, *Trace Elements in Coal: Occurrence and Distribution*, Illinois State Geological Survey Circular 499.

Goodarzi, F., 1987, Concentration of elements in lacustrine coals from zone A Hat Creek deposit No. 1, British Columbia, Canada, *Int. J. Coal Geol.* **8**:247–268.

Goodarzi, F., 1988, Elemental distribution in coal seams at the Fording coal mine, British Columbia, Canada, *Chem. Geol.* **68**: 129–154.

Gorham, E., and Santelmann, M. V., 1984, *Peatland Bibliography*, University of Minnesota Press, Minneapolis.

Graham, J. W., 1907, *The Destruction of Daylight*, G. Allen, London.

Guin, J. A., Tarrier, A. R., Lee, J. M., Lo, L., and Curtis, C. W., 1979, Further studies of the catalytic activity of coal minerals in coal liquefaction: 1. Verification of catalytic activity of mineral matter by model compound studies, *Ind. Eng. Chem. Process Des. Dev.* **18**(3):371–376.

Hart, R. J., and Leahy, R. M., 1983, The geochemical characterization of coal seams from the Witbank Basin, *Spec. Publ. Geol. Soc. S. Afr.* **7**:169–174.

Hatch, J. R., 1983, Geochemical processes that control minor and trace element composition of United States coals, in: *Unconventional Mineral Deposits* (W. C. Shanks, III, ed.), Society of Mining Engineers, New York, pp. 89–98.

Hatch, J. R., Avcin, M. J., Wedge, W. K., and Brady, L. L., 1976a, Sphalerite in Coals from Southeastern Iowa, Missouri, and Southeastern Kansas, U.S. Geological Survey Open-File Report 76-796.

Hatch, J. R., Gluskoter, H. J., and Lindahl, P. C., 1976b, Sphalerite in coals from the Illinois basin, *Econ. Geol.* **71**:613–624.

Helble, J. J., Srinivasachar, S., and Boni, A. A., 1990, A fundamental study of ash particle adhesion, in: *Proceedings of the Seventh Annual International Pittsburgh Coal Conference*, University of Pittsburgh, Pittsburgh, pp. 52–61.

Honea, F. I., Montgomery, G. G., and Jones, M. L., 1982, Recent research on ash fouling in combustion of low rank coals, in: *Technology and Use of Lignites, Vol. 1, Proceedings, Eleventh Biennial Lignite Symposium* (W. R. Kube, E. A. Sondreal, and D. M. White, compilers), Grand Forks Energy Technology Center IC-82/1, Grand Forks, North Dakota, pp. 504–545.

Jenkins, R. G., and Walker, P. L., Jr., 1978, Analysis of mineral matter in coal, in: *Analytical Methods for Coal and Coal Products*, Vol. 2 (C. Karr, Jr., ed.), Academic Press, New York, pp. 265–292.

Jenny, W. P., 1903, The chemistry of ore-deposition, *Am. Inst. Min. Eng. Trans.* **33**:445–498.

Karner, F. R., Benson, S. A., Schobert, H. H., and Roaldson, R. G., 1984, Geochemical variation of inorganic constituents in a North Dakota lignite, in: *The Chemistry of Low-Rank Coals* (H. H. Schobert, ed.), ACS Symposium Series No. 264, American Chemical Society, Washington, D.C., pp. 175–193.

Keystone Coal Industry Manual, 1989, McGraw-Hill, New York.

Kuellmer, F. J., Kendrick, D. T., and Baker, L., 1987, Trace Element Distributions in Some New Mexico Coals, New Mexico Research and Development Institute Report 2-74-4321.

Lindahl, P. C., and Finkelman, R. B., 1986, Factors influencing major, minor, and trace element variations in U.S. coals, in: *Mineral Matter and Ash in Coal* (K. S. Vorres, ed.), ACS Symposium Series No. 301, American Chemical Society, Washington, D.C., pp. 61–69.

Lyons, P. C., Outerbridge, W. F., Evans, Jr., H. T., and Triplehorn, D. M., 1990, Carboniferous (Westphalian) volcanic ash deposits, Appalachian Basin (U.S.A.), in: *13th International Sedimentological Congress*, International Association of Sedimentologists, Utrecht, p. 320.

McCarthy, T. S., McIver, J. R., Cairncross, B., Ellery, W. N., and Ellery, K., 1989, The inorganic chemistry of peat from the Maunachia channel-swamp system, Okavango Delta, Botswana, *Geochim. Cosmochim. Acta* **53**:1077–1089.

Miller, R. T., and Given, P. H., 1978, A Geochemical Study of the Inorganic Constituents in Some Low-Rank Coals, U.S. Department of Energy Report FE-2494-TR-1.

Nichols, C. L., and D'Auria, J. M., 1981, Seam and location differentiation of coal specimens using trace element concentrations, *Analyst* **106**:874–882.

Noble, E. A., 1973, Uranium in coal, in: *Mineral and Water Resources of North Dakota*, North Dakota Geological Survey Bulletin 63, pp. 80–83.

O'Connor, J. T., 1988, The Campbell Creek/No. 2 Gas/Peerless/Powellton coal bed correlation from the middle of the Kanawha Formation of the central Appalachian Basin, in: *USGS Research on Energy Resources—1988 Program and Abstracts* (L. M. H. Carter, ed.), U.S. Geological Survey Circular 1025, p. 39.

O'Gorman, J. V., 1971, Studies on mineral matter and trace elements in North American coals, Ph.D. Dissertation, The Pennsylvania State University.

Oman, C. L., Finkelman, R. B., Coleman, S. L., and Bragg, L. J., 1988, Selenium in coals from the Powder River Basin, Wyoming and Montana, in: *USGS Research on Energy Resources—1988 Program and Abstracts* (L. M. H. Carter, ed.), U.S. Geological Survey Circular 1025, pp. 16–17.

Pevear, D. R., Williams, V. E., and Mustoe, G. E., 1980, Kaolinite, smectite, and K-rectorite in bentonites: Relation to coal rank at Tulameen, British Columbia, *Clays Clay Miner.* **28**(4):241–254.

Raymond, R., Jr., Bish, D. L., and Cohen, A. D., 1990, Inorganic contents of peats, in: *Mineral Matter and Ash Deposition from Coal* (R. W. Breyers and K. S. Vorres, eds.), Engineering Foundation Conference, New York, pp. 23–37.

Reid, W. T., 1981, Coal ash—its effects on combustion systems, in: *Chemistry of Coal Utilization*, 2nd supplementary volume (M. A. Elliot, ed.). John Wiley & Sons, New York, pp. 1389–1445.

Robbins, E. I., D'Agostino, J. P., Carter, V., Fanning, D. S., Gamble, C. J., Ostwald, J., Van Hoven, R. L., and Young, G. K., 1990a, Manganese nodules and microbial fixation of oxidized manganese in the Huntly Meadows wetland, Fairfax County, Virginia, *U.S. Geol. Surv. Circ.* **1060:**69–70.

Robbins, E. I., Zielinski, R. A., Otton, J. K., Owen, D. E., Schumann, R. R., and McKee, J. P., 1990b, Microbially mediated fixation of uranium, sulfur, and iron in a peat-forming montane wetland, Larimer County, Colorado, *U.S. Geol. Surv. Circ.* **1060:**70–71.

Roscoe, B. A., and Hopke, P. K., 1982, Analysis of mineral phases in coal utilizing factor analysis, in: *Atomic and Nuclear Methods in Fossil Energy Research* (R. H. Filby, ed.), Plenum Press, New York, pp. 163–174.

Schnable, R. W., 1975, Uranium, in: Mineral and Water Resources of South Dakota, Report prepared by the U.S. Geological Survey for the Committee on Interior and Insular Affairs, United States Senate, pp. 172–176.

Severson, R. C., and Shacklette, H. T., 1988, Essential elements and soil amendments for plants: Sources and use for agriculture, *U.S. Geol. Surv. Circ.* **1017**.

Shepard, M., 1987, Toxic resources and the real world, *EPRI J.* **1987** (Sept):17–21.

Swaine, D. J., 1983, Geological aspects of trace elements in coal, in: *The Significance of Trace Elements in Solving Petrogenetic Problems and Controversies* (S. S. Augustithis, ed.), Theophrastus Publications, Athens, pp. 521–532.

Swaine, D. J., 1990, *Trace Elements in Coal*, Butterworths, London.

Swanson, V. E., Medlin, J. H., Hatch, J. R., Coleman, S. L., Wood, G. H., Jr., Woodruff, S. D., and Hildebrand, R. T., 1976, Collection, chemical analysis, and evaluation of coal samples in 1975, U.S. Geological Survey Open-File Report 76-468.

Triplehorn, D. M., and Finkelman, R. B., 1989, Replacement of glass shards by aluminum phosphates in a middle Pennsylvanian tonstein from eastern Kentucky, *Geological Society of America Abstracts with Programs*, Vol. 21, No. 6, p. PA 52.

U.S. EPA, 1976, Quality Criteria for Water, U.S. Environmental Protection Agency, Washington, D.C.

U.S. EPA, 1987, Wastes from the Combustion of Coal by Electric Utility Power Plants, U.S. Environmental Protection Agency Report to Congress, June 1987.

U.S. National Committee for Geochemistry, 1980, *Trace-Element Geochemistry of Coal Resource Development Related to Environmental Quality and Health*, National Academy Press, Washington, D.C.

Valkovic, V., 1983, *Trace Elements in Coal*, Vol. 1, CRC Press, Boca Raton, Florida.

Wood, G. H., Jr., Kehn, T. M., Carter, M. D., and Culbertson, W. C., 1983, Coal resource classification system of the U.S. Geological Survey, *U.S. Geol. Surv. Circ.* **891**.

Zubovic, P., 1966, Minor element distribution in coal samples of the Interior coal province, in: *Coal Science* (R. F. Gould, ed.), Advances in Chemistry Series No. 55, American Chemical Society, Washington, D.C., pp. 232–247.

VII

The Precambrian

Chapter 29

Evolution of the Earth's Atmosphere and Hydrosphere

Hadean to Recent

JAMES F. KASTING

The evolution of the Earth's atmosphere and hydrosphere is a long and interesting story—one that is much too complex to be told in any detail in a single book chapter. The discussion that follows outlines the essential elements of the story, with particular emphasis on matters relating to the prebiotic synthesis of organic matter and the early evolution of life. More complete treatments of this subject can be found in Walker (1977), Schopf (1983), Holland (1984), and Schopf and Klein (1992).

Although the title of this chapter implies that the discussion will extend from Hadean to recent times, the reader will find that most of the events that I describe took place during the Precambrian. This is not because the atmosphere and oceans have ceased to evolve during the Phanerozoic. Significant changes in atmospheric pO_2 and pCO_2 and in various trace gas concentrations have probably occurred in the relatively recent past and will likely continue to occur in the future. The changes that have occurred during the last 600 million years, however, are much smaller than those that took place during the preceding 4 billion years. Indeed, the largest transformations in atmospheric and oceanic composition occurred in the first half of the Precambrian, and particularly in the Hadean Era (4.6 to 3.8 Ga B.P.), prior to the beginning of the rock record. So, if it seems that I have devoted an inordinate amount of space to such a poorly understood time period, it is because much of the Earth's subsequent history, including the history of life, was determined by what happened during those early years.

1. Formation of the Atmosphere and Ocean

Theories of how the Earth's atmosphere and ocean originated have changed markedly within the last 10 to 15 years. Before that time, it was generally assumed that the Earth accreted as a cold, airless body and that the atmosphere and ocean were formed by outgassing from the interior (e.g., Rubey, 1951, 1955; Holland, 1962). The evidence used to support this hypothesis was the low noble-gas content of the atmosphere, which implied that the atmosphere could not have been captured gravitationally from the solar nebula. While this latter inference remains reasonably secure, the concept of an initially airless planet does not. Studies by Benlow and Meadows (1977), Lange and Ahrens (1982), and others have shown that once the proto-Earth had reached about 10% of its present size, in-falling planetesimals should have been degassed on impact and released their

JAMES F. KASTING • Department of Geosciences, The Pennsylvania State University, University Park, Pennsylvania 16802.
Organic Geochemistry, edited by Michael H. Engel and Stephen A. Macko. Plenum Press, New York, 1993.

volatiles directly into the atmosphere. For presently accepted accretionary time scales, 10 to 100 million years (Safronov, 1969), the heat released by the infalling material would have been sufficient to vaporize any water that was present at the surface, creating a dense steam atmosphere (Matsui and Abe, 1986a, b; Zahnle *et al.*, 1988). The surface temperature of this atmosphere would have been of the order of 1500 K, near the solidus for typical silicate rocks. Thus, the Earth's surface should have been covered by a magma ocean that could, in principle, have extended well down into its interior (Zahnle *et al.*, 1988).

The idea that the Earth was originally enveloped in a steam atmosphere has been criticized by Stevenson (1987), who has pointed out that the latter stages of accretion were probably highly sporadic, with most of the material arriving in large, moon-sized chunks (Wetherill, 1985). Between impacts, the Earth's surface may have cooled sufficiently for a thin crust to have formed and for liquid water to have condensed. If this hypothesis is correct, the ocean may have formed not just once, but numerous times during the accretion process. It seems likely, however, that a steam atmosphere would have existed during at least part of this time. This idea is supported by data on neon isotopes, which show that the $^{20}Ne/^{22}Ne$ ratio in the atmosphere is lower than that in either the mantle or the solar wind (Craig and Lupton, 1976; Ozima and Igarashi, 1989). This observation is most easily explained if ^{20}Ne was lost during rapid, hydrodynamic escape of hydrogen from a primordial steam atmosphere (Zahnle *et al.*, 1990).

Even if it did exist, the initial steam atmosphere may have been largely removed by a giant moon-forming impact toward the end of the main accretion period (Hartmann *et al.*, 1986). If so, much of the Earth's volatile inventory may have been brought in at a later time by comets or asteroids that impacted the Earth during the period of heavy bombardment, 4.5 to 3.8 Ga B.P. (Chyba, 1987, 1989). As discussed below, the impacting bodies may also have affected atmospheric composition during this period. For this and other reasons, it is difficult to predict atmospheric density and composition during the first several hundred million years of Earth history.

2. Early Atmospheric Composition and Climate

Once the main part of accretion had ended, most of the water vapor that was in the atmosphere would have condensed to form an ocean, leaving an atmosphere dominated by carbon compounds (mostly CO_2 and CO) and N_2. Molecular hydrogen (H_2) would have been an important trace constituent; an H_2 mixing ratio (i.e., mole fraction) of $\sim 10^{-3}$ could have been maintained by the balance between volcanic outgassing and escape to space (Walker, 1977). Free oxygen (O_2) would have been essentially absent, except for a thin layer formed at high altitudes from the photodissociation products of CO_2 and H_2O (Kasting *et al.*, 1984, and references therein).

Perhaps the most interesting question about this early atmosphere concerns the abundance and oxidation state of carbon. Current models (e.g., Walker, 1977; Holland, 1984; Kasting *et al.*, 1984; Walker, 1986; Kasting and Ackerman, 1986) predict a CO_2-rich atmosphere containing trace amounts of methane and slightly greater amounts of CO. The reasoning is as follows. CO_2 is the dominant gas released by volcanoes today and should have been so in the past if the oxidation state of the upper mantle was about the same as today. Furthermore, both methane and carbon monoxide can be oxidized by the by-products of water vapor photolysis. For example, one possible reaction pathway is

$$H_2O + h\nu \rightarrow H + OH \tag{1a}$$

$$CH_4 + OH \rightarrow CH_3 + H_2O \tag{1b}$$

$$CH_3 + OH \rightarrow H_2CO + H_2 \tag{1c}$$

$$H_2CO + h\nu \rightarrow H_2 + CO \tag{1d}$$

$$CO + OH \rightarrow CO_2 + H \tag{1e}$$

The hydrogen produced in these reactions can escape to space, leaving carbon in its fully oxidized form.

Exactly how much CO_2 was present in the early atmosphere is uncertain. The estimated crustal abundance of carbon, most of which is stored in carbonate rocks, is about 10^{23} g (Ronov and Yaroshevsky, 1967; Holland, 1978). This is enough to produce an atmospheric pressure of 60 bars, were it all present as gaseous CO_2. Holland (1984) has estimated that as much as one-third of this CO_2, or 20 bars, may have been present in the atmosphere at the close of accretion. How long this dense CO_2 atmosphere would have lasted depends on how fast it was removed by the conversion of silicate minerals into carbonates. This rate, in turn, would have depended on the surface temperature of the early Earth and on the amount of continental area exposed to weathering. Walker (1986) has argued that a 10-bar CO_2 atmosphere could have persisted for several hundred million years if the early Earth was entirely covered by oceans. Taking into account an estimated 30% decrease in solar luminosity at that time compared to today (Gough,

1981), this 10-bar atmosphere would have produced a mean global surface temperature of approximately 85°C (Kasting and Ackerman, 1986). Estimated surface temperatures for different CO_2 partial pressures are shown in Fig. 1.

The existence of such a warm, dense CO_2 atmosphere can be neither demonstrated nor ruled out by geologic evidence, since the earliest sedimentary rocks are only 3.8 Ga. One argument against it is that it may not have been a favorable setting for the origin of life. As discussed below and in Chapter 30, the abiotic synthesis of organic compounds proceeds most readily in a more reducing environment. Ideally, carbon should have been present as methane. However, methane, as explained above, is photochemically oxidized even in an anoxic atmosphere. Since there is no obvious source for methane on the prebiotic Earth, its concentration should have been very low. Carbon monoxide, on the other hand, may have been present in higher abundances. Although it, too, can be photochemically oxidized, CO could have been produced in copious quantities by late-arriving planetesimals (Kasting, 1990). Reduced minerals (mainly metallic iron) in the partially vaporized impactors would have reacted with atmospheric carbon dioxide and water vapor, according to

$$Fe + CO_2 \rightarrow FeO + CO \quad (2)$$

$$Fe + H_2O \rightarrow FeO + H_2 \quad (3)$$

Most of the hydrogen produced in this manner would have escaped to space. The CO, however, was too heavy to escape and would have remained in the atmosphere. Additional CO might have been produced by oxidation of organic carbon in carbonaceous chondrite or cometary impactors. If the rate of impacts was high, it is possible that a large fraction of the atmospheric CO_2 could have been converted to CO.

The reason that the atmospheric oxidation state is so important is that it affects the types and yields of organic compounds that can be formed from natural processes. The two most obvious ways of synthesizing organic compounds on the early Earth involve atmospheric photochemistry and lightning discharges. Photochemistry in CO_2-rich atmospheres has been shown to be an effective mechanism for producing formaldehyde (H_2CO), one of the basic building blocks for amino acids and sugars (Pinto *et al.*, 1980; Kasting *et al.*, 1984). However, another key precursor for amino acids, hydrogen cyanide (HCN), is more difficult to create in such an environment. Zahnle (1986) has shown that HCN can be formed photochemically if atomic nitrogen atoms from the ionosphere react with trace amounts (several parts per million) of atmospheric methane. However, this mechanism fails if methane was not present. Alternatively, HCN could have been formed in lightning discharges if the atmospheric C/O ratio was unity or higher

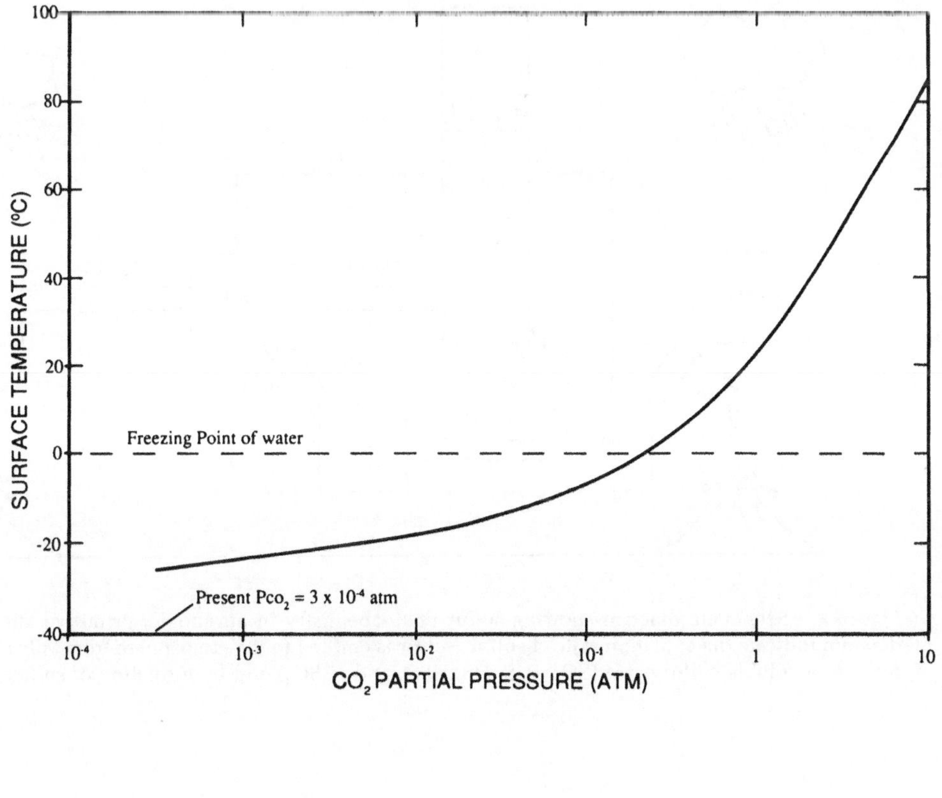

Figure 1. Mean surface temperature of the early Earth as a function of atmospheric CO_2 partial pressure. A 30% reduction in solar luminosity relative to the present-day level is assumed. The results are based on calculations by Kasting and Ackerman (1986).

(Chameides and Walker, 1981). (At lower C/O ratios, nitrogen atoms produced in the shock combine with oxygen to form NO.) Thus, HCN could have been produced quite efficiently in a CO-dominated atmosphere but not in one that consisted mostly of CO_2. More work should be done to examine the suggestion that the early atmosphere was rich in CO (Kasting, 1990) and to further reconcile theories of atmospheric and biological evolution.

3. Sulfur Gases and Ultraviolet Radiation

Another important environmental parameter on the early Earth was the flux of solar ultraviolet (UV) radiation. UV radiation between 200 and 300 nm is presently absorbed within the atmosphere by stratospheric ozone. If the early atmosphere was anoxic, ozone would have been absent as well (Levine *et al.*, 1980; Kasting and Donahue, 1980), and solar UV radiation in this wavelength region may have reached the ground virtually unattenuated. This might have posed a serious problem for early organisms and could also have affected the prebiotic synthesis of organic compounds. Although UV energy may have facilitated some synthetic reactions, its most likely effect would have been to inhibit the formation of large, complex molecules, since such molecules are generally destroyed by the absorption of UV radiation.

A possible solution to the UV problem is offered by sulfur photochemistry (Kasting *et al.*, 1989). Sulfur would have been emitted to the atmosphere from volcanoes, mostly as H_2S or SO_2. As indicated in Fig. 2, these gases could have been photochemically converted to a variety of other sulfur species. Most of the sulfur that entered the atmosphere should have either dissolved in the ocean as SO_2 or been oxidized to sulfate by reactions with O and OH radicals. However, a substantial fraction of the outgassed H_2S and SO_2 could have been converted into elemental sulfur vapor, consisting of S_8 along with other gaseous sulfur molecules S_n, $2 \leqslant n \leqslant 12$. S_8 is a ring molecule which absorbs well into the near UV, yet is thought to be relatively resistant to photolysis. (Absorption of a

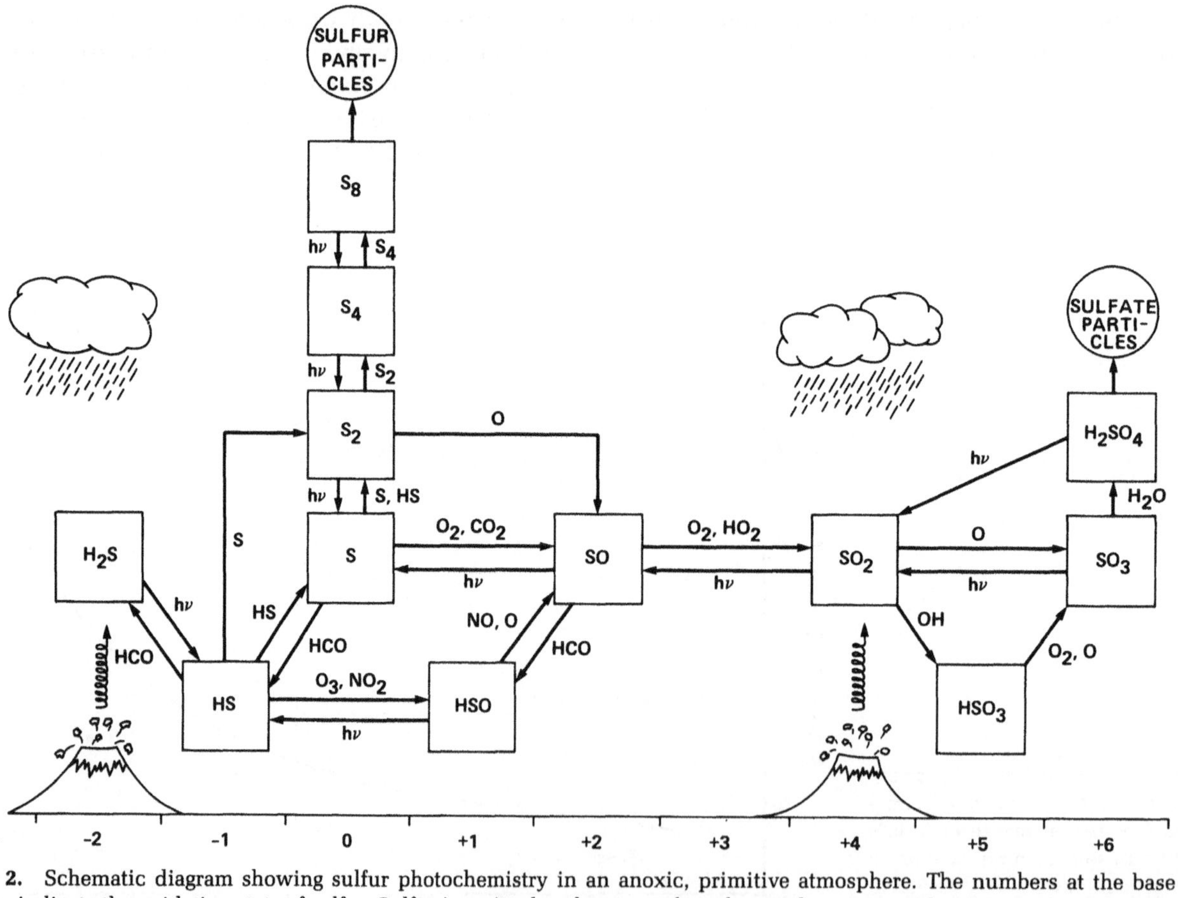

Figure 2. Schematic diagram showing sulfur photochemistry in an anoxic, primitive atmosphere. The numbers at the base of the diagram indicate the oxidation state of sulfur. Sulfur is emitted to the atmosphere from volcanoes as either SO_2 or H_2S. It is removed by rainout of soluble sulfur gases (SO_2, H_2S, HS, HSO, and H_2SO_4) and by formation of sulfate and elemental sulfur particles.

photon cleaves the ring. If the ring recloses before another photon is absorbed, the original molecule is preserved.) Its UV absorption coefficient is comparable to or higher than that of ozone (Fig. 3). The concentration of sulfur vapor in the primitive atmosphere would have been limited by its saturation vapor pressure, which depends strongly on temperature. If the surface temperature of the early Earth was the same as today or cooler, the amount of sulfur vapor in the atmosphere would have been negligible. If, however, the Earth was significantly (30° or more) warmer than today, sulfur vapor could have been present in sufficient concentrations to shield the surface from solar UV. From this standpoint, a warm, CO_2-rich atmosphere may have been beneficial to early life, whereas the conversion of CO_2 into CO by impacts could actually have been detrimental.

The presence of sulfur gases could have affected atmospheric photochemistry in a variety of other ways. Perhaps most importantly, they may have inhibited the photolysis of ammonia (NH_3). Ammonia is another compound that is considered important for prebiotic synthesis (Bada and Miller, 1968; Wigley and Brimblecombe, 1981). It absorbs strongly below about 225 nm (Fig. 3); hence, in the absence of shielding, it would have been rapidly converted to N_2 plus H_2 (Kuhn and Atreya, 1979; Kasting, 1982). However, both ammonia and its photodissociation product hydrazine (N_2H_4) could have been shielded from photolysis by SO_2, H_2S, and S_8. A warm, early atmosphere containing appreciable amounts of sulfur gases might thus have had a relatively high ammonia concentration; and this, too, could have facilitated the origin of life.

4. Ocean Composition and pH

The ocean, as pointed out earlier, was probably formed along with the planet and could have been close to its present volume even at 4.5 Ga B.P. Its composition and pH, however, must have been quite different from today. The present average pH of the deep ocean is about 8 (Broecker and Peng, 1982). The major cations and anions are shown in Table I. At present, the positive charge from Na^+, Mg^{2+}, Ca^{2+}, and K^+ is almost entirely balanced by the negative charge from Cl^- and SO_4^{2-}.

An ocean underlying an anoxic, CO_2-rich, primi-

TABLE I. Present Ocean Composition

Species	Concentration (mmol kg^{-1})
Na^+	468
Mg^{2+}	53.2
Ca^{2+}	10.2
K^+	10.2
Cl^-	545
SO_4^{2-}	28.2
HCO_3^-	2.38
CO_3^{2-}	0.18

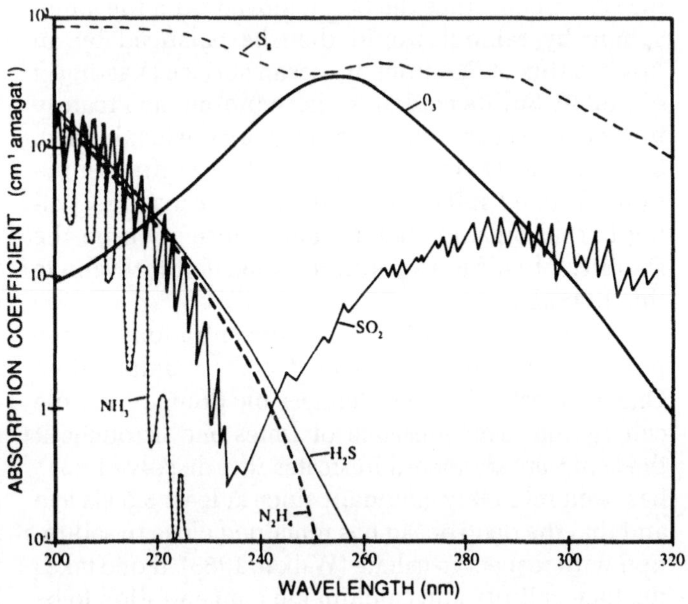

Figure 3. Absorption coefficients of various gases in the near ultraviolet. The S_8 absorption spectrum is that of sulfur dissolved in ethyl alcohol (Bass, 1953). The N_2H_4 curve is an extrapolation by Kasting (1982) of measurements at shorter wavelengths.

tive atmosphere would necessarily have had a very different chemical composition. The absence of atmospheric O_2 implies that sulfur-bearing minerals, such as pyrite, would not have been oxidized during continental weathering as they are today. The remaining source of sulfate is photochemical oxidation of sulfur gases, primarily SO_2 and H_2S, emitted from volcanoes. Even if this oxidation process were 100% efficient (which it is not), the rate at which sulfate was supplied to the early oceans would have been some 30 times less than the present rate (Holland, 1978; Walker and Brimblecombe, 1985). If the residence time of sulfate in the oceans was the same as today, the sulfate concentration would have been lower by this same factor. Thus, sulfate should have played only a minor role in the charge balance of the early oceans.

An interesting possibility, suggested by Skyring and Donnelly (1982), is that the primitive ocean may have contained appreciable concentrations of sulfite and bisulfite—the dissociation products of dissolved SO_2. Although sulfite is thermodynamically unstable with respect to disproportionation to sulfate and either elemental sulfur or sulfide,

$$3SO_3^{2-} + 2H^+ \rightarrow 2SO_4^{2-} + H_2O + S \quad (4)$$

$$4SO_3^{2+} \rightarrow 3SO_4^{2-} + S^{2+} \quad (5)$$

the mechanism by which these reactions would occur in dilute solution has not been established. If sulfite was indeed metastable, it might have accumulated in the oceans until its concentration approached saturation. This, in turn, would have greatly enhanced the likelihood of developing an effective sulfur UV screen, since the SO_2 removed from the atmosphere by rainout would then be balanced by an upward flux of SO_2 from the ocean surface (Kasting *et al.*, 1989). Sulfite could also have presented a readily utilizable energy source for early organisms; indeed, Skyring and Donnelly suggested that sulfite reduction preceded sulfate reduction as a means of recycling buried organic matter. More information on the stability of sulfite in solution is needed to evaluate this hypothesis.

Higher atmospheric CO_2 concentrations in the past would also have affected ocean composition. One can estimate the changes by doing a simple calculation. The presence of limestone throughout the sedimentary record indicates that dissolved Ca^{2+} has been relatively abundant since at least 3.5 Ga ago and that the deep ocean has remained close to saturation with respect to calcite (Walker, 1983). If one takes the total salinity and calcium ion concentration to be the same as today and assumes chemical equilibrium between the ocean and the atmosphere, then both $[H^+]$ and $[HCO_3^-]$ should be proportional to $pCO_2^{1/2}$. This result is expressed graphically in Fig. 4. An ocean in equilibrium with a CO_2-rich atmosphere should have been more acidic and had much higher bicarbonate ion concentrations than today. This prediction does not change greatly if the calcium concentration was somewhat different in the past or if the ocean was not completely saturated with respect to calcite.

An additional constraint imposed by the geologic record stems from the observation that gypsum generally follows carbonate minerals in evaporite sequences dating back to the late Precambrian. As the precipitation of calcite by the reaction

$$Ca^{2+} + 2HCO_3^- \rightarrow CaCO_3 + CO_2 + H_2O \quad (6)$$

has always left some Ca^{2+} in solution, it follows that the bicarbonate ion concentration has not exceeded twice the calcium ion concentration (shown by the dashed line in Fig. 4) during this time (Holland, 1972). Hence, the right-hand portion of Fig. 4 ($pCO_2 \geqslant 0.3$ atm) is excluded during the last 600 million years of the Earth's history. This result is not particularly useful, since the CO_2 partial pressure has probably been lower than 0.3 atm since at least the late Archean (see below). The constraint is worth bearing in mind, though, as it has occasionally been violated in model calculations of the long-term carbon cycle (Berner *et al.*, 1983; Kasting, 1984).

It is interesting to examine the effect on ocean composition of the dense, 10-bar CO_2 atmosphere alluded to earlier. According to Fig. 4, the bicarbonate ion concentration in equilibrium with such an atmosphere would be equal to ~130 mmol kg^{-1}, or over 20% of the present-day salinity. An ocean in which the bicarbonate concentration was comparable to that of chloride might be termed a "soda ocean," following Kempe and Degens (1985). Kempe and Degens, however, proposed that the primitive ocean had a pH comparable to that of modern soda lakes (pH 9–11). This would have been possible only if atmospheric CO_2 had been reduced to very low concentrations by the formation of carbonates. (Simultaneous high ocean pH and high atmospheric pCO_2 would imply enormous bicarbonate concentrations and, hence, an ocean that was much saltier than the present one.) However, low atmospheric CO_2 levels would, in turn, have left the primitive Earth without any obvious way of compensating for the lower solar luminosity at that time. Hence, Kempe and Degens's soda ocean would have turned into a gigantic soda ice cube unless some

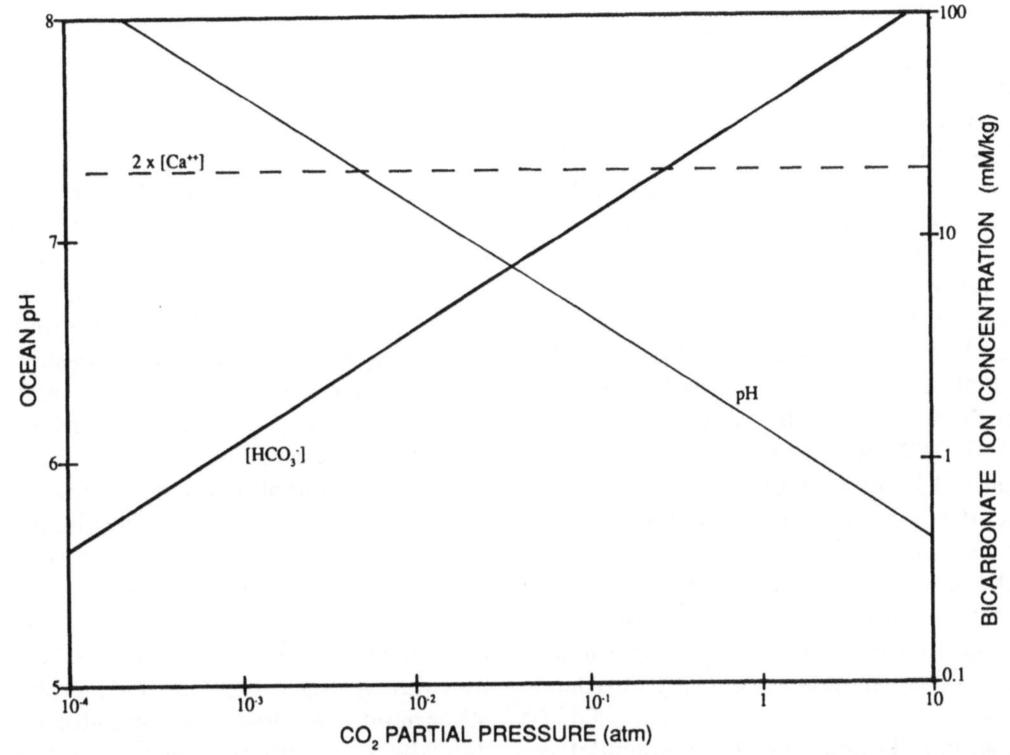

Figure 4. Ocean pH and bicarbonate ion concentration as functions of the atmospheric CO_2 partial pressure. Equilibrium at 25°C between the atmosphere and dissolved carbonate species in the ocean is assumed. Total salinity and calcium ion concentration are assumed to be the same as today.

greenhouse gas other than CO_2 was able to keep the Earth's surface warm. The question of how the Earth managed to maintain a temperate climate is discussed further below.

5. The Faint Young Sun Problem and the Decline in Atmospheric CO_2

Astronomers have known for many years that the sun has been gradually increasing in luminosity during its lifetime on the main sequence. As hydrogen is converted into helium, the density of the sun's core increases. This increase in density must be countered by an increase in temperature and pressure in order to keep the sun from collapsing on itself. Higher temperatures lead to an increase in the rate of thermonuclear burning, which, in turn, causes an increase in the rate of energy emission. Detailed numerical calculations indicate that solar luminosity was about 30% lower than today 4.6 Ga ago and has increased more or less linearly with time up to the present (Gough, 1981).

Sagan and Mullen (1972) were the first to point out the effects of changing solar luminosity on the long-term evolution of the Earth's climate. If the planetary albedo and the greenhouse effect of the atmosphere had remained constant over time, the mean surface temperature of the Earth would have been below freezing prior to about 2 Ga ago. The present mean temperature, by comparison, is 15°C. Since the geologic record contains no indication that the Earth was ever completely ice-covered, it is clear that either the planetary albedo was lower or the atmospheric greenhouse effect was larger in the past.

A variety of solutions to the problem have been suggested; these are reviewed elsewhere (Kasting *et al.*, 1984; Kasting, 1987, 1989). The most promising one involves a gradual decrease in atmospheric CO_2 levels with time. This idea, which was originally suggested by Hart (1978), was first investigated with a sophisticated, radiative–convective climate model by Owen *et al.* (1979). Their calculations have since been repeated with at least five different radiative–convective climate models, all of which are presently in good agreement (Kasting, 1989). The conclusion is that atmospheric CO_2 levels must have been at least several hundred times the present-day atmospheric level (PAL), or a few tenths of a bar, during the early Archean to keep the mean surface temperature above freezing. If the early Archean was warmer than today, as seems likely, pCO_2 was probably somewhat higher than this value.

The same type of climate calculation can be performed for different times in the Earth's history, taking into account the gradual increase in solar output. The results are most useful if one chooses periods during which there is evidence for glaciation, because the climate during these times could not have been much warmer than today. Using this approach, Kasting (1992) has estimated that pCO_2 during the Huronian glaciation, ~2.5 Ga ago, was 0.02 to 0.25 bar and the pCO_2 during the late Precambrian, 0.8 Ga ago, was 10^{-4} to 0.025 bar. These estimates are crude because we do not know how other factors, such as cloud cover and surface albedo, have changed over time. The uncertainties are somewhat mitigated by the fact that the present climate is glacial in the sense considered here (i.e., there are ice caps at the poles). Thus, the assumption that other climate forcing factors have remained constant is not entirely unreasonable.

It is amusing, although admittedly speculative, to try to plot atmospheric pCO_2 versus time based on considerations such as those above. My own "best guess" for how the concentrations of CO_2 and other gases evolved is shown in Fig. 5. The CO_2 curve is constrained, albeit not very tightly, at the two glacial periods and at the present. I have assumed, for reasons discussed earlier in this chapter, that the initial CO_2 partial pressure was 10 atm. This would, of course, imply that the earth was warm during the Hadean and early Archean (Fig. 1). The slope of the CO_2 curve during this time period is arbitrary, since we do not know how the actual climate changed during this time. In contrast, the predicted decrease of pCO_2 between 2.5 Ga and the present is much more secure, since a change of about this magnitude is required to keep the Earth's surface temperature within the observed bounds.

Why should atmospheric CO_2 have decreased in just such a way as to counteract the increase in solar output? Actually, the negative correlation between these two factors has not always been that good. The early Archean seems, if anything, to have been warmer than today, based on the lack of evidence for glaciation during that time. The Late Proterozoic, on the other hand, was close to being a global icebox; glacial deposits are found on nearly all of the continents at paleolatitudes as low as 15° (Frakes, 1979; Schopf and Klein, 1992). Nonetheless, there is a good reason why pCO_2 should have declined as the sun warmed up

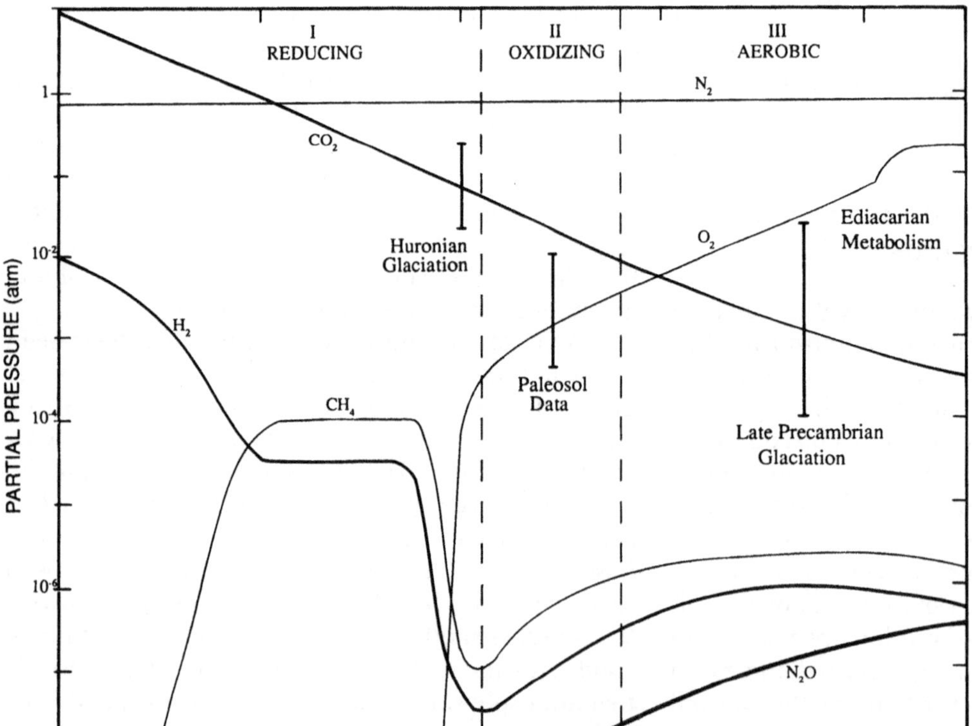

Figure 5. Possible changes in atmospheric composition with time. The two error bars on the CO_2 curve show upper and lower limits on pCO_2 during glacial periods calculated with a 1-D climate model (Kasting *et al.*, 1992). The error bar on the O_2 curve shows upper and lower limits on pO_2 between 1.5 Ga and 2.5 Ga, based on a combination of paleosol data and the theoretically derived pCO_2 curve. The curves for H_2, CH_4, and N_2O are based entirely on theoretical expectations (see text).

(Walker *et al.*, 1981). The loss process for atmospheric CO_2 over long time scales involves the weathering of silicate rocks—a process that depends on both temperature and rainfall. Cooler climates should cause a reduction in the rate of silicate weathering, and a corresponding increase in the concentration of atmospheric CO_2. Conversely, warmer climates should speed up the weathering rate and should draw down atmospheric CO_2. This negative feedback between CO_2 levels and surface temperature is probably the main reason for the long-term stability of the Earth's climate (Kasting, 1989).

6. The Rise of Oxygen and Ozone

Perhaps the most dramatic transformation that has taken place on the Earth's surface during its entire history has been the development of our present oxygen-rich atmosphere. We are reasonably certain that very little free O_2 was present in the atmosphere prior to the development of oxygenic photosynthesis. Walker (1977) was the first to calculate the prebiotic O_2 concentration correctly, taking into account the atmospheric hydrogen budget and the concept of diffusion-limited flux (Hunten, 1973). The ground-level O_2 concentration that Walker obtained from a simple, back-of-the-envelope calculation, 2×10^4 O_2 molecules cm^{-3}, or about 10^{-14} PAL, is within the range of estimates derived from more elaborate photochemical models (Kasting *et al.*, 1979; Canuto *et al.*, 1982; Kasting *et al.*, 1984).

How long oxygen remained at these extremely low levels is uncertain. One view (Walker *et al.*, 1983; Kasting, 1987; Kasting *et al.*, 1992) is that this weakly reducing, "Stage I" atmosphere persisted until 2.4–2.0 Ga ago. This view is based on the scarcity of Archean redbeds and on the preservation during the Archean of detrital minerals (uraninite and pyrite) that today are oxidized during weathering. In contrast, Holland (1984) suggested that free oxygen appeared much earlier than this, based on data obtained from paleosols (ancient soils). In my view, the bulk of the evidence favors the first interpretation. (Only three Archean paleosols have been identified, none of which is as firmly established as paleosols of more recent origin.) A relatively late date for the appearance of free oxygen is also consistent with data on sulfur isotopes, which exhibit a much broader range of $\delta^{34}S$ values during the Proterozoic than during the Archean (Schidlowski, 1979; Lambert and Donnelly, 1986). This pattern is most easily explained by low sulfate levels in the Archean oceans, which, in turn, imply an atmosphere lacking in free oxygen (Walker and Brimblecombe, 1985). An alternative explanation, proposed by Ohmoto and Felder (1987), is that Archean sulfate concentrations were high, but that the isotopic fractionation produced by bacterial sulfate reducers was diminished by higher ocean temperatures.

The next stage ("Stage II") in the evolution of atmospheric oxygen lasted from approximately 2.4 Ga to 1.7 Ga B.P. During this time, the atmosphere appears to have been oxidizing (based, again, on redbeds and detrital mineral deposits), but the deep ocean remained reduced. The evidence for an anoxic deep ocean comes from the occurrence of massive banded iron formations (BIFs), which are thought to require a deep ocean rich in dissolved ferrous iron (Holland, 1973; Walker *et al.*, 1983). The idea that the deep ocean was anoxic at this time can be used to derive an upper limit of 0.03 PAL on atmospheric pO_2 prior to 1.7 Ga B.P. (Kasting, 1987). The assumptions made in deriving this limit are that the rate of organic carbon burial and the rate of deep ocean mixing were the same as today. [Since this chapter was written, the estimated duration of Stage II has been shortened considerably, based on reevaluation of the redbed and BIF data (Kasting *et al.*, 1992). The most likely time limits now appear to be 2.0 Ga for its beginning and 1.85 Ga for its end. This period is so short that it is no longer clear that this stage ever occurred.]

An independent check on atmospheric O_2 concentrations during the early Proterozoic can be obtained from paleosol data. An analysis of paleosols ranging in age from 1.5 Ga to 2.5 Ga indicates that the ratio pO_2/pCO_2 during this time was approximately 0.03 ± 0.01 (Holland and Zbinden, 1986; Holland and Grandstaff, 1992). Combining this estimate with the limits on CO_2 partial pressure presented earlier (0.02–0.25 atm) yields 4×10^{-4} atm $\leq pO_2 \leq 0.01$ atm. This result is shown as a vertical "error bar" in Fig. 5. Encouragingly, the O_2 concentration obtained in this manner is consistent with the value derived from theoretical considerations of ocean mixing.

The disappearance of the BIFs at ~1.7 Ga signaled the end of Stage II and the beginning of the third and final stage of atmospheric O_2 evolution. Walker (1977) argued that, once the deep oceans were cleared of ferrous iron, atmospheric O_2 would rapidly have approached its present concentration, since the carbon cycle would then have been operating more or less as it does today. This argument is not very convincing, however, since many aspects of carbon cycling have changed since that time. In particular, the evolution of multicellular organisms during the late

Proterozoic and the appearance of land plants during the late Silurian and early Devonian should each have facilitated the burial of organic carbon (Kump, 1988; Kasting *et al.*, 1992). However, the geologic record indicates that the rate of carbon burial has remained roughly constant (Holland, 1984), except for transient excursions to higher values during the late Proterozoic and the late Paleozoic (Holser, 1984; Knoll *et al.*, 1986). DesMarais *et al.* (1992) have argued for changes in organic carbon burial rates during the early Proterozoic as well, based on new isotopic evidence. One way of countering such evolutionary advances would have been for atmospheric oxygen to have increased. It follows that O_2 concentrations were probably lower prior to these developments. The idea that atmospheric O_2 levels were still low (~0.1 PAL) during the late Proterozoic is consistent with ideas advanced by Nursall (1959) and Berkner and Marshall (1965), who suggested that increases in atmospheric O_2 may have triggered the Cambrian radiation in the biota. Although this hypothesis remains unproven, it is consistent with information from carbon isotopes and with theoretical models of the carbon cycle.

Accompanying the rise in atmospheric oxygen was an increase in atmospheric ozone and a corresponding decrease in the solar UV flux reaching the Earth's surface. The development of the ozone layer has now been studied by several different modelers (e.g., Levine *et al.*, 1979; Kasting and Donahue, 1980, and references therein; Kasting, 1987). The results are in reasonably good agreement. The UV screen should have started to be effective at an O_2 level of 10^{-3} PAL and was firmly established by the time pO_2 had reached 10^{-2} PAL. Thus, it seems likely that an efficient UV screen has been in place since sometime in the early Proterozoic. This is consistent with the observed proliferation of phytoplankton beginning around 2 Ga B.P. (Schopf *et al.*, 1983).

7. Other Trace Gases

Other important trace gases whose concentrations should have been much different in the past include hydrogen (H_2), methane (CH_4), and nitrous oxide (N_2O). H_2 produced from the reactions in Eqs. (1c) and (1d) above may have been a major constituent (perhaps 50% by volume) of an accretionary steam atmosphere (Zahnle *et al.*, 1988). Its concentration would have rapidly declined to more modest levels, however, as the flux of impactors decreased. Atmospheric H_2 mixing ratios expected during the early Archean are of the order of 10^{-4} to 10^{-3}, depending on the volcanic outgassing rates of hydrogen and carbon monoxide at that time (Walker, 1977; Kasting *et al.*, 1984). These estimates are derived by balancing the production rate of H_2 with its diffusion-limited escape rate (Hunten, 1973).

As the early biosphere became established, H_2 was probably used by organisms to reduce CO_2 and produce organic matter (Walker, 1977). Much of this organic matter may have been recycled by fermentation followed by methanogenesis. Schematically,

$$2H_2 + CO_2 \rightarrow CH_2O + H_2O \quad (\times 2)$$
$$\underline{2CH_2O \rightarrow CO_2 + CH_4}$$
$$4H_2 + CO_2 \rightarrow CH_4 + 2H_2O \tag{7}$$

The net effect of this process would have been to convert hydrogen into methane. Additional methane could have been generated from organic matter produced by oxygenic photosynthesis. Hence, the methane concentration in the Archean atmosphere could have been considerably higher than its present value of 1.6 ppm. A mixing ratio of $\sim 10^{-4}$, or 100 ppm, is suggested in Fig. 5 and in studies by Kasting *et al.* (1983) and Zahnle (1986). Indirect evidence that methane was abundant around 2.8 Ga B.P. is provided by the extremely low $\delta^{13}C$ values (down to −50‰) of many kerogens of this age (Schidlowski *et al.*, 1983). One explanation, proposed by Hayes (1983), is that these light kerogens represent organic matter that was doubly fractionated in a biological cycle involving methane transport either through the atmosphere or through the water column.

The appearance of free O_2 would have caused a dramatic decrease in the abundance of H_2 and CH_4 and any other reduced gases that may have been present in the Archean atmosphere. It has even been suggested that the elimination of methane could have triggered the Huronian glaciation in the early Proterozoic by reducing the magnitude of the greenhouse effect (Kasting *et al.*, 1983; Lovelock, 1988). After an initial drop to levels that were even lower than today, it is expected that H_2 and CH_4 concentrations would have recovered somewhat as pO_2 continued to increase (Kasting, 1987). The stability of both gases maximizes at an O_2 level of about 10^{-1} PAL; hence, they may have reached a secondary peak in concentration sometime during the late Precambrian. This prediction is based on the assumption of a constant methane flux from the biota. As there is no way of checking this assumption, or of measuring the past abundances of these gases, the curves shown in Fig. 5 are merely suggestive of how these gases may have varied with time.

Finally, nitrous oxide—the main source of odd nitrogen in the present stratosphere—is shown as monotonically increasing during the last half of the Earth's history. This is what one would expect if the biospheric N_2O flux has remained constant with time (Kasting, 1987). The N_2O concentration increases along with that of O_2 because the oxygen helps to shield it from photolysis.

8. Summary

The Earth's atmosphere and oceans formed along with the planet from impact degassing of in-falling planetesimals. The ocean was completely vaporized during at least part of the accretion period, creating a dense steam atmosphere that was hot enough to partially melt the Earth's surface. Evidence for the existence of this steam atmosphere is provided by neon isotopes.

The steam atmosphere collapsed once the main accretion process was over, leaving an atmosphere composed mostly of CO_2 or CO. CO_2 should have been the major constituent if photochemistry was the most important process; CO might have dominated if impacts continued to have a significant effect. A CO-rich atmosphere might have been a more promising environment for prebiotic synthesis and the origin of life.

Sulfur gases released by volcanoes into an anoxic early atmosphere could have created a layer of UV-absorbing elemental sulfur vapor. If the early atmosphere was substantially warmer than today, this sulfur layer could have shielded the surface from harmful solar ultraviolet radiation and could thereby have facilitated the origin and early evolution of life. The prospects for the existence of a sulfur screen are greatly enhanced if the primitive ocean was saturated with dissolved S(IV) species, so that rainout of SO_2 could be balanced by an upward flux of SO_2 from the ocean surface.

The early atmosphere must have contained at least a few tenths of a bar of CO_2 in order to counteract the climatic effects of the faint young sun. The primitive ocean must therefore have been more acidic than today and contained large amounts of bicarbonate ion. Atmospheric carbon dioxide levels would have declined with time as CO_2 was converted to carbonates during weathering. The negative feedback between weathering rates and temperature was probably an important stabilizing influence on the long-term evolution of the Earth's climate.

Free oxygen was essentially absent from the atmosphere until about 2.0 Ga B.P. The major reactive gases prior to this time were hydrogen and methane. A significant increase in pO_2 at 2.0 Ga is indicated by the disappearance of detrital uraninite deposits and the widespread appearance of redbeds. Low O_2 concentrations (0.03 PAL or below) during the early Proterozoic gave way to higher concentrations by 1.85 Ga, as evidenced by the disappearance of the BIFs. Atmospheric pO_2 remained appreciably lower than today throughout the Proterozoic, however, and probably did not reach its modern value until after the proliferation of land plants during the Devonian.

Much of the foregoing discussion is, of course, extremely speculative. Since the earliest history of the Earth is not recorded in the rocks, our ideas of what Hadean surface conditions were like are based totally on theoretical models. Matters improve somewhat in the Archean, but it remains exceedingly difficult to determine the nature of either the atmosphere or the ocean based on direct evidence. Hence, theoretical models are useful here as well. Future advances in our understanding of atmospheric and hydrospheric evolution will depend both on improvements in the models and on refinements in the geologic data base and its interpretation.

References

Bada, J. L., and Miller, S. L., 1968, Ammonium ion concentration in the primitive ocean, *Science* **159**:423–425.

Bass, A. M., 1953, The optical absorption of sulfur, *J. Chem. Phys.* **21**:80–86.

Benlow, A., and Meadows, A. J., 1977, The formation of the atmospheres of the terrestrial planets by impact, *Astrophys. Space Sci.* **46**:293–300.

Berkner, L. V., and Marshall, L. C., 1965, On the origin and rise of oxygen concentration in the Earth's atmosphere, *J. Atmos. Sci.* **22**:225–261.

Berner, R. A., Lasaga, A. C., and Garrels, R. M., 1983, The carbonate–silicate geochemical cycle and its effect on atmospheric carbon dioxide over the past 100 million years, *Am. J. Sci.* **283**:641–683.

Broecker, W. S., and Peng, T.-H., 1982, *Tracers in the Sea*, Lamont-Doherty Geological Observatory, Palisades, New York.

Canuto, V. M., Levine, J. S., Augustsson, T. R., and Imhoff, C. L., 1982, UV radiation from the young sun and oxygen and ozone levels in the prebiological paleoatmosphere, *Nature* **296**: 816–820.

Chameides, W. L., and Walker, J. C. G., 1981, Rates of fixation by lightning of carbon and nitrogen in possible primitive atmospheres, *Origins Life* **11**:291–302.

Chyba, C. F., 1987, The cometary contribution to the oceans of primitive Earth, *Nature* **330**:632–635.

Chyba, C. F., 1989, Impact delivery and erosion of planetary oceans, *Nature* **343**:129–132.

Craig, H., and Lupton, J. E., 1976, Primordial neon, helium, and hydrogen in oceanic basalts, *Earth Planet. Sci. Lett.* **31:** 369–385.

DesMarais, D. J., Strauss, H., Summons, R. E., and Hayes, J. M. 1992, Carbon isotopic evidence for the stepwise oxidation of the Proterozoic environment, *Nature* **359:**605–609.

Frakes, L. A., 1979, *Climates throughout Geologic Time*, Elsevier, New York.

Gough, D. O., 1981, Solar interior structure and luminosity variations, *Sol. Phys.* **74:**21–34.

Hart, M. H., 1978, The evolution of the atmosphere of the earth, *Icarus* **33:**23–39.

Hartmann, W. K., Phillips, R. J., and Taylor, G. J., 1986, *Origin of the Moon*, Lunar and Planetary Institute, Houston.

Hayes, J. M., 1983, Geochemical evidence bearing on the origin of aerobiosis, a speculative hypothesis, in: *The Earth's Earliest Biosphere: Its Origin and Evolution* (J. W. Schopf, ed.), Princeton University Press, Princeton, New Jersey, pp. 291–301.

Holland, H. D., 1962, Model of the evolution of the Earth's atmosphere, in: *Petrologic Studies: A Volume to Honor A. F. Buddington* (A. E. J. Engle, H. L. James, and B. F. Leonard, eds.), Geological Society of America, New York, pp. 447– 477.

Holland, H. D., 1972, The geologic history of sea water—an attempt to solve the problem, *Geochim. Cosmochim. Acta* **36:**637–651.

Holland, H. D., 1973, The oceans: A possible source of iron in ironformations, *Econ. Geol.* **68:**1169–1172.

Holland, H. D., 1978, *The Chemistry of the Atmosphere and Oceans*, John Wiley & Sons, New York.

Holland, H. D., 1984, *The Chemical Evolution of the Atmosphere and Oceans*, Princeton University Press, Princeton, New Jersey.

Holland, H. D., 1992, Distribution and paleoenvironmental interpretation of Proterozoic paleosols, in: *The Proterozoic Biosphere: A Multidisciplinary Study* (J. W. Schopf and C. Klein, eds.), Cambridge University Press, New York, 153–155.

Holland, H. D., and Zbinden, E. A., 1986, Paleosols and the evolution of the atmosphere, Part I, in: *Physical and Chemical Weathering in Geochemical Cycles* (A. Lerman and M. Meybeck, eds.), NATO ASI Series, Reidel, Dordrecht, pp. 61–82.

Holser, W. T., 1984, Gradual and abrupt shifts in ocean chemistry during Phanerozoic time, in: *Patterns of Change in Earth Evolution* (H. D. Holland, and A. F. Trendall, eds.), Springer-Verlag, Berlin, pp. 123–143.

Hunten, D. M., 1973, The escape of light gases from planetary atmospheres, *J. Atmos. Sci.* **30:**1481–1494.

Kasting, J. F., 1982, Stability of ammonia in the primitive terrestrial atmosphere, *J. Geophys. Res.* **87:**3091–3098.

Kasting, J. F., 1984, Comments on the BLAG model: The carbonate–silicate geochemical cycle and its effect on atmospheric carbon dioxide over the past 100 million years, *Am. J. Sci.* **284:** 1175–1182.

Kasting, J. F., 1987, Theoretical constraints on oxygen and carbon dioxide concentrations in the Precambrian atmosphere, *Precamb. Res.* **34:**205–229.

Kasting, J. F., 1989, Long term stability of the earth's climate, *Palaeogeogr. Palaeoclimatol. Palaeoecol. (Global and Planetary Change 1)* **75:**83–95.

Kasting, J. F., 1990, Bolide impacts and the oxidation state of carbon in the Earth's early atmosphere, *Origins Life* **20:**199–231.

Kasting, J. F., 1992, Proterozoic climates: The effect of changing atmospheric carbon dioxide concentrations, in: *The Proterozoic Biosphere: A Multidisciplinary Study* (J. W. Schopf and C. Klein, eds.), Cambridge University Press, New York, 165–168.

Kasting, J. F., and Ackerman, T. P., 1986, Climatic consequences of very high CO_2 levels in Earth's early atmosphere, *Science* **234:** 1383–1385.

Kasting, J. F., and Donahue, T. M., 1980, The evolution of atmospheric ozone, *J. Geophys. Res.* **85:**3255–3263.

Kasting, J. F., Liu, S. C., and Donahue, T. M., 1979, Oxygen levels in the prebiological atmosphere, *J. Geophys. Res.* **84:**3097–3107.

Kasting, J. F., Zahnle, K. J., and Walker, J. C. G., 1983, Photochemistry of methane in the earth's early atmosphere, *Precamb. Res.* **20:**121–148.

Kasting, J. F., Pollack, J. B., and Crisp, D., 1984, Effects of high CO_2 levels on surface temperature and atmospheric oxidation state on the early earth, *J. Atmos. Chem.* **1:**403–428.

Kasting, J. F., Zahnle, K. J., Pinto, J. P., and Young, A. T., 1989, Sulfur, ultraviolet radiation, and the early evolution of life, *Origins Life* **19:**95–108.

Kasting, J. F., Holland, H. D., and Kump, L. R., 1992, Atmospheric evolution: The rise of oxygen, in: *The Proterozoic Biosphere: A Multidisciplinary Study* (J. W. Schopf and C. Klein, eds.), Cambridge University Press, New York, 159–163.

Kempe, S., and Degens, E. T., 1985, An early soda ocean?, *Chem. Geol.* **53:**95–108.

Knoll, A. H., Hayes, J. M., Kaufman, A. J., Swett, K., and Lambert, I. B., 1986, Secular variations in carbon isotope ratios from Upper Proterozoic successions of Svalbard and East Greenland, *Nature* **321:**832–838.

Kuhn, W. R., and Atreya, S. K., 1979, Ammonia photolysis and the greenhouse effect in the primordial atmosphere of the earth, *Icarus* **37:**207–213.

Kump, L. R., 1988, Terrestrial feedback in atmospheric oxygen regulation by fire and phosphorus, *Nature* **335:**152–154.

Lambert, I. B., and Donnelly, T. H., 1986, The paleoenvironmental significance of trends in sulfur isotope compositions in the Precambrian: A critical review, in: *Stable Isotopes and Fluid Processes in Mineralisation* (H. K. Herbert, ed.), Special Publication, Geological Society of Australia, 328–348.

Lange, M. A., and Ahrens, T. J., 1982, The evolution of an impact-generated atmosphere, *Icarus* **51:**96–120.

Levine, J. S., Hays, P. B., and Walker, J. C. G., 1979, The evolution and variability of atmospheric ozone over geological time, *Icarus* **39:**295–309.

Levine, J. S., Boughner, R. E., and Smith, K. A., 1980, Ozone, ultraviolet flux, and the temperature of the paleoatmosphere, *Origins Life* **10:**199–207.

Lovelock, J. E., 1988, *The Ages of Gaia*, W. W. Norton, New York.

Matsui, T., and Abe, Y., 1986a, Evolution of an impact-induced atmosphere and magma ocean on the accreting earth, *Nature* **319:**303–305.

Matsui, T., and Abe, Y., 1986b, Impact-induced oceans on Earth and Venus, *Nature* **322:**526–528.

Nursall, J. R., 1959, Oxygen as a prerequisite to the origin of the metazoa, *Nature* **183:**1170–1172.

Ohmoto, H., and Felder, R. P., 1987, Bacterial activity in the warmer, sulphate-bearing Archean oceans, *Nature* **328:**244–246.

Owen, T., Cess, R. D., and Ramanathan, V., 1979, Early Earth: An enhanced carbon dioxide greenhouse to compensate for reduced solar luminosity, *Nature* **277:**640–642.

Ozima, M., and Igarashi, G., 1989, Terrestrial noble gases: Constraints and implications on atmospheric evolution, in: *Origin and Evolution of Planetary and Satellite Atmospheres* (S. K. Atreya, J. B. Pollack, and M. S. Matthews, eds.), University of Arizona Press, Tucson, pp. 306–327.

Pinto, J. P., Gladstone, C. R., and Yung, Y. L., 1980, Photochemical production of formaldehyde in the earth's primitive atmosphere, *Science* **210**:183–185.

Ronov, A. B., and Yaroshevsky, A. A., 1967, Chemical structure of the earth's crust, *Geochemistry* **11**:1041–1066 [translated from *Geokhimiya* **11**:1285–1309 (1967)].

Rubey, W. W., 1951, Geological history of seawater. An attempt to state the problem, *Geol. Soc. Am. Bull.* **62**:1111–1147.

Rubey, W. W., 1955, Development of the hydrosphere and atmosphere, with special reference to probable composition of the early atmosphere, in: *Crust of the Earth* (A. Poldervaart, ed.), Geological Society of America, New York, pp. 631–650.

Safronov, V. S., 1969, *Evolution of the Protoplanetary Cloud and Formation of the Earth and the Planets*, Nauka, Moscow [in Russian, translation NASA Report TTF-677, 1972].

Sagan, C., and Mullen, G., 1972, Earth and Mars: Evolution of atmospheres and surface temperatures, *Science* **177**:52–56.

Schidlowski, M., 1979, Antiquity and evolutionary status of bacterial sulfate reduction: Sulfur isotope evidence, *Origins Life* **9**: 299–311.

Schidlowski, M., Hayes, J. M., and Kaplan, I. R., 1983, Isotopic inferences of ancient biochemistries: Carbon, sulfur, hydrogen, and nitrogen, in: *The Earth's Earliest Biosphere: Its Origin and Evolution* (J. W. Schopf, ed.), Princeton University Press, Princeton, New Jersey, pp. 149–186.

Schopf, J. W. (ed.), 1983, *The Earth's Earliest Biosphere: Its Origin and Evolution*, Princeton University Press, Princeton, New Jersey.

Schopf, J. W., and Klein, C. (eds.), 1992, *The Proterozoic Biosphere: A Multidisciplinary Study*, Cambridge University Press, New York.

Schopf, J. W., Hayes, J. M., and Walter, M. R., 1983, Evolution of the Earth's earliest ecosystems: Recent progress and unsolved problems, in: *The Earth's Earliest Biosphere: Its Origin and Evolution* (J. W. Schopf, ed.), Princeton University Press, Princeton, New Jersey, pp. 361–384.

Skyring, G. W., and Donnelly, T. H., 1982, Precambrian sulfur isotopes and a possible role for sulfite in the evolution of biological sulfate reduction, *Precamb. Res.* **17**:41–61.

Stevenson, D. J., 1987, Steam atmospheres, magma oceans, and other myths, Paper presented at the American Geophysical Union Meeting, San Francisco, Dec. 3–7.

Walker, J. C. G., 1977, *Evolution of the Atmosphere*, Macmillan, New York.

Walker, J. C. G., 1983, Possible limits on the composition of the Archean ocean, *Nature* **302**:518–520.

Walker, J. C. G., 1986, Carbon dioxide on the early Earth, *Origins Life* **16**:117–127.

Walker, J. C. G., and Brimblecombe, P., 1985, Iron and sulfur in the pre-biologic ocean, *Precamb. Res.* **28**:205–222.

Walker, J. C. G., Hays, P. B., and Kasting, J. F., 1981, A negative feedback mechanism for the long-term stabilization of Earth's surface temperature, *J. Geophys. Res.* **86**:9776–9782.

Walker, J. C. G., Klein, C., Schidlowski, M., Stevenson, D. J., and Walter, M. R., 1983, Environmental evolution of the Archean–Early Proterozoic Earth, in: *The Earth's Earliest Biosphere: Its Origin and Evolution* (J. W. Schopf, ed.), Princeton University Press, Princeton, New Jersey, pp. 260–290.

Wetherill, G., 1985, Occurrence of giant impacts during the growth of the terrestrial planets, *Science* **228**:877–879.

Wigley, T. M. L., and Brimblecombe, P., 1981, Carbon dioxide, ammonia, and the origin of life, *Nature* **291**:213–215.

Zahnle, K. J., 1986, Photochemistry of methane and the formation of hydrocyanic acid (HCN) in the earth's early atmosphere, *J. Geophys. Res.* **91**:2819–2834.

Zahnle, K. J., Kasting, J. F., and Pollack, J. B., 1988, Evolution of a steam atmosphere during Earth's accretion, *Icarus* **74**: 62–97.

Zahnle, K. J., Kasting, J. F., and Pollack, J. B., 1990, Mass fractionation of noble gases in diffusion-limited hydrodynamic hydrogen escape, *Icarus* **84**:502–527.

Chapter 30

The Prebiotic Synthesis of Organic Compounds on the Early Earth

STANLEY L. MILLER

In the past three decades there have been a wide variety of experiments designed to simulate conditions on the primitive Earth and to demonstrate how the organic compounds that made up the first living organisms were synthesized. This chapter will review this work and indicate the status of such syntheses. There is too much material to review in detail, and the reader is directed to a number of more complete discussions (Miller and Orgel, 1974; Kenyon and Steinman, 1969; Lemmon, 1970).

1. The Composition of the Primitive Atmosphere

There is no agreement on the constituents of the primitive atmosphere. It is to be noted that there is no geological evidence concerning the conditions on the Earth from 4.5×10^9 years to 3.8×10^9 years, since no rocks older than 3.8×10^9 years are known. Even the 3.8×10^9-year-old Isua Rocks in Greenland are not sufficiently well preserved to infer details of the atmosphere at that time. Proposed atmospheres and the reasons given to favor them will not be discussed here. As shown in Sections 2, 3, and 4, the more reducing atmospheres favor the synthesis of organic compounds both in terms of yields and in terms of the variety of compounds obtained. Some of the organic chemistry can give explicit predictions about atmospheric constituents. Such considerations cannot prove that the Earth had a certain primitive atmosphere, but the prebiotic synthesis constraints should be a major consideration.

2. Energy Sources

A wide variety of energy sources have been utilized with various gas mixtures since the first experiments using electric discharges. The importance of a given energy source is determined by the product of the energy available and its efficiency for organic compound synthesis. Even though both factors cannot be evaluated with precision, a qualitative assessment of the energy sources can be made. It should be emphasized that a single source of energy of a single process is unlikely to account for all the organic compounds on the primitive Earth (Miller *et al.*,

This chapter was adapted in part from a paper by the author in Cold Spring Harbor Symposium on Quantitative Biology ***52****:17–27 (1987).*

STANLEY L. MILLER • Department of Chemistry, University of California–San Diego, La Jolla, California 92093-0317.
Organic Geochemistry, edited by Michael H. Engel and Stephen A. Macko. Plenum Press, New York, 1993.

1976). An estimate of the sources of energy on the Earth at the present time is given in Table I.

The energy from the decay of radioactive elements was probably not an important energy source for the synthesis of organic compounds on the primitive Earth since most of the ionization would have taken place in silicate rocks rather than in the reducing atmosphere. The shock wave energy from the impact of meteorites on the Earth's atmosphere and surface and the larger amount of shock waves generated in lightning bolts have been proposed as energy sources for primitive earth organic synthesis. Very high yields of amino acids have been reported in some experiments with shock waves (Bar-Nun *et al.*, 1970), but it is doubtful whether such yields would be obtained in natural shock waves. Cosmic rays are a minor source of energy on the Earth at present, and it seems unlikely that any increase in the past could have been so great as to make them a major source of energy.

The energy in the lava emitted at the present time is a significant but not a major source of energy. It is generally supposed that there was a much greater amount of volcanic activity on the primitive Earth, but there is no evidence to support this. Even if the volcanic activity was a factor of 10 greater than at present, it would not have been the dominant energy source. Nevertheless, molten lava may have been important in the pyrolytic synthesis of some organic compounds.

Ultraviolet light was probably the largest source of energy on the primitive Earth. The wavelengths absorbed by the atmospheric constituents are all below 2000 Å except for ammonia (<2300 Å) and H_2S (<2600 Å). Whether ultraviolet light was the most effective source of organic compounds is not clear. Most of the photochemical reactions would occur in the upper atmosphere, and the products formed would, for the most part, absorb the longer wavelengths, and so be decomposed before reaching the protection of the oceans. The yield of amino acids from the photolysis of CH_4, NH_3, and H_2O at wavelengths of 1470 and 1294 Å is quite low (Groth and von Weyssenhoff, 1960), probably due to the low yields of hydrogen cyanide. The synthesis of amino acids by the photolysis of CH_4, C_2H_6, NH_3, H_2O, and H_2S mixtures by ultraviolet light of wavelengths greater than 2000 Å (Sagan and Khare, 1971; Khare and Sagan, 1971) is also a low-yield synthesis, but the amount of energy is much greater in this region of the sun's spectrum. Only H_2S absorbs the ultraviolet light, but the photodissociation of H_2S results in a hydrogen atom having a high kinetic energy, which activates or dissociates the methane, ammonia, and water. This appears to be a very attractive prebiotic synthesis. However, it is not clear whether a sufficient partial pressure of H_2S could be maintained in the atmosphere since H_2S is photolyzed rapidly to elemental sulfur and hydrogen. The same applies to other molecules that might generate hot hydrogen atoms.

A photochemical source of HCN using the very short wavelengths of the Lyman continuum (796 to 912 Å) has been proposed by Zahnle (1986). The N atoms produced diffuse lower into the atmosphere and react with CH_2 and CH_3, producing HCN. The process depends on N atoms reacting with nothing else before the CH_2 and CH_3. Whether this is valid remains to be determined.

The most widely used sources of energy for laboratory syntheses of prebiotic compounds are electric discharges. These include sparks, semicorona, arc, and silent discharges, with the spark being the most frequently used type. The ease of handling and high efficiency of electric discharges are factors favoring their use, but the most important reason is that electric discharges are very efficient energy sources for the synthesis of hydrogen cyanide, whereas ultraviolet light is not. Hydrogen cyanide is a central intermediate in prebiotic synthesis, being needed for amino acid synthesis from the Strecker reaction, or by self-polymerization to amino acids, and, most importantly, for the prebiotic synthesis of adenine and guanine.

An important feature in the use of all these energy sources is that the activation of molecules in a

TABLE I. Present Sources of Energy Averaged over the Earth

Source	Energy	
	cal cm^{-2} yr^{-1}	J cm^{-2} yr^{-1}
Total radiation from sun	260,000	1,090,000
Ultraviolet light		
<3000 Å	3,400	14,000
<2500 Å	563	2,360
<2000 Å	41	170
<1500 Å	1.7	7
Electric discharges	4[a]	17
Cosmic rays	0.0015	0.006
Radioactivity (to 1.0-km depth)	0.8	3.0
Volcanoes	0.13	0.5
Shock waves	1.1[b]	4.6

[a]3 cal cm^{-2} yr^{-1} of corona discharge + 1 cal cm^{-2} yr^{-1} of lightning.
[b]1 cal cm^{-2} yr^{-1} of this is in the shock wave of lightning bolts and is also included under electric discharges.

local area is followed by quenching of this activated mixture and that the organic compounds are then protected from further influence of the energy source. The quenching and protective steps are critical because the organic compounds will be destroyed if subjected continuously to the energy source.

3. Prebiotic Synthesis of Amino Acids

Atmospheres containing CH_4, NH_3, and H_2O with or without added H_2 are considered strongly reducing. The atmosphere of Jupiter contains these species with the H_2 in large excess relative to CH_4. The first successful prebiotic amino acid synthesis was carried out using an electric discharge as an energy source (Miller, 1953, 1955). The result was a large yield of amino acids (the yield of glycine alone was 2.1% based on carbon), together with hydroxy acids, short aliphatic acids, and urea (Table II). One of the surprising results of this experiment was that the products were not a random mixture of organic compounds, but rather a relatively small number of compounds were produced in substantial yield. In addition, the compounds were, with a few exceptions, of biological importance.

TABLE II. Yields from Sparking a Mixture of CH_4, NH_3, H_2O, and H_2

Compound	Yield µmoles	Yield %[a]
Glycine	630	2.1
Glycolic acid	560	1.9
Sarcosine	50	0.25
Alanine	340	1.7
Lactic acid	310	1.6
N-Methylalanine	10	0.07
α-Amino-n-butyric acid	50	0.34
α-Aminoisobutyric acid	1	0.007
α-Hydroxybutyric acid	50	0.34
β-Alanine	150	0.76
Succinic acid	40	0.27
Aspartic acid	4	0.024
Glutamic acid	6	0.051
Iminodiacetic acid	55	0.37
Iminoacetic-propionic acid	15	0.13
Formic acid	2330	4.0
Acetic acid	150	0.51
Propionic acid	130	0.66
Urea	20	0.034
N-Methylurea	15	0.051
Total		15.2

[a]The percent yields are based on carbon; 59 mmol (712 mg) of carbon were added as CH_4.

The mechanism of synthesis of the amino and hydroxy acids was investigated (Miller, 1957). It was shown that the amino acids were not formed directly in the electric discharge but were the result of solution reactions of smaller molecules produced in the discharge, in particular, hydrogen cyanide and aldehydes. The reactions are given in Scheme 1.

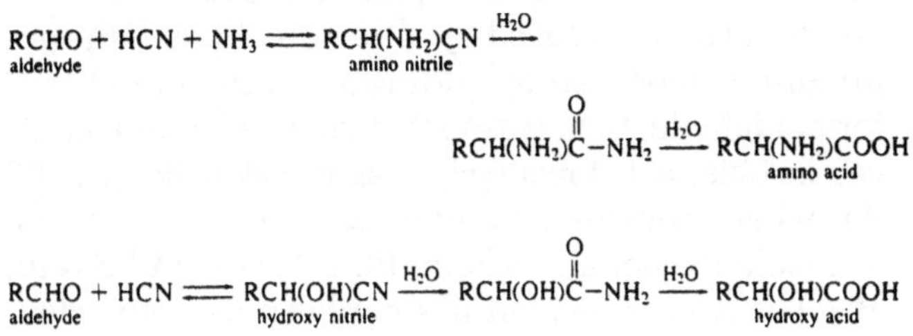

(1)

These reactions were studied subsequently in detail, and their equilibrium and rate constants were measured (Miller and Van Trump, 1981). These results show that amino and hydroxy acids can be synthesized at high dilutions of HCN and aldehydes in a primitive ocean. It is also to be noted that the rates of these reactions were rather rapid. The half-lives for the hydrolysis of the amino and hydroxy nitriles are less than 10^3 years at 0°C.

This synthesis of amino acids, called the Strecker synthesis, requires the presence of NH_4^+ (and NH_3) in the primitive ocean. On the basis of the experimental equilibrium and rate constants, it can be shown (Miller and Van Trump, 1981) that equal amounts of amino and hydroxy acids are obtained when the NH_4^+ concentration is about 0.01M at pH 8 and 25°C, with this NH_4^+ concentration being insensitive to temperature and pH. This translates into a pNH_3 in the atmosphere of 2×10^{-7} atm at 0°C and 4×10^{-6} atm at 25°C. This is a low partial pressure, but it would seem to be necessary for amino acid synthesis. A similar estimate of the NH_4^+ concentration in the primitive ocean can be obtained from the equilibrium decomposition of aspartic acid (Bada and Miller, 1968). Ammonia is decomposed by ultraviolet light, but mechanisms for resynthesis are available. The details of the ammonia balance on the primitive Earth remain to be worked out.

In a typical electric discharge experiment, the partial pressure of CH_4 is 0.1 to 0.2 atm. This pressure is used for convenience, and it is likely, but has never been demonstrated, that organic compound synthesis would work at much lower partial pressures of methane. There are no estimates available for pCH_4 on the primitive Earth, but 10^{-5} to 10^{-3} atm seems plausible. Higher pressures are not reasonable because the sources of energy would convert the CH_4 to organic compounds in the oceans too rapidly for higher pressures of CH_4 to build up.

Ultraviolet light acting on this mixture of gases is not effective in producing amino acids except at very short wavelengths (<1500 Å), and even then the yields are very low (Groth and von Weyssenhoff, 1960). The low yields are probably due to the low yields of HCN produced by ultraviolet light. If the gas mixture is modified by adding gases such as H_2S or formaldehyde, then reasonable yields of amino acids can be obtained at relatively long wavelengths (<2500 Å), where considerable energies from the sun are available (Sagan and Khare, 1971; Khare and Sagan, 1971). It is possible, but has not been demonstrated, that HCN and other molecules are produced, which then form amino acids in the aqueous part of the system.

Pyrolysis of CH_4 and NH_3 gives very low yields of amino acids. The pyrolysis conditions are temperatures from 800 to 1200°C with contact times of a second or less (Lawless and Boynton, 1973). However, the pyrolysis of CH_4 and other hydrocarbons gives good yields of benzene, phenylacetylene, and many other hydrocarbons. It can be shown that phenylacetylene would be converted to phenylalanine and tyrosine in the primitive ocean (Friedmann and Miller, 1969). Pyrolysis of the hydrocarbons in the presence of NH_3 gives substantial yields of indole, which can be converted to tryptophan in the primitive ocean (Friedmann *et al.*, 1971).

A mixture of CH_4, N_2, and H_2O with traces of NH_3 is a more realistic atmosphere for the primitive Earth because large amounts of NH_3 would not have accumulated in the atmosphere since the NH_3 would have dissolved in the ocean. It is still, however, a strongly reducing atmosphere. This mixture of gases is quite effective with an electric discharge in producing amino acids (Ring *et al.*, 1972; Wolman *et al.*, 1972). The experimental apparatus used for this type of synthesis is shown in Fig. 1. The yields are somewhat lower than with higher partial pressures of NH_3, but the products are more diverse (Table III). Hydroxy acids, short aliphatic acids, and dicarboxylic acids are produced along with the amino acids (Peltzer and Bada, 1978; Peltzer *et al.*, 1984). Ten of the 20 amino acids that occur in proteins are produced directly in this experiment. Counting asparagine and glutamine, which are formed but hydrolyzed before analysis, and methionine, which is formed when H_2S is added (Miller and Van Trump, 1972), one can say that 13 of the 20 amino acids in proteins can be formed in this single experiment. Cysteine was found in the photolysis of CH_4, NH_3, H_2O, and H_2S (Khare and Sagan, 1971). The pyrolysis of hydrocarbons, as discussed above, leads to phenylalanine, tyrosine, and tryptophan. This leaves only the basic amino acids: lysine, arginine, and histidine. Thus far, no prebiotic syntheses of these amino acids have been established. There is no fundamental reason why the basic amino acids cannot be synthesized, and this problem may be solved before too long.

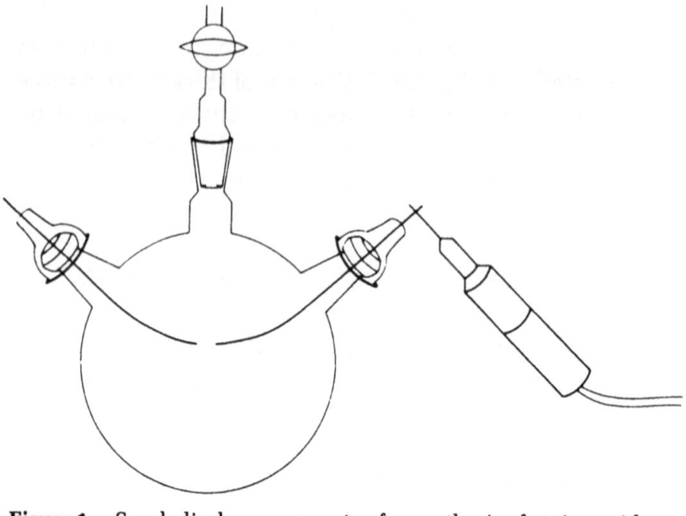

Figure 1. Spark discharge apparatus for synthesis of amino acids in a strongly reducing atmosphere containing traces of NH_3. The three-liter flask is shown with the two tungsten electrodes and a spark generator. The second electrode is usually not grounded. In the experiments described in Table III, the flask contained 100 ml of 0.05 *M* NH_4Cl brought to pH 8.7, giving pNH_3 of 0.1 torr. The pCH_4 was 200 torr, and pN_2 was 80 torr. Since the temperature was about 30°C during the sparking, pH_2O was 32 torr.

4. Mildly Reducing and Nonreducing Atmospheres

There has been less experimental work with gas mixtures containing CO and CO_2 as carbon sources instead of CH_4. Spark discharges have been the source of energy most extensively investigated (Abelson, 1965; Schlesinger and Miller, 1983a, b; Stribling and Miller, 1987). Figure 2 compares amino acid yields using CH_4, CO, and CO_2 as a carbon source with various amount of H_2. Separate experiments were performed with and without added NH_3. In the case of CH_4 without added NH_3, the yield of amino acids is 4.7% at $H_2/CH_4 = 0$ and drops to 1.4% at $H_2/CH_4 = 4$.

TABLE III. Yields from Sparking CH_4 (336 mmol), N_2, and H_2O with Traces of NH_3

Compound	μmoles
Glycine	440
Alanine	790
α-Amino-*n*-butyric acid	270
α-Aminoisobutyric acid	~30
Valine	19.5
Norvaline	61
Isovaline	~5
Leucine	11.3
Isoleucine	4.8
Alloisoleucine	5.1
Norleucine	6.0
tert-Leucine	<0.02
Proline	1.5
Aspartic acid	34
Glutamic acid	7.7
Serine	5.0
Threonine	~0.8
Allothreonine	~0.8
α,γ-Diaminobutyric acid	33
α-Hydroxy-γ-aminobutyric acid	74
α,β-Diaminopropionic acid	6.4
Isoserine	5.5
Sarcosine	55
N-Ethylglycine	30
N-Propylglycine	~2
N-Isopropylglycine	~2
N-Methylalanine	~15
N-Ethylalanine	<0.2
β-Alanine	18.8
β-Amino-*n*-butyric acid	~0.3
β-Aminoisobutyric acid	~0.3
γ-Aminobutyric acid	2.4
N-Methyl-β-alanine	~5
N-Ethyl-β-alanine	~2
Pipecolic acid	0.05

[a]Yields based on the carbon added as CH_4: glycine, 0.26%; alanine, 0.71%; total yield of amino acids in the table, 1.90%.

With CO and no added NH_3, the amino acid yield is 0.05% at $H_2/CO = 0$ and rises to a maximum of 2.7% at $H_2/CO = 3$. With CO_2 and no added NH_3, the amino acid yield is 7×10^{-4}% at $H_2/CO_2 = 0$. This is close to the level of reagent contamination and is so low that this possibility cannot be considered as a significant source of amino acids on the primitive Earth. At higher H_2/CO_2 ratios, however, the yield rises to about 2%. With both CO and CO_2, the presence of added NH_3 increases the yield of amino acids by a factor of ~10 at low H_2/CO and H_2/CO_2 ratios. The amino acids produced in the CH_4 experiments were similar to those shown in Table III. With CO and CO_2, glycine was the predominant amino acid with little else besides some alanine being produced (Table IV).

A mixture of CO + H_2 is used in the Fischer–Tropsch reaction to make hydrocarbons in high yields. The reaction requires a catalyst, usually Fe or Ni supported on silica, a temperature of 200–400°C, and a short contact time. Depending on the conditions, aliphatic hydrocarbons, aromatic hydrocarbons, alcohols, and acids can be produced. If NH_3 is added to the CO + H_2, then amino acids, purines, and pyrimidines can be formed (Hayatsu and Anders, 1981). The intermediates in these reactions are not known, but it is likely that HCN is involved together with some of the intermediates postulated for the electric discharge processes.

Passage of electric discharges through a mixture of CO + H_2O is not particularly effective in organic compound synthesis, but irradiation with ultraviolet light that is absorbed by the water (<1849 Å) results in the production of formaldehyde and other aldehydes, alcohols, and acids in fair yield (Bar-Nun and Hartman, 1978; Bar-Nun and Chang, 1983). The mechanism seems to involve splitting the H_2O to H + OH with the OH converting CO to CO_2 and the H reducing another molecule of CO.

Electric discharges and ultraviolet light do not produce substantial amounts of organic compounds from a mixture of CO_2 + H_2O. Ionizing radiation (e.g., 40-MeV helium ions) gives small yields of formic acid and formaldehyde (Garrison *et al.*, 1951).

Calculations using one-dimensional photochemical models of CO_2 + H_2O atmospheres show that substantial amounts of H_2CO can be produced in these atmospheres by solar ultraviolet light (Pinto *et al.*, 1980; Kasting *et al.*, 1984).

The action of γ rays on an aqueous solution of CO_2 and ferrous ion gives fair yields of formic acid, oxalic acid, and other simple products (Getoff, 1962). Ultraviolet light gives similar results. In these reactions, the Fe^{2+} is a stoichiometric reducing agent rather than a catalyst. Nitrogen in the form of N_2 does not react, and experiments with NH_3 have not been tried.

The implication of these results in considering the composition of the primitive Earth is that CH_4 is the best carbon source for prebiotic synthesis, especially for amino acid synthesis. Although glycine was essentially the only amino acid synthesized in the spark discharge experiments with CO and CO_2, other amino acids (e.g., serine, aspartic acid, alanine) would probably have been formed from this glycine, H_2CO, and HCN as the primitive ocean matured. Since we do not know which amino acids made up the first organism, we can only say that CO and CO_2 are less favorable than CH_4 for amino acid synthesis,

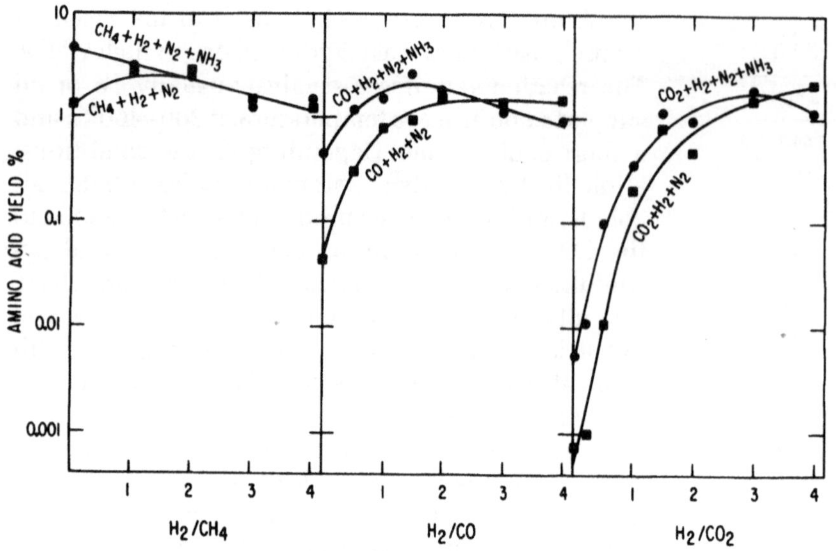

Figure 2. Amino acid yields, based on initial carbon, from sparking gas mixtures. In all experiments, pN_2 = 100 torr and pCH_4, pCO, or pCO_2 = 100 torr with 100 ml of H_2O. The flask was kept at room temperature, and the spark generator was operated continuously for 48 h.

but these amino acids may have been adequate. The synthesis of purine and sugars described below would not be greatly different with CH_4, CO, or CO_2 with adequate H_2. Although the spark discharge yields of amino acids, HCN, and H_2CO are about the same with CH_4 and with $H_2/CO > 1$ and $H_2/CO_2 > 2$, it is not clear how such high H_2/carbon ratios could have been maintained in the primitive atmosphere since H_2 escapes from the Earth's atmosphere into outer space. These problems are poorly understood and beyond the scope of this chapter.

5. Purine and Pyrimidine Synthesis

Hydrogen cyanide is used in the synthesis of purines as well as amino acids. This is illustrated in a remarkable synthesis of adenine. If concentrated solutions of ammonium cyanide are refluxed for a few days, adenine is obtained in up to 0.5% yield along with 4-aminoimidazole-5-carboxamide and the usual cyanide polymer (Oro and Kimball, 1961, 1962).

The mechanism of adeninesynthesis in these experiments is probably as shown in Scheme 2.

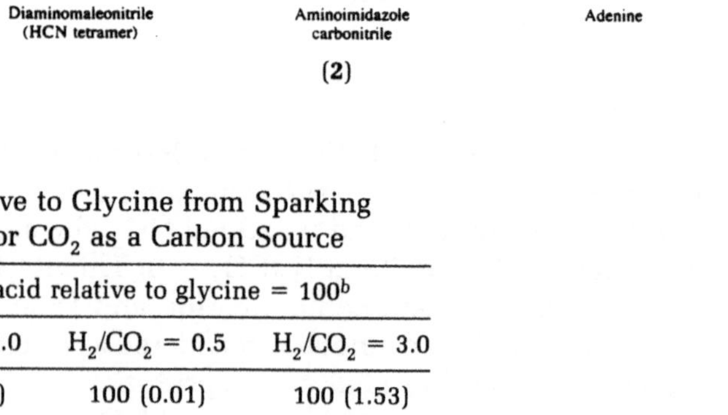

(2)

TABLE IV. Amino Acid Yields Relative to Glycine from Sparking Gas Mixtures Containing CH_4, CO, or CO_2 as a Carbon Source

	Mole ratio of amino acid relative to glycine = 100[b]			
Amino acid	H_2/CH_4 = 3.0	H_2/CO = 3.0	H_2/CO_2 = 0.5	H_2/CO_2 = 3.0
Glycine	100 (0.40)	100 (1.42)	100 (0.01)	100 (1.53)
Alanine	101	2.4	7.0	0.87
α-Aminobutyric	30	0.04	<0.001	0.09
Valine	1.2	0.005	<0.001	<0.001
Norvaline	1.3	0.01	<0.001	<0.001
Aspartic acid	1.6	0.09	0.22	0.14
Glutamic acid	1.5	0.01	0.06	<0.001
Serine	3.1	0.15	0.40	0.23

[a]Adapted from Schlesinger and Miller (1983a).
[b]The percent yield of glycine based on carbon is given in parentheses.

The difficult step in this synthesis of adenine is the reaction of tetramer with formamidine. This step may be bypassed by the photochemical rearrangement of tetramer to aminoimidazole carbonitrile (Scheme 3), a reaction that proceeds readily in contemporary sunlight (Sanchez *et al.*, 1967, 1968).

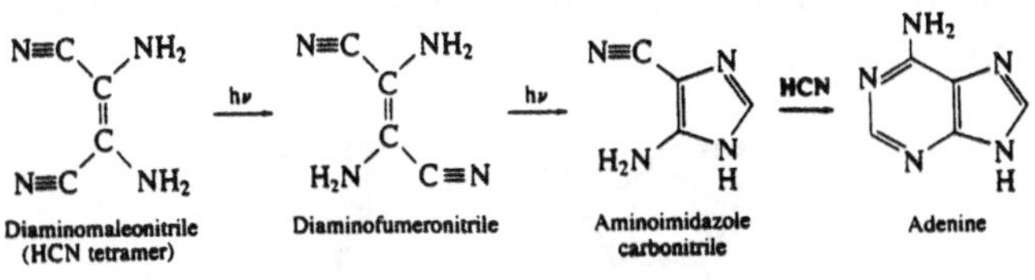

(3)

A further possibility is that tetramer formation may have occurred in a eutectic solution. High yield of tetramer (>10%) can be obtained by cooling dilute cyanide solutions to between −10 and −30°C for a few months (Sanchez *et al.*, 1966b). The cyanide polymerization reaction to adenine is even more complex than indicated here, with the synthesis proceeding through 8-cyanoadenine and 2-cyanoadenine under some conditions (Voet and Schwartz, 1983).

Guanine, hypoxanthine, xanthine, and diaminopurine would have been synthesized by variations of the adenine synthesis, as shown in Scheme 4, using aminoimidazole carbonitrile and aminoimidazole carboxamide.

The prebiotic synthesis of the pyrimidine cytosine involves cyanoacetylene, which is synthesized in good yield by sparking mixtures of CH_4 + N_2. Cyanoacetylene reacts with cyanate to give cytosine

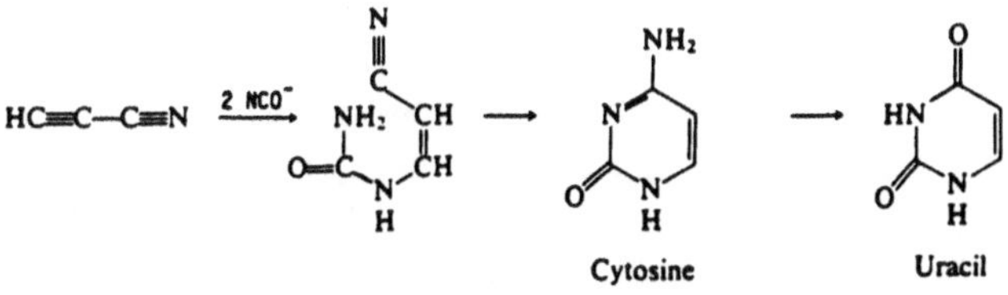

(5)

(Sanchez *et al.*, 1966a; Ferris *et al.*, 1968), and the cytosine can be converted to uracil (Scheme 5). Cyanate can form from cyanogen or by the decomposition of urea.

A related synthesis (Scheme 6) starts with cyanoacetaldehyde, from the hydration of cyanoacetylene, which reacts with guanidine to give diaminopyrimidine. This is then hydrolyzed to cytosine and uracil (Ferris *et al.*, 1974).

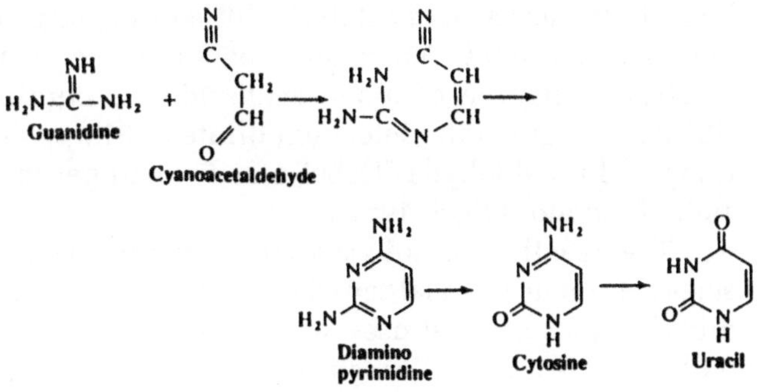

(6)

Another prebiotic synthesis of uracil (Scheme 7) starts from β-alanine and cyanate and proceeds with ultraviolet light (Schwartz and Chittenden, 1977).

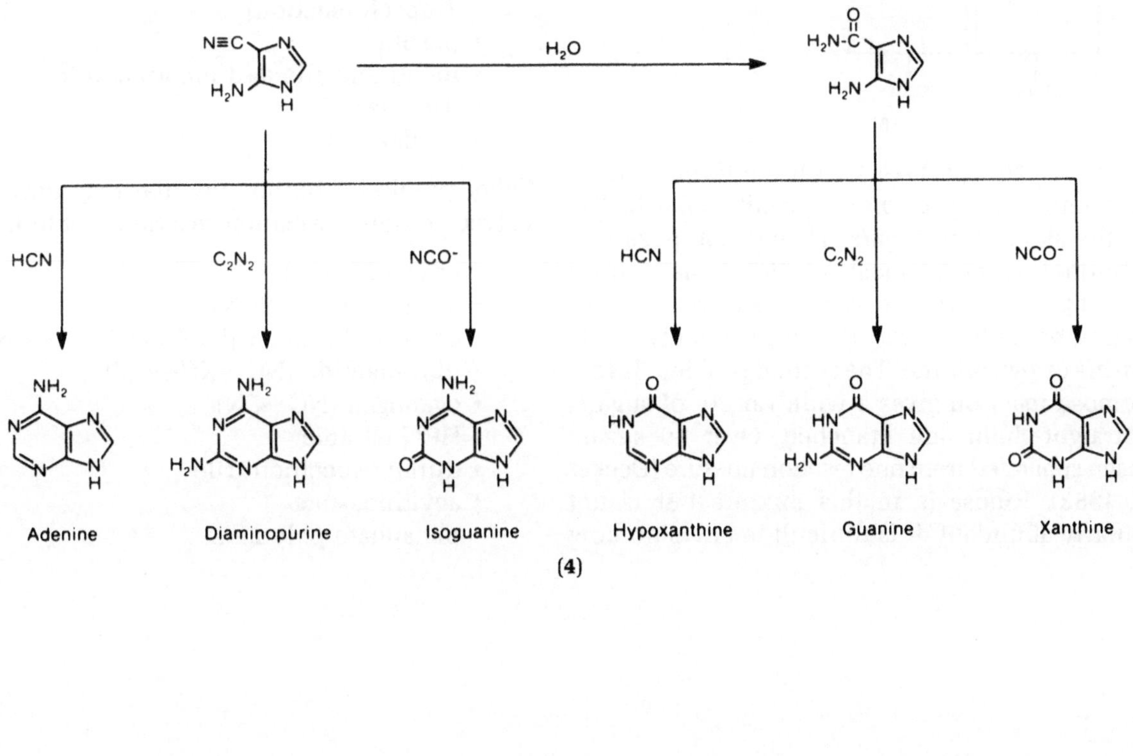

(4)

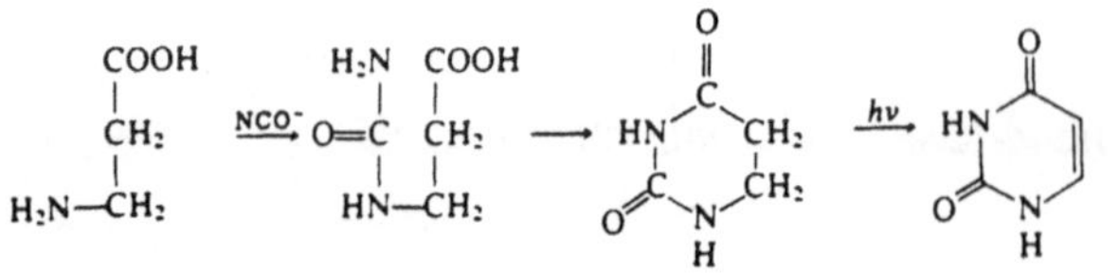

(7)

6. Sugars

The synthesis of reducing sugars from formaldehyde under alkaline conditions was discovered long ago (Butlerow, 1861). However, the Butlerow or formose reaction is very complex and incompletely understood. It depends on the presence of a suitable catalyst, with calcium hydroxide and calcium carbonate being the most popular heterogeneous catalysts. In the absence of catalysts, little or no sugar is obtained. At 100°C, clays such as kaolin serve to catalyze formation of monosaccharides, including ribose, in significant yield from dilute (0.01*M*) solutions of formaldehyde (Gabel and Ponnamperuma, 1967; Reid and Orgel, 1967).

The reaction is autocatalytic and proceeds in stages through glycolaldehyde, glyceraldehyde and dihydroxyacetone, tetroses, and pentoses to give finally hexoses, including glucose and fructose. One proposed reaction sequence is given in Scheme 8.

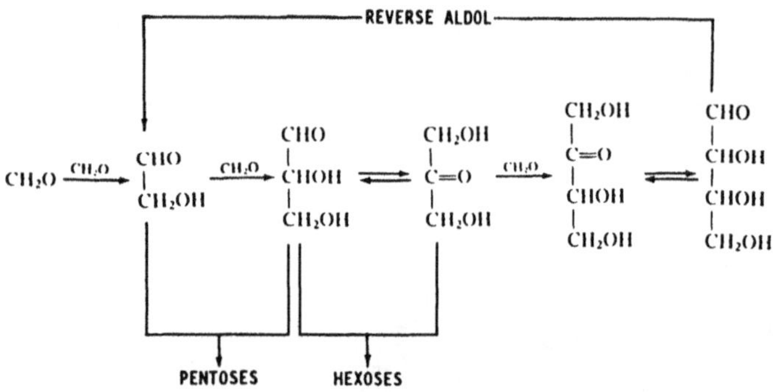

(8)

There are two problems with the formose reaction as a source of sugars on the primitive Earth. The first is the stability of sugars. They decompose in a few hundred years, at most, at 25°C. There are a number of possible ways to stabilize sugars, the most interesting being to convert the sugar to a glycoside of a purine or pyrimidine. The second problem is that the formose reaction gives a wide variety of sugars, both straight-chain and branched. Over 40 sugars have been separated from one reaction mixture (Decker *et al.*, 1982). Ribose is in this mixture but is not particularly abundant. It is difficult to envision how the relative yield of ribose could be greatly increased in this reaction or how any prebiotic reaction of sugars could give mostly ribose. It has therefore become apparent that ribonucleotides could not have been the first components of the prebiotic nucleic acids (Shapiro, 1988). A number of alternatives have been proposed (Joyce *et al.*, 1987), including the following compounds:

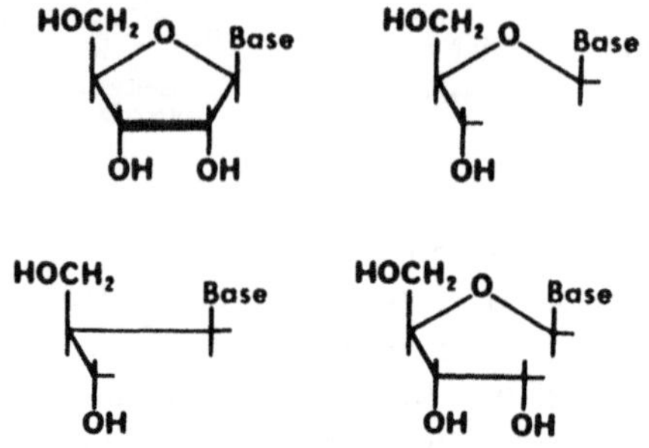

There are many other possibilities, but the ones shown are attractive because they are open chain, flexible, and prochiral. The prebiotic synthesis of these compounds has not yet been demonstrated.

7. Other Prebiotic Compounds

There are a number of compounds that have been synthesized under primitive Earth conditions, but space does not permit an adequate discussion. These include:

- dicarboxylic acids
- tricarboxlic acids
- fatty acids, C_2–C_8 (branched and straight)
- fatty alcohols (straight chain via Fischer–Tropsch reaction)
- porphin
- nicotinonitrile and nicotinamide
- triazines
- imidazoles

Other prebiotic compounds that may have been involved in polymerization reactions include:

- cyanate (NCO)
- cyanamide (H_2NCN)
- cyanamide dimer [$H_2NC(NH)NH$—CN]
- dicyanamide (NC—NH—CN)
- cyanogen (NC—CN)
- HCN tetramer
- diiminosuccinonitrile
- acylthioesters
- phosphate polymers

8. Compounds That Have Not Been Synthesized Prebiotically

It is a matter of opinion as to what constitutes a prebiotic synthesis. In some cases the conditions are so forced (e.g., the use of anhydrous solvents) or the concentrations so high (e.g., 10*M* formaldehyde) that they could not have occurred extensively on the primitive Earth. Reactions under these and other extreme conditions cannot be considered prebiotic.

There have been many claimed prebiotic syntheses in which the compound has not been properly identified. The best method for unequivocal identification these days is gas chromatography/mass spectrometry of a suitable derivative, although melting points and mixed melting points can sometimes be used. The amino acid analyzer alone or chromatography in multiple solvent systems does not prove the identification of a compound.

Some of the compounds which do not yet have adequate prebiotic syntheses are:

- arginine
- lysine
- histidine
- straight-chain fatty acids ($>C_8$)
- porphyrins
- pyridoxal
- thiamine
- riboflavin
- folic acid
- lipoic acid
- biotin

It is probable that prebiotic syntheses will be available before too long for some of these compounds. In other cases the compounds may not have been synthesized prebiotically, so their occurrence in living systems started after the origin of life.

9. Organic Compounds in Carbonaceous Chondrites

On 28 September 1969 a type II carbonaceous chondrite fell in Murchison, Australia. Surprisingly, large amounts of amino acids were found in the chondrite by Kvenvolden *et al.* (1970, 1971) and Oro *et al.* (1971). The first report identified 7 amino acids (glycine, alanine, valine, proline, glutamic acid, sarcosine, and α-aminoisobutyric acid), of which all but valine and proline had been found in the original electric discharge experiments (Miller, 1953, 1955, 1957). The most striking are sarcosine and α-aminoisobutyric acid. The second report identified 18 amino acids of which 9 had previously been identified in the original electric discharge experiments, but the remaining 9 had not.

At that time we had identified the hydrophobic amino acids from the low-temperature electric discharge experiments described above, and therefore we examined the products for the nonprotein amino acids found in Murchison. We were able to find all of them (Ring *et al.*, 1972; Wolman *et al.*, 1972).

There is a striking similarity between the products and relative abundances of the amino acids produced by electric discharge and the meteorite amino acids and their relative abundances. The results are compared in Table V. The most notable difference between the meteorite and the electric discharge amino acids is in the yield of pipecolic acid, which is extremely low in the electric discharge. The amount of α-aminoisobutyric acid is greater than that of α-amino-*n*-butyric acid in the meteorite, but the reverse is the case in the electric discharge. Reasonable differences in ratios of amino acids do not detract from the overall picture. Indeed, the ratio of α-aminoisobutyric acid to glycine is quite different in two meteorites of the same type, being 0.4 in the Murchison meteorite and 3.8 in the Murray meteorite (Cronin and Moore, 1971).

TABLE V. Relative Abundances of Amino Acids in the Murchison Meteorite and in an Electric-Discharge Synthesis[a]

Amino acid	Murchison meteorite	Electric discharge
Glycine	****	****
Alanine	****	****
α-Amino-n-butyric acid	***	****
α-Aminoisobutyric acid	****	**
Valine	***	**
Norvaline	***	***
Isovaline	**	**
Proline	***	*
Pipecolic acid	*	<*
Aspartic acid	***	***
Glutamic acid	***	**
β-Alanine	**	**
β-Amino-*n*-butyric acid	*	*
β-Aminoisobutyric acid	*	*
γ-Aminobutyric acid	*	**
Sarcosine	**	***
N-Ethylglycine	**	***
N-Methylalanine	**	**

[a]Mole ratio to glycine (= 100): *, 0.05–0.5; **, 0.5–5; ***, 5–50; ****, >50.

A similar comparison has been made between the dicarboxylic acids in the Murchison meteorite (Lawless *et al.*, 1974) and those produced by an electric discharge (Zeitman *et al.*, 1974), and the product ratios are quite similar.

The close correspondence between the amino acids found in the Murchison meteorite and those produced by an electric discharge synthesis, in terms of both the amino acids produced and their relative ratios, suggests that the amino acids in the meteorite were synthesized on the parent body by means of an electric discharge or analogous processes. A quantitative comparison of the amino acid and hydroxy acid abundances (Peltzer and Bada, 1978) shows that these compounds can be accounted for by a Strecker–cyanohydrin synthesis on the parent body (Peltzer *et al.*, 1984). Electric discharges appear to be the most favored source of energy, but sufficient data are not available to make realistic comparisons with other energy sources.

Our ideas on the prebiotic synthesis of organic compounds are based largely on the results of experiments in model systems. Accordingly, it is extremely gratifying to see that such synthesis really did take place on the parent body of the meteorite, and so it becomes plausible but not proved that they took place on the primitive Earth.

10. Interstellar Molecules

In the past twenty years a large number of organic molecules have been found in interstellar dust clouds, mostly by emission lines in the microwave region of the spectrum [for a summary, see Mann and Williams (1980)]. The concentration of these molecules is very low (a few molecules per cubic centimeter at the most), but the total amount in a dust cloud is large. The molecules found include formaldehyde, hydrogen cyanide, acetaldehyde, and cyanoacetylene. These are important prebiotic molecules, and this immediately raises the question of whether the interstellar molecules played a role in the origin of life on the Earth. In order for this to have taken place, it would have been necessary for the molecules to have been greatly concentrated in the solar nebula and to have arrived on the Earth without being destroyed by ultraviolet light or pyrolysis. This appears to be unlikely. In addition, it is necessary for some molecules to have been continuously synthesized (unless life started very quickly) because of their instability, and an interstellar source could not be responsible for these.

For these reasons, it is generally felt that the interstellar molecules played at most a minor role in the origin of life. However, the presence of so many molecules of prebiotic importance in interstellar space, combined with the fact that their synthesis must differ from that on the primitive Earth, where the conditions were very different, indicates that some molecules are particularly easily synthesized when radicals and ions recombine. Another way of saying this is that there appears to be a universal organic chemistry, which shows up in interstellar space, in the atmospheres of the major planets, and in the reducing atmosphere of the primitive Earth.

11. Production Rates and Concentrations of Hydrogen Cyanide, Formaldehyde, and Amino Acids in the Primitive Ocean

The amount of quantitative data available for the synthesis of HCN and H_2CO by various energy sources is limited, but a preliminary calculation of production rates can be made (Stribling and Miller, 1987).

Figure 3 shows the corona discharge synthesis of HCN and H_2CO at various H_2/CH_4, H_2/CO and H_2/CO_2 ratios (Stribling and Miller, 1987). For CH_4 atmospheres the energy yield for HCN is about 5 nmol J^{-1}. Taking the corona discharge energy as 12.6 J cm^{-2} yr^{-1} gives about 60 nmol cm^{-2} yr^{-1}. The calculated lightning yields in CH_4 + N_2 atmospheres are about 500 nmol J^{-1} (Chameides and Walker, 1981), but prelimi-

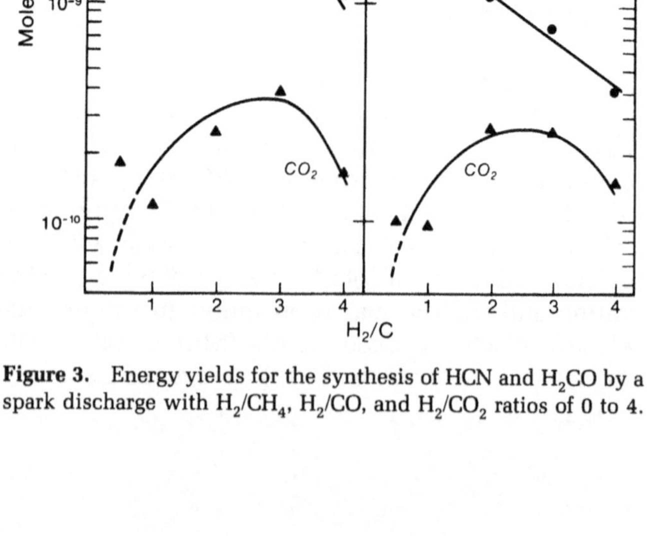

Figure 3. Energy yields for the synthesis of HCN and H_2CO by a spark discharge with H_2/CH_4, H_2/CO, and H_2/CO_2 ratios of 0 to 4.

nary experiments indicate that much lower yields of about 30 nmol J^{-1} are obtained. The calculated yields are very low for CO and CO_2 atmospheres.

A photochemical source of HCN has been proposed by Zahnle (1986), as discussed earlier. The yields of HCN range between a maximum of 520 nmol cm^{-2} yr^{-1} when the production of N atoms is limiting and 5 to 50 nmol cm^{-2} yr^{-1} at lower CH_4 fluxes.

We will assume a combined production rate of HCN of 100 nmol cm^{-2} yr^{-1} for a CH_4 atmosphere. The production rates would be not much lower in CO and CO_2 atmospheres if the Zahnle N atom process were important. In the absence of this mechanism, the HCN production rates in these atmospheres would be one or two orders of magnitude lower.

The production rates of H_2CO from corona discharge are also shown in Fig. 3. This is an effective energy source, especially with a CO-containing atmosphere. Yields from photochemical experiments of about 3 nmol J^{-1} from CH_4 + H_2O (Ferris and Chen, 1975) and about 8 nmol J^{-1} from CO + H_2O have also been reported (Bar-Nun and Hartman, 1978; Bar-Nun and Chang, 1983).

The production rate of H_2CO in a model CO_2 atmosphere has been calculated by Pinto *et al.* (1980). The rainout rate is about 15 nmol cm^{-2} yr^{-1}. The assumed atmosphere is relatively reducing since H_2/CO_2 = 3. Another calculation by Kasting *et al.* (1984) gives 1.6 to 45 nmol cm^{-2} yr^{-1}. An important part of these calculations is the rainout rate of the H_2CO, since H_2CO absorbs in the ultraviolet at 3200 Å. It is not clear whether this factor has been adequately treated. As in the case of HCN, the H_2CO production rates from the various sources, including the results of our experiment, are generally comparable.

The production rates of HCN estimated above allow a calculation of this steady-state concentration in the primitive ocean. We assume that all the HCN produced in the atmosphere enters the ocean and remains there and that the only pathway for the loss of HCN is hydrolysis to formamide and then to formic acid. At steady state, the production rate (S_{HCN} in mol cm^{-2} yr^{-1}) equals the rate of hydrolysis:

$$S_{HCN} = \frac{d[HCN]}{dt} V_o$$

where [HCN] is the molar concentration, and V_o is the volume of the ocean in liters per square centimeter, now 300 liter cm^{-2}. Writing the hydrolysis rate of HCN as a pseudo-first-order reaction,

$$-\frac{d[HCN]}{dt} = k_1[\Sigma HCN]$$

where k_1 is the pseudo-first-order rate constant in yr^{-1} and depends on temperature and pH, and $\Sigma HCN = HCN + CN^-$.

The hydrolysis of HCN is both acid- and base-catalyzed and has been discussed by Stribling and Miller (1987). The half-lives for hydrolysis and the concentrations of HCN in the primitive ocean for several pH values and temperatures are given in Table VI. The concentrations of HCN are proportional to the production rate and the H^+ concentration and inversely proportional to the volume of the ocean.

The results of Table VI make it clear that low temperatures and pH favor higher concentrations of HCN, but even the concentration at pH 7 and 0°C (3.5×10^{-5}M) is much lower than that in the usual prebiotic experiment. Concentrations of HCN of 10^{-6}M are sufficient to make amino acids by the Strecker synthesis (Miller and Van Trump, 1981), but adenine synthesis would require a concentration mechanism, with freezing being the most likely (Sanchez *et al.*, 1967).

The yields for the spark discharge synthesis of HCN would have been only 1 nmol cm^{-2} yr^{-1} for H_2/CO_2 = 1 and 4 nmol cm^{-2} yr^{-1} for H_2/CO_2 = 2, which is considerably lower than the 100 nmol cm^{-2} yr^{-1} assumed in this calculation. This would lead to HCN concentrations 25 to 100 times lower than given in Table VI, which might raise problems for amino acid and purine synthesis. If the Zahnle N atom synthesis were as efficient on the primitive Earth as calculated, then this problem would not arise.

12. The Concentration of Amino Acids in the Primitive Ocean

The production rates of HCN and H_2CO estimated in Section 11 permit us to calculate the rate of buildup of amino acids in the primitive ocean and their steady-state concentration. Provided the con-

TABLE VI. Half-Lives for Hydrolysis of HCN and the HCN Concentrations in the Primitive Ocean

Temperature (°C)	pH	$t_{1/2}$ (yr)	M_{HCN}
0	8	7,000	4×10^{-6}
	7	70,000	4×10^{-5}
25	8	40	2×10^{-8}
	7	400	2×10^{-7}
50	8	0.5	3×10^{-10}
	7	5	3×10^{-9}

[a]Based on an HCN production rate of 100 nmol cm^{-2} yr^{-1} and an ocean of 300 liter cm^{-2}.

centrations of HCN and aldehydes do not drop too low, the Strecker synthesis will be effective in the primitive ocean (Miller and Van Trump, 1981). We will assume an amino acid yield of 10% based on the HCN production. This is the approximate yield reported with CH_4, CO, and CO_2 atmospheres (Schlesinger and Miller, 1983a, b). The yield is much higher (~90%) with the Strecker synthesis, while the cyanide polymerization gives only about 1% (Lowe *et al.*, 1963; Ferris *et al.*, 1978).

Taking a combined HCN production rate of 100 nmol cm^{-2} yr^{-1}, the steady-state amino acid production rate would then have been 10 nmol cm^{-2} yr^{-1}. Taking the volume of the ocean as 300 liter cm^{-2}, the increase in amino acid concentration would have been 3.3×10^{-11}M yr^{-1}. Assuming no losses, this gives 3.3×10^{-4}M in 10 million years. In a low-temperature ocean the losses from thermal decomposition should be low for amino acids such as glycine, threonine, and methionine. The losses from adsorption on clays, ionizing radiation, and ultraviolet light are much more difficult to estimate. The most important loss mechanism was probably heating in the submarine vents, in which the seawater reaches at least 350°C, thereby decomposing all the amino acids. Since the entire ocean on the average passes through the vents in 10 million years (Edmond *et al.*, 1982). the amino acid concentration could not have risen higher than 3×10^{-4}M with the above assumptions. Increasing the production rate of HCN or decreasing the size of the ocean would increase proportionally the steady-state concentration of amino acids.

It is more difficult to estimate the concentrations of adenine and other purines because the details of the different pathways from HCN are not known. In the Murchison meteorite the ratio of purines to amino acids is about 1:20. A figure for the primitive ocean of 1% of the amino acid concentration (within an order of magnitude) is in accord with general experience in prebiotic chemistry. An adenine concentration of 3×10^{-5}M seems low, and a means to concentrate it may have been needed.

As discussed for the HCN production, CO_2 atmospheres give considerably lower HCN yields from spark discharges. Without the HCN production from the Zahnle N atom mechanism, the amino acid concentrations would be 25 to 100 times lower than the 3×10^{-4}M estimated above. There is no way at the present time to estimate the amino acid and purine concentrations needed for life to arise. A detailed mechanism of the process leading to the origin of life is needed to place constraints on the amino acid concentration as well as the atmosphere conditions and composition required for the synthesis of necessary amino acids and purines.

ACKNOWLEDGMENT. This work was supported by NASA grant NAGW-20.

References

Abelson, P. H., 1965, Abiogenic synthesis in the Martian environment, *Proc. Natl. Acad. Sci. U.S.A.* **54:**1490–1497.

Bada, J. L., and Miller, S. L., 1968, Ammonium ion concentration in the primitive ocean, *Science* **159:**423–425.

Bar-Nun, A., and Chang, S., 1983, Photochemical reactions of water and carbon monoxide in earth's primitive atmosphere, *J. Geophys. Res.* **88:**6662–6672.

Bar-Nun, A., and Hartman, H., 1978, Synthesis of organic compounds from carbon monoxide and water by UV photolysis, *Origins Life* **9:**93–101.

Bar-Nun, A., Bar-Nun, N., Bauer, S. H., and Sagan, C., 1970, Shock synthesis of amino acids in simulated primitive environments, *Science* **168:**470–473.

Butlerow, A., 1861, Formation synthétique d'une substance sucrée, *C. R. Acad. Sci.* **53:**145–147.

Chameides, W. L., and Walker, J. C. G., 1981, Rates of fixation by lightning of carbon and nitrogen in possible primitive atmospheres, *Origins Life* **11:**291–302.

Cronin, J. R., and Moore, C. B., 1971, Amino acid analyses of the Murchison, Murray, and Allende carbonaceous chondrites, *Science* **172:**1327–1329.

Decker, P., Schweer, H., and Pohlmann, R., 1982, Identification of formose sugars, presumable prebiotic metabolites, using capillary gas chromatography/gas chromatography–mass spectrometry of *n*-butoxime trifluoroacetates on OV-225, *J. Chromatogr.* **225:**281–291.

Edmond, J. M., Von Damm, K. L., McDuff, R. E., and Measures, C. I., 1982, Chemistry of hot springs on the east Pacific Rise and their effluent dispersal, *Nature* **297:**187–191.

Ferris, J. P., and Chen, C. T., 1975. Chemical evolution. XXVI. Photochemistry of methane, nitrogen, and water mixtures as a model for the atmosphere of the primitive earth. *J. Am. Chem. Soc.* **97:**2962–2967.

Ferris, J. P., Joshi, P. C., Edelson, E. H., and Lawless, J. G., 1978. HCN: A plausible source of purines, pyrimidines and amino acids on the primitive earth. *J. Mol. Evol.* **11:**293–311.

Ferris, J. P., Sanchez, R. A., and Orgel, L. E., 1968, Studies in prebiotic synthesis. III. Synthesis of pyrimidines from cyanoacetylene and cyanate, *J. Mol. Biol.* **33:**693–704.

Ferris, J. P., Zamek, O. S., Altbuch, A. M., and Frieman, H., 1974, Chemical evolution. XVIII. Synthesis of pyrimidines from guanidine and cyanoacetaldehyde, *J. Mol. Evol.* **3:**301–309.

Friedmann, N., and Miller, S. L., 1969, Phenylalanine and tyrosine synthesis under primitive earth conditions, *Science* **166:** 766–767.

Friedmann, N., Haverland, W. J., and Miller, S. L., 1971, Prebiotic synthesis of the aromatic and other amino acids, in: *Chemical Evolution and the Origin of Life* (R. Buvet and C. Ponnamperuma, eds.), North-Holland, Amsterdam, pp. 123–135.

Gabel, N. W., and Ponnamperuma, C., 1967, Model for origin of monosaccharides, *Nature* **216:**453–455.

Garrison, W. M., Morrison, D. C., Hamilton, J. G., Benson, A. A., and Calvin, M., 1951, Reduction of carbon dioxide in aqueous solutions by ionizing radiation, *Science* **114**:416–418.

Getoff, N., 1962, Über die bildung organischer substanzen aus kohlensäure in wässeriger lösung mittels ^{60}Co-gamma-strahlung, *Z. Naturforsch.* **17B**:751–757.

Groth, W., and von Weyssenhoff, H., 1960, Photochemical formation of organic compounds from mixtures of simple gases, *Planet. Space Sci.* **2**:79–85.

Hayatsu, R., and Anders, E., 1981, Organic compounds in meteorites and their origins, *Top. Curr. Chem.* **99**:1–37.

Joyce, G. F., Schwartz, A. W., Miller, S. L., and Orgel, L. E., 1987, The case for an ancestral genetic system involving simple analogues of the nucleotide, *Proc. Natl. Acad. Sci. U.S.A.* **84**:4398–4402.

Kasting, J. F., Pollack, J. B., and Crisp, D., 1984, Effects of high CO_2 levels on surface temperature and atmospheric oxidation state of the early earth, *J. Atmos. Chem.* **1**:403–428.

Kenyon, D. H., and Steinman, G., 1969, *Biochemical Predestination*, McGraw-Hill, New York.

Khare, B. N., and Sagan, C., 1971, Synthesis of cystine in simulated primitive conditions, *Nature* **232**:577–578.

Kvenvolden, K. A., Lawless, J. G., and Ponnamperuma, C., 1971, Nonprotein amino acids in the Murchison meteorite, *Proc. Natl. Acad. Sci. U.S.A.* **68**:486–490.

Kvenvolden, K., Lawless, J. G., Pering, K., Peterson, E., Flores, J., Ponnamperuma, C., Kaplan, I. R., and Moore, C., 1970, Evidence for extraterrestrial amino-acids and hydrocarbons in the Murchison meteorite, *Nature* **228**:923–926.

Lawless, J. G., and Boynton, C. G., 1973, Thermal synthesis of amino acids from a simulated primitive atmosphere, *Nature* **243**:405–407.

Lawless, J. G., Zeitman, B., Pereira, W. E., Summons, R. E., and Duffield, A. M., 1974, Dicarboxylic acids in the Murchison meteorite, *Nature* **251**:40–42.

Lemmon, R. M., 1970, Chemical evolution, *Chem. Rev.* **70**:95–109.

Lowe, C. U., Rees, M. W., and Markham, R., 1963. Synthesis of complex organic compounds from simple precursors: Formation of amino acids, amino-acid polymers, fatty acids and purines from ammonium cyanide. *Nature* **199**:219–222.

Mann, A. P. C., and Williams, D. A., 1980, A list of interstellar molecules, *Nature* **283**:721–725.

Miller, S. L., 1953, Production of amino acids under possible primitive earth conditions, *Science* **117**:528–529.

Miller, S. L., 1955, Production of some organic compounds under possible primitive earth conditions, *J. Am. Chem. Soc.* **77**: 2351–2361.

Miller, S. L., 1957, The formation of organic compounds on the primitive earth, *Ann. N. Y. Acad. Sci.* **69**:260–274 [Also in: *The Origin of Life on the Earth* (A. Oparin, ed.), Pergamon Press, Oxford, 1959, p. 123–135].

Miller, S. L., and Orgel, L. E., 1974, *The Origins of Life on the Earth*, Prentice-Hall, Englewood Cliffs, New Jersey.

Miller, S. L., and Van Trump, J. E., 1981, The strecker synthesis in the primitive ocean, in: Origin *of Life* (Y. Wolman, ed.), Reidel, Dordrecht, p. 135–141.

Miller, S. L., Urey, H. C., and Oro, J., 1976, Origin of organic compounds on the primitive earth and in meteorites, *J. Mol. Evol.* **9**:59–72.

Oro, J., and Kimball, A. P., 1961, Synthesis of purines under possible primitive earth conditions. I. Adenine from hydrogen cyanide, *Arch. Biochem. Biophys.* **94**:221–227.

Oro, J., and Kimball, A. P., 1962, Synthesis of purines under possible primitive earth conditions. II. Purine intermediates from hydrogen cyanide, *Arch. Biochem. Biophys.* **96**:293–313.

Oro, J., Nakaparksin, S., Lichtenstein, H., and Gil-Av, E., 1971, Configuration of amino acids in carbonaceous chondrites and a Precambrian chert, *Nature* **230**:107–108.

Peltzer, E. T., and Bada, J. L., 1978, α-Hydroxy carboxylic acids in the Murchison meteorite, *Nature* **272**:443–444.

Peltzer, E. T., Bada, J. L., Schlesinger, G., and Miller, S. L., 1984, The chemical conditions on the parent body of the Murchison meteorite: Some conclusions based on amino, hydroxy and dicarboxylic acids, *Adv. Space Res.* **4**(12):69–74.

Pinto, J. P., Gladstone, G. R., and Yung, Y. L., 1980, Photochemical production of formaldehyde in earth's primitive atmosphere, *Science* **210**:183–185.

Reid, C., and Orgel, L. E., 1967, Synthesis of sugar in potentially prebiotic conditions, *Nature* **216**:455.

Ring, D., Wolman, Y., Friedmann, N., and Miller, S. L., 1972, Prebiotic synthesis of hydrophobic and protein amino acids, *Proc. Natl. Acad. Sci. U.S.A.* **69**:765–768.

Sagan, C., and Khare, B. N., 1971, Long-wavelength ultraviolet photoproduction of amino acids on the primitive earth, *Science* **173**:417–420.

Sanchez, R. A., Ferris, J. P., and Orgel, L. E., 1966a, Cyanoacetylene in prebiotic synthesis, *Science* **154**:784–786.

Sanchez, R. A., Ferris, J., and Orgel, L. E., 1966b. Conditions for purine synthesis: Did prebiotic synthesis occur at low temperatures? *Science* **153**:72–73.

Sanchez, R. A., Ferris, J. P., and Orgel, L. E., 1967, Studies in prebiotic synthesis. II. Synthesis of purine precursors and amino acids from aqueous hydrogen cyanide, *J. Mol. Biol.* **30**:223–253.

Sanchez, R. A., Ferris, J. P., and Orgel, L. E., 1968, Studies in prebiotic synthesis. IV. The conversion of 4-aminoimidazole-5-carbonitrile derivatives to purines, *J. Mol. Biol.* **38**:121–128.

Schlesinger, G., and Miller, S. L., 1983a, Prebiotic synthesis in atmospheres containing CH_4, CO, and CO_2. I. Amino acids, *J. Mol. Evol.* **19**:376–382.

Schlesinger, G., and Miller, S. L., 1983b, Prebiotic synthesis in atmospheres containing CH_4, CO, and CO_2. II. Hydrogen cyanide, formaldehyde and ammonia, *J. Mol. Evol.* **19**:383–390.

Schwartz, A. W., and Chittenden, G. J. F., 1977, Synthesis of uracil and thymine under simulated prebiotic conditions, *Biosystems* **9**:87–92.

Shapiro, R., 1988. Prebiotic ribose synthesis: A critical analysis. *Origins of Life Evol. Bioshpere* **18**:71–85.

Stribling, R., and Miller, S. L., 1987, Energy yields for hydrogen cyanide and formaldehyde syntheses: The HCN and amino acid concentrations in the primitive ocean, *Origins Life Evol. Bioshpere* **17**:261–273.

Van Trump, J. E., and Miller, S. L., 1972, Prebiotic synthesis of methionine, *Science* **178**:859–860.

Voet, A. B., and Schwartz, A. W., 1983, Prebiotic adenine synthesis from HCN—Evidence for a newly discovered major pathway, *Bioinorg. Chem.* **12**:8–17.

Wolman, Y., Haverland, W. J., and Miller, S. L., 1972, Nonprotein amino acids from spark discharges and their comparison with the Murchison meteorite amino acids, *Proc. Natl. Acad. Sci. U.S.A.* **69**:809–811.

Zahnle, K. J., 1986, The photochemistry of hydrocyanic acid (HCN) in the earth's early atmosphere, *J. Geophys. Res.* **91**:2819–2834.

Zeitman, B., Chang, S., and Lawless, J. G., 1974, Dicarboxylic acids from electric discharge, *Nature* **251**:42–43.

Chapter 31

The Initiation of Biological Processes on Earth

Summary of Empirical Evidence

MANFRED SCHIDLOWSKI

1. Introduction

With the terrestrial rock record constituting the only source of geological and paleontological information, empirical evidence as to the initiation of life processes on the ancient Earth necessarily cannot predate the appearance of the oldest sediments about 3.8 Gyr ago (1 Gyr = 10^9 yr). Accordingly, the beginnings of biology in the geologically undocumented "Hadean" era that preceded the onset of the sedimentary record are likely to remain shrouded in mystery. However, several lines of indirect and circumstantial evidence tend to constrain relevant speculations, permitting plausible inferences as to the sequence of events that led to the establishment of life on the surface of this planet.

It is reasonable conjecture that life (or protein chemistry) appeared at a certain stage of either cosmic or planetary evolution as an intrinsically new property of matter based on a set of specific characteristics. Foremost among these are (1) the buildup of living matter primarily from a rather limited number of chemical elements (C, O, H, N, S, P), all of which figure among the most abundant elements in the cosmos, (2) the existence of living systems as dynamic states removed from thermodynamic equilibrium to markedly decreased entropy levels, (3) their conspicuous structural differentiation ("compartmentalization"), culminating in the establishment of the cell as the basic morphological and functional entity of life, and (4) the capability of identical reproduction. There is no doubt that an extended period of *chemical evolution* must have preceded the emergence of the first self-replicating life-like systems ("protobionts") before *biological* or *"Darwinian" evolution* took over and consequently gave rise to the abundance and morphological variety of life as we know it today.

As for the probable beginnings of prebiological evolution in space, radio astronomical molecular spectroscopy has, of late, permitted the tabulation of an impressive list of organic molecules that have been identified in the interstellar medium (cf. Irvine and Knacke, 1989). These include species that have been shown to function as essential intermediates in the prebiotic synthesis of sugars, proteins, and nucleic acids [e.g., formaldehyde (CH_2O), hydrogen cyanide (HCN), acetaldehyde (CH_3CHO), cyanoacetylene (HC_2CN), and several others]. Accordingly, it is safe to assume that the principal building blocks of living matter have been continually synthesized in the space environment since the universe came into being some 2 × 10^{10} yr ago.

Moreover, it has been demonstrated that the myriads of submicron-sized frozen dust particles that make up the bulk of matter in interstellar molecular clouds are sites of a complex organic chemistry driven by photochemical reactions (cf. Greenberg, 1984, 1985; Agarwal *et al.*, 1985). According to recent estimates, about half of the O-, C-, and N-containing

MANFRED SCHIDLOWSKI • Max-Planck-Institut für Chemie (Otto-Hahn-Institut), D-55020 Mainz, Germany.
Organic Geochemistry, edited by Michael H. Engel and Stephen A. Macko. Plenum Press, New York, 1993.

gaseous constituents of intergalactic clouds (mostly CO, H_2, and CH_4) are converted to high-molecular-weight organic polymers as a result of photochemical processing of both the primary gases and the resulting organic condensates, which form mantles around the silicate cores of the carrier grains. Photoprocessing by cosmic ultraviolet (UV) radiation of the interstellar medium in both the gas phase and the solid state thus gives rise to a sizable intergalactic reservoir of organic substances, estimated to account for about 1‰ of the total mass of the Milky Way (Greenberg, 1984), the bulk being tied up in the refractory organic mantles of common interstellar grains.

In the course of time, all interstellar dust clouds are destined to be consumed by star formation, with a conspicuous fraction of their carbonaceous component ending up—after extensive high-temperature processing—in reconstituted form as part of the chondritic constituent of newly formed planets (cf. Chapter 33). As a rule, some material from the remote marginal reaches of the primary dust cloud escapes this fate, leaving a thin veil of pristine material from the parent nebula around the newly formed solar system. These relics of the primary protostellar dust cloud are the birthplaces of comets, which are made up of material that apparently represents a state of "arrested" development of interstellar dust as it existed some 4.5 Gyr ago (Greenberg, 1985). The organic mantles of cometary dust particles have, meanwhile, been found to consist of highly unsaturated compounds (Kissel and Krueger, 1987), apt to be immediately involved in vigorous organic reactions when encountering aqueous environments, with carbohydrates and nucleoside bases as the most probable reaction products. Accordingly, cometary material may have served as a convenient organic molecular seeding for the initiation of prebiotic evolution on a previously sterile planetary surface, and its potential role in the origin of life on Earth still awaits proper evaluation.

Apart from the prebiological organic chemistry proceeding in the cosmic environment, the early degassing products released from the Earth's mantle immediately after the formation of the planet could also have provided convenient reactants for Earth-based prebiotic syntheses energized by electric discharges, solar UV radiation, or thermal energy (cf. Miller *et al.*, 1976). Since the pioneering investigation by Miller (1955), it has been well known that exposure of mixtures of reducing gases (such as are present in the reduced gas fraction of volcanic emanations) to electric spark discharges gives rise to an impressive suite of amino acids, with yields of up to 5% of the carbon originally added to the system in the form of methane (CH_4). Moreover, electric discharge reactions have been shown to produce high yields of hydrogen cyanide and cyanoacetylene, which figure, *inter alia*, as essential reactants in the prebiotic synthesis of nucleoside bases (adenine, guanine) and pyrimidines, respectively (see also Chapter 30). Along with a set of related complementary syntheses powered by either solar UV or local sources of telluric thermal energy (e.g., Fischer–Tropsch type processes), or both, the above reactions appear to be potentially capable of providing an efficient base for the chemical evolution at or near the planetary surface that must have preceded the appearance of biological processes on the early Earth.

Since the terrestrial rock record has hitherto failed to furnish unequivocal evidence of prebiotic organic chemistry, early chemical evolution was almost certainly confined to the "Hadean" era bracketed by the time of the Earth's formation (~4.5 Gyr) and the onset of the sedimentary record about 3.8 Gyr ago (cf. Fig. 1). As this time segment is devoid of geological documents, current inferences about the beginnings of organic chemistry on the juvenile planet are largely extrapolations from cosmic (notably cometary) scenarios as well as from model experiments simulating prebiotic key reactions under possible primitive Earth conditions. On the other hand, students of early life on Earth have been stunned since long by the steadily increasing evidence of biological activity already in Archean times (Fig. 1). Specifically, the isotopic composition of the carbon constituents of the oldest sediments (in the form of carbonate and kerogenous materials) has prompted the conclusion that biological modulation of the global carbon cycle had been established already by the time of the deposition of these rocks, with a fully developed biogeochemical carbon cycle thus dating back to ~3.8 Gyr (Schidlowski *et al.*, 1979, 1983). Also, the morphological record of microbial (prokaryotic) life has been shown to proceed over at least 3.5, if not 3.8, b.yr of geological history. Summing up, we may state with fair confidence that not only chemical evolution, but also the beginnings of biological evolution most probably predated the start of the presently known sedimentary record, whose oldest manifestations give fair testimony as to the presence of a full-fledged microbial biosphere on the terrestrial surface as from about 3.8 Gyr.

In the following, a synopsis will be presented of the currently available empirical evidence related to the antiquity of life on Earth, with special reference to (1) morphological remnants (microfossils, biosedi

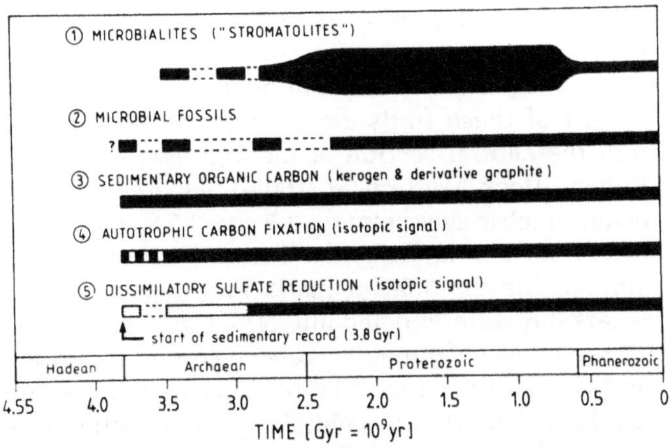

Figure 1. Record of fossil microbial life (1,2) and selected categories of biogeochemical evidence (3–5) within the temporal framework of 4.55 Gyr of Earth history. The Hadean, Archean, and Proterozoic eons are conventionally grouped together as the "Precambrian," with the Hadean (>3.8 Gyr) lacking a sedimentary record and thus being virtually devoid of geological documents. The stromatolite diagram (1) reflects the exuberant proliferation of microbial ecosystems after the establishment of extended stable marine shelves during the Proterozoic as well as their sudden decline following the rise of metazoan grazers at the dawn of the Phanerozoic. The isotopic signature of autotrophic carbon fixation (4) as revealed by sedimentary organic carbon quite uniformly throughout Earth history (see also Fig. 7) exhibits a well-defined metamorphic overprint for $t > 3.5$ Gyr (broken line). A corresponding isotopic index line for dissimilatory sulfate reduction (5) apparently fades between 2.7 and 2.8 Gyr (see also Fig. 9), but it seems most likely that the underlying biologically mediated isotope effect has been camouflaged (stippled signature) by either geochemical or environmental controls that were selectively operative during the Earth's earliest history. For further explanations, see the text.

mentary structures) and (2) selected biogeochemical vestiges (such as the carbon and sulfur isotope records) left by the oldest microbial ecosystems. For other parts of the biogeochemical evidence preserved in Precambrian rocks, the reader is referred to Chapter 32.

2. Origins of Life on Earth: Evidence and Constraints from the Geological Record

Most commonly, organisms leave a distinct morphological and chemical record of their existence in sedimentary rocks. Though in part highly selective, this record may persevere through billions of years under favorable circumstances (cf. Fig. 1) before being finally destroyed during the metamorphic and anatectic reconstitution of the host rock. This holds for both higher multicellular life forms (Metaphyta and Metazoa) and for microorganisms, which held dominion over the Earth during the first 3 b.yr of recorded geological history. Here, an overview will be given of relevant paleontological and biogeochemical evidence that bears on the presence of life during the time of formation of the oldest terrestrial (Archean) rocks between about 3.3 and 3.8 Gyr.

2.1. The Paleontological Record

Known as the age of microorganisms, the Precambrian was in fact dominated by prokaryotic and eukaryotic microbial ecosystems until the rise of the first Ediacaran-type metazoan faunas some 0.6–0.7 Gyr ago (cf. Glaessner, 1983, 1984). In particular, the Proterozoic segment (2.5–0.6 Gyr) of the Precambrian era has been shown to be characterized by prolific microbial communities of both the benthic and planktonic type that have left marvelous records of cellularly preserved microbiota, notably in the siliceous (cherty) facies of Proterozoic sequences such as the remarkably differentiated microfloras of the ~0.9-Gyr-old Bitter Springs Formation of Australia (Schopf, 1968) and the ~2.0-Gyr-old Gunflint iron formation of Canada (Tyler and Barghoorn, 1954; Barghoorn and Tyler, 1965; Cloud, 1965). Also, an abundance of biosedimentary structures ("stromatolites") preferentially preserving the matting behavior of benthic prokaryotes in the form of impressive sequences of lithified microbial carpets has been recorded from this time interval (cf. Hofmann, 1973; Walter, 1983). Evolution of the early microbial world culminated in the emergence of the eukaryotic (nucleated) cell probably some 1.4 Gyr ago (cf. Cloud, 1976; Schopf and Oehler, 1976); dissent in favor of a considerably earlier appearance of eukaryotes around 2 Gyr ago has been occasionally voiced by other investigators (e.g., Kazmierczak, 1979; Pflug and Reitz, 1985). With the advent of eukaryotic organization, marking one of the most important discontinuities in biological evolution, the stage was actually set for all subsequent differentiation and diversification of life, which, after a lengthy and largely enigmatic retardation period, resulted in the so-called "Cambrian explosion" with the sudden rise of virtually all major animal phyla close to the dawn of the Phanerozoic.

2.1.1. The Early Microfossil Record

As we go back to the Archean (3.8–2.5 Gyr), the fossil record progressively fades (Awramik, 1982;

Schopf and Walter, 1983). This is particularly true for the preservation of microbial fossils of mostly cyanobacterial affinity (Fig. 1) which had been so excellently documented in Proterozoic cherty series. A reexamination by Schopf and Walter (1983) of some 40 assemblages of putative microfossils and microfossil-like objects from 28 Archean stratigraphic units showed that only few of these stood the test of a critical reappraisal, the overwhelming majority representing, in all likelihood, mineralogical phenomena, preparation artifacts, or, at best, dubiofossils of highly variable or even questionable confidence levels. The conspicuous preponderance of this latter category among the Archean finds can be readily explained by the increasing diagenetic alteration and the incipient metamorphic overprint to which the delicate primary microstructures had been exposed in progressively older rocks, with a concomitant large-scale loss of contours and other crucial morphological detail.

Comparative studies performed on differentially altered sediments have, for instance, clearly demonstrated that the dominant filamentous and coccoid morphotypes of well-preserved Proterozoic cyanobacterial communities are bound to end up as "sticks and balls" when subject to increasing degrees of diagenetic and metamorphic alteration (Knoll *et al.*, 1988). Accordingly, we may reasonably infer that a large proportion of the Archean record of putative dubiomicrofossils characterized by "hazy" or severely impaired cellular morphologies (e.g., Muir and Grant, 1976; Knoll and Barghoorn, 1977) are, in fact, authentic microfossils in advanced stages of obliteration within progressively older (and hence more reconstituted) host rocks. Apart from relevant documentations given by Golubic and Barghoorn (1977) and Knoll *et al.* (1988), enlightening examples of such alteration series have been recently submitted by Pflug and Reitz (1992). Thus, if assessed in the light of these facts, the apparent impoverishment of Archean microfossil assemblages in both number and morphological complexity may be primarily due to a poor preservation of the record rather than a reflection of a reduction in both activities and diversification of the oldest microbial ecosystems. As has been argued persuasively by Knoll *et al.* (1988), any diversified Gunflint-type Proterozoic microflora would ultimately grade into a depauperate and poorly preserved microfossil assemblage of the kind found in Archean rocks if exposed to an extended period of low-grade metamorphic reconstitution.

In spite of the blatant empoverishment of the oldest record (Fig. 1), there are at least a few reports of singularly well preserved microbial communities which may be accepted as conclusive cellular evidence of Archean life. Prominent among the most ancient of these finds are microfossil assemblages from the central section of the Warrawoona Group (Pilbara Block, Western Australia) that are bracketed by radiometric ages between 3.3 and 3.5 Gyr, with the lower fossiliferous sequence (Towers Formation) well approaching the 3.5-Gyr mark. First reports of this remarkably differentiated microflora (Awramik *et al.*, 1983) had been disputed largely on grounds of imprecisely constrained rock relationships involving a putatively secondary (diagenetic) chert host believed to occur as fissure fillings (Buick, 1984). Subsequently, however, the respective lithology has proved fossiliferous also when intercalated in normally bedded series that leave little doubt about the primary character of the enclosing chert facies (Schopf and Packer, 1987). Most conspicuous among the Warrawoona microfossil assemblages hitherto described are both the coccoidal and septate tubular (or filamentous) morphotypes (Fig. 2) that were so abundantly present in fossil cyanobacterial communities of Proterozoic age, suggesting a remarkable degree of evolutionary conservatism of prokaryotic morphologies and ecosystem composition through time. The septate filaments described by both Awramik *et al.* (1983) and Schopf and Packer (1987) can be most readily explained as fossil trichomes derived from either filamentous cyanobacteria or more primitive prokaryotes of related morphologies (e.g., Chloroflexaceae), whereas the recently discovered sheath-enclosed coccoidal unicell aggregates have been claimed to exclude other than strictly cyanobacterial (notably chroococcalean) affinities (Schopf and Packer, 1987). With cyanobacteria carrying out the water-splitting variant of the photosynthetic process [cf Section 2.2.1., Eq. (1)], a well-proven cyanobacterial link of the Warrawoona microflora would have crucial implications for the onset of oxygenic photosynthesis, lending support to proposals of a very early emergence of this process based on various geochemical inferences (e.g., Schidlowski, 1978, 1984; Walker *et al.*, 1983). In sum, the body of evidence provided by the Warrawoona assemblage testifies to the occurrence of prolific fossil microbial communities characterized by a high degree of morphological and bioenergetic diversification in ancient sedimentary environments approaching 3.5 Gyr. Complementary evidence for this conclusion is provided by the intercalation in this sequence of stromatolites (see below).

As coeval microfossil assemblages, notably those from the Swaziland Supergroup, South Africa (cf.

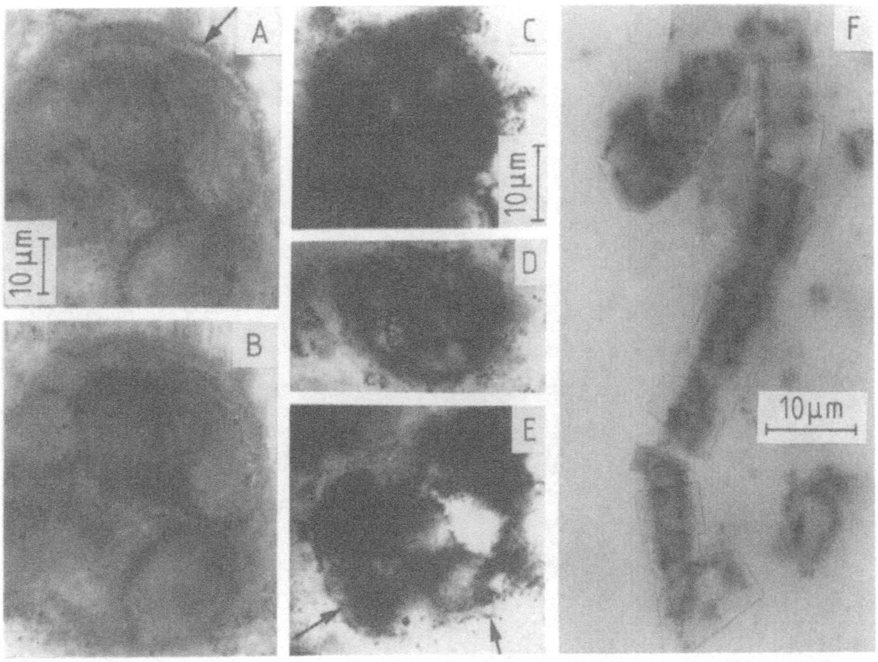

Figure 2. Early Archean (3.3–3.5 Gyr) microfossils from the Warrawoona Group (Pilbara Block, Western Australia). (A)–(E) Sheath-enclosed cell colonies suggestive of cyanobacterial (chroococcalean) affinity in different stages of preservation (C–E represent poor preservation states; arrows point to relics of primary sheaths). (F) Filamentous microfossil (photomontage) resembling trichomes of prokaryotic microorganisms of either cyanobacterial or flexibacterial affinities. Scales in panels A and C apply to panels A and B and panels C–E, respectively. [Panels A–E from Schopf and Packer (1987); panel F by courtesy of S. M. Awramik.]

Muir and Grant, 1976; Knoll and Barghoorn, 1977; Walsh and Lowe, 1985), are largely dominated by evidence on the dubiofossil level, the Warrawoona microflora constitutes a crucial benchmark in the evolution of Archean life. In view of both its degree of diversification and the evolutionary complexity of the prokaryotic cell in general, we may reasonably assume that the lineages of the principal microbial genera of the Warrawoona community must extend well beyond 3.5 Gyr. Accordingly, it seems almost certain that precursor floras had existed in older times where the rock record tends to be dramatically attenuated and progressively metamorphosed. In this context, assemblages of cell-like morphologies described from the ~3.8-Gyr-old amphibolite-grade metasediments from Isua, West Greenland [see illustrations in Pflug (1978, 1987), Pflug and Jaeschke-Boyer (1979), Bridgwater *et al.* (1981), and Roedder (1981)] were bound to arouse considerable interest.

A most articulate advocate of the biogenicity of the Isua microstructures has been H. D. Pflug, who observed a globular to elongate, supposedly sheath-enclosed cell-like morphotype in a quartzite (metachert) sample collected during the 1976 Isua field campaign of the Max Planck Institute for Chemistry, Mainz (cf. Schidlowski *et al.*, 1979). Described as *Isuasphaera isua* (Pflug, 1978), the biogenicity of this morphotype (Fig. 3) has been subsequently disputed on a variety of grounds, *inter alia* the improbability of survival of cellular microstructures during the amphibolite-grade reconstitution of the host rock (Bridgwater *et al.*, 1981). Meanwhile, however, there is ample evidence that microfossils may indeed preserve their identity through low- to medium-grade metamorphism (notably the greenschist and lower amphibolite facies) to a degree permitting even a palynostratigraphic approach to the study of metamorphosed sediment series (Pflug and Reitz, 1992). Hence, objections to the proposed biogenicity of the Isua objects can be based *solely* on a critique of the microstructures *per se* and not on attendant petrological or related criteria.

Since the structurally more differentiated septate filaments and coccoid colonial unicells characterizing the younger record (cf. Fig. 2) are virtually absent from the Isua assemblage, the morphological evidence is markedly empoverished and decidedly less unequivocal than, for instance, in the case of the Warrawoona microflora. Moreover, much of the electron microscopic data hitherto submitted (Pflug, 1987; Pflug and Reitz, 1992) may be debatable as this particular approach has been shown to be most prone to the introduction of preparation artifacts and other ambiguities (Schopf and Walter, 1983). Part of the depauperation of the Isua record can be readily attributed to the amphibolite-grade metamorphic overprint of the host rock that was bound to result in a large-scale loss of informative morphological detail. However, in spite of the obvious reduction of the morphotype inventory and the conspicuous absence (or

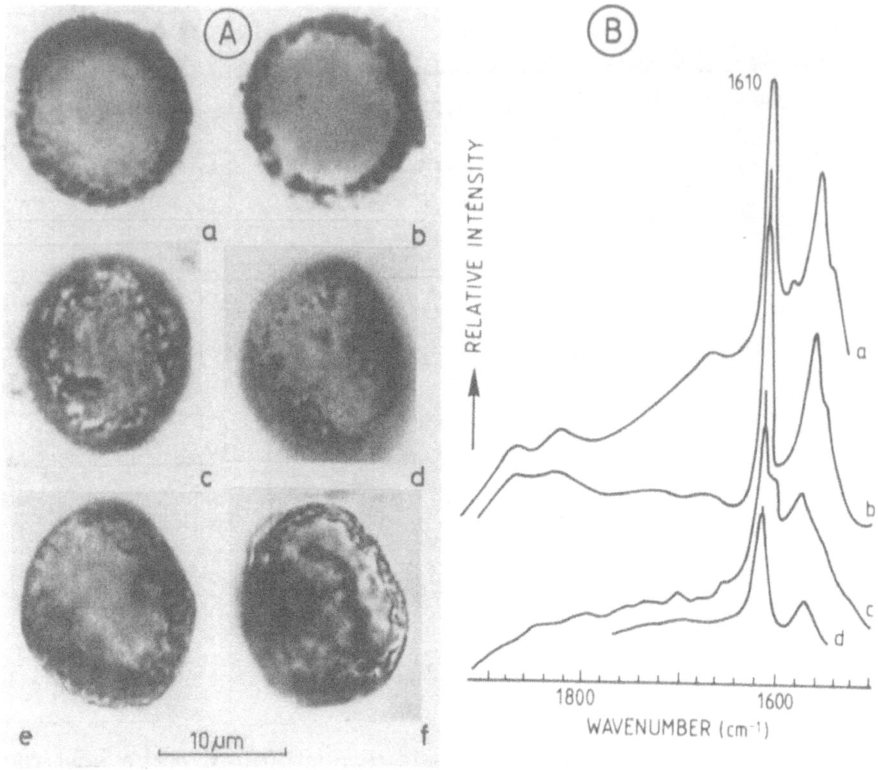

Figure 3. Comparison of *Huroniospora* sp. from the Proterozoic (~2.0 Gyr) Gunflint iron formation, Ontario (a–c) with *Isuasphaera* sp. from the ~3.8-Gyr-old Isua metasedimentary suite, West Greenland (d–f). The optically distinctive marginal rim may be explained as a relic of the original cell wall. (B) Laser Raman spectra obtained from *Huroniospora* sp. as an isolated particle (a) and in thin section (b) compared to those from *Isuasphaera* sp. (c, d) obtained under the same conditions. The close resemblance of the spectra suggests similarities in the composition of the residual organic component of the two types of microstructures (the prominent peak close to 1610 cm^{-1} is indicative of aromatic double bonds among the carbon atoms of the molecular structure). [Adapted from Pflug (1987).]

secondary obliteration) of structural detail, there is a striking resemblance of the preponderant Isua morphotype to a possible counterpart in the Archean–Proterozoic record, known as *Huroniospora* sp. (Fig. 3A), whose biogenic interpretation appears to be well founded and generally accepted. It is interesting to note that the near-perfect morphological analogy between the two forms is matched by their microchemical characteristics (Fig. 3B). Hence, even with due application of the critical standards called for in this case, selected microstructures from the Isua metasedimentary sequence appear to qualify for at least dubiofossil status as they stand in the continuity of the younger record. If confirmed by subsequent work, the recent description from the Isua iron formation of possible aggregate-forming iron bacteria of the *Siderocapsa* type with intricate external cell wall structures (Robbins, 1987) would lend additional support to the existence of a flourishing microbial life in Isua times.

Summing up the available evidence, the statement seems warranted that at least some of the microstructures reported from the 3.8-Gyr-old Isua metasedimentary suite may suggest microbial affinities irrespective of the extraordinary degree of uncertainty surrounding individual morphotypes and of occasional convergences with purely mineralogical features (such as limonite-coated dissolution cavities; cf. Roedder, 1981). While there may be differences of opinion as to whether or not the degree of structural differentiation, and the quality of preservation, of the observed morphotypes justify their formal description as genera and species, it probably cannot be excluded at this stage that the inventory as such contains at least some elements of a structurally degenerated "sticks and balls" assemblage that would result from an intense metamorphic alteration of any Warrawoona type Archean microfossil community. As will be pointed out below, the existence of prolific microbial ecosystems in Isua times would be decidedly consistent with the conspicuous organic (reduced) carbon content of the Isua rocks and with their carbon isotope geochemistry.

2.1.2. The Early Stromatolite Record

A second category of paleontological evidence attesting to the presence of life on the ancient Earth is provided by organosedimentary structures of the "stromatolitic" type. In fact, stromatolites constitute the most glaring morphological (macroscopic) expression in the rock record of the Earth's microbial benthos, being fairly common and widespread already in Archean formations (Fig. 1). Incidentally, these structures had long been regarded as the only undisputed manifestations of microbial life in the Archean, their geological record as a whole reflecting the evolutionary history of mat-forming bacterial and/or algal benthos over a time span of some 3.5 Gyr (Fig. 1).

Stromatolites in general (Fig. 4) represent lithified stacks of successively superimposed laminated microbial (mostly prokaryotic) communities that had originally thrived as organic films at the sediment–water interface. Accordingly, they may be aptly referred to as "microbialites" (Burne and Moore, 1987) since they ultimately derive from the interaction of a benthic microbial layer with the ambient sedimentary environment, the subsequent lithification of the structures resulting from either trapping, binding, or biologically mediated precipitation of selected mineral constituents (cf. Hofmann, 1973; Walter, 1977). The conspicuous lamination typical of most stromatolites preserves a record of transient stages in the growth of the parent microbial mat community, the complex interplay with the physical environment giving rise to a wealth of accretionary morphologies (stratiform, domal, columnar, branched, etc.; cf. Fig. 4).

There is by now ample evidence of the presence of fossil microbial mat communities in Archean sedimentary sequences approaching ages of 3.5 Gyr. Most prominent among their geologically oldest manifestations are finds from the 3.3–3.5-Gyr-old Warrawoona Group in the Pilbara Block of Western Australia (Fig. 5B) which have been painstakingly documented from several localities (Dunlop *et al.*, 1978; Lowe, 1980; Walter *et al.*, 1980) and thus leave little doubt that laminated prokaryotic ecosystems were indeed common in sedimentary environments of this age. Occurring in a cherty facies, the morphological inventory of these stromatolite occurrences seems virtually modern, attesting to the fact that fossilized microbial mats and related biosedimentary structures figure among the most conservative and persistent features of the paleontological record covering 3.5 Gyr of geological history. As pointed out above, selected chert members of the stromatolite-hosting sequence have been shown to also harbor a distinctive microbial flora of cyanobacterial and flexibacterial affinities (cf. Fig. 2).

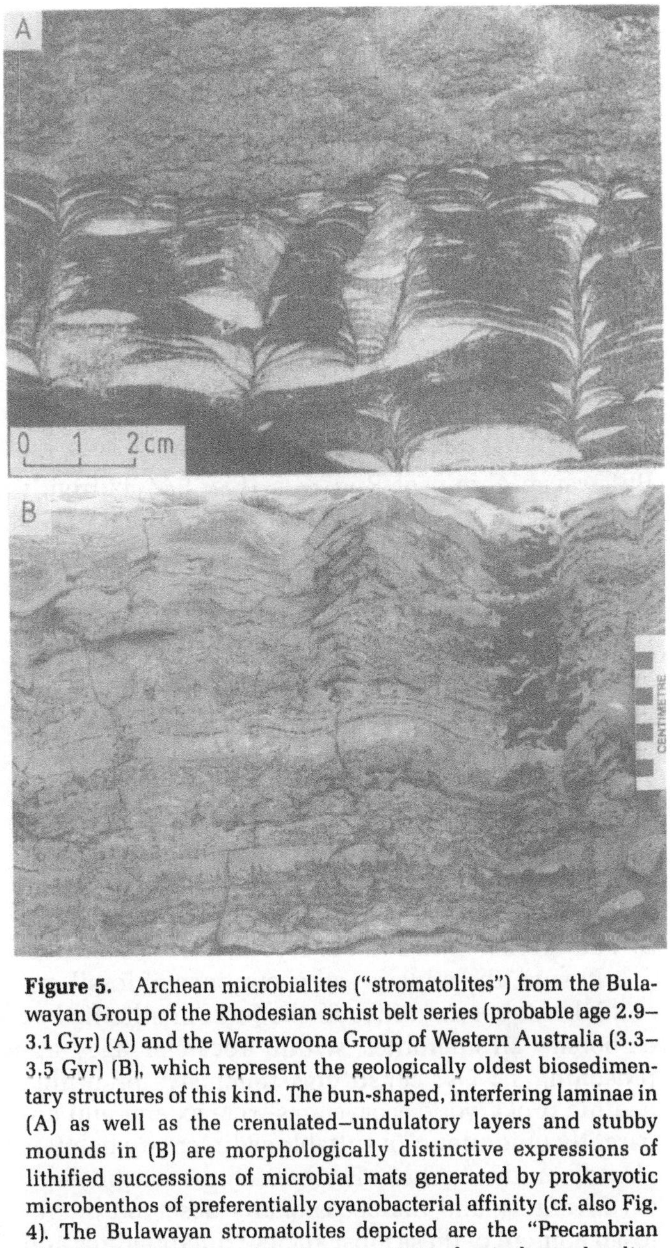

Figure 5. Archean microbialites ("stromatolites") from the Bulawayan Group of the Rhodesian schist belt series (probable age 2.9–3.1 Gyr) (A) and the Warrawoona Group of Western Australia (3.3–3.5 Gyr) (B), which represent the geologically oldest biosedimentary structures of this kind. The bun-shaped, interfering laminae in (A) as well as the crenulated–undulatory layers and stubby mounds in (B) are morphologically distinctive expressions of lithified successions of microbial mats generated by prokaryotic microbenthos of preferentially cyanobacterial affinity (cf. also Fig. 4). The Bulawayan stromatolites depicted are the "Precambrian algal limestones" from MacGregor's (1940) classical type locality. [Panel A from Schidlowski (1978); panel B by courtesy of S. M. Awramik.]

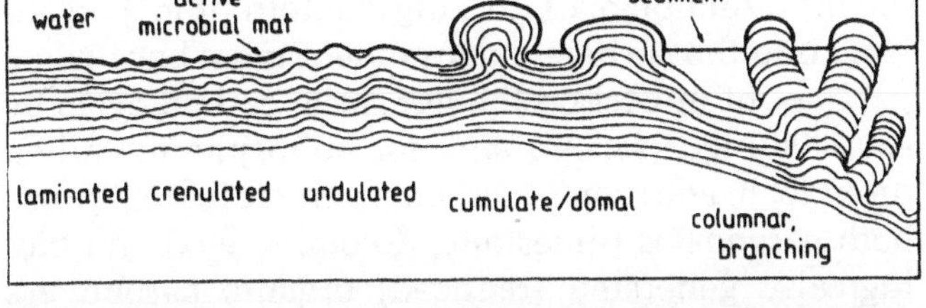

Figure 4. The principal morphologies of microbially active interfaces established between solid underground and the overlying water column that may subsequently lend themselves to lithification in the form of stromatolites. For comparison with the structural inventory displayed by the geologically oldest stromatolites, see Fig. 5.

After the discovery of the Warrawoona stromatolites, approximately coeval finds were reported from various other early Archean sequences such as the ~3.5-Gyr-old Sebakwian Group of the Rhodesian greenstone belt series (Orpen and Wilson, 1981) and the Fig Tree Group of the Barberton Mountain Land, South Africa (Byerly *et al.*, 1986), roughly bracketed between 3.3 and 3.5 Gyr. Corresponding documentations from younger Archean terranes are accumulating at a rapid rate and will probably be legion before long. Worthy of special mention among the reports from Middle and Late Archean rocks are the stromatolite occurrences in the ~3.0-Gyr-old Pongola Group of South Africa (Mason and von Brunn, 1977) as well as the classical description by MacGregor (1940) of what he termed "algal limestones" from the Bulawayan Group of Rhodesia (Fig. 5A), currently assigned a most probable age between 2.9 and 3.1 Gyr (Bond *et al.*, 1973). Incidentally, MacGregor's seminal paper served as a source of inspiration for a whole generation of Precambrian paleontologists during the emergence of this discipline between 1950 and 1970. With the establishment of extensive marine shelf facies during the Proterozoic, stromatolite development reached its peak (cf. Hofmann, 1973; Monty, 1984; see also Fig. 1), being followed by a conspicuous decline at the dawn of the Phanerozoic (0.6–0.7 Gyr) due to the takeover by the oldest Ediacaran-type metazoan faunas, whose advent marks the end of the dominion held over the Earth for three billion years by both prokaryotic and eukaryotic microorganisms.

In sum, the widespread occurrence of fossil microbial mat communities in the form of silicified or carbonate stromatolites within rocks of ages approaching 3.5 Gyr constitutes conclusive proof that benthic prokaryotic ecosystems were extant, and apparently ubiquitous, in suitable sedimentary habitats already on the Archean Earth. Both the morphological inventory and the occasionally observed microfossil content of these most ancient microbialites and their host rocks permit a detailed reconstruction of the Earth's earliest microbial communities, indicating specifically that Archean stromatolites were not markedly different from their counterparts occurring in the younger geological record, and that their morphologies can be well interpreted in terms of modern (Holocene) analogs (Walter, 1983; Knoll and Awramik, 1984). Moreover, it seems established beyond reasonable doubt that the principal microbial mat builders in Archean times were filamentous and unicellular prokaryotes capable of both photic responses ("phototropism") and photoautotrophy (cf. Walter, 1983). The almost unbroken stromatolite record from these times onward would, accordingly, attest to a blatant uniformity in the physiology and social organization of prokaryotic microbenthos of preferentially cyanobacterial affinity over at least 3.5 Gyr of geological history. As a whole, these findings are decidedly consistent with the micropaleontological evidence set out above.

2.2. The Geochemical Record

The advent of life on the ancient Earth and the subsequent establishment of the oldest microbial (prokaryotic) ecosystems were bound to have profound geochemical consequences. With living systems possessing an intrinsic property to proliferate until they encounter externally imposed limits, we may reasonably expect that the surficial environment was, before long, covered by a thin but fairly continuous veneer of living matter constituting a reactive film at, or close to, the Earth's surface, Being based, for the most part, on photosynthesis as an endergonic (or thermodynamic "uphill") reaction, life was apt to consequently figure as a powerful geochemical agent, its influence ultimately stemming from the marked accumulation of negative entropy in all living systems; in fact, life processes may be conveniently defined in terms of their entropy-deferring character. The negative entropy accumulated in the biosphere subsequently gives rise to a thermodynamic gradient that, in turn, acts as a driving force for a number of geochemically important reactions in the weathering and erosion cycles (prominent examples being the biologically imposed redox imbalance at the terrestrial surface or the large-scale release of bacteriogenic hydrogen sulfide in the marine realm). Since the products of biologically mediated geochemical transformations are commonly preserved in sediments, it is possible to trace the underlying biological processes back into the distant geological past. Specifically, the geochemical cycles of carbon, sulfur, and oxygen have been modulated by the global biosphere over most of geologic time.

With carbon constituting the key element of life, the introduction of CO_2-fixing ("autotrophic") organisms into the surficial environment must have had a paramount impact on the terrestrial carbon cycle by establishing a second sink for atmospheric carbon dioxide, in addition to carbonate precipitation. Since both carbonates (limestone, dolomite, etc.) and biologically generated (reduced) organic carbon are abundantly preserved in sedimentary rocks, relevant evidence should be encoded in the geologic record from its very onset. Here, a summary will be given of the principal biogeochemical and carbon isotope data supporting the existence as early as 3.8 Gyr of autotrophic carbon fixation, and specifically of photosynthesis as the quantitatively most important CO_2-fixing process. The following argument draws extensively on previous papers on the subject by the author (Schidlowski *et al.*, 1975, 1979, 1983; Schidlowski, 1987, 1988) and is consonant with the paleontological evidence outlined above.

2.2.1. Organic Carbon through Geologic Time

Carbon occurs in the Earth's sedimentary shell in two different forms, namely, (1) as *oxidized* or carbonate carbon (C_{carb}) and (2) as *reduced* or organic carbon (C_{org}), the latter constituting the fossil residue of primary biological substances. In the average sediment, organic carbon is present in amounts between 0.5 and 0.6% (Ronov, 1980), mostly in the form of kerogen, the acid-insoluble, polycondensed end product of the diagenetic alteration of organic debris derived from organisms and the products of their metabolism (cf. Durand, 1980). With the mass of sedimentary rocks totaling some 2.4×10^{24} g (Garrels and Mackenzie, 1971), the above C_{org} content of 0.5–0.6% translates into a sedimentary C_{org} reservoir of between 1.2×10^{22} and 1.4×10^{22} g. Carbonate carbon (C_{carb}), as the other sedimentary carbon species, is about four times more abundant than C_{org} and, for the most part, is concentrated in carbonate rocks such as limestones and dolomites.

While carbonates are formed by precipitation from aqueous solution driven, in essence, by the equilibria of the heterogeneous $CO_{2(gas)}$–$HCO^-_{3(aq)}$–$CO^{2-}_{3(solid)}$ system, the bulk of terrestrial organic matter is primarily generated by a limited number of photoautotrophic pathways involving the light-powered reduction of carbon dioxide to the carbohydrate level (CH_2O) by suitable reductants such as water or hydrogen sulfide, i.e.,

$$2H_2O + CO_2 \xrightarrow{h\nu} CH_2O + H_2O + O_2 \qquad \Delta G'_0 = +470.7 \text{ kJ} \qquad (1)$$

$$0.5H_2S + H_2O + CO_2 \xrightarrow{h\nu} CH_2O + H^+ + 0.5SO_4^{2-} \qquad \Delta G'_0 = +117.2 \text{ kJ} \qquad (2)$$

We may reasonably assume that H_2S-based (bacterial) photosynthesis (Eq. 2) preceded the water-splitting variant (Eq. 1) in the evolution of photoautotrophy because of its markedly lower energy requirements. On the other hand, the very fact that most autotrophs have subsequently come to capitalize on the energetically more demanding H_2O-splitting process constitutes eloquent testimony that there must have been a strong incentive to overcome the existing energy threshold by the integration into the assimilatory pathway of a second light reaction ("photosystem II"). There is little doubt that the utilization of water for the reduction of CO_2 was a necessary prerequisite for any major proliferation of biomass on Earth as it permitted the exploitation of a virtually inexhaustible pool of readily accessible reducing power.

Both carbon species, and specifically organic carbon (Fig. 6), can be traced back in sedimentary rocks to the very beginning of the record 3.8 Gyr ago (Allaart, 1976; Schidlowski *et al.*, 1979). Systematic assays for organic carbon carried out on Phanerozoic rocks (Trask and Patnode, 1942; Ronov, 1958) as well as observed oscillations of the carbon isotope age functions (see Fig. 7) indicate that C_{org} deposition rates have just moderately oscillated around a mean of perhaps 0.5% in the average sediment over this time span. Although both the preserved rock record and the currently available data base are progressively attenuated as we proceed into the geologic past, it is fairly evident that the scatter of C_{org} in Precambrian sediments is virtually the same as in Phanerozoic rocks (Fig. 6). Hence, the average content of organic carbon in the oldest sediments does not appear to be basically different from that of geologically younger formations. It is worth noting that the largely graphitized organic carbon burden of the 3.8-Gyr-old Isua metasediments of West Greenland marking the beginning of the record amounts to >0.6% (cf. 1 in Fig. 6) in the case of carbon-rich members of the suite.

Although astounding at first sight, the surprisingly modern C_{org} content of Archean sedimentary rocks may not be unusual at all. It is by now firmly established that microbial communities figure among the biosphere's most productive ecosystems, with benthic prokaryotes capable of sustaining prodigious rates of primary productivity, between 8 and 12 g C_{org} m^{-2} day^{-1} (Krumbein and Cohen, 1977; Cohen *et al.*, 1980). As pointed out before, the Precambrian was unquestionably the "Golden Age" of prokaryotic ecosystems. If such high rates of primary production can be sustained by microbial photoautotrophs operating at the prokaryotic level, a good case can be made that global production of biomass per time (and the pro rate concomitant C_{org} flux into sediments) probably did not increase very much during subsequent evolution of photosynthesis. As any generation of biomass is finally nutrient-limited, it seems most reasonable to envision for the Earth's earliest biosphere a scenario characterized by a state of global plenitude or "biotic saturation" (Stanley, 1981) in which microbial ecosystems had proliferated in existing aquatic habits to ultimate limits set by the availability of environmental resources, notably phosphorus and nitrogen as critical nutrients.

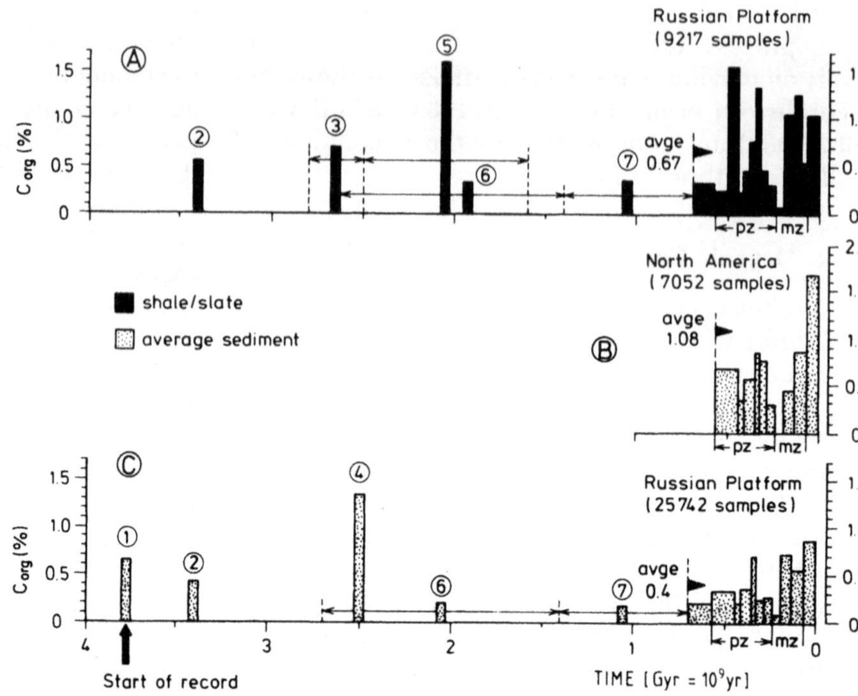

Figure 6. Organic carbon content of shales (A) and average sediments (B, C) through geologic time. Note that the data base for $t > 0.6$ Gyr is still deplorably scanty as compared to the younger record, but the scatter of C_{org} in Precambrian sediments is apparently the same as in Phanerozoic ones [Phanerozoic record after Trask and Patnode (1942) and Ronov (1958, 1980)]. Numbered Precambrian occurrences: (1) Early Archean metasediments from Isua, West Greenland; (2) Swaziland System, South Africa; (3) Archean pelitic sediments, Canadian Shield; (4) Hamersley Group, Australia; (5) Proterozoic pelitic sediments, Canadian Shield; (6) Proterozoic 1–2 of Russian Platform; (7) Proterozoic 3 of Russian Platform. [From Schidlowski (1982); see also references therein.]

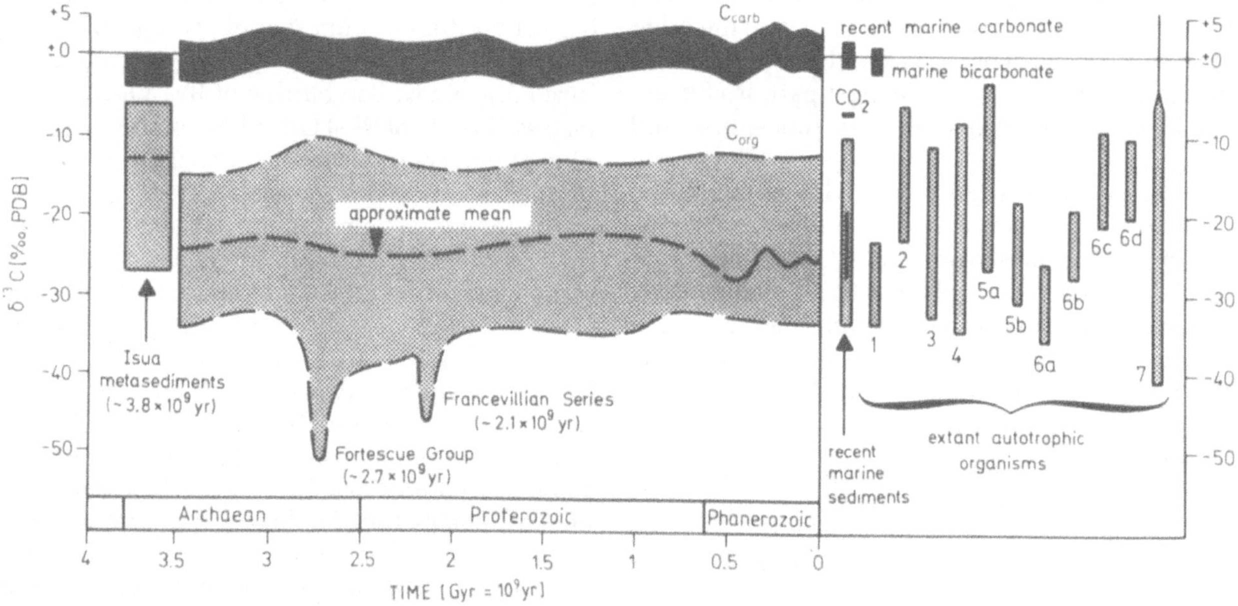

Figure 7. Isotope age functions of organic carbon (C_{org}) and carbonate carbon (C_{carb}) over 3.8 Gyr of recorded Earth history as compared with the isotopic compositions of their progenitor substances in the present environment (marine bicarbonate and biogenic matter of various parentage; cf. right panel). Isotopic compositions are given as $\delta^{13}C$ values, indicating either a relative increase (+) or decrease (−) in the $^{13}C/^{12}C$ ratio of the respective substance (in per mil difference) as compared to that of the Peedee belemnite (PDB) standard with $^{12}C/^{13}C = 88.99$, which defines the zero per mil line on the δ-scale. Note that the $\delta^{13}C_{org}$ spread of the extant biomass is basically transcribed into recent marine sediments and the subsequent record back to 3.8 Gyr (the Isua values have been moderately reset by amphibolite-grade metamorphism; cf. Fig. 8). The envelope shown for fossil organic carbon is an update of the data base presented by Schidlowski *et al.* (1983), comprising the means of some 150 Precambrian kerogen provinces as well as the currently available information on the Phanerozoic record (Degens, 1969; Veizer *et al*, 1980; and others). The negative spikes at 2.7 and 2.1 Gyr indicate a large-scale involvement of methane in the formation of the respective kerogen precursors. Extant autotrophs contributing to the contemporary biomass are C_3 plants (1), C_4 plants (2), crassulacean acid metabolism (CAM) plants (3), eukaryotic algae (4), natural and cultured cyanobacteria (5a, 5b), groups of photosynthetic bacteria other than cyanobacteria (6), and methanogens (7). $\delta^{13}C_{org}$ range in recent marine sediments is that reported by Deines (1980) based on some 1600 data points (black line within bar covers >90% of the data base). [From Schidlowski (1988).]

Summing up the evidence presently available, we may state that both organic carbon and carbonate are common constituents of sedimentary rocks over the whole of the hitherto known record, with the graphitic derivatives of primary kerogenous materials present in the 3.8-Gyr-old Isua suite attesting to the operation of life processes already during the time of formation of the Earth's oldest sediments. Moreover, currently available assays indicate that the C_{org} content of the average sediment has probably stayed fairly uniform from the Archean to the Recent, lying mostly between 0.4 and 0.6%.

2.2.2. Sedimentary Carbon Isotope Record: Implications for the Antiquity of Autotrophic Carbon Fixation

Additional proof for the biogenicity of the reduced carbon fraction of common sedimentary rocks, including their oldest occurrences, comes from observed $^{13}C/^{12}C$ fractionations between the two sedimentary carbon species. As has been known since the early investigations by Nier and Gulbransen (1939), Murphey and Nier (1941), and Rankama (1948), transformation of inorganic carbon into biological substances entails a marked bias in favor of the light isotope (^{12}C), with the heavy isotope (^{13}C) preferentially retained in the inorganic feeder pool and finally precipitated in the form of carbonate. Subsequent work by numerous investigators (e.g., Craig, 1953; Park and Epstein, 1960; O'Leary, 1981) has confirmed that all common pathways of autotrophic carbon fixation, and notably the quantitatively important photosynthetic ones, discriminate against ^{13}C, primarily as a result of a kinetic isotope effect imposed on the first irreversible enzymatic CO_2-fixing carboxylation reaction (see Chapter 3 for a detailed discussion). This reaction is responsible for the incorporation of CO_2 into the COOH group of an organic acid, which, in turn, is further metabolized in ensuing metabolic pathways. This has consequently given rise to a pronounced enrichment of ^{12}C in all forms of biogenic (reduced) carbon as compared with the inorganic (oxidized) carbon reservoir of the surficial environment, consisting mainly of dissolved marine bicarbonate ion (HCO_3^-) and atmospheric carbon dioxide. In terms of the conventional δ-notation, the $\delta^{13}C$ values of average biomass are usually between 20 and 30‰ more negative than those of marine bicarbonate, the prevalent oxidized carbon species of the environment (Fig. 7, right panel).

The isotopic difference thus established between biologically derived (organic) carbon and carbonate is largely retained when both carbon species are incorporated in newly formed sediments. Although diagenesis has been shown to cause distinct shifts in the primary isotopic compositions of both C_{org} and C_{carb}, the secondary overprint thus imposed is usually moderate, with diagenetically altered $\delta^{13}C$ values rarely differing from the original ones by more than 2–3‰ (for a detailed discussion, see, *inter alia*, Hayes *et al.*, 1983; Schidlowski, 1987; Schidlowski *et al.*, 1983). Consequently, the original isotope fractionations between reduced and oxidized carbon as established in the surficial environment appear to be largely "frozen" after the entry of the respective carrier phases into the sedimentary realm. Over geologic time, continuous biological processing of carbon in the exogenic compartment of the geochemical cycle has thus brought about a conspicuous isotopic disproportionation of the planet's primordial carbon endowment into an isotopically "light" (C_{org}) and an isotopically "heavy (C_{carb}) reservoir, both stored in the Earth's sedimentary shell.

Students of the sedimentary carbon record have long been impressed by the relative constancy on the gigayear scale of both the $\delta^{13}C_{org}$ and $\delta^{13}C_{carb}$ age functions. As can be inferred from Fig. 7, both the carbon isotope spreads of extant primary producers and those of recent marine bicarbonate and carbonate are virtually superimposable, with minor oscillations, on the record back to 3.5, if not 3.8, Gyr, constituting evidence that organic carbon and carbonate carbon have been transferred from the surficial exchange reservoir to sedimentary rocks with little change in their isotopic compositions.

The narrow spread of the C_{carb} age function suggests that the $\delta^{13}C$ values of the bicarbonate precursor of ancient marine carbonates were closely tethered to the zero per mil line over 3.8 Gyr of Earth history, allowing little departure in either direction. As for the corresponding C_{org} function, it is obvious that the $\delta^{13}C_{org}$ range observed in recent marine sediments (Fig. 7, right panel) spans the respective spreads of the $\delta^{13}C$ values of the principal contributors to the contemporary biomass with just the extremes eliminated. This shows that the influence of a later diagenetic overprint on the primary values is rather limited in extent and, for the most part, gets lost within the broad scatter of the primary values that has consequently come to be projected into the geologic past over billions of years (cf. envelope for C_{org} in Fig. 7). Over the 3.5 Gyr of the unmetamorphosed rock record, isotopic fractionations between organic carbon and carbonate carbon thus appear to be demonstrably the same as in the present world, with the characteristic differences in $\delta^{13}C$ of some 20 to 30‰ and a corresponding enrichment of light carbon (^{12}C) in the biogenic phase. Quite obviously, the kinetic isotope effect inherent in autotrophic (photosynthetic) carbon fixation has been propagated from the biosphere into the rock section of the carbon cycle almost unaltered, which opens up the possibility of tracing the isotopic signature of the underlying process far back into the geologic past.

With these relationships established, the implications of both carbon isotope age functions for the antiquity of photoautotrophy as the quantitatively most important biochemical process seem to be straightforward. There is little doubt that the conspicuous ^{12}C enrichment displayed by the data envelope for fossil organic carbon (Fig. 7) constitutes a coherent signal of autotrophic carbon fixation over 3.8 Gyr of geological history as it ultimately originated with the process that gave rise to the biological precursor materials. Moreover, the long-term uniformity of this isotopic signal attests to an extreme degree of conservatism of the basic biochemical mechanisms of carbon fixation. In fact, the mainstream of the envelope for $\delta^{13}C_{org}$ shown in Fig. 7 can be most readily explained as the geochemical manifestation of the isotope-discriminating properties of one single enzyme, namely, ribulose-1.5-bisphosphate (RuBP) carboxylase, the key enzyme of the Calvin cycle.

It is well established today that the carbon transfer from the inorganic to the organic world basically proceeds via the RuBP carboxylase reaction that feeds CO_2 directly into the Calvin cycle as a three-carbon compound (phosphoglycerate). Most autotrophic microorganisms and all green plants operate this pathway of carbon assimilation; higher plants relying on it entirely are termed C_3 plants. As a result, the bulk of the Earth's biomass (both extant and fossil) bears the isotopic signature of C_3 (or Calvin cycle) photosynthesis characterized by the sizable fractionations of the RuBP carboxylase reaction that assign a mean $\delta^{13}C_{org}$ range of -26 ± 7‰ to the bulk of terrestrial organic matter. Accordingly, the $\delta^{13}C_{org}$ age function presented in Fig. 7 may be aptly referred to as an *index line of autotrophic carbon fixation over almost 4 Gyr of Earth history*. It is worth noting that the first important conjectures about the antiquity of photosynthesis on Earth were based on simple precursor curves of the currently established carbon isotope age function (Eichmann and Schidlowski, 1975; Junge *et al.*, 1975).

Occasional negative offshoots from this long-term average (Fig. 7) are commonly restricted to the Precambrian (Schoell and Wellmer, 1981; Hayes *et al.*, 1983; Schidlowski *et al.*, 1983; Weber *et al.*, 1983) and suggest the involvement of methanotrophic pathways in the formation of the respective kerogen precursors. Basically, these excursions appear to be oddities confined to regional side stages of the carbon cycle that have hardly affected the economy of the global cycle as a whole.

The only glaring discontinuity in the sedimentary carbon isotope age functions is the break between the ~3.8-Gyr-old Isua metasediments from West Greenland and the whole of the geologically younger record. While the long-term averages for $\delta^{13}C_{carb}$ and $\delta^{13}C_{org}$ may be given as $+0.5 \pm 2.5$‰ and -26.0 ± 7.0‰, respectively (Schidlowski *et al.*, 1975, 1983; Veizer *et al.*, 1980), and interpreted as being indicative of a biologically mediated carbon isotope fractionation over the last 3.5 Gyr, the distinctly displaced means for the Isua rocks ($\delta^{13}C_{carb} = -2.3 \pm 2.2$‰, $\delta^{13}C_{org} = -13.0 \pm 4.9$‰; cf. Schidlowski *et al.*, 1983) could raise doubts as to whether these values constitute equally unequivocal evidence of biological activity.

An early proposal that the Isua anomaly is due to a metamorphic overprint (Schidlowski *et al.*, 1979) has been subsequently borne out by a wealth of corroborative evidence. The observed isotope shifts are, in fact, fully consistent with the predictable effects of a high-temperature reequilibration between coexisting sedimentary carbonate and organic carbon of "normal" isotopic composition in response to the amphibolite-grade reconstitution of the host rock. Specifically, it has been demonstrated that $^{13}C/^{12}C$ exchange between kerogenous rock constituents and isotopically heavy CO_2 released from carbonate rocks as a result of metamorphic decarbonation reactions starts already in the greenschist facies (300–450°C) and becomes more pronounced in higher metamorphic grades (Valley and O'Neil, 1981), with isotopic reequilibration being governed by a fractionation factor that decreases with increasing temperature. The overall result is a marked diminution in the magnitude of fractionation between carbonate and organic carbon (including its graphitic derivatives) as a function of metamorphic temperature (Fig. 8). These findings are consonant with thermodynamic data on high-temperature carbon isotope exchange as well as with observational evidence from geologically younger metamorphic terranes and, accordingly, permit a fairly conclusive interpretation of those parts of the record that bear a metamorphic overprint. Therefore, it appears virtually certain that the "normal" sedimentary $\delta^{13}C_{org}$ and $\delta^{13}C_{carb}$ records originally extended back to Isua times (3.8 Gyr), suggesting that biological modulation

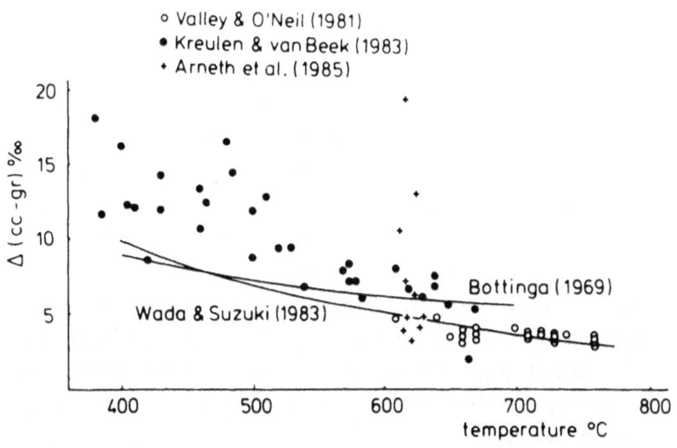

Figure 8. Decrease of isotopic fractionation $\Delta(cc - gr) = \delta^{13}C_{cc} - \delta^{13}C_{gr}$ between coexisting calcite (cc) and graphite (gr) as a function of increasing temperature of metamorphism (sedimentary organic carbon preferentially occurs as graphite in higher metamorphic grades). Observed Δ(cc − gr) values reported by several authors from various metamorphic terranes are shown to scatter around both Bottinga's (1969) function of thermodynamically calculated isotope equilibria and an empirical fractionation curve from Wada and Suzuki (1983) calibrated by dolomite–calcite solvus temperatures. Note that the markedly decreased average fractionation of 10–11‰ between organic carbon and carbonate in the 3.8-Gyr-old Isua metasediments (cf. Fig. 7) falls into the temperature range of the lower amphibolite facies (450–550°C).

of the global carbon cycle had already been established at the time of formation of the Earth's oldest sedimentary rocks (Schidlowski *et al.*, 1979, 1983).

2.2.3. Sedimentary Sulfur Isotope Record and Antiquity of Dissimilatory Sulfate Reduction

Although not immediately relevant for the beginnings of life processes on Earth, the emergence of dissimilatory sulfate reduction certainly was an important quantum step in the early evolution of bioenergetic processes (cf. Broda, 1975; Trudinger, 1979; Trüper, 1982). As H_2S-based bacterial photosynthesis (cf. Section 2.2.1., Eq. 2) surely preceded the H_2O-splitting variant of the photosynthetic process in the evolution of photoautotrophy, sulfate ion (SO_4^{2-}) as a mild oxidant must have appeared in the ancient environment well before the advent of free oxygen. The availability of an oxidized sulfur species was, in turn, a prerequisite for the rise of dissimilatory sulfate reduction, an energy-yielding adaptive reversal of the primary SO_4^{2-}-generating photosynthetic process that couples the reduction of sulfate to hydrogen sulfide with the oxidation of organic substances. A simplified version of these cyclic transformations, ultimately powered by solar energy, may be written as

$$H_2S + 2HCO_3^- \underset{\text{dissimilation}}{\overset{\text{photosynthesis}}{\rightleftharpoons}} 2\,CH_2O + SO_4^{2-}$$

Since sulfate instead of oxygen serves as an oxidant for organic "foodstuffs" (CH_2O) in the above back reaction, the dissimilatory process may be regarded as a form of anaerobic respiration ("sulfate respiration"). Although utilized by only a few bacterial genera presently restricted to the anaerobic layer of the littoral seafloor, this reaction is of crucial geochemical importance as it is responsible for a large-scale conversion of sulfate to sulfide that fully depends on biological mediation under the low-temperature conditions prevailing in terrestrial near-surface environments.

Bacterial sulfate reduction has been shown to release geochemically significant quantities of H_2S whose $\delta^{34}S$ values are shifted by about 40‰ in the negative direction as compared to the parent seawater sulfate. Part of this isotopically light (^{32}S-enriched) hydrogen sulfide is precipitated in the marine realm in the form of metal sulfides (mostly pyrite) which basically inherit the isotopic composition of the bacteriogenic H_2S precursor, thereby preserving the isotopic signature of the underlying biochemical process. With marine sulfate likewise preserved in sulfate evaporites, the isotopic fractionation between sulfide and sulfate can be traced back into the sedimentary record, the respective isotope age function providing an index of dissimilatory sulfate reduction through geologic time. Utilizing the presently available isotopic evidence (Fig. 9) for pinning down the emergence of this process, it would appear that bacteriogenic sulfur isotope patterns in sedimentary sulfides arose about 2.8 Gyr ago, which is astoundingly late in view of the fact that photosynthesis almost certainly emerged one billion years earlier (cf. Fig. 1). It is, moreover, well known that sulfate was available in the Archean oceans as of at least 3.5 Gyr ago (cf. Perry *et al.*, 1971; Vinogradov *et al.*, 1976). Therefore, it is most likely that early trends to establish a biologically mediated isotopic differentiation of marine sulfur were either suppressed or camouflaged by geochemical processes or environmental controls operative on the early Earth that are as yet poorly understood. As has been argued elsewhere (Schidlowski, 1989), any large-scale isotopic reequilibration between marine sulfur and the primordial sulfur reservoir of the early oceanic crust could have possibly tethered the $\delta^{34}S$ values of both sulfide and sulfate to a mean not far from zero per mil, as is actually shown by the Early

Archean record (1–5 in Fig. 9). With extended areas of newly formed oceanic crust exposed to the Archean seas, and a substantially higher contemporaneous heat flow, basalt–seawater interactions should have been at a maximum in those times, probably culminating in an essentially mantle-dominated geochemistry of the ancient seas (Veizer *et al.*, 1982). Alternatively, Ohmoto and Felder (1987) have invoked a combination of increased rates of bacterial sulfate reduction in the surficial sedimentary layer of a warmer Archean ocean and a system closed or semiclosed to sulfate (due to the absence of a burrowing fauna) to account for the markedly reduced fractionations between sulfide and sulfate observed in the oldest sediments (Fig. 9).

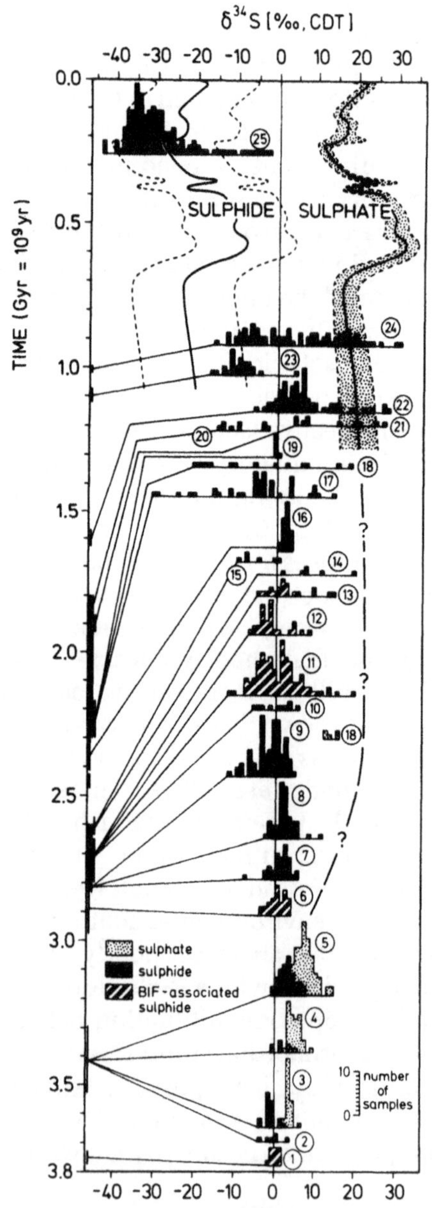

Summing up, we may reasonably suspect that the oldest presumably bacteriogenic sulfur isotope patterns appearing in the record ~2.8 Gyr ago just give a *minimum age* for the operation of dissimilatory sulfate reduction as a bioenergetic process. With evidence in favor of the existence of both prokaryotic life and photoautotrophy predating the appearance of these patterns by approximately one billion years, it seems most likely that the isotopic signature of this process either remained cryptic due to environmental controls specific to the Archean, or was obliterated by a mantle-dominated geochemistry of the Earth's oldest oceans.

3. Summary and Conclusions

Based on the paleontological evidence set out above and moderate extrapolations thereof, it can be stated with reasonable confidence that the Earth was inhabited by life from about the start of the sedimentary record 3.8 Gyr ago. Both fossil biosedimentary structures ("stromatolites") attributable to the matting behavior of prokaryotic microbenthos and coeval microbiota of mainly cyanobacterial affinity become significant in the record as of ~3.5 Gyr ago, attesting to the existence of prolific microbial ecosystems in suitable habitats of the Early Archean Earth. The degree of diversification displayed by the Warrawoona microflora representative of this age suggests, furthermore, that the lineages of the principal microbial morphotypes might well extend into the older geological past. This, in turn, adds credence to the biogenicity of more debatable morphologies on the dubiofossil level from the 3.8-Gyr-old Isua suite,

Figure 9. Isotopic evolution of sedimentary sulfide and sulfate over 3.8 Gyr of geological history. The microbially generated isotopic differentiation patterns between the two sulfur species are progressively blurred from about $t > 1.3$ Gyr, mainly as a result of the low preservation potential of marine sulfate evaporites that tend to fade from the record with increasing geologic age. Therefore, the oldest isotopic evidence of bacterial sulfate reduction primarily rests on the typical $\delta^{34}S$ distribution patterns of bacteriogenic sulfides that are characterized by extended spreads coupled with a "Rayleigh tail" at the positive end, reflecting a reservoir effect (such as exemplified by the Permian Kupferschiefer of Central Europe shown by no. 25). Applying these criteria to the early record, the presumably oldest bacteriogenic sulfide patterns appear around 2.7–2.8 Gyr [Deer Lake Greenstone Belt, Minnesota (no. 8); Michipicoten, Woman River, and Lumby–Finlayson Lakes banded iron formations, Canada (nos. 11–13)]. Note that the $\delta^{34}S$ values of the sulfide constituents of the Isua banded iron formation (no. 1) cluster closely around the zero per mil line defined by the isotope ratio of the Canyon Diablo troilite (CDT) standard with $^{32}S/^{34}S$ = 22.22. BIF, banded iron formation. [From Schidlowski (1989); for respective data bases of nos. 1–25, see references therein.]

which, at least in part, may represent members of a structurally degenerated microfossil assemblage such as would result from intense metamorphic impairment of any Warrawoona-type microfossil community.

The occurrence of sedimentary organic carbon in the form of kerogen and its graphite derivatives over 3.8 Gyr of recorded Earth history is fully consistent with the conclusions drawn from the paleontological record. Moreover, with the currently documented isotope age functions of sedimentary carbon at hand, a convincing case can be made that $^{13}C/^{12}C$ fractionations between kerogenous materials and carbonates are diagnostic of the isotope-discriminating properties of ribulose-1,5-bisphosphate (RuBP) carboxylase as the principal CO_2-fixing enzyme of the photosynthetic pathway which demonstrably channels the bulk of the carbon transfer from the inorganic to the living world. Accordingly, the isotope age curve of sedimentary organic carbon can be best explained as an index line of autotrophic carbon fixation. Allowing for a metamorphic overprint of the oldest (>3.5 Gyr) record, enzymatically produced carbon isotope fractionations have persisted rather uniformly over the past ~3.8 Gyr, indicating an extraordinary degree of evolutionary conservatism in the biochemistry of autotrophic carbon fixation.

Summarizing the facts pertinent to the advent of biological processes on the ancient Earth, we may state that, irrespective of the blatant depauperation and structural impoverishment of the earliest paleontological record and of the metamorphic overprint on the carbon isotope record prior to 3.5 Gyr, evidence for the existence of life during Early Archean times is so firmly established as to be virtually unassailable. As can be specifically inferred from the isotope age function of sedimentary organic carbon, autotrophy—and notably photoautotrophy as the quantitatively most important process of biological CO_2 fixation—has been extant as both a biochemical process and as a geochemical agent from at least 3.8 Gyr, implying that biologically mediated photoinduced "uphill" electron transfer as a principal means of creating reducing power for assimilatory reactions had already evolved in much older times that are not covered by the sedimentary record.

References

Agarwal, V. K., Schutte, W., Greenberg, J. M., Ferris, J. P., Briggs, R., Connor, S., Van de Bult, C. P. E. M., and Baas, F., 1985, Photochemical reactions in interstellar grains: Photolysis of CO, NH_3 and H_2O, *Origins Life* **16**:21–40.

Allaart, J. H., 1976, The pre-3760-Myr old supracrustal rocks of the Isua area, central West Greenland, and the associated occurrence of quartz-banded ironstone in: *The Early History of the Earth* (B. F. Windley, ed.), John Wiley & Sons, New York, pp. 177–189.

Arneth, J. D., Schidlowski, M., Sarbas, B., Goerg, U., and Amstutz, G. C., 1985, Graphite content and isotopic fractionation between calcite–graphite pairs in metasediments from the Mgama Hills, Southern Kenya, *Geochim. Cosmochim. Acta* **49**:1553–1560.

Awramik, S. M., 1982, The pre-Phanerozoic fossil record, in: *Mineral Deposits and Evolution of the Biosphere* (H. D. Holland and M. Schidlowski, eds.), Springer-Verlag, Berlin, pp. 67–81.

Awramik, S. M., Schopf, J. W., and Walter, M. R., 1983, Filamentous fossil bacteria from the Archean of Western Australia, in: *Developments and Interactions of the Precambrian Atmosphere, Lithosphere and Biosphere* (B. Nagy, R. Weber, J. C. Guerrero, and M. Schidlowski, eds.), *Developments in Precambrian Geology*, Vol. 7, Elsevier, Amsterdam, pp. 249–266.

Barghoorn, E. S., and Tyler, S. A., 1965, Microorganisms from the Gunflint chert, *Science* **147**:563–577.

Bond, G., Wilson, J. F., and Winnall, N. J., 1973, Age of the Huntsman limestone (Bulawayan) stromatolites, *Nature* **244**:275–276.

Bottinga, Y., 1969, Calculated fractionation factors for carbon and hydrogen isotope exchange in the system calcite–carbon dioxide–graphite–methane–hydrogen–water vapor, *Geochim. Cosmochim. Acta* **33**:49–64.

Bridgwater, D., Allaart, J. H., Schopf, J. W., Klein, C., Walter, M. R., Barghoorn, E. S., Strother, P., Knoll, A. H., and Gorman, B. E., 1981, Microfossil-like objects from the Archean of Greenland: A cautionary note, *Nature* **289**:51–53.

Broda, E., 1975, *The Evolution of the Bioenergetic Processes*, Pergamon Press, Oxford.

Buick, R., 1984, Carbonaceous filaments from North Pole, Western Australia: Are they fossil bacteria in Archean stromatolites? *Precamb. Res.* **24**:157–172.

Burne, R. V., and Moore, L. S., 1987, Microbialites: Organosedimentary deposits of benthic microbial communities, *Palaios* **2**:241–254.

Byerly, G. R., Lowe, D. R., and Walsh, M. M., 1986, Stromatolites from the 3,300–3,500-Myr Swaziland Supergroup, Barberton Mountain Land, South Africa, *Nature* **319**:489–491.

Cloud, P. E., 1965, Significance of the Gunflint (Precambrian) microflora, *Science* **148**:27–45.

Cloud, P. E., 1976, Beginnings of biospheric evolution and their biogeochemical consequences, *Paleobiology* **2**:351–387.

Cohen, Y., Aizenshtat, Z., Stoler, A., and Jorgensen, B. B., 1980, The microbial geochemistry of Solar Lake, Sinai, in: *Biogeochemistry of Ancient and Modern Environments* (J. B. Ralph, P. A. Trudinger, and M. R. Walter, eds.), Springer-Verlag, Berlin, pp. 167–172.

Craig, H., 1953, The geochemistry of stable carbon isotopes, *Geochim. Cosmochim. Acta* **3**:53–92.

Degens, E. T., 1969, Biogeochemistry of stable carbon isotopes, in: *Organic Geochemistry* (G. Eglinton and M. T. J. Murphy, eds.), Springer-Verlag, Berlin, pp. 304–329.

Deines, P., 1980, The isotropic composition of reduced organic carbon, in: *Handbook of Environmental Isotope Geochemistry, Vol. 1* (P. Fritz, and J. C. Fontes, eds.), Elsevier, Amsterdam, pp. 329–406.

Dunlop, J. S. R., Muir, M. D., Milne, V. A., and Groves, D. I., 1978, A new microfossil assemblage from the Archaean of Western Australia, *Nature* **274**:676–678.

Durand, B. (ed.), 1980, *Kerogen—Insoluble Organic Matter from Sedimentary Rocks*, Editions Technip, Paris.

Eichmann, R., and Schidlowski, M., 1975, Isotopic fractionation between coexisting organic carbon–carbonate pairs in Precambrian sediments, *Geochim. Cosmochim. Acta* **39**:585–595.

Garrels, R. M., and Mackenzie, F. T., 1971, *Evolution of Sedimentary Rocks*, Norton, New York.

Glaessner, M. F., 1983, The emergence of Metazoa in the early history of life, in: *Developments and Interactions of the Precambrian Atmosphere, Lithosphere and Biosphere* (B. Nagy, R. Weber, J. C. Guerrero, and M. Schidlowski, eds.), Elsevier, Amsterdam, pp. 319–333.

Glaessner, M., 1984, *The Dawn of Animal Life*, Cambridge University Press, Cambridge.

Golubic, S., and Barghoorn, E. S., 1977, Interpretation of microbial fossils with special reference to the Precambrian, in: *Fossil Algae* (E. Flügel, ed.), Springer-Verlag, Berlin, pp. 1–14.

Greenberg, J. M., 1984, Chemical evolution in space, in: *Proceedings of the 27th International Geological Congress*, Volume 19 (N. A. Bogdanov, ed.), VNU Science Press, Utrecht, pp. 209–228.

Greenberg, J. M., 1985, The chemical and physical evolution of interstellar dust, *Phys. Scr.* **T11**:14–26.

Hayes, J. M., Kaplan, I. R., and Wedeking, K. W., 1983, Precambrian organic geochemistry: Preservation of the record, in: *Earth's Earliest Biosphere: Its Origin and Evolution* (J. W. Schopf, ed.), Princeton University Press, Princeton, New Jersey, pp. 93–134.

Hofmann, H. J., 1973, Stromatolites: Characteristics and utility, *Earth Sci. Rev.* **9**:339–373.

Irvine, W. M., and Knacke, R. F., 1989, The chemistry of interstellar gas and grains, in: *Origin and Evolution of Planetary and Satellite Atmospheres* (S. K. Atreya, J. B. Pollack, and M. S. Matthews, eds.), University of Arizona Press, Tucson, pp. 3–34.

Junge, C. E., Schidlowski, M., Eichmann, R., and Pietrek, H., 1975, Model calculations for the terrestrial carbon cycle: Carbon isotope geochemistry and evolution of photosynthetic oxygen, *J. Geophys. Res.* **80**:4542–4552.

Kazmierczak, J., 1979, The eukaryotic nature of Eosphaera-like ferriferous structures from the Precambrian Gunflint Iron Formation, Canada: A comparative study, *Precamb. Res.* **9**:1–22.

Kissel, J., and Krueger, F.R., 1987, The organic component in dust from comet Halley as measured by the PUMA mass spectrometer on board Vega 1, *Nature* **326**:755–760.

Knoll, A. H., and Awramik, S. M., 1984, Ancient microbial ecosystems, in: *Microbial Geochemistry* (W. E. Krumbein, ed.), Blackwell Scientific, Oxford, pp. 287–315.

Knoll, A. H., and Barghoorn, E. S., 1977, Archean microfossils showing cell division from the Swaziland System of South Africa, *Science* **198**:396–398.

Knoll, A. H., Strother, P. K., and Rossi, S., 1988, Distribution and diagenesis of microfossils from the Lower Proterozoic Duck Creek Dolomite, Western Australia, *Precamb. Res.* **38**:257–279.

Kreulen, R., and van Beek, P. C. J. M., 1983, The calcite–graphite isotope thermometer; data on graphite-bearing marbles from Naxos, Greece, *Geochim. Cosmochim. Acta* **47**:1527–1530.

Krumbein, W. E., and Cohen, Y., 1977, Primary production, mat formation and lithification chances of oxygenic and facultative anoxygenic cyanophytes (cyanobacteria), in: *Fossil Algae* (E. Flügel, ed), Springer-Verlag, Berlin, pp. 37–56.

Lowe, D. R., 1980, Stromatolites 3,400-Myr old from the Archean of Western Australia, *Nature* **284**:441–443.

MacGregor, A. M., 1940, A Precambrian algal limestone in Southern Rhodesia, *Trans. Geol. Soc. S. Afr.* **43**:9–15.

Mason, T. R., and von Brunn, V., 1977, 3-Gyr-old stromatolites from South Africa, *Nature* **266**:47–49.

Miller, S. L., 1955, Production of some organic compounds under possible primitive Earth conditions, *J. Am. Chem. Soc.* **77**: 2351–2361.

Miller, S. L., Urey, H. C., and Oro, J., 1976, Origin of organic compounds on the primitive Earth and in meteorites, *J. Mol. Evol.* **9**:59–72.

Monty, C., 1984, Stromatolites in Earth history, *Terra Cognita* **4**: 423–430.

Muir, M. D., and Grant, P. R., 1976, Micropaleontological evidence from the Onverwacht Group, South Africa, in: *The Early History of the Earth* (B. F. Windley, ed.), Wiley, London, pp. 595–604.

Murphey, B. F., and Nier, A. O., 1941, Variations in the relative abundance of the carbon isotopes, *Phys. Rev.* **59**:771–772.

Nier, A. O., and Gulbransen, E. A., 1939, Variations in the relative abundance of the carbon isotopes, *J. Am. Chem. Soc.* **61**: 697–698.

Ohmoto, H., and Felder, R. P., 1987, Bacterial activity in the warmer, sulphate-bearing, Archaean oceans, *Nature* **328**:244–246.

O'Leary, M. H., 1981, Carbon isotope fractionation in plants, *Phytochemistry* **20**:553–567.

Orpen, J. L., and Wilson, J. F., 1981, Stromatolites at 3,500 Myr and a greenstone–granite unconformity in the Zimbabwean Archaean, *Nature* **291**:218–220.

Park, R., and Epstein, S., 1960, Carbon isotope fractionation during photosynthesis, *Geochim. Cosmochim. Acta* **21**:110–126.

Perry, E. C., Monster, J., and Reimer, T., 1971, Sulfur isotopes in Swaziland System barites and the evolution of the Earth's atmosphere, *Science* **171**:1015–1016.

Pflug, H. D., 1978, Yeast-like microfossils detected in the oldest sediments of the Earth, *Naturwissenschaften* **65**:611–615.

Pflug, H. D., 1987, Chemical fossils in early minerals, *Top. Curr. Chem.* **139**:1–55.

Pflug, H. D., and Jaeschke-Boyer, H., 1979, Combined structural and chemical analysis of 3,800-Myr-old microfossils, *Nature* **280**: 483–486.

Pflug, H. D., and Reitz, E., 1985, Earliest phytoplankton of eukaryotic affinity, *Naturwissenschaften* **72**:656–657.

Pflug, H. D., and Reitz, E., 1992, Palynostratigraphy in Phanerozoic and Precambrian metamorphic rocks, in: *Early Organic Evolution: Implications for Mineral and Energy Resources* (M. Schidlowski, S. Golubic, M. M. Kimberley, D. M. McKirdy,and P. A. Trudinger, eds.), Springer-Verlag, Berlin, pp. 509–518.

Rankama, K., 1948, New evidence of the origin of Pre-Cambrian carbon, *Geol. Soc. Am. Bull.* **59**:389–416.

Robbins, E. I., 1987, *Appelella ferrifera*, a possible new iron-coated microfossil in the Isua iron-formation, southwestern Greenland, in: *Precambrian Iron-Formations* (P. W. U. Appel and G. L. LaBerge, eds.), Theophrastus Publications, Athens, pp. 141–154.

Roedder, E., 1981, Are the 3,800-Myr-old Isua objects microfossils, limonite-stained fluid inclusions, or neither? *Nature* **293**: 459–462.

Ronov, A. B., 1958, Organic carbon in sedimentary rocks (in relation to the presence of petroleum), *Geochemistry* **1958**:510–536.

Ronov, A. B., 1980, *Osadochnaya Obolochka Zemli* (Earth's Sedimentary Shell), 20th Vernadsky Lecture, Izdatelstvo Nauka, Moscow [in Russian].

Schidlowski, M., 1978, Evolution of the Earth's atmosphere: Current state and exploratory concepts, in: *Origin of Life* (H. Noda, ed.), Center for Academic Publications, Japan, Tokyo, pp. 3–20.

Schidlowski, M., 1982, Content and isotopic composition of reduced carbon in sediments, in: *Mineral Deposits and the Evolution of the Biosphere* (H. D. Holland and M. Schidlowski, eds.), Springer-Verlag, Berlin, pp. 103–122.

Schidlowski, M., 1984, Early atmospheric oxygen levels: Constraints from Archean photoautotrophy, *J. Geol. Soc. London* **141**:243–250.

Schidlowski, M., 1987, Application of stable carbon isotopes to early biochemical evolution on Earth, *Annu. Rev. Earth Planet. Sci.* **15**:47–72.

Schidlowski, M., 1988, A 3,800-million-year isotopic record of life from carbon in sedimentary rocks, *Nature* **333**:313–318.

Schidlowski, M., 1989, Evolution of the sulphur cycle in the Precambrian, in: *Evolution of the Global Biogeochemical Sulphur Cycle*, SCOPE Volume 39 (P. Brimblecombe and A. J. Lein, eds.), Wiley, Chichester, England, pp. 3–19.

Schidlowski, M., Eichmann, R., and Junge, C. E., 1975, Precambrian sedimentary carbonates: Carbon and oxygen isotope geochemistry and implications for the terrestrial oxygen budget, *Precamb. Res.* **2**:1–69.

Schidlowski, M., Appel, P. W. U., Eichmann, R., and Junge, C. E., 1979, Carbon isotope geochemistry of the 3.7×10^9 yr old Isua sediments, West Greenland: Implications for the Archean carbon and oxygen cycles, *Geochim. Cosmochim. Acta* **43**: 189–199.

Schidlowski, M., Hayes, J. M., and Kaplan, I. R., 1983, Isotopic inferences of ancient biochemistries: Carbon, sulfur, hydrogen and nitrogen, in: *Earth's Earliest Biosphere: Its Origin and Evolution* (J. W. Schopf, ed.), Princeton University Press, Princeton, New Jersey, pp. 149–186.

Schoell, M., and Wellmer, F. W., 1981, Anomalous ^{13}C depletion in Early Precambrian graphites from Superior Province, Canada, *Nature* **290**:696–699.

Schopf, J. W., 1968, Microflora of the Bitter Springs Formation, Late Precambrian, Central Australia, *J. Paleontol.* **42**:651–688.

Schopf, J. W., and Oehler, D. Z., 1976, How old are the eukaryotes? *Science* **193**:47–49.

Schopf, J. W., and Packer, B. M., 1987, Early Archean (3.3-billion to 3.5-billion-year-old) microfossils from Warrawoona Group, Australia, *Science* **237**:70–73.

Schopf, J. W., and Walter, M. R., 1983, Archean microfossils: New evidence of ancient microbes, in: *Earth's Earliest Biosphere: Its Origin and Evolution* (J. W. Schopf, ed.), Princeton University Press, Princeton, New Jersey, pp. 214–239.

Stanley, S. M., 1981, *The New Evolutionary Timetable*, Basic Books Inc., New York.

Trask, P. D., and Patnode, H. W., 1942, *Source Beds of Petroleum*, American Association of Petroleum Geologists, Tulsa, Oklahoma.

Trudinger, P. A., 1979, The biological sulfur cycle, in: *Biogeochemical Cycling of Mineral-Forming Elements* (D.J. Swaine and P. A. Trudinger, eds.), Elsevier, Amsterdam, pp. 293–313.

Trüper, H. G., 1982, Microbial processes in the sulfur cycle through time, in: *Mineral Deposits and the Evolution of the Biosphere* (H. D. Holland and M.Schidlowski, eds.), Springer-Verlag, Berlin, pp. 5–30.

Tyler, S. A., and Barghoorn, E. S., 1954, Occurrence of structurally preserved plants in pre-Cambrian rocks of the Canadian shield, *Science* **119**:606–608.

Valley, J. W., and O'Neil, J. R., 1981, $^{13}C/^{12}C$ exchange between calcite and graphite: A possible thermometer in Grenville marbles, *Geochim. Cosmochim. Acta* **45**:411–419.

Veizer, J., Holser, W. T., and Wilgus, C. K., 1980, Correlation of $^{13}C/^{12}C$ and $^{34}S/^{32}S$ secular variations, *Geochim. Cosmochim. Acta* **44**:579–587.

Veizer, J., Compston, W., Hoefs, J., and Nielsen, H., 1982, Mantle buffering of the early oceans. *Naturwissenschaften* **69**:173–180.

Vinogradov, V. I., Reimer, T. O., Leites, A. M., and Smelov, S. B., 1976, The oldest sulfates in Archaean formations of the South African and Aldan Shields and the evolution of the Earth's oxygenic atmosphere, *Lithol. Miner. Resour. (USSR)* **11**:407–420.

Wada, H., and Suzuki, K., 1983, Carbon isotopic thermometry calibrated by dolomite–calcite solvus temperatures, *Geochim. Cosmochim. Acta* **47**:697–706.

Walker, J. C. G., Klein, C., Schidlowski, M., Schopf, J. W., Stevenson, D. J., and Walter, M. R., 1983, Environmental evolution of the Archean–Early Proterozoic Earth, in: *Earth's Earliest Biosphere: Its Origin and Evolution* (J. W. Schopf, ed.), Princeton University Press, Princeton, New Jersey, pp. 260–290.

Walsh, M. M., and Lowe, D. R., 1985, Filamentous microfossils from the 3,500-Myr-old Onverwacht Group, Barberton Mountain Land, South Africa, *Nature* **314**:530–532.

Walter, M. R., 1977, Interpreting stromatolites, *Am. Sci.* **65**:563–571.

Walter, M. R., 1983, Archean stromatolites: Evidence of the Earth's earliest benthos, in: *Earth's Earliest Biosphere: Its Origin and Evolution* (J. W. Schopf, ed.), Princeton University Press, Princeton, New Jersey, pp. 187–213.

Walter, M. R., Buick, R., and Dunlop, J. S. R., 1980, Stromatolites 3,400–3,500 Myr old from the North Pole area, Western Australia, *Nature* **284**:443–445.

Weber, F., Schidlowski, M., Arneth, J. D., and Gauthier-Lafaye, F., 1983, Carbon isotope geochemistry of the lower Proterozoic Francevillian Series of Gabon (Africa), *Terra Cognita* **3**:220.

Chapter 32

Organic Geochemistry of Precambrian Sedimentary Rocks

SCOTT W. IMBUS and DAVID M. McKIRDY

1. Introduction

Recent decades have witnessed a revolution in our understanding of the geological events and nature of life in Precambrian time. A period of Earth history once described simply as "the Azoic" (e.g., Dana, 1866) is now recognized as an interval of great complexity in terms of biological systems and their influence on the evolution of the hydrosphere and atmosphere. Characterization of Precambrian tectonic processes and climatic regimes, and their consequent sedimentological responses (e.g., transgressive–regressive sequences, glaciation, euxinic oceans), has demonstrated that, as in the Phanerozoic, biological radiations and extinctions probably reflect selective pressures exerted by such events. The evolution of new metabolic pathways and the advent of biogenic sedimentation transformed the character of sedimentary rocks. Accumulation of atmospheric oxygen, often attributed to increased levels of photoautotrophy, led to the deposition of iron formations and redbeds as well as enabling the evolution of later life forms (i.e., eukaryotic algae, metazoans, and metaphytes), with their profound sedimentological influences. Geochemical inquiry into the nature of the composition of organic materials in Precambrian sedimentary rocks continues to contribute to our understanding of these processes and events. Renewed interest in the economic resources of Precambrian basins, and the availability of advanced analytical techniques, has opened the Precambrian rock record to detailed organic geochemical investigations of the type traditionally reserved for Phanerozoic sediments.

McKirdy (1974) presented a comprehensive review of the earlier goals, strategies, problems, and accomplishments of organic geochemical studies of Precambrian sedimentary and metamorphic rocks. Since then, several papers reviewing various aspects of Precambrian organic geochemistry have appeared (e.g., McKirdy and Hahn, 1982; Hayes *et al.*, 1983; Summons *et al.*, 1988a; Summons and Walter, 1990; Summons, 1992; Strauss *et al.*, 1992a). The present chapter seeks to summarize and (where appropriate) update the information and insights provided by these and other important works.

2. The Precambrian: Geological and Paleontological Aspects

A broad consensus has emerged concerning the timing and significance of major landmarks in the physical, chemical, and biological evolution of the Earth. The evolution of tectonic and sedimentological

SCOTT W. IMBUS • School of Geology and Geophysics, The University of Oklahoma, Norman, Oklahoma 73019. DAVID M. McKIRDY • Department of Geology and Geophysics, The University of Adelaide, Adelaide, South Australia 5001, Australia. *Present address for Scott W. Imbus:* E & P Technology Department, Texaco, Inc., Houston, Texas 77042.

Organic Geochemistry, edited by Michael H. Engel and Stephen A. Macko. Plenum Press, New York, 1993.

processes has been examined in a volume edited by Kröner (1981) and by Windley (1983, 1984). Aspects of the composition of the Precambrian oceans and atmosphere, and in particular their interaction with biological systems, have been reviewed by Holland (1984) and in volumes edited by Trudinger *et al.* (1980), Holland and Schidlowski (1982), Schopf (1983), Holland and Trendall (1983), Sundquist and Broecker (1985), Campbell and Day (1987), and Gregor *et al.* (1988). In lieu of a detailed treatment of the aforementioned papers, Fig. 1 is presented as a summary of their pertinent findings.

The principle of uniformitarianism, or "the present is the key to the past," is a basic tenet of geology that invokes present-day processes to interpret the rock record. Because the processes affecting the Earth's lithosphere, hydrosphere, atmosphere, and biosphere have changed with time, this principle is of decreasing utility as increasingly ancient rocks are examined. In an elegant treatment of this topic, Veizer (1988) applied the concept of biological population dynamics to describe the cyclic nature of terrestrial exogenic systems. These interdependent systems or "populations" (with rates of recycling increasing in

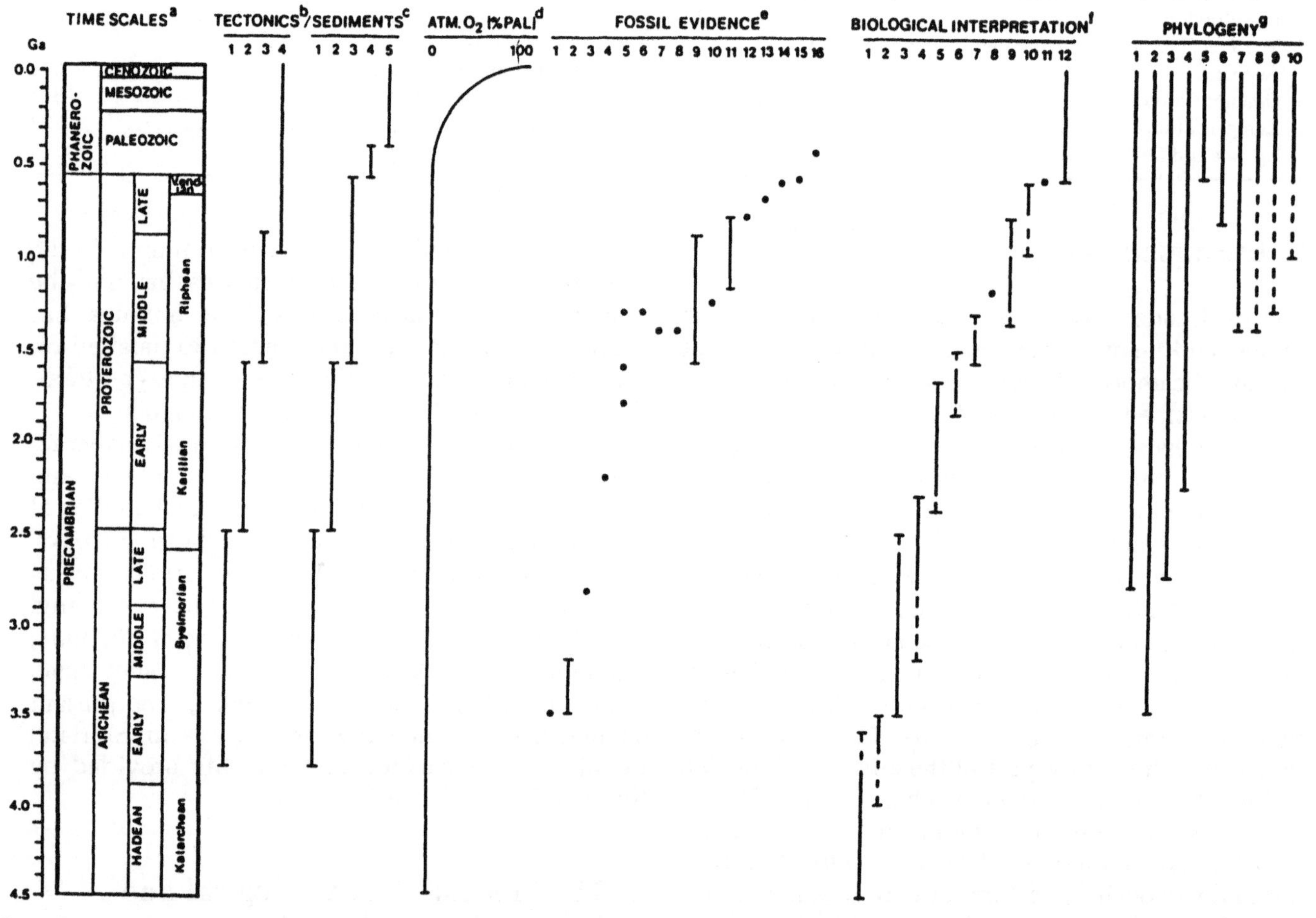

[a]TIMESCALES (in part after Salop, 1983; Plumb and James, 1986).

[b]TECTONICS (after Windley, 1984). 1. <u>Permobile tectonics</u>: granite-greenstone/granulite-gneiss associations. 2. <u>Rigid continental lithosphere</u>: aulacogens, rift systems with alkaline complexes and extensive, thick sedimentary basins; first orogenic belts developed along stable continental margins. 3. <u>Widespread fragmentation attempts</u>. 4. <u>New continental fragmentation, drift and collision</u>.

[c]SEDIMENTATION (after Veizer, 1988). 1. <u>Granite-greenstone belts</u>: characterized by immature assemblages of graywackes, paraconglomerates, volcanoclastic cycles capped by chemical sediments. 2. <u>Rapid growth of sedimentary mass</u>: mature miogeoclinal platform sedimentary assemblages; mature and fractionated clastics and (bio)chemical sediments. 3. <u>Detrital pyrite and uraninite in fluvial-deltaic deposits, increase in redbeds and redbed copper and Clinton type ores</u>. 4. <u>Increased role of (bio)chemical sediments</u>: important formation of shallow marine limestones/dolomites/phosphates. 5. <u>Shift in organic deposition to terrestrial environments (Paleozoic), pelagic $CaCO_3$ sedimentation shifts carbonate deposition to deeper marine environments (Mesozoic), increase in unstable tectonic settings (Phanerozoic in general)</u>.

[d]LEVELS OF ATMOSPHERIC OXYGEN (after Cloud, 1983; 1987). Expressed in percent of present atmospheric level.

[e]FOSSILS - DESCRIPTION (after Schopf et al., 1983). 1. Oldest stromatolites, filamentous microbial fossils. 2. Spheroidal carbonaceous dubiomicrofossils. 3. "Cyanobacterium-like" stromatolitic fabric and fossil filaments. 4. Oldest abundant microfossils. 5. Oldest possible millimetric (Chuaria-like) fossils. 6. Oldest established millimetric (Chuaria-like) fossils. 7. Size increase of coccoid microfossils. 8. Oldest established thick-walled acritarchs. 9. Increase in diversity of microfossils. 10. Diverse cyanobacterial filaments and coccoids. 11. Oldest body fossils of the higher algae. 12. Oldest acritarchs with pylomes and/or spines. 13. Oldest traces of invertebrate animals. 14. Oldest invertebrates with exoskeletons.

[f]BIOLOGIC INTERPRETATION (after Schopf et al., 1983; Hayes, 1983). 1. Abiotic chemical evolution; origin of anaerobic life. 2. Diversification of early anaerobes; origin of anaerobic photosynthetic bacteria. 3. Continued diversification (e.g., of methanogenic and sulphate reducing bacteria). 4. Origin/dispersal of protocyanobacteria; first widespread amphiaerobes; methylotrophs-locally abundant; extinction of some anaerobes. 5. Diversification of amphiaerobes; increase of prokaryote plankton. 6. Diversification and dispersal of obligate anaerobes; increased importance of plankton. 7. Origin of unicellular mitotic eukaryotes. 8. Cyanobacteria diverse. 9. Diversification of microscopic eukaryotes; origin of meiosis/sexuality. 10. Origin and diversification of metazoans and metaphytes. 11. Decrease in abundance of stromatolite biocoenesis. 12. Diversification of large eukaryotes.

[g]METABOLIC PATHWAYS (after Walter, 1987). <u>Archaebacteria</u> 1. Methanogens. <u>Eubacteria</u> 2. Anaerobic photoautrophs. 3. Cyanobacteria. 4. Sulphate reducers. 5. Actinomycetes. 6. <u>Protista</u>. 7. <u>Plantae</u> 8. Chlorophyta. 9. Phaeophyta. 10. <u>Animalia</u>.

Figure 1. Synopsis of documented geological and paleontological findings from Precambrian sedimentary rocks and their possible paleobiogeochemical significance.

the order geotectonic realms, oceans, atmospheric constituents, living systems) form a nested hierarchy wherein the larger slow-cycling "populations" generally establish the limits and existential basis of smaller, more rapidly cycling "populations." Two evolutionary phenomena, superimposed on this recycling process, reflect the process of self-organization of the Earth: (1) exponential decrease in heat flux from the interior of the Earth during the course of terrestrial differentiation and (2) the evolution of life. Based on this hypothesis, the first phenomenon can be seen as influencing tectonic style (and consequently sedimentary processes) and affecting, in the early phase of Earth history, the chemistry of the atmosphere and of the hydrosphere and its precipitates (Cloud, 1976). Important consequences of the second phenomenon include the deposition of biogenic sediments and the indirect, but profound, influence of this evolution on exogenic cycles through photosynthetically produced oxygen.

The Precambrian paleontological record may be examined from two perspectives: (1) association with lithology and depositional environment (facies) and (2) taxonomically, with respect to the origin of organisms and associations of organisms forming ecosystems (Hofmann, 1987; Knoll, 1989; Knoll *et al.*, 1989). Awramik (1982) recognized three main categories of Precambrian fossils: stromatolites, microfossils, and algal megafossils. Microfossils, commonly occurring as organic-walled microstructures of varying size, shape, and organization, may be stromatolitic or nonstromatolitic. Stromatolites and associated microfossils typically occur in carbonate–chert lithofacies, whereas nonstromatolitic microfossils (e.g., acritarchs, cryptarchs) and algal megafossils (carbonaceous disks and ribbons) are found in fine-grained siliciclastics (shale, siltstone). Walter (1987) examined the geological and geochemical evidence for the first appearance of the Eubacteria, Archaebacteria, eukaryotes, fungi, metaphytes, and metazoans (Fig. 1).

The oldest documented stromatolites are from the 3.3–3.5-Ga Warrawoona Group of northwestern Australia (Walter *et al.*, 1980; Lowe, 1983; Schopf and Packer, 1987). However, it is not until the Early Proterozoic (~2.3 Ga), when extensive shallow marine shelf areas developed, that the stromatolite record becomes conspicuous (Schopf, 1975). Stromatolitic microfossils are usually preserved by silicification and appear to be morphologically analogous to cyanobacteria and bacteria (Awramik, 1982). Nonstromatolitic, or shale-facies microbiota, are common in Proterozoic sedimentary rocks and comprise both benthonic and planktonic forms. The latter include acritarchs (solitary spheroidal to polyhedral vesicles) and cryptarchs (vesicles in clusters, colonies, or filaments; Diver and Peat, 1979) of probable eukaryotic algal affinity and thick-walled, vase-shaped microfossils (melanocyrillids), considered to be encystment stages of algae (Bloeser, 1985), but which also resemble chitinozoans (Hofmann, 1987) or tintinnid protozoans (Vidal and Knoll, 1983) and therefore are possible early protists.

Carbonaceous films occur as macroscopic ribbons (Walter *et al.*, 1976; Horodyski and Bloeser, 1978; Horodyski, 1980) and as spheroids (e.g., *Chuaria*; Hofmann and Aitken, 1979) in Proterozoic sediments up to 1.8 Ga (Hofmann, 1985, 1987).* *Chuaria* is commonly associated with a sausage-shaped form (*Tawuia*) in Late Proterozoic rocks (0.7–0.9 Ga). The biostratigraphic potential of the *Chuaria–Tawuia* assemblage and other carbonaceous fossils is limited by their sporadic occurrence. These problematic carbonaceous films may represent the fossil remains of both prokaryotic (cyanobacterial) and eukaryotic (algal and ?metazoan) biota (Hofmann, 1987). Hence, together with acritarchs and cryptarchs, they are of great interest for documenting the origin of the eukaryotic cell. Other examples of possible and probable eukaryotic microfossils in the Proterozoic record include so-called "spot cells" and tetrahedral tetrads (Knoll, 1983).

Acritarchs are potentially useful for the stratigraphic correlation of Late Proterozoic and early Cambrian clastic sequences (e.g., Vidal and Knoll, 1982, 1983; Knoll and Swett, 1987). In addition, acritarchs allow a distinction between inshore and moderately shallow-to-offshore marine depositional settings on the basis of simple structures of low diversity and highly diverse, more complex structures, respectively (Vidal and Knoll, 1983; Butterfield *et al.*, 1988; Zang and Walter, 1989; Jenkins *et al.*, 1992). The prospect for similar biostratigraphic correlations and environmental determinations in pre-Late Proterozoic clastic sequences is at present not promising. Early acritarchs occur sporadically and do not possess the distinctive ornamentation necessary for correlation purposes (e.g., Peat *et al.*, 1978).

Recent studies on extant organisms (e.g., biochemical pathways, ultrastructural studies, chemical analysis of cell wall components, nucleic acid and protein sequencing; Walter, 1987) reveal that genetic distinctions among prokaryotes can now be made. Hence, the traditional classification of life, which

*Recently, Han and Runnegar (1992) reported "megascopic eukaryotic algae" from the 2.1-Ga Negaunee Iron Formation of Michigan.

recognized five kingdoms (Monera, Protista, Fungi, Plantae, and Animalia), has been superseded by *inter alia* the Archaebacteria–Eubacteria ("true bacteria")–Eukaryote scheme of Fox *et al.* (1980) and the Parkaryote (Eubacteria, Methanogens, Halobacteria)–Karyote (Eocytes, Eukaryotes) scheme of Lake (1988). Although the Precambrian paleontological record is extensively documented (e.g., Walter, 1976, 1987; Schopf, 1983; Hofmann, 1985, 1987; Knoll, 1989), the taxonomic affinities of fossil biota are typically decided on the basis of rather generalized criteria such as size or morphology. The distinction between fossil prokaryotes and eukaryotes is tenuous until the Late Proterozoic, and recognition of different fossil prokaryote taxa is difficult under any circumstances. Distinctions between groups of fossil organisms ideally require a combined paleontologic–organic geochemical approach (Glaessner and Foster, 1992). Such investigations of Phanerozoic samples have met with considerable success (e.g., Comet and Eglinton, 1987; Hoffmann *et al.*, 1987), but analogous Precambrian studies remain highly speculative (e.g., Hayes, 1983; Schopf *et al.*, 1983; Summons *et al.*, 1988a, b; Summons and Walter, 1990).

It is also important to mention the role of stable isotope geochemistry in bridging the gap between Precambrian paleontologic findings and the results of biochemical studies on extant organisms. Other authors in this volume consider in detail the mechanisms and attendant isotopic fractionation involved in the incorporation of biologically significant elements into living organisms (Summons, Chapter 1; Fogel and Cifuentes, Chapter 3) and into sediments (de Leeuw and Largeau, Chapter 2). Schidlowski (Chapter 31) shows how secular trends in the isotopic composition of carbon and sulfur in Precambrian sediments provide evidence for the development of, first, photoautotrophy (at 3.5 Ga or earlier) and, subsequently, dissimilatory sulfate reduction (at ~2.3 Ga).

3. Organic Carbon

Processes that influenced the accumulation and preservation of organic matter [conveniently measured as total organic carbon (TOC)] in ancient sediments are known to have varied in intensity, at least through Phanerozoic time. Higher-than-average organic preservation in Phanerozoic marine basins has been variously ascribed to high sedimentation rates in the Holocene (Bralower and Thierstein, 1987); climate, transgressive and regressive events, and continental drift in the Cretaceous (de Graciansky *et al.*, 1987); marine transgressions in the Mesozoic (Hallam, 1987); and climate, high stands of sea level, paleogeography, and, to a lesser extent, primary productivity and sedimentation rate in the early Paleozoic (Thickpenny and Leggett, 1987). The reasons for the variability of organic carbon preservation in Precambrian sequences are poorly known and may be difficult to determine, given the substantial portion of the Archean and Early Proterozoic rock record that is missing, and the possibility that many representative environments may have been preferentially destroyed (Walter, 1987). Fundamental parameters such as average TOC in the sediments deposited during successive periods of the Precambrian, and the corresponding primary productivities, can only be estimated indirectly using the geochemical methods outlined below.

Studies of secular trends in the stable isotopic composition of reduced carbon (C_{org}) and coeval carbonate carbon (C_{carb}), and isotopic mass-balance considerations, reveal that C_{org} probably has accounted for approximately 20% of the total carbon in the exogenic exchange reservoir since the deposition of the Isua Supracrustal sequence ~3.8 Ga (Schidlowski *et al.*, 1979; Schidlowski, 1982, 1987, 1988). This relationship supports the concept of "biogenic plenitude" which is based on the intrinsic property of life to proliferate to the ultimate limits set by nutrient supply (Schidlowski, 1987, 1988). As land surfaces were virtually barren of life during the entire Precambrian, critical nutrients such as nitrogen and phosphorus were not retained by terrestrial biota. Accordingly, the degree of eutrophication of the Precambrian seas was probably greater (by a factor of at least 2; Schidlowski, 1988) than that of the present ocean. Huc (1980) has estimated that the present relative primary productivity of the land and the oceans is at least 1:1 and may be as high as 1:6. This finding lends strong support to the hypothesis that primary productivity in the Precambrian was at least as high as it is today. However, high primary productivity does not necessarily translate into enhanced organic preservation (cf. Demaison and Moore, 1980).

Des Marais *et al.* (1992) recently challenged the concept of C_{org}:C_{carb} (20:80) stasis over time by proposing that stable carbon isotope data, corrected for metamorphic overprint, indicate an increase of the C_{org} relative to the C_{carb} reservoir over the Proterozoic (2.5–0.54 Ga). Concomitant with the increase in the C_{org} reservoir, was an increase in oxidizing power released to the environment.

In view of the important role of organic carbon in the Earth's exogenic carbon exchange reservoir through time, it is of interest to examine actual TOC levels in the sedimentary record. Table I is a compilation of TOC determinations on selected suites of Precambrian and Phanerozoic sedimentary rocks. Although the actual number of Precambrian TOC measurements probably exceeds 10,000, only a few studies have been designed to systematically evaluate organic carbon levels through contiguous vertical sequences, or on a regional basis.

In the classic TOC survey of the Russian Platform by Ronov and Migdisov (1971), average weighted organic carbon levels in Paleozoic sediments were found to be nearly twice that of Proterozoic metasediments (Table I). This finding does not necessarily imply a significant difference in original organic carbon levels between these rock sequences because the Proterozoic metasediments are likely to have experienced significant carbon loss. Based on isotopic evidence, Hayes *et al.* (1983) speculated that low measured TOC values may represent metagenetic carbon loss rather than low initial organic carbon contents. Given the long periods of time during which many Precambrian rocks were exposed to tectonic deformation and erosion, it is also possible that, in certain circumstances, less resistant (but more organic-rich) units such as shales were preferentially destroyed.

A mean weighted TOC value of 0.43% was determined by Reimer *et al.* (1979) for Early Archean

TABLE I. Summary of Selected Total Organic Carbon (TOC) Studies on Precambrian Sediments/Metasediments and Phanerozoic Sediments

Rock unit or sequence	Age (Ga)	Rock type	TOC n	TOC Mean	TOC Range	Reference
Canadian Archean	>2.4	Shales	14[a]	1.04[a]	0.1–7.6[a]	Cameron and Jonasson, 1972
Canadian Aphebian	2.4–1.6	Shales	9[b]	1.61[b]	0.1–5.1[b]	Cameron and Jonasson, 1972
Russian Proterozoic 1–2	(2.5–1.6)	Metasediments[c]	1229	0.25[c]	—	Ronov and Migdisov, 1971
Russian Proterozoic 3	(1.6–0.6)	All sediments	71	0.18[d]	—	Ronov and Migdisov, 1971
Russian Paleozoic	(0.6–0.2)	All sediments	1575	0.34[d]	—	Ronov and Migdisov, 1971
Russian Mesozoic–Cenozoic	(0.2–0.0)	All sediments	801	0.64[d]	—	Ronov and Migdisov, 1971
Swaziland Sequence	3.4	All sediments	10	0.43[d]	0.1–2.1	Reimer *et al.*, 1979
PPRG survey (worldwide)						Hayes *et al.*, 1983
Archean	3.8–2.5	Carbonates	26	0.07	0.0–0.2	
		Banded iron formations	9	0.48	0.0–2.5	
		Cherts	60	0.24	0.0–1.9	
		Shales/sandstones	46	2.55	0.0–16.5	
		Schist/gneiss	4	0.00	0.0	
Proterozoic	2.5–0.8	Carbonates	61	0.07	0.0–0.9	
		Banded iron formations	3	0.01	0.0	
		Cherts	17	0.05	0.0–0.1	
		Shales/sandstone	31	0.44	0.0–2.7	
		Schist/gneiss	—	—	—	
Prospective oil and gas basins						
McArthur Basin	1.4–1.7	Mostly clastics				Jackson *et al.*, 1988
McMinn Formation			—	—	0.7–2.9	
Velkerri/Lansen Creek Shale			—	—	0.9–7.2	
Yalco Formation			—	—	0.8–5.4	
Lynott Caranbirini Member			—	—	0.2–3.4	
Barney Creek Formation			—	—	0.2–10.4	
Amadeus Basin			—	—	0.1–1.0	Jackson *et al.*, 1984
Southeastern Arabian Platform	0.6–0.7	Clastics/carbonates	4	3.98	2.2–6.2	Grantham *et al.*, 1988
Mid-Continent Rift System						
Nonesuch Formation	1.1	Euxinic silts and shales	183	0.38	0.0–3.0	Imbus *et al.*, 1988
Solor Church Formation	1.1	Mudstones, siltstones, fine sandstones	25	0.24	0.0–1.7	Hatch and Morey, 1985

[a]Values listed were compiled by site (n = 14), for which individual formations represented were not listed. A total of 406 analyses were performed.
[b]Values listed were compiled by formation (n = 9). A total of 396 analyses were performed.
[c]Includes sandstones and quartzites; paragneiss, phyllites, and schists; carbonates; and ferruginous rocks. Mean value is a weighted average taking into account the proportion of each (meta)sediment type.
[d]Mean values are weighted averages (excluding tuffs and effusives).

sediments in the Swaziland Sequence of South Africa. These authors suggested that, even if a low rate of sedimentation was assumed for all sediments in the sequence, the rate of production of organic carbon was still at least one-fourth the present rate. The mean weighted TOC values of shale (0.56%), graywacke (0.76%), and chert (0.14%) in the Sheba Formation (part of the 3.4-Ga Fig Tree Group of the Swaziland Sequence) were interpreted to indicate an organic production rate of 0.32 g m^{-2} yr^{-1}, which is of the same order of magnitude as organic productivity in several modern aquatic settings. This study indicates that, although organic-rich sediments may not have been as widespread in the Archean as in the Phanerozoic (e.g., because of the smaller areal extent of stable marine shelves), significant organic production has persisted at relatively high levels since the Early Archean.

In a more comprehensive survey by the Precambrian Paleobiology Research Group (PPRG) of UCLA, Hayes *et al.* (1983) presented the results of analyses of TOC and stable isotope composition of kerogen carbon ($\delta^{13}C_{ker}$) of approximately 315 Archean and Proterozoic rocks. Considerable variation in organic carbon levels (TOC) was reported: <0.5%, 210 samples; 0.5–1%, 55 samples; 1–5%, 43 samples; 5–10%, 6 samples; >10%, 1 sample. Although the TOC values for the Archean samples were surprisingly high [compared to the data of Reimer *et al.* (1979), Cameron and Garrells (1980), and Ronov (1980)], it was acknowledged that these samples were not collected with the purpose of representing all the rock types in a given unit and that there was probably a bias toward lithologies rich in organic carbon. Middle and Late Proterozoic carbon levels equivalent to those found in some analogous Phanerozoic settings have been reported by several workers (e.g., McKirdy, 1977; Jackson *et al.*, 1984, 1988; Imbus *et al.*, 1988; Table I) and recently confirmed by a second PPRG survey (Strauss *et al.*, 1992b).

4. Kerogen Studies

Biological, depositional, and maturity factors influencing the chemical composition of kerogens have been reviewed by Hunt (1979) and Tissot and Welte (1984). "Bulk" techniques such as elemental analysis (McIver, 1967), Rock-Eval pyrolysis (Espitalié *et al.*, 1977), and optical microscopy (Alpern, 1980; Hutton *et al.*, 1980; Stach *et al.*, 1982; Crick *et al.*, 1988) have led to the fundamental concept of kerogen type. In the kerogen typing scheme first proposed by Tissot *et al.* (1974), the vast majority of Phanerozoic kerogens analyzed were found to plot into three distinct regions on a van Krevelen diagram (i.e., atomic H/C versus O/C). On this basis, kerogens were designated as types I, II, or III. A fourth variety of kerogen (type IV) is now widely recognized (e.g., Harwood, 1977; McKirdy *et al.*, 1980; Horsfield, 1984). Kerogens falling into each of these categories have predictable "coalification tracks" (i.e., changes of elemental composition with rank, particularly H/C atomic ratios). Jones (1987) has proposed an "organic facies" classification along the same lines, but in greater detail. In Table II the various types and chemical characteristics of microbial kerogens recognized and their relationship to petrographically classified coal macerals and maceral groups are summarized. The selected examples of Proterozoic and Cambrian kerogens attest to the chemical variability of the kerogens found in pre-Devonian sediments.

Another "bulk" analytical technique applicable to kerogen analysis is stable isotope mass spectrometry (especially of carbon and hydrogen; Hoering, 1977; Schidlowski *et al.*, 1983). In general, kerogen carbon isotopic values ($\delta^{13}C_{ker}$) are of limited use in distinguishing between specific biological source materials: most values fall in the range -26 ± 7‰ for both the Precambrian and Phanerozoic records, consistent with ubiquitous biological carbon fixation via photosynthesis (Schidlowski, 1988). Notable excursions to anomalously negative values (-40 to -53‰) have been recorded in various late Archean (2.6–2.8 Ga) and early Proterozoic (~2.1 Ga) carbonaceous sediments (Schoell and Wellmer, 1981; Hayes *et al.*, 1983; Strauss, 1986; Cortial *et al.*, 1990), where they implicate methanogenic and methylotrophic bacteria as kerogen precursors. Accurate interpretation of the isotopic composition of Precambrian kerogens must also take into account their degree of preservation. Thermal alteration and hydrogenation beyond a threshold atomic H/C ratio of 0.20–0.15 is accompanied by an increase in $\delta^{13}C_{ker}$ of 10–12‰ (McKirdy and Powell, 1974; McKirdy *et al.*, 1975; Hayes *et al.*, 1983; Strauss *et al.*, 1992a). Stable isotopic analysis of kerogen pyrolysates (Burwood *et al.*, 1988) is another potentially useful technique in Precambrian studies, although few such studies have yet been undertaken.

The high-resolution techniques of combined pyrolysis–gas chromatography (Py–GC)) and pyrolysis–gas chromatography/mass spectrometry (Py–GC/MS) are commonly employed to characterize Phanerozoic kerogens (e.g., van de Meent *et al.*, 1980; Larter, 1984). These techniques and various chemical degradation methods (e.g., ozonolysis, KOH fusion, and HI reduction) have also been used to analyze Precambrian kerogens. Hayes *et al.* (1983) compiled the results of many such studies up to 1980 but concluded that little had been learned from this work. Many samples analyzed were from highly mature or metamorphosed sedimentary sequences, and the lack of reported yields precluded an assessment of how representative the compounds detected are of the overall kerogen composition. Nevertheless, several interesting compounds (chemical fossils) indicative of Archean and Early Proterozoic biochemistries were detected in these pioneering studies. For example, furan derivatives, possible remnants of cyanobacterial carbohydrates, were detected in the ozonolysis and pyrolysis products of kerogen from the 2.2–Ga Transvaal stromatolitic carbonates (Zumberge and Nagy, 1975; Nagy, 1976). In the case of kerogen from the 2.7-Ga Rupemba–Belingwe stromatolites, C_9–C_{20} *n*-alkanes detected by Sklarew and Nagy (1979) were interpreted as being derived from fatty acids.

For the sake of simplicity and relevance to recent and continuing studies, the ensuing discussion will deal mainly with the results of kerogen characterization by elemental analysis (or Rock-Eval pyrolysis) and high-resolution analyses by Py–GC and Py–GC/MS. It is important to be aware that, with high-resolution pyrolysis techniques, abundant unsaturated hydrocarbons such as alkenes are generated. These compounds largely represent pyrolytic remnants of their saturated counterparts in the kerogen macromolecular structure. This problem can be obviated by the innovations of hydrous pyrolysis (Hy/Py) with subsequent GC or GC/MS analysis (Lewan *et al.*, 1979) and by pyrolysis–hydrogenation–GC (Py/H/GC; McHugh *et al.*, 1976).

One of the earliest "high-resolution" pyrolysis studies of a substantial suite of organic-rich Precambrian rocks, and unfortunately the most discouraging, was that of Leventhal *et al.* (1975). In this study, samples of 30 rocks (ranging in age from 0.7 to 38 Ga) were analyzed using a "stepwise" Py–GC technique. Only molecules with fewer than eight carbon atoms (sensitivity level of 10^{-9} g organics/10 mg rock sample) were detected. It was concluded that little useful biochemical information could be expected from pre-Phanerozoic kerogens, except those rare samples that display low CH_4/C_2H_4 or CH_4/C_2H_6 ratios in their pyrolysate and hence have H/C atomic ratios greater than 1. Even with more sensitive techniques, many of the samples analyzed in this study may indeed yield no "useful biochemical information." However, because the kerogens analyzed by Leventhal and coworkers were not isolated from their rock matrices prior to pyrolysis and multiple pyrolysis steps were employed, it is clear that sensitivity levels were insufficient to detect compounds other than low-molecular-weight hydrocarbons.

Work done on Precambrian and Early Paleozoic "microbial" (i.e., algal-or bacterial-derived) kerogens by McKirdy (1976, 1977) and McKirdy *et al.* (1980) was reviewed by McKirdy and Hahn (1982) and is summarized here in Table II. The approach taken in these studies involved the correlation of Py/H/GC and stable carbon isotopic data with kerogen type as determined by elemental composition (H/C and O/C atomic ratios) and, where possible, petrographic identification of macerals. The kerogens are surprisingly diverse and include those with chemical compositions characteristic of type III and type IV organic matter in younger rocks. Type III kerogens, for example, are commonly thought to originate from terrestrial higher plant debris and therefore to occur only in post-Silurian sediments. For microbial type III kerogens occurring in low concentrations in many Precambrian and Early Paleozoic sedimentary rocks, a primary or secondary origin may be proposed (Table II). Primary type III microbial kerogen may represent geochemically gelified algal humic acids (incorporating some bacterial lipids) in clastic setting or may originate from cyanobacterial mucilage and algal and bacterial pectic tissue (cell walls) in stromatolitic carbonates and cherts. Secondary type III kerogen may form in a manner analogous to the genesis of micrinite from bituminite in bituminous coals (i.e., hydrogen disproportionation during oil generation; cf. Teichmüller, 1974). Examples of such kerogen have been documented in Late Proterozoic sediments of the Amadeus Basin, central Australia (McKirdy, 1977; Jackson *et al.*, 1984), and in the Adelaide Geosyncline (Belperio, 1987). In the case of type IV microbial kerogens, which are typically associated with columnar-branching stromatolites, partial oxidation during early diagenesis may have been a contributing factor to their high O/C ratios. The distribution of aliphatic moieties in the pyrolysates of both type III and type IV microbial kerogens is suggestive of a bacterial contribution (McKirdy *et al.*, 1980).

Kerogens from eight Precambrian units (with ages in the range 1.0–3.1 Ga) were analyzed by Philp and van de Meent (1983) using Py–GC and Py–GC/MS. Two groups of kerogens were established on the basis of pyrolysate composition. Pyrograms of the group 1 kerogens (Soudan Iron Formation, northeastern Minnesota; Onverwacht and Daspoort groups, South Africa; Kalgoorlie Shale, Western Australia) are dominated by normal alkanes, alkenes, and alkadienes in

TABLE II. Origin(s), Chemical Characteristics, Typical Associated Depositional Environments, and Cambrian and Precambrian Examples of Principal Types of Microbial Kerogens[a]

Kerogen type	Maceral group	Equivalent individual macerals	Source	Chemical characterization[b]	Depositional environment	Precambrian and Cambrian examples (sediment type)
I	Liptinite	Alginite A (Telalginite) Alginite B (Lamalginite)	Lipid-rich chlorophytes and cyanophytes (e.g., *Tasmanites*, *Gloeocapsomorpha*) Bloom and mat-forming cyanophytes; acritarchs; archaebacteria	Hydrogen-rich (alginite coalification track), *n*-alkanes up to at least C_{25} are dominant in pyrolysate; simple aromatic hydrocarbons (benzene, toluene, ethylbenzene, and the xylenes) occur in very low concentrations relative to aliphatic moieties of equivalent carbon numbers.	Reported from Ordovician and older rocks for which anoxic lacustrine or marine environments may be inferred	Beetle Creek Fm., Georgina Basin, Queensland (Cambrian, phosphorite restricted marine shelf) Observatory Hill Fm., Officer Basin, South Australia (Cambrian, dolomitic argillaceous mudstone; alkaline playa lake, evaporitic) Barney Creek Fm., McArthur Basin, Northern Territory (Proterozoic, dolomitc shale; lacustrine)
II	Liptinite	Bituminite Liptodetrinite	Partially degraded (i.e., sapropelized) algae and zooplankton; bacterial lipids Acritarchs	Intermediate coalification track, complex chemical composition; pyrolysates differ little in aromaticity from type I kerogen but are considerably more naphthenic; *n*-alkyl chains prominent but shorter and present in lower concentration relative to branched and cyclic alkyl moieties.	Various anoxic to suboxic environments, including lacustrine and marine sabkha to outer shelf	Ouldburra Fm., Officer Basin, South Australia (Cambrian, micritic carbonate; shallow marine to evaporitic) Nonesuch Fm., Keweenawan Trough, Michigan (Proterozoic shale; lacustrine) McMinn Fm., McArthur Basin, Northern Territory (Proterozoic, shale and siltstone; outer marine shelf) Velkerri Fm., McArthur Basin, Northern Territory (Proterozoic, pyritic mudstone and siltstone; ?barred marine basin)

III[c]	Vitrinite	Desmocollinite	Geochemically gelified algal humic acids incorporating some bacterial lipids. In the case of stromatolites, most probably derived from cyanobacterial mucilage and from algal and bacterial pectic tissue (cell walls).	Oxygen-rich, hydrogen-poor; pyrolysate commonly displays an odd-carbon-number predominance in the C_{10+} range which may be a biological marker for bacterial cell wall lipids; simple aromatic hydrocarbons prominent relative to short-chain *n*-alkyl moieties.	Suboxic, shallow to deep marine	Pertatataka Fm., Amadeus Basin, Northern Territory (Proterozoic, shale and siltstone; outer marine shelf) Bitter Springs Fm., Amadeus Basin, Northern Territory (Proterozoic, dolomite, in part anhydritic; shallow marine to evaporitic)
	(Inertinite)[d]	Rank micrinite	Coalification residue of bituminite			
IV	Inertinite	Degradofusinite	Partially oxidized algal cellulose and sheath mucilage; bacterial pectic tissue	As above but with lower pyrolysate yield	Oxic, shallow marine. Typically found in very low concentrations in certain late Proterozoic and early Paleozoic stromatolitic carbonates and cherts for which an inter- to supratidal environment may be inferred	Wilkawillina Limestone, Arrowie Basin, South Australia (Cambrian, stromatolitic limestone) Wonoka Fm., Adelaide Geosyncline, South Australia (Proterozoic, stromatolitic limestone)

[a]Compiled from McKirdy (1976), McKirdy *et al.* (1980, 1984), McKirdy and Hahn (1982), Jackson *et al.* (1984), and Crick *et al.* (1988).
[b]Pyrolysis–hydrogenation–GC, pyrolysis–GC, Rock-Eval pyrolysis, elemental analysis.
[c]May include partially oxidized liptodetrinite.
[d]Although micrinite has traditionally been classified as a member of the inertinite group, it "is neither inert nor especially carbon-rich, since it contains relatively large amounts of hydrogen and volatile matter" (Teichmüller, 1975; cf. McKirdy *et al.*, 1980).

the C_8–C_{35} range. In contrast, those of the group 2 kerogens (Fig Tree Group, South Africa; Nonesuch Formation, Michigan; Moodies Group, South Africa; Soudan Iron Formation graphitic lens, northeastern Minnesota) have as dominant components methyl-branched alkenes occurring at every third carbon number between C_9 and C_{33}. Among these branched alkenes, which were tentatively attributed to isoprenoid precursors, C_9H_{18} and $C_{15}H_{30}$ are the most and second most abundant, respectively, and there is a slight but significant prevalence of odd-carbon-numbered homologs. Additionally, a complex mixture of branched hydrocarbons at every carbon number, and substituted aromatics at lower carbon numbers, was observed in the group 2 pyrograms. The ratio of aromatic to aliphatic compounds was higher in the group 2 pyrolysates than in those of group 1. It was suggested that this difference may reflect the presence of "a 'loose,' easily thermodegradable kerogen structure" in the case of the group 2 samples, rather than a condensed one (i.e., higher "aromaticity" or rank). High-rank kerogens comprise highly condensed aromatic structures, but these will yield only small amounts of aromatic hydrocarbons upon pyrolysis. In the absence of information on pyrolysate yields and independent data on the degree of thermal alteration of the kerogens analyzed (e.g., atomic H/C ratios were not cited), the significance of the two kerogen groups remains unclear.

A search for "molecular fossils," including branched and isoprenoid alkanes, steranes, and triterpanes, as undertaken by Hoering and Navale (1987) using GC/MS to analyze extracts and hydrous pyrolysates obtained from 16 Precambrian and Cambrian shales and kerogen concentrates. Of the 12 Precambrian samples analyzed, *n*-alkanes were detected in pyrolysates of the Nonesuch Shale (1.1 Ga, Michigan), the Muhos Shale (1.3 Ga, Finland), and the McArthur Shale (~1.7 Ga, Northern Territory, Australia). Monomethyl-branched alkanes, including 2-methyl- and 3-methylalkanes, were not found in any of the Precambrian kerogens. There were questionable occurrences of isoprenoid and triterpenoid hydrocarbons in the Bethel Shale (0.76 Ga, South Australia) and of triterpenoids in the Cardup Shale (0.7 Ga, Western Australia). Likewise, trace amounts of possible steranes were detected in an extract of the Tindelpina Shale (0.75 Ga, South Australia). Steranes and triterpanes were positively identified only in a gilsonite from the 0.5-Ga Nama Group, Namibia. Dehydrogenation reactions involving cross linking, cyclization, and aromatization of organic structures in the kerogens of these ancient shales appear to have resulted in the virtual obliteration of all molecular fossils. This conclusion may be somewhat pessimistic because even less altered kerogens of this age are likely to have already generated and expelled their most labile (biomarker) hydrocarbon moieties, and, in any case, yields from hydrous pyrolysis (at ~330°C; 72 h) may be substantially lower than those from methods involving anhydrous pyrolysis (flash pyrolysis) at temperatures above 600°C. Nevertheless, the absence of intact polycyclic biomarker hydrocarbons in the Tindelpina Shale extract and kerogen pyrolysate is hardly surprising in view of the previously reported graphitic rank of its kerogen at the locality sampled (atomic H/C = 0.05–0.06; McKirdy *et al.*, 1975). Future success in finding such compounds in Precambrian rocks will depend on the discovery of sediments in which the kerogen is much better preserved (i.e., less thermally altered).

Among the Precambrian sequences presently known to contain kerogen that is not extensively thermally altered (atomic H/C > 0.75; Rock-Eval T_{max} < 450°C) are several Late Proterozoic units, including the 0.85-Ga Walcott Member, Chuar Group, Grand Canyon, Arizona (Summons *et al.*, 1988b); the 0.85-Ga Bitter Springs and 0.61-Ga Pertatataka Formations, Amadeus Basin, Northern Territory, Australia (McKirdy, 1977; Jackson *et al.*, 1984); the 0.61-Ga Rodda Beds, Officer Basin, South Australia (D. M. McKirdy, S. D. Pell, and M. Smyth, unpublished data); and the ~0.6-Ga Huqf Group, Oman (Grantham *et al.*, 1988). Precambrian, organic-rich sediments containing kerogen of exceptional preservation include the lacustrine Barney Creek Formation (1.69 Ga, McArthur Group) and marine Velkerri and McMinn Formations (1.43 Ga, Roper Group) of the Middle Proterozoic McArthur Basin, Northern Territory, Australia (McKirdy, 1977; Peat *et al.*, 1978; Crick *et al.*, 1988; Jackson *et al.*, 1988). Kerogen in the former unit is of type I and type II composition (H/C = 0.70–1.66; HI = 150–740), whereas in the latter units it is mostly type II (Velkerri: H/C = 0.76–1.01; HI = 100–641) and type II/III (McMinn: H/C = 0.90–1.17; HI = 100–484). A diverse array of microbial biomarker hydrocarbons has been identified in extracts of these rocks (Summons *et al.*, 1988a). In all cases, lamalginite [alginite B of Hutton *et al.* (1980)] is the dominant and primary petrographic component of the kerogen. In thermally immature to marginally mature samples (i.e., with respect to petroleum generation), lamalginite appears as discontinuous sheets or lamellae of fluorescent organic matter, commonly stacked in bundles up to 0.05–1 mm thick and aligned parallel to bedding. Fluorescence colors change from bright yellow through orange and dull orange-brown with increasing maturity. "Nonfluorescent" lamalginite occurs in mature to overmature samples (Crick *et al.*, 1988). Bacterial degradation (sapropelization) of the primary algal biota (Peat *et al.*, 1978), and perhaps its partial oxidation in the inner marine shelf environment, may explain the trend to type III kerogen compositions in the McMinn Formation.

Py–GC/MS was employed in a study of kerogens from the ~1.1-Ga Nonesuch Formation, a rift-associated, primarily lacustrine deposit in Michigan and Wisconsin (Imbus *et al.*, 1988; Elmore *et al.*, 1989). Different pyrolysate compound distributions were found to correlate with two "organic petrographies" defined using incident white light and blue light fluorescence microscopy. As in the McArthur Basin sequences, differential preservation of a common algal precursor, rather than different source inputs or thermal regimes, appears to have been the primary cause of the two distinct chemical compositions of the kerogens. Further studies on the variety of Nonesuch Formation kerogen yielding abundant aliphatic pyrolysates (organic petrography 1; Imbus *et al.*, 1988) resulted in the detection of hopanoid homologs (Imbus, 1990). The three-stage Py–GC/MS technique employed was conclusive in demonstrating the indigineity of these compounds as (a) pyrolysate yields were most abundant in the highest "temperature slice" (450–615°C), (b) unsaturated homologs were evolved, and (c) compound redistribution was noted between the middle (330–450°C) and highest temperature slices.

Argillaceous dolostone from the fossiliferous Late Proterozoic Walcott Member (Chuar Group, Arizona) was the subject of a novel and elegant study by Summons *et al.* (1988b). Hydrocarbon distributions and relative yields of compound classes in solvent extracts were compared to those derived by chemical and thermal degradation of kerogen. Degradation of the kerogen isolate involved solvent extraction (to remove "trapped hydrocarbons") and BBR_3 treatment to release ether-linked compounds as alkyl bromides, which were then reduced by $LiAlD_4$ deuteration to produce deuterated alkanes, and sealed-tube (hydrous) pyrolysis of purified kerogen at 330°C for 72 h. The compound classes detected in the extracted hydrocarbons were largely analogous to, although typically greater in diversity and abundance than, those obtained from kerogen degradation procedures. This experiment was unique for several reasons. First, it established the syngenicity of the solvent extract. Second, it also demonstrated what yields and distributions can be expected when the products of chemical and thermal kerogen degradation are compared to those of solvent extraction. Finally, it proved that adequate biomarker characterization of thermally mature kerogens (Rock-Eval T_{max} = 446–449°C) is possible when the presence of nonsyngenetic compounds is suspected in solvent extracts.

5. Extract and Oil Studies

Documentation of a broad range of biological marker compounds in Precambrian rock extracts and oils is a recent development enabled by advanced instrumentation and new analytical strategies. Prior to the papers of Jackson *et al.* (1986) and Summons *et al.* (1988a) on the geochemistry of Middle Proterozoic oil and source rocks in the McArthur Basin, uncontested reports of terpane and sterane biomarkers were known only from oils of latest Precambrian age from the Siberian Platform (e.g., Arefev *et al.*, 1980; Fowler and Douglas, 1987). At present, these and other biological marker compounds have been documented in oils and extracts from several units ranging in age from the late Early Proterozoic (~1.7 Ga) to Vendian (~0.6 Ga).

The identification, distribution, and origin of biological marker compounds in Phanerozoic oils and sediments, and their uses in petroleum exploration studies, have been reviewed by Seifert and Moldowan (1978), Mackenzie (1984), Philp (1985), and Philp and Lewis (1987). Molecular criteria for recognizing marine, terrestrial, and microbial organic materials have been outlined by Tissot and Welte (1984). Specific molecular product–precursor correlations are usually difficult to establish. Somewhat easier is the use of multiple parameters to infer depositional environments. Relative levels of anoxia may be ascertained using the methods of Powell and McKirdy (1975), Hussler *et al.* (1984), and Moldowan *et al.* (1986). Criteria for distinguishing organic materials deposited in carbonate and siliciclastic sediments have been proposed by several contributors to a volume edited by Palacas (1984). More specific environments, such as hypersaline (McKirdy *et al.*, 1984; ten Haven *et al.*, 1986; Philp and Zhaoan, 1987; Sinninghe-Damsté and de Leeuw, 1987) and lacustrine (Moldowan and Seifert, 1980; Shi *et al.*, 1982; McKirdy *et al.*, 1986) settings, may also be amenable to characterization by molecular parameters.

The problem of finding definitive biological marker compounds in Precambrian organic materials is more acute. Many of the more highly diagnostic biological marker compounds in routine use appear to be derived from higher plants (e.g., certain bicyclic sesquiterpanes and diterpanes and the distinctive lupanes and oleananes found in many post-Paleozoic sediments), whereas most of the remaining compounds are nonspecific microbial derivatives considered to be ubiquitous in sedimentary organic matter of all ages. In Table III the various biological markers (compound classes, individual compounds, and homologous series) that have been identified in Precambrian and Early Paleozoic sediments are listed, together with their probable precursor biota. Given the great antiquity of Precambrian organic matter, questions arise concerning the applicability of Phanerozoic findings in interpreting the Precambrian record. Although a basic understanding of problems related to syngenicity (e.g., Hoering, 1967; McKirdy,1974; Oehler, 1977; McKirdy and Hahn, 1982; Summons and Hayes, 1992), thermal maturity (Hayes *et al.*, 1983; Wedeking and Hayes, 1983; Hoering and Navale, 1987), uniformitarian considerations of evolutionary conservatism (Hayes, 1983; Schidlowski, 1987; Summons and Walter, 1990), and the relative constancy of depositional processes (Veizer, 1988) has been addressed, caution is still warranted in the collection and interpretation of Precambrian organic geochemical data, particularly where solvent extracts are concerned.

From data compiled by Hayes *et al.* (1983) and

TABLE III. Hydrocarbons with Potential as Precambrian Biological Markers and Their Provable Precursor Biota[a]

Biological marker	Precursor biota
n-Alkanes	
C_{12}–C_{20} with marked odd/even predominance	Bacteria, cyanobacteria
C_{17}	Cyanobacteria
C_{21}–C_{35}	Bacteria (e.g., sulfate reducers), eukaryotic algae, ?graptolites
Methylalkanes	
C_{12}–C_{22}, *iso* (2-Me) and *anteiso* (3-Me)	Bacteria
C_{20}–C_{26}, mid-chain Me	?Bacteria
Isoprenoid alkanes	
C_{15}–C_{20} regular	Photosynthetic algae
C_{15}–C_{25} regular	Archaebacteria (methanogens, halophiles)
C_{30}–C_{40} irregular	Archaebacteria (methanogens, halophiles)
Pristane/phytane	
<1	Anoxic
1–2	Suboxic
>2	Oxic
Aryl isoprenoids	
C_{13}–C_{31} irregular	Green sulfur bacteria (meta- to hypersaline environments)
Alkylcyclohexanes	
C_{13}–C_{21} *n*-alkyl with odd/even predominance	Cyanobacteria, bacteria (including thermoacidophiles)
C_{14}–C_{20} Me-*n*-alkyl with even/odd predominance	Cyanobacteria, bacteria (including thermoacidophiles)
Tricyclic terpanes	
C_{19}–C_{33}	?Eukaryotic algae, bacteria
Tetracyclic terpanes	
C_{24} dominant	?Bacteria
Hopanes	
C_{27}–C_{30} dominant	Cyanobacteria, bacteria
C_{31}–C_{35} dominant	Cyanobacteria, bacteria
28,30-bisnorhopane	Anaerobic bacteria
C_{27}–C_{35} neohopanes	?Bacteria
Methylhopanes	
C_{28}–C_{36} 2-methyl	Bacteria (carbonate environments)
Gammacerane	??Protozoa
Steranes	
C_{26}–C_{30} desmethyl	Eukaryotic algae
C_{28} dominant	?Halophilic algae (evaporitic environments)
Methylsteranes	
C_{28}–C_{30} 2-,3-, and 4-methyl	Eukaryotic algae

[a]Modified from McKirdy and Hahn (1982).

Tissot and Welte (1984), kerogen H/C atomic ratios of 0.6–0.7 (approximately equivalent to vitrinite R_0 = 1.2–1.3% and Rock-Eval T_{max} = 460°C) likely represent a maturity limit beyond which preservation of intact biological marker compounds would not be expected such that, if compounds associated with lower kerogen H/C atomic ratios are found, they can be reasonably considered to be adventitious (i.e., not syngenetic). The factor of time in time–temperature equations describing thermal maturation (e.g., Waples, 1984, and references therein) also suggests that, even in cases where Precambrian organic materials have never been exposed to thermal regimes warmer than the "oil window," many of the more diagnostic biological marker compounds may be destroyed by cracking or profoundly altered by spontaneous dehydrogenation, as described by Hoering and Navale (1987). A further consideration, recognized in Phanerozoic studies, is the phenomenon whereby the hydrocarbon products of kerogen generation change according to the stage of catagenesis reached. Comet and Eglinton (1987) recognized preferential incorporation of lipid compounds (by functionality) into protokerogens and their subsequent preferential release during catagenesis as limitations on the use of geolipids as facies indicators. Despite these complications, syngenetic hydrocarbons diagnostic of various specific and general classes of organisms will continue to be detected in Precambrian extracts and oils that have eluded extensive thermal alteration.

A substantial body of biological marker data on a variety of Precambrian oils and extracts is now available for interpretation. A summary of the major compounds found in several such studies, and their likely paleobiological significance, is discussed below.

The saturated hydrocarbon fractions of Precambrian oils and extracts are typically dominated by normal alkanes encoding classic "marine signatures." This "signature" comprises a prevalence of C_{12}–C_{20} *n*-alkanes, commonly with a predominance of odd- over even-carbon-numbered homologs in the vicinity of *n*-C_{15} and *n*-C_{17} (Tissot and Welte, 1984). This distribution of *n*-alkanes was interpreted by Tissot *et al.* (1977) to be the result of decarboxylation of the corresponding normal fatty acids (especially the *n*-C_{16} and *n*-C_{18} homologs) which are important lipid constituents of planktonic and benthic algae. A notable exception to the aforementioned *n*-alkane pattern was reported by Summons *et al.* (1988b) in extracts of argillaceous dolostones from the Walcott Member of the Kwagunt Formation, Chuar Group, Arizona. In these sediments, a slight even-carbon-number predominance is observed in the *n*-alkane distribution, which has a maximum at C_{20}. This feature was interpreted to be indicative of a nonbacterial input to an intermittently hypersaline, carbonate environment (cf. ten Haven *et al.*, 1988).

A wide variety of isoprenoid hydrocarbons have been reported in Precambrian oils and extracts. The C_{12}–C_{20} "regular" (i.e., head-to-tail linked) isoprenoid alkanes are the most frequently recorded isoprenoid compounds. Regular isoprenoids with up to 20 carbon atoms have several possible origins (Volkman and Maxwell, 1986). Although these compounds are generally thought to be derived primarily from the phytol side chain of chlorophyll *a* (photoautotrophs), other origins include chlorophyll *b* and bacteriochlorophyll *a* (Gillan and Johns, 1980) and α- and γ-tocopherols (Brassell *et al.*, 1981). Lipids derived from Archaebacteria, such as the extreme halophiles (Kates, 1978) and methanogens (Risatti *et al.*, 1984), also may be important contributors in ancient sedimentary environments (e.g., McKirdy *et al.*, 1984). The pristane/phytane ratio (Pr/Ph), a classic indicator of redox conditions in ancient sedimentary environments, varies from an average of ~4.1 in Nonesuch Formation bitumens (Hieshima *et al.*, 1989) to around 0.6 in some Omani oils and Huqf Group rock extracts (Grantham, 1986) and in several McArthur Basin extracts (Summons *et al.*, 1988a). Given the recent recognition of multiple sources of $\leq C_{20}$ acyclic isoprenoids, the validity of the Pr/Ph ratio as an index of oxicity (e.g., values less than 1 being taken to indicate anoxic or highly reducing conditions) may be called into question. However, according to Volkman and Maxwell (1986), reducing or oxidizing conditions can indeed by inferred if the Pr/Ph ratio is significantly below unity or much greater than unity, respectively. On this basis, most of the Precambrian oils and extracts so far analyzed were derived from kerogens formed under marginally reducing conditions.

Homologous series of C_{21}–C_{35} extended regular isoprenoid alkanes were found in varying abundance by Fowler and Douglas (1987) in several Siberian Platform oils. Summons *et al.* (1988a) detected regular extended isoprenoid alkanes (C_{20}–C_{25}) in the Barney Creek Formation. The C_{29} and C_{30} (squalane) "irregular" (tail-to-tail) isoprenoid alkanes were also detected in these extracts. The C_{20}–C_{25} and C_{30} compounds indicate deposition under hypersaline conditions with important inputs of organic matter by Archaebacteria, particularly the extreme halophiles. The C_{35}–C_{40} irregular isoprenoids, markers of methanogenic Archaebacteria, were not detected. Finally, a homologous series of C_{14}–C_{20} aryl isoprenoids, probably derived from the carotenoids of fossil Chlorobiaceae (green sulfur bacteria), were identified in the aromatic hydrocarbon fraction of Siberian Platform oils analyzed by Summons and Powell (1990).

Among the most distinctive compounds found in Precambrian oils and extracts are the monomethyl-branched alkanes. Both *iso*- and *anteiso*-alkanes (2- and 3-methyl-substituted, respectively) are common components of most oils and extracts (e.g., Kissin, 1987) and widely attributed to (eu)bacterial inputs (Tissot and Welte, 1984). Other monomethyl alkanes such as the 4-, 5-, and 6-methyl-substituted alkanes and the midchain methyl-substituted alkane isomers [the so-called "X compounds" first reported in Omani oils by Klomp (1986)], which are found in unusual abundances in Precambrian oils and extracts (Fowler and Douglas, 1987; Summons and Powell, 1990), have a less definitive origin. Hoering (1981) found that branched alkane species (2- and 6-substituted) increase in abundance in oils with increasing geologic age. This phenomenon was attributed to rearrangements and cracking of carbon skeletons during long periods of kerogen maturation. The midchain methyl-substituted alkanes, apparently found exclusively in Precambrian and Early Paleozoic oils and extracts, were believed by Summons *et al.* (1988a) to be of biological (perhaps bacterial) derivation; possible inorganic (thermal and/or catalytic) origins were considered unlikely. Alkylcyclohexyl hydrocarbons, comprising homologous series of *n*-alkylcyclohexanes and methyl-*n*-alkylcyclohexanes, occur in oils and extracts of all ages, including the Precambrian. These compounds are primarily derived from the cyclization of normal fatty acids upon maturation (Summons *et al.*, 1988a).

Polycyclic hydrocarbons, such as terpanes (tricyclic, tetracyclic, and pentacyclic) and steranes, are ubiquitous components of Phanerozoic oils and extracts (Mackenzie, 1984; Tissot and Welte, 1984). Earlier reports of these compounds in Precambrian materials (e.g., 2.7-Ga Soudan Shale extracts; Johns *et al.*, 1966) were commonly dismissed as representing examples of nonsyngenetic organic matter (i.e., contamination). Indeed, after the pessimistic results of studies by Leventhal *et al.* (1975) and Hoering (1976), many workers were probably reluctant to search for, much less report, occurrences of such compounds. Several Precambrian extracts and oils are now known to contain concentrations of these compounds varying from mere traces (e.g., steranes and triterpanes in the Nonesuch Shale and associated oil; Summons, 1990) to levels comparable to those found in their Phanerozoic counterparts. As outlined in Table III, there is strong evidence that the terpanes have a dominantly microbial source, whereas steroidal compounds appear to originate from the membrane lipids of eukaryotic organisms [Taylor, 1984; see discussion by Summons *et al.* (1988a)].

Tricyclic terpane homologs comprising at least the C_{20}–C_{30} series have been recorded in the Siberian Platform oils (Fowler and Douglas, 1987), the Omani Huqf and "Q" oils (Grantham *et al.*, 1988), in various sediments from the McArthur basin (Summons *et al.*, 1988a), and in the Chuar Group Walcott Member (Summons *et al.*, 1988b). Little effort has been made in the above studies to interpret the significance of tricyclic terpane distributions. As is also the case in Phanerozoic studies, these compounds are apparently considered to be poorly diagnostic of specific source biota. Tetracyclic terpane compounds of uncertain (but possibly bacterial) affinity also were reported in the studies of Fowler and Douglas (1987) and Summons *et al.* (1988b).

Various pentacyclic triterpanes of the hopanoid type are known to occur in the oils and extracts of Precambrian age. The prevalence of hopanes over steranes observed in the McArthur Basin sediment extracts and the Siberian Platform oils has been cited as evidence of a major contribution by prokaryotic biota, and a subordinate eukaryotic input, to their respective source materials (Fowler and Douglas, 1987; Summons *et al.*, 1988a). The 17α(*H*),21β(*H*) hopanes constitute the dominant hopane series in the Siberian oils. The prevalence of C_{35} hopane over C_{34} hopane and of Tm [17α(*H*)-22,29,30-trisnorphopane] over Ts [18α(*H*)-22,29,30-trisnorhopane] was regarded by Fowler and Douglas (1987) as indicative of a carbonate source rock (cf. McKirdy *et al.*, 1983). This interpretation is consistent with other features of the triterpenoid alkanes in a suite of Siberian oils analyzed by Summons and Powell (1992), viz., high abundances of 29,30-bisnorhopane and norhopane relative to hopane, and of 28,30-bisnorhopane and a series of 2α-methylhopanes. Grantham (1986) ascribed the contrasting tricyclic terpane, hopane, and sterane distributions in Omani oils "A" and "B" [the Huqf and "Q" oils, respectively, of Grantham (1968)] to the different membrane lipids of the microbial ecosystems which populated the depositional environments of two separate and distinct source rocks. Summons *et al.* (1988a) reported that the McArthur Basin sediments contain "the usual C_{27+} pseudohomologous series of pentacyclic triterpanes" (i.e., distributions similar to those in Phanerozoic materials). In addition, a series of "putative" C_{29}–C_{34}

18α(H)-neohopanes* was detected in the McMinn and Barney Creek Formations. No definitive origin for these neohopanes has yet been determined. However, in discussing a similar series of C_{29}–C_{35} neohopanes in the Walcott Member (Chuar Group), Summons *et al.* (1988b) favored a bacterial origin. Like the mid-chain methyl-branched hydrocarbons described previously, neohopanes appear to be restricted to sediments of Proterozoic and Early Paleozoic age. For example, they also occur in marine mudstones of the Late Proterozoic Rodda Beds, Officer Basin, and in shales of the Nonesuch Formation remote from the White Pine Mine locality of the famous Nonesuch oil seep (Summons, 1992). Both 2α- and 3β-methylhopanes were detected in the McMinn and Barney Creek Formations (Summons *et al.*, 1988a). The dominant 2α-methyl series appear to be derived from the 2β-methyl hopanoids found in certain methylotrophic bacteria and cyanobacteria, whereas the 3β-methyl series have been tentatively linked with bacteria of the *Acetobacter* type (Summons and Jahnke, 1990). A substantial concentration of gammacerane in the Walcott Member extract was interpreted by Summons *et al.* (1988b) to represent a contribution by lipids of primitive Protozoa via the reduction of tetrahymanol. This compound replaces steroids in cellular membranes of the extant protozoan *Tetrahymena pyriformis* (cf. Ourisson *et al.*, 1987). Hence, there may be a causal relationship between gammacerane and the vase-shaped microfossil *Melanocyrillium* (described elsewhere in the Chuar Group; Bloeser, 1985), which some workers (e.g., Vidal and Knoll, 1983) interpret as a heterotrophic protist similar to tintinnid protozoans. It is also significant that gammacerane is commonly enriched in younger sediments which were deposited in hypersaline environments (cf. Brassell *et al.*, 1987; ten Haven *et al.*, 1988). The two Walcott Member dolostone samples analyzed come from an evaporitic carbonate sequence.

The first report of possible steroid hydrocarbons in the Nonesuch Formation seep oil was that by Barghoorn *et al.* (1965). Although Hoering (1976) failed to detect either steroid or terpenoid compounds in this oil, eventually Summons *et al.* (1988a) confirmed the presence of both C_{27}–C_{29} steranes and C_{27}–C_{35} triterpanes. Recently, distributions and ratios of these compounds have been used to infer maturity levels of the Nonesuch Formation (Hieshima *et al.*, 1991). Grantham and Wakefield (1988) have suggested that the abundance of C_{29} steranes (relative to the C_{27} and C_{28} homologs) in sedimentary organic matter increases with increasing geologic age (i.e., Cenozoic to late Precambrian). The C_{27} homologs, however, are dominant in the C_{27}–C_{29} series of desmethylsteranes in the Omani "Q" oils (cf. Grantham and Wakefield, 1988) and in extracts of the McMinn and Barney Creek Formations (McArthur Basin) and of the Walcott Member (Chuar Group, Arizona) (Summons *et al.*, 1988a, b). The question of the origin and significance of C_{27} or C_{29} sterane prevalences in Precambrian materials has therefore not yet been resolved. The C_{29} steranes have been attributed to various precursors, including cyanobacteria (Fowler and Douglas, 1987) and primitive higher algae (Grantham, 1986), whereas the C_{27} steranes may have a marine algal origin (e.g., Rhodophyta; Grantham, 1986). However, as pointed out by Summons *et al.* (1988b), evidence that extant cyanobacteria can synthesize sterols is at best equivocal.

Diasterane abundance varies from low (<10% of total steranes) in the apparently carbonate-sourced Siberian Platform (Fowler and Douglas, 1987) and Omani Huqf oils (Grantham *et al.*, 1988) to concentrations exceeding those of the unrearranged steranes [e.g., in the McMinn Formation, McArthur Basin (Summons *et al.*, 1988a) and the Walcott Member, Chuar Group (Summons *et al.*, 1988). Unusual ring-methylated steroid compounds, such as the C_{28} and C_{29} methylsteranes and C_{29} dimethylsterane, occur in the Walcott Member (Summons *et al.*, 1988b). The sites of methylation in the monomethylsteranes (i.e., at C-2 and C-3) are analogous to those observed in the monomethyl hopanoids of various prokaryotes, and thus the monomethylsteranes may have a similar biosynthetic origin. The small quantities of C_{30} 4-methylsteranes detected in the Siberian Platform oils were suggested by Fowler and Douglas (1987) to be indicative of an unknown eukaryotic organism (other than dinoflagellates, the commonly proposed precursor organisms for these compounds in Phanerozoic sediments) that used 4-methylsterols as membrane constituents.

The most important conclusion concerning the presence of steranes in Precambrian oils and sediments [as pointed out by all authors reporting them and discussed in detail by Summons *et al.* (1988a, b) and Summons (1990)] is that they provide definitive evidence of eukaryotic inputs to their precursor organic matter. Based on the presence of steranes in the ~1.7-Ga Barney Creek Formation of the McArthur

*The 18α(H)-neohopanes reported by Summons *et al.* (1988a, b) are in fact 17α(H)-diahopanes (R. E. Summons, personal communication to D. M. McKirdy, 1991; Pratt *et al.*, 1991).

Basin, Summons *et al.* (1988a) favored an early origin of the eukaryotes. This is contrary to well-established fossil evidence which indicates the presence of eukaryotes back to only ~1.4 Ga (e.g., Walter, 1987), although carbonaceous films (spheroids, ribbons) of possible eukaryotic affinity do occur in Proterozoic sediments as old as 1.8 Ga (Hofmann, 1987).

To date, apparently the oldest porphyrins detected in sedimentary rocks are those (including a vanadyl porphyrin) found in the Nonesuch Formation by Barghoorn *et al.* (1965). As porphyrins are reasonably stable (cf. Baker and Louda, 1986), their lack of documentation in Precambrian organic matter probably reflects the emphasis placed on other compounds.

The concept that distinct metabolic pathways associated with particular classes of extant organisms have persisted for long periods of geologic time is well accepted. Arguably, the best example is photoautotrophy, which the sedimentary carbon isotopic record indicates has been operative since the early Archean (i.e., at least ~3.8 Ga ago; Hayes *et al.*, 1983; Schidlowski *et al.*, 1983; Schidlowski, 1988).

Major (negative) excursions in the same isotopic record also provide evidence for methylotrophy and methanogenesis in sediments as old as 2.8 Ga. By reference to appropriate modern analogs, varying lipid structures and/or isotopic signatures preserved in Precambrian sedimentary organic matter should therefore allow an assessment of the biochemical processes active at the time of deposition. The recent application of combined GC and isotope-ratio mass spectrometry (GC/IRMS) to acquire $\delta^{13}C$ values on individual compounds shows great promise in efforts to distinguish among such processes (e.g., Freeman *et al.*, 1989; Hieshima, 1991). In the rare cases where thermal alteration of the organic matter is minimal, it may even be possible to reconstruct Precambrian ecosystems and depositional settings. In the absence of land-plant cover, mechanical weathering and erosion of Precambrian terrains may have been more efficient than was the case on the post-Silurian Earth, resulting in higher rates of burial and enhanced preservation of aquatic organic matter (cf. Bralower and Thierstein, 1987). Lower free oxygen levels in the Precambrian atmosphere and hydrosphere may have markedly reduced the rate of bacterial degradation of organic matter (microbial activity may be higher under aerobic conditions; cf., Demaison and Moore, 1980), thus allowing for more faithful preservation of both autotrophic organic matter and bacterial remains. Furthermore, post Silurian sediments are often characterized by considerable allochthonous organic inputs whereas Precambrian sequences, at least those preferentially studied because of their high organic contents, are more likely to be sites where primary productivity coincided with organic accumulation (e.g., rift-related marine and lacustrine settings and marine shelves).

6. Economic Geology

6.1. Oil and Gas

The petroleum prospectivity of Precambrian sediments has now passed from the realm of speculation (e.g., Murray, 1965; Vassoyevich *et al.*, 1971; Murray *et al.*, 1980) to that of practical assessment by active exploration programs in parts of the Middle East, Eurasia, China, Australia, and North America. Vassoyevich *et al.* (1971) outlined the general tectonic basis for the development of Proterozoic basins that may have served as suitable sites for the accumulation of organic-rich sediments. Potentially petroliferous basins of Proterozoic age where identified and classified. Of these, the Angara–Tunguska Basin (Siberian Platform), the Potwar–Punjab Basin (Pakistan–India), and the Amadeus Basin (central Australia) were known at the time to contain oil or gas deposits. At present, the only large commercial oil and gas fields with proven or suspected Late Proterozoic (Vendian) source beds are those in the Lena–Tunguska region of East Siberia (Meyerhoff, 1980; Kontorovich *et al.*, 1990), central Oman (Hughes-Clarke, 1988), and the Sichuan Basin, southern China (Fu Jia Mo *et al.*, 1984). Outlined below is a brief history of developments in Precambrian petroleum geology. For a more comprehensive summary, the reader is referred to McKirdy and Imbus (1992).

The famous Markovo field in Siberia, discovered in 1962 and located at the southern end of the north-northeast-trending Angara–Lena Trough (part of the Baikalian foredeep), produces gas and light oil from Vendian sandstones and Lower Cambrian carbonates. Total proven and probable reserves are 0.62 Tcf of gas and 26 MMbbl of oil, whereas Vendian sandstones in the adjacent Yaraktin field may contain up to 210 MMbbl of oil (Meyerhoff, 1980). Some 600 km away of the northern flank of the same trough, two giant fields have been discovered in which production is mostly from Vendian siliciclastic reservoirs: Sredne–Botuobin (149 MMbbl of oil and 17.3 Tcf of gas recoverable) and Verkhne–Vilyuchanskoye (260 MMbbl of oil and 10.3 Tcf of gas recoverable) (Carmalt and St. John, 1986). The total resource potential of the Lena–Tunguska petroleum province is estimated by Meyerhoff (1982) to be 2000 MMbbl of oil and 83 Tcf of gas. Although no specific source horizon has yet been identified for these Siberian hydrocarbons, they could be derived from deeply buried Riphean black shales (TOC = 3–5%) or organic-rich marls (TOC = 1–10%) in the Vendian to Early Cambrian carbonate sequence (Summons and Powell, 1992; T. G. Powell, personal communication to D. M. McKirdy, 1988). The latter are the most likely source of the oils since aspects of their biomarker geochemistry (e.g., Pr < Ph; virtual absence of diasteranes; high abundance of 29,30-bisnorhopane and norhopane relative to hopane) are strongly indicative of a carbonate source affinity (Fowler and Douglas, 1987; Summons and Powell, 1992). The gas, on the other hand, may be of Riphean origin.

Since 1960, petroleum exploration in Oman has been spectacularly successful, leading to the discovery and development of some 50 fields with total daily production currently in excess of 500,000 bbl of oil, mostly from Cretaceous, Jurassic, and Permian reservoirs. A detailed geochemical study by Grantham *et al.* (1988) identified potential source rocks ranging in age from Late Proterozoic (Huqf Group) to mid-Tertiary and five geochemically distinct end-member oil types (two of Proterozoic origin). The Huqf Group comprises a sequence of shallow marine clastics and carbonates which pass upward into a major evaporite unit (Ara Formation). Within and beneath these thick, basin-wide salt deposits, rich source beds occur as micritic carbonates and claystones (typically containing 2–9% TOC; G. W. M. Lijmbach, personal communication to D. M. McKirdy, 1988). These oil-prone Proterozoic source rocks are richer and more extensive than the younger marine Paleozoic and Mesozoic source beds and therefore could have generated the bulk of Oman's identified oil and gas reserves. Using biomarker and carbon isotopic evidence, Grantham *et al.* (1988) distinguished two families of Proterozoic-sourced oils, one from Carboniferous, Permian, and Cretaceous reservoirs in south Oman (the "Huqf" oils) and the other from a Permian reservoir in central Oman (the "Q" oil). Definitive oil-source correlations were established only in the case of the "Huqf" crudes. The "Q" oil is believed to have originated from a second source interval (as yet unsampled) deeper in the Huqf Group of the South Oman Salt Basin.

The giant Weiyuan gas field is the largest commercial gas accumulation in China. It is one of several medium to large gas fields in southern Sichuan Basin, where total annual production from Triassic, Permian, Cambrian, and Sinian (Late Proterozoic) carbonate reservoirs is 205 Bcf (Xu *et al.*, 1984). The main natural gas reserves of the Weiyuan field (80–87% methane,7% nitrogen, 5% carbon dioxide) are located in Upper Sinian (0.7–0.8 Ga) dolomites and, on the basis of geochemical criteria outlined by Xu *et al.* (1992), are considered to have been derived from thermally mature to overmature source beds (dark gray to black, micritic limestones, and dolomites) within the same sequence.

In all three case examples (Siberia, Oman, and China), organic geochemical analyses have provided definitive evidence for the Proterozoic origin of commercial quantities of petroleum hydrocarbons.

The existence of large areas of unmetamorphosed and structurally little disturbed Proterozoic and early Paleozoic strata in Australia (Jackson *et al.*, 1988) has made that continent a particularly attractive region in which to search for Precambrian oil and gas. Exploration has centered on three areas: the Amadeus, Officer, and McArthur basins. Geochemical and organic petrographic studies of Late Proterozoic units from the Amadeus Basin (central Australia) indicate that the Gillen Member of the Bitter Springs Formation and the Pertatataka Formation in places have fair to good source potential (Jackson *et al.*, 1984; Summons and Powell, 1991). However, the available geochemical data indicate a prevalence of gas-prone kerogen (type III). The contrasting carbonate–evaporite and shale facies of the above units are reflected in differences in their biomarker alkane distributions (Summons, 1992; Summons and Walter, 1990). The Dingo gas field, in which the Late Proterozoic–Cambrian Arumbera Formation is the productive unit, is an example of recent exploration success in the Amadeus Basin.

Despite early optimism regarding the oil and gas potential of the Officer Basin, based partly on the discovery of Cambrian oil shows and source rocks in its eastern part (McKirdy and Kantsler, 1980; McKirdy *et al.*, 1983, 1984), recent stratigraphic and exploration drilling in the Officer Basin has failed to locate economic occurrences of hydrocarbons in either the Cambrian or underlying Adelaidean (Late Proterozoic) sequences (e.g., Townson, 1985; Brewer *et al.*, 1987). Nevertheless, thin potential source beds (TOC = 0.4–1.5%; Summons, 1990; D. M. McKirdy and M. Smyth, unpublished results) have been identified in the Rodda Beds, a Proterozoic marine unit >1.2 km thick. Moreover, additional oil shows have been recorded in this unit, in the underlying Murnaroo Formation (Proterozoic), and in the overlying Relief Sandstone (basal Cambrian). Preliminary biomarker data on these oils (D. M. McKirdy, unpublished results) reveal differences between their isoprenoid alkane, sterane, and triterpane distributions and those of the previously analyzed Cambrian oils and extracts from the eastern Officer Basin. This is consistent with the intra-Proterozoic origin of the new oil shows and with the reported similarity between the alkane patterns of Rodda Beds extracts and those of Late Proterozoic oils from Oman and Siberia (notably a dominance of C_{29} steranes, Summons, 1992).

An extensive collaborative effort to characterize and assess the oil and gas potential of Middle Proterozoic sediments in the McArthur Basin has been made by Crick *et al.* (1988), Jackson *et al.* (1988), and Summons *et al.* (1988a). As summarized by Jackson *et al.* (1988), five potential source rocks at various stages of maturity have been identified. Most prospective among these are the lacustrine or lagoonal Barney Creek Formation and the marine (outer shelf) Velkerri Formation (and its stratigraphic equivalent, the Lansen Creek Shale). The Barney Creek Formation, the oldest potential source rock in the basin, contains oil-prone type I–II kerogens varying in maturity from marginally mature to overmature. Where mature, this unit has TOC values in the range 0.8 to 7.6% and hydrocarbon genetic potentials (assessed by Rock-Eval pyrolysis) of up to 40 kg $tonne^{-1}$. The Velkerri Formation is similarly oil prone and contains type II kerogen with a genetic potential of up to 33 kg $tonne^{-1}$. Live oil which bled from the core of the Velkerri Formation was analyzed by Jackson *et al.* (1986). Other, less prolific units are the lacustrine Caranbirini Member of the Lynott Formation, the lacustrine or lagoonal Yalco Formation, and the marine (inner shelf) McMinn Formation. Although organic-rich sediments in the McArthur Basin may indeed have generated significant quantities of oil and gas, variable development of reservoir facies, early hydrothermal fluid movement, and lack of structure at the time of peak hydrocarbon generation appear to be the major factors which limit the accumulation and preservation of oil and gas in this ancient basin.

Vassoyevich *et al.* (1971) cited North America, South America, and Antarctica as continents with little or no Precambrian oil and gas potential. The situation in North America, at least, now appears somewhat more promising. Here, exploration for Proterozoic-generated oil or gas has focused on the Mid-Continent Rift, which extends 1300 km from the Lake Superior area to northeastern Kansas. There is little question that TOC values of up to 3% and appropriate levels of thermal maturity make the rift an attractive exploration target in the Lake Superior region, provided suitable reservoirs and traps can be delineated (Imbus *et al.*, 1988; Elmore *et al.*, 1989; Imbus *et al.*, 1990). In Minnesota, southwest of Lake Superior, organic-rich Proterozoic units are overmature (Hatch and Morey, 1985). Drilling in northeastern Kansas has neither yielded gas or oil nor revealed the presence of possible source rocks (Dickas, 1988). Nevertheless, the rift is still considered prospective, as recently released results of drilling in Iowa indicate small gas shows and the presence of organic-rich lithologies (albeit overmature) similar to those reported in the Lake Superior region (R. M. McKay, personal communication to S. W. Imbus, 1990).

The Proterozoic oil or gas potential or other areas in North America is unknown. Proterozoic strata in western North America (e.g., Lower Belt Supergroup) are indeed overmature for oil generation and preservation, although dry gas accumulations may exist in certain structurally favorable settings. Recently, a strong case has been made for the source potential of the Chuar Group (~0.9 Ga) of northern Arizona and adjacent Utah (Fritz, 1990; Rauzi, 1990). Finally, Hagen (1988) recognized Precambrian rift trends throughout Canada (notably in the Northwest Territories, Yukon Territory, British Columbia, and Alberta) as a possible new petroleum exploration frontier. Organic geochemical data on Proterozoic rocks in this area are apparently not available, and drill holes are rare. However, gas flowed from sandstones of the lower Cambrian Old Fort Island Formation in a well drilled in the Great Bear Basin near Tedji Lake, Northwest Territories (Petersen, 1974; Meyerhoff, 1982). These hydrocarbons may have originated from shales in the underlying thick Proterozoic sequence.

6.2. Economic Metals

Although the present interest in the oil and gas potential of Proterozoic sedimentary rocks may prove to be transitory due to unfilled expectations, there can be little doubt that the genesis of Precambrian ore deposits associated with carbonaceous or bituminous organic matter will continue to be a topic of active research. The temporal distribution of many mineral deposits (notably sedimentary iron and uranium ores and phosphorites) was influenced by the evolving biosphere and its effect on the chemistry of the ocean–atmosphere system (Folinsbee, 1982; Veizer, 1988). Increasingly, the origin of ore deposits is being examined from the perspective of processes that take account of the observed (or former) presence of organic materials (e.g., Powell *et al.*, 1975; Williams, 1978; Zumberge *et al.*, 1978; Connan, 1979; Macqueen and Powell, 1983; Ilchik *et al.*, 1986; Gize and Barnes, 1987; Hoffmann *et al.*, 1988; Puttmann *et al.*, 1988; Disnar and Sureau, 1990).

Among the roles envisaged for organic matter and living organisms in the genesis of "low-temperature" (<200°C) sedimentary and hydrothermal ore deposits are modification of chemical environments, oxidation or reduction of metals, catalysis of reactions, and sequestration and mobilization of metals (Giordano, 1985, and references cited therein). In the particular case of Mississippi Valley-type (MVT) carbonate-hosted Pb–Zn deposits, organic matter appears to have played a key role in the production of sulfide (Powell and Macqueen, 1984) and in the transportation and precipitation of metals (Giordano, 1985). Theoretical aspects of the involvement of organic matter in the metal transport processes which gave rise to these and other low-temperature mineral deposits (e.g., sandstone-hosted Cu–U ores) have been reviewed by Manning (1986) and Manning and Gize (this volume, Chapter 25). Of particular interest is the role which migrating oils and associated brines might have played in the precipitation of ore metals (Macqueen and Powell, 1983; Etminan and Hoffmann, 1989). Migration of formation fluids consequent upon basinal sediment compaction (e.g., Connan, 1979) is just one of the various mechanisms invoked to explain the genesis of MVT ore deposits. Arguments for the syngenetic, diagenetic, or epigenetic origins of these important orebodies, all of which are associated with organic matter, have been summarized by Guilbert and Park (1986).

Element enrichment involving metal–organic interactions may occur at any stage from the living biota or ecosystem, through sedimentation and burial of organic matter, to metamorphism of carbonaceous sedimentary rocks. Saxby (1976) recognized three possible modes of element concentration during sedimentation and early diagenesis: (1) complex formation whereby suitable dissolved organic ligands react with metals in aqueous solution, forming insoluble metal–organic compounds that may be incorporated directly into sediments or be absorbed onto minerals with high surface areas, particularly clays; (2) physical adsorption and chemisorption processes that occur when metallic ions interact with soluble, colloidal, or particulate organic matter; and (3) chemical precipitation of metals or metal complexes present in percolating groundwaters or hydrothermal fluids upon encountering carbonaceous materials. Postdepositional maturation of organic-rich sediments during catagenesis and metamorphism may act to further concentrate metals through the release of gases (e.g., methane, hydrogen sulfide, and ammonia) which create the pH and Eh conditions necessary for the solubilization or precipitation of metals and hence modification of the primary ore mineralogy. Although now containing little or no trace of organic carbon, some metamorphic orebodies may have formed by a process (or processes) in which organic matter played a major role (Saxby, 1976).

The association of biogenic carbon with a wide variety of Precambrian ore deposits was noted by McKirdy (1974). Since then, the involvement of organic matter in the genesis of selected Archean and Proterozoic Au, U, Cu, and Pb–Zn–(Ag) deposits has been the subject of increasingly sophisticated organic geochemical research.

Stratiform organic matter (kerogen) is intimately associated with Archean Au–U ores of the Witwatersrand Sequence, South Africa (Zumberge *et al.*, 1978, 1981) and with Lower Proterozoic U ores in the Elliot Lake–Blind River region of Ontario, Canada (Willingham *et al.*, 1985; Mossman and Dyer, 1985). These kerogens are generally considered to be remnants of Precambrian cyanobacterial mats which trapped detrital uranium minerals and gold prior to burial. Prolonged exposure to ionizing radiation has converted this presumably aliphatic microbial organic matter (initial H/C atomic ratio >1) into a highly aromatic kerogen (H/C = 0.4–0.6) rich in free radicals. Analysis by Py–GC/MS shows the kerogen pyrolysate to comprise mainly alkyl-substituted aromatic hydrocarbons, low-molecular-weight aliphatic hydrocarbons, and lesser amounts of aromatic sulfur compounds (thiophenes). At Elliot Lake a second variety of organic matter occurs. This so-called "globular kerogen" contains only 3–5 ppm U (cf. a U content of 20–32% in stratiform kerogen) and appears to be a resolidified expulsion product of the primary stratiform kerogen (Nagy and Mossman, 1992).

Of strikingly different origin is the vein-type U mineralization which occurs in strongly altered quartzofeldspathic and graphitic pelitic gneisses of probable Archean age in the Cluff Lake area of Saskatchewan, Canada. Blebs and pods of crystallographically disordered organic matter host grains of uraninite and subordinate coffinite in veins of the Claude deposit. Leventhal *et al.* (1987) attributed the origin of this isotopically light organic matter ($\delta^{13}C = -43$ to -45%) to the radiation-induced degradation of methane, which (contrary to their suggestion) was probably derived from graphite-bearing country rocks.

Cortial *et al.* (1990) demonstrated a convincing

genetic relationship between solid bitumens derived from the thermal alteration of crude oil and uraninite ore in Lower Proterozoic sandstones of the Franceville Basin, Gabon. The oil originated in contiguous carbonaceous shales (with TOC contents of up to 20%) which are now thermally overmature and, in places, host to economic Mn mineralization (e.g., at Moanda; B. R. Bolton, personal communication to D. M. McKirdy, 1988). Likewise, liquid and solid hydrocarbons may have controlled the main phase of copper mineralization at White Pine, Michigan, where there are close temporal and spatial associations of oil with Cu–Fe sulfides and native Cu in carbonaceous shale and sandstone of the 1.1-Ga Nonesuch Formation (Kelly and Nishioka, 1985).

Diagenetic formation of pyrite in recent anoxic, organic-rich sediments involves microbial sulfate reduction and gives rise to unimodal and symmetrical distributions of sulfide-S/organic-C ratios (Goldhaber and Kaplan, 1974). The recognition by Williams (1974) that fine-grained, well-mineralized Middle Proterozoic rocks comprising the concordant Pb–Zn–Ag sulfide deposits at McArthur River, Northern Territory, Australia, have positively skewed S/C frequency distributions firmly constrains the syngenetic or epigenetic processes which led to the formation of their second-generation pyrite, galena, and sphalerite. At Mount Isa, Queensland, Australia, similar Middle Proterozoic stratiform Pb–Zn–Ag sulfide ores have undergone a higher degree of metamorphism (viz., lower greenschist facies). Mycke *et al.* (1988) used mild catalytic hydrogenation to release sulfide-entrapped organic matter from samples of Mount Isa pyritic black shale and massive ore. Analysis by GC and GC/MS of the saturated hydrocarbons revealed the presence of *n*-alkanes, monomethylalkanes (notably 3- and 5-methyl), cycloalkanes, and a series of unassigned C_{17}–C_{31} odd-carbon-numbered branched alkanes. Major cyclic components were the *n*-alkylcyclohexanes (C_{12}–C_{21}, with marked odd/even predominance only in the massive sulfide). Among the polycyclic alkanes, steranes (C_{29} dominant) were more abundant than hopanes. This array of biomarkers was taken to indicate the contribution of thermoacidophilic bacteria and primitive algae to the 1.5–1.6-Ga environment of early diagenetic sulfide formation. The slight dominance of pristane over phytane (Pr/Ph = 1.25–2.0) and the predominance of even-carbon-numbered homologs in the C_{14}–C_{24} *n*-alkanes isolated from the massive sulfide are consistent with the primary sabkha-type lithofacies of these sulfide-bearing metasediments (Neudert and Russell, 1981).

Tectonic reconstruction of the Precambrian with respect to ore genesis (e.g., Garson and Mitchell, 1981; Hutchison, 1981; Windley, 1984) and documentation of those metal deposits which can be genetically linked to fundamental changes in the chemistry of the atmosphere and hydrosphere that accompanied early biospheric evolution (e.g., Folinsbee, 1982; Cloud, 1983; Meyer, 1988) are now sufficiently advanced for systematic organic geochemical study of Precambrian mineralization to proceed apace.

An important first step in this endeavor is the worldwide literature compilation by Sozinov and Gorbachev (1987) of anomalous metal concentrations associated with carbonaceous sedimentary rocks dating back to the Early Archean. From a complex array of ores formed in specific geological settings throughout Earth history, they have identified four distinct types which are enriched in organic matter:

1. terrigenous–carbonaceous
2. siliceous–carbonaceous
3. carbonate–carbonaceous
4. volcanic–siliceous (–carbonate)–carbonaceous.*

The terrigenous–carbonaceous type is marked by occurrences of Ni, Co, Au, and P ore (manganese and sulfide ores are typical) with concentrations of Ni, Au, and Co reaching their zenith during the Lower–Middle Proterozoic. The overall importance of this mineralization diminished as shallow localized Precambrian basins with rigid basement gave way to extensive marine platforms in the Phanerozoic. The siliceous–carbonaceous ores are enriched in V, Mo, Ag, and Fe. Temporally, the character of these ores evolved from banded iron formations (BIF) in the Archean to maximal development of V-, Mo-, and Ag-bearing carbonaceous shales in the Early Paleozoic, particularly throughout Eurasia. Carbonate–carbonaceous ores are typically enriched in Pb, Cu, and Zn sulfides and had their maximum development during the Late Proterozoic.

Sozinov and Gorbachev (1987) also studied the enrichment of metallic elements in organic-rich black shales (>10% TOC) through time. Black shales, regardless of age, were found to be enriched in Co, Ni, Cr, Mo, Cu, Ag, Zn, V, Pb, U, P, Sr, and Ba. In unmetamorphosed sediments, the elements V, U, Ag, and Mo (among others) were found to be quantitatively related to organic matter. Several elements (Ni, Co, Cr, Cu, Ag, Th, V) maintain their high concentrations during metamorphism. In Precambrian black shales

*As type 4 is often polyformational, only types 1–3 were considered further in this study.

associated with economic ore deposits, particular elements have an affinity for one or more of the aforementioned depositional settings: terrigenous—Cu, Au, Mn, P; siliceous—V, W, U, Au; carbonate—Pb, Zn, P; and volcanogenic—Au, U, W, Mn, pyritic Fe.

Notwithstanding the aforementioned studies, little is yet known about the nature and significance of the organic matter preserved in certain Precambrian ore deposits. How, and to what extent, did organic materials participate in the processes of ore genesis? Is the presence of organic matter essential to the origin of the mineralization, or merely fortuitous? Difficulties are to be expected in tackling any multidisciplinary problem which involves aspects of inorganic and organic chemistry, organic and ore petrography, sedimentology, and plate tectonics. The Precambrian age and, in many cases, metamorphic alteration of these ores further complicates the reconstruction of the syngenetic and epigenetic events leading to their formation. Nevertheless, the rewards of such research are potentially great for both mineral and petroleum exploration. In some instances, oil genesis and ore genesis in Precambrian rocks appear to have been intimately related.

7. Summary and Conclusions

Problems encountered in early organic geochemical investigations of Precambrian sedimentary rocks, notably advanced thermal alteration and the difficulty of distinguishing indigenous geolipids from contamination, are now well understood and may be circumvented. Important scientific and economic reasons justify continued attention to this and related fields of inquiry.

The nonuniformitarian view of Earth history holds that tectonic and sedimentological processes have varied in kind and intensity through time. In shaping the surface of the lithosphere, these processes influenced the course of biospheric evolution of life by providing appropriate ecological settings for biota of increasing diversity. The early development of contemporary metabolic pathways, in particular photoautotrophy, has in turn influenced the character of sedimentary rocks as well as creating the atmospheric and hydrospheric conditions required for more advanced eukaryotic organisms, notably metaphytes and metazoans. The Precambrian paleontological record has been extensively documented in morphologically descriptive terms. Organic geochemical (including isotopic) techniques have demonstrated their usefulness in providing diagnostic evidence from Phanerozoic sediments of ancient ecosystems which employed a variety of metabolic pathways. The application of these techniques to selected Precambrian sediments containing well-preserved organic matter has already met with considerable success. In the future, collaboration between paleontologists and organic geochemists may lead to chemical distinctions between the fossil remains of individual Precambrian organisms (or groups of organisms) and thereby permit recognition of their earliest occurrence in the geological record.

The mean total organic carbon contents of Precambrian sedimentary rocks are lower than those of comparable Phanerozoic lithofacies. However, there are numerous instances where Precambrian TOC values are equivalent to (or even greater than) those in analogous Phanerozoic settings. The sedimentary record of many representative Precambrian environments may have been lost to erosion. Moreover, low TOC values in some metamorphosed sediments reflect metagenetic carbon loss rather than poor initial organic productivity and preservation. Indeed, stable carbon isotopic mass-balance calculations show that organic preservation (and hence productivity) during the Precambrian may have been as high as that of today, back to at least 3.5 Ga.

Precambrian kerogens, despite their predominantly microbial origin, display as much chemical variability as kerogens in the Phanerozoic. Pyrolysis studies (Py–GC and Py–GC/MS) have been useful in assessing both kerogen type and maturity.

Remarkable progress has been made in detecting biologically significant hydrocarbons (biological markers) in late Precambrian rock extracts and crude oils. Distinctive C_{21}–C_{30} midchain monomethylalkanes, 17α(*H*)-diahopanes, and 3β- and 2α-methylsteranes appear to be indicative of Proterozoic or Early Paleozoic organic matter. The presence of significant quantities of steroid hydrocarbons in extracts as old as 1.7 Ga suggests an origin for eukaryotes earlier than that indicated by the traditional paleontological record of organic-walled microfossils (N.B., footnote on p. 659). The oldest known 4-methylsteranes and C_{30} desmethylsteranes occur in marine carbonates and evaporites of the 0.85-Ga Bitter Springs Formation. The prevalence of hopanoid over steroid hydrocarbons in well-preserved sediments dating back to the Middle Proterozoic highlights the importance of prokaryotes as a source of sedimentary organic matter throughout Earth history.

Increasingly, the rationale of organic geochemical investigations of Precambrian sediments is being viewed in economic terms. Commercial production of Precambrian-derived hydrocarbons occur in Oman (oil and gas), Siberia (oil and gas), and southern China (gas). The McArthur and Amadeus basins of central Australia contain thick sequences of Proterozoic sediments which currently are the focus of active petroleum exploration. Precambrian oil and gas may also occur elsewhere in Asia and in Africa and North America. Biogenic organic carbon is commonly found in association with major Precambrian stratiform Au, U, Cu, Pb, Zn, and Ag ore deposits. The ability of organic matter to concentrate these and other metals has been linked to every step of the biosynthesis–*post mortem* accumulation–diagenesis–metamorphism–weathering continuum. However, little is yet known concerning the precise role(s) or organic matter in the genesis of these orebodies.

ACKNOWLEDGMENTS. We wish to thank Roger Summons for preprints of his recent publications. Richard Jenkins, Paul Philp, and John Zumberge reviewed various versions of the manuscript and provided helpful suggestions. Alison McKirdy and Marjorie Starr helped with word processing. S. W. I. acknowledges the patience of Mike Engel during the preparation of this manuscript and his industry (Texaco Inc.) and government (U.S. Department of Energy and National Science Foundation) sponsors for financial support.

References

Alpern, B., 1980, Pétrographie du kérogène, in: *Kerogen—Insoluble Organic Matter from Sedimentary Rocks* (B. Durand, ed.), Editions Technip, Paris, pp. 339–371.

Arefev, O. A., Zabrodina, M. N., Makushina, V. M., and Petrov, A. A., 1980, Relic tetra- and pentacyclic hydrocarbons in ancient petroleums of the Siberian Platform, *Izv. Akad. Nauk SSR Ser. Geol.* **2**:135–140 [in Russian].

Awramik, S. M., 1982, The Pre-Phanerozoic fossil record, in: *Mineral Deposits and the Evolution of the Biosphere* (H. D. Holland and M. Schidlowski, eds.), Springer-Verlag, Berlin, pp. 67–81.

Baker, E. W., and Louda, J. W., 1986, Porphyrins in the geological record, in: *Biological Markers in the Sedimentary Record* (R. B. Johns, ed.), Elsevier, Amsterdam, pp. 125–225.

Barghoorn, E. S., Meinschein, W. G., and Schopf, J. W., 1965, Paleobiology of a Precambrian shale, *Science* **148**:461–472.

Belperio, A. P., 1987, Hydrocarbon potential of Late Proterozoic graben sediments, *Aust. J. Earth Sci.* **34**:403–404.

Bloeser, B., 1985, *Melanocyrillium*, a new genus of structurally complex Late Proterozoic microfossils from the Kwagunt Formation (Chuar Group), Grand Canyon, Arizona, *J. Paleontol* **59**: 741–765.

Bralower, T. J., and Thierstein, H. R., 1987, Organic carbon and metal accumulation rates in Holocene and mid-Cretaceous sediments: Paleooceanographic significance, in: *Marine Petroleum Source Rocks* (J. Brooks and A. J. Fleet, eds.), Blackwell Scientific, Oxford, pp. 345–369.

Brassell, S. C., Wardroper, A. M. K., Thomson, I. D., Maxwell, J. R., and Eglinton, G., 1981, Specific acyclic isoprenoids as biological markers of methanogenic bacteria in sediments, *Nature* **290**:693–696.

Brassell, S. C., Eglinton, G., and Howell, V. J., 1987, Palaeoenvironmental assessment for marine organic-rich sediments using molecular organic geochemistry, in: *Marine Petroleum Source Rocks* (J. Brooks and A. J. Fleet, eds.), Blackwell Scientific, Oxford, pp. 79–98.

Brewer, A. M., Dunster, J. N., Gatehouse, C. G., Henry, R. L., and Weste, G., 1987, A revision of the stratigraphy of the eastern Officer Basin, *Q. Geol. Notes Geol. Surv. S. Aust.* **102**:2–15.

Burwood, R., Drozd, R. J., Halpern, H. I., and Sedivy, R. A., 1988, Carbon isotopic variations of kerogen pyrolyzates, *Org. Geochem.* **12**:195–204.

Butterfield, N. J., Knoll, A. H., and Swett, K., 1988, Exceptional preservation of fossils in an Upper Proterozoic shale, *Nature* **334**:424–427.

Cameron, E. M., and Garrells, R. M., 1980, Geochemical compositions of some Precambrian shales from the Canadian Shield, *Chem. Geol.* **28**:181–197.

Cameron, E. M., and Jonasson, I. R., 1972, Mercury in Precambrian shales of the Canadian Shield, *Geochim. Cosmochim. Acta* **36**:985–1005.

Campbell, K. W. S., and Day, M. F. (eds.), 1987, *Rates of Evolution*, Allen and Unwin, London.

Carmalt, S. W., and St. John, B., 1986, Giant oil and gas fields, in: *Future Petroleum Provinces of the World* (M. T. Halbouty, ed.), American Association Petroleum Geologists Memoir 40, pp. 11–53.

Cloud, P. E., Jr., 1976, Beginnings of biospheric evolution and their biogeochemical consequences, *Paleobiology* **2**:351–387.

Cloud, P. E., Jr., 1983, Aspects of Proterozoic biogeology, *Geol. Soc. Am. Mem.* **161**:245–251.

Cloud, P. E., Jr., 1987, Trends, transitions, and events in Cryptozoic history and their calibration. Apropos recommendations by the subcommission on Precambrian stratigraphy, *Precamb. Res.* **37**:257–267.

Comet, P. A., and Eglinton, G., 1987, The use of lipids as facies indicators, in: *Marine Petroleum Source Rocks* (J. Brooks and A. J. Fleet, eds.), Blackwell Scientific, Oxford, pp. 99–117.

Connan, J., 1979, Genetic relation between oil and ore in some Pb–Zn–Ba ore deposits, *Geol. Soc. S. Afr. Spec. Publ.* **5**:264–274.

Cortial, F., Gauthier-Lafaye, F., Lacrampe-Couloume, G., Oberlin, A., and Weber, F., 1990, Characterization of organic matter associated with uranium deposits in the Francevillian Formation of Gabon (Lower Proterozoic), *Org. Geochem.* **15**:73–85.

Crick, I. H., Boreham,, C. J., Cook, A. C., and Powell, T. G., 1988, Petroleum geology and geochemistry of Middle Proterozoic McArthur Basin, Northern Australia II: Assessment of source rock potential, *Am. Assoc. Petrol. Geol. Bull.* **72**:1495–1514.

Dana, J. D., 1866, *A Textbook of Geology*, Theodore Bliss and Co., Philadelphia.

de Graciansky, P. C., Brosse, E., Deroo, G., Herbin, J. P., Montadert, L., Muller, C., Sigal, J., and Schaaf, A., 1987, Organic-rich sediments and palaeoenvironmental reconstructions of the Cretaceous North Atlantic, in: *Marine Petroleum Source Rocks* (J. Brooks and A. J. Fleet, eds.), Blackwell Scientific, Oxford, pp. 317–344.

Demaison, G. J., and Moore, G. T., 1980, Anoxic environments and oil source bed genesis, *Am. Assoc. Petrol. Geol. Bull.* **64:**1179–1209.

Des Marais, D. J., Strauss, H., Summons, R. E., and Hayes, J. M., 1992, Carbon isotope evidence for the stepwise oxidation of the Proterozoic environment, *Nature* **359:**605–609.

Dickas, A. B., 1988, The control of extrusion and sedimentary patterns within Keweenawan age structures of the Lake Superior Basin by taphrogenesis, in: *Upper Keweenawan Rift-Fill Sequence, Mid-Continent Rift System, Michigan* (M. S. Wollensak, ed.), Michigan Basin Geological Society (1988 Fall Field Trip Guidebook), East Lansing, Michigan, pp. 81–94.

Disnar, J. R., and Sureau, J. F., 1990, Organic matter in ore genesis: Progress and perspectives, *Org. Geochem.* **16:**577–600.

Diver, W. L., and Peat, C. J., 1979, On the interpretation and classification of Precambrian organic-walled microfossils, *Geology* **7:**401–404.

Elmore, R. D., Milavec, G. J., Imbus, S. W., and Engel, M. H., 1989, The Precambrian Nonesuch Formation of the North American Mid-Continent Rift, sedimentology and organic geochemical aspects of lacustrine deposition, *Precamb. Res.* **43:**191–213.

Espitalié, J., Laporte, J. L., Medec, M., Marquis, F., Leplat, P., Paulet, J., and Boutejeu, A., 1977, Méthode rapide de caractérisation des roches mères, de leur potentiel pétrolier et de leur degré d'évolution, *Rev. Inst. Fr. Petr.* **32:**23–42.

Etminan, H., and Hoffmann, C. F., 1989, Biomarkers in fluid inclusions: A new tool in constraining source regimes and its implications for the genesis of Mississippi Valley-type deposits, *Geology* **17:**19–22.

Folinsbee, R. E., 1982, Variations in the distribution of mineral deposits with time, in: *Mineral Deposits and the Evolution of the Biosphere* (H. D. Holland and M. Schidlowski, eds.), Springer-Verlag, Berlin, pp. 219–236.

Fowler, M. G., and Douglas, A. G., 1987, Saturated hydrocarbon biomarkers in oils of Late Precambrian age from Eastern Siberia, *Org. Geochem.* **11:**210–213.

Freeman, K. H., Hayes, J. M., Trendel, J. M., and Albrecht, P., 1989, Evidence from GC–MS carbon-isotopic measurements for multiple origins of sedimentary hydrocarbons, *Nature* **353:** 254–256.

Fritz, M., 1990, Chuar inspires quest in the west, *Am. Assoc. Petrol. Geol. Explor.* **11**(10):8–9.

Fox, G. E., Stackebrante, E., Hespell, R. B., Gibson, J., Maniloff, J., Dyer, T. A., Wolfe, R. S., Balch, W. E., Tanner, R. S., Magrum, L. J., Zaben, L. B., Blakemore, R., Gupta, R., Bonen, L., Lewis, B. J., Stahl, D. A., Luehrsen, K., Chen, N., and Woese, C. R., 1980, The phylogeny of prokaryotes, *Science* **209:**457–463.

Fu Jia Mo, Liu De Han, Jia Rong Fen, and Dai Yong Ding, 1984, Distribution and origin of hydrocarbons in carbonate rocks (Precambrian to Triassic) in China, in: *Petroleum Geochemistry and Source Rock Potential of Carbonate Rocks* (J. G. Palacas, ed.), AAPG Studies in Geology No. 18, American Association of Petroleum Geologists, Tulsa, Oklahoma, pp. 1–12.

Garson, M. S., and Mitchell, A. H. G., 1981, Precambrian ore deposits and plate tectonics, in: *Precambrian Plate Tectonics* (A. Kröner, ed.), Elsevier, Amsterdam, pp. 689–731.

Gillan, F. T., and Johns, R. B., 1980, Input and early diagenesis of chlorophylls in a temperate intertidal sediment, *Mar. Chem.* **9:**243–253.

Giordano, T. H., 1985, A preliminary evaluation of organic ligands and metal–organic complexing in Mississippi Valley-type ore solution, *Econ. Geol.* **80:**96–106.

Gize, A. P., and Barnes, H. L., 1987, The organic geochemistry of two Mississippi Valley-type lead–zinc deposits, *Econ. Geol.* **82:**457–470.

Glaessner, M. F., and Foster, C. B., 1992, Palaeontology and biogeochemical research: A powerful synergy in: *Early Organic Evolution and Mineral and Energy Resources* (M. Schidlowski, S. Golubic, M. M. Kimberley, D. M. McKirdy, and P. A. Trudinger, eds.), Springer-Verlag, Berlin, pp. 193–202.

Goldhaber, M. B., and Kaplan, I. R., 1974, The sulfur cycle, in: *The Sea, Vol. 5, Marine Chemistry* (E. D. Goldberg, ed.), Wiley, New York, pp. 569–655.

Grantham, P. J., 1986, The occurrence of unusual C_{27} and C_{29} sterane predominances in two types of Oman crude oil, *Org. Geochem.* **9:**1–10.

Grantham, P. J., and Wakefield, L. L., 1988, Variations in the sterane carbon number distributions of marine source rock derived crude oils through geological time, *Org. Geochem.* **12:**61–73.

Grantham, P. J., Lijmbach, G. W. M., Posthuma, J., Hughes-Clark, M. W., and Willink, R. J., 1988, Origin of crude oils in Oman, *J. Petrol. Geol.* **11:**61–80.

Gregor, C. B., Garrells, R. M., Mackenzie, F. T., and Maynard, J. B. (eds.), 1988, *Chemical Cycles in the Evolution of the Earth*, John Wiley & Sons, New York.

Guilbert, J. M., and Park, C. F., Jr., 1986, *The Geology of Ore Deposits*, W. H. Freeman and Co., New York.

Hagen, D. W., 1988, Precambrian rift trends, pre-Devonian strata: A realistic exploration frontier (part 1); Tathlina High: Greatest concentration of well control (part 2); Southern N.W.T.: Realistic exploration area (part 3); Buried rifts likely places to find oil, gas (part 4), *Oil Gas J.* **82**(July 4):52–56; **82**(July 11):114–118; **82**(July 18):59–64; **82**(July 25):103–109.

Hallam, A., 1987, Mesozoic marine organic-rich shales, in: *Marine Petroleum Source Rocks* (J. Brooks and A. J. Fleet, eds.), Blackwell Scientific, Oxford, pp. 251–261.

Han, T.-M., and Runnegar, B., 1992, Megascopic eukaryotic algae from the 2.1-billion-year-old Negaunee Iron-Formation, Michigan, *Science* **257:**232–235.

Harwood, R. J., 1977, Oil and gas generation by laboratory pyrolysis of korogon, *Am. Assoc. Petrol. Geol. Bull.* **61:**2082–2102.

Hatch, J. R., and Morey, G. B., 1985, Hydrocarbon source rock evaluation of Middle Proterozoic Solor Church Formation, North American Mid-Continent Rift System, Rice County, Minnesota, *Am. Assoc. Petrol. Geol. Bull.* **69:**1208–1216.

Hayes, J. M., 1983, Geochemical evidence bearing on the origin of aerobiosis, a speculative hypothesis, in: *Earth's Earliest Biosphere* (J. W. Schopf, ed.), Princeton University Press, Princeton, New Jersey, pp. 291–301.

Hayes, J. M., Kaplan, I. R., and Wedeking, K. M., 1983, Precambrian organic geochemistry, preservation of the record, in: *Earth's Earliest Biosphere* (J. W. Schopf, ed.), Princeton University Press, Princeton, New Jersey, pp. 93–134.

Hieshima, G. B., 1991, Source and thermal maturity of bitumen in Precambrian Nonesuch Formation inferred from isotopic compositions of individual normal alkanes [abstract], *Am. Assoc. Petrol. Geol. Bull.* **75:**594.

Hieshima, G. B., Simmons, R. E., and Pratt, L. M., 1991, Thermal maturity of the 1.05 Ga Nonesuch Formation, North American Mid-Continent Rift: Biomarker ratios and hopane and sterane stereoisomers [abstract], *Am. Assoc. Petrol. Geol. Bull.* **75:**595.

Hieshima, G. B., Zoback, D. A., and Pratt, L. M., 1989, Petroleum potential of Precambrian Nonesuch Formation, Mid-Continent Rift System [abstract], *Am. Assoc. Petrol. Geol. Bull.* **73:**363.

Hoering, T. C., 1967, The organic geochemistry of Precambrian rocks, in: *Researches in Geochemistry vol. 2*. (P. H. Abelson, ed.) John Wiley & Sons, New York, pp. 87–111.

Hoering, T. C., 1976, Molecular fossils from the Precambrian Nonesuch Shale, *Carnegie Inst. Washington Yearb.* **75**:806–813.

Hoering, T. C., 1977, The stable isotopes of hydrogen in Precambrian organic matter, in: *Chemical Evolution of the Early Precambrian* (C. Ponnamperuma, ed.), Academic Press, New York, pp. 81–86.

Hoering, T. C., 1981, Monomethyl, acyclic hydrocarbons in petroleum and rock extracts, *Carnegie Inst. Washington Yearb.* **80**: 389–393.

Hoering, T. C., and Navale, V., 1987, A search for molecular fossils in the kerogen of Precambrian sedimentary rocks, *Precamb. Res.* **34**:247–267.

Hoffmann, C. F., Foster, C. B., Powell, T. G., and Summons, R. E., 1987, Hydrocarbon biomarkers from Ordovician sediments and the fossil alga *Gloeocapsomorpha prisca* Zalessky 1917, *Geochim. Cosmochim. Acta* **51**:2681–2697.

Hoffmann, C. F., Henley, R. W., Higgins, N. C., Solomon, M., and Summons, R. E., 1988, Biogenic hydrocarbons in fluid inclusions from the Aberfoyle tin–tungsten deposit, Tasmania, Australia, *Chem. Geol.* **70**:287–299.

Hofmann, H. J., 1985, Precambrian carbonaceous megafossils, in: *Paleolgology: Contemporary Research and Applications* (D. F. Toomey and N. H. Nitechi, eds.), Springer-Verlag, Berlin, pp. 20–33.

Hofmann, H. J., 1987, Precambrian biostratigraphy, *Geosci. Can.* **14**: 135–154.

Hofmann, H. J., and Aitken, J. D., 1979, Precambrian biota from the Little Dal Group, Mackenzie Mountains, northwestern Canada, *Can. J. Earth Sci.* **16**:150–166.

Holland, H. D., 1984, *The Chemical Evolution of the Atmosphere and Oceans*, Princeton University Press, Princeton, New Jersey.

Holland, H. D., and Schidlowski, M. (eds.), 1982, *Mineral Deposits and the Evolution of the Biosphere*, Springer-Verlag, Berlin.

Holland, H. D., and Trendall, A. F. (eds.), 1983, *Patterns of Change in Earth Evolution*, Springer-Verlag, Berlin.

Horodyski, F. J., 1980, Middle Proterozoic shale facies microbiota from the Lower Belt Supergroup, Little Belt Mountains, Montana, *J. Paleontol.* **54**:649–663.

Horodyski, R. J., and Bloeser, B., 1978, 1400 million-year-old shale facies microbiota from the Lower Belt Supergroup, Montana, *Science* **199**:682–684.

Horsfield, B., 1984, Pyrolysis studies and petroleum exploration, *Adv. Petrol. Geochem.* **1**:247–298.

Huc, A. Y., 1980, Origins and formation of organic matter in recent sediments and in relation to kerogen in: *Kerogen—Insoluble Organic Matter from Sedimentary Rocks* (B. Durand, ed.), Editions Technip, Paris, pp. 445–474.

Hughes-Clarke, M. W., 1988, Stratigraphy and rock unit nomenclature in the oil producing area of interior Oman, *J. Petrol. Geol.* **11**:5–60.

Hunt, J. M., 1979, *Petroleum Geochemistry and Geology*, W. H. Freeman and Co., San Francisco.

Hussler, G., Connan, J., and Albrecht, P., 1984, Novel families of tetra- and hexacyclic aromatic hopanoids predominant in carbonate rocks and crude oils, *Org. Geochem.* **6**:39–49.

Hutchison, R. W., 1981, Metallogenic evolution and Precambrian tectonics, in: *Precambrian Plate Tectonics* (A. Kröner, ed.), Elsevier, Amsterdam, pp. 733–759.

Hutton, A. C., Kantsler, A. J., Cook, A. C., and McKirdy, D. M., 1980, Organic matter in oil shales, *Aust. Petrol. Explor. Assoc. J.* **20**: 44–67.

Ilchik, R. P., Brimhall, G. H., and Schull, H. W., 1986, Hydrothermal maturation of indigenous organic matter at the Alligator Ridge gold deposits, Nevada, *Econ. Geol.* **81**:113–130.

Imbus, S. W., 1990, Organic petrologic and geochemical studies on the Oronto Group Nonesuch Formation (Middle Proterozoic) of the Midcontinent Rift System, northern Wisconsin and Upper Peninsula Michigan, Ph.D. Thesis, The University of Oklahoma, Norman.

Imbus, S. W., Engel, M. H., Elmore, R. D., and Zumberge, J. E., 1988, The origin, distribution and hydrocarbon generation potential of the organic-rich facies in the Nonesuch Formation, Central North American Rift system: A regional study, *Org. Geochem.* **13**:207–219.

Imbus, S. W., Engel, M. H., and Elmore, R. D., 1990, Organic geochemistry and sedimentology of the Middle Proterozoic Nonesuch Formation: Hydrocarbon source rock assessment of a lacustrine rift deposit, in: *Lacustrine Basin Exploration: Case Studies and Modern Analogs* (B. J. Katz, ed.), American Association of Petroleum Geologist Memoir 50, pp. 197–208.

Jackson, K. S., McKirdy, D. M., and Dechelman, J. A., 1984, Hydrocarbon generation in the Amadeus Basin, Central Australia, *Aust. Petrol. Explor. Assoc. J.* **24**:42–65.

Jackson, M. J., Powell, T. G., Summons, R. E., and Sweet, I. P., 1986, Hydrocarbon shows and petroleum source rocks in sediments as old as 1.7×10 years, *Nature* **322**:727–729.

Jackson, M. J., Sweet, I. P., and Powell, T. G., 1988, Petroleum geology and geochemistry of the Middle Proterozoic, McArthur Basin, Northern Australia. I: Petroleum potential, *Aust. Petrol. Explor. Assoc. J.* **28**:283–302.

Jenkins, R. J. F., McKirdy, D. M., Foster, C. B., O'Leary, T., and Pell, S. D., 1992, The record and stratigraphic implications of organic-walled microfossils from the Ediacaran (terminal Proterozoic) of South Australia, *Geol. Mag.* **129**:401–410.

Jones, R. W., 1987, Organic facies, *Adv. Petrol. Geochem.* **2**:1–90.

Johns, R. B., Belsky, T., McCarthy, E. D., Burlingame, A. L., Haug, P., Schnoew, H. K., Richter, W. J., and Calvin, M., 1966, The organic geochemistry of ancient sediments, II, *Geochim. Cosmochim. Acta* **30**:1191–1222.

Kates, M., 1978, The phytanyl ether linked polar lipids and isoprenoid neutral lipids of extremely halophilic bacteria, *Prog. Chem. Fats Lipids* 15:301–342.

Kelly, W. C., and Nishioka, G. K., 1985, Precambrian oil inclusions in late veins and the role of hydrocarbons in copper mineralization at White Pine, Michigan, *Geology* **13**:334–337.

Kissin, Y. V., 1987, Catagenesis and composition of petroleum: Origin of *n*-alkanes and isoalkanes in petroleum crudes, *Geochim. Cosmochim. Acta* **51**:2445–2457.

Klomp, U. C., 1986, the chemical structure of a pronounced series of isoalkanes in south Oman crudes, in: *Advances in Organic Geochemistry 1985* (D. Leythaeuser and J. Rullkötter, eds.), Pergamon Press, Oxford, pp. 807–814.

Knoll, A. H., 1983, Biological interactions and Precambrian eukaryotes, in: *Biotic Interactions in Recent and Fossil Benthic Communities* (M. J. S. Tevesz and P. C. McCall, eds.), Plenum Press, New York, pp. 251–281.

Knoll, A. H., 1989, The paleomicrobiological information in Proterozoic rocks, in: *Microbial Mats—Physiological Ecology of Microbial Communities* (Y. Cohen and E. Rosenberg, eds.), American Society of Microbiology, Washington, D.C., pp. 469–484.

Knoll, A. H., and Swett, K., 1987, Micropaleontology across the

Precambrian–Cambrian boundary in Spitsbergen, *J. Paleontol.* **61:**898–926.

Knoll, A. H., Swett, K., and Burkhardt, E., 1989, Paleoenvironmental distribution of microfossils and stromatolites in the Upper Proterozoic Backlundtoppen Formation, Spitsbergen, *J. Paleontol.* **63:**129–145.

Kontorovich, A. E., Mandel'baum, M. M., Surkov, V. S., Trofimuk, A. A., and Zolotov, A. N., 1990, Lena–Tunguska Upper Proterozoic–Palaeozoic petroleum superprovince, in: *Classic Petroleum Provinces* (J. Brooks, ed.), Geological Society of London Special Publication 50, pp. 473–489.

Kröner, A. (ed.), 1981, *Precambrian Plate Tectonics*, Elsevier, Amsterdam.

Lake, J. A., 1988, Origin of the eukaryotic nucleus determined by rate-invariant analysis of rRNA sequences, *Nature* **331:**184–186.

Larter, S. R., 1984, Application of analytical pyrolysis techniques to kerogen characterizations and fossil fuel exploration/exploitation, in: *Analytical Pyrolysis* (K. J. Voorhees, ed.), Butterworths, London, pp. 212–275.

Leventhal, J., Suess, S. E., and Cloud, P., Jr., 1975, Nonprevalence of biochemical fossils in kerogen from pre-Phanerozoic sediments, *Proc. Natl. Acad. Sci. U.S.A.* **72:**4706–4710.

Leventhal, J. S., Grauch, R. I., Threlkeld, C. N., Lichte, F. E., and Harper, C. T., 1987, Unusual organic matter associated with uranium from the Claude deposit, Cluff Lake, Canada, *Econ. Geol.* **82:**1169–1176.

Lewan, M. D., Winters, J. C., and McDonald, J. H., 1979, Generation of oil-like pyrolysates from organic-rich shales, *Science* **203:** 897–899.

Lowe, D. R., 1983, Restricted shallow-water sedimentation of Early Archean stromatolitic and evaporitic strata of the Strelley Pool Chert Pilbara Block, Western Australia, *Precamb. Res.* **19:** 239–283.

Mackenzie, A. S ., 1984, Applications of biological markers in petroleum geochemistry, in: *Adv. Petrol. Geochem.* **1:**115–214.

Macqueen, R. W., and Powell, T. G., 1983, Organic geochemistry of the Pine Point lead–zinc ore field and region, Northwest Territories, Canada, *Econ. Geol.* **78:**1–25.

Manning, D. A. C., 1986, Assessment of the role of organic matter in ore transport processes in low-temperature base-metal systems, *Trans. Inst. Min. Metall.* **95:**B195–B200.

McHugh, D. J., Saxby, J. D., and Tardif, J. W., 1976, Pyrolysis–hydrogenation–gas chromatography of carbonaceous material from Australian sediments, Part I: Some Australian coals, *Chem. Geol.* **17:**243–259.

McIver, R. D., 1967, Composition of kerogen: Clue to its role in the origin of petroleum, in: *Proceedings of the 7th World Petroleum Congress*, Mexico City, Vol. 2, Elsevier, London, pp. 25–36.

McKirdy, D. M., 1974, Organic geochemistry in Precambrian research, *Precamb. Res.* **1:**75–137.

McKirdy, D. M., 1976, Biochemical markers in stromatolites, in: *Stromatolites* (M. R. Walter, ed.), Elsevier, Amsterdam, pp. 163–191.

McKirdy, D. M., 1977, Diagenesis of microbial organic matter: A geochemical classification of its use in evaluating the hydrocarbon-generation potential of Proterozoic and Lower Paleozoic sediments, Amadeus Basin, central Australia, Ph.D., Thesis, Australian National University, Canberra.

McKirdy, D. M., and Hahn, J. D., 1982, The composition of kerogens and hydrocarbons in Precambrian rocks, in: *Mineral Deposits and the Evolution of the Biosphere* (H. D. Holland and M. Schidlowski, eds.), Springer-Verlag, Berlin, pp. 123–154.

McKirdy, D. M., and Imbus, S. W., 1992, Precambrian petroleum: A decade of changing perceptions, in: *Early Organic Evolution and Mineral and Energy Resources* (M. Schidlowski, S. Golubic, M. M. Kimberley, D. M. McKirdy, and P. A. Trudinger, eds.), Springer-Verlag, Berlin, pp. 176–192.

McKirdy, D. M., and Kantsler, A. J., 1980, Oil geochemistry and potential source rocks of the Officer Basin, South Australia, *Aust. Petrol. Explor. Assoc. J.* **20:**68–86.

McKirdy, D. M., and Powell, T. G., 1974, Metamorphic alteration of carbon isotopic composition in ancient sedimentary organic matter: New evidence from Australia and South Africa, *Geology* **2:**591–595.

McKirdy, D. M., Sumartojo, J., Tucker, K. H., and Gostin, V., 1975, Organic, mineralogic and magnetic indications of metamorphism in the Tapley Hill Formation, Adelaide Geosyncline, *Precamb. Res.* **2:**345–373.

McKirdy, D. M., McHugh, D. J., and Tardif, J. W., 1980, Comparative analysis of stromatolitic and other microbial kerogens by pyrolysis–hydrogenation–gas chromatography (PH-GC), in: *Biogeochemistry of Ancient and Modern Environments* (P. A. Trudinger, M. R. Walter, and B. J. Ralph), Springer-Verlag, Berlin, pp. 187–200.

McKirdy, D. M., Aldridge, A. K., and Ypma, P. J. M., 1983, A geochemical comparison of some crude oils from pre-Ordovician carbonate rocks, *Advances in Organic Geochemistry 1981* (M. Bjorøy *et al.*, eds.), Wiley–Heyden, Chichester, England, pp. 99–107.

McKirdy, D. M., Kantsler, A. J., Emmett, J. K., and Aldridge, A. K., 1984, Hydrocarbon genesis and organic facies in Cambrian carbonates of the eastern Officer Basin, South Australia, in: *Petroleum Geochemistry and Source Rock Potential of Carbonate Rocks* (J. G. Palacas, ed.), AAPG Studies in Geology No. 18, American Association of Petroleum Geologists, Tulsa, Oklahoma, pp. 13–31.

McKirdy, D. M., Cox, R. E., Volkman, J. K., and Howell, V. J., 1986, Botryococcane in a new class of Australian non-marine crude oils, *Nature* **320:**57–59.

Meyer, C., 1988, Ore deposits as guides to geologic history of the Earth, *Annu. Rev. Earth Planet. Sci.* **16:**147–171.

Meyerhoff, A. A., 1980, Geology and petroleum fields in Proterozoic and Lower Cambrian Strata, Lena–Tunguska Petroleum Province, Eastern Siberia, in: *Giant Oil and Gas Fields of the Decade* (M. T. Halbouty, ed.), American Association of Petroleum Geologists Memoir 30, pp. 225–252.

Meyerhoff, A. A., 1982, Hydrocarbon resources in arctic and subarctic regions, in: *Arctic Geology and Geophysics* (A. F. Embry and H. R. Balkwill, eds.), Canadian Society of Petroleum Geologists Memoir 8, pp. 451–552.

Moldowan, J. M., and Seifert, W. K., 1980,First discovery of Botryococcane in petroleum, *J. Chem. Soc., Chem. Commun.* **1980:** 912–914.

Moldowan, J. M., Sundararaman, P., and Schoell, M., 1986, Sensitivity of biomarker properties to depositional environment and/or source input in the Lower Toarcian of SW-Germany, *Org. Geochem.* **10:**915–926.

Mossman, D. J., and Dyer, B. D., 1985, The geochemistry of Witwatersrand-type gold deposits and the possible influence of ancient prokaryotic communities on gold dissolution and precipitation, *Precamb. Res.* **30:**303–319.

Murray, G. E., 1965, Indigenous Precambrian petroleum, *Am. Assoc. Petrol. Geol. Bull.* **49:**3–21.

Murray, G. E., Kaczor, M. J., and McArthur, R. E., 1980, Indigenous Precambrian petroleum revisited, *Am. Assoc. Petrol. Geol. Bull.* **64:**1681–1700.

Mycke, B., Michaelis, W., and Degens, E. T., 1988, Biomarkers in sedimentary sulfides of Precambrian age, *Org. Geochem.* **13:** 619–625.

Nagy, B., 1976, Organic chemistry on the young Earth: Evolutionary trends between 3,800 m.y. and 2,300 m.y. ago, *Naturwissenschaften* **63:**499–505.

Nagy, B., and Mossman, D. J., 1992, Stratiform and globular organic matter in the Lower Proterozoic metasediments at Elliot Lake, Ontario, Canada, in: *Early Organic Evolution and Mineral and Energy Resources* (M. Schidlowski, S. Golubic, M. M. Kimberley, D. M. McKirdy and P. A. Trudinger, eds.), Springer-Verlag, Berlin, pp. 224–231.

Neudert, M. K., and Russell, R. E., 1981, Shallow water and hypersaline features from the Middle Proterozoic Mt. Isa Sequence, *Nature* **293:**284–286.

Oehler, J. H., 1977, Irreversible contamination of Precambrian kerogen by ^{14}C-labelled organic compounds, *Precamb. Res.* **4:** 221–227.

Ourisson, G., Rohmer, M., and Prolla, K., 1987, Prokaryotic hopanoids and other polyterpenoid sterol surrogates, *Annu. Rev. Microbiol.* **41:**301–333.

Palacas, J. G. (ed.), 1984, *Petroleum Geochemistry and Source Rock Potential of Carbonate Rocks*, AAPG Studies in Geology No. 18, American Association of Petroleum Geologists, Tulsa, Oklahoma.

Peat, C. J., Muir, M. D., Plumb, K. A., McKirdy, D. M., and Norvick, M. S., 1978, Proterozoic microfossils from the Roper Group, Northern Territory, *Bureau of Mineral Resources J. Aust. Geol. Geophys.* **3:**1–17.

Petersen, E. V., 1974, Promising new play gaining momentum in the Great Bear Lake area of NWT, *Oilweek* **1974**(Nov. 25):14–15.

Philp, R. P., 1985, Biological markers in fossil fuel production, *Mass. Spectrom. Rev.* **4:**1–54.

Philp, R. P., and Lewis, C. A., 1987, Organic geochemistry of biomarkers, *Annu. Rev. Earth Planet. Sci.* **15:**363–395.

Philp, R. P., and van de Meent, D., 1983, Characterization of Precambrian kerogens by analytical pyrolysis, *Precamb. Res.* **20:** 3–16.

Philp, R. P., and Zhaoan, F., 1987, Geochemical investigation of oils and source rocks from Qianjiang Depression in Jianghan Basin, a terrigenous saline basin, China, *Org. Geochem.* **11:** 549–562.

Plumb, K. A., and James, H. L., 1986, Subdivisions of Precambrian time: Recommendations and suggestions by the subcommission on Precambrian stratigraphy, *Precamb. Res.* **20:**3–16.

Powell, T. G., and Macqueen, R. W., 1984, Precipitation of sulfide ores and organic matter: Sulfate reactions at Pine Point, Canada, *Science* **224:**63–66.

Powell, T. G., and McKirdy, D. M., 1975, Crude oil composition in Australia and Papua–New Guinea, *Am. Assoc. Petrol. Geol. Bull.* **59:**1176–1197.

Powell, T. G., Cook, P. J., and McKirdy, D. M., 1975, Organic geochemistry of phosphorites: Relevance to petroleum genesis, *Am. Assoc. Petrol. Geol. Bull.* **59:**618–632.

Pratt, L. M., Summons, R. E., and Hieshima, G. B., 1991, Sterane and triterpane biomarkers in the Precambrian Nonesuch Formation, North American Midcontinent Rift, *Geochim. Cosmochim. Acta* **55:**911–916.

Puttmann, W., Hagemann, H. W., Merz, C., and Speczik, S., 1988, Influence of organic material on mineralization processes in the Permian Kupferschiefer Formation, Poland, *Org. Geochem.* **13:**357–363.

Rauzi, S. L.,1990, Distribution of Proterozoic Hydrocarbon Source Rock in Northern Arizona and Southern Utah, *Arizona Oil and Gas Conservation Committee Special Publication.*

Reimer, T. O., Barghoorn, E. S., and Margulis, L., 1979, Primary production in an Early Archean microbial ecosystem, *Precamb. Res.* **9:**93–104.

Risatti, J. B., Rowland, S. J., Yon, D. A., and Maxwell, J. R., 1984, Stereochemical studies of acyclic isoprenoids—XII. Lipids of methanogenic bacteria and possible contributions to sediments, in: *Advances in Organic Geochemistry 1983* (P. A. Schenck, J. W. de Leeuw, and G. W. M. Lijmbach, eds.), Pergamon Press, Oxford, pp. 105–114.

Ronov, A. B., 1980, Osadochnaya Obolochka Zemli Kolichestvennyye Zakonomernosti Stroyeniya Sostavai, Izdatel'stvo Nauka, Moscow; *Int. Geol. Rev.* **24:**1313–1388 (1982).

Ronov, A. B., and Migdisov, A. A., 1971, Geochemical history of the crystalline basement and the sedimentary cover of the Russian and North American platforms, *Sedimentology* **16:**137–185.

Salop, L. J., 1983, *Geological Evolution of the Earth during the Precambrian*, Springer-Verlag, New York.

Saxby, J. D., 1976, The significance of organic matter in ore genesis, in: *Handbook of Strata-Bound and Stratiform Ore Deposits* (K. H. Wolf, ed.), Elsevier, Amsterdam, pp. 111–133.

Schidlowski, M., 1982, Content and isotopic composition of reduced carbon in sediments, in: *Mineral Deposits and the Evolution of the Biosphere* (H. D. Holland and M. Schidlowski, eds.), Springer-Verlag, Berlin, pp. 103–122.

Schidlowski, M., 1987, Application of stable carbon isotopes to early biochemical evolution of Earth, *Annu. Rev. Earth Planet. Sci.* **15:**47–72.

Schidlowski, M., 1988, A 3,800-million-year isotopic record of life from carbon in sedimentary rocks, *Nature* **333:**313–318.

Schidlowski, M., Appel, P. W. U., Eichmann, R., and Junge, C. E., 1979, Carbon isotope geochemistry of the 3.7 billion year-old Isua sediments, West Greenland: Implications for the Archean carbon and oxygen cycles, *Geochim. Cosmochim. Acta* **43:** 189–200.

Schidlowski, M., Hayes, J. M., and Kaplan, I. R., 1983, Isotopic inferences of ancient biochemistries: Carbon, sulfur, hydrogen, and nitrogen, in: *Earth's Earliest Biosphere* (J. W. Schopf, ed.), Princeton University Press, Princeton, New Jersey, pp. 149–186.

Schoell, M., and Wellmer, F. W., 1981, Anomalous ^{13}C depletion in early Precambrian graphites from Superior province, Canada, *Nature* **290:**696–699.

Schopf, J. W., 1975, Precambrian paleobiology: Problems and perspectives, *Annu. Rev. Earth Planet. Sci.* **3:**213–249.

Schopf, J. W. (ed.), 1983, *Earth's Earliest Biosphere*, Princeton University Press, Princeton, New Jersey.

Schopf, J. W., and Packer, B. M., 1987, Early Archean (3.3-billion to 3.5-billion-year-old) microfossils from Warrawoona Group, Australia, *Science* **237:**70–73.

Schopf, J. W., Hayes, J. M., and Walter, M. R., 1983, Evolution of Earth's earliest ecosystems: Recent progress and unsolved problems, in: *Earth's Earliest Biosphere* (J. W. Schopf, ed.), Princeton University Press, Princeton, New Jersey, pp. 361–384.

Seifert, W. K., and Moldowan, J. M., 1978, Applications of steranes, terpanes and monoaromatics to the maturation, migration and source of crude oils, *Geochim. Cosmochim. Acta* **42:**77–95.

Shi, Ji-Yang, Mackenzie, A. S., Alexander, R., Eglinton, G., Gowan, A. P., Wolff, G. A., and Maxwell, J. R., 1982, A biological marker investigation of petroleums and shales from the Shengli oilfield, The People's Republic of China, *Chem. Geol.* **35:**1–31.

Sinninghe-Damsté, J. S., and de Leeuw, J. W., 1987, The origin and fate of isoprenoid C_{20} and C_{15} sulfur compounds in sediments and oils, *Int. J. Environ. Anal. Chem.* **28**:1–19.

Sklarew, D. S., and Nagy, B., 1979, 2,5-Dimethyl furan from 2.7 × 10^9 year old Rupemba–Belingwe stromatolite, Rhodesia: Potential evidence for remnants of carbohydrates, *Proc. Natl. Acad. Sci. U.S.A.* **76**:10–14.

Sozinov, N. A., and Gorbachev, O. V., 1987, Precambrian carbonaceous formations: Their evolution and metal content, in: *Proterozoic Lithospheric Evolution* (A. Kröner, ed.), American Geophysical Union Series 17, American Geophysical Union, Washington, D.C., pp. 35–42.

Stach, E., Mackowsky, M.-Th., Teichmuller, R., Taylor, G. H., Chandra, D., and Teichmuller, R. (eds.), 1982, *Stach's Textbook of Coal Petrography*, Gebruder Borntraeger, Berlin.

Strauss, H., 1986, Carbon and sulfur isotopes in Precambrian sediments from the Canadian Shield, *Geochim. Cosmochim. Acta* **50**:2653–2662.

Strauss, H., Des Marais, D. J., Hayes, J. M., and Summons, R. E., 1992a, Proterozoic organic carbon—its preservation and isotopic record, in: *Early Organic Evolution and Mineral and Energy Resources* (M. Schidlowski, S. Golubic, M. M. Kimberley, D. M. McKirdy, and P. A. Trudinger, eds.), Springer-Verlag, Berlin, pp. 201–211.

Strauss, H., Des Marais, D. J., Summons, R. E., and Hayes, J. M., 1992b, Proterozoic biogeochemistry: Concentrations of organic carbon and maturities and elemental compositions of kerogens, in: *The Proterozoic Biosphere: A Multidisciplinary Study* (J. W. Schopf and C. Klein, eds.), Cambridge University Press, Cambridge, pp. 95–99.

Summons, R. E., 1992, Proterozoic biogeochemistry: Abundance and composition of extractable organic matter, in: *The Proterozoic Biosphere: A Multidisciplinary Study* (J. W. Schopf and C. Klein, eds.), Cambridge University Press, Cambridge, pp. 101–115.

Summons, R. E., and Hayes, J. M., 1992, Proterozoic biogeochemistry: Principles of molecular and isotopic biogeochemistry, in: *The Proterozoic Biosphere: A Multidisciplinary Study* (J. W. Schopf and C. Klein, eds.), Cambridge University Press, Cambridge, pp. 83–93.

Summons, R. E., and Jahnke, L. L., 1990, Identification of the methylhopanes in sediments and petroleum, *Geochim. Cosmochim. Acta* **54**:247–251.

Summons, R. E ., and Powell, T. G., 1991, Petroleum source rocks of the Amadeus Basin, in: *Geological and Geophysics Studies in the Amadeus Basin, Central Australia* (R. J. Korsch and J. M. Kennard, eds.), Bureau of Mineral Resources, Australia, Bulletin 236, pp. 511–524.

Summons, R. E., and Powell, T. G., 1992, Hydrocarbon composition of the Late Proterozoic oils of the Siberian Platform: Implications for the depositional environment of source rocks, in: *Early Organic Evolution and Mineral and Energy Resources* (M. Schidlowski, S. Golubic, M. M. Kimberley, D. M. McKirdy, and P. A. Trudinger, eds.), Springer-Verlag, Berlin, pp. 296–307.

Summons, R. E., and Walter, M. R., 1990, Molecular fossils and microfossils of prokaryotes and protists from Proterozoic sediments, *Am. J. Sci.* **290-A**:212–244.

Summons, R. E., Powell, T. G., and Boreham, C. J., 1988a, Petroleum geology and geochemistry of the Middle Proterozoic McArthur Basin, northern Australia: III. *Geochim. Cosmochim. Acta* **52**: 1747–1763.

Summons, R. E., Brassell, S. C., Eglinton, G., Evans, E., Horodyskik, R. J., Robinson, N., and Ward, D. M., 1988b, Distinctive hydrocarbon biomarkers from fossiliferous sediment of the Late Proterozoic Walcott Member, Chuar Group, Grand Canyon, Arizona, *Geochim. Cosmochim. Acta* **52**:2625–2637.

Sundquist, E. T., and Broecker, W. S. (eds.), 1985, *The Carbon Cycle and Atmospheric CO_2: Natural Variations Archean to Present*, Geophysical Monograph 32, American Geophysical Union, Washington, D.C.

Tannenbaum, E., Ruth, E., and Kaplan, I. R ., 1986, Steranes and triterpanes generated from kerogen pyrolysis in the absence and presence of minerals, *Geochim. Cosmochim. Acta* **50**: 805–812.

Taylor, R. F., 1984, Bacterial triterpenoids, *Microbiol. Rev.* **48**:181–198.

Teichmüller, M., 1974, Generations of petroleum-like substances in coal seams as seen under the microscope, in: *Advances in Organic Geochemistry 1973* (B. Tissot and F. Bienner, eds.), Editions Technip, Paris, pp. 379–407.

Teichmüller, M., 1975, Origin of the petrographic constituents of coal, in: *Stach's Textbook of Coal Petrology* (E. Stach, M. Th. Mackowski, M. Teichmüller, G. H. Taylor, D. Chandra, and R. Teichmuller, eds.), Gebruder Borntraeger, Berlin, pp. 176–238.

ten Haven, H. L., de Leeuw, J. W., Peakman, T. M., and Maxwell, J. R., 1986, Anomalies in steroid and hopanoid maturity indices, *Geochim. Cosmochim. Acta* **50**:853–855.

ten Haven, H. L., de Leeuw, J. W., Sinninghe-Damsté, J. S., Schenck, P. A., Palmer, S. E., and Zumberge, J. E., 1988, Application of biological markers in the recognition of palaeohypersaline environments, in: *Lacustrine Petroleum Source Rocks* (K. Kelts, A. Fleet, and M. Talbot, eds.), Blackwell Scientific, Oxford, pp. 123–130.

Thickpenny, A., and Leggett, J K., 1987, Stratigraphic distribution and palaeo-oceanographic significance of European early Palaeozoic organic-rich sediments, in: *Marine Petroleum Source Rocks* (J. Brooks and A. J. Fleet, eds.), Blackwell Scientific, Oxford, pp. 231–247.

Tissot, B. P., and Welte, D. H., 1984, *Petroleum Formation and Occurrence*, 2nd ed. Springer-Verlag, Berlin.

Tissot, B., Durand, B., Espitalié, J., and Combaz, A., 1974, Influence of nature and diagenesis of organic matter in formation for petroleum, *Am. Assoc. Petrol. Geol. Bull.* **58**:499–506.

Tissot, B., Pelet, R., Roucache, J., and Combaz, A., 1977, Utilisation des alcanes comme fossiles géochimiques indicateurs des environnements géologiques, in: *Advances in Organic Geochemistry 1975* (R. Campos, and J. Goni, eds.), Madrid, ENADIMSA, pp. 117–156.

Townson, W. G., 1985, The subsurface geology of the western Officer Basin—results of Shell's 1980–1984 petroleum exploration campaign, *Aust. Petrol. Explor. Assoc. J.* **25**:34–51.

Trudinger, P. A., Walter, M. R., and Ralph, B. J. (eds.), 1980, *Biogeochemistry of Ancient and Modern Environments*, Springer-Verlag, Berlin.

van de Meent, D., Brown, S. C., and Philp, P., 1980, Pyrolysis high resolution gas chromatography–mass spectrometry of kerogens and kerogen precursors, *Geochim. Cosmochim. Acta* **44**: 999–1013.

Vassoyevich, V. B., Vysotskiy, I. V., Sokolov, B. A., and Tatarenko, Y. I., 1971, Oil–gas potential of Late Precambrian deposits, *Int. Geol. Rev.* **13**:407–418.

Veizer, J., 1988, The evolving exogenic cycle, in: *Chemical Cycles in the Evolution of the Earth* (C. B. Gragor, R. M. Garrells, F. T. MacKenzie, and J. B. Maynard, eds.), John Wiley & Sons, New York, pp. 175–220.

Vidal, G., and Knoll, A. H., 1982, Radiations and extinctions of plankton in the late Proterozoic and early Cambrian, *Nature* **297**:57–60.

Vidal, G., and Knoll, A. H., 1983, Proterozoic plankton, *Geol. Soc. Am. Mem.* **161:**265–277.

Volkman, J. K., and Maxwell, J. R., 1986, Acyclic isoprenoids as biological markers, in: *Biological Markers in the Sedimentary Record* (R. B. Johns, ed.), Elsevier, Amsterdam, pp. 1–42.

Walter, M. R. (ed.), 1976, *Stromatolites*, Elsevier, Amsterdam.

Walter, M. R., 1987, The timing of major evolutionary innovations from the origins of life to the Metaphyta and Metazoa: The geological evidence, in: *Rates of Evolution* (K. W. S. Campbell and M. F. Day, eds.), Allen and Unwin, London, pp. 15–33.

Walter, M. R., Oehler, J. H., and Oehler, O. Z., 1976, Megascopic algae 1300 million years old from the Belt Supergroup, Montana: A reinterpretation of Walcott's Helminthoidichnites, *J. Paleontol.* **50:**872–881.

Walter, M. R., Buick, R., and Dunlop, J. S. R., 1980, Stromatolites 3.4–3.5 billion years old from the North Pole area, Pilbara Block, Western Australia, *Nature* **441:**443–445.

Waples, D. W., 1984, Thermal models for oil generation, *Adv. Petrol. Geochem.* **1:**7–67.

Wedeking, K. W., and Hayes, J. M., 1983, Carbonization of Precambrian kerogens, in: *Advances in Organic Geochemistry 1981* (M. Bjorøy *et al.*, eds.), Wiley, Chichester, England, pp. 546–553.

Williams, N., 1974, Epigenetic processes in the stratiform lead–zinc deposits and McArthur River, Northern Territory, Australia [abstract], *Geological Society of America Abstracts with Programs*, Vol. 6, pp. 1006–1007.

Williams, N., 1978, Studies of the base metal sulfide deposits at McArthur River, Northern Territory, Australia: II. The sulfide-S and organic-C relationships of the concordant deposits and their significance, *Econ. Geol.* **73:**1036–1056.

Willingham, T. O., Nagy, B., Nagy, L. A., Krinsley, D. H., and Mossman, D. J., 1985, Uranium-bearing stratiform organic matter in paleoplacers of the Lower Huronian Supergroup, Elliot Lake–Blind River region, *Can. J. Earth Sci.* **22:**1930–1944.

Windley, B. F., 1983, A tectonic review of the Proterozoic, *Geol. Soc. Am. Mem.* **161:**1–10.

Windley, B. F., 1984, *The Evolving Continents*, 2nd ed. John Wiley & Sons, New York.

Xu, Y., Shen, P., and Li, Y., 1992, Natural gas in Sinian reservoirs of the Weiyuan area, Sichuan Province: The oldest gas in China, in: *Early Organic Evolution and Mineral and Energy Resources* (M. Schidlowski, S. Golubic, M. M. Kimberley, D. M. McKirdy, and P. A. Trudinger, eds.), Springer-Verlag, Berlin, pp. 317–323.

Zang, W. L., and Walter, M. R., 1989, Latest Proterozoic plankton from the Amadeus Basin in central Australia, *Nature* **337:** 642–645.

Zumberge, J. E., and Nagy, B., 1975, Alkyl substituted cyclic ethers in 2,300 M year-old Transvaal algal stromatolite, *Nature* **255:** 695–696.

Zumberge, J. E., Sigleo, A. C., and Nagy, B., 1978, Molecular and elemental analyses of the carbonaceous matter in the gold and uranium bearing Vaal Reef carbon seams, Witwatersrand Sequence, *Miner. Sci. Eng.* **10:**223–246.

Zumberge, J. E., Sigleo, A. C., and Nagy, B., 1981, Some aspects of Vaal Reef uranium–gold carbon seams, Witwatersrand Sequence, U.S. Geological Survey Professional Paper 1161–0, pp. 1–7.

Chapter 33

The Organic Geochemistry of Carbonaceous Meteorites

Amino Acids and Stable Isotopes

MICHAEL H. ENGEL, STEPHEN A. MACKO, and BARTHOLOMEW NAGY

1. Introduction

The stable carbon isotope record of bulk organic matter (i.e., kerogen) in ancient rock samples indicates that living systems may have existed on Earth 3.5 billion years ago (e.g., Schidlowski, 1987, 1988, 1991). It is generally accepted that life originated on the Earth and that organic compounds deemed essential for its origin, for example, amino acids, formed via prebiotic mechanisms prior to this event. Thus, an appreciation of the distribution, stereochemistry, and stable isotope compositions of amino acids in ancient terrestrial and extraterrestrial systems is of fundamental importance with respect to (1) evaluating mechanisms for prebiotic organic synthesis at the time of (and perhaps prior to) the formation of the solar system (e.g., Ferris, 1984; Cronin *et al.*, 1988; Oró *et al.*, 1990), (2) determining the potential contribution of extraterrestrial organic matter to Earth via comet and meteorite bombardment (e.g., Nagy, 1985; Nagy *et al.*, 1987; Zhao and Bada, 1989; Zahnle and Grinspoon, 1990; Chyba *et al.*, 1990; Chyba and Sagan, 1992), and (3) understanding the nature of the organic compounds available for the origin of the first living systems on Earth (e.g., Mullie and Reisse, 1987; Miller, 1992, and this volume, Chapter 30).

The unambiguous detection of prebiotic amino acids in Precambrian sedimentary rocks has been unsuccessful. This is the result of the scarcity and metamorphic history of these crustal remnants prior to 3.5 Ga and the fact that amino acids are ubiquitous in all living systems on Earth. It has not been possible, by conventional criteria, to distinguish indigenous, Precambrian amino acids from those entering sedimentary systems via fluid migration at later times (e.g., Kvenvolden *et al.*, 1969; Nagy *et al.*, 1981). While, as will be discussed below, the recent development of new methods for direct stable isotope analyses of individual amino acid enantiomers in geologic materials may provide an alternative approach for the study of these compounds in Precambrian sedimentary rocks, the current appreciation of amino acid distributions on the Earth prior to the origin of life is based largely on indirect evidence obtained by analyses of carbonaceous meteorites and by laboratory simulation experiments. Attempts have been made to

MICHAEL H. ENGEL • School of Geology and Geophysics, The University of Oklahoma, Norman, Oklahoma 73019. STEPHEN A. MACKO • Department of Environmental Sciences, The University of Virginia, Charlottesville, Virginia 22903. BARTHOLOMEW NAGY • Department of Geosciences, The University of Arizona, Tucson, Arizona 85721.

Organic Geochemistry, edited by Michael H. Engel and Stephen A. Macko. Plenum Press, New York, 1993.

characterize organic matter in carbonaceous meteorites for over a century (Nagy, 1975). Most recent investigations of amino acids have focused on stones from the Murchison meteorite, an organic-rich (~2.% total organic carbon), type CM2 meteorite that fell in Australia in 1969. Amino acid analyses of other carbonaceous meteorites that fell prior to 1969 are problematic. This is because earlier methods of sample handling and prolonged residence times on Earth enhance the possibility for terrestrial overprints, i.e., contamination (Nagy, 1975).

In this chapter a brief review of the distribution, stereochemistry, and stable isotope composition of amino acids in the Murchison meteorite is presented. These data are discussed within the context of laboratory experiments that have attempted to simulate the prebiotic synthesis of amino acids in the early solar system and the accumulation of an amino acid inventory on the early Earth that would have been necessary for the evolution of life. Finally, it must be emphasized that although amino acids are certainly essential "building blocks" of life, they do not, in themselves, provide the complete organic inventory necessary for the formation of a living cell (Miller, 1992), nor do they reflect the bulk composition of organic matter in carbonaceous meteorites (Chang, 1977; Mullie and Reisse, 1987). The reader is referred to other general reviews for additional information concerning the classes of organic compounds that have been detected in carbonaceous meteorites and that have been synthesized under prebiotic conditions in the laboratory (see, e.g., Nagy, 1975; Chang, 1977; Mullie and Reisse, 1987; Cronin *et al.*, 1988; Miller, 1992; Miller, this volume, Chapter 30).

2. Distribution of Amino Acids in the Murchison Meteorite

Compared to that in living systems and the terrestrial rock record, the distribution of amino acids in the Murchison meteorite is exotic, consisting of numerous nonprotein amino acids, many of which have never or only rarely been associated with living systems on earth (Kvenvolden *et al.*, 1971; Engel and Nagy, 1982; Cronin *et al.*, 1988), and characterized by relatively high concentrations of specific protein amino acids, e.g., glycine, alanine, aspartic acid, and glutamic acid, and the noticeable absence or extremely diminished concentrations of others, e.g., serine, phenylalanine, tyrosine, threonine, lysine, histidine, and arginine (Kvenvolden *et al.*, 1970; Engel and Nagy, 1982, 1983; Cronin *et al.*, 1988). In general, individual amino acid concentrations range from subnanomole to several tens of nanomoles per gram of stone (Cronin *et al.*, 1988). As will be discussed below, it is this exotic distribution that has been used as one of the standard criteria for determining the indigeneity of amino acids in meteorites.

Differences in amino acid concentrations for individual stones of the Murchison meteorite have been commonly attributed to varying levels of terrestrial contamination and differences in laboratory extraction and analytical methods (Bada *et al.*, 1983; Engel and Nagy, 1983; Cronin *et al.*, 1988). The possibility of minor inhomogeneities in amino acid distributions in individual stones (Cronin *et al.*, 1988) as well as the decomposition of amino acids with prolonged residence time on Earth also cannot be entirely excluded at this time. Nevertheless, it is important to note that, with few exceptions, the distributions and absolute concentrations of amino acids reported for the different stones are in fairly good agreement. For example, the amino acid abundances reported by Engel *et al.* (1990) and Silfer (1991) for the nonhydrolyzed water extract of a Murchison stone are almost identical to those previously reported by Cronin (1976a, b) (Table I).

Shock and Schulte (1990) have suggested that practically all of the minor differences in amino acid distributions and abundances reported for Murchison stones may, based on solubility data, simply reflect differences in extraction procedures. To date, no extraction procedure exists for the complete recovery of all of the amino acids in a meteorite sample. In

TABLE I. Amino Acid Abundances in Nonhydrolyzed Water Extracts of the Murchison Meteorite[a]

	Abundance (nmol g^{-1})		
Amino acid	Engel *et al.* (1990) and Silfer (1991)	Cronin (1976a)	Cronin (1976b)
Glycine	28.1	31.0	25.5
α-Aminoisobutyric acid	16.0	19.9	15.0
Alanine	12.9	17.1	15.9
β-Alanine	8.1	6.9	5.7
Isovaline + valine	7.5	4.6	6.0
Glutamic acid	4.6	3.5	1.9
Aspartic acid	3.9	1.0	1.2
α-Amino-*n*-butyric acid	3.7	5.3	4.6
Serine	3.3	1.7	1.9
Threonine	2.0	1.2	0.8
Isoleucine	1.8	0.9	1.4
Leucine	1.6	0.8	1.2

[a]Amino acid abundances were determined by high-performance liquid chromatography.

support of this suggestion, it is well known that the recovery of amino acids from a water extract of the Murchison meteorite is enhanced by acid hydrolysis of the water residue (Cronin, 1976a) and that repeated extractions of the same meteorite sample lead to the recovery of additional amino acids (Engel and Nagy, 1982; Silfer, 1991). While the recovery of amino acids via acid digestion of the meteorite stone subsequent to water extraction (Engel and Nagy, 1982) has been suggested to reflect the release of amino acid constituents of terrestrial protein contaminants (Bada *et al.*, 1983), the distribution of amino acids released by acid digestion does not resemble a common protein contaminant. Engel and Nagy (1983) have suggested that this amino acid fraction is more likely representative of components incorporated into the kerogen matrix of the stone that are not "released" by the initial, milder water reflux. Becker and Epstein (1982) have also reported the occurrence of intractable organic matter in the Murchison meteorite that is not released until dissolution of the mineral phase. Engel and Nagy (1982) also observed that some of the more abundant (and exotic) amino acids (e.g., α-aminoisobutyric acid, isovaline) appear to be the most easily extracted with water. Perhaps this reflects alternative, diagenetic pathway(s) of formation (e.g., Peterson *et al.*, 1991) subsequent to the initial synthesis of these compounds.

With respect to potential mechanisms for the prebiotic synthesis of amino acids in carbonaceous meteorites, Strecker synthesis (Miller, 1992, and this volume, Chapter 30) and, to a lesser extent, Fischer–Tropsch synthesis (Hayatsu and Anders, 1981) have received attention, as both pathways lead to the formation of amino acids that have been detected in the Murchison meteorite. It should be stressed, however, that alternative mechanisms have also been proposed (e.g., Sagan and Khare, 1971; Matthews *et al.*, 1977; McPherson *et al.*, 1987) and that the possibility also exists that a percentage of the amino acids in the Murchison meteorite may in fact reflect a contribution of precursors from interstellar space (Cronin *et al.*, 1988; Kerridge, 1991).

By comparing the amino acids in the Murchison meteorite with those formed in laboratory simulation experiments, the assumptions are made that only one of the proposed mechanisms is valid and that the Murchison amino acids reflect a "primary" distribution at the time of formation, with minimal, subsequent alteration. Given the information currently available, there is no compelling reason to accept either of these assumptions. In fact, the organic matter in the Murchison meteorite may reflect a diverse history of prebiotic synthetic processes, including components derived from the presolar nebula (Robert and Epstein, 1982; Yang and Epstein 1983; Epstein *et al.*, 1987; Cronin *et al.*, 1988; Kerridge, 1991). Also, the potential alteration of organic matter by, for example, cosmic radiation (Roessler *et al.*, 1991) and shock-induced reactions during impact (Peterson *et al.*, 1991) leaves open the question of whether the organic compounds in carbonaceous meteorites truly represent the original distribution at the time of synthesis.

3. Amino Acid Stereochemistry

Amino acids that comprise living systems on Earth consist almost exclusively of the L-enantiomer. How this "handedness" came to be has been the subject of intense investigation for many years (Bonner, 1991, and references therein). Central to this problem has been the ongoing debate of whether optical activity was a necessary precondition for the origin of life (Bonner, 1991; Goldanskii and Kuzmin, 1991), a natural consequence of processes during the early stages of formation of the first primitive cells (Miller, 1992), or a random event during the early stages of the origin of life (Ferris, 1984). The development of routine gas-chromatographic methods for resolving the D- and L-enantiomers of amino acids (Gil-Av *et al.*, 1965; Pollock *et al.*, 1965) permitted the establishment of the stereochemical apportionment of these compounds in ancient terrestrial and extraterrestrial materials.

3.1. Terrestrial Systems

Hare and Mitterer (1966) were the first to observe that, upon death, amino acid constituents of organisms undergo racemization with the passage of time, eventually approaching racemic mixtures of the L-reactant and D-product. This led to the assumption that the occurrence of nonracemic (L-enantiomer predominating) amino acids in significantly older (i.e., greater than several hundred thousand to several million years) fossils and sediments is likely to reflect contamination from the surrounding environment (e.g., Williams and Smith, 1977). As a result, earlier observations of nonracemic amino acids in Precambrian sedimentary rocks (Schopf *et al.*, 1968) were attributed to contamination (Abelson and Hare, 1969; Kvenvolden *et al.*, 1969; Nagy *et al.*, 1981).

Nonracemic amino acids have, however, been reported in well-preserved fossil mollusks ranging in age from late Cretaceous (Buchardt and Weiner, 1981) to Miocene (Hoering, 1980). More recently, Rafalska *et al.* (1991) have reported that diagenetic processes resulting in the incorporation of amino acids into humic-like materials retard the racemization process. Thus, it is possible that a portion of the nonracemic amino acids in ancient sedimentary systems are actually indigenous. The problem that remains is how to isolate this indigenous component from more recent overprints. Alternative, indirect approaches for the assessment of the stereochemistry of amino acids on the ancient Earth have consisted of investigations of carbonaceous meteorites and the products of laboratory simulation experiments.

3.2. Carbonaceous Meteorites

Whereas the distribution of amino acids in Murchison meteorite stones has been investigated in detail (Cronin *et al.*, 1988, and references therein), there are surprisingly few published accounts concerning the stereochemistry of amino acids isolated from these stones. In their initial studies, Kvenvolden *et al.* (1970, 1971) reported that several of the protein (e.g., alanine, valine, proline, glutamic acid) and nonprotein (e.g., α-amino-*n*-butyric acid, pipecolic acid) amino acids in the Murchison meteorite were approximately racemic. Pollock *et al.* (1975) later isolated and resolved the D- and L-enantiomers of isovaline from a Murchison stone. This amino acid was shown to be racemic. Based on the assumption that prebiotic synthesis should result in racemic amino acids, it was concluded that the amino acids in this Murchison stone were indeed extraterrestrial and that the slight excesses of the L-enantiomers were a consequence of terrestrial contamination, which might have occurred, for example, upon impact or during sample handling and/or preparation (Kvenvolden *et al.*, 1970).

More recently, Engel and Nagy (1982) and Engel *et al.* (1990) reported that the enantiomeric compositions of amino acid constituents of the Murchison meteorite were quite varied, ranging from approximately racemic (e.g., isovaline, α-amino-*n*-butyric acid) to only partially racemized (e.g., glutamic acid, aspartic acid). Alanine was not entirely racemic, with D/L values ranging from 0.67 in hydrolyzed water extracts to 0.85 in nonhydrolyzed water extracts. Also, it was observed that while the overall amino acid distribution was similar, the extent of racemization was less in acid extracts of residual meteorite stone than in the initial nonhydrolyzed and hydrolyzed water extracts of the same samples (Engel and Nagy, 1982; Silfer, 1991).

The occurrence of nonracemic amino acids in the Murchison meteorite has been commonly attributed to the introduction of an excess of the L-enantiomer from terrestrial sources. It is, however, interesting that the overall amino acid distributions and abundances in extracts of these stones were quite similar to those reported by others in previous studies (Engel and Nagy, 1983; Shock and Schulte, 1990) (Table I). If the stones were contaminated by, for example, microbes, it is difficult to account for the low to below detection limit levels of many of the common amino acids that comprise the proteins of living systems, e.g., lysine, histidine, arginine, threonine, serine, phenylalanine, and tyrosine (Engel and Nagy, 1982, 1983; Engel *et al.*, 1990). Also, if it is eventually demonstrated that nonracemic amino acids are indigenous to the Murchison stone, the question arises as to how they came to be this way. As will be discussed below, new analytical methods for determining the stable isotope composition of individual amino acid enantiomers in extracts of Murchison stones may provide a solution to the indegeneity question. The question concerning the origin of optical activity is also briefly discussed below.

3.3. Simulation Experiments

The distribution of amino acids resulting from laboratory simulation experiments has been studied in detail (e.g., Cronin *et al.*, 1988; Miller, 1992, and references therein). Although there exists little documentation on the stereochemistry of the amino acid products of earlier, prebiotic experiments, the results of more recent work indicate that the amino acids are racemic. For example, Khare *et al.* (1986) reported that common protein amino acids isolated from synthetic tholins were racemic. Similarly, Hennet *et al.* (1992) have reported the detection of racemic alanine, aspartic acid, and glutamic acid resulting from the prebiotic synthesis of amino acids under hydrothermal vent conditions. Spark discharge experiments performed in our laboratory (unpublished results) have also resulted in the formation of racemic amino acids.

Recent simulation studies in which L-amino acids were added to powdered samples of the Allende meteorite and subsequently shocked (to simulate impact) resulted in partial racemization and the secondary formation of protein and nonprotein amino acids (Peterson *et al.*, 1991). While it is unclear how primarily L-amino acids could initially occur in a meteorite via stereoselective synthesis or alteration, a similarity exists between the results of this experiment and the distribution and stereochemistry reported by Engel and Nagy (1982) and Engel *et al.* (1990) for Murchison amino acids.

4. Stable Isotopes

In the formulation of hypotheses concerning the origin(s) of organic matter in carbonaceous meteorites, studies of organic compound distributions and stereochemistry have been augmented by investigations of noble gases and the stable isotope ($\delta^{13}C$, δD, $\delta^{15}N$) compositions of insoluble organic matter (i.e., kerogen), carbonate, bulk organic extracts, and, more recently, individual organic compounds, i.e., alkanes, carboxylic acids, and amino acids. Isotopic anomalies of noble gases (e.g., Anders, 1988, and references therein) and the extreme enrichment of deuterium (>+1000‰) (Robert and Epstein, 1982; Yang and Epstein, 1983; Kerridge *et al.*, 1987; Epstein *et al.*, 1987; Zinner, 1988) and carbon (e.g., Swart *et al.*, 1983; Zinner, 1988) in some Murchison components indicate that at least a portion of this material is likely to be interstellar (and perhaps circumstellar) in origin. Kerridge (1991) has recently proposed a model whereby interstellar components (e.g., cyanide, ammonia, aldehydes, ketones) reacted within the aqueous medium of the meteorite parent body via Strecker synthesis to form a more complex suite of organic acids. More detailed information on this topic may be found in Cronin *et al.* (1988) and Kerridge (1991).

A limiting factor with respect to using the currently available stable isotope data for individual organic compounds as the basis for defining prebiotic synthetic pathways is that the present data set is quite small, with little in the way of independent confirmations of the original reports of Yuen *et al.* (1984) for the alkanes and carboxylic acids and Engel *et al.* (1990) and Pizzarello *et al.* (1991) for the amino acids. Also, these initial reports present data for only a few of the components of each compound class. This is a consequence of the challenges encountered when attempting to perform these types of analyses. What follows is a brief discussion of new technology that may lead to the expansion of this initial data set and a summary of the stable isotope data currently available for amino acids isolated from the Murchison meteorite.

4.1. Analytical Procedures

It must be recalled that the concentrations of individual amino acids in extracts of the Murchison meteorite are on the order of nanomoles per gram. The isolation of individual amino acids by liquid chromatography, therefore, requires the processing of very large amounts of stone (50 grams or more) to reach the limits of detection for stable isotope analyses of individual amino acids by conventional methods. Also, chromatographic procedures for the isolation of organic compounds for stable isotope analyses are complicated by column bleed (Pizzarello *et al.*, 1991) and the potential for isotopic fractionation if the recovery of an individual component is not quantitative (Macko *et al.*, 1987, 1990). Given these constraints, it is not surprising that there have been very few reports of individual amino acids isolated by chromatographic methods for stable isotope analysis (see, e.g., Abelson and Hoering, 1961; Engel and Macko, 1984, 1986; Macko *et al.*, 1986, 1987; Serban *et al.*, 1988; Hare *et al.*, 1991; Pizzarello *et al.*, 1991).

One of the most significant advances in the stable isotope analysis of organic matter has been the development of combined gas chromatography/isotope-ratio mass spectrometry (GC/IRMS). This method permits the direct analysis of individual organic compounds at subnanomole levels (Freedman *et al.*, 1988; Hayes *et al.*, 1989), provided that they are sufficiently volatile for separation by GC.

Hydrocarbons are particularly attractive targets for GC/IRMS analysis in that they are volatile and hence do not require derivatization. Numerous applications of this approach for determining the $\delta^{13}C$ compositions of hydrocarbons in terrestrial (Kennicutt and Brooks, 1990; Freeman *et al.*, 1990; Hayes *et al.*, 1990; Bjorøy *et al.*, 1991; Rieley *et al.*, 1991) and extraterrestrial materials (Franchi *et al.*, 1989) have been recently reported.

The application of GC/IRMS for amino acid analysis is complicated by the fact that, unlike hydrocarbons, amino acids require derivatization to achieve adequate volatility for analysis by GC. Although the derivatization procedure introduces additional carbon and apparent fractionations during esterification and acylation (Silfer *et al.*, 1991), carbon isotope compositions of individual amino acids and their respective D- and L-enantiomers can be computed through analysis of standards prepared in a similar fashion (Silfer *et al.*, 1991). An ion trace (mass 44) for the continuous combustion of a standard mixture of trifluoroacetyl isopropyl esters of D- and L-amino acids is shown in Fig. 1. The concentration of each amino acid was approximately 2 to 3 nmole. Additional GC/IRMS procedures for the analysis of compounds that require derivatization, e.g., fatty acids (Abrajano *et al.*, 1992; Goodman and Brenna, 1992) and carbohydrates (Macko *et al.*, 1992), have recently been reported.

An obvious limitation of GC/IRMS is that it is currently only possible to analyze carbon isotopes by this method. However, modifications of this method for the analysis of δD (Fenwick *et al.*, 1992) and $\delta^{15}N$ (Brand and Tegtmeyer, 1992) are currently under investigation. With respect to the analysis of organic constituents of carbonaceous meteorites, GC/IRMS is the method of choice because the amount of meteorite sample required for analysis is significantly reduced and potential problems associated with column bleed and isotope fractionation by liquid-chromatographic methods can be avoided.

4.2. Stable Isotope Analyses of Amino Acids in the Murchison Meteorite

As indicated above, stable isotope values of amino acid constituents of the Murchison meteorite may provide new insights into their mechanisms of formation (Cronin *et al.*, 1988; Kerridge, 1991). Also, the fact that the amino acids are significantly enriched in ^{13}C, D, and ^{15}N relative to terrestrial organic matter may provide a means of assessing the extent to which Murchison stones have been compromised by contamination.

The initial $\delta^{13}C$ values for amino acids in Murchison stones were reported for fractions eluted from paper-chromatographic bands (Chang *et al.*, 1978) and a bulk extract (Epstein *et al.*, 1987). While all of the values were enriched in ^{13}C (Table II), the value reported by Epstein *et al.* (+23.1‰) is on the depleted end of the range reported by Chang *et al.* (+23–+44‰). This discrepancy can likely be attributed to differences in separation procedures rather than sample inhomogeneities. It must be recalled that bulk amino acid extracts of complex samples such as those from the Murchison meteorite may contain over 100 individual components, each having a distinct $\delta^{13}C$ value. Slight differences in extraction methods and/or chromatographic separations may result in differences in recovery, which in turn could shift the bulk isotope values. Also, as discussed above, it is possible to introduce an isotope fractionation during chro-

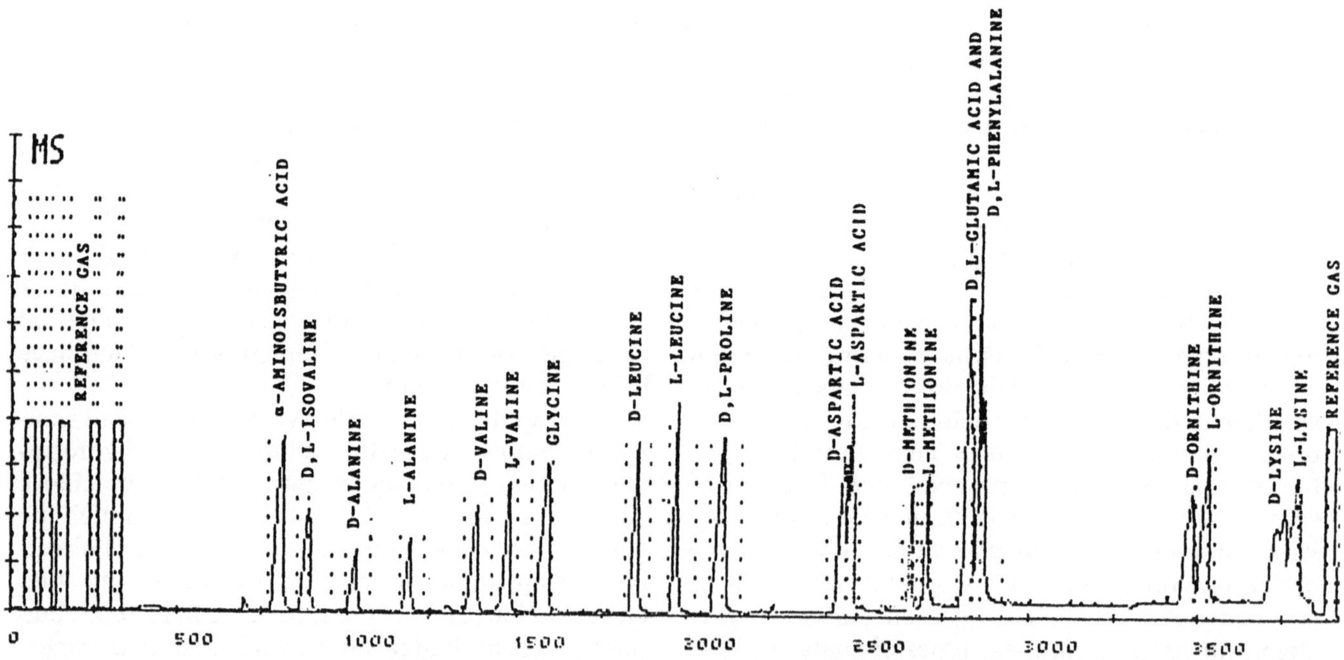

Figure 1. Mass-44 ion trace for the continuous combustion of a standard mixture of trifluoroacetyl isopropyl esters of amino acid enantiomers. The GC/IRMS system consists of a Hewlett-Packard 5890 gas chromatograph (GC) interfaced to a VG PRISM isotope-ratio mass spectrometer (IRMS) via a combustion furnace/water trap. The GC was equipped with a 50-m × 0.25-mm-i.d. fused-silica capillary column coated with an optically active stationary phase (Chirasil-Val; Alltech Associates, Deerfield, Illinois). Details of the method are reported in Silfer *et al.* (1991).

TABLE II. Stable Carbon Isotope Values ($\delta^{13}C$) Reported for Amino Acids in the Murchison Meteorite

	$\delta^{13}C$ (‰ PDB)			
Component	Chang *et al.* (1978)[a]	Epstein *et al.* (1987)[b]	Engel *et al.* (1990)[c]	Pizzarello *et al.* (1991)[d]
Total amino acid extract	+23–+44	+23.1	n.d.[e]	+26
α-Aminoisobutyric acid	n.d.	n.d.	+5	−1
L-Glutamic acid	n.d.	n.d.	+6	n.d.
Aspartic acid	n.d.	n.d.	n.d.	+4
Valine + isovaline	n.d.	n.d.	n.d.	+30
D,L-Isovaline	n.d.	n.d.	+17	n.d.
Glycine	n.d.	n.d.	+22	n.d.
D-Alanine	n.d.	n.d.	+30	n.d.
L-Alanine	n.d.	n.d.	+27	n.d.
Glycine + alanine	n.d.	n.d.	n.d.	+41

[a]The $\delta^{13}C$ values are for a nonhydrolyzed water extract. The range of values are for discrete amino acid fractions (valine, alanine, glycine, glutamic acid) eluted from paper-chromatographic bands. Specific values for individual amino acids were not reported.
[b]The $\delta^{13}C$ value is for the acid-hydrolyzed water extract, corrected for column bleed.
[c]The $\delta^{13}C$ values are for the nonhydrolyzed water extract. Values were determined by gas chromatography/isotope-ratio mass spectrometry (GC/IRMS) and reflect background correction.
[d]The $\delta^{13}C$ values are for the hydrolyzed water extract. Individual amino acids were isolated by high-performance liquid chromatography (HPLC) and reflect background correction for column bleed.
[e]n.d., Not determined.

matographic separation unless the recovery of each component in the mixture is quantitative.

More recently, Pizzarello *et al.* (1991) have attempted a liquid-chromatographic separation of several fractions of amino acids from the Murchison meteorite for stable carbon and hydrogen isotope analyses (Table II). The limitations of this approach are the need for a large meteorite sample (25 g) for analysis, the potential interference from citrate buffers and column bleed, and, more importantly, the fact that each individual fraction, although containing one or two major amino acid components, may also contain additional components with unique isotope signatures. Whereas our initial approach for isolating individual amino acid enantiomers from less complicated mixtures encountered in terrestrial samples involved liquid chromatography employing volatile buffers (Engel and Macko, 1984, 1986; Serban *et al.*, 1988), efforts have been initiated to circumvent some of the inherent problems of this approach by employing GC/IRMS for the direct stable carbon isotope analysis of amino acid constituents of the Murchison meteorite at nanomole levels.

Recent analyses of a 3.8-g interior sample of a 2Murchison stone (Engel *et al.*, 1990; Silfer, 1991) revealed a bulk isotopic composition similar to that previously reported by Kerridge (1985) and an amino acid distribution similar to that previously reported by Engel and Nagy (1982). As in the previous study (Engel and Nagy, 1982), it was observed that amino acids in the extracts of this stone were partially racemized but not entirely racemic (Engel *et al.*, 1990).

It was initially suggested by Pillinger (1982) that one possible way to assess whether the L-enantiomer excess reported by Engel and Nagy (1982) was indigenous or the result of contamination would be to compare the stable isotope compositions of the D- and L-enantiomers of the individual amino acids. Given the ^{13}C enrichments of bulk amino acid extracts for this stone, it was hypothesized that terrestrial contributions of L-amino acid enantiomers would result in marked ^{13}C depletions of the L-enantiomers relative to the D-enantiomers.

The $\delta^{13}C$ values for individual free amino acids isolated from this new Murchison stone were determined by GC/IRMS (Table II). It is readily apparent that the individual components are indeed enriched in ^{13}C relative to terrestrial organic matter. Alanine was not racemic (D/L = 0.85). Whereas L-alanine is slightly depleted in ^{13}C relative to D-alanine, mass-balance considerations indicate that if the excess of the L-enantiomer was contributed from a terrestrial source, it would have had to have a stable carbon isotope composition of about +10‰ (Engel *et al.*, 1990). Thus, a terrestrial overprint appears unlikely. Clearly, GC/IRMS will eventually permit the acquisition of the larger data set that will facilitate evaluations of the origin and indigeneity of organic constituents in terrestrial and extraterrestrial materials.

In addition to establishing indigeneity, the acquisition of $\delta^{13}C$ values for individual amino acid constituents of the Murchison meteorite may provide the key to understanding their respective mechanism(s) of formation. In a previous study, Yuen *et al.* (1984) reported that the $\delta^{13}C$ values of monocarboxylic acids (C_2 to C_5) and *n*-alkanes (C_1 to C_5) decreased with increasing carbon number (Fig. 2). The depletion in ^{13}C was interpreted as a kinetic effect, resulting from the synthesis of higher molecular weight compounds from lower molecular weight homologs (Yuen *et al.*, 1984). A similar kinetic isotope effect has been reported for low-molecular-weight hydrocarbons synthesized from methane in a spark discharge reaction (Chang *et al.*, 1983).

For comparison, the $\delta^{13}C$ values reported by Engel *et al.* (1990) for individual Murchison amino acids are plotted with the hydrocarbon data of Yuen *et al.*

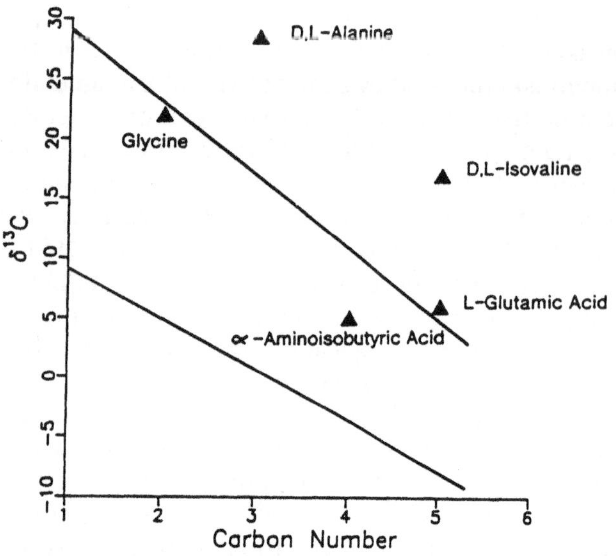

Figure 2. Variation of $\delta^{13}C$ values for amino acids in the nonhydrolyzed water extract of the Murchison meteorite with carbon number (after Silfer, 1991). The value shown for alanine is the average value of the D- and L-enantiomers. Trends for the monocarboxylic acids (upper line) and saturated hydrocarbons (lower line) (after Yuen *et al.*, 1984) are shown for comparison.

(1984) in Fig. 2. While the lower carbon number amino acids (glycine, alanine) are enriched in ^{13}C relative to higher carbon number amino acids (α-aminoisobutyric acid, isovaline, glutamic acid), an assessment of possible kinetic effects during synthesis will require additional $\delta^{13}C$ values for amino acids of increasing carbon number that comprise a homologous series (Silfer, 1991). Finally, the acquisition of $\delta^{13}C$ values for individual amino acids from laboratory experiments (e.g., spark discharge, Fischer–Tropsch reaction) will help to resolve the question as to which, if any, of these mechanisms best accounts for the prebiotic synthesis of amino acids in the Murchison meteorite.

5. Implications for the Origin of Life on Earth

5.1. Sources of Organic Matter

Two fundamental questions concerning the origin of life on Earth are (1) what were the sources of organic matter prior to the origin of life and (2) was optical activity a necessary precondition for life's origin or an early consequence of it? With respect to the first question, two schools of thought exist concerning the accumulation of organic matter on the early Earth. One proposes that the early Earth's organic inventory was a consequence of prebiotic organic synthesis primarily on Earth (e.g., Miller, 1992, and this volume, Chapter 30). The alternative scenario has been that the Earth's initial organic matter was derived via comet/asteroid/meteorite bombardment (e.g., Oró, 1961; Oró *et al.*, 1980; Chyba *et al.*, 1990). The relative importance of either model is dependent to a large extent on the composition of the Earth's early atmosphere (Chyba and Sagan, 1992; Kasting, this volume, Chapter 29), which mediated the types of reactions that could occur on Earth. More importantly, early hydrosphere conditions would have certainly influenced the decomposition rates of organic compounds, irrespective of their initial sources (e.g., Miller, 1982; Bada, 1991).

As stated at the beginning of this chapter, the search for amino acids in the early Precambrian terrestrial rock record has been unsuccessful. The inability to distinguish indigenous amino acids from more recent contaminations (e.g., Kvenvolden *et al.*, 1969; Nagy *et al.*, 1981) is further complicated by the paucity of the sedimentary rock record prior to 3.5 Ga. Also, the oldest crustal remnants on Earth have been metamorphosed (e.g., Nagy *et al.*, 1981; Bowring *et al.*, 1989), thus making the possibility for survival of indigenous amino acids fairly remote. Although the stable isotope composition of kerogens associated with some of these rocks may hold clues as to the antiquity of living systems (Schidlowski 1987, 1988, 1991, and this volume, Chapter 31), the only viable alternative for exploring the prebiotic organic inventory that may have been available on the early Earth is by studying comets, meteorites, interstellar dust, and planetary and lunar samples.

It is important to bear in mind that inferences from meteorites concerning the composition of organic matter on the early Earth assume that this extraterrestrial material reflects the original composition at the time of formation. The validity of this assumption is unknown. It is suggested that, in addition to possible alteration during impact (Peterson *et al.*, 1991), diagenetic alteration via processes such as prolonged cosmic ray bombardment may have a significant impact on the isotopic (e.g., Jull *et al.*, 1989) and chemical (e.g., Rössler *et al.*, 1984; Roessler *et al.*, 1991) composition of extraterrestrial organic matter. Clearly, this type of information needs to be considered when attempting to reconstruct the prebiotic organic inventory that preceded life's origin.

5.2. Chirality

Central to any discussion of life's origin is the question of chirality and whether its occurrence preceded life or was, perhaps, simply a consequence of chance, the latter being ". . . almost impossible to test experimentally, because it is based on the premise of a random event" (Ferris, 1984). In a comprehensive review of this problem, Bonner (1991) concluded that it is not possible for chirality to have developed on Earth, either prebiotically or via some early biological discrimination process. This point of view has been questioned by some (e.g., Bada and Miller, 1987) and supported by others (e.g., Goldanskii and Kuzmin, 1991).

The problem of chirality is unlikely to be resolved in the immediate future. However, it is interesting to speculate that if a significant portion of the early Earth's organic inventory was extraterrestrial in origin (e.g., Oró, 1961; Chyba and Sagan, 1992) and the L-enantiomeric excess reported for amino acids in Murchison stones is indeed extraterrestrial (Engel and Nagy, 1982; Engel *et al.*, 1990; Macko and Engel, 1991), then an explanation may exist for the introduction of chirality to the Earth. These, of course, are big ifs. Also, the problem of chirality is not resolved, but transferred to space. If it is assumed that extraterrestrial, prebiotic synthesis resulted in racemic mixtures of amino acids, alternative pathways must be sought for the preferential destruction of D-enantiomers in space (e.g., Goldanskii and Kuzmin, 1991).

6. Conclusions

The search for a reliable record of the organic precursors of life on Earth is hampered by the lack of appropriate terrestrial samples, a paucity of extraterrestrial materials, and the haunting question of whether past attempts to simulate prebiotic synthesis truly reflect conditions at that time. Nevertheless, the origin of life will continue to be one of the fundamental questions addressed by future generations. It is possible that recent advances in stable isotope geochemistry at the molecular level will provide some new information toward the solution of this problem.

ACKNOWLEDGMENTS. M.H.E. and S.A.M. wish to acknowledge NASA, Division of Exobiology (grant NAGW–2765), for support of our meteorite research and the National Science Foundation, Division of Earth Sciences (grant EAR 9118011), for support of the development of GC/IRMS for amino acid analyses. We acknowledge J. A. Silfer and Yaorong Qian for their contributions to this work.

References

Abelson, P. H., and Hare, P. E., 1969, Recent amino acids in the Gunflint chert, *Carnegie Inst. Washington Yearb.* **67**:208–210.

Abelson, P. H., and Hoering, T. C., 1961, Carbon isotope fractionation in formation of amino acids by photosynthetic organisms. *Proc. Natl. Acad. Sci. U.S.A.* **47**:623–632.

Abrajano, T. A., Fang, J., Comet, P. A., and Brooks, J., 1992, Compound-specific carbon isotope analysis of fatty acids, 203rd National Meeting of the American Chemical Society, San Francisco, GEOC Abstract 104.

Anders, E., 1988, Circumstellar material in meteorites: Noble gases, carbon and nitrogen, in: *Meteorites and the Early Solar System* (J. F. Kerridge and M. S. Matthews, eds.), University of Arizona Press, Tucson, pp. 927–955.

Bada, J. L., 1991, Amino acid cosmogeochemistry, *Philos. Trans. R. Soc. London, Ser. B* **333**:349–358.

Bada, J. L., and Miller, S. L., 1987, Racemization and the origin of optically active organic compounds in living organisms, *BioSystems* **20**:21–26.

Bada, J. L., Cronin, J. R., Ho, M.-S., Kvenvolden, K. A., Lawless, J. G., Miller, S. L., Oró, J., and Steinberg, S., 1983, On the reported optical activity of amino acids in the Murchison meteorite, *Nature* **301**:494–496.

Becker, R. H., and Epstein, S., 1982, Carbon, hydrogen and nitrogen isotopes in the solvent-extractable organic matter from carbonaceous chondrites, *Geochim. Cosmochim. Acta* **46**:97–103.

Bjorøy, M., Hall, K., Gillyon, P., and Jumeau, J., 1991, Carbon isotope variations in *n*-alkanes and isoprenoids of whole oils, *Chem. Geol.* **93**:13–20.

Bonner, W. A., 1991, The origin and amplification of biomolecular chirality, *Origins Life* **21**:59–111.

Bowring, S. A., Williams, I. S., and Compston, W., 1989, 3.98 Ga gneisses from the Slave province, Northwest Territories, Canada, *Geology* **17**:971–975.

Brand, W. A., and Tegtmeyer, A. R., 1992, Isotope ratio monitoring: Extended toward ^{15}N and ^{18}O, 203rd Annual Meeting of the American Chemical Society, San Francisco, GEOC Abstract 89.

Buchardt, B., and Weiner, S., 1981, Diagenesis of aragonite from Upper Cretaceous ammonites: A geochemical case study, *Sedimentology* **28**:423–438.

Chang, S., 1977, Comets: Cosmic connections with carbonaceous meteorites, interstellar molecules and the origin of life, in: *Space Missions to Comets* (M. Neugebauer, D. K. Yeomans, J. C. Brandt, and R. W. Hobbs, eds.), NASA Publication 2089, pp. 59–111.

Chang, S., Mack, R., and Lennon, K., 1978, Carbon chemistry of separated phases of Murchison and Allende meteorites, *Lunar Planet. Sci.* **IX**:157–159.

Chang, S., Des Marais, D., Mack, R., Miller, S. L., and Strathearn, G. E., 1983, Prebiotic organic synthesis and the origin of life, in: *Earth's Earliest Biosphere: Its Origin and Evolution* (J. W. Schopf, ed.), Princeton University Press, Princeton, New Jersey, pp. 53–92.

Chyba, C., and Sagan, C., 1992, Endogenous production, exogenous delivery and impact-shock synthesis of organic molecules: An inventory for the origins of life. *Nature* **355**:125–132.

Chyba, C. F., Thomas, P. J., Brookshaw, L., and Sagan, C., 1990, Cometary delivery of organic molecules to the early Earth, *Science* **249**:366–373.

Cronin, J. R., 1976a, Acid-labile amino acid precursors in the Murchison meteorite. I. Chromatographic fractionation, *Origins Life* **7**:337–342.

Cronin, J. R., 1976b, Acid-labile amino acid precursors in the Murchison meteorite II. A search for peptides and amino acyl amides, *Origins Life* **7**:343–348.

Cronin, J. R., Pizzarello, S., and Cruikshank, D. P., 1988, Organic matter in carbonaceous chrondites, planetary satellites, asteroids and comets, in: *Meteorites and the Early Solar System* (J. F. Kerridge and M. S. Matthews, eds.), University of Arizona Press, Tucson, pp. 819–857.

Engel, M. H., and Macko, S. A., 1984, Separation of amino acid enantiomers for stable nitrogen and carbon isotopic analyses, *Anal. Chem.* **56**:2598–2600.

Engel, M. H., and Macko, S. A., 1986, Stable isotope evaluation of the origins of amino acids in fossils, *Nature* **323**:531–533.

Engel, M. H., and Nagy, B., 1982, Distribution and enantiomeric composition of amino acids in the Murchison meteorite, *Nature* **296**:837–840.

Engel, M. H., and Nagy, B., 1983, Authors' reply, *Nature* **301**:496–497.

Engel, M. H., Macko, S. A., and Silfer, J. A., 1990, Carbon isotope composition of individual amino acids in the Murchison meteorite, *Nature* **348**:47–49.

Epstein, S., Krishnamurthy, R. V., Cronin, J. R., Pizzarello, S., and Yuen, G. U., 1987, Unusual stable isotope ratios in amino acid and carboxylic acid extracts from the Murchison meteorite, *Nature* **326**:477–479.

Fenwick, C. S., Kempe, K., and Jumeau, E. J., 1992, On-line measurement of HD/H_2 ratios by GC–C–IRMS, 203rd Annual Meeting of the American Chemical Society, San Francisco, GEOC Abstract 91.

Ferris, J. P., 1984, The chemistry of life's origin, *Chem. Eng. News* **1984**(Aug. 27):22–35.

Franchi, I. A., Exley, R. A., Gilmour, I., and Pillinger, C. T., 1989, Stable isotope and abundance measurements of solvent extractable compounds in Murchison, *14th Symposium on Antarctic Meteorites*, June 1989, Tokyo, National Institute of Polar Research, Tokyo, Japan, pp. 40–42.

Freedman, P. A., Gillyon, E. C. P., and Jumeau, E. J., 1988, Design and application of a new instrument for GC–isotope ratio MS, *Am. Lab.* **1988**(June):114–119.

Freeman, K. H., Hayes, J. M., Trendel, J.-M., and Albrecht, P., 1990, Evidence from carbon isotope measurements for diverse origins of sedimentary hydrocarbons, *Nature* **343**:254–256.

Gil-Av, E., Charles, R., and Fischer, G., 1965, Resolution of amino acids by gas chromatography, *J. Chromatogr.* **17**:408–410.

Goldanskii, V. I., and Kuzmin, V. V., 1991, Chirality and the cold origin of life, *Nature* **352**:114.

Goodman, K. J., and Brenna, T. J., 1992, High sensitivity tracer detection using high-precision gas chromatography–combustion isotope ratio mass spectrometry and highly enriched [U-^{13}C]-labeled precursors, *Anal. Chem.* **64**:1088–1095.

Hare, P. E., and Mitterer, R. M., 1966, Nonprotein amino acids in fossil shells, *Carnegie Inst. Washington Yearb.* **65**:362–364.

Hare, P. E., Fogel, M. L., Stafford, T. W., Mitchell, A. D., and Hoering, T. C., 1991, The isotopic composition of carbon and nitrogen in individual amino acids isolated from modern and fossil proteins. *J. Arch. Sci.* **18**:277–292.

Hayatsu, R., and Anders, E., 1981, Organic compounds in meteorites and their origins, in: *Cosmo- and Geochemistry*, Topics in Current Chemistry, Vol. 99 Springer-Verlag, Berlin, pp. 1–37.

Hayes, J. M., Freeman, K. H., Ricci, M. P., Studley, S. A., Merritt, D. A., Brzuzy, L., Brand, W. A., and Habfast, K., 1989, Isotope ratio monitoring gas chromatography mass spectrometry, Paper presented at the 37th American Society of Mass Spectrometry Conference on Mass Spectrometry and Allied Topics, Miami Beach, Florida, pp. 33–34.

Hayes, J. M., Freeman, K. H., Popp, B. N., and Hoham, C. H., 1990, Compound specific isotope analysis: A novel tool for reconstruction of ancient biogeochemical processes, *Org. Geochem.* **16**:1115–1128.

Hennet, R. J.-C., Holm, N. G., and Engel, M. H., 1992, Abiotic synthesis of amino acids under hydrothermal conditions and the origin of life: A perpetual phenomenon? *Naturwissenschaften* **79**:361–365.

Hoering, T. C., 1980, The organic constituents of fossil mollusc shells, in: *Biogeochemistry of Amino Acids* (P. E. Hare, T. C. Hoering, and K. King, Jr., eds.), John Wiley & Sons, New York, pp. 193–201.

Jull, A. J. T., Donahue, D. J., and Linick, T. W., 1989, Carbon-14 activities in recently fallen meteorites and Antarctic meteorites, *Geochim. Cosmochim. Acta* **53**:2095–2100.

Kennicutt, M. C., and Brooks, J. M., 1990, Unusual normal alkane distributions in offshore New Zealand sediments, *Org. Geochem.* **15**:193–197.

Kerridge, J. F., 1985, Carbon, hydrogen and nitrogen in carbonaceous chondrites: Abundances and isotopic compositions in bulk samples, *Geochim. Cosmochim. Acta* **49**:1707–1714.

Kerridge, J. F., 1991, A note on the prebiotic synthesis of organic acids in carbonaceous meteorites, *Origins Life* **21**:19–29.

Kerridge, J. F., Chang, S., and Shipp, R., 1987, Isotopic characterization of kerogen-like material in the Murchison carbonaceous chondrite, *Geochim. Cosmochim. Acta* **51**:2527–2540.

Khare, B. N., Sagan, C., Ogino, H., Nagy, B., Er, C., Schram, K. H., and Arakawa, E. T., 1986, Amino acids derived from Titan Tholins, *Icarus* **68**:176–184.

Kvenvolden, K. A., Peterson, E., and Pollock, G. E., 1969, Optical configuration of amino acids in Pre-Cambrian Fig Tree chert, *Nature* **221**:141–143.

Kvenvolden, K. A., Lawless, J., Pering, K., Peterson, E., Flores, J., Ponnamperuma, C., Kaplan, I. R., and Moore, C., 1970, Evidence for extraterrestrial amino acids and hydrocarbons in the Murchison meteorite, *Nature* **228**:923–926.

Kvenvolden, K. A., Lawless, J. G., and Ponnamperuma, C., 1971, Nonprotein amino acids in the Murchison meteorite, *Proc. Natl. Acad. Sci. U.S.A.* **68**:486–490.

Macko, S. A., and Engel, M. H., 1991, Assessment of indigeneity in fossil organic matter: Amino acids and stable isotopes, *Philos. Trans. R. Soc. London, Ser. B* **333**:367–374.

Macko, S. A., Estep, M. L. F., Engel, M. H., and Hare, P. E., 1986, Kinetic fractionation of stable nitrogen isotopes during amino acid transamination, *Geochim. Cosmochim. Acta* **50**:2143–2146.

Macko, S. A., Estep, M. L. F., Hare, P. E., and Hoering, T. C., 1987, Isotopic fractionation of nitrogen and carbon in the synthesis of amino acids by microorganisms, *Isot. Geosci.* **65**:79–92.

Macko, S. A., Helleur, R., Hartley, G., and Jackman, P., 1990, Diagenesis of organic matter—a study using stable isotopes of individual carbohydrates, *Org. Geochem.* **16**:1129–1137.

Macko, S. A., Leskey, T., Ryan, M., and Engel, M. H., 1992, Carbon isotopic analysis of individual carbohydrates by GC/IRMS, 203rd National Meeting of the American Chemical Society, San Francisco, GEOC Abstract 107.

Matthews, C., Nelson, J., Varma, P., and Minard, R., 1977, Deuterolysis of amino acid precursors: Evidence for hydrogen cyanide polymers as protein ancestors, *Science* **198**:622–625.

McPherson,D. W., Rahman, K., Martinez, I., and Shevlin, P. B., 1987, The formation of amino acid precursors in the reaction of atomic carbon with water and ammonia at 77 K, *Origins Life* **17**:275–282.

Miller, S. L., 1982, Prebiotic synthesis of organic compounds, in: *Mineral Deposits and the Evolution of the Biosphere* (H. D. Holland and M. Schidlowski, eds.), Springer-Verlag, Berlin, pp. 155–176.

Miller, S. L., 1992, The prebiotic synthesis of organic compounds as a step toward the origin of life, in: *Major Events in the History of Life* (J. W. Schopf, ed.), Jones and Bartlett, Boston, pp. 1–28.

Mullie, F., and Reisse, J., 1987, Organic matter in carbonaceous meteorites, *Top. Curr. Chem.* **139**:83–117.

Nagy, B., 1975, *Carbonaceous Meteorites*, Elsevier, Amsterdam.

Nagy, B., 1985, New aspects of early organic evolution, *Terra Cognita* **5**:128–129.

Nagy, B., Engel, M. H., Zumberge, J. E., Ogino, H., and Chang, S. Y., 1981, Amino acids and hydrocarbons in the ~3,800-Myr old Isua rocks, southwestern Greenland, *Nature* **289**:53–56.

Nagy, B., Burke, M. F., and Boynton, W. V., 1987, Possibilities and constraints of the presence of amino acids in cometary nuclei, *Bull. Am. Astron. Soc.* **19**:894.

Oró, J., 1961, Comets and the formation of the biochemical compounds on the primitive Earth, *Nature* **190**:389–390.

Oró, J., Holzer, G., and Lazcano, A., 1980, The contribution of cometary volatiles to the primitive Earth, *Life Sci. Space* **18**: 67–82.

Oró, J., Miller, S. L., and Lazcano, A., 1990, The origin and early evolution of life on Earth, *Annu. Rev. Earth Planet. Sci.* **18**: 317–356.

Peterson, E., Horz, F., Haynes, G., and See, T., 1991, Fate of amino acids during simulations of impacts at ≤5 km/s, *Comets and the Origins and Evolution of Life*, Proceedings of the symposium held at the University of Wisconsin, Eau Claire, Sept. 30–Oct. 2, 1991, p. 31.

Pillinger, C. T., 1982, Not quite full circle? Non-racemic amino acids in the Murchison meteorite, *Nature* **296**:802.

Pizzarello, S., Krishnamurthy, R. V., Epstein, S., and Cronin, J. R., 1991, Isotopic analyses of amino acids from the Murchison meteorite, *Geochim. Cosmochim. Acta* **55**:905–910.

Pollock, G. E., Oyama, V. I., and Johnson, R. D., 1965, Resolution of racemic amino acids by gas chromatography, *J. Gas Chromatogr.* **3**:174–176.

Pollock, G. E., Cheng, C.-N., Cronin, S. E., and Kvenvolden, K. A., 1975, Stereoisomers of isovaline in the Murchison meteorite, *Geochim. Cosmochim. Acta* **39**:1571–1573.

Rafalska, J. K., Engel, M. H., and Lanier, W. P., 1991, Retardation of racemization rates of amino acids incorporated into melanoidins, *Geochim. Cosmochim. Acta* **55**:3669–3675.

Rieley, G., Collier, R. J., Jones, D. M., Eglinton, G., Eakin, P. A., and Fallick, A. E., 1991, Sources of sedimentary lipids deduced form stable carbon isotope analyses of individual compounds, *Nature* **352**:425–427.

Robert, F., and Epstein, S., 1982, The concentration and isotopic composition of hydrogen, carbon and nitrogen in carbonaceous meteorites, *Geochim. Cosmochim. Acta* **46**:81–95.

Roessler, K., Atwa, S. T., Kaiser, R., Mahfouz, R. M., and Sauer, M., 1991, Formation and destruction of complex organic matter in space: A steady state, *Comets and the Origins and Evolution of Life*, Proceedings of the symposium held at the University of Wisconsin, Eau Claire, Sept. 30–Oct. 2, 1991, p. 33.

Rössler, K., Jung, H.-J., and Nebeling, B., 1984, Hot atoms in cosmic chemistry, *Adv. Space Res.* **4**:83–95.

Sagan, C., and Khare, B. N., 1971, Long-wavelength ultraviolet photoproduction of amino acids on the primitive earth, *Science* **173**:417–420.

Schidlowski, M., 1987, Application of stable carbon isotopes to early biochemical evolution on Earth, *Annu. Rev. Earth Planet. Sci.* **15**:47–72.

Schidlowski, M., 1988, A 3,800-million-year isotopic record of life from carbon in sedimentary rocks, *Nature* **333**:313–318.

Schidlowski, M, 1991, Organic carbon isotope record: Index line of autotrophic carbon fixation over 3.8 Gyr of Earth history, *J. Southeast Asian Earth Sci.* **5**:333–337.

Schopf, J. W., Kvenvolden, K. A., and Barghoorn, E. S., 1968, Amino acids in Precambrian sediments: An assay, *Proc. Natl. Acad. Sci. U.S.A.* **59**:639–648.

Serban, A., Engel, M. H., and Macko, S. A.,1988, The distribution, stereochemistry and stable isotopic composition of amino acid constituents of fossil and modern mollusc shells, *Org. Geochem.* **13**:1123–1129.

Shock, E. L., and Schulte, M. D., 1990, Summary and implications of reported amino acid concentrations in the Murchison meteorite, *Geochim. Cosmochim. Acta* **54**:3159–3173.

Silfer, J. A., 1991, Ph.D. Thesis, The University of Oklahoma.

Silfer, J. A., Engel, M. H., Macko, S. A., and Jumeau, E. J., 1991, Stable carbon isotope analysis of amino acid enantiomers by conventional isotope ratio mass spectrometry and combined gas chromatography–isotope ratio mass spectrometry, *Anal. Chem.* **63**:370–374.

Swart, P. K., Grady, M. M., Pillinger, C. T., Lewis, R. S., and Anders, E., 1983, Interstellar carbon in meteorites, *Science* **220**:406–410.

Williams, K. M., and Smith, G. G., 1977, A critical evaluation of the application of amino acid racemization to geochronology and geothermometry, *Origins Life* **8**:1–144.

Yang, J., and Epstein, S., 1983, Interstellar organic matter in meteorites, *Geochim. Cosmochim. Acta* **47**:2199–2216.

Yuen, G., Blair, N., Des Marais, D. J., and Chang, S., 1984, Carbon isotopic composition of low molecular weight hydrocarbons and monocarboxylic acids from the Murchison meteorite, *Nature* **307**:252–254.

Zahnle, K., and Grinspoon, D., 1990, Comet dust as a source of amino acids at the Cretaceous/Tertiary boundary, *Nature* **348**: 157–160.

Zhao, M. X., and Bada, J. L., 1989, Extraterrestrial amino acids in Cretaceous/Tertiary boundary sediments at Stevns Klint, Denmark, *Nature* **339**:463–465.

Zinner, E., 1988, Interstellar cloud material in meteorites, in: *Meteorites and the Early Solar System* (J. F. Kerridge and M. S. Matthews, eds.), University of Arizona Press, Tucson, pp. 956–983.

VIII

Applications of Organic Geochemistry for Quaternary Research

Chapter 34

Applications of Biomarkers for Delineating Marine Paleoclimatic Fluctuations during the Pleistocene

SIMON C. BRASSELL

1. Introduction

1.1. Influences on Climate and the Climate Record

The Earth's climate is a dynamic system governed by many factors including the complex, constantly changing interactions between the atmosphere and the ocean, the configuration of the continent, and external variables such as solar luminosity and periodicities in perturbations of the Earth's orbit. Evidence of increasing levels of CO_2 (Keeling *et al.*, 1976) attributed to anthropogenic burning of fossil fuels and the effects of deforestation has heightened concerns about future global warming of a few degrees, together with an associated reduction in ice volume and a consequent rise in sea level. However, investigations of the stratigraphic record have revealed dramatic changes in the Earth's climate of comparable or greater magnitude in the geological past (e.g., Lamb, 1977; Tarling, 1978; Frakes, 1979; Barron, 1989a, b; Crowley and North, 1991; Bowen, 1991). Such variation in the ancient climatic history of the Earth has been unraveled and elucidated using a diverse range of approaches to which new methodologies continue to be added (e.g., Hecht, 1985).

Studies in paleoclimatology vary in scope; they may be focused on the description and interpretation of climate at a given time in the geological past (e.g., during the Cretaceous; Barron, 1983, 1989a) or concerned with the nature and rate of climatic change during specific time intervals (e.g., Pleistocene ice ages; CLIMAP, 1984). Many investigations also tend to have a geographical frame of reference; for example, they may be focused on climatic variability on a global scale (CLIMAP, 1984) or on events that appear restricted to a specific area of the Earth (e.g., the North Atlantic; Shackleton *et al.*, 1984; Ruddiman *et al.*, 1986) or a particular climatic regime, such as the polar or equatorial regions (Shackleton and Kennett, 1975; Kennett, 1977). Such differences in scale reflect the geographical influences on climate and its dependency on both regional and global factors, especially those influencing oceanic and atmospheric CO_2 levels and their circulation patterns (Barnett, 1978; Sundquist and Broecker, 1985; Broecker and Denton, 1989).

Climatic variations are recognized to operate over a vast range of time scales; some are short-lived

SIMON C. BRASSELL • Department of Geology, Stanford University, Stanford, California 94305-2115. *Present address:* Biogeochemical Laboratories, Department of Geological Sciences, Indiana University, Bloomington, Indiana 47405-5101.

Organic Geochemistry, edited by Michael H. Engel and Stephen A. Macko. Plenum Press, New York, 1993.

phenomena (e.g., Philander, 1990), some represent long-term trends on a geological time scale (Frakes, 1979), whereas others occur cyclically or periodically (e.g., Imbrie *et al.*, 1989). The ability to identify and recognize past climatic events depends on historical accounts or on the preservation of their signals in the lithological, biological, paleontological, elemental, isotopic, or molecular characteristics and compositions of sediments. For example, in many areas of temperate zones the seasonal fluctuations in rainfall and primary photosynthetic production can exert a major influence on sediment composition, perhaps best documented in varved or laminated sequences. Phenomena associated with weather patterns and systems, such as storm deposits and erosional surfaces, can also be recorded in sediments (e.g., Wells, 1987; Brandt and Elias, 1989). However, the climatic features best documented in the sedimentary record are probably ice ages. These significant events do not necessarily represent continuous glaciation but appear as cycles of alternating glacial and interglacial periods controlled by orbital forcing factors apparent from time series analysis (e.g., Imbrie and Imbrie, 1986; Imbrie *et al.*, 1989). They exert a major influence on biological populations and, hence, on both sedimentary fossil and biomarker records. In addition, the restriction of glaciation sequences to particular eras of geological time has prompted the conjecture that there are larger scale cycles or "supercycles" which are a function of sea level and atmospheric concentrations of CO_2 and water vapor (Fischer and Arthur, 1977; Fischer, 1982). In this assessment Earth's climate is thought to oscillate between conditions with and without permanent ice caps, referred to as cooler "icehouse" and warmer "greenhouse" states, respectively.

1.2. Assessment of Paleoclimates

The most comprehensive means for evaluation of past climates is examination of historical records of meteorological observations, such as air temperatures and rainfall. However, such direct measurements of these climatic variables only span, at best, the last few hundred years and are limited to those areas where such records have been kept (e.g., Gribbin and Lamb, 1978; Landsberg, 1985). Significant additional evidence, commonly referred to as parameteorological phenomena, can be drawn from historical documents which report weather-dependent events, either in terms of their annual variability or in recording exceptional incidents (e.g., Lamb, 1977; Bradley, 1985; Grove, 1988). Many examples of well-documented climatic variability exist in the annals of ancient civilizations, with the annual levels of flooding of the River Nile in ancient Egypt perhaps the oldest and most comprehensive record (e.g., Lamb, 1977). Other historical records of climate include commentaries on ancient floods and droughts, some of which can be described as catastrophic events from which recovery was slow, long delayed, or never complete. Perhaps the most significant and widely reported events are those in accounts of the biblical flood and in similar documentation of a great flood in the legends of Babylon and other ancient civilizations. It remains an open question whether these accounts represent a human description of the eustatic rise in sea level associated with the waning of the last glaciation (Lamb, 1977).

The sequences of major changes in Earth's climate are also punctuated by smaller climatic oscillations. The best documented event in recent historical times is the Little Ice Age, which lasted from ca. 1450 to 1850 A.D. (e.g., Lamb, 1977; Grove, 1988). During this period, contemporary records report markedly harsher winters than those experienced and recorded in historical annals of earlier times. The marked change in climate is well illustrated in paintings and drawings depicting glaciers in the Alps and the freezing of various waterways, including Dutch canals and the River Thames. The generally cooler climate during the Little Ice Age is also apparent in terms of its impact on agriculture, for example, the frequency and duration of failures in crop harvests in Europe and the demise of medieval vineyards in many parts of England (Lamb, 1977; Grove, 1988). Examination and evaluation of moraines show that glaciers were expanded in the Alps, Scandinavia, Iceland, Greenland, Spitsbergen, Alaska, the Himalayas, New Zealand, and South America and on Mount Jaya in Indonesia at this time, revealing the global scale of this minor climatic event (Grove, 1988). The Little Ice Age was immediately preceded by a less well documented period (ca. 1150–1300 A.D.) of warmer climatic conditions, known as the medieval warm epoch or little optimum. The impact of this climatic optimum has been documented in North America, Europe, and the North Atlantic (Lamb, 1982, 1984; Pfister, 1988, and references therein). For example, the limits of grain cultivation expanded into the Arctic Circle, and there is evidence of raised sea levels in the 13th century (Tooley, 1978). Evaluation of the limits of vine cultivation enables estimates for summer temperatures in England and central Europe to be made, which suggests that these areas were 0.7–1.0°C and 1.0–1.4°C warmer, respectively, than 20th century average values (Lamb, 1982). In North America the event is recognized in the retreat of glaciers in Alaska (Denton and Karlén, 1977) but best detailed in the record of annual variations in tree ring widths of bristlecone pines from the White Mountains of Southern California (LaMarche, 1974).

Climate change can have a dramatic effect on vegetation. The success and survival of floral and faunal populations, their geographic distributions and relative abundances are all influenced by climatically dependent factors such as temperature, rainfall, and humidity, irrespective of human intervention. Thus, the stratigraphic record of pollen and spores in lake sediments can record changes in forestation and vegetation affected by climate (e.g., Lamb, 1977; Wijmstra, 1978). The variations in pollen associated with different plant populations enable the elucidation and classification of zones which correspond to specific climatic conditions (e.g., Flenley, 1979). Interestingly, such pollen zones often appear to remain largely unaltered for substantial periods of time and then to change rapidly. Although such dramatic transformations can sometimes be attributable to forest fires, they appear, more typically, to reflect an abrupt environmental change (e.g., Lamb, 1977). Another biological record of climatic change lies in the annual growth rings of trees. Studies in dendroclimatology have demonstrated the capability to interpret variations in ring width and character as a measure of seasonal rainfall (e.g., LaMarche, 1974, Stockton *et al.*, 1985).

In the oceanic environment the populations and abundance of many organisms are highly sensitive to environmental changes, whether in nutrient supply, water temperature, or salinity (e.g., Bukry,1974). Hence, variations in both planktonic and benthonic organisms can represent a response to a changing environment. Such changes can be interpreted from shifts in the predominant species, or in the dominant class of organism, or in the balance between different organisms. Alternatively, they may find expression in the appearance of dwarf forms or in the direction of coiling of various foraminifera (Bandy, 1959). Open-ocean records, in contrast to those of lacustrine sedimentary environments, which may be subject to local perturbation, typically provide an indication of events that are global in scale, reflecting changes in ocean circulation patterns, ice volume, or temperature. Thus, at a given oceanic site the variations in individual species observed, for example, in a sediment core may record events affecting the entire planet. The evidence of biological variability can also provide a comparison of the importance of a given event between different regions or its intensity over separate time periods. In particular, the populations of a number of indicator fossils—notably species of foraminifera—fluctuate in their abundance in sediment cores, apparently in response to climate change. In an early application of this approach the foraminifer *Globorotalia menardii* was found to vary in downcore sections of a Caribbean sediment; horizons with abundant *G. menardii* were related to warmer, interglacial periods (Ericson *et al.*, 1961). This concept has been developed and expanded and its applicability well demonstrated in subsequent years. It is now possible to decipher and interpret the proxy records of climate change apparent in paleontological data based on various evidence from the physical remains of organisms, notably changes in their individual morphology or in their community structure and species distribution. In many instances, the qualitative character of such paleontological investigations concerned with the presence/absence and the abundance of a given species can distinguish warm and cold climatic episodes but does not furnish a precise temperature record without some reference to external criteria. However, it becomes possible to attain a more quantitative assessment of temperatures through the combination of the distributions and abundances of numerous species with statistical data treatments, such as factor analysis (Imbrie and Kipp, 1971).

Probably the most powerful and widely used tool in the assessment of paleoclimates is the study of the isotopic composition of the calcareous skeletons of foraminifera (e.g., Emiliani, 1955; Savin, 1977; Duplessy, 1978; Berger, 1981). The value of this technique lies in the fact that the fractionation of oxygen and carbon isotopes in skeletal carbonate is a function of the sources of CO_2 and temperature (e.g., Shackleton, 1982), and the wide applicability of the approach is aided by the extensive occurrence of foraminiferal tests in marine sediments. The uptake of oxygen isotopes in the production of calcareous tests is sensitive to water temperature, is species-dependent, and reflects the isotopic composition of the CO_2 or HCO_3^- available for carbonate formation. The $\delta^{18}O$ values of target species provide a record of oceanic water temperatures and variations in ice volume. Such records have been extensively explored on the carbonate tests of planktonic and benthic foraminifera (e.g., Savin, 1977; Williams *et al.*, 1988), but the populations of calcareous nannoplankton collected by centrifugation have also been studied (Margolis *et al.*, 1975). In addition, the differences between planktonic and benthic foraminifera have been studied as measures of the geological history of variance between surface and deep-water temperatures in different oceanic water masses (e.g., Savin, 1977). The cumulative data sets for foraminiferal $\delta^{18}O$ values have led to the acceptance of an isotopic stratigraphy for glacial/interglacial events which divides them into stages, where stage 1 is the present interglacial, stage 2 the last glacial maximum, etc. (Fig. 1). Major boundaries between the various stages are also given discrete names; the rapid warming event denoted as the division of stages 1 and 2 is called termination I, corresponding to the end of the last glacial maximum. However, the record of $\delta^{18}O$ changes in foraminiferal carbonate is not the sole source of such climatic information. More recently, excellent records of temporal variations in temperature have been obtained from the values of oxygen isotopes of ice cores collected from both the Arctic and Antarctic ice caps (e.g., Lorius *et al.*, 1985; Genthon *et al.*, 1987; Lorius, 1989). In addition, the $\delta^{13}C$ values of foraminiferal carbon bear on ocean productivity levels, and their variations have been related to changes in atmospheric levels of CO_2 (Shackleton *et al.*, 1983). Other measures also help describe and define ocean variability. For example, the cadmium/calcium ratios of benthic foraminifera reflect nutrient distributions that, in turn, are influenced by changes in oceanic circulation patterns (e.g., Boyle, 1988). Thus, there are varied paleoceanographic records contained in the chemical and isotope signatures of fossils in marine sediments.

Antarctic ice cores have provided a record of atmospheric CO_2 partial pressures (pCO_2) over the last 160 kyr, revealing values of ca. 200 ppm and ca. 300 ppm during glacial and interglacial episodes, respectively (Barnola *et al.*, 1987; Jouzel *et al.*, 1987). These data demonstrate a clear coupling between climate and pCO_2 and, hence, surface water dissolved CO_2 concentrations [CO_2 (aq)], which can be calculated directly from pCO_2 using Henry's law (e.g., Rau *et al.*, 1989, 1991). However, this link between atmospheric and oceanic CO_2 levels is a function of temperature because of the greater solubility of CO_2 at lower water temperatures, as observed at high latitudes (Rau *et al.*, 1989). Variations in [CO_2 (aq)] influence the isotopic fractionation associated with the uptake of carbon during photosynthesis in marine environments (e.g., Rau *et al.*, 1989). Similarly, the isotopic fractionation achieved in photosynthetic fixation of carbon is dependent on a variety of factors—physiological, ecological, and environmental (Jasper and Hayes, 1990), including latitudinal variations in planktonic $\delta^{13}C$ attributable to variations in [CO_2(aq)] (Rau *et al.*, 1989). Despite these constraints, it is apparent that the carbon isotope signatures of planktonic organisms may preserve a proxy record of CO_2 fluctuations both in their bulk organic matter (e.g., Rau *et al.*, 1991) and in individual molecular components (Jasper and Hayes, 1990).

All of the data described above help to define and validate general circulation models of the atmosphere (GCM) and their coupling with oceanic circulation patterns (Schlesinger, 1988). The role of GCMs includes both the assessment and interpretation of past climatic events, but they are primarily focused on modeling the present-day climate system and seeking to predict future climate change.

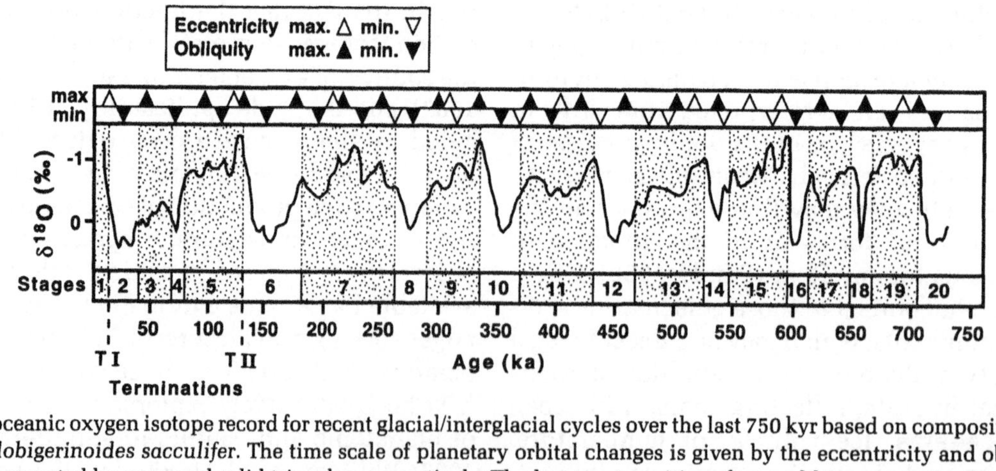

Figure 1. The oceanic oxygen isotope record for recent glacial/interglacial cycles over the last 750 kyr based on composite $\delta^{18}O$ data from four cores for *Globigerinoides sacculifer*. The time scale of planetary orbital changes is given by the eccentricity and obliquity maxima and minima, represented by open and solid triangles, respectively. The last two transitions from cold to warm episodes are indicated as termination I (TI) between stages 2 and 1 and termination II (TII) between stages 6 and 5. [Redrawn and modified from Bowen (1991).]

2. Climatic Changes during the Pleistocene

2.1. Glacial/Interglacial Cycles

The principal global climatic changes occurring during the Pleistocene have been the series of ice ages associated with the present glacial age. The geological record reveals that similar episodes of glaciation have occurred at other times in Earth history, including several major glacial events in the Precambrian, Ordovician, Carboniferous, and late Cenozoic (e.g., Frakes, 1979). The present phase of glaciation can be taken to have begun approximately ten million years ago with the advent of the formation of an Antarctic ice sheet and glaciers in the northern hemisphere (e.g., Imbrie and Imbrie, 1986). These events were followed by the appearance of ice sheets in the Northern Hemisphere in the late Pliocene; these ice sheets have expanded and contracted several times in the last few million years and may appear again at the end of the present interglacial. Initially, the existence of widespread glaciations was inferred from field evidence, such as erratic boulders, moraines, striations, and tills at various sites (e.g., Geikie, 1894). The concept of a series of ice ages in the geological past was first successfully championed by Louis Agassiz (see, e.g., Imbrie and Imbrie, 1986). In more recent times, this information has been augmented and extended by extensive studies of the oxygen isotopic record of foraminifera from oceanic sediment cores (e.g., Savin, 1977; Ruddiman, 1985; Ruddiman *et al.*, 1986; Shackleton 1988). Specifically, the variations in $\delta^{18}O$ values of planktonic foraminiferal species reveal a characteristic "sawtooth" pattern of glaciations, with episodes of gradual cooling and ice buildup being followed by a rapid warming and decrease in ice volume (Fig. 1; e.g., Imbrie and Imbrie, 1986). The timing of these events first became apparent from the determination of magnetic reversals associated with glacial deposits and subsequently has been refined and better calibrated using a wide range of floral and faunal biostratigraphic data, coupled with radioactive dating procedures.

The principal influence on the periodicity of glacial episodes is the variations in the Earth's orbit first suggested in the 1860s (Croll, 1875) and later quantitated in detail as the Milankovitch cycles (Fig. 2; e.g., Imbrie and Imbrie, 1986). These fluctuations in the orbit of the Earth are the frequencies of its precession, obliquity, and eccentricity, which have periodicities of 19 and 23, 41, and 100 thousand years, respectively, and occur in combination to create a

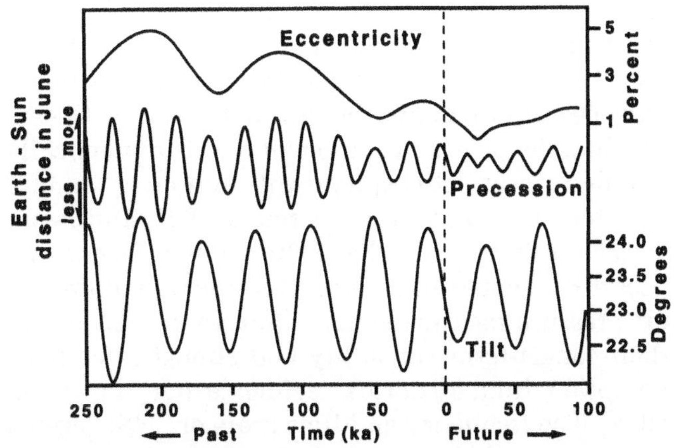

Figure 2. Past and predicted future variations in the eccentricity, the precession, and the tilt (obliquity) of the Earth's orbit. Note the difference in units for the three curves. [After Berger (1978).]

complex pattern of variability in the incident solar radiation with time at different latitudes of the Earth. Convincing confirmation of the correspondence between ice volume and variations in the Earth's orbit has emerged from detailed oxygen isotope studies on foraminifera from oceanic sediment cores (Hays *et al.*, 1976). Recent detailed examination of multiple cores using spectral analysis techniques has revealed that the 23-kyr and 41-kyr cycles show significant phase shifts between the hemispheres, with the southern responses leading (Imbrie *et al.*, 1989). Also, the 41-kyr cycle is primarily observed at high latitudes. In the North Atlantic the 100-kyr cycle appears in phase with ice volume, whereas the records elsewhere are more varied (Imbrie *et al.*, 1989). In general, the climatic events of the Pleistocene can be attributed to orbital forcing factors, with the exception of specific episodes, notably the Younger Dryas cooling event (Kennett, 1990), which appear to be independent of these controls (Broecker and Denton, 1989). Thus, an explanation of such phenomena requires other mechanisms to be invoked, for example, differences in the salinity balance of the ocean arising from changing meltwater influences on individual oceanic water bodies (Broecker *et al.*, 1989).

2.2. El Niño Southern Oscillation (ENSO)

A major signal of modern short-term climatic variability is that derived from the El Niño Southern Oscillation (ENSO), which represents a major, frequent perturbation of the Earth's climatic system (Philander, 1990). The precise trigger of ENSO fluctuations is not well understood, but the events are associated with changes which affect atmospheric and oceanic circulation patterns in the Pacific Ocean. El Niño, the oceanic component of this dynamically coupled ocean–atmosphere interaction, is chiefly manifest as a large-scale warming of Pacific Basin surface waters (Philander, 1990). El Niño events are best documented in records of sea temperature and air pressure measurements, although an account of their longer-term variability and strength has been compiled from historical evidence (Quinn *et al.*, 1987). The frequency and time scale of ENSO events vary, as does their severity. In Peru, field evidence of catastrophic floods in arid regions caused by exceptional rainfall has demonstrated that severe El Niño events have been occurring at frequent intervals for several thousands of years (Wells, 1987).

3. Biomarkers as Climatic Indicators

3.1. Principles

Many organisms have been found to possess the ability to regulate their biosynthetic production of their constituent lipids and thereby retain their viability under changing environmental conditions, such as fluctuations in temperature, light, or salinity (e.g., Harwood and Russell, 1984). The effects of these external influences are manifest in the constituent lipids of the organism as it strives to survive in its changing environment. Clearly, the development of an adaptive biochemistry that can respond to changing environmental conditions gives a species an evolutionary advantage through this greater tolerance to climatic fluctuations. Such flexibility in biosynthesis can be held to increase the competitiveness of the organism and is a necessary prerequisite for eurythermal species. The possibility exists for compounds indicative of climatic variations to be preserved in the molecular sedimentary record, although their survival is also a function of their chemical resilience to early diagenetic alteration. Hence, the primary objective of molecular investigations of paleoclimates is to explore the biological signals in the sedimentary record of those diagnostic marker compounds that are preserved.

A key factor in the potential use of biomarkers as climatic indicators is therefore the survival of their signal in sediments. Thus, refractory biomarkers are preferable candidates for evaluation and use as paleoclimatic indicators. Unfortunately, the more diagnostic components are sometimes the least stable. For example, the biological specificity of carotenoids is high—higher than that of most other lipid classes—but their lability severely restricts their applicability in studies of the sedimentary record of biomarkers (e.g., Brassell *et al.*, 1987). In addition, many components of potential value in paleoclimatic assessment are too widespread in biota (e.g., Brassell and Eglinton, 1986) and bear little diagnostic information since there is no guarantee that they originate from particular organisms within a given environmental niche. For example, the possibility of long-range transportation of aeolian dusts (e.g., Simoneit *et al.*, 1977) generally reduces the value of their constituent *n*-alkanes as measures of local climatic variations, although they can record prevailing wind patterns (e.g., Gagosian and Peltzer, 1986).

In the examination of short-term climatic trends, it is important that the pool of marker compounds is generated and buried in a relatively short period of time to provide a suitably time-dependent signal. Thus, allochthonous lipids from terrigenous higher plants are less appropriate, given the longevity of their transportation and durability in soils, although the pattern of their occurrence can aid the description of aeolian sources and river runoff. Ideally, target compounds should be biologically specific, preferably originating from organisms that inhabit and/or are restricted to a particular zone of the water column or underlying sediment. For example, the most useful components are those that occur in specific environmental settings, and display characteristics analogous to organisms such as foraminifera whose occurrences tend to be restricted to particular temperature, salinity, and bathymetric zones. A further desirable quality is that the compounds should occur widely and therefore provide information that is generally applicable to a variety of sedimentary regimes, preferably over an extended period of geologic time. Preservation is another important factor; ideally, the compounds should survive quantitatively during sediment deposition and subsequent burial. Hence, anoxic or dysaerobic environments are the preferred sites of study, given the reduced extent of degradation of organic matter where infaunal sediment reworking is minimized.

The range and variety of biomarkers in sediments is extensive. For simplicity, the following discussion focuses first on the distribution of members of a few significant compound types, namely *n*-alkanes, carboxylic acids, alkenones and their related alkyl alkenoates and alkenes, and sterols, which have demonstrable potential or proven applicability as climatic indicators. Second, evidence of climatic change provided by variations in the abundances of these compounds in the sedimentary record is considered.

3.2. *n*-Alkanes and *n*-Alkanols

The *n*-alkane distributions of organisms show significant variation; lower *n*-alkane homologs, notably C_{17}, tend to be predominant in many algae whereas the leaf waxes of higher plants are typically dominated by odd-numbered higher *n*-alkanes (i.e., C_{27}, C_{29}, and C_{31}; e.g. Brassell *et al.*, 1978). Recently, another source of *n*-alkanes has been recognized in the aliphatic biopolymers of algae (e.g., Goth *et al.*, 1988); these biopolymers may be preserved and become incorporated into kerogen and subsequently act as an algal source for higher *n*-alkanes in waxy crude oils (Tegelaar *et al.*, 1989). It is the physical properties of the higher *n*-alkanes—low solubility in water, low volatility—that provides vascular plants with some protection from dessication and immunity to bacterial attack. These same properties ensure that *n*-alkanes can survive aeolian transportation over oceans (Fig. 3); they can be carried great distances and can be deposited at sites remote from their origin (Fig. 4; e.g., Gagosian and Peltzer, 1986). Thus, it is the refractory nature of the organic matter in aeolian material that makes it possible for the organic contents of oceanic sediments to be dominantly derived from terrestrial sources though they lie thousands of kilometers from significant land masses.

The importance of aeolian transportation of organic matter is partly dependent on climatic variables, specifically wind strength and direction, which determines the effectiveness of winds in carrying aeolian particles for long distances (e.g., Simoneit *et al.*, 1977; Chesselet *et al.*, 1981; Zafiriou *et al.*, 1985; Gagosian and Peltzer, 1986) (Fig. 4). The exposure of land areas following the retreat of glaciers and changes in rainfall patterns are perhaps the most climatically dependent controls on the formation of dust deposits amenable to aeolian transportation. The existence of the two major present-day sources of aeolian particulates, namely, Chinese loess deposits and Saharan dusts, are directly related to these factors. Both pollen and molecular records of aeolian fallout can provide paleoclimatic evidence of wind direction or of changing environmental conditions (Hooghiemstra *et al.*, 1987; Hooghiemstra and Agwu, 1988; Cox *et al.*, 1982). Recent studies of sediments from the equatorial eastern Atlantic Ocean have revealed populations of *n*-alkanes related to different sources on the mainland. They show a correspondence between average carbon chain length of *n*-alkanes and climatic variation and an increase in the flux of higher plant *n*-alkanes in sediments deposited during cold stages, attributable to an intensification of trade winds during glacial times (Poynter *et al.*, 1989b). Hence, perturbations in climate appear to be reflected by variations in the abundance and distribution of *n*-alkanes derived from higher plant sources linked to changes in wind strength and direction. Overall, the refractory character of *n*-alkanes aids their survival during aeolian transportation and the preservation of such molecular signatures in the sedimentary record.

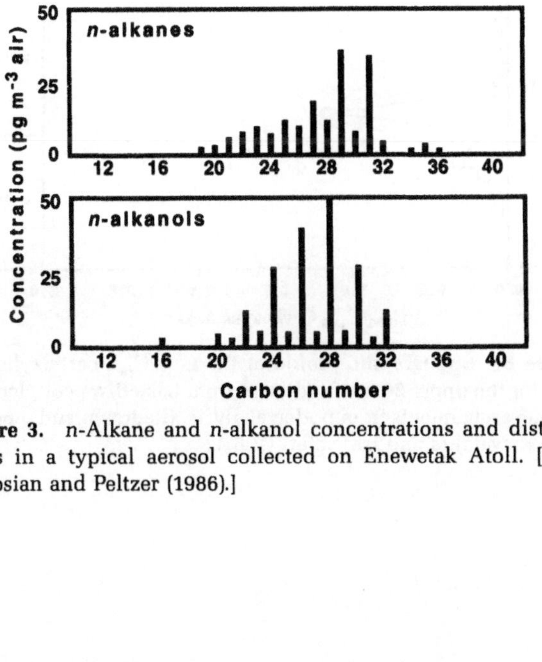

Figure 3. *n*-Alkane and *n*-alkanol concentrations and distributions in a typical aerosol collected on Enewetak Atoll. [After Gagosian and Peltzer (1986).]

The distributions of long-chain *n*-alkanols in aerosols over the Pacific Ocean (Fig. 3; Gagosian and Peltzer, 1986) and in sediments from the eastern equatorial Atlantic Ocean (Poynter *et al.*, 1989b), like *n*-alkane distributions, provide evidence for aeolian transportation of organic matter from terrigenous sources and can be similarly related to seasonal or longer-term variations in wind patterns.

3.3. Carboxylic Acids

Carboxylic acids are among the most widespread lipid components of recent sediments (e.g., Volkman and Johns, 1977; Smith *et al.*, 1983a, b), yet their diagnostic value in interpreting the sedimentary record is limited by their ubiquity in organisms both as free components and as integral parts of larger molecules such as triacylglycerides (e.g., Harwood and Russell, 1984). Two aspects of the distributions of carboxylic acids in sediments provide information pertinent to climatic evaluation. First, the abundances of long-chain *n*-alkanoic acids with prominent even-over-odd predominance (EOP), which can be assigned to terrestrial sources, may record aeolian contributions to marine sediments and therefore variations in wind strength (Prahl and Muehlhausen, 1989). The approach is similar to that described above

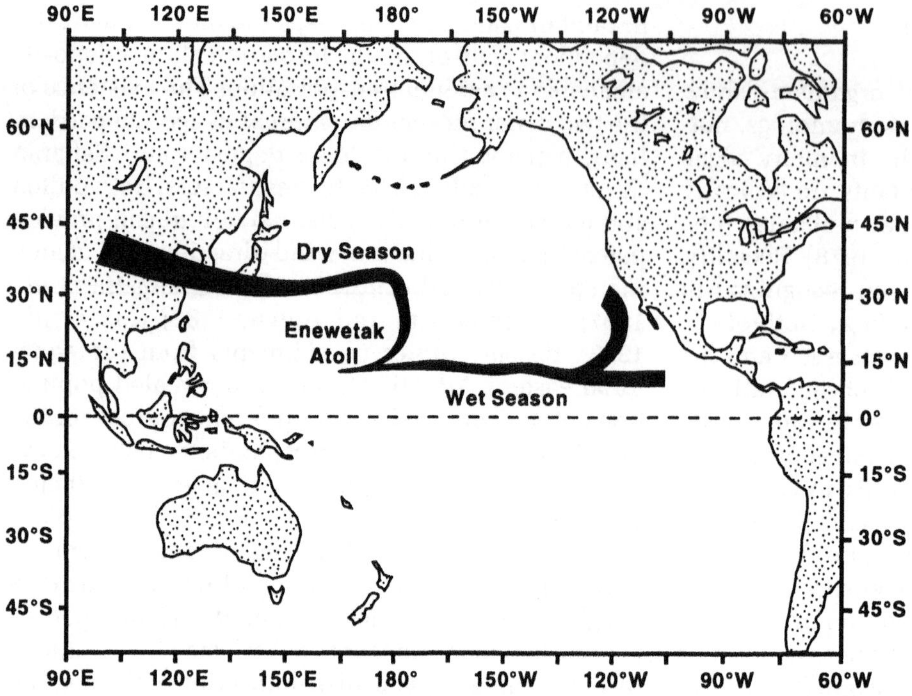

Figure 4. Typical air mass trajectories for aeolian material collected on Enewetak Atoll. Note the difference between the wet and dry seasons. [After Gagosian and Peltzer (1986).]

for terrestrial *n*-alkanes. Second, the abundances and the degree of unsaturation of alkenoic acids, which are regulated by biota as a means to maintain the fluidity of their membranes (Marr and Ingraham, 1962), can reflect temperature or climate changes if preserved. In one geochemical investigation this characteristic has been exploited to derive a climatic record for a lake sediment. Specifically, the downcore variation in the ratio of alkenoic to alkanoic acids (e.g., $C_{18:2}/C_{18:0}$; Fig. 5) for the upper 20 m of sediments from Lake Biwa reveals increases at four intervals over the last 20,000 years which are generally coincident with evidence from pollen analyses of cooler climatic conditions (Fig. 5; Kawamura and Ishiwatari, 1981). The ease of degradation of alkenoic acids (e.g., Rhead *et al.*, 1971) in most environments, however, effectively precludes the use of such labile lipids as paleoclimatic indicators in the majority of sediment sequences. For example, the only sediments found to contain significant abundances of polyunsaturated alkenoic acids are shallow cores from areas of high productivity and excellent preservation of organic matter, such as the upwelling areas of Walvis Bay (Boon *et al.*, 1975) and Peru (e.g., Smith *et al.*, 1983a, b). In addition, the widespread occurrence of alkenoic acids (e.g., Smith *et al.*, 1983a) creates an intractable problem in calibrating such responses, given the improbability of a uniform temperature dependence in alkenoic acid production by

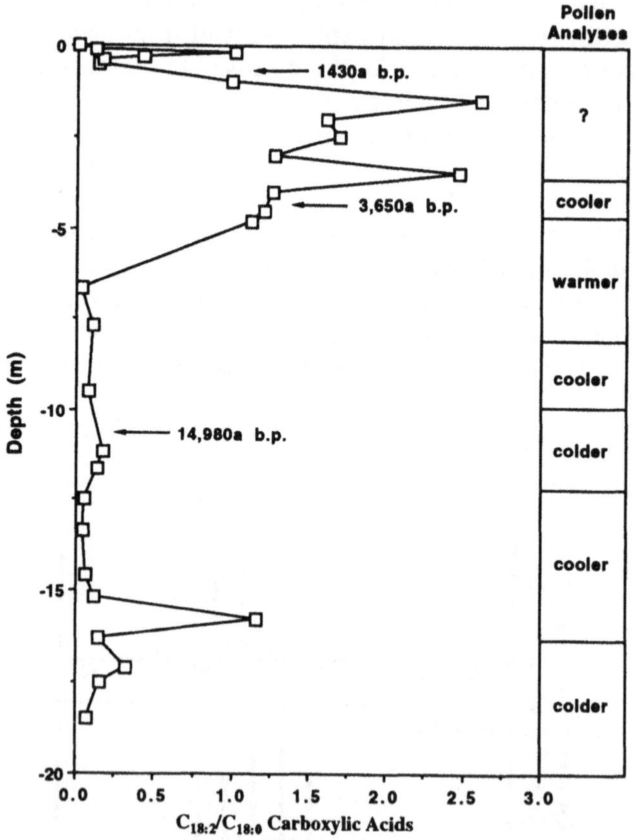

Figure 5. Stratigraphic profile of the $C_{18:2}/C_{18:0}$ carboxylic acid ratio for the upper 20 m of sediment from Lake Biwa coupled with climate assignments from pollen analysis. [Redrawn and modified from Kawamura and Ishiwatari (1981).]

different organisms. Furthermore, the lack of biological specificity in the synthesis of alkenoic acids means that they can be produced in various water masses, for example, by plankton in the photic zone and by bacteria in bottom sediments. Hence, correlations such as that between the abundances of alkenoic acids in sediments from Lake Biwa and colder episodes in the past have value in climatic studies of restricted lacustrine regimes, but the abundances of alkenoic acids are unlikely to be generally applicable as a molecular measure of climate change in the marine sedimentary record.

3.4. Alkenones

One major family of temperature-sensitive lipids comprises series of C_{37}–C_{39} di-, tri-, and tetraunsaturated methyl and ethyl ketones (Fig. 6). These components show demonstrable promise and application in paleoclimatic assessment (Brassell *et al.*, 1986a,b; Prahl and Wakeham, 1987). Herein, they are discussed in terms of the history of their recognition, their occurrence, and the development of their use as paleoclimatic, specifically paleotemperature, indicators.

3.4.1. Occurrence and Identifications of Alkenones

Series of alkenones were first observed as abundant components of Miocene to Pleistocene sediments from Walvis Ridge recovered during Leg 40 of the Deep Sea Drilling Project (DSDP; Boon *et al.*, 1978). In addition, reexamination of published gas-chromatographic traces suggests that they were certainly present, though unrecognized, in earlier analyses of DSDP sediments (cf. Table I; Simoneit and Burlingame, 1972, 1973; Simoneit *et al.*, 1973). Since their initial discovery in marine sediments, their range of occurrence has been found to encompass all the oceans. Indeed, they occur in marine sediments

Figure 6. Selected structures, abbreviated notations (cf. Marlowe *et al.*, 1984a, b; Sinninghe Damsté *et al.*, 1989), and chemical names of major long-chain alkenones, alkyl alkenoates, alkenes, and dialkylthiolanes and dialkylthianes found in sediments (e.g., de Leeuw *et al.*, 1980; Volkman *et al.*, 1980a, b; Marlowe *et al.*, 1984 a, b; Sinninghe Damsté *et al.*, 1989). Compound assignments are made as follows: *, Structures verified by synthesis (Rechka and Maxwell, 1988a, b); †, double-bond positions supported by mass spectra of trimethylsilyloxy derivatives (de Leeuw *et al.*, 1980); §, structures based on mass spectral characteristics and gas-chromatographic retention times, with the positions (i.e., ω-15, ω-22, ω-29) and configurations (i.e., E) of double bonds in alkenones, alkyl alkenoates, and alkenes assigned by analogy with standards (de Leeuw *et al.*, 1980; Volkman *et al.*, 1980a, b; Marlowe *et al.*, 1984a; Sinninghe Damsté *et al.*, 1989). Other related long-chain constituents are homologs of these compounds with extra carbon atoms and/or additional positions of unsaturation (i.e., ω-8) for the alkenones, alkyl alkenoates, and alkenes (e.g., Marlowe *et al.*, 1984a, b; Farrimond *et al.*, 1986; Volkman *et al.*, 1989) or with extra carbon atoms and/or different alkyl substituents in the dialkylthiolanes and dialkylthianes (Sinninghe Damsté *et al.*, 1989).

TABLE I. Occurrences of Long-Chain Alkenones, Alkyl Alkenoates, and Alkenes in Marine and Lacustrine Sediments, in Sediment-Trap and Particulate Material, and in Aerosols

Location	Site/code[a]	Latitude	Longitude	No. of samples	Geological age	Approximate age
Equatorial East Pacific	9-84	5°45′N	82°53′W	1	Pliocene	NR[c]
Ionian Basin	13-128	35°42′N	22°28′E	2	Pleistocene	NR
Levantine Basin	13-130	33°36′N	27°52′E	1	Pleistocene	NR
Cap Blanc	14-138	25°55′N	25°33′W	1	Oligocene	35 Ma
Cariaco Trench	15-147	10°43′N	65°10′W	6	Pleistocene	NR
Cariaco Trench	15-147	10°43′N	65°10′W	2	Pleistocene	2/100 ka
Cariaco Trench	SPM	10°22′N	65°18′W	2	NA[e]	NA
Porcupine Seabight	D51704	NR	NR	1	Holocene	NA
Iceland–Faeroe Ridge	38-336	63°21′N	07°47′W	3	Oligocene	27–35 Ma
Iceland–Faeroe Ridge	38-336	63°21′N	07°47′W	2	Eocene	37/39 Ma
St. Lawrence Gulf	81-025-12	47°45′N	60°08′W	8	Holocene	<13 ka
St. Lawrence Gulf	81-025-3	42°50′N	61°45′W	1	Holocene	NR
Walvis Ridge	40-362	19°45′S	10°32′E	1	Pleistocene	NR
Walvis Ridge	40-362	19°45′S	10°32′E	1	Pliocene	NR
Walvis Ridge	40-362	19°45′S	10°32′E	2	Miocene	NR
Walvis Ridge	75-532	19°45′S	10°31′E	1	Pleistocene	0.5 Ma
Walvis Ridge	75-532	19°45′S	10°31′E	2	Pliocene	2/3.5 Ma
Walvis Ridge	75-532	19°45′S	10°31′E	20	Pliocene	2 Ma
Messina Abyssal Plain	42A-374	35°51′N	18°12′E	1	Pleistocene	NR
Mediterranean	42A-376	34°52′N	31°48′E	1	Pliocene	2.5 Ma
Cretan Basin	42A-378	35°56′N	25°07′E	1	Pleistocene	NR
Mediterranean	T83-7	33°40′N	25°59′E	5	Pleistocene	160/180 ka
Mediterranean	T83-30	33°48′N	28°37′E	2	Pleistocene	8/125 ka
Vrica	NR	NR	NR	1	Pliocene	NR
Japan Trench	56-434	39°45′N	144°06′E	1	Pliocene	3 Ma
Japan Trench	56-435	39°44′N	143°48′E	1	Pliocene	2.5 Ma
Japan Trench	56-436	39°56′N	145°33′E	1	Pliocene	2 Ma
Japan Trench	57-438	40°38′N	143°14′E	1	Pliocene	3 Ma
Japan Trench	57-438	40°38′N	143°14′E	1	Miocene	13 Ma
Japan Trench	57-440	39°44′N	143°56′E	4	Pleistocene	0.5–1 Ma
Japan Trench	57-440	39°44′N	143°56′E	2	Pliocene	2/4 Ma
Japan Trench	57-440	39°44′N	143°56′E	1	Miocene	7 Ma
San Miguel Gap	63-467	33°51′N	120°45′W	1	Pleistocene	0.2 Ma
San Miguel Gap	63-467	33°51′N	120°45′W	2	Pliocene	2.5/4.5 Ma
San Miguel Gap	63-467	33°51′N	120°45′W	1	Miocene	7 Ma
Santa Catalina Basin	NR	NR	NR	8	Holocene	NR
Gulf of California	64-479	27°51′N	111°37′W	3	Pleistocene	NR
Gulf of California	64-481	27°15′N	111°30′W	1	Pleistocene	NR
Middle America Trench	66-487	15°51′N	99°11′W	1	Pleistocene	NR
Middle America Trench	66-491	16°02′N	98°58′W	1	Pleistocene	NR
Middle America Trench	67-493	16°23′N	98°55′W	1	Pleistocene	NR
Middle America Trench	67-493	16°23′N	98°55′W	1	Pliocene	3.5 Ma
Middle America Trench	67-493	16°23′N	98°55′W	2	Miocene	7/22 Ma
Middle America Trench	67-494	12°43′N	90°56′W	1	Pleistocene	NR
Middle America Trench	67-494	12°43′N	90°56′W	1	Eocene	45 Ma
Middle America Trench	67-495	12°30′N	91°02′W	1	Pleistocene	0.6 Ma
Middle America Trench	67-496	13°04′N	90°48′W	1	Pleistocene	NR
Middle America Trench	67-499	12°40′N	90°57′W	1	Pleistocene	NR
Falkland Plateau	71-511	51°00′S	46°58′W	4	Oligocene	32–34 Ma
Falkland Plateau	71-511	51°00′S	46°58′W	2	Eocene	39–41 Ma
South Atlantic	71-514	46°03′S	26°51′W	1	Pleistocene	NR
South Atlantic	71-514	46°03′S	26°51′W	1	Pliocene	3.2 Ma
Blake–Bahama Basin	76-534	28°21′N	75°23′W	2	Cretaceous	95/105 Ma
Sierra Leone Rise	M13519	5°40′N	19°51′W	3	Pleistocene	NR
Kane Gap	M16415-2	9°34′N	19°06′W	72	Holocene	0–1 Ma
Cap Blanc	108-658	20°45′N	18°35′W	86	Pleistocene	0–575 ka
Cap Blanc	108-658	20°45′N	18°35′W	2	Pliocene	NR

[a]ST, Sediment trap; SPM, suspended particulate material; NZAS, New Zealand aerosol sample.
[b]MS, Compound(s) evident in published mass spectra; GC, compound(s) evident in published gas chromatogram; Y, compound(s) present; N, compound(s) absent; NR, not reported.
[c]NR, Not reported.

TABLE I. *(Continued)*

Water depth (m)	Sediment depth (m)	Alkenones[b]	Alkenoates[b]	Alkenes[b]	U^K_{37} Index	Reference(s)
3096	146	MS	NR	NR	ND[d]	Simoneit and Burlingame, 1972
4640	85	GC	NR	NR	ND	Simoneit and Burlingame, 1973
2980	16	GC	NR	NR	ND	Simoneit and Burlingame, 1973
5288	118	GC	NR	NR	ND	Simoneit *et al.*,1973
892	0.3–171	Y	N	N	0.89	Marlowe *et al.*, 1984a
892	2/50	Y	NR	NR	ND	de Leeuw *et al.*, 1980
50/150	NA	Y	NR	NR	0.88/0.89	Prahl and Wakeham, 1987
NR	0–0.065	Y	Y	N	0.54	Marlowe, 1984
830	184/239	Y	Y	Y	0.51–0.53	Marlowe, 1984
830	250/271	Y	N	N	0.57/0.84	Marlowe, 1984
507	0–7.6	Y	GC	NR	ND	Nichols and Johns, 1986
980	0–0.002	Y	NR	NR	ND	Nichols and Johns, 1986
1325	62	Y	NR	N	ND	Boon *et al.*, 1978, de Leeuw *et al.*, 1980
1325	157	Y	NR	N	ND	Boon *et al.*, 1978; de Leeuw *et al.*, 1980
1325	263/453	Y	NR	N	ND	Boon *et al.*, 1978, de Leeuw *et al.*, 1980
1331	19	Y	Y	N	0.56	Brassell *et al.*, 1986a
1331	83/173	Y	Y	N	NR	Brassell and Eglinton, 1983a
1331	93–94	Y	Y	N	0.61–0.69	Marlowe, 1984
4078	208	Y	N	Y	0.47	Comet, 1982; Brassell *et al.*, 1986a
2101	41	Y	N	Y	0.87	Comet, 1982; Brassell *et al.*, 1986a
1835	86	Y	N	Y	0.66	Comet, 1982; Brassell *et al.*, 1986a
2480	1.8–3.2	Y	NR	NR	0.48–0.71	ten Haven *et al.*, 1987
2810	NR	Y	NR	NR	0.64/0.80	ten Haven *et al.*, 1987
NA	NA	Y	NR	NR	0.87	ten Haven *et al.*, 1987
6000	209	Y	NR	NR	0.65	Brassell, 1980; Brassell *et al.*, 1980a, b
3413	115	Y	NR	NR	0.66	Brassell, 1980; Brassell *et al.*, 1980a, b
5248	99	Y	Y	Y	0.64	Brassell, 1980; Brassell *et al.*, 1980a, b
1552	87	Y	Y	N	0.39	Marlowe, 1984
1552	674	Y	N	N	0.92	Marlowe, 1984
4517	42–212	Y	Y	Y	0.51–0.55	Brassell, 1980; Brassell *et al.*, 1980a, b
4517	354/504	Y	Y	Y	0.51/0.61	Brassell, 1980; Brassell *et al.*, 1980a, b
4517	779	Y	Y	Y	0.80	Brassell, 1980; Brassell *et al.*, 1980a, b
2128	21	Y	NR	Y	0.58	McEvoy, 1983
2128	165/333	Y	NR	NR	0.83/0.85	McEvoy, 1983
2128	494	Y	NR	NR	1.00	McEvoy, 1983
1297	0–0.18	Y	NR	NR	0.50–0.58	Shaw and Johns, 1986
747	15–153	Y	Y	N	0.88–0.95	Marlowe, 1984
1998	8	Y	Y	N	0.86	Brassell *et al.*, 1986a
4777	5	Y	N	N	0.97	Brassell *et al.*, 1981
2877	7	Y	N	N	0.98	Brassell *et al.*, 1981
676	4	Y	N	N	0.78	Marlowe, 1984
676	153	Y	N	N	0.99	Marlowe, 1984
676	297/620	Y	N	N	0.99/1.00	Marlowe, 1984
5529	18	Y	N	N	0.92	Brassell *et al.*, 1987
5529	270	Y	N	N	1.00	Marlowe, 1984
4150	23	Y	N	N	0.93	Brassell *et al.*, 1987
2064	12	Y	N	N	0.88	Brassell *et al.*, 1987
6127	9	Y	N	N	0.99	Brassell *et al.*, 1987
2589	20–139	Y	Y	N	0.31–0.54	Marlowe, 1984
2589	158/178	Y	Y	N	0.63–0.65	Marlowe, 1984
4318	4	Y	Y	N	0.33	Marlowe, 1984
4318	109	Y	Y	N	0.24	Marlowe, 1984
4971	766–832	Y	N	N	1.00	Farrimond *et al.*, 1986
2860	0.32–1.88	Y	N	N	0.88–0.97	Marlowe, 1984
3851	0.05–13	Y	Y	N	0.76–0.94	Brassell *et al.*, 1986a, b
2264	0–5.6	Y	Y	N	0.60–0.82	Poynter *et al.*, 1989a
2264	183/213	Y	NR	NR	ND	ten Haven *et al.*, 1989

[d]ND, Not determined.
[e]NA, Not applicable.
[f]Range of values reported for all five samples from Washington Coast and Astoria Fan sites.

(continued)

TABLE I. *(Continued)*

Location	Site/code[a]	Latitude	Longitude	No. of samples	Geological age	Approximate age
NE Cape Verde	108-659	18°05′N	21°02′W	3	Pleistocene	NR
NE Cape Verde	108-659	18°05′N	21°02′W	2	Pliocene	NR
NE Sierra Leone Rise	108-660	10°01′N	19°15′W	25	Pleistocene	0–250 ka
Peru: Lima Basin Ridge	112-679	11°04′S	76°16′W	7	Pleistocene	NR
Peru: Lima Basin Ridge	112-679	11°04′S	76°16′W	3	Pliocene	NR
Peru: Salaverry Basin	112-681	10°59′S	77°57′W	10	Pleistocene	NR
Peru: Trujillo Basin	112-684	8°59′S	79°54′W	3	Pleistocene	NR
Peru: Trujillo Basin	112-684	8°59′S	79°54′W	4	Pliocene	NR
Peru: West Pisco Basin	112-686	13°29′S	76°54′W	84	Pleistocene	0–350 ka
Owen Ridge	117-721	16°41′N	59°52′E	7	Pleistocene	NR
Oman Margin	117-723	18°03′N	57°37′E	18	Pleistocene	0–80 ka
Oman Margin	117-728	16°40′N	57°50′E	30	Pleistocene	0–275 ka
Owen Ridge	117-731	17°28′N	59°42′E	26	Pleistocene	0–275 ka
Orca Basin	96-618	27°01′N	91°16′W	1	Holocene	NR
Pigmy Basin	96-619	27°12′N	91°25′W	33	Pleistocene	1.8–96 ka
Peruvian Shelf	SC4	NR	NR	18	Holocene	NR
Peruvian Shelf	SC6	NR	NR	13	Holocene	0–125 a
Peruvian Shelf	BC7	15°02′S	75°31′W	5	Holocene	NR
Peruvian Shelf	ST	NR	NR	4	NA	NA
Peruvian Shelf	110	12°02′S	77°29′W	1	Holocene	NR
North Sea	IGS SLN33	58°4′N	0°23′E	1	Holocene	NR
North Sea	IGS SP333CS	58°4′N	0°34′E	1	Holocene	NR
Walvis Bay	Cruise 1/75	22°35′S	13°45′E	2	Holocene	NR
Santa Barbara Basin	Spoul	34°14′N	120°01′W	133	Holocene	0–80 a
Guinea Basin	10529	00°21′E	4°41′S	1	Holocene	NR
Namibian Shelf	NR	22°15′S	14°16′E	1	Holocene	1–1.5 ka
Namibian Shelf	NR	22°52′S	14°15′E	1	Holocene	NR
Black Sea	Unit 1	42°41′N	29°40′E	1	Holocene	0–3 ka
Black Sea	Unit 2	41°39′N	30°44′E	1	Holocene	3–7 ka
Ionian Sea	10103	36°10′N	20°29′E	5	Holocene	2.5–10 ka
Gulf of Arabia	NR	NR	NR	1	Holocene	NR
Corner Inlet	NR	38°49′S	146°20′E	1	Holocene	NR
Lake Burland	NR	NR	NR	2	Holocene	50 ka
Coniston	NR	NR	NR	4	Holocene	9–14.5 ka
Windermere	NR	NR	NR	1	Holocene	<3 a
Crose Mere	NR	NR	NR	2	Holocene	<30–400 a
Upton Broad	NR	NR	NR	2	Holocene	<8–450 a
Ace Lake	NR	68°28′S	78°11′E	1	Holocene	NR
Madeira Abyssal Plain	86P5	32°03′N	24°13′W	3	Pleistocene	140 ka
Madeira Abyssal Plain	86P25	30°44′N	25°23′W	3	Pleistocene	140 ka
North Atlantic	29GGC	56°13′N	12°37′W	23	Pleistocene	0–17 ka
N.W. Africa	2BC5	14°41′N	17°50′W	2	Holocene	NR
E. eq. Atlantic	4BC14	9°14′N	17°00′W	2	Holocene	NR
E. eq. Atlantic	5BC17	8°49′N	17°16′W	2	Holocene	NR
Eq. Atlantic	13BC55	2°06′S	23°03′W	1	Holocene	NR
Eq. Atlantic	12PC51	0°00′N	23°00′W	18	Holocene	NR
W. eq. Atlantic	110KBC84	04°22′N	43°31′W	2	Holocene	NR
North Atlantic	0173	31°54′N	64°08′W	1	Holocene	NR
Peru	MWSC2	11°04′S	78°03′W	2	Holocene	NR
Peru	MWSC7	14°56′S	75°37′W	2	Holocene	NR
Peru	SC3	15°06′S	75°42′W	50	Holocene	0-300 a
Peru	SC3 Subcore	15°06′S	75°42′W	15	Holocene	0-20 a
Peru	SC3 Floc	15°06′S	75°42′W	1	Holocene	0
Peru	SC2	11°04′S	78°03′W	12	Holocene	NR
Peru	SC7	14°57′S	75°37′W	15	Holocene	NR
Gulf of Alaska	PAR87A	54°13′N	148°49′W	1	Holocene	NR
N.W. Pacific	RNDB24GG	51°17′N	167°27′E	2	Holocene	NR
N.E. Pacific	PAPA C4	49°26′N	136°39′W	1	Holocene	NR
California	BC207	35°40′N	121°37′W	2	Holocene	NR
California	BC238	35°57′N	122°59′W	2	Holocene	NR
Black Sea	K134BC34	42°16′N	37°32′E	2	Holocene	NR
Black Sea	134KB53	42°41′N	37°37′E	1	Holocene	NR

TABLE I. *(Continued)*

Water depth (m)	Sediment depth (m)	Alkenones[b]	Alkenoates[b]	Alkenes[b]	U^{K}_{37} Index	Reference(s)
3070	2–41	Y	NR	NR	ND	ten Haven *et al.*, 1989
3070	123/276	Y	NR	NR	ND	ten Haven *et al.*, 1989
4327	0.4–5.6	Y	Y	N	0.77–0.94	Poynter *et al.*, 1989a
461	0.35–61	Y	NR	NR	ND	ten Haven *et al.*, 1990
461	68/74	Y	NR	NR	ND	ten Haven *et al.*, 1990
161	0.35–87	Y	NR	NR	ND	ten Haven *et al.*, 1990
437	0.35–13	Y	NR	NR	ND	ten Haven *et al.*, 1990
437	18–36	Y	NR	NR	ND	ten Haven *et al.*, 1990
447	0–160	Y	Y	NR	0.66–0.92	Farrimond *et al.*, 1990a, b
1945	2.5–22	Y	NR	NR	0.87–0.98	ten Haven and Kroon, 1991
808	0.5–16	Y	NR	NR	0.80–0.89	ten Haven and Kroon, 1991
1428	0.14–14	Y	NR	NR	0.69–0.92	ten Haven and Kroon, 1991
2366	0.35–10	Y	NR	NR	0.65–0.89	ten Haven and Kroon, 1991
2412	14	Y	NR	NR	0.84	Brassell *et al.*, 1986a
2259	1.6–188	Y	NR	NR	0.62–0.89	Jasper and Gagosian, 1989
90	0–0.68	Y	NR	NR	0.65–0.88	Farrington *et al.*, 1988
268	0–0.52	Y	NR	NR	0.64–0.78	Farrington *et al.*, 1988
85	0–0.19	Y	NR	Y	0.23–0.70	Volkman *et al.*, 1983
14/52	NA	Y	NR	NR	ND	Volkman *et al.*, 1983
145	0.001–0.002	Y	NR	NR	0.87	Smith *et al.*, 1983a
155	0–4.3	Y	NR	NR	0.51	Volkman *et al.*, 1980a
155	0–0.9	Y	Y	Y	0.40	Volkman *et al.*, 1980a
127	0.4–0.8	Y	NR	NR	0.52	Volkman *et al.*, 1980a
NR	0–0.3	Y	Y	N	0.47–0.55	Kennedy and Brassell, 1992a, b
4735	1	Y	NR	NR	ND	Brassell *et al.*, 1986a; Morris *et al.*, 1984
NR	0.4	Y	NR	N	ND	de Leeuw *et al.*, 1980
NR	0.7	Y	NR	N	ND	de Leeuw *et al.*, 1980
NR	0.1	Y	NR	Y	ND	de Leeuw *et al.*, 1980
2150	0.5	Y	NR	N	ND	de Leeuw *et al.*, 1980
2880	0–1.1	Y	NR	NR	0.44–0.79	Smith *et al.*, 1986
NR	NR	Y	NR	NR	ND	Marlowe *et al.*, 1984a
NR	0–0.005	Y	NR	NR	0.56	Volkman *et al.*, 1981
NA	28/29	Y	NR	NR	ND	Cranwell, 1988
8	0.90–1.6	Y	NR	NR	0.16–0.37	Cranwell, 1985
38	0–0.01	Y	NR	NR	−0.49	Cranwell, 1985
9.2	0–0.60	Y	NR	NR	−0.57/−0.60	Cranwell, 1985
1.4	0–0.65	Y	NR	NR	−0.41/−0.58	Cranwell, 1985
23	0–0.05	Y	Y	N	ND	Volkman *et al.*, 1988
5400	0.70–0.78	Y	Y	NR	0.68–0.70	Prahl *et al.*, 1989b
5400	0.78–0.92	Y	Y	NR	0.70–0.72	Prahl *et al.*, 1989b
2626	0.01–1.9	Y	NR	NR	0.32–0.77	Sikes, 1990
1445	0	Y	NR	NR	1.00	Sikes *et al.*, 1991
1011	0	Y	NR	NR	0.96/1.00	Sikes *et al.*, 1991
3749	0	Y	NR	NR	1.00	Sikes *et al.*, 1991
5065	0	Y	NR	NR	0.83	Sikes *et al.*, 1991
3875	0.01–0.43	Y	NR	NR	0.86–0.98	Sikes, 1990; Sikes *et al.*, 1991
2947	0	Y	NR	NR	1.00	Sikes *et al.*, 1991
4469	0	Y	NR	NR	0.92	Sikes *et al.*, 1991
255	0	Y	NR	NR	0.78	Sikes *et al.*, 1991
105	0	Y	NR	NR	0.64/0.68	Sikes *et al.*, 1991
253	0–1	Y	NR	NR	0.68–0.75	McCaffrey *et al.*, 1991
253	0–0.05	Y	NR	NR	0.69–0.73	McCaffrey *et al.*, 1991
253	0	Y	NR	NR	0.70	McCaffrey *et al.*, 1991
255	0–0.08	Y	NR	NR	0.73–0.84	McCaffrey *et al.*, 1991
105	0.01–0.14	Y	NR	NR	0.65–0.79	McCaffrey *et al.*, 1991
3446	0	Y	NR	NR	0.42	Sikes *et al.*, 1991
2323	0	Y	NR	NR	0.52/0.83	Sikes *et al.*, 1991
3740	0	Y	NR	NR	0.30	Sikes *et al.*, 1991
795	0	Y	NR	NR	0.54/0.56	Sikes *et al.*, 1991
3590	0	Y	NR	NR	0.64/0.69	Sikes *et al.*, 1991
2060	0	Y	NR	NR	0.34/0.44	Sikes *et al.*, 1991
2154	0	Y	NR	NR	0.51	Sikes *et al.*, 1991

(continued)

TABLE I. *(Continued)*

Location	Site/code[a]	Latitude	Longitude	No. of samples	Geological age	Approximate age
New Zealand	NZAS	34°06′S	172°08′W	9	NA	NA
Washington Coast	NR	46°50′N	124°26′W	1	Holocene	0
Washington Coast	NR	46°50′N	124°36′W	2	Holocene	0/200 a
Washington Coast	NR	46°45′N	125°01′W	1	Holocene	0
Astoria Seafan	NR	46°04′N	125°05′W	1	Holocene	0
Manop C	W8420A-14GC	0°57′N	138°57′W	11	Pleistocene	0.9–26.2 ka
Manop C	W8402A-14GC	0°57′N	138°57′W	88	Pleistocene	0.8–253 ka
Manop C	ST	0°57′N	138°57′W	1	NA	434 days
Vertex I	ST	35°50′N	124°40′W	2	NA	NA
Vertex II	ST	18°00′N	108°00′W	3	NA	NA
Vertex II	SPM	18°00′N	108°00′W	2	NA	NA
Vertex III	SPM	15°47′N	108°18′W	2	NA	NA
VertexIII	ST	15°47′N	108°18′W	3	NA	NA
Vertex VA	SPM	33°10′N	138°37′W	3	NA	NA
Vertex VB	SPM	35°22′N	126°10′W	1	NA	NA
Vertex VX	SPM	35°31′N	124°59′W	1	NA	NA
Vertex VC	SPM	36°06′N	122°38′W	2	NA	NA
Parflux E	ST	13°30′N	54°00′W	2	NA	NA
Parflux P1	ST	15°21′N	151°28′W	2	NA	NA
Sea of Japan	128-798	37°02′N	134°48′W	56	Pleistocene	0–750 ka
Guaymas Basin	AII125-8	27°47′N	111°43′W	39	Holocene	NA
Midland Valley	139-855	48°27′N	128°38′W	19	Pleistocene	NR
Midland Valley	139-856	48°26′N	128°41′W	8	Pleistocene	NR
Midland Valley	139-857	48°26′N	128°43′W	17	Pleistocene	NR
Midland Valley	139-858	48°27′N	128°43′W	27	Pleistocene	NR
Black Sea	SPM	NR	NR	12	NA	NA
Black Sea	ST, BSK-2	NR	NR	1	NA	NA
Black Sea	NR	NR	NR	1	Holocene	0
Eastern North Atlantic	D183 SPM:SAP	59°25′N	20°59′W	1	NA	NA
Eastern North Atlantic	D183 SPM:CTD	59°42′N	21°15′W	5	NA	NA
Eastern North Atlantic	D183 SPM:SAP	59°48′N	21°10′W	3	NA	NA
Eastern North Atlantic	D183 SPM:CTD	55°59′N	19°49′W	3	NA	NA
Eastern North Atlantic	D183 SPM:CTD	55°59′N	19°56′W	1	NA	NA
Eastern North Atlantic	D183 SPM:CTD	51°57′N	19°51′W	1	NA	NA
Eastern North Atlantic	D183 SPM:SAP	51°48′N	19°26′W	1	NA	NA
Eastern North Atlantic	D183 SPM:SAP	46°55′N	20°02′W	1	NA	NA
Eastern North Atlantic	D183 SPM:CTD	46°50′N	20°02′W	2	NA	NA
Eastern North Atlantic	D183 SPM:SAP	46°45′N	19°52′W	1	NA	NA
Eastern North Atlantic	CD46 SPM:SAP	49°51′N	18°32′W	1	NA	NA
Eastern North Atlantic	CD46 SPM:SAP	48°54′N	17°01′W	1	NA	NA
Eastern North Atlantic	D191 SPM:CTD	48°30′N	17°48′W	1	NA	NA
Eastern North Atlantic	D191 SPM:CTD	48°29′N	17°48′W	1	NA	NA
Eastern North Atlantic	D191 SPM:CTD	48°28′N	17°34′W	1	NA	NA
Eastern North Atlantic	D191 SPM:CTD	48°27′N	17°30′W	1	NA	NA
Eastern North Atlantic	CD47 SPM:SAP	48°26′N	17°29′W	1	NA	NA
Eastern North Atlantic	CD47 SPM:SAP	48°17′N	17°19′W	2	NA	NA
Eastern North Atlantic	CD47 SPM:SAP	47°56′N	16°49′W	1	NA	NA
Eastern North Atlantic	CD53 SPM:UW	47°04′N	19°44′W	1	NA	NA
Eastern North Atlantic	CD53 SPM:UW	44°39′N	20°34′W	1	NA	NA
Eastern North Atlantic	CD53 SPM:CTD	47°48′N	19°26′W	2	NA	NA
Eastern North Atlantic	CD53 SPM:UW	42°09′N	21°28′W	1	NA	NA
Eastern North Atlantic	CD53 SPM:UW	41°08′N	20°42′W	1	NA	NA
Eastern North Atlantic	CD53 SPM:CTD	42°10′N	21°28′W	1	NA	NA
Eastern North Atlantic	CD53 SPM:UW	38°48′N	19°53′W	1	NA	NA
Eastern North Atlantic	CD53 SPM:UW	35°07′N	19°14′W	1	NA	NA
Eastern North Atlantic	CD53 SPM:UW	32°54′N	18°57′W	1	NA	NA
Eastern North Atlantic	CD53 SPM:UW	32°52′N	19°17′W	1	NA	NA
Eastern North Atlantic	CD53 SPM:CTD	32°52′N	20°28′W	1	NA	NA
Eastern North Atlantic	CD53 SPM:CTD	24°27′N	20°24′W	1	NA	NA
Eastern North Atlantic	CD60 SPM:CTD	60°10′N	20°00′W	1	NA	NA
Eastern North Atlantic	CD60 SPM:CTD	60°40′N	22°35′W	1	NA	NA

TABLE I. *(Continued)*

Water depth (m)	Sediment depth (m)	Alkenones[b]	Alkenoates[b]	Alkenes[b]	U^K_{37} Index	Reference(s)
NA	NA	Y	GC	GC	0.45–0.55	Sicre *et al.*, 1990; Poynter *et al.*, 1989b
73	0.01	Y	Y	NR	0.42–0.46*	Prahl *et al.*, 19888
110	0.01/0.29	Y	Y	NR	0.42–0.46*	Prahl *et al.*, 1988
700	0.01	Y	Y	NR	0.42–0.46*	Prahl *et al.*, 1988
2235	0.01	Y	Y	NR	0.42–0.46*	Prahl *et al.*, 1988
4287	0.15–0.44	Y	Y	NR	0.92–0.96	Prahl *et al.*, 1989a; Lyle *et al.*, 1992
4287	0.15–3.25	Y	Y	NR	0.88 to 0.97	Lyle *et al.*, 1992
3900	NA	Y	Y	NR	0.94	Prahl *et al.*, 1989a
100/750	NA	Y	NR	NR	0.48/0.50	Prahl and Wakeham, 1987
100/470/1500	NA	Y	NR	NR	0.99–1.00	Prahl and Wakeham, 1987
5/200	NA	Y	NR	NR	0.95–0.96	Prahl and Wakeham, 1987
5/200	NA	Y	NR	NR	0.95–1.00	Prahl and Wakeham, 1987
100/470/1500	NA	Y	NR	NR	0.95–0.96	Prahl and Wakeham, 1987
10/10/100	NA	Y	NR	NR	0.56–0.70	Prahl and Wakeham, 1987
10	NA	Y	NR	NR	0.49	Prahl and Wakeham, 1987
10	NA	Y	NR	NR	0.39	Prahl and Wakeham, 1987
10/60	NA	Y	NR	NR	0.37/0.40	Prahl and Wakeham, 1987
389/5068	NA	Y	NR	NR	0.94/0/95	Prahl and Wakeham, 1987; Jasper *et al.*, in press
389/5582	NA	Y	NR	NR	0.93/0.95	Prahl and Wakeham, 1987
906	1.1–84.6	Y	Y	NR	0.26 to 0.73	Kheradyar and Brassell, unpublished data
745	NA	Y	NR	NR	0.83 to 0.89	Kennedy and Brassell, unpublished data
2457	0.7–108.5	Y	NR	NR	0.13 to 0.46	Dans *et al.*, 1992; Simoneit *et al.*, in press
2406	0.01–55.3	Y	NR	NR	0.26 to 0.36	Dans *et al.*, 1992; Simoneit *et al.*, in press
2433	1.9–82.2	Y	NR	NR	0.15 to 0.31	Dans *et al.*, 1992; Simoneit *et al.*, in press
2420	0.03–15.8	Y	NR	NR	0.17 to 0.63	Dans *et al.*, 1992; Simoneit *et al.*, in press
10 to 2000	NA	Y	NR	NR	0.01 to 0.40	Freeman and Wakeham, 1992
NR	NA	Y	NR	NR	0.02	Freeman and Wakeham, 1992
NR	0–0.01	Y	NR	NR	0.37	Freeman and Wakeham, 1992
20	NA	Y	Y	NR	0.24	Conte *et al.*, 1992; Conte and Eglinton, in press
4 to 25	NA	Y	Y	NR	0.30 to 0.46	Conte *et al.*, 1992; Conte and Eglinton, in press
20/30/50	NA	Y	Y	NR	0.25 to 0.51	Conte *et al.*, 1992; Conte and Eglinton, in press
4/10/30	NA	Y	Y	NR	0.46 to 0.52	Conte *et al.*, 1992; Conte and Eglinton, in press
30	NA	Y	Y	NR	0.40	Conte *et al.*, 1992; Conte and Eglinton, in press
30	NA	Y	Y	NR	0.50	Conte *et al.*, 1992; Conte and Eglinton, in press
40	NA	Y	Y	NR	0.39	Conte *et al.*, 1992; Conte and Eglinton, in press
30	NA	Y	Y	NR	0.43	Conte *et al.*, 1992; Conte and Eglinton, in press
4/15	NA	Y	Y	NR	0.45/0.46	Conte *et al.*, 1992; Conte and Eglinton, in press
10	NA	Y	Y	NR	0.45	Conte *et al.*, 1992; Conte and Eglinton, in press
15	NA	Y	Y	NR	0.34	Conte and Eglinton, in press
15	NA	Y	Y	NR	0.39	Conte and Eglinton, in press
10	NA	Y	Y	NR	0.47	Conte *et al.*, 1992; Conte and Eglinton, in press
10	NA	Y	Y	NR	0.51	Conte *et al.*, 1992; Conte and Eglinton, in press
10	NA	Y	Y	NR	0.50	Conte *et al.*, 1992; Conte and Eglinton, in press
10	NA	Y	Y	NR	0.47	Conte *et al.*, 1992; Conte and Eglinton, in press
10	NA	Y	Y	NR	0.38	Conte and Eglinton, in press
10/25	NA	Y	Y	NR	0.37/0.38	Conte and Eglinton, in press
10	NA	Y	Y	NR	0.37	Conte and Eglinton, in press
4	NA	Y	Y	NR	0.44	Conte and Eglinton, in press
4	NA	Y	Y	NR	0.54	Conte and Eglinton, in press
10/30	NA	Y	Y	NR	0.50/0.51	Conte and Eglinton, in press
4	NA	Y	Y	NR	0.64	Conte and Eglinton, in press
4	NA	Y	Y	NR	0.69	Conte and Eglinton, in press
10	NA	Y	Y	NR	0.70	Conte and Eglinton, in press
4	NA	Y	Y	NR	0.69	Conte and Eglinton, in press
4	NA	Y	Y	NR	0.80	Conte and Eglinton, in press
4	NA	Y	Y	NR	0.84	Conte and Eglinton, in press
4	NA	Y	Y	NR	0.82	Conte and Eglinton, in press
10	NA	Y	Y	NR	0.85	Conte and Eglinton, in press
10	NA	Y	Y	NR	0.88	Conte and Eglinton, in press
8	NA	Y	Y	NR	0.27	Conte and Eglinton, in press
8	NA	Y	Y	NR	0.29	Conte and Eglinton, in press

(*continued*)

TABLE I. *(Continued)*

Location	Site/code[a]	Latitude	Longitude	No. of samples	Geological age	Approximate age
Eastern North Atlantic	CD60 SPM:CTD	60°25′N	14°50′W	1	NA	NA
Eastern North Atlantic	CD61 SPM:CTD	61°05′N	19°60′W	1	NA	NA
Eastern North Atlantic	CD61 SPM:SAP	61°17′N	19°47′W	1	NA	NA
Eastern North Atlantic	CD61 SPM:UW	60°33′N	20°09′W	1	NA	NA
Eastern North Atlantic	CD61 SPM:UW	60°33′N	19°33′W	1	NA	NA
Eastern North Atlantic	CD61 SPM:UW	60°07′N	19°41′W	1	NA	NA
Eastern North Atlantic	CD61 SPM:UW	56°04′N	19°59′W	1	NA	NA
Eastern North Atlantic	CD61 SPM:UW	55°56′N	20°01′W	1	NA	NA
Eastern North Atlantic	CD61 SPM:UW	55°37′N	20°00′W	1	NA	NA
Eastern North Atlantic	191'A'	~59°N	~20°W	20	Holocene	NR
Eastern North Atlantic	191'E'	~48°N	~19°W	21	Holocene	NR
Eastern North Atlantic	191'A' SPM	~59°N	~20°W	13	NA	NA
Eastern North Atlantic	191'E' SPM	~48°N	~19°W	16	NA	NA
Northeast Pacific	Nearshore ST	~42°N	~132°W	6	NA	NA
Northeast Pacific	Midway ST	~42°N	~128°W	6	NA	NA
Northeast Pacific	Gyre ST	~42°N	~126°W	5	NA	NA
Northeast Pacific	Nearshore	~42°N	~132°W	1	Holocene	NA
Northeast Pacific	Midway	~42°N	~128°W	1	Holocene	NA
Northeast Pacific	Gyre	~42°N	~126°W	1	Holocene	NA
Southern Oceans	FR9/86 SPM	~43°–57°S	~155°E	19	NA	NA
Southern Oceans	AA 1.1/91 SPM	~44°–62°S	~138°–146°E	18	NA	NA
Northern Indian Ocean	MD 900963	5°04′N	73°53′E	NR	Pleistocene	0–170 ka
E. eq. Pacific	138-846	3°06′S	90°49′W	45	Pleistocene	0–700 ka
Northeast Pacific	W8709A-8TC	42°N	NR	NR	Pleistocene	NR
Chile Trench	141-861	45°51′S	75°41′W	1	Pleistocene	NR
Chile Trench	141-861	45°51′S	75°41′W	2	Pliocene	NR
Cap Blanc	108-658	20°45′N	18°35′W	>250	Pleistocene	0–730 ka

on a worldwide basis (Table I and Fig. 7) and have also been observed in lacustrine sediments (Cranwell, 1985, 1988; Volkman *et al.*, 1988). Thus, their presence as globally significant constituents of contemporary sediments fulfills the criterion that biomarkers proposed as paleoclimatic indicators should occur widely. In addition, the record of their presence in pre-Pleistocene sediments (Table I; Marlowe *et al.*, 1990) demonstrates that this global prevalence is long-standing.

The chain lengths of the alkenones and their positions of unsaturation (Fig. 6) were elucidated (de Leeuw *et al.*, 1980) at the same time as the compounds were first recognized in an organism, namely, *Emiliania huxleyi*, an abundant marine unicellular coccolithophorid alga belonging to the Prymnesiophyceae (Volkman *et al.*, 1980a). Subsequently, the alkenones have been identified in other members of this family, and they appear to be biochemically restricted to it (Marlowe *et al.*, 1984a, b). In addition, their unsaturation has been confirmed as the biologically less common *E* configuration (Rechka and Maxwell, 1988a, b), a factor that may aid their preservation in sediments through resistance to bacterial attack.

3.4.2. Temperature Dependence in Unsaturation

Examination of the distributions of alkenones in sediments from two disparate environmental regimes, namely, the Japan Trench and the Middle America Trench, revealed a major difference in their proportion of di- and triunsaturated alkenones (Brassell *et al.*, 1981; Brassell and Eglinton, 1983a; Fig. 8). Specifically, the proportion of alkatrienones was found to be markedly higher in sediments from the Japan Trench than in those from the Middle America Trench. The sources of organic matter, and hence extractable lipids, in these two environments show major similarities because both environments receive contributions from autochthonous and allochthonous sources (e.g., Brassell *et al.*, 1980a, b, 1981, 1987). Hence, the factor determining their long-chain alkenone compositions was unlikely to be a function of their sources of organic matter. Rather, it seemed that the alkenones might be revealing a relationship that reflected the different climates of the two environments (Brassell and Eglinton, 1984), specifically a link between the growth temperature of their source algae and the degree of unsaturation in the alkenones—behavior comparable to that seen in the biological

TABLE I. *(Continued)*

Water depth (m)	Sediment depth (m)	Alkenones[b]	Alkenoates[b]	Alkenes[b]	U^K_{37} Index	Reference(s)
5	NA	Y	Y	NR	0.40	Conte and Eglinton, in press
10	NA	Y	Y	NR	0.46	Conte and Eglinton, in press
15	NA	Y	Y	NR	0.39	Conte and Eglinton, in press
4	NA	Y	Y	NR	0.43	Conte and Eglinton, in press
4	NA	Y	Y	NR	0.45	Conte and Eglinton, in press
4	NA	Y	Y	NR	0.39	Conte and Eglinton, in press
4	NA	Y	Y	NR	0.36	Conte and Eglinton, in press
4	NA	Y	Y	NR	0.37	Conte and Eglinton, in press
4	NA	Y	Y	NR	0.36	Conte and Eglinton, in press
3070	NR	Y	Y	NR	NR	Conte *et al.*, 1992
4105	NR	Y	Y	NR	NR	Conte *et al.*, 1992
0–2000	NA	Y	Y	NR	NR	Conte *et al.*, 1992
0–4000	NA	Y	Y	NR	NR	Conte *et al.*, 1992
1000	NA	Y	Y	NR	0.33 to 0.42	Prahl *et al.*, in press
1000	NA	Y	Y	NR	0.33 to 0.44	Prahl *et al.*, in press
1000	NA	Y	Y	NR	0.33 to 0.44	Prahl *et al.*, in press
2712	0–0.01	Y	Y	NR	0.42	Prahl *et al.*, in press
3111	0–0.01	Y	Y	NR	0.42	Prahl *et al.*, in press
3680	0–0.01	Y	Y	NR	0.44	Prahl *et al.*, in press
10–250	NA	Y	NR	Y	0.06 to 0.32	Sikes and Volkman, 1993
3/10	NA	Y	NR	Y	0.03 to 0.34	Sikes and Volkman, 1993
2450	0–10	Y	NR	NR	~0.89 to 0.99	Rostek *et al.*, 1992
3307	NR	Y	NR	NR	0.75–0.93	Mayer *et al.*, 1992; Emeis *et al.*, 1992
NR	NR	Y	NR	NR	~0.30 to 0.45	Prahl *et al.*, 1992
1652	45	Y	GC	NR	ND	Behrmann *et al.*, 1992
1652	215/391	Y	GC	NR	ND	Behrmann *et al.*, 1992
2264	0–100	Y	NR	NR	0.58 to 0.88	Eglinton *et al.*, 1992

production of carboxylic acids. This characteristic prompted an initial test of this hypothesis through evaluation of the distribution of alkenones in *E. huxleyi* grown in laboratory cultures at a variety of different temperatures (Marlowe, 1984; Brassell *et al.*, 1986a). These preliminary experiments and subsequent, independent investigations (Prahl and Wakeham, 1987; Prahl *et al.*, 1988) have verified a temperature dependence in the biosynthetic production of unsaturation in the alkenones, with higher proportions of the more unsaturated components (i.e., $C_{37:3}$ and $C_{37:4}$) at lower temperatures (Fig. 9). It is noteworthy that the relationship between the alkenones and growth temperature also appears to be independent of culture media (Marlowe, 1984). In order to permit the extent of unsaturation in the alkenones to be directly compared between different culture experiments and sediment samples, the term U^K_{37}, defined as the alkenone unsaturation index, was introduced (Brassell *et al.*, 1986a). This index is calculated from the relative concentrations of the di-, tri-, and tetra-unsaturated C_{37} alkenones ($C_{37:2}$, $C_{37:3}$, and $C_{37:4}$, respectively) in either organism or sediment according to the formula

$$U^K_{37} = [C_{37:2}] - [C_{37:4}]/([C_{37:2}] + [C_{37:3}] + [C_{37:4}])$$

which simplifies to

$$U^{K'}_{37} = [C_{37:2}]/([C_{37:2}] + [C_{37:3}])$$

in the absence of the tetraunsaturated component.

Results from series of culture experiments with *E. huxleyi* have revealed an apparent linear relationship between U^K_{37} and growth temperature (Fig. 10a; Brassell *et al.*, 1986a, Strains 92 and 92d with $R^2 = 0.791$; Prahl and Wakeham, 1987, $R^2 = 0.977$; Prahl *et al.*, 1988; $R^2 = 0.994$ for $U^{K'}_{37}$), enabling estimation of sea surface water temperatures from U^K_{37} values—but see below. The original relationship [$U^K_{37} = 0.040T - 0.110$ from data of Marlowe (1984) or $U^K_{37} = 0.033T + 0.043$ from Prahl and Wakeham (1987)] was subsequently revised to yield the equation $U^{K'}_{37} = 0.034T + 0.039$ (Prahl and Wakeham, 1987; Prahl *et al.*, 1988), which has become the standard calibration for conversion of U^K_{37} values into water temperatures. The alkenone composition of water column particulate matter and sediment-trap material collected from various sites, principally in the tropical Pacific, bears a similar linear relationship to the measured water tem-

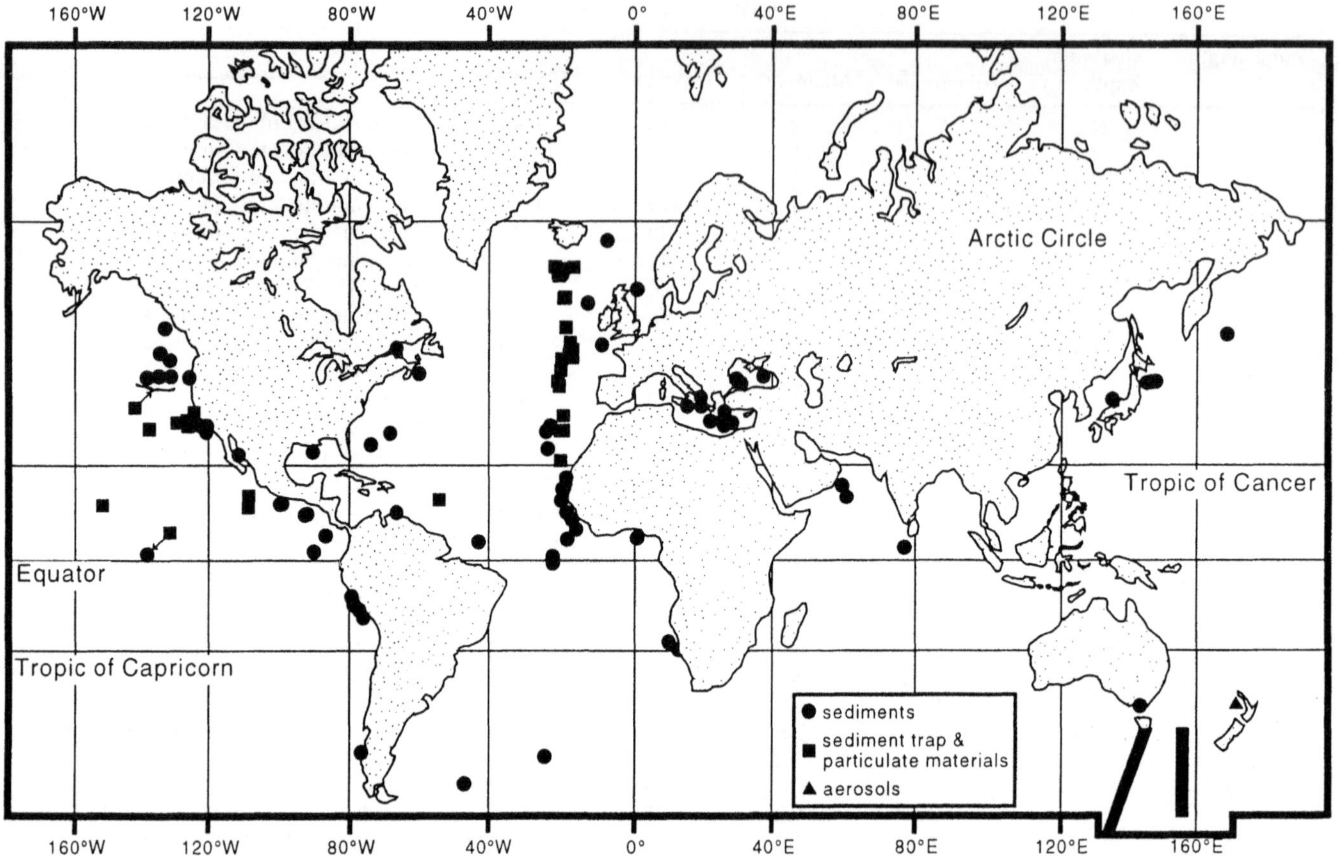

peratures (Prahl and Wakeham, 1987), confirming that this behavior is a characteristic of natural algal populations and substantiating the validity of the temperature calibration for warmer waters in the Pacific.

However, U^{K}_{37} (or $U^{K'}_{37}$) values and oceanic water temperatures at other locations show different correlations betwen U^{K}_{37} and temperature (Fig. 10b). The various equations calculated from data for cultures of *E. huxleyi* and field measurements are summarized in Table II; although there are some significant differences in the slopes of these linear correlations, they tend to follow generally similar trends. Further evaluation of the various measurements for particulate organic matter and sediment traps shows that an excellent correlation exists between U^{K}_{37} and temperature when data for some of the colder (<4°C) and warmer (>25°C) waters are excluded. For example, this more restricted data set provides a marked improvement in the Southern Ocean (i.e., $r^2 = 0.808$ versus 0.720; Table II, data from Sikes and Volkman, 1993). Similarly, the best correlation ($r^2 = 0.980$) for data from the North Atlantic is found for a series of determinations made for warmer waters (>16°C). These considerations are considered further in Section 3.4.4b.

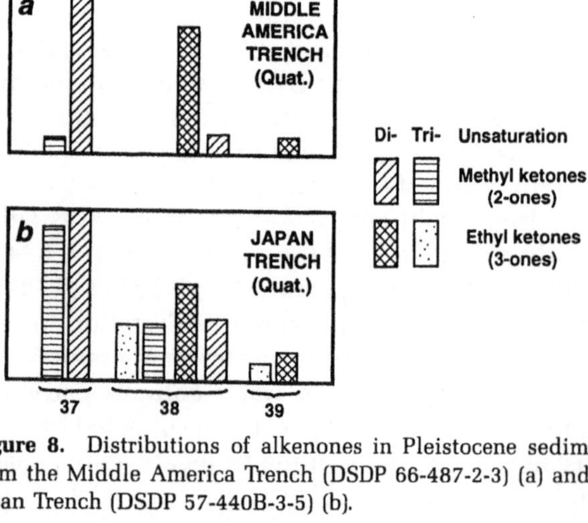

Figure 8. Distributions of alkenones in Pleistocene sediments from the Middle America Trench (DSDP 66-487-2-3) (a) and the Japan Trench (DSDP 57-440B-3-5) (b).

The production of alkenones by *E. huxleyi* also responds rapidly to changes in water temperatures. Specifically, when cultures of *E. huxleyi* acclimated to a growth temperature of 20°C were subjected to an abrupt 5°C drop in temperature, their production of

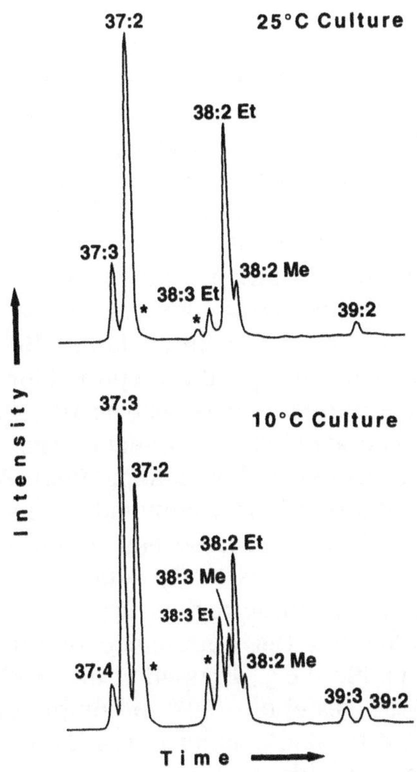

Figure 9. Gas chromatograms of alkenones in cultures of *Emiliania huxleyi* grown at 25°C and at 10°C. [Redrawn from Prahl and Wakeham (1987) and Prahl *et al.* (1988).]

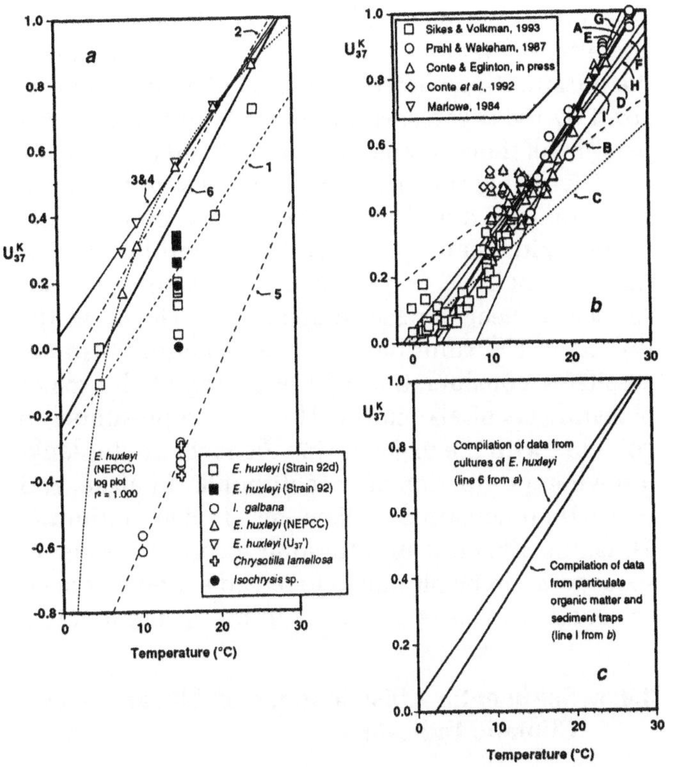

Figure 10. Relationships between U^{K}_{37} and temperature. (a) Data from cultures of *Emiliania huxleyi*, *Isochrysis galbana*, *Chrysotila lamellosa*, and *Isochrysis* sp. (b) Data for field samples of particulate organic matter and sediment traps. (c) Comparison of linear correlations for culture and field data. The equations of the lines and corrolation coefficients are discussed in Table II and in the text.

TABLE II. U^{K}_{37} (or $U^{K'}_{37}$) vs Temperature: Culture and Field Samples

Line[a]	Relationship	r^2	n	Strain(s)/location(s)	Reference(s)[b]
Cultures					
1.	$U^{K}_{37} = 0.036T - 0.282$	0.764	15	*E. huxleyi* (92 and 92d)	(Marlowe, 1984; Volkman *et al.*, 1980b)
2.	$U^{K}_{37} = 0.040T - 0.110$	0.989	20	*E. huxleyi* (NEPCC[c])	Prahl and Wakeham, 1987; Prahl *et al.*, 1988
3.	$U^{K'}_{37} = 0.033T + 0.043$	0.997	20	*E. huxleyi* (NEPCC)	Prahl and Wakeham, 1987; Prahl *et al.*, 1988
4.	$U^{K'}_{37} = 0.034T + 0.039$	0.994	22	*E. huxleyi* (NEPCC)	Prahl *et al.*, 1988
5.	$U^{K}_{37} = 0.052T - 1.120$	0.791	6	*I. galbana*	(Marlowe, 1984)
6.	$U^{K}_{37} = 0.044T - 0.255$	0.722	37	*E. huxleyi*	This compilation
Particulate organic matter and sediment traps					
A.	$U^{K'}_{37} = 0.037T - 0.070$	0.983	25	mainly Pacific	Prahl and Wakeham, 1987
B.	$U^{K}_{37} = 0.018T + 0.205$	0.260	25	North Atlantic	Conte *et al.*, 1992
C.	$U^{K'}_{37} = 0.022T + 0.017$	0.720	37	Southern Ocean	(Sikes and Volkman, 1993)
D.	$U^{K'}_{37} = 0.033T - 0.090$	0.808	23	Southern Ocean >4° C	(Sikes and Volkman, 1993)
E.	$U^{K'}_{37} = 0.0414T - 0.156$	0.958	40	Combined D and A <25°C	Sikes and Volkman, 1993
F.	$U^{K}_{37} = 0.033T - 0.009$	0.799	51	North Atlantic	(Conte and Eglinton, in press)
G.	$U^{K}_{37} = 0.056T - 0.469$	0.980	14	North Atlantic (>16°C)	Conte and Eglinton, in press
H.	$U^{K}_{37} = 0.030T + 0.042$	0.720	76	North Atlantic	(Conte *et al.*, 1992; Conte and Eglinton, in press)
I.	$U^{K'}_{37} = 0.037T - 0.083$	0.874	116	All data >4° C and <25°C	This compilation

[a]Line for cultures and particulate organic matter and sediment traps are shown in Figs. 10a and 10b, respectively.
[b]References in parentheses are the source of data used to derive the U^{K}_{37} versus temperature relationship, where the publications themselves do not contain this specific calculation.
[c]NEPCC = North East Pacific Culture Collection.

alkenones, as calculated from cell densities before and after the temperature change, demonstrated that the unsaturation characteristics of alkenones synthesized by the organisms immediately reflect the temperature of their new environment (Prahl *et al.*, 1988).

Given that the algae generating alkenones, such as *E. huxleyi*, are phototrophic, living, and actively synthesizing their lipid constituents within the photic zone, their U^K_{37} values should approximate to sea surface or near surface temperatures. However, the linking of U^K_{37} values to water temperatures also necessitates consideration of the timing of the major biosynthesis of alkenones. This can be presumed to coincide with the major spring bloom of phytoplankton when production of coccolithophorid algae, and hence their constituent alkenones, is at an optimum. Thus, the alkenone unsaturation record in sediment sequences can be presumed to record a profile of near-sea-surface water temperature in the spring months.

3.4.3. Sedimentary Distributions of Alkenones as Climatic Indicators

3.4.3a. Contemporary Marine Sediments and Particulates. A preliminary appraisal of alkenones as potential indicators of water sea surface temperatures (SST) was made by examining a compilation of U^K_{37} values for a number of surface or Pleistocene sediments from different latitudes (Brassell *et al.*, 1986a), revealing a general decrease in U^K_{37} values from the warmer equatorial regions toward the cooler polar regions. Subsequent analyses tend to confirm this pattern, both for sediments and for water column particulates (Prahl and Wakeham, 1987). In cold-water regimes, triunsaturated alkenones dominate their diunsaturated counterparts; for example, in the St. Lawrence Estuary U^K_{37} is estimated as 0.32 (Nichols and Johns, 1986). Similar results have been observed for higher latitudes (>40°N) in the North Pacific (Sikes *et al.*, 1991) and South Atlantic (Brassell *et al.*, 1986a). The lowest $U^{K'}_{37}$ values reported for an oceanic settings are those for an extensive suite of samples of particulate organic matter collected from the Southern Ocean (Sikes and Volkman, 1993). All possess alkenone distributions dominated by the $C_{37:3}$ alkenone, with $U^{K'}_{37} \ll 0.1$ in several of the more southerly samples. In contrast, U^K_{37} values for bottom sediments from equatorial regions are typically close to unity, reflecting the dominance of the C_{37} alkadienone (Table I; e.g., Brassell *et al.*, 1986a; Prahl *et al.*, 1989a; Sikes *et al.*, 1991). A compilation of published reports on alkenones in marine sediments reveals similar trends on a global basis [Fig. 11, updated and expanded from Brassell *et al.* (1986a)], with the envelope of values in the general shape of a chevron, suggesting a broad correspondence between U^K_{37} and surface oceanic temperatures. Direct comparisons between measured oceanic temperatures and U^K_{37} values

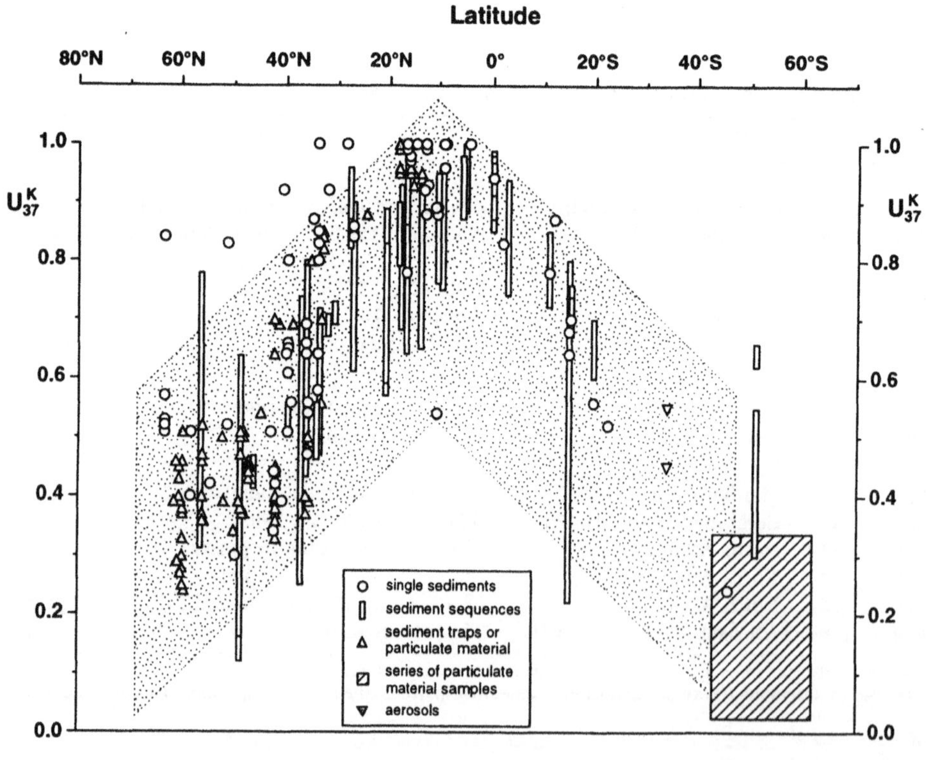

Figure 11. Latitudinal variation in U^K_{37} values of sediments and particulate samples listed in Table I. Individual samples are represented by the symbols identified on the figure, and the ranges of U^K_{37} values for sequences of sediments (often encompassing glacial/interglacial cycles) are designated by the vertical lines. The shaded, chevron-shaped area depicts the general trend in the data, showing higher values at lower latitudes.

for sediments (e.g., Sikes *et al.*, 1991) and particulate matter (Prahl and Wakeham, 1987; Sikes and Volkman, 1993; Conte *et al.*, in press) have been made at various locations. Some discrepancies exist between the different sets of determinations and their calibration curves, but, in general, $U_{37}^{K'}$ values show a consistent and coherent relationship to water temperatures in the contemporary oceanic environment, prompting and supporting their use in the assessment of sea surface temperatures.

3.4.3b. Contemporary Lacustrine Sediments. The values estimated for U_{37}^{K} in sediments from published histograms of their comparative abundances in a number of English lakes [(Table I; estimated from data of Cranwell (1985)] include negative values unprecedented in oceanic sediments (e.g., Cranwell, 1988). Similarly, the alkenone unsaturation of sediments from an Antarctic lake (Volkman *et al.*, 1988) is consistent with the cold temperatures anticipated for this regime, but the dominance of the $C_{37:4}$ alkenone suggests that there may be a major distinction between the alkenone compositions of marine and lacustrine systems. Indeed, lacustrine sediments appear to contain higher concentrations of $C_{37:4}$ than oceanic sediments at similar latitudes (e.g., Cranwell, 1985) and $C_{37:4}$ is a minor alkenone even in cold polar waters (<5°C; Sikes and Volkman, 1993). These observations suggest a different calibration between U_{37}^{K} and water temperature in lacustrine and marine settings. Since *E. huxleyi* is a restricted marine species and cannot be the biological source of alkenones in lacustrine sediments, another biological source must be sought. Given the recognized difference between the U_{37}^{K} values of the pyrmnesiophyte species *E. huxleyi* and *Isochrysis galbana* in 15°C culture experiments [0.20 versus −0.33, respectively, calculated from data in Marlowe (1984); Fig. 10a], it seems appropriate to attribute the discrepancy in U_{37}^{K} values between marine and lacustrine sediments to differences in the production of alkenones by the algal species that are contributing them to the respective environmental settings. The presence of C_{40} alkenones in the lake settings (Cranwell, 1985; Volkman *et al.*, 1988) further confirms that alkenone characteristics in lakes are distinct from those in the oceanic realm. Such differences suggest that the use of U_{37}^{K} values in lacustrine environments should be viewed with caution until an appropriate calibration is achieved based on a proven biological source for alkenones (a lacustrine prymnesiophyte?) in lakes.

3.4.3c. Glacial/Interglacial Cycles and Relationship to Foraminiferal $\delta^{18}O$. The application of alkenone unsaturation to the study of ancient climate records was pioneered in sediments from the eastern equatorial Atlantic Ocean (Brassell *et al.*, 1986a,b). The specific study site in the Kane Gap was selected because several characteristics of a Kasten core (M16415-2) collected from this area in 1983 were appropriate for the proposed investigation of U_{37}^{K} values as measures of glacial/interglacial climatic variability. The chosen core spanned a 1-Myr sediment record with little disturbance associated with bioturbation, and it contained alkenones in sufficient abundance for their determination in closely spaced sediment samples of small size (i.e., a few grams). Also, species of planktonic and benthic foraminifera suitable for $\delta^{18}O$ measurements were present throughout most of the core, accompanied by alkenones as major constituents of the lipid extracts (Fig. 12). Initial results revealed a general correspondence between U_{37}^{K} and fluctuations in $\delta^{18}O$ values for planktonic foraminifera over the past 700,000 yr (Fig. 13; Brassell *et al.*, 1986a). These trends were reinforced and confirmed by subsequent analyses (Brassell et *al.*, 1986b; Poynter *et al.*, 1989b). More recently, similar climate records in U_{37}^{K} signals have been further documented in other sediment cores from the same region, specifically those collected during Leg 108 of the Ocean Drilling Project (ODP) (Poynter et *al.*, 1989a, b; Eglinton *et al.*, 1992), and in several other oceanographic regions (e.g., Gulf of Mexico; Jasper and Gagosian, 1989). The major results from the Kane Gap study of glacial/interglacial U_{37}^{K} records included (i) recognition of the characteristic "sawtooth" pattern in the inferred temperature profiles, attributable to a sequence of gradual coolings from interglacial to glacial conditions and rapid warmings associated with terminations of glacial periods (Fig. 13), (ii) a general relationship between U_{37}^{K} values and the $\delta^{18}O$ record for foraminifera, including a correspondence to individual isotopic stages where the $\delta^{18}O$ record was incomplete (notably for stage 10 in the Kane Gap core; Fig. 13), (iii) time series variations in U_{37}^{K} related to Milankovitch cycles (Brassell *et al.*, 1986b), and (iv) a deterioration in the correspondence between the U_{37}^{K} and $\delta^{18}O$ profiles below ca. 8 m. The precise cause of the lack of correspondence below ca. 8 m is unclear. It may be attributable to a variety of factors (Brassell *et al.*, 1986a) including diagenetic alteration of alkenone distributions or foraminiferal $\delta^{18}O$ values (rather unlikely explanations) or changes in the speciation of phytoplankton contributing alkenones or is perhaps evidence of difference in the influence of ice volume and water temperatures on the two measures. However, the upper portion of the record to ca. 575 kyr is excellent and is perhaps best illustrated by the broad correspondence between the U_{37}^{K} profile for ODP Hole 658A and stacked profiles for foraminiferal $\delta^{18}O$ from

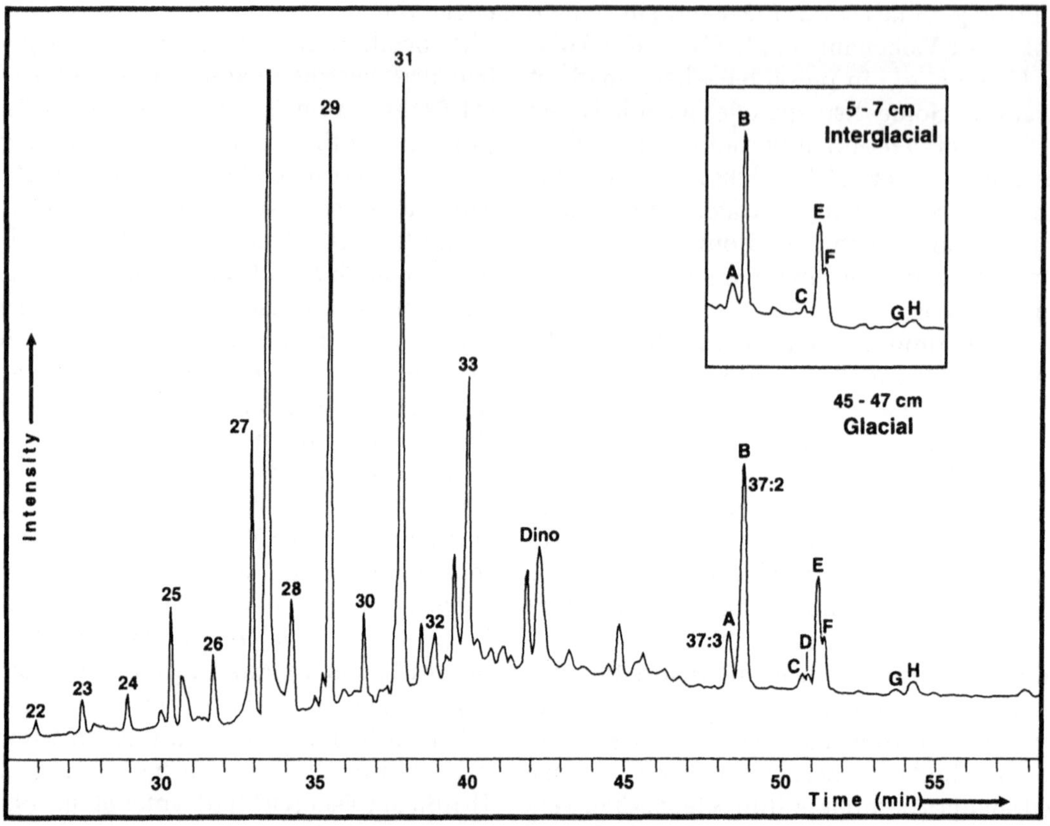

Figure 12. Gas chromatogram of the total lipid extract of a sediment (*Meteor* core 16415-2, 45–47 cm core depth) from the Kane Gap deposited during the last glaciation (U^K_{37} = 0.78). The inset shows the alkenone distribution in a sediment (5–7 cm depth) from the present interglacial deposited at the same site (U^K_{37} = 0.89). Numbered peaks designate *n*-alkanes of corresponding carbon number. The peak labeled "Dino" represents the sterol dinosterol. Lettered peaks are alkenones; their designations in shorthand notation (cf. Fig. 6) are as follows: A, $C_{37:3}$Me; B, $C_{37:2}$Me; C, $C_{38:3}$Et; D, $C_{38:3}$Me; E, $C_{38:2}$Et, F, $C_{38:2}$Me; G, $C_{39:3}$Et; H, $C_{39:2}$Et. [Redrawn from Brassell *et al.* (1986a).]

five low-latitude cores (Poynter *et al.*, 1989a). The close resemblance between glacial/interglacial trends in alkenone unsaturation and in established $\delta^{18}O$ profiles indicates that U^K_{37} values are valid molecular measures of climatic variability.

The U^K_{37} profile for the Kane Gap sediments shows a closer match to the $\delta^{18}O$ trends for the two planktonic species *Globigerinoides sacculifer* and *G. ruber* than for the benthic species *Cibiddoides wuellerstorfi* (Fig. 13). This observed correspondence is to be expected, given that alkenones derive from planktonic organisms inhabiting the photic zone. Following the investigations of U^K_{37} values in the eastern equatorial Atlantic, down-hole trends in this parameter for other sediments sequences have been determined and have reinforced the value of alkenone unsaturation as a molecular indicator of climate change. For example, the alkenone unsaturation has been determined in samples from a sediment core taken from the Pigmy Basin in the Gulf of Mexico (Jasper and Gagosian, 1989). The sequence spanned ca. 190 kyr and revealed a U^K_{37} profile that showed broad correspondence to glacial/interglacial stages recognized in $\delta^{18}O$ trends for *G. sacculifer* from the Caribbean Sea (Fig. 14). In contrast, the $\delta^{18}O$ record for the planktonic foraminifer *G. ruber* in the Pigmy Basin cores was characterized by a series of minima attributed to freshwater runoff. Also, the temperature difference between glacial and interglacial times estimated from U^K_{37} was 8° ± 1°C, a value significantly greater than that (ca. 0.8–2.0°C) indicated by foraminiferal $\delta^{18}O$ signals or foraminiferal assemblages in the northern Gulf of Mexico. A satisfactory resolution of this discrepancy seems to lie in a better understanding of the factors influencing alkenone unsaturation.

U^K_{37} values in sediments from the Peru margin covary with the $\delta^{18}O$ signal of benthic foraminifera (Farrimond *et al.*, 1990b), although the latter record is incomplete due to poor preservation in the upper portion (<55 m, ca. 350 kyr) of the sediment section. The lower core section (>55 m) provides a reasonable agreement between the values for temperature changes from glacial to interglacial periods afforded by ben-

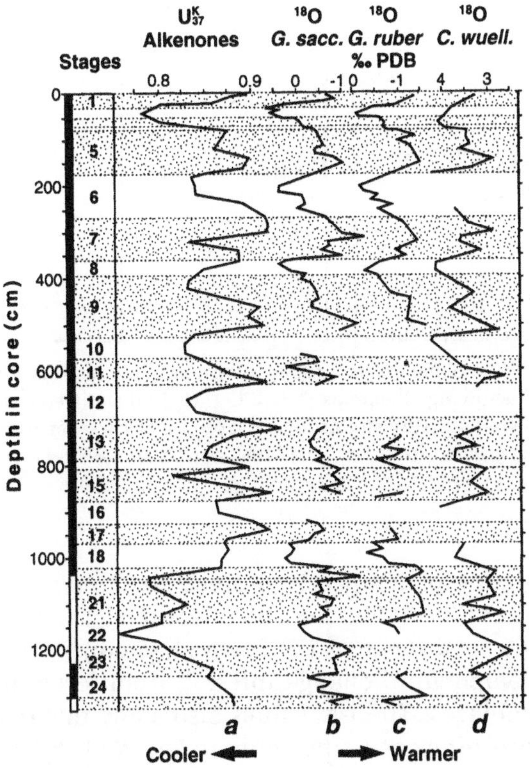

Figure 13. Records for U^K_{37} and for the $\delta^{18}O$ values of selected foraminifera in *Meteor* core 16415-2 from Kane Gap. (a) Alkenone unsaturation index (U^K_{37})values. (b)–(d) $\delta^{18}O$ values of $CaCO_3$ of *Globigerinoides sacculifer* and *G. ruber* and benthic *Cibiddoides wuellerstorfi*, respectively. Gaps in the $\delta^{18}O$ data indicate complete dissolution of $CaCO_3$. Details of the oxygen isotope stages and the magnetic stratigraphy (thickening of the ordinate) are given in Brassell *et al.* (1986a). The general trends reflecting cooler and warmer temperatures are indicated. [Redrawn and modified from Brassell *et al.* (1986a).]

thic foraminiferal assemblages and U^K_{37} (Farrimond *et al.*, 1990b). However, the upper core sequence displays significant variability in its U^K_{37} record, which may be attributable to the influence of changes in the intensity of coastal upwelling on water temperatures. Also, these major fluctuations in U^K_{37} values cannot be related to the 100-kyr cycles observed in other records.

Investigation of alkenone unsaturation as a measure of sea surface temperatures from several sites offshore Oman (ten Haven and Kroon, 1991) reveals values for contemporary sediments that match present-day water temperatures and show down-hole fluctuations related to glacial/interglacial cycles through isotopic stage 8 (Fig. 15; ca. 250 kyr). The lack of a direct correspondence between U^K_{37} and $\delta^{18}O$ values in all the sediment sequences may possibly reflect the dependence of the latter on both local water temperatures and ice volume. The particularly low U^K_{37} values observed for glacial isotopic stages 3 and 8 are somewhat surprising, given that biotic assemblages suggest that upwelling was stronger during interglacials. Thus, it would appear that variations in the intensity of monsoonal upwelling may not be the only influence on water temperatures in the Arabian Sea. For example, changes in wind strength may also be important when coupled with expanded ice sheets on the Tibetan Plateau.

Studies of four different sapropelic horizons (designated S_1, S_5, S_6, and S_7) from the Mediterranean ranging from 8 to 175 ka have revealed that their U^K_{37} values broadly correspond to the $\delta^{18}O$ record for *G. ruber* in the eastern Mediterranean (Fig. 16; ten Haven *et al.*, 1987). The higher U^K_{37} values are those for the sapropels deposited during warmer interglacial intervals, and the lower values were determined for the S_6 sapropel laid down during the penultimate glacial maximum. Interestingly, the glacial/interglacial temperature variations inferred from the U^K_{37} index reflect a less extreme range (ca. 8°C) than that implied by the $\delta^{18}O$ values, which are affected by the existence of low-salinity surface waters. Thus, U^K_{37} values may be independent of salinity constraints.

The trend in U^K_{37} values in sediments from the eastern tropical Pacific Ocean depicts a warming following the last glacial maximum and the relative

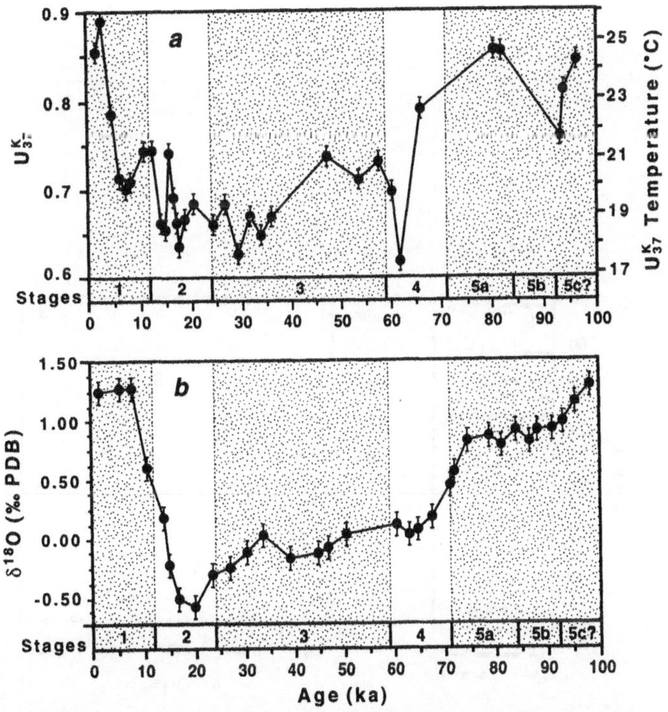

Figure 14. U^K_{37} values for sediments from the Pigmy Basin (DSDP Site 619) (a) compared with the $\delta^{18}O$ profile of *Globigerinoides sacculifer* from the Caribbean Sea (b). Alkenone temperatures are calculated from U^K_{37} using the relationship $U^K_{37} = 0.033T + 0.043$. [After Jasper and Gagosian (1989).]

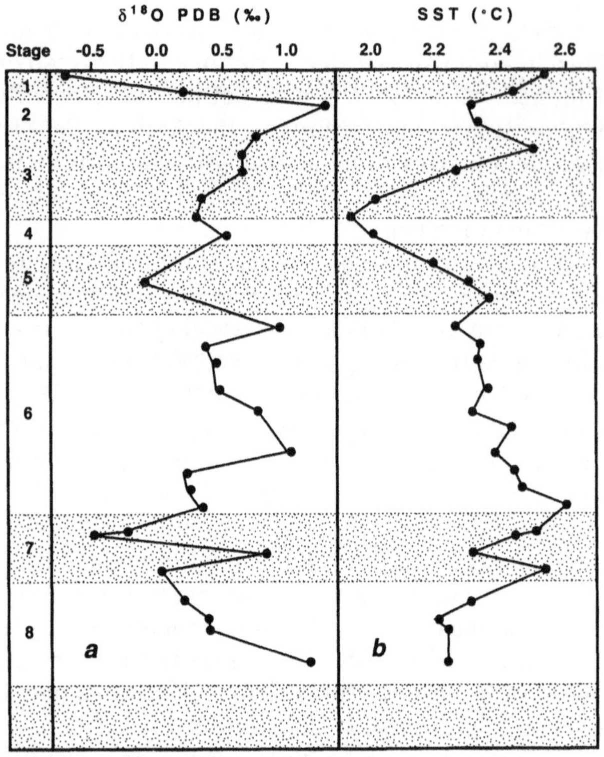

Figure 15. Depth profiles for $\delta^{18}O$ values in *Neogloboquadrina dutertrei* (a) and sea surface temperatures (SST) calculated from U^K_{37} values at Ocean Drilling Program (ODP) Hole 728A (b). [After ten Haven and Kroon (1991).]

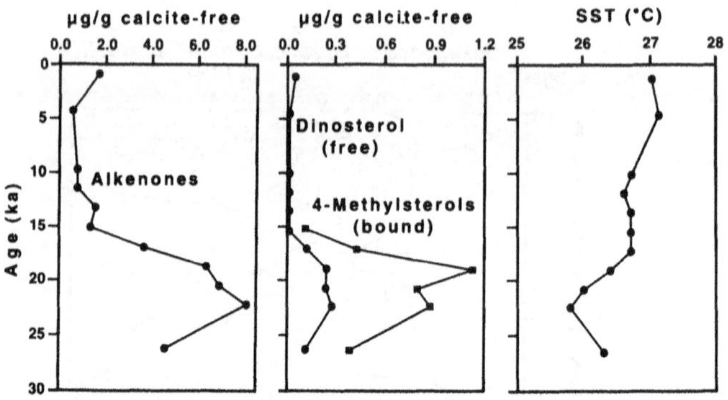

Figure 17. Time series for concentrations of biomarkers at MANOP site C, showing alkenones (Σ C_{37}, C_{38}, C_{39}) (a) and free dinosterol and bound 4-methylsterols (b) and a profile of sea surface temperature calculated from U^K_{37} (c). [After Prahl *et al.* (1989a).]

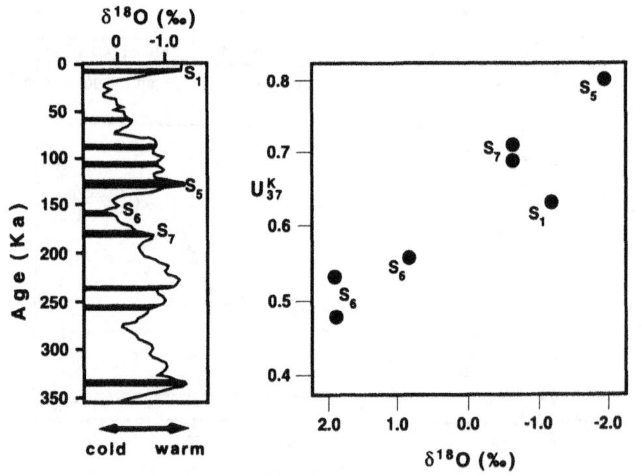

Figure 16. *Left*: The timing of sapropel deposition in the Mediterranean, shown as shaded intervals superimposed on a generalized open-ocean oxygen isotope record; *right*: a plot of U^K_{37} vs. $\delta^{18}O$ for *G. ruber* (b). [After ten Haven *et al.* (1987).]

constancy of temperature since that time (Fig. 17; Prahl *et al.*, 1989a). Specifically, the U^K_{37} index indicates a total range of 1.3°C, which is consistent with the temperature change inferred from foraminiferal assemblages. The U^K_{37} values also show that 0.9°C of the warming occurring between 22.1 and 16.9 ka and that subsequently the temperature has been more uniform, varying by ±0.4°C. The variation in these values is comparatively small and the sampling density is low, yet the systemic warming trend from the last glacial maximum to the present is clearly discernible.

Longer-term records (*ca.* 250 kyr) from the same site reveal SST values derived from $U^{K'}_{37}$ that vary slightly (<2°C) but coherently with Milankovitch insolation cycles (Lyle *et al.*, 1992). These data were also used, in conjunction with a simple model of heat balance to estimate upwelling rates in the equatorial Pacific, demonstrating the potential of alkenone paleotemperatures to assist in assessing records of paleoclimate and paleoproductivity.

In a sediment sequence from the Japan Sea, SST fluctuations based on U^K_{37} depict glacial/interglacial cycles (Kheradyar *et al.*, unpublished data) in combination with changes in the *N. pachyderma* coiling ratio (Kheradyar, 1992), which appear comparable to the climatic record shown by a composite oxygen isotope profiles for foraminifers (Prell *et al.*, 1986). The downhole profiles for alkenones provide more detail of the smaller scale SST fluctuations, whereas the foraminifers offer strong evidence for the warmest intervals. The parallelism between the curves is most apparent above the late Brunhes isotope stage 12 where temperature values based on U^K_{37} show maxima and minima which appear to correspond to glacial and deglacial stages. The late Pleistocene isotope stage 12 (at 423 kyr) represents the coldest period and the best defined boundary, characterized by the sharpest rise (>10°C) in temperature. Comparison of the two methods permits evaluation of the reliability of the depicted events and facilitates stratigraphic correlations. The combined evidence from U^{K}_{37} and the distributions of planktonic foraminifera, coupled with additional $\delta^{18}O$ values for foraminifers (Dunbar *et al.*, 1992), may lead to adjustment of stage boundaries previously based on foraminiferal data which, in turn, directly influences sedimentation rates and the potential resolution required for examination of the periodicities in climatic fluctuations within the northwest Pacific.

The various studies described above illustrate the potential of the alkenone unsaturation index as a measure of sea surface water temperatures, but they also show that a number of unresolved problems exist in the use of U^{K}_{37} values. For example, some sediment sequences reveal discrepancies in the temperature ranges deduced from U^{K}_{37} and $\delta^{18}O$ values, perhaps linked to the differences in the response of these measures to temperature and ice volume effects. Also, concerns have been expressed that the temperature calibration of U^{K}_{37} may be dependent on the speculation of the phytoplankton contributing alkenones, and the possibility has been raised that U^{K}_{37} values are affected by diagenesis. These matters are further considered below, but they do not appear to detract from the general utility of alkenone unsaturation as a sedimentary record of glacial/interglacial climatic cycles.

3.4.3d. Other Climatic Events, Including El Niño. In addition to the application of U^{K}_{37} values to climate variations occurring over glacial/interglacial cycles (i.e., 100 kyr), it is appropriate to explore the applicability of the approach to events occurring over much shorter time periods. Specifically, the apparent speed of response to a change in temperature of alkenone biosynthesis by *E. huxleyi* (Prahl *et al.*, 1988) demonstrates that U^{K}_{37} values possess the potential to reflect climatic variability over time scales shorter than those typically resolvable in the sedimentary record.

El Niño events represent significant perturbations of global climate which occur at somewhat irregular intervals (ca. 3–10 yr). They are suitable candidates for molecular investigations of short-term climatic variations, given the organic richness of the sediments off the coast of Peru which underlie the region of coastal upwelling where the impact of El Niño events is most striking. An initial study (Farrington *et al.*, 1988) of sediments from this area revealed that the general temperature range calculated from U^{K}_{37} values corresponds to that expected for oceanic waters off Peru and showed broad trends for the bulk samples which were consistent with the frequency of El Niño events. However, the sampling frequency chosen was not suitable for evaluation of individual El Niño events, and the poor preservation of carbonate tests due to remineralization precluded any comparison of U^{K}_{37} and $\delta^{18}O$ signals. Evaluation of more closely spaced samples (McCaffrey *et al.*, 1990) revealed that U^{K}_{37} signals corresponding to individual strong El Niño events were less apparent than evidence for periods (e.g., 1870–1891) when the events were more prevalent. The molecular signal attributable to a specific El Niño event may be less discernible in the sedimentary U^{K}_{37} record due to a decrease in alkenone production by prymnesiophytes. It is to be expected that the abundance of prymnesiophytes, like that of other phytoplankton, will be severely curtailed during the catastrophic decline in oceanic productivity off Peru associated with strong El Niño conditions. The potential for alkenones to provide a sedimentary record of major climatic perturbations may therefore be limited when the conditions also create a drastic reduction in the productivity of their source organisms. Under these circumstances, it seems appropriate to explore the molecular signals of such climatic events in areas more distant from the focal point of the phenomenon, yet still affected by it. For example, the El Niño signal off southern California is not accompanied by the major decline in productivity observed off Peru because the water temperature change, though significant, is not prejudicial to the viability of the phytoplankton community. Thus, the U^{K}_{37} record for laminated sediments from the Santa Barbara Basin provides evidence of water temperature increases associated with the El Niño phenomenon throughout the 20th century (Fig. 18; Kennedy and Brassell, 1992, a, b). Such data demonstrate that molecular indicators possess the potential to resolve annual events in the sedimentary record under suitably favorable conditions of high sedimentation rates and high productivity.

3.4.4. Validation and Refinement of Alkenone Stratigraphic Approaches

The further development of alkenone unsaturation as a molecular signal of climatic change requires its validation, calibration, and extension back in time and its further integration with other climatic measures. One potential major advance in the value of U^{K}_{37} signals lies in the exploration and application of their relationship to $\delta^{18}O$ profiles; the two independent

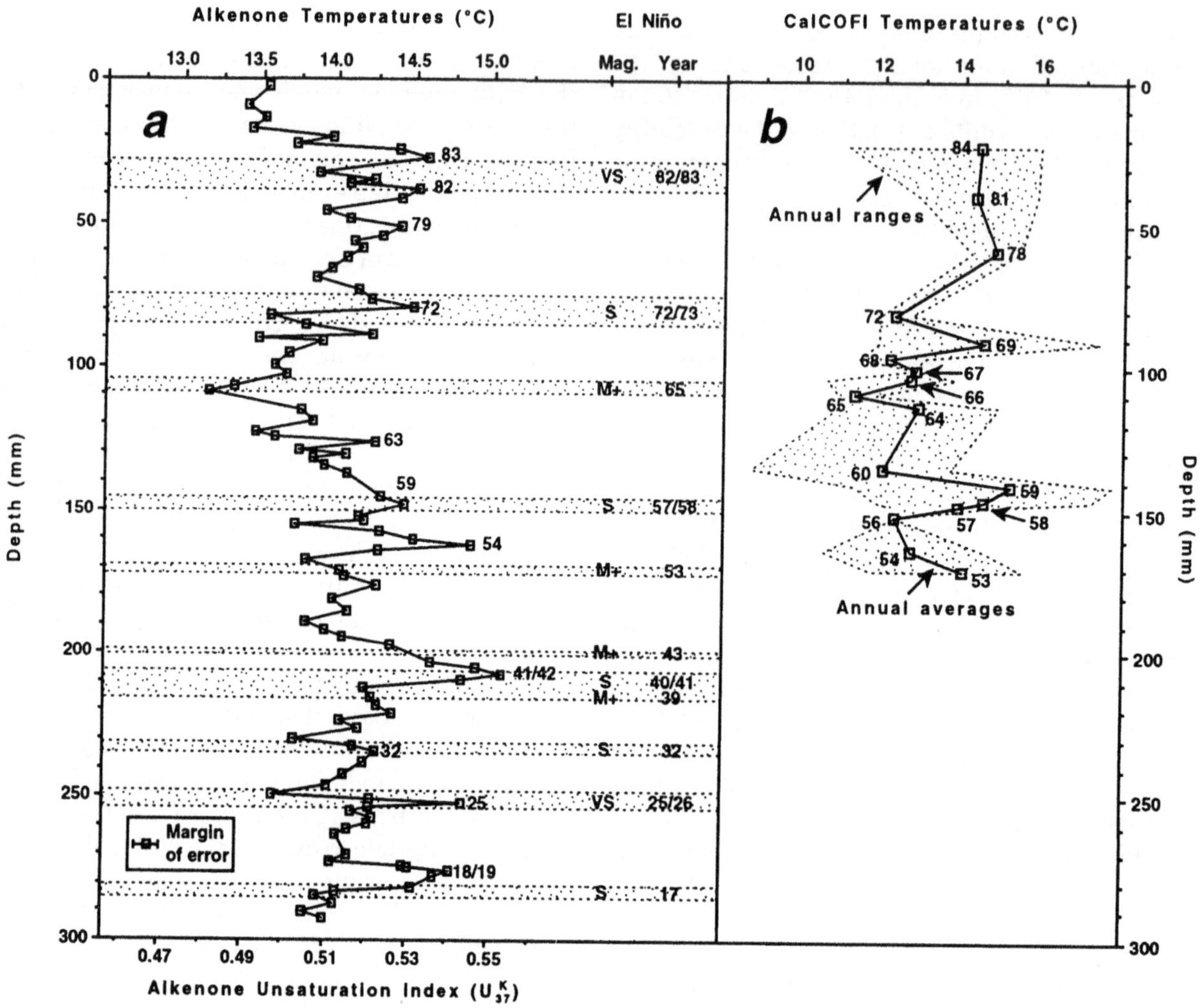

Figure 18. (a) Profile of 20th century variability in alkenone temperatures derived (using the relationship $U^K_{37} = 0.034T + 0.039$) from the alkenone unsaturation indices (U^K_{37} values) for a composite of three cores of laminated sediments from the Santa Barbara Basin (SBB). The margin of error (±0.003) in determination of the U^K_{37} values corresponds to an error in temperature determinations of ±0.09°C. The age assignments for specific depths are based on the stratigraphy for the SBB established at SIO. Magnitudes and dates of equatorial El Niño events (after Quinn *et al.*, 1987) are shown with their intensities assessed as "very strong" (VS), "strong" (S), and "moderate to strong" (M+) and with their approximate duration represented by the shaded intervals. Note the general agreement between warm excursions in the molecular temperature record and El Niño events, allowing for the one-year time lag in their appearance offshore California relative to their manifestation in equatorial regions. (b) Annual averages of all CalCOFI temperatures (irrespective of season) measured at 20-m water depth plotted against age-equivalent sediment depths. The shaded area represents the range of values recorded in each given year. [Data redrawn from Kennedy and Brassell (1992a, b).]

measures offer the possibility of separating the effects of ice volume and temperature in the climatic record (Brassell *et al.*, 1986a). In addition, there are instances, for example, in sediments deposited below the lysocline, when molecular signatures can survive whereas carbonate records are not preserved (e.g., Brassell *et al.*, 1980b). Similarly, alkenones occur in, and may derive from, species of nannofossils that do not bear coccoliths throughout their life cycle (Brassell *et al.*, 1987).

To varying extents, the use of alkenones as molecular indicators of climatic variability is hindered by limitations in the current understanding of their biological and geochemical behavior. Specifically, the broader application of alkenones require an improved knowledge both of their origins, in particular, the specificity of the biological sources of alkenones at the present day and in the geological past, and of their sedimentary fate. Other restrictions that hinder a broader and more definitive use of the alkenone unsaturation index include concerns relating to the temperature calibration of U^K_{37}, especially in sediments that predate the evolution of *E. huxleyi* (Brassell *et al.*, 1986a, b; ten Haven *et al.*, 1987). Also, it remains an open question whether there is diagenetic overprinting of the alkenone signature arising from the preferential degradation of more unsaturated components (i.e., $C_{37:3} > C_{37:2}$). The recognition of organosulfur compounds related to the alkenones (Fig. 6; Sinninghe Damsté *et al.*, 1989) confirms that diagenetic products of the C_{37}–C_{39} lipids of prymnesiophytes exist, although it seems probable that they are derivatives of alkenes rather than alkenones (cf. Sinninghe Damsté *et al.*, 1990). The use of U^K_{37} also presumes that dissolution and storage effects exert no significant influence on the index—an assumption that requires testing. Such issues potentially affecting the integrity of sedimentary U^K_{37} signals are considered below.

3.4.4a. Speciation of Algae Contributing Sedimentary Alkenone. A major potential influence on alkenone compositions is variation in the speciation of the algae from which the alkenones derive. Result from the Kane Gap sediments reveal that the correspondence between U^K_{37} and $\delta^{18}O$ of planktonic foraminifera does not change significantly at 250 kyr, the timing of the first appearance of *E. huxleyi*. This observation lends credence to the conclusion that U^K_{37} is, to some degree, species-independent and that the alkenone–temperature relationship of the extant alga *E. huxleyi* is apparently a characteristic inherited from its ancestors. Comparison of alkenone and nannofossil assemblages in oceanic sediments suggests that members of the family Gephyrocapsaceae have possessed the capability for alkenone biosynthesis for at least 45 Myr (Marlowe *et al.*, 1990). However, it is not possible to determine an absolute, or even relative, calibration for U^K_{37} without regard for the biological origin of sedimentary alkenones at the species level, and this problem is complicated by the production of alkenones by species that are no longer extant. Furthermore, differences may exist in the alkenone unsaturation and its temperature response between morphotypes of the same species; specifically, no comparison of the characteristics of the "cold" and "warm" morphotypes of *E. huxleyi* (Winter, 1985) has been made. The use of the terms "cold" and "warm" morphotypes, however, is now discouraged (Young and Wesbroek, 1991) because variations in calcification attributed to temperature effects are not systematic. Differences in the culture strains of *E. huxleyi*, especially their genotypic variation, enable their division into types A and B which appear to represent two groups of a restricted number of naturally occurring varieties. Interestingly, strains 92 and 92d of *E. huxleyi* which have a distinctive size difference, are members of type A and type B, respectively, yet the relationships between their alkenone compositions and growth temperatures appear comparable (Fig. 10a), arguing that this behavior may be uniform within the species. Also, it has yet to be demonstrated and verified that the changes in alkenone unsaturation observed in the natural system are a function of the biosynthesis in a single species, rather than a variation in the speciation of the principal contributors of alkenones. However, the temperature behavior of culture experiments with *E. huxleyi* appears to be representative of typical signals for phytoplankton in the Pacific (Prahl *et al.*, 1988), but not in the North Atlantic (Conte and Eglinton, in press). Other potential contributors of alkenones in contemporary sediments include *Gephyrocapsa oceanica* (Marlowe *et al.*, 1990). Ideally, a chemotaxonomic basis for species assignments is desirable; a potential conflict arising from the morphological basis for nannofossil speciation is illustrated by the existence of dimorphic coccospheres composed of placoliths assigned to different species (e.g., *E. huxleyi* and *G. oceanica*; Clochiatti, 1971). In general, the precise assessment of the biological sources of alkenones at the species level and the comparative influence of mixed assemblages on alkenone compositions seem likely to remain intractable problems best addressed through empirical determinations of U^K_{37} in circumstances which provide the means to test it with other temperature indicators.

3.4.4b. Temperature Calibration. A further concern is the ability to calculate sea surface water temperatures from U^K_{37} values. Evidence from culture experiments supports a linear relationship between alkenones and temperature (Fig. 10a), although there is little inherent reason why such a relationship should necessarily be linear unless dictated by biosynthetic and biophysiological grounds. Interestingly, the logarithmic correlation between U^K_{37} and the growth temperatures of one series of *E. huxleyi* cultures (Prahl *et al.*, 1988) provides a perfect fit (R^2 = 1.000; Fig. 10a).

Indeed, other data for alkenones recovered from sediment traps in the North Atlantic (Conte and Eglinton, in press) and from filtered seawater from the Southern Ocean (Sikes and Volkman, 1993) are best fit by nonlinear correlations between U^K_{37} and $U^{K'}_{37}$, respectively, and water temperature. However, in both instances linear relationships do yield good correlations over narrow temperature ranges, for example, exclusion of low temperature values (<4°C) in the Southern Ocean (Table II). The combination of reports from the Pacific, Atlantic, and Southern Oceans provides a well-correlated (r^2 = 0.874) relationship ($U^{K'}_{37} = 0.037T - 0.083$) when colder and warmer values are excluded (Table II) which broadly corresponds (Fig. 10c) with the temperature relationship compiled from culture experiments ($U^K_{37} = 0.044T - 0.255$). Hence, it appears that a linear approximation of the U^K_{37} versus temperature relationship is valid over a broad temperature range, though not for the extremes of oceanic conditions.

In the application of U^K_{37} values to oceanic settings, it is presumed, with substantial supporting evidence (Prahl *et al.*, 1988; Sikes and Volkman, 1993), that data for *E. huxleyi* are representative of the marine phytoplankton which are the contributors of sedimentary alkenones. This correlation tends to suggest that the alkenone behavior is characteristic of a single group of organisms and strengthens the arguments for prymnesiophyte algae (e.g., Marlowe *et al.*, 1984a, b) as the biological source of alkenones. However, alkenones may be produced at various depths (and hence various temperatures) within the water column by populations of a single species with a uniform relationship between alkenone composition and growth temperature. In addition, the timing of alkenone production is undoubtedly seasonal as shown by differences in their abundance within time series of sediment trap samples collected at bimonthly (F. Prahl *et al.*, personal communication) or biweekly (J. Kennedy and S. Brassell, unpublished data) intervals. This natural variability in the depth and timing of alkenone biosynthesis can also help explain differences between measured temperatures and U^K_{37} values, and anomalies in the carbon isotopic compositions of $C_{37:2}$, $C_{37:3}$, and $C_{37:4}$ alkenones in the Black Sea (Freeman and Wakeham, 1992). Specifically, mixed contributions from two (or more) separate sources for alkenones offer an explanation more in keeping with general observations of alkenone characteristics than the possibility of preferential degradation of specific alkenones. Despite these considerations, there is no guarantee that all organisms producing alkenones possess similar temperature responses. Indeed, laboratory culture experiments with *E. huxleyi* and *I. galbana* demonstrate that this is not the case (Fig. 10a; Marlowe, 1984). The temperature dependence of alkenone production shown by *I. galbana* reveals a relationship between U^K_{37} and growth temperature for this prymnesiophyte alga which possesses a different slope and intercept from that of *E. huxleyi* (Fig. 10a; *I. galbana* yields the relationship $U^K_{37} = 0.052T - 1.12$ with $R^2 = 0.791$). It should be stressed that *I. galbana* is highly unlikely to be a significant source of alkenones in marine sediments and that its temperature dependence appears to better correspond to the range of U^K_{37} values found in lacustrine sediments (Table I; Cranwell, 1985). The significant difference in the temperature calibration of U^K_{37} for *E. huxleyi* and *I. galbana* does not preclude the existence of smaller discrepancies in the alkenone characteristics of other marine prymnesiophyte species at the present day or in the geological past. For example, differences in the unsaturation–temperature response of prymnesiophyte species containing alkenones have been suggested as one possible cause of the disparity between U^K_{37} and $\delta^{18}O$ profiles in deeper (>8 m) sediments of the Kane Gap (Brassell *et al.*, 1986a). However, this explanation remains speculative and is only one of a number offered to account for the observed discrepancy between the molecular and isotopic climatic indicators. At present, it is expedient to utilize the temperature calibration for U^K_{37} described above ($U^K_{37} = 0.037T - 0.083$) based primarily on data from series of particulate samples (Prahl and Wakeham, 1987; Sikes and Volkman, 1993; Conte and Eglinton, in press) and generally supported by results from laboratory cultures of *E. huxleyi* unless conclusive evidence demonstrates that it is invalid for a particular sediment core (e.g., in polar regimes with $T < 5°C$; Sikes and Volkman, 1993). The further development of alkenone stratigraphy undoubtedly requires a calibration for U^K_{37} which is valid and applicable for older sedimentary sequences and for ancestral alkenone production. Such future modifications of the temperature calibration seem likely to arise from comparisons of alkenone compositions with other climatic measures.

3.4.4c. Integrity of Alkenone Unsaturation during Sedimentation, Diagenesis, and Core Storage. The validity and applicability of alkenone unsaturation as a measure of water temperature requires that it remain unaltered during sediment deposition and diagenesis, during core collection and storage, and during alkenone extraction and analysis. Of these, the most likely cause for modification of U^K_{37} values is diagenetic degradation of the various unsaturated alkenones at different rates. For example, if double-bond degradation was more prevalent close to the carbonyl group (cf. Fig. 6), the tetra- and triunsaturated alkenones would be more rapidly transformed than their diunsaturated counterpart. To date, there is no convincing evidence to substantiate this possibility. Rather, it appears that alkenones do experience degradation, especially under oxic conditions, but that such diagenetic degradation has no discernible effect on their U^K_{37} values (Prahl *et al.*, 1989b) and must therefore affect all the alkenones in a comparable manner. It is also important that U^K_{37} values be immune to bacterial alteration in the sediment and remain unaffected by subsequent diagenetic processes. A recent investigation of the effects of oxidation on alkenones was undertaken for turbiditic sediments containing both oxic and anoxic zones (Prahl *et al.*, 1989b). The results of this study showed that their absolute abundance was significantly reduced but that their distributions, and hence U_{37}^{K} values, appeared to be unaffected by oxidative degradation process. The rate of microbial degradation of alkenones may be significantly slower than that of other unsaturated lipids because their double bonds—the most likely positions for initial microbial attack—possess the E (*trans*) rather than the more common Z (*cis*) configuration (Rechka and Maxwell, 1988a, b) so that many organisms may not possess the enzymatic capability to degrade them. Second, as long-chain molecules, alkenones may be expected to possess rather low solubilities in aqueous media, which may further restrict the possibilities for their microbial degradation. However, these characteristics may inhibit but cannot exclude the degradation of the alkenones in oxic sediments where there is a net decrease in their abundance. Third, the alkenones may be packaged in a manner that makes them less accessible to degradation. Perhaps they reside in membraneous material that is protected by an enveloping layer, although this possibility is entirely speculative; the precise location of alkenones within the algal cell is unknown.

In the oceanic water column, many fates can befall the alkenones. For example, the prymnesiophyte algae which biosynthesize alkenones may be ingested by zooplankton; however, it appears that alkenones survive well in fecal pellets (Volkman *et al.*, 1980c) and through food web processes (Rowland and Volkman, 1982). Similarly, carbonate dissolution should not effect U_{37}^{K}, unlike $\delta^{18}O$ values for carbonate tests, because the alkenones are chemically inert to such changes and occur in sediments deposited below the lysocline (e.g., Brassell *et al.*, 1980b). Recent laboratory experiments confirm the apparent immunity of alkenones to changes in unsaturation under conditions of carbonate dissolution (Sikes *et al.*, 1991).

The survival of alkenones and preservation of U_{37}^{K} values may depend on factors other than oxidation or carbonate dissolution. For example, it appears that sulfur can be introduced into these molecules, or more probably their alkene counterparts, to form dialkylthiophenes and dialkylthiolanes (Fig. 6; Sinninghe Damsté *et al.*, 1989). The diagenetic stage at which this process takes place is unclear. If such organosulfur components are indeed the diagenetic products of associated C_{37} and C_{38} alkenes rather than alkenones, then U_{37}^{K} values may be unaltered in sediments within which significant amounts of organosulfur compounds are formed. Enhanced abundances of C_{37} and C_{38} *n*-alkanes have been recognized in various sediments (e.g., McEvoy *et al.*, 1981), in hydrous pyrolysates of sediments (Sofer, 1988), and among the products of selective chemical treatments of sediments (e.g., Sinninghe Damsté *et al.*, 1988, 1989; Sinninghe Damsté and de Leeuw, 1990; Kohnen *et al.*, 1990; Rullkötter and Michaelis, 1990) and petroleums (Zumberge, 1987). The occurrence of these components demonstrates that the C_{37} and C_{38} alkyl skeletons can survive significant diagenetic alteration, although the specific organosulfur compounds and their alkane counterparts do not preserve evidence of the extent of unsaturation of their precursors. The presence of alkenones in Albian shales from the Blake–Bahama Basin illustrates that they can survive a mild thermal history, although the components recognized were all alkadienones (Farrimond *et al.*, 1986) and therefore can only be taken to suggest water temperatures of $\geq 29°C$, presuming that the temperature calibration derived from contemporary algae and particulate matter is valid in the Cretaceous.

A further consideration is the possibility that U_{37}^{K} values are altered during core (or sample) storage. Present evidence suggests that the alkenones remain unaltered during sediment storage at temperatures slightly above freezing. For example, the U_{37}^{K} values for sediment samples from the Kane Gap in the eastern equatorial Atlantic which were taken after 34 months' storage were comparable to those obtained from analysis of samples collected immediately upon core recovery (Brassell *et al.*, 1986b). A more recent study (Sikes *et al.*, 1991) has shown that U_{37}^{K} values for pairs of cores, one frozen and the other nonfrozen, are generally comparable; they differed significantly at only one site in the northwest Pacific where the alkenone concentrations were low (<50 ng g^{-1}). Overall, frozen storage of cores does not appear to be a prerequisite for preservation of alkenone unsaturation.

3.4.5. Carbon Isotopic Composition of Sedimentary Alkenones and pCO_2

Recent studies comparing the $\delta^{13}C$ values of individual alkenones and co-occurring skeletal carbonate from foraminifera in sediments from the Pigmy Basin demonstrate the possibility of using these data to calculate variations in atmospheric pCO_2 levels (Fig. 19; Jasper and Hayes, 1990). The principle of the approach depends on the link between isotopic fractionation associated with carbon fixation and dissolved CO_2 concentrations, selecting compounds that

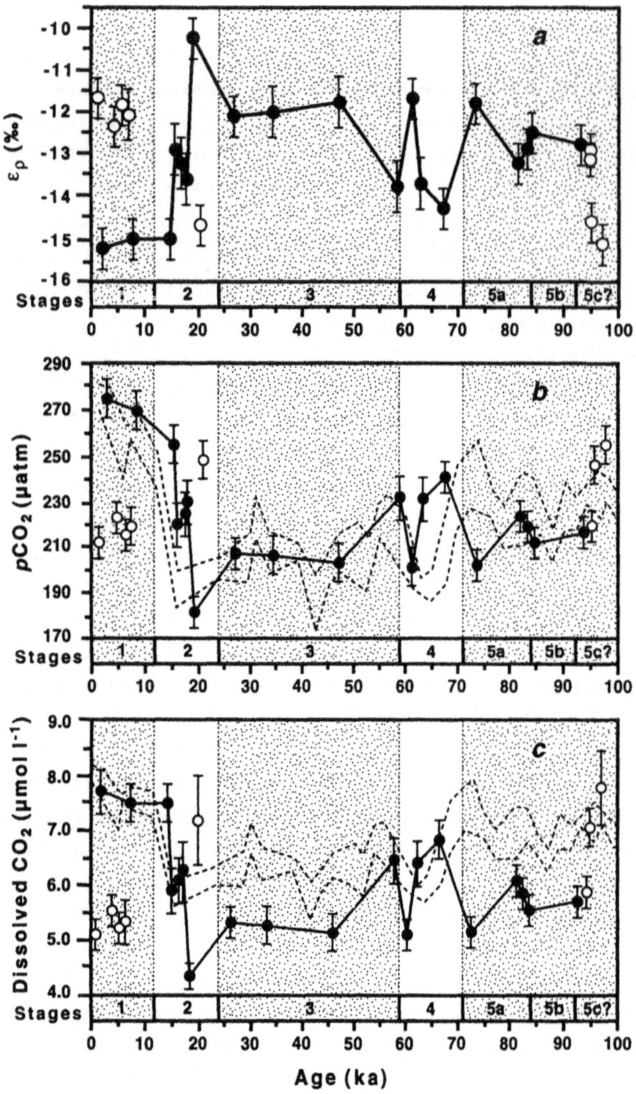

Figure 19. (a) Isotope fractionation between C_{37} alkadienones and *G. ruber* calcite expressed in terms of the fractionation factor ϵ_p. (b) The record of equilibrium atmospheric pCO_2 based on the concentrations of dissolved CO_2 calculated from the ϵ_p values in (a). The envelope (mean $\pm$ 2σ) of the atmospheric CO_2 levels measured in the Vostok ice core (dashed lines) is shown for comparison. (c) The isotopically based record of sea surface concentrations of dissolved CO_2 compared with a hypothetical record (dashed lines) derived from the Vostok results based on assumption of ocean–atmosphere equilibration. [After Jasper and Hayes (1990).]

derived from organisms which are primary photosynthetic producers. The values for CO_2 concentrations calculated from the carbon isotopic signals of alkenones in the Pigmy Basin show a correspondence to those measured in ice cores (Barnola *et al.*, 1987), demonstrating the potential of the method. However, the precision of this relationship depends on equilibrium between atmospheric CO_2 and oceanic CO_2 in the waters inhabited by the phytoplankton-generating alkenones. In the central equatorial Pacific it appears that this condition may not be met. The prymnesiophyte algae grow most abundantly at depths of 60–80 m, hence the $\delta^{13}C$ of the alkenones that they produce records subsurface rather than surface CO_2 levels. This characteristics appears to reduce the amplitude of the glacial/interglacial signals, though not their phasing (Jasper *et al.*, in press). The carbon isotopic composition of individual compounds can also be pursued in sediment samples that predate ice core records. Interestingly, the isotopic fractionation associated with the synthesis of alkenones is better defined and controlled than the factors involved in production of the foraminiferal carbonate tests. Hence, future efforts to enhance and develop this approach need to explore the isotopic dependence associated with the formation of carbonate tests from CO_2 and/or HCO_3^- and to improve the calibration between isotopic fractionation and atmospheric CO_2 levels. In addition, the range of molecular constituents of sediments that are recognized as biosynthetic products of primary producers offers the possibility for investigation of several measures of pCO_2 within a single sediment sequence and establishing an excellent proxy record for atmospheric CO_2 concentrations.

3.5. Alkyl Alkenoates and Alkenes

Both the alkyl alkenoate and alkene families of long-chain compounds are structurally related to alkenones (Fig. 6) and are biosynthesized by the same species of prymnesiophyte algae (Marlowe *et al.*, 1984a, b), although production of the alkenes is not evident in all strains of *E. huxleyi* (Marlowe et *al.*, 1984a, 1990). In the sedimentary record the alkyl alkenoates are as prevalent as the alkenones, whereas there are fewer reported occurrences of the alkenes (Table I; e.g., de Leeuw *et al.*, 1980). The alkyl alkenoates (Fig. 6) vary in the chain length of their acid moiety (C_{36}, C_{37}), their degree of unsaturation (di-, tri-, and tetraunsaturated), and their alkylation (methyl or ethyl; Marlowe *et al.*, 1984a). In cultures of *E. huxleyi* the relative abundance of the $C_{38:2}$ ethyl alkenoate relative to the total C_{37} alkenones (expressed as EE/K_{37}) decreased with increasing growth temperature (Prahl *et al.*, 1988). However, this laboratory observation is borne out by few sedimentary sequences such as Pliocene cores from Walvis Ridge (Marlowe, 1984), although recent data on these compounds in water column particulates have shown that the proportion of the $C_{37:2}$ methyl and $C_{38:2}$ ethyl alkenoates relative to alkenones is temperature-dependent (Conte *et al.*, 1992). Specifically an index AA_{36} was defined (Conte *et al.*, 1992) as follows:

$$AA_{36} = [38{:}3\ Et]/([36{:}3\ FAME] + [36{:}2\ FAME] + [38{:}3\ Et])$$

where 38:3 Et denotes the $C_{38:3}$ ethyl alkenoate and FAME refers to the methyl esters of the respective C_{36} alkenoic acids. AA_{36} shows a better correspondence to temperature than U^{K}_{37} for the samples and its slope suggests a greater role for alkyl alkenoates at lower growth temperatures, similar to the relationship first noted by Marlowe (1984). Alkyl alkenoates, however, are minor components of particulate organic matter from the Southern Ocean where it appears that an increase in the relative abundance of alkenes with decreasing water temperature (Sikes and Volkman, 1993) may reflect the inability of alkenones to provide the needed response to maintain membrane fluidity (?) at such low temperatures. This observation raises the possibility that the nonlinearity of U^{K}_{37} at lower temperatures may result from changes in the proportion of the differently functionalized long-chain lipids that are biosynthesized. In general, the potential for changes in many different compounds with growth temperature suggests that multivariate calibration of temperature trends based on a number of molecular characteristics may offer a more comprehensive approach to interpretation of such signals (Conte and Eglinton, in press).

The production of long-chain alkenes by prymnesiophyte algae is reflected in the sporadic occurrence of these compounds both in algal cultures and in the sedimentary record (Table I). Although structurally and biosynthetically related to the long-chain alkenones and alkyl alkenoates, there appear to be no systematic relationships that link the C_{37} and C_{38} alkenes to the growth temperatures of their source organisms. Thus, they appear to possess limited potential as paleoclimatic measures.

3.6. Sterols

A diverse range of sterols has been recognized in marine sediments on a global basis, including components with various side chain methylations (e.g., methyl, ethyl, propyl, methylidene, ethylidene, or propylidene substituents at C-24; 23,24-dimethyl or 22,23-methylene-23,24-dimethyl; e.g., Brassell and Eglinton, 1983a, b), positions of unsaturation (e.g., Δ^{5}, $\Delta^{5,22}$, Δ^{7}, $\Delta^{5,7}$, $\Delta^{5,7,22}$, $\Delta^{8(14)}$) ring methylation (e.g., at C-4) and stereochemistry (e.g., at C-3, C-5, C-24; e.g., Brassell and Eglinton, 1981; Mackenzie *et al.*, 1982; Brassell, 1992; Fig. 20). The complexity in sedimentary sterol distributions stems primarily from the variety of their biological sources—they originate from a multiplicity of eukaryotic organisms, including phytoplankton (e.g., diatoms, coccolithophorids, dinoflagellates), vascular plants, and zooplankton (Mackenzie *et al.*, 1982; Volkman, 1986)—and also include the products of diagenetic transformations. In several cases sterols are restricted to specific families of biota and are therefore diagnostic indicators for sedimentary contributions from their specific biological sources; for example, dinosterol is recognized as a marker for dinoflagellate inputs to sediments (e.g., Boon *et al.*, 1979; Robinson *et al.*, 1984), although it has recently also been reported in cultured diatoms (Volkman *et al.*, 1993).

In one instance a variation in the proportion of two sterols has been found to preserve a climatic record. A study of sapropels from the Mediterranean showed a correlation between the ratio of specific bound sterols [cholesta-5,22-dien-3β-ol/(cholesta-5,22-dien-3β-ol + 27-nor-24-methylcholesta-5,22-dien-3β-ol)], defined as the sterol temperature index (STI), and U^{K}_{37} profiles in the same core (Fig. 21; ten Haven *et al.*, 1987). Hence, higher STI values can be attributed to higher temperatures. However, in the case of STI, unlike U^{K}_{37}, no culture experiments nor systematic analysis of particulate matter and sediment samples from specific temperature regimes have been conducted which verify its relationship with temperature. Other temperature variations in sterol production are known; one phytoplankton, the phytoflagellate *Ochromonas danica*, biosynthesizes a higher proportion of sterols at lower culture temperatures (Betouhim-El *et al.*, 1977). However, this isolated observation relates to sterol concentrations rather than distributions and therefore cannot be used as a geochemical measure given the natural variability in phytoplankton abundance. Also, it is unclear whether the STI reflects changes in phytoplankton assemblages or in the production of sterols by a given species. The lack of a precise rationale for the temperature dependence in STI further brings into question its general applicability. Interestingly, the STI was determined for bound sterols, and the same components in the "free" sterol fraction do not show similar behavior, suggesting that only esterified sterols may exhibit this behavior. Additional data on the composition of bound sterols in marine sediments are comparatively rare. The published concentrations of sterols in Black Sea Unit I sediments (de Leeuw *et al.*, 1983) give an STI value of ca. 0.35 which, taken with

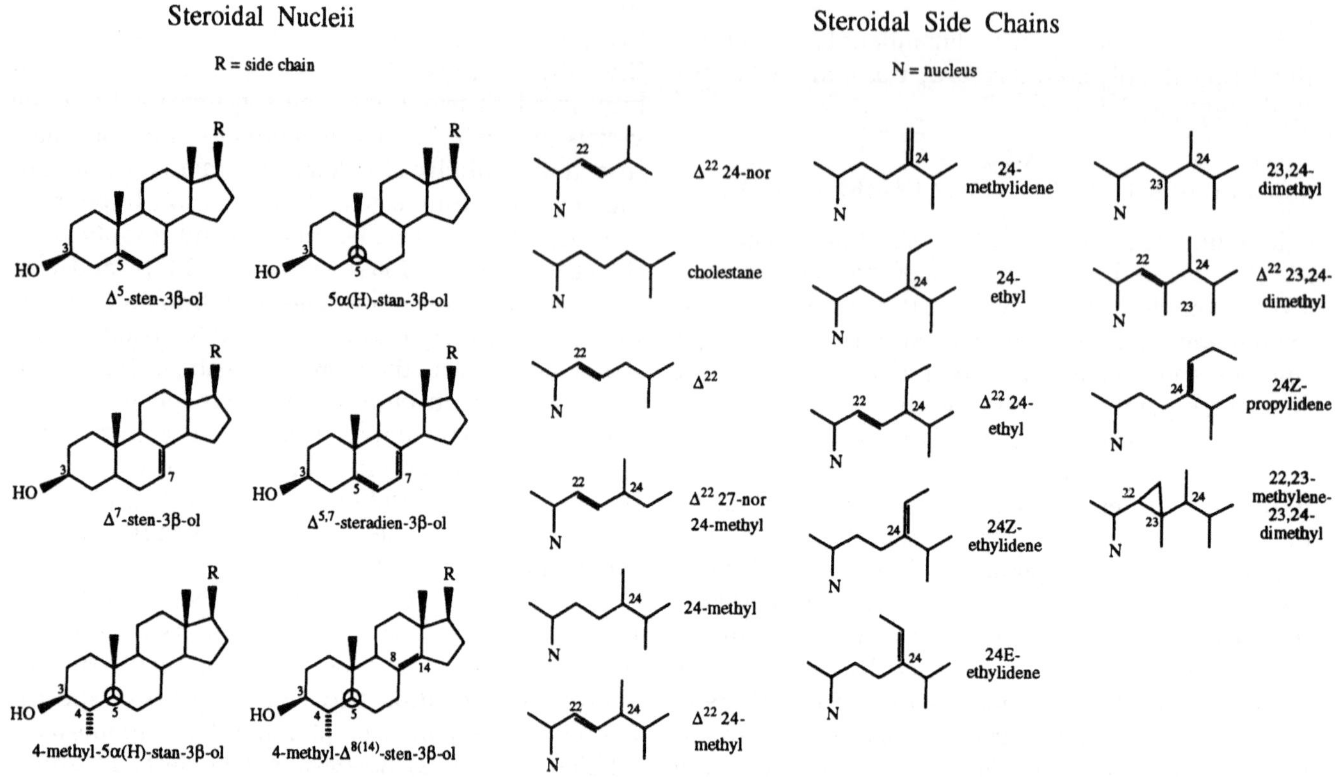

Figure 20. Structures of steroidal nuclei and side chains described in the text.

U_{37}^{K} values for other Black Sea sediments [0.338–0.511 (Sikes *et al.*, 1991)], compares favorably with the correlation seen in the Mediterranean sapropels. The recent discovery of pyropheophorbide steryl esters in the Black Sea (King and Repeta, 1991) raises the possibility that the production of such compounds containing esterified sterols may in some way be climatically controlled and thereby influence the STI. Overall, the STI constitutes an interesting characteristic of a single suite of organic-rich sediments which perhaps warrants further investigation as a climatic tool, but its significance cannot be assessed at present in the absence of multiple data sets providing a global basis for its evaluation.

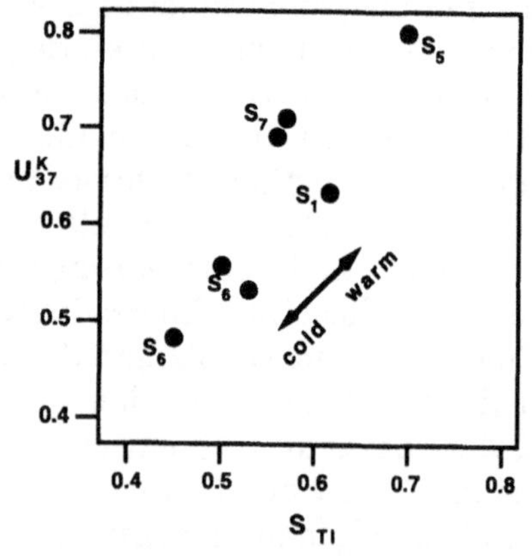

Figure 21. Plot of U_{37}^{K} vs. the sterol temperature index (S_{TI}) for Mediterranean sapropels (see Fig. 16). [Redrawn from ten Haven *et al.* (1987).]

3.7. Biomarker Abundances as Climatic Indicators and Their Statistical Evaluation

The preceding sections have primarily concerned the ratios of specific molecular components within sediment sequences as measures of climatic variability. Another approach lies in the potential of the abundances of individual diagnostic marker compounds to preserve a record of climate change. Specifically, fluctuations in molecular abundances can reflect climatically induced fluctuations in the productivity of the oceanic environment, especially in organic-rich sequences where the preservation of such signatures is unlikely to be a major influence. The choice of components is important; such studies are typically focused on marker compounds for specific phytoplankton which can be interpreted in terms of the productivity of particular source organisms, or a specific family of biota (e.g., prymnesiophyte algae, dinoflagellates, diatoms).

Two different, but related, approaches have been adopted in the investigation and evaluation of trends in marker compound abundances. First, profiles of single compounds have been examined to determine whether their variability can be related to climatic factors, such as glacial/interglacial cycles or El Niño events (e.g., Brassell *et al.*, 1986b; Poynter *et al.*, 1989a, b; Prahl *et al.*, 1989a; McCaffrey *et al.*, 1990; Kennedy and Brassell, 1992a). In these investigations, further assessment of the productivity signals recorded in the sedimentary abundances of the source-specific components can be made by comparison to the sedimentary abundance of marker fossils such as calcareous nannoplankton or diatom tests. Alternatively, contemporaneous measurements of phytoplankton productivity, when available, can provide a direct and independent verification of the molecular signals. For example, there is a correspondence between the downcore profile of dinosterol abundances in the Santa Barbara Basin and observations of dinoflagellate blooms, typically as "red tides," offshore southern California (Kennedy and Brassell, 1992b). Second, statistical methods such as factor or principal component analysis have been used to explore the interrelationships in the abundances of series of components (e.g., Fig. 22; Brassell *et al.*, 1986b; Poynter *et al.*, 1989b; McCaffrey *et al.*, 1991) and can reveal evidence of diagenetic trends. The value of this technique in investigating the application of biomarker abundances as climatic indicators lies in its ability to tackle multivariate data. Such complexity is a fundamental characteristic of the study of sedimentary biomarkers, which may typically comprise determinations of several tens of different components in many tens of individual samples within a selected sediment sequence. However, it remains the role of the organic geochemist to interpret whether a given trend is attributable to climatic or diagenetic factors.

Down-hole variations in the distributions and abundances of individual marker compounds have been investigated for many years, but most of the early studies of such trends in recent sediments examined comparatively few samples and were primarily targeted toward an evaluation of the biological sources of the components or toward an exploration of their diagenetic transformation pathways (e.g., Gaskell and Eglinton, 1976; Brassell and Eglinton, 1983b; Shaw and Johns, 1986; ten Haven *et al.*, 1987). It is only in the last decade that the potential of molecular compositions as measures of climate has

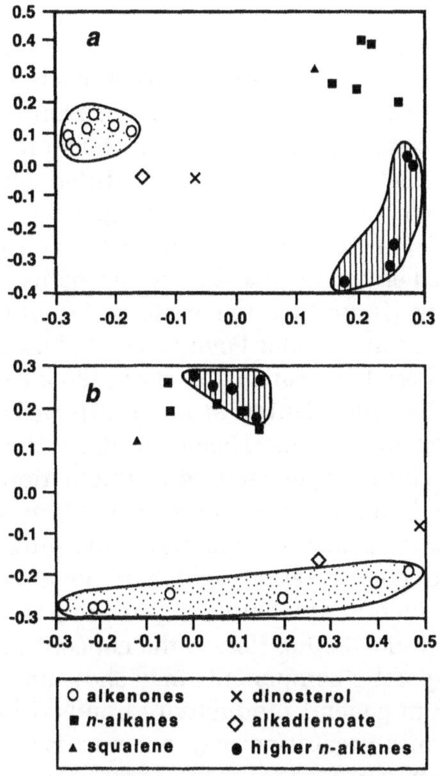

Figure 22. Principal component eigenvector plots. The principal components (PC) for 71 samples from the 0–705-cm section of a core from the Kane Gap have been calculated using the amounts of 20 compounds: 10 individual *n*-alkanes, including 5 terrigenous (higher) alkanes; 7 alkenones; methyl alkadienoate; dinosterol; and squalane (see key). The principal component eigenvectors are plotted for the 20 compounds as follows: (a) PC2 vs. PC1, for scaled data (scaled to constant weight per sample); (b) PC1 vs. PC3 for scaled data. [Redrawn from Brassell *et al.* (1986b).]

been recognized and evaluated (e.g., Brassell and Eglinton, 1986; Brassell *et al.*, 1986b; ten Haven *et al.*, 1987; Prahl *et al.*, 1989a; Poynter *et al.*, 1989a), specifically through the examination of series of many samples that can provide profiles of compounds which are markers for contributions from specific organisms, such as alkenones (Fig. 6) for prymnesiophyte algae (Volkman *et al.*, 1980a, b; Marlowe *et al.*, 1984a, b) and dinosterol (Fig. 20) for dinoflagellates (Boon *et al.*, 1979; Robinson *et al.*, 1984). Examination of the downcore trends for marker compounds in the eastern equatorial Atlantic reveals a general correspondence between their concentrations and glacial/intergalacial cycles (e.g., Brassell *et al.*, 1986b; Poynter *et al.*, 1989a), with their maximum abundances coincident with glacial maxima (e.g., alkenones) or preceding them (e.g., dinosterol). Dinosterol occurs in higher abundance in glacial sediments in other regions, including the eastern tropical Pacific Ocean, and the concentrations of bound 4α-methylsterols and alkenones show parallel trends (Fig. 17; Prahl *et al.*, 1989a). The precise cause of fluctuations in the abundances of individual components or groups of components cannot be deduced from their profiles, but such fluctuations may arise from differences in the productivity levels of competing organisms, in nutrient supply, in biological speciation, or in climatically influenced environmental factors. In regions of coastal upwelling, including the eastern Atlantic Ocean (Brassell *et al.*, 1986b; Poynter *et al.*, 1989a,b) and offshore Peru (e.g., Farrington *et al.*, 1988; Farrimond *et al.*, 1990b; McCaffrey *et al.*, 1991), variations in upwelling intensity have been inferred as the cause of productivity changes leading to fluctuations in the individual components. In sediments from the Peru margin, two subsurface maxima are recognized in the dinosterol concentrations which do not appear to be mirrored by other planktonic lipid components (Farrimond *et al.*, 1990b). Hence, the concentration profiles of specific components may show an independence from general productivity trends, illustrating that specific organisms may not necessariiy be favored during episodes when overall phytoplankton production is enhanced.

Statistical treatments of molecular data can reveal a covariance in related components, such as alkenones and alkyl alkenoates, and differences in the responses of other constituents attributed to different sources (e.g., Poynter *et al.*, 1989a; McCaffrey *et al.*, 1991). The primary result from such investigations has been the confirmation or refutation of the inferred biological origins of specific components. Time series variations in the molecular data have been related to Milankovitch cycles (Brassell *et al.*, 1989b), most notably the 100-kyr cycle, which reflects changes in the eccentricity of the Earth's orbit (Fig. 2). Overall, the use of statistical measures in assessing molecular characteristics offers the capability of recognizing trends in the analytical data that are not apparent from visual inspection for various reasons, including occasions where there is an ignorance of the potential correlation and other instances where correlations are obscured by the volume of data.

4. Future Potential and Scope for Biomarkers in Paleoclimatic Assessment

Climatic variations can have a profound influence on the abundance and assemblages of organisms, which is reflected in their biomarker compositions. Hence, the distribution and abundances of many different classes of biomarkers in oceanic sediments preserve a record of climatic change, providing independent measures of palaeoclimates and records of annual or longer term changes in the productivity of different species or classes of organism. The exploration of biomarker records in sediments as measures of climatic change is still in its infancy. The most promising approaches include: (i) the use of molecular constituents, specifically the degree of unsaturation in alkenones codified as U^K_{37}, in water column particulates and in sediments as measures of sea surface water temperatures, (ii) variations in compound abundances, notably alkenones and dinosterol, that can be related to changes in productivity over time, or to differences in wind strength (e.g., long-chain *n*-alkanes), and (iii) the ability of carbon isotope composition of individual biomarkers biosynthesized by photosynthetic organisms to record pCO_2 variations. Data from many different oceanic sites demonstrate the widespread applicability of these methods, although significant refinements of these approaches are required to expand their scope to a more general scale. Undoubtedly, the use of alkenone unsaturation indices as measures of sea surface water temperatures has opened new horizons for molecular data in climatological studies, though the results from U^K_{37} determinations are occasionally puzzling, and some questions regarding the applicability of the index remain unanswered. Specifically, the temperature calibration of the alkenone response is critical to its widespread use and value. Also, an explanation for the differences between U^K_{37} values in marine and lacustrine environments should be sought, an objective that requires the identification of the source of alkenones in lake settings.

Several studies have demonstrated that changes in the abundances of individual biomarkers in sediment sequences can reflect differences in the contributions of organic components from specific sources at particular times and can correlate to variations in phytoplankton productivity associated with climatic perturbations, including glacial/interglacial sequences. The concentrations of marker compounds can also denote changes related to fluctuations in upwelling intensity or, for terrestrially sourced components derived from aeolian dusts, to variations in the strength or direction of prevailing winds. However, the capability to elucidate and interpret such molecular signals within the sedimentary record is also a function of stratigraphic resolution and preservation of organic matter. Hence, the best oppportunities for unravelling these molecular signals lie in the examination of organic-rich, finely laminated sequences, such as the Santa Barbara basin. An additional important component is the acquistion of relevant data sets concerning the factors influencing the production of molecular components in the modern ocean. Investigations that seek to characterize the depths, temperatures, and seasonal fluctuations in molecular compositions provide invaluable information that can constrain the application of such parameters.

The molecular compositions of sediments are highly complex, especially in regions where contributions of organic matter originate from many different sources. Hence, the deconvolution of signals of climatic significance requires statistical treatments—a situation analogous to that encountered in the interpretation of the time dependence in $\delta^{18}O$ profiles which was resolved by spectral analysis. An important objective in the use of statistical packages should be to extend the range of biomarkers and aid the search for other diagnostic components that reveal links to environmental variables. Clearly, a major limitation lies in the ease of degradation of many of the most informative biomarkers, especially carotenoids, although the survival of selective degradation products of such components, for example, loliolides, may still preserve a record of the source organisms (e.g., ten Haven *et al.*, 1987).

In the future, a major task lies in the recognition of relationships between the variations in the molecular constituents of sediments and broader issues of global change. It is becoming increasingly apparent that there is a wealth of information contained in the compositions, concentrations, and isotopic characteristics of biomarkers in the sedimentary record. Ultimately, the independent nature of the organic and inorganic signals may offer the potential to resolve the various factors represented in other paleoclimatic signals, notably the $\delta^{18}O$ values for foraminifera, that reflect a complex web of interplay between ice volume, water temperature, and the CO_2 uptake. Clearly, the recently acquired capability to determine $\delta^{13}C$ values of individual biomarkers and thereby hindcast pCO_2 concentrations offers significant potential for exploring and interpreting important facets of geochemical cycles throughout Earth history.

ACKNOWLEDGMENTS. I would like to thank my former colleagues at the University of Bristol, especially Geoffrey Eglinton and Ian Marlowe, for their collaboration in the development of alkenones as molecular stratigraphic tools. I am also grateful to the many other organic geochemists who have shared their thoughts on the potential of biomarkers as measures of climatic variability, including John Farrington, Bob Gagosian, Lo ten Haven, John Hayes, John Jasper, Julie Kennedy, Jan de Leeuw, James Maxwell, Mark McCaffrey, John Poynter, Fred Prahl, John Volkman, and Stuart Wakeham. I acknowledge support for organic geochemical research from several sources, including a Fellowship in Science and Engineering from the David and Lucile Packard Foundation, NSF (OCE88-22583), and contributions from Chevron Oil Field Research Company and Unocal Science and Technology Division to the Stanford Organic Geochemistry Affiliates Program.

References

Bandy, O. L., 1959, The geological significance of coiling ratios in the foraminifer *Globigerina pachyderma* (Ehrenberg), *Geol. Soc. Am. Bull.* **70**:1708.

Barnett, T. P., 1978, The role of the oceans in the global climate system, in: *Climate Change* (J. Gribben, ed.), Cambridge University Press, Cambridge, pp. 157–177.

Barnola, J. M., Raynaud, D., Korotkevich, Y. S., and Lorius, C., 1987, Vostok ice core provides 160,000 year record of atmospheric CO_2, *Nature* **329**:408–414.

Barron, E. J., 1983, A warm, equable Cretaceous: The nature of the problem, *Earth Sci. Rev.* **19**:309–338.

Barron, E. J., 1989a, Studies of Cretaceous climate, in: *Understanding Climate Change* (A. Berger, R. E. Dickinson, and J. W. Kidson, eds.), American Geophysical Union, Washington, D.C., pp. 149–157.

Barron, E. J., 1989b, Pre-Pleistocene climates: Data and models, *Climate and Geo-Sciences* (A. Berger, S. Schneider, and J. C. Duplessy, eds.), Kluwer, Dordrecht, pp. 3–19.

Behrmann, J. H., Lewis, S. D., Musgrave, R. J., *et al.*, 1992, *Proc. ODP Init. Repts.* **141**, ODP, College Station.

Berger, A., 1978, Long-term variations in caloric insolation resulting from the Earth's orbital elements, *Quat. Res.* **9**:139–167.

Berger, W. H., 1981, Oxygen and carbon isotopes in foraminifera: An introduction, *Palaeogeogr. Palaeoclimatol. Palaeoecol.* **33**: 3–7.

Betouhim-El, T., Kahan, D., and Eckstein, B., 1977, Influence of temperature on the sterols of *Ochromonas danica*, *Comp. Biochem. Physiol.* **58B**:243–248.

Boon, J. J., de Leeuw, J. W., and Schenck, P. A., 1975, Organic geochemistry of Walvis Bay diatomaceous ooze—I. Occurrence and significance of the fatty acids, *Geochim. Cosmochim. Acta* **39**:1559–1565.

Boon, J. J., van der Meer, F. W., Schuyl, P. J. W., de Leeuw, J. W., Schenck, P. A., and Burlingame, A. L., 1978, in: *Initial Reports of the Deep Sea Drilling Project*, Vols. 38, 39, 40, and 41, Supplement, Walvis Ridge DSDP Leg 40, U.S. Government Printing Office, Washington, D.C., pp. 627–637.

Boon, J. J., Rijstra, R. I. C., de Lange, F., de Leeuw, J. W., Yoshioka, M., and Shimizu, Y., 1979, The Black Sea sterol—a molecular fossil for dinoflagellate blooms, *Nature* **277**:125–127.

Bowen, R., 1991, *Isotopes and Climates*, Elsevier Applied Science, London.

Boyle, E. A., 1988, Cadmium: Chemical trace of deepwater paleoceanography, *Paleoceanography* **3**:471–489.

Bradley, R. S., 1985, *Quaternary Paleoclimatology*, Allen and Unwin, Boston.

Brandt, D. S., and Elias, R. J., 1989, Temporal variations in tempestite thickness may be a geologic record of atmospheric CO_2, *Geology* **12**:614–618.

Brassell, S. C.,1980, The lipids of deep sea sediments: Their origin and fate in the Japan Trench, Ph.D. Thesis, University of Bristol.

Brassell, S. C., 1992, Biomarkers in recent and ancient sediments: The importance of the diagenetic continuum, in: *Organic Matter: Productivity, Accumulation, and Preservation in Recent and Ancient Sediments* (J. K. Whelan and J. W. Farrington, eds.), Columbia University Press, New York.

Brassell, S. C., and Eglinton, G., 1981, Biogeochemical significance of a novel C_{27} stanol, *Nature* **290**:579–582.

Brassell, S. C., and Eglinton, G., 1983a, The potential of organic geochemical compounds as sedimentary indicators of upwelling, in: *Coastal Upwelling, Its Sediment Record, Part A* (E. Suess and J. Thiede, eds.), Plenum Press, New York, pp. 545–571.

Brassell, S. C., and Eglinton, G., 1983b, Steroids and triterpenoids in deep sea sediments as environmental and diagenetic indicators, in: *Advances in Organic Geochemistry 1981*, (M. Bjorøy *et al.*, eds.) Wiley, Chichester, England, pp. 684–697.

Brassell, S. C., and Eglinton, G., 1984, Lipid indicators of microbial activity in marine sediments, in: *Heterotrophic Activity in the Sea* (J. E. Hobbie, and P. J. leB. Williams, eds.), Plenum Press, New York, pp. 481–503.

Brassell, S. C., and Eglinton, G., 1986, Molecular geochemical indicators in sediments, in: *Marine Organic Geochemistry* (M. Sohn, ed.), ACS Symposium Series No. 305, American Chemical Society, Washington, D.C., pp. 10–32.

Brassell, S. C., Eglinton, G., Maxwell, J. R., and Philp, R. P., 1978, Natural background of alkanes in the aquatic environment, in: *Aquatic Pollutants: Transformation and Biological Effects* (O. Hutzinger, L. H. van Lelyveld, and B. C. J. Zoeteman, eds.), Pergamon Press, Oxford, pp. 69–86.

Brassell, S. C., Comet, P. A., Eglinton, G., Isaacson, P. J., McEvoy, J., Maxwell, J. R., Thomson, I. D., Tibbetts, P. J. C., and Volkman, J. K., 1980a, Preliminary lipid analyses of Sections 440A-7-6, 440B-3-5, 440B-8-4, 440B-68-2, and 436-11-4; Legs 56 and 57, in: *Initial Reports of the Deep Sea Drilling Project*, Vol. 56/57(2), U.S. Government Printing Office, Washington, D.C., pp. 1367–1390.

Brassell, S. C., Comet, P. A., Eglinton, G., Isaacson, P. J., McEvoy, J., Maxwell, J. R., Thomson, I. D., Tibbetts, P. J. C., and Volkman, J. K., 1980b, The origin and fate of lipids in the Japan Trench, in: *Advances in Organic Geochemistry 1979* (A. G. Douglas and J. R. Maxwell, eds.), Pergamon Press, Oxford, pp. 375–391.

Brassell, S. C., Maxwell, J. R., and Eglinton, G., 1981, Preliminary lipid analyses of two Quaternary sediments from the Middle America Trench, southern Mexico transect, in: *Initial Reports of the Deep Sea Drilling Project*, Vol. 66, U.S. Government Printing Office, Washington, D.C., pp. 557–580.

Brassell, S. C., Eglinton, G., Marlowe, I. T., Pflaumann, U., and Sarnthein, M., 1986a, Molecular stratigraphy: A new tool for climatic assessment, *Nature* **320**:129–133.

Brassell, S. C., Brereton, R. G., Eglinton, G., Grimalt, J., Leibezeit, G., Marlowe, I. T., Pflaumann, U., and Sarnthein, M., 1986b, Palaeoclimatic signals recognized by chemometric treatment of molecular stratigraphic data, in: *Advances in Organic Geochemistry 1985* (D. Leythaeuser and J. Rullkötter, eds.), Org. Geochem. **10**:649–660.

Brassell, S. C., Eglinton, G., and Howell, V. J., 1987. Palaeoenvironmental assessment for marine organic-rich sediments using molecular organic geochemistry, in: *Marine Petroleum Source Rocks* (A. J. Fleet and J. Brooks, eds.), Geological Society Special Publication 26, Blackwell Scientific, London, pp. 79–98.

Brasseur, G., and Verstraete, M. M., 1989, Atmospheric chemistry—climate interactions, in: *Climate and Geo-Sciences* (A. Berger, S. Schneider, and J. C. Duplessy, eds.), Kluwer, Dordrecht, pp. 279–302.

Broecker, W. S., and Denton, G. H., 1989, The role of ocean–atmosphere reorganizations in glacial cycles, *Geochim. Cosmochim. Acta* **53**:2465–2501.

Broecker, W. S., Kennet, J. P., Teller, J., Trumbone, S., Bonani, S. G., and Wolfi, W., 1989, Routing of meltwater from the Laurentine Ice Sheet during the Younger Dryas cold episode, *Nature* **341**: 318–321.

Bukry, D., 1974, Coccoliths as palaeosalinity indicators—evidence from the Black Sea, in: *The Black Sea—Geology, Chemistry and Biology* (E. T. Degens and D. A. Ross, eds.), American Association of Petroleum Geologists Memoir 20, pp. 353–363.

Chesselet, R., Fontugne, M., Buat-Menard, P., Ezat, U., and Lambert, C. E., 1981, The origin of particulate organic carbon in the marine atmosphere as indicated by its stable carbon isotopic composition, *Geophys. Res. Lett.* **8**:345–348.

CLIMAP Project Members, 1984, The last interglacial ocean, *Quat. Res.* **21**:123–224.

Clochiatti, M., 1971, Sur l'existence de coccosphères portant des coccolithes de *Gephyrocapsa oceanica* et de *Emiliania huxleyi* (Coccolithophoroidés), *C. R. Acad. Sci., Sér. D.* **273**: 318–321.

Comet, P. A., 1982, The use of lipids as facies indicators. Ph.D. Thesis, University of Bristol.

Comet, P. A., and Eglinton, G., 1987, The use of lipids as facies indicators, in: *Marine Petroleum Source Rocks* (A. J. Fleet and J. Brooks, eds.), Blackwell Scientific, London, pp. 99–117.

Conte, M. H., Eglinton, G., and Madureira, L. A., 1992, Long-chain alkenones and alkyl alkenoates as paleotemperature indicators: Their production, flux and early diagenesis in the eastern North Atlantic, in: *Advances in Organic Geochemistry 1991* (C. B. Eckardt, S. R. Larter, D. A. Manning, and J. R. Maxwell, eds.), *Org. Geochem.* **19**:287–298.

Conte, M. H., and Eglinton G., in press, Alkenone and alkenoate distributions within the Euphotic zone of the eastern North Atlantic: Correlation with production temperature. *Deep-Sea Research*.

Cox, R. E., Mazurek, M. A., and Simoneit, B. R. T., 1982, Lipids in Harmatten aerosols of Nigeria, *Nature* **296**:848–849.

Cranwell, P. A., 1985, Long-chain unsaturated ketones in recent lacustrine sediments, *Geochim. Cosmochim. Acta* **49**:1545–1551.

Cranwell, P. A., 1988, Lipid geochemistry of Late Pleistocene lacustrine sediments from Burland, Cheshire, U.K., *Chem. Geol.* **68**:181–197.

Croll, J., 1875, *Climate and Time*, Appleton, New York.

Crowley, T. J., and North, G. R., 1991, *Paleoclimatology*, Clarendon, Oxford.

Davis, E. E., Mohl, M. J., Fisher, A. T., *et al.*, 1992, *Proc. ODP Init. Repts.* **139**, ODP, College Station.

de Leeuw, J. W., van der Meer, F. W., and Rijpstra, W. I. C., 1980, On the occurrence and structural identification of long-chain unsaturated ketones and hydrocarbons in sediments, in: *Advances in Organic Geochemistry 1979* (A. G. Douglas and J. R. Maxwell, eds.), Pergamon Press, Oxford, pp. 211–217.

de Leeuw, J. W., Rijpstra, W. I. C., Schenck, P. A., and Volkman, J. K., 1983, Free, esterified and residual bound sterols in Black Sea Unit 1 sediments, *Geochim. Cosmochim. Acta* **47**:455–465.

Denton, G. H., and Karlén, W., 1977, Holocene glacial and tree-line variations in the White River Valley and Skolai Pass, Alaska and Yukon Territory, *Quat. Res.* **7**:63–111.

Dunbar, R. B., deMenocal, P. B., and Burckle, L., 1992, Late Pliocene–Quaternary biosiliceous sedimentation at Site 798, Japan Sea, in: *Proc. ODP Scientific Results*, 127/128, Pt. 1, 439–448.

Duplessy, J.-C., 1978, Isotope studies, in: *Climate Change* (J. Gribben, ed.), Cambridge University Press, Cambridge, pp. 46–67.

Eglinton, G., Bradshaw, S. A., Rosell, A., Sarnthein, M., Pflaumann, U., and Tiedemann, R., 1992, Molecular record of secular sea surface temperature changes on 100-year timescales for glacial terminations I, II and IV. *Nature* **356**:423–426.

Emeis, K. C., Bull, D., and Doose, H., 1992, Alkenone sea-surface temperatures and productivity at ODP Site 846 (eastern equatorial Pacific) in the Late Quaternary, *Fourth International Conference on Paleoceanography*, *Abs.*, p. 105.

Emiliani, C., 1955, Pleistocene temperatures, *J. Geol.* **63**:538–578.

Ericson, D. B., Ewing, M., Wollin, G., and Heezen, B. C., 1961, Atlantic deep-sea sediment cores, *Geol. Soc. Am. Bull.* **72**: 193–286.

Farrimond, P., Eglinton, G., and Brassell, S. C., 1986, Alkenones in Cretaceous black shales, Blake–Bahama Basin, western North Atlantic, in: *Advances in Organic Geochemistry 1985* (D. Leythaeuser and J. Rullkötter, eds.), *Org. Geochem.* **10**:897–903.

Farrimond, P., Poynter, J. C., and Eglinton, G., 1990a, Molecular composition of sedimentary lipids off the Peru Margin, Leg 112, in: *Proceedings of the Ocean Drilling Project*, Vol. 112B, pp. 539–546.

Farrimond, P., Poynter, J. G., and Eglinton, G., 1990b, A molecular stratigraphic study of Peru Margin sediments, Hole 686B, Leg 112, in: *Proceedings of the Ocean Drilling Project*, Vol. 112B, pp. 547–553.

Farrington, J. W., Davis, A. C., Sulanowski, J., McCaffrey, M. A., McCarthy, M., Clifford, C. H., Dickinson, P., and Volkman, J. K., 1988, Biogeochemistry of lipids in surface sediments of the Peru upwelling area at 15°S, in: *Advances in Organic Geochemistry 1987*, (L. Mattavelli and L. Novelli, eds.), *Org. Geochem.* **13**:607–617.

Fischer, A. G., 1982, Long-term climatic oscillations recorded in stratigraphy, in: *Climate in Earth History* (W. H. Berger and J. C. Crowell, eds.), National Academy, Washington, D.C., pp. 97–104.

Fischer, A. G., and Arthur, M. A., 1977, Secular variations in the pelagic realm, *Soc. Econ. Paleontol. Mineral.* **25**:19–50.

Flenley, J. R., 1979, The Quaternary vegetational history of the equatorial mountains, *Prog. Phys. Geogr.* **3**:488–509.

Frakes, L. A., 1979, *Climates throughout Geological Time*, Elsevier, Amsterdam.

Freeman, K. H., and Wakeham, S. G., 1992, Variations in the distribution and isotopic compositions of alkenones in Black Sea particles and sediments, in: *Advances in Organic Geochemistry 1991* (C. B. Eckardt, S. R. Larter, D. A. Manning, and J. R. Maxwell, eds.), *Org. Geochem.* **19**:277–285.

Gagosian, R. B., and Peltzer, E. T., 1986, The importance of atmospheric input of terrestrial organic material to deep sea sediments, in: *Advances in Organic Geochemistry 1985* (D. Leythaeuser and J. Rullkötter, eds.), *Org. Geochem.* **10**:661–669.

Gaskell, S. J., and Eglinton, G., 1976. Sterols of a contemporary lacustrine sediment, *Geochim. Cosmochim. Acta* **40**:1221–1228.

Geikie, J., 1894, *The Great Ice Age*, 3rd ed., Stanford, London.

Genthon, C., Barnola, J. M., Raynaud, D., Lorius, C., Jouzel, J., Barkov, N. I., Korotkevich, Y. S., and Kotlaykov, V. M., 1987, Vostok ice core: Climatic response to CO_2 and orbital forcing changes over the last climatic cycle, *Nature* **329**:414–418.

Goth, K., de Leeuw, J. W., Püttmann, W., and Tegelaar, E. W., 1988, Origin of Messel Oil Shale kerogen, *Nature* **336**:759–761.

Gribbin, J., and Lamb, H. H., 1978, Climate change in historical times, in: *Climate Change* (J. Gribben, ed.), Cambridge University Press, Cambridge, pp. 68–82.

Grove, J. M., 1988, *The Little Ice Age*, Methuen, New York.

Harwood, J. L., and Russell, N. J., 1984, *Lipids in Plants and Microbes*, Allen and Unwin, London.

Hays, J. D., Imbrie, J., and Shackleton, N. J., 1976, Variations in the Earth's orbit: Pacemaker of the ice ages, *Science* **194**:1121–1132.

Hecht, A., 1985, *Paleoclimatic Analysis and Modeling*, John Wiley & Sons, New York.

Hooghiemstra, H., and Agwu, C. O. C., 1988, Changes in the vegetation and trade winds in equatorial northwest Africa 140,000–70,000 yr B.P. as deduced from two marine pollen records, *Palaeogeogr. Palaeoclimatol. Palaeoecol.* **66**:173–213.

Hooghiemstra, H., Bechler, A., and Beug, H.-J., 1987, Isopollen maps for 18,000 years B.P. of the Atlantic offshore of Northwest Africa: Evidence for paleowind circulation, *Paleoceanography* **2**:561–582.

Imbrie, J., and Imbrie, K. P., 1986, *Ice Ages: Solving the Mystery*, Harvard University Press, Cambridge.

Imbrie, J., and Kipp, N. G., 1971, A new micropaleontological method for quantitative paleoclimatology: Application to a late Pleistocene Caribbean core, in: *The Late Cenozoic Glacial Ages* (K. K. Turekian, ed.), Yale University Press, New Haven, pp. 71–181.

Imbrie, J., McIntyre, A., and Mix, A., 1989, Oceanic response to orbital forcing in the late Quaternary: Observational and experimental strategies, in: *Climate and Geo-Sciences* (A. Berger, S. Schneider, and J. C. Duplessy, eds.), Kluwer, Dordrecht, pp. 121–164.

Jasper, J. P., 1988, An organic geochemical approach to problems of glacial–interglacial climatic variability, Ph.D. Thesis, Massachusetts Institute of Technology/Woods Hole Oceanographic Institution.

Jasper, J. P., and Gagosian, R. B., 1989, Alkenone molecular stratigraphy in an oceanic environment affected by glacial freshwater events, *Paleoceanography* **4**:603–614.

Jasper, J. P., and Hayes, J. M., 1990. A carbon isotope record of CO_2 levels during the late Quaternary, *Nature* **347**:462–464.

Jasper J. P., Mix, A. C., Prahl, F. G., and Hayes, J. M., in press, Estimated CO_2 levels from photosynthetic ^{13}C fractionation in the Central Equatorial Pacific over the last 255,000 years. *Paleoceanography*.

Jouzel, J., Lorius, C., Petit, J. R., Genthon, C., Barkov, N. I., Kotlyakov, V. M., and Petrov, V. N., 1987, Vostok ice core: A continuous isotope temperature record over the last climatic cycle (160,000 years), *Nature* **329**:403–409.

Kawamura, K., and Ishiwatari, R.,1981, Polyunsaturated fatty acids in a lacustrine sediment as a possible indicator of paleoclimate, *Geochim. Cosmochim. Acta* **45**:149–155.

Keeling, C. D., Bacastow, R. B., Bainbridge, A. E., Ekdahl, C. A., Guenther, P. R., Waterman, L. S., and Chin, J. S., 1976, Carbon dioxide variations at Mauna Loa Observatory, Hawaii, *Tellus* **28**:538–551.

Kennedy, J. A., and Brassell, S. C., 1992a, Molecular records of 20th century El Niño events in laminated sediments from the Santa Barbara Basin, California, *Nature* **357**:62–64.

Kennedy, J. A., and Brassell, S. C., 1992b, Molecular stratigraphy of the Santa Barbara Basin: Comparison with historical records of annual climate change, in: *Advances in Organic Geochemistry 1991* (C.B. Eckardt, S. R. Larter, D. A. Manning, and J. R. Maxwell, eds.), *Org. Geochem.* **19**:235–244.

Kennett, J. P., 1977, Cenozoic evolution of Antarctic glaciation, the circum-Antarctic Ocean, and their impact on global paleoceanography, *J. Geophys. Res.* **82**:3843–3860.

Kennett, J. P., 1990, The Younger Dryas cooling event: An introduction, *Paleoceanography* **5**:891–895.

Kheradyar, T., 1992, Pleistocene planktonic foraminiferal assemblages and paleotemperature fluctuations in Japan Sea, Site 798. In: *Proc.ODP, Scientific Results*, 127/128, Pt. 1, 457–470.

King, L. L., and Repeta, D. J., 1991, Novel pyropheophorbide steryl esters in Black Sea sediments, *Geochim. Cosmochim. Acta* **55**: 2067–2074.

Kohnen, M. E. L., Sinninghe Damsté, J. S., Rijpstra, W. I. C., and de Leeuw, J. R., 1990, Alkylthiophenes as sensitive indicators of palaeoenvironmental changes, in: *Geochemistry of Sulfur in Fossil Fuels* (W. Orr and C. White, eds.), ACS Symposium Series No. 429, *American Chemical Society*, Washington, D.C., pp. 444–485.

LaMarche, V. C., Jr., 1974, Paleoclimatic inferences from long treering records, *Science* **183**:1043–1048.

Lamb, H. H., 1977, *Climate: Present Past and Future, Vol. 2, Climatic History and the Future*, Methuen, London.

Lamb, H.H., 1982, *Climate, History and the Modern World*, Methuen, London.

Lamb, H. H., 1984, Climate in the last thousand years: Natural climatic fluctuation and change, in: *The Climate of Europe: Past, Present and Future* (H. Flohn and R. Fantechi, eds.), Reidel, Amsterdam, pp. 25–64.

Landsberg, H. E., 1985, Historic weather data and early meteorological observations, in: *Paleoclimate Analysis and Modeling* (A. D. Hecht, ed.), John Wiley & Sons, New York, pp. 27–70.

Lorius, C. J., 1989, Polar ice cores and climate, in: *Climate and Geo-Sciences* (A. Berger, S. Schneider, and J. C. Duplessy, eds.), Kluwer, Dordrecht, pp. 77–103.

Lorius, C. J., Jouzel, J., Ritz, C., Merlivat, L., Barkov, N. I., Korotkevich, Y. S., and Kotlyakov, V. M., 1985, A 150,000 year climatic record from Antarctic ice, *Nature* **316**:591–596.

Lyle, M. W., Prahl, F. G., and Sparrow, M. A., 1992, Upwelling and productivity changes inferred from a temperature record in the central equatorial Pacific. *Nature* **355**:812–815.

Mackenzie, A. S., Brassell, S. C., Eglinton, G., and Maxwell, J. R., 1982. Chemical fossils: The geological fate of steroids, *Science* **217**:491–504.

Margolis, S. V., Kroopnick, P. M., Goodney, D. E., Dudley, W. C., and Mahoney, M. E., 1975, Oxygen and carbon isotopes from calcareous nannofossils of paleoceanographic indicators, *Science* **189**:555–557.

Marlowe, I. T., 1984, Lipids as paleoclimatic indicators, Ph.D. Thesis, University of Bristol.

Marlowe, I. T., Brassell, S. C., Eglinton, G., and Green, J. C., 1984a, Long chain unsaturated ketones and esters in living algae and marine sediments, in *Advances in Organic Geochemistry 1983* (P. A. Schenck, J. W. de Leeuw, and G. M. W. Lijmbach, eds.), *Org. Geochem.* **6**:135–141.

Marlowe, I. T., Green, J. C., Neal, A. C., Brassell, S. C., Eglinton, G., and Course, P. A., 1984b, Long chain (n-C_{37}–C_{39}) alkenones in the Prymnesiophyceae. Distribution of alkenones and other lipids and their taxonomic significance, *Br. Phycol. J.* **19**: 203–216.

Marlowe, I. T., Brassell, S. C., Eglinton, G., and Green, J. C.,1990, Long-chain alkenones and alkyl alkenoates and the fossil coccolith record of marine sediments, *Chem. Geol.* **88**:349–375.

Marr, A. G., and Ingraham, J. L., 1962, Effect of temperature on the composition of fatty acids in *Escherichia coli*, *J. Bacteriol.* **84**: 1260–1267.

Mayer, L. A., Pisias, N. G., Janecek, T. R., *et al.*, 1992. *Proc. ODP Init. Repts.* **138**, ODP, College Station.

McCaffrey, M. A., Farrington, J. W., and Repeta, D. J., 1990, The organic geochemistry of Peru margin surface sediments—I. A comparison of the C_{37} alkenone and historical El Niño records, *Geochim. Cosmochim. Acta* **54**:1671–1682.

McCaffrey, M. A., Farrington, J. W., and Repeta, D. J., 1991. The organic geochemistry of Peru margin surface sediments—II. Paleoenvironmental implications of hydrocarbon and alcohol profiles, *Geochim. Cosmochim. Acta* **55**:483–498.

McEvoy, J., 1983, The origin and diagenesis of organic lipids in sediments from the San Miguel Gap, Ph.D. Thesis, University of Bristol.

McEvoy, J., Eglinton, G., and Maxwell, J. R., 1981, Preliminary lipid analyses of sediments from Sections 467-3-3 and 467-97-2, in: *Initial Reports of the Deep Sea Drilling Project*, Vol. 63, U.S. Government Printing Office, Washington, D.C., pp. 763–774.

Morris, R. J., McCartney, M. J., and Weaver, P. P. E., 1984, Sapropelic deposits in a sediment from the Guinea Basin, South Atlantic, *Nature* **309**:611–614.

Nichols, P. D., and Johns, R. B., 1986, The lipid chemistry of sediments from the St. Lawrence estuary. Acyclic unsaturated long chain ketones, diols and ketone alcohols, *Org. Geochem.* **9**:25–30.

Pfister, C., 1988, Variations in the Spring–Summer climate of central Europe from the high Middle Ages to 1850, in: *Long and Short Term Variability of Climate* (H. Wanner and U. Siegenthaler, eds.), Lecture Notes in Earth Sciences, Vol. 16, Springer-Verlag, Berlin, pp. 57–83.

Philander, S. G., 1990, *El Niño, La Niña, and the Southern Oscillation*, Academic Press, San Diego.

Poynter, J. G., Farrimond, P., Brassell, S. C., and Eglinton, G., 1989a, Molecular stratigraphic study of sediments from Holes 658A and 660A, Leg 108, in: *Proceedings of the Ocean Drilling Program, Scientific Results* (W. Ruddiman, M. Sarnthein *et al.*, eds.), Vol. 108, U.S. Government Printing Office, Washington, D.C., pp. 387–394.

Poynter, J. G., Farrimond, P., Robinson, N., and Eglinton, G., 1989b, Aeolian-derived higher plant lipids in the marine sedimentary record: Links with palaeoclimate, in: *Paleooclimatology and Paleometerology: Modern and Past Patterns of Global Atmospheric Transport* (M. Leinen and M. Sarnthein, eds.), Kluwer, Dordrecht, pp. 435–462.

Prahl, F. G., and Muehlhausen, L. A., 1989, Lipid biomarkers as geochemical tools for paleoceanographic study, in: *Productivity of the Oceans: Present and Past* (W. H. Berger, V. S. Smetacek, and G. Wefer, eds.), John Wiley & Sons, pp. 271–289.

Prahl, F. G., and Wakeham, S. G., 1987, Calibration of unsaturation patterns in long-chain ketone compositions for palaeotemperature assessment, *Nature* **330**:367–369.

Prahl, F. G., Muehlhausen, L. A., and Zahnle, D. L., 1988, Further evaluation of long-chain alkenones as indicators of paleoceanographic conditions, *Geochim. Cosmochim. Acta* **52**: 2303–2310.

Prahl, F. G., Muehlhausen, L. A., and Lyle, M., 1989a, An organic geochemical assessment of oceanographic conditions at MANOP site C over the past 26,000 years, *Paleoceanography* **4**:495–510.

Prahl, F. G., de Lange, G. J., Lyle, M., and Sparrow, M. A., 1989b, Post-depositional stability of long-chain alkenones under contrasting redox conditions, *Nature* **341**:434–437.

Prahl, F. G., Sparrow, M. A., and Eversmeyer, B., 1992, Biomarker perspective on oceanographic changes in the northeast Pacific during the last glacial–interglacial transition, *Fourth International Conference on Paleoceanography*, *Abs.*, p. 231.

Prell, W. L., Imbrie, J., Martinson, D. G., Morley, J. J., Pisias, N. G., Shackleton, N. J., and Streeter H. F., 1986, Graphic correlation of oxygen isotope stratigraphy application to the late Quaternary. *Paleoceanography* **1**:137–162.

Quinn, W. H., Neal, V. T., and Antunez de Mayolo, S. E., 1987, El Niño occurrences over the past four and a half centuries, *J. Geophys. Res.* **92**:14449–14461.

Rau, G. H., Takahashi, T., and Des Marais, D. J., 1989, Latitudinal variations in plankton $\delta^{13}C$: Implications for CO_2 and productivity in past oceans, *Nature* **341**:516–518.

Rau, G. H., Froelich, P. N., Takahashi, T., and Des Marais, D. J., 1991, Does sedimentary organic $\delta^{13}C$ record variations in Quaternary ocean [CO_2(aq)]? *Paleoceanography* **6**:335–347.

Rechka, J. A., and Maxwell, J. R., 1988a, Unusual long chain ketones of algal origin, *Tetrahedron Lett.* **29**:2599–2600.

Rechka, J. A., and Maxwell, J. R., 1988b, Characterization of alkenone temperature indicators in sediments and organisms, in: *Advances in Organic Geochemistry 1987* (L. Mattavelli and L. Novelli, eds.), *Org. Geochem.* **13**:727–734.

Rhead, M. M., Eglinton, G., Draffan, G. H., and England, P. J., 1971, Conversion of oleic acid to saturated fatty acids in Severn Estuary sediments, *Nature* **232**:327–330.

Robinson, N., Eglinton, G., Brassell, S. C., and Cranwell, P. A., 1984, Dinoflagellate origin for sedimentary 4α-methylsteroids and 5α(*H*)-stanols, *Nature* **308**:439–442.

Rostek, F., Ruhland, G., Bassinot, F., Müller, P. J., Bard, E., Labeyie, L., and Lancelot, Y., 1992, 170,000 year sea-surface temperature (SST) and organic carbon record from the northern Indian Ocean, *Fourth International Conference on Paleoceanography*, *Abs.*, p. 243.

Rowland, S. J., and Volkman, J. K., 1982, Biogenic and pollutant aliphatic hydrocarbons in *Mytilus edulis* from the North Sea, *Mar. Environ. Res.* **7**:117–130.

Ruddiman, W. F., 1985, Climate studies in ocean cores, in: *Paleoclimate Analysis and Modeling* (A. D. Hecht, ed.), John Wiley and Sons, New York, pp. 197–257.

Ruddiman, W. F., Shackleton, N. J., and McIntyre, A., 1986, North Atlantic sea-surface temperatures for the last 1.1 million years, in: *North Atlantic Palaeoceanography* (C. P. Summerhayes and N. J. Shackleton, eds.), Geological Society Special Publication 21, Blackwell Scientific, London, pp. 155–173.

Rullkötter, J., and Michaelis, W., 1990, The structure of kerogen and related materials. A review of recent progress and future trends, in: *Advances in Organic Geochemistry 1989* (B. Durand and F. Behar, eds.), *Org. Geochem.* **16**:829–852.

Savin, S. M., 1977, The history of the earth's surface temperature during the past 100 million years, *Annu. Rev. Earth Planet. Sci.* **5**:319–335.

Schlesinger, M. E. (ed.), 1988, *Physically-Based Modelling and Simulation of Climate and Climate Change*, Kluwer, Dordrecht.

Shackleton, N. J., 1982, The deep sea sediment record of climate variability, in: *Prog. Oceanogr.* **11**:199–218.

Shackleton, N. J., 1988, Oxygen isotopes, ice volume, and sea level, *Quat. Sci. Rev.* **6**:183–190.

Shackleton, N. J., and Kennett, J. P., 1975, Late Cenozoic oxygen and carbon isotope changes at DSDP Site 284: Implications for glacial history of the northern hemisphere and the Antarctic, in: *Initial Reports of the Deep Sea Drilling Project*, Vol. 29, U.S. Government Printing Office, Washington, D.C., pp. 801–807.

Shackleton, N. J., Hall, M. A., Line, J., and Shuxi, C., 1983, Carbon isotope data in core V19–30 confirm reduced carbon dioxide concentration of the ice age atmosphere, *Nature* **307**:620–623.

Shackleton, N. J., Backman, J., Zimmerman, H., Kent, D. V., Hall, M. A., Robert, D. G., Schnitker, D., Baldauf, J. G., Desprairies, A., Homrighausen, R., Huddlestun, P., Keene, J. B., Kaltenback, A. J., Krumsiek, K. A. O., Morton, A. C., Murray, J. W., and Westberg-Smith, J., 1984, Oxygen isotope calibration of the onset of ice-rafting and history of glaciation in the North Atlantic region, *Nature* **306**:319–322.

Shaw, P. M., and Johns, R. B., 1986, The identification of organic input sources of sediments from the Santa Catalina Basin using factor analysis, in: *Advances in Organic Geochemistry 1985* (D. Leythaeuser and J. Rullkötter, eds.), *Org. Geochem.* **10**: 951–958.

Sicre, M.-A., Gagosian, R. B., and Peltzer, E. T., 1990, Evaluation of the atmospheric transport of marine derived particles using long-chain unsaturated ketones, *J. Geophys. Res.* **95**(D2):1789–1795.

Sikes, E. L., 1990, Refinement and application of a new paleotemperature estimation technique, Ph.D. Thesis, Massachusetts Institute of Technology/Woods Hole Oceanographic Institution.

Sikes, E. L., and Volkman, J. K., 1993, Calibration of alkenone unsaturation ratios $U^{K'}_{37}$s for palaeotemperature estimation in cold polar waters. *Geochim. et Cosmochim. Acta* **57**:1883–1889.

Sikes, E. L., Farrington, J. W., and Keigwin, L. D., 1991, Use of the alkenone unsaturation ratio U^{K}_{37} to determine past sea surface temperatures: Core-top SST calibrations and methodology considerations, *Earth Planet Sci. Lett.* **104**:36–47.

Simoneit, B. R., and Burlingame, A. L., 1972, Further preliminary results on the higher weight hydrocarbons and fatty acids in the DSDP cores, Leg 9, in: *Initial Reports of the Deep Sea Drilling Project*, Vol. 9, U.S. Government Printing Office, Washington, D.C., pp. 859–901.

Simoneit, B. R., and Burlingame, A. L., 1973, Preliminary organic analyses of DSDP cores, Legs 12 and 13, in: *Initial Reports of the Deep Sea Drilling Project*, Vol. 17, U.S. Government Printing Office, Washington, D.C., pp. 561–590.

Simoneit, B. R., Scott, E. S., and Burlingame, A. L., 1973, Preliminary organic analyses of DSDP cores, Leg 14, Atlantic Ocean, in: *Initial Reports of the Deep Sea Drilling Project*, Vol. 16, U.S. Government Printing Office, Washington, D.C., pp. 575–600.

Simoneit, B. R. T., Chester, R., and Eglinton, G., 1977, Biogenic lipids in particulates from the lower atmosphere over the eastern Atlantic, *Nature* **267**:682–685.

Sinninghe Damsté, J. S., and de Leeuw, J. W., 1990, Analysis, structure and geochemical significance of organically-bound sulphur in the geosphere: State of the art and future research, in: *Advances in Organic Geochemistry 1989* (B. Durand and F. Behar, eds.), *Org. Geochem.* **16**:1077–1101.

Sinninghe Damsté, J. S., Rijpstra, W. I. C., de Leeuw, J. W., and Schenck, P. A., 1988, Origin of organic sulphur compounds and sulphur-containing high molecular weight substances in sediments and immature crude oils, in: *Advances in Organic Geochemistry 1987* (L. Mattavelli and L. Novelli, eds.), *Org. Geochem.* **13**:593–606.

Sinninghe Damsté, J. S., Rijpstra, W. I. C., Kock-van Dalen, A. C., de Leeuw, J. W., and Schenck, P. A., 1989, Quenching of labile functionalised lipids by inorganic sulphur species: Evidence for the formation of sedimentary organic sulphur compounds at the early stages of diagenesis, *Geochim. Cosmochim. Acta* **53**:1343–1355.

Sinninghe Damsté, J. S., Kohnen, M. E. L., and de Leeuw, J. W., 1990, Thiophenic biomarkers for palaeoenvironmental assessment and molecular stratigraphy. *Nature* **345**:609–611.

Smith, D. J., Eglinton, G., and Morris, R. J., 1983a, The lipid chemistry of an interfacial sediment from the Peru continental shelf: Fatty acids, alcohols, aliphatic ketones and hydrocarbons, *Geochim. Cosmochim. Acta* **47**:2225–2232.

Smith, D. J., Eglinton, G., and Morris, R. J., 1983b, Interfacial sediment and assessment of organic input from a highly productive water column, *Nature* **304**:259–262.

Smith, D. J., Eglinton, G., and Morris, R. J., 1986, The lipid geochemistry of a recent sapropel and associated sediments from the Hellenic Outer Ridge, eastern Mediterranean Sea, *Philos. Trans. R. Soc. London, Ser. A* **319**:375–415.

Sofer, Z., 1988, Hydrous pyrolysis of Monterey asphaltenes, in: *Advances in Organic Geochemistry 1987* (L. Mattavelli and L. Novelli, eds.), *Org. Geochem.* **13**:939–945.

Stockton, C. W., Boggess, W. R., and Meko, D. M., 1985, Climate and tree rings, in: *Paleoclimate Analysis and Modeling* (A. D. Hecht, eds.), John Wiley & Sons, New York, pp. 71–151.

Sundquist, E. T., and Broecker, W. S. (eds.), 1985, *The Carbon Cycle and Atmospheric CO_2: Natural Variations Archean to Present*, Geophysical Monograph 32, American Geophysical Union, Washington, D.C.

Tarling, D. H., 1978, The geological–geophysical framework of ice ages, in: *Climate Change* (J. Gribben, ed.), Cambridge University Press, Cambridge, pp. 3–24.

Tegelaar, E. W., Matthezing, R. M., Jansen, J. B. H., Horsfield, B., and de Leeuw, J. W., 1989, Possible origins of *n*-alkanes in high-wax crude oils, *Nature* **342**:529–531.

ten Haven, H. L., and Kroon, D., 1991, Late Pleistocene sea surface water temperature variations off Oman as revealed by the distribution of long-chain alkenones in: *Proceedings of the Ocean Drilling Program*, Vol. 117B, U.S. Government Printing Office, Washington, D.C., pp. 445–452.

ten Haven, H. L., Baas, M., Kroot, M., de Leeuw, J. W., Schenck, P. A., and Ebbing, J., 1987, Late Quaternary Mediterranean sapropels III—Assessment of source of input and paleotemperature as derived from biological markers, *Geochim. Cosmochim. Acta* **51**:803–810.

ten Haven, H. L., Rullkötter, J., and Stein, R., 1989, Preliminary analysis of extractable lipids in sediments from the eastern North Atlantic (Leg 108): Comparison of a coastal upwelling area (Site 658) with a nonupwelling area (Site 659), in: *Proceedings of the Ocean Drilling Program*, Vol. 108B, pp. 351–360.

ten Haven, H. L., Littke, R., Rullkötter, J., Stein, R., and Welte, D. H., 1990, Accumulation rates and composition of organic matter in Late Cenozoic sediments underlying the active upwelling area off Peru, in: *Proceedings of the Ocean Drilling Program*, Vol. 112B, pp. 591–606.

Thomson, I. D., Brassell, S. C., Eglinton, G., and Maxwell, J. R., 1982, Preliminary lipid analysis of Section 481-22-2 in: *Initial Reports of the Deep Sea Drilling Project*, Vol. 64, No. 2, U.S. Government Printing Office, Washington, D.C., pp. 913–919.

Tooley, M. J., 1978, Interpretation of Holocene sea level changes, *Geol. Fören. Stockholm Förh.* **100**:203–212.

Volkman, J. K., 1986, A review of sterol markers for marine and terrigenous organic matter, *Org. Geochem.* **9**:83–99.

Volkman, J. K., and Johns, R. B., 1977, The geochemical significance of positional isomers of unsaturated acids from an intertidal zone sediment, *Nature* **267**:693–694.

Volkman, J. K., Eglinton, G., Corner, E.D. S., and Sargent, J. R., 1980a, Novel unsaturated straight-chain C_{37}–C_{39} methyl and ethyl ketones in marine sediments and a coccolithophore *Emiliania huxleyi*, in: *Advances in Organic Geochemistry 1979* (A. G. Douglas and J. R. Maxwell, eds.), Pergamon Press, Oxford, pp. 219–227.

Volkman, J. K., Eglinton, G., Corner, E. D. S., and Forsberg, T. E. V., 1980b, Long-chain alkenes and alkenones in the marine coccolithophorid *Emiliania huxleyi*, *Phytochemistry* **19**:2619–2622.

Volkman, J. K., Corner, E. D. S., and Eglinton, G., 1980c, Transformations of biolipids in the marine food web and in underlying bottom sediments, in: *Biogéochemie de la Matière Organique à l'Interface Eau–Sédiment Marin* (R. Daumas, ed.), Éditions du CNRS, Paris, pp. 185–197.

Volkman, J. K., Gillan, F. T., Johns, R. B., and Eglinton, G., 1981, Sources of neutral lipids in a temperate intertidal sediment, *Geochim. Cosmochim. Acta* **45**:1817–1828.

Volkman, J. K., Farrington, J. W., Gagosian, R. B., and Wakeham, S. G., 1983, Lipid composition of coastal marine sediments from the Peru upwelling region, in: *Advances in Organic Geochemistry 1981* (M. Bjorøy *et al.* eds.), Wiley, Chichester, England, pp. 228–240.

Volkman, J. K., Farrington, J. W., and Gagosian, R. B., 1987, Marine and terrigenous lipids in coastal sediments from the Peru upwelling region at 15°C: Sterols and triterpene alcohols, *Org. Geochem.* **11**:463–477.

Volkman, J. K., Burton, H. R., Everitt, D. A., and Allen, D. I., 1988, Pigment and lipid compositions of algal and bacterial communities in Ace Lake, Vestfold Hills, Antarctica, *Hydrobiologia* **165**:41–57.

Volkman, J. K., Jeffrey, S. W., Nichols, P. D., Rogers, G. I., and Garland, C. D., 1989, Fatty acid and lipid composition of 10 species of microalgae used in mariculture, *J. Exp. Mar. Biol. Ecol.* **128**:219–240.

Wells, L. E., 1987, An alluvial record of El Niño events from northern coastal Peru, *J. Geophys. Res.* **92**:14463–14470.

Wijmstra, T. A., 1978, Palaeobotany and climate change, in: *Climate Change* (J. Gribben, ed.), Cambridge University Press, Cambridge, pp. 25–45.

Williams, D. F., Lerche, I., and Full, W. E., 1988, *Isotope Chronostratigraphy: Theory and Methods*, Academic Press, San Diego.

Winter, A., 1985, Distribution of living coccolithophores in the California Current system, Southern California Borderland, *Mar. Micropaleontol.* **9**:385–393.

Young, J. R., and Westbroek, P., 1991, Genotypic variation in the coccolithophorid species *Emilliania huxleyi*. *Marine Micropaleontology* **18**:5–23.

Zafiriou, O. C., Gagosian, R. B., Peltzer, E. T., and Alford, J. B., 1985, Air-to-sea fluxes of lipids at Enewetak Atoll, *J. Geophys. Res.* **90**:2409–2423.

Zumberge, J. E., 1987, Terpenoid biomerker distributions in low maturity crude oils, *Org. Geochem.* **11**:479–496.

Chapter 35

The Diagenesis of Proteins and Amino Acids in Fossil Shells

RICHARD M. MITTERER

1. Introduction

Proteins are one of the major classes of biopolymers in organisms. These nitrogen-containing macromolecules, comprised of amino acid building blocks, constitute more than 50 percent of the dry weight of most animal tissues. Proteins fulfill a variety of biochemical roles including strengthening of connective tissues, muscles, and membranes and regulation of metabolic activities and immunological functions. Geological interest in these compounds focuses on the role of proteins in invertebrate biomineralization and on the application of amino acid racemization reactions in fossil shells and bones to chronostratigraphy.

Abelson's (1954) discovery of amino acids in fossil shells was the first demonstration that original biochemical constituents of a specific organism were preserved intact millions of years after death of the animal. Hare and Mitterer (1967, 1969) and Hare and Abelson (1968) subsequently discovered that amino acids in fossil shells undergo epimerization and racemization reactions and that these reactions can be used to estimate the age of fossils or their burial temperature. Bada (1972a) and Bada *et al.* (1973) first applied racemization reactions to the determination of ages and burial temperatures of bone. Early work on amino acid geochemistry was reviewed by Hare (1969). Later research, with emphasis on the geochemistry of amino acid racemization reactions, has been summarized by Bada and Schroeder (1975), Kvenvolden (1975), Schroeder and Bada (1976), Williams and Smith (1977), and Bada (1982). An extensive compilation of papers, emphasizing amino acid racemization reactions and their geological applications, appeared in a volume edited by Hare *et al.* (1980).

This chapter focuses on the diagenetic changes that organic matrices of fossil shells and tests experience during their burial history and especially on the racemization/epimerization reactions that amino acids in fossil shells undergo as part of the diagenetic changes.

2. Diagenesis of Proteins and Amino Acids in Fossil Shells

Biominerals, such as shells, bones, and teeth, are composed of a mineral structure and an intimately associated protein matrix, which serves as a substrate

RICHARD M. MITTERER • Department of Geosciences, University of Texas at Dallas, Richardson, Texas 75083-0688.
Organic Geochemistry, edited by Michael H. Engel and Stephen A. Macko. Plenum Press, New York, 1993.

on and within which the mineral phase nucleates. In the case of bones and teeth, the mineral phase is hydroxyapatite. Mollusk shells, as well as most other mineralized structures in invertebrates and plants, are comprised of one or both of two polymorphs of calcium carbonate, calcite and aragonite. A few groups of organisms, notably diatoms and radiolaria, produce opaline tests. After burial of the shell, the mineral structure provides a generally protective medium for the organic matrix and effectively isolates it from microbial reactions. Significant nonmetabolic changes occur, however, over time in the concentration and composition of the organic matrix due to physical processes and chemical reactions (Fig. 1). These processes and reactions are interdependent and generally proceed simultaneously.

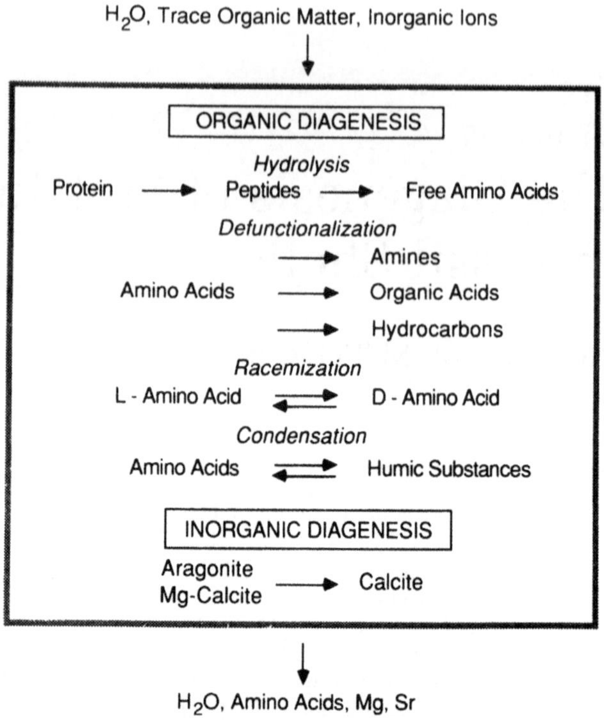

Figure 1. Summary of diagenetic changes in organic and inorganic phases of a fossil shell originally composed of aragonite or high-magnesium calcite. Dissolved organic matter and inorganic ions seep into the shell from the surrounding sediment, and shell constituents are leached or modified by the added water. In addition to mineralic stabilization and loss of Mg and Sr, other changes, such as carbon and oxygen isotopic modifications, occur under specific conditions. Shells composed initially of low-magnesium calcite may experience less modification of the mineral structure. Due to leaching and diffusion, the concentration of organic constituents decreases with increasing age of the fossil. [After Hoering (1980), Wehmiller (1982), and Bada (1985).]

2.1. Physical Processes

Leaching and contamination are the major physical processes affecting amino acid content and composition in fossils. Leaching is broadly used here to include physical transport as well as diffusion of amino acids out of shells. An aqueous phase is necessary for both processes. Water gradually infiltrates fossil shells at a rate that depends on shell structure, amount of organic matrix, conditions of burial, porosity and permeability of surrounding sediment, and local environmental effects (e.g., rainfall, temperature). Except under extraordinary conditions of preservation, such as occur when fossils are embedded in a tarry medium or in sediments containing a very high organic carbon content, water probably cannot be excluded from fossil shells.

Penetration of water into the internal shell structure may cause leaching of original organic components and introduction of contaminants. Leaching leads to loss of soluble portions of the organic matrix, usually free amino acids and small peptides that have formed by hydrolysis of the larger protein molecules (Müller, 1984). Extraneous amino acids can be introduced from the surrounding sediment or from organisms growing on or near the shells. The two processes, leaching and contamination, have opposite effects on amino acid concentration and composition. Leaching causes a decrease in amino acid concentration, while contamination introduces additional compounds. Although leaching may occur without adding contaminants, it is unlikely that the introduction of contaminants occurs without leaching. The net effect of both processes is loss of original amino acids and their gradual replacement, at much lower concentrations, by amino acids originating in the surrounding environment (Hare and Mitterer, 1969; Mitterer, 1972). Contaminants consist of amino acids such as serine that are major by-products of metabolic processes (Hamilton, 1965; Hare, 1965). Consequently, contamination of an older fossil with new amino acids results in an amino acid spectrum resembling that of modern organisms or younger fossils. Contaminating amino acids may also be derived from older sediments or fossils that surround (e.g., in archeological burial sites), or that lie stratigraphically beneath, younger samples.

In addition to affecting amino acid composition and concentration, leaching and contamination can modify the extent of amino acid racemization observed in fossils (Wehmiller *et al.*, 1976). Preferential loss of free amino acids, one of the most highly racemized fractions, leads to a decrease in the overall extent of racemization in a sample (Müller, 1984). Similarly, preferential addition of L-amino acids from modern contaminants will reduce the extent of amino acid racemization in a sample. Both effects tend to make apparent racemization ages younger than true ages of fossils. Contamination from older sediments or fossils will increase the observed extent of racemization.

2.2. Chemical Reactions

Four types of chemical reactions occur in fossil shells (Fig. 1): (1) hydrolysis, (2) racemization, (3) defunctionalization, and (4) condensation (Hare and Hoering, 1977; Hoering, 1980). These reactions generally occur continuously and simultaneously, although there is probably a temporal trend in relative importance of each one. Hydrolysis and racemization proceed rapidly during early diagenesis whereas the latter two reactions may predominate after the protein has been largely hydrolyzed.

Hydrolytic reactions break peptide bonds and convert the original protein initially to a mixture of large and small peptides and free amino acids and ultimately to a mixture of only free amino acids. As hydrolysis proceeds, smaller, more readily leachable molecules are produced. Free amino acids are the smallest and most easily leached amino compounds in fossils. Thus, hydrolysis and subsequent leaching lead to a preferential loss of free amino acids and a decrease in the concentration of amino acids in fossils (Müller, 1984). Because of its importance in affecting the extent of amino acid racemization in fossils, hydrolysis is treated in greater detail in the next section.

Racemization of the optically active L-amino acids that are initially present in calcified proteins of modern shells begins after formation of the shell. Assuming no preferential gain or loss of either the D- or the L-isomer, the extent of racemization is related to the age of the fossil and its temperature history after deposition. This reaction, which has seen extensive applications in geochronology and chronostratigraphy, will be discussed in more detail in Section 4.

Defunctionalization involves loss of amino acid functional groups by decarboxylation, deamination, and other reactions. Although detailed studies of specific pathways of amino acid defunctionalization have not been conducted, initial products are probably organic acids and amines (Hare and Hoering, 1977; Hoering, 1980), with low-molecular-weight hydrocarbons the ultimate products of complete defunctionalization (Philippi, 1977). In some cases, partial defunctionalization leads to other amino acids. These newly formed compounds may be nonprotein amino acids, such as α-aminobutyric acid, cysteic acid, methionine sulfoxides, and ornithine, or they may be other common protein amino acids such as glycine and alanine (Vallentyne, 1964; Hare and Mitterer, 1967). Thus, some portion of common protein amino acids in fossil shells can be diagenetic products of other amino acids.

Condensation reactions convert amino acids and other reactive compounds that may be present, such as amino sugars, into insoluble humic- or kerogen-like substances (Hare and Hoering, 1977; Hoering, 1980). Some compounds incorporated into the condensed substances may be contaminants that have been introduced from the surrounding sediment, although no evidence for contamination was found in a shell as old as Miocene (Hare and Hoering, 1977; Hoering, 1980).

As a consequence of these chemical reactions, the amino acid composition of the remnant organic matrix of mollusk shells, despite significant differences in original composition, converges toward a more limited suite of stable, racemic, free amino acids. Acidic, basic, hydroxy, and sulfur amino acids, which contain more than two functional groups, are especially reactive and disappear relatively rapidly (Abelson, 1959; Vallentyne, 1964; Hare and Mitterer, 1967). Neutral amino acids are more stable and dominate the suite of amino acids in older fossils. As a result of the loss of amino acids by defunctionalization, condensation, and leaching, the concentration of amino acids in shells decreases with increasing age of the fossil (Abelson, 1954; Hare and Mitterer, 1967, 1969). The relative importance of these various processes in contributing to the decrease in amino acid concentration in fossils has not been firmly established. A decrease in concentration of isoleucine in deep-sea foraminifera with increasing age and extent of isoleucine epimerization is illustrated in Fig. 2. Müller (1984) attributed this decrease primarily to leaching of free amino acids, a process that is dependent on the rate of hydrolysis.

Addition of new amino acids by natural contamination, combined with loss of original amino acids due to defunctionalization, condensation, and leaching, becomes an increasingly important process in older fossils. Amino acids in these older samples are only partially contemporaneous with the shells. At low to mid-latitudes in continental locations, fossils much older than Neogene probably have retained few of their original amino acids. For example, a Miocene

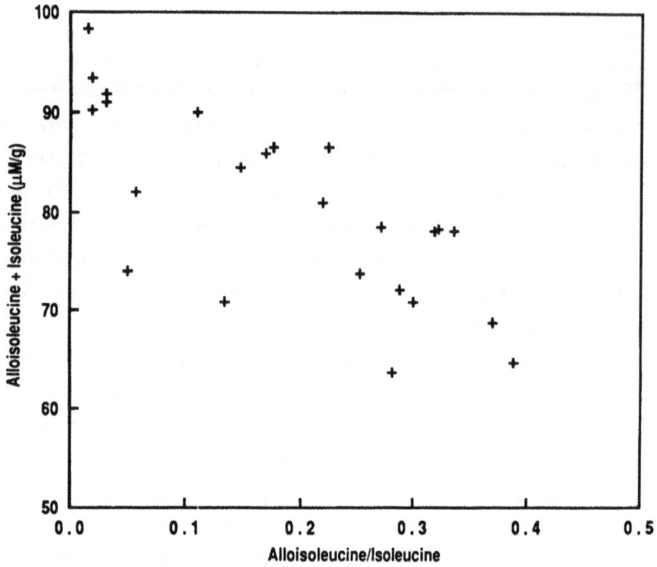

Figure 2. Concentration of isoleucine + alloisoleucine in foraminiferal tests of *Orbulina universa* as a function of alloisoleucine/isoleucine. The decrease in concentration with increasing alloisoleucine/isoleucine (increasing age) is attributed to leaching of free amino acids from the foraminiferal tests. [Data from Müller (1984).]

bivalve, *Mercenaria* sp., was reported to have about 2% of its original amino acid content (Hare and Mitterer, 1967). At higher latitudes and in deep-sea cores, lower temperatures retard hydrolytic and leaching reactions sufficiently to permit preservation of proteins and amino acids in shells for longer time periods. Eventually, all fossils will lose their original amino acids.

3. Hydrolysis of Proteins and Peptides in Fossil Shells

Hydrolytic reactions gradually convert the original protein in shells into a complex mixture of free amino acids and smaller peptides of different molecular weights and finally into free amino acids entirely (Fig. 1). The total amino acid population in a fossil shell, as a consequence, is comprised of a number of analytical fractions (e.g., free, bound, insoluble, total) for which amino acid composition and extent of racemization can be determined individually (Fig. 3). The *total amino acid* composition consists of the entire suite of amino acids in a sample and is determined by ion-exchange chromatographic analysis after decalcification and laboratory hydrolysis. The extent of racemization of a particular amino acid in this preparation (e.g., "total isoleucine") is the measurement generally used for age estimations. The total amino acids are comprised of the following subset fractions:

1. The *free fraction* consists of free amino acid molecules that were formed by natural hydrolysis *in situ* during the burial history of the fossil. The amino acid composition and extent of racemization in the free fraction are established by analyzing a sample that has not been hydrolyzed in the laboratory.
2. The *bound fraction* consists of amino acids remaining in peptide linkages of various lengths. Because this fraction cannot be separated easily from free amino acids, the amino acid composition and extent of racemization in the bound fraction are determined as the difference between quantitative analyses of the *total* and *free* amino acids.
3. The *insoluble fraction* is a component of the bound fraction of very high molecular weight, rendering it insoluble during decalcification. This fraction can be physically isolated by filtration or centrifugation and analyzed after laboratory hydrolysis.

In addition to these fractions, advances in ultrafiltration membrane technology allow different *molecular weight fractions* of soluble peptides (e.g., <500, 500–1000, 1000–10,000 daltons, etc.) to be isolated and analyzed (Kriausakul and Mitterer, 1980a, b). Although these separations are time-consuming and therefore are not routine procedures in amino acid diagenetic studies, important information has been gained from this effort.

During gradual hydrolysis of a protein, an amino acid may exist in five different positional configurations (Mitterer and Kriausakul, 1984): (1) as an internal amino acid in a peptide chain with other amino acids linked to it on each side; (2) and (3) as a terminal amino acid at either end of a protein or peptide

FRACTION	ANALYTICAL METHOD
Bound	Difference between Free and Total
Insoluble	Filtration or centrifugation
Large peptides Small peptides	Ultrafiltration
Free amino acids	Analysis of unhydrolyzed sample
Total amino acids	Analysis of hydrolyzed sample

Figure 3. Amino acid-containing fractions comprising the total amino acids in fossil shells and the method of isolation or analysis of each of the fractions as well as the total amino acids.

molecule (C-terminal and N-terminal, respectively); (4) as a cyclic structure formed by joining the two terminal functional groups of small peptides to form a diketopiperazine; and (5) as a free amino acid (Fig. 4).

The rate of hydrolysis of peptide bonds depends on many factors, including length of peptide chains and nature of adjacent amino acids (i.e., amino acid sequence) (Hill, 1965; Hare *et al.*, 1975; Hare, 1976; Kriausakul and Mitterer, 1978; Wehmiller, 1980). Peptide bonds in larger molecules are generally more labile than those in shorter chains. As protein composition and amino acid sequence are genetically determined, calcified proteins in organisms that are taxonomically distinct will have different rates of hydrolysis (Wehmiller, 1980).

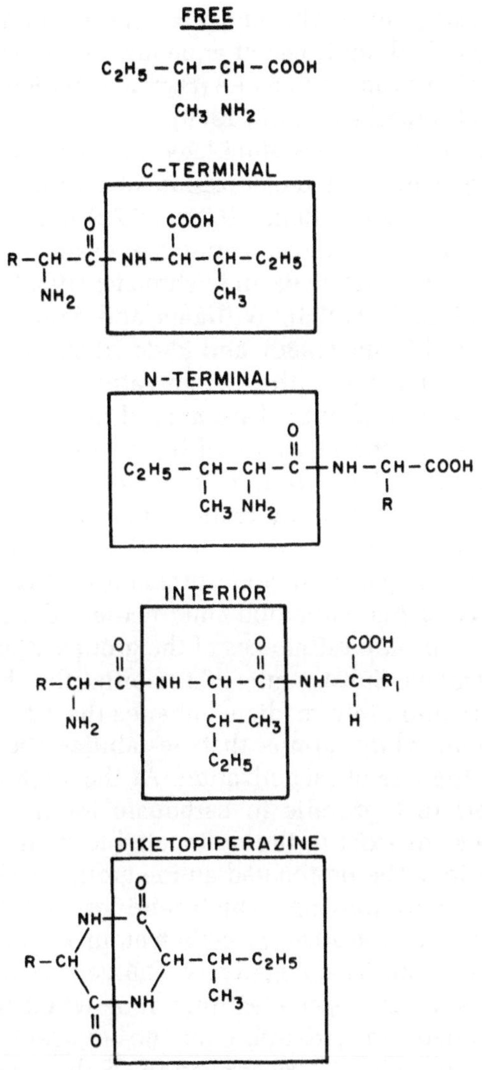

Figure 4. Five positional configurations in which isoleucine and many other amino acids may exist. The peptide bond connects adjacent amino acids. An interior amino acid in a peptide chain is bonded through the α-amino group to one amino acid and through the α-carboxyl group to another amino acid. Terminal amino acids, at the end of a peptide chain, have either a free amino group (NH_2-terminal or N-terminal) or a free carboxyl group (COOH-terminal or C-terminal) depending on at which end of the molecule they are located. Diketopiperazine is formed when the free α-COOH and free α-NH_2 groups of a dipeptide join to form a cyclic structure. [From Mitterer and Kriausakul (1984).]

4. Stereochemistry and Racemization of Amino Acids

All protein amino acids except glycine have at least one asymmetric or chiral carbon atom. Because of this molecular asymmetry, protein amino acids (except glycine) exist in two nonsuperimposable, mirror-image forms termed optical isomers or enantiomers. At a different rate for each amino acid, enantiomers interconvert or racemize in accordance with the equation

$$\text{L-amino acid} \underset{k_2}{\overset{k_1}{\rightleftarrows}} \text{D-amino acid} \qquad (1)$$

At equilibrium, the amino acid is fully racemized, and the concentrations of the two enantiomers are equal (i.e., D-amino acid/L-amino acid, or D/L, = 1). Because both enantiomers have equal free energies of formation, chemical synthesis of amino acids also generates racemic mixtures. A few common protein amino acids, such as isoleucine and threonine, contain a second asymmetric carbon atom. When more than one chiral center is present in an amino acid molecule, inversion occurring only at the α-carbon results in formation of diastereomers or epimers (which are not mirror-image forms) rather than enantiomers; the process of interconversion of epimers is epimerization rather than racemization. Epimerization is analogous to racemization (both occur by abstraction of a proton to form a carbanion intermediate), and the two reactions usually are considered simply as "racemization" in the geochemical literature. The relationship between enantiomers (optical isomers) and epimers (diastereomers) is illustrated for isoleucine in Fig. 5. Both L- and D-isoleucine and L- and D-alloisoleucine are enantiomeric pairs because the L- and D-forms are nonsuperimposable mirror images. Isoleucine and alloisoleucine are epimers.

Most research on racemization of amino acids in mollusk shells has focused on epimerization of isoleucine and racemization of leucine and aspartic acid. (In this chapter the term "epimerization" is used

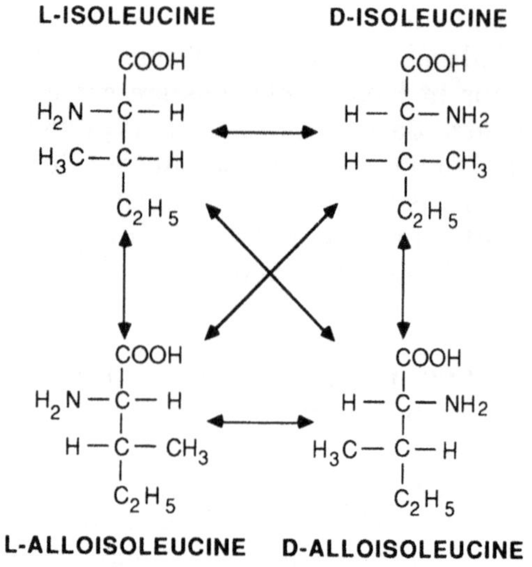

Figure 5. Stereochemical configurations of isoleucine and alloisoleucine (Fischer projections). Horizontal arrows represent interconversion of enantiomers (optical isomers); diagonal arrows represent interconversion of epimers (diastereomers). The major reaction pathway of L-isoleucine is to D-alloisoleucine. [After Kvenvolden (1975).]

in reference to interconversion of L-isoleucine and D-alloisoleucine specifically, while the term "racemization" is used in a generic sense to refer to interconversion of L- and D-isomers of an unspecified amino acid or of amino acids in general.) Because epimers have slightly different physical and chemical properties, they are more easily separated by standard high-pressure liquid-chromatographic (HPLC) techniques. Interconversion of the epimers L-isoleucine and D-alloisoleucine is a reversible reaction analogous to the reaction in Eq. (1):

$$\text{L-isoleucine} \underset{k_2}{\overset{k_1}{\rightleftarrows}} \text{D-alloisoleucine} \qquad (2)$$

Biosynthetic pathways, which are enzymatically stereospecific, lead to proteins containing amino acids almost entirely in the L-configuration. This stereochemical uniformity of L-amino acids in proteins is necessary for these biopolymers to maintain an ordered structure and to fulfill metabolic functions. D-Amino acids, however, are present in some peptides, in bacterial cell membranes, and as free amino acids in animal fluids and tissues (Corrigan, 1969). After death of the organism, or after formation of shell proteins and other inactive proteins that are not metabolically replaced during life, the thermodynamically unstable assemblage of optically active L-amino acid molecules spontaneously and gradually converts to an equilibrium mixture of L- and D-isomers in accordance with the reactions in Eqs. (1) and (2). For enantiomeric forms such as L- and D-leucine, $k_1 = k_2$, and the relative amount of D-amino acid gradually increases from zero in newly synthesized proteins of organisms to a 1:1 mixture with the L-enantiomer in fossil shells (i.e., D/L increases from 0 to 1). Unlike enantiomers, non-mirror-image epimers such as L-isoleucine and D-alloisoleucine do not have identical stabilities ($k_1 \neq k_2$), and their equilibrium D/L values may differ from unity. The equilibrium ratio of D-alloisoleucine to L-isoleucine (A/I), as measured in a fossil bivalve (Miocene *Mercenaria* sp.), a heated modern shell, and heated aqueous solutions of free isoleucine, is about 1.25–1.3 (Hare and Mitterer, 1967, 1969; Nakaparksin *et al.*, 1970).

Neuberger (1948) elucidated the current view of the mechanism of racemization of amino acids in aqueous solution. Bada (1971, 1972b) expanded on the mechanism of racemization of free amino acids at various pH values. Bada and Schroeder (1975), Schroeder and Bada (1976), Williams and Smith (1977), Smith and Evans (1980), and Bada (1982) have more recently discussed the racemization mechanism. Racemization of optical isomers of amino acids involves abstraction by base of the proton attached to the chiral carbon, forming an intermediate planar carbanion, followed by reattachment of a proton to either side of the planar configuration, forming the D- or L-enantiomer with equal probability (Fig. 6). The stability of the carbanion intermediate depends on the electrostatic influences of the groups attached to the α-carbon. Substituents that stabilize the charge on the carbanion intermediate enhance the rate of racemization, while groups that destabilize the charge retard the rate of racemization. At the slightly alkaline pH that prevails in carbonate fossils, neutral amino acids exist in the protonated form and as the zwitterion. The protonated amine group ($—NH_3^+$) is electron-withdrawing, which tends to stabilize the carbanion on the adjacent carbon atom and enhances racemization. The negatively charged carboxylate group ($—COO^-$) is an electron donor, which tends to destabilize the carbanion on the adjacent carbon atom and inhibits racemization. Substituents attached to the amine and carboxylate groups modify their stabilizing and destabilizing capacities. In peptides and proteins the attached substituents are other amino acids. Consequently, in peptides the presence and nature of adjacent amino acids have a significant influence on the stability of the carbanion intermediate and on the rate of racemization of an individual amino acid.

By analogy with the mechanism for free amino

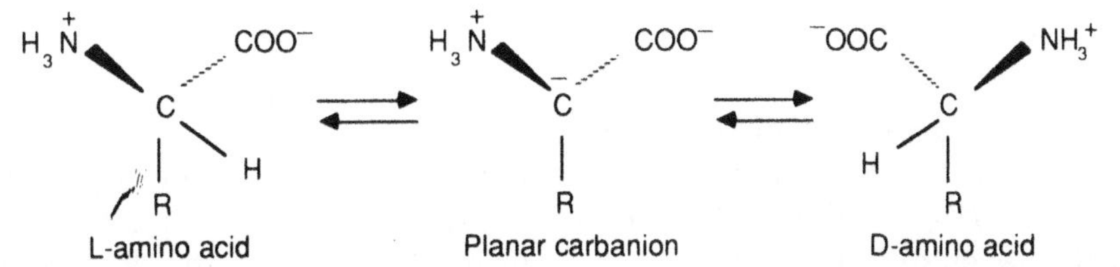

Figure 6. Mechanism of racemization of free and peptide-bound amino acids via a carbanion intermediate. The reaction proceeds by abstraction of the α-proton by base to form the planar carbanion; in this configuration readdition of a proton to form either the L-enantiomer or the D-enantiomer occurs with equal probability. [Based on Neuberger (1948) and after Bada and Schroeder (1975) and Smith and Evans (1980).]

acids, a tentative mechanism can be proposed for racemization of amino acids in various positional configurations in peptides (e.g., terminal, interior; Bada, 1982).

Individual amino acids in various peptides have different rates of racemization (e.g., Kriausakul and Mitterer, 1978; Smith and Sol, 1980; Mitterer and Kriausakul, 1984). These differences are consistent with observations of relative rates of amino acid racemization reactions with various degrees of modification of the amine and carboxyl groups and with variations in the ionic state of amino acids (Neuberger, 1948; Bada and Shou, 1980; Smith and Evans 1980; Bada, 1982). At a slightly alkaline pH, dipeptides and tripeptides, by analogy with free amino acids, are present as fully protonated species and as zwitterions. Racemization of amino acids at the C-terminal position (the end of a peptide molecule having the free carboxyl group) should be slower due to the presence of a negative charge on the free carboxyl group and neutralization of the electron-withdrawing protonated amine group by attachment to an adjacent amino acid (Mitterer and Kriausakul, 1984). Conversely, the presence of the protonated amine group at the N-terminal position (the other end of a peptide molecule, having the free amine group) and neutralization of the charge on the carboxyl group by attachment to an adjacent amino acid should enhance racemization at the N-terminal position. Overall, general theory predicts that the relative amino acid racemization rate in dipeptides is N-terminal > C-terminal (Smith and Sol, 1980). Experimental data for isoleucine, described in Section 6, are in agreement with this prediction.

5. Amino Acid Racemization in Fossil Shells

The increase in the proportion of D-amino acid to L-amino acid in fossil shells and bones of increasing age forms the basis of a relative and an absolute dating method (Wehmiller, this volume, Chapter 36). Because racemization is a chemical reaction, its rate is a function of temperature. Consequently, the relative amount of D-amino acid in a fossil is a measure not only of a fossil's age (i.e., the time since reaction began) but also of the temperature to which the fossil has been subjected since deposition (i.e., the rate of reaction). In practice it is generally assumed that all fossils of the same genus within a limited geographic region have experienced virtually the same thermal history and thus have identical rates of racemization for a particular amino acid. With this assumption, the extent of amino acid racemization in fossils of the same genus within a region is then related to their ages. Without some knowledge of temperature history, it is possible only to assign relative ages to fossils and to make stratigraphic correlations of deposits in which they are found. Absolute ages can be calculated from calibration samples of known age from the region of interest (Bada and Protsch, 1973) or from kinetic models (Wehmiller and Belknap, 1978).

Much of our knowledge of racemization reaction kinetics is based on detailed analyses of foraminifera from deep-sea sediment cores and on laboratory experimentation with heated shell fragments and peptide solutions. Deep-sea cores provide a continuous fossil record that has been incubated at low and nearly uniform temperatures in a stable environment. In laboratory experiments with heated shells and peptide solutions, the entire progress of the reaction can be observed within a few days or weeks under a variety of conditions and with various reactants. The assumption with this approach is that the same types of reactions observed in the laboratory at high temperatures over short time periods operate at lower *in situ* temperatures in fossils over long time periods. Good agreement between laboratory results and analyses of fossils gives confidence that this assumption is reasonably justified, at least qualitatively. Analysis of a suite of radiocarbon-dated Holocene shells has recently provided detailed information on the kinetics of racemization of seven amino acids based on land snail shells with excellent age control (Goodfriend, 1991). The results demonstrate a strong correlation between radiocarbon age and extent of racemization for each of the seven amino acids.

5.1. Isoleucine Epimerization in Fossil and Heated Shells

Experiments designed to study racemization reactions that most nearly duplicate conditions in fossils involve the heating of modern shells under an inert atmosphere or in vacuum in sealed tubes in the presence of water. Water is necessary to permit hydrolytic reactions. Experimental results using fragments of a modern mollusk (*Mercenaria* sp.) are illustrated in Fig. 7a for isoleucine, the amino acid most commonly used in racemization/epimerization studies. The ratio of D-alloisoleucine to L-isoleucine (A/I), which represents the progress of the reaction, increases with heating time. When mostly L-isoleucine is present, reaction progress is initially rapid as the overall reaction rate is controlled by the rate of the forward reaction. Reaction progress slows as the concentration of the product, D-alloisoleucine, increases, thereby causing the rate of the reverse reaction to increase and the rate of the net reaction to decrease as equilibrium is approached (Eq. 2). At equilibrium, no further change in A/I occurs.

Racemization and epimerization are reversible first-order reactions, and the extent of reaction for isoleucine epimerization can be expressed in the first-order kinetic form [see Bada and Schroeder (1972) and Williams and Smith (1977) for derivations]:

$$\ln\left[\frac{1 + (A/I)}{1 - K'(A/I)}\right] - \text{constant} = (1 + K')\cdot k_1 t \quad (3)$$

where $K' = 1/K_{eq}$; for $K_{eq} = 1.3$, $K' = 0.769$. The constant of integration is determined from A/I values in modern specimens. Results from fossil foraminifera in deep-sea cores and from heating experiments with modern shells establish that linear first-order kinetics are followed for only a portion of the reaction range. In heated *Mercenaria* and *Chione* shells, isoleucine epimerization follows reversible first-order kinetics up to an A/I value of about 0.6 (Masters and Bada, 1977; Kriausakul and Mitterer, 1980b) (Fig. 7b). At higher ratios the reaction rate slows considerably and deviates from the reversible first-order rate law. The overall rate curve for isoleucine epimerization in *Mercenaria* sp. can be interpreted as consisting of two sequential linear segments reflecting an initial rapid rate of epimerization followed by a much slower rate (Kriausakul and Mitterer, 1980b).

Foraminifera from deep-sea cores show a similar general pattern of isoleucine epimerization to that observed for heated *Mercenaria* sp. (Wehmiller and Hare, 1971; Bada and Schroeder, 1972; King and Hare, 1972; King and Neville, 1977; Bada and Man, 1980; Müller, 1984). Isoleucine epimerization in foraminifera can also be interpreted in terms of two sequential linear rate segments (Müller, 1984). The epimerization rate is rapid in the first portion but decreases by about an order of magnitude beyond an A/I value of about 0.5 to 0.6 (Fig. 8a). The first portion of the rate curve, examined in detail, consists of two or three shorter linear segments of decreasing rates with increasing age, indicating that isoleucine epimerization in foraminifera does not follow first-order kinetics beyond A/I values of about 0.30 (Fig. 8b; Bada and

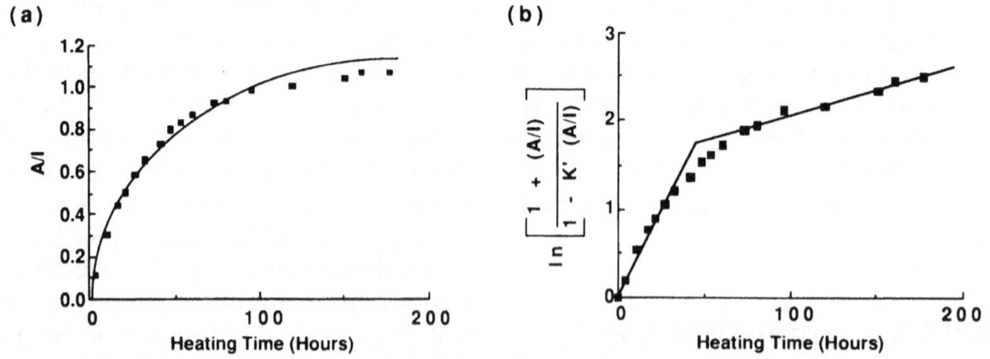

Figure 7. (a) Extent of isoleucine epimerization, plotted as the ratio of D-alloisoleucine to L-isoleucine (A/I), in the modern bivalve *Mercenaria sp.* heated at 152°C in sealed tubes with 1 ml of water. The line is a parabolic curve fitted to the data empirically. (b) Data from (a) plotted according to the first-order reversible rate law. The reaction kinetics are modeled as two sequential first-order reversible reactions with different rates (straight lines).

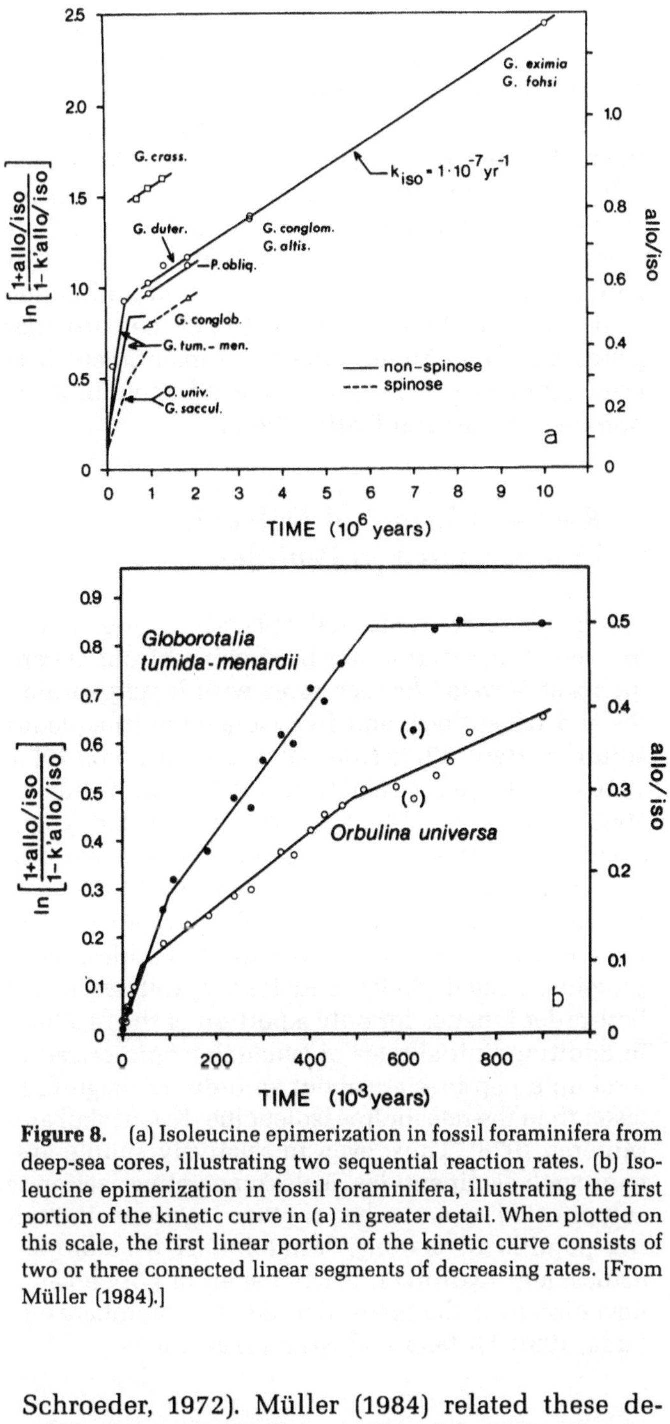

Figure 8. (a) Isoleucine epimerization in fossil foraminifera from deep-sea cores, illustrating two sequential reaction rates. (b) Isoleucine epimerization in fossil foraminifera, illustrating the first portion of the kinetic curve in (a) in greater detail. When plotted on this scale, the first linear portion of the kinetic curve consists of two or three connected linear segments of decreasing rates. [From Müller (1984).]

Schroeder, 1972). Müller (1984) related these decreases in rate in the first portion of the epimerization curve to leaching of free amino acids and to a decreasing rate of hydrolysis of the protein. In addition, as observed for mollusks (Wehmiller, 1980), rate curves for isoleucine epimerization in foraminifera are genus specific (King and Neville, 1977; Müller, 1984).

Comparable decreases in the rate of isoleucine epimerization are not seen in heating experiments with mollusk shells until A/I values of about 0.6 are reached, probably because of the more rapid reaction rates at the much higher temperatures used in laboratory experiments. Well-dated fossils are not available to give precise details of isoleucine epimerization kinetics in mollusks at earth-surface temperatures beyond an A/I value of about 0.3 (Goodfriend, 1991).

Not all experimental results conform to the patterns depicted in Figs. 7 and 8. Kimber *et al.* (1986) and Kimber and Griffin (1987) examined the racemization trend for several amino acids in heated bivalves of two genera. While some amino acids exhibited the expected pattern of an initial rapid increase in D/L values followed by a slower increase, other amino acids, including isoleucine, displayed the opposite pattern (i.e., an initial slow increase in D/L followed by a more rapid increase). Two amino acids, however, showed a significant reversal in D/L values at longer heating times. These results indicate that not all genera follow the expected racemization trend, and therefore some genera may not be suitable specimens for use in estimation of racemization ages (Kimber *et al.*, 1986; Kimber and Griffin, 1987).

5.2. Extent of Racemization in Different Amino Acid Fractions

Earlier studies showed that significant differences exist in the extent of isoleucine epimerization in various amino acid fractions of fossil shells and foraminifera (Hare and Mitterer, 1967; Wehmiller and Hare, 1971). Subsequent analyses by many investigators have confirmed these results. In Fig. 9, from Müller (1984), for example, the extent of epimerization of total isoleucine is compared with the extent of isoleucine epimerization in the bound and free fractions from foraminifera in a deep-sea core. Free isoleucine is more extensively epimerized than bound isoleucine. Similar results are obtained from fossil *Mercenaria*; isoleucine in the free fraction, as well as that in small peptides and in the C-terminal position, is more extensively epimerized than isoleucine in large peptides (Fig. 10) (Hare and Mitterer, 1967; Hare, 1976; Kriausakul and Mitterer, 1980a, 1983). In aqueous solution, however, free isoleucine epimerizes at a very slow rate compared to rates in peptide-bound isoleucine (Hare and Mitterer, 1969; Hare, 1971; Kriausakul and Mitterer, 1978). The differences in extent of isoleucine epimerization in various fractions in fossils, coupled with results of laboratory investigations described below, provide important clues to an understanding of some of the factors affecting rates of epimerization.

Several amino acids in the higher molecular

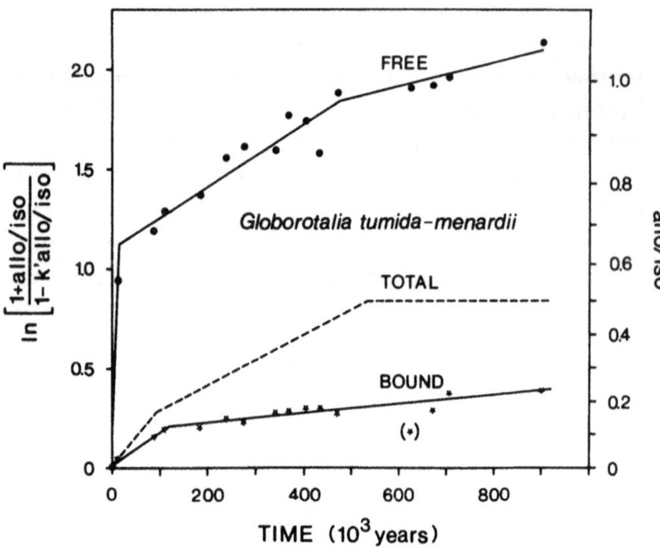

Figure 9. Extent of epimerization of total isoleucine compared to the extent of epimerization of bound and free isoleucine in foraminiferal tests from a deep-sea core. [From Müller (1984).]

weight (>1000 daltons) fraction from bivalves studied by Kimber and Griffin (1987) display reversals in D/L values at longer heating times. Evidently, a portion of the larger peptides in these shells is more resistant to hydrolysis, and racemization proceeds more slowly in this fraction because many of the amino acids remain in more slowly racemizing internal positions (Kriausakul and Mitterer, 1978, 1983). After the more labile peptide bonds, and more highly racemized amino acids, are hydrolyzed and lost from the higher molecular weight fraction, the stable peptide core is composed of amino acids with lower D/L values. Thus, this larger peptide fraction exhibits a reversal in D/L values at longer heating times. This pattern may be due to the composition and amino acid sequence of the organic matrix of the particular genus studied, although there are insufficient data from different genera to evaluate this possibility at present (Kimber and Griffin, 1987).

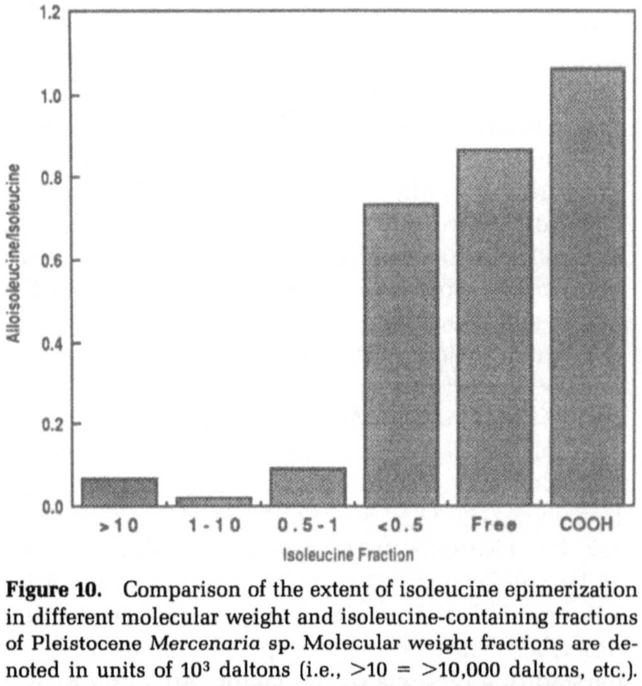

Figure 10. Comparison of the extent of isoleucine epimerization in different molecular weight and isoleucine-containing fractions of Pleistocene *Mercenaria* sp. Molecular weight fractions are denoted in units of 10^3 daltons (i.e., >10 = >10,000 daltons, etc.). Free: Free isoleucine; COOH: C-terminal isoleucine. Data were obtained from two different specimens with total A/I = 0.49 and 0.56. [Modified from Kriausakul and Mitterer (1983).]

6. Rate and Extent of Isoleucine Epimerization in Peptides

Insight into positional effects on rates of isoleucine epimerization has been gained from experiments at elevated temperatures with large peptides, di- and tripeptides, and free isoleucine in aqueous solution (Hare, 1971; Kriausakul and Mitterer, 1978, 1980a, b, 1983; Smith and Evans, 1980; Smith and Sol, 1980; Steinberg and Bada, 1981; Mitterer and Kriausakul, 1984). In aqueous solution, epimerization of free isoleucine obeys the reversible first-order rate law in accordance with Eq. (3). By comparison, isoleucine epimerization in di- and tripeptides, as in proteins, heated shells, and fossils, follows linear first-order kinetics for only a portion of the reaction. In addition, initial rates of isoleucine epimerization in simple peptides are about an order of magnitude faster than the rate for free isoleucine (Kriausakul and Mitterer, 1978). Thus, even in relatively simple systems such as dipeptides, isoleucine epimerization is complicated by a number of other reactions, including peptide inversions, formation of diketopiperazines, and hydrolytic reactions, all of which cause deviation from the first-order rate law (Steinberg and Bada, 1981; Mitterer and Kriausakul, 1984).

6.1. Products of Heated Peptide Solutions

In elevated temperature experiments with a typical dipeptide [e.g., isoleucylglycine (ile-gly)], isoleucine species present or formed during the experiment are: (1) the remainder of the original dipeptide, (2) diketopiperazine (DKP), a cyclic dipeptide formed by intramolecular reaction, (3) inverted dipeptide (gly-ile), and (4) free isoleucine (Steinberg and Bada, 1981; Mitterer and Kriausakul, 1984). An equilibrium relationship exists between original and inverted dipeptides and DKP:

$$\text{ile-gly} \rightleftarrows \text{DKP} \rightleftarrows \text{gly-ile} \quad (4)$$

A typical dipeptide molecule in this experiment rapidly cyclizes to form DKP. Subsequently, DKP partially hydrolyzes to either the original dipeptide or its inverted form depending on which peptide bond is broken. A molecule may toggle back and forth between the two dipeptide configurations through the DKP intermediate. If no other reactions occurred, equilibrium concentrations eventually would be established among the three species. Equilibrium is not attained because peptide concentrations decrease due to their eventual hydrolysis to free amino acids.

Heating of tripeptides produces a more complex suite of products than occurs with heated dipeptides. Isoleucine-containing products present or formed during heating experiments with the tripeptide ile-gly-gly are: (1) the remainder of the original tripeptide (ile-gly-gly), (2) a dipeptide (ile-gly) formed by partial hydrolysis of the tripeptide, (3) DKP, (4) inverted dipeptide (gly-ile), and (5) free isoleucine (Mitterer and Kriausakul, 1984). After formation by partial hydrolysis, the dipeptide ile-gly undergoes intramolecular reaction to form DKP. Partial hydrolysis of DKP leads to regeneration of the dipeptide (ile-gly) or its inverted form (gly-ile). The major isoleucine-containing species present after short heating times are the original tripeptide and DKP. With longer heating times, the proportion of DKP and dipeptide products increases and the relative concentration of tripeptide decreases due to gradual hydrolysis. Eventually, all peptides are completely hydrolyzed to free amino acids. The sequence of reactions for the isoleucine-containing molecules is illustrated as (Mitterer and Kriausakul, 1984):

$$\text{ile-gly-gly} \rightarrow \{\text{ile-gly} \rightleftarrows \text{DKP} \rightleftarrows \text{gly-ile}\} \rightarrow \text{ile} \quad (5)$$

Steinberg and Bada (1983) have shown that diketopiperazines are also major products from the heating of higher molecular weight tri- and hexapeptides.

6.2. Isoleucine Epimerization in Heated Proteins, Peptides, and Peptide Products

Epimerization of isoleucine occurs before and after hydrolysis and at widely different rates in solutions of heated proteins, peptides, and peptide products depending on positional configuration (Kriausakul and Mitterer, 1978; Steinberg and Bada, 1981; Mitterer and Kriausakul, 1984). Highest rates of epimerization occur in the N-terminal position and in DKP. Rates of epimerization in the C-terminal position and in interior positions of peptides and proteins are very slow and are comparable to the rate of epimerization of free isoleucine in aqueous solution (Table I).

Despite a slow rate of epimerization, the extent of epimerization observed for C-terminal and free isoleucine is relatively high (Kriausakul and Mitterer, 1983). This is because some of the more highly epimerized isoleucine is transferred from faster to slower epimerizing forms as a consequence of the equilibrium relationships among initial dipeptides, inverted dipeptides, and DKP. Inverted dipeptides, for example, can only form by partial hydrolysis of DKP, and the extent of isoleucine epimerization in the inverted state is affected by prior epimerization both in DKP and in the original dipeptide. Partial hydrolysis of DKP can also produce more highly epimerized isoleucine in the original dipeptide as well. In fact, the high rate of epimerization in DKP enhances the extent of isoleucine epimerization in all dipeptides and dipeptide products. Similarly, prior isoleucine epimerization in peptides enhances the extent of epimerization of isoleucine in DKP.

The rate of hydrolytic release of isoleucine plays an important role in the kinetics of total isoleucine epimerization (Kriausakul and Mitterer, 1978; Wehmiller, 1980). The proportion of isoleucine in various positions and the location of the transition zone between the two linear segments of the rate curve (e.g., Fig. 7b) are controlled by the rate of formation of free isoleucine. If hydrolysis of a protein to free amino acids occurs very rapidly compared to the rate of epimerization in the N-terminal position and in DKP, the total isoleucine epimerization rate curve would be comparable to that of free isoleucine. Conversely, if proteins experience a very slow rate of hydrolysis, the initial isoleucine epimerization rate curve would also be comparable to that of free isoleucine due to slow epimerization occurring in interior positions.

TABLE I. Rate and Extent of Epimerization of Isoleucine in Different Positional Configurations

Positional configuration	Rate of ile epimerization	Extent of ile epimerization
Interior	Slow	Low
C-terminal	Slow	High
N-terminal	Fast	High
Diketopiperazine	Fast	High
Free	Slow	High

Neither of these extremes occurs because the activation energy for hydrolysis of dipeptide-bound isoleucine is only slightly less (by about 9 kJ) than the activation energy for isoleucine epimerization in dipeptides (Kriausakul and Mitterer, 1980a). In actuality, therefore, hydrolysis occurs at a sufficient rate to allow transfer of interior isoleucine to terminal positions but not so fast as to convert all isoleucine rapidly to the free state. Thus, an isoleucine molecule occupies a terminal position long enough for epimerization to occur before it is released to the free state. In summary, results of heating experiments with solutions of isoleucine-containing peptides and proteins lead to the following conclusions: (1) reversible transfer of isoleucine residues in dipeptides between N- and C-terminal positions through the DKP intermediate (Eq. 4) has a cumulative effect on the extent of isoleucine epimerization observed in each position; (2) hydrolytic transfer of isoleucine from faster to slower epimerizing positions imparts a high degree of epimerization to the product; (3) the high degree of epimerization exhibited by C-terminal isoleucine in dipeptides and by free isoleucine is due to prior epimerization in faster epimerizing positions (i.e., DKP and N-terminal isoleucine in dipeptides) and subsequent inversion to the C-terminal position or hydrolysis to free isoleucine.

Based on experimental results with proteins and peptides and on analyses of fossils, it is clear that the relative rate of isoleucine epimerization differs from the relative extent of epimerization in the five different positional configurations (Mitterer and Kriausakul, 1984). Relative rates of isoleucine epimerization are (Table I)

$$\{\text{N-terminal, DKP}\} \gg \{\text{C-terminal, interior, free}\}$$

while relative extents of isoleucine epimerization are

$$\text{DKP} > \text{N-terminal} > \text{C-terminal} > \text{free} \gg \text{interior}$$

7. A Model for Isoleucine Epimerization in Proteins

Mitterer and Kriausakul (1984) proposed a general model for isoleucine epimerization reactions in protein systems undergoing hydrolysis (Fig. 11). The following discussion is based on this model. Epimerization of isoleucine in each of its five positional configurations is a reversible reaction that obeys the first-order rate law. Rates of epimerization in individual positions differ, however, with isoleucine in the N-terminal position and in cyclic dipeptides epimerizing relatively rapidly while isoleucine in interior and C-terminal positions and free isoleucine epimerize more slowly. Initially, in a protein or polypeptide, isoleucine may exist entirely as an internal amino acid. Subsequent partial hydrolysis of the protein or polypeptide creates a mixture of products with isoleucine present in up to five different positional configurations, in each of which isoleucine epimerizes at different rates. As hydrolysis proceeds, the proportion of isoleucine in the various positions continually changes. In such a mixture the rate of total isoleucine epimerization deviates from the first-order rate law as new hydrolytic products, each with different epimerization rates, form. Only very slow epimerization occurs while isoleucine is in interior and C-terminal positions, whereas rapid epimerization occurs in the N-terminal position. As dipeptides are produced, they readily convert to cyclic diketopiperazines, in which further rapid epimerization occurs. Hydrolysis of highly epimerized terminal isoleucine yields highly epimerized free isoleucine. As the rate of epimerization of free isoleucine is very slow, continued epimerization of isoleucine in the free amino acid fraction occurs at a greatly reduced rate.

The rate of production of terminal isoleucine during partial hydrolysis depends on the strength of

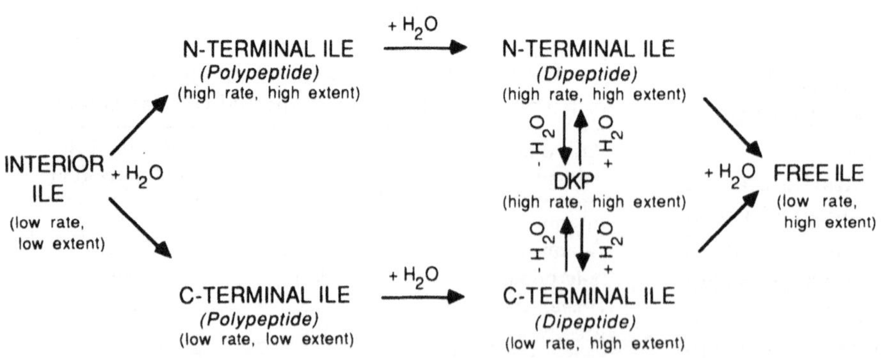

Figure 11. General model to explain the differences in rate and extent of isoleucine epimerization during hydrolysis of a protein. [Modified from Mitterer and Kriausakul (1984).]

isoleucine peptide bonds (Hill, 1965; Hare, 1976; Kriausakul and Mitterer, 1978). Some isoleucine peptide bonds are more labile than others, depending on the nature of the adjacent amino acid. Wehmiller (1980) and Müller (1984) showed that differences in rates of release of isoleucine from proteins can explain the different total isoleucine epimerization rates found for different genera of mollusks and foraminifera. Thus, differences in rate of production of terminal and free isoleucine probably account for the separate epimerization rate curves reported for taxonomically distinct organisms (e.g., the "generic effect") (King and Neville, 1977; Lajoie *et al.*, 1980; Wehmiller, 1980; Müller, 1984).

The model in Figure 11 can be invoked to explain the nonlinear kinetics of isoleucine epimerization in fossil shells (Kriausakul and Mitterer, 1980b; Mitterer and Kriausakul, 1984). The total isoleucine epimerization reaction in fossils can be interpreted as two consecutive first-order reactions, with the second having a rate about an order of magnitude slower than the first (Fig. 7b). Hydrolytic reactions rapidly convert the original high-molecular-weight protein into a mixture of smaller peptides in which some isoleucine is in the N-terminal position and in DKP. These smaller peptides hydrolyze more slowly. The rate-controlling step of the first linear portion of the kinetic curve is the rate of isoleucine epimerization in the N-terminal position and in DKP. With further hydrolysis, the proportion of slowly epimerizing free isoleucine increases and the overall rate of epimerization decreases. When free isoleucine is predominant or when all bound isoleucine is fully epimerized, the rate-controlling step is the rate of epimerization of free isoleucine. This is the second linear portion of the rate curve in Fig. 7b.

8. Summary

Leaching is the most important physical process affecting the amino acid concentration in fossil shells. Although the rate of leaching is a complex function of many variables, the eventual consequence is loss of original amino acids in older fossils. Hydrolysis, racemization, and defunctionalization are the most important chemical reactions affecting proteins in shells. Racemization rate is primarily a function of diagenetic temperature. Because of hydrolytic reactions, amino acid positional configurations change during diagenesis, leading to the complex kinetics of racemization seen in fossil shells. Isoleucine epimerization in shells can be modeled as two sequential, reversible, first-order reactions. The first linear portion of the rate curve represents faster epimerization in the N-terminal position and in diketopiperazines; the second linear portion is controlled by the slower rate of epimerization of free isoleucine.

ACKNOWLEDGMENTS. I thank G. A. Goodfriend for his review of an earlier draft. Some of the research described herein and preparation of this manuscript were supported by NSF grants EAR-8508298 and EAR-8816391. This is contribution 634 of the Department of Geosciences, University of Texas at Dallas.

References

Abelson, P. H., 1954, Organic constituents of fossils, *Carnegie Inst. Washington Yearb.* **53**:97–101.

Abelson, P. H., 1959, Geochemistry of organic substances, in: *Researches in Geochemistry* (P. H. Abelson, ed.), John Wiley & Sons, New York, pp. 79–103.

Bada, J. L., 1971, Kinetics of the non-biological decomposition and racemization of amino acids in natural waters, in: *Nonequilibrium Systems in Natural Water Chemistry* (J. D. Hem, ed.), ACS Advances in Chemistry Series No. 106, American Chemical Society, Washington, D.C., pp. 309–311.

Bada, J. L., 1972a, The dating of fossil bones using the racemization of isoleucine, *Earth Planet. Sci. Lett.* **15**:223–231.

Bada, J. L., 1972b, Kinetics of racemization of amino acids as a function of pH, *J. Am. Chem. Soc.* **95**:1371–1373.

Bada, J. L., 1982, Racemization of amino acids in nature, *Interdiscip. Sci. Rev.* **7**:30–46.

Bada, J. L., 1985, Amino acid racemization dating of fossil bones, *Annu. Rev. Earth Planet. Sci.* **13**:241–268.

Bada, J. L., and Man, E. H., 1980, Amino acid diagenesis in Deep Sea Drilling Project cores: Kinetics and mechanisms of some reactions and their applications in geochronology and in paleotemperature and heat flow determinations, *Earth Sci. Rev.* **16**:21–55.

Bada, J. L., and Protsch, R., 1973, Racemization reaction of aspartic acid and its use in dating fossil bones, *Proc. Natl. Acad. Sci. U.S.A.* **70**:1331–1334.

Bada, J. L., and Schroeder, R. A., 1972, Racemization of isoleucine in calcareous marine sediments: Kinetics and mechanism, *Earth Planet. Sci. Lett.* **15**:1–11.

Bada, J. L., and Schroeder, R. A., 1975, Amino acid racemization reactions and their geochemical implications, *Naturwissenschaften* **62**:71–79.

Bada, J. L., and Shou, M.-Y., 1980, Kinetics and mechanisms of amino acid racemization in aqueous solution and in bones, in: *Biogeochemistry of Amino Acids* (P. E. Hare, T. C. Hoering, and K. King, Jr., eds.), John Wiley & Sons, New York, pp. 235–255.

Bada, J. L., Protsch, R., and Schroeder, R. A., 1973, The racemization reaction of isoleucine used as a paleotemperature indicator, *Nature* **241**:394–395.

Corrigan, J. J., 1969, D-Amino acids in animals, *Science* **164**:142–149.

Goodfriend, G. A., 1991, Patterns of racemization and epimerization of amino acids in land snail shells over the course of the Holocene, *Geochim. Cosmochim. Acta* **55**:293–302.

Hamilton, P. B., 1965, Amino acids on hands, *Nature* **205**:284–285.

Hare, P. E., 1965, Amino acid artifacts in organic geochemistry, *Carnegie Inst. Washington Yearb.* **64**:232–235.

Hare, P. E., 1969, Geochemistry of proteins, peptides, and amino acids, in: *Organic Geochemistry—Methods and Results* (G. Eglinton and M. T. J. Murphy, eds.), Springer-Verlag, Berlin, pp. 438–463.

Hare, P. E., 1971, Effect of hydrolysis on the racemization rate of amino acids, *Carnegie Inst. Washington Yearb.* **70**:256–258.

Hare, P. E., 1976, Relative reaction rates and activation energies for some amino acid reactions, *Carnegie Inst. Washington Yearb.* **75**:801–806.

Hare, P. E., and Abelson, P. H., 1968, Racemization of amino acids in fossil shells, *Carnegie Inst. Washington Yearb.* **66**:526–528.

Hare, P. E., and Hoering, T. C., 1977, The organic constituents of fossil mollusc shells, *Carnegie Inst. Washington Yearb.* **76**: 625–631.

Hare, P. E., and Mitterer, R. M., 1967, Nonprotein amino acids in fossil shells, *Carnegie Inst. Washington Yearb.* **65**:362–364.

Hare, P. E., and Mitterer, R. M., 1969, Laboratory simulation of amino acid diagenesis in fossils, *Carnegie Inst. Washington Yearb.* **67**:205–210.

Hare, P. E., Miller, G. H., and Tuross, N. C., 1975, Simulation of natural hydrolysis of proteins in fossils, *Carnegie Inst. Washington Yearb.* **74**:609–612.

Hare, P. E., Hoering, T. C., and King, K., Jr. (eds.), 1980, *Biogeochemistry of Amino Acids*, John Wiley & Sons, New York.

Hill, R. L., 1965, Hydrolysis of proteins, *Adv. Protein Chem.* **20**: 37–107.

Hoering, T. C., 1980, The organic constituents of fossil mollusc shells, in: *Biogeochemistry of Amino Acids* (P. E. Hare T. C. Hoering, and K. King, Jr.), John Wiley & Sons, New York, pp. 193–201.

Kimber, R. W. L., and Griffin, C. V., 1987, Further evidence of the complexity of the racemization process in fossil shells with implications for amino acid racemization dating, *Geochim. Cosmochim. Acta* **51**:839–846.

Kimber, R. W. L., Griffin, C. V., and Milnes, A. R., 1986, Amino acid racemization dating: Evidence of apparent reversal in racemization with time of shells of *Ostrea*, *Geochim. Cosmochim. Acta* **50**:1159–1161.

King, K., Jr., and Hare, P. E.,1972, Species effects in the epimerization of L-isoleucine in fossil plankton foraminifera, *Carnegie Inst. Washington Yearb.* **71**:596–598.

King, K., Jr., and Neville, C., 1977, Isoleucine epimerization for dating marine sediments: Importance of analyzing monospecific foraminiferal samples, *Science* **195**:1333–1335.

Kriausakul, N., and Mitterer, R. M., 1978, Isoleucine epimerization in peptides and proteins: Kinetic factors and application to fossil proteins, *Science* **201**:1011–1014.

Kriausakul, N., and Mitterer, R. M., 1980a, Comparison of isoleucine epimerization in a model dipeptide and fossil protein, *Geochim. Cosmochim. Acta* **44**:753–758.

Kriausakul, N., and Mitterer, R. M., 1980b, Some factors affecting the epimerization of isoleucine in peptides and proteins, in: *Biogeochemistry of Amino Acids* (P. E. Hare, T. C. Hoering, and K. King, Jr., eds.), John Wiley & Sons, New York, pp. 283–296.

Kriausakul, N., and Mitterer, R. M., 1983, Epimerization of COOH-terminal isoleucine in fossil dipeptides, *Geochim. Cosmochim. Acta* **47**:963–966.

Kvenvolden, K. A., 1975, Advances in the geochemistry of amino acids, *Annu. Rev. Earth Planet. Sci.* **3**:183–212.

Lajoie, K. R., Wehmiller, J. F., and Kennedy, G. L., 1980, Inter- and intrageneric trends in apparent racemization kinetics of amino acids in Quaternary mollusks, in: *Biogeochemistry of Amino Acids* (P. E. Hare, T. C. Hoering, and K. King, Jr., eds.), John Wiley & Sons, New York, pp. 305–340.

Masters, P. M., and Bada, J. L., 1977, Racemization of isoleucine in fossil mollusks from Indian middens and interglacial terraces in southern California, *Earth Planet. Sci. Lett.* **37**:173–183.

Mitterer, R. M., 1972, Calcified proteins in the sedimentary environment, in: *Advances in Organic Geochemistry* (H. R. van Gaertner and H. Wehner, eds.), Pergamon Press, Oxford, pp. 441–452.

Mitterer, R. M., and Kriausakul, N., 1984, Comparison of rates and degrees of isoleucine epimerization in dipeptides and tripeptides, *Org. Geochem.* **7**:91–98.

Müller, P. J., 1984, Isoleucine epimerization in Quaternary planktonic foraminifera: Effects of diagenetic hydrolysis and leaching, and Atlantic–Pacific intercore correlations, *"Meteor" Forschungsergeb. Reihe C* **38**:25–47.

Nakaparksin, S., Gil-Av, E., and Oró, J., 1970, Study of the racemization of some neutral alpha-amino acids in acid solution using gas chromatographic techniques. *Anal. Biochem.* **33**:374–382.

Neuberger, A., 1948, Stereochemistry of amino acids, *Adv. Protein Chem.* **4**:297–383.

Philippi, G. T., 1977, Proteins as a possible source material of low molecular weight petroleum hydrocarbons, *Geochim. Cosmochim. Acta* **41**:1083–1086.

Schroeder, R. A., and Bada, J. L., 1973, Glacial–postglacial temperature difference deduced from aspartic acid racemization in fossil bones, *Science* **182**:479–482.

Schroeder, R. A., and Bada, J. L., 1976, A review of the geochemical applications of the amino acid racemization reaction, *Earth Sci. Rev.* **12**:347–391.

Smith, G. G., and Evans, R. C., 1980, The effect of structure and conditions on the rate of racemization of free and bound amino acids, in: *Biogeochemistry of Amino Acids* (P. E. Hare, T. C. Hoering, and K. King, Jr., eds.) John Wiley & Sons, New York, pp. 257–282.

Smith, G. G., and Sol, B. S., 1980, Racemization of amino acids in dipeptides shows COOH $>$ NH_2 for nonsterically hindered residues, *Science* **207**:765–767.

Steinberg, S. M., and Bada, J. L., 1981, Diketopiperazine formation during investigations of amino acid racemization in dipeptides, *Science* **213**:544–545.

Steinberg, S. M., and Bada, J. L., 1983, Peptide decomposition in the neutral pH region via the formation of diketopiperazines, *J. Org. Chem.* **48**:2295–2298.

Vallentyne, J. R., 1964, Biogeochemistry of organic matter—II. Thermal reaction kinetics and transformation products of amino compounds. *Geochim. Cosmochim. Acta* **28**:157–188.

Wehmiller, J. F., 1980, Intergeneric differences in apparent racemization kinetics in mollusks and foraminifera: Implications for models of diagenetic racemization, in: *Biogeochemistry of Amino Acids* (P. E. Hare, T. C. Hoering, and K. King, Jr., eds.), John Wiley & Sons, New York, pp. 341–355.

Wehmiller, J. F., 1982, A review of amino acid racemization studies in Quaternary mollusks; stratigraphic and chronologic applications in coastal and interglacial sites, Pacific and Atlantic coasts, United States, United Kingdom, Baffin Island, and tropical islands, *Quat. Sci. Rev.* **1**:83–120.

Wehmiller, J. F., and Belknap, D. F., 1978, Alternative kinetic models for the interpretation of amino acid enantiomeric ra-

tios in Pleistocene mollusks: Examples from California, Washington and Florida, *Quat. Res.* **9**:330–348.
Wehmiller, J. F., and Hare, P. E., 1971, Racemization of amino acids in marine sediments, *Science* **173**:907–911.
Wehmiller, J. F., Hare, P. E., and Kujala, G. A., 1976, Amino acids in fossil corals: Racemization (epimerization) reactions and their implications for diagenetic models and geochronological studies, *Geochim. Cosmochim. Acta* **40**:763–776.
Williams, K. M., and Smith, G. G., 1977, A critical evaluation of the application of amino acid racemization to geochronology and geothermometry, *Origins Life* **8**:91–144.

Chapter 36

Applications of Organic Geochemistry for Quaternary Research

Aminostratigraphy and Aminochronology

JOHN F. WEHMILLER

1. Introduction

Stratigraphic correlation can be accomplished using a variety of physical, chemical, and paleontological methods. The success of different approaches depends upon the nature of the preserved geologic record, the details required by the stratigraphic analysis, and the resolving power of the technique(s) employed. Classical stratigraphic methods include lithostratigraphy and biostratigraphy, which employ properties of the rocks or fossils that usually remain unchanged since burial. Chemical stratigraphic methods rely upon properties that may or may not change during burial. Current uses of stable isotope data (e.g., carbon, oxygen, strontium) rely upon the assumption (usually testable) that paleochemical information is preserved since the time of burial. At the other extreme, however, unstable isotope data are employed as time-stratigraphic tools *because* of their known time-dependent properties. Because of their firmly calibrated radioactive half-lives, these techniques are considered as the most reliable correlation or numerical dating tools (Colman *et al.*, 1987), though their accuracy in specific cases must always be evaluated by consideration of possible diagenetic effects on the observed parent or daughter abundances. A third category of chemical stratigraphic tool is represented by those molecular chemical properties that do change with time but in a manner that, in principle, can be calibrated well enough to be useful for stratigraphic correlation and age estimation. The major characteristic of these methods, when compared to those involving radioactive decay, is that their reaction rates vary with temperature and often with time. Beyond this single distinguishing trait, many of the same tests or assumptions regarding the chemical integrity of isotopic parent–daughter pairs are required for reactant–product systems. The application of these molecular chemical stratigraphic tools combines litho- and chronostratigraphic strategies, each approach being modified somewhat to account for temperature factors, which are responsible for alteration of material properties at different rates.

In the field of Quaternary research, several chemical or isotopic chronostratigraphic techniques have been developed over the past 10 to 15 years. The

JOHN F. WEHMILLER • Department of Geology, University of Delaware, Newark, Delaware 19716.
Organic Geochemistry, edited by Michael H. Engel and Stephen A. Macko. Plenum Press, New York, 1993.

principles and characteristics of most of these methods are summarized in a number of recent publications (Mahaney, 1984; Rutter, 1985; Hurford *et al.*, 1986; Pierce, 1986; Colman *et al.*, 1987; Easterbrook, 1988). The nomenclature proposed by Colman *et al.* (1987), summarized in Table I, is a particularly useful approach for representing the qualitative and quantitative properties of various Quaternary dating methods.

Amino acid racemization (AAR) is a good example of a chemical dating method that has been used for both relative and calibrated age estimation (see Table I). Amino acids found in fossil mineralized skeletons represent the diagenetically altered remains of the original mineralization proteins. During the fossilization process, these proteins are hydrolyzed into lower molecular weight polypeptides and free amino acids. A labile fraction (perhaps as much as 50%) of the modern amino acid population is lost from the fossil during early diagenesis (10^0 to 10^4 y), but the portion that remains is usually quite stable. Racemization is the process of conversion of the "left-handed" (L-) amino acids commonly found in the preserved skeleton into a mixture of "right-handed" (D-) and L-amino acids. In most sample types that have been studied, the initial (zero-age) abundance of D-amino acids is negligible, so the D/L value of any amino acid can be expected to increase from 0.0 (living samples) to some equilibrium value (usually 1.0) in samples of "infinite" age. Both temperature and sample type affect the time to equilibrium, which can range from a few thousand to a few million years.

TABLE I. Classification of Quaternary Dating Methods[a]

Type of method	Example	Type of result[b]
Sidereal	Dendrochronology	*Numerical*
Isotopic	^{14}C, U-series	*Numerical*, calibrated
Radiogenic	Fission track	*Numerical*
	Electron spin resonance	*Calibrated*, numerical
	Thermoluminescence	*Calibrated*, numerical
Chemical or biological	Amino acid racemization	*Calibrated, relative*
	Obsidian hydration	*Calibrated, relative*
Geomorphic	Soil profiles	*Relative*
	Rock and mineral weathering	*Relative*
	Geomorphic position	*Relative*
Correlation	Lithostratigraphy	*Correlated*
	Tephrochronology	*Correlated*
	Paleomagnetism	*Correlated*
	Fossils	*Correlated*
	Stable isotopes	*Correlated*

[a] Modified from Colman *et al.* (1987).
[b] Italicized term represents most common type of result. Definitions [modified from Colman *et al.* (1987)] as follows: Numerical results produce quantitive age estimates (often cited as "absolute" ages); relative age methods establish an age sequence and perhaps also estimates of age differences; calibrated results are those that can produce a sequence of relative ages, with at least some results directly calibrated by a numerical method; correlated results are those that demonstrate equivalence to independently dated units.

For purposes of the discussion to follow, two approaches to the use of amino acid racemization will be referred to as *aminostratigraphy* [a term first proposed by Miller and Hare (1980)] and *aminochronology*. Aminostratigraphy is based on the simple arrangement of sites or samples into a relative age sequence based on observed clusters of D/L values (aminozones, or aminogroups). Aminochronology uses independent age control for one or more aminozones and kinetic modeling to estimate ages for samples in a particular aminostratigraphic region. In some cases, aminochronologic approaches produce what Colman *et al.* (1987) refer to as a numerical-age result, though the most common type of aminochronologic product is the calibrated-age result (see Table I). Because of various chemical and thermal factors that can affect amino acid diagenesis, aminochronologic results are best qualified as *age estimates* to avoid misrepresentation of the inherent uncertainties of the dating method. Often aminostratigraphic and aminochronologic approaches complement each other, because differences in D/L values can be used to estimate age differences between aminozones even if no independent age control is available for either of these zones.

The literature on the geochemistry and geological applications of racemization is extensive. A major book on the biogeochemistry of amino acids (Hare *et al.*, 1980) is a useful background for more recent references. Papers by Wehmiller (1982, 1984, 1986, 1990), Miller (1985), Miller and Mangerud (1985), Bowen *et al.* (1985), Hearty *et al.* (1986), Wehmiller *et al.* (1988a), Bowen and Sykes (1988), Miller and Brigham-Grette (1989), and Kaufman and Miller (1992) present new data or review developments in the field. These papers demonstrate that marine mollusks are the most commonly studied sample type. Many other fossil types have been studied, including bones, teeth, wood, corals, terrestrial mollusks, and foraminifera. Bada (1985a, b) reviewed the chronological applications of racemization for fossil bone and tooth samples from African and North American sites. Müller's (1984) study of foraminifera built upon earlier studies (Wehmiller and Hare, 1971; King and Neville, 1977) and serves as an excellent review of current models of diagenetic mechanisms (Mitterer and Kriausakul, 1984; Mitterer, this volume, Chapter 35) and taxonomic differences in racemization rates.

Many of the analytical methods frequently used for amino acid geochemical studies were reviewed by Hare *et al.* (1985) and Engel and Hare (1985).

2. Foraminiferal Aminostratigraphy and Aminochronology: Examples of the Relationships among Enantiomeric Ratios, Stratigraphic Position, and Taxonomy

Because of the many variables (particularly temperature and sample type) that affect the relationship between D/L values and the geologic age of a sample, it is generally accepted that the least complicated use of racemization is as a simple stratigraphic tool. This approach avoids any assumptions about racemization kinetics or temperature dependence, instead using D/L values measured in samples with similar or identical present temperatures to infer an apparent relative age sequence. In order to document the general trends that are expected in geological racemization, it is always useful to measure D/L values in samples where clear stratigraphic relationships can be recognized. Samples of this type serve as tests of the reliability of racemization as a relative dating tool and provide insight into the resolving power of different stratigraphic tools.

The aminostratigraphic results with the clearest demonstration of the relation between D/L values and stratigraphic age are those obtained from foraminifera from deep-sea sediment cores. The relatively stable thermal environment of these samples is critical for understanding of the actual kinetics of diagenetic racemization. Often these studies are conducted using sediment cores with well-defined radiometric chronologies, so the observed D/L values can be plotted as a function of time, thereby becoming calibrated racemization reactions. A good example of this approach is the study by Müller (1984), who reported the quantitative abundance of L-isoleucine and D-alloisoleucine (produced by the epimerization, or racemization about the α-carbon, of L-isoleucine) in the samples of the foraminiferal taxa *Orbulina universa* and *Globorotalia tumida-menardii* (complex) from a core representing a 900-ka sequence of marine sediments from the eastern North Atlantic ocean. Müller's data for the total sample hydrolysates of these two taxa are plotted in Fig. 1, with D/L values plotted against time, a known function of depth given the independent chronologic information for the core. Curves shown in Fig. 1 are logarithmic least-squares best fits to the data for each taxon. The general trends of these curves are similar to those observed in other studies of foraminifera samples from this range of ages (Wehmiller and Hare, 1971; King and Neville, 1977; Jansen and Sejrup, 1986).

The curves in Fig. 1 are also good examples of well-known taxonomic effects on observed racemization reactions. In every case, samples of *G. tumida*

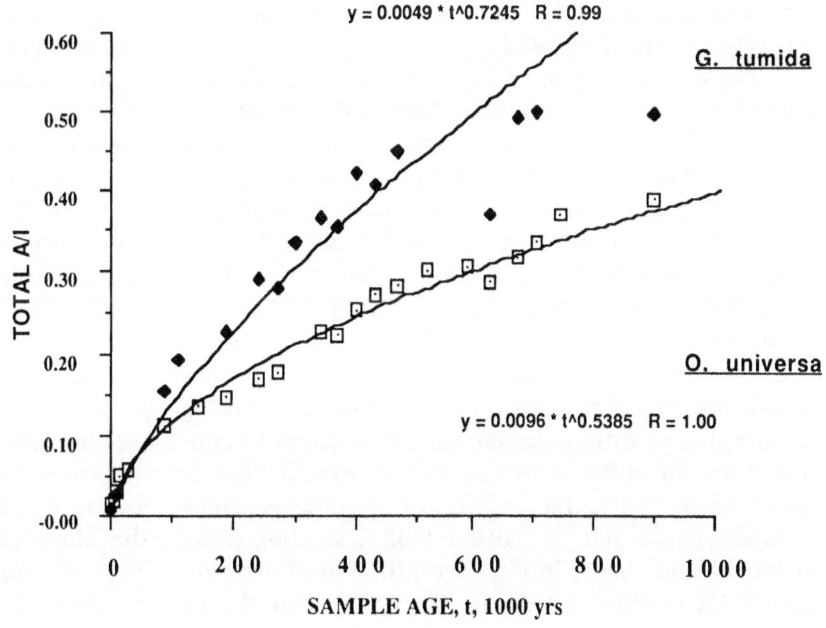

Figure 1. Epimerization of L-isoleucine (I) to D-alloisoleucine (A) in the total sample hydrolysates of *Orbulina universa* and *Globorotalia tumida* samples in eastern Atlantic Ocean core 13519-2 (Müller, 1984).

have higher D/L values than do coeval samples of *O. universa*. The observed differences in D/L values among different genera of samples are referred to as intergeneric differences in *apparent rates of racemization* (Lajoie *et al.*, 1980; Wehmiller, 1990) rather than as intergeneric differences in *racemization rates*, because it is clear that D/L values measured in the total amino acid population of any sample are the result of a complex combination of diagenetic reactions, each with independent rates that probably vary during diagenesis. Müller (1984) and Wehmiller (1990) discussed these intergeneric differences in terms of the diagenetic model proposed by Mitterer and Kriausakul (1984; Mitterer, this volume, Chapter 35). That model indicates that the position of peptide-bound amino acids changes during diagenetic hydrolysis, from interior to terminal positions, then to free amino acid configurations. The actual rates of racemization of terminally bound amino acids are relatively large (compared to the racemization rates of interior or free amino acids), but the extent of racemization in the total amino acid hydrolysate is a function of the relative amount of terminal amino acids and their residence time at that position. The commonly observed high degree of racemization of free amino acids in carbonate samples is apparently caused by the production of these free amino acids from fast-racemizing terminal amino acids, not because of an inherently rapid rate of racemization of free amino acids themselves (Mitterer and Kriausakul, 1984). In the case of the taxa shown in Fig. 1, the observed extents of racemization are consistent with measured extents of hydrolysis: the fast-racemizing *G. tumida* has a smaller proportion of its amino acids in bound form at any selected time than does *O. universa* (Müller, 1984; Wehmiller, 1990).

The curves shown in Fig. 1 demonstrate the expected relation between D/L values and sample age, but they also demonstrate the amount of scatter that can be observed in these trends. These deviations from the smooth trend of D/L versus time probably occur because of any or all of the following factors: analytical complications, sample contamination (lowering the D/L value from its true value) during diagenesis (or analysis), diagenetic alteration of the amino acid mixture (thereby affecting specific racemization rates), leaching of free amino acids, or natural mixing of samples of different ages during sediment accumulation. In spite of the scatter in results that is observed at particular depths for specific samples [usually explained by Müller (1984) as diagenetic artifacts], the curves in Fig. 1 could be used as independently calibrated tools for the estimation of ages of samples from other cores where no independent chronologic information exists.

Intergeneric differences in apparent rates of racemization are known for several foraminiferal and molluscan taxa (King and Neville, 1977; Lajoie *et al.*, 1980; Müller, 1984; Miller and Mangerud, 1985; Bowen *et al.*, 1985; Hearty *et al.*, 1986; Bowen and Sykes, 1988; Wehmiller *et al.*, 1988a). These differences are related to taxonomic differences in amino acid composition (King, 1980; Wehmiller, 1980; Andrews *et al.*, 1985; Kaufman *et al.*, 1992). Though these intergeneric differences may complicate any geochronological or stratigraphic application of aminostratigraphy because no single genus might be found at all the sites of interest, they provide useful insights into models of diagenetic racemization (Wehmiller, 1980, 1990) and permit the use of multiple independent "clocks" in stratigraphic or chronologic applications. Different apparent racemization rates can be applied as needed, to problems of short-term, high-resolution stratigraphy or longer term, lower resolution stratigraphy, just as unstable isotopes with different half-lives are applied to chronologic problems with different time scales.

Figure 2 shows how the calibrated aminostratigraphic data in Fig. 1 can be used in a manner that is similar to the use of "concordia" curves for relating two isotope ratio pairs (concordance) to each other via one curve which represents time. The curve in Fig. 2 represents the D/L values predicted by the equations in Fig. 1 for 50,000-yr increments. The spacing between the points diminishes with increasing sample age because of the decreased overall rate of racemization inherent to older samples. Age estimates for samples between the Holocene and about 250,000 yr B.P. have an uncertainty of about 20,000 yr, given typical analytical uncertainties. For older samples the resolving power diminishes in proportion to the spacing of the plotted points, but even the poorest resolving power should be on the order of 75,000 yr for early Pleistocene samples. These estimates of precision are based upon analytical issues only and do not take into account diagenetic or reworking factors that might distort these intergeneric relationships. Other examples of covarying aminostratigraphic data include comparisons of D/L values in both free and total amino acid populations of specific samples (Miller, 1985) or comparisons of D/L values for different amino acids in the total sample hydrolysate (Lajoie *et al.*, 1980). These latter examples differ from that shown in Fig. 2 in that they deal with the covariance of amino acid D/L values within samples of a single genus (*intrageneric* comparisons), rather than

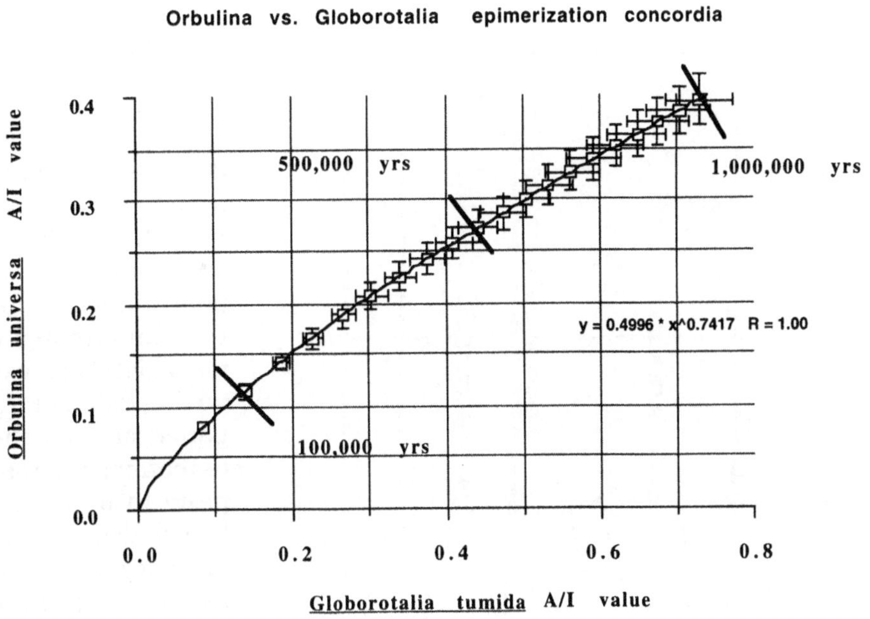

Figure 2. A concordia curve for the covariance of D-alloleucine/L-isoleucine (A/I) values in *Orbulina universa* and *Globorotalia tumida*, derived from the equations shown in Fig. 1. Each data point represents an increment of 50,000 years, plotted with an analytical uncertainty of 8%.

the covariance of results for different genera (*intergeneric* comparisons). Figure 2 represents one of the best-calibrated intergeneric concordia figures because of the stable thermal environment and the relatively close time spacing of the samples.

3. Molluscan Aminostratigraphy: General Depositional Framework Models

Many aminostratigraphic applications involve the study of marine mollusks found in coastal sequences formed during successive relative sea level maxima (ice-volume minima) of Quaternary interglacials. The nature of the preserved record of these sea levels precludes the availability of samples representing the continuum of ages found in deep-sea sediments, and samples from these coastal environments have been exposed to greater Quaternary temperature fluctuations than found in the deep sea. Consequently, it is more difficult to construct calibrated racemization curves, like those in Figs. 1 and 2, for Quaternary mollusk samples.

Figure 3 presents a conceptual model of how the timing of ice-volume minima and molluscan racemization kinetics could combine, in certain coastal depositional situations, to produce a particular record of D/L values. The oxygen isotope record of ice-volume fluctuation is shown in this figure [modified from Imbrie *et al.* (1984) and Prell *et al.* (1986)]. This curve is often used to predict the positions (relative to present sea level) of various interglacial or interstadial sea levels during the middle and late Quaternary [see, e.g., Chappell and Shackleton (1986) and Shackleton (1987)]. Although inference of precise sea level positions from the marine istopic record is debated, these records can be used to predict the ages of Quaternary shoreline deposits that *might* be preserved today at or above present sea level in stable or slowly uplifting coastal regions. The plot of D/L values versus time in Fig. 3 represents a *hypothetical* curve, similar to that shown for *O. universa* in Fig. 1, that might be appropriate for a certain molluscan genus in a particular coastal region. This curve does not represent the racemization kinetics that samples would follow if exposed to *present* temperatures [hereinafter referred to as current mean annual temperature (CMAT)], but rather represents the integrated kinetic effect of all glacial–interglacial temperature fluctuations for the region of interest. This kinetic averaging results in what is known as the effective temperature [often referred to as the effective diagenetic temperature (EDT) or effective Quaternary temperature (EQT)], which represents the one temperature that would be consistent with a particular racemization curve (kinetic model) and D/L values for known-age samples in the region of study.

The curves in Fig. 3 show that even the most completely preserved coastal record of all the interglacial (odd-numbered) isotopic stages would yield clusters of enantiomeric ratios rather than the continuum that can be observed in the deep-sea record. For the older deposits, however, the differences in D/L

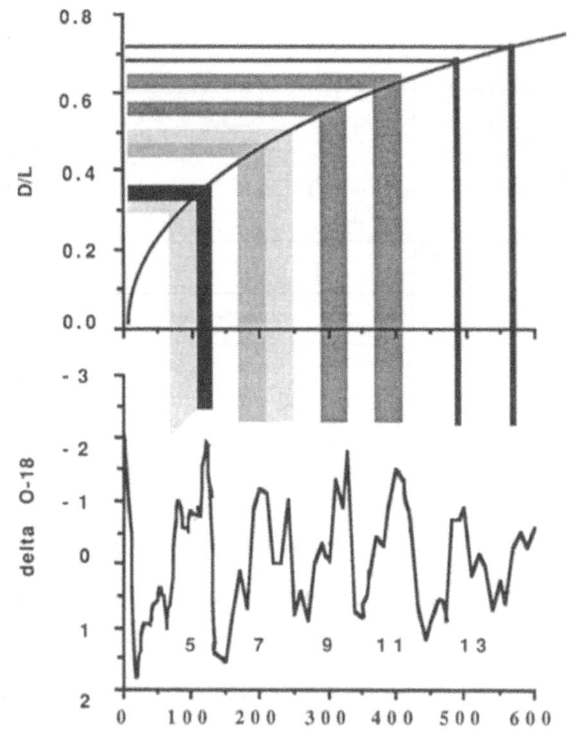

Figure 3. Conceptual model for the expected clustering of amino acid D/L values in a coastal record of late Quaternary sea level maxima inferred from the marine oxygen isotope record. The upper curve shows a representative relation between D/L values and time (see Fig. 1); the lower curve shows the time scale for ice-volume and sea level fluctuations, with stages 5, 7, 9, 11, and 13 identified (Prell *et al.*, 1986). The shaded bands represent the approximate intervals during which deposits of these sea level maxima might have formed and how these times of deposition might be recorded by racemization kinetics that slow with increasing sample age. The heavily shaded band represents the commonly observed "last interglacial" ($\delta^{18}O$ substage 5e) portion of the record, which is often used for calibration of local aminostratigraphic sections.

values would be relatively small, a fundamental consequence of the general slowing of overall racemization rates in carbonate fossils. Though it is often possible to achieve analytical precision of 5% or better when a single shell is analyzed, it is quite common to observe a coefficient of variation of 10% or more for multiple shells from a single outcrop (or from discontinuous sites within a single aminozone). The observed variations in D/L values for single aminozones are probably caused by a combination of analytical limitations, slight variation in the thermal and diagenetic environments of coeval units, and slight but significant age differences between deposits that formed during different phases of a particular interglacial stage (Murray-Wallace and Kimber, 1987; Cann *et al.*, 1988; Wehmiller *et al.*, 1988a; Miller and Brigham-Grette, 1989). These observations, combined with the trends depicted in Fig. 3, indicate that racemization techniques can be used most easily to resolve deposits of late Quaternary age. In most cases, deposits of middle Quaternary age can be resolved from both early and late Quaternary units, and some resolution *within* either the middle or early Quaternary is usually possible. However, the overall uncertainty inherent to the geochemical record implies that aminostratigraphic identification of multiple interglacial stages would be possible only in those cases where detailed lithostratigraphic sequences supported such a high-resolution interpretation. For taxa or thermal environments where the racemization appears to be more rapid than depicted in Fig. 3, a higher degree of age resolution is possible for late Pleistocene units. Additional factors, such as reoccupation of depositional regions by multiple transgressions to similar elevations, reworking of older samples into a younger deposit, and incomplete preservation of the depositional sequence, combine with the analytical and geochemical factors listed above to complicate the interpretation of aminostratigraphic data in coastal sequences. Consequently, studies such as these require numerous analyses of samples from unambiguous local stratigraphic sections so that data from regionally widespread sites can be related to well-constrained aminostratigraphic sections, as is discussed further in Section 4.

4. Molluscan Aminostratigraphy: Examples of the Relationship between Enantiomeric Ratios, Stratigraphic Position, and Taxonomy

Quaternary coastal records that offer unambiguous demonstration of the relationship between molluscan D/L values and lithostratigraphic or morphologic sequences are recognized in areas of tectonic uplift, such as southern California [reviewed by Wehmiller (1990)], the Mediterranean Basin (Hearty *et al.*, 1986), New Guinea (Hearty and Aharon, 1988), and along the South American coast from 15° to 30°S (Hsu, 1988; Hsu *et al.*, 1989). In regions such as these, interglacial high sea levels are recorded by marine terrace deposits that generally increase in age with elevation, though lateral variation in uplift and incomplete terrace preservation preclude correlation based on elevation alone. Vertical terrace sequences provide excellent tests of the fundamentals of aminostratigraphy.

Figure 4 shows the location of several of the coastal southern California aminostratigraphic rec-

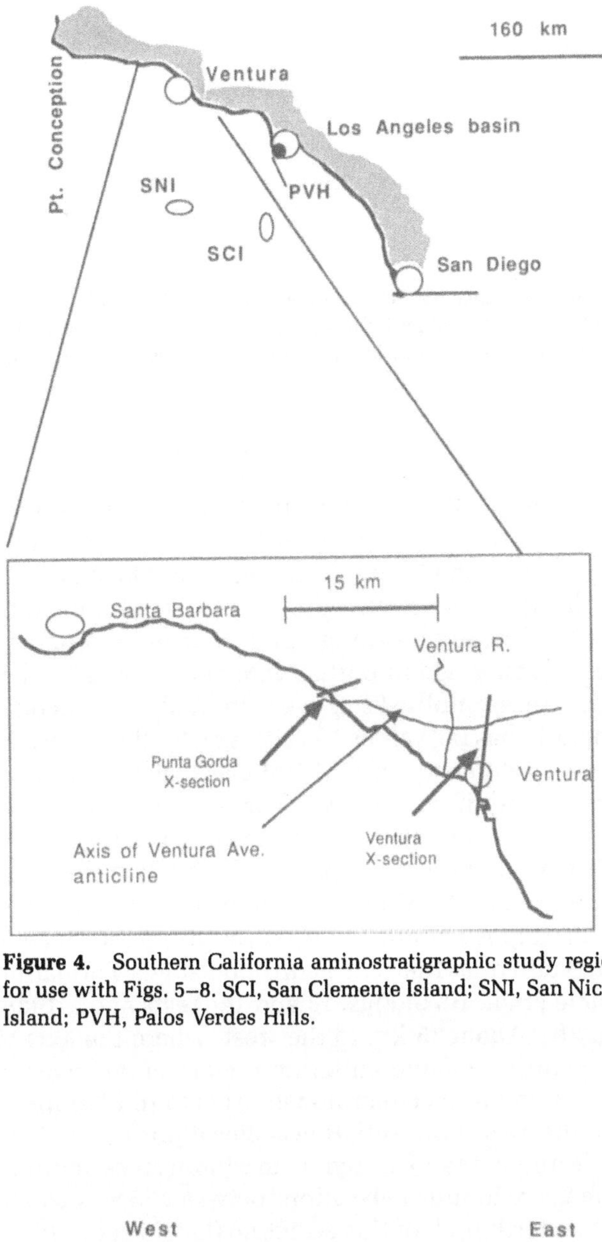

Figure 4. Southern California aminostratigraphic study regions, for use with Figs. 5–8. SCI, San Clemente Island; SNI, San Nicolas Island; PVH, Palos Verdes Hills.

ords that have been obtained. Sequences of multiple terraces (10 or more) are found on the Palos Verdes Peninsula, San Clemente Island, and San Nicolas Island, and molluscan aminostratigraphic data conform to the relative ages inferred from each of these sequences (Mitterer and Hare, 1967; Wehmiller *et al.*, 1977; Muhs, 1985), though extensive sampling of all the preserved terraces has not been possible. U–Th dating of solitary corals provides independent calibration for the San Nicolas Island and San Clemente Island aminostratigraphic sequences (Wehmiller *et al.*, 1977; Kennedy *et al.*, 1982; Muhs, 1985; Muhs *et al.*, 1988).

Sections in southern California (Fig. 4) where more extensive sampling has been possible include the Los Angeles Basin and Point Loma, at San Diego. Figure 5 shows the relationship of D/L leucine values in the bivalve *Protothaca staminea* to a sequence of terrace deposits on Point Loma. This section is one of the key late Pleistocene aminostratigraphic reference sections for the North American Pacific coast, because there is a clear relation between the stratigraphy, U-series chronology, and aminostratigraphy. The Nestor and Bird Rock terraces (upper and lower, respectively) are used as the early and late stage 5 (125 and 80 ka, respectively) calibration points for the aminostratigraphic data that are available for these deposits. In addition, the molluscan faunal data from the Nestor and Bird Rock terrace deposits indicate a trend from warmer to cooler water temperatures, consistent with the isotopic record for this time interval (Kennedy *et al.*, 1982; Kennedy, 1988).

The section in the Los Angeles Basin includes

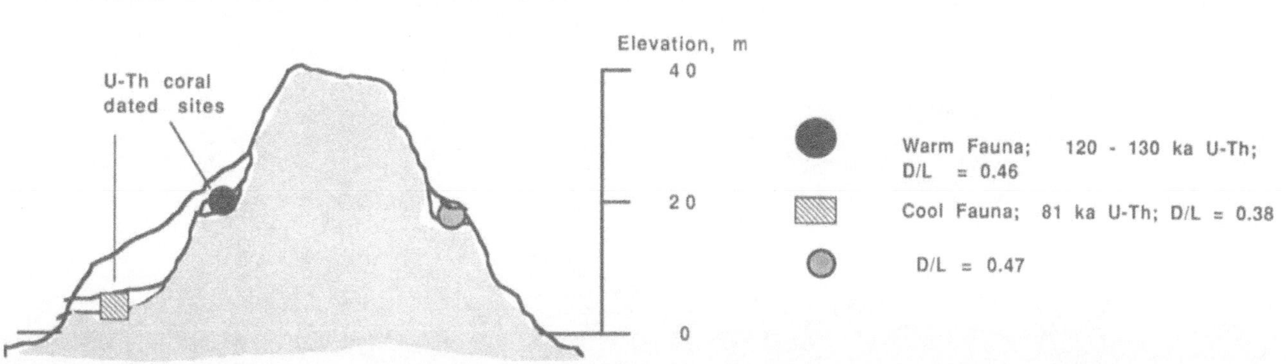

Figure 5. East–west cross section of the southern end of Point Loma, San Diego Harbor (Fig. 4). On the west side of the peninsula, the upper (Nestor) and lower (Bird Rock) terraces are morphologically resolved and have faunal, isotopic, and aminostratigraphic data that permit correlation of these two terraces with substages 5e and 5a of the marine isotopic record, respectively. The terrace site on the east side of the peninsula has no radiometric data, but aminostratigraphic data suggest that this terrace remnant is correlative with the Nestor Terrace. [Redrawn from Lajoie *et al.* (1979) with the addition of data from Emerson *et al.* (1981) and Muhs *et al.* (1988).]

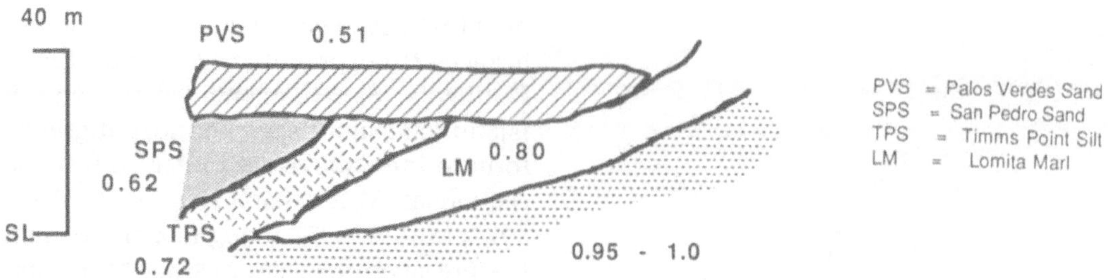

Figure 6. Schematic cross section of Pleistocene deposits in the Los Angeles Basin area. SL, Sea level; PVS, Palos Verdes Sand; SPS, San Pedro Sand; TPS, Timms Point Silt; LM, Lomita Marl. The numbers on the figure are D/L values for leucine in the bivalve *Saxidomus*. The Palos Verdes Sand is considered to be 125,000 years old, based on aminostratigraphic and faunal evidence (Kennedy *et al.*, 1982). [Redrawn from Lajoie *et al.* (1979) and Wehmiller (1984, 1990).]

superposed early and middle Pleistocene basin deposits and late Pleistocene marine terrace deposits (Fig. 6). At several outcrops, these mapped units can be found with clear stratigraphic relationships that, when combined with the D/L data for the units, demonstrate at least five clearly resolved aminozones that conform to the local stratigraphic control. The leucine data shown in Fig. 6 are for *Saxidomus*, the molluscan genus most commonly sampled at these particular sites. Enantiomeric ratios for *Saxidomus* and *Protothaca*, the genus reported in Fig. 5, are generally within 5% (*Protothaca* ≤ *Saxidomus*) when both taxa are found at the same site (Lajoie *et al.*, 1980). The D/L values shown in Fig. 6 for the Palos Verdes Sand, a late Pleistocene marine terrace deposit, are interpreted to represent the same age (early stage 5, ca. 125 ka) as those shown for the Nestor Terrace in Fig. 5 (Kennedy *et al.*, 1982). Ponti *et al.* (1987) reported AAR age estimates for this region that appear consistent with the summary in Fig. 6.

Northwest of Los Angeles, in the area of Ventura, California, a remarkable sequence of fossiliferous Quaternary basin and terrace deposits is exposed as a consequence of uplift along the Ventura Avenue anticline (Yeats *et al.*, 1988). Figure 7, which shows aminozones established from *Protothaca* and *Macoma* data (Table II), demonstrates the relation of aminostratigraphic data to the stratigraphic control provided by this sequence. An important intergeneric relation between these mollusks is seen in Table II: *Macoma* generally has D/L values 10–20% greater than those in coeval *Protothaca*, and this intergeneric relationship can be used, if necessary, to convert D/L data for one genus into equivalent data for the other (Lajoie *et al.*, 1980). At Ventura, a south-to-north transect along the south limb of the Ventura Avenue anticline reveals a terrace deposit (aminozone 3) overlying a Pleistocene unit known as the San Pedro or Saugus Formation (Lajoie *et al.*, 1979, Figs. 13 and 14; Yeats *et al.*, 1988, Fig. 2B). About 15 km to the west, where the axis of the Ventura Avenue anticline intersects the coast, a low marine terrace (aminozone 1) is found "draped" over the axis of the anticline at elevations up to 20 m, while remnants of a higher marine terrace (aminozone 2) are found at elevations between 75 and 200 m on the north limb of this anticline (Lajoie *et al.*, 1979,

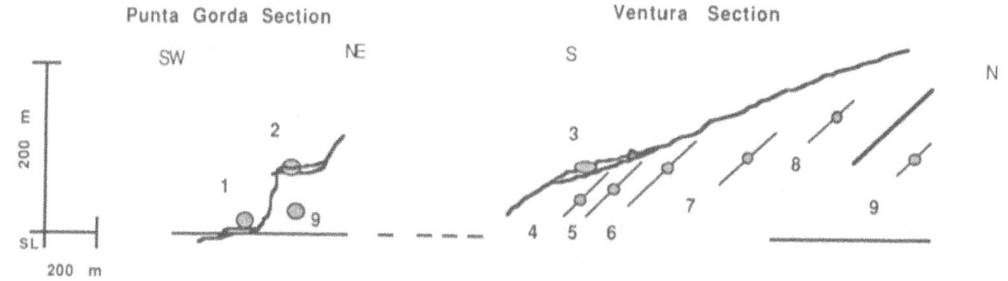

Figure 7. Combined litho- and aminostratigraphic section for the Punta Gorda–Ventura region, southern California (Fig. 4). The Punta Gorda section is a short east–west section on the north limb of the Ventura Avenue anticline; the Ventura section is a longer, north–south section on the south limb of the same structure. Aminozones 1–9 are identified, in proper stratigraphic order, based on the information in Table II. The diagonal lines sloping to the south in the Ventura section represent the regional dip of marine units in this structure. The heavy line between aminozone 8 and 9 collection points represents a volcanic ash that is approximately 700,000 years old (Sarna-Wocicki *et al.* 1987).

TABLE II. Ventura–Punta Gorda Aminozones: Mean D/L Leucine Values for Nine Aminozones, Based on Results for *Macoma* and *Protothaca* Samples

	D/L	
Aminozone	*Macoma*	*Protothaca*
1	n.d.[a]	0.10
2	0.38	0.30
3	0.50	0.41
4	0.65	(0.51)[b]
5	0.71	(0.56)
6	0.77	(0.60)
7	0.81	0.75
8	n.d.	0.84
9	n.d.	0.96[c]

[a]n.d., No data.
[b]Values in parentheses are predicated *Protothaca* values based on intergeneric conversions (Lajoie *et al.*, 1980).
[c]From a sample of *Tivela*, thought to have racemization kinetics similar to those of *Protothaca* (Lajoie *et al.*, 1980).

Fig. 13). The combined aminostratigraphic section for this region (Fig. 7) identifies nine separate aminozones, each consistent with its physical stratigraphic position. The Ventura–Punta Gorda aminostratigraphic section is unusual because so many aminozones (representing nearly the complete range of D/L values) are found in a small region. The exposure of these multiple aminozones is an apparent consequence of both rapid deposition and tectonic deformation during the late Cenozoic in this region (Morton and Yerkes, 1987; Yeats *et al.*, 1988).

The Ventura–Punta Gorda aminostratigraphic section shown in Fig. 7 has some independent age control, providing calibration for conversion of the D/L values into numerical-age estimates. Aminozone 1, with the lowest D/L values, is represented by samples on the low (≤20 m) terrace at Punta Gorda and is Holocene in age (≤6 ka) based on ^{14}C dates on shells (Lajoie *et al.*, 1979; Yerkes and Lee, 1987). The terrace deposit at Ventura (aminozone 3) has *Protothaca* D/L values that are about 15% lower than those of *Protothaca* from the Nestor Terrace (125 ka calibration site) in San Diego (Fig. 5). This difference, considered in light of molluscan faunal data and the 1°C difference in CMAT values for Ventura and San Diego, has resulted in the conclusion that the Ventura terrace is about 80,000 yr in age, or late-stage 5 correlative (Lajoie *et al.*, 1979; Kennedy *et al.*, 1982). Though a precise age estimate for the Ventura terrace may remain elusive in the absence of nearby calibration sites, a stage 5 estimate (100 ± 25 ka) for this terrace seems secure. Aminozone 2, found on the second terrace at Punta Gorda, has D/L values between those of the calibrated aminozones discussed above and is estimated to be about 45,000 y in age (stage 3 correlative) (Lajoie *et al.*, 1979). Interestingly, Kaufman *et al.* (1971) obtained U-series isotopic dates of 5 ka and 50 ka for mollusks from the sites represented by aminozones 1 and 2 but dismissed the dates as being inaccurate, because of a combination of geochemical evidence from the analyzed shells and assumptions about the tectonic history of the region (W. S. Broecker, personal communication, 1981). Consequently, it appears that these molluscan U-series dates can also be used as independent calibration of the younger portion of aminostratigraphic section shown in Fig. 7. Age control for the older part of the section also exists, because a mid-Pleistocene (ca. 700 ka) volcanic ash is found stratigraphically below most of the units analyzed for aminostatigraphic data (Sarna-Wocicki *et al.*, 1987). Further discussion of the use of the Ventura area data in age estimation is deferred to Section 5.

The leucine enantiomeric ratio data shown in Table II conform well to the local stratigraphic control. Figure 8 shows how two other amino acids, valine and glutamic acid, relate to this control. Both

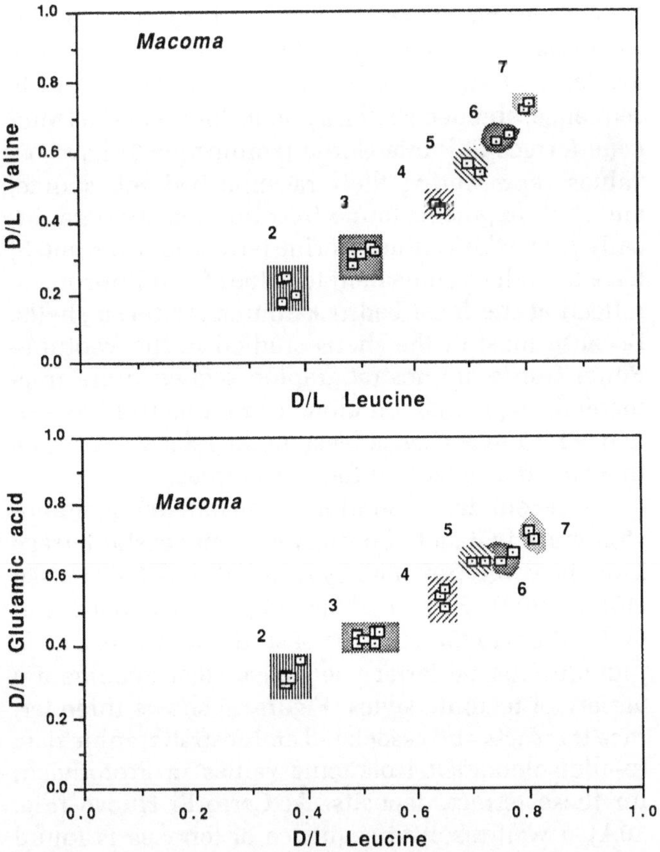

Figure 8. Intrageneric plots of D/L values for leucine, valine, and glutamic acid in the *Macoma*-based aminozones shown in Fig. 7.

of these amino acids are "slower racemizers" than leucine, in that they generally have lower D/L values than those for leucine when all three are measured in the same sample. Valine D/L values are resolved in the same manner as leucine D/L values; the resolution of the glutamic acid D/L values is not quite as good, especially for the older, more extensively racemized deposits. The data shown in Fig. 8 indicate that D/L values for multiple amino acids may be particularly valuable in those situations where the number of analyzed samples is relatively small.

The Ventura–Punta Gorda section is unusual in many respects, including its demonstration of reworking of older shells into younger deposits. This is a phenomenon that is often suspected in biostratigraphic analysis, and for which aminostratigraphic data may prove useful (Mitterer, 1975; Miller and Brigham-Grette, 1989). At issue, however, is the method of *objective* interpretation of aminostratigraphic data when a few aberrant (higher) D/L values are encountered among results for a specific site. It is easy to interpret these anomalous ratios as indicative of reworking, but, in the absence of other evidence, geochemical or thermal effects must also be considered as possible factors. Ideally, reworking should be invoked in those cases where *evidence obtained at the time of sample collection* suggested that the samples might have been reworked. Such field evidence was available at two sites in the Ventura–Punta Gorda sequence: abraded shell fragments found in the Holocene terrace at Punta Gorda (aminozone 1) have D/L values representing their racemic bedrock source, and shell fragments found in colluvium stratigraphically above the Ventura marine terrace (aminozone 3) have D/L values equivalent to values for aminozone 6, reflecting the local bedrock source for these shells. Because most of the shells studied in the Ventura–Punta Gorda aminostratigraphic sequence are fragmented, fragmentation alone cannot be used as evidence for reworking, at least on the relative age scale that was recognized in these examples.

A recent demonstration of the relation of aminostratigraphic data to South American coastal terrace sequences was provided by Leonard *et al.* (1987), Hsu (1988; 1992), Hsu *et al.* (1989), and Leonard and Wehmiller (1992). The area of study, shown in Fig. 9, includes coastal terrace sequences that document a variety of tectonic styles. Figure 10 shows three terrace transects and associated aminostratigraphic data (D-alloisoleucine/L-isoleucine values in *Protothaca*) for these terrace deposits. At Cerro El Huevo (Fig. 10A), a well-resolved sequence of terraces is found where comparatively rapid Quaternary uplift is occurring in the region of subduction of the Nazca Ridge

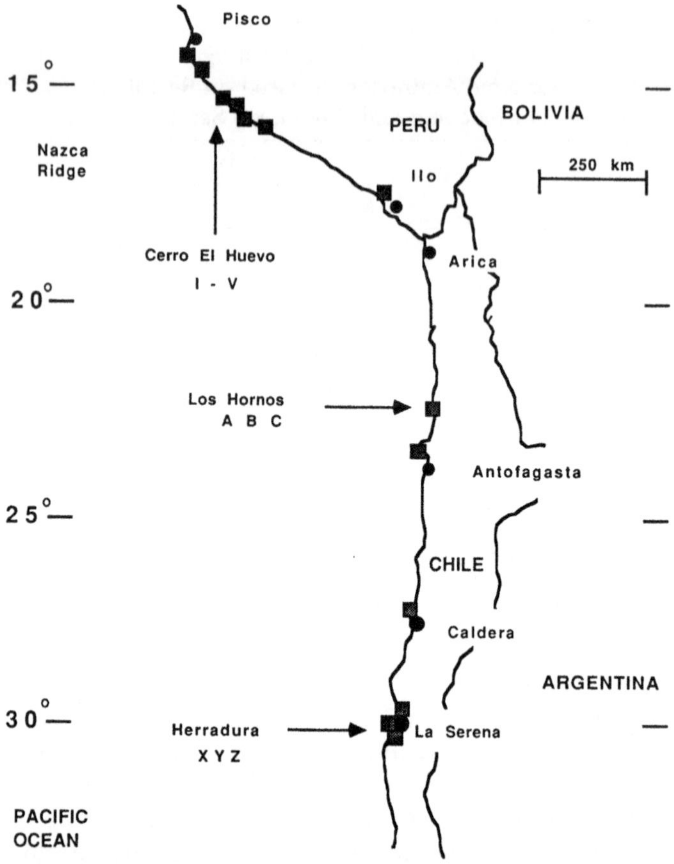

Figure 9. Map of South American Pacific coast sites studied for aminostratigraphy (Hsu et *al.*, 1989). Aminostratigraphic sections for Cerro El Huevo, Los Hornos, and Herradura are shown in Fig. 10. The local aminozone designations for these terrace transects are I–V, A, B, and C, and X, Y, and Z, respectively. Aminozone C is poorly defined because of few samples and a wide range of D/L values (see Fig. 10).

(Hsu, 1988). Aminozones I, IIa, IIb, III, IV, and V are clearly identifiable and conform to the local relative age sequence of these terraces. Elsewhere in central and southern Peru, in areas of lower uplift rates and poorer terrace resolution, data representing aminozones IIa and IIb are found on what appears as the same terrace (Hsu *et al.*, 1989). At Ilo (Fig. 9), aminozones IIa, IIb, and III are found in stratigraphic superposition in a single thick (ca. 30 m) terrace deposit, implying low and/or oscillatory uplift in the region (Hsu, 1988). Independent calibration of the Cerro El Huevo aminostratigraphic section is available in the form of electron spin resonance (ESR) dates, obtained by U. Radtke (on portions of the *same shell* analyzed for amino acid D/L values), as summarized by Hsu (1988) and Hsu *et al.* (1989). Numerical dates derived from ESR dating (See Table I) require calibration procedures and assumptions similar to those for amino acid racemization, but a variety of geochemical and

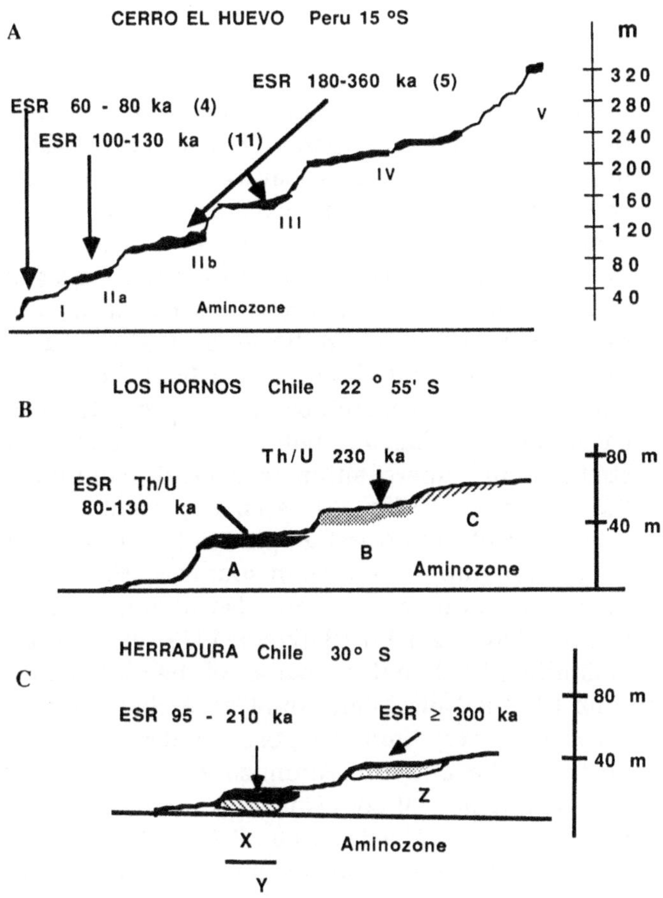

Figure 10. Peru–Chile terrace transects and aminostratigraphy, with associated independent chronologic control (Radtke, 1987; Hsu, 1988). Mean D-alloisoleucine/L-isoleucine values in *Protothaca thaca* samples from each aminozone are as follows: (A) I, 0.272; IIa, 0.441; IIb, 0.55; III, 0.625; IV, 0.78; V, 1.02; (B) A, 0.40; B, 0.57; C, ca. 0.80 (wide range, few samples; (C) X, 0.40; Y, 0.61; Z, 0.80 (X and Y are bimodal—0.38, 0.46, and 0.57, 0.67, respectively; see Hsu et *al.*, 1989).

stratigraphic information supports the general validity of the ESR dates shown in Fig. 10A, especially the 11 dates constraining aminozone IIa to the interval between 100 and 130 ka.

Chilean terrace transects at Los Hornos and Herradura (Fig. 10B and 10C) also demonstrate clear relations between aminostratigraphic data and local stratigraphic control. Independent age control is less clear in the case of these terraces, because mollusk samples for ESR and U–Th dating were not collected in conjunction with the aminostratigraphic collection (Radtke, 1987). Aminozones A and B, identified in the Los Hornos section, are thought to be correlative with stages 5 and 9 (Hsu *et al.*, 1989). Aminozone C, found on higher terraces at Los Hornos, spans a broad range of D/L values and is best interpreted as "pre-stage 11." Similar correlations of aminozones X, Y, and Z, at Herradura, are proposed. The Herradura section is particularly significant because stratigraphic, amino acid, and ESR data collectively imply reoccupation of the lowest terrace by transgressions during stages 9, 7, and 5 (Hsu *et al.*, 1989; Leonard and Wehmiller, 1992).

Coastal areas lacking the obvious terrace sequences described above often have more complex stratigraphic records because most of the Quaternary interglacial transgressions reached to the same approximate relative elevation, resulting in combinations of vertically and laterally superposed sequences with frequent channeling and destruction of preexisting units. The U.S. Atlantic Coastal Plain (Fig. 11) is a region with this type of homotaxial history. The aminostratigraphic data for this region were reviewed recently by Wehmiller *et al.* (1988a); data from segments of the Coastal Plain have generated significantly different models for age assignments (Corrado *et al.*, 1986; Wehmiller *et al.*, 1988a; Colman and Mixon, 1988; Hollin and Hearty, 1990). Only a limited number of critical localities in the Atlantic Coastal Plain provide the kind of unambiguous stratigraphic resolution that is seen in the lithostratigraphic sections described above. Instead, the vast majority of localities for which aminostratigraphic data exist represent isolated sites within thin or discontinuous units that may not be preserved at a scale that permits simple one-to-one correlation with the isotopic model of Quaternary sea level fluctuations (Fig. 3).

In Dare County, northeastern North Carolina (Fig. 11), a vertical and lateral sequence of coastal deposits demonstrates an unequivocal relation between aminostratigraphic data and physical stratigraphic position. This section is shown in Fig. 12, with D/L values for *Mulinia* and *Mercenaria* samples from various positions in the sequence given in Table III (York *et al.*, 1989; York, 1990; Riggs *et al.*, 1992). Four distinct aminozones (designated IIIa, IIIb, IIIc, and IIId) are recognized. The nomenclature of these aminozones was established by Wehmiller *et al.* (1988a), who labeled aminozones with letters (a, b, c, ...) from lowest to highest D/L values for different geographic regions (I, II, III, IV, and V, from north to south) of the Coastal Plain. The results shown in Fig. 12 demonstrate the lateral and vertical stratigraphic integrity of the observed amino acid D/L values. Aminozone IIIc is regionally bimodal, but D/L values assigned to this aminozone cluster tightly at specific sties. Aminozone IIId, the oldest of the units shown in Fig. 12, also represents a distribution of D/L values that are usually site specific and which probably reflect units of at least two ages (Wehmiller *et al.*, 1988a; York *et al.*, 1989). The section at Stetson Pit (Fig. 12) has both

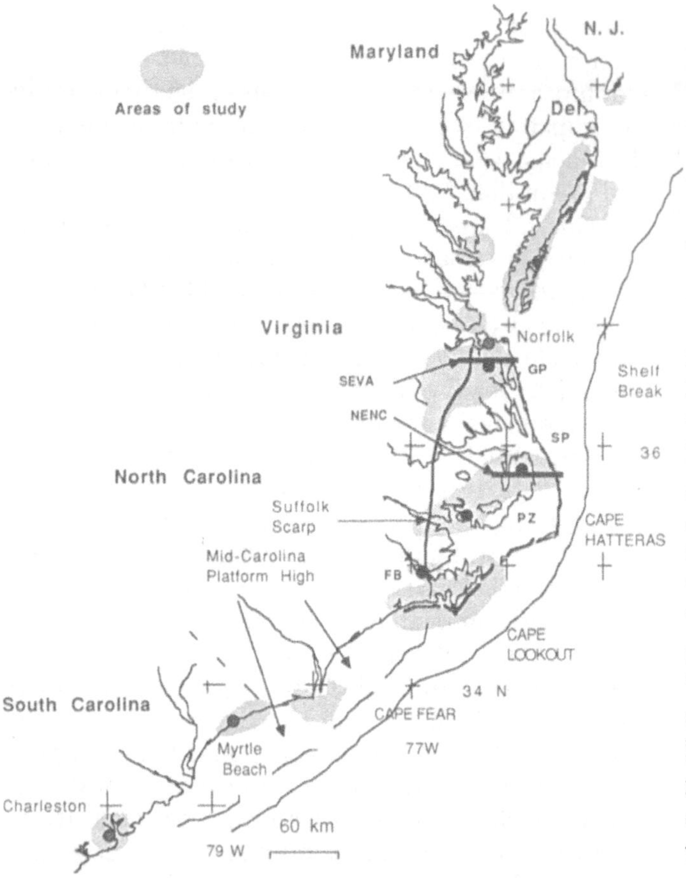

Figure 11. Map of the Atlantic Coastal Plain from New Jersey to South Carolina, showing important regions of study and particular sites discussed in text. Solid dots identify Gomez Pit (GP), Stetson Pit (SP), Ponzer (PZ), and Flanner Beach (FB). The locations of two important cross sections (Figs. 12 and 13) are shown as NENC and SEVA. The Suffolk Scarp, a feature mappable from the southern Chesapeake Bay region to near Cape Lookout, represents the inland limit of middle to late Quaternary deposits (Peebles *et al.*, 1984). The offshore projection of the Suffolk Scarp, southwest of Cape Lookout, appears to represent the offshore limit of pre-Quaternary outcrops on the inner shelf, common features of the Mid-Carolina Platform High. The nondepositional character of the inner shelf south of Cape Lookout is both a consequence and an expression of the Cape Fear Arch (Riggs and Belknap, 1988; Gohn, 1988).

faunal and floral data that permit subdivision of the record into five or six climatic cycles (York *et al.*, 1989). Independent calibration for the Dare County cross section is provided from two sites in the region (Szabo, 1985): (1) a U-series date of 75 ka on a solitary coral from a depth equivalent to aminozone IIIb; and, (2) a U-series coral date of about 200 ka was obtained at Ponzer, North Carolina (Fig. 11), an aminozone IIIc site. These relationships were summarized by Wehmiller *et al.* (1988a, Fig. 13).

Southeastern Virginia (Fig. 11) provides an opportunity (rare in the Coastal Plain) to examine *outcropping* middle and late Pleistocene units, often in stratigraphic superposition. There is a long history of study of these units (Oaks *et al.*, 1974; Peebles *et al.*, 1984; Spencer and Campbell, 1987) that has benefited from extensive (but temporary) borrow pit activity in the region. Three such excavation sties for which abundant amino acid data exist are the New Light, Gomez, and Yadkin pits, all south–southeast of Norfolk, Virginia (Fig. 11). The data from these sites are summarized in Fig. 13; further discussion of these results is found in Wehmiller *et al.* (1988a, Fig. 10) Mirecki (1990), and Groot *et al.* (1990). Gomez Pit, where two distinct aminozones (IIa and IIc, each based on more than 75 shell analyses) are found in stratigraphic superposition in a nearly continuous 300-m exposure (Fig. 13), is an important link between recent aminostratigraphic research and previous stratigraphic studies in southeastern Virginia. These two aminozones were also identified in the nearby New Light Pit (Belknap, 1979; Belknap and Wehmiller, 1980), but the reality of the older aminozone (IIc) at New Light could never be confirmed because of a very limited exposure and rapid deterioration of the outcrop. Aminozone IIa is independently calibrated at both New Light and Gomez pits, with U-series coral dates of about 75 ka (Szabo, 1985). At Gomez Pit, ESR dates for mollusks from aminozones IIa and IIc cluster around 110 and 220 ka, respectively, suggesting that these two units represent deposition during portions of isotope stages 5 and 7 (Mirecki *et al.*, 1989). The Gomez and Yadkin pit data carry the regional aminostratigraphy still further back in time, because at both sites an older unit (aminozone IIe) is found stratigraphically below aminozone IIc. The shells that yielded aminozone IIe D/L values (in the range of 0.90–1.0) were collected from units mapped as the Yorktown Formation (at Yadkin Pit) and the Chowan River Formation (at Gomez Pit) (see Wehmiller *et al.*, 1988a, Fig. 10), but the placement of these two formations in the same aminozone does not necessarily imply age equivalence of the units, because of the decreased age-resolution capability of racemization data in this numerical range (e.g., Fig. 1–3).

The record at Gomez Pit provides numerous insights into the complexity of the Coastal Plain chronostratigraphy. The distinct temporal difference between aminozones IIa and IIc at Gomez Pit, where their stratigraphic relationship is clear, requires revision of previous stratigraphic interpretations of the region, because the IIa and IIc shell beds were interpreted to represent a single transgression (Peebles *et al.*, 1984; Spencer and Campbell, 1987; Wehmiller *et*

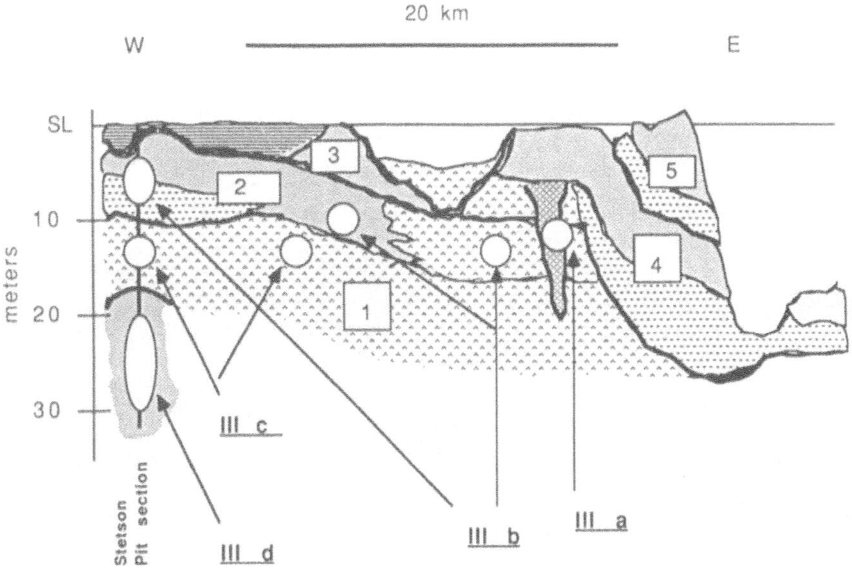

Figure 12. Cross section of Quaternary sediments in Dare County, northeastern North Carolina (NENC in Fig. 11). Numbered boxes represent depositional units identified from lithostratigraphic evidence (Eames, 1983). Aminozones (IIIa, IIIb, etc.) follow the nomenclature proposed by Wehmiller *et al.* (1988a) with data for this cross section summarized in Table III.

al., 1988a). Aminozones IIa and IIIb (Stetson Pit, above) are calibrated as being essentially equal in age (75 ka), so the differences in D/L values (20–30%) for these closely spaced sites (CMAT difference of about 0.5°C) must be interpreted to reflect either local diagenetic effects on racemization at particular sites or greater age differences than implied by the calibration U-series data. The comparison of IIa and IIIb data indicates that the variability in D/L values for multiple sites within an aminozone may be greater than the variability seen at individual sites.

Evidence of reworking of older shells into younger deposits is seen at Gomez Pit, though the field evidence for reworking is very subtle and could be confirmed only after frequent reexamination of the outcrop during a four-year period of field and laboratory study (Mirecki, 1985, 1990). Along one 30-m exposure of the basal contact of the aminozone IIc shell-bearing unit (Fig. 13), whole but disarticulated *Mercenaria* valves are found within a gravel and cobble lag deposit. Of the shells analyzed from this localized section, 75% have D/L values 50% greater than those of aminozone IIc shells, and they are interpreted as having been reworked from an older unit. No in-place shells with these higher D/L values have been found within Gomez Pit, but shells with these same D/L values (designated as aminozone IId) have been collected at other sites in the central and southern Chesapeake Bay region (Wehmiller *et al.*, 1988a), indicating that units of this age could have served as the source for the reworked shells found at Gomez Pit.

TABLE III. Mean D-Alloisoleucine/L-Isoleucine (A/I) and D/L-Leucine (D/L) Values for Dare County Cross Section (Fig. 12) Used to Relate This Section to the Aminostratigraphic Terminology of Wehmiller *et al.* (1988a)[a]

	Mulinia		*Mercenaria*	
Aminozone	A/I	D/L	A/I	D/L
IIIa	0.10	—	—	—
IIIb	0.162	0.28	0.23	0.33
IIIc	0.25	—	0.47	—
IIId	0.48	0.51	0.76	0.74

[a]Data from York *et al.* (1989) and York (1990).

A third Coastal Plain region of combined aminostratigraphic and lithostratigraphic studies is in northeastern South Carolina, where extensive outcrops of Pleistocene units are seen along the Intracoastal Waterway, near Myrtle Beach (Fig. 11). The data from this region have been summarized by McCartan *et al.* (1982) and Wehmiller *et al.* (1988a, Fig. 15). A threefold lithostratigraphic nomenclature (Socastee–Canepatch–Waccamaw formations, for the youngest to oldest Pleistocene units) is used by most workers in this region, modified from DuBar *et al.* (1980) as discussed by McCartan *et al.* (1982) and Wehmiller *et al.* (1988a). While there is general agreement among workers that amino acid D/L values conform to local stratigraphic control at specific outcrops along the Intracoastal Waterway, there is debate about aminostratigraphic and lithostratigraphic correlations from section to section along the Waterway (McCartan *et*

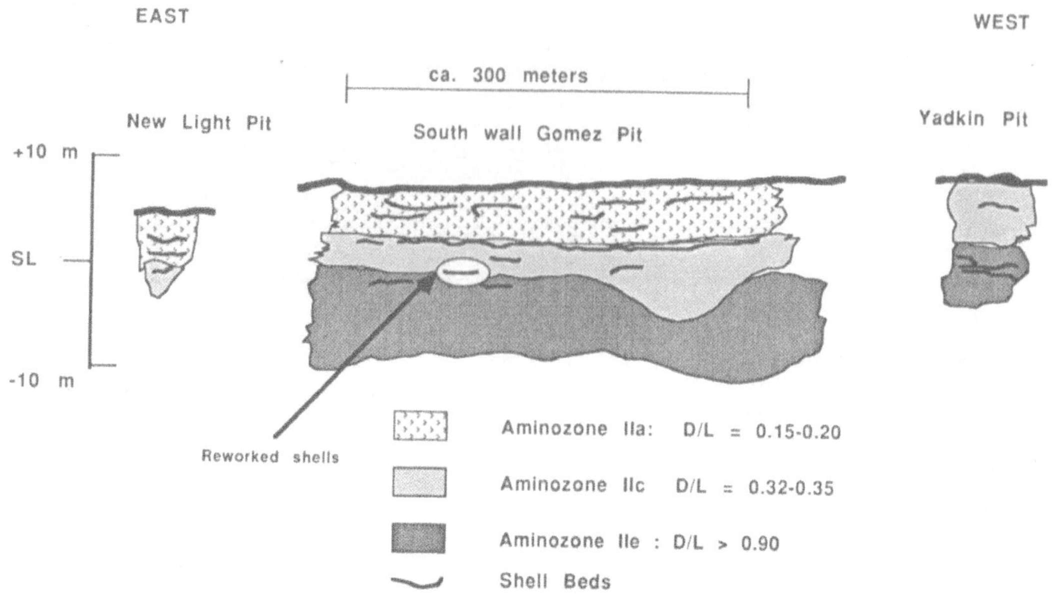

Figure 13. Schematic cross section of units in southeastern Virginia (SEVA in Fig. 11), showing aminozones (based on *Mercenaria* D-alloisoleucine/L-isoleucine values) in stratified outcrops at New Light, Gomez, and Yadkin pits, all south–southeast of Norfolk, Virginia.

al., 1982; Wehmiller *et al.*, 1988a; Hollin and Hearty, 1990). Some of this disagreement hinges on variable application of lithostratigraphic terms to specific outcrops or portions of outcrops, some is caused by the fact that four aminozones are identified where only three formations have been mapped (Wehmiller *et al.*, 1988a), and some results from ambiguous U-series data for the region (Szabo, 1985). Consequently, in spite of the excellent outcrops seen along the Intracoastal Waterway, a clear relation between *regional* aminostratigraphic and lithostratigraphic data cannot be documented, even though aminostratigraphic data from specific outcrops are internally consistent. The aminostratigraphic placement of the units from the Intracoastal Waterway is a vigorously debated topic (Wehmiller *et al.*, 1988a, 1992; York and Wehmiller, 1992), with significant implications for general principles of aminostratigraphic correlation. This issue is discussed further in Section 6.

Formerly glaciated coastal regions often contain *in situ* marine sections and glacially or fluvially transported mollusks derived from these deposits. Andrews *et al.* (1983) used aminostratigraphic methods to evaluate the chronostratigraphy of the southern Hudson Bay lowlands. The D/L data for in-place shells were used to establish the local aminostratigraphic range, representing last interglacial (?) and Holocene marine incursions. The D/L values in mollusk fragments found in fluvial and glacial sediments in the area clustered into three pre-Holocene groups, the oldest (highest D/L values) of which is equivalent to measured last interglacial (?) shells. Lower D/L groups apparently represent younger marine incursions in the source region for the reworked shells, depositional events not yet identified in local lithostratigraphic sections. This study demonstrates the value of local reference sections for the time-stratigraphic interpretation of samples with less control.

The examples cited above demonstrate the fundamental ability of molluscan amino acid racemization data to conform to a relative age sequence that is firmly established by the local lithostratigraphy. This ability has to be documented wherever possible if the stratigraphic utility of amino acid racemization is to be accepted. Local aminostratigraphic reference sections are essential for the interpretation of regionally widespread results from isolated outcrops because they define aminozone boundaries that are then used to "map" the other sites in the region of study. Other coastal areas for which extensive molluscan aminostratigraphic data exist include the Mediterranean Basin (Hearty *et al.*, 1986), the British Isles and northwestern Europe (Bowen *et al.*, 1985; Miller and Mangerud, 1985; Bowen and Sykes, 1988), high-latitude sites in Alaska, Canada, and Greenland (Brigham, 1983; Miller, 1985), and Australia (Murray-Wallace and Kimber, 1987). In almost every case, it is possible to establish at least one local aminostratigraphic reference section which provides structure for the regional data and demonstrates the stratigraphic coherence of the method. Recent research demonstrates that such reference sections are possible for terrestrial mollusks as well (Miller *et al.*, 1979; Miller *et al.*, 1987; McCoy, 1987; Goodfriend and Mitterer, 1988; Vacher and Hearty, 1989; Bowen *et al.*, 1989; Clark *et al.*, 1989).

5. Molluscan Aminochronology: Examples of Calibrated Racemization Kinetics in Sequences of Coastal Deposits

Aminochronology represents the application of calibrated amino acid D/L values to the estimation of ages for samples with uncalibrated D/L values. Inherent to any aminochronologic strategy is the choice of a kinetic model (or model options) that describes D/L values as a function of time, for the particular genus and effective temperature of interest. In many cases, rigorous calibration may not exist but inferred ages for certain samples are chosen as calibrations, with resulting age estimates for uncalibrated sites being evaluated in light of any independent stratigraphic information for the region of study.

The only kinetic model of racemization that has a mathematical expression based on mechanistic theory is a first-order reversible model that is often used in the interpretation of diagenetic racemization data (Bada, 1985b). The relation between D/L values and time for this model is

$$\ln\{[1 + (\text{D/L})]/[1 - K(\text{D/L})]\} - C = (1 + K)(k_L)t \quad (1)$$

where D/L is the enantiomeric ratio at any time t, and $K = k_D/k_L$, the ratio of the reverse and forward rate constants. C is the value of the ln function on the left-hand side of Eq. (1) at zero time; this value is usually less than 0.03. For amino acids with one center of asymmetry, $K = 1.0$.* These kinetics are often decribed as "linear" because the slope of $\ln\{[1 + (\text{D/L})]/[1 - (\text{D/L})]\}$ versus time would be linear (Fig. 14). However, kinetics of this type have not been observed in foraminifera from deep-sea sediments (Fig. 1), where samples representing a continuum of ages (and D/L values) and nearly constant temperatures are available.

*The expected consequence of $K = 1.0$ is that the equilibrium D/L value equals 1.0; however, for various analytical (chromatographic) reasons, or because of diagenetic factors, actual measured D/L values in "infinitely old" samples may not quite reach 1.0.

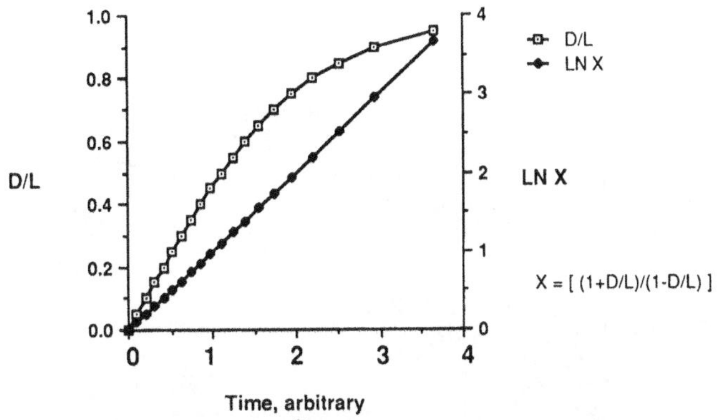

Figure 14. Comparison of curves predicted by first-order reversible kinetics [Eq. (1) with $K = 1$] plotted on an arbitrary time scale. The ln function is linear; the D/L function is nonlinear, though the nonlinearity becomes pronounced only above D/L values of about 0.50. These curves are to be compared with those in Figs. 1, 15, and 16.

Because of the nonlinearity of foraminifera racemization kinetics (epimerization, in the case of the isoleucine results shown in Fig. 1), several predictive nonlinear kinetic models have been developed (Wehmiller and Belknap, 1978, 1982; Miller and Mangerud, 1985; Wehmiller *et al.*, 1988a; Hearty and Aharon, 1988; Mitterer and Kriausakul, 1989). These models have evolved from "two-linear-stage" models (Wehmiller and Hare, 1971; Wehmiller, 1980) to parabolic or logarithmic curves, shown in Fig. 1 as curvilinear regressions to observed data. Figures 15 and 16 show kinetic model curves for leucine racemization and isoleucine epimerization, respectively, derived from the equations given by Wehmiller *et al.* (1988a, Appendixes I–III). These two amino acids are used most frequently in aminostratigraphic and aminochronologic studies. These model curves, like all other kinetic models proposed by Wehmiller and colleagues (Wehmiller *et al.*, 1977; Wehmiller and Belknap, 1978), rely upon the assumption that the overall kinetics observed in foraminifera (Fig. 1; Wehmiller and Hare, 1971; King and Neville, 1977) represent the best analog for diagenetic racemization in mollusks. Logarithmic curves are plotted in Figs. 15 and 16, each derived by curve-fitting routines which extrapolate the nonlinear kinetics observed in foraminifera at low temperatures (Wehmiller *et al.*, 1988a) to the higher effective temperatures typical of mid-latitude coastal regions. The temperature values assigned to each curve are implicit to the model used for curve generation (Wehmiller *et al.*, 1988a) and are not necessarily valid for all molluscan taxa unless suitably calibrated. The temperature differences be-

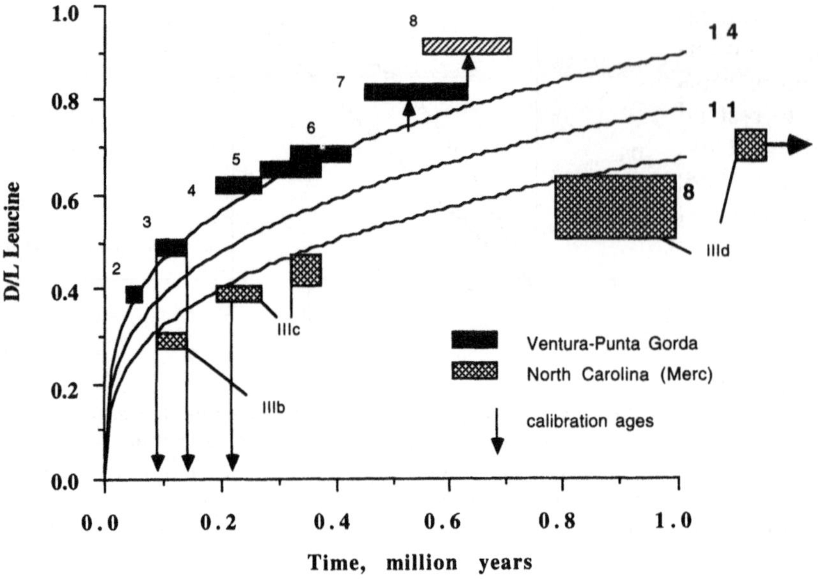

Figure 15. D/L-Leucine racemization kinetic model, derived from the equations of Wehmiller *et al.* (1988a). The curves represent the extrapolation of foraminifera kinetic data to effective temperature of 8, 11, and 14°C. D/L-Leucine data are shown for Ventura–Punta Gorda aminozones 2–8 (Table II) and for North Carolina aminozones IIIb–IIId (Figs. 12 and 13). Aminozones 3 and IIIb are used as the calibration aminozones, with assigned ages of 75–125 ka based on available chronostratigraphic control. A portion of IIIc is also calibrated at about 200 ka. Ages of other aminozones can be estimated by plotting the D/L data on the curve appropriate to the calibration samples.

tween the curves represent the generally observed temperature dependence of racemization. Because these curves are best used where local calibration is available, the curve appropriate for a particular taxon in a given region can be selected once a single calibration point is known, without consideration of any temperature value assigned to that curve.

The stratified and calibrated molluscan aminostratigraphic records described in the previous section can be used to demonstrate the application and testing of kinetic models. This particular form of kinetic modeling is based on temperature assumptions similar to those of aminostratigraphy in that sites with similar CMAT values are assumed to have had similar effective temperatures. Figure 15 shows that if the Ventura–Punta Gorda (Fig. 7) aminozone 3 established from *Macoma* data is used as a stage 5 calibration aminozone and assigned an age of between 80 and 125 ka (an age consistent with available southern California chronostratigraphic control), then age estimates for older aminozones 4–7 can be made using the assumption that these aminozones fall on the same effective temperature curve as the calibration aminozone. The midpoints of the resulting age estimates [cited by Yeats *et al.* (1988)] for these older aminozones are approximately 210, 310, 375, 500 ka, respectively, with uncertainties of about 25%. These uncertainties represent the combined effect of ana-

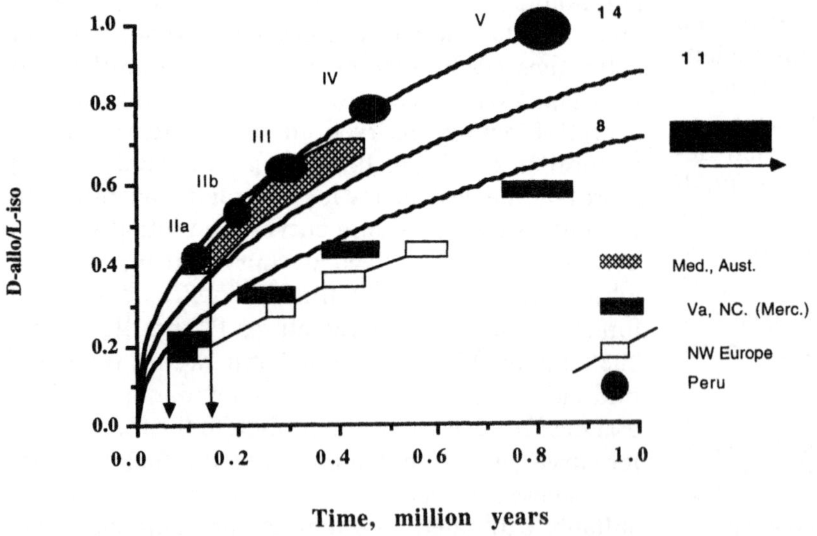

Figure 16. Isoleucine epimerization model curves, derived as in Fig. 15. Data and inferred kinetics for the Mediterranean Basin (Hearty *et al.*, 1986), Australia (Murray-Wallace and Kimber, 1987), Peru (Fig. 10A), the Virginia–North Carolina Coastal Plain (Figs. 12 and 13), and intermediate racemizing taxa from northwest Europe (Miller and Mangerud, 1985; Bowen and Sykes, 1988) are superimposed on these curves, using the indicated aminozones as calibrations.

lytical variability within the aminozone, a range of possible ages for the calibration aminozone, and a range of possible effective temperatures. The age estimates for aminozones 4–7 are consistent with available chronologic control for the deeper part of the section (Fig. 7). Aminozones 7 and 8 are constrained to ages less than 700 ka (Fig. 7); hence, the results for these aminozones suggest that geothermal heating of these deeply buried samples has caused the observed D/L values to be greater than expected from the calibrated curve.

The ages of North Carolina aminozones are estimated in the same manner. Aminozones IIIc and IIId are bimodal, with the lower D/L values in IIIc being calibrated by a 200-ka date and aminozone IIIb being calibrated to between 75 and 125 ka. The ages of the older portion of IIIc and IIId appear to be about 400 ka and about 1 Ma, respectively (Fig. 15) (York *et al.*, 1989).

Figure 16 shows the model curves for isoleucine epimerization and their application to age estimation for the Peruvian samples shown in Fig. 10A. Age estimates of approximately 210, 280, 470 and 900 ka for aminozones IIb, III, IV, and V, respectively, result from Peruvian aminozone IIa being calibrated at 125 ka. Also shown in Fig. 16 are the curves of D-alloisoleucine/L-isoleucine versus time proposed for the molluscan aminochronology of northwest Europe (Miller and Mangerud, 1985; Bowen and Sykes, 1988), the Mediterranean basin (Hearty *et al.*, 1986), and southern Australia (Murray-Wallace and Kimber, 1987; 1993). Note that these latter curves were proposed strictly on the basis of available local chronostratigraphic control, yet their similarity to the curves derived from extrapolated (low-temperature) foraminifera kinetics is striking. One can conclude from these observations that while the details of any kinetic model are dependent on specific genera, there is a consensus among aminostratigraphers about the general trends versus time of molluscan diagenetic racemization or epimerization. Given suitable calibration, or even reasonable inferences about the age of any one aminozone in a particular region, these curves can be used to propose *model age estimates* for comparison with other stratigraphic or chronologic information. The more rigorous the calibration, the more confident one will be in the age estimates derived from the kinetic model. Ideally, the curves used in these models would be radiometrically calibrated at a minimum of two points within the Pleistocene portion of the curve (to reduce the contrast in effective temperatures), but such calibration has been rare in most studies, where it is more common to have last interglacial samples (some or all of isotope stage 5) as the only Pleistocene calibration point.

6. Correlation

Correlation between aminostratigraphic sections is one of the primary goals of regional studies, because local calibration sites are rarely available for all sections of interest and because not all such sections contain complete records of Quaternary interglacials (Fig. 3). Accurate correlation of aminostratigraphic data between regions with different CMATs must demonstrate the principle that D/L values in samples of equal age will increase in proportion to the effective temperature of the sample. There are several approaches to the quantification of this function.

The simplest approach, referred to as the *calibrated correlation method*, relies upon the availability of enantiomeric ratio data for equal-age samples (independently calibrated) from a set of localities with different CMATs, but from within a region that has probably experienced simultaneous temperature changes of comparable magnitude during the Quaternary. The typical format for presentation of results is a graph of D/L values versus CMAT for the region of study, with isochrons drawn to connect D/L values for samples of equal age (calibrated or inferred) at different temperatures (Wehmiller *et al.*, 1977; Wehmiller, 1982, 1984; Miller and Mangerud, 1985; Hearty *et al.*, 1986; Hearty and Miller, 1987; Wehmiller *et al.*, 1988b; Bowen and Sykes, 1988; Hsu *et al.*, 1989). Calibrated isochrons for 125-ka samples are the most commonly constructed isochrons because of the frequent occurrence of samples of this age (last interglacial) in the late Quaternary record. Other samples in the region of interest can be assigned ages greater than, less than, or equal to the calibration isochron based upon their plotted position in this format.

The 125-ka isochron equations for D-alloisoleucine/L-isoleucine values in *Protothaca* for the Pacific coasts of North and South America (Wehmiller *et al.*, 1988b) were compared with a similar isochron derived from data for intermediate-rate racemizing mollusks from Arctic, European, and western Pacific sites (Hearty and Miller, 1987). These equations are as follows:

North America $\quad D/L = 0.171[10^{(0.0247\times CMAT)}] \quad (2)$

South America $\quad D/L = 0.233[10^{(0.0148\times CMAT)}] \quad (3)$

Hearty and Miller $\quad D/L = 0.077[10^{(0.042\times CMAT)}] \quad (4)$

Although the isochrons predicted by Eqs. (2)–(4)

have similar trends, they differ enough in detail (probably because of local differences in temperature history) to prevent their application in regions other than those in which they are calibrated (Wehmiller *et al.*, 1988b; Hsu *et al.* 1989).

Calibrated isochrons (Eqs. 2–4) serve as the basis for *derived isochrons*, those isochrons that can be calculated using the model curves in Fig. 15, fixed to local calibration points. Although derived isochrons are model dependent, they are often the only quantitative correlation tools available for estimation of sample ages. Their accuracy is limited by the implicit assumptions of aminostratigraphy, namely, that effective temperatures are equal for all sites with the same CMAT and that CMAT differences serve as proxies for EQT difference.

The Pacific coast of North America represents a good test of this approach, because of available calibration and a wide latitude/temperature range for localities that have not been exposed to large temperature fluctuations during the late Quaternary. Figure 17 shows an array of calibrated and derived isochrons for D/L leucine values in *Protothaca staminea* from sites between 34° and 23°N (Emerson *et al.*, 1981; Kennedy *et al.*, 1982; Wehmiller, 1984; Wehmiller *et al.*, 1988a, b). The calibration isochron is fixed to 125-ka independently dated sites, and the derived isochrons are shown for selected older ages. The assumptions inherent to age estimates from these derived isochrons can be tested by superimposing an array of isochrons like those in Fig. 17 on a regional data set and comparing the resulting age estimates with any available stratigraphic information (Wehmiller, 1982, Fig. 6).

Figure 17 shows an additional isochron ("Modeled 125,000") for specific EQTs (plotted on the same temperature axis as the calibrated and derived isochrons, which are plotted as a function of CMAT). This isochron represents the locus of D/L leucine values for 125-ka samples, predicted by the kinetic model curves in Fig. 15. Comparison of this isochron and the calibrated isochron permits an evaluation of the kinetic model and an estimate of late Quaternary temperature fluctuations in the region. The vertical spacing between the calibrated and model isochrons for 125 ka represents the reduced extent of racemization caused by the cooler temperatures that samples of this age have experienced. The horizontal spacing between the two isochrons represents the difference between current and effective temperatures needed to account for the observed difference in D/L values. The *Protothaca* data (D/L leucine values of about 0.46) for

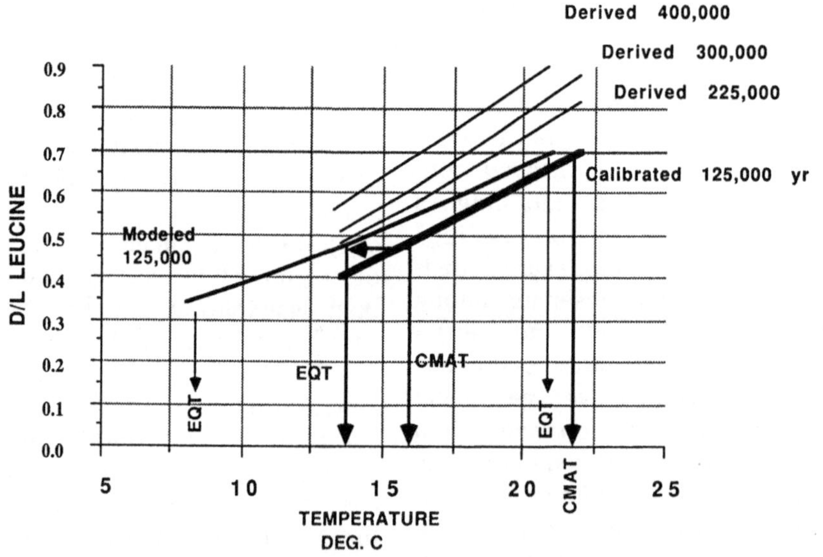

Figure 17. Plot of calibrated, derived, and modeled isochrons, using the kinetic model curves shown in Fig. 15. The model isochron predicts the locus of D/L-leucine values in 125-ka samples at the effective temperatures indicated on the temperature axis. This isochron is a direct consequence of the relation between D/L values, time, and effective temperature depicted in Fig. 15. The calibrated 125-ka isochron is plotted against current temperatures (CMAT) and is the trend observed in mollusk samples of this age from the Pacific coast of North America (Kennedy *et al.*, 1982; Wehmiller, 1982, 1984; Wehmiller *et al.*, 1988b). The derived isochrons are calculated using the curves of Fig. 15, fixed to the D/L values of the calibration isochron. The Nestor Terrace calibration data (Fig. 5) are used here as an example. These D/L values, plotted vs. CMAT (16°C), can be explained by the kinetic model if the effective temperature (EQT) for this region was about 13.5°C. If the EQT and CMAT were both 16°C, the kinetic model predicts that the Nestor Terrace samples would have D/L values about 20% greater than observed.

the 125-ka Nestor Terrace (Fig. 5) demonstrate this principle. These samples are predicted by the kinetic model to have D/L values of about 0.56 if their EQT had been 16°C (equal to CMAT). Instead, their lower D/L value is consistent with an EQT of about 13.5°C (Fig. 17). For samples of this age, the difference between EQT and CMAT represents about 60% of the full-glacial temperature reduction (Wehmiller, 1982, Appendix). The convergence of the modeled and calibrated isochrons in Fig. 17 implies that these full-glacial temperature reductions were less at lower latitudes, consistent with paleoclimatic records for the eastern Pacific (Imbrie *et al.*, 1983). The EQT value derived from the approach outlined in Fig. 17 is dependent on both the kinetic model (Fig. 15) and the genus for which the data are obtained. However, the general reliability of the kinetic model for a particular genus can be tested by comparing these modeled EQT values with estimates of late Quaternary temperature change for selected regions (Wehmiller, 1982, Appendix). Though the number of variables in an exercise such as this may appear excessive, it is possible to reduce these variables to a set of reasonable combinations of kinetic models and temperature histories if enough calibrated samples are available. Regions with moderate late Quaternary temperature excursions are especially helpful for developing these insights. In the absence of any calibrated data, model isochrons like the one shown in Fig. 17 (and others derived from Fig. 15) can be used for age estimation *if* it can be shown that the isochrons are appropriate for the genera being analyzed and if independent estimates of EQT can be made, using paleoclimatic data for the region.

Figure 17 demonstrates a fundamental property (*not* model dependent) of temperature-dependent reactions, namely, that isochrons (representing the extent of the reaction) must rise and diverge toward higher temperatures. Any exercise in correlation of aminostratigraphic data, whether qualitative or quantitative, must be consistent with this principle. Furthermore, it can be expected that the relation of D/L values to latitude will be represented by a function similar to that describing CMAT versus latitude, unless dramatic latitudinal variations in EQT have affected a region.

The aminostratigraphy of Quaternary marine deposits of the Atlantic Coastal Plain of the United States (Fig. 11), reviewed by Wehmiller *et al.* (1988a), presents a particularly interesting challenge to the principles of regional correlation outlined above. Amino acid D/L values exist for over 150 localities between Nova Scotia and Florida, a region characterized by a smooth and nearly linear relation between CMAT (range from 7 to 25°C) and latitude. Coastal Plain U-series coral dates are concentrated at sites near Charleston and Myrtle Beach, South Carolina, and near Norfolk, Virginia (Fig. 11). Two interpretations [presented as Figs. 19 and 21 in Wehmiller *et al.* (1988a)] are available for correlating the regional aminostratigraphic data in terms of these U-series results. These options, which conflict in a few areas of the Coastal Plain, are summarized in Fig. 18. One option, often referred to as the "empirical approach," fixes *all* the amino acid data to the existing U-series dates, thereby assuming that all these dates are correct and must be used as rigorous calibrations for the aminostratigraphy (Szabo, 1985; Corrado *et al.*, 1986; Cronin, 1987; Wehmiller *et al.*, 1987; Hollin and Hearty, 1990; Wehmiller *et al.*, 1992). This empirical approach *requires* the isochrons of D/L values to follow trends that *would not be predicted* by current latitudinal temperature trends, instead being extremely steep in southeastern North Carolina and perhaps even inverted in Virginia (Wehmiller and Belknap, 1982; Wehmiller *et al.*, 1988a). The other option, more theoretical, questions the accuracy of *some* of the U-series dates and instead relies upon the aminostratigraphic strategy (Fig. 17) that assumes that isochrons of D/L values should be smoothly varying functions of latitude or CMAT. The empirical approach requires little or no similarity between the kinetic pathways or temperature histories for the Virginia and South Carolina regions (Wehmiller *et al.*, 1987, 1988a), thereby violating some of the basic assumptions of aminostratigraphy. The theoretical approach requires that the South Carolina stage 5 (85- to 120-ka) U-series coral dates be increased, by factors of 2–3, in order to maintain the integrity of these fundamental principles. The older ages (ca. 200 to 500 ka) for the Myrtle Beach aminozones are supported by a combination of biostratigraphic and isotopic data (McCartan *et al.*, 1982). Both the theoretical and empirical approaches require that *some* of the U-series dates be rejected as calibrations for portions of the Coastal Plain aminostratigraphy, because of either coral reworking or geochemical alteration of the dated corals.

Although there are several detailed issues that remain to be resolved in the Atlantic Coastal Plain research, the essential conflict defined by the results shown in Fig. 18 has to do with the presence or absence of emergent stage 5 (ca. 125–80 ka) deposits in northeastern South Carolina. The empirical approach outlined above assumes that these deposits are present, because deposits of this age (the most

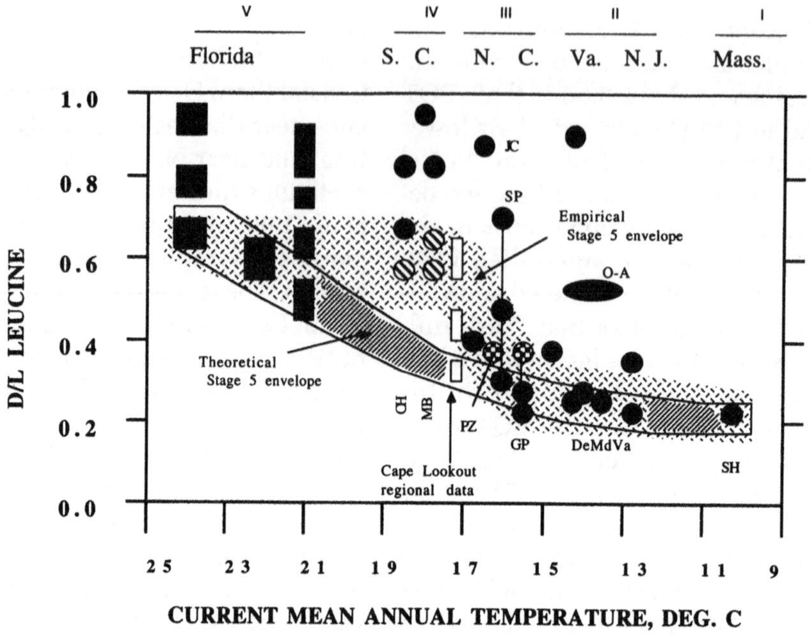

Figure 18. Summary of the U.S. Atlantic Coastal Plain aminostratigraphic data, based on *Mercenaria* D/L leucine values. Roman numerals identify aminostratigraphic regions as presented by Wehmiller *et al.* (1988a). Shaded circles represent regional aminozones between South Carolina and Massachusetts (Wehmiller *et al.*, 1988a). Shaded rectangles represent Florida aminozones inferred by Mitterer (1975). The "stage 5 empirical" and "stage 5 theoretical" correlation envelopes represent the two conflicting interpretations of the Coastal Plain aminostratigraphy (Wehmiller *et al.*, 1988a, 1992). The data points connected by vertical lines represent important superposed sections at Stetson Pit, North Carolina (SP), and Gomez Pit, Virginia (GP). The black squares and circles represent data points with "stage 5" calibration by U–Th dating of solitary corals (at least 80% of these dates are between 70 and 90 ka; the remainder are between 100 and 130 ka). The checkered circles labeled Ponzer (PZ) and Gomez Pit (GP) represent data points with U–Th or ESR calibrations, respectively, of "stage 7." Open boxes depict the aminostratigraphic data obtained from the Cape Lookout, North Carolina, region (Fig. 11). Other important aminostratigraphic reference units are SH (Sankaty Head, Massachusetts), O-A (Omar-Accomac Formation, Chesapeake Bay region), and JC (James City Formation). Diagonally shaded circles at Charleston (CH) and Myrtle Beach (MB), South Carolina, emphasize the magnitude of the difference between the theoretical and empirical isochrons. These sites are interpreted by some workers (e.g., Hollin and Hearty, 1990) as representing stage 5, perhaps even early and late stage 5, because of selected U–Th coral dates from the Myrtle Beach region and additional U–Th coral dates from the Charleston region. The Myrtle Beach units represented by these data points are the Socastee and Canepatch Formations, previously considered as "middle Pleistocene" (older than "Sangamon" and perhaps as old as 600 ka, respectively, based on a combination of U–Th and biostratigraphic data).

recent Pleistocene interglacial) are commonly encountered elsewhere (Hollin and Hearty, 1990). Two U-series coral dates in this age range have been obtained from this region, but the majority of the U-series dates from northeastern South Carolina are greater than 200 ka (Szabo, 1985), and the younger dates show evidence of open-system behavior, and hence are considered suspect (Szabo, 1985). The theoretical approach, involving interpolation of aminostratigraphic data from regions to the north and south, based on present latitudinal gradients of CMAT, indicates that the predominant aminozones found in emergent deposits in central and northeastern South Carolina are all older than stage 5 (Fig. 18) and that stage 5 material is missing from the sampled record. Recently obtained aminostratigraphic data (Wehmiller *et al.*, 1988a, 1989, 1992; York and Wehmiller, 1992) from subsurface and offshore sections from the Cape Lookout–Cape Fear region, North Carolina (Fig. 11), combined with offshore stratigraphic data (Mixon and Pilkey, 1976; Meisburger, 1979; Hine and Snyder, 1985), provide strong geomorphic evidence in favor of the theoretical approach to Coastal Plain aminostratigraphic correlation. It appears that the inner shelf area south–southwest of Cape Lookout (west of the southward projection of the Suffolk Scarp; Fig. 11) has been nondepositional during the middle and late Pleistocene, because this region of the shelf is dominated by Tertiary and early Pleistocene outcrops [see Riggs and Belknap (1988) for a review of the processes affecting this region)]. The thin-to-absent Quaternary record represents the ongoing lack of deposition over the Cape Fear Arch (Fig. 11) during the entire Cenozoic (Gohn, 1988, Fig. 2). Aminostratigraphic data for inner shelf sections around Capes Fear and Lookout reflect these nondepositional processes: south and west of Cape Lookout, samples from the inner shelf and from subsurface portions of the barrier islands indicate that the *youngest* pre-Holocene aminozone is of IIId "age" (which is interpreted to represent middle or early Pleistocene samples see Fig. 12 and Table III), while just to the north and northeast of Cape Lookout, samples from aminozone IIId are found in thicker Quaternary sections (≥15 m), in clear stratigraphic relation to younger aminozone IIIc, IIIb, and IIIa (middle to late Pleistocene) samples. Consequently, Wehmiller *et al.* (1992) concluded that the "offset" or steep gradient of D/L values that is required by the empirical correlation model (Fig. 18) is actually an artifact of the assumption that the youngest aminozone found along the Coastal Plain should everywhere be last interglacial (stage 5) in age. Instead, it appears that the "missing" young aminozones southwest of Cape Lookout simply represent a continuation into the Quaternary of a long (ca. 75 Myr) history of nondeposition in the Cape Fear Arch region (Gohn, 1988). Though these results are preliminary and require extensive investigation, they are critically important because they will confirm or deny the theoretical basis for aminostratigraphic correlation between sections at different latitudes or CMATs.

7. Geochemical Techniques for the Recognition of Anomalous Results

Beyond simple stratigraphic tests, there are several geochemical criteria that can be used (with partial success) to evaluate the quality of amino acid enantiomeric ratio data. Murray-Wallace and Kimber (1987) and Miller and Brigham-Grette (1989) reviewed some of the criteria that others have proposed (see also Wehmiller, 1982; Wehmiller and Belknap, 1982; Wehmiller *et al.*, 1988a). The precision of results for samples from a specific site or aminozone is often used as one of the basic criteria for acceptability. Precision is a function of sample quality and the number of analyses and determines the ability of AAR methods to define different aminozones. In some cases, the available precision does not permit resolution of physically mappable lithostratigraphic units (perhaps because they are not distinctly different in age), while in other cases high-precision aminostratigraphic data suggest separate ages where only one lithostratigraphic unit can be mapped (Wehmiller *et al.*, 1988a; Miller and Brigham-Grette, 1989). Although instrumental precision is often as good as 2%, most single-site, single-genus precisions [expressed as coefficients of variation (CV)] are between 5 and 8% and most aminozone CVs are between 8 and 12%.

Other tests of sample or data integrity often rely upon a general diagenetic model that includes leaching (loss of amino acids) and contamination (gain of amino acids); both of these processes are expected to lower a sample's D/L values from the true values, but they produce opposite effects on the sample's amino acid content. Graphs relating total amino acid content (or abundances of particular amino acids such as thermally unstable serine) to D/L values are often used to depict the covariance of these independent diagenetic parameters and to identify the possible effects of leaching or contamination on samples. King (1980) used quantitative amino acid data as a measure of sample leaching and attempted to relate the extent of leaching to the level of randomness in observed amino acid data. Corrado *et al.* (1986) used a plot of total amino acid content versus D/L values in *Mulinia lateralis* shells to evaluate sample alteration, working with the model that leaching would remove extensively racemized free amino acids, lowering both the amino acid content and total hydrolysate D/L values in altered samples. Corrado *et al.* (1986) also plotted histograms of either nmol/g or D/L data (in the same samples) in an attempt to identify different apparent ages, but these plots showed that the relation between these two variables is not monotonic (Wehmiller *et al.*, 1987). In addition, data presented in Fig. 19 (from Müller, 1984) show how difficult it can be to use these effects as objective criteria for sample evaluation. Figure 19 shows total isoleucine (D-allo- + L-isoleucine) versus time and D/L value, for *O. universa* and *G. tumida-menardii*. For these two genera, the only substantial deviation in the racemization curve is seen for the oldest portion (≥600 ka) of the *G. tumida-menardii* data (Fig. 1). These results are accompanied by an apparent increase (contamination?) in amino acid content when plotted versus time but no obvious deviation in amino acid content when plotted versus D/L value, which is the only objective test (of those presented) that can be used to evaluate the chronological significance of D/L data. For *O. universa*, there is substantial scatter in quantitative results that has little or no parallel in the racemization data, further indicating the difficulty of using plots such as these as criteria for evaluation of data integrity.

Other co-varying relations are useful for evaluation of internal consistency. Examples include intrageneric plots, which represent the trends of different D/L values in either the total or free amino acid population, as seen in Fig. 8 (see also Lajoie *et al.*, 1980;

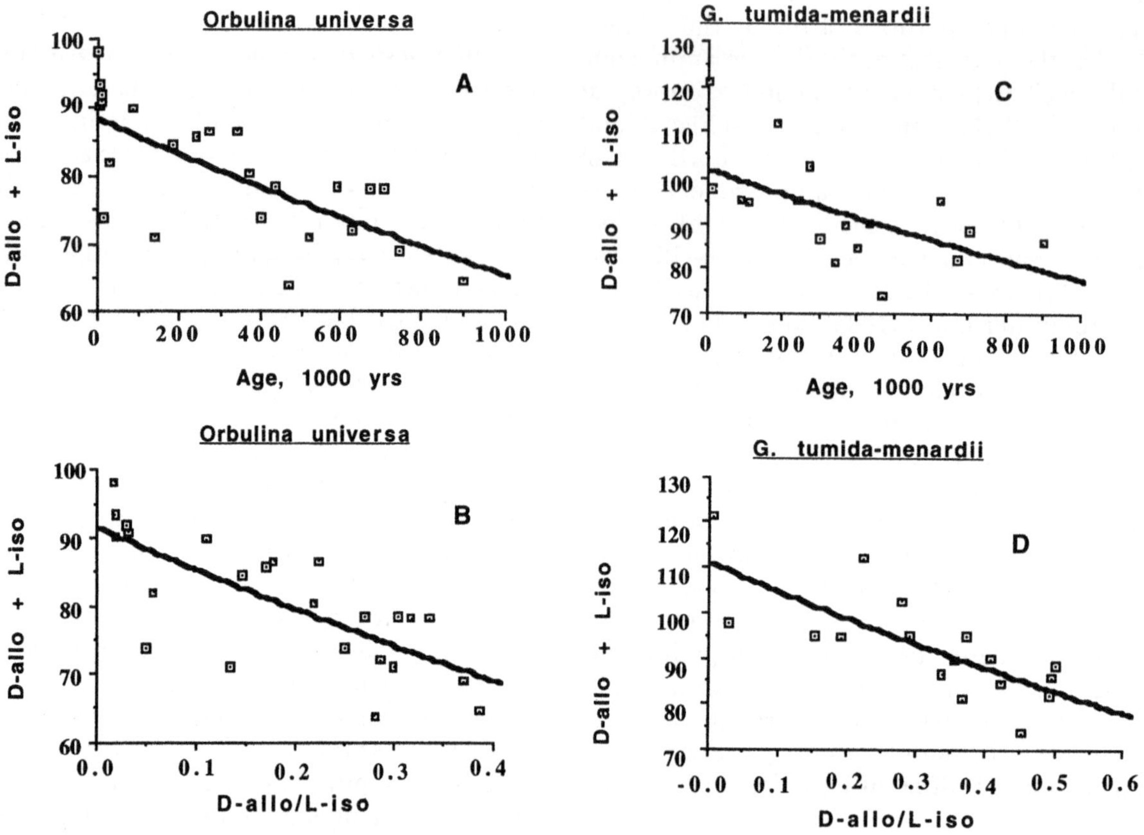

Figure 19. Quantitative amino acid abundance data, and epimerization data, for the foraminifera *O. universa* and *G. tumida-menardii*, from Müller (1984). See Fig. 1 for related plots. Equations describing the curves are as follows: (A) D-allo + L-iso = $88.1(10^{(-0.000013)t})$, $R = 0.71$; (B) D-allo + L-iso = $91.3(10^{([-0.31(A/I)])})$, $R = 0.74$; (C) D-allo + L-iso = $101.4(10^{(-0.00012)(t)})$, $R = 0.58$; (D) D-allo + L-iso = $110.7(10^{[-0.26(A/I)]})$, $R = 0.75$, where t is the sample age in 1000 years, and A/I = D-alloisoleucine/L-isoleucine.

Miller, 1985). Intergeneric enantiomeric ratio data (such as seen in Fig. 2) are especially valuable tests of internal consistency, because aminostratigraphic conclusions must be supported by data for more than one genus. Because examples are known where either intergeneric or intrageneric data do not yield identical aminostratigraphic conclusions (Wehmiller *et al.*, 1977; Kennedy, 1978; Wehmiller *et al.*, 1988a; Rutter *et al.*, 1989; Muhs *et al.*, 1990), aminozones constructed from data for a single genus or a single amino acid should be based on conservative estimates of aminozone precision (assumed CV no less that 10%) unless there is clear geologic evidence for better chronostratigraphic resolution in the data. Preservation characteristics (visual, mineralogic, etc.) of analyzed shells can often be used as qualitative guides prior to analysis, but examples can be cited in which these criteria do not specifically predict observed enantiomeric ratios (Wehmiller, unpublished results; Mirecki, 1990). Advancing analytical capabilities permit the study of AAR in smaller samples. Such improving sensitivity allows the quality of large samples to be evaluated by serial or multiple-fragment analyses. Conversely, however, enhanced analytical sensitivity, when applied to very small samples (with proportionately large surface areas) produced by natural fragmentation processes, requires that the chemical integrity of the samples be evaluated as fully as possible.

We are in the process of evaluating an alternative approach to the comparison of enantiomeric ratio and quantitative amino acid data. This method is derived from standard isochron plots of parent–daughter isotope ratio pairs, particularly the Rb–Sr system, where $^{87}Sr/^{86}Sr$ values (daughter) are graphed against $^{87}Rb/^{86}Sr$ values (parent). In the application of this model to amino acid data, we use the relative abundance of D-alloisoleucine (normalized to the abundance of amino acids such as valine, leucine, or alanine) as a measure of the "daughter" and plot these parameters against the same relative abundance of the "parent" (L-isoleucine). A graph with representative values is produced if the daughter and parent are each compared to three stable amino acids with substantially different abundances (usually leucine, valine, and alanine), so that the numerical values of these ratios will have a range that is statistically useful in defining the "isochron." Figure 20 shows the general format for presentation of data derived from typical liquid-chromatographic amino acid analyses. One of the advantages of this approach is that ratios of amino acids are used rather than quantitative abundances, which are not easily determined with precisions better than about 15%.

The basic premise of the approach shown in Fig. 20 is that zero-age (modern) samples start with D-alloisoleucine/valine, D-alloisoleucine/alanine, and D-alloisoleucine/leucine values of zero and with finite and rather narrowly defined L-isoleucine/valine, L-isoleucine/alanine, and L-isoleucine/leucine values [see, e.g., Andrews *et al.* (1985)]. As diagenesis proceeds, the ratios involving D-alloisoleucine will increase at the expense of those involving L-isoleucine. In a perfectly closed system, each unit decrease in the L-isoleucine ratios would be matched by an equivalent increase in the ratios involving D-alloisoleucine.

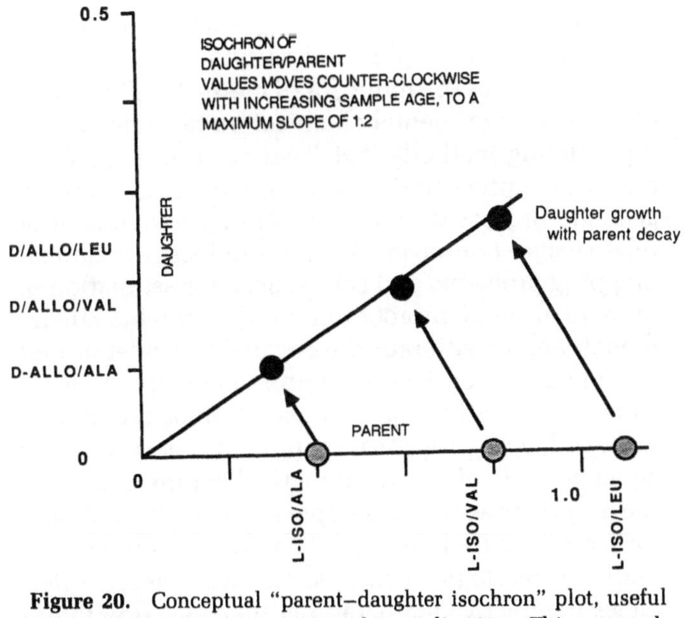

Figure 20. Conceptual "parent–daughter isochron" plot, useful in high-resolution aminostratigraphic applications. This approach is applicable when liquid-chromatographic data for D-alloisoleucine (D-allo) and L-isoleucine (L-iso) can be normalized to the abundances of other amino acids in the sample. In this example, alanine, valine, and leucine are used. Ratios involving L-isoleucine decrease with time as the ratios involving D-alloisoleucine increase from zero in modern samples. The slope of the best-fit isochron should represent the D-alloisoleucine/L-isoleucine value in the sample.

Normal scatter observed in these ratios indicates that sample systems are not closed. For a group of samples from a single deposit, all analyzed in this manner, the slope of the best-fit line through all the data points should be the same as the average D-alloisoleucine/L-isoleucine ratio determined directly from all the samples. The intercept of this line should be the origin. In a few detailed studies of 80- and 125-ka (stage 5) marine terrace deposits in central California (Kelson *et al.*, 1987; Wehmiller, 1992 and references therein), we have observed that the intercept value can deviate from the origin while the precision of multiple D/L values appears to remain acceptable. Furthermore, the intercept value deviates most from the origin when the linear least-squares regression coefficient for the daughter–parent ratio plot drops below 0.985. The relative age assignments for samples with low R values (<0.985) conflict with the available stratigraphic and age control, though they are still consistent with a stage 5 age estimate. Consequently, it is possible to establish as a standard of acceptability a minimum correlation coefficient of 0.990 and to reject data which cause this coefficient to drop below this value, if high-resolution aminostratigraphic interpretations are to be accomplished. Analytical strategies such as these require multiple analyses of the same genus from all sites in question.

Thermal factors are potentially the greatest sources of uncertainty in AAR applications. The kinetic and correlation curves presented above (Figs. 15–17) demonstrate how postulated differences in effective temperatures between calibrated and uncalibrated samples result in various age estimates. Without rigorously calibrated samples from a wide range of ages for sites within a region of identical CMATs, it is almost impossible to determine if different effective temperatures are appropriate for these sites. Consequently, aminostratigraphers are forced to assume the equality of effective temperatures for all samples within a region of identical CMATs. The coefficient of variation of any inferred regional aminozone is directly linked to the uncertainties inherent to this assumption about effective temperature equalities.

This assumption has been proven incorrect in a few cases, most notably where comparatively young samples (Holocene) have been shallowly buried (<1 m) in deposits where significant temperature excursions (daily or seasonal) occur (Wehmiller, 1977; Miller and Brigham-Grette, 1989). In these cases, effective temperatures are substantially greater (often 3–6°C) than the CMAT, so ages inferred from comparison of D/L values to the CMAT can appear incorrectly old. This effect is most important for samples that have been buried in the thermally "active" (large-amplitude daily and seasonal temperature variations) soil zone for a substantial proportion (ca. 25% or more) of their geological age. It can be an important factor for Pleistocene samples in arid continental environments but is probably not a factor for most coastal or marine deposits, where deeper burial and smaller temperature excursions are the general rule. In southern California (CMAT 15.5–16.0°C), shallowly buried late Holocene mollusks from archeological sites have D/L values almost as high as those of samples of the same genus from late Pleistocene marine terrace deposits (Wehmiller, 1977; Wehmiller *et al.*, 1977), demonstrating the importance of understanding the depositional context of all analyzed samples. The magnitude of effects of thermal factors on D/L values in Pleistocene samples can be estimated by comparing results from surface samples with those from deep within the outcrop. Tests at several U.S. Atlantic and Pacific coast sites have demonstrated that these effects are not statistically detectable (Wehmiller, unpublished results).

One example of probable combined chemical and thermal effects on enantiomeric ratio data is found in mollusk samples from the Sankaty Sand, a shallow marine deposit exposed between glacial units at Sankaty Head, Nantucket Island, Massachusetts (Oldale *et al.*, 1982). The Sankaty Sand is divided into upper and lower units, separated by a pronounced unconformity. The lower unit (1–3 m thick) is a fine indurated sand with abundant serpulid worm tubes and articulated valves of *Crassostrea virginica* and *Mercenaria mercenaria*, many in growth position. The upper Sankaty Sand, generally less than 0.5 m thick, consists of coarser sand and gravel with abundant disarticulated and fragmented shells, including *Mercenaria* and *Crassostrea*. The two units are interpreted as transgressive deposits (lower) overlain by a reworked beach sand (upper) (Oldale *et al.*, 1982).

A peculiar stratigraphic inversion of enantiomeric ratios in *Mercenaria* samples from the upper and lower Sankaty Sand exists (Oldale *et al.*, 1982). For several amino acids D/L values in upper Sand samples were generally 8–15% greater than those in lower Sand samples of the same genus (Oldale *et al.*, 1982; Wehmiller, unpublished results). Because the upper Sand mollusk samples most likely represent reworked lower Sand material, the difference in D/L values between the two units is an apparent consequence of different thermal or diagenetic histories. Rahaim (1986), using Sankaty Sand samples collected specifically to evaluate the effects of these differences, confirmed the observations made by Oldale *et al.* (1982). Rahaim (1986) also showed that no statistically significant difference in D/L values was observed between the upper and lower Sand when only articulated, well-preserved (lustrous) samples were analyzed from the two units (four samples from each unit). The samples from the upper Sand that yield anomalously high D/L values were chalkier than those from the lower Sand, probably as a consequence of surf-zone abrasion with associated shallow-burial thermal effects. The shells in the upper Sand that yield these anomalous D/L values occur at the base of a thick (25 m) sequence of porous sand units, where groundwater flow is perched on the indurated lower Sankaty Sand. Consequently, the upper Sand samples are exposed to a diagenetic environment that differs both chemically and thermally from the environment of the lower Sand samples. Recognition of effects such as these is often difficult when D/L data for isolated localities are compared, but the Sankaty Sand example serves as a useful guide for understanding the level of precision that is realistically achievable in aminostratigraphy (Wehmiller, 1987).

8. Summary and Conclusions

In the past 15 years, AAR has found abundant use as a Quaternary stratigraphic and chronologic tool, especially in studies involving fossil mollusks. This is one of several experimental chemical, radiogenic, or isotopic dating methods that have been developed recently to improve understanding of many Quaternary geologic processes. AAR is useful for assignment of relative ages (aminostratigraphy) in local or regional mapping problems and can be used for estimation of numerical ages (aminochronology) if suitable kinetic models and/or calibration are available. Under proper circumstances of data or sampling density, AAR results have a stature equal to that of lithostratigraphic data, and, with proper calibration, they have a stature equal to that of many numerical dating methods. Assumptions about the temperature histories of samples are required for both aminostratigraphic and aminochronologic methods. In most cases, *differences* in effective temperatures are equated with differences in current mean annual temperatures, the former being the kinetically important variable but the latter being "real data." Numerical values of effective temperatures can be estimated independently if paleoclimatic data for a region are available, or if an appropriate kinetic model and calibrated Pleistocene samples are available.

Because of the experimental nature of AAR, the method must be evaluated repeatedly using field and laboratory data. Tests of the "local reliability" of AAR are best done using well-constrained lithostratigraphic sequences to establish aminostratigraphic reference sections. Examples from uplifted, stable, and subsiding coastal records in North and South America were cited here, and many other examples from Europe, Australia, and the Arctic could be used to expand this documentation. Further field testing can be made using data from samples with well-established radiometric dates and by correlation between local aminostratigraphic sections. Age estimates derived by extrapolation of model kinetics from calibrated to uncalibrated samples have an inherent uncertainty that may appear unacceptably large to those more familiar with the uncertainties quoted for radiometric dating methods. Nevertheless, extrapolated AAR age estimates are often more readily available than other chemical or isotopic data (because of sample availability or analytical advantages of AAR), thereby being more easily tested against all independent stratigraphic information. Laboratory studies of amino acid diagenetic reactions provide insight into natural reactions, but the real utility of AAR in chronostratigraphic applications must be established using field samples that have been exposed to natural diagenesis. The aminostratigraphic records reviewed here are clear examples of the fact that tests of AAR data in *local* stratigraphic exercises (deep-sea and coastal sediment sequences) almost always demonstrate the stratigraphic coherence of the data. Correlation between these local aminostratigraphic sections (often required in order to relate calibrated and uncalibrated data) is attempted using isochrons (based on models or calibrated data) and assumptions about relative effective temperatures. Problematic correlations demonstrate the potential inapplicability of these assumptions about effective temperature, but also identify issues about the accuracy of the available control used to test proposed aminostratigraphic correlations. Proposed correlations using data for rate-equivalent samples must be internally consistent with appropriate kinetic and temperature history models, if the fundamentals of the method are to be considered regionally (or universally) applicable.

The temperature uncertainties inherent to any chemical dating method usually affect the precision of a stated age estimate more than analytical uncertainties. Aside from the temperature uncertainty factor, chemical dating methods are similar to radiometric dating methods because both types of methods rely upon changing proportions of products (daughters) and reactants (parents), proportions that can be affected by diagenetic reactions other than the one with direct geochronological application. Because the development of objective analytical criteria for recognition of altered samples remains an elusive goal for many dating methods, evaluation of these methods usually relies on frequent testing of chemical-stratigraphic data within evolving frameworks of time-, bio-, and lithostratigraphic information.

ACKNOWLEDGMENTS. Many of the results reviewed here were obtained with support from the U.S. Geological Survey and the U.S. National Science Foundation (grants EAR-8407024 and EAR-8915747). I have benefitted from collaboration with many who were involved in the work reported here, including D. F. Belknap, A. L. Bloom, B. S. Boutin, T. M. Cronin, J. T. Hsu, E. M. Keenan, G. L. Kennedy, K. R. Lajoie, E. M. Leonard, J. E. Mirecki, S. D. Rahaim, and L. L. York. Special gratitude for the use of unpublished results goes to D. F. Belknap, J. T. Hsu, E. M. Leonard, J. E. Mirecki, S. D. Rahaim, and L. L. York.

References

Andrews, J. T. Shilts, W. W., and Miller, G. H., 1983, Multiple deglaciations of the Hudson Bay Lowlands, Canada, since deposition of the Missinaibi (last interglacial?) Formation, *Quat. Res.* **19**:18–37.

Andrews, J. T., Miller, G. H., Davies, D. C., and Davies, K. H., 1985, Generic identification of fragmentary Quaternary molluscs by amino acid chromotography: A tool for Quaternary and paleontological research, *Geol. J.* **20**:1–20.

Bada, J. L., 1985a, Amino acid racemization dating of fossil bones, *Ann. Rev. Earth Planet. Sci.* **13**:241–268.

Bada, J. L., 1985b, Racemization of amino acids, in: *Chemistry and Biochemistry of the Amino Acids* (G. C. Barrett, ed.), Chapman and Hall, London, pp. 399–414.

Belknap, D. F., 1979, Application of amino acid geochronology to stratigraphy of the late Cenozoic marine units of the Atlantic coastal plain, Ph.D. Dissertation, University of Delaware, Newark.

Belknap, D. F., and Wehmiller, J. F., 1980, Amino acid racemization in Quaternary mollusks: Examples from Delaware, Maryland, and Virginia, in: *Biogeochemistry of Amino Acids* (P. E. Hare, T. C. Hoering, and K. King, Jr., eds.), John Wiley & Sons, New York, pp. 401–414.

Bowen, D. Q., and Sykes, G. A., 1988, Correlation of marine events and glaciations on the northeast Atlantic margin, *Philos. Trans. R. Soc. London, Ser. B* **318**:619–635.

Bowen, D. Q., Sykes, G. A., Reeves, A., Miller, G. H., Andrews, J. T., Brew, J. W., and Hare, P. E., 1985, Amino acid geochronology of raised beaches in southwest Britain, *Quat. Sci. Rev.* **4**:279–318.

Bowen, D. Q., Hughes, S., Sykes, G. A., and Miller, G. H., 1989, Land-sea correlations in the Pleistocene based on isoleucine epimerization in non-marine molluscs, *Nature* **340**:49–51.

Brigham, J. K., 1983, Stratigraphy, amino acid geochronology, and correlation of Quaternary sea level and glacial events, Broughton Island, Arctic Canada, *Can. J. Earth Sci.* **20**:577–598.

Cann, J. H., Belperio, A. P., Gostin, V. A., and Murray-Wallace, C. V., 1988, Sea-level history, 45,000 to 30,000 yr B.P., inferred from benthic foraminifera, Gulf St. Vincent, South Australia, *Quat. Res.* **29**:153–175.

Chapell, J., and Shackleton, N. J., 1986, Oxygen isotopes and sea level, *Nature* **324**:137–140.

Clark, P. U., Nelson, A. R., McCoy, W. D., Miller, B. B., and Barnes, D. K., 1989, Quaternary aminostratigraphy of Mississippi Valley loess, *Geol. Soc. Am. Bull.* **101**:918–926.

Colman, S. M., and Mixon, R. B., 1988, The record of major Quaternary sea-level changes in a large coastal plain estuary, Chesapeake Bay, eastern United States, *Palaeogeogr. Palaeoclimatol. Paleaeoecol.*, **68**:99–116.

Colman, S. M., Pierce, K. L., and Birkeland, P. W., 1987, Suggested terminology for Quaternary dating methods, *Quat. Res.* **28**: 314–319.

Corrodo, J. C., Weems, R. E., Hare, P. E., and Bambach, R. K., 1986, Capabilities and limitations of applied aminostratigraphy as illustrated by analyses of *Mulinia lateralis* from the late Cenozoic marine beds near Charleston, South Carolina, *S. C. Geol.* **30**:19–46.

Cronin, T. M., 1987, Quaternary sea level studies in the eastern United States: A methodological perspective, in: *Sea Level Changes* (M. J. Tooley and I. Shennan, eds.), Blackwell Scientific, Oxford, pp. 225–248.

DuBar, J. R., DuBar, S. S., Ward, L. W., Blackwelder, B. W., Abbott, W. H., and Huddleston, P. F., 1980, Cenozoic biostratigraphy of the Carolina outer coastal plain, in: *Excursions in Southeastern Geology: Geological Society of America 1980 Annual Meeting Field Trip Guidebook*, Vol. 1 (R. W. Frey, ed.), Geological Society of America, Boulder, Colorado, pp. 179–236.

Eames, G. B., 1983, The Late Quaternary seismic stratigraphy, lithostratigraphy and geologic history of a shelf–barrier–estuarine system, Dare County, North Carolina, Master of Science Thesis, East Carolina University.

Easterbrook, D. J. (ed.), 1988, *Dating Quaternary Sediments*, Geological Society of America Special Paper 227.

Emerson, W. K., Kennedy, G. L., Wehmiller, J. F., and Keenan, E. M., 1981, Age relations and zoogeographic implications of late Pleistocene marine invertebrate faunas from Turtle Bay, Baja California Sur, Mexico, *Nautilus* **95**:105–116.

Engel, M. H., and Hare, P. E., 1985, Gas liquid chromatographic separation of amino acids and their derivatives, in: *Chemistry and Biochemistry of Amino Acids* (G. C. Garrett, ed.), Chapman and Hall, London, pp. 462–479.

Gohn, G. S., 1988, Late Mesozoic and early Cenozoic geology of the Atlantic Coastal Plain: North Carolina to Florida, in: *The Atlantic Continental Margin: U.S.* (R. E. Sheridan and J. A. Grow, eds.), Geological Society of America Decade of North American Geology Vol. 1–2, pp. 107–130.

Goodfriend, G. A., and Mitterer, R. M., 1988, Late Quaternary land snails from the north coast of Jamaica: Local extinctions and climatic change, *Palaeogeogr. Palaeoclimatol. Palaeoecol.* **63**: 293–311.

Groot, J. J., Ramsey, K. W., and Wehmiller, J. F., 1990, Ages of the Bethany, Beaverdam, and Omar Formations of southern Delaware, Delaware Geological Survey Report of Investigations No. 47, pp.1–19.

Hare, P. E., Hoering, T. C., and King, K., Jr. (eds.), 1980, *Biogeochemistry of Amino Acids*, John Wiley & Sons, New York.

Hare, P. E., St. John, P. A., and Engel, M. H., 1985, Ion exchange separation of amino acids, in: *Chemistry and Biochemistry of Amino Acids* (G. C. Barrett, ed.), Chapman and Hall, London, pp. 415–425.

Hearty, P. J., and Aharon, P., 1988, Amino acid chronostratigraphy of late Quaternary coral reefs: Huon Peninsula, New Guinea and the Great Barrier Reef, Australia, *Geology* **16**:579–583.

Hearty, P. J., and Miller, G. H., 1987, Global trends in isoleucine epimerization: Data from the circum-Atlantic, the Mediterranean, and the South Pacific, in: *Geological Society of America Abstracts with Programs*, Vol. 19, p. 698.

Hearty, P. J., Miller, G. H., Stearns, C. E., and Szabo, B. J., 1986, Aminostratigraphy of Quaternary shorelines in the Mediterranean Basin, *Geol. Soc. Am. Bull.* **97**:850–858.

Hine, A. C., and Snyder, S. W. P., 1985, Coastal lithosome preservation: Evidence from the shoreface and inner continental shelf off Bogue Banks, North Carolina, *Mar. Geol.* **63**:307–330.

Hollin, J. T., and Hearty, P. J., 1990, South Carolina interglacial sites and Stage 5 sea levels, *Quat. Res.* **33**:1–17.

Hsu, J. T., 1988, *Emerged Quaternary marine terraces in southern Peru: Sea level changes and continental margin tectonics over the subducting Nazca Ridge*, Ph.D. Dissertation, Cornell University.

Hsu, J. T., 1992, Quaternary uplift of the Peruvian coast related to subduction of the Nazca Ridge: 13.5 to 15.6 degrees south latitude, *Quat. Int.* **15/16**:87–98.

Hsu, J. T., Leonard, E. M., and Wehmiller, J. F., 1989, Aminostratigraphy of Peruvian and Chilean Quaternary marine terraces, *Quat. Sci. Rev.* **8**:255–262.

Hurford, A. J., Jager, E., and Ten Cate, J. A. M. (eds), 1986, *Dating Young Sediments*, United Nations CCOP Publication 16, Bangkok, pp. 1–393.

Imbrie, J., McIntyre, A., and Moore, T. C., Jr., 1983, The ocean around North America at the last glacial maximum, in: *Late Quaternary Environments of the United States, The Late Pleistocene*, Vol. 1 (S. Porter, ed.), University of Minnesota Press, Minneapolis, pp. 230–236.

Imbrie, J., Hays, J. D., Martinson, D. G., McIntyre, A., Mix, A. C., Morley, J. J., Pisias, N. G., Prell, W. L., and Shackleton, N. J., 1984, The orbital theory of Pleistocene climate: Support from a revised chronology of the marine ^{18}O record, in: *Milankovitch and Climate* (A. Berger, J. Imbrie, J. D. Hays, G. Kukla, and B. Saltzman, eds.), Reidel, Dordrecht, pp. 269–305.

Jansen, E., and Sejrup, H. P., 1986, Stable isotope stratigraphy and amino acid epimerization for the last 2.4 m. y. at site 610, holes 610 and 610A in: *Initial Reports of the Deep Sea Drilling Project*, Vol. 54, (W. F. Ruddiman, R. B. Kidd, E. Thomas, *et al.*, eds.), U.S. Government Printing Office, Washington, D.C., pp. 879–888.

Kaufman, A., Broecker, W. S., Ku, T. L., and Thruber, D. S., 1971, The status of U-series methods of mollusk dating, *Geochim. Cosmochim. Acta* **35**:1155–1183.

Kaufman, D. S., and Miller, G. H., 1992, Overview of amino acid geochronology, *Comp. Biochem. Physiol.* **102B(2)**:199–204.

Kaufman, D. S., Miller, G. H., and Andrews, J. T., 1992, Amino acid composition as a taxonomic tool for molluscan fossils: an example from Pliocene-Pleistocene Arctic marine deposits, *Geochim. Cosmochim. Acta* **56**:2445–2453.

Kelson, K. I., Lettis, W. R., Weber, G. E., Kennedy, G. L., and Wehmiller, J. F., 1987, Amount and timing of deformation along the Wilmar Avenue, Pismo, and San Miguelito Faults, Pismo Beach, California, in: *Geological Society of America Abstracts with Programs*, Vol. 19, p. 394.

Kennedy, G. L., 1978, Pleistocene paleoecology, zoogeography and geochronology of marine invertebrate faunas of the Pacific Northwest Coast (San Francisco Bay to Puget Sound), Ph.D. Thesis, University of California at Davis.

Kennedy, G. L., 1988, Zoogeographic discordancy in Late Pleistocene northeastern pacific marine invertebrate distributions explained by astronomical theory of climatic change, in: *Geological Society of America Abstracts with Programs*, Vol. 20, p. A207.

Kennedy, G. L., Lajoie, K. R., and Wehmiller, J. F., 1982, Aminostratigraphy and faunal correlations of late Quaternary marine terraces, Pacific coast U.S.A., *Nature* **299**:545–547.

King, K., Jr., 1980, Applications of amino acid biogeochemistry for marine sediments, in: *Biogeochemistry of Amino Acids* (P. E. Hare, T. C. Hoering, and K. King, Jr., eds), John Wiley & Sons, New York, pp. 377–392.

King, K., Jr., and Neville, C., 1977, Isoleucine epimerization for dating marine sediments: Importance of analyzing monospecific foramiferal samples, *Science* **195**:1333–1335.

Lajoie, K. R., Kern, J. P., Wehmiller, J. F., Kennedy, G. L., Mathieson, S. A., Sarna-Wojcicki, A. M., Yerkes, R. F., and McCrory, P. A., 1979, Quaternary shorelines and crustal deformation, San Diego to Santa Barbara, California, in: *Geological Excursions in the Southern California Area*, (P. L. Abbott, ed.), Department of Geology, San Diego State University, San Diego, California, pp. 3–15.

Lajoie, K. R., Wehmiller, J. F., and Kennedy, G. L., 1980, Inter- and intrageneric trends in apparent racemization kinetics of amino acids in Quaternary mollusks, in: *Biogeochemistry of Amino Acids*, (P. E. Hare, T. C. Hoering, and K. King, Jr., eds.), John Wiley & Sons, New York, pp. 305–340.

Leonard, E. M., and Wehmiller, J. F., 1992, Low uplift rates and terrace reoccupation inferred from mollusk aminostratigraphy, Coquimbo Bay area, Chile, *Quat. Res.* **38**:246–259.

Leonard, E. M., Mirecki, J. E., and Wehmiller, J. F., 1987, Amino acid age estimates for late Pleistocene marine terraces in northern Chile: Implications for uplift rates, in: *Geological Society of America Abstracts with Programs*, Vol. 19, p. 744.

Mahaney, W. C. (ed.), 1984, *Quaternary Dating Methods*, Elsevier, Amsterdam.

McCartan, L., Owens, J. P., Blackwelder, B. W., Szabo, B. J., Belknap, D. F., Kriausakul, N., Mitterer, R. M., and Wehmiller, J. F., 1982, Comparison of amino acid racemization geochronometry with lithostratigraphy, biostratigraphy, uranium-series coral dating, and magnetostratigraphy in the Atlantic coastal plain of the southeastern United States, *Quat. Res.* **18**:337–359.

McCoy, W. D., 1987, Quaternary aminostratigraphy of the Bonneville Basin, western United States, *Geol. Soc. Am. Bull.* **98**: 99–112.

Meisburger, E. P., 1979, Reconnaissance Geology of the Inner Continental Shelf, Cape Fear Region, North Carolina, U.S. Army, Corps of Engineers, Coastal Engineering Research Center, Technical Paper no. 79-3.

Miller, B. B., McCoy, W. D., and Bleur, N. K., 1987, Stratigraphic potential of amino acid ratios in Pleistocene terrestrial gastropods: An example from west-central Indiana, *Boreas* **16**: 133–138.

Miller, G. H., 1985, Aminostratigraphy of Baffin Island Shell-bearing deposits, in: *Quaternary Environments, the Eastern Canadian Arctic, Baffin Bay, and West Greenland* (J. T. Andrews, ed.), Allen and Unwin, Winchester, Massachusetts, pp. 394–427.

Miller, G. H., and Brigham-Grette, J., 1989, Amino acid geochronology: Resolution and precision in carbonate fossils, *Quat. Int.* **1**: 111–128.

Miller, G. H., and Hare, P. E., 1980, Amino acid geochronology: Integrity of the carbonate matrix and potential of molluscan fossils, in: *Biogeochemistry of Amino Acids* (P. E. Hare, T. C. Hoering, and K. King, Jr., eds.), John Wiley & Sons, New York, pp. 415–444.

Miller, G. H., and Mangerud, J., 1985, Aminostratigraphy of European marine interglacial deposits, *Quat. Sci. Rev.* **4**:215–278.

Miller, G. H., Hollin, J. T., and Andrews, J. T., 1979, Aminostratigraphy of U.K. Pleistocene deposits, *Nature* **281**:539–543.

Mirecki, J. E., 1985, Amino acid racemization dating of coastal plain sites, southeastern Virginia and northeastern North Carolina, Master of Science Thesis, University of Delaware, Newark.

Mirecki, J. E., 1990, Aminostratigraphy, geochronology, and geochemistry of fossils from late Cenozoic marine units in southeastern Virginia, Ph.D. Dissertation, University of Delaware, Newark.

Mirecki, J. E., Skinner, A., and Wehmiller, J. F., 1989, Resolution of a depositional hiatus in the late Pleistocene record, southeastern Virginia, using amino acid racemization and electron spin resonance dating methods, Geological Society of America Southeastern Sectional Meeting, Abstracts, Vol. 21, p. 51.

Mitterer, R. M., 1975, Ages and diagenetic temperatures of Pleistocene deposits of Florida based upon isoleucine epimerization in *Mercenaria*, *Earth Planet. Sci. Lett.* **28**:275–282.

Mitterer, R. M., and Hare, P. E., 1967, Diagenesis of amino acids in fossil shells as a potential geochronometer, Geological Society of America 1967 Annual Meeting, Abstracts, p. 152.

Mitterer, R. M., and Kriausakual, N., 1984, Comparison of rates and degrees of isoleucine epimerization in dipeptides and tripeptides, *Org. Geochem.* **7**:91–98.

Mitterer, R. M., and Kriausakul, N., 1989, Calculation of amino acid racemization ages based on apparent parabolic kinetics, *Quat. Sci. Rev.* **8**:353–357.

Mixon, R. B., and Pilkey, O. H., 1976, Reconnaissance Geology of the Submerged and Emerged Coastal Plain Province, Cape Lookout Area, North Carolina, U.S. Geological Survey Professional Paper 859.

Morton, D., and Yerkes, R. F., 1987, Recent Reverse Faulting in the Transverse Ranges, California, U.S. Geological Survey Professional Paper 1339.

Muhs, D. R., 1985, Amino acid age estimates of marine terraces and sea levels on San Nicolas Island, California, *Geology* **13**:58–61.

Muhs, D. R., Kennedy, G. L., and Rockwell, T. K., 1988, Uranium-series ages of corals from marine terraces, Pacific coast of North America: Implications for the timing and magnitude of late Pleistocene sea-level changes, American Quaternary Association Meeting, Abstracts, p. 140.

Muhs, D. R., Kelsey, H. M., Miller, G. H., Kennedy, G. L., Whelan, J. F., and McInelly, G. W., 1990, Age estimates and uplift rates for late Pleistocene marine terraces: Southern Oregon portion of the Cascadia forearc, *J. Geophys. Res.* **95(B5)**:6685–6698.

Müller, P. J., 1984, Isoleucine epimerization in Quaternary planktonic foraminfera: Effects of diagenetic hydrolysis and leaching, and Atlantic–Pacific intercore correlations, *"Meteor" Forschungs Ergebnisse, Reihe C.* **38**:25–47.

Murray-Wallace, C. V., and Kimber, R. W. L., 1987, Evaluation of the amino acid racemization reaction in studies of Quaternary coastal and marine sediments in South Australia, *Aust. J. Earth Sci.* **34**:279–292.

Murray-Wallace, C. V., and Kimber, R. W. L., 1993, Further evidence for apparent "parabolic" racemization kinetics in Quaternary molluscs, *Aust. J. Earth Sci.* **40**:313–317.

Oaks, R. Q., Jr., Coch, N. K., Sanders, J. E., and Flint, R. F., 1974, Post-Miocene shorelines and sea levels, southeastern Virginia, in: *Post-Miocene Stratigraphy, Central and Southern Atlantic Coastal Plain* (R. Q. Oaks, Jr. and J. R. DuBar, eds.), Utah State University Press, Logan, Utah, pp. 53–87.

Oldale, R. N., Cronin, T. M., Valentine, P. C., Spiker, E. C., Blackwelder, B. W., Belknap, D. F., Wehmiller, J. F., and Szabo, B. J., 1982, The stratigraphy, structure, absolute age, and paleontology of the upper Pleistocene deposits at Sankaty Head, Nantucket Island, Massachusetts, *Geology* **10**:246–252.

Peebles, P. C., Johnson, G. H., and Berquist, C. R., 1984, The middle and late Pleistocene stratigraphy of the outer coastal plain, southeastern Virginia, *Va. Miner.* **30(2)**:14–22.

Pierce, K. L., 1986, Dating methods, in: *Active Tectonics*, National Academy of Sciences, Washington, D.C., pp. 195–214.

Ponti, D. J., Lajoie, K. R., and Appel, S. H., 1987, Aminostratigraphic classification of the "type" Quaternary marine formations from the San Pedro–Palos Verdes area, Los Angeles Basin, California, in: *Geological Society of America Abstracts with Programs*, Vol. 19, pp. 440–441.

Prell, W. L., Imbrie, J., Martinson, D., Morley, J., Pisias, N. G., Shackleton, N. J., and Streeter, H. F., 1986, Graphic correlation of oxygen isotope stratigraphy: Application to the late Quaternary, *Paleoceanography* **1**:137–162.

Radtke, U., 1987, Palaeo sea levels and discrimination of the last and the penultimate interglacial fossiliferous deposits by absolute dating methods and geomorphological investigations, *Berl. Geogr. Stud.* **25**:313–342.

Rahaim, S. D., 1986, The aminostratigraphy of the Sankaty Sand, Nantucket Island, Massachusetts, Master of Science Thesis, University of Delaware, Newark.

Riggs, S. R., and Belknap, D. F., 1988, Upper Cenozoic processes and environments of continental margin sedimentation: Eastern United States, in: *The Atlantic Continental Margin: U.S.* (R. E. Sheridan and J. A. Grow, eds.), Geological Society of America Decade of North American Geology Vol. 1–2, pp. 131–176.

Riggs, S. R., York, L. L., Wehmiller, J. F., and Snyder, S. W., 1992, Depositional patterns resulting from high frequency Quaternary sea-level fluctuations in northeastern North Carolina, in: *Quaternary Coasts of the United States: Marine and Lacustrine Systems* (C. H. Fletcher III and J. F. Wehmiller, eds.), Society of Economic Paleontologists and Mineralogists Special Publication No. 48, pp. 141–153.

Rutter, N. W. (ed.), 1985, *Dating Methods of Pleistocene Deposits and Their Problems*, Geoscience Canada Reprint Series 2, Geological Association of Canada, Toronto, pp. 1–87.

Rutter, N., Schnack, E. J., de Riio, J., Fasano, J. L., Isla, F. I., and Radtke, U., 1989, Correlation and dating of Quaternary littoral zones along the Patagonian coast, Argentina, *Quat. Sci. Rev.* **8**: 213–234.

Sarna-Wocicki, A. M., Morrison, S. D., Meyer, C. E., and Hillhouse, J. W., 1987, Correlation of upper Cenozoic tephra layers between sediments of the western United States and eastern Pacific ocean and comparison with biostratigraphic and magnetostratigraphic age data, *Geol. Soc. Am. Bull.* **98**:207–233.

Shackleton, N. J., 1987, Oxygen isotopes, ice volumes, and sea level, *Quat. Sci. Rev.* **6**:183–190.

Spencer, R. S., and Campbell, L. D., 1987, The fauna and paleoecology of the late Pleistocene marine sediments of southeastern Virginia, *Bull. Am. Paleontol.* **92**(327):1–124.

Szabo, B. J., 1985, Uranium-series dating of fossil corals from marine sediments of the United States Atlantic coastal plain, *Geol. Soc. Am. Bull.* **96**:398–406.

Vacher, H. L., and Hearty, P. J., 1989, History of stage-5 sea level in Bermuda: Review with new evidence of a rise to present sea level during substage 5a, *Quat. Sci. Rev.* **8**:159–168.

Wehmiller, J. F., 1977, Amino acid studies of the Del Mar, California midden site: Apparent rate constants, ground temperature models, and chronological implications, *Earth Planet. Sci. Lett.* **37**:184–196.

Wehmiller, J. F., 1980, Intergeneric differences in apparent racemization kinetics in molluscs and foraminifera: Implications for models of diagenetic racemization, in: *Biogeochemistry of Amino Acids* (P. E. Hare, T. C. Hoering, and K. King, Jr., eds.), John Wiley & Sons, New York, pp. 341–355.

Wehmiller, J. F., 1982, A review of amino acid racemization studies in Quaternary mollusks: Stratigraphic and chronologic applications in coastal and interglacial sites, Pacific and Atlantic coasts, United States, United Kingdom, Baffin Island, and tropical islands, *Quat. Sci. Rev.* **1**:83–120.

Wehmiller, J. F., 1984, Relative and absolute dating of Quaternary mollusks with amino acid racemization: Evaluation, application, questions, in: *Quaternary Dating Methods* (W. C. Mahaney, ed.), Elsevier, Amsterdam, pp. 171–193.

Wehmiller, J. F., 1986, Amino acid racemization geochronology, in: *Dating Young Sediments* (A. J. Hurford, E. Jager, and J. A. M. Ten Cate, eds.), United Nations CCOP Technical Publication No. 16, Bangkok, pp. 139–158.

Wehmiller, J. F., 1987, Aminostratigraphic resolving power for last interglacial marine molluscs in coastal sequences, in 12th International Congress, International Quaternary Association, Ottawa, Abstracts, p. 286.

Wehmiller, J. F., 1990, Amino acid racemization applications in chemical taxonomy and chronostratigraphy of Quaternary fossils, in: *Metazoan Biomineralization*, 1989 International Geological Congress Workshop (J. Carter, ed.), Van Nostrand Reinhold, New York. p. 583–608.

Wehmiller, J. F., 1992, Aminostratigraphy of southern California marine terraces, in: *Quaternary Coasts of the United States: Marine and Lacustrine Systems*, Soc. for Sedimentary Geology (SEPM) Spec. Pub. 48, pp. 317–321.

Wehmiller, J. F., and Belknap, D. F., 1978, Alternative kinetic models for the interpretation of amino acid enantiomeric ratios in Pleistocene mollusks: Examples from California, Washington, and Florida, *Quat. Res.* **9**:330–348.

Wehmiller, J. F., and Belknap, D. F., 1982, Amino acid age estimates, Quaternary Atlantic coastal plain: Comparison with U-series dates, biostratigraphy, and paleomagnetic control, *Quat. Res.* **18**:311–336.

Wehmiller, J. F., and Hare, P. E., 1971, Racemization of amino acids in marine sediments, *Science* **173**:907–911.

Wehmiller, J. F., Lajoie, K. R., Kvenvolden, K. A., Peterson, E. A., Belknap, D. F., Kennedy, G. L., Addicott, W. O., Vedder, J. G., and Wright, R. W., 1977, Correlation and Chronology of Pacific Coast Marine Terraces of Continental United States by Amino Acid Stereochemistry: Technique Evaluation, Relative Ages, Kinetic Model Ages, and Geologic Implications, U.S. Geological Survey Open-File Report 77-680.

Wehmiller, J. F., Belknap, D. F., and York, L. L., 1987, Comments on Corrado et al., 1986, Capabilities and limitations of applied aminostratigraphy as illustrated by analyses of *Mulinia lateralis* from late Cenozoic marine beds near Charleston, South Carolina, *S. C. Geol.* **31**:109–118.

Wehmiller, J. F., Belknap, D. F., Boutin, B. S., Mirecki, J. E., Rahaim,

S. D., and York, L. L., 1988a, A review of the aminostratigraphy of Quaternary mollusks from United States Atlantic Coastal Plain sites, in: *Dating Quaternary Sediments* (D. L. Easterbrook, ed.), Geological Society of America Special Paper 227 pp. 69–110.

Wehmiller, J. F., Hsu, J. T., and Leonard, E. M., 1988b, Latitudinal isochrons of amino acid enantiomeric ratios in Quaternary molluscs: Implications of North and South American Pacific Coast data for aminostratigraphy and effective temperature gradients, in: *Geological Society of America Abstracts with Programs*, Vol. 20, p. A53.

Wehmiller, J. F., York, L. L., and Belknap, D. F., 1989, Aminostratigraphic offsets in the Atlantic Coastal Plain Quaternary: Relation to North Carolina lithostratigraphy and coastal geomorphology, in: *Geological Society of America Abstracts with Programs*, Vol. 21, p. A283.

Wehmiller, J. F., York, L. L., Belknap, D. F., and Snyder, S. W., 1992, Aminostratigraphic discontinuities in the U.S. Atlantic Coastal Plain and their relation to preserved Quaternary stratigraphic records, *Quat. Res.*, **38**:275–291.

Yeats, R. S., Huftile, G. J., and Grigsby, F. B., 1988, Oak Ridge fault, Ventura fold belt, and the Sisar decollement, Ventura Basin, California, *Geology* **16**:1112–1116.

Yerkes, R. F., and Lee, W. H. K., 1987, Late Quaternary deformation in the western Transverse Ranges, in: U. S. Geological Survey Professional Paper 1339, pp. 71–82.

York, L. L., 1990, Aminostratigraphy of U.S. Atlantic Coast Pleistocene deposits: Maryland Continental Shelf and North and South Carolina Coastal Plain, Ph.D. Dissertation, University of Delaware, Newark.

York, L. L., and Wehmiller, J. F., 1992, Aminostratigraphic results from Cape Lookout, N.C. and their relation to the preserved Quaternary marine record of SE North Carolina, *Sediment. Geol.* **80**:279–291.

York, L. L., Wehmiller, J. F., Cronin, T. M., and Ager, T. A., 1989, Stetson Pit, Dare County, North Carolina: An integrated chronologic, faunal, and floral record of subsurface coastal sediments, *Palaeogeogr. Palaeoclimatol. Palaeoecol.* **72**:115–132.

Chapter 37

Sources and Cycling of Organic Matter within Modern and Prehistoric Food Webs

PEGGY H. OSTROM and BRIAN FRY

1. Introduction

Over the past two decades, analysis of the stable isotopes of carbon, nitrogen, and sulfur has become a widely used and important technique for tracing the flow of organic matter in modern and prehistoric food webs (Miyake and Wada, 1967; DeNiro and Epstein, 1978; Fry, 1981a; van der Merwe, 1982). This approach is based on the observation that the carbon, nitrogen, and sulfur isotopic compositions ($\delta^{13}C$, $\delta^{15}N$, and $\delta^{34}S$,* respectively) of an organism are similar to or vary predictably from those of its food source. As a measure of assimilable material, this type of data offers a unique opportunity to delineate pathways of energy transfer.

Recent studies have focused on determining the $\delta^{13}C$ and $\delta^{15}N$ of individual compound classes (Macko *et al.*, 1987; Tuross *et al.*, 1988; Hare *et al.*, 1991). The isotopic composition of single molecules is related to the analogous signal from the diet and shifts, or fractionations, that occur during metabolism (Macko *et al.*, 1986; Hare *et al.*, 1991). Consequently, in addition to furthering an understanding of feeding relationships, this approach provides a basis for documenting elemental transfer during metabolism.

*$\delta^{I}E = [(R_{sample}/R_{standard}) - 1] \times 10^3$, where I is the heavy isotope of element E, and R is the abundance ratio of the heavy to the light isotope.

The use of isotopic techniques to evaluate past food webs is limited by the ability to isolate a remnant of the original organic material from fossils. Evaluation of indigeneity has been based on ^{14}C dating techniques, C/N values, the distribution and relative abundance of amino acids, and the extent to which individual amino acids have undergone racemization (DeNiro, 1985; Engel and Macko, 1986; Gurfinkel, 1987; Ostrom *et al.*, 1990). Additional evidence can be obtained by comparing the isotopic composition of the bulk organic component or individual compounds isolated from fossils to that of a modern analog (Tuross *et al.*, 1988; Ostrom *et al.*, 1990; Hare *et al.*, 1991). Recovery of indigenous geochemical signals has provided an indication of the paleoecology of Recent to Late Cretaceous organisms (Nelson *et al.*, 1986; Tuross *et al.*, 1988; Ostrom *et al.*, 1990).

Stable isotope analyses of food web components provide a unique opportunity for delineating feeding relationships among organisms and transfers of energy through a food web. This information is particularly relevant to prehistoric systems where such relationships may not be preserved in the fossil record. This chapter will concentrate on the use of C, N, and S isotopes as indicators of diet and metabolism. An understanding will initially be developed with regard to (1) isotopic variability among primary producers and consumers, (2) fractionation effects, and

PEGGY H. OSTROM • Department of Geological Sciences, Michigan State University, East Lansing, Michigan 48824-1115. BRIAN FRY • The Ecosystems Center, Marine Biological Laboratory, Woods Hole Oceanographic Institution, Woods Hole, Massachusetts 02543.

Organic Geochemistry, edited by Michael H. Engel and Stephen A. Macko. Plenum Press, New York, 1993.

mates of the relative importance of dietary sources. These concepts are fundamental for reconstruction of prehistoric diets from analyses of bulk materials and individual molecules isolated from fossils.

2. Isotopic Variation among Primary Producers

Global distributions of isotope values for primary producers have broad ranges. For example, marine autotrophs have values of −4 to −30 for δ^{13}C, −2 to +19‰ for δ^{15}N, and −13 to +20‰ for δ^{34}S (Table I). This variability is a consequence of two factors: (1) differences in the isotopic composition of inorganic nutrient sources and (2) the magnitude of discrimination against the heavy isotope during nutrient uptake and subsequent fixation by the plant. The relationship between these factors and the δ^{13}C, δ^{15}N, and δ^{34}S distributions among primary producers will be briefly discussed here. Fogel and Cifuentes (this volume, Chapter 3) present a more thorough treatment of this subject.

TABLE I. Ranges of δ^{13}C, δ^{15}N, and δ^{34}S Values of Autotrophs and Particulate Organic Matter within Terrestrial and Marine Environments

Source	δ^{13}C (‰)	δ^{34}S (‰)	δ^{15}N (‰)	References[a]
Terrestrial C_3 plants	−30–−23	−7–14	−7–6	1, 4, 6, 10, 12, 22, 23, 26
Marsh C_3 plants	−29–−23	6	3–5	5, 6, 7, 20, 24, 26
Marsh C_4 plants	−15–−12	−10–6	1–8	4, 5, 13, 15, 22, 23, 24, 26
Seagrasses	−16–−4	−13–15	0–6	8, 13, 14, 15, 23, 29, 31
Macroalgae	−27–−10	15–20	−1–10	2, 8, 13, 14, 15, 17, 23, 25, 29, 30, 31
Mangroves	−29–−25	5–14	6–7	13, 29, 31
Temperate marine POM	−24–−18	18–19	−2–10	3, 9, 19, 27
Temperate estuarine POM	−30–−15	18–20	2–19	11, 15, 16, 18, 21, 27, 28, 29, 30

[a]References: 1, Hoering, 1957; 2, Miyake and Wada, 1967; 3, Deuser, 1970; 4, Smith and Epstein, 1971; 5, Johnson and Calder, 1973; 6, Troughton et al.; 1974; 7, Haines, 1976a, b; 8, Mekhityeva et al., 1976; 9, Wada and Hattori, 1976; 10, Shearer and Kohl, 1978; 11, Spiker and Schemel, 1979; 12, Chukhrov et al., 1980; 13, Fry, 1981a; 14, Macko, 1981; 15, Fry et al., 1982; 16, Sherr, 1982; 17, Schell, 1983; 18, Tan and Strain, 1983; 19, Gearing et al., 1984; 20, Sternberg et al., 1984; 21, Owens, 1985; 22, Peterson et al., 1985; 23, Peterson et al., 1986; 24, DeLaune and Lindau, 1987; 25, Sealy, 1987; 26, Peterson and Howarth, 1986; 27, Fry, 1988; 28, Cifuentes et al., 1988; 29, Harrigan et al., 1989; 30, Ostrom and Macko, 1992; 31, Zieman, unpublished data.

Carbon fixed by autotrophs is derived from the pool of dissolved inorganic carbon in the immediately surrounding environment. Equilibrium fractionation during the exchange reaction between CO_2 and $HCO_3{}_-$ within aquatic systems results in a difference between the δ^{13}C of the atmospheric CO_2 (−8‰) and oceanic (0‰) carbonate reservoir. As a consequence of isotopic shifts during diffusion of CO_2 into the cell and subsequent fixation, photoautotrophs are depleted in ^{13}C by approximately 10 to 20‰ with respect to their inorganic carbon source (Craig, 1953; Peters et al., 1978).

The magnitude of the depletion in ^{13}C of the plant relative to the inorganic carbon source depends on the degree of fractionation during fixation and diffusion of CO_2 into the cell (O'Leary, 1981). For example, differences in primary carbon-fixing enzymes and associated fractionations result in a difference of at least 5‰ between C_3 and C_4 photosynthetic types (Table I; Smith and Epstein, 1971). In comparison to terrestrial plants, phytoplankton are more commonly diffusion-limited. Under these circumstances, a large portion of the CO_2 pool within the cell is enzymatically incorporated. This effectively reduces the enzymatic fractionation and results in the δ^{13}C enrichment in marine particulate organic matter (POM) relative to terrestrial C_3 plants (Table I; O'Leary, 1981).

Autotrophs often have nitrogen and sulfur isotopic compositions approximating those of their nutrient source (Tables II and III). However, there is a wide range of δ^{15}N values for phytoplankton that utilize NH_4^+ or NO_3^- and large variability in the δ^{34}S

TABLE II. Nitrogen Isotope Values for Some Marine Primary Producers and Their Inorganic Nutrient Sources

Primary producer	δ^{15}N (‰)	Inorganic nutrient source	δ^{15}N (‰)	Reference(s)[a]
N_2-fixing phytoplankton	−2–2	N_2	−1–2	1, 2
Phytoplankton[b]	−3–12	NH_4	−4–42	2–4
		NO_3^-	5–19	
Phytoplankton[c]	5[d]	NO_3^-	2–4	5

[a]References: 1, Wada, 1980; 2, Miyake and Wada, 1967; 3, Cline and Kaplan, 1975; 4, Velinsky et al., 1989; 5, Wada et al., 1987a.
[b]These phytoplankton represent a mixed seston sample composed primarily of phytoplankton believed to be assimilating the nutrient source indicated
[c]Phytoplankton sorted by hand.
[d]Average value.

TABLE III. Sulfur Isotope Values for Some Marine Primary Producers and Their Inorganic Nutrient Sources

Primary producer	$\delta^{34}S$ (‰)	Inorganic nutrient source	$\delta^{34}S$ (‰)	Reference(s)[a]
Plankton	19[b]	Seawater sulfate	20[b]	1, 2
Upland plants	5[b]	Sulfate in rainwater	2–8	1, 3
Spartina	−6–9	Pore-water sulfate	28[b]	4
		Pore-water sulfide	−22[b]	4
Seagrass				
Leaves	−11–15	Seawater sulfate	20[b]	5
Roots	−12–12	Pore-water sulfate	16–17	5
		Pore-water sulfide	−23–−25	5

[a]References: 1, Peterson *et al.*, 1985; 2, Kaplan et *al.*, 1963; 3, Nriagu and Coker, 1970; 4, Peterson *et al.*, 1986; 5, Fry *et al.*, 1982.
[b]Average value.

of *Spartina* and seagrass (Tables II and III). The $\delta^{15}N$ of phytoplankton depends on the extent of fractionation during uptake. When assimilation is not limited by availability of nutrients, a large fractionation effect can occur (Saino and Hattori, 1987; Macko et *al.*, 1987). In contrast, when the supply of inorganic nitrogen is low, isotopic shifts during uptake are small. Values of $\delta^{34}S$ for rooted plants can change as a function of the relative utilization of sulfate versus ^{34}S-depleted sulfides or sulfide oxidation products (Fry et *al.*, 1982; Carlson and Forrest, 1982).

Despite the variability observed in global ranges of $\delta^{13}C$, $\delta^{15}N$, and $\delta^{34}S$ of primary producers, isotope tracing techniques have provided important information regarding the base of nutritional support to food webs. Differentiation among primary producers is often assisted through the analysis of multiple isotopes. The successful use of the method also requires careful documentation of the isotopic compositions of primary producers within a study site.

3. Isotopic Composition of Consumers

A similarity in $\delta^{13}C$ values between animals and plants from the same environment was documented in the first carbon isotope studies in marine environments (Craig, 1953; Parker, 1964). Subsequently, stable carbon, nitrogen, and sulfur isotope measurements have been widely used to trace sources of organic matter that are important to consumers (Thayer et *al.*, 1978; Fry et *al.*, 1987, Fry, 1988; Harrigan et *al.*, 1989). This methodology requires that autotrophs can be differentiated based on their isotope values and that these values are passed on to consumers in a predictable manner.

Many field and laboratory studies have documented small shifts of 1 to 2‰ in $\delta^{13}C$ and smaller shifts in $\delta^{34}S$ between the muscle tissue or whole body of a consumer and its food source (DeNiro and Epstein, 1978; Fry and Sherr, 1984; Peterson and Fry, 1987). The difference between the $\delta^{13}C$ of bone collagen and that of the food source appears to be close to 5‰ (DeNiro and Epstein, 1978; Price *et al.*, 1985). The precise magnitude of the difference in $\delta^{13}C$ between the diet and a particular tissue type depends on the extent to which the heavy isotope is incorporated or lost during synthesis and catabolism.

Owing to the relative fidelity in $\delta^{13}C$ of muscle tissue and large differences in plant values, carbon isotope signatures are valuable tracers in food web studies. For example, carbon isotope values can be used to distinguish marine and terrestrial dietary sources (Rodelli et *al.*, 1984; McConnaughey and McRoy, 1979a, b). An increase in the influence of marine relative to terrestrial carbon is reflected in the $\delta^{13}C$ of filter feeders living along a nearshore to offshore transect within an estuary (Stephenson and Lyon, 1982).

Similarly, terrestrial organisms dependent on C_3 plants have $\delta^{13}C$ values that are distinct from those of animals which derive their carbon from C_4 autotrophs (Fry et *al.*, 1978). Within food webs of salt marshes, the large difference in $\delta^{13}C$ values between C_3 and C_4 plants is complicated by the presence of other primary producers having intermediate carbon values. However, the relative contributions of upland C_3 plants, C_4 plants of the genus *Spartina*, and phytoplankton can be traced to consumers through the combined use of carbon and sulfur isotopes (Peterson et *al.*, 1985; Peterson and Howarth, 1986). The combined use of $\delta^{13}C$, $\delta^{15}N$, and $\delta^{34}S$ can often provide excellent separation among multiple primary producers (Fig. 1).

In contrast to $\delta^{13}C$ and $\delta^{34}S$ values, the $\delta^{15}N$ of muscle tissue, bone collagen, or a whole organism shows approximately a 3 to 3.5‰ enrichment relative to the food source (Miyake and Wada, 1967; Minigawa and Wada, 1984; Schoeninger and DeNiro, 1984; Harrigan et *al.*, 1989). Nitrogen isotope values are a good indicator of dietary source when this fractionation is accounted for. In addition, the 3 to 3.5‰ shift occurs with each trophic level along the food chain and, thus, provides a basis for establishing trophic structure. As a consequence of trophic enrichments, plots of $\delta^{15}N$ versus $\delta^{13}C$ for food web components show a linear relationship with a positive slope. This trophic continuum is consistent among several subtropical to near-Arctic food webs (Dickson, 1987; Harrigan et *al.*, 1989; Wada et *al.*, 1987b).

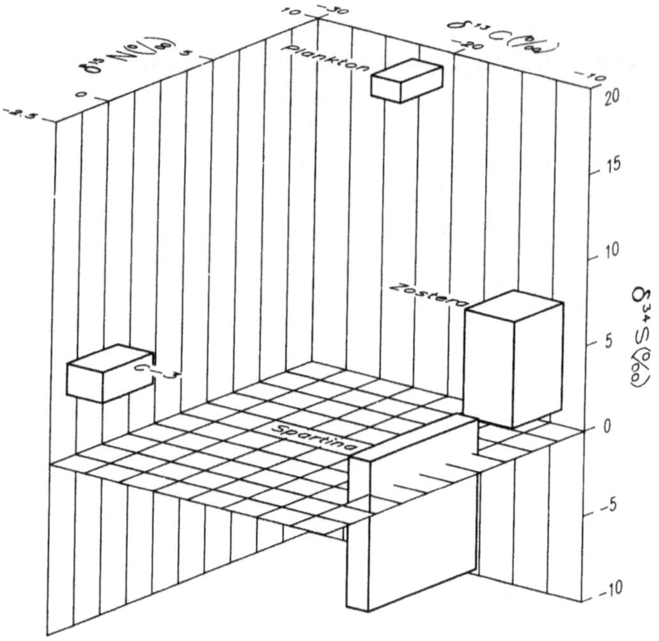

Figure 1. Carbon, nitrogen, and sulfur isotope values of primary producers from northeastern Atlantic salt marshes (Peterson *et al.*, 1985; J. C. Zieman, unpublished data). Boxes represent one standard deviation about the mean.

The use of nitrogen isotopes in conjunction with carbon is also valuable in differentiating among primary producers that could not be separated based on $\delta^{13}C$ alone, e.g., macroalgae versus seagrass, marine POM versus estuarine POM, and algae versus mangrove (Fry, 1981a; Schoeninger and DeNiro, 1984; Zieman *et al.*, 1984; Dickson, 1987). The dual-tracer method often supplies mutually supporting data for distinction between terrestrial and marine food sources (Schoeninger and DeNiro, 1984). Analyses of collagen from more than 100 animals from various locations showed clear differences in $\delta^{13}C$ and $\delta^{15}N$ between marine and terrestrial inhabitants (Schoeninger and DeNiro, 1984). The average $\delta^{13}C$ and $\delta^{15}N$ of marine organisms are −13.7 and +18.8‰, respectively. In contrast, average $\delta^{13}C$ and $\delta^{15}N$ compositions of terrestrial organisms are −18.8‰ and 5.9‰, respectively. Ultimately, the choice of isotopes will depend on the complexity of the system and the isotopic values of the primary producers.

Whereas the majority of applications indicate that $\delta^{13}C$ and $\delta^{15}N$ provide excellent data for dietary tracing, recent studies in terrestrial ecosystems indicate that, in addition to diet, nitrogen isotopic compositions may be affected by climate, physiology, and/or microhabitat (Ambrose and DeNiro, 1986a; Sealy *et al.*, 1987; Ambrose and DeNiro, 1989). In two cases, the collagen $\delta^{15}N$ values of herbivores from arid locations were much greater than those of similar animals from moist habitats (Heaton *et al.*, 1986; Sealy *et al.*, 1987). An inverse relationship between $\delta^{15}N$ of the animals and rainfall was also apparent (Heaton *et al.*, 1986; Sealy *et al.*, 1987). These results show that inhabitants of arid terrestrial environments can have $\delta^{15}N$ values that are similar to those of marine consumers.

4. Quantitative Analysis through Mixing Equations

Tracing diets with stable isotopes has several advantages over more traditional methods of food web analysis. In comparison to stomach content data, which identify material that has been ingested, $\delta^{13}C$, $\delta^{15}N$, and $\delta^{34}S$ reflect the time-integrated average of the assimilated diet. Consequently, stable isotope data can be used to assess the nutritional dependence of an animal on a food source. Quantitative estimates of the relative importance of food sources to the diet can be obtained from mixing models (Dunton and Schell, 1987; Harrigan *et al.*, 1989). The general expression for the mixing equation is

$$\delta_{\text{diet}} = \Sigma f_j \delta_j$$

where $\delta_{\text{diet}} = \delta_{\text{consumer}}$ − estimate of trophic fractionation, f_j is the fractional contribution of the food item, such that $\Sigma f_j = 1$, and δj is the isotopic composition of the food item.

A unique mixing equation can be written in terms of each of the isotopic compositions, $\delta^{13}C$, $\delta^{15}N$, and $\delta^{34}S$. If the isotopic compositions of the food sources and the consumer are measured and trophic fractionation is known, the mixing equations can be solved simultaneously for f_j. The use of C, N, and S isotopes provides a solution for the fractional contributions of up to four end members. Most applications of mixing models have been based on $\delta^{13}C$ and $\delta^{15}N$ or $\delta^{13}C$ and $\delta^{34}S$ analyses.

Mixing equations have been used to estimate the relative contribution of C, N, or S from various primary producers to a consumer (Dunton and Schell, 1987; Harrigan *et al.*, 1989). Such evaluations of the reliance of consumers on primary producers are a direct indication of energy transfer in food webs and can be used to address long-standing questions regarding flow of nutrients and energy transfers in ecosystems (Odum and de la Cruz, 1967; Teal, 1962; Odum and Heald, 1972). Isotopic measurements show strong linkages between seagrass habitats and feeding by commercially important shrimp and fish (Fry, 1981b; Zieman, 1982; Harrigan *et al.*, 1989). This provides a conservation rationale to protect these important coastal ecosystems.

Despite the need for quantitative estimates of primary source contributions, there are only a few studies that have utilized this technique (Dunton and Schell, 1987; Harrigan *et al.*, 1989). In the natural environment, the presence of multiple sources can complicate the use of mixing models. To deal with this complexity, additional measures of food web structure are needed. This allows constraints to be placed on possible outcomes. Stable isotopes can be combined with biomarkers to assist in discriminating among sources (Saliot *et al.*, 1988; Des Marais *et al.*, 1980). Studies of plant abundance and productivity may also allow the elimination from mixing models of certain plant sources that are unlikely to significantly contribute to the nutrient pool.

The use of mixing models relies on precise measurements of trophic fractionation and of the isotopic composition of consumers and their dietary sources. Owing to variability in literature values, it is most accurate to evaluate fractionation within a particular food web by comparing isotope values of a consumer and that of its diet. The application of mixing models to upper level carnivores is complicated because any errors associated with an estimate of fractionation are compounded at higher trophic positions.

A stepwise approach involving a combination of stomach content and stable isotope analysis provides an alternative method for evaluating the relative importance of primary producers to an upper level carnivore (Fig. 2). First, the relative contributions of prey items are quantified through stomach content analysis, and the isotopic compositions of the prey, predator, and dominant primary producers are determined. With this information, mixing equations are used to estimate the dependence of each prey on various primary producers. The extent to which the upper level consumer is dependent on these same primary producers is based on the fractional dietary contribution of each prey item to the consumer and the relative importance of each primary producer to the prey.

This approach has several advantages over more traditional methods of food web analysis. Unlike stomach content analysis and documentation of feeding strategies, the stable isotope approach identifies material that has been assimilated. By tracing transfers of carbon, nitrogen, and sulfur to each member in a food chain, pathways of energy flow are identified. This type of analysis provided a quantitative estimate of the amount of C or N contributed by

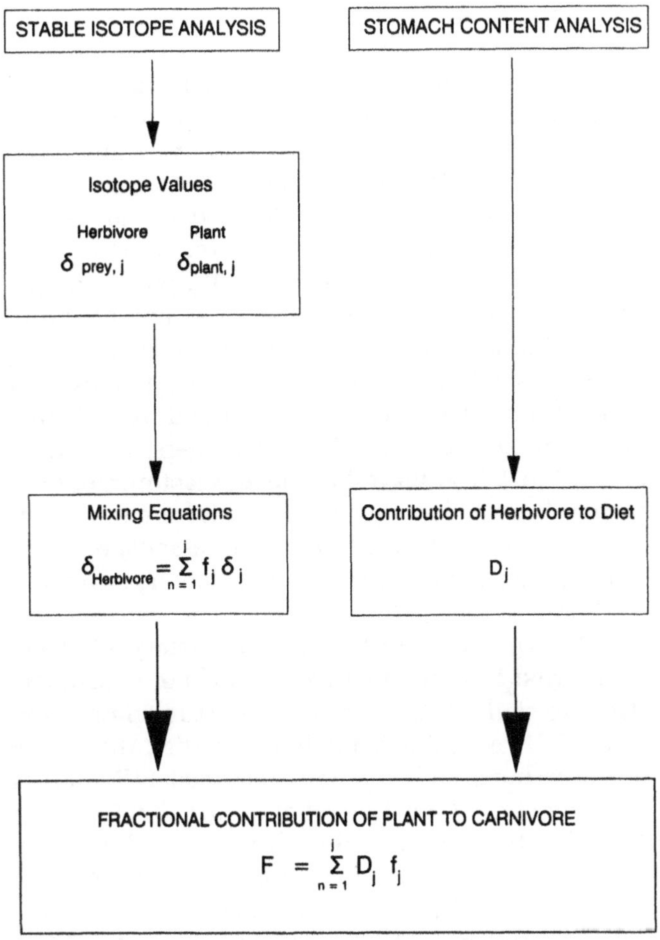

Figure 2. Stepwise approach for the quantitative assessment of the relative contribution of carbon, nitrogen, or sulfur from different plants to a carnivore.

seagrass to a south Florida coastal food web (Harrigan *et al.*, 1989).

5. Isotopic Compositions of Individual Compounds

With the recent coupling of gas chromatography and isotope-ratio mass spectrometry, information on isotopic compositions of specific molecules will rapidly expand. These data will assist in interpreting feeding relationships and evaluating biosynthetic pathways for modern organisms. The ability to distinguish differences in the synthesis of compounds among various taxa provides a basis for comparative biochemistry.

5.1. Food Web Analysis

The δ¹³C values of individual molecules such as alkanes and single amino acids from a consumer largely reflect the isotopic composition of their diet. For example, differences in δ¹³C values between C_3 and C_4 plants appear to be passed on to certain classes of consumer alkanes and amino acids (Des Marais *et al.*, 1980; Hare *et al.*, 1991). Such isotopic differences are clearly observed in a comparison of the δ¹³C of amino acids between consumers fed C_3 and C_4 diets (Fig. 3). Serine, threonine, glycine, and valine from collagen of a consumer and from its diet have similar δ¹³C values (Fig. 3; Winters, 1971; Hare *et al.*, 1991). The magnitude of the isotopic enrichments observed for other amino acids from the organism relative to its food source is associated with fractionations that occur during metabolism (refer to Section 5.2).

As with δ¹³C, differences in dietary δ¹⁵N are transferred to individual amino acids of consumers (Gaebler *et al.*, 1966; Hare *et al.*, 1991). In comparison to the δ¹⁵N of amino acids from the diet, the amino acids or bone collagen have enriched δ¹⁵N values (Gaebler *et al.*, 1966; Hare *et al.*, 1991). Fractionation of individual amino acids during metabolism is likely to be more pronounced for δ¹⁵N than δ¹³C. As a consequent of the higher abundance of carbon relative to nitrogen in an amino acid, typically 3:1, a greater number of metabolic fractionations of similar magnitude are required before a change is observed for δ¹³C than for δ¹⁵N. The study of individual reactions is required to assess where such isotopic shifts occur.

Despite variations in dietary δ¹³C or δ¹⁵N, a similar pattern of isotopic compositions among nonessential amino acids exists among consumers (Hare *et al.*, 1991). This can be observed by comparing the δ¹³C values of the amino acids from the collagen of two pigs that were raised on different food sources (Fig. 3). Further studies will be necessary to determine if the pattern of isotopic compositions of essential amino acids is similar among consumers or if these shifts will differ among organisms as a consequence of variations in metabolism. The study of essential amino acids may be fruitful for assessing the effects of dietary stresses at a molecular level.

5.2. Comparative Biochemistry

With new data emerging on isotopic compositions of amino acids, hydrocarbons, and other compound classes, a clearer understanding of the factors that control the isotopic compositions of individual molecules is needed. For example, this information will be essential for distinguishing whether diagenetic or metabolic processes control the δ¹³C or δ¹⁵N values of individual amino acids isolated from fossils. Factors that govern the δ¹³C or δ¹⁵N of an individ-

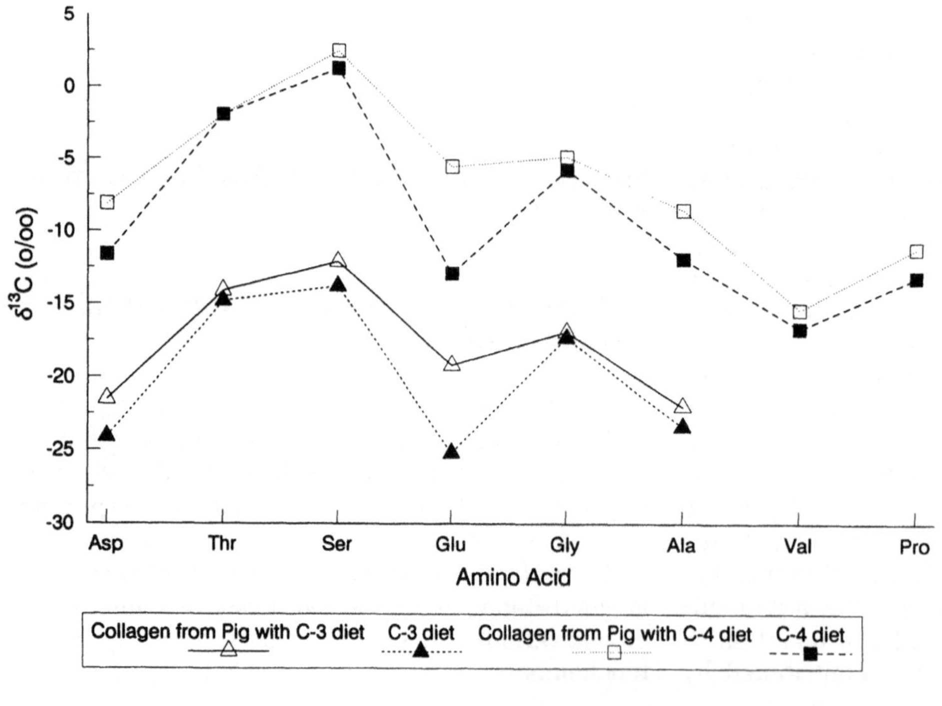

Figure 3. Carbon isotope values of amino acids from consumers and diets. Data are given for one pig fed a controlled diet derived from plants which had C_3 metabolism and for another pig fed a controlled diet derived from plants which had C_4 metabolism. [From Hare *et al.* (1991).]

ual molecule include the signature of its ultimate precursor, the degree of fractionation during rate-determining steps and at major branching points of metabolic pathways, and the number of reactions in which the compound is involved. Insights into the biochemical controls of the isotopic composition of amino acids and fatty acids are presently being developed.

It is well known that lipids are depleted in ^{13}C relative to the whole cell (Park and Epstein, 1961). A consistency in the pathway of fatty acid synthesis among autotrophs was suggested by early investigations that showed a similarity in the extent of ^{13}C depletion among individual fatty acids with respect to CO_2 (Parker, 1962). Fractionation effects occur during fatty acid synthesis at the carbonyl position of pyruvate during the formation of acetyl coenzyme A (acetyl-CoA) by pyruvate dehydrogenase and during the hydrolysis of the thiol ester bond associated with the acyl chain of the carrier protein (Fig. 4; DeNiro and Epstein, 1977; Monson and Hayes, 1980, 1982). Isotopic measurements at specific atomic positions have shown that, during the latter step, fatty acids with light carboxylic acid groups are more rapidly incorporated in long-chain lipids (Monson and Hayes, 1980, 1982).

While the metabolism of lipid formation appears similar among organisms, intraspecies differences in isotopic discrimination could arise due to many factors such as the magnitude of the effect for specific enzymes and the degree to which a substrate is converted to product. For example, if all the pyruvate were oxidized to acetyl-CoA, no isotope distinction would be observed between pyruvate and the acetyl group. The magnitude of the ^{13}C depletions increases as smaller portions of the total pyruvate are utilized. Further consideration of interspecies differences in the degree and process of ^{13}C segregation during fatty

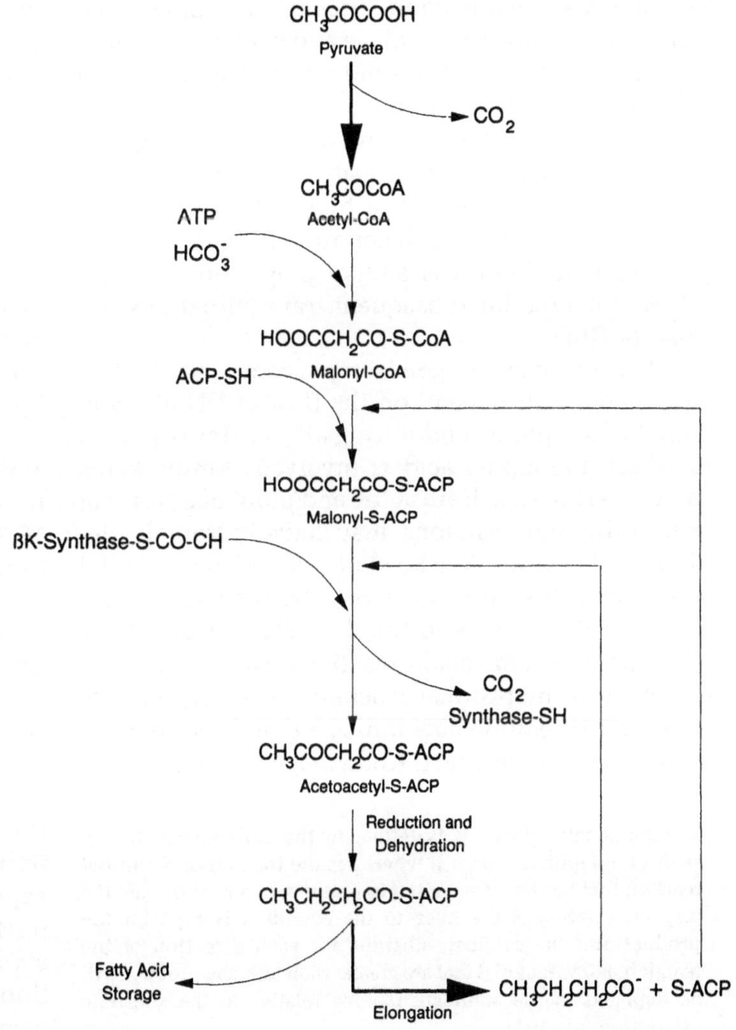

Figure 4. Simplified mechanism of fatty acid synthesis, emphasizing reaction steps in which discrimination against ^{13}C occurs. ACP, Acyl carrier protein. Fractionation occurs at the steps indicated by enlarged arrows (Monson and Hayes, 1980, 1982; DeNiro and Epstein, 1977).

acid synthesis will provide insight into the prevalence of this metabolism in nature.

The $\delta^{15}N$ of a whole cell is dependent on the degree of fractionation during uptake and enzymatic incorporation of nitrogen into the cell (Macko *et al.*, 1987). However, the broad partitioning of ^{15}N among individual amino acids is a consequence of fractionation within metabolic pathways during synthesis (Macko *et al.*, 1987). Transamination is a main pathway of nitrogen transfer involving at least 13 amino acids. An example of a reaction that involves transfer of nitrogen from one amino acids to an α-keto acid is the formation of aspartic acid (Asp) from glutamic acid (Glu) (Fig. 5). The experimentally determined fractionation factor, β,* for the forward reaction, 1.0083, reflects approximately an 8‰ depletion in ^{15}N in the product, Asp, relative to the substrate, Glu (Macko *et al.*, 1986). Analysis of amino acid isolates from bacterial cultures shows a similar discrimination against ^{15}N of 9‰ in Asp relative to Glu (Macko *et al.*, 1987). Observed ^{15}N enrichments of Glu relative to valine, isoleucine, leucine, tyrosine, and phenylalanine in some prokaryotic algae and bacteria are probably linked to this transamination isotope effect (Macko *et al.*, 1986, 1987).

The isotope effect for the reverse reaction, β = 1.0012, results in an approximate 2‰ depletion in ^{15}N in Glu relative to Asp. The higher $\delta^{15}N$ of Asp in comparison to Glu within certain algal cultures suggests that, in these cases, Asp plays the role of a nitrogen source for subsequent transaminations, instead of Glu.

The ultimate nitrogen isotopic composition of an amino acid will depend on the flow of ^{15}N at metabolic branch points and during all transfer reactions in which the amino acid is involved. Amino acids such as serine which are at the end point of a series of transamination reactions may have extremely depleted $\delta^{15}N$ values (Macko *et al.*, 1987). The essential amino acid threonine appears to be consistently depleted in ^{15}N relative to the diet (Hare *et al.*, 1991). Whereas most enzymatic reactions result in the enrichment of the residual reactant, the depleted $\delta^{15}N$ values for threonine may indicate that threonine has an inverse isotope effect. Knowledge of the diversity

*The fractionation factor, β, is defined by the Rayleigh equation, $\beta = [\ln(1-f)]/[\ln(1-fr_p/r_s)]$ where f is the fraction of the initial reactant that has been converted to product, and r_p and r_s are the respective ratios of the trace to the abundant isotope for the product and the original substrate for each direction of the reaction. For values of β that are greater than one, the product will be depleted in the abundant isotope relative to the substrate (Mariotti *et al.*, 1981).

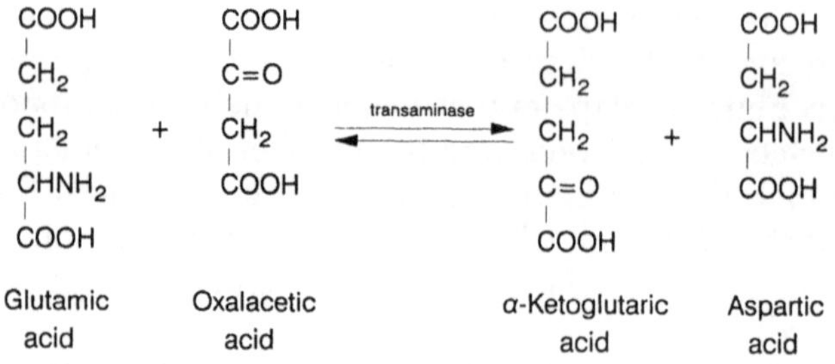

Figure 5. Molecular transformations during transamination.

of fractionations that occur during synthesis will enhance the interpretation of ^{15}N partitioning among amino acids.

Biosynthetic pathways and branching ratios also influence the $\delta^{13}C$ distribution of amino acids. Owing to the high carbon-to-nitrogen ratio of this compound class, carbon isotope fractionation is more complex than that of nitrogen. The $\delta^{13}C$ values of amino acids in bacteria and algae reflect the origin of the α-carboxyl group. This moiety is generally enriched in ^{13}C relative to the total amino acid (Abelson and Hoering, 1961). Distributions of $\delta^{13}C$ in both bacteria and algae show ^{13}C enrichments in the acidic amino acids and ^{13}C depletions among the basic and neutral fractions (Winters, 1971; Macko *et al.*, 1987). The $\delta^{13}C$ of the acidic amino acids may be related to the presence of a secondary carboxyl group (Macko *et al.*, 1987). Enzymatic fractionation associated with the formation of acetyl-CoA may result in the observed ^{13}C depletion in lysine, isoleucine, and leucine (DeNiro and Epstein, 1977; Macko *et al.*, 1987).

In addition to the fractionation factors for transamination, those for a large number of decarboxylation reactions have been determined (O'Leary *et al.*, 1970; Abell and O'Leary, 1988a, 1988b). The loss of CO_2 from an amino acid or its precursor could influence the $\delta^{13}C$ amino acid pattern. Future $\delta^{13}C$ and $\delta^{15}N$ analyses of the primary amine decarboxylation product and amino acid precursor could provide a better understanding of ^{13}C and ^{15}N abundances of amino acids.

The relative $\delta^{13}C$ or $\delta^{15}N$ values among amino acids are similar between autotrophic and heterotrophic organisms (Macko *et al.*, 1987). This similarity begins to shape a perspective as to what may be expected when comparing modern organisms and tracking biosynthetic pathways through the fossil record. A more complete understanding of fractionation effects during metabolism and diagenesis of compound classes will assist in the understanding of the isotopic composition of the molecular component of the paleontological record.

6. Prehistoric Dietary Reconstruction

Numerous archaeological studies have used the information from stable isotope analysis of collagen extracted from fossil bones to reconstruct the diet of prehistoric human populations (Schoeninger and DeNiro, 1982; Chisholm *et al.*, 1983; Hobson and Collier, 1984; Ambrose and DeNiro, 1986b). Just as in modern food webs, carbon isotope values differentiate between prehistoric consumption of C_3 and C_4 plants, and both $\delta^{13}C$ and $\delta^{15}N$ distinguish between marine and terrestrial food sources (Chisholm *et al.*, 1982, 1983; Schoeninger *et al.*, 1983; Schoeninger and DeNiro, 1985; Schwarcz *et al.*, 1985; Nelson *et al.*, 1986). Animals whose diet is predominantly marine-based show $\delta^{13}C$ and $\delta^{15}N$ values enriched by approximately 5 and 7‰, respectively, relative to terrestrial C_3 consumers. As in modern systems, the nitrogen isotope distinction may be obscured in arid environments where climatic effects result in enrichments in $\delta^{15}N$ among terrestrial animals (Ambrose and DeNiro, 1989).

The mineral phase of bones and teeth, hydroxyapatite, provides an alternative material for $\delta^{13}C$ analyses. Carbon exists as carbonate ions substituted in the apatite crystal or adsorbed onto crystal surfaces and hydration layers (Betts *et al.*, 1981). It has been argued that bone apatite is an unsuitable substrate for isotopic analyses on the basis that diagenetic changes such as exchange reactions involving carbonate may alter the indigenous isotopic signal retained within the apatite (Tamers and Pearson, 1965; Schoeninger and DeNiro, 1982). However, pretreatment to remove diagenetic carbonate can provide a substrate that is suitable for accurate isotopic evaluation for materials which are older than 50,000 years (Krueger and Sullivan, 1984).

In contrast to the large amount of work that has been done on humans, there has been little effort on analysis of feeding relationships among other prehistoric organisms (Katzenberg, 1989; Ostrom *et al.*, 1990). The potential of this type of work is emphasized by the striking resemblance in trophic structure that is observed between a modern and an ancient food web (Fig. 6). Nitrogen isotope ratios are strong trophic indicators and, as such, can play an important role in establishing positions of ancient organisms within the hierarchy of a food web. This is particularly valuable in cases where certain faunal elements are not represented in the fossil record.

Isotopic compositions of individual molecules can be another source of information about ancient diets. Applications of this new approach have been limited to three studies (Hare and Estep, 1983; Tuross *et al.*, 1988; Hare *et al.*, 1991). The $\delta^{13}C$ and $\delta^{15}N$ values of amino acids in fossil bones mirror those of modern analogs (Fig. 7). Shifts in $\delta^{13}C$ and $\delta^{15}N$ of fossil

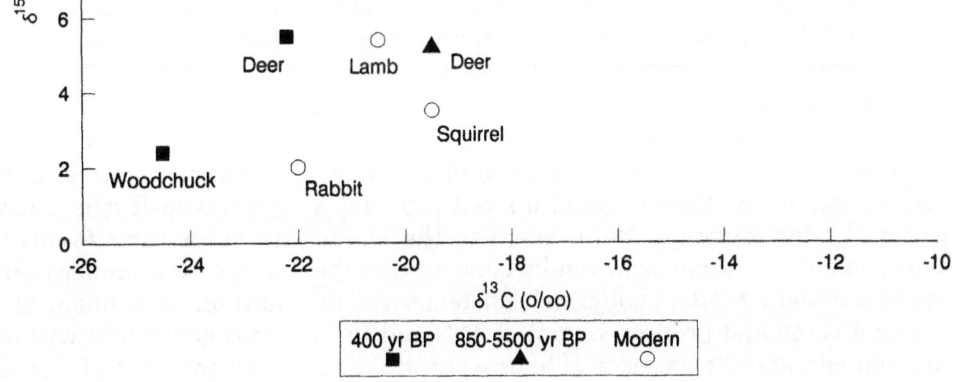

Figure 6. Carbon and nitrogen isotopic compositions of bone collagen from modern and prehistoric terrestrial organisms (Schoeninger and DeNiro, 1984; DeNiro, 1985; Katzenberg, 1989). Modern bones were obtained from southern California. Fossils were obtained from southern Ontario (400 yr B.P.) and southern California (850–5500 yr B.P.).

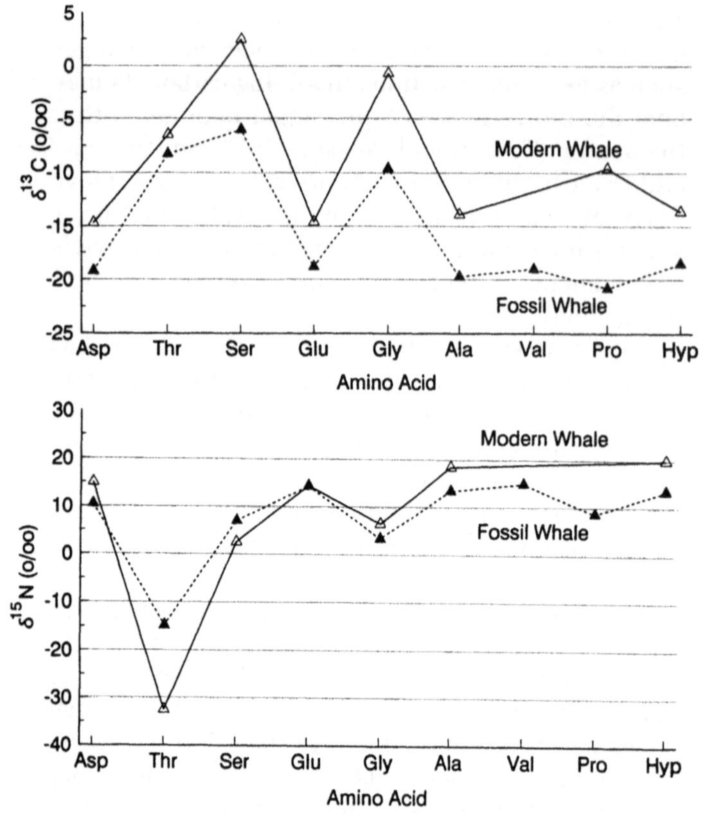

Figure 7. Carbon and nitrogen isotope distributions of amino acids obtained from fossil and modern whale collagen (Tuross *et al.*, 1988).

collagen could result if isotopically distinct amino acids such as glycine, threonine, and serine are preferentially lost during diagenesis (Tuross *et al.*, 1988).

The use of isotopes as tracers in prehistoric food webs depends on isolation of indigenous organic material and the retention of a dietary signal in the isolate. In the past, four types of data have been used to assess the integrity of fossil collagen: (1) radiocarbon dating, (2) amino acid patterns, (3) the ratio of amino acid enantiomers, and (4) elemental ratios of carbon to nitrogen. The measurement of ^{14}C is useful for materials which are less than, approximately, 75,000 years old. Exogenous substances such as humic acids frequently have a different ^{14}C composition from the indigenous organic component (Stafford *et al.*, 1988). Consequently, a comparison of the radiocarbon date to the known age of a fossil provides a gauge of contamination. With regard to the amino acid content, fossil collagen can be compared to the modern protein. Modern collagen is characterized by a high glycine and proline content of 35% and 12%, respectively, and the presence of hydroxyproline and hydroxylysine (Lehninger, 1979; Hare, 1980). These characteristic amino acid patterns have been observed in isolates from Recent to Pleistocene bones (Wycoff, 1972; Tuross *et al.*, 1988).

The ratio of amino acid stereoisomers has also been used as an indicator of contamination [for a detailed review, see Schroeder and Bada (1976)]. Whereas amino acids in living organisms are primarily in the L configuration, postmortem racemization results in a mixture of D- and L-amino acids in fossils. Equilibrium mixtures of amino acids are reached in approximately the first two to three million years. Consequently, materials that are older than this and still contain a high abundance of L-amino acids are likely to have been contaminated.

It has also been suggested that C/N values between 2.9 and 3.5, the reported range for modern collagen, are a good indication of preservation (DeNiro, 1985). Collagen extracted from fossils whose values fall in this range have $\delta^{13}C$ and $\delta^{15}N$ values comparable to those of bone collagen from modern analogs.

The inability to detect collagen does not, unequivocally, indicate contamination. Laboratory experiments indicate that, over time, collagen is progressively leached, and during the later stages of diagenesis the amino acid composition of bones is completely unlike that of collagen (Hare, 1980). Noncollagenous proteins (NCP) have been successfully identified by immunological techniques in fossil bone material (Tuross, 1987). High concentrations of aspartic acid, serine, and glutamic acid are typical of some NCP (Masters, 1985; Takagi and Veis, 1984).

Chromatographic studies show that hydroxyapatite, the main mineral component of bone, adsorbs synthetic proteins and peptides containing carboxyl groups (Bernardi and Kawasaki, 1968). High concentration of the dicarboxylic amino acids, aspartic acid and glutamic acid, may enhance the preservation of NCP within the hydroxyapatite matrix.

Analysis of an organic fraction contained in fused aggregates of hydroxyapatite crystals of modern bones showed an amino acid composition similar to that of NCP (DeNiro and Weiner, 1988). Isolates of this material from fossil bone were similar in organic content, $\delta^{13}C$, and $\delta^{15}N$ to a modern analog. These observations were consistent for both well and poorly preserved fossils. In contrast, the collagen fraction from the same fossils was not always comparable to that of the modern analog. Further characterization and an assessment of postmortem alteration effects on this material will provide a stronger basis for the interpretation of isotopic analyses of NCP.

The presence of an organic remnant in fossil material is not always an indication that a biogeochemical signal has been retained. The present criteria for indigeneity are limited. Fossils are not closed systems and are subject to exchange of amino acids with the surrounding environment. In addition, the ratio of D- to L-amino acids may be low if the amino acids are derived from a higher molecular weight material (Kimber and Griffin, 1987; Serban *et al.*, 1988). Perhaps the best evaluation of indigeneity can be made from isotopic comparisons of amino acid enantiomers. Whereas diagenetic alteration is likely to result in isotopic differences between the D- and the L-stereoisomer, low-temperature racemization reactions do not cause a significant shift in the stable carbon or nitrogen isotopic composition of the resultant amino acid enantiomer produced (Engel and Macko, 1986). Therefore, a comparison of enantiomers separated from fossil organic material may prove to be the best criterion for establishing indigeneity.

7. Conclusions

Stable isotope measurements provide a means to trace the flow of energy within natural systems. Perhaps the greatest limitations of this method are the potential for overlap in isotope values among sources of organic matter and loss of the signal during diagenesis. Attempts at overcoming these difficulties have prompted new avenues of research.

In many situations, multidisciplinary approaches have assisted in interpretations of isotope data. For example, knowledge of the ecology of past and present organisms can be used to place constraints on the number of food sources that a consumer is likely to encounter. The recently expanding knowledge of isotopic compositions of individual compounds, in addition to more traditional analyses of bulk materials, will aid in evaluating present and past food webs. It may also be possible to differentiate between a contaminant and an indigenous material through analyses of specific molecules.

Geochemists studying cycling of organic matter in modern and prehistoric environments must continue to ask questions and develop approaches that will expand current understandings of pathways of energy transfer and transformations that occur during diagenesis. Such knowledge is essential for distinguishing among sources of organic matter in the existing environment and determining the origin of postdepositional substances based on isotopic analyses.

References

Abell, L. M., and O'Leary, M. H., 1988a, Nitrogen isotope effects on glutamate decarboxylase from *Escherichia coli*, *Biochemistry* **27**:3325–3330.

Abell, L. M., and O'Leary, M. H., 1988b, Isotope effect studies of the pyridoxal 5′-phosphate dependent histidine decarboxylase from *Morganella morganii*, *Biochemistry* **27**:5927–5933.

Abelson, P. H., and Hoering, T. C., 1961, Carbon isotope fractionation in formation of amino acids by photosynthetic organisms, *Proc. Natl. Acad. Sci. U.S.A.* **47**:623–632.

Ambrose, S. H., and DeNiro, M. J., 1986a, Stable carbon and nitrogen isotope analysis of human diet in Africa, *J. Hum. Evol.* **15**:707–731.

Ambrose, S. H., and DeNiro, M. J., 1986b, Reconstruction of African diet using bone collagen carbon and nitrogen isotope ratios, *Nature* **319**:321–323.

Ambrose, S. H., and DeNiro, M. J., 1989, Climate and habitat reconstruction using stable carbon and nitrogen isotope ratios of collagen in prehistoric herbivore teeth from Kenya, *Quat. Res.* **31**:407–422.

Bernardi, G., and Kawasaki, T., 1968, Chromatography of polypeptides and proteins on hydroxyapatite columns, *Biochim. Biophys. Acta* **160**:301–310.

Betts, F., Blumenthal, N. C., and Posner, A. S., 1981, Bone mineralization, *J. Cryst. Growth* **53**:63–73.

Carlson, P. R., Jr., and Forrest, J., 1982, Uptake of dissolved sulfide by *Spartina alterniflora*: Evidence from natural sulfur isotope abundance ratios, *Science* **216**:633–635.

Chisholm, B. S., Nelson, D. E., and Schwarcz, H. P., 1982, Stable carbon isotope ratios as a measure of marine versus terrestrial protein in animal diets, *Science* **16**:1131–1132.

Chisholm, B. S., Nelson, D. E., and Schwarcz, H. P., 1983, Marine and terrestrial protein in prehistoric diets on the British Columbia coast, *Curr. Anthropol.* **24**:396–398.

Chukhrov, F. V., Ermilova, L. P., Churikov, V. S., and Nosik, L. P., 1980, The isotopic composition of plant sulfur, *Org. Geochem.* **2**:69–75.

Cifuentes, L. A., Sharp, J. H., and Fogel, M. L., 1988, Stable carbon and isotope biogeochemistry in the Delaware Estuary, *Limnol. Oceanogr.* **33**:1102–1115.

Cline, J. D., and Kaplan, I. R., 1975, Isotopic fractionation of dissolved nitrate during denitrification in the eastern tropical North Pacific Ocean, *Mar. Chem.* **3**:.271–299.

Craig, H., 1953, The geochemistry of the stable carbon isotopes, *Geochim. Cosmochim. Acta* **3**:53–92.

DeLaune, R. D., and Lindau, C. W., 1987, $\delta^{13}C$ signature of organic carbon in estuarine bottom sediment as an indicator of carbon export from adjacent marshes, *Biogeochemistry* **4**:225–230.

DeNiro, M. J., 1985, Postmortem preservation and alteration of *in vivo* bone collagen isotope ratios in relation to palaeodietary reconstruction, *Nature* **317**:806–809.

DeNiro, M. J., and Epstein, S., 1977, Mechanism of carbon isotope fractionation associated with the lipid synthesis, *Science* **197**: 261–263.

DeNiro, M. J., and Epstein, S., 1978, Influence of diet on the distribution of carbon isotopes in animals, *Geochim. Cosmochim. Acta* **42**:495–506.

DeNiro, M. J., and Weiner, S., 1988, Organic matter within crystalline aggregates of hydroxyapatite: A new substrate for stable isotopic and possibly other biogeochemical analyses of bone, *Geochim. Cosmochim. Acta* **52**:2415–2423.

Des Marais, D. J., Mitchell, J. M., Meinschein, W. G., and Hayes,

J. M., 1980, The carbon isotope biogeochemistry of the individual hydrocarbons in bat guano and the ecology of insectivorous bats in the region of Carlsbad, New Mexico, *Geochim. Cosmochim. Acta* **44**:2075–2086.

Deuser, W. G., 1970, Isotopic evidence for diminishing supply of available carbon during diatom bloom in the Black Sea, *Nature* **225**:1069–1071.

Dickson, M. L., 1987, A comparative study of the pelagic food chains in two Newfoundland fjords using stable carbon and nitrogen isotope tracers, M.Sc. Thesis, Memorial University of Newfoundland.

Dunton, K. H., and Schell, D. M., 1987, Dependence of consumers on macroalgal (*Laminaria solidungula*) carbon in an arctic kelp community: $\delta^{13}C$ evidence, *Mar. Biol.* **93**:615–625.

Engel, M. H., and Macko, S. A., 1986, Stable isotope evaluation of the origins of amino acids in fossils, *Nature* **323**:531–533.

Fry, B., 1981a, Tracing shrimp migrations and diets using natural variations in stable isotopes, Ph.D. Thesis, University of Texas at Austin.

Fry, B., 1981b, Natural stable isotope tag traces Texas shrimp migrations, *Fish. Bull.* **79**:337–345.

Fry, B., 1988, Food web structure on Georges Bank from stable C, N, and S isotopic compositions, *Limnol. Oceanogr.* **3**:1182–1190.

Fry, B., and Sherr, E. B., 1984, $\delta^{13}C$ measurements as indicators of carbon flow in marine and freshwater ecosystems, *Contrib. Mar. Sci.* **27**:13–47.

Fry, B., Jeng, W.-L., Scalan, R. S., and Parker, P. L., 1978, $\delta^{13}C$ food web analysis of a Texas sand dune community, *Geochim. Cosmochim. Acta* **42**:1299–1302.

Fry, B., Scalan, R. S., Winters, J. K., and Parker, P. L., 1982, Sulfur uptake by salt grasses, mangroves, and seagrasses in anaerobic sediments, *Geochim. Cosmochim. Acta* **46**:1121–1124.

Fry, B., Macko, S. A., and Zieman, J. C., 1987, Review of stable isotopic investigations of food webs in seagrass meadows, Florida Marine Research Publications, No. 42, pp. 190–209.

Gaebler, O. H., Vitti, T. G., and Vukmirovich, R., 1966, Isotope effects in metabolism of ^{14}N and ^{15}N from unlabeled dietary proteins, *Can. J. Biochem.* **44**:1249–1257.

Gearing, J. N., Gearing, P. J., Rudnick, D. T., Requejo, A. G., and Hutchins, M. J., 1984, Isotopic variability of organic carbon in a phytoplankton-based, temperate estuary, *Geochim. Cosmochim. Acta* **48**:1089–1098.

Gurfinkel, D. M., 1987, Comparative study of the radiocarbon dating of different bone collagen preparations, *Radiocarbon* **29**:45–52.

Haines, E. B., 1976a, Relation between the stable carbon isotope composition of fiddle crabs, plants, and soils in a salt marsh, *Limnol. Oceanogr.* **21**:880–883.

Haines, E. B., 1976b, Stable carbon isotope ratios in the biota, soils, and tidal water of a Georgia salt marsh, *Estuarine Coastal Shelf Sci.* **4**:609–616.

Hare, P. E., 1980, Organic geochemistry of bone and its relation to the survival of bone in natural environments, in: *Fossils in the Making: Vertebrate Taphonomy and Paleoecology* (A. K. Behrensmeyer and A. P. Hill, eds.), University of Chicago Press, Chicago, pp. 208–219.

Hare, P. E., and Estep, M. F., 1983, Carbon and nitrogen isotopic composition of amino acids in modern and fossil collagens, *Carnegie Inst. Washington Yearb.* **82**:410–414.

Hare, P. E., Fogel, M. L., Stafford, T. W., Jr., Mitchell, A. D., and Hoering, T. C., 1991, The isotopic composition of carbon and nitrogen in individual amino acids isolated from modern and fossil proteins, *J. Arch. Sci.* **18**:277–292.

Harrigan, P., Zieman, J. C., and Macko, S. A., 1989, The base of nutritional support for the gray snapper (*Lutjanus griseus*): An evaluation based on a combined stomach content and stable isotope analysis, *Bull. Mar. Sci.* **44**:65–77.

Heaton, T. H. E., Vogel, J. C., Chenvallarie, G., and Collett, G., 1986, Climate influence on the isotopic composition of bone nitrogen, *Nature* **322**:822–823.

Hobson, K. A., and Collier, S., 1984, Marine and terrestrial protein in Australian Aboriginal diets, *Curr. Anthropol.* **25**:238–240.

Hoering, T. C., 1957, The isotopic composition of the ammonia and the nitrate in rain, *Geochim. Cosmochim. Acta* **12**:97–102.

Johnson, R. W., and Calder, J. A., 1973, Early diagenesis of fatty acids and hydrocarbons in a salt marsh environment, *Geochim. Cosmochim. Acta* **37**:1943–1955.

Kaplan, I. R., Emery, K. O., and Rittenberg, S. C., 1963, The distribution and isotopic abundance of sulphur in recent marine sediments off southern California, *Geochim. Cosmochim. Acta* **27**:297–331.

Katzenberg, M. A., 1989, Stable isotope analysis of archaeological faunal remains from southern Ontario, *J. Archaeol. Sci.* **16**: 319–329.

Kimber, R. W. L., and Griffin, C. V., 1987, Further evidence of the complexity of the racemization process in fossil shells with implications for amino acid dating, *Geochim. Cosmochim. Acta* **51**:839–846.

Krueger, H. W., and Sullivan, C. H., 1984, Models for carbonate isotope fractionation between diet and bone, in: *Stable Isotopes in Nutrition* (J. R. Turnlund and P. E. Johnson, eds.), American Chemical Society, Washington, D.C., pp. 205–220.

Lehninger, A. L., 1979, *Biochemistry*, Worth Publishers, New York.

Macko, S. A., 1981, Stable nitrogen isotope ratios as tracers of organic geochemical processes, Ph.D. Dissertation, University of Texas at Austin.

Macko, S. A., Estep, M. F., Engel, M. H., and Hare, P. E., 1986, Kinetic fractionation of stable nitrogen isotopes during amino acid transamination, *Geochim. Cosmochim. Acta* **50**:2143–2146.

Macko, S. A., Estep, M. F., Hare, P. E., and Hoering, T. C., 1987, Isotopic fractionation of nitrogen and carbon in the synthesis of amino acids by microorganisms, *Chem. Geol.* **65**:79–92.

Mariotti, A., Germon, J. C., Hubert, P., Kaiser, P., Letolle, R., Tardieux, A., and Tardieux, P., 1981, Experimental determination of nitrogen kinetic isotope fractionation: Some principles; illustration for the denitrification and nitrification processes, *Plant Soil* **62**:413–430.

Masters, P. M., 1985, *In vivo* decomposition of phosphoserine and serine in noncollagenous protein from human dentin, *Calcif. Tissue Int.* **37**:236–241.

McConnaughey, T., and McRoy, C. P., 1979a, ^{13}C label identifies eelgrass (*Zostera marina*) carbon in an Alaskan estuarine food web, *Mar. Biol.* **53**:263–265.

McConnaughey, T., and McRoy, C. P., 1979b, Food web structure and the fractionation of carbon isotopes in the Bering Sea, *Mar. Biol.* **53**:257–262.

Mekhityeva, V. L., Pankina, R. G., and Gavrilov, Y. Y., 1976, Distributions and isotopic compositions of forms of sulfur in water, animals and plants, *Geochem. Int.* **13**:82–87.

Miyake, Y., and Wada, E., 1967, The abundance ratio of $^{15}N/^{14}N$ in marine environments, *Rec. Oceanogr. Works Jpn.* **9**:176–192.

Minigawa, M., and Wada, E., 1984, Stepwise enrichments of ^{15}N along food chains: Further evidence and the relation between $\delta^{15}N$ and animal age, *Geochim. Cosmochim. Acta* **48**:1135–1140.

Monson, K. D., and Hayes, J. M., 1980, Biosynthetic control of the natural abundance of carbon 13 at specific positions within fatty acids in *Escherichia coli*, *J. Biol. Chem.* **55:**11435–11441.

Monson, K. D., and Hayes, J. M., 1982, Carbon isotopic fractionation in the biosynthesis of bacterial fatty acids. Ozonolysis of unsaturated fatty acids as a means of determining the intramolecular distribution of carbon isotopes, *Geochim. Cosmochim. Acta* **46:**139–149.

Nelson, B. K., DeNiro, M. J., Schoeninger, M. J., De Paolo, D. J., and Hare, P. E., 1986, Effects of diagenesis on strontium, carbon, nitrogen and oxygen concentration and isotopic composition of bone, *Geochim. Cosmochim. Acta* **50:**1941–1949.

Nriagu, J. O., and Coker, R. D., 1983, Sulphur in sediments chronicles past changes in lake acidification, *Nature* **303:**692–694.

Odum, E. P., and de la Cruz, A. A., 1967, Particulate organic detritus in a Georgia salt marsh–estuarine ecosystem, in: *Estuaries* (G. H. Lauff, ed.), American Association for the Advancement of Science, New York, pp. 383–388.

Odum, W. E., and Heald, E. J., 1972, Trophic analysis of an estuarine mangrove community, *Bull. Mar. Sci.* **22:**671–738.

O'Leary, M. H., 1981, Carbon isotope fractionation in plants, *Phytochemistry* **20:**553–567.

O'Leary, M. H., Richards, T. D., and Hendrickson, D. W., 1970, Identification of rate limiting step in the chymotrypsin catalyzed hydrolysis of N-acetyl-L-tryptophanamide, *J. Am. Chem. Soc.* **92:**4435.

Ostrom, N. E., and Macko, S. A., 1992, Sources, cycling and distribution of water column particulate and sedimentary organic matter in northern Newfoundland fjords and bays: A stable isotope study, in: *Organic Matter Productivity, Accumulation, and Preservation in Recent and Ancient Sediments* (J. K. Whelan and J. W. Farrington, eds.), Columbia University Press, New York, pp. 55–81.

Ostrom, P. H., Macko, S. A., Engel, M. H., Silfer, J. A., and Russell, D. A., 1990, Geochemical characterization of high molecular weight material isolated from Late Cretaceous fossils, in: *Advances in Organic Geochemistry 1989* (F. Behar and B. Durand, eds.), Org. *Geochem.* **16:**1139–1144.

Owens, N. J. P., 1985, Variations in the natural abundance of ^{15}N in estuarine suspended particulate matter: A specific indicator of biological processing, *Estuarine Coastal Shelf Sci.* **20:** 505–510.

Park, R., and Epstein, S., 1961, Metabolic fractionation of C^{13} and C^{12} in plants, *Plant Physiol.* **36:**133–138.

Parker, P. L., 1962, The isotopic composition of the carbon of fatty acids, Annual Report of the Director, Geophysical Laboratory, *Carnegie Inst. Washington Yearb.* **1962–1963:**187–190.

Parker, P. L., 1964, The biogeochemistry of the stable isotopes of carbon in a marine bay, *Geochim. Cosmochim. Acta* **28:**1155–1164.

Peters, K. E., Sweeney, R. E., and Kaplan, I. R., 1978, Correlation of carbon and nitrogen stable isotope ratios in sedimentary organic matter, *Limnol. Oceanogra.* **23:**598–604.

Peterson, B. J., and Fry, B., 1987, Stable isotopes in ecosystem studies, *Annu. Rev. Ecol. Syst.* **18:**293–320.

Peterson, B. J., and Howarth, R. W., 1986, Sulfur, carbon, and nitrogen isotopes used to trace organic matter flow in the salt-march estuaries of Sapelo Island, Georgia, *Limnol. Oceanogr.* **32:**1195–1213.

Peterson, B. J., Howarth, R. W., and Garrit, R. H., 1985, Multiple stable isotopes used to trace the flow of organic matter in estuarine food webs, *Science* **227:**1361–1363.

Peterson, B. J., Howarth, R. W., and Garritt, R. H., 1986, Sulfur and carbon isotopes of salt-marsh organic matter flow, *Ecology* **67:** 865–874.

Price, T. D., Schoeninger, M. J., and Armelagos, G. J., 1985, Bone chemistry and past behavior: An overview, *J. Hum. Evol.* **14:** 419–447.

Rodelli, M. R., Gearing, J. N., Gearing, P. J., Marshall, N., and Sasekumar, A., 1984, Carbon sources used by organisms in Malaysian mangrove swamps and nearshore waters as determined by ^{13}C values, *Oecologia (Berlin)* **61:**326–333.

Saino, T., and Hattori, A., 1987, Geographical variation of the water column distribution of suspended particulate organic nitrogen and its ^{15}N natural abundance in the Pacific Ocean and its marginal seas, *Deep Sea Res.* **34:**807–827.

Saliot, A., Tronczynski, J., Scribe, P., and Letolle, R., 1988, The application of isotopic and biogeochemical markers to the study of the biochemistry of organic matter in a macrotidal estuary, the Loire, France, *Estuarine Coastal Shelf Sci.* **27:** 645–669.

Schell, D. M., 1983, Carbon-13 and carbon-14 abundances in Alaskan aquatic organisms: Delayed production from peat in Arctic food webs, *Science* **219:**1068–1071.

Schoeninger, M. J., and DeNiro, M. J., 1982, Carbon isotope ratios of apatite from fossil bone cannot be used to reconstruct diets of animals, *Nature* **297:**577–578.

Schoeninger, M. J., and DeNiro, M. J., 1984, Nitrogen and carbon isotopic composition of bone collagen from marine and terrestrial animals, *Geochim. Cosmochim. Acta* **48:**625–639.

Schoeninger, M. J., DeNiro, M. J., and Tauber, H., 1983, $^{15}N/^{14}N$ ratios of bone collagen reflect marine and terrestrial components of prehistoric human diet, *Science* **220:**1831–1383.

Schroeder, R. A., and Bada, J. L., 1976, A review of the geochemical applications of the amino acid racemization reaction, *Earth Sci. Rev.* **12:**347–391.

Schwarcz, H. P., Melbye, J., Katzenberg, M. A., and Knyf, M., 1985, Stable isotopes in human skeletons of southern Ontario: Reconstruction of palaeodiet, *J. Archaeol. Sci.* **12:**187–206.

Sealy, J. C., van der Merwe, N. J., Thorp, J. A. L., and Lanham, J. L., 1987, Nitrogen isotopic ecology in southern Africa: Implications for environmental and dietary tracing, *Geochim. Cosmochim. Acta* **51:**2707–2717.

Serban, A., Engel, M. H., and Macko, S. A., 1988, The distribution, stereochemistry and stable isotopic constituents of fossil and modern mollusk shells, *Adv. Org. Geochem.* **13:**1123–1129.

Shearer, G., and Kohl, D. H., 1978, ^{15}N abundance in N-fixing and non-N-fixing plants, in: *Recent Developments in Mass Spectrometry in Biochemistry and Medicine: Volume 1* (A. Frigeria, ed.), Plenum Press, New York, pp. 605–622.

Sherr, E. B., 1982, Carbon isotope composition of organic seston and sediments in a Georgia salt marsh estuary, *Geochim. Cosmochim. Acta* **36:**1227–1232.

Smith, B. N., and Epstein, S., 1971, Two categories of $^{13}C/^{12}C$ ratios for higher plants, *Plant Physiol.* **47:**380–384.

Spiker, A. C., and Schemel, L. E., 1979, Distribution and stable isotope composition of carbon in San Francisco Bay, in: *San Francisco Bay: The Urbanized Estuary* (T. J. Conomos, ed.), Pacific Division, American Association for the Advancement of Science, San Francisco, pp. 195–212.

Stafford, T. W., Jr., Brendel, K., and Duhamel, R. C., 1988, Radiocarbon, ^{13}C, and ^{15}N analysis of fossil bone: Removal of humates with XAD-2 resin, *Geochim. Cosmochim. Acta* **2:**2257–2267.

Stephenson, R. L., and Lyon, G. L., 1982, Carbon-13 depletion in estuarine bivalve: Detection of marine and terrestrial food sources, *Oecol* **55:**110–113.

Sternberg, L., DeNiro, M. J., and Koeley, J. E., 1984, Hydrogen, oxygen, and carbon isotope ratios of cellulose from submerged aquatic crassulacean acid metabolism and noncrassulacean acid metabolism plants, *Plant Physiol.* **76**:68–70.

Takagi, Y., and Veis, A., 1984, Isolation of phosphophoryn from human dentin organic matrix, *Calcif. Tissue Int.* **36**:259–265.

Tamers, M. A., and Pearson, F. J., 1965, Validity of radiocarbon dates on bone, *Nature* **208**:1053–1055.

Tan, F. C., and Strain, P. M., 1983, Sources, sinks and distribution of organic carbon in the St. Lawrence Estuary, Canada, *Geochim. Cosmochim. Acta* **47**:125–132.

Teal, J. M., 1962, Energy flow in the salt marsh ecosystem of Georgia, *Ecology* **43**:614–624.

Thayer, G. W., Adams, S. W., and LaCroix, M. W., 1978, The stable carbon isotope ratio of some components of an eelgrass, *Zostera marina* bed, *Oecologia (Berlin)* **35**:1–12.

Troughton, J. H., Card, K. A., and Hendy, C. H., 1974, Photosynthetic pathways and carbon isotope discrimination by plants, *Carnegie Inst. Washington Yearb.* **73**:768–780.

Tuross, N., 1987, Molecular preservation in human bones from the Windover archeological site, *Geological Society of America Abstracts with Programs*, pp. 872–873.

Tuross, N., Fogel, M. L., and Hare, P. E., 1988, Variability in the preservation of the isotopic composition of collagen from fossil bone, *Geochim. Cosmochim. Acta* **52**:929–935.

van der Merwe, N. J., 1982, Carbon isotopes, photosynthesis and archaeology, *Am. Sci.* **70**:596–606.

Velinsky, D. J., Pennock, J. R., Sharp, J. H., Cifuentes, L. A., and Fogel, M. L., 1989, Determination of the isotopic composition of ammonium-nitrogen at the natural abundance level from estuarine waters, *Mar. Chem.* **26**:351–361.

Wada, E., 1980, Nitrogen isotope fractionation and its significance in biogeochemical processes occurring in marine environments, in: *Isotope Marine Chemistry* (E. D. Goldberg, Y., Horibe, and J. K. Saruhashi, eds.), Uchida Rokakuho, Tokyo, pp. 375–398.

Wada, E., and Hattori, A., 1976, Natural abundance of ^{15}N in particulate organic matter in the North Pacific Ocean, *Geochim. Cosmochim. Acta* **40**:249–251.

Wada, E., Miagawa, M., Mizatani, H., Takashi, T., Imaizumi, R., and Karasawa, K., 1987a, Biogeochemical studies on the transport of organic matter along the Otsuchi River watershed, Japan, *Estuarine Coastal Shelf Sci.* **25**:321–326.

Wada, E., Terazaki, M., Kabaya, Y., and Nemoto, T., 1987b, ^{15}N and ^{13}C abundances in the Antarctic ocean with emphasis on the biogeochemical structure of the food web, *Deep Sea Res.* **5/6:** 829–841.

Winters, J. K., 1971, Variations in the natural abundance of ^{13}C in proteins and amino acids, Ph.D. Dissertation, University of Texas at Austin.

Wycoff, R. W. G., 1972, *The Biochemistry of Animal Fossils*, Williams and Wilkins Company, Baltimore.

Zieman, J. C., 1982, The Ecology of the Seagrasses of South Florida: A Community Profile, U. S. Fish and Wildlife Service, Office of Biological Services, Washington, D. C., Report FWS/OBS-82/25.

Zieman, J. C., Macko, S. A., and Mills, A. L., 1984, Role of seagrasses and mangroves in estuarine food webs: Temporal and spatial changes in stable isotope composition and amino acid content during decomposition, *Bull. Mar. Sci.* **35**:380–392.

Chapter 38

Macromolecules from Living and Fossil Biominerals

Implications for the Establishment of Molecular Phylogenies

L. L. ROBBINS, G. MUYZER, and K. BREW

1. Introduction

1.1. Foraminifera and Mollusks

The study of fossils as a means of establishing the geologic age and long-distance correlation of strata has been recognized as one of the most precise and reliable instruments of stratigraphy. Relatively rapid rates of evolution and easily fossilizable shells are two important elements which account for this fact. Two different invertebrate taxa whose stratigraphic utility has been extensively exploited are the microfossil foraminifera and the macrofossil mollusks. These taxa will be discussed in this chapter in terms of their application in molecular paleontology.

1.1.1. Classification

Taxonomy of fossil organisms has traditionally been based on classifications of the fossilized hard parts, but often this presents a number of different problems. Shelled organisms often demonstrate a wide range of phenotypic variability, which is often a manifestation of environmental variability (Kennett, 1976; Malmgren and Kennett, 1972; Robbins, 1988). Such phenotypic plasticity can create difficulty in discerning whether the morphologic changes are attributable to either genetic or environmental influences. The genetic makeup of an organism is known as the genotype. In living organisms the genotype can be studied by analyzing the DNA. Unfortunately, in fossils, the cellular material is usually completely degraded except in a few rare cases of exceptional preservation, and so retrieval of DNA is not typically possible. Although not providing the genotype, amino acid sequences of preserved homologous proteins which may be preserved in the shell will give a genetically related signal.

1.1.2. Use of Proteins to Obtain Genetic Signals

The rates of evolution of different proteins have varied through time, and comparative biochemistry of homologous proteins is a vast potential source of

L. L. ROBBINS • Department of Geology, University of South Florida, Tampa, Florida 33620-5200. G. MUYZER • Department of Biochemistry, Leiden University, 2333 CC Leiden, The Netherlands. *Present address*: Molecular Ecology Unit, Max-Planck-Institute for Marine Microbiology, D-2800 Bremen 33, Germany. K. BREW • Department of Biochemistry and Molecular Biology, University of Miami School of Medicine, Miami, Florida 33101-9990.

Organic Geochemistry, edited by Michael H. Engel and Stephen A. Macko. Plenum Press, New York, 1993.

historical information (Runnegar, 1986). Because of the enormous amount of information in an amino acid sequence, proteins give substantially more phylogenetic information than can be obtained from morphology alone. However, one problem in the study of fossil remnant proteins is the limited amounts available for study. Fortunately, recent technological improvements in molecular biology provide paleontologists with tools that allow the retrieval and use of trace amounts of proteins from shells of fossils such as foraminifera and mollusks (Muyzer, 1988; Curry, 1987a, b; Robbins, 1987; Robbins and Healy-Williams, 1991; Robbins *et al.*, in press) to provide information for molecular taxonomy.

A powerful framework of taxonomic relationships can be built by the integration of data from a number of techniques, such as biochemical, immunological, morphological, and other geochemical methods, for example, isotope analysis (Robbins, 1987; Robbins and Healy-Williams, 1991). Proteins which have been encapsulated in the crystals of the calcium carbonate shell can provide the basis for establishing biochemical taxonomic relationships. The location of the organics varies from taxon to taxon, but in most cases they are located between and within the carbonate crystals of the shell, often forming layered structures. For example, the foraminifera possess a series of organic layers within the calcium carbonate test (Fig. 1A, B). These layers are records of the periodic growth and chamber formation of the organism (Towe and Cifelli, 1967; Towe, 1971; Hemleben *et al.*, 1977, 1989). During the biomineralization process, the organism lays down a continuous primary organic membrane (POM) in the shape that the calcium carbonate structure will ultimately reflect. Calcite crystals subsequently nucleate on the proximal and distal sides of the POM (Fig. 1). As a new chamber is added, an additional organic layer is deposited, and crystals once again proceed to nucleate on the membrane. The induction of calcite on the surface of the newly formed organic "chamber" is also accompanied by additional organic deposition on the surface of older chambers, resulting in alternating layers of organic membrane and calcium carbonate (Towe, 1971; Hemleben *et al.*, 1977, 1989). This bilamellar crystal/inorganic structure encloses and tends to protect the organic/protein sandwich from a relatively rapid decay (Curry, 1987a, b; Robbins, 1987; Robbins and Brew, 1990). During growth, some of the organics are also trapped within the crystals. These intracrystalline organics have been likened to fluid inclusions preserved in rocks; they are relatively protected from microbial degradation and contamination by organic matter from extraneous sources (Curry, 1987a). Most mollusk shells also have layered organic and inorganic structures with both intra- and extracrystalline organics, which are thought to play an important role in biomineralization (e.g., Crenshaw,

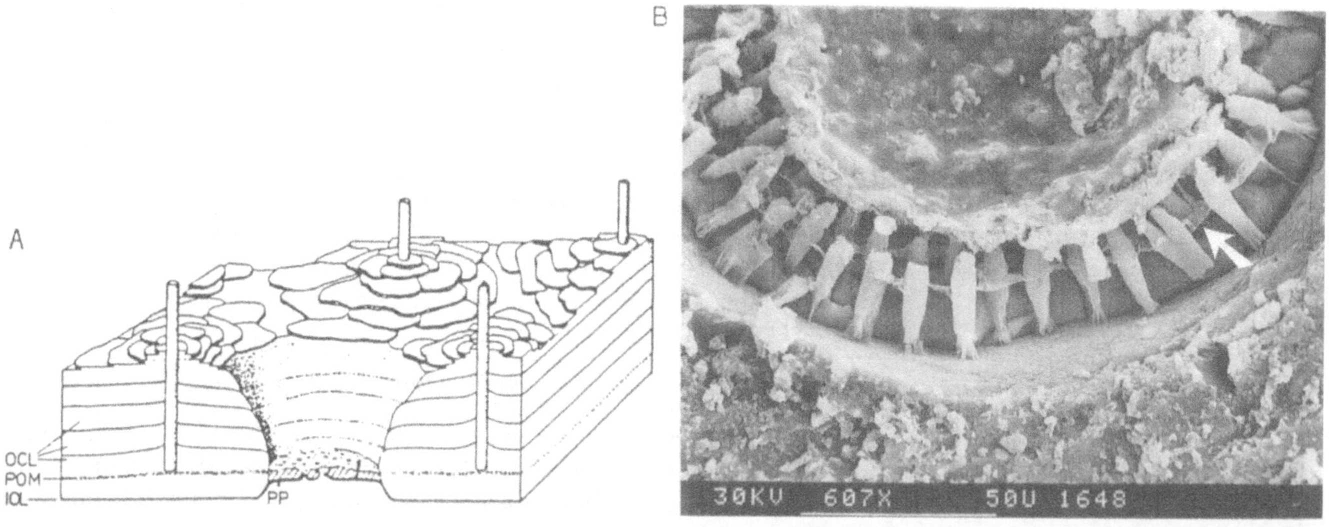

Figure 1. (A) Schematic diagram of the location of the organic membranes within the foraminiferal shell. OCL, Outer calcite layer; POM, primary organic membrane; IOL, inner organic lining; PP, pore plate. [From Hemleben *et al.* (1989).] (B) Scanning electron microscopic (SEM) photomicrograph of a cross section through a shell of a 300,000-year-old *Orbulina universa*. Sample has been embedded in resin and subsequently decalcified. Pillars represent resin infilled pores. The arrow points to the preserved primary organic membranes. By itself, the presence of this organic membrane does not guarantee that a biogeochemical signal will be preserved (Towe, 1980) but does reveal the excellent preservational conditions of the organic matrix.

1972; Weiner *et al.*, 1983; Krampitz *et al.*, 1983; Weiner and Traub, 1984; Wheeler *et al.*, 1987, 1988; see also Lowenstam and Weiner, 1989).

Research on fossil organics from a wide variety of taxa such as mollusks and vertebrate bone and teeth has indicated compositional similarities to living counterparts. Certain amino acids, however, may disappear quickly during the breakdown of the organics (Abelson, 1955; Wyckoff and Doberenz, 1965; Hare, 1969; Gillespie, 1970; Totten *et al.*, 1972; Weiner *et al.*, 1979). Particularly well protected macromolecular fractions have been shown to escape destruction and have part of their original structure preserved over long periods of geologic time (Muyzer, 1988). Despite the inevitability of diagenetic alteration of the protein through time, a surprising number of studies have shown that, under certain conditions, proteins are remarkably stable and may survive periods of several million years (e.g., Matter *et al.*, 1969; Gillespie, 1970; Dungworth *et al.*, 1975; Jope, 1973, 1979; Prager *et al.*, 1980; Weiner and Lowenstam, 1981). For instance, Fig. 2A demonstrates that while a rapid decrease of the amino acids in fossil scallops occurs in the first 2 million years, the rate of decay reaches a plateau at approximately 1–2% survival at 100 Ma and remains relatively constant thereafter (Akiyama, 1971; Wyckoff, 1972; Curry, 1988). Apparently, amounts of the insoluble proteins remain approximately the same (Fig. 2B). However, very little is known about these insoluble proteins and their ability to solubilize over time.

Use of Recent and fossil proteins from two invertebrate calcareous shelled organisms, planktonic foraminifera and mollusks, will be discussed in this chapter. Explanations for protein preservation and two approaches for extracting phylogenetic information from Recent and fossil proteins (viz., biochemical analysis and immunological detection) will be addressed.

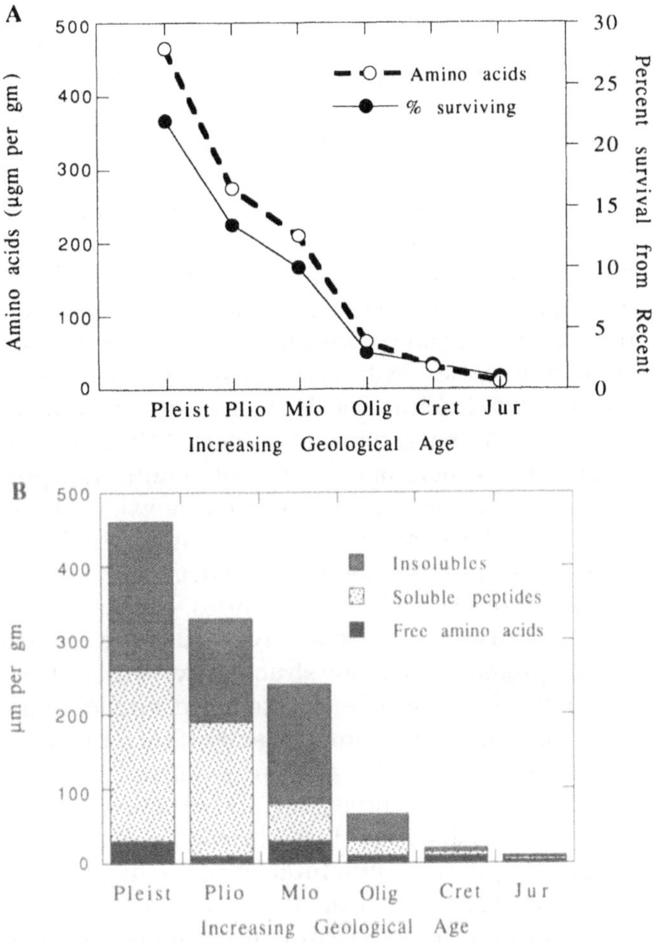

Figure 2. Preserved organic composition of fossil scallops. (A) Time-dependent decrease of amino acids in fossil scallops. (B) Relative proportions of free amino acids, soluble peptides, and insoluble residues in Pleistocene, Pliocene, Miocene, Oligocene, Cretaceous, and Jurassic scallops. [Data collated from Akiyama (1971) and Wyckoff (1972) by Curry (1988).]

1.2. Recent Proteins

1.2.1. Structure and Functions of Proteins

The basic structure of a protein is the polypeptide chain, a linear polymer formed by joining α-amino acids (Fig. 3) through peptide linkages. There are 20 different amino acids which typically comprise a protein. Regardless of function and origin, every protein is distinguished by a specific and unique arrangement of amino acids along its polypeptide chain. The term peptide refers to one or a small number of amino acids linked together, whereas polypeptide refers to longer chains, although in both cases neither the length nor the sequences need be defined. In general, there are four levels of protein structure. The basic covalent structure of the protein is called the primary structure or amino acid se-

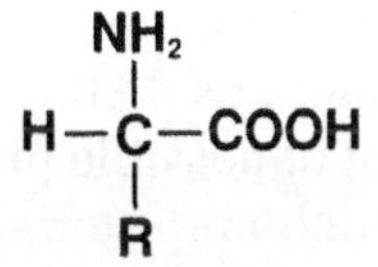

Figure 3. Structural formula of α-amino acids, the basic building blocks of protein. R is a side chain which can vary in size, shape, charge, hydrogen binding capacity, and chemical reactivity.

quence. The local arrangement of the polypeptide backbone is termed the secondary structure, while the three-dimensional structure of the polypeptide is the tertiary structure. Quaternary structure is the aggregation of the polypeptides according to specific interactions.

An understanding of the structures and roles of proteins is one of the bases of modern molecular biology but may be used in the study of ancient organisms as well. Proteins represent an interface through which genetic information is translated into cellular activities. With very few exceptions, the activities and structures of living systems are a direct reflection of the actions of proteins. Examples of these actions include muscle contraction, transport across cell membranes, enzymatic catalysis and mediation of the regulation of metabolic activities of cells, control of cell shape and movement, and the formation of extracellular structures, such as bones and shells. Despite the diversity of activities of individual proteins, they are structurally similar, their functional differences being associated with architectural variations on common themes. Since the basic chemical structure of a protein is derived from the translation of a structural gene, the structures of proteins are directly inherited. Mutational changes in genes can produce structural changes in proteins, sometimes resulting in modifications in their functional properties. Structural differences between equivalent proteins from different organisms tend to reflect their degree of temporal divergence so that species variations in individual proteins provide phylogenetic information that is close to the level of the genome.

The primary structure of a protein is a reservoir of biological information. Insight into its information content can be obtained by considering the number of alternative similar structures from which it has been selected. Since proteins are composed of 20 different amino acids, for a dipeptide (two amino acids) there are 400 (20×20) different sequences, and for a tripeptide, 8000. This number increases exponentially so that for a small protein containing 61 amino acids there are $20^{61} = 2.3 \times 10^{79}$ alternative sequences. The immensity of this number can be conceived by relating it to the Einstein model of the universe, which contains 0.88×10^{79} atoms. Clearly, only a small subset of possible protein structures can exist, many of which are interrelated. The information content in protein sequences is both historical and functional. Sequences reflect the evolutionary history of individual proteins; quite frequently, significant similarities are seen between the sequences of proteins with different biological functions, indicating that their structural genes are derived from a common ancestor.

In addition to the primary structure (the basic covalent structure), every protein has a distinct, three-dimensional structure or conformation. This is stabilized by noncovalent interactions between elements of the polypeptide chain and can be disrupted by treatments that do not disturb the primary structure, a process that is known as denaturation.

1.2.2. Proteins and Evolution

1.2.2a. Relationship between Protein Structure and Common Ancestry. Despite the incredible diversity that is seen at the macro- and microscopic level, there is a certain amount of biochemical unity to macromolecules (Creighton, 1983). A particular protein present today is merely one of the evolutionary products of many millennia, its particular amino acid sequence being primarily determined by its function and ancestry. Proteins that have descended from a common ancestry are called homologous. We can infer homology in proteins by structural conformation or by amino acid sequence and composition. Furthermore, the inference of evolution in proteins is often based on such information; structural conformation tends to be conserved, while composition tends to be more variable. Importantly, homologous proteins may differ significantly in sequence, while nonhomologous proteins may be very similar due to convergence of similar functions. Inferences of homology, therefore, are based on a statistically significant level of similarity in sequence and structure, implying that the genes for the proteins (and therefore the species) have descended from a common ancestor. A high level of dissimilarities in sequence and in length as a result of amino acid insertions and deletions imply widely separated species. One type of homologous proteins is orthologous proteins. Orthologous proteins have very similar tertiary structures or conformations as a result of gene duplication. Two or more proteins having equivalent conformations but characteristic differences in composition are called isologous. For example, horse and human hemoglobin are strikingly different in amino acid composition but have exactly the same folding of the protein, indicating a distant but common ancestry. However, it is important in this respect that comparison of species is based on orthologous or isologous proteins, since an individual organism will contain homologous proteins with divergent sequences. The comparison of homologous but not orthologous proteins from two species may give very misleading phylogenetic information.

During the evolution of a protein, in addition to insertions and deletions of amino acids, general replacements of amino acids will also occur. These replacements of amino acids tend to occur with other amino acids with similar side chains, such as gly/ala, ala/ser, ser/thr, ile/val/leu, asp/glu, lys/arg, and try/phe. The rate of change varies widely in different proteins (Kimura, 1977). A comparison of rates of evolution expressed as single base-pair mutations per position as a function of time, for two types of proteins, fibrous and globular proteins, is shown in Table I. Clearly, fibrous structural proteins show a relatively slow rate of mutation as compared to other proteins, a feature which makes these proteins attractive to study in evolutionary terms.

The structural proteins found in the shells of many calcitic organisms may be functionally related as a result of their implied involvement in the biomineralization of the shell. The specific differences in amino acid sequences will give an indication of the relationships between the different species. Unfortunately, as of yet there are very few data on sequences of calcareous shelled organisms owing to technical difficulties involving extraction, purification, and sequencing (Donachy *et al.*, 1992).

1.2.2b. Determination of Phylogenetic Relationships Using Electrophoresis. A widely used method to determine evolutionary relationships between organisms and to analyze mixtures of proteins is electrophoresis. This technique and immunological techniques are considered indirect because they do not exploit the primary structure (the amino acid sequence) of a protein. In general, electrophoresis is the migration of charged biopolymers, such as proteins, in a semisolid supporting medium (gel) under the influence of an electric field. Many biopolymers, and all proteins, have a particular size and differ in charge characteristics. As a result, they will have different electrophoretic mobilities. The electrophoretic mobility is therefore used as an indicator of similarity.

TABLE I. Rates of Protein Evolution[a]

Protein	Accepted point mutations per 100 residues per 10^8 yr
Fibrous proteins	
Collagen (α-1)	2.8
Crystallin (αA)	4.5
Other proteins	
Parvalbumin	20
Fibrinopeptide A	59
Casein κ chain	71

[a]Data adapted from Wilson *et al.* (1977).

Allozyme electrophoresis has been used in many phylogenetic studies, the enzymes being separated in starch gels on the basis of charge and size. Other electrophoretic techniques use only one of these two properties to separate biopolymers.

Isoelectric focusing (IEF), for instance, is used to separate biopolymers solely on the basis of their charge. A protein mixture is applied to a gel in which a pH gradient is established. The gel has wide pores, and therefore the size of the proteins has no effect on the migration. Each protein will migrate to a location in the gel where the pH is equal to its isoelectric point (pI). At this location the protein has no net charge and will move no further. The positions of the separated proteins are then determined by using various general staining techniques, such as Coomassie brilliant blue (CBB) or silver stain. As is often the case for polyanionic proteins, shell matrix proteins do not stain well using CBB. However, one method that has proved successful for detecting small amounts of foraminiferal shell proteins is autoradiography, whereby proteins are labeled with ^{125}I prior to electrophoresis (see Tolan *et al.*, 1980; Robbins, 1987) and then run on a gel. The gel is subsequently exposed to X-ray film to visualize the location of the protein bands. Using antibodies, biopolymers can be detected specifically.

To separate biopolymers on the basis of their size, polyacrylamide gel electrophoresis (PAGE) is used. The gels act as molecular sieves. The anionic detergent sodium dodecyl sulfate (SDS) is added to the sample to give the proteins a large negative charge, masking the protein's intrinsic charge. The result is a separation based solely on molecular weight.

Two-dimensional electrophoresis (2D electrophoresis) is a combination of IEF and SDS-PAGE. The technique is performed in three steps. First, the proteins are separated on the basis of their charge. The gel is then placed on an SDS-containing polyacrylamide gel, and the proteins are run perpendicularly to their movement in the focusing gel, resulting in a separation on the basis of their molecular weight. Finally, the position of the separated proteins in the gel is determined, using the various detection techniques outlined above.

Two-dimensional electrophoresis is a very powerful technique; it can resolve hundreds of individual proteins from a mixture, and it has been successfully applied to the study of molecular phylogenetics. Spicer (1988), for example, determined the evolutionary and phylogenetic relationships among seven *Drosophila* species groups using 2D-electrophoresis on whole-body extracts.

While electrophoretic techniques have been successfully applied to analyze matrix proteins from planktonic foraminifera isolated from a core top sediment and from a 300,000-year-old sample (Robbins and Brew, 1990), their application to more ancient samples may be limited because of degradation and modification of original biopolymers and contamination with organic matter from extraneous sources, a recurrent problem in the study of fossil organic matter. An alternative to determine sequence homology between proteins from Recent and fossil biominerals may be through peptide mapping, as suggested by Armstrong *et al.* (1983). Another alternative may be epitope mapping, involving proteolytic digestion followed by 2D-electrophoresis and the immunological identification of original antigenic fragments.

1.2.2c. Determination of Phylogenetic Relationships Using Immunology. Probably the oldest of all the approaches employed for the indirect retrieval of phylogenetic information about biopolymers is the use of immunological methods. As early as 1901, Nuttall and Dinkelspiel (1901) used immunological assays to determine systematic relationships between organisms. From then on, other investigators have used this approach in a great number of taxonomic studies. The principle of serotaxonomy is simple and is explained below in general terms. A more detailed description is given in Chapter 39.

When a laboratory animal, such as a rabbit, is injected with a protein isolated from a shell, the immune system of the rabbit responds to this foreign substance and produces large amounts of antibodies. These antibodies, which can be obtained from the blood of the animal, will recognize and bind to small domains of the biopolymer, typically four to eight amino acids in the case of a protein (Goodman, 1980). Usually, several different antibodies are produced that bind to different portions of the same protein. Evolutionary changes in these domains, which are named antigenic determinants or epitopes, will alter the antibody response to the protein, a principle employed in measures of immunological distance (Wilson *et al.*, 1977). The intensity of the immune reaction of an antiserum raised against a protein from species X (homologous antigen) and tested with the same protein from species Y (heterologous antigen) is a measure of the degree of similarity between the proteins, and thus of the kinship between the two species.

The use of antibodies to determine phylogenetic relationships has great advantages over other molecular approaches. Immunological assays are extremely sensitive and highly specific. Furthermore, the antibodies can be applied to mixtures to assay for one particular compound, which omits the time-consuming and difficult task of purification. An additional advantage of this approach is that antibodies can be used to retrieve phylogenetic information from fossil biopolymers, because they will recognize and bind to small parts of the biopolymer, which have a greater chance of survival than the entire protein (Lowenstein, 1980a; Muyzer, 1988). In addition, specific antibodies will not react with degradation products or contaminating organic matter.

In addition, when coupled to inert beads, antibodies can be used to isolate specific proteins or fragments thereof from complex mixtures of organic matter. These isolated biopolymers can then be further characterized by other biochemical techniques.

1.2.2d. Determination of Phylogenetic Relationships Using Protein Sequences. Sequencing of the amino acids in proteins remains one of the most direct methods of obtaining the maximum amount of phylogenetic information from an organism. One current method for determining amino acid sequences uses Edman degradation, which sequentially cleaves off amino acids from the amino terminus of the polypeptide. The amino acids are then identified by liquid chromatography. Comparisons of two or more polypeptide sequences are aided by computer software designed to align the sequences and statistically compare them. The divergences of proteins may be traced by reconstruction of the phylogenetic tree from sequences of similar proteins. Cladograms, a reflection of phylogenetic relationships, may be constructed from a variety of different types of evolutionary evidence, including morphological and molecular data; some of the strongest evidence is provided by the comparison of multiple protein sequences (Penny *et al.*, 1982).

1.2.2e. Construction of Phylogenetic Trees by Computer Analysis of Molecular Data. To construct phylogenetic trees from sequence data requires the use of sophisticated computer programs to help align several different sequences. This is important because, for example, over two million trees are possible using only ten species. Two main methods are used: cladistic and phenetic (Joysey and Friday, 1982). The phenetic method uses a distance matrix to compare the overall similarity between protein or base (DNA or RNA) sequence differences among species (Fitch and Margoliash, 1967; Elwood *et al.*, 1985; Olsen, 1988). This method makes no assumptions about homogeneity of the rate of evolution (see Farris, 1972). A cladistic approach actually infers the ancestral sequences of the proteins that occurred at branch points and reconstructs how mutations have proceeded along lines of descent from ancestor to living species. Each ancestral sequence is based on a minimum number of mutational events that have occurred throughout the tree.

There are a number of methods and software programs that are available for protein and base sequence comparisons and cladogram construction. These widely used programs include PHYLIP, PAUP, HENNIG 86, Maximum-Parsimony (MP), Fitch–Margoliash (FM), Maximum-Likelihood (ML), and Unweighted Pair Group Mathematic Average (UPGMA); some of these programs can compare sequences directly, while others need a distance matrix for the construction of a phylogenetic tree. Each of these programs or methods contains particular constraints. For example, PHYLIP does not count all base substitutions but only those substitutions which change amino acids. Moreover, constant and varying rates of substitutions should be considered when reconstructing the threes.

2. Fossil Proteins

2.1. Spontaneous Chemical Changes That Can Occur in Proteins

A number of spontaneous chemical changes can occur in both newly formed proteins and preserved proteins through the processes of degradation. Many of these have been discussed in earlier chapters. This section will focus on the correlation of degradation with molecular properties of the protein. Therefore, knowledge of the original molecular composition and structure of the protein is critical as a predictive tool.

On the average, acidic proteins will degrade more rapidly than neutral or basic ones. Specifically, aspartyl bonds are known to break down rapidly (Hill, 1967), and glutamic and aspartic acids are susceptible to deamidation, which in turn will accelerate degradation. Because the organic shell matrix of varied organisms contains abundant quantities of these acidic amino acids (Mann, 1988), deamidation can play a significant role. As a consequence, more acidic shell proteins degrade first in the rock record (Hoering, 1980). On the other hand, it is interesting to note that formation of β-aspartyl linkages, a phenomenon occurring in peptides exposed to acidic conditions for long periods of time, may actually inhibit hydrolysis (Schroeder *et al.*, 1963).

Rates of protein degradation in living systems are related to stabilities of protein 3D structures and strength of peptide bonds (Creighton, 1983). Therefore, some factors that are significant in influencing the rate of degradation are the unequal strength of peptide bonds, unfolding of the protein, dissociation of a stabilizing ligand, and the susceptibility to protease digestion. Similar principles might be expected to affect the decomposition of proteins in fossils. An important factor that may contribute to inhibition of degradation is the tenaciously enduring relationship between organic phases and crystals (Abelson, 1956; Weiner and Traub, 1984), which may be partially attributable to the binding of the protein to the crystal matrix (Sikes and Wheeler, 1986; Addadi and Weiner, 1985).

The complexity of protein degradation is exemplified by the fact that there are over 200 posttranslational modifications that proteins can undergo, some of which are manifestations of old age (Creighton, 1983). Proteolytic cleavage, hydrolysis, deamidation, and oxidation of polypeptides are just some of the many forces that drive the process toward the breakdown to smaller polypeptides and eventually to individual amino acids. Some of these and other chemical changes known or expected to occur in proteins are listed in Table II. Often, the end products of such processes are ammonia and acidic byproducts. These products in themselves may in turn accelerate degradative processes. Moreover, certain amino acids may be modified into other amino acids. The epimerization of isoleucine to alloisoleucine is one of the more well documented changes occurring in fossils. This process and other L to D conversions which are important in the postmortem transformation of proteins are discussed further in Chapters 35 and 36 of this volume.

One particularly troublesome interconversion is serine to alanine (Table II). Such an interconversion may mask the original amino acid sequence of a fossil protein. This problem may be confounded by the fact that some shell proteins are silk fibroin-like structural proteins (Weiner and Erez, 1984; Weiner *et al.*, 1983) which may originally contain substantial amounts of alanine. Therefore, it is critical to know the original protein composition or sequence prior to evaluating fossil proteins.

TABLE II. Some Chemical Changes of Proteins

Hydrolysis
- Some peptide bonds are more susceptible, such as the aspartyl bonds

Peptide bond rearrangements
- Asparagine bonds rearrange to produce isopeptide linkages, i.e.,

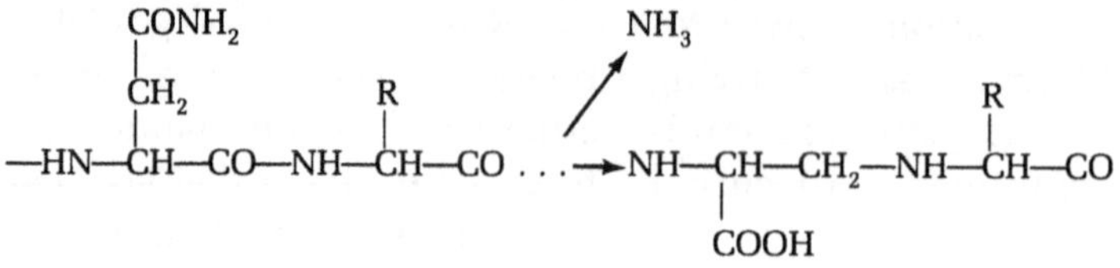

Cross-linking
- Tyrosine may form di-tyrosine cross links as the result of exposure to UV light
- Glutamine–lysine side chains

Oxidation
- Methionine, tryptophan, tyrosine, histidine, and cysteine are particularly susceptible

Deamidation
- Aspargine forms aspartic acid
- Glutamine forms glutamic acid

Other chemical changes
- β-Elimination of O-modified serines to form alanine
- Dehydration of unmodified serines

2.2. Conditions That Might Favor Preservation

2.2.1. Environment of Hardparts

As discussed earlier, preserved proteins may owe their longevity to a number of contributing factors. One such factor is their enclosure within a calcite shell, limiting accessibility to solvent and slowing the rate of hydrolysis. For example, in the case of the mollusk or planktonic foraminifera, unless the shell is broken or until the shell is recrystallized, the intracrystalline protein will have only limited exposure to seawater. Once fully exposed to seawater, however, the proteins will undergo a complex set of degradative reactions including hydrolysis (Hoering, 1980).

2.2.2. Environment of Deposition

Also playing an important role in the preservation of shell proteins is the environment of deposition. For instance, once planktonic foraminifera die or shed their gametes, the empty test (shell) is deposited onto the ocean floor as much as 2–4 km below. The temperature of deposition often ranges between 2 and 4°C for globigerina oozes, a temperature which is ideal for preserving and inhibiting thermally induced protein decomposition. However, after burial the temperature may rise and initiate protein hydrolysis and degradation. Typically, initial conditions are oxidative, but the shell may experience less aerobic conditions once burial commences. It is likely that the rapidity of burial and subsequent dewatering of the sediment partially determines the extent of preservation. The more "desiccated" the environment, the more protected the organics are from hydrolysis. If the shell is deposited in an anaerobic environment, such as an anoxic basin, protein oxidation and subsequent degradation may be fairly slow, and even cellular material, such as DNA, may be preserved. Fossils preserved in this specialized environment are likely candidates for other types of molecular studies such as outlined in Section 3.

2.2.3. Protein Structure

Protein structure (primary, secondary, tertiary, and quaternary) plays an important role in the ability to withstand degradation. In particular, structural proteins, such as those found in shells (Weiner *et al.*, 1983; Mann, 1988), have repetitive amino acid sequences and regular, linear structures which are especially resistant to degradation. This fact is obvious from the preservation of ancient fabrics made from silks (a fibrous protein, silk fibroin) and wool (predominantly the protein keratin).

2.3. General Considerations and Specific Problems of Protein Decomposition

One of the major limitations of using proteins for taxonomic studies of fossils continues to be the breakdown and diagenetic alteration of the organics. The question arises, how does the original protein relate to the residual left after millions of years and how will this affect our assessment of phylogenies? This question of indigeneity cannot be answered without the complete characterization of the actual components of the organic phases of both the Recent and the fossil shells. Isotopic analysis of individual amino acids from high-molecular-weight fractions and individual proteins and polypeptides has been suggested as ways to provide criteria in assessing indigeneity (Tuross *et al.*, 1988; Hare *et al.*, 1991; Ostrom *et al.*, in press).

2.4. Detection of Proteins in Recent and Fossil Foraminifera

Despite the fact that the majority of proteins degrade over geologic time, small amounts of the original macromolecule can often be detected in well-preserved fossils (Muyzer *et al.*, 1984; Robbins, 1987; Muyzer, 1988; Stathoplos, 1989; Robbins and Brew, 1990). By isolating and characterizing trace amounts of fossil proteins, a clearer understanding of phylogenetic relationships and how individual proteins degrade over geologic time can be gained. Such studies require sophisticated biochemical technology in order to purify the biopolymers. Fortunately, recent advances in instrumentation have dramatically improved the chance of overcoming the problems of using fossil proteins that appeared insurmountable just ten years ago.

For instance, it is not known to what extent amino acid sequences survive fossilization. However, one goal of these studies is to sequence these biopolymers. A recent study by Robbins and Donachy (1991) reported a partial sequence of the N-terminus of one of the mineralization regulation proteins found in the shell matrix of two species of core-top planktonic foraminifera. This study revealed a hydrophilic domain at the N-terminus. As of yet, the phylogenetic implications of this domain are not known. It is possible that such sequence data will also lead to a better understanding of some of the degradational processes that take place in the organic matrix.

Because of the relative ease in collecting core-top sediment containing thousands of foraminifera versus collecting living foraminifera from plankton tows, the former has been studied almost exclusively. Well-preserved shell material less than 4000 years old has been shown to contain up to seven matrix proteins (Robbins and Brew, 1990). Of these, a subset has been shown to survive at least 300,000 years (Fig. 4). However, it is clear that an understanding of the organic constituents and how they degrade in any system will benefit from the characterization of the original components in the living organism.

The methods used in isolating and characterizing fossil planktonic foraminifera are similar to those employed in the study of other proteins. Cleaning of the shell is critical because of the variety of debris and contaminants that adhere to both the outside and the inside of the foraminferal chambers. There is also the possibility of free amino acids and humic acids adhering to the calcite test (Mitterer and Cunningham, 1985; Robbins and Brew, 1990). Effective methods of removing the debris from the shell exterior include repeated sonic cleaning in distilled water and guanidine–HCl or NaOH rinses. The latter two methods may also remove some of the proteins associated with the inner organic lining (IOL), an organic membrane lying adjacent to the outside of the calcite shell, and

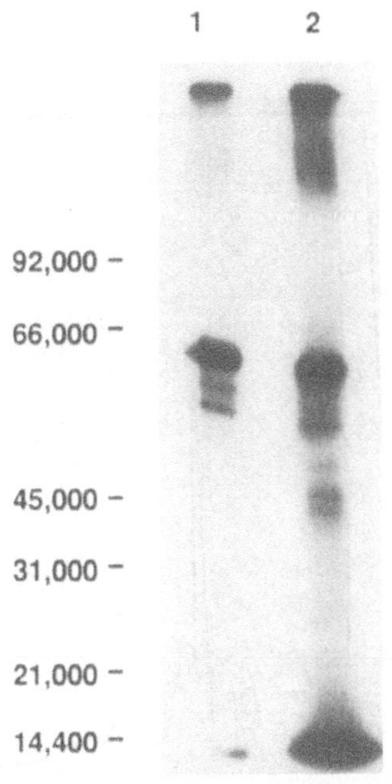

Figure 4. Autoradiograph of 10% polyacrylamide gel of high-molecular-weight (HMW) material from *Globigerinoides sacculifer* and mixed assemblage tests. Lane 1: 300,000-year-old *Globigerinoides sacculifer*; Lane 2: Recent *G. sacculifer*. Bands indicate the presence of proteinaceous material. Molecular weight standards ran at the designated locations on the left.

therefore should be used cautiously. After cleaning, samples are checked by scanning electron microscopy (SEM) to ensure all contaminants have been removed from the shell. The calcium carbonate is then dissolved using an acid treatment, such as cold 50% formic acid (Robbins and Brew, 1990), or a chelating reagent, such as EDTA (Sikes and Wheeler, 1983; Robbins and Donachy, 1991). Several researchers (L. L. Robbins and J. Donachy, unpublished data; C. W. Sikes, 1988, personal communication; Lowenstam and Weiner, 1989) have found that the use of cold dilute acid does not tend to hydrolyze the peptides and that these two methods produce similar results on polyacrylamide gels and on reverse-phase high-pressure liquid chromatography (RPHPLC), although the formic acid treatment may tend to precipitate some of the more acidic proteins.

After dissolution of the calcium carbonate and solubilization of the protein, the sample is centrifuged at 20,000 × g for 20 min. A supernatant comprised of what will be designated as soluble matrix (SM) and a pellet comprised of insoluble matrix (IM) remains. The SM is extracted and dialyzed (Spectrapor 6000–8000 nominal pore cutoff) exhaustively against water and freeze-dried. Later, the SM is reconstituted in buffer and either applied to an RPHPLC column or run on a polyacrylamide gel.

RPHPLC has been shown to be one of the most powerful methods for protein and peptide purification (Shively, 1986) and has been used extensively for isolation of numerous types of Recent and fossil shell proteins (Weiner, 1982; Weiner and Erez, 1984; Rusenko, 1988; Robbins and Brew, 1990; Robbins and Donachy, 1991). The principle of RPHPLC is the separation of proteins based on hydrophobicity, which is related to the content of nonpolar amino acid residues. Reverse-phase columns, composed of silica to which hydrocarbon chains containing between 3 and 18 carbons have been attached, are used with a variety of elution buffers. Lyophilized samples are dissolved in a buffer and injected into the column equilibrated with the buffer. A gradient of increasing concentrations of an organic solvent (e.g., acetonitrile) in water or a salt buffer (Weiner, 1982; Robbins, 1987) is used to elute proteins of increasing hydrophobicity. In general, the solvents are acidified with a low concentration of trifluoracetic acid (0.1%), whereas the salt buffer may be only slightly acidic (Weiner, 1982). Eluted proteins are detected by monitoring the effluent at a wavelength (210–220 nm) that is sensitive to peptide-bond ultraviolet (UV) absorption and are collected. A chromatogram of extracted protein from the shell of *Globigerinoides ruber* is presented in Fig. 5.

Gel filtration has been used in a number of

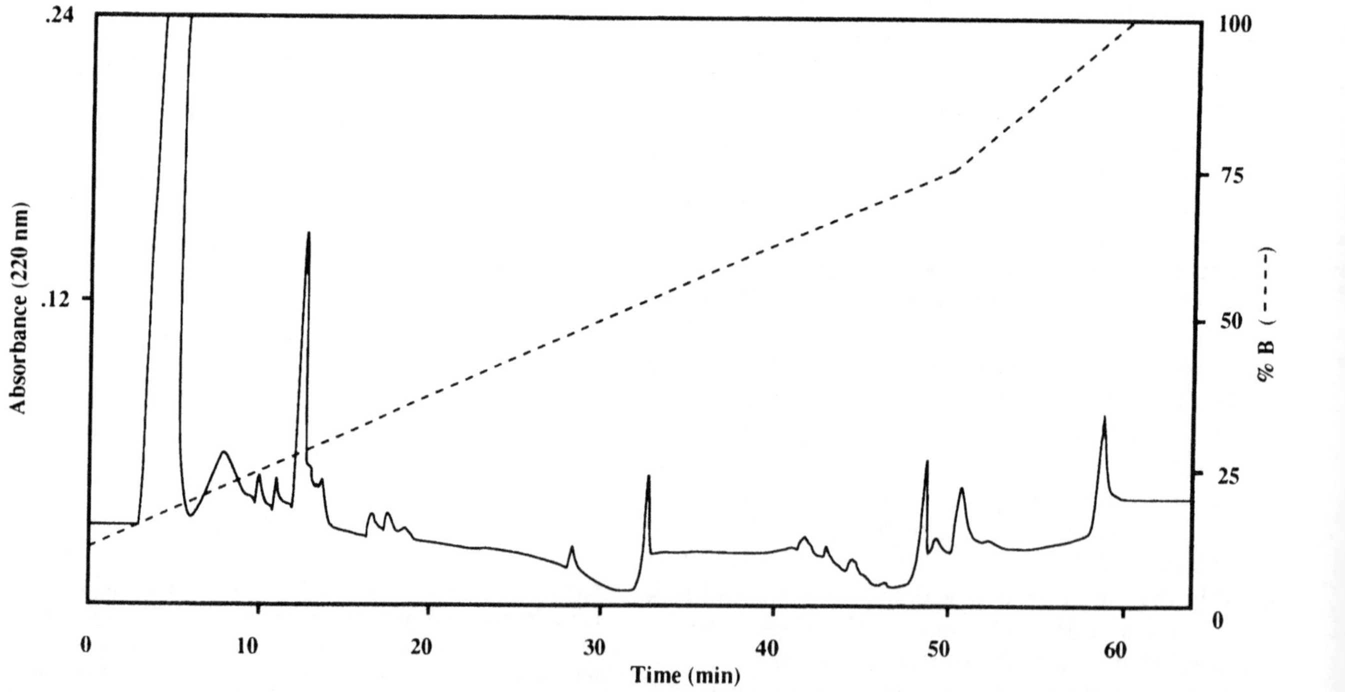

Figure 5. Chromatogram of the organic matrix of *Globigerinoides ruber*. The isolation was carried out at 25°C with an initial linear gradient (10–75%) of acetonitrile in 10m*M* ammonium acetate (pH 6.5) for 50 min and an increase (75–100%) of acetonitrile for 10 min. Flow rate was 1 ml min^{-1}. Fractions were monitored at 214 nm.

studies to show that the organic fraction of fossils contains major protein fractions. This is a suitable method for the preliminary cleanup of samples and to separate material into general size fractions if sufficient material is available, but it has lower resolving power than is often needed and is insufficient for separating specific components of the major fractions (Rusenko, 1988) and therefore has not been used in our studies. Weiner (1982) has indicated that ion-exchange chromatography, another method widely used in separating proteins, has been applied to shell proteins with varying success.

In our work, freeze-dried fractions from RPHPLC containing foraminiferal polypeptides and proteins were hydrolyzed in 200 μl of 6*N* HCl at 110°C *in vacuo* for 18 h. The hydrolysates were vacuum dried to remove residual HCl, and the amino acids were converted to their phenylthiocarbamyl (PTC) derivatives by reaction with phenylisothiocyanate (PITC) (Heinrickson and Meridith, 1984). The PTC amino acids were analyzed by a Hewlett-Packard 1090 HPLC equipped with a Waters PICO-TAG column (Robbins, 1987). Another way to monitor amino acids is by their reaction with *ortho*-phthalaldehyde (OPA). This has been successful in identification of various enantiomers of amino acids (Hare, 1969) but may be unsuitable for analyzing proteins with significant amounts of proline, such as found in the structural protein collagen.

Although little is known about the shell matrix proteins within a foraminiferal test, in recent years a few studies have addressed this topic (Weiner and Erez, 1984; Robbins, 1987; Robbins and Brew, 1990; Robbins and Donachy, 1991; Toler, 1993). Two broad groups of proteins, an aspartic acid- and serine-rich fraction and a glycine-, serine-, and alanine-rich fraction, were identified in benthic foraminifera (Weiner and Erez, 1984) and planktonic foraminifera (Robbins, 1987; Robbins and Brew, 1990) although the distinction between these two groups of proteins in planktonic foraminifera is unclear because of an apparent gradation. For example, the amino acid composition of protein fractions from the organic matrix of *Globigerinoides ruber* obtained by RPHPLC has demonstrated that two initial hydrophilic fractions are high in aspartic and glutamic acids and serine while a later fraction has high amounts of glycine, glutamic acid, and serine. Other fractions, however, have characteristics of both groups of proteins (Table III).

TABLE III. Amino Acid Composition (mol %) of Four Proteins from Shells of *Globigerinoides ruber* (red)

	HPLC—C_{18} fractions			
Amino acid	1	2	8	16
Asp + Asn	32.0	50.7	11.8	8.3
Glu + Gln	15.1	11.3	9.5	10.9
Ser	11.4	10.7	9.5	11.6
Gly	5.7	4.5	13.9	19.2
His	0.9	0.8	Trace	0.6
Arg	2.0	7.7	3.1	Trace
Thr	9.9	5.1	5.7	8.1
Ala	7.6	2.7	11.5	8.0
Pro	3.2	2.2	5.5	3.6
Tyr	2.7	0.3	1.6	1.3
Val	1.2	0.6	7.4	6.1
Met	3.1	0.8	5.4	5.0
Ile	1.6	0.8	3.0	3.7
Leu	1.7	0.9	5.4	6.5
Phe	1.1	0.6	3.2	2.6
Lys	0.4	0.3	3.1	4.4

At least one protein fraction has been suggested to be species specific, because of its reproducible retention time and its compositional differences from species to species (Robbins and Healy-Williams, 1991). Additional data from morphology and stable isotopes gave a powerful indication of generic and species differences (Robbins and Healy-Williams, 1991). Species-specific trends in planktonic foraminifera were first suggested by the pioneering work of King and Hare (1972) on the total amino acid composition of planktonic foraminiferal shells. Similar trends were also shown for benthic foraminifera (Haugen *et al.*, 1989). These studies, however, implicitly assumed that the shell matrix was composed of only one protein. This assumption was shown to be incorrect as Recent shell matrix from foraminifera was observed to contain at least seven polypeptides (Robbins, 1987; Robbins and Brew, 1990; Toler, 1993). The total amino acid analysis of the shell therefore averaged together the contributions from these different proteins. Moreover, problems associated with using total amino acid analysis include the possibility of analyzing free amino acids, such as aspartic acid, and humic acids adsorbed to the test (Mitterer and Cunningham, 1985; Robbins and Brew, 1990). This can be problematic since taxonomic distinctions have been based on aspartic acid, one of the most abundant amino acids (King and Hare, 1972; Robbins and Brew, 1990).

Based on composition, a similarity matrix of one prominent protein, FP8, from different species (Table IV) indicated the degree of relatedness between species (Robbins and Healy-Williams, 1991). This type of information provides the basis for establishing taxonomic classifications based on molecular data if we

TABLE IV. Amino Acid Composition (mol %) of Protein FP8 from Six Different Species

	Species[a]					
Amino acid(s)	1	2	3	4	5	6
Asp + Asn	9.1	11.8	10.6	3.4	3.9	5.9
Glu + Gln	12.6	9.5	8.0	8.9	7.3	14.7
Ser	9.9	9.5	10.4	9.6	8.8	8.9
Gly	14.0	13.9	14.2	16.3	15.4	18.2
His	6.9	Trace	0.7	2.6	Trace	2.2
Arg	3.5	3.1	3.2	3.2	3.8	0.9
Thr	5.4	5.7	4.6	5.5	5.4	3.4
Ala	10.5	11.5	11.8	11.9	10.7	9.7
Pro	5.0	5.5	4.5	7.5	7.6	3.7
Tyr	1.1	1.6	1.7	4.5	1.0	3.4
Val	3.2	7.4	8.3	7.2	5.8	8.1
Met	3.9	5.4	3.2	2.0	5.5	2.2
Ile	3.7	3.0	5.1	4.5	3.7	2.5
Leu	4.3	5.4	6.3	6.8	7.0	7.1
Phe	4.1	3.2	3.7	3.0	8.3	2.8
Lys	2.4	3.1	3.7	3.1	5.8	6.4

[a]1, *Globigerinoides sacculifer*; 2, *Globigerinoides ruber* (red); 3, *Orbulina universa*; 4, *Globorotalia menardii*; 5, *Globorotalia tumida*; 6, *Pulleniatina obliquiloculata*.

assume that the degree of relatedness corresponds to mutation distances. Cornish-Bowden (1983) indicated that this is valid for homologous proteins, although such assumptions may be dangerous in situations when the proteins are not actually homologous. Cluster analysis of the data also portrayed species relationships similar to those based on morphology alone (Fig. 6).

FP8 and additional proteins can be found in 300,000-years-old and older samples, suggesting that these may allow tracing of evolutionary trends on a molecular level. A crucial aspect to this type study is determining whether indigenous proteins are present and the amount and specific type of diagenetic alteration that the proteins encounter over time (Ostrom *et al.*, 1993). While sequencing will be able to partially address these problems, other approaches include use of monoclonal antibodies (Robbins *et al.*, in press) and geochemical criteria (Ostrom *et al.*, 1993).

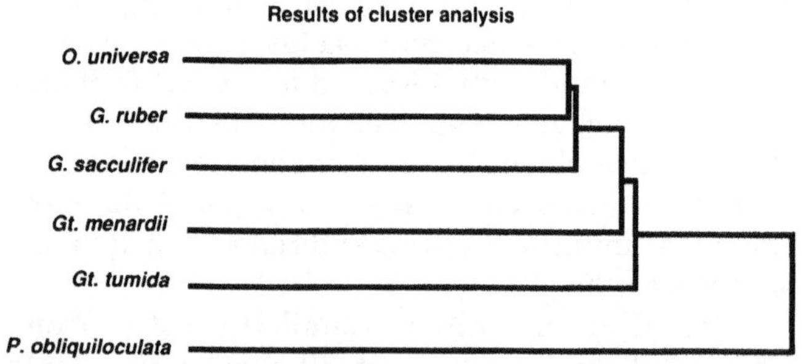

Figure 6. Cluster analysis of the amino acid composition data of protein FP8 isolated from shells of different foraminifera. The results demonstrate that the relationship between species based on the amino acid data are in good agreement with those based on morphological characters.

2.5. Detection of Proteins in Recent and Fossil Mollusks

Although immunological techniques have been used to compare proteins from less ancient, but extinct species (Lowenstein, 1980a, b; Lowenstein *et al.*, 1981), they have not been used extensively to determine phylogenetic relationships of geologically old samples. An exception to this is the work of Westbroek and co-workers (de Jong *et al.*, 1974; Westbroek *et al.*, 1979), who showed by immunological means that original macromolecular information was preserved in fossils more than 70 million years old. More recent studies by this group on the preservation and diagenesis of biopolymers in fossils have been conducted on well-characterized mollusk shells collected by J. F. Wehmiller of the University of Delaware from five Pleistocene localities along the Atlantic seaboard of the United States.

An antiserum produced against the EDTA-soluble organic matrix from Recent shells of the bivalve mollusk *Mercenaria mercenaria* was used to study phylogenetic relationships between Recent and fossil mollusks (Muyzer *et al.*, 1988). In addition, the serum was applied to organic matter from fossil *Mercenaria* shells to determine the diagenetic stability of the antigenic determinants (Muyzer *et al.*, 1988), and used to localize the antigenic material *in situ* (Muyzer and Westbroek, 1989).

The taxonomic specificity of the antiserum was determined with shell fragments of a variety of Recent bivalves (Fig. 7) using an enzyme-linked immunosorbent assay (ELISA). Strong positive reactions were obtained with representatives of the Paleoheterodonta and of the Heterodonta to which *Mercenaria* belongs. An exception was a lack of reaction with the aberrant genus *Teredo*, which may implicate that this genus is incorrectly grouped. No or only weak reactions were found with members of the other three bivalve subclasses: The Palaeotaxodonta, the Cryptodonta, and the Pteriomorphia. The serum can therefore distinguish between shell macromolecules from (paleo)heterodonts and those from other bivalve subclasses. A more specific antibody preparation was obtained after absorption of the antiserum with etched shell powders of various (paleo)heterodonts. Anti-

SUBCLASS	ORDER	SUPERFAMILY	GENUS
Paleotaxodonta			*Nucula*
Cryptodonta			*Solemya*
Pteriomorphia	Arcoida	Arcacea	*Arca*
		Limopsacea	*Glycymeris*
	Mytiloida	Mytilacea	*Mytilus*
			Brachidontes
		Pinnacea	*Pinna*
	Pteroida	Pteriacea	*Pteria*
			Pinctada
			Isognomon
		Pectinacea	*Pecten*
			Chlamys
		Anomiacea	*Anomia*
		Ostreacea	*Ostrea*
Paleoheterodonta	Unionida		*Unio*
			Anodonta
Heterodonta	Veneroida	Cardiacea	*Cardium*
			Cerastoderma
			Laevicardium
		Tridacnacea	*Tridacna*
		Mactracea	*Mactra*
			Spisula
		Solenacea	*Solen*
			Ensis
			Pharus
		Tellinacea	*Macoma*
			Donax
			Scrobicularia
		Corbiculacea	*Polymesoda*
		Veneracea	*Venus*
			Meretrix
			Dosinia
			Venerupis
			Mercenaria
		Myacea	*Mya*
	Myoida	Pholadacea	*Zirfaea*
			Teredo

Figure 7. Classification of the species used in antibody research. All species belong to the class Bivalvia, phylum Mollusca.

bodies in the serum which recognize similar antigenic determinants on biopolymers of other species will be absorbed and therefore removed from the serum; antibodies which do not recognize similar antigenic determinants are not absorbed and will be retained in the serum. The preparations reacted only with members of the Veneracea, although *Meretrix*, which is classified with the Veneracea, behaved as if it were a nonveneracean heterodont.

The phylogenetic specificity was also apparent in tests with fossil shells (see Fig. 8). The heterodont bivalve *Dinocardium* (80,000 to 250,000 years old) gave a significant positive reaction with the antiserum, while the nonheterodont bivalves and the gastropod *Busycon* did not. After absorption of the antiserum with etched shell powders of Recent *Cardium* (closely related to *Dinocardium*), all cross-reactivity with *Dinocardium* was removed, although the absorbed antibody preparation still reacted with the *Mercenaria* samples.

To determine the potential preservation of original biopolymers in fossils, the antiserum was applied to a series of Pleistocene *Mercenaria* shells for which amino acid racemization, radiometric (U–Th), and biostratigraphic data were available. The immunological reactivity decreased with sample age as expected. However, small amounts of antigenic determinants could still be detected in fossils over 1 M year old (Fig. 8). Figure 9 shows the relationship between the immunological reactivity and the degree of leucine racemization in modern and fossil *Mercenaria* shells. The plot is virtually linear, suggesting parallel diagenetic alteration of the antigenic material. These results are consistent with the ELISA results of shell fragments of *Mercenaria mercenaria* which were heated at a high temperature for different numbers of hours to mimic the process of diagenesis. The immunological reactivity decreased with increasing heating times; however, significant positive reactions were still found with shell fragments which were incubated for 80 h at 140°C (Fig. 10). Similar results were found in a previous study on the preservation of biopolymers from *Mercenaria* shells (Muyzer *et al.*,

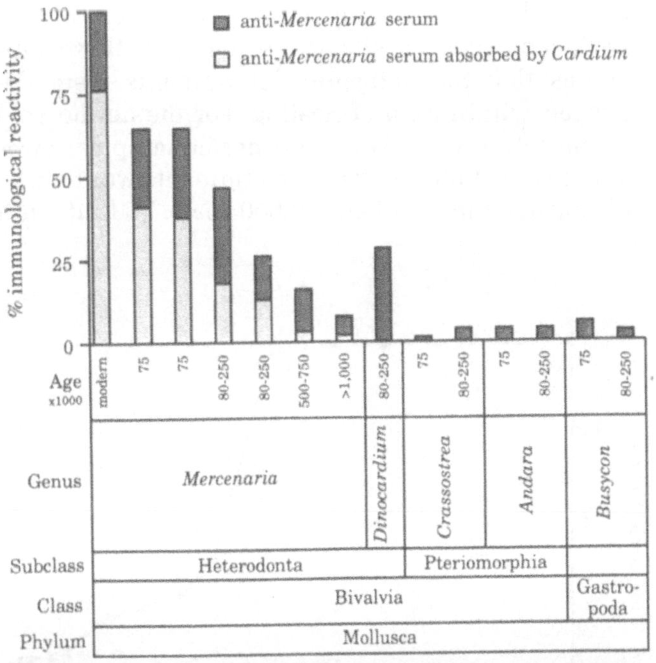

Figure 8. Immunological reactivity with fossil mollusk shells. Note the phylogenetic specificity of the antiserum, reacting only with *Mercenaria* and the other heterodont bivalve *Dinocardium*. The intensity of the immunological reaction is expressed as the percentage of the reaction between the crude antiserum and the homologous antigen *Mercenaria mercenaria*.

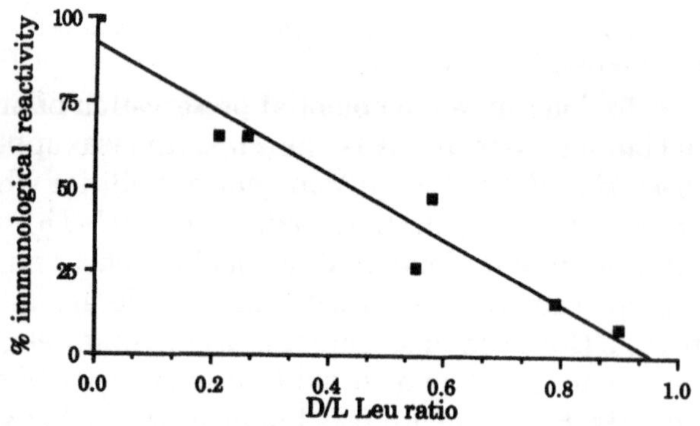

Figure 9. Plot of the relationship between the immunological reactivity and the extent of racemization of leucine in Recent and Pleistocene *Mercenaria* shells.

1984). In an artificial diagenesis experiment carried out with shell fragments of *Mercenaria mercenaria*, the degradation of individual antigenic determinants was followed using monoclonal antibodies. In contrast to a conventional antiserum, which contains a large variety of antibodies raised against different biopolymers and different antigenic determinants, monoclonal antibodies (Köhler and Milstein, 1975) are homogeneous antibody preparations, reacting only with one particular antigenic determinant. In this study, monoclonal antibodies were produced against the EDTA-soluble matrix from Recent *Mercenaria mercenaria* shells. It was found for these antibodies that the antigenic determinants were destroyed within 32 h of heating. For the amino acid racemization analysis on these heated samples, it was assumed that the antigenic determinants would have disappeared in less than 120,000 years. To find out if

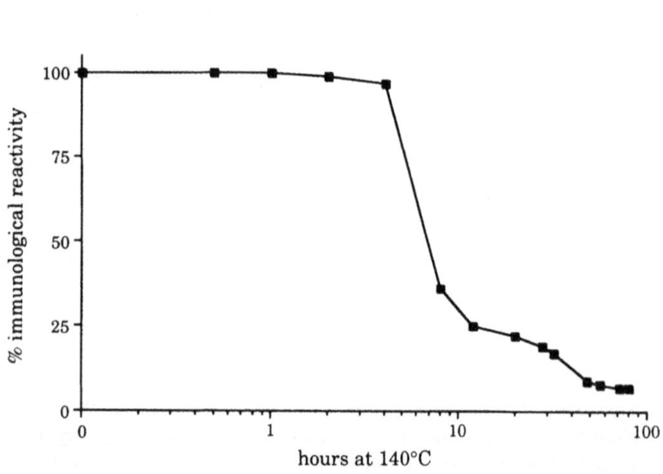

Figure 10. ELISA results of an anti-*Mercenaria* serum with shell fragments of Recent *Mercenaria mercenaria* which were heated at 140°C for several hours to mimic the process of diagenesis.

this assumption was acceptable, the results were compared with ELISA results of the antibodies tested with fossil *Mercenaria* shells for which the age was known from amino acid racemization and radiometric data. One monoclonal antibody, viz., M5, reacted only with fossil samples not older than 75,000 years old, while two other monoclonal antibodies only reacted with shell fragments of Recent *Mercenaria*. This indicates that, with the help of antibodies, the preservation of biopolymers can be determined and that the process of degradation of organic matter over geologic time can be followed.

Although the shells were scrubbed clean and incubated overnight in a solution of EDTA to remove the outermost shell material containing contaminating organic matter, nothing was known of possible contamination sources such as endoliths. In principle, this could mean that an antiserum produced against the EDTA-soluble matrix could also have antibodies against these organisms. Therefore, to further characterize the antigenic material, an immunohistochemical method was applied to shell fragments of Recent and fossil *Mercenaria*. Shells were sliced, polished, and etched with an EDTA solution to expose the antigenic determinants. Subsequently, the fragments were incubated with the anti-*Mercenaria* serum and thereafter with a goat anti-rabbit antibody conjugated with an enzyme to visualize the bound antibody. The enzyme converts a noncolored substrate into a colored product, which will precipitate in the immediate vicinity of the enzyme. The samples were examined by light microscopy with incident illumination. Strong staining reaction were obtained with organic matter from Recent shells. The staining reaction with a 75,000-year-old sample was more blurred, although the matrix surrounding the secondary prisms could still be observed. No cross-reactions were found with possible contaminating organisms, such as bacteria or algae. Figure 11 is an illustration of the immunohistochemical staining of antigenic macromolecules in a longitudinal section of a Recent *Mercenaria* shell.

Although this immunohistochemical technique has been applied to Recent and Pleistocene shells only, it could be envisaged that with other markers (viz., gold labels or fluorescent compounds) and high-resolution devices, such as SEM or laser scanning microscopy, much older material can be studied. The technique is therefore a welcome extension to those already available in the study of original organic remains in the fossil record.

From these results, the following conclusions can be drawn: (1) biopolymers from mollusk shells

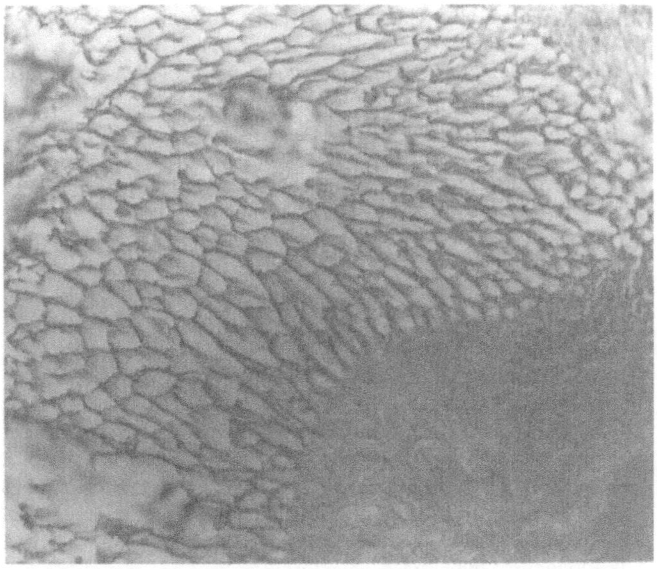

Figure 11. *In situ* localization of antigenic determinants in a longitudinal section of a Recent *Mercenaria mercenaria* shell.

contain phylogenetic information which can be retrieved with immunological methods, and (2) original biopolymers or fragments thereof can be preserved in biominerals over long periods of geologic time.

Recent studies by one of us (G. M.) on the retrieval of phylogenetic information of biopolymers from fossil brachiopod shells support these conclusions. Systematic specific results were found with EDTA extracts from a variety of fossil brachiopod shells (Collins *et al.*, 1989), using antibodies produced against biopolymers from shells of different Recent brachiopod taxa (Collins *et al.*, 1988).

The fact that small traces of original determinants are present in well-preserved fossils of considerable age raises hopes for the independent immunological establishment of the systematic position of fossil taxa of uncertain taxonomic position, such as the stromatoporoids.

3. Future Directions

3.1. Sequencing of Proteins Using New Methods

Separation of individual protein fractions from the shell matrix of fossils is in its early stages. However, there are exciting prospects for future directions of this research. For instance, in the near future amino acid sequencing (either partial or complete) of the fossil proteins will be achieved. These data ultimately will allow comparison of sequences between species and will establish more reliable phylogenetic schemes. However, diagenetic alteration of sequences may produce problems in liquid-phase sequencing, and, when problems do arise, other means of sequencing should be explored. New approaches such as mass spectrometry (MS), fast atom bombardment (FAB) mass spectrometry, and tandem MS/MS directly ionize peptides and small proteins. These methods may be used to define molecular weights of amino acids and therefore help determine primary structure of proteins/peptides of a molecular weight less than 6000 (Biemann, 1988). It appears likely that mass spectrometry will be most useful for proteins which have blocked N-termini or modified (phosphorylated, sulfated, or glycosylated) amino acids or have had amino acid replacements (Biemann, 1988)—an important step forward for modified fossil proteins which have been subjected to oxidative processes, including intermolecular cross-linking.

3.2. DNA from Fossils

The relatively new field of molecular evolutionary genetics is applicable to foraminifera, mollusks, and other fossilizable organisms. DNA sequencing of living and ancient representatives has exciting implications in the establishment of phylogenies and the study of evolution and is a likely research direction for the future (Pääbo, 1989). Comparative studies of the nucleotide sequences of ribosomal RNA in living organisms in which limited amounts can be obtained may be accomplished using the polymerase chain reaction (PCR) to amplify the small amounts available (only 0.1–10 ng of DNA are required). rRNA can then be directly sequenced. Recently, these techniques have been successfully applied to living benthic and planktic foraminifera (Langer *et al.*, 1992; Lipps *et al.*, 1992) and have exciting prospects for being extended to fossilized material. This method may also be used on DNA found in exceptionally well preserved fossils in which protoplasm may still be partially preserved. By sequencing DNA, phylogenetic relationships and possible speciation events can be established.

ACKNOWLEDGMENTS. We are greatly indebted to Drs. Peter Westbroek and Matthew Collins (Leiden University, The Netherlands) for critically reading the manuscript and to Drs. J. Donachy (University of South Alabama) and W. Allmon (University of South Florida) for helpful discussions. Financial support to G. M. was given by the Richard Lounsbery Foundation (New York) and to L. L. R. by the National Science Foundation (grant OCE-8817343).

References

Abelson, P. H., 1955, Organic constituents of fossils, *Carnegie Inst. Washington Yearb.* **54**:107–109.

Abelson, P. H., 1956, Paleobiochemistry, *Sci. Am.* **195**(1):2–7.

Addadi, L., and Weiner, S., 1985, Interactions between acidic proteins and crystals: Stereochemical requirements in biomineralization, *Proc. Natl. Acad. Sci. U.S.A.* **82**:4110–4114.

Akiyama, M., 1971, The amino acid composition of fossil scallop shell proteins and non-proteins, *Biominer. Res. Rep.* **3**:65–70.

Armstrong, W. G., Halstead, L. B., Reed, F. B., and Wood, L., 1983, Fossil proteins in vertebrate calcified tissue, *Philos. Trans. R. Soc. London, Ser. B* **301**:301–343.

Biemann, K., 1988, Contributions of mass spectrometry to peptide and protein structure, *Biomed. Environ. Mass Spectrom.* **16**: 99–111.

Collins, M. J., Curry, G. B., Quinn, R., Muyzer, G., Zomerdijk, T., and Westbroek, P., 1988, Sero-taxonomy of skeletal macromolecules in living terebratulid brachiopods, *Hist. Biol.* **1**: 207–224.

Collins, M., Curry, G., Muyzer, G., and Quinn, R., 1989, The prospects for molecular paleontology, *Terra Abstracts* **1**:194.

Cornish-Bowden, A., 1983, Relating proteins by amino acid composition, *Methods Enzymol.* **91**:60–75.

Creighton, T. E., 1983, *Proteins: Structures and Molecular Principles*, W. H. Freeman and Co., New York.

Crenshaw, M. A., 1972, The soluble matrix from *Mercenaria mercenaria* shell, *Biomineralization* **6**:6–11.

Curry, G., 1987a, Molecular palaeontology: New life for old molecules, *Trends Ecol. Evol.* **2**:161–165.

Curry, G., 1987b, Molecular palaeontology, *Geol. Today* **1987**(Jan.): 12–16.

Curry, G., 1988, Molecular evolution and the fossil record, in: *Short Courses in Paleontology*, Vol. 1 (Thomas W. Broadhead, ed.), Paleontological Society, Knoxville, TN, pp. 20–33.

de Jong, E. W., Westbroek, P., Westbroek, J. F., and Bruning, J. W., 1974, Preservation of antigenic properties in macromolecules over 70 Myr old, *Nature* **252**:63–64.

Donachy, J. E., Drake, B., and Sikes, C. S., 1992, Sequence and atomic-force microscopy analysis of a matrix protein from the shell of the oyster *Crassostrea virginica*, *Marine Biology* **114**:423–428.

Dungworth, G., Vincken, J. A., and Schwartz, A. W., 1975, Amino acid compositions of Pleistocene collagens, *Comp. Biochem. Physiol. B* **51**:331–335.

Elwood, H. J., Olsen, G. J., and Sogin, M. L., 1985, The small-subunit ribosomal RNA gene sequences from the hypotrichous ciliates *Oxytricha nova* and *Stylonchia pustulosa*, *Mol. Biol. Evol.* **2**:399–410.

Farris, J. S., 1972, Estimating phylogenetic trees from distance matrices, *Am. Nat.* **106**:645–668.

Felsenstein, J., 1988, Phylogenies from molecular sequences: Inference and reliability, *Annu. Rev. Genet.* **22**:521–566.

Fitch, W. M., and Margoliash, E., 1967, Construction of phylogenetic trees: A method based on mutational distances as estimated from cytochrome c sequences is of general applicability, *Science* **155**:279–284.

Gillespie, J. M., 1970, Mammoth hair: Stability of α-keratin structure and constituent proteins, *Science* **170**:1100–1101.

Goodman, J. W., 1980, Immunogenicity and antigenic specificity, in: *Basic and Clinical Immunology* (H. H. Fundenberg, D. P. Sites, J. L. Caldwell, and J. V. Wells, eds.), Lange Medical Publications, Los Altos, California, pp. 44–52.

Hare, P. E., 1969, Organic geochemistry of proteins, peptides, and amino acids, in: *Organic Geochemistry: Methods and Results* (G. Eglinton and M. Murphy, eds.), Springer-Verlag, New York, pp. 438–463.

Hare, P. E., Fogel, M. L., Stafford, T. W., Jr., Mitchell, A. D., and Hoering, T. C., 1991, The isotopic composition of carbon and nitrogen in individual amino acids isolated from modern and fossil proteins, *J. Archaeol. Sci.* **18**:277–292.

Haugen, J.-E., Sejrup, H.-P., and Vogt, N. B., 1989, Chemotaxonomy of Quaternary benthic foraminifera using amino acids, *J. Foraminiferal Res.* **19**(1):38–51.

Heinrickson, R. L., and Meridith, S. C., 1984, Amino acid analysis by reverse-phase high-performance liquid chromatography: Precolumn derivatization with phenylisothiocyanate, *Anal. Biochem.* **136**:65–74.

Hemleben, C., Be, A., Anderson, O. R., and Tuntivate, S., 1977, Test morphology, organic layers, and chamber formation of the planktonic foraminifera *Globorotalia menardii* (d'Orbigny), *J. Foraminiferal Res.* **7**:1–25.

Hemleben, C., Spindler, M., and Anderson, O. R., 1989, *Modern Planktonic Foraminifera*, Springer-Verlag, New York.

Hill, R. L., 1967, Hydrolysis of proteins, in: *Advances in Protein Chemistry* (C. B. Anfinsen, Jr., M. L. Anson, J. T. Edsall, and F. M. Richards, eds.), Academic Press, New York, pp. 37–107.

Hoering, T. C., 1980, The organic constituents of fossil mollusc shells, in: *Biogeochemistry of Amino Acids* (P. E. Hare, T. C. Hoering, and K. King, Jr., eds.), John Wiley & Sons, New York, pp. 193–201.

Jope, M., 1973, The protein of brachiopod shell—V. N. terminal end groups, *Comp Biochem. Physiol. B* **45**:17–24.

Jope, M., 1979, The protein of the brachiopod shell—VI. C-terminal end groups and sodium dodecylsulphate-polyacrylamide gel electrophoresis: Molecular constitution and structure of the protein, *Comp. Biochem. Physiol. B* **63**:163–173.

Joysey, K. A., and Friday, A. E. (eds.), 1982, *Problems of Phylogenetic Reconstruction*, Systematics Association Special Volume 21, Academic Press, New York.

Kennett, J. P., 1976, Phenotypic variation in some Recent and Late Cenozoic planktonic foraminifera, in: *Foraminifera*, Vol. 2 (R. H. Hedley and C. D. Adams, eds.), Academic Press, London, pp. 111–169.

Kimura, M., 1977, Preponderance of synonymous changes as evidence for the neutral theory of molecular evolution, *Nature* **267**:275–276.

King, K., Jr., and Hare, P. E., 1972, Amino acid composition of the test as a taxonomic character for living and fossil planktonic foraminifera, *Micropaleontology* **18**:285–293.

Köhler, G., and Milstein, C., 1975, Continuous cultures of fused cells secreting antibodies of pre-defined specificity, *Nature* **256**:495–497.

Kramptiz, G., Drolshagen, H., Hausle, J., and Hoflrmscher, K., 1983, Organic matrices of mollusc shells, in: *Biomineralization and Biological Metal Accumulation* (P. Westbroek and E. W. de-Jong, eds.), Reidel, Dordrecht, pp. 231–247.

Langer, M. R., Lipps, J. H., and Simison, W. B., 1992, Testing the molecular clock of evolution with planktic foraminifera, Abstract in *American Geophysical Union EOS 73* (43, supplement):273.

Lipps, J. H., Langer, M. R., Piller, W. E., Simison, W. B., Berbee, M., Lo Buglio, K., and Taylor, J., 1992, Molecular phylogeny of Foraminifera and Radiolaria Abstract, in *American Geophysical Union EOS 73* (43, supplement):319.

Lowenstam, H. A., and Weiner, S., 1989, *On Biomineralization*, Oxford University Press, New York.

Lowenstein, J., 1980a, Immunospecificity of fossil collagens, in: *Biogeochemistry of Amino Acids* (P. E. Hare, T. C. Hoering, and K. King, eds.), John Wiley & Sons, New York, pp. 41–51.

Lowenstein, J. M., 1980b, Species-specific proteins in fossils, *Naturwissenschaften* **67**:343–346.

Lowenstein, J. M., Sarich, V. M., and Richardson, B. J., 1981, Albumin systematics of the extinct mammoth and Tasmanian wolf, *Nature* **291**:409–411.

Malmgren, G., and Kennett, J. P., 1972, Biometric analysis of Phenotypic variation of *Globigerina pachyderma* (Ehrenberg) in the S. Pacific Ocean, *Micropaleontology* **8**(2):241–248.

Mann, S., 1988, Molecular recognition in biomineralization, *Nature* **32**:119–124.

Matter, P. F., Davidson, D., and Wyckoff, R. W. G., 1969, The composition of fossil oyster shell proteins, *Proc. Natl. Acad. Sci. U.S.A.* **64**:970–972.

Mitterer, R. M., and Cunningham, R., Jr., 1985, The interaction of natural organic matter with grain surfaces: Implications for calcium carbonate precipitation, In *Carbonate Cements* (N. Schneidermann and P. M. Harris, eds.) *Soc. Econ. Paleontol. Mineral. Spec. Publ.* **36**:17–31.

Muyzer, G., 1988, Immunological approaches in geological research, Thesis, Leiden University, The Netherlands.

Muyzer, G., and Westbroek, P., 1989, An immunohistochemical technique for the localization of preserved biopolymeric remains in fossils, *Geochim. Cosmochim. Acta* **53**:1699–1702.

Muyzer, G., Westbroek, P., deVrind, J. P. M., Tanke, J., Vrijheid, T., de Jong, E. W., Bruning, J. W., and Wehmiller, J. F., 1984, Immunology and organic geochemistry, *Org. Geochem.* **6**:847–855.

Muyzer, G., Westbroek, P., and Wehmiller, J. F., 1988, Phylogenetic implications and diagenetic stability of macromolecules from Pleistocene and Recent shells of *Mercenaria mercenaria* (Mollusca, Bivalva), *Hist. Biol.* **1**:135–144.

Nuttall, G. H. F., and Dinkelspiel, E. M., 1901, On the formation of specific anti-bodies in the blood following upon treatment with the sera of different animals, together with their use in legal medicine, *J. Hyg.* **1**:357–387.

Olsen, G. J., 1988, Phylogenetic analysis using ribosomal RNA, *Methods Enzymol.* **164**:793–812.

Ostrom, P. H., Zonneveld, J.-P., and Robbins, L. L., Organic geochemistry of hard parts: Assessment of isotopic variability and indigeneity, *Palaeogeography, Palaeoclimatology, Palaeoecology*, in press.

Pääbo, S., 1989, Ancient DNA: Extraction, characterization, molecular cloning and enzymatic amplification, *Proc. Natl. Acad. Sci. U.S.A.* **86**:1939–1943.

Penny, D., Foulds, L. R., and Hendy, M. D., 1982, Testing the theory of evolution by comparing phylogenetic trees constructed from five different protein sequences, *Nature* **297**:197–200.

Prager, E. M., Wilson, A. C., Lowenstein, J. M., and Sarich, V. M., 1980, Mammoth albumin, *Science* **209**:287–289.

Robbins, L. L., 1987, Morphologic variability and protein isolation and characterization of recent planktonic foraminifera, Ph.D. Dissertation, University of Miami.

Robbins, L., 1988, Environmental significance of morphologic variability in open-ocean versus ocean-margin assemblages of *Orbulina universa*, *J. Foraminiferal Res.* **18**(4):326–333.

Robbins, L., and Brew, K., 1990, Proteins from the organic matrix of Recent and fossil planktonic foraminifera, *Geochim. Cosmochim. Acta* **54**:2285–2292.

Robbins, L. L., and Donachy, J., 1991, Mineralization regulating proteins in fossil planktonic foraminifera, in: *Commodity Polypeptides* (C. S. Sikes and A. P. Wheeler, eds.), ACS Books, Washington, D.C.

Robbins, L. L., and Healy-Williams, N., 1991, Towards a classification of planktonic foraminifera based on biochemical, geochemical, and morphological criteria, *J. Foraminiferal Res.* **21**(2):159–167.

Robbins, L. L., Toler, S. K., and Donachy, J. E., Immunological and biochemical analysis of shell matrix proteins in living and fossil foraminifera, *Lethaia*, in press.

Runnegar, B., 1986, Molecular palaeontology, *Palaeontology* **29**(1): 1–24.

Rusenko, K. W., 1988, Studies on the structure and function of shell matrix proteins from the american oyster, *Crassostrea virginica*, Ph.D. Dissertation, Clemson University.

Schroeder, W. A., Shelton, J. R., Shelton, J. B., Cormick, J., and Jones, R. T., 1963, The amino acid sequence of the γ chain of human fetal hemoglobin, *Biochemistry* **2**:992–1008.

Shively, J. E. (ed.), 1986, *Methods of Protein Microcharacterization: A Practical Handbook*, Humana Press, Clifton, New Jersey.

Sikes, C. S., and Wheeler, A., 1983, A systematic approach to some fundamental questions of carbonate calcification, in: *Biomineralization and Biological Metal Accumulation* (P. Westbroek and E. W. de Jong, eds.), Reidel, Dordrecht, pp. 285–289.

Sikes, C. S., and Wheeler, A., 1986, The organic matrix from oyster shell as regulator of calcification *in vivo*, *Biol. Bull.* **170**: 494–505.

Spicer, G. S., 1988, Molecular evolution among some *Drosophila* species groups as indicated by two-dimensional electrophoresis, *J. Mol. Evol.* **27**:250–260.

Stathoplos, L., 1989, Amino acids in planktonic foraminiferal tests, Ph.D. Dissertation, University of Rhode Island.

Tolan, D., Lambert, S. M., Boileau, G., Fanning, T. G., Kenny, J. W., Vassos, A., and Traut, R. R., 1980, Radioiodination of microgram quantities of ribosomal proteins from polyacrylamide gels, *Anal. Biochem.* **103**:101–109.

Toler, S. K., 1993, Characterization of shell soluble matrix proteins from six genera of Soritacea Foraminifera, M.S. Thesis, University of South Florida.

Totten, D. K., Davidson, F. D., and Wyckoff, R. W. G., 1972, Amino acid composition of heated oyster shells, *Proc. Natl. Acad. Sci. U.S.A.* **69**:784–785.

Towe, K. M., 1971, Lamellar wall construction in planktonic foraminifera, in: *Proceedings of the IInd Planktonic Conference, Rome* (A. Farinacci, ed.), pp. 1213–1218.

Towe, K. M., 1980, Preserved organic ultrastructure: An unreliable indicator for Paleozoic amino acid biogeochemistry, in: *Biogeochemistry of Amino Acids* (P. E. Hare, T. C. Hoering, and K. King, Jr., eds.), John Wiley & Sons, New York, pp. 65–74.

Towe, K. M., and Cifelli, R., 1967, Wall structure in the calcareous foraminifera: Crystallographic aspects and a model for calcification, *J. Paleontol.* **41**(3):742–762.

Tuross, N., Fogel, M. L., and Hare, P. E., 1988, Variability in the preservation of isotopic composition of collagen from fossil bone, *Geochim. Cosmochim. Acta* **52**:929–935.

Weiner, S., Lowenstam, H. A., Taborek, B. and Hood, L., 1979, Fossil mollusk shell organic matrix components preserved for 80 million years, *Paleobiology* **5**:144–150.

Weiner, S., 1982, Separation of acidic proteins from mineralized tissues by reversed phase high performance liquid chromatography, *J. Chromatogr.* **245**:148–154.

Weiner, S., and Erez, J., 1984, Organic matrix of the shell of the foraminifer, *Heterostegina depressa*, *J. Foraminiferal Res.* **14** (3):206–212.

Weiner, S., and Lowenstam, H. A., 1981, Well preserved fossil mollusk shells: Characterization of mild diagenetic processes, in: *Biogeochemistry of Amino Acids* (P. E. Hare, T. C. Hoering, and K. King, Jr., eds.), John Wiley & Sons, New York, pp. 95–119.

Weiner, S., and Traub, W., 1984, Macromolecules in mollusc shells and their function in biomineralization, *Philos. Trans. R. Soc. London, Ser. B* **304**:425–434.

Weiner, S., Traub, W., and Lowenstam, H., 1983, Organic matrix in calcified exoskeletons, in: *Biomineralization and Biological Metal Accumulation* (P. Westbroek and E. W. de Jong, eds.), Reidel, Dordrecht, pp. 205–224.

Westbroek, P. van der Meide, P. H., van der Wey-Kloppers, J. S., van der Sluis, R. J., de Leeuw, J. W., and de Jong, E. W., 1979, Fossil macromolecules from cephalopod shells: Characterization, immunological response and diagenesis, *Paleobiology* **5**: 151–167.

Wheeler, A. P., Rusenko, K. W., George, J. W., and Sikes, C. S., 1987, Evaluation of calcium binding by oyster shell soluble matrix and its role in biomineralization, *Comp. Biochem. Physiol., B* **87**:953–960.

Wheeler, A. P., Rusenko, K. W., Swift, D. M., and Sikes, C. S., 1988, Regulation of *in vitro* and *in vivo* $CaCO_3$ crystallization by fractions of oyster shell organic matrix, *Mar. Biol.* **98**:71–80.

Wilson, A. C., Carlson, S. S., and White, T. J., 1977, Biochemical evolution, *Annu. Rev. Biochem.* **45**:573–639.

Wyckoff, R. W. G., 1972, *The Biochemistry of Animal Fossils*, Scientechnica, Bristol.

Wyckoff, R. W. G., and Doberenz, A. R., 1965, Electron microscopy of Rancho La Brea bone, *Proc. Natl. Acad. Sci. U.S.A.* **53**: 230–233.

Chapter 39

Immunospecificity of Fossil Proteins

Implications for the Establishment of Evolutionary Trends

JEROLD M. LOWENSTEIN

1. Introduction

1.1. The Molecular Basis of Evolution

The atomic structure of matter has been generally accepted by physicists and chemists for less than 100 years, and the molecular basis of genetics, the DNA double helix, has been understood only within the past half century. During recent decades, molecular evidence for the timing and topology of evolutionary relationships among bacteria, fungi, plants, and animals has been accumulating rapidly (Nei, 1987).

The bacterial genome consists of DNA dispersed in single cells. The eukaryotic genome includes nuclear DNA, which consists of about 3 billion base pairs, plus mitochondrial DNA and, in the case of plants, chloroplast DNA, each of which consists of some thousands of base pairs. Mitochondria and chloroplasts apparently originated as eubacterial and cyanobacterial invaders, respectively, of the early eukaryotic cell (Margulis, 1981).

DNA is transcribed as messenger RNA, which encodes the chains of amino acids that make up proteins, the main structural molecules of all life. The genome determines all the genetic characteristics of an organism, and mutations in the DNA message, accumulating over many generations, are the stuff of organic evolution.

One of the great surprises coming out of genomic exploration in the past few years is the discovery that only 1 to 5% of the genomic DNA is actually "expressed" as proteins. This means that 95–99% is "silent" or "junk" or "selfish" DNA, which "goes along for the ride," replicating from generation to generation without contributing to the structure or behavior of the organism (Doolittle and Sapienza, 1980; Orgel and Crick, 1980). Not being constrained by natural selection, this silent DNA usually evolves more rapidly than the "exons," the expressive DNA, and can speak eloquently on the subject of evolutionary comparisons among different taxa (Kimura, 1983; Nei, 1987).

JEROLD M. LOWENSTEIN • Department of Medicine, University of California–San Francisco, San Francisco, California 94143-0568.

Organic Geochemistry, edited by Michael H. Engel and Stephen A. Macko. Plenum Press, New York, 1993.

1.2. Molecules versus Morphology in Constructing Family Trees

All life on Earth speaks the same chemical language of DNA, RNA, and proteins, and so, in principle, any two organisms, no matter how far removed from their common ancestor in evolutionary time, can be subject to molecular comparisons. Because the number of mutable elements is so large—about 3 billion for eukaryotes—the probability of convergence within a single sequence is very low. Only one case of apparent convergence, in the digestive enzymes of plant-fermenting primates and bovids, has been reported (Stewart and Wilson, 1987).

Comparison of such conservative proteins as cytochrome *c* makes it possible to estimate the degree of relatedness of organisms as different from each other as bacteria, yeasts, plants, and animals—a task obviously impossible with morphological criteria (Margoliash, 1980). Furthermore, comparison of ribosomal sequences has revealed that all bacteria are not the same and that a group called the Archaebacteria are so different from the ordinary Eubacteria as to constitute a third form of life, equal in status to the Eubacteria and eukaryotes (Woese, 1987).

Morphological convergence, on the other hand, is commonplace. The parallels between Australian marsupials and placental mammals, which separated from a common ancestor more than 100 million years ago, are well known. Morphology represents an inextricable mixture of evolutionary history and function. Animals that have a similar way of life in similar environments may come to resemble each other, no matter how distantly related they are.

Though molecular methods are playing a larger and larger role in establishing relationships among extant species of plants, animals, fungi, and microorganisms, the fossil record is replete with the remains of extinct species no longer available for DNA, RNA, or protein sequencing. Until recently, we had to rely entirely on morphology for our understanding of the relations of fossil species to each other and to living taxa. In the past decade, molecular methods have begun to be applied to fossils too.

1.3. Molecular Genetic Analyses of Fossil Material

A generation ago, Abelson (1954) demonstrated the presence of amino acids in many fossils. Unfortunately, isolated amino acids are not very useful for determining genetic relationships, though occasionally the amino acid profiles would indicate the presence of a particular protein, such as collagen, which is unique in being one-third glycine and about 10% hydroxyproline (Fietzek and Kuhn, 1976; Wyckoff, 1972).

The problems in identifying specific protein or DNA sequences in fossils are that these compounds break down with time, so that at best the remaining molecules are changed and fragmented, and at worst they are gone, degraded and leached out by time and the elements (Wyckoff, 1972; Hare *et al.*, 1980).

2. Radioimmunoassay

In 1976 I began applying the very sensitive immunological technique of radioimmunoassay (RIA) to the detection and identification of proteins in fossils (Lowenstein, 1980). The advantage in this case of immunology over the "more precise" methods of protein or DNA sequencing is that antibodies bind to a sequence of some six to ten adjacent amino acids, and antiserum to a particular protein such as collagen or albumin contains dozens of different antibodies (Benjamin *et al.*, 1984). Therefore, the immunological approach is particularly suited to detecting fragmentary proteins.

The RIA that I use is a solid-phase double-antibody technique carried out in several steps:

1. A protein solution or fossil extract is placed in some of the wells of a 96-well polyvinyl microtiter plate. Some protein binds to the plastic as the "solid phase."
2. Unbound protein is washed out of the wells with soy protein (0.25‰), which blocks any remaining binding sites on the plastic.
3. Rabbit antisera to various albumins or collagens are added in 0.02-ml aliquots to the wells and allowed to remain for 24 h. These antisera bind best to their strict homologs but also cross-react with related species, in proportion to the degree of relatedness.
4. A second soy protein wash removes unbound antisera.
5. Radioactive (^{125}I-labeled) goat anti-rabbit gamma globulin (GARGG) is added and allowed to remain for 24 h. This antibody binds very specifically with the rabbit antibody already bound to the plastic—hence the term "double antibody."
6. Unbound GARGG is removed by running tap water.

7. The individual wells are cut from the plate and their radioactivity measured in a scintillation counter. Total radioactivity in each well depends on the amount of bound antibody, which is a function of the species specificity of the protein.
8. From the count data, quantitative immunological distance (ID) is obtained:

$$ID = 100 \log A/B$$

where A and B are different antiserum concentrations that yield identical radioactivities for homologous and heterologous proteins, respectively.

ID has been found to be linearly correlated with the amino acid sequence differences between compared proteins (Prager and Wilson, 1971). From matrices of these distances, phylogenetic trees may be constructed.

This assay can measure nanogram quantities of protein and distinguish between closely related species such as the African and Asian elephants, whose albumins react identically in other immune tests (Lowenstein, 1981, and 1985; Lowenstein *et al.*, 1981).

3. Applications of RIA to Fossils

3.1. An Experimental Model of Diagenesis: Heated Bone

In an experimental simulation of the processes of diagenesis, in which fossil bones are subjected to weathering, with denaturation and leaching out of their proteins, modern cattle bone fragments were heated to 80°C in a moist environment for a period of 10 weeks. At intervals, extracts of these bones were tested by RIA for the presence of bovine-specific albumin and collagen (Fig. 1). Though the concentrations of both proteins decreased rapidly with time, the extracts continued to give a stronger reaction with antisera to bovine albumin and collagen than with antisera to human albumin and collagen. This indicated the retention of species-specific molecular configurations in the heated bone proteins.

There are two classes of antigenic determinants on a protein—*sequential* and *conformational* (Benjamin *et al.*, 1984). Sequential determinants depend only on the amino acid sequence of a segment about 6 to 10 amino acids in length. Such segments will continue binding to their corresponding antibody even after the original protein is broken down by time, enzymes, bacterial action, or heat. *Conformational* determinants are adjacent amino acids on the protein surface which may consist either of a sequential segment or residues far apart in the amino acid sequence but brought together by the folding of the protein. These determinants are unlikely to survive diagenesis and denaturation.

3.2. Fossil Mammoths, Mastodons, and Sea Cows

Traditionally, evolutionary relationships among members of the order Proboscidea have been inferred

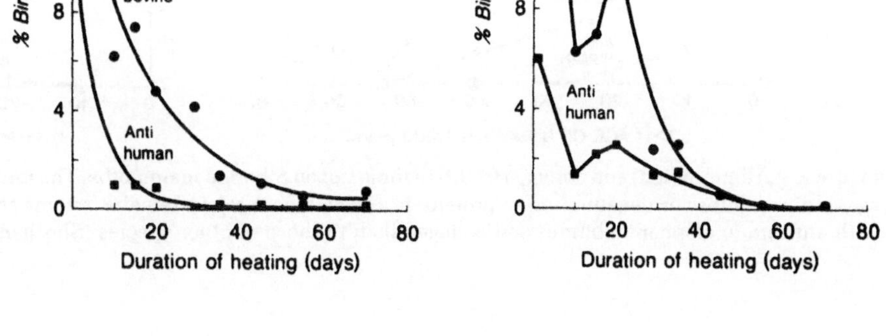

Figure 1. An experimental model for diagenesis. Modern cattle bone was heated at 80°C for 10 weeks, and samples were tested at intervals during this period for albumin (left) and collagen (right). Immunoreactivity decreased with time but species specificity persisted, as shown by the relative binding of antisera to bovine and human proteins. The "hump" on the collagen curve is due to increased collagen solubility with heating. The heated specimens were obtained from Dr. P. E. Hare.

almost exclusively from bones and teeth. The family Elephantidae includes 6 genera and 26 species, of which the African elephant (*Loxodonta africana*) and the Asian elephant (*Elephas maximus*) are the only living representatives. One extinct species, the woolly mammoth (*Mammuthus primigenius*), is ideal for fossil protein studies because a number of frozen specimens have been recovered from the Siberian permafrost, and one would expect the proteins in these specimens to be particularly well preserved (Lowenstein *et al.*, 1981; Prager *et al.*, 1980).

Albumin and collagen were extracted from three frozen Siberian mammoths, 10, 40, and 50 thousand years old and analyzed by RIA (Shoshani *et al.*, 1985) (Fig. 2). Though the amount of protein is considerably reduced with geological age, the albumins and collagens have retained species specificity and immunologically are nearly identical to those of living elephants. Mammoth teeth have been interpreted as being more like those of *Elephas* than those of *Loxodonta*, but the albumins of the three species are equidistant from each other, so that no pairing is favored over any other (Lowenstein *et al.*, 1981).

The American mastodon (*Mammut americanum*) is classified in the family Mammutidae, which has a fossil record extending back to the Oligocene. In 1968 a partial skeleton ("Elmer") was unearthed in a construction site in southeast Michigan and found to have a ^{14}C age of 10,000 years. This provided the opportunity to see whether molecular genetic information could be obtained from bones preserved for millennia at ambient temperatures. Protein was extracted from the bone dust drilled for mounting the skeleton. This extract was tested with antisera to the albumins and collagens of the living elephant species. Also, the extract was injected into rabbits to raise anti-mastodon antisera (Shoshani *et al.*, 1985).

Anti-mastodon antisera reacted highly specifically with elephant serum and elephant collagen and only weakly with cow serum and collagen. This

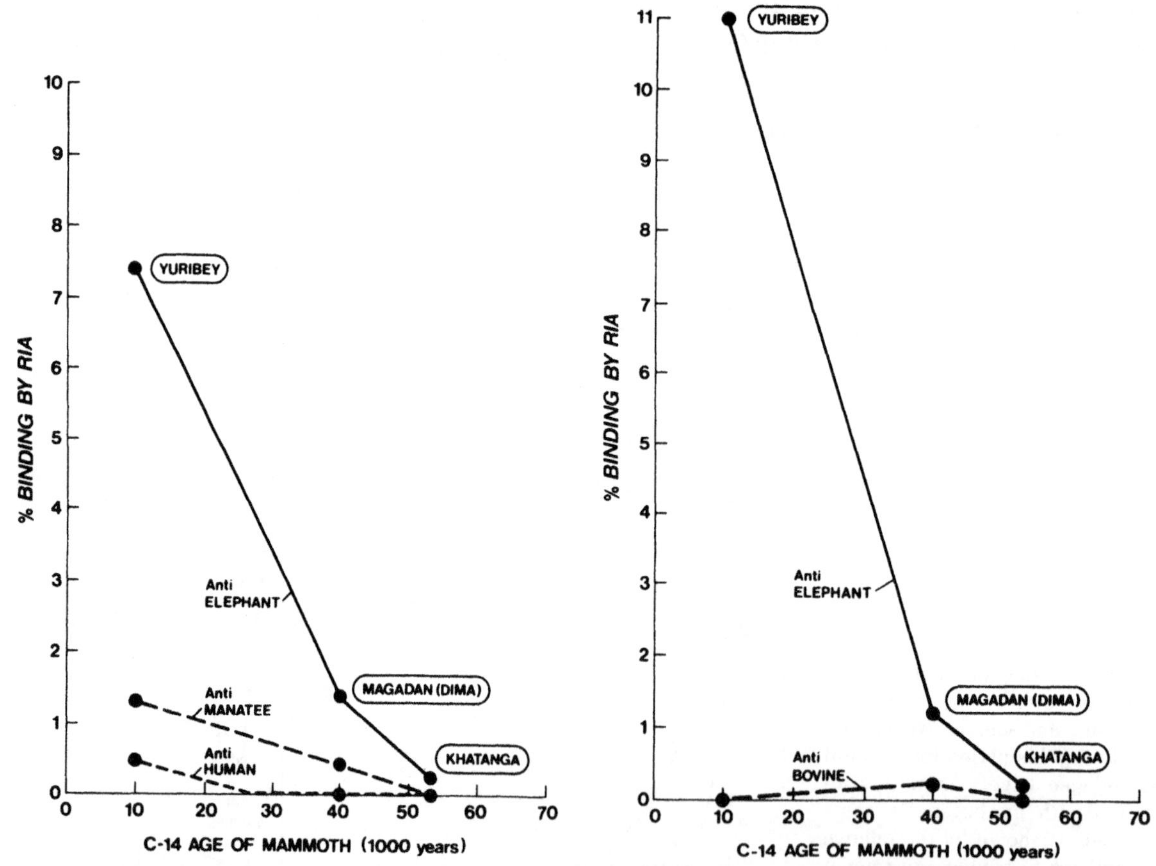

Figure 2. Albumin (left) and collagen (right) in three frozen Siberian mammoths. The immunological reactions decrease with increasing age, indicating loss or denaturation of protein. However, what protein remains retains some species specificity, binding more strongly with antisera to elephant albumin and collagen than to those of other species (Shoshani *et al.*, 1985).

showed that species-specific proteins were preserved in the mastodon skeleton. Unlike the frozen mammoth tissues, mastodon extracts did not react with anti-elephant albumin and collagen. This asymmetrical pattern of reactions suggests that injection of fossil material into a rabbit amplifies the residual protein determinants and makes it possible to detect them.

Sea cows are considered the closest living relatives of the elephants. Four species of sirenians survive, the Pacific dugong and three Atlantic manatees. The now extinct Steller's sea cow (*Hydrodamalis gigas*) was discovered in 1741 in the Commander Islands off the Kamchatka Peninsula. By 1768 this immense herbivorous marine mammal had been exterminated by sailors and fur hunters. My colleagues and I made extracts from a skeleton in a museum collection in order to establish the molecular affinities of Steller's sea cow with its living relatives. Bone extracts were tested with antisera to the albumins of all extant sea cows, as well as of hyrax and aardvark (Rainey *et al.*, 1984).

As expected from its anatomy and geography, Steller's sea cow is more closely related to the Pacific dugong than to the three Atlantic manatees. Paleontologically, the divergence time between Steller's sea cow and the dugong has been estimated at 20–30 Ma, but the RIA results suggest a much more recent split, in the 4–8-Ma range. Likewise, the dugong–manatee divergence has been thought to date from the middle Eocene (about 45-Ma), but RIA results indicate 17–20 Ma. Immunologically, elephant, sea cow, hyrax, and aardvark are about equidistant and form a monophyletic unit relative to other mammals (Fig. 3).

3.3. Taxonomic Controversies: The Extinct Quagga and Tasmanian Wolf

During the past century, human activities have exterminated numerous mammalian species, among them the quagga (*Equus quagga*) and the Tasmanian wolf (*Thylacinus cynocephalus*). The taxonomic position of each species has been disputed, and in each case RIA of fossil tissue supports one of the contesting opinions.

Huge herds of quaggas roamed the plains of Cape Province in South Africa early in the last century, but their habitat was destroyed and they were hunted to extinction by the end of the century. The quagga was striped on its face and forequarters like a zebra but was chestnut-toned on its body and hindquarters like a horse. From this striping pattern and from features of the bones and teeth, three distinct opinions have emerged about quagga relationships:

1. Its closest relative is the domestic horse.
2. It was a subspecies of the plains zebra.
3. It was equally distant from the three surviving zebra species—plains, mountain, and Grevy's.

Both RIA (Lowenstein and Ryder, 1985) and analysis of mitochondrial DNA (Higuchi *et al.*, 1987) from quagga skin showed quagga proteins to be much more like those of the plains zebra than those of other members of the horse family. Thus, the second hypothesis, that the quagga was a subspecies of the plains zebra, is supported by molecular evidence (Fig. 4).

The Tasmanian wolf was shot out after the turn of the century as a pest that preyed on cattle. The last

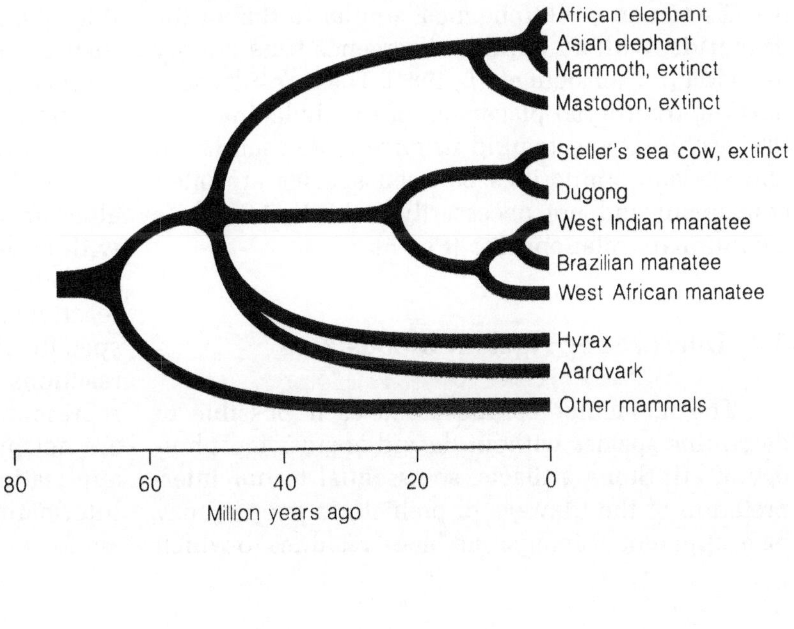

Figure 3. Phylogeny of the "paenungulates"—elephants, sea cows, hyraxes and aardvark—from RIA of their serum albumins. This molecular family tree is unique in including three extinct species—mammoth, mastodon, and Steller's sea cow.

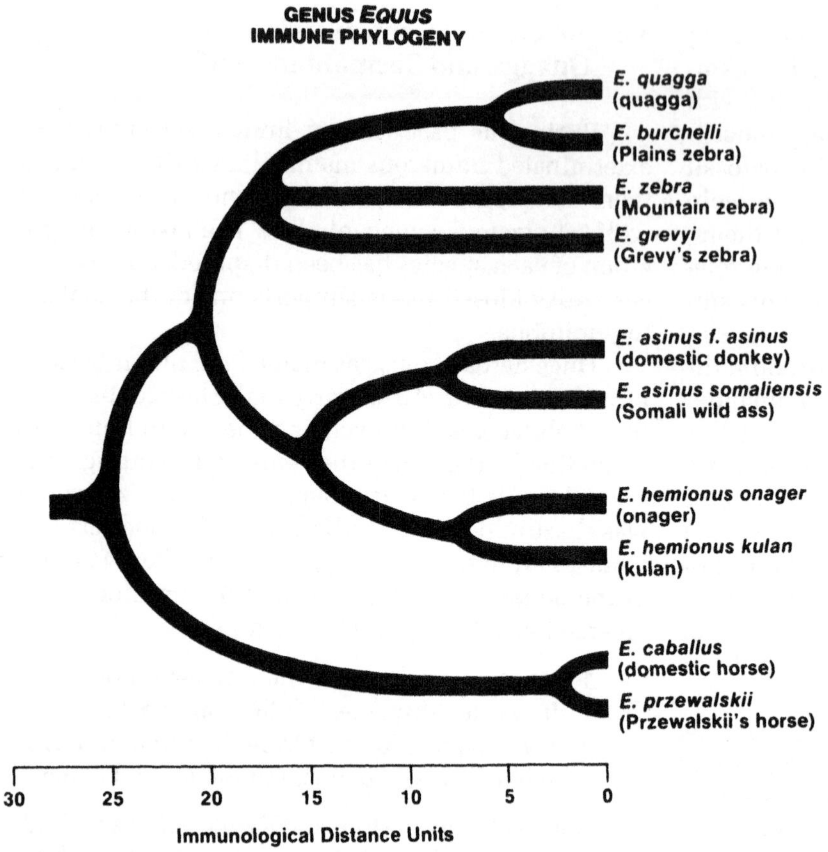

Figure 4. Family tree of the horse family, including the extinct quagga. Proteins extracted from quagga skin react most like those of the plains zebra. The domestic horse, *Equus caballus*, which has also been suggested as the quagga's nearest relative, is actually the most distant by RIA analysis.

thylacine died in a zoo in 1933. This species has been placed by most systematists with the other Australian marsupial carnivores, the dasyurids. However, its remarkable dental and skeletal resemblance to the extinct South American marsupial carnivore *Borhyaena* has raised the question of whether it might be a relict borhyaenid (Archer, 1976). RIA of skin and muscle specimens preserved in museums showed albumin of the Tasmanian wolf to be most similar to that of the dasyurids, with an implied divergence time of only 6–10 Ma (Lowenstein *et al.*, 1981). These results support the traditional placement of the thylacine with the dasyurids and remind us once more that dental and skeletal similarities between species are often convergent and not necessarily indicative of close evolutionary relationships (Fig. 5).

3.4. Interpreting Ancient Bloodstains

The molecular approach makes it possible to determine species without the aid of gross morphology at all. Stone artifacts, so essential to our interpretation of the lifeways of prehistoric people, may have apparent or inapparent blood residues to which analytical techniques can be applied. Loy (1983) reported the identification of animal species from bloodstains on stone tools by the method of hemoglobin recrystallization. Subsequently, he and I collaborated in examining a collection of stone spear points by the more sensitive RIA technique (Lowenstein, 1988).

Each tool was inspected for the presence of visible bloodstains or discolorations. Next, a drop of distilled water was placed on the stain, which was scraped gently with a wooden stick to bring it into solution. The drop was then transferred for RIA analysis (Lowenstein, 1988).

Of the ten implements tested, four gave specific albumin reactions: two with anti-elk albumin, one with anti-bison, and one with anti-human (Table I). The other six gave either no significant albumin-like reactions, or the reactions were weak and not species-specific. As the tools originated in British Columbia, reactions with elklike albumin could equally well represent moose or reindeer. Specific anti-albumins are not presently available for those species. Wider application of this method could provide fascinating information about the uses of tools and weapons among ancient peoples.

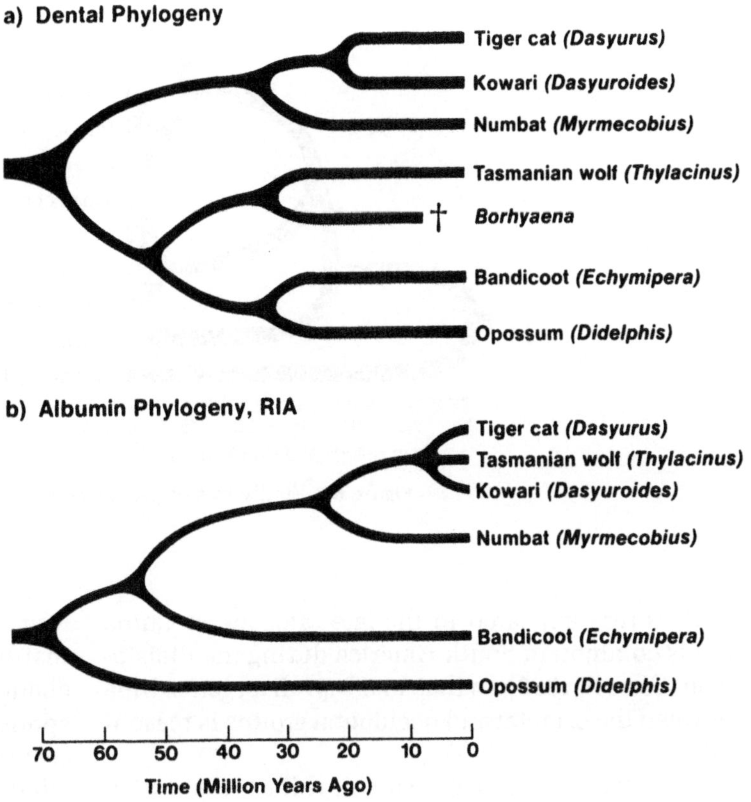

Figure 5. Dental vs. albumin phylogenies of the extinct Tasmanian wolf. Archer (1976) deduced from similarities of tooth structure that the Tasmanian wolf was close to the extinct South American marsupial carnivore *Borhyaena*. However, albumin, extracted from skin and muscle in museums, shows the Tasmanian wolf to be a dasyurid.

4. RIA Phylogenies, Biogeography, and Molecular Clocks

RIA detects such smaller amounts of protein than standard immunological tests, by several orders of magnitude. This sensitivity makes it possible to distinguish very closely related subspecies, such as the quagga and the plains zebra, and to measure cross-reactions between very distantly related groups such as mammals, birds, reptiles, amphibians, and fish.

Perhaps the most vigorously disputed topic in molecular evolution is that of "molecular clocks." Do particular molecules such as albumin evolve at fairly steady rates over long stretches of geological time, as Wilson *et al.* (1977, 1988) claim, or do the rates fluctuate too wildly to be useful for timing evolutionary events (Goodman, 1985)?

TABLE I. RIA of Bloodstains from 3000-Year-Old Spear Points

Specimen	Reaction[a] with antisera against albumin of: Elk	Bison	Horse	Human	Rat	Fish
FST1-4	6.7	1.8	1.4	1.3	0.8	0.0
FST1-6	5.0	1.3	0.7	0.4	0.5	0.0
GSK7-31	0.8	0.1	0.5	5.5	1.1	0.0
GSK7-32	0.4	6.0	2.2	2.8	0.7	0.2

[a]The numbers in the table are percent uptake of radioactive second antibody (GARGG). The first two specimens give positive reactions for elk serum albumin, the third for human, and the fourth for bison. Measurable cross-reactions are commonly observed between mammalian albumins but almost none between mammals and fish.

4.1. Perissodactyls

Perissodactyls originated in the earliest Eocene, according to the fossil record, and only six genera of living tapir, rhino, and horses remain of the many lineages that existed in the past. An RIA phylogeny was constructed by making antisera in rabbits to whole serum and measuring immunological distances between the genera. In such sera, 70% of the antibodies are directed against albumin, and the remainder against a variety of other serum proteins.

The resulting family tree of perissodactyls (Fig. 6) conforms well with current taxonomy and the fossil record, if one ID unit is taken to represent one million years. The tapir and rhino are slightly closer to each other than either is to *Equus*, and the basic split occurred around 55 Ma, with the tapir–rhino split following around 49 Ma. The modern genus

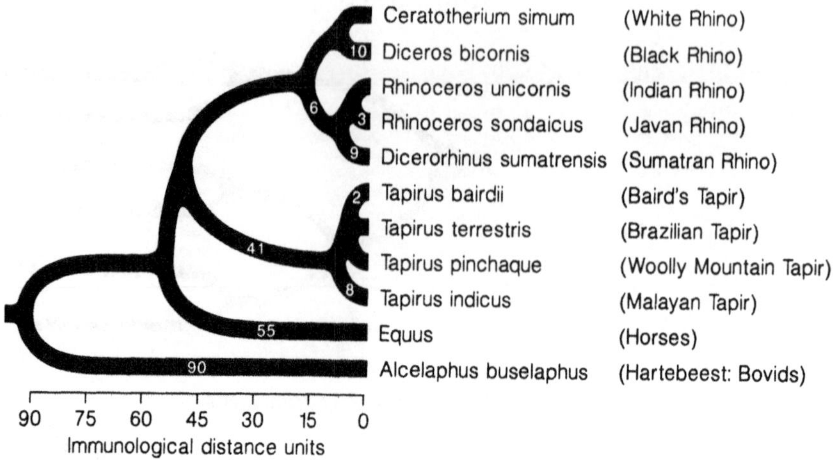

Figure 6. Family tree of perissodactyls, by RIA of serum proteins.

Tapirus first appeared in the late Miocene of China and is common in South America during the Pleistocene (Carroll, 1988), so that an 8-Myr divergence time between the oriental and occidental groups is reasonable.

The white and black African rhinos, *Ceratotherium simum* and *Diceros bicornis*, are thought to have diverged around 12 Ma, not far from the date of 10 Ma indicated by RIA. *Dicerorhinus*, now restricted to southeast Asia and Indonesia, was supposedly present in Europe and Africa at 25–20 Ma, an interpretation not in accordance with the RIA split between *Dicerorhinus* and *Rhinoceros* at about 9 Ma. *Rhinoceros* is only known from the late Pliocene, consistent with the 3-Myr RIA divergence between the two extant *Rhinoceros* species.

Perissodactyls and artiodactyls are assumed to have diverged from a common ancestor around 65 Ma (Carroll, 1988). The 90 ID units between the hartebeest *Alcelaphus buselaphus* and the perissodactyls, almost 50% greater than expected, suggests two possibilities: (1) that the average rate of serum protein evolution along the artiodactyl lineage has been considerably more rapid than the average along the perissodactyl lineages; or (2) that RIA immunological distances are nonlinear with time, the smaller values underestimating evolutionary divergence times and the larger values overstimating them. I suspect that both mechanisms are at work here to some extent.

4.2. Reptiles and Birds

Albumin distances between reptiles, birds, and mammals reveal significant differences in rates of change in different lineages during the past 100–200 Myr. Prager *et al.* (1974) have previously reported that bird albumins and transferrins have apparently changed at only about one-third the rate of the corresponding mammalian molecules. Turtle albumins seem to have evolved at an even slower rate than bird albumins (Fig. 7). The factors determining these rate differences are not well understood (Doolittle, 1979).

4.3. Amphibians

Amphibian albumins, on the other hand, seem to have changed at about the same rate as mammalian albumins (Wilson *et al.*, 1977). Maxson *et al.* (1975) have correlated albumin immunological differences between frogs on different landmasses with the estimated separation times of those landmasses. It is believed that Australia broke away from South America around 70 Ma, and the albumin "clock" between Australian and South American frogs indicates an appropriate interval of difference.

With RIA it is possible to estimate albumin-based divergence times of even more ancient groups of amphibians: primitive frogs, salamanders, and caecilians (Fig. 8). The phylogeny is consistent with the known biogeography of these organisms. The three subclasses of Amphibia apparently diverged at about the same time, as no pair shows a closer immunological relationship than any other.

5. Future Prospects

The extraction of fossil DNA from a 100-year old quagga skin (Higuchi *et al.*, 1984) raised considerable excitement and speculation in the popular press

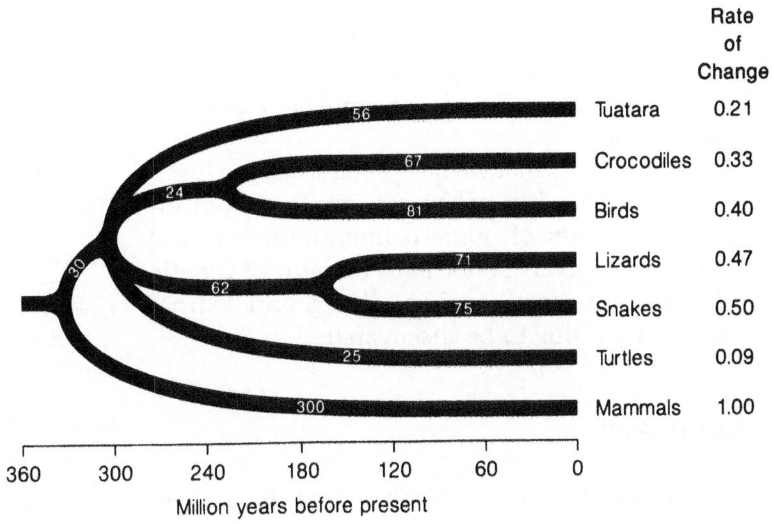

Figure 7. Relations among reptiles, birds, and mammals, by RIA of serum albumin. The numbers on the branches are immunological distances. Rates of change have differed along the different lineages, being notably slow in turtles.

about the prospect of being able to reconstruct the genomes of extinct species such as dinosaurs. Unfortunately, the reality is that the DNA recovered was just a small fraction of the mitochondrial genome, which consists of some 15,000 base pairs, and none of the nuclear genome of three billion base pairs was retrieved. Although DNA was recovered and cloned from an ancient Egyptian mummy (Pääbo, 1985) and a 40,000-year-old mammoth (Higuchi and Wilson, 1984), most of the latter appeared to be microbial rather than proboscidean DNA.

The new polymerase chain reaction (PCR), which can amplify DNA sequences *in vitro* (Mullis and Faloona, 1987), may improve the recovery of DNA from fossils. With PCR one can obtain sufficient DNA for analysis from a single hair (Higuchi *et al.*, 1988). Whether or not enough native DNA survives in most fossils for meaningful genetic comparisons is the question to be answered.

Proteins do survive in many fossils for thousands and sometimes millions of years (Lowenstein, 1985; Rowley *et al.*, 1986; de Jong *et al.*, 1974). The sensitivity of our present methods for detecting proteins could undoubtedly be improved. One possibility for enhancement is to use monoclonal rather than polyclonal antibodies. It may be that some parts of common protein molecules such as collagen and albumin are less subject to change and breakdown with time than others. In making monoclonal antibodies, one produces unlimited amounts of a single antibody directed against a particular short molecular sequence. If such "high probability" sequences can be identified, the monoclonal approach might increase the yield of useful fossil proteins.

Another immunological technique for improving sensitivity of detection is affinity chromatography. Antigens or antibodies are bound to particles on a column, and the corresponding antisera or antigens

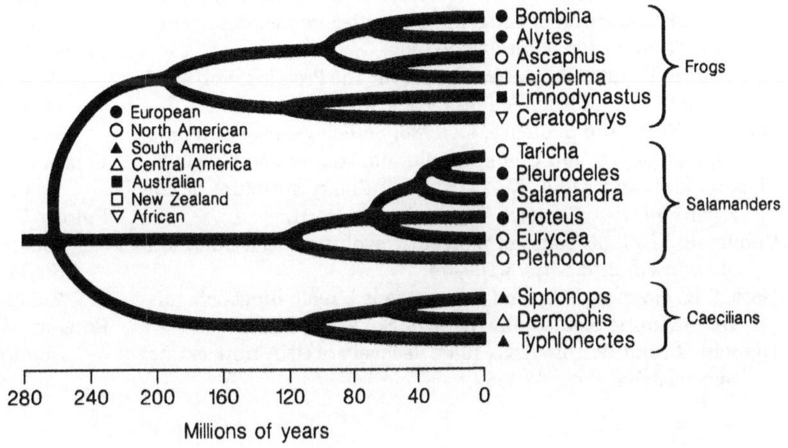

Figure 8. Albumin family tree of amphibians, by RIA. Distances between these genera are too great to be measured by the standard immunological technique of microcomplement fixation. The albumins and antisera were provided by Dr. Linda Maxson.

are passed through the column, where they bind to their "mirror images." In this way, the concentrations of antigens such as fossil proteins, or of weak antibodies to fossil materials, can be amplified as much as thousandfold.

In the few years that we have been studying fossil molecules, we have barely begun to apply the powerful armamentarium of modern immunological and analytical methods. An enormous amount of genetic evolutionary information about living and extinct organisms is waiting to be discovered.

6. Summary

The molecular biology revolution of the past 30 years has had a profound effect on evolutionary studies. In addition to the morphological resemblances between related organisms, we now have DNA, RNA, and protein sequences and immunological comparisons for quantitating genetic evolutionary relationships. Until recently, these molecular approaches could only be applied to living species. A sensitive radioimmunoassay technique has made it possible to study the proteins of extinct species such as mammoths, mastodons, the quagga, and the Tasmanian wolf and has helped to resolve controversies about their evolutionary affinities.

References

Abelson, P. H., 1954, Organic constituents of fossils, *Carnegie Insti. Washington Yearb.* **53**:97–101.

Archer, M., 1976, The dasyurid dentition and its relationships to that of the didelphids, thylacynids, borhyaenids (Marsupicarnivora) and peramelids (Peramelina: Marsupialia), *Aust. J. Zool. Suppl.* **39**:1–24.

Benjamin, D. C., *et al.*, 1984, The antigen structure of proteins, *Annu. Rev. Immunol.* **2**:67–101.

Carroll, R. L., 1988, *Vertebrate Paleontology and Evolution*, W. H. Freeman, New York.

de Jong, E. W., Westbroek, P., Westbroek, J. F., and Bruning, J. W., 1974, Preservation of antigenic properties in macromolecules over 70 myr old, *Nature* **252**:63–64.

Doolittle, R. F., 1979, Protein evolution, in: *The Proteins*, Vol. 4, (H. Neurath, ed.) Academic Press, New York, pp. 1–118.

Doolittle, W. F., and Sapienza, C., 1980, Selfish genes, the phenotype paradigm and genome evolution, *Nature* **284**:601–603.

Fietzek, P. P., and Kuhn, K., 1976, The primary structure of collagen, *Int. Rev. Connect. Tissue Res.* **7**:61–101.

Goodman, M., 1985, Rates of molecular evolution: The hominoid slowdown, *BioEssays* **3**(1):9–14.

Hare, P. E., Hoering, T. C., and King, K. (eds.), 1980, *Biogeochemistry of Amino Acids*, John Wiley & Sons, New York.

Higuchi, R., and Wilson, A. C., 1984, Recovery of DNA from extinct species, *Fed. Proc.* **43**:1557.

Higuchi, R., Bowman, B., Freiberger, M., Ryder, O. A., and Wilson, A. C., 1984, DNA sequences from the quagga, an extinct member of the horse family, *Nature* **312**:282–284.

Higuchi, R., Wrischnik, L. A., Oakes, E., George, M., Tong, B., and Wilson, A. C., 1987, Mitochondrial DNA of the extinct quagga: Relatedness and extent of post-mortem change, *J. Mol. Evol.* **25**:283–287.

Higuchi, R., von Beroldingen, C., Sensabaugh, G. F., and Erlich, H. A., 1988, DNA typing from single hairs, *Nature* **332**:543–546.

Kimura, M., 1983, *The Neutral Theory of Molecular Evolution*, Cambridge University Press, New York.

Lowenstein, J. M., 1980, Species-specific proteins in fossils, *Naturwissenschaften* **67**:343–346.

Lowenstein, J. M., 1981, Immunological reactions from fossil material, *Philos. Trans. R. Soc. London, Ser. B* **292**:143–149.

Lowenstein, J. M., 1985, Radioimmunoassay and molecular phylogeny, *BioEssays* **2**:60–62.

Lowenstein, J. M., 1988, Immunological methods for determining phylogenetic relationships, in: *Molecular Evolution and the Fossil Record* (B. Runnegar and J. W. Schopf, eds.), The Paleontological Society, Knoxville.

Lowenstein, J. M., and Ryder, O. A., 1985, Immunological systematics of the extinct quagga (Equidae), *Experientia* **41**:1192–1193.

Lowenstein, J. M., Sarich, V. M., and Richardson, B. J., 1981, Albumin systematics of the extinct mammoth and Tasmanian wolf, *Nature* **291**:409–411.

Loy, T. H., 1983, Prehistoric blood residues: Detection on tool surfaces and identification of species of origins, *Science* **120**: 1269–1271.

Margoliash, E., 1980, Evolutionary adaptation of mitochondrial cytochrome *c* to its functional milieu, in: *The Evolution of Protein Structure and Function* (D. S. Sigman and M. A. B. Brazier, eds.), Academic Press, London.

Margulis, L., 1981, *Symbiosis in Cell Evolution*, W. H. Freeman and Co., San Francisco.

Maxson, L. R., Sarich, V. M., and Wilson, A. C., 1975, Continental drift and the use of albumin as an evolutionary clock, *Nature* **255**:397–399.

Mullis, K. B., and Faloona, F. A., 1987, Specific synthesis of DNA *in vitro* via a polymerase-catalysed chain reaction, *Methods Enzymol.* **155**:335–350.

Nei, M., 1987, *Molecular Evolutionary Genetics*, Columbia University Press, New York.

Orgel, L. E., and Crick, F. H. C., 1980, Selfish DNA: The ultimate parasite, *Nature* **284**:604–607.

Pääbo, S., 1985, Molecular cloning of ancient Egyptian mummy DNA, *Nature* **314**:644–645.

Prager, E. M., and Wilson, A. C., 1971, The dependence of immunological cross-reactivity upon sequence resemblance among lysozomes, *J. Biol. Chem.* **246**:5978–5989, 7010–7017.

Prager, E. M., Brush, A. H., Nolan, R. A., Nakanishi, M., and Wilson, A. C., 1974, Slow evolution of transferrin and albumin in birds according to micro-complement fixation analysis, *J. Mol. Evol.* **3**:243–262.

Prager, E. M., Wilson, A. C., Lowenstein, J. M., and Sarich, V. M., 1980, Mammoth albumin, *Science* **209**:287–289.

Rainey, W. E., Lowenstein, J. M., Sarich, V. M., and Magor, D. M., 1984, Sirenian molecular systematics—including the extinct Steller's sea cow (*Hydrodamalis gigas*), *Naturwissenschaften* **71**:586–588.

Rowley, M. J., Rich, P. V., Rich, T. H., and Mackay, I. R., 1986, Immunoreactive collagen in avian and mammalian fossils, *Naturwissenschaften* **73**:620–623.

Shoshani, J., Lowenstein, J. M., Walz, D. A., and Goodman, M., 1985, Proboscidean origins of mastodon and woolly mammoth demonstrated immunologically, *Paleobiology* **11**:429–437.

Stewart, C.-B., and Wilson, A. C., 1987, Sequence convergence and functional adaptation of stomach lysozymes from foregut fermenters, *Cold Spring Harbor Symp. Quant. Biol.* **52**:891–899.

Wilson, A. C., Carlson, S. S., and White, T. J., 1977, Biochemical evolution, *Annu. Rev. Biochem.* **46**:573–639.

Wilson, A. C., Ochman, H., and Prager, E. M., 1988, Molecular time scale for evolution, in: *Molecular Evolution and the Fossil Record* (B. Runnegar and J. W. Schopf, eds.), The Paleontology Society, Knoxville.

Woese, C. R., 1987, Bacterial evolution, *Microbiol. Rev.* **51**:221–271.

Wyckoff, R. W. G., 1972, *The Biochemistry of Animal Fossils*, Scientechnica, Bristol.

IX

Summary

Chapter 40

Postscript

GEOFFREY EGLINTON

Twenty-four years have been added to the Earth's history since the publication of the text *Organic Geochemistry: Methods and Results*, comprising chapters written by those then actively engaged in the subject and edited by myself and Sister Mary Murphy (Springer-Verlag, New York, 1969). The editors of the present volume, which has a similar format, kindly invited me to comment on the current state of the subject. How have things changed—is there anything left to do?! Simply put, the answers are "enormously" and "yes, indeed."

Organic geochemists continue to extend their activities into fields well beyond the classic topic of petroleum geochemistry. Certain major problem areas need increased analytical study by chemists, whether they be termed organic geochemists, biogeochemists, petroleum geochemists, marine chemists, or whatever. One important area is the characterization of the very complex mixtures of individual organic components present in natural and polluted waters and sediments. Substantial progress has, of course, been made, but minor components and trace components are still largely uncharacterized. New methods are required which will bring about high-efficiency molecular fractionations, based on size, shape, and functional groups.

The combination of molecular separation techniques with isotope measurement is burgeoning, and future studies may be expected to include elements other than carbon, such as hydrogen, oxygen, nitrogen, and sulfur. Stable isotope determinations have been plowed as lonely furrows (using total organic fractions) for many years by determined pioneers. The new compound-specific measurements must surely enhance their efforts and open up completely new vistas.

Data processing is another major technology, important for the ability to handle multiple analyses in an automated fashion. Continued development of data processing is a high priority, especially in those fields requiring quantitative data for the construction of budgets and the testing of models.

Next, there is the challenge of polymeric organic material. Thus, one major carbon pool of truly global significance is the dissolved organic carbon (DOC) in the oceans. A hot debate continues as to the validity of the different methods for measuring this DOC, but satisfactory quantitation will be merely a beginning, since what we really need to know is what the DOC is made of, how it is assembled, and how it is taken apart and utilized. Also, what isotopic signals does it carry and just how good a tracer is it for original water masses and their subsequent history? Comprehensive studies of DOC will be an essential part of developing our understanding of the planetary carbon cycle and will remain a challenge for years to come. However, it is only one illustration of the enormous lack of detailed knowledge concerning the organic chemistry of the environment. True "natural product chemistry" has barely begun to be studied. Biochemistry and biological organic chemistry need to be seen as parts of the wider biogeochemical whole. It is not a view held by most chemists, and we need to continue to seek chemical approbation for our field! Indeed, we need to develop organic geochemical participation in environmental and ecological studies, such as those relevant to paleoclimatology and paleoceanography. We have a part to play in the current projects aimed at documenting and assessing the recent past and in predicting the future. Claims to environmental credibility by government, industry, and funding agencies are becoming increas-

GEOFFREY EGLINTON • Organic Geochemistry Unit, School of Chemistry, University of Bristol, Bristol BS8 1TS, England.
Organic Geochemistry, edited by Michael H. Engel and Stephen A. Macko. Plenum Press, New York, 1993.

ingly common, and we should contribute our services appropriately.

Finally, education—the teaching of organic geochemistry. The widespread adoption of teaching of organic geochemistry in universities and institutes is now timely, especially the development of suitable high-level short courses, such as spring or summer schools, of one or two weeks' duration. My personal view is that instruction should be given by experts drawn from wherever appropriate but that the courses should demand as a prerequisite a sound basis in organic and natural product chemistry. This chemical knowledge may then be built on by describing the analytical methods and techniques employed in the study of complex organic materials in natural environments and in geological samples—thus giving immediate scope for exploring both processes and applications. Such courses, then, would have a different role from and not overlap with those for educating petroleum geologists and others in the assessment of maturity and petroleum potential of source rocks.

Index

GPSR Compliance

The European Union's (EU) General Product Safety Regulation (GPSR) is a set of rules that requires consumer products to be safe and our obligations to ensure this.

If you have any concerns about our products, you can contact us on ProductSafety@springernature.com

In case Publisher is established outside the EU, the EU authorized representative is:

Springer Nature Customer Service Center GmbH
Europaplatz 3
69115 Heidelberg, Germany

Batch number: 10723888

Printed by Printforce, the Netherlands